BUILDING STRUCTURAL DESIGN HANDBOOK

BUILDING STRUCTURAL DESIGN HANDBOOK

Edited by

RICHARD N. WHITE
Professor of Civil and Environmental Engineering
Cornell University

CHARLES G. SALMON
Professor of Civil and Environmental Engineering
University of Wisconsin

A Wiley-Interscience Publication
JOHN WILEY & SONS
New York · Chichester · Brisbane · Toronto · Singapore

Library of Congress Cataloging in Publication Data:

Building structural design handbook.

"A Wiley-Interscience publication."
1. Structural design—Handbooks, manuals, etc.
I. White, Richard N. II. Salmon, Charles G.

TA658.3.B85 1986 624.1′771 86-15808
ISBN 0-471-08150-7

Printed in the United States of America

10 9 8 7 6 5 4 3 2 1

CONTRIBUTORS

David L. Adler, Senior Associate, Simpson Gumpertz & Heger Inc., Consulting Engineers, Arlington, Massachusetts

M. S. Agbabian, President, Agbabian Associates, Engineers and Consultants, El Segundo, California, and Fred Champion Professor of Engineering and Chairman, Department of Civil Engineering, University of Southern California, Los Angeles, California

Horatio Allison, Allison, McCormac & Nickolaus, Structural Engineers, Rockville, Maryland

Horst Berger, Dipl. Ing., P.E., Principal, Horst Berger Partners, New York, New York

Omer W. Blodgett, Consultant, The Lincoln Electric Company, Cleveland, Ohio

Richard E. Chambers, P.E., Principal, Simpson Gumpertz & Heger Inc., Arlington, Massachusetts

John V. Christiansen, Consulting Structural Engineer, 7799 Hansen Road, Bainbridge Island, Washington

Joseph P. Colaco, CBM Engineers, Inc., Houston, Texas

Stephen J. Condren, Formerly, Staff Engineer, Simpson Gumpertz & Heger Inc., Consulting Engineers, Arlington, Massachusetts

Daniel A. Cuoco, Vice President, Lev Zetlin Associates, Inc. & Thornton-Tomasetti, P.C., New York, New York

Bruce Ellingwood, Research Professor, Department of Civil Engineering, The Johns Hopkins University, Baltimore, Maryland

Melvin I. Esrig, Vice-President, Woodward-Clyde Consultants, Wayne, New Jersey

Robert D. Ewing, President, Ewing & Associates, Engineering Consultants, 28907 Doverridge Drive, Rancho Palos Verdes, California

David H. Geiger, Ph.D., P.E., Principal, Geiger Associates, P.C., New York, New York

Jacob S. Grossman, Vice-President, Robert Rosenwasser Associates, P.C., Consulting Engineers, New York, New York

Robert W. Haussler, Consulting Structural Engineer, P.O. Box 669, Templeton, California

Frank J. Heger, Senior Principal, Simpson Gumpertz & Heger Inc., Consulting Engineers, Arlington, Massachusetts

ROBERT J. HOYLE, JR., Mechanical & Structural Engineer, Lewiston, Idaho

JEROME S. B. IFFLAND, Iffland Kavanagh Waterbury, P.C., New York, New York

MOHAMMAD IQBAL, Chief Structural Engineer, Walker Parking Associates, Elgin, Illinois

SRINIVASA H. IYENGAR, General Partner, Skidmore, Owings & Merrill, Chicago, Illinois

LEONARD M. JOSEPH, Lev Zetlin Associates, Inc. & Thornton-Tomasetti, P.C., New York, New York

MILO S. KETCHUM, Ketchum, Konkel, Barrett, Nickel, Austin, Denver, Colorado

FRITZ KRAMRISCH, Dr. Eng., P.E., Retired Chief Civil Engineer, Albert Kahn Associates, Inc., Architects/Engineers, Detroit, Michigan

H. S. LEW, Leader, Structural Evaluation Group, Center for Building Technology, National Bureau of Standards, Gaithersburg, Maryland

PAUL T. LOVEJOY, Supervisor, Materials Processing, Allegheny Ludlum Steel Corporation, Research Center, Brackenridge, Pennsylvania

LEROY A. LUTZ, Vice-President, Computerized Structural Design, Milwaukee, Wisconsin

ROBERT F. MAST, President, ABAM Engineers Inc., Federal Way, Washington

JAMES L. NOLAND, Atkinson-Noland & Associates, Inc., Consulting Engineers, Boulder, Colorado

PARTNERS, Jaros, Baum & Bolles, Consulting Engineers, New York, New York

CLARKSON W. PINKHAM, S. B. Barnes and Associates, Consulting Structural Engineers, Los Angeles, California

LESLIE E. ROBERTSON, Chief Designer and Chairman of the Board, Leslie E. Robertson & Associates, New York, New York

CHARLES G. SALMON, Professor of Civil and Environmental Engineering, University of Wisconsin, Madison, Wisconsin

SAW-TEEN SEE, Partner and Vice-President, Leslie E. Robertson & Associates, New York, New York

REXFORD L. SELBE, Senior Consultant, Wiss, Janney, Elstner Associates, Northbrook, Illinois

EMIL SIMIU, Structural Engineer, Center for Building Technology, National Bureau of Standards, Gaithersburg, Maryland

JERRY G. STOCKBRIDGE, Vice-President, Wiss, Janney, Elstner Associates, Northbrook, Illinois

RICHARD L. TOMASETTI, Senior Vice President, Lev Zetlin Associates, Inc. & Thornton-Tomasetti, P.C., New York, New York

RICHARD N. WHITE, Professor of Civil and Environmental Engineering, Cornell University, Ithaca, New York

GEORGE WINTER, Professor of Engineering, Emeritus (Deceased 1982), Cornell University, Ithaca, New York

LORING A. WYLLIE, JR., H. J. Degenkolb Associates, Engineers, San Francisco, California

PREFACE

This handbook contains up-to-date information on preliminary design, analysis, and construction of a variety of *buildings*. The book has two primary purposes: (1) to enable practicing structural engineers to make the best decisions in the crucial early stages of building design when alternative schemes are being formulated and evaluated, and (2) to provide information needed for the final design which is not readily available in standard textbooks or other traditional references.

Structural engineering is a combination of art and science. This book documents the ideas of many experts regarding the decisions that must be made in the "real world" of structural engineering. The authors of the chapters are practicing engineers, and the chapters represent the experience of these authors in making the planning and design decisions based on that extensive accumulated experience. While the book does contain some traditional "fact" material, it is primarily an integration of fundamental engineering concepts, code requirements, and experience. It is the intent of the book to provide engineers with this "experience," in order to improve their designs and to minimize the long period of trial and error associated with becoming an expert in building structural design.

During the planning of chapters, selection of authors, and production of the manuscript, we always received enthusiastic response. Each author developed their own chapter in light of their own design experience, so that the scope of chapters tends to vary. We attempted to include all major topics of importance in the design of buildings; inevitably, since there are 30 chapters and even more authors, gaps in coverage will occur. We have tried to coordinate chapters so that no large gaps occur and that no excessive overlap occurs. Cross-referencing between chapters is extensive but not 100% complete, and we encourage the user to consult the index to best guarantee that all pertinent information on a given topic is found. A good example is the effect of lateral loads on buildings, including sway, the P–Δ effect, and related topics; these subjects appear in many chapters in the book.

While the major intended users will be those in design offices, the book will be found useful in advanced undergraduate and graduate level structural design courses in universities. It should also be useful to architects and other members of the building design team who want to better understand the concerns and approaches of the structural engineer.

The first three chapters provide treatment of the general design process, loads, and design philosophies. In particular, Chapter 1 details how the remaining chapters relate to the overall design process for buildings. Chapters 4 and 5 on mechanical and electrical system requirements and on vertical transportation show how these all-important nonstructural items interrelate with the structure itself. The properties and special characteristics of construction materials used in buildings are presented in the various chapters devoted to structures made of those materials. However, since welding affects material properties and is such a common method of connecting parts of many buildings, Chapter 6 is devoted to welding-related considerations. Structural walls and diaphragms are covered in Chapter 7. Function and form for buildings are treated in Chapter 8.

The largest portion of the book, and perhaps its most unique feature, is the group of chapters (9 through 16) devoted to preliminary design. Included are such topics as preliminary design of low-rise buildings, high-rise buildings, single-story open-space buildings, shells and folded plates, space trusses and frames, membrane and air supported structures, and building foundations. This information should prove to be valuable in helping structural designers arrive at better solutions more quickly.

Chapter 17 relates the computer to design practice. Chapter 18 provides basic information regarding dynamic loading and its effects on buildings. Chapters 19 through 23 on structural steel, cold-formed steel, prestressed concrete, reinforced concrete, and composite construction extend traditional textbook coverage of each of these topics into the real design world. Special situations not covered in the classroom, such as choice of structural system, constructability, efficiency and economy, stiffness assumptions, deflections, and long-term deformations are discussed in these chapters. It is assumed that the reader has previously acquired basic concepts relating to these five chapters; thus, such material is covered lightly if at all.

Other special features of this book are the more complete treatment on subjects not so readily available to the typical structural engineer. Chapters 24 and 25, on masonry and wood structures, present the subject assuming that the reader does not have previous knowledge. Chapter 26 on roofs and roofing and Chapter 27 on building facades are on critical subjects relating to building design and performance; this unique treatment is not usually covered in books. Chapters 28 on aluminum structures, 29 on stainless steel, and 30 on structural plastics provide basic concepts for design in these materials.

We gratefully acknowledge the effort (and in many cases the patience) of the chapter authors, all of them busy prominent consulting engineers who made their contributions to this handbook because they were convinced it was a worthy task. Their generous sharing of experience is in the highest tradition of professional engineering, and the cooperation we received from each of the authors in bringing this project to fruition is sincerely appreciated.

Many people at John Wiley & Sons played key roles in the development, editing, and production of the book, including Thurman Poston (who helped inspire us to undertake the project), Valda Aldzeris, Robert Golden, Everett Smethurst, Rose Ann Campise, and Glenn Curry. We are indebted to these individuals as well as to the many others "behind the scenes" at John Wiley. We also acknowledge the understanding and encouragement of our wives, Marge White and Bette Salmon, without which the project would not have been completed.

RICHARD N. WHITE
CHARLES G. SALMON

Ithaca, New York
Madison, Wisconsin
October 1986

CONTENTS

BUILDING STRUCTURAL DESIGN HANDBOOK

CHAPTER 1

INTRODUCTION TO THE HANDBOOK

CHARLES G. SALMON

Professor of Civil and Environmental Engineering
University of Wisconsin
Madison, Wisconsin

RICHARD N. WHITE

Professor of Civil and Environmental Engineering
Cornell University
Ithaca, New York

1.1 PURPOSE

This handbook has two primary purposes: (1) to enable the practicing structural engineer to make the best decisions in the crucial early stages of building design when alternative designs are being explored; and (2) to provide guidance and information needed for final design which are not readily available elsewhere.

This book does *not* cover member design topics usually available in standard textbooks on structural steel, reinforced concrete, and prestressed concrete. Instead the focus is on preliminary design and the special concerns that are usually beyond the scope of traditional academic classroom courses.

The authors of the various chapters were selected because of their expertise as designers in the "real" world. The documentation of how the design decisions are actually made, from the preliminary design stage to the final construction, was the objective of each of the authors. It is hoped that a sufficient variety of building design topics has been treated so that the relative novice can gain the needed experience by studying the appropriate chapters of this book.

1.2 STRUCTURAL DESIGN

Structural design has been defined* as

> *. . . a mixture of an art and science, combining the experienced engineer's intuitive feeling for the behavior of a structure with a sound knowledge of the principles of statics, dynamics, mechanics of materials, and structural analysis, to produce a safe economical structure which will serve its intended purpose.*

In the early years before 1850, structural design was essentially an art, with reliance on previous experience and intuition to determine the size and arrangement of the structural elements. Early man-made structures essentially conformed to those that could be observed in nature, such as beams and arches. As the principles governing the behavior of structures and structural materials have become better understood, design procedures have become more scientific.

Computations involving scientific principles should usually be viewed as a guide to decision making and not be followed blindly. The art or the intuitive ability of the experienced engineer is used to make decisions, *guided* by computational results.

Design is an iterative process the end objective of which is an optimum solution; in the context of this book, that solution is the appropriate building structure. In a building design, criteria must be established to ascertain whether or not an optimum is obtained. Typical criteria may be: (1) minimum cost; (2) minimum weight; (3) minimum construction time; (4) minimum maintenance (or maximum life); (5) minimum labor cost; (6) minimum cost of manufacture of owner's products; and (7) maximum efficiency of operation to owner, including energy efficiency.

Usually several criteria are involved, each of which may require weighting. Having an understanding of the many possible design criteria, the experienced designer understands that setting precise, clearly measurable criteria (such as weight or cost) for establishing an optimum is difficult, perhaps even impossible. Thus, *experience* becomes the crucial attribute to evaluate whether or not an optimum is obtained.

If an objective criterion can be mathematically expressed, optimization techniques may be used to obtain a maximum or minimum for the objective function. Optimization procedures and techniques comprise an entire subject that is outside the scope of this handbook. Here the "optimum" is that which each author of a chapter has determined from "experience." Since many design criteria are not easily quantifiable, the documentation of how eminent consulting engineers actually determine what is the "optimum" is intended to supplement the "fact" treatment available in standard textbooks.

The cost of the structural system in a modern building is usually but a small fraction of the total project cost, and the prudent (and successful) structural engineer always keeps this fact of life in mind. For example, design decisions that reduce the cost of the structure but greatly increase the costs of providing utilities are not in keeping with overall optimization. Similarly, choosing a structural system that stretches out the construction time may increase project costs even if the structure itself has the lowest net cost. Hence the structural engineer must be fully aware of the other requirements for the total building, and work closely with the owner, the architect, mechanical and electrical engineers, and potential constructors, in arriving at the final design of the structural system.

---

* Charles G. Salmon and John E. Johnson, *Steel Structures: Design and Behavior,* 2nd ed. New York: Harper & Row, 1980, p. 1.

1.3 DESIGN PROCEDURE

The design procedure may be considered as two parts—functional design and structural framework design. Functional design is the design that ensures that the intended results are achieved, such as: (1) adequate and properly arranged working areas and clearances; (2) proper ventilation and/or air conditioning; (3) adequate transportation facilities, such as elevators, stairways, and cranes or materials-handling equipment; (4) adequate lighting; (5) architectural attractiveness; and (6) energy efficiency.

Structural framework design is determining the structural form, establishing the loads acting on the structure, making preliminary selection of member sizes, establishing the mathematical model of the structure and doing the structural analysis, and selecting the final arrangement and sizes of the structural elements and connections, so that the structure and its elements will have adequate strength and be properly serviceable (i.e., no problems with items such as large deflections, cracking of nonstructural elements, excessive noise transmission, and/or unacceptable vibration).

The iterative procedure of design may be outlined as follows:

1. *Planning*. Establishment of the functions for which the structure must serve. Set criteria against which to decide whether or not the resulting design is an optimum.
2. *Preliminary configuration of structure*. Arrange the structural elements (and establish the model for structural analysis) to serve the functions in step 1. Material in Chapter 8 (Structural Form) should be of substantial help in arriving at the overall structural configuration; this phase of the design is normally done in close collaboration with the architect.
3. *Model the loads*. Establish the loads to be carried and identify the model to be used in structural analysis.
4. *Preliminary member selection*. Based on the decisions of steps 1, 2, and 3, select the member sizes to satisfy the objective criteria as established in step 1.
5. *Structural analysis*. Perform the analysis of the structural model acted upon by the modeled loads.
6. *Refined member design*. Check the adequacy of the members selected in the preliminary design; check the strength and serviceability.
7. *Evaluation and redesign*. Are all the requirements satisfied and is the result optimum? Compare the result with the predetermined criteria. If not optimum, repeat any of the previous steps.
8. *Final decision*. Evaluate whether or not the optimum has been achieved.

1.4 LOADS

Many of the loads to be used in building design are prescribed by regional legal requirements with which the designer must become familiar. Chapter 2 contains a brief treatment of typical requirements for live loads in buildings. However, a major feature of Chapter 2 is the extensive treatment of wind and snow loads. Often the design of low-rise buildings is such that a simplified treatment of wind using a statically equivalent lateral force is permitted by codes. However, when the shape of the building is unusual, the building is tall, or the roof system or building cladding is susceptible to wind damage, the treatment of wind as a dynamic force causing pressure, suction, uplift, and drag must be considered. Chapter 2 contains detailed treatment to permit designing buildings for this dynamic action, and the special considerations on wind loadings for tall buildings are treated in Chapters 10 and 11.

Similarly to wind, snow is often treated in a simplified fashion as a uniformly distributed load on the roof, exactly as a dead load. However, the deposition of snow rarely is uniformly distributed; thus, the factors affecting the magnitude and distribution of snow loading are treated in detail in Chapter 2.

1.5 DESIGN PHILOSOPHIES

Since design in a given material must conform to the local code, and such codes usually require design in accordance with a "Standard" as developed by a recognized national organization, such as the American Concrete Institute, the American Institute of Steel Construction, or the National Forest Products Association, the design philosophy must be as set forth in the Standard.

Some of the codes (specifications) adopted by local or state ordinance provide alternative philosophies of design, such as the working stress method using service loads or the strength design method using factored loads. These design philosophies are given different names by the national organizations; however, they generally may be divided into those two categories. The recognized design philosophies are discussed in Chapter 3.

1.6 LATERAL LOAD RESISTANCE

Designing a building for lateral load resistance is one of the most important functions performed by the structural engineer, particularly for high buildings and in regions of substantial wind or seismicity. The goal of the designer is to minimize the additional cost of the framing needed to resist lateral loads, above and beyond the framing needed to carry gravity loads.

This design requires considerable insight into the stiffness of different system configurations as well as a keen understanding of how framing, walls, and diaphragms function in complicated three-dimensional assemblages. The importance of lateral load resistance is indicated by the large number of chapters containing material on lateral loads, on behavior of lateral load resisting systems, and on preliminary design of these systems: included are Chapters 2, 3, 7, 8, 10, 11, 18, 19, 21, 22, 23, and 25. Interaction of the lateral load resisting framing and facades is discussed in Chapter 27, and the influence of lateral drift on design of vertical transportation is treated in Chapter 5.

1.7 STRUCTURAL STEEL BUILDINGS

Steel structures are treated in several places throughout the handbook. Chapter 19 (Steel Structures—Special Considerations) should be considered the next higher level of topics after a traditional textbook* used in, say, two semesters of academic study. In that chapter general economic considerations, serviceability criteria, and stability and design of frames are emphasized. The AISC (American Institute of Steel Construction) Specification is used as the referenced set of design rules and brief treatment is provided of member design, including laterally unbraced, cantilever, and composite beams; compression members, including beam-columns; and connections.

1.7.1 Cold-Formed Steel Structures

While the steel used for the major supporting elements is usually hot-rolled steel covered under the AISC Specification, many building components consist of cold-formed steel shapes covered under the American Iron and Steel Institute (AISI) Specification. The special concerns and behavior unique to the use of elements thinner than those covered by AISC are treated in Chapter 20. Though cold-formed steel may have a thickness of up to 1 in. (25 mm), the common use of this material is for material less than $\frac{3}{16}$ in. (8 mm) thick, referred to as "light gage" steel.

The development of the AISI Specification and supporting documents occurred largely through the effort and guidance of the author of Chapter 20, the late George Winter of Cornell University. The various AISI design guides are considered to be indispensable in designing cold-formed steel structures and it is assumed that the reader has access to these references. Cold-formed steel decks used as roofing systems are treated in Chapter 26 (Roofs and Roofing), and composite metal–concrete decks are discussed in Chapter 23 (composite construction).

1.7.2 Preliminary Design

Preliminary design of steel structures is discussed in Chapter 19. In addition, Chapter 9 (Preliminary Design of Low-Rise Buildings), Chapter 10 (Tall Buildings—Load Effects and Special Design Considerations), Chapter 11 (Preliminary Design of High-Rise Buildings), Chapter 12 (Preliminary Design of Single-Story Open-Space Buildings), Chapter 14 (Preliminary Design of Space Trusses and Frames), and Chapter 23 (Composite Construction) all contain preliminary design methods for steel structures.

1.7.3 Braced and Unbraced Frames

The reader is advised to use Chapter 19 as the initial source for designing multistory braced and unbraced frames in steel. Of major concern are the determination of the effective length factor to use in evaluating frame stability, and for unbraced frames, the secondary bending effect, known as the P–Δ effect. These are treated in Chapter 19. The chapters relating to high-rise buildings, Chapters 10 and 11, both treat frame stability and the P–Δ effect.

---

* Charles G. Salmon and John E. Johnson, *Steel Structures: Design and Behavior,* 2nd ed. New York: Harper & Row, 1980.

1.7.4 Special Considerations

The design of steel structures relies on the proper and adequate treatment of connections. High-strength bolted connections, both bearing-type and friction-type, are treated in Chapter 19 as well as in Chapter 12. Welded connections give rise to metallurgy-related concerns. A special feature of this handbook is the detailed treatment of welded connections in Chapter 6 (Welding-Related Considerations in Building Design). The treatment of lamellar tearing which occurs in heavily restrained welded connections is covered in Chapter 19 and in more detail in Chapter 6.

1.8 REINFORCED CONCRETE BUILDINGS

Chapter 22 (Reinforced Concrete Design) emphasizes the special concerns of reinforced concrete buildings; problems relating to excessive deformation, including constructability and aesthetic integrity; simplified approach to stiffness assumptions, particularly related to lateral load resistance; and time-dependent deformation considerations. The chapter also includes discussion of economic considerations, compliance with codes, professional liability, and office practices relating to design of concrete buildings. Additional information on reinforced concrete also appears in the chapter on composite construction (Chapter 23).

1.8.1 Preliminary Design

Preliminary design treatment occurs primarily in Chapter 22, with lesser coverage in Chapter 7 (Structural Walls and Diaphragms—How They Function), Chapter 9 (Preliminary Design of Low-Rise Office-Type Buildings), Chapter 11 (Preliminary Design of High-Rise Buildings), Chapter 13 (Preliminary Design of Shells and Folded Plates), and Chapter 16 (Foundations—Preliminary Design). Cast-in-place concrete decks and precast-prestressed concrete decks used as roof systems are treated in Chapter 26 (Roofs and Roofing).

1.8.2 Deformations

A major feature of this book is the extensive practical treatment of deflections, relating to serviceability, constructability, and aesthetic integrity. The special coverage of construction-related deformation problems will be particularly useful. Treatment of dimensional changes, cladding-related considerations, and horizontal joints are topics rarely treated in the literature. All of these have detailed discussion in Chapter 22. A minimum thickness equation is presented to aid the designer in satisfying deflection limitations.

1.8.3 Stiffness Assumptions

Analysis of concrete structures requires an assumption of the stiffnesses of the members. Cracking and time-dependent deformations always affect the stiffness of members. Chapter 22 provides unique coverage of this important topic.

1.9 PRESTRESSED CONCRETE BUILDINGS

Chapter 21 reviews the fundamentals of prestressing and then moves on quickly into special topics of critical importance to the designer of prestressed concrete buildings, including framing scheme selection, lateral load resistance of precast systems, connections, and composite construction. Statically indeterminate members, curved beams, and special problems in post-tensioning are also treated. The coverage emphasizes both the differences and similarities between reinforced and prestressed concrete.

1.10 COMPOSITE CONSTRUCTION

The basic principle underlying composite construction is to utilize the most desirable attributes of several materials (usually structural steel and concrete) to their best advantage so that the combination may result in a higher order of structural efficiency and cost-effectiveness. Composite construction is treated in Chapter 23 as the interactive and integral behavior of concrete and steel components and systems. Composite members involve interactive behavior of concrete and steel in single, isolated, and specific elements of a structure (beams, columns, walls, trusses, and slabs). A composite system is an assemblage of reinforced concrete, structural steel, and/or composite members, resulting in a combination building system that is often called a "mixed steel–concrete system." Such systems have become particularly popular for high-rise buildings; coverage of four specific mixed systems is included in Chapter 23, along with connections for these systems.

1.11 WOOD BUILDINGS

All aspects of the design of wood structures are treated in Chapter 25. Included are not only the principal structural aspects of solid lumber, but also plywood and glued laminated members. Detailed treatment of the design of members and connections is included. Further, design of special wood structures such as arches, domes, diaphragms and shear walls, folded plates, hyperbolic paraboloids, and curved roof panels, is included. The design of wood structures is treated in the context of the design rules promulgated by the National Forest Products Association (NFPA) and the American Plywood Association (APA). Wood planks and plywood decks used as roofs are treated in Chapter 26 (Roofs and Roofing).

1.12 MASONRY BUILDINGS

The basic concepts of masonry construction are treated in Chapter 24 (Masonry). The chapter begins with the differences between concrete and masonry, and follows with the definition of the specialized terminology, the types of masonry units, the standard tests, and the basic properties relating to masonry. The design of masonry beams, walls, and columns is treated in detail.

Both the "empirical" design and the "engineered" design of masonry are included. In addition, masonry-related considerations are included in Chapter 27 (Building Facades), such as special design aspects of clay unit masonry, concrete unit masonry, and natural stone masonry.

1.13 OTHER MATERIALS IN BUILDINGS

In addition to structural steel, reinforced and prestressed concrete, wood, and masonry, buildings often contain other materials used for structural or partially structural elements. The structural design aspects of aluminum are treated in Chapter 28; stainless steel is treated in Chapter 29; and structural plastics are covered in Chapter 30.

1.13.1 Aluminum

Aluminum is used for building hardware, such as for sash on glass door frames and jambs, as well as for partially structural exterior cladding, and structural residential patio covers. Aluminum is also used for structural purposes when high strength is needed but the magnetic properties of steel cannot be tolerated. The properties of the more common structural aluminum alloys are presented in Chapter 28, as well as the structural design requirements, in accordance with the Specifications of the Aluminum Association.

1.13.2 Stainless Steel

Stainless steel is a material choice in many of the same situations where the alternative might be aluminum. Corrosion resistance and appearance are the principal reasons for using stainless steel. Chapter 29 contains general information about the common stainless steel alloys, including the factors affecting corrosion resistance, mechanical properties, joining techniques, surface appearance, and fabrication techniques.

1.13.3 Structural Plastics

The use of plastics for structural or semistructural purposes occurs for interior drain waste and vent lines, water supply, sewer, foundation drainage, and electrical conduits. Vessels for containment of industrial corrosive products and structural plastic foam cores for structural sandwich panels faced with metal skins are two of the more strength-related uses of plastics. Chapter 30 presents data on structural reinforced and unreinforced plastics, and treats the strength and stiffness properties of some of the structural plastics. In addition, the design of plastic thin-walled members, structural sandwich members, box sections, and rings and shell structures is included.

1.14 SHELL STRUCTURES

Analysis of shell structures is complex and time consuming, even when done with the aid of a computer; hence, reasonably accurate preliminary design procedures are particularly important to the structural designer. Chapter 13 contains a number of these practical design approaches for proportioning folded plates, barrel shells (both long and short), several types of hyperbolic paraboloid (hypar) shells, domes of revolution (including ring beam design), translational shells, and funicular shells. The material in this chapter also provides the engineer with considerable insight into shell behavior.

Design of folded plates, hypars, and curved panels constructed of wood is treated in Chapter 25.

1.15 CABLE SUPPORTED AND AIR SUPPORTED SYSTEMS

Chapter 15 treats the preliminary design of air supported and cable supported structures. The structural principles involved are presented along with the loads and materials typically used. The presentation includes one-way and two-way cable systems. In addition, the chapter includes saddle surface tension structures, arch supported membrane structures, and point-supported membrane structures.

1.16 BUILDING SERVICE FACILITIES

1.16.1 Mechanical and Electrical Systems

The design of buildings must include the structurally-related aspects of the interior facilities, such as the heating, ventilating, air conditioning, and vertical transportation systems. Chapter 4 treats the mechanical and electrical systems. Included are air and water heating and cooling systems, the loads and support system design requirements for the mechanical rooms or areas, and the duct systems and their supports. In addition, special treatment is given to isolating from the user areas of the structure the vibration caused by mechanical equipment. The relationship of expansion and control joints in the structure to the mechanical system requirements is also included.

1.16.2 Vertical Transportation

Chapter 5 presents the vertical transportation requirements in terms of both human and structural effects. Included are moving walks, accelerating moving walks, escalators, and elevators.

1.17 BUILDING CLADDING

1.17.1 Facades

The structural requirements for building facades, including aesthetics, loadings, thermal resistance, moisture resistance, fire resistance, durability, volumetric changes, and facade/structure interaction, are presented in Chapter 27. In addition, masonry-related considerations are included, such as special design aspects of clay unit masonry, concrete unit masonry, and natural stone masonry. Design considerations are also included for concrete facades, metal/glass facades, stucco facades, and metal stud back-up walls. A discussion of joints and sealants is also provided.

1.17.2 Roofs

Of all problems faced by owners of buildings, leaky roofs are perhaps the most common. There has been a lack of information on how to successfully design, detail, and construct a roof that will give satisfactory performance for the specified design life. The detailed treatment of all aspects of building roofs given in Chapter 26 is intended to provide the structural engineer a firmer basis for design. The chapter contains treatment of the design considerations affecting roofing, energy conservation principles and related insulation procedures, various types of roof decks and support systems, roof system selection principles, and detail of various types of roofing. Materials used in roofs are presented in detail, including polymeric and modified bituminous sheet roofing, liquid-applied polymeric roofing, shingles, slate, and tile roofing. Procedures for construction are presented including contract documents to be used, actual construction, inspection, testing, and finally maintenance and long-term performance.

1.18 DYNAMIC BEHAVIOR OF BUILDINGS

Chapter 18 (Dynamic Loading—Concepts and Effects) contains the basic principles relating to the response of buildings to dynamic loads, such as wind, earthquake, or vibration-inducing equipment. The chapter contains basic concepts of the dynamic characteristics of structures and structural components, including degrees of freedom, damping, natural frequency, and natural period. Treatment is presented of how to model the structure to perform a dynamic analysis. Finally, the procedure of performing an analysis is presented, including what to look for, and how to interpret the results. The specifics relating to wind and earthquake loading are presented in Chapter 2 (Loads).

1.19 COMPUTER-AIDED AND INTERACTIVE DESIGN

Chapter 17 (Computer-Aided Analysis and Design) contains guidance on how the computer should be incorporated into the design of buildings, including use of computers for preliminary design, final analysis, and evaluating the output. Discussion is also provided on what is good design practice relating to the computer, incorporating computer results into design calculations, and documenting computer analysis and design.

1.20 FOUNDATION DESIGN

The foundation of a building may be invisible to the public and to the owner, but it is a critical part of the total building system and deserves careful attention in any building design. Chapter 16 (Foundations—Preliminary Design) is written to aid the structural engineer in selecting, for purposes of preliminary design, a foundation system appropriate for a particular structure. The chapter begins with the development of an understanding of the many factors affecting the choice of a foundation system. This is followed by a description of the many available foundation systems and their suitability for each class of subsurface conditions. A formal protocol is presented to aid the engineer in the preliminary selection of a foundation system. This protocol is followed with information necessary for the preliminary design of the selected system or systems. Costs, construction sequencing, and the influence of construction on adjacent structures and facilities are incorporated into the selection process.

CHAPTER 2
LOADS

H. S. LEW

Leader, Structural Evaluation Group
Center for Building Technology
National Bureau of Standards
Gaithersburg, Maryland

EMIL SIMIU

Structural Engineer
Center for Building Technology
National Bureau of Standards
Gaithersburg, Maryland

BRUCE ELLINGWOOD

Research Professor
Department of Civil Engineering
The Johns Hopkins University
Baltimore, Maryland

Formerly Leader, Loads and Analysis Group
Center for Building Technology
National Bureau of Standards
Gaithersburg, Maryland

The accurate determination of the loads* that a structure or structural element may carry is not always possible. Even if loads are well known at one location in a structure, the distribution of load from element to element throughout the structure usually requires assumptions and approximations. This chapter is intended to help the designer establish the loads for which structures should be designed. Dead loads, live loads, wind loads, snow loads, earthquake loads, and load combinations are discussed.

2.1 DEAD LOADS

Dead load is a fixed-position gravity load, so-called because it acts permanently toward the earth when the structure is in service. The weight of the structure is a dead load, as are attachments to the structure such as pipes, electrical conduits, air-conditioning and heating ducts, lighting fixtures, floor covering, roof covering, and suspended ceilings; that is, all the items intended to remain in place throughout the life of the structure. Under normal circumstances, the dead load may contribute a significant fraction of the total load effect.

The dead load may be assumed to remain essentially constant throughout the life of the structure. However, when the original structure is altered, such as by making changes in floor, wall, and ceiling surfaces, and adding partitions, the dead load on the structure will change.

Determination of the dead load is not complicated once a set of drawings is available. In early stages of design, a reasonable estimate of the total weight of the structure can be made from a preliminary set of drawings. This information is needed for the foundation design and for preliminary calculation of the earthquake loads. However, experience has shown that contributions of nonstructural items to dead load frequently are not estimated accurately at the preliminary design stage. A correction to the initially estimated dead loads should be made in the final design with correct unit weights of the materials used for structural and nonstructural members.

Estimates of the dead loads on structural members usually are made on the basis of tributary areas appropriate to simply supported framing. The actual distribution of the dead load through the structural system is dependent on the continuity in framing as well as on the sequence of construction. A conservative approach in estimating the weights of trial members, such as using centerlines instead of clear dimensions in estimating volumes, could counterbalance errors that might be introduced in dead load estimates in the preliminary design.

2.2 LIVE LOADS

Gravity loads acting when the structure is in service, but varying in magnitude and location, are termed *live loads*. Examples of live loads are human occupants, furniture, movable equipment, vehicles, and stored goods. Snow and wind loads are excluded from this definition and are considered separately. Some live loads may be practically permanent, such as book stack loads in libraries; others may be highly transient. Because of the unknown nature of the magnitude, location, distribution, and density of live load items, realistic magnitudes and the positions of such loads are difficult to determine.

Because of the concern for public safety, live loads used in design are usually prescribed by state and local building codes. These loads are generally conservative and are selected on the basis of experience and practice; however, during the past 15 years, several live load surveys have been conducted to improve the data base [2.6, 2.27]. Whenever local codes do not apply, or do not exist, the provisions from one of several regional and national building codes and standards may be used. One such widely recognized standard is the *American National Standard Minimum Design Loads for Buildings and Other Structures* of the American National Standards Institute (ANSI) [2.6], from which some typical live loads are presented in Table 2.1.

---

* Throughout this chapter the SI units are the primary ones with the Inch–Pound units in parentheses.

Table 2.1 Minimum Uniformly Distributed Live Loads (Adapted from Ref. 2.6)

Occupancy or Use	Live Load psf	Live Load Pa*
1. Hotel guest rooms School classrooms Private apartments Hospital private rooms	40	1900
2. Offices	50	2400
3. Assembly halls, fixed seat Library reading rooms	60	2900
4. Corridors, above first floor in schools, libraries, and hospitals	80	3800
5. Assembly areas; theater lobbies Dining rooms and restaurants Office building lobbies Main floor, retail stores Assembly hall, movable seats	100	4800
6. Wholesale stores, all floors Manufacturing, light Storage warehouses, light	125	6000
7. Armories and drill halls Stage floors Library stack rooms	150	7200
8. Manufacturing, heavy Sidewalks and driveways subject to trucking Storage warehouses, heavy	250	12,000

* SI values are approximate conversions. 1 lb/sq ft = 47.9 Pa.

Live load when applied to a structure should be positioned to give the maximum effect, including partial loading, alternate span loading, or full span loading as may be necessary. The simplified assumption of full uniform loading everywhere should be used only when it agrees with reality or is an appropriate approximation. The probability of having the prescribed loading applied uniformly over an entire floor, or over all floors of a building simultaneously, is almost nonexistent. Most codes recognize this by allowing for some percentage reduction from full loading. For instance, for live loads of 100 psf or less ANSI [2.6] allows members having an influence area of 400 sq ft or more to be designed for a reduced live load according to Eq. (2.2.1), as follows:

$$L = L_0\left(0.25 + \frac{15}{\sqrt{A_I}}\right) \tag{2.2.1}$$

where L = reduced live load per sq ft of area supported by the member
L_0 = unreduced live load per sq ft of area supported by the member (from Table 2.1)
A_I = influence area, sq ft

The influence area A_I is four times the tributary area for a column, two times the tributary area for a beam, and is equal to the panel area for a two-way slab. The reduced live load L shall not be less than 50% of the live load L_0 for members supporting one floor, nor less than 40% of the live load L_0 otherwise.

The live load reduction referred to above is not permitted in areas to be occupied as places of public assembly and for one-way slabs, when the live load L is 100 psf or less. Reductions are permitted for occupancies where L_0 is greater than 100 psf and for garages only under special circumstances (ANSI-4.7.2) [2.6].

2.3 WIND LOADS

Wind loads depend on the wind environment and on the aerodynamic, dynamic, and aeroelastic behavior of the structure or element being designed. The material of this chapter includes a description of the wind environment and a general discussion concerning aerodynamic, dynamic, and aeroelastic behavior. In addition, information is presented on: (1) design wind pressures on building facades (see also Chapter 27) and roofs (see also Chapter 26); (2) the dynamic response of tall buildings (see also Chapters 10, 11, and 18); (3) the behavior of slender towers and stacks; (4) the behavior of low-rise buildings (see also Chapters 9 and 12); (5) the behavior of cooling towers; (6) the behavior of trussed frameworks and bridges; (7) the behavior of offshore structures; and (8) the behavior of nuclear power plants.

Wind Environment

Wind speeds within any given storm vary both in time and location. The patterns that govern this variability on the micrometeorological scale are referred to as the *structure of wind.* Wind speeds also vary from storm to storm. This variability can be described statistically to provide information on the *extreme wind climate* at any one location.

2.3.1 Structure of Wind

Dependence Of Wind Speeds Upon Averaging Time

The variation of wind speeds with time during any given storm is illustrated by the wind speed record of Fig. 2.1. Note that in Fig. 2.1 the highest wind speed averaged over a period of about 2 sec is approximately 29 mph, whereas the highest wind speed averaged over about 1 hr is only approximately 19 mph. This example shows that a statement concerning the highest wind speed in a storm is meaningful only if the corresponding average time interval is specified.

Most principal weather stations in the United States use equipment that automatically records wind speeds averaged over the time interval required for the passage over the anemometer of a column of air with a horizontal length of 1 mile. The highest wind speed recorded by that equipment during any one storm is known as the *fastest-mile* wind speed for that storm. The averaging time t,

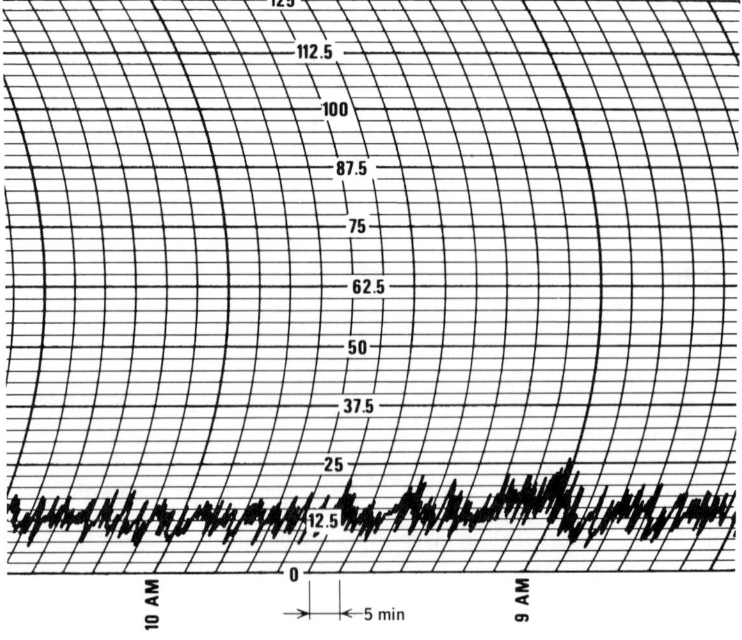

Fig. 2.1 Typical wind speed record.

Table 2.2 Coefficient $c(t)$ for Eq. (2.3.2)

$t(s)$	1	10	20	30	50	100	200	300	600	1000	3600
$c(t)$	3.00	2.32	2.00	1.73	1.35	1.02	0.70	0.54	0.36	0.16	0.00

in sec, for a recorded fastest-mile wind speed u_f, in mph, is

$$t = \frac{3600}{u_f} \tag{2.3.1}$$

Wind speeds are specified in the ANSI A58 Standard [2.6] in terms of fastest miles. The following approximate relation holds between the wind speed at elevation z, averaged over t sec, $u_t(z)$, and the corresponding wind speed $\bar{u}(z)$ averaged over 1 hr:

$$u_t(z) = \bar{u}(z)\left(1 + \frac{\sqrt{\beta}\, c(t)}{2.5 \ln \dfrac{z}{z_0}}\right) \tag{2.3.2}$$

where z_0 = roughness length. The coefficient $c(t)$ in Eq. (2.3.2) is listed in Table 2.2. The parameters z_0 and β are listed in Table 2.3 [2.25, 2.71]. For brevity, the hourly mean wind speed $\bar{u}(z)$ is commonly referred to simply as the mean wind speed.

Dependence of Mean Wind Speed on Height Above Ground

The following relationship holds over horizontally homogeneous terrain:

$$\bar{u}(z) = \bar{u}(z_{\text{ref}}) \frac{\ln \dfrac{z}{z_0}}{\ln \dfrac{z_{\text{ref}}}{z_0}} \tag{2.3.3}$$

Table 2.3 Micrometerological Parameters (From Refs. 2.6 and 2.11)

	Terrain Category			
	A*	B†	C‡	D§
z_0	2.50 m (10.76 ft)	1.00 m (3.28 ft)	0.07 m (0.23 ft)	0.005 m (0.016 ft)
β	4.00	4.85	6.00	6.00
r	0.41	0.62	1.00	1.27
α	1/3	1/4.5	1/7	1/10

Note: Depending upon the actual features of the terrain surface (e.g., extent to which surface is built up, type of vegetation, and so forth), actual parameter values may differ from the conventional, standard values listed in this table.

* Large city centers with at least 50% of the buildings having a height in excess of 20 m (70 ft).

† Wooded terrain and urban and suburban areas with numerous closely spaced obstructions having the size of single family dwellings or larger.

‡ Open terrain with scattered obstructions having heights generally less than 10 m (30 ft). This category includes flat, open country and grasslands.

§ Flat, unobstructed coastal areas directly exposed to wind flowing over large bodies of water.

where z_{ref} = reference height [commonly z_{ref} = 10 m (approximately 33 ft)], and z_0 = roughness length (see Table 2.3). Equation (2.3.3), a cornerstone of micrometeorology, is known as the *logarithmic law*. The following approximation to Eq. (2.3.3), known as the *power law*, is used traditionally in engineering applications:

$$\bar{u}(z) = \bar{u}(z_{ref}) \left(\frac{z}{z_{ref}} \right)^{\alpha} \tag{2.3.4}$$

where α is an exponent listed in Table 2.3 [2.6]. One may note that this approximation is close at some elevations, but less satisfactory at others.

Relation Between Mean Wind Speeds in Different Roughness Regimes

Mean wind speeds at a specified elevation are weaker over built-up terrain, and stronger over water, than those occurring within the same storm over open terrain. The following relationship holds

$$\bar{u}(10) = r\bar{u}^C(10) \tag{2.3.5}$$

where $\bar{u}(10)$ = mean wind speed at 10 m (33 ft) elevation over terrain with roughness length z_0; $\bar{u}^C(10)$ = mean wind speed at 10 m (33 ft) elevation over open (Category C) terrain; and r = coefficient listed in Table 2.3 [2.11].

Mean Wind Profiles Downwind of a Terrain Roughness Change

Downwind of a roughness change from smooth to rougher terrain mean wind speeds may be reduced as follows. Above a height $\delta_i \simeq x/12.5$, where x = downward distance from roughness change, the wind profile is approximately the same as upwind of the roughness change. Below elevation δ_i the profile is logarithmic, with zero speed at ground surface and a speed at elevation δ_i equal to the speed at that elevation upwind of the roughness change. For a distance of approximately $x < 500$ m ($\approx$ 1600 ft), it is prudent to assume $\delta_i = 0$. At distances $x > 5$ km ($\approx$ 3 miles), it may be assumed that a logarithmic profile is established which corresponds to the roughness length prevailing downwind of the roughness change [2.11].

Atmospheric Turbulence—Integral Scales of Turbulence.

Information on the features of atmospheric turbulence is required for the purpose of (1) estimating fluctuating loads due to wind gustiness, and (2) verifying the adequacy of laboratory simulations of atmospheric flows.

The integral scales of turbulence are measures of the average size of a turbulent eddy. Integral scales tend to increase with height above ground and appear to be larger over smooth than over rough terrain [2.20]. Data on integral scales of turbulence are listed in Counihan [2.20]. For example, at about 30 m ($\approx$ 100 ft) elevation near a shoreline, the longitudinal integral scale of turbulence L_u^x corresponding to the along-wind velocity fluctuations was found by measurement to be about 100 to 300 m (328 to 985 ft).

Failure to reproduce correctly the turbulence scale in wind tunnel tests may result in distorted modeling of the wind pressures. To ensure that such distortions are kept within acceptable bounds, the ANSI A58 Standard [2.6] requires the ratio between the longitudinal scales of turbulence in the wind tunnel and in the atmosphere to be at least one-third times the geometric scale for the model being tested.

Turbulence Intensity

The turbulence intensity is defined as the ratio between the root mean square (rms) of the turbulent wind speed fluctuations and the mean wind speed. The longitudinal turbulence intensity at elevation z may be expressed as

$$I(z) \simeq \frac{\sqrt{\beta}}{2.5 \ln \dfrac{z}{z_0}} \tag{2.3.6}$$

in which β and z_0 are listed in Table 2.3. The ANSI A58 Standard [2.6] requires that the intensity of turbulence be accounted for when modeling the natural wind flow in the laboratory.

Turbulence Spectra

Turbulence spectra provide a representation of the contributions of turbulence components with various frequencies to the mean square value of the turbulent fluctuations. They are of interest in situations where the turbulence induces fluctuating forces that cause significant resonant amplification effects on flexible structures.

The spectrum $S_u(z, n)$ of the longitudinal velocity fluctuations may be divided into two parts. In the higher frequency range

$$\frac{nS_u(z, n)}{\left[\dfrac{\bar{u}(z)}{2.5 \ln \dfrac{z}{z_0}}\right]^2} \simeq 0.26 f^{-2/3} \tag{2.3.7}$$

where n = frequency, and the nondimensional frequency $f = nz/\bar{u}(z)$. Equation (2.3.7) holds for $f >$ 0.2 or so. In the lower frequency range no description of the spectra based on first principles is available. However, spectral curves in this range can be developed [2.71] on the basis of the following requirements

$$S_u(z, 0) = 4[I(z)]^2 L_u^x \, \bar{u}(z) \tag{2.3.8}$$

and

$$\int_0^\infty S_u(z, n)\, dn = [\bar{u}(z) I(z)]^2 \tag{2.3.9}$$

Cross-Spectra of Longitudinal Velocity Fluctuations

The cross-spectrum of the velocity fluctuations at two points in space provides a measure of the extent to which the various frequency components of the velocities at those two points are correlated. The modulus of the cross-spectrum, also referred to as the square root of the coherence function, has the following empirical expression

$$\sqrt{C(r, n)} = e^{-\tilde{f}} \tag{2.3.10}$$

where

$$\tilde{f} = \frac{n[C_z^2(z_1 - z_2)^2 + C_y^2(y_1 - y_2)^2]^{0.5}}{\frac{1}{2}[\bar{u}(z_1) + \bar{u}(z_2)]} \tag{2.3.11}$$

where z_i, y_i $(i = 1, 2)$ are the coordinates of the two points in the vertical direction and the horizontal direction normal to the wind speed, respectively, and $C_z \simeq 10$, $C_y \simeq 16$ are empirical coefficients [2.22, 2.84].

2.3.2 Extreme Wind Climate

Traditional descriptions of the extreme wind climate do not take wind direction into consideration. In recent years, owing to the increased use of wind tunnel tests, data have become available, or can be obtained, on the dependence of aerodynamic effects on direction. To utilize these data, information is needed on the directional characteristics of the extreme wind climate.

Extreme Wind Speeds in Zones Not Subjected to Hurricane Winds—Winds Blowing from Any Direction

Extreme wind speeds $u_t^N(z)$, corresponding to an N-year mean recurrence interval (or, equivalently, to a probability of occurrence in any one year equal to $1/N$) may be estimated as follows [2.71] for $N \geq 20$ years:

$$u_t^N(z) \approx a + b \ln N \tag{2.3.12}$$

where

$$a = \bar{X} - 0.45s \tag{2.3.13a}$$

$$b = 0.78s \tag{2.3.13b}$$

[For the definitions of t and z, see Eq. (2.3.2).] In Eqs. (2.3.12) and (2.3.13), $\bar{X}$ and s are the sample mean and sample standard deviation, respectively, of the wind speed data, $u_{t,1}(z)$, $u_{t,2}(z)$, . . . , $u_{t,m}(z)$, representing the largest yearly wind speed during the first, second, . . . , mth year of record. The error in the estimation of $u_t^N(z)$ may be calculated as a function of sample size m [2.70]. It is desirable that $m > 20$ years.

In the absence of long-term records, it is possible to obtain respectable estimates of $u_t^N(z)$ from wind speed data, $u_{t,1}^m(z)$, $u_{t,2}^m(z)$, . . ., $u_{t,r}^m(z)$, representing the largest monthly wind speed during the first, second, . . ., rth month of record, where the length of the record can be as small as 3 years (i.e., 36 months). The expression for $u_t^N(z)$ is in this case

$$u_t^N(z) \approx a_m + b_m \ln(12N) \tag{2.3.14}$$

$$a_m = \bar{X}_m - 0.45s_m \tag{2.3.15a}$$

$$b_m = 0.78s_m \tag{2.3.15b}$$

where $\bar{X}_m$ and s_m are the sample mean and standard deviation, respectively, of $u_{t,i}^m(z)$ ($i = 1, 2, . . . , r$) [2.65]. Equations (2.3.12) through (2.3.15) represent Extreme Value Type I distributions with parameters estimated by the method of moments. (The latter is sufficiently precise for design purposes, as shown by Simiu and Scanlan [2.71].)

Extreme Wind Speeds in Zones Not Subjected to Hurricane Winds—Directional Extreme Wind Speeds

The probabilistic modeling of extreme wind speeds blowing from any direction, Eq. (2.3.12), is well established. However, no validated models of the joint probability distribution of the extreme wind speeds and direction are currently available. Nevertheless, procedures have been developed, which are described in Section 2.3.4, by which the needed design information can be obtained simply and rigorously from extreme yearly wind speed data recorded for each octant over a sufficiently large number m of consecutive years ($m > 20$). Such a procedure, applicable to structures or elements that do not exhibit dynamic amplification or aeroelastic effects, is described by Simiu [2.64] (see also Section 2.3.4). For a procedure applicable to structures that experience such effects, see Ref. 2.65. Directional data are listed for a number of U. S. principal weather stations in Ref. 2.15. For stations not included in Ref. 2.15, directional extreme wind speeds can be obtained from local climatological data (LCD) summaries issued by the National Climatic Center, Asheville, North Carolina. It was shown in Ref. 2.65 that errors resulting from use of LCD data are negligible for practical design purposes.

Extreme Wind Speeds in Hurricane-Prone Regions

The number of hurricane occurrences at any one station over a period of, say, 50 years is usually too small to allow statistical inferences to be made confidently from the corresponding hurricane wind speeds. For this reason, an approach was developed in Ref. 2.57 for estimating extreme wind speeds in hurricane-prone regions which makes use of statistics of meteorological and climatological characteristics of hurricanes, including atmospheric pressure defect, radii of maximum wind speeds within hurricane, storm translation speeds, and rate of occurrence of hurricanes in the entire region surrounding the station of concern. These characteristics can be related to the hurricane wind speeds at the site being investigated and to their probabilities of occurrence. Probabilistic estimates of hurricane wind speeds in the United States incorporated in the ANSI A58 Standard [2.6] were developed and are listed in Ref. 2.10 (in knots) on the basis of this approach. Corresponding hurricane directional wind speed data covering the entire Gulf and Atlantic Coasts are listed on tape available at the National Technical Information Service [2.47].

Climatology of Tornadoes

For most buildings, the cost of providing sufficient strength to withstand tornado effects is considered to be significantly higher than the expected loss associated with the risk of a tornado strike. For this reason, tornado-resistant design requirements are not included in building codes and standards, and the possibility of occurrence of tornadoes is not reflected in the respective design wind maps. However, the effects of a tornado strike must be explicitly taken into account in the design of structures important to the safety of nuclear power plants.

The probability that a tornado with maximum wind speeds higher than some specified value will strike a location in any one year is estimated on the basis of (1) the number of tornado occurrences within some specified geographical area surrounding that location, (2) data or models (inferred mostly from observations of damage) on tornado path areas, (3) maximum wind speeds, and (4) velocity distributions. According to estimates presented in Ref. 2.38, maximum tornado wind speeds corresponding to a 10^{-7} probability of occurrence in any one year vary from 400 mph in Oklahoma and Nebraska to 240 mph in Northern California and Oregon. According to Ref. 2.81, these wind speeds could be reduced significantly. Reference 2.5 specifies tornado speeds generally lower than those of Ref. 2.38 though higher than those proposed in Ref. 2.81.

Nuclear power plant design data on tornado wind speeds, pressure drops, and pressure drop rates in the United States are summarized in Ref. 2.38.

2.3.3 Aerodynamic, Dynamic, and Aeroelastic Behavior of Structures

Engineering aerodynamics is concerned with the study of loads arising from the relative motion of air with respect to a structure. Because aerodynamic loads induced by atmospheric flows are generally time-dependent, they may cause certain flexible structures (e.g., high-rise buildings, long-span bridges) to experience resonant amplification effects that must be determined by using the methods of structural dynamics (see also Chapter 18).

If the structure is sufficiently flexible, the relative motion of the fluid with respect to the body is due not only to the motion of the fluid, but to the flow-induced motion of the structure as well. A complex feedback mechanism thus occurs, whereby the aerodynamic forces that cause the structure to move are themselves modified by the motion of the structure. If this modification is such that the aerodynamic forces—and the motion they produce—increase strongly with time, an aeroelastic instability is said to occur. Aeroelasticity is concerned with the study of aerodynamic forces and structural motions significantly influenced by the feedback mechanism just described.

In the present state-of-the-art the wind flow around a bluff (nonstreamlined) body in the presence of turbulence cannot be determined from first principles. For this reason, virtually all information on wind-induced loads used in the design of buildings or other structures is based on measurements obtained in the wind tunnel. The question arises of the extent to which the various types of wind tunnel currently in use are acceptable for the purpose of modeling the full-scale (prototype) wind loads.

The answer to this question is not clear-cut. Certain tests have, even in recent years, been conducted with satisfactory results in virtually turbulence-free flows, for example, for the purpose of estimating mean loads on trussed frameworks [2.30], or the flutter velocity of suspended-span bridge decks. On the other hand, mean loads on buildings, towers, or stacks, buffeting effects due to incident flow fluctuations, across-wind effects due to vorticity shed in the wake of bodies, torsional effects, and local pressures, can be grossly misrepresented if based on smooth flow tests. When such effects are investigated it is therefore appropriate to test structures in wind tunnels in which the mean speed profiles and the features of turbulence are similar to those typical of atmospheric flows. Requirements concerning the longitudinal scale and the intensity of turbulence in the wind tunnel are specified in Section 6.4 of the ANSI A58 Standard [2.6] as mentioned previously. However, as evidenced by studies presented in Ref. 2.71, the extent to which various types of turbulent flow simulation are appropriate for various types of tests is not yet established conclusively.

Another question arises in connection with the fact that geometric model scales commonly used in wind tunnel tests are of the order of 1/100 or less. Inherent in such scales is a distortion of the Reynolds number by two or even three orders of magnitude, that is, the ratio between viscous and inertia forces is considerably larger in the wind tunnel than in the full-scale flow. It is believed that such differences are in most cases not consequential in the case of bodies with sharp edges. However, the violation of the Reynolds number may pose difficult problems concerning the modeling of loads on bodies with curved shapes, particularly when such loads are controlled by vortices shed in the wake of the body or are associated with aeroelastic effects. Some of these problems appear to have been solved satisfactorily, for example, in the case of hyperbolic cooling tower tests. However, unless interpreted with extreme care, wind tunnel tests of structures such as slender stacks with circular cross-section can lead to misleading representations of the wind loads [2.86].

Finally, the quality of wind tunnel load simulations depends on whether or not the instrumentation used is adequate. Instrumentation problems require special attention for fluctuating pressure measurements. Attempts to develop standard criteria for the validation of pressure measurement systems are currently in progress.

2.3.4 Wind Pressures

Wind pressures p measured in a wind tunnel may be expressed as

$$p(\alpha) = \frac{1}{2}\rho C_p(\alpha)\bar{u}^2(h, \alpha) \tag{2.3.16}$$

where ρ = air density; $\bar{u}(h, \alpha)$ = mean wind speed at a reference elevation h (usually h is the height of the building) corresponding to wind blowing from direction α; and $p(\alpha)$ and $C_p(\alpha)$ = pressure and pressure coefficient, respectively, corresponding to wind blowing from direction α. If $\bar{u}$ is expressed in mph and p in psf, then for Eq. (2.3.16), $p = 0.00256\ C_p\bar{u}^2$.

The following example shows how aerodynamic and climatological data can be combined to provide estimates of design wind pressures on a given element belonging to a structure with known orientation.

EXAMPLE 2.3.1 Illustrate how a designer may use data on largest yearly wind speeds and pressure coefficients to obtain largest yearly pressures.

Solution The largest mean wind speeds $\bar{u}(h, \alpha)$ recorded during a given year, and the pressure coefficients $C_p(\alpha)$ for a given element, have the values listed on lines 1 and 2 of Table 2.4.

The values of $p(\alpha)$ obtained by using these values in Eq. (2.3.16) are listed on line 3 of Table 2.4, and represent the largest pressures induced on the element by winds blowing from the direction α during the year being considered.

It is seen from Table 2.4 that the largest pressure induced during the year of concern by winds blowing from any direction is

$$p = \max[p(\alpha)] = 12.38 \text{ psf}$$

In traditional engineering practice, the nominal largest yearly wind pressure $p^{(n)}$ is assumed to be given by

$$p^{(n)} = \frac{1}{2}\rho \max_{\alpha}[C_p(\alpha)]\{\max_{\alpha}[u^2(h, \alpha)]\} \tag{2.3.17}$$

For the data of Table 2.4, $p^{(n)} = 0.00256(2.90)(61)^2 = 27.62$ psf, that is, the nominal value is considerably higher than the actual value of the largest yearly wind pressure, $p = 12.38$ psf.

If largest yearly wind speed data such as those listed on line 1 of Table 2.4 are available for a sufficient number of consecutive years (say, 20 years or more), then a set of largest yearly pressures, $p_1, p_2, \ldots, p_m$ can be obtained by using the procedure illustrated in Table 2.4. The estimated mean and standard deviation of the variates $\sqrt{p_1}, \sqrt{p_2}, \ldots, \sqrt{p_m}$ are denoted by $\bar{X}_{eq}$ and s_{eq}, respectively. The pressure p^N corresponding to an N-year mean recurrence interval (or with a probability of exceedance equal to $1/N$ per year) may then be estimated by applying Eqs. (2.3.12) and (2.3.13) to the variate $\sqrt{p}$, that is,

$$p^N \approx (a_{eq} + b_{eq} \ln N)^2 \tag{2.3.18}$$

where

$$a_{eq} = \bar{X}_{eq} - 0.45 s_{eq} \tag{2.3.19}$$

$$b_{eq} = 0.78 s_{eq} \tag{2.3.20}$$

Table 2.4 Largest Yearly Wind Speeds, Pressure Coefficients, and Largest Yearly Pressures

		Direction, α							
		N	NE	E	SE	S	SW	W	NW
1	$\bar{u}(h, \alpha)$ (mph)	38	28	27	31	25	61	42	34
2	$C_p(\alpha)$	0.42	1.21	2.90	0.72	0.79	1.30	1.47	0.41
3	$p(\alpha)$ (psf)	1.55	2.43	5.41	1.77	1.26	12.38	6.63	1.21

A similar procedure applicable to structures in hurricane-prone zones was developed in Ref. 2.66. Equations (2.3.18) through (2.3.20) can in principle also be used for the estimation of design pressures for buildings with unknown orientation [2.64]. The estimation of directional effects on design pressures is also treated in Refs. 2.79, 2.23, and 2.92. For a critique of the approaches used in Refs. 2.23 and 2.92, see Ref. 2.65.

2.3.5 Tall Buildings

Tall buildings subject to wind action respond in the along-wind direction, that is, the direction of the mean wind, in the across-wind direction, and in torsion.

Tall buildings are referred to as isolated if they are not affected significantly by the flow in the wake of other tall structures. Otherwise they are referred to as buildings subjected to interference effects. The latter are in many cases designed on the basis of ad hoc wind tunnel tests. Useful information on the response of such buildings can be found in Refs. 2.12, 2.49, 2.56, and 2.59.

Occupant comfort depends on the acceleration experienced by the building under the action of fluctuating loads. Various tentative occupant comfort criteria are summarized in Ref. 2.71.

The remainder of this section presents information on the dynamic response of isolated tall buildings. See also Chapters 10, 11, and 18 for more on the dynamic response of tall buildings.

Along-Wind Response

Experiments on square buildings indicate that the largest along-wind response occurs when the mean speed is normal to a building face. Computer programs are available for estimating the along-wind response of tall buildings [2.46]. The simplified procedure consisting of Eqs. 1 through 23 of Table 2.5 [2.72] may be used in lieu of a computer program, if modes higher than the fundamental one contribute negligibly to the response, and the fundamental mode shape is nearly linear.

Table 2.5 Equations for Estimating the Along-Wind Response of Buildings Having a Linear Modal Shape (from Ref. 2.72)

(1) $$Q = 2 \ln \frac{h}{z_0} - 1$$

(2) $$J = 0.78Q^2$$

(3) $$B = \frac{6.71Q^2}{1 + 0.26 \dfrac{b}{h}}$$

(4) $$u_* = \frac{\bar{u}(10)}{2.5 \ln \dfrac{10}{z_0}}$$

(5) $$\tilde{f}_1 = \frac{n_1 h}{u_*}$$

(6) $$C(x) = \frac{1}{x} - \frac{1}{2x^2}(1 - e^{-2x})$$

(7) $$x_1 = 12.32 \frac{\tilde{f}_1}{Q} \frac{d}{h}$$

(8) $$N(\tilde{f}_1) = C(x_1)$$

(9) $$C_{Df}^2(\tilde{f}_1) = C_w^2 + 2C_w C_\ell N(\tilde{f}_1) + C_\ell^2$$

(10) $$x_2 = 3.55 \frac{\tilde{f}_1}{Q}$$

Table 2.5 (*Continued*)

(11)	$M(z) = bd\rho_b(z)$
(12)	$R = 0.59 \dfrac{Q^2}{\zeta} \left[\dfrac{Q}{\tilde{f}_1}\right]^{2/3} \left[\dfrac{C_{Df}^2(\tilde{f}_1)}{C_D^2}\right] C(x_2) \left[\dfrac{1}{1 + 3.95 \dfrac{\tilde{f}_1 b}{Qh}}\right]$
(13)	$M_1 = \dfrac{1}{h^2} \int_0^h M(z) z^2 dz$
(14)	$q_* = \dfrac{1}{2} \rho u_*^2$
(15)	$\bar{X} = \left[\dfrac{C_D bhq_*}{M_1(2\pi n_1)^2}\right] J$
(16)	$\sigma_x = \left[\dfrac{C_D bhq_*}{M_1(2\pi n_1)^2}\right]\left[\dfrac{\beta B}{6} + R\right]^{1/2}$
(17)	$\nu_x = n_1 \left[\dfrac{R}{\dfrac{\beta B}{6} + R}\right]^{1/2}$
(18)	$g_x = [1.175 + 2 \ln (\nu_X T)]^{1/2}$
(19)	$G = 1 + g_x \dfrac{\sigma_x}{\bar{X}}$
(20)	$X = G\bar{X}$
(21)	$\sigma_{\ddot{x}} = \dfrac{C_D bhq_*}{M_1} R^{1/2}$
(22)	$g_{\ddot{x}} = [1.175 + 2 \ln (n_1 T)]^{1/2}$
(23)	$\ddot{X} = g_{\ddot{x}} \sigma_{\ddot{x}}$

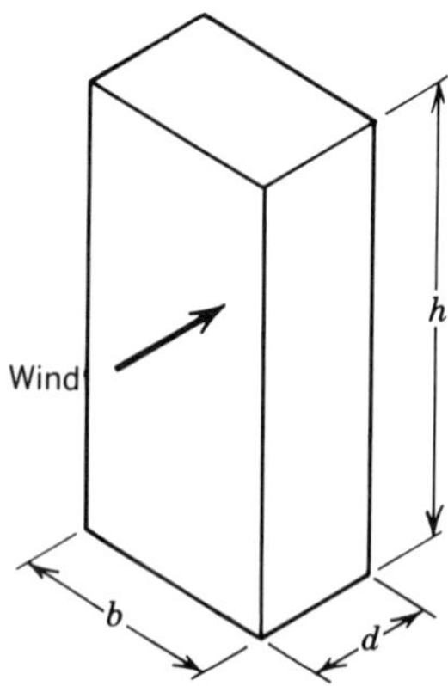

Fig. 2.2 Dimensions used in Table 2.5.

In Table 2.5, the terms are as follows:

b = across-wind dimension of the structure (Fig. 2.2)
C_D = drag coefficient ($C_D = C_w + C_\ell$)
C_w = average pressure coefficient on windward face of the building
C_ℓ = average pressure coefficient on leeward face of the building
d = along-wind dimension of the structure (Fig. 2.2)
G = gust response factor
h = the vertical dimension (shown in Fig. 2.2)
$M(z)$ = mass of building per unit height
n_1 = natural frequency in fundamental mode of vibration
T = duration of storm ($T \simeq 3600$ sec)
$\bar{u}(10)$ = hourly mean speed at 10 m (33 ft) above ground upwind of building
X = peak displacement at top of structure
$\bar{X}$ = mean displacement at top of structure
$\ddot{X}$ = peak acceleration at top of structure
z = height above ground
z_0 = roughness length upwind of building (in m) (see Table 2.3)
β = coefficient given in Table 2.3
ζ = damping ratio
ρ = mass of air per unit volume
$\rho_b(z)$ = bulk mass of building per unit volume
σ_x = rms of fluctuating part of displacement at top of structure
$\sigma_{\ddot{x}}$ = rms acceleration at top of structure

EXAMPLE 2.3.2 Evaluate the along-wind response of a building having $h = 200$ m (660 ft); $b = 35$ m (120 ft); $d = 35$ m (120 ft); $n_1 = 0.175$ Hz; $\zeta = 0.01$; $\rho_b = 200$ kg/m^3; $C_w = 0.8$; $C_\ell = 0.5$; and $C_D = 1.3$. The building is located in a town* (terrain Category B, see Table 2.3). It is assumed $\rho = 1.25$ kg/m^3, and that the fastest-mile wind speed at 10 m (35 ft) above ground in open (Category C) terrain is $u_f(10) = 78$ mph.

Solution The averaging time for the fastest-mile wind speed is 3600/78 = 46 sec from Eq. (2.3.1); from Table 2.2, $c(t) \simeq 1.42$, and from Eq. (2.3.2), the hourly mean speed at 10 m (33 ft) above ground in open (Category C) terrain is

$$\bar{u}^C(10) = \frac{78}{1 + \dfrac{1.42\beta^{1/2}}{2.5 \ln \dfrac{z}{z_0}}} \tag{2.3.21}$$

or $\bar{u}^C(10) = 62.3$ mph (27.8 m/s). From Eq. (2.3.5) and Table 2.3, the mean hourly wind speed at 10 m (33 ft) above ground in Category B terrain ($z_0 = 1$ m) is $\bar{u}(10) = 27.8(0.62) \simeq 17.2$ m/s. Then, referring to the equations of Table 2.5, $Q = 9.60$ (Eq. 1); $\underline{J} = 71.83$ (Eq. 2); $B = 591$ (Eq. 3); $u_* = 2.98$ m/s (Eq. 4); $\bar{f}_1 = 11.74$ (Eq. 5); $x_1 = 2.62$ (Eq. 7); $N(\bar{f}_1) = 0.31$ (Eqs. 6 and 8); $C^2_{Df}(\bar{f}_1) = 1.14$ (Eq. 9); $x_2 = 4.34$ (Eq. 10); $M(z) = 245{,}000$ kg (Eq. 11); $R = 353$ (Eq. 12); $M_1 = 16{,}333{,}300$ kg (Eq. 13); $q_* = 5.55$ kg/m/s^2 (Eq. 14); $\bar{X} = 0.184$ m (Eq. 15); $\sigma_x = 0.074$ m (Eq. 16); $\nu_x = 0.114$/s (Eq. 17); $g_x = 3.63$ (Eq. 18); $G = 2.46$ (Eq. 19); $X = 0.452$ m (Eq. 20); $\sigma_{\ddot{x}} = 0.058$ m/s^2 (Eq. 21); $g_{\ddot{x}} = 3.75$ (Eq. 22); $\ddot{X} = 0.218$ m/s^2 (Eq. 23).

Across-Wind Response

The across-wind motion, that is, the horizontal motion normal to the mean wind direction, is due primarily to vorticity shed in the wake of the structure. It may be assumed in practice that if the calculated rms of the across-wind oscillations at the tip of the structure exceeds a critical value σ_{ycr}, then aeroelastic effects become significant and cause a considerable increase of the across-wind response. Wind tunnel experiments [2.35, 2.54] suggest that it is prudent to assume $\sigma_{ycr}/b \simeq 0.015$, 0.025, and 0.045 in open, town, and city center, respectively (b = across-wind dimension of building).

---

* It is assumed that the terrain roughness is uniform over a distance upwind of at least 12.5h (see Section *Mean Wind Profiles Downwind of a Terrain Roughness Change*).

Table 2.6 Values $10^2\tilde{Y}$

		$\frac{n_1 b}{\bar{u}(h)}$						
Terrain	b/h	0.04	0.05	0.06	0.07	0.08	0.09	0.10
Urban*	1/9	4.7	5.7	7.2	9.7	12.0	17.0	23.0
	1/8.33	5.3† 5.3§	6.3† 6.5§	7.8† 8.1§	11.0† 11.0§	13.0† 13.0§	17.0† 14.0§	23.0† 12.0§
	1/6	8.4	9.7	11.0	13.0	15.0	19.0	21.0
	1/4	6.2† 7.9§	7.6† 9.9‡	9.3†	11.0† 16.0‡	12.0† 17.0‡	17.0† 17.0‡	15.0† 12.0‡
	1/3	3.9	5.5	7.1	8.2	9.1	14.0	10.0
Suburban*	1/18	3.6	4.1	5.0	5.7	8.4	11.0	22.0
	1/9	5.1	6.0	7.5	10.0	13.0	19.0	28.0
	1/6	5.3	6.9	8.9	12.0	16.0	21.0	23.0
	1/4	5.0†	6.2†	8.0†	11.0†	15.0†	18.0†	19.0†
	1/3	4.6	5.4	7.1	10.0	13.0	15.0	15.0
Open	1/4	6.2‡	7.1‡	8.7‡	13.0‡	17.0‡	28.0‡	29.0‡
All Terrain¶	1/3.4–1/7.1	56.0	39.0	30.0	24.0	20.0	16.0	14.0‡
All Terrain‖	1/3.4–1/7.1	29.0	22.0	17.0	14.0	12.0	10.0	8.9

* Results reported in Refs. 2.35 and 2.58, unless other noted.
† Values obtained by interpolation between results reported in Ref. 2.58 for $b/h = 1/6$ and $b/h = 1/3$, and $b/h = 1/9$ and $b/h = 1/6$.
‡ Results reported in Ref. 2.34.
§ Results reported in Ref. 2.53.

If $\sigma < \sigma_{ycr}$, the across-wind response corresponding to the case where the mean wind is normal to a building face can be estimated as

$$\sigma_y(h) \simeq \frac{\pi^{1/2}}{2\zeta^{1/2}} \frac{1}{(2\pi n_1)^2 M_1} \frac{1}{2} \rho \bar{u}^2(h) bh \tilde{Y} \tag{2.3.22}$$

and

$$\sigma_{\ddot{y}}(h) = (2\pi n_1)^2 \sigma_y \tag{2.3.23}$$

where the notations are similar to those of Table 2.5, $\bar{u}(h)$ = mean wind speed at elevation h, and $\tilde{Y}$ = factor determined empirically. For buildings with a uniformly distributed mass, a square shape in plan, and a fundamental modal shape that may be approximated by a straight line,

$$M_1 = \frac{1}{3} \rho_b bh^2 \tag{2.3.24}$$

and

$$\sigma_y(h) \simeq 0.0337 \left[\frac{\bar{u}(h)}{n_1 b}\right]^2 \frac{\rho}{\rho_b} \left(\frac{1}{\zeta^{1/2}}\right) b\tilde{Y} \tag{2.3.25}$$

Values of $\tilde{Y}$ based on measurements reported in Refs. 2.34, 2.35, 2.53, and 2.58, and implicit in Refs. 2.45 and 2.85, are given in Table 2.6. Note that considerable differences exist among values of $\tilde{Y}$ obtained from various sources (e.g., for urban terrain, $nb/\bar{u}(h) = 0.105$, and $b/h = 1/8.33$, $\tilde{Y} = 10$ according to Ref. 2.53, and $\tilde{Y} = 23$ on the basis of data from Ref. 2.58). Studies reported in Ref. 2.53 indicate that, for buildings having a square shape in plan, the across-wind response corresponding to cases where the wind is not normal to a building face is less than that given by Eq. (2.3.22).

EXAMPLE 2.3.3 Compute the across-wind response for the building of Example 2.3.2 when the building is assumed to be acted on by wind corresponding to a fastest-mile speed at 10 m (33 ft) above ground in open (Category B) terrain, $u_f(10) = 78$ mph.

Solution As shown previously, the mean hourly wind speed at 10 m (33 ft) above ground in Category B terrain is $\bar{u}(10) = 17.2$ m/s, or $\bar{u}(200) = 17.2[\ln(200/1.00)]/[\ln(10/1.00)] = 39.4$ m/s from Eq. (2.3.3). Then $n_1 b/u(h) = 0.155$; $b/h = 1/5.7$; $\tilde{Y} \simeq 0.075$ from Table 2.6; $\sigma_y = 0.23$ m from Eq. (2.3.25); $\sigma_{\ddot{y}} = 0.28$ m/s^2 from Eq. (2.3.23). Assuming that the peak factors are $g_y = 3.5$, $g_{\ddot{y}} = 4.0$, it

				$\frac{n_1 b}{\bar{u}(h)}$				
0.105	0.115	0.125	0.14	0.16	0.18	0.20	0.25	0.30
24.0	19.0	13.0	9.0‡	6.8	5.7	5.2	4.5	3.9
23.0† 10.0§	19.0† 7.7§	14.0† 6.3§	10.0† 5.3§	6.9† 4.7§	6.3† 3.9§	5.7† 3.6§	4.7† 3.2§	3.8† 2.9§
21.0	20.0	18.0	15.0	12.0	10.0	8.2	5.6	3.5
15.0† 11.0‡	15.0† 9.5§	13.0† 8.8‡	12.0† 7.8‡	9.6† 6.8‡	13.0† 5.9‡	6.8† 5.7‡	4.9† 4.9‡	3.4† 4.0‡
10.0	9.5	9.0	8.4	7.2	6.5	5.5	4.2	3.3
29.0	22.0	10.0	6.0	3.6	2.8	2.4	2.0	1.7
29.0	13.0	7.8	6.0	4.5	3.5	2.8	2.6	2.3
23.0	20.0	14.0	10.0	7.7	6.1	4.5	2.8	1.7
18.0†	16.0†	12.0†	9.5†	7.4†	5.9†	4.7†	3.7†	2.1†
13.0	12.0	10.0	8.9	7.1	5.7	4.9	3.5	2.6
28.0	11.0‡	8.7‡	7.5‡	6.6‡	5.9‡	5.5‡	4.8‡	4.0‡
13.0	11.0	10.0	8.5	7.0	5.8	5.0	3.6	2.7
8.3	7.1	6.6	5.9	4.9	4.1	3.6	2.8	2.1

¶ Values implicit in Ref. 2.85. Experimental values may differ from tabulated values by ±40%. Application restricted to buildings with density of the order of 200 kg/m^3 and damping ratio of the order of 1%. According to Ref. 2.85, these values may be applied to building with cross-sections other than square, i.e., triangular, rectangular, and circular.

‖ Values implicit in Ref. 2.45.

follows that the peak across-wind response and acceleration are $Y = 0.805$ m and $\ddot{Y} \simeq 1.12$ m/s^2. These values are larger than the corresponding values of the along-wind response calculated previously, that is, $X \simeq 0.452$ m and $\ddot{X} \simeq 0.22$ m/s^2.

Torsional Response

Wind-induced torsional effects occur because the center of mass and/or the elastic center do not coincide with the instantaneous point of application of the aerodynamic loads. Although ad hoc tests simulating torsional effects have been conducted in wind tunnels for a number of years, it is only recently that work has been performed toward the development of design information and procedures [2.71]. According to Ref. 2.53, wind-induced torsional moments on a square building are largest when the mean wind is normal to a building face, and may cause column displacements in slender buildings on the order of 50% of those associated with the along-wind response.

Tuned Mass Dampers

A tuned mass damper (TMD) is a system consisting essentially of a mass, a spring, and an energy dissipation device (dashpot). Its function is to reduce the amplitude of response oscillations of a body to which it is attached (see Fig. 2.3).

The effect of the TMD may be viewed as changing the damping of the structure from the value ζ_1, which would prevail in the absence of the TMD, to an increased effective damping value ζ_e. To the extent that the reliability of the TMDs is uncertain, TMD effects are counted on for serviceability (occupant comfort) purposes only, and are disregarded in structural design.

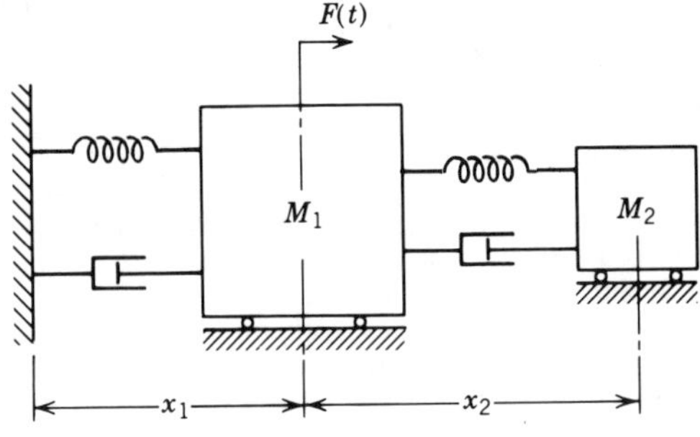

Fig. 2.3 Tuned mass damper.

The generalized mass of a building, the natural frequency, and the damping ratio in the fundamental mode of the building are denoted by M_1, n_1, and ζ_1, respectively, and the mass, natural frequency, and damping ratio of the TMD system are denoted by M_2, n_2, and ζ_2, respectively. Using results from Refs. 2.21 and 2.40, it follows that

$$\zeta_e = \frac{1}{2}\frac{\alpha_1(\alpha_2\alpha_3 - \alpha_1) - \alpha_0\alpha_3^2}{\dfrac{\beta_0^2}{\alpha_0}(\alpha_2\alpha_3 - \alpha_1) + \alpha_3(\beta_1^2 - 2\beta_0) + \alpha_1} \tag{2.3.26}$$

where

$$\alpha_0 = f^2 \tag{2.3.26a}$$

$$\alpha_1 = 2f(\zeta_1 f + \zeta_2) \tag{2.3.26b}$$

$$\alpha_2 = 1 + f^2(1 + \mu) + 4f\zeta_1\zeta_2 \tag{2.3.26c}$$

$$\alpha_3 = 2\zeta_1 + 2\zeta_2 f(1 + \mu) \tag{2.3.26d}$$

$$\beta_0 = f^2 \tag{2.3.26e}$$

$$\beta_1 = 2\zeta_2 f \tag{2.3.26f}$$

$$\mu = M_2/M_1 \tag{2.3.26g}$$

$$f = n_2/n_1 \tag{2.3.26h}$$

For example, if $\mu = 0.01$, $f = 0.98$, $\zeta_1 = 0.01$, and $\zeta_2 = 0.0515$, then $\zeta_e = 0.03226$. From Eq. (2.3.26) it is possible to obtain the optimum (largest) value of ζ_e, denoted by ζ_e^{opt}, consistent with any given set of values ζ_1, f, μ. Corresponding to ζ_e^{opt} there is a value of ζ_2 denoted by ζ_2^{opt}.

Let the displacement of the TMD mass with respect to mass M_1 be denoted by x_2. It can be shown that the mean square value of x_2 can be expressed as

$$\overline{x_2^2} = \frac{2\zeta_1\alpha_1}{\alpha_1(\alpha_2\alpha_3 - \alpha_1) - \alpha_0\alpha_3^2}\overline{x_{1,0}^2} \tag{2.3.27}$$

where $\overline{x_{1,0}^2}$ = variance of wind-induced displacement of mass M_1 *due to resonant amplification effects only*. It is shown in Ref. 2.40 that the order of magnitude of x_2 may be as high as 1 m (3 ft) in practical situations.

2.3.6 Slender Towers and Stacks

Slender towers and stacks are designed to withstand the effects of both along-wind and across-wind loads.

Along-Wind Response

The along-wind response can be estimated by using the procedures referred to in Section 2.3.5 (Tall Buildings, Along-Wind Response), in which appropriate values for the pressure coefficients C_ℓ and C_w are used. In the case of structures having a circular shape in plan, it may be assumed for design purposes that $C_\ell \simeq 1.2$, $C_w \simeq 0$ (for Reynolds numbers Re $\approx < 3 \times 10^5$), and $C_\ell \simeq 0.8$, $C_w \simeq 0$ (Re $\approx > 3 \times 10^5$), where Re $\simeq 67{,}000\,\bar{u}D$, $\bar{u}$ = wind speed in m/s, and D = diameter in m.

Across-Wind Response

The character of the across-wind response depends on whether or not the across-wind deflections are sufficiently large for aeroelastic (lock-in) effects to occur. If the rms σ_y of the across-wind deflections is less than a critical value σ_{ycr}, it may be assumed that aeroelastic effects are negligible. For towers having square cross-section the criteria concerning σ_{ycr} given in Section 2.3.5 (Tall Buildings, Across-Wind Response) may be used.

As noted in Ref. 2.86, aeroelastic wind tunnel tests of chimneys having circular cross-section cannot in general provide a useful representation of full-scale behavior unless carefully interpreted within the framework of an aeroelastic theory complemented by full-scale data or data obtained in special tests conducted at a relatively large scale.

Rumman's Procedure for Estimating Across-Wind Response [2.39, 2.55]

In this procedure it is assumed that towers or stacks having a circular cross-section are subjected to a sinusoidal force per unit length

$$F_0(t) = \frac{1}{2} C_L \rho [\bar{u}(z_{e_i})]^2 D(z) \tag{2.3.28}$$

where ρ = air density
$\bar{u}(z_{e_i})$ = mean wind speed at elevation z_{e_i}
C_L = lift coefficient
$D(z)$ = diameter of structure at elevation z

Reference 2.55 suggests $z_{e_i} = h$, where h = height of structure. Other practitioners suggest $z_{e_i} = \frac{2}{3}h$ to $\frac{5}{6}h$ [2.50], or $z_{e_i} = \frac{2}{3}h$ [2.62]. The wind speeds being considered are those producing at elevation z_{e_i} vortex shedding with frequencies equal to the natural frequencies of the structure, that is,

$$u(z_{e_i}) = \frac{1}{S} n_i D(z_{e_i}) \qquad (i = 1, 2, \ldots) \tag{2.3.29}$$

For Reynolds numbers based on $u(z_{e_i})$ and $D(z_{e_i})$ such that Re $> 2 \times 10^6$ or so, it is usually assumed that the Strouhal number is $S \simeq 0.20$ to 0.25. The peak response for the structure excited in the ith mode may be written as

$$y_{pk}(z) = \frac{\rho C_L}{16\pi^2 \zeta_i} D^2(z_{e_i}) \frac{\int_0^h D(z_1) x_i(z_1)\, dz_1}{M_i S^2} x_i(z) \tag{2.3.30}$$

where $x_i(z)$ = ith, h = normal mode of vibration, ζ_i = damping in ith mode, and M_i is the generalized mass (Eq. 13, Table 2.5). According to Ref. 2.55, many tall reinforced concrete chimneys were designed using a ratio $C_L/\zeta_1 = 13$ to 16.

Equation (2.3.30) shows that for a structure having a uniform mass m per unit length, and having a uniform circular cross-section, the ratio of the tip deflection to the diameter is inversely proportional to the parameter (referred to as the Scruton number)

$$c = 2m\zeta/\rho D^2 \tag{2.3.31}$$

On the basis of empirical studies of the behavior of full-scale chimneys of wind tunnel tests conducted in uniform flow [2.62] it has been suggested that the structure may not be expected to experience lock-in effects if $c > 25$ [2.50].

It is generally agreed that the loading and response models inherent in Rumman's procedure are not entirely consistent with advances made over the last two decades in the fields of micrometeorology, aerodynamics, and structural dynamics. This could lead, in certain situations, to the underestimation of the across-wind response. For example, according to Ref. 2.88, Rumman's procedure appears in certain cases to underestimate significantly the response in the second mode of vibration. Research aimed at developing more realistic design procedures is reported in Refs. 2.9, 2.43,and 2.87.

2.3.7 Low-Rise Buildings

A large body of data on the aerodynamic behavior of low-rise buildings in boundary layer flows was obtained recently at the University of Western Ontario [2.24]. On the basis of these data, improved design criteria for low-rise buildings were incorporated in the 1982 ANSI A58 Standard [2.6].

2.3.8 Hyperbolic Cooling Towers

Information on wind pressures on hyperbolic cooling towers obtained by wind tunnel and full-scale measurements was reported by Niemann [2.42], Armitt [2.7], Sollenberger, Scanlan, and Billington [2.73], Sun and Zhou [2.74], and Propper and Welsch [2.51]. Measurements clearly show that wind effects on cooling towers can be reduced significantly by providing the outside surface of the towers with properly designed ribs [2.42, 2.74].

Niemann [2.42] has shown that dynamic amplification effects on typical cooling towers are negligible. However, it is necessary to account for the quasi-static effects of fluctuating loads acting on the tower at any instant [2.1, 2.51]. Results presented by Niemann [2.42] show that for typical towers the total effects of the mean and fluctuating loads are about twice as large as the effects of the mean load alone. Wind loads on individual towers belonging to a group can be significantly larger than on isolated towers. In the case of the Ferrybridge, U. K. towers that collapsed on November 1, 1965, the wind load affecting the towers had been intensified by channeling effects [2.7].

2.3.9 Trussed Frameworks

Trussed frameworks exposed to wind have routinely been used in structural engineering applications for more than a century. Nevertheless, the state of knowledge concerning the effects of wind on this type of structure is still imperfect, and provisions concerning such effects included in various standards, codes, and design guides are in some cases mutually inconsistent and in disagreement with experimental data [2.30]. It appears that the potential for improving the aerodynamic design criteria of trussed frameworks such as antenna towers and towers for transmission lines still exists. Recent measurements of loads on trussed frameworks were reported by Refs. 2.30 and 2.93 for structures having sharp-edged members, and by Ref. 2.89 for structures having circular members. Research on the nonlinear response of guyed towers to turbulent winds—a problem that has not yet been solved—is in progress in Italy [2.13].

2.3.10 Bridges

Leading centers of research on bridge behavior under wind loads and on the improvement of deck shapes include, among others, the Federal Highway Administration, U. S. Department of Transportation [2.60], the University of Tokyo [2.32], and the National Research Council of Canada [2.90]. Summaries of methods for the evaluation of bridge behavior under wind loads are given by Simiu and Scanlan [2.71] and Ito and Nakamura [2.33].

2.3.11 Offshore Structures

Offshore structures for which the effects of wind may be a major design consideration are primarily of the compliant type (semi-submersibles, tension leg platforms, guyed towers, and articulated towers). Since these structures have nominal natural frequencies of vibration that are very low (on the order of 0.01 Hz), it is important to represent the wind spectra in the low-frequency region as correctly as possible [2.69, 2.71]. A procedure for estimating turbulent wind effects on compliant offshore structures in the presence of current and waves, based on the integration of the equations of motion in the time domain, is presented by Simiu and Leigh [2.69]. For information and references concerning (a) wind effects on semi-submersibles and jack-up platforms, and (b) wind flow considerations for the design of helicopter landing pads on offshore structures, see *Wind Effects on Structures* [2.71].

2.3.12 Nuclear Power Plants

Nuclear power plant structures that are essential from a safety viewpoint are designed in the United States against tornado effects. These include the direct aerodynamic action of the tornado winds, effects of the difference between the relatively low atmospheric pressure at the center of the tornado and the normal atmospheric pressure within an enclosed space in the tornado path, and effects of objects (missiles) hurled by the tornado [2.67, 2.82].

2.4 SNOW LOADS

2.4.1 Basic Determination of Roof Snow Loads

Snow loads provide the governing load requirements for the structural design of roofs in many northern or mountainous regions. Long-span roof systems are especially vulnerable to the effects of snow. During the severe winters of the late 1970s, numerous roofs in the northern United States suffered severe damage or collapsed because of heavy snow. These failures naturally have focused the attention of architects, engineers, and the public on the snow load requirements in codes and standards that are used for structural design.

Conceptual mathematical models have been developed to predict snow accumulation on roofs. One such model is that of Isyumov and Davenport [2.31),

$$S_r(t) = \sum_{i=1}^{N(t)} \Delta R_i - \int_0^t \Delta r(\tau)\, d\tau \qquad (2.4.1)$$

in which $S_r(t)$ is the roof snow load expressed as a function of time t; ΔR_i is the increment of snow load caused by the ith snowfall; $N(t)$ describes the number of snowfalls during time t; and $\Delta r(\tau)$ is the rate of snow removal due to wind, melting, and evaporation. The term ΔR_i depends on the rate and magnitude of snowfall, wind speed, temperature, shape of roof, and its exposure. The function $\Delta r(\tau)$ depends on the current depth and consolidation of snow on the roof, temperature, humidity, and solar radiation. However, at present, the mechanisms of snow deposition and removal are not understood well enough to evaluate ΔR_i and $\Delta r(\tau)$ in terms of the basic variables that describe the local climate and the roof. Accordingly, simpler methods must be used to develop practical loading requirements for structural codes.

The current practice is to calculate the roof snow load S_r as

$$S_r = CS_g \tag{2.4.2}$$

in which S_g is the ground snow load and C is a dimensionless ground-to-roof conversion factor. The ground snow load depends on the geographical location of the building and is determined from meteorological data provided at weather stations by agencies such as the National Oceanic and Atmospheric Administration (formerly the U. S. Weather Bureau). Many of these stations are located at airports and thus are fully exposed to the scouring effects of wind. The conversion factor C relates the weight and distribution of snow on the roof to the ground snow. Factor C has been established from field measurements of snow accumulation on roofs [2.48, 2.75]. The exposure of the roof to the effects of wind, its geometry, and the thermal properties of the materials used in its construction are known to be important factors in determining factor C. Equation (2.4.2) has been the basis of the snow load calculations in the *National Building Code of Canada* [2.44] since 1965, in the ANSI A58 Standard [2.6] since 1972, and reportedly, in the Soviet Union since the mid-1930s. Using C and S_g as defined subsequently, Eq. (2.4.2) provides an approximation to Eq. (2.4.1) that is sufficient for checking the strength of most roof structural systems.

2.4.2 Ground Snow Loads

Basic meteorological data on snow accumulation is reported at first-order weather stations as water-equivalent of snow, in inches. The water-equivalents reflect moderate rainfalls that may occur occasionally during the snow season as well. The water-equivalents are converted easily to ground snow load, in psf, by multiplying by 62.4/12. However, relatively few weather stations report data in this form. Many more stations report data simply in terms of depth of snow. These data on snow depth then must be converted to snow loads using a relationship between depth and density of snow. It has been common to assume a specific gravity for snow of 0.19 (equivalent to 12 pcf) to convert depth measurements to loads [2.75]. However, the use of one specific gravity for calculating load from depth at all sites is not appropriate, as the depth–density relationship varies considerably from region to region. The specific gravity may range from 0.05 to 0.1 for fresh snow to 0.3 or more if consolidation or wind packing has occurred or if the snowpack is the result of several snowfalls over a period of time. At many sites, particularly in colder climates, the depth and water-equivalent may not attain their maximum values during the winter at the same time.

The design-basis ground snow load is determined from the probability distribution for the annual extreme ground snow load as that load with a small probability of being exceeded in any given year. For most permanent structures, this probability is taken as 0.02 in the ANSI A58 Standard [2.6] and 0.033 in the *National Building Code of Canada* [2.44]. Equivalently, the design ground snow load is said to have a mean recurrence interval (MRI) equal to the inverse of this probability—50 years in the ANSI A58 Standard [2.6] and 30 years in the Canadian Code [2.44]. The ground snow loads in the 1972 edition of the ANSI A58 Standard were developed from an analysis of annual extreme water-equivalents for 10 years of record at 140 weather stations [2.77]. The ground snow loads in A58-1982 [2.6] have been determined from a statistical analysis of annual extreme water-equivalents for up to 28 years at 184 first-order stations operated by the National Weather Service (NWS), augmented by snow depth data at approximately 9000 sites. Snow depths were converted to loads through a depth–density relationship developed at the 184 NWS sites.

The annual extreme ground snow load S_g at a majority of sites in the northeast United States can be described by a lognormal cumulative probability distribution [2.28]. The lognormal distribution is given by

$$F_{S_g}(x) = \Phi\left[\frac{\ln x - \lambda_g}{\zeta_g}\right] \tag{2.4.3}$$

Parameters λ_g and ζ_g^2 are the mean and variance of $\ln S_g$, defined, for a sample of size M, as

$$\lambda_g = \left(\sum_i \ln x_i\right)\Big/M \tag{2.4.4a}$$

$$\zeta_g^2 = \left[\sum_i (\ln x_i - \lambda_g)^2\right] \Big/ (M - 1) \tag{2.4.4b}$$

The function Φ[ ] is the standard normal probability integral which is tabulated in most statistical textbooks. The parameters λ_g and ζ_g^2 are site dependent. The N-year mean recurrence interval ground snow load S_{gN} is that value having a probability $1/N$ of being exceeded in any year. It may be obtained from Eq. (2.4.3):

$$S_{gN} = \exp\left[\lambda_g + \zeta_g \Phi^{-1}\left(1 - \frac{1}{N}\right)\right] \tag{2.4.5}$$

in which $\Phi^{-1}(\,)$ is the percent point function of the standard normal distribution. This percent point function equals 1.751, 2.054, and 2.326 for $N = 25$, 50, and 100 years, respectively. Table 2.7 presents the results of a statistical analysis of annual extreme water-equivalents at sites in the northern United States. Estimates of the 25-year, 50-year, and 100-year MRI water-equivalents are obtained from probability plots such as the one shown in Fig. 2.4, or by direct substitution of λ_g and ζ_g into Eq. (2.4.5)[2.26, 2.28].

The ANSI A58 Standard [2.6] specifies that most permanent structures should be designed using a 50-year MRI ground snow load. Certain buildings presenting an unusually high hazard to human life must be designed using a 100-year MRI ground snow instead. Such buildings and structures would include hospitals and other emergency care facilities, fire and police stations, disaster mitigation centers, primary communication facilities, and utility structures whose continued operations are required in an emergency. Other buildings and structures that present essentially no hazard to human occupants may be designed using a 25-year MRI ground snow load; these would include buildings used for agricultural and temporary storage purposes.

The regional variation of ground snow loads is presented in the ANSI A58 Standard [2.6] by a snow map, obtained by solving Eq. (2.4.5) at each site for $N = 50$ years and drawing a smooth contour map consistent with the calculated data. This mapped snow load P_g is approximately equal to S_{g50} at each site. Loads P_g are listed in Table 2.7 for several sites. Small differences between S_{g50} and P_g are because of smoothing while drawing the snow map. Snow loads in areas where rough terrain or bodies of water create extreme local variation in snow accumulation are not included on the map. Loads for these special areas should be determined by analyzing local weather bureau data using the statistical techniques described by Eqs. (2.4.3) through (2.4.5).

Maps corresponding to $N = 100$ and/or $N = 25$ years can also be constructed. However, it has been noted [2.6, Appendix] that the average value of the ratio S_{g100}/S_{g50} is about 1.2, while a similar average of S_{g25}/S_{g50} is about 0.8. The scatter about these averages is quite small. Therefore, the 1982 ANSI A58 Standard [2.6] provides only one ground snow map, which is for a 50-year MRI. The design-basis ground snow load is then given by IP_g, in which I is an *importance factor*. The 100-year and 25-year ground snow loads are computed by multiplying the map values by I of 1.2 or 0.8, respectively.

Table 2.7 Basic Water-Equivalent Ground Snow Data for the Years 1952 through 1980 (psf)

	Sample Statistics							Mapped Values P_g	
Site	M	Maximum Observed	λ_g	ζ_g	S_{g25}	S_{g50}	S_{g100}	A58 1982	A58 1972
Concord, NH	28	36	2.71	0.71	52	65	78	60	37
Boston, MA	27	25	2.23	0.59	26	31	36	30	30
Rochester, NY	28	33	2.46	0.59	33	40	47	40	35
Grand Rapids, MI	28	32	2.12	0.75	31	39	48	35	25
Alpena, MI	19	34	2.76	0.60	45	54	64	50	42
Chicago, IL	26	37	1.64	0.72	18	23	28	25	19
Milwaukee, WI	28	34	1.96	0.76	27	34	41	35	25
Green Bay, WI	28	37	2.06	0.78	31	39	48	40	28
Des Moines, IA	28	22	1.57	0.76	18	23	28	25	27
Minneapolis, MN	28	34	2.22	0.85	41	53	67	50	42
Duluth, MN	28	55	3.20	0.49	58	67	76	70	45
Bismarck, ND	28	27	1.72	0.75	21	26	32	30	10

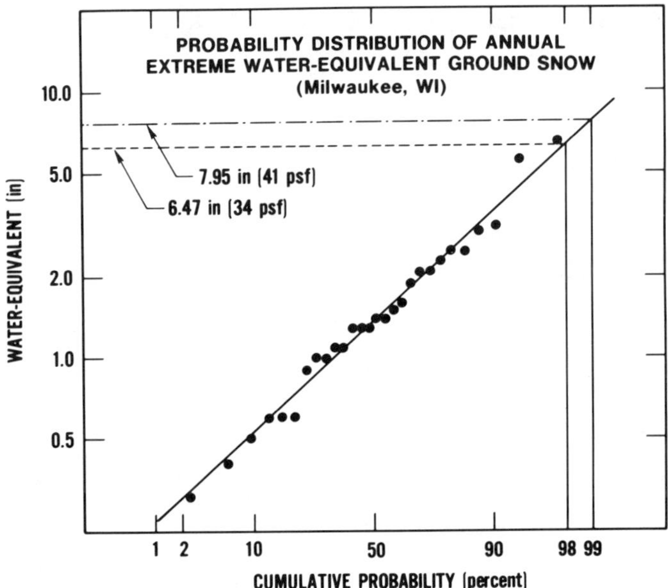

Fig. 2.4 Probability distribution of annual extreme water-equivalent ground snow at Milwaukee, Wisconsin.

As an example for Rochester, New York, statistical analysis of 28 years of annual extreme water-equivalents shows that the parameters for the lognormal distribution describing S_g are λ_g = 2.46 and ζ_g = 0.59. Using these parameters in Eq. (2.4.5), the 25-, 50-, and 100-year MRI ground loads are 32.9 psf (1.6 kPa), 39.8 psf (1.9 kPa), and 46.8 psf (2.2 kPa). For comparison, the mapped load P_g for Rochester is 40 psf (1.9 kPa); multiplied by the importance factor I = 1.2, this becomes 48 psf (2.3 kPa), while multiplied by I = 0.8, it is 32 psf (1.6 kPa).

Ground snow loads in the 1972 ANSI A58 Standard are shown in the last column of Table 2.7. The changes in the design loads have occurred as a result of the greatly expanded data base used in preparing the 1982 ANSI A58 Standard snow map.

2.4.3 Uniform or Balanced Roof Snow Loads

Snow accumulation on both the ground and on the roofs of structures is influenced by prevailing meteorological conditions at the site. The pattern of snow accumulation on a roof depends on whether or not the wind is blowing while the snow is falling, the exposure of the roof to wind, the roof configuration, and the heat loss through the roof. In general, structural designers should make provision both for uniformly distributed snow loads and for the effects of drifting snow and unbalanced load conditions.

The uniform flat-roof design snow load P_f is specified as the product of the design-basis ground snow load IP_g described in Section 2.4.2, and a snow load coefficient C that transforms the ground snow load to a design roof snow load [see Eq. (2.4.2)],

$$P_f = C(IP_g) \tag{2.4.6}$$

Equation (2.4.2) was adopted for convenience in relating the design-basis ground and roof snow loads. In climates typified by infrequent snowfall, the annual maximum ground and roof snow loads may occur at about the same time. However, when there are many snowstorms during the winter and the snow cover is nearly continuous, the roof load usually reaches its yearly maximum value *before* the ground snow load reaches its maximum value [2.48]. The snow load on a well-insulated flat roof on a building that is perfectly sheltered from the effects of wind would be essentially the same as the ground snow load, in which case C would equal 1.0. Most buildings are not so situated, however, and the effects of wind, melting, or sublimination cause C to have a lesser value. A survey of roof snow loads was initiated in Canada in 1956 to determine C for different roof configurations and exposures [2.75]. The Canadian studies indicated that C might be reduced to 0.8 for roofs that are sheltered from the wind and to about 0.6 for roofs that are exposed completely to the effects of wind on all sides. Further reductions in load are permitted for roofs having slopes greater than 30° (a pitch slightly greater than 6 on 12). These snow load coefficients have been used in the *National Building Code of Canada* since 1965 and were adopted by the ANSI A58 Standard in its 1972 edition.

More recent studies have confirmed that roof snow loads are affected by regional exposure to wind and temperature as well as roof thermal characteristics and geometry. Snow load criteria developed recently for Alaska [2.78] took the form

$$P_f = C_r C_e C_t (I P_g) \tag{2.4.7}$$

in which C_r is a ground-to-roof factor which takes regional climatology into account, C_e is a factor depending on the exposure of the roof, and C_t is a factor relating to the thermal characteristics of the roof. The design snow load for sloped roofs was computed as

$$P_s = C_s P_f \tag{2.4.8}$$

in which C_s depends on the roof slope and composition of the roof surface. Additional surveys of roof snow loads were conducted over several winters beginning in 1975 to establish appropriate design values of the coefficients in Eqs. (2.4.7) and (2.4.8) [2.48]. These surveys gathered information on the depth and density of snow on the ground and on roofs, and on patterns of snow accumulation on roofs. The majority of roofs surveyed had slopes less than 30°, at which slope effects begin to become significant; thus, little information on slope effects was obtained. Moreover, the focus of the study was on the uniform or area-averaged snow loads on roofs, and while significant drifts were encountered, no detailed analysis of drift loads was performed.

The basic flat-roof design snow load of the 1982 ANSI A58 Standard [2.6] is calculated using Eq. (2.4.7), taking C_r as 0.7 in the contiguous United States and 0.6 in Alaska. Thus, the snow load coefficient C in Eq. (2.4.6) for flat roofs is $0.7C_eC_t$. The factors C_e and C_t have been set so as to equal 1.0 for heated buildings in normal settings, where wind cannot be counted on to remove snow from the roof. C_e varies from 0.8 for roofs exposed on all sides in windy areas to 1.2 for roofs that are sheltered in densely forested areas. However, the snow accumulation on a very large roof may be greater than on a smaller roof having the same basic exposure. The accelerated flow over the leading edge of any roof has less overall effect on a large roof [2.31]; as a consequence, the average wind velocity over a large expanse of roof is less and more snow tends to accumulate. Designers should be cautious about using a C_e less than unity in calculating snow loads for flat roofs that are larger than about 400 ft (120 m) in either plan dimension.

The factor C_t for thermal effects varies from 1.0 for continuously heated buildings to 1.2 for unheated structures. Although heat loss through the roof of a heated building may cause some melting of snow on the roof, this melting should not be counted on to reduce significantly the snow load on the roof.

The factor C_s in Eq. (2.4.8) depends on the slope of the roof, the slipperiness of the roof surface, and the thermal characteristics of the roof. For most roof surfaces, C_s is reduced from 1.0 at a slope of 30° and less to zero at slopes of 70° and more. Load P_s is considered to act on the horizontal projection of the sloped surface.

2.4.4 Drifting and Unbalanced Roof Snow Loads

Wind is the most important factor in determining the deposition of snow around buildings and on roofs. Snow accumulation patterns are a direct reflection of the turbulence and eddies in the windflow. While the area-averaged uniform snow loads on roofs are less than the ground snow loads, drifting and unbalanced accumulations of snow can cause extreme load intensities over certain roof areas. Approximately 75% of all roof failures related to snow occur because of drifting [2.48]. Thus, drifting and unbalanced snow load conditions warrant careful attention in design.

Basic patterns of snow accumulation and drifting can be predicted in many cases from principles of fluid mechanics and aerodynamics. For complicated roof geometries, wind tunnel or water flume tests might be desirable for determining snow accumulation patterns. With the volume of snow in the drift defined by its geometry, the drift load can be calculated if the density of snow in the drift is known.

Two fundamental principles are useful in predicting the pattern of snow accumulation on roofs [2.76]: (1) Snow is picked up in areas where wind speeds are high and the snow is deposited where wind speeds are low, and (2) snow cannot move from a lower surface to a higher surface unless there is a continuous upward slope. The growth of a snow drift around an obstruction according to these principles is shown in Fig. 2.5. Snow particles may be picked up from fresh snow cover and transported at windspeeds greater than about 4 m/s (12 mph). A turbulent wake region in which the particle velocities are low exists in the lee of the obstruction, which causes snow to accumulate. In

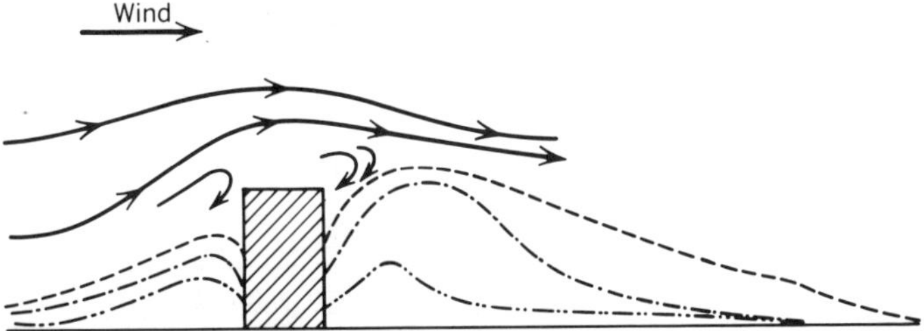

Fig. 2.5 Typical pattern of snow accumulation adjacent to an obstruction.

a frigid climate, a drift grows until its profile is so streamlined that the wind flow is not disturbed and subsequent accumulation takes place at the same rate as in the surrounding snow field.

Building height is important to the analysis of drifts because most of the snow that forms drifts is transported horizontally very close to the ground [2.76]. Therefore, for buildings of normal height having vertical walls, the amount of snow available for drifting usually is limited to the snow that falls on the roof and on higher adjacent roofs. The total volume of available snow on the upper roof limits the snow that can drift or slide from an upper roof to an adjacent lower roof.

Substantial drifts can form on a low roof that is in the wind shadow of a building having a higher roof. Drift load intensities in excess of 100 psf (4.5 kPa) are common [2.16], and intensities as high as 240 psf (11.5 kPa) have been measured on such roofs [2.61]. However, the scouring action of the wind can cause drifting and nonuniform accumulation of snow even on open flat roofs. Figure 2.6 illustrates a typical pattern of snow accumulation on a large flat roof having a parapet. The wake region downwind from and adjacent to the parapet causes the deposition of additional snow. Similarly, a drift may form upwind of the parapet on the opposite side of the roof, as wind scours the snow off the middle of the roof. The height of the drift is limited to the height of the parapet, which seldom exceeds 3 to 4 ft (0.9 to 1.2 m).

The snow load coefficients in the ANSI A58 Standard [2.6] take into account localized heavy loading due to drifting or unbalanced conditions for a number of common roof geometries. Drift surcharges are defined for the lower level of multilevel roofs, for roofs having vertical projections, and for roofs within 20 ft (6 m) of a higher structure or terrain feature that could lead to drifting conditions. Figure 2.7 shows a typical drift surcharge load that might occur at an abrupt change in roof elevation. Surveys of snow drifts on roofs have determined that the drift surcharge can be represented adequately by a triangle in which the horizontal leg W is approximately four times the vertical leg h_d [2.6, Appendix]. The height of the drift need not exceed the clear distance between the top of the uniform load and the upper roof. The maximum intensity of the drift surcharge then is $P_d = h_d\gamma$, in which γ is the density of snow in the drift.

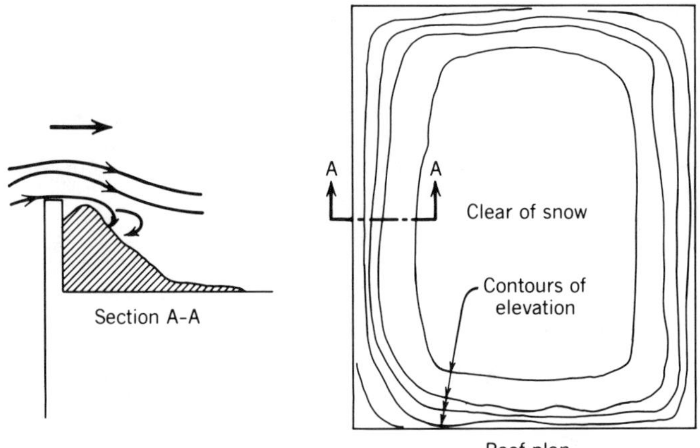

Fig. 2.6 Typical pattern of snow accumulation on large flat roofs with parapets.

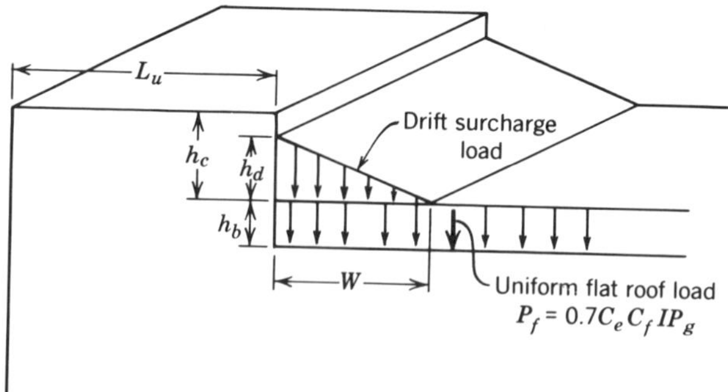

Fig. 2.7 Uniform and drift surcharge snow loads for design.

The measured densities of snow in drifts on roofs are consistently higher than the densities of snow on the ground. The higher density is caused by compaction by wind, heat loss through the roof (which may cause a mixture of ice and snow in the drift), and poor drainage and capillary action of the snow, which cause meltwater to be retained in the snowpack. The ANSI A58 Standard [2.6] provides a tabulation of drift densities that increase from 15 pcf (240 kg/m^3) when P_g is 30 psf (1.4 kPa) or less to 25 pcf (400 kg/m^3) when P_g is 60 psf (2.9 kPa) or more. For comparison, surveys of drift intensities on multilevel flat industrial roofs in Ottawa, Ontario, where the design ground snow is 60 psf (2.9 kPa) show a continuous increase in average density during the winter from 12 pcf (190 kg/m^3) in mid-December to about 20 pcf (320 kg/m^3) in mid-March [2.75]. An analysis of densities obtained from 30 samples removed from snow drifts adjacent to 10 collapsed roofs in Chicago, Illinois, during the winter of 1978–1979 revealed an average drift density of 17 pcf (270 kg/m^3), a minimum of 11 pcf (180 kg/m^3), and a maximum of 24 pcf (390 kg/m^3). These measured drift densities are consistent with the densities in the ANSI A58 Standard [2.6]. However, they tend to be much larger than the 12 pcf (190 kg/m^3), which has been assumed frequently when converting snow depth to load.

Figure 2.8 illustrates how an unbalanced snow load can accumulate on the leeward slope of a common gabled roof. The windward slope is blown clear, while the leeward slope in the wake region picks up additional load. This load pattern is provided for by increasing the balanced load by $1.5/C_e$ on the leeward slope while removing the load on the windward slope, as shown in Fig. 2.8. Further unbalanced load conditions are defined in the ANSI A58 Standard [2.6] for hip, arched, multiple folded plate, sawtooth, and barrel vault roofs. In addition, an unbalanced load condition that is caused by removing half the balanced snow load from any area on the roof is included for all roof types.

2.4.5 Other Snow Considerations

The previous discussion covers a majority of the basic snow load situations for the structural design of roofs. Additional items should also be considered in designing certain roofs. The possibility of a heavy early spring rain on top of an existing heavy snowfall should be considered in temperate climates. The snowmass can retain a considerable amount of rainwater by capillary action [2.19]. The melting of snow by the rain may be negligible unless the rain is very heavy. The practice in Canada [2.75] is to surcharge the ground snow load with the load that results from the maximum 24-hr rainfall during the winter months. There is no such requirement in the United States; however, a conservative designer may wish to add a uniform 5 psf (240 Pa) surcharge if the roof slope is less than about 0.5 in. (12 mm) per foot of span.

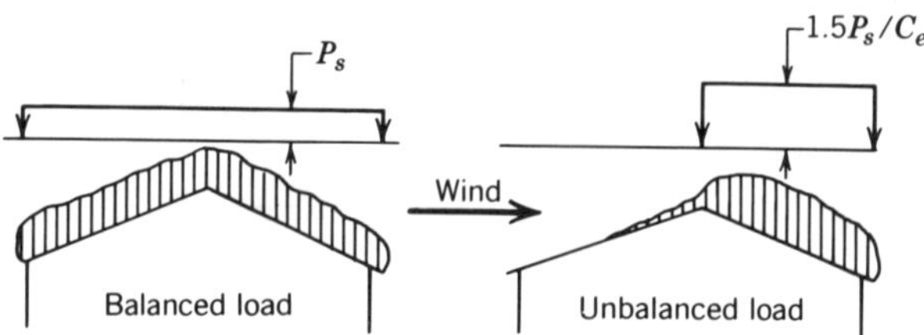

Fig. 2.8 Snow load on the leeward slope of a gable roof.

Thin ice layers frequently are observed intermixed with snow on some roofs because of poor thermal characteristics of the roof and the resulting heat loss. It is unusual for ice to block interior roof drains because they generally are heated sufficiently from the interior of the building to prevent freezing over. However, drifting snow and freezing may be sufficient to block peripheral drains and scuppers at the base of parapets, which may lead to ponding problems. Moreover, ice dams may form as meltwater freezes at the eaves of sloped roofs, blocking flow to gutters and causing substantial loads on roof overhangs and damage to moisture-resistant roofing membranes [2.37]. Ice formation is an especially critical problem with warm, poorly insulated structures in cold climates.

Sliding snow can be a problem with roofs having slopes greater than about 10°. The sliding snow may increase the load on lower adjacent roofs or may endanger pedestrians or block entryways. There are little data for predicting sliding snow loads. As a guide, it may be assumed that 50% of the snow on the upper roof can slide off onto the lower roof [2.6].

Snow loads may be important for designing outdoor parking decks or garages. When designing such structures, local accumulations of snow due to snow clearing operations and the localized weight of snow removal equipment also should be taken into account.

2.5 EARTHQUAKE LOADS

2.5.1 General

When a building or other structure is subjected to earthquake motion, its base moves with the ground. The ground shaking during an earthquake occurs in all directions simultaneously. The motion may be treated in terms of its horizontal, vertical, and rotational components. Generally, the horizontal motion is the component of major concern in design. The vertical component is usually not considered in design because it is considered that ordinary gravity load provisions would be sufficient to accommodate the dynamic effects of the vertical component. The rotational component is simply not considered in design because partly its effect on the structure is not well understood at this time and partly its effect on the structure is not significant as compared to the other two components. Discussion on design philosophy for seismic loads is given in Section 3.7.

When the ground suddenly moves horizontally under an object (structure) having mass, the inertia of the mass tends to resist the movement, as shown in Fig. 2.9. A shear force is developed between the ground and the mass. For a building where the mass is primarily located at the floor levels, and each floor mass is connected to the adjacent floor mass by columns which are essentially elastic springs, the relative horizontal motion induced between each floor by the ground motion causes shears in the columns at the floor level. Thus, the earthquake-induced forces are interdependent with the structural characteristics and configuration of the structure.

Design of tall buildings for earthquake should include consideration of performing a dynamic analysis of the entire structure. Without the aid of a computer it is impractical to determine the detailed dynamic behavior of structures under earthquake conditions. Such an analysis could include both the horizontal and vertical components of ground motion. Structural dynamics, which includes how the structure should be modeled for such an analysis, as well as how the dynamic loading should be modeled, is treated in Chapter 18 as well as in several textbooks [2.17, 2.41, and 2.94] and is outside the scope of this chapter.

2.5.2 Equivalent Lateral Force

In order to simplify the design process, most building codes contain an equivalent lateral force procedure for designing to resist earthquakes. One of the most widely used design recommenda-

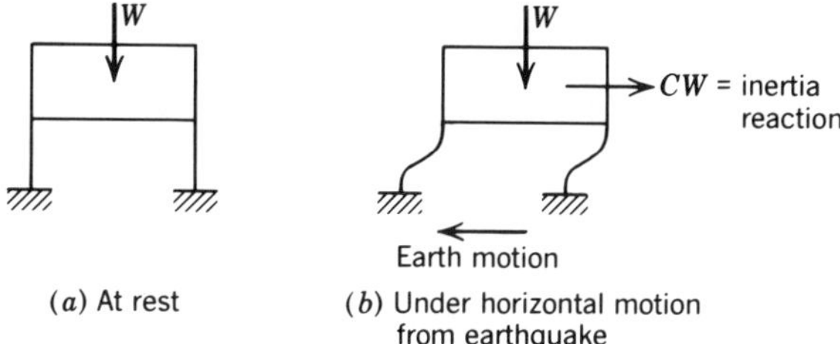

Fig. 2.9 Force developed by an earthquake.

tions is that of the Structural Engineers Association of California (SEAOC), the latest version of which is 1974 [2.63]. Recently, the Applied Technology Council (ATC) prepared a set of design provisions [2.8]. The ATC provisions were prepared by 85 leading professionals, including engineers, seismologists, geologists, and building officials, The ATC provisions can serve the designer as a useful source for earthquake loads and their treatment. The most recent rules for the equivalent lateral force procedure are those given by the ANSI A58 Standard [2.6], which have been based on the 1979 *Uniform Building Code* [2.83] and certain aspects of the ATC study. The lengthy ATC provisions needed refining and testing before being incorporated in codes of practice; thus, the 1982 ANSI A58 Standard should be regarded as being an interim standard for the transition between one generation of standards and another.

The total lateral seismic forces assumed to act nonconcurrently in the direction of each of the main axes of the structure are given [2.6] by

$$V = ZIKCSW \tag{2.5.1}$$

where Z = seismic zone coefficient, varying from ⅛ for the zone of lowest seismicity, to 1 for the zone of highest seismicity

I = occupancy importance factor, varying from 1.5 for buildings designated as "essential facilities," 1.25 for buildings where the primary occupancy is for assembly for greater than 300 persons, to 1.0 for usual buildings

K = horizontal force factor, varying from 0.67 to 2.5, indicating capacity of the structure to absorb plastic deformation (low values indicate high ductility)

$$C = \frac{1}{15\sqrt{T}} \leq 0.12, \text{ the seismic coefficient, equivalent to the maximum} \tag{2.5.2}$$

acceleration in terms of acceleration due to gravity

T = fundamental elastic period of vibration of the building or structure in the direction of motion being considered, in sec

S = soil profile coefficient, varying from 1.0 for rock to 1.5 for soft- to medium-stiff clays and sands

W = total dead load of the building, including interior partitions

In application of Eq. (2.5.1), the product CS need not exceed 0.14 in general, nor 0.11 for the soft to medium-stiff clay ($S = 1.5$) soil profile in seismic zones 3 and 4.

The fundamental elastic period of the structure may be computed "using the structural properties and deformational characteristics of the resisting elements in a properly substantiated analysis" [2.6]. In the absence of such an analysis, the value of T for buildings may be obtained as follows:

1. For shear walls or exterior concrete frames utilizing deep beams or wide piers, or both,

$$T = \frac{0.05h_n}{\sqrt{D}} \tag{2.5.3}$$

where D is the dimension of the structure in the direction of the applied forces, in ft, and h_n is the height of the building.

2. For isolated shear walls not interconnected by frames, and for braced frames,

$$T = \frac{0.05h_n}{\sqrt{D_s}} \tag{2.5.4}$$

where D_s is the longest dimension of a shear wall or braced frame in a direction parallel to the applied forces, in ft.

3. For buildings in which the lateral force-resisting system consists of moment-resisting space frames capable of resisting 100% of the required lateral forces, and where such system is not enclosed by or attached to rigid elements that would tend to prevent the frame from deflecting under lateral forces,

$$T = C_T h_n^{0.75} \tag{2.5.5}$$

where C_T = 0.035 for steel frames, and 0.025 for concrete frames.

Once the base shear V has been determined, the lateral force must be distributed over the height of the building. For structures having regular shapes or framing systems, the total base shear V is considered to be the sum of a concentrated lateral force F_t at the top of the structure and the lateral forces F_i applied at each intermediate level i, according to the following:

$$V = F_t + \sum_{i=1}^{n} F_i \tag{2.5.6}$$

The concentrated force F_t should be determined as

$$F_t = 0.07TV \tag{2.5.7}$$

However, F_t need not exceed $0.25V$ and may be taken as zero when the period T is 0.7 sec or less. The top force F_t is introduced to accommodate the higher modes of vibration by increasing the design shears in the upper stories where the high modes have the greater effect. Higher modes are important for long-period tall buildings.

The remaining portion of the total base shear V [i.e., the ΣF_i in Eq. (2.5.5)] must be distributed over the height of the structure. At any level x, including the top level n, the force F_x is given by the following:

$$F_x = \frac{(V - F_t)w_x h_x}{\sum_{i=1}^{n} w_i h_i} \tag{2.5.8}$$

At each level designated as x, the force F_x shall be applied over the area of the building in accordance with the mass distribution on that level.

The overturning moment is determined as the static effect of the lateral forces,

$$\sum_{i=1}^{n} F_i h_i + F_t h_n \tag{2.5.9}$$

where h_i is the height above the base to level i and h_n is the height to the roof.

2.5.3 Dynamic Analysis

If the structure has an unusual plan configuration and framing systems and/or is unusually tall, the equivalent lateral load approach may be inadequate. An analysis would then be carried out considering the dynamic characteristics of the structure following the principles discussed in Chapter 18. Actual earthquake records may be used to determine the dynamic force acting on the mathematical model of the structure. Ground acceleration records are the usual basic records available, from which ground velocity and ground displacement can be computed. The record, known as an accelerogram, is continuous during the earthquake (as well as before and after). However, it should be recognized that there are important uncertainties in a dynamic analysis, such as what expected base movements would occur at the building site because of an earthquake, and what damping coefficients and ductility factors to use in the dynamic analysis.

A dynamic analysis of a structure can be performed by subjecting a mathematical model of the structure to a particular input motion. Because earthquake ground motions are erratic in nature, they cannot be represented by a single continuous function. Therefore, the load function is discretized and the numerical integration of the equation of motion is carried out. This gives the time history of the response of the structure to a particular earthquake motion. This procedure, which is known as the step-by-step integration method, is generally based on linear (i.e., elastic) theory. Although nonlinear analysis is possible, the procedure is time consuming and costly. In general, the cost of implementing the analysis is directly proportional to the size of the time step used in the integration. A balance between cost and accuracy of the analysis is a principal concern, particularly in the case of large structures.

Another dynamic analysis procedure is a modal analysis method based on the mode-superposition concept. In this approach, the response of a complex system is obtained by combining the responses of the individual modes of vibration. A structure may vibrate with as many mode shapes and periods as it has degrees-of-freedom.

For a multistory building, the weight is assumed to be concentrated at the floor levels. Such a lumped-mass model is used to determine the required periods and mode shapes by established methods of mechanics for the fixed-base condition. The response of each mode of vibration is calculated as though it were a single mass system.

The design values for each of the story shears, moments, and drifts are determined by combining each modal value by taking the square root of the sum of the squares of each of the modal values. A modal analysis procedure is given in Ref. 2.8.

2.5.4 General Comments

Detailed treatment of earthquake loads and corresponding specific design procedures are not included in this book because most of the United States does not require special design for earthquake loads; that is, the provisions for wind will supersede any requirements that may be earthquake related.

2.6 COMBINATIONS OF LOADS

Most structural loads vary with time. Calculations of structural performance that are used in checking safety and serviceability require the maximum combined load effect occurring during some suitable period of reference. This period T might be on the order of 50 to 100 years for safety-related performance criteria, and approximately 1 to 10 years for serviceability. If the structure or structural component is subjected to a set of N loads $Q_i(t)$, $i = 1, \ldots, N$, which vary in time, the designer would then be concerned with determining an appropriate value of the maximum combined load effect Q_{max},

$$Q_{max} = \max_T [Q_1(t) + Q_2(t) + \cdots + Q_N(t)] \tag{2.6.1}$$

which can be used for design and detailing purposes.

As has been shown in earlier sections of this chapter, structural loads are random in space and time. During the past decade, load combinations have been analyzed using the theory of stochastic processes to take into account the spatial and temporal correlation and variation of the loads [2.36, 2.80, 2.91]. The discussion herein uses an intuitive approach to arrive at similar conclusions. Additional discussion on load combinations is presented in Chapter 3.

Figure 2.10 gives typical sample functions that show the structural effects of several common loads considered in codes and standards [2.6]. Permanent load effects such as those due to dead load or prestressing change slowly with time and may be assumed to remain essentially constant during the period of reference. Occupancy live loads, temperature effects, and snow loads vary with time, but their effects on the structure are essentially static. Finally, wind loads that are structurally significant and earthquake loads occur relatively infrequently and their durations are much smaller than the durations of permanent and other variable loads described above.

The load at any point in time is described by a probability density function $f(x)$ shown at the left side of Fig. 2.10. This point-in-time load could be determined by means of a load survey. The mean point-in-time load usually is much less than the nominal load Q_n that is specified for design purposes in the ANSI A58 Standard [2.6] and in other regulatory documents. For example, a survey [2.27] of live loads in office buildings showed that the mean survey load is about 12 psf (0.6 kPa) while the nominal design load is 50 psf (2.4 kPa). The maximum load during the period T also is described by a probability density $f_{max}(x)$, as shown in Fig. 2.10.

If the structural component is subjected to only one variable load in addition to the permanent load, the load combination analysis is relatively straightforward, since the maximum combined load effect during T occurs when the variable load attains its maximum value. In the design of large interior slabs or long-span roofs, for example, the maximum occupancy live load on the slab or the maximum snow load on the roof during T would be combined with the dead load. However, the analysis is not so simple when more than one variable load acts on the structure at any given time. One may observe from Fig. 2.10 that the likelihood is small that two or more of the variable loads in the combination will attain their maximum values simultaneously. Accordingly, one would expect that

$$Q_{max} < [\max_T Q_1(t) + \max_T Q_2(t) + \cdots + \max_T Q_N(t)] \tag{2.6.2}$$

in which $\max_T Q_i(t)$ is the maximum value of $Q_i(t)$ to occur during T. Consequently, structural members may be designed for a total load that is less than the sum of the peak loads. Current standards recognize this by prescribing procedures for reducing the effect of the combination of peak load values for design purposes.

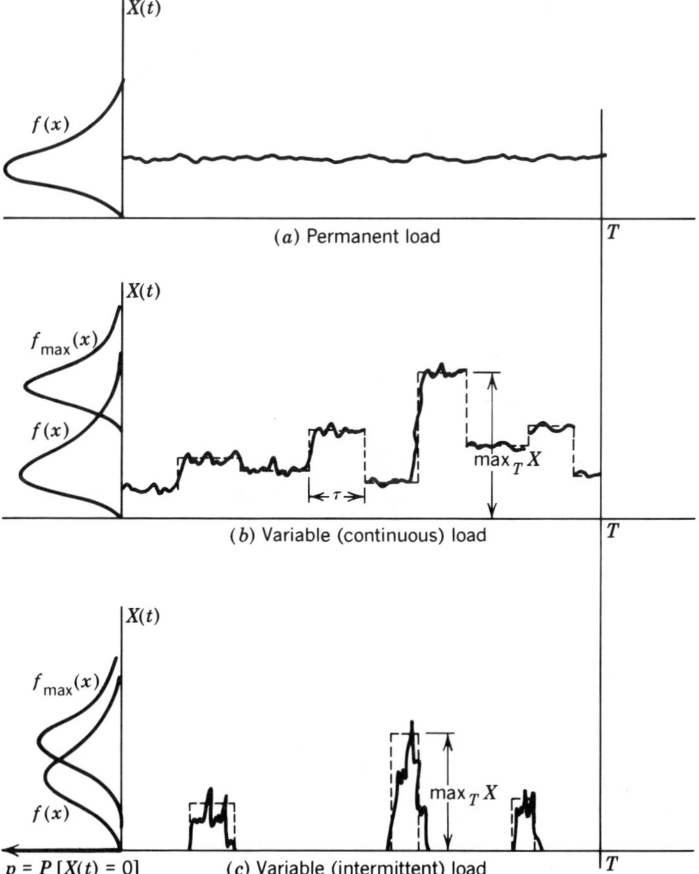

Fig. 2.10 Stochastic process models of structural loads.

In allowable or working stress design, the total load effect may be multiplied by a load combination probability factor ψ less than unity to obtain a design load Q_d as

$$Q_d = \psi(Q_{n1} + Q_{n2} + \cdots + Q_{nN}) \tag{2.6.3}$$

In the ANSI A58 Standard [2.6], the load combination probability factor is taken as 0.75 when the combination includes two time-varying loads and 0.66 when the combination includes three time-varying loads. For example, analysis of load combinations involving dead load D, live load L, and wind load W, would require the following combinations to be considered:

$$Q_d = \max \begin{cases} D + L & (2.6.4a) \\ D + W & (2.6.4b) \\ 0.75(D + L + W) & (2.6.4c) \end{cases}$$

Alternatively, in some allowable stress specifications, the allowable stress is increased for some load combinations. For example, the *AISC Specification* [2.3] permits the allowable stress to be increased by 33% for load combinations involving wind load (i.e., both the $D + W$ and $D + L + W$ combinations), but the combined load effects are not adjusted. This approach is consistent with the treatment of $D + L + W$ in the ANSI A58 Standard [2.6] since $0.75 = 1/1.33$; however, it is not consistent for $D + W$, where ANSI A58 Standard makes no adjustment because W is the only variable load in the combination. Apparently, the 33% increase in allowable stress permitted by

AISC [2.3] for combinations involving wind represents more than simply a way to account for the low probability that the peaks of the variable loads coincide.

In strength or plastic design (see also Chapter 3), the small likelihood of peak loads coinciding is treated by reducing the combined factored loads. This adjustment usually has been consistent with the treatment of load combinations in allowable stress design, which predated the strength design procedures. In the *ACI Code* [2.2], the combinations for dead, live, and wind loads are

$$Q_d = \max \begin{cases} 1.4D + 1.7L & (2.6.5a) \\ 0.75(1.4D + 1.7L + 1.7W) & (2.6.5b) \end{cases}$$

Note that the combination involving factored wind load is multiplied by the factor 0.75, just as it had been done in allowable stress design. Similarly, Part 2 (Plastic Design) of the *AISC Specification* [2.3] gives

$$Q_d = \max \begin{cases} 1.7(D + L) & (2.6.6a) \\ 1.3(D + L + W) & (2.6.6b) \end{cases}$$

in which it might be observed that $1.3/1.7 \approx 0.75$.

The Canadian Limit States Design Standard [2.18] takes a slightly different approach to load combinations, using the following load requirement,

$$Q_d = 1.25D + \psi(1.5L + 1.5W + 1.25T) \tag{2.6.7}$$

in which ψ is a load combination factor equal to 1.00 when one of L, W, or T (temperature effect) act; 0.7 when two of L, W, or T act; and 0.6 when all three act. Unlike the load combinations in Eqs. (2.6.4) through (2.6.6), the load combination factor is not applied to the permanent load D, which is always present. Therefore, Eq. (2.6.7) provides a better description of how structural loads actually combine in practice than do Eqs. (2.6.4) through (2.6.6).

More recent load combination rules draw upon results of the probabilistic load combination studies referred to earlier. Such studies have led to the observation that the maximum effect of a combination of loads usually occurs when one of the loads reaches its maximum value during time period T while the other loads are equal to their point-in-time values (see Fig. 2.10). The maximum load effect then is approximated by [2.80],

$$Q_{\max} \simeq \max_{i}^{N} \left[\max_{T} Q_i(t) + \sum_{j \neq i} Q_j(t) \right] \tag{2.6.8}$$

The term $\left(\max_{T} Q_i\right)$ is denoted the principal variable load, while the Q_j are denoted companion actions. It is necessary to consider M distinct load combinations in order to compute the maximum load effect for structural design purposes, that is, each time-varying load must assume, in turn, the position of the principal variable load in Eq. (2.6.8). Equation (2.6.8) neglects the possibility that two or more variable loads will reach their respective maximum values at the same time, or that the maximum effect will occur when two of the loads attain "near-maximum" values. This introduces a degree of unconservatism which may be significant when two of the loads are positively correlated as a consequence of arising from the same underlying physical phenomenon. However, studies have shown that Eq. (2.6.8) is a good approximation for most practical cases involving building structures. It is also consistent with the observation that structural failures (at least those not due to human error or willful abuse) usually occur as a consequence of one load attaining an extreme value.

Several practical load combination rules in limit states design standards (see Chapter 3) now being proposed in the United States, Canada, and Europe have evolved from Eq. (2.6.8). The probability-based load combinations in the 1982 ANSI A58 Standard [2.6] have an appearance that is similar to what already is used in the *ACI Code* [2.2] for reinforced concrete design. The same load combinations have been used in the proposed AISC Load and Resistance Factor Design (LRFD) [2.4, 2.52] for steel buildings. For combinations involving dead, live, and wind loads, for example,

$$Q_d = \max \begin{cases} 1.2D + 1.6L & (2.6.9a) \\ 1.2D + 1.3W + 0.5L & (2.6.9b) \\ 0.9D + 1.3W & (2.6.9c) \end{cases}$$

The load factor of 0.5 on the nominal live load in Eq. (2.6.9b) is a consequence of the load combination analysis represented by Eq. (2.6.8). The factored load $0.5L$ is equivalent to a factored point-in-time load; recall that the average point-in-time load is much less than the nominal live load.

An alternative load combination rule has been proposed for several European model limit states design codes, such as CEB [2.14], namely

$$Q_d = \max_i \left[(0.9D \text{ or } 1.2D) + \gamma_Q \left(Q_{ki} + \sum_{j \neq i} \psi_{oj} Q_{kj} \right) \right] \tag{2.6.10}$$

in which D is permanent load, Q_{ki} is the characteristic (nominal) value of the principal variable load, and ψ_{oj} are termed companion action factors. In the proposed load requirements, γ_Q is about 1.4 to 1.5 while ψ_{oj} are on the order of 0.4 to 0.6 for the strength or safety-related limit states. When the dead load has a beneficial stabilizing effect on the structure, the load factor on D in Eqs. (2.6.9) and (2.6.10) is 0.9 rather than 1.2.

It should be noted that in all recent load combination proposals, no adjustment is ever made to the dead load when it is combined with other variable or transient loads. This is a conceptual improvement over existing load combination rules in which the dead load is multiplied by 0.75 when combining it with wind or earthquake loads.

The magnitudes of the variable and transient loads are much more unpredictable than the magnitude of the permanent load. It is not possible to achieve uniform reliability and performance if the same factor of safety or load factors are applied to both the permanent and the variable or transient loads, as is done in Eqs. (2.6.3), (2.6.4), and (2.6.6)[2.29]. Assigning different load factors to dead load and time-varying loads, as in Eqs. (2.6.9) and (2.6.10), makes it possible to achieve more uniform reliability and performance for different combinations of loads.

SELECTED REFERENCES

2.1 Salman H. Abu-Sitta and Mahmoud G. Hashish. "Dynamic Wind Stresses in Hyperbolic Cooling Towers," *Journal of the Structural Division,* ASCE, **99,** September 1973 (ST9), 1823–1935.

2.2 ACI. *Building Code Requirements for Reinforced Concrete* (ACI 318-83). Detroit: American Concrete Institute, 1983.

2.3 AISC. *Specification for the Design, Fabrication and Erection of Steel Buildings.* Chicago: American Institute of Steel Construction, 1978.

2.4 AISC. *Proposed Load and Resistance Factor Design Specification for Structural Steel Buildings.* Chicago: American Institute of Steel Construction, September 1, 1983.

2.5 ANS. *American Nuclear Society Standard for Estimating Tornado and Extreme Wind Characteristics at Nuclear Power Sites,* (contained in American National Standard, ANSI/ANS-2.3-1983). La Grange, Illinois: American Nuclear Society, 1983.

2.6 ANSI. *American National Standard Minimum Design Loads for Buildings and Other Structures* (A58.1-1982). New York: American National Standards Institute, 1982. (1430 Broadway, New York, NY 10018)

2.7 John Armitt. "Wind Loading on Cooling Towers," *Journal of the Structural Division,* ASCE, **106,** March 1980 (ST3), 623–641.

2.8 Applied Technology Council (ATC), Associated with the Structural Engineers Association of California. *Tentative Provisions for the Development of Seismic Regulations for Buildings (ATC 3-06)* (NBS Special Publication 510). Washington, D.C.: Center for Building Technology, National Bureau of Standards, June 1978.

2.9 R. I. Basu and B. J. Vickery. "Across-Wind Vibrations of Structures of Circular Cross-Section, Part 2, Development of a Mathematical Model for Full Scale Application," *Journal of Wind Engineering and Industrial Aerodynamics,* **13,** 1983.

2.10 Martin E. Batts, Larry R. Russell, and Emil Simiu. "Hurricane Wind Speeds in the United States," *Journal of the Structural Division,* ASCE, **106,** October 1980 (ST10), 2001–2016.

2.11 Jacque Bietry, Christian Sacre, and Emil Simiu. "Mean Wind Profiles and Changes of Terrain Roughness," *Journal of the Structural Division,* ASCE, **104,** October 1978 (ST10), 1585–1593.

2.12 J. Blessman and J. D. Pierce. "Interaction Effects in Neighboring Tall Buildings," *Proceedings of the Fifth International Conference on Wind Engineering,* Fort Collins, Colorado, July 1979 (Vol. 2). Oxford, New York: Pergamon Press, 1980.

2.13 H. A. Bucholdt and P. Spinelli. *Static and Dynamic Analysis of Guyed Masts,* Progress Report UFIST/04/1983, University of Florence, Department of Civil Engineering, Florence, Italy, 1983.

2.14 CEB. *Common Unified Rules for Different Types of Construction and Material* (Bulletin D'Information No. 124E). Paris: Comite Euro-International du Beton (CEB), April 1978.

2.15 M. E. Changery et al. *Directional Extreme Wind Speed Data for Buildings and Other Structures* (Building Science Series 160). Washington, D.C.: National Bureau of Standards, 1984.

2.16 I. Chin, A. Gouwens, and J. Hanson. "Review of Roof Failures in the Chicago Area Under Heavy Snow Load," ASCE Convention and Exposition, Portland, Oregon, April 1980, Preprint 80–145.

2.17 A. K. Chopra. *Dynamics of Structures—A Primer.* Berkeley, CA: Earthquake Engineering Research Institute, 1981.

2.18 CISC. *Limit States Design Steel Manual.* Willowdale, Ontario, Canada: Canadian Institute of Steel Construction, 1984.

2.19 Samuel G. Colbeck. "Roof Loads Resulting from Rain-On-Snow," U.S. Army Cold Regions Research and Engineering Laboratory Report 77-12, May 1977.

2.20 J. Counihan. "Adiabatic Atmospheric Boundary Layers: A Review and Analysis of Data from the Period 1880–1972," *Atmospheric Environment,* **9,** October 1975 (10).

2.21 S. H. Crandall and W. D. Mark. *Random Vibrations in Mechanical Systems.* New York: Academic Press, 1963.

2.22 A. G. Davenport. "The Dependence of Wind Load Upon Meteorological Parameters," *Proceedings of the International Research Seminar on Wind Effects on Buildings and Structures.* Toronto, Canada: University of Toronto Press, 1968, pp. 19–82.

2.23 A. G. Davenport. "The Prediction of Risk Under Wind Loading," *Proceedings of the Second International Conference on Structural Safety and Reliability,* Munich, West Germany, September 1977, pp. 511–538.

2.24 A. G. Davenport, D. Surry, and T. Stathopoulos. "Wind Loads on Low Rise Buildings: Final Report on Phases I and II—Parts 1 and 2," BWLT Report SSB-1977, The University of Western Ontario, London, Canada, November 1977.

2.25 C. S. Durst. "Wind Speeds Over Short Periods of Time," *Meteorological Magazine,* **89,** 1960, 181–186.

2.26 Bruce Ellingwood. "Wind and Snow Statistics for Probabilistic Design," *Journal of the Structural Division,* ASCE, **107,** July 1981 (ST7), 1345–1350.

2.27 Bruce Ellingwood and Charles Culver. "Analysis of Live Loads in Office Buildings," *Journal of the Structural Division,* ASCE, **103,** August 1977 (ST8), 1551–1560.

2.28 Bruce Ellingwood and Robert Redfield. "Ground Snow Loads for Structural Design," *Journal of Structural Engineering,* ASCE, **109,** April 1983 (ST4), 950–964.

2.29 Bruce Ellingwood, James G. MacGregor, Theodore V. Galambos, and C. Allin Cornell. "Probability-Based Load Criteria: Load Factors and Load Combinations," *Journal of the Structural Division,* ASCE, **108,** May 1982 (ST5), 978–996.

2.30 P. N. Georgiou and B. J. Vickery. "Wind Loads on Building Frames," *Proceedings of the Fifth International Conference on Wind Engineering,* Fort Collins, Colorado, July 1979. New York: Pergamon Press, 1980, pp. 421–443.

2.31 N. Isyumov and A. G. Davenport. "A Probabilistic Approach to the Prediction of Snow Loads," *Canadian Journal of Civil Engineering,* **1,** 1974, 28–49.

2.32 M. Ito. "On the Wind-Resistant Design of Truss-Stiffened Suspension Bridges," *Proceedings of the Second USA Japan Research Seminar on Wind Effects on Structures,* Kyoto, 1974. Tokyo: University of Tokyo Press, 1976, pp. 285–296.

2.33 Manabu Ito and Yasuharu Nakamura. "Aerodynamic Stability of Structures in Wind," *IABSE Surveys* S-20/82, International Association for Bridge and Structural Engineering, ETH-Honggerberg, Zurich, Switzerland, May 1982, 33–56.

2.34 Ahsan Kareem. "Across-Wind Response of Buildings," *Journal of the Structural Division,* ASCE, **108,** April 1982 (ST4), 869–887.

2.35 Kenny C. S. Kwok and William H. Melbourne. "Wind-Induced Lock-In Excitation of Tall Structures," *Journal of the Structural Division,* ASCE, **107,** January 1981 (ST1), 57–72.

2.36 Richard D. Larrabee and C. Allin Cornell. "Combinations of Various Load Processes," *Journal of the Structural Division,* ASCE, **107,** January 1981 (ST1), 223–239.

2.37 I. Mackinlay, "The Neglected Hazards of Snow and Cold," *AIA Journal,* February 1983, 52–59.

2.38 E. H. Markee, J. G. Beckerley, and K. E. Sanders. *Technical Basis for Interim Regional Tornado Criteria* [WASH-1300 (UC-11)]. Washington, D.C.: Nuclear Regulatory Commission, 1974.

2.39 Lawrence C. Maugh and Wadi S. Rumman. "Dynamic Design of Reinforced Concrete Chimneys," *ACI Journal, Proceedings,* **64,** September 1967, 558–567.

2.40 Robert J. McNamara. "Tuned Mass Dampers for Buildings," *Journal of the Structural Division,* ASCE, **103,** September 1977 (ST9), 1785–1798.

2.41 N. M. Newmark and E. Rosenblueth. *Fundamentals of Earthquake Engineering.* Englewood Cliffs, NJ: Prentice-Hall, 1970.

2.42 Hans-Jurgen Niemann. "Wind Effects on Cooling-Tower Shells," *Journal of the Structural Division,* ASCE, **106,** March 1980 (ST3), 643–661.

2.43 M. Novak and H. Tanaka. "Pressure Correlations on a Vibrating Cylinder," *Proceedings of the Fourth International Conference, Wind Effects on Buildings and Structures,* September 1975. Cambridge, U.K.: Cambridge University Press, 1976.

2.44 NRCC. *National Building Code of Canada, Supplement No. 4.* Ottawa, Canada: National Research Council of Canada, 1985.

2.45 NRCC. *Commentaries on Part 4 of the National Building Code of Canada 1975,* Supplement No. 4 of the National Building Code of Canada (NRCC No. 13989). Ottawa, Canada: Associate Committee of the National Building Code, National Research Council of Canada, 1975.

2.46 NTIS. *Wind Load Program,* Computer Program for Estimating Along-Wind Response (NTIS Accession Number PB 294 757/AS). Springfield, VA: National Technical Information Service, 1979.

2.47 NTIS. *Hurricane Induced Wind Loads* (NTIS PB 821 32 259). Springfield, VA: National Technical Information Service, 1982.

2.48 Michael J. O'Rourke, Robert Redfield, and Peter von Bradsky. "Uniform Snow Loads on Structures," *Journal of the Structural Division,* ASCE, **108,** December 1982 (ST12), 2781–2798.

2.49 Jon A. Peterka and Jack E. Cermak. "Adverse Wind Loading Induced by Adjacent Buildings," *Journal of the Structural Division,* ASCE, **102,** March 1976 (ST3), 533–548.

2.50 G. M. Pinfold, *Reinforced Concrete Chimneys and Towers.* Flushing, New York: Viewpoint Publications, Scholium International, Inc., 1975.

2.51 H. Propper and J. Welsch. "Wind Pressures on Cooling Tower Shells," *Proceedings of the Fifth International Conference on Wind Engineering,* Fort Collins, Colorado, July 1979 (J. E. Cermak, Ed.). New York: Pergamon Press, 1980.

2.52 Mayasandra K. Ravindra and Theodore V. Galambos. "Load and Resistance Factor Design for Steel," *Journal of the Structural Division,* ASCE, **104,** September 1978 (ST9), 1337–1353.

2.53 T. A. Reinhold and P. R. Sparks. "The Influence of Wind Direction on the Response of a Square-Section Tall Building," *Proceedings of the Fifth International Conference on Wind Engineering,* Fort Collins, Colorado, July 1979 (Vol. 2). New York: Pergamon Press, 1980, pp. 685–699.

2.54 P. A. Rosati. *An Experimental Study of the Response of a Square Prism to Wind Load* (BLWT II-68), Faculty of Graduate Studies, University of Western Ontario, London, Ontario, Canada, 1968.

2.55 Wadi S. Rumman. "Basic Structural Design of Concrete Chimneys," *Journal of the Power Division,* ASCE, June 1970 (PO3), 309–318.

2.56 H. Ruscheweyh. "Dynamic Response of High-Rise Buildings Under Wind Action with Interference Effects from Surrounding Buildings of Similar Size," *Proceedings of the Fifth International Conference on Wind Engineering,* Fort Collins, Colorado, July 1979 (Vol. 2). New York: Pergamon Press, 1980, pp. 725–735.

2.57 Larry R. Russell. "Probability Distributions for Hurricane Effects," *Journal of Waterways, Harbors, and Coastal Engineering Division,* ASCE, **97,** February 1971 (WW1), 139–184.

2.58 J. W. Saunders. *Wind Excitation of Tall Buildings with Particular Reference to the Cross-Wind Motion of Tall Buildings of Constant Rectangular Cross-Section,* thesis presented to the Department of Mechanical Engineering, Monash University, Victoria, Australia, 1975, in fulfillment of the requirement for the degree of Doctor of Philosophy.

2.59 J. W. Saunders and W. H. Melbourne. "Buffeting Effects of Upstream Buildings," *Proceedings of the Fifth International Conference on Wind Engineering,* Fort Collins, Colorado, July 1979 (Vol. 1). Oxford, New York: Pergamon Press, 1980, pp. 593–607.

2.60 R. H. Scanlan. *Recent Methods in the Application of Test Results to the Wind Design of Long Suspended-Span Bridges,* Report No. FHWA-RD-75-115. Washington, D.C.: Federal Highway Administration, Office of Research and Development, 1975.

2.61 W. R. Schriever. "Estimating Snow Loads on Roofs," *Canadian Building Digest No. 193,* Division of Building Research, National Research Council of Canada, February 1978, 4 pp.

2.62 C. S. Scruton and A. R. Flint. "Wind Excited Oscillations of Structures," *Proceedings of the Institution of Civil Engineers,* April 1964, pp. 673–702.

2.63 SEAOC. *Recommended Lateral Force Requirements and Commentary.* San Francisco: Seismology Committee, Structural Engineers Association of California, 1974.

2.64 Emil Simiu. "Aerodynamic Coefficients and Risk-Consistent Design," *Journal of Structural Engineering,* ASCE, **109,** May 1983, 1278–1289.

2.65 E. Simiu et al. "Multivariate Distributions of Directional Wind Speeds," *Journal of Structural Engineering,* ASCE, **111,** April 1985, 939–946.

2.66 Emil Simiu and Martin E. Batts. "Wind-Induced Cladding Loads to Hurricane-Prone Regions," *Journal of Structural Engineering,* ASCE, **109,** January 1983, 262–266.

2.67 E. Simiu and M. Cordes. "Probabilistic Assessment of Tornado-Borne Missile Speeds," NBSIR 80-2117, National Bureau of Standards, Washington, D.C., September 1980.

2.68 Emil Simiu, James J. Filliben, and James R. Shaver. "Short-Term Records and Extreme Wind Speeds," *Journal of the Structural Division,* ASCE, **108,** November 1982 (ST11), 2571–2577.

2.69 E. Simiu and S. D. Leigh. *Turbulence Wind Effects on Tension Leg Platform Surge* (Building Science Series 151). Washington, D.C.: National Bureau of Standards, March 1983.

2.70 Emil Simiu and Stefan D. Leigh. "Turbulent Wind and Tension Leg Platform Surge," *Journal of Structural Engineering,* ASCE, **110,** April 1984, 785–802.

2.71 E. Simiu and R. H. Scanlan. *Wind Effects on Structures,* 2nd ed. New York: Wiley-Interscience, 1985.

2.72 Giovanni Solari. "Along-Wind Response Estimation: Closed Form Solution," *Journal of the Structural Division,* ASCE, **108,** January 1982 (ST1), 225–244.

2.73 Norman J. Sollenberger, Robert H. Scanlan, and David P. Billington. "Wind Loading and Response of Cooling Towers," *Journal of the Structural Division,* ASCE, **106,** March 1980 (ST3), 601–621.

2.74 T. F. Sun and L. M. Zhou. "Wind Pressure Distribution on a Ribless Hyperbolic Cooling Tower," *Journal of Wind Engineering and Industrial Aerodynamics,* **13,** 1983, 181–182.

2.75 D. A. Taylor, "Roof Snow Loads in Canada," *Canadian Journal of Civil Engineering,* **7,** March 1980, 1–18.

2.76 J. T. Templin and W. R. Schriever. "Loads Due to Drifted Snow," *Journal of the Structural Division,* ASCE, **108,** August 1982 (ST8), 1916–1925.

2.77 H. C. S. Thom. "Distribution of Maximum Annual Water-Equivalent of Snow on the Ground," *Monthly Weather Review,* **94,** April 1966 (No. 4), 265–271.

2.78 W. Tobiasson and R. Redfield. "Alaskan Snow Loads," Presented at the 24th Alaskan Science Conference, University of Alaska, August 1973, 28 pp.

2.79 Bjanni V. Tryggvason, David Surry, and Alan G. Davenport. "Predicting Wind-Induced Response in Hurricane Zones," *Journal of the Structural Division,* ASCE, **102,** December 1976 (ST12), 2333–2350.

2.80 Carl J. Turkstra and Henrik D. Madsen. "Load Combinations in Codified Structural Design," *Journal of the Structural Division,* ASCE, **106,** December 1980 (ST12), 2527–2543.

2.81 Lawrence A. Twisdale. "Tornado Data Characterization and Wind Speed Risk," *Journal of the Structural Division,* ASCE, **104,** October 1978 (ST10), 1611–1630.

2.82 L. A. Twisdale and W. L. Dunn. *Tornado Missile Simulation and Design Methodology,* EPRI NP-2005. Palo Alto, CA: Electrical Power Research Institute, August, 1981.

2.83 UBC. *Uniform Building Code.* Whittier, CA: International Conference of Building Officials, 1985.

2.84 B. J. Vickery. "On the Reliability of Gust Loading Factors," *Proceedings of the Technical Meeting Concerning Wind Loads on Buildings and Structures* (Building Science Series 30). Washington, D.C.: National Bureau of Standards, 1970, pp. 93–104.

2.85 B. J. Vickery. "Notes on Wind Forces on Tall Buildings," Annex to *Australian Standard 1170, Part 2-1973, SAA Loading Code Part 2—Wind Forces.* Sydney, Australia: Standards Association of Australia, 1973.

2.86 B. J. Vickery. "The Aeroelastic Modeling of Chimneys and Towers," *Proceedings of the Wind Tunnel Modeling for Civil Engineering Applications,* Gaithersburg, MD, April 1982. Cambridge, New York: Cambridge University Press, 1983.

2.87 B. J. Vickery and R. I. Basu. "Across-Wind Vibrations of Structures of Circular Cross-Section, Part I, Development of a Two-Dimensional Model for Two-Dimensional Conditions," *Journal of Wind Engineering and Industrial Aerodynamics,* **13,** 1983.

2.88 Barry J. Vickery and Arthur W. Clark. "Lift of Across-Wind Response of Tapered Stacks," *Journal of the Structural Division,* ASCE, **98,** January 1972 (ST1), 1–20.

2.89 H. B. Walker, Ed. *Wind Forces on Unclad Tubular Structures.* (Constrado Publication 1/75). Croydon, U.K.: Constructional Steel Research and Development Organization, 1975.

2.90 R. L. Wardlaw. *Static Force Measurement of Six Deck Sections for the Proposed New Burrard Inlet Crossing,* Report No. LTR-LA-53, NAE. Ottawa, Canada: National Research Council, 1970.

2.91 Yi-Kwei Wen. "Statistical Combination of Extreme Loads," *Journal of the Structural Division,* ASCE, **103,** May 1977 (ST5). 1079–1095.

2.92 Yi-Kwei Wen. "Wind Direction and Structural Reliability," *Journal of Structural Engineering,* ASCE, **108,** April 1983, 1028–1041.

2.93 R. E. Whitbread. "The Influence of Shielding on the Wind Forces Experienced by Arrays of Lattice Frames," *Proceedings of the Fifth International Conference on Wind Engineering,* Fort Collins, Colorado, July 1979. New York: Pergamon Press, 1980, pp. 405–420.

2.94 R. L. Weigel, Ed. *Earthquake Engineering.* Englewood Cliffs, NJ: Prentice-Hall, 1970.

CHAPTER 3
DESIGN PHILOSOPHIES

CLARKSON W. PINKHAM

S. B. Barnes and Associates
Consulting Structural Engineers
Los Angeles, California

3.1 STRUCTURAL DESIGN

The structural design of a building is the process by which adequate strength, rigidity, and toughness are obtained. In this case "adequate" implies that the structure throughout its usable life will provide satisfactory service to its owners and occupants when natural or man-made loads or motions are imposed on it. In order to perform this design, the effects that various loads or combination of loads produce on the structure are determined by structural load analysis. These load effects are then compared to the capability of the structure so that the adequacy of the structure or its components can be assessed.

The initial part of structural design is a collaborative effort with the other members of the design team (architect, planner, and the mechanical, electrical, and acoustical engineers) to develop the optimum form of the structural system. Due consideration to these other requirements is essential to the structural performance of the building. For a thorough discussion of form as a part of the structural design process, see Chapter 8.

The structural design engineer determines the level of approximation (i.e., the appropriate structural "model") to be used in the load analysis and also in the member strength analysis. This level varies depending on the particular structure being designed. Frequently, the crudest approximations are all that are needed with little, if any, effect on the overall cost of the project. As the building increases in size and importance, or if members are repetitive, it becomes appropriate to analyze both loads and strength in more detail. Thus while most structural analysis textbooks emphasize the complex and detailed analytical procedures, it is just as important to understand the methods and limitations of approximations. The determination of the precision required is frequently the most important decision the designer will be required to make.

An important part of preliminary structural design is the selection of the structural system with consideration given to its relationship to construction economics. Material, fabrication, and erection costs need to be correlated with the cost related to the time and speed of erection, loan repayment schedules, and the maintenance cost during the life of the structure so that the optimum system can be chosen.

3.2 STRUCTURAL SAFETY

The excess of calculated strength over calculated load effect is a measure of implied safety. An excess is necessary to provide for the uncertainties in predicting strength by calculation, the uncertainties of determining the exact magnitude of load that will be imposed, and the uncertainty in the determination of the effect on the structure by the imposed loading.

Many studies into the concepts and formulation of structural safety have been made within the last 25 years [3.6, 3.8, 3.13, 3.15, 3.16]. The main concern of these studies is to determine by various probabilistic methods the chances of having a failure occur in a member, connection, or building system. The use of the word "failure" usually implies only that the member, connection, or system is not performing satisfactorily to the loading condition imposed on it. A thorough analysis of all the uncertainties that might influence the "failure" becomes very involved. The current trend of finding a simplified method using a probability-based assessment of structural safety is termed a *first-order, second-moment* procedure [3.10]. It is assumed that the load Q and resistance R are random variables in the design equation. Structural failure can then be defined in two ways as shown in Fig. 3.1. This figure shows the probability density functions using two different methods of comparing R and Q. In both cases the cross-hatched area indicates the area in which failure is defined. The location of the failure line from the mean value of each function [$R - Q$ or $\ln(R/Q)$] is defined as a given multiple of the standard deviation σ of the function. The multiplier is termed a reliability (or safety) index β. This index becomes useful in several ways:

1. It can be used to give an indication of the consistency of safety for various components of current design methods.
2. It can be used to determine design methods having fairly consistent factors of safety.
3. It can be used to vary the margin of safety for those members, connections, or systems having a greater or lesser need for safety than those used for normal design situations.

A general expression for the structural design process has been given by Ravindra and Galambos [3.18] as

$$\phi R \geq \Sigma \gamma Q \tag{3.2.1}$$

in which ϕR represents the resistance or strength of a member and $\Sigma\gamma Q$ represents the effects (such as bending moment, shear, or axial force) at any section from the loads applied to the structure. This expression will be used to show the relationship between the different strategies used in design.

Loads on a building producing the load effects Q can be categorized into several types [3.10]:

1. *Permanent* loads remain relatively constant over the life of the building even though random minor variations do occur. The dead load of a structure is a loading of this type.
2. *Sustained* loads remain relatively constant between times of major change and can be absent entirely at times.
3. *Transient* loads are loads of relatively short duration that usually occur infrequently.

Live load on a building can be the sum of all three types of loads. The magnitude of live load (or other loads composed of a combination of sustained and transient loads) that is likely to occur at

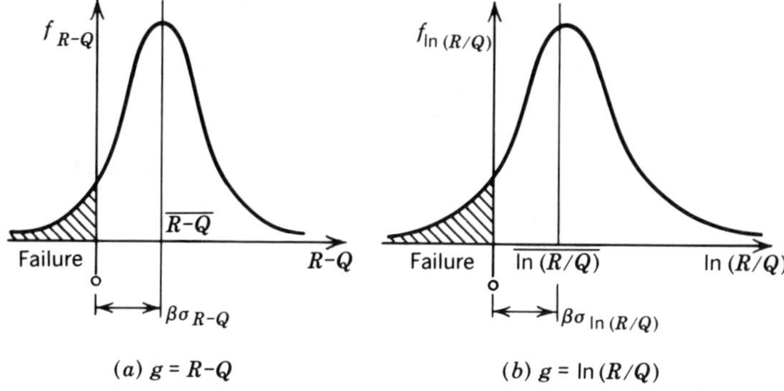

Fig. 3.1 Illustrative examples of definitions of "failure" and reliability index β.

any time a load evaluation is made is called an "arbitrary-point-in-time" load. This load is considerably less than the maximum live load which will occur at relatively rare intervals. In any case, live loads have a higher level of uncertainty than dead loads, and this fact is explicitly accounted for when load factors are used.

Wind loads are a combination of sustained and reversing transient loads while earthquake loads are solely reversing loads of the transient type.

Impact loads as covered by building design specifications are modeled as a percentage of the transient live load.

Horizontal earth loads may be a combination of permanent and sustained loads. They have a high level of uncertainty so that when load factors are used, the factor is similar to that used for live loads. Horizontal fluid pressure is a sustained loading which can be predicted with a greater certainty, hence the load factor for fluid pressure is similar to that for dead load.

When considering floor live loads, the transient loads are usually heavy objects or stored material occurring over a relatively small area. Thus when members support large floor areas, the average maximum floor load per unit area is reduced. In the past these reductions have been based on a method using "tributary areas" [3.14] to the member being designed. In ANSI A58.1 [3.4] the reductions for area use an "influence area" of the member. As explained in Chapter 2, the influence area for beams is twice and for columns it is four times the tributary area. The resulting values are not significantly different but the new method uses the actual area that would influence the loading to a member and results in consistent loading to beams and columns. Restrictions and limitations are placed on the use of floor live load reductions by the governing building code and these should be verified before using them.

When the load effects from several different load sources are combined, the probability is extremely low that the maximum effect from each load source will occur simultaneously. The way that the different design methods handle this problem is discussed later.

Structural safety does not only relate to strength mechanisms of failure. Unsatisfactory performance of a member or system can also occur by exceeding what is called *serviceability limit states*. These limit states can be categorized as:

1. Local damage (local yielding, buckling, slipping, or creeping) requiring excessive maintenance or local conditions conducive to corrosion.
2. Vibrations induced by transient live loads or wind which adversely affect the comfort of occupants of the structure or the satisfactory operation of mechanical equipment.
3. Deflection or rotation that significantly affects the appearance, use, or drainage of a structure, or may cause damage to the nonstructural components or their attachments.

Some serviceability limit states are specifically found in building design requirements, such as deflection or drift limitations, span-to-depth ratio limitations, and crack control (in concrete). Other provisions are hidden by using high factors of safety for design or by using thickness limitations. Devising specification provisions for controlling these many serviceability limit states is difficult because the limits are dependent on the type of structure and its intended use. In most cases, therefore, the serviceability of the structure is left to the experience and ingenuity of the design team.

3.3 ELASTIC DESIGN PROCEDURES

Design experience has shown that design assumptions using a completely elastic stress–strain relationship provide the simplest means of determining the distribution of load effects on the structural system and for analyzing the strength of members and connections. Use of the linear equations resulting from these assumptions not only is simple but it also permits the independent analysis of different loading conditions with direct superposition for load combinations. Unfortunately no material deforms precisely in an elastic manner and the strengths of the members in the structural system are often significantly in excess of those that would be predicted by elastic methods. Thus it has been found that in order to closely assess the performance of the structural system, modifications to the design assumptions are needed. It is these design modifications that result in many of the added design complexities decried by the design professions in recent years. Examples of some of these modifications are redistribution of moments in continuous members (such as steel beams and concrete slabs), plastic strength of steel members, nonlinear compression stress distribution in concrete flexural members, and allowable stress modifications based on load duration in wood members.

In design, the distribution of loads to determine their effect on members within the structural system is usually performed on a member-by-member basis using the summation of loads tributary to the member under design consideration. When the structural system becomes complex, such as when moment-resisting frames are used to resist lateral loads in multistory buildings, elastic two-dimensional frame analyses are often used. The advent of the computer has made practical three-dimensional elastic analyses of structural frames of buildings of modest complexity. It is important to consider the limitations of these elastic analyses so that changes can be made to either correct weaknesses or to redistribute load effects where appropriate.

3.4 LIMIT STATES DESIGN

Limit states design has been defined as a process that involves [3.10]:

1. Identification of all modes of failure or ways in which the structure might fail to fulfill its intended purpose (limit states).
2. Determination of acceptable levels of safety against occurrence of each limit state.
3. Consideration by the designer of the significant limit states.

When considering the usual design situation, items 1 and 2 are in general predetermined for the designer by building code requirements and material standards. Item 3 is the step performed by the designer.

The usual limit states that must be considered in item 1 are:

1. Collapse.
2. Instability.
3. Deformation (elastic and plastic, static, and dynamic).
4. Durability.
5. Fatigue.
6. Brittle rupture.

Limit states can also be categorized by the way in which failure is manifested. Thus there are strength limit states and serviceability limit states. In many cases elastic analysis of serviceability limit states provides a good estimation of the behavior. Some, however, such as long-term deflection, creep, and shrinkage in wood, masonry, or concrete are actually nonlinear effects, and even though elastic methods are used, modifications are needed to account for the nonlinearity. In most cases, the bulk of the designer's time is spent in consideration of the strength limit state.

In the design of concrete structural systems, an elastic load analysis is normally performed. Some redistribution of moments is permitted for continuous systems. Research is continuing on additional modifications of elastic methods so that a closer representation of actual conditions found in buildings can be made. In general, the strengths of members are found by using nonlinear stress–strain relationships with limitations placed on concrete strains. Elastic stress–strain relationships continue to be used by many designers, however, and modifications to elastic procedures have been made so that the final design is approximately the same as when using nonlinear assumptions.

In steel design both elastic and nonlinear (plastic) design procedures are used for load analysis. The determination of strength can be made also using elastic or plastic procedures. In the design specifications for steel buildings, the elastic strength limits are modified so that there is little difference between the two methods in the final design.

Aluminum structures are usually designed only by elastic methods.

Special design approaches are needed for structural plastics because the plastics are sensitive to load duration, environmental conditions, and other factors (Chapter 30).

3.5 ALLOWABLE (WORKING) STRESS METHODS

The general expression, Eq. (3.1.1), for structural design may be modified to become

$$\phi R/\gamma \geq \Sigma Q \tag{3.5.1}$$

This places all the variability of loads and strengths on the strength side of the equation. In order to simplify the design procedure it can be assumed that all loads have a common (or average) variability. This then is the essence of the *allowable stress method* (or *working stress method*) of design. The $\phi R/\gamma$ values are the allowable stresses assigned by codes or specifications and the ΣQ is the sum of the nominal load effects at the section being designed.

This is the design method most frequently used over the past 75 years by designers for common structural materials. The greatest attribute has been its simplicity since consideration of the variable risks and probability of failure need not be part of design concern. In most design situations, this method can be used to produce reasonably safe and usable structures as long as the designer remains cognizant of the shortcomings so that when appropriate, other methods of design would be used.

In addition to the inclusion of the variability of load and strength into the ratio ϕ/γ, nonlinear behavior such as the formation of plastic hinges and redistribution of moment in flexural members can also be included. This is accomplished by simply modifying the allowable stress of the member.

The allowable stresses are also increased by one-third for load combinations involving wind or seismic forces. This factor is viewed in two ways, both resulting from the short duration of these types of loadings. First, some materials (such as wood and concrete) do exhibit higher strength for short-time loading vs. long-time loading. The short-term strength of concrete members occurs for stress consideration which is then modified by creep. The second view is that the simultaneous occurrence of maximum load effects from several load sources is very remote so that under the temporary condition a reduced safety margin can be tolerated. To recognize this, the allowable stresses are increased. Some building codes (such as ICBO [3.14]) permit this increase to be used to resist stresses resulting from wind or seismic forces alone or in combination with other load sources. Other building codes (such as ANSI [3.4]) have a double reference. One is that load reductions are permitted when loads from more that two sources are considered. If load reductions are used, then the increase in stresses stipulated in each of the material sections would not be used. As all of the current allowable stress material specifications do permit the stress increase, the net effect is the same as for ICBO [3.14].

For the design of wood members, allowable stresses depend on load duration. Increases are permitted for loads with duration less than two months, and a reduction is prescribed when the member is continuously loaded for many years.

The allowable stress modifications based on the probability of occurrence of load combinations is a simplified design approach which can lead to variable real margins of safety as opposed to the nominal safety factors assumed to be used in development of the method. This is particularly so when a combination of loads oppose each other such as the combination of dead load with uplift wind load effects. The dead load usually has only a small variation during the life of a building, whereas wind is a highly uncertain type of loading. When the design situation is such that the dead load significantly (but not entirely) reduces the wind load effect, the margin of safety can be very low. This can be seen by a simple example, shown in Fig. 3.2.

In Fig. 3.2a assume the dead load effect on a column is 100 kips ± 10 kips (compression), and the wind load effect is 150 kips ± 100 kips (tension). The allowable stress design load would be 150 − 100 = 50 kips (tension). The actual load for which a margin of safety should be provided would be 150 + 100 − 100 + 10 = 160 kips (tension). The required margin of safety would be 160/50 = 3.2. Assuming as in Fig. 3.2b the dead load was only 10 kips ± 1 kips (compression), and the wind load was 60 kips ± 40 kips (tension), the design load would be 60 − 10 = 50 kips (tension). This is the same as before, thus the member would be sized to resist the same tension in both cases. The real

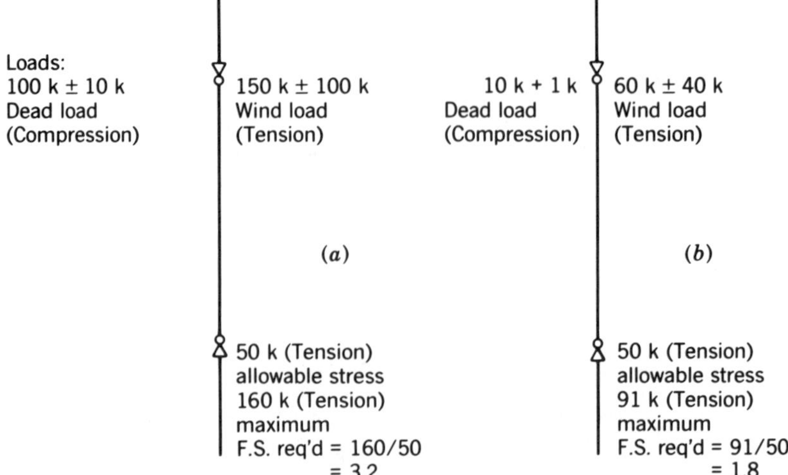

Fig. 3.2 Inconsistency of margins of safety.

load would be 60 + 40 − 10 + 1 = 91 kips (tension). This indicates that the margin of safety of would be 91/50 = 1.8. Thus on any member or structure of high importance, a close look at the implied margin of safety should be made when using this method of design.

3.6 LOAD AND RESISTANCE FACTOR DESIGN

The *load and resistance factor design* (LRFD) method is one in which the general design equation, Eq. (3.1.1), is used with a factor ϕ representing the variability of the resistance R_n function and a factor γ representing the variability of the load Q_n function. The influence of the safety index β is included in both ϕ and γ. The American Concrete Institute (ACI) Code and Commentary [3.1, 3.2] have been available in this format since 1956. Canadian specifications (NRCC) [3.17] are also available in LRFD format. The American Institute of Steel Construction (AISC) also has a proposed specification [3.3] in LRFD format.

The methods used in the development of both load and resistance factors have had several variations. In general, the design equation has been made slightly more explicit. Thus

$$\phi R_n \geq \sum_{k=1}^{j} \gamma_k Q_{km} \tag{3.6.1}$$

In this equation R_n is the nominal resistance of the member or connection and Q_{km} is the mean load effect for the k type load. The total number of concurrent loads is j.

Some of the variations on this approach propose to use split load factors for load combinations while others provide load combination modifiers. Some have tried to establish as near as possible a constant margin of safety in which the level of safety was selected from current design practices [3.18]. Still others have determined the range of safety that current practice yields and then selected appropriate values to minimize the variations [3.16].

The nominal resistance R_n is a generalized force such as bending moment, shear, or axial force associated with the particular limit state of strength or serviceability under consideration. The determination would be based on the methods outlined in the materials specification or code using nominal material and cross-sectional properties. Interaction equations between the different forces (such as bending moment and axial load, bending moment and shear, etc.) may also be represented by R_n.

In code development for the steel industry [3.13, 3.18], Eq. (3.6.1) was further refined to obtain the final LRFD format. Thus,

$$\phi R_n \geq \gamma_E(\gamma_D C_D D_m + \gamma_L C_L L_m) \tag{3.6.2}$$

assuming that only dead and live loads are considered. In this case, ϕ is a factor representing the variability of resistance, whereas γ_D and γ_L are load factors representing the variability of the dead and live loads; D_m and L_m are the mean dead and live loads; C_D and C_L represent the factors converting load to the appropriate load effect (flexure, shear, or axial load) on the structure. γ_E in this equation is the connecting load factor representing the other uncertainties of analysis vs. strength that would provide an appropriate margin of safety. In the latest development of LRFD [3.3, 3.10], γ_E has been combined with the other load factors and does not appear in the design equation.

In the past, load and resistance factors have frequently been chosen with a heavy reliance on subjective judgment. Current studies in the development of both factors have been based on the use of probabilistic methods of limit states procedures. These methods have been used to form a guide in the selection of the design factors in order to account for the variabilities of loads and resistances so that a more uniform overall level of safety can be achieved.

In order to perform this analysis, the following expression has been used (considering the dead plus live load case only),

$$\exp(-\alpha\beta V_R)R_m \geq \exp(\alpha\beta V_E)[(1 + \alpha\beta\sqrt{V_A^2 + V_D^2})C_D D_m + (1 + \alpha\beta\sqrt{V_B^2 + V_L^2})C_L L_m] \tag{3.6.3}$$

where D and L = the random variables representing dead and maximum live load intensities

A and B = random variables representing the uncertainty of transforming loads to load effects

E = random variable representing the uncertainty of structural analysis

V_D, V_L, V_A, V_B, V_E = coefficients of variation of the above variables

α = factor producing a linear approximation separating load and resistance variables (a factor of 0.55 was found to provide reasonable approximations within the area of interest [3.18])

Comparing Eq. (3.6.3) with Eq. (3.6.2) it can be seen that

$$\phi = \exp(-\alpha\beta V_R)R_m/R_n \tag{3.6.4a}$$

$$E = \exp(\alpha\beta V_E) \tag{3.6.4b}$$

$$D = 1 + \alpha\beta\sqrt{V_A^2 + V_D^2} \tag{3.6.4c}$$

$$L = 1 + \alpha\beta\sqrt{V_B^2 + V_L^2} \tag{3.6.4d}$$

Placing numerical values into these equations, the design equation for plastic design of a simply supported steel beam becomes [3.18]

$$0.86ZF_y \geq 1.1(1.1C_D D_m + 1.4C_L L_m) \tag{3.6.5}$$

Modifying this equation to the values consistent with the later development of ANSI A58.1 [3.4], this equation becomes

$$0.86ZF_y \geq 1.2C_D D_m + 1.6C_L L_m \tag{3.6.6}$$

The method of combining other load types used in ANSI A58.1 [3.4] is based on the use of the dead load plus the maximum of the principal load type being considered plus the "arbitrary-point-in-time" loads of the other relevant load types. For opposing load cases, the live load is assumed to be zero and the maximum effect of the opposing loads is considered.

The final load combinations and load factors selected for ANSI A58.1 [3.4] are:

$$\left.\begin{aligned}
&1.\ 1.4(D + T)\\
&2.\ 1.2D + 1.6L + 0.5(L_r \text{ or } S \text{ or } R)\\
&3.\ 1.2D + 1.6(L_r \text{ or } S \text{ or } R) + (0.5L \text{ or } 0.8W)\\
&4.\ 1.2D + 1.3W + 0.5G\\
&5.\ 1.2D + 1.5E + (0.5L \text{ or } 0.2S)\\
&6.\ 0.9D - (1.3W \text{ or } 1.5E)
\end{aligned}\right\} \tag{3.6.7}$$

The symbols used in these load combinations are defined as follows:

D = dead loads consisting of
- (a) the weight of the member itself
- (b) the weight of all materials of construction incorporated into the building to be permanently supported by the member, including built-in partitions
- (c) the weight of permanent equipment
- (d) net forces due to prestressing except that the prestress effect should be taken as 0.75 of the calculated effect if the net prestress is beneficial to the limit state considered

G = variable gravity load consisting of
- (a) L = live loads due to intended use and occupancy (including loads due to movable objects and partitions, impact, traveling cranes and loads temporarily supported by the structure during maintenance)
- (b) L_r = roof live loads
- (c) S = snow loads
- (d) R = rain loads

For roofs and members supporting only roof loads,

$$G = L_r \text{ or } S \text{ or } R$$

For members subjected to roof loads and live loads from other portions of a structure, live loads from the other portions shall be considered to act in combination with snow, rain, or roof live loads as follows:

$$G = L + L_r$$

$$G = L + S$$

$$G = L + R$$

T = loads, forces, and effects due to contraction or expansion resulting from temperature changes, shrinkage, moisture changes, creep in component materials, movement due to differential settlement, soil and hydrostatic pressure, or combinations thereof
W = wind load
E = earthquake load

Serviceability limits are usually placed on the performance of the building under normal service load conditions. Thus the loads have a load factor of 1 and the response of the structure is normally elastic. Long-term effects, however, such as creep, are frequently part of the serviceability behavior.

The appropriate resistance factors ϕ for all types of members are being developed for both hot-rolled and cold-formed steel members and should be available as an alternate design procedure by the late 1980s. In order to use the ANSI load factors in the design of concrete members, adjustments are necessary to the factors given in the *ACI Code* [3.1]. Proposals for the modifications are being reviewed and should be available in future codes as an alternative design method.

It has been noted by Ellingwood et al. [3.10] that the main advantages of load and resistance factor design are:

1. More consistent reliability is attained for different design situations because the different variabilities of various strengths and loads are considered explicitly and independently.
2. The reliability level can be chosen to reflect consequences of failure.
3. The designer has a better understanding of the fundamental structural requirements and of the behavior of the structure in meeting those requirements.
4. The design process is simplified by encouraging the same design philosophy and procedures to be adopted for all materials of construction.
5. It is a tool for exercising judgment in nonroutine situations.
6. It provides a tool for updating standards in a rational manner.

The main drawback of the LRFD procedures is that they do produce a somewhat more complex system of design than the allowable stress procedures. This can be eased by multiplying the nominal strength R_n by ϕ/γ_{avg}, in which ϕ/γ_{avg} is an average factor for the material and load combinations being considered. The design equation then reverts to Eq. (3.5.1) and thus is the same as the allowable stress procedure. This modification, however, should be used conservatively and only when accuracy can by sacrificed to gain simplicity.

3.7 EARTHQUAKE DESIGN CONCEPTS

Design of buildings to resist earthquake motions got its impetus during the rebuilding of San Francisco after the 1906 earthquake. This was accomplished by requiring a horizontal force of 30 psf (1.44 kN/m^2) applied to the exterior of the building [3.20]. This force was to provide for both wind and earthquake design. This was considered proper for providing earthquake resistance because it had been noticed that the collapse of buildings during an earthquake normally occurred in a horizontal direction.

The concept that the horizontal force to be used would be proportional to the mass of a building finally appeared in PCBOC [3.19] of 1927. These design provisions were introduced therein by the following statement:

> *The design of buildings for earthquake shocks is a moot question but the following provisions will provide adequate additional strength when applied in the design of buildings or structures.*

This early code also contained as part of its requirements:

> *All buildings shall be firmly bonded and tied together as to their parts and each one as a whole in such a manner that structure will act as a unit.*

These two statements are still relevant to earthquake design even though much more thorough, explicit, and detailed provisions are currently used.

The next basic change in design concepts came in 1941 when the City of Los Angeles in its building code provisions permitted a gradation of seismic horizontal force factors for multistory buildings. The factor was heaviest on the top story and became gradually smaller as the lower stories were considered. The only variable other than mass (weight) was the number of stories. It should be noted that at the time buildings in the city were limited to 13 stories or 150 ft (45 m) in height (not for seismic concerns but for zoning).

It had long been recognized that a principal parameter affecting the response of a building to earthquake motions was the stiffness of the building. This was first recognized in design provisions by the report of the Joint Committee in 1952 [3.5]. It was accomplished through the use of a base shear directly proportional to the building mass and inversely proportional to the fundamental period of vibration. In order to predetermine the period for design purposes, a simplified equation based on building geometry was provided. The base shear was distributed as a series of forces which were in the shape of an inverted triangle for a uniformly stiff building with uniformly distributed masses. These concepts have been the model for most of the seismic design code provisions since 1952. In the most recently developed provisions, this method is referred to by the Applied Technology Council (ATC) [3.7] as the "Equivalent Lateral Force Procedure."

In general, it is intended that the forces used in design be placed on an elastic model of the building. Load effects from the combination of dead, live, and earthquake loads are determined and members sized to be capable of producing the required strengths.

Starting in 1959, the Structural Engineers Association of California (SEAOC) has periodically issued its Recommendations for use in seismic design. In the Commentary of the 1967 edition of these Recommendations SEAOC outlined its goals in determining earthquake design criteria:

> *The SEAOC Code is intended to provide criteria to fulfill the purposes of building codes generally. More specifically with regards to earthquakes, structures designed in conformance with the provisions and principles set forth therein should be able to:*
>
> 1. *Resist minor earthquakes without damage;*
> 2. *Resist moderate earthquakes without structural damage, but with some nonstructural damage;*
> 3. *Resist major earthquakes, of the intensity or severity of the strongest experienced in California, without collapse, but with some structural as well as nonstructural damage.*

It can be noted that in referring to the size of earthquakes, reference is made to the intensity of shaking, not the magnitude. This means that an earthquake of "medium" magnitude can produce local shaking that is very severe.

In order to achieve the capability of performing satisfactorily in "major" earthquakes, a number of detailed requirements and limitations are given in specifications. These are based on a concept of providing for nonbrittle, stable elements when the system is required to perform beyond its elastic limit.

Various forms of "dynamic" analyses can also be used in design (see also Chapter 18). These are permitted by general clauses in building codes and are seldom mentioned in detail. These analyses can be described as:

1. Elastic modal analyses using a design spectra (either specified in the code or specific for the site).
2. Elastic modal analyses using earthquake time-histories.
3. Integrated elastic time-history analyses.
4. Integrated nonlinear time-history analyses.

Dynamic methods as well as the equivalent lateral force method can be modified by considering the interaction of the building with its foundation. This effect is normally not included in the analysis as a rigid base is the usual assumption.

In the four types of dynamic analyses mentioned, the important ingredient of design is in the selection of the "design" earthquake. First, either through seismicity studies or by studies of the effects of previous earthquakes in the area, a group of earthquake motions having a reasonable chance of occurring during the life of the building are chosen. Reasonableness is not a definitive way to describe a quantity. In ATC [3.7] it was assumed to be an earthquake that would not be exceeded within 50 years with a probability of about 90%. From the group of motions chosen, the response spectra can be derived. A response spectrum is found by determining the response of all elastic single-degree-of-freedom vibrators having periods within the range of interest. When plotted, these become a response spectrum. A composite smoothed spectrum can be drawn which considers the spectra of the chosen earthquakes. This becomes the design spectrum. Similar spectra can be derived assuming a multistory building with uniform stiffness and mass distribution. This can be used to represent the response of multistory buildings in general but would only be a crude approximation in some cases.

In the case of either the equivalent force method or the elastic dynamic methods, a reduction in response is introduced to account for nonlinear behavior, soil–structure interaction, and past observed reliability of the structural system being used. If the allowable stress method of design is used, a further reduction is made so that the working stress level forces would be achieved. In the case of the design spectra for the equivalent lateral force method [3.7, 3.14] and the modal analysis with design spectrum [3.7], a single design spectrum was chosen modified for different foundation conditions. For the other types of "dynamic" analyses, design spectra are usually based on site specific studies or on four or more time-history plots.

Thus for earthquake design, it is anticipated that the response to the specified forces in building codes will produce elastic response. However, it is to be anticipated that the building during its useful life will be required to undergo motions very much in excess of these elastic responses. The satisfactory performance of the building to these motions will be determined by attention to detail in design, elimination of brittle response characteristics, and by how well the design drawings are actually implemented during the construction.

If the design concept of the building calls for essentially elastic response to the design earthquake, some relaxation of the arbitrary provisions for producing ductility in the system would be reasonable. But at the same time the design earthquake should not be reduced for nonlinear response. In order to accomplish this, elastic design forces would be from four to eight times those factors given in the ATC [3.7]. Thus in areas of high seismicity, this method of design would in general not be practical.

An additional requirement has been added in recent years to earthquake design which stipulates that some critical buildings shall remain operational after a "major" earthquake. This is accomplished by maintaining a limit on deformations the building will undergo during the "major" earthquake and by close attention to the bracing of nonstructural systems within the building.

3.8 BUILDING CODES AND THEIR IMPACT ON STRUCTURAL DESIGN

Building codes are the laws generated by a governing agency that deal with the planning, design, construction, and use of buildings within their jurisdiction. Despite the many efforts to provide standardized or uniform building regulations, there are literally thousands of jurisdictions within the United States which in one way or another vary. Three major organizations have made efforts to provide uniform regulations. These are the Southern Building Code Congress International (SBCCI), Building Officials and Code Administrators International (BOCA), and the International Conference of Building Officials (ICBO). These three groups have also jointly established an agency called the Council of American Building Officials (CABO) to try to establish uniform provisions for model code implementation where it is possible. In addition to these three, many of the larger jurisdictions maintain independent building codes with varying success in obtaining uniformity with the model codes. Another organization active in the effort to provide uniform regulations is the National Conference of States on Building Codes and Standards (NCSBCS). This organization is composed of appointees from individual states. It has made progress in the area of factory built housing.

The zoning, fire, and panic regulations within a building code do have a strong impact on building and structural design, often dictating what can and what cannot be done. These regulations often directly place limitations on some materials and systems. The floor area, number of stories, fire zone, combustibility, and high temperature material strength all have a direct effect on the appropriate system to be used. The limits imposed by various building codes vary with the jurisdiction. Thus determination of the governing code or codes is necessary even prior to the conceptual design of a building.

Most building codes use as standard references the code provisions or specifications published by the technical trade associations for each basic structural building material. As these provisions or specifications are not inherently legal documents, they require adoption by legislative bodies. Thus, they are frequently not current. The appropriate year of the provisions or specifications to be used must be stated in the design documents. In addition to the minimum legal requirements delineated in a code, it is necessary for a designer to determine whether or not the general requirements contained therein are appropriate in each particular design. If not appropriate, then variances or modifications may be required from the building official prior to the start of design.

SELECTED REFERENCES

3.1 ACI Committee 318. *Building Code Requirements for Reinforced Concrete* (ACI 318-83). Detroit: American Concrete Institute, 1983.

3.2 ACI Committee 318. *Commentary on Building Code Requirements for Reinforced Concrete* (ACI 318-83). Detroit: American Concrete Institute, 1983.

3.3 AISC. *Proposed Load & Resistance Factor Design Specification for Structural Steel Buildings*. Chicago: American Institute of Steel Construction, September 1, 1983.

3.4 ANSI Committee A58. *Building Code Requirements for Minimum Design Loads in Buildings and Other Structures* (ANSI A58.1). Washington, D.C.: American National Standards Institute, 1982.

3.5 Arthur W. Anderson, John A. Blume, Henry J. Degenkolb, Harold B. Hammill, Edward M. Knapik, Henry L. Marchand, Henry C. Powers, John E. Rinne, George A. Sedgewick, and Harold O. Sjoberg. "Lateral Forces of Earthquake and Wind," *Transactions,* ASCE, **117,** 1952, 716–780.

3.6 ASCE-STD. Task Committee on Structural Safety of the Administrative Committee on Analysis and Design of the Structural Division. "Structural Safety—A Literature Review," *Journal of the Structural Division,* ASCE, **98,** April 1972 (ST4), 845–884.

3.7 ATC. *Tentative Provisions for the Development of Seismic Regulations for Buildings* (ATC 3-06), NBS SP510, NSF 78-8. Washington, D.C.: Applied Technology Council, June 1978.

3.8 J. Bantanero. "Theme Report for Technical Committee No. 19—Load Factor (Limit States) Design," *Proceedings of ASCE-IABSE International Conference,* II-19. New York: American Society of Civil Engineers, August 1972, pp. 1–9.

3.9 C. Allin Cornell, Luis Esteva, and Roberto Meli. "Structural Safety and Probabilistic Methods," *Monograph on Planning and Design of Tall Buildings,* Vol. CL, *Tall Building Criteria and Loading.* New York: American Society of Civil Engineers, 1980, pp. 535–643.

3.10 Bruce Ellingwood, Theodore V. Galambos, James G. MacGregor, and C. Allin Cornell. *Development of a Probability Based Load Criterion for American National Standard A58* (NBS Special Publication 577). Washington, D.C.: U.S. Department of Commerce, National Bureau of Standards, June 1980.

3.11 Alfred M. Freudenthal, Jewell M. Garrelts, and Masanobu Shinozuka. "The Analysis of Structural Safety," *Journal of the Structural Division,* ASCE, **92,** February 1966 (ST1), 267–325.

3.12 Alfred M. Freudenthal. Safety and the Probability of Structural Failure," *Transactions,* ASCE, **121,** 1956, 1337–1375.

3.13 Theodore V. Galambos and Wei-Wen Yu. "Load and Resistance Factor Design of Cold-Formed Steel. Tentative Recommendations—Load and Resistance Factor Design Criteria for Cold-Formed Steel Structural Members and Commentary Thereon," *Civil Engineering Study 80-1,* Structural Series, University of Missouri-Rolla, Rolla, MO, March, 1980.

3.14 ICBO. *Uniform Building Code.* Whittier, CA: International Conference of Building Officials (5360 South Workman Mill Road), 1985.

3.15 O. G. Julian. "Synopsis of First Progress Report of Committee on Factors of Safety," *Journal of the Structural Division,* ASCE, **83,** July 1957 (ST4), 1316–1322.

3.16 Niels C. Lind. "Consistent Partial Safety Factors," *Journal of the Structural Division,* ASCE, **97,** June 1971 (ST6), 1651–1669.

3.17 NRCC. *National Building Code of Canada* (NRCC NO. 17303). Ottawa: National Research Council of Canada, Associate Committee on the National Building Code, 1980.

3.18 Mayasandra K. Ravindra and Theodore V. Galambos. "Load and Resistance Factor Design for Steel," *Journal of the Structural Division,* ASCE, **104,** September 1978 (ST9), 1337–1353.

3.19 PCBOC. *Uniform Building Code.* Los Angeles, CA: Pacific Coast Building Officials Conference, 1927.

3.20 SEAOC Seismology Committee. *Recommended Lateral Force Requirements and Commentary,* 4th ed. San Francisco, CA: Structural Engineers Association of California, 1980.

CHAPTER 4

MECHANICAL AND ELECTRICAL SYSTEMS

PARTNERS OF JAROS, BAUM & BOLLES

Consulting Engineers
New York, New York

4.1 INTRODUCTION

The design of mechanical and electrical systems for buildings is not possible without an understanding of how these systems will relate to the structural elements in the building. A symmetrical observation is possible concerning the design of structures, to wit:. It is not possible to properly develop a structural design without understanding the requirements of the mechanical and electrical systems, and what impact these needs will have on the building's structure.

It is hoped that this chapter will permit some understanding of the complex alternative solutions possible for the mechanical and electrical designs, and more important how these alternatives affect the structural design. Sections 4.3 through 4.9 on mechanical systems include a general discussion of heating, ventilating, and air conditioning (HVAC) alternatives, including the equipment used to achieve these alternatives; a general discussion of the plumbing system, including fire protection; and then a discussion of how the equipment, piping and ductwork included for the various systems, affects the structure.

On the electrical side (Sections 4.10–4.14), alternative power from typical utilities is discussed. This is followed by a review of how the electrical system is distributed throughout the building, and what the importance of alternative distribution means on the structural design.

4.2 CONCEPTUAL COALS

The design process of the mechanical and electrical systems must be conditioned by the goals, both stated and implied, which are inherent in the building and its function. One need only reflect on the fact that mechanical and electrical systems for a building can constitute anywhere from 25 to 45% of the total building cost to appreciate the need to provide proper system designs. Moreover, the mechanical and electrical systems affect not only the building's first cost but also its operating costs. The mechanical and electrical systems selected for the building are the points of energy consumption in the building. By this we mean the components that make up the various mechanical and electrical systems in a sense are the users of energy in a building. The fact that they are the means of delivering or transferring energy from one point to another must be recognized. This marks a fundamental distinction with structural systems. The structural systems constitute a significant portion of the construction costs of a project. They, however, have no impact on operating costs for the building after it has been completed.

Generally all mechanical and electrical systems that are selected for a building must be responsive to a series of considerations, including, among others, the following:

1. Building type and usage
2. Occupancy requirements (present and future)
3. Architectural objectives
4. Structural design concepts
5. Initial cost
6. Available energy alternatives

7. Energy consumption
8. Annual operating and maintenance costs

4.3 AIR CONDITIONING AND HEATING SYSTEMS

4.3.1 Reference Material on Air Conditioning

The design technology for air conditioning systems has resulted in numerous books, magazines, and periodicals which address the issue of design and operation of air conditioning systems. We have utilized, for purposes of this text, a primary authoritative reference. This is the American Society of Heating, Refrigerating and Air Conditioning Engineers, Inc. (ASHRAE). This Society publishes a *Handbook* series on a four-year cycle which is a comprehensive source of information on HVAC system design. The *Handbooks* currently in the series are:

1. *Handbook of Fundamentals*
2. *Applications Handbook*
3. *Equipment Handbook*
4. *Systems Handbook*

These handbooks [4.1, 4.2, 4.3, and 4.4] will be used as a common reference throughout this discussion by reference to the handbook volume and, where appropriate, to the particular chapter in the handbook. The four-year cycle for these handbooks will be modified starting in 1986 by a redistribution of the material into new volumes titled *Refrigeration Handbook* (1986), *Systems and Application Handbook* (1987), *Equipment Handbook* (1988), and *Handbook of Fundamentals* (1989). Other ASHRAE sources and other references, where applicable, are also cited herein.

4.3.2 Calculation of Heating and Cooling Loads

The design of air conditioning systems must be started by the calculation of the heating and cooling loads for the building. There are a number of alternative sources for determining these loads; a primary reference for load calculations is the ASHRAE *Handbook of Fundamentals* [4.1]. Another is the ASHRAE *Cooling and Heating Load Calculation Manual*, published in 1979 [4.5]. A number of computer programs have been developed to use the procedures outlined in these references. In determining loads, particular attention must be paid to applicable codes which may affect the loads indirectly (by, e.g., controlling in a prescriptive sense the exterior envelope or lighting of the building) or directly (by, e.g., mandating the amount of outside air that must be brought into a building for ventilation purposes). In any case, the heating and cooling load for the building must be determined.

The loads that are calculated as the basis of the design are maximums or design loads. As an additional matter, the question of relative energy consumption must be reviewed early in the design procedure. Load determination and energy consumption, while clearly interrelated, are significantly different concepts. The first (i.e., load determination) is concerned with the maximum load on the building for both heating and cooling at a particular time during the year. The second (i.e., energy consumption) is a function of the varying weather and the varying occupancy to which a building is subjected over the 8760 hr of a year.

The two concepts are interrelated in that they are both a function of the siting and massing of the building, the envelope or outer skin as developed by the architect, the ventilation or outside air requirements, the lighting system selected, the air conditioning system selected, the energy conversion components (e.g., boilers and refrigeration machines) and energy delivery system (including fans and pumps) used, the domestic hot water load, and any heating or cooling process load that exists in the building.

They differ in that one is concerned with instantaneous maximum requirements including peak efficiencies of system components, while the other is concerned with the climate and occupancy (hours of usage), annual average efficiencies of system components, and how these variables interrelate with the building and its energy systems over an entire year. Clearly, for example, the energy consumed by lighting systems, domestic hot water, and process loads is almost entirely determined by the type of occupancy and the hours of operation of the building.

It is not intended to dwell on the determination of energy consumption of alternative solutions in a given building in a particular location. Suffice to say that several alternative means of calculating annual energy usage are available. They vary from manual calculations using weather data that has been developed over time, to computer programs available from several public and private sources. The standard reference for manual and computer techniques is Chapter 43 of Ref. 4.4.

4.3.3 Alternative Air Conditioning Systems

The selection of an air conditioning system type to handle a particular building involves a number of factors. The selection procedure is one of the most difficult tasks faced by the designer. Clearly the system selected should respond to the overall mechanical and electrical conceptual goals outlined earlier. The system selected must be responsive to the building, its usage or occupancy, as well as to the constraints developed by the owner, the architect, and the structural engineer. The system must respect, within the initial cost budget of the project, the energy design criteria (both peak loads and annual usage) for the project.

While the available alternative systems for buildings in a sense would seem to vary widely, it is possible to simmer the multiplicity of systems down to a small number of generic types that are categorized by the means of delivering energy to the conditioned space. Reference 4.4 categorizes the systems as described in the following Sections 4.3.3.1 through 4.3.3.4.

4.3.3.1 All-Air Systems

All-air systems provide full air conditioning cooling capacity by the air delivered by the system. Heating can be accomplished by a number of alternatives including the delivered air or by a separate water, steam, or electric heating function. Systems that can be considered all-air systems are discussed below.

In a sense they all present structural problems since the delivery system, being limited to ductwork, will have larger space requirements than would be true with the air–water or all-water systems discussed below. Buildings designed in the 1960s had high internal cooling loads and large glass areas, resulting in high air conditioning loads and high air quantities with all-air designs. As a result, such designs did not find wide building application at that time. With the recent development of buildings that use less glass, are better insulated, and have lower internal cooling loads, an inherent reduction in total cooling loads and the resultant air quantities has occurred. As a result, the problems associated with all-air systems, while not eliminated, have been mitigated, and the all-air systems, particularly in the variable air volume form, have found wide usage.

The simplest all-air system variant would be the *single zone system*. In this system, a single supply air duct would be extended from an air conditioning supply unit to the point of usage. A return air duct would collect air in the conditioned space and return it to the air conditioning supply unit where it mixes with outside air. The mixture of return air and outside air would, in turn, be cooled and redirected to the supply duct and then the cooled space. This is one of the simplest systems available. It finds wide usage in, for example, industrial spaces or large department stores.

If the single zone system is altered to the extent of including a reheat coil in the duct as it enters individual spaces, the system is categorized as a *reheat system*. The reheat coil can use hot water, steam, or electric energy. In this system, the heating coil is controlled by a space thermostat. These systems have found wide usage in spaces with fluctuating loads or in large spaces that have small variations but are easily developed as part of a larger duct system such as the interior spaces of an office floor. They have also found application in projects with a critical need for close control on temperature and humidity, such as laboratories, research facilities, and hospitals. The system, which thermally provides excellent control, has been limited recently in application due to its inherent high energy usage.

Variable air volume (VAV) systems, a third all-air variant, adjust through thermostatic control the amount of cool air being delivered to maintain the condition in the space. While available in a large number of configurations, the VAV system is one of the most popular systems employed over the past 10 years. This is a reflection in part of the energy concerns of the building community and in part the improvement in the technology of the available products. The system is widely applied in office buildings and many other occupancies, particularly where relatively small control zones are required.

Dual duct and *multizone systems* have been used and are all-air systems. The dual duct system requires two ducts (one hot and one cold) to be extended to the space where they mix the air from both ducts to the required temperature to maintain space conditions. Where the air is mixed at the fan system with a single distribution duct to the space, the system would fall into the multizone type system. The systems have found wide use in the past for many applications but are finding more limited usage today since they are, in a thermal sense, similar to a reheat system. They do find usage in small commercial applications such as small office buildings.

4.3.3.2 Air–Water Systems

Air–water systems provide part of the air conditioning cooling capacity through a ducted air system and part through a piped chilled water network. Usually, a majority of the cooling (as much as 70 or 75%) is through the piping system. In addition, the air or water can be elevated in temperature to provide heating as required by season. Typically, the widest application of this generic system type has been with induction systems. The induction system, in general, has been limited to exterior spaces with multiple thermostatically controlled spaces such as office buildings and patient rooms in hospitals. It utilizes one or more small air conditioning units within each space. Structurally, because of the reduced duct size requirement resulting from the use of water for a substantial part of the cooling load, this system is more easily coordinated than is an all-air system. These systems have found wide use in the exterior portion of office buildings having considerable glass and high internal cooling loads because of the more limited distribution space such as required with all-air systems. With the design of less energy intensive buildings, the application of this system type has been virtually eliminated when used with an induction unit. It has continued to find application in conjunction with fan coil units of the type described in Section 4.3.3.3 as a hybrid system between an all-water and air–water system.

4.3.3.3 All-Water Systems

All-water systems provide virtually all of the air condition cooling (and usually heating) through a piping network. A small amount of cooling (or heating) can be provided through a relatively small amount of air brought to the space for ventilation purposes. The system has found its widest application with fan coil units. These units located within each space of an occupancy, such as hotels or offices, are largely limited to exterior spaces. In hotels, with limited floor-to-floor height and small internal shafts, even the distribution of the water piping and limited ventilation air can present serious coordination problems. The units have the advantages of being capable of providing small control zones and having on–off control on an individual basis in periods of vacancy. The system is energy efficient in many applications. Another variant of this system is unit ventilators (widely used at one time in schools). This alternative structurally requires piping only and limited ductwork coordinations.

4.3.3.4 Multiple Unit or Unitary Systems

These systems use products that directly contain the means for space cooling. This requires each unit installed within or adjacent to the point of application to include the cooling compressor and direct expansion coil within the unit or space. It can include window units, self-contained water-cooled or through-the-wall heat pumps, or even packaged air conditioning units with ductwork to the space. Usage is wide but usually limited to small commercial installations or hotels. It has minor impact on structure (except in some cases at the outside wall where penetrations may be required) since the cooling effect is self-contained and ductwork and piping extensions are limited.

4.3.4 Heating, Ventilating, and Air Conditioning Equipment

The several systems discussed require, within their respective configurations, various pieces of equipment. Some of this equipment is used to convert one energy source into another more usable source. This includes hot water or steam boilers, hot air furnaces, refrigeration machines of a number of alternative types, and air-cooled condensers or cooling towers to handle the heat of rejection from the refrigeration equipment.

Additional equipment can be considered to be part of the energy distribution system which transmits the converted energy to the point of application. The transmittal is through the piping and ductwork systems discussed below. This grouping of equipment includes the filters, cooling and heating coils and fans that make up the air conditioning supply units, heat exchangers to change the form of energy (e.g., from steam to hot water) or the temperature level of the heating and cooling fluid used in the building and pumps.

There are multiple sources of information on the various pieces of equipment that are employed as part of the systems employed in the air conditioning of a building. Introductory details on most equipment are included in Ref. 4.3. In addition, each chapter in that volume includes additional reference material that can be used to obtain further information on any type of equipment.

4.4 MECHANICAL EQUIPMENT ROOMS

4.4.1 General

Generally, the air conditioning supply systems for multistory buildings fall into two major categories—those of a *central station* variety (serving multiple floors) and those of a *local* type (serving, usually, the floor upon which the equipment is located). Central or local air conditioning distribution systems may be either of the factory-assembled type or of the field-fabricated variety where the unit is actually built in place in the building. Draw-through system arrangements where the air is pulled through the filters and coils or blow-through system arrangements where the fan is located before the coils are also possible. Generally, the blow-through arrangement requires more space but, in cold climates, will provide fewer problems related to stratification of outside air with respect to recirculated air. The using of either a draw-through or blow-through arrangement, or for that matter, a built-up as opposed to a factory-assembled system can only be resolved after a thorough examination of space conditions, cost, local labor practices, and project quality objectives.

4.4.2 Location

The location of mechanical equipment rooms is a major decision that can affect the structural design of a building. Usually specific equipment would be located in the basement (e.g., domestic plumbing service equipment and hot water heaters, fire protection equipment, electrical transformers and switchgear, etc.) although this is not always the case. In very tall buildings, for example, hot water heaters may be required in the basement as well as on upper levels to keep water operating pressures on equipment at manageable levels.

Alternatively, cooling towers are usually located at the highest level in the building. This minimizes the noise generated by the tower from being carried into the building and also keeps any intermediate season cooling tower discharge plume or summertime carryover from the tower from being carried onto the building facade.

The remaining equipment (boilers, chillers, fans, etc.) can be located at virtually any building level. Location is a function of costs of piping, ductwork, building structure, available space, and construction phasing. Quite frequently mechanical equipment rooms are located at midheight (or slightly lower) in a multistory building. The advantages of such a location are that one equipment room can frequently handle an entire building, outside air for ventilation purposes is readily available, and the phasing of the construction is simplified since major pipe and duct runs, as well as complicated equipment installations, can be handled early in the construction, thereby tightening the entire construction phase.

The inclusion of significant quantities of equipment in upper portions of buildings, which is a post-World War II phenomenon, has been accompanied by two structural changes that have complicated the process of vibration control of fans, pumps, and other equipment. The structural changes are (1) the trend toward erecting buildings that are much lighter and (2) buildings that include increased floor spans to provide unobstructed floor space for better space layouts. The concurrent movement of equipment to more sensitive areas of a building and the above structural modifications have resulted in considerable refinements of the design of vibration isolation systems to eliminate structure-borne noise from being carried to occupied portions of the building.

4.4.3 Vibration Isolation

The key to the design of systems that will keep a definable percentage of equipment vibration out of the building structure is to install the machinery that causes vibration on a system of vibration isolators that will resonate at a frequency which is much lower than the frequency of the vibrating mass. When the vibrating frequency of the equipment (usually a function of the rotating speed of the equipment) is three times the natural frequency of the isolation system, 90% of the vibration is theoretically eliminated. Since the natural frequency is inversely proportional to the square root of the static deflection of the isolation material, it is clear that the greater the deflection, the smaller the natural frequency and the greater the elimination of potential vibration problems.

The elaboration of the overly simple arithmetic discussed above has resulted in a better understanding of the key nature of floor deflections and the need for mass under the equipment. Today, vibration isolation system deflections are based on theoretical considerations that have been tem-

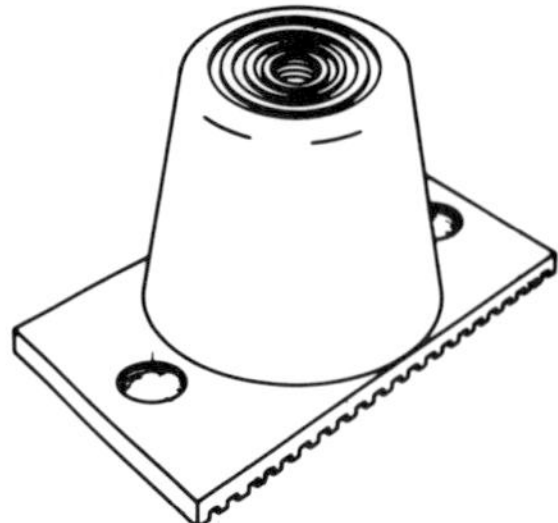

Fig. 4.1 Double deflection vibration isolation. Courtesy of Mason Industries, Inc.

pered by the empirical results observed in actual installations. Acoustical and vibration consultants recommend isolation systems that add mass under mechanical equipment, frequently utilize high deflection materials, and have successfully coped with the reduction, if not elimination, of structure-borne equipment vibration to occupied portions of a building. Reference 4.4 discusses in detail the material contained in Sections 4.4.3.1 through 4.8.3.

4.4.3.1 Pad Materials

Pad materials, including neoprene, cork, combinations of cork and neoprene, fiber glass and other material, have limited deflection of between 10 and 20% of the pad thickness. They are acceptable for high-frequency noise isolation but, since their deflections are small (0.2–0.5 in.) (5–12 mm) when related to potential upper floor deflections, they are limited, in general, to basement areas or less critical installations. A typical double deflection neoprene mount is shown in Figure 4.1.

4.4.3.2 Steel Spring Isolation

Steel spring isolation material under fans, pumps, chillers, and other equipment is the most common solution for vibration control. Steel springs are available with static deflections of 5 in. (128 mm) or more and, accordingly, can be provided to isolate properly the disturbing frequencies. A typical spring mount in combination with neoprene is shown in Figure 4.2. Supplementary steel to tie together equipment and maintain alignment (e.g., between a pump and a motor) is often used.

4.4.3.3 Supplementary Steel Bases

Supplementary steel or concrete equipment bases are frequently used to integrate the base under two pieces of apparatus that form a single unit, for example, a fan and motor to drive the fan or a large motor-driven centrifugal compressor. In these cases, the base is a function of keeping the pieces of apparatus in alignment. Where the base has a primary purpose of integration, it is common to make it of steel rather than concrete since the lighter weight of the steel members, as compared to concrete, lessens the floor load and minimizes the need for strengthening of the floor slab. The key issue with the steel base is to make it sufficiently rigid to provide proper support and not resonate at the frequency of the equipment it is supporting.

4.4.3.4 Floating Concrete Bases

Floating concrete bases are often employed to provide stiffness for equipment and to resist unbalanced forces that exist in the equipment. In large, low-speed (350 rpm) air compressors, balancing of internal forces is not possible. A typical concrete base is shown in Fig. 4.3. Accordingly, a concrete base, weighing six times the weight of the air compressor, is included to bring possible motion down to controllable, acceptable levels. Similarly, in the case of refrigeration equipment or certain fans, concrete bases are frequently included to reduce potential problems. These bases can weigh between one and two times the equipment weight, substantially changing the structural loading of the equipment. Finally, concrete bases are usually installed for pumps because pump bases are designed to be grouted to concrete floors. The concrete base provides a grouting surface as well as providing required stiffness.

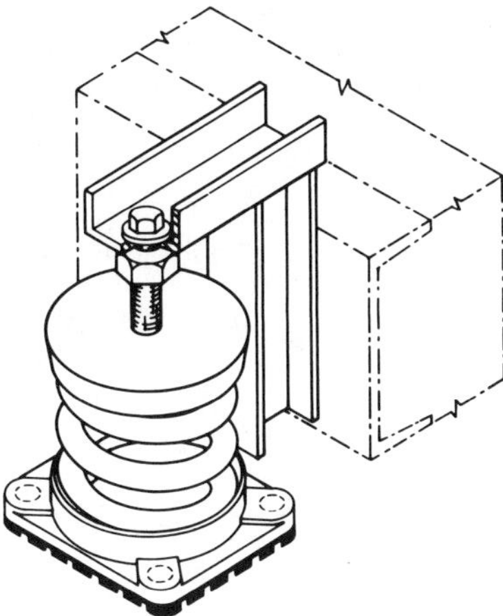

Fig. 4.2 Combination spring with neoprene vibration isolation. Courtesy of Mason Industries, Inc.

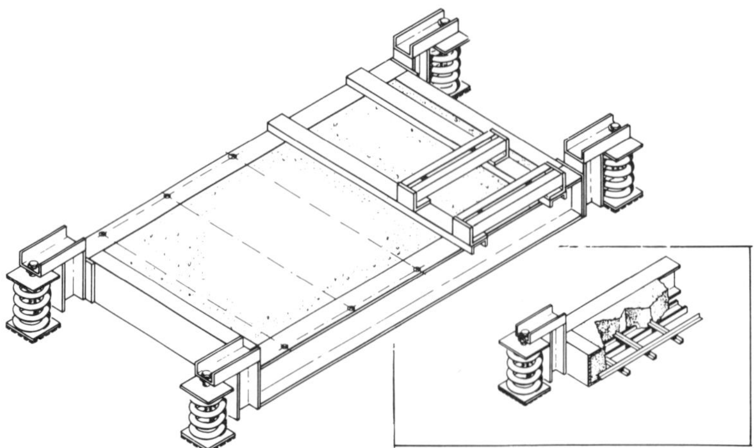

Fig. 4.3 Concrete base for fan. Courtesy of Mason Industries, Inc.

4.4.3.5 Other Base Considerations

Wherever equipment is located, but particularly on upper story levels, air and water systems must be designed to control the structure-borne noise and vibration problem. As noted, alternative mounting systems may require concrete inertia blocks with spring mountings to minimize potential transmission into the structure. In all cases 4-in. (100-mm) thick housekeeping pads should be provided for all equipment. This will provide some mass below the equipment (with the benefits outlined above), keep the isolation material off the floor, and protect the equipment from possible water damage and rust due to leaks or washing down of adjacent equipment. Housekeeping pads are normally an integral part of the floor slab, being placed at the same time.

4.4.4 Floor Loads and Design Requirements

The floor on which mechanical equipment is installed, as a general rule, should be a minimum of 8 in. (200 mm) thick and of high density concrete to isolate (both structurally and acoustically) the floor from the building. The minimum depth of 8 in. (200 mm) must be observed. In the case of steel deck type construction, the measurement must be taken from the top of the slab to the highest point in the deck flute. Mechanical equipment is by its very nature noisy. Therefore the containing of this noise is critical. The floor should be designed to handle a structural live load of 150–200 psf (7.2–9.6 kN/m^2) with allowance for the concentrated live load of particularly heavy pieces of equipment. Certain large pieces of equipment (e.g., boilers and chillers) produce floor loads of as much as 600–800 psf (30–40 kN/m^2). A final observation is that the mechanical equipment room floor must be waterproofed because of the use of the space.

4.4.5 Overhead Structure

The overhead structure should be designed to handle the piping and duct loads as discussed hereinafter. For preliminary purposes in a typical building, the overhead structure should be designed to handle loads of 40–100 psf (2–5 kN/m^2) for these items. This is only for schematic design and actual loads should be developed for final structural loadings. Frequently the overhead structure will be provided with trolley beams to permit servicing of equipment by removing major subsections at required intervals of time. Even if the trolley beams are not provided, the overhead structure must be capable of accepting the loading of the serviced equipment since it will be done by using chains or cables attached temporarily to the structure.

4.4.6 Seismic Considerations

As detailed in Chapter 18 , the possibility of an earthquake is a known condition in specific parts of the world. The *Uniform Building Code* [4.6] includes earthquake zones for the United States. The zones are described numerically from 0 to 3, where Zone 0 is an area expecting no potential damage from an earthquake; Zone 1, minor damage; Zone 2, moderate damage; and Zone 3, major damage.

Computer programs have been developed by public and private sources which provide the loads placed on the equipment as a function of the seismic response curves. These response curves are available for Zone 3 locations. The computer output will detail, for an input system configuration, the maximum displacement and force load at each equipment component. This will allow the development of proper seismic specifications for equipment.

In general, the potential for equipment being torn loose from foundation bolts by the acceleration forces that occur in an earthquake has resulted in the development of systems that, as a function of expected loads, will minimize the possible damage. These designs are passive and maintenance free, and retain the isolation system provided to handle normal vibration concerns. Many designs include snubbers installed below or on the sides of equipment, as well as piping isolators to prohibit the isolation equipment from moving beyond a certain limit regardless of the force to which it is subjected. Seismic restraint systems are widely used in code-mandated areas (usually Zone 3 sections of the United States).

4.5 IMPACT OF STRUCTURAL EXPANSION JOINTS

Expansion joints, which are designed to permit controlled horizontal and/or vertical movement of the structure, place design restraints on the mechanical systems. Ductwork and piping systems that traverse these structural joints must be designed to accept the horizontal and vertical displacements established by the structural design.

In the case of duct systems, several techniques can be utilized. Attachments to ducts may incorporate springs whose deflections correspond to the anticipated structural vertical movement. In certain applications, flexible ducts can be introduced to account for both horizontal and vertical deflections; and, in still other instances, combinations of techniques employing resilient supports and flexible materials may be invoked.

Similar techniques can be used for piping systems. Spring hangers have applications for vertical movement; however, piping that crosses structural expansion joints should, in most cases, be designed to account for horizontal and vertical movement associated with both the piping and structural systems. In addition, the required pitch of the piping systems may further complicate the solution.

In general, all of the above parameters can be satisfied with the proper introduction of offsets, loops, mechanical joints (ball, bellows, slip, etc.), spring hangers, anchorages, and guides in various combinations to suit the design condition.

4.6 DUCT SYSTEMS

4.6.1 Design Considerations and Types of Materials Used

In the various heating, ventilating, and air conditioning systems that have been outlined the importance of the network of ductwork is clearly established. The ducts have the requirement to deliver hot or cold conditioned air to the point of application. This should be done in a proper design, as directly as possible. It is a point of major interface between the structure and the air conditioning system. Accordingly, duct network design requires full coordination between the mechanical and structural engineer.

The use of the term ductwork can be construed to include supply ductwork, return air ductwork, exhaust ductwork, as well as related items normally installed by the sheet metal trade such as plenums and apparatus casings. While a substantial majority of ductwork is undoubtedly field fabricated of galvanized sheet metal, other alternatives or materials have found extensive use. These include fibrous glass board, factory-fabricated spiral sheet metal ductwork (either round or oval in shape) or aluminum, stainless steel, copper, and black iron ductwork. In the case of industrial exhaust where corrosive gases may be involved, iron and other materials find use as a function of the chemical composition of the gases being exhausted.

4.6.2 Insulation

These alternative materials, for a given air quality and air velocity, will vary widely in their weight per running foot. Also, the insulation of supply ductwork is a requirement under most energy codes and many return or exhaust ducts must be insulated in the interest of energy conservation or for alternative reasons. For example, proper procedure would indicate that the exhaust ductwork from kitchen ranges, often made of 10-gage to 16-gage black iron, be insulated with calcium silicate or other materials that effectively make the duct a low-temperature chimney. This is due to the potential of fire in that duct with its exhausting of grease laden air. The point is that the weight per foot of duct will vary widely as a function of the system involved, as well as the size of the duct.

4.6.3 Classification—Pressure and Velocity

Supply ductwork is classified as a function of the pressure within the ductwork (low, medium, or high pressure), the velocity (low, medium, or high velocity), and, in the case of fibrous ducts, as a function of flame spread and smoke developed classifications as defined by NFPA 90A [4.7] and the Underwriters Laboratories. The ability to use high-velocity ductwork as contrasted with low-velocity ductwork results in a significant decrease of duct size for a given air quantity since the velocity is increased to 2500–3000 ft (760–915 m) per minute from 1200–1500 ft (365–460 m) per minute. This clearly simplifies structural clearances but results in a different variation of the system involved. The high-velocity ductwork, if rectangular, is constructed to a different standard (it is heavier and when referenced to air leakage, tighter), requires an alternative supply fan construction and classification, has a greater potential for noise generation, is, in general, more expensive and uses more energy per delivered cubic foot of air per minute (cfm) than low-velocity ductwork. But high-velocity ductwork also can be factory-fabricated spiral ductwork which is lightweight, of an efficient shape, and meets high standards in strength and rigidity. The difficulty with round ductwork is that the structural design often places a limit on the depth of the duct and round ductwork, all other things being equal, tends to be deeper than rectangular ductwork.

The decision as to which duct classification to use for velocity and pressure typically extends beyond any structural tradeoff. Nevertheless, the ductwork must be coordinated with the structure. The rectangular ductwork permits the designer to adapt most readily the duct system to the space allowed structurally and architecturally in the building for that purpose. This adjustment is limited by design requirements on duct aspect ratios. The most efficient rectangular duct is a square one (as judged by weight of duct per cfm of delivered air). It is possible to go to aspect ratios (width of duct to depth of duct) of 6 to 1, or even 8 to 1, but not without added cost. The alternative duct shapes and layouts may require alternative structural designs, penetrations of the steel, or revisions to both duct and structure by both engineering disciplines. This is not a simple issue. The use of deeper, lighter, more economical structural sections with rectangular openings (often reinforced) or with many round unreinforced openings must be studied as an option to shallower, heavier structural members, with the ductwork running underneath. The key issue is the need to coordinate both designs.

4.7 DUCT SUPPORTS AND HANGERS

Once designed and coordinated, the duct system must be supported and tied into the building structure. The ductwork is usually relatively light so it does not normally present a structural problem. The hanging of ductwork requires the integration of an attachment at the duct, a hanger, and finally, an attachment to the building itself.

4.7.1 Hanger Types

A number of alternative means have been developed for supporting ducts as a function of the location of the duct (e.g., horizontal ducts on a wall, vertical ducts penetrating a floor, or overhead ducts). To illustrate the general nature of several support hangers see Fig. 4.4, taken from materials supplied by the Sheet Metal and Air Conditioning Contractors' National Association, Inc. (SMACNA). SMACNA has specific recommendations on the size of the hanger straps and their spacing as a function of the duct size. The trapeze hangers illustrated are recommended for larger, heavier ducts.

4.7.2 Concrete Inserts

The support hangers, in turn, must be attached to the building itself to ensure that the duct is firmly in place. This can be attained with concrete inserts installed prior to placing the concrete. This approach is limited to simple duct layouts where the ductwork routing is established early and without possibility of change. The inserts are available from a number of different sources. There are manufactured inserts available that will permit minor adjustment of the ductwork, but these adjustments are small, being measured in inches.

4.7.3 Concrete Fasteners

An alternative to the restriction of relatively accurate duct location required with concrete inserts involves the use of concrete fasteners. These are located after the concrete is placed. They can be expanding concrete anchors which require the insertion of the expansion shield into a hole drilled in the concrete. Alternatively, powder-actuated fasteners are available although their use is limited to

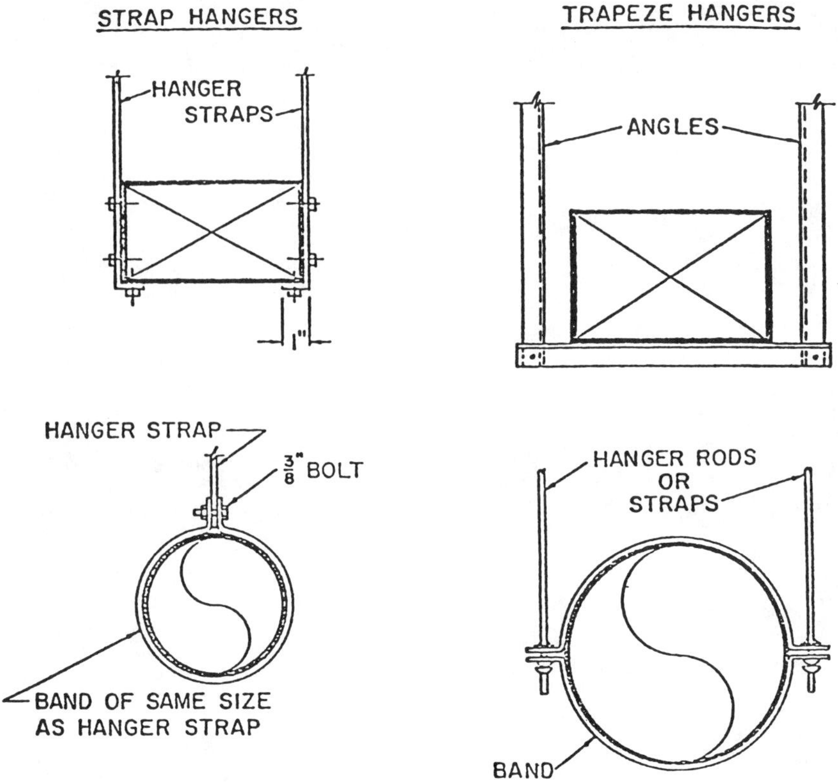

Fig. 4.4 Hangers for ducts. Courtesy of Sheet Metal and Air Conditioning Contractors National Association, Inc.

slabs at least 4 in. (100 mm) thick. Moreover, the powder-actuated fasteners should not be used in certain lightweight aggregate concretes.

4.7.4 Structural Steel Fasteners

A number of different clamps are available which can be employed to attach the hanger to the flange of the building structural steel.

4.7.5 Cellular Panel Support

As is discussed in Section 4.12.2 of this chapter, many office buildings utilize a cellular deck which is used to carry various power wiring and communication wiring. Cellular panels are placed on framing steel and then a finish or topping coat of concrete integrates the panel into the building floor. In this alternative, the duct support cannot be resolved with fasteners since they could penetrate the electrical deck. The manufacturers of these decks provide hanging systems or, alternatively, it is required that the hangers be installed before the concrete is placed. In any case any attachment should be made prior to the application of fireproofing materials to ensure the fire integrity of the deck as is required by applicable codes.

4.7.6 Examples

Examples of alternative hangers and means of attachment to the buildings are included in Figure 4.5, also prepared by SMACNA. These are selected examples; the range of alternative products is vast. Whatever the ultimate choice, major coordination between the structural engineer and the mechanical engineer is required concerning the support of ductwork systems in a building.

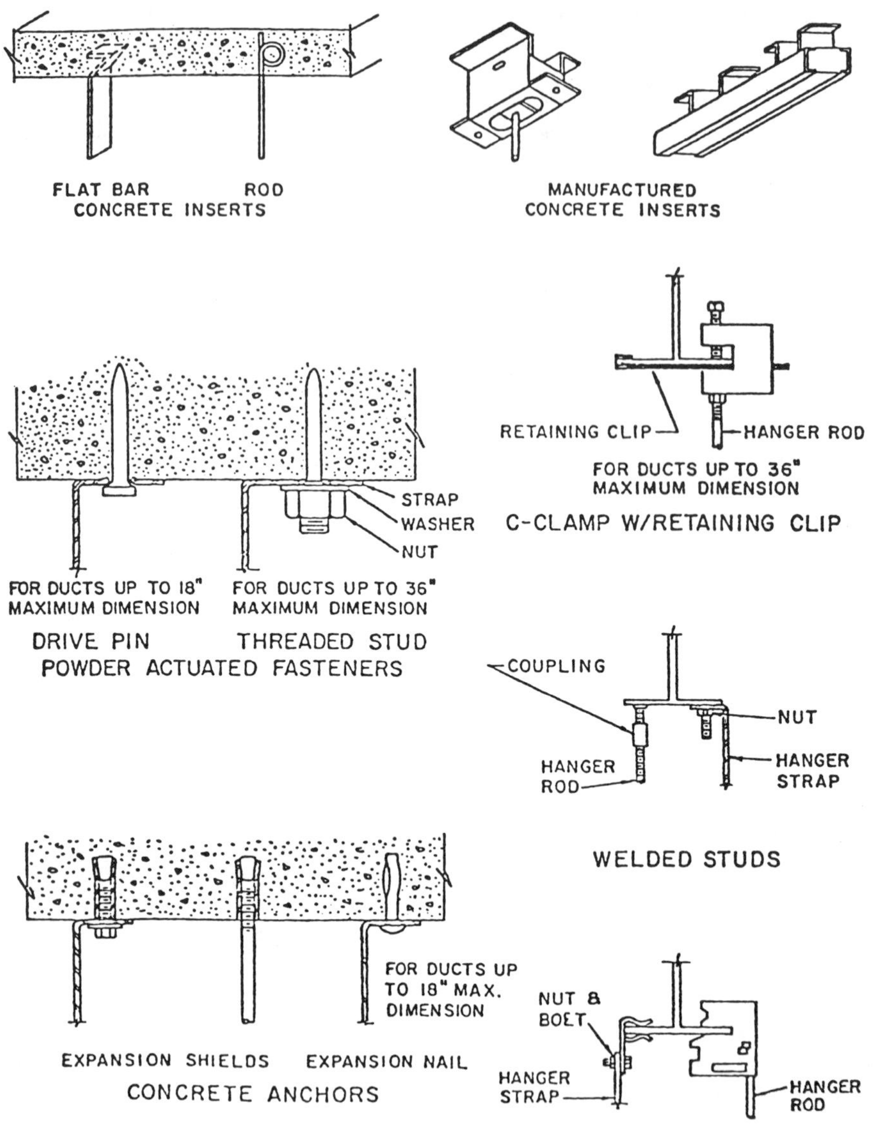

Fig. 4.5 Hangers for ducts—upper attachments. Courtesy of Sheet Metal and Air Conditioning Contractors National Association, Inc.

4.8 PIPING SYSTEMS

4.8.1 Design Considerations

The routing of HVAC plumbing and fire protection piping throughout a building requires coordination with the structural engineers. The penetrations required will be less than for ductwork (since the dimensions of the pipes are usually smaller) but the weight of pipes is much more than that of ducts.

The piping system presents several unique problems to the structural engineer. The problems are imposed by the weight of the piping (including such items as all pipe fittings, valves, the fluid in the piping system, and insulation), temperature changes that occur within the piping system with a resultant dimensional change in the pipe element, and dynamic or shock loads placed on the piping by the fluid flowing through the pipe or through seismic loads. It should be established that the internal stresses placed on the piping are not a matter of concern to the structural engineer. That is clearly the mechanical engineer's responsibility. What is of concern is the effect of these problems on the building's structure.

4.8.2 Piping Layout and Support Systems

All piping in a building should be properly supported or suspended on stands. The support system should be designed to permit free expansion and contraction of the piping while minimizing vibration. Often it is necessary to anchor the piping to the structure to control its possible motion. The mechanical engineer should prepare a layout of the building equipment and the piping system that interconnects the equipment to permit an analysis of the dynamic nature of the piping system. This first step of layout will permit a determination of where pipe hangers that will transmit piping loads overhead to the building structure should be located, or, alternatively, whether it would be preferable to include independent support stanchions and transmit the load downward to the structure. Finally, the preliminary layout will permit the engineer to determine the degree to which the piping must be anchored or restrained to control the direction of the movement of the pipe and its ultimate load on the hangers, supports, and structure. This ability to control the potential direction that a piping system would tend to move, if unrestrained, is a major means of allowing the determination of hanger location and ultimately of the stress that will occur in the pipe itself. Both establish a degree of predictability of the direction of motion (and resultant pipe stress) and also allow control of that motion. It is therefore an important matter to the piping designer.

4.8.3 Location of Supports and Hangers

Once the preliminary design has been completed and anchor locations are set, then the supports or hangers should be located. In general, the smaller the pipe the shorter the span between hangers. The hanger locations and spacing are not subject to a firm set of rules, these being largely an exercise in judgment. Where practical, hangers should be located close to concentrated weights (e.g., valves or heavy fittings) and, if possible, adjacent to the point in which a change in piping direction occurs.

4.8.4 Types of Hangers and Attachments to Building

Hangers are available in a large number of configurations. The type of hanger selected will be a function of the position of the support point relative to the supporting structure, any interferences that may exist and must be cleared, the nature of the piping, and the potential movement that the piping may experience due to its expansion and contraction. Several manufacturers provide full catalogue information of available alternatives.

The hanger that is selected, as in the case of the ductwork, must then be attached to the building. While marginally different, the attachment techniques are similar to those described for ductwork. The key difference is the loadings on the structure are generally substantially greater for piping as contrasted with most ductwork. As a result, for example, all main pipe lines and any large or heavy pipes should be carried by piping supported by beam clamps or cast-in-place attachments.

4.9 PLUMBING AND FIRE PROTECTION SYSTEMS

The selection of plumbing systems for a building involves several disparate systems. These include distinct functions such as storm drainage systems, sanitary drainage systems, as well as the domestic water, both hot and cold, which is required for the building.

4.9.1 Storm Drainage System

The *storm drainage system* is provided to carry precipitation in the form of rainwater or melted snow from the roofs, parking lots, and other paved areas, or any other point where its continued accumulation could be hazardous to the public or the structure or could cause damage to the public or building. The rate of removal of rainwater (or melted snow) is determinable from climatological data that is usually readily available. The design of roof drainage systems often is not handled on an instantaneous basis. Under this design approach, the water runoff system is designed to handle substantial rainfall but not the highest rainfall ever recorded in the area. This controlled flow design means that there will be a buildup of water levels on a roof that will add to the structural load. The system and loads should be reviewed early in the design process by the structural and the plumbing engineer. In addition, the structural engineer must allow for a proper amount of snow loading on the roof, as discussed in Chapter 2.

4.9.2 Sanitary Drainage System

The *sanitary drainage system* is concerned only with sewage. If a public sanitary system is available, the drainage will connect into that system. If public sewers are not installed at the building, then it would be necessary to include a sewage disposal capability for the building. In any event raw sewage should not be discharged into the ground or into a storm sewer system until it has been treated to an acceptable level. Sewage treatment plants are available as factory-packaged units, constructed primarily of steel; however, due to size or geographical location, they are sometimes built in place. In these cases, the structural engineer must work closely with the plumbing or sanitary engineer to provide the required structure.

4.9.3 Domestic Water System

The *domestic water systems* are routed through the building to the terminal water use fixtures installed in toilets and in other areas that require hot or cold water. The piping must be coordinated with the structure. There are many methods of delivering the water to the fixtures and equipment requiring it. Pumping systems can be designed to meet instantaneous, fluctuating demands or, with the incorporation of elevated storage tanks, they can be designed for average consumption. In either event, steps must be taken to prevent transmission of vibration to the structure and, in the gravity tank system, the structure must be designed for the weight of the stored water. In addition, the hot water system would include heat transfer equipment usually installed in mechanical equipment rooms to heat the water. As in the case of the cold water supply, storage tanks are used to minimize the rate at which energy is consumed; in these cases, additional allowance must be made for the weight of the stored water.

4.9.4 Fire Protection System

The *fire protection system* can include both a fire standpipe and sprinkler system. Buildings more and more are being full sprinklered since it is recognized by code officials that this is the single most effective way of limiting the size of a fire and the generation of smoke. The virtue of a sprinkler system is that extinguishment occurs before a fire can spread throughout the building.

As fire protection technology has developed, it has become more and more apparent that application of a relatively small amount of water at the point of an incipient conflagration is the best means of providing life and property protection against fire. The ideal means of applying the water required to suppress or extinguish a fire is an automatic sprinkler system, installed in the interest of economy as part of the fire standpipe system.

The sprinkler system in an office building, hospital, or residential building, for example (considered by NFPA as "light hazard" occupancy), is usually hydraulically designed to deliver water at a density of 0.1 gpm/sq ft (4.07 L per min/m^2), over an application area of 1500 sq ft (140 m^2). The design of the system is such that areas considered as "ordinary hazard," requiring higher densities of water application, can also be accommodated. Use of hydraulic design minimizes the amount of water that can be "dumped" needlessly when heads are fused, thus reducing the need for excessive water supplies, equipment, and pipes.

Combining the sprinkler system and the fire standpipe system results in a reduction of common purpose items, such as pumps and pipes, needed to apply the extinguishing agent (water) to a fire. Of course, the water sources for the suppression–extinguishing system must be reliable, with sufficient redundancy of tanks and pumps to assure that water will be available in adequate quantity and at the required pressure at all items. The type of system selected will reflect on the structure differently, depending on whether the water supply is primarily pumped or stored in elevated tanks.

The cost impact of the standpipe–sprinkler system can be minimized to some degree by tradeoffs where they are permitted by the authorities having jurisdiction. These tradeoffs can take the form of increased travel distances to exits, elimination of compartmentation and areas of refuge, reduced egress requirements, reduced fire ratings of shaft enclosures, and a decrease in the amount of structural fireproofing. The possibility of these potential tradeoffs and the advisability of their acceptance should be addressed for each building that is in design.

In a real sense the requirements of the plumbing and fire protection are similar to the piping systems for the heating, ventilating, and air conditioning system. Certainly the means of anchoring and support systems are similar. The discussion contained in Section 4.7 should be referred to for this discipline as well. The structural engineer and plumbing engineer must also carefully coordinate the plumbing shafts so that the piping fits and can be serviced. The need for penetration out of the shafts and the horizontal distribution on a floor, particularly for sprinkler piping, will require close coordination between the structural engineer and fire protection engineer.

4.10 ELECTRICAL DESIGN CONCEPTS

4.10.1 General Considerations

Electrical service is provided for buildings by various means depending on the locations, dimensions of the structure, type and voltage of service available from the utility company, magnitude of electrical load required by the particular usage, degree of reliability necessary to fulfill the basic mission of occupancy, and need for future growth.

Generally, rate differentials exist for service purchased at certain voltage levels as well as variations in the division of responsibility between equipment furnished and installed by the utility company and the user. When low voltage (600 volts and below) is supplied, the utility company furnishes all transformation and primary distribution. The customer usually furnishes everything downstream of the secondary connections of the power transformers. Medium voltage, from 2300 to 13,800 volts, is generally supplied by the utility company to the property line beyond which all equipment and installation, including such items as primary feeders, transformers, and connections, are in most jurisdictions the responsibility of the customer.

4.10.2 Transformer Interface Between Utility and Customer

In rural as well as certain urban areas, the transformer or substation interface between utility company and customer is constructed outdoors on a concrete pad on-grade, thereby having no impact on the building. Frequently, however, due to architectural, space, environmental, economic, utility company, or local code requirements, the transformers must be located in the building. Adequate space will be necessary to house the equipment while providing work space, ventilation to remove heat produced by transformer losses, and a route to remove the transformer in the event it has to be replaced.

One of the most reliable methods of providing service is by means of a network arrangement which requires the utility to provide a primary feeder for each transformer, the secondaries of which are paralleled after passing through network protectors which protect the utility company supply grid from power aberrations occurring within the building. Sufficient transformers are provided such that, if one or, in some instances, two combinations of transformers and incoming primary feeder fail, the remaining transformers will have adequate capacity for the entire building. The transformers are located in vaults, one per transformer in certain cities and a common vault for all or groups of transformers in other cities. The network protectors are usually mounted on the secondary end of the transformers, but, in certain jurisdictions, separate compartments are required for these devices.

4.10.3 Utility Transformers (Oil Cooled)

Utility company transformers generally utilize oil as a cooling medium which is circulated around the windings contained within a heavy gage metal tank. The flammable quality of the oil requires the vault in which the transformer is located to be constructed according to the requirements of the *National Electrical Code* [4.8]. The walls, floor, and ceiling must be constructed of reinforced concrete or masonry designed to carry the weight of the transformer. In certain cities, it is permissible to locate the vaults under sidewalks having gratings to provide natural ventilation while permitting the transformers to be lifted out in the event a replacement is required. Other cities require the transformers to be within the building, frequently in the basement, in which case they would require mechanical ventilation. However, in tall structures, there could be financial and electrical distribution advantages in locating the vaults on an upper floor. The preferable location on the floor would be at the perimeter of the building so that, by means of louvers or other similar devices, the transformer could be naturally ventilated. Another advantage of the perimeter location is as a removal path for the transformer in the event it has to be replaced (which does happen occasionally). By removing the perimeter louver panels, the transformer can be rigged out the side of the building by use of a roof-mounted boom, a street crane, or a telescoping trolley beam that can sling the transformer outside from which point it can be winched down.

4.10.4 Customer Transformers (Air Cooled or Submerged)

As noted previously, when power is purchased at primary voltage, it is incumbent on the customer to provide his own transformers. These transformers need not be oil immersed, but can be either dry, air cooled, or submerged in a nonflammable liquid, thereby eliminating the need for vault type construction of the transformer room. Although these two alternate types are heavier than the oil cooled, the overall weight of the installation, including the enclosed space, would represent a net reduction from utility-owned oil-filled transformers. The customer's primary protective and secondary distribution equipment is frequently close-coupled to the transformer in this case, which makes for a rather compact assembly averaging approximately 200–300 psf (9.6–14.1 kN/m^2). The same problems of removal exist for noninflammable transformers as with oil filled; however, dry type dimensions are somewhat more compact since the core and coil assembly, the only part requiring removal, can be rigged out separately from the enclosure.

4.10.5 Location and Impact on Structure

Locating power transformers on upper floors of the building requires extension of primary feeders to their location. At times, electrical protection for these feeders does not occur at the point they enter the building, in which event the raceway in which they are installed will require a concrete encasement to within 5 ft (1.5 m) of the transformer throat-mounted primary switch. Some local codes require cable supports be provided at prescribed spacings depending on the size and type of cable. The cable support chambers may have to be concrete structures with vault type Class A doors. Special supports are required for the concrete-encased raceway, generally at each floor level, as well as the support chamber.

The impact on the structure from service entrance electrical equipment, including primary and secondary distribution, transformers, and primary and secondary feeders, is essentially limited to support of equipment weight, not only in the permanent locations but also, especially in the case of transformers, the removal route. It might be necessary, in order to create a large enough opening for the transformer to be rigged through the exterior wall, to remove a structural member in addition to the venting panel. Although these measures are temporary, the structure must be designed accordingly.

4.11 DISTRIBUTION AND FEEDER ROUTING

The service entrance and distribution equipment discussed above is the heart of the electrical system from which point electrical feeders of relatively large amperage radiate to subdistribution equipment. The feeders consist of cable in conduit or bus duct which is a prefabricated assembly of aluminum or copper bars separated by insulating material and enclosed in sheet metal weighing as much as 100 lb/ft (1.5 kN/m). When bus duct is installed as a vertical riser running through slots in the floor, generally in an electric closet, it is supported on each floor slab by direct bolting or on spring mounts to permit expansion of the housing. Since the sheet steel housing has a different coefficient of expansion than the aluminum or copper bars, an expansion joint is included in long runs to take up the differential. The slab openings, through which the bus duct passes, are generally framed to carry the supports.

Conduit and cable feeders, when vertically oriented, can be installed through sleeves cast in the concrete slab or through slots. In either case, each conduit is supported by resting the conduit clamp on the concrete floor. Cable support boxes are utilized to support the cable by means of split inserts wedged into the ends of the conduits at the point they enter the box. The vertical distance between supports is determined by the size and weight of cable involved.

Major feeder groupings leaving the building main switchgear room represent an appreciable weight and dimension factor and are most frequently routed along the same general path as major piping and ventilating ducts on their way to vertical riser shafts or equipment rooms. Slabs and beams must be designed to include these loads.

4.12 FLOOR DISTRIBUTION SCHEMES

4.12.1 Branch Cabling

Branch cables (cables from final distribution points), be they lighting and receptacle panels, telephone strip terminals, television splitter cabinets or communication closets to the using equipment, such as a task light on a desk, an electric typewriter, a telephone, or a CRT or television set, are run concealed within finished areas. In office occupancies, the areas of concealment are the space between the hung ceiling and the underside of slab above (either on the particular floor or the floor below), within the floor or ceiling slab, or within any other cavities created for the purpose.

Generally, the space between hung ceiling and slab above is used as a plenum to return ventilating or conditioned air from the occupied area back to the supply air or exhaust air equipment.

Cables installed in plenums used for environmental air must, in most jurisdictions, comply with Article 300-22c of the *National Electrical Code* [4.8], which requires them to be in approved raceways or have insulation, such as Teflon, approved for the purpose.

4.12.2 Branch Circuit Distribution Schemes

The following branch circuit distribution schemes are commonly used in office buildings and have applications in most commercial types of structures.

4.12.2.1 Cellular Metal Floor Raceways

Cellular metal floor raceways are fluted panels of sheet steel fabricated in standard 10- or 20-ft (approx. 3–6 m) lengths, either 2 or 3 ft (approx. 0.6 or 1 m) wide, containing two or three cells, each used for a different service (telephone, electric, communication), complete with a bottom closure plate. The panels are butted end to end to form a continuous raceway and interlocked with adjacent panels which, if they do not have a bottom plate, will not have wiring capability. The resulting assembly is then used as a concrete form on which concrete is placed to a depth generally 2 to $3\frac{1}{4}$ in. (51–83 mm) above the top of the cell. Since the cells are obtainable 2 or 3 in. (51 or 77 mm) deep, the overall floor thickness can vary from 4 to $6\frac{1}{4}$ in. (102 to 160 mm) (see Chapter 20 and Fig. 4.6).

Branch cable originating at final distribution points is fed into a header system that runs on top of and at right angles to the cells for the full length of the floor. The header system is segregated by service, that is, one compartment for power, one for telephone, and one (if there are three) for communications. The cable can be routed through the header until it reaches the cell to be activated, then into the cell through a hole cut into the cell for that purpose. To gain access to the cell and provide an exit for the cable, a hole is drilled through the concrete into the cell, and a monument type outlet containing either a receptacle or a bushed hole is installed on the floor and attached to the cell opening with appropriate fittings to which the cable can be connected or pulled through, as shown in Fig. 4.6.

In some installations, boxes are set in a predetermined pattern on the metal deck prior to the concrete placement, each box having access to two or three adjacent cells via prefabricated openings, thereby eliminating the necessity of coring the floor into the cell and providing the means to contain all electrical outlets below the finished floor in the preset box.

The structural design of the floor system, including framing steel, metal deck, and concrete, can be considerably influenced and conversely can influence the direction of the cellular floor and location of the header. The best arrangement electrically might result in the most disadvantageous and expensive structural solutions. However, this type of floor distribution system combines flexibility in location of furniture and equipment with ease of making changes after the space has been occupied.

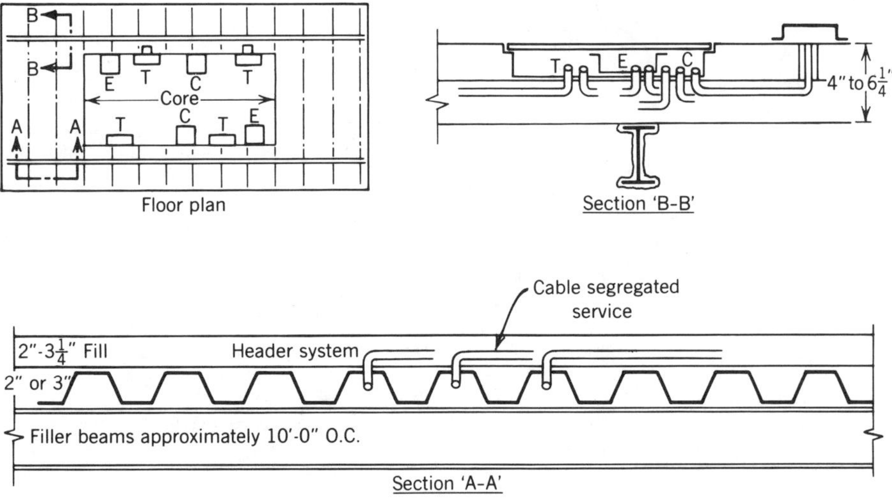

Fig. 4.6 Electrical cellular floor system.

4.12.2.2 Underfloor Duct Raceways

An underfloor duct raceway system consists of a grid of hollow rectangular steel ducts with trunks running the length of the floor and branches intersecting at right angles at the same elevation as the trunk. A barriered junction box is installed at the intersection of the trunk and branch ducts, the barriers serving to keep the high- and low-tension cables isolated one from the other. Single, double, and triple duct systems are available, depending on the number of services or voltage levels to be accommodated (see Fig. 4.7).

Most frequently, these systems are installed on top of a structural slab and require $2\frac{1}{2}$–3 in. (64–77 mm) of fill which is the minimum depth of the junction box. The addition of fill and finish to slab thickness imposes a structural penalty for the additional weight as well as impacting the overall building height if a given floor-to-ceiling height is to be maintained.

Whereas the cross-sectional area of a cellular floor cell is from 12 to 17 sq in. (7800 to 11,000 mm^2) depending on cell height and manufacture, the largest underfloor duct has an area of slightly more than 8 sq in. (5200 mm^2). The differential is most critical in the trunks and headers which, because of their limited capacity, represent a major bottleneck in the use of underfloor duct which can be partially offset by augmenting the headers with conduit from the junction boxes to the final distribution points in the concrete fill.

4.12.2.3 Raised Floors in Computer Rooms

Raised floors in computer rooms are used as raceways for power and low-voltage cable associated with electronic data processing equipment and frequently as a supply air plenum to provide cooling for the equipment installed thereon. Chilled water and/or condenser water piping required for air conditioning and water-cooled computer mainframes is also installed within the cavity.

More recently, in institutional type occupancies, where large densities and varieties of communication systems, minicomputers, cathode ray tubes (CRTs), etc., have resulted in masses of cable oriented horizontally on many floors of a building to be collected in riser closets for interfloor connections, raised floors have been used as cable raceways in noncomputer areas. Generally, these floors do not act as a means of transporting environmental air and need be high enough only for cable, approximately 6 in. (150 mm) above finished floor slab, whereas in computer areas, the recommended height might be 12–24 in. (300–600 mm).

If core areas, including toilets and elevator lobbies, are at finished floor level and the raised floor is 6–24 in. (150–600 mm) higher, a series of ramps would be required to move material and personnel from one level to another. To avoid ramps, which are space consuming, the floor slab on which the raised floor will be installed can be dropped to make the top of the raised floor level with the core and other areas where the cavity is not required. Dropping a slab on one floor will impact

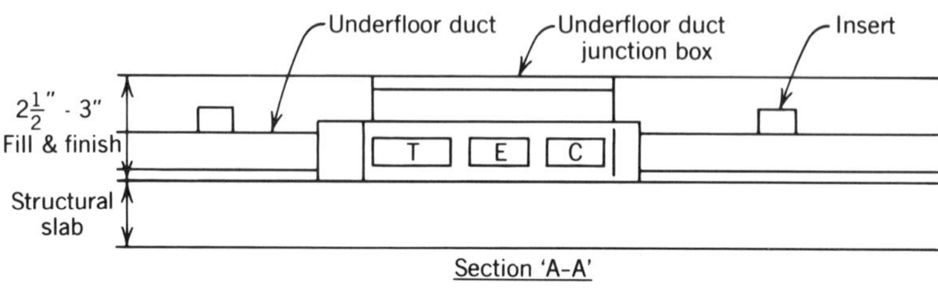

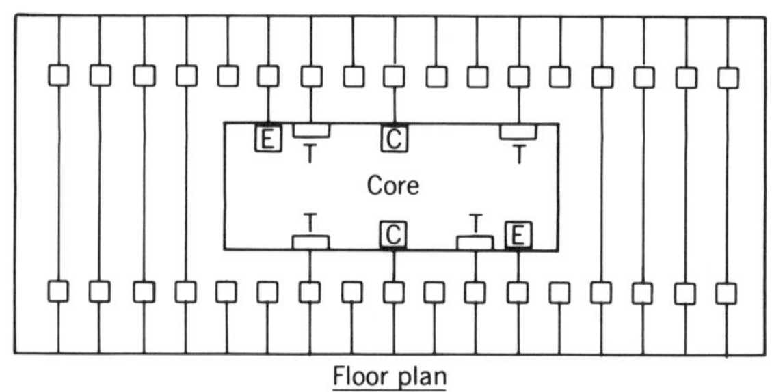

Fig. 4.7 Underfloor duct system.

on the floor-to-floor height of the floor below, all of which should be considered in the structural design.

A raised floor system is one of the most flexible of floor distribution systems; however, its major disadvantage is its potential impact on floor-to-floor height and therefore on the overall height of the building.

4.12.2.4 Poke-Through Distribution Systems

A poke-through distribution system is one in which cable is installed in the ceiling cavity of the floor below and pokes up into the working floor through a hole core drilled for that purpose. If the ceiling cavity is used as a plenum for environmental air, raceways will be required for the cable or an approved insulation must be used. In most jurisdictions, special fittings are mandated to be installed through the cored hole to maintain the fire rating integrity of the floor slab.

4.12.2.5 Cast-in-Slab Conduits

A floor distribution system can be developed by casting conduits in the floor slab to feed a grid of floor boxes embedded in the concrete with their tops flush with the finish. Gang boxes are available which provide for segregation of services, one compartment for electric, one for telephone, and the third for communications. Multiple service boxes are between 3 and 4 in. (75 and 100 mm) minimum overall height which could require a greater slab thickness than need otherwise be provided.

The major disadvantage to this system is its total lack of flexibility. If an outlet is required any place on the floor other than at the location of a floor box, the poke-through method discussed above would be the only solution other than chopping the floor to a partition or column in order to access the hung ceiling space and a distribution point.

An alternative approach to this system, which will improve the flexibility, is the installation in the ceiling slab of homerun conduits from a final distribution point to various locations in the space, at which locations the conduit will be turned down out of the slab and terminate in an outlet box mounted tight to the underside of slab. These boxes could be located one per bay per service, as an example, each supplied by a separate homerun conduit. When the occupancy requirements are known, conduit can be extended in the hung ceiling from the appropriate box to the partition or column on which the outlet to be wired is located.

4.12.3 Structural Design Considerations

The structural design for these distribution systems must cope with a multitude of conduits in the slab which tend to bunch up as they approach the final distribution point. It is conceivable that a considerable percentage of the slab thickness in the congested areas is conduit rather than concrete. Structural supervision during construction must be alert when a significant number of conduits is cast in the slab if the design is sensitive to the position of the conduits in the slab and their relationship to reinforcing bars.

4.13 LIGHTING TRENDS AND SYSTEM TYPES

4.13.1 Efficiency

In today's climate of very high electrical energy charges, the objective of most lighting designs is to maximize production of high-quality illumination with the least expenditure of energy. Development of more efficient light sources and lighting fixtures has aided in achieving more for less. A fluorescent tube will produce as much as seven times more light than an incandescent bulb. Not only will the energy to achieve a given illumination level be less, the air conditioning burden will be decreased as a function of lower heat production. Recessed fluorescent lighting fixtures are available today with efficiencies as high as 85% designed to minimize direct glare and veiling reflections which, in a sense, are a measure of unusable light. Lighting intensity can be reduced if the quality of light is improved and the relationship between the lighting fixtures and the work task is optimized to produce good contrast.

4.13.2 Task Ambient Lighting

A modern trend in lighting design is task ambient espousing the use of desk- or furniture-mounted fixtures to provide a substantial percentage of light at the work task combined with fixtures also mounted on the furniture, the floor, or in the ceiling to illuminate circulation areas and soften shadows created by the task light. On–off control of the fixtures is local, at the desk, enabling at least the task light to be turned off when not needed.

4.13.3 Indirect Lighting

Certain architectural solutions for interior spaces preclude the use of false ceilings, thereby exposing the underside of slab above. Frequently, the structure in these designs is coffered or might consist of precast concrete members, such as tee or double tee. Given a reasonably good quality concrete finish of the exposed surface, the coffer or tee can be converted into a luminous element by mounting an indirect lighting fixture within the confines of the structural element, causing the light to be diffused and reflected down to the working surface. The quality of light produced in this manner is relatively good and essentially shadow free.

4.13.4 Natural Lighting

The ultimate energy conservative lighting scheme is the reduction of artificial lighting by substituting natural daylighting when possible. The goal is to enhance the quantity and quality of daylight entering the occupied space and balancing it by modulating the artificial light to produce a desired illumination level at the task. Direct sunlight, because of its great intensity, is a detriment for this purpose and also has an undesirable effect on air conditioning systems in summer. Indirect daylight is the best for this purpose.

4.13.5 Structural Considerations

Various lighting designs impact structural design by affecting floor-to-ceiling height, the depth of hung ceiling cavity, and the quality of concrete finish when the exposed surface is to be used as a reflective element. Efficient recessed fluorescent fixtures with photometrically designed louvers rather than plastic lenses tend to be 2–3 in. (50–75 mm) deeper, thereby requiring corresponding space in the hung ceiling cavity. Task ambient lighting having the ambient component produced by a desk- or floor-mounted fixture directed toward and reflecting from the ceiling is more effective with higher ceilings, again possibly affecting floor-to-floor heights. The quantity of daylight entering occupied spaces can be greatly increased by including a structural member at the outside perimeter of the building which, if properly oriented to the window, can reflect daylight through the window into the space that otherwise would not have reached the task.

4.14 EMERGENCY POWER

4.14.1 Requirements

Building codes throughout the country mandate an auxiliary power supply for many types of structures. High-rise office buildings and hospitals have very specific requirements as to which systems and equipment should remain operable when the normal source of electricity is interrupted. Emergency power must be made automatically available within 10 sec for critical equipment and systems, while less essential equipment can be manually transferred. To accomplish this, the normal supply failure is sensed by an automatic transfer switch which will transmit a signal to the engine control causing the engine to start. When the engine–generator set achieves rated speed, voltage, and frequency output, the automatic transfer switch will connect the essential loads to the generator output.

4.14.2 Engine–Generator Sets

Engine–generator sets are most frequently used to provide standby and emergency services for buildings although batteries better serve the purpose for a limited number of applications. Engine–generator sets consist of a generator driven by a direct coupled prime mover which can be a reciprocating engine utilizing various grades of fuel oil or natural gas, or a turbine fired by diesel fuel oil or natural gas under relatively high pressure. Generator selection is basically unaffected by choice of prime mover; however, turbines require a geared coupling to the generator to reduce rotational speed of the generator.

The advantages of the turbines are more compact size and lighter weight. The reciprocating engine is less expensive and consumes much less fuel. The major benefit derived from diesel fuel over natural gas is that the diesel fuel can be stored on site in sufficient quantities for whatever operating period is desired without dependency on an external fuel source which, in a disaster, may not be available.

4.14.3 Batteries

Batteries as a source of emergency power are generally used when an interruption of service, no matter how short, cannot be tolerated. It might be crucial that certain lighting systems be on continuously during the performance of certain building or occupancy functions. An example might be the money counting and sorting area of a bank or the operating room light or respirator in a hospital for which a 10-sec outage would be unacceptable. Batteries for these applications are small. However, when used to provide uninterruptible power supplies for computer centers where a power disturbance of millisecond duration could cause equipment failure, battery capacity can be very large since they would supply the total power requirement for the computers.

4.14.4 Structural Considerations

The impact of engine–generators on the structure is a function of the weight of the machine and amount of the fuel oil piping, frequently required to be in a masonry-enclosed riser shaft, and the fuel oil storage tanks required by most codes to be located, if in the building, in the lowest level. The structural design, along with acoustical and vibration eliminators, must also consider the rotational forces or vibration of the turbine or reciprocating engine, respectively, to prevent the resulting noise from being carried by the structure to other areas of the building. Concrete mass and isolated mounting bases will assist in minimizing the problems.

Large battery systems, primarily because of the lead content of the cells, are very heavy, as much as 400 psf (20 kN/m^2)—a characteristic that presents the major impact on structural design.

SELECTED REFERENCES

4.1 ASHRAE. *Handbook of Fundamentals*. Atlanta: American Society of Heating, Refrigerating and Air Conditioning Engineers, Inc., 1981. (1791 Tullie Circle N.E., Atlanta, GA 30329)

4.2 ASHRAE. *Applications Handbook*. Atlanta: American Society of Heating, Refrigerating and Air Conditioning Engineers, Inc., 1982.

4.3 ASHRAE. *Equipment Handbook*. Atlanta: American Society of Heating, Refrigerating and Air Conditioning Engineers, Inc., 1983.

4.4 ASHRAE. *Systems Handbook*. Atlanta: American Society of Heating, Refrigerating and Air Conditioning Engineers, Inc., 1980.

4.5 ASHRAE. *Cooling and Heating Load Calculation Manual*. Atlanta: American Society of Heating, Refrigerating and Air Conditioning Engineers, Inc., 1979.

4.6 ICBO. *Uniform Building Code*. Whittier, CA: International Conference of Building Officials, 1985. (5360 South Workman Mill Road, Whittier, CA)

4.7 NFPA. *Standard for the Installation of Air Conditioning and Ventilation Systems (90A)*. Quincy, MA: National Fire Protection Association, 1981. (Batterymarch Park, Quincy, MA 02269)

4.8 NFPA. *National Electrical Code*. Quincy, MA: National Fire Protection Association, 1981.

4.9 IES. *Lighting Handbook*. New York: Illuminating Engineering Society, 1981.

CHAPTER 5

VERTICAL TRANSPORTATION —STRUCTURAL ASPECTS

PARTNERS OF JAROS, BAUM & BOLLES

Consulting Engineers
New York, New York

5.1 TYPES OF VERTICAL TRANSPORTATION—AN OVERVIEW

5.1.1 Introduction

This chapter identifies the interfaces between the building structural systems and the vertical transportation systems in major high-rise commercial, institutional, and residential buildings, as well as discusses, in general, the concepts of both pedestrian and materials movement which result in the various system components. Special emphasis will be placed on the discussion of building dynamics as a result of changing environmental conditions. It has only been relatively recently, particularly with the construction of the World Trade Center and the Sears Tower, that human factors permit a relatively dynamic building environment. This had not then been anticipated by the vertical transportation industry. The rapid technological changes with respect to the structural aspects of a major building resulted in increased dynamics of the building which were evaluated within the human response and comfort criteria. The advent of high-strength steel, sprayed-on fireproofing, and dry wall construction, coupled with the optimization capability with computer-aided design, now results in buildings of substantially less mass and, as a result, these buildings are more dynamic. This can seriously affect the vertical transportation systems if not accommodated by some elevator structural considerations as well as special dampening or detuning considerations on the part of the elevator subcontractor. The objective of this chapter is to establish a level of understanding between these two previously separate disciplines.

The national advisory safety code relating to all the vertical transportation systems, except conveyors, is the ANSI/ASME A17.1 *Safety Code for Elevators and Escalators* [5.1]. This code includes requirements for elevators, escalators, dumbwaiters, moving walks, material lifts and dumbwaiters with automatic transfer devices, inclined lifts, and an Appendix F, Recommended Elevator Safety Requirements for Seismic Risk Zone 3 or Greater. This Code is referred to in this chapter with the understanding that it is an annually revised advisory code that may not be the actual code used by the regulatory authorities having jurisdiction over a specific project.

5.1.2 Passenger Elevators

Passenger elevators are batch conveyors of people in a substantially vertical orientation to transport passengers among a multiplicity of levels in both directions. A passenger elevator by Code must have horizontally sliding doors for passenger safety as opposed to freight elevators, which may have either horizontally sliding doors or vertically sliding doors.

The usual connotation for an elevator with a passenger classification to be used primarily for freight movement and the incidental transport of passengers is called a service elevator. This classification, however, is not currently recognized by the Code and, hence, all elevators must comply with either the passenger elevator classification requirements or the freight elevator classification requirements. The use of freight elevators by general passengers is prohibited except in emergencies and, hence, the transport of materials in combination with passengers should utilize the passenger elevator classification. As in the case of health-care facilities or major office building facilities, the shape of the service elevator platform car interior should be related to the freight, materials, or vehicles carried rather than to the most efficient use of that interior space by passengers. In passenger elevators, the emphasis is on the people who must frequently move from the rear of the car through a group of remaining passengers standing in the front of the elevator car in order to exit the elevator. The platform design of the passenger elevator attempts to achieve the shortest time for passenger transfer.

Single-Deck Elevators

The most common elevator configuration is a single-deck elevator which involves a single passenger compartment with car and landing doors at the front of the elevator car serving a multiplicity of levels. The usual configuration is that the car platform is wider than it is deep to permit ease of passenger transfer in a crowded elevator as well as providing a wide center-opening door for the shortest door operating times. The service elevator variant of the passenger elevator is usually narrow and deep to accommodate materials movement and only incidentally passenger movement. It usually is of such a construction as to provide a higher interior space for large outsized pieces of furniture or equipment and usually has the highest and widest door openings available commensurate with the hoistway construction and the architectural considerations of ceiling height.

Double-Deck Elevators

In recent years, there has been a demand for vertical transportation concepts that most efficiently utilize the volume of elevator hoistways in a building. One such concept involves the double-deck elevator, which has two separate passenger compartments, vertically connected structurally to each other, and stopping simultaneously on adjacent floors. This concept results in a reduction of approximately 25% in the number of elevator shaftways required and an increase in the net rentable and net usable floor area. The amount of saving is a function of the gross area per floor and the number of floors served by each group of double-deck elevators. Double-deck elevators have a very rigid structural requirement in that the floors served by double-deck elevators must be constructed with a closely controlled floor-to-floor tolerance so that the simultaneous stopping can be accommodated without generating a tripping hazard in the landing-sill to car-sill relationship. The Sears Tower in Chicago utilizes double-deck shuttle or express elevators to serve two Sky Lobbies. The Standard Oil (Indiana) Building in Chicago utilizes double-deck local elevators serving local floors in a conventionally zoned system. Both buildings require the passenger to identify his destination floor at the two-level lobby entrances prior to boarding the elevator system for the up trip.

Oil Hydraulic Elevators

The least expensive and slowest speed elevator is the oil hydraulic elevator utilizing a direct-acting hydraulic piston which is pumped under pressure by a dedicated hydraulic system to raise the elevator (see Fig. 5.1). The elevator travels downward by gravity. This motion in both directions, and the leveling, are controlled by a complex hydraulic oil control valve which gradually accelerates an elevator car, travels it at a constant speed in both directions, and decelerates it when approaching a landing for an accurate stop. Hydraulic elevators have the advantage of low overhead heights and shallow pit depths as well as remote machine rooms in some local jurisdictions. The Code permits hydraulic elevators to be installed without car safeties which are required on electric elevators. This deletion concentrates the hydraulic elevator's structural reaction in the elevator pit under all operating conditions. Hydraulic elevators can operate at speeds up to 200 ft/min (1.0 m/s), but most often operate at speeds of 150 ft/min (0.75 m/s) or less. The maximum vertical rise is limited to approximately 70 ft (21 m) or approximately seven served landings.

Geared Elevators

Electric geared elevators are used for intermediate speed ranges and for higher travels than permitted by hydraulic elevators (see Fig. 5.2). A geared elevator with a counterweight is called a traction elevator, which has a current upper speed limitation of 400–450 ft/min (2.0–2.3 m/s) with a maximum travel of from 40 to 50 floors depending on the building use and on the necessity of special machine design. The geared elevator is a cost-effective design which uses a high-speed, low-torque electric motor with a worm-to-worm gear or helical gear speed reduction between the motor shaft and the driving sheave. It is usually installed with a maximum vertical acceleration and deceleration of from 3.0 to 3.5 ft/s^2 (0.9 to 1.1 m/s^2).

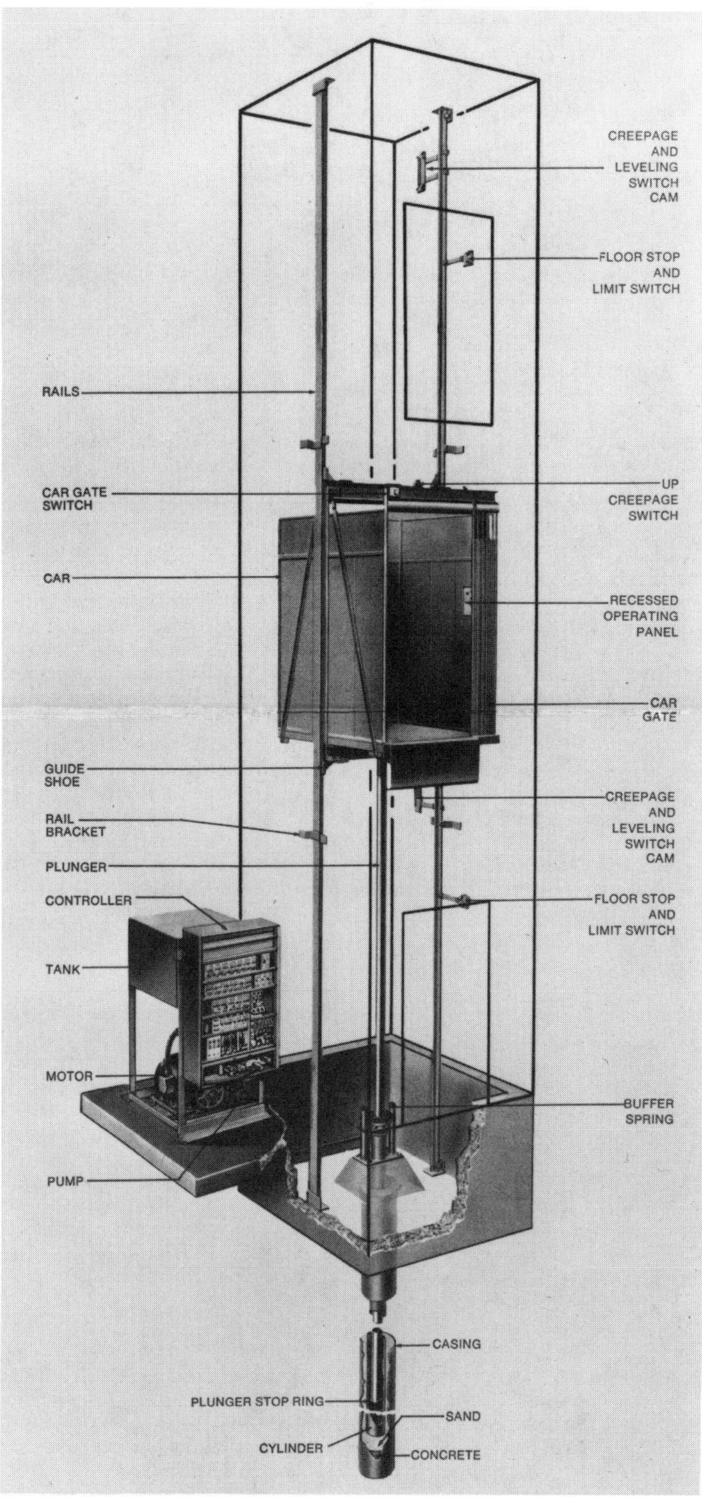

Fig. 5.1 Hydraulic plunger elevator. Courtesy of Otis Elevator Co.

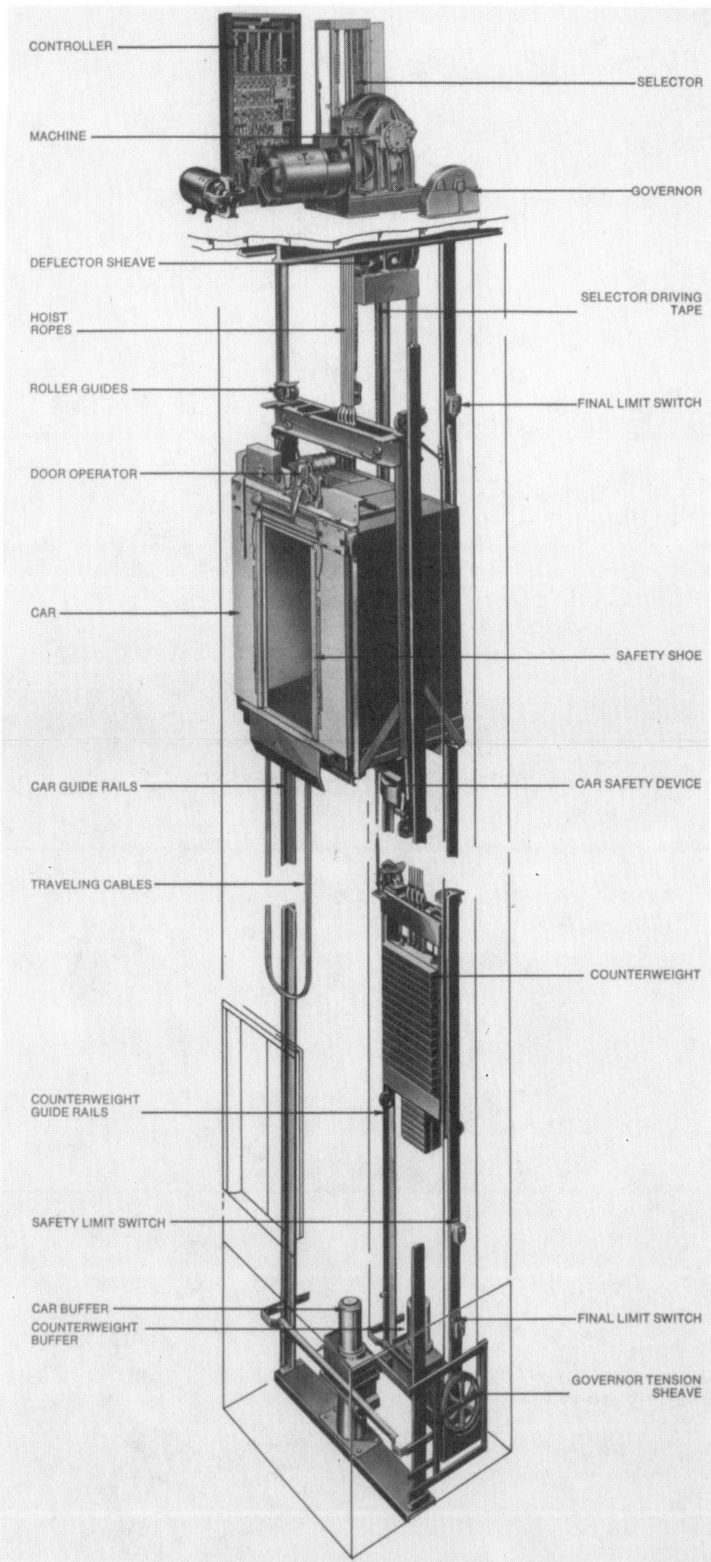

Fig. 5.2 Geared traction elevator. Courtesy of Otis Elevator Co.

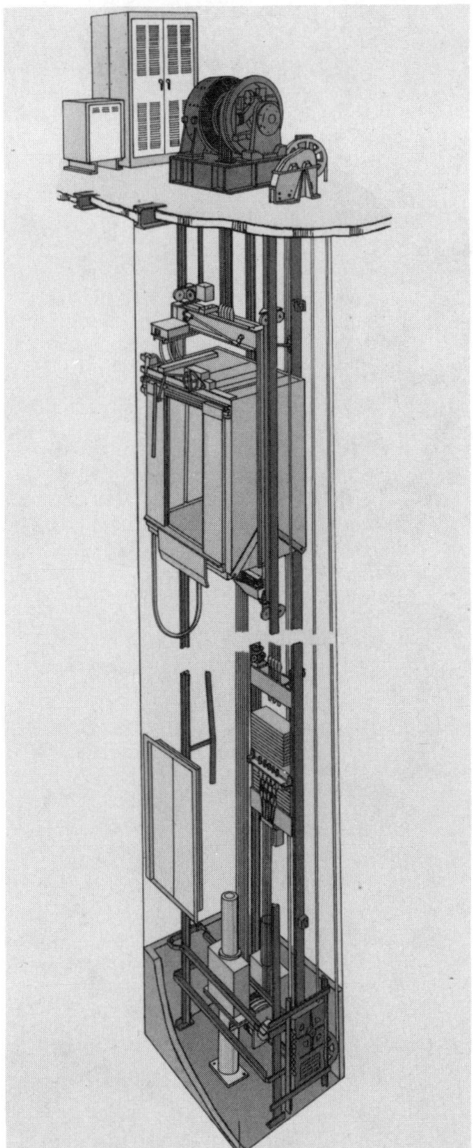

Fig. 5.3 Gearless traction elevator. Courtesy of Otis Elevator Co.

Gearless Elevators

A gearless elevator is a traction elevator with a counterweight using a low-speed, high-torque electric motor with the drive sheave mounted directly on the motor armature shaft (Fig. 5.3). It is installed at speeds from 400 to 2000 ft/min (2 to 10 m/s). The vertical rise is currently limited only by the height of the building, approximately 1500 ft (450 m) in the Sears Tower in Chicago and the World Trade Center in New York City. It is usually installed with a maximum vertical acceleration and deceleration of from 4.0 to 4.5 ft/s^2 (1.2 to 1.4 m/s^2).

5.1.3 Escalators

Escalators (see Fig. 5.4) are a continuous unidirectional flow conveyor of people which are boarded when the step treads are in a horizontal plane, traveled on an incline, with the steps

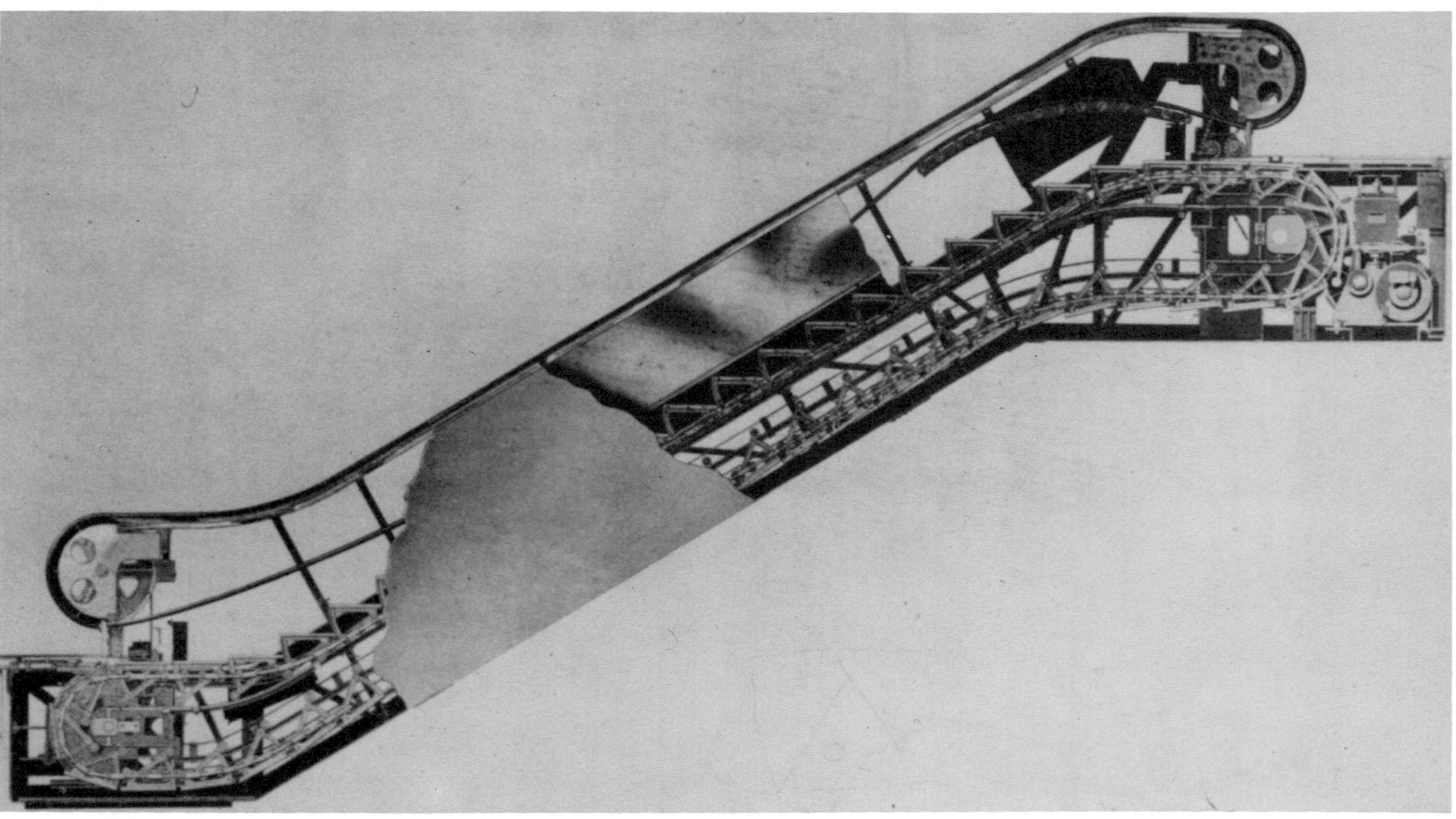

Fig. 5.4 Escalator. Courtesy of Otis Elevator Co.

configured a conventional step–riser relationship, and exited at the end of the travel with the step treads realigned in a horizontal plane.

Step Widths

The width of an escalator step is prescribed by code minima and maxima in various jurisdictions. In the United States, Europe, and Canada, the minimum tread width permitted is approximately 24 in. (600 mm) and the maximum tread width permitted is 40 in. (1000 mm). A variant in Europe and other countries permits a step width of 37 in. (800 mm). For practical purposes, the 24-in. (600-mm) and 37-in. (800-mm) step widths permit a passenger on every other step for dynamic considerations. The nominal 40-in. (1000-mm) step width will permit one passenger on every step for through-put considerations. The step depth in the traveled direction is limited currently in all widths to $15\frac{3}{4}$ in. (400 mm).

Escalator Lengths

The length of an escalator is not limited by its basic design but is only limited by the ability of the building to accommodate, in the case of some manufacturers, an external machine room outside of the truss. In the case of a modular escalator, the length is theoretically limitless because the escalator machinery is located inside the truss and in multiples required for the load requirements. In cases where the rise exceeds the normal floor height, special consideration should be given to the requirement for intermediate truss supports which vary among the manufacturers involved. A special consideration under all circumstances should be given to the exterior cladding of the truss with respect to the weight of the material used as it may affect the need for and location of such intermediate supports, and their sizes.

Code Incline Limitations

The codes in the United States and Canada permit an escalator design with a maximum angle of incline of 30°. Other jurisdictions relating to European codes permit an incline of 28.5, 30, or 35°. These limitations are based on the human engineering factor relating to using the stopped escalator as a stairway as a means of access or egress. The European codes permit the 35° escalator only for relatively short rises, 20 ft (6 m), as even the 30° escalator has an uncomfortable tread-riser relationship when the stopped escalator must be negotiated for an extended distance.

5.1.4 Moving Walks

A moving walk is another continuous flow conveying device which is essentially a flattened "escalator" in that it may have flat steps or pallets which have no riser between them. Moving walks are usually installed as an alternative to long horizontal traverses in densely traveled pedestrian ways to reduce fatigue since the horizontal speed on a moving walk of a standing passenger is not as fast as a walking pedestrian. A walking passenger on a moving walk can approximately double his net horizontal velocity.

Moving Walk Widths

Moving walk widths are essentially unrestricted at slow speeds but restricted at higher speeds to encourage the use of the handrail. At inclines up to 15° from the horizontal, the widths are restricted to 40 in. (1000 mm). The minimum width permitted is 16 in. (400 mm).

Moving Walk Lengths

The length of a moving walk is unrestricted; however, it probably represents a mechanical limitation of from 300 to 600 ft (90 to 180 m) depending upon installation costs and other factors relating to the slope of the treadway and the availability of truss and machine room space.

Code Restrictions

The maximum slope of the moving walk treadway is limited to the human engineering factor of 15°, above which fatigue in the Achilles' tendon in the up direction and discomfort in the down direction are generated. The Code restriction separates the speed and width based on a 0–5° slope, a 5–8° slope, and an 8–15° slope.

5.1.5 Combined Systems

Combined systems in any complex building may configure elevator, escalator, and moving walk systems to respond to the peak arrival rates and departure rates that represent the traffic in a continuous flow context by escalators and by a series of batch elevators that approximate the continuous flow during a traffic peak. Since the practical maximum number of elevators in a single group is eight, with four elevators facing four elevators, the handling capacity of that group should not be exceeded without another group of elevators or groups of elevators being added to accommodate the traffic demand of the total arrival or departure rates. The building is thus zoned so that each zone is served by only one group of elevators. It is this limitation that separates conventional buildings into various rises of elevators serving 10 to 14 floors each. The speeds of each rise increase as each express run increases in order to provide qualitatively an approximately equal level of service related to waiting time for each group even though the vertical travel distance is different for each group.

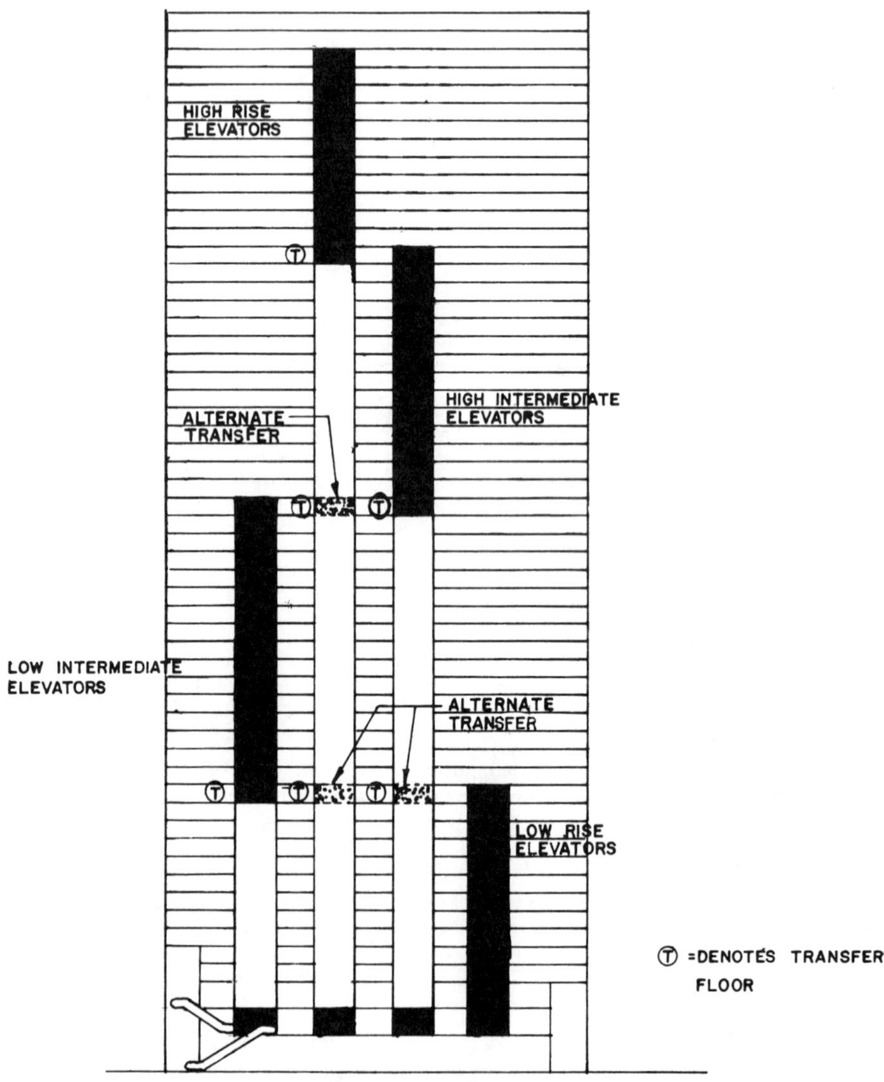

Fig. 5.5 Conventional concept—single-deck local elevators. Courtesy of Jaros, Baum & Bolles.

Conventional Elevator System

The conventional elevator system utilizes single-deck elevators originating in a single lobby at the street entry and separating incoming passengers into lobbies serving a designated portion of the building. Usually these elevator groups have exchange or transfer floors so that interfloor traffic among the groups can be accommodated without travel to the main lobby and generating another up trip. The limitation of this system varies with the total building population and type of occupancy; however, generally, it is restricted to 40 to 50 occupied floors, as shown in Fig. 5.5.

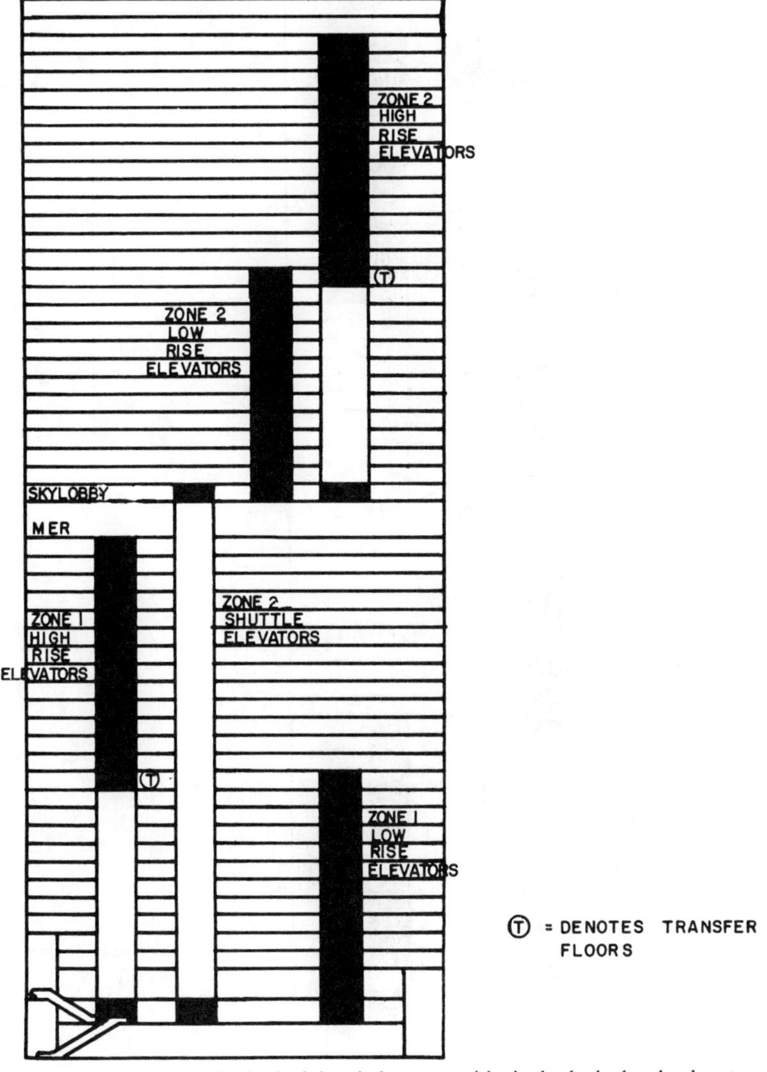

Fig. 5.6 Skylobby concept—single-deck local elevators with single-deck shuttle elevators. Courtesy of Jaros, Baum & Bolles.

Sky Lobby Systems

The sky lobby system essentially divides the building into two or more office building sections served from the sky lobby by a conventionally zoned elevator system. The sky lobbies use dedicated shuttle elevators which may be either single-deck or double-deck serving the main lobby or lobbies at street level and the sky lobby or sky lobbies above. The purpose of this configuration is to reduce the amount of square footage in a core dedicated to elevator service that cannot be leased. While the sky lobby approach was an essential concept for the World Trade Center or Sears Tower, it can be applied to lower-rise buildings under special conditions, as illustrated in Figs. 5.6 and 5.7.

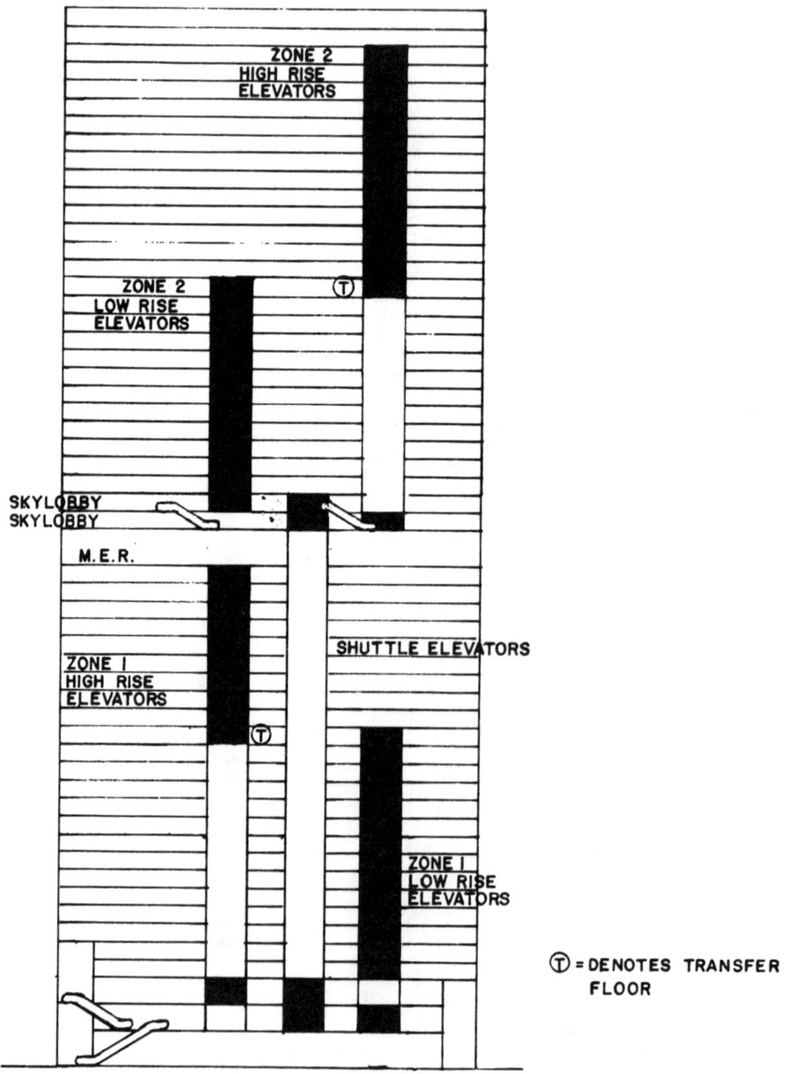

Fig. 5.7 Skylobby concept—single-deck local elevators with double-deck shuttle elevators. Courtesy of Jaros, Baum & Bolles.

Double-Deck Elevators

Double-deck elevators serving local floors or sky lobbies require the use of a double level entrance generated at the street level with either natural multilevel access or escalator access equal to both the upper and lower decks since they will be assigned to serve, in up-peak conditions, only "odd" or "even" floors from a specific main lobby. When double-deck local or shuttle elevators interact with a single level street lobby, it is necessary to provide an escalator core that handles the up-peak or down-peak traffic from the upper deck or the lower deck or, in the case of when this lobby is split between the two levels, to handle separately the upper deck and the lower deck traffic simultaneously (see Fig. 5.8).

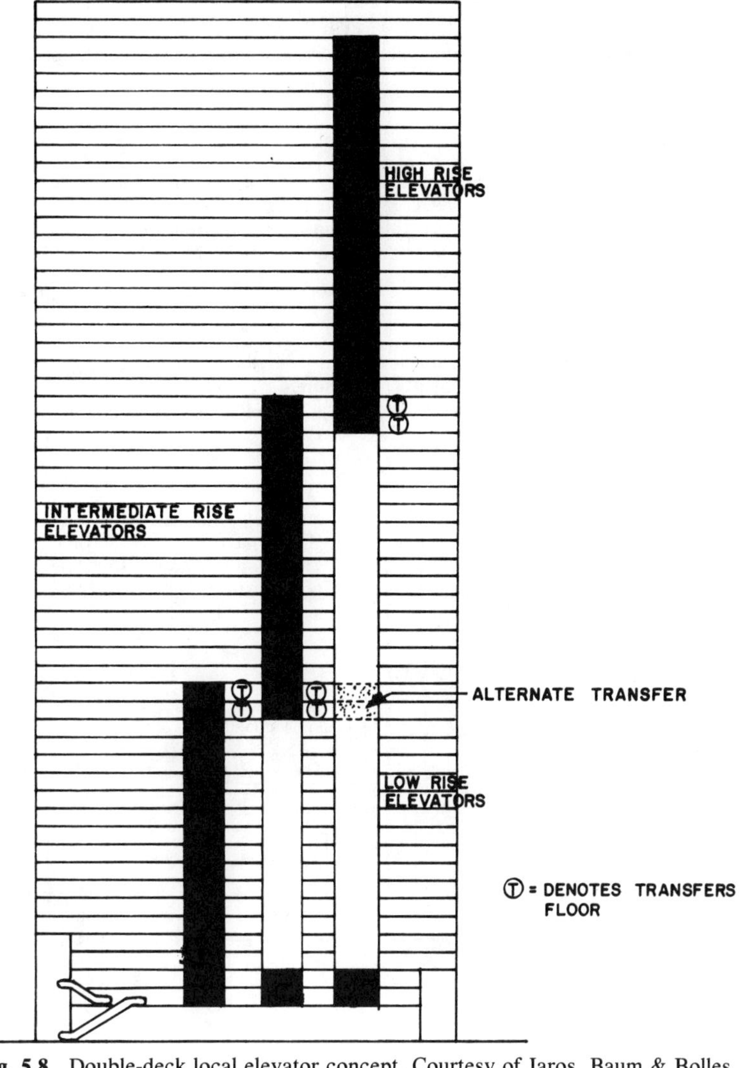

Fig. 5.8 Double-deck local elevator concept. Courtesy of Jaros, Baum & Bolles.

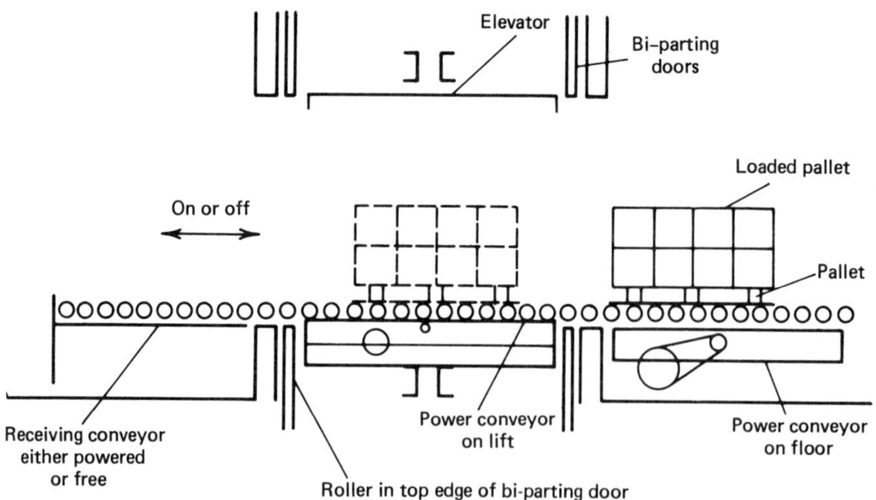

Fig. 5.9 Automated pallet lift with roller conveyor. Courtesy of Jaros, Baum & Bolles.

5.1.6 Freight Elevators

Freight elevators are recognized as a separate category in the Code to permit vertically sliding counterbalanced doors since often they accommodate the use of heavy fork trucks carrying loaded pallets and, as such, the doors must be as wide as the car enclosure for convenience of loading. A freight elevator usually is narrow and deep since the efficiency of simultaneous loading and unloading is not important as in passenger elevators.

Standard Freight Elevators

The standard freight elevator can be propelled either by hydraulic, geared, or gearless machines, depending upon speed and rise requirements. It is limited by Code to permit only freight-handling personnel from riding on the elevator and, depending upon its use during loading and unloading, may transport fork trucks for that purpose.

Special Freight Lifts

The automated or semiautomated handling of freight on pallets or in carts can be accommodated in the vertical portion of their movement with freight lifts that use powered roller conveyors for pallets, "power-and-free" overhead conveyors for carts, or "in-floor" tow line conveyors for carts, as shown in Figs. 5.9, 5.10, and 5.11. These lifts have special Code restrictions and requirements.

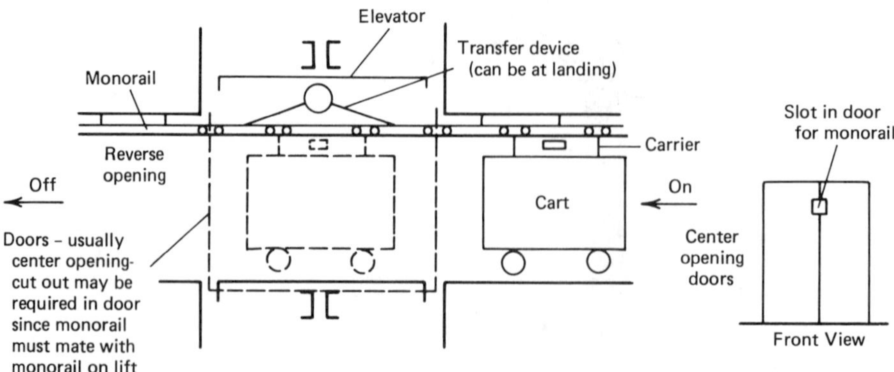

Fig. 5.10 Automated cart lift with overhead "power-and-free" conveyor. Courtesy of Jaros, Baum & Bolles.

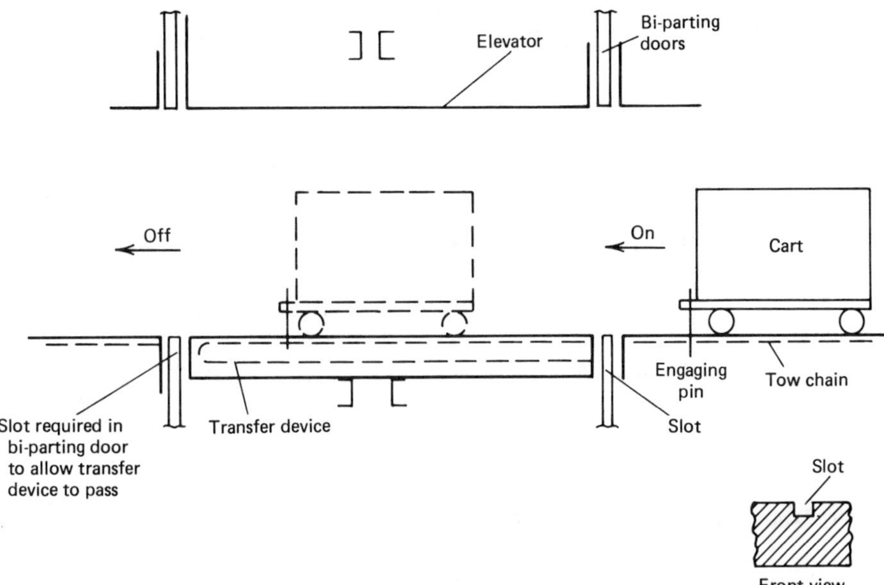

Fig. 5.11 Automated cart lift with "in-floor" tow line conveyor. Courtesy of Jaros, Baum & Bolles.

5.1.7 Cart Lifts

Cart lifts with automated inject–eject transfer devices have special consideration in the Code since they are designed to accommodate only the cart to be transported without an accompanying passenger or freight handler.

The basic cart lift is essentially a vertical materials-handling device for carts, which automatically injects at an origin floor and automatically ejects a loaded cart at a destination floor or, in its reverse program, injects a cart at a floor and returns it to the main floor where it is ejected. These lifts are used primarily in health-care facilities and office buildings to transport carts vertically which will be subsequently manually handled on each floor for horizontal travel. They also can carry an automated guided vehicle (AGV) that self-loads and self-unloads automatically (see Figs. 5.12 and 5.13).

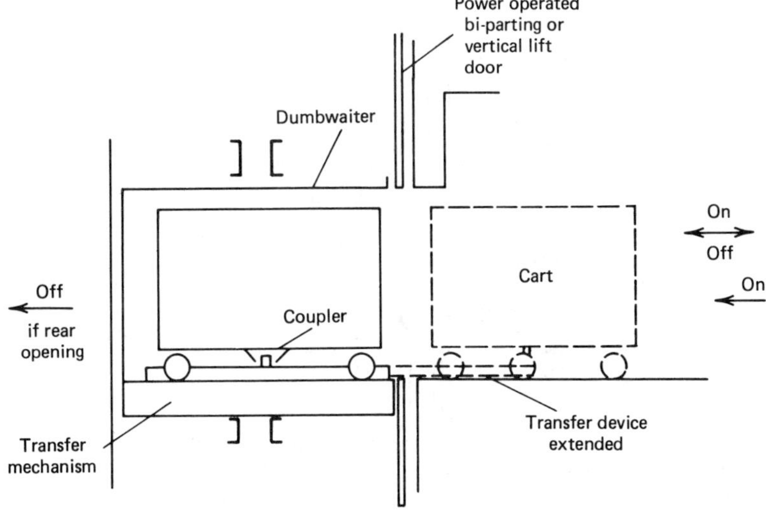

Fig. 5.12 Automated cart lift with inject–eject transfer device. Courtesy of Jaros, Baum & Bolles.

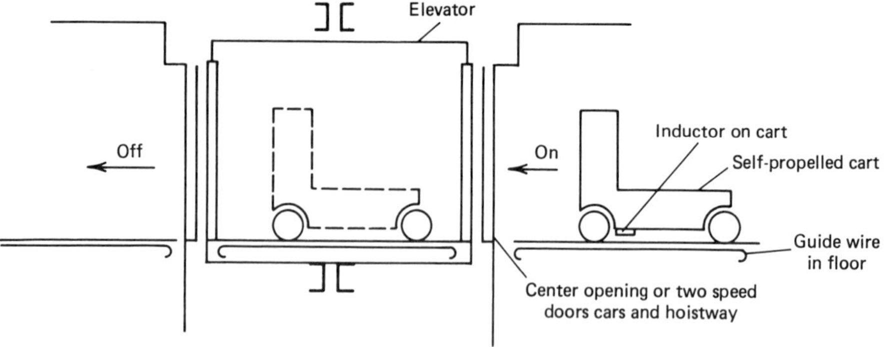

Fig. 5.13 Automated vehicle lift with automated guided vehicle. Courtesy of Jaros, Baum & Bolles.

Vertical Cart Lifts and Horizontal Conveyors

In many health-care facilities, the density of traffic is sufficient to warrant the installation of either "overhead" or "in-floor" conveyors to self-load the cart lift without an attendant (Fig. 5.10). Most of these installations are configured around the overhead "power-and-free" conveyor system on high traffic floors with automated loading and unloading at typical floors.

5.1.8 Dumbwaiters

Dumbwaiters are essentially small lifts limited by Code to 9 sq ft (0.84 m^2) in area and 48 in. (1220 mm) in door height for the transport of tote boxes, trays, or carts. The size is limited to that which will not comfortably accommodate a passenger or freight handler.

The usual dumbwaiter configuration is a counterstopping unit serving a multiplicity of floors for the transport of trays, tote boxes, or other containers which are manually loaded and manually unloaded at both the origin and destination. They can be configured to automatically inject a tote box at an origin floor and eject it at a destination floor (Fig. 5.14). They can also stop at floor level with a resulting pit projection if carts are to be transported.

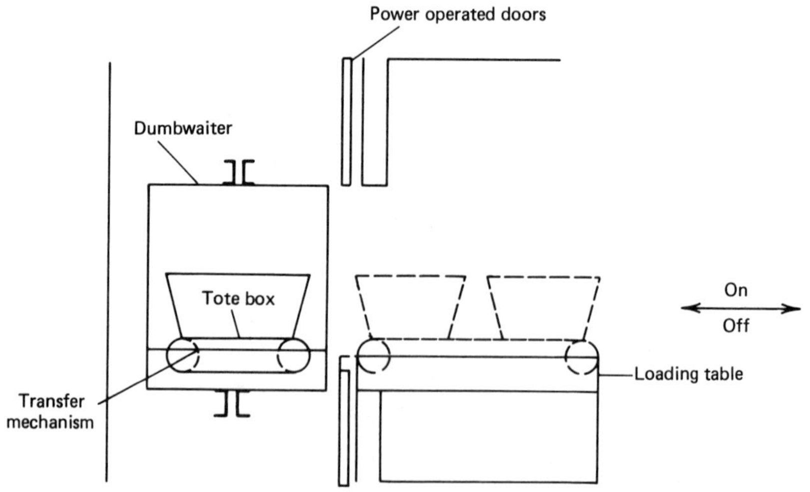

Fig. 5.14 Automated dumbwaiter with inject–eject device for tote boxes. Courtesy of Jaros, Baum & Bolles.

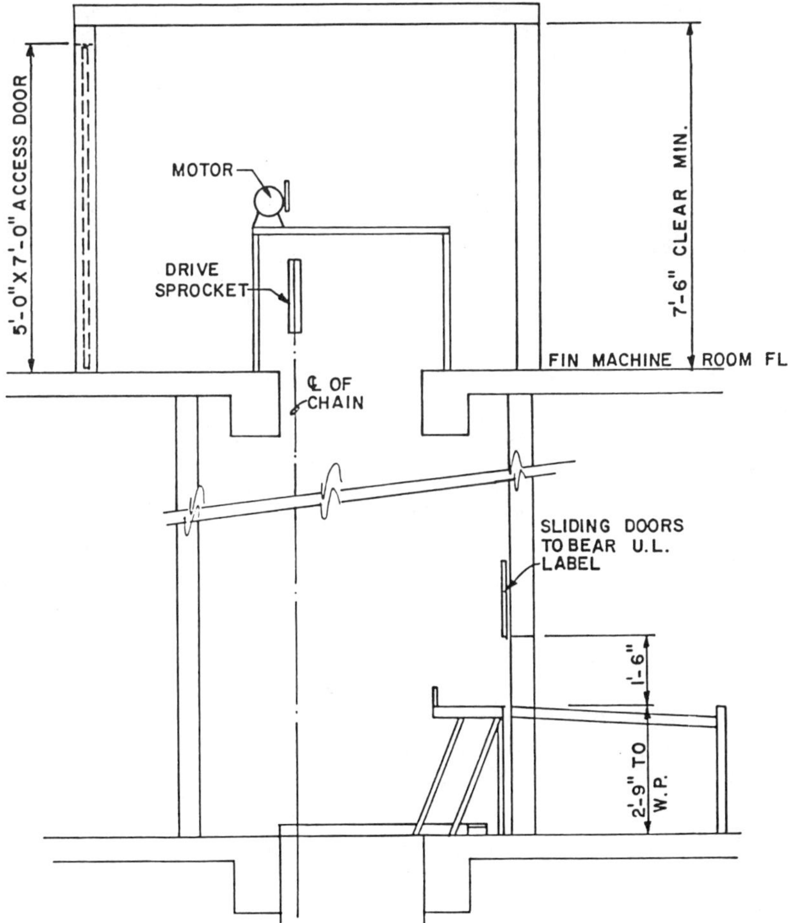

Fig. 5.15 Selective vertical conveyor—overhead section. Courtesy of Jaros, Baum & Bolles.

5.1.9 Vertical Conveyors

Vertical conveyors are continuous flow devices (Fig. 5.15) which have a high through-put capacity that cannot be accommodated by batch dumbwaiters or cart lifts in that the device will carry trays, tote boxes, or pallets in multiple on an endless chain.

The vertical conveyor in its elementary form requires the manual loading at an origin floor of a standardized container with the insertion of an address for its destination floor. At that floor, the vertical conveyor self-unloads the container onto a receiving gravity conveyor.

The vertical conveyor can be combined with a horizontal powered belt conveyor system to respond to the requirements of both vertical and horizontal movement with escort address on the container so that it can be dispatched from and received at a point remote from the vertical shaft. The horizontal portion along with the vertical portion are a combination of two unidirectional conveyors.

5.2 HUMAN FACTORS IN DESIGN CONCEPTS

5.2.1 Horizontal and Vertical Acceleration

The human engineering factors that make up the parameters for an acceptable environment differ widely based on whether the individual had a preconceived understanding of whether or not his immediate environment is a moving environment, either horizontally or vertically, or whether it is a stationary environment. The limits of acceptable horizontal and vertical acceleration are based on experience and subject to an arbitrary but usually accepted level for design. While the horizontal motion of the building is linear about the primary axes or torsional, the linear motion is the portion that imparts detrimental movement to the ropes and traveling cables in an elevator system. The horizontal and vertical elevator accelerations will usually mask any building sway motion perception by passengers and, hence, they are design factors within the elevator discipline.

The maximum horizontal deceleration component affecting escalator or moving walk travel is that which would be experienced by a traveling passenger during an emergency stop when the escalator or moving walk is lightly loaded. This stop would be the most abrupt and, hence, have the tendency to cause unbalance by a passenger if the person was not holding on to the handrail. In order to stop an escalator or moving walk safely from a remote location, such as a fire command center, or automatically by smoke detectors, it has been necessary to establish two levels of stopping criteria, one for the "emergency stop" required by Code and another as a passenger safe "controlled stop" from remote locations.

5.2.2 Building Sway Motion

A proposed lower-limit criterion for design has been suggested by Reed, Hansen, and Vanmarcke [5.2] at 0.005 g (0.112 m/s^2) rms acceleration under a condition when 2% or more of the occupants would object to the motion based on a lower limit of 6 years for the return period of storms. In this referenced paper two 40-story buildings were instrumented. One building in its most sensitive response had a peak-to-peak acceleration of 0.010 g (0.245 m/s^2). During the instrumented storm, approximately 40% of the people noted an abnormal scraping or slapping condition while riding in the elevators. This condition represents a sensitive interaction between building sway and the elevator systems from the standpoint of perceived passenger discomfort. As can be seen from the specifications for elevator ride characteristics in the following paragraph, a poor elevator ride in conjunction with building sway can create an uncomfortable, though short, ride as an elevator traverses to or from the upper portion of a high-rise building.

5.2.3 Elevator Ride Characteristics

The experience with the quality of elevator ride at the Sears Tower in Chicago and at the World Trade Center in New York City has resulted in industry-acknowledged standards for acceptable and commercially achievable ride characteristics based on elevator speed. These or more stringent standards, as experience permits, will, in the future, become part of "Performance Specifications" for quality vertical transportation in first-class buildings. The current specifications given in Table 5.1 have been accepted by all major manufacturers in the speeds listed. It is anticipated that experience will modify the criteria and also extend the speed range.

The recording accelerometer tests are conducted prior to final acceptance on each elevator traveling at its contract speed for the full length of the hoistway between the terminal landings in both the up and down directions with the maximum noneccentric load of 500 lb (2.2 kN). Recordings are taken on the platform in the plane of the car guide rails and perpendicular to the plane of the car guide rails. The accelerometer sensing unit is placed in the center of the elevator platform. The accelerometer must be sensitive in a bandwidth from 0.25 to 2.0 Hz. The maximum horizontal acceleration in each plane is determined from the charts and evaluated against the following criteria. If the test results exceed the maxima permitted, the guide rail alignment, the guide rail joints, and the roller guides must be adjusted to correct the ride characteristic in each plane separately to fall below the maximum permitted. The adjacent peak-to-peak values at instantaneous acceleration are used to determine the zero reference line for peak-to-peak values.

Table 5.1 Permissible Accelerations

Contract Speed in ft/min (m/s)	Adjacent Peak-to-Peak Acceleration in *g*'s [ft/s^2]
300 (1.5) and below	0.020 (0.65)
400 (2.0)	0.025 (0.82)
500 (2.5)	0.025 (0.82)
800 (4.0)	0.025 (0.82)
1000 (5.0) and above	0.030 (1.0)

As can be seen from Table 5.1, the adjacent peak-to-peak acceleration permitted in the ride of an elevator is approximately twice that of the acceleration acceptable in building sway conditions.

5.2.4 Escalator Stopping Characteristics

The life safety control of escalators, until recently, had been left to individuals using pushbuttons at the top and bottom of each escalator unit to initiate an emergency stop condition. Under certain loading conditions, the escalator could stop so abruptly as to cause passengers to lose their balance and fall down the escalator in a cascade reaction. Newer buildings have equipment that shut down escalators remotely and safely by smoke detector or pushbutton from a fire command center. It is anticipated that the Code-required "emergency" stop will be supplemented by a "controlled stop" which will decelerate the escalator at a maximum of 1.5 ft/s^2 (0.45 m/s^2) and between 18 in. (0.5 m) minimum and 48 in. (1.2 m) maximum stopping distance from no load to full load on a 40-in. (1000-mm) wide escalator. Similar specifications have been developed for other sizes of escalators. Passengers may be able to return to areas of safety using the stopped escalator as a stairway as a result of emergency instructions that are given over the public address system so that they will travel away from a fire or smoke-involved floor.

5.2.5 Pedestrian Planning

The pedestrian planning for a complex vertical transportation system involving a combination of elevators and escalators requires that an adequate study of anticipated arrival and departure rates be conducted in order to establish the proper queuing areas and circulation paths that will result in the peak traffic periods. In general, this should start outside the building and continue up to the destination floor for up-peak traffic periods and the reverse for down-peak traffic periods to assure that there will be adequate spaces or transfer from one system to another. As a general rule, it is accepted that the minimum lobby width between elevator banks be 10 ft (3 m) and that the maximum width should not exceed 13 ft (4 m) for convenience in access as well as exiting the elevator group. In a similar way, the queuing distance at the end of an escalator run is an essential dimension which permits passengers to select alternate paths and should be, under heavy traffic condition, a minimum of 12 ft (3.7 m) at each landing from the newel of the escalator to the nearest obstruction in the path of traffic (see Section 5.13).

5.3 ELEVATORS, DUMBWAITERS, LIFTS, AND CONVEYORS—THE STRUCTURAL INTERFACE

The basic design and function of the various categories have been described in Section 5.1. The limits of the design function of each and the proper application of the equipment should be left to competent consulting engineers to accommodate properly the pedestrian and materials flow within a major building commensurate with its use and the proper separation of the circulation pattern. The following paragraphs in this section are intended to cover the scope of the structural and related architectural considerations to establish the proper interfaces among the various components of a complex vertical transportation system. In general, except as noted, the structural interface for a group of similar units can be handled on the basis of designing the interface for a single unit and repeating it wherever a similar unit is installed.

5.3.1 Elevator Pits

Elevator pits of the proper depth must be provided with various structural considerations. The elevator main and counterweight guide rails must be bracketed to the building construction below the lowest landing served and above the pit to accommodate the load transfer reactions that will occur on the lower guides of the elevator car in conjunction with the rails when it is at its bottom terminal landing. Buffers are located under each elevator car, regardless of whether or not the elevator is hydraulic, geared, or gearless, to absorb the impact of the car when it overtravels the bottom terminal landing (see Figs. 5.16 and 5.17). The hydraulic elevator buffers and low-speed geared elevators are spring buffers and generally are of minor consequence in considering structural reactions due to the low kinetic energy level involved. The buffers of high-speed geared elevators and all gearless elevators are hydraulic oil buffers intended to retard an elevator car at the governor tripping speed should the safety not set until the car reaches the run-by at the bottom terminal landing. The buffers then are intended to retard the car over its stroke and an average retardation of 1.0 g and a maximum of 2.5 g as a peak. This reaction is immediately absorbed by a steel beam in the pit provided by the elevator subcontractor which spans between the two main

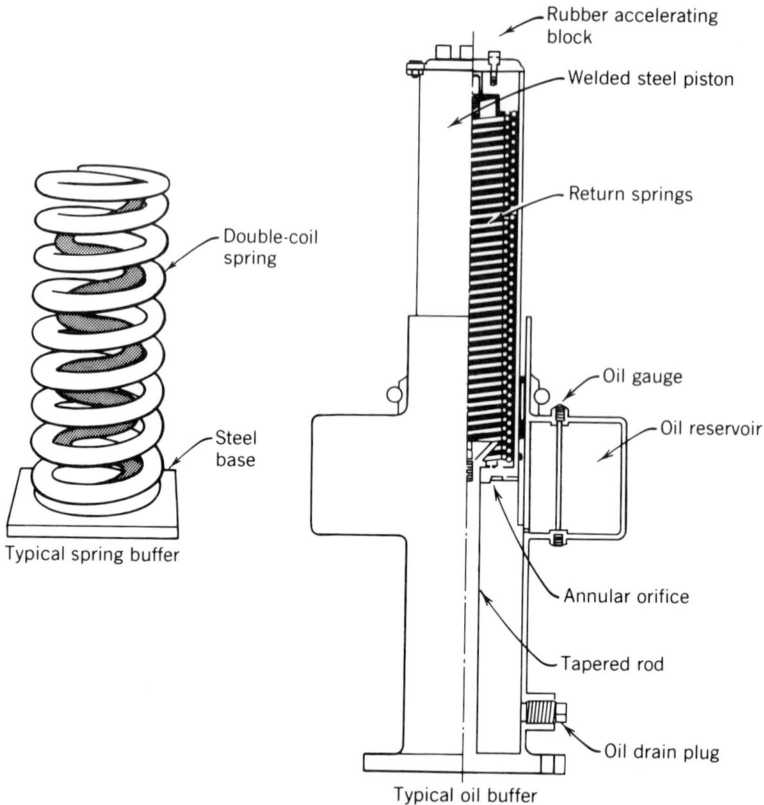

Fig. 5.16 Typical spring and oil buffers. Courtesy of Westinghouse Elevator Co.

guide rails. Similar buffers with similar construction and reactions are placed under the counterweight for all traction elevators to accomplish the same purpose should the counterweight strike the buffers just prior to the car being shut down in the up direction at the governor tripping speed. It should be noted that these two reactions can physically never occur simultaneously. Hence, this should be taken into consideration with the structural design of the pit. Under all conditions, the main car guide rails of electric traction elevators will be used as columns to retard an elevator car that may overspeed in the down direction by the application of the safety device that has been tripped by the governor. The rail reactions normally occur in the pit at the pit floor with a reaction which has been doubled for impact, but which is modified by the location of the safety application in the hoistway and the sliding rail clips at each floor bracket interface. The worst condition is assumed to occur if the car safety applies directly above the buffers where the pit reaction will see most of the safety application forces.

When there is space that may be occupied or used beneath a pit floor, the counterweight must be equipped with a governor-operated safety so that it will not free-fall to the pit floor. The safety application of the counterweight is similarly imposed on the pit floor. In the physical application of car safety or counterweight safety, it should be noted that these two safety applications are similar to buffer applications and again cannot occur simultaneously. This should be taken into consideration in a structural design of the pit. Forces are shown in Figs. 5.18 and 5.19.

The weight compensation of the hoist ropes and the traveling cables is usually configured by adding a set of compensating ropes between the underside of the counterweight and the underside of the car when necessary for traction considerations. When the speed exceeds 700 ft/min (3.5 m/s), this compensating sheave, which normally would not impose any uplift reaction on the pit floor, must be restrained after a certain travel dimension has been exceeded to prevent the counterweight from momentarily jumping free of its hoist ropes when the car has a safety application in the down direction or vice versa when the counterweight has a safety application with the car in the up

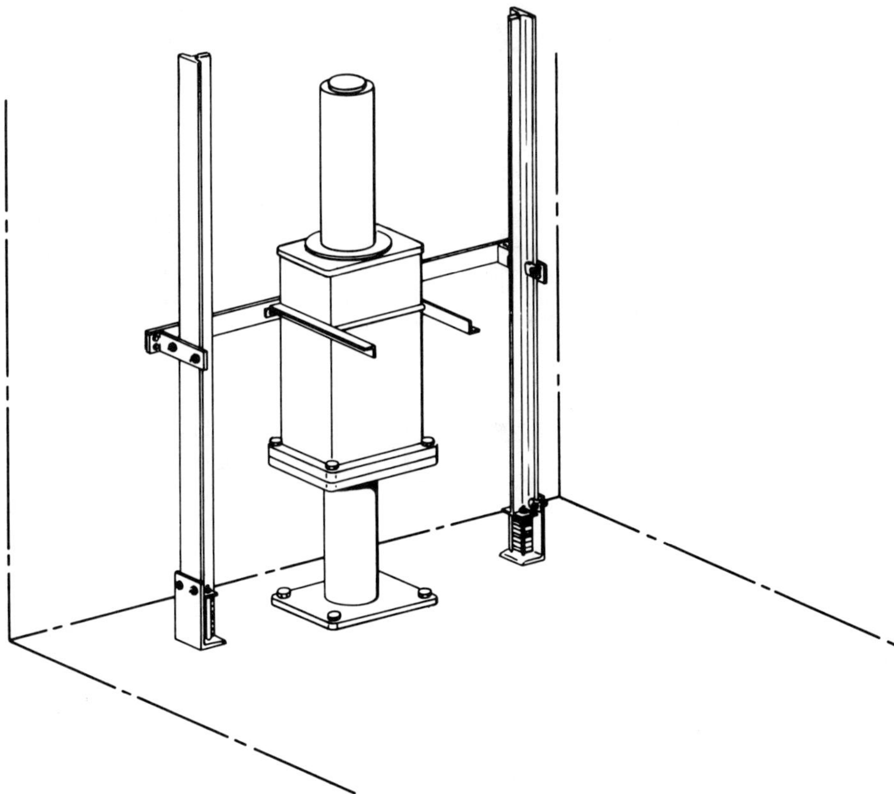

Fig. 5.17 Typical oil buffers located between car rails and counterweight rails. Courtesy of Otis Elevator Co.

direction. This restraining action occurs as an uplift in the pit, approximately equidistant between the car buffer and the counterweight buffer located in the centerline of each set of main and counterweight guide rails, as shown in Figs. 5.20 and 5.21. It should be noted that the pit uplift reaction from tie-down compensation, along with either the safety application from the car or the safety application from the counterweight, may occur almost simultaneously and should be taken into consideration with the structural design. The normal construction relationship requires that, in tie-down compensation, the structural engineer must provide the beams to which the counterweight and car buffers are attached and to which the channels that support the tie-down compensating sheaves that run between the car buffers and counterweight buffers are attached. The steel support beam system should have a minimum flange width of 8 in. (200 mm) and should be located at least 4 in. (100 mm) above the pit floor so that through-bolt connections can be used.

The usual configuration among most manufacturers for the bottoming of the car or counterweight guide rails with safeties is to install a jack screw arrangement with a minimum of $\frac{1}{4}$-in. (7-mm) steel shims that can be used to effectively bottom the car rails or the counterweight rails when counterweight safeties are used on the pit floor without excessive rail slide in the sliding rail clips at each bracket. Excessive rail slide by one or both rails involved in a safety application would have an unequal slide dimension between them and, as a result, rack the car frame or counterweight frame with possible extensive damage. When a safety test is conducted with a loaded car for the car safety or with an empty car with a counterweight safety, the jack screws are temporarily bottomed to provide a rigid connection between the building structure and the underside of the car or counterweight rails. At the conclusion of the test, the jack screws are backed off to provide a clearance of approximately $\frac{3}{16}$ in. (5 mm) so that any ensuing building compression and subsequent rail slide would not cause misalignment of the car rails because of a compression force exerted on the rail alignment from the pit floor. During the initial occupancy and fitting period and extending through the first heating season, the clearance between the jack screws and the underside of the rail

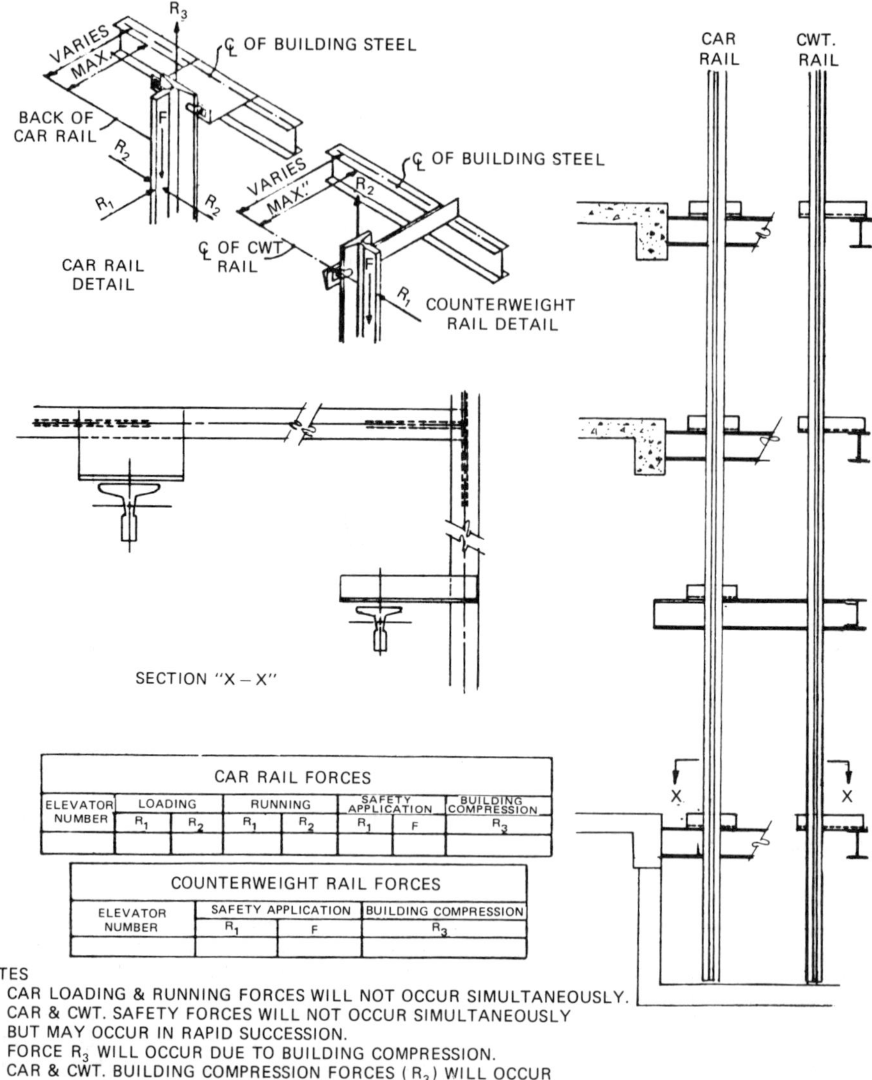

CAR RAIL FORCES							
ELEVATOR NUMBER	LOADING		RUNNING		SAFETY APPLICATION		BUILDING COMPRESSION
	R_1	R_2	R_1	R_2	R_1	F	R_3

COUNTERWEIGHT RAIL FORCES			
ELEVATOR NUMBER	SAFETY APPLICATION		BUILDING COMPRESSION
	R_1	F	R_3

NOTES
1– CAR LOADING & RUNNING FORCES WILL NOT OCCUR SIMULTANEOUSLY.
2– CAR & CWT. SAFETY FORCES WILL NOT OCCUR SIMULTANEOUSLY BUT MAY OCCUR IN RAPID SUCCESSION.
3– FORCE R_3 WILL OCCUR DUE TO BUILDING COMPRESSION.
4– CAR & CWT. BUILDING COMPRESSION FORCES (R_3) WILL OCCUR SIMULTANEOUSLY. FORCES EXPECTED TO BE LIMITED BY SLIDING RAIL CLIPS DURING & AFTER CONSTRUCTION
5– MAXIMUM ALLOWABLE DEFLECTION 1/8 IN.

Fig. 5.18 Forces of car rail and counterweight rail (side counterweight). Courtesy of Jaros, Baum & Bolles.

assembly must be frequently monitored so that, as the building compresses, this clearance is maintained. Rail misalignment and subsequent rough-riding elevator cars are often a result of the failure to accommodate the building compression with jack screws or with properly working sliding rail clips, or both.

5.3.2 Elevator Hoistways

The hoistway for each elevator contains, in addition to the car and counterweight, the guide rails that are supported at each floor level for guiding the car and counterweight and the electrical wire trough carrying the conductors to the landing signals, pushbutton stations, and the wiring to the halfway box which terminates the flexible traveling cable to the elevator car. Under normal condi-

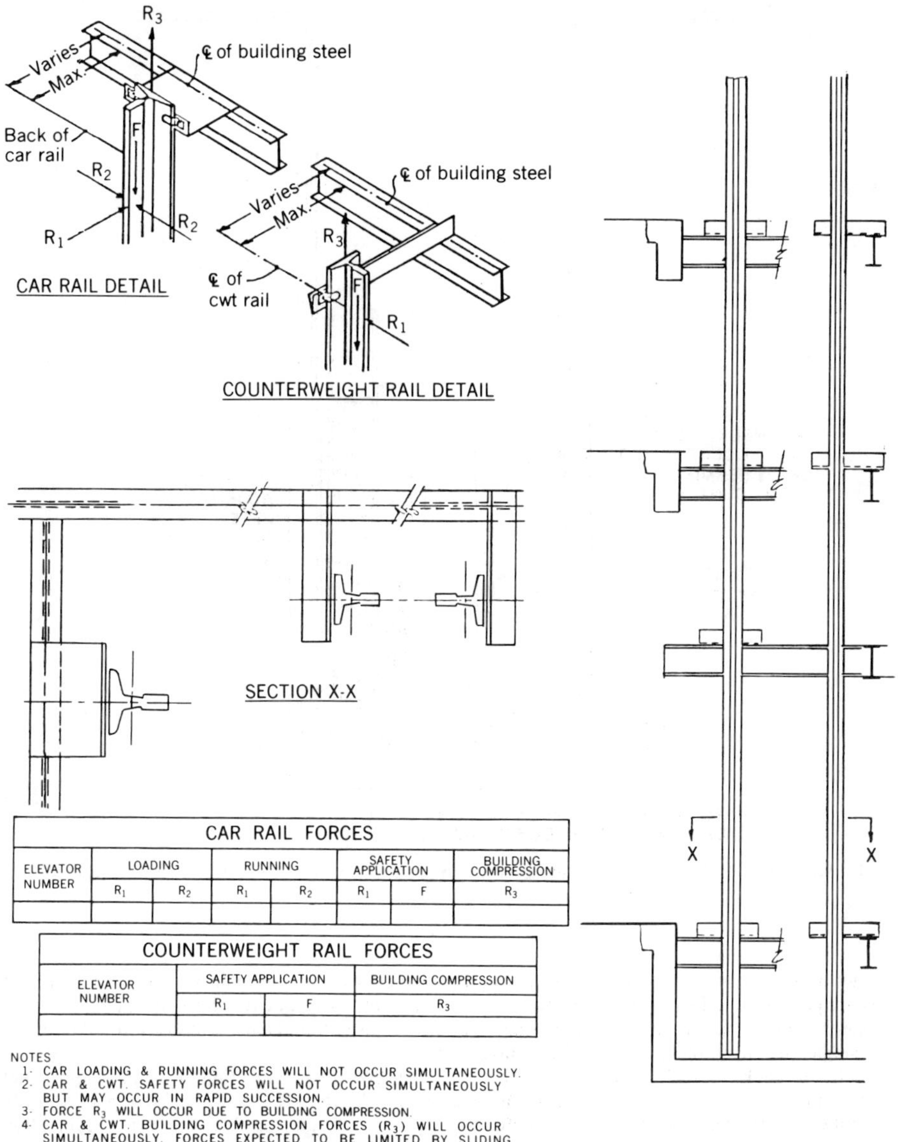

ELEVATOR NUMBER	LOADING		RUNNING		SAFETY APPLICATION		BUILDING COMPRESSION
	R_1	R_2	R_1	R_2	R_1	F	R_3

CAR RAIL FORCES

ELEVATOR NUMBER	SAFETY APPLICATION		BUILDING COMPRESSION
	R_1	F	R_3

COUNTERWEIGHT RAIL FORCES

NOTES
1- CAR LOADING & RUNNING FORCES WILL NOT OCCUR SIMULTANEOUSLY.
2- CAR & CWT. SAFETY FORCES WILL NOT OCCUR SIMULTANEOUSLY BUT MAY OCCUR IN RAPID SUCCESSION.
3- FORCE R_3 WILL OCCUR DUE TO BUILDING COMPRESSION.
4- CAR & CWT. BUILDING COMPRESSION FORCES (R_3) WILL OCCUR SIMULTANEOUSLY. FORCES EXPECTED TO BE LIMITED BY SLIDING RAIL CLIPS DURING & AFTER CONSTRUCTION.
5- MAXIMUM ALLOWABLE DEFLECTION $\frac{1}{8}$ INCH.

Fig. 5.19 Forces on car rail and counterweight (rear counterweight). Courtesy of Jaros, Baum & Bolles.

tions, the Code maximum car rail bracket spacing interval should not exceed 14 ft (4.2 m) and the maximum counterweight rail bracket spacing interval should not exceed 16 ft (4.9 m). The Code requirements are specific with respect to the load on the rails, the size or weight of the rails, and the bracket interval connecting the rail system to the building structure. Under special conditions, it is necessary to provide intermediate (between floors) rail bracket supports at the perimeter of an elevator hoistway to accommodate an extra high floor height at the main lobby or elsewhere. Under special infrequent conditions, the guide rails may be reinforced with rail backing provided by the elevator subcontractor to accommodate this extended bracket interval. In order to prevent subsequent postcontract extras, the rail bracket spacing of 14 ft (4.2 m) and 16 ft (4.9 m) should be adhered to in the design of the structural support system. The intermediate steel required by this

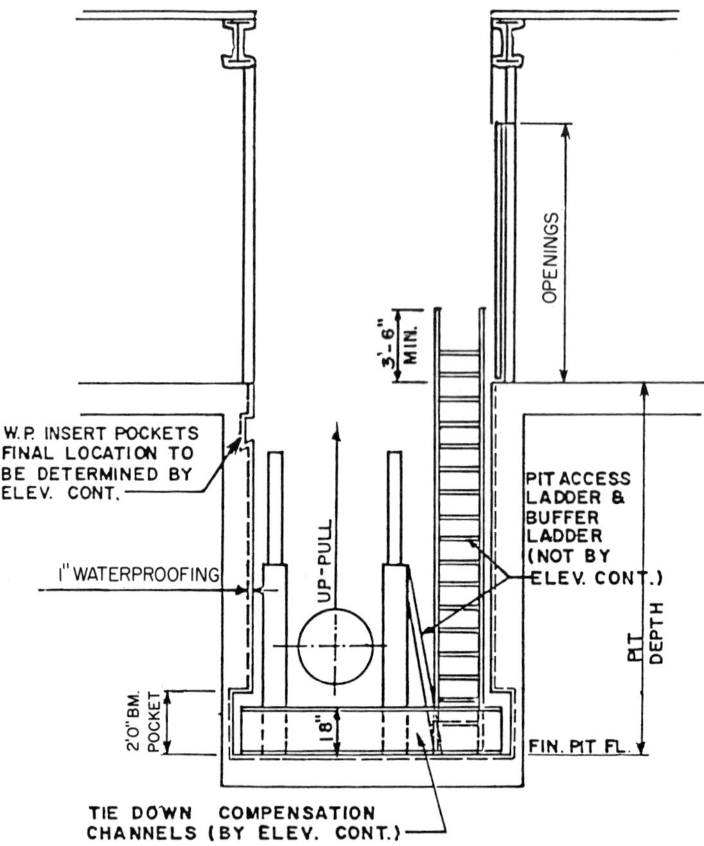

Fig. 5.20 Pit uplift forces from tie-down compensation. Courtesy of Jaros, Baum & Bolles.

restriction need not be provided between multiple hoistways, as the main and/or counterweight rails are permitted to be bracketed together to accommodate this Code requirement. The intermediate steel needs to be provided in the perimeter wall structure around a bank of elevators that occur in a single shaftway made up of multiple hoistways.

Each rail bracket should be provided with a pair of sliding clips connecting the rail to the bracket which will permit vertical slide between the bracket and the rail. These should be applied to both car and counterweight rails to accommodate building compression as well as the safety application on rails with safeties. The reaction between the rail and the bracket, if not properly maintained, can exert a substantial force at the rear (flange) of the rail. The maximum force anticipated from a single rail in a properly aligned and maintained condition is approximately 1500 lb (6.7 kN) per bracket as the maximum slide-through force. In considering the loading forces, running forces, safety application forces, and building compression forces, the ANSI A17.1 Code [5.1] is specific in its requirement under Rule 200.5B (fastenings and supports), as follows:

> *The guide rail brackets, their fastenings and supports, such as building beams and walls, shall be capable of resisting the horizontal forces imposed by the class of loading with a total deflection at the point of support not in excess of 1/8 in. (3.2 mm).*

(See Figs. 5.18, 5.19, and 5.22.)

At least one subcontractor may prefer to pin the guide rails at approximately two-thirds of the distance up the hoistway to the structural building system instead of bottoming the rails in the pit with jack screws or by other means. This preference comes about as a result of observation of the movement of elevator guide rails with sliding clips in high-rise buildings. It appears that a neutral zone exists at approximately the two-thirds point in the hoistway above which the rails slide up

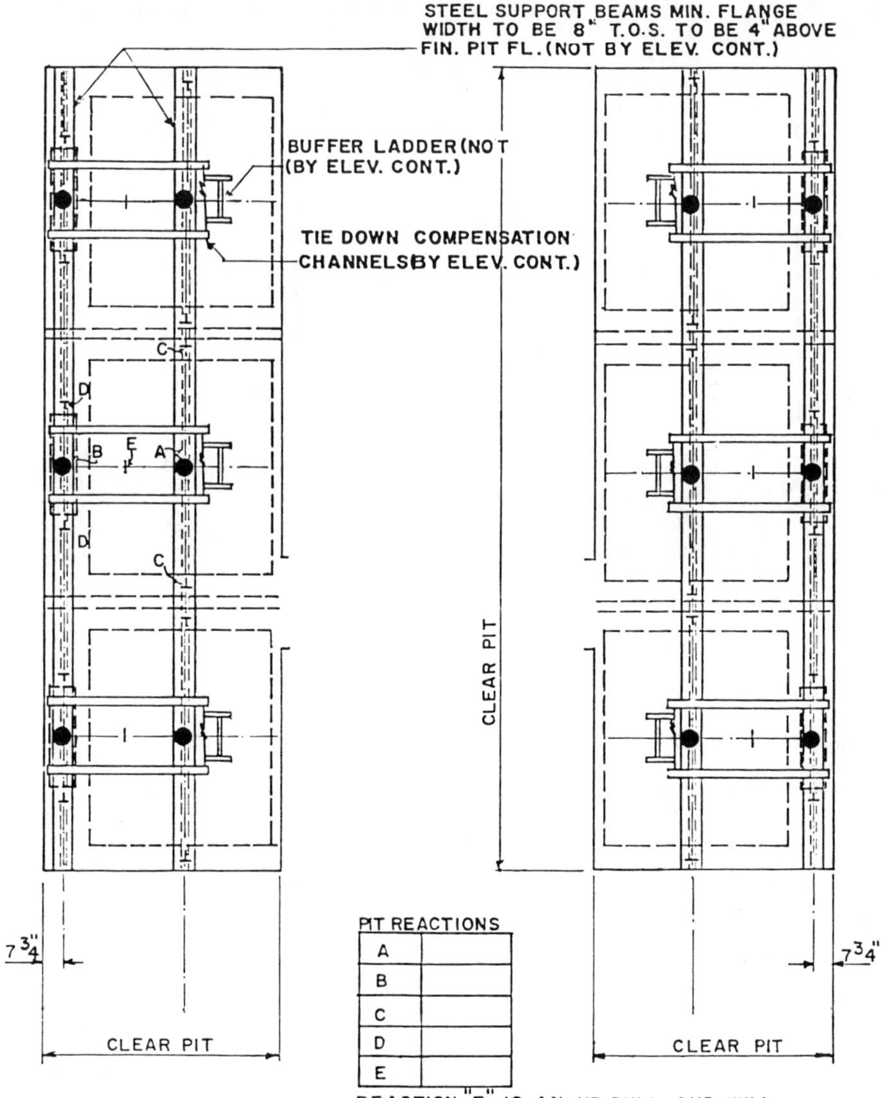

STEEL SUPPORT BEAMS MIN. FLANGE WIDTH TO BE 8" T.O.S. TO BE 4" ABOVE FIN. PIT FL. (NOT BY ELEV. CONT.)

BUFFER LADDER (NOT BY ELEV. CONT.)

TIE DOWN COMPENSATION CHANNELS (BY ELEV. CONT.)

CLEAR PIT

$7\frac{3}{4}$"

CLEAR PIT

CLEAR PIT

$7\frac{3}{4}$"

PIT REACTIONS

A	
B	
C	
D	
E	

REACTION "E" IS AN UP PULL AND WILL OCCUR SIMULTANEOUSLY WITH RAIL REACTIONS "C" OR "D"

Fig. 5.21 Pit plan—tie down compensation scheme. Courtesy of Jaros, Baum & Bolles.

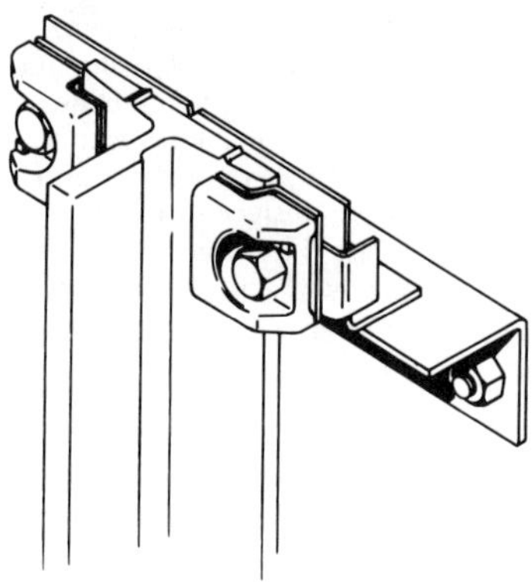

Fig. 5.22 Typical rail clip. Courtesy of Otis Elevator Co.

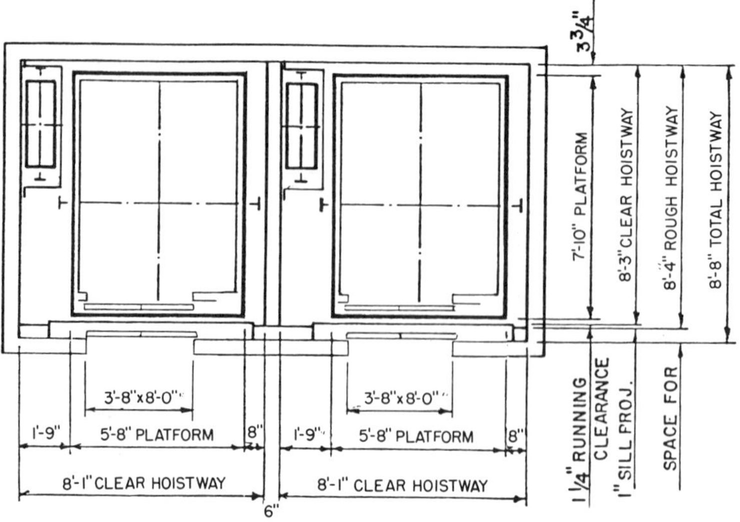

Fig. 5.23 Hoistway plan for earthquake zone, passenger elevators. Courtesy of Jaros, Baum & Bolles.

through the floor brackets and below which they slide down through the floor brackets. In some cases, pinning of the rails may result in extra cost to support the system at that point since most elevator consulting engineers prefer to support the guide rails for a safety application in the elevator pit. In high-rise buildings above 50 stories with high-speed elevators, this can become a serious elevator problem without specially designed fishplates and "body-fit" fishplate bolts since a safety application below the pinning points (normally at two contiguous floors) may tend to open the guide rail joints if the fishplates and rails are not fastened with "body-fit" bolts. In the pinning configuration, the rails with safeties are cut off at least 6–8 in. (150–200 mm) above the pit floor bottoming steel.

The design of the divider beams between adjacent elevator hoistways should take into consideration the common main guide rail fastening as well as the counterweight guide rails fastenings if the counterweight is located at the side of the elevator car rather than at the rear. A side-mounted counterweight with counterweight safety can exert a substantial movement on the divider beam during a counterweight safety application. Usually there is not a counterweight with its rails from the adjacent elevator located on the same beam which would help to overcome this movement by bracketing the rails themselves together at intermediate points (see Fig. 5.23).

5.3.3 Elevator Machine Rooms

The elevator machine rooms in high-rise buildings are located over the elevator hoistway. The space required in the machine room is usually of the order of 200% of the area of the hoistway with the elevator hoisting machine located directly above the hoistway and supported by perimeter framing associated with each hoistway at a Code-dictated level above the topmost landing and above the secondary level when required. The support of the elevator machinery is always a divided responsibility. The elevator machinery is supported directly by elevator machine beams or sheave beams, the design and provision of which are the responsibilities of the elevator subcontractor. The bearing points of these beams on the supporting structure must be designed to withstand the reactions shown on the elevator layout. Section 105 of ANSI A17.1 [5.1] indicates the design conditions for the elevator machine beams, sheave beams, and the associated supporting structure (see Figs. 5.24, 5.25, and 5.26). The pertinent rules are quoted as follows:

> *Rule 105.1 Beams and Supports Required: Machines, machinery, and sheaves shall be so supported and maintained in place as to effectually prevent any part from becoming loose or displaced under the conditions imposed in service.*
>
> *Supporting beams, if used, shall be of steel or reinforced concrete. Beams are not required under machines, sheaves, and machinery or control equipment which are supported*

on floors provided such floors are designed and installed to support the load imposed thereon.

Rule 105.2 Loads on Machinery and Sheave Beams, Floors or Foundations and Their Supports: 105.2a Overhead Beams, Floors and Their Supports: Overhead beams, floors and their supports shall be designed for not less than the sum of the following loads: 1) The load resting on the beams and their supports which shall include the complete weight of the machine, sheaves, controller, governor and any other equipment together with that portion, if any, of the machine room floor supported thereon. 2) Two (2) times the sum of the tensions in all wire ropes supported by the beams with the rated load in the car.

MACHINE ROOM PLAN
PASSENGER ELEVATORS

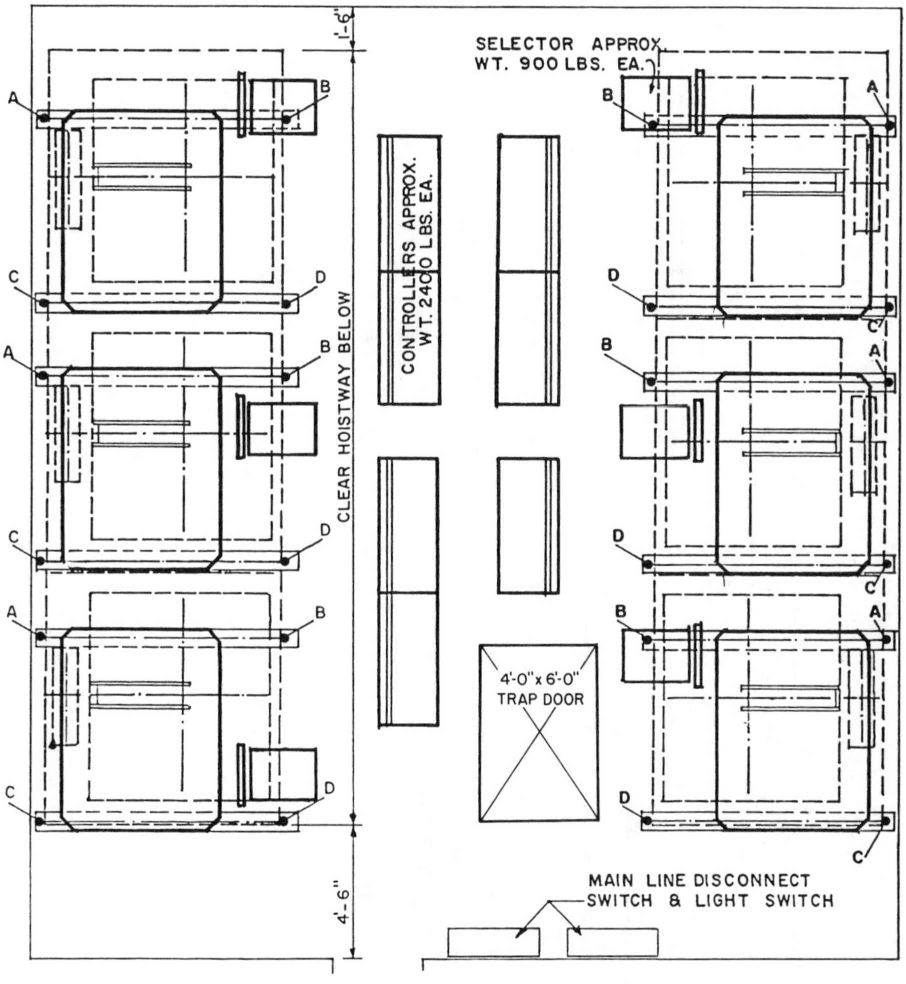

MACHINE REACTIONS

A	30.0^k	C	17.0^k
B	21.0^k	D	12.0^k

NOTE: ALL LIVE LOADS HAVE BEEN DOUBLED FOR IMPACT.

3500 LBS @ 700, 800, 1000 & 1200 F.P.M.

Fig. 5.24 Machine room plan, passenger elevators. Courtesy of Jaros, Baum & Bolles.

Note: These tensions are doubled to take care of impact, accelerating stresses, etc.

Rule 105.4 Allowable Stresses for Machinery and Sheave Beams or Floors and Their Supports: The unit stresses for all machinery and sheave beams and floors and their supports, based on the loads computed as specified in Rule 105.2 shall not exceed eighty (80) percent of those permitted for static loads by the following standards: a) Structural Steel: AISC Specifications for the design, fabrication and erection of structural steel for buildings. b) Reinforced Concrete: Building Code requirements for reinforced concrete, ANSI A89.1.

When the stresses due to loads other than elevator loads supported on the beams of floors exceed those due to elevator loads 100% of the permitted stresses may be used.

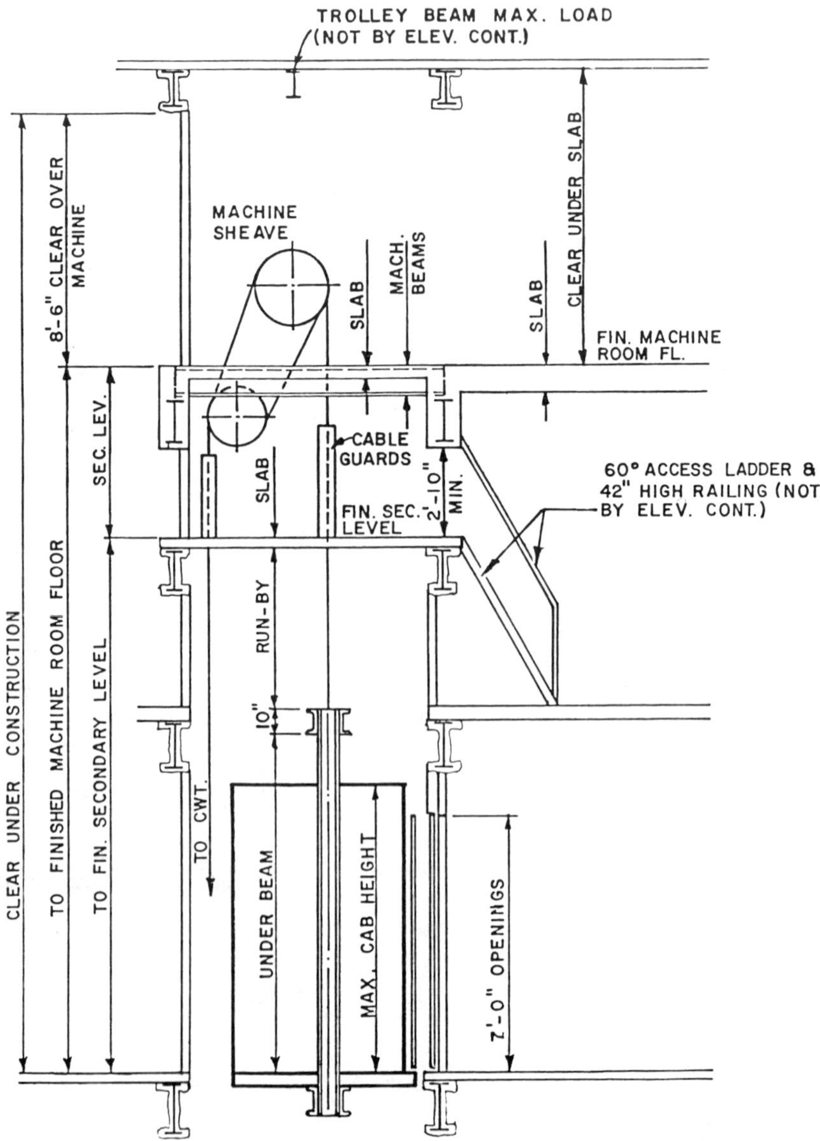

Fig. 5.25 Sheave beams for gearless elevator (overhead section). Courtesy of Jaros, Baum & Bolles.

Rule 105.5 Allowable Deflection of Machinery and Sheave Beams and Their Supports: The allowable deflections of machinery and sheave beams and their immediate supports under static loads, shall not exceed 1/1666 of the span.

In machine rooms for geared or gearless elevators, it is important to ascertain whether or not the roping between the machine car and counterweight is 1 : 1 or 2 : 1. In a 1 : 1 configuration (see Fig. 5.27), the dead-end hitches terminate on the car and the counterweight and, hence, there are no rope tensions supported by the machine room construction other than that imposed by the ropes through the elevator machine. However, when a 2 : 1 configuration exists (see Fig. 5.28), the dead-end hitches of the ropes terminate either on the secondary level below the machine room or on the machine room level. When such occurs, it is necessary to make sure that the reaction imposed on the supporting steel by the machine beams directly related to the machinery and the dead-end hitch beams related to both dead-end hitches be included in the calculations for deflection and stress. It is often the current practice to eliminate the secondary level which previously contained the dead-end hitch beams and the governor and now relocate those reactions on the machinery support beams in addition to the machine reaction.

The elevator machine rooms, in addition to the floor reactions of controllers and motor generator sets, if provided, should be equipped with hoist beams above each elevator hoistway so that the machinery can be easily installed and properly aligned during construction. Subsequently the major components can be easily removed for repair or replacement. In major installations, the hoist

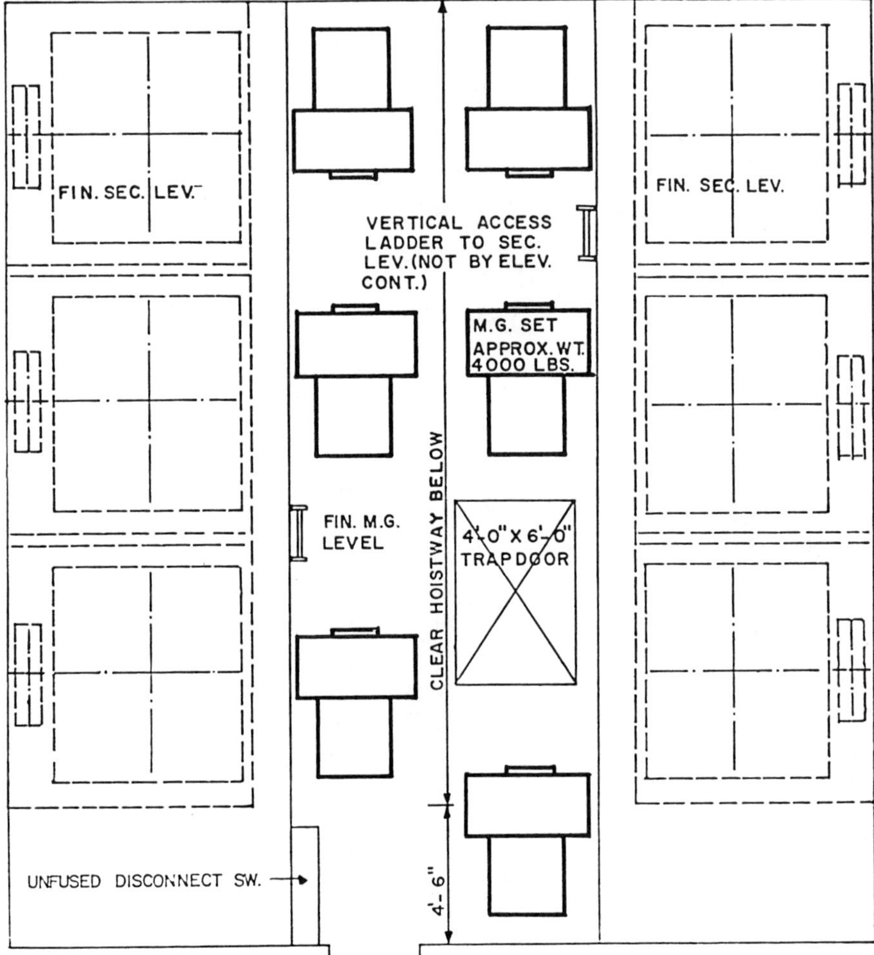

Fig. 5.26 Mezzanine grade and secondary level passenger elevators. Courtesy of Jaros, Baum & Bolles.

beams should be configured as trolley beams with a portion of the trolley beam passing over each elevator machine and arranged to travel over a trap door in the elevator machine room floor slab so that the gearless armatures may be lowered to the top terminal landing of either the service elevator or passenger elevator system. This may entail a second trap door in a secondary level when it is provided.

5.3.4 Secondary Level Access Space

A secondary level is provided underneath a machine room to provide access for lubrication service and inspection of a secondary sheave beneath a gearless machine since the elevator car at its topmost position cannot adequately provide access to the secondary sheave when it is located below the machine room slab (see Fig. 5.29). When this slab is provided, it usually will contain the governor and any dead-end hitches with 2 : 1 roping. It should be pointed out that the governor does provide a reaction when it has tripped at approximately 1000 lb (4.5 kN) or as much as 2000 lb (9 kN) as a pull-through of the governor rope as the car continues to slide after the application of the safety. This reaction may or may not be important, depending on the roping conditions and depending upon the manufacturer since some manufacturers bracket the governor from the guide rail. When this occurs, the penetration of the guide rail through the secondary level must be free of the slab in order that it may move with the rail system during building compression.

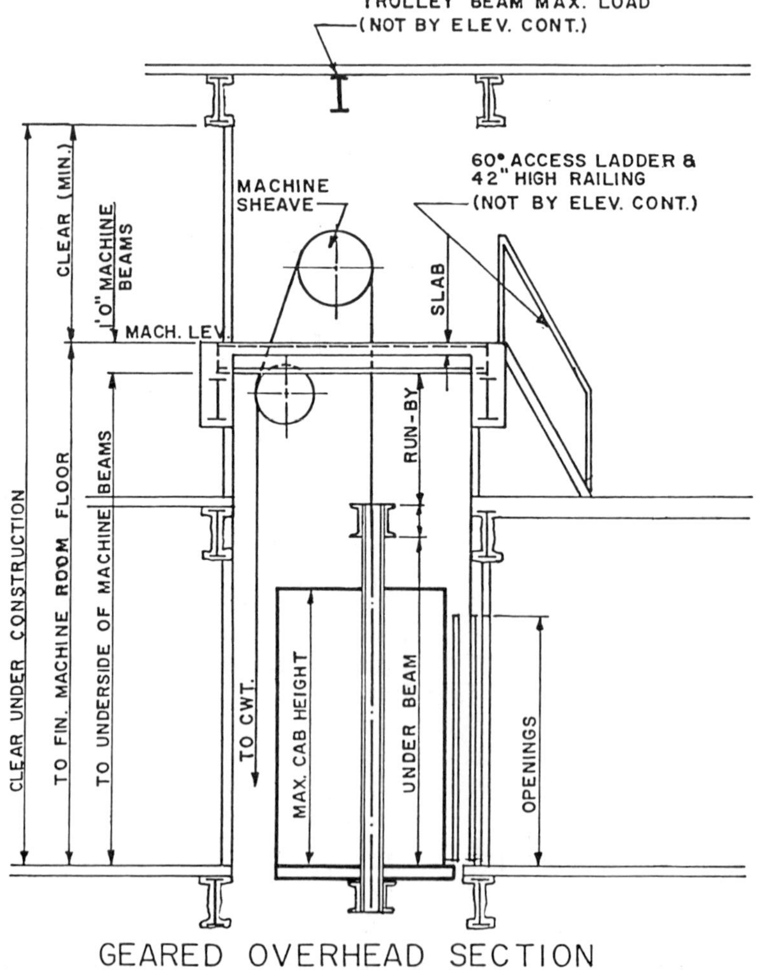

Fig. 5.27 1 : 1 roping among machine, car, and counterweight. Courtesy of Jaros, Baum & Bolles.

5.3.5 Special Structural Support for Freight Elevators

Freight elevators above 6000 lb (27 kN) with Class C Loading in each of the classes (Class C1, Class C2, and Class C3) must take into consideration special loading and unloading reactions imposed on the guide rails and the associated building structure during the loading and unloading process. The special structural requirement for this loading usually takes the form of two (2) vertical columns with beams at each side of the hoistway located inside the hoistway walls and fastened appropriately at floor levels to the main structural framing. The columns in the hoistway at each main rail are required to permit the insertion of beams between the two columns which are bolted directly to the flange of the main guide rails wherever the elevator guide shoe will bear on the guide rail during the loading and unloading impact. Each rail during the loading and unloading process has two guide shoes bearing on it at the top of the "cross-head" above a car and at the bottom of the "safety plank" beneath the platform. It is the usual practice to provide two beams for each guide shoe location or a total of eight beams per landing so that the loading and unloading forces will be immediately transferred from the guide rail to the supporting structure. Figure 5.30 shows the typical arrangement for elevators with a capacity of 6000 lb (27 kN) or more where this truck loading may occur. For more specific information concerning the loading classifications and the

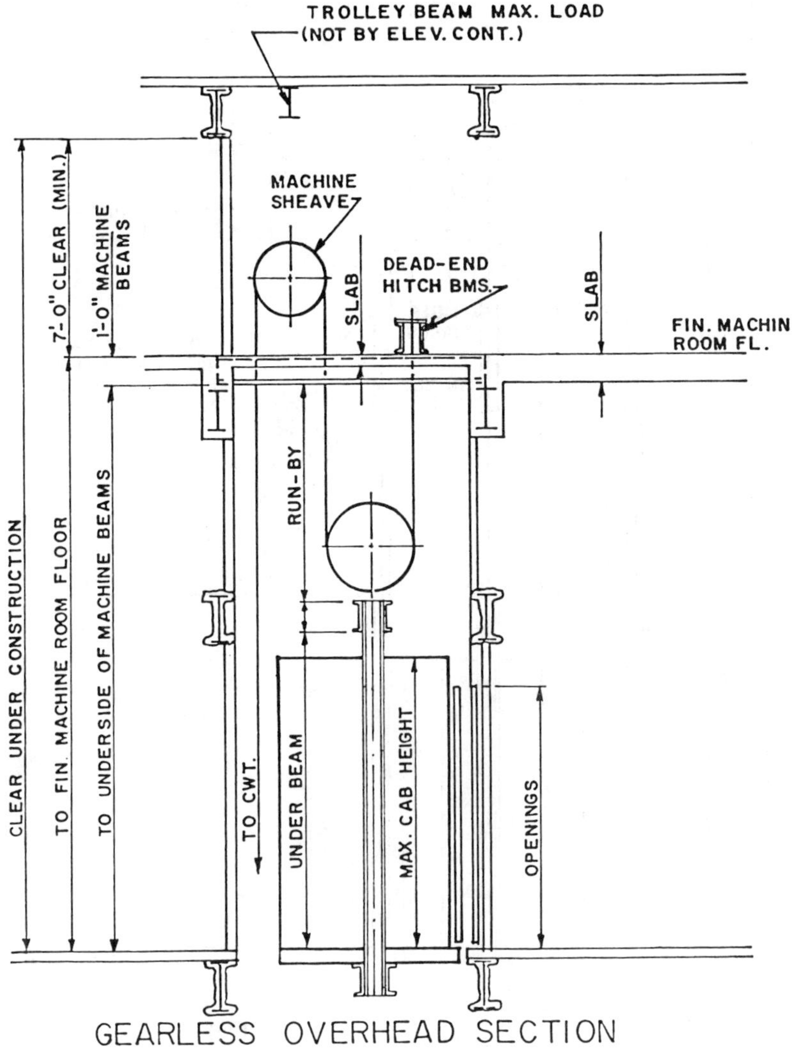

Fig. 5.28 2:1 roping among machine, car, and counterweight. Courtesy of Jaros, Baum & Bolles.

special constructions necessary to accommodate them, refer to Rule 207.2, Minimum Rated Load for Freight Elevators, in the Code.

5.3.6 Emergency Stop Reactions

The actuation of the emergency stop switch in the elevator car in each elevator pit and machine room or secondary level will result in an electrical stop using the machine brake as a retardation device rather than only a holding device when used in normal operation in conjunction with the dynamic braking feature. While this stop is a severe stop, it is not as severe with respect to building reactions as the application of either the counterweight or the car safety. As such, no special structural considerations are necessary when all the other considerations have been accommodated.

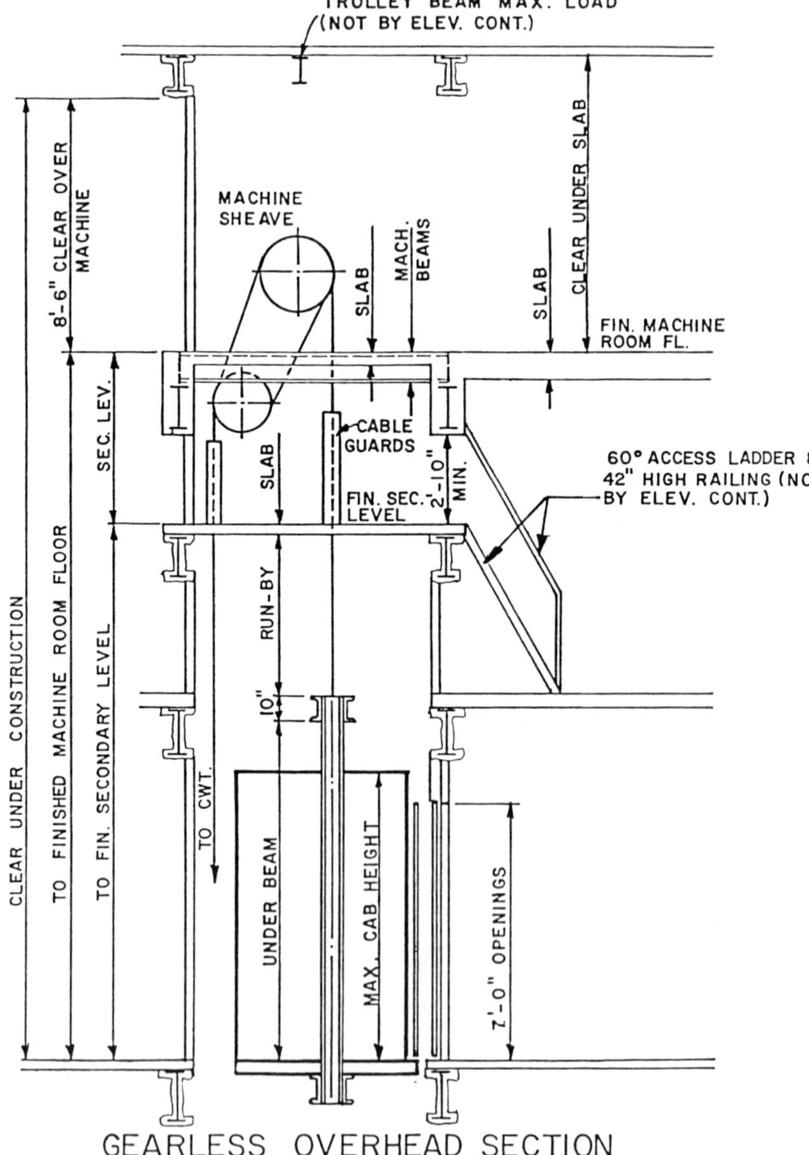

Fig. 5.29 Secondary level access space. Courtesy of Jaros, Baum & Bolles.

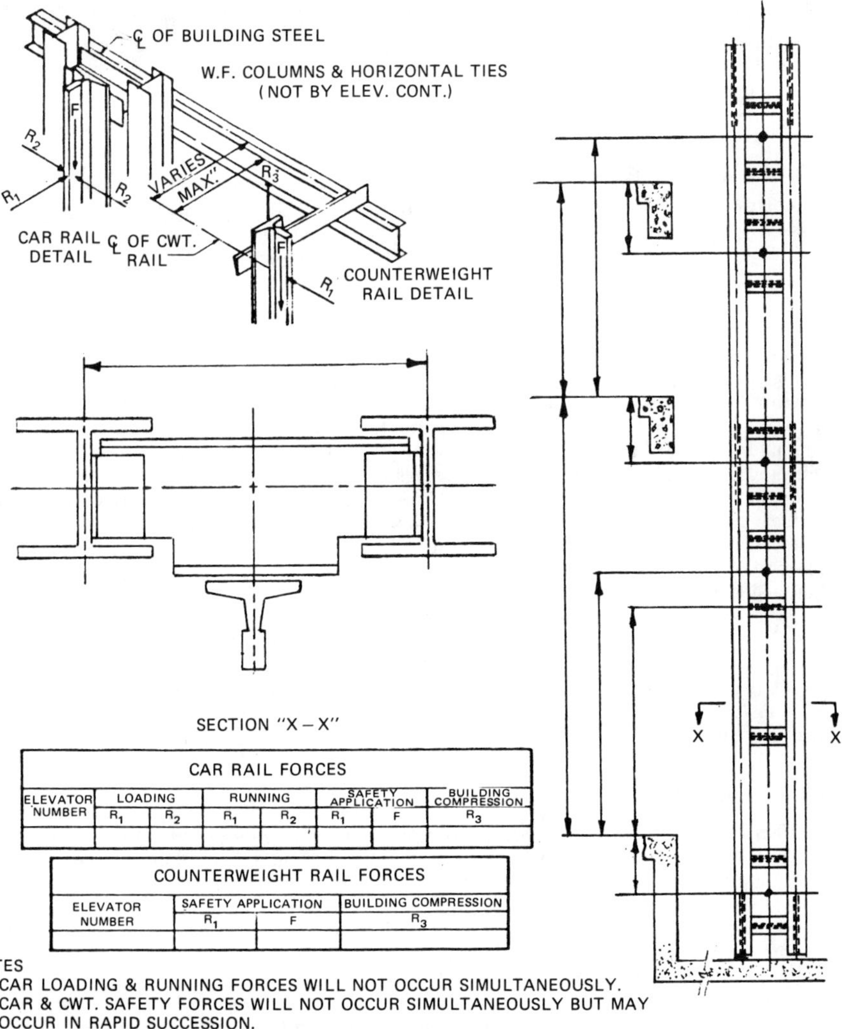

CAR RAIL FORCES							
ELEVATOR NUMBER	LOADING		RUNNING		SAFETY APPLICATION		BUILDING COMPRESSION
	R_1	R_2	R_1	R_2	R_1	F	R_3

COUNTERWEIGHT RAIL FORCES			
ELEVATOR NUMBER	SAFETY APPLICATION		BUILDING COMPRESSION
	R_1	F	R_3

NOTES

1– CAR LOADING & RUNNING FORCES WILL NOT OCCUR SIMULTANEOUSLY.
2– CAR & CWT. SAFETY FORCES WILL NOT OCCUR SIMULTANEOUSLY BUT MAY OCCUR IN RAPID SUCCESSION.
3– FORCE R_3 WILL OCCUR DUE TO BUILDING COMPRESSION.
4– CWT. BUILDING COMPRESSION FORCES EXPECTED TO BE LIMITED BY SLIDING RAIL CLIPS.
5– RAIL BACKING COLUMN CONNECTIONS TO BUILDING STRUCTURE MUST ACCOMODATE BUILDING COMPRESSION.

Fig. 5.30 Column backing for car rails with side counterweight—freight elevators. Courtesy of Jaros, Baum & Bolles.

5.4 SPECIAL STRUCTURAL SUPPORT REQUIREMENTS

5.4.1 Special Hoistway Construction

When special hoistway construction is envisioned which exceeds the usual elevator hoistway plumb tolerance of ±1 in. (±25 mm), consideration needs to be given to the hoistway running clearances to evaluate their adequacy for the conditions envisioned. If the hoistway is to be erected intentionally out of plumb, such as the observation elevator hoistways at the Sears Tower in Chicago, with a 7.6-in. (190-mm) permanent lean to the west, special advice from experienced consultants or subcontractors should be sought with respect to the specific capacities and speeds involved. If cast-in-place concrete construction is envisioned, it is suggested that the formwork for each floor be checked by the site engineer to comply with reference marks permanently located in each elevator pit so that the runout of the forms is not cumulative in the hoistway. Special consideration should be given to the front hoistway walls when they are cast-in-place concrete with recesses for the subsequent installation of entrance frames and door panels. Additional clearances must be maintained, depending on the design of the frame and the operation of the door panel with similar control from permanent markings in the pit.

5.4.2 Earthquake Restraints

The A17.1 Code and the California Elevator Safety Orders prescribe the design criteria for earthquake restraints on elevators located within Seismic Risk Zone 3 and within the State of California, respectively. It is recommended that, whenever a building is under design which requires structural consideration of a seismic risk zone, consideration for elevator restraints be given when the seismicity lies between Seismic Risk Zones 1 and 2. Earthquake restraints should be employed for elevators in all structures in Seismic Risk Zone 2 and above but with discretion in a less active location. It is suggested that this consideration be given to all major office buildings and to all buildings with sleeping quarters such as apartment houses, hotels, and hospitals. Unfortunately, at this writing, there are no established design criteria for elevator restraints less than Seismic Risk Zone 3, but it is recommended that, where restraints are deemed desirable, the Seismic Risk Zone 3 restraints be specified as the expense involved is minimal and these restraints have already been engineered in a number of forms for installations in California by all elevator subcontractors. The earthquake restraints should be on all major components in the elevator machine room, such as machines, controllers, selectors, and motor generator sets if provided. The elevator guide rails and the brackets for both car and counterweight will be specially designed to deflection criteria and consideration should be taken to support these brackets appropriately to the building framing. Additional space between the car and the counterweight may be necessary to accommodate the installation of the counterweight tie or spreader brackets between the car and the counterweight as opposed to the usual location between the counterweight and the hoistway wall. Such an arrangement will contain the counterweight in the counterweight guide rails. Such a configuration may preclude the necessity of a seismic switch in the machine room to shut down the elevator when a seismic disturbance occurs. Under some circumstances, the compensation in the pit may be "tied down" as opposed to "free" at the election of the elevator subcontractor.

5.4.3 Building Sway

The effect of building sway on vertical transportation systems is serious in a modern high-rise building that is relatively limber with elevators that travel from the main lobby to the topmost part of the building. The building displacement during each cycle imparts a rope deflection at the topmost fixed point (machine room) that cumulatively builds up until it is dampened by the inherent characteristics of the ropes or by striking a fixed portion of the building or elevator equipment in the hoistway. Since the hoist and compensation ropes of modern high-speed elevators are conventionally fixed at the bottommost landing by a tensioning compensating sheave and at the topmost landing by the machine room, free space is available in the hoistway for the buildup of substantial amplitudes, which can cause a destructive interaction with traveling cables. In the past, where building sway conditions existed, it was conventionally thought that reducing the elevator speed would permit continued operation without danger of destructive amplitudes. However, this has proved only partially correct if the elevators were kept in constant motion as opposed to being stopped at their terminal landings.

This condition was observed by the author at the Sears Tower in a wind storm gusting to 87 mph and a constant velocity of 60 mph from the west. The observation elevators were operating at their normal 1800 ft/min (9 m/s) with some slapping of the $\frac{3}{4}$-in. (19-mm) hoist ropes against the car. The car speed was reduced to 900 ft/min (4.5 m/s) and the elevators kept in constant motion from top to bottom with rest periods only during the passenger loading and unloading transfer times. At the peak of the wind storm, the three $1\frac{1}{4}$-in. (32-mm) compensating ropes became tangled with the three traveling cables pulling them out of the halfway box anchorage. Immediately after the shutdown, the dynamic action of the hoist ropes and compensating ropes were observed after the disabled elevator was moved to and parked at the top terminal landing. Approximately one-third of the distance down the shaftway from the top landing the hoist ropes were moving approximately 60 in. (1500 mm) peak-to-peak amplitude when the building was moving in a north–south direction and approximately 30 in. (750 mm) when the building was moving in the east–west direction because of the ropes slapping the hoistway walls and self-dampening. The compensating ropes with the tangled traveling cables were describing a Lissajous series of figures with a maximum amplitude of approximately 40–50 in. (1000–1300 mm). Subsequent observations in other buildings of 30 to 40 stories have indicated similar dynamics with respect to the hoist ropes or compensating ropes striking the car enclosure and failure of the traveling cables.

The problem with elevator hoist and compensating ropes is associated with the inherent safety factors and loading of these rope systems themselves as dictated by Code and by their basic function with an elevator car and counterweight. The hoist ropes continuously travel between the car and the counterweight propelled by the driving machine so that their length above the car and above the counterweight during travel is constantly changing. The compensating ropes are attached to the underside of the car and to the underside of the counterweight to compensate for the shifting weight of the hoist ropes so that the elevator hoisting machine has a minimum unbalance related to the capacity load of the car. The hoist ropes are configured on the car side to be attached to the car near its center of gravity. The ropes above the car hence travel in a large free space limited only by the hoistway walls without any fixed guiding elements possible between the extremities of the elevator travel. The hoist ropes are restrained at the hitch point of the counterweight, the hitch point of the elevator car, and by the intermediate elevator machine in the machine room. Similarly, the compensating ropes are restrained at the compensating rope hitch on the underside of the car, the compensating rope hitch on the underside of the counterweight, and by the intermediate compensating sheave tensioning this system in the elevator pit.

When the building moves horizontally due to wind action, the topmost fixed points in the elevator system move with the building. The ropes, acting as tension members with negligible dampening and with distributed mass, develop a wave action in the vertical ropes. The actual wave pattern and its propagation rate are functions of the distributed mass and the tension in the rope system. Since tensions in elevator hoist ropes are not equal and constant at all points when the system is moving, the curve shape at resonance is not symmetrical about the midpoint of the rope system.

The governor rope system is essentially the same system on a much smaller scale. The governor rope system, however, permits its free running side to be restrained at the antinodal points in accordance with its calculated resonance so that its movement is dampened.

The traveling cables connected between the halfway point in the hoistway and the underside of the car are essentially free as the traveling cable has no tension at the bottom of the loop and, hence, its natural frequency is a direct function of its length and distributed mass.

In recognizing the building sway effect on the elevator system, it is essential to understand that as a building's stiffness increases, the period of oscillation becomes shorter and the magnitude of the imposed horizontal disturbances generally tends to reduce. A short-period building is less likely to cause rope resonance problem than a long-period building. The critical condition occurs when an elevator system is installed whose full rope length has a natural period longer than the natural building period. At some position in the hoistway, the car can have a resonance match which will impair the normal operation of the elevator when the building displacement becomes substantial. It is the responsibility of the structural design engineers to inform the elevator subcontractor accurately through the appropriate contract documents of not only any unusual static conditions of the building, but also of the normal dynamic conditions anticipated for the building in its initial configuration as well as when the building has reached full occupancy with its population, office equipment, furniture, and partitions. It is suggested that a specification insertion, similar to the following, be placed in Division 14, Conveying Systems (Construction Specification Institute [CSI] Format):

> *The fundamental periods of the building for the two principal directions of motion have been calculated. The period in the North-South direction is initially _____ seconds; this will increase to _____ seconds at full occupancy. The corresponding periods in the East-West direction are _____ and _____ seconds. From the computer analysis, the maximum dynamic displacements to be expected are _____ inches North-South (half-amplitude off plumb) and _____ inches East-West (half-amplitude off plumb).*

Such advice in the contract documents during the bidding stage will permit the subcontractor to detune the rope systems appropriately as much as possible by installing restraining brackets on the counterweight guide rails so as to restrain the hoist ropes and compensating ropes from hitting the car, and to consider the installation of "following carriages" beneath the cars so that the compensating ropes and traveling cables are permanently separated just above the traveling cable loop by roping the carriage 2:1 with respect to the elevator car. Similar carriages installed on sensitive elevators at the Sears Tower and the World Trade Center have prevented subsequent damage among the rope and cable systems of the high-speed, high-rise elevators. Elevators mounted exterior to the structure, similar to the Canadian National Tower in Toronto, require special considerations from the standpoint of structure dynamics and dampening of the rope and traveling cable systems, not only for building sway but also for additional reaction to gusting winds. In the design of any such structure, it is essential that the structural design parameters for the elevator interfaces be developed with the aid of an elevator consultant or elevator subcontractor experienced with building sway and with elevators exposed to weather.

5.5 VIBRATION ISOLATION, EMERGENCY ACCESS, AND AIR TRANSFER

5.5.1 Multiple Unit Elevator Installations

The multiple elevator installation presents the need for special consideration in developing a structural system which, by its design, minimizes the transfer through the structure of vibrations that cannot be completely eliminated in respect to the elevator hoistway, pit, or machine room. This usually takes the configuration of vertically coupling the supporting beams of multiple units together with columns to as many floors as possible to eliminate horizontally structure-borne vibrations through occupied spaces. This is particularly sensitive in buildings where occupied spaces surround internal elevator machine rooms, pits, and secondary levels.

The source of structure-borne vibrations in the elevator pit can originate from the lay of the strands in the compensating ropes and from the movement of the tensioned compensating sheave in the pit caused by acceleration and deceleration of the elevator car.

The structure-borne noise from the elevator hoistway occurs primarily because of the motion of the elevator car and counterweight as they travel along the guide rails. The noise is caused by the imperfections of roller guide bearing on the rails, misalignment of the elevator guide rails [predominantly at the rail joints that occur every 16 ft (4.9 m)], or under adverse building sway conditions, the rubbing or slapping of the hoist ropes, compensating ropes, or compensating chains.

The structure-borne vibrations from the elevator machine room originate from the lay of the strands in the hoist ropes reacting with the elevator hoisting machine and/or secondary sheave along with the operation of the brake, associated roller bearings, electric contactor operation, and the solid-state motor drive's 360-Hz silicon controlled rectifier (SCR) waveform irregularities with a 60-Hz power supply.

In multiple unit installations, the hoistways in an elevator core may have to be divided so that no more than four hoistways are in a common shaftway. This requirement stems from building codes and may require special consideration for the design of divider beams between hoistways which must accommodate not only the elevator running reactions and building compression reactions, but also the dead weight of an intermediate separating wall. Some consideration should be given in the design of these walls so that the air pressures developed between moving elevator cars can be accommodated. The considerations for these intermediate walls are identical to the considerations necessary to design the perimeter hoistway walls, depending upon the shaft wall construction selected.

5.5.2 Single Unit Elevator Installations

Emergency Access

In high-speed, high-rise elevator installations involving single elevators in single blind hoistways, special consideration must be given to the installation of emergency access doors at every third floor or every 36 ft (11 m) in accordance with Rule 110.1. Extensive construction may be necessary with respect to exterior elevators and with respect to elevators traveling through express zones with a shear wall core tightly designed around the elevator shaft which would inhibit this emergency access.

This rule with respect to the operation of an elevator in a single blind hoistway can have a significant impact on the operation of a building that has two elevators operating in a common shaftway without the emergency access doors. High-rise, high-speed observation elevators, conventionally installed as two units side by side operating between street lobby floors and observation deck floors, usually do not comply with the emergency access door requirement of the Code. However, when one of the units is not available for rescue purposes within a very short period of time, the second elevator must be shut down since it is, in fact, under these temporary circumstances, operating in a single blind hoistway and hence does not comply with a strict interpretation of the Code requirements for emergency access of rescue personnel. The failure to recognize this subtle requirement may necessitate special consideration for access to these floors during any such interim period.

Air Transfer

The air transfer between the building spaces and the hoistway of a single elevator due to its high ascending or descending speed will cause a buffeting of the elevator car which will increase its linear, horizontal, and torsional oscillations as well as increasing the objectionable noise from sharp edged elevator equipment or projections in an otherwise smooth hoistway. This condition increases with the size of the elevator platform and the velocity of travel as it relates to the volume of air displaced and the available area at the perimeter of the platform for it to dissipate. Special vent consideration may be necessary to permit this transfer to be relieved at both the top of the hoistway and the bottom of the hoistway to accommodate both directions of travel. This accommodation should take the form of a substantial area for relief, if possible, to permit a low-velocity air transfer among a multiplicity of similarly designed hoistways, particularly at the pit level, if Code requirements prohibit it at the top of the hoistway. Special arrangements may be necessary with respect to the venting required by Code in consideration of this transfer.

5.5.3 Vibration Isolation of Elevator Equipment

Hydraulic Elevators

The vibration isolation with respect to hydraulic elevators usually is limited to vibration isolation pads under the pump and controller unit or between the motor-pump drive and the frame of the pump unit. This vibration may achieve objectionable levels if occupied space surrounds the hydraulic elevator pump room since it occurs when the elevator is traveling only in the up direction. A different and equally objectionable noise occurs when the elevator travels in the down direction, generating hydraulic turbulence in the tank above the pump. Consideration should be given to the vibration isolation of the piping between the pump unit and the hydraulic cylinder when it is supported from above or as it penetrates walls. Usually quality hydraulic elevator installations include a vibration- and sound-dampening muffler in the "to-and-from" piping located as close to the pump unit as possible. Additional mufflers should be considered if there are sensitive structural considerations with respect to the pump room and adjacent occupancies.

Geared Elevators

The vibration isolation with respect to geared elevators relates primarily to the elevator machine room with its hoisting machine, controller, motor generator set or solid-state controller, and any associated deflector sheave. Usually the vibration isolation in the hoistway is of minor consideration due to the relatively low velocities involved. Since the resulting live load reaction on the elevator machine is made up of a primary vertical and a secondary horizontal component, vibration isolation with respect to the machine room is needed to accommodate the vertical major component and in the horizontal minor component when there is a deflector sheave. Because of the level of forces involved, this can generally be easily accommodated with the installation of vibration pads for the vertical component and the combination of vibration pads and earthquake type horizontal structural restraints for the horizontal component. Conventional treatment for controllers and motor generator sets accommodates these units.

Gearless Elevators

The vibration isolation with respect to gearless units increases proportionately to the capacity and speed of the elevator as it relates to the masses and the kinetic energy involved. The arrangement of gearless elevators usually involves the driving sheave and a deflecting sheave which spans the rope drop between the centerline of the car rails and the centerline of the counterweight rails, but may often also increase its traction by doubling the wrap and the traction on the gearless machine drive sheave. This combination of deflecting and doubling the traction changes the appellation from a "deflector sheave" to a "secondary sheave." This doubling may increase substantially the horizontal component with respect to the building structure. The introduction of vibration-absorbing material between the secondary sheave and the driving sheave is discouraged as it could introduce vibrations of an amplitude and frequency that would destroy the traction between the ropes and the drive sheave. Hence, the relationship between the drive sheave and the secondary sheave should be a fixed dimensional relationship. In cases where occupied space surrounds the elevator machine room or the secondary level, or any special occupancies such as training centers, auditoria, or board rooms, occur directly below or above these machinery spaces, the elevator equipment should be specified so that the secondary sheave is mounted directly to the gearless machine base or bedplate, resulting in a structural interaction with the supporting machine beams that has only a vertical component. Under these circumstances, with horizontal safety restraints, vibration isolation will be the most effective. This configuration has been the standard of one manufacturer for many years, but has not yet been adopted universally by all others. As the substitution of a static silicon controlled rectifier (SCR) drive for the rotating converter drive of a motor generator set continues, the introduction of 360-Hz electrical vibrations between the SCR controller and the machine will appear as a 360-Hz structure-borne mechanical vibration if that vibration between the machine and the supporting structure is not accommodated. The usual introduction of a reactor between the SCR controller and the machine may or may not satisfactorily reduce the transmitted mechanical vibrations. The responsibility for minimizing the 360-Hz electromechanical vibrations of the elevator component and the electrical power feeder conduits, as well as the mechanical vibration of the hoist ropes, should be that of the elevator subcontractor through a delineation of responsibility in the contract documents and the installation of isolation transformers and line filters.

Guide Rail Isolation for Adjacent Sleeping Quarters

Special vibration consideration should be given to the connection of the elevator car and/or counterweight guide rails to the building structure through the rail brackets when sensitive occupied office spaces or sleeping quarters are immediately adjacent to the elevator shaft walls. In the cases of sleeping quarters in apartment houses, hotels, or hospitals immediately surrounding an elevator shaft, consideration should be given to require that the car guide rails or the counterweight guide rails or both have vibration isolation at each car and counterweight rail bracket. It is impossible to maintain roller guides with their resilient tired wheels in a condition that is ideal. Early consideration of the sensitivity of the combined architectural and structural design of the core should take into account the necessity of this feature and its relative cost effectiveness. The vibration isolation of both the car rails and counterweight rails has been demonstrated in combination with rail brackets and rail clips that also satisfy the most stringent Seismic Risk Zone 3 requirements.

Underslung Elevators

An underslung elevator is a special design of geared or gearless elevator that has its machine room at the side or rear of the hoistway so as to reduce the overhead requirements to a minimum for architectural or structural considerations. The machine exerts an up-pull in the basement or pit machine room (see Fig 5.31). The elevator car and counterweight are roped 2 : 1 so as to place the hoist ropes at the side of the hoistway. The car is thus able to overtravel the side and rear sheave and dead-end hitch beams at the top of the hoistway to keep the hoistway height to a minimum above the topmost floor served. This arrangement requires perimeter structural supports below the top of the hoistway which is unique to this configuration (see Figs. 5.32 and 5.33). The pit depth is usually deeper by 24 in. (600 mm) to accommodate the "underslung" sheaves beneath the car platform.

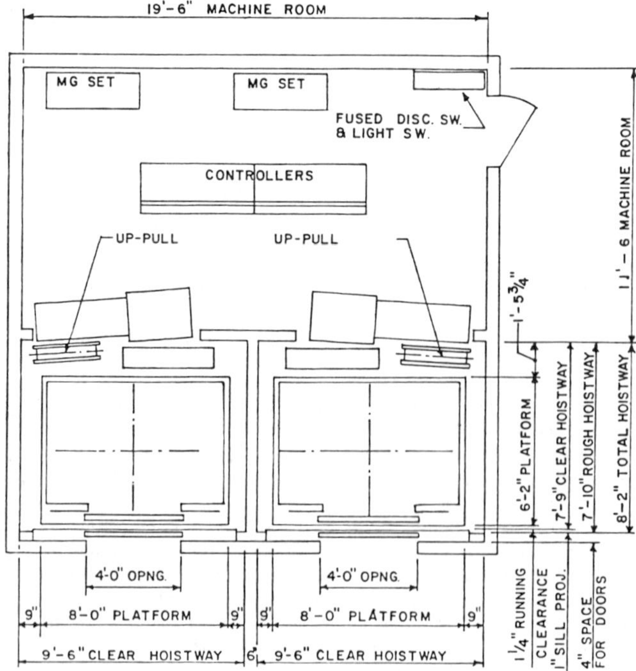

Fig. 5.31 Underslung elevator—up-pull forces in machine room. Courtesy of Jaros, Baum & Bolles.

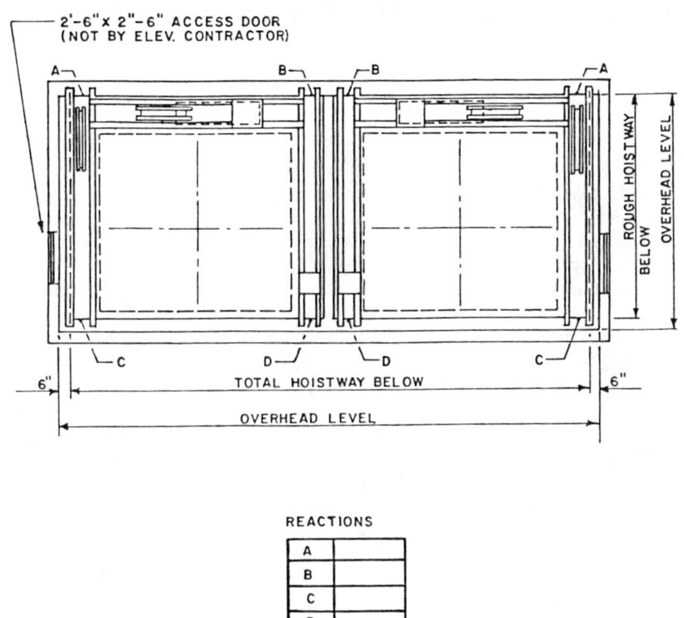

REACTIONS

A	
B	
C	
D	

NOTE: ALL LIVE LOADS HAVE BEEN DOUBLED FOR IMPACT.

Fig. 5.32 Underslung elevator—structural supports at overhead level above top landing. Courtesy of Jaros, Baum & Bolles.

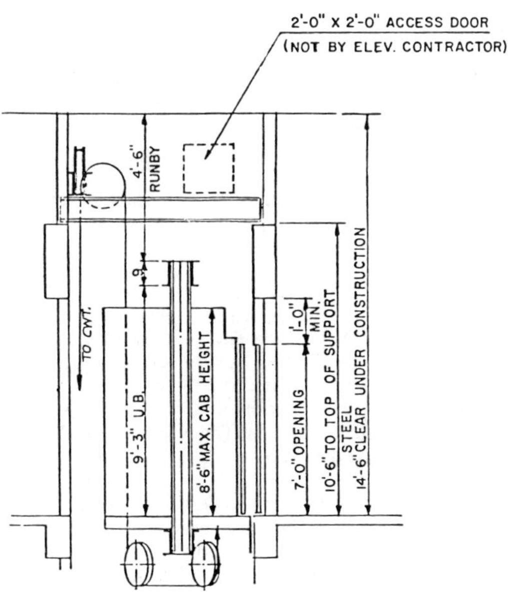

Fig. 5.33 Underslung elevator—overhead section above top landing. Courtesy of Jaros, Baum & Bolles.

5.6 SPECIAL STRUCTURAL REQUIREMENTS DURING CONSTRUCTION

5.6.1 Tolerances

If there are special architectural features to be controlled with respect to the elevator entrances and lobby conditions, the hoistway construction may require special consideration, depending upon whether it is a steel frame building or a concrete building with shear wall hoistway construction. The entrances may demand special alignment tolerances so that they collectively properly respond to lobby conditions using cut stone, metal, or other special features. It is usually an accepted practice where special control is necessary that additional tolerances in the running clearance be factored into the design of the structure so that essentially plumb elevator hoistways can occur with respect to the entrance sills and the associated entrance frames. With respect to cast-in-place concrete walls in the front wall of the elevator hoistway which must accommodate the sliding doors, it is essential that an additional running clearance [at least 1 in. (25 mm)] be included so that the runout usually experienced in the erection of the forms does not necessitate cutting or chipping. In such cases, it may be essential to insert in the specification section on forming the concrete, that special tolerances be exercised in conjunction with the forming of the hoistway walls that may require a continuing floor-to-floor plumb relationship with the elevator pit. The specification also should require that the elevator subcontractor provide the necessary control points at the sensitive location in the elevator pits that will permit the forming above to proceed in a plumb relationship on a floor-by-floor basis. In this manner, the control of the forming can be accurate and checked prior to the placement of the concrete.

It is also desirable to exercise special erection tolerances with respect to steel construction that relates to the front hoistway walls or other close tolerances with moving elevator equipment. With the spray-on fireproofing in current practice, the construction tolerances are just as sensitive as if the steel structure were fireproofed with concrete.

Special attention should be given to the detailing of floor construction with respect to the location of elevator landing entrance sills if carpet or other thick materials are to be applied directly to the cast-in-place floor construction. The sills should be raised appropriately to accommodate the floor material that will be installed after the flour slab is cast. In general, this requires notching the entrance side jambs to accommodate simultaneously the structural, architectural, and fire rating requirements of the entrances since the final finish floor will be installed at a substantially later time in the construction process.

5.6.2 Fireproofing

The spray-on fireproofing occurring in elevator machine rooms and hoistways which can in any way adversely affect the operation of electric contacts must be stabilized to prevent erratic operation. This requires that the fireproofing be hardened after installation in a secondary sprayed-on process. Failure to harden the fireproofing will result in continuing problems with electric contacts in the interlocks on each hoistway door and on the relays and contactors in the elevator machine room. These are affected by the airborne particles dislodged by the air movement and turbulence caused by moving elevator and winter stack effect. The hardening is best accomplished late in the construction sequence so that all corrective fireproofing has been completed. In order to assure the thorough application of the hardener, it is suggested that a visible color pigment be added to the spray so that visual inspection is easily accomplished from the machine room and from the top of the elevator car by traveling the hoistway.

5.6.3 Guide Rail Installation

There are no special structural requirements necessary during construction for the installation of guide rails since they are usually installed in 16-ft (4.9-m) sections, and it is the responsibility of the elevator subcontractor to accommodate hoisting from the existing structure. It is necessary, however, to ascertain early in the project if any pair of guide rails will be pinned at an upper floor at the shop drawing stage by the elevator subcontractor. The pinning of guide rails that will experience a safety application at approximately the two-thirds point in the travel of an elevator imposes unusual and special structural impact reactions on the building, which may not have been anticipated in the original design (see Section 5.3.2, Elevator Hoistways). The fact that most elevator consultants and elevator subcontractors do not pin the rails probably predicts that most structural design data will indicate to the structural engineer that the rails will not be pinned and that the reaction of a safety application will occur in the pit along with the comparable buffer reactions. It must be made the responsibility of the elevator subcontractor to advise the designer of pinning intentions. The elevator subcontractor is also responsible for the cost of the reevaluation of the interface structure, along with the cost of any necessary additional structural reinforcement required by his specific requirements.

5.6.4 Hoistway Requirements

The special construction requirements of installing the elevator machines, controllers, and motor generator sets occur primarily with respect to the final machine location and involve the necessity of hoisting from the street level to the machine room by a hoist beam or other structural support provided in the machine room. In gearless installations, it is important to design the hoist beam not just for maintenance purposes for removing of armature and the heavy field components, but also to design the hoist beam so that it may be used for construction in hoisting the machine and, usually, supporting it during the final location process of the elevator machine beams and the subsequent forming and pouring of the machine room floor. It should be noted that Rule 100.3 requires that the machine room floor slab be above or level with the top of the machine beams under all conditions when the elevator machine is located above the hoistway.

5.7 SPECIAL STRUCTURAL REQUIREMENTS FOR REPAIR PROCEDURES

5.7.1 General Requirements

In general, there are no special structural requirements to accommodate the repair procedures involved during the life of an elevator installation. However, there are certain repair procedures that should be understood which involve structural support during the removal and replacement of secondary sheaves, hoist ropes, compensating ropes or chains, and traveling cable. In general, these requirements are met by the structural support involved with the elevator; however, it is essential to understand the sequence of procedures when high-speed, high-rise elevators that require large and heavy equipment are involved.

5.7.2 Sheave Replacement

The replacement of a secondary sheave with associated shaft and bearings, while infrequent, must be considered. With elevators above 700 ft/min (3.5 m/s), the run-by distances may become substantial which should dictate the installation of a secondary slab below the machine room slab to provide a working surface for the replacement of the secondary sheave, shaft, or bearing. Under those circumstances, it is necessary to relieve the tension in the hoist ropes by supporting the car at the top of the hoistway and by bottoming the counterweight in the pit. The car is subsequently raised by chain fall to relieve the rope tension so that the ropes may be removed and the sheave dropped to the secondary slab level for replacement of the shaft, the bearings, or the sheave as an assembly. Usually the secondary slab will at best only absorb the reaction imposed upon it by the governor in its pull-through of the governor rope during a safety application; however, it is desirable that it additionally accommodate the dead weight of the secondary sheave assembly and associated tools and workmen.

5.7.3 Hoist Rope Replacement

The replacement of the hoist ropes and compensating ropes will also require that the counterweight be bottomed in the hoistway and the car lifted to the top of the hoistway to such a distance that will relieve the tension in the ropes and permit their replacement. The replacement of any single hoist rope or compensating rope usually dictates the replacement of the entire set since the tension in each rope must be equal at all times and the rope stretch in each must proceed an equal rate. The compensating ropes are replaced with a similar procedure.

5.7.4 Traveling Cable Replacement

The replacement of the electrical traveling cables usually occurs individually on a failure basis rather than the change of a complete set. The replacement of traveling cable usually is a very simple procedure since one end of it is fastened to the car when it is positioned at the lowest level and then raising the car subsequently unreels the traveling cable with the other end fastened to the halfway box in the hoistway. The traveling cable provides the electrical energy and control logic circuits to the elevator car for its safe operation.

5.8 SPECIAL STRUCTURAL REQUIREMENTS

5.8.1 Double-Deck Elevators

Double-deck elevators are essentially two, single-deck elevator cars contained in a structural frame serving two adjacent floors simultaneously. The special structural requirements for these elevators involve a precise control of all floor levels served by the double-deck elevator system. The tolerance between the finished floors must not exceed $\pm\frac{1}{4}$ in. (6 mm) about the design floor height dimension, so that the sills can be finally leveled to a tolerance of $\pm\frac{1}{8}$ in. (3 mm). It is essential that the placing of each final floor slab be controlled to accommodate these tolerances particularly when stone or concrete hoistway walls are involved at typical floors.

Increased side running clearances in the hoistways are necessary with double-deck elevators since the side supporting structures are usually fabricated trusses rather than rolled shapes as usual in single-deck elevators. In the larger double-deck installations, a suitable special structural support may be required in order to support the elevator car for the reroping conditions discussed in Section 5.7.3, if the machine beams are not accessible. This would include the installation of a beam beneath the secondary slab which would support the dead weight of the elevator car and its hoist ropes, compensating ropes and traveling cables, or slab access through the slab to the machine beams.

The desirability of having all passenger elevator compartments serve all floors under certain conditions may necessitate that the top elevator deck overtravels the top floor or the bottom elevator deck undertravels the bottom floor. In either case, the location of the elevator machine room or the elevator pit equipment must be extended to accommodate this overtravel which does not occur in single-deck elevators. Such over- and undertravel conditions usually necessitate the installation of emergency access doors for each elevator hoistway. These doors may be similar to the hoistway doors at the typical floors or may be special doors that comply with the requirements of emergency access as defined in Rule 110.1.

5.8.2 Sky Lobby Buildings

The only special structural requirements that may be required in sky lobby buildings, apart from the multiple group arrangement and the usual installation of shuttle elevators serving the sky lobby of either the single- or double-deck variety, are directed toward any pits with occupied areas beneath them requiring additional horizontal hoistway space for counterweight safeties and the structural accommodation of the counterweight safety application on the counterweight rails. This most often occurs in a stacked configuration over heating, ventilating, and air conditioning mechanical equipment rooms or above elevator machine rooms. Sufficient structure should be provided to accommodate the vertical buffer reactions, safety application reactions, and compensating rope uplift reactions in these pits.

5.8.3 Lifts and Dumbwaiters

Separate categories are established in the Code to recognize that there is a special requirement for vertical transportation of materials alone and the configuration of the enclosure is such as to discourage the use of the conveyance by people. Such equipment has essentially the safety features of a passenger or freight elevator, but they are less stringent. In high-rise buildings, this equipment generally is used for the transport of containerized items for the horizontal on-floor movement such as carts or tote boxes. Dumbwaiters are often used to transport by tray or other kitchen container bulk food to and from an eating facility if the preparation and consumption facilities are vertically related.

Dumbwaiters

Dumbwaiters are limited in the Code to an inside car enclosure defined by a maximum of 9 sq ft (0.84 m^2) in area and a vertical door height of 4 ft (1.2 m). In this configuration, the dumbwaiter may stop either at counter height, which does not necessitate a floor penetration to accommodate the pit requirements, or it may stop at floor level for cart or vehicle transport which does necessitate a floor penetration to accommodate a pit. In general, the reactions of the dumbwaiter equipment can be accommodated by the perimeter framing of each floor slab. Often low-rise dumbwaiter machinery and controller can be accommodated at the top of the hoistway beneath the overhead floor slab so that the space involved for equipment construction is limited to the topmost floor served. Under special conditions, it may be desirable to require that the dumbwaiter be provided by the dumbwaiter subcontractor as a self-supporting angle-iron structure that imposes its reactions on the hoistway perimeter floor slab. On high-rise high-speed dumbwaiters, it may be necessary that the dumbwaiter machine room be a normal machine room which permits "standup" maintenance from inside the room so that the equipment can meet the intensive use and reliability requirements with a minimum maintenance and repair downtime.

Cart Lifts

A cart lift is a "junior elevator" that has a platform area and door height exceeding that of a dumbwaiter. The dumbwaiter size limitation essentially addresses a compartment which discourages, while it may not prevent, the transportation of people. The cart lift size, even though larger than a dumbwaiter, specifically requires the installation of a transfer device located in the floor so as to discourage its use for passengers or freight handling other than that for which it was designed and for which the transfer device is made compatible with the carts. The size of the cart lift is such as to prohibit a person from riding in the lift when the cart is being transported. The cart lift has all of the accessories and safety devices of an elevator, including rope compensation, governor, and safety applications of an elevator. It may travel at speeds up to 1000 ft/min (5 m/s).

The only special consideration structurally to accommodate cart lifts is an out-of-level tolerance with respect to the floor in front of the cart lift openings. A level area is necessary in order to permit the proper engagement of the "on-lift" transfer device to engage with an "on-cart" coupler on the underside of each cart. The floor level requirement is difficult to obtain in practice and the solution requires that it should be supplemented by tracks located in the floor to accommodate the casters and should be provided by the cart lift subcontractor so as to provide the level floor tolerances under their subcontract responsibility. The floor is placed and grouted to the tracks as a final step in the procedure. The finished flooring is then approximately level with the top of the tracks. This requires a typical floor depth that permits a recess of approximately 2 in. (50 mm) to accommodate the tracks with their leveling jack screws. The recess encompasses an area in front of each cart lift opening approximately the size of the lift platform area.

5.8.4 Vertical Conveyors

The installation in a high-rise building of a vertical conveyor requires the usual fire-resistant shaft construction and may take the form of either a continuously running "selective vertical conveyor" for "passive" tote boxes on an endless moving chain or of "active" tote boxes that are self-propelled in a fixed bidirectional "track/conveyor" system. The advantage of these conveyors versus dumbwaiters or lifts is that they provide a continuous bidirectional flow system in that each has a dedicated up path and a dedicated down path, resulting in a high "through-put" capability. This is opposed to the batch system represented by the dumbwaiters and cart lifts in Section 5.8.3.

The usual structural configuration necessary to accommodate either form of conveyor system is to support the chain conveyor or the track guideway at each floor level to spread the load proportionately (see Figs. 5.15 and 5.34). The chain conveyor will impose a small additional requirement at the topmost landing where a drive motor and gear reduction unit is located. The self-propelled tote box system does not have any special requirements at the topmost landing.

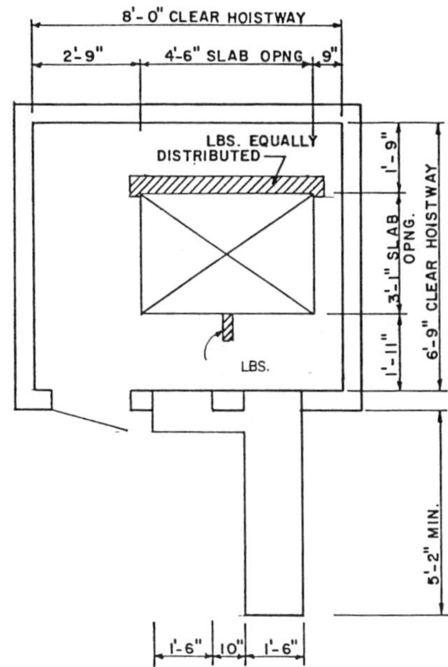

Fig. 5.34 Selective vertical conveyor—typical floor plan. Courtesy of Jaros, Baum & Bolles.

5.9 ESCALATORS

5.9.1 Basic Design Parameters

The basic structural design parameters for escalators require that the location of the supporting structure be horizontally adjustable until such time as the escalator subcontractor has been selected and shop drawings submitted to accommodate appropriately the architectural requirements for the location of newels at the landings. This flexibility is required because of the variation in escalator drive and truss designs available from among the major manufacturers. These variations are exacerbated by the extended newel arrangements required by various architectural treatments. In the structural design to support an escalator, it is necessary only to recognize on the structural drawings the necessity of holding the truss-supporting steel members at a movable location until the escalator subcontractor has been selected. On fast-track construction projects, it may dictate that an early bid package be prepared for escalators in order to accommodate the erection schedule.

The A17.1 escalator has a Code-mandated usual tread–step relationship installed at a 30° angle. The standard worm–worm-gear driving machine is located at the topmost landing with an associated controller. A bottom landing pit to accommodate the reversal of the handrail and the steps is of a minimum dimension. Such design is standard among most manufacturers which provides an unequal dimension from the supporting steel at the bottom landing of the escalator to the vertical portion of the newel as compared to an increased comparable dimension at the top landing (see Fig. 5.35).

One manufacturer's design is modular, which incorporates in the inclined portion of the truss and between the steps the drive unit or units in multiple so that the newels at the top and the bottom have an equal dimension with respect to the supporting steel. The modular units more easily accommodate the newel alignment in multiple floor escalator installations where the escalators are arranged in a criss-cross relationship (see Fig. 5.36).

5.9.2 Truss Design

The truss design of an escalator supports the machinery, steps, chains, and balustrade, with its handrail. A number of factors that are dictated from the architectural considerations may affect the structural support requirements at each end of the truss or, in addition, require an intermediate support as a result of either truss-cladding considerations or the extended rise of the escalator itself.

The truss design as a standard from all manufacturers is based on the assumption that the exterior cladding is installed as an architectural treatment not provided by the escalator subcon-

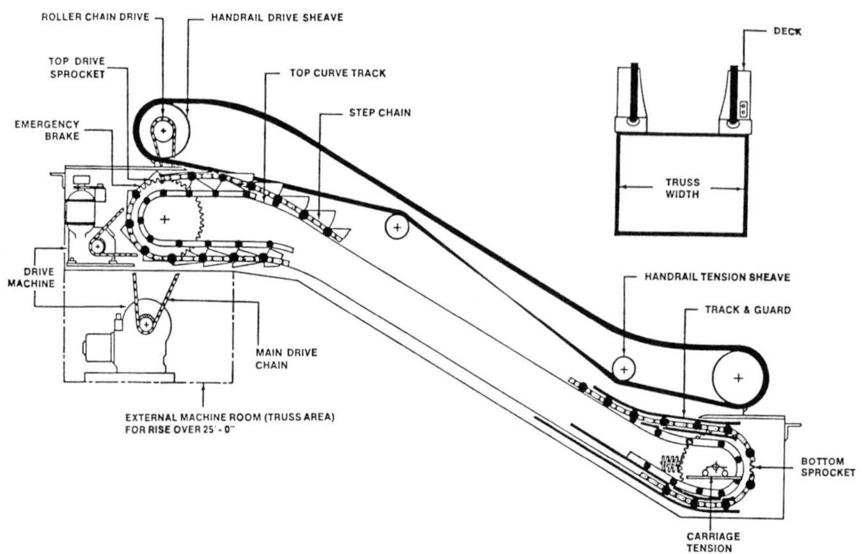

Fig. 5.35 Conventional escalator drive showing both internal and external drives. Courtesy of Otis Elevator Co.

tractor. It must be of relatively lightweight material such as a metal cladding or a decorative plastic or gypsum plaster construction. If special cladding is selected, such as stone or decorative features that exceed the weight of 1 in. (25 mm) of plaster on both sides and soffit of the escalator truss, special consideration must be given in the design of the escalator truss to accommodate the additional dead weight which increases the reactions on the supporting structure.

Additional Flat Steps

If additional flat steps are specified in addition to the usual 1.3 to 1.5 flat steps at the top and bottom, the truss will be extended at the top and at the bottom by the additional flat steps specified above the standard. The purpose of increasing the flat steps above the standard is to accommodate, more easily and hence more safely, the transfer of passengers to and from the horizontal treadway. It has been determined through pedestrian movement evaluation that 2.3 to 2.5 flat stops in a configuration at both the top and the bottom of an escalator provide a safer boarding and exiting condition than the standard, particularly at the speed of 120 ft/min (0.6 m/s). In most high traffic installations, this flat step arrangement will become a standard in the future to accommodate this feature. This parameter relates to the maximum stride length of a person as he passes the comb plate in the boarding or exiting process. The addition of one flat step at the top and the bottom will extend the truss at least twice the front-to-back step dimension of approximately 16 in. (400 mm) for each step.

Newel Arrangement for Multiple Escalators

The architectural requirement for the alignment of newels among a multiplicity of escalators serving more than two floors requires careful consideration in the initial layout of the structural supports to accommodate all escalator manufacturers. This will dictate extensions of the trusses so that, when the newels line up appropriately and depending upon whether they are of a conventional design or of a modular design, the truss supporting steel will be common to both the up and the down escalators. Such an arrangement must be retained as changeable, until such time as the manufacturer is selected, so that the field dimensioning can be made final. An alternative procedure would be to fix the steel to the worst condition among all of the manufacturers and indicate to each manufacturer through the escalator contract documents that the steel cannot be moved and that truss extensions appropriate to each manufacturer be included by them in their bidding process and in the construction of the truss. This construction condition may adversely affect some manufacturers since it dictates different truss extensions depending on the manufacturer rather than accommodating each in the least costly manner. Assessment of the impact of such a construction restriction should be made in each case, depending upon the time table for obtaining shop drawings from both the structural and escalator disciplines (see Figs. 5.37 and 5.38).

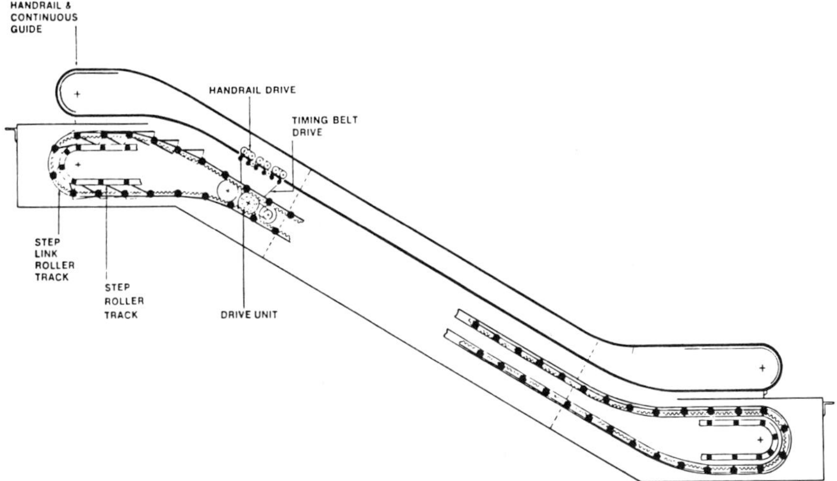

Fig. 5.36 Modular escalator drive. Courtesy of Westinghouse Elevator Co.

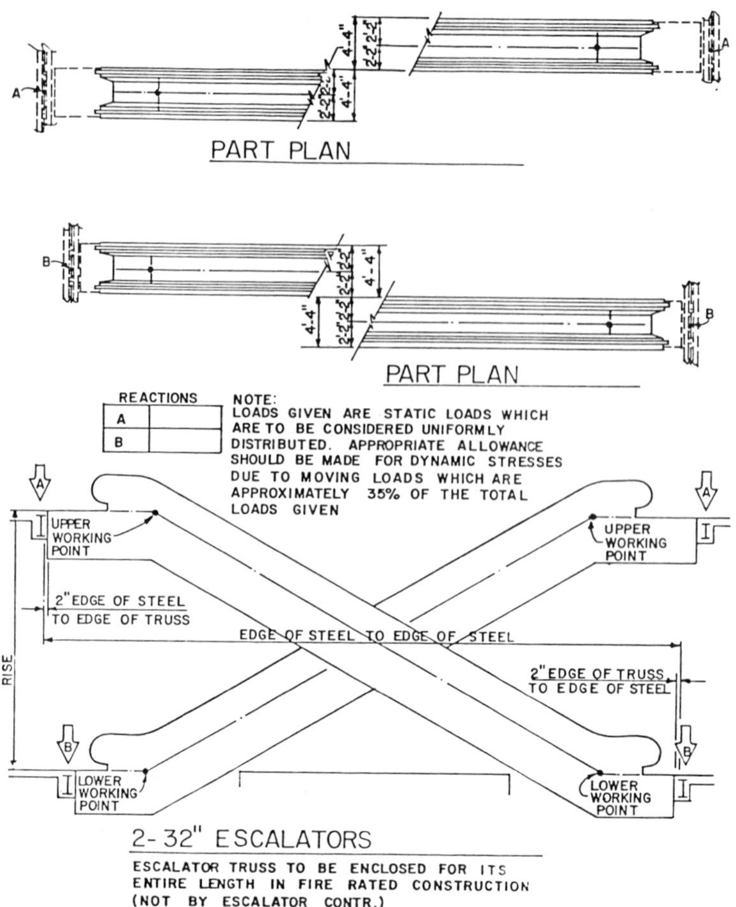

Fig. 5.37 Supports for escalator truss—criss-cross arrangement. Courtesy of Jaros, Baum & Bolles.

5.9.3 Truss Supports and Reactions

In escalator vertical rises over 20 ft (6 m), it is essential to consider the necessity of intermediate supports in addition to those provided at the end of each truss in conjunction with the floor framing.

The standard configuration indicates that 40-in. (1000-mm) step width escalators with a rise over 20 ft (6 m) and 24-in. (600-mm) step width escalators with a rise above 25 ft (7.5 m) should require the installation of intermediate supports. This intermediate support requirement may be modified by extended truss designs, truss cladding, or for newel alignment considerations. Therefore, it is recommended that the escalator structural reactions be a "worst-case" design, depending upon the cladding, truss extension, and newel alignment considerations, involving extended trusses in the vicinity of the rises indicated.

When a moderate or extended rise condition occurs, it is necessary to consider as a first step an intermediate support located at the apex of the 30° angle so that it can bear on supplementary truss framing at the lower floor. Rises that indicate more than one intermediate support and a more central location of the support must be accommodated in conjunction with the architectural design of the building. Long extended rises or intensive heavy duty operation with a normal rise may indicate that the space in the upper truss is not sufficiently large to contain the machine from certain manufacturers. In this case, it is necessary to provide outside of the truss and the truss-supporting steel, a remote machine room to accommodate the larger machine and control equipment. This requires special access considerations and may require the necessity of penetrating the upper steel member supporting the truss for the silent chain drive from the machine to the sprockets driving the steps and handrails.

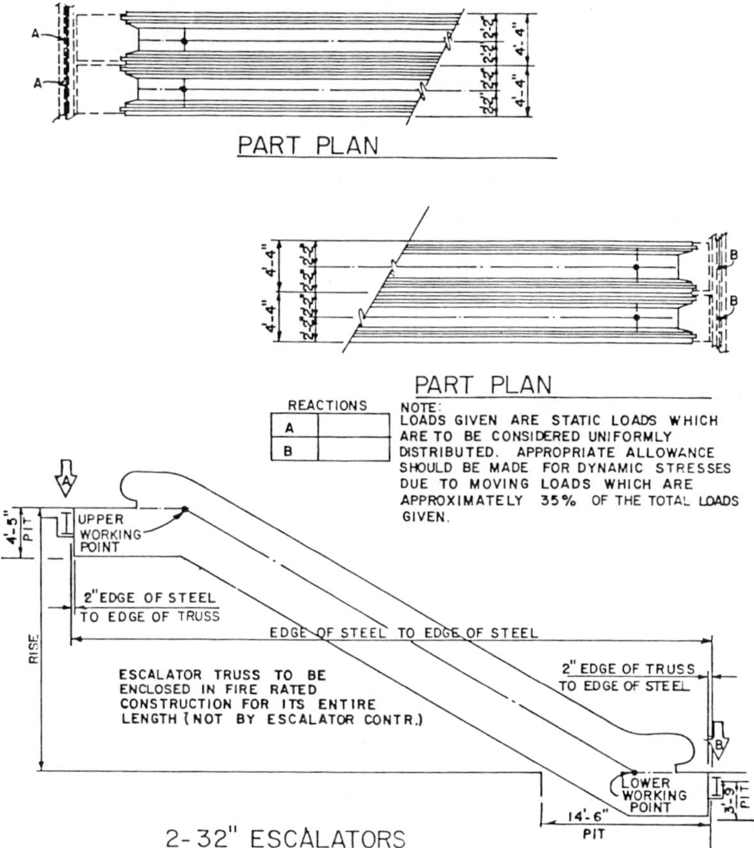

Fig. 5.38 Supports for escalator truss—parallel arrangement. Courtesy of Jaros, Baum & Bolles.

5.9.4 Vibration Isolation

In installations where structure-borne noise may be a sensitive consideration, vibration isolation between the escalator truss and the structural angle at the top, bottom, or intermediate supports can be incorporated to dampen a major portion of the vibrations if the associated floor interface construction will accommodate the movement amplitude. The usual isolation is a canvas-impregnated multilayer neoprene pad which is placed under the bearing supports involved (see Fig. 5.39). In some instances, this may not be feasible because of architectural considerations.

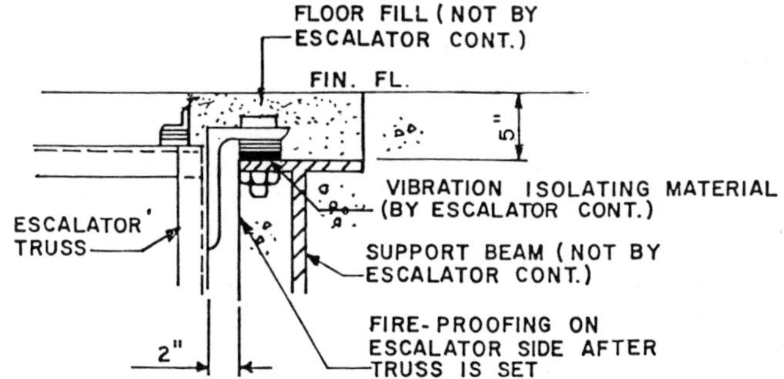

Fig. 5.39 Escalator truss support with vibration isolation. Courtesy of Jaros, Baum & Bolles.

5.9.5 Fire-Resistive Enclosures

The Escalator Safety Code has interpreted Rule 801.1 as requiring that the sides and undersides of the escalator truss and as well as the machinery and pit spaces shall be enclosed in fire-resistive materials. This indicates that the upper or machinery space and the lower pit space must be enclosed in a fire-resistive enclosure meeting the requirements of the local building codes. It is anticipated that this requirement will be more rigidly enforced by local authorities in the future than it has been in the past due to the increased scrutiny for smoke control and for strict code compliance. In conjunction with this compliance, it is possible, under some building codes, to permit an up traveling escalator to be an accredited means of egress, as well as the down traveling escalator, if the escalators are stopped with a controlled stop by smoke detector or water flow (New York City Building Code C26-604.11).

5.9.6 Special Construction Requirements

The angle of inclination for an escalator is limited by Rule 802.1 to 30° from the horizontal. It has been interpreted that this rule will permit an angle of inclination based on a construction variation increase to 31° from the horizontal and still be in compliance with the rule. The erection of structural supporting members should have tolerances that do not exceed this limitation.

5.10 MOVING WALKS

5.10.1 Basic Design

The basic design of moving walks consists of a belt or pallet type treadway with an inclination limitation of 15° from the horizontal based on speed and width considerations established by the Code. The basic unit can take the form of a belt treadway which is a structural belt with escalator-type tread grooves in the direction of travel that is edge supported along the length of the treadway. The pallet type treadway is essentially a "flattened" escalator step construction which may be either a pallet of single or double depth tread with escalator step grooving in the direction of travel. The drive units may be similar to the escalator drive unit configuration in that the standard units would involve driving machines at one end or at both ends for extended length units. The modular design would involve equally spaced drive units along the treadway as in the modular escalator design.

5.10.2 Load Support Requirements

The load support structural requirements are specified in Rule 900.2 which states that the structural support design be based on a live load rating equal to or greater than 100 psf (4.8 kN/m^2) of exposed treadway area between the skirt panels at each side and between the comb plates at each end. This is in addition to any special requirements based on a concentrated dead load for end-mounted driving machines.

5.10.3 Fire-Resistive Construction

The sides and undersides of the supporting structure and the machinery and pit spaces must be enclosed in fire-resistive construction similar to the escalator requirement. This has been similarly interpreted with respect to moving walks under Rule 903.1 as that required for escalators under Rule 801.1.

5.11 ACCELERATING MOVING WALKS

The state of the current art for demonstration purposes in accelerating moving walks has been explored by the Port Authority of New York and New Jersey [5.4, 5.5]. The structural design associated with each proposed moving walk system varies among them and no general comments can be made until a successful demonstration has been achieved by a sponsoring authority or imported from an approved foreign demonstration.

The proposed demonstration suggested by Urban Mass Transportation Administration (UMTA) has not yet achieved favorable consideration for funding and hence the demonstrations probably will have to be privately funded. The current assessment of the demand for such equipment in the major activity centers such as intermodal terminals has not been sufficient to encourage the private sector to build a full-scale demonstration with the accompanying construction and insurance costs.

5.12 INCLINED ELEVATORS

The installation of inclined elevators has been successfully demonstrated in conjunction with escalators to provide safe transport of the handicapped to and from a major subway system in Stockholm, Sweden. The suitability for installation in conjunction with escalator transport has been proven and reported on in UMTA [5.6]. It is apparent in that report that special structural considerations for the inclined hoistway must be considered completely separate from any of the special considerations necessary for the accompanying and parallel escalators serving the same landings and at the same 30° angle of inclination from the horizontal. The A17.1 Code now recognizes and covers this type of elevator.

5.13 DESIGN PARAMETERS FOR ELEVATOR AND ESCALATOR CORES

5.13.1 Elevator Core

Passenger Elevators

Passenger elevators should be arranged in a group not exceeding eight cars with a physical arrangement of no more than four elevators in-line. This arrangement is a function of the human response time to an elevator system that, for practical purposes, limits the effective number of elevators in a line to four. This optimizes the travel time between the directional lantern indication of the arriving elevator and the ability of the passenger to traverse the intervening distance from a central pushbutton to the specific opening within the announcement time and the door-open dwell time. The eight-elevator restriction is based on the standard group automatic operation available from all the major manufacturers with that as the upper limit of design. The elevator lobby at the main landing should be a minimum of 10 ft (3 m) in width clear when normal traffic is expected and wider if the lobby is used for other pedestrian traffic in addition to traffic waiting for the elevator group involved. With eight elevators, the main lobby should be capable of loading from either end to efficiently handle all eight elevators with the potential for simultaneous multiple elevator loading. The lobby of a four-car elevator group may be arranged for a single-ended loading in either the two-facing-two or four-in-a-line configuration. Other arrangements must be subject to a value judgment in each installation for applicability to this optimum. In general, the location of passenger elevator cores should provide vertical transportation with an "on-floor" horizontal walking distance on typical floors in good practice of 150–180 ft (45–55 m). Only in special cases should this be extended to 200 ft (60 m) or more.

The negative result of structural columns or pilasters that project into an elevator lobby cannot be overemphasized when such projections block the view of the directional lanterns from any point in the elevator lobby. Confusion of waiting passengers and the attendant delays in boarding result in "missed" elevators and a reduction in the actual handling capacity.

Service Elevators

The installation of a service elevator which is dedicated to the movement of materials and not a combination passenger/service elevator is based usually on the building design that relates to the typical floor area in an office building with its type of occupancy, along with the possible installation of supplementary materials-handling systems such as a vertical conveyor or cart lift system for handling mail and packages. A dedicated service elevator is usually considered necessary in an office building when the gross area approaches 500,000 sq ft (46,000 m^2). A second service elevator is considered when the gross area approaches 800,000–1,000,0000 sq ft (75,000–90,000 m^2), depending upon occupancy and special material movement considerations.

5.13.2 Escalator Cores

Escalators are installed usually in pairs to accommodate the simultaneous up and down traffic associated with high-density pedestrian circulation. If the escalators are the major access and egress means, it is essential to consider a third escalator or adjacent stairway to provide for the condition when an escalator is shut down and it may not be available for use as a down stairway. The tread–riser relationship of a stopped escalator is uncomfortable for many people and, as a result, the stairway adjacent with its normal tread–riser relationship should be immediately available as an alternative means of down transportation. Under these circumstances, it is assumed that the remaining operational escalator will operate usually in the up direction.

In design of escalator cores, it is essential to have a sufficient dimension at the top and bottom of an escalator core to provide an unobstructed platform that is at least 10 ft (3 m) and, preferably, 12 ft (3.6 m) directly in front of the newels so that, during peak periods in both directions, the passengers can accommodate their horizontal travel to the platform and exit promptly without being crowded by the "conveyor belt of people" behind them. If there are other obstructions such as swing or revolving doors immediately in the pedestrian path, an additional dimension with additional queuing area should be provided to accommodate surges or temporary blockages in the existing doors.

If the escalator core is arranged to have an intermediate dedicated platform without building or floor egress, it is necessary to provide the platform with a dimension similar to the consideration for the boarding or exiting platform with the additional feature that the escalators must be interlocked directionally so that they do not feed people to a blind platform from which there is only egress up or down a stopped escalator. Under these circumstances, the controlled stop feature currently being installed on some escalators should be employed to stop a loaded escalator safely without upsetting the traveling pedestrians.

Whenever escalator transport is provided between occupied floors, it is essential to provide elevator access for the handicapped that is reasonably accessible from the immediate area of the escalator core. This also will accommodate those people who hesitate to use escalators because of impaired motor or vision functions.

SELECTED REFERENCES

5.1 ANSI/ASME. *Safety Code for Elevators and Escalators* (A17.1). New York: American Society of Mechanical Engineers, 1984.

5.2 John W. Reed, Robert J. Hansen, and Erik H. Vanmarcke. "Human Response to Tall Buildings Wind-Induced Motion," State-of-Art Report 6, *Proceedings of the International Conference on Planning and Design of Tall Buildings*, August 21–26, 1972, Vol. II-17, ASCE, 1973, p. 687.

5.3 ANSI/ASME. *Handbook on A17.1 Elevators and Escalators, 1984 Edition*. New York: American Society of Mechanical Engineers, 1985.

5.4 J. Fruin, R. Marshall, and O. Perilla. *Accelerating Moving Walkway Systems. Market Attributes, Applications, Benefits, Report D* (UMTA IT-06-0126-78-4, PB-287 083 /OGA), Port of New York and New Jersey Authority, Engineering Research and Development Division, March 1978.

5.5 J. Fruin, R. Marshall, and M. Zeigen. *Accelerating Moving Walkway Systems. Technology Assessment. Report B*. (UMTA IT-06-0126-78-2, PB-287 082 /2GA), Port of New York and New Jersey Authority, Engineering Research and Development Division, April 1978.

5.6 T. B. Hansen, J. S. Worrell, J. King, R. E. Reinsel, and T. O. O'Brien. *Assessment of the Inclined Elevator and Its Use in Stockholm*. (UMTA IT-06-0172-79-1, PB-294 854 /5GA), DeLeuw, Cather & Co., Washington, D.C., September 1978.

CHAPTER 6

WELDING-RELATED CONSIDERATIONS IN BUILDING DESIGN

OMER W. BLODGETT

Consultant
The Lincoln Electric Company
Cleveland, Ohio

6.1 INTRODUCTION

Architects and engineers having responsibility for the design and specification of materials and processes for welded steel structures have found it difficult to keep up with the technological developments affecting their work and interests. Not only has there been substantial progress in welding processes, the design of arc-welded joints applicable to building construction, and knowledge of stresses and their transfer through materials, but also many changes in codes and standards. It has been said that what was good practice 10 years ago in welded steel construction is almost certain to be obsolete, or partially obsolete, today, as a result of the pace of development in the technology. And with this development continuing—spurred by the use of computers to make possible more precise analysis of the engineering factors affecting structural design—it would appear decades will pass before engineers can be certain that they are dealing with the optimum methods of joining steel with fused weld metal.

As an example of the intricacies of the advanced welding technology, two years of extensive computer-based analysis had led to the establishment of formulas to accurately assess the cooling rate of welded steel plate. Offhand, this may seem unimportant to the design engineer, but the mathematics involved considers such heretofore imponderables as changes in thermal conductivity and specific heat with temperature change, plate chemistry, the geometry of joint, plate thickness, and welding heat input. Knowing with high precision the cooling rate in any welding assembly, a preheat can be established to give crack-resistant welds and the proper rate of diffusion of absorbed gases from the molten weld pool.

Structural steel members may be connected with welds or bolts; welding can give more complete continuity to the structure. Once the framing for a multistory rigid frame building has been completely arc welded, the entire framing acts as one unit. With good welding practice, the joints that hold the many columns and beams together are stronger than the base metal. No other method of joining can give such continuity.

To create such a "one-piece" frame efficiently, however, one does not "select" welding as the joining method. The designer must start with the concept of a welded design, and then size and specify the materials, plan the connections, determine the erection methods, and decide on procedures and processes of fabrication at the drawingboard stage. The full advantages of welded steel construction in competition with other methods of attachment, or other materials, will only be realized when the structure is conceived and created as a welded design.

6.2 THE ARC-WELDED JOINT

The arc-welded joint is, of course, the key to welded design. It holds structural elements together and in doing so, must be able to transfer all forces and resist all stresses to which it will be subjected. The joint is made by melting the surface to be joined and adding molten metal from the welding electrode to give adequate material to fill the joint. While molten, the base metal and filler metal in the joint mix, establishing metallurgical continuity. After solidification of the mixture in the joint, the two pieces being welded become essentially one. Many welding processes have been developed for joining steel sections; they are described in Section 6.10.

Figure 6.1 shows the joint and weld types. Specifying a joint does not by itself describe the type of weld to be used. Thus, 10 types of welds are shown for making a butt joint. Although all but two welds are illustrated with butt joints here, some may be used with other types of joints. Thus, a single-bevel weld may also be used in a T or corner joint (Fig. 6.2), and a single-V weld may be used in a corner or butt joint.

The fillet weld shown in Fig. 6.3, requiring no groove preparation, is one of the most commonly used welds. It is the weld by which fabricators join plate material to make girders and beams. In the United States and Canada, a fillet weld is measured by the leg size (Fig. 6.3) of the largest right triangle that may be inscribed within the cross-sectional area. The throat, a better index to strength, is the shortest distance between the root of the joint and the face of the diagrammatical weld.

With thick plates, such as in fabrication of a column, a partial-penetration groove joint, as shown in Fig. 6.4, is often used. This requires beveling, which adds to the cost but this weld uses less weld metal than full-penetration corner joints, which also require beveling. The size of the weld should always be designed with reference to the thinner member, as shown in Fig. 6.5. The joint cannot be made stronger by using the thicker member for weld size, which would use more weld metal. A combination of a partial-penetration groove weld and a fillet weld (Fig. 6.6) is used for many joints in building construction. The AWS (American Welding Society) prequalified, single-bevel groove T joint is reinforced with a fillet weld.

Full-strength welds are not always required in the design, and economies can often be achieved by using partial-strength welds where these are applicable and permissible. Referring to Fig. 6.7, it can be seen that on the basis of an unreinforced 1-in. (25-mm) throat, a 45° partial-penetration,

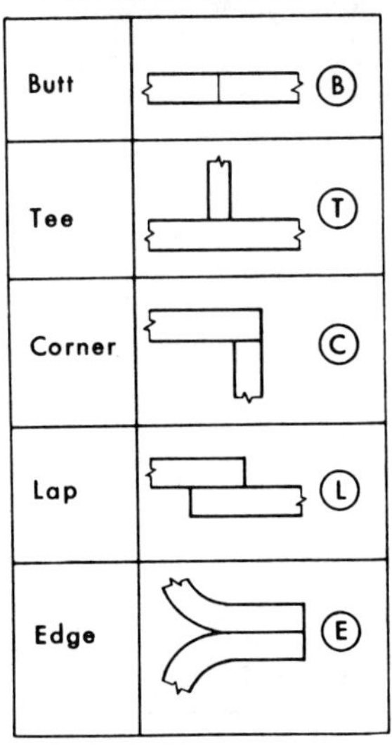

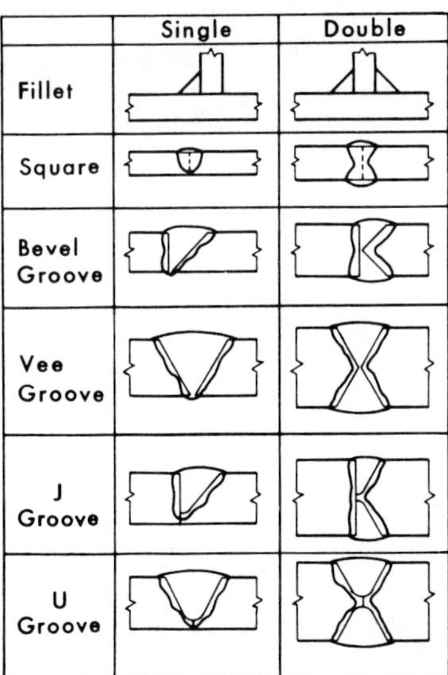

Fig. 6.1 Types of joints and types of welds.

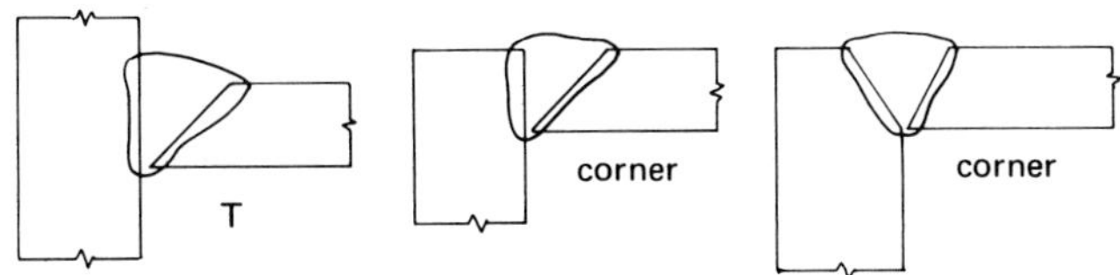

Fig. 6.2 Single-bevel weld used in T joint (left); corner joint (center); single-V weld in corner joint (right).

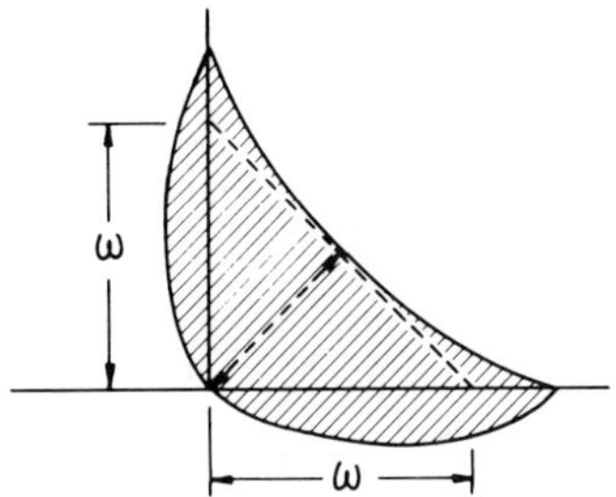

Fig. 6.3 Leg size ω of a fillet weld.

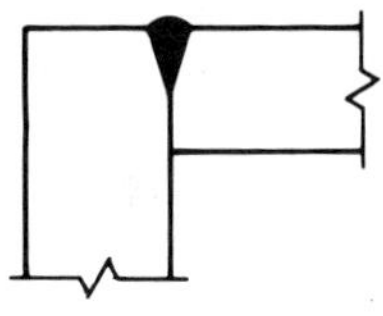

Fig. 6.4 A partial-penetration groove joint, such as used in column fabrication.

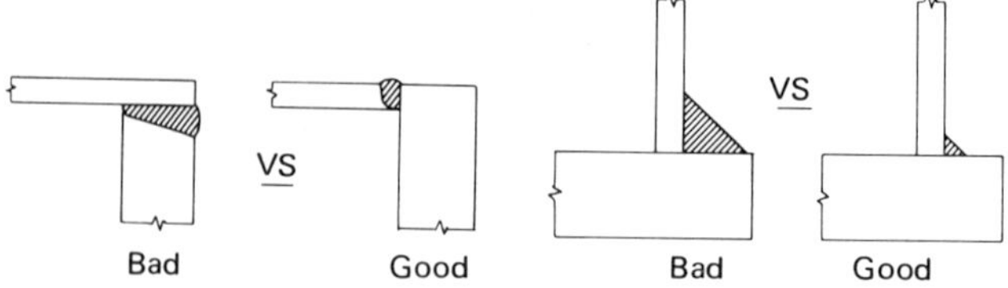

Fig. 6.5 Size of weld should be determined with reference to thinner member.

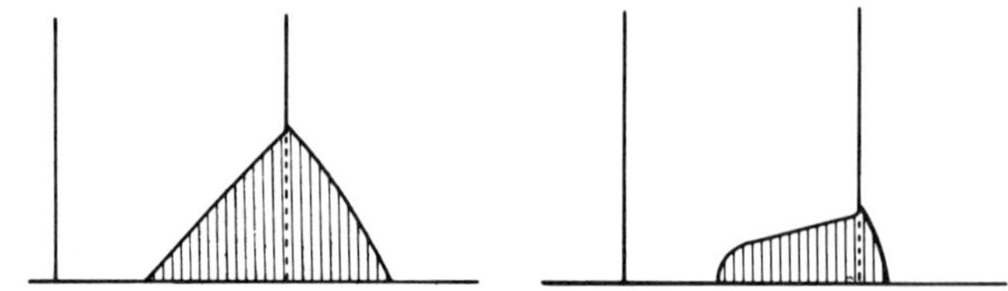

Fig. 6.6 Combinations of partial-penetration groove weld and a fillet weld.

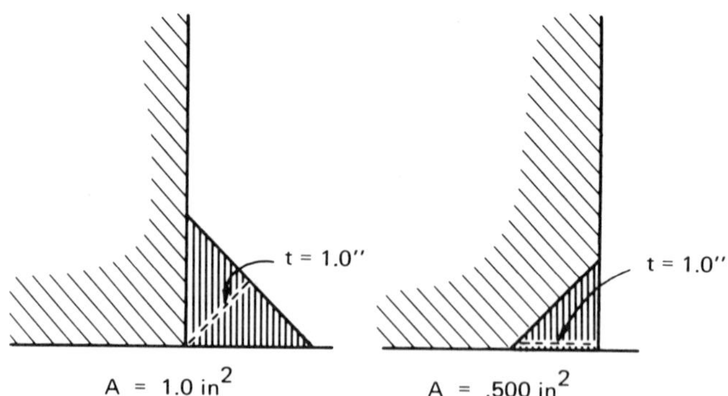

Fig. 6.7 Comparison of weld joints having equal throats.

single-bevel groove weld requires just one-half the weld area needed for a fillet weld. Such a weld may not be as economical as the same strength fillet weld, however, because of the cost of edge preparation and need to use a smaller electrode and lower current on the initial pass.

When a partial-penetration groove weld is reinforced with a fillet weld, the minimum throat is used for design purposes, just as a minimum throat of a fillet or partial-penetration groove weld is used. However, as Fig. 6.8 shows, the allowable for this combination weld is not the sum of the allowable limits for each portion of the combination weld. This would result in a total throat larger than the actual.

Groove joints are used with structural assemblies where full-strength welds are mandatory. Since they require relatively large amounts of weld metal and can present complications in placement, cost is frequently a factor in selecting the type of groove joint. A root opening is used with groove joints to give electrode accessibility. Figure 6.9 shows the root opening (R) with various types of groove joints. The smaller the angle of bevel, the larger the root opening must be to get good fusion at the root. If the root opening is too small, root fusion is more difficult to obtain, and smaller electrodes must be used, thus slowing down the welding process. If the root opening is too large, weld quality does not suffer but more weld metal is required; this increases welding cost and will tend to increase distortion.

Figure 6.10 indicates how the root opening must be increased as the included angle of the bevel is decreased. Backup strips are used on larger root openings. All *three* preparations are acceptable; all are conducive to good welding procedure and good weld quality. Selection, therefore, is usually based on cost.

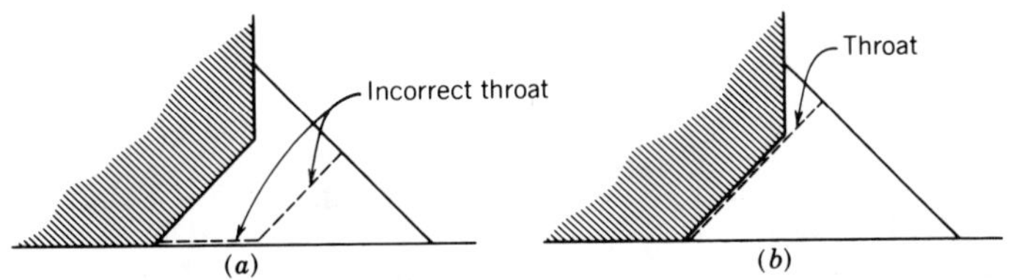

Fig. 6.8 Determining minimum throat: (*a*) incorrect result; (*b*) correct result.

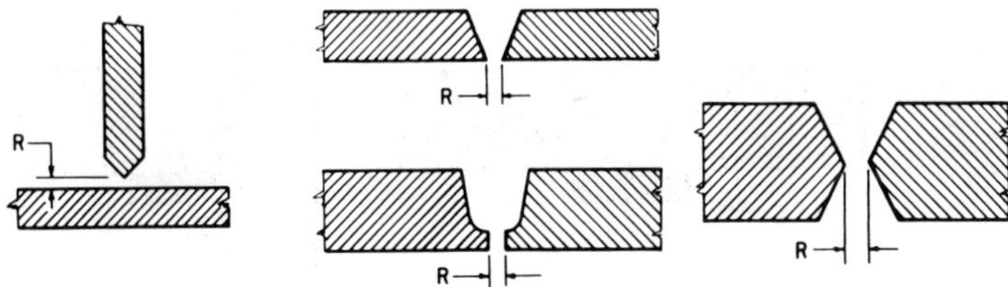

Fig. 6.9 Root opening (*R*) with various types of groove joints.

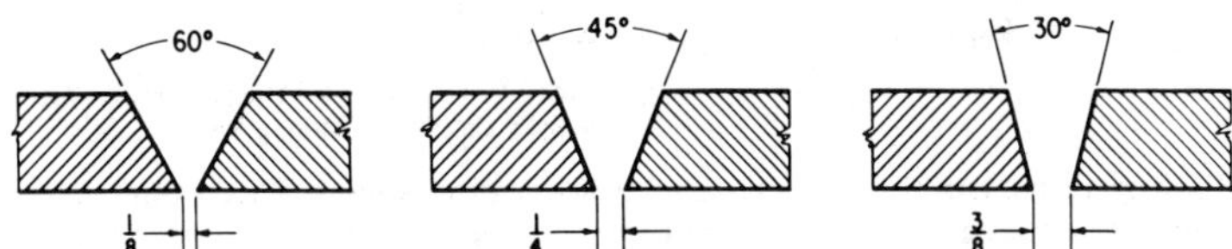

Fig. 6.10 Root opening related to bevel angle and backup strips.

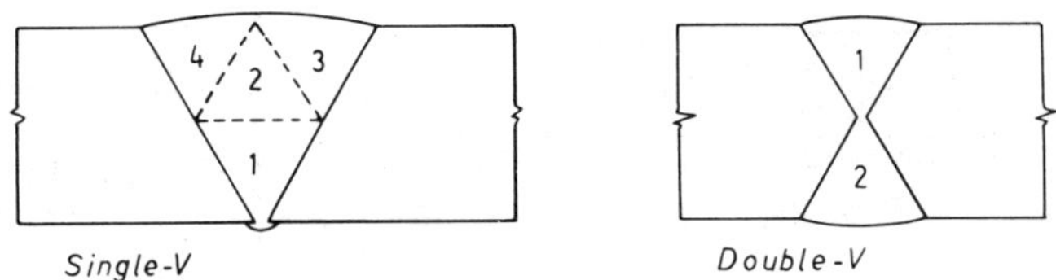

Fig. 6.11 Double-groove joint used in place of single-groove joint reduces amount of welding.

Root opening and joint preparation will directly affect weld cost (pounds of metal required), and choice should be made with this in mind. Joint preparation involves the work required on plate edges prior to welding and includes beveling and providing a root face.

Using a double-groove joint in preference to a single-groove (Fig. 6.11) cuts the amount of welding in half. This reduces distortion and makes possible alternating the weld passes on each side of the joint, again reducing distortion. However, the double-groove joint requires the ability to turn the assembly to avoid out-of-position welding.

Designers are not expected to know all the intricacies of shop or erection welding, although in specifying welds some basic understanding of shop and field problems will be helpful. The aim should always be to facilitate production and minimize costs.

Figure 6.12*b* shows how proper joint preparation and procedure will produce good root fusion and will minimize back-gouging. In Fig. 6.12*c*, a large root opening will result in burn-through. A spacer strip may be used, in which case the joint must be back-gouged. Backup strips are commonly used when all welding must be done from one side, or when the root opening is excessive.

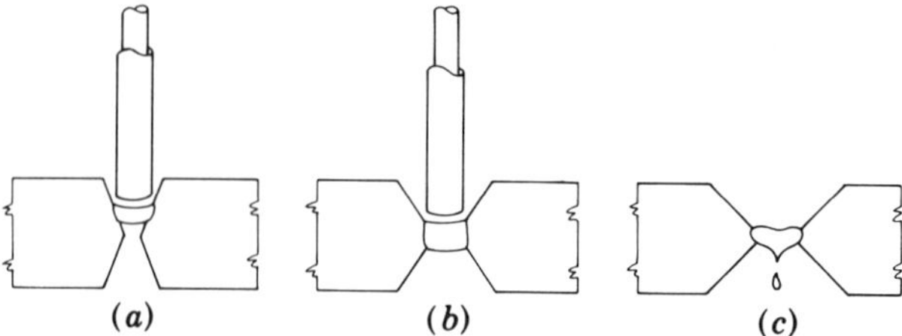

Fig. 6.12 (*a*) If the gap is too small, the weld will bridge the gap, and may leave slag at the root; (*b*) a proper joint preparation; and (*c*) a root opening too large will result in burn-through.

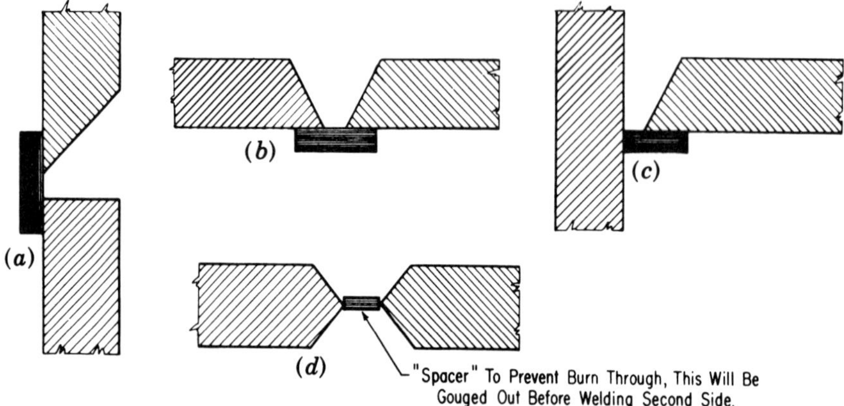

Fig. 6.13 Backup strips: (*a*), (*b*) and (*c*) are used when all welding is done from one side or when the root opening is excessive. A spacer (*d*) to prevent burn-through will be gouged out before welding the second side.

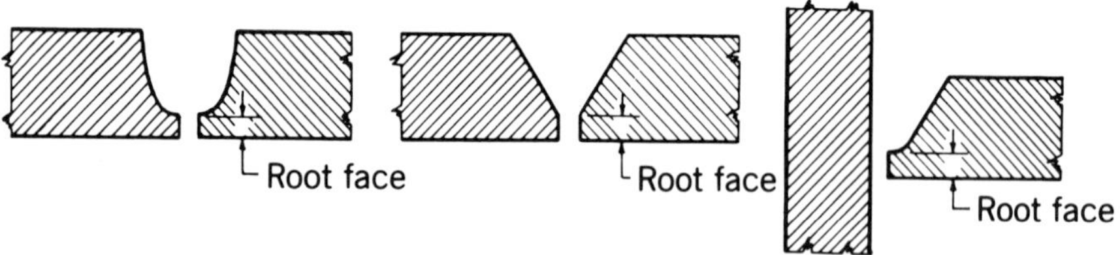

Fig. 6.14 A root face minimizes tendency to burn-through.

Backup strips, shown in Fig. 6.13*a*, 6.13*b*, and 6.13*c*, are generally left in place and become an integral part of the joint. Spacer strips may be used especially in the case of double-V joints to prevent burn-through. The spacer in Fig. 6.13*d* to prevent burn-through will be gouged out before welding the second side. A groove joint will require edge preparation, which not only affects cost in terms of flame cutting or machining but also the ease or difficulty of welding.

The main purpose of a root face (Fig. 6.14) is to provide an additional thickness of metal, as opposed to a feather edge, in order to minimize any burn-through tendency. A feather-edge preparation is more prone to burn-through than a joint with a root face, especially if the gap gets a little too large (Fig. 6.15).

A root face is not as easily obtained as a feather edge. A feather edge is generally a matter of one cut with a torch, while a root face will usually require two cuts or possibly a torch cut plus machining or grinding. A root face usually requires back-gouging if a 100% weld is required. A root face is not recommended when welding into a backup strip, since some slag may be trapped.

Plate edges are beveled to permit accessibility to all parts of the joint and ensure good fusion throughout the entire weld cross-section. Accessibility can be gained by compromising between

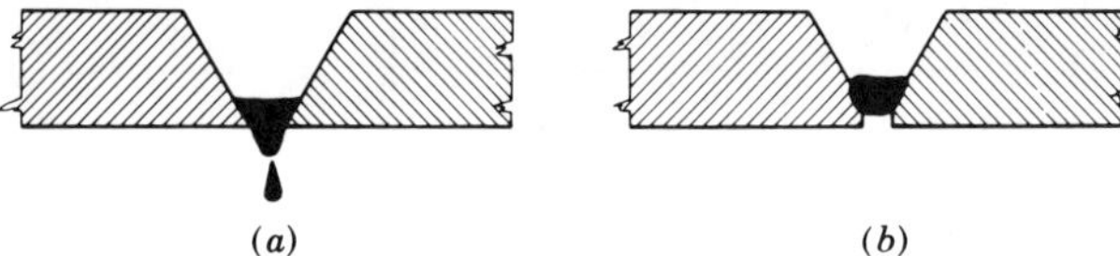

Fig. 6.15 A feather edge (*a*) is more prone to burn-through than a joint with a root face (*b*).

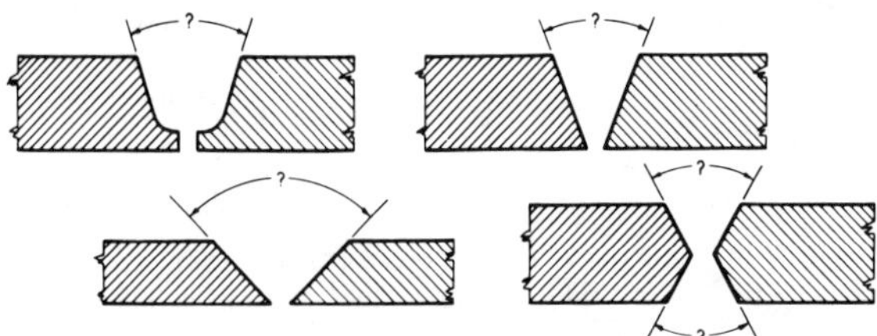

Fig. 6.16 Accessibility is gained by compromising between bevel and root opening.

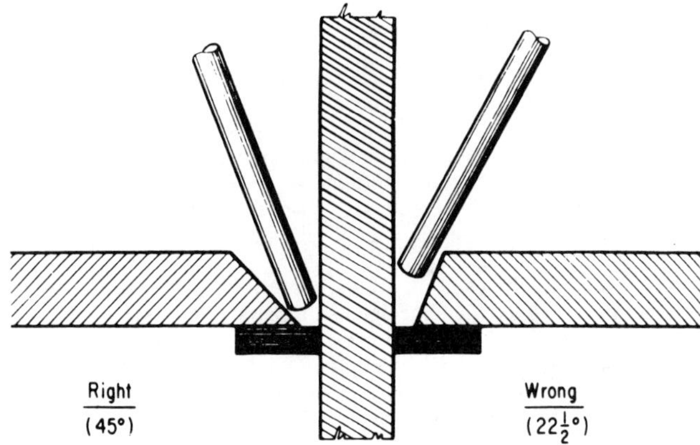

Fig. 6.17 Degree of bevel may be dictated by the need for maintaining proper electrode angle.

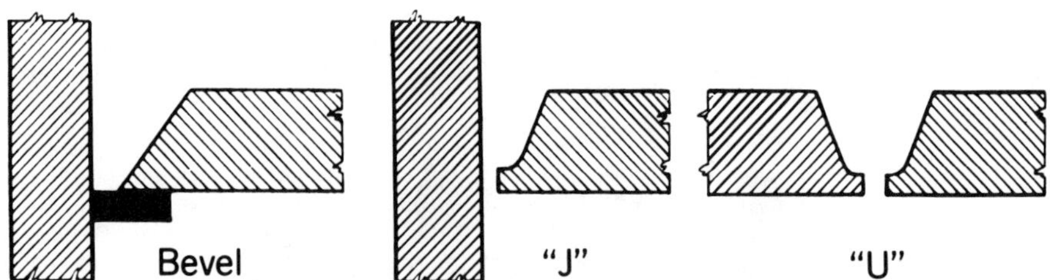

Fig. 6.18 A bevel preparation with a backup strip may be more economical than a J- or U-groove and it eliminates back-gouging.

maximum bevel and minimum root opening (Fig. 6.16). Degree of bevel may be dictated by the importance of maintaining proper electrode angle in confined quarters (Fig. 6.17). For the joint illustrated, the minimum recommended bevel is 45°.

J and U preparations (Fig. 6.18) are excellent to work with, but require air carbon arc gouging or machining as opposed to simple torch cutting. Also a J or U groove requires a root face and thus back-gouging.

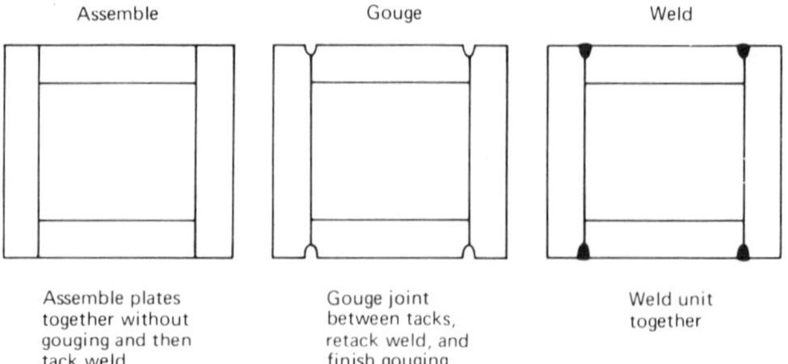
Assemble
Gouge
Weld
Assemble plates together without gouging and then tack weld
Gouge joint between tacks, retack weld, and finish gouging
Weld unit together

Fig. 6.19 Partial-penetration submerged-arc groove welds. U-groove joint preparation may be done either prior to or after fitting.

In the past, there has been minimal use of the technique of U-joint preparation after assembly of the materials making up the structural element. However, clarification of the AWS *Structural Welding Code* [6.1] has made this method more acceptable. The AWS now notes that U-groove joints for complete-penetration and partial-penetration welds may be made "prior to or after fitting." Figure 6.19 illustrates how this applies to the assembly, joint gouging, and welding of the plates making up a building column.

6.3 HOW GOOD ARE WELDS?

As noted in the introduction, the complete metallurgical unity is the most appealing aspect of the welded rigid frame structure. And it is proclaimed that the weld is as strong or stronger than the metal it joins.

Two factors support these statements on the characteristics of welded steel construction. The first factor relates to the amount of pretesting done on welded joints and the amenability of the welded joint to radiographic and other methods of nondestructive inspection after it has been finished. Secondly, there are strict codes defining what types of joints, weld metal, procedures, and processes must be used in structural welding.

Extensive, severe testing is imposed on welds, merely because such tests are possible. For example, as shown in Fig. 6.20, welded specimens prepared with a selected welding process and electrode are subjected to severe bending to prove ductility of the weld. Similar tests for other methods of joining are difficult if not impossible. Nor would the structure have any useful value if it were bent this amount. Also, weld joints in a structure may be radiographed or ultrasonically inspected to assure freedom from cracks, inclusions, or porosity.

Granting that the weld is amenable to tests where it almost always shows up well, the engineer might ask whether or not these tests correlate with the performance of the weld in a finished structure. What goes first when the structure is subjected to destructive loads?

The common beam-to-column connection in building framing illustrates the strength of welded joints. It is usually believed that the weakest portion in such a connection is the groove weld on the tension flanges, and, if column stiffeners are not provided, there will be an uneven transfer of tensile forces through the weld. This is logical, since the column flange is very rigid where it joins the web, becoming more flexible toward its outer edges. Thus, less force is transferred through the

Fig. 6.20 Welded specimens prepared with a selected welding process and electrode are subjected to severe short radius bending to prove ductility of the weld. Photo courtesy of Lincoln Electric Co., Cleveland, Ohio.

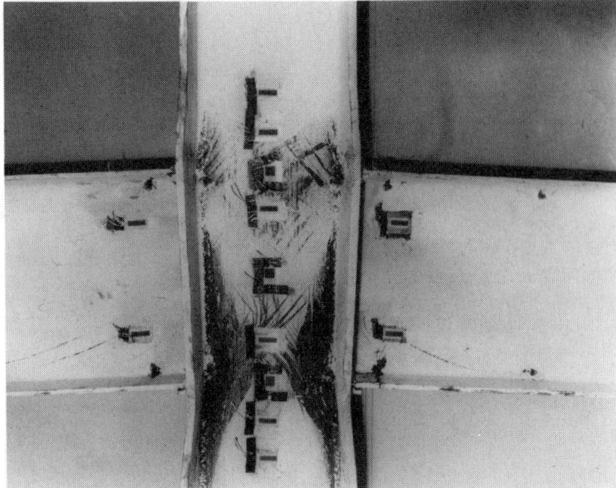

Fig. 6.21 When a beam-to-column connection is loaded to destruction, the unexpected rather than the expected happens: The column web usually buckles under compression loading rather than the tension weld failing. Photo courtesy of Fritz Engineering Laboratory, Lehigh University, Bethlehem, Pa.

welds at the outer edges, with a corresponding increase in the transfer at the center of the weld. From this analysis, one would expect the center of the weld to give way if the beams on such a connection were loaded to the point of connection destruction.

In Fig. 6.21, a connection loaded to destruction is shown. The weld that appears vulnerable is still intact, whereas the column web has buckled under compression. Even with horizontal column flange stiffeners added, as shown in Fig. 6.22, the tension weld remains intact and the buckling occurs in the compression flanges of the beams.

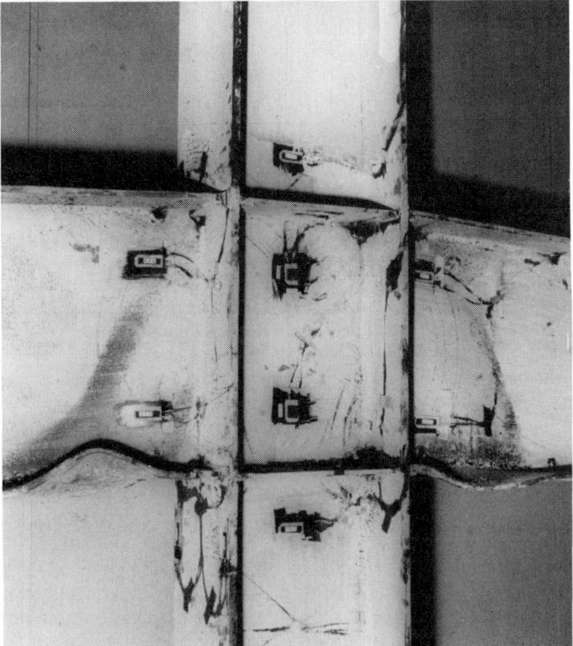

Fig. 6.22 Even when column flange stiffeners are added, the tension weld remains intact and the buckling occurs in the compression flanges of the beams. Photo courtesy of Fritz Engineering Laboratory, Lehigh University, Bethlehem, Pa.

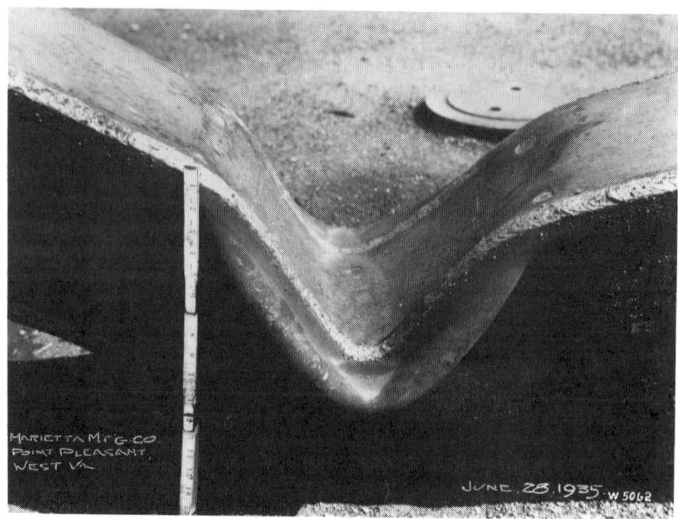

Fig. 6.23 When a box section was subjected to destructive impact load it collapsed in this manner—but none of the weld cracked. Photo courtesy Lincoln Electric Co., Cleveland, Ohio.

Figure 6.23 illustrates what happens when a welded box section is tested to destruction by impact load. It shows the condition of the weldment after an 1800-lb (8000-N) load had been dropped repeatedly 100 times onto it, from heights ranging from 3 to 75 ft (1 to 23 m). The section squashed under the severe treatment, but none of the welds cracked. It is inconceivable that a beam or column in any building could ever be subjected to such shock loading without the entire building, and probably its immediate environment, being destroyed.

Failure is unlikely in any weld that has been properly done by a qualified weldor, when the conditions conducive to brittle fracture (triaxial residual stresses and low service temperature) and fatigue are not present. Design for fatigue loadings is treated in Section 6.6. Preheating and procedures needed to prevent embrittlement are discussed in Section 6.8.

6.4 ALLOWABLES FOR WELDS

To make sure that a weld of proper strength is used in a joint in structural steel, various "allowable" weld strengths are specified by the American Welding Society (AWS), the American Institute of Steel Construction (AISC), and various other engineering and governmental organizations. So-called "allowables" are designated for various types of welds under steady and fatigue loads.

In Figs. 6.24*a* and 6.24*b*, complete joint-penetration groove welds are illustrated. These are considered full-strength welds, since they are capable of transferring the full strength of the members they connect. In calculations, such welds are allowed the same stress as the plate, provided the proper strength level of weld metal is used. In such complete joint-penetration welds, when loaded in tension transverse to the axis of the weld, the mechanical properties of the weld metal must at least match the strength of the base material. If the plates joined are of different strengths, the weld metal must at least match the strength of the weaker plate.

Figures 6.24*c–f* illustrate partial joint-penetration groove welds, which are widely used in the economical welding of very heavy plate. A partial joint-penetration groove weld in such heavy material will usually accomplish a savings in weld metal and welding time, while giving the required strength at the joint. The faster cooling and increased restraint, however, justify establishment of a minimum effective throat t_e.

Various factors must be considered in determining the allowable stresses on the throat of partial joint-penetration groove welds. Joint configuration is one. If a V, J, or U groove is specified, it is assumed that the weldor can easily reach the bottom of the joint, and the effective weld throat t_e equals the depth of the groove. If a bevel groove with an included angle of 45° or less is specified and the manual shielded metal-arc process is used, $\frac{1}{8}$ in. (3 mm) is deducted from the depth of the prepared groove in defining the effective throat. This does not apply to the submerged-arc welding process because of its deeper penetration properties. In the case of gas metal-arc welding or flux-cored arc welding, the $\frac{1}{8}$-in. (3-mm) deduction in throat only applies to bevel groove with an included angle of 45° or less in the vertical or overhead positions.

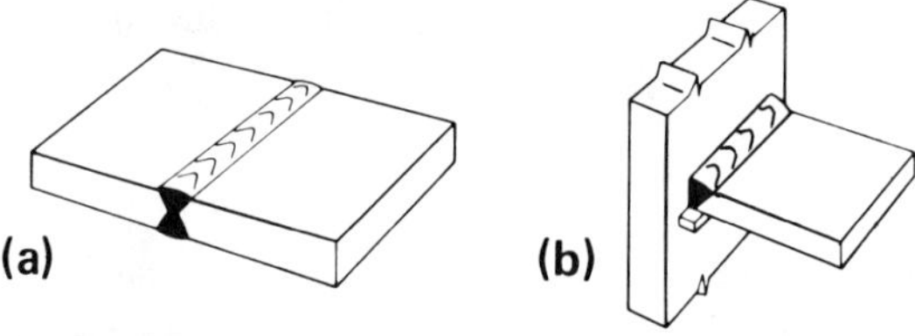

TYPICAL COMPLETE PENETRATION GROOVE WELDS

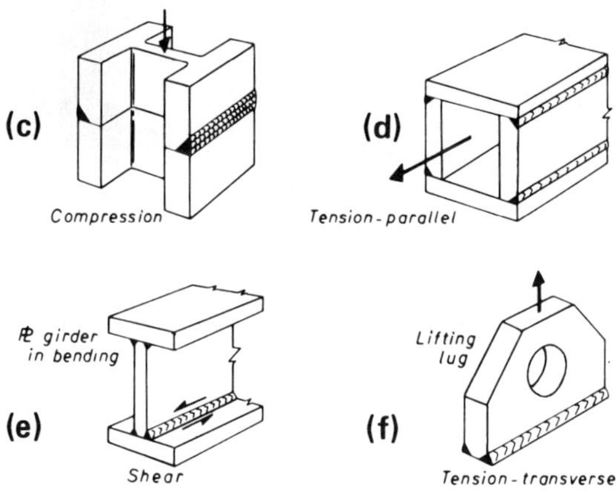

TYPICAL PARTIAL PENETRATION GROOVE WELDS

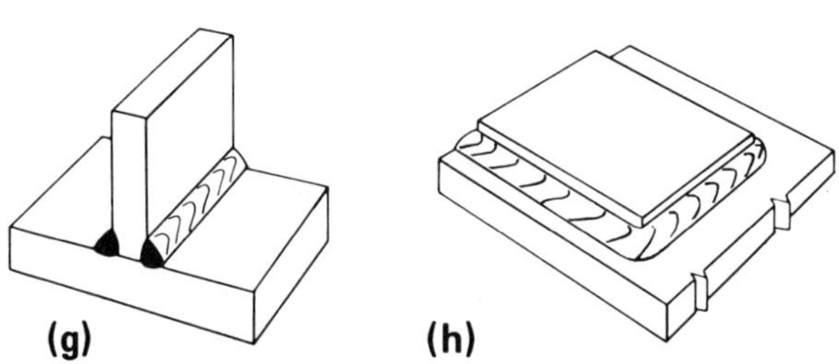

TYPICAL FILLET WELDS

Fig. 6.24 Typical welded joints: (*a*) and (*b*), complete joint-penetration groove welds; (*c*) to (*f*), partial joint-penetration groove welds; and (*g*) and (*h*), fillet welds. From Lincoln Electric Co., Publication D412, used by permission.

Weld metal subjected to compression in any direction or tension parallel to the axis of the weld, Figs. 6.24*c* and 6.24*d*, will have the same allowable as the base material. It is not necessary to use matching weld metal.

The existence of tension forces transverse to the axis of the weld or shear in any direction, Figs. 6.24*e* and 6.24*f*, requires the use of weld metal allowables that are the same as those used for fillet welds. The selected weld metal may have mechanical properties equal to or lower than those of the material being joined. If the weld metal has lower strength, however, its allowable must be used for calculating weld size or maximum allowable weld stress. The allowable used shall not exceed the allowable of the plate.

The AWS [6.1] has established the allowable shear value for weld metal in a fillet or partial-penetration bevel groove weld for buildings as

$$\tau = 0.30(\text{electrode nominal tensile strength})$$

The validity of this equation was proven [6.3] with a series of fillet weld tests conducted by a special Task Committee of AISC and AWS.

Table 6.1 presents the allowable shear values for various weld metal strength levels and the more common fillet weld sizes. These values are for equal-leg fillet welds where the effective throat t_e equals 0.707 times the leg size ω. With the table, one can calculate the allowable unit force f per linear inch for a weld size made with a particular electrode type. For example, calculating the allowable unit force f per inch for the $\frac{1}{2}$-in. (12.8-mm) fillet weld made with an E70 electrode,

$$\begin{aligned} f &= 0.707\omega\tau \\ &= 0.707\omega(0.30)(\text{EXX}) \\ &= 0.707\ (\tfrac{1}{2}\text{ in.})(0.30)(70\text{ ksi}) \\ &= 7.42\text{ kips/in. } (66.7\text{ N/mm}) \end{aligned}$$

The minimum allowable sizes for fillet welds, Figs. 6.24*g* and 6.24*h*, are given in Table 6.2. Where materials of different thickness are joined, the minimum fillet weld size is governed by the thicker material, but this size does not have to exceed the thickness of the thinner material unless required by the calculated stress.

6.5 PRIMARY AND SECONDARY WELDS

There are two types of welds, primary and secondary. *Primary welds* are those that transfer the entire load at the point where they are located. The weld must have the same property as the member at this point, and, if the weld fails, the member fails. *Secondary welds* are those that simply hold the parts together to form a built-up member. In most cases, the forces on these welds are relatively low.

If a full-strength weld (primary weld) is required for tension, the weld metal is selected to match the mechanical properties of the plate. Generally it is unnecessary to match the chemistry, since the weld will undergo a faster rate of cooling, which increases its strength. An exception to this might be when the weldment is to be heat treated and it is required that the weld metal have the same mechanical properties as the steel after heat treatment. Although high-strength quenched-and-tempered plate cools much faster during manufacture than hot-rolled plate because of the quench, the cooling rate is limited somewhat by the large amount of heat contained in the mass. The weld, on the other hand, is quenched by the mass of the plate on which it is deposited, and this rate is very fast—as fast or faster than the cooling rate during the initial quenching of tempered plate. Of course, any preheating and a high welding heat input will decrease the cooling rate of the weld metal by decreasing the temperature gradient.

In welding high-strength alloy steels, one should not use matching strength welds unless they are required. High-strength steel may require additional preheat and greater care in welding because there is an increased tendency for cracking, especially if the joint is restrained. Non-full-strength welds or secondary welds can be made with lower-strength weld metal—E70, E80, or E90 electrodes. Since the E70 electrode would be easier to use, it would be the preferable choice. The only design concern is that the weld be sized to give sufficient strength to the joint.

Figure 6.25 shows a full-strength and non-full-strength weld in A514 (110,000 psi tensile strength) (760 MPa) steel plate. One weld is transverse to the plate; the other parallel.

In the transverse welded joint, both the plate and the weld will be stressed together, and their behavior can be followed by the stress–strain curve shown in the figure. The plate and weld are fairly well matched. If a test specimen were pulled, it is probable that the plate would neck down and fail first, because of the reinforcement of the weld and slightly higher strength developed in the weld and heat-affected zone due to the rapid cooling following welding.

In the parallel welded joint, both the plate and the weld would be strained together. As the member is loaded, the strain increases from point 1 to point 2, with a corresponding increase in the stress in both the plate and weld from point 1 to point 3. At this point, the E7018 weld metal has

Table 6.1 Allowable Load for Various Sizes of Fillet Welds†

	Strength Level of Weld Metal (EXX)						
	*60	*70	80	*90	100	*110	120
Allowable Shear Stress on Throat ksi (1000 psi) of Fillet Weld or Partial Penetration Groove Weld							
τ =	18.0	21.0	24.0	27.0	30.0	33.0	36.0
Allowable Unit Force on Fillet Weld kips/linear in.							
f =	12.73ω	14.85ω	16.97ω	19.09ω	21.21ω	23.33ω	25.45ω
Leg Size ω, in.	Allowable Unit Force for Various Sizes of Fillet Welds kip/linear in.						
1	12.73	14.85	16.97	19.09	21.21	23.33	25.45
7/8	11.14	12.99	14.85	16.70	18.57	20.41	22.27
3/4	9.55	11.14	12.73	14.32	15.92	17.50	19.09
5/8	7.96	9.28	10.61	11.93	13.27	14.58	15.91
1/2	6.37	7.42	8.48	9.54	10.61	11.67	12.73
7/16	5.57	6.50	7.42	8.35	9.28	10.21	11.14
3/8	4.77	5.57	6.36	7.16	7.95	8.75	9.54
5/16	3.98	4.64	5.30	5.97	6.63	7.29	7.95
1/4	3.18	3.71	4.24	4.77	5.30	5.83	6.36
3/16	2.39	2.78	3.18	3.58	3.98	4.38	4.77
1/8	1.59	1.86	2.12	2.39	2.65	2.92	3.18
1/16	.795	.930	1.06	1.19	1.33	1.46	1.59

* Fillet welds actually tested by the joint AISC-AWS Task Committee.
† From The James F. Lincoln Arc Welding Foundation, Publication D412, "New Stress Allowables Affect Weldment Design." Used by permission.

Table 6.2 Minimum Fillet Weld Size ω or Minimum Throat *t* of Partial Penetration Groove Weld†

Material Thickness of Thicker Part Joined (in.)		in.
** to 1/4 incl.		1/8
over 1/4	to 1/2	3/16
over 1/2	to 3/4	1/4
* over 3/4	to 1-1/2	5/16
over 1-1/2	to 2-1/4	3/8
over 2-1/4	to 6	1/2
over 6		5/8

Not to exceed the thickness of the thinner part.

* For minimum fillet weld size, Table does not go above 5/16″ fillet weld for over 3/4″ material.
** Minimum size for bridge application does not go below 3/16″.
† From The James F. Lincoln Arc Welding Foundation, Publication D412, "New Stress Allowables Affect Weldment Design." Used by permission.

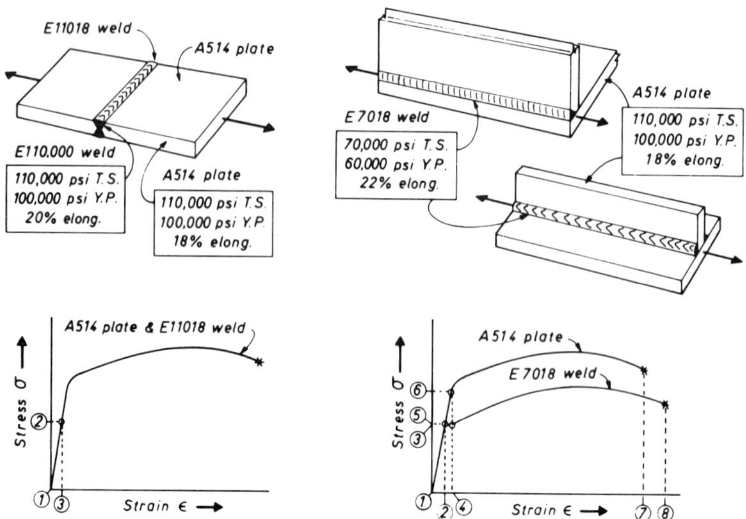

Fig. 6.25 Stress–strain relationships of transverse and parallel welds. From Lincoln Electric Co., Publication D412, used by permission.

reached its 60-ksi (414-MPa) yield stress. On further loading, the strain is increased to point 4. The weld is now stressed beyond its yield point (point 5) and is plastically yielding. The plate, however, is still below its yield point, point 6. With still further loading, the strain will reach point 7, at which time the elongation of the plate may be exhausted. It will probably fail before the weld metal, which has a higher elongation. Finally at point 8, the weld will also fail when reaching the limit of its elongation.

From the stress–strain curve for the parallel weld, it can be seen that nothing would be gained by using an electrode with a strength higher than E70. In this situation, the weld will take any parallel load that the 110-ksi (760-MPa) plate can withstand.

An AISC [6.2] provision (Section 1.14.6.2) gives limited credit for penetration beyond the root of a fillet weld made with the submerged-arc process, as shown in Fig. 6.26. Since penetration increases the effective throat thickness of the weld, the provision permits an increase in this value when calculating weld strength. For fillet welds $\frac{3}{8}$ in. (9.5 mm) and smaller, the effective throat t_e is now assumed to be equal to the leg size ω of the weld:

$$\text{When } \omega \le \tfrac{3}{8} \text{ in. (9.6 mm), then } t_e = \omega$$

For submerged-arc fillet welds larger than $\frac{3}{8}$ in. (9.6 mm), the effective throat of the weld is obtained by adding 0.11 to 0.707 ω:

$$\text{When } \omega > \tfrac{3}{8} \text{ in. (9.6 mm), then } t_e = 0.707\omega + 0.11 \text{ in.}$$

The cost reduction of these increases in effective throat over those used in earlier AWS codes is substantial. The 41% increase (using ω instead of 0.707 ω) in effective weld throat t_e for fillets up to and including $\frac{3}{8}$ in. (9.6 mm) means that the allowable strength of these welds is increased 41%. Or the area of the weld can be cut nearly in half and still have the same allowable unit force per inch.

For example, compare the allowable unit force for a $\frac{1}{2}$-in. E70 weld, 8 in. long, *without* recognizing penetration and a $\frac{3}{8}$-in E70 weld, 8 in. long, recognizing penetration:

$$f\,(\text{for } \tfrac{1}{2}) = \tfrac{1}{2}(0.707)(21{,}000)8 = 59{,}400 \text{ lb}$$

$$f\,(\text{for } \tfrac{3}{8}) = \tfrac{3}{8}(21{,}000)8 = 63{,}000 \text{ lb}$$

The penetration adjustment permits a designer to obtain a given weld strength with about half the weld metal previously required. Obviously, costs are substantially reduced. The benefit from penetration is not as large if the weld size is greater than ⅜ in. but it is still substantial. Note that allowance for penetration applies only to fillet welds made by the submerged-arc welding process. Electrode positive polarity will provide this penetration.

Table 6.3 summarizes the AWS [6.1] and AISC [6.2] stress allowables for weld metal.

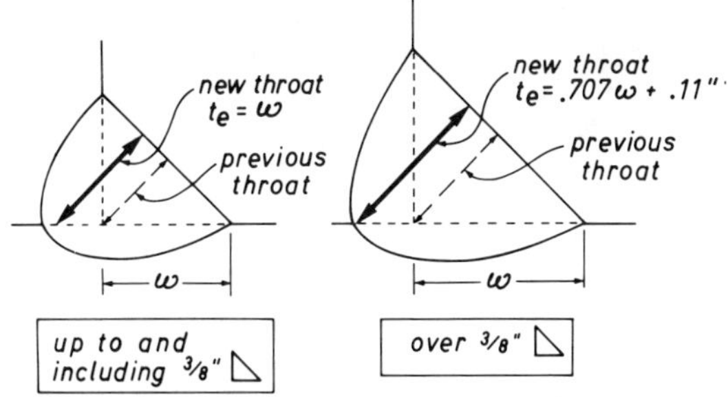

Fig. 6.26 AISC [6.2] gives credit for penetration beyond the root of fillets made with the submerged-arc process.

Table 6.3 Stress Allowables for Weld Metal*

Type of Weld Stress	Permissible Stress	Required Strength Level (1)(2)
	COMPLETE PENETRATION GROOVE WELDS	
Tension normal to the effective throat.	Same as base metal.	Matching weld metal must be used. See Table below.
Compression normal to the effective throat.	Same as base metal.	Weld metal with a strength level equal to or one classification (10 ksi) less than matching weld metal may be used.
Tension or compression parallel to the axis of the weld.	Same as base metal.	Weld metal with a strength level equal to or less than matching weld metal may be used.
Shear on the effective throat.	.30 x Nominal Tensile strength of weld metal (ksi) except stress on base metal shall not exceed .40 x yield stress of base metal.	
	PARTIAL PENETRATION GROOVE WELDS	
Compression normal to effective throat.	Designed not to bear — .50 x Nominal Tensile strength of weld metal (ksi) except stress on base metal shall not exceed .60 x yield stress of base metal. Designed to bear. Same as base metal.	Weld metal with a strength level equal to or less than matching weld metal may be used.
Tension or compression parallel to axis of the weld. (3)	Same as base metal.	
Shear parallel to axis of weld.	.30 x Nominal Tensile strength of weld metal (ksi) except stress on base metal shall not exceed .40 x yield stress of base metal.	
Tension normal to effective throat. (4)	.30 x Nominal Tensile strength of weld metal (ksi) except stress on base metal shall not exceed .60 x yield stress of base metal.	
	FILLET WELDS (3)	
Stress on effective throat, regarless of direction of application of load.	.30 x Nominal Tensile strength of weld metal (ksi) except stress on base metal shall not exceed .40 x yield stress of base metal.	Weld metal with a strength level equal to or less than matching weld metal may be used.
Tension or compression parallel to axis of weld.	Same as base metal.	
	PLUG AND SLOT WELDS	
Shear parallel to faying surfaces.	.30 x Nominal Tensile strength of weld metal (ksi) except stress on base metal shall not exceed .40 x yield stress of base metal.	Weld metal with a strength level equal to or less than matching weld metal may be used.

(1) For matching weld metal, see AISC Table 1.17.2 or AWS Table 4.1.1 or table below.
(2) Weld metal, one strength level (10 KSI) stronger than matching weld metal may be used when using alloy weld metal on A242 or A588 steel to match corrosion resistance or coloring characteristics (Note 3 of Table 4.1.4 or AWS D1.1).
(3) Fillet welds and partial penetration groove welds joining the component elements of built up members (ex. flange to web welds) may be designed without regard to the axial tensile or compressive stress applied to them.
(4) Cannot be used in tension normal to their axis under fatigue loading (AWS 2.5). AWS Bridge prohibits their use on any butt joint (9.12.1.1), or any splice in a tension or compression member (9.17), or splice in beams or girders (9.21), however, are allowed on corner joints parallel to axial force of components of built up members (9.12.1.2 (2). Cannot be used in girder splices (AISC 1.10.8).

MATCHING WELD METAL AND BASE METAL

Weld Metal	60 or 70	70	80	100	110
Type of Steel	A36; A53, Gr. B; A106, Gr. B; A131, Gr. A, B, C, CS, D, E; A139, Gr. B; A381, Gr. Y35; A500, Gr. A, B; A501; A516, Gr. 55, 60; A524, Gr. I, II; A529; A570, Gr. D, E; A573, Gr. 65; A709, Gr. 36; API 5L, Gr. B; API 5LX, Gr. 42; ABS, Gr. A, B, D, CS, DS, E	A131, Gr. AH32, DH32, EH32, AH36, DH36, EH36; A242; A441; A516, Gr. 65; 70; A537, Class 17; A572, Gr. 42, 45, 50, 55; A588 (4 in. and under); A595, Gr. A, B, C; A606; A607, Gr. 45, 50, 55; A618; A633, Gr. A, B, C, D (2-1/2 in. and under); A709, Gr. 50, 50W; API 2H; ABS Gr. AH32, DH32, EH32, AH36, DH36, EH36.	A572, Gr. 60, 65; A537, Class 2; A63, Gr. E	A514 [over 2-1/2 in. (63 mm)]; A709, Gr. 100, 100W [2-1/2 to 4 in. (63 to 102 mm)]	A514 [2-1/2 in. (63 mm) and under]; A517; A709, Gr. 100, 100W [2-1/2 in. (63 mm) and under]

* For bridges, reduce these allowable values by 10%.
From AWS D1.1 [6.1] and AISC [6.2].

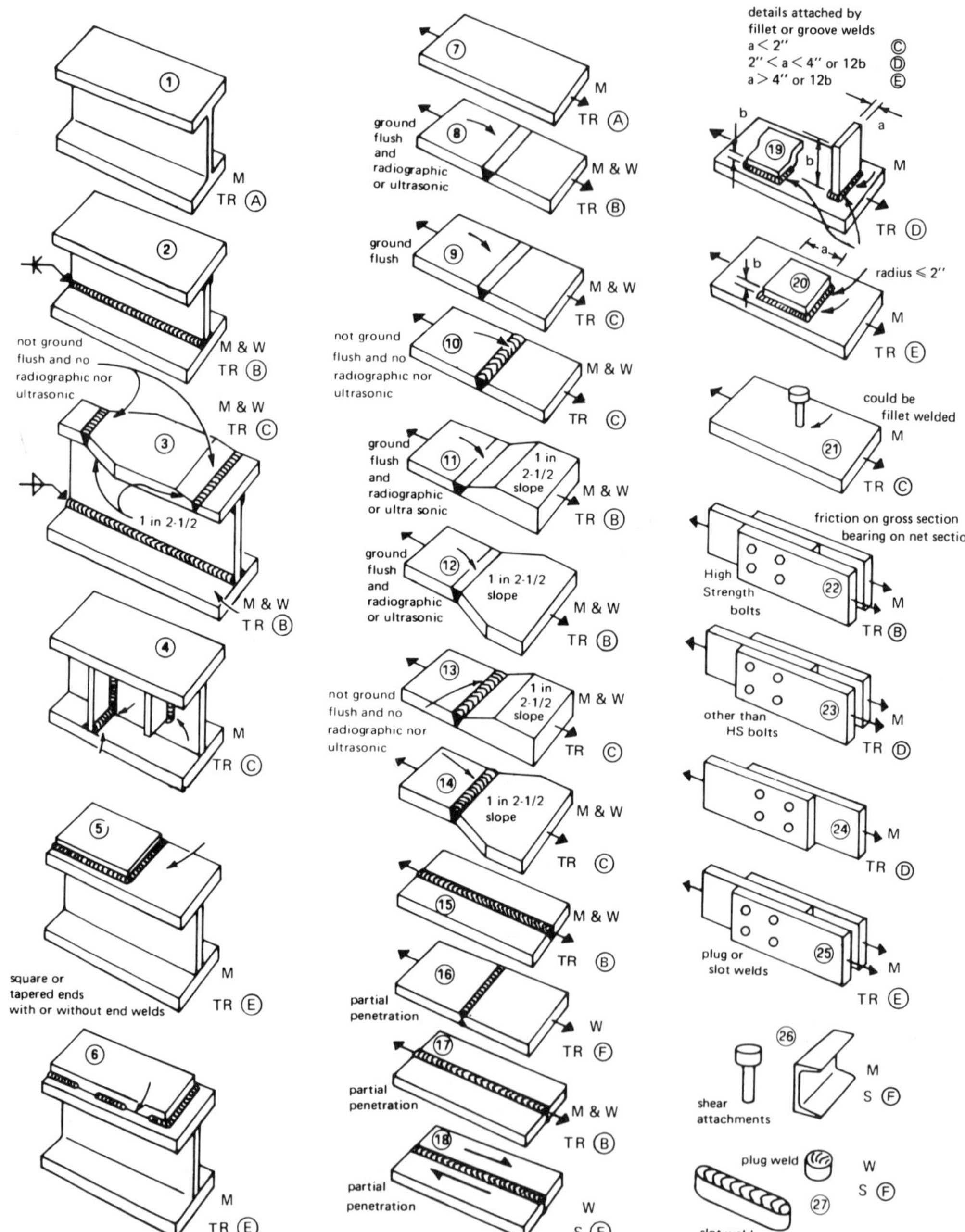

Fig. 6.27 Allowable fatigue stress (σ_{sr} and τ_{sr}) as used by AISC [6.2] and AWS [6.1]. From Lincoln Electric Co., Publication D412, used by permission.

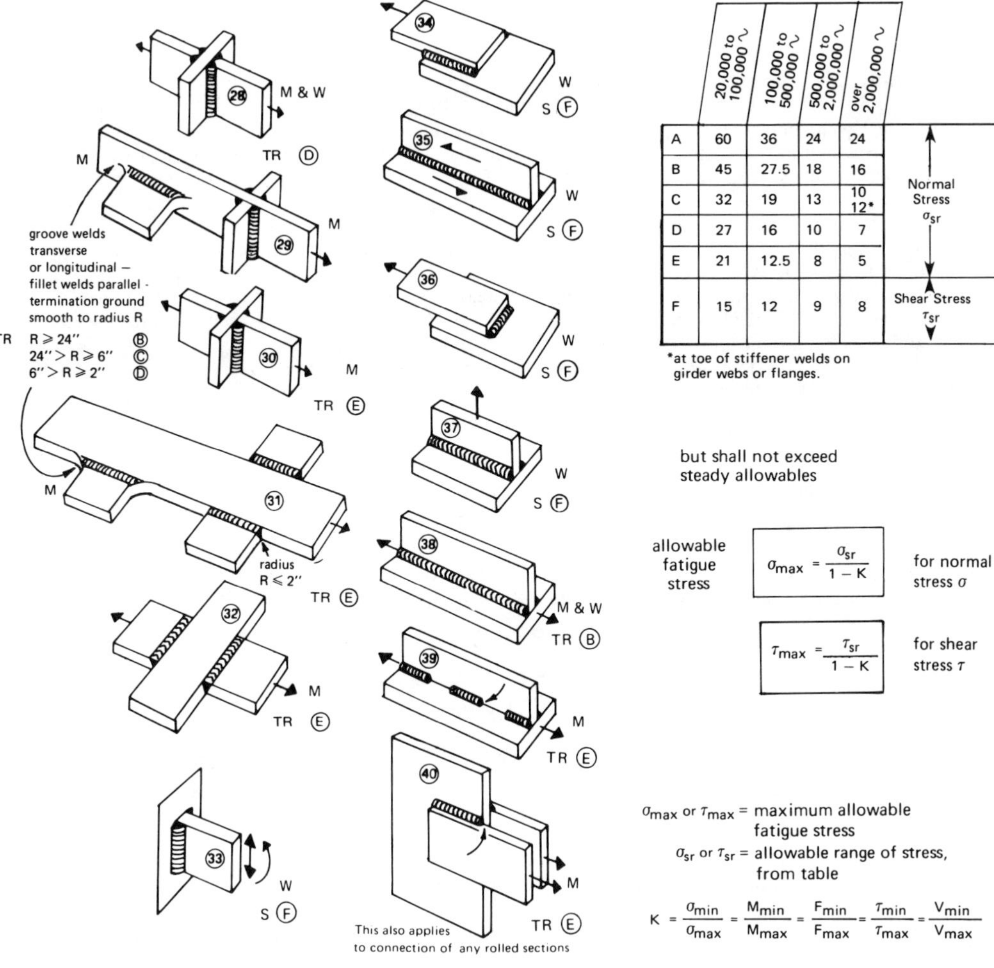

	20,000 to 100,000 ~	100,000 to 500,000 ~	500,000 to 2,000,000 ~	over 2,000,000 ~	
A	60	36	24	24	Normal Stress σ_{sr}
B	45	27.5	18	16	
C	32	19	13	10 12*	
D	27	16	10	7	
E	21	12.5	8	5	
F	15	12	9	8	Shear Stress τ_{sr}

*at toe of stiffener welds on girder webs or flanges.

but shall not exceed steady allowables

allowable fatigue stress

$$\sigma_{max} = \frac{\sigma_{sr}}{1 - K}$$ for normal stress σ

$$\tau_{max} = \frac{\tau_{sr}}{1 - K}$$ for shear stress τ

σ_{max} or τ_{max} = maximum allowable fatigue stress

σ_{sr} or τ_{sr} = allowable range of stress, from table

$$K = \frac{\sigma_{min}}{\sigma_{max}} = \frac{M_{min}}{M_{max}} = \frac{F_{min}}{F_{max}} = \frac{\tau_{min}}{\tau_{max}} = \frac{V_{min}}{V_{max}}$$

Curved arrow indicates region of application of fatigue allowables

Straight arrows indicate applied forces

Grind in the direction of stressing only

(when slope is mentioned (ex. 1 in 2-1/2) this is always the maximum value. Less slope is permissable.

S = shear
T = tension
R = reversal
M = stress in metal
W = stress in weld
$\underline{\tau}$ = allowable steady shear stress

Fig. 6.27 *(Continued)*

6.6 ALLOWABLE FATIGUE STRENGTHS OF WELDS

Some failures in welded structures, other than those caused by old age, corrosion, and lack of maintenance, are attributed to fatigue. The weld must be designed so it does not contribute to fatigue failure. Thus, the performance of a weld under cycles of stress is an important consideration, and specifications relating thereto have been developed after extensive research by the AISC.

Although sound weld metal has about the same fatigue strength as unwelded plate, the change in section that a weld induces may lower the fatigue strength. In the case of a complete-penetration groove weld, the reinforcement and any undercut, lack of penetration, or a crack will act as a notch, which in turn acts as a stress raiser and may result in a lower fatigue strength. The very nature of a fillet weld, because it is used in lap and tee joints, provides an abrupt change in section which lowers fatigue strength.

The *AISC Specification* [6.2, Appendix B] provisions for fatigue cover a wide range of welded joints and also take into consideration the fatigue strength of members attached by welds. Figure 6.27 shows various types of joints and contains a chart that tabulates the allowable range of stress. Covered are steels having a yield strength of 36,000 psi (250 MPa) (A36) up to 110,000 psi (760 MPa) (A514) and weld metal from E60XX to E120XX. The fatigue allowables for members are designated as M and the fatigue allowables for welds as W. A tensile load is T, reversal is R, and shear is S. In the chart used for determining values for allowable range of stress (Fig. 6.27), there are four groups representing fatigue life:

1. 20,000–100,000 cycles
2. 100,000–500,000 cycles
3. 500,000–2,000,000 cycles
4. Over 2,000,000 cycles

and six different categories (A–F) representing type of joint and member loading. The chart provides the allowable *range* in stress (normal stress σ_{sr} or shear stress τ_{sr}), which value may be used in the conventional fatigue formulas,

$$\sigma_{\max} = \frac{\sigma_{sr}}{1 - K} \quad \text{or} \quad \tau_{\max} = \frac{\tau_{sr}}{1 - K} \tag{6.6.1}$$

$$\text{where } K = \frac{\text{min stress}}{\text{max stress}} \text{ or } \frac{\text{min force}}{\text{max force}} \text{ or } \frac{\text{min moment}}{\text{max moment}} \text{ or } \frac{\text{min shear}}{\text{max shear}} \text{ or } \frac{\text{min torque}}{\text{max torque}} \tag{6.6.2}$$

(K will vary from +1, through zero, to −1)

$\sigma_{\max}$ = maximum allowable fatigue stress in tension or compression (normal stress)
$\tau_{\max}$ = maximum allowable fatigue stress in shear

In every case, the allowable should not exceed the steady (i.e., static) stress allowable for the plate and weld. Equation (6.6.1) is used to reduce the fatigue allowable because of the fatigue condition. As the range of fatigue stress increases, this will provide a lower and lower fatigue allowable. When a high value for K is inserted (almost 1), it may give a value higher than the steady value, which, of course, must not be exceeded.

Using the information in Fig. 6.27, it can be seen that the fatigue allowable of a welded member in bending, as shown by case 2, is really determined by the allowable of the plate when connected by fillet welds parallel to the direction of the applied stress. M and W are equal, and the applicable category is B in the chart.

If stiffeners are used in the weldment, as in case 4, the fatigue allowable of the web or flange is determined by the allowable in the member at the termination of the weld or adjacent to the weld, Category C.

The fatigue allowable of the flange plate at the termination of a cover plate, either square or tapered end, is represented by configuration 5. The applicable category to use in the chart is E.

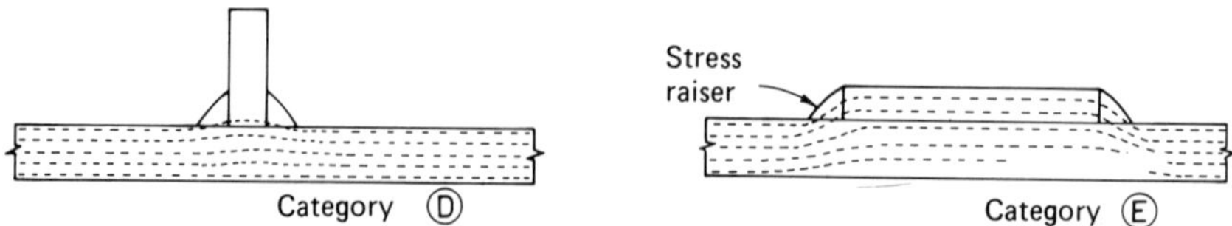

Fig. 6.28 Effect of attachments.

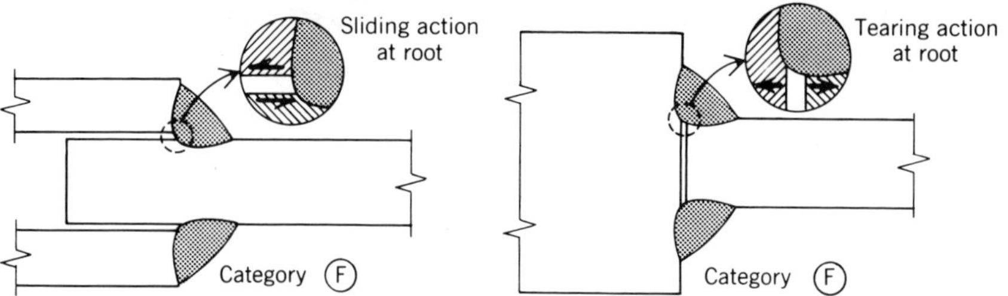

Fig. 6.29 Fatigue strength of transverse fillet welds.

If intermittent fillet welds are used on members, the fatigue allowable of the plate adjacent to the termination of the weld is Category E, as in cases 6 and 39.

Fatigue allowables for partial-penetration groove welds may be determined by reference to configurations 16 to 18.

According to cases 19 and 20, the fatigue allowable for a member with a transverse attachment increases as the length of the attachment decreases, measured parallel to the axis of the load (see Fig. 6.28). Although there may be a similar geometrical notch effect or abrupt change in section in both, it is the stress raiser that is important. The transverse bar in case 19 is so short, insofar as far as the axis of the member and the load are concerned, that very little of the force is able to swing up and into the bar and then back down again. Consequently, the stress raiser is not as severe. The longer bar attachment in case 20, however, is sufficiently long to provide a path for the force through it and the connecting welds. Because of this force transfer into the bar and through the welds (Fig. 6.28) there will be a higher stress raiser and, as a result, a reduction of the fatigue strength of the member.

Configuration 30 in Fig. 6.27 falls into Category E, and should not be confused with case 37, which is Category F. Both depict transverse fillet welds, but case 30 provides a fatigue allowable for the member adjacent to the fillet weld, while case 37 provides a fatigue shear allowable for the throat of the fillet weld.

Since the static strength of a transverse fillet is about one-third stronger than a parallel fillet, one might question why the fatigue allowable for a parallel fillet, as in cases 34 and 35, Category F, is the same as the transverse fillet in cases 36 and 37, Category F. The actual fatigue strength of the transverse fillet 36 is only slightly higher than the parallel fillet 34, and both have been placed in the same range covered by Category F. Although there might be a slightly lower fatigue strength for a transverse fillet-welded tee joint in case 37 than a transverse fillet-welded lap joint in case 36, both are given Category F as detailed in Fig. 6.29.

6.7 PREHEATING BEFORE WELDING

The architect or designer can hardly be expected to be expert in the precise shop or field practices that lead to a high-quality weld, but background knowledge can be helpful in specifying welds.

Preheating the assembly before welding is used for one or more of the following reasons:

1. To reduce shrinkage stresses in the weld and adjacent base metal—especially important with highly restrained joints.
2. To provide a slower rate of cooling through the critical temperature range (about 1600 to 1330°F) (870 to 720°C), preventing excessive hardening and lower ductility of both the weld and heat-affected area of the base plate.
3. To provide a slower rate of cooling through the 400°F (204°C) range, allowing more time for any hydrogen that is present to diffuse away from the weld and adjacent plate to avoid underbead cracking.

As suggested by the above, a main purpose of preheat is to slow down the cooling rate—to allow more "Time at Temperature," as illustrated in Fig. 6.30. As the insets show, there is a greater temperature drop in 1 sec at a given temperature (T_1) when the initial temperature of the plate is 70°F (21°C) than when the initial temperature is 300°F (150°C). In other words, the cooling rate (°F or °C/sec) is slower when preheat is used. Thus, the amount of heat in the weld area as well as the temperature is important. A thick plate could be preheated quickly to a specified temperature in a localized area and the heating be ineffective because of rapid heat transfer, the reduction of heat in the welding area, and, thus, no marked effect on slowing the cooling rate. Having a thin surface

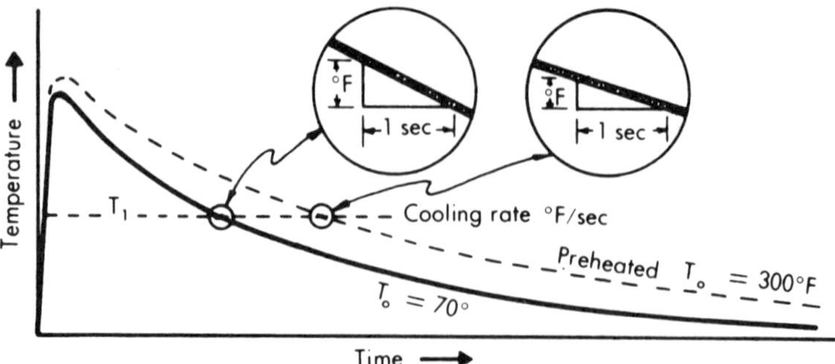

Fig. 6.30 Effect of preheat in slowing down the cooling rate.

area at a preheat temperature is not enough if there is a mass of cold metal beneath it into which the heat can rapidly transfer.

Because of the heat-absorption capacity of a thick plate, the heat-affected zone and the weld metal after cooling may be in a highly quenched condition unless sufficient preheat is provided. What really matters is how long the weld metal and adjacent base metal are maintained in a certain temperature range during the cooling period. This, in turn depends on the amount of heat in the assembly, the heat transfer properties of the material, and its configuration. Without adequate preheat, the cooling could be rapid and intolerably high hardness and brittleness could occur in the weld or adjacent area.

Welding at low ambient temperatures or on steel brought in from outside storage on cold winter days greatly increases the need for preheat. It is true that preheating rids the joint of moisture, but preheating is usually not specified for that purpose.

6.7.1 Amount of Preheat Required

The amount of preheat required for any application depends on such factors as base metal chemistry, plate thickness, restraint and rigidity of the members, and heat input of the process. Unfortunately, there is no method for metering the amount of heat put into an assembly by a preheat torch. The best shop approach for estimating the preheat input is a measure of the temperature at the welding area by temperature-indicating crayon marks or pellets. These give approximate measures of temperature at the spots where they are placed, and they are taken as indices to the heat input and are correlated with thickness of metal and chemistry of metal in tables specifying minimum preheat temperatures. Thus, temperature is the gage to preheat inputs, and preheating to specified temperatures is the practical method of obtaining the amount of preheat needed to control the cooling rate after welding.

There are various guides for use in estimating preheat temperatures, including the recommendation of the suppliers of special steels. No guide, however, can be completely and universally applicable because of the varying factors of rigidity and restraint in assemblies. (*Note:* As indicated in the introduction, formulas are available which give information on the cooling rate, whatever the configuration. Correlated with experimental data, it is now possible to assess the optimum preheat.) Recommendations are, thus, presented as "minimum preheat recommendations," and they should be accepted as such. However, the quenched-and-tempered steels can be damaged if the preheat is too high and the precautions necessary for these steels are discussed later.

The AWS *Structural Welding Code* [6.1] has established minimum preheat and interpass temperature requirements for common weldable steels, as shown in Table 6.4. While material thickness, ranges of metal chemistry, and the welding process are taken into account in the minimum requirements, some adjustments may be needed for specific steel chemistry, welding heat input, joint geometry, and other factors.

Generally, the higher the carbon content of a steel, the lower the critical cooling rate and the greater the necessity for preheating and using low-hydrogen electrodes.

Carbon, however, is not the only element that influences the critical cooling rate. Other elements in the steel (discussed in Section 6.11.1) are responsible for the hardening and loss of ductility that occur with rapid cooling. Total hardenability is thus a factor to be considered when determining preheat requirements. Total hardenability can be expressed in terms of a "carbon equivalent", and this common measure of the effects of carbon and other alloying elements on hardening can be the basis for preheat and interpass temperature estimates.

Carbon equivalents (C_{eq}) are empirical values, determined by various carbon-equivalent formulas that represent the sum of the effects of various elements in steel on its hardenability. One of these (International Institute of Welding) is

$$C_{eq} = C + \frac{Mn}{6} + \frac{Cr + Mo + V}{5} + \frac{Ni + Cu}{15} \tag{6.7.1}$$

Theoretically, it is possible to reduce the preheat temperature requirement below the value listed in preheat tables when using welding currents in the high range of the procedures for semiautomatic and automatic processes. The justification for this is that the welding heat input is likely to be much higher than anticipated by the preheat recommendations. In such cases, heat losses from the assembly might more than be balanced by the welding heat input, bringing the affected metal up to or beyond the minimum preheat and interpass recommendations before it starts to cool.

The heat input during welding for a specific welding procedure is readily calculated by the formula

$$KJ = \frac{60EI}{1000V} \tag{6.7.2}$$

where KJ = heat input in kilojoules/inch or kilowatt sec/inch
E = arc voltage in volts
I = welding current in amperes
V = arc speed in inch/minute

The method of preheating depends on the thickness of the plate, the size of the weldment, and the heating equipment available.

One method is torch heating, using natural gas premixed with compressed air. This produces a hot flame and burns clean. Torches can be connected to convenient gas and compressed-air outlets around the shop. Acetylene, propane, and oil torches can also be used. On large weldments, banks and heating torches may be used to bring the material up to temperature quickly and uniformly.

Electrical strip heaters are used on longitudinal and girth seams on plate. The heaters are clamped to the plate parallel to the joint and about 6 in. (150 mm) from the seam. After the plate reaches the proper preheat temperature, the heaters may remain in place to add heat if necessary to maintain the proper interpass temperature.

Although it is important that the work be heated to a minimum temperature, no harm is done if this temperature is exceeded by about 100°F (about 40°C). This is not true, however, for quenched-and-tempered steels, since welding on an overheated plate may cause damage in the heat-affected zone. For this reason, the temperature should be measured as accurately as possible with such steels.

Temperature-indicating crayons and pellets are available for a wide range of temperatures. A crayon mark for a given temperature on the work will melt suddenly when the work reaches that temperature. Two crayon marks, one for the lower limit and one for the upper limit of temperature, show clearly when the work is heated to the desired temperature range.

Several types of portable pyrometers are available for measuring surface temperature. Properly used, these instruments are sufficiently accurate, but must be periodically calibrated to ensure reliability.

Thermocouples may be attached to the work and used to measure temperature. Thermocouples, of course, are the temperature-sensing devices in various types of ovens used for preheating small assemblies.

6.7.2 Interpass Temperatures

Usually a steel that requires preheating to a specified temperature also must be kept at this temperature between weld passes. With many weldments, the heat input during welding is adequate to maintain the interpass temperature. On a massive weldment, it is not likely that the heat input of the welding process will be sufficient to maintain the required interpass temperature. If this is the case, torch heating between passes may be required.

Once an assembly has been preheated and the welding begun, it is desirable to finish the welding as soon as possible so as to avoid the need for interpass heating.

Since the purpose of preheating is to reduce the quench rate, it logically follows that the same slow cooling should be accorded all passes. This can only be accomplished by maintaining an interpass temperature which is at least equal to the preheat temperature. If this is not done, each individual bead will be subjected to the same high quench rate as the first bead of a nonpreheated assembly.

Table 6.4 Minimum Preheat and Interpass Temperature[3,4] (from AWS D1.1 [6.1], Table 4.2)

Group	Steel Specification	Welding Process	Thickness of Thickest Part at Point of Welding		Minimum Temperature	
			in.	mm	°F	°C
I	ASTM A36[2] ASTM A53 Grade B ASTM A106 Grade B ASTM A131 Grades A, B, CS, D, DS, E ASTM A139 Grade B ASTM A381 Grade Y35 ASTM A500 Grade A Grade B ASTM A501 ASTM A516 Grades 55 & 60 ASTM A524 Grades I & II ASTM A529 ASTM A570 All grades ASTM A573 Grade 65 ASTM A709 Grade 36[2] API5L Grade B API5LX Grades X42 ABS Grades A, B, D, CS, DS Grade E	Shielded metal arc welding with other than low hydrogen electrodes	Up to 3/4	19 incl.	None[1]	
			Over 3/4 thru 1-1/2	19 38 incl.	150	66
			Over 1-1/2 thru 2-1/2	38 64	225	107
			Over 2-1/2	64	300	150
II	ASTM A36[2] ASTM A53 Grade B ASTM A106 Grade B ASTM A131 Grades A, B, CS, D, DS, E AH 32 & 36 DH 32 & 36 EH 32 & 36 ASTM A139 Grade B ASTM A242 ASTM A381 Grade Y35 ASTM A441 ASTM A500 Grade A Grade B ASTM A501 ASTM A516 Grades 55 & 60 65 & 70 ASTM A524 Grades I & II ASTM A529 ASTM A537 Classes 1 & 2 ASTM A570 All grades ASTM A572 Grades 42, 50 ASTM A573 Grade 65 ASTM A588 ASTM A595 Grades A, B, C, ASTM A606 ASTM A607 Grades 45, 50, 55 ASTM A618 ASTM A633 Grades A, B Grades C, D ASTM A709 Grades 36, 50, 50W API5L Grade B API5LX Grade X42 API Spec. 2H ABS Grades AH 32 & 36 DH 32 & 36 EH 32 & 36 ABS Grades A, B, D, CS, DS Grade E	Shielded metal arc welding with low hydrogen electrodes, submerged arc welding, gas metal arc welding, flux cored arc welding	Up to 3/4	19 incl.	None[1]	
			Over 3/4 thru 1-1/2	19 38 incl.	50	10
			Over 1-1/2 thru 2-1/2	38 64 incl.	150	66
			Over 2-1/2	64	225	107

III	ASTM A572 ASTM A633 API5LX	Grades 60 & 65 Grade E Grade X52	Shielded metal arc welding with low hydrogen electrodes, submerged arc welding, gas metal arc welding, flux cored arc welding	Up to 3/4	19 incl.	50	10
				Over 3/4 thru 1-1/2	19 38 incl.	150	66
				Over 1-1/2 thru 2-1/2	38 64 incl.	225	107
				Over 2-1/2	64	300	150
IV	ASTM A514 ASTM A517 ASTM A709	 Grades 100 & 100W	Shielded metal arc welding with low hydrogen electrodes, submerged arc welding with carbon or alloy steel wire, neutral flux, gas metal are welding or flux cored arc welding	Up to 3/4	19 incl.	50	10
				Over 3/4 thru 1-1/2	19 38 incl.	125	50
				Over 1-1/2 thru 2-1/2	38 64 incl.	175	80
				Over 2-1/2	64	225	107

Notes:

A. For modification of preheat requirements for submerged arc welding with parallel or multiple electrodes, see AWS D1.1 [6.1], Sec. 4.10.6 or 4.11.6.

B. Zero° F (−18° C) does not mean the ambient environmental temperature but the temperature in the immediate vicinity of the weld. The ambient environmental temperature may be below 0° F, but a heated structure or shelter around the area being welded could maintain the temperature adjacent to the weldment at 0° F or higher.

[1] When the base metal temperature is below 32° F (0° C), the base metal shall be preheated to at least 70° F (21° C) and this minimum temperature maintained during welding.

[2] Only low hydrogen electrodes shall be used when welding A36 or A709 Grade 36 steel more than 1 in. thick for bridges.

[3] Welding shall not be done when the ambient temperature is lower than 0° F (−18° C). When the base metal is below the temperature listed for the welding process being used and the thickness of material being welded, it shall be preheated (except as otherwise provided) in such manner that the surfaces of the parts on which weld metal is being deposited are at or above the specified minimum temperature for a distance equal to the thickness of the part being welded, but not less than 3 in. (76 mm) in all directions from the point of welding. Preheat and interpass temperature must be sufficient to prevent crack formation. Temperature above the minimum shown may be required for highly restrained welds. For A514, A517, and A709 Grades 100 and 100W steel, the maximum preheat and interpass temperature shall not exceed 400° F (205° C) for thickness up to $1\frac{1}{2}$ in. (38 mm) inclusive, and 450° F (230° C) for greater thickness. Heat input when welding A514, A517, and A709 Grades 100 and 100W steel shall not exceed the steel producer's recommendations.

[4] In joints involving combinations of base metals, preheat shall be as specified for the higher strength steel being welded.

6.8 PREHEATS FOR QUENCHED-AND-TEMPERED STEELS

Since the low-alloy quenched-and-tempered steels are already in a heat-treated condition, any heating beyond a certain temperature will affect the properties developed in them by the manufacturing process. Some assemblies must be preheated before welding to prevent cracking on rapid cooling, but the preheat must be controlled so as not to destroy throughout the mass of material that high yield strength and toughness that characterize these steels and give them special applications. Yet, during welding the heat-affected zone will be heated far above the allowable preheat temperatures. This zone must then cool rapidly enough so as to reestablish the original properties and avoid a brittle structure. As a consequence, preheat temperatures and welding heat inputs must be closely controlled. Narrow limits are thus placed on the procedures.

Through research, welding procedures have been developed that assure high strength and good toughness, ductility, and impact properties in the welded joints. The recommended heat inputs and preheat temperatures are intended to allow sufficiently fast cooling rates to avoid brittle structure. In general, this means a cooling rate of 6°F (3°C) or more per second through the 900°F (480°C) temperature range. The chemistry of these steels is such that the carbon equivalent is low enough to minimize the preheat.

In welding quenched-and-tempered steels, the proper low-hydrogen welding process is selected. Next, the required preheat temperature is determined, based upon the chemistry of the weld metal and plate thickness. Knowing the preheat temperature and the plate thickness, the maximum permissible welding heat input per pass can be found. A welding procedure is then selected that will stay below this maximum value. Welding heat input may be reduced by decreasing the welding current or increasing the arc travel speed. Either change will decrease the amount of weld metal deposited per pass and will result in more passes being used for a given joint. For this reason, stringer beads are used frequently in welding quenched-and-tempered steels.

Steel manufacturers publish tables suggesting minimum preheat temperatures and maximum welding heat inputs for their quenched-and-tempered steels in various thicknesses. Sometimes, the procedures most desirable from the economic standpoint in welding these steels will lead to a total heat input—preheat plus welding heat—that exceeds the steel manufacturer's recommendations. In such cases, one might question whether the weldment needs maximum notch toughness as well as high yield strength. If it does, the procedures should be modified to reduce the total heat input—not the preheat. Reducing preheat would be too risky, since such action might lead to weld cracking, and maximum toughness in the heat-affected zone would then be of no value. If maximum notch toughness is not required, total heat input limits can be exceeded somewhat without materially reducing the yield strength but there is little information for fatigue and impact properties.

As mentioned in the introduction, the use of computers in solving complex problems with many interrelated variables has given what appears to be a breakthrough in the engineer's ability to assess cooling rates in an assembly being welded, no matter what the configuration. When cooling rates have been assessed by this mathematical method and correlated with known rates that prevent weld cracking, the long-time problem of devising optimum preheat treatments to give crack-resistant welds appears to have been mastered.

6.9 ARC WELDING CONSUMABLES

Since the designer specifies both the joint and the weld, knowledge about arc-welding consumables is important—especially the electrodes that supply molten filler metal. Arc-welding consumables are the materials consumed or used up during welding, such as electrodes, filler rods, fluxes, and externally applied shielding gases. All of the commonly used consumables are covered by AWS specifications.

Twenty specifications in the AWS A5.x series prescribe the requirements for welding electrodes, rods, and fluxes. Below are reviewed some of the important requirements of the A5.x series, with the intent of providing a guide to the selection of the proper specification. When detailed information is required, the actual AWS specification should be consulted.

As described in detail in Table 6.5, the classifications of mild and low-alloy steel electrodes for shielded metal-arc welding (SMAW) are based on an "E" prefix and a four- or five-digit number. The first two digits (or three, in a five-digit number) indicate the minimum required tensile strength in thousands of pounds per square inch. For example, 60 = 60,000 psi, 70 = 70,000 psi, and 100 = 100,000 psi. The next to the last digit indicates the welding position in which the electrode is capable of making satisfactory welds: 1 = all positions, flat, horizontal, vertical, and overhead; 2 = flat and horizontal fillet welding. The last two digits indicate the type of current to be used and the type of covering on the electrode.

Table 6.5 AWS A5.1 and A5.5 Designations for Manual Electrodes

a. The prefix "E" designates arc-welding electrode.

b. The first two digits of four-digit numbers and the first three digits of five-digit numbers indicate minimum tensile strength.

E60XX	60,000-psi minimum tensile strength
E70XX	70,000-psi minimum tensile strength
E110XX	110,000-psi minimum tensile strength

c. The next-to-last digit indicates position:

EXX1X	All positions
EXX2X	Flat position and horizontal fillets

d. The suffix (Example: EXXXX-A1) indicates the approximate alloy in the weld deposit:

-A1	0.5% Mo
-B1	0.5% Cr, 0.5% Mo
-B2	1.25% Cr, 0.5% Mo
-B3	2.25% Cr, 1% Mo
-B4	2% Cr, 0.5% Mo
-B5	0.5% Cr, 1% Mo
-C1	2.5% Ni
-C2	3.25% Ni
-C3	1% Ni, 0.35% Mo, 0.15% Cr
-D1 and D2	0.25–0.45% Mo, 1.75% Mn
-G	0.5% min. Ni, 0.3% min. Cr, 0.2% min. Mo, 0.1% min. V, 1% min. Mn (only one element required)

Table 6.6 AWS A5.1 Minimum Mechanical Property and Radiographic Requirements for Covered Arc Welding Electrode Metal

AWS Classification	Tensile Strength, min, psi	Yield Point, min, psi	Elongation in 2 in., min, percent	Radiographic Standard [a]	V-Notch Impact [d]
			E60 Series [b]		
E6010	62,000	50,000	22	Grade II	20 ft/lb at −20°F
E6011	62,000	50,000	22	Grade II	20 ft/lb at −20°F
E6012	67,000	55,000	17	Not required	Not required
E6013	67,000	55,000	17	Grade II	Not required
E6020	62,000	50,000	25	Grade I	Not required
E6027	62,000	50,000	25	Grade II	20 ft/lb at −20°F
			E70 Series [c]		
E7014	72,000	60,000	17	Grade II	Not required
E7015			22	Grade I	20 ft/lb at −20°F
E7016			22	Grade I	20 ft/lb at −20°F
E7018			22	Grade I	20 ft/lb at −20°F
E7024			17	Grade II	Not required
E7028			22	Grade II	20 ft/lb at 0°F

a See AWS A5.1-69, Fig. 3

b For each increase of one percentage point in elongation over the minimum, the yield point or tensile strength, or both, may decrease 1,000 psi to a minimum of 60,000 psi for the tensile strength and 48,000 psi for the yield point for all classifications of the 60-series except E6012 and E6013. For the E6012 and E6013 classifications the yield point and tensile strength may decrease to a minimum of 65,000 psi for the tensile strength and 53,000 psi for the yield point.

c For each increase of one percentage point in elongation over the minimum, the yield point or tensile strength, or both, may decrease 1,000 psi to a minimum of 70,000 psi for the tensile strength and 58,000 psi for the yield point.

d The extreme lowest value and the extreme highest value obtained in the test shall be disregarded. Two of the three remaining values shall be greater than the specified 20 ft/lb energy level; one of the three may be lower but shall not be less than 15 ft/lb. The computed average value of the three remaining values shall be equal to or greater than the 20 ft/lb energy level.

Table 6.7 Composition Requirements of Low-Alloy Weld Metal AWS A5.5

Electrode Classification	Composition (%)								
	C	Mn	P	S	Si	Ni	Cr	Mo	V
					Carbon-Molybdenum Steel				
E7010-A1 E7011-A1 E7015-A1 E7016-A1 E7018-A1 E7020-A1 E7027-A1	0.12	0.60 0.60 0.90 0.90 0.90 0.60 1.00	0.03	0.04	0.40 0.40 0.60 0.60 0.80 0.40 0.40	...	...	0.40 to 0.64	...
					Chromium-Molybdenum Steel				
E8016-B1 E8018-B1	0.12	0.90	0.03	0.04	0.60 0.80	...	0.40 to 0.65	0.40 to 0.65	...
E8015-B2L	0.05	0.90	0.03	0.04	1.00	...	1.00 to 1.50	0.40 to 0.65	...
E8016-B2 E8018-B2	0.12	0.90	0.03	0.04	0.60 0.80	...	1.00 to 1.50	0.40 to 0.65	...
E8018-B2L	0.05	0.90	0.03	0.04	0.80	...	1.00 to 1.50	0.40 to 0.65	...
E9015-B3L	0.05	0.90	0.30	0.04	1.00	...	2.00 to 2.50	0.90 to 1.20	...
E9015-B3 E9016-B3 E9018-B3	0.12	0.90	0.03	0.04	0.60 0.60 0.80	...	2.00 to 2.50	0.90 to 1.20	...
E9018-B3L E8015-B4L E8016-B5	0.05 0.05 0.07 to 0.15	0.90 0.90 0.40 to 0.70	0.03 0.03 0.03	0.04 0.04 0.04	0.80 1.00 0.30 to 0.60	... ...	2.00 to 2.50 1.75 to 2.25 0.40 to 0.60	0.90 to 1.20 0.40 to 0.65 1.00 to 1.25	... ... 0.05
					Nickel Steel				
E8016-C1 E8018-C1	0.12	1.20	0.03	0.04	0.60 0.80	2.00 to 2.75	...	...	...
E8016-C2 E8018-C2	0.12	1.20	0.03	0.04	0.60 0.80	3.00 to 3.75	...	...	...
E8016-C3 E8018-C3	0.12	0.40 to 1.25	0.030	0.030	0.80	0.80 to 1.10	0.15	0.35	0.05
					Manganese-Molybdenum Steel				
E9015-D1 E9018-D1	0.12	1.25 to 1.75	0.03	0.04	0.60 0.80	...	...	0.25 to 0.45	...
E10015-D2 E10016-D2 E10018-D2	0.15	1.65 to 2.00	0.03	0.04	0.60 0.60 0.80	...	...	0.25 to 0.45	...
					Other Low-Alloy Steel				
EXX10-G EXX11-G EXX13-G EXX15-G EXX16-G EXX18-G E7020-G	...	1.00 min	...	...	0.80 min	0.50 min	0.30 min	0.20 min	0.10 min
E9018-M E10018-M E11018-M E12018-M	0.10 0.10 0.10 0.10	0.60 to 1.25 0.75 to 1.70 1.30 to 1.80 1.30 to 2.25	0.030 0.030 0.030 0.030	0.030 0.030 0.030 0.030	0.80 0.60 0.60 0.60	1.40 to 1.80 1.40 to 2.10 1.25 to 2.50 1.75 to 2.25	0.15 0.35 0.40 0.30 to 1.50	0.35 0.25 to 0.50 0.30 to 0.55 0.30 to 0.55	0.05 0.05 0.05 0.05

Note: Single values shown are maximum percentages except where otherwise specified.

6.9.1 Mild Steel Covered Arc Welding Electrodes, AWS A5.1

The scope of this specification prescribes requirements for covered mild steel electrodes for shielded metal-arc welding of carbon and low-alloy steels. The minimum mechanical property requirements are shown in Table 6.6. Radiographic standard Grade I has less and smaller porosity than Grade II. The actual standards are not contained herein, and, if a comparison is required, the standard in AWS A5.1 should be used.

6.9.2 Low-Alloy Steel Covered Arc Welding Electrodes, AWS A5.5

This specification prescribes covered electrodes for shielded metal-arc welding of low-alloy steel. The same classification system is used as for mild steel covered electrodes, with an added suffix that indicates the approximate chemistry of the deposited weld metal (see Table 6.5). The chemistry composition of the deposited weld metal is shown in Table 6.7. The electrodes with the suffix G need have only one alloy above the minimum to qualify for the chemical requirements. Electrodes with the suffix M will meet or be similar to certain military requirements.

Table 6.8 shows the tensile strength, yield strength, and elongation requirements. The preheat, interpass temperature, and postheat treatments are not the same for all electrodes. For this reason, the complete AWS A5.5 specification should be consulted before conducting any tests. Table 6.9 shows the impact requirements. The impact test specimens receive the same heat treatment as the tension test specimens.

Table 6.8 AWS A5.5 Tensile Strength, Yield Strength, and Elongation Requirements for All-Weld-Metal Tension Test[a]

AWS Classification	Tensile Strength, min, psi	Yield Strength at 0.2 percent offset, psi	Elongation in 2 in., min, percent
E7010-X E7011-X E7015-X E7016-X E7018-X E7020-X E7027-X	70,000	57,000	22 22 25 25 25 25 25
E8010-X E8011-X E8013-X E8015-X E8016-X E8018-X	80,000	67,000	19 19 16 19 19 19
E8016-C3 E8018-C3	80,000	68,000 to 80,000	24
E9010-X E9011-X E9013-X E9015-X E9016-X E9018-X	90,000	77,000	17 17 14 17 17 17
E9018-M	90,000	78,000 to 90,000	24
E10010-X E10011-X E10013-X E10015-X E10016-X E10018-X	100,000	87,000	16 16 13 16 16 16
E10018-M	100,000	88,000 to 100,000	20
E11015-X E11016-X E11018-X	110,000	97,000	15
E11018-M	110,000	98,000 to 110,000	20
E12015-X E12016-X E12018-X	120,000	107,000	14
E12018-M	120,000	108,000 to 120,000	18

a For the E8016-C3, E8018-C3, E9018-M, E10018-M, E11018-M, and E12018-M electrode classifications the values shown are for specimens tested in the as-welded condition. Specimens tested for all other electrodes are in the stress-relieved condition.

Table 6.9 AWS A5.5 Impact Requirements

AWS Classification	Minimum V-Notch Impact Requirement [a]
E8016-C3 E8018-C3	20 ft/lb at −40°F[b]
E9015-D1 E9018-D1 E10015-D2 E10016-D2 E10018-D2	20 ft/lb at −60°F[c]
E9018-M E10018-M E11018-M E12018-M	20 ft/lb at −60°F[b]
E8016-C1 E8018-C1	20 ft/lb at −75 F[c]
E8016-C2 E8018-C2	20 ft/lb at −100°F[c]
All other classifications	Not required

a The extreme lowest value obtained together with the extreme highest value shall be disregarded for this test. Two of the three remaining values shall be greater than the specified 20 ft/lb energy level; one of the three may be lower but shall not be less than 15 ft/lb. The computed average value of the three remaining values shall be equal to or greater than the 20 ft/lb energy level.

b As-welded impact properties.

c Stress-relieved impact properties.

Table 6.10 AWS A5.17 Chemical Composition Requirements for Submerged-Arc Electrodes

AWS Classification	Chemical Composition, percent						
	Carbon	Manganese	Silicon	Sulfur	Phos-phorus	Cop-per [a]	Total other Ele-ments
Low Manganese Classes							
EL8	0.10	0.30 to 0.55	0.05				
EL8K	0.10	0.30 to 0.55	0.10 to 0.20				
EL12	0.07 to 0.15	0.35 to 0.60	0.05				
Medium Manganese Classes							
EM5K[b]	0.06	0.90 to 1.40	0.40 to 0.70	0.035	0.03	0.15	0.50
EM12	0.07 to 0.15	0.85 to 1.25	0.05				
EM12K	0.07 to 0.15	0.85 to 1.25	0.15 to 0.35				
EM13K	0.07 to 0.19	0.90 to 1.40	0.45 to 0.70				
EM15K	0.12 to 0.20	0.85 to 1.25	0.15 to 0.35				
High Manganese Class							
EH14	0.10 to 0.18	1.75 to 2.25	0.05				

[a] The copper limit is independent of any copper or other suitable coating which may be applied to the electrode.

[b] This electrode contains 0.05 to 0.15% titanium, 0.02 to 0.12% zirconium, and 0.05 to 0.15% aluminum, which is exclusive of the "Total Other Elements" requirement.

Note 1—Analysis shall be made for the elements for which specific values are shown in this table. If, however, the presence of other elements is indicated in the course of routine analysis, further analysis shall be made to determine that the total of these other elements is not present in excess of the limits specified for "Total Other Elements" in the last column of the table.

Note 2—Single values shown are maximum percentages.

6.9.3 Bare Mild Steel Electrodes and Fluxes for Submerged-Arc Welding, AWS A5.17

Most of the heavy welding in shop fabrication of structural elements is done with the submerged-arc process, using automatic equipment. Bare electrode wire and separate granular fluxes are required with this process. Since the electrode and flux are two separate consumable items, they are classified separately. Electrodes are classified on the basis of chemical composition, as shown in Table 6.10. In the classifying system, the letter E indicates an electrode, as in the other classifying systems, but here the similarity stops. The next letter L, M, or H, indicates low-, medium-, or high-manganese, respectively. The following number or numbers indicate the approximate carbon content in hundredths of 1%. If there is a suffix K, this indicates a silicon-killed steel.

Fluxes are classified as shown in Table 6.11 on the basis of the mechanical properties of the weld deposit made with a particular electrode. The classification designation given to a flux consists of a prefix F (indicating a flux) followed by a two-digit number representative of the tensile strength and impact requirements for test welds made in accordance with the specification. This is then followed by a set of letters and numbers corresponding to the classification of the electrode used with the flux.

Test welds are radiographed and must meet the Grade 1 standard of AWS A5.1 specification.

6.9.4 Mild Steel Electrodes for Flux-Cored Arc Welding, AWS A5.20

Flux-cored arc welding is used extensively in shop fabrication of structural elements. The semiautomatic process using the self-shielded electrode wire has become a favorite method for field and erection welding. AWS A5.20 prescribes requirements for mild steel composite electrodes for flux-cored arc welding of mild and low-alloy steels.

Electrodes are classified on the basis of single- or multiple-pass operation, chemical composition of the deposited weld metal, mechanical properties, and whether or not carbon dioxide is required

Table 6.11 AWS A5.17 Mechanical Property Requirements for Submerged-Arc Flux Classification

AWS Flux[a] Classification	Tensile Strength psi	Yield Strength at 0.2% Off-set, min, psi	Elongation in 2 in., min, %	Charpy V-Notch Impact Strength[b]
F60-XXXX F61-XXXX[c] F62-XXXX[c] F63-XXXX[c] F64-XXXX[c]	62,000 to 80,000	50,000	22[d]	Not required 20 ft/lb at 0°F 20 ft/lb at −20°F 20 ft/lb at −40°F 20 ft/lb at −60°F
F70-XXXX F71-XXXX[c] F72-XXXX[c] F73-XXXX[c] F74-XXXX[c]	72,000 to 95,000	60,000	22[e]	Not required 20 ft/lb at 0°F 20 ft/lb at −20°F 20 ft/lb at −40°F 20 ft/lb at −60°F

a The letters "XXXX" as used in this table stand for the electrode designations EL8, EL8K, etc. (see Table 4-11).

b The extreme lowest value obtained, together with the extreme highest value obtained, shall be disregarded for this test. Two of the three remaining values shall be greater than the specified 20 ft/lb energy level; one of the three may be lower but shall not be less than 15 ft/lb. The computed average value of the three values shall be equal to or greater than the 20 ft/lb energy level.

c Note that if a specific flux-electrode combination meets the requirements of a given F6X-xxxx classification, this classification also meets the requirements of all lower numbered classifications in the F6X-xxxx series. For instance, a flux-electrode combination meeting the requirements of the F63-xxxx classification, also meets the requirements of the F62-xxxx, F61-xxxx, and F60-xxxx classifications. This applies to the F7X-xxxx series also.

d For each increase of one percentage point in elongation over the minimum, the yield strength or tensile strength, or both, may decrease 1000 psi to a minimum of 60,000 psi for the tensile strength and 48,000 psi for the yield strength.

e For each increase of one percentage point in elongation over the minimum, the yield strength or tensile strength, or both, may decrease 1000 psi to a minimum of 70,000 psi for the tensile strength and 58,000 psi for the yield strength.

Table 6.12 AWS A5.20 Mechanical Property Requirements for Flux-Cored Arc Welding Weld Metal*

AWS Classification	Shielding Gas [b]	Current and Polarity [c]	Tensile Strength min.[f], psi	Yield Strength at 0.2% Offset, min.[f], psi	Elongation in 2 inches, min.[f], psi
E60T-7	None	DC, straight polarity	67,000	55,000	22
E60T-8	None	DC reverse polarity	62,000	50,000	22
E70T-1	CO_2		72,000	60,000	22
E70T-2			72,000	Not required	
E70T-3	None		72,000	Not required	
E70T-4	None		72,000	60,000	22
E70T-5[g]	CO_2 None		72,000	60,000	22
E70T-6	None		72,000	60,000	22
E70T-G	not spec.	not spec.	72,000 [d]	Not required	
			72,000 [e]	60,000 [e]	22 [e]

a As-welded mechanical properties.
b Shielding gases are designated as follows:
CO_2 = carbon dioxide
None = no separate shielding gas
c Reverse polarity means electrode is positive; straight polarity means electrode is negative.
d Requirement for single-pass electrodes.
e Requirement for multiple-pass electrodes.
f For each increase of one percentage point in elongation over the minimum, the minimum required yield strength or the tensile strength, or both, may decrease 1000 psi, for a maximum reduction of 2000 psi in either the required minimum yield strength or the tensile strength, or both.
g Where CO_2 and None are indicated as the shielding gases for a given classification, chemical analysis pads and test assemblies shall be prepared using both CO_2 and no separate shielding gas.

as a separate shielding gas. Tables 6.12 and 6.13 show the minimum mechanical property requirements.

Gas-shielded flux-cored electrodes are available for welding the low-alloy high-tensile strength steels. Self-shielded flux-cored electrodes are available for all-position welding, as in building construction. Fabricators that use or anticipate using the flux-cored arc welding processes should keep in touch with the electrode manufacturers for new or improved electrodes not included in the present specifications.

Table 6.13 AWS A5.20 Impact Property Requirements for Flux-Cored Arc Welding Weld Metal

AWS Classification	Minimum V-Notch Impact Requirement[a]
E70T-5	20 ft-lb at −20°F
E70T-8 E70T-1 E70T-6	20 ft-lb at 0°F
E70T-7 E70T-2 E70T-3 E70T-4 E70T-G	Not required

[a] The extreme lowest value obtained, together with the extreme highest value obtained, shall be disregarded for this test. Two of the three remaining values shall be greater than the specified 20 ft-lb energy level; one of the three may be lower but shall not be less than 15 ft-lb. The computer average value of the three values shall be equal to or greater than the 20 ft-lb energy level.

Table 6.14 AWS A5.18 Mechanical Property Requirements for Gas Metal-Arc Welding Weld Metal*

AWS Classification	Shielding Gas[b]	Current and Polarity[c]	Tensile Strength min., psi	Yield Strength at 0.2% Offset, min.	Elongation in 2 inches, min. %
GROUP A – MILD STEEL ELECTRODES					
E70S-1	AO	DC reverse polarity	72,000[e,f]	60,000[e,f]	22[e,f]
E70S-2 E70S-3	AO & CO_2[d]				
E70S-4 E70S-5 E70S-6	CO_2				
E70S-G	not spec.	not spec.			
GROUP B – LOW-ALLOY STEEL ELECTRODES					
E70S-1B E70S-GB	CO_2 not spec.	DC, reverse polarity not spec.	72,000[e,f] 72,000[e,f]	60,000[e,f] 60,000[e,f]	17[e,f] 22[e,f]
GROUP C – EMISSIVE ELECTRODE					
E70U-1	AO & A[d]	DC, straight polarity	72,000[e]	60,000[e]	22[e]

a As-welded mechanical properties
b Shielding gases are designated as follows:
AO = argon, plus 1 to 5 percent oxygen
CO_2 = carbon dioxide
A = argon
c Reverse polarity means electrode is positive; straight polarity means electrode is negative.
d Where two gases are listed as interchangeable (that is, AO and CO_2 and AO & A) for classification of a specific electrode, the classification tests may be conducted using either gas.
e Mechanical properties as determined from an all-weld-metal tension-test specimen.
f For each increase of one percentage point in elongation over the minimum, the yield strength or tensile strength, or both, may decrease 1,000 psi to a minimum of 70,000 psi for the tensile strength and 58,000 psi for the yield strength.

Table 6.15 AWS A5.18 Impact Property Requirements for Gas Metal-Arc Welding Weld Metal

AWS Classification	Minimum V-Notch Impact Requirement*
E70S-2 E70S-6 E70S-1B E70U-1	20 ft-lb at −20°F
E70S-3	20 ft-lb at 0°F
E70S-1, E70S-4, E70S-5, E70S-G, E70S-GB	Not required

* The extreme lowest value obtained, together with the extreme highest value obtained, shall be disregarded for this test. Two of the three remaining values shall be greater than the specified 20 ft-lb energy level, one of the three may be lower but shall not be less than 15 ft-lb. The computed average value of the three values shall be equal to or greater than the 20 ft-lb energy level.

Table 6.16 AWS A5.18 Chemical Composition Requirements for Gas Metal-Arc Welding Electrode

AWS Classification	Chemical Composition, percent											
	Carbon	Manganese	Silicon	Phosphorus	Sulfur	Nickel [a]	Chromium [a]	Molybdenum [a]	Vanadium [a]	Titanium	Zirconium	Aluminum
GROUP A – MILD STEEL ELECTRODES												
E70S-1	0.07 to 0.19		0.30 to 0.50									
E70S-2	0.06		0.40 to 0.70							0.05 to 0.15	0.02 to 0.12	0.05 to 0.15
E70S-3	0.06 to 0.15	0.90 to 1.40	0.45 to 0.70	0.025	0.035							
E70S-4	0.07 to 0.15		0.65 to 0.85									
E70S-5	0.07 to 0.19		0.30 to 0.60									0.50 to 0.90
E70S-6	0.07 to 0.15	1.40 to 1.85	0.80 to 1.15									
E70S-G	no chemical requirements [b]											
GROUP B – LOW-ALLOY STEEL ELECTRODES												
E70S-1B	0.07 to 0.12	1.60 to 2.10	0.50 to 0.80	0.025	0.035	0.15		0.40 to 0.60				
E70S-GB	no chemical requirements											
GROUP C – EMISSIVE ELECTRODE												
E70U-1	0.07 to 0.15	0.80 to 1.40	0.15 to 0.35	0.025	0.035							

Note—Single values shown are maximums.

[a] For Groups A and C these elements may be present but are not intentionally added.

[b] For this classification there are no chemical requirements for the elements listed with the exception that there shall be no intentional addition of Ni, Cr, Mo or V.

6.9.5 Mild Steel Electrodes for Gas Metal-Arc Welding, AWS A5.18

This specification prescribes requirements for mild steel solid electrodes for gas metal-arc welding of mild and low-alloy steel. The electrodes are classified on the basis of their chemical composition and the as-welded mechanical properties of the deposited weld metal (see Tables 6.14 and 6.15). For the chemical composition requirements of the deposited weld metal, see Table 6.16.

Table 6.16 includes a Group B classification, entitled "Low-Alloy Steel Electrodes." The alloy additions here do not meet the accepted definitions of mild steel. The basis for including this classification in a mild steel specification is that the alloy additions are for deoxidation and usability improvement and not for the purpose of upgrading the mechanical properties.

6.10 ARC WELDING PROCESSES

The process used in making a weld has an effect on suitability for the joint and a still larger effect on welding economics. Welding processes, just as the electrode filler metal or the procedure, are qualified for structural work. The engineer or designer needs at least a fundamental understanding of the various processes used in structural welding in order to specify professionally to get the quality welds needed—and especially to minimize shop and field welding costs. As an example of the latter consideration, in the joining of a beam to a column in erection welding, the use of the semiautomatic self-shielded flux-cored electrode process can often reduce the cost of such field welding by as much as one-fourth the cost with manual electrodes.

6.10.1 Shielded Metal-Arc (Manual) Process

The shielded metal-arc process—commonly called "stick-electrode" welding or "manual" welding—is the most widely used of the various arc-welding processes. It is characterized by application versatility and flexibility and relative simplicity in the equipment.

With this process (Fig. 6.31), an electric arc is struck between the electrically grounded work and a 9–18 in. length of covered metal rod—the electrode. The electrode is clamped in an electrode

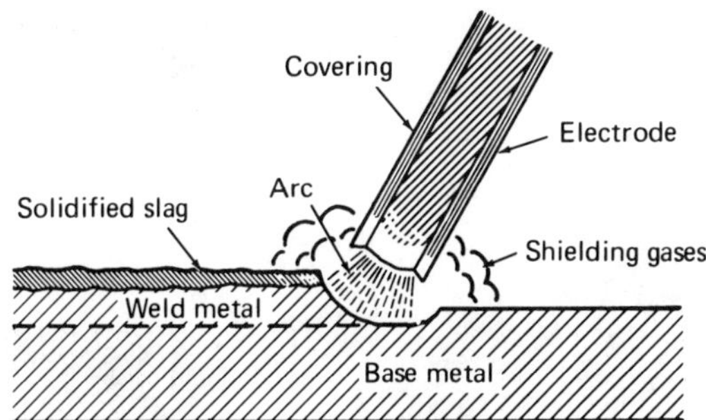

Fig. 6.31 Schematic representation of shielded metal-arc welding.

holder, which is joined by a cable to the power source. The weldor grips the insulated handle of the electrode holder and maneuvers the tip of the electrode with respect to the weld joint. When the tip of electrode is touched against the work, and then withdrawn to establish the arc, the welding circuit is completed. The heat of the arc melts the base metal in the immediate area, the electrode's metal core, and any metal particles that may be in the electrode's covering. It also melts, vaporizes, or breaks down chemically nonmetallic substances incorporated in the covering for arc shielding, metal protection, or metal conditioning purposes. The mixing of molten base metal and filler metal from the electrode provides the coalescence required to effect joining.

Gases generated by the decomposition and vaporization of materials in the electrode covering—including vaporized slag—provide a dense shield around the arc stream and over the molten puddle. Molten and solidified slag above the newly formed weld metal protects it from the atmosphere while it is hot enough to be chemically reactive with oxygen and nitrogen.

As welding progresses, the covered rod becomes shorter and shorter. Finally, the welding must be stopped to remove the stub and replace it with a new electrode. This periodic changing of electrodes decreases the operating factor, or the percent of the weldor's time spent in the actual operation of laying weld beads.

There is a limitation placed on the current that can be used. High amperages, such as those used with semiautomatic guns or automatic welding heads, are impractical because of the long (and varying) length of electrode between the arc and the point of electrical contact in the jaws of the electrode holder. The welding current is limited by the resistance heating of the electrode. The electrode temperature must not exceed the "breakdown" temperature of the covering.

A solid wire core is the main source of filler metal in electrodes for the shielded metal-arc process. However, the so-called iron powder electrodes also supply filler metal from iron powder contained in the electrode covering or within a tubular core wire. Iron powder in the covering increases the efficiency of use of the arc heat and thus the deposition rate. With thickly covered iron powder electrodes, it is possible to drag the electrode over the joint without the electrode freezing to the work or shorting out. Even though the heavy covering makes contact with the work the electrical path through the contained powder particles is not adequate in conductivity to short the arc, and any resistance heating that occurs supplements the heat of the arc in melting the electrode. Because heavily covered iron powder electrodes can be dragged along the joint, less skill is required in their use.

Now that semiautomatic self-shielded flux-cored arc welding has been developed to a similar (or even superior) degree of versatility and flexibility in welding, there is less justification for adhering to stick-electrode welding in steel fabrication and erection wherever substantial amounts of weld metal must be placed. In fact, the replacement of shielded metal-arc welding with semiautomatic processes has been a primary means by which steel fabricators and erectors have met the cost squeeze in their welding operations.

6.10.2 Submerged-Arc Process

Submerged-arc welding (Fig. 6.32) differs from other arc-welding processes in that a blanket of fusible, granular material—commonly called flux—is used for shielding the arc and the molten metal. The arc is struck between the workpiece and a bare wire electrode, the tip of which is submerged in the flux. Since the arc is completely covered by the flux, it is not visible and the weld is run without the flash, spatter, and sparks that characterize the open-arc processes. The nature of the flux is such that no smoke or visible fumes are developed.

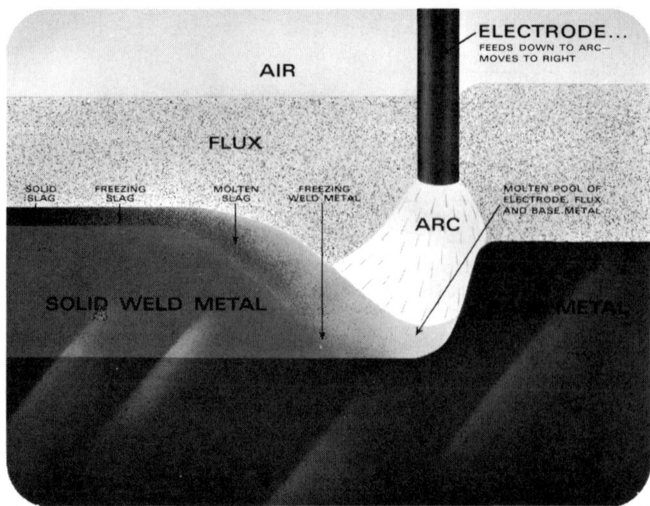

Fig. 6.32 The mechanics of the submerged-arc process. Photo courtesy Lincoln Electric Co., Cleveland, Ohio.

The process is either semiautomatic or full-automatic, with electrode fed mechanically to the welding gun, head, or heads. In semiautomatic welding, the weldor moves the gun, usually equipped with a flux-feeding device, along the joint. Flux feed may be by gravity flow through a nozzle concentric with the electrode from a small hopper atop the gun, or it may be through a concentric nozzle tube connected to an air-pressurized flux tank. Flux may also be applied in advance of the welding operation or ahead of the arc from a hopper run along the joint. In fully automatic submerged-arc welding, flux is fed continuously to the joint ahead of or concentric to the arc, and full-automatic installations are commonly equipped with vacuum systems to pick up the unfused flux left by the welding head or heads for cleaning and reuse.

During welding, the heat of the arc melts some of the flux along with the tip of the electrode as illustrated in Fig. 6.32. The tip of the electrode and the welding zone are always surrounded and shielded by molten flux, surmounted by a layer of unfused flux. The flux protects the weld metal from contamination and concentrates the heat into the joint. The electrode is held a short distance above the workpiece, and the arc is between the electrode and the workpiece. As the electrode progresses along the joint, the lighter molten flux rises above the molten metal in the form of a slag. The weld metal, having a higher melting (freezing) point, solidifies while the slag above it is still molten. The slag then freezes over the newly solidified weld metal, continuing to protect the metal from contamination while it is very hot and reactive with atmospheric oxygen and nitrogen. Upon cooling and removal of any unmelted flux for reuse, the slag is readily peeled from the weld.

High currents can be used in submerged-arc welding and extremely high heat is developed. Because the current is applied to the electrode a short distance above its tip, relatively high amperages can be used on small-diameter electrodes. This results in extremely high-current densities on relatively small cross-sections of electrode. Currents as high as 600 amp can be carried on electrodes as small as 5/64 in., giving a density in the order of 125,000 amp/sq in.—six to ten times that carried on stick electrodes.

Because of the high current density, the melt-off rate is much higher for a given electrode diameter than with stick-electrode welding. The melt-off rate is affected by the electrode material, the flux, type of current, polarity, and length of wire beyond the point of electrical contact in the gun or head.

The insulating blanket of flux above the arc prevents rapid escape of heat and concentrates it in the welding zone. Not only are the electrode and base metal melted rapidly, but the fusion is deep into the base metal. The deep penetration allows the use of small welding grooves, thus minimizing the amount of filler metal per unit length of joint and permitting fast welding speeds. Fast welding, in turn, minimizes the total heat input into the assembly and, thus, tends to prevent problems of heat distortion. Even relatively thick joints can be welded in one pass by submerged-arc welding.

Welds made under the protective layer of flux have good ductility and impact resistance and uniformity in bead appearance. Mechanical properties at least equal to those of the base metal are consistently obtained. In single-pass welds, the fused base material is large compared to the amount of filler metal used. Thus, in such welds the base metal may greatly influence the chemical and mechanical properties of the weld. For this reason, it is sometimes unnecessary to use electrodes of the same composition as the base metal for welding many of the low-alloy steels. The chemical composition and properties of multipass welds are less affected by the base metal and depend to a greater extent on the composition of the electrode, the activity of the flux, and the welding conditions.

Through regulation of current, voltage, and travel speed, the operator can exercise close control over penetration to provide any depth ranging from deep and narrow with high-crown reinforcement, to wide, nearly flat beads with shallow penetration. Beads with deep penetration may contain on the order of 70% melted base metal, while shallow beads may contain as little as 10% base metal. In some instances, the deep penetration properties of submerged-arc can be used to eliminate or reduce the expense of edge preparation.

Multiple electrodes may be used, two side by side or two or more in tandem, to cover a large surface area or to increase welding speed. If shallow penetration is desired with multiple electrodes, one electrode can be grounded through the other (instead of through the workpiece) so that the arc does not penetrate deeply.

Deposition rates are high—up to 10 times those of stick-electrode welding. Figure 6.33 shows approximate deposition rates for various submerged-arc arrangements, with comparable deposition rates for manual welding with covered electrodes.

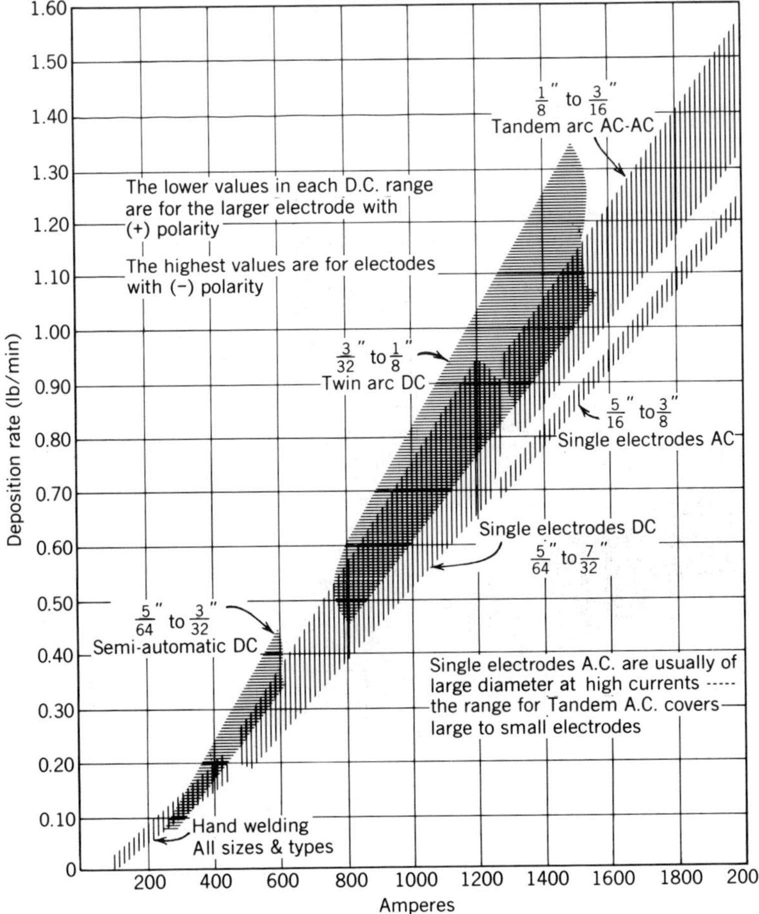

Fig. 6.33 Approximate deposition rates of various submerged-arc arrangements, compared with the deposition rates of stick-electrode welding.

6.10.3 Self-Shielded Flux-Cored Process

The self-shielded flux-cored arc welding process (Fig. 6.34) is an outgrowth of shielded metal-arc welding. The versatility and maneuverability of stick electrodes in manual welding stimulated efforts to mechanize the shielded metal-arc process. The thought was that if some way could be found for putting an electrode with self-shielding characteristics in coil form and feeding it mechanically to the arc, welding time lost in changing electrodes and the material lost as electrode stubs would be eliminated. The results of these efforts were the development of the semiautomatic and full-automatic processes for welding with continuous flux-cored tubular electrode "wires". The electrode may be viewed as an "inside-out" construction of the stick electrode used in shielded metal-arc welding. Putting the shield-generating materials inside the electrode allows the coiling of long, continuous lengths of electrode and gives an outside conductive sheath for carrying the welding current from a point close to the arc.

In essence, semiautomatic welding with flux-cored electrodes is manual shielded metal-arc welding with an electrode many feet long instead of just a few inches long. By the press of the trigger completing the welding circuit, the operator activates the mechanism that feeds the electrode to the arc. A gun is used instead of an electrode holder, but it is similarly light in weight and easy to maneuver. The only other major difference is that the weld metal of the electrode surrounds the shielding and fluxing chemicals, rather than being surrounded by them.

One of the limitations of the stick electrode is the long and varying length of electrode between the point of electrical contact in the electrode holder and the electrode tip. This limits the current that can be used because of electrical resistance heating. High currents, capable of giving high deposition rates, in passing though an electrode length more than a few inches long would develop enough resistance heating to overheat and damage the covering. But when electrical contact can be made close to the arc, as with the inside-out construction of tubular electrodes, relatively high currents can be used even with small-diameter electrode wires.

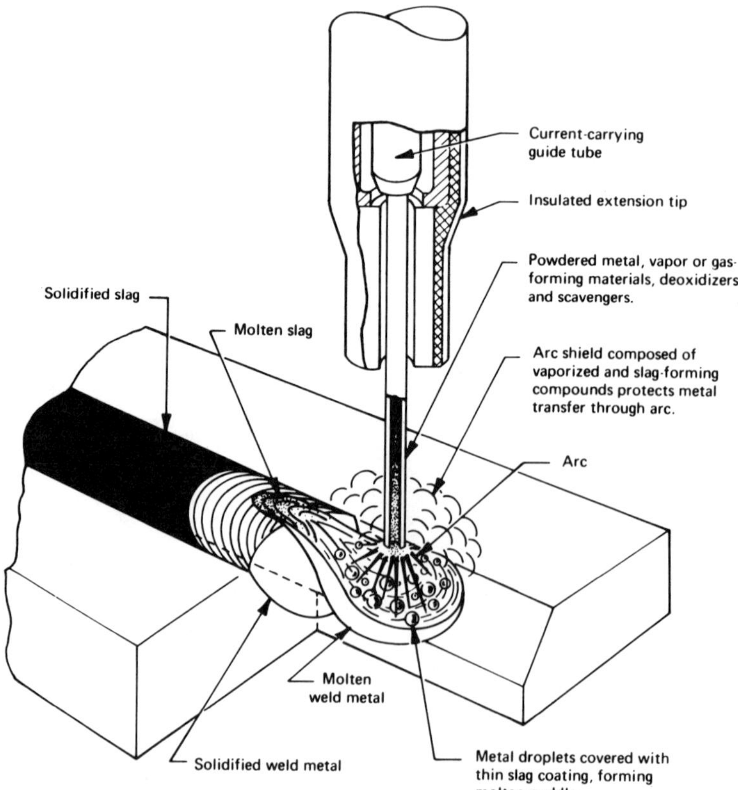

Fig. 6.34 Principles of the self-shielded flux-cored arc welding process. Courtesy of Lincoln Electric Co., Cleveland, Ohio.

The inside-out construction of the self-shielded electrode brought to manually manipulated welding the possibility of using higher-amperage currents than feasible with stick-electrode welding. As a result, much higher deposition rates are possible with the hand-held semiautomatic gun than with the hand-held stick-electrode holder.

Higher deposition rates, plus automatic electrode feed and elimination of lost time for changing electrodes, have resulted in substantial production economies wherever the semiautomatic process has been used to replace stick-electrode welding. Decreases in welding costs as great as 50% have been common, and in some production welding deposition rates have been increased as much as 400%.

The intent behind the development of self-shielded flux-cored electrode welding was to mechanize and increase the efficiency of manual welding. The semiautomatic use of the process does just that—it serves as a direct replacement for stick-electrode welding. The full-automatic use of the process, on the other hand, competes with other fully automatic processes and is used in production where it gives the desired performance characteristics and weld properties, while eliminating problems associated with flux or gas handling. Although the full-automatic process is important to a few industries, the semiautomatic version has the wider application possibilities. In fact, semiautomatic self-shielded flux-cored welding has potential for substantially reducing welding costs when working with steel wherever stick-electrode welding is used to deposit other than minor volumes of weld metal.

Since the semiautomatic process can be used any place stick electrodes can be used, it makes possible one-process welding in the erection of structural steel in building framing. As proved on major high-rise projects, this factor is possibly as important as the higher deposition rates of semiautomatic welding in reducing erection costs. One process substantially reduces the amount of equipment needed on the job and allows every welding operator to be qualified for every joint. This, in turn, permits the more systematic deployment of workers, equipment, and materials and reduces delays and minimizes equipment handling. One-process welding—from tack welding to column splice and beam-to-column welding—gives erectors the opportunity to take a "systems approach" to erection logistics, with the result that cost savings above those attributable to the speed of the semiautomatic process are realizable.

6.10.4 Gas-Shielded Arc Welding Processes

As noted in the preceding, the shielding metal-arc process (stick-electrode) and self-shielded flux-cored electrode process depend in part on gases generated by the heat of the arc to provide arc and puddle shielding. In contrast, the gas-shielded arc-welding processes use either bare or flux-cored filler metal and gas from an external source for shielding. The gas is impinged upon the work from a nozzle that surrounds the electrode. It may be an inert gas—such as argon or helium—or carbon dioxide (CO_2), a much cheaper gas that is suitable for use in the welding of steels. Mixtures of the inert gases, oxygen, and carbon dioxide also are used to produce special arc characteristics.

There are three basic gas-shielded arc-welding processes that have broad application in industry. They are the gas-shielded flux-cored process, the gas tungsten-arc (TIG) process, and the gas metal-arc (MIG) process.

The *gas-shielded flux-cored process* may be looked upon as a hybrid between self-shielded flux-cored arc welding and gas metal-arc welding. Tubular electrode wire is used (Fig. 6.35), as in the self-shielded process, but the ingredients in its core are for fluxing, deoxidizing, scavenging, and sometimes alloying additions, rather than for these functions plus the generation of protective vapors. In this respect, the process has similarities to the self-shielded flux-cored electrode process, and the tubular electrodes used are classified by the AWS along with electrodes used in the self-shielded process. On the other hand, the process is similar to gas metal-arc welding in that a gas is externally applied to act as arc shield.

The guns and welding heads for semiautomatic and full-automatic welding with the gas-shielded process are of necessity more complex than those used in self-shielded flux-cored welding. Passages must be included for the flow of gases. If the gun is water cooled, additional passages are required for this purpose. This complexity makes the equipment more difficult to handle than self-shielded equipment in structural erection work, and in high-rise framing there can also be a problem with wind blow, requiring shelters to prevent loss of shielding gas.

The gas-shielded flux-cored process is used for welding mild and low-alloy steels. It gives high deposition rates, high deposition efficiencies, and high operating factors. Radiographic-quality welds are easily produced, and the weld metal with mild and low-alloy steels has good ductility and toughness. The process is adaptable to a wide variety of joints and gives the capability for all-position welding.

Gas metal-arc welding as shown in Fig. 6.36, popularly known as MIG welding, uses a continuous solid electrode for filler metal and an externally supplied gas or gas mixture for shielding. The shielding gas—helium, argon, carbon dioxide, or mixtures thereof—protects the molten metal from

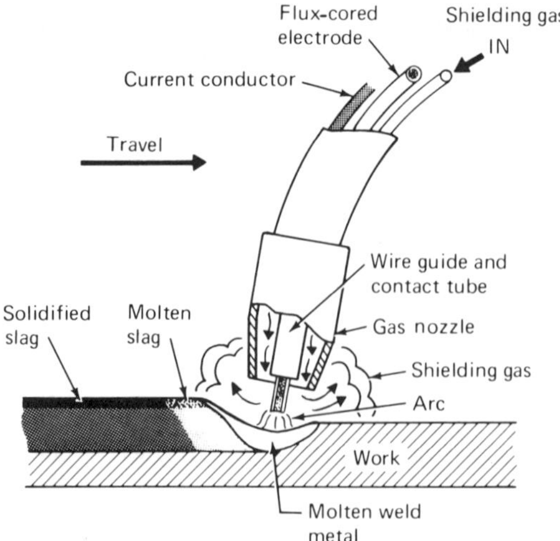

Fig. 6.35 Principles of the gas-shielded flux-cored process. Gas from an external source is used for the shielding; the core ingredients are for fluxing and metal-conditioning purposes.

reacting with constituents of the atmosphere. Although the gas shield is effective in shielding the molten metal from the air, deoxidizers are usually added as alloys in the electrode. Sometimes light coatings are applied to the electrode for arc stabilizing or other purposes. Lubricating films may also be applied to increase the electrode-feeding efficiency in semiautomatic welding equipment. Reactive gases may be included in the gas mixture for arc-conditioning functions. Figure 6.36 illustrates the method by which shielding gas and continuous electrode are supplied to the welding arc.

Metal transfer with the MIG process is by one of two methods: "spray-arc" or short circuiting. With spray-arc, drops of molten metal detach from the electrode and move through the arc column to the work. With the short-circuiting technique—often referred to as short-arc welding—metal is transferred to the work when the molten tip of the electrode contacts the molten puddle.

MIG welding and TIG welding—the other gas-shielded arc welding process—are widely used in industry, but are of lesser significance in structural steel welding than the processes discussed.

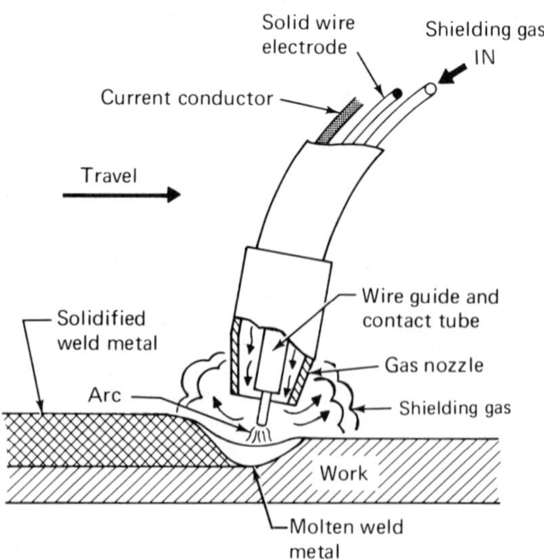

Fig. 6.36 Principle of the gas metal-arc process. Continuous solid-wire electrode is fed to the gas-shielded arc.

Table 6.17 Preferred Analyses for Steels To Be Arc Welded

Element	Composition (%)	
	Preferred	High*
Carbon	0.06 to 0.25	0.35
Manganese	0.35 to 0.80	1.40
Silicon	0.10 or less	0.30
Sulfur	0.035 or less	0.05
Phosphorus	0.030 or less	0.04

Additional care is required in welding of steels containing these amounts of the elements listed.

6.11 WELDABILITY OF STRUCTURAL STEELS

In a welded structure, consideration must be given to the weldability of the steels selected. Carbon and low-alloy steels are the *weldable* steels used mostly in building construction.

A metal is considered to have good weldability if it can be welded without excessive difficulty or the need for special and costly procedures and the weld joints are equal in all necessary respects to a similar piece of solid metal. Weldability varies with the grade, chemistry, and mechanical properties of the steel, and, when weld joining is to be a major factor in the attachment of steel parts, weldability should be given proper attention in specifying and ordering materials for the job.

Several methods are used to identify and specify steels. These are based on chemistry, on mechanical properties, on an ability to meet a standard specification or industry-accepted practice, or on an ability to be fabricated into a certain type of product.

6.11.1 Specifying by Chemistry

A desired composition can be produced in one of three ways: to a maximum limit, to a minimum limit, or to an acceptable range.

For economical, high-speed welding of carbon-steel plate, the composition of the steel should be within the "preferred-analysis" ranges indicated in Table 6.17. If one or more elements varies from the ranges shown, cost-increasing methods are usually required to produce good welding results. Thus, steels within these ranges should be used whenever extensive welding is to be done unless their properties do not meet service requirements. Published welding procedures generally apply to normal welding conditions and to the more common preferred-analysis mild steels. Low-hydrogen electrodes and processes will generally tolerate a wider range of the elements than shown in Table 6.17.

If the chemical specification of a steel falls outside the preferred-analysis range, it is usually not necessary to use special welding procedures based on the extremes allowed by the specification. The chemistry of a specific heat, under average mill-production conditions, may be considerably below the top limits indicated in the specification. Thus, for maximum economy, welding procedures for any type of steel should be based on actual rather than allowed chemistry values. A mill test report* can be obtained which gives the analysis of a heat of steel. From this information, a welding procedure can be established that ensures production of quality welds at lowest possible cost.

Some of the commonly specified elements and their effects on weldability and other characteristics of steels follow:

Carbon is the principal hardening element in steel. As carbon content increases, hardenability and tensile strength increase, and ductility and weldability decrease. In steels with a carbon content over 0.25%, rapid cooling from the welding temperature may produce a hard, brittle zone adjacent to the weld. Also, if considerable carbon is picked up in the weld puddle through admixture from the metal being welded, the weld deposit itself may be hard. Addition of small amounts of elements

---

* A mill test report is usually based on a ladle analysis and is an average for an entire heat. Most low-carbon steels are rimmed steels, widely used because of their excellent forming and deep-drawing properties but not used in structures. The analysis of a rimmed steel varies from the first ingot to the last ingot of a single heat and also from the top to the bottom of a single ingot. Thus, a mill test report is an average and should be interpreted as such.

Fig. 6.37 Sulfur segregations. Dark lines in etched section indicate areas of high sulfur concentration. Photo courtesy of Lincoln Electric Co., Cleveland, Ohio.

other than carbon can produce high tensile strengths without a detrimental effect on weldability. In general, carbon content should be low for best weldability.

Manganese increases hardenability and strength, but to a lesser extent than carbon. Properties of steels containing manganese depend principally on carbon content. Manganese content of less than 0.30% may promote internal porosity and cracking in the weld bead.

For good weldability, the ratio of manganese to sulfur should be at least 10 to 1. If a steel has a low manganese content in combination with a low carbon content, it may not have been properly deoxidized. In steel, manganese combines with sulfur to form MnS, which is not harmful. However, a steel with a low Mn/S ratio may contain sulfur in the form of FeS, which can cause cracking (a "hot-short" condition) in the weld.

In general, manganese increases the rate of carbon penetration during carburizing and is beneficial to the surface finish of carbon steels.

Sulfur increases the machinability of steels, but reduces transverse ductility, impact toughness, and weldability. Sulfur in any amount promotes hot shortness in welding, and the tendency increases with increased sulfur. It can be tolerated up to about 0.035% (with sufficient Mn); over 0.050% it can cause serious problems. Sulfur is also detrimental to surface quality in low-carbon and low-manganese steels.

A common cause of poor welding quality that is not apparent from analyses made in the usual way is segregated layers of sulfur in the form of iron sulfide. These layers, which cause cracks or other defects at the fusion line of an arc-welded joint, can be detected by examination of a deep-etched cross-section as illustrated in Fig. 6.37.

Silicon is a deoxidizer that is added during the making of steel to improve soundness. Silicon increases strength and hardness, but to a lesser extent than manganese. It is detrimental to surface quality, especially in the low-carbon, resulfurized grades. If carbon content is fairly high, silicon aggravates cracking tendencies. Amounts up to 0.30% are not as serious as high sulfur or phosphorus content.

Phosphorus, in large amounts, increases strength and hardness, but reduces ductility and impact strength, particularly in the higher-carbon grades. In low-carbon steels, phosphorus improves machinability and resistance to atmospheric corrosion. As far as welding is concerned, phosphorus is an impurity, and should be kept as low as possible. Over 0.04% makes welds brittle and increases the tendency to crack. Phosphorus also lowers the surface tension of the molten weld metal, making it difficult to control.

Copper improves atmospheric corrosion resistance when present in excess of 0.15%. (A minimum of 0.20% is usually specified for this purpose.) Most carbon steels contain some copper as a "tramp element", up to about 0.15%. Copper content up to about 1.50% has little or no effect on the weldability of a steel. Copper content over 0.50% may reduce mechanical properties, however, if the steel is heat treated. Copper content is detrimental to surface quality, particularly in high-sulfur grades.

6.11.2 Specifying by Mechanical Properties

The producer of steels specified by mechanical properties is free to alter the chemistry of the steel (within limits) to obtain the required properties. Mechanical tests are usually specified under one of these conditions:

1. Mechanical test requirements only, with no limits on chemistry.
2. Mechanical test requirements, with limits on one or more elements.

Generally, these tests have been set up according to practices approved by ASTM (American Society for Testing and Materials) or to the requirement of other authorized code-writing organizations, such as the ASME (American Society of Mechanical Engineers) or the API (American Petroleum Institute).

The most common tests are bend tests, hardness tests, and a series of tensile tests that evaluate modulus of elasticity, yield strength, and tensile strength. In some cases, impact tests are required.

6.11.3 Metallurgy of a Weld Bead

The heat of welding brings about certain changes, both in the structure of the steel being welded and in the weld metal. Some of these changes occur during welding; others, after the metal has cooled.

During welding, the temperature of the molten weld metal reaches 3000°F (1650°C) or higher. A short distance from the weld, the temperature of the plate may be only about 600°F (315°C). When the steel reaches or exceeds certain critical temperatures between these values, changes occur which affect grain structure, hardness, and strength properties. These changes and the temperatures at which they occur are illustrated by Fig. 6.38, a schematic diagram of a section through a weld bead.

The schematic diagram represents a strip cut vertically through the weld shown. Significance of the four numbered zones are as follows: (1) Metal that has been melted and resolidified; the grain

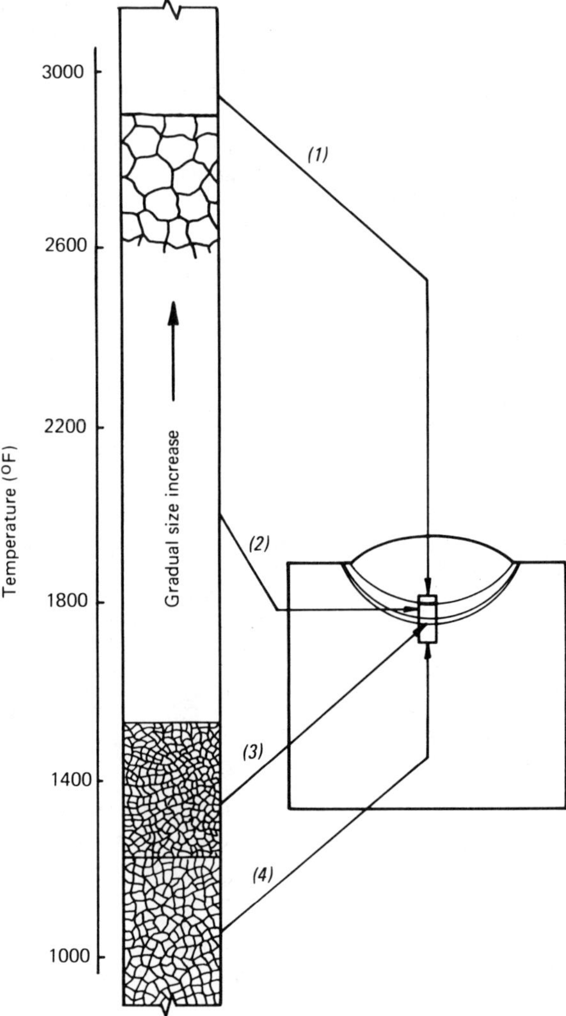

Fig. 6.38 Effect of welding heat on hardness and microstructure of an arc-welded 0.25% carbon steel plate.

structure is coarse. (2) Metal that has been heated above the upper critical temperature (1525°F)(830°C) for 0.25% carbon steel but has not been remelted. This area of large grain growth is where underbead cracking can occur. (3) Metal that has been heated slightly above the critical temperature of 1333°F (723°C) but not to the upper critical temperature. Grain refinement has taken place. (4) Metal that has been heated and cooled, but not to a high enough temperature for a structural change to occur.

The extent of change in structure depends on the maximum temperature to which the metal is subjected, the length of time the temperature is sustained, the composition of the metal, and the rate of cooling. The principal factor that controls these changes is the amount of heat that is put into the plate—both from preheating and from the welding process.

Cooling rate affects properties along with grain size. Rapid cooling rates produce stronger, harder, and less ductile steels; slow cooling rates produce the opposite properties. With low-carbon steels, the relatively small differences in cooling rates in normal practice have negligible effects on these properties. However, with steels of higher carbon contents or those with an appreciable amount of alloying elements, the effect can be significant.

Holding the plate material at a high temperature (above the upper critical temperature) for a long time produces a structure with large grain size. During welding, however, the metal adjacent to the weld (Zone 3 in Fig. 6.38) is at the high temperature for a very short time. The result is a slight decrease in grain size and an increase in strength and hardness, compared with the base metal.

In multipass weld joints, each bead produces grain-refining action on the preceding bead as it is reheated. However, this refining is not likely to be uniform throughout the joint.

6.12 CRACKING IN STRUCTURAL WELDS

Cracking can occur either in the deposited metal or in the heat-affected zone of the base metal adjacent to the weld. The major cause of cracking in the base metal or in the weld metal is a high carbon or alloy content that increases the hardenability. High hardenability, combined with a high cooling rate, produces the brittle condition that leads to cracking. Other causes of weld cracking are: joint restraint that produces high stresses in the weld, improper shape of the weld bead, hydrogen pickup, and contaminants on the plate or electrode.

6.12.1 Factors Causing Underbead Cracking

Subsurface cracks in the base metal, under or near the weld, are known as underbead cracks. Underbead cracking in the heat-affected base metal is caused by:

1. A relatively high-carbon or alloy-content steel that is allowed to cool too rapidly from the welding temperature.
2. Hydrogen pickup during welding.

Underbead cracking seldom occurs with the preferred-analysis steels (Table 6.17). With carbon steels above 0.35% carbon content and with the low-alloy structural grade steels, underbead cracking can be minimized by using a low-hydrogen welding process. The problem is most severe with materials such as the heat-treated structural steels having tensile strengths of 100 ksi (689 MPa) and higher.

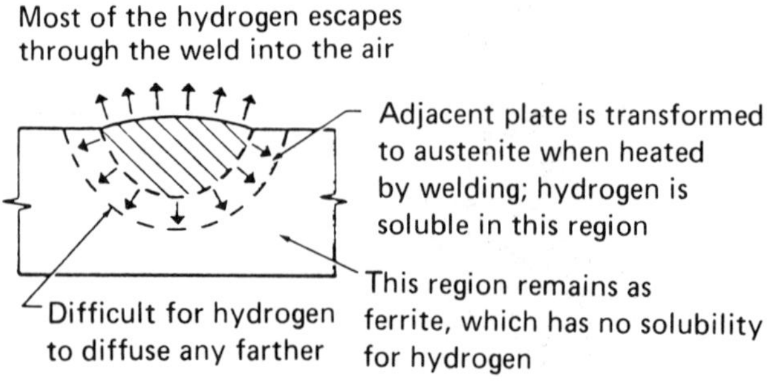

Fig. 6.39 Austenitic heat-affected zone of a weld has high solubility for hydrogen. Upon cooling, the hydrogen builds up pressure that can cause underbead cracking.

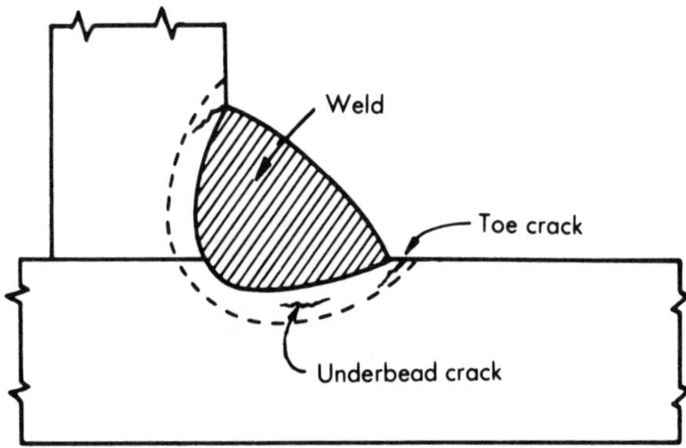

Fig. 6.40 Underbead cracking and toe cracks caused by hydrogen pickup in heat-affected zone of plate.

The second factor that promotes underbead cracking—the pickup and retention of hydrogen—is also influenced by the cooling rate from the welding temperature. During welding, some hydrogen—a decomposition product of moisture from the air, electrode coating, wire, flux, shielding gas, or the surface of the plate—can dissolve into the molten weld metal and from there into the extremely hot (but not molten) base metal. If cooling occurs slowly, the process reverses, and the hydrogen has sufficient time to escape through the weld into the air. But if cooling is rapid, some hydrogen may be trapped in the heat-affected zone next to the weld metal, as illustrated by Fig. 6.39. The hydrogen is absorbed and produces a condition of low ductility known as hydrogen embrittlement.

One theory suggests that the hydrogen produces a pressure, which, combined with shrinkage stresses and any hardening effect from the chemistry of the steel, causes tiny cracks in the metal immediately under the weld bead (Fig. 6.40). Similar cracks that appear on the plate surface adjacent to the weld are called "toe cracks."

Slower cooling (by welding slower, or by preheating) allows more of the hydrogen to escape and helps control the problem. In addition, the use of low-hydrogen welding materials eliminates the major source of hydrogen and usually eliminates underbead cracking.

Rapid cooling rates occur when the arc strikes on a cold plate—at the start of a weld with no previous weld bead to preheat the metal. The highest cooling rates occur on thick plate and in short tack welds. The effect of weld length on cooling rate can be illustrated by the time required to cool welds from 1600 to 200°F (870 to 93°C) on a ¾-in. (19-mm) steel plate:

2½-in. (61-mm) weld	1.5 min.
4-in. (102-mm) weld	5 min.
9-in. (230-mm) weld	33 min.

A 9-in. (230-mm) long weld made on plate at 70°F (21°C) has about the same cooling rate as a 3-in. (76-mm) weld on a plate that has been preheated to 300°F (150°C).

Welds with large cross-sections require greater heat input than smaller ones. High welding current and slow travel rates reduce the rate of cooling and decrease the likelihood of cracking.

6.12.2 Effects of Section Thickness

In a steel mill, billets are rolled into plate or shapes while red hot. The rolled members are then placed on finishing tables to cool. Because a thin plate has more surface area in proportion to its mass than a thick plate, it loses heat faster (by radiation) and cools more rapidly.

If a thick plate has the same chemistry as a thin one, its slower cooling rate results in lower tensile and yield strength, lower hardness, and higher elongation. In very thick plates, the cooling rate may be so low that the properties of the steel may not meet minimum specifications. Thus, to meet specified yield-strength levels, the mill increases the carbon or alloy content of the steels that are to be rolled into thick sections.

In welding, cooling rates of thin and thick plates are just the opposite. Because of the larger mass of plate, the weld area in a thick plate cools more rapidly than the weld area in a thin one. The heat input at the weld area is transferred, by conduction, to the large mass of relatively cool steel, thus cooling the weld area relatively rapidly (heat is transferred more rapidly by conduction than by

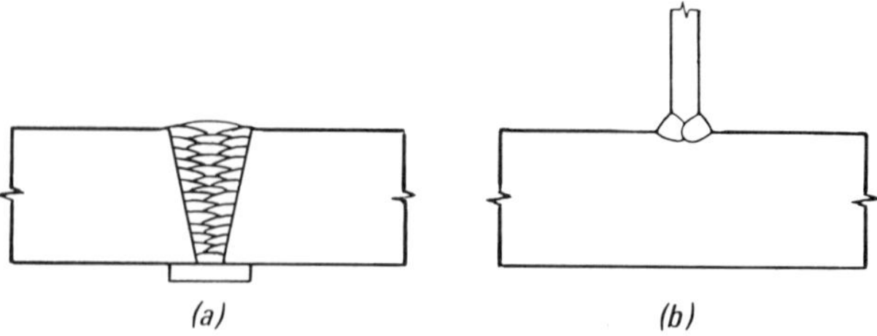

Fig. 6.41 A groove-welded butt joint in thick plate (*a*) might require a higher preheat because of joint restraint, than a fillet-welded joint (*b*) of a thin member and a thick plate. For the minimum size weld required by AWS, see Ref. 6.1, Sec. 2.7.

radiation.) The thin plate has less mass to absorb the heat, and it cools at a slower rate. The faster cooling of the thicker plate produces higher tensile and yield strengths, higher hardness, and lower elongation.

Welds in structural steel shapes and plate under ½ in. (13 mm) thick have less tendency toward cracking than welds in thicker plate. In addition to the favorable (slower) cooling rate of thinner members, two other factors minimize causes of cracking:

1. Thinner plate weldments usually have a good ratio (high) of weld-throat-to-plate thickness.
2. Because they are less rigid, thinner plates can flex more as the weld cools, thus reducing restraint on the weld metal.

Thicker plates and rolled sections do not have these advantages. Because a weld cools faster on a thick member, and because the thick member probably has a higher carbon or alloy content, welds on a thick section have higher strength and hardness but lower ductility than similar welds on thin plate. If these properties are unacceptable, preheating (especially for the more critical root pass) may be necessary to reduce the cooling rate.

Because it increases cost, preheating should be used only when needed. For example, a thin web to be joined to a thick flange plate by fillet welds may not require as much preheat as two highly restrained thick plates joined by a multiple-pass butt weld (Fig. 6.41).

6.12.3 Effect of Joint Restraint

If metal-to-metal contact exists between thick plates prior to welding, the plates cannot move—the joint is restrained. As the weld cools and contracts, all shrinkage stress must be taken up in the weld, as illustrated in Fig. 6.42. This restraint may cause the weld to crack, especially in the first pass on the second side of the plate.

Joint restraint can be minimized by providing a space of 1/32 to 1/16 in. (0.8–1.6 mm) between the two members to allow movement during cooling. Such spaces or gaps can be incorporated by several simple means:

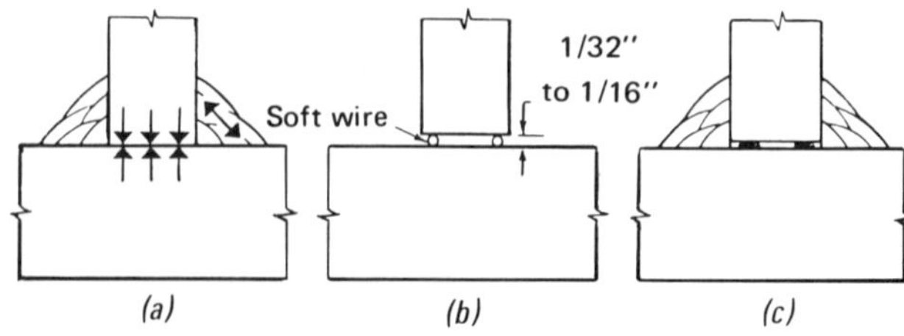

Fig. 6.42 In a restrained joint in thick plates (*a*), all shrinkage stress must be taken up in the weld. Separating the plates with soft wires (*b*) allows the plates to move slightly during cooling. The wires flatten (*c*) and remove most of the stress from the weld metal.

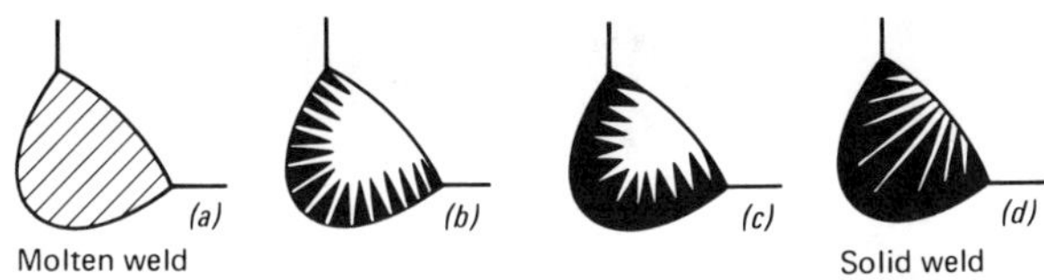

Fig. 6.43 A molten fillet weld (*a*) starts to solidify along the sides next to the plate (*b*). Solidification proceeds as shown in (*c*) and (*d*).

1. Soft steel wire spacers may be placed between the plates, as in Fig. 6.42. The wire flattens out as the weld shrinks. (Copper wire should not be used because it may contaminate the weld metal.)
2. Rough flame-cut edges on the plate. The peaks of the cut edge keep the plates apart, yet can deform and flatten out as the weld shrinks.
3. Upsetting the edge of the plate with a heavy center punch. Results are similar to those on the flame-cut edge.

Fillet Welds

A molten fillet weld starts to solidify, or freeze, along the sides of the joint, as in Fig. 6.43, because the heat is conducted to the adjacent plate, which is at a much lower temperature. Freezing progresses inward until the entire weld is solid. The last material to freeze is that at the center, near the surface of the weld.

Although a concave fillet weld may appear (Fig. 6.44) to be larger than a convex weld, it may have less penetration into the welded plates and a smaller throat than the convex bead. Thus the convex weld may be the stronger of the two, even though it appears to be smaller.

In the past, the concave weld has been preferred by designers because of the apparent smoother stress flow it offers to resist a load on the joint. Experience has shown, however, that single-pass concave fillet welds have a greater tendency to crack during cooling than do convex welds. This disadvantage usually outweighs the effect of improved stress distribution, especially in steels that require special welding procedures.

When a concave bead cools and shrinks, the outer surface is in tension and may crack. A convex bead has considerably reduced shrinkage stresses in the surface area, and the possibility of cracking during cooling is slight.

When design conditions require concave welds for smooth flow of stresses in thick plate, the first bead (usually three or more passes are required) should be slightly convex. The others are then built up to the required shape.

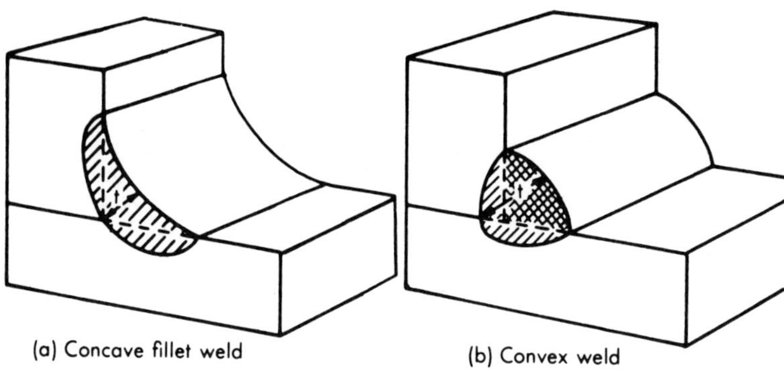

Fig. 6.44 The leg size and the surface of a concave fillet weld (*a*) may be larger than that of a convex bead (*b*), but its throat *t* may be considerably smaller.

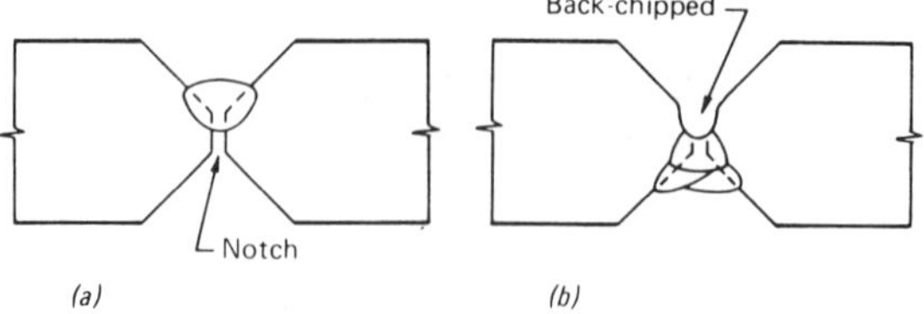

Fig. 6.45 The root pass of a double-V joint is susceptible to cracking because of the notch effect (*a*). On high quality work the notch is minimized by back-chipping (*b*).

Groove Welds

The root pass of a groove weld in heavy plate usually requires special welding procedures. For example, the root pass on the first side of a double-V joint is susceptible to cracking because of the notch, as illustrated in Fig. 6.45, which is a crack starter. On high-quality work, this notch is back-chipped in order to:

1. Remove slag or oxides from the bottom of the groove.
2. Remove any small cracks that may have occurred in the root bead.
3. Widen the groove at the bottom so that the first bead of the second side is large enough to resist the shrinkage that it must withstand due to the rigidity of the joint.

The weld metal tends to shrink in all directions as it cools, and restraint from the heavy plates produces tensile stresses within the weld. The metal yields plastically while hot to accommodate the stresses; if the internal stresses exceed the strength of the weld, it cracks, usually along the centerline.

The problem is greater if the plate material has a higher carbon content. If this is the case, the weld metal usually picks up additional carbon through admixture with the base metal. Under such conditions, the root bead is usually less ductile than subsequent beads.

A concave root bead in a groove weld, as shown in Fig. 6.46, has the same tendency toward cracking as it does in a fillet weld. Increasing the throat dimension of the root pass, as in Fig. 6.46, helps to prevent cracking. Electrodes and procedures should be used which produce a convex bead shape. A low-hydrogen process usually reduces cracking tendencies; if not, preheating may be required.

Centerline cracking can also occur in subsequent passes of a multiple-pass weld if the passes are excessively wide or concave. This can be corrected by putting down narrower, slightly convex beads, making the weld two or more beads wide, as in Fig. 6.47.

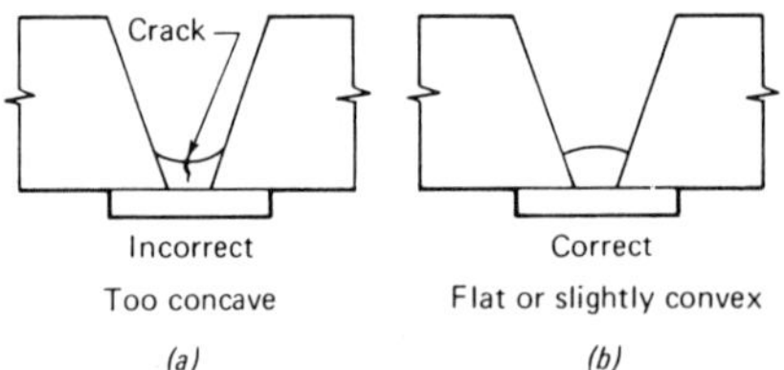

Fig. 6.46 A concave root pass (*a*) may crack because tensile stresses exceed the strength of the weld metal. A slightly convex root-pass bead (*b*) helps prevent cracking.

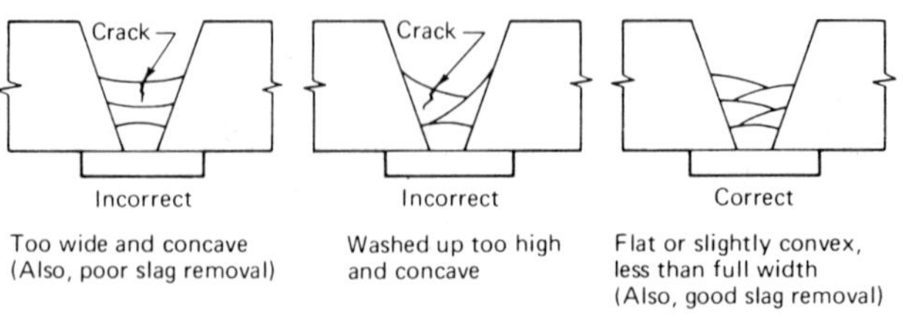

Fig. 6.47 Wide, concave passes (*a*) and (*b*) in a multiple-pass weld may crack. Slightly convex beads (*c*) are recommended.

Width-to-Depth Ratio

Cracks caused by joint restraint or material chemistry usually appear at the face of the weld. In some situations, however, internal cracks occur which do not reach the surface. These are usually caused by improper joint design (narrow, deep grooves or fillets) or by misuse of a welding process that can achieve deep penetration.

If the depth of fusion is much greater than the width of the weld face, the surface of the weld may freeze before the center does. When this happens, the shrinkage forces act on the almost-frozen center (the strength of which is lower than that of the frozen surface) and can cause a crack that does not extend to the surface, as shown in Fig. 6.48*a*.

Internal cracks can also be caused by improper joint design or preparation. Results of combining thick plate, a deep-penetrating welding process, and a 45° included angle are shown in Fig. 6.48*b*. A similar result on a fillet weld made with deep penetration is shown in Fig. 6.48*c*. A too-small bevel and arc-gouging a groove too narrow for its depth on the second pass side of a double-V groove weld, can cause the internal crack shown in Fig. 6.48*d*.

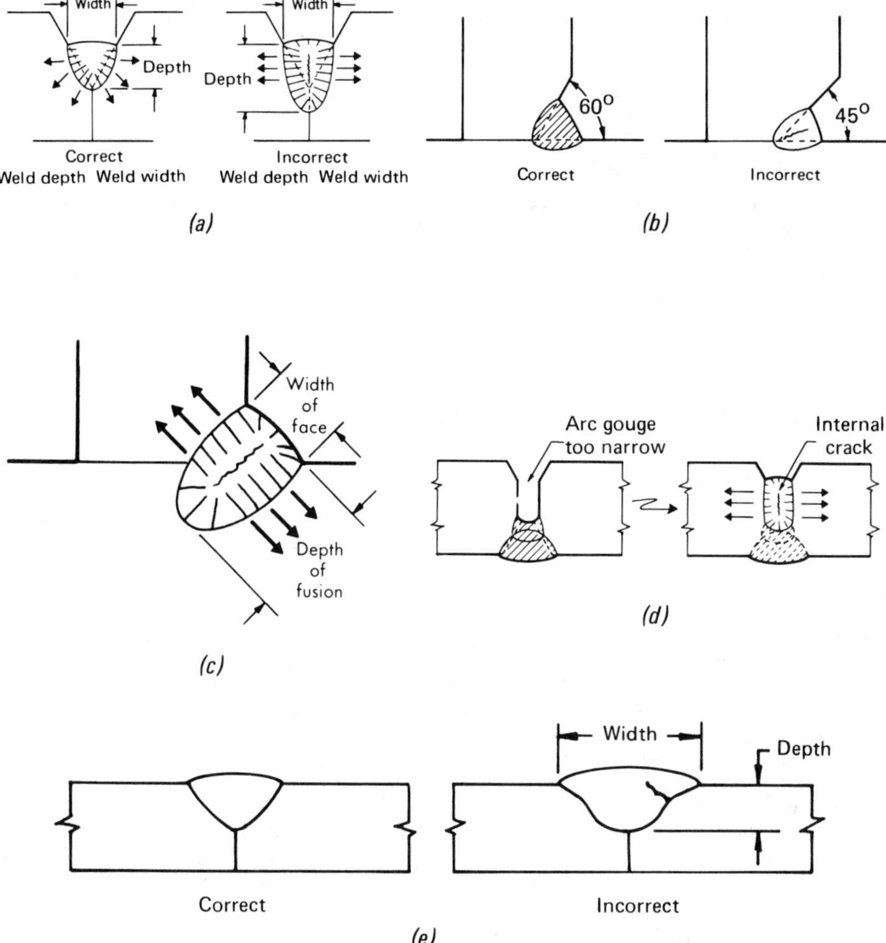

Fig. 6.48 Internal cracking can occur when weld penetration is greater than width. Correct and incorrect proportions are shown in (*a*), (*b*), and (*c*). Arc-gouging a groove too narrow for its depth can cause a similar internal crack (*d*). Width of a weld should not exceed twice its depth (*e*).

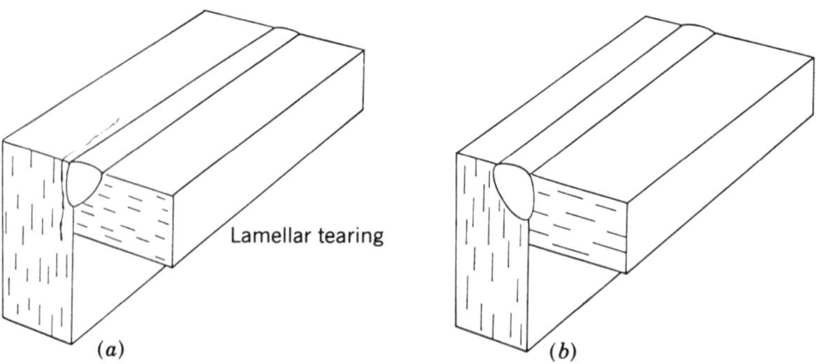

Fig. 6.49 Lamellar tearing (*a*) and a suggested solution (*b*).

Internal cracks are serious because they cannot be detected by visual inspection methods. But they can be eliminated if preventive measures are used. Penetration and volume of weld metal deposited in each pass can be controlled by regulating welding speed and current and by using a joint design which establishes reasonable depth-of-fusion requirements. Recommended ratios of width of each individual bead to depth of fusion are between 1.2 to 1 and $1\frac{1}{2}$ to 1.

A different type of internal crack occurs in submerged-arc welding when the width-to-depth ratio is too large as in Fig. 6.48*c*. Cracks in these so-called "hat-shaped" welds are especially dangerous because radiographic inspection may not detect them. The width-to-depth ratio of any individual bead should not exceed 2 : 1.

Lamellar Cracking or Tearing

This is illustrated in Fig. 6.49. In Fig. 6.49*a*, the shrinkage forces on the upright member are perpendicular to the direction in which the plate was rolled at the steel mill. The inclusions within the plate are strung out in the direction of rolling. If the shrinkage stress should become high enough, a lamellar tear might occur by the progressive cracking from one inclusion to the next. One way to prevent this is illustrated in Fig. 6.49*b*. Here the bevel has been made in the upright plate. The weld now cuts across the inclusions, and the shrinkage forces are distributed, rather than applied to a single plane of inclusions.

6.13 WELDABLE STRUCTURAL STEELS

6.13.1 Low-Carbon Steels

In general, steels with carbon contents to 0.30% are readily joined by all common arc welding processes. These grades account for the greatest tonnage of steels used in welded structures.

6.13.2 AWS Structural Steels

The American Welding Society does not write specifications for structural steel, but does recognize many steels specified by ASTM, API, and ABS as suitable for welded structures with the various arc welding processes. Table 6.18 shows a list of these steels with the mechanical property requirements and the proper filler metals for welding. Since the table does not contain the complete mechanical property or chemical requirements, it is suggested that the reader consult the original specification for further information.

In general, these steels have maximum limits on carbon, sulfur, and phosphorus. Manganese may be specified as a range or in a maximum amount. Small amounts of other alloys may be added in order to meet the mechanical property requirements. All the steels listed in Table 6.18 have satisfactory weldability characteristics but some may require special procedures or techniques, such as limited heat input or minimum preheat and interpass temperatures.

6.13.3 High-Strength Low-Alloy Structural Steels

Higher mechanical properties and, usually, better corrosion resistance than the structural carbon steels are characteristics of the high-strength low-alloy (HSLA) steels. These improved properties are achieved by additions of small amounts of alloying elements (see Section 6.11.1). Some of the

Table 6.18 Minimum Mechanical Properties for ASTM HSLA Steels Approved for Use by AISC Specification for the Design, Fabrication, and Erection of Structural Steel for Buildings and AWS Building Code D1.0

A36 ASTM GRADE and Descriptive Information	Mechanical Properties			Material Shape	Thickness Group or Grade	Chemical Requirements (Ladle) Percent						
	Tensile Strength† (ksi min.)	Yield Point (ksi min.)	Elongation in 2 in. % min.			C Max.	Mn† Max.	P Max.	S Max.	Si †	Cu Min.	V Min.
Structural Steel	58 to 80	36	23	Shapes		0.26		0.04	0.05		0.20*	
	58 to 80	36	23	Plate	to 3/4" incl.	0.25		0.04	0.05		0.20*	
	58 to 80	36	23		over 3/4" to 1-1/2" incl.	0.25	0.80–1.20	0.04	0.05		0.20*	
	58 to 80	36	23		over 1-1/2" to 2-1/2" incl.	0.26	0.80–1.20	0.04	0.05	0.15–0.40	0.20*	
	58 to 80	36	23		over 2-1/2" to 4" incl.	0.27	0.85–1.20	0.04	0.05	0.15–0.40	0.20*	
	58 to 80	36	23		over 4" to 8" incl.	0.29	0.85–1.20	0.04	0.05	0.15–0.40	0.20*	
	58 to 80	36	23	Bars & Bar Shapes	to 3/4" incl.	0.26		0.04	0.05		0.20*	
	58 to 80	36	23		over 3/4" to 1-1/2" incl.	0.27	0.60–0.90	0.04	0.05		0.20*	
	58 to 80	36	23		over 1-1/2" to 4" incl.	0.28	0.60–0.90	0.04	0.05		0.20*	
A53 GRADE B Welded & Seamless Steel Pipe electric resistance or seamless (only chemistry limit is phosphorus)	60 min	35	**		B			**				

Table 6.18 *(Continued)*

	Mechanical Properties					Chemical Requirements (Ladle) Percent*						
	Tensile Strength† (ksi min.)	Yield Point (ksi min.)	Elongation in 2 in. % min.	Material Shape	Thickness Group or Grade	C Max.	Mn† Max.	P Max.	S Max.	Si †	Cu Min.	V Min.
A242 High-Strength Low-Alloy Structural Steel	70 min.	50	**	Plates & Bars	to 3/4" incl.	0.20	1.35	0.04	0.05		0.20	
	67 min.	46	21		over 3/4" to 1-1/2" incl.	0.20	1.35	0.04	0.05		0.20	
(other alloying elements may be added.)	63 min.	42	21		over 1-1/2" to 4" incl.	0.20	1.35	0.04	0.05		0.20	
† For Type 2	70 min.	50	**	Struc. Shapes	I	0.20	1.35	0.04	0.05		0.20	
	67 min.	46	**		II	0.20	1.35	0.04	0.05		0.20	
	63 min.	42	21		III	0.20	1.35	0.04	0.05		0.20	
A441 High-Strength Low-Alloy Structural Manganese Vanadium Steel	70 min.	50	**	Plates and Bars	to 3/4" incl.	0.22	.85–1.25	0.04	0.05	0.40	0.20	0.02
	67 min.	46	**		over 3/4" 1-1/2" incl.	0.22	.85–1.25	0.04	0.05	0.40	0.20	0.02
	63 min.	42	24		over 1-1/2" to 4" incl.	0.22	.85–1.25	0.04	0.05	0.40	0.20	0.02
	60 min.	40	24		over 4" to 8" incl.	0.22	.85–1.25	0.04	0.05	0.40	0.20	0.02
	70 min.	50	**	Struc. Shapes	I	0.22	.85–1.25	0.04	0.05	0.40	0.20	0.02
	67 min.	46	**		II	0.22	.85–1.25	0.04	0.05	0.40	0.20	0.02
	63 min.	42	24		III	0.22	.85–1.25	0.04	0.05	0.40	0.20	0.02

A500 Cold-Formed Welded and Seamless Carbon Steel Structural Tubing in Rounds and Shapes	45 min.	33	25	Round Structural Tubing	A	0.26		0.04	0.05		0.20*
	58 min.	42	23		B	0.26		0.04	0.05		0.20*
	62 min.	46	21		C	0.23	1.35	0.04	0.05		0.20*
	45 min.	39	25	Shaped Structural Tubing	A	0.26		0.04	0.05		0.20*
	58 min.	46	23		B	0.26		0.04	0.05		0.20*
	62 min.	50	21		C	0.23	1.35	0.04	0.05		0.20*
A501 Hot-Formed Welded and Seamless Carbon Steel Structural Tubing.	58 min.	36	23			0.26		0.04	0.05		0.20*
A529 Structural Steel with 42 ksi Minimum Yield Point (1/2 in. Maximum Thickness)	60 to 85	42	19			0.27	1.20	0.04	0.05		0.20*
A570 GRADES 40 and 45 Steel, Sheet and Strip, Carbon, Hot-Rolled, Structural Quality	55 min.	40	**		Grade 40	0.25	0.90	0.04	0.05		0.20*
	60 min.	45	**		Grade 45	0.25	1.35	0.04	0.05		0.20*

* When specified.
** See ASTM Standards for details.
† Where two figures are given this is a min-max range.

ASTM GRADE and Descriptive Information	Material Shape	Thickness Or Group	Mechanical Properties: Tensile Strength† (ksi min.)	Yield Point (ksi min.)	Elonga. in 2 in. % min.	Grade	Chemical Requirements (Ladle) Percent: C† Max.	Mn† Max.	P Max.	S Max.	Si Max.	Ni† Max.	Cr† Max.	Mo† Max.	Cu† Max.	V† Max.	Cb Max.	Ti† Max.	Others
A572 High-Strength Low-Alloy Columbium–Vanadium Steels of Structural Quality	Shapes And Plates		60 min.	42	24	42	0.21	1.35	0.04	0.05	(5)					(2)	(1)(3)		N (4)
			65 min.	50	21	50	0.23	1.35	0.04	0.05	(5)					(2)	(1)(3)		(4)
			75 min.	60	18	60	0.26	1.35	0.04	0.05	(5)					(2)	(1)(3)		(4)
Alloy content shall be in accordance with one of the following —			80 min.	65	17	65	0.26	1.35	0.04	0.05	(5)					(2)	(1)(3)		(4)

(1) Cb 0.005–0.05
(2) V 0.01–0.15
(3) Cb (.05 max)+V 0.02–0.15
(4) N (with V) 0.015 max

(4) N(.015 max) when added as a supplement to V shall be reported, and the minimum ratio of V to N shall be 4 to 1.
(1)(3) Cb when added either singly or in combination with V shall be restricted for Grades 42 and 50 to plate or bar thickness of ½ in. max and to Group I (Table A, ASTM A6) shapes; and for Grades 60 and 65 to plate or bar thickness of ¾ in. max and to Groups I and II shapes.
(5) Si (0.40 max) for shapes and plates to $1\frac{1}{2}''$ in. For plates over $1\frac{1}{2}''$ in., Si range 0.15–0.40.

Table 6.18 (*Continued*)

A588 High-Strength Low-Alloy Structural Steel with 50 ksi Minimum Yield Point to 4-in. Thick

Material Shape	Thickness Or Group	Tensile Strength† (ksi min.)	Yield Point (ksi min.)	Elonga. in 2 in. % min.
Plates And Bars	to 4″ incl.	70 min.	50	21
Plates And Bars	over 4″ to 5″ incl.	67 min.	46	21
Plates And Bars	over 5″ to 8″ incl.	63 min.	42	21
Shapes	All Groups	70 min.	50	21

Chemical Requirements (Ladle) Percent

Grade	C† Max.	Mn† Max.	P Max.	S Max.	Si Max.	Ni† Max.	Cr† Max.	Mo† Max.	Cu† Max.	V† Max.	Cb Max.	Ti† Max.	Others
A	0.19	0.90–1.25	0.04	0.05	0.30–0.65	0.40	0.40–0.65		0.25–0.40	0.02–0.10			
B	0.20	0.75–1.25	0.04	0.05	0.15–0.50	0.50	0.40–0.70		0.20–0.40	0.01–0.10			
C	0.15	0.80–1.35	0.04	0.05	0.15–0.40	0.25–0.50	0.30–0.50		0.20–0.50	0.01–0.10			
D	0.10–0.20	0.75–1.25	0.04	0.05	0.50–0.90		0.50–0.90		0.30		0.04		Zr 0.05–0.15
E	0.15	1.20	0.04	0.05	0.30	0.75–1.25		0.08–0.25	0.50–0.80	0.05			
F	0.10–0.20	0.50–1.00	0.04	0.05	0.30	0.40–1.10	0.30	0.10–0.20	0.30–1.00	0.01–0.10			
H	0.20	1.25	0.035	0.04	0.25–0.75	0.30–0.60	0.10–0.25	0.15 Max.	0.20–0.35	0.02–0.10		0.005–0.030	

A514 High-Yield-Strength, Quenched-and-Tempered Alloy Steel Plate, Suitable for Welding

Material Shape	Thickness Or Group	Tensile Strength† (ksi min.)	Yield Point (ksi min.)	Elonga. in 2 in. % min.
Plate	to 3/4″ incl.	110 to 130	100	18
Plate	over 3/4″ to 2-1/2″ incl.	110 to 130	100	18
Plate	over 2-1/2″ to 6″ incl.	100 to 130	90	16

Chemical Requirements (Ladle) Percent

Grade	C† Max.	Mn† Max.	P Max.	S Max.	Si Max.	Ni† Max.	Cr† Max.	Mo† Max.	Cu† Max.	V† Max.	Cb Max.	Ti† Max.	Others
E	0.12–0.20	0.40–0.70	0.035	0.04	0.20–0.40		1.40–2.00	0.40–0.60				0.01–0.10	B 0.001–0.005
F	0.10–0.20	0.60–1.00	0.035	0.04	0.15–0.35	0.70–1.00	0.40–0.65	0.40–0.60	0.15–0.50	0.03–0.08			B 0.0005–0.006

† Where two figures are given this is a min-max range.

HSLA types are carbon–manganese steels; others contain different alloy additions, governed by requirements for weldability, formability, toughness, or economy. Strength of these steels is between those of structural carbon steels and the high-strength quenched-and-tempered steels.

High-strength low-alloy steels are usually used in the as-rolled condition. These steels are produced to specific mechanical property requirements as well as to chemical compositions. Minimum mechanical properties available in the as-rolled condition vary among the grades and, within most grades, with thickness. Ranges of properties available in this group of steels are:

1. Minimum yield stress from 42 to 70 ksi (290 to 480 MPa).
2. Minimum tensile strength from 60 to 85 ksi (410 to 590 MPa).
3. Resistance to corrosion, classed as: equal to that of carbon steels, twice that of carbon steels, or four to six times that of carbon steels.

The HSLA steels are used extensively in products and structures that require higher strength-to-weight ratios than the carbon structural steels offer. Typical applications are buildings, bridge decks, and similar structures.

The high-strength low-alloy steels should not be confused with the high-strength quenched-and-tempered alloy steels. Both groups are sold primarily on a trade-name basis, with different letters or numbers being used to identify each. The quenched-and-tempered steels are full-alloy steels that are heat treated at the mill to develop optimum properties. They are generally martensitic in structure, whereas the HSLA steels are mainly ferritic steels; this is the clue to the metallurgical and fabricating differences between the two types. In the as-rolled condition, ferritic steels are composed of relatively soft, ductile constituents; martensitic steels have hard, brittle constituents that require heat treatment to produce their high-strength properties.

Strength in the HSLA steels is achieved instead by relatively small amounts of alloying elements dissolved in a ferritic structure. Carbon content rarely exceeds 0.28% and is usually between 0.15 and 0.22%. Manganese content ranges from 0.85 to 1.60%, depending on grade and other alloy additions—chromium, nickel, silicon, phosphorus, copper, vanadium, columbium, and nitrogen—are used in amounts less than 1%. Welding, forming, and machining characteristics of most grades do not differ markedly from those of the low-carbon steels.

To be weldable, the high-strength steels must have enough ductility to avoid cracking from the rapid, cooling inherent in welding processes. Weldable HSLA steels must be sufficiently low in carbon, manganese, and all "deep-hardening" elements to ensure that appreciable amounts of martensite are not formed upon rapid cooling. Superior strength is provided by solution of the alloying elements in the ferrite of the as-rolled steel. Corrosion resistance is also increased in certain of the HSLA steels by the alloying additions.

Addition of a minimum of 0.20% copper usually produces steels with about twice the atmospheric corrosion resistance of structural carbon steels. Steels with four to six times the atmospheric corrosion resistance of structural carbon steels are obtained in many ways, but, typically, with additions of nickel and/or chromium, often with more than 0.10% phosphorus. These alloys are usually used in addition to the copper.

ASTM Specifications

Four ASTM specifications cover the high-strength low-alloy structural steels. They are: A242, A441, A572, and A588. Table 6.18 lists the mechanical properties of these steels. Specifications A374 and A375 cover similar steels in sheet and strip form.

ASTM A242 covers HSLA structural steel shapes, plates, and bars for welded or bolted construction. Maximum carbon content of these steels is 0.24%, typical content is from 0.09 to 0.17%. Materials produced to this specification are intended primarily for structural members where light weight and durability are important.

Some producers can supply copper-bearing steels (0.20% minimum copper). Steels meeting the general requirements of ASTM A242 but modified to give four times the atmospheric corrosion resistance of structural steels are also available. These latter grades—sometimes called "weathering steels"—are used for architectural and other structural purposes where it is desirable to avoid painting for either aesthetic or economic reasons.

Welding characteristics vary according to the type of steel; producers can recommend the most weldable material and offer welding advice if the conditions under which the welding will be done are known.

ASTM A441 covers the intermediate-manganese HSLA steels that are readily weldable with proper procedures. The specification calls for additions of vanadium and a lower manganese content (1.25% maximum). Minimum mechanical properties are the same as A242 steels except that plates and bars from 4 to 8 in. (102 to 205 mm) thick are covered in A441.

Atmospheric corrosion resistance of this steel is approximately twice that of structural carbon steel. Another property of ASTM A441 steel is its superior toughness at low temperatures. Only shapes, plates, and bars are covered by the specification, but weldable sheets and strip can be supplied by some producers with approximately the same minimum mechanical properties.

ASTM A572 includes four grades of high-strength low-alloy structural steels in shapes, plates and bars. These steels offer a choice of strength levels with yield points ranging from 42 to 65 ksi (290 to 450 MPa) (Table 6. 18). Proprietary HSLA steels of this type with 70 and 75 ksi (480 and 520 MPa) yield points are also available. Increasing care is required for welding these steels as strength level increases. A572 steels are distinguished from other HSLA steels by their columbium, vanadium, and nitrogen content. Copper additions above the minimum of 0.20% may be specified for atmospheric corrosion resistance about double that of structural carbon steels.

A supplementary requirement is included in the A572 Specification that permits designating the specific alloying elements required in the steel. Examples are the Type 1 designation, for columbium; Type 2, for vanadium; Type 3, for columbium and vanadium; and Type 4, for vanadium and nitrogen. Specific grade designations must accompany this type of requirement.

ASTM A588 provides for a steel similar in most respects to A242 weathering steel, except that the 50-ksi (345-MPa) yield point is available in thicknesses to at least 4 in. (102 mm).

6.13.4 High-Yield-Strength Quenched-and-Tempered Alloy Steels

The high-yield-strength quenched-and-tempered construction steels are full-alloy steels that are treated at the steel mill to develop optimum properties. Unlike conventional alloy steels, these grades do not require additional heat treatment by the fabricator.

These steels are generally low-carbon grades (upper carbon limit of about 0.20%) that have minimum yield strengths from 80 to 125 ksi (550 to 860 MPa).

ASTM Specifications

Two plate specifications, ASTM A514 and A517 for welded structures, allow for the effect of section size on yield strength, tensile strength, and ductility. ASTM A514 requires a minimum yield strength of 100 ksi (690 MPa) for material up to 2½ in. (64 mm) thick and 90 ksi (620 MPa) for material from 2½ to 6 in. (64 to 152 mm) thick. ASTM A517 requires uniform yield strengths of 100 ksi (690 MPa) for all material up to 2½ in. (65 mm) thick. Representative trade names of the A514 and A517 steels are given in Table 6.19.

Weldability

If suppliers' recommendations are followed for controlling welding procedures, 100% joint efficiency can be expected in the as-welded condition for the 90 and 100 ksi (620 and 690 MPa) yield-strength grades.

If the heat-affected zone cools too slowly, the beneficial effects of the original heat treatment (particularly notch toughness) are destroyed. This can be caused by excessive preheat temperature, interpass temperature, or heat input. On the other hand, if the heat-affected zone cools too

Table 6.19 Representative ASTM A514/517 Steels Designated According to Trade Name

Producer	Trade Name
Armco Steel Corp.	SSS-100 SSS-100A SSS-100B
Bethlehem Steel Corp.	RQ-100A, RQ-100 RQ-100B
Great Lakes Steel Corp. and Phoenix Steel Corp.*	N-A-XTRA 100 N-A-XTRA 110
Jones & Laughlin Steel Corp.	Jalloy-S-100 Jalloy-S-110
United States Steel Corp. and Lukens Steel Corp.*	T-1 T-1 Type A T-1 Type B

* Licensee

Table 6.20 Maximum Suggested Heat Units* Input for USS T-1 Steel per Inch of Weld†

Preheat and Interpass Temperature (°F)	Plate Thickness (in.)							
	3/16	1/4	1/2	3/4	1	1 1/4	1 1/2	2
70	27	36	70	121	Any	Any	Any	Any
200	21	29	56	99	173	Any	Any	Any
300	17	24	47	82	126	175	Any	Any
350	15	21.5	43.5	73.5	109.5	151	Any	Any
400	13	19	40	65	93	127	165	Any

* Heat units usually called kilojoules/in.
† From the *Welding Heat Input Calculator* by the United States Steel Corporation.

rapidly, it can become hard and brittle and may crack. This is caused by insufficient preheat or interpass temperature or insufficient heat input during welding. Producer's recommendations should be followed closely.

The quenched-and-tempered steels can be welded by the shielded metal-arc, submerged-arc, flux cored, and gas-shielded-arc processes. Weld cooling rates for these processes are relatively rapid and mechanical properties of the heat-affected zones approach those of the steel in the quenched condition. Reheat treatment, such as quenching and tempering after welding, is not recommended.

Because of the desirability of relatively rapid cooling after welding, thin sections of these materials can usually be welded without preheating. When preheating is required, both maximum and minimum temperatures are important. If the sections to be welded are warm as a result of preheating and heat input from previous welding passes, it may be necessary to reduce current or increase arc travel speed for subsequent passes, or to wait until the metal cools somewhat. Interpass temperature is just as important as preheat temperature and should be controlled with the same care.

In the ASTM specifications A514 and A517, there are several grades of quenched-and-tempered constructional steels listed. Welding procedures for all of these steels are similar, but no one procedure is right for all grades. When in doubt, consult the steel manufacturer for the proper procedure.

The following is a general shielded metal-arc procedure for one of the popular grades of quenched-and-tempered constructional steels and can be used as a guide for all grades or other welding processes.

Use only low-hydrogen electrodes; usually the electrode specified for A514 and A517 steels is E11018. Under some conditions, a lower tensile strength electrode may be used, as will be discussed later. Make sure electrodes are dry. Under normal conditions of humidity, electrodes should be returned to the drying ovens after an exposure of 4 hr maximum. If the humidity is high, reduce the exposure time. Electrodes are shipped in hermetically sealed containers and the contents of any damaged container should be redried before using.

Clean the joint thoroughly. Remove all rust and scale, preferably by grinding. If the base metal has been exposed to moisture, preheat to drive off the moisture. On thin sections, allow the plate to cool, if necessary, before starting to weld.

The amount of preheat and the amount of welding heat put into the weld must be kept within definite boundaries during the actual welding.* Usually preheating is not necessary or desirable on thin sections, but in order to avoid cracks preheating is necessary if: (1) the joints are highly restrained, (2) the structure is very rigid, and (3) the weld joint is on thick sections.

Whether or not the base metal is preheated, it is necessary to approximate the heat input before starting to weld. Calculation by Eq. (6.7.2) is only approximate because the heat losses can be large. Also, there are many variables that affect the heat distribution and the maximum temperature of the base metal at the joint, but the formula is sufficiently accurate to predict the maximum allowable heat input for a given set of conditions.

In industry, the term "heat unit" is used and is equal to the watt-sec/in. of weld divided by 1000, usually called kilojoules/in.

Maximum suggested heat units input for welding one type of quenched-and-tempered steel per linear inch of weld are available from the steel producers. The suggested values for USS T-1 steel are given in Table 6.20.

* A calculator is available from the United States Steel Corporation for quickly determining heat units. Also available are tables for maximum heat units when welding these steels.

Usually the electrodes used are the E11018 type, but lower strength electrodes may be specified where the stress does not require the high yield strength of E11018. A good example is the lower stress in the web-to-flange fillet welds. However, if lower strength electrodes are used, the same limitations apply as to heat input and interpass temperature.

6.14 SOME DESIGN IDEAS FOR HEAVY WELDED COLUMNS

In the construction of multistory welded joint buildings in the recent decades, designers and engineers have developed techniques that could well become standardized and applied to future structures.

Representative of such techniques are the innovations in the design and fabrication of welded steel columns. Since structures are reaching 100 or more stories in the air and have greater spans between columns, thus requiring heavier columns than formerly fabricated by arc welding, questions arise as to the configuration of column sections, the transfer of moments created by beam or wind loads, the weld sizes, the effects of residual stresses, and other factors.

As Fig. 6.50 illustrates, the development of such heavier columns has come about by an evolutionary procedure. In the past when engineers needed steel columns of heavier section than available commercially, they riveted steel cover plates to the flanges of 14-in. (360-mm) wide-flange sections, as in Fig. 6.50*a*. The next step was to tack-weld the cover plates to the rolled section so the drilling and reaming could be done while the parts were locked together. Eventually, the tack welds were replaced with fillets as shown in Fig. 6.50*b* and the rivets were eliminated entirely.

The latter practice produced the same column section, but presented a problem in efficiently transferring tensile force from the beam flange through the cover plate and into the column without pulling the cover plate away from the flange. As noted in Fig. 6.51, the cover plate, being attached near its outer edges only, tends to bow, resulting in uneven distribution of forces on the beam-to-column weld.

A logical next step was to eliminate the rolled component and build the column up completely from steel plate, Fig. 6.50*c*. This permits the exact section dimensions required with no increase in welding. It also eliminates the problem of transfer of tensile forces from the beam flange through cover plate of a column.

Columns of H design fabricated from steel plate with fillet welds have been engineered on a sound and economical basis. Figure 6.52 shows a column section used in framing the C.I.L. House in Montreal. This column weighed 1700 lb/ft (2530 kg/m); others in the structure weighed as much as 2000 lb/ft (2980 kg/m). The heavy flange plates were welded to the web with single-pass $\frac{3}{4}$-in. (19-mm) fillets, made with a triple tandem submerged-arc welding head.

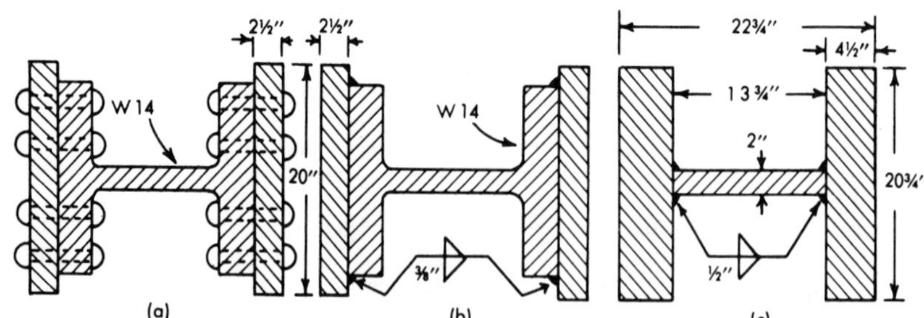

Fig. 6.50 Attachment of cover plates to flanges of 14-in. wide-flange sections and evolution of the heavy H column section.

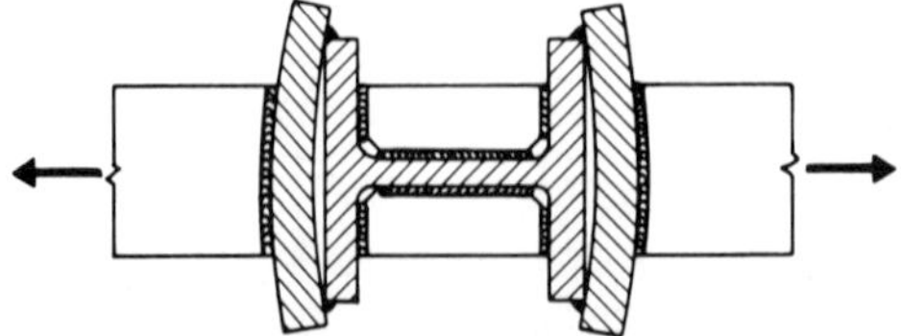

Fig. 6.51 Bowing of cover plates following attachment to flanges near outer edges of plates.

The trend toward taller and taller structures, with the need for heavier and heavier columns, led to increased use of the welded box section. These can be made with or without a web, as indicated in Fig. 6.53. In addition to providing a means of fabricating a column of any size, the scheme permits the lightening of the column in its upper reaches without changing outside section dimensions merely by first omitting the web to form a box, then omitting a plate to form an H section, and finally going to a rolled wide-flange section.

A variation of the box-section column is shown in Fig. 6.54. Here a 14-in. (360-mm) wide-flange rolled section (now called W shape) is used as a web between two massive rolled flange plates. The

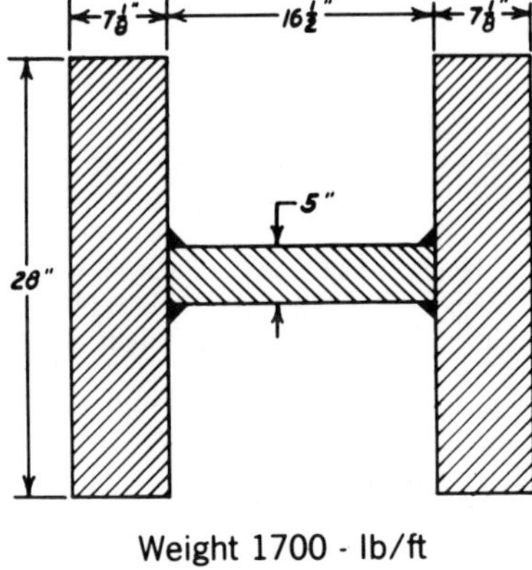

Fig. 6.52 Schematic of column section used in framing in C.I.L. House in Montreal, Canada.

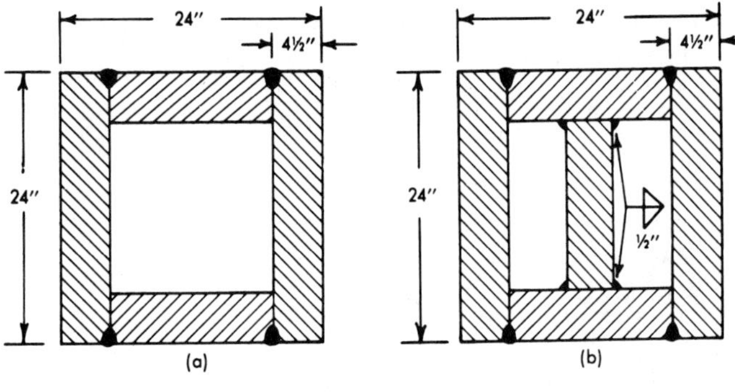

Fig. 6.53 Fabrication of heavier columns with and without webs.

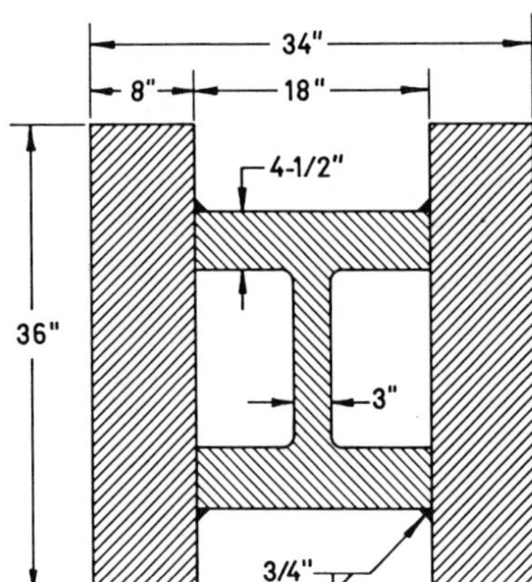

Fig. 6.54 Cross-section of 2700 lb/ft column used in construction of Toronto Dominion Bank Tower.

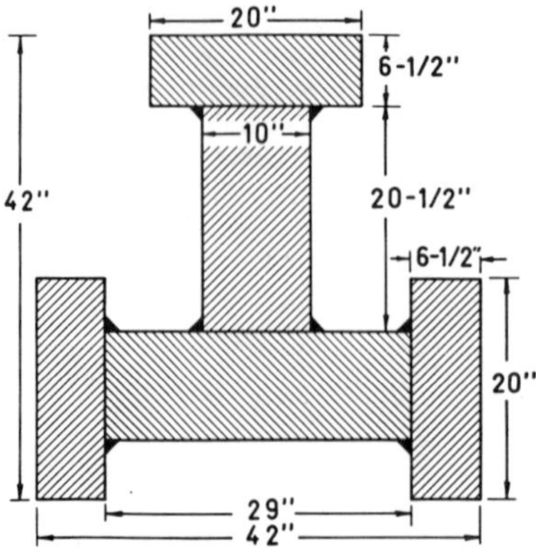

Fig. 6.55 Section of welded column for Commerce Court Building, Toronto, Canada.

W shape component is fillet welded to the plates the length of the column, except in the region of a beam-to-column connection, where a 45° groove weld is used beneath the fillet. The section shown in Fig. 6.54 was used in the construction of the Toronto Dominion Bank Tower. The weight of the column is 2700 lb/ft (4020 kg/m).

Some of the heaviest columns on record were fabricated for the Commerce Court Building in Toronto, Canada. These are combinations of T and H sections, as shown in Figs. 6.55 and 6.56. Single-pass $\frac{5}{8}$ and $\frac{3}{4}$ in. (16 and 19 mm) fillet welds were used to join the heavy plates. The column in Fig. 6.55 weighed 3000 lb/ft (4460 kg/m) and the column in Fig. 6.56, 3540 lb/ft (5270 kg/m).

6.14.1 Requirements for the Welds in Column Fabrication

There are two main requirements for the longitudinal welds used in column fabrication:

1. The welds must be able to withstand any longitudinal shear resulting from moments applied to the column by wind or beam loads as shown in Fig. 6.57. (Note from the Fig. 6.57 moment diagram that the change in moment is small along most of the column length.)

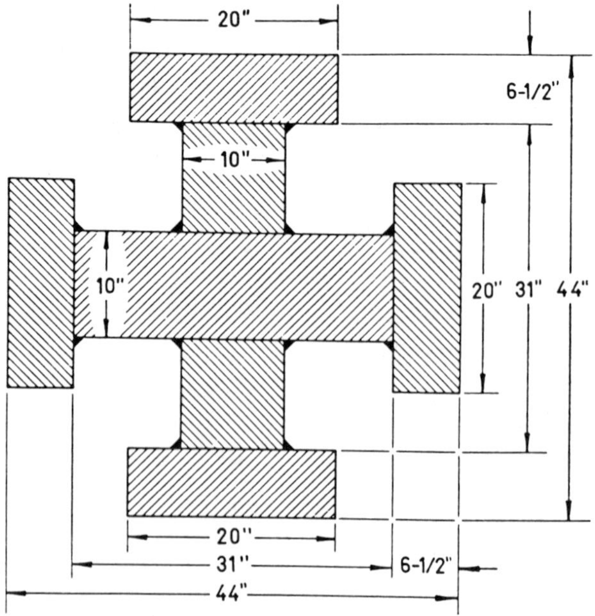

Fig. 6.56 Section of welded column for Commerce Court Building, Toronto, Canada.

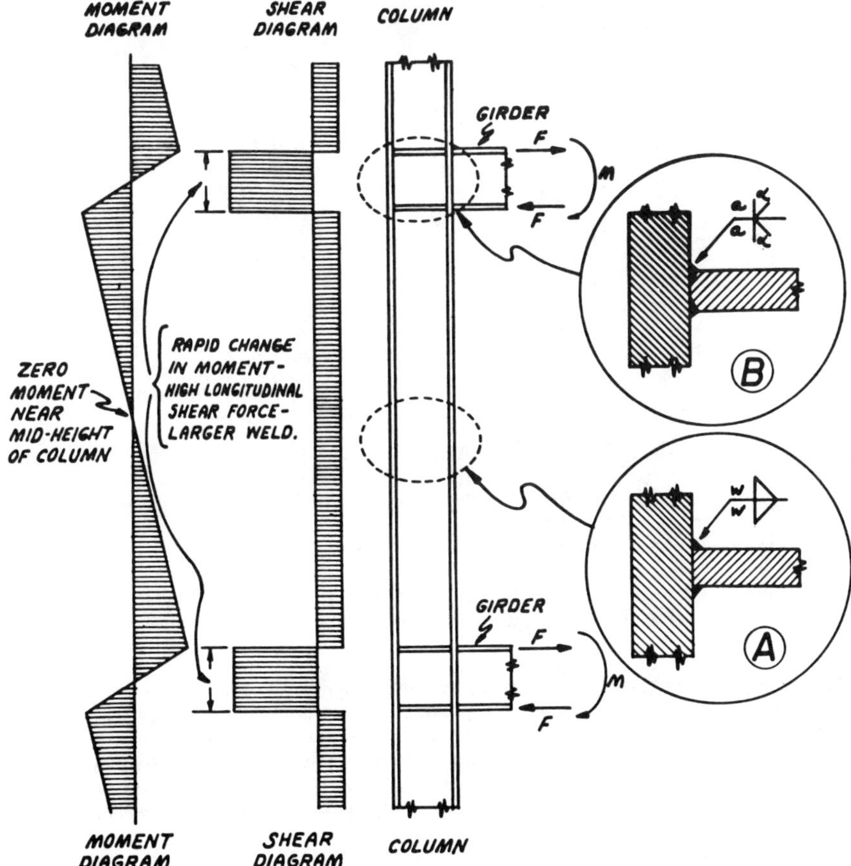

Fig. 6.57 Moment and shear diagrams for welded columns.

2. The welds in the region where the beams connect to the column must be able to withstand the longitudinal shear and tensile forces to which they are subjected.

Since there is a rapid change in moment in the region of the beam connection, the longitudinal shear is much greater. Also, the tensile force from the beam is transferred through a portion of this weld. Thus, the welds in this area must be larger than at other points along the column.

The formula to determine the unit shear force on the welds is

$$f = \frac{Vay}{I} \tag{6.14.1}$$

where V = horizontal shear acting on the column
a = area held by the weld
y = distance from the center of gravity of this area to the neutral axis of the whole section
I = moment of inertia of the whole section

In an unsymmetrical column, any twisting action applied to it must be considered. In the case of wind loading (Fig. 6.58), an unbalanced moment is applied to the column from the beams attached to it. One way to analyze this problem is to assume that the unbalanced moment is resisted by the wide-flange section and the outer flange plate in proportion to their relative moments of inertia. The portion of the moment resisted by the outer flange plate must be transferred as torque from the column by an arrangement called a torque box (Fig. 6.59).

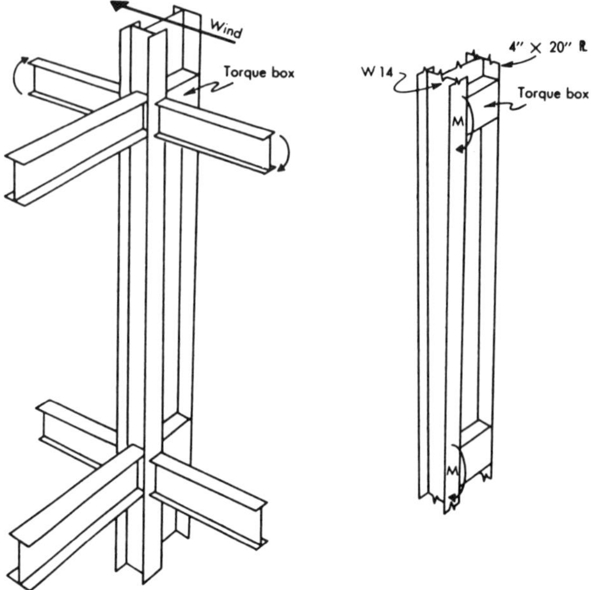

Fig. 6.58 Wide-flange section through a torque box.

Another method of checking this twisting action is to consider the end moments of the beams as applying torque to the built-up column section. The applied torque may be considered as two forces from the upper and lower flanges of the spandrel beam, but in opposite directions (see Fig. 6.60). Since these forces are not applied at the "shear center" of the column, a twisting action will be applied to the column about its longitudinal axis within the region of the beam connection where these forces are applied. There is no twisting action along the length of the column between these regions.

The only other concern about this built-up column detail is the sharp reentrant corners at points "d" and "f" in Fig. 6.61. In structural steel, any stress concentration in these areas would probably be relieved through plastic flow and should be neglected, unless fatigue loading were an

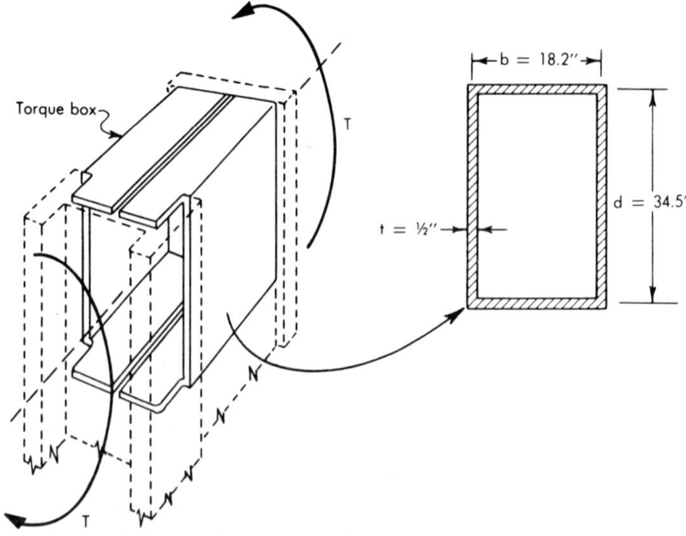

Fig. 6.59 Details for a torque box, as shown in Fig. 6.58.

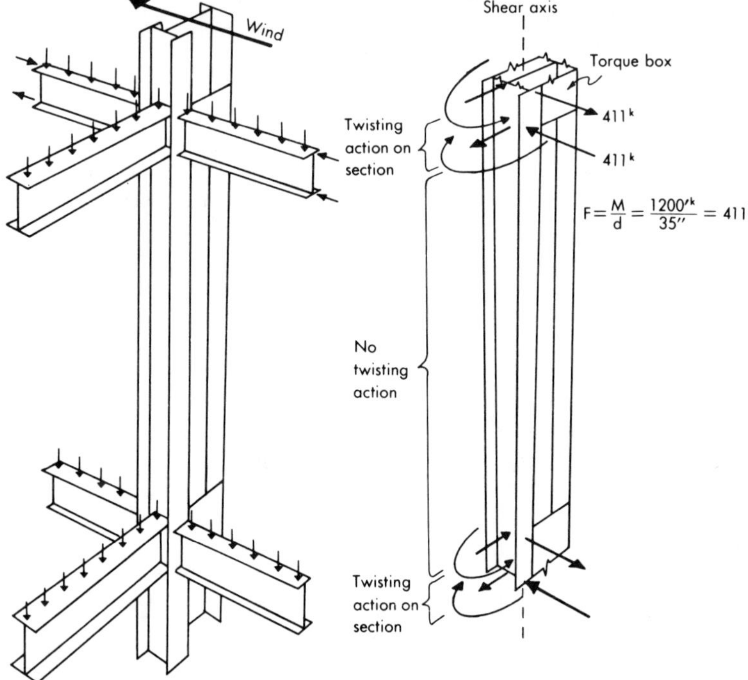

Fig. 6.60 Built-up column with an "open" section.

important factor. Of course, if a fillet weld could be made on the inside corner, it would eliminate the problem (see inset in Fig. 6.61). This might be possible, since the plates for the "torque box" are not very long, and the welding operator might reach in from each end to make the inside weld.

6.14.2 Types of Welds Used in Column Fabrication

Three common types of welds are used in column fabrication—fillets, bevel and V-groove, and J- and U-groove.

Fillet welds, as shown in Fig. 6.62, require no plate preparation. Their size can be increased simply by making more passes. Fillets are usually used for nominal-size welds. As weld size increases, they become less desirable because of the excessive amount of weld metal required. When the weld size makes fillets uneconomical, they are replaced by some type of groove weld.

Single-bevel and V-groove welds, illustrated in Fig. 6.63, require beveling of joint edges, usually by the oxygen cutting process. The preparation cost on large welds is offset by savings in weld metal over what would be required with fillets. A $\frac{1}{8}$-in. (3-mm) deduction is allowed in the depth of chamfer of a partial-penetration bevel groove weld having an included angle of 45° when made by shielded metal-arc welding in any position, or by gas metal-arc or flux-cored arc welding in the vertical or overhead positions.

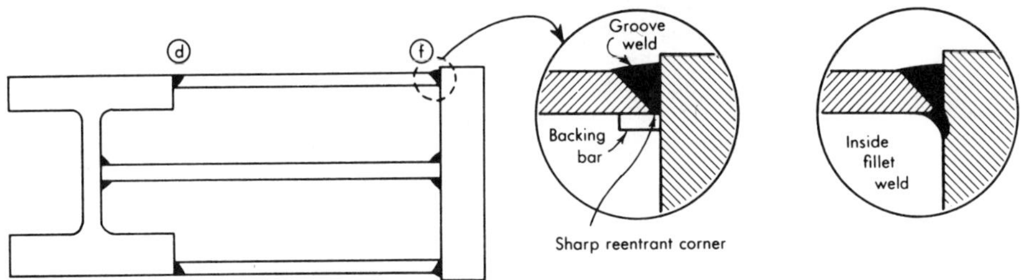

Fig. 6.61 Built-up column with sharp reentrant corners at torque box.

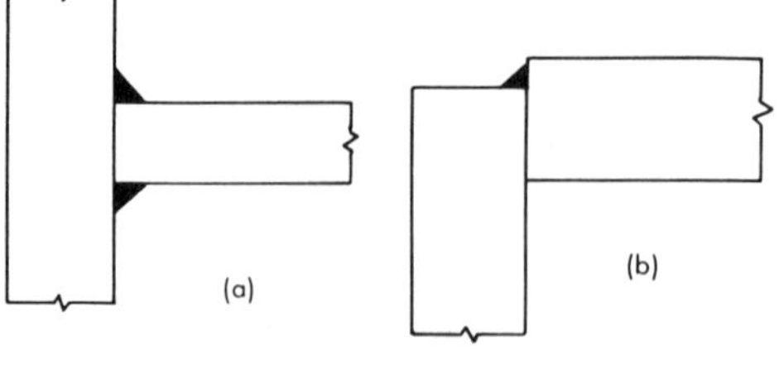

Fig. 6.62 Fillet welds.

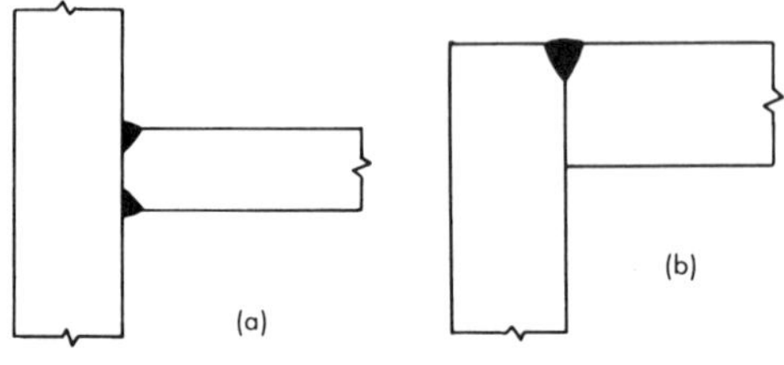

Fig. 6.63 (*a*) Double-bevel groove weld; (*b*) single-V groove weld.

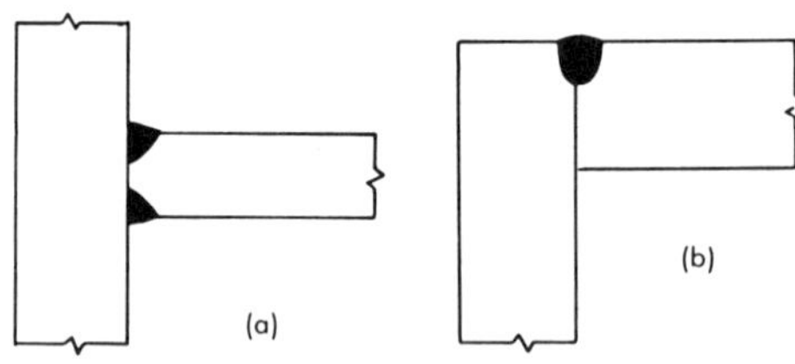

Fig. 6.64 (*a*) J-groove weld; (*b*) U-groove weld.

J- and U-groove welds are illustrated in Fig. 6.64. These require the plate edges to be gouged or machined. Arc carbon arc gouging is becoming a preferred method for forming such grooves with heavy structural material.

Engineering drawings must indicate the effective throat required. Shop drawings indicate the actual depth of groove, taking into consideration any possible $\frac{1}{8}$-in. (3-mm) increase in the throat.

With partial-penetration groove welds, tension applied parallel to the weld axis, or compression in any direction, has the same allowable stress as the plate. Tension applied transverse to the weld axis, or shear in any direction, has an allowable stress equal to that for the throat of a fillet weld.

Just as fillet welds joining thick plate must have a minimum size because of fast cooling and the need to prevent excessive restraint, so must partial-penetration groove welds have a minimum effective throat t_e. Such minimum throats, based on the thickness of the thicker plate, are given in Table 6.21.

Table 6.21 Minimum Throat t_e of Partial-Penetration Groove Weld*

Material Thickness of Thicker Part Joined (in.)		t_e (in.)
To 1/4 incl.		1/8
Over 1/4	to 1/2	3/16
Over 1/2	to 3/4	1/4
Over 3/4	to 1-1/2	3/16
Over 1-1/2	to 2-1/4	3/8
Over 2-1/4	to 6	1/2
Over 6		5/8

* Not to exceed the thickness of thinner part.

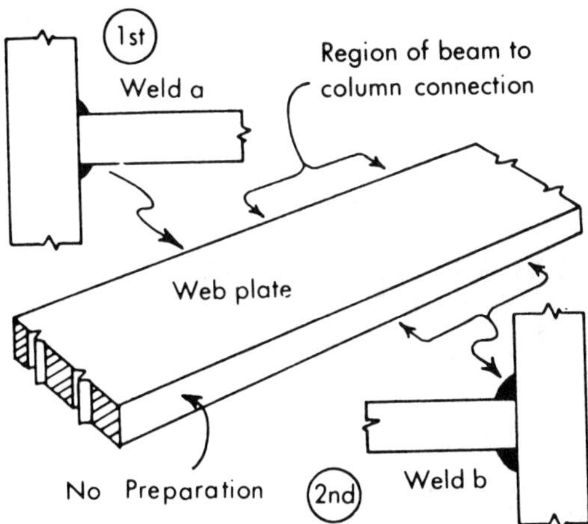

Fig. 6.65 Schematic of one method to enable heavier welds to withstand higher longitudinal shear.

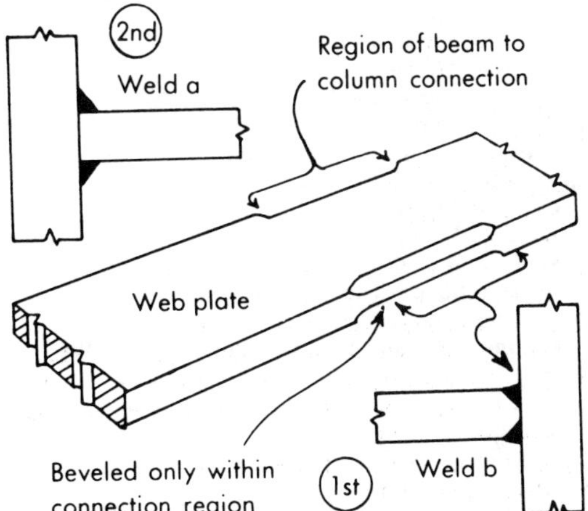

Fig. 6.66 Technique of beveling web at connection region.

In building up columns from steel plate, there are several ways for getting the heavier welds needed in the beam-connection regions to withstand the higher longitudinal shear produced by abrupt change in moment and to carry the tensile force from the beam flanges. Figure 6.65 illustrates one of the simplest methods. If the weld sizes are not too large, the column may be fillet welded with weld "a" along its entire length. Additional passes are then made in the beam-connection region to bring the fillet up to the proper size for weld "b". No joint preparation is required.

In Fig. 6.66, the web is beveled at the connection region only. Groove weld "b" is first made flush with the surface, after which fillet weld "a" is made along the entire length of the column.

SELECTED REFERENCES

6.1 AWS. *Structural Welding Code,* D1.1-84. Miami, Fla.: American Welding Society, 1984.

6.2 AISC. *Specification for the Design, Fabrication & Erection of Structural Steel for Buildings* (Effective November 1, 1978). Chicago: American Institute of Steel Construction, 1978.

6.3 T. R. Higgins and F. R. Preece. "Proposed Working Stresses for Fillet Welds in Building Construction," *Engineering Journal,* AISC, **6,** 1 (January 1969), 16–20.

CHAPTER 7

STRUCTURAL WALLS AND DIAPHRAGMS—HOW THEY FUNCTION

LORING A. WYLLIE, JR.

H. J. Degenkolb Associates, Engineers
San Francisco, California

7.1 INTRODUCTION

This chapter treats two commonly used structural elements—walls and diaphragms and how they function within a structure. Many buildings contain structural walls and virtually all buildings have floor and roof diaphragms which provide important and essential links to a stable structural system. However, since these elements are usually present in the average building and perform their structural role effectively without sophisticated analysis or design, building designers tend to overlook their importance. It is the intent of this chapter to review the basic functions of structural walls and diaphragms, how they work when made of different building materials, and how they can be properly analyzed and designed.

Everyone involved in building design or construction knows what a structural wall is, but the term defies an accurate, comprehensive definition in a concise manner. Traditionally, walls were essentially solid vertical elements built of masonry, concrete, stone, or wood with occasional openings to serve as doors and windows. These elements supported the gravity loads of the building as well as providing a measure of lateral stability. Modern construction, however, frequently emphasizes open areas with extensive glass at the perimeter and minimal structural walls. Interior structural walls sometimes complicate the relocation of interior partitions as building functions or uses change. Thus, many structural walls of modern buildings are isolated elements which seem to defy the basic definition of a wall, which is to enclose or separate spaces. It frequently becomes impossible to identify when a structural wall actually becomes a column in today's use. This may seem like an academic concern, but is a practical one since building codes generally have somewhat different requirements for columns than for walls. Thus, it is important for designers to understand how walls function so they may be properly designed to perform their intended role.

Structural diaphragms as used in building design usually consist of the floor and roof construction which provide in-plane stiffness and strength to the structure. The diaphragms tie the building together, allowing it to function as a unit and to distribute lateral forces to vertical resisting elements. Diaphragms also provide stabilizing forces to slender columns from the stiffer walls or frames that provide lateral bracing and stability for the structure. Although diaphragms are usually horizontal, they can be curved or sloped as is frequently the case in roof construction. They can also have vertical offsets at abrupt changes in floor grades. Diaphragms are often overlooked in design because in many buildings they are inherently suitable for their structural role without extensive analysis or design. However, in many sophisticated buildings, buildings with discontinuous wind bracing elements, industrial buildings with large floor openings, or buildings designed to be earthquake resistant, diaphragms must be given proper attention in design and construction.

This chapter first treats structural walls, their purpose, and how they function to fulfill those purposes when constructed of different building materials. A similar discussion of building diaphragms follows. Design considerations for walls and diaphragms are presented as they relate to building design as well as the significance of their elements. A section on general analysis considerations is provided to give guidance in modeling walls and diaphragms for computer or hand analysis calculations.

7.2 PURPOSES OF STRUCTURAL WALLS

Structural walls serve many purposes in a building, some structural and some nonstructural. Walls perform these various functions simultaneously and proper design must recognize these different conditions. Some of these effects are occasionally overlooked by designers and the results can lead to either structural distress or failure or to an unserviceable situation. The structural purposes of a wall are summarized in Fig. 7.1.

7.2.1 Resisting Gravity Loads

Resisting gravity loads is perhaps the most basic structural purpose of a wall. Walls resisting gravity loads are commonly called bearing walls. The loads result from dead loads, live loads, snow loads, and any other items that must be supported by the building. These loads cause axial forces and bending moments in a wall. The axial force part of the load is easily recognized and determined. However, the important bending moments resulting from gravity loads can be overlooked if the designer has not fully considered the details of construction and how the wall works in the structure.

Bending moments from gravity loads come from many sources. In continuous structures such as monolithic, cast-in-place concrete construction, the structural wall resists moments induced from the continuity with the floor and roof slabs and/or beams. In precast concrete or masonry construction, bending moments often arise in the structural wall because of an eccentric bearing of the floor or roof system from the center of gravity of the wall cross-section. For example, precast concrete floor or roof elements may be supported on a corbel or on one edge of a wall section, and the bearing load times the eccentricity of that bearing load from the center of gravity of the wall cross-section is a concentrated moment per unit length applied to the wall. That moment causes shears

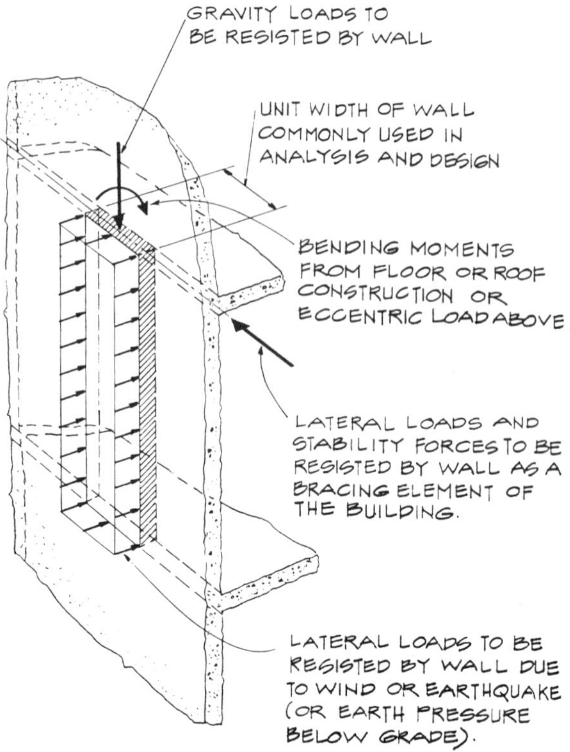

Fig. 7.1 Forces acting on a segment of a wall in a building.

and moment gradients in the wall, depending on the restraint conditions of the wall at the floors. Interior bearing walls also have bending moments resulting from unbalanced live loads, different span lengths of the floor or roof framing, or unsymmetrical support details for the framing.

Another source of bending moments in a wall is eccentric bearing from the wall above. This can be due to a change in thickness of the wall, an offset in the wall for some reason, or an eccentric bearing which sometimes occurs in precast or prefabricated construction. For example, when an exterior concrete or masonry wall changes in thickness from an 8-in. (200-mm) wall to a 12-in. (300-mm) wall at a floor and the outside faces are aligned for aesthetic considerations, a moment equal to the axial load times the eccentricity of the centers of gravity of the wall sections, 2 in. (50 mm), is applied at the joint. This moment must be distributed to the members depending on the continuity within the structure and the stiffness of the members.

7.2.2 Resisting Perpendicular Lateral Loads

Walls, both structural and nonstructural, resist lateral forces perpendicular to the face of the wall. Perhaps the most common of these lateral loads is wind on the exterior wall of a building. The wind pressure or suction causes a lateral load and the wall resists that load spanning vertically between floors or horizontally between cross walls or plasters. The spanning of the wall results in bending moments in the wall which must be considered in design in conjunction with the axial load and bending moments due to gravity loads.

Walls in buildings subjected to earthquake ground shaking also develop lateral forces perpendicular to the faces of the walls. For design purposes, coefficients are given in the earthquake-resisting design provisions of building codes for an equivalent lateral force which can be analyzed similar to a wind pressure or suction. It is extremely important that walls subjected to earthquakes are properly tied into the floors and roof. A major reason for failure of structural walls in earthquakes is an improper connection to the horizontal diaphragms.

Retaining walls or walls below grade resisting earth pressures are another form of the structural wall resisting lateral loads perpendicular to the face of the wall. Retaining walls braced at their top by a floor of the building or similar restraint are commonly called basement walls and can be designed to resist earth pressures in the same fashion as the above-grade wall is designed to resist wind pressures. Retaining walls not braced at their top which cantilever from their footing are the classic retaining wall and are discussed more fully in Section 7.9.

7.2.3 Resisting Parallel Lateral Loads

Structural walls resist parallel or in-plane lateral loads as bracing elements of a building. These in-plane lateral loads are generally due to wind or earthquake forces acting on the building as a whole although they can also be due to blast or other sources. The lateral forces are delivered to the wall by the floor and roof diaphragms or by other horizontal bracing systems. These in-plane lateral loads result in shear stresses within the wall as well as tension and compression stresses due to in-plane moments. These in-plane moments are sometimes called overturning moments because in typical walls they are the moments tending to overturn the wall or building.

The amount of lateral force resisted by a specific wall within a building is a function of the magnitude of the lateral loads acting on the building, the stiffness of the wall, and the location of the wall within the building. After the loads to be resisted by the wall at each floor and roof are determined, the wall can be analyzed to determine its in-plane shears and moments in each story. Structural walls come in all shapes and sizes from simple, solid rectangular shaped walls to walls of irregular elevation punctured with a regular or irregular pattern of openings for windows, doors, and building services. The wall must be analyzed by an appropriate procedure to determine the in-plane shears, moments, and axial forces in each element. These forces can than be combined with gravity loads for the design of the wall for adequate strength. These aspects of design and analysis are discussed more fully in Sections 7.7 and 7.8.

Another function of walls resisting in-plane forces as bracing walls, is to provide sufficient strength and stiffness to provide stability to the building. Columns or frames that comprise the remainder of the vertical support of the building are commonly designed as frames or columns braced against sidesway, relying on the walls and horizontal diaphragms for stability. This basic concept is expanded on in Section 7.4.1.

Walls resisting in-plane lateral forces have commonly been called shear walls. The term shear wall is sometimes confusing as shear forces may not be the critical force to be resisted by the wall. The term shear wall dates to the 1930s or earlier when the exterior walls of most buildings were either cast-in-place concrete or masonry. The term "shear wall" probably infers that these walls resist the building's "shears" or lateral loads, or else it comes from early wood diaphragm tests where "shears" were delivered from the diaphragm to the resisting wall elements. There is an emerging trend in the profession to call shear walls "structural walls" to avoid potential confusion as to the nature of forces to be resisted. Shear walls are treated in Section 11.3.7 and in other chapters of this book.

7.2.4 Other Functions of Structural Walls

In addition to the structural functions described, walls also serve a significant architectural and functional purpose in buildings. Walls often provide the exterior closure of the building, providing both protection from weather, wind, and the elements, as well as the aesthetic finish of the building. A structural wall as the exterior closure of a building provides these several functions at a single cost. Thus, it is often possible to gain structural benefit from the exterior cladding or enclosure at a cost comparable with the cost of a nonstructural closure such as complete glazing or other non-structural finishes. Exterior walls can be treated with various finishes to provide architectural interest to the building, such as special forming for concrete walls, decorative block or brick for masonry construction, and special plywood or lumber sidings for wood construction. These walls, properly detailed, will protect the building from weather, and provide insulation, although windows and other penetrations are always essential. Walls can easily accommodate such penetrations, especially in a regular pattern.

Structural walls of reinforced concrete or masonry can also provide protection against blast and other severe pressure exposures. Hardened structures to resist the effects of nuclear weapons almost universally have substantial perimeter walls of reinforced concrete or masonry. The mass of the walls also provides shielding against radiation from nuclear explosions. For further information on this subject, the reader is referred to Ref. 7.1.

7.3 EXAMPLES OF STRUCTURAL WALL SYSTEMS

This section is intended to illustrate the basic principles of how various structural wall systems work to meet the various functions described in the previous section. This section is not intended to be a design guide for various structural walls although references are given to other chapters of this handbook and other documents which should prove useful to the structural designer.

7.3.1 Concrete Walls

Concrete walls are a widely used building element. Many cast-in-place concrete buildings have concrete walls acting as both bearing and shear walls. Buildings utilizing precast concrete elements usually incorporate precast concrete wall panels which also serve as bearing walls and as shear walls. Many industrial buildings in the United states, known as tilt-up buildings, utilize precast concrete wall panels cast flat at the site on the slab on grade to minimize forming costs. The walls are then lifted into place to serve as bearing walls. Below-grade construction also widely utilizes reinforced concrete walls.

The design criteria for reinforced concrete walls are included in the *ACI Code* [7.2]. Concrete walls are treated as solid elements of unit length and designed separately for their axial load and out-of-plane bending and for their in-plane shear, bending, and axial force. Stiffness assumptions for shear walls are discussed in Section 22.4.

In the design for axial load and out-of-plane bending, a unit length of wall can be designed as a concrete column subjected to axial load and moment, considering the effects of moment magnification as appropriate for a slender column. Chapter 14 of the *ACI Code* does contain an empirical design method for reinforced concrete walls, but this method requires that for all conditions of loading, the combined axial load plus out-of-plane bending results in no tension at the various critical sections of the wall. This is usually described as requiring the resultant factored axial load to be within the middle third, or kern, of the cross-section. This empirical design method is valid only when the resultant force is within the kern of the cross-section for *all* conditions of loading, including wind or seismic forces perpendicular to plane of the wall.

While most cast-in-place concrete walls are generally of uniform thickness, precast walls can vary in cross-section when ribbed panels or those with architectural relief are utilized. Precast concrete wall panels often are subjected to their greatest design forces when they are transported, lifted, and erected, and the reinforcement in these panels must be designed considering these conditions. Connection details also require special attention for precast concrete wall elements as they are generally always weaker than comparable cast-in-place construction. Reference 7.3 gives some guidance for the analysis and design of such connections.

In non-seismic areas, concrete walls are sometimes designed as plain concrete walls without reinforcement. The design of such plain concrete walls must contain control joints and consider the possibility of cracking from shrinkage of concrete or other sources. Stresses from combined axial load and flexure are limited to values less than the modulus of rupture of the concrete. Design criteria for plain concrete walls can be found in Ref. 7.4. It is advisable in multistory construction to design concrete walls as reinforced walls as such walls have a great increase in ductility and integrity at nominal additional cost, which is highly beneficial in resisting unforeseen loads and conditions.

7.3.2 Masonry Walls

Masonry walls are historically the first walls used in building construction. In earliest times, stones would be stacked to form a building wall, sometimes using weak mortars or mud to fill the voids and sometimes using carved stones that fit neatly without mortar or filler. These walls were often sufficiently thick to provide adequate stability in spite of weak mortars and lack of steel reinforcement. Bricks and other prefabricated masonry elements were also made in early times to become a form of wall construction. Adobe bricks, made of clay soil with straw added as binder and baked in the sun or in ovens, are another early form of masonry.

Modern masonry construction consists of a wide range of masonry units assembled in a variety of methods. Most common are brick or other solid masonry units and hollow unit masonry such as concrete blocks. From a design or performance viewpoint, the most significant differences in types of masonry wall construction are whether the masonry wall is reinforced or unreinforced and whether the wall is a bearing or nonbearing wall.

Masonry walls are traditionally designed similar to concrete walls considering a unit length of wall and designing for the loads and moments applied. Empirical methods have been developed to aid in wall design and are contained in most building codes. If walls are unreinforced, it is essential that tension forces due to bending are prevented or sufficiently below the tension capacity of the wall. Lateral loads causing in-plane forces require consideration of the entire wall system for distribution of forces to the piers or elements in proportion to their rigidities. For more information on the design of masonry walls and additional references, see Chapter 24.

Another form of masonry wall construction is the infilled masonry wall, where a frame, probably of reinforced concrete, is built and then masonry is infilled between the columns to form a masonry wall. Such walls are non-load-bearing although they will be effective in resisting lateral loads. The masonry must be doweled or adequately fastened to the frame around its perimeter to maintain stability for wind or seismic forces perpendicular to its plane.

7.3.3 Walls in Timber Construction

Walls in timber construction (see Chapter 25) are inherently different from the essentially solid concrete or masonry walls. The basic gravity loads in a timber wall are carried by studs, nominal 2 × 4s or heavier, spaced at perhaps 16 in. (400 mm) on center. Where heavier loads occur due to a large floor beam or similar, a double or triple stud, or a post of the same nominal width as the studs, can be provided.

Figure 7.2 illustrates the key elements of a wood wall. A bottom plate (2 × 4 or heavier) is installed flat on the floor construction and nailed to the construction below. The bottom plate provides a bottom support for the sheathing and a bearing for the studs. Studs [2 × 4 or heavier at 16 or 24 in. (400 or 600 mm) on center] are erected with toenails to the bottom plate. The studs support the gravity loads, span vertically to support lateral loads perpendicular to the plane of the wall, and provide an attachment for the sheathing. On top of the studs, a double top plate is installed, consisting of two 2 × 4s or heavier which are nailed together and which have staggered splice locations with extra nails added at the splices. The purpose of the double top plate is to provide a continuous chord or tie for the structure plus a bearing and header for the joists above which might not align with the studs below. Sheathing material is applied to the face of the studs on one or both sides as desired. The purpose of the sheathing is to resist lateral loads in the plane of the wall as in a shear wall and to give bracing to the studs in their weak direction. The sheathing can also serve as a wall finish, or finish material can be applied over the sheathing.

Sheathing can be of many different materials. In modern construction when lateral forces are significant, plywood sheathing is the usual selection with appropriate thickness and nailing patterns to resist the design shears. Before plywood was readily available, wood sheathing consisted of straight or diagonally placed boards, 1 × 4, 1 × 6 or 1 × 8, nailed at all bearings. The diagonally sheathed boards were considerably stronger than the straight sheathed boards as a truss action was established. Gypsum board sheathing is commonly used in residential and light wood framing and is acceptable where in-plane shear forces are small. Wire lath and cement plaster, commonly called stucco, is another frequently used sheathing material. Other wood products such as particle board and proprietary materials can also be used for sheathing.

Figure 7.2 also illustrates a header which is a timber beam spanning over an opening in a wall. The header is usually selected not just for its strength but more for its stiffness so that deflections will be minimal and wall surfaces above will not be cracked or buckled. Headers are commonly supported on a cripple as shown in the illustration.

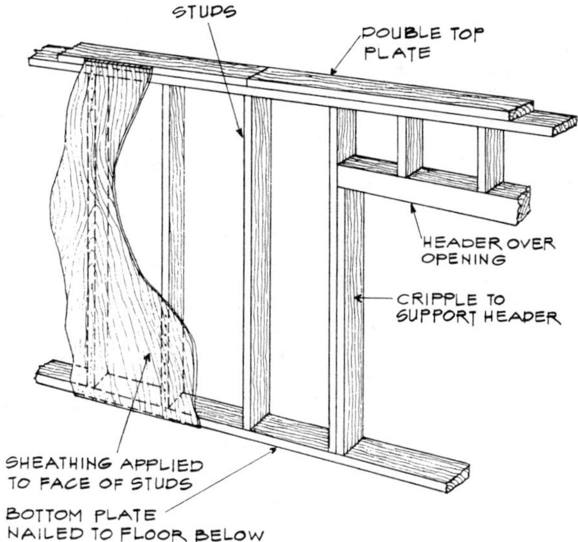

Fig. 7.2 Typical wood wall construction.

When timber walls act as shear walls, it is frequently the situation that analysis will show that an uplift or tension situation will occur at the end of a shear wall or shear panel. This is because of the relatively low gravity dead loads that are carried by the studs, combined with the usual situation that solid wall lengths are relatively short because of openings in the wall. Thus, it is necessary to provide a holddown to the wall or foundation below. A holddown is attached to the stud or post at the end of the wall panel and passes through the floor with a similar attachment or anchorage below. The holddown can be a metal strap or a rod attached to brackets or a similar detail. All elements of the holddown construction must be detailed to resist the design uplift. Holddowns are generally needed at both ends of a wall panel since lateral loads are reversible, except for earth pressure.

7.3.4 Walls in Light Metal Construction

Light gage metal construction (see Chapter 20) is becoming more common and is essentially patterned after timber construction. The elements of a wall in light metal construction are essentially identical to wood construction, with metal studs of various gages replacing the wood stud, and top and bottom metal plates replacing their wood counterparts. The pieces are generally tack welded together and sheathing material is fastened with self-tapping screws. Diagonal metal straps are often added both to provide erection stability and provide additional strength to resist lateral loads. Section 20.8 has additional discussion on metal stud walls.

7.4 PURPOSES OF STRUCTURAL DIAPHRAGMS

Structural diaphragms are one of the most important structural building elements but probably the least understood by the average design engineer. Diaphragms are frequently taken for granted, as in most buildings they are present in the form of floor and roof slabs or framing and they work fine without any special design considerations. However, in buildings with discontinuous lateral force bracing elements, large or numerous floor openings as in many industrial structures, or in buildings designed for seismic resistance, the proper design of diaphragms is essential for satisfactory structural performance.

7.4.1 Bracing of Columns, Walls, and Beams

Structural diaphragms generally have tremendous strength and stiffness in their plane. Even when floors or roofs are somewhat flexible to walk on, they are extremely stiff in their own plane. This inherent stiffness and strength allows them to tie the vertical elements, columns and walls, at each level together and provide lateral bracing to those vertical elements that require bracing. Also, the diaphragm braces the top or compression side of floor beams against lateral buckling.

Perhaps several examples will best illustrate this function of a structural diaphragm. Consider a one-story warehouse structure, perhaps built as a "tilt-up" building with concrete wall panels around the perimeter and a wood-framed roof system supported by pipe columns. The plywood of the roof system is the diaphragm. The plywood is stabilized against locally buckling up or down by nailing to the roof joists and beams. Meanwhile, nailing the plywood to the joists and beams provides lateral bracing to the top edge of the joist and beams, which is the compression portion of the beams generally requiring bracing to prevent lateral buckling. The plywood diaphragm is attached to the exterior concrete walls which provide the lateral bracing for the building. The in-plane stiffness and strength of the plywood diaphragm thus extends the lateral bracing of the walls throughout the roof structure and provides lateral support or restraint against translation to the top of the interior pipe columns. Obviously, some local bracing may be needed if the pipe column terminates at the bottom of the roof beam to prevent local rotation of the beam and column end.

Another example that can easily be visualized is a high-rise steel-framed building with its perimeter frame providing lateral bracing for wind or seismic loadings. The interior beam-to-column connections might consist only of beam web connections, or simple connections. Since these interior frames are not moment resisting, they rely on the stiffness and strength of each floor diaphragm, either concrete slab or metal deck with concrete fill, to provide lateral support or bracing to the interior columns so they will function properly. This steel structure relies on the perimeter frames for bracing and on the floor diaphragms to extend that bracing to the interior columns. Forces on floor diaphragms are discussed in Section 11.2.

7.4.2 Transferring Lateral Forces at Each Level

One of the most important functions of a diaphragm is the transferring of lateral forces at each level to the vertical bracing elements within the building. Consider the simple building in Fig. 7.3, with

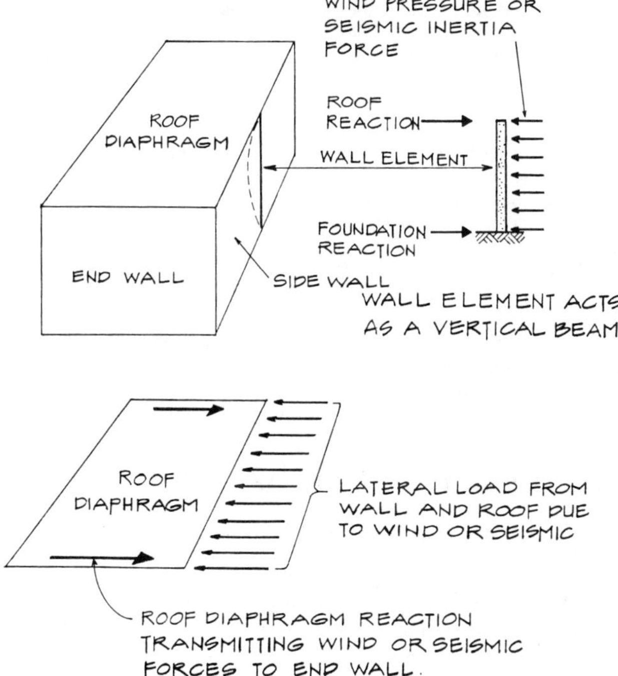

Fig. 7.3 Schematic diagram of a simple building with wind or seismic forces transmitted to end walls by roof diaphragm.

exterior perimeter walls and a roof diaphragm. Wind pressure or seismic inertia forces act on the side wall and the wall acts as a vertical beam, transferring lateral loads to the foundation and to the roof. Those loads transferred to the roof are resisted by the roof diaphragm. The roof diaphragm in this simple structure then acts as a deep horizontal beam and transfers the lateral loads to the end walls, which in turn brace the building in this direction and transfer these forces to the ground. This same basic concept acts in both directions of the building as the diaphragm works both ways, transmitting lateral forces to the vertical bracing elements.

In real buildings, this concept of each diaphragm transferring wind or seismic forces to the vertical bracing elements (walls, moment-resisting frames, diagonally braced frames, etc.) is conceptually treated the same in both directions of the building as the simple structure in Fig. 7.3. At each floor and at the roof, the diaphragm can be analyzed for forces it must transfer, combined with any transfers of shears from one bracing element to the next as discussed in the next subsection, and then designed to resist adequately the design forces that may be imposed.

7.4.3 Transferring Lateral Forces at Discontinuous Vertical Bracing Elements

When structural walls or frames are discontinuous at one level in a building, a significant redistribution of lateral forces to vertical bracing elements occurs. Such redistribution can occur when selected walls terminate at a midheight within the building or when stiff, rigid perimeter frames change to an open, flexible frame at a high lobby entry to a building. Changes in stiffness of vertical bracing elements at a floor level can also cause redistribution of lateral forces.

When such redistribution of force to vertical bracing elements occurs, it is the function of the diaphragm at that level to provide the necessary stress transfer. The diaphragm must be strong enough at that level to resist the diaphragm forces that result from the load transfer. The diaphragm must also be stiff enough to allow the forces to be transferred. For example, if one compares the stiffness of thick reinforced concrete shear walls to a thin diaphragm of concrete or metal deck with fill, the diaphragm transferring load may be the most flexible element in the continuous stress path of lateral forces down the building to the foundations. In such a case, the flexibility of the diaphragm should be considered in the analysis. This will frequently require three-dimensional computer analysis with finite element representation of the floor diaphragm. See Section 7.8.2 for more discussion on diaphragm analysis.

At levels where large transfers of lateral force between vertical bracing elements occur in a floor diaphragm, it is not uncommon for the floor at that level to be of special construction. A concrete slab may be thicker and it may also contain additional continuous reinforcement, as well as additional chord and "drag" steel, as discussed in Section 7.5. A metal deck floor system may have a thicker concrete fill. In steel-framed buildings, it is not uncommon to add diagonal steel members beneath the floor system to form a truss with the regular framing to transfer large diaphragm forces. Special steel framing connections that transmit diaphragm tension or compression forces in addition to the usual vertical shear at beam ends are frequently required. In concrete framed buildings, additional continuous reinforcing steel may be added to beams or slabs to resist similar diaphragm tension forces. In wood diaphragms, steel straps and ties are common in the properly designed diaphragm.

Even when vertical bracing elements are not discontinuous, diaphragms generally transfer forces between vertical bracing elements. In any multistory building, the relative stiffness of the various vertical bracing elements may vary significantly at different heights within a building. The diaphragm must transfer these relatively small forces back and forth between vertical bracing elements to allow the building to function consistent with the stiffness and strength of all of its elements.

7.5 ELEMENTS OF STRUCTURAL DIAPHRAGMS

Before discussing structural diaphragms further, it is important to understand the various elements of a diaphragm and their purpose. Many of these elements and concepts have been developed for seismic-resistant design requirements, where diaphragm failures have led to increased damage in buildings subjected to strong ground shaking. However, these same concepts are valid for buildings in nonseismic regions or zones of low seismicity where severe wind storms will create similar demands on structural diaphragms for strength.

7.5.1 Shear Web

The shear web is the key element of a diaphragm. If one considers a diaphragm as a deep beam in a horizontal position, the beam web is synonymous with the shear web of the diaphragm. The shear web resists the shear forces within the diaphragm. In a truss-type diaphragm without a solid slab element, the diagonal and other members act like web members of a truss to provide the shear web.

7.5.2 Diaphragm Chord

The diaphragm chord is a boundary member or added reinforcement capable of resisting tension or compression at the perimeter of a diaphragm. The chord is similar to the chord of a truss or the flange of a beam. Chords can be beams or framing members which are connected together for tension and compression forces or chords can simply be continuous reinforcement as in the edge of a concrete slab.

While diaphragm chords are always required at diaphragm boundaries, they are also often needed at large openings in floor or roof slabs or at reentrant corners.

Obviously, diaphragm chords must all be capable of resisting both tension or compression, as lateral loads can act in either direction on both axes of the building. Figure 7.4 illustrates diaphragm chords for one axis of the building. Obviously, they are appropriate in both axes of the building.

The calculation of chord force usually ignores such complexities of deep beam action and shear lag which might be theoretically acting in a diaphragm. The usual procedure consists of calculating a maximum moment within the diaphragm and then determining the force simply by dividing the moment by the distance between chords. For example, in Fig. 7.4 assuming a 100 by 200 ft diaphragm with shear walls only at the ends and a uniform force of 300 lb/ft applied to the diaphragm, the maximum diaphragm moment would be

$$M = \frac{wl^2}{8} = \frac{0.300(200)^2}{8} = 1500 \text{ ft-kips}$$

The corresponding chord force would be

$$F = \frac{M}{d} = \frac{1500}{100} = 15 \text{ kips}$$

This force acts at the midspan of the chord and reduces toward the corners of the building.

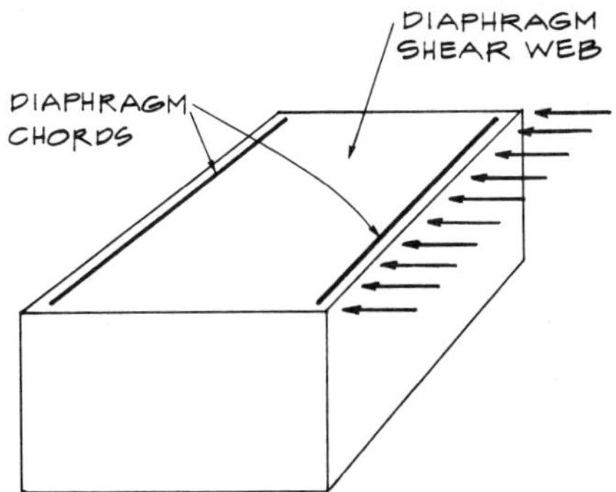

Fig. 7.4 Diaphragm chords resist tension at edges of diaphragm.

7.5.3 Drag or Collector Members

Drag or collector members are added to a diaphragm along lines where vertical bracing elements exist for a portion of the diaphragm length. The purpose of the drag or collector member is to collect the approximately uniform shear from the diaphragm shear web and drag or transfer that shear to the vertical bracing element. Figure 7.5 shows a simple diaphragm with three concrete shear walls and the drag or collector members.

Sometimes the shear web will have sufficient strength without adding additional strength for this collection of forces, but the designer should consider that the strength present may be needed for other building purposes. For example, a concrete slab with shrinkage or temperature reinforcement in the direction of a needed collector is required to control cracking from concrete shrinkage and temperature changes and may be already at its yield stress when collector reinforcement acts to resist lateral forces. In such a case, continuous reinforcement can easily be added to the slab for the drag or collector member.

Drag or collector reinforcement is usually calculated assuming a simple uniform shear distribution in the shear web and calculating the force that needs to be collected and dragged. For example, consider in Fig. 7.5 the left edge shear wall which is 20 ft (6.1 m) long at the end of a 100-ft (30 m) diaphragm. If 50 kips (220 kN) is to be transferred to the shear wall, the drag reinforcement can be designed to resist a force of

$$\frac{80}{100}\,(50 \text{ kips}) = 40 \text{ kips}$$

7.5.4 Tension Ties

The concept of a diaphragm providing a tension tie holding the walls into the building and extending across the building is a common-sense building technique known for centuries. Historically, buildings in Europe of masonry wall and timber floor construction would have steel or iron rods

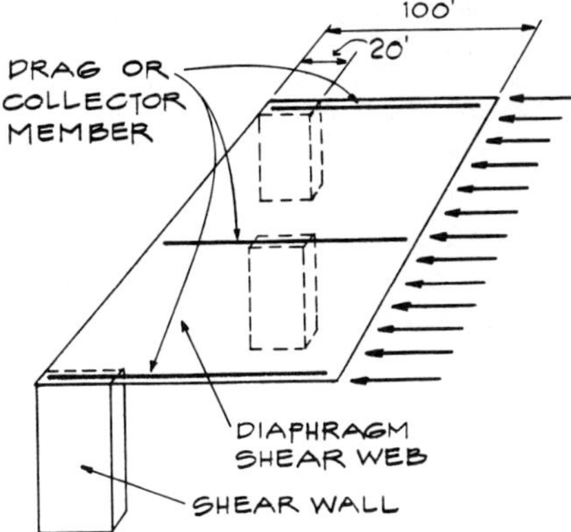

Fig. 7.5 Drag or collector members transfer forces from diaphragm to bracing element.

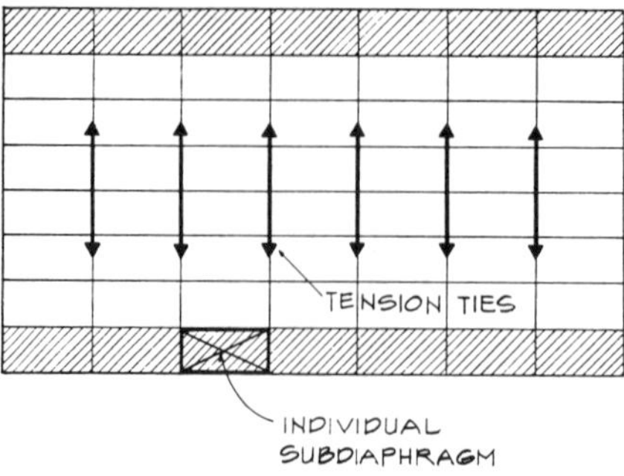

Fig. 7.6 Diaphragm subdiaphragm braces exterior wall between tension ties across diaphragm.

extending the width and length of the building and connected to decorative rosettes or steel bars on the exterior face of the building. These ties hold the structure together and greatly minimize the tendency of walls to fall out during an earthquake or severe windstorm.

Requirements for tension ties in building diaphragms are not clearly outlined in current building codes. Codes frequently require that walls be anchored to floors and walls for some nominal amount, but seldom state a requirement for tension capacity across the entire building. However, the importance of providing such a tension tie cannot be overemphasized.

Tension ties are easy to provide in building diaphragms and are often naturally present. Concrete slab or beam systems frequently have continuous reinforcement which provides a tie. Steel-framed buildings have an inherent tie when steel framing members align and are positively connected as supports. Timber framing usually requires light metal straps or similar details as plywood edge nailing cannot be relied on for significant tension strength.

7.5.5 Subdiaphragms

The concept of subdiaphragms has been expressed in recent years to explain the concept of locally transferring loads in a portion of a diaphragm. It is particularly appropriate in wood diaphragms

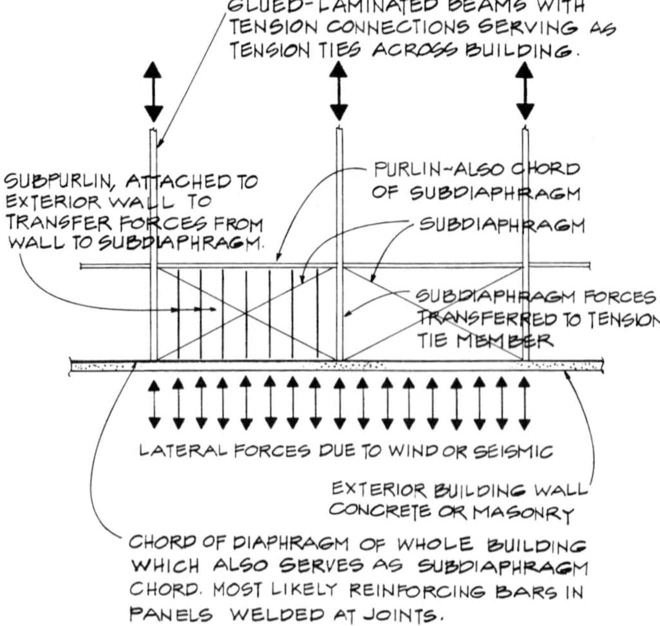

Fig. 7.7 Detail of a subdiaphragm in a wood roof system bracing a concrete or masonry exterior wall.

connected to concrete or masonry walls or in steel framed structures where the diaphragm consists of steel framing members acting as a horizontal truss.

Figure 7.6 illustrates the concept of a subdiaphragm for the roof of a one-story building. Lateral forces due to wind or seismic loads acting on the exterior wall are transferred at close spacings, perhaps every 4 ft (1.2 m), to the subdiaphragm. The subdiaphragm transfers these forces to the major girder lines which serve as tension ties across the building and into the roof diaphragm. Obviously, this same concept must be executed in both directions of the building construction, as the same concept obviously applies and is appropriate in both directions.

Figure 7.7 illustrates the concept in more detail in a "tilt-up" type building with concrete exterior wall panels and a panelized wood roof system. While this is a particular example, it illustrates the principles involved. Lateral forces acting on the concrete wall (pressure or suction due to wind or seismic inertia forces) are transmitted to subpurlins by positive wall to subpurlin ties. The subpurlin transfers the forces to the girders, which serve as tension ties, as a "mini" diaphragm with its own chords. The diaphragm as a whole then transmits these lateral forces to the end walls or other bracing elements.

Many diaphragms do not have subdiaphragms. For example, the cast-in-place concrete slab simply works as a diaphragm for the whole area. Subdiaphragms are needed to rationalize force transfer only when unitized construction results in the need for numerous stress transfers to provide a continuous stress path for lateral forces.

7.6 EXAMPLES OF STRUCTURAL DIAPHRAGMS

This section is intended to illustrate how structural diaphragms act when constructed of the various materials common to construction. This section is not intended to be a design guide for various materials, but a summary of how various building materials fulfill the important role of a structural diaphragm.

7.6.1 Diaphragms in Concrete Structures

Cast-in-place reinforced concrete slabs or slab-beam systems are one of the easiest types of diaphragms to understand. The slab acts as the shear web of the diaphragm and chords and collectors can easily be accomplished by adding reinforcing bars in beams or in the slab as appropriate. The continuous reinforcement inherent in monolithic concrete floor and roof systems provides a structural tie across the building. Chapter 22 has more detailed discussion on concrete floor systems.

Precast floor systems offer more of a challenge to provide a sound structural diaphragm. When diaphragm stresses are significant, a cast-in-place topping slab is used over the precast elements. This topping slab can be treated in the same manner as the cast-in-place slab regarding diaphragm forces and stresses, while the precast elements simply prevent the relatively thin topping slab from buckling under high in-plane stresses. Precast concrete diaphragms without a topping slab must be interconnected by adequate fasteners to transmit all shear, tension, and compression forces at precast element boundaries. Fasteners generally consist of welding steel plates or bars between inserts in adjacent panels. The inserts must be well anchored back into the precast element so that tension forces are transferred to the reinforcement of the precast element. Another form of fastener between precast elements is a cast-in-place closure strip with reinforcement extending from adjacent panels into the cast-in-place concrete. Such cast-in-place closures are particularly beneficial at diaphragm boundaries or other locations where cords or drag members are required, as continuous reinforcing bars can be added in the closure, thus eliminating the need to weld chord and drag reinforcement.

Cast-in-place post-tensioned slabs can also serve as effective diaphragms. Analysis of such diaphragms is similar to cast-in-place conventionally reinforced slabs. When shear stresses are high in the diaphragm, it may be necessary to add continuous bonded reinforcement. Bonded reinforcement is also necessary for chord and drag reinforcement.

7.6.2 Diaphragms in Steel Structures

Floor diaphragms in steel buildings are generally a concrete slab or metal decking with concrete fill. The concrete slab diaphragm is identical to the slab of a concrete building except that proper bond or attachment must be provided to the steel floor beams, both to stabilize the compression flange of the steel beam as well as to interconnect the diaphragm to the steel framing to transfer diaphragm forces to the steel frame. This generally requires encasing the entire steel beam with concrete or providing some welded stud connectors on the top flange of the steel beam.

Metal deck diaphragms with concrete fill (see Section 20.11) rely on the concrete to provide the stiffness while tensile strength is provided by the metal decking itself, added reinforcing in the concrete fill, and steel beams of the floor framing. A key requirement of a metal deck and concrete diaphragm is the proper interconnection of all elements. The metal deck must be securely fastened to the steel beams or joists of the frame, which is usually accomplished by puddle welds through the decking. Proper welding procedures are necessary to prevent burning holes in the deck and providing an ineffective weld and stress transfer. The concrete fill must be thoroughly bonded to the metal decking, which is most effectively done by lugs or embossments in the metal deck which dramatically increase the bond. The development of composite metal deck systems for gravity loadings has greatly improved this stress transfer for diaphragm loads. Oil, grease, or sawdust spilled on the decking during construction and not thoroughly cleaned can destroy bond if such lugs or embossments are not present. The important thing to remember is that diaphragm forces must constantly be transferred between the steel framing and the concrete fill and that a complete stress path must be provided. For more information on metal deck diaphragms, see Porter and Greimann [7.5].

Metal deck roof diaphragms without concrete fill are also used on many structures. These diaphragms are considerably weaker than those with concrete fill. Manufacturer's test data should be consulted for allowable diaphragm stresses in such applications.

In many steel-framed buildings, particularly industrial or warehouse structures with long span trusses, the roof diaphragm is provided by diagonal steel bracing and struts. Such a system is analyzed and designed as a horizontal steel truss, utilizing purlins and vertical truss chords as members of the horizontal truss bracing system. Again, the most important consideration is to provide a complete stress path and then to adequately resisting moments caused by eccentricities which may result in connections.

7.6.3 Diaphragms in Wood-Framed Structures

In modern wood-framed construction, diaphragms consist of plywood or similar sheathing material applied over the floor framing. The plywood or sheathing acts as the shear web while floor framing with proper connections forms long continuous elements which can resist tension from the chords, ties, and drag elements. For example, in light wood-framed construction, the double top plate of wood stud walls can serve as an effective chord if the splices in the two plates are sufficiently staggered and the plates are well nailed together. In other cases, joists are end-lapped at bearings and nailed or bolted together or end-butted with metal connectors to resist tension and sometimes compression as appropriate. Shear strength of the plywood diaphragm depends on panel layout and edge nailing while appropriate chord strength must be provided in the framing. For more information on wood-framed diaphragms, see Section 25.15 and Ref. 7.6.

In older wood-framed structures built before plywood was readily available, diaphragms were constructed of straight or diagonally sheathed boards. Diagonal sheathing is considerably stiffer and stronger than straight sheathing as the triangular arrangement of the framing joists and trusses or beams and flat sheathing boards at about 45° provides considerable strength. Greater strength yet was historically provided by double diagonal sheathing, or two layers of diagonal sheathing installed with the boards of one layer at an angle of 90° to the boards in the other layer. Diagonally sheathed diaphragms require careful detailing of all framing components in order to provide a complete stress path and resist all components of force. For more information on straight and diagonally sheathed diaphragms, see Ref. 7.7.

7.6.4 Other Diaphragm Materials

Several other diaphragm materials have been developed in recent years for roof construction on single or low-rise construction. Many of the systems are proprietary or involve proprietary materials and specific information should be obtained from the manufacturer or supplier.

Diaphragms utilizing various types of lightweight insulating concretes have been used applied over light gage metal decking in industrial and low-rise office buildings. The metal decking generally provides strength for roof gravity loads while the insulating concrete provides stiffness for diaphragm action. The combined metal and insulating concrete system is relied on for diaphragm strength. Such systems have the obvious advantage of including insulation material in the structural system. Perhaps the greatest difficulty with such systems is providing sufficient chord strength and edge anchorage and interconnection between diaphragm and framing materials.

Gypsum diaphragms have also been used for roof construction and have similar advantages to diaphragms of insulating concrete. Small metal subpurlins constructed similar to open web joists are installed between steel beams or joists. A gypsum panel is installed between supports on the lower flange of the metal subpurlin to serve as a form material. Light mesh similar to poultry fabric is placed over the subpurlins and liquid gypsum is poured to provide a continuous diaphragm also capable of supporting limited gravity loads such as normal roof loadings.

7.7 DESIGN CONSIDERATIONS

Structural design is an art rather than a science, although it relies considerably on science to achieve satisfactory solutions. Structural design relies on experience, professional judgment, a thorough understanding of structural performance, a thorough understanding of construction techniques and procedures, and a determination to express strongly the needs of the structural system to the other design professionals. Buildings can be designed in many different ways, but the successful building from a structural point of view contains materials, bracing elements, and diaphragm layouts that complement the functional and aesthetic considerations of the building without compromising the professional judgment of the structural designer.

Good design must be coupled with proper analysis and comes in several stages. First, a preliminary design is established, based largely on professional judgment and experience, usually utilizing limited rough calculations to determine the order of magnitude of forces and stresses. Then the analysis phase follows to determine accurately the forces that must be resisted. The analysis phase is followed by design, or the final design of the details of the structural members and their connections to resist satisfactorily the forces and stresses determined by the analysis. Both the preliminary design, or layout, and the final design, or detailing, must be well conceived to achieve a sound structural design.

Several important factors must always be considered in the design of walls and diaphragms. These factors are related to the purposes of walls and diaphragms as discussed in Sections 7.2 and 7.4, respectively. When walls are incorporated in a building, their layout to support gravity loads is a key design issue. Perhaps, even more important is their layout to resist lateral loads. For this purpose, they should be adequately distributed throughout the building to minimize building torsion, of adequate length for stability, and of proper proportion to accommodate required openings. Diaphragms must be of adequate strength and stiffness to fulfill their purpose, and the layout of walls or other vertical bracing elements dictates the design of the diaphragm elements. Building setbacks or large floor openings are major design factors. Volume change considerations, such as thermal changes, are also a prime consideration in the layout and design of structural walls and diaphragms.

Material selection is largely a function of building type, economics, and fire-resistive requirements. The materials selected for walls and diaphragms must have adequate strength and stiffness to fulfill all their functions. For example, material selection for floor diaphragms is usually based on an economical floor system to support dead and live loads, and after that selection has taken place, the designer considers what modifications might be necessary for proper diaphragm performance.

The most important element in design is attention to detail and ensuring that a complete stress path is always provided. It is relatively easy to design the basic elements of building construction. It is much more difficult to design properly the construction details to provide a complete and adequate stress path while also permitting structural elements to easily interconnect, which leads to an economic design. The design of details is an important and imperative task of the structural designer.

Since good design is so dependent on sound judgment, it is difficult to give any exact prescription to good design of walls and diaphragms. Each building is unique with its own specific design needs and requirements. Examples 7.10.1 through 7.10.6 illustrate various lessons in design of walls and diaphragms leading toward sound structural solutions.

7.8 GENERAL ANALYSIS CONSIDERATIONS

Analysis consists of manual or computer-aided calculations to determine the forces in structural members. The design process involves analysis combined with each stage of design. Analysis should be appropriate to the design stage and the degree of accuracy required. For many phases of design, manual calculations based on appropriate assumptions provides an adequate answer. For other phases of design, detailed computer analysis may be required to define adequately the stresses within the structural members. Detailed analysis considerations for walls and diaphragms generally pertain to the lateral force resisting system of buildings. Gravity load analysis of walls in conventional structures is relatively straightforward, provided discontinuities or large openings do not occur. However, structural walls and diaphragms are always the major portion of the wind- or earthquake-resisting system, and appropriate analysis is always required.

Analysis is a function of the assumptions made by the analyst. Once the analysis assumptions are determined, the remainder of the analysis is simple (or maybe not so simple) mathematics to arrive at the results desired. The assumptions thus become the key elements of the analysis. If the assumptions are crude, the results will be crude, despite the apparent precision of the mathematical process that occurs. If the assumptions involve detailed and precise models of material performance, the mathematics may become tedious but results may be of high precision. If the assumptions are inappropriate, the results will be inappropriate and misleading. The assumptions must be carefully considered so the results will be of sufficient accuracy and not misleading. The following sections will attempt to discuss some of those assumptions.

7.8.1 Analysis Assumptions for Walls

The analysis of structural walls generally consists of two phases. The first is obtaining a proper rigidity of the structural wall so the lateral forces of the building can be properly distributed to the resisting elements. The second phase then involves analyzing the wall for the applied lateral loads to determine the internal stresses within the wall. Both of these phases involve assumptions which must be properly considered.

Appropriate assumptions to determine wall rigidity depend on the type of building and proportions of walls involved. In most cases, the actual wall rigidity is unimportant but the *relative* rigidity is the important quantity needed for proper analysis. For example, if a building is of constant height with all walls of equal height, length, and thickness, all walls have an equal relative rigidity and no more analysis of this type is needed. However, real buildings are not this simple and relative rigidity calculations are usually required. Factors that are frequently considered in some detail are the shear and bending deformations in the wall, assuming a uniform section without openings.

The effects of openings in the wall are sometimes ignored or sometimes modeled to varying degrees of success. A regular pattern of openings may permit the use of a computer frame program that includes joint rigidity, or the rigidity of the wall portion connecting the piers and spandrels which must be considered with this type of analysis. In walls with openings, it is generally inappropriate to consider the relative rigidity of the piers between openings alone, assuming the pier fixed against rotation at top and bottom. Proper boundary conditions must be considered, as they are usually most significant. For random opening patterns in walls where a reasonable degree of accuracy is desired, it may be necessary to use finite element representation of the wall elements to obtain proper rigidity and stress distribution within the wall. Clearly, the assumptions used in analysis must be consistent with the accuracy desired and the mathematics the analyst is willing to perform.

The fixity at the base of a wall is another assumption that is often overlooked. It is not uncommon to assume that the base of a wall is fixed against rotation. This is particularly common for tall slender walls of constant proportion as the wall is analyzed as a vertical cantilever and cantilevers usually are fixed against rotation at their support. However, this vertical cantilevered wall is supported on soil or piles or some foundation system which will deform under load. Under lateral loads, the base moment of the wall exerts uneven foundation pressure resulting in a rotation of the base of the wall. Although this rotation is small, it frequently can be very significant in determining the relative rigidity of various walls. Example 7.10.6 compares two 30-ft (9.1 m) high walls, one 10 ft (3 m) long and one 30 ft (9.1 m) long, under differing analysis assumptions. This example illustrates the significant influence of foundation rotation on relative wall rigidity.

Figure 7.8 also illustrates the effect of foundation fixity as well as wall height. Assume that three 40-foot long walls of equal thickness and construction exist in a two-story building. Two walls extend to the roof and are two stories high. The third wall only extends up one story. Ignoring building torsion due to the location of the walls within the building, their relative rigidities will be determined by varying assumptions under lateral loads as shown. In the upper floor there is no question of shear distribution with each of the two-story walls resisting one-half of the 50-kip (220 kN) shear, or 25 kips (110 kN) each. However, the shear distribution in the lower story depends on the assumptions of analysis. Figure 7.8*a* illustrates the ordinary assumption by length or story shear deformation which would result in 33 kips (150 kN) per wall or equal force to each wall of equal length.

If a similar analysis is performed by computer with foundations fixed against rotation, a shear distribution as shown in Fig. 7.8*b* is obtained with 28 kips (120 kN) in the two-story walls and 44 kips (200 kN) in the single story wall. This reflects the deformations in the second story of the two-story walls which influence deformations of the vertical cantilevered wall in the lower story. Another assumption that could be made is that the walls are rigid with respect to the foundations and that the base of each wall will rotate an equal amount. Since equal base rotation means equal base moments, statics will give the results in Fig. 7.8*c* with 25 kips (110 kN) in the lower story of the two-story walls and 50 kips (220 kN) being resisted by the one-story wall. However, the two-story wall will have more vertical load to support and will most likely have a wider footing at its base. If it is assumed that the two-story walls have footings twice as wide as the single story wall and that the walls are rigid with respect to the soil and thus all footings rotate an equal amount, then the base moment resisted by the two-story walls is twice the base moment of the one-story wall, resulting in shear distributions as shown in Figure 7.8*d* with 35 kips (160 kN) in the high walls and 30 kips in the low wall. Note, by coincidence, the similarity of these results to the shear length assumption in Fig. 7.8*a*, which is probably the least accurate of the four assumptions studied.

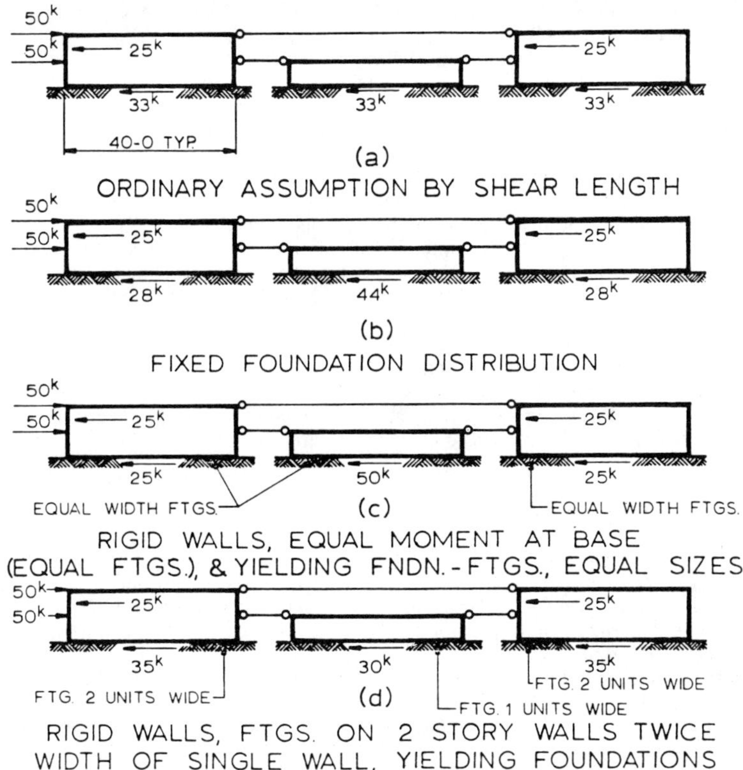

Fig. 7.8 Shear resisted by three walls based on different analysis assumptions.

The purpose of the above examples has not been intended to indicate a need for involved analysis or sophisticated assumptions for all structural walls. Rather, it is an attempt to illustrate that factors in analysis usually neglected by assumption may be more significant than those considered in detail. In many structures it is unimportant as there is adequate strength or redundancy to resist the applied forces. However, in some structures these varying assumptions may be appropriate to consider. But, most important is the need for structural engineers and designers to understand the significance of the assumptions they make in analysis, so that the analysis and resulting design are appropriate and adequate to perform satisfactorily.

7.8.2 Analysis Assumptions for Diaphragms

Diaphragm analysis is an often overlooked facet of building design, and the assumptions used in analyzing diaphragms can make a tremendous difference in the results obtained from the analysis. A common assumption used in building analysis is that diaphragms are infinitely rigid. This is a convenient assumption for computer analysis as it eliminates many factors which can complicate analysis. However, particularly in low-rise buildings, the diaphragms are usually more flexible than the walls which are analyzed with great care, and more appropriate results can be obtained considering the flexibility or rigidity of the diaphragm. Consider the wood or light metal diaphragm or even a thin concrete slab acting as a diaphragm spanning perhaps 100 ft between rigid concrete walls. The diaphragm is orders of magnitude more flexible than the walls, and a proper analysis should reflect this fact.

Example 7.10.7 illustrates two similar buildings with an interior wall B, located in a different position in each building. The example requires the calculation of force distribution to the walls assuming both a rigid and a flexible diaphragm system. Note in both cases that the diaphragm assumption makes a substantial difference in the loads assigned to each wall, particularly in building II with the greater torsion or eccentricity between center of load and center of rigidity. Note also that walls perpendicular to the applied loads resist substantial torsional forces inherent in the rigid diaphragm assumption.

In major structures where bracing walls or elements are discontinuous at a particular level or the plan dimensions of the building change substantially, a careful structural analysis properly modeling the stiffness of the diaphragms is essential if the proper interpretation of structural performance is to be determined. Many computer programs exist today where the walls or bracing elements can be located in a three-dimensional model connected by diaphragms with their stiffness modeled by finite elements. With both vertical bracing elements and horizontal diaphragms modeled with elastic properties reasonably representing their stiffness in a single building representation, an accurate assessment of load distribution can be obtained in analysis.

In seismic-resistant design, the discontinuity of major structural walls at an intermediate level should be avoided if at all possible, as the discontinuity of stiffness accentuates dynamic response. In such cases, an appropriate model of the structure should be subjected to a dynamic analysis to judge properly the effect of the discontinuity in the building's bracing system.

Major openings in a floor or roof diaphragm can also affect the distribution of forces to bracing elements. This is particularly true in some industrial buildings where major pieces of process equipment may extend through openings in several floors, leaving the structural diaphragm resembling a slice of Swiss cheese. In such a case, careful computer modeling is necessary or a detailed set of free body diagrams are needed to analyze properly the force transfers.

7.9 RETAINING WALLS

Retaining walls that retain earth are a somewhat unique type of wall and, thus, they will be briefly discussed in this separate section. Retaining walls, as discussed here, are cantilever retaining walls constructed of reinforced concrete or masonry. Walls retaining earth which are braced at their top, commonly called basement walls, are not cantilevered walls and are designed following the basic principles discussed in Sections 7.2 and 7.3.

The pressure that the earth exerts on a retaining wall should be based on the properties of soil and backfill, the amount of positive drainage provided the soil behind the wall, plus any surcharge effects due to loads on the soil being retained. These loads should be obtained from a geotechnical consultant or based on long experience with similar soil conditions. Positive drainage by weepholes or drain tile in the backfill plus permeable backfill cannot be overemphasized, as a tremendous increase in pressure can result from rainwater or other trapped water which cannot flow from behind the wall.

There are two basic types of footing design for cantilevered retaining walls. The gravity wall or footing with a large heel, as illustrated in Fig. 7.9*a*, relies on the gravity weight of the earth on top of the heel for stability. Alternatively, a retaining wall with a footing having a large toe, as illustrated in Fig. 7.9*b*, places the foundation forward to intersect the natural thrust line of the forces applied. Both types of wall foundations have their advantages. The retaining wall with a large toe requires less excavation into the soil to be retained while the retaining wall with a large heel has less footing interfering with the downhill uses beneath the wall.

Conventional construction practice for retaining walls results in the concrete foundation being cast first with dowels extending upward for the wall. This causes a lap splice in the reinforcing bars

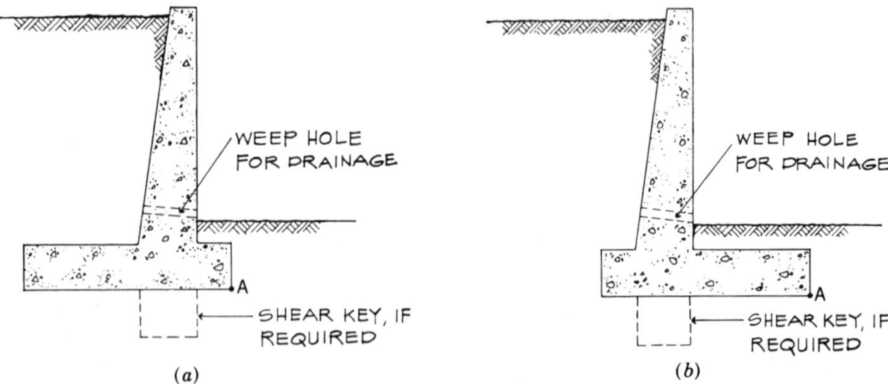

Fig. 7.9 Typical retaining walls configurations. (*a*) Retaining wall with large heel small toe. (*b*) Retaining wall with small heel large toe.

at a point of maximum moment under a permanent load condition. Since lap splices are discouraged at points of maximum moment, the designer is encouraged to either extend the dowels full height, thus eliminating the lap splice, or provide a splice length considerably longer than the minimum length required by code. Furthermore, the bars being spliced should be wired together to ensure positive stress transfer.

Analysis of a cantilever retaining wall requires a static summation of moments about point A at the toe to ensure stability, that is, the weight of the wall plus the backfill on the heel exceeds the overturning moment of the active soil pressure. Once a stable configuration is achieved, the soil pressure distribution on the footing must be calculated to ensure that the bearing pressures are within allowable values for the soil involved. Finally, a check must be made to determine that there is no problem with sliding of the base. Passive pressure on the toe and friction on the foundation can be utilized and, if necessary, a shear key below the footing can be constructed to provide additional passive pressure resistance against sliding. Shear keys are shown dashed in Fig. 7.9.

7.10 EXAMPLES

EXAMPLE 7.10.1

A five-story building with a single basement is being designed to resist wind forces. Based on discussions with the architect and because of aesthetic and functional limitations, it has been tentatively decided that a 20-ft long shear wall should resist the wind loads shown in Fig. 7.10*a*. Determine if the design is feasible and if additional space requirements are required. Tributary gravity dead loads at each end of the 20-ft long wall immediately above the first floor are 125 kips and the dead load tributary to a basement wall including footings is 4 kips/ft.

Solution

Determine the shear and moment in the wall at the first floor level.

$$V = 23 + 42 + 36 + 27 + 30 = 158 \text{ kips}$$

$$M = 23(63) + 42(51) + 36(39) + 27(27) + 30(15) = 6174 \text{ ft-kips}$$

Assuming for initial comparison that the wall is a 12-in. reinforced concrete wall, the average unfactored horizontal shear stress in the wall is

$$\frac{158{,}000}{(20)(12)(12)} = 55 \text{ psi}$$

This is an acceptable value, indicating that the wall can be designed for shear on either reinforced concrete or reinforced masonry.

The overturning force at each end of the wall will be about 6174/20 = 309 kips. Since this is more than twice the dead load in the end of the wall (125 kips), it will be necessary to extend the wall in the basement to mobilize additional dead load for stability.

As a first trial, extend the wall in the basement 20 ft to one side, as shown in Fig. 7.10*b*. This engages a column with a dead load of 75 kips at the first floor. If moments are taken about point A, the stability moment is

$$M_{\text{stability}} = 125(20) + 75(40) + 40(4)(20) = 8700 \text{ ft-kips}$$

and the overturning moment is

$$M_{\text{overturn}} = 6174 + 158(12) = 8070 \text{ ft-kips}$$

While the stability moment is greater than the overturning moment, most building codes require a stability moment 1.5 times the wind overturning moment. Therefore, the wall must be longer in the basement.

As a second trial, extend the wall in the basement 20 ft in each direction for a total length of 60 ft, as shown in Fig. 7.10*c*. This engages a column with a dead load of 75 kips at the first floor at each end. If moments are taken about point B, the stability moment is

$$M_{\text{stability}} = 125(20) + 125(40) + 75(60) + 60(4)(30) = 19{,}200 \text{ ft-kips}$$

The overturning moment is 8070 ft-kips as before, and this trial is a successful solution.

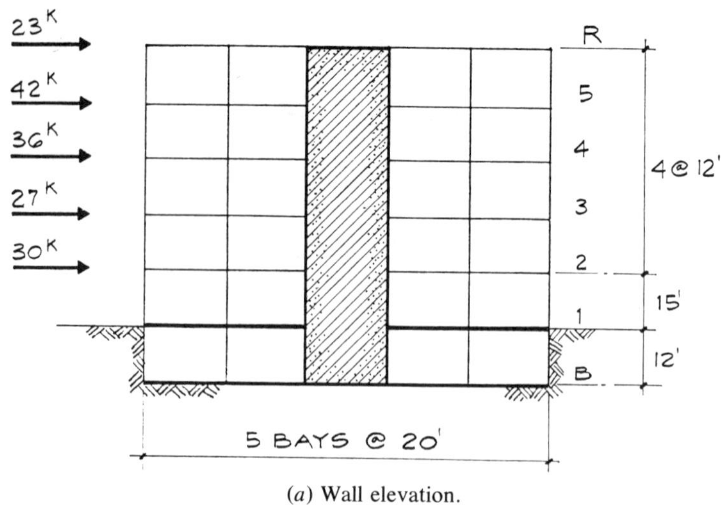

(*a*) Wall elevation.

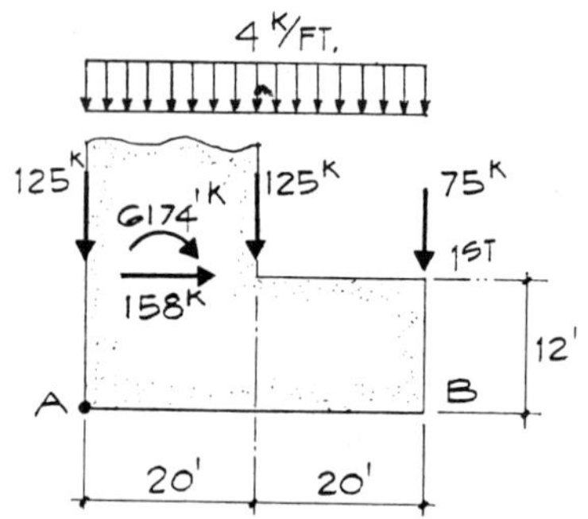

(*b*) First trial with a 40-ft long basement wall showing applied loads (foundation reactions not shown).

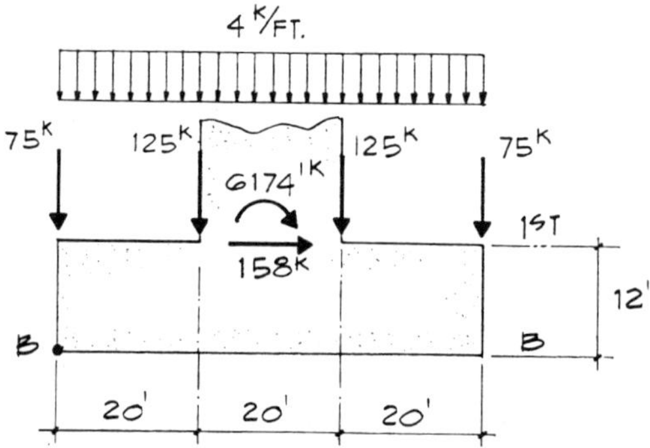

(*c*) Second trial with a 60-ft long basement wall showing applied loads (foundation reactions not shown).

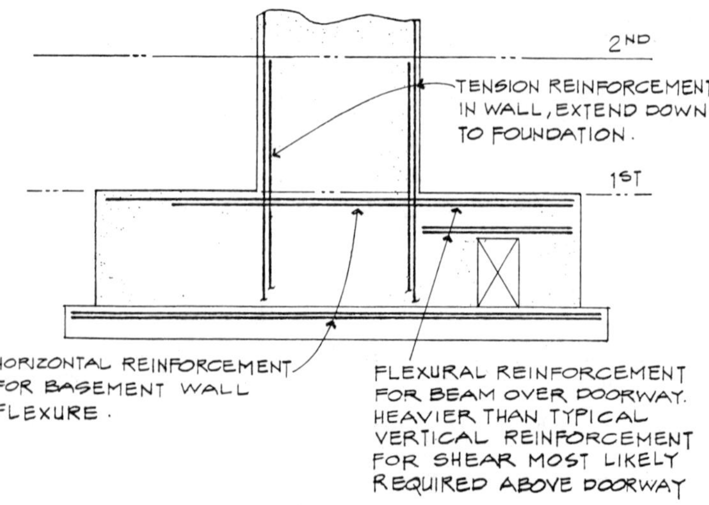

(*d*) Schematic basement wall elevation.

Fig. 7.10 Data for Example 7.10.1.

Discussion with the architect indicates that the basement wall 60 ft long is acceptable, but a 5-ft wide double door 7 ft high must be provided through the 20-ft wall extension. To determine the vertical shear and bending moment in the 5-ft deep spandrel section above the doorway, it is necessary to distribute the bearing pressures from the overturning moment to the footings based on appropriate assumptions. Then free body diagrams through the two ends of the spandrel beam can be analyzed to determine shears and moments in that portion of the wall. Likewise, a free body diagram drawn at the face of the 20-ft wall above will provide a design moment at the face of the wall to size horizontal reinforcing at the first floor and basement levels in the 20-ft wall extension.

Figure 7.10*d* is a schematic elevation of the wall in the basement showing the wall and the major reinforcing required. Note that wall boundary reinforcement has been well extended and anchored more than a single development length, which is appropriate in flexural members of this size.

EXAMPLE 7.10.2

A building is located on a corner lot with two solid concrete or masonry walls on the property lines, as shown in Fig. 7.11. If the typical floor framing cannot effectively resist lateral forces in frame action, is this building adequately designed? If not, what design solutions might be implemented?

Solution

The building is *not* adequately designed, as there is no way to resist building torsion. Lateral forces due to wind or seismic are centered near the center of the building, while the center of rigidity is at the intersection of the two walls. Statics of a diaphragm will show that there is no way to resist the torsion that results.

Solutions for this design involve adding portions of structural wall or adding moment-resisting frames along or near the other two exterior lines of the building. Analysis can then be performed similar to Example 7.10.7 to determine the force applied to each wall or moment-resisting frame. This is an illustration of a building with a lot of wall but not enough for stability and an adequate structural solution.

EXAMPLE 7.10.3

A large one-story industrial building is to be built for two separate tenants. The building is approximately 300 by 900 ft in plan with concrete or masonry walls and a lightweight roof of steel or wood construction. Local practice indicates that thermal expansion joints should be provided to prevent structural dimensions greater than 300 ft, which would indicate two expansion joints breaking the building into three separate structures, as shown in Fig. 7.12*a*. Recognizing that the masonry or concrete walls are effective bracing elements and that lightweight roofs have limited diaphragm shear strength, discuss solutions to brace adequately this building for wind or seismic forces.

Solution

One solution is to utilize only the walls provided for bracing and design the diaphragms for the large forces involved. For lateral forces in the north–south direction, the diaphragm for each unit

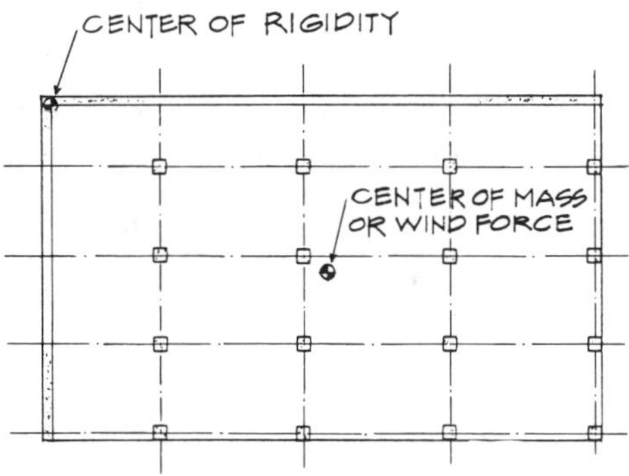

Fig. 7.11 Example 7.10.2. Building plan with two solid masonry or concrete walls on property lines.

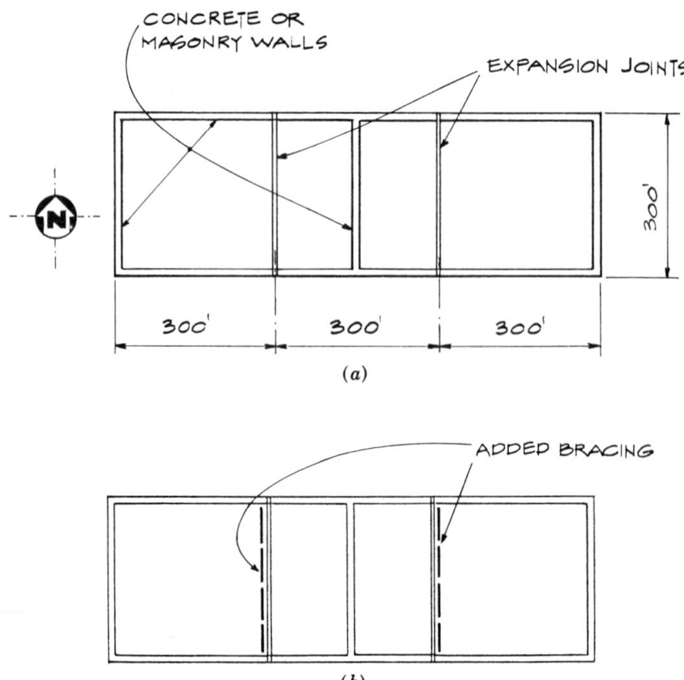

Fig. 7.12 Example 7.10.3. (*a*) Plan of single story industrial building. (*b*) Solution to high diaphragm stresses and flexibility by adding bracing.

transfers lateral forces to the end or center wall, and torsional tendencies are resisted by the north and south exterior walls. High diaphragm stress will result in the end units with a 300-ft diaphragm span, possibly requiring special design solutions such as added horizontal steel bracing truss in the roof framing. Diaphragm deflections at the expansion joint for north–south lateral forces should be checked to ensure building stability under extreme lateral forces. Shear distortion in the roof diaphragm may be excessive in the end units.

A second solution is shown in Fig. 7.12*b* where bracing is added as shown at the expansion joints in the end units. Ideally, this would consist of a partial length concrete or masonry wall, which would have stiffness comparable with the other walls. However, this may not be possible while the functional usage of the building space is maintained. Alternatively, moment-resisting rigid or braced steel frames could be added on these lines. While their rigidity will be considerably less than the concrete or masonry exterior walls, they may be quite effective bracing elements when the designer considers the flexibility of the roof diaphragm. In this example, the walls are quite rigid while the diaphragm is quite flexible.

EXAMPLE 7.10.4

During preliminary design of a concrete building, it is determined that the concrete walls around a stairway at the corner of the building must be fully utilized in the north–south direction as bracing walls for lateral forces. Figure 7.13*a* shows the typical floor framing plan with an architectural reveal in the building line adjacent to the stairway. Stair construction is to be metal stairs inside the concrete shaft. A problem exists because there is no easy way to transfer lateral forces from the floor diaphragm into the wall on line *B*, leaving only the wall on line *C* effective. Wall C is overstressed by analysis. How can the wall on line *B* become equally effective?

Solution

Several solutions are possible, depending somewhat on the magnitude of the forces involved. Figure 7.13*b* shows the most direct solution by adding an exterior beam across the reveal in the building exterior. If this beam matches patterns in the concrete wall or has appropriate finish

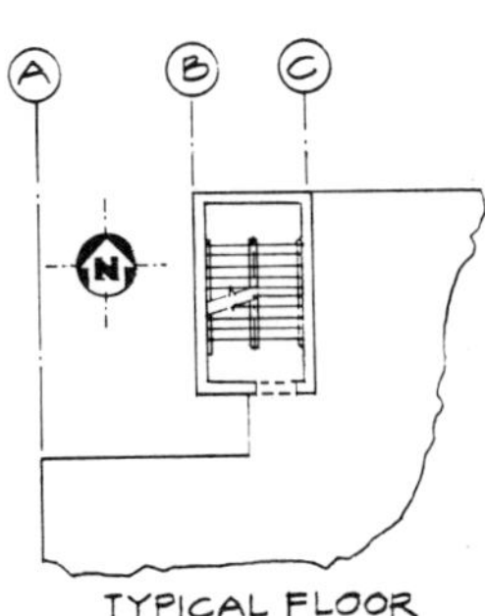

(*a*) Design problem.

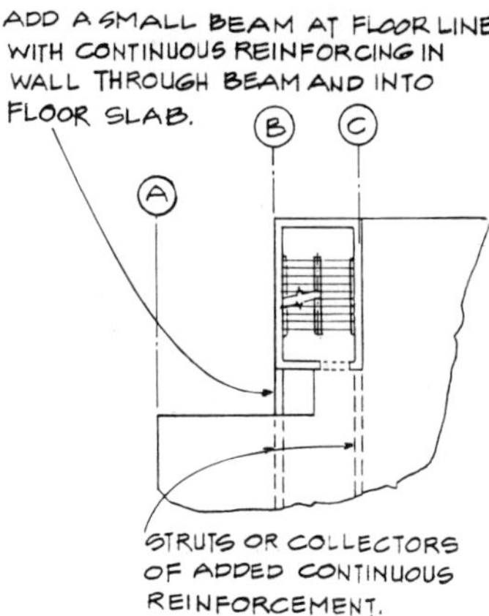

(*b*) Solution by adding beam for collector.

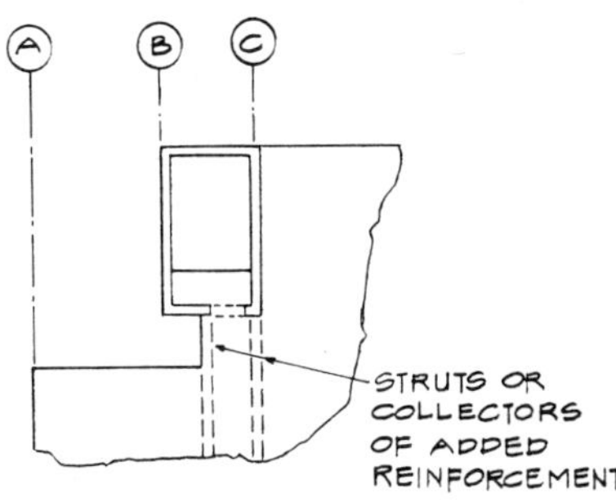

(*c*) Solution using stair landing to transfer lateral forces to wall B.

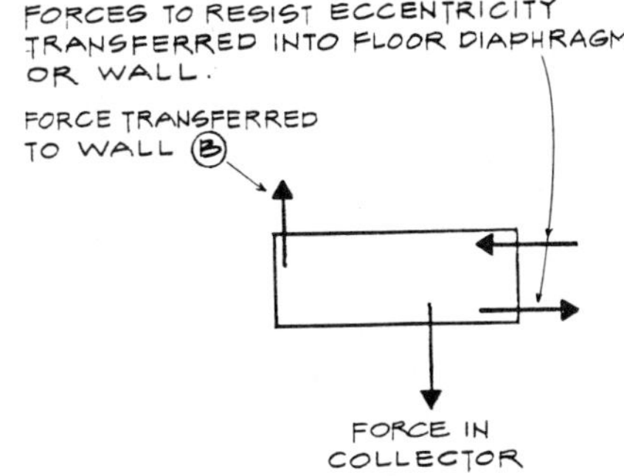

(*d*) Free body diagram of stair landing.

Fig. 7.13 Example 7.10.4.

treatment, it can blend well with the building's aesthetics. Continuous reinforcement in the wall can extend through the beam and into the slab or a beam which aligns with the wall. If desired, for functional or aesthetic purposes, a slab can be added in the reveal behind the beam to form an exterior balcony.

A second solution is shown in Fig. 7.13*c* where the architectural reveal in the building is not affected. The collector in the slab is located at the edge of the reveal and the stair landing at the floor level is made of reinforced concrete and designed to transfer the lateral forces to wall B. Figure 7.13*d* is a free body diagram of the stair landing showing the forces applied. The forces to resist the eccentricity are determined by statics and must be extended into the floor slab and considered in the design of the diaphragm.

EXAMPLE 7.10.5

A large hospital complex, Fig. 7.14*a*, is to be constructed of precast concrete structural elements in an area where large temperature changes can affect the structure. The building has four service cores with cast-in-place concrete walls which can serve as structural bracing walls, but it is necessary to divide the building into three separate structures to prevent excessive stress buildup due to thermal expansion. Precast framing details indicate that frame action to resist lateral forces is not feasible for this system. No additional bracing walls can functionally be allowed within the central building area. Furthermore, preliminary analysis shows that the service cores are suffi-

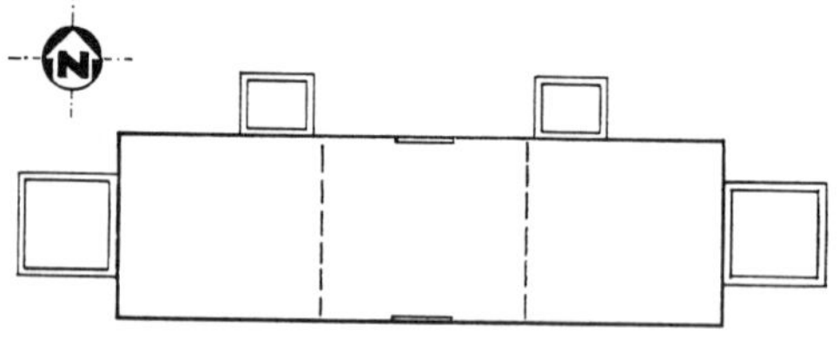

(*a*) Plan layout of building.

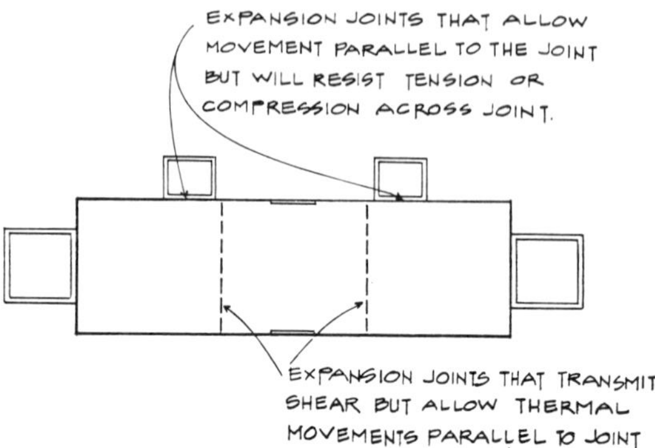

(*b*) Expansion joint requirements for building.

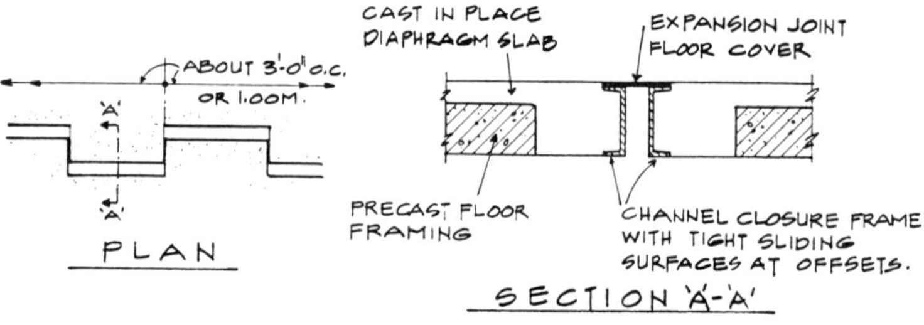

(*c*) Schematic expansion joint detail that transmits diaphragm shear but allows thermal expansion.

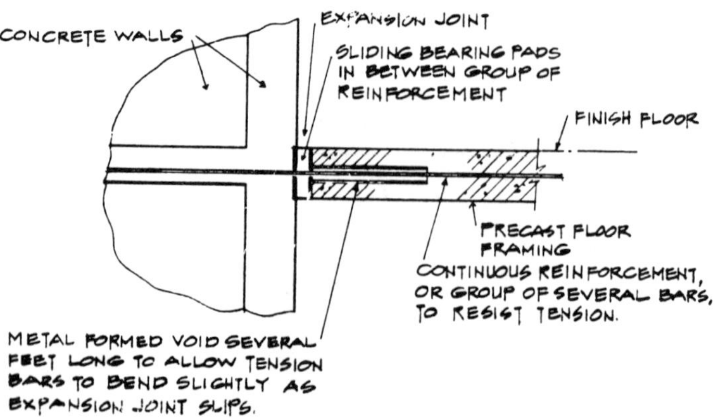

(*d*) Schematic section at expansion joint that allows parallel movement while transmitting tension or compression.

Fig. 7.14 Example 7.10.5.

ciently rigid that diaphragm stresses will be excessive if two adjacent cores are connected by the same diaphragm. How might the lateral forces be resisted in this building?

Solution

With virtually no alternative but to use the four service cores as structural wall complexes, it is necessary to devise a system of expansion joints which will allow the building to move with thermal changes but be adequately braced for lateral loads due to wind or earthquake. A solution is to design the expansion joints to permit the required thermal movements but still transmit shear or tension forces as necessary. This is schematically shown in Fig. 7.14*b*. For lateral forces in the north–south direction, the diaphragm of the central portion of the building spans to the two expansion joints and transfers lateral forces to the diaphragms of the end units. The diaphragms of the end units in turn resist their lateral loads and the force applied at the expansion joint and span the loads to the structural wall complexes. In the east–west direction, each unit has some wall in that direction requiring proper collector elements of reinforcing in the topping slab.

Figure 7.14*c* illustrates schematically the expansion joint in the floor which transmits shear but allows longitudinal thermal movement. The joint is keyed for bearing at the ends of each key with sufficient gaps for structural expansion and contraction based on anticipated thermal extremes and analysis of movements possible within the structure. Bearings at the ends of the keys must be long enough to function under maximum contraction of the building and should be well greased or provided with sliding surfaces.

Figure 7.14*d* shows schematically the expansion joint detail that allows movement parallel to the joint while resisting tension or compression. Tension is resisted by continuous reinforcement across the joint which passes through a void long enough to permit a slight offset of the reinforcement without excessive stress. Compression is resisted by sliding bearing pads in the joint. Proportions and actual details for both expansion joint types must suit the actual building configuration and the analysis.

The author wishes to add that more direct methods of resisting lateral forces in each segment of the building are recommended. This example, although based on an actual building, is provided to illustrate the principles of diaphragm action rather than encourage similar design.

EXAMPLE 7.10.6

Given: Assume lateral forces due to wind or earthquake are to be resisted by two 30-ft-high solid concrete walls. Wall A is 10 ft long and wall B is 30 ft long and both walls are 8 in. thick. See Fig. 7.15. Determine the distribution of lateral force to the two walls assuming:

(a) Relative rigidities of walls are proportional to the shear deformations of the walls.

(b) Relative rigidities of walls are proportional to shear and flexural deformations of the walls.

(c) Relative rigidities of walls are proportional to shear and flexural deformations of the walls plus foundation deformations assuming the soil deforms $\frac{1}{8}$ in. under 6000 psf and footings are 2 ft wide by the wall length.

Solution

(a) *Relative rigidities proportional to shear deformations.* These deformations can be computed as follows, assuming an elastic cantilever wall,

$$\Delta = \frac{1.2Vh}{AG}$$

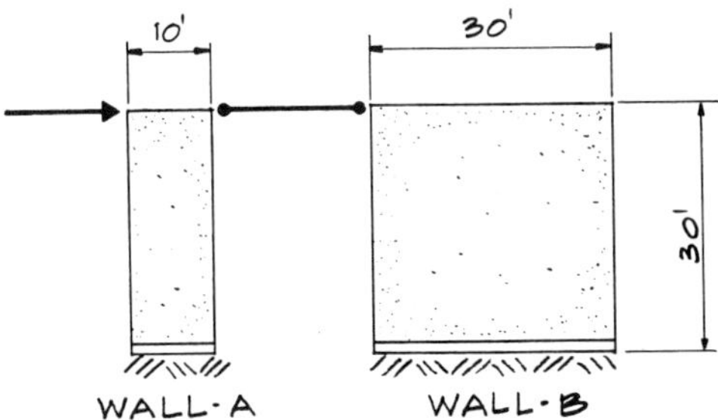

Fig. 7.15 Example 7.10.6.

where V is the lateral load, h is the wall height, A is the wall area, and G is the shear modulus. For an equal load applied to each wall, the wall deflection will be inversely proportional to the wall length, since the height, wall thickness, and G are constant for both walls. Wall rigidity is the inverse of the deflection under equal load, so rigidity is thus proportional to length. Therefore,

$$\begin{aligned} \text{rigidity of wall A} &= 10 \\ \text{rigidity of wall B} &= \underline{30} \\ \text{total} &= 40 \end{aligned}$$

$$\text{percentage of load to wall A} = \frac{10}{40}(100\%) = 25\%$$

$$\text{percentage of load to wall B} = \frac{30}{40}(100\%) = 75\%$$

(b) *Relative rigidities proportional to flexural and shear deformations.* Assuming an elastic cantilever wall, these can be calculated by

$$\Delta = \frac{Vh^3}{3EI} + \frac{1.2Vh}{AG}$$

where V is the lateral load, h is the wall height, I is the moment of inertia of the wall, E is the modulus of elasticity, A is the wall area, and G is the shear modulus, which can be taken as $0.4E$. Assuming a lateral load of 10 kips to each wall and 3000 psi concrete with $E = 3150$ ksi,

$$\text{wall A } \Delta = \frac{(10)(30)^3}{(3)(3150)(144)\dfrac{(0.67)(10)^3}{12}} + \frac{(1.2)(10)(30)}{(0.67)(10)(0.4)(3150)(144)}$$

$$= 0.00355 + 0.00030 = 0.00385$$

$$\text{wall B } \Delta = \frac{(10)(30)^3}{(3)(3150)(144)\dfrac{(0.67)(30)^3}{12}} + \frac{(1.2)(10)(30)}{(0.67)(30)(0.4)(3150)(144)}$$

$$= 0.00013 + 0.00010 = 0.00023$$

$$\text{rigidity of wall A} = \frac{1}{0.00385} = 260$$

$$\text{rigidity of wall B} = \frac{1}{0.00023} = \underline{4348}$$

$$\text{total} = 4608$$

$$\text{percentage of load to wall A} = \frac{260}{4608}(100\%) = 5.6\%$$

$$\text{percentage of load to wall B} = \frac{4348}{4608}(100\%) = 94.4\%$$

Thus, in this particular example, the added calculation of flexural deformations has considerable effect.

(c) *Relative rigidities proportional to shear and flexural deformations plus a foundation deformation.* The bearing pressure due to moment at the foundation for wall A is approximately

$$\frac{10(30)}{(2)[(10)^2/6]} = 9 \text{ ksf}$$

Thus, foundation vertical deformation $= \dfrac{9}{6}(0.125) = 0.1875$ in.

wall A horizontal deflection due to foundation deformation

$$= 0.1875 \left(\frac{30}{5}\right)\left(\frac{1}{12}\right) = 0.09375 \text{ ft}$$

$$\text{wall A deflection in part (b)} = \underline{0.00385}$$

$$\text{total deflection wall A} = 0.09760$$

Likewise,

$$\text{bearing pressure, wall B} = \frac{10(30)}{2[(30)^2/6]} = 1.0 \text{ ksf}$$

$$\text{foundation deformation} = \frac{1}{6}(0.125) = 0.0208 \text{ in.}$$

wall B horizontal deflection due to foundation deformation

$$= 0.0208 \left(\frac{30}{5}\right)\left(\frac{1}{12}\right) = 0.00347$$

$$\text{wall B deflection in part (b)} = \underline{0.00023}$$

$$\text{total deflection wall B} = 0.00370$$

$$\text{rigidity of wall A} = \frac{1}{0.09760} = 10.2$$

$$\text{rigidity of wall B} = \frac{1}{0.00370} = \underline{270.3}$$

$$\text{total} = 280.5$$

$$\text{percentage of load to wall A} = \frac{10.2}{280.5}(100\%) = 3.6\%$$

$$\text{percentage of load to wall B} = \frac{270.3}{280.5}(100\%) = 96.4\%$$

Note the foundation consideration is not a very significant factor in determining relative wall rigidity in this example, although it does reduce the percentage of load in wall A from 5.6 to 3.6%, which is a reduction of one-third.

EXAMPLE 7.10.7

Given: Two commercial buildings of identical size are illustrated in Fig. 7.16, with a sales area to the right behind a glass curtain wall (not shown) and one interior wall dividing the sales area from the work or storage area behind. Lateral load on the roof diaphragm due to wind or earthquake design is 300 lb/ft as shown. Each building has four walls with relative rigidities determined by a separate analysis as shown. Columns and framing are not shown. The only difference in the two buildings is that wall B is located in a different position. Determine the loads to each wall for the forces shown, assuming both a rigid and a flexible diaphragm.

Solution

(a) *Building I*

Rigid Diaphragm Assumption: The center of rigidity for the direction walls A and B, measured from wall A, is

$$\frac{6(150)}{6 + 10} = 56.25 \text{ ft}$$

The center of rigidity in the other direction is the centerline of the building by symmetry.

With a rigid diaphragm, the walls in both directions resist the torsion due to the eccentricity between the center of load and the center of rigidity. The force from the torsion equals the torsion

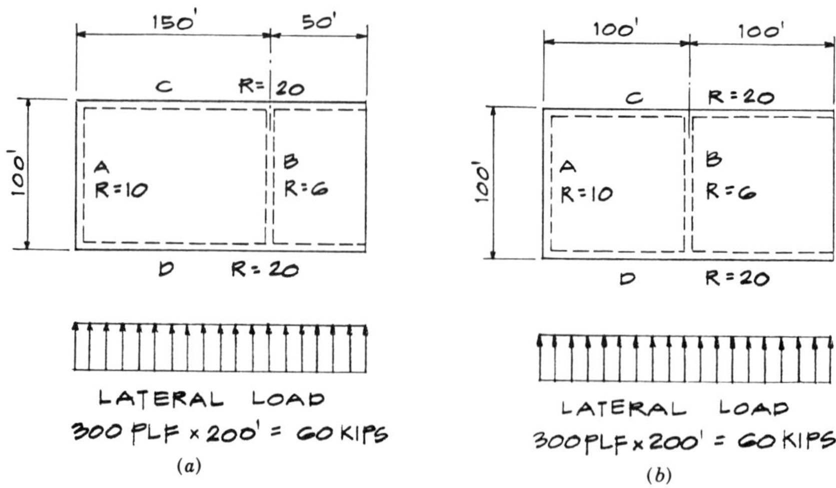

Fig. 7.16 Example of 7.10.7. (*a*) Building I. (*b*) Building II.

times the relative rigidity times the distance from the center of gravity divided by the polar moment of inertia of the wall rigidities. Thus, in tabular form:

Wall	R	d	Rd^2	Rd/J
A	10	56.25	31,640	0.00305
B	6	93.75	52,734	0.00305
C	20	50	50,000	0.00542
D	20	50	50,000	0.00542
		$J = \Sigma Rd^2 =$	184,374	

Thus, total forces to the walls due to translation and rotation are as follows:

$$\text{force to wall A} = \left(\frac{10}{16}\right)(60) - 0.00305(60)(43.75) = 37.5 - 8.0 = 29.5 \text{ kips}$$

$$\text{force to wall B} = \left(\frac{6}{16}\right)(60) + 0.00305(60)(43.75) = 22.5 + 8.0 = 30.5 \text{ kips}$$

$$\text{force to walls C and D} = 0.00542(60)(43.75) = 14.2 \text{ kips}$$

Statics can be used to check these results.

Flexible Diaphragm Assumption: Forces are distributed to walls by a tributary area basis without any participation of perpendicular walls under a flexible diaphragm assumption. Thus,

$$\text{force to wall A} = \frac{75}{200}(60) = 22.5 \text{ kips}$$

$$\text{force to wall B} = \frac{125}{200}(60) = 37.5 \text{ kips}$$

(b) *Building II.* Calculations are similar to building I and are omitted for brevity. Results are:

Rigid Diaphragm Assumption:

force to wall A = 27.3 kips
force to wall B = 32.7 kips
forces to walls C and D = 27.3 kips

Flexible Diaphragm Assumption:

force to wall A	= 15.0 kips
force to wall B	= 45.0 kips

SELECTED REFERENCES

7.1 R. E. Crawford, C. J. Higgins, and E. H. Bultmann. *The Air Force Manual for Design and Analysis of Hardened Structures,* AFWL-TR-74-102. Air Force Weapons Laboratory, Kirtland AFB, New Mexico, October 1974.

7.2 ACI. *Building Code Requirements for Reinforced Concrete* (ACI 318-83). Detroit: American Concrete Institute, 1983.

7.3 PCI. *PCI Design Handbook, Precast and Prestressed Concrete, 3d ed.* Chicago: Prestressed Concrete Institute, 1985.

7.4 ACI. *Building Code Requirements for Structural Plain Concrete* (ACI 318.1-83). Detroit: American Concrete Institute, 1983.

7.5 M. L. Porter and L. F. Greimann. *Seismic Resistance of Composite Floor Diaphragms.* Ames, Iowa: Ames Engineering Research Institute, May 1980, 187 pp. (NTIS PB 81-102774).

7.6 ATC. *Guidelines for the Design of Horizontal Wood Diaphragms.* Berkeley, CA: Applied Technology Council, September 1981.

7.7 H. J. Degenkolb and L. A. Wyllie, Jr. "Design—Lateral Forces," *Western Woods Use Book.* Portland, OR: Western Wood Products Association, 1973.

CHAPTER 8
STRUCTURAL FORM

JOHN V. CHRISTIANSEN

Consulting Structural Engineer
Bainbridge Island, Washington

8.1 DEFINITIONS OF STRUCTURAL FORM

Reference to a standard dictionary [8.1] produces the following definitions for the word "form":

> *(1) The shape and structure of something as distinguished from its material. (2) The essential nature of a thing as distinguished from its matter. (3) An orderly method of arrangement, a manner of coordinating elements, or a particular kind or instance of such arrangement.*

The same dictionary produces the following definitions for the word "structure":

> *(1) The action of building. (2) Something that is constructed. (3) Something arranged in a definite pattern of organization. (4) An arrangement or interrelation of parts as dominated by the general character of the whole. (5) The aggregate of elements of an entity and their relationships to each other.*

Thus, "structure" and "form" are complementary descriptive words and, in the context of this book, structural form is seen as a man-constructed orderly pattern of organization or arrangement of elements. In the strict engineering sense, a structural form is a structural system, defined as a three-dimensional system of interconnected planar, linear, and (sometimes) curved and warped structural elements arranged in such a way as to serve the architectural and service engineering functions of the building and to survive a host of gravity-induced and environmental conditions.

Form has been called "the basic ingredient of the art of architecture." Form is often considered the major element of the various elements of design: form, line, space, texture, light, and color.

Those with an interest in the visual arts have developed a vocabulary for the principles of design, expressed in such words as emphasis, continuity, rhythm, balance, equilibrium, dominance, contrast, repetition, alternation, progression, transition, symmetry, asymmetry, radiation, and proportion [8.2].

Down through history, whatever the culture and philosophy of a particular society or time, the structural systems for buildings have played a prominent role in the form that buildings have taken. The remainder of this chapter attempts to outline briefly those significant factors that have determined and continue to determine structural form, and then to outline the current vocabulary of structural form.

The forms discussed here all receive more detailed treatment in subsequent chapters.

8.2 DETERMINANTS OF STRUCTURAL FORM

The structural form of a building develops as a result of a creative design process which develops a design in response to a wide range of conditions and constraints, including technical, material, utilitarian, and aesthetic considerations. The following are major determinants of structural form.

8.2.1 Building Function

A major determinant of structural form is the plan for use of the structure, that is, its program and its functional and social requirements. The forms of the buildings of a particular culture or era are largely determined by the functional and social requirements for which the structure was built. Dramatically different functional uses of buildings exist in our present society, and these various building functional types may be listed as follows:

1. *Buildings for Business Activities*. High-rise office buildings often have a distinctive structural form, and the proliferation of such buildings is changing the face of our cities.
2. *Housing*. Single family residences, apartments, hotels, dormitories.
3. *Buildings for Commercial Activities*. Retail stores, shopping centers, banks, transportation terminals, and similar facilities.
4. *Institutional Buildings*. Buildings for educational purposes, medical and health care, and religious services and other activities.
5. *Entertainment and Convention Center Facilities*. Exhibition halls, theaters, gymnasiums, and stadiums. Because of the very long-span requirements of such buildings, structural form often dominates the design of such facilities.
6. *Industrial Buildings*. Light and heavy manufacturing, utility plants, and cargo-handling facilities. The social use of these facilities is minimal. The heavy loads and other special requirements often result in unusual structural forms.
7. *Storage and Maintenance Facilities*. Warehouses, parking garages, aircraft hangars, and maintenance shops. Again, the social use of these facilities is minimal, but the long-span requirements often dictate structural solutions which produce unusual forms.

8.2.2 Building Loads

A principal determinant of structural form is the loads to which the structural system is subjected. These loads may be thought of in two principal groups: gravity loads and lateral loads. Extensive information on load intensities is given in Chapter 2.

Gravity loads are described as dead or live loads. Dead loads act throughout the entire life of a structure. The principal dead load is usually the weight of the structure itself.

Live loads are thought of as gravity loads of a variable nature. A live load may only occur for a small portion of the life of a structure.

The lateral loads of wind and earthquake are often thought of as equivalent static loads, but actually they are imposed environmental conditions that cause the building structure to respond dynamically, as discussed in Chapters 2, 10, 11, and 18.

The relative influence of the various types of loads on the structural form is primarily a function of the physical overall scale of the structural system and the weight and strength properties of the structural material. For a large-scale and/or long-span structure constructed of relatively heavy-weight material (concrete, steel, wood, or masonry), the gravity dead load, principally the structure self-weight, may be the primary determinant of structural form, whereas for a very small or short-span structure using the same materials, the primary determinant may be the live load of a piece of heavy equipment or truck, or some other movable load.

On the other hand, for an equally large-scale structural system, but constructed of very lightweight materials (such as cable-stayed fabric), the short-term live and wind load conditions may be major structural form determinants.

Linear and planar structures may carry loads by developing flexural and normal shear stresses. Such systems generally consist of slabs, beams, and joists and are normally arranged in horizontal planes to serve as either roof or floor systems (or in vertical planes as walls). The material requirements are largely determined by the loads and the span, and it can be shown that such structures do not represent the most efficient use of materials.

By contrast, structures may be arranged to carry loads by direct axial forces in linear members, or direct inplane axial and shear forces in planar structures. It can be demonstrated that such structures are far more efficient in the use of materials. Such structures include, in two dimensions, the truss, arch, and tension cable and, in three dimensions, space frames, shells, and tension structures.

8.2.3 Site Geography

The geography of the building site, both natural and man-made, may significantly impact the structural form. The topography of the site may require special engineering solutions (retaining walls, buttresses, etc.). The form may be significantly affected by its relation to the transportation and utility network and the forms, materials, and purposes of the neighboring buildings, as well as the city code and zoning ordinances governing the use of the building site. The configuration (and structural form) of office buildings in major metropolitan centers has been largely determined by zoning ordinances and planning regulations.

8.2.4 Environmental Conditions

Building and structural form are greatly influenced by external environmental conditions, that is, the climatic conditions of sun, light, temperature and humidity, precipitation, wind, fire, and seismic.

With respect to sun and light, a structure meets fundamentally different conditions on each of its different sides. The effects of temperature, humidity, and precipitation are significant. Thermal forces may play a principal role in the selection of structural form; witness the contrast between the igloo of the Eskimo and the tent of the desert Bedouin.

Natural wind has played a major role in the development of structural systems and forms. The micro wind climate around a building structure is substantially affected by the structural form of the building and the surrounding buildings. The wind climates of entire cities are dramatically influenced by the structural form of their buildings.

In seismically active areas, earthquakes may dramatically affect the structural form. Fire has played a major role through history, literally changing the course of architecture. The fire codes and regulations of today play a major role in determining the structural form of our buildings.

8.2.5 Mechanical, Electrical, and Service Systems

The internal mechanical, electrical, and service systems of buildings are playing an ever-increasing role in determining building form. The whole concept of energy conservation required by our dwindling supply and rising cost of fossil fuels may be expected to change dramatically the structural form of our buildings.

The cost of these internal systems in modern buildings runs from 20 to as much as 50% of the total cost of construction. When thought of in terms of life-cycle cost, the percentage may be higher.

Large areas and volumes of modern buildings are required to house the heating, ventilating, and air conditioning (HVAC) equipment; electrical equipment and distribution; plumbing and fire protection; and the internal transportation systems. The impacts of these systems on the structural design process are presented in Chapters 4 and 5.

The HVAC system may be at scattered local sites throughout the building, but it is often centrally located in a large mechanical room, or it may even occupy entire floors. In major buildings, these required systems are often concentrated in cores which are often used as portions of a lateral bracing system.

Internal transportation systems, such as the traditional people-moving systems of stairs and ramps, have always provided structural engineers the opportunity to stiffen and brace building structures. People-moving systems now also include escalators, moving walks, and even personal rapid transit (PRT) systems.

The combination of high land cost and lack of sites in our central cities has sent buildings upward, and elevatoring systems have become significant building components. Passenger and service elevators and their paraphernalia, such as sky lobbies and sky bridges, are impacting the structural form of our buildings.

For commercial and industrial buildings, material-handling systems play an ever-increasing role in design.

8.2.6 Structural Materials

The properties of structural materials have traditionally been a major determinant of structural form. Strength, durability, stiffness and flexibility, and fire resistance, as well as availability and cost, have been primary factors in the development of architectural and engineering design down through the centuries.

Four principal structural materials have been and continue to be used in our building structures: masonry, timber, concrete, and ferrous metals. The earliest structures were of timber and stone; they were the principal materials used by the master builder who designed by intuition, rule of thumb, and trial and error.

More recently, concrete reinforced with both conventional reinforcing bars and high-strength prestressed steel, and structural steel have developed as our principal structural materials for large structures.

The nature of these structural materials plays a principal role in the development of structural form. Timber grows (generally) straight and thus lends itself to structural systems that are assemblies of linear members. The use of timber has been expanded beyond the linear member by the development of curved glued laminated lumber and has been extended to planar structures by the development of plywood.

Masonry consists of relatively small units laid up by hand and bonded together with portland cement paste mortar, which suggests its use for walls and piers, vaults, and arches, where the stresses are principally compressive.

Fresh concrete is a perfectly plastic material. Thus, the structural form applications of concrete are almost without limit, especially with the addition of reinforcement to overcome the relative lack of tensile strength. There is a tendency for the form of a concrete structure to be limited somewhat by the cost of its formwork and the nonplastic nature of formwork materials.

Structural steel and other metals are produced by rolling the material into straight linear members or flat sheets; thus, structural forms using assemblies of linear members (such as trusses, posts, and beams) are appropriate, as well as flat sheet structures (such as metal decks and corrugated iron). The structural designer works always with both the overall and detailed nature of the structural materials.

8.2.7 Construction Methods

The construction methods used cannot help but influence structural form. The quality and quantity of the labor force in the field, on the job site, and in the fabricating plant (if any), may significantly affect the choice of structural system and construction method.

The availability and capability of construction equipment and the use of off-site or plant-prefabricated building elements, components, or systems, may significantly affect form. For example, off-site precast concrete components assembled on the site may assume a dramatically different structural form than site-cast concrete work would assume.

Standard manufactured products are often used in structural systems, and the use of such products usually results in a modular structural system.

8.2.8 Scale

Scale or size plays a major role in the determination of structural form. For very small-scale structures, such as a small residence or utility building, the structural requirements can be economically met by simple structural systems in any material, and the structural design is a trivial task.

However, for very large-size structures, the forces of nature; loads, both gravity and lateral; environmental conditions—wind, seismic, fire, and temperature; limitations of the structural materials; and limitations of construction methods begin to dominate the concept of the structural system.

Hence, the structural form of long-span structures and high-rise buildings is dominated by the scale of the structure.

8.2.9 Economy

Our society demands economical buildings. The cost of our buildings involves two components: (1) the construction cost and (2) the operating cost.

The construction cost may be thought of as a blend of three primary cost items: (1) labor, (2) materials, and (3) equipment. The total cost is a function of all three.

In the early history of construction, material costs were generally low, labor costs were low to medium, and the little equipment available was probably high priced for its time. Today, the cost of materials varies greatly—some high and some low. However, the labor costs are high.

Labor costs have steadily increased, and, as a result, equipment has been developed to replace labor. Some very high-cost materials have been developed. However, even though some material costs are very high, it has been possible to achieve cost-competitive designs in those materials. An example is the air supported, cable-stayed fabric roof (see Chapter 15) which utilizes very expensive structural materials (fabric and cables), but reduces the labor by the use of sophisticated equipment such as large cranes and helicopters, and reduces the material quantities by the use of mechanically generated (and energy-consuming) internal air pressure.

Operating cost is of much greater interest to building owners and designers today than it has been in the recent past. Rapidly escalating energy costs will most likely dramatically change the character of our buildings and will certainly impact structural form.

Maintenance cost may play an increasing role as the relatively high cost of building forces owners and designers to look to more permanent and long-lived systems.

8.2.10 The Nature and Mood of Man

The ancient soldier-philosopher Vitruvius wrote that the three goals of architecture are "commodity, firmness, and delight." To translate this into the more popular words of today, the goals of architecture would be "function, durability, and aesthetics."

The nine determinants of structural form preceding this have dealt principally with the first two goals—commodity and firmness (function and durability). For buildings designed for human use or the storage of goods or services for human use, the nature and mood of the owner, user, and designer all play an important role in the form that the design takes.

It has been suggested [8.3] that "delight" has three sources: moral, sensual, and intellectual. The moral source may be examined in two parts:

1. The morality of the project as a whole—its relation to the culture and philosophy of a particular society at a particular time.
2. The morality of its execution—this refers to such commonly espoused design principals as "honesty of design", "being true to the nature of the materials", and "form follows function". Thus, a designer can take a moral position that the structural form of a building should be expressed.

The second source of delight is sensual. The primary sense related to building architecture is vision, where the viewer observes shape, texture, line, light and shadow, color, and contrast. The question of whether or not a variety of viewers respond identically to the same visual stimulus is of interest and has been studied extensively by scientists in the field.

A third source of delight is intellectual, an ability to transmit the impressions received through the senses into an understandable pattern. Thus, this depends on the level of knowledge and understanding of the viewer. A case can be made for simplicity and directness in design in order to have the widest intellectual appeal.

Visually apparent structural clarity has a universal appeal. In a building, this expresses itself in a natural and orderly flow of loads from the top of the building down through the building structure to the foundations.

8.3 THE VOCABULARY OF STRUCTURAL FORM

The structural system of a building is a complex, three-dimensional assemblage of interconnected linear, planar, curved, and warped structural elements designed in response to a host of conditions and constraints, the "determinants of form", such as those listed in Section 8.2.

The following vocabulary of structural form can only be general in nature. The structural forms illustrated represent general categories only (although some illustrations may be recognized as specific constructed works). Time and space does not permit a discussion of every detailed type of structural member or system.

For clarity, the elements of structural form will be treated in three categories: (1) floors (which are flat) and flat roofs, (2) roofs and enclosures, and (3) bracing systems.

8.3.1 Floors and Flat Roofs

Floors and flat roofs may be thought of as assemblies of linear members or as planar systems.

Linear Members

1. *Columns (Posts, Piers, Pilasters, Caissons, etc.).* Linear, generally vertical members of compact shape which must be designed to carry loads imposed by roof and floor systems above. Their size (cross-sectional area) is determined by the load carried. They are limited by the material strength and stiffness properties and their slenderness. They may be of any shape but, where slender, certain shapes (forms) are most efficient, notably the hollow, circular cylinder.

2. *Hangers.* Slender linear, generally vertical, members arranged to carry floor loads. They may be of any shape, and they generally are relatively small.

3. *Beams.* Horizontal linear members that carry loads principally by developing flexural stresses. The size and shape, and hence the form, of the beams are determined by the load, span, and nature of the material. The traditional, most-efficient flexural member for uniaxial bending is the I beam. The distribution of moment varies throughout the length of a beam and, if the section properties of the beam (principally its depth) are varied to achieve a constant level of extreme fiber stress, some interesting beam forms can be developed.

4. *Trusses.* Assemblages of straight linear members in a plane connected together at common joints in order to carry imposed loads by direct axial forces in members. As a result of the virtual elimination of flexural stresses, they are efficient and economical in use of material. Figure 8.1 illustrates some common truss types. If the trusses are shaped to follow the funicular line of the loads (i.e., for constant chord stress), the web members of the truss have little or no stress. Refer to Fig. 8.2 for examples of trusses shaped for constant chord stress: (1) the common bow-string truss, an efficient truss for single span construction; (2) the cantilever bridge, an often-used bridge type in steel and iron; and (3) a transit shed roof designed by the early twentieth century Swiss engineer Robert Maillart [8.4].

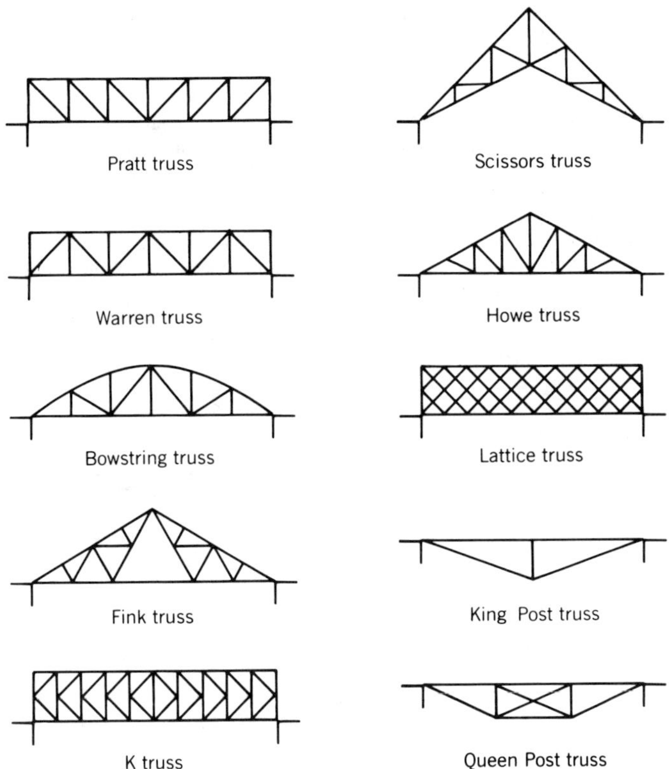

Fig. 8.1 Simple building trusses.

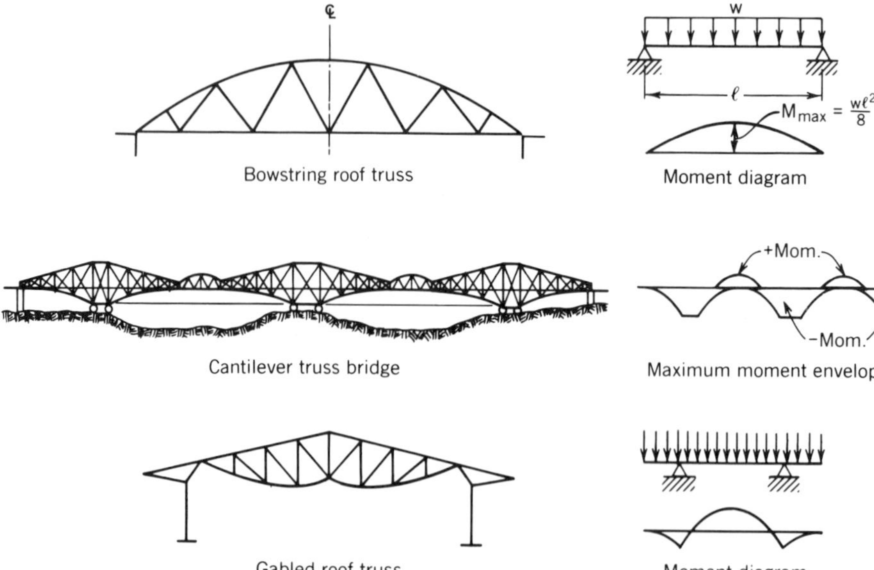

Fig. 8.2 Trusses having uniform chord stress.

5. *Frames*. If beams are rigidly connected to posts or columns in a common plane, the column will contribute to the flexural strength and stiffness of the beam, and "frame action" is developed. The form of the frame, that is, its configuration and relative stiffness (determined principally by depth), tends to determine the magnitude and distribution of flexural stresses throughout the frame.

Planar Systems

1. *Slabs and Plates*. Slabs or plates may be supported by linear members such as beams or joists. In this case, the slab or plate may span in one direction or in two or more directions and may consist of a reinforced concrete slab, corrugated metal deck, laminated or tongue-and-groove wood decking, plywood, or a corrugated metal deck composite with a concrete slab. Such systems are often constant in thickness.

Slabs or plates may be supported directly by columns, piers, or hangars. In this case, they are generally of concrete and called flat slabs. When supported directly by the column without any transition, they are called flat plates.

The transition zone between the flat plate slab and the column is a highly stressed one. Often this zone is reinforced with a drop panel and/or capital. Figure 8.3 illustrates common types of capitals often used and also illustrates a more sophisticated curved transition capital. In general, slab

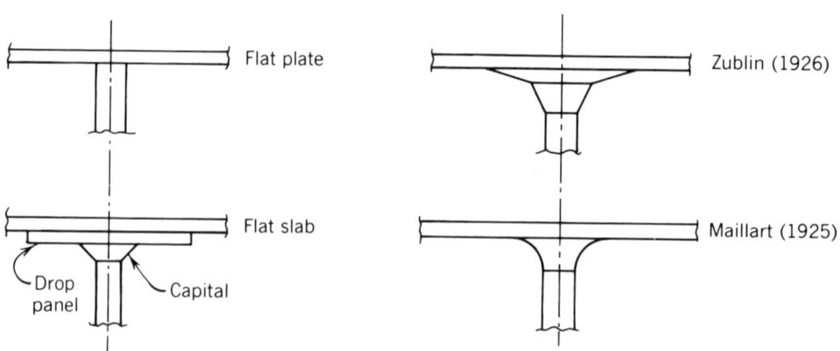

Fig. 8.3 Column capitals.

construction uses minimum structural depth, thus presenting the designer the opportunity of achieving minimum story heights.

Slab systems are adaptable to irregular support layouts but have application to relatively short-span construction.

2. *Ribbed Slabs and Plane Grids*. These systems consist of very thin slabs (concrete, corrugated metal, wood decking, or plywood) cast integrally with, or fastened to, closely spaced ribs which may be arranged in a one-way pattern or a two- (or more-) way pattern. For rectangular framed panels, two-way ribs may be oriented normal and parallel to the panel edges; however, greater structural efficiency may be achieved by skewing the ribs. The ribs may be precast or cast-in-place concrete, light-gage steel, steel bar joists, or solid or fabricated wood joists. The cost of connecting together and fabricating a two- (or more-) way system in steel or wood discourages its use.

One- and two-way systems in concrete are common and are called pan joist systems. Some very appealing structural forms have been developed utilizing rib slabs where the ribs are aligned along lines of principal stress (isostatic lines). Figure 8.4 illustrates three such applications.

3. *Space Frames*. In this planar context, space frames are three-dimensional trusses with top chords and (most likely) bottom chords in horizontal planes. Figure 8.5 illustrates several geometries that have been used for such space frames. The economy and erectability of these space frames is very much dependent on the joint connector design.

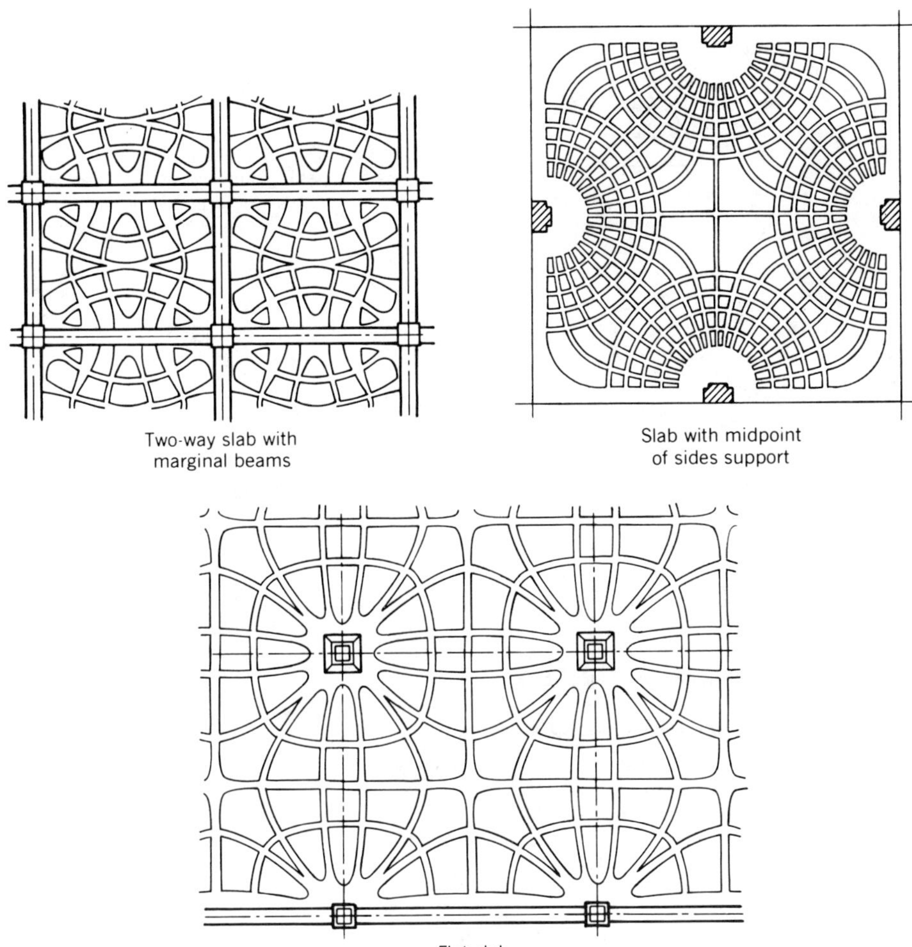

Fig. 8.4 Ribbed slabs.

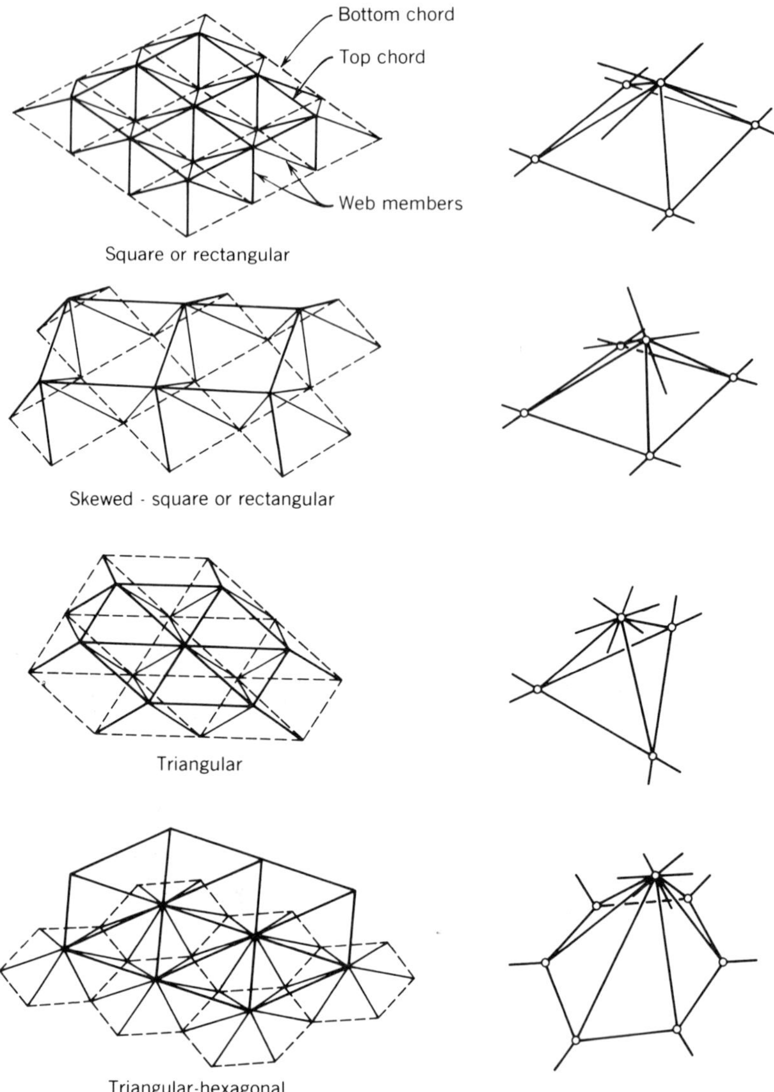

Fig. 8.5 Space frame geometry.

8.3.2 Roofs and Enclosures

Linear Member Assemblies (Two-Dimensional)

Linear members may be assembled and connected together in a plane in a wide variety of configurations to carry roof and wall enclosures. Most loads imposed on such enclosures, such as snow and wind, are distributed over the surface. Thus, to reduce the impact of flexural stresses, structures need be shaped to the ''funicular'' curve of the loads where the funicular curve is the natural shape of a structure to carry loads by direct axial tension and compression alone.

In practice, because of the linear nature of most structural components and the rectilinear volumetric character of most enclosed spaces, the forms most used are a compromise between an idealized funicular curve and a practical assembly of linear members.

Figure 8.6 illustrates some structural form possibilities for a simple frame carrying a load uniformly distributed in a horizontal plane. Note the relative magnitude of the bending moments,

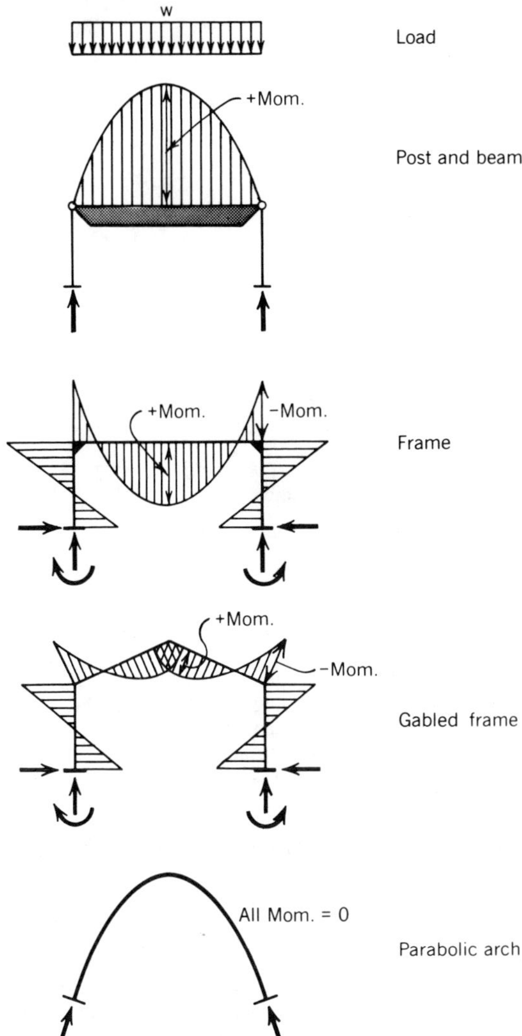

Fig. 8.6 Single span frames.

remembering that bending moments are a principal determinant of member size. This illustrates a range of structural forms from a least-efficient (post and beam) to a most-efficient (arch).

Further refinements of structure or form may be accomplished by varying the member section properties proportional to the strength requirements. Thus, frame columns and beams may be haunched at or near their connection joints. Figure 8.7 illustrates the forms of commonly used single span frames and some common and historical arch forms.

Surface Systems, Three-Dimensional

1. *Shell Structures*. Shell structures are curved and warped or folded surfaces whose thicknesses are small compared to their other dimensions. Shells are characterized by their three-dimensional load-carrying behavior determined by their geometry, boundary and support conditions, and the nature of the applied loads. Shells are usually bounded by supporting members and edge members, provided to stiffen the shell and distribute or carry loads in composite action with that shell. Approximate methods for preliminary design of shells are given in Chapter 13.

Thin shells are generally constructed in concrete, but other materials have been used, such as

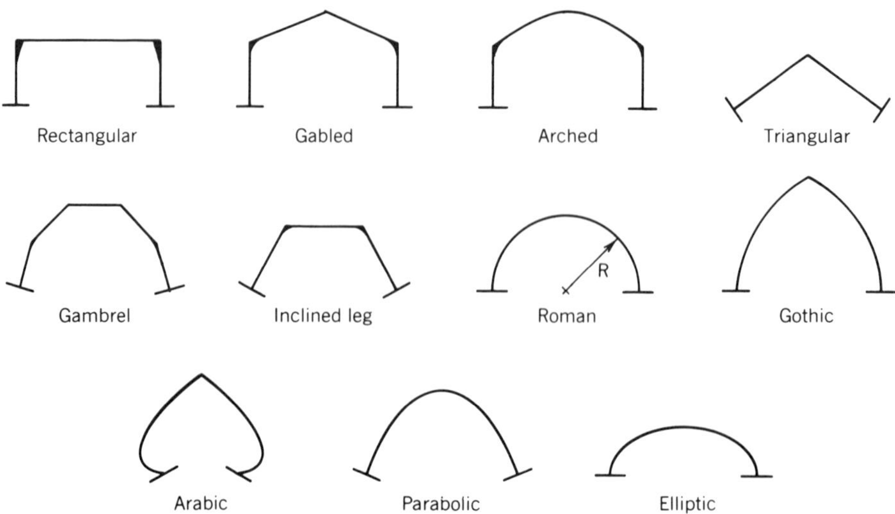

Fig. 8.7 Frames and arches.

welded steel plates, metal decking, plywood, multiple-layer timber decks, and fiberglass-reinforced plastic.

Shell structures are called "form-resistant structures" because of the nature of their design, which is to achieve maximum efficiency by shaping the curvature of the structure to carry the loads most efficiently. The design process selects a shape that will transfer the applied loads into direct forces in the middle surface of the shell. Such a surface is often called a membrane surface, and the state of stress is a membrane state of stress, that is, a state of stress that does not include bending about any axis in the shell surface.

In contrast to a tension membrane structure, a shell structure will most likely have both in-plane compressive and shear forces, and thus the shell must have some stiffness in order to remain stable and retain its shape.

Shell structures have been categorized according to their geometrical shape as shown in Table 8.1. The principal classifications deal with the curvature of the shell. Shells may have single curvature or double curvature. Doubly curved shells may have synclastic or anticlastic curvature.

Synclastically curved shells have centers of radial curvature on only one side of the surface, whereas anticlastically curved shells have their centers of radial curvature on opposite sides of the surface (see Fig. 8.8). Single curved shells are developable, that is, their surface may be flattened onto a plane. Doubly curved surfaces are nondevelopable.

Shell surfaces may be generated by rotating a plane curve about an axis lying in the plane of the curve. Such shells are called shells of revolution. Shell surfaces also may be generated by translating one plane curve over another plane curve. Such shells are called shells of translation. Shells are ruled surfaces if they can be described by a series of straight lines in the surface. Thus, ruled surfaces may be generated by moving a straight line (1) always through a fixed point, (2) parallel to itself, and (3) parallel to a plane. A conical shell is a ruled surface generated by moving a straight line through a fixed point, a cylindrical shell is a ruled surface generated by moving a straight line parallel to itself, and a conoid is a ruled surface generated by moving a straight line parallel to a plane. Doubly curved, synclastic shells are not ruled surfaces, whereas certain anticlastic shells (hyperbolic paraboloids, conoids, and hyperboloids of revolution) are ruled surfaces.

Ruled surfaces are advantageous for concrete shells because they are readily formed. A number of commonly used shell surfaces are illustrated in Figs. 8.9 and 8.10.

1. *Conical shell.* Generated by rotating a straight line through a point and around a closed curve (circular, elliptical, or other).
2. *Cylindrical shell.* Generated by translating a convex curved segment along a straight line or by moving a straight line parallel to itself along a convex curve.
3. *Dome.* Formed by rotating a convex plane curve about an axis lying in the plane of the curve. Domes may be spherical, parabolic, elliptical, or another shape as determined by the plane curve.

Table 8.1 Shell Structure Classification by Geometry

General Curvature	Single Curved (Developable)		Double Curved (Nondevelopable)					
Type of Curvature			Synclastic		Anticlastic		General	
Derivation	Shell of Revolution	Shell of Translation	Shell of Revolution	Shell of Translation	Shell of Revolution	Shell of Translation	Alternately Synclastic and Anticlastic	Membrane Condition
Examples of ruled surfaces	Conical shells	Cylindrical shells Vaults			Hyperboloid of revolution	Hyperbolic paraboloid	Corrugated shells	
Examples of nonruled surfaces		Folded plates	Domes	Elliptic paraboloid				Funicular shells

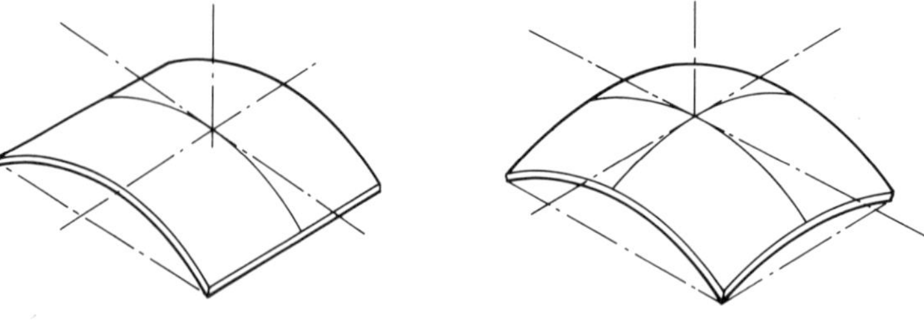

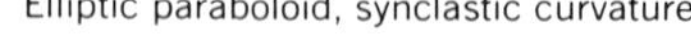

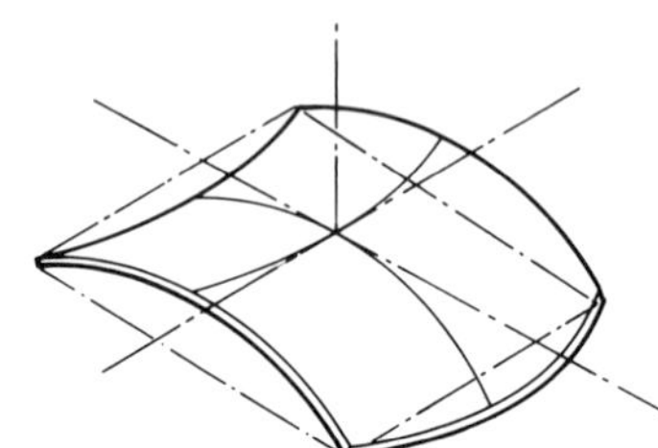

Fig. 8.8 Examples of shells of translation.

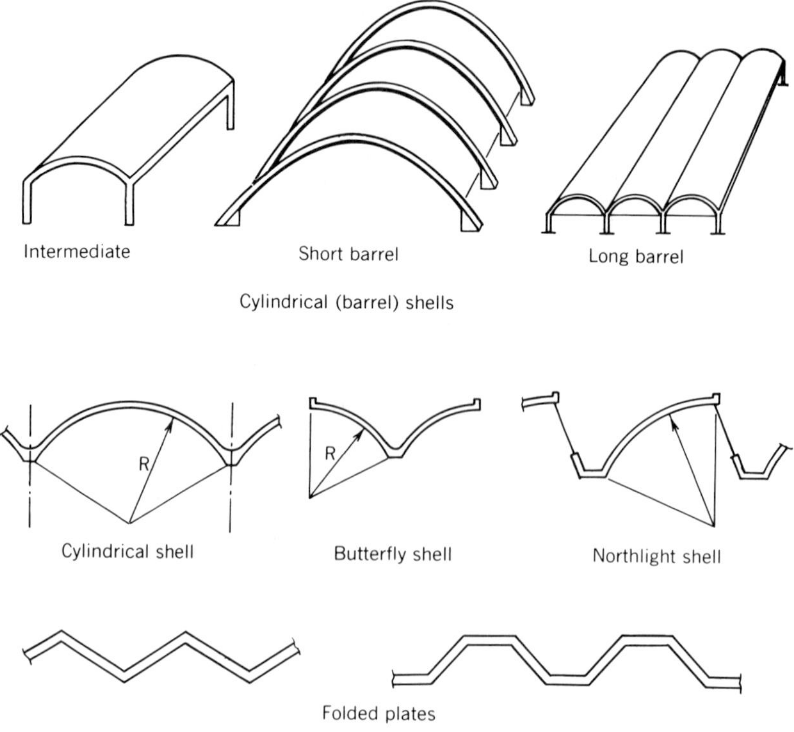

Fig. 8.9 Singly curved shells of translation.

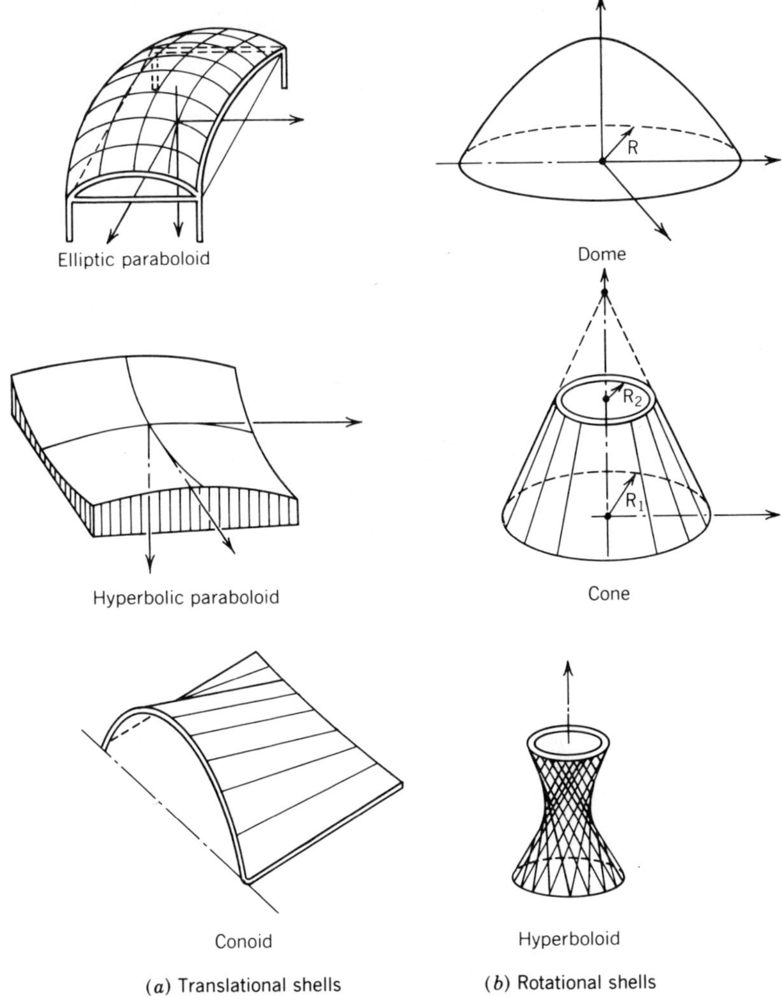

Fig. 8.10 Shell forms.

4. *Elliptic paraboloid.* Generated by translating a convex parabola over a convex parabola.
5. *Hyperboloid of revolution.* Generated by rotating an hyperbola about its y axis. The hyperboloid of revolution is also a ruled surface.
6. *Hyperbolic paraboloid.* A ruled surface, which can be generated by translating a straight line parallel to itself in a horizontally projected plane and at a constantly changing slope in a vertical plane. Hyperbolic paraboloids can also be generated by translating a convex parabola over a concave parabola, thus creating a saddle surface. Hyperbolic paraboloid panels can be combined with appropriate edge and stiffener members and support points to create a wide variety of structural forms (see Fig. 8.11).
7. *Conoid.* A surface generated by a straight line moving parallel to a fixed plane and intersecting one straight line and one curved line.

All of these surfaces are readily defined geometrically, but they may or may not be ideal surfaces with respect to certain loading conditions. For example, parabolic domes, paraboloids of revolution, and hyperbolic paraboloids are idealized surfaces to carry loads uniformly distributed over the surface in a horizontal plane. This derives from the parabolic nature of the curvature throughout these surfaces.

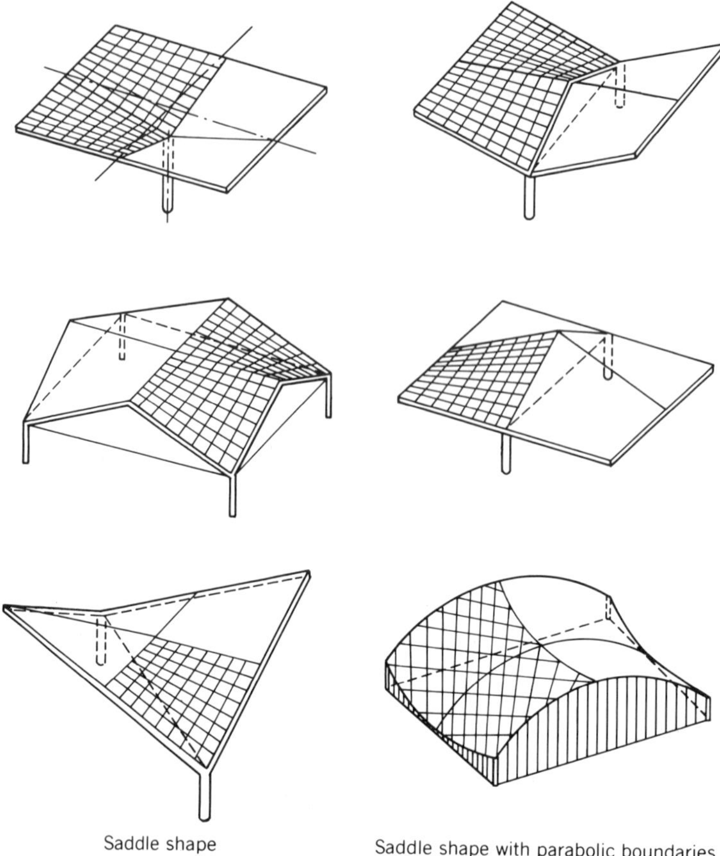

Fig. 8.11 Hyperbolic paraboloid shell forms.

However, they are unable to carry their self-weight, and other load types, such as concentrated loads or line loads, by membrane action alone.

For a given set of loads and support conditions, there are a number of idealized surface structures which can carry these loads by generating only in-plane (membrane) stresses: tensile, compressive, or shear. Such a surface is called funicular, and it may not be readily geometrically defined.

Such funicular shells have been developed analytically and (principally) experimentally by such techniques as hanging weights from tensile networks or by inflating plastic materials with air pressure. There is no limit to the variety of shell surfaces that can be developed.

Folded plates may be thought of as a special class of shell structure where the surface is approximated by an interconnected group of plane plate structures. "Single curve" folded plate structures have been used to approximate the behavior of cylindrical shells. Folded plates have also been developed to dome-like shapes.

2. *Tension Structures.* Tension structures may be defined as thin membranes or networks with a thickness that is small compared to their other dimensions, and characterized by a three-dimensional, load-carrying behavior determined by their geometrical shape and level of prestress such that all stresses are in the surface of the structure and are entirely tensile. Tension structures are thin-stretched flexible membranes or networks of tensile materials such as steel cables and rods. Preliminary design of both cable supported and air supported structures is treated in Chapter 15.

a. *Cable Networks.* The surface of tension structures may be framed by a series of parallel or intersecting curved cables, and where cables occur in two or more directions, the structure is called a cable net. Cable nets may be single curved or double curved (synclastic or anticlastic).

Stability under other than a single-load condition requires that cable nets be stiffened: (1) by having a stiff, shell-like surfacing; (2) by external support provided by other structural members or cables; (3) by having two or more sets of intersecting cables connected together and pretensioned; or (4) by mechanically generated differential air pressure.

Figure 8.12 illustrates several types of cable nets that have been used.

1. A single curvature roof tension structure supported by columns and tied-down anchor columns.
2. A circular axisymmetric tension roof with radial cables framing between a circular outside-edge compression ring and a circular centrally located tension ring.
3. A double curved tension structure with sagging primary cables and arching secondary cables supported around its edges by a compression-flexural ring (a saddle-shaped surface).

Cable nets form skeletal structures, and sheathing is required to provide a total building enclosure. Rigid sheet materials, such as flat and corrugated metal decks, precast concrete planks, wood decking, and plywood have been used. Fabric and film materials have also been used, in which case the material either must be stretched into a sufficiently taut membrane or supported by internal air pressure.

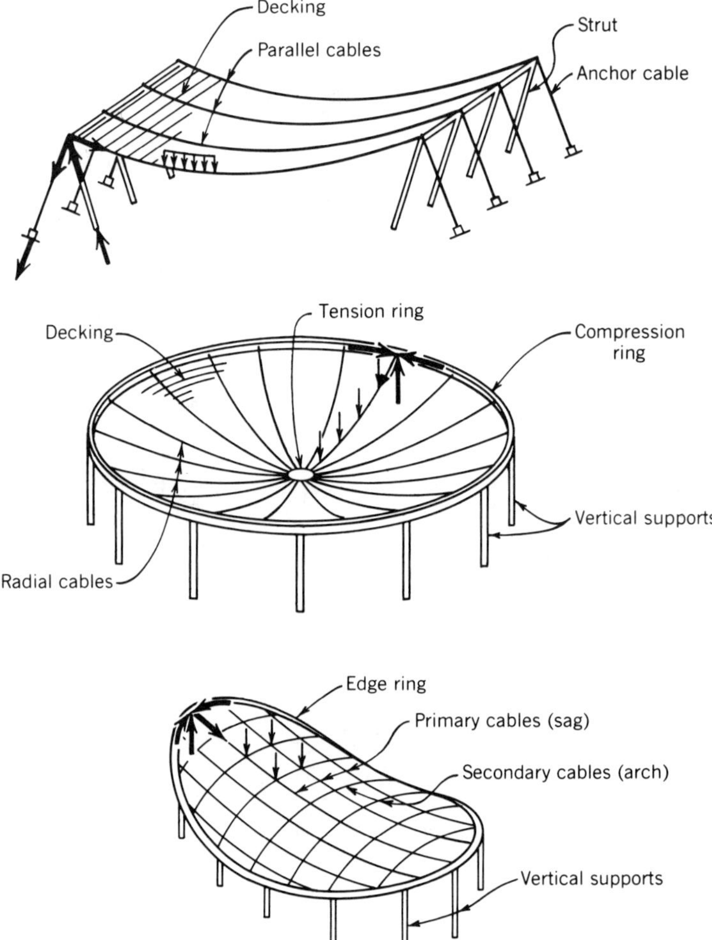

Fig. 8.12 Cable net structures.

b. *Stretched Membranes.* Stretched membranes have little or no flexural stiffness but must have substantial tensile strength. They may be homogenous films or reinforced membranes, either laminates or coated fabrics. In structures of any size, they require support by a primary structural system, which may be a rigid structure but is often a cable net.

Stretched membranes may support loads by either being tightly stretched and tensioned (usually in a doubly curved shape) between supporting members or supported by internal air pressure. Air supported or pressure-preloaded structures may be classified as either air supported or air inflated.

An air supported structure is a membrane envelope enclosing a pressurized occupied space. The air inflated structure consists of self-enclosed membranes which are inflated to form members capable of transmitting applied loads to their supports, as described in chapter 15.

Air inflated structures may take any of the forms of structure that are suitable for rigid materials, such as frames, arches, beams and slabs. The shapes of air supported structures, on the other hand, are limited by the nature of the shape-determining load—the air pressure. Differential air pressure provides a uniformly distributed pressure normal to the surface of the structure, and thus produces surfaces, all areas of which approximate some spherical surface. The variation of the spherical radius over the surface is a function of the restraining structure and edge support conditions.

Several common forms for air supported structures are shown in Figure 8.13.

1. *Low-profile dome.* An economical form for an air supported structure. At large scale, the dome fabric is stayed by a two-way cable network.
2. *Cylindrical segments.* The application of the air supported enclosure to a rectangular plan produces cylindrical shapes with half-dome ends.

3. *Space Frames (Three-Dimensional).* In the context of three-dimensional surface systems, space frames are thought of as skeletal structural systems consisting of linear members connected together to form a three-dimensional structural system with sufficient strength and stiffness to support a nonstructural skin subjected to all gravity and environmental load conditions.

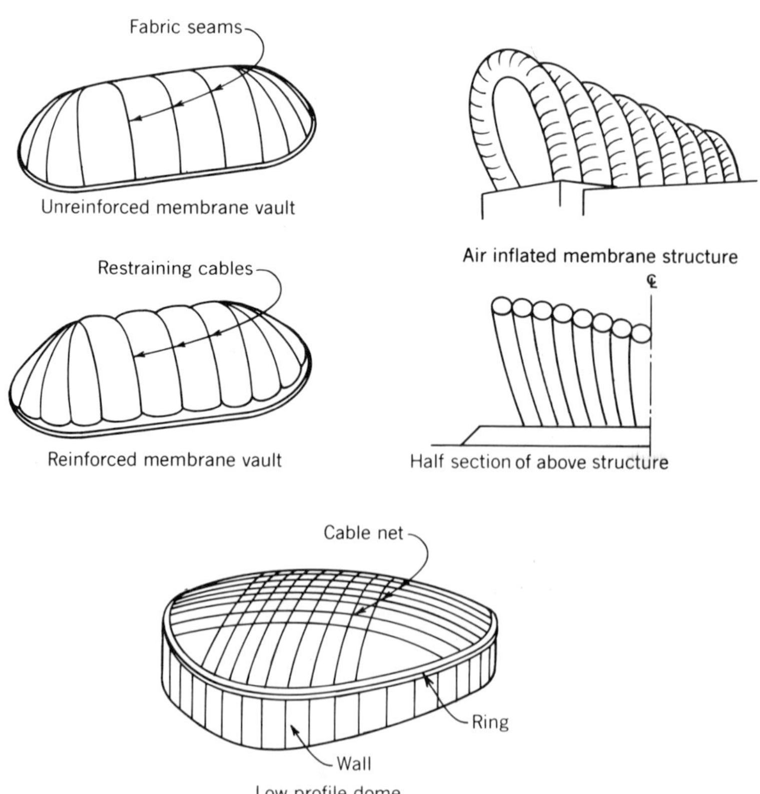

Fig. 8.13 Air supported membrane structures.

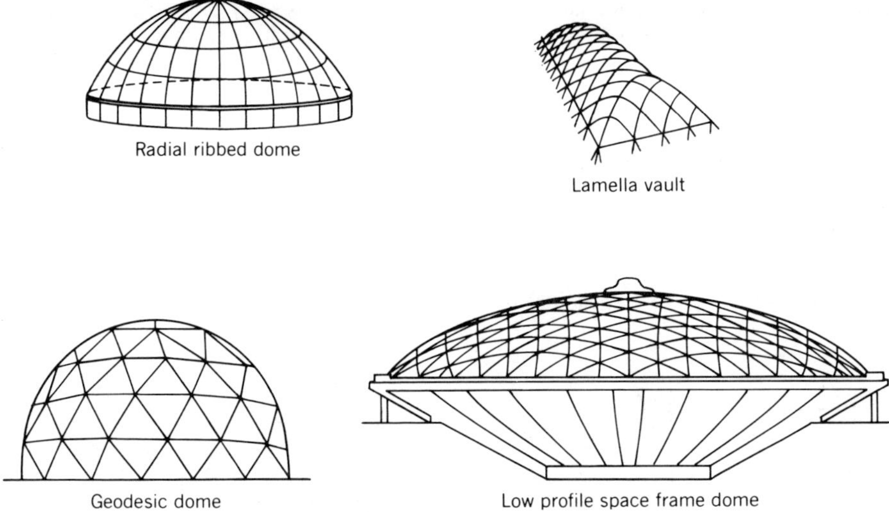

Fig. 8.14 Dome configurations.

All the geometrical forms discussed under the categories of shell and tension structures may be adapted to space frame structures. Domes, vaults, and doubly curved surfaces have been designed and constructed. Figure 8.14 illustrates a number of dome configurations that have been used.

8.3.3 Bracing Systems

A building must have a structural system capable of carrying its gravity loads, and it must be laterally stabilized and supported by a bracing system capable of preventing lateral instability of the gravity load-carrying elements and resisting any combination of environmental conditions, such as wind, earthquake, earth pressure, and temperature. The required bracing system may have a significant impact on the form of a structure. Bracing systems and requirements are treated in many places in this book, including Chapters 9, 10, 11, and 12. In the broadest sense, there are three fundamental types of lateral bracing systems: (1) moment-resisting frame, (2) braced frame, and (3) shear wall.

Moment-Resisting Frames

Moment-resisting frames consist of floor or roof members in plane with, and connected to, column or pier members with rigid or semi-rigid joints. The strength and stiffness of a frame is proportional to the beam and column sizes, and inversely proportional to the column's unsupported height and spacing. A moment-resisting frame may be internally located within the building, or it may be in the plane of the exterior walls or facade.

Cast-in-place concrete or precast concrete with cast-in-place joints provides the rigid or semi-rigid monolithic joints required. Frames may consist of beams and columns, flat slabs and columns, and slabs and bearing walls. Structural steel beams and columns may be connected together to develop moment frame action by means of welding or high-tensile bolting. Frames may be constructed of other materials, but with difficulty. The normal action of a moment-resisting frame produces significant bending moments in the beam at the face of the column and in the column at the face of the beam, with inflection points (points of zero moment) near the midlengths of the beams and columns.

This has led to designs with fillets at beam/column connections and tapered beam/column configurations.

Braced Frames

Braced frames consist of beam/column frameworks infilled with diagonal bracing. A system composed entirely of linear members, it has found extensive application in wood- and steel-framed buildings.

A braced frame may be located internally within the building or it may be placed in the exterior facade. It may be concealed in walls or partitions, or it may be exposed to view, in which case it establishes a strong structural form. A great variety of bracing systems are in use, and Fig. 8.15 illustrates a number of these.

1. Single diagonal bracing.
2. Double diagonal bracing.
3. K-bracing, either vertical or horizontal.
4. Lattice bracing.
5. Knee bracing.
6. Eccentric bracing.

Both knee and eccentric bracing produce a structure somewhere between a braced frame and a moment-resistant frame.

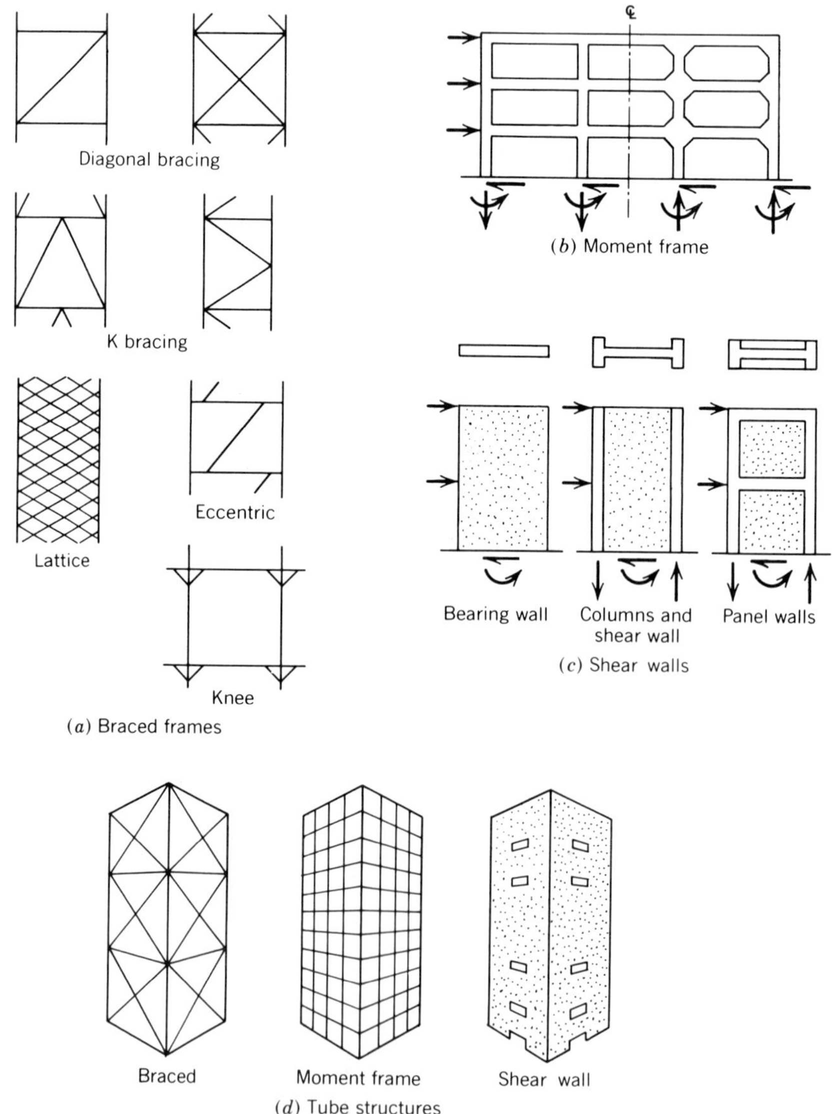

Fig. 8.15 Bracing systems for lateral loads.

Shear Walls

Shear walls are planar, generally vertical elements which are relatively thin and long. Shear walls generally have few openings or penetrations. If two or more shear wall elements are connected together in plane with relatively rigid members, they are called coupled shear walls. The shear wall may or may not carry substantial gravity loads, and it may be a single-bearing wall, a wall connecting two or more columns, or a panel wall filling the openings of a beam-column frame.

Shear walls may be organized in plan and connected together at their edges to form boxlike cellular structures. The obvious location for the introduction of shear walls into buildings is in and around utility cores which, in high-rise buildings, include elevators, stairs, mechanical rooms and shafts, and often toilets and janitor closets.

Shear walls may be developed in the exterior facade of a building, where they may contribute substantially to the structural form of a building. Often, a bearing and/or shear wall may have regular penetrations such that its structural action is intermediate between a shear wall and a moment-resisting frame. Some tube systems fall in this category.

Cast-in-place concrete, precast concrete, masonry, plywood and diagonally sheathed wood, and steel plate may function as shear walls. Additional treatment of wall design is given in Chapters 7, 24, and 25.

Combination Systems

In practice, bracing systems for buildings may not fall into one of the standard categories mentioned. Often, the variety of conditions that determine structural form dictate that the bracing system contain frames, walls, and bracing, all interacting together.

Several current examples of combination systems are illustrated in Fig. 8.16.

1. *Tube Structures*. The tube structure has been defined as a three-dimensional space structure composed of three, four, or (possibly) more frames; braced frames; or shear walls joined at or near their edges to form a vertical tubelike structural system capable of resisting lateral forces in any direction by cantilevering from the foundation. Thus, the tube may be a framed tube with moment-resistant frames where the columns are closely spaced, a braced tube (three-dimensional diagonally braced), or a shear wall tube formed by joining three, four, or more shear walls at their edges to form a true tubular structure. Centrally located cores and cores in other locations have been developed into substantially stiff tubes.

For very tall buildings, say over 20 to 30 stories, a tubular structural frame is a must, and, for office buildings with their rather limited plan dimensions, the tube must be placed in the exterior wall. Thus, the lateral bracing system for a high-rise office building is often dramatically expressed as a structural form.

2. *Multiple Tubes*. More than a single tube may be placed in any one building. Where a core wall tube is placed within an exterior facade tube, the structural system has been called *tube in tube*. For buildings with large floor areas, tubes have been grouped together to form a *bundled tube* system, a number of vertical tubular elements fitted together and sharing common side frames. Such tubes have been terminated at various levels of a building, thus producing variable floor areas.

3. *Core Interaction Systems*. Systems have been developed that connect a core tube to the exterior columns of the building at one or more locations in the building height where beams or

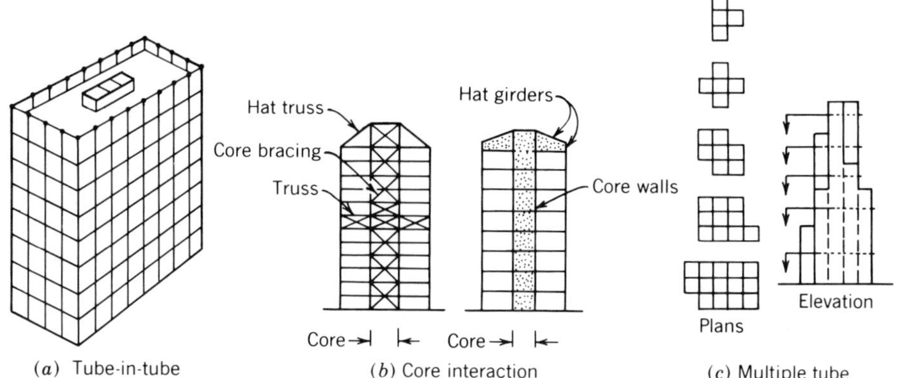

Fig. 8.16 Combination bracing systems.

cross bracing may not interfere with the building function. Thus, the core is used to fully develop horizontal shears, but the connection with the exterior wall effectively develops the full overturning capacity of the full building dimension. Figure 8.16 illustrates this principle in both steel and concrete construction.

SELECTED REFERENCES

8.1 *Webster's New Collegiate Dictionary*. Springfield, Massachusetts: G. C. Merriam Co., 1973.

8.2 Ray Faulkner and Edwin Ziegfeld. *Art Today; An Introduction to the Visual Arts,* 5th ed. New York: Holt, Rinehart & Winston, 1969.

8.3 Stefan J. Medwadowski. "Conceptual Design of Shells," *Concrete Thin Shells,* ACI SP-28. Detroit: American Concrete Institute, 1971, pp. 29–32.

8.4 David P. Billington. *Robert Maillart's Bridges—The Art of Engineering*. Princeton, New Jersey: Princeton University Press, 1979, pp. 64–67.

8.5 Council on Tall Buildings. *Tall Buildings Systems and Concepts,* Vol. SC, Monograph on Planning and Design of Tall Buildings, "Structural Systems," Chapter SC-1. New York: American Society of Civil Engineers, 1980, pp. 23–53.

CHAPTER 9

PRELIMINARY DESIGN OF LOW-RISE BUILDINGS

JOSEPH P. COLACO

CBM Engineers, Inc.
Houston, Texas

9.1 DEFINITION

In the context of this chapter, a low-rise building is defined as one in which the design of the structural system is predominantly governed by vertical (gravity) loads. Lateral loads due to wind or earthquake have a small effect on the sizing of members, but they cannot be ignored. In designing and detailing the structure, all forces that act on it must be considered. Building codes that used working stress design generally allowed an increase of up to 33% in allowable stress for the design of members subjected to wind forces in addition to gravity load. In strength design of concrete structures and in load and resistance factor design for steel structures, the load factors are lower for members subjected to combined gravity and lateral loads. This reduction in load factor (or increase of allowable stress) for the wind load included case minimizes the impact of wind load on low-rise buildings. Consequently, at most, a few members in the structure will be increased in size from the effects of lateral forces except in severe seismic loading situations. Wind design criteria and snow load intensities are given in Chapter 2.

9.2 BUILDING OCCUPANCY

There are innumerable uses for low-rise buildings. Several broad categories (by no means all-encompassing) can be distinguished by occupancy:

1. *Office buildings*. These buildings cover the range from regular commercial office structures to specially designed institutional buildings. Both single and multi-tenant occupancies are common.
2. *Residential buildings*. These include hotels, motels, dormitories, and low-rise apartment buildings.
3. *Hospitals and medical facilities*. Hospital construction ranges from single purpose to general occupancy hospitals.
4. *Parking garages*. These range from single story to approximately 15 stories in height.
5. *Shopping facilities*. Occupancies such as shopping malls, department stores, and grocery stores fall within this category.
6. *Educational facilities*. Most schools, university buildings, and libraries are low-rise structures.
7. *Other structures*. Several other types are identifiable, such as warehouses, industrial and manufacturing facilities, and penal institutions. Warehouses and industrial buildings are also treated in Chapter 12, Preliminary Design of Single-Story Open-Space Buildings.

9.3 DESIGN REQUIREMENTS

9.3.1 Loads

The vertical loads are generally specified by the applicable building code for each occupancy. Lateral loads are also specified by the building code for the location. A thorough discussion of loads is given in Chapter 2.

9.3.2 Fire Requirements

The local building code or the *National Building Code* of the National Board of Fire Underwriters generally specify fire-rating requirements for buildings and their components. The fire ratings of structural elements are given in terms of hours and refer to the time that a structural element is subjected to a standard fire test (ASTM E329). Generally, low-rise structures have less stringent fire-rating requirements than taller structures. Most fire codes will specify minimum fire ratings for structures less than 75 ft (23 m) in height. For most structures lower than 75 ft (23 m) in height, for normal commercial occupancy, the fire-rating requirements are 1 hr for floors and 2 hr for the structural frame. Moreover, most codes do not require sprinklers for office buildings of this type.

9.3.3 Flexibility

With the rapid changes taking place in society, considerations of future uses of the building play a part in the design. The main features that require structural consideration are future occupancy, span length, and mechanical systems.

Future Occupancy Loads

The use of a given space often changes. For example, in a commercial office building, the occupancy of a given space may change from normal office use to computer use or file storage space. Live load requirements may increase from 70 psf (3400 N/m^2) to 125 psf (6000 N/m^2) or more. Where large increases in loading are anticipated and full flexibility to make this change over large areas of the building is desirable, steel structures are generally favored. Other options are to design predesignated areas in the building for heavier live loads with minimal live load flexibility in other areas.

Spans

Longer spans obviously allow for more freedom in future changes. The initial cost of large-span column-free spaces must be weighed against future benefits. Moreover, longer spans require greater structural depths and generally, larger floor-to-floor heights. This increases the building height and further increases the cost.

Mechanical Systems

Experience with remodeling of older structures indicates changes do occur in mechanical, electrical, telecommunication, and vertical transportation systems more rapidly than in the structure. Some consideration should be given to accommodating these changes within the useful life of the structure. Two possibilities are additional floor heights for duct and pipe runs above the ceiling and the use of knock-out panels in the floor for future vertical penetrations, as discussed in Chapter 4.

9.3.4 Design Criteria

Above and beyond the vertical and lateral loads established by code, certain other criteria need to be considered.

Structural Depth

The depth of the structure (floor system) is governed by spans, loads, and choice of construction materials. The structural depth becomes critical in areas where the height of the building is restricted by zoning ordinances, such as limits to building heights in the vicinity of airports, and restrictions by city ordinances of building heights in certain zones. In general, the shallowest structural depth is obtained by concrete construction. In some cases, the overall depth can be reduced by integration of mechanical and structural systems to occupy the same volume in the ceiling instead of being in separate volumes.

Floor Vibration Under Pedestrian Traffic

In some building types, floor vibration becomes important. Several research programs, especially those done by Lenzen [9.7], Murray [9.8], and Allen, Rainer, and Pernica [9.1], have contributed greatly to our knowledge of the phenomena associated with floor vibration. The problem arises primarily in long-span, low-weight construction having low damping. It is most often found in typical bar joist construction for spans in the 28–42 ft (8.5–13 m) range in open lease spaces. In regular steel construction vibration is a consideration where noncomposite design is used with shallow beams. Cast-in-place concrete construction very seldom experiences the problem of floor vibration under pedestrian traffic, but precast construction may need investigation.

9.4 FLOOR SYSTEMS—MATERIALS

9.4.1 Structural Steel

Open-Web Steel Joist or Bar Joist Construction

Bar joist construction is common throughout the United States in low-rise construction. The fabrication and design of bar joist members are governed by the Steel Joist Institute. Bar joists consist of open-web or "trusslike" members that are fabricated to standard depths and member sizes. Catalogs showing load-carrying capacities for different spans and different joist designations are published by joist manufacturers. Typical connection details are also provided. It is difficult to fire proof bar joists by conventional sprayed-on fire proofing. Where fire rating is required, it is provided by placing gypsum board of appropriate thickness directly below the bar joist, or by a fire-rated ceiling. In the latter case, all penetrations through the ceiling need to have fire dampers.

Wide-Flange Steel Beam Construction

Rolled steel beams are used to frame floors. Common spacings of floor beams vary from 7 ft (2.1 m) to 15 ft (4.6 m) centers. The floor beams are supported by rolled steel girders which frame between columns. Corrugated metal deck (either galvanized or phos-painted) is used to span between floor beams. The assembly is completed by the placement of a concrete slab on the metal deck. The thickness of the slab and the type of concrete (either normal-weight or lightweight) depend both on the structural design of the slab as well as the fire-rating requirements. In general, lightweight concrete slabs can be thinner than normal-weight concrete slabs to meet a given fire rating. A minimum of $3\frac{1}{4}$-in. (83-mm) thickness (above the deck) is required in lightweight concrete to meet a 2-hr fire rating. In some cases, it is advantageous to make the slab composite with the steel beams. This reduces the weight of steel beams and also reduces the deflection under live loads. The technique used to develop composite action is the placement of shear studs on the top flange of the steel beams which extend into the concrete slab. Extensive tables are available from metal deck manufacturers on the load-carrying capacities of composite construction. Chapters 20 and 23 contain material on composite decks.

Stub-Girder Systems

The stub-girder system is a Vierendeel beam system where the concrete slab is the top chord, and steel sections form the verticals and the bottom chord. A detailed description is given in the article by Colaco [9.3]. The objective of the system is to reduce the steel quantities and to integrate the mechanical ducts into the structure in a natural way instead of through costly beam penetrations.

The stub-girder system can be conventionally fireproofed with sprayed-on fireproofing. One further advantage of the system is the reduction in floor-to-floor height since the mechanical ducts are integrated into the structure.

Plastic–Composite Design

The use of this design technique has advanced in the last 10 years. Fundamentally, it is a design technique whereby a girder is moment-connected to a column for fixity and the slab is made composite with the girder. By carefully balancing the negative and positive moment capacities, the girder weight is substantially reduced. As a further refinement, floor beams framing into the girders can also develop continuity by appropriate detailing. The cost of moment connections minimizes some of the cost savings. However, continuity greatly reduces the deflection of the floor system. A study of the system is given by LeMessurier [9.6].

Composite Bar Joist Systems

These are systems where the concrete slab is made composite with the bar joist by suitable connection. Composite bar joist systems minimize the floor vibration characteristics of regular bar joist systems and further reduce the steel weight. The diagonal web members are extended above the top chord of the joist to anchor into the concrete slab, thereby providing composite action. The drawback of the system is that the metal deck can no longer be placed to run continuously over the tops of adjacent bar joists since the web extensions are in the way. Conventional plywood decking or single span metal decking can be used.

9.4.2 Concrete

Concrete floors can be divided into the following types:

1. All cast-in-place construction.
2. Precast structural members with cast-in-place concrete topping.
3. All precast floor framing members.

From a structural behavior standpoint, construction is normally split into two parts, namely, one-way construction which includes one-way slabs and pan-joist construction, or two-way construction which includes flat plate, flat slab, waffle slab, and two-way slab systems. A detailed discussion of these systems is given by Colaco [9.4] and detailed material and cost estimates have been prepared by the Portland Cement Association in their *Material and Cost Estimating Guide* [9.9].

Pan-Joist System

By far the most common method of concrete construction uses pan forms made of steel or fiberglass. In essence, it gives rise to a ribbed slab with the ribs being spaced anywhere from 2 ft (0.6 m) to approximately 9 ft (2.7 m) on centers. In general, the joists are carried by a girder which most commonly is the same depth as the joist.

Haunched-Girders

This is a system similar to the one described above except the girders are not of uniform depth. The girders are generally the same depth as the joist in the center two-thirds of the girder span but the girder depth is increased toward the column. This has the advantage of providing the largest depth at the supports where the moments and shears are maximum. This, therefore, gives rise to economy in concrete construction. A description of the advantages of the system is given by Colaco [9.5].

Flat Slabs

This mode of construction with uniform thickness of slab is used to span between columns. In cases where the spans are long (over 25 ft) (approximately 7.5 m) and loads are heavy or post-tensioning is not used, then generally the slab is thickened at the columns to take care of shear and moment requirements. A prime example of utilization of this kind of structure is in hotels, apartments, warehouses, short-span office buildings, and parking garages. The structural design criteria, concrete and reinforcing bar quantities, and costs are contained in the two papers by Colaco [9.4, 9.5].

The major drawback of this system in low-rise structures is the lack of repetition of formwork. On the other hand there is a reduction in the structural depth and the elimination of fireproofing. The structure is extremely rigid.

9.4.3 Precast Construction

Several kinds of precast members are generally used for low-rise buildings. The type of members used depends upon the occupancy and the spans. Chapter 21 treats precast prestressed members.

Hollow Cored Slabs

Hollow cored slabs are precast elements (so-called planks) with a flat top and bottom containing round holes within the thickness of the slab running parallel to the span. The slabs are generally pretensioned and are used in spans ranging from 15 to 35 ft (5 to 11 m). Most common use of these planks is where a flat ceiling is needed. Such uses are common in hotel and motel construction and low-rise apartments.

Double Tees and Channels

Double tees and channels are generally precast, pretensioned elements that range in span from 30 ft (9 m) to 65 ft (20 m). The depth, rib spacing, and overall width of the elements vary among different manufacturers. Generally, the most common use of this construction is where long spans and light weight is essential. This type of construction is most prevalent in parking garages, shopping malls, some warehouses, and some office buildings. It is occasionally seen in apartment construction.

Rigid Frames

Several attempts have been made to obtain precast rigid frames and several of these are available in the market place under trade names. Most of them utilize either welding of reinforcing steel or post-tensioning in the field in order to achieve continuity. Such trade names as Dynaframe and Omega-frame are available for low-rise industrial buildings and office and apartment buildings.

9.4.4 Masonry

The use of load-bearing masonry in low-rise structures goes back to antiquity. Modern versions include the use of load-bearing block and brick masonry for various kinds of structures. Brick and block masonry is found in building exteriors both in the load-bearing and veneer modes, and sometimes, on the inside of a structure where regular modules of structure are contemplated. This type of construction is generally found in hotel and motel construction, some school buildings, low-rise office structures, and medical clinics. The advantage of this kind of structure is that it provides common construction with excellent fireproofing and sound attenuation. The floors are generally made of lightweight steel or precast construction. Excellent wind resistance can be provided in both directions because of the rigidity of the walls and with suitable detailing, seismic conditions can also be met. Chapter 24 provides detailed discussion of masonry structures.

9.4.5 Mixed Construction

The economics of the project sometimes dictate the use of mixed construction. Special care must be taken because of different trades and scheduling problems that may arise.

Steel and Precast Concrete

In some conditions, load-bearing precast exterior members are used in conjunction with steel construction for the remainder of the building. The load-bearing precast members could be either special architectural precast panels or standard shapes such as double tees. The floor system consists of either bar joist framing or regular wide-flange steel sections. Many buildings have been built using this construction and range in height up to approximately six stories.

Masonry and Precast Concrete

Load-bearing masonry and hollow cored precast planks are fairly common in low-rise hotel and motel construction. The advantages of each material are utilized in this process to provide a rapid construction technique with a low floor-to-floor height and cost.

9.5 SELECTION OF FRAMING SYSTEM

The selection of the framing system for a low-rise structure depends on a careful evaluation of all the factors that are relevant for the project. The following are factors that need to be considered:

1. Economics.
2. Speed of construction.
3. Integration of the architectural and structural elements.
4. Integration of the structural and mechanical elements.
5. Flexibility for future use.
6. Spans and floor loads.
7. Foundation considerations.

There are several other factors that are also important but generally, to a lesser degree than those mentioned above. These additional factors include labor and materials availability, building codes, and the availability of certain trades. Every project needs to have a detailed investigation made of different systems in order to obtain the one that is most desirable for the project. In general, the process is called "value engineering" and can only be conducted by a group of professionals operating as a team. The professionals involved consist of the client, the architect, structural and mechanical engineers, and a cost consultant. This group of individuals will study different ideas and alternatives which are developed by each professional in the team.

With regard to the economics, several excellent references are available. For steel construction in low-rise structures, an article by Ruddy [9.10] and a series of Bethlehem Steel Tables [9.2] for parking structures provide good source material. Also, trade manuals by different steel fabricators such as Strand, Butler, etc. show proprietary building systems.

A key element in obtaining cost-effective structures is the consideration of integration of different functional elements. For example, the exterior wall of the building could be both the architectural feature and the load-bearing structure, as described in Chapter 27. The advent of joist-girders (or truss-girders) provides an element which can both carry the floor and provide wind resistance, while at the same time be a carrying element for masonry or granite facades. "Flame-shielded" girders have been used on several projects whereby the girders of weathering steel are the finish element to the structure, while at the same time providing the load-bearing element for the floor system.

Another important integration is of the structure and the mechanical system (see Chapter 4). The objective is the development of a single element mechanical and the structural system. Such a system, called "duct-beam" has been developed but at the present time has not been widely used. At the present state-of-the-art, most of the structural–mechanical systems consist of taking advantage of the knowledge of the two disciplines that are vying for the same space in the ceiling plenum and integrating them efficiently. Such systems as the "stub-girder," "haunched-girder," and "joist-girder" are well suited for this purpose. The final choice of the framing system can only be made by consideration of the alternatives for each project.

9.6 DESIGN EXAMPLES

9.6.1 Case Study 1: Low-Rise Office Building

A specific example of a low-rise office building is an eight-story office building in Houston that was studied in 1978 utilizing four different types of structural systems:

Scheme 1

Haunched-girder and pan-joist system: This system is shown in Fig. 9.1. It consists of pan joists spaced at 3 ft (0.9 m) on center and running in the long direction of the building. The pans are 30 in. (760 mm) wide and 16 in. (400 mm) deep. A $4\frac{1}{2}$-in. (114-mm) concrete slab was used to maintain the required fire rating. Haunched-girders running in the narrow direction of the building were used to support the pan joists. The haunched-girder system also provided for the lateral wind resistance for wind acting on the broad face of the building. In the narrow direction, two short shear walls were used by the elevator core. General information was given to the contractor for column sizes, foundations, and other dimensions. The cost of this scheme was $7.37/sq ft.

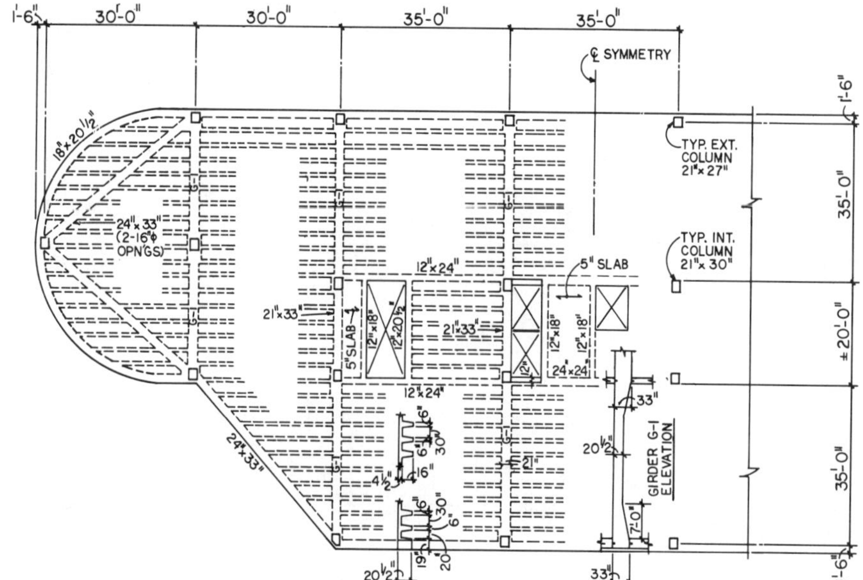

Typical floor plan (8 story building)

Notes: 1.) Concrete: hardrock 4,000 psi except columns 5,000 psi; 2.) Rebar: grade 60. Total (including footings) 350 tons; 3.) Mesh in all slabs 4×4-W3.5×W3.5 additional; 4.) Bottom of footings: 10′-0″ below grade. Footings designed to allow 4,500 psf; 5.) Footings (3,000 psi stone), interior columns combined 37′-0″×16′-0″×3′-6″, rebar 5,500 lbs; exterior columns 13′-6″×13′-6″×3′-0″, rebar 1,050 lbs.

RELATIVE COST SCHEME 1 $7.37/SQ.FT.

Fig. 9.1 Scheme 1—concrete haunched-girder and pan-joint system.

Scheme 2

A regular rolled steel beam, fireproofed system, as shown in Fig. 9.2: The scheme consists of steel beams spaced at up to 10 ft (3 m) on centers with a 2-in. (50-mm) metal deck and 3¼ in. (83 mm) of lightweight concrete on top of the deck. Once again, all the beams, girders, bracing, and foundation were shown. The cost of this scheme was $7.91/sq ft.

Scheme 3

A standard bar joist system shown in Fig. 9.3: The bar joists were spaced approximately at 2 ft—6 in. (0.8 m) centers. Standard galvanized corruform metal deck was placed on top of the bar joists and a 3-in. (75-mm) concrete slab was used on top in order to minimize the floor-vibration problems. Joist sizes, girder sizes, column sizes, and bracing were also indicated. Foundations were sized for load. The cost of this scheme was $6.34/sq ft.

Scheme 4

A proprietary composite bar joist system with bracing similar to scheme 3: The bar joists were spaced at 4 ft (1.2 m) on centers to suit the proprietary composite joist system. The cost of this scheme was $7.06/sq ft.

Considerable effort was spent to be sure that the prices included such items as the floor-to-floor height differentials, fireproofing differentials, and mechanical duct work difficulties. The net result of this analysis indicated that the bar joist system (scheme 3) was the most economical way to build the building on an overall basis and that was the one selected for the building. The building design was completed in October 1978 and the construction then proceeded.

9.6.2 Case Study 2: Low-Rise Condominium

In June 1979, a six-story condominium was to be designed. Seven schemes were studied in order to obtain the most cost-effective structure for these condominiums. A typical wing of the condominium was studied based on a 26-ft (8-m) module. The schemes are shown in Fig. 9.4 and Fig. 9.5.

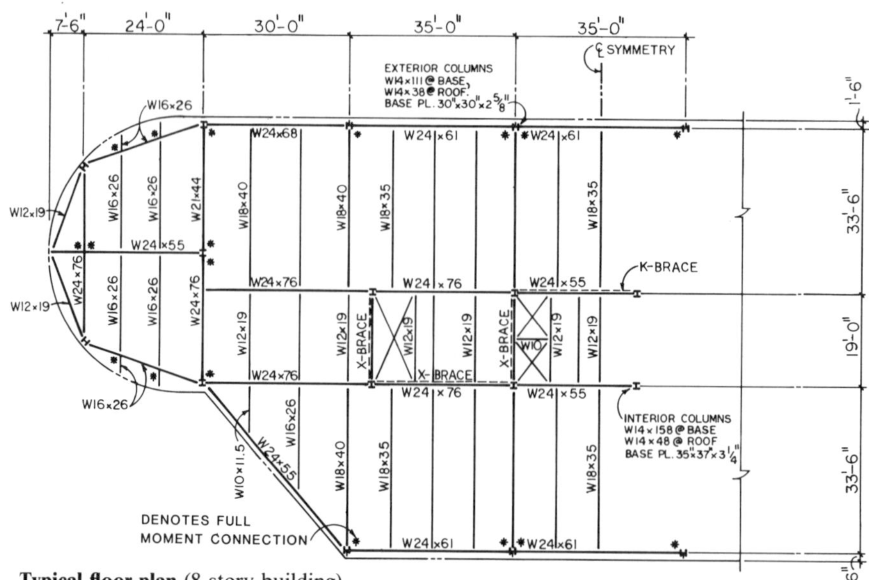

Typical floor plan (8 story building)

Notes: 1.) All steel A-36; 2.) Bracing 2 angles 6″×4″×3/4″ average; 3.) All beams & girders are composite assume 3,000 studs 3/4″×3½″/floor; 4.) Metal deck 2″ Mac-Lok 19 ga./equal; 5.) Slab 3,000 psi light weight 5¼″ thick; 6.) Mesh 6×6-6/6; 7.) Allow total 15^T steel for misc.; 8.) Footings: (3,000 psi stone) interior columns 14′×14′×3′-3″, rebar 1,300 lbs; exterior columns 11′-6″×11′-6″×2′-9″, rebar 700 lbs; 9.) Bent plate at exterior edges, and all openings.

RELATIVE COST
SCHEME 2
$7.91/SQ.FT.

Fig. 9.2 Scheme 2—regular rolled steel beam, fire-proofed system.

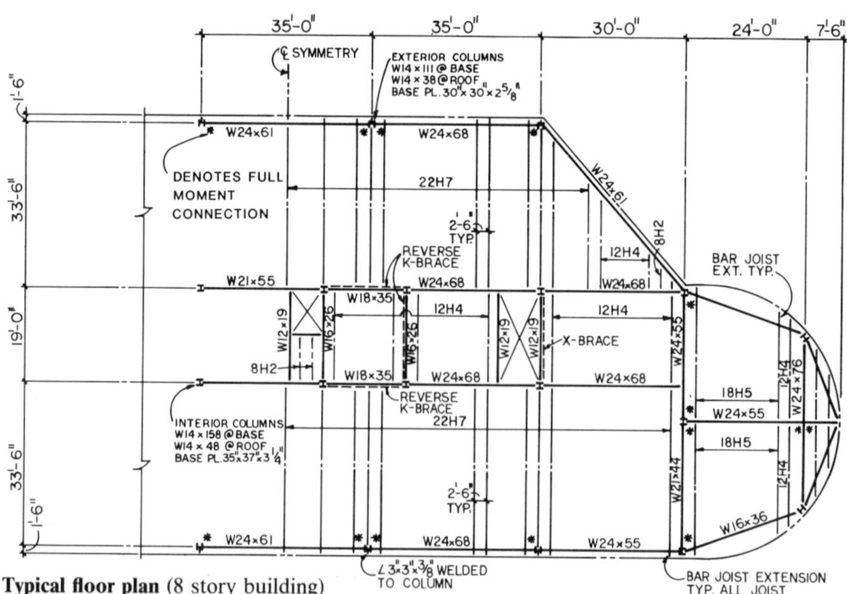

Typical floor plan (8 story building)

Notes: 1.) All steel A-36; 2.) Bracing 2 angles 6″×4″×3/4″ average; 3.) Bridging as per SJI standards; 4.) Metal deck galvanized Corruform; 5.) Slab 3,000 psi hardrock 3″ thick; 6.) Mesh 6×6-10/10; 7.) Allow total 15^T steel for misc.; 8.) Footings: (3,000 psi stone) interior columns 14′×14′×3′-3″; rebar 1,300 lbs; exterior columns 11′-6″×11′-6″×2′-9″, rebar 700 lbs. 9.) Provide bent plate at exterior edges and all openings.

RELATIVE COST
SCHEME 3
$6.34/SQ.FT.

Fig. 9.3 Scheme 3—standard bar joist system.

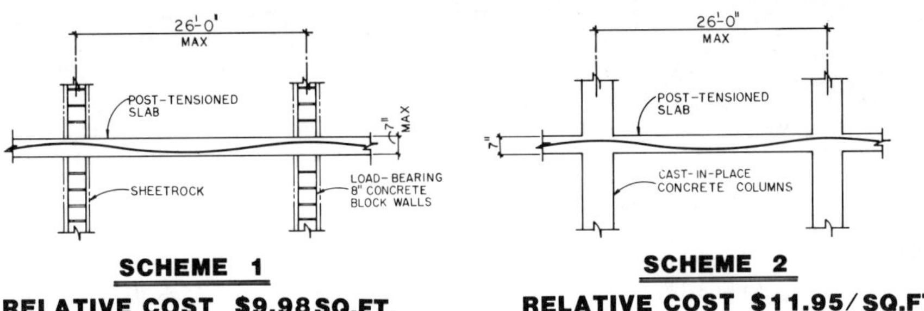

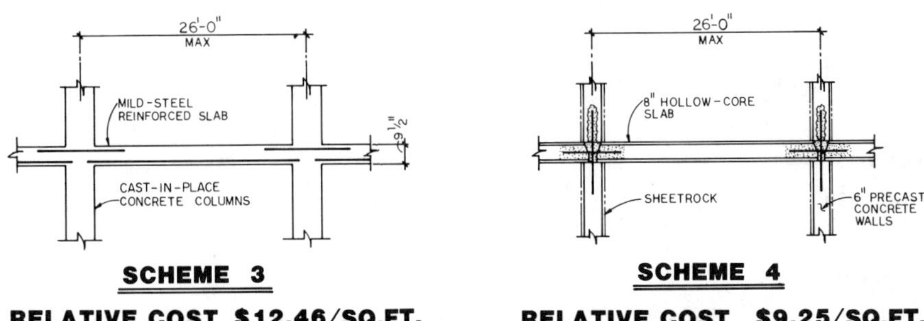

Fig. 9.4 Floor schemes for condominium case study.

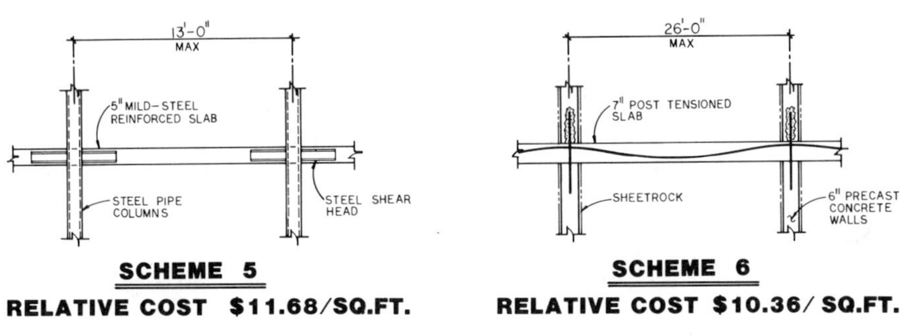

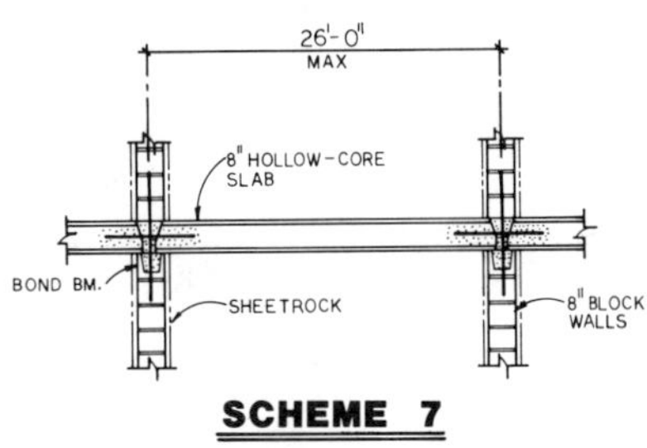

Fig. 9.5 Floor schemes for condominium case study (continued).

Scheme 1

A post-tensioned, flat slab on load-bearing concrete masonry block walls: The total cost of construction was \$9.98/sq ft. This scheme assumes that the block walls had furring and a layer of drywall to provide a finish for the wall construction. If the block walls could be painted and the drywall eliminated, the price would have been \$8.95/sq ft.

Scheme 2

A post-tensioned flat plate and cast-in-place concrete columns: This scheme cost \$11.95/sq ft.

Scheme 3

A mild-steel reinforced slab with cast-in-place concrete columns: This scheme was priced at \$12.46/sq ft.

Scheme 4

A precast concrete wall with hollow-core precast plank system: This scheme was priced at \$9.25/sq ft.

Scheme 5

A 5-in. (130-mm) flat plate on steel pipe columns: This scheme was priced at \$11.68/sq ft. It should be noted that the span was reduced to 13 ft (4 m).

Scheme 6

A precast concrete wall system and post-tensioned slab: This scheme was priced at \$10.36/sq ft. Once again, if the precast wall could be painted in lieu of furring and drywall, the price would be \$9.33/sq ft.

Scheme 7

A load-bearing concrete masonry with hollow-cored precast planks: This scheme was priced at \$8.47/sq ft. If the furring of the precast walls was removed, the cost was \$7.44/sq ft.

9.6.3 Case Study 3: Parking Garage

A 12-story parking garage having approximately 1,000,000 sq ft (93,000 m^2) was studied. The parking studies indicated that a spacing of 50 ft (15 m) in one direction and 19 ft–8 in. (6 m) in the other direction would be an optimum for both the layout of the parking garage and the parking of automobiles. Based on this layout, several alternate structural systems were studied and priced in August, 1981. These systems are indicated in Figs. 9.6 through 9.11. The prices are for the floor system only and do not include columns and foundations.

Scheme 1

Mild steel haunched-girders spanning the 50-ft (15-m) direction and mild steel reinforced pan-joists at 6 ft (1.8 m) on centers spanning 19 ft–8 in. (6 m): The relative cost of this floor system only was \$6.00/sq ft.

Scheme 2

This scheme was the same as scheme 1 except that the haunched-girder was post-tensioned. The cost of this scheme was \$5.67/sq ft. The effective depth of the haunched-girder at midspan was only 16 in. (400 mm), thereby further reducing floor-to-floor heights.

Scheme 3

This system was the same as scheme 1 except that a constant depth girder was used. The cost of this scheme was \$6.55/sq ft. Moreover, there is a 10-in. (250-mm) floor height penalty when compared to scheme 1.

Scheme 4

A beam and slab system with post-tensioned beams 30 in. (760 mm) deep spanning the 50-ft (15-m) direction and a post-tensioned slab 5 in. (130 mm) thick spanning the 19 ft–8 in. (6 m) direction: The cost of this scheme was \$5.97/sq ft.

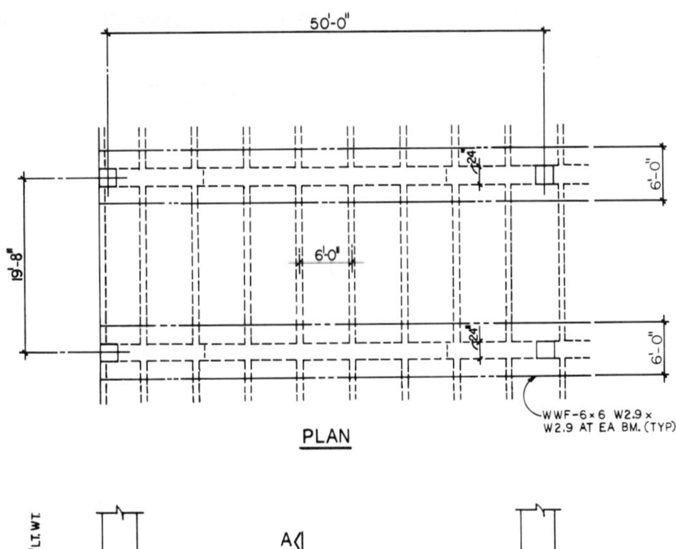

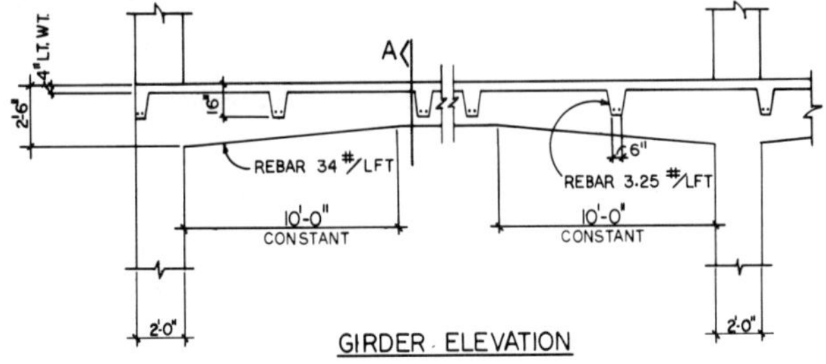

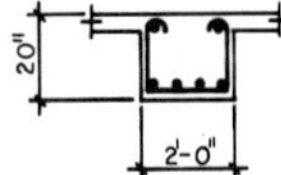

SECTION – A **RELATIVE COST $6.00/SQ. FT.**

Fig. 9.6 Parking garage case study (scheme 1)—haunched-girder (mild steel reinforced).

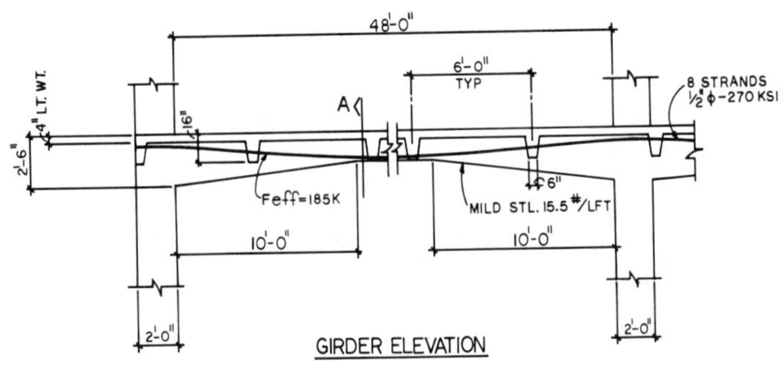

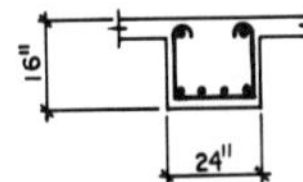

SECTION – A **RELATIVE COST $5.67/SQ. FT.**

Fig. 9.7 Parking garage case study (scheme 2)—post-tensioned haunched-girder.

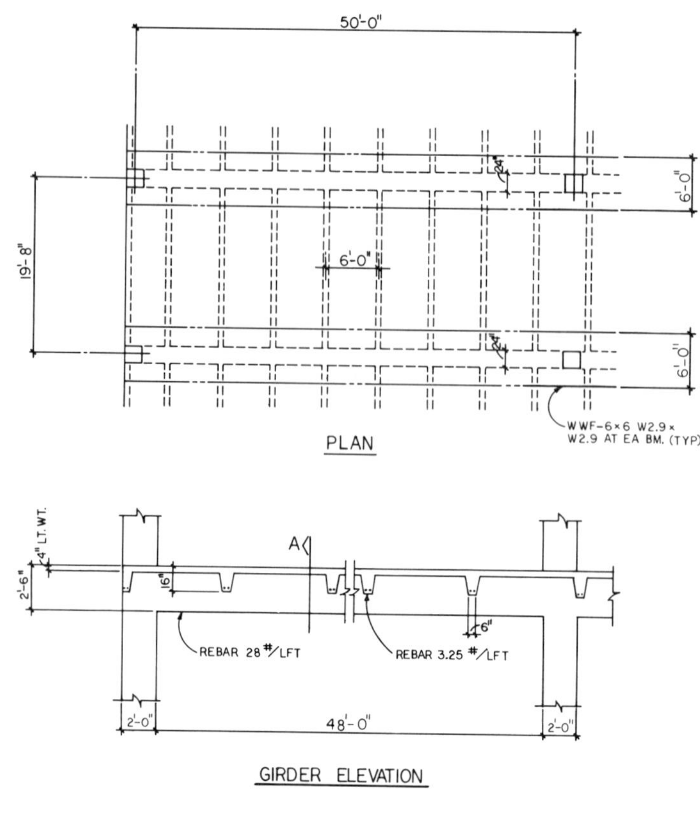

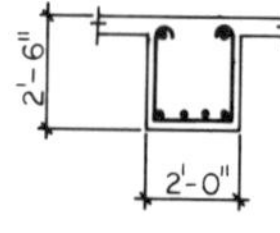

SECTION - A **RELATIVE COST $6.55/ SQ. FT.**

Fig. 9.8 Parking garage case study (scheme 3)—prismatic girder (mild steel reinforced).

As an extension to these systems and to obtain cost comparisons only, several short span schemes were studied where the bay spacing was changed to 33 ft (10 m) in one direction and 26 ft (8 m) in the other direction.

Scheme 5

16-in. (400-mm) deep pan-joists spanning the 33-ft (10-m) direction with a girder of the same depth as the pans spanning the 26-ft (8-m) direction: Cost of this scheme was $5.54/sq ft.

Scheme 6

9-in. (230-mm) flat slab with a 14-in. (360-mm) thick drop-panel which measured 10 × 11 ft (3 × 3.4 m). The cost of this scheme was $6.31/sq ft.

9.7 CONCLUSIONS

The design for low-rise buildings utilizes methodology similar to that used in the design of any height of building; however, the effects of gravity loads are substantially more significant than the effects of wind loads or moderate seismic forces. The process needs the ingenuity of each and every individual in the design team, namely, the engineer, the architect, the contractor, and the owner.

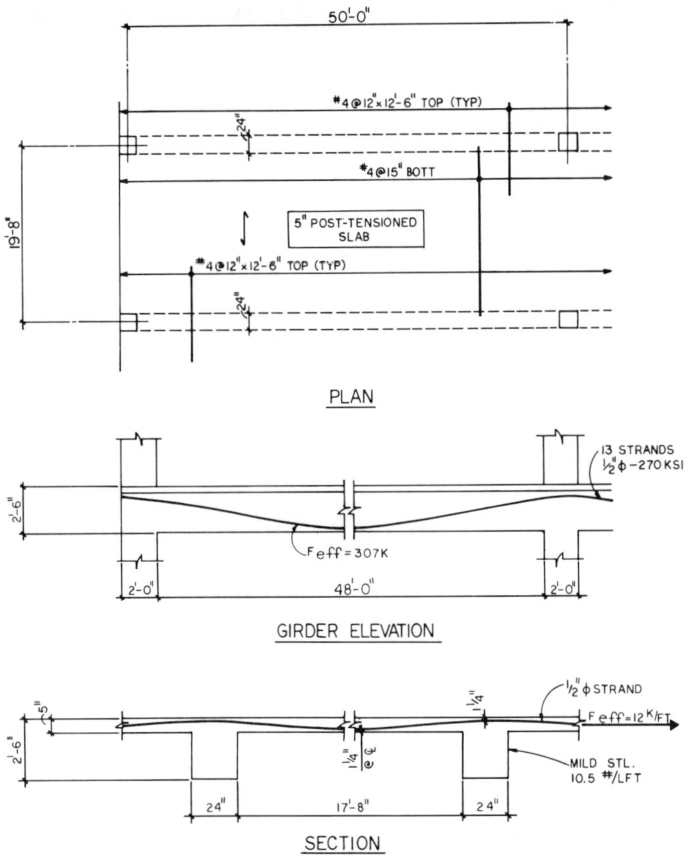

RELATIVE COST $5.97/SQ.FT

Fig. 9.9 Parking garage case study (scheme 4)—post-tensioned beam and slab.

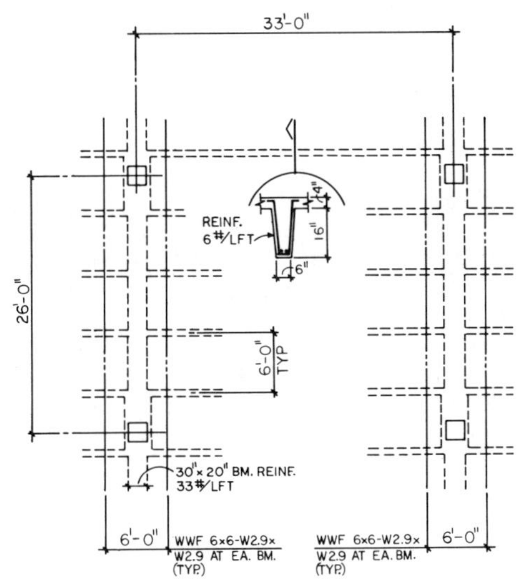

RELATIVE COST $5.54/SQ.FT.

Fig. 9.10 Parking garage case study (scheme 5)—short span pan-joist system.

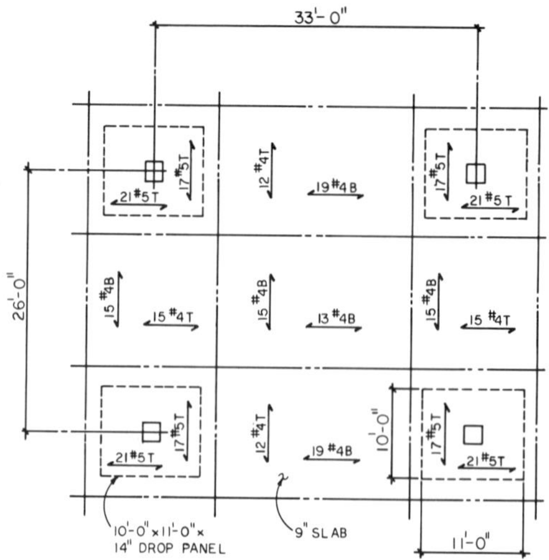

RELATIVE COST $6.31/SQ. FT.

Fig. 9.11 Parking garage case study (scheme 6)—short span flat slab system.

SELECTED REFERENCES

9.1 D. E. Allen, J. H. Rainer, and G. Pernica. "Vibration Criteria for Long-Span Concrete Floors," DBR Paper, No. 858, Division of Building Research, National Research Council of Canada, September 1975.

9.2 "Cost Optimization Tables for Steel-Framed Automobile Parking Structures," Booklet 2870, Bethlehem Steel Corporation.

9.3 Joseph P. Colaco. "A Stub-Girder System for High-Rise Buildings," *Engineering Journal,* AISC, **9,** July 1972, 89–95.

9.4 J. P. Colaco. "Concrete Floor Systems," State of Art Report No. 6, Technical Committee No. 21, *Elastic Analysis—Strength of Members and Connections,* Structural Design of Tall and Masonry Buildings, ASCE-IABSE International Conference Preprints, Reports Vol. III-21, August 21–26, 1972.

9.5 Joseph P. Colaco. "Haunched Girder Concept for High Rise Office Buildings in Reinforced Concrete," *Significant Developments in Engineering Practice and Research,* SP-72. Detroit: American Concrete Institute, 1981, pp. 305–317.

9.6 William J. LeMessurier. "Plastic—Composite Design Cuts Steel Tonnage in Johns-Manville's New Headquarters Building," *Architectural Record,* September 1977, 127–128.

9.7 Kenneth H. Lenzen. "Vibration of Floor Systems of Tall Buildings," ASCE-IABSE International Conference Preprints, Vol. II-17, Lehigh University, Bethlehem, PA, August, 1972.

9.8 Thomas M. Murray. "Design to Prevent Floor Vibrations," *Engineering Journal,* AISC, **12,** Third Quarter 1975, 82–87.

9.9 PCA. *Material and Cost Estimating Guide for Concrete Floor and Roof Systems.* Skokie, Illinois: Portland Cement Association, 1974.

9.10 John L. Ruddy. "Economics of Low-Rise Steel-Framed Structures," *Engineering Journal,* AISC, **20,** Third Quarter 1983, 107–118.

CHAPTER 10

TALL BUILDINGS—LOAD EFFECTS AND SPECIAL DESIGN CONSIDERATIONS

RICHARD L. TOMASETTI

Senior Vice President
Lev Zetlin Associates, Inc.
& Thornton-Tomasetti, P.C.
New York, New York

LEONARD M. JOSEPH

Associate
Lev Zetlin Associates, Inc.
& Thornton-Tomasetti, P.C.
New York, New York

DANIEL A. CUOCO

Vice President
Lev Zetlin Associates, Inc.
& Thornton-Tomasetti, P.C.
New York, New York

The same basic principles of design apply for high-rise buildings as for any other type of construction. The individual members and overall structure must be designed for adequate strength under gravity and lateral loads. They must have enough stiffness to restrict deflections to acceptable levels. They must be checked for dynamic response under applied loads so that the structure is serviceable. They must be stable. Their selection should take into consideration the present and potential future uses of the structure. And they should accommodate the other components of the building, that is, architectural, mechanical, and electrical, as required to satisfy the user.

What is "special" about these principles when designing a high-rise building is that, because of the tremendous wind exposure and typically large height-to-width aspect ratios, the provisions for lateral forces in strength, drift control, dynamic behavior, and stability overshadow the provisions for gravity load-carrying ability. Indeed, the proper choice of a lateral load-resisting system can make or break a high-rise project in terms of economics, constructability, and usefulness.

This chapter highlights the special nature of high-rise buildings and their response to load. Chapter 11 builds on this material and extends the discussion to types of load-resisting forms and how to determine the approximate behavior of tall buildings under gravity and lateral load systems.

10.1 LOAD CONSIDERATIONS AND EFFECTS

Proper determination of gravity and lateral loads during the preliminary design phase is critical in design of high-rise buildings. Gravity loads include both dead and live loads, and lateral loads include wind and/or seismic loads, depending upon building site. In some cases, especially for buildings having fully or partially exposed structural elements, thermal loadings must also be considered in the preliminary design phase. The material given here on loads complements the basic treatment of loads in Chapter 2.

10.1.1 Dead Loads

Dead loads are defined as fixed, nonmovable loads of a permanent nature which can be divided into two categories: (1) self-weight of the structure and (2) superimposed dead loads. The self-weight of the structure includes all beams, girders, columns, slabs, walls, bracing, and any other structural elements. Concrete framing systems are typically heavier than steel framing systems, which can sometimes be an advantage from a wind overturning standpoint but can also be a disadvantage in terms of seismic and foundation considerations.

Superimposed dead loads consist of partitions, hung ceilings, hung mechanical/electrical loads (e.g., sprinklers, lights, etc.), special floor fills and finishes, facade weight, and any other dead load which acts in addition to the weight of the structural elements. Many building codes stipulate that an allowance for partition loads equal to 20 psf (1 kN/m^2) over the floor area must be considered unless calculations based upon partition layouts and weights demonstrate that lesser loadings are applicable.

The lightweight partition types that are utilized in modern high-rise buildings usually result in an equivalent uniform superimposed dead load ranging between 8 and 12 psf (400 and 600 N/m^2). For preliminary design, however, it is prudent to include a 20-psf (1-kN/m^2) allowance for partitions since the actual partition types, weights, and locations are often not defined at this stage.

Suspended ceiling weights and mechanical/electrical loadings vary from project to project, depending upon the type of occupancy and the type of structural system utilized, and usually range between 2 and 10 psf (100 and 500 N/m^2). Floor fills or other special floor finishes can result in significant dead loads. Examples are brick or stone pavers in lobby areas, concrete fill over a waterproofing membrane for mechanical room floors, concrete housekeeping pads for mechanical equipment rooms in order to control the transmission of noise and vibration, fill over floor slabs in order to accommodate electrification grids, and similar special loadings.

The facade weight can vary significantly, depending upon the type of facade to be used, for example, glass curtain wall, precast concrete, masonry, or stone. The weight of glass curtain wall systems usually ranges between 8 and 12 psf (400 and 600 N/m^2), but the weight of precast or masonry facades can be 40 to 80 psf (2 to 4 kN/m^2) or more.

If the type of facade has not yet been determined at the preliminary design stage, and if the options being considered vary widely in their weights, it may be necessary to consider these facade options separately when evaluating the various structural framing systems. Another factor that must be considered in evaluating framing systems is whether the type of facade requires that the supporting structural elements meet special stiffness criteria, such as a limitation of live load deflection for spandrel beams. Chapter 27 contains detailed information on facades.

10.1.2 Live Loads

Live loads are nonpermanent in nature and vary depending upon the usage of the building floor area in question. For example, most building codes specify a minimum design live load of 50 psf (2.4 kN/

m^2) for typical office areas. Increased live loads for special usage areas that are known at the time of preliminary design should be taken into account, such as lobbies, restaurants, mechanical equipment rooms, cooling towers, landscaped planting areas, computer rooms, and places of assembly.

Localized areas to be used for storage or heavy filing loads are often unknown at the time of preliminary design and, therefore, must be taken into account during final design or, as sometimes is necessary, during or after construction. Roof live loads, which will be a very small portion of total gravity load, include snow loads with due consideration given to drifting, for example, at vertical surfaces of parapets, penthouses, setbacks, and adjacent structures (see Section 2.4). Allowable reductions of live load in accordance with applicable building code provisions should be applied during the preliminary design phase.

10.1.3 Wind Loads

Outside of high-risk seismic zones, wind is the force that most affects the design of high-rise buildings. For this reason, a brief summary of current approaches for determining wind force is presented here, to supplement the coverage of wind loading in Chapter 2 (Section 2.3.5, in particular). Additional comments on wind loads on tall buildings are provided in Section 11.3.2.

For preliminary design, wind loads are usually derived from the wind pressures specified in the governing building codes. Most building codes differentiate between the wind pressures acting on the overall structure and those acting on the facade or secondary wall framing elements. The latter values are higher in magnitude because of the localized effects of gusts and minimal structural damping characteristics.

The wind loads acting on the overall structure are generally given as stepped increments of wind pressure along the height of the building, with the pressures increasing in magnitude as the height above ground level increases. These wind loads are taken to act normal to the vertical surfaces of the building with consideration also given to the effect of quartering wind loads. For certain configurations of the structural bracing system, quartering wind loads can be much more severe than normal wind loads.

The inherent stiffness and damping characteristics of most high-rise buildings preclude the possibility of wind-induced resonance and aeroelastic instability. For very flexible structures, however, dynamic wind effects must be considered during the preliminary phase.

Early Approach

Historically, wind force has been presented in building codes as a quasi-static lateral force, that is, assuming steady-state flow and ignoring building motions. Code writers estimated the wind speed at various elevations for the worst storm that a building would be likely to experience, and the wind velocities and average air density were used to equate the kinetic energy of a given windspeed with the potential energy of wind pressure on a building face. This gave a stagnation pressure (Fig. 10.1).

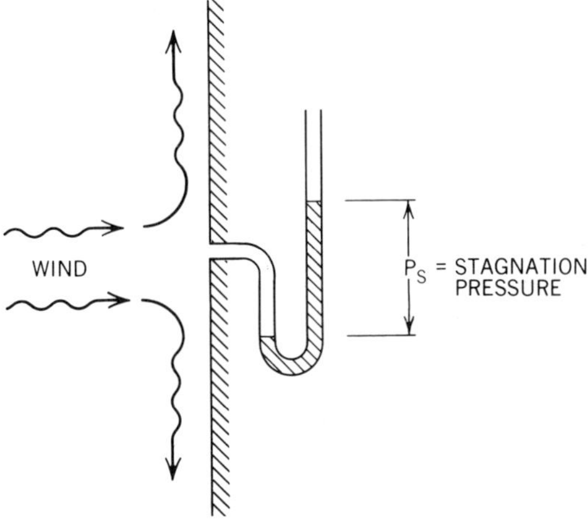

Fig. 10.1 Stagnation pressure.

Pressure Coefficients

With the advent of aeronautical engineering and the use of uniform-flow wind tunnels, it was recognized that solid surfaces could simultaneously experience both pressure and suction. Pressure coefficients were introduced in some codes as a factor on the stagnation pressure, which could result in a wind pressure in excess of the stagnation pressure when windward and leeward coefficients were considered simultaneously (Fig. 10.2). Overall wind action, however, was still regarded as quasi-static. Special, higher coefficients have been established for cladding, especially at edges and corners of walls and roofs (Fig. 10.3).

Variation with Height

As model codes attempted to be universally applicable, the effect of surface roughness had to be considered. This was done by considering open area wind speed as a standard, and using power laws to develop different wind speed profiles for open, suburban, and urban areas (as explained in Chapter 2).

Gust Factors

Recent research into wind fluctuations, in the field and in state-of-the-art boundary-layer wind tunnels designed to simulate scale surface roughness, has revealed consistent and recognizable classes of turbulence or gusts with associated frequencies.

When the frequency of turbulence is much higher than the natural frequency of a building, the building does not have time to respond fully to the gust, but the cladding does. When the gust frequency is much lower, the building responds in a quasi-static fashion, that is, the duration of the gust is long enough for the building to deflect downwind and stay in the deflected position. But when gust frequency and building frequency are similar, building response can significantly exceed that expected using the peak windspeed quasi-static fashion, similar to a dynamic load factor or resonant response. In addition, flexible buildings can develop significant cross-wind motion. In order to consider these effects, gust factors have been developed. Along-wind and cross-wind response calculation methods are illustrated in Section 2.3.5.

Differences in U.S. and Canadian Factors

In the United States, National Weather Service stations record wind in terms of "fastest mile of wind" passing their anemometers. This is, in effect, providing wind speeds averaged over about $\frac{1}{2}$ to 1 min. The American National Standards Institute (ANSI) [10.3] provides in its Standard A58.1-1982 fastest-mile wind speeds for various return periods (see p. 256), and tables presenting pressure coefficients, wind pressure profiles for quasi-static wind, and wind pressure profiles which include average gust factors for the overall structure and for cladding.

The factors vary inversely with height. The overall factor is for relatively stocky ($H/D = 5$) stiff buildings and considers only downwind action. It specifically excludes the potentially significant effects of vortex shedding and galloping, which can occur with flexible, towerlike structures.

ANSI [10.3] states that when height H divided by minimum width D exceeds 5, a separate gust factor must be calculated, which depends on building frequency. For preliminary design, of course, frequency is still unknown. ANSI suggests using $f = 10/(\text{number of stories})$ as a rough approximation, but the actual building frequency should later be established and back-checked, since the

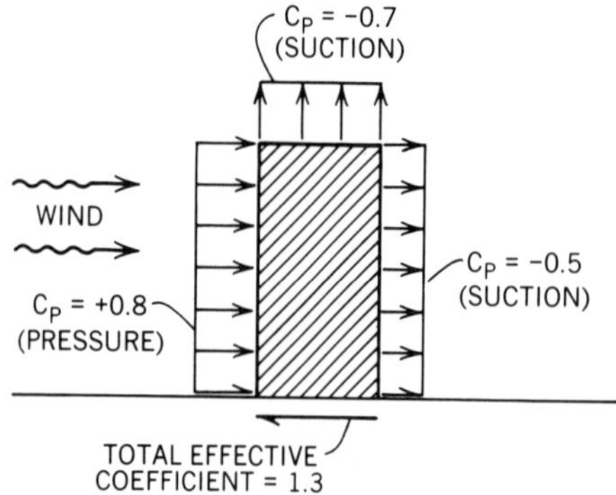

Fig. 10.2 Sample overall pressure coefficients, C_p. Wind variation with height is not shown here.

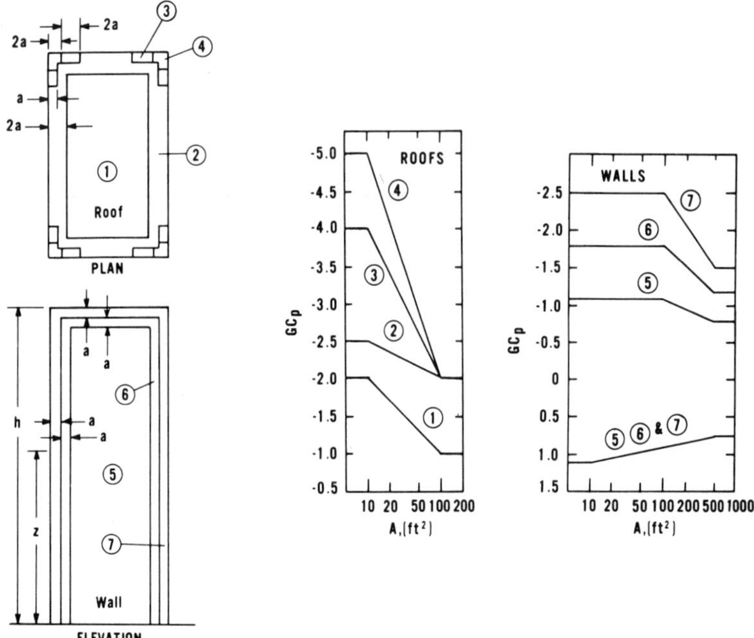

Notes:
(1) Vertical scale denotes GC_p to be used with appropriate q_z or q_h.
(2) Horizontal scale denotes tributary area A, sq. ft.
(3) Use q_h with negative values of GC_p and q_z with positive values of GC_p.
(4) Each component shall be designed for maximum positive and negative pressures.
(5) If a parapet is provided around the roof perimeter, Zones ③ and ④ may be treated as Zone ②.
(6) Not applicable at roof for roofs with slope of more than 10°.
(7) Plus and minus signs signify pressure acting toward and away from the surfaces, respectively.
(8) Notation:
a: 5% of minimum width or $0.5h$, whichever is smaller.
h: Mean roof height, ft.
z: Height above ground, ft.
This material is reproduced with permission from *American National Standard Minimum Design Loads for Buildings and Other Structures,* ANSI A58.1-1982, copyright 1982 by the American National Standards Institute. Copies of this standard may be purchased from the American National Standards Institute at 1430 Broadway, New York, N.Y. 10018.

Fig. 10.3 External pressure coefficients, GC_p, for loads on building components and cladding for buildings with mean roof height h greater than 60 ft.

calculated gust factor is quite sensitive to the frequency used. For preliminary design, the gust pressures should be no less than those shown in the main tables of ANSI A58.1.

Canadian weather data are in the form of miles of wind per hour, which automatically smooths out gusts over a 60-min period. Since this value is used as the base wind speed, the Canadian gust factors are necessarily larger than those in ANSI, as they must account for wind variations in the range of several minutes in addition to gusts lasting several seconds. As a minimum, gust factors of 2.0 for structure and 2.5 for cladding were established. The Canadian Code [10.15] stipulates that a dynamic calculation or aeroelastic wind tunnel test is required to establish gust factors for buildings taller than 394 feet (120 m), or with height-to-least-width ratios greater than 4, or with special properties which make them susceptible to vibration.

In practice in the United States, sufficient experience in wind tunnel tests is available to permit a good estimate of suitable gust factors on a preliminary basis.

Some researchers [10.5] have noted that the use of 1-hr data, rather than fastest-mile data, may tend to underestimate the frequency of severe storms and short-term winds, such as thunderstorms.

Return Periods

In tall buildings there is a premium paid in structural costs for providing stiffness above that provided when minimum strength requirements are met. One rule of thumb is that buildings taller than 15 stories are typically drift controlled [10.7, pp. 346–394]. Different probabilities of exceeding the design force may be used when designing against structural collapse, local cladding failure, and excessive or annoying deflections and/or accelerations. ANSI [10.3] provides importance factors I depending on nature of building use. This is similar to the previous ANSI Code A58.1-1972 which specified using a 50-yr wind for the design of typical structural framing and cladding, with a 100-yr wind used for critical buildings and a 25-yr wind for low-risk buildings. The Canadian Code [10.15] specifies a 10-yr wind for cladding, deflection, and vibration, a 30-yr wind for framing strength, and a 100-yr wind for critical buildings.

Factors of Safety

The Council on Tall Buildings [10.5] points out that many buildings over the years have been designed using low, quasi-static wind forces, and that no tall buildings have collapsed from wind loadings after construction. Therefore, it may be rational to use new, higher wind forces with gust factors when checking for deflection, but to use a reduced factor of safety when checking for strength to maintain a factor of safety consistent with historical experience.

The ANSI Code [10.3] requires a 1.5 safety factor against overturning and no particular safety factor against sliding. The Canadian Code requires a 2.0 factor against both overturning and sliding. When checking for overturning or uplift, it is important to use reasonable minimum dead loads, rather than the usual maximum dead and live loads normally used in design. Both the *ACI Code* [10.1] and the *AISC Specification* [10.2] reduce dead load when checking for possible load reversals.

Wind Tunnel Testing

Although the above wind criteria are sufficient for preliminary design purposes, wind tunnel testing is important in order to verify the actual wind pressures acting on the building and to determine nonuniform wind effects. The latter can result from unsymmetrical features of the building and the effects of adjacent structures. Even perfectly symmetrical buildings are subject to wind-induced torsional forces, as shown by wind tunnel tests, because of nonuniform wind pressure distributions. In addition, crosswind effects can be determined by wind tunnel testing. Since the actual wind tunnel testing often does not occur during the initial stages of a project, some of the above factors can be approximated analytically for preliminary design (see Section 10.2.2 for crosswind effects and Section 10.2.3 for torsion effects).

From the standpoint of wind pressures acting on the building facade, it is common for peak pressures measured during wind tunnel testing to be significantly in excess of code-specified values (Fig. 10.4).

10.1.4 Seismic Loads

For high-rise buildings located in active seismic zones, the lateral forces induced by seismic accelerations must be considered. Comparisons between wind and earthquake forces are given in Section 11.3.3, where it is shown that wind loads, in general, are more critical than seismic loads for typical tall buildings.

For preliminary design, these dynamic forces can be taken as an equivalent static system of lateral forces (see Section 2.5) which can be calculated in accordance with the provisions of most building codes, such as the *Uniform Building Code* [10.12]. In this design process, torsional effects due to the lack of coincidence between the center of mass of the building and the center of rigidity of the lateral bracing system must be considered.

It is important for the designer to note that the choice of structural framing system will have a very strong influence on the design seismic forces. For example, "box" systems stiffened by "X" bracing must be designed for greater lateral loads than frame systems.

When necessary, a dynamic analysis can be subsequently performed once the structural components have been selected based upon the preliminary design, as explained in Chapter 18. This is recommended for critical occupancies, such as hospitals and schools, for very flexible or unusually shaped buildings, and as required by local building codes.

10.1.5 Thermal Loads

As for other structures, temperature changes create strains in high-rise members. When members are free to move, length change occurs. When members are restrained, axial stress is generated.

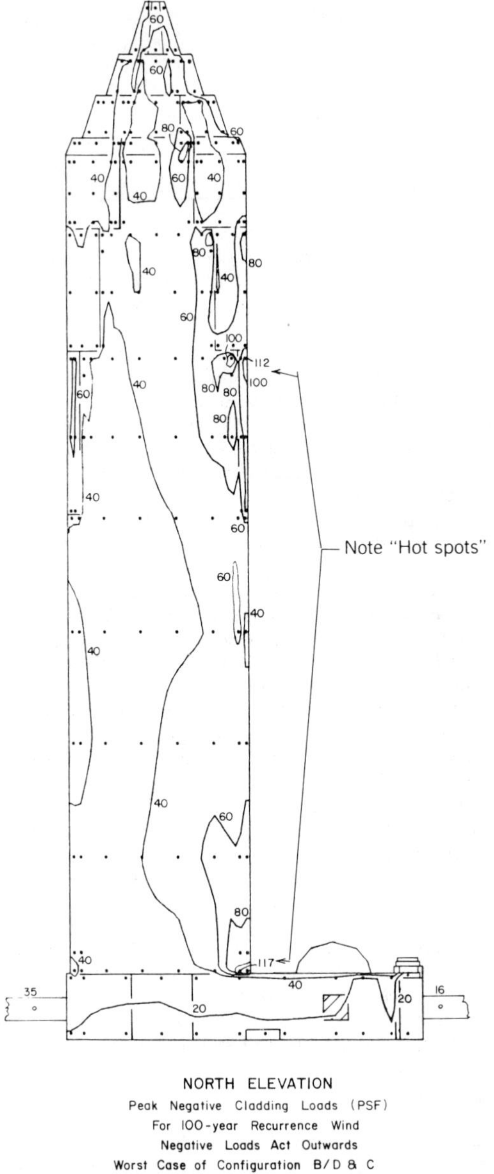

Note "Hot spots"
112
117
NORTH ELEVATION
Peak Negative Cladding Loads (PSF)
For 100-year Recurrence Wind
Negative Loads Act Outwards
Worst Case of Configuration B/D & C

Fig. 10.4 Peak-pressure distributions on a building for cladding loads.

Because of the extreme length of high-rise members, the length change can be significant. Because the members are generally heavy, the forces generated when motion is restrained can be large. Both effects must be considered and can be minimized by thoughtful design (see Section 10.2.5).

10.2 SPECIAL DESIGN CONSIDERATIONS

10.2.1 Building Drift

As stated earlier, high-rise building design is usually governed by deflection, not by strength. Historically, limits on overall building deflection and interstory drift were thought to accomplish three things: (1) limit stresses on, and damage to, nonstructural partitions and finishes within the building, (2) ensure overall building stability, and (3) hold perception of motion to acceptable levels.

Point 1, avoiding damage to nonstructural items, is still a valid objective.

Point 2, building stability, can be addressed more directly by a second-order analysis which includes the *P*–Δ effect. This will be described in Section 10.2.4 and in Chapter 11. At this point, suffice it to say that application of the usual sway limitations, coupled with the usual wind loads, should result in a trial structure which will converge to a satisfactory final solution in one or two design cycles. If no sway limitation were imposed, the resulting trial structure could indeed prove to be unstable.

Point 3, motion perception, has less relationship to sway limitations than was originally believed. Laboratory studies have shown that humans perceive acceleration and changes in acceleration ("jerk") rather than displacement or velocity. Perception of motion will be discussed in Section 10.2.2. However, it will be seen that acceleration is difficult to determine accurately in the preliminary stages of design, and that acceleration cannot be easily changed by modifying structural stiffness. On a preliminary basis, designers usually start by assuming that the generally accepted drift limits which have proven to give acceptable results for similar buildings will be satisfactory for their building. Acceleration would then be checked as a later design step.

Engineers differ on the amount of drift which is permissible, and on the proper return period to be used for drift calculations. A review of the literature and polls of designers [10.9] shows that, for buildings that would be considered high-rise, a drift of 1/400 to 1/500 of the height would be a reasonable limit under working loads.

Most building codes in the United States do not specify drift limitations, but do note the need to protect nonstructural elements and hold perceived motion to tolerable levels. The Canadian Code [10.15] specifies an interstory drift of 1/500 of the story height, except for industrial structures. Note that this is applied at *each* story, and is the sum of shear (racking) drift and bending (cantilever) drift. As such, it is considerably more conservative than limits based upon overall building drift. On the other hand, the stiffness required to satisfy this limitation on a 10-yr return period, as per the Canadian Code, and that required to satisfy less stringent limits on a 50-yr return period, such as ANSI wind forces, may well be similar.

When writers refer to the amount of drift calculated for a given building, they generally do not state whether it is the first-order (wind only) deflection or the second-order (wind plus P–Δ) deflection. To be consistent, the latter should be used. It must also be kept in mind that these drift limitations assume along-axis deflection. Torsion coupled with along-axis deflection can cause unacceptable movement at building corners. Torsion is addressed further in Section 10.2.3.

Another consideration regarding drift is the effect of rotation and torsion on microwave or satellite stations which may be mounted on the building roof.

Drift due to seismic loading is a separate issue from that due to wind loading. First, occupant comfort is not an issue, as anyone experiencing tremors is uncomfortable. Second, deflections based upon an elastic analysis are meaningless once the structure is loaded into the range of plastic deformations. Third, depending on the design, it may be beneficial to stiffen or soften various parts of the building in order to achieve adequate ductility and energy absorption. And fourth, in seismic areas, nonstructural partitions should be detailed for free motion or designed to harmlessly crush in the event of a major earthquake. Therefore, drift control to protect the partitions is unnecessary. The applicable seismic codes should be consulted for drift limitations under this special condition.

10.2.2 Perception of Motion

As stated above, laboratory tests [10.7, pp. 346–394] have shown that humans are sensitive to acceleration and changes in acceleration ("jerk") rather than displacement or velocity.

Tests by several experimenters have shown that, under laboratory conditions, the threshold of acceleration perception under cyclic motion similar to that of a building (period of several seconds) is about 0.005*g*. The accelerations become annoying when they reach 0.015 to 0.020*g*. Note, however, that perception and annoyance also depend on cues, activity, and nature of motion.

Wind whistling through the facade or bolts creaking in the frame will trigger awareness of a storm to occupants, who will then be receptive to detecting motion. Occupants will be more sensitive when in an apartment or hotel than in a busy office. (Note that this runs counter to the tendency to place apartments at the top of multiuse towers.) Torsional motion will be obvious to an occupant who can sight distant objects through the window and note relative motion of the mullions past the backdrop. Floor tilt due to cantilever action may be similarly visible.

With this information, a designer should attempt to get the building owner to agree to a set of accelerations and probabilities of occurrence. It is uneconomical to create a building designed to have no perceptible or objectionable motions at any time in its useful life. The economic tradeoff between structural costs and loss of "goodwill" from occupants must be made by the owner.

Opinions on acceptability vary widely [10.9]. The World Trade Center in New York City was designed to limit accelerations on occupied floors to 0.01*g* on a one-month return period (equivalent to 12 times per year, on the average). Others have suggested values ranging from $0.05g$ once every 6 yr to 0.01–$0.03g$ once every 10 yr, where $0.01g$ applies to apartments and $0.03g$ to offices. A check on the World Trade Center showed that the 10-yr return wind would create accelerations of $0.028g$, which is similar to the latter recommendation.

Calculation of Acceleration

Acceleration limits that should be checked for occupant comfort were described above. However, the calculation of acceleration is very sensitive to the building frequency, which is not accurately available for preliminary design. Acceleration is also strongly affected by building damping. For steel structures acting under wind loads, damping is normally taken as 0.01 of critical, and for concrete as 0.02 of critical [10.3, 10.8, 10.16]; see Fig. 10.5. During earthquakes, damping would be much higher.

To add to the difficulty, acceleration is very difficult to change, as can be seen from the following expressions:

$$a = A(2\pi f)^2 \tag{10.2.1}$$

where a is acceleration, A is fluctuating amplitude (deflection), and f is frequency. Also

$$2\pi f = \sqrt{\frac{k}{m}} \tag{10.2.2}$$

and

$$A \approx \frac{1}{k} \tag{10.2.3}$$

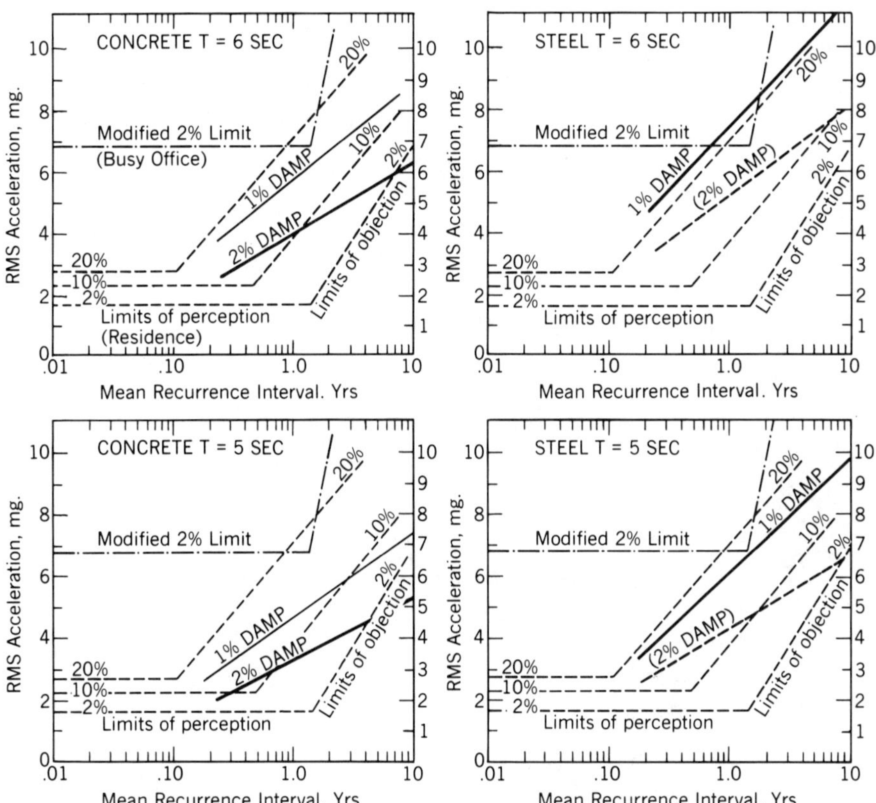

Fig. 10.5 Study of acceleration vs. return period for 66-story tower—varied mass (concrete vs. steel tube), stiffness, damping. Percentages of population perceiving or objecting to motion shown.

where k is stiffness and m is mass. Thus, if stiffness is doubled, the square of frequency is doubled while deflection is halved. The end product is the same, except that, at the higher frequency, the gust factor for wind will be slightly different. To make a significant change in acceleration, either mass or damping could be varied. Adding mass is very expensive, since it has to be supported by the structure. Adding damping directly has been tried on several projects, but sometimes the dampers absorb so much energy that they fail or the damping effect is less than was hoped for.

Tuned-mass dampers (TMD) have been used, most notably at the Citicorp Building in New York. There, when the TMD is activated it boosts damping of this steel-framed building from its natural 1% level to 4%. The TMD is used only to improve occupant comfort in moderate winds.

Crosswind Acceleration

When wind strikes a prismatic object, a downwind or drag force is generated, which is the conventional building behavior usually associated with drift, gust factors, and maximum displacements. When the building rebounds from a gust, it is moving upstream against the wind, creating a helpful damping effect. Depending upon the angle of attack, a crosswind force can also be generated as one side of the building experiences more suction, or "lift", than the other. This crosswind force causes lateral oscillations which are not damped by the wind. Indeed, at certain wind speeds, crosswind vibration can continue to build.

Although the amount of force and drift generated by the crosswind effect will often be much less than for the downwind case, the crosswind motion is all fluctuating, while the downwind motion consists of a quasi-static deflection plus a smaller fluctuation (Fig. 10.6). Since acceleration is related to the fluctuating deflection, it is not unusual for the crosswind acceleration to exceed the downwind acceleration.

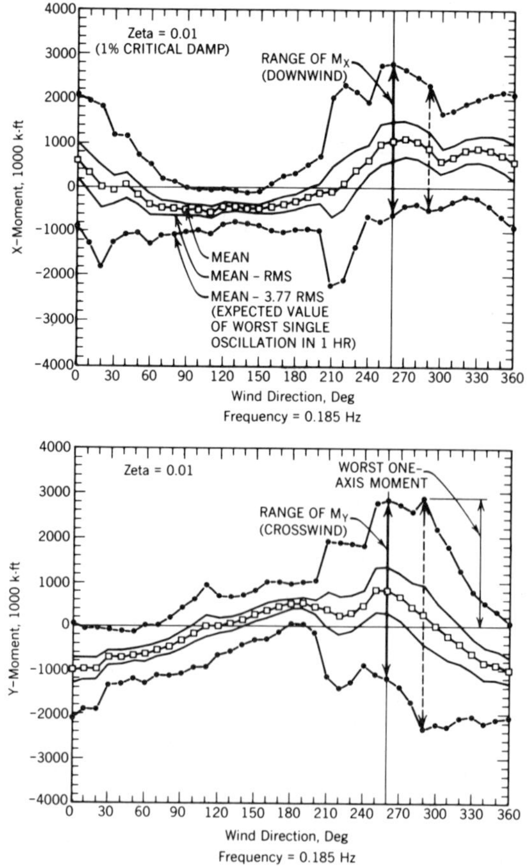

Fig. 10.6 Wind tunnel force–balance results—critical quartering wind at 260°.

Crosswind-motion has been analytically rather intractable. It appears to be most strongly influenced by the turbulence initiated in the wake of an object upwind. Therefore, where similarly tall buildings exist or are planned nearby, the prudent course of action is to perform an aeroelastic wind tunnel test, complete with the possibly troublesome neighbors. Unfortunately, construction of an aerolastic model requires knowledge of building geometry, mass and stiffness—attributes unavailable until a design is well underway.

Another recent approach is to use a force–balance model. As soon as building geometry is known, a stiff, light model is placed in the wind tunnel and strain gages on the beams supporting the model record the variations of overturning moments and torsion over time due to wind gusts. In this way a wind forcing function can be inferred. Using this forcing function, dynamic building response for a wide range of mass distributions, damping values, and stiffnesses (periods) can be calculated by computer (Fig. 10.6). Frequently, at least some preliminary design work is desired before a force–balance model can be prepared.

As an interim measure, the Canadian Code Supplement [10.16] offers the following approximate formulas:

Crosswind acceleration may exceed downwind acceleration if

$$\frac{\sqrt{WD}}{H} < \frac{1}{3} \tag{10.2.4}$$

where W, D, H are width (across wind), length (along wind), and height, in the same units.

Crosswind acceleration may be expressed as

$$a_w = n_w^2 g_p \sqrt{WD}\left(\frac{a_r}{\rho_B\sqrt{\beta_w}}\right) \qquad (10.2.5)\dagger$$

and along-wind accceleration as

$$a_D = 4\pi^2 n_D^2 g_p \sqrt{\frac{KsF}{C_e\beta_D}}\left(\frac{\Delta}{C_g}\right) \qquad (10.2.6)\dagger$$

where† W, D = across-wind and along-wind building dimensions, m
a_w, a_D = peak acceleration in across-wind and along-wind directions, m/s²
$a_r = 78.5 \times 10^{-6}[V_H/(n_w\sqrt{WD})]^{3.3}$, kN/m³
ρ_B = average density of the building, kg/m³
β_w, β_D = fraction of critical damping in across-wind and along-wind directions
n_w, n_D = fundamental natural frequencies in across-wind and along-wind directions, Hz
Δ = maximum wind-induced lateral deflection at the top of the building in along-wind direction, m
K = a factor related to the surface roughness coefficient of the terrain
= 0.08 for Exposure A (city)
= 0.10 for Exposure B (suburban, rolling, wooded)
= 0.14 for Exposure C (exposed, grassland)
C_e = exposure factor obtained from Fig. 10.7
B = background turbulence factor, obtained from Fig. 10.8 as a function of height, H, and width, W, of the windward face of the structure
s = size reduction factor, obtained from Fig. 10.9 as a function of the ratio of width, W, to height, H, of the windward face of the structure and the reduced frequency
F = gust energy ratio at the natural frequency of the structure, obtained from Fig. 10.10 as a function of the wave number (natural frequency, Hz, divided by mean wind speed (m/s) at height, H, of structure)
β = critical damping ratio (0.01 for steel frames, 0.02 for reinforced concrete frames; prestressed concrete with no microcracking may have very low damping)
V_H = wind velocity at top of building, m/s
g_p = peak factor of total loading effect (varies from 3.4 for flexible to 4.2 for stiff buildings—use 3.8 for typical high rise)

$$C_g = 1 + g_p\sqrt{\frac{K}{C_e}\left(B + \frac{sF}{\beta}\right)} = \text{gust effect factor}$$

† Equations (10.2.5) and (10.2.6) and definition list are reprinted with permission from the *Supplement to the National Building Code of Canada* 1980, NRCC No. 17724, Copyright 1980 by the National Research Council of Canada, Ottawa, Ontario, Canada.

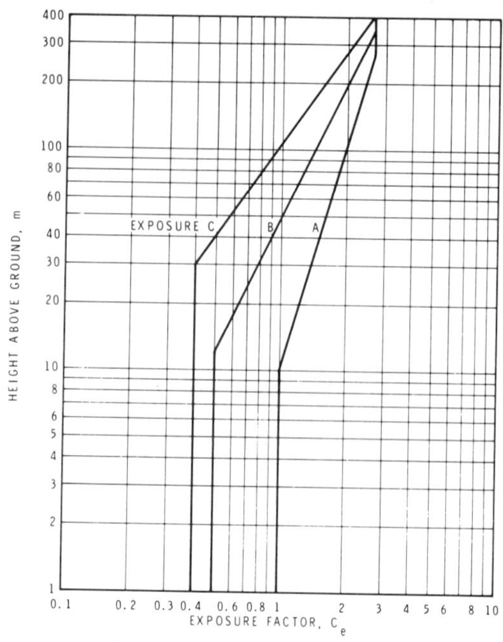

Fig. 10.7 Exposure factor as a function of terrain roughness and height above ground. Reprinted with permission from the *Supplement to the National Building Code of Canada* 1980, NRCC No. 17724, copyright 1980 by the National Research Council of Canada, Ottawa, Ontario, Canada.

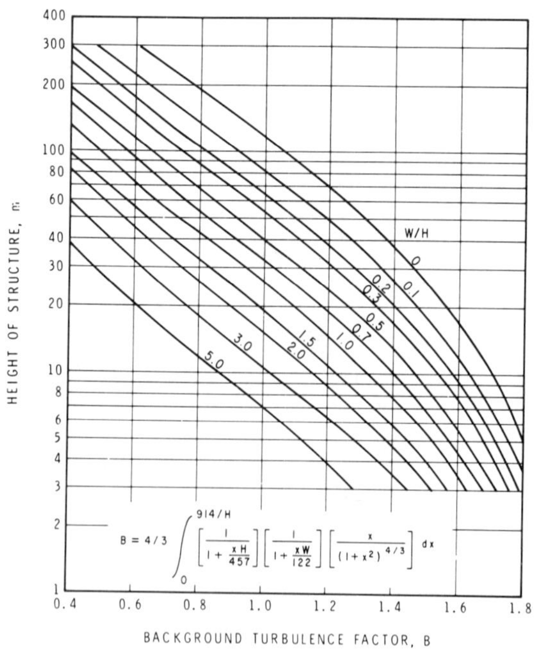

$$B = 4/3 \int_0^{914/H} \left[\frac{1}{1+\frac{xH}{457}}\right]\left[\frac{1}{1+\frac{xW}{122}}\right]\left[\frac{x}{(1+x^2)^{4/3}}\right] dx$$

Fig. 10.8 Background turbulence factor as a function of width and height of structure. Reprinted with permission from the *Supplement to the National Building Code of Canada* 1980, NRCC No. 17724, copyright 1980 by the National Research Council of Canada, Ottawa, Ontario, Canada.

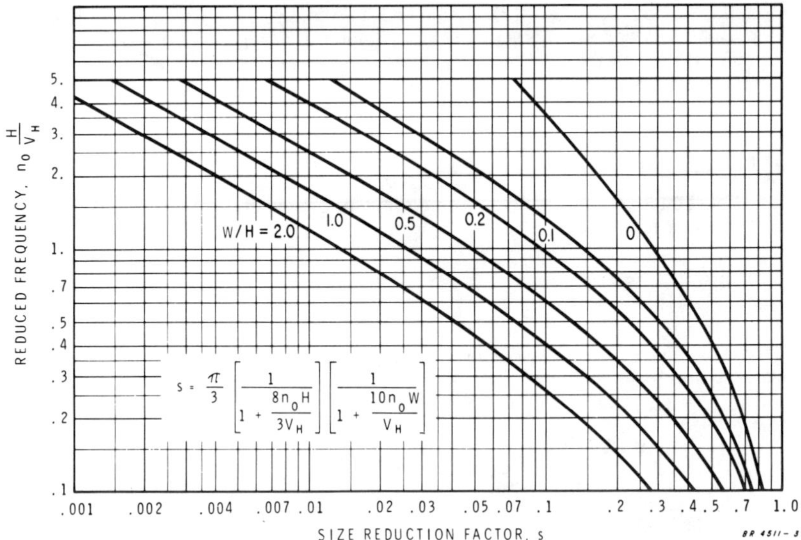

Fig. 10.9 Size reduction factor as a function of width, height, and reduced frequency of structure. Reprinted with permission from the *Supplement to the National Building Code of Canada* 1980, NRCC No. 17724, copyright 1980 by the National Research Council of Canada, Ottawa, Ontario, Canada.

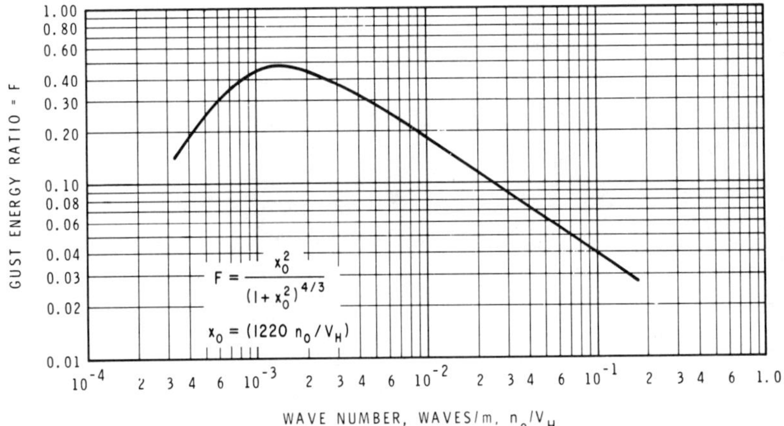

Fig. 10.10 Gust energy ratio as a function of wave number. Reprinted with permission from the *Supplement to the National Building Code of Canada* 1980, NRCC No. 17724, copyright 1980 by the National Research Council of Canada, Ottawa, Ontario, Canada.

In addition, the Canadian Code [10.15] recommends designing a structure for 75% of wind simultaneously on each of its two axes. Note that this is similar to the quartering wind case, where 71% would be applied to each face. In any case, quartering wind should be checked for its effect on corner columns (Fig. 10.11).

Natural Frequencies

For tall, flexible buildings (e.g., moment frame, tube, or braced core), a very approximate formula for natural frequency, f (in Hz), is

$$f = \frac{10}{N} \tag{10.2.7}$$

where N is the number of stories. For use in preliminary design, once a trial structure has been developed, the approximate method presented by Sandhu [10.17] can be used.

$$T_1 = 4 \sum_{i=1}^{n} \sqrt{\frac{m_i}{k_i}} \tag{10.2.8}$$

where T_1 = mode 1 period,
m_i = mass of story i
k_i = stiffness of story i including effects of flexure, shear, and column strain (overturning)
n = total number of stories

Another, less refined approach considers a tall, relatively uniform tower as a cantilever beam with uniform mass distribution and uniform stiffness, setting stiffness to give the calculated tip

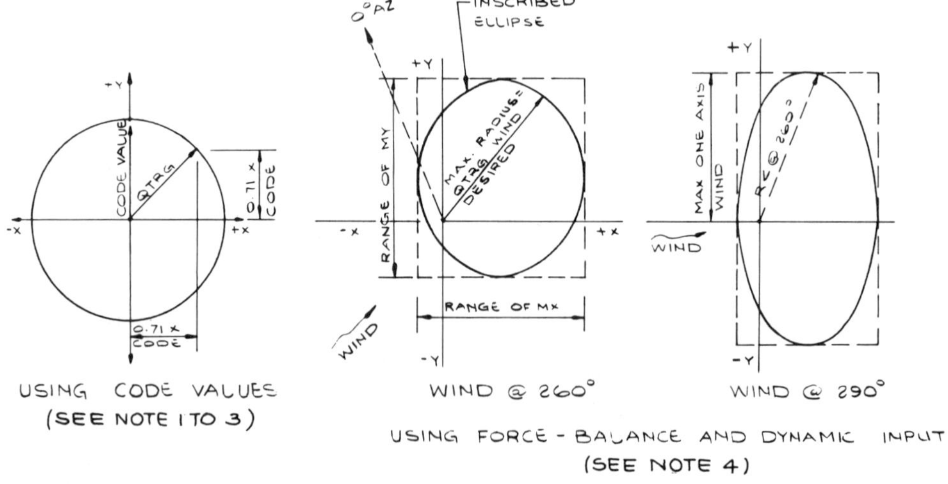

Notes:

1. Apply maximum one-axis wind to model in each direction separately.
2. For quartering wind, use sum of two winds × 0.707 × (quartering wind radius/maximum one-axis wind).
3. If symmetrical structure, can use one wind, add windward + corresponding sidewall member forces and factor per step 2.
4. Work with Fig. 10.6. when using force–balance and dynamic input.

Fig. 10.11 Simplified approach to quartering wind.

deflection under wind load. Then the classic formula is applied:

$$T_1 = 1.78H^2 \sqrt{\frac{m}{hEI}} \tag{10.2.9}$$

where T_1 = mode 1 period
H = building height, ft
m = mass per floor, lb-sec^2/ft
h = floor height, ft
EI = equivalent building stiffness as cantilever, lb-ft^2

A variation on this approach is to apply a unit load at the building top to find a cantilever spring constant, k. The pattern of deflection and the mass distribution are used to define the generalized mass, m^*. For an assumed straight line mode shape,

$$f = \frac{1}{2\pi} \sqrt{\frac{k^*}{m^*}} \tag{10.2.10}$$

where $k^* = kH^2$ (10.2.11)

$$m^* = \sum_{i=1}^{n} m_i(L_i)^2 \tag{10.2.12}$$

L_i = distance from base to floor i

Again, since acceleration is sensitive to frequency these formulas can offer no more than rough estimates. The final building frequency can be determined by using dynamic analysis computer programs which are readily available.

10.2.3 Torsional Effects

While torsion may be present in structures of any height, it can be particularly troublesome in high-rise buildings. First, motions associated with torsion can add to motion along building axes, creating unacceptable translations and accelerations. Second, the turbulent wind associated with "gustiness" does not strike a building uniformly. Differences in pressures across the building face can easily create torsion in even the most symmetrical building layout. Third, due to the extreme height of high-rise buildings, a certain story torsion and associated story rotation, which would normally be considered acceptable, accumulates over many stories to cause an unacceptable total rotation for the building.

Torsion Due to Layout

In the preliminary planning stage, every effort should be made to provide a lateral resistance system which is relatively well-positioned with respect to the applied lateral forces. Where wind governs, this would mean ensuring that the center of rigidity of the bracing system is close to the center of wind pressure, as determined by the building sail area. In seismic design, the center of rigidity should coincide as closely as possible with the center of gravity of the floors. The latter would usually be determined by using the weight of the floor slabs plus any heavy, permanent partitions such as masonry walls. Note that, for any plan other than a uniform, fully clad circle or rectangle, the center of wind force and center of gravity for seismic force can be in very different locations.

While a well-positioned bracing system is preferable in theory, it often cannot be accepted as practical in reality. Where a building is of moderate height and virtually fills its lot, at least one building face may be blank to allow for abutting buildings, and this is usually an ideal location for elevator banks and service cores from a space planning point of view. When a building is of extreme height, it often stands clear of the property lines and can be glazed on all sides. In this case, a central core is preferred as it maximizes the amount of high rental perimeter space.

Regardless of core location, the preferred lateral resistance system is of the closed type, with bracing or frame action forming a complete tube. Examples of this are (Fig. 10.12): tubular framed towers, with continuously moment-connected spandrels and columns all around the building perimeter; braced cores, with all core sides stiffened by diagonals or knee braces; and structural concrete

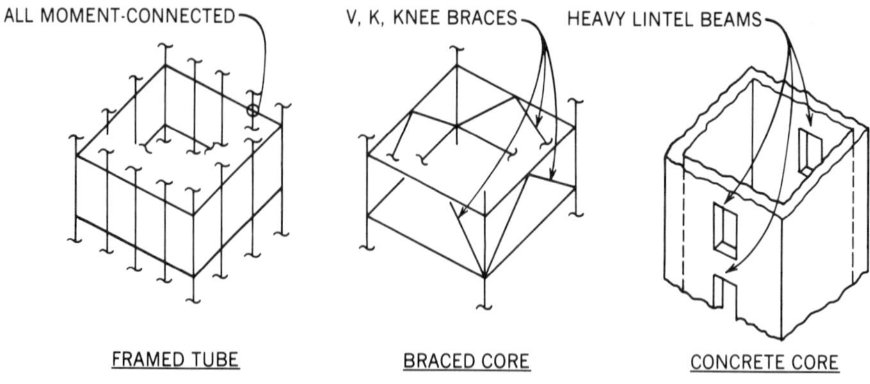

Fig. 10.12 Closed bracing systems.

cores, with heavily reinforced lintel beams over doorways acting as links between wall segments. Closed forms are preferred because of their inherent torsional stiffness.

Open forms of bracing are inherently weak in torsional stiffness and should be avoided. "L", "T", and "X" plan arrangements are the worst in this respect (Fig. 10.13*a*). "C" and "Z" shapes are slightly better (Fig. 10.13*b*). Where such forms occur, torsional stiffness must be provided independently, by adding moment frames, braces, or shear walls in a symmetric fashion perpendicular to, and at the greatest distance possible from, the center of rigidity (Fig. 10.13*c*). Some designers automatically provide spandrel frames on towers having a central braced core. As an alternative, a pair of cores, one at each end of the building, could provide overall torsional rigidity.

Accidental Torsion

Regardless of the bracing arrangement used and the theoretical alignment between the center of force and the center of resistance, some torsion should be specifically provided for. First, the bracing may not actually be as favorable as assumed. For example, during final design, braced bays may change from "K" braces to knee braces, member sizes may change, and design assumptions

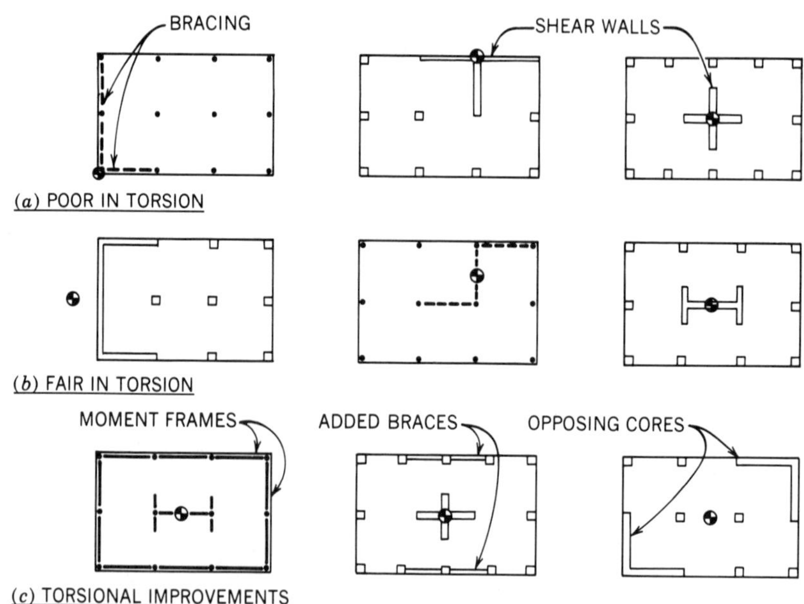

Fig. 10.13 Open bracing systems. ◐ = center of building stiffness.

such as concrete cracked-section properties may not hold true. Second, the applied force will certainly differ from the ideal case. The Canadian Code [10.15] specifies that 25% of the applied (wind) load should be removed from selected portions of the building in order to reflect nonuniform pressures and generate torsions. Wind tunnel tests have shown that torsions can be even larger than this, but it is a reasonable starting point. The model seismic code developed by the Structural Engineers Association of California and adopted by the UBC [10.12] recommends an accidental eccentricity of building center of gravity of at least 5% of the building plan dimension.

If the resisting system has a large torsional moment of inertia (e.g., tube or very generous braced core), the effect of accidental eccentricity will be small and easily handled. If the resisting system has a small torsional moment of inertia (e.g., small core or open section), the additional forces generated by torsion can be several times larger than the eccentricity of force might indicate. Thus, the designer must be aware during the preliminary stage that such a system will require additional material and will be less efficient than a simple analysis with little or no nominal allowances for torsion would indicate.

Dynamic Effects of Torsion

Torsion will have a major effect on building behavior if the first torsional mode frequency and lateral mode frequency are similar [10.5]. Then cyclic motion at building corners will be a combination of the two motions, and may prove unacceptable. For preliminary design, approximations using distributed building mass and stiffness may be adequate to indicate if a problem exists. If so, more exact analyses using individual story stiffnesses or three-dimensional dynamic computer programs are warranted.

Generally speaking, interaction between torsional and lateral modes is not a problem when the resisting systems are located around the perimeter, as in tube systems and braced exterior walls. This is because the forces required to move two walls in the same direction (lateral mode) and four walls in opposing directions (torsional mode) are very different. When the lateral resisting system is not distributed along the perimeter, the difference between the two modes is not as obvious.

Stability Effects of Torsion

Where deflection of building corners under torsion is large, this deflection must be added to the lateral deflection in order to ensure that local column stability is not more critical than the overall floor stability addressed in the usual lateral stability checks. Further stability considerations are discussed in the following section.

10.2.4 Building Stability

Traditionally, buildings have been designed member by member, and design for stability was no exception. Using that philosophy, a designer would find the axial, shear, and bending forces acting on a particular one-story high column; establish an equivalent unbraced length KL for that individual column, using empirical rules, judgment, or design aids, such as the AISC nomograph [10.2]; and design the column using the applicable code-specified interaction equations.

In high-rise design this can lead to misleading and unconservative results because it ignores overall building behavior, overall story behavior, and the fact that, although only a few columns may act as part of the lateral bracing system at each floor, they must be sufficiently stiff and strong to brace adequately all the columns at that floor against buckling. (The traditional approach above considers each column to be acting on its own.) This concern can be addressed by considering the force generated by the sum of all column axial forces at a story P times the story drift Δ (Fig. 10.14). The traditional approach also ignores the additional bracing forces and beam moments required to resist the P–Δ effect, and overstates the actual building stiffness.

A high-rise building can be viewed about each axis as a cantilever column with varying axial load (floor loads accumulate with distance from roof), significant lateral load (wind or seismic forces), varying moment of inertia (as column areas change), and a relatively soft web (racking or story shear distortions). It must be checked for overall buckling as a cantilever column (rarely a problem for buildings of normal height-to-width ratios and lateral drift limitations), buckling of individual stories in a shear mode, and a combination of the two [10.6], as

$$\frac{1}{P_{cr}} = \frac{1}{P_{o,\text{flex}}} + \frac{1}{P_{o,\text{shear}}} \tag{10.2.13}$$

(See Fig. 10.15.)

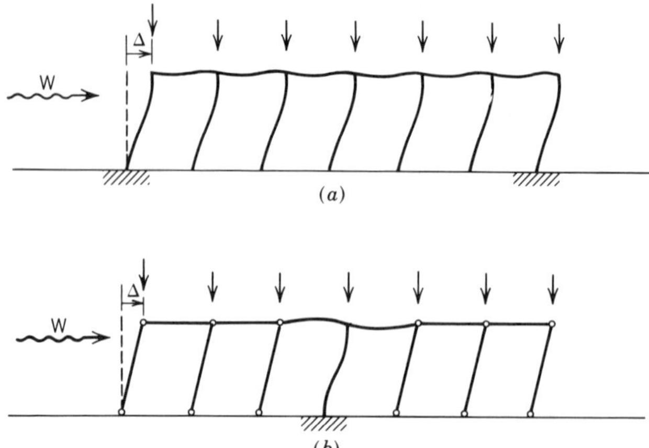

Fig. 10.14 *P*–Δ cases. (*a*) Case 1: all columns carry load and resist sway. Reasonable to use nomograph. (*b*) Case 2: only one column resists the sway effects of many columns. *P*–Δ should be explicitly considered.

Overall Flexural Stability

Considering the tower as a cantilever column with linearly varying load and linearly varying moment of inertia,

$$P_{o,\text{flex}} = \left(\frac{7.83EI_o}{H^2}\right)(1 - 0.3B) \tag{10.2.14}$$

where P_o = critical total building load
H = tower height
B = fraction such that EI_o applies at base and $EI_o(1 - B)$ applies at top (i.e., if top $EI = EI_o/3$, then $B = 2/3$)
E = modulus of elasticity in consistent units
I_o = effective overall moment of inertia

Where frames or braced bays are used and column wind forces seem to follow the plane section assumption well,

$$I_o = \Sigma A_i d_i^2 \tag{10.2.15}$$

where A_i = column i area
d_i = distance to neutral axis of building from column i

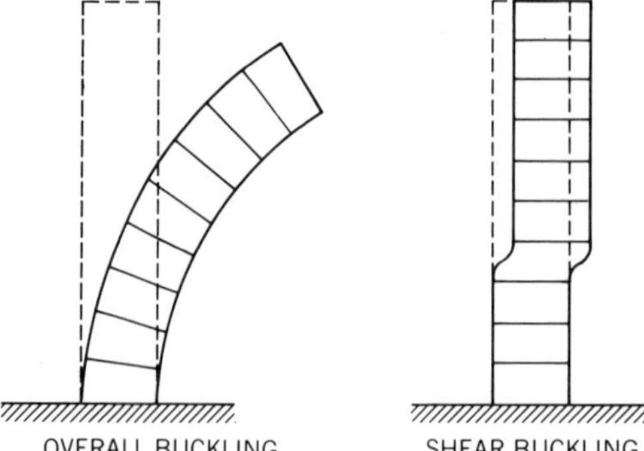

Fig. 10.15 Building buckling modes.

When shear lag makes the assumption of plane sections unlikely, such as in tube systems [10.7, pp. 242–308],

$$I_o = \Sigma n_i A_i d_i^2 \tag{10.2.16}$$

where n_i = fraction indicating relative effectiveness of the column i. This fraction can be established by finding the relative column forces which would exist if an ideal $I = \Sigma\, Ad^2$ situation existed, comparing them to the actual results of analysis, and setting the column with the highest $(F_{\text{actual}}/F_{\text{ideal}})$ as $n = 1.0$. Thus, for all other columns, $n_i = (F_{i,\text{actual}}/F_{i,\text{ideal}})$ (best $F_{\text{ideal}}/F_{\text{actual}}$) and n_i will be less than 1.0. Generally the corner columns will have $n = 1.0$ or nearly so.

As stated above, overall stability should not be a problem except for unusually slender and/or flexible structures, but should be checked.

Shear Stability

For the column sizes normally used in high-rise buildings, slenderness ratios are small enough that *elastic* buckling on a one-story basis should not be a problem. However, *inelastic* beam-column action (i.e., combined axial plus flexure approaching yield stress) is critical and requires consideration of lateral forces and stiffness.

Because all columns in a high-rise building are connected by a relatively rigid floor diaphragm at each floor, no single column can buckle in a sidesway mode independently (Fig. 10.14). One way to account for this, used in the *ACI Code* and British Standard CP, is to use the sum ΣP of the floor loads and the sum $\Sigma\, P_u$ of the bracing column Euler strengths in the moment magnification equation, and to use an average story slenderness ratio KL/r for all columns in the story. Another method [10.7, pp. 242–308] is to use an effective length factor K_n defined as

$$K_n = K_o\sqrt{1 + n} \tag{10.2.17}$$

where K_o = normally assumed K
n = load at pinned columns/load at bracing columns

These methods, however, still do not explicitly cover the P–Δ effect.

There are numerous factors which affect story stiffness, but the two most significant are the P–Δ effect and the loss of bending stiffness with increasing axial load. The loss of bending stiffness is related to the ratio of column load to Euler buckling load. Because the column sections used in lateral resisting systems tend to be stocky, the Euler buckling loads are very high, the ratio P/P_{cr} is very small [10.15], and the effect on lateral stiffness is usually ignored. The P–Δ effect, on the other hand, cannot be ignored.

In low-rise buildings, the lateral forces applied are generally small, column loads are small, and the bracing design is relatively stiff, such as X-braced bays or moment-connected frames sized for combined vertical and lateral loads, where a design sufficient for strength is more than amply stiff to limit deflection. Therefore, the story drift Δ experienced is relatively small, the column forces P are small, and the P–Δ effect is minor.

In high-rise buildings, the lateral resisting system is sized for deflection, the lateral forces are large, and column loads are large. The P–Δ effect becomes significant and must be explicitly considered.

The P–Δ method [10.13, 10.14] is iterative. A first-order analysis including lateral loads is performed, and lateral deflections Δ are determined. Then a second analysis is made with the original forces plus fictitious P–Δ forces, such that an additional story shear at level i occurs, equal to $\Sigma P_i\,(\Delta_{i+1} - \Delta_i)/h_i$. If deflections changed significantly between the first and second analysis, a third analysis is performed using deflections obtained from the second analysis. If deflections do not seem to converge after one or two cycles, the structure is too flexible and could buckle because of the P–Δ effect after any initial lateral disturbance.

The following points should be noted regarding the P–Δ method:

1. The P used is the total column load present at the story of interest, including both pinned (braced) and lateral resisting (bracing) columns.

2. The converged deflections resulting from second or third analyses are the deflections that should be used when evaluating drift limits and establishing building stiffness.

3. Deflections including the P–Δ effect are greater than those without the effect, but the externally applied loads are unchanged. Thus, the P–Δ method reflects a reduced lateral stiffness of the building under vertical load.

4. The P–Δ method can be used equally well with moment frame, X-braced, and shear wall systems. The fictitious forces applied will, of course, assume different distributions depending upon the deflected shapes that are characteristic of the different systems.

5. The P–Δ method considers overall story lateral action, but does not consider story torsion.

6. Where it is desired to speed convergence, the Δ used in establishing the P–Δ forces can be factored up as [10.6]

$$\Delta_{2i} = \frac{\Delta_{1i}}{1 - \dfrac{\Sigma P_i \Delta_{1i}}{F_i h_i}} \tag{10.2.18}$$

where F_i = story shear
h_i = story height
Δ_{1i} = first-order deflection in story i

7. The P–Δ method is applicable only where inelastic stability (strength of materials) governs, not where elastic buckling governs [10.6]. For limber columns, where elastic (Euler) buckling is a concern, the loss of bending stiffness discussed above becomes significant and leads to additional deflection which the P–Δ method neglects. Elastic stability problems begin at allowable drift ratios of about $H/300$ (working loads) and $H/200$ (ultimate loads). Since high-rise buildings are typically stiffer than this, elastic stability is usually not a problem.

8. For ultimate strength design (concrete) or LRFD design (steel), factored gravity and lateral loads should be used. For determining building lateral stiffness, working gravity and working lateral loads should be used. For allowable stress design, it has been argued [10.11] that the P–Δ sidesway forces should be increased by a factor of safety.

Once deflections, axial forces, shears, and moments satisfying the P–Δ effect are established, the arbitrary K (length) factors and unrestrained C_m (bending) factors used to cover this effect are not needed. Use $K = 1.0$ and C_m as appropriate for the applied moments, and design members in the usual fashion.

One suggested check for shear mode stability [10.6] is

$$P_{ci} = \left(\frac{F_i}{\gamma}\right)\left(\frac{h_i}{\Delta_{1i}}\right) \tag{10.2.19}$$

where γ = 1.0 for flexible beams
= 1.22 for rigid beams
P_{ci} = critical load at story i
F_i = story shear
h_i = story height

Other Stability Checks [10.6, 10.7, pp. 78–100]

A model European code states that, for buildings with first-order deflection less than $H/500$, simply add horizontal forces to create additional lateral shears equal to $\Sigma P_i/80$. If deflection is still acceptable then the structure has adequate stiffness for stability.

Dutch and Danish codes call for story shears equal to wind or $0.015\Sigma P_i$, whichever is larger.

The 1972 West German Code considers column miserection and misalignment, and gives potential offsets of sloping, kinked, and corkscrew forms.

Design procedures for towers with multistory X-bracing (such as the Hancock Tower in Chicago) are still unclear. Panel points of the giant X's are obviously good brace points, but such points occur several floors apart. Obviously an unbraced length of one story is too small, but a length equal to the height between panel points is too conservative. Careful judgment is required in such cases. Addition of light core bracing, suitable to provide bracing forces at intermediate floors and tie them to the panel point floors, can prove beneficial. Suitable treatment for design of staggered-truss systems is also not firmly established.

10.2.5 Thermal Effects

Motion and framing systems are interrelated. If a column is framed with all pin-connected members, it can change length freely and the beams framing into it will simply tilt. If the column is part of a bracing system or moment frame, its length change will affect the rest of the system. An example of this is the difficulty erectors experience in plumbing up a high-rise frame on a sunny day. The sun-heated columns expand and cause the tower to point away from the sun. As the sun moves, the tower tilt follows, so the tower tip describes a semicircular path each day.

Control Methods

In this age of energy consciousness and central climate control, the simplest way to minimize thermal problems is to include the structural frame within the insulated building envelope. Once the building is occupied, temperature variations will be limited by the occupants' desire for comfort.

Where enclosing the entire frame is considered unacceptable, the design must accommodate anticipated thermal motion due to exposed columns as described below.

If the building uses a shear wall or braced core, the free perimeter motion is best achieved by ensuring that core-to-perimeter beams are simple span. Recognizing that thermal motion will cause beam tilting, short core-to-perimeter beams should be avoided as slab cracking may result. In monolithic concrete structures, a long span and thin slab may be flexible enough to permit such motion. If so, the shears and moments generated by motion must be considered in the floor design. At upper floors where the motion is largest, forces generated may be excessive. In that case, creating a hinge detail at one or both ends of the spans may provide enough flexibility (Fig. 10.16*a*).

While outrigger trusses, girders, "hats," or "brackets" which engage perimeter columns are usually helpful in stiffening the core and resisting overturning, they must be avoided when thermal motion is anticipated. Unequal thermal changes will cause tilt of the overall building, rather than just the floor beams. This can be allowed for. But uniform thermal changes, such as seasonal variations, can cause all the perimeter columns to pull down or push up on the core via the outrigger trusses (Fig. 10.16*b*). With heavy column sections and stiff trusses, the forces generated can be very large. Even if one desired to account specifically for these forces in designing the perimeter columns, core columns, and outriggers, the resulting frame would be quite inefficient.

Exposing one side of a braced core to the elements is also undesirable, as it will exaggerate the tilting caused by unequal thermal motion. For any given nonuniform temperature distribution, unequal strains result. Difference in strain divided by distance over which the difference occurs defines curvature. Curvature times building height provides building top lateral movement (Fig. 10.16*c*). Thus, the motion induced by unequal temperature in a 50-ft (15-m)-wide concrete core will be about four times larger than for a 200-ft (60-m)-wide tube-type concrete building.

If exposed core walls are desired, and one core is placed at each end of a building, the curvatures will tend to act in opposition. The structure between the cores must be designed to carry the tension or compression required to force the cores back to a vertical position (Fig. 10.16*d*), and the cores themselves must be designed for the additional stresses induced by these straightening forces. Again, the structure will be less than fully efficient.

A tube-type building will exhibit overall tilting under nonuniform thermal conditions, since the relatively stiff perimeter frame forces compatibility of motion. The magnitude of motions can be conservatively estimated by considering the tower as a cantilever column with a variation in

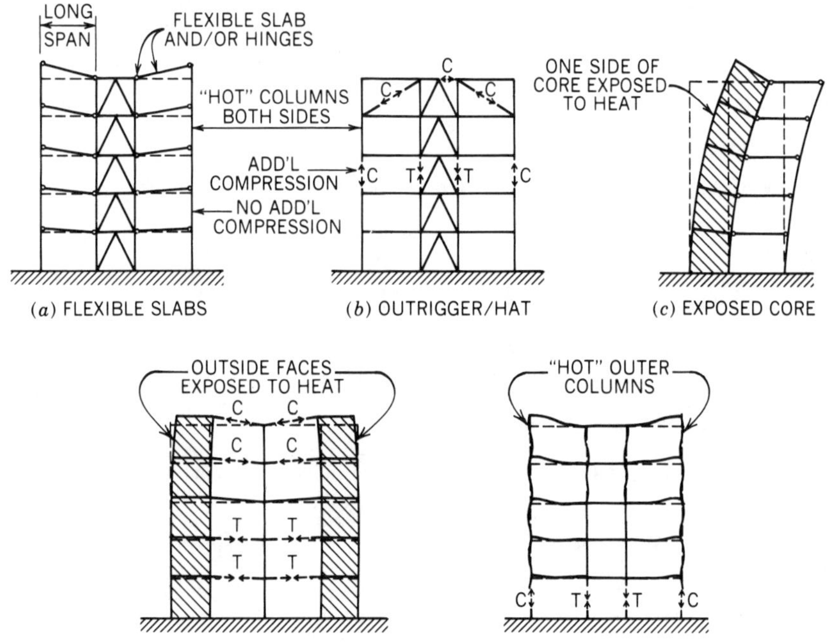

Fig. 10.16 Thermal behavior.

column strain, yielding a corresponding curvature. Rate of curvature times building height gives lateral motion. As for the braced-core schemes, if floor beams between interior and exterior columns are reasonably long and pin-ended, seasonal variation in column length is no problem. Slight tilting of the floor will be the only effect. Monolithic slabs should be checked for induced forces and detailed to provide the required flexibility.

Where conventional frames are used and include both interior and exterior columns, thermal effects must be specifically designed for by considering imposed length changes. For moment-connected girders of normal depth spanning typical 30- to 40-ft (10–12 m) bays, the moments and shears generated will be significant but not overwhelming as in the case of outrigger trusses above. Lateral motion due to unequal temperature would be somewhat less than in the idealized cantilever column model, since interior columns would resist the thermal strains.

Vertical strains produce the most dramatic effects in high-rise buildings, but exposed spandrels will also cause lateral thermal strains. The large area of the constant-temperature interior floor slabs produces horizontal shear between the spandrels and the floor. If the framing permits, a joint between the two could relieve the problem. More commonly, the framing is monolithic and must be reinforced for the anticipated shear and tension or compression which can result.

Load Combination Factors

When considering thermal forces in combination with other loads, some building codes permit the use of a reduction factor. For example, ANSI [10.3] and the *National Building Code of Canada* [10.15] use a reduction factor of 0.75 for a combination of dead load plus live load plus thermal load, and 0.66 for a combination of dead load plus live load plus wind (or seismic) load plus thermal load.

10.2.6 Axial Shortening

As was the case for thermal movement discussed in Section 10.2.5, axial shortening occurs in all structures, but is particularly troublesome in high-rise buildings because of the cumulative nature of the motion, which appears most strongly on the upper floors.

Axial shortening can be a problem in two ways. First, absolute axial shortening causes story heights to decrease, and elements that span between floors, such as partitions, window frames, and facades, must be detailed to accommodate this motion by use of expansion joints, relief angles, soft joints, and the like. Otherwise the elements will be crushed. This topic will be further discussed in Section 10.2.7.

Second, differential shortening can occur between axial members on the same floor. This can cause sloping and bending of slabs and generate large internal forces in stiff structures. Preliminary design of the structure should take this effect into consideration, as will be discussed below.

Shortening can be due to three major mechanisms. Elastic strain, which occurs when any material is stressed, is straightforward. Creep is the additional time-dependent strain which occurs when concrete is stressed. Shrinkage is the additional strain, dependent upon time and humidity, which occurs when concrete dries out from its original, saturated state.

Elastic Shortening

For engineering purposes, elastic shortening at any point along an axially stressed member, such as a column, can be considered as following Hooke's law, using a generally accepted modulus of elasticity for the material in question. For structural steel, published values for the cross-sectional area would be used. For reinforced concrete, the concrete area plus transformed area of vertical reinforcing would be used.

If all vertical members were equally stressed, the entire building would shorten uniformly. If desired, the vertical members could be fabricated or erected slightly long, so that all floors of the completed building would be at the design elevations. In actual practice, however, these corrections are rarely implemented for the uniform shortening case because occupants seldom care if the stories are slightly shorter than designed, or if, sitting on the 50th Floor, they are a few inches closer to the street than specified in the design.

Because of the critical nature of wind resistance and wind drift, columns acting as part of the lateral resistance system can be subjected to combined axial load, biaxial bending, and shear, and will be designed very differently than columns carrying only vertical load. If all vertical members are part of the lateral resistance system, such as in a three-dimensional frame where all columns act to resist wind in each direction, uniform shortening may occur. In all other cases, the participating members will be acting at lower axial stresses than the nonparticipating members.

In a braced-core building, core columns will be large to carry the additional overturning loads from wind or earthquake and to limit the amount of cantilever drift. Shear walls are similarly large to carry cantilever bending and wind shear, and to provide weight to minimize uplift problems. Thus, for these schemes, the core will have lower dead and live load axial stresses and will tend to shorten less than the other columns, unless compensation is provided.

In a perimeter frame or perimeter tube building, the perimeter columns will be relatively large, less axially stressed, and will tend to shorten less than the interior columns. An additional consideration for the tube case is the extreme stiffness of the perimeter frames. Because of this stiffness, relatively uniform axial strains of adjacent perimeter columns are enforced. If strains based upon tributary loading are different, the frame will tend to shift load from the more-strained column to the less-strained one. Therefore, the frame must be designed with adequate strength to perform this redistribution of load, and the column receiving more load must be designed for it.

Creep and Shrinkage

Considerable theoretical, laboratory, and field research has been conducted on creep [10.4, 10.9]. Only a general discussion is presented here.

Both creep and shrinkage represent volumetric changes in concrete which result from changes in the pores and interfaces between coarse aggregate, fine aggregate, and cement particles. When the concrete is first cast, these pores are filled with water. As hydration of the cement proceeds, some water chemically reacts with the cement and cannot be driven off unless the concrete is heated above the boiling point. Another fraction of the water is adsorbed to the hydrophyllic ("water-loving") cement particles in a boundary layer several Ångstroms deep. This will only evaporate at low levels of ambient relative humidity.

The remainder of the water is loosely held by capillary action and can evaporate relatively easily. When this "free" water evaporates to reach equilibrium with the surrounding atmosphere, the pores tend to become smaller. This is shrinkage, and is strongly dependent upon time, the surrounding relative humidity, and the concrete configuration (since a high volume-to-surface area ratio will slow the evaporation process). If concrete has experienced shrinkage due to drying out, reintroduction of humidity can cause some recovery.

When the concrete is stressed in compression, creep begins. Pore size reduction can again occur. In addition, calcium silicate hydrate "flowers", which form as part of the products of cement hydration, can be crushed, squeezing out the adsorbed water layer. A third mechanism is microcracking and slippage at the cement paste/aggregate interface. Because these processes are not reversible, creep rebound is limited.

Creep is strongly affected by the strength of the concrete, curing method (steam or field), applied axial stress as a percentage of strength, concrete age at the time of load application, volume-to-surface area ratio, and percentage of longitudinal steel reinforcing. The last factor is significant because, as concrete creeps, more load is redistributed from the concrete to the steel reinforcing. Thus, heavily reinforced columns creep much less than lightly reinforced ones.

Compensation for Shortening

Where it is anticipated that all axial members will shorten similarly, special corrections are generally not required. For cast-in-place structures, some adjustment may be inherent in the construction process as the contractor attempts to hold proper elevations at ever-shortening columns. Where differential shortening is expected, however, it should be specifically compensated for in detailing and construction.

For steel structures, differential shortening can be offset by providing a length adjustment for the columns having higher stresses. Each tier or every few tiers of these columns are fabricated slightly long (cambered), so that under a specified axial loading each floor beam connecting adjacent columns will be approximately level (Fig. 10.17). Note that the correction is uniformly spread throughout the column length, assuming that the differential stress is relatively constant over the building height.

For a typical office building, the actual dead load (including permanent partitions) would be an appropriate basis for the column length adjustment. Since the full load will not yet be on the columns during construction, the floor beams between adjusted and unadjusted columns will be sloping. Concrete slabs should be finished with reference to this slope, rather than with reference to a laser level. Otherwise, when construction is finished, the columns will shorten to the proper elevations, the beams will be level, but the concrete surface will be sloped.

Compensation for differential shortening in concrete structures is much more difficult to achieve because of the effects of creep and shrinkage. During construction, ambient relative humidity will vary with the weather. Even after construction is complete and surface sealers are applied to exposed concrete, interior relative humidity will vary with the seasons. Much of this effect is beyond the control of the designer, but if all columns are concrete the differential effect of shrinkage will be small. If the building has a combination of concrete and steel framing, an estimate of the final shrinkage should be made.

Creep is strongly influenced by concrete strength, stress level, and age at loading. In high-rise construction it is common to use different concrete strengths for columns at different floors. Therefore, a careful stepwise analysis of the construction sequence and schedule should be performed, as a cooperative designer/contractor effort.

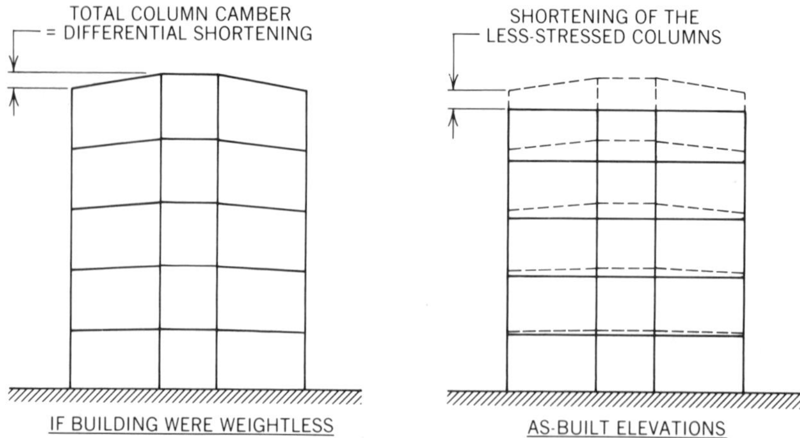

Fig. 10.17 Column cambering. Note: framed-tube case shown.

For each story, estimated one-story column deflection vs. time can be plotted based on the incremental column loading during construction. For each load increment, elastic shortening occurs and a certain total creep and shrinkage can be assumed to occur in a relatively standard time-dependent way, with the creep rate slowing over time. By superposition of these incremental elastic and creep curves, overall shortening of the column with time is estimated (Fig. 10.18].

Then the contractor can deliberately set forms high an amount equal to the total shortening at a floor expected at some time after initial occupancy, minus the shortening already expected to have occurred when that floor is cast (Fig. 10.19). The slab should also be appropriately sloped between column lines. Of course, if the schedule assumed in design is not followed fairly closely (as when a work stoppage occurs), the adjustments as calculated could be off somewhat.

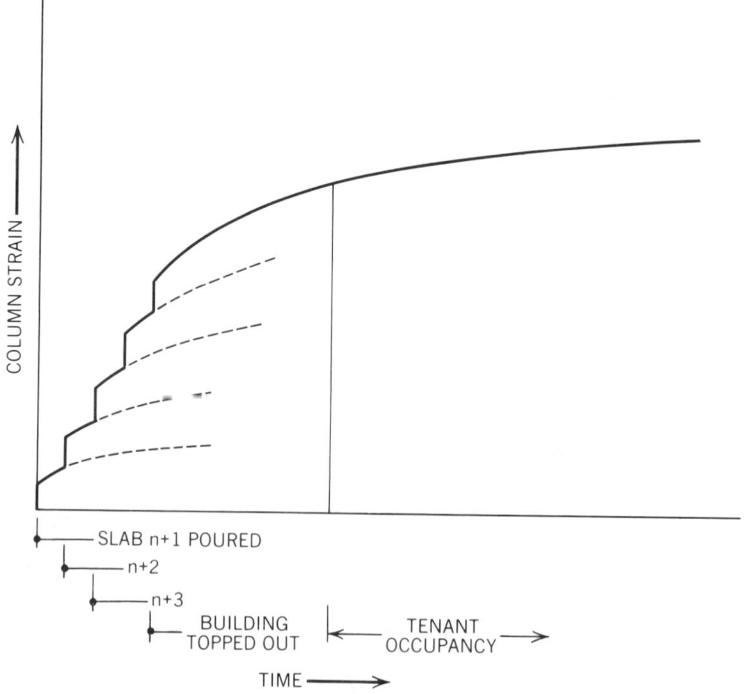

Fig. 10.18 Typical strain on column at floor *n*.

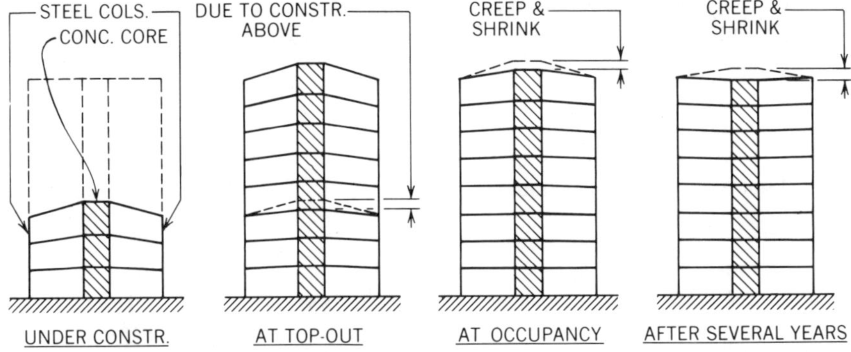

Fig. 10.19 Creep compensation.

Where both core and perimeter are concrete, and the combined anticipated differential creep, shrinkage, and axial shortening is minor, the use of published prediction methods such as the Bazant–Panula model [10.4] may be sufficient. However, where perimeter and core columns are of dissimilar materials (composite construction), the differential shortening and its timing will be critical. A full laboratory testing program of the actual concrete mixes and cement suppliers proposed for use is essential to establish elastic, creep, and shrinkage parameters. The testing should begin as soon as composite construction is found to be a viable option, so that meaningful time-dependent data will be available to guide the contractor from the start of construction.

Control Methods for Shortening

The paragraphs above make it clear that compensating for elastic, creep, and shrinkage effects in a cast-in-place concrete building can be time consuming and only approximate at best. To minimize the number of calculations and the number of variations in field adjustments required, the preliminary design should strive to use axial members of similar concrete design strength, axial stress, volume-to-area ratio, and percentage of reinforcement wherever possible on a floor. Thus, the calculation and adjustment for one member would suffice for all. The standardization and simplicity resulting from this may offset whatever minor savings might otherwise have resulted from trimming each individual column to the minimum.

A similar situation occurs for both steel and concrete tube systems, as stated earlier. Some small inefficiency in column sizing may be acceptable to provide uniform axial strains, avoid placing load-redistribution demands on the framing system, and permit the use of standardized column sections.

Where differential column shortening is anticipated, adequate flexibility must be provided in the framing to accept this differential motion. For example, where interior columns are more highly stressed than perimeter columns, the span between them should be as long as possible to minimize slab warping.

For monolithic cast-in-place concrete, the moments and shears in floor beams resulting from differential motion occurring since the date of casting should be added to the other design moments. However, the forces generated are not as large as one would calculate using short-term elastic beam stiffness with long-term anticipated differential motion, since the beam will creep (and thus relax) at the same time that the columns are creeping (and generating the differential motion). Some researchers [10.9] have found that, if the differential motion occurs over a 1-yr time span, creep within the beam will cause a 50% relaxation of peak moments and shears. If the differential motion from elastic, creep, and shrinkage (and thermal) effects causes excessive moments and shears, the slab and beams must be provided with hinges to improve flexibility. Usually hinging one end and leaving the other end monolithic provides sufficient flexibility.

Hat trusses, outrigger trusses, and brackets connecting differently stressed core and perimeter elements can be used successfully [10.10] if the differential motion is well defined. The preferable method of handling this motion is to delay truss installation until the major portion of the load is applied, as when a steel building has all slabs placed. In steel construction this is easily accomplished by delaying the final bolt-up of such trusses as long as possible. The truss will then attempt to redistribute only the differential live load stresses, which are generally small, and any thermal stresses.

In concrete buildings, or composite concrete and steel buildings, where creep and shrinkage motions occur slowly and are of large magnitude, delay of outrigger installation will not be as helpful. The differential motion anticipated to follow outrigger erection will induce very large forces in the outriggers. Conservatively, the forces generated are those required to enforce compatible deflections. A more precise analysis would use compatibility of deflection while considering elastic and creep deflection (relaxation) of the outrigger, increased creep on the newly loaded columns, and reduced creep on the relieved columns. Loaded columns must be designed for the redistributed force.

10.2.7 Wall/Structure Interaction

Building deflections are the most significant items to be considered in wall design for high-rise buildings.

Because of the relatively flexible nature of modern construction, racking between floors can be significant. If partition material is rigid and brittle, such as masonry, the top of the wall should be detailed with a cap channel or similar device to permit the floor above to slip along the wall axis. Such walls should not be built tight against columns since the columns will tilt as the building racks. Soft joints should also be provided at right-angle intersections between walls, as one wall will try to tilt as the other tries to stay vertical.

Facade joint details (see Chapter 27) and methods of support should also consider racking deflections, as determined for the building under lateral wind load and an allowance for torsion. Precast concrete panels are usually horizontally oriented, with two supports on the lower floor providing vertical and horizontal support, and two upper-floor tiebacks which are stiff against wind pressure but flexible in lateral motion (Fig. 10.20*a*). Joints should be generous, so that the racking between upper and lower panels does not create unacceptably high shear strains in the sealant.

Where facade panels are tall and narrow, racking can be accommodated by pivoting on a central

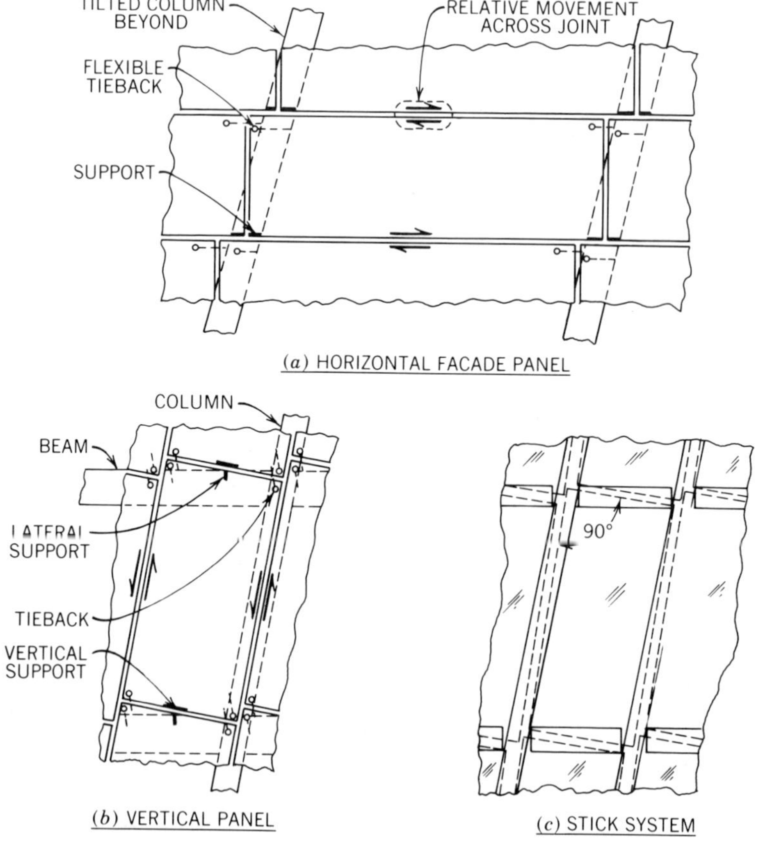

Fig. 10.20 Facade system movement patterns.

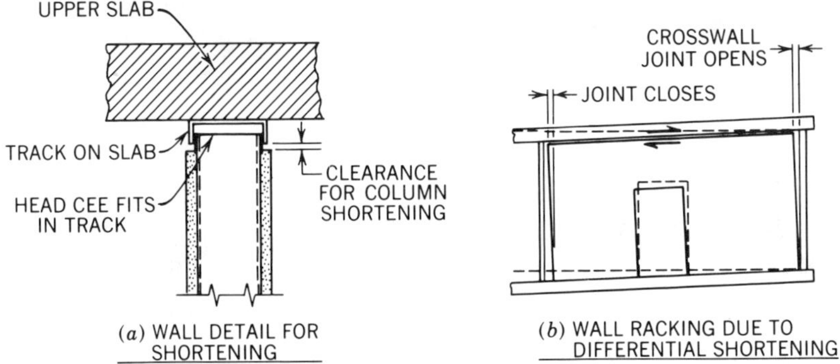

Fig. 10.21 Interior partition considerations.

support point, with solid connections to upper and lower floors. In that orientation, joint shearing is critical along the panel sides (Fig. 10.20*b*).

Shearing is not a serious problem for glass and metal curtain walls which are assembled piece by piece. Each "stick" tilts to follow building racking, and the glass vision or spandrelite rectangular panels "float" as their frames form parallelograms (Fig. 10.20*c*). Generous frame allowances and placement of setting and antiwalk blocks near panel centers rather than at corners are essential for a proper float.

Axial shortening due to progressive loads, thermal changes, creep, and shrinkage must be accommodated. Thermal motion is well understood, and the amount of axial shortening under load for a one-story steel column is rather small. But the cumulative effects of creep and shrinkage in concrete can be several times that of the elastic strain alone. And for the high-strength highly stressed columns currently used in the lower floors of tall buildings, even the elastic strain itself can be significant. In order to accommodate these movements, continuous vertical lines of partitions should be interrupted with soft joints under the floor slabs, and facade joints must provide enough room so that the sealant will not be crushed or extruded under the worst conditions (Fig. 10.21*a*). Numerous facade failures have occurred because of these movements, and allowances for them should be calculated and set on the generous side.

Differential axial shortening should also be considered. This type of movement subjects partitions to a vertical racking, and the details mentioned above for racking should also be applied here (Fig. 10.21*b*).

In areas of high seismic risk, special isolation details and large lateral motions are specified by codes and must be accommodated.

As a final point, the wall design and the structural design should be compatible. As one example, if a precast facade is to be used, setting column spacing consistent with the spanning capabilities of the facade may permit support directly from the columns. Spandrel beams can then be lighter, as they carry only floor load and have no special deflection criteria. Another example is the significant economies obtained by using 10-ft (3-m)-wide steel facade panels structurally, attaching them to columns on a 10-ft (3-m) spacing (Fig. 10.22) [10.20]. A third example is the inherent conflict between an architectural design calling for floor-to-ceiling windows and a structural design calling for upset spandrels. Coordination between the facade designers and the frame designers is essential.

10.2.8 Interface with Mechanical and Electrical Design, and Future Flexibility

A major fallacy in the design of structures is the emphasis on the cost of isolated systems, while ignoring their impact on other systems. For example, minimizing floor heights will reduce wind forces and save on wall height, but may require special structural and mechanical details which prove to be very expensive. The impact of mechanical systems on structural requirements is explored in detail in Chapter 4, as well as in Chapter 11.

Also, although current real estate theory assumes that buildings have a rather short useful economic life, and that changing rental needs must lead to demolition and new construction, recent experience points in a different direction. Renovation and restoration are becoming more and more popular. Major buildings are stripped to their frames and given new facades, cores, and mechanical plants. Reinsulation of facades and reworking of HVAC systems is common, as owners seek to

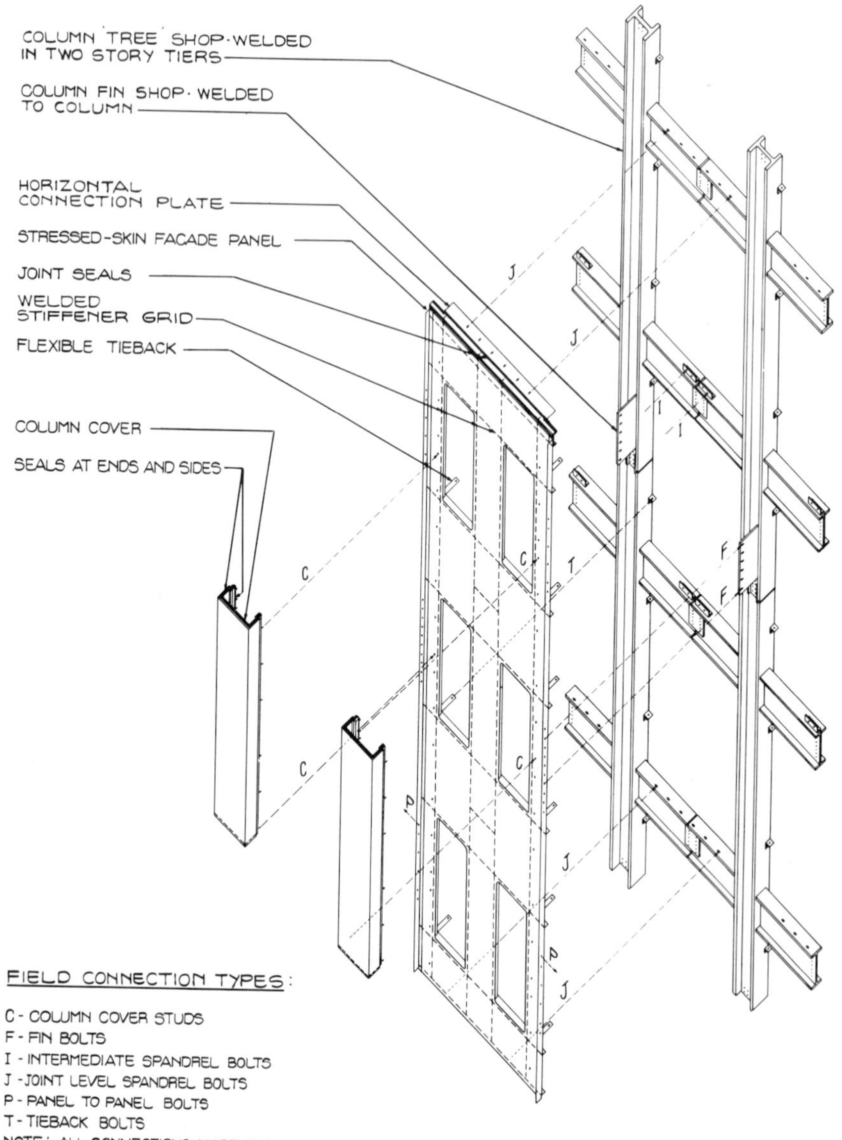

Fig. 10.22 One Mellon Bank center: stressed skin–tube interface.

reduce energy costs. The buildings with the longest useful life will be those that have enough flexibility to accept such changes.

Structural modification can be accomplished for structural steel systems by cutting, reinforcing, welding, and bolting. For cast-in-place concrete, the location of existing reinforcement within the concrete must be considered. For prestressed and post-tensioned concrete, an additional consideration is the existence of prestressing tendons which must be avoided during structural modification work.

The flexibility of the mechanical and electrical distribution systems at each floor is dependent upon the type of floor construction and the floor-to-floor height. A slab system comprised of cellular metal deck with concrete fill can be used to provide electrical distribution to all points on an office floor without disturbing occupants below or incurring the fire hazard of "poke-through"

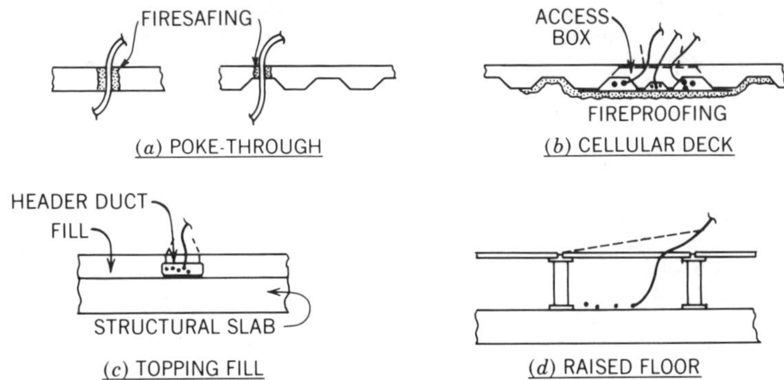

Fig. 10.23 Slab electrification schemes.

installation. On the other hand, the cells and trenches will fill quickly unless they are sized generously and provided on a close module (Fig. 10.23*a* and 10.23*b*).

Similar distribution can be provided in a topping fill over a concrete slab, at some penalty in weight and floor thickness (Fig. 10.23*c*). Electrical distribution systems can also be provided within the thickness of a concrete structural slab, thereby giving rise to additional structural, coordination, and scheduling considerations. The provision of an electrical distribution system within the floor slab is a minor consideration in residential buildings. Note that "poke-through" installation in post-tensioned slabs must consider the location of existing prestressing tendons.

Recently, developers anticipating the all-electronic office have found that providing raised "computer floors" throughout a building can make sense. Complete tenant wiring flexibility is provided, while the builder may offset the raised-floor cost by using the most economical floor system, be it post-tensioned, precast, flat or waffle slab, or metal deck (Fig. 10.23*d*).

The floor framing system which provides the most flexible mechanical and electrical distibution is one which has adequate space below the beams and above the desired ceiling (Fig. 10.24*a*). Next in flexibility are open-web systems, such as joists, trusses, and castellated beams (Fig. 10.24*b*). Large ducts may have to be split to fit, at some penalty in cost. Truss schemes have also been built in concrete. Stub-girders can be used with shallow continuous beams, or open-web beams, for two-way flexibility (Fig. 10.24*c*).

Next, and providing less flexibility for distribution, are wide-flange beam systems with individual web penetrations, which require detailed coordination between the trades for the original design openings, a time-consuming and error-prone process (Fig. 10.24*d*). New openings can be cut and reinforced in the future, but at considerable cost and inconvenience. Following that, reinforced concrete beams with predetermined openings are feasible, but very difficult to modify after erec-

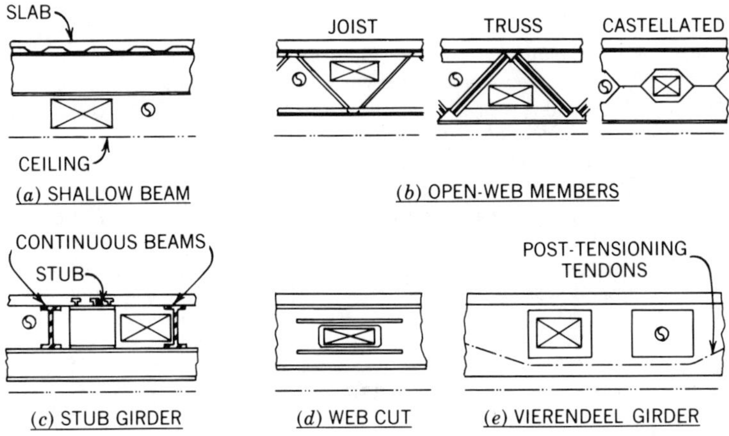

Fig. 10.24 Floor framing for mechanical distribution.

tion. Some schemes have used multiple web slots to ensure future flexibility, but the resulting beams are similar to Vierendeel trusses and are not very efficient. For post-tensioned and prestressed beam systems, tendons must be specially harped to clear original openings and must be avoided for any future openings (Fig. 10.24*e*).

Flexibility in locating core walls, elevator shafts, HVAC shafts, plumbing risers, and electrical risers is desirable in designing a new building, and essential in the successful renovation of an existing frame. At each floor, freedom of distribution from the core to the main floor area for pedestrian traffic, ductwork, and piping is also desirable. For this criterion, schemes using the perimeter for lateral resistance are preferred. Tubes and perimeter "X" braces or shear walls leave the core completely open. Structural steel can be modified relatively easily. Reinforced concrete framing within the core could be removed and replaced in its entirety if required.

A braced steel core using X, K, or knee braces is next in flexibility, since ductwork can be snaked around the bracing members. In some cases, shafts can be located immediately outside the structural core, thus enabling distribution to the main floor area without interfering with the bracing members. If circumstances warrant, the bracing system could even be modified in the future to suit new doorways and the like. Beams at the bracing lines are horizontal struts which cannot be moved, but other framing could be modified to accommodate new shafts.

The least flexible lateral system in the core area, from the standpoint of distribution, is the concrete shear wall core. Lintel beams across doorways are usually heavily reinforced in the original design, thereby inhibiting the possible relocation of doorways. The core walls, being concrete, are also fixed. Wall penetrations for mechanical services require special reinforcement unless they are small and well separated. Therefore, later openings are also limited in size, quantity, and location.

Freedom in facade design runs counter to the above. Narrow perimeter columns on 30-ft (10-m) centers, set back from the building face, and with heavy spandrel beams between them, can accept virtually any architectural facade desired. Tubular framing has heavier columns and shorter bay spacing, while braced perimeters express a strong architecture of their own. Thus, these schemes limit future facade variations. An exposed concrete perimeter frame virtually eliminates variations. Upset spandrels, while structurally advantageous, will limit the height of future windows and make access to future abutting buildings or connecting bridges difficult or impossible.

Selection of a suitable structural scheme is the first, and most critical, decision in the development of a high-rise building. Careful analysis of all factors and thoughtful discussion of various possible systems, tradeoffs, and compromises by all members of the design team and the owner are essential to achieve a satisfactory final design.

SELECTED REFERENCES

10.1 ACI. *Building Code Requirements for Reinforced Concrete* (ACI 318-83). Detroit: American Concrete Institute, 1983, pp. 32–33.

10.2 AISC. *Manual of Steel Construction,* 8th ed. Chicago: American Institute of Steel Construction, 1980, Chap. 5.

10.3 ANSI. *American National Standard Minimum Design Loads for Buildings and Other Structures* (A58.1-1982). New York: American National Standards Institute, 1982.

10.4 Zdenek P. Bazant and Liisa Panula. "Creep and Shrinkage Characterization for Analyzing Prestressed Concrete Structures," *PCI Journal,* **25,** May/June 1980, 86–122.

10.5 Council on Tall Buildings. "Wind Loadings and Wind Effects," *Planning and Design of Tall Buildings,* **CL-3.** New York: American Society of Civil Engineers, 1980, pp. 151–200.

10.6 Council on Tall Buildings. *Structural Design of Tall Concrete and Masonry Buildings.* New York: American Society of Civil Engineers, 1978.

10.7 Council on Tall Buildings. *Structural Design of Tall Steel Buildings.* New York: American Society of Civil Engineers, 1980.

10.8 Alan G. Davenport. "Gust Loading Factors," *Journal of the Structural Division,* ASCE, **93,** June 1967 (ST3), 11–34.

10.9 Mark Fintel and Fazlur R. Khan. "Effects of Column Creep and Shrinkage in Tall Structures—Analysis for Differential Shortening of Columns and Field Observation of Structures," *Designing for Effects of Creep, Shrinkage, Temperature in Concrete Structures* (SP-27). Detroit: American Concrete Institute, 1971, pp. 95–119.

10.10 J. F. Fleming. "Lateral Truss Systems in Highrise Buildings," *Regional Conference on Tall Buildings.* Bangkok, Thailand: IABSE, ASCE, AIA, AIP, IFHP, VIA, January 1974, pp. 33–47.

10.11 Jerome S. B. Iffland. "High Rise Building Column Design," *Journal of the Structural Division,* ASCE, **104,** September 1978 (ST 9), 1469–1483.

10.12 ICBO. *Uniform Building Code.* Whittier, CA: International Conference of Building Officials, 1980, pp. 131–151.

10.13 Bruce G. Johnston, ed. *Guide to Stability Design Criteria for Metal Structures,* 3rd ed., Structural Stability Research Council. New York: Wiley, 1976, pp. 410–454.

10.14 L. Lu and E. Ozer. "Lateral Deflection and Stability of Tall Buildings," *Conference on Tall Buildings,* Kuala Lumpur, December 1984, pp. 4-1 to 4-8.

10.15 NRCC. *National Building Code of Canada* (NRCC No. 17303). Ottawa, Canada: National Research Council of Canada, 1980, pp. 137–147.

10.16 NRCC. *Supplement to the National Building Code of Canada* (NRCC No. 17724). Ottawa, Canada: National Research Council of Canada, 1980, pp. 144–175.

10.17 Balbir S. Sandhu. "Dynamic Analysis of Multistory Buildings," *Engineering Journal,* AISC, **11,** 3rd quarter, 1974, 67–72.

10.18 Richard Tomasetti, Abraham Gutman, Leonard Joseph, and David Beer. "One Mellon Bank Center: Skin is More Than Beauty-Deep!," *Modern Steel Construction,* AISC, **24,** 3rd quarter, 1984, 5–10.

CHAPTER 11

PRELIMINARY DESIGN OF HIGH-RISE BUILDINGS

LESLIE E. ROBERTSON

Senior Partner, Director of Design and Construction
Leslie E. Robertson Associates
New York, New York

SAW-TEEN SEE

Partner
Leslie E. Robertson Associates
New York, New York

Over the last decades the methodologies employed by structural engineers in the design of tall buildings have shown a marked increase in complexity. The reasons for this change are to be found in the increased variety of materials available to the engineer, in the increased acceptance by the profession of modern computing devices, in the substantial increase in the complexity and the cost of building service systems (air conditioning, vertical transportation, communications systems and the like) and in the development of more sophisticated structural systems by the structural engineer.

In order to consider the subject of high-rise buildings, one need first arrive at an understanding of the term "high-rise." Quoting from the Monograph *Planning and Design of Tall Buildings* [11.1]:

> *A tall building is not defined by its height or number of stories. The important criterion is whether or not the design is influenced by some aspect of "tallness." It is a building in which "tallness" strongly influences planning, design, and use. It is a building whose height creates different conditions in the design, construction, and operation from those that exist in "common" buildings of a certain region and period.*

From the point of view of the structural system, tallness can be related to dynamic behavior. That is, buildings that display a dominance of dynamic behavior under wind or seismic excitation can be considered "high-rise." Another aspect, from the point of view of the wind engineer, is the reaching of the building above its surroundings. That is, a 30-story building in a field of 30-story buildings is not "tall," but, in a field of 20-story buildings, it becomes "tall," this, because the turbulent wind is shielded from the one, but the wind dominates the design of the other "tall" building.

The Monograph [11.1] goes on to define "building" as a structure that is designed essentially for residential, commercial, or industrial purposes and allows also for structures for institutional, public assembly, and multiple-use functions. Following the lead of the Monograph, this chapter does not treat specialty structures such as towers, monuments, and "space needles."

The materials of construction of high-rise buildings are varied and are used often in combination: structural steel, reinforced and prestressed concrete, precast concrete, and both brick and block masonry. While heavy timber has been used in high-rise buildings, the application is rare and is not considered herein.

The structural systems for high-rise buildings seem enormously varied. However, under closer scrutiny, it is seen that the basic tools of the structural designer are applied in mixes and combinations that, while appearing endless, are rather more finite than might be supposed. Indeed, there is very little that is truly "new" in the structures of contemporary high-rise buildings, except, perhaps, the willingness of designers to apply the many and varied structural systems of the past to the new and the exciting architectural forms of the present.

Surprisingly, the *Encyclopaedia Britannica* [11.2] has devoted significant space to the new spirit of cooperation between the architect and the structural engineer and speaks also of the opportunities to combine current technology with historically evolved forms. Probably the first lessons that any aspiring designer need learn are those associated with the accomplishments of other designers of other buildings . . . and, if the designer is to be successful, the lessons need be relearned over and over again. It has been said that those who do not study and learn from the past are condemned to relive it. Indeed, in nearly every high-rise building, there are "mistakes" to be observed and to be avoided. Often, however, it takes the combination of an uncanny perception and an acute eye to observe them.

11.1 BUILDING FUNCTION

Both the materials of construction and the structural system itself are dependent largely on the use of the finished building. Interestingly, this stems more from differences in architectural and service systems (heating, ventilating and air conditioning, plumbing and the like) than from the appropriateness of a given structural system to a given function. Examples include the following.

11.1.1 Hotels

Hotels and patient rooms of hospitals have rooms of relatively small size which are fixed in position, with the room boundaries largely unchanged over long periods of time. This type of discipline allows and even encourages the use of relatively short-span construction.

Service systems are restricted largely to the room corridors, to the baths, and to the entrance hallway; lowered ceilings, then, are found commonly in those areas. The major room area does not require ceilings if the structure floor above can be inexpensively treated for this purpose. Floor construction tends to be of cast-in-place concrete or of precast concrete, neither requiring hung ceilings. For reasons associated with acoustical separation, floor slabs tend to be relatively thick and heavy, perhaps 6–8 in. (150–200 mm) or thicker. Piping and air shafts (toilet exhausts) tend to fall in good vertical alignment, but down-going elements often need be offset above the ground floor where the rigor of the room discipline ceases.

Recognizing that nonstructural walls need not be movable, but need be durable and provide a high level of acoustical separation, such walls are constructed generally of concrete block, 4–8 in. (100–200 mm) in thickness.

11.1.2 Apartment Buildings

Apartment buildings have many similarities to hotels, except that dropped ceilings are seldom used at any location because service systems can be organized to preclude their need. Again, floor construction tends to be of cast-in-place or of precast concrete 6–8 inches (150–200 mm) or thicker.

11.1.3 Office Buildings

Office buildings usually require air-carrying ductwork at all locations. Particularly in the United States, dropped ceilings are used generally in all public and rentable areas and even in work areas, except for mechanical rooms. Acoustical separation is of minor importance (again, except for mechanical rooms) and slabs as thin as 2 in. (50 mm) over steel deck can be employed, although 2½ in. (60 mm) or thicker is more common. Mechanical rooms generally require slabs weighing 100 psf (490 kg/m^2), accomplished often with 8 in. (200 mm) of normal-weight concrete, in order to achieve the needed acoustical separation from contiguous office areas.

The distribution of electricity (see also Chapter 4) for low-voltage systems, such as communication wiring, and for high-voltage systems for lighting and equipment (110 volts in the U.S., but commonly 220 volts in other parts of the world), requires that access be available at relatively closely-spaced locations. Three basic systems are used:

Poke-through systems, wherein a hole is drilled through the floor so as to allow wiring to be pulled through from conduits or trays in the ceiling below. In order to maintain the required fire separation, fittings with intumescent coatings are provided through the floor. An example of this type of system is shown in Fig. 11.1. After passing through the floor fitting, wires are carried in trays or in conduits in the ceiling space below. The details of the system are highly dependent on the regulations of the governing jurisdiction.

Cellular distribution systems, such as those used in metal deck construction, are found where the inconvenience of poke-through systems warrants the higher cost of in-floor systems. In metal deck construction, the need for trench headers, which collect wiring from the various distribution cells, provides a particular inconvenience to the structural design in that it cuts completely through the concrete floor slab. Steel deck is generally 2 or 3 in. (50 or 75 mm) in depth, usually composite with the concrete slab, 2½ in. (60 mm) of normal-weight concrete with spray-applied fireproofing, or 3¼ in. (80 mm) of lightweight concrete without other fireproofing. These assemblies provide the needed 2-hr fire separation. The system is shown in Fig. 11.2.

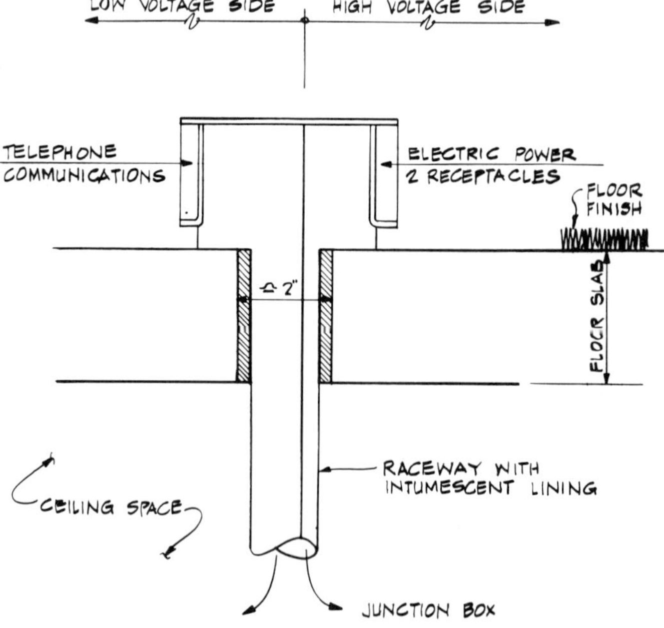

Fig. 11.1 Poke-through electrical system.

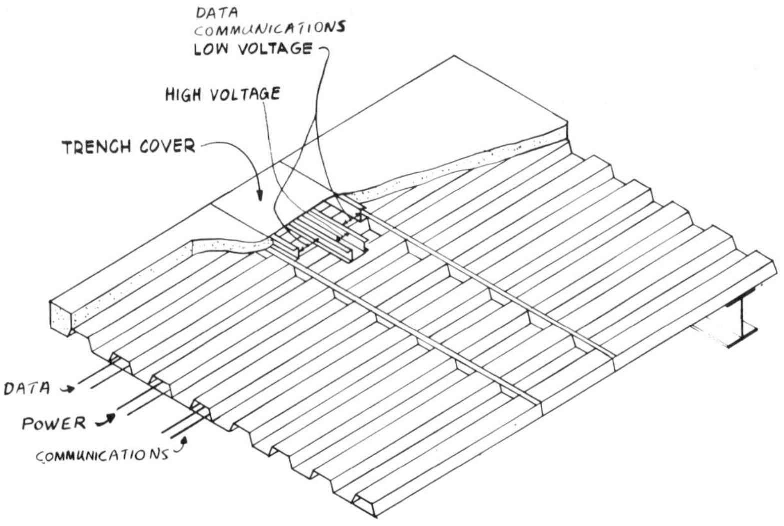

Fig. 11.2 Trench header system.

Fig. 11.3 Ducted distribution system.

Ducted distribution systems, either in a concrete topping slab or in the structural slab are a common alternative to cellular deck systems, particularly in reinforced concrete buildings. The systems are generally proprietary. One such system is shown in Fig. 11.3.

Recognizing that the direction of span of the deck and the required thickness of the concrete slabs are all dependent on the chosen electrical distribution system, it should be clear that floor systems cannot be selected until the electrical system is established.

Partition systems need not provide the relatively rigorous acoustical barrier required in residential buildings, but fire separation is required for most permanent nonstructural partitions. Accordingly, partitions are generally lighter in weight, being of metal stud and gypsum wallboard; fire-rated walls generally use a system commonly called "shaftwall."

11.2 STRUCTURAL SYSTEMS TO RESIST GRAVITY LOADS

The classical approach to the design of structural systems for high-rise buildings is in a three-part format [11.3]:

1. Systems to resist gravity loads.
2. Systems to resist lateral loads.
3. Energy-dissipation systems.

It should be noted immediately that, except for the need for high in-plane (diaphragm) strength, gravity load systems differ little from those found in low-rise buildings; accordingly, the subject is covered herein with some brevity. The reader is referred to other chapters of this handbook for more detailed information.

Eccentricities can seldom be tolerated in the gravity load system of taller buildings because the structural inefficiency becomes overwhelming. For example, an eccentricity of but one-sixth of the depth of an axially loaded rectangular section doubles the maximum stress in that member; larger concrete columns may carry reinforcing steel weighing as much as 20,000 pounds per foot (30,000 kg/m) of column, and cannot possibly afford such eccentricities. It follows that column centerlines need be held in a rigorous fashion.

11.2.1 Floor Diaphragms

In high-rise buildings, column axial forces tend to be high. Figure 11.4 shows heavy steel column sections designed for use in the United States Steel Headquarters in Pittsburgh. Such columns carry 40,000 kips (180,000 kN) and more. For the Bank of China Headquarters in Hong Kong (Fig. 11.25), a 72-story building, gravity loads are supported largely on its four corners, with columns as large as 5 m (16 ft–5 in.) to a side, and the largest carrying more than 440,000 kN (100,000 kips). These are probably the largest column loads ever used in building construction.

Such columns, of course, are not perfectly straight. At points of change in slope, the floor (gravity) system must resist forces in-plane to provide the required restraining force. This process, with deformations exaggerated, is depicted in Fig. 11.5 for columns of structural steel.

The AISC *Code of Standard Practice* [11.4] defines the acceptable plumbness of steel columns as $H/500$ for individual pieces of length H. Specific tolerances for out-of-straightness and total deviation from a true and plumb vertical are provided also, but are not of importance here.

The ACI *Standard Tolerances for Concrete Construction and Materials* [11.5] cites an acceptable out-of-plumbness of $\frac{1}{4}$ in. (6 mm) in 10 ft (3 m) ($H/480$), for columns. This standard discusses also tolerances for a range of concrete members.

For purposes of evaluating the required strength of the floor diaphragm, let us assume that a plumbness of 1/500 is achieved. From Fig. 11.5, it can be seen that the required resisting force, for a theoretical column, is $P/250$ or 0.4% of the axial force in the column. Codes have specified much higher loads [11.6, 11.7], as high as 2% of the column load, but most codes do not specify any load at all, which is a clear and important omission of a substantial gravity-induced force.

It can be argued that all columns will not slope the same amount in a given floor, so that the global content of some of the restraining forces may cancel out. However, the nature of building construction, particularly buildings of structural steel, tends to produce more or less equal slopes in all columns of a given story. For purposes of preliminary design, then, the floor diaphragm could be designed to provide for the lateral support of all columns at something more than 0.4% of their axial loads. In doing so, it is important to recall that these loads are directly associated with the axial forces in columns so that the dead load plus live load component of these stabilizing forces is directly additive to the dead load plus live load gravity load on impacted elements. Specifically, it is not proper to allow an increase in allowable stresses for these stabilizing forces.

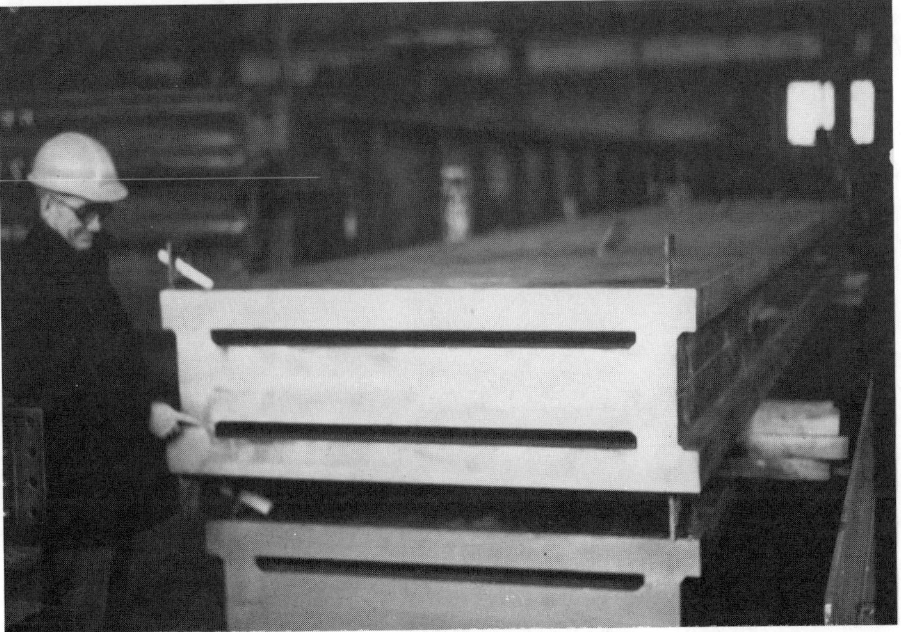

Fig. 11.4 Heavy steel columns.

A free-body diagram of an idealized floor diaphragm, wherein basic building lateral forces from wind and earthquake are carried in the outside wall, is given in Fig. 11.6. The applied forces, then, are associated with the out-of-plumbness of individual columns, while the resisting forces are supplied by the basic lateral force system of the building. ΣP, as used in Fig. 11.6, is the summation of the gravity loads in all columns at the given floor.

The structural system required to transmit these forces can be idealized as a stiffened membrane as shown in Fig. 11.7. The stiffeners or flanges can be actual beams, whether steel or concrete, or they can be steel reinforcement embedded in the slab, or both.

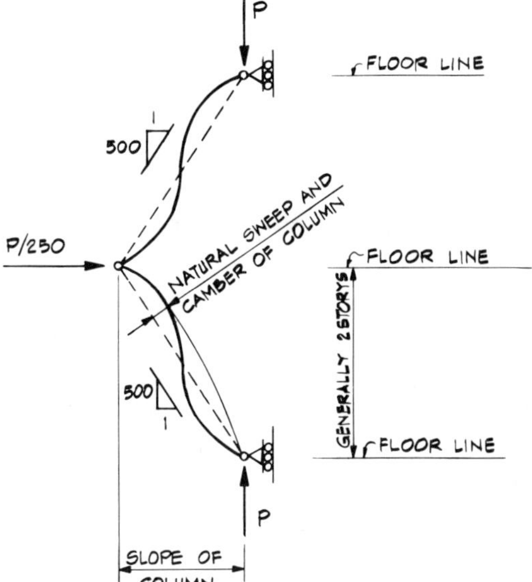

Fig. 11.5 Restraining forces on columns.

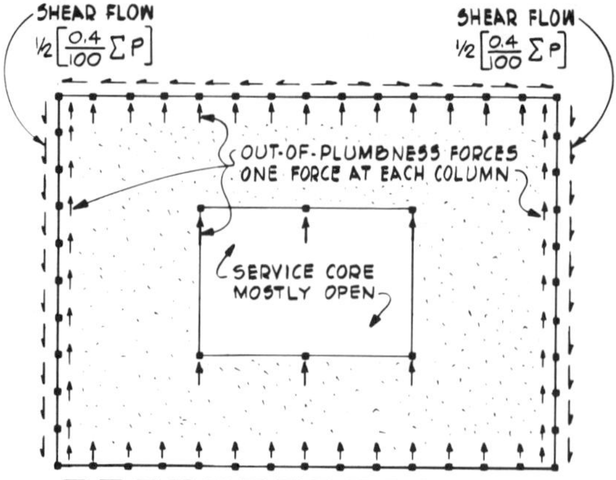

Fig. 11.6 Idealized floor diaphragm.

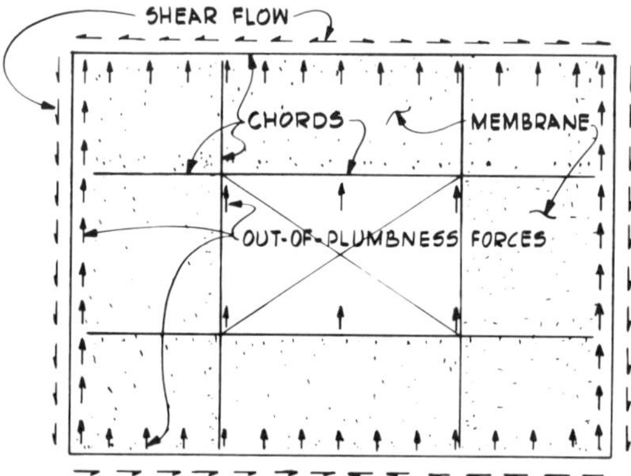

Fig. 11.7 Stiffened membrane.

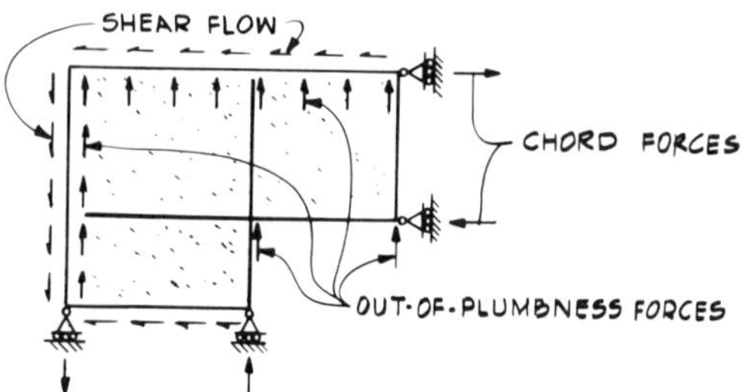

Fig. 11.8 Symmetrical membrane.

From concepts of symmetry, the structural system can sometimes be further idealized as shown in Fig. 11.8. From this model it is a more or less trivial operation to determine slab shear stresses and chord forces within realistic limits. Of course, for the final design of more complex floor diaphragms, the designer can use finite element methods or can resort to truss-analogy to obtain forces and stresses at higher precision.

For columns located at the building perimeter, or at openings, a buckling tendency toward the open side (i.e., away from the concrete diaphragm) could pull a column right off the building. It is important, then, that a beam, girder, or other tensile tie be framed directly to the column, and that it both be adequately connected to the column and be rooted into the diaphragm in sufficient depth to allow the tensile force to be resisted by diaphragm shear. This concept is illustrated in Fig 11.9; while structural steel is depicted, the need exists also for columns of reinforced concrete. These

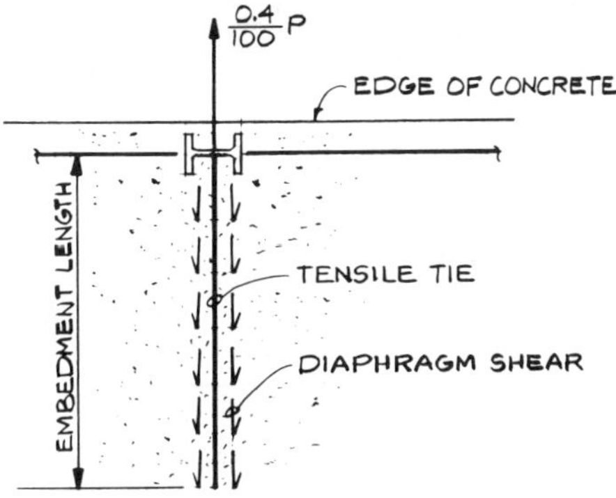

Fig. 11.9 Restraint of individual columns.

local shear stresses are additive to the global shear stresses as developed under the analysis depicted in Fig. 11.8. It should be noted that at least one major high-rise building was reinforced to correct this deficiency.

11.3 STRUCTURAL SYSTEMS TO RESIST LATERAL LOADS

In high-rise buildings, the structural system tends to be dominated by the need to resist lateral forces. In the not-too-distant past, the engineer could rely heavily on the massive nature of non-structural cladding and partitions to resist lateral forces [11.8]. The *Commentary to the AISC Specification* [11.9], in bolstering this philosophy, has revised the AISC posture here in nearly every edition of that *Commentary*. The 1978 edition states in part that "ordinarily the existence of masonry walls provides enough lateral support for tier building frames to control lateral deflection . . . ," a posture clearly not consistent with the design of most contemporary high-rise buildings. In such buildings, masonry is seldom used and, when used, is almost always deliberately cut free from the structural frame with the goal of minimizing cracking and distress in the masonry associated with the lateral deflection of the structural frame. For most high-rise buildings, then, the structural frame alone need resist all lateral forces.

The loss of the stiffening influence of walls and partitions has perplexed many engineers and has led to the construction of some tall buildings with excessive swaying motion. The root cause of the problem can be traced often to an incomplete understanding of the P–Δ phenomenon.

11.3.1 The P–Δ Phenomenon

This critically important effect was introduced in Section 10.2.4 of the previous chapter. It will be given additional treatment here.

As tall structures are displaced from a true and plumb position, the weight of the structure, now displaced from the center, contributes an additional overturning moment. The magnitude of this additional moment is commonly of the order of 10% of the moment creating the original displacement—and has reached 20% and even 50% in some building structures. Because it is not possible to design safe structures without a thorough understanding of this phenomenon, known as P–Δ, the subject is considered in some detail.

The principle has long been known (and often neglected) by most engineers [11.10]. Here, the P–Δ phenomenon is considered through the vehicle of the magnification factor, or MF, as it is reflected in the natural period of oscillation of the structure. The MF is that factor which relates the overturning moment, uncorrected for P–Δ, to the true, corrected overturning moment. That is,

$$M_T = M(MF)$$

where M is the overturning moment not corrected for P–Δ (i.e., the overturning moment as determined by first order, elastic analysis) and M_T is the corrected overturning moment.

Simple methods can be developed for calculating the MF and can relate also the magnitude of the P–Δ phenomenon (i.e., the MF) to commonly used deflection criteria used for the design of high-rise buildings [11.10]. The natural period of oscillation can be shown to be a convenient yardstick for evaluating the importance of the P–Δ phenomenon.

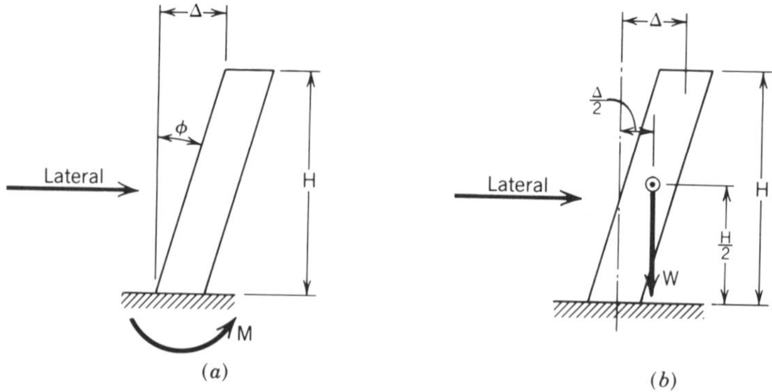

Fig. 11.10 Building drift.

The Magnification Factor

Consider a structure of stiffness K (Fig. 11.10a), where K is defined as that overturning moment producing unit rotation (slope) of the building

$$\tan \phi = \frac{\Delta}{H} \cong \phi \tag{11.3.1}$$

so that

$$K \cong \frac{M}{\phi} \tag{11.3.2}$$

Further, consider a building of linear mode shape; that is, the deflection per unit of height is constant throughout the height. The center of mass is taken at midheight as shown in Fig. 11.10b. Both approximations are generally satisfactory for preliminary design, but can be corrected easily should the need exist.

The initial P–Δ overturning moment, taken at the base, is $W(\Delta°/2)$, where $\Delta°$ is the uncorrected displacement.

But $W(\Delta°/2)$ produces Δ', thus further increasing the P–Δ overturning moment by $W(\Delta'/2)$. But $W(\Delta'/2)$ produces Δ'', and so forth.

Recognizing that the ratio of a Δ to the overturning moment creating that Δ is constant (within the elastic range), it is easy to see that

$$\frac{\Delta°}{M} = \frac{\Delta'}{0.5W\Delta°} = \frac{\Delta''}{0.5W\Delta'} = \frac{\Delta'''}{0.5W\Delta''}, \quad \text{etc.} \tag{11.3.3}$$

and that

$$M_T = M + 0.5W\Delta° + 0.5W\Delta' + 0.5W\Delta'' + \cdots \tag{11.3.4}$$

It follows that

$$M_T = M\left[\frac{1}{1 - \dfrac{W\Delta}{2M}}\right]$$

Substituting $M = K\phi$ and $\Delta = H\phi$,

$$M_T = M\left[\frac{1}{1 - \dfrac{WH}{2K}}\right] \tag{11.3.5}$$

The magnification factor MF is then seen to be

$$\mathrm{MF} = \frac{1}{1 - \dfrac{WH}{2K}} = \frac{1}{1 - \dfrac{\text{total weight (total height)}}{2\text{ (building stiffness)}}} \tag{11.3.6}$$

The MF developed here is taken from concepts associated with the building or structure as a whole. The expression for the MF can be derived also from concepts associated with individual columns in specific structural systems.

The Period of Vibration

The period of vibration is often calculated for structural systems, particularly for those with significant dynamic response to input forces such as wind or earthquake. The period can be developed easily with a hand-calculator by using Rayleigh's method or it can be determined readily by most applicable computer programs. While the period determined by these methods is often not corrected for the P–Δ phenomenon, a simple method for accomplishing that correction is shown here.

Consider a building or structure with a linear mode shape and with W, H, K, Δ, and M as before; g is gravity. Also, rotary energy is neglected. I is the mass moment of inertia about the base:

$$I = \frac{WH^2}{3g}$$

(for uniform cylindrical structures such as a square or rectangular tower of constant cross-section.)

If P–Δ is neglected, the natural period of oscillation, T_o, can be calculated from elementary principles:

$$T_o = 2\pi \sqrt{\frac{I}{K}}$$

Substituting I,

$$T_o = 2\pi H \sqrt{\frac{W}{3gK}} \tag{11.3.7}$$

When considering P–Δ, the building stiffness has to be corrected for the magnification factor MF. That is, the period of vibration, T, corrected for P–Δ, is given by

$$T = 2\pi H \sqrt{\frac{W(\mathrm{MF})}{3gK}} \tag{11.3.8}$$

Therefore,

$$T = \frac{T_o}{\sqrt{1 - \dfrac{WH}{2K}}} \tag{11.3.9}$$

or

$$T = T_o\sqrt{\mathrm{MF}} \tag{11.3.10}$$

It follows that the magnification factor can be stated in terms of the period of oscillation, both corrected and uncorrected for the P–Δ effect. Combining Eqs. (11.3.6), (11.3.7), and (11.3.10),

$$\mathrm{MF} = \frac{1}{1 - \dfrac{3gT_o^2}{8\pi^2 H}} = \frac{1}{1 - \dfrac{1.22T_o^2}{H}} \quad \text{(ft, lb, sec)} \tag{11.3.11}$$

and, including P–Δ,

$$\mathrm{MF} = \frac{1}{1 - \dfrac{3gT^2}{8\pi^2 H(\mathrm{MF})}}$$

or

$$\mathrm{MF} = 1 + \frac{3gT^2}{8\pi^2 H} = 1 + \frac{1.22T^2}{H} \qquad \text{(ft, lb, sec)} \tag{11.3.12}$$

Commonly Used Stiffness and Deflection Parameters

It is common for the designers of tall buildings to find that the uncorrected period of vibration, T_o, is about equal to one-tenth the number of floors. That is,

$$T_o \cong \frac{n}{10} \qquad (n = \text{number of floors})$$

This expression is given commonly in various building codes and standards such as the *Uniform Building Code*.

Taking the story height as 12.2. ft, H becomes $12.2n$. Substituting in Eq. (11.3.11),

$$\mathrm{MF} \cong \frac{1}{1 - \dfrac{1.22T_o^2}{H}} \qquad \text{(ft, lb, sec)}$$

or

$$\mathrm{MF} \cong \frac{1}{1 - \dfrac{n}{1000}} \tag{11.3.13}$$

which shows that the following of this guideline will result in structures wherein the P–Δ effect is small, even for buildings with a large number of floors—say up to 100 stories.

A more direct approach is to limit the period of vibration so that

$$T_o \leq \sqrt{n}$$

Substituting in Eq. (11.3.11) and again taking the story height at 12.2 ft,

$$\mathrm{MF} = \frac{1}{1 - \dfrac{1.22n}{12.2n}} = 1.11 \tag{11.3.14}$$

It is seen that this rule will limit the P–Δ amplification to 11%—a level consistent with good engineering practice.

Some designers establish structure stiffness by limiting the lateral displacement, or drift, under design wind or earthquake loading to some fraction of the structure height [11.10]. These "drift ratios" R are commonly set at $H/400$ or $H/500$.

For a building of height H and width B (perpendicular to the wind pressure w), the base moment is given by $M = wBH^2/2$ and the drift ratio R is $R = \Delta/H \cong \phi$ from Fig. 11.10*a*. Since the structure stiffness K is given by $K = M/\phi$,

$$K = \frac{wBH^2}{2R} \quad \text{or} \quad R = \frac{wBH^2}{2K} \tag{11.3.15}$$

which provides a direct relationship between the drift ratio R and the structure stiffness K.

It is possible, then, to calculate the structure period of oscillation from a known drift ratio, building height, and building weight. From Eq. (11.3.7),

$$T_o = 2\pi H \sqrt{\frac{W}{3gK}} = 2\pi H \sqrt{\frac{WR}{3g\,\dfrac{wBH^2}{2}}} \tag{11.3.16}$$

where $wBH^2/2$ is the base overturning moment due to the lateral pressure w. Correcting for P–Δ, Eq. (11.3.9) gives

$$T = \frac{T_o}{\sqrt{1 - \frac{WH}{2K}}} = 2\pi H \sqrt{\frac{WR}{3g\left(\frac{wBH^2}{2} - \frac{WRH}{2}\right)}} \tag{11.3.17}$$

where $wBH^2/2$ is as defined above.

It is clear, then, that the MF can be developed from a known drift ratio, building height, and building weight. Using Eq. (11.3.6)

$$\text{MF} = \frac{1}{1 - \frac{WH}{2K}} = \frac{1}{1 - \frac{WH}{2}\left(\frac{2R}{wBH^2}\right)}$$

or

$$\text{MF} = \frac{1}{1 - \frac{WR}{wBH}} \tag{11.3.18}$$

The use of the drift ratios, particularly in association with wind loads in the long direction (against the narrow face) of tall buildings, is now examined.

Taking the building density at 12 lb/ft^3 (190 kg/m^3), the wind load at an average of 40 psf (1.9 kPa), and a drift ratio of $H/500$, B = 90 ft (27.4 m), length of building = 300 ft (91.4 m), and H = 800 ft (243.8 m).

From Eq. (11.3.15),

$$K = \frac{40(90)(800)^2}{2(1/500)} = 5.76 \times 10^{11} \text{ lb-ft } (7.81 \times 10^8 \text{ kN·m})$$

$$W = 300(90)(800)(12) = 2.59 \times 10^8 \text{ lb } (1.17 \times 10^8 \text{ kg})$$

$$\text{MF} = \frac{1}{1 - \frac{WH}{2K}} = \frac{1}{1 - \frac{(2.59 \times 10^8)(800)}{2(5.76 \times 10^{11})}} = 1.22 \qquad \underline{\text{MF} = 1.22}$$

With a wind load of 30 psf (1.4 kPa) and a drift ratio of $H/400$,

$$\text{MF} = \frac{1}{1 - \frac{2.59 \times 10^8(800)}{2(3.46 \times 10^{11})}} = 1.43 \qquad \underline{\text{MF} = 1.43}$$

That is, overturning moments due to applied lateral forces are 22 and 43%, respectively, larger than would be calculated if P–Δ were neglected.

Magnification factors of this magnitude, not recognized in design, could result in disastrous consequences, both in structure performance or structure safety, or both. Indeed, at least one major high-rise building was substantially reinforced for precisely this reason.

It is interesting to note that, for structures where the length-to-width ratio is large (10 or more), and using only drift ratio criteria under wind load for establishing structure stiffness, the structure is likely to prove to be unstable under gravity load alone.

As a final word of caution, while most column design methodologies provide some recognition of the P–Δ phenomena, these commonly used methods may not be applicable to cases where the MF exceeds 10 to 12% and will likely not provide at all for moment magnification into girders.

In summary, proper recognition of P–Δ is essential in all structures, but is absolutely vital in tall buildings.

11.3.2 Wind Engineering

Most older high-rise buildings have been designed by intuitive means to resist wind forces, relying heavily on building code requirements and using concepts limited to steady-state forces and response. Wind effects on tall buildings are treated in Chapter 2, and the true complexity of the random and turbulent action of wind on structures was described in Chapter 10. Some additional comments on wind engineering of high-rise buildings are provided here.

The tests of Jensen and Franck [11.11] first demonstrated the importance of the ground roughness in the consideration of the magnitude and distribution of wind pressures on buildings. Designers of more contemporary buildings have taken advantage of these new concepts in wind engineering. A "classical" program in wind engineering involves the following coordinated phases:

1. Environmental study—the gathering and analysis of pertinent data related to wind velocity, turbulence, and the like (and to air temperature, if applicable).
2. Simulation by wind tunnel models—the topographic model to establish the structure of the wind appropriate for the testing of building models; the aeroelastic model to obtain both the static and the dynamic response of the proposed structure; the pressure model to examine both steady-state and fluctuating local pressures and for cross-checking of results; and the plaza and the roof-top models to examine the characteristics of the air flow in order to evaluate human adaptability to the plaza winds and the suitability of the shrubs and trees to be used in the plaza, and in order to evaluate the several problems associated with the operation of aircraft from the roof top.
3. Synthesis of environmental data and the results of the wind tunnel measurements.
4. Recognition in the structural concept of the problems uncovered in the previous phases and incorporation of the capability to accommodate the deflections and motions uncovered in these phases into the structural design.

The reader is referred to Fig. 11.11 which shows something of the flow of turbulent air around a high-rise building as developed from model simulation.

For unusual structures, it is important to recognize that these tests need be carried out in a preliminary manner during the preliminary phases of design.

Detailed descriptions of the techniques involved in the completion of these studies are available [11.12]; see also Chapter 2. Not as well documented is the establishment of criteria of various kinds—particularly for limitations on the steady-state and the fluctuating components of structure motion. In establishing such criteria it need be recognized that the "windiness" of various locations throughout the world is not constant.

Wind speeds in New York, for example, are seen in Fig. 11.12 to be perhaps one-third higher than those of Los Angeles. If identical buildings were constructed in each of these locations, both their steady-state and their dynamic components of motion would vary widely. A "good" structure in Los Angeles may not perform adequately in New York, even if strengthened to resist the higher loads. This follows because the stiffness characteristics of a building are not changed merely by adding strength.

Considering the dynamic components of structure motion, it is useful to reiterate some of the points made in Section 10.2.2 and to recognize that the human perception of that motion is affected by the frequency of vibration, the amplitude of the motion, and the accompanying accelerations [11.13]. The concept of occupant sensitivity quotient [11.14] can be used to relate the human response factor to anticipate structure motions. The probability distribution of peak acceleration is not in itself useful without weighing these factors against the sensitivity of building occupants to these same accelerations.

In conclusion, the acceptance of simple "drift" criteria of the form Δ/H is known to be misleading as an indicator of the expected performance of the structural system in the light of human response to the lateral oscillation under wind excitation. Further, some caution must be followed in assuming that a "proven" building system, found to be satisfactory in one geographic environment, will perform as well in other, windier, geographical areas.

11.3.3 Earthquake Engineering

The importance of seismically induced loads to the design of tall buildings is very dependent on the height of the building. Figures 11.13 and 11.14 demonstrate that, for taller steel-frame buildings, the forces induced by earthquake are substantially smaller than those induced by wind. Here, both base shear and base overturning moment are compared for a building 1000 ft (305 m) in height. It is seen that, for taller buildings, wind loads will dominate both the strength and performance characteristics of the structural frame. This is true for all areas of the world, including the highly seismic areas of the Pacific rim.

Despite the fact that loads are smaller, the basic requirements of good earthquake-resistant design need to be followed in high-rise buildings. The most important aspects are those associated with the need for ductility, the need to properly locate the formation of earthquake-induced plastic hinges and so forth. Fortunately, most of these requirements flow smoothly from considerations of strength.

For lower buildings, the importance of seismically induced loads increases until, for low-rise buildings, they dominate.

Fig. 11.11 Flow around tall buildings.

11.3.4 Commonly Used Structural Systems

With loads measured in tens of thousands of kips, there is little room in the design of high-rise buildings for excessively complex thoughts. Indeed, the better high-rise buildings carry the universal traits of simplicity of thought and clarity of expression.

It does not follow that there is no room for grand thoughts. Indeed, it is with such grand thoughts that the new family of high-rise buildings has evolved. Perhaps more important, the new concepts of but a few years ago have become commonplace in today's technology.

Omitting some concepts that are related strictly to the materials of construction, the most commonly used structural systems used in high-rise buildings can be categorized as follows:

1. Moment-resisting frames.
2. Braced frames, including eccentrically braced frames.

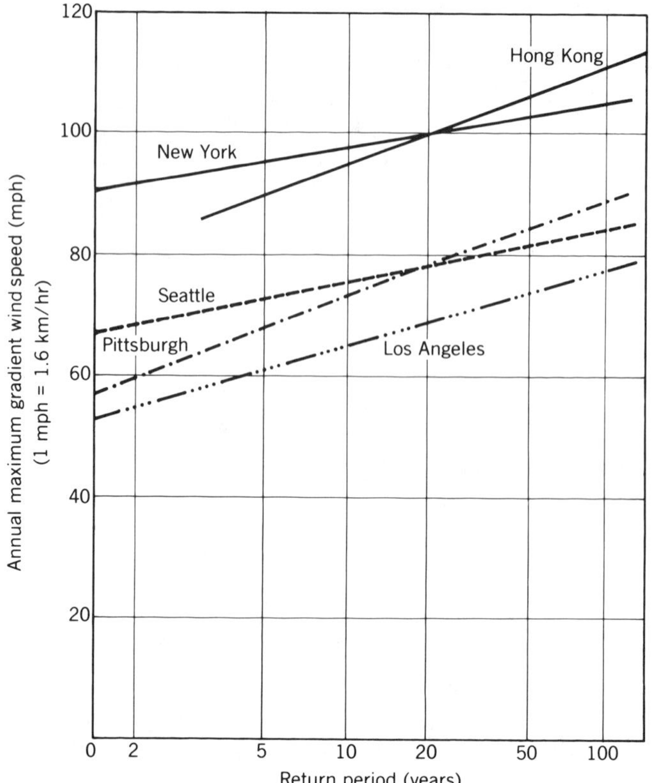

Fig. 11.12 Wind speeds.

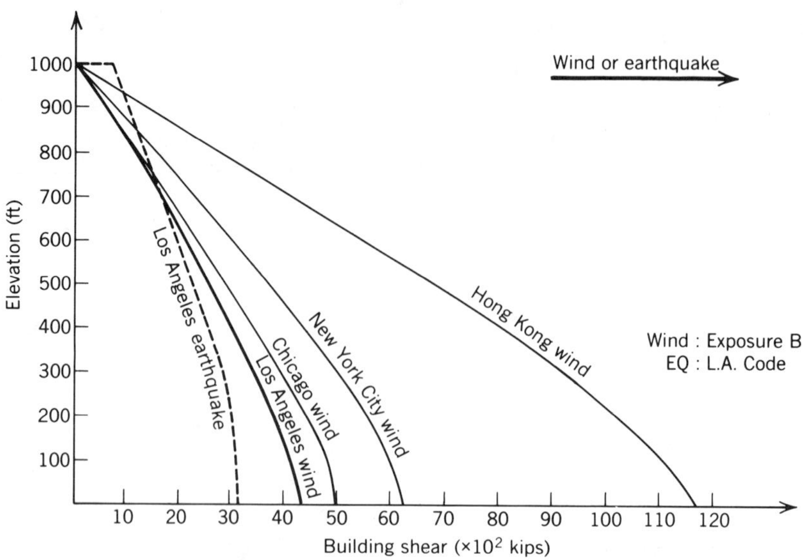

Fig. 11.13 Comparative base shears.

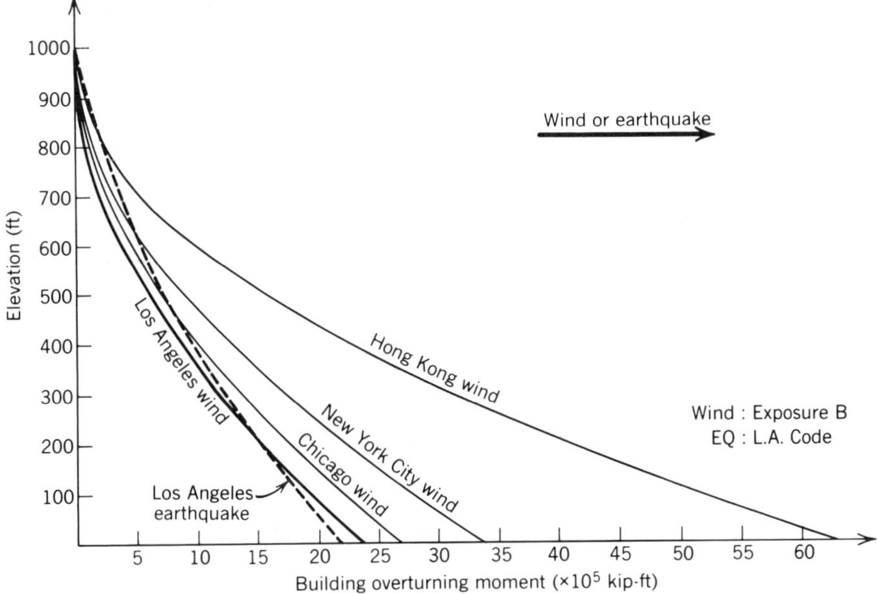

Fig. 11.14 Comparative base overturning moments.

3. Shear walls, including steel plate shear walls.
4. Tubular frames.
5. Tube-in-tube structures.
6. Core-interactive structures.
7. Cellular or bundled-tube systems.

Megastructures and bridging structures are considered later in this chapter.

Particularly with the recent trend toward more complex architectural forms, but in response also to the need for increased stiffness to resist the forces from wind and earthquake, most high-rise buildings have structural systems built up of combinations of frames, braced bents, shear walls, and related systems. Further, for the taller buildings, the majority are composed of interactive elements in three-dimensional arrays.

The method of combining these elements is the very essence of the design process for high-rise buildings. These combinations need evolve in response to environmental, functional, and cost considerations so as to provide efficient structures that provoke the architectural development to new heights. This is not to say that imaginative structural design can create great architecture. To the contrary, many examples of fine architecture have been created with only moderate support from the structural engineer, while only fine structure, not great architecture, can be developed without the genius and the leadership of a talented architect. In any event, the best of both is needed to formulate a truly extraordinary design of a high-rise building.

While comprehensive discussions of these seven systems are generally available in the literature, further discussion is warranted here. The essence of the design process is distributed throughout the discussion.

11.3.5 Moment-Resisting Frames

Perhaps the most commonly used system in low- to medium-rise buildings, the moment-resisting frame, is characterized by linear horizontal and vertical members connected essentially rigidly at their joints. Such frames are used as a stand-alone system or in combination with other systems so as to provide the needed resistance to horizontal loads. In the taller of high-rise buildings, the system is likely to be found inappropriate for a stand-alone system, this because of the difficulty in mobilizing sufficient stiffness under lateral forces.

Analysis can be accomplished by STRESS, STRUDL, or a host of other appropriate computer programs; analysis by the so-called portal method or the cantilever method has no place in today's technology.

Because of the intrinsic flexibility of the column/girder intersection, and because preliminary designs should aim to highlight weaknesses of systems, it is not unusual to use center-to-center dimensions for the frame in the preliminary analysis. Of course, in the latter phases of design, a realistic appraisal of in-joint deformation is essential.

11.3.6 Braced Frames

The braced frame, intrinsically stiffer than the moment-resisting frame, finds also greater application to higher-rise buildings. The system is characterized by linear horizontal, vertical, and diagonal members, connected simply or rigidly at their joints. It is used commonly in conjunction with other systems for taller buildings and as a stand-alone system in low- to medium-rise buildings.

On a larger scale, wherein the bracing work points are several floors apart, the system falls closer to that discussed later as a "megastructure." While the use of structural steel in braced frames is common, concrete frames are more likely to be of the larger-scale variety.

Of special interest in areas of high seismicity is the use of the eccentric braced frame. This system is discussed in Section 11.4.

Again, analysis can be by STRESS, STRUDL, or any one of a series of two- or three-dimensional analysis computer programs. And again, center-to-center dimensions are used commonly in the preliminary analysis.

11.3.7 Shear Walls

The shear wall is yet another step forward along a progression of ever-stiffer structural systems. The system is characterized by relatively thin, generally (but not always) concrete elements that provide both structural strength and separation between building functions.

In high-rise buildings, shear wall systems tend to have a relatively high aspect ratio, that is, their height tends to be large compared to their width. Lacking tension in the foundation system, any structural element is limited in its ability to resist overturning moment by the width of the system and by the gravity load supported by the element. Limited to a narrow width, shear walls need be

Fig. 11.15 AT&T steel plate shear walls.

augmented in some way so as to provide the needed resistance to overturning. One obvious use of the system, which does have the needed width, is in the exterior walls of buildings, where the requirement for windows is kept small.

Structural steel shear walls, generally stiffened against buckling by a concrete overlay, have found application where shear loads are high. The system, intrinsically more economical than steel bracing, is particularly effective in carrying shear loads down through the taller floors in the areas immediately above grade. The system has the further advantage of having high ductility—a feature of particular importance in areas of high seismicity. A typical installation is shown in Fig. 11.15.

The analysis of shear wall systems is made complex because of the inevitable presence of large openings through these walls. Preliminary analysis can be by truss-analogy, by the finite element method, or by making use of a proprietary computer program designed to consider the interaction, or coupling, of shear walls. Chapter 7 contains additional material on structural walls and on shear walls.

11.3.8 Framed or Braced Tubes

The concept of the framed or braced tube erupted into the technology with the IBM Building in Pittsburgh Fig. 11.16, but was followed immediately with the twin 110-story towers of the World Trade Center, New York and a number of other buildings. The system is characterized by three-dimensional frames, braced frames, or shear walls, forming a closed surface more or less cylindrical in nature, but of nearly any plan configuration. Because those columns that resist lateral forces are placed as far as possible from the centroid of the system, the overall moment of inertia is increased and stiffness is very high.

The analysis of tubular structures is done using three-dimensional concepts, or by two-dimensional analogy, where possible. Whichever method is used, it must be capable of accounting for the effects of shear lag.

Fig. 11.16 IBM Pittsburgh.

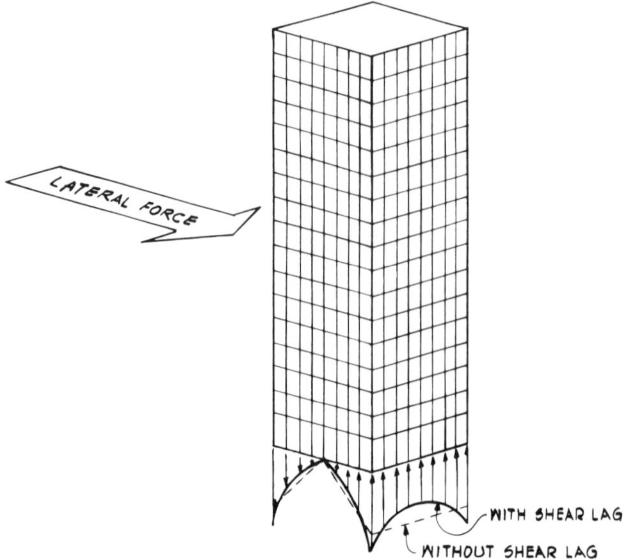

Fig. 11.17 Shear lag.

The presence of shear lag, detected first in aircraft structures, is a serious limitation in the stiffness of framed tubes. The concept, illustrated in Fig. 11.17, has limited recent applications of framed tubes to the order of 60 stories. Designers have developed various techniques for reducing the effects of shear lag, most noticeably the use of belt trusses. This system, shown in Fig. 11.18, finds application in buildings perhaps 40 stories and higher. However, except for possible aesthetic considerations, belt trusses interfere with nearly every building function associated with the outside wall; the trusses are placed often at mechanical floors, much to the disapproval of the designers of the mechanical systems. Nevertheless, as a cost-effective structural system, the belt truss works well and will likely find continued approval from designers. Numerous studies have sought to optimize the location of these trusses, with the optimum location very dependent on the number of trusses provided. Experience would indicate, however, that the location of these trusses is provided by the optimization of mechanical systems and by aesthetic considerations, as the economics of the structural system is not highly sensitive to belt-truss location.

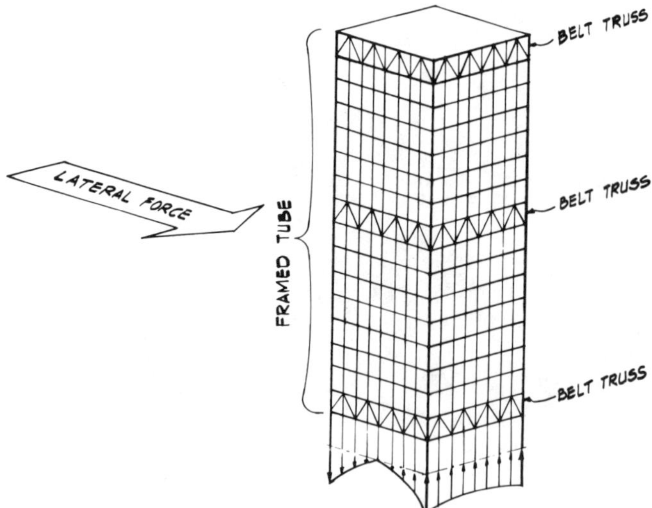

Fig. 11.18 Belt trusses.

11.3.9 Tube-in-Tube Structures

The tubular framing system mobilizes every column in the exterior wall in resisting overturning and shearing forces. The term "tube-in-tube" is largely self-explanatory in that a second ring of columns, the ring surrounding the central service core of the building, is used as an inner framed or braced tube. The purpose of the second tube is to increase resistance to overturning and to increase lateral stiffness. The tubes need not be of the same character; that is, one tube could be framed, while the other could be braced.

In considering this system, it is important to understand clearly the difference between the shear and the flexural components of deflection, the terms being taken from beam analogy. In a framed tube, the shear component of deflection is associated with the bending deformation of columns and girders (i.e., the webs of the framed tube) while the flexural component is associated with the axial shortening and lengthening of columns (i.e., the flanges of the framed tube). In a braced tube, the shear component of deflection is associated with the axial deformation of diagonals while the flexural component is associated with the axial shortening and lengthening of columns.

Following beam analogy, if plane surfaces remain plane (i.e., the floor slabs), then axial stresses in the columns of the outer tube, being farther from the neutral axis, will be substantially larger than the axial stresses in the inner tube. However, in the tube-in-tube design, when optimized, the axial stresses in the inner ring of columns may be as high, or even higher, than the axial stresses in the outer ring. This seeming anomaly is associated with differences in the shearing component of stiffness between the two systems. This is easiest to understand where the inner tube is conceived as a braced (i.e., shear-stiff) tube while the outer tube is conceived as a framed (i.e., shear-flexible) tube. For purposes of illustration, consider two beams as shown in Fig. 11.19:

Beam #1—deep, but with low shear stiffness.

Beam #2—shallow, but with high shear stiffness.

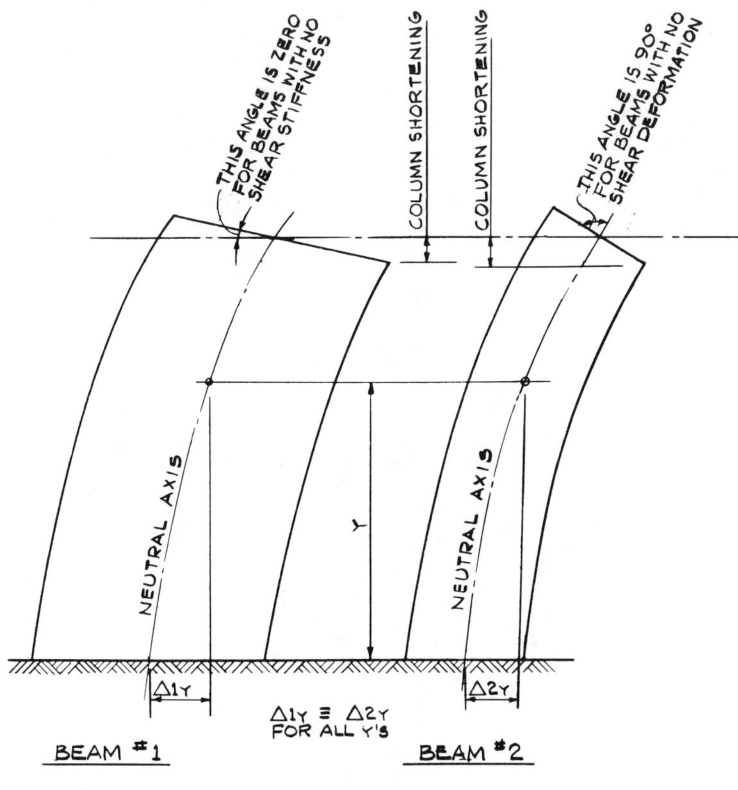

Fig. 11.19 Tube-in-tube concept.

Both beams follow the same elastic curve; the stresses in the columns are proportional to the change in length of the columns. Because of the larger shear deformation of the deeper beam, the flange axial stress is reduced and can be lower than that of the shallower beam. In order to mobilize the inner tube effectively, it is essential to take full advantage of this phenomenon.

Methods of analysis are common with those used for framed and braced tubes . . . STRESS, STRUDL, and the like.

11.3.10 Core Interactive Structures

Core interactive structures are a special case of a tube-in-tube wherein the two tubes are coupled together with some form of three-dimensional space frame. Indeed, the system is used often wherein the shear stiffness of the outer tube is zero. The United States Steel Building, Pittsburgh, Fig. 11.20, illustrates the system very well. Here, the inner tube is a braced frame, the outer tube has no shear stiffness, and the two systems are coupled with a space frame or "hat" structure. The system is illustrated in Fig. 11.21. Note that the exterior columns would be improperly modeled if they were considered as systems passing in a straight line from the "hat" to the foundations; these columns are perhaps 15% stiffer as they follow the elastic curve of the braced core. Note also that the axial forces associated with the lateral forces in the inner columns change from tension to compression over the height of the tube, with the inflection point at about $\frac{5}{8}$ of the height of the tube. The outer columns, of course, carry the same axial force under lateral load for the full height of the columns because the shear stiffness of the system is close to zero.

The space structures or outrigger girders or trusses, that connect the inner tube to the outer tube, are located often at several levels in the building. The AT&T Headquarters [11.15], Fig. 11.22,

Fig. 11.20 U.S. Steel Building.

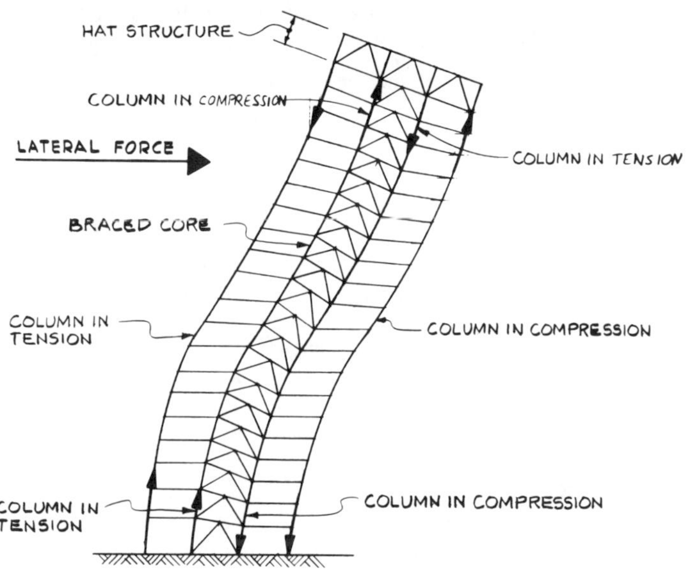

Fig. 11.21 Core-interactive structure.

Fig. 11.22 AT&T Building.

is an example of an astonishing array of interactive elements:

1. The structural system is 94 ft (28.6 m) wide, 196 ft (59.7 m) long, and 601 ft (183.3 m) high.
2. Two inner tubes are provided, each 31 ft (9.4 m) by 40 ft (12.2 m), centered 90 ft (27.4 m) apart in the long direction of the building.
3. The inner tubes are braced in the short direction, but with zero shear stiffness in the long direction.
4. A single outer tube is supplied, which encircles the building perimeter.
5. The outer tube is a moment-resisting frame, but with zero shear stiffness for the center 50 ft (15.2 m) of each of the long sides.
6. A space-truss hat structure is provided at the top of the building.
7. A similar space truss is located near the bottom of the building.
8. The entire assembly is laterally supported at the base on twin steel-plate tubes, because the shear stiffness of the outer tube goes to zero at the base of the building.

11.3.11 Cellular Structures

A classic example of a cellular structure is the Sears Tower, Chicago, a bundled tube structure of nine separate tubes, Fig. 11.23. While the Sears Tower contains nine nearly identical tubes, the basic structural system has special application for buildings of irregular shape, as the several tubes need not be similar in plan shape. It is not uncommon that some of the individual tubes stop at one or more points over the height of the building. Indeed, this characteristic is both one of the strengths and one of the weaknesses of the system.

This special weakness of this system, particularly in framed tubes, has to do with the concept of differential column shortening. The shortening of a column under load is given by the expression

$$\Delta = \sum \frac{fL}{E}$$

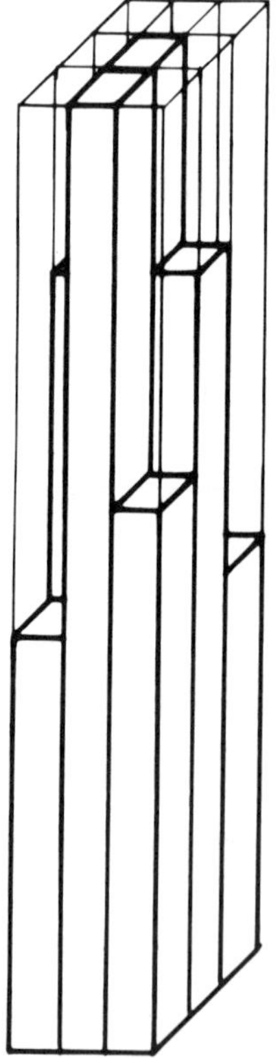

Fig. 11.23 Sears Tower.

For buildings of 12 ft (3.66 m) floor-to-floor distances and an average compressive stress of 15 ksi (138 MPa), the shortening of a column under load is 15(12)(12)/29,000 or 0.074 in. (1.9 mm) per story. At 50 stories, the column will have shortened to 3.7 in. (94 mm) less than its unstressed length. Where one cell of a bundled tube system is, say, 50 stories high and an adjacent cell is, say, 100 stories high, those columns near the boundary between the two systems need to have this differential deflection reconciled.

Major structural work has been found to be needed at such locations. In at least one building, the Rialto Project, Melbourne, the structural engineer found it necessary to vertically prestress the lower height columns so as to reconcile the differential deflections of columns in close proximity with the post-tensioning of the shorter column simulating the weight to be added on to adjacent, higher columns.

11.3.12 Megastructures

The concept of the megastructure was perhaps best popularized by Paolo Soleri [11.16]. The development of such structures has been relatively slow. The evolutionary process can be seen in three buildings:

1. The United States Steel Building, Pittsburgh [11.17], wherein the braced frame is on a three-story module, with the two secondary floors between posted down onto the primary floor, Fig. 11.20.
2. The John Hancock Center, Chicago [11.18], constructed with a braced tube of megastructure proportion, but with the floors supported in a conventional manner, Fig. 11.24.

Fig. 11.24 John Hancock Center. (Courtesy of Ezra Stoller Associates, Inc.)

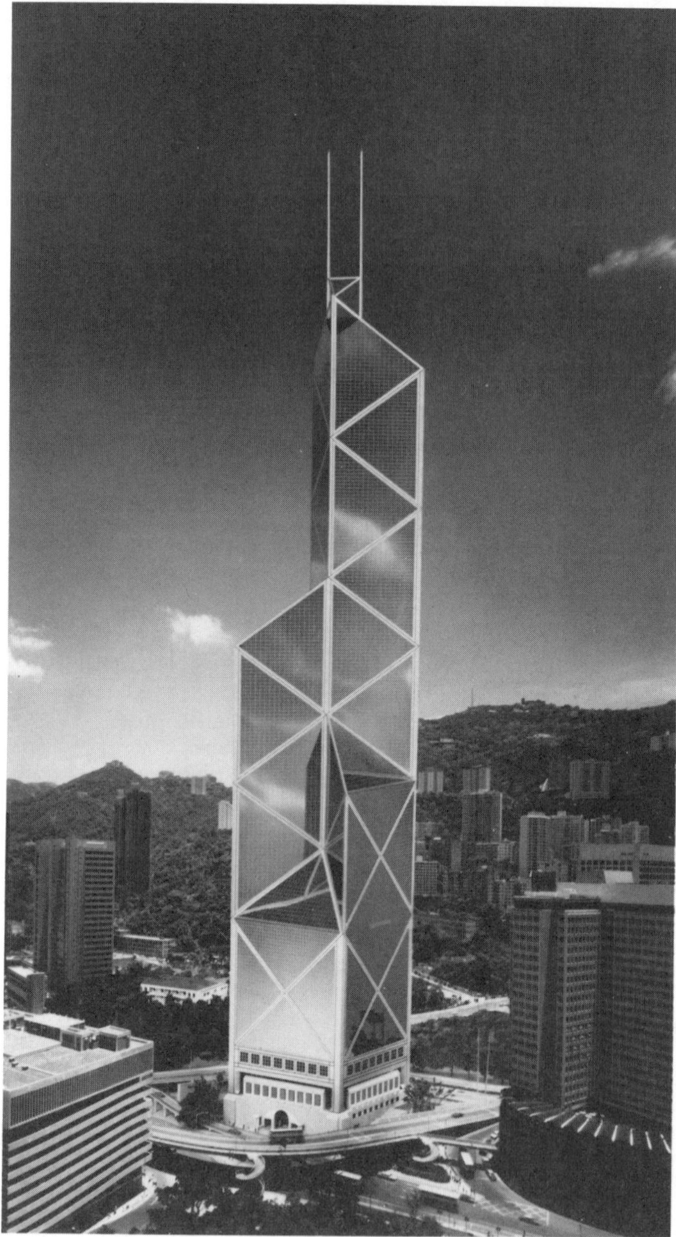

Fig. 11.25 Bank of China megastructure.

3. The Bank of China Building, Hong Kong [11.19], a space-frame braced tube, has the internal loads transferred to the building corners at 13-story intervals, so that almost the entire weight of the building is carried on the four corner columns, Fig. 11.25.

Professor Soleri considered such structures in the framework of enormous building-cities wherein a single "structure" houses all of the activities of a human population center; such structures have yet to be built. The current generation of megastructure responds to functional, cost, and environmental factors that have led to the development of these unusual structures.

Fig. 11.26 Federal Reserve Bank.

Important considerations include:

1. For the United States Steel Building in Pittsburgh [11.17] an unfireproofed steel frame, positioned outside of the building envelope, was placed on a three-story module in order to resolve differences in construction tolerance between structural and nonstructural elements.
2. For the John Hancock Center in Chicago, the vertical stacking of apartments, office, and parking.
3. For the Bank of China, the need to mobilize nearly all of the dead weight of the building in order to remain stable under the overturning action of wind. Here, design wind pressures are in excess of 100 psf (5 kPa) for the structural system.

11.3.13 Spanning Structures

The Federal Reserve Bank in Minneapolis [11.20] , shown in Fig. 11.26, is surely the best known example of a structure spanning a significant distance. Here, twin slip-formed concrete towers support a roof-top box stiffening truss and post-tensioned catenary supporting steel. In the short direction, lateral forces are carried in the twin, cantilevering concrete tubes. In the long direction, lateral forces are carried in the complex interaction of towers, truss, catenaries, and floor construction.

11.4 ENERGY DISSIPATION SYSTEMS

The total dynamic response of building structures stimulated by wind or earthquake is associated with a host of factors but with structural damping high on the list of importance. All structural systems carry some level of intrinsic damping, generally less than 1% of critical damping so long as the structure is deformed in the elastic range. Artificial damping has been achieved in a few buildings.

The World Trade Center in New York Fig. 11.27, was the first building to incorporate damping devices as a part of the structural system. Each of the two buildings uses some 10,000 viscoelastic

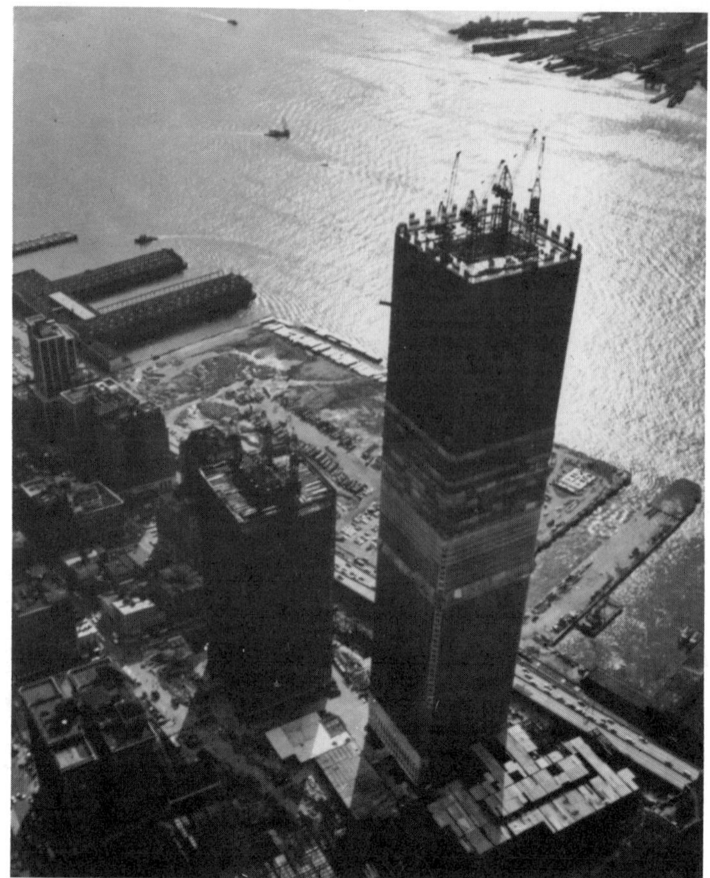

Fig. 11.27 World Trade Center.

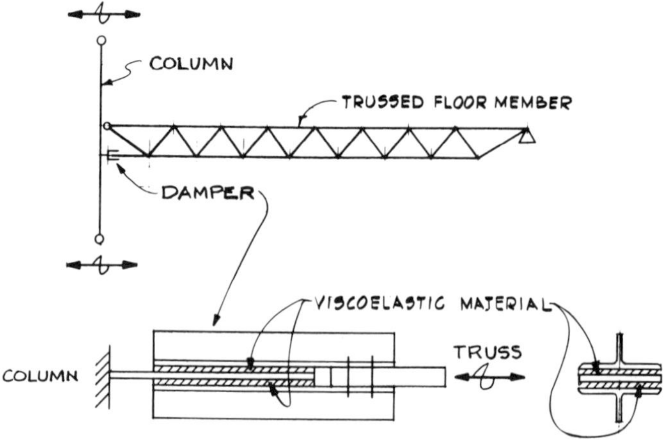

Fig. 11.28 World Trade Center damper.

damping units, making use of materials in shear, manufactured by 3M Company. Columbia Center in Seattle uses similar devices incorporated into a braced tube concept.

These devices operate in a manner not unlike that of a door closer designed to prevent the slamming of simple screen doors. However, in order to dissipate energy, viscoelastic materials stressed in shear are substituted for the air plunger and outlet venturi of the door closer. An example of such a damper is shown in Fig. 11.28.

The viscoelastic materials may be as thin as 0.05 in. (1.2 mm) and axial forces in the damper may be on the order of 50 to 500 kips (200 to 200 kN) or larger. These devices can add 0.5% of critical damping to that which is intrinsic in the building system. That is, with an intrinsic damping of, say, 0.75%, these devices can raise the damping level to 1.25% or higher.

As an added feature, viscoelastic damping units in the floor construction add to floor system damping so as to reduce floor tremor under footfall and other dynamic loading.

A tuned mass damper was used first in the Citicorp Center, New York, Fig. 11.29. The equipment is designed to reduce the swaying motion under wind excitation so as to reduce human perception of that motion. Here an 820,000-lb (372,000-kg) mass, floated on oil bearings, is laterally supported (or driven) by nitrogen springs. Similar dampers were retrofit to the John Hancock Building in Boston, following the addition of core bracing and other work into the completed building. A schematic drawing of such a damper is given in Fig. 11.30.

The design of, or even a good description of these devices, is too complex to include in this handbook. It should be clear, however, that if they are to be used, provision need be made in the preliminary design for the incorporation of such dampers into the building system.

Fig. 11.29 Citicorp Building.

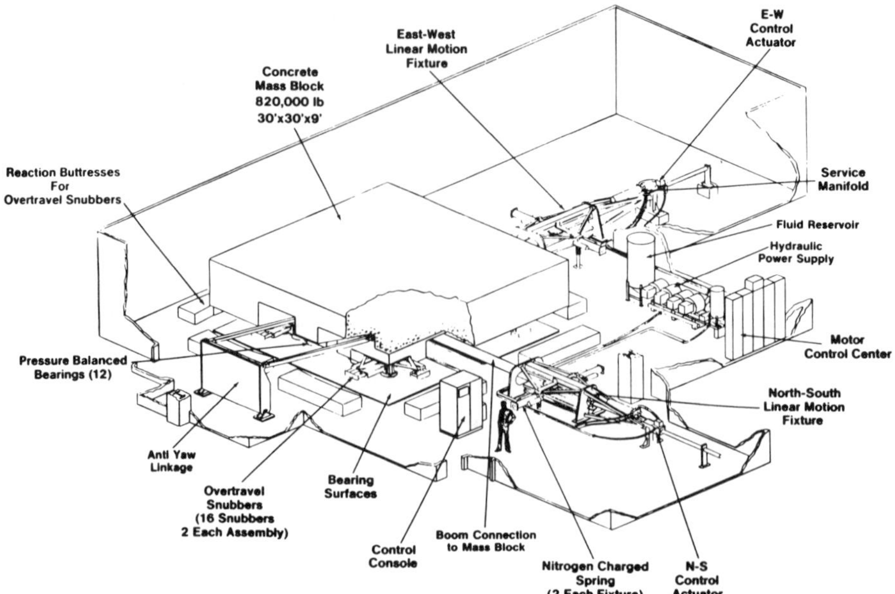

Fig. 11.30 Citicorp Building tuned-mass damper.

Building shape has been modified so as to increase the contribution of aerodynamic damping to that of structural damping. Further, various approaches have been suggested to break up vortex excitation [11.21] so as to reduce the dynamic input from the wind.

Probably the best example of the use of building shape to reduce wind-induced building motion is the United States Steel Building, Pittsburgh [11.17]. For this project, a comprehensive aerodynamic comparison of alternative building shapes was accomplished before a notched corner triangular shape was selected for the design.

Figure 11.31 shows the aerodynamic performance of alternative building shapes. In this study the area, the height, and the density of the various shapes are held constant.

The eccentrically braced frame has been developed to provide increased structural damping under earthquake excitation. Eccentric bracing, of course, is not new, having been incorporated into bracing systems for many years, largely to deal with conflicts between bracing and functional considerations (Fig. 11.32). Here, the eccentricity is an imposition on the structural system. It was tolerated by structural designers because there was no other choice; the resulting bending was considered detrimental to both stiffness and strength.

Contemporary designs now take advantage of these bending stresses in carefully designed configurations. Here, the eccentric joint is designed to deform plastically under the extremes of earthquake loading so as to introduce damping and controlled deformation into the structural system.

11.5 METHODS OF ANALYSIS

As noted at the outset of this chapter, unlike lower buildings, structural systems appropriate to high-rise buildings need be relatively simple and pure in concept. It follows that appropriate analytical systems are likely to be less complex than might be supposed. However, both because of the large number of members involved and because of the need to consider volumetric effects, analytical models are expensive to formulate and can be expensive to execute. Firms designing such buildings need have ready access to electronic computing systems for large-scale analysis, for data handling, and for component design.

11.5.1 Basic Assumptions

In order to make practical the repeated and the iterative preliminary analysis of large-scale systems, it is essential to develop simple and easy-to-modify computer models.

(*a*) Alternative building shapes.

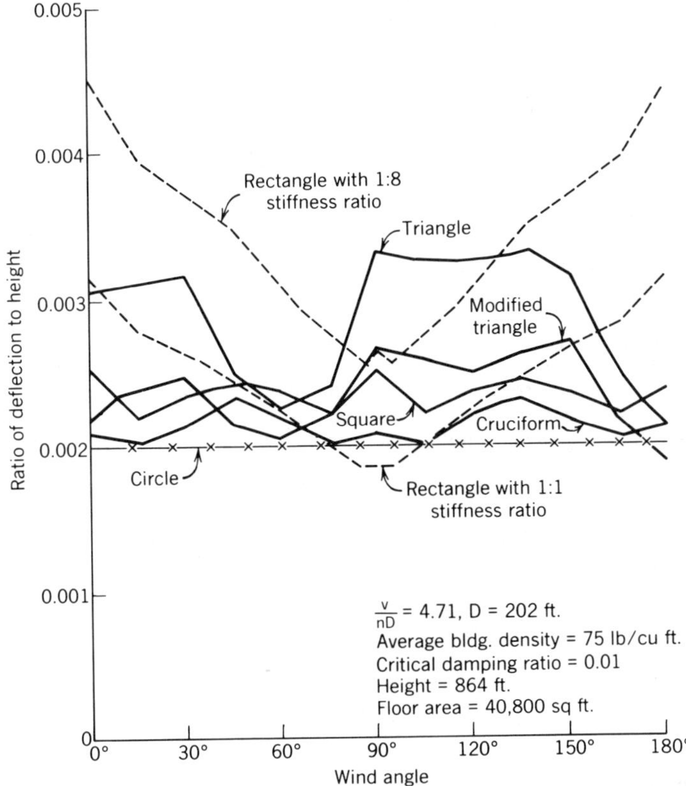

(*b*) Comparison of peak resultant deflections of various building shapes with equal floor area.

Fig. 11.31 Wind tunnel model study for United States Steel Building.

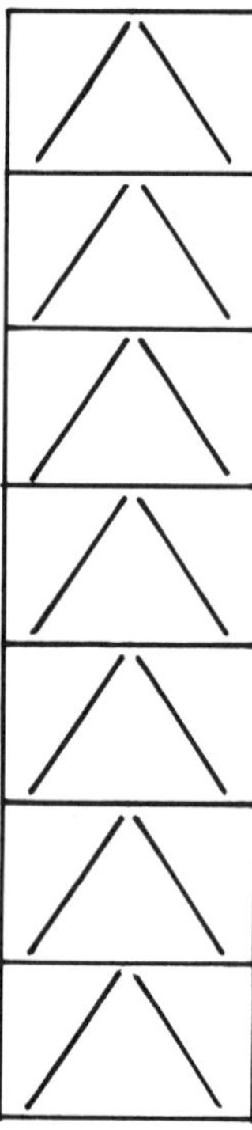

Fig. 11.32 Eccentric bracing.

1. By limiting the impact of P–Δ forces, it is proper to assume that linear, elastic behavior prevails.
2. Consistent with the first assumption, out-of-plane behavior of frames, slabs, walls, and the like can likely be neglected in preliminary design.
3. The deformations of horizontal diaphragms can be considered small and may be neglected but, where the results of analyses indicate to the contrary, the analysis should be corrected.
4. As noted earlier, nonstructural elements do not participate in considerations of strength and, only where specifically designed to participate, do not contribute to structure stiffness.
5. Other normal assumptions, where they will contribute to the speed of data formulation, can usually be accepted.

The classical analytical approaches, making use of the cantilever and of the portal methods, have little meaning in the present-day technology. Except for the educational process, the design of major structural systems, not done by computer, has no place in the technology of today's high-rise buildings.

This does not imply that there is no need for intuitive inputs or that only computer analysis is appropriate. Indeed, many (perhaps most) of the initial calculations will be made by most skilled designers by making use of "back-of-the-envelope" type calculations. Computer simulations, then, look at highly simplified models of the building structure, but use precise (i.e., computer) analyses of those simplified systems.

The need for simplification of analytical models cannot be overemphasized. For example, despite the "apparent" symmetry in the structural systems of the twin towers of the World Trade Center, Fig. 11.27, the two buildings are far from identical and each of the quadrants of any one tower differs from the next. A computer model recognizing the as-built structure needs to consider about 300,000 separate members. For preliminary designs, the designers were able to simplify the system adequately so as to consider only one-hundredth of the total number of components.

11.5.2 Symmetrical Framed and Braced Tubes

In the design of framed and braced tubes, it is important to recognize that primary stresses are associated with warped, two-dimensional elements. It is often possible, then, to resort to two-

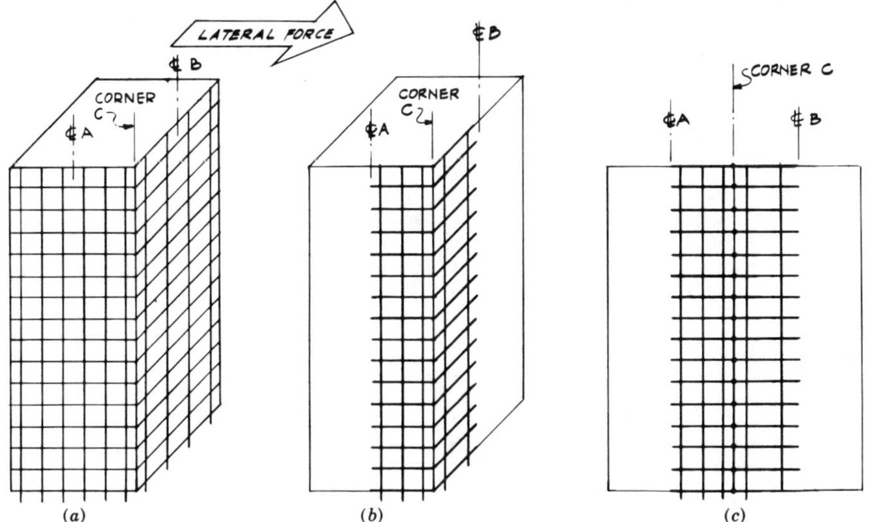

Fig. 11.33 Framed tube: (*a*) full structure, (*b*) three-dimensional model, (*c*) two-dimensional model.

dimensional analogies *without* resorting to approximations associated with the two-dimensional nature of the work. In being able to make use of this two-dimensional analysis, a host of advantages ensue:

1. The likelihood of the introduction of human error is reduced because of the reduction in input data.
2. The cost of the development of the computer program is reduced.
3. The size and the cost of execution of the computer program are reduced.

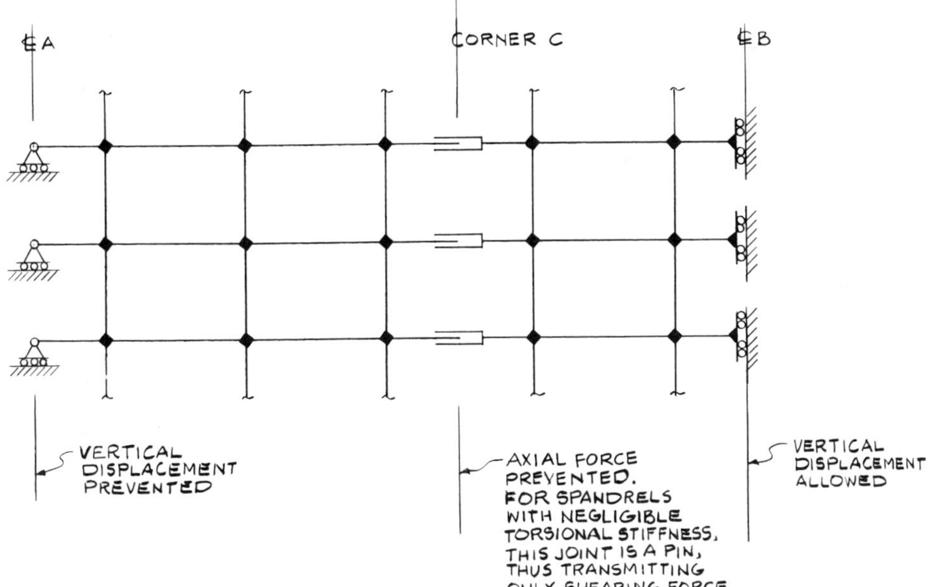

Fig. 11.34 Mechanism at building corner.

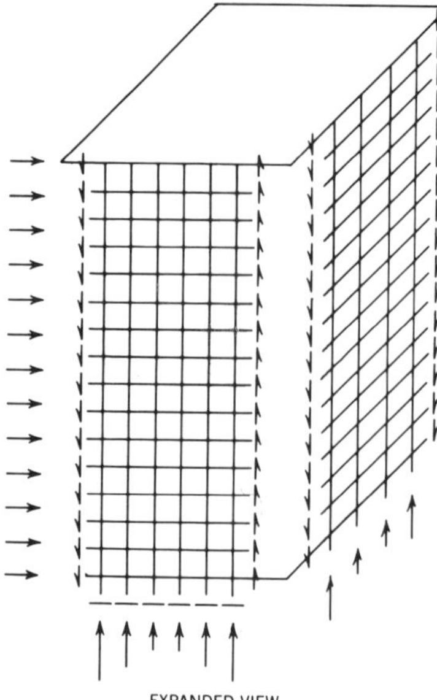

Fig. 11.35 Expanded view.

For purposes of this chapter, a framed tube with two axes of symmetry is examined. The example used has no column in the corner of the tube, but the method works equally well for frames with corner columns and is directly applicable to braced tubes. The basic structural system is shown in Fig. 11.33*a*, a three-dimensional model taking advantage of the planes of symmetry is shown in Fig. 11.33*b*, and an equally precise two-dimensional model is shown in Fig. 11.33*c*. The mechanism assumed in the building corner is shown in Fig. 11.34. A simplified expanded view, showing the load transfer between planes of the tube is shown in Fig. 11.35.

It should be noted that the applied loads will arrive from the floor diaphragm, whether from earthquake or wind, and are *not* applied at the windward side of the building. The structure loading, then, should be by a uniform shear load applied at each node of the model. This method eliminates the buildup of unrealistic axial forces in the spandrels which would occur if the applied loads were taken as shown (literally) in Fig. 11.35.

While only one example of techniques suitable for the simplification of analysis is presented, the ingenuity of designers is endless in conceiving others. In attempting these simplifications, designers are cautioned that the cost of computer time is generally modest when compared to the cost of engineer's hours required to formulate and to reformulate the needed computer programs. Cost-saving measures, then, need focus on the cost of engineers and not the cost of computers.

SELECTED REFERENCES

11.1 Council on Tall Buildings. *Planning and Design of Tall Buildings,* Monograph in 5 volumes. New York: American Society of Civil Engineers, 1978–1981.

11.2 "Science and the Future, Architecture and Civil Engineering," *Encyclopaedia Britannica,* 1984.

11.3 Council on Tall Buildings. "Structural Systems," *Planning and Design of Tall Buildings,* SC-11 (John V. Christiansen, Chairman). New York: American Society of Civil Engineers, 1980.

11.4 AISC. *Code of Standard Practice for Steel Buildings and Bridges.* Chicago: American Institute of Steel Construction, 1976.

11.5 ACI. *Standard Tolerances for Concrete Construction and Materials* (ACI 117-81). Detroit: American Concrete Institute, 1981.

11.6 *Building Code of the City of New York.* New York: Van Nostrand Reinhold, 1970.

11.7 NRCC. *National Building Code of Canada* (NRCC No. 17303). Ottawa, Canada: National Research Council of Canada, 1980.

11.8 J. Charles Rathbun. ''Wind Forces on a Tall Building,'' *Transactions ASCE,* **105,** 1940, 1–41.

11.9 AISC. *Commentary on the Specification for the Design, Fabrication and Erection of Structural Steel for Buildings.* Chicago: American Institute of Steel Construction, 1978.

11.10 Leslie E. Robertson, ''Le Phénomene P–Δ Évalué Sur la Base de le Période Fondamentale D'Oscillation,'' *Construction Métallique,* No. 3, 1981.

11.11 M. Jensen and N. Franck. *Model-Scale Tests in Turbulent Winds,* Part II. Copenhagen: The Danish Technical Press, 1965.

11.12 Leslie E. Robertson and Peter W. Chen. ''Effects of Environmental Loads on Tall Buildings,'' *Proceedings of U.S.-Japan Research Seminar,* Honolulu, Hawaii, 1970.

11.13 Peter W. Chen and Leslie E. Robertson. ''Human Perception Thresholds of Horizontal Motion,'' *Journal of the Structural Division,* ASCE, **98,** August 1972 (ST8), 1681–1695.

11.14 ''Limitations on Swaying Motion of Tall Buildings Imposed by Human Response Factors,'' Sydney, Australia: *Proceedings of the Australian and New Zealand Conference on the Planning and Design of Tall Buildings,* 1973.

11.15 AT&T Corporate Headquarters, *Architectural Record,* October 1980.

11.16 Paolo Soleri. *Arcology, The City in the Image of Man.* Cambridge, Massachusetts: MIT Press, 1969.

11.17 *The Steel Triangle.* United States Steel Corporation, ADUSS 88-4431-0, 1969.

11.18 John Hancock Center, *AIA Journal,* October 1980.

11.19 The Bank of China, *Architectural Record,* September 1985.

11.20 Leslie E. Robertson. ''Una Cellula Urbana del Futuro,'' *L'Industria Delle Construzioni,* 1970.

11.21 Leslie E. Robertson, ''Heights We Can Reach,'' *AIA Journal,* January 1973.

CHAPTER 12

PRELIMINARY DESIGN OF SINGLE-STORY OPEN-SPACE BUILDINGS

FRITZ KRAMRISCH, DR. ENG., P.E.

Retired Chief Civil Engineer
Albert Kahn Associates, Inc.
Architects/Engineers
Detroit, Michigan

Former Lecturer, Lawrence Institute of Technology
Southfield, Michigan

Former Lecturer, Arizona State University
Tempe, Arizona

12.1 INTRODUCTION

Many designers consider it practical to copy the structural design of an existing structure built for a similar purpose. Such practices have led to the repeated use of old-fashioned design schemes and have contributed greatly to earning the engineer the reputation of being over conservative. Designers must keep in mind that technology is continuously advancing, that economy is always changing, and that the expectations regarding the functions of a building, that is, the requirements for strength, safety, and overall performance, are becoming more and more pretentious. If one wants, therefore, to have a reasonably up-to-date design, an old scheme, as inviting as it may appear, cannot be merely copied; instead the more important design criteria must be reevaluated and most of the other ones must be reviewed every time to keep abreast with the latest developments. An actual design is the end result of a multitude of decisions and, therefore, quite a complex thing. It can, however, in most cases be broken down in such a way that minor decisions can be made ahead of time to prepare the basis for making the major decisions on the selection of the most suitable building material and the choice of the most appropriate design scheme.

This is not such a difficult task, as formidable as it may appear at the beginning. It needs an analytical mind, a good knowledge of the engineering sciences, a practical sense, an overall feeling of responsibility for the safety of everyone connected with the structure, and an understanding of economy.

In the exact solution of some of the more involved design problems and especially in the speedy execution of comparative estimates, a computer can be of great help. It will, however, never replace the intuition and ingenuity of the individual designer whose creative mind will again and again successfully come through with flying colors.

12.2 SELECTION OF MATERIAL

One of the first decisions an architect or engineer has to make is the selection of the type of material to be used for the construction of the building to be designed; that is, whether structural steel, reinforced concrete, or wood is to be used for the structural skeleton. This decision, which is of paramount importance for the proper functioning of the structure, is not always made properly. One should think that the expected degree of permanency, the intended purpose, the architectural appearance, the simplicity of fabrication and speed of erection, the geographic location and cost of transportation, the condition of the subsoil, and the prevailing economy (all items that are discussed later) should be the governing factors in making such an important decision; but this is not always true. Often it is simple personal preference for the one or the other material which makes the designer think (perhaps subconsciously) in one direction only. Whether it is the material that is most familiar to the designer, whether it is the material in which the designer has the greatest practice and experience, or whether it is the material in whose promotion the designer has some special interest; whatever it may be, it may constitute the basis for a preference which sometimes may even degenerate into a prejudice. Such an approach is inappropriate, not only for the successful outcome of the design, but also for the designer's personal reputation, and for that of the whole profession. Many structures can be built satisfactorily from more than one material and some even from all three materials under consideration. However, the use of any one of the three materials will unavoidably be associated with certain advantages or disadvantages. It is the duty of the architect or design engineer to inform the owner of these conditions, and only after a proper evaluation of these items, the designer may feel justified in the recommendation to the owner.

12.2.1 Degree of Permanency

It is primarily up to the owner to decide if the structure should serve only on a temporary basis or if it must be used for many years. Wood construction tends to be easier to assemble and to disassemble and is perhaps better suited for temporary structures than other materials. For this reason it is often used for such purposes, especially in areas where it is plentiful. This does not, however, mean that wood cannot be used for permanent structures as well.

All permanent structures require maintenance. Reinforced concrete requires probably the least amount of maintenance after completion of construction; structural steel requires regular rust removal and painting of exposed surfaces. Where steel is embedded in concrete, it requires no maintenance; but where it remains exposed in an inaccessible location, it should receive an asphaltic coating or be galvanized for protection from corrosion. Wood requires occasional repainting and sometimes replacement of certain members that have been distorted or cracked from excessive moisture or dryness.

12.2.2 Purpose of Structure

The intended use and purpose of a building has an important role in the decision regarding the type of material to be used. The following list gives a few typical examples of structures that should preferably be made of a particular material. There are, however, no definite lines of demarcation and there is probably not a single structure in these groups which has not already been built of one or the other materials. The purpose of this list, therefore, is to serve primarily as a guide, but not as a limitation. All structures are of the single-story, open-space type.

1. *Structural steel.* Buildings for light, medium, and heavy industrial work, with and without such items as hoists, hanging loads, and craneways; assembly buildings; warehouses; larger stores and office buildings; large enclosures; aircraft hangars; and convention halls.
2. *Reinforced concrete.* Buildings for light and medium industrial work, mostly with floor conveyors and with few hanging loads; recreational buildings; and utility structures.
3. *Wood.* Sawn timber for workshops, utility structures, and residential buildings; glued laminated for churches, schools, recreational buildings, offices, and stores; shear-connector trusses and rafters for workshops, assembly halls, recreational buildings, aircraft hangars, and residential buildings.

12.2.3 Architectural Appearance

For factory buildings, warehouses, and utility structures, the architectural appearance will be integrally related to the structural characteristics of the building; but where the building is a church, a recreational building, or a store, the architectural expression may be of prime importance and the structural design will have to be adjusted to suit architectural requirements. At any rate, the most successful overall design will be obtained if the architect and structural engineer collaborate from the beginning, to produce a building that is at the same time structurally functional and architecturally pleasing.

12.2.4 Simplicity of Fabrication and Speed of Erection

Simplicity of fabrication and speed of erection are two characteristics that have a decisive influence on the selection of the building material, although in a somewhat indirect way. In selecting a particular structural design scheme, the designer is often induced to use also the material that is most suitable for it. In other words, competitive groups for the various materials will utilize the claimed superiority of certain design schemes to promote the use of their material.

Presently, the cost of the field work constitutes an important part of the overall cost of the structure. Designers have done their best to simplify fabrication and speed up erection by shop fabricating as much of the structure as practical and thus reduce the field work to mere assembly of the prefabricated items.

Structural steel has led the way in this respect by fabricating items as large as railroad or highway clearances would permit. At the same time erection equipment has been developed to such a degree that crane-lifting capacity seldom controls. Trusses are shop fabricated to depths of 12 ft (3.7 m) and to lengths such that a 120-ft (37-m) span requires only one field splice. Welding is used in the fabricating shop and high-strength bolts are used in the field, primarily because quality control is easier to obtain for field bolting than for field welding. Steel joists are available in a variety of depths, weights, and spans. Since they are manufactured on a standardized basis, they are becoming more and more competitive with deep rolled beam sections. Although structural steel is paid for by the pound, it is frequently economical to waste some material if simplification of details or a speed up in erection can be achieved. This is particularly true in the United States and in most countries of Western Europe; however, it may not be applicable elsewhere where the material-to-labor cost ratio is different. In each case, the estimator's advice should be sought before a decision is made.

In reinforced concrete design, schemes employing formwork that can be rented and reused, such as flat slabs and joist floors, are more frequently used than beam-slab construction that requires new individual forms for each job. In many cases, metal decking used as permanent slab form is competitive with removable wood forms. Here as well, the trend to prefabrication has brought about a great development not only in the use of secondary building components, but also for principal items of construction. Entire buildings consisting of columns, beams, roofs, and wall panels can be prefabricated and erected. The limits to such endeavors are set by the means and methods of connecting the various separate pieces and transforming them into one monolithic structure. This is not always necessary and in some cases designers are satisfied with having the prefabricated roof units simply rest on masonry walls or on similar supports. However, where monolithic action is mandatory, it may sometimes be quite tricky not only to develop a detail that is able to transfer all particular forces, but also to execute it properly; it has, however, successfully been done in many instances.

Another item that has to be given close attention by the designer is the use of high-early-strength cement. Contractors are understandably interested in the early removal of forms in order to reuse them elsewhere. The cost of the entire structure can be influenced greatly by such an undertaking. Climate, time of year, mix design, type of construction, eventual reshoring, and other items have to be taken into careful consideration because the entire structure and all people working on it may depend on its safety.

Timber design and construction practices have also managed to develop shop fabrication to a high degree even though skilled carpentry in the field is in many places still available, especially where timber construction is more commonly employed. Trussed rafters and trusses using shear connectors are predominantly fabricated in shops to sizes as large as shipping limits permit. Glued laminated beams and arches must be made in shops where all necessary jigs are available and where strict moisture control, which is of great importance for this kind of fabrication, can be enforced. All that remains to be done in the field is to erect the large pieces and to install the smaller interconnecting members, such as beams, joists, and sheathing.

Summarizing, in order to produce an economical building construction, the designer must strive to arrive at a design concept that is most simple in its configuration and most economical in its requirements for fieldwork, regardless of the selected kind of material.

12.2.5 Geographic Location and Cost of Transportation

The geographic location of a building, and in certain cases its particular location or lack of accessibility, can influence the selection of a building material greatly. In areas where a certain building material is plentiful, there will be also a greater number of shops experienced in the fabrication and erection of this material; consequently, there will be also a sufficient amount of properly trained labor available. All these factors will contribute to make a structure using this particular material less costly, and thus more competitive. At the same time, the structure will probably be of a higher overall quality because of the better workmanship available in this location. This makes timber construction more commonly used in the vicinity of forested areas, structural steel in the vicinity of rolling mills and large industrial plants, and reinforced concrete in areas that are somewhat remote from heavy industrial areas but close to sources of good aggregate and cement. The natural reason for this distribution lies not only in the availability of the material itself, but also in the extra costs involved in bringing other materials to this location. In large cities or commercial centers there will, in all probability, be all kinds of materials available; but not necessarily on a competitive basis. In places with difficult access, such as in mountainous terrain or in underdeveloped areas, availability and transportation may be the determining factors. Using these factors, the evaluation of the economy of a structure has to be done with care by people who are experienced in this field and who will give conscientious attention even to the smallest detail. Miscalculation in the cost of transportation is an easily made mistake, which is hard to correct, and has driven many architects and contractors into costly debacles.

12.2.6 Condition of Subsoil

The selection of the most suitable building material is influenced indirectly by the type of the subsoil and by the choice of the design scheme that is most appropriate for the particular soil encountered at the site. Firm soils have practically no influence on the selection of the building material and design scheme; they all will work satisfactorily. Softer soils, however, or sites with variable subsoil conditions present a greater chance for differential settlements and require more flexible design schemes. Under such conditions, timber and structural steel are the preferred materials because they permit a better use of flexible connections. In other words, materials and design schemes that lend themselves to use simply supported beams or seated connections will be best. Industrial buildings are most affected by such restrictions because industry usually acquires large parcels of cheap, leftover land that could not be used for any better purpose.

12.3 SELECTION OF STRUCTURAL DESIGN SCHEME

Every structure was originally designed and built to serve a certain purpose. This does not necessarily mean that each building is presently serving the purpose for which it was originally designed, because buildings change ownership, and even if still in the hands of the original owner, the use of the building may have changed many times since it was built. This should, however, not discourage the engineer from designing a new building to suit the very purpose for which it is intended. To this end, certain decisions as discussed later must be made, usually in collaboration with the architect and/or owner.

One of the basic decisions to be made is whether the total area of the building or structure has to be unobstructed or whether intermediate columns can be tolerated. After this decision has been made, all further decisions depend primarily on the maximum span (distance between supports) in

Table 12.1 Relation Between Type of Structure, Span and Design Scheme in Direction of Main Span

No.	Type of Structure	Main span L		Beams	Joists	Continuous Beams or Cantilever Beams	Frames	Trusses		Special Design
		ft	m					Shop Fabricated	Field Assembled	
1	Apartment houses, office buildings, warehouses, stores; no suspended loads	≤40	≤12	X	X					
2	Workshops, light manufacturing buildings with suspended loads	≤40	≤12	X						
3	Garages, stores, public buildings, gymnasium courts, warehouses; no suspended loads	40 to 60 60 to 100	12 to 18 18 to 30		X X		X			
4	Medium heavy manufacturing buildings with suspended loads	40 to 100	12 to 30			X		X		
5	Heavy manufacturing buildings, machine shops; with suspended loads and craneways	40 to 120	12 to 36					X		
6	Airplane hangars, convention halls, assembly plants	120 to 300	36 to 90						X	X
7	Tall buildings, stacker housings, special structures	Varies	Varies				X	X		

the main direction of the building, whether the supports are continuous bearing walls or isolated columns. This problem, of course, can be fundamentally different if the designer has a structure that is covering only one bay or if the structure extends over several bays of approximately equal width. The magnitude of spans and loads are of greatest importance for the selection of the type of structural design to be used. This is not only structurally true but also economically important, because the magnitude of the column spacing has probably the greatest influence on the cost of the structural system. The most common types of structures are divided into several basic types, given in Table 12.1. The list is not complete, but it should enable the designer to locate a type or group that is close enough to be applicable to the type of structure about to be designed. The various spans indicated in the table are by no means intended to represent "limitations" but simply magnitudes that past experience has shown to be safe and appropriate for the various types of structural designs.

Table 12.2 is intended to be used primarily as an aid in selecting the proper type of member in the direction perpendicular to the main span; that is, the type of member to be used for girders, carrying members, and similar.

12.3.1 Column Spacing or Span

Rolled Wide-Flange Sections

For column or girder spacings of up to 40 ft (12 m), rolled wide-flange beam sections (W shapes) will unquestionably provide the most economical roof framing for all single-story open-space buildings. The *AISC Manual* [12.5] contains standard sections having various weights and depths from which the designer can select the most economical section for each particular span and loading condition. Toward the upper span limit of 35–40 ft (10–12 m), heavy sections are indicated, but may still provide reasonable economy. For column spacings in a rectangular array it is usually advantageous to place the girders in the direction of the larger span as shown in Fig. 12.1*a*. The transverse beams can either be set on top of the girders or be framed into the girder webs when headroom is limited. This is also done where greater stiffness is desired, as for machine- or fan room floors. The first method is less expensive and is preferred wherever possible. In some cases it can be advantageous to reverse the direction of the framing and place the girders in the direction of the shorter span and the beams in the direction of the longer span as shown in Fig. 12.1b. If flexibility is not an essential requirement, continuity or the cantilever beam design can be applied to the design of the girders to reduce the structural steel tonnage and, consequently, the total cost. If also applied to the roof beams, the savings are often less significant because of the extra costs involved for the great number of field splices or hinges.

Table 12.2 Type of Support in Direction Perpendicular to Main Span

Type of Structure from Table 12.1	Span Perpendicular to Main Direction		Continuous Walls	Beams	Joists	Continuous Beams or Cantilever Beams	Frames	Trusses		Special Designs
	ft	m						Shop Fabricated	Field Assembled	
1, 2	≤40	≤12		X						
1, 2			X			X				
3, 4, 5	40–60	12–18				X		X		
3, 4, 5	60–120	18–36						X		
6	≥120	≥36							X	
7	Varies	Varies					X	X		X

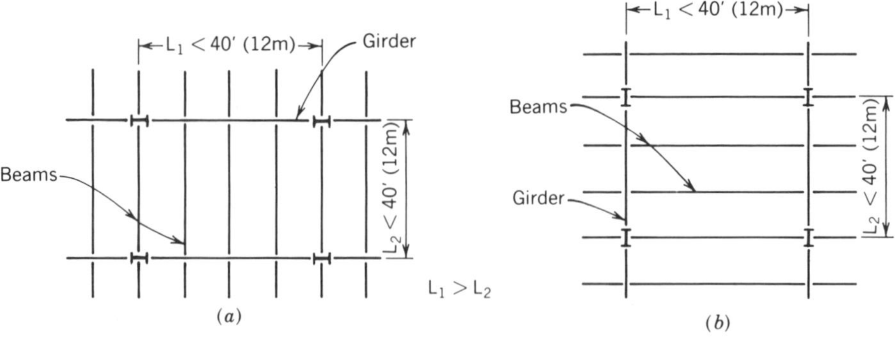

Fig. 12.1 Framing plan using rolled wide-flange beam sections spanning (*a*) the short and (*b*) the long direction.

Steel Joists

When the column spacing in one direction exceeds about 40 ft (12 m), the selection of rolled wide-flange beam sections becomes uneconomical because overly deep and heavy members are needed. Such column spacings can be framed by using steel joists in the direction of the longer span and rolled wide-flange girders in the direction of the shorter span, as shown in Fig. 12.2*a*. Joists can also be used economically for lesser spans down to a minimum of about 25 ft (7.5 m) if the loads are small and uniformly distributed. Using joists with underslung ends will further help to reduce the depth of construction (see Fig. 12.2*b*). In such cases, joists with bottom chords extending over the full length are only used at the column lines for lateral stiffness. Joists can also be placed directly on masonry walls or other continuous supports as indicated in Fig. 12.3.

Roof structures made of steel joists are well suited for a support of uniformly distributed roof loads, suspended ceilings, ducts, and similar; but they should not be used for the support of hoists, conveyor lines, and similar devices because the shear capacity of joists is often insufficient to sustain concentrated loadings and their lateral stability is small in spite of the frequent lines of bridging required by the *AISC Specification* [12.6]. It is good practice to check the shear capacity of a joist which is given in the tables when selecting a joist section from the *AISC Manual* [12.5] or from any other applicable trade catalog.

Where large concentrated loads or heavier uniformly distributed loads must be provided for, the designer should use trusses at the column lines placed in the direction of the longer span as shown in Fig. 12.4. Where such trusses are used to support beams or joists, they require either a panel point at each location where a cross member is supported or a flexurally stiff top chord section to transfer all intermediate loads into the panel points.

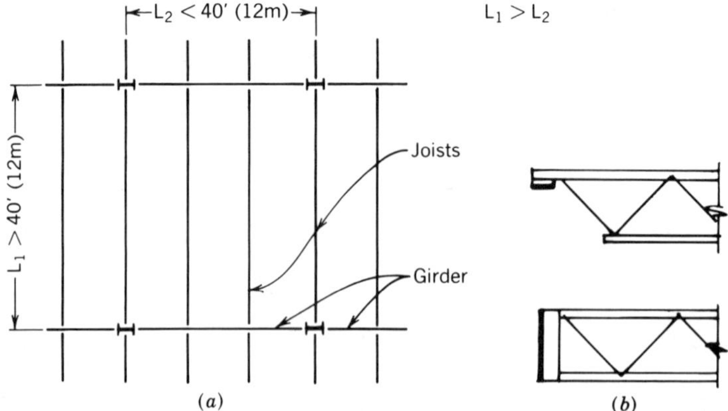

Fig. 12.2 Framing plan using rolled girders and steel joists. (*a*) Plan. (*b*) Joists.

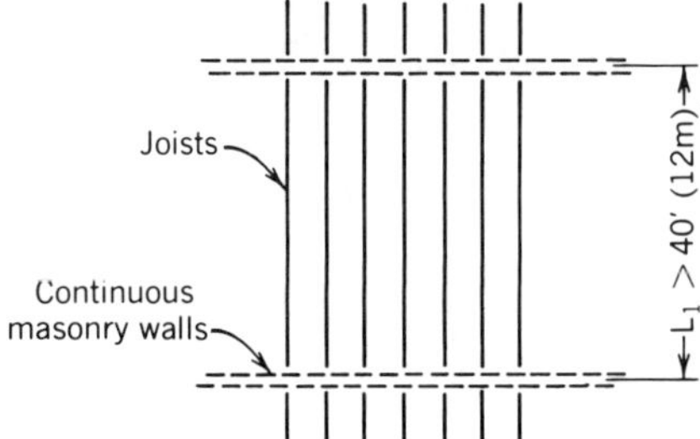

Fig. 12.3 Joists framing between masonry walls.

Shop-Fabricated Trusses

If the column spacing in both directions exceeds about 40 ft (12 m), but is not larger than 120–140 ft (36–42 m), shop-fabricated trusses can be used to best advantage to support the roof beams. The reasons for these span limitations are as follows: The truss depth at midspan is usually selected to vary between 1/8 to 1/12 of the span, most commonly using about 1/10. Applying this rule to the lower limit of span length, the overall depth (also called out-to-out or back-to-back) is about $3\frac{1}{2}$–4 ft (1–$1\frac{1}{4}$ m) which is considered to be the minimum for practical construction. The upper limit is governed by railroad clearance restrictions which do not permit shipments of units deeper than 12 ft (3.7 m). Highway clearances are not so tightly controlled and some overpasses provide greater clearance than others. Special routing has often helped to ship larger units by truck than are ordinarily permitted by railroad. Cases where the tires of low-boy trailers had to be partly deflated to let a unit move under an overpass are not uncommon. For very large trusses, all possibilities of transportation (including by barge, if feasible) must be carefully investigated before the actual design is undertaken.

The framing scheme shown in Fig. 12.5 subdivides the shorter column spacing L_2 into two or more bays to reduce the span and the weight of the roof beams. The intermediate roof trusses are supported by carrying trusses (also called jack trusses), which adds a third type of member to the framing system. These trusses have a dual purpose: first, they carry the intermediate roof trusses, but in addition, they act also as roof purlins.

Since the weight of the long-span purlins would constitute a sizable portion of the total steel weight of the roof construction, it is advantageous to introduce intermediate roof trusses to reduce the span of the roof beams. It is customary, therefore, to place roof trusses at intervals of 18–25 ft (5.5–7.5 m) for which spans the rolling mills (see *AISC Manual* [12.5]) offer a wide selection of economical wide-flange sections. To this end the shorter column spacing is subdivided in such a way that each space is approximately within the above limits. Subsequently, the panel-point spacing of the carrying trusses has to be correlated with the spacing of the roof trusses in a way similar to the panel-point spacing of roof trusses being governed by the desirable purlin spacing. The decision on the most economical spacing of the roof trusses usually depends on a comparative estimate. Sometimes one roof truss more per bay with shorter span and lighter weight purlins may be more advantageous than the saving of a roof truss at the expense of making the purlins and trusses longer and heavier. Sometimes the opposite is true. In every case, the steel tonnage, number of units, amount of shop fabrication, transportation, field work, and ease of erection have to be taken into consideration to arrive at the most economical result. The same evaluation procedure has to be repeated every time because the results are not consistent.

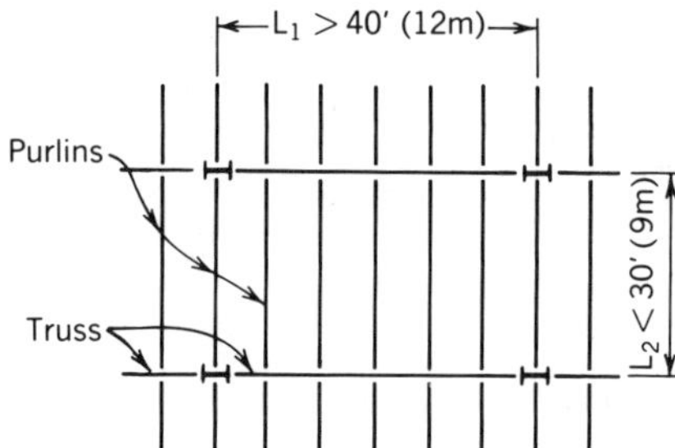

Fig. 12.4 Rolled beams framing between trusses.

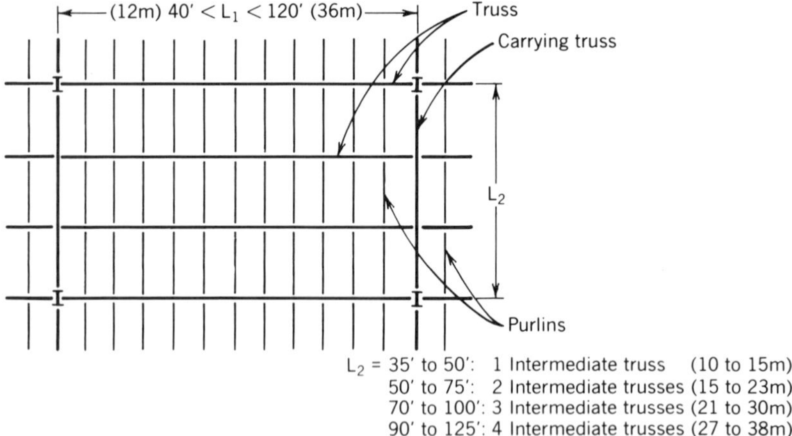

Fig. 12.5 Typical beam framing between intermediate roof trusses.

If the structure in the direction of the long span consists of several bays of about equal size, and no particular requirements for flexibility in the future exist, trusses can be designed to be continuous. If a continuous structure is to be utilized, the possibility of large magnitude or nonuniform soil settlement must be investigated. The latter can create a major obstacle to this design scheme, because differential settlement will cause additional (sometimes quite substantial) secondary stresses in the continuous structure. The bending moment and shear distribution in the continuous truss is the same as in a continuous beam of the same span and loading. The trusses used in such a design scheme are usually of the parallel chord type, for which the internal forces are relatively easy to determine. Since the deflection of a continuous truss is much smaller than that of a simply supported one, the continuity enables the designer to hold the overall depth of such trusses to 12 ft (3.7 m) for spans exceeding 120 ft (36.5 m); that is, factory-made trusses can still be used and the only work that has to be done in the field consists in splicing of the chords, as required by the maximum permitted shipping length.

Where soil conditions are not favorable and differential settlements are a possibility, the designer may use instead of the continuous beam scheme a cantilever type beam, also called the "Gerber beam." This scheme utilizes a series of simply supported cantilever beams interconnected by short simply supported beam links. Since the location of the hinges in this scheme governs the distribution of the positive and negative moments, it is up to the designer to find the proper hinge locations to keep the maximum moments approximately balanced for all conditions of loading. Figure 12.6

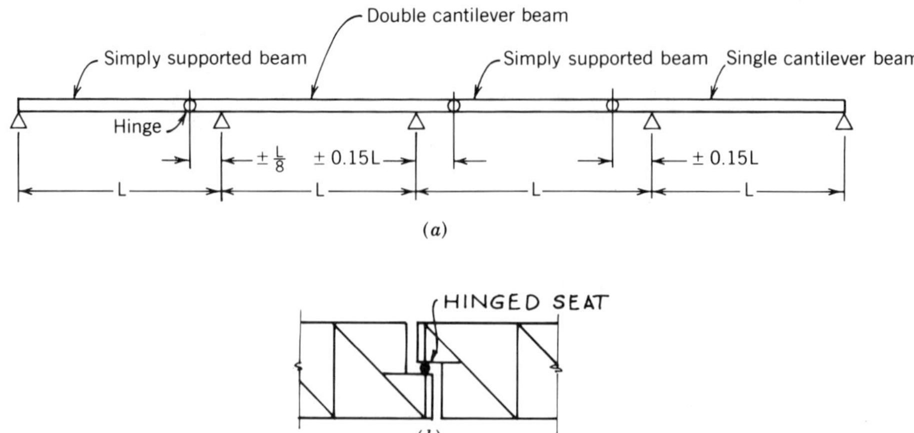

Fig. 12.6 Basic elements of cantilever beam design. (*a*) Location of hinges along the span. (*b*) Detail of hinge.

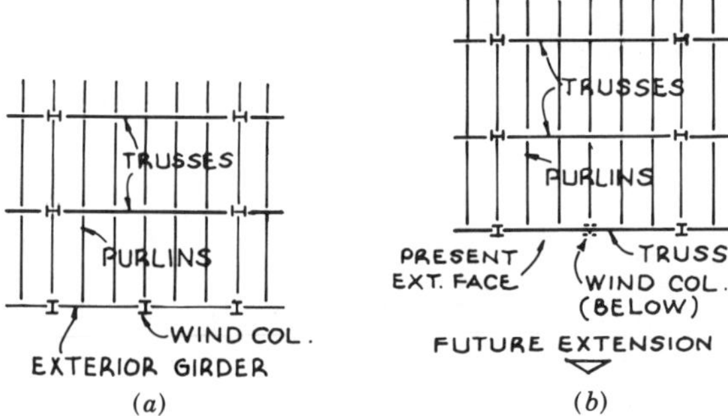

Fig. 12.7 Placement of exterior columns. (*a*) Without future extension. (*b*) With future extension.

shows the various basic elements of a cantilever beam design, including the general idea of a truss hinge (Fig. 12.6*b*). When the moments and shears are known, the design of the members and connections can proceed as for any ordinary truss. Particular attention must be given to the design, detailing, and practical execution of the hinges, because the simplicity of erection as well as the general functioning of the whole system depend on the hinges. Cantilever beams are more flexible than continuous beams. They can take differential settlements without causing any secondary stresses, but these settlements can be detrimental to the roof unless allowances for movements are provided for in the design.

If a structure has a fixed overall size, that is, if its exterior walls are in a permanent position, it would be wasteful to provide a roof truss or a carrying truss alongside an exterior wall face. Secondary columns (so-called wind columns, or wind posts) are usually installed along the building face at 18–25 ft (5.5–7.5 m) intervals to provide lateral as well as vertical support for the girts and other wall framing. It is practical and economical to utilize these posts also for the vertical support of the roof framing as shown in Fig. 12.7*a*. However, where the present face of the building is only temporary, that is, where a future extension of the building is contemplated or feasible, then it is advisable to provide the trusses at the time of construction, as shown in Fig. 12.7*b*. These trusses should be designed strong enough to support the future roof and permit removal of the wall below with least interference with the owner's operations. It is advisable also to check the design of these trusses for the present loading, because in some cases the present wall construction, which is sometimes hung from the trusses, may be heavier than the future roof load.

Frames

When the column spacing in the direction of the longer span exceeds the maximum length [about 40 ft (12 m)] for which a rolled beam framing is economical, and the building is only one bay wide, neither continuity nor cantilever beam design can be utilized to reduce the flexural moments. In this case designers may mobilize the stiffness of the columns and introduce stiff corner connections to create frame action. Such frames, whether made of regular rolled wide-flange sections, of tapered wide-flange sections, or of welded plate girders, can extend the upper limits of the economical span length considerably and provide lateral wind resistance at the same time. Welding has played a big role in the promotion of moment-resisting frames and deserves to be given credit for improvement in their quality. Welding has not only simplified details and reduced girder cross-sections, but has also opened up a new field for tapered or otherwise built-up sections. Figure 12.8 indicates how extra-deep sections of parallel flanged girders can be made by inserting a plate, or how tapered girders can be obtained by cutting at an angle the web of rolled sections, turning one portion around and rewelding the sections as indicated. Such frames are commonly built for spans from 50 to 150 ft (15 to 45 m) and are used mostly where heavy loads have to be carried. Figure 12.9 shows a frame of this type. A disadvantage of this design scheme is that most longitudinal services, such as pipes and ducts, have to be placed either below the bottom flanges or holes have to be cut through their webs. Such holes often must be reinforced by welding on rings or plates to replace the lost material. Since these frames are usually designed as two- or three-hinged arches with a tie rod below the floor slab, they cannot be used where floor conveyor trenches have to be provided. In some cases, ties have been dropped below the trenches, and piers have been designed to resist the horizontal thrust similar to the action of a buttress.

For preliminary design calculations, the formulae contained in Kleinlogel's *Rahmenformeln* [12.9] are practical to use.

Similar schemes are widely applied to the design and construction of prefabricated building

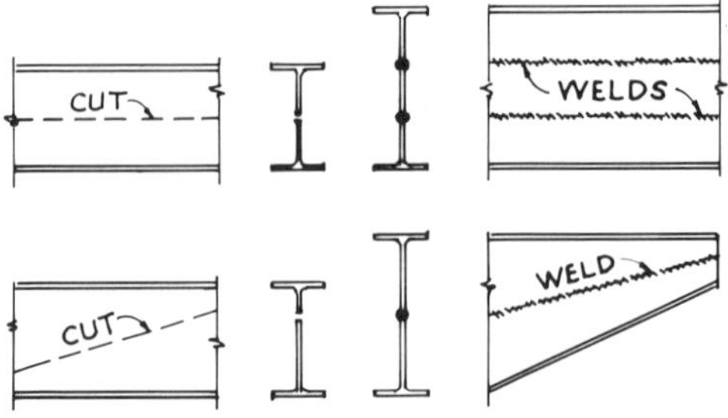

Fig. 12.8 Fabrication of tapered girders and columns.

frames. Standardization and mechanized shop fabrication have transformed prefabricated frames into a highly competitive manufactured product. Such frames are usually made either of special hot-rolled thin sections, or more commonly of cold-formed sections made of high-strength [$F_y = 50$ ksi (350 MPa)] steel plates bent into special shapes suitable for construction. The use of high-strength steels permits reduction in the size of their cross-section, that is, smaller weight, less expensive transportation, and easier erection. Here as well as with other prefabricated designs, economy depends to a great degree on whether or not the units can be used the way they come, because alterations and deviations from the basic detail can easily ruin their economy. Like any other system, they have also their disadvantages. The shear capacity of the thin lightweight sections is small, which may make them not suitable for the support of heavier concentrated loads, such as hanging loads of any appreciable magnitude. Other disadvantages are the same as for other solid-web frames. The large spaces between the frames are essentially lost for the use of utilities because pipes and ducts have to be placed beneath them. Holes for penetrating pipes have to be drilled (or cut) and reinforced to make up for the lost material. Pipes placed under a slope need careful and skillful layout work in order that the holes do not have to be unnecessarily oversized.

The greatest field of application for prefabricated moment-resisting frames is for warehouses, assembly buildings, gym courts, and similar, where no concentrated loads of any magnitude occur, where the spans are relatively large but the live load (roof load only) is small, and where most of the services and utilities can be buried beneath the floor slab.

Frame action can also be created by combining roof trusses and trussed columns into a stiff moment-resisting frame. Frame action may be desired where the column spacing in the direction of the longer span exceeds about 120 ft (37 m), but continuity or cantilever design cannot be applied to keep the overall depth of trusses below the maximum shipping limit. This may help to reduce the maximum moment in the roof truss sufficiently to design it with a maximum overall depth of about 12 ft (3.7 m); the general intention is to create a trussed frame with stiff corners. A tie rod can be provided beneath the floor slab and restraint at the column bases can be utilized where soil conditions permit. Such a design is not only used for trusses over large openings, but can also be applied to structures with lesser span but greater height, where lateral resistance to wind can create sizable problems. To this category belong the so-called stacker buildings, a recent kind of warehouse construction, where automatic hoists, operating from a moving column shaft, distribute the load and carry it to, or discharge it from, a particular stacker shelf. These buildings are basically of two types. In the first kind, as shown in Fig. 12.10, the building represents not much more than an

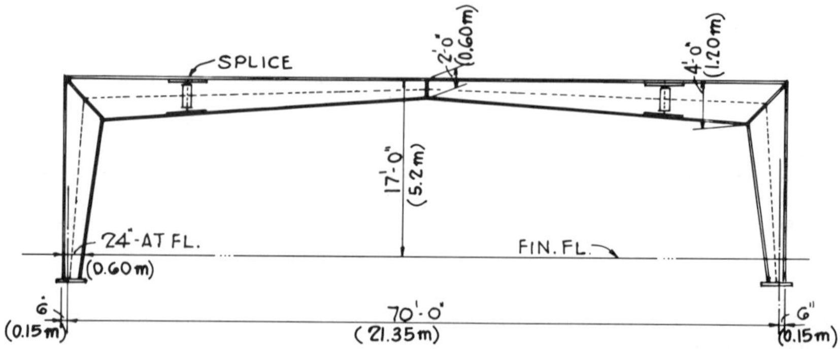

Fig. 12.9 Tapered girder rigid frame. Courtesy of Albert Kahn Associates, Detroit.

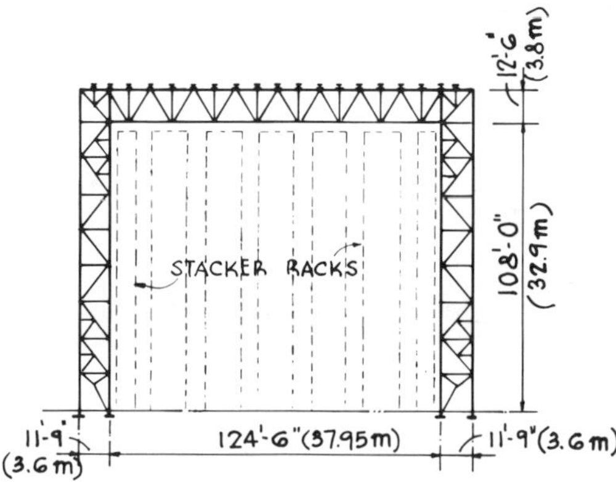

Fig. 12.10 Stacker building frame. Courtesy of Albert Kahn Associates, Detroit.

enclosure, entirely separated from the stackers, where the various hoist-columns are guided to their particular destinations. In the second type, the stacker columns themselves provide intermediate vertical support for the roof trusses, and the trussed exterior columns provide mainly the lateral resistance against the wind.

Field-Assembled Trusses

Field assembly of trusses is used frequently in connection with airplane hangars, convention halls, concert auditoriums, and similar structures where spans often reach 300 ft (90 m) or more, as shown in Fig. 12.11. In such a case it is most economical to fabricate as much as possible in the shop and ship the units separately to the site, to be assembled and fully connected in the field. It is common practice to have the trusses preassembled in the shop, to ascertain that all connections will fit, and to take them apart again before shipment. Whenever possible, reassembling in the field is done flat on the ground on a wooden platform or on wood blockings. After the assembly has been completed, the whole truss is hoisted into place. Where length or weight make it difficult to erect the truss in one piece, it is subdivided into suitable sections and each section is then hoisted into place and set on temporary intermediate supports. After all sections are in position, they are fully connected to each other to form one unit. It is easy to realize that this kind of fieldwork is expensive, time consuming, and sometimes difficult, considering adverse climatic conditions such as wind, cold, or sleet, that may have to be contended with. From a design standpoint, there is no difference between a truss that was shop fabricated and shipped in one piece, and a truss whose members alone were shop fabricated but which was otherwise finally assembled and connected in the field. Such trusses usually have large forces in their chords and web members and also large distances between panel points, because of their extreme size. Wide-flange sections with the webs in horizontal position are often used for the truss chords and also for the web members, because they combine a large cross-sectional area with large lateral stiffness (low L/r). They offer also the possibility to use two planes of gusset plates, which reduces the length of each connection. Where the depths of the various members do not match, they have to be adjusted by means of preconnected filler plates to suit. Shop fabrication of the various members is mostly done by welding; however, the field connections are almost invariably done with high-strength bolts to take advantage of the easier supervision and control.

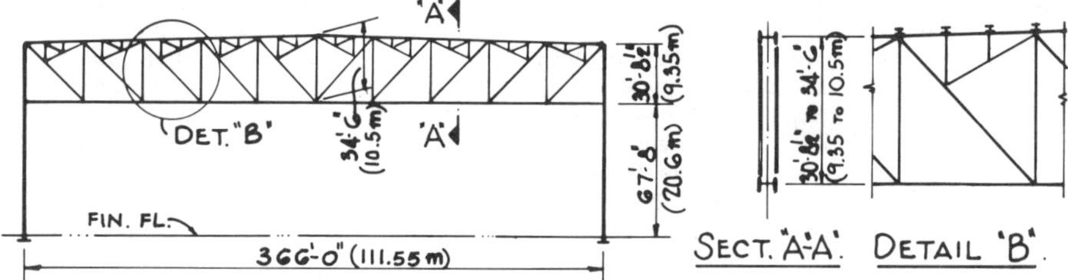

Fig. 12.11 Building using field-assembled trusses. Courtesy of Albert Kahn Associates, Detroit.

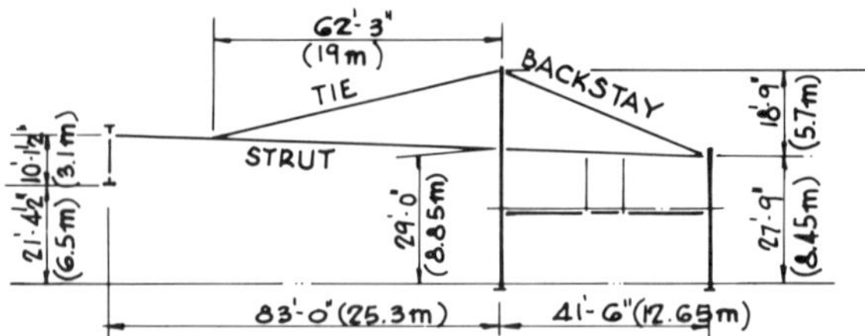

Fig. 12.12 Cantilever hangar. Courtesy of Albert Kahn Associates, Detroit.

Special Systems

Considering the great costs involved in the field assembly and erection of long deep trusses, it is not surprising that designers have tried to develop other schemes that require less fieldwork and permit easy erection. Typical examples in this respect are suspension systems. Such systems can be used in many different ways. One of the most commonly used prototypes in this respect is that of a cantilever hangar as shown in Fig. 12.12. This hangar design has the additional advantage of being laterally extendable without special provisions. In this design, the cantilevering strut whose protruding end is held up by a cable or tie provides the main member for the support of the roof construction. The necessary construction depth of the large cantilever portion is not created by a truss, which would require large field assembly, but by the combined action of a flexurally stiff compression strut at the bottom and a tension tie at the top. The bottom strut can be a rolled beam, a plate girder, or a shallow truss, and the top tie can be a cable, a pipe, a tube, or any other shape that can develop not only the required tension, but has also sufficient stiffness to develop a small compression force under certain uplift conditions. The necessary construction depth is obtained by raising the tie above the compression strut until the required design depth is attained. Since strut and tension tie are two independent pieces, they can be shipped and erected as separate units until incorporated into the final structure. The downward force at the backstay has to be provided by the weight of the structure itself or by a foundation block acting as counterweight. It is important to note that a hangar structure with open front doors represents a dangerous wind catch and that specific provisions have to be made to resist the upward wind force acting against the underside of the roof. Such provisions may include any of the following: a sufficiently large dead weight; a prestressing of the suspension cable; or the stiffness of the tie itself designed to act not only as a tie, but also as compression strut resisting uplift forces.

12.3.2 Flexibility of Space Utilization

The importance of flexibility of utilization of space in industrial buildings cannot be overemphasized. In many manufacturing processes, the required layout or the provided flowchart of production represents only a solution to the present requirements. This makes the chance for a change in the future not only possible but probable. If flexibility to accommodate structural changes without excessive damage to the existing structure and with the least possible interference with the owner's daily operations is one of the owner's requirements, then the designer must select a material and a structural system that lends itself to such an undertaking. Relocation of web members in a truss to let ducts or conveyors pass through; raising of roof portions to provide necessary clearance for mechanical equipment or operations; reinforcing of trusses, beams, and columns to support heavier loads; introduction of a new mezzanine to provide extra floor space; cutting openings through the roof construction, varying in size from small holes for the passage of pipes or ducts, to openings large enough for a whole fan room; and even entire removal or relocation of existing columns, are just a few of the typical alterations that confront an industrial building designer. Most of these alterations can be made with relative ease as long as the structure was designed with the possibility of such alterations in mind, that is, with a built-in flexibility. For this reason, design for continuity in girders, beams, or trusses, and the use of moment connections and similar design approaches should be avoided. They will not only complicate any sizable alteration and therefore be costly, but they may also cause considerable interference with the owner's operations. The use of simply supported members, whether they are trusses, purlins, or beams, will serve flexibility best. Simple details and ample (neither wasteful nor skimpy) design of members and connections are desirable. In general, intricate designs should preferably be avoided; even if they offer certain economical advantages at the present time, in the long run they usually reduce flexibility by making the inevitable modifications both difficult and costly.

In this respect, steel structures have a definite advantage over cast-in-place reinforced concrete structures. Structures using precast (often prestressed) concrete units have a greater degree of flexibility than cast-in-place structures, because entire members can be replaced if necessary. Timber structures have a high degree of flexibility; however, they are, for other reasons, not too commonly used for industrial buildings.

Provisions for future extensions should also be considered. If extensions are likely, then the affected girders, roof trusses, carrying trusses, columns, and footings along the exterior building face should be designed for the future loads. Figure 12.7*b* shows a condition where interior trusses are used at the exterior building face to provide for extension of the building in the future. Steel structures have a definite advantage here because welding makes the attachment of seats and ledges simple. In reinforced concrete structures, steel plates (weld plates) have to be embedded and anchored back in the particular faces of beams or columns to permit later welding on of seat angles, clips, and other devices. In exposed concrete structures, the designer must be aware of the fact that heat from welding may discolor the concrete in the vicinity of the weld plate. To establish continuity across the joint, dowels can be extended from the present structure or threaded couplers can be embedded instead. Similar provisions can also be made if an upward extension of a presently single-story structure is contemplated.

12.3.3 Simplicity of Detail and Speed of Erection

Both factors are important in making any type of structure more practical and economical. It was their application to industrial work that gave them the significance which they deservedly have today. When an industrial management group decides on the construction of a new plant, they want it ready for operation in the shortest possible time. Since the investment is large and during the time of construction completely unproductive (from the owner's point of view), the selection of a design whose fabrication and erection requires the least amount of time will always be attractive. This is the major factor behind the widespread development of prefabricated units. The final answer to this problem, however, does not lie in the development of prefabrication alone, because it does not provide the necessary overall stiffness and stability of the structure. To transform a structure consisting of prefabricated parts into a monolithic unit is a formidable task.

In most cases the owner has a certain preference for the type of structure; that is, whether it should be regularly constructed or prefabricated. When the designer has to compete with a structure partly or entirely made up of prefabricated units, the only way to succeed lies in a careful selection of the right kind and strength of construction material, in the clarity and simplicity of the detail, and in an austere treatment of expensive items, such as field welding, shear connectors, and formwork. There is some money available, because the manufacture, transportation, and erection of the ready-made structure is not exactly inexpensive. The real outcome, however, can only be decided by a close design and by a sharp estimate.

12.3.4 Prevailing Economy

This is not the place to discuss the art and value of making comparative estimates. The designer must be reminded, in general, to exercise caution in applying the many cost comparisons and promotional discussions that are found in trade publications and professional journals. Accepting without question the accuracy of the statements, there still remain the variations in time, location, cost fluctuation, design concept, and execution of detail, just to mention a few, which can create enough differences in costs to make these comparisons seldom exactly applicable. What was practical and economical in one situation may not necessarily be practical and economical in the next. The designer should, therefore, review all recommendations with a critical and analytical scrutiny and select from them only those applicable. Approaching the decision with an open, unprejudiced mind will provide the best chance of obtaining the solution that is structurally best suited, as well as most functional and economical.

12.4 STEEL STRUCTURES—PARAMETERS FOR PRELIMINARY DESIGN

For design of steel structures one must consider the roof slope, the roof decking material, the grade of steel to be used, the structural arrangement of the main supporting members, the roof framing plan, column design, provision to support materials-handling equipment, and the bracing of the entire system. In addition to the detailed discussion given here on steel structures, Chapter 19 contains treatment of many special considerations.

12.4.1 Roof Slope

Roof slopes may be divided into three categories:

1. Steep roofs, having a slope greater than 1 : 25 (1/2 in. per foot of length)
2. Flat roofs, having a slope of 1 : 200 to 1 : 25 (1/16 to 1/2 in. per foot of length)
3. Level roofs, having no slope.

Steep Roofs

Steep roofs tend to be heavy and expensive. They originated many years ago in colder climates, when carrying snow represented a major problem. As explained in Chapter 26, roofing materials for steep roofs include shingles and slate or clay tiles, each of which function by overlapping. Some of the units are anchored back to the supporting framing to keep them from getting lifted by wind suction on the leeward side of the roof. The only lighter materials that can be used on a steeply sloped roof are sheet metal, copper, and terne. Such metal roofs have to be held down continuously by seam clips nailed to a complete wooden sheathing. For a steep roof, the area of the roof decking as well as that of the roof construction is much larger than the absolute minimum for a flat roof. All these factors add up to an expensive way of roofing, such that it is presently used mostly for architectural expression.

Somewhat flatter roofs can be made with less expensive conventional asphalt tiles nailed to a plywood sheathing. This kind of roof is commonly used for residential housing. For a more detailed discussion on roofing materials and their weights to be used for preliminary design purposes, see Chapter 26.

The old "mill" buildings, shed roofs, and similar designs required steep slopes for ventilation and lighting. Since power ventilation did not exist, natural draft had to be utilized as much as possible. Similarly, since artificial lighting was costly and of inferior quality, extensive use of daylight was highly desirable. Steep slopes, preferably with exposure to the north, seemed ideal in this respect. Unhealthy dust accumulations, difficult maintenance (painting), and an occasional leak in the roof seemed to have been of lesser importance in those days. An entirely different set of conditions and requirements are currently met in the satisfactory functioning of a building.

Flat Roofs

Flat (not level) roofs are presently most commonly used to cover single-story open-space buildings, the majority of which belongs to the so-called type of "industrial buildings." These roofs are flat enough to prevent the runoff of bitumastic material in hot weather, but also steep enough to get rid of rainwater in a short time without creating local puddles or extensive ponding. Such roofs are usually sloped in two directions; the primary slope, perpendicular to the ridge line, is about 1 : 100 to 1 : 50 (1/8 to 1/4 in. per foot of length). The secondary slope which usually has a magnitude of about 1 : 200 (1/16 in. per foot of length), runs parallel to the ridge line and is sometimes provided only locally near the sumps; the primary slope is created by the slope of the top chord of the roof trusses and the secondary slope either by varying the elevations of the end connections of the purlins or by building up the valley with filler material and roofing, as detailed in Chapter 26. Flat roofs are not wasteful in area, they are relatively light in weight, reliably waterproof, and can easily be repaired if they have to be penetrated by vents, shafts, or other kinds of openings. They provide easy access to existing installations protruding above the roof. Though they are strong enough to resist the careful walking of single maintenance personnel, they are not suitable for regular pedestrian traffic. In case heavier units or building materials have to be moved across the roof, a strip of wooden planking should be placed on the roof perpendicular to the line of traffic to spread the load and protect the roofing. It is not economical to over-design the roof decking, whatever type is used, for any such future occurrence, because it is difficult to foresee the magnitude of any overloading and the exact location where it may take place. It is important, however, to impress upon maintenance engineers and similar personnel to report potential overloads or unusual roof uses to the designer or engineer. The condition can be investigated and the engineer or designer can recommend any necessary measures to protect the existing roof construction.

For preliminary design purposes, the weight (mass/unit area) of composition roofing can be assumed to be 6 psf (30 kg/m^2).

Level Roofs

The simplicity of a level roof is intriguing, but it is more vulnerable to defects than any other type of roof. Malfunctioning or accidental clogging of scuppers (drains) and overflows can easily lead to excessive ponding. The *AISC Manual* [12.5] contains stiffness requirements to prevent unforeseen ponding caused by the deformation of a roof that is designed to be level. The best protection in the author's opinion is to avoid the level roof.

12.4.2 Roof Decking

Years ago, the most commonly used roof deckings for industrial buildings were of the cement tile type. They had the advantage of furnishing a reasonably substantial, fireproof deck and providing a good stiff base for insulation and roofing. These advantages, however, could not offset the extra cost due to the weight of the decking. The most commonly used cement tiles, made of lightweight (low-density) concrete, weigh about 10.5 psf (50 kg/m^2). In addition, there was always the disadvantage that the tiles did not provide lateral support for the purlins. Lighter weight insulation board and gypsum-type roofs have been developed to try to overcome some of these difficulties and preserve simultaneously most of the advantages of cement tiles. Most of these lighter-weight materials provide lateral support to the compression flange of the purlins through the use of special ⊥-bars. These ⊥-bars span between purlins and are welded to them. They act as subpurlins and carry the insulation or planking. Some of these decking types, especially those that are cast-in-place, are influenced by climatic conditions, since adverse weather during installation can affect their quality and the quality of the roofing material placed over them.

Presently, metal decking (see Chapter 20) is the most commonly used type of roof decking for industrial buildings. Its relatively low cost, light weight, and ease of installation have given it a highly competitive status. It requires, of course, good insulation and reliable fire protection, because it is lacking in both of these qualities. There are many different types of metal roof decking with slightly varying shapes, depths, and configurations. They consist usually of a single sheet metal plate bent in a corrugated or other regular pattern with a locking device along its two long edges to permit pinching adjacent strips into large, almost monolithic-acting, panels. The most commonly used type is $1\frac{1}{2}$ in. (38 mm) deep and carries a live load of 30 psf (1.4 kN/m^2) at a maximum span of about 8 ft (2.5 m). There are also 3-in. (76-mm) and $4\frac{1}{2}$-in. (114-mm) deep sections available that can be used on greater spans, but with less economy. In general, panel thicknesses vary from 22 to 12 gage, with dead loads from 2 to 6 psf (0.1 to 0.3 kN/m^2); they can be used for maximum spans from 7 to 11 ft (2 to 3.5 m). Metal decking is fastened to the supporting roof beams by intermittent welding. Properly attached to the purlins, the decking can create a large diaphragm of high quality and stiffness. The fastening of the metal decking requires good workmanship and care during the welding operation. Insufficient welding or excessive heat that has burned through the thin metal, can permit high winds to lift the decking off the purlins and roll it back like a carpet.

Where metal decking spans purlins, the valleys of its configuration run parallel to the fall line of the roof, which gives it a natural drainage for any condensation that may occur. However, when metal decking spans between rafters, its valleys will be in horizontal position (perpendicular to the fall line) and condensation can accumulate in the grooves and cause rusting.

Stiff insulation boards are used in connection with metal roof decking to bridge over the gaps (valleys) of the corrugated configuration. Most roof deckings are not symmetrical in shape, that is, the valleys at one side of the x–x axis are wider than at the other side. When placed with the wider opening up, some insulations will not be stiff enough to bridge the gap.

12.4.3 Grade of Steel

Along with the selection of the appropriate design scheme, the designer has to decide on the quality or strength of the structural steel material to be used. In general, there are steels having yield stresses from 36 to 100 ksi (250 to 700 MPa) available. However, only two of them are commonly used. With a few exceptions, as noted below, structural steel with the ASTM designation A36, having a minimum yield stress 36 ksi (250 MPa), is by far the most commonly used material for steel structures. It has good welding qualities and is presently the most economical steel material available. For extra long spans requiring field assembly or for special designs where it becomes essential to reduce the cross-section in order to save weight, A572 Grade 50 steel (similar to A242 or A441 formerly used) having a minimum yield stress F_y of 50 ksi (350 MPa) is frequently substituted. Similarly, where structural steel is to remain exposed to the weather, corrosion-resistant steels such as ASTM A588, whose minimum yield stress F_y is 50 ksi (350 MPa), are used. Steel joists are manufactured and available for F_y of both 36 and 50 ksi (250 and 350 MPa). Structural shapes of A36 are available from warehouses in a wide variety of sections even in small quantities; sections rolled from higher strength materials usually have to be mill ordered. Table 12.3 gives typical allowable stresses for the two main yield stress levels of structural steel as given by the *AISC Specification* [12.6]. Whenever a steel is used having yield stress higher than A36, one should be aware that the reduction in cross section, assuming constant span, means a reduction in moment of inertia, and therefore an increased deflection of the member.

Table 12.3 Allowable Steel Stresses*

ASTM Designation				A36		F_y = 50 ksi†	
Yield Stress		$F_y =$		36 ksi	250 MPa	50 ksi	345 MPa
Tension		$F_t =$	$0.60F_y$	22.0	150	30.0	205
Flexure	Compact	$F_b =$	$0.66F_y$	24.0	165	33.0	225
	Others	$F_b =$	$0.60F_y$	22.0	150	30.0	205
Shear		$F_v =$	$0.40F_y$	14.5	100	20.0	140
Compression							
	L/r = 1	$F_c =$	Tables 3-36 and 3-50 from Ref. 12.6	21.6	150	29.9	205
	= 40	$F_c =$		19.2	130	25.8	180
	= 80	$F_c =$		15.4	105	19.0	130
	= 120	$F_c =$		10.3	70	10.4	70
Bearing		$F_p =$	$0.90F_y$	33.0	230	45.0	310

* From *AISC Specification* [12.6].
† ASTM Specifications A242, A441, A572.

12.4.4 Structural Framework

Direction of Roof Beams (Purlins)

It is most practical to run the top chord of a roof truss parallel to the fall line (slope) of the roof surface. This permits setting the roof beams on top of the roof trusses. This method is simple for fabrication and erection and is least expensive. Such roof beams (purlins) simply rest on top of the roof trusses and are secured to them with two ordinary machine bolts. Where two purlins meet, a simple web plate with two bolts on either side is provided to keep the two adjacent webs in line. At roof sumps, the purlins are partly lowered by coping their bottom flange or they may be framed-in completely, if so required. Placing the purlins on top of the roof trusses has another advantage; it permits placing these purlins in two-bay lengths. Since the maximum moment (positive at midspan) for a simply supported beam is the same as the maximum moment (negative over interior support) for a two-span continuous beam, there is no loss in flexibility connected with it in case the purlin has to be cut at a later date. On the other hand, as long as the two-bay length is intact, the deflection of the purlin is greatly reduced, and the stiffness of the whole roof is increased at no extra cost.

The lateral inclination of a purlin on a flat roof is usually so small that the effect of bending about its y–y axis (weak axis) can be disregarded. However, if purlins are used on steeper roof slopes, this influence on the resultant stress has to be taken into consideration. Another way to overcome this difficulty is by providing sufficient bracing with sag rods for intermediate lateral support.

On steep roofs, the roof framing can also be turned by 90°. This approach will place the roof trusses parallel to the ridge line and the roof beams perpendicular to it, that is, parallel to the fall line of the roof surface. Placing the roof beams (rafters) this way will permit full contact between the roof decking and the supporting beams; but the rafters themselves will have to be supported from the level top chords of the roof trusses by means of clip angle seats.

Spacing of Roof Beams

The maximum spacing of the roof beams (purlins) can be found in information available from roof decking manufacturers. Strength is not always the governing factor; deflections can control the recommendation for a maximum spacing of the roof beams. Increasing the depth and gage of the units in order to obtain a larger purlin spacing has often been tried, but is usually not economical. Any designer who has walked across a metal deck roof with large purlin spacing will remember the feeling; however, this may not reflect on the strength of the decking but instead indicates its flexibility. Such roofs are more vulnerable to defects in their roofing and will normally require greater maintenance.

Stiff roof decking will protect the roof covering better but is not necessarily safer. Because of the apparent stiffness of such decking, workers have been induced to store or drop construction materials on it. Such actions have often led to dangerous accidents.

Conservatism is recommended in the selection of the purlin spacing. The extra cost for another line of purlins is not so great when the reduced load is accounted for by selecting a lighter section. In addition, it will probably pay for itself through improved serviceability and reduced maintenance.

Design of Roof Beams

In most cases roof beams are designed as simply supported beams, which is also correct from the standpoint of maximum moment in roof beams furnished in two-bay lengths, that is, continuous over two bays. In some foreign countries it is common practice to design purlins as continuous members and in some areas to base the design on the plastic theory. All these approaches are based purely on economics, realizing the great influence the purlin weight has on the total steel tonnage. There is nothing wrong with either one of the above design methods, if the following items are taken into consideration:

1. Purlins designed as continuous beams or by the plastic theory require full-capacity moment splices at each splice. Such splices have to be done in the field. Under the prevailing labor-to-material cost ratio in the United States, the cost for the fieldwork usually exceeds the savings that can be achieved with the lesser material.
2. The continuous design unfavorably affects the future flexibility of the building; no bay can be cut out or be altered without either reestablishing the continuity or strengthening the adjacent bays.
3. Continuous designs are sensitive to differential settlements and since industrial complexes are often forced to use cheap land with nonuniform soil conditions, simply supported designs are usually preferred.

To save on the expensive field splices, cantilever (also known as Gerber) designs have been used (more often for girders than for purlins) in several cases instead of the continuous designs—with variable economical success. (See also Fig. 12.6 with accompanying text.)

The greatest economy in the design of a roof beam will probably be achieved through proper bracing; that is, the compression flanges of the sections must be braced laterally well enough that the maximum allowable flexural stress can be used in their design.

At any rate, the selected section should provide the desired depth-to-span ratio, and must have a section modulus satisfying the allowable stress. The section must also satisfy the "compact section" requirements if the allowable stress is for that condition.

For spans varying from 15 to 30 ft (4.5 to 9.0 m), and for total loads of 25 to 50 psf (1.2 to 2.4 kN/m^2), excluding the weight of girders or trusses, the weight of the purlins can, for preliminary design purposes, be approximated as 1½ to 3 psf (70 to 140 N/m^2), depending on span and load.

For ordinary roof loadings excessive deflections can usually be avoided if the beam depth is selected within the limits given by the Commentary to the *AISC Specification* [12.6]. Since this value is only a recommendation (the only stipulated maximum live load deflection in the *AISC Specification* [12.6] is that for beams or purlins supporting a plastered ceiling), shallower members can be used if the deflection is not excessive. Maximum live load deflections varying from $L/180$ to $L/240$, depending on the sensitivity of the structure, are commonly used. Level roofs must be checked for ponding.

For preliminary design purposes the following formulas are suitable to determine the approximate deflection at midspan,

$$\Delta = \frac{K_d f L^2}{1000h} \tag{12.4.1}^*$$

* For SI, with f in MPa, L in m, and h in mm,

$$\Delta = \frac{K_d f L^2}{h} \tag{12.4.1}$$

where Δ = deflection at midspan, in.
f = extreme fiber stress at midspan due to the investigated loading condition, ksi
L = span, ft
h = total depth of beam, in.
K_d = a dimensionless coefficient, as defined below

1. For uniformly distributed loads:

for simple spans: $K_d = 1.0$

For continuous spans over at least four bays, with all bays loaded:

for interior bays: $K_d = 0.5$

for exterior bays: $K_d = 0.9$

2. For a single concentrated load applied at midspan:

for simple spans: $K_d = 0.8$

For continuous spans over at least four bays, with all bays loaded:

for interior bays: $K_d = 0.35$

for exterior bays: $K_d = 0.65$

The modulus of elasticity E used in the development of Eq. (12.4.1) was assumed to be 29,000 ksi (200,000 MPa).

For other or combined loading conditions, the deflection at midspan can be approximated by interpolation between the given values.

Roof Framing Plan

Once the design of the roof beams has been completed and the designer has decided on the layout of the various roof bracings, the drawing of the typical roof framing plan can practically be completed. At this time the designer will probably have become aware of the fact that it is much easier to make a theoretical design than to execute it in practice. All kinds of ducts and openings will interfere with the typical layout. During the necessary alterations, relocations, and eventual redesigns, the designer must not get frustrated and discouraged, but should try again and again to restore the continuity and integrity of the structure. This especially applies to bracing systems, because in their proper functioning lies the secret of a safe structure.

The roof framing plan of an industrial building resembles in many ways a geometric pattern. This layout is a rather regular one, as long as special conditions do not overweigh the typical ones, as it is sometimes the case. Because of the basic regularity, it is recommended to start the drafting of the typical roof framing plan at the beginning of the conceptual design. It is advisable to show the regular pattern over the entire area of the building and to mark on this layout not only the typical members but all deviations from and additions to the typical design at their proper locations. One by one, as soon as a particular member has been designed, its size should be entered in the material schedule, until every member is called for. This makes the roof framing plan a kind of a record sheet on which not only the members but all installations, whether they are hanging from or supported by the roof framing, are shown or called for. This drawing should always be kept up to date and any changes should immediately be recorded on it; otherwise the designer will very soon lose control over the special conditions one has to contend with.

At this time, before the final layout of the purlin framing and the truss geometry is completed, the designer should also give thought to the type and layout of the roof bracing that will be required.

The following list contains items usually involved in the design of a roof framing. As an aid to the designer they are given in their logical sequence. The list provides also the sections of this chapter in which each particular item is treated in detail:

1. Column layout (Sections 12.3 and 12.3.1).
2. Spacing of trusses or girders (Sections 12.3.1 and 12.4.5).
3. Positioning of columns (Section 12.4.6).
4. Spacing of roof beams (purlins) (Sections 12.3.1 and 12.4.4).

5. Design of roof beams (purlins) (Section 12.4.4).
6. Provisions for materials-handling equipment (Section 12.4.7).
7. Design of trusses (Sections 12.4.1 and 12.4.5).
8. Design of bracing (Section 12.4.8).
9. Design of columns and base plates (Section 12.4.6).

12.4.5 Design of Main Supporting Members

Spacing of Girders and Trusses

Following a reasonable design approach, the selection of the spacing for the main members, regardless of whether they are girders or trusses, should immediately follow the selection of the column spacing and be either equal to it or a fraction thereof, such as $L/2$ or $L/3$. If the girders and trusses are too closely spaced, the purlin size will be the practical minimum section, and therefore unnecessarily heavy. If the spacing is too far apart, the purlins will have to be excessive in size and weight in order to satisfy strength and deflection requirements. For most practical design purposes, spacings of 18–25 ft (5.5–7.5 m) will offer the most economical design. Where a larger spacing cannot be avoided or where it is architecturally desirable, it becomes reasonable to replace the rolled wide-flange sections by joists. Such a solution is often preferred for roofs over apartment buildings, warehouses, and stores. In most of these cases the spans exceed 25 ft (7.5 m) and the units are rather closely spaced, about 4 ft (1.2 m) apart. They span either from wall to wall without any intermediate support or with an intermediate small girder, along a passageway, supported by closely spaced columns. In all these cases there is enough headroom for the joists; however, the excess space that would have been needed to accommodate trusses would have been considered wasteful. Joists have also their limitations and disadvantages as discussed in Section 12.3.1. Poundwise, they are more expensive than rolled sections. Because of their limited shear capacity, joists are not good for concentrated or moving loads of any magnitude. They are therefore not well suited for industrial work, unless it is a warehouse roof, a storage area enclosure, or the roof of a building to be used for light manufacturing without provisions for hanging loads.

Joists may also be used advantageously to provide for extra headroom where an intermediate truss has to be eliminated. A similar approach is often used, where the installation of a tall machine, like a press, requires local raising of the roof to provide needed headroom.

Truss Type

The selection of the type and layout of a steel truss is a purely functional operation that can be performed by every designer without having to resort to copying an existing example. It is unfortunate that some designers are of the opinion trusses have to be of a certain shape and that it is good practice to copy exactly old designs, without giving them the necessary scrutiny.

A typical case in this respect is the so-called "mill building" truss shown in Fig. 12.13 which has at the present time hardly any structural justification except for certain types of short-span timber truss buildings. Nevertheless, it is still repeated in many textbooks and used in actual design cases. This old truss type had its origin in England or France where steel trusses were used shortly after the rolling mills started to turn out the first Z and L shapes. The French "Polonceau" and the British "Fink" trusses were the forerunners of this once successful type. Their steeply sloped top chord, once so desirable for natural ventilation, is now redundant; their pointed ends always created structurally undesirable stress concentrations; and they offered no lateral stiffness to the building unless knee braces were provided which represented a nuisance for ducts and pipes. There is really no reason to repeat such an outmoded design, especially when it is so easy to design a truss

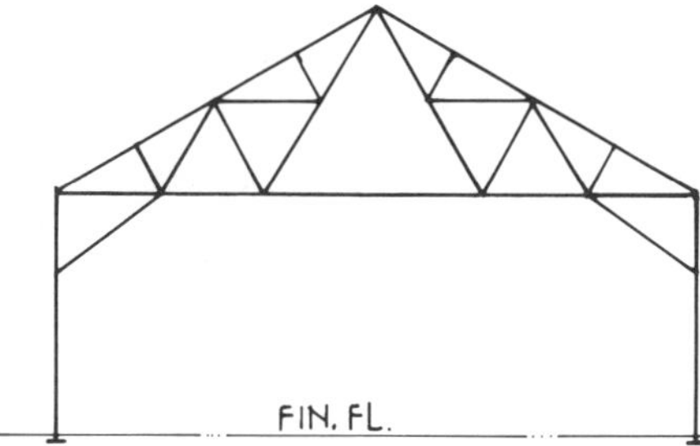

Fig. 12.13 Old type mill building.

functionally to serve our present-day requirements by the following steps:

1. The slope of the top chord is determined architecturally or by the type of roofing to be used. See Section 12.4.1.

2. The depth at midspan of an ordinary roof truss is generally assumed to be about 1/10 of its span. This is no fixed rule but it is known that this depth-to-span ratio will produce stiff trusses with reasonable member sizes and connections. Extra heavy loads, as in the case of carrying trusses, provisions for the passage of large ducts, matching the depth of cross-trusses, or architectural requirements may be good reasons for using deeper trusses. Deepening of a truss beyond the recommended depth will not necessarily make it less expensive. Although the chord stresses will be reduced, the shear forces in the verticals remain unchanged and the sizes of the compression web members will increase because of their greater buckling length. On the other hand, light loading, continuity, or cantilever action will reduce the required moment capacity and permit a reduction in depth. This will, in turn, increase the size of the web members. In questionable cases it is advisable to ascertain that a reduction in depth is not accompanied by an increase in deflection beyond the permissible limits (see Section 12.4.5).

3. Since all loads must be applied at panel points to have no bending in members of a truss, the panel point spacing along the top chord is governed by the assumed spacing of the roof beams (purlins). Since it is also desirable (but not absolutely necessary for the economy of the roof decking) that the purlins be spaced equally, the panel point spacing is made a fraction of the top chord length, such as $L/8$ or $L/12$ or whatever is required. In case a truss is peaked at the center, it is advisable to select an even number of purlin spacings; for a truss without peak, the number of spacings may as well be an odd one. In general, an even number of purlin spacings will often simplify the layout of the top chord bracing.

4. The panel point loads consist primarily of the purlin reaction made up by the dead and live load of the roof, the purlin weight, and an allowance for the dead weight of the truss which is commonly applied at the top chord. The panel point load may also include a load from a piece of equipment above the roof, such as an air conditioning unit, fan platform, or housing.

The fact that purlins are often specified to be installed in double lengths (continuous over two bays) will increase their reaction at the interior support; this is however disregarded because it is difficult to predict or control their placement in the field. This condition is usually alleviated by placing the purlins in such a way that interior supports and end supports are alternating. For reasons of simplicity all purlin reactions are, therefore, considered like those of simply supported beams.

Ordinarily, panel point loads are applied as vertical loads; only for very steep trusses, where the purlins may have to be designed for biaxial bending, it may be advantageous to apply them as two components (parallel and perpendicular to the top chord).

Reactions from conveyer lines, hoists, ceiling, or other hanging loads are applied as concentrated loads at the panel points of the bottom chord.

For preliminary design purposes, the dead weight of a truss per unit roof area can be approximated from the following formula:

$$W_T = \frac{wL}{30F_y} \quad \text{(for Inch-Pound units)}$$

$$W_T = \frac{3}{4}\frac{wL}{F_y} \quad \text{(for SI units)} \tag{12.4.2}$$

where W_T = weight of truss, psf (kN/m^2), of serviced area
w = total load of roof carried by truss ($w_D + w_L$, but excluding truss weight), psf (kN/m^2)
L = truss span, ft (m)
F_y = yield stress of steel, ksi, (MPa)

The total weight of the truss can be found from $W_T sL$, where s is the spacing between adjacent trusses in ft (m).

5. Theoretically, the centroid of each cross-section should be placed at the centerline of each truss member. For reasons of simplicity, however, the gage lines* (inner gage for sections having more than one gage line) are substituted for the centroid in most structural truss designs. For symmetrical cross-sections, the centroid is always used. For practical purposes, the theoretical depth of a truss is often assumed to be 6 in. (150 mm) less than its out-to-out depth.

---

* Standard locations for bolts or rivets.

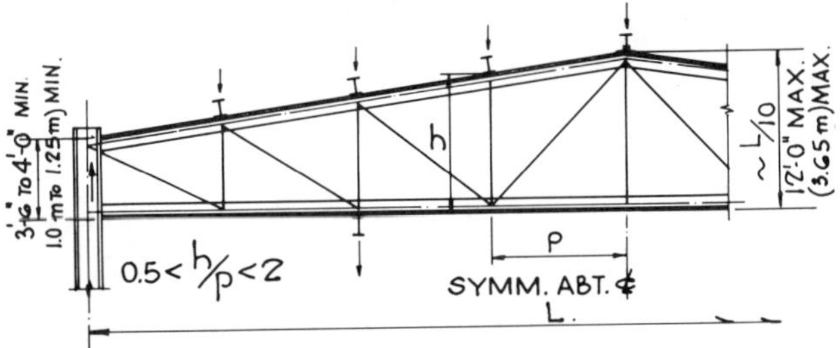

Fig. 12.14 Typical truss layout.

6. It is good practice to start the end diagonal of a truss at the intersection of the top or bottom chord centerline with the centerline of the supporting column. This will keep the column concentrically loaded and free of any bending moments; the end connection of the truss, however, if connected to the column flange, will have to be designed not only for the truss reaction R, but also for an additional bending moment $M = Rh/2$, where h is the depth of the column section. To simplify the connection and reduce the number of bolts required, the end diagonal is sometimes started at the intersection of the top or bottom chord centerline with the column face. This will remove the additional bending moment from the end connection, but exert a bending moment of the same magnitude on the column itself, unless it is an interior column and assumed to be symmetrically loaded. The first method is considered preferable to minimize unaccounted bending moments in the columns due to accidental differences in the two adjacent truss reactions.

7. Since any sequence of triangles can result in the formation of a feasible truss, there is no theoretical rule about the placing of the panel point locations along the bottom chord of a truss. Unless certain locations have to be used because of conveyor lines, hanging loads, or other architectural requirements, it is still good economy to follow the Pratt truss idea and put compression verticals at every top chord panel point and tension diagonals in between (see Fig. 12.14), because this arrangement will make the compression members as short as possible.

It is good practice to make the ends of a truss not less than 3.5–4 ft (1.0–1.2 m) deep.

The optimum ratio (depth h of shorter side to length p) of any truss panel (Fig. 12.14) should be about 1 : 1, but not greater than 1 : 2. In deep trusses it is advisable to place the main verticals at every second or third panel point, as shown in Fig. 12.15, and use a secondary framing for the support of the intermediate panel points.

If one follows these few steps, as described above, a truss layout presents itself almost automatically and does not require the copying of an existing design.

The following exceptions to the above guidelines must be considered:

1. In peaked roof trusses the center diagonals at either side of the peak will usually be in compression (Fig. 12.16*a*) unless their inclination is reversed from that of the other tension diagonals on their side of the truss as shown in Fig. 12.16*b*.

2. Where large ducts are expected to run through the roof construction, two provisions may facilitate their placement: (a) The depth of the trusses may be increased beyond the optimum, especially near the end supports. (b) A change from the "Pratt" to a "Warren" type web arrangement (Fig. 12.17), using alternate tension and compression diagonals, will permit omission of certain zero-stress verticals and provide for larger openings.

Where entire panels have to be kept clear of any intersecting diagonals, designers may have to resort to a Vierendeel truss design (rigid frame) which consists only of chords and posts. In order to prevent parallelogrammatic deformation of the panels, not only the chords and posts themselves, but also their connections at the panel points have to be designed flexurally stiff. If, for preliminary

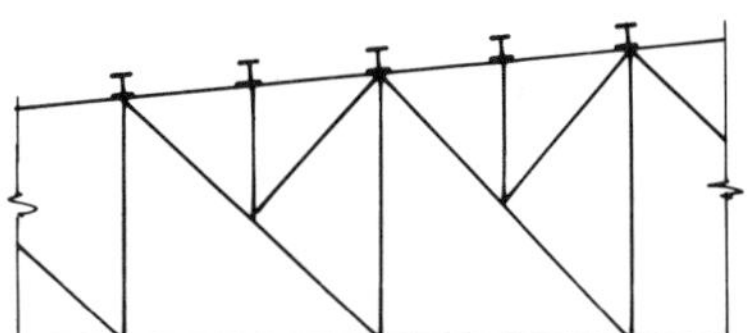

Fig. 12.15 Subdivided panels for deep trusses.

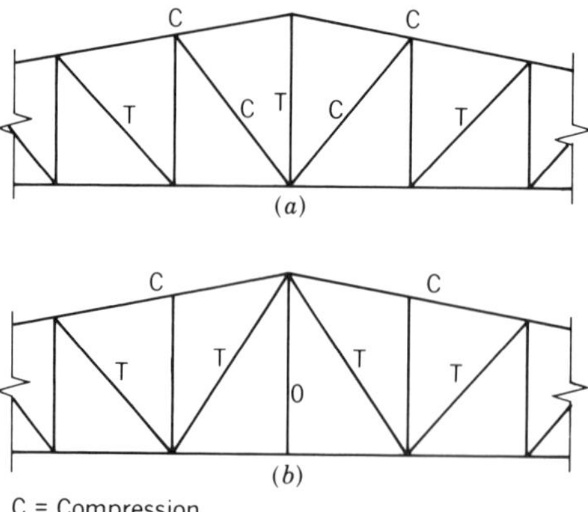

Fig. 12.16 Truss layout for peaked roof trusses.

design purposes, the points of contraflexure are assumed to be at midspan of each panel and at midheight of each post, their design can be approximated as follows: The chords have to be designed in addition to their axial force, for a bending moment of $M_{CH} = VL_p/4$, and the posts have to be designed in addition to their axial force, for a bending moment of $M_p = VL_p/2$ at each panel point. In the above equations, V is the maximum shear at each particular location and L_p is the length of each particular panel. Vierendeel trusses are uneconomical in material and costly in labor, and should, therefore, be used only when there is no other way to achieve the desired result. Since they have less stiffness than ordinary trusses, they often require overdesign of members and connections to keep their deformation within tolerable limits. To reduce costs and improve stiffness, the Vierendeel design is sometimes restricted only to certain panels of an otherwise ordinarily designed truss. In such a case, particular attention must be paid to the design and detailing of the points of transition.

3. To simplify erection, local concentration of details should be avoided. At truss intersections it is advisable to have the diagonals of the carrying truss meet at the bottom, and those of the carried truss meet at the top as shown in Fig. 12.18. Where the roof truss frames in at the center of the carrying truss, this condition is usually automatically obtained; however, where the roof trusses frame in at the 1/3 or 1/4 points of the carrying trusses, some diagonals may have to be reversed in their direction and be designed as compression members. The same reasoning applies where large girders frame in.

4. To simplify the connection of the bottom chord of a roof truss it should be lifted about 1–2 in. (25–50 mm) above the bottom chord of the carrying truss. This will permit bringing the bottom chord of the roof truss closer to the center of the carrying truss bottom chord, without having to cope the chord member. See Fig. 12.18*a*.

5. Truss verticals are ordinarily designed as double-angle members with the long legs back-to-back as is usual for web members in compression. Where purlins or beams frame in, the connections would be somewhat eccentric if the beams were fastened to the outstanding legs. Where larger reactions have to be accommodated, as in the case where trusses or girders frame in, it is advisable to substitute a two- or four-angle star section for the double-angle section in order to permit concentric connection of the load.

Force Evaluation

Before the evaluation of the forces in the various truss members is started, it is advisable to check whether the truss layout is internally statically determinate. This is the case if the equation $2p - m = \Sigma R$ is satisfied, where p is the number of panel points, m is the number of members in the truss, and R is the number of reactions. For ordinary simply supported conditions $\Sigma R = 3$.

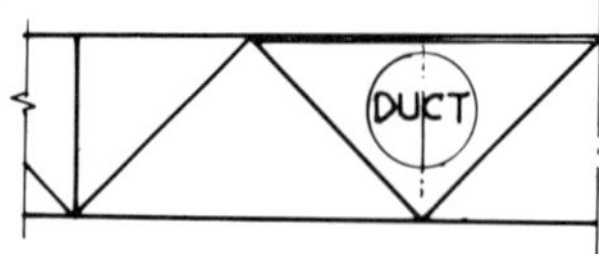

Fig. 12.17 Truss layout to permit passage of ducts.

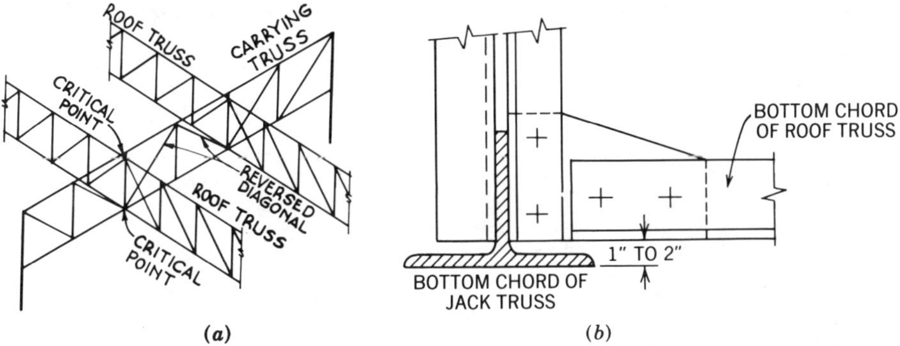

Fig. 12.18 Special conditions influencing truss layout. (*a*) Truss layout. (*b*) Detail of bottom chord intersection.

If a reliable computer program for the evaluation of the forces in the truss members is available, then it is, of course, the right thing to use. Some programs not only evaluate the forces, but also design the members, calculate the total weight of the truss, and find the maximum deflection at midspan. If such a program is not available, the easiest approach should be followed to obtain the desired information.

If the truss has parallel, or almost parallel, chords, the analytical beam approach is most practical, as shown in Fig. 12.19. The chord forces, T and B, in each panel of the top or bottom chord can easily be found from M/d, where M is the bending moment at the opposite panel point and d is the distance of the panel point from the centerline of the chord under consideration. The forces in the web members are determined by equating the vertical component of the force in the web member to the beam shear at the appropriate location.

If the truss chords are not parallel to each other, the most practical approach is a force diagram, also known under the name "Cremona plan" or "Maxwell diagram" [12.11, 12.12]. As a matter of fact, the more unusual the truss layout becomes, the more advantageous becomes this method. Using "Bow's" notation, it is simple and almost foolproof as long as the basic procedure is adhered to and no shortcuts are attempted. It also offers an easy way to determine the sign (+ tension, − compression) of the force in each member without cluttering up the diagram with all kinds of arrows. No particularly large scale is needed, because it is sufficiently exact to scale the forces to the nearest kip (kN) and disregard any decimals.

If a force diagram does not close, check the equilibrium of loads and reactions (force polygon). A wrong beginning can never lead to a correct ending.

Member Design

As far as trusses for ordinary building structures are concerned, the truss chords are always designed for the maximum force, or force combination, wherever it may occur along the whole length, and are executed as one continuous cross-section. Only in very heavy building structures, with trusses having long spans, or in bridge design, do designers deviate from this practice, because the waste would become excessive. In such cases the chords are designed for an average chord force, equal to one that occurs about at the quarter point, and are strengthened by addition of plates

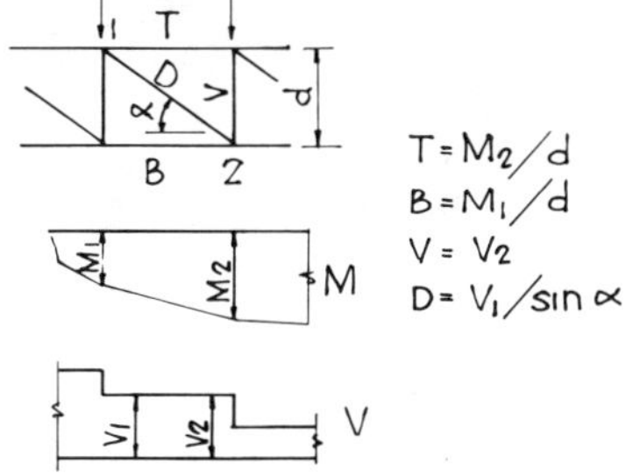

$T = M_2/d$
$B = M_1/d$
$V = V_2$
$D = V_1/\sin\alpha$

Fig. 12.19 Forces in parallel chord truss.

to satisfy increased requirements at midspan. The basic cross-section, however, runs (like a backbone) continuous over the whole length of the truss.

Two back-to-back angles with a gap for a gusset plate between them, represent the most commonly used cross-section for a truss chord, where high-strength bolts are used for connections, as is the case with most field-assembled and also with some shop-assembled trusses. If necessary, channels or flange plates can be added for strengthening. For a welded, shop-fabricated design, T-sections (usually half wide-flange sections) prove to be more economical and practical, because they permit the use of their vertical web portion as a gusset plate. It is advisable to use relatively deep, thin sections to gain the advantage of good lateral stiffness, and to have enough web section available for the connection of the web members. If necessary, the vertical portion of the T's can be extended by welding on extension gussets having the same thickness as the T webs. It is also advisable to select for both chords T-sections with about the same web thickness so that the web members will fit flat to the web or its extension; otherwise filler plates have to be added.

Although the bottom chord of the truss is usually a tension member and may never experience any stress reversal, it is good practice to make it laterally stiff. The *AISC Manual* [12.6] recommends a maximum slenderness $L/r = 240$.

Monorails, conveyor lines, hoists, and other hanging loads should be supported from panel points only. Where full freedom of locating and relocating such loads is desired, the bottom chords should be designed, in addition to their tensile capacity, flexurally stiff to transfer any intermediate load by beam action to the nearest panel points.

In some factories it is common practice to connect all hanging loads with clamps to the bottom chords, thus avoiding holes used for attaching bolts. If attachment with clamps can be enforced, it will contribute to an economical design of all truss bottom chords. Otherwise, it is good (safe) practice to deduct an additional hole from each outstanding bottom chord leg (a minimum total of two holes per chord) to permit fastening of hanging loads.

Before the design of the top chord section is started, the designer should have developed a complete concept of the top chord bracing. Since the design of the top chord section will be affected by the spacing of the lateral supports a decision must be made on whether every one or every second or third panel point will form a part of the lateral bracing system. Not every type of roof decking will stiffen a purlin sufficiently to make it an adequate lateral support. For ordinary roof construction, metal decking is usually considered sufficient to act as bracing because the decking is continuously welded to every purlin. Some designers are concerned by the fact that the bracing (roof decking), which is placed on top of the roof trusses, is not in the plane of the top chord but offset by the depth of the purlins. Other designers consider the roof decking as too flimsy to act as bracing for heavy trusses. Openings in the roof and other interferences may also reduce the effectiveness of the roof decking as top chord bracing. In such cases the designer may add independent top chord bracing. Such bracing is usually provided in the plane of the truss top chord, that is, either right below the bottom flanges of the purlins or level with their top flanges, depending on whether the purlins rest on top of the trusses or frame into them. In most cases it is an inexpensive addition that will provide the designer not only peace of mind, but will constitute also a worthwhile stiffening of the entire roof system. It is often possible to combine top chord bracing with the upper wind bracing, in which case it can be utilized for the purpose of top chord bracing at no extra cost.

Since compression governs the design of the top chord section of simply supported trusses, the most economical section will be one for which the slenderness ratios in each direction (about the x–x and y–y axes) are approximately the same. For buckling in the vertical direction (about the x–x axis), the panel point spacing of the top chord will provide the critical slenderness ratio; for buckling about the y–y axis (laterally), the distance between braced purlins will provide the critical slenderness ratio.

Double-angle sections are used mostly where gusset plates occur, as is the case with high-strength bolted trusses; T-sections are used mostly for welded trusses, and horizontal wide-flange sections for heavy large-span trusses with one or two planes of gussets. Unequal angles with their longer legs back-to-back or T-sections having about the same dimension in horizontal as in vertical direction, have about the same slenderness ratio in both directions. Equal angles back-to-back, unequal angles with their short legs back-to-back, or T-sections with a greater width than depth, have greater stiffnesses (smaller slenderness ratios) about their y–y axis, and are used where the lateral supports are spread farther apart than one panel point spacing.

Top chords are normally designed in one continuous piece for maximum stress combination, wherever it may occur. The waste connected with such a design is easily offset by the saving of connections and splices that would have been necessary otherwise.

For very large spans of 250–300 ft (75–90 m) it is often practical to use wide-flange sections for the top and bottom chords; the sections are placed with webs in the horizontal position as an H shape. To connect the web members, either T-sections are attached to the chord webs to be used as gussets or double gussets are connected to their vertical flanges. Different compressive forces in adjacent bays can be accommodated by using sections of the same nominal depth, but of different cross-sectional area, and by butting them through milling to reduce the size of the connection. Another way to accommodate higher forces without excessive waste is by designing the section for a force occurring at about the 1/4 point of the span and reinforcing it with plates where higher forces occur.

It is customary to use double-angle sections for the web members of ordinary trusses regardless of the type of connections used. For web members in compression, sections having about the same slenderness ratio in both directions are the most economical. Unequal leg angles, with their long legs back-to-back satisfy this requirement best and are the most commonly used sections. For tension members, any combination of sections, including all angles, can be used as long as it is symmetrical and connected in such a way that the centroid of the section coincides with the theoretical centerline of the web member. The only exception from the practice of using double angles, back-to-back, for web members is for verticals, where cross members such as beams or trusses frame in. In this case it is better to place the two angles in star fashion or to use four angles back-to-back to permit a concentric connection of the cross member.

Large trusses with wide-flange sections for the chords may also use wide-flange sections for the web members, preferably of the same nominal depth, with filler plates at the ends to adjust their depths to the width of the chords. Such filler plates should be preconnected to assure their action as part of the web member.

Connections

Structural steel connections are made by welding in the fabricating shop and by high-strength bolting in the field. The reason for this separation lies in the relative ease with which the proper tightening of a high-strength bolt can be checked, in contrast to the involved methods still required to check the quality and workmanship of a field weld, especially when it has to be done under difficult physical conditions.

High-strength bolts are available in two qualities. A325 bolts are generally used for the connection of A36 steel and steel having $F_y = 50$ ksi (345 MPa), and the A490 bolts for higher-strength steels.

Originally, high-strength bolts were tightened with calibrated impact wrenches which stopped their impounding automatically when the specified bolt tension (proof load) was attained. The proof load was a sufficient torque to produce a bolt tension of 0.7 of the tensile strength, 60 to 75 ksi (410 to 520 MPa) for A325 bolts and 75 to 80 ksi (520 to 550 MPa) for A490 bolts, computed for the nominal bolt cross-sectional area. Since the calibration was difficult to maintain, this method has been replaced generally by the *turn-of-the-nut method,* which is easier to perform in the field and of at least equal quality. This method still uses impact wrenches, but without calibration. It requires about a 2/3 to 3/4 turn of the nut, after all nuts in the connection have been brought to a "snug tight" position.

To become independent of the calibration of the impact wrenches, special types of bolts, such as those with twistoff shanks, or special washers, have been developed to provide for tightening by means of a *direct tension indicator.*

Because of the fact that it is bothersome to use threads of exact length to keep threads out of the shear plane, three conditions for the use of high-strength bolts are recognized by the *AISC Specification* [12.6]:

1. *Friction-type connection* (identified in Table 12.4 by an F). This type of connection has the highest factor of safety. The load is transferred entirely by friction, with a factor of safety against slip. The computation for design, however, uses a nominal allowable shear stress computed on the bolt cross-sectional area. The friction-type connection requires contact surfaces to be kept free of oil, paint, lacquer, and galvanizing, and the use of washers to ascertain that the threads will remain outside of the shear plane. This method is used primarily where fatigue, vibration, and similar conditions make its use essential, such as for crane girder connections, brackets, hoist beam connections, and machine platforms.

2. *Bearing-type connection* having threads *not* excluded from the shear plane (identified in Table 12.4 by an N). For bearing-type connections the allowable shear capacity of the bolt is to be compared with its bearing capacity against the connected metal to determine which one (the smaller one) should govern.

This most commonly used connection permits the contact surfaces to be painted and requires no washers. However, where field checking of the attained bolt tension is specified, washers should be used to facilitate the use of manual torque wrenches for checking.

3. *Bearing-type connection* having threads *ex*cluded from the shear plane (identified in Table 12.4 by an X). This connection utilizes the highest allowable shear values but requires strict control over the length of the threads, and the compulsory use of washers to ascertain that the threads will be kept out of the shear plane.

Table 12.4 Allowable Stresses in ksi (MPa) for Steel Fasteners*

		Unfinished Bolts		Rivets and Turned Bolts in Reamed Holes With A36		High-Strength Bolts					
						A325			A490		
		A307	A36	A502-1	A502-2	F	N	X	F	N	X
Shear		10.0 (70)	9.9 (70)	17.5 (120)	22.0 (150)	17.5 (120)	21.0 (145)	30.0 (210)	22.0 (150)	28.0 (195)	40.0 (280)
Bearing	$F_y = 36$ (250)	87.0 (600)		87.0 (600)		—	87.0 (600)		—	—	
	$F_y = 50$ (345)	—		—		—	—		—	97.5 (670)	
Tension		23.0 (160)	29.0 (200)	20.0 (140)	22.0 (150)	44.0 (340)			54.0 (370)		

* From *AISC Specification* [12.6].

Table 12.4 gives the AISC [12.6] allowable stresses for A325 and A490 bolts, as well as those for rivets and unfinished bolts.

Where high-strength bolts are provided for strength, they are seldom smaller than 3/4 in. (19 mm) in diameter, nor larger than 7/8 in. (22 mm) in diameter. On one job, they should preferably be all of the same size to prevent a smaller bolt from getting into a larger hole. As a guide for the selection of the proper bolt diameter, the designer should examine the largest connection (usually that of the end diagonal) in the truss and see that the number of bolts required in one row (in direction of loading) is not excessive. Seven bolts in one row are considered to be about the maximum to avoid overstressing of the extreme bolt in each row. Where more bolts are required, either a larger bolt diameter should be selected, or lug angles provided to increase the number of parallel rows of bolts. To avoid double slippage, the number of bolts connecting the lug angle to the web member should be increased by 25 to 30% over the number required to connect the lug angle to the gusset.

For spacing of bolts, end distance of bolts, and similar placement requirements, see *AISC Specification* [12.6].

Steels having a yield stress of 36 ksi (250 MPa) are usually welded using E60XX or E70XX electrodes. E70XX electrodes should be used for steels having a yield strength of 50 ksi (350 MPa). Higher-strength electrodes E80XX, E100XX, and E110XX are obtainable for higher strength steels. In each case, the particular number gives the tensile strength f_{ts} of the electrode material in ksi or in MPa if multiplied by 6.895. The allowable shear stress on a fillet weld is specified as $0.3f_{ts}$; that is, 18 ksi (125 MPa) for E60XX and 21 ksi (145 MPa) for E70XX electrodes. The stress on a fillet weld is measured through its "throat," but the size of the fillet weld is measured along its "leg" (here in the United States, but not necessarily elsewhere). The basic capacities of fillet welds can be calculated as follows. For E60XX electrodes: $0.707(0.125)(18) = 1.59$ kips per 1/8-in. weld size, 1 in. long [$0.707(1)(125) = 88.4$ N per 1-mm weld size, 1 mm long]; for E70XX electrodes: $0.707(0.125)(21) = 1.86$ kips per 1/8-in. weld size, 1 in. long [$0.707(1)(145) = 102.5$ N per 1-mm weld size, 1 mm long].

For rules regarding minimum weld sizes, maximum weld sizes, intermittent welds, and other requirements, see *AISC Specification* [12.6]. Chapter 6 also contains extensive information on welding.

For axially loaded members, such as all truss members, the centroid of the weld connection should coincide with the centerline of the member.

For the details of calculations for bolted and welded connections according to the *AISC Specification* [12.6] the reader is referred to standard textbooks, such as Salmon and Johnson [12.11].

Deflection and Camber

Where trusses have to be made shallower than ordinarily recommended, or when trusses are highly stressed because high-strength steel is used, or when the conditions are such that only a certain maximum deflection can be tolerated, then the deflection of the truss should be calculated to see whether or not its stiffness is adequate. In such a case the following approximate method is recommended because it is easily done and easily repeated during the design process. It can be checked by a more accurate method in the final design. If computer programs are available for this purpose, they should be used; but if no program is available and time is short, the method below will furnish an adequate answer.

For quick approximation it is expedient to consider the truss to act like a beam having a fictitious moment of inertia I_T, which can be expressed

$$I_T = A_{CH} h^2 K_I \tag{12.4.3}$$

where A_{CH} = gross area of the chord (average of top and bottom)
h = back-to-back dimension of the truss
K_I = dimensionless factor varying between 0.15 for flat panels to 0.30 for steep panels

The deflection Δ at midspan can then be approximated by the equation

$$\Delta = \frac{WL^3}{K_D I_T} \text{ in. (mm)} \tag{12.4.4}$$

where W = total load carried by the truss for the loading condition under consideration, kips (kN)
L = span, ft (m)
K_D = deflection factor varying from 800 (0.095) when W is applied as a concentrated load at midspan, to 1300 (0.155) when W is applied uniformly distributed over the entire span
I_T = moment of inertia of the truss evaluated from Eq. (12.4.3), in.^{4} (cm^4)

The modulus of elasticity for steel has been included in Eq. (12.4.4).

For a more accurate method of computing the deflection of a truss, the reader may use any elastic analysis method, such as the "unit load method" (that is, "virtual work method") as described in standard texts [12.12, 12.13] on structural analysis.

In calculating the total deflection on a roof structure, one should realize that all contributing deflections have to be superimposed on one another. In the case of a structure employing carrying trusses, for example, the total deflection of a purlin is obtained by adding the deflection of the carrying truss to the deflection of the roof truss and the deflection of the purlin itself.

Roof trusses seldom have attached to them structural elements that are sensitive to deflections. No definite rules have been established in this respect, but it is recommended that deflections of ordinary roof construction under the action of live load should not exceed 1/180 of the span. Where elements sensitive to deflections are connected to them, the maximum deflection under live load should not exceed 1/240 of the span, and in case a suspended plaster ceiling is supported the live load deflection should not exceed 1/360 of the span. This last limitation is the only one which occurs in regulations, such as the *AISC Specification* [12.6]; it has been in use for a long time and seems to work. In most cases the calculated deflection of a truss will be less than the limitations given above. This is good because trusses function best when they are stiff.

Trusses with level, or almost level, top chords must be checked for ponding. The *AISC Specification* [12.6] provides a formula for this purpose. As already discussed in Section 12.4.1 the structural advantage of a level truss is small; it should be used, therefore, only where accumulation of water at all times is specifically desired.

Members having spans exceeding 40 ft (12 m) should be cambered. Since trusses are used only when the span exceeds this dimension, it means that all trusses need to be cambered. Where the camber is to be calculated, it is usually made equal to a deflection under a load totaling (DL + 0.5LL); for this reason the truss should deflect slightly below the horizontal only under full design load, which seldom occurs.

For ordinary structural work the magnitude of the camber is seldom calculated, but usually assumed to be 1/8 in. per 10 ft of span (3 mm per 3 m), which is about equivalent to 1/1000 of the span. As little as this seems to be, it furnishes satisfactory results.

Camber must be built-in by properly changing the length of the web members and by careful bending of the involved chord sections to suit. Changes in the length of a chord are usually so small that they can ordinarily be disregarded.

The purpose of a camber is solely architectural or visual. There is no structural advantage gained by cambering, nor is the actual deflection affected by it in any way. Since it is only done for reasons of appearance, designers often content themselves with the cambering of the bottom chord alone.

It is common practice to give the cambered chord a parabolic shape with the maximum rise at midspan. To obtain this distortion, all web members have to be shortened by calculated amounts and the bottom chord itself must be bent to follow the curve. Where heavier chord sections are involved, they have to be pre-bent either by running them through a bending machine (set of adjustable rollers) or by preheating and bending. Where top and bottom chords are to be cambered by the same amount, the tension diagonals have to be shortened and the compression diagonals have to be lengthened, but the lengths of the verticals remain constant.

Camber usually diminishes with time, however, but at a rate that is sometimes difficult to predict. Where large span trusses are adjacent to doors, as it is the case in large warehouses or airplane hangars, and where they are supposed to carry some of the door guides, collateral steel, and other items, an erratic relief of the camber can be troublesome. In conditions like that, it is advisable to provide ample space and means for vertical adjustment, in both up *and* down directions, in order to become independent of the trusses own time schedule.

12.4.6 Columns

Loading

Columns must have sufficient strength to carry the maximum combination of dead and live load plus any vertical resultant caused by wind, earthquake, or other forces, as well as any bending moments caused by them. If the calculation of loadings is made by accumulating the various reactions of members framing into a particular column, mistakes will frequently occur. With the exception of special conditions, where large concentrated loads are present, the best way to arrive at the proper load is to establish the tributary area of loading and multiply it by the applicable area loads. For this purpose all loadings should be converted into loads per sq ft (m^2) of area. This method will also permit easy checking. Where future extension is anticipated, it is advisable to design at the beginning for the future load on the columns.

Positioning

The positioning of the columns should not be random because of its influence on the lateral stability of the structure.

Exterior Columns. The column webs of exterior columns should be placed perpendicular to the face of the wall, unless uniformity with other columns (in case of future extension) requires a different positioning. Exterior columns require, in addition to their vertical load capacity, flexural capacity to resist any moments created by wind. For this reason, the columns should be placed with their strong axis parallel to the face of the building. Since future extension columns have to be overdesigned for the future loading, they may have enough reserve strength to resist presently the additional wind moments about their weak axis. Extreme conformity, however, may create unreasonable requirements and costs, which do not always appear warranted.

Interior Columns. Column webs of interior columns should be positioned in the direction of the stiffer member (truss or girder), except where wind or other lateral forces influence the stability of the whole building. In such cases, column webs should be positioned in the direction in which the columns will provide the required lateral stiffness. In free-standing buildings of long rectangular plan, the column webs of all interior columns, or a predominant part of them, may have to be placed perpendicular to the long direction of the building or perpendicular to the direction in which the greater number of columns occur. For free-standing buildings of approximately square plan, or having about an equal number of columns in each direction, it may be desirable to turn columns alternately in order to obtain about the same lateral stiffness in each direction.

In structures using stiff welded frames or trusses, the columns, which are usually welded sections or built-up truss legs, are always positioned in the direction of the frame. The same is true for large span openings, such as airplane hangars. In such buildings, lateral stiffness perpendicular to the large span is not obtained by the columns alone but by bracing systems between the frame columns.

Shape

The most commonly used cross-section for a building column is a wide-flange section. Theoretically, the most economical section for axial compression is a tube or pipe, because its slenderness ratio (KL/r) is the same in all directions and the least in magnitude per unit area of cross-section. Practical difficulties in providing simple connections restrict the use of tubes or pipes to stores and residential buildings, where a simple cap plate can serve as support for beams or joists. Framed connections to pipe columns require fabrication such as slotting, cutting, and welding, which make such connections expensive. For ordinary conditions, involving primarily axial forces, column sizes having a depth of 8–14 in. (200–350 mm) and a width of not less than 8 in. (200 mm) are used. Strengthwise, these columns cover a wide scope of applications because of the many different weights that are rolled from every nominal depth, especially for W14. All wide-flange sections are ideally suited for splicing, because within one nominal depth all sections have a constant inside clearance and require reinforcement plates to match heavier sections only on the outside of the flanges. Where large bending moments exist in addition to the axial forces, deeper (> 14 in.) wide-flange sections (beam-columns) may be used. Where rolled wide-flange sections or sections reinforced with flange plates are not strong enough to carry the axial forces and bending moments, double-wall or box sections can be built-up, using a rolled section as the core and adding plates around the core section. Where tapered legs are desired, such sections can be obtained as described under "Frames" in Section 12.3.1. If none of these simplifications seem to be applicable, any shape, size, and thickness can be produced by welding the column section entirely from plates. For airplane hangars, assembly halls, stacker buildings, and similar tall structures, where it is not axial load but lateral stiffness that requires the greatest attention, trussed columns are often employed. Such columns can either be built up from two spaced column sections, usually having their webs placed perpendicular to the plane of the frame, and be laced or interconnected by web members; when the flexural capacity is of predominant importance, they may also be designed like regular trusses with chords and web members as discussed under truss design, in Section 12.4.5.

Design

Before the design of the building columns is undertaken, the designer has to investigate whether the building can be designed under the assumption that the columns are laterally supported at the top, that is, sidesway is prevented, or whether the design of the building columns has to be governed by a condition in which sidesway is uninhibited.

Since the first of the two conditions is by far the more economical one and also much more desirable because of its superior all-around structural quality, the designer should investigate first the layout of the structure to see whether it lends itself to fit into this category. If the conditions do not seem to satisfy the requirements, the designer should see what changes could be introduced to make the structure fit into this category. If it has been determined that nothing can be done to improve the lateral stiffness, the columns of the structure should be designed under the assumption that sidesway is uninhibited.

Basically, there are two conditions which can be utilized to design the structure with prevented sidesway:

1. The building is of limited size and wall bracings are architecturally permissible and are also located close enough to be called upon for lateral support of the upper column end.
2. There are shaft enclosures, like elevators, stairs, or similar within the building area, which either permit installation of wall bracing or are of stiff reinforced concrete or masonry construction so that the building may be tied to them for lateral support.

In either of the above cases the designer can use the most simple and most economical design approach of a column that is hinged and restrained from motion at the top and bottom. The section can then be selected for an allowable compressive stress governed by the slenderness ratio KL/r, using an effective length factor $K = 1$ (i.e., a pin-end column).

It is commonly assumed that sidesway in a structure is prevented if the sum of all moments of inertia of the bracing units (wall bracings, shear walls, masonry enclosures and similar) is about six times the sum of all moments of inertia of the braced members (columns).

If, however, the above described conditions, or other means of lateral support, are not available, and if the building extends over a large area with neither sufficient shaft enclosures, nor exterior walls in the vicinity to be utilized for a bracing, then the building will have to be designed under the assumption that sidesway is uninhibited.

In this case the evaluation of the K-factor depends on the ratio of the degree of restraint that can be mobilized at the top of the column and the degree of restraint that can be afforded by the stiffness of the column base and footing. The restraint at the top can be expressed by

$$G_T = \frac{\Sigma(I_C/L_C)}{\Sigma(I_G/L_G)}$$

where I_C and L_C are the moments of inertia and lengths of the column sections and I_G and L_G are the moments of inertia and lengths of the girders or beams framing in at the top, all in the direction under consideration. The restraint at the bottom is not so easy to define for one-story constructions. The Commentary to the *AISC Specification* [12.6] recommends the following extreme values: $G_B = 1$ where the soil is stiff and column base and footing are designed for full restraint, and $G_B = 10$ where the column rests on a regular base plate without restraint; in other words, where the condition approaches that of a hinge. Deviations may be taken care of by interpolation between these extremes. After having obtained the degree of restraint at top and bottom, a so-called Alignment Chart may be used to obtain an appropriate K value. The Alignment Chart can be found in the *AISC Manual* [12.5] under Commentary on AISC Specification. Examination of this Alignment Chart will show that if the top of a column is not laterally supported, that is, if sidesway is not prevented by special structural elements, a sizable restraint at top and bottom is essential to arrive at a reasonable economical design solution. If a building column is fully fixed to the foundation and stiffly framed to the roof construction consisting of trusses or girders, then a K value of 1.2 to 1.3 can usually be obtained, which is considered reasonable; but, of course there are cases where the K value is higher. Designers are advised to assume a low restraint at the column base, if possible, because just a slight yielding of the subsoil can make the column lose most of its restraint.

Column Bases

Where a column is carrying only an axial load, the base is not much more than a means of spreading the concentrated load to the somewhat softer pier material (concrete). The design method for the evaluation of the base plate thickness, as described in the *AISC Manual* [12.5], furnishes reasonable thicknesses for such plates, though their actual behavior is probably quite different. To transfer the load from the column to the base plate, the bottom end of the column should be either milled or at least square-cut to be reasonably smooth; the top of the base plate should be straightened and/or flattened to such a degree that full bearing of the column on the plate can be established by contact, not by welding. For this type of column, welding is used only for fastening, and is applied mostly only to one side of the flange or web, to save turning the column in the fabricating shop.

The anchor bolts are of secondary importance unless they are provided to resist uplift due to wind, as in the case of the deadman (right exterior column in Fig. 12.12) or overturning (force T in Fig. 12.20). In this case the anchor bolts have to be designed as tension members for the smallest cross-section within their threaded portion. In most ordinary cases, however, they are primarily needed during erection to keep the column in upright position while it is waiting for the installation of the trusses and beams. The anchor bolts should not, however, be too small, but in some relation to the size and height of the column. The nuts of the anchor bolts should be kept in reasonably tight condition to prevent accidental knocking over of a column when hit by a moving hoist tackle or cart.

Quite different is the condition when the column base is supposed to develop a restraining moment. In this case the anchor bolts and the stiffness of the base assume a definite purpose and require careful design. One of the most practical and least sophisticated approaches to such a design is shown in Fig. 12.20, where the stress distribution across the pier is assumed to be comparable to that of a reinforced concrete beam. In this method the compressive stress (bearing) under the base plate is assumed to be uniformly distributed over 1/4 of the depth of the plate and counteracted by the tension of the anchor bolts. Simple equilibrium equations can be set up, as shown in Fig. 12.20, to determine the necessary size of the base plate and the force in the anchor bolts. This approach is probably as good as any of the more sophisticated methods and is generally applicable regardless of the stiffness of the base plate, or whether stiffeners, channels, or wing-plates are needed. If channels are used, ship channels are preferred because of their wider flanges which will permit the use of larger anchor bolts without losing too much of the flange material. Welded stiffeners should be used on both sides of the bolts in this commonly used approach to make a base plate flexurally stiff [12.5]. There are other ways to achieve similar results.

The long-accepted common practice of filling the space between base plate and column pier with

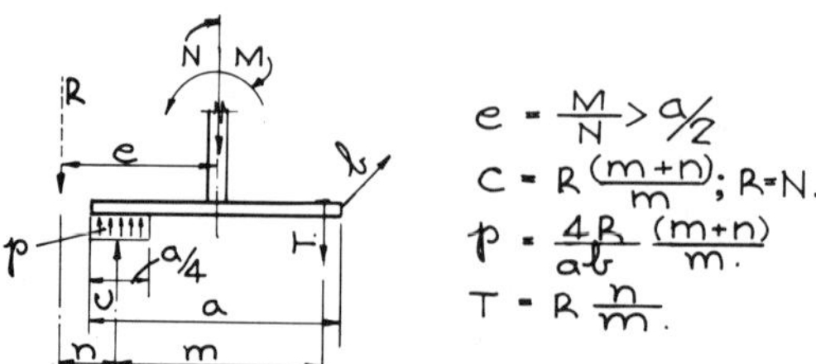

Fig. 12.20 Evaluation of forces acting on column base under overturning conditions.

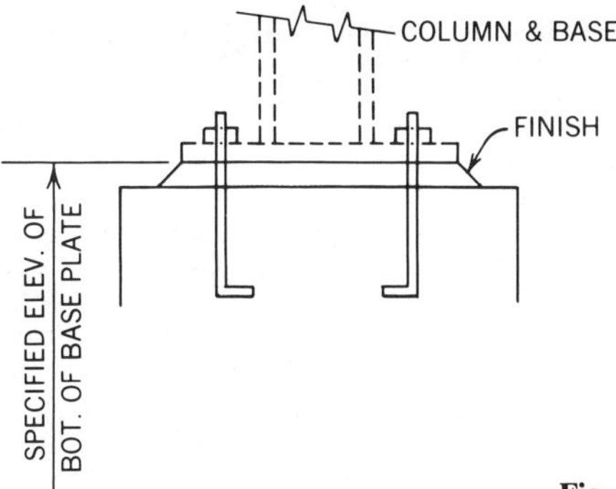

Fig. 12.21 Typical detail of finish below column base.

a stiff cement grout to assure complete transfer of the load over the rough concrete surface is still used in some parts of the United States. The thickness of this grout varies between 1 and 2 in. (25 and 50 mm), depending on the size of the base plate. Ordinarily 1/4-in. (6-mm) thick steel plates (masonry plates) are used; they are shimmed up at the four corners and leveled to the proper elevation on the anchor bolts which were placed when the concrete pier was cast. The masonry plate has oversized (bolt diameter + 5/16 in.) holes to fit over the protruding anchor bolts. The masonry plate forms the actual base on which the column base plate is set. In special cases (bridges, towers), where the anchor bolts are also used to transfer shear, the masonry plates are omitted and small shafts are kept open during the casting of the piers to receive the anchor bolts. In such cases, the actual base plate has the holes for the anchor bolts drilled to exact size and location. The whole base assembly, with the anchor bolts dangling from it, is lowered into place and the shafts as well as the space between pier and plate are filled with grout.

Presently such methods are considered to be too slow, too expensive, and unnecessary for structural purposes, and have been replaced by faster methods which seem to work satisfactorily. The grout is replaced by a finish of similar thickness, except that it is applied open ended, directly to the pier, and struck to a smooth level surface at the correct elevation as shown in Fig. 12.21. Care must be exercised to keep the protruding anchor bolts clean and unharmed. Small wooden frames are often used to keep the finish from spreading. In this way the quality of the finish as well as its top elevation can easily be checked by the steel fabricator before erection and can, if necessary, be corrected. The steel bases, with columns, are then carefully set on top of the finish and secured by tightening the nuts of the anchor bolts. This method is not only faster and less expensive, but it also provides a firm base for the steel column to sit on. Furthermore, it avoids common faulty conditions where the grout did not fill the entire space, or the corner shims were forgotten to be removed (see Fig. 12.21).

The niche formed between the flanges and the web of a building column is usually utilized for the placing of rainwater conductors and other vertical services which have to be turned on or off below the first floor elevation, as shown in Fig. 12.22. The largest space for such a turnoff is required by

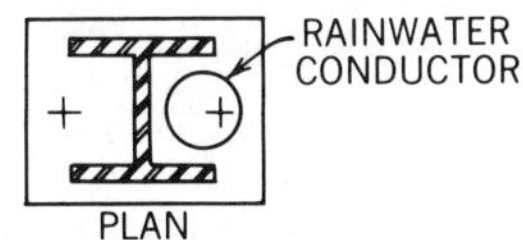

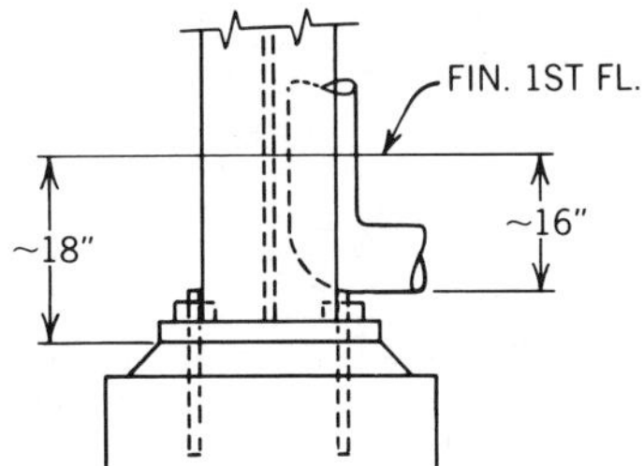

Fig. 12.22 Typical detail of down spout turnoff.

the rainwater conductor, approximately 16 in. (400 mm). It has been common practice, therefore, to place the bottom elevation of ordinary column base plates 18 in. (450 mm) below the finished first floor elevation, so that the turnoff can be installed on top of the base plate. Since the exact location of the columns affected by this condition is not known during the preliminary design stage, and even later subject to changes, it is advisable to place the bottom elevations of all columns at least 18 in. (450 mm) below the finished first floor elevation. Where rainwater conductors or other services occur along column bases stiffened by channels, wingplates, or other elements, the bases must be set low enough not to cause any interference with the turnoffs to be placed above.

12.4.7 Provisions to Support Material-Handling Equipment

Modern industrial buildings must have extensive provisions for lifting and moving loads of various magnitudes. Aside from floor-supported conveyors, the equipment used can generally be divided into three categories:

1. Monorails and suspended conveyors, belts, and similar.
2. Stationary and movable hoists.
3. Bridge cranes.

Monorails

Monorails, suspended conveyors, belts, and similar are widely used as means of transportation for bulk materials, single units, and parts.

Since the rail along which the units move is usually a part of their mechanism, the structural designer is only concerned with the reaction the equipment will exert on the structure. This reaction depends on the magnitude of the load itself, the weight of the conveyor, the spacing of the loads along the conveyor, and the distance between conveyor supports. The rails are normally fastened to the bottom chord of the trusses or to the bottom flange of the purlins or roof beams. In addition to the vertical reaction at these points, conveyors or monorails may also exert lateral forces, especially at curves and anchor points. Buildings that house such conveyors should have bottom chord bracing to transmit these forces into the columns. The vertical reactions from the monorails and conveyors should be applied to the truss bottom chords only at panel point locations. Where such a panel point does not occur at the particular location and the unit cannot be moved to meet one, either a secondary panel point as shown in Fig. 12.23*a* has to be created, or the bottom chord has to be given enough flexural stiffness to act as a header beam and transmit the load to the adjacent panel points. Where a possibility for changing the location can be foreseen, the entire bottom chord should be designed to be flexurally stiff. Where neither of these provisions has been incorporated in the design, headers will have to be added later, but, of course, at greater cost. Figure 12.23*b* shows two ways to change a regular bottom chord section into a flexurally stiff header. In any case, the stiffening reinforcements have to be extended at either side until they reach, or slightly extend beyond, the nearest panel point; there, they must be welded to the gusset or truss member for the full shear they are designed to transmit.

Stationary and Movable Hoists

Stationary and movable hoists are employed to lift larger loads and move them over relatively short (usually straight) distances. Such hoists normally run on, or are supported by, level beams (hoist beams) which either span between two adjacent trusses or are otherwise supported. Such hoists

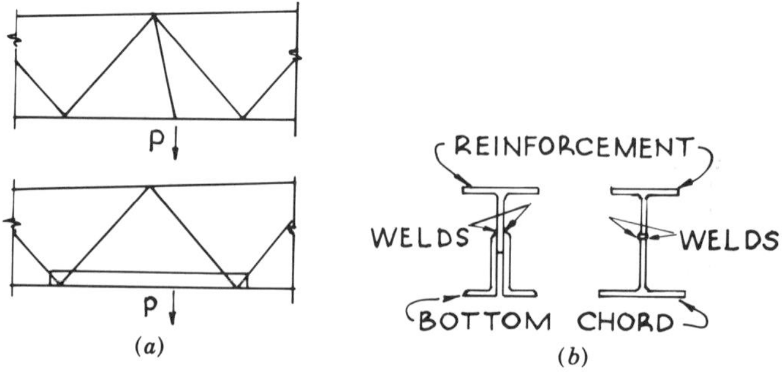

Fig. 12.23 Secondary framing to transfer concentrated loads to adjacent panel points.

consist of the lifting mechanism, that is, tackle and chain or electric motor, depending on the way they are operated. They usually hang from the beam on a small carriage that runs along the bottom flange of the hoist beam. Standard I-beam sections (so-called S shapes) are preferably used for such hoist beams because they have thicker webs (for greater shear resistance) and a sloping top of the bottom flange that permits toeing in of the running wheels of the hoist to make them self-aligning. The design of the hoist beam itself should be based on the critical positions causing maximum shear or maximum moment in the beam. Where the exact weight of the hoist is not known (which is usually the case at the time of a preliminary design), the designer may assume the total reaction of an electrically operated hoist to be about 1.6 times its required capacity. This value consists of W (the capacity), $0.35W$ (for the weight of the hoist), and $0.25W$ (for impact). The impact may be reduced to $0.15W$ if the hoist is manually operated. Reactions from hoists may amount to several tons and should only be applied at panel point locations.

Bridge Cranes

Bridge cranes service large areas with their capacity to lift heavy loads and to move them to any place in the building. Bridge cranes move along runways (crane rails) placed on top of welded or built-up beams (crane girders) which form part of the structural design. The bridge itself is furnished by the crane manufacturer and carries the main hoist and usually a smaller auxiliary hoist, both moving on tracks carried by the crane bridge. Since it is often difficult to fasten the pick-up loop of a machine or load precisely over its centroid, the auxiliary hoist is used to balance and stabilize the load during lifting and transport. At each end of the bridge there is a carriage with two to four wheels running on top of the crane rail.

The minimum weight of the rail is given in lb/yd and is specified by the crane manufacturer, who also furnishes the physical properties (dimensions and weights) of the various parts and the maximum reaction (wheel load) of the crane bridge with the hoist in its extreme position, that is, closest to the crane rails.

The maximum wheel load and other pertinent dimensions and weights for a bridge crane of desired capacity and span can be obtained from a manufacturer's catalog. For preliminary design purposes, however, if no such catalog is available yet, the maximum wheel load can roughly be approximated from the formulas given below. Because of the great number of manufacturers and variations in design, the error in wheel loads can be as much as ±20%.

For two-wheel carriages:

$$W = (7 + 1.4C)\left(\frac{40 + L/2}{60}\right) \qquad \text{(Inch-Pound units)} \qquad (12.4.3)$$

For four-wheel carriages:

$$W = (5 + 0.6C)\left(\frac{40 + L/2}{60}\right) \qquad \text{(Inch-Pound units)} \qquad (12.4.4)$$

where W = the maximum wheel load, kips
C = crane capacity, tons
L = span of bridge, ft

The design of crane girders is based on the maximum wheel load of the carriage as determined above, placed in the positions to give the maximum bending moment and shear. An impact factor of 25% is recommended to be used in the design.

In large buildings two or more bridge cranes may use the same crane rails. For physical reasons there is a certain minimum distance which has to be maintained between two bridge cranes to assure their free movability. The manufacturer not only furnishes this information but also provides spacers (bumpers) protruding from the carriages to ascertain that this distance is always maintained. Where the possibility exists that two or more cranes could simultaneously be working on the same runway, then the wheel loads from all cranes have to be considered in the design of the crane girders. This results sometimes in very heavy crane girders. Owners do not always need such close spacing of the bridge cranes and are often willing to accept a larger distance between two adjacent bridge cranes in order to economize on the crane girder design. In such a case, the manufacturer can be requested to extend the spacers to the specified length.

The crane rails rest directly on the top flange of the crane girders and are fastened to them with adjustable clamps. These clamps should be carefully aligned to keep the travel of the carriage as shockfree as possible. To give the crane bridge the necessary lateral guidance, the wheels of the carriage are double rimmed. Large bridge cranes often have double rimmed wheels only at one side

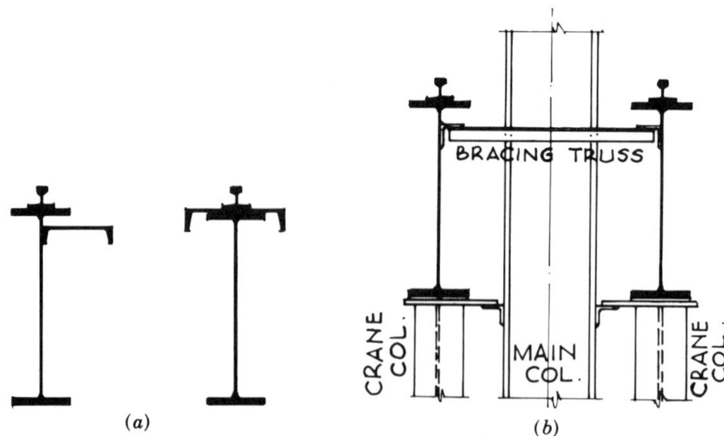

Fig. 12.24 Lateral supports for the compression flange of crane girders. (*a*) Crane girder cross-sections. (*b*) Bracing arrangement.

of the bridge, whereas the wheels on the opposite side are plain rollers without rims. This is done to prevent binding caused by changes in the length of the bridge due to temperature. Where rolled sections are not sufficient, their capacity can be increased by inserting a piece of web plate or by adding some extra flange plates. Crane girders may also entirely be welded up from plates. Deep plate girders require web stiffeners.

The top flange of a crane girder requires lateral stiffness not only to develop its full flexural capacity but also to resist lateral forces caused by the starting and stopping of the crane trolley (hoist carriage) on the crane bridge. An allowance for somewhat inclined load pickups is also needed. AISC specifies a horizontal force of 20% of the sum of the lifted load (capacity) plus the weight of the trolley itself. The horizontal force is applied at the top of the crane rail, one-half on either side of the crane bridge. For bridge cranes having rimmed wheels only at one side of the bridge, the total horizontal force should be applied at either side of the crane bridge, since it is hard to predict on which side the rimmed wheels will be placed. The bending moment created by the horizontal force can either be taken by the top flange of the crane girder in biaxial bending or it can be resisted independently by a horizontal channel, fastened to the web of the crane girder, right below the top flange, as shown in Fig. 12.24*a*. For large cranes a small horizontal truss brace may be used in the same location. Where craneways are located in two adjacent building bays, the brace may be placed between the adjacent crane girders and serve both craneways at the same time as shown in Fig. 12.24*b*. By some means the horizontal bracing has to be connected to the building columns in order to transfer all horizontal shear forces ultimately down to the ground.

At one time it was common practice to support crane girders from the main building columns by means of brackets. In this case, the entire cross section of the building, consisting of truss and columns, acted as a frame to resist the moments created by the vertical and horizontal reactions of the crane girders. Such designs resulted in heavy building columns and often caused undesirable structural conditions such as vibrations, cracking, or roof leaks. Not only was every impact on the bracket transmitted to the entire building, but the bracket connections themselves were very unfavorably stressed and plagued by fatigue. Such bracket connections in riveted construction were subject to failure by having the rivets work loose, and frequently had to be replaced by using turned bolts in reamed holes. Presently, column brackets for the support of crane girders are used primarily for small crane loads and for short spans.

Another way to provide support for a crane girder is by using step-columns. Such columns have a larger cross-section in their lower portion, below the crane girder, and a smaller column section above, supporting the roof. As before, the whole building becomes a frame consisting of stepped columns and a stiffly connected roof truss. Tables have been worked out to determine the corner and base moments of such building frames for various conditions of loading. The design results in somewhat lighter building columns, because the upper part is greatly reduced in size, but has otherwise similar disadvantages as the design using crane girder brackets.

For heavy crane loads and long crane girder spans the most practical approach is to use separate crane columns, set in front of the main building columns, as shown at the right support in Fig. 12.25. These crane columns are placed with their webs perpendicular to the direction of the webs of the main columns in order to utilize their stiffness for lateral bracing. This way, the crane column

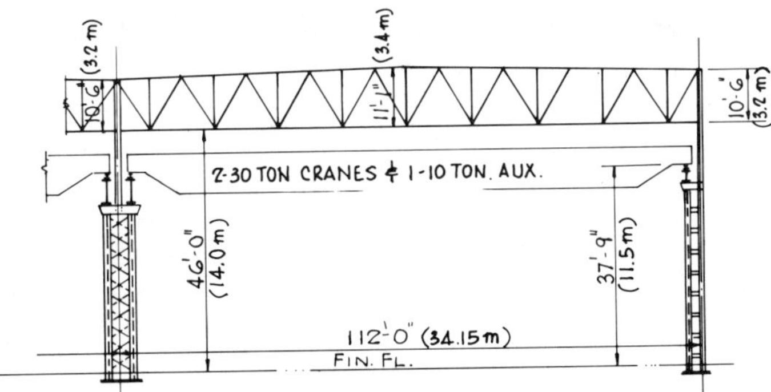

Fig. 12.25 Factory building having bridge cranes. Courtesy of Albert Kahn Associates, Detroit.

can be designed just strong enough to carry the reaction of the crane girder, but remains otherwise unaffected by the building design. Such columns are usually topped by a cap plate to provide a seat for the crane girder which is laterally braced to the main column. For very long spans and heavy loads, intermediate crane columns can be provided which have to be laterally supported by an A-frame or similar, because the lateral support of the main column is missing.

In order not to have three parallel columns below the elevation of the crane girders, the design is sometimes modified by stopping the center (building) column somewhat below the crane girders and transferring its load by means of a header beam into the crane columns as shown at the left support in Fig. 12.25. As a substitute for the missing stiffness of the center (building) column, the crane columns are usually laced for lateral stiffness. Crane girders and columns have also to resist horizontal forces in the longitudinal direction produced by the moving of the crane bridge. These forces must be applied longitudinally at the top of the rail with a magnitude of 10% of the maximum wheel load, according to the *AISC Specification* [12.6]. Bracing in one of the bays (preferably near the middle of the run) or frame action by means of knee braces is usually utilized to transmit this force to the ground. To assure that these knee braces will only create frame action for the longitudinal horizontal force, but will not create any bending moments in the relatively small crane columns due to any vertical load, they should be provided at one side of the bay only, as in Fig. 12.26*a*. To avoid unsymmetrical arrangement or, for better appearance, they can be placed symmetrically at every second crane column, as shown in Fig. 12.26*b*.

Underslung Cranes

Where the entire floor area of a building does not have to be serviced by the crane, or where the lifting capacity is not very large, underslung cranes are often used. In this type of crane, the carriages for the crane bridge run on the bottom flanges of runway beams, as shown in Fig. 12.27. The runway beams themselves are suspended from the bottom chords of the trusses similar to hoist beams. Since the crane bridge represents a major portion of the truss load, the truss must be laid out in such a way that a bottom chord panel point occurs at the location where the runway beam is to be connected. For longer spans an intermediate runway may be provided at midspan of the roof trusses. This design penalizes the truss design to some degree, but not excessively, because the

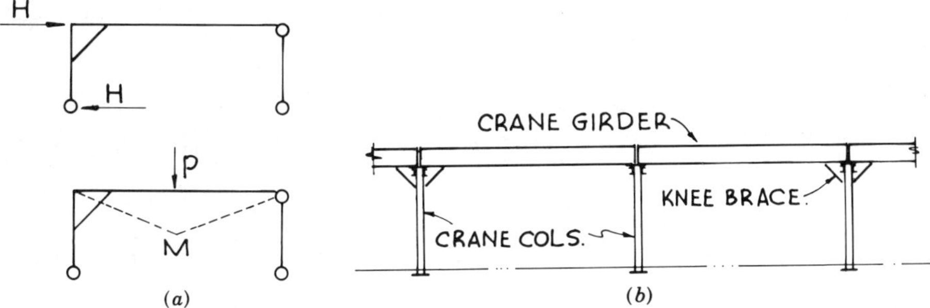

Fig. 12.26 Longitudinal stiffness of crane girders.

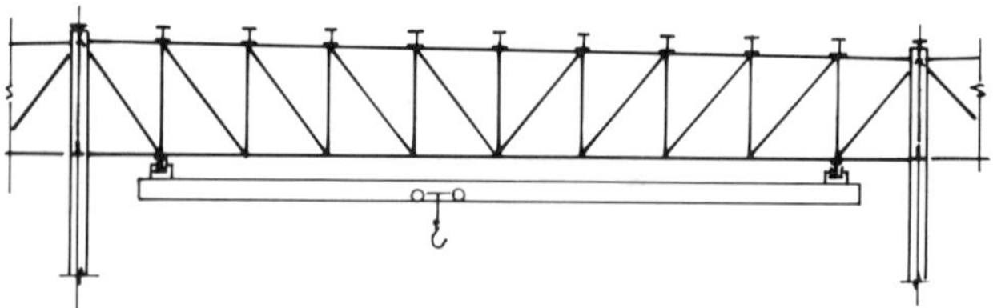

Fig. 12.27 Truss with underslung crane (schematic).

runway locations are kept close to the supporting columns. Manufacturer's catalogs provide all necessary data regarding lifting capacities, weights, and clearances. Standard I-beams (S shapes) are preferably used for runways because they provide a self-aligning capability as discussed above under hoist beams.

Crane girder columns must always be designed with full restraint at their base whether they carry the crane girder on a bracket or on a separate small column placed in front of the main building column. Where the main column and the crane girder column are separate, they should both use the same base plate and be anchored to the footing. Channels or welded wing plates give the base plate the required stiffness and anchor bolts transmit the moment capacity into the footing. Base plates are considered a part of the column and superstructure and any impact from a load above should be considered in their design. In the evaluation of the soil pressure beneath the footing, however, effects of impact may be disregarded.

12.4.8 Bracing

The designer should think about bracing schemes throughout the design procedure. Proper bracing means bracing in the right amount and at the right location. Materialwise and costwise, bracing is a small item, but an item of paramount importance on the overall stability and functional behavior of a structure. This fact is often not sufficiently well recognized by some designers.

Bracing, in general, is defined as any member or arrangement of members (such as a truss) that either resists forces not directly related to gravity, or provides stability and/or stiffness necessary for the proper functioning of the structure. For this reason, it may be divided into two broad categories, namely *strength bracing* and *stiffness bracing*.

Strength Bracing—General

To the category of strength bracing belong the following bracing systems whose members can be designed for a certain known force.

1. Wind bracing, whether in horizontal position as a part of the roof framing, or in vertical position, such as wall bracing or a knee brace.
2. Crane runway bracing in longitudinal as well as in transverse direction (see Section 12.4.7).
3. Wales, shores, or other members to resist earth and water pressures.
4. Bracing to resist seismic forces or blast loading.

Such bracing systems may assume the appearance of beams, struts, ties, or that of a whole truss assembly, depending on the magnitude of the force, its location within the structure, and the size of area over which it acts. In every case strength bracing resists forces that either exist continuously or are assumed to exist under certain conditions. At any rate, the forces are real at certain times and the stresses caused by them are calculable. Since these forces usually act horizontally, somewhere above the ground, and since the most effective way to deal with them is to absorb them by gravity and friction, designers have to find a reasonable, uninterrupted way to direct the forces to the ground and react them with the foundation of the building. If the design is lacking in this respect, nature will seek its own path to carry these forces to the ground, sometimes in a way not compatible with the design.

Strength Bracing—Wind

The most common type of strength bracing is wind bracing. This bracing must resist the lateral forces caused by wind action on a building and transmit them safely to the ground by way of the foundation. In this respect columns act like vertical beams and girts like floor members. The upper ends of the beam-like columns are laterally supported either by frame action of the whole building cross-section or by a horizontal truss (wind bracing) located either at the level of the top or bottom chord of the roof trusses. Such a wind truss usually finds its reaction at the two opposite ends of the building in the form of wall bracing, which carries the reactions to the ground. This combination of wall bracing and wind truss (at roof level) actually forms what may be called a *bracing belt*. Such a bracing belt is especially useful in resisting wind in the direction perpendicular to the plane of the main trusses, because there is no frame action available in this direction. If such bracing belts were to be provided near each end of the building, the building between them would be held too firmly and could not respond to movements such as those due to temperature variation. It is therefore advisable to place bracing belts in the middle ⅓ of the building, if the building is about 100 ft (30 m) long, and let the wall girts on either side act as struts. For longer buildings two bracing belts are indicated, one at approximately each ⅓ point of the building.

Wind columns are usually provided to reduce the span of the girts. They are designed like vertical beams and should have vertical-slotted-hole connections at the top. Such connections can provide all necessary lateral support but, will at the same time, prevent interference with the vertical deformation of the roof construction.

Other strength bracing systems are treated in a similar manner. The general treatment of bridge cranes in Section 12.4.7 discussed crane runway bracing.

Stiffness Bracing—General

The second category of bracing systems is usually even less understood than strength bracing, because it is not governed by any force but by stability or stiffness considerations. To this category belong such items as purlin bracing, top chord bracing, lateral support of compression flanges, column lacing and battens, and sway bracing. For members forming a part of such bracing systems, strength is not the most important factor in their design, because the least cross-section, as governed by slenderness considerations, will usually suffice. The *AISC Manual* [12.5] specifies a minimum slenderness ratio L/r of 300 for bracing members. Their mere existence and readiness for action when needed is of utmost importance, not their strength. A designer must realize that the compression flange of a flexural member, whether it is a truss, a girder, or a beam, has the tendency to buckle laterally, unless it is prevented from doing so. Furthermore, any series of parallel compression members, be it a set of purlins or a group of building columns, cannot perform as expected without definite provisions to prevent the possibility of lateral swaying and buckling.

Such considerations have to be present in the mind of an engineer in all stages of design. Recognition of this omnipresent danger, and expertise in counteracting it constitutes a main portion of the designer's responsibility. There are usually many ways to take proper care of such bracing requirements; some designers do it more simply than others. The crucial point is that it be done.

There is no reason why a bracing system cannot serve two or more purposes at the same time; for example, wall bracing may serve simultaneously as wind bracing and as sway bracing. Another example of bracing that can be used for a dual purpose is a bottom chord bracing. Where trusses are placed parallel and adjacent to an end wall of a building, as is the case with buildings earmarked for future extension, the intermediate wind columns will stop at the bottom chord of the trusses and should be connected with vertical slotted holes as discussed under *strength bracing*. To take the horizontal thrust exerted at the upper column ends, bracing is usually provided in the first bay. In this case, the bottom chords of the first two trusses serve also as chords of the wind bracing; the tension diagonals are made up of double-angle sections and the compression posts are of light wide-flange sections.

Where conveyor lines are supported from the bottom chords of the trusses, lateral stiffness is required to resist all anchor-forces caused by the action of the conveyor chain. This is achieved by introducing bottom chord bracing at each end of the building, and by providing lateral support to all intermediate trusses by means of sag rods. The sag rods are placed at the locations of the panel points of the wind-bracing truss.

Sag rods, in general, are used with many kinds of bracing. They are lightweight, easy to install, and therefore economical. The advantage of easy installation, however, incorporates also the danger of easy removal by anyone without permission. Engineers should be aware of that and use angles instead of rods, because the cutting off of an angle will, in all probability, not be undertaken without special permission.

Sag rods are useful only if they are straight. They do not have to be stressed but they must be taut. They are usually installed loosely and then tightened after erection has been completed. A loosely hanging sag rod, as is often seen, is of no use whatsoever.

Top Chord Bracing

This is used to provide lateral stiffness to the compression chord of a truss. Since the panel points of the top chord represent the locations for the lateral support of the top chord, its layout should be conceived simultaneously with the design of the top chord section of the truss. There is little effort

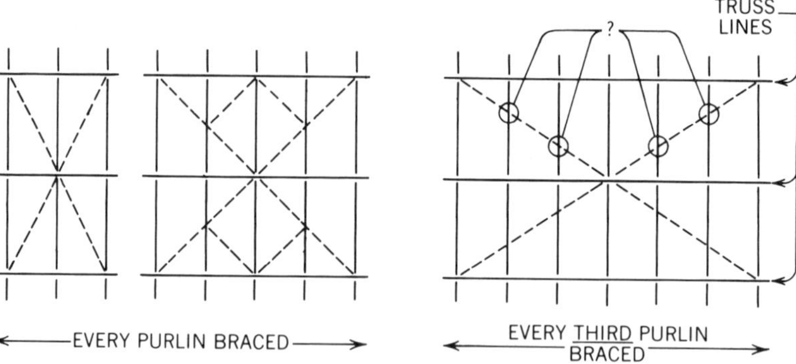

Fig. 12.28 Various roof bracing arrangements.

connected with the provision of top chord bracing. The chords of two adjacent trusses can be utilized to serve as chords for the bracing truss and the purlins can be used as posts. The only extra cost lies in the diagonals which are usually crossed to act in tension in either direction. Where purlins are placed on top of the roof trusses, the diagonals of the top chord bracing will be located below the purlin's bottom flange; where purlins frame into the verticals of the roof truss, the diagonals of the bracing will have to be placed between the purlins in the plane of their top flanges. This arrangement requires more pieces and more intermediate connections than the first type. In both kinds of bracing the purlins become not only the compression posts of the bracing system, but also the struts or ties for the lateral support of adjacent trusses. This way, top chord bracing does not have to be placed in every bay, and rows of parallel trusses may be braced at the same time by the lateral stiffness of the braced bay. Because of unavoidable "give" in the connections, accompanied by some lateral deformations of the purlins, some effectiveness of the bracing is lost with increasing distance from the bracing system. It is, therefore, advisable to repeat top chord bracing about every 100 to 150 ft (30 to 45 m). For shorter buildings it is often enough to use one top chord bracing system about in the middle of the building. In cases where it is possible to utilize the top chord bracing system also as wind bracing, it may become a part of the bracing belt discussed above as strength bracing.

Where the number of purlin spaces crossed by one bracing diagonal is one, two, or four, the designer has a chance to make every intermediate purlin location braced depending on whether or not the brace is connected to the bottom flange of the intermediate purlin. However, where the diagonal crosses, say, three purlin spaces, the intermediate purlin location is not braced even if the brace is connected to the purlins because they are held only by the flexural stiffness of the bracing angle, which is negligible. See Fig. 12.28.

Purlin Bracing

Where the roof decking does not provide sufficient lateral support to use the maximum allowable flexural stress in the design of a purlin, for instance, where cement tiles or insulating planks are used, additional lateral support has to be provided by means of a bracing system. Such points of support are usually provided at midspan, or at closer spacing, depending on the span of the purlin. The bracing itself is often started in the second bay from the face of the building to avoid interference with rainwater conductors and other standpipes which are usually located along the face of the building. Such a braced bay is provided at either end of the building and all other purlins between these two braced bays are laterally held by sag rods from either side, as shown in Fig. 12.29. The plane of such a purlin bracing is about 3 in. (75 mm) below the top flange of the purlins. This distance is considered close enough to give the top flange adequate lateral support and at the same time large enough to permit easy installation.

Sway Bracing

Any series or row of columns or compression struts has to be prevented from lateral swaying in either direction; if swaying cannot be controlled, it will invariably lead to failure. This is not a matter of strength but of stability and applies not only to columns but also to shores supporting formwork for reinforced concrete, to purlins acting as struts in a top chord bracing system, and to others. Generally speaking, such bracing controls swaying by preventing the rectangular spaces between the members from changing into parallelograms. The most common way to prevent such a

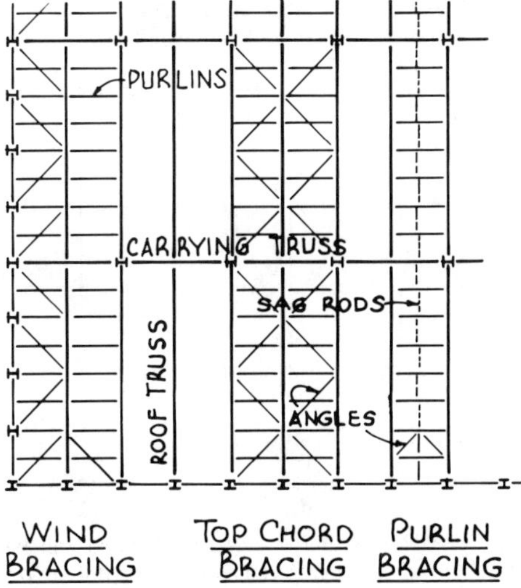

Fig. 12.29 Part of typical roof framing plan.

deformation is by providing X-bracing in some of the bays (not necessarily in all of them). Braced bays should preferably be provided about five to six bays apart. Sway bracing does not have to occur in every row as long as a truss or a diaphragm is provided to let intermediate rows partake in the stiffness of the braced rows. Another way of providing lateral stability is through transforming the columns and beams into stiff frames by means of kneebraces or rigid end connections. Such frames do not have to be provided in every row as long as specific provisions are made that the top of every compression member is laterally held in each direction.

Walls and shaft enclosures for stairs, elevators, and ducts built of cast-in-place concrete or masonry can similarly act as shear walls or stiffeners as long as the steel structure is properly tied to them.

12.5 CONCRETE STRUCTURES—PARAMETERS FOR PRELIMINARY DESIGN

12.5.1 General

Reinforced concrete is not as often used for single-story open-space buildings as is structural steel, though the first automobile factories built in this country in the early 1920s, in and around Detroit, were constructed mainly of reinforced concrete. Reinforced concrete was a relatively new material which produced buildings that were strong, robust, fireproof, and economical. They were also capable of sustaining high levels of accidental overloading and other abuses, which is borne out by the fact that some of them are still standing and in use after more than 60 years of continuous service. Battered and cracked as most of them now are, and inadequate in many respects, they still behave like old work horses, with little glamour but with lots of stamina.

These factory buildings were in general not designed as "single-story open-space" buildings, as would presently be done, but rather as two- or three-story buildings as explained below. Changed circumstances, such as the lack of steel for structural purposes during World War II, induced engineers to develop design schemes that provided large, single-story open-space factory areas for the war effort, with minimum construction time and with a minimum amount of reinforcing steel.

Reinforced concrete is a heavy building material and needs a sizable cross-section just to carry its own weight. In the early decades of reinforced concrete design and construction, before shell roofs were developed, the flat roofs that were used consisted mostly of joist construction, beam-slabs, or flat slabs. Consequently, it was obvious that just a little extra reinforcement could make this type of construction carry a reasonable floor live load, and without any extra cost for formwork, which has always been expensive. The economical advantage gained from repeated use of formwork was recognized early and taken into consideration. These were probably the main reasons for the use of more than one story in the early factory buildings constructed of reinforced concrete.

At the same time, availability of a highly suitable design concept, namely that of a flat slab system, was a strong contributing factor for the success of reinforced concrete. The flat slab system provided the engineer with a floor of high load capacity and simple construction. It permitted not only the use of plain boards (later plywood) for the bottom forms of the slab, but allowed also a standardization of the columns and capitals so that the forms could be rented and reused on other jobs. In addition, the overall simplicity in positioning the steel reinforcement and placing the concrete on a flat working surface made the whole design very practical. Since there were not protruding beams or girders to contend with, the overall equal clearance, so characteristic for this design, provided almost ideal working conditions.

Most of the manufacturing processes in those days were performed on the floor itself and the materials and parts were moved on floor-supported conveyors; the most important part of the structure, therefore, was the floor slab, which had to be stiff, strong and reliable, as provided by a flat slab design. Later on, when gasoline-powered lift trucks became available, the apparently "unlimited" stiffness of the floor slabs was soon exhausted when lift trucks were indiscriminately used on floors that were never designed for loads of such magnitude. Consequently, cracks and spalls occurred in the floor slabs which, however, rarely affected the strength of the building.

Reinforced concrete is a fireproof construction, but this does not mean the prolonged intense heat would not cause the protective concrete cover to crack and to spall. After removal of the damaged portions, however, the structure can be restored to its original condition. Since open fires used to play a much greater role in the early days of manufacturing, and sprinklers were not yet available, a building that could effectively withstand fire was highly desirable.

Use of reinforced concrete for industrial buildings proceeded at a much slower pace, however, than use of structural steel, even though reinforced concrete did not require the long lead time for mill ordering and fabrication, and needed a shorter time for finishing after the concrete was placed and had hardened.

The early manufacturing buildings constructed of reinforced concrete incorporated all the above advantages as well as one deficiency which ultimately turned out to be decisive; they did not have enough "flexibility." In those days flexibility of building use was not needed as much as it is needed now. Once a product design process has been established, it stayed essentially the same for a long period. The overall stiff floor slab and the clearance up to the ceiling provided all the user flexibility that was needed then.

At the time of World War II, steel framing systems had already replaced reinforced concrete for heavy and medium-heavy manufacturing buildings. When the war broke out, and the use of steel for structural purposes was curtailed to the very minimum, industry resurrected the design of reinforced concrete buildings and utilized all possible advantages. This endeavor resulted in the construction of some of the largest plants in the United States, built in record time and using just a fraction of the steel reinforcement that is normally used for plants of equal column spacing. Since the plants were built for the manufacture of war materials, they were considered to be of "temporary" nature and permitted to be designed for stresses 20% higher than those permitted by the existing building codes. Some of these "temporary" plants are still in existence and have been in continuous use for 40 years. A more detailed description of one of these plants, which is a true single-story open-space building made of reinforced concrete, is given here. This facility was one of the largest plants ever built, not only of reinforced concrete but of any material. The principal reason for including this information is not because of the size of the plant, but because its design concept provides an excellent example of what *can* be done under trying conditions.

The plant, Chrysler Corporation's Dodge Machine and Assembly Building, was designed by Albert Kahn Associates in Detroit and constructed in Chicago in 1942. It covered, including all ancillary buildings, an area of about $5\frac{1}{2}$ million sq ft (500,000 m^2). The main building alone had a size of 2340 × 1520 ft (713 × 463 m), which is more than $3\frac{1}{2}$ million sq ft (300,000 m^2) under one roof. With the exception of some smaller areas, the building was designed as a single-story building with a column spacing of 30 × 38 ft (9.1 × 11.6 m) and a clear height under the roof construction of 21.83 ft (6.65 m). The roof construction (as shown in the typical cross section of Fig. 12.30) consisted of a segmental barrel-shaped slab, thickened toward the supports, spanning in the long direction of the bay. Each barrel was considered to act primarily as a plain concrete arch, reinforced only with a light wire mesh. To take the horizontal thrust, ties were provided at 15 ft (4.55 m) on center. The ties also form the basis for the stiffening ribs, which had several openings to permit passage of pipes and conduits. Each intermediate rib was supported by a continuous girder, haunched at the columns, and made large enough not to require any shear reinforcement. The few stirrups that were provided were used only to hold the longitudinal reinforcement in place. The columns were designed large enough to require only minimum steel reinforcement. The footings were either spread footings or stepped footings as required by the conditions of the subsoil. All footings were made of structural plain concrete. The formwork consisted of mobile units, one bay wide and four bays long, sitting on jacks and running on wheels. The wheels supporting the formwork carriage ran on rails set on the concrete floor slab, which was always placed ahead of the roof construction. A total of 60 such units were used at the same time. Starting at each end and working toward the middle, in alternate bays, they were moved and reset every seven days. Since work proceeded at 24 hr a day,

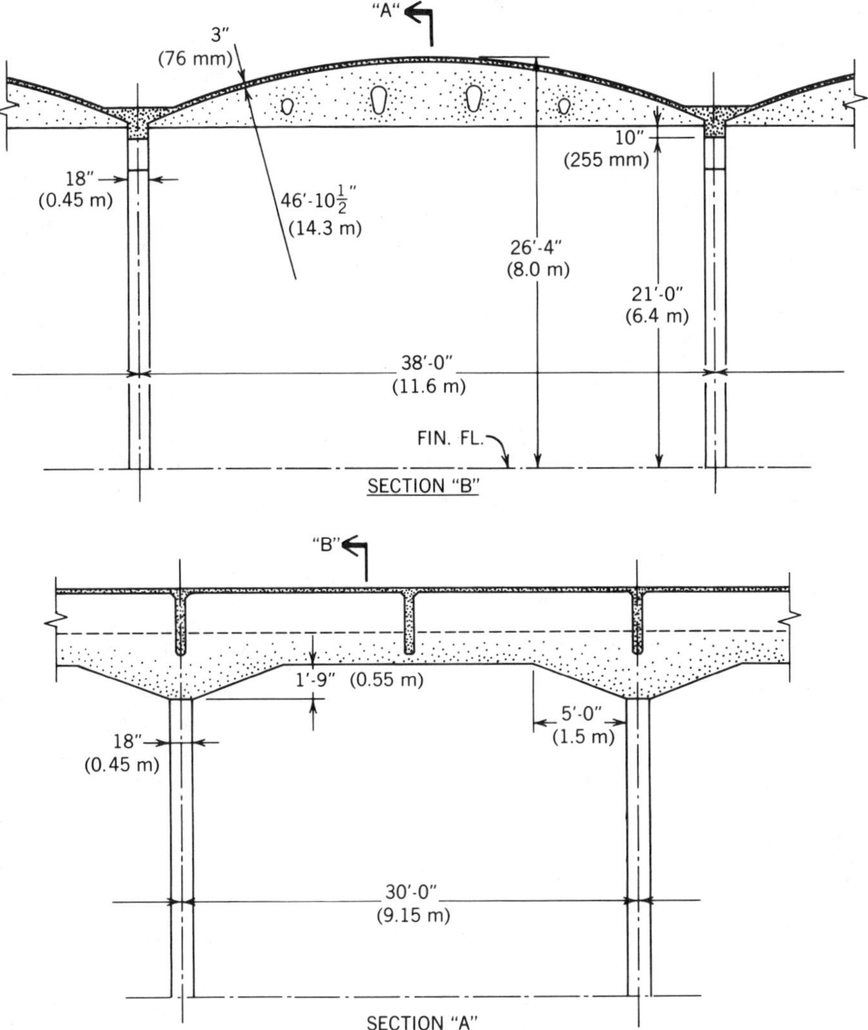

Fig. 12.30 Typical cross sections of Chrysler Corporation Dodge Chicago Plant. Courtesy of Albert Kahn Associates, Detroit.

it is no wonder that the plant was completed in record time, substantially faster than a mill order, fabrication, and erection of structural steel would have permitted. In addition, as soon as the formwork had left a position, the area was practically ready for installation of machines and actual work.

All this was accomplished at a tremendous saving in steel. The amount of reinforcing steel used for the entire construction was less than 2.6 lb/sq ft (12.7 kg/m^2), a fraction of what is normally required. The concrete thickness used for the superstructure amounted to a little over 6 in. (150 mm), which is also quite thin. This plant was not the only one of its kind; there were several other ones built of similar design and even larger column spacing, but of somewhat "smaller" area.

Should a designer in the 1980s try to design a building using minimum steel a similar design would probably be obtained. If steel reinforcement is not at a premium and high strength material can be used, a more economical design with smaller cross-sections or larger column spacings could probably be produced.

Chapters 21 and 22 should also be consulted for preliminary design of concrete structures.

12.5.2 Selection of Materials

Before one starts with the actual design of a cast-in-place reinforced concrete structure the concrete strength and the steel reinforcement grade must be selected, because they influence the successful outcome of the design. Presently such a variety of strengths is available to choose from that for many designers it may be more confusing than helpful. It is, therefore, necessary that a designer acquire a certain know-how to decide when a greater strength will be of help or when it might just involve some extra costs without particular gain. The designer must also ascertain that what is selected is obtainable at the site and controllable by a laboratory. Last, but not least, a contractor must be available who is capable of placing and curing with all the knowledge and workmanship required. Extra concrete strength is expensive; not only because of the extra amount of cement that is needed, but because the water/cement ratio will have to be lowered to obtain the increased strength. Because of the reduction in water, more admixtures will have to be added to disperse the cement and to retard the setting. Extra time and effort will have to be expended for placing, consolidating, vibrating, and curing. All this will find its expression in the final cost per cubic yard of concrete.

The most commonly available concrete mix has a 28-day compressive strength f'_c of 3000 psi (20 MPa). On large jobs some cement can be saved by using an f'_c = 2500 psi (17 MPa) for ordinary foundations and mass concrete. A lesser quality is used only for lean concrete fills. Foundations, such as mats, grillages, and caissons, in general use concrete with f'_c = 3000 psi (20 MPa). Superstructures for industrial buildings, with which we are mostly concerned in this chapter, ordinarily use 3000–4000 psi (20–28 MPa) concrete if they are cast-in-place. Higher strength concrete having an f'_c = 5000 psi (35 MPa) is primarily used for precast and prestressed elements made at a fabricator's shop. Under shop conditions, mixing, placing, and curing can better be controlled; the resulting smaller cross-sections will favorably affect the cost of transportation and the ease of erection.

Industrial buildings made of cast-in-place reinforced concrete will predominantly use stone (regular density) aggregate for the foundation as well as for the superstructure. The use of lightweight (low-density) aggregate should only be taken into consideration where large spans are contemplated; in this case it might be used for the roof construction, but preferably not for the columns.

In summary, the designer should carefully consider whether the use of a higher strength concrete is warranted for the design and construction of a single-story open-space building of reinforced concrete.

The selection of the grade of the steel reinforcement is much easier to make. Theoretically, there are four types of steel reinforcement available, identified according to their minimum specified yield stress f_y as 40, 50, 60, and 75 ksi (275, 345, 415, and 515 MPa) grades.

Most commonly used are Grade 40 having an f_y of 40,000 psi (275 MPa), and Grade 60 having an f_y of 60,000 psi (415 MPa). Either one can be used equally well, though Grade 60 is most readily available. Designers who still adhere to the ACI *alternate design method,* previously called *working stress design method* or *service load design method,* are better off to use Grade 40, because the *ACI Code* [12.1] does not permit full utilization of the strength of Grade 60. If the designer, however, uses the *strength design method* which is presently the basic method used and recommended by the *ACI Code* [12.1], it is advantageous to use Grade 60 steel, because the savings in reinforcement will usually well exceed the extra costs for the high quality steel. Two design handbooks are published by the American Concrete Institute, one for working stress design [12.2] and one for the strength design method [12.3]. These contain the necessary tables and graphs to expeditiously prepare a reinforced concrete design. Attention is also drawn to the Step-by-Step Design Procedures* [12.4] which simplify design by the strength design method.

12.5.3 Structural Framework Scheme

After having decided on the type and strength of the concrete and reinforcing steel to be used, the designer has to determine the type of the structure or the design scheme to be used. The most probable scheme for a "single-story open-space" cast-in-place reinforced concrete structure will in all probability consist of a series of stiff frames of any feasible configuration, spaced about 18 to 30 ft (5.5 to 9.0 m) apart, with joist-slabs, beam-slabs, or two-way slabs covering the area between them. The moments and shears caused in the various parts of the frames under the particular loading conditions can easily be evaluated either by an available computer program or by using the formulas in Kleinlogel's *Rahmenformeln* [12.9] prepared for a multitude of different frame shapes

---

* Originally published in the *ACI Journal,* they have been revised and expanded and form a part of the *Strength Design Handbook* [12.3].

and loading conditions. The formulas are easy to use once one has become familiar with the correct meaning of each particular letter used in the tables.

The design of the members should be done preferably by the strength design method. This method unquestionably offers the engineer the widest choice in the selection of the size of the members and their required reinforcement. However, the designer should not be misled by this freedom of choice and select sections which are too small in size and which will require a high percentage of reinforcement. Small sized sections will be more costly and will at the same time introduce a certain limberness into the structure that may lead to cracks and movements. Design of flexural members should preferably be based on reinforcement ratios ρ between 1/3 and 2/3 of the maximum reinforcement ratio ρ permitted by code. With regard to shear, beam sizes should be selected such that the maximum stirrup spacing, where needed, does not have to be closer than $d/2$, at least for the predominant part of the beam length. Designers should always keep in mind that the greatest economy is obtained through savings in reinforcing steel and simplicity of formwork, and not usually through the saving of concrete. Where provisions for hoists or craneways are to be incorporated in the design, brackets must be given special attention. This can be done using the *ACI Code* [12.1] which has specific requirements for the design of brackets.

Present-day engineering has tried hard to utilize the inherent advantages of reinforced concrete design and has strained at the same time to overcome certain difficulties in order to make it more economical and more flexible. Of special importance in this respect are four items:

1. The use of lightweight aggregate to reduce the density of the final concrete without affecting its strength.
2. The use of admixtures to produce higher strength concrete that is still manageable enough to permit handling and placement.
3. Precasting of structural units to enable the engineer to shop-fabricate members large enough to still permit transportation and erection. Such precasting contributes much to reduce the time for construction as well as to increase the flexibility of the structure.
4. Prestressing (pretensioning as well as post-tensioning) to utilize higher concrete qualities as well as greater steel strength; it is used to raise the capacity of members without increasing their dead weight and to improve their resistance to cracking (see Chapter 21).

In unison, these improvements permit the engineer to exercise better control over materials and fabrication. It is becoming recognized that the correct way to produce reliable quality concrete is to produce it in a stationary plant. Using the same type of material over and over, handling it under controlled conditions, curing and storing the finished material in a proper manner, will all contribute to achieve a quality product that is difficult to be equaled if done in the field.

When designing precast units, the engineer must keep aware of the fact that the construction is not monolithic reinforced concrete. Since the structure is made up of reinforced concrete elements, special attention must be given to the connections between them. If monolithic behavior of the assembled structure is expected, connections must be designed to transfer all forces necessary to achieve it; this is not always easy.

Prefabricated structures are more flexible than cast-in-place monolithic concrete structures. This is true because certain elements can be removed, if necessary, to make space or be replaced by stronger ones, without interference with the rest of the structure. Prefabricated structures are also more sensitive to impact and vibrations, not because of the members, which can be designed ductile and strong, but because of their connections. Structures consisting entirely of precast (conventionally reinforced or prestressed) members are not as common as structures using precast (usually prestressed) units, such as roof members spanning between walls. These units, whether in double-T, single-T, or giant-T shape are treated like steel joists. They are usually handled and erected, depending on their size and weight, by steel erectors or by the general contractor.

Precast prestressed units used over large spans require a camber which is provided by the manufacturer during the prestressing operations. Since it is common that the camber will vary from unit to unit in spite of all proper care by the manufacturer, it is advisable to survey and record the actual cambers of the units after delivery and to install them in such sequence that the cambers will gradually vary from one extreme to the other.

Prefabrication is not restricted to roof units, but it is applied also to columns, walls, slabs, stairs, and practically all building components. Buildings can, therefore, be constructed entirely of precast elements, just as they can be built from prefabricated steel units. Concrete seats and ledges are preferred over the use of steel seat angles, because the concrete seats are fireproof, whereas the steel angles have to be made fireproof through the provision of a protective cover. Simple seat connections will not add much to the lateral stiffness of a building, regardless of the material of which they are constructed. To obtain lateral stiffness for the entire structure, shear walls, and enclosures for stairs, elevators, and other shafts have to be made an integral part of the structure.

In their absence, either bracing has to be used or other provisions have to be made in the connections to transfer safely all the lateral forces down to the foundations.

Reinforced concrete structures are more often used in areas farther away from steel mills and steel fabricators, and where adverse climatic conditions place the maintenance (painting) of structural steel at a premium; they also may be preferred where better insulation and energy conservation in the building is desired.

12.6 WOOD STRUCTURES—PARAMETERS FOR PRELIMINARY DESIGN

12.6.1 General

Wood, used since earliest times, is one of the oldest building materials used by man to construct a roof over an open space. No wonder, therefore, that design with wood has influenced considerably the design with other materials, particularly structural steel. The enrichment, however, was mutual. Solid timber trusses unquestionably influenced early steel truss designs, but on the other hand, welded steel frames probably served as prototypes for many glued-laminated arches and frames. Similarly, the invention of shear connectors (split-rings) enabled wood designers to deviate from the old timber truss design and fashion their trusses along the lines of modern steel trusses.

Chapter 25 has extensive coverage of timber structures.

12.6.2 Selection of Material

The sawn timber beams of large cross-section and long length have become architectural exhibition pieces and are nowadays used more for their beauty and value than for their strength. The old time-proven treatment of soaking, drying, and storing has become too expensive and too time consuming to be employed for commercial purposes. The pieces used in most commercial wood construction are of small cross-section so that they can be kiln-dried and used on relatively short notice. Such small cross-section members are beams, such as joists, studs, boards, and planks, which are only obtainable in relatively small cross-sections. These members used mainly for short spans of about up to 25 ft (7.5 m) or as basic parts for larger members such as trusses, trussed rafters, and other built-up members. For preliminary design purposes, the allowable stresses given in Table 12.5 are conservative. Better qualities of wood for construction purposes are usually used in actual design and the allowable stresses given in the *National Design Specification* [12.10] should be used for final design.

Designers must consider the *load duration factors* as given in the *National Design Specification* [12.10] which permit an increase of the allowable stresses, in case one or the other portion of the loads included in the total design load, is temporary.

To bolster the insufficient strength of smaller cross-sections, steel plates or rolled sections are sometimes used in composite action with the wood beams or columns. These so-called *Flitch-beams* are often used by architects because the stiffening steel portions can completely be encased in the wood sections to make them more visually pleasing.

Table 12.5 Approximate Allowable Unit Stresses* for Preliminary Design with Wood

Stresses / Material	Flexure	Horizontal Shear	Compression Parallel to Grain	Compression Perpendicular to Grain	Modulus of Elasticity
Solid sawn lumber	1200 (8.30)	85 (0.60)	1000 (6.90)	385 (2.65)	1,600,000 (11,030)
Glued-laminated members	1800 (12.40)	165 (1.15)	1500 (10.30)	385 (2.65)	1,700,000 (11,700)
Trusses and trussed rafters	1400 (9.65)	85 (0.60)	1200 (8.30)	455 (3.15)	1,700,000 (11,700)

* Inch-Pound values are in psi; SI values are in (MPa).

The market for larger wood sections was taken over by glued-laminated members; they are available in a wide variety of sizes and lengths (see Chapter 25) and if not readily available, they can be made to order as required. They can be manufactured in vertically as well as in horizontally laminated fashion; they can be prismatic or tapered; they can be straight, slightly cambered, or bent to any curvature; they resist higher stresses than ordinary solid wood, and can be rough on the outside or polished, as desired. In other words, they are very versatile, useful, and good looking. In design they are treated like ordinary solid beams. For preliminary design purposes the stresses in Table 12.5 can be used. Better quality material is usually available and can be substituted in the actual design.

12.6.3 Structural Framework Scheme

Beams having small cross-sections are used for intermediate members having spans of up to about 25 ft (7.5 m). To this group belong joists, planking, columns, and studs, which are still made of solid sawn lumber. They are designed as solid sections, under consideration of all particular requirements for the design of solid timber sections, using the allowable stresses given in the *National Design Specification* [12.10] (or Table 12.5 for preliminary purposes).

A highly successful application of glued-laminated wood that is of particular interest for the enclosure of single-story open-space structures is the arch or frame (see Section 25.14). Though little used for industrial purposes because of the greater cost, its beauty, as expressed by shape, color, and texture, is often desired for architectural reasons and used for churches, gym courts, and assembly halls. Spans from 50 to 250 ft (15 to 75 m) have been successfully used. For design they are usually calculated as three-hinged arches with a steel tie embedded in the floor slab. Two-hinged arches are less frequently used, because of greater shipping difficulties. Although the ideal shape to resist a uniformly distributed loading is a parabola, better results, especially for an unsymmetrical loading, have been obtained with arches that follow more closely the line of a semicircle. For preliminary designs, the minimum rise of an arch above the springline should be assumed to be about 1/6 to 1/7 of its span. See Table 12.6.

The introduction of split-rings and other related shear connectors has opened up a vast field of application for trusses and trussed rafters. They can either be manufactured entirely in the shop or, in extreme cases, be built or assembled at the site from shop-fabricated parts. With the help of shear connectors several smaller or thinner members can be combined to produce a member having a large cross-section. Similarly as with structural steel, the possible variations seem unlimited and the final product is an economical unit that can be manufactured with local material and local labor. Widely used for roofs of residential buildings, smaller stores, and work shops, they can also be used for larger span roofs over gym courts, assembly halls, and similar. Strengthwise good enough to be applied for large design projects, they have successfully been used for airplane hangars of medium size and similar structures. Alternating absorption of moisture and drying out under the sun have caused some of them to distort through swelling, shrinking, or warping. Such undesirable deformations can only be controlled and retarded by careful maintenance and strict humidity control. They are therefore used to best advantage under interior, weather protected, conditions.

Appearance-wise, there is no difference between a truss and a trussed rafter; the difference lies in the load carried, that is, in the strength of their members and connections. If the units are placed about 2–4 ft (0.6–1.2 m) apart, they are called *trussed rafters*. In this case the roof sheathing [usually less than 1 in. (25 mm) thick] is nailed directly to the rafters either perpendicular to the direction of the rafters or inclined at 45° to improve its bracing effect. If the units are placed farther apart, up to a maximum span of about 20 ft (6 m), they are called *trusses*. For closer spaced trusses, planking thicker than 1 in. (25 mm) may be used as decking directly nailed to the rafters. Otherwise roof beams have to be introduced as intermediate members to carry the sheathing. When placed about 16–24 in. (0.4–0.6 m) on center, the roof beams are called *joists*. They run perpendicular to the direction of the trusses and carry the sheathing that is nailed to them. In this case the top chord of the trusses has to be designed flexurally stiff to make the positioning of the joists independent of the panel point spacing. If the roof beams are placed farther apart, they become purlins and should be placed at panel point locations. On steeper roofs, the position of joists or purlins can be inclined and it may become necessary to support some of them in the lateral direction. The remainder of the roof beams can then be laterally supported by means of rods or struts. However, other methods can also be used to achieve the same thing. If the Belgian-type of web member configuration is used, that is, some of the web diagonals are placed perpendicular to the inclination of the top chord, these diagonals can be extended beyond the top chord to give the purlin adjacent to them good lateral support. See Table 12.6.

Similarly, where monitors, fan platforms, and other housings are to be supported above the roof, Pratt and Howe web configurations can be used to advantage, because their verticals can be extended beyond the roof line to form an integral part of the housings and lend them considerable lateral stiffness, as shown in Table 12.6.

Table 12.6 Characteristics and Range of Span for Various Types of Timber Roof Constructions

Type of Construction	Pitch	h/l	Name	Typical Shape	Optimum Span	
					ft	m
Glued-laminated	Varies	~1/7 to 1/6	Arches; frames	h, l	50 to 250	15 to 75
Timber trusses	~1/50 to 1/10	~1/8 to 1/6	Howe	h, $\frac{l}{2}$; h, $\frac{l}{2}$	30 to 100	9 to 30
Pitched roof trusses	~1:4 to 1:1 1/2	~1/8 to 1/3	Belgian	h, l	20 to 60	6 to 18
			Pratt	h, l		
			Howe	h, l		
			Fink	h, l		
	~1:1	~1/2	Scissor	h, l	30 to 50	9 to 15

Bowstring trusses	Varies	~1/8 to 1/6	Curved	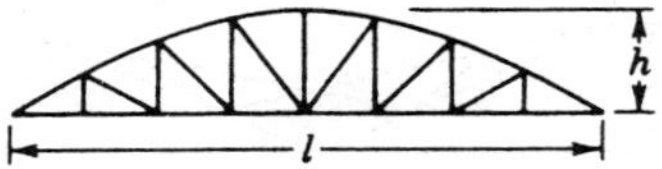 h, l	80 to 200	24 to 60
			Segmental	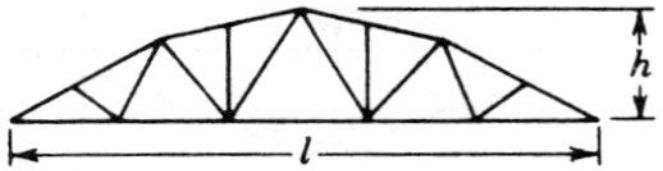 h, l		
Flat trusses	~1/50 to 1/10	~1/10 to 1/8	Howe	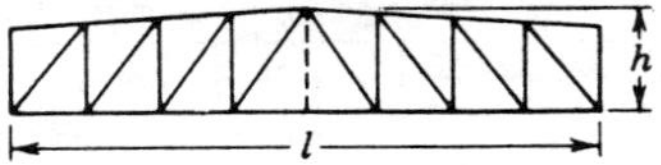 h, l	~40 to 80	~12 to 24
			Pratt	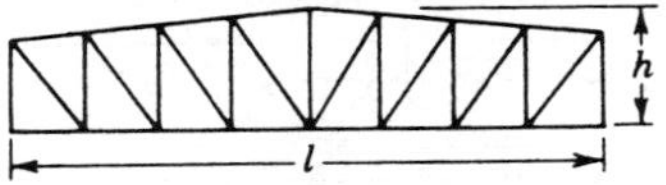 h, l		
			Warren	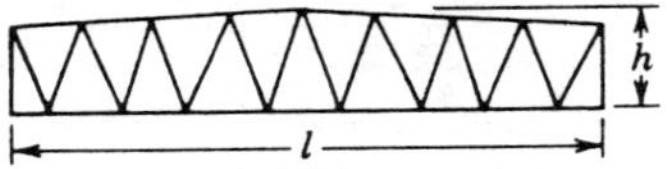 h, l		

Table 12.6 (*Continued*)

Type of Construction	Pitch	h/l	Name	Typical Shape	Optimum Span	
					ft	m
Special type trusses	Varies	~1/10 to 1/8	Sawtooth	h, l	~20 to 50	~6 to 15
		~1/6	Shed	h, l	~20 to 40	~6 to 12
		~1/10 to 1/8	Lank-Teco	h, l	~30 to 50	~9 to 15
Pole type	~1/10 to 1/4	Varies	Any type roof truss	l, Restr.	~40 to 60	~12 to 18

The Fink configuration is often used for steep trusses, because it helps reduce the length of the compression diagonals perpendicular to the roof line. A rod can be provided at the center of a Fink truss to reduce sagging of the bottom chord section.

Where multiple bays of pitched trusses or rafters are used, roof drains have to be provided in the valleys between adjacent bays.

Where vertical clearance or a feeling for space is desired, Scissor trusses are sometimes used to enclose open spaces. But similar to the use of glued-laminated arches their application is primarily for churches, gym courts, and assembly halls, where they are used for architectural expression.

For relatively long spans the bowstring truss is structurally one of the most effective and economical types of trusses that can be used. Bowstring trusses have been designed and built for spans of 200 ft (60 m) or more. They are, therefore, particularly useful for the roofs of aircraft hangars and assembly halls. The curved top chord has been built as a vertically laminated segmental arch section or as a horizontally laminated curved section. The bottom chord can be constructed in a similar fashion, although its curvature is much less, just as much as required by the camber. Because of their great height, bowstring trusses often can not be shop-fabricated and shipped in one piece, but have to be assembled at the site. Curved chord sections, however, are usually shop-fabricated to facilitate their assembly in the field. To simplify this procedure, bowstring trusses have also been designed as segmental bowstring trusses, that is, with a polygonal shaped top chord, usually straight over two to three panel point spacings. See Table 12.6.

The three types of flat trusses shown in Table 12.6 are not economical for small spans such as residential buildings. Such trusses are, however, frequently used for industrial buildings with greater clearance requirements below the bottom chord, because they provide a far superior end restraint. If, for example, a Pratt-type web configuration is used, it is no problem to design the truss in such a way that the end verticals are left out and the truss chords are slipped over the columns and connected to them. In multiple bay plan layouts, every alternate bay may be so designed. Trusses in the intermediate bays can, for example, have Howe-type web configurations which lend themselves to be supported on corbels (wood blocks) bolted to the building columns. Such trusses would have to be constructed for a total length fitting into the clear opening between the columns. Flat trusses usually carry composition type roofing. Their open spaces between the web members (especially if a Warren system is used) lend themselves best for the passage of such things as ducts and pipes.

Sawtooth trusses are actually flat trusses with the triangular framings superimposed on them, and shed roof trusses can be treated like unsymmetrically pitched trusses.

All pitched roof trusses, including the bowstring truss, have a common design deficiency, the pointed ends, which prevent them from developing an end restraint unless knee braces are provided. A similar condition has been discussed in Section 12.4.5 with regard to steel trusses for mill buildings, but in the case of a wood structure, knee braces seem to be more natural and less objectionable. Knee braces, of course, can be avoided if the wall sheathing of the building can be utilized as bracing, as it is commonly done in residential buildings. For larger structures, bracing systems that are required to transfer the lateral forces to the exterior walls become too formidable. In case the sheathing is secured directly to the rafters or trusses, its diaphragm action represents undoubtedly an excellent bracing system. But if the sheathing is fastened to the joists or purlins then its plane will be several inches (depth of roof beams) away from the top flange of the trusses. In such a case, some designers prefer to provide extra bracing members directly at the top chord level, that is, below the bottom of the purlins. Wood runners (bottom chord struts) should be provided for the full length of larger buildings. Their spacing should be about the same as the spacing of the trusses, and their locations should be coordinated with that of the sway frames.

Vertical sway frames are preferably placed in the end bays of the structure and for larger structures also at intermediate locations. Sway frames are required at the center of the span and at the column lines, unless a cross truss is already there. The need for additional sway frames is up to the judgment of the designer.

The evaluation of the forces in the various members of a wood truss or trussed rafter is exactly the same as that for a steel truss and a graphical solution by means of a Maxwell force diagram can be similarly used. When solid timber is used for the construction of the truss, the forces must be transferred from one member to the other through carpenter connections; all dabs and notches have to be cut skillfully and be made to fit. For such designs it is practical to let the wood diagonals take the compression and use steel rods to take the tensile forces in the verticals. This is the way the old Howe trusses were built. If shear connectors are used, such considerations are not necessary and the wood truss can be designed just like any steel truss.

Table 12.6 provides an overview of the various types of roof trusses and their web configurations, as well as their critical depth-to-span ratios and optimum range of spans. Obviously, the values given in the table are commonly used values, rather than absolute limits. Their purpose is primarily to simplify the concept and selection of what is structurally feasible and commonly available.

Another large group of timber structures come under the collective term of pole structures (Table 12.6). Basically, a pole structure is a structure where lateral resistance such as to sidesway or wind is provided by restraining the bottom (butt) end of the column pole in the ground, in a similar way as it is done with any ordinary utility pole. The degree of restraint can be varied depending on whether the pole is simply embedded in the ground by tamping the backfill or whether the annular space around the butt end is filled with compacted sand or concrete. On soft soils with low bearing capacity, it may even be necessary to provide a small footing with a steel base plate below the butt end, to keep the pole from sinking into the ground.

The lateral restraint of the whole building can be greatly augmented by providing in addition to the bottom fixity some additional restraint at the top of the pole. This can be achieved by utilizing the stiffness of the roof structure and framing it securely to the pole type column. The roof structure can be a truss, a deep beam (such as a glued-laminated member) with knee braces, a king- or queen-post framing, a scissorlike framing, or any other combination of beams, struts, and ties. The members may be built up of round poles, roughly hewn timber, sawn lumber, or of mixed units. If the upper restraint is used along with the lower embedment, the column stiffness greatly increases and the point of inflection may be considered to be about 2/3 of the free length of the column; this will render the column design much more reasonable. In some cases, the top restraint is developed to such a degree that it permits elimination of the bottom embedment. In this case only a simple concrete footing with a steel base plate is needed.

Where the embedment of the butt end of the pole is utilized, it is usually about 6 ft (1.8 m) long. All round poles or sawn timber posts must have received a preservative treatment according to Standards C1 and C4 of the American Wood Preservers Association, [12.8].

The *Timber Construction Manual* [12.7] contains various tables giving the allowable concentric loads and moment capacities for various diameters and classes of poles as well as a guide to evaluate the restraining capacity of the soil.

Greatly expanded treatment of the design of wood structures is provided in Chapter 25, Wood Structures Design.

12.7 DESIGN EXAMPLE

Design the structural frame for an industrial building, approximately 20,000 sq ft in plan area. Provide for a future extension to the south.

12.7.1 Selection of Material

Consideration of the following factors (See Section 12.2) dictates the decision to use structural steel:

1. Degree of permanency—permanent.
2. Purpose of structure—medium to heavy manufacturing.
3. Architectural appearance—functional as well as pleasing.
4. Simplicity of fabrication and speed of erection—reasonable.
5. Geographic location and cost of transportation—industrialized area in one of the North Central states.
6. Condition of subsoil—average.

12.7.2 Selection of Structural Design Scheme

The following factors as discussed in Section 12.3 are considered:

1. Column spacing or span—from manufacturing point of view a 40 × 60 ft unobstructed area appears satisfactory.
2. Flexibility—as much as possible.
3. Simplicity of detail and speed of erection—as much as economically feasible.
4. Prevailing economy—average.

Based on the above considerations a building plan is selected having three large bays in the long direction and three short bays in the short direction,

$$3(60) \times 3(40) = 180(120) = 21{,}600 \text{ sq ft} > 20{,}000 \text{ sq ft}$$

The selected framing scheme using shop-fabricated trusses and rolled wide-flange beams is as follows:

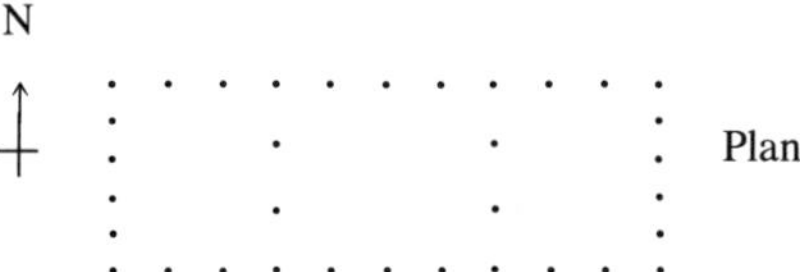

12.7.3. Roof Framing Arrangement and Selection of Purlins

As discussed in Section 12.4.1 through 12.4.4, the following parameters are established prior to computing roof loads:

1. Roof slope—use flat roof, with main slope for drainage approximately 1/8 in./ft, and secondary slope approximately 1/16 in./ft.
2. Roof decking—use $1\frac{1}{2}$-in. metal decking.
3. Grade of steel—use A36 (F_y = 36 ksi).
4. Purlin arrangement—place perpendicular to roof trusses at a spacing of 60/8 = 7.5 ft.

The dead load to be carried is:

composition roofing and insulation	6.0 psf
metal decking	1.5 psf
purlins (estimated)	2.0 psf
bracing	0.5 psf
ducts, pipes, etc.	2.0 psf
total dead load	12.0 psf

The live load from snow is 30 psf. Then,

$$w = (12.0 + 30)7.5 = 315 \text{ lb/ft}$$

$$M = \frac{0.315(20)^2}{8} = 15.75 \text{ ft-kips}$$

$$\text{minimum depth,} \qquad h_{\min} = \frac{L}{28} = \frac{20(12)}{28} = 8.6 \text{ in.}$$

Use 10-in. Purlins

The selection of the purlin section for F_y = 36 ksi is made by assuming a "compact section" and using an allowable bending stress F_b of 24 ksi. If a lighter "noncompact" section is available, check whether it may be satisfactory with a reduced allowable stress. The purlin arrangement is shown in Fig. 12.31.

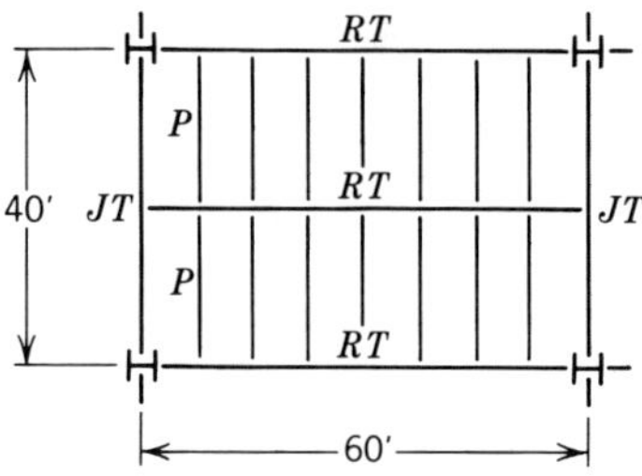

Fig. 12.31 Roof framing plan for design example. Typical framing for one bay.

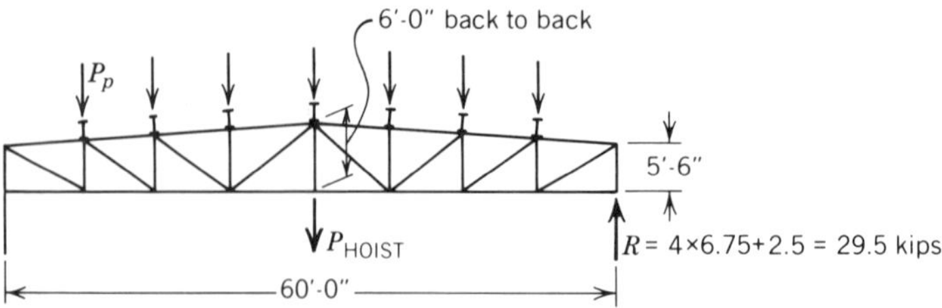

Fig. 12.32 Roof truss dimensions and loads for design example.

12.7.4 Design of Main Supporting Member

The truss dimensions and loading arrangement are shown in Fig. 12.32. The panel point loads are then computed,

$$\frac{wL}{30F_y} = \frac{42(60)}{30(36)} = 2.3 \text{ psf} \qquad \text{(estimated truss dead load)}$$

The total loading of the roof (psf) is 12.0 + 30.0 + 2.3 = 44.3 psf, say 45 psf. The panel point load P_p is

$$P_p = 45(7.5)(20) = 6750 \text{ lb} \qquad (6.75 \text{ kips})$$

Since the manufacturer requires a 1.5-ton capacity hoist on monorail at midspan of roof trusses (see Section 12.4.7),

$$P_{\text{hoist}} = 1.5(2)1.6 = 4.8 \text{ kips}, \qquad \text{say 5 kips}$$

The truss analysis is next performed (omitted here) to determine the axial forces in all the members. The members are then selected using design criteria of the *AISC Specification* [12.6].

12.7.5 Design of Jack Truss

The jack truss dimensions and loading arrangement are shown in Fig. 12.33. The reaction P_R from the roof truss is

$$P_R = 2R = 2(29.5) = 59 \text{ kips}$$

The load from the roof tributary to the jack truss plus the estimated weight of the jack truss

from roof, 45(10)7.5	3.5 kips
truss weight (estimated)	1.0 kips
	4.5 kips

Thus, the total load P_R is

$$P_R = 59 + 4.5 = 63.5 \text{ kips}$$

The truss analysis and design of members according to AISC are omitted here.

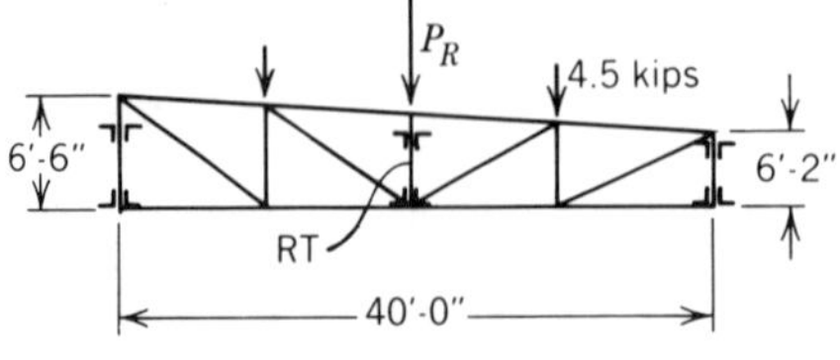

Fig. 12.33 Jack truss dimensions and loads for design example.

12.7.6 Design of Columns and Column Bases

The loading for an interior column is

roof	46(40)60/1000	= 110.4 kips
hoists	2(5)	= 10.0 kips
column		= 0.6 kips
total for interior column		= 121.0 kips

The exterior columns are positioned with the web perpendicular to the building face. The interior columns are positioned with the web parallel to the direction of the roof trusses (see Fig. 12.34, the E–W direction). The columns are to be wide-flange sections. The clear height to the bottom chord of trusses (as required by manufacturer) is 14 ft. The column design (not given here) is for axial load and wind moments. *AISC Specification* [12.6] allows a $33\frac{1}{3}\%$ overstress for loading including wind, but the sections must also be satisfactory for gravity loads without the $33\frac{1}{3}\%$ overstress.

The column bases are designed for concentric load. Provide four anchor bolts for lateral stiffness of the building, even if the moment is taken entirely at the top.

Wind posts are placed at 20 ft on centers between main columns (See Fig. 12.34). These are to be extended up to the bottom chord of trusses and connected with vertical slotted hole connection.

12.7.7 Strength Bracing for Wind

Wind in the N–S Direction

Provide at bottom chord level a bracing truss across entire building in the N–S direction, and use wall bracing in the east and west walls to carry wind reactions down to the ground (see Fig. 12.34). Bracing will also provide lateral stiffness against hoist racking.

The wind load H_{p1} is next evaluated. Assuming the parapet to be 8 ft high above bottom chord elevation,

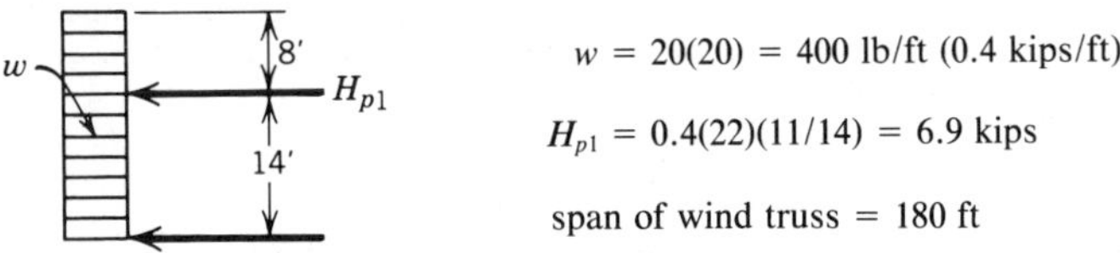

$$w = 20(20) = 400 \text{ lb/ft } (0.4 \text{ kips/ft})$$

$$H_{p1} = 0.4(22)(11/14) = 6.9 \text{ kips}$$

$$\text{span of wind truss} = 180 \text{ ft}$$

The forces in the members of the wind truss (bracing) are evaluated by structural analysis and the axially loaded members are designed according to the *AISC Specification* [12.6].

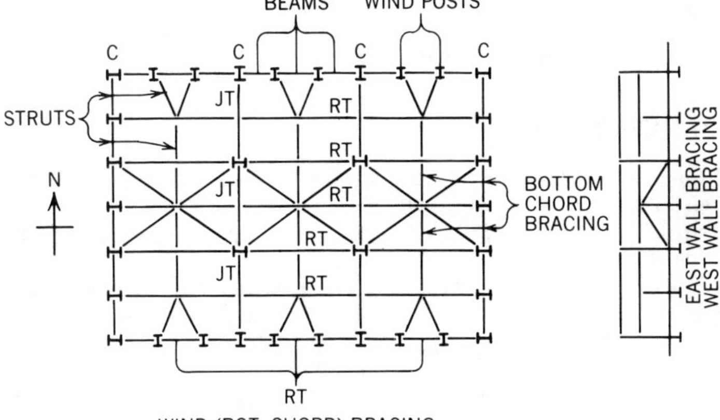

Fig. 12.34 Wind bracing for design example.

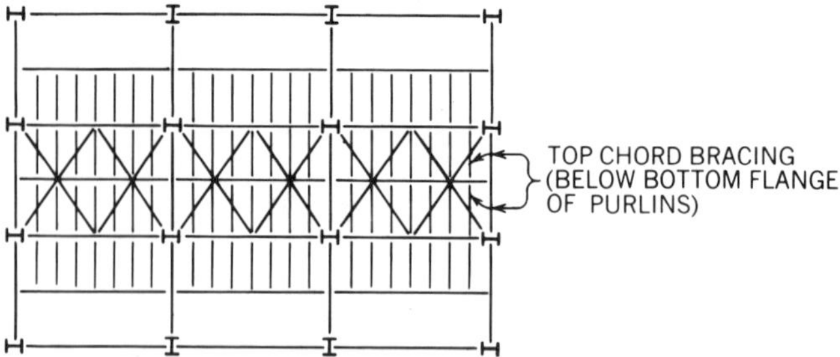

Fig. 12.35 Stiffness bracing for design example.

Wind in the E–W Direction

Since provisions for future expansion to the south are to be made, no wall bracing can be used in the south wall. Wind must be resisted, therefore, by frame action. Moments are taken about the upper end of the column. The wind load H_{p2} is next evaluated,

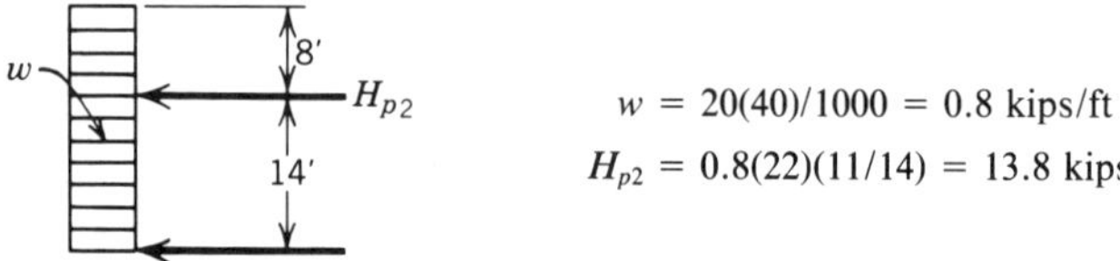

$$w = 20(40)/1000 = 0.8 \text{ kips/ft}$$
$$H_{p2} = 0.8(22)(11/14) = 13.8 \text{ kips}$$

Assuming the stiffness of the exterior columns to be one-half of that of the interior columns, the wind load for one interior column is 13.8/4 = 3.45 kips. The interior columns are then designed for an additional wind moment of

$$M_{wi} = 3.45(14) = 48.3 \text{ ft-kips}$$

This moment is assumed to vary from M_{wi} at the top of column to zero at the base. Exterior columns are designed for an additional moment at the top of column equal to one-half that assigned to the interior columns, thus

$$M_{we} = 24.15 \text{ ft-kips acting at the top}$$

12.7.8 Stiffness Bracing

Provision must be made for the lateral support of the compression chord of the roof trusses. Some designers may consider the metal decking to provide this lateral support. However, since the decking is 10 in. away from the plane where it is needed, it is advisable to provide bracing at the elevation of the top chord (below the purlin flanges), arranged as shown in Fig. 12.35.

SELECTED REFERENCES

12.1 ACI Committee 318. *Building Code Requirements for Reinforced Concrete* (ACI 318-83). Detroit: American Concrete Institute, 1983.

12.2 ACI. *Reinforced Concrete Design Handbook* (SP-3), 2nd ed. Detroit: American Concrete Institute, 1965.

12.3 ACI Committee 340. *Design Handbook,* Vol. 1, *Beams, One-Way Slabs, Brackets, Footings, and Pile Caps* [SP-17(84)], 4th ed. Detroit: American Concrete Institute, 1984.

12.4 ACI Committee 340. "Step-by-Step Design Procedures in Accordance with the Strength Design Method of ACI 318-71," *ACI Journal, Proceedings,* **74,** August 1977, 333–360.

12.5 AISC. *Manual of Steel Construction,* 8th ed. Chicago: American Institute of Steel Construction, 1980.

12.6 AISC. *Specification for the Design, Fabrication and Erection of Structural Steel for Buildings, Effective November 1, 1978, With Commentary*. Chicago: American Institute of Steel Construction, 1978.

12.7 AITC. *Timber Construction Manual,* 3rd ed. New York: Wiley, 1986.

12.8 American Wood-Preservers' Association. *Book of Standards*.

12.9 A. Kleinlogel. *Rigid Frame Formulas* (*Rahmenformeln*), 2nd American ed. (Translated from 12th German ed.). Frederick Ungar Publishing Company, 1958.

12.10 NFPA. *National Design Specification for Wood Construction*. Washington, DC: National Forest Products Association, 1986.

12.11 Charles G. Salmon and John E. Johnson. *Steel Structures, Design and Behavior,* 2nd ed. New York: Harper & Row, 1980.

12.12 Richard N. White, Peter Gergely, and Robert G. Sexsmith. *Structural Engineering, Combined Edition, Volumes 1 & 2*. New York: Wiley, 1976.

12.13 Chu-Kia Wang and Charles G. Salmon. *Introductory Structural Analysis*. Englewood Cliffs, NJ: Prentice-Hall, 1984.

CHAPTER 13

PRELIMINARY DESIGN OF SHELLS AND FOLDED PLATES

MILO S. KETCHUM

Ketchum, Konkel, Barrett, Nickel, Austin
Denver, Colordo

13.1 INTRODUCTION

The preliminary design of shells and folded plates* serves a number of purposes. Among them may be listed the following:

1. To study the structural action of the shell so that a thorough understanding of the analysis and design problems can be obtained.

---

* Coverage in this chapter is restricted to concrete shells and folded plates. See Chapter 25 for timber shells and folded plates.

2. To arrive at preliminary sizes for the structural elements so that important architectural decisions can be made.
3. To make quantity surveys and cost estimates of possible alternates for other shell structures or for more conventional systems of construction.

Preliminary design involves both the analysis of structural systems and the selection of member sizes and reinforcing. The methods of analysis involve a wide variety of methods ranging from the simple application of the equations of statics, to methods just short of precise analysis and are often very useful, if not indispensable, for the final checking of completed calculations. The accuracy of approximate methods is highly dependent on the form and basic details of the structure, since they are based on concepts of structural behavior which must be thoroughly understood. The methods to be described here do not involve elaborate equations or numerical procedures but rather are methods based on statics and deflections which are intended to give insight into the structural action of the shell or folded plate.

Shells cannot be designed independently of the architectural considerations, and it is important for the engineer to avoid making decisions on the shapes, forms, and details without the sanction of the architect. In the same way the architect also must not make decisions that are not compatible with structural integrity. It may be possible to design the shell so it will stand, but the cost of the structure may be all out of reason.

It is important that both the preliminary and the final precise analysis be started as early as possible during the project planning. The lead time to develop methods of analysis, to understand their limitations, and either to write or to obtain access to computer programs is vital to the success of any design.

Design of a serviceable shell depends on the proper design of the supporting structure and the stiffening elements. These elements provide the shell with boundary conditions which influence its behavior. Thus, if the supporting structure is inappropriate, acceptable performance of the shell will be difficult to achieve. The interaction of the shell and the structure must be considered during both the conceptual and the final design. The design of the supporting structure is often more difficult and critical than the design of the shell elements. For example, end supports of cylindrical shells are usually arches having loads from the shell that are tangent to the slope of the arch. The use of a plane frame computer program is virtually a necessity and the results require careful interpretation.

Types of shell structures are described and illustrated in Chapter 8, Structural Form, and this information will not be given here. It is important for the reader to study that section carefully.

To cover all of the methods for the preliminary analysis of shell structures would require much more space than is available here. In this chapter, the writer has tried to accomplish the following:

1. Describe the structural action so the potential designer will know how the shell performs.
2. Give one or more simple methods to determine dimensions and forces.
3. Show a simple, practical example of the structure to illustrate the methods proposed.

These shells are generally not completely designed, including steel reinforcement details. It is presumed that the reader is a competent structural engineer familiar with codes and methods for the design of reinforced concrete. The reader should also study the requirements of the current *ACI Code* [13.18] sections for shells.

The literature of shell structures is very extensive and only a limited number of references are given here. Many of them have additional bibliographies that should be studied by the serious reader. Unfortunately, many of the references may not be available in technical libraries.

It should be cautioned that the methods shown are normally suitable only for preliminary design to obtain member sizes and reinforcing bar quantities for cost and feasibility studies. The final design should be made using the more precise tools described in texts on shell design. In particular, analysis of long-span hypars should be made by the finite element method.

In the next sections, the various types of folded plates and shells are examined for their structural action and simple methods for the preliminary analysis and design are devised. In the final section, the economics of shells is discussed.

13.2 FOLDED PLATES

A folded plate structure is made from thin slabs folded to form beamlike elements. A typical folded plate is show in Fig. 13.1. The principal structural elements are: (1) A slab spanning transversely between folds; (2) slanting plates spanning longitudinally from support to support; (3) edge members such as beams or ribs at the junction of plates or at unsupported edges; (4) supports, which

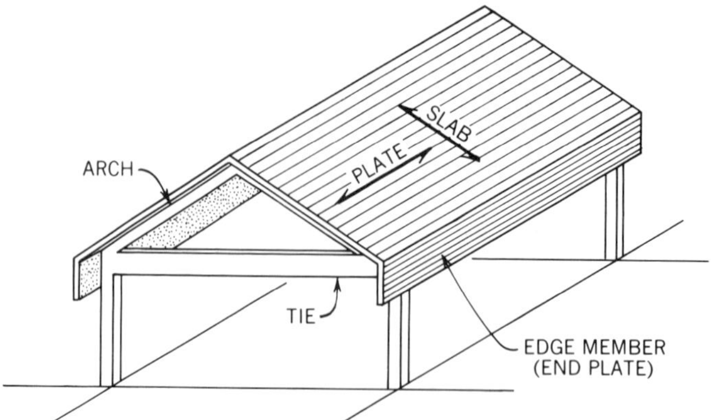

Fig. 13.1 Structural elements of a folded plate.

may be either triangular beams (in elevation), walls, or tied arches. The latter is shown in the sketch. There may be any number of folds from valley to valley. The sketch shows a two-element folded plate. Profiles of several other types with their names are shown in Fig. 13.2. At the edges of a series of plates, an additional plate or rib is required to end the series. This will be called an end plate to distinguish it from interior plates. It is usually not as deep, and therefore not as stiff, as the interior plates, especially if it is turned upward rather than downward. The dashed line is the effective neutral axis of the section.

13.2.1 Structural Action

Folded plates may be designed as beams spanning longitudinally and as slabs spanning transversely between folds. Reactions from the slabs are resolved into forces in the plates and the stresses are determined from these forces. However, the stresses on each side of the joint will not necessarily be the same so analysis must be made to adjust these stresses. The next factor that must be taken into account is the relative movement of the joints between folds. If all of the plates are of the same depth and configuration, and have the same load, then there will be no relative deflection. There-

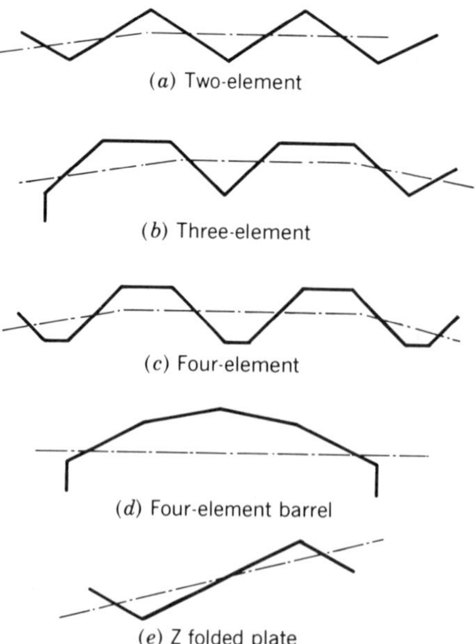

Fig. 13.2 Types of folded plate sections.

fore the center plates of a series of two- or three-element folded plates can be designed as a beam without having to resolve the forces into components.

For the end plates, however, the relative deflection of the valleys and ridges may be so great that both the longitudinal stresses and the transverse moments are affected. A precise analysis will take these deflections into account. The task of the preliminary analysis, then, is to predict these stresses and moments without having to go through the more elaborate analysis.

Effective Section of End Plates

A key to the understanding of the action of the end plates is to draw the neutral axis through the end plate and the first interior plate as illustrated in Fig. 13.2. In (*a*), (*c*), and (*e*), the effective sections are quite small because the neutral axis must bisect the end plate. In (*b*) and (*d*), however, the effective section is quite large. The least effective pattern for an edge member is a horizontal plate which has a very small effective depth.

An analysis of the longitudinal stresses can be made by assuming a neutral axis, and resolving all loads in the direction perpendicular to the neutral axis. However, the most important factors are the bending moment at the crown of the first interior plate element and the deflection of the edge plate, neither of which is easy to determine by this method.

Unless absolutely necessary, large deflections of small end plates are to be avoided. Usually the valley at an end plate is at a wall or can be supported by intermediate columns. This support will greatly reduce the relative deflections, will avoid the excessive stresses and deflections at the end plates, and will require less reinforcement. A simple method of analysis is to treat the intermediate columns as redundants. The forces in the redundants are determined, and the stresses from the redundants are added to those from the basic structure. The reinforcement required is greatly reduced for both the longitudinal stresses and the transverse moments.

For general guidelines for dimensions, the span-to-depth ratio for the vertical dimension of the plate elements should be from 10 to 12. If the depth is too small, the deflections, including those due to creep and shrinkage, will be high. Prestressing will, of course, permit shallower structures. The width-to-depth ratio (the transverse width) normally should be lower but the two ratios are interlocked, depending on the type of structure.

The maximum slab span for a typical 4-in. (100-mm) thick plate should be not more than 12 to 14 ft (3.7 to 4.2 m) unless it is haunched at the valley and the crown. If the slab span gets too large, then more elements should be used. The end spans are the most critical.

13.2.2 Methods of Preliminary Analysis

The interior plates of two- and three-element folded plates are essentially beam elements with little relative joint movement and may be analyzed as ordinary beams. The important task of preliminary analysis is to predict accurately the longitudinal stresses and the slab moments in the edge plate and the first interior plates.

For a two- or three-element plate, with a downturned edge plate, Fig. 13.1, a simple analysis can be made that will give accurate results for preliminary design because the stiffness of the edge plate combined with the slanting plate is large and the relative deflection of the joints is small. This method follows the example to be described later.

If the edge plate is upturned, then the stresses and moments can be made by one iteration for the additional stresses caused by deflection as shown by Ramaswamy [13.13]. For the four-element folded plate, Fig. 13.2*c*, it is more difficult to predict the effect of an end span on stresses and slab moments, but the effective depth of the edge plates is fairly large, so for preliminary design, the same thickness can be used as in other parts of the structure. The Z folded plate of Fig. 13.2*e* is statically determinate and the deflections of the joints do not affect the stresses.

13.2.3 Intermediate Edge Supports

Because of the importance of intermediate columns, a simple analysis for obtaining the stresses in edge-supported folded plates will be presented. It follows the normal initial steps for manual calculations as presented in standard texts [13.13] as follows:

1. Determine longitudinal stresses and transverse bending moments in the structure due to dead and live loads, assuming no joint movement.
2. Calculate the vertical deflection of the edge at the location of the column.
3. Apply a unit upward load at the point where the column is placed.
4. Determine the upward deflection.

5. The force in the column will be the deflection due to the dead and live load divided by the deflection due to the upward force.
6. Add the stresses due to the dead and live loads to those due to the actual upward force.

The resulting stresses, deflections, and reinforcing are much reduced from the structure without intermediate columns.

13.2.4 Example: Intermediate Edge Supports

A sketch showing the dimensions of a folded plate with intermediate columns supporting the exterior valleys is shown in Fig. 13.3. The structure is a two-element folded plate having a span of 50 ft (15 m). There are two sets of these elements each 20 ft (6 m) wide so the horizontal slab span is 10 ft (3 m). The structure is symmetrical about the center so only one 20-ft (6-m)-wide element need be considered. The edge plate is 4 ft (1.2 m) wide, horizontal, and is turned upward at the same slope as the other plates. It is supported by a edge column at the center of the 50-ft (15-m) span.

The calculations in the following section are all made using ordinary methods for stresses and deflections in beams. Moments are distributed from initial fixed-end moments by the moment distribution method. Stresses produced by continuity of adjacent plates are determined from the initial stresses in disconnected plates by stress distribution which is analogous to moment distribution. The short BASIC program shown in Fig. 13.4 was used to determine both moments and stresses. The resolution of forces can be made with simple graphic solutions and the deflection by construction of graphic Williot diagrams.

Following is a list of the steps required to make the preliminary analysis. Many of the steps are more fully explained in the example calculations.

1. Select the appropriate thickness of the slab, the weight of roofing, and the appropriate live load.
2. Determine the dead and live load per horizontal foot of slab and analyze the slab as a horizontal, continuous beam. Determine the vertical reactions on this beam.
3. Resolve the reactions into plate loads.
4. Determine the plate stresses in individual plates from the plate loads.
5. Distribute these plate stresses to obtain the stresses in the continuous plates.
6. Determine the downward deflection of the plates at the center of the 50-ft (15-m) span where the auxiliary column will be placed.
7. Apply an upward unit load at the center of the edge plate and determine the stresses and the deflection as for the previous step.
8. The force on the column will be the deflection due to dead and live load (step 6) divided by the deflection due to the unit load (step 7).
9. Add to the original stresses (step 5) the stresses from step 7, multiplied by the ratio of the column load to the unit load. These must be plotted both transversely and longitudinally. The result will be the stresses in the structure.

The calculations and their description follow:

Dimensions and Loads

The live load is 30 psf of horizontal surface. The slab thickness chosen is 4 in., based on the thickness required for placing the layers of reinforcement. The allowance for roofing is 5 psf so the total weight on the slope is 4(150)/12 + 5 = 55 psf. The first step, shown in Fig. 13.5, is to apply the

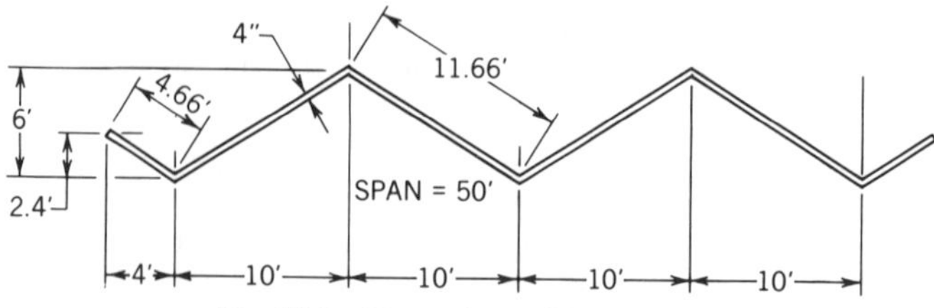

Fig. 13.3 Dimensions of folded plate.

```
10 CLS:PRINT "MOMENT DISTRIBUTION":PRINT
20 INPUT "ENTER NUMBER OF SUPPORTS = ";N
30 INPUT "ENTER NUMBER OF DISTRIBUTIONS ";DI
40 N1=N-1:N=N+1:N2=N*2:PRINT
50 DIM D(N2,2),M1(N2,2)
60 FOR M=2 TO N
70 INPUT "TWO K VALUES, EACH SUPPORT ";K1,K2
80 K3=K1+K2:D(M,1)=K1/K3:D(M,2)=K2/K3:NEXT
90 FOR I=1 TO N2:M1(I,1)=0:M1(I,2)=0:NEXT
100 PRINT:PRINT "OPERATIONS LIST "
110 PRINT "1    STRESS DISTRIBUTION"
120 PRINT "2    MOMENT DISTRIBUTION, UNIFORM LOADS "
130 INPUT "   ORDER BY NUMBER ";Z
140 ON Z GOTO 150,220
150 CLS:PRINT
160 PRINT "STRESS DISTRIBUTION"
170 PRINT "FORCE TO THE RIGHT IS POSITIVE, BOTH ENDS""
180 FOR M=2 TO N
190 INPUT "STRESS TO EACH SIDE OF JOINT ";M1(M,1),M1(M,2)
200 NEXT M
210 GOTO 320
220 CLS:PRINT
230 PRINT "ENTER CANTILEVER MOMENTS "
240 PRINT "MOMENT CLOCKWISE ABOUT JOINT IS POSITIVE"
250 INPUT "LEFT END ";M1(2,1)
260 INPUT "RIGHT END ";M1(N,2)
270 PRINT "UNIFORM LOADS, EACH SPAN"
280 FOR M=2 TO N-1
290 INPUT" W, SPAN ";W,S
300 M1(M,2)=W*S*S/12:M1(M+1,1)=-M1(M,2)
310 NEXT M
320 PRINT "DISTRIBUTIONS":FOR K=1 TO DI
330 PRINT K;:FOR M=2 TO N
340 U=M1(M,1)+M1(M,2):M1(M,1)=M1(M,1)-U*D(M,1)
350 M1(M-1,2)=M1(M-1,2)-U*D(M,1)/2
360 M1(M,2)=M1(M,2)-U*D(M,2)
370 M1(M+1,1)=M1(M+1,1)-U*D(M,2)/2
380 NEXT M,K
390 PRINT:PRINT
400 PRINT "MOMENTS OR STRESSES AT JOINTS"
410 FOR M=2 TO N:L=M-1
420 PRINT USING "JOINT ## #####.##   #####.## ";L,M1(M,1),M1(M,2)
430 NEXT M:PRINT "END"
```

Fig. 13.4 Program for moment and stress distribution.

dead and live load to a continuous beam of 10-ft (3-m) spans. The dead load per horizontal foot is $w = 55(11.66)/10 = 64.1$, so the total load is $64.1 + 30 = 94.1$ lb/ft.

Stresses

The structure is symmetrical about the center, so it is necessary to consider only the cantilever and two spans. The stiffness factors, fixed-end moments, and distributed moments are shown in Fig. 13.5. The reactions considering the internal moments are shown on the next line. These reactions are resolved in an equilibrium diagram to give the components of the reactions on the plates. In this case, it was easiest to calculate the reactions rather than draw force diagrams because the slopes of the adjacent plates are the same. In the case of varying slopes, it may be quicker to draw force diagrams to scale. The plate loads are the sum of the forces at the joints.

The next step is to determine the plate stresses from the plate loads.

The section modulus of plate A is

$$S = \frac{4(4.66)^2(12)^2}{6} = 2085 \text{ in.}$$

For plates B and C,

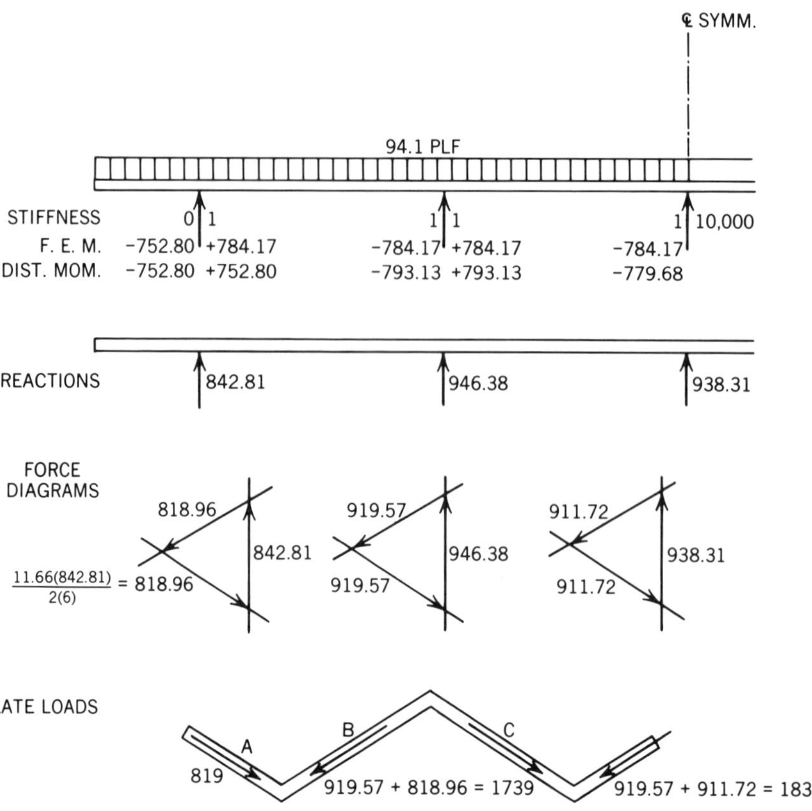

Fig. 13.5 Slab moments and plate loads.

$$S = \frac{4(11.66)^2(12)^2}{6} = 13052 \text{ in.}$$

The stresses in plates A, B, and C are

$$\text{in A,} \quad f = \frac{819(50)^2(12)}{8(2085)} = 1473 \text{ psi}$$

$$\text{in B,} \quad f = \frac{1739(50)^2(12)}{8(13052)} = 500 \text{ psi}$$

$$\text{in C,} \quad f = \frac{1831(50)^2(12)}{8(13052)} = 526 \text{ psi}$$

These stresses are not the same at each side of the joint, so it is necessary to distribute them in a procedure analogous to moment distribution [13.13]. The stresses are analogous moments, and the sign convention is the initial stresses are positive if the forces that cause them are pointed to the right. Free ends must be treated as fixed ends, and the stiffness of a plate is $K = 4/tL$ where t is the thickness and L is the length of the plate. Because only relative values are required for uniform depth, it is easiest to use the reciprocal of the length times any convenient constant. The stiffness, initial stresses, and the distributed stresses are summarized in Fig. 13.6. The stresses are plotted in this figure.

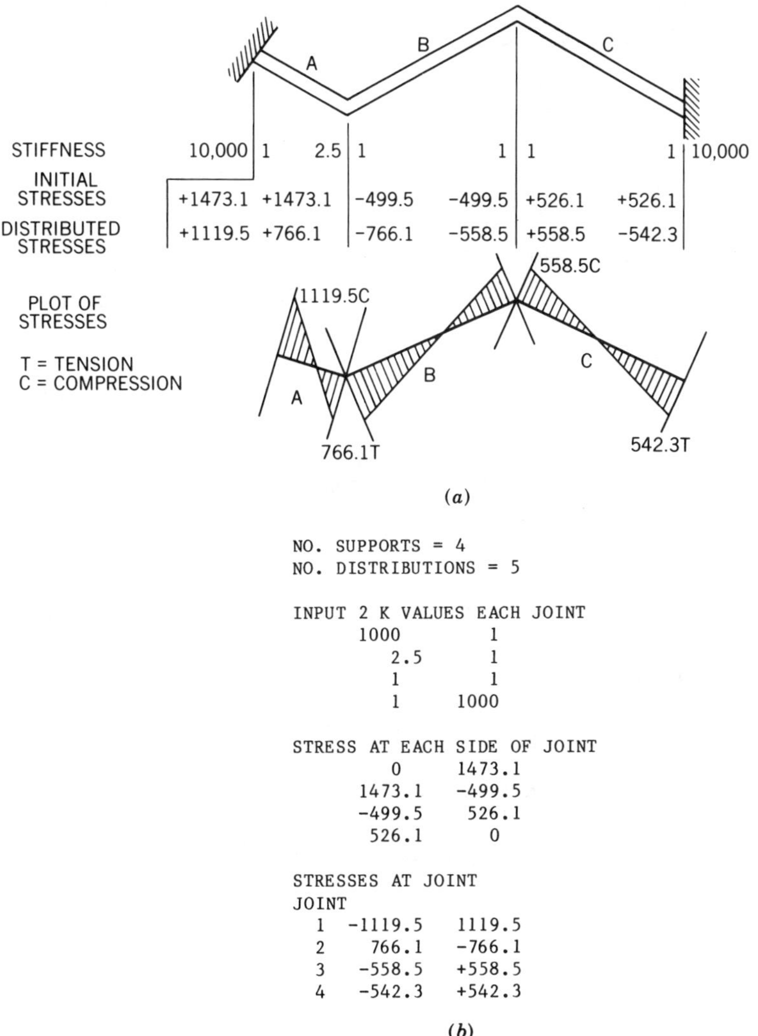

Fig. 13.6 Distributed stresses, dead + live load. (*a*) Stresses in plates. (*b*) Computer output for stress distribution.

Deflections

Deflection in the plates at the center of the outside span is determined by the equation

$$D = \frac{5fL^2}{24Ed}$$

where f is the average stress (compression and tension), L is the span, d is the depth, and E is the modulus of elasticity. The average stresses in plates A and B are

$$\text{in A,} \qquad f = \frac{(1119.5 + 766.1)}{2} = 943 \text{ psi}$$

$$\text{in B,} \qquad f = \frac{(766.1 + 558.5)}{2} = 662 \text{ psi}$$

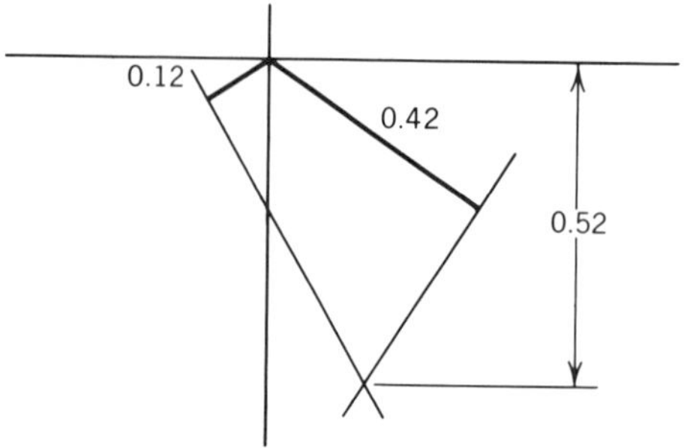

Fig. 13.7 Deflections from member loads.

The deflections are, using $E = 3(10)^6$ psi

$$\text{in A,} \qquad D = \frac{5(943)(50)^2(12)^2}{24(3)(10)^6(4.66)(12)} = 0.42 \text{ in.}$$

$$\text{in B,} \qquad D = \frac{5(662)(50)^2(12)^2}{24(3)(10)^6(11.66)(12)} = 0.12 \text{ in.}$$

The vertical deflection can be obtained by a graphic Williot diagram or by calculation. The diagram is shown in Fig. 13.7. The actual diagram was drawn to a much larger scale.

Unit Upward Load

This completes the determination of the vertical deflection at the center of the outside span produced by dead and live loads. The next step is to determine the deflection produced by unit upward load.

The value of unit load has been chosen as $P = 10{,}000$ lb, so the deflection will be scaled for easy calculation. The same steps are repeated as for uniform dead and live load, except that the reaction is already known, $R = 10{,}000$ lb, at the center of the span. The plate loads (a single concentrated load at the center of the span) are shown in Fig. 13.8 and are each 19433 lb. Only plates A and B are affected.

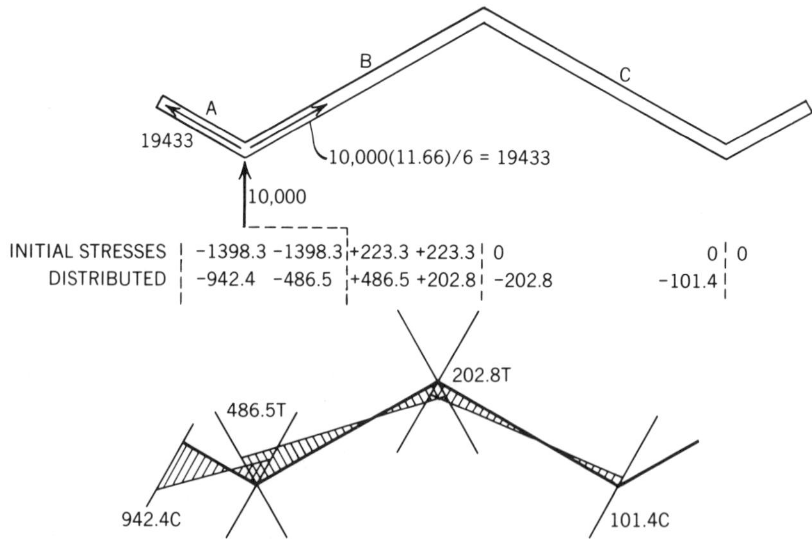

Fig. 13.8 Distributed stresses, unit upward load.

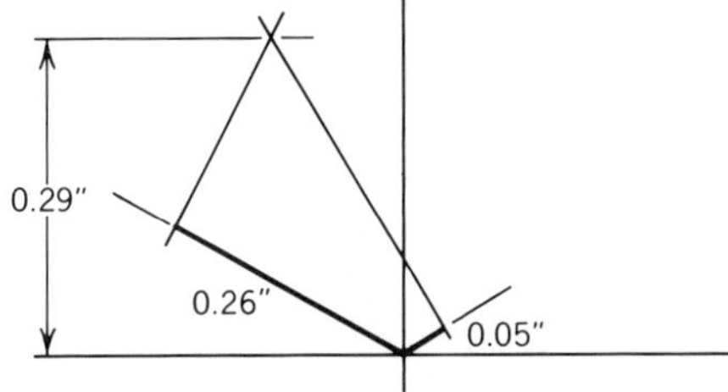

Fig. 13.9 Deflections from upward load.

$$\text{in A,} \quad f = \frac{PL}{4S} = \frac{19{,}433(50)(12)}{4(2085)} = 1398 \text{ psi}$$

$$\text{in B,} \quad f = \frac{19{,}433(50)(12)}{4(13052)} = 223 \text{ psi}$$

The initial and the distributed stresses are shown in Fig. 13.8. The average stresses are

$$\text{in A,} \quad f = \frac{(942.4 + 486.5)}{2} = 714 \text{ psi}$$

$$\text{in B,} \quad f = \frac{(486.5 + 202.8)}{2} = 345 \text{ psi}$$

The deflections are those due to a concentrated load at the center of a span:

$$\text{in A,} \quad D = \frac{fL}{6Ed} = \frac{714(50)^2(12)^2}{6(3)(10)^6(4.66)(12)} = 0.26 \text{ in.}$$

$$\text{in B,} \quad D = \frac{345(50)^2(12)^2}{6(3)(10)^6(10.66)(12)} = 0.05 \text{ in.}$$

The deflection diagram is shown in Fig. 13.9.

Column Load

The force in the column is equal to the deflection due to the dead and live loads divided by the deflection due to the unit upward load.

$$P = \frac{10{,}000(0.52)}{0.29} = 17{,}900 \text{ lb}$$

Multiply all stresses for the upward load of 10,000 lb by the ratio 17,900/10,000 = 1.79.

The final step is to determine the combined stresses. These are the original stresses less the stresses due to the upward force. These stresses have been drawn for each plate in Fig. 13.10. It may be seen that the stresses are considerably reduced by providing an extra column at the edge plate.

Reinforcement in the slanting slab span is designed as for an ordinary slab span. The moments at the valley and the crown should not approach the maximum strength design values. Tests indicate that for tensile forces on an inside corner, the ultimate strength may be less than 50% of that for a straight section.

Reinforcing for the plate elements of folded plates is very similar to that for barrel shells, Section 13.3.

This method has further advantages. For a two-span structure, the horizontal transverse ties at the center can be completely eliminated by using as redundants the additional edge columns and the interior columns.

Timber folded plates are also popular. Their design is covered in Section 25.16.

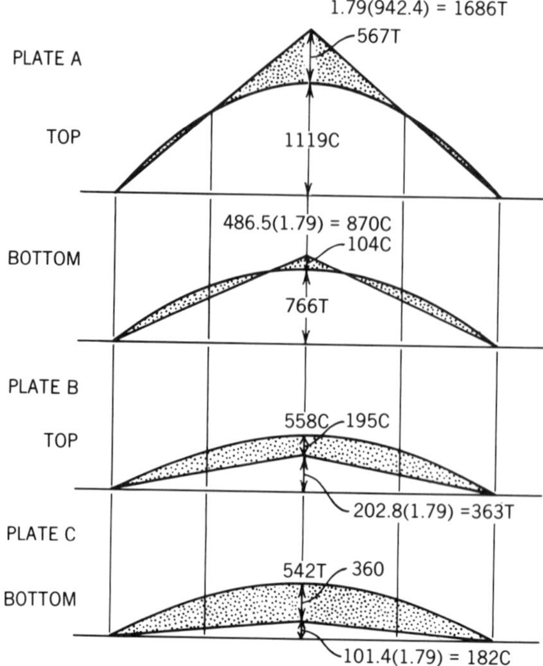

Fig. 13.10 Combined stresses.

13.3 BARREL SHELLS

Barrel shells may be classified as short, intermediate, or long, depending on the ratio of the length along the axis compared to the radius of the circle. The long barrel, if properly supported, tends to have a linear distribution of longitudinal stress as in a beam. Short and intermediate barrels have a distinctly nonlinear distribution.

Profiles of several configurations of long barrel shells are sketched in Fig. 13.11. The first is a circular barrel with no edge beams and with upturned end plates. This shape is easier to form than the barrel with edge beams and downturned interior edge beams, Fig. 13.11*b*. However, due to the greater width required to obtain the same effective depth, the transverse bending moments are greater, and the thickness of the shell may be greater as well as the quantity of reinforcing. The system with edge beams may also require more concrete. It has been used extensively for very long spans with prestressed edge beams.

The profile in Fig. 13.11*c* is for a barrel with reverse curvature at the valleys. It is an interesting architectural shape both externally and internally. It has another advantage for continuous spans; the valleys may be increased in thickness where additional compression capacity is required.

The last profile, Fig. 13.11*d*, is for a north light shell, similar to the Z folded plate, Fig. 13.2*e*, in function. The individual shells are separated by windows. Also the shells mutually support each

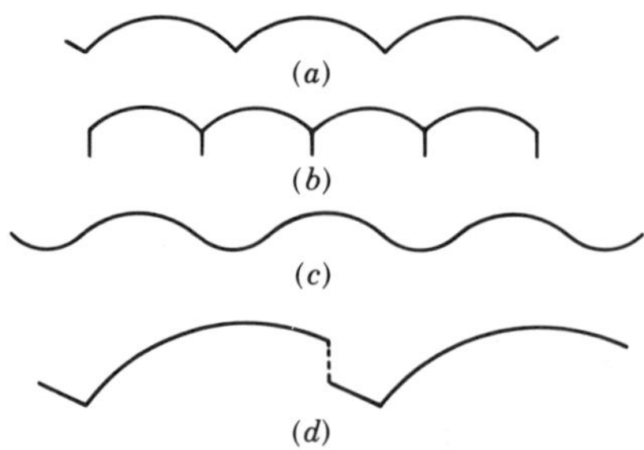

Fig. 13.11 Profiles for long barrel shells.

other by the window mullions. This shell is not suitable for long spans because the effective depth is quite small for a circular curve of such large diameter.

13.3.1 Structural Action

Barrel shells carry load in the longitudinal direction as a beam and in the transverse direction as an arch. The arch, however, is supported not at its ends as for an ordinary arch, but by the internal shear forces associated with the beam. A long barrel shell may be analyzed by this method using elementary beam and arch theory. As in the case of a folded plate, long barrel shells will depart from the mechanics of ordinary beam theory when there is considerable difference between the deflections of the crown and the valley edge. The transverse bending moments in the barrel result from the equilibrium of the vertical loads and shear forces. In an internal barrel of a series of shells, these forces are nearly in equilibrium so the moments are low, but at an exterior bay, the forces may produce rather high transverse bending moments that may be too large for a thin barrel. Also these high moments lead to a more nonlinear longitudinal stress distribution.

The key to the design of long barrels is to avoid these excessive edge deflections. As in the case of folded plates, intermediate columns should be used at edge plates. If this is done, then the longitudinal stress distribution tends to become linear, and simpler methods of analysis can be used for preliminary and sometimes final design.

Short barrels are also nonlinear, but part of the reason is that the depth of the internal, longitudinal, beam is large in comparison to the span, and the short shell acts as a deep beam. An approximate analysis can be made assuming that the depth of the beam at the valley is half of the span. These internal beams are loaded by the arch above the beams. The moments in the arch are very low and can be disregarded for preliminary design. In Fig. 13.12 are shown the longitudinal stresses in a typical short shell. They were taken from Parme and Conner [13.11].

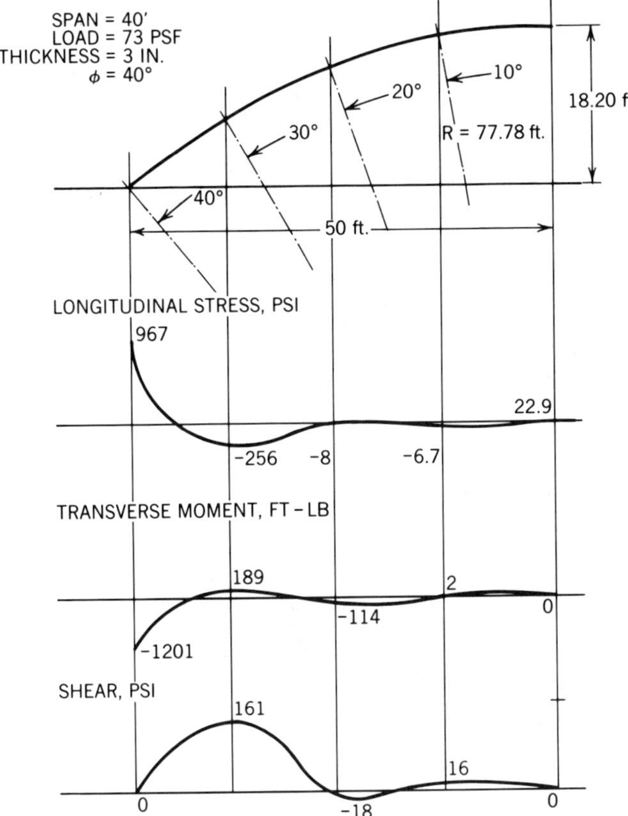

Fig. 13.12 Distribution of stresses in a short shell.

13.3.2 Preliminary Analysis of Long Barrel Shells

There are a number of tables of coefficients or formulas for the design of long barrel shells [13.2, 13.5, 13.6]. Because these shells come in so many different shapes and spans, the tables may not always fit the design problem, but they give useful information if properly interpreted. The longitudinal and shear reinforcing can be estimated from purely statical considerations, but the transverse moments are more difficult to estimate.

Dimensions for long barrel shells are similar to those for folded plates. A span-to-depth, or width-to-depth ratio of 10 or 12 is optimum. Again the ratio can be increased for prestressing. Thickness of barrel shells is more a function of the number of layers of reinforcing than the level of stresses. Usual depths are from 3 to 4 in (75 to 100 mm).

In Fig. 13.13, coefficients are shown for transverse moments in an interior barrel of a series. They were derived from tables for stresses by the beam method given in Chinn [13.5]. They are sufficiently accurate for preliminary design for a wide range of central angles. The shear stresses and the longitudinal reinforcing can be obtained directly from statics by assuming the arm length of the internal tension–compression couple. Although these coefficients were selected for barrels without edge beams, they will give a very good estimate for barrels with interior edge beams as in Fig. 13.11*b*.

A coefficient for the transverse moment in an exterior barrel is more difficult to estimate, but the moments may not govern the design thickness, so for a preliminary estimate of reinforcing, an amount of transverse reinforcing that is 50% larger than an interior barrel will probably suffice. It is important, on the end barrel, to use either an intermediate support or a downturned edge member.

Example of Design by Coefficients

An example for the design of an interior barrel of a long barrel shell having a span of 50 ft (15 m) follows. Steps in the design are:

1. Select dimensions and loads.
2. Calculate transverse bending moments from coefficients and select transverse reinforcing for an internal barrel.
3. Estimate the depth of the internal tension–compression couple and determine the longitudinal force and the longitudinal reinforcing required.
4. Determine the shear at the neutral axis from equilibrium of the longitudinal force with the shear forces, and select the shear reinforcing.
5. Complete the design, estimating sizes of edge members and reinforcing and support structure.
6. Estimate quantities for a feasibility study of the structure.

Dimensions and Loads

The barrel is simply supported and the center-to-center span is $L = 50$ ft. The width to the center lines of the barrels is $S = 30$ ft. The rise $H = 5$ ft and the radius $R = 25$ ft. The half-angle $\phi = 36.87°$ and the length of the half-arc is 16.1 ft. The thickness of the shell is assumed to be 3.5 in. based on the minimum thickness required to place the reinforcing and not on strength considerations.

The design live load is 30 psf and the roof load includes 5 psf for roofing and insulation. The weight of concrete is 45 psf. For convenience, the total load is assumed to be the sum of the dead and live loads: $w = 30 + 5 + 45 = 80$ psf.

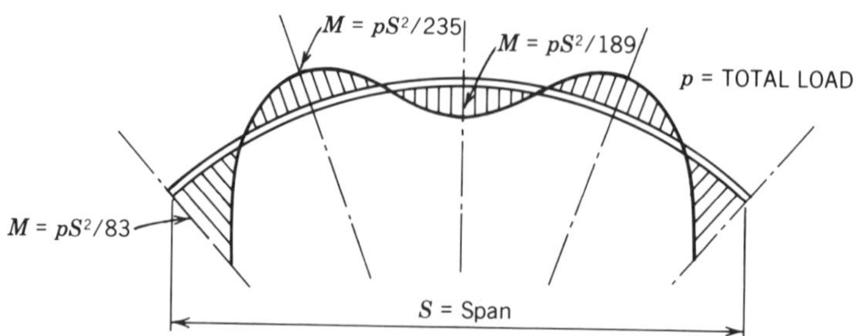

Fig. 13.13 Transverse moment coefficients.

Transverse Moments and Reinforcement

The transverse moments, using the coefficients from Fig. 13.13 are

$$\text{moment in valley} \quad \frac{80(30)^2}{83} = 867 \text{ ft-lb}$$

$$\text{moment in crown} \quad \frac{80(30)^2}{189} = 381 \text{ ft-lb}$$

$$\text{moment at quarter point} \quad \frac{80(30)^2}{235} = 306 \text{ ft-lb}$$

The required area of reinforcement will be based on the working stress method using a cover of steel of 0.75 in. or effectively 1 in. to the center of the bar. The value of j is assumed to be 0.875 and the allowable stress is 24,000 psi.

At the valley, the shell will be thickened locally to 5 in. so the required area is

$$A = \frac{M}{jdf} = \frac{867(12)}{4(0.875)(24000)} = 0.124 \text{ sq in.}$$

The corresponding areas for the crown, using an effective depth of 3.5 in. is $A = 0.087$ sq in. At the quarter point the minimum requirement will hold; $A = 2.5(0.0025)(12) = 0.075$ sq in. The following bars will be used: at valley #3 @ 10 in., at crown and quarter point, #3 @ 12. The reinforcement is shown in the preliminary sketches in Fig. 13.14.

Longitudinal Moments and Reinforcement

The load on a half-shell is the load per foot multiplied by the length of the half-arc: $w = 80(16.1) = 1288$ lb/ft. In addition a weight of 60 lb/ft will be added for the valley where it is thickened so the total is $1288 + 60 = 1348$ lb/ft. The moment at the center of the simple span is

$$M = \frac{1.35(50)^2}{8} = 422 \text{ ft-kips}$$

The location of the center of tension for the compression–tension couple is selected by estimating the center of a likely number of reinforcing bars and in this case it was assumed to be 12 in. above the valley. The center of compression was estimated conservatively by using a moment arm of 0.92 times the distance from the center of tension to the crown. The effective depth is $0.92(60 - 12) = 44.1$ in. The longitudinal force is $422(12)/44.1 = 114.8$ kips. The area of reinforcement is $A = 114.8/24 = 4.78$ sq in. Use 11 #6 bars, $A = 4.84$ sq in.

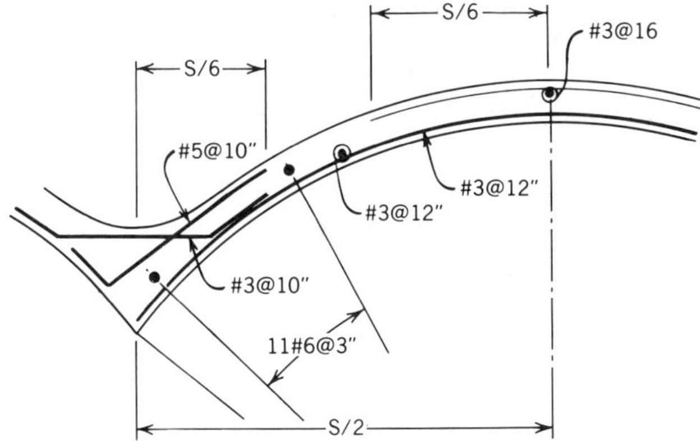

Fig. 13.14 Reinforcement in barrel.

Shear

The longitudinal shears at the neutral axis at the end of the barrel were obtained by assuming that the total shear force must equal the tensile (or compressive) force, and that the shear varies linearly from a maximum at the support to zero at the center of the span. A precise analysis will not give much more information than this approximate method.

The shear S is two times the tensile force divided by half the area of the shell from the center to the support:

$$S = \frac{(2)(114.8)}{(3.5)(25)(12)} = 0.219 \text{ ksi} = 219 \text{ psi}$$

At this point it is interesting to compare the stresses from this preliminary analysis with the values obtained from a computer analysis using the solution of the eighth order differential equation.

	Approximate	Computer
Moment at valley, ft-lb	867	900
Quarter point, ft-lb	306	331
Crown, ft-lb	381	418
Shear stress, psi	219	183

The diagonal shear reinforcement is obtained by assuming that on a small element there is no force perpendicular to the axis of the shell, in which case the shear is numerically equal to diagonal tension. Actually, there is usually some compressive force so this assumption is conservative. At the support, the area of bars required is the shear times the area, divided by the allowable tension stress. No allowance is made for using the concrete as part of the tension force. The area is

$$A = \frac{219(3.5)(12)}{24{,}000} = 0.383 \text{ sq in.}$$

The spacing along the diagonal at the support using #6 bars would be 11 in. Toward the center, where the shear is less, the spacing can be increased or the bar size reduced.

13.3.3 The Beam Method for Long Barrels

A very useful method for the design of long barrel shells is the beam method [13.5, 13.10]. The longitudinal stresses are determined assuming that the barrel is a beam. The usual assumption is that there is a straight line distribution of longitudinal stress, but other nonlinear distributions are possible. One of the main advantages is that the method can be used for shells of any shape or with varying thickness. Given the longitudinal stress distribution, the transverse moments and shear can be determined by statics. A typical transverse element of a barrel is shown in Fig. 13.15. The vertical forces from dead load must be in equilibrium with the components of the change in shear between the two sections. If the structure is a single barrel, the arch has no reactions at the springing, but if it is one of a multiple set of barrels, there is a horizontal force and a moment but no vertical force. The reactions, shears, and transverse moments may be obtained by a typical statically indeterminate arch analysis.

Algebraic methods have been given [13.5] for circular cross-sections of uniform thickness, but for other shapes and for varying thicknesses, the cross-section can be divided into segments and the properties of these segments used in the analysis [13.14]. The beam method has also been used

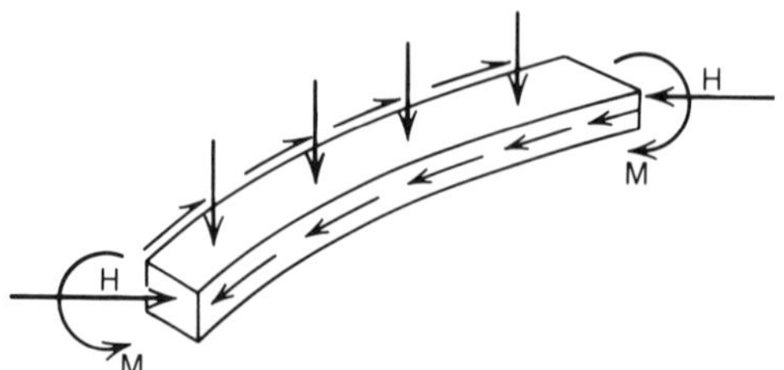

Fig. 13.15 Element of a barrel shell.

for unsymmetrical sections such as north light shells [13.13], but may be inaccurate because of the relative deflections of the elements of the shell.

The beam method is a very useful tool for preliminary design of barrel shells, especially if nonlinear distribution of stress is required. A short computer program can be developed for this purpose.

13.3.4 Short Barrel Shells

The distribution of stresses in a short barrel shell is considerably different from that of a long barrel shell as evidenced by the plot of stresses for an interior barrel in Fig. 13.12. The longitudinal tension stress is concentrated at the bottom of the shell, and the maximum compression is not at the top of the shell. The transverse bending moment is a maximum at the support if the edge is fixed as in this case, and is very small at the crown. There are many types of members used to support the edge of short shells but the usual solution is a horizontal or downturned beam. A short shell acts virtually as a membrane shell at the crown supported by the sloping beam at the valley.

The most important structural element of a short barrel shell is the arch required to support the shell, and the design of this element requires considerable care. It should be noted that the loads from the shell are delivered not as vertical loads but as tangent to the curve of the circle. Another confusing factor is that unbalanced live loads tend to be distributed throughout the arch rather than going directly to the arch as for a normal arch with beams and a roof deck.

Preliminary Design

The tables in Parme and Conner [13.11] are very useful for preliminary design of short shells, especially if the shells are multiple and not single barrels. However, most long span short barrel shells are single barrels so that the tables are only a general guide.

An approach that will give sizes of elements that are satisfactory for preliminary design is to assume that the shell is a membrane structure at the crown, and a beam element near the valley. A rule of thumb for the inclined depth of this beam is that the depth should be half the span. This beam can be thicker than the membrane element at the crown to accommodate the reinforcement required for longitudinal stress, transverse bending, and shear. At the crown, only two layers of reinforcing bars are required. There will still be no key to the amount of transverse bending reinforcement required but normally it is quite small and can be estimated for preliminary costs.

Example of Preliminary Design

The dimensions of the shell are span between arches, $L = 40$ ft, half-central angle $= 40°$. The rise then becomes 18.20 ft from the springing to the crown, and the radius is 77.78 ft. The length of the cylindrical curve from the springing to the crown is $S = 54.3$ ft. The effective beam is 20 ft deep on the slant, (half of the span L), and the thickness will be assumed to be 4 in. for this effective beam. The thickness above the beam is 2.5 in. A sketch is shown in Fig. 13.16. The barrels are multiple (continuous spans).

The loads are

2.5-in shell	=	32 psf
roofing	=	5 psf
live load	=	30 psf
total	=	67 psf

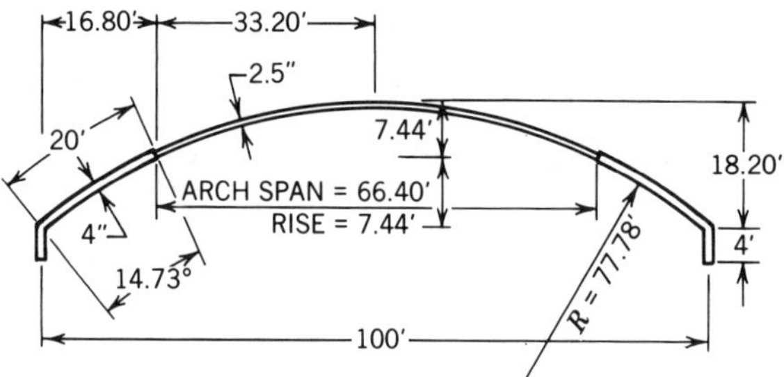

Fig. 13.16 Design example, short shell.

The horizontal thrust H and the vertical reaction R on the central arch above the effective beam for a span $L = 66.4$ ft and a rise $h = 7.44$ ft are

$$H = \frac{wL^2}{8h} = \frac{67(66.4)^2}{8(7.44)} = 4963 \text{ lb/ft}$$

$$R = \frac{wL}{2} = \frac{67(66.4)}{2} = 2224 \text{ lb/ft}$$

The inclined thrust in the shell at the springing of the interior arch is:

$$T = \sqrt{H^2 + R^2} = \sqrt{(4963)^2 + (2224)^2} = 5438 \text{ lb/ft} = 5.44 \text{ kips/ft}$$

These thrusts, in turn, are applied to the inclined beam which has a depth of 20 ft. If the barrels are continuous, then the maximum moment in this beam at the support (the main arches), would be $M = wL^2/12$. The precise effect of continuity is unknown and there are other factors to be considered. For example, how will the shell be constructed? Will each shell be independently supported as it is sprung? Will there be expansion joints in every other span as is frequently done? All of these questions must be answered before the final design.

For the preliminary estimate, a moment of $M = wL^2/10$ for both the top and the bottom area of reinforcement will be used.

$$M = 5.44(40)^2/10 = 870 \text{ ft-kips}$$

The stress in a beam 4 in. wide and 20 ft (240 in.) deep, assuming a prismatic section is:

$$\frac{I}{c} = \frac{bd^2}{6} = \frac{4(240)^2}{6} = 38400 \text{ in.}^3$$

$$f = \frac{Mc}{I} = \frac{870(12)}{38400} = 0.272 \text{ ksi}$$

The stress on the concrete is quite low so the thickness of 4 in. is satisfactory.

Assuming an effective depth of 215 in., the area A of steel required for each face, for an allowable steel stress of 24 ksi, and an effective moment arm ratio of $j = 0.875$ is:

$$A = \frac{M}{jdf} = \frac{870(12)}{215(0.875)(24)} = 2.31 \text{ sq in.}$$

This will require 6-#6 bars.

The shear reinforcement must also be selected for these beams. It is best to assume that the concrete takes no shear. Stirrups may either be vertical or preferably on a diagonal.

The transverse bending reinforcement must also be selected. The tables in Parme and Conner [13.11] will serve as a general guide to the level of moments.

13.4 HYPARS

Hyperbolic paraboloids, or hypars, the commonly used shortened name, have the distinct advantage that they can be formed from straight form boards and can be built in a variety of configurations. The principal types are saddles, gables, inverted umbrellas, and groined vaults. These structures can be designed on a preliminary basis with the use of the membrane theory [13.13, 13.4], and many successful shells have been built using only this theory plus considerable common sense. Recent studies [13.15, 13.16] have indicated that the actual behavior of the shell is not well represented by the membrane theory and these studies, using the results from the analysis by the finite element method, will be described here for each of the types. One of the problems in the design of hypars is the tendency of the designer to make the shells too shallow. Shells obtain their strength through form and not by using higher stresses. Many of the failures or partial failures of hypars have occurred in shallow structures [13.7]. Hypars may also be constructed from timber; see Section 25.17.

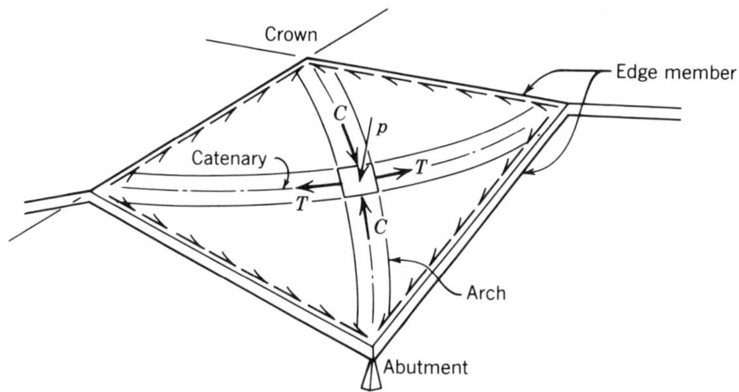

Fig. 13.17 Element of hypar surface.

13.4.1 Membrane Theory

The membrane theory is well covered in many texts [13.13, 13.14], so only a brief description will be given here. The forces on an element of a hypar surface are shown in Fig. 13.17. The top edges are assumed to be horizontal. From the crown to the abutment, the surface is curved upward in a parabolic arch. In the other direction, it sags downward in a parabolic catenary. Assuming a statically determinate structure at 45° to the orientation of the diagonal element shown, there will be pure shear forces. If the shell is assumed to be shallow, these shear forces are

$$S = \frac{pab}{2f}$$

where a and b are the horizontal dimensions and f is the vertical distance of the crown above the abutment. Also, from statics, the compression or tension on the element as shown in the sketch is numerically equal to the shear so that

$$T = \frac{pab}{2f}$$

In the example shown, the shear forces are transferred to the edge members which in turn carry them to the abutment. The force in an edge member, disregarding arch action, is assumed to vary from zero at the top to a maximum at the abutment. The level top edge members, in this case, are in compression and are assumed to be in equilibrium with each other.

13.4.2 Saddle Hypars

The saddle hypar shown in Fig. 13.18 may be square or diamond shaped and can be considered as a beam cantilevering from a line through the abutments and the crown to the tips. The total moment at a section through the abutments and the crown must be resisted by a force couple formed by compression forces C in the edge beams and the tension forces T in the shell. For uniformly distributed loads on a relatively deep shell, the reinforcing requirements can be estimated using a stress distribution similar to the membrane theory as sketched in Fig. 13.19*a*. That is, the tensile stresses are constant and the compressive forces are all concentrated at the edge beam. The internal moment is actually greater than that predicted by the membrane theory and is more nearly the distribution represented by Fig. 13.19*b* because there is also some compression in the shell. For dead load, the membrane theory gives satisfactory results for preliminary design.

The design of the shell for the weight of the edge beam should be based on a much shorter moment arm similar to the distribution of stresses shown in Fig. 13.19*c*. About 90% of the loads on the edge beam will be resisted by the compression-tension couple and the other 10 to 20% are carried by direct bending in the edge beam. Therefore reinforcing for bending moments in the edge beam will be required near the support. For hypars square in plan up to a span of 75 ft (23 m), a constant-depth edge beam is satisfactory but it should not be made heavier than necessary because any extra load must be carried on a short moment arm and the deflections will be greater. For

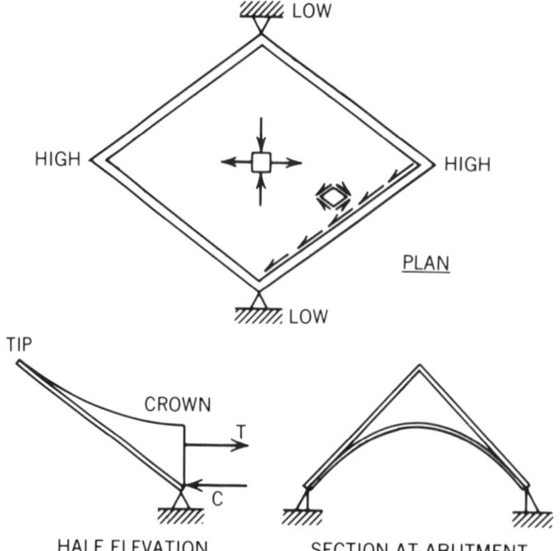

Fig. 13.18 Elements of a saddle shell.

longer shells, the edge beam should be increased in size near the support. A tapered edge beam will reduce the the bending moment and the direct tension forces in the shell.

A saddle shell is not, by itself, stable against unbalanced live loads as is evident from the sketch in Fig. 13.18. Therefore additional columns must be place under the edge beams, at least on one side of the structure, to make it stable. It is not easy to predict the moments in the edge beam from unbalanced live load, but the design of the reinforcing should be conservative for this detail. Also the edge beams are subjected to considerable torsion from the restraint of the edge beam to the bending in the shell slab. The larger and more nearly square the edge beam, the greater the torsional moments. These are difficult to estimate on a preliminary basis but in any event the edge beam should have considerable stirrup reinforcing and bar splices should be well over the minimum requirements.

Slab depths for saddle shells are normally from 2.5 to 3.5 in (60 to 90 mm). The tendency to increase the depth to be conservative should be resisted. Calculations for the preliminary design of a saddle shell using some of the considerations described above follow.

Example, Saddle Hypar

The shell is 100 ft square in plan. The rise from the abutment to the crown is $f = 15$ ft, so the total height is 30 ft from the abutment to the tips. The thickness is 2.5 in. and the live load is 30 psf; including an allowance of 5 psf for roofing, the total live and dead load is $p = 67$ psf.

The membrane stress is

$$S = T = \frac{pab}{2f} = \frac{67(50)^2}{2(15)} = 5583 \text{ lb/ft}$$

The tensile reinforcing in the shell, using a unit stress of 24 ksi, is

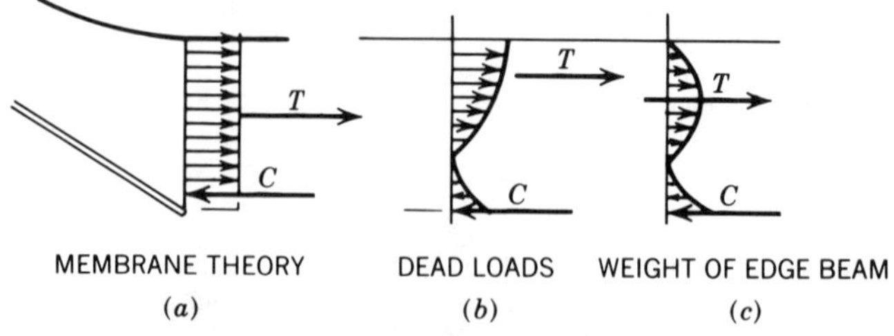

Fig. 13.19 Internal forces in a saddle shell.

$$A = \frac{5583}{24{,}000} = 0.23 \text{ sq in/ft}$$

Use #4 bars at 10 in. each way.

Assume an edge beam, 18 in. square, at the abutment which tapers to 12 in. at the tip. The thrust in the beam at the abutment, using $L = 100$ ft as the length of the beam is

$$C = SL = 5583(100) = 558{,}300 \text{ lb} = 558 \text{ kips}$$

The load per foot on the edge beam, using a unit weight of 150 pcf for the concrete and an average size of 15 in. square is, approximately, 0.234 kips/ft. The total moment is

$$M = \frac{0.234(100)^2}{2} = 1170 \text{ ft-kips}$$

Using 15% of the moment, the design moment is

$$M = 0.15(1170) = 175 \text{ ft-kips}$$

A column size at the abutment of 20 × 20 in. was selected by column formulas for strength design based on an average load factor of 1.6 times the total load. Three #9 bars were required in each face. The taper should be increased near the support to obtain the 20 in. square section.

Another important factor is the thrust of the shell at the abutments. It tends to be less than the statical value obtained from the membrane theory. If there is a prestressed tie for the abutments, then the thrust will influence the design.

13.4.3 Gabled Hypars

Analysis of gabled hypars by the finite element method [13.16] has demonstrated that the force system in a gabled hypar system, Fig. 13.20, is essentially membrane (small bending stress in the shell except at the edge members), but the usual assumption of equal tension and equal compression in an interior element does not necessarily hold. If the abutments are rigid, that is, not held by steel tie bars on columns, then the stiffest path for the forces is not through the catenary (tension) elements to the edge members but is directly over the crossed parabolic arches (compression elements). In this case, the membrane compression stress may be nearly double, and the tension stress is greatly reduced or nearly zero. On the other hand, if the abutments are allowed to move outward as will be the case if the abutments are held by horizontal steel ties to take the thrust, then the tensile forces are larger and may approach the classical membrane theory. Because the compression stresses are still very low, even if doubled, the thickness of the shell need not be increased. On the other hand, the tensile stresses require a minimum of reinforcing and the movement of the abutment is possible, so it is prudent for preliminary design to provide the full capacity required by the classical theory.

A considerable portion of the self-weight of the exterior edge member is carried by the shell, and the edge member is supported at both ends. The largest moments are concentrated near the abutments because the compressive shell forces must be transferred to the edge members. If the shell is made thicker near the abutment, then these high moments in the edge beams can be reduced. Several gabled hypars have failed because of excessive settlement of the crown [13.7].

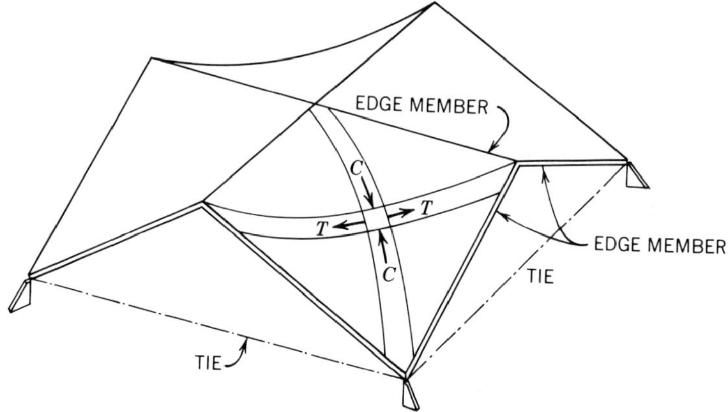

Fig. 13.20 Gabled hypar.

The edge members at the ridges should be made as small as possible consistent with the compressive stresses, and should be given some camber, say 4 to 6 in. (100 to 150 mm). In shell design, conservative does not always mean making the structural element bigger. An example for the preliminary design of a gabled shell follows.

Design Example, Gabled Hypar

The hypar shell roof is 100 ft square in plan and has a rise of f = 15 ft from the abutment to the crown. The shell thickness is 3 in. except near the abutment where it will be tapered for distance of 10 ft from the corner from 3 in. to the 10 in. at the abutment. The live load is 30 psf and the roofing load is 5 psf so the total load is 73 psf.

The membrane stress is

$$S = T = \frac{pab}{2f} = \frac{73(50)^2}{2(15)} = 6083 \text{ lb/ft}$$

The area of reinforcement requirement in tension is

$$A = \frac{6083}{24{,}000} = 0.25 \text{sq in.} \qquad (\#4 \text{ bars @ 9 in. on center})$$

The edge beam thrusts are

$$C = 6083(50)/1000 = 304 \text{ kips}$$

The moments in the sloping beam will be assumed to be 25% of the simple beam moment for the weight of the rib on the 50-ft span. Assuming a 14 in. square member weighing 0.21 lb/ft, the moment is

$$M = \frac{0.25(0.21)(50)^2}{8} = 16 \text{ ft-kips}$$

Using a load factor of 1.6, a 14 in. square edge member is satisfactory. Use 4-#7 bars.

The remaining element that must be selected is the top edge member. It should be proportioned so that the compression stress on the rib and a portion of the shell is reasonable. It should be cambered about 4 in.

13.4.4 Inverted Umbrella Hypars

The inverted umbrella hypar, Fig. 13.21, may also be analyzed by the membrane theory with reasonable accuracy because the internal forces are well contained within the primary structure. Note that multiple repeating umbrellas may also be considered as multiple repeating gables. Single inverted umbrellas should be analyzed as slabs in the regions near the tips. Shell curvature is very slight in this region, so that the tips have a propensity to droop and the edge beams must be reinforced in bending. Prestressing will help this condition.

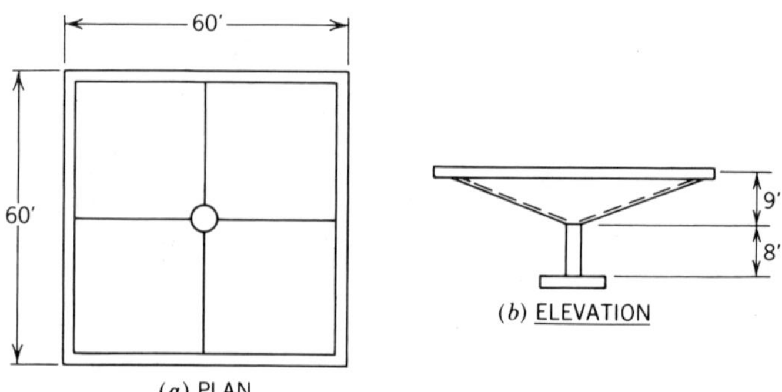

Fig. 13.21 Umbrella hypar.

Experience and theory has shown that the exterior edge beams of singly supported umbrella hypars should be upturned so that the tensile stress will cause the end of the beam to move upward. The preliminary design of an inverted umbrella hypar, square in plan follows.

Design Example, Inverted Umbrella Hypar

A sketch of the structure is shown in Fig. 13.21. The spans L each way are 60 ft. The rise is 9 ft. The height from the base to the lowest point on the shell is 8 ft. The thickness will be assumed as 2.5 in. The structure is part of a series of shells but will be designed as a free standing element. The live load is 20 psf of horizontal surface. For the design of the pedestal, the unbalanced live load will be assumed as half the full load on half of the structure. The dead load is 2.5(150)/12 = 32 psf plus an additional roofing load of 5 psf for a total dead load of 37 psf and live plus dead of 37 + 20 = 57 psf.

The membrane stress on a 30 × 30 ft element due to a load of p = 57 psf with a rise f = 9 ft is

$$T = \frac{pab}{2f} = \frac{57(30)^2}{2(9)} = 2850 \text{ lb/ft}$$

The compressive or tensile unit stress is

$$S = \frac{2850}{12(2.5)} = 95 \text{ psi}$$

Using an allowable unit stress of 24 ksi, the area of tension reinforcement is

$$A = \frac{2850}{24{,}000} = 0.119 \text{ sq in./ft}$$

This requires #3 bars @ 12 in. They should be placed diagonally and lapped at the splices and into edge members.

The top edge member will have a tensile force of

$$T = 2850(30)/1000 = 86 \text{ kips}$$

Using an allowable stress of 24 ksi, the required area at the center of the edge member is

$$A = \frac{86}{24} = 3.6 \text{ sq in.}$$

This would indicate a member with 3-#7 bars in the top and 3-#7 bars in the bottom so a member 12 × 12 in. is required. However, another consideration is the possible bending in the edge beam. At the center the effective depth including a portion of the shell is quite large. An educated guess for determining the bending moments is to design the edge member at the quarter point for the full cantilever bending of the edge member plus a small width of shell, say 3 ft. The moment at this point, assuming an edge member 12 in. square, will be

$$M = \frac{[150 + 3(57)](15)^2}{2} = 36{,}112 \text{ ft-lb} = 36.11 \text{ ft-kips}$$

The tensile force at this point is

$$T = 2850(15)/1000 = 42.7 \text{ kips}$$

The area of reinforcement can be selected as sketched in Fig. 13.22, assuming that the concrete is not included. The edge beam is upturned. In this example it is assumed to be at the same vertical location as the bottom layer of bars so the tensile force is at the bottom of the section. A free body diagram of the forces shows that the tensile force in the top layer is 54.16 kips, and the bottom layer is in compression. Therefore in the top, 3-#8 bars are used in place of the 3-#7 bars previously selected for the centerline of the edge beam. Although not required in this analysis, the same 3 #7 bars in the bottom are used.

The next element to be considered is the pedestal to support the shell. It is assumed that each hypar unit is free standing and does not receive lateral support from adjacent units. It is not clear what unbalanced load will be required to design the connection between the top of the column and

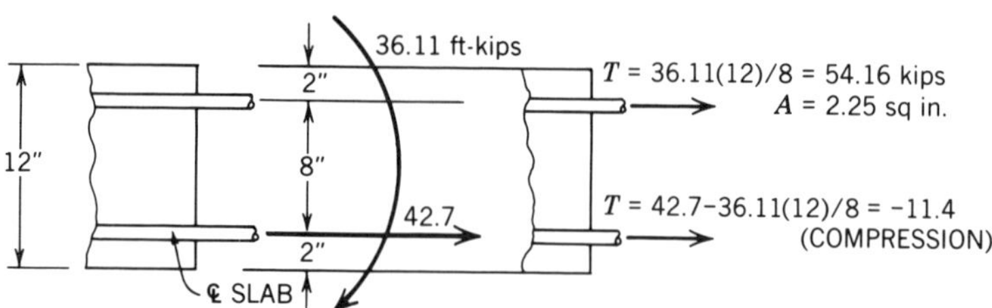

Fig. 13.22 Reinforcement in upturned edge beam.

the shell. A conservative value based on an ice load would seem to be half of the live load, that is, 10 psf.

The analysis of stresses in the connection between shell and column is a very difficult problem to solve precisely. Tests of large-scale umbrella hypars [13.17] have indicated that the beam element embedded in the shell at the top of the column should be designed for half of the bending moment at the top of the column. The column below the shell would be designed for the full moment plus the vertical load which will help to reduce the reinforcement.

The moment due to a 10-psf load on a 30-ft cantilever for a 60-ft width is

$$M = \frac{wL^2}{2} = \frac{60(10)(30)^2}{2(1000)} = 270 \text{ ft-kips}$$

The vertical load, including the weight of the shell is

$$P = [40(60)^2 + 10(30)(60)]/1000 = 162 \text{ kips}$$

Using an average load factor of 1.6, a 20 in. square column with 8-#8 bars is required.

13.5 DOMES OF REVOLUTION

Domes of revolution are built in spans from 50 to 300 ft (15 to 90 m), and are often used for covers of water tanks or sewage digesters as well as roofs for sports facilities. A typical dome of revolution is a portion of a sphere supported by an exterior ring beam which, in turn, is supported either by a circular wall or by columns. It is a very efficient structure because the stresses are generally membrane, except for small bending moments at the junction of the sphere and the ring beam where the outward movement of the ring will cause an eccentricity of the thrust from the shell. The shell thickness varies from 2 to 4 in. (50 to 100 mm) and may be governed, for large spans, by deflection and stability, rather than by stress. Most of the stresses are compression except for the possibility of some tension in the lower areas of the shell. Timber domes are discussed in Section 25.14.7.

13.5.1 Membrane Stresses in Domes

For the preliminary and for the final analysis, a dome of revolution is assumed to be axisymmetric for dead load and for vertical live load. For the final analysis on large structures, it may be necessary to consider other loading patterns for wind or seismic loads but this is not generally necessary for preliminary analysis except for a general study of the structural action. Of course the rule applies that the larger the structure, the greater the analysis.

The literature [13.1, 13.13] abounds with formulas for membrane stresses on many types of axisymmetric shells and loading configurations which will not be repeated here. However, two simple rules are sufficient to obtain the membrane stresses for any configuration of a dome of revolution for axisymmetric loading:

1. The membrane stresses T_1, Fig. 13.23*a*, may be obtained by statics. The sum of the vertical components of stress V must be equal to the total vertical force P. The stress T_1 must be tangent to the curve of the shell so $T_1 = V/\cos\phi$.
2. The membrane stress T_2, Fig. 13.23*b*, may be obtained by statics by adding the vector sums of the stresses T_1 and T_2 in the direction of the applied load Z perpendicular to the surface of

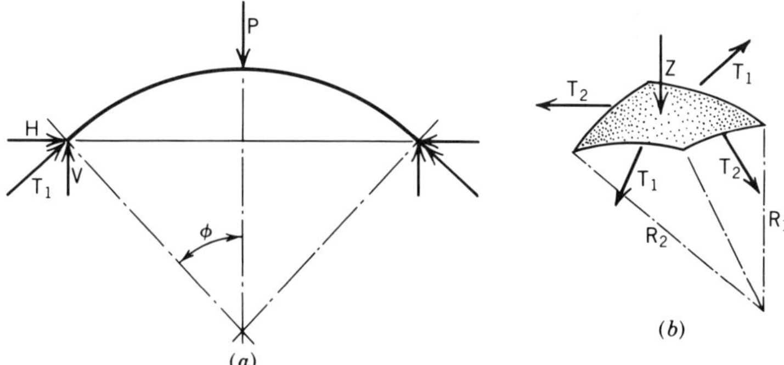

Fig. 13.23 Elements of a circular dome.

the element. The relationship between T_1 and T_2 may be obtained from the cylinder formula

$$\frac{T_1}{R_1} + \frac{T_2}{R_2} = Z$$

where R_1 and R_2 are the radii corresponding to each stress.

It should be noted that the formulas shown in texts do not necessarily give the maximum T_2 stress. If on an element, there is only dead load but the T_1 stress is based on both dead and live load, then T_2 must be computed for the Z force due to both dead and live load. This is further illustrated in the design example.

The ring beam tension T can be obtained from H, the horizontal component of T at the ring, by using the cylinder formula $T = Hr$, where r is the horizontal radius of the ring beam.

13.5.2 Moments in the Shell Near the Ring

The analysis of the bending moments in the shell near the ring is extensively covered in the literature [13.1, 13.13, 13.14]. These moments usually do not govern the preliminary design and it is sufficient merely to thicken the shell near the support, and to place two layers of radial bars to resist possible bending moments. These moments are secondary and any cracking will merely relieve the moments and the structure will not fail from this cause. If the ring beam is to be prestressed, the moments in the shell will be completely changed, generally to relieve the bending moments.

13.5.3 Design of the Ring Beam

If the ring beam is supported by the wall of a tank, then the ring beam will participate in the water loads from the tank, and a more sophisticated analysis may be necessary for the preliminary design. If the ring is supported on columns, the ring beam and shell must be designed as a beam to carry the structure between columns. The total depth of the internal beam may generally be estimated as half of the span between columns rather than just the depth of the ring beam. These considerations may require extra tension reinforcement in the shell near the columns. The forces are balanced so that the curvature of this beam can be disregarded. Any assumptions for preliminary analysis should, of course, be verified in the final analysis.

13.5.4 Arrangement of Reinforcement

Although most of the forces in a dome of revolution are compression, a nominal amount of reinforcement is required in both directions. The inclined reinforcement is usually radial, but there is no reason why all of the reinforcement could not be placed in a grid pattern.

For small structures, the ring beam is reinforced with ordinary reinforcing bars but care must be taken to splice the bars properly and extra ties should be used. For larger structures, it is customary to use prestressing, either by tendons or by wire wrapping, especially in the case of large tanks.

Design Example, Dome of Revolution

The dome will have a diameter of 94 ft and a height of 18 ft. The thickness will be 3 in., the radius of the sphere is $R = 70.36$ ft, and the half-central angle is $\phi = 41.9°$. The structure will be designed for a roofing load of 7 psf so the total dead load is 45 psf. The live load is 30 psf.

The area of the sphere is

$$A = 2\pi Rh = 2\pi(70.36)(18) = 7958 \text{ sq ft}$$

The area in plan is

$$\pi r^2 = \pi(47)^2 = 6940 \text{ sq ft}$$

The total load of the shell above the ring beam is

$$P = [7958(45) + 6940(30)]/1000 = 566.3 \text{ kips}$$

The vertical reaction per foot length of the ring beam is

$$V = \frac{566.3}{\pi(94)} = 1.92 \text{ kips/ft}$$

The horizontal component is

$$H = \frac{1.92}{\tan(41.91)} = 2.14 \text{ kips/ft}$$

The inclined component is

$$T_1 = \frac{1.92}{\sin(41.91)} = 2.87 \text{ kips/ft}$$

The next step is to determine the value of T_2 at the ring for full live and dead load. From Fig. 13.23,

$$Z = 30 \cos^2(41.91) + 45 \cos(41.49) = 50.1 \text{ psf}$$

$$T_2 = -ZR + T_1 = -50.1(70.36)/1000 + 2.87 = -0.66 \text{ kips/ft, compression}$$

The value of T_2 for dead load only is

$$Z = 45 \cos(41.91) = 33.5 \text{ psf}$$

$$T_2 = -33.5(70.36)/1000 + 2.87 = 0.51 \text{ kips/ft, tension}$$

Thus the sign of T is completely reversed. In some designs, this might make a difference in the reinforcing.

The reinforcing in the shell is a minimum, say #3 bars at 12 in. on centers, both ways. A double layer of radial bars of the same size will be used at the ring and the shell will be increased in thickness from 3 to 5 in. for a distance of 5 ft from the edge of the ring.

The ring tension is

$$T = \frac{Hd}{2} = \frac{2.14(94)}{2} = 100.6 \text{ kips}$$

The area of reinforcing in the ring, using a unit stress of 24 ksi, is 4.2 sq in., so 8-#7 bars are required.

13.6 TRANSLATION SHELLS

A translation shell is formed by a vertical curve moving along another curve as illustrated by the shell shown in Fig. 13.24. If the curve is a circle, it looks, superficially, like a spherical dome trimmed to square in plan. However, all of the vertical radii are the same, instead of varying, as for

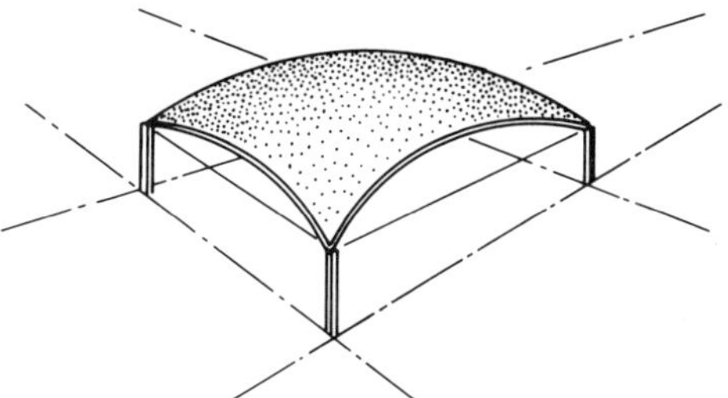

Fig. 13.24 Translation shell.

a sphere. A translation shell is easy to form and is favored for square plan domes. Large areas can be covered with a series of square or rectangular multiple domes, which is not possible with domes circular in plan. The curves that generate the shape may be circles, parabolas, or other conic sections. The shell is supported by arches on each of its sides, which in turn are supported by columns, or in some cases by walls. For multiple elements, the arches may be part of the shell. Thickness of the shell varies from 2 to 4 in. (50 to 100 mm), except near the arches where it is thickened and extra reinforcing added as in the case of domes of revolution.

13.6.1 Preliminary Analysis

It is a general rule of shell analysis that the greater the curvature, the less important the actual stress in the shell, and the more important the method of support and the design of the supports. A translation shell follows this rule. There are small membrane compression stresses over most of the surface except for tension stresses near the corners. There are local bending moments in the shell adjacent to the arch supports as for circular domes.

Excellent design tables for membrane stresses are given by Parme [13.12] and Candela [13.4] and they are recommended for preliminary design.

An insight into the structural action of translation shells may be obtained by considering that a translation shell, square in plan, acts essentially like a dome of revolution on top of an assemblage of arches, and a preliminary design may be made on this basis. For the tension in the element of the shell represented by a ring at the top of the arches, the dome theory may be used. The thrust in the side arches may be estimated by assuming that each arch takes a quarter of the total load. Bending moment in the arches is due mostly to unbalanced live load. The shell tends to spread the load over a considerable area and to make the loads uniform on the arches. A reasonable and conservative estimate of the moments can be made by assuming that unbalanced load is 50% of the live load. As a guide, the moment on a two-hinged arch due to a live load on half of the span is $M = wL^2/64$, where w is the live load, and L is the span of the arch. This occurs at the quarter point of the arch.

Example, Translation Shell

The translation shell is 64 ft square in plan. The rise from the lowest point is 8 ft to the top of the arches and 16 ft to the crown. The shell thickness is 3 in. Loads are roofing = 5 psf, total dead load = 43 psf. The live load is 30 psf so the total load is 73 psf.

The first step is to analyze the dome above the top of the arches as a spherical dome. This has already been described under the design of domes of revolution, Section 13.5. From the dimensions given above, the vertical radius $R = 68$ ft, and the radius of the ring at the top of the arches is $r = 32$ ft. The half-central angle is $\phi = 28.07°$. The area of the shell above the ring is $2\pi Rh = 2\pi(68)(8) =$ 3418 sq ft. The area of the dome in plan is $A = \pi r^2 = \pi(32)^2 = 3217$ sq ft.

The weight of the dome above the ring is $W = [3418(43) + 3217(30)]/1000 = 243.5$ kips. The vertical reaction of the ring per foot is $243.5/(64)(\pi) = 1.21$ kips/ft. $H = 1.21/\tan(28.07) = 2.26$ kips/ft. The inclined thrust $T_1 = 1.21/\sin(28.07) = 2.57$ kips/ft. The ring force in the ring at the top of the arches is $T = Hr = 2.26(32) = 72.3$ kips. This thrust is to be distributed over a wide band around the dome.

Next estimate the thrusts in the arch at the springing. The total load on the arch is approximately $W = 73(64)^2/1000 = 299$ kips. The total load per foot of arch is $w = 299[4(64)] = 1.168$ kips/ft. The horizontal thrusts from this arch can be estimated by the equation $H = wL^2/8h$, where h is the rise of the arch. $H = 1.168(64)^2/[8(8)] = 74.8$ kips. The vertical reaction at the springing is $R = wL/2 =$ $1.168(64)/2 = 37.4$ kips. The approximate thrust is the square root of the sum of the squares for the reaction and the horizontal thrust, $T = \sqrt{74.8^2 + 37.4^2} = 83.6$ kips.

The design moment in the arch is $M = wL^2/64$ where w is half of the live load on the arch; thus $w = 30(64)^2/[64(2)(4)(1000)] = 0.24$ kips/ft. The moment is $M = 0.24(64)^2/64 = 15.36$ ft-kips. The reinforcement for this moment should be used for all points on the arch.

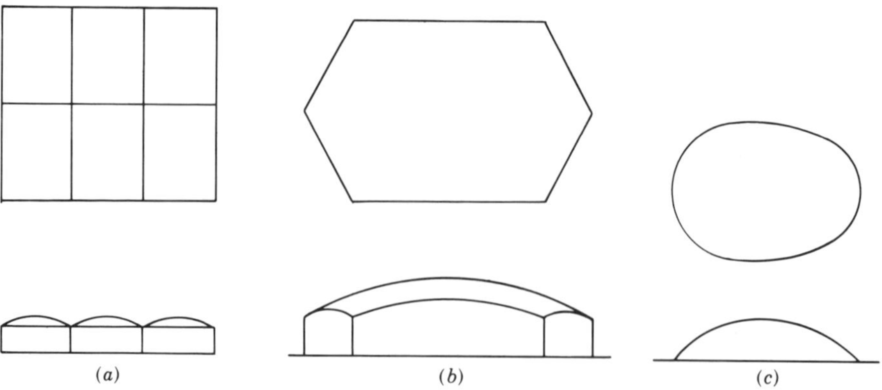

Fig. 13.25 Shapes of funicular shells.

13.7 FUNICULAR SHELLS

The traditional shell types already described are suitable for roofs over areas rectangular or circular in plan with the exception of hypars which may have diamond or polygonal shapes. A funicular shell can be placed over a building which has any shape in plan or in the elevation of the supports. Some simple examples are shown in Fig. 13.25.

The shape of the shell is based on either the solution of the differential equation for a shallow membrane by the finite difference method [13.8] for any shape, or the Valeria equation [13.8] for a rectangular area that gives parabolic sections in both the X and Y directions.

13.7.1 The Valeria Equation

A plan of a rectangular area is shown in Fig. 13.26. The vertical height z at any point is

$$z = h - \frac{(h - f)x^2}{a^2} - \frac{(h - g)y^2}{b^2} + \frac{(h - g - f)x^2y^2}{a^2b^2}$$

This shape is similar to the translation shell, but the height of the sides is not fixed and may be varied. Often the edges of the shell are horizontal with f and g equal to zero.

13.7.2 The Membrane Equation and Solution

The differential equation for a shallow membrane is

$$\frac{\partial^2 z}{\partial x^2} + \frac{\partial^2 z}{\partial y^2} = pN$$

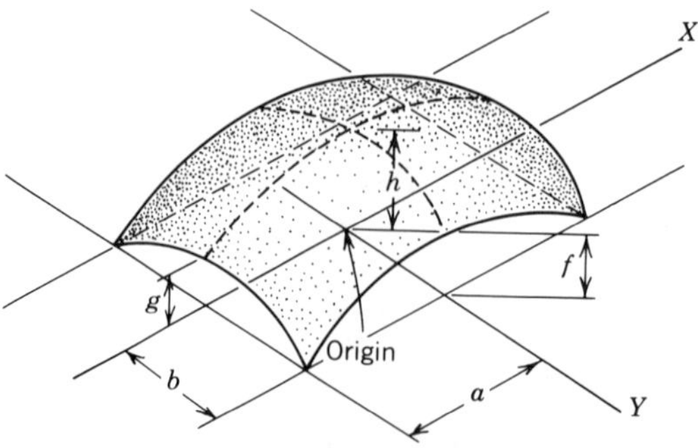

Fig. 13.26 Dimensions of a Valeria shell.

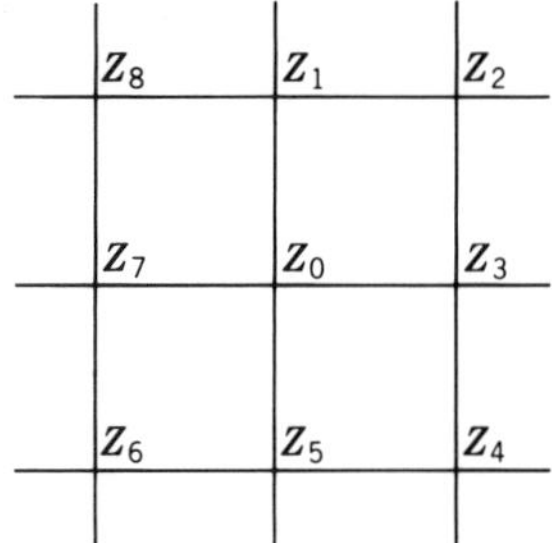

Fig. 13.27 Grid for finite difference equation for a membrane.

where x, y, and z are the ordinates to the surface, p is the pressure, and N is the tension in the membrane.

An element of a uniform square grid with a spacing h is shown in Fig. 13.27. The finite difference equation is

$$-z_1 - z_3 + 4z_0 - z_5 - z_7 = \frac{ph^2}{N}$$

The concentrated load, ph^2, can be varied at each node to shape the ordinates of the shell for functional or aesthetic reasons. A larger vertical load at a point will cause the shell to hump upward. A negative value will cause a dimple in the surface. Techniques for the determination of shapes for other floor plans and elevations are described by this writer [13.8] and Ramaswamy [13.13].

13.7.3 Shell Shapes from Models

It is possible to develop shapes from models by stretching a membrane, loading it with vertical loads, and measuring the resulting shape. This technique has resulted in some incredibly beautiful shells by Heinz Isler in Switzerland and is described by Billington [13.3].

13.7.4 Analysis

These shells are essentially funicular. That is, the dead load stresses in the shell are membrane if the shell is properly supported. For the shell shown in Fig. 13.25*b*, however, thrusts must be transferred to the side walls so the bending moments in the shell may become large and ribs may be required across the surface. For the preliminary design, the thickness of the shell and the amount of reinforcement are minimal, so the problems in design are in the support systems such as walls, beams, and arches, or elements of adjacent shells. Final analysis of the shell for larger structures may be made with the finite element method.

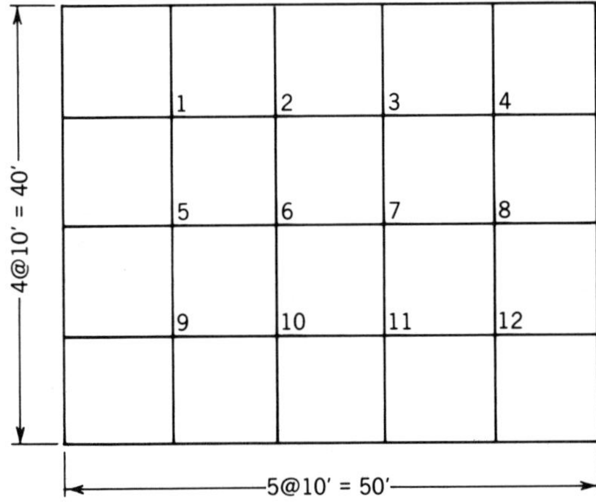

Fig. 13.28 Rectangular funicular shell.

	1	2	3	4	5	6	7	8	9	10	11	12	ph^2/N	Solution
1	4	−1			−1								8	7.15
2	−1	4	−1			−1							8	10.76
3		−1	4	−1			−1						8	10.76
4			−1	4				−1					8	7.15
5	−1				4	−1			−1				8	9.86
6		−1			−1	4	−1			−1			20	17.13
7			−1			−1	4	−1			−1		20	17.13
8				−1			−1	4				−1	8	9.86
9					−1				4	−1			8	7.15
10						−1			−1	4	−1		8	10.76
11							−1			−1	4	−1	8	10.76
12								−1			−1	4	8	7.16

Fig. 13.29 Equations for membrane surface and solution.

13.7.5 Example: Solution of Membrane Equations

A rectangular plan for a shell is shown in Fig. 13.28 with a grid of four elements in one direction and five elements in the other. The edges of the shell are horizontal. The nodes are numbered from 1 to 12 but the edges are not numbered. The finite difference equations for the membrane surface are shown in Fig. 13.29. The values on the right side of the equations essentially represent the load at each element and on the elements 6 and 7 the load has been increased to make the surface hump at the middle. The values used are arbitrary so at points 1 through 5, and 8 through 12, a value of ph^2/N of 8 was used and at points 6 and 7 a value of 20 was used. These equations can be solved with a simultaneous equation program. The solution is shown in the last column of Fig. 13.29. Several trials with load sizes and patterns may be required to get the required shape. An isometric of the surface is shown in Fig. 13.30. It is obvious that if more points had been used, the surface would have been defined better. Advantage can be taken of the property that the surface is symmetrical about both the X and Y axes. In writing the equations, if points have the same elevation, their number will be the same. If the edges of the shell had not been horizontal, then the elevations of the edges above the base would be added to the existing values of ph^2/N on the right side of the equations.

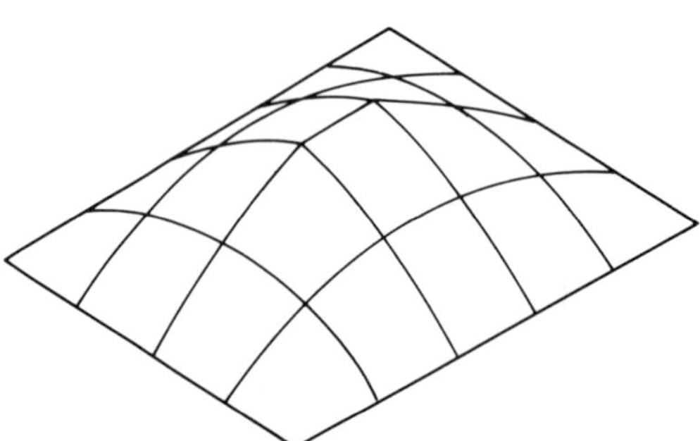

Fig. 13.30 Elevation of surface.

13.8 QUANTITY SURVEYS AND COSTS

One of the most important functions of preliminary design is to prepare a list of quantities of materials in the structure for a preliminary cost estimate so that the economy of the project can be determined.

Surprisingly, it is often easier to estimate the costs of the structure of a shell roof building than it is to estimate the costs of a building built with other structural materials. A steel frame must be designed accurately, piece by piece, and the weight of steel per sq ft varies greatly for various bay widths, spans, and roofing systems.

The quantities for shells, on the other hand, show little variation with the span. The thickness of the shell can accurately be guessed for all of the common shell systems. Folded plates are generally 4 in. (100 mm) thick for slab spans up to 12 ft (3.7 m). Barrel shells are 3.5 to 4 in. (90 to 100 mm), hypars and domes of revolution are 2.5 to 3 in. (40 to 80 mm). The area of forming can be estimated quickly from a simple sketch. The quantity of reinforcing steel does not vary greatly with larger spans, and even so, it is a small part of the total cost. If the designer does not have information to estimate quantities of reinforcement then the preliminary design will give a good estimate.

As a general guide, the following formulas were developed from studies by the writer's firm [13.9]. They include columns but not footings and these shells were designed for a live load of 30 psf.

folded plates to 60 ft	$w = 1.25 + 0.05L$
barrel shells, 50 to 100 ft	$w = 1.8 + 0.046L$
umbrella hypars to 60 × 60 ft	$w = 1.4 + 0.023L$
gabled hypars to 100 × 100 ft	$w = 1.43 + 0.42L$

where L = span in ft
w = weight of reinforcement per sq ft of building

These shells were designed for a live load of 30 psf.

Having obtained the quantities of materials, it is next necessary to apply unit prices for the concrete, steel reinforcement, and forming. Those for concrete and steel are not difficult to determine, and local contractors can give fairly accurate unit prices. Forming costs are a different matter because there are so many different methods of forming combined with the ideas of many contractors that almost any answer is possible. Published costs of forming are no real guide. It is necessary for the engineer to do a considerable amount of work to get forming costs under control. He should go back to fundamentals with a study of the design of the possible forming systems, working with contractors to come up with forms of reasonable cost that will fit the budget.

13.8.1 Types of Forms

A general classification of forms is: (1) single-use forms, (2) intermediate use, two to four times, and (3) repetitive use, over five uses. In the writer's experience, the intermediate type will seldom give an economical solution. Unless a forming system can be devised that will permit a pour of reasonable size with five to six reuses, it is better to stick with a single-use form. A pour of reasonable size is one that will keep the placing crew busy for at least 6 hr. However, the initial pour always seems to take longer to get the crew organized and productive. In terms of the projected area of the roof, this is probably about 4000 to 6000 sq ft.

If a single-use form is used, it is not always necessary to place all of the concrete in one operation. The advantages of a single-use form are:

1. Each trade can completely finish its assigned task and be off the job before the next crew arrives.
2. After concrete has been placed, it can be allowed to cure the required length of time.
3. The forms can be taken down by a smaller crew of laborers, working without the high pressure which may lead to inefficiency and therefore to extra cost.

The trade secret of using single-use forms is very simple. Persuade the contractor that he should not write off all the cost of the form material against this single job and to design the forms so that there is a minimum of cutting of plywood so that most of it can be salvaged.

Repetitive forms may be of two types: (1) demountable forms that have separate components such as panels, beams, and posts that can be disassembled piece by piece and reerected for the next shell; (2) movable forms that are moved all in one piece. The advantage of the demountable form is that the form panels can be left in place longer than the beams and posts and the elements to be moved are light in weight. It may not require as large a crew to move the forms.

SELECTED REFERENCES

13.1 E. H. Baker, L. Kovalevsky, and F. L. Rish. *Structural Analysis of Shells*. New York: McGraw-Hill, 1972.

13.2 J. D. Bennett. "Emperical Design of Symmetrical Cylindrical Shells," *Colloquim on Simplified Calculation Methods*, Brussels, 1961, pp. 314–332.

13.3 David P. Billington. *Thin Shell Concrete Structures*, 2nd ed. New York: McGraw-Hill, 1982.

13.4 Felix Candela. "General Formulas for Membrane Stresses in Hyperbolic Paraboloid Shells," *ACI Journal, Proceedings*, **57**, October 1960, 353–371.

13.5 James Chinn. "Cylindrical Shells Simplified by Beam Method," *ACI Journal, Proceedings*, **55**, May 1959, 1183–1192; Disc. **55**, December 1959, 1583–1604.

13.6 M. L. Kalra. "Emperical Formulae for Designing Long Multi-Barrel Cylindrical Shells," *Bulletin*, International Association for Shell Structures, **41**, March 1970, 7–32.

13.7 M. S. Ketchum. "Lessons Learned from Experience with Hypars," Preprint 3490, ASCE, Boston Meeting, April 2–6, 1979.

13.8 M. S. Ketchum. "Funicular Frameworks," Preprint 81-119, ASCE Convention, New York, May 11–15, 1981.

13.9 M. S. Ketchum. "Economic Factors in Shell Roof Construction," *Proceedings, World Conference on Shell Structures*, October 1–4, 1962, San Francisco, CA. Washington, D.C.: National Academy of Sciences, Publication 1187.

13.10 Andrew R. Nasser and Carl B. Johnson. "Semigraphical Analysis of Long Prestressed Concrete Vaulted Shells," *ACI Journal, Proceedings*, **59**, May 1962, 659–672.

13.11 A. L. Parme and H. W. Conner. "Design Constants for Interior Cylindrical Concrete Shells," *ACI Journal, Proceedings*, **58**, July 1961, 83–106.

13.12 Alfred L. Parme. "Shells of Double Curvature," *Transactions*, ASCE, **123**, 1958, 989–1025.

13.13 G. S. Ramaswamy. *Design and Construction of Concrete Shell Roofs*. New York: McGraw-Hill, 1968.

13.14 W. C. Schnobrich. "Thin Shell Structures," *Handbook of Concrete Engineering*, 2nd ed., Mark Fintel, ed. New York: Van Nostrand Reinhold, 1984.

13.15 A. C. Scordelis and M. A. Ketchum. "Structural Behavior and Design of Saddle HP Shells," International Association for Shell Structures, World Congress on Shell and Spacial Structures, Madrid, 1979, 4.239–4.254.

13.16 Ahmed Shaaban and Milo S. Ketchum. "Design of Hipped Hypar Shells," *Journal of the Structural Division*, ASCE, **102**, November 1976 (ST11), 2151–2161.

13.17 C. W. Yu and L. B. Kriz. "Tests of Hyperbolic Paraboloid Reinforced Concrete Shells," *Proceedings, World Conference on Shell Structures*, October 1–4, 1962, San Francisco, CA. Washington, D.C.: National Academy of Sciences, Publication 1187.

13.18 ACI. *Building Code Requirements for Reinforced Concrete* (ACI 318-83). Detroit: American Concrete Institute, 1983.

CHAPTER 14

PRELIMINARY DESIGN OF SPACE TRUSSES AND FRAMES

JEROME S. B. IFFLAND

Iffland Kavanagh Waterbury, P.C.
New York, New York

14.1 TYPES OF SPACE TRUSSES AND FRAMES

This chapter is restricted to the preliminary planning of metal flat-surfaced space trusses and frames. With this restriction, the flat-surfaced space truss is a special case of a latticed structure defined as a "structural system in the form of a network of elements" [14.8]. Since curved structures are not being considered, nor are folded or combinations of flat surfaces, the space trusses and frames under consideration here are usually termed plane latticed space grids or frameworks. Makowski [14.6], who is one of the pioneers in the design of these types of structures, defines a plane grid framework "as a two-dimensional structure consisting of two or more sets of beams or latticed girders, intersecting each other at right or oblique angles. The beams are interconnected at all intersections and are loaded by forces perpendicular to the plane of the grid or by moments whose vectors lie in the plane of the grid."

Space structures of this type have been used for all kinds of buildings or portions of buildings including small canopies, enclosures as art forms, floors, roofs, and exterior walls. The number of such structures in existence runs into the hundreds. To illustrate the types of buildings where plane latticed space frameworks have been used, Table 14.1 has been developed. This table has been prepared from a review of available literature and is by no means a substantially complete listing of buildings. It does illustrate that space trusses are widely used for sports arenas, recreational facilities, and other buildings where wide open areas are functional requirements. In general, space trusses can be classified as proprietary systems or nonproprietary systems. The large number of proprietary systems has led to many designs being prepared for design–build contracts with the basic parameters established by the designer and the bidders providing solutions with their own systems. Because of this, a review of some of the proprietary systems is in order.

14.1.1 Proprietary Systems

The possibility of repetitive use of both members and joints for a large variety of structures of different shapes and areas has led to the development of standardized proprietary systems by a number of manufacturers. Plane space grids are ideally suited to industrialization and new types

Table 14.1 List of Representative Space Truss and Frame Projects

Project	Location	Size	Depth, ft	Weight, psf	Type*
Towson State College Physical Education Building	Towson, Maryland	200 × 230 ft 25 × 26 ft panels	2.2	18	
Metro. Atlanta Rapid Transit Authority Grand Street Station	Atlanta, Georgia	50,000 ft^2 5-ft module			PG square tube
AT&T Long Lines Data Center	Atlanta, Georgia	34 × 220 ft 4.25-ft module			PG square tube
Popular Creek Music Theatre	Hoffman Estates, Illinois	54,000 ft^2 160-ft span 50-ft cantilever 10-ft module			PG square tube
Office building	Caracas, Venezuela	7200 ft^2 84.6-ft span 7.1-ft module			PG pipe
Grand Hyatt Hotel	New York, New York	154-ft long cantilever structure			PG
New York Exposition and Convention Center	New York, New York	860,000 ft^2 5- and 10-ft modules			PG
General Motors Assembly Division Tech. Ctr. Administration Bldg.	Warren, Michigan	150-ft long 7.5-ft module			PG pipe
Sheraton Park Hotel Building Entrances	Washington, D.C.	4500 ft^2 6.25-ft module			PG pipe
North Shore Hospital	Long Island, New York	53 × 53 ft 7.5-ft module		3.5	PG pipe
Crowtree Leisure Centre	Sunderland, England	482 × 256 ft 157 × 118 ft clear span	9.8	14.2	NODUS
Fil Exhibition Palace	Coronmeuse, Belgium	190,500 ft^2 34- and 69-ft spans	3.9 and 5.7	5.8	

Dusseldorf Industrial Fair Exhibition Halls	Dusseldorf, Germany	94.5 × 94.5 ft	8.2	8.0	OKTAPLATTE
State Government Store, New South Wales	Alexandria, Australia	150,000 ft^2 105 × 98 ft bays	6.5		Triodetic
Brown-Boveri Storage Building	Bornem, Belgium	17,200 ft^2 66 × 66 ft bays	4.0	5.1	Bureau d'Etudes Daniel
International Exhibition and Congress Center-Amstel Hall	Amsterdam, Netherlands	203 × 643 ft 71.5 × 132 ft bays	5.9	8.1	
Hydrographic Research Centre	Madrid, Spain	54,450 ft^2 48 × 60 ft bays (ave)	9.8	10.3	
Philips-Halle	Dusseldorf, Germany	54,000 ft^2 217-ft span	14.5	11.9	MERO
Summerlands Sports and Leisure Centre	Isle of Man, U.K.	126 × 226 ft	4.0		NODUS
Habitat Warehouse	Wallingford, England	100 × 120 ft	6.0		NODUS
N.Y. State University Health & Physical Education Building	Potsdam, New York	118,300 ft^2 148 × 220 ft 13-ft module	7.7	10	PG
Reunion Arena	Dallas, Texas	174,000 ft^2 412-ft span	20 max.	22	PG
Eastern Airlines B747 & L1011 Overhaul & Maintenance Facility	Miami, Florida	280 × 440 ft	20	19	Takenaka
Museum of Natural History "Can Man Survive" Exhibition	New York, New York	110 × 62 ft			Takenaka
East Carolina University Minges Coliseum	Greenville, North Carolina	224 × 230 ft	13		Grid
Gymnasium	Kudamatsu City, Japan	149 × 178 ft	6.6	14.3	
Currigan Exhibition Hall	Denver, Colorado	Four 170 × 240 ft 10 × 10 ft	14.5	12	

(continued)

Table 14.1 (continued) List of Representative Space Truss and Frame Projects

Project	Location	Size	Depth, ft	Weight, psf	Type*
University of California Davis Campus Recreational Center	Davis, California	252 × 315 ft 21 × 21 ft	14	18	Grid
Hartford Civic Center	Hartford, Connecticut	300 × 360 ft	21.2	25.9	
Crosby Kemper Memorial Sports Arena	Kansas City, Missouri	325 × 425 ft	27		
Roissy-Charles de Gaulle Airport, Air France Freight Terminal	Paris, France	102,250 ft^2 141.7 × 141.7 center zone	7.87 center zone	5.5	TUBACORD
Sports Centre	Alton, England	117 × 226 ft	4.9	6.8	NODUS
Zurich-Kloten Airport Swissair Maintenance Hangar	Zurich, Switzerland	421 × 423 ft (four supports)	38.2	35.6	Grid
Exhibition Pavilion for Evers & Company	Copenhagen, Denmark	82 × 82 ft 65.6 × 65.6 ft span	2.3	7	
Royal Flemish Academy of Music	Antwerp, Belgium	102 × 132 ft	5.7		
Coquets Shopping Center	Rouen, France	80,700 ft^2 42.6 × 42.6 ft spans	3.9	5.1	
Heathrow Airport BOAC Car Park	London, England	43,000 ft^2 49.2-ft span	2.7		Unibat
All Sports Theatre	Nantes, France	Hexagon with 141-ft sides 262-ft span	8.2	7.2	
New Printing Works—Main Shop, Imprimerie Nationale Francaise	Douvai, France	426 × 157 ft 131-ft span	6.2	7.2	TUBACORD
B & B Italia Upholstery Factory	Novedrate, Italy	157 × 98 ft	5.2		One-way
National Exhibition Centre	Birmingham, England	91.6 × 91.6 ft 93 units			NODUS

University of Surrey Structural Engineering Laboratory	Guildford, England	100 × 100 ft	5.0	4.1	NODUS
SIG Canteen	Neuhausen, Switzerland	72.25 × 58.5 ft	5.4	4.1	NODUS
Harrow Leisure Centre	London Borough of Harrow, England	142 × 142 ft	6.6		NODUS
James Buchanan and Co., Ltd. Office Building	Stepps, Scotland	150 × 63.5 ft	3.5	5.1	NODUS
Harland and Wolf Belfast Shipyard Shot Blasting and Paint Shop	Belfast, Ireland	115 × 138 ft	9.8		NODUS
Stratford Shopping Precinct	Stratford, London, England	106 × 130 ft	5.9		NODUS
Herringthorpe Leisure Centre	Rotherham, England	105 × 126 ft	5.3	5.3	NODUS
Gruze Garage	Winterthur, Switzerland	51,700 ft^2 41.6-ft main span	5.3		NODUS
National Gallery	West Berlin, Germany	220 × 220 ft	7.0		Grid
University of South Carolina Coliseum	Columbia, South Carolina	356 × 356 ft 29.33 × 29.33 ft panels	20 25	18.3	
Heathrow Airport BOAC Hanger	London, England	257 × 560 ft	27 48	16.7	
Dallas Convention Center Addition	Dallas, Texas	330 × 600 ft	20	22.2	
Buffalo Memorial Auditorium	Buffalo, New York	260 × 275 ft			
Narita International Airport Japan Airline Hangar	Tokyo, Japan	627 × 300 ft	26		
Larkspur Ferry Terminal	San Francisco, California	Triangular 192-ft sides	10		
Brussels Trade Mart	Brussels, Belgium	152 × 152 ft	6.6		Grid
Market	Marmande, France	114.8 × 114.8 ft	4.1		

(continued)

Table 14.1 (continued) List of Representative Space Truss and Frame Projects

Project	Location	Size	Depth, ft	Weight, psf	Type*
Market Square	Storcentret City, Denmark	265 × 353 ft	19.7	12	R & H
Reception Hall of the "Agora"	Evry, France	27,000 ft^2 100-ft spans	4.9	5.4	Unibat
Municipal Swimming Pool	Hilversum, Netherlands	126 × 126 ft 78.7 max. span	4.6	3.2	NODUS
Alfa-Romeo Alfasud Factory Buildings (3)	Pomigliano, d'Arco, Italy	226,000 ft^2 65.6 × 39.4 ft bays			Grid
Doha Stadium	Doha, Emirate of Qatar	14.7-ft module	8.2		MERO
Dierdorf Swimming Pool	Dierdorf, Germany	7.9-ft module columns: 15.7 to 31.5 ft	6.6		MERO
Thermae Museum	Heerlen, Netherlands	157 × 181 ft 86.8 × 118 ft columns	5.2		NODUS
Hyatt Hotel Atrium	Washington, D.C.	76,000 ft^2 128-ft triangular span	8.0	1.5	ORBA-HUB
Federal Express Pavilion	Knoxville, Tennessee	6500 ft^2	4.25	1.5	OCTA-HUB II
Hartz Mountain Office Atrium	Meadowlands, New Jersey	7200 ft^2 hexagonal 90-ft span	6.71	1.7	OCTA-HUB III
St. Lukes Hospital Entryway	Ft. Thomas, Kentucky	6200 ft^2	3.53	1.5	OCTA-HUB II
Aramco School Gymnasiums	Dhahran, Saudi Arabia	7200 ft^2 60 × 120 ft	3.53	1.5	OCTA-HUB II

* See Sections 14.1.1 and 14.1.2 for definitions.

continually appear. Since many of the structures built utilize a proprietary system it is important to give consideration to these for use in a specific design. Most of the available systems provide a complete "kit" of members and nodal joints and many provide for changes in both member and nodal joint sizes depending upon the grid size and depth selected, or upon the stresses in the structure. The resulting structures are of fairly low weight per unit surface area. The nonproprietary systems, usually constructed of rolled steel shapes with fabricated steel joints, are often used for larger grids with fewer members. These systems can require a greater amount of steel than the smaller grid structures although total weight is only one component of the cost.

A brief description of a number of proprietary systems follows. The list is by no means complete but was compiled from readily available manufacturers' catalogs. These descriptions are adapted from the catalogs. Table 14.1 indicates additional types not described herein. There is no significance to the order of presentation. The key component to each of these systems is the type of nodal joint utilized and these are illustrated in Figs. 14.1*a* through 14.1*k*.

*Figure 14.1*a*: Space Structures—ORBA-HUB*

The joint is a solid sphere tapped for attachment of each member by a single bolt. The elements are round tubular members. The system is manufactured by Space Structures International Corporation.

*Figure 14.1*b*: Space Structures—OCTA-HUB*

The system is similar to the previous one and is made by the same manufacturer. The difference is the use of square tubular members instead of round members, resulting in a more complicated joint. Each member is connected by two or four bolts. The basic hub consists of one, two, three, or more sections of the OCTA-HUB extrusion joined together by a high-strength center bolt which is pretensioned.

*Figure 14.1*c*: Space Frame Flat Plate Connectors*

For this system manufactured by Power-Strut Space Frame, the chord and web members are joined together by a universal flat plate with specially designed nuts and bolts. The flat plate connector is common to both top and bottom chords. Members are cold-formed tubes having square cross-sections.

*Figures 14.1*d*: Butler Triodetic Hub*

The primary elements include hubs and structural tubes. The hub is a cylindrical shaped extrusion, with six to nine serrated slots or keyways, which serve as the joint for the structural tubes. The tubes are press-formed to fit the key and lock into position. The joint is completed when the members are attached to the hub and held in place by end washers and a screw bolt. Multilayer grids can be constructed by extending the hub. The system is manufactured by Butler Manufacturing Company.

*Figure 14.1*e*: Synestructics Nodeless Multihinge*

This system, manufactured by Synestructics, Inc., which is also known as the Superstructures System, does not use a nodal element. Instead, the ends of connecting elements are formed with flanges so that the components can be bolted to each other and the entire bolted assembly forms the nodal joint. As many as 26 members can be connected together at the joint. Some joint eccentricity exists depending upon the number of members.

*Figure 14.1*f*: MERO JOINT*

The MERO building system is manufactured by MERO-Raumstruktur Gmbh & Company. It consists of spherical shaped nodes with provisions for 18 connections. Nodes are provided in seven different sizes as are the round tubular member elements. Each element is attached to the node by a single pin bolt which is fixed by a locking pin. The tubular members have cone-shaped steel forgings welded to the ends to accommodate the connection bolts. There is no eccentricity in the joints.

*Figure 14.1*g*: Space Deck NODUS*

The NODUS system is manufactured by Space Decks Limited. The connector joint is made of two hemispherical shaped shells that are shaped to receive the member elements and are bolted together. Eight members can frame into a joint. These elements can be either round or square tubular shapes. There is some joint eccentricity inherent in the system.

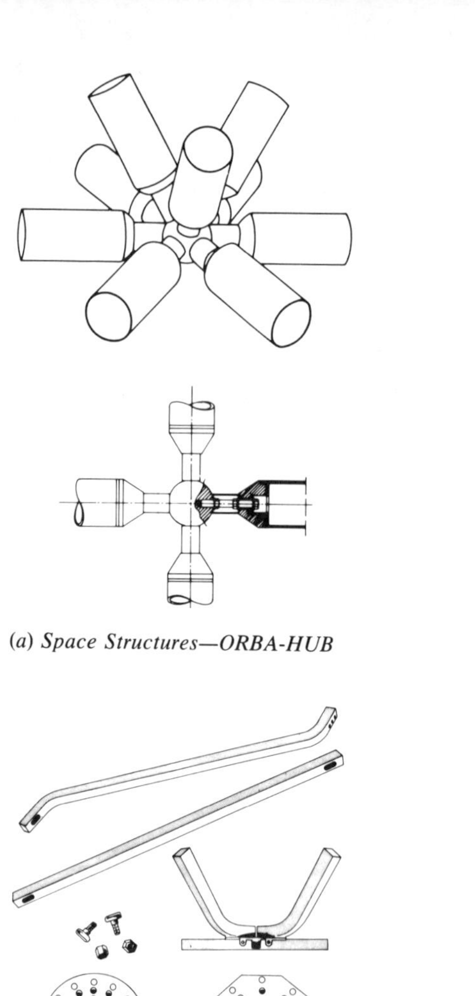

(*a*) *Space Structures—ORBA-HUB*

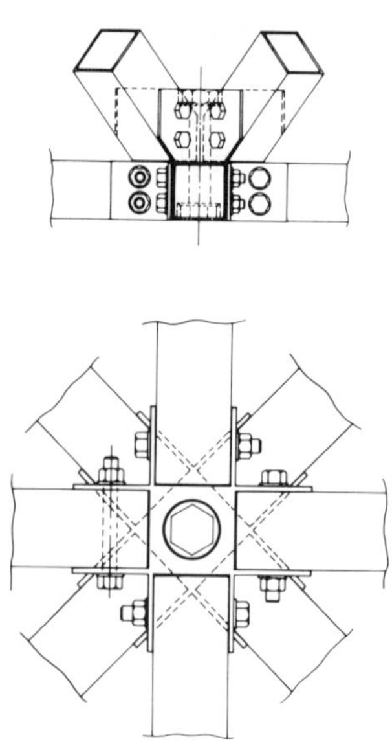

(*b*) *Space Structures—OCTA-HUB*

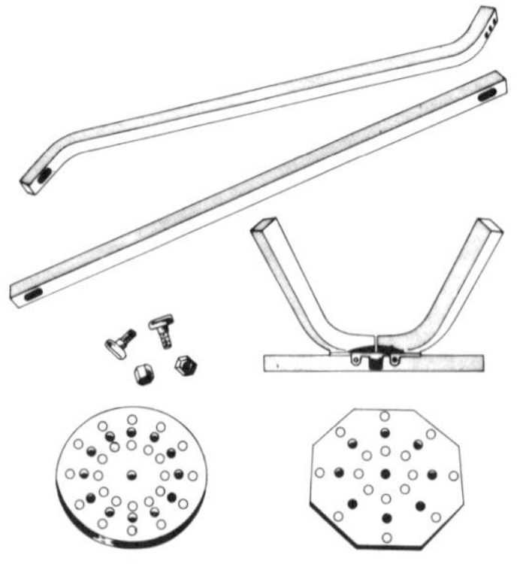

(*c*) *Space Frame Flat Plate Connectors*

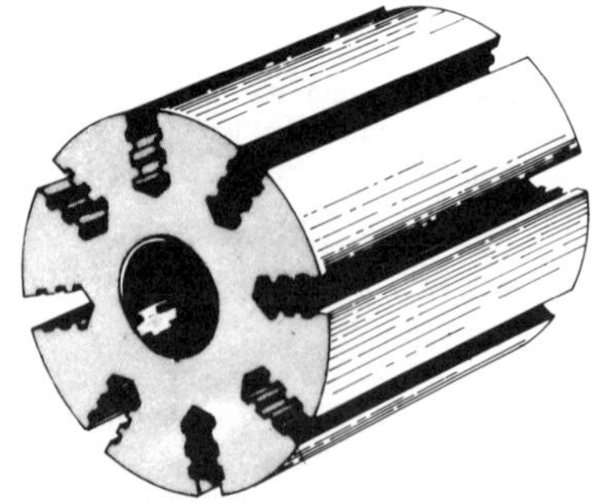

(*d*) *Butler Triodetic Hub*

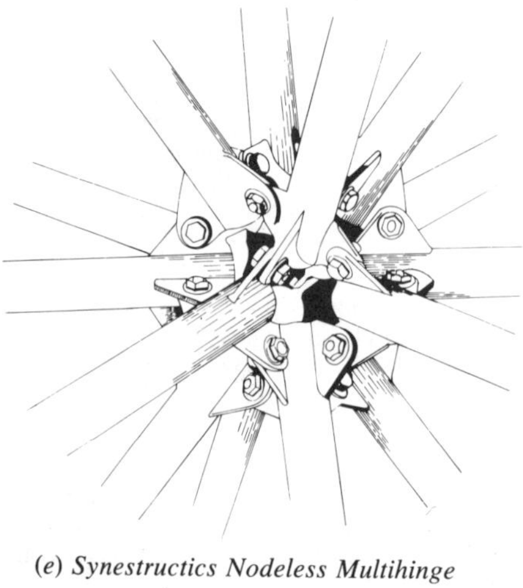

(*e*) *Synestructics Nodeless Multihinge*

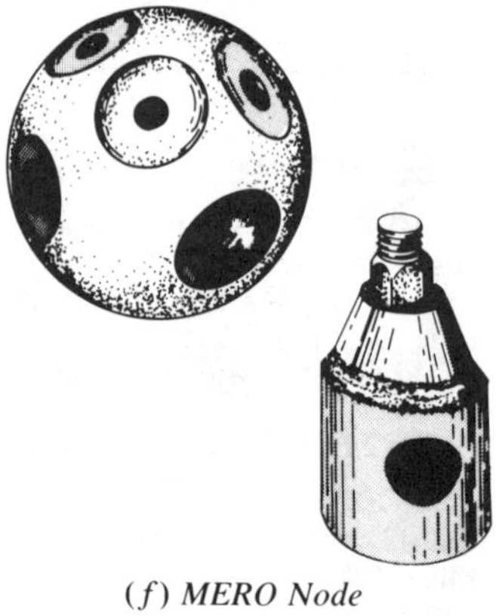

(*f*) *MERO Node*

Fig. 14.1 Representative proprietary space frame nodal joint details.

(*g*) *Space Deck NODUS Joint*

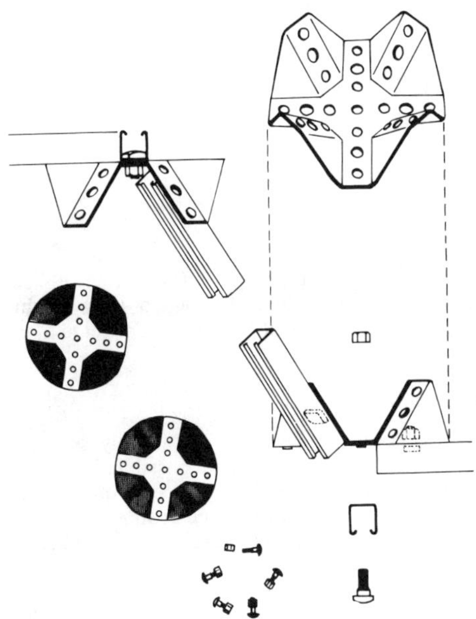

(*h*) *Moduspan Unistrut Connectors—Instruct and Outstrut*

(*i*) *PG Space Structures Spherical Nodes*

Fig. 14.1 (continued).

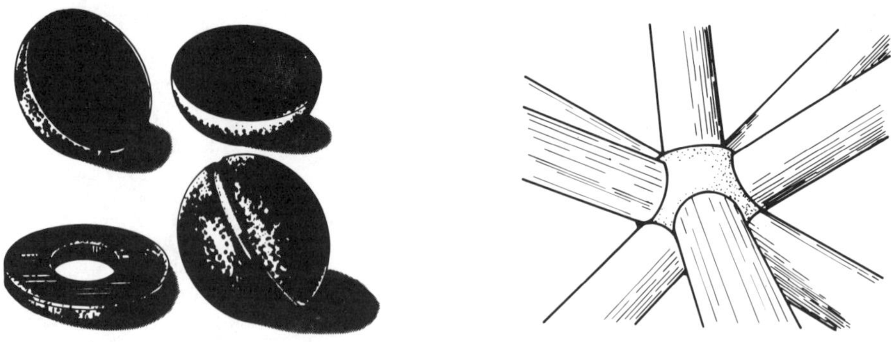

(*j*) *OKTAPLATTE Nodes*

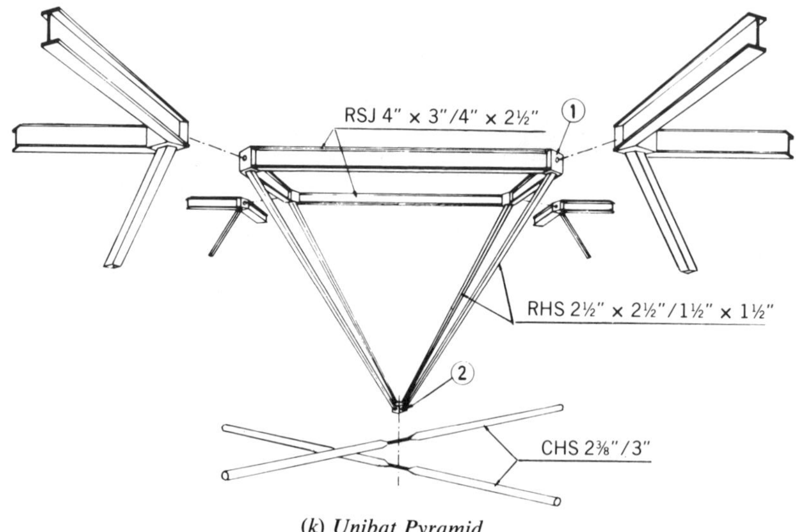

(*k*) *Unibat Pyramid*

Fig. 14.1 (continued).

*Figure 14.1*h: *Moduspan Unistrut Connector*

The Moduspan space frame is manufactured by the Unistrut Corporation. The system utilizes instrut and outstrut connectors which are plate connectors shaped to fit the angle of the attaching members. The member elements are channel-shaped cold-formed sections and are connected to the connector plate by bolts.

*Figure 14.1*i: *PG Space Structures Spherical Nodes*

This system is manufactured by PG Structures Incorporated. The nodes are usually spherical connectors that receive eight member elements. Other joint configurations are possible. The members may be either round or square tubular shapes. The connection is achieved by mechanical means with no welding. A high-strength steel rod, which is located within each member for its full length, passes through a predrilled hole in the connector. A removable portion of the spherical connector provides access for placing the nuts on the ends of the steel rods.

*Figure 14.1*j: *OKTAPLATTE Nodes*

The OKTAPLATTE system utilizes hollow steel spheres and round tubular member elements that are connected by fillet welding. The nodes consist of two hemispherical shells and a diaphragm. The system is manufactured by the Mannesmann Company.

Figure 14.1k: Unibat Pyramid

The Unibat Pyramid is jointly furnished by Stephane du Chateau and H. Ron Taylor. There are no joint components. Prefabricated pyramids are bolted directly together at the joints through split castings using high-tensile bolts. The top chord members are beam shaped, the lower chord members are round tubular shaped, and the diagonals are square tubular shaped. The top chords are set on a diagonal grid relative to the bottom chord.

14.1.2 Nonproprietary Systems

When a proprietary system is not used, a fabricated joint must be designed. This component of the design usually puts restrictions on the geometry of the framing system selected. Larger modules are utilized since the frame members are selected from standard available sizes. The number of members framing into a node is held to a minimum with simple geometry to make the connection detail possible. Two examples are given to show how engineers have utilized nonproprietary plane grid systems.

The first example is a structure designed by the author. It is the roof for the Capital Plaza Sports and Convention Center in Frankfort, Kentucky. The roof is a square on square offset grid similar to Fig. 14.5*c*. The basic module is 19 ft-8 in. (6 m) square and the roof is 275 ft.-4 in. (81 m) square with 19 ft-8 in. (6 m) cantilevers on all sides. The truss is 9 ft-3 in. (2.8 m) deep at the perimeter and 11 ft-6 in. (3.5 m) deep at the center to provide for drainage as well as increased depth where needed structurally. The typical node is shown in Fig. 14.2. The cylindrical bar used for the connection was 3 in. (76 mm) in diameter. Gusset plates were butt welded to this bar. The joint assembly was prefabricated and all field connections were high-strength bolts. A number of full-scale tests were conducted to verify the strength of the joint assembly.

The second example is of a special truss called the Takenaka truss [14.5]. The truss is a diagonal system grid overlaying a square grid similar to Fig. 14.5*h* except that the directions of the chord system are reversed. The resulting truss has six members meeting at a node in the top chord and eight members meeting at a node in the bottom chord. The truss has been studied in detail by Geiger [14.3] and has been used on a number of occasions. The example shown is for the Eastern Airlines

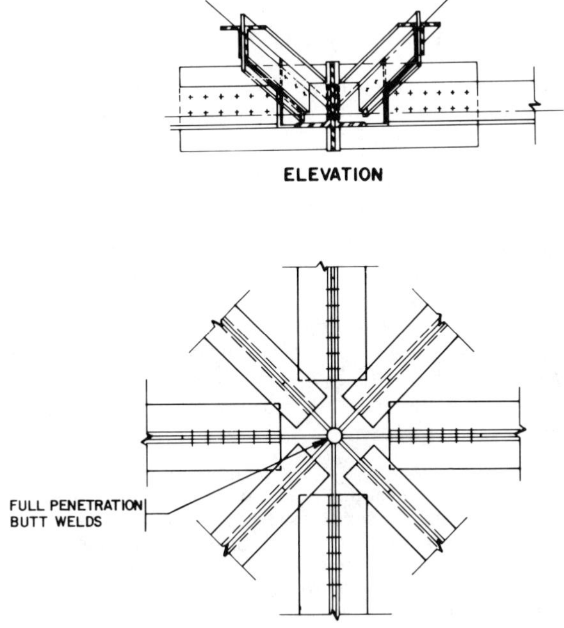

Fig. 14.2 Joint used for roof truss of Capital Plaza Sports and Convention Center, Frankfort, Ky.

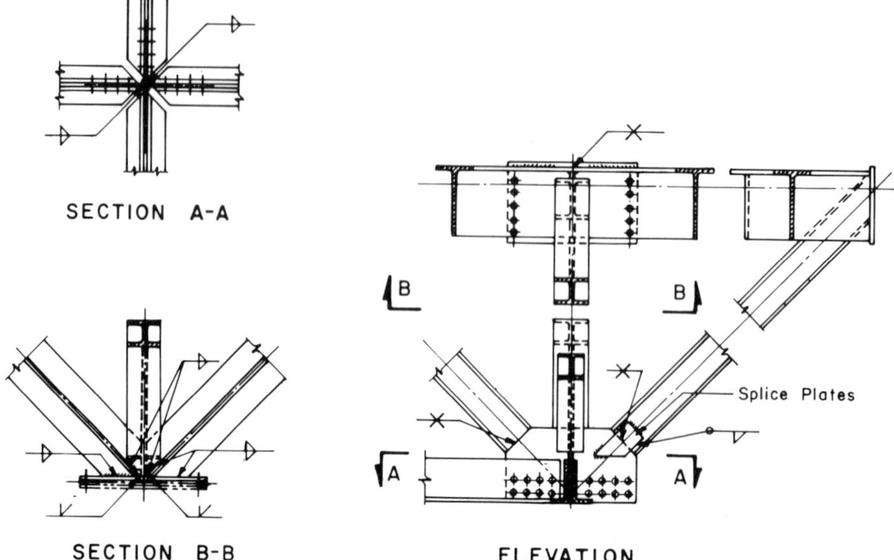

Fig. 14.3 Joint used for roof truss of Eastern Airlines B-747 and L-1011 Overhaul and Maintenance Facilities, Miami, Fl.

B747 and L-1011 Overhaul and Maintenance Facilities in Miami, Florida [14.4]. Details of construction are shown in Fig. 14.3. The roof is 20 ft (6.1 m) deep and 280 × 440 ft (85 × 134 m) in plan.

14.2 PRELIMINARY PLANNING GUIDELINES

In planning a structure to support either a floor or roof over a specific area, a number of parameters must be studied and evaluated before proceeding to the structural analysis and design. These include the possible use of a proprietary system, the geometry of the framing system, type of nodal joint, locations and methods of support and use of cantilevers, and depth of the truss. All of these will influence the least cost of the structure.

While plane truss grids are ideal structures for covering large areas without using internal supports, each case must be judged on its own merit, considering the loads to be carried, the shape of the area to be spanned, and the architectural configuration desired. In general, grids should only be used for fairly long spans—over approximately 130 ft (40 m)—and when the geometric arrangement is suitable to the volume to be constructed. This implies boundary conditions and supports corresponding to the grid truss nodal points. For rectangular areas, plane truss grids have the same limitations as plate or slab structures. A one-way or unidirectional system should always be used for a length-to-width ratio of 2.0 or more.

14.2.1 Geometry of Framing Systems

Plane grid frameworks have the capacity of multidirectional strength. Depending upon the system utilized, this capacity will vary between that of an isotropic plate and an orthogonal grid system. The orthogonal grids have negligible torsional resistance. For other systems, the torsional strength will vary depending upon the basic element used to construct the truss. These basic elements are the rectangular and triangular prisms used for the two-way and three-way grids, a pyramid with a square or rectangular base which is essentially part of an octahedron, and a pyramid with a triangular base (tetrahedron). These basic elements used for the various framing systems are shown in Fig. 14.4. The rectangular and triangular prisms used for the two-way and three-way grids do not utilize space diagonals while the rectangular base and triangular base pyramids used for the other grid systems utilize space diagonals.

A number of framing systems are shown in Figs. 14.5*a* through Fig. 14.5*l*. These systems are developed by varying the directions of the top and bottom chords with respect to each other and also by the positioning of the top chord nodal points with respect to the bottom chord nodal points.

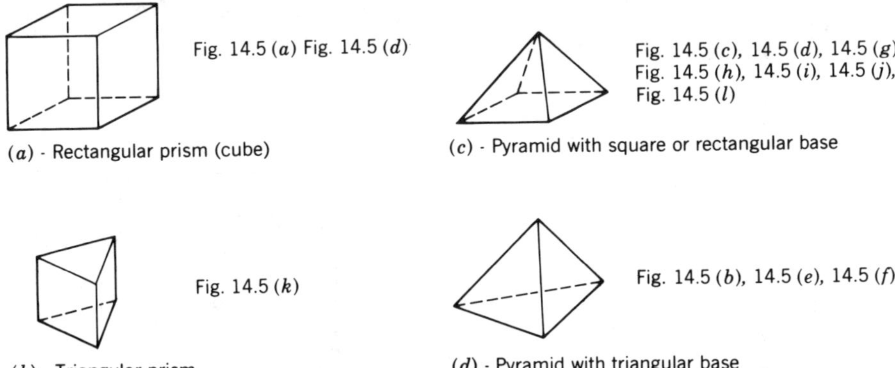

Fig. 14.4 Basic elements of plane grid frameworks.

Additional variations can be introduced by changing the size of the top chord grid with respect to the bottom chord grid. All of the systems shown in Fig. 14.5, except for the three-way grid, have the consistent features of having the components of each chord orthogonal. Top chord members are depicted with heavy solid lines, bottom chord members are depicted with heavy dashed lines, and chord members, where visible, are shown by light solid lines. The areas and number of panels have been selected arbitrarily so that the systems can be illustrated.

Figure 14.5*a* shows a simple or orthogonal grid. The basic element is a cube. Because of its negligible torsional strength, it is structurally inefficient with relatively poor load distribution characteristics. However, it has one major advantage that has resulted in its selection as the optimum design for a number of structures. This advantage is the simplicity of the joint detail. All members (chords and diagonals) lie in two planes that intersect at right angles to each other.

Figure 14.5*b* shows a system with the top chords in a square pattern over bottom chords in a square pattern offset on a diagonal. The basic element is a pyramid with a triangular base. The load distribution characteristic is excellent and the system has a high torsional strength. A drawback is that the number of members and the complicated joints result in higher costs.

Figure 14.5*c* shows one of the most commonly utilized framing patterns of a top chord square grid offset over a bottom chord square grid. All grid lines are mutually parallel and perpendicular with the basic element being a pyramid with a square base.

Figure 14.5*d* is a combined system formed from both pyramids with square bases and cubes as basic elements. It is a combination of the orthogonal grid system in Fig. 14.5*a* and the square on square offset shown in Fig. 14.5*c*.

Figure 14.5*e* shows a variation of Fig. 14.5*c*. This pattern has both chord systems on diagonal grid lines and offset from each other. The basic unit is still a pyramid with a square base. The major advantage here is that the grid lines vary in length and therefore the stiffness of the system varies. Loads in the corner regions are transmitted by the shortest distance to the supports resulting in an increased structural efficiency compared to the system shown in Fig. 14.5*c*.

Figure 14.5*f* is a variation of Fig. 14.5*b*. The basic element is still the tetrahedron. Certain members have been eliminated so that square skylights can be utilized. The system illustrates the flexibility of grid systems to meet architectural requirements and also the inherent strength of the type shown in Fig. 14.5*b*. Even with the removal of a large number of members, the system is still a stable efficient grid.

Figure 14.5*g* is a similar modification to the system shown in Fig. 14.5*c*. Skylights have been provided. The basic element is still a pyramid with a square base.

Figure 14.5*h* shows a square upper chord system over a diagonal lower chord system while Fig. 14.5*i* shows the reverse pattern of a diagonal upper chord system over a square lower chord system. Square skylights can be used with both of these systems. Both of these grids have the pyramids with the square base as the basic element.

Figure 14.5*j* shows how Fig. 14.5*i* can be modified by using a larger grid module for the lower chord system to increase the size of the skylight areas. The basic element is the pyramid with a square base.

Figure 14.5*k* shows a three-way grid. Like the two-way grid of Fig. 14.5*a*, it is not basically a space truss since the diagonal web members lie in the vertical planes containing the top and bottom chords. However, it is an extremely stiff and efficient system which is adaptable to spanning odd shaped areas. The basic element is a triangular prism.

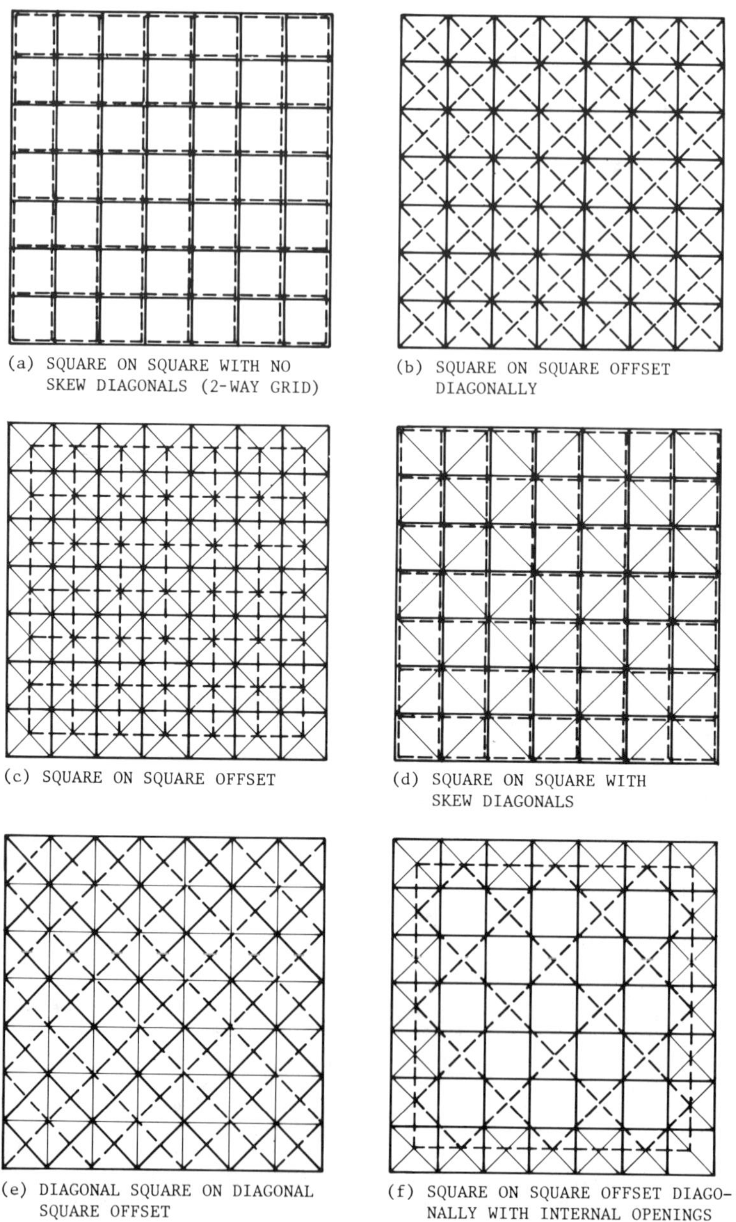

(a) SQUARE ON SQUARE WITH NO SKEW DIAGONALS (2-WAY GRID)
(b) SQUARE ON SQUARE OFFSET DIAGONALLY
(c) SQUARE ON SQUARE OFFSET
(d) SQUARE ON SQUARE WITH SKEW DIAGONALS
(e) DIAGONAL SQUARE ON DIAGONAL SQUARE OFFSET
(f) SQUARE ON SQUARE OFFSET DIAGONALLY WITH INTERNAL OPENINGS

Fig. 14.5 Typical types of framing systems.

Figure 14.5*l* shows how rectangular grid patterns can be utilized. This system uses a rectangular top chord grid offset over an equal size bottom chord grid. It is identical to Fig. 14.5*c* except that the basic element is a pyramid with a rectangular base rather than a square base.

14.2.2 Planning Parameters

Nodal Joints

It should be remembered that anywhere from 25 to 75% of the cost of the truss is associated with the joints. This "connection" cost is a major part of the fabrication and erection costs. The actual

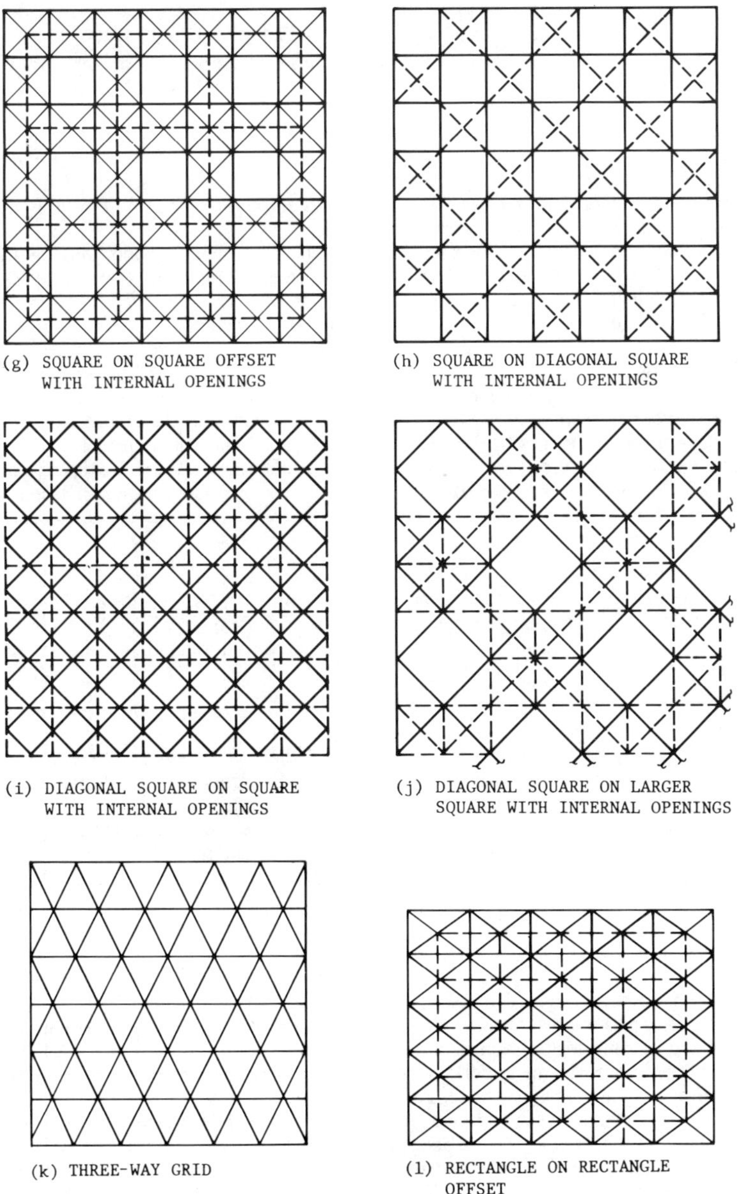

Fig. 14.5 (continued).

weight of the truss can be of secondary importance. In general, the finer the grid, the less the member weight, but the penalty is more joints. This is satisfactory for the proprietary systems but if one of these systems is not used, serious consideration and study must be given to the number and details of the nodal joint.

Many engineers use a rule of thumb of one joint for every 500 sq ft (46.5 m^2) of truss area as an upper boundary on the number of joints; however, there are exceptions to the rule and each space frame must be judged on its own merit. An economically designed joint that can be used repetitively permits more joints and a finer grid and will result in an overall cost savings. This is why the proprietary systems are so competitive. In general, they result in space frames weighing 25 to 30% less than the nonproprietary designs.

Methods of Support

Plane grid frameworks may have a variety of support conditions. These vary from individual columns with or without structural cantilevers to continuous point supports along the periphery of the structure, also, with or without structural cantilevers. It should be noted that plate structures tend to lift up at the corners under dead load. In addition, if a roof structure is being considered, wind suction forces must be included as one of the loading conditions. It is, therefore, important to consider the effect of uplift forces at all supports. As on any large roof structure, not only should wind forces be considered, but also variable live load due to drifting snow and water ponding loads. These will result in nonuniform support conditions.

One of the most important considerations in developing the support conditions for a grid structure is the need to provide stability to the system. The termination of the structural system should be examined at all locations, in both the top and bottom chords, to verify that all forces are accounted for and that the structure has been adequately braced.

Most space frames are large, running to several hundred feet in length. This means that expansion and contraction movements require consideration. While a steel roof can be constructed for a heated and air conditioned building up to a length well over 600 ft (180 m) without expansion joints, the supports must be designed to provide for the thermal movements in all directions. Transmittal of the thermal and other forces through these support details must be considered.

Supports for large structures should also have provisions for rotational capacity to allow the structure to deflect without restraint if this restraint is not considered in the analysis.

Cantilevers

Use of cantilevers has two major advantages. Since long spans are always involved, they reduce positive moments and the corresponding truss forces with economy resulting. They also simplify support conditions. Their use implies support at interior nodal points where the major forces to be carried at the support are vertical without any complicated rotational characteristics required. In general, it is usually better to avoid continuity between the truss grid and the supports.

Truss Depths

Truss depths vary from 1/20 to 1/60 of the span, depending on the system utilized. The diagonal torsionally stiff grids usually are shallower as are most of the proprietary systems. The rectangular grid systems with fabricated steel joints usually result in deeper structural systems. Once a module size has been selected, the depth is limited by geometry since the angle of the diagonals from the plane of the truss should be at an efficient slope. For many of the proprietary systems, this angle is fixed and the truss depth is an exact function of the grid size.

14.3 DESIGN LOADS

Design loads have been covered in Chapter 2. For space trusses and frames, because of their usually large size, it is important to give adequate consideration to forces resulting from thermal movements and to consider differential live and wind loadings on roof areas. Uplift forces due to negative wind pressure are an important consideration since the normal bottom chord tension members can have compressive stresses for this loading condition.

For preliminary design, truss dead loads must be estimated. To provide assistance in estimating dead loads, truss weights per sq ft have been tabulated in Table 14.1 where available. These weights should be used cautiously since every structure has its own particular properties. These include method and position of supports, grades of steel, dead and live loads being carried, type of system being used, and geometry. All of these affect the unit weight.

14.4 PRELIMINARY DESIGN

In today's computer-oriented society, the discussion of methods for preliminary design is almost unnecessary. However, there are various procedures available and a few of these will be discussed and referenced.

As part of a University of Michigan Research Institute project, Coy [14.1] prepared a complete design procedure for Unistrut space frames. The method developed is a strip analysis which is suitable for pocket calculator use. The method considers variable depth trusses and trusses interacting with support structures of variable stiffness. Many examples and convenient tables are provided. The method is applicable to other types of systems and is not restricted to the space frame system.

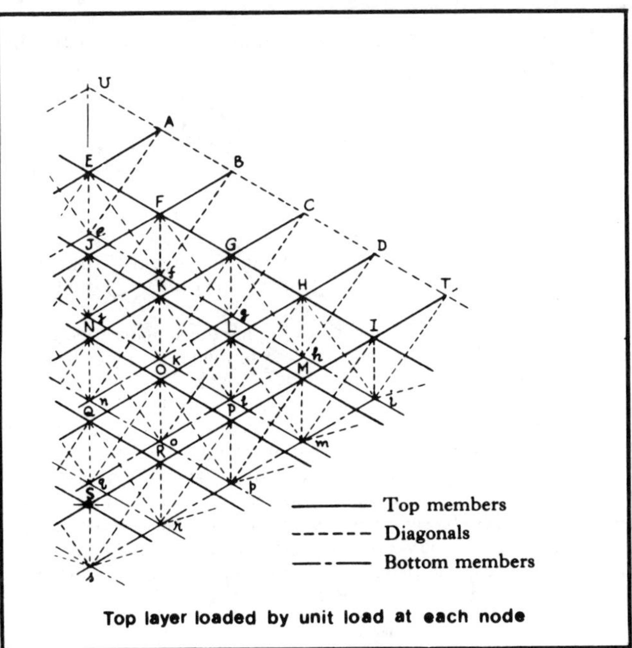

LATTICE GRID

Member Mkd.		Influence Coeff.	Plate Analogy	Plane Grid Analogy
Top-layer Members	AE	−1·513	−1·487	−1·492
	BF	−2·594	−2·652	−2·646
	CG	−3·375	−3·470	−3·436
	DH	−3·895	−3·890	−3·899
	TI	−4·248	−4·055	−4·053
	EF	−2·359	−2·393	−2·317
	FG	−2·789	−2·804	−2·801
	FJ	−4·270	−4·403	−4·301
	GH	−2·923	−3·073	−3·061
	GK	−5·712	−5·858	−5·762
	HI	−2·868	−3·201	−3·144
	HL	−6·645	−6·740	−6·643
	IM	−7·053	−7·039	−6·937
	JK	−5·280	−5·309	−5·290
	KL	−5·804	−6·099	−5·818
	KN	−7·202	−7·251	−7·214
	LM	−5·970	−6·497	−5·985
	LO	−8·420	−8·458	−8·415
	MP	−8·857	−8·864	−8·822
	NO	−8·024	−8·050	−8·000
	OP	−8·295	−8·348	−8·247
	OQ	−9·414	−9·454	−9·392
	PR	−9·870	−9·934	−9·870
	QR	−9·742	−9·816	−9·701
	RS	−10·203	−10·324	−10·203
Bottom-layer Members	ef	+1·513	+1·487	+1·492
	fg	+2·359	+2·393	+2·317
	fj	+2·594	+2·652	+2·646
	gh	+2·789	+2·804	+2·801
	gk	+3·375	+3·470	+3·436
	hi	+2·923	+3·073	+3·061
	hl	+3·895	+3·890	+3·899
	im	+4·248	+4·055	+4·053
	jk	+4·270	+4·403	+4·301
	kl	+5·280	+5·309	+5·290
	kn	+5·712	+5·858	+5·762
	lm	+5·804	+6·099	+5·818
	lo	+6·645	+6·740	+6·643
	mp	+7·053	+7·039	+6·937
	no	+7·202	+7·251	+7·214
	op	+8·024	+8·050	+8·000
	oq	+8·420	+8·458	+8·415
	pr	+8·857	+8·864	+8·822
	qr	+9·414	+9·454	+9·392
	rs	+9·870	+9·934	+9·870

Member Mkd.		Influence Coeff.	Plate Analogy	Plane Grid Analogy
Vertical Members	Ee	−2·269	−2·425	−2·238
	Ff	−2·580	−2·696	−2·604
	Gg	−2·854	−2·988	−2·939
	Hh	−3·022	−3·088	−3·120
	Ii	−3·104	−3·116	−3·163
	Jj	−2·515	−2·533	−2·483
	Kk	−2·510	−2·528	−2·485
	Ll	−2·456	−2·485	−2·454
	Mm	−2·353	−2·431	−2·414
	Nn	−2·234	−2·224	−2·178
	Oo	−1·949	−1·953	−1·918
	Pp	−1·759	−1·811	−1·785
	Qq	−1·491	−1·496	−1·465
	Rr	−1·250	−1·277	−1·250
	Ss	−1·000	−1·000	−1·000
Inclined Members	Ae	+1·891	+2·021	+1·865
	Bf	+3·242	+3·462	+3·308
	Cg	+4·219	+4·365	+4·295
	Dh	+4·868	+4·816	+4·874
	Ti	+5·310	+4·988	+5·066
	Ef	+1·058	+1·032	+1·032
	Fg	+0·538	+0·615	+0·604
	Fj	+2·095	+2·111	+2·069
	Gh	+0·168	+0·331	+0·325
	Gk	+2·922	+2·950	+2·907
	Hi	−0·069	+0·103	+0·103
	Hl	+3·438	+3·466	+3·429
	Im	+3·506	+3·629	+3·606
	Jk	+1·262	+1·264	+1·236
	Kl	+0·655	+0·675	+0·661
	Kn	+1·862	+1·853	+1·815
	Lm	+0·208	+0·211	+0·208
	Lo	+2·219	+2·253	+2·215
	Mp	+2·255	+2·391	+2·356
	No	+1·028	+1·002	+0·982
	Op	+0·338	+0·314	+0·310
	Oq	+1·243	+1·247	+1·220
	Pn	+1·265	+1·337	+1·309
	Qr	+0·409	+0·396	+0·387
	Rs	+0·417	+0·430	+0·417

+ Denotes Tension
− Denotes Compression

Fig. 14.6 Final forces in members—lattice grid.

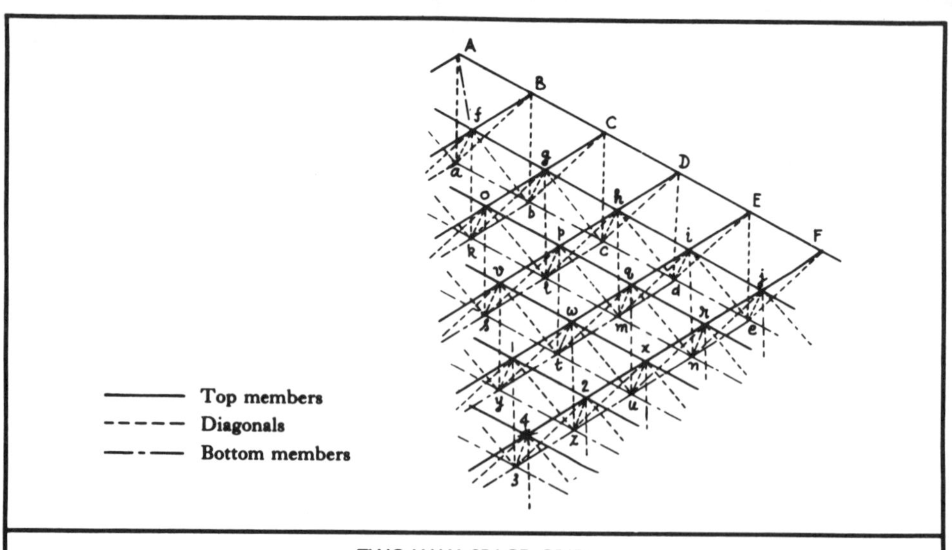

TWO-WAY SPACE GRID

	Member Mkd.	Methods of Analysis		
		Influence Coeff.	Plate Analogy	Plane Grid Analogy
Top-layer Members	AB	+0·023		
	BC	+0·104		
	Bf	−0·604	−0·637	−0·639
	CD	+0·299		
	Cg	−1·118	−1·137	−1·134
	DE	+0·392		
	Dh	−1·605	−1·487	−1·473
	EF	+0·392		
	Ei	−1·609	−1·667	−1·671
	Fj	−1·723	−1·738	−1·737
	fg	−1·487	−1·663	−1·633
	gh	−2·059	−2·227	−2·193
	go	−3·042	−3·024	−2·978
	hi	−2·413	−2·519	−2·512
	hp	−4·070	−3·998	−3·942
	ij	−2·608	−2·689	−2·659
	iq	−4·509	−4·556	−4·518
	jn	−4·705	−4·755	−4·710
	op	−4·177	−4·162	−4·111
	pq	−4·845	−4·889	−4·761
	pv	−5·689	−5·618	−5·561
	qr	−5·151	−5·398	−5·059
	qw	−6·497	−6·513	−6·454
	rx	−6·789	−6·816	−6·754
	vw	−6·647	−6·558	−6·520
	wx	−7·090	−7·028	−6·962
	wl	−7·708	−7·677	−7·631
	x2	−8·075	−8·056	−8·011
	l-2	−8·273	−8·258	−8·183
	2-4	−8·680	−8·682	−8·603

	Member Mkd.	Methods of Analysis		
		Influence Coeff.	Plate Analogy	Plane Grid Analogy
Inclined Members	aA	−0·051		
	aB	+0·771	+0·730	+0·719
	af	−1·492	−1·460	−1·439
	bB	+0·588	+0·730	+0·719
	bC	+1·478	+1·294	+1·276
	bf	−0·144	−0·352	−0·321
	bg	−1·922	−1·672	−1·674
	cC	+1·039	+1·294	+1·276
	cD	+1·909	+1·659	+1·657
	cg	−0·885	−1·078	−1·043
	ch	−2·062	−1·874	−1·890
	dD	+1·701	+1·659	+1·657
	dE	+1·810	+1·858	+1·880
	dh	−1·355	−1·552	−1·531
	di	−2·157	−1·964	−2·006
	eE	+1·809	+1·858	+1·880
	eF	+1·938	+1·924	+1·954
	ei	−1·748	−1·825	−1·840
	ej	−1·999	−1·957	−1·994
	fk	+0·494	+0·756	+0·796
	gk	+0·635	−0·418	−0·400
	gl	+0·886	+1·012	+1·031
	hl	+1·021	+0·912	+0·888
	hm	+1·110	+1·022	+1·246
	im	+1·294	+1·221	+1·197
	in	+1·326	+1·360	+1·363
	jn	+1·356	+1·360	+1·351
	ko	−1·765	−1·592	−1·596
	lo	−0·155	−0·318	−0·321
	lp	−1·752	−1·606	−1·598
	mp	−0·711	−0·889	−0·866
	mq	−1·693	−1·566	−1·578
	nq	−1·188	−1·244	−1·242
	nr	−1·494	−1·476	−1·471
	os	+0·790	+0·955	+0·953
	ps	+0·358	+0·219	+0·224
	pt	+0·820	+0·935	+0·955
	qt	+0·706	+0·607	+0·600
	qu	+0·889	+0·929	+0·935
	ru	+0·851	+0·816	+0·828
	sv	−1·505	−1·393	−1·400
	tv	−0·216	−0·318	−0·321
	tw	−1·310	−1·224	−1·233
	uw	−0·695	−0·733	−0·735
	ux	−1·045	−1·012	−1·028
	vy	+0·651	+0·756	+0·757
	wy	+0·168	+0·086	+0·092
	wz	+0·550	+0·577	+0·590
	xz	+0·402	+0·391	+0·385
	yl	−0·988	−0·929	−0·942
	zl	−0·291	−0·332	−0·321
	z2	−0·662	−0·637	−0·654
	l-3	+0·283	+0·265	+0·299
	2-3	+0·019	+0·017	+0·023
	3-4	−0·322	−0·299	−0·322

	Member Mkd.	Methods of Analysis		
		Influence Coeff.	Plate Analogy	Plane Grid Analogy
Bottom-layer Members	ab	+0·640	+0·623	+0·568
	bc	+1·035	+1·020	+0·957
	bk	+1·836	+1·727	+1·596
	cd	+1·172	+1·213	+1·176
	cl	+2·620	+2·532	+2·382
	de	+1·480	+1·326	+1·293
	dm	+3·121	+3·050	+2·901
	ee	+1·534	+1·374	+1·330
	en	+3·331	+3·317	+3·159
	kl	+2·840	+2·891	+2·729
	lm	+3·490	+3·502	+3·394
	ls	+4·315	+4·353	+4·148
	mn	+3·844	+3·848	+3·748
	mt	+5·258	+5·339	+5·119
	nn	+3·966	+3·984	+3·859
	nu	+5·714	+5·842	+5·609
	st	+5·335	+5·407	+5·238
	tu	+5·872	+5·990	+5·825
	ty	+6·615	+6·748	+6·541
	uu	+6·260	+6·198	+6·011
	uz	+7·261	+7·434	+7·213
	yz	+7·343	+7·536	+7·324
	zz	+7·574	+7·807	+7·573
	z3	+8·107	+8·331	+8·107
	3-3	+8·376	+8·635	+8·393

Top layer loaded by unit load at each node

+ Denotes Tension
− Denotes Compression

Reprinted by permission from:
"Steel Space Structures" by Z.S. Makowski
Michael Joseph Ltd., London, 1965

Fig. 14.7 Final forces in members—two-way space grid.

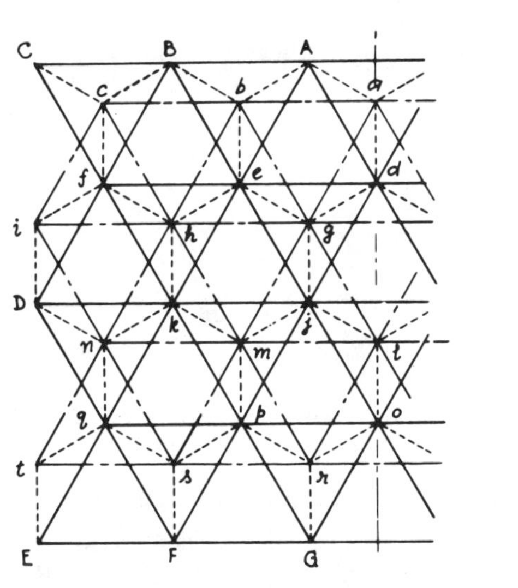

Top layer loaded by unit load at each node

THREE-WAY SPACE GRID

Member Mkd.	Methods of Analysis		Experimental Results	Member Mkd.	Methods of Analysis		Experimental Results
	Influence Coeff.	Plane Grid Analogy			Influence Coeff.	Plane Grid Analogy	
AA	−0·364			ej	−0·668	−0·678	
AB	−0·290			ek	−0·666	−0·757	
Aa	+0·790	+0·731		fh	+0·491	+0·447	
Ab	+0·848	+1·033		fi	−0·843	−0·745	
Ad	−0·170	−0·365		fk	−0·479	−0·332	−0·430
Ae	−0·376	−0·517		gg	+0·221	+0·318	
BC	−0·056			gh	+0·214	+0·149	
Bb	+0·592	+0·338		gj	−0·531	−0·967	
Bc	+0·705	+0·745		gl	+0·922	+1·018	+1·045
Be	+0·074	−0·169		gm	+0·975	+1·018	+0·955
Bf	−0·507	−0·372	−0·330	hi	+0·130		
Cc	+0·167	+0·183		hk	−0·354	+0·255	
Cf	−0·056	−0·091		hm	+0·567	+0·481	+0·315
Df	−0·548	−0·372	−0·560	hn	+0·811	+0·481	+0·950
Di	+0·843	+0·745	+0·805	in	−0·130		
Dk	−0·241	−0·513		jj	−0·391	−0·316	−0·570
Dn	+1·523	+1·770	+1·305	jk	−0·404	−0·670	
Dq	−0·493	−0·372		jl	−0·213	−0·287	
EF	+0·129			jm	−0·482	+0·030	
Eq	−0·258	−0·091		jo	−0·877	−0·874	+0·700
Et	+0·387	+0·183	+0·315	jp	−0·669	−0·678	−1·010
FG	−0·0001			km	+0·157	−0·157	
Fp	−0·274	−0·169		kn	−1·028	−1·323	
Fq	−0·532	−0·372		kp	−0·881	−0·757	−0·715
Fs	+1·209	+1·083	+1·075	kq	−0·210	−0·332	
GG	−0·042			lm	+0·531	+0·316	+0·590
Go	−0·461	−0·365		lo	+0·425	+0·574	
Gp	−0·544	−0·517		lr	+0·710	+0·731	+0·760
Gn	+1·507	−1·764	+1·440	mn	+0·604	+1·025	+0·445
ab	+0·024			mp	+0·324	+0·127	
ad	−1·581	−1·462	−1·200	mr	+0·797	+1·033	+0·690
ag	+0·790	+0·731	+0·860	ms	+0·421	+0·338	
bc	+0·008			nq	−0·495	−0·447	
be	−1·441	−1·371	−1·330	ns	+0·789	+0·745	+0·750
bg	+0·576	+0·338		nt	+0·387	+0·183	
bh	+0·864	+1·033	+0·640	op	−0·090	−0·234	
cf	−0·872	−0·927	−0·850	or	−0·825	−0·900	
ch	+0·160	+0·183		pq	−0·027	−0·075	
ci	+0·713	+0·745	+0·810	pr	−0·682	−0·865	
de	−0·495	−0·234	−0·400	ps	−0·867	−0·487	
dg	+0·178	+0·118		qs	−0·342	−0·595	
dj	−0·756	−0·874	−0·680	qt	−0·387	−0·183	
ef	−0·476	−0·075	−0·640	rs	+0·446	+0·149	
eg	+0·353	+0·849		rr	+0·561	+0·318	
eh	−0·137	−0·702		st	0		

Reprinted by permission from
"Steel Space Structures" by Z.S. Makowski
Michael Joseph Ltd., London, 1965

Fig. 14.8 Final forces in members—three-way space grid.

Makowski [14.6, 14.7] has developed a method of analysis for which pocket calculators can be used. He provides examples and design aids. Reference [14.7] includes influence coefficients for lattice grids, three-way space grids, and two-way space grids. These tables also compare computed influence coefficients to those that would be obtained utilizing a plate analogy and a plane orthotropic grid analogy. These have been reproduced by permission as Figs. 14.6, 14.7, and 14.8.

A third method has been suggested by Dauner [14.2]. He has developed lattice characteristics which can be applied to known elastic isotropic solutions. He has developed these lattice characteristics for a grid of orthogonal beams parallel at the edges, a three-way grid, and an orthogonal grid of diagonal beams.

14.5 STABILITY REQUIREMENTS

With a number of failures of major space frame roof structures occurring in recent years, stability requirements demand some attention. In general, space trusses are highly redundant with ample capabilities for redistribution of loads should failure of individual members or connections occur. When a major failure does occur, it results from a progressive loss of stiffness until zero stiffness is approached and instability results.

As part of the final design, a stability analysis should be included. This should be a post-buckling analysis. This post-buckling analysis, which is a second-order analysis, should at least include deformations, inelastic behavior, and the reduction in stiffness in compression members due to axial force in the members. Some designers have included the effect of residual stresses resulting from manufacture and fabrication of the members, but this is extremely difficult and generally should not be considered necessary.

14.6 ROOF CONSTRUCTION—SPECIAL PROBLEMS

There are several items connected with the roof construction which require consideration in design of space trusses and frames.

Roof purlins span between truss top chord members. The span of these purlins must be alternated in a checkerboard pattern so that all nodal joints as well as top chord members receive equal distribution of loads. For unsymmetrical grids or irregular shapes, this unidirectional requirement of purlins may result in some special requirements along the edges of the system.

While through-structural expansion joints are not generally required in space frames and although provision for the resulting thermal forces have been included in designing the supports, provision must be made for expansion in the roof material spanning between purlins. This presents a problem because usually the roof deck is utilized as a structural diaphragm to transmit horizontal forces. There are proprietary products available that allow for both panel movement and transmittal of forces to supports. This is usually accomplished by crimped or upset joints that permit the roof deck to breathe but still provide positive attachment for force transmittal. Roofing details are treated in Chapter 26.

Roof deck should be sized for deflection requirements as well as for strength. The deflection under the construction loads, including concentrated weights of personnel, should be limited so that the resulting deflections will not damage the connections.

Space trusses are generally sufficiently stiff that camber is often not required. In any case, attention should be directed to proper roof drainage, including provision for overflow should drains become clogged which usually happens since maintenance is difficult. The possibility of ponding and progressive load buildup has to be considered as in all roof structures.

SELECTED REFERENCES

14.1 P. H. Coy. *Structural Analysis of Unistrut Space-Frame, Part a and Part b*. Ann Arbor, MI: University of Michigan Press, 1959.

14.2 H-G. Dauner. "Reflections on the Choice, Design, and Calculation of Plane Steel Space Frame," *Acier-Stahl-Steel,* **42,** 3, March 1977, 107–112.

14.3 David H. Geiger. "A Cost Evaluation of Space Trusses of Large Span," *Engineering Journal,* AISC, **5,** 2(April 1968), 49–61.

14.4 D. H. Geiger. "Skew Chord Truss Used in L-1011 Widebody Jet Facilities," *Acier-Stahl-Steel,* **37,** 6, June 1972, 294–297.

14.5 B. Kato, K. Takanashi, Y. Tsushima, and T. Hirata. "The Analysis of a Space Truss Composed of Square Pyramid Units," *Proceedings of Space Structures,* R. M. Davis, ed. New York: Wiley, 1967, pp. 201–212.

14.6 Z. S. Makowski. "Modern Grid Structures," *Architectural Science Review,* Academic Press Proprietary Limited, **3,** July 1960, 52–65.

14.7 Z. S. Makowski. *Steel Space Structures*. London: Michael Joseph Ltd., 1963.

14.8 Donald R. Sherman, Chairman. "Bibliography on Latticed Structures," by the Subcommittee on Latticed Structures of the Task Committee on Special Structures of the Committee on Metals of the Structural Division, *Journal of the Structural Division,* ASCE, **98,** July 1972 (ST7), 1545–1566.

CHAPTER 15

PRELIMINARY DESIGN OF LIGHTWEIGHT MEMBRANE STRUCTURES INCLUDING AIR SUPPORTED AND STRUCTURALLY SUPPORTED SYSTEMS

DAVID H. GEIGER, Ph.D., P.E.

Principal
Geiger Associates, P.C.
New York, New York

HORST BERGER, Dipl.Ing., P.E.

Principal
Horst Berger Partners
New York, New York

15.1 INTRODUCTION

15.1.1 Structural Principles

The structures discussed in this chapter consist of thin, lightweight structural membrane surfaces of compound curvature supported by edge rings, masts, frames, or arches with or without internal air pressure. Having only negligible weight and thickness, the structural behavior of lightweight membrane structures is radically different from all other structural systems [15.4]. Conventional structures achieve their stability generally by a combination of gravity and rigidity. In some structures, such as arches and shells, curvature is used as an additional means of carrying load and achieving stability, resulting in structures of lighter weight and more efficient use of materials. In other systems, such as prestressed concrete, internal stresses are deliberately introduced for the purpose of improving the structural capacity and efficiency of the system, which might further reduce the reliance on gravity and rigidity. A prestressed concrete shell structure would go farthest in this direction. However, weight and especially bending and shear stiffness still play a significant role in making such a structure stable. This is not the case for membrane structures.

The weight of membrane structures is so small as to have little or no impact on stability and is quite often negligible with regard to the stresses produced by it. A fabric membrane may have a weight of 0.25 psf (1.2 kg/m^2); a fabric membrane on a cable net may vary between 0.5 and 1.5 psf (2.4 and 7.3 kg/m^2); a metal deck on a cable net may weigh as much as 15 psf (73 kg/m^2). Furthermore, the components of a membrane structure have no rigidity at all with regard to their span, thus providing no capacity to carry load in bending or in shear stress. Stability has to be generated by totally different means, using curvature and prestress only. The use of curvature and prestress, therefore, becomes the dominant feature of membrane structures.

The structural principles of lightweight membrane structures can most readily be illustrated in the case of a two-way orthogonal cable net (Fig. 15.1*a*). For an air supported structure, the intersection point of two cables (node) will be subjected to the loads and stresses shown in Figure 15.1*b*.

If there is no air support (tension structure) the load configurations are those in Fig. 15.1*c*. The discussion of these configurations contains the basic principles underlying the behavior of lightweight membrane structures. The weight of the structure has been neglected in this discussion. If in a particular structure the self-weight needs to be considered, its effect is similar to an external downward load.

In Fig. 15.1*b* the air pressure represented by the force F_a is in equilibrium with the membrane forces F_p. Since the air pressure is consistently directed outward, the curvatures and stress levels of the two stress lines must be so proportioned as to produce an inward reaction equal and opposite to F_p. This can obviously be achieved with various curvatures.

The most efficient configuration producing the lowest maximum stress in the system is achieved by equal curvature in both directions (synclastic shape). Having no curvature in one of the two directions is a possible choice resulting, however, in the least stable system. Even the use of curvatures of opposite Gaussian sign (anticlastic shape) is conceivable and might be considered in a system where stability must be sustained in case of loss of air pressure.

As Fig. 15.1*b* also shows, the superimposed load is again most efficiently carried by the system with synclastic shape and equal curvature in both directions. The anticlastic shape has the potential advantage that under superimposed loads one of the two directions can be allowed to go slack, provided that other considerations regarding the stability of the overall system and its dynamic behavior are not violated.

Referring to Fig. 15.1*c*, it is noted that the absence of the air pressure force F_a requires that the two stress lines be of opposite Gaussian curvature since otherwise there can be no equilibrium in

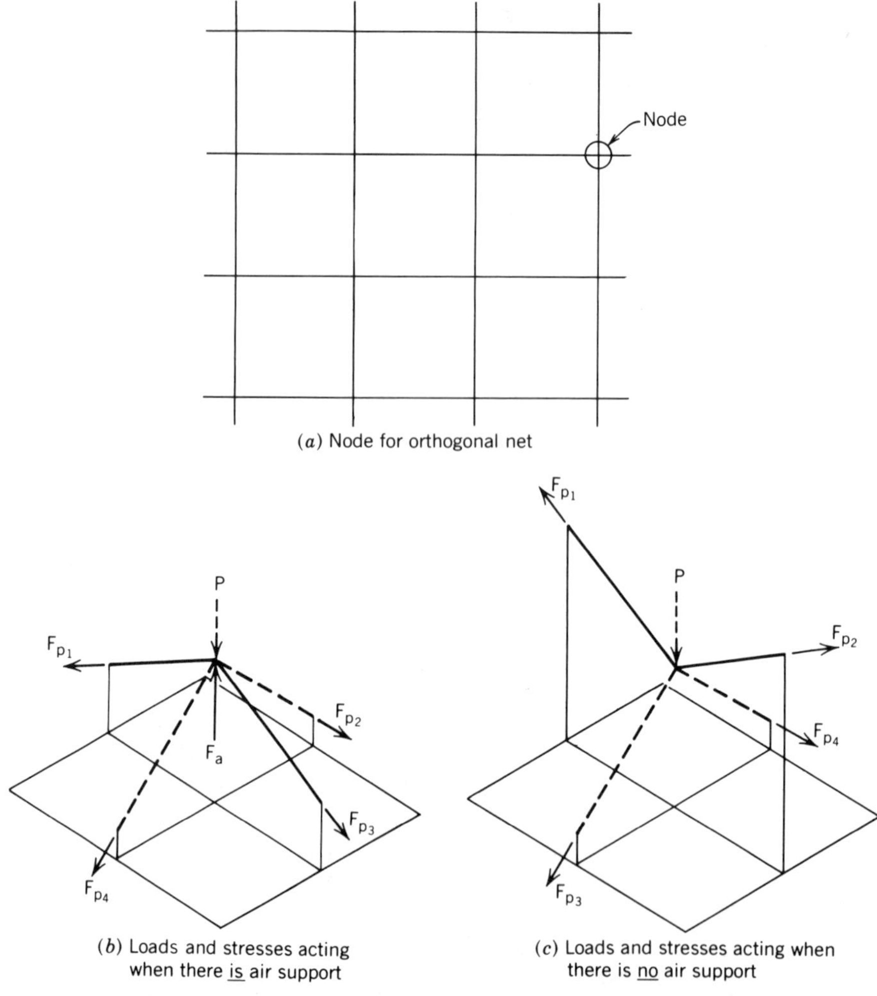

Fig. 15.1 Lightweight membrane structure having orthogonal cable net.

the system. Tension structure membranes, therefore, must have anticlastic shape. As the illustration also shows, they are capable of carrying superimposed loads. Stability is always achieved if both directions stay in tension under all load cases. However, if certain requirements for overall stability of the system and its dynamic behavior are satisfied, loss of stress in one of the two directions under certain load cases might be permissible in limited areas of the membrane.

15.1.2 Design Considerations

In a real building the membrane of an air supported structure will generally have a synclastic configuration as shown in Fig. 15.2*a*, and a tension structure membrane will have an anticlastic configuration as shown in Fig. 15.2*b*. In order to have economical structures, the proper choice of curvature is essential. Finding the proper shape of the membrane, therefore, becomes a critical part of the design.

The actual form of a structural system utilizing lightweight membranes follows from the selection of the membrane shape and the choice of its support system. As a result, an almost unlimited variation of forms is possible. But since form and structure are identical, and since shape is directly related to stress, the design process must begin with a rational geometric configuration in which the internal forces and the distance they travel are kept to a minimum. Because of the three-dimensional nature of membrane structures, design aids must be three-dimensional, such as physical models (stretch fabric, soap film, etc.) or based on computer graphics (Fig. 15.3). Once a basic

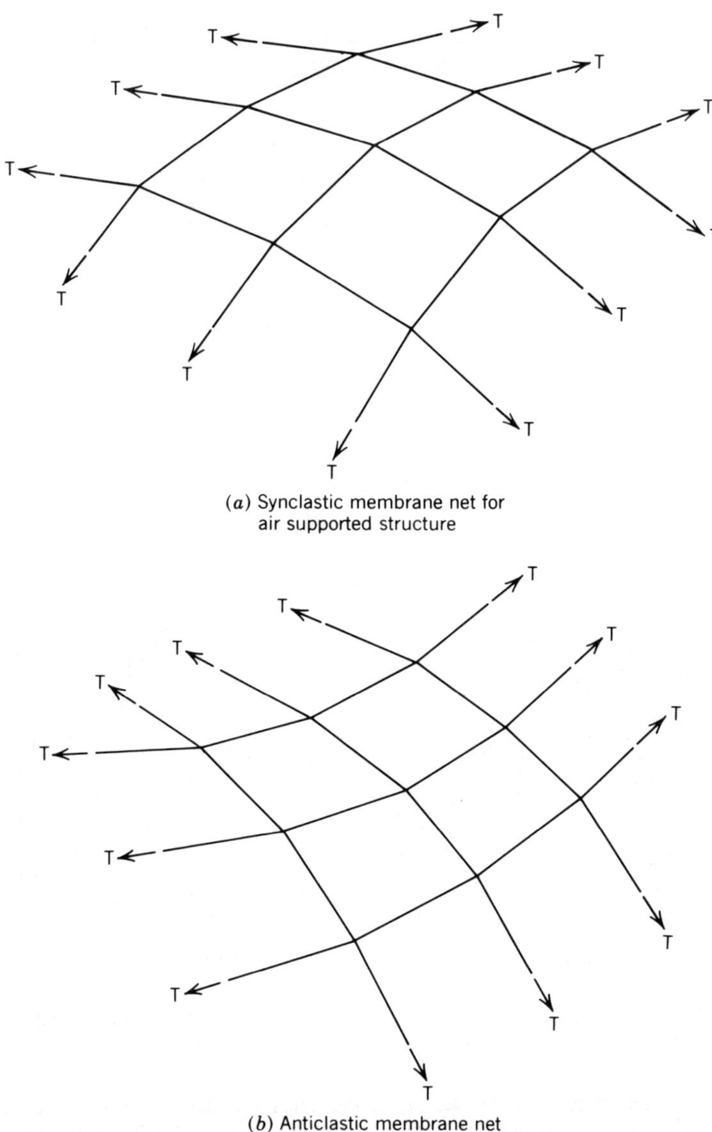

(*a*) Synclastic membrane net for air supported structure

(*b*) Anticlastic membrane net for tension structure

Fig. 15.2 Membrane net configurations.

structural configuration has been adopted, it is often advantageous to input a preferred stress pattern and find the shape that results. Correction of shape for functional, economic, and aesthetic reasons can then be achieved by changing the input stress pattern until the desirable final shape has been achieved.

A further characteristic of lightweight membrane structures is their behavior with regard to deformation. Because of the absence of rigidity and the use of high-strength materials, membrane structures have relatively large deformations under superimposed load. Since deformations in tensile systems tend to increase their capacity to carry load, this is in principle a desirable system. However, it has the result that certain conventional assumptions underlying the design of structures no longer apply. The analysis of these structures requires methods that take into account the deformation of the structure, generally called "nonlinear" methods. While the behavior of membrane structures has been understood in general terms for a long time, it is to a large degree the complexity of the design and analysis process which has prevented the use of membranes in permanent buildings long after their inherent advantages were known.

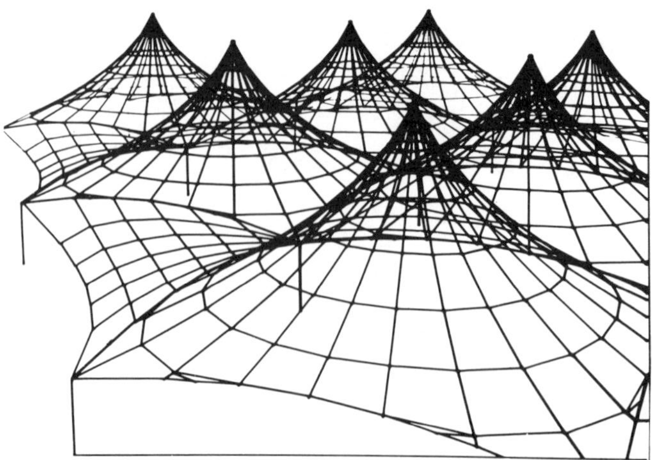

Fig. 15.3 Model of multiple tents—computer graphics drawing. Courtesy of Geiger Associates, New York.

15.1.3 Historical Background

Tents are among the oldest forms of man-made dwellings, dating back as far as 28,000 years, according to a recent discovery in the Soviet Union. The flexible nature of its components has made the tent the ideal transportable shelter which has been used extensively throughout history [15.7, 15.11]. The vela, a shade structure system used in many Roman arenas, could be regarded as the forerunner of today's covered stadium structures. Medieval armies were housed in tents of excellent design and efficiency. The nineteenth century circus tent was the first low-cost movable enclosure for large assembly spaces. Like most of the other historic applications, it was made of woven natural fibers. It was the development of synthetic materials which made the fabric membrane suitable for use in major engineered structures.

The suspension bridge, deriving from historic applications using nature's materials, made its entry into modern construction with the invention of the high-strength steel cable and its extensive application by John Roebling. While many forms of hanging roof structures were attempted in the last hundred years, an essential breakthrough was achieved by Novicki and Severud in 1950 using high-strength steel cables forming a cable net of hyperboloid surface configuration for the design of the 60,000-sq ft (5600-m^2) livestock pavilion in Raleigh, North Carolina (Fig. 15.4). This project led to numerous derivative applications and established a structural system with great future potential. Frei Otto [15.12] developed the idea of tensile architecture systematically into a species of tremendous structural and sculptural variety, with the roof of the Munich Olympic Stadium (1972), an

Fig. 15.4 Livestock Pavilion in Raleigh, North Carolina. Courtesy of Severud Perrone Sregetly Sturm Engineers, New York.

Fig. 15.5 Munich Olympic Stadium, Germany. Courtesy of Horst Berger, New York.

acrylic-sheathed cable net of great complexity and structural sophistication, as his major achievement (Fig. 15.5).

Balloons were no doubt the inspiration behind the idea of the air-inflated and air supported structure. With the advent of vinyl-coated fiberglass fabrics, Walter Bird built the first air structures during World War II [15.6]. The low cost and ease of erection of air structures led to their enormous popularity in use for pool covers, tennis halls, warehouses, and many other uses. The low-profile, cable-reinforced roof structure patented by David Geiger* and used in the design for the U.S. Pavilion in Osaka, Japan in 1970 (Fig. 15.6) opened a totally new potential leading to the record

Fig. 15.6 U.S. Pavilion in Osaka, Japan. Courtesy of Geiger Associates, New York.

* Patents by David Geiger referred to in this chapter have since been sold to Frederick Lang of Landenburg, PA, and are available for use at a 3% royalty fee.

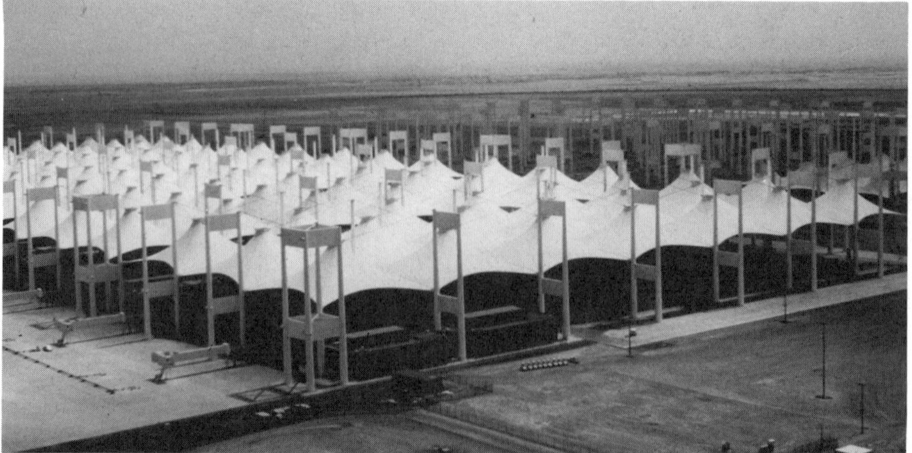

Fig. 15.7 Haj Pavilion, Jeddah Airport, Saudi Arabia. Courtesy of Geiger Associates, New York.

breaking stadium roof covers of recent years, and suggesting the possible encapsulation of entire communities [15.9, 15.10].

The success of the low-profile air supported structures using noncombustible and durable fabrics encouraged the development of a new generation of tension structures. The development of computer-assisted design of radial tent shapes by Horst Berger led to a series of novel structures including the 105-acre (425,000-m^2) roof system for the Haj Pavilion of the Jeddah Airport (Fig. 15.7) [15.5]. These developments, including computerized determination of the membrane shape, nonlinear stress analysis, wind tunnel testing, new materials development, invention of new, more efficient configurations, improved detailing, simplified construction, and experience with a growing number of completed buildings form the basis for the increased use of lightweight membrane structures today. Environmental and energy considerations, within the framework of total design concepts, are further reasons for their use in many circumstances. The following sections will provide a basis for the preliminary design of some of the more common forms of lightweight membrane structures.

15.1.4 Common Structural Systems

By far the most successful membrane structures are *air supported* systems. Low-profile designs are best suited for economical, long-span applications, because their cost per unit area varies little with spans ranging from a few hundred to several thousands of feet.

The basic configuration is the synclastic net, which is best arranged in a diagonal pattern and restrained by a funicular edge ring as shown in Fig. 15.8*c*. Other configurations are discussed in Section 15.2. The economy of the air supported structure derives from its shape with low edge supports, its small ratio of surface area to plan area, its low stresses, and its elimination of interior supports. It requires on the other hand that the building be airtight, that exits are specially de-

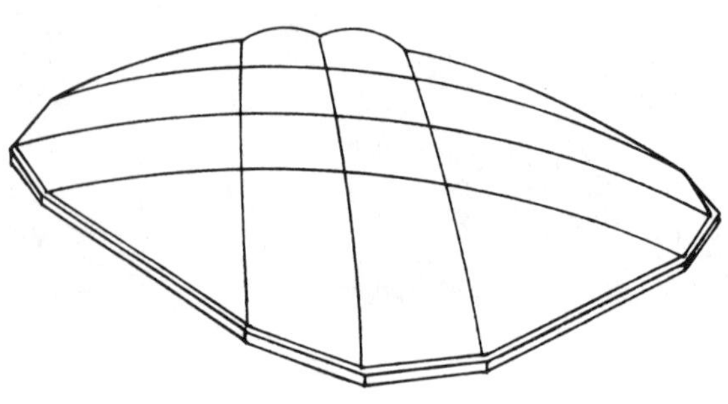

(*a*) Low profile roof

Fig. 15.8 Common membrane structural systems.

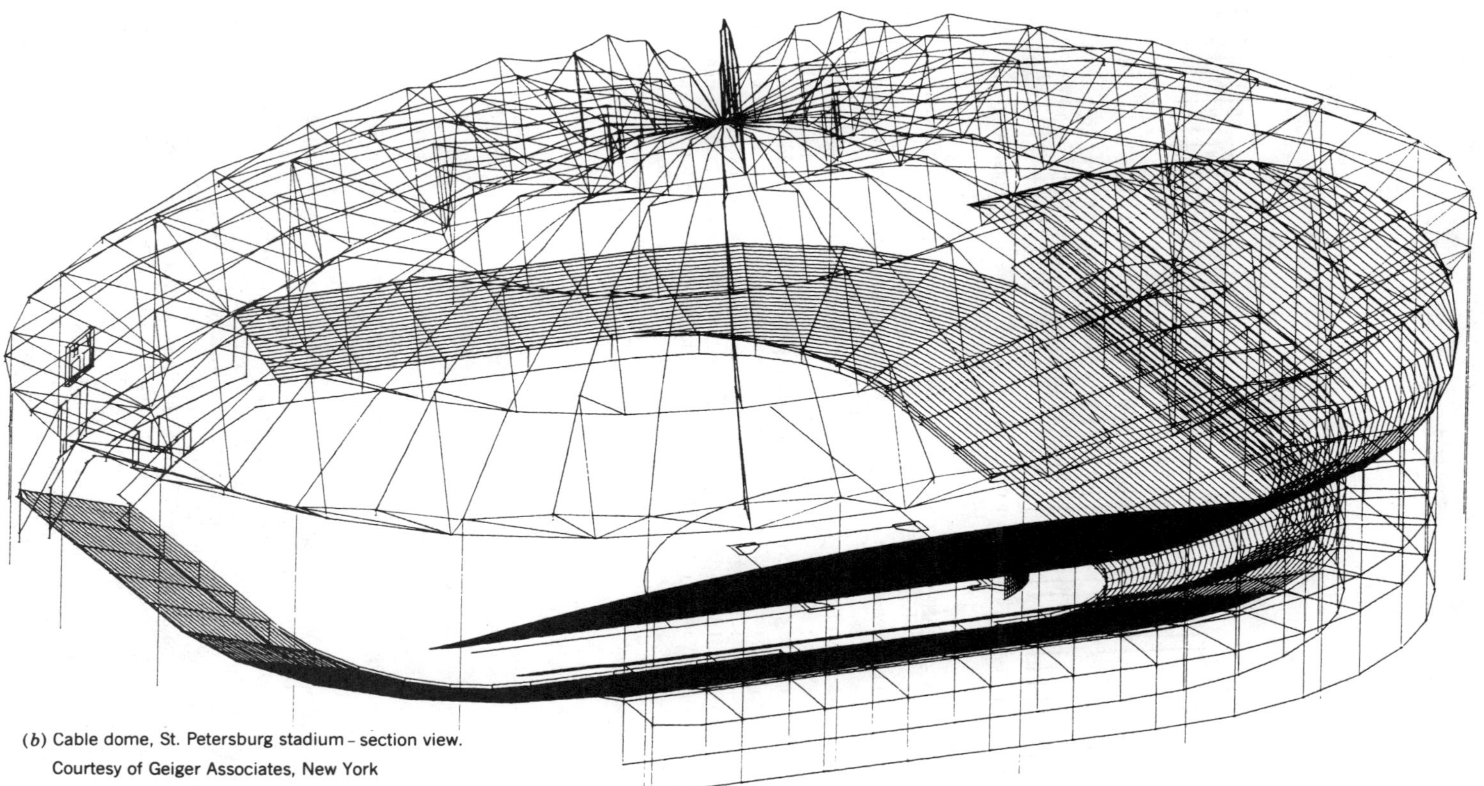

(*b*) Cable dome, St. Petersburg stadium – section view.
Courtesy of Geiger Associates, New York

Fig. 15.8 (continued) Common membrane structural systems.

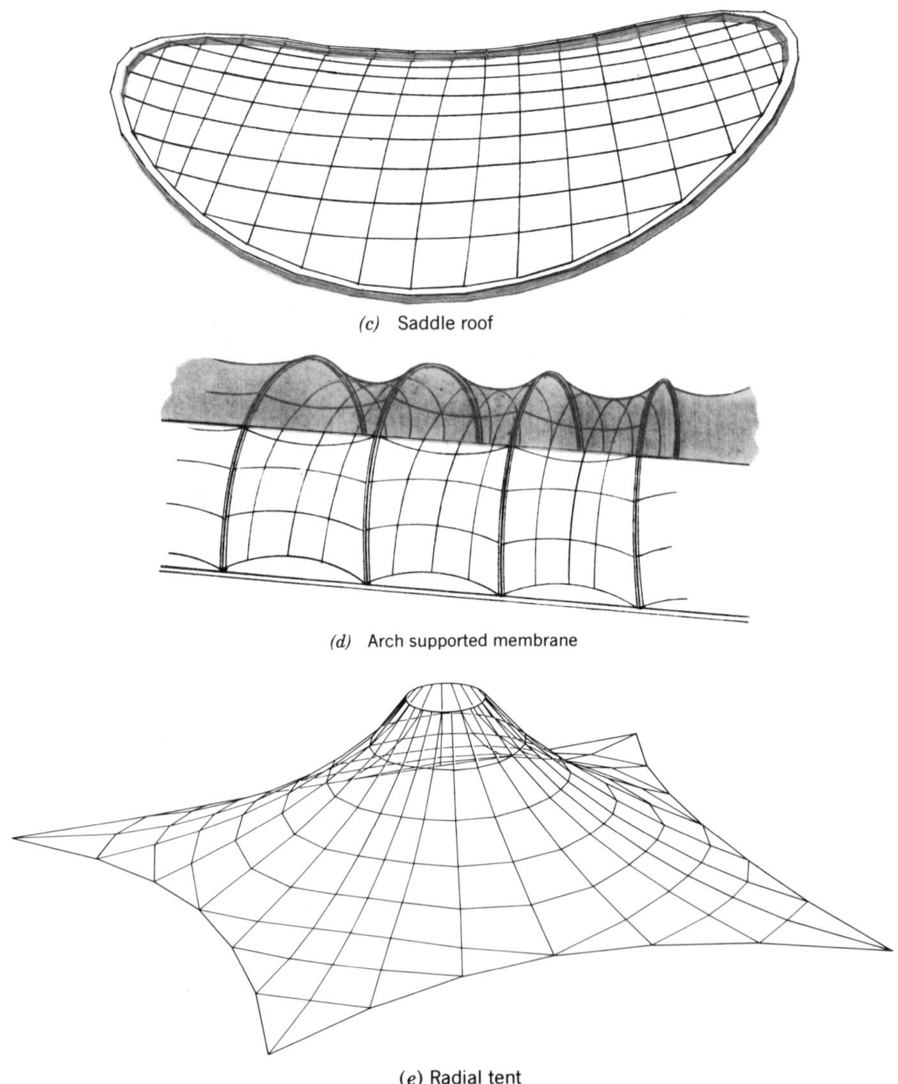

(*c*) Saddle roof

(*d*) Arch supported membrane

(*e*) Radial tent

Fig. 15.8 (continued) Common membrane structural systems.

signed, and that air-handling equipment be responsibly maintained, since the building's stability depends on air pressure sustained by machines.

A recent development in membrane structures that incorporates into a tension structure the advantages of the air structure is the cable dome developed and patented by David Geiger. Herein a synclastic cable net of ridge cables is supported by a system of diagonal cables, floating tension rings, and posts (Fig. 15.8*b*). The fabric membrane spans between ridge cables, it is pulled down by the valley cables, and thus is formed into an anticlastic surface. Since the cable dome, as with the air structure, is a tension dome, the rise can be kept to a minimum given consideration of drainage; hence the ratio of surface area to plan area can be kept to a minimum. By constructing the cable dome of prestressing strand rather than bridge cables, the total cable costs can be made similar to that of the air structure, even though the total cable tonnage required is nearly double that of the air structure. Taken together these attributes result in structural costs similar to that of the air structure with additional savings resulting from elimination of the air support and consequent savings in mechanical equipment and architectural elements.

Another structure without air support that is similar to the low-profile air roof is the *saddle surface* structure. In this case an anticlastic membrane is restrained by an edge ring of varying elevation whose shape derives directly from the shape of the saddle surface (Fig. 15.8*c*). Many

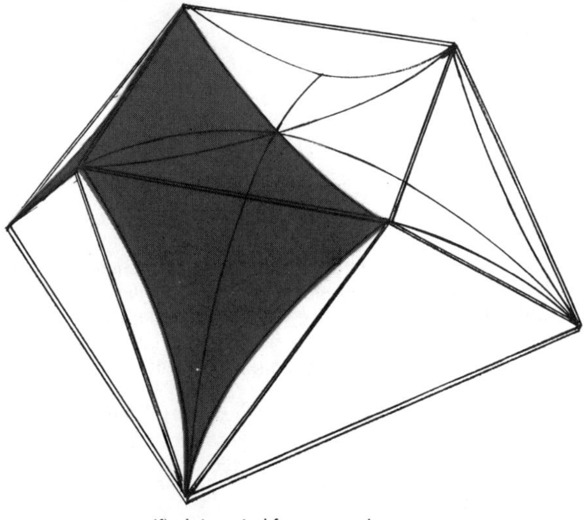

(f) Integrated frame membrane

Fig. 15.8 (continued) Common membrane structural systems.

saddle surface structures have been built using concrete, metal deck, wood, fabric, and other materials supported by cable net systems. (See Section 15.3.)

The introduction of arches to subdivide the roof surface permits a level edge configuration and reduced stresses in the membrane. A typical portion of an *arch supported membrane* system is shown in Fig. 15.8*d*. Other configurations are indicated in Section 15.4.

While tension structures with continuous supports as described above have a membrane net based on the Cartesian coordinate system, point supported structures lead to radial nets based on polar coordinates. The *radial tent* structure in Fig. 15.8*e* typically has a membrane supported by radial cables terminating at a central pole. Other tent configurations are described in Section 15.5.

Replacing the mast of the point supported system by members that form a trusslike frame in balanced equilibrium with the stresses of the tent membrane, results in a new type of structure called "integrated frame membrane structure" (Inframembrane). Its simplest form is shown in Fig. 15.8*f*.

15.1.5 Loads

Since the gravity loads of membrane structures are minimal, the proper evaluation of superimposed loads is of particular importance. Stresses due to wind, snow, and temperature change govern the design. Their evaluation needs more than the normal effort. As explained in Chapter 2, building codes are often lacking in detailed information regarding wind loads. The requirements might not be sufficient or might be so conservative as to prohibit economical design. The most comprehensive treatment of wind loads in the United States appears in ANSI A58.1 [15.2], which contains maps on expected maximum wind velocities and methods to develop pressure coefficients for common structural shapes. The wind load distribution for a specific membrane structure shape requires evaluation from wind load tests either derived from exact models of the project or by interpretation of existing wind tunnel information by experienced designers. For larger structures tests are best performed in a boundary layer tunnel, as described in Chapter 2. In some cases dynamic testing might be required.

Snow loads require equal care in evaluation. Air supported structures generally are designed with snow-melting equipment, though this may need to be augmented by manual removal of snow. In steeper parts of tension structures snow will slide off. It may, however, accumulate locally because of such sliding or wind effects. Again, ANSI A58.1 [15.2] has good information to supplement that given in Chapter 2.

Temperature effects must consider the various coefficients of the materials involved and the differential temperatures due to exposure to sun or location inside or outside of the membrane.

15.1.6 Materials

Except for structural fabric membranes, all the materials used are common to other structural systems. High-strength steel cable properties are given by AISI [15.1].

The following structural materials are available:

Vinyl-coated polyester fabric or scrim
Vinyl-coated polyester scrim laminate with or without Teflon film
Vinyl-coated fiberglass fabric
Vinyl-coated Kevlar fabric
Teflon-coated fiberglass fabric
Silicon-coated fiberglass fabric

Of these materials, vinyl-coated polyester is the most common material for temporary structures (lifespan up to 10 yr). Teflon-coated and silicon-coated fiberglass are the only noncombustible durable materials usable in permanent buildings. Teflon-coated fiberglass to date is the only one of these with a long experience record in buildings (first structures built in 1972). The following table (Table 15.1) is an excerpt from the fabric specifications for this material by Owens-Corning Fiberglas Corporation.

Practically identical materials are available from Chemfab/Birdair. Silicon-coated fiberglass materials with similar properties have been developed. For purposes of preliminary design, the allowable stress should not exceed 10% of the dry strip tensile strength for sustained loads, and 20% for temporary loads (wind). Final design requires expert experience with regard to acceptable stress

Table 15.1 Properties of Teflon-Coated Fiberglass Fabric as Provided by Owens-Corning Fiberglas Corporation

	Exterior Fabric			
	Structo-Fab 450	Structo-Fab 375	Interior Fabric Structo-Fab 120	Method or Standard
Coated fabric				
Weight, oz/sq yd				
a. Minimum average across width	45	37.5	12.0 approx.	ASTM D1910
b. Minimum single value	42	34.5	10.0	
Coating distribution, approximate				
a. Face, back, %	50,50	50,50	50,50	ASTM D1910
b. Mils thickness each side	4	4	—	O.C.F. W-01C,G
Thickness, approximate, in.	0.038	0.032	0.012	Method #5030
Strip tensile, lb/in. (rate of separation 2 in./min)				ASTM D1682
Dry, warp, min.	800	520	320	
Dry, fill, min.	700	430	230	
Wet, warp, min.	700	440	—	
Wet, warp, min.	600	360	—	O.C.F. Lab. Proc. S-26C
Strip tensile, after flexfold, lb/in.				O.C.F. Lab. Proc. S-26B
Dry, warp, min.	700	440	270	
Dry, fill, min., average	600	360	200	
Min. single value	540	300	180	
Trapezoid tear, lb				O.C.F. Lab. Proc. S-94Ad
Warp, min.	60	35	20	
Fill, min.	80	38	18	

Table 15.1 (*Continued*)

	Exterior Fabric		Interior Fabric	
	Structo-Fab 450	Structo-Fab 375	Structo-Fab 120	Method or Standard
Coating adhesion, lb/in.				O.C.F. Lab. Proc. S-08Q1
Dry or wet min., average	13	13	4	
Min. single value	10	10	4	
Burst drum tear, lb/in. stress for no propagation	80	60		O.C.F. Lab. Proc. CF17
Surface burning characteristics				ASTM E84-79a
Flame spread	5	5	5	
Smoke developed	5	5	5	
Fuel contributed	0	0	0	
Optical properties				
Available translucencies				
a. High trans., %	13 ± 3	13 ± 3	21 ± 5	ASTM E424-71, Method A
b. Low trans., %	6 ± 2	6 ± 2	—	
Reflectance, % min.	65–72	65–72	65	
Color	White	White	White	
Sound absorption, noise reduction coefficient (NRC), approximate	—		0.55–0.65	ASTM C423-77

1 in. = 25.4 mm; 1 oz/yd^2 = 0.034 kg/m^2; 1 lb/in. = 0.175 N/mm.

distribution and detailing of the fabric to avoid local overstress, damage, or tear, and should be based on the fabricator's recommended maximum stress loads.

15.1.7 Thermal Properties of Translucent Fabrics

Beyond their structural capabilities fabric membranes now available have additional properties which are of importance in the overall design of the building. These properties include translucency, reflectivity, and low heat absorbance. The highest translucency achieved to date in a completed structure with Teflon-coated fiberglass is 16%. A "translucent sky" enclosure as in Figs. 15.9 and 15.51 creates a sense of outdoors and permits even the growth of palm trees. The elimination of artificial light in the daytime results in substantial energy savings. The high reflectivity of the material reduces heating load in summer days and allows lower artificial light levels in the night.

New approaches to the design of climate control were required for structures made of these new materials. Dr. Karl Beitin [15.3] has done pioneering work in this area with the result that for some fabric structures, including the Pontiac Silverdome, air conditioning could be eliminated. New developments with silicone fabrics indicate potential translucencies of up to 80%. Since a 6% translucency provides adequate illumination in normal conditions, this opens the potential for translucent insulated roof covers with very major energy savings. In certain cases air conditioning may be eliminated entirely, and heating reduced to local radiant sources.

Combining various membranes and insulation components in a static configuration or even moving them relative to each other to decrease or increase the intake of light and heat at various times in the form of a "thermally active roof"* can further optimize the energy consumption of the building. Specific examples of thermally active roofs are discussed by Geiger [15.9].

* Patent pending by David Geiger.

Fig. 15.9 University of Santa Clara Student Activities Center, Santa Clara, California. Courtesy of Geiger Associates, New York.

15.2 AIR SUPPORTED STRUCTURES

15.2.1 Ground Mounted Air Structures

Ground mounted air structures are characterized by a shape that permits the structure to meet the ground vertically, thus permitting the membrane tension to be resisted by gravity loads. These anchorage systems may be continuous systems (concrete grade beams for permanent installations or water bags for temporary installations) or discrete systems such as screw-in earth anchors with cable or webbing catenaries built into the fabric which distribute the fabric stresses to point loads. The rise of the roof is typically nearly ½ of the span. The resultant semicircular cross-section has a sufficiently large radius of curvature that the environmental forces can generally be carried by fabric alone without the aid of a cable system. If films or lightweight fabrics are to be used then cables or webbing may be required in conjunction with the fabric. Webbing is typically sewn into the fabric seams forming a one-way system, whereas cables are incorporated into pockets in a one-way system or are formed into a cable net harness placed over the fabric in a two-way system. See Figs. 15.10 and 15.11. This type of structure typically spans less than 150 ft (45 m) and is used extensively for the covering of tennis courts and other moderate span enclosures.

Fig. 15.10 Air supported structure having one-way cable or webbing reinforcement.

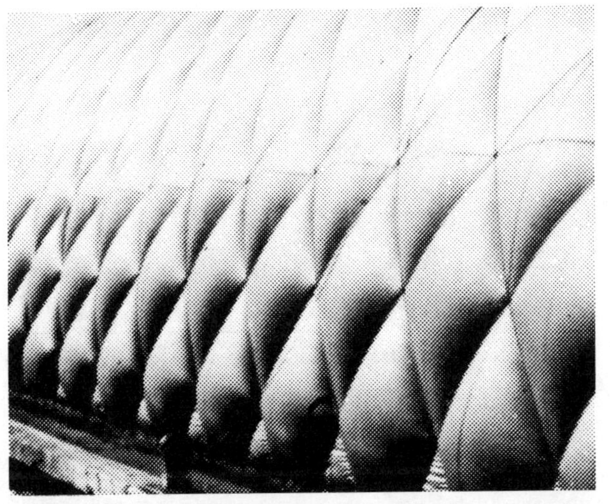

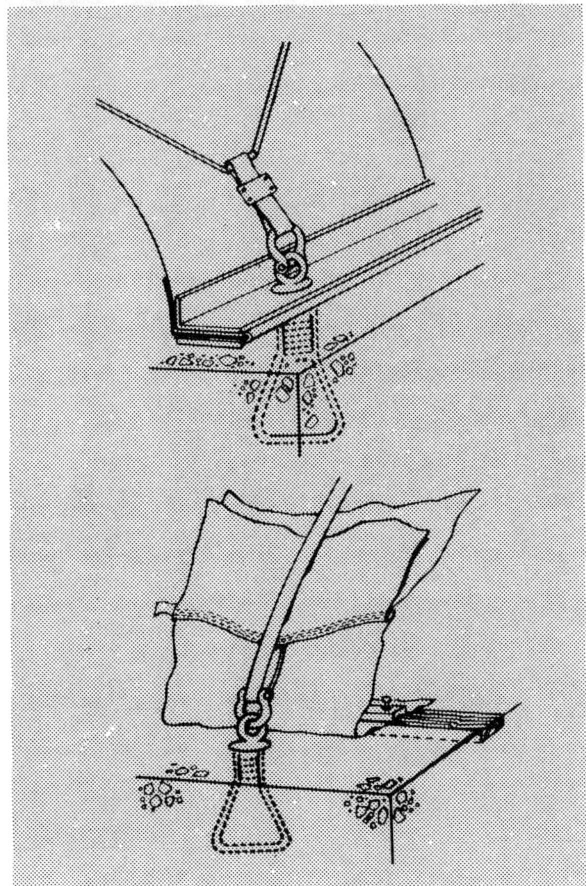

Fig. 15.11 Air supported structure having cable harness system reinforcement.

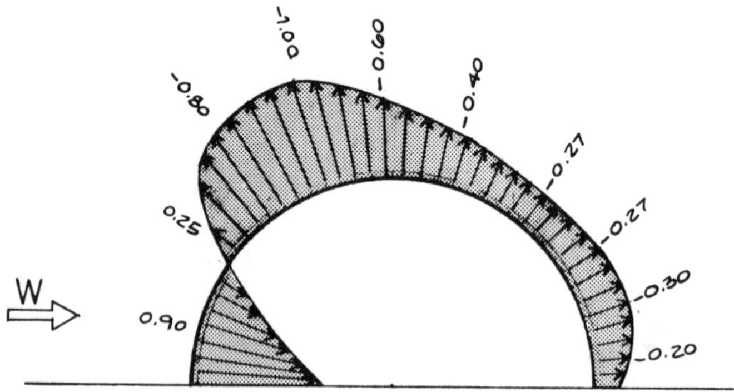

Fig. 15.12 Wind forces acting on the semicircular cross-section of an air structure.

As shown in Fig. 15.12, the internal pressure required to keep this structure inflated must be greater than the stagnation pressure due to the direct force of the wind against the nearly vertical face of the structure [15.8]. As a consequence the pressure must increase with wind velocity and the structure is vulnerable to deflation from tears, openings, etc., under high winds. Snow loads slide off most of the roof area with snow melting being required primarily for the center third of the structure. This is typically accomplished through heating the total space but in some cases a fabric plenum has been created to deliver maximum heat directly to the center of the roofs. The operations of the structure require that the snow sliding off the roof not be allowed to collect at the base of the structure as the weight of the snow and ponded water against its side could cause the collapse of the structure.

The equations that relate to fabric stress and the combination of wind suction and internal pressure are similar to those for cylindrical pressure vessels as shown in Fig. 15.13.

The moderate spans and the inexpensive temporary fabric used in ground mounted fabric structures result in low-cost systems. However, they can only be used when a certain degree of risk is acceptable under extreme load conditions.

15.2.2 Low-Profile Air Supported Structures—One-Way Cable Systems

When the rise of the roof is significantly reduced the wind pressure distribution becomes one of suction alone on the fabric membrane. This can result from either mounting the structure on a berm (Fig. 15.14*a*) where the positive pressure is taken by the berm or mounting the structure on a wall (Fig. 15.14*b*) where the positive wind pressure is resisted by the wall and the edge of the roof. In the latter case the positive pressure is best carried by a rigid roof at the edge, which in cross-section continues the curvature of the fabric roof. If this is done the separation point in the air flow will

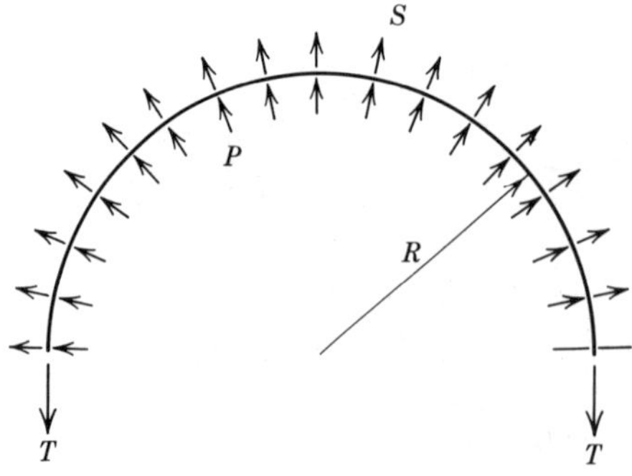

From vertical equilibrium: $T = (P + S)R/12$

P = internal pressure
= 10 psf (0.5 kN/m^2) minimum
S = maximum wind suction, psf
R = radius of the cylinder, ft
T = fabric tension, lb/in.

Fig. 15.13 Wind suction and internal pressure.

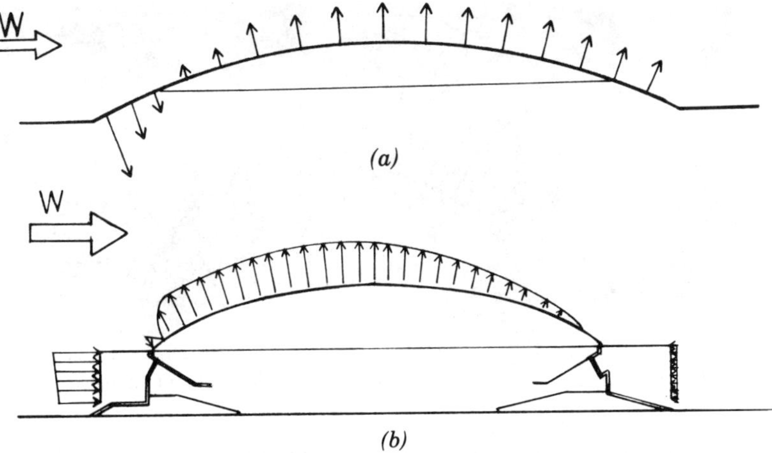

Fig. 15.14 Wind forces on a low-profile air structure. (*a*) Mounted on a berm. (*b*) Mounted on a wall.

occur at the outer edge of the building and snow drifting will occur on the rigid edge roof and not on the fabric membrane. For the low-profile air structure, wind suction alone acts on the roof; consequently a low internal pressure may be used, and with a roof weight less than 1 psf (4.9 kg/m^2) an internal pressure of 3 to 4 psf (0.14 to 0.19 kN/m^2) is possible.

Additionally, the low rise permits the roof to be designed so that in the deflated position it can hang free over the occupied space, thus minimizing the potential of damage in the deflated position. (See Fig. 15.15). Moreover at the low point roof vents can act as roof drains, thus minimizing the potential of roof damage due to the water ponding. Well-placed interior fountains and drains can carry away drainage from the interior space in the event of deflation. If a central spine is used (see Fig. 15.16) drainage in the deflated state can be made to flow to the perimeters.

Supporting the membrane on berms or walls may permit the roof structure to satisfy codes for permanent construction. This normally requires that the roof membrane be noncombustible and at a required minimum distance from the occupied space—typically 20 to 25 ft (6 to 8 m).

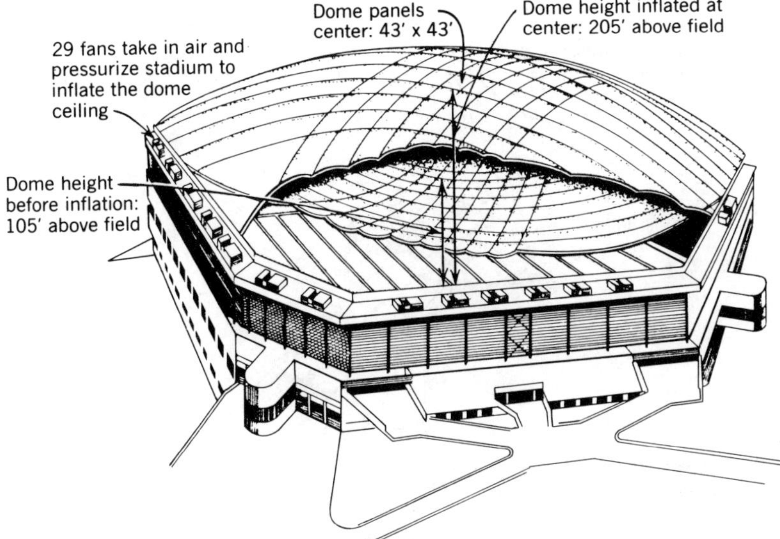

Fig. 15.15 Section through The Silverdome (Pontiac, Michigan) showing roof in deflated position. Courtesy of Owens-Corning Fiberglas Corp., Toledo, Ohio.

Fig. 15.16 Air supported roof carried by a central spine (proposed Jacksonville Junior College project, Jacksonville, Florida, by Geiger Associates).

Clamp

Clamp

A

Typ solar cell

Upper membrane

Middle membrane

Clamp assembly

Insulation

Roped edge

Lower membrane assembly

Absorptive membrane

Detail 'A'

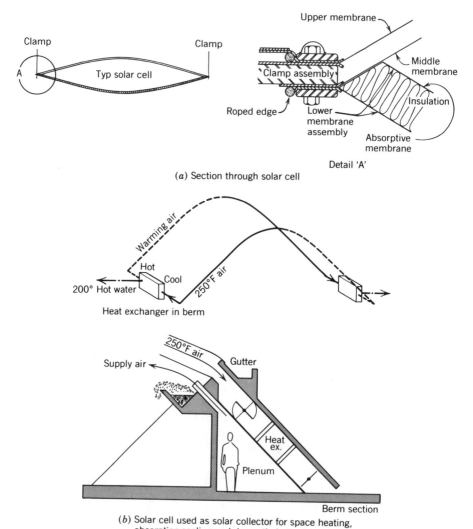

(*a*) Section through solar cell

Warming air

Hot

Cool

200° Hot water

250°F air

Heat exchanger in berm

250°F air

Supply air

Gutter

Heat ex.

Plenum

Berm section

(*b*) Solar cell used as solar collector for space heating, absorption cooling and domestic hot water

Fig. 15.17 Air-inflated and air supported membrane solar cell.

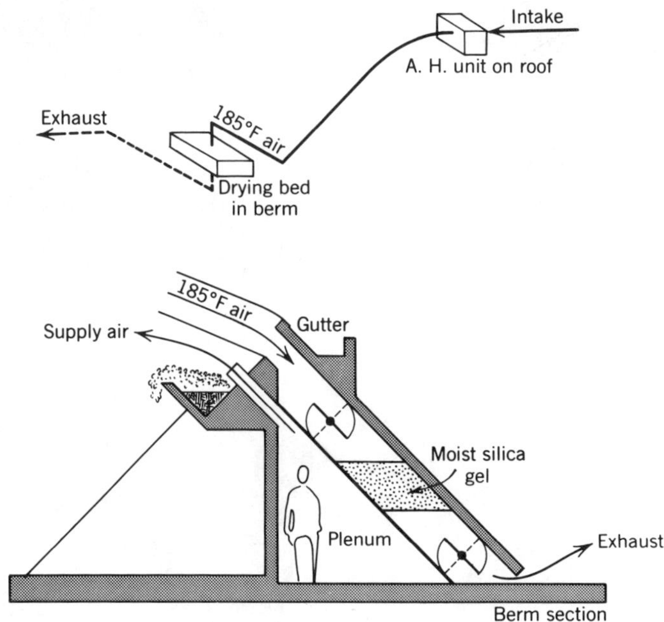

(*c*) Solar cell used to dry dehumidifying desiccants

Fig. 15.17 (continued).

One-way cable systems permit the incorporation of thermally active roof concepts as proposed in the Jacksonville Junior College project (Figs. 15.16 and 15.17) and the GSA Megastructure project (Figs. 15.18 and 15.19) [15-9].

Using these design techniques and highly translucent fabrics the roof may be made to function as either an active or passive solar collector (Figs. 15.16–15.19).

The major drawback to the low-rise one-way cable roof system is that the horizontal force that must be resisted at the cable ends may be significant and much of the economy of the roof system may be lost in the anchorage system. If these systems are integrated with the building structure then these added costs may be less significant. These horizontal and vertical forces may be calculated as indicated in Fig. 15.20.

One-way cable systems require that snow melting occur over the total roof surface. This is accomplished for a single membrane structure and for a multiple membrane structure by heating the total space or passing heated air between the upper cells. Typically the structure in the deflated position is required to carry a minimum of 60% of the code snow load on the deflated roof. This may exceed the value of $P + S$ given by Fig. 15.20 and when this occurs the snow load on the deflated roof may govern the roof design. Note that it is this case that causes the greatest downward force on the perimeter support system.

Fig. 15.18 GSA megastructure project—exterior.

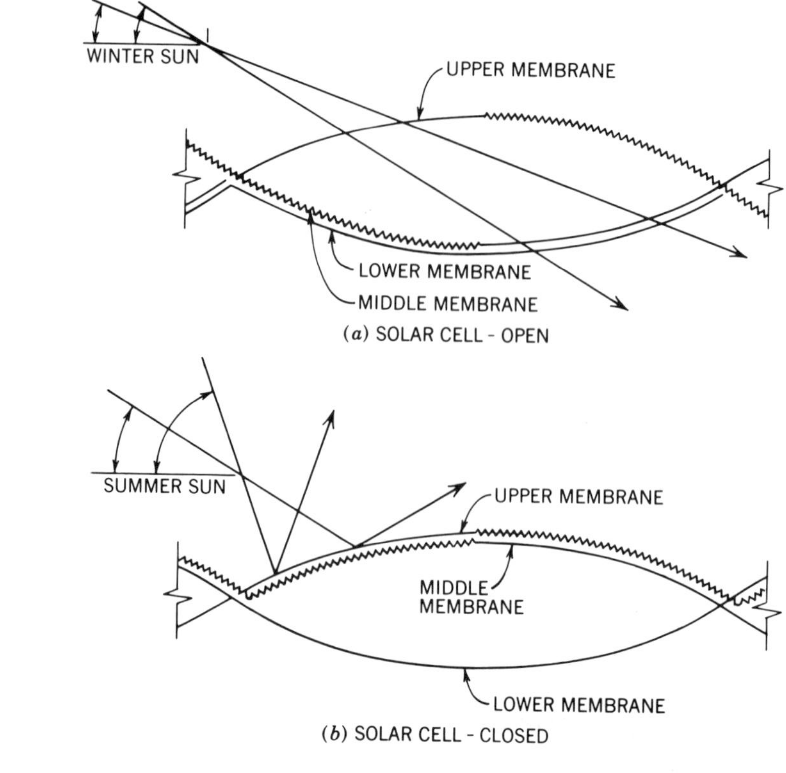

Fig. 15.19 Cross-section of passive heating solar cell. Cell runs east–west allowing for maximum penetration of winter sun. (*a*) Solar cell—open. (*b*) Solar cell—closed.

15.2.3 Low-Profile Air Supported Structures—Two-Way Cable Grids

For the low-profile air supported roof the problem of the anchorage of the end cable force may be overcome by using a two-way cable grid and anchoring the cables to a funicular compression ring. Examples of funicular compression rings with the corresponding ring loading are shown in Fig. 15.21. The corresponding cable layouts that load a polygonal ring with vertices on the previously defined funicular ring curves are shown in Fig. 15.22. The funicular ring is defined as a closed curve which for a given distribution of loads carries those loads without bending moments or shear forces in the ring, that is, only direct axial compressive forces are required for equilibrium. Funicular rings are also characterized by the fact that if they are loaded by a cable grid spanning the space, the funicular ring will not buckle globally in the ring plane since the action of buckling will cause a redistribution of cable forces so as to restore the ring to its original shape. This is not to imply that buckling of the ring locally between cable anchorage points or out of plane is not to be investigated. The fact that the loadings shown in Fig. 15.21 are funicular can be demonstrated through the use of free bodies of the ring and simple equations of statics.

Although the circular funicular ring (Fig. 15.21*a*) and the resultant radial cable (or rib) pattern (Fig. 15.22*a*) have been known since antiquity it is not a cable layout that lends itself to use in an air supported roof. The reason is that the cables collect at a tension ring at the center of the roof and the weight of the tension ring will result in the roof dimpling at the center, thus creating a place for water ponding. On the other hand, the cable patterns shown in Figs. 15.22*b* and *c* result in a uniform roof weight that can easily be carried by a uniform internal pressure and the roof drains readily to the perimeter. If it is assumed that the roof rise is sufficiently small so that a vertical force can be used to approximate the direction of the internal pressure (actually normal to the surface), then it can be shown that for an elliptical ring with a uniform cable spacing all the cables would be of the same size and each cable would conform to a parabolic curve in its elevation. Note that since a

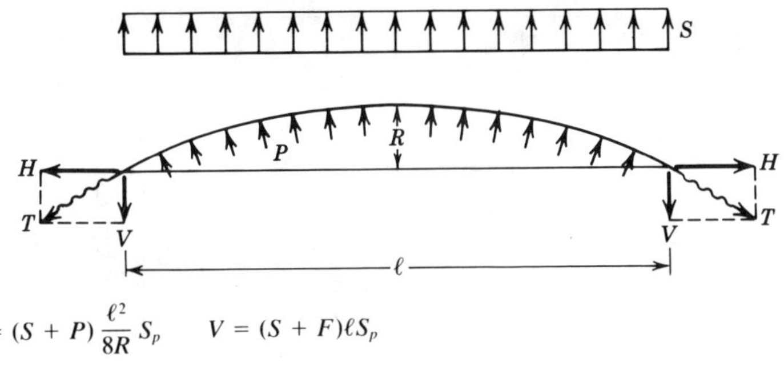

$$H = (S + P)\frac{\ell^2}{8R} S_p \qquad V = (S + F)\ell S_p$$

$$T = \sqrt{H^2 + V^2}$$

S_p = cable spacing
S = maximum wind suction
ℓ = cable length
P = maximum pressure [3 psf (0.14 kN/m^2) minimum, 6 psf (0.29 kN/m^2) maximum]
H = horizontal component of cable tension
V = vertical component of cable tension
T = cable tension
R = roof rise

Fig. 15.20 End reaction on low-profile one-way cable system.

circular ring is a specific case of the elliptical ring these conclusions and cable directions apply to a circular ring as well where the spanning cables intersect in plan at 90°.

It may be desirable for architectural reasons to have a super-elliptical (Fig. 15.23) rather than an elliptical plan configuration. If a super-elliptical ring shape is desired the load distributions shown in Figs. 15.21*b* and *c* do not apply. It can be demonstrated that with the super-ellipse functioning as a funicular ring the load distribution is as shown in Fig. 15.24.

With $M = 2$ this loading becomes the uniform loading of the ellipse as in Fig. 15.21*b* and for M equal to infinity the curve becomes a rectangle where the funicular loads are point loads at the corners coincident with the edges of the rectangle. For the super-elliptical polygonal ring and a

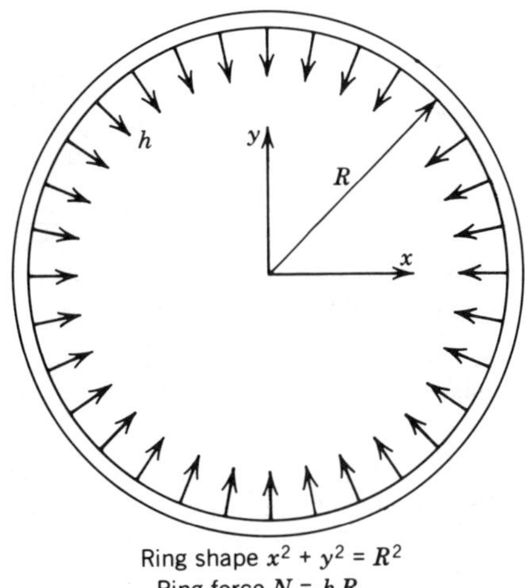

Fig. 15.21 Funicular compression rings and loading.

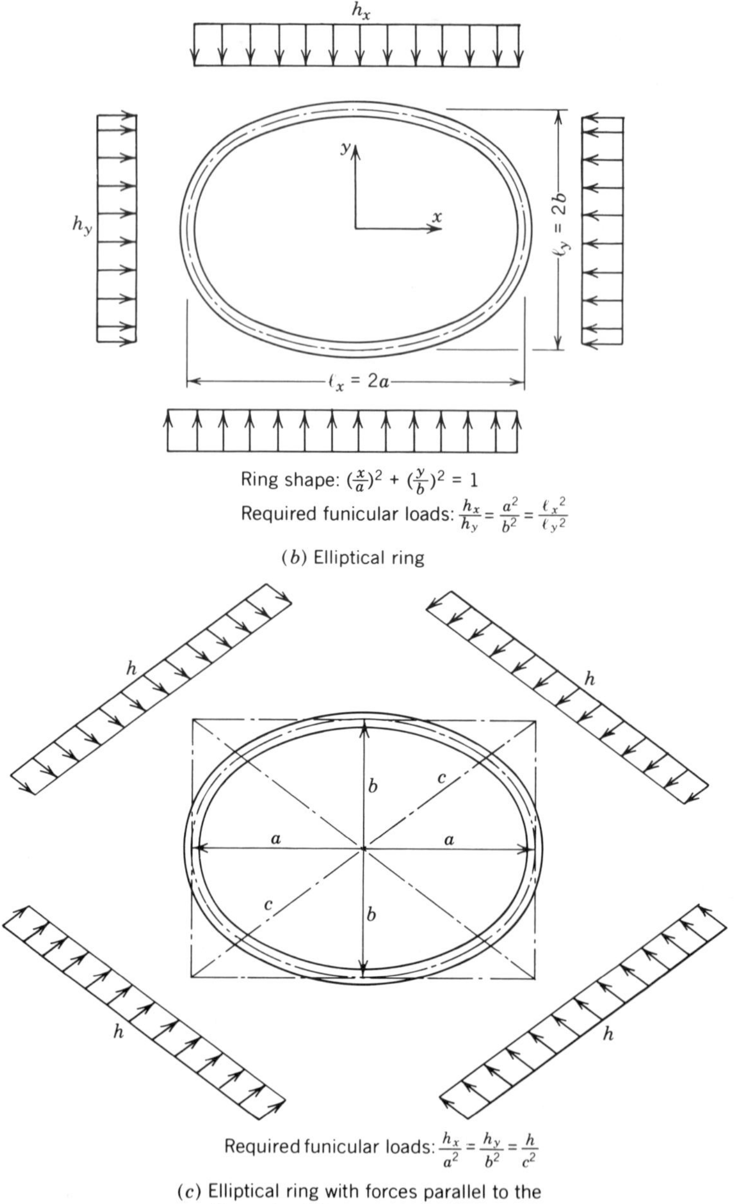

(*b*) Elliptical ring

(*c*) Elliptical ring with forces parallel to the diagonals of the superscribed rectangle

Fig. 15.21 (continued). Funicular compression rings and loading.

given cable spacing, discrete loads acting on the polygonal ring are obtained. If the roof shape for the super-ellipse is then derived, it may be shown that the cable forces spanning across points of the ring with greater curvature are significantly greater than those cable forces spanning across points of the ring with less curvature; consequently, the higher cable forces preload those cables that carry less force with the resultant cross-section along the major or minor diameter of the roof as shown in Fig. 15.25. This condition is undesirable in that it results in an increase in the cable weight for the roof and also may result in possible drainage problems at the flat areas near the perimeter.

An alternate cable direction for the super-elliptical ring is parallel to the diagonals of the superscribed rectangle. In this case the longest cables span from points of maximum curvature of the ring

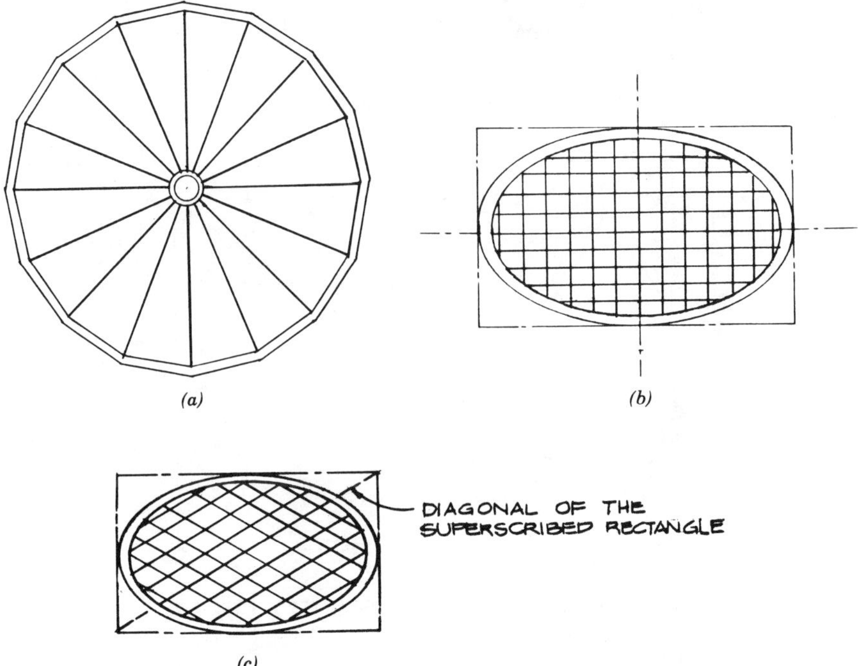

Fig. 15.22 Cable patterns for funicular rings. (*a*) Circular ring. (*b*) Elliptical ring. (*c*) Elliptical ring.

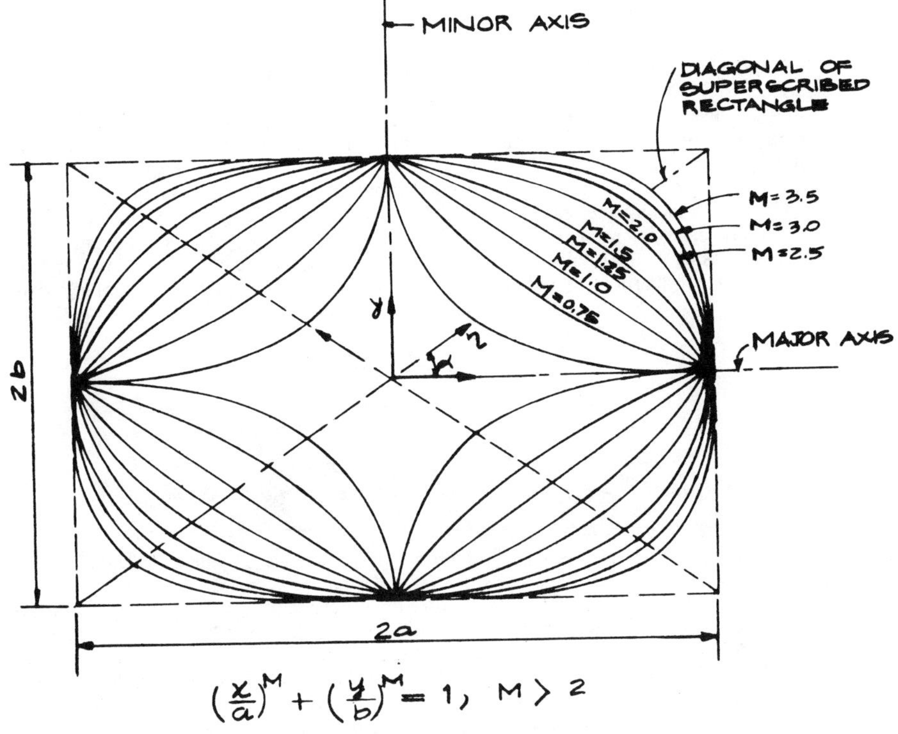

Fig. 15.23 Super-elliptical curves.

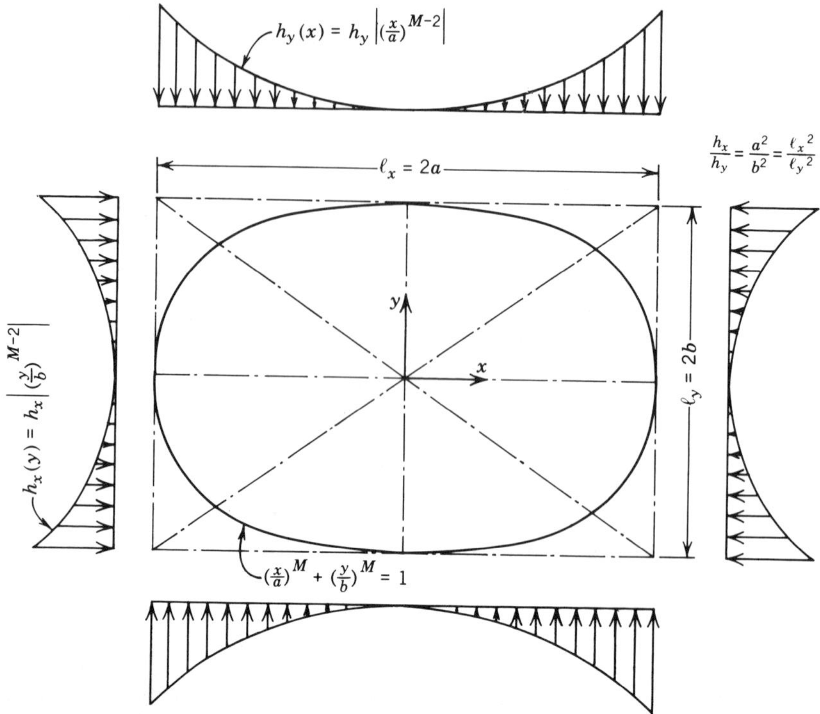

Fig. 15.24 Load distribution required for a super-elliptical funicular ring.

and in finding the roof shape the cross-section is more like that shown in Fig. 15.25*b*.

The procedure used to actually determine the cable forces and the roof shape is summarized here. With the ring shape of the super-ellipse given (i.e., a, b, and m are specified), draw the cable pattern parallel to the diagonals of the superscribed rectangle. Then isolate a free body of the ring as shown in Fig. 15.26 and establish the horizontal component of the cable tensions in terms of the internal compressions C by taking successive free bodies and solving the moment equations about points 2, 3, 4, etc. In this manner each of the unknown horizontal components of the cable tensions H_1, H_2, H_3, etc. can be determined in terms of the internal compressing force C. The curious point is that if the free body at the other ends of the cables were taken, the results would be the same. This is true because if the roof is acted upon by vertical or skewed symmetric loads, the horizontal

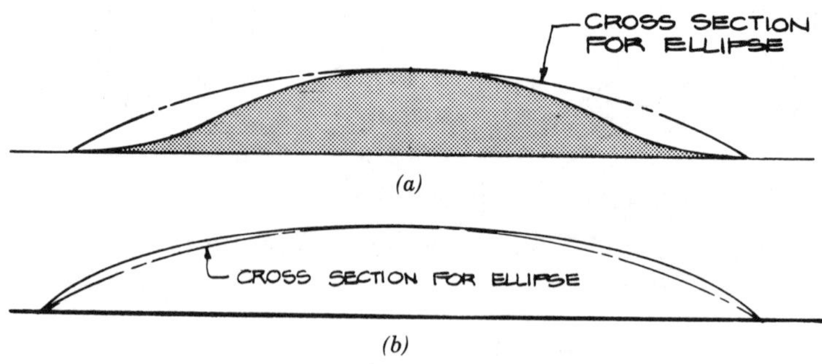

Fig. 15.25 Cross-sections for super-ellipse plan. (*a*) Orthogonal cable grid parallel to major and minor axes. (*b*) Skewed cable grid parallel to the diagonals of the superscribed rectangle.

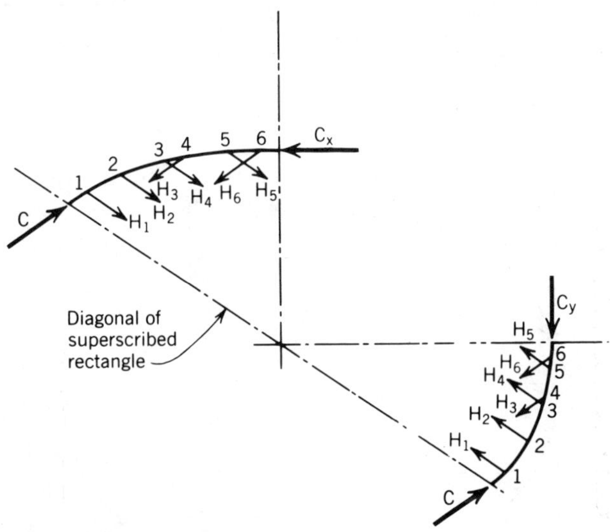

Fig. 15.26 Free body of ring segment showing horizontal components of cable tensions H_i and internal ring forces, C, C_x, and C_y.

component of the cable tension at the ends or at symmetric points on a cable is unchanged. But this does not explain why this in fact occurs when the free bodies of this ring are considered. It occurs because the tangent of the curve at the point where it intersects the diagonal of the superscribed rectangle is parallel to the other diagonal and the offsets from this tangent to the curve are equal for each of the curved segments. This principle is shown in Fig. 15.27. These equal offsets result in equal moment arms to the internal force C for the free bodies at each end of the cable. Thus the super-ellipse obeys the property of "skewed symmetry" expressed by the relationship

$$\eta = \pm f(\pm\xi)$$

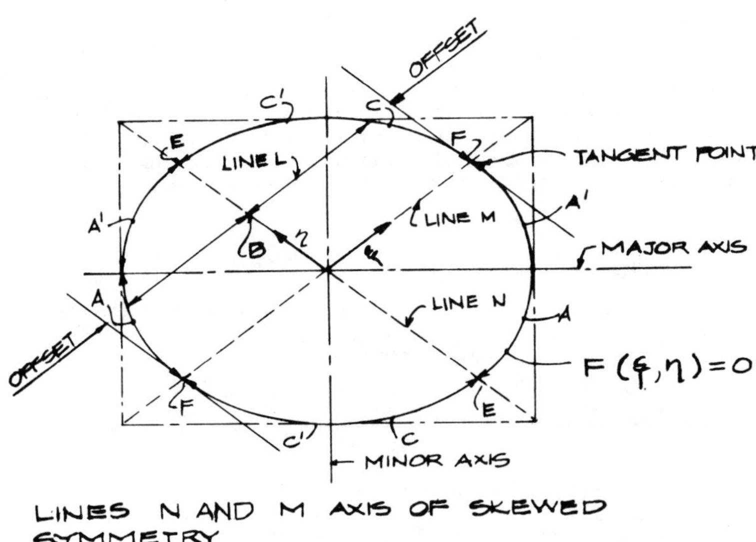

Fig. 15.27 Principle of skewed symmetry.

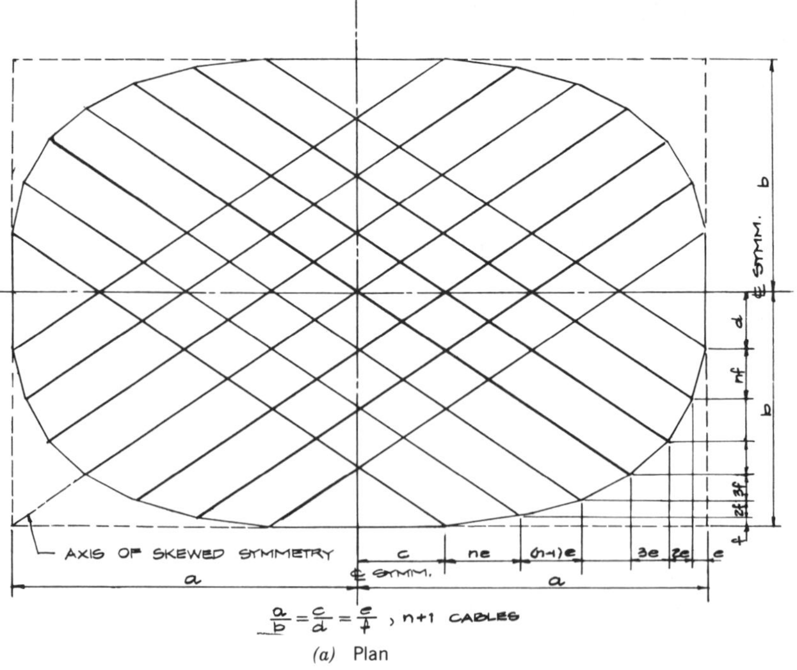

(a) Plan

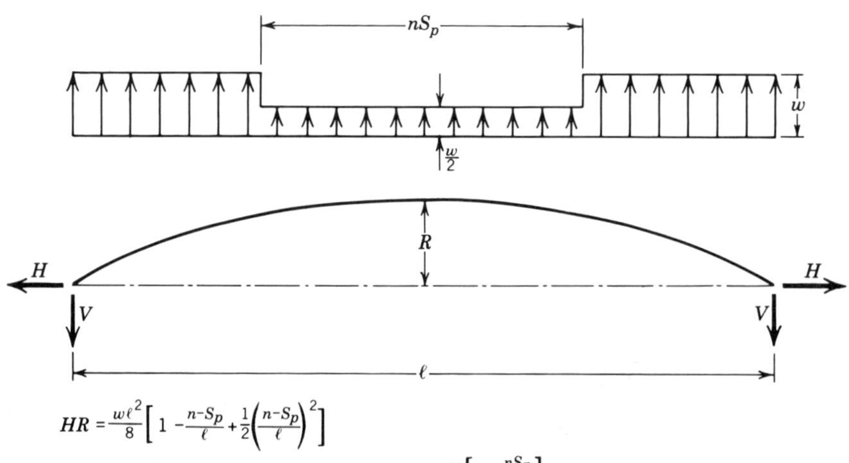

$$HR = \frac{w\ell^2}{8}\left[1 - \frac{n\text{-}S_p}{\ell} + \frac{1}{2}\left(\frac{n\text{-}S_p}{\ell}\right)^2\right]$$

V = vertical component of cable tension $= \frac{w}{2}\left[\ell - \frac{nS_p}{2}\right]$

T = cable tension $= \sqrt{H^2 + V^2}$

n = number of cables spanning in one direction (half of total number of cables)

ℓ = cable span

S_p = cable spacing

R = roof rise

$w = (p + S)\,S_p$

p = initial pressure

S = average wind suction

H = horizontal component of cable tension

(*b*) Loading

Fig. 15.28 Funicular ring with sparsest cable grid.

for the ξ and η axis shown. Furthermore, it can be shown that any ring that is skewed symmetic is funicular for any cable system that is parallel to the skewed symmetric axis.

Under these conditions it is possible to derive a funicular ring shape that satisfies certain ideal conditions as to cable forces and cable spacing. By postulating the sparsest cable grid with cables of equal spacing and carrying equal horizontal components of cable tension, it is possible to obtain a roof with the minimum number of cable intersections, the minimum attachments of fabric to cables, the minimum number of cable anchorages, and most likely the lowest cable weight. This would lead to the lowest roof cost. Postulating this condition leads to the plan configuration shown in Fig. 15.28*a*. The cross-section of Fig. 15.25*b* would apply in this case as well, and it is quite easy to calculate the roof shape under uniform load for purposes of preliminary design. Since at the points where the cables cross, half the uniform load acting on the roof is carried by each of the cable systems, taking the longest cable for which the rise is either given or is to be determined for the horizontal component H of cable tension, then the loading on the cable is as shown in Fig. 15.28*b*. Using the beam analogy for cables where the bending moment at the center of a simply supported beam is equal to the horizontal cable tension times the rise for a similarly loaded cable, the relationship shown in Fig. 15.28*b* is obtained and the rise R or the horizontal component H of the cable tension can be solved for accordingly.

With the relationship between the horizontal components of the cable and the ring shape established so that the ring is funicular, then the roof shape may be established once the roof loading is given. The only requirement is that the roof loading be skewed symmetric so that the skewed symmetric relation of the cable grid is not disturbed.

The vertical ordinates of the cable intersection points are determined as follows. With vertical roof load and if the cables intersect in plan as straight lines then the horizontal equilibrium of the cables is identically satisfied. If the cable grid is continuous over the whole area, or is made so by the addition of parallel almost zero tension dummy cables (required for Fig.15.28*a*) and if the vertical loads are applied to each of the cable intersection points, then the equations of statics for vertical equilibrium will provide the vertical roof ordinates Z through the solution of a system of simultaneous equations. There are as many unknown roof ordinates as there are equations of statics (Fig. 15.28*b*).

15.2.4 Low-Profile Air Structures—Consideration of Structural Design

In the foregoing discussion, once the plan configuration is established, the rise must be set in order to establish the maximum cable tension and thus the required cable size. The roof rise should be a minimum of 1/20 of the span for reasons of drainage and maximum of approximately 1/10 of the span based on wind loads and roof dynamics.

With the cable tensions determined the cable sizes can be established using the design criteria set by the American Iron and Steel Institute [15.1].

Cable clamps must be designed so as to prevent cable slippage at cable intersections. Split pipe details (Fig. 15.29) have been used effectively to this end.

Cable anchorage details must take into account the fact that the cables must be free to move from the inflated to the deflated positions without damage to the cable. This is best handled by either open or closed pin connected sockets. Allowance must also be made to prevent damage to the fabric due to movement of the cable around its pin. This is best handled by bringing the axis of rotation of the fabric to the same center as the axis of rotation of the pin. This detail is accomplished through use of an eccentric eliminator frame (Fig. 15.30). Drainage considerations from the axis of rotation of the eccentric eliminator frame to the edge of the structure requires that the center of the pin be higher than the top of the ring. The resultant eccentricity of the cable force relative to the ring centroid results in a torsional moment being imparted to either the ring support frames or

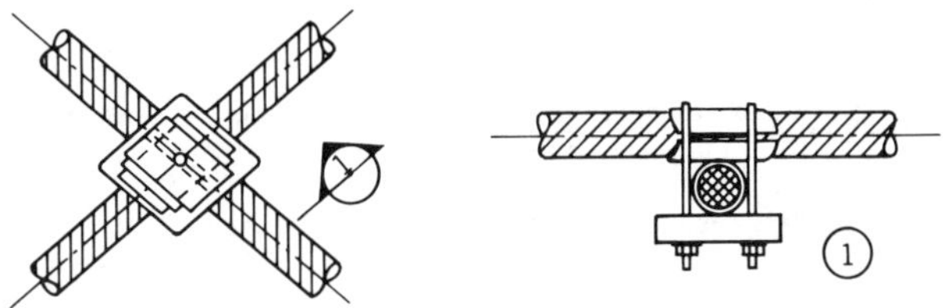

Fig. 15.29 Cable clamp details.

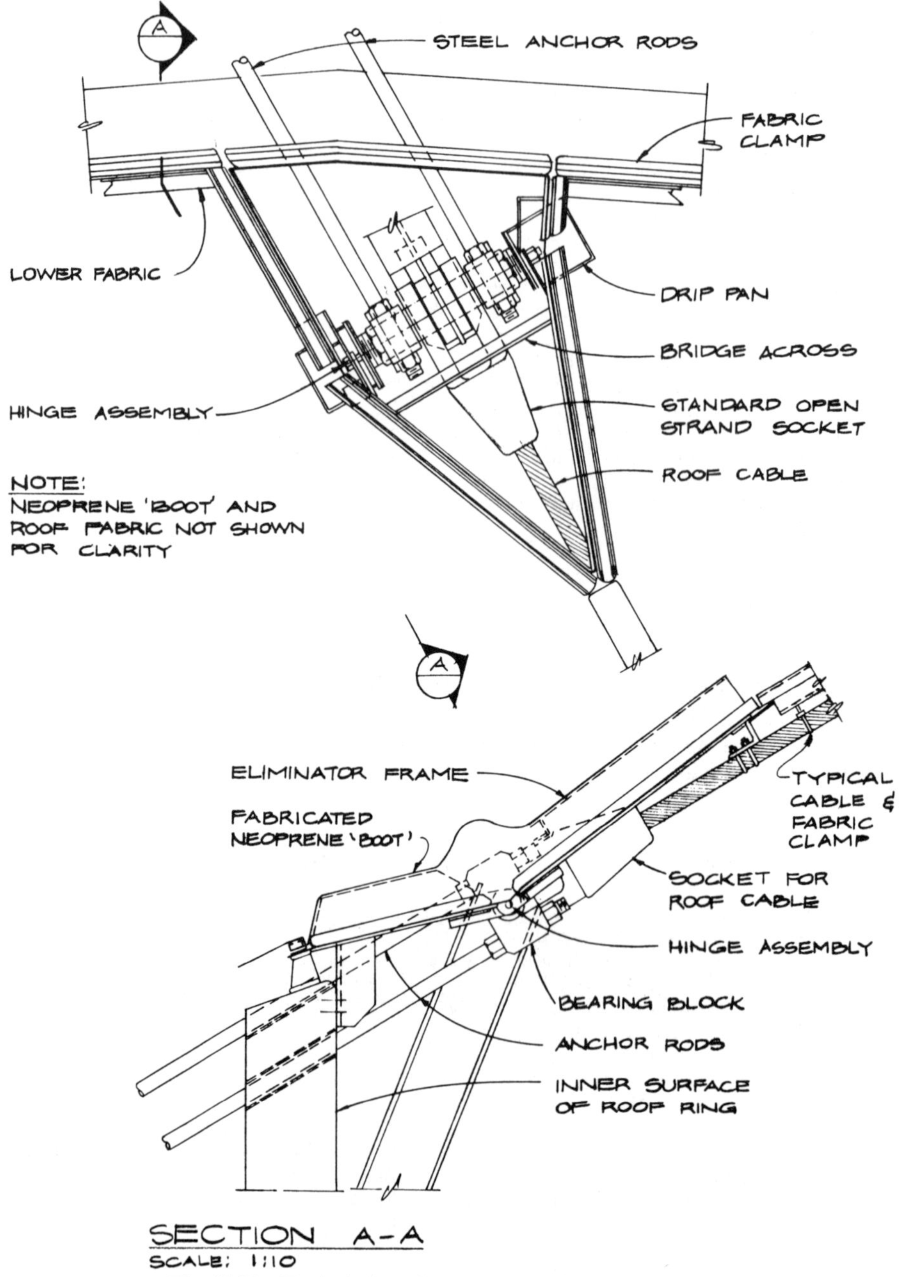

Fig. 15.30 Typical plan of eccentric eliminator frame.

columns. Note that in any case these moments cannot be avoided since also they generally occur because of the change in the line of action of the cable force in moving from the inflated to the deflated position.

With the horizontal component of the cable tensions established the axial compression force in the ring for the funicular load case can be determined (Fig. 15.26). For load cases other than the funicular load case, the redistribution of cable tensions will result in nonfunicular ring loading with consequent ring bending. Since the nonlinear analysis of the cable net required to find these nonfunicular cable tensions is beyond the scope of this book, certain approximate rules are required for the purpose of preliminary design: If the ring is supported on a support where it is free to slide and the only restraint against movement is friction, then the bending moments on the ring may be considered to be sufficient to reduce the compressive stresses to zero on the one face and to the maximum on the opposite face. Consequently, if the centroid is at the center of the ring, it can be sized by using the maximum compressive force for the funicular load case and a concrete stress equal to ½ of the maximum allowable. Bear in mind that the ring cross-section should have enough mass to resist the uplift forces on the ring with a factor of safety of 1.5.

If the ring is supported by a column or frame system where that system not only takes out the local eccentricities due to cable force introduction but also offers horizontal restraint to ring movements, it is possible that the stiffness of this system may take out almost all this nonfunicular loading of the ring. In this case the allowable concrete stress can approach the code value. The supporting structure can also be used to resist the uplift forces on the ring; consequently the ring weight need not be governed by consideration of uplift. Generally ring reinforcing will be between ½ and 1% of the ring cross-sectional area.

Considerations of fabric design and shipping will limit cable spacing to a maximum of 45 ft (14 m). Low-profile air structures are normally of such a large span as to require the fabric to be shop-fabricated and shipped in panels that are clamped to the cables in the field. These clamp details, which are costly both in terms of material and erection labor, must be kept to a minimum consequently the minimum cable spacing considered economically feasible is 35 ft (11 m) on center.

From consideration of fabric stress the base dimension of the four triangular panels should not exceed 200 ft (60 m).

Fabric cable clamp details (Fig. 15.31) have evolved into a set of recognized details that perform adequately and permit recaulking as required to prevent roof leakage. Fabric patterning is generally done by the roof contractor although this does not relieve the engineer of checking the work because of the importance of the performance of the fabric panels in the overall roof performance.

15.2.5 Low-Profile Air Structures—Consideration of Mechanical Design

Ground mounted air structures are typically of such a short span that their inflation needs as an air structure exceed the ventilation requirements for purposes of occupancy. On the other hand for large-scale low-profile air structures the ventilation requirements generally exceed the blower

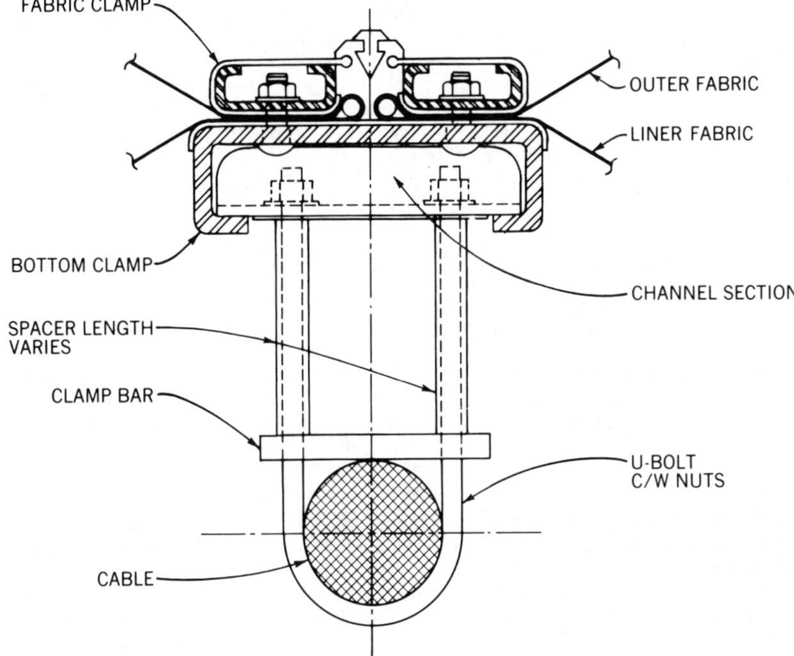

Fig. 15.31 Fabric cable clamp detail.

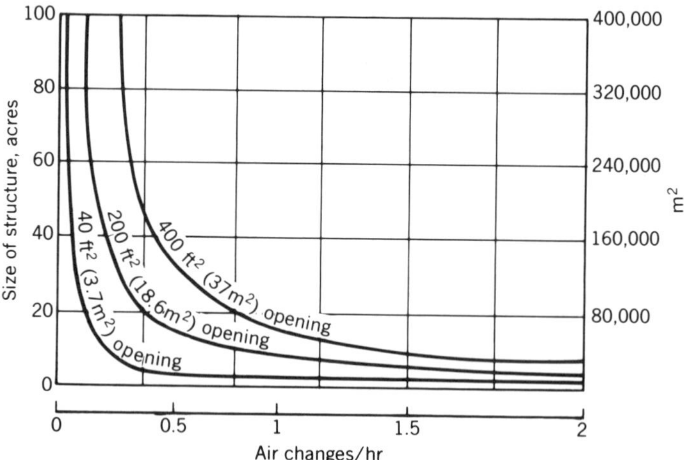

Fig. 15.32 Air changes per hour required to prevent deflation.

requirements to prevent deflation (Fig. 15.32) [15.8]. These calculations are based on an assumed roof weight of 1 psf (4.9 kg/m²).

15.2.6 Low-Profile Air Structures—Consideration of Scale

As the size of the low-profile air structure increases in span its reliability increases. The ultimate catastrophe would be simultaneous failure of the mechanical system, including the standby generator, coupled with a large opening in the structure. For large span structures there will, however, be sufficient time to evacuate people from the structure and there should be sufficient time to close openings, and/or to restart the blowers before full deflation occurs (Fig. 15.33). All of the above considerations do not take into account the fact that as the span—and consequently the height—of the structure increases, the buoyancy due to the thermal gradient of the stratified air within the structure, plus lift due to wind forces integrated over the roof area, begin to exert forces such that, for very large roofs, the pressure differential supplied by the mechanical system is not required. Figure 15.33 was determined based on a roof weight of 1 psf (4.9 kg/m²). Through the introduction of new materials and the elimination of clamp plates through new techniques of field assembly it is envisioned that in the very near future roof weights may be reduced to less than ¼ psf (1.2 kg/m²). At this weight structures of over 2000-ft (600-m) span may be kept aloft by thermal buoyancy alone. Structures of 8000-ft (2400-m) span have been considered in terms of the practical architectural implications [15.8].

15.3 SADDLE SURFACE TENSION STRUCTURES

15.3.1 Design Considerations

As discussed in Section 15.1, pure tension structures (without air support) require anticlastic surface shapes to create stable structural systems and to carry the superimposed loads. The simplest membrane surface satisfying this requirement is the four point structure shown in Fig. 15.34. Locating one of the four points outside the plane created by the other three results in an orthogonal anticlastic net with edge catenaries taking the surface stresses to the support points.

Extending the surface to meet a circumferential ring generates a single surface membrane tension structure most similar to the low-profile air supported system (Fig. 15.35*b*). While identical in plan configuration the two structures are different in elevation: The saddle-shaped tension structure has a ring of varying elevation containing both high points and low points of the system, and the carrying cables and tie-down cables are oriented to obtain maximum curvature (Fig. 15.35*a*).

The form of the structure is determined by the shape of the edge ring. Many such shapes are possible; however, of particular interest are continuous ring forms of circular, super-elliptic, or polygonal shapes.

The economy of these systems depends on the sag-to-span ratios for the two membrane directions and the design and restraint of the ring. The cost of the ring is generally governed by the

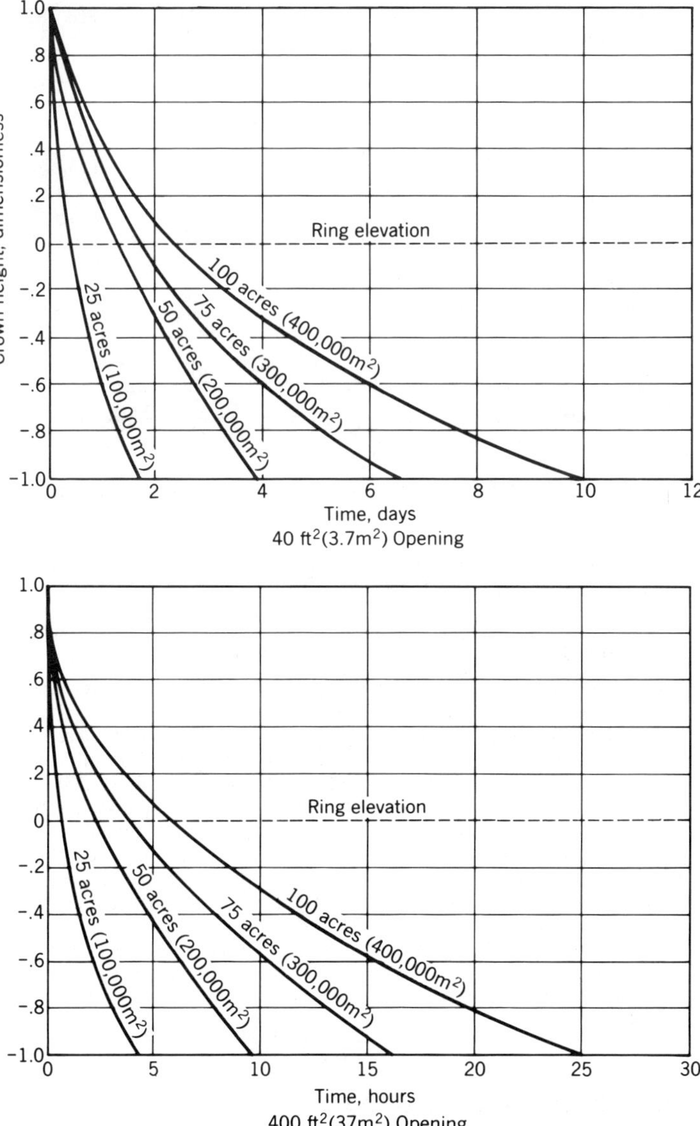

Fig. 15.33 Deflation curves when there are no blowers. Weight of roof 1 psf (4.9 kg/m²) assumed.

amount of horizontal bending to be carried. Shallow membranes will have high membrane stresses, but low bending moments on the ring; deep membranes, the opposite. The curvature available for the membrane is further limited by the height of the building available without costly increase in exterior supports and wall enclosure.

The bending moments in the ring can be further reduced by restraining the ring by its supports or by adding straight prestressing cables. Restraining supports must be elastic to permit movement of the ring. Prestressing cables are most effective if placed parallel to the load-carrying cables above and below the roof surface. This method was used in the Capital Center Arena, which is shown during construction in Fig. 15.36. The roof has a circular plan of 400 ft (120 m) diameter. The skin is metal deck on bar joists supported on galvanized bridge strand cables. The bar joists are welded and stressed against the ring to form the tie-down system.

A fabric membrane may be used in saddle roofs of large span if the fabric can be given a curvature greater than that of the saddle roof. In the Lindsay Park Athletic Centre in Calgary,

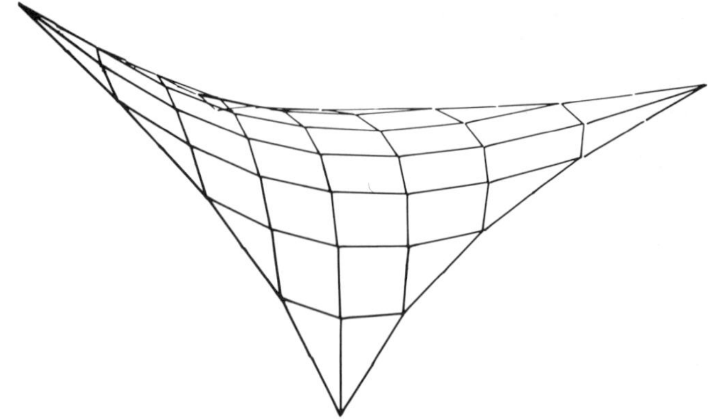

Fig. 15.34 An orthogonal anticlastic saddle surface.

A

B

ROOF PLAN

SECTION 'B'

SECTION 'A'

(a)

(b)

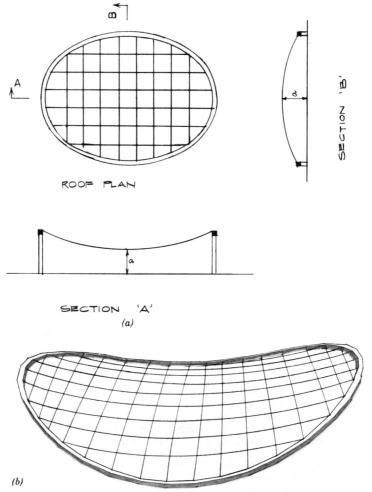

Fig. 15.35 Single surface membrane tension structures. (*a*) Plan and elevation. (*b*) Saddle-shaped structure.

Fig. 15.36 Roof system for Capital Bullets Arena near Washington, D.C. Courtesy of Geiger Associates, New York.

Fig. 15.37 Saddle roof for Lindsay Park Athletic Centre, Calgary, Canada. (*a*) Actual structure. Courtesy of Geiger Associates, New York.

(*continued on next page*)

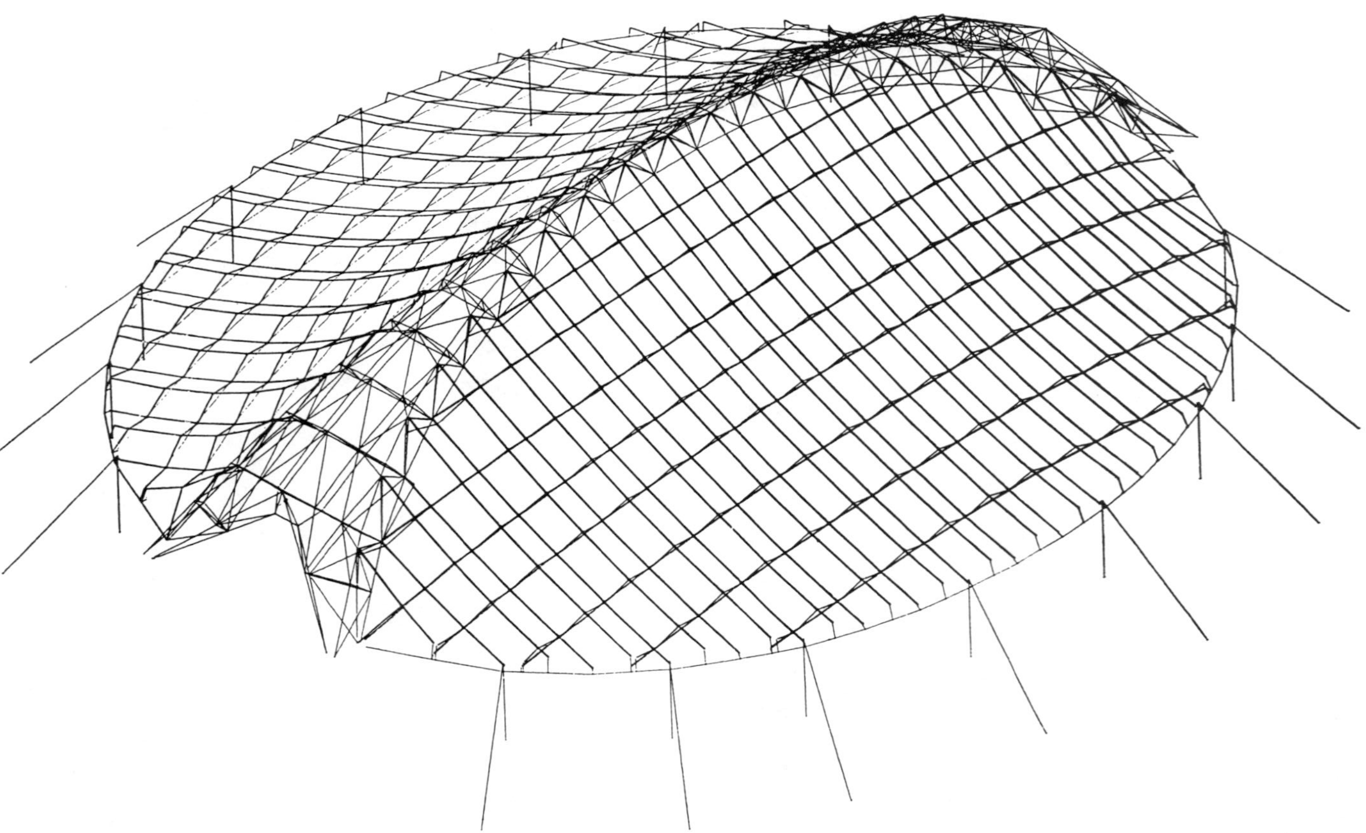

Fig. 15.37 (continued) (*b*) Computer model. Courtesy of Geiger Associates, New York.

Canada, this was accomplished by clamping fabric rolls to nearly parallel draped cables attached to a two-way cable net formed into a saddle surface. The fabric was then pulled upward by a parallel draped cable at the midpoint of the fabric roll, thus forming a saw toothed surface (Fig. 15.37).

This project also gave rise to the first insulated fabric roof with a resultant translucency of 4% and a thermal resistance of R16. The method of construction patented by David Geiger involves the use of a translucent upper membrane, fibrous translucent insulation, and a transparent vapor barrier of low enough mass as to allow sound vibrations to be transmitted to the fibrous insulation.

Architecturally, closed ring saddle-shaped roofs are particularly advantageous for arenas or similar structures where the varying height of the roof matches the space requirements (See Fig. 15.35*a*). Many saddle-shaped roof structures have been built using concrete, both precast and cast-in-place, as roof surface materials. If monolithically executed, especially if post-tensioned, the roof structure then becomes a concrete shell. Metal deck, wood, fabric, or acrylic panels are more appropriate skin materials for true lightweight membrane structures.

15.3.2 Loading

Downward loads are mainly created by snow loads or minimum code requirements. The effect of slopes has to be evaluated for potential sliding and accumulation of snow. Chapter 2 and ANSI A58.1 [15.2] provide useful guidelines.

For lightweight roofs, wind uplift is a major load case. In the direction of the upward bow of the surface the roof has great similarity with vaulted roof shapes for which ANSI has formulas for wind load distribution. In the direction of downward sag additional information may be necessary to evaluate the wind effects properly. For final design of large span structures of this type, wind tunnel tests may be necessary as a basis of establishing wind loading.

15.3.3 Membrane Design

The preliminary design of the tension membrane requires two main steps: the determination of the surface shape and the evaluation of stresses under maximum live loads, both for selecting the membrane components and for determining the resulting loads on the ring.

The process is illustrated in the example of an elliptical roof shown in Fig. 15.38. As shown in Fig. 15.21*b* a funicular ring requires the following relationship of the horizontal components of the membrane stresses:

$$\frac{h_x}{h_y} = \frac{a^2}{b^2} = \frac{\ell_x^2}{\ell_y^2} \tag{15.3.1}$$

The horizontal membrane stresses along the center strip in the x and y directions are as follows:

$$h_x = \frac{(w + p)\ell_x^2}{8f_x} \tag{15.3.2}$$

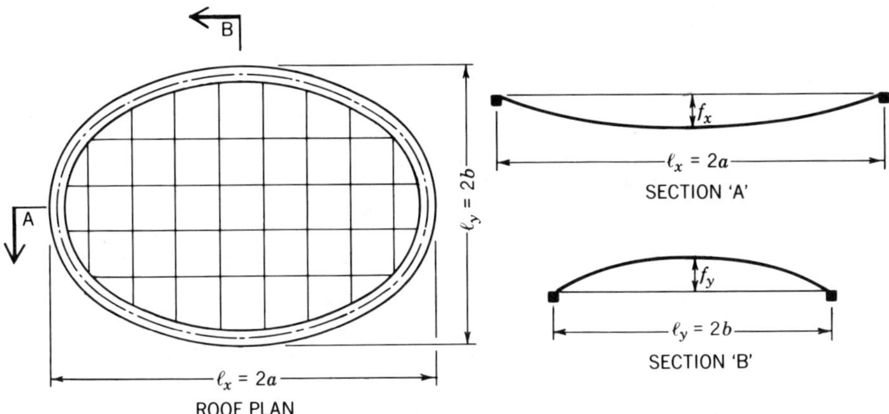

Fig. 15.38 Elliptical saddle surface roof.

$$h_y = \frac{p\ell_y^2}{8f_y} \tag{15.3.3}$$

where w = dead load of membrane
p = vertical reaction of internal prestress force acting between the two main directions of the membrane

Therefore,

$$\frac{f_x}{f_y} = \frac{w + p}{p} \tag{15.3.4}$$

Since $f_x + f_y = f$, this expression will determine the curvature of the center strips if p is given.

The choice of the prestress load p is to be made so that membrane and ring design become optimal. Therefore, the magnitude of upward and downward loads, the spans, the material of the membrane, and the shape and support of the ring must be considered.

The variables are too many to reduce their influence to a simple equation. However, a preferred ratio of sags can be derived from the maximum load conditions. For the case of the elliptical ring and a membrane or cable net with equal strength in both directions, the ratio becomes

$$\frac{f_x}{f_y} = \frac{w + p_x}{p_y} \tag{15.3.5}$$

where p_x = downward live load
p_y = upward live load

The internal prestress force should be chosen so that the actual f_x/f_y ratio deviates as little as possible from this preferred ratio. For preliminary purposes the following minimum levels of prestress might be practical:

$$\frac{w + p}{p_x} \geqq 0.20; \qquad \frac{p}{p_y} \geqq 0.20 \tag{15.3.6}$$

The selection of the membrane members or determination of stresses for a given membrane or cable net follows from these equations for membrane stresses t_x and t_y:

$$\max t_x = \frac{1.10(w + p_x + 0.2p)\ell_x^2}{8f_x} \tag{15.3.7}$$

$$\max t_y = \frac{1.10(p_y + 0.2p)\ell_y^2}{8f_y} \tag{15.3.8}$$

15.3.4 Design of the Ring

The loads on the ring are highly indeterminate since under the load the ring will deflect in a fashion which will change the membrane stresses. As a result, the loads acting on the ring will decrease in areas of high loads, and will increase in areas of low loads. With the addition of restraining cables or supports with lateral stiffness the determination of the ring bending forces is even more complex. Therefore, no simple equations can be established for preliminary computation. However, the following suggestions will lead to good results with reasonable effort.

If cable support points are located over columns, consider bending in the horizontal plane only. For cable support points occurring between columns, compute the vertical bending moments separately on a curved continuous beam system; this again reduces the ring into a system loaded in the horizontal plane only.

As horizontal loads on the ring, use full static membrane stresses max t_x and max t_y as determined above. These stresses vary along the arch linear to the respective membrane span.

The configuration, loading, and behavior of the membrane on cable net result in a pattern of ring stresses which can be assumed to have double symmetry. Therefore the analysis of the ring can be reduced to one-quarter of the roof structure, as shown in Fig. 15.39. If there is no horizontal restraint by columns or prestressing cables the curved beam can be analyzed as a first degree indeterminate system, requiring only the computation of simple cantilever moments as in Fig. 15.39*c*. With the notation of Fig. 15.39 the statically indeterminate moment M_1 is computed:

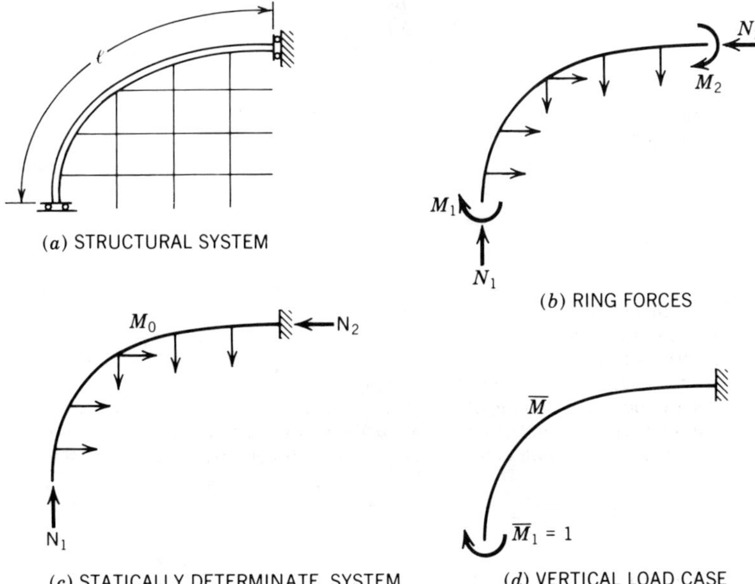

Fig. 15.39 Design of the ring for doubly symmetric saddle surface tension structure.

a. For the general case, since $\overline{M}$ = constant = 1

$$M_1 = \frac{\int \frac{M_0 \overline{M}}{EI} ds}{\int \frac{\overline{M}^2}{EI} ds} = \frac{\int \frac{M_0}{EI} ds}{\int \frac{1}{EI} ds} \tag{15.3.9}$$

b. For EI = constant

$$M_1 = \frac{\int M_0 \overline{M}\, ds}{\int ds} = \frac{\int M_0\, ds}{1} \tag{15.3.10}$$

c. For a polygon ring, with EI = constant

$$M_1 = \frac{\sum_0^n (M_{0,i} + M_{0,k}) S_{ik}}{2 \sum_0^n S_{ik}} \tag{15.3.11}$$

The final ring moments are

$$M_r = M_0 + M_1 \tag{15.3.12}$$

In case of elastic horizontal restraint by columns or prestressing cables the system becomes highly indeterminate and can no longer be analyzed by simplified methods.

15.4 ARCH SUPPORTED MEMBRANE STRUCTURES

15.4.1 Design Considerations

Arch supported membrane structures are another means of enclosing spaces that require free span, a low edge, and a high center. Since curved compression members are costly, the economy of the

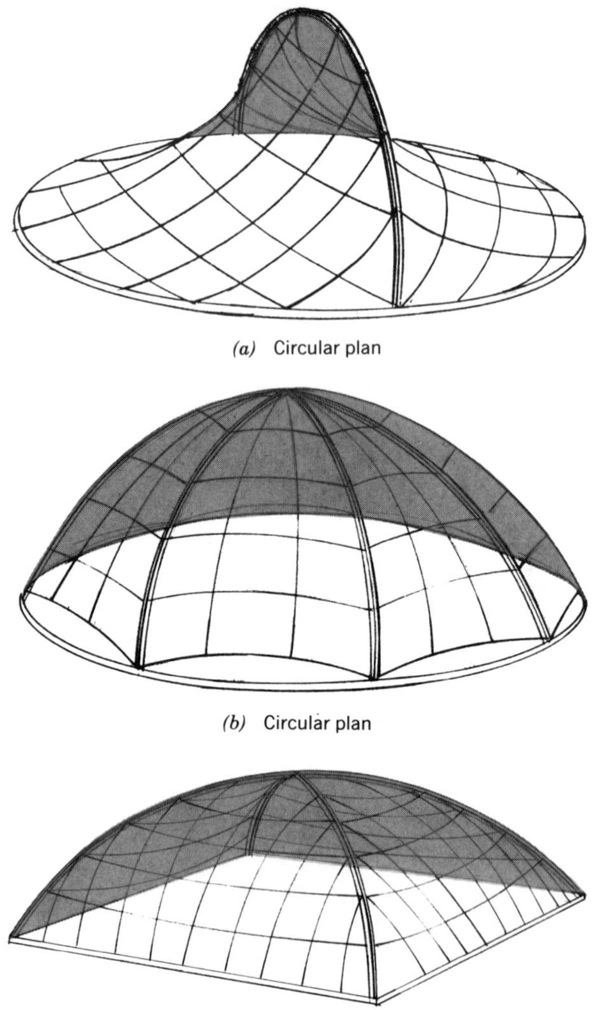

(a) Circular plan

(b) Circular plan

(c) Square or rectangular plan

Fig. 15.40 Geometrical arrangements for arch supported membrane structures.

system depends largely on the design and erection method of the arches. Configuration and spacing of the arches have to be selected in such a way as to minimize the cost of the membrane. Large membrane panels will require cable nets which can be utilized to stabilize the arch against buckling. Small membrane panels can consist of fabric only, in which case the arches must be stable by themselves. The fabric can be attached or ride free over the arches. The resulting lateral forces on the arches from unequal membrane loads or friction must be properly considered in the arch analysis.

A large number of geometrical arrangements is possible. Figure 15.40 shows typical arrangements for circular, square, and rectilinear plan configurations.

15.4.2 Loading

Arch supported membrane structures have characteristics similar to domes and vaults, both of which have been explored in great detail with regard to wind and snow loads. For purposes of preliminary design ANSI A58.1 [15.2] provides sufficient information. In locations with no snow load the superimposed load will be almost entirely suction, thus relieving the arches. The effect of

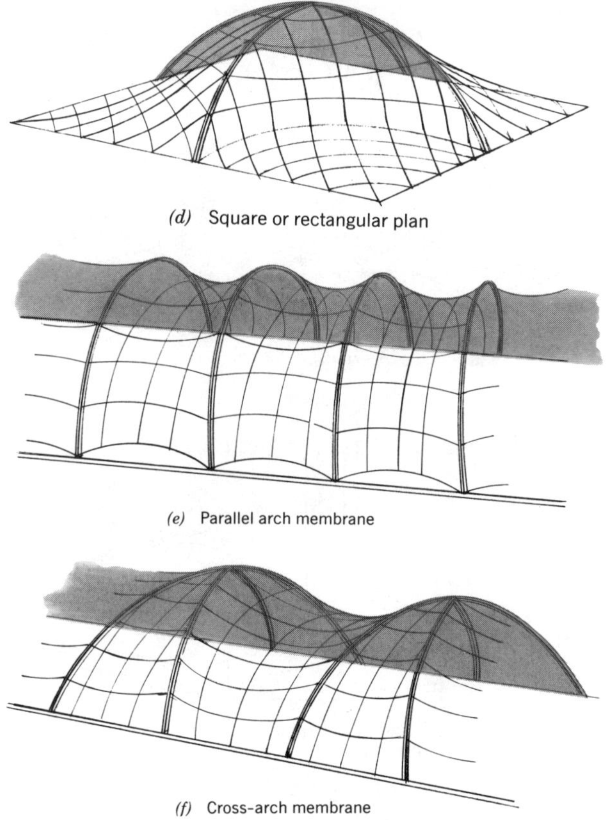

(d) Square or rectangular plan

(e) Parallel arch membrane

(f) Cross-arch membrane

Fig. 15.40 (continued).

uneven snow load on the arches will usually be greatly diminished by the distributing effect of the membrane action.

15.4.3 Membrane Design

The preliminary design of the membrane requires again two major steps: (a) the determination of the basic shape and (b) the checking of stresses produced by superimposed loads. In the example of Fig. 15.41 which shows a membrane panel between two parallel arches, the process is as follows:

1. Assumptions: There is no cable reinforcement in the membrane, and the fabric stresses under prestress are to be equal and constant in both directions.
2. The assumption of equal stress requires equal curvature in the two main directions expressed as $r_x = r_y$. Using parabolic curves for mathematical simplification and an internal prestress force p, the fabric tensions t_x and t_y are

$$t_x = \frac{p\ell_x^2}{8f_x} \qquad t_y = \frac{p\ell_y^2}{8f_y} \tag{15.4.1}$$

When $t_x = t_y$,

$$f_x = \left(\frac{\ell_x^2}{\ell_y^2}\right) f_y \tag{15.4.2}$$

The total rise of the arch is

$$f_a = f_x + f_y \tag{15.4.3}$$

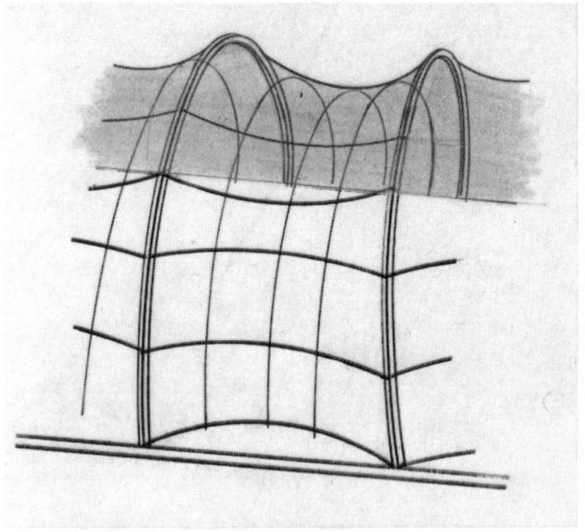

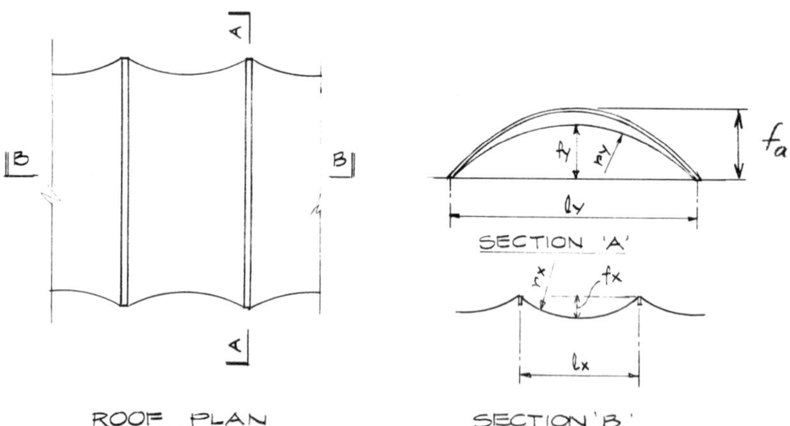

Fig. 15.41 Membrane panel between two parallel arches.

3. For unequal stress in the two directions of the fabric, or with cables in only one direction, the same equations can be used by introducing an appropriate multiplier. For instance, if $t_x = 2t_y$ then

$$f_x = 2\left(\frac{\ell_x^2}{\ell_y^2}\right) f_y \tag{15.4.4}$$

4. Determine maximum fabric stresses under superimposed loads. If under downward load w_d in the x-direction the fabric goes slack in the y-direction, the fabric stress is

$$t_x = \frac{w_d \ell_x^2}{8f_y} \tag{15.4.5}$$

Similarly if under imposed load w_u in the y-direction, the fabric goes slack in the x-direction, the fabric stress is

$$t_y = \frac{w_u \ell_y^2}{8f_y} \tag{15.4.6}$$

Allow a 15 to 20% margin against the allowable stress for the residual prestress from the other direction and for the influence of the vertical component.

15.4.4 Design of Edge Catenary

The lower edge of the fabric can be formed by an edge catenary or by clamping the fabric against a rigid edge member. The use of an edge catenary has the advantage of permitting the stressing of the fabric by jacks against the arches, thereby closing the stress system before it gets anchored against the supports. Stresses from the prestress load therefore remain virtually internal with negligible impact on the supports. A good method for closing a roof system is to prestress by means of an edge catenary and then clamping a cover flap against a rigid edge using a low, locally applied prestress in that area.

The maximum catenary cable force F_{max} can be determined by assuming the catenary to be in a plane and loaded with the full membrane stresses. The horizontal and vertical components F_h and F_v of F_{max} are as follows:

$$F_h = \frac{t_y \ell_x^2}{8f_c} \tag{15.4.7}$$

$$F_v = \frac{t_y \ell_x}{z} \tag{15.4.8}$$

and

$$F_{max} = \sqrt{F_h^2 + F_v^2} \tag{15.4.9}$$

In case of a cover flap, the catenary force will be smaller since the flap will take a percentage of the load directly to the edge which has to be designed for it.

15.4.5 Design of the Arches

In order to minimize lateral loads on the arches, the membranes on both sides must be designed properly. If the membrane is connected to the arch, the horizontal components of the two membranes normal to the plane of the arch should be equal under prestress (Fig. 15.42*a*). If the membrane is allowed to slide over the arch, the total membrane stress on both sides has to be equal (Fig. 15.42*b*). For purposes of fabrication the arches are preferably of circular shape. The loading of the arches is complex, especially in the case of cross-arches and end-arches. Preliminary design of the arches can be facilitated by establishing a number of simplified loading situations derived from the membrane stresses and analyzing the arches as three-hinged systems. If the membrane consists of a two-way cable system connected to the arches, the membrane can be used to stabilize the arches and reduce their bending stresses. Fabric membranes should not be relied upon for the stability of arches. In that case it might be economical to brace the arches against each other.

15.4.6 Support Loads

Membrane structures, because of their weight, are often governed by uplift loads. Arch supported membranes generally have uplift forces greater than the weight of the arches and the membrane.

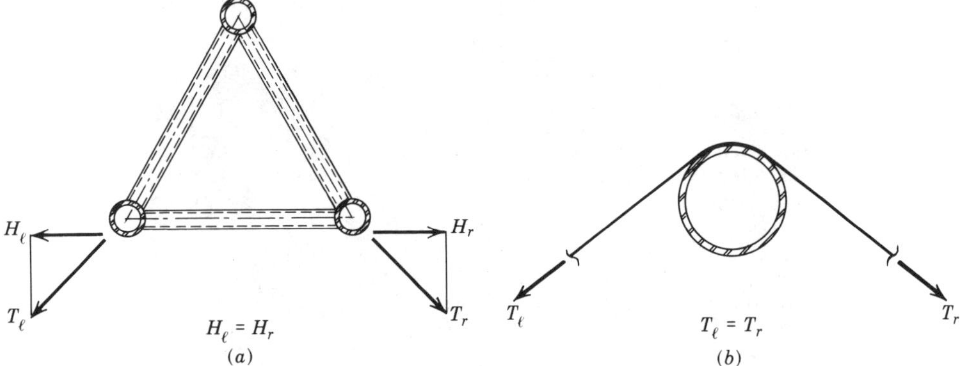

Fig. 15.42 Membrane attachment to arches. (*a*) Horizontal components of two membranes normal to plane of arch equal under prestress. (*b*) Equal total membrane stress on each side.

This results in inward thrusts of the arches. Tension ties are therefore not always practical. The arches have to be restrained at their supports against inward and outward forces in addition to the vertical reactions which might be directed up or down. If the system is closed above the support points, which means the internal stresses are balanced as in the case of edge catenaries, the reactions on the support system are those of the overall structure, as would result from any rigid dome or vault. Care must be exercised to avoid any unbalanced catenary loads on the support.

15.4.7 Final Design Considerations

The final shape should be determined by a computerized shape-determination program which begins with a given stress field and part of the geometry, and outputs the final shape in exact equilibrium. Details along arches and edges must be carefully worked out to avoid stress risers, and permit easy erection and stressing. Large roofs must come in sections for shipment, erection, and potential replacement. Arches can be placed on the outside or the inside. Membranes can have single or double skin. Figure 15.43 shows a double skin, parallel arch arrangement for the translucent roof at the University of Florida Student Center in Gainsville, Florida, and Fig. 15.44 shows a double skin, cross-arch arrangement for a translucent roof over a Bullock's department store in San Jose, California.

15.5 POINT SUPPORTED MEMBRANE STRUCTURES (TENT)

15.5.1 Design Considerations

Curved compression members are expensive to fabricate, ship, and erect. They need to be restrained against buckling. Tension members require no stiffness and therefore follow curved pro-

Fig. 15.43 Double skin, parallel arch arrangement for translucent roof, University of Florida, Gainesville, Florida. Courtesy of Geiger Associates, New York.

Fig. 15.44 Double skin, cross-arch arrangement for translucent roof, Bullock's Department Store, San Jose, California. Courtesy of Geiger Associates, New York.

files readily. Therefore, point supported structures are advantageous in many circumstances especially when the space covered does not need to be free of interior supports.

The simplest form of the point supported membrane structure is the radial tent (Fig. 15.45). The illustrated shape has a central mast and four corner anchor points. Radial cables carry the membrane stresses directly to the center support point. Along the periphery they are connected to edge catenaries which transmit the loads to the corner column points. The ring direction can be a uniform fabric membrane, a set of ring cables, or a combination.

A uniform two-way membrane without radial cables is not a good arrangement, since it results in excessively nonuniform stress patterns with a high stress concentration at the neck. Although under certain circumstances other arrangements might be more economical, the following discussion will be restricted to the radial pattern of point supported structures.

Point supported membrane structures of the radial tent type can take on a variety of shapes. Figure 15.46 shows a computer drawing of a single tent with rigid edges as built in 1974 for Great Adventure in New Jersey. The configuration shown in Fig. 15.47 represents a multiple tent mem-

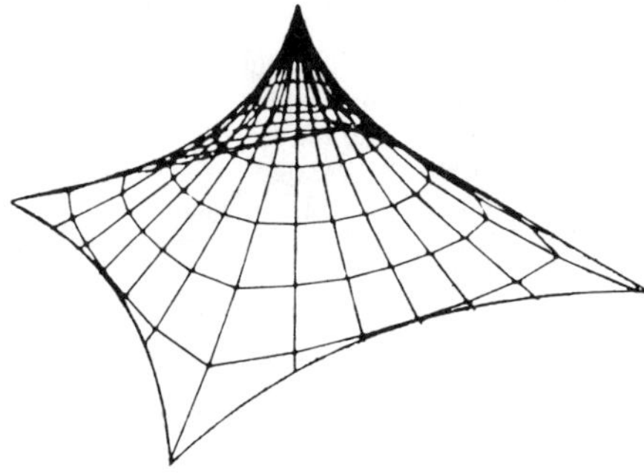

Fig. 15.45 Radial tent. Courtesy of Geiger Associates, New York.

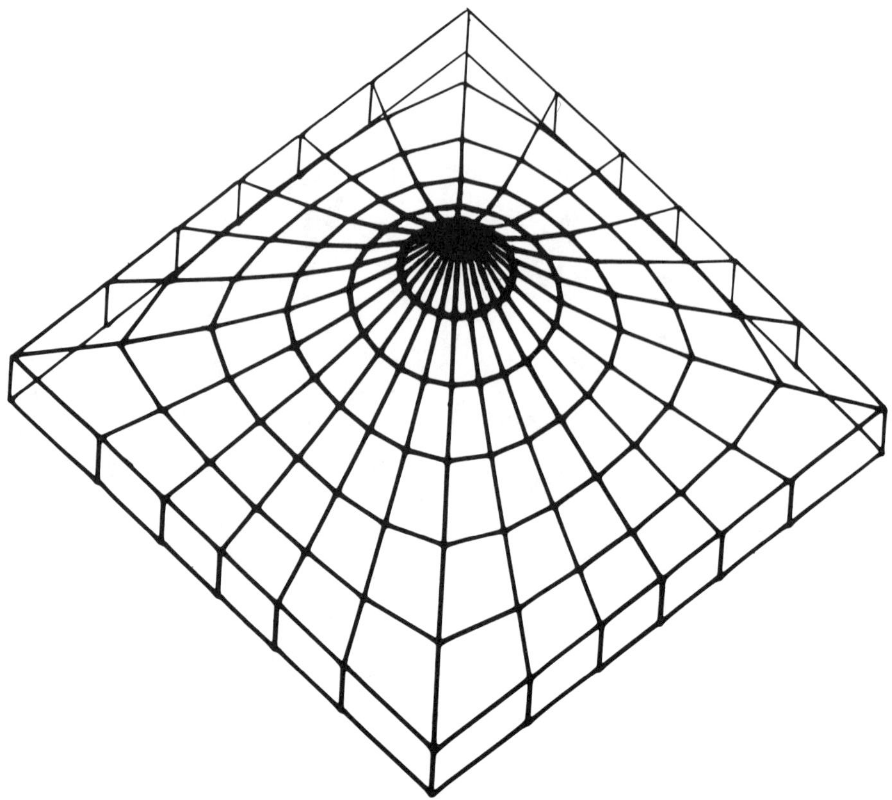

Fig. 15.46 Single tent with rigid edges, as built for Great Adventure Amusement Park in New Jersey. Courtesy of Geiger Associates, New York.

brane. A similar arrangement was used in the Haj Pavilion of the Jeddah International Airport (see Fig. 15.7). Reversed tent shapes are shown in Fig. 15.48. And finally a combination of tent shapes is shown in Fig. 15.49 delineating the membrane shape for the Florida Festival structure at Seaworld in Orlando Florida. Figures 15.50 and 15.51 show exterior and interior views of this structure, completed in 1979.

When mast supports under the tent points are not desirable, the interior mast can be replaced by an exterior A-frame. Figure 15.52 shows the Queeny Park structure in St. Louis County with a 100-ft (30-m) high main A-frame support. A-frames can be arranged in a trusslike structural system which interconnects all the support points in such a fashion as to eliminate or at least substantially reduce the impact of internal membrane stresses on the support system.

These "integrated frame membrane structures" ("inframembranes") are illustrated in Figs. 15.53 and 15.54.

Variations of the membrane shape permit structures that suit special purposes such as the cantilever roof design for the Interama Amphitheater (Fig. 15.55) and the 950-ft (285-m) diameter roof structure for the Riyadh International Stadium (Fig. 15.56). These illustrations give an idea of the variety of structural shapes which can be created using the radial tent concept.

In addition to stress flow, other important design considerations include fabrication, erection, and the introduction of prestress. Small curvatures, regular shapes, and repetition of patterns are considerations in reducing cost of fabrication. Avoidance of scaffolding and minimizing field splices and assemblies are essential to keep erection costs minimal. And finally, the capability of introducing the prestress by simple but accurate methods is absolutely essential. Being able to jack the peaks of the tent shapes with calibrated equipment is the basic approach.

Membrane areas not influenced by the jacking at the peaks (see Fig. 15.57) need careful consideration, and if necessary, a means of local adjustment.

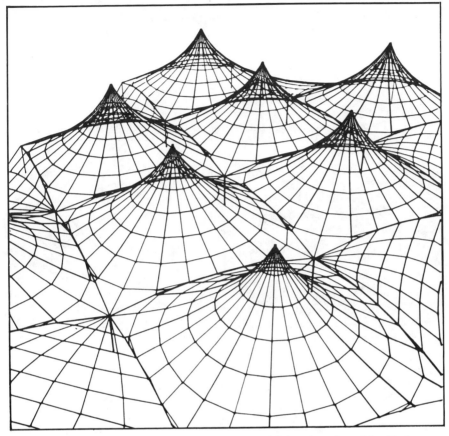

Fig. 15.47 Multiple tent membranes. Courtesy of Geiger Associates, New York.

15.5.2 Loading

Point supported membrane structures are new structural forms. Therefore, only limited information is found in the literature on wind and snow loads. Again, for preliminary design ANSI A58.1 [15.2] provides basic data. In the vicinity of a valley between two tent shapes, values derived from arch-shaped structures are useful for preliminary evaluation of wind loads. The cone-shaped portions of the tent experience inward pressures for the windward quadrant, and suction for the other three. The magnitude is somewhat smaller than what a vertical cylinder would experience. These

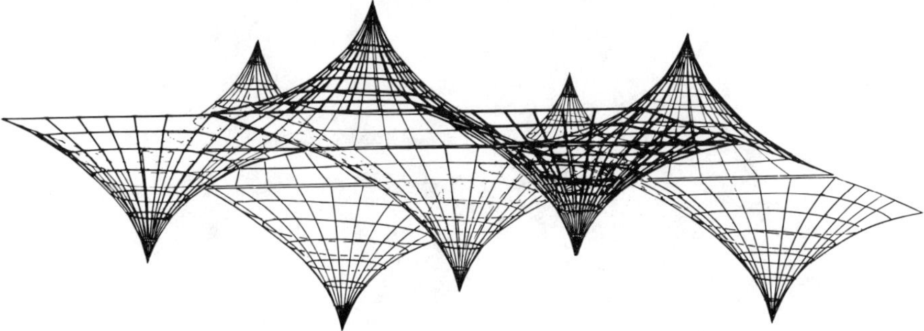

Fig. 15.48 Reversed tent shapes. Courtesy of Geiger Associates, New York.

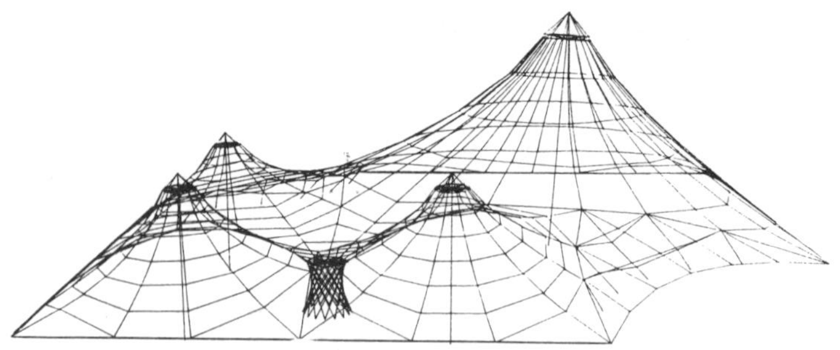

Fig. 15.49 Membrane shape (computer model) for the Florida Festival structure, Seaworld, Orlando, Florida. Courtesy of Geiger Associates, New York.

Fig. 15.50 Florida Festival structure, Seaworld, Orlando, Florida. Courtesy of Geiger Associates, New York.

Fig. 15.51 Interior view of Florida Festival structure, Seaworld, Orlando, Florida. Courtesy of Geiger Associates, New York.

Fig. 15.52 Exterior A-frame supports tent structure, Queeny Park, St. Louis County, Missouri. Courtesy of Geiger Associates, New York.

observations apply to closed structures. Open roofs (with no sidewalls or very large openings in sidewalls) are subject to additional internal loads which may change the picture substantially. For larger structures of this kind wind tunnel tests might be necessary to evaluate the loads properly. In general the cost of the wind tunnel tests pays off for these structures since in many cases the load intensities found are considerably lower than anticipated.

Snow will again tend to accumulate in the valleys and low points, which have to be able to carry the additional load without ponding.

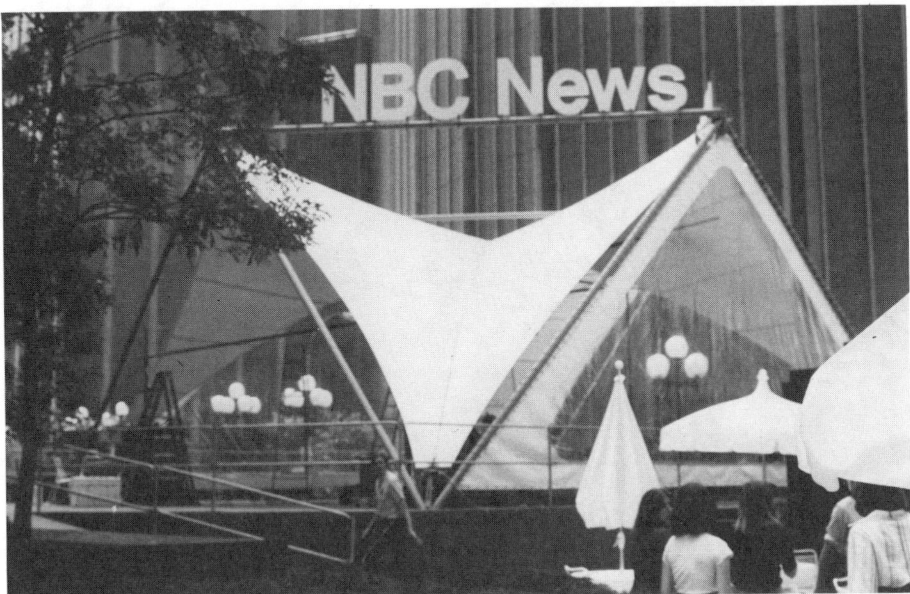

Fig. 15.53 Integrated frame membrane structure. Courtesy of Geiger Associates, New York.

Fig. 15.54 Integrated frame membrane structure. Courtesy of Geiger Associates, New York.

15.5.3 Shape Generation

Choice of support points, edge members, and membrane surface construction determine the configuration of point supported membrane structures. Finding the proper shape is the most important step in the design process. Since the resulting structural shape is also the architectural shape, the design process must include functional and aesthetic space considerations.

Because of the three-dimensional nature of membrane tension structures, drawings are generally not sufficient as a means of exploring the form of the system. An excellent aid is a study model using stretch fabric to simulate the membrane. With some experience such models can be very helpful in the preliminary engineering design, yielding measurable curvatures, slopes, and dimensions which are difficult to develop analytically. In combination with sketches and simple computations, stretch fabric models can be an excellent design tool, having the additional advantage of doubling as visual models for presentation of the design to a client.

For regular tent shapes analytical methods can be used to approximate the shape of the membrane. For the tent shape of Fig. 15.58 the following information is given:

1. Location of radial cables in plan
2. Location of ring cables in elevation

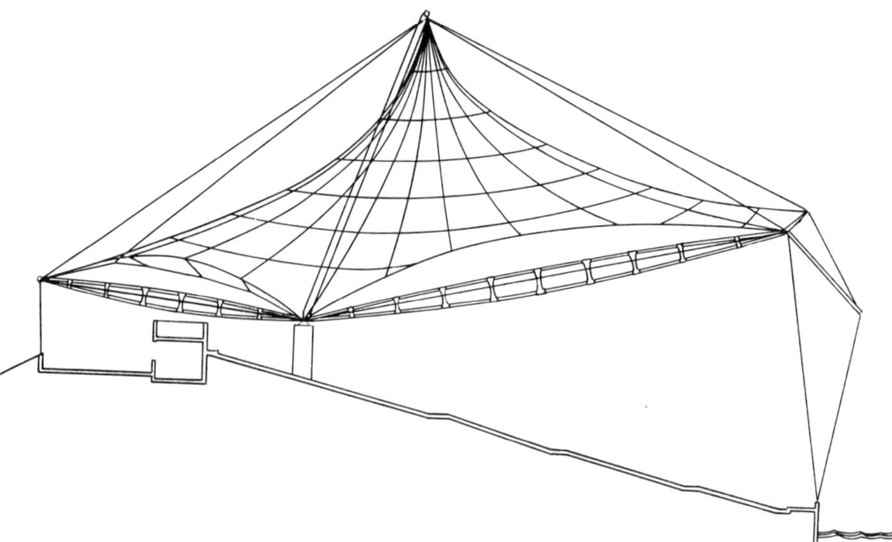

Fig. 15.55 Cantilever roof design for Interama Amphitheater. Courtesy of Geiger Associates, New York.

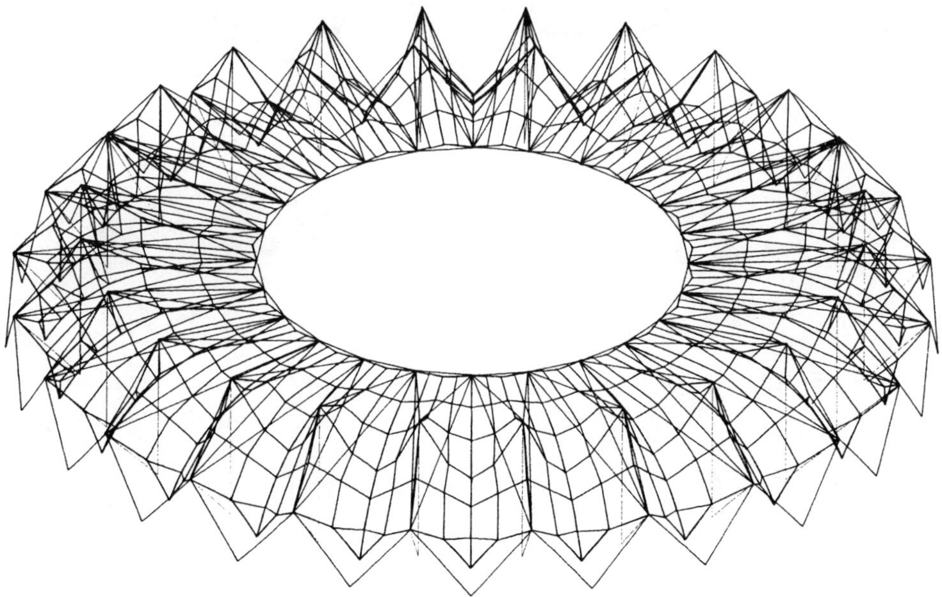

Fig. 15.56 Roof for Riyadh International Stadium, Saudi Arabia. Courtesy of Geiger Associates, New York.

3. Vertical component of radial cable force (prestress stage)
4. Horizontal component of ring cable force (prestress stage)

Assuming for the first approximation that the rings are all circular polygons, a set of horizontal inward prestress forces p acting on the radial cables is established.

With given loads p any radial cable shape can be quickly determined following the process indicated in Fig. 15.58. The resulting values for the distances x of the intersection points between rings and radials result in ring shapes that can be used to determine a corrected loading p on the radials following the process in Fig. 15.58. This method can very quickly lead to a good approximation of membrane shapes.

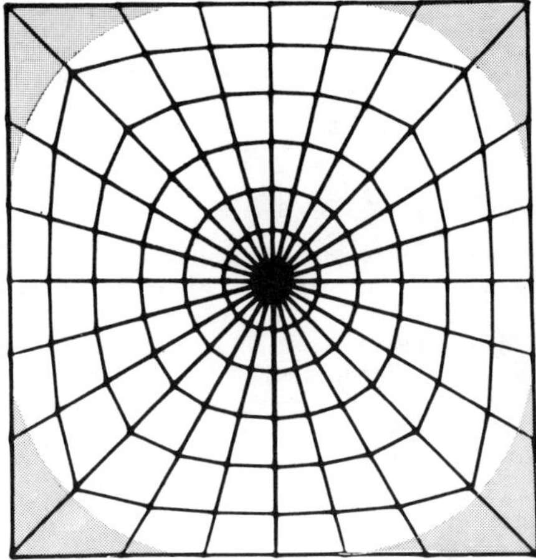

Fig. 15.57 Membrane areas not influenced by jacking.

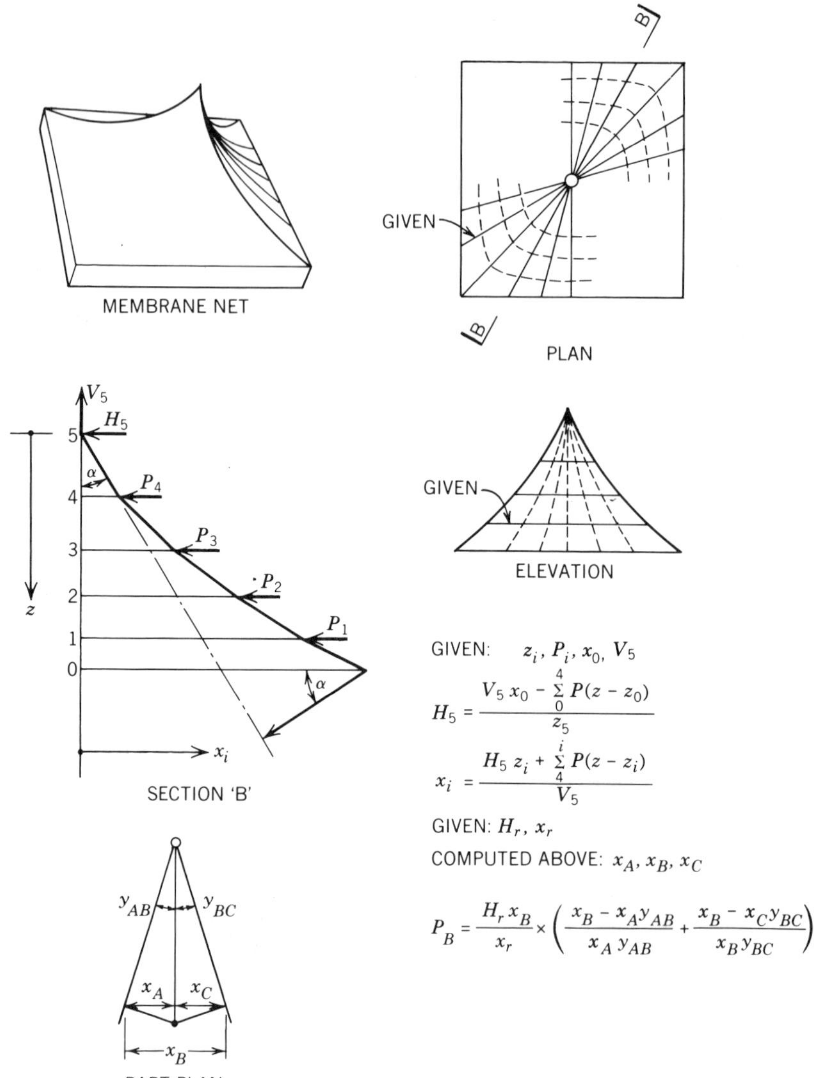

Fig. 15.58 Analytical data for regular tent-shaped membrane structure.

15.5.4 Stress Due to Superimposed Loads

The evaluation of these stresses requires some imagination to avoid complex nonlinear computer analysis for preliminary purposes. A few suggestions will help to find useful values.

The force on the radial cables at the top ring, the load on the ring, and the load of the mast can be found by assuming the maximum downward load to occur uniformly around the entire tent shape and using the resulting reactions allowing for a small portion of the prestress to remain. Similarly the radial cable force at the lower edge, or the load on the edge beam or edge catenary, can be found by taking the maximum uplift uniformly, adding a small fraction of the mast prestress force, and using the resulting reactions.

The radial cables can be checked, assuming that they carry the direct inward load of the related membrane strip with the sag found for the prestress shape. A simple universal method of checking stresses in the ring direction is not available. However, reversing the process used for shape generation by loading the rings with the radials might give an idea of the order of magnitude of the ring stresses. Because of the beneficial effect of redistribution of stresses produced by change of shape, approximate methods for membrane design will generally be conservative.

The final analysis of these structures must use mathematical models that reflect the real behavior of the materials and the system with sufficient accuracy. Unless the methods used have been tried and carefully checked the results cannot be assumed to be accurate enough for safe construction of permanent large span membrane structures.

SELECTED REFERENCES

15.1 AISI. *Manual for Structural Applications of Steel Cables for Buildings*. New York: American and Iron Steel Institute, 1973.

15.2 ANSI. *Building Code Requirements for Minimum Design Loads in Buildings and Other Structures* (ANSI A58.1). Washington, D.C.: American National Standards Institute, 1982.

15.3 Karl I. Beitin. "Alternative Solar Systems," 2nd Passive Solar Conference, Philadelphia, PA, March 1978.

15.4 Horst Berger. "The Engineering Discipline of Tent Structures," *Architectural Record,* February 1975, 81–88.

15.5 Horst Berger. "Shaping Cable Supported Lightweight Tensile Structures," *Bulletin of the International Association for Shell and Spatial Structures,* **XVII-3,** 62, 1977.

15.6 Walter W. Bird and Ronald A. Kinnius. *The History of the Air Structures in the USA*. Buffalo, NY: Birdair Structures, Inc., 1976.

15.7 Philip Drew. *Tensile Architecture*. London: Granada Publishing Ltd., 1979.

15.8 David H. Geiger. "Pneumatic Structures," *Progressive Architecture,* August 1972.

15.9 David H. Geiger. "Development in Incombustible Fabrics and Low Profile Air Structures Including Those with Thermally Active Roofs," *Bulletin of the International Association of Shell and Spatial Structures,* **XVIII-2,** 64, August 1977.

15.10 David H. Geiger. "U.S. Pavilion at Expo '70 Features Air Supported Cable Roof," *Civil Engineering,* March 1970, 48.

15.11 E. M. Malton. *The Tent Book*. Boston: Houghton Mifflin, 1979.

15.12 Frei Otto. *Tensile Structures,* Vols. 1 and 2. Cambridge, MA: MIT Press, 1973.

CHAPTER 16

FOUNDATIONS—PRELIMINARY DESIGN

MELVIN I. ESRIG

Vice-President
Woodward-Clyde Consultants
Wayne, New Jersey

16.1 INTRODUCTION

The information provided in this chapter is directed at aiding the structural engineer in selecting, for purposes of preliminary design, a foundation system appropriate for a particular structure. The chapter is organized to first provide an understanding of the many factors affecting the choice of a foundation system. This is followed by a description of the foundation systems available to choose among, and the subsurface conditions for which each is appropriate. A protocol is presented to aid the engineer in the preliminary selection of a foundation system. It is followed by a presentation of the information necessary for preliminary design of the system selected.

The selection of a foundation system for a particular structure is the culmination of a decision process that requires compromises to be made among conflicting and often interdependent needs. Ordinarily, the goal is to select a low-cost system that will perform the required function without excessive risk to the owner. Therefore, consideration is given in the selection process to the following:

1. The predicted performance of the foundation system for the loads to be imposed by the structure, considering the stiffness of the structure and its tolerance for displacement.
2. The likely effects of construction of the foundation system on adjacent or nearby structures or facilities.
3. The cost of construction of the foundation system, including the cost associated with the time necessary for its construction.
4. The level of risk that is considered acceptable by the owner, the community, and the design team.

Many of the significant decisions in the selection process are based on qualitative, rather than quantitative, information; judgments must be made about risk, about the effects of the rigidity of the structure on predicted displacements, and about some of the effects on nearby structures of the construction of the foundation system. Calculations of the total and differential settlement of structures supported by different foundation systems and bearing on different foundation materials can be made. However, judgment and experience must be used to assess the reliability of the calculated values and to assess the ability of the structure to tolerate the predicted displacements. Thus, as described in the sections that follow, the selection of a suitable foundation system for a particular structure in a particular location requires the application of engineering science, tempered by judgment and an understanding of the limitations of available engineering "tools."

16.2 FACTORS AFFECTING SELECTION OF FOUNDATION SYSTEM

16.2.1 Selection Affected by Prediction of Foundation Performance

Foundations must be safe against bearing capacity failure (failure in shear) of the supporting materials, and they must not settle during the lifetime of the structure more than some acceptable amount. In addition, they must not adversely affect existing nearby structures or structures that may legally be constructed nearby in the future.

In order to predict the safety of a foundation system, it is necessary to estimate the loads that will be applied to the foundations and to understand the behavior of the materials beneath the foundation system when subjected to these loads. Dead loads can be calculated and distributed to the foundation with a reasonably high degree of accuracy. For most structures (other than tanks, silos, and the like), live loads and their distribution are not known as well as dead loads, but are probably better defined than extreme loadings from winds, waves, and earthquakes.

The properties of small samples of foundation materials can be measured in the laboratory or in the field. The repeatability of these measurements and their accuracy in reflecting the properties needed for design are related to the natural variability of the materials, the ability of the engineer to obtain even relatively undisturbed soil or rock specimens or, for that matter, to recover any samples of the material that controls the behavior of a particular foundation system, and the degree to which the laboratory test reproduces the stress system in the ground. Heterogeneity far exceeds homogeneity in natural soil deposits. The belief that some of the parameters obtained from laboratory or field tests (e.g., the undrained shear strength of clays) are appropriate for use in analyses to predict the performance of foundations is based on empirical evidence obtained from case studies in which post hoc predictions of foundation behavior are compared with the behavior observed. Efforts to predict foundation behavior prior to making detailed observations of actual behavior have not proven to be very successful despite the eminence of the predictors. Therefore, it is evident that there is significant uncertainty associated with the prediction of foundation behavior and that the uncertainty arises from uncertainty in the loads imposed on the foundation, in the material properties needed for engineering analyses, and in the applicability of the analytic tools available.

In order to reduce (but not eliminate) the need to predict foundation behavior with accuracy and to still provide "safe" foundations, it is common practice for foundation systems to be required to provide at least twice the bearing capacity that is required to support the *maximum possible loads,* and three times the capacity required to support *maximum probable loads* (i.e., safety factors of 2 and 3, respectively). Simultaneous with the requirement for adequate bearing capacity is the requirement for safe performance when settlements occur. Settlement predictions for foundations bearing on clay soils, which are expected to settle (consolidate) over a period of years, are commonly based on the *average probable loads* to be applied to the structure. Settlement predictions for foundations bearing on sand, which settles rapidly on application of loads, are based on the *maximum probable loads*.

It is of more than passing interest to note that a disproportionately large number of foundation failures have occurred for structures for which the loads are well known, such as tanks and silos, or where dead load is 80 or 90% of total load. This might be interpreted to suggest that the margin for error in the assessment of the safety of a structure is significant and that the commonly recommended safety factors are not excessively conservative.

For most buildings, the bearing capacity of the soils supporting shallow foundations is normally of controlling concern only for a limited range of materials. These materials are generally clays or weak rocks that are hard, brittle, and frequently fissured either macroscopically or microscopically. For the remaining materials, soft to stiff clays, sands of all densities, and the many intermediate materials occurring in nature, protection against excessive settlement usually controls the design of the foundation system.

It is increasingly common for engineers assessing the stability of a foundation system to consider the accuracy of each component of load and of resistance and to apply to those components partial safety factors related to the uncertainty of their estimates. For example, gravity always acts; therefore, an appropriate partial safety factor for dead load should be close to or equal to unity. Significant uncertainty with live load computations for buildings or the need to design for live loads mandated by building codes or regulatory agencies may suggest use of a partial safety factor of 1.5, while for grain silos or water tanks, a factor of 1.2 may be more appropriate. The use of partial safety factors applied to loads will produce about the same overall safety factors as suggested earlier (2 or 3) if the partial safety factor applied to material strength (usually the unconfined compressive strength) for calculation of bearing capacity of a foundation on clay is assumed as 2.0, or if the partial safety factor for calculation of bearing capacity of a foundation on granular soil is applied to the coefficient of sliding friction, tan ϕ', and is taken as 1.2.

Interestingly, the partial safety factor approach to prediction of the safety of foundations can be shown to be a special case of a more general probabilistic approach to foundation design. The probabilistic approach is not yet in wide use, in part because of philosophical difficulties associated with the statement that every foundation has some finite probability of failure, and also in recognition of the difficulty engineers have with selecting an acceptable value of a probability of failure for use in design.

Significant advances have been made in recent years in the ability of engineers to predict the deformations of soil-structure systems under working stresses and under stresses approaching failure. Predictions of the deformations of structures due to soil consolidation (settlement) are routinely made, while predictions of the lateral deformations of the sides of excavations or of retaining structures are only made when there is an unusual circumstance requiring them. The former predictions are relatively inexpensive, straightforward, and needed for design, while the latter are expensive, difficult, and only contribute to design under unusual circumstances mainly associated with new construction in well-established urban areas.

The statement that predictions of the settlement of structures are straightforward and routinely made must be qualified. The prediction methods commonly in use tend to overpredict, by a factor of as much as 2, settlements of structures on hard clay and dense sand and to underpredict, also by a factor that may be as large as 2, the settlement of structures or facilities on weak, compressible foundation materials. The causes of this level of inaccuracy in the prediction of the settlement of foundations are not well understood. They are believed to be the result of inaccuracy in the average loads estimated as applied by the building columns to the foundations, inaccuracy in the distribution of stress through the soil strata supporting the foundations, as well as inaccuracy in the selection of the material properties appropriate for use in the analyses. Therefore, there is an obvious need to temper calculated settlements with experience and there is an obvious difficulty in assessing, in many cases, whether or not the structure is likely to undergo displacements that will be considered tolerable.

16.2.2 Selection Affected by Effects of Foundation Installation on Adjacent or Nearby Structures

The selection of a foundation system for a structure in an urban area is often dictated by the need to avoid the potential adverse effects on existing structures of installation of the system. Those effects might include inducing movements of existing structures (a) in response to adjacent or nearby excavations, (b) in response to installation of a pile foundation or retaining structure, or (c) in response to dewatering. A summary of some of the effects of foundation construction on nearby structures is provided in Table 16.1, together with a summary of possible avoidance measures. Each of the potential causes of movements is discussed in greater detail below and, where possible, the likely extent of these movements is indicated. A more extensive discussion is given by D'Appolonia [16.5].

It should be recognized that, on occasion, it may not be appropriate for a contractor or owner to take measures designed to prevent the movement of adjacent or nearby structures. For example, when movements are likely to be small and unlikely to endanger the structural integrity of a building, it may be more reasonable and economical to permit the movements induced by the new construction to occur and, once construction is completed, to repair the damage, than to prevent the movements in the first place.

16.2.3 Movements in Response to Excavations

When an excavation is made, ground movements are inevitable. In general, heave of the bottom of the excavation will occur in response to unloading; settlement of the ground surface will occur adjacent to the excavations, and the cut face moves inward toward the center of the excavation, sometimes causing stretching of the ground adjacent to its perimeter. The shape of these movements is shown schematically in Fig. 16.1.

Heave

Figure 16.2*a* is based on observations of heave within the boundaries of excavations reported in the literature and obtained from engineering practitioners. The empirical data suggest that the maximum heave in the center of a large excavation underlain by clay is approximately 0.06 in./ft (5 mm/m) of depth of excavation. Thus, the center of an excavation about 33 ft (10 m) in depth can be expected to heave about 2 in. (50 mm). However, in order to heave upon excavation, there must be a sufficient thickness of soil beneath the bottom of the excavation to expand upward in response to the unloading. The same data as in Fig. 16.2*a* are replotted in Fig. 16.2*b* where heave, as a percentage of the thickness of the compressible soil layer beneath the excavation, is plotted against depth of excavation. This curve suggests that heave of the center of an excavation will be about 0.2% of the thickness of the compressible layer beneath the bottom of the excavation. Thus, an excavation to a depth of 33 ft (10 m), underlain by about 50 ft (15 m) of compressible soil, can be expected to heave about 1.2 in. (30 mm). The use of the empirical relations in Figs. 16.2*a* and 16.2*b* are believed to provide a reasonable guide to the amount of heave to be expected in an excavation. Conflicting predictions of heave obtained when these relations are used, as in the example given, must be resolved by judgment or supplemental computations.

The heave resulting from an excavation is believed associated with the unloading of soil weight, rather than the result of deformations of the retaining structure around the rim of the excavation. Therefore, this heave is believed to have no effect on adjacent structures. However, the building to be constructed in the excavation can be expected to undergo vertical settlement that is at least as much as the heave resulting from the unloading. This settlement (reconsolidation of the soil) normally occurs early in the period of construction, when the weight of the soil removed from the excavation has been equaled by the weight of the structure, and before architectural finishes are applied within the structure. Therefore, it is ordinarily not noticed nor is it usually significant for the structure.

Settlement of Ground Surface

Settlement of the ground surface adjacent to and nearby an excavation is related to the depth of the excavation, the care with which the internal bracing is constructed, the type of soil through which the excavation is dug, and the control exercised over groundwater levels during and after construction. Groundwater is discussed separately and is assumed, for purposes of the discussion in this section, to be under complete control at all times.

Peck [16.17] prepared a figure, reproduced as Fig. 16.3, suggesting, for different materials and average workmanship, the approximate distribution of ground settlement (as a percent of depth of excavation) with distance from the face of the excavation (also expressed as a percent of the depth

Table 16.1 Summary of Effects of Foundation Construction on Nearby Structures

Cause	Effect	Avoidance Measures
Excavations		
Granular soils (groundwater assumed under control)	a. Ground settlement potentially affecting structures within a distance equal to twice the depth of excavation b. Lateral displacement (stretching) potentially affecting structures within a distance equal to depth of excavation	a. Some displacement inevitable. Reduce displacements by better-than-average workmanship b. Avoid loss of ground through opening in sheeting system
Soft to hard clay (compressive strength q_u greater than about 40 kN/m^2 (0.8 ksf))	Same as granular soils above	a. Same as granular soils above b. Special attention needed for deep excavations ($D_m > 0.1q_u$ kN/m^2 or $D_{ft} > 15q_u$ ksf)
Very soft to soft clay ($q_u < 40$ kN/m^2 (0.8 ksf))	a. Ground settlement potentially affecting structures within a distance equal to four or more times the depth of excavation b. Lateral displacement (stretching) potentially affecting structures within a distance equal to the depth of excavation c. Potential failure suspected when depth of excavation D in meters exceeds about 10% of compressive strength q_u in kN/m^2, or D in ft exceeds $15q_u$ in ksf	a. Special attention to deep excavations ($D_m > 0.1q_u$ kN/m^2 or $D_{ft} > 15q_u$ ksf) b. Require better-than-average workmanship c. Install braces early and use berms for stability d. Consider appropriateness of requiring "stiff" retaining walls, such as concrete wall constructed in slurry trench and braced as excavation proceeds
Deep Foundation Installation		
Pile Driving		
Granular soils Mainly for loose to very loose deposits	Vibration-induced settlement (vibratory pile drivers affect larger areas than impact pile hammers)	Reduce pile hammer energies Use nondisplacement pile system in loose to very loose granular deposits
Clay and silt Potentially a problem when large pile groups and many piles are to be driven	Ground heave which can lift piles off bearing stratum or fail concrete piles in tension Lateral displacement of adjacent piles, bulkheads, and retaining walls Development of large excess pore water pressures that can lead to failure of adjacent slopes or excavations	Use low displacement piles (H piles) Preauger for portion of pile length before inserting pile Prescribe pile driving sequence

Bored or Augered Piles		
Granular soils	Loss of ground into augered hole causing settlement of adjacent structure Inadequate grout injection pressure and grout volume as auger is removed causing "necking" of pile section	Construction control by knowledgeable personnel essential to success
Soft clay or peat	Loss of ground resulting from unstable sides of prebored hole causing pile necking Excessive grout injection pressure causing hydrofracture of ground and tension failure of adjacent piles	Use cased hole leaving casing in ground rather than removing casing as concrete is placed Monitor injection pressure, ground displacements, and displacements of completed piles near injection point
Stiff to hard clay	Ground heave due to excessive grout injection pressure	Monitor injection pressure, ground displacements, and displacements of completed piles near injection point
Dewatering		
Granular soils	Fine sand and/or silt can be drawn from the ground into the dewatering system causing ground subsidence Inadequate drawdown produces boiling of subgrade and loss of soil-bearing capacity	Use properly designed filters around dewatering wells Deepen well system or change system to increase drawdown
Silt	Inadequate dewatering by conventional systems is common, leading to failure of excavations	Consult experts for advice and for recommendations on nonconventional dewatering systems or cutoff walls
Clay or organic soils	Dewatering *above* clay or organic soil increases effective stresses and, therefore, settlement of nearby structures	Limit the extent of drawdown by installation of cutoff walls. Use of recharge systems not always effective

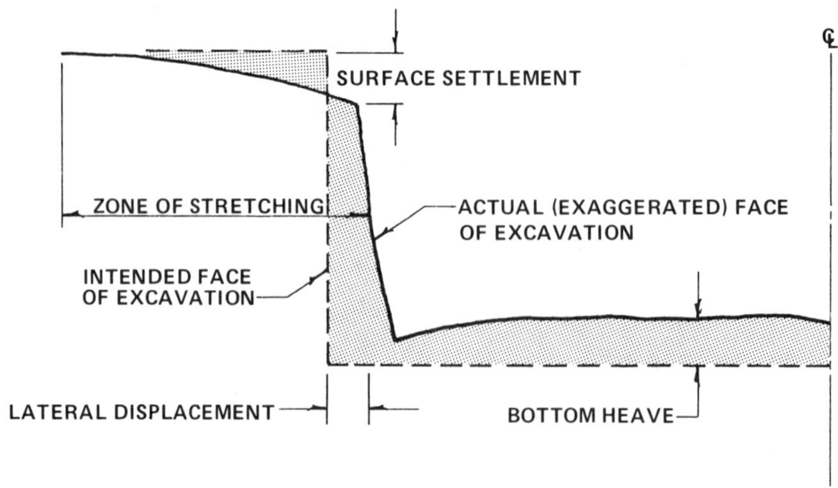

Fig. 16.1 Exaggerated schematic of soil movement due to braced excavation.

of excavation). The figure is appropriate for excavations that are internally braced or tied back and whose exterior walls are either of steel sheet piles or soldier piles with timber lagging. Several investigators have verified independently the appropriateness of the relations shown in this figure subsequent to its first publication.

The relationships of Fig. 16.3 indicate that the ground surface will settle least when the soils are similar to those in Zone I and are granular (sand, gravel, etc.) or soft to hard clays [clays in Zone I exhibit unconfined compressive strengths in excess of about 1000 psf (50 kN/m^2)]. The behavior shown in Zones II and III results when excavations are made through soft to very soft clays, and where the depth of the excavation is such that significant plastic flow begins to occur. Zone II is

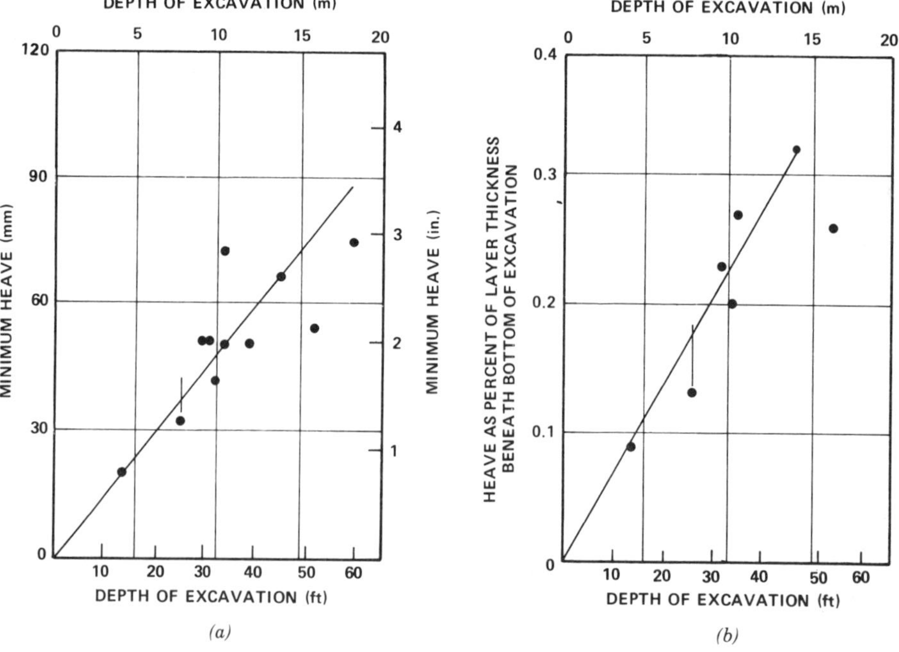

Fig. 16.2 Heave of the bottom of excavations in clay.

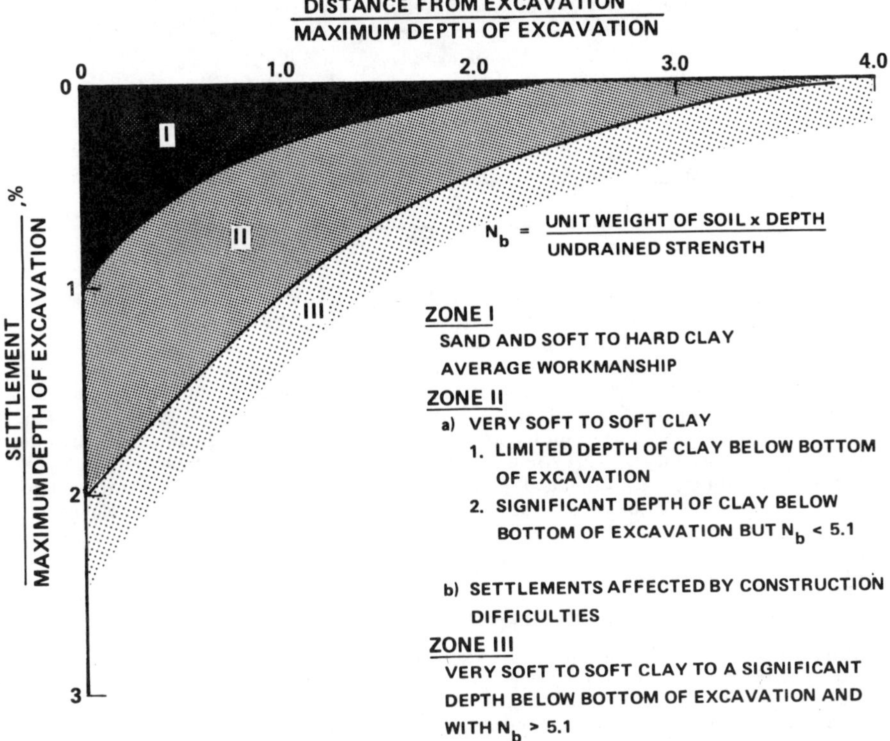

Fig. 16.3 Distribution of ground settlement adjacent to excavations (from Peck [16.17]).

also characteristic of conditions in which construction difficulties (such as failure of the braced excavation) have caused increases in building settlement.

The movements associated with Zone I are rarely a problem. Those associated with Zones II and III have, in the extreme, resulted in adjacent buildings falling into the excavation. Potential problems with failure of the excavation and the associated adverse effect on adjacent structures should be suspected when the depth D of the excavation in ft exceeds about 1.5% of the compressive strength of the soil expressed in psf (or the depth D in m exceeds about 10% of the compressive strength of the soil expressed in kN/m^2).

Lateral Movement of Retaining Walls

Retaining wall design is discussed in Section 7.9. Experience has shown that these earth retaining structures can be expected to deflect inward at least the minimum amount necessary to fully mobilize the active pressure against the wall. The inward movement will "stretch" the ground adjacent to the wall some small amount. Sometimes this stretching is enough to damage utilities or structures within the affected zone. A discussion of the tolerable settlement of a structure adjacent to an excavation is included in Section 16.3. Of interest for this discussion is the magnitude of the stretching and the distance from the excavation that might be expected to be affected.

There are few "hard" data available from which to define the lateral movement of the face of a retaining structure and the zone of stretching. It can be safely assumed that the *minimum* deflection of a retaining wall subjected to average construction practices will be about equal to that necessary to develop active earth pressures on the wall. The average deflections necessary to develop active earth pressures are given in Table 16.2 (assuming a linear distribution of deflection from top to bottom of wall, these are the displacements at its midheight).

Thus, the deflection of the midheight of the face of a retaining structure surrounding an excavation 10 ft (3 m) deep, subjected to average workmanship, can be expected to be about 0.06 in. (1.5 mm) in dense sand, 0.12 in. (3 mm) in loose sand, 0.6 in. (15 mm) in stiff clay, and 1.2 in. (30 mm) in soft clay. The distribution of displacement along the wall is related to the type of bracing system used and the details of construction of the retaining structure. A typical retaining wall that is unbraced at the top can be expected to deflect about twice these amounts at the top and zero at the

Table 16.2 Deflections to Develop Active Pressure

Material	Approximate Displacement of Midheight of Retaining Wall of Height H
Dense sand	$0.0005H$
Loose sand	$0.001H$
Stiff clay	$0.005H$
Soft clay	$0.01H$

base. A typical braced excavation with a brace provided near the ground surface *before* the excavation is made can be expected to exhibit near zero displacement at the top and about twice the average displacement indicated above at the base.

Observations of ground movements adjacent to braced excavations that have moved laterally, but for which groundwater was continually under control and, therefore, loss of ground did not have a significant effect on the ground movements, suggest that the zone of ground stretching is limited to a distance from the face of the excavation less than about the depth of the excavation. Earth pressure theory suggests that stretching should only occur within a lateral distance less than 70% of the depth of the excavation and, perhaps, as small as 20% of the depth of excavation.

The displacements described above show clearly that excessive lateral movement (elongation) of a structure immediately adjacent to a deep excavation is possible unless substantial care is taken to brace and, perhaps, to prestress the retaining system *as it is constructed.*

16.2.4 Movements in Response to Deep Foundation Installation

The installation of a deep foundation by driving piles or by drilling shafts or piers can cause distress in nearby structures as a result of (1) the vibrations associated with pile driving, (2) the ground heave or settlement associated with the installation of piles that displace soil as they penetrate the ground, and the loss of ground that may be associated with the installation of drilled shafts or piers, and (3) the buildup of water pressure in the pores of the soil which accompanies the installation of displacement piles in fine-grained soils (clay or silt).

Densification and Ground Settlement Due to Vibrations

Vibrations associated with driving piles into deposits of sand cause the sand to become denser and result in settlements of the ground surface. No similar effects occur when piles are driven into cohesive soils. Measurements suggest that increases in density of sands as a result of driving piles with *impact* hammers normally occur within a distance of a few pile diameters of the pile shaft and pile top, except when the sand is loose to very loose. Loose sands have been observed to densify to distances of 60 ft (20 m) from the driven pile.

Observations of surface settlement of sand adjacent to H-piles and sheet piles driven with a *vibratory* hammer suggest that the sustained vibrations produced by the vibratory hammer cause densification to a substantially greater distance from the pile than the impact hammer. As a tentative conclusion that could change substantially as more becomes known, it appears that the vibratory pile driver causes significant compaction of sand in all areas where the ground acceleration exceeds about 0.05 to 0.1 g. Thus, damage to structures adjacent to areas in which piles are to be driven into sand strata is possible and this damage must be guarded against. Ground acceleration due to pile driving can be predicted with some uncertainty and can be measured with ease either in a test section or at the beginning of pile driving operation. Damage due to densification of sands beneath a structure caused by vibrations must be considered independently from damage due to the response of the structure to vibrations.

Further consideration of the effects of the response of structures to vibrations from all sources is included in Section 16.3.7.

Ground Heave, Lateral Displacement, and Settlement

The fact that driving piles into the ground can produce heave and lateral displacement in areas adjacent to the piles is well documented. Heave and lateral displacement are of concern when many piles are driven through or into clay or silt. When driving piles into saturated clays and some silts, the total volume of soil that is heaved vertically, or moved laterally, for example, in a waterfront area where lateral movement can readily occur, is approximately equal to the volume of pile material inserted into the ground. Therefore, when large numbers of piles that displace significant volumes of soil (such as pipe piles) are being driven into saturated clay adjacent to an existing structure, the possibility that damage may occur should be examined. When large numbers of piles are being driven, movement of structures has been observed at distances in excess of 150 ft (50 m). Within 30 ft (10 m) of the edge of pile groups installed in soft to medium clays in Chicago, Illinois, heave of 0.1–0.2 in. (2–4 mm) was reported by Ireland [16.10]. The heave was followed by net settlements of 0.2–0.4 in. (4–8 mm).

Lateral displacements of the ground surface due to piles driven through clay for the Tokyo Telephone Exchange were reported as 0.2–0.4 in. (5–10 mm) within 40 ft (12 m) of the area of pile driving and were accompanied by heave of the ground surface of 0.1–0.4 in. (3–9 mm). These displacements, which not only affected adjacent structures but also moved piles driven for the building's foundation system, caused the change of the foundation system from driven displacement piles to drilled (nondisplacement) piers. Alternatively, the magnitude of the lateral displacement and heave could have been reduced, but probably not eliminated, by preaugering for part of the length of each pile before driving it and accepting the reduced capacity that results from the associated reduction in skin frictional support.

The fact that lateral displacement was observed at a distance of 40 ft (12 m) from the edge of the area of pile driving indicates that the driving of a large number of piles in an urban area might also apply very high lateral stresses to the walls of adjacent basements, perhaps causing them to crack. Thus, it is clear that driven piles are no panacea for foundation problems, nor are drilled piles or piers. Their installation has also been known to produce settlements of adjacent structures.

For example, the movement of soil into a drilled shaft (caving of the sides) for any of a number of possible reasons will result in settlements near the area of lost ground. For one case of a building constructed on auger cast piles bearing in loose sand adjacent to the Mississippi River, the flowing of sand (beneath the water table) into several holes augered to a depth of about 80 ft (25 m), but from which the auger had carelessly been removed, caused an adjacent building to settle about 1 in. (25 mm). The objective of the augered piles was to avoid the settlements of this same building which had been expected if driven piles were used.

The soil heave associated with the driving of large numbers of piles in relatively small areas often causes the vertical displacement of some or all of the piles in the group being driven. When more than about six piles are being driven in a group and penetrate silt or clay deposits, measurements of pile heave should be made. The designer should be aware of the following possible consequences of vertical pile displacement and include provisions in the construction contract to correct any measured heave:

1. Bearing piles that heave and lose contact with the bearing stratum must be redriven if future settlement of the building is to be avoided. However, friction piles that develop little or none of their load-carrying capacity in end bearing need not be redriven.
2. Pile heave can cause unreinforced or lightly reinforced concrete piles to stretch and break. Therefore, heaving of concrete piles is particularly dangerous and should be avoided. If concrete piles are found to have heaved, then the designer must institute a testing program to determine that they have not failed in tension and will perform in a satisfactory manner when subject to the structural loads.

Pore Water Pressures Due to Pile Driving

Driving displacement piles into soils that are not free draining generates very high water pressures in the ground. These water pressures are in response to increases in total stress resulting from the displacement of soil by the piles. A potentially serious effect of these pore water pressures, which effect is independent of the effect of the changes in total stress that have led to the pore water pressures, occurs when piles are driven on slopes. The driving of a large number of piles on or adjacent to a slope, say for the abutment of a bridge, can produce pore water pressures in clay or silt soils large enough to cause the slope to become unstable. In such a case, the foundation system installed to avoid instability causes instability.

This potential problem should be considered when the following conditions exist:

1. A large number of displacement piles are to be driven on or adjacent to the crest of a slope or adjacent to a marine bulkhead or a braced excavation.
2. The piles are to penetrate clay or silt materials that are beneath the groundwater table.
3. The slope inclination exceeds about 4 horizontal : 1 vertical (15°).

The problem may be avoided or its importance reduced by use of a drilled pier foundation rather than driven piles, by predrilling at each pile location or, perhaps, by using an H-pile that displaces a relatively small volume of soil as it penetrates the ground.

16.2.5 Movements in Response to Dewatering

Excavating below the water table will ordinarily lower the groundwater level for some distance from the face of the excavation. This distance is dependent on the material through which the excavation is made, the depth of the excavation, and the time the excavation is open. Groundwater levels are affected for distances of 1000 ft (300 m) or more from the face of the excavation when the excavation is deep and in sand. The effect occurs over distances 10 to 20 times smaller when the excavation is for a finite, brief time and in clay.

The dewatering that results when an excavation is made can cause vertical movements due to soil consolidation (compression that manifests itself as settlement) or, if the dewatering is not controlled, can cause vertical movements in response to the flow of soil particles into the excavation (loss of ground).

Movements Due to Soil Consolidation

Lowering the groundwater level can be viewed as increasing the weight of the material in the zone of groundwater lowering by eliminating the "buoyancy" provided by the water. This increase in weight per unit area, technically termed an increase in effective stress, causes compression of all materials that feel the stress increase. Significant settlements resulting from groundwater lowering frequently occur when compressible deposits underlie an area in which groundwater levels are lowered. Such settlement occurs slowly and over a long period of time. It is rare that lowering groundwater levels in a deposit of sand not underlain by a compressible material will cause significant settlements of the ground or of adjacent structures.

The settlement of streets and buildings adjacent to an excavation because of dewatering and the resultant compression of peat or clay layers at depth occur with some regularity. The adverse effects of this compression are easiest to discern in areas occupied by high-rise buildings. For example, a ⅜-in. (10-mm) difference in settlement between the opposite faces of a 65-ft (20-m) wide building can be magnified to ¾ in. (20 mm) at a height of 130 ft (40 m) (about 13 stories) above ground surface, and to 1.5 in. (40 mm) at a height of 260 ft (80 m) (about 26 stories). Interior cracks would be expected in the upper stories of such a building as a result of the "stretching" produced by the settlement.

Movements Due to Poor Control of Groundwater

Uncontrolled flow of water into an excavation can erode particles of sand and silt from the surrounding ground, deposit these particles in the excavation where they are removed in the normal course of routine dewatering, and, as a consequence, produce undetected voids beneath adjacent areas. This process of erosion, sometimes referred to as *piping* because channels or pipes seem to be created in the ground, frequently occurs when excavations are made through silt strata that are below groundwater level, and only occurs under unusual circumstances when clay soils are adjacent to the excavation. The loss of ground due to erosion can lead to settlement of streets and adjacent buildings.

For soil erosion to occur, the velocity of groundwater flow must be high. Therefore, erosion always begins at the face of the excavation where some opening in the soil retaining system occurs, such as a space between wooden lagging or between sheet piles driven out of interlock. It then progresses backward, up water gradient, as a function of time. It is unlikely that this form of soil erosion can occur at distances from the face of the excavation greater than about the depth of the excavation. Avoiding this form of soil erosion requires that careful attention be paid to points of leakage in the retaining system surrounding the excavation and that filters be installed at points of leakage as soon as they are discovered. Salt hay is a common filter material used in many urban areas, although graded filters or filter fabrics are also used.

Another form of soil erosion all too frequently occurs and sometimes adversely affects the building to be constructed in the excavation. This erosion results from heave or boiling of the bottom of the excavation in response to the upward flow of inadequately controlled groundwater. The consequence of heave and boiling is the loosening of the stratum on which the building is to be founded, resulting in excessive and unanticipated settlement of the structure, and, all too often, the failure of the retaining system surrounding the excavation.

Heave or boiling occurs when an inappropriate system (or no system) is used for the control of groundwater. Most groundwater control on construction projects is done by pumping from sumps. Sumps are adequate when excavations in sand or nonplastic silt are only to be less than about 3 ft (1 m) below groundwater level, or when the excavations are through clay soils. For excavations that must penetrate sand or silt deposits to a depth in excess of about 3 ft (1 m) below the level of groundwater, other systems are available.

Groundwater may be controlled in excavations through sand that must extend to about 16 ft (5 m) below the water level by pumping from shallow, closely spaced wells, known as well points.

Excavations through sand that extend to depths of 60–100 ft (say 20–30 m) below groundwater level may be controlled by pumping from closely spaced wells of low efficiency called ejectors or eductors, or from widely spaced, high-volume, deep wells. At times, it has proven necessary to avoid dewatering entirely by creating a subsurface wall of relatively impervious material around the excavation. This is sometimes done when the soil is very permeable, when the wall can be "keyed into" a relatively impervious bottom material, and when it is believed that wells cannot handle the volume of water expected to flow to the excavation efficiently, or when the drawdown from dewatering is expected to extend sufficiently far that it has the potential for causing damage to other buildings. These walls are constructed in trenches held open by a bentonite slurry and are commonly referred to as slurry trench walls. The wall itself may be constructed continuously or in panels and may be of concrete, cement/bentonite, or soil/bentonite, depending on the required depth, strength, available space, and other factors.

It is most difficult to avoid heave and boiling, as well as piping, when making deep excavations through deposits of nonplastic silt below the water table. Designers should be alert to identify the occurrence of these materials and seek special advice on potential problems. It is not sufficient to attempt to avoid the problem by placing all responsibility on contractors. Control of groundwater in excavations through nonplastic silt may require the use of vacuum well points, a slurry wall, or another retaining structure that will prohibit the movement of soil particles as water flows into the excavation from the sides and the bottom.

Movements Due to Rotting Timbers

Foundations of many old, historic structures are constructed of timber. Timber treated with creosote has been used for less than about 100 years to avoid rotting when foundations are exposed to alternate wetting and drying and to many organisms. Therefore, structures constructed prior to about 1900 are frequently founded on untreated timber piles, on untreated timber grillages, or on masonry bearing on an untreated timber plate. When temporary dewatering occurs near such buildings, the question often arises about the length of time that the water level may be lowered without causing the timber to rot excessively. Little is known about this. However, it is evident that the effects of drying in response to dewatering are mitigated by the ability of capillary forces to raise water considerable distances in soil. Capillary rise of several feet (1 m) is common in sand, and of more than 10 feet (3 m) is to be expected in clay.

It is clear that the older the timber, the more rapidly it will rot when dried. Fresh untreated timber can sustain the effects of exposure to air as a result of dewatering for many months and, perhaps, for 1 or more years. The limited information available seems to suggest that timber which is about 100 yr old and is now to be subjected to substantial water drawdown may deteriorate excessively when exposed for 3–6 months. An extreme case of rapid deterioration of timber occurred in Appleton, Wisconsin. A caisson was drilled through trees buried, according to radio carbon dating analyses, about 10,000 years ago. When inspected in the ground where they had always been in a fully saturated environment, the trees appeared to be strong and fresh. The timber appeared green where it had been broken by the drilling equipment. However, the pieces of wood brought to the surface by the drilling equipment and permitted to dry turned into brown wood fibers that could easily be pulled apart within 2 hr of exposure to air. Qualitatively, therefore, it is evident that there is an inverse relationship between the age of the timber and the time it will take to rot when exposed to air and be unable to support structural loadings.

The example cited above indicates quite clearly that exposure to air accelerates the deterioration of wet timber. Therefore, designers should be aware that the act of inspecting the integrity of timber piles supporting an aged, historic structure may lead to accelerated deterioration of the timber if the material surrounding the piles which is excavated in order to make the inspection is not replaced as a saturated soil.

16.2.6 Selection Affected by Cost of Construction Including Time

The cost of a foundation system is the sum of the costs of at least the following:

1. The intrinsic cost of the foundation system when installed at a particular site.
2. The monetary value associated with the time required to complete the installation of the foundation system (e.g., inflation or labor increases resulting from periodic renegotiation of union contracts).
3. The increase in contract price required by the constructor to protect himself against uncertainties associated with its installation. This protection may be needed because of uncertainty about the need for a dewatering system, about the length of piles to be driven, about the quantity of rock to be excavated, or for other reasons related to the subsurface conditions that may be encountered during the process of construction.

4. The cost arising from the effects of construction on adjacent or nearby structures or facilities.
5. The cost of money and/or other financial penalties in the event that construction is delayed because of unexpected foundation conditions.

It is self-evident that not all foundation systems that may be satisfactory to support a particular structure at a particular site are equally expensive to construct. Selecting the least expensive, satisfactory foundation system often requires considerable judgment about the costs associated with future risks. For example, if a driven pile foundation is considered for use in close proximity to existing structures, someone must assess the risk of damage to those structures as a result of the vibrations produced, or as a result of the soil displacements resulting from pile installation. If damage is believed likely, the cost of constructing the system should be increased by the estimated cost of repairs. In some cases, it has proven most economical to purchase and tear down structures damaged by nearby pile driving rather than to repair them.

An alternate system to the driven pile foundation may also be considered and its cost assessed. A drilled pile foundation is one alternative. The capacity of such a system to support axial loads is often more difficult to predict then that of a driven pile system. This may require that conservative estimates be made of the allowable load-carrying capacity of each drilled pile. However, the cost of this needed conservatism may be smaller than the cost assumed necessary to repair or purchase adjacent structures.

Each uncertainty about the likelihood of some problem during construction adds to the cost of the contract; the occurrence of one of these problems during construction usually delays completion and has financial implications. Therefore, an assessment must also be made of the potential for particular problems developing during the construction period and delaying the project. The financial impact of such a delay must be evaluated and a decision made to proceed with the foundation system being evaluated, or to select another system with, perhaps, higher intrinsic costs of construction but lower risks of delay due to unexpected foundation conditions. It is now possible to base decisions and choices on the results of a formal analysis of the risks and costs of delays resulting from specific events which can be postulated as occurring (see, e.g., *Acceptable Risk* by Fischhoff et al. [16.6]).

16.3 TOLERABLE DISPLACEMENTS OF STRUCTURES

16.3.1 General

The ability to predict the total and differential displacements of a structure for the period of its lifetime is not useful unless the tolerance of the structure to displacements is understood. Those displacements that are associated with vertical movements of the foundations, as contrasted with the displacements of the internal structural members under applied loads, are called settlement. Both total settlement of structural supports and differential settlement between supports must be considered as inevitable for every structure. These settlements occur in response to the loads applied by foundation elements to the foundation materials, as well as in response to the construction procedures followed.

Settlements are time related. Some settlement occurs during construction, immediately upon application of building loads. Almost all the settlement of a structure founded on granular soils that is expected in response to static loads, will have occurred by the time the building is completed and occupied. Settlements occurring in response to building loads applied to foundation materials composed of fine-grained soils (clay) frequently continue at an ever-decreasing rate for many years after construction has been completed. Empirical evidence indicates clearly that settlement occurring slowly produces less damage to a structure than settlement occurring rapidly. This is believed to be the result of gradual stress readjustment as a structure settles slowly. Therefore, in discussing settlement, consideration must be given to the magnitude of the initial settlements and to the time required for the total settlement to occur.

16.3.2 Definitions of Terms

The definitions of terms used in this section are shown in Fig. 16.4. These terms include maximum settlement, ρ_{max}, which is also the total settlement of the point of maximum settlement; the maximum differential settlement, δ_{max}; the maximum net slope of the curve of settlement between supports, $(\delta/\ell)_{max}$ (also known as the maximum angular distortion); the total uniform settlement, ρ_{unif}, of the structure; and the displacement between adjacent supports, ρ_{tilt}. It is believed that the critical parameters in assessing whether actual or predicted displacements are tolerable for a

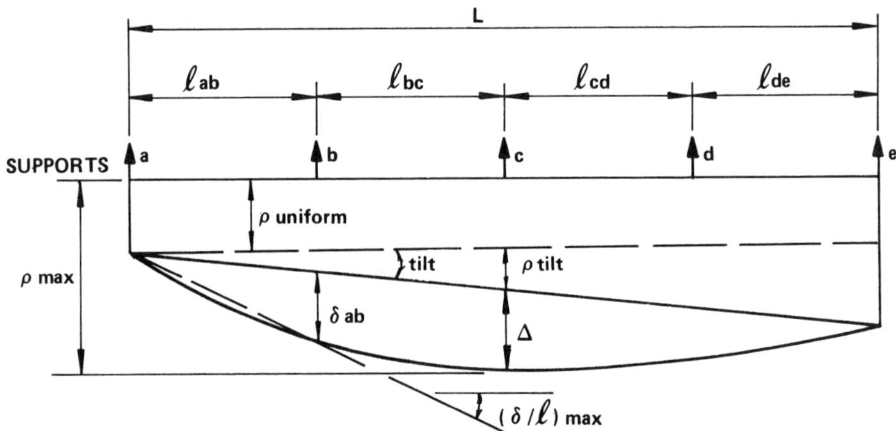

Fig. 16.4 Definition of terms for tolerable settlement.

structure are, in order or importance, (1) the angular distortion (corrected for tilt) (δ/ℓ), (2) the maximum displacement, ρ_{max}, and (3) the acceptable tilt, ρ_{tilt}.

It is important to emphasize that the maximum angular distortion or maximum net slope of the settlement curve, $(\delta/\ell)_{max}$, is *not* usually equal to the maximum tilt per unit length. Tilt per unit length is the difference in displacement between supports, ρ_{tilt}, divided by the distance, ℓ, between these supports. However, the slope of the settlement curve, δ/ℓ, can only be determined after a complete survey of a structure has been made and plotted, while the displacement between adjacent supports may be measured directly and the tilt per unit length computed. Criteria tor tolerable settlement are commonly couched in terms of the angular distortion computed by considering three adjacent columns or measurement points, eliminating the tilt associated with these columns to find the differential settlement, δ, and then computing the angular distortion, δ/ℓ.

16.3.3 Definitions of Tolerance

Whether or not the settlement of a structure is tolerable is dependent on the judgement of what is tolerable, as well as the type of distress the settlement induces. Engineers are normally concerned with three categories of distress: (1) functional distress, (2) architectural distress, and (3) structural distress.

Functional Distress

Functional distress occurs when displacements are (1) large enough to interfere with the intended purpose of the structure, (2) require continuous maintenance, or (3) produce hazards to safety. Although portions of the Auditorium Building (now Roosevelt College) in Chicago settled as much as 2.5 ft (0.75 m) between 1890 and about 1959, and exhibited differential settlements of about one-half the total settlement, its function was impaired only slightly. More than 6 ft (2 m) of settlement is reported for the Palace of Fine Arts in Mexico City, yet its usefulness and beauty remain. In contrast, displacements that produce angular distortions as small as 1 in 50,000 can make precision-tracking radar systems sufficiently inaccurate to be considered inoperative. Continuous maintenance would be required if settlements of a building periodically produced significant cracks in its exterior wall permitting rain to damage the interior walls. These cracks could also lead to a hazard to safety if they weakened the exterior brickwork or other veneer so that sections might come loose and fall to the ground.

Thus, criteria for functional distress, which are related primarily to the uses of structures, vary between widely separated extremes.

Architectural Distress

Architectural distress is defined as distress that interferes with the visual, aesthetic, and psychological value of the structure. It, rather than structural distress, is most often the criterion that defines tolerable settlement. Typical examples of architectural distress are discernible cracks in the exterior or interior walls, floors, and finishes, perhaps accompanied by some malfunctioning of doors and windows. Permissible limits for this type of damage are variable, depending on such factors as the purpose of the facility and the personalities involved with its design and use. Minor cracks in walls that may be acceptable in a warehouse are the source of strong criticism if they occur in "show piece" structures like an opera building, city hall, or a convention center. Acceptance of architectural distress, which often occurs from factors not related to distortion in the foundations, may depend upon the cost of remedial measures and the frequency that remedial measures are needed. Architectural distress can result from settlement of structures. More frequently, however, it is the inevitable consequence of normal shrinkage of materials, of temperature changes that subject building materials to cyclic loadings and lead to failure in fatigue, of vibrations from internal or external sources, of weathering, or even of sonic boom.

Structural Distress

Structural distress is said to occur when the distortions of a structure result in (1) the overstressing of some of its structural members, (2) the cracking of bearing walls, or (3) the excessive tilting of tall buildings. When structural distress occurs, questions about the safety of the structure follow close behind, and costly repairs are usually necessary. It is of considerable interest to note that it has not yet proven possible to compute the stresses in structural elements resulting from foundation displacements and, from these computations, arrive at realistic conclusions about the magnitude of tolerable displacements. This inability to compute stresses in members is ascribed to many factors, such as the indeterminate nature of the problem, variable and unknown stiffness of joints, the details and sequence of construction operations, the effects of internal partitions and external walls on the stiffness of portions of the structure, material creep and load redistribution as settlements occur slowly, and many other such factors. For example, Skempton and MacDonald [16.24] reported on a series of attempts to correlate calculated and measured stresses in structures, and also on an effort by G. G. Meyerhof to estimate by calculation the point at which distress due to differential settlement would occur. They reported in 1956, and their conclusions have not been altered by subsequent researchers or practitioners, "that it is difficult to predict an allowable differential settlement from stress calculations in the framework of building," perhaps for the following reasons:

1. Live loads are commonly estimated incorrectly and generally conservatively.
2. Composite action of the frame, floors, wall panels, etc. which will lower stress levels, are difficult or impossible to consider in the calculations.
3. Stress redistribution due to time effects (i.e., creep) tends to reduce the level of stress at critical locations.

It is, therefore, conventional practice to use empirical criteria to estimate whether or not settlements are likely to be excessive, rather than make computations that always result in overestimates of the levels of stress in critical members.

16.3.4 Magnitude of Tolerable Settlements

A comprehensive review of documented experience with the tolerable settlement of structures was presented by Grant et al. [16.7]. These authors followed the approach of Skempton and MacDonald [16.24] and added new observations to the literature. Burland and Wroth [16.4] presented a paper on tolerable settlements at about the time that the Grant et al. [16.7] paper appeared. Burland and Wroth [16.4] introduced the concept that it is usually architectural distress that one seeks to avoid, that architectural distress occurs at some critical strain that is a function of the architectural materials, and that there is more tolerance of a structure to settlements producing sag at its center than to settlements producing "hogging" (stretching, as in the top fibre of a cantilever beam). This is discussed separately later.

Grant et al. [16.7], as well as others before them, recognized that there was a necessary relationship between total settlement, ρ_{max}, and the maximum angular distortion, $(\delta/\ell)_{max}$, and that this relationship depended upon such factors as soil type, thickness of deposit, structural rigidity, and uniformity of loadings. The suggested limiting values for settlement of structures are discussed below.

Total Settlement

There is general agreement that a limitation on total settlement of structures is wise because of the relationship between total and differential settlement. Relationships between total settlement, ρ_{max}, and the maximum slope of the settlement curve, $(\delta/\ell)_{max}$, were developed by Skempton and MacDonald [16.24] and by Grant et al. [16.7]. Those relationships lead to the following limitations on settlement if a maximum slope of the settlement curve of $1/300$ is assumed acceptable:

Foundation	Maximum Settlement (in.)	(mm)
Isolated foundation on clay	4	100
Isolated foundation on sand	2	50
Raft on clay (believed conservative)	4¼	110
Raft on sand (believed conservative)	2½	65
Raft on clay related to width (B) of structure:		
B = 65 ft (20 m)	3	75
B = 130 ft (40 m)	6	150
B = 200 ft (60 m)	9	220
Stiff foundations for silos, stacks, etc.	12	300

The above limitations on total settlement are essentially identical to those suggested by all other investigators. However, they do not take into account the rigidity of the structure or the thickness of any compressible strata beneath the structure.

Attempts to investigate the effects of rigidity of the structure on allowable settlements have not been successful; the studies only suggest, but do not show conclusively, that the more rigid the entire structure, the more uniform the settlement and, perhaps, the greater the allowable total settlement. Several examples of buildings that have successfully tolerated total settlements larger than those indicated above have been given earlier. Others have been described in the engineering literature (e.g., Ref. 16.28). In principal, the magnitude of acceptable total settlement is unlimited, provided that excessive differential settlement does not develop and that the problems of utilities entering the structure, access to the structure, and drainage around the structure have been solved.

Differential Settlement

Tolerable differential settlements are normally expressed as acceptable angular distortion (corrected for tilt). A summary of limiting values of angular distortion suggested by a number of investigators is provided in Table 16.3. Most of the values in this table do not differentiate between foundations on sand or clay, for which materials the time for the development of the differential settlement will be different, nor do they differentiate between settlements that take place during or after construction.

It is believed that angular distortions, δ/ℓ, of a frame structure of 1/500 will not produce cracking of architectural finishes, that cracks in load bearing or panel walls of frame structures are likely when the angular distortion exceeds 1/300, and that structural damage is likely when the angular distortion is as high as 1/150.

The depth to and thickness of compressible strata beneath a structure have long been recognized as affecting both its total settlement and its pattern of settlement (differential settlement). Large total settlements resulting from founding structures on compressible deposits cause smaller differential settlements when the compressible deposit is found at great depth than when the deposit is at shallow depth. This is believed to be the result of the distribution of stress beneath a structure which tends to make the stress increases felt by deep soil deposits more uniform than the increases felt by near-surface soil deposits. The fact that differential settlement is mainly associated with compression of the soils near foundation level has been confirmed by observations of the settlements of buildings during construction; the maximum differential settlement has been found to develop early in the life of most structures as the loads are applied (the time when settlements are mainly the result of compression of near-surface soils), and this differential settlement has been found not to change appreciably with time after the completion of the building (as soils at great depth are compressed).

16.3.5 Concept of Critical Strain

Burland and Wroth [16.4] introduced the concept that architectural damage, which frequently is the basis of acceptability or unacceptability of structural deflections, can be related to a "critical tensile strain" of the structure. They considered, in their initial efforts, a simple model of a structure, assuming that it could be represented as a beam, and correlated initial cracking with a critical strain of 0.035% and structural damage with a critical strain of 0.075%. Their work is described in a convenient and accessible form by Wahls [16.26]. The validity of their concept is slowly being confirmed as additional information on building damage become available [16.25].

Critical strain is related to the relative sag, defined by the ratio Δ/L, where Δ is the maximum deflection, corrected for uniform settlement and tilt, and L is the length of the deflected portion of the structure, as defined in Fig. 16.4. The tolerable relative sag, in turn, depends on the ratio L/H,

Table 16.3 Criteria, in Terms of Angular Distortion, to Evaluate Extent of Differential Settlement a Structure Can Tolerate

Angular Distortion, δ/ℓ	Conditions
1/50–1/100	Difficulty with drainage of floors
1/100	Difficulty in controlling hand trucks in warehouse
1/150	Structural damage to beams likely to occur
	Safe limit for flexible brick walls where (length, L)/(height, H) ≥ 4
	Considerable cracking in panel walls and brick walls
1/200	Suggested maximum for simple steel frame
1/250	Tilting of high, rigid buildings might become visible
	Suggested maximum if settlement occurs during construction, or over a period in excess of 10 years
1/300	Cracking of panel walls likely in steel or reinforced concrete frame buildings without diagonal bracing
	Operation of overhead cranes affected
	Suggested maximum for reinforced concrete building curtain walls
1/500	*Safe limit* for buildings where cracking is not permissible
	Suggested maximum for continuous steel frame structures
1/600	Suggested maximum for frames with diagonals
1/750	Operation of sensitive machines may be affected
1/1000	Plaster (gypsum) cracking
	Suggested maximum for brick walls of one-story industrial buildings
1/1500	Suggested maximum for brick walls of multistory buildings with footings on plastic clay, $L/H \geq 5$
1/2000	Suggested maximum for brick walls of multistory buildings with footings on dense sand or hard clay, $L/H \geq 5$
1/2500	Suggested maximum for brick walls of multistory buildings with footings on plastic clay, $L/H \leq 3$
1/3000	Suggested maximum for brick walls of multistory buildings with footings on dense sand or hard clay, $L/H \leq 3$
1/5000	Operation of turbogenerator affected

where H is the height of the building; the L/H ratio of high-rise buildings is less than unity and may, for industrial buildings, exceed 5 or 10.

Frame structures for which damage data are available exhibit L/H ratios less than 3. In this range, there appears to be no difference between the design/damage criteria proposed by Skempton and MacDonald [16.24] (included in Table 16.3) and those suggested by the critical strain concept. However, the Skempton and MacDonald criteria have been found to be unconservative when applied to structures supported on load-bearing walls. Damage criteria for such structures are dependent on the L/H ratio. The concept of damage occurring at some critical strain, as proposed by Burland and Wroth produces damage criteria similar to those given below from the building code of the USSR, as reported by Polshin and Tokar [16.20]:

	Damage Criteria, Δ/L	
	Sand and Hard Clay	Plastic Clay
Relative deflection of plain brick walls:		
For multistory dwellings and civil buildings		
$L/H \leq 3$	0.003	0.004
$L/H \geq 5$	0.0005	0.0007
For one-story mill buildings	0.0010	0.0010

A major contribution of the concept of critical strain is the recognition that buildings whose edges settle more than their center, producing a settlement pattern that is concave downward, will be damaged by much smaller angular distortion than buildings exhibiting a settlement pattern that is concave upward. Damage to structures with a concave downward settlement pattern, termed *hogging,* occurs at about one-half the values suggested by Polshin and Tokar [16.20]. Hogging frequently occurs to existing structures adjacent to new excavations, sometimes occurs when an addition is put on an existing building, and has been known to occur as a result of tunneling adjacent to one wall of a structure, or from activities that cause land subsidence, such as pumping of oil or water from the ground.

For the present, the concept of critical strain only produces a better understanding of why damage may occur in response to settlement. It does not alter in any significant way the existing damage criteria given in Table 16.3 or above, except that it identifies clearly that hogging is more critical for a structure than is sag. It holds promise, however, of providing a useful analytical tool sometime in the near future.

16.3.6 Effects of Time and Sequence of Construction

The few data available on stress measurements in structures appear to confirm that the slower the settlement of a structure occurs, the greater the opportunity for internal stress redistribution and the greater the tolerable angular distortions. This is suggested, but not proven, by the tabulation of maximum total settlement given earlier in which it is suggested that the limitation on total settlement of a foundation on clay, which is expected to settle slowly, is about twice that of the same foundation on sand, which compresses rapidly. Observations of the settlement of structures also suggest that, in response to inhomogeneity of the compressible strata and variations in the distribution of stresses beneath the foundations, the differential settlement of a flexible structure bearing on clay is expected to be about one-half to two-thirds of its total settlement (rigid structures, one-third to one-half). Similar foundations on sand are expected to exhibit differential settlements that are 80 to 100% of their total settlement.

It is also obvious (and should not be forgotten) that those settlements occurring during the construction of a building only affect the members in place when they occur. Furthermore, many secondary members may not be in place when construction-induced settlements occur, reducing the rigidity of portions of the structure and reducing the stresses induced by the displacements. Thus, the relationship between the details and sequence of construction and the time of settlement affect the consequences of the settlement.

16.3.7 Effects of Vibrations

Construction-related vibrations produced by blasting, pile driving, and vibratory compactors have been known to induce the cracking of structures. The cracking is related to the properties of the materials and their age, as well as to the vibration characteristics.

The occurrence of vibration-induced cracking of structures has been related to measurable indices, the most common of which is peak particle velocity. The maximum peak particle velocity normally considered safe for structures in sound condition is 2 in./sec (50 mm/s); at most, minor damage is expected if peak particle velocities are maintained below about 4 in./sec (100 mm/s). However, lower values of peak particle velocity may be appropriate for structures that are old and/or in poor condition. During the many years of their existence, these structures probably have been subjected to cyclic stresses induced by extreme temperature changes over long periods of time, to stresses arising from minor, long-term-related, differential settlements resulting from external causes, to stresses associated with the drying of the materials used to construct the structure or, perhaps, to stresses induced by earlier (and forgotten), nearby construction. Residual strains associated with these earlier stresses now make these old structures more susceptible to cracking in response to new vibrations. Consequently, the safe level of vibration (maximum peak particle velocity) for old or historic structures is frequently established at 0.5 in./sec (12 mm/s) and, sometimes, as low as 0.3 in./sec (7 mm/s).

However, a study by Hendron and Dowding [16.8] suggested that the use of peak particle velocity as an indicator of potential structural damage is only appropriate for structures (or portions of structures) with natural frequencies between about 5 and 20 Hz. Damage to facilities with natural frequencies below about 5 Hz appears to be related to peak particle displacement rather than peak particle velocity, and damage to facilities with natural frequencies above about 20 Hz appears to be related to peak particle acceleration. As a result, the use of a response spectrum approach to damage from vibrations, as suggested by Hendron and Dowding, is increasing and will probably supplant the more common use of peak particle velocity.

It is also important to recognize, when developing specifications for the construction of buildings, that vibrations associated with peak particle velocities in excess of about 0.3 in./sec (7 mm/s) are likely to be very disturbing to people. Vibrations associated with peak particle velocities of 0.04 in./sec (1 mm/s) are still strongly perceptible by humans. It is quite clear that establishing acceptable vibrations at levels that are not disturbing to people will, for example, increase the cost of blasting considerably and, at times, may make blasting impossible. Therefore, complaints from people who perceive construction vibrations are a social (environmental?) issue and are almost inevitable; the engineering issue is associated with designing the construction so that damage due to vibrations does not occur and the legal issue is associated with establishing and documenting whether damage did or did not occur.

16.4 FOUNDATION SYSTEMS—TYPES AND WHEN TO CONSIDER

16.4.1 General

The foundation system selected for a particular structure depends on the size, importance and use of the structure, the subsurface conditions at the site, the design of the structure, and the cost of the foundation system. These factors are interrelated, but often one will dominate the choice.

A tall building with heavy loads can readily be supported on inexpensive footing foundations if they bear on rock. The same building constructed at a site underlain by clay may require pile support if the clay is soft; it may be founded most economically on a mat if the use of the building necessitates the construction of deep basements or if deep basements are found advantageous. It may be supported by drilled piers bearing on rock or a hard stratum or by caissons drilled into rock if the building is designed with a small number of highly loaded main columns and competent rock is within economical reach. A low building founded in an area of weak soils may require a deep foundation if it is an important structure, such as a museum or a convention center. However, if it is a flexible steel frame warehouse structure, it may be appropriate to support it on footings and to allow tolerable settlement to occur.

With experience, the selection of a foundation system appropriate for a particular structure can be made rapidly from among a relatively few foundation types. The final choice is commonly dependent upon the results of the following types of studies:

1. Studies of the estimated settlements, both total and differential, that the structure might undergo if one or more of the "suitable" foundation systems is adopted.
2. Comparison of the probable cost of each of these foundation systems, including the hidden costs associated with such uncertainties in the subsurface conditions as the presence of large numbers of boulders or man-made obstructions, or uncertainties about the actual quantity of water to be controlled in excavations below the groundwater table.
3. Estimates of the risks of damage to adjacent structures resulting from the installation of each of the systems.
4. Estimates of the time necessary to construct each of the systems, of the possibility that other work can begin before the foundation system is completely installed, and of the risks of delay associated with construction of each of the systems.

A brief description of the foundation systems commonly considered as support for buildings is provided in the remainder of this section. Some of the factors that influence the selection of each system are discussed.

16.4.2 Shallow Foundations

Shallow foundations may be defined as foundations that obtain their support on soil or rock just below the bottom of the structure, even if the structure has one or more basements. Normally the ratio of the depth D of the foundation below basement or grade to the least width B of the foundation is less than about 1.0 for shallow foundations.

Footing Foundations

Footings are normally (1) individual foundations carrying a single column, (2) combined foundations carrying more than one column, or (3) strip footings carrying a wall. They are usually constructed of reinforced concrete and are sized to distribute the column loads to the underlying materials at contact pressures that will not produce bearing failure or excessive settlement. When the total area of footing foundations for a particular structure exceeds about 50% of the plan area of the building, it is often advantageous, from a cost viewpoint, to consider substituting a mat or raft foundation for the footings.

Mat Foundations

A mat foundation covers the entire plan area of the building and may, if necessary, extend beyond the limits of the walls of the building. It has several advantages:

1. A mat is stiffer than a series of individual footings, permitting some bridging of localized weak zones of soil beneath foundation level and, thereby, reducing differential settlements.

2. The increased bearing area of the mat, as compared to footings, reduces the average contact pressure, thereby reducing the settlements associated with the compression of materials just below foundation level. It is the material just below foundation level that is most likely to produce intolerable differential building displacements.

3. The excavation for basements may remove (or may be made to remove) a weight of soil equal to the weight of the building, thereby limiting building settlement to the recompression of material that heaved during the excavation process. This is not true beneath a footing foundation established below even a deep basement. Footing foundations will ordinarily produce a soil contact pressure exceeding the average weight of the excavated material. This stress, which attenuates with depth below the foundation, will cause compression (consolidation) of the soil directly beneath the footings resulting in differential settlements.

4. If a "pressure slab" is required in the basement of a building to resist upward hydrostatic pressure, it is sometimes more economical to construct a mat of slightly greater thickness than the pressure slab than it is to construct individual footings covered by a pressure slab. The pressure slab must be designed with water stops at every column or other penetrations, resulting in construction difficulties (water stops are not always installed properly) and, frequently, leakage.

16.4.3 Deep Foundations

The function of a deep foundation is to carry building loads beneath a stratum of material deemed unsatisfactory (for whatever reason) to a satisfactory bearing stratum. A foundation is considered to be deep when the ratio of its depth D to its least width B exceeds about $2\frac{1}{2}$ to 5; typically, this ratio actually exceeds 10 for most deep foundations. Contrary to popular belief, structures on deep foundations will settle, although the magnitude of the settlement can almost always be expected to be smaller than that of the same structure founded on a shallow foundation.

Deep foundations include piers, caissons, and piles installed in a variety of ways, and without a clear distinction possible (or necessary) among them. For example, there is normally no difference between a drilled caisson and a drilled pier and, most often, only a modest difference in diameter between a drilled caisson or pier, and what the British call a bored pile.

Piers

Most often piers are installed in small, braced, hand-dug or machine-excavated excavations extending to a shallow bearing stratum. Very deep, hand-dug piers have been installed in Chicago and elsewhere. However, they require great care and favorable subsurface conditions. Piers have a constant cross-section from the bearing stratum to some level within the basement of a structure where a column will bear on the pier. Under favorable conditions, some widening or belling of the bottom of the pier at the bearing stratum is possible. The ratio of depth D to width B for piers generally ranges between $2\frac{1}{2}$ and 5; that is, a pier 4×4 ft (1.2×1.2 m) in plan dimensions will frequently be founded at depths of 10–20 ft (3–6 m) below basement level. However, the ratio D/B for hand-dug piers can be 10 or more.

Piers are used to carry structural loads beneath a near-surface layer of unsuitable bearing material to a nearby competent stratum. The cost of hand-dug piers relative to other foundation systems may be high because of the need for a braced excavation and control of groundwater. The sheeting and bracing is commonly left in place when concrete is placed. Hand-dug piers are used infrequently in most of the United States; drilled piers or caissons are used in their place.

Drilled and Belled Piers

The excavation for drilled piers is made most often using truck-mounted rotating augers or buckets fitted with cutting tools. Under favorable conditions in stable cohesive soils (clay), the excavation of a 3–6 ft (1–2 m) shaft proceeds rapidly to a bearing stratum at great depth. There, the contact pressure on the bearing stratum can be reduced substantially by forming a bell. The diameter of the base of the bell can be as much as $2\frac{1}{2}$ times the diameter of the shaft, which reduces the end-bearing stress below that in the shaft by a factor of 6. The bell and shaft may then be filled with concrete, ordinarily without the need for much steel reinforcing.

Problems arise in the construction of drilled pier foundations and, therefore, costs increase substantially, when boulders must be excavated to reach the bearing stratum and when water-bearing strata, such as sand or gravel, must be penetrated. Furthermore, bells are not possible in most silt deposits that are beneath groundwater level and in sand. In addition to the added cost of excavating through sand or silt beneath groundwater level, the inability to form a bell increases the cost of the drilled pier foundation significantly because the total number of piers required must be increased to make up for the loss in increased bearing area provided by the bell.

The technology exists to construct drilled piers safely through sand strata that are below groundwater level. One of the best techniques makes use of a slurry of drilling mud. The sides of the excavation through sand are stabilized by maintaining the hole filled with a slurry of drilling mud (bentonite) at all times. If desired, steel reinforcing cages can be placed in the slurry-filled excavation prior to filling with concrete. Concrete is tremie-placed in the excavation when it is completed, displacing the drilling mud from the hole and from around the reinforcement. There is substantial evidence that the slurry does not affect the steel-to-concrete bond and, therefore, the performance of the system.

This technology is far more advanced outside the United States than within. Therefore, the American designer should be certain that the many details associated with the successful use of this system are understood by the prospective specialty contractor before this method of pier installation is adopted, and also be certain that the owner is willing to accept, in return for lower cost, the added risk of tremie-placement of concrete in an uninspected excavation.

Casing has also been used to seal a sand stratum beneath groundwater which is penetrated by a drilled pier excavation. The success of this system depends upon the success the contractor has in sealing the bottom of the casing. Sealing the bottom is simple if the casing penetrates into an impervious layer, such as a hard clay. It is quite difficult if the casing penetrates to rock because the surface of the rock is almost always fractured and is almost never horizontal. Therefore, a complete seal is difficult to achieve. Leakage into the bottom of the casing as a result of an ineffective seal will, almost always, cause loss of ground and could endanger the stability of nearby structures or other facilities.

There are also occasions when a drilled pier is to penetrate soft clay or silt strata that may squeeze into an open excavation for the shaft before or while the concrete is being placed. Such strata must be protected against squeezing, sometimes by the use of drilling mud, but most often by a steel pipe casing. There is a substantial risk to the integrity of the pier associated with attempts to remove this casing as concrete is filling the excavation. In general, removal should be avoided, although procedures have been developed to reduce this risk to the integrity of the pier.

In assessing the suitability and economics of a drilled pier foundation, the following should be considered:

1. The presence of water-bearing strata that must be controlled.
2. The presence of boulders or other obstructions that must be penetrated.
3. Whether the soils just above the bearing stratum are cohesive and strong enough for the formation of a bell.
4. Whether the ground that is to be penetrated will squeeze into the shaft and, therefore, must be protected by a casing or by drilling mud.
5. Whether there are contractors available who are experienced in installing drilled piers in the subsurface conditions at the specific site to be constructed on.

Drilled-in (Rock) Caissons

When column loads are high, exceeding about 4500 kips (20 MN), and when sound rock is at reasonable depth, drilled-in caissons are sometimes appropriate and economical. They are installed by successively driving and cleaning out a steel pipe casing until it reaches rock, where it is seated. Rock is then excavated by one of several available methods for some depth below the tip of the casing. Rock drilling methods include rotary drilling and percussion drilling using a down-hole air hammer or a churn drill. The depth of rock excavation is frequently two or more times the inside diameter of the casing. A steel "core" is then set in the center of the shaft and held in place while concrete is tremie-placed to fill the rock socket and casing.

The loads applied to the caisson are assumed distributed by shear to the side of the rock socket and by bearing to the base of the socket. The allowable end bearing and shear stresses in the rock socket are frequently established by code. They are, in reality, related to the compressive strength of the rock and to the spacing and frequency of fractures and other characteristics of its structure.

The infrequent use of drilled-in caissons is a reflection of the relatively small number of occasions when they prove to be economical. Most often, they are selected for use in response to other than economic concerns. For example, they have been used to provide a rigid resistance to tension loads resulting from wind or other environmental forces, to deliver loads deep into a rock mass to reduce the possibility of overstressing an adjacent buried structure, to reduce the number of foundation elements within a given area, or to reduce the differential settlement between columns supporting a structure whose tolerance to differential settlement is less than 0.05 in. (1 mm).

Piles

Piles are, by far, the most common deep foundation system. They are used when the near-surface materials cannot support the building loads without unacceptable deformations. Trees were the first piles. The use of brute force to pound a tree into the ground dates at least to Roman times. Throughout history, when it has appeared that the efficiency of existing piles or the ability of existing pile hammers to drive piles to the required depth was in doubt, new pile materials were developed or new advances in pile hammers were made. Today, it is possible to drive underwater piles 450 ft (150 m) long and 80 in. (2.1 m) in diameter using a hammer delivering an energy of 1000 ft-kips (1400 kN·m).

The selection of pile types available for use in buildings is large. The designer may choose among timber, concrete (both precast and cast-in-place), steel (pipe, heavy-wall shell, and H-section), and a variety of specialty systems. Design capacities vary between about 25 tons (250 kN) and 200 tons (roughly 2000 kN), depending on the material selected and the subsurface conditions.

Selection of the type of pile to be used to support a particular structure is, after considering costs, related to the magnitude of the loads to be supported and the minimum number of piles needed in a pile group for stability and redundancy. Thus, a light building will not be supported on piles that provide 200 tons (roughly 2000 kN) of capacity each because one pile or less would be needed to support the column loads, leading to an inefficient system that requires the use of lateral bracing. However, if this same building is already laterally braced at foundation level because of earthquake considerations, high-capacity piles may yield a cost savings.

A summary of pile systems and commonly accepted load-carrying capacities in the United States is provided in Table 16.4. When comparing the economics of various pile types as part of the selection procedure, it has been found useful to estimate the total cost of each of the piles under consideration, including the cost of the associated pile cap, and divide the estimated cost by the useful capacity to determine the cost of each pile type per ton (or say, per kN) of useful capacity. Such a comparison necessitates that estimates be made of the probable length of different pile types when driven at a specific site. Those estimates are generally difficult to make. However, some rough guides are available:

1. Low displacement piles, such as H-piles, will probably have driven lengths that are longer than displacements piles, such as pipe piles, assuming that they both are not to be driven to bearing on sound rock.
2. Tapered piles, such as the Monotube pile, can be expected to be shorter for the same capacity than H-piles, pipe piles, and cast-in-place shell piles when driven into granular soils, and may be marginally shorter when driven into clay.
3. Specialty piles, such as compacted concrete or tapered tip piles, can be installed at roughly (or specifically) predetermined tip elevations, reducing or eliminating the need to estimate lengths. Augercast and other drilled and grouted piles can be installed to predetermined depths in an almost vibration-free manner.

The designer, in selecting a pile for use, also should be wary of specifying both the pile and the pile hammer to be used to drive the pile unless he knows, from experience in the industry, which

Table 16.4 Summary of Pile Systems and Commonly Accepted Load-Carrying Capacities in the United States

Type	Common Capacity		Usual Length	Taper	
	tons	kN		Yes	No
Timber	10– 35	100– 350	25–50 ft common (7–13 m)	X	
Concrete					
Cast-in-place	50–100	500–1000	60 ft common (20 m) 100–125 ft with pipe extension (30–40 m)		X
Precast	30–100	300–1000	15–90 ft (5–30 m)	X	X
Steel					
Pipe, concrete filled	30–100	300–1000	15–90 ft, splice to 150 ft (5–30 m, splice to 50 m)		X
H-section	50–150	500–1500	15–90 ft, splice to 150 ft (5–30 m, splice to 50 m)		X
Monotube	50–100	500–1000	20–60 ft (7–20 m)	X	
Specialty					
Compacted concrete	50–150	500–1500	15–50 ft (5–15 m)		X
Tapered tip	40–150	400–1500	15–75 ft (5–25 m)	X	
Auger cast	30– 60	300– 600	15–75 ft (5–25 m)		X

hammer should be associated with which pile type, or he has performed the appropriate dynamic analysis to match the pile to the hammer. It is preferable to have the contractor specify which hammer to be used and to substantiate its appropriateness by performing the dynamic analysis. Alternatively, the analysis can be performed by some independent agency or performance specifications can be written, placing the total responsibility on the contractor.

16.4.4 Special Foundation Treatment or Systems

The purpose of the discussion in this section is to bring to the attention of building designers that methods exist, and often are found to be very economical, to improve subsurface conditions so that less expensive foundation systems can be used than originally believed necessary. Not all methods available are described here. The discussion is limited to those methods deemed to have relatively frequent application to building construction. Many of these methods have been understood and used for many years; some are relatively recent developments. Little is said in this section on preliminary design about the details of design necessary to make use of many of these special systems. Their use requires special knowledge not appropriate for inclusion in this book. An overview of the applicability of foundation soil improvement for different soil conditions so that shallow foundations may be used was prepared by Mitchell [16.14] and is presented as Table 16.5.

Foundation treatment methods appropriate for use in design of buildings fall into two general categories: (1) replacement of material of inadequate strength or excessive compressibility by material suitable to construct upon; and (2) improvement in place of the critical properties of the foundation materials.

Replacement of Weak Materials

1. *Total Replacement.* The replacement of weak materials by strong materials has been practiced for many years by highway designers and constructors. Most replacement materials have been granular soils although, in recent years, there have been examples of cement or lime-stabilized soils being used successfully as replacement materials. Total replacement of weak materials beneath buildings usually can only be accomplished with confidence if it is possible to inspect the excavated area visually to be certain that all unsuitable material has been removed before the replacement material is installed.

In order to inspect the bottom of an excavation completely, it almost always must be dry. Thus, some method of controlling water flowing into the excavation is usually required. Excavation "in the wet", followed by underwater inspection of the bottom by divers, has been used successfully at several building sites. However, the risks associated with this procedure are greater (by some unquantifiable amount) than those associated with excavation under dry conditions. There are known examples of weak materials, inadvertently left in wet excavations, that have affected adversely the performance of the structure. In addition, because of problems related to compaction of the backfill materials, the magnitude of the building loads that can be applied to the replacement materials will usually be smaller if replacement is done in a wet excavation than if replacement is done in a dry excavation.

Replacement granular soils can be placed in a dry excavation in layers 8–12 in. (200–300 mm) thick, each layer can be compacted to a predetermined density, and the degree of compaction achieved can be verified by tests. Densification of granular soils is best achieved using a vibratory compactor. Footing foundations bearing on compacted fill that has been properly controlled can often be used in conjunction with high bearing pressures to support structural loads without risk of excessive deformation. A wide range of mainly granular materials, most often available in the vicinity of the proposed structure, can be used to fill the excavation.

Compaction of replacement materials placed in a wet excavation is best done by vibratory means, or by other methods which also require the rapid drainage of water from the soil as it is compacted. Therefore, these materials must be relatively clean, coarse granular soils. Replacement materials normally must be placed in a wet excavation to about a height of 2–3 ft (600–900 mm) above the water level before compaction by surface vibratory compactors can be begun. Surface compaction is rarely effective to depths of more than 5–6.5 ft (1.5–2 m), even if very heavy vibratory compactors are used (Moorhouse and Baker, Ref. 16.16). This limits the usefulness of surface compaction and generally requires that the allowable bearing pressure of the foundation elements be fairly small, unless only about 3 ft (1 m) of soil is to be placed below water level. Furthermore, because the material necessary for successful compaction in a wet excavation must be coarse grained, of relatively uniform gradation, and essentially free of silt-size or smaller particles, the cost of the material is frequently high.

When backfilling must be done under wet conditions and to depths beyond those that can be compacted satisfactorily by surface compactors, methods of in-situ compaction must be utilized before the structure can be built. Several of these methods are discussed below in the section on soil improvement in place.

Table 16.5 Summary of Soil Improvement Methods*

Method	Principle	Most Suitable Soil Conditions/ Types	Maximum Effective Treatment Depth	Special Materials Required	Special Equipment Required	Properties of Treated Material	Special Advantages and Limitations	Relative Cost
				In-Situ Deep Compaction of Cohesionless Soils				
Blasting	Shock waves and vibrations cause liquefaction and displacement, with settlement to higher density	Saturated, clean sands; partly saturated sands and silts (collapsible loess) after flooding	>30 m	Explosives, backfill to plug drill holes, hole casings	Jetting or drilling machine	Can obtain relative densities to 70–80% may get variable density; time-dependent strength gain	Rapid, inexpensive, can treat any size areas; variable properties, no improvement near surface, dangerous	Low
Vibratory probe	Densification by vibration; liquefaction-induced settlement under overburden	Saturated or dry clean sand	20 m (ineffective above 3–4 m depth)	None	Vibratory pile driver and 750-mm diameter open steel pipe	Can obtain relative densities of up to 80%; ineffective in some sands	Rapid, simple, good underwater; soft underlayers may damp vibrations, difficult to penetrate, stiff overlayers, not good in partly saturated soils	Moderate
Vibrocompaction	Densification by vibration and compaction of backfill material	Cohesionless soils with less than 20% fines	30 m	Granular backfill, water supply	Vibroflot, crane, pumps	Can obtain high relative densities, good uniformity	Useful in saturated and partly saturated soils, uniformity	Moderate
Compaction piles	Densification by displacement of pile volume and by vibration during driving	Loose sandy soils; partly saturated clayey soils; loess	>20 m	Pile material (often sand or soil plus cement mixture)	Pile driver, special sand pile equipment	Can obtain high densities, good uniformity	Useful is soils with fines, uniform compaction, easy to check results; slow, limited improvement in upper 1–2 m	Moderate
Heavy tamping (dynamic consolidation)	Repeated application of high-intensity impacts at surface	Cohesionless soils, waste fills, partly saturated soils	30 m	None	Tampers of up to 200 tons, high-capacity crane	Can obtain good improvement and reasonable uniformity	Simple, rapid, suitable for some soils with fines; usable above and below water; requires control, must be away from existing structures	Low

(continued)

Table 16.5 (*Continued*)

Method	Principle	Most Suitable Soil Conditions/ Types	Maximum Effective Treatment Depth	Special Materials Required	Special Equipment Required	Properties of Treated Material	Special Advantages and Limitations	Relative Cost
				Precompression				
Preloading	Load is applied sufficiently in advance of construction so that compression of soft soils is complete prior to development of the site	Normally consolidated soft clays, silts, organic deposits, completed sanitary landfills	—	Earth fill or other material for loading the site; sand or gravel for drainage blanket	Earth moving equipment; large water tanks or vacuum drainage systems sometimes used; settlement markers, piezometers	Reduced water content and void ratio, increased strength	Easy, theory well developed, uniformity; requires long time (vertical drains can be used to reduce consolidation time)	Low (moderate if vertical drains are required)
Surcharge fills	Fill in excess of that required permanently is applied to achieve a given amount of settlement in a shorter time, excess fill then removed	Normally consolidated soft clays, silts, organic deposits, completed sanitary landfills	—	Earth fill or other material for loading the site; sand or gravel for drainage blanked	Earth moving equipment; settlement markers, piezometers	Reduced water content, void ratio, and compressibility; increased strength	Faster than preloading without surcharge, theory well developed; extra material handling; can use vertical drains to reduce consolidation time	Moderate
Electro-osmosis	DC current causes water flow from anode toward cathode where it is removed	Normally consolidated silts and silty clays		Anodes (usually rebars or aluminum), cathodes (wellpoints or rebars)	DC power supply, wiring, metering systems	Reduced water content and compressibility, increased strength, electrochemical hardening	No fill loading required, can use in confined area, relatively fast, nonuniform properties between electrodes, no good in highly conductive soils	High

				Injection and Grouting				
Particulate grouting	Penetration grouting-fill soils pores with soil, cement, and/or clay	Medium to coarse sand and gravel	Unlimited	Grout, water	Mixers, tanks, pumps, hoses	Impervious, high strength with cement grout, eliminate liquefaction danger	Low-cost grouts, high strength; limited to coarse grained soils, hard to evaluate	Lowest of the grout systems
Chemical grouting	Solutions of two or more chemicals react in soil pores to form a gel or a solid precipitate	Medium silts and coarser	Unlimited	Grout, water	Mixers, tanks, pumps, hoses	Impervious, low to high strength, eliminate liquefaction danger	Low viscosity, controllable gel time, good water shutoff; high cost, hard to evaluate	High to very high
Pressure-injected lime	Lime slurry injected to shallow depths under high pressure	Expansive clays	Unlimited but 2–3 m usual	Lime, water, surfactant	Slurry tanks, agitators, pumps, hoses	Lime encapsulated zones formed by channels resulting from cracks, root holes, hydraulic fracture	Only effective in narrow range of soil conditions	Competitive with other solutions to expansive soil problems
Displacement grout	Highly viscous grout acts as radial hydraulic jack when pumped in under high pressure	Soft, fine grained soils; foundation soils with large voids or cavities	Unlimited, but a few m usual	Soil, cement, water	Batching equipment, high-pressure pumps, hoses	Grout bulbs within compressed soil matrix	Good for correction of differential settlements, filling large voids; careful control required	Low material, high injection
Electrokinetic injection	Stabilizing chemicals moved into soil by electro-osmosis or colloids into pores by electro-phoresis	Saturated silts, silty clays (clean sands in case of colloid injection)	Unknown	Chemical stabilizer colloidal void fillers	DC power supply, anodes, cathodes	Increased strength, reduced compressibility, reduced liquefaction potential	Existing soil and structures not subjected to high pressures; no good in soil with high conductivity	Expensive
Jet grouting	High speed jets at depth excavate, inject, and mix stabilizer with soil to form columns or panels	Sands, silts, clays		Water, stabilizing chemicals	Special jet nozzle, pumps, pipes, and hoses	Solidified columns and walls	Useful in soils that can not be permeation grouted, precision in locating treated zones	

(continued)

Table 16.5 (*Continued*)

Method	Principle	Most Suitable Soil Conditions/ Types	Maximum Effective Treatment Depth	Special Materials Required	Special Equipment Required	Properties of Treated Material	Special Advantages and Limitations	Relative Cost
				Admixtures				
Remove and replace	Foundation soil excavated, improved by drying or admixture, and recompacted	Inorganic soils	10 m (?)	Admixture stabilizers	Excavating, mixing, and compaction equipment, dewatering	Increased strength and stiffness, reduced compressibility	Uniform, controlled foundation soil when replaced; may require large area dewatering	High
Structural fills	Structural fill distributes loads to underlying soft soils	Use over soft clays or organic soils, marsh deposits	—	Sand, gravel, flyash, bottom ash, slag, expanded aggregate, clam shell or oyster shell, incinerator ash	Mixing and compaction equipment	Soft subgrade protected by structural load-bearing fill	High strength, good load distribution to underlying soft soils	Low to high
Mix-in-place piles and walls	Lime, cement, or asphalt introduced through rotating auger or special in-place mixer	All soft or loose inorganic soils	>20 m	Cement, lime, asphalt, or chemical stabilizer	Drill rig, rotary cutting and mixing head, additive proportioning equipment	Solidified soil piles or walls of relatively high strength	Uses native soil, reduced lateral support requirements during excavation, difficult quality control	Moderate to high

				Thermal				
Heating	Drying at low temperatures; alteration of clays at intermediate temperatures (400–600°C), fusion at high temperatures (>1000°C)	Fine grained soils, especially partly saturated clays and silts, loess	15 m	Fuel	Fuels tanks, burners, blowers	Reduced water content, plasticity, water sensitivity; increased strength	Can obtain irreversible improvements in properties; can introduce stabilizers with hot gases	High
Freezing	Freeze soft, wet ground to increase its strength and stiffness	All soils	Several m	Refrigerant	Refrigeration system	Increased strength and stiffness; reduced permeability	No good in flowing groundwater, temporary	High
				Reinforcement				
Vibro-replacement stone and sand columns	Hole jetted into soft, fine grained soil and backfilled with densely compacted gravel or sand	Soft clays and alluvial deposits	20 m	Gravel or crushed rock backfill	Vibroflot, crane or vibrocat, water	Increased bearing capacity, reduced settlements	Faster than precompression, avoids dewatering required for remove and replace; limited bearing capacity	Moderate to high
Root piles, soil nailing	Inclusions used to carry tension, shear, compression	All soils	?	Reinforcing bars, cement grout	Drilling and grouting equipment	Reinforced zone behaves as a coherent mass	In-situ reinforcement for soils that cannot be grouted or mixed in place with admixtures	Moderate to high
Strips, grids, and membranes	Horizontal tensile strips, membranes buried in soil under embankments, gravel base courses and footings	Cohesionless, soils, some $c-\phi$ soils	Can construct earth structures to heights of several tens of meters	Metal or plastic strips, geotextiles	Excavating, earth-handling, and compaction equipment	Self-supporting earth structures, increased bearing capacity, reduced deformations	Economical, earth structures coherent, can tolerate deformations; increased allowable bearing pressure	Low to moderate

* From J. K. Mitchell "Soil Improvement—State-of-the-Art Report," *Proceedings of the Tenth International Conference on Soil Mechanics and Foundation Engineering*, Stockholm, 1981, Vol. 4 [16.14].

The method of total excavation and replacement is often most appropriate when weak materials near the ground surface can be excavated with little need to be concerned about control of groundwater. Thus, areas in which unsuitable materials (fill, peat, or other marsh deposits) overlie clay or some other material of low permeability are particularly suitable for the use of this technique.

A major advantage of the excavate-and-replace technique is not only that it provides increased bearing pressures for shallow foundations, but also that it permits the ground floor slab to be constructed as a slab-on-grade, rather than as a structural slab spanning between columns. Eliminating the need to support the ground floor slab can produce major cost savings.

2. *Partial Replacement.* On occasion, it is possible to excavate only the portion of the building area beneath the footings and replace the excavated material with controlled, compacted granular fill. This most often is appropriate when the unsuitable material is an uncontrolled fill that is too variable in composition and compressibility to support the loads of the columns, but is satisfactory, after a small amount of near-surface compaction or material replacement, to carry the loads of a slab-on-grade.

For convenience of construction, excavation of the material to be replaced is done in strips along column lines. Its width at the bottom of the excavation is two to three times the least width of the footings to be established on the fill placed in the excavation. To minimize the width of the excavated strip, the column footings are frequently combined into strips or are designed to be rectangular rather than square.

A major advantage of the strip excavation is that it reduces the volume of material that must be excavated and wasted, and requires much less backfill material than if the entire building were excavated. In addition, excavation in narrow strips reduces the quantity of groundwater to be controlled during the period of excavation and backfilling, making groundwater control more manageable. In fact, there are circumstances under which the excavation and replacement of material in an entire building area can be accomplished with reduced problems of groundwater control by excavating the entire area as a series of individual strips, each of which is backfilled before the adjacent area is excavated.

Soil Improvement In-Place

Soil may be improved in-place, that is, without removal and reworking or replacement, by some method of densification (reduction in volume), or by injection of additives. An overview of the materials that may be treated by the different methods that are available is provided in Fig. 16.5.

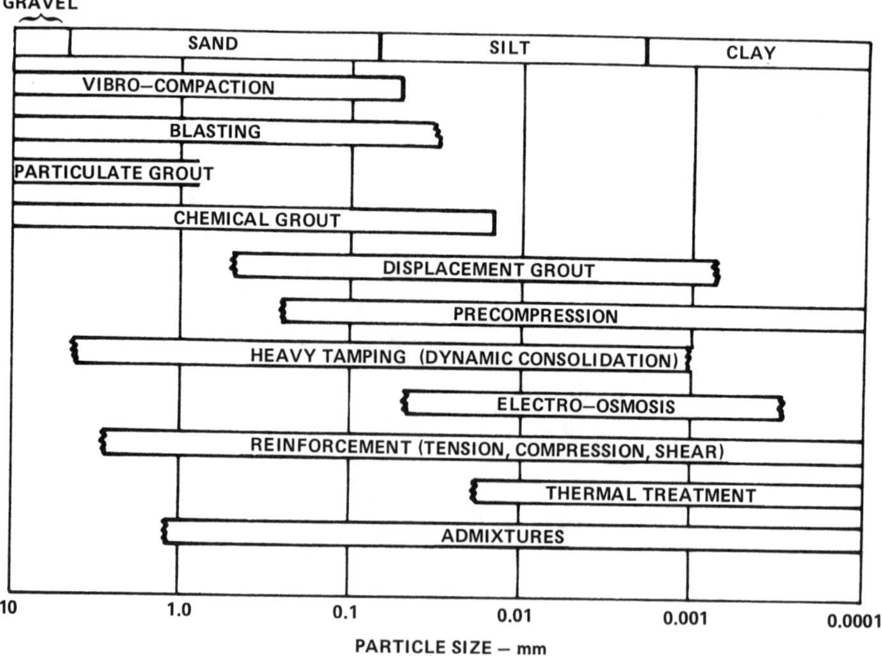

Fig. 16.5 Applicable grain-size ranges for different stabilization methods.

Only densification has broad application to building design. This discussion is limited, therefore, to the techniques available for soil improvement through densification.

Granular materials can be densified by vibratory compactors, vibroflotation, driving compaction piles, dynamic compaction, and blasting. Both granular and cohesive materials can be densified by surcharging. The densification methods are described as follows:

1. *Vibratory Compactors*. Vibratory compactors, which are effective in compacting sands, are a poor tool for compacting clays. They are available in sizes varying from hand-operated plate compactors, suitable for use in confined areas and to compact layers of soil 2–4 in. (50– 100 mm) thick, to heavy, self-propelled and towed units capable of compacting to depths of 6 ft (2 m). Thus, vibratory compaction is only suitable for surface compaction to shallow depths.

The results of compaction are affected by such factors, in addition to soil type, as the weight of the compactor, the thickness of the layer being compacted, the frequency of vibration of the compactor, and the number of times the compactor passes over a particular point. Soil improvement by compaction, when applied appropriately, has been shown to be an inexpensive and effective method for reducing the cost of building construction. For example, the compaction of the near-surface materials of an uncontrolled fill, when the problem soils are within the fill and not below it, can at times permit the substitution of a slab-on-grade for a structural slab at a substantial cost savings.

Care must be taken if compaction is to be done adjacent to an existing building. Heavy compactors should not be allowed within about 6 ft (2 m) of an existing foundation wall for fear of causing excessive lateral force on the wall. Furthermore, vibrations can produce densification of granular material beneath the wall, causing it to settle and producing cracks in the existing structure.

2. *Vibroflotation*. Densification by vibroflotation to depths of 60–100 ft (20–30 m) is achieved by jetting into loose sand and/or gravel deposits a cylindrical vibrator unit about 15 in. (380 mm) in diameter and 6 ft (1.8 m) long and activating the unit fully as it is withdrawn from the ground. A combination of the vibrations produced by a rotating eccentric weight and the water injected into the ground as the vibrator is withdrawn at a rate of about 1 ft/min (0.3 m/min), densifies the granular deposit and creates a depression at the ground surface. This depression is filled with 6 to 18 ft^3/ft (0.5 to 1.5 m^3/m) of sand to take up the space resulting from the soil densification. Vibroflotation is normally done using a triangular pattern of points spaced about 5–8 ft (1.5–2.5 m) apart.

Vibroflotation is commonly effective in sands and/or gravels containing less than about 20% silt and clay particles (material finer than about 0.06 mm). The finer the soil and the greater the proportion of silt, the closer the required spacing of the vibroflotation points.

Foundations on sand densified by vibroflotation may be footing foundations, if the inevitable variation in density between points of vibrofloat application is deemed not to be a problem, or a mat foundation, if it is considered necessary to bridge across materials of differing density. Allowable bearing pressures after successful vibroflotation vary between about 3 and 8 ksf (about 150 and 400 kN/m^2), but should be determined by test once compaction has been completed.

3. *Compaction Piles*. Compaction piles produce densification by displacement of material equal to the volume of the pile and from the effects of the vibrations produced by pile driving. Those piles, which are pipes or casings with a detachable end plate, are first driven into the sand to be densified. The pipes are then filled with sand before being withdrawn, leaving the sand behind to fill the hole. Pile spacings of 3–5 ft (1–1.5 m) are common for piles with diameters between 12 and 18 in. (300 and 450 mm).

Densification by compaction piles is frequently cost competitive with vibroflotation, but is likely to produce a somewhat less dense final product. Either footing or mat foundations can be used on sand densified in this manner, as described in the section on vibroflotation. They should be designed for allowable bearing pressures that are defined by field tests once compaction is completed.

4. *Dynamic Compaction*. In the late 1970s, a process called dynamic compaction was introduced to densify soils. Considerable skepticism was expressed by the engineering community about early claims of major improvements to cohesive soils. It has now been established that some improvement can be made to some cohesive soils at the cost of two or more cycles of compaction and the increased time associated with repeating the process. However, the method is clearly effective in compacting fill and other materials that are primarily granular, to depths of 30–60 ft (about 10–20 m).

Compaction is achieved by raising a 10- to 20-ton (about 100- to 200-kN) weight 50 to 130 ft (15 to 40 m) using a crane, and dropping the weight at points several feet apart across a site. The amount of compaction that can occur is related to the energy of the weight, and to the rate at which water can drain from the soil being compacted.

The vibrations associated with dynamic compaction can be felt at substantial distances from the point of impact. Therefore, care must be exercised in using this method in urban areas, where building damage might result and where the complaints of residents could delay a project. Dynamic compaction has been used successfully to increase the density of sand deposits beneath several major structures, to improve the behavior of rubble fill when subjected to the loads of low-rise buildings, and for other landfill improvements.

5. *Blasting.* Densification by blasting of loose sands, generally with less than 20% silt or clay size particles contained within the sand, has been done a number of times. Successive detonations of small explosive charges are required to improve the uniformity of the densification. However, detailed design procedures are not available.

Densification by blasting is economical when compaction of deep deposits of sand is required. The available data and experience indicate that it is unlikely that the high relative densities or the uniformity of densities that can be achieved by vibroflotation can be obtained by blasting, yet the system is occasionally found to be useful. Relative densities of 70 to 80% have been reported, although variability in relative density should be expected throughout a deposit compacted by blasting.

Clearly, the effects of the vibrations associated with blasting on adjacent structures and on people must be considered before selecting this method for use.

6. *Surcharging.* The term surcharging (or preloading) refers to the temporary placement of an overload of soil over the area to be covered by a building or other structure. This load is at least equal to, but is normally greater than, the weight of the facility and serves to compress the soils beneath the area before construction begins. The load, or surcharge, is removed when sufficient compression or densification of compressible soils is deemed to have occurred. Surcharging prepares the ground to receive the structural loadings without the occurrence of excessive displacements.

Surcharging is effective for compressing either sand or clay. However, the time needed to complete the compression (consolidation) process is orders of magnitude greater for cohesive deposits than for granular deposits of similar thickness. Because the time necessary for surcharging increases with the square of the thickness of the layer to be compressed, it is rare for most commercial applications that time is available to prepare *thick* layers of clay by surcharging, even if special drainage wicks or sand drains are used to speed up the process. It is used frequently when relatively thin layers [3–6 ft (1–3 m) thick] of compressible materials are found.

Surcharging is commonly achieved by heaping fill, generally needed for or obtained from other portions of the site being developed, on the building area. About 18 in. (500 mm) or more of fill per building floor is used as a surcharge. Thus, a 30-story building may require a pile of fill 50 ft (15 m) or more high covering the entire area of the structure. However, fill this high could cause a bearing failure in the near-surface soils to be improved by the surcharge and in causing a bearing failure, could move adjacent or nearby structures laterally. Furthermore, it must be recognized that the vertical movements produced by the surcharge extend beyond the limits of the filled area and could also cause settlement of nearby structures. Thus, considerable care is necessary in designing a surcharge program.

Surcharges may be required to remain in place one to three months if shallow organic deposits and underlying sands, such as are found in South Florida, are being compressed. Six months to several years could be necessary to compress thick deposits of clay, depending on the thickness of the compressible deposit and whether or not sand drains or wicks are used to speed up the consolidation process.

Although the proper design of surcharge programs requires special expertise, they should be considered frequently. They are often very effective in eliminating, at low cost, the need for a pile-supported ground floor slab or eliminating excessive settlements of roadways, parking areas, and buried utility lines. They are effective in eliminating the need for an expensive deep foundation for an entire building often enough that they should not be forgotten.

16.5 SELECTION OF A FOUNDATION SYSTEM

16.5.1 Need for Subsurface Investigation

The need for an adequate program of subsurface investigation goes substantially beyond the designer's need for subsurface information before a foundation system can be selected. The owner of the proposed structure needs to have the information obtained from the subsurface investigation so that he can gain confidence in the safety of the structure once constructed, and confidence that it can be constructed without damaging neighboring property; public officials, charged with responsibility for the safety of the public, need the information in order to be confident of the adequacy of the design of the proposed structure and confident that damage to surrounding property will not occur during or subsequent to construction; contractors need to know, at the time they are preparing their bids for construction of the structure, what subsurface conditions they can expect to encounter.

The primary objectives of the subsurface investigation are to provide information specific to the building site on the subsurface stratigraphy to some depth of interest, and to provide information on the physical properties of the subsurface materials, such as strength, compressibility, and permeability, so that a safe and economical foundation system can be selected and designed. Furthermore, information should be obtained as part of the subsurface investigation to permit design of appropriate earth-retaining (sheeting and bracing) and/or dewatering systems that may be required during construction. Frequently, it is during the progress of the subsurface investigation that information is gathered on the foundations of adjacent structures, on the location of nearby buried facilities (utilities, vaults, abandoned foundations, etc.), and on other facilities that might affect or be affected by construction activities.

The subsurface investigation must be designed to develop the information necessary to design for local conditions. For example, special attention must be given in some cities to the presence beneath a particular site of abandoned mines that might collapse at some future date, or to a history of sinkhole formation in an area and the likelihood that future sinkholes will appear beneath the proposed structure. The stability of slopes adjacent to the property being investigated must be studied to be confident that they will not fail during or subsequent to construction and affect the structure under design, and/or the stability of the slope on which the proposed building is to be constructed must be verified. For example, a design engineer must be particularly cautious about constructing on or adjacent to all slopes inclined steeper than about 1 vertical : 4 horizontal.

Construction costs can escalate dramatically if subsurface conditions disclosed during construction differ materially from those anticipated by the contractor from the results of the subsurface investigation. Furthermore, important design assumptions may prove to be invalid, with serious consequences for the structure, if subsurface conditions differ materially from those expected by the designer. Therefore, it is not only necessary that a comprehensive program of subsurface exploration be completed prior to final design, but also that the information on subsurface conditions and on the design assumptions for the foundation system be made available to those who supervise and inspect the construction activities. These individuals must verify, on a continuing basis, that the subsurface conditions assumed by the designer actually do exist, and that the subsurface conditions also conform to those that the contractor should have expected.

It is not considered appropriate to include in this chapter a (necessarily) long discussion of the design of a program for subsurface exploration. Those who require detailed information are referred to the 1972 ASCE report "Subsurface Investigation for Design and Construction of Foundations of Buildings" [16.1]. It should also be recognized that there are frequently specific requirements for subsurface investigation programs that are included in the various building codes governing structures in different areas. Normally, engineers must comply with the requirements of those codes, in addition to obtaining sufficient subsurface information of the appropriate type and quality to permit them to produce an economical design of the structure.

The ASCE report [16.1] on subsurface investigations states some general guidelines that may be valuable to the designer, as follows:

> *Thus, foundation investigations for design should proceed through four phases:*
>
> 1. *Initial studies and explorations to determine soil stratification and characteristics required for design, considering both the structure under design and possible adjoining structures. These investigations should be planned considering structural requirements, depths of excavation, available knowledge of soil conditions, and possible requirements for dewatering and other construction problems.*
> 2. *Amplification, if necessary, of specific portions of the initial investigation to obtain more complete information, as desirable, during the design phase and for preparation of contract documents.*
> 3. *Verification of anticipated foundation conditions during construction so that changes may be made, if necessary, to ensure proper performance and control for assurance of compliance with design.*
> 4. *Observation of structure and soil performance following construction.*
>
> *Items 1 and 3 are essential. Items 2 and 4 may be of limited or of substantial scope, depending upon the nature of the project.*

16.5.2 Suggested Protocol for Foundation Selection

The foundation system chosen for a specific building is usually that system believed by the owner and designer to provide the lowest construction cost, to provide the performance necessary for satisfactory behavior of the structure it supports, and that which is believed unlikely to damage adjacent or nearby structures or other facilities. The effects on adjacent structures are, most frequently, concerns when constructing in urban areas, but also must be considered when adding to an existing structure. A protocol for the selection of a foundation system is provided below for consideration by the design engineer.

1. Define clearly the characteristics of the building such as height, foundation loading, and number of basements.

2. Establish realistic performance criteria for the foundation system based upon structural requirements, architectural requirements, and the use of the structure. In establishing these criteria, recognition should be given to several fundamental facts:

a. Zero settlement is an unacceptable criterion; the imposition of load on every material causes displacement. The question that must be addressed is how much settlement is tolerable. This has been discussed in some detail in Section 16.3.

b. Architectural finishes, which are usually most sensitive to differential movements, are installed in all structures late in the construction sequence, well after much of the dead load has been applied to the foundations and therefore, after some portion of the anticipated total and differential settlement has already occurred. Often, essentially all the settlement of a building has occurred before the sensitive architectural finishes are in place.

c. Settlement of some structures, for example, some warehouses, does not affect their usefulness for their intended purpose. Therefore, some significant settlement [which may at times be as much as 12 in. (300 mm)] may be acceptable to the user of the building. The decision on acceptibility of significant settlement should be the owners; it is the owner who receives the benefit of the low-cost foundation system that will permit settlements to occur rather than paying for a higher-cost system that will avoid and resist building settlement. In exchange for this benefit of lower cost, it is equitable that the owner assume the risks of cracking or other distress associated with settlement.

3. Summarize the information on subsurface conditions (soil, rock, groundwater) at the building location. Selection of foundations most frequently begins with an assessment of the suitability of a shallow foundation system, commonly the least expensive system of support.

a. Assess the likely total and differential settlements of the structure under the average structural loads if the foundation bears on or above clay soils, or under the maximum probable loads if the foundation bears on sand. Settlement is ordinarily not a problem if all foundation elements bear on rock, but might be a problem if some of the elements bear on rock or some other relatively rigid material while others bear on soil.

b. Assess the bearing capacity of the subsurface materials and compare them with the maximum probable and maximum possible loads. Sometimes it is necessary to consider the frequency of application of the maximum loads and, in the extreme, to also consider the time of application of the loads.

c. If either the predicted settlements are intolerable for the structure or the bearing capacity is inadequate, assess the potential for increasing the depth of the basements or improving the subsurface conditions so that shallow foundations might be utilized.

d. If increased basement depth is uneconomical or not useful, compare the cost of soil improvement with the cost of the least expensive deep foundation system. There are many geographic areas in which a low-capacity timber pile foundation is less expensive than soil improvement, and there are other areas in which moderate-capacity short, precast concrete piles are less expensive than soil improvement.

e. If a shallow foundation system proves to be inappropriate, assess the costs and appropriateness of deep foundation systems to support the foundation loads. Recognize that supporting the main columns on a deep foundation does not imply the need to support the ground floor slab on a similar foundation. Frequently, the necessary excavation (soil unloading) to provide even a single basement level will make possible the use of a slab-on-grade. Otherwise, the use of a low surcharge to precompress the area of the lowest floor slab should be considered. Frequently, pile-supported structural floor slabs can be avoided at low cost, yielding substantial savings.

4. Compare the costs of the various deep foundation systems considered suitable for the particular structure. The cost estimates for each system should provide a range within which the actual cost can be expected to fall. The range provides an indication of the engineers confidence in his estimate, as well as an indication of construction uncertainties that tend to increase costs.

5. List, for each of the acceptable foundation systems, the following:

a. Uncertainties associated with construction of the foundation system that could increase its cost and time for construction. For example, a dewatering system may be necessary if soil improvement by excavation and replacement is chosen, while it may not be necessary if a pile foundation is utilized; required rock excavation may be greater than that assumed when rock elevations are inferred between widely spaced borings; the foundation system that appears to be most economical may not have been installed before under similar conditions, perhaps leading to construction delays.

b. Estimates of the effects of the installation of each foundation system on adjacent or nearby structures and the cost associated with avoiding these effects or correcting the damage that is caused. Consider the effects of vibrations due to pile driving or blasting, the settlements produced by dewatering or by excavating adjacent to an existing structure, and the effects of the settlement of the new structure under its building loads on the settlement of adjacent structures.

c. Estimates of the time for construction of each acceptable foundation system. These estimates cannot consider explicitly delays produced by litigious neighbors, but can take into account, in a qualitative way, the potential for delays resulting from third party litigation.

d. Estimates of the relative reliability of each of the foundation systems. For example, boulders may deflect H-piles during installation, perhaps reducing their load-carrying ability; therefore, a pipe pile that can be inspected for verticality after installation may be preferable. The construction of a pier or caisson for which the bearing surface can be inspected before concrete is placed may be considered more reliable than a similar system for which inspection of the bearing surface must be done below water (or not at all) and concrete must be tremie-placed.

e. Estimates of the reliability of predictions of the performance of each of the foundation systems being considered. Commonly, the reliability of predictions of performance increases as the cost of the system increases. Therefore, the level of risk associated with the choice of a particular foundation system should be identified to the owner of the building, the person who benefits from all reductions in the cost of construction.

6. Select a foundation system from the above information. It should be clear that the selection is based on a set of relative values and judgments. The value placed on initial estimated construction cost, on uncertainties that may lead to delays in construction and increased cost, on the level of confidence that is possible in predicting the performance of foundation systems, and on the risk acceptable to the owner, are all included in making a selection of a foundation system for a particular structure.

16.6 PRELIMINARY DESIGN OF SHALLOW FOUNDATIONS

16.6.1 Introduction

The information provided in this section is designed to permit the structural engineer to prepare preliminary designs for several foundation systems so that their likely costs and performance can be estimated. Preliminary design, for purposes of this discussion, includes the proportioning of footings and mats and the identification of pile type and probable length, but not the detailed design of reinforcing steel or pile caps and pile connections.

This discussion of preliminary design has deliberately been placed late in the chapter in an effort to emphasize that design of the foundation elements is only one small part of the process of selecting a foundation system. Design is the simplest and most clear-cut part of engineering; it must follow the clear definition of the problems that the design must take into account. The fundamental challenge faced by engineers is that of identifying the critical engineering problems that need to be addressed in design; it is not the problem of the design itself.

16.6.2 Shallow Foundations

The design of shallow footing or spread footing foundations bearing on sand or on clay is best thought of as two individual problems. Although footings bearing on both sand and clay must be safe against bearing capacity failure (failure in shear of the material supporting the footing) and also against excessive settlement, the bearing capacity of footings on sand is rarely in question and the critical time in the life of foundations bearing on sand is quite different from that of foundations bearing on clay. Foundations on sand respond rapidly to the maximum load that is applied to them even for a short time. This is because sand is permeable, water can drain readily from between the sand grains, and volume reductions (soil consolidation), producing settlement of foundations, take place rapidly. Clay, on the other hand, does not drain rapidly. Therefore, volume changes (settlements) occur under average loads, over long periods of time. Furthermore, the increase in bearing capacity that is associated with soil consolidation occurs slowly in clay, so slowly, in fact, that usually no strength increase may be assumed for purposes of design. Bearing capacity failures of foundations bearing on clay do occur, usually on first loading, and must be guarded against. The selection of loadings for use in design has been discussed in Section 16.2.1.

The design procedures described in this section are valid for static loads only. Vibrations resulting from equipment, transient repeated loads from wind or waves, and earthquake-induced stresses will often cause additional settlement of foundations on sand and may cause additional settlement of foundations on clay. Settlements from these sources are not considered in the discussions that follow.

16.6.3 Shallow Foundations on Sand

Bearing Capacity

The *ultimate* bearing capacity of foundations may be estimated with sufficient accuracy for design, as follows:

$$q = \frac{Q}{B'L'} = d_c i_c s_c c N_c + d_q i_q s_q \gamma D N_q + \frac{1}{2} d_\gamma i_\gamma s_\gamma \gamma B' N_\gamma \tag{16.6.1}$$

The terms in Eq. (16.6.1) are defined below and in Fig. 16.6:

q = unit soil contact pressure
Q = total column load

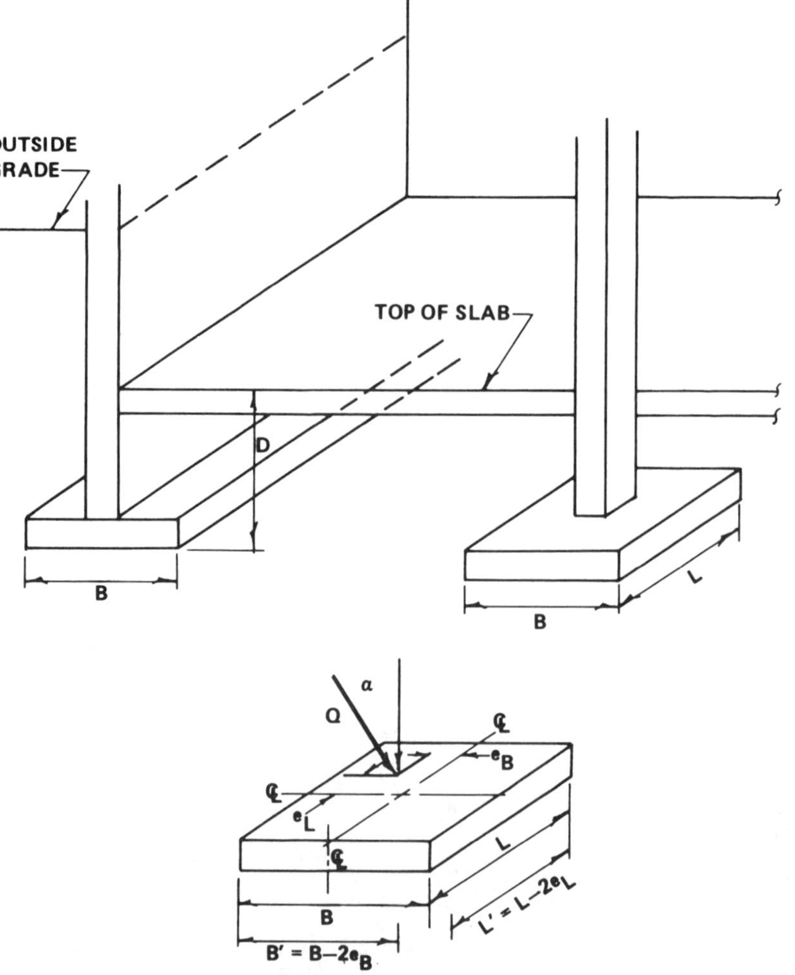

Fig. 16.6 Definition of terms in Eq. (16.6.1).

B = least width of foundation
B' = effective width of foundation = $B - 2e_B$
e = eccentricity of resultant load R in B and L directions
L = length of foundation
L' = effective length of foundation = $L - 2e_L$
c = apparent cohesion of soil
D = *least* depth from top of adjacent ground to bottom of footing
γ = effective unit weight of soil (see further discussion)

N_c, N_q, N_γ = bearing capacity factors that depend mainly on the angle of internal friction or angle of shearing resistance ϕ of the soil, as shown in Fig. 16.7 and defined below.

d_c, d_q, d_γ = factors associated with the depth of the foundation
i_c, i_q, i_γ = factors associated with the inclination of the applied load,
s_c, s_q, s_γ = factors associated with the shape of the foundation

The bearing capacity factors for a shallow horizontal strip foundation under vertical load, where the depth D of the foundation is less than or equal to its width ($D/B \leq 1$) and ϕ is greater than 10° are

$$N_c = (N_q - 1) \cot \phi \qquad \text{(theoretical)} \qquad (16.6.2a)$$

$$N_q = e^{(\pi \tan \phi)} N_\phi \qquad \text{(theoretical)} \qquad (16.6.2b)$$

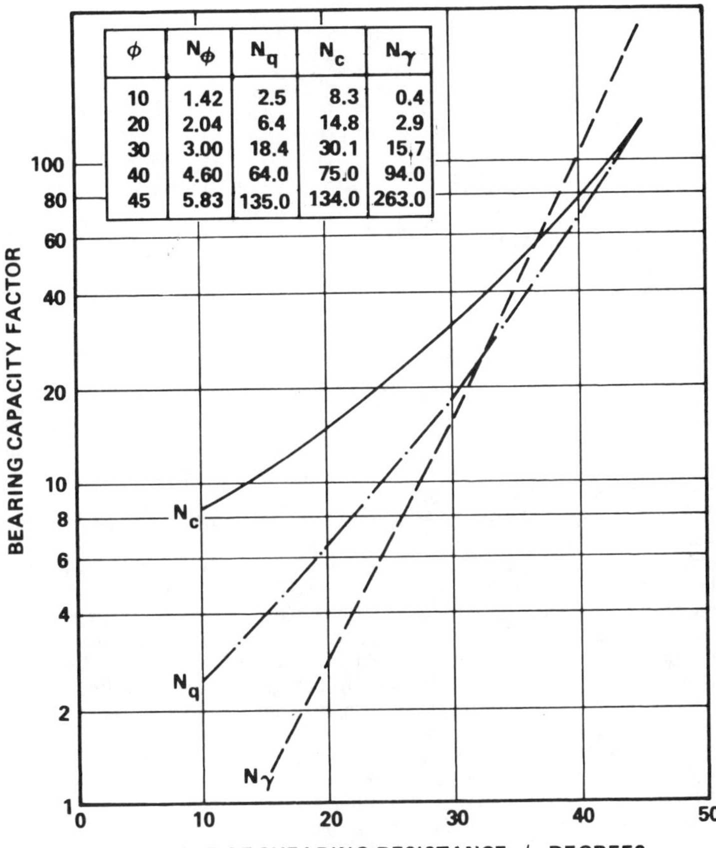

ϕ	N_ϕ	N_q	N_c	N_γ
10	1.42	2.5	8.3	0.4
20	2.04	6.4	14.8	2.9
30	3.00	18.4	30.1	15.7
40	4.60	64.0	75.0	94.0
45	5.83	135.0	134.0	263.0

Fig. 16.7 Bearing capacity factors.

$$N_\gamma = (N_q - 1)\tan(1.4\phi) \qquad \text{(approximate)} \tag{16.6.2c}$$

$$N_\phi = \tan^2\left(\frac{\pi}{4} + \frac{\phi}{2}\right) \qquad \text{(theoretical)} \tag{16.6.2d}$$

The depth factors for a depth of foundation less than its width ($D/B \leq 1$) are

$$d_c = 1 + (0.2\sqrt{N_\phi})\frac{D}{B} \tag{16.6.3a}$$

$$d_q = d_\gamma = 1 \qquad \text{when} \quad \phi = 0 \tag{16.6.3b}$$

$$d_q = d_\gamma = 1 + (0.1\sqrt{N_\phi})\frac{D}{B} \qquad \text{when} \quad \phi > 10° \tag{16.6.3c}$$

The inclination factors are

$$i_c = i_q = \left(1 - \frac{\alpha}{90°}\right)^2 \tag{16.6.4a}$$

$$i_\gamma = \left(1 - \frac{\alpha}{\phi}\right)^2 \tag{16.6.4b}$$

where α is the inclination from the vertical, as illustrated in Fig. 16.6.

The shape factors are approximately

$$s_c = 1 + (0.2N_\phi)\frac{B}{L} \tag{16.6.5a}$$

$$s_q = s_\gamma = 1 \qquad \text{when} \quad \phi = 0 \tag{16.6.5b}$$

$$s_q = s_\gamma = 1 + (0.1N_\phi)\frac{B}{L} \qquad \text{when} \quad \phi > 10° \tag{16.6.5c}$$

Equation (16.6.1) is a general bearing capacity equation, applicable to both sands and clays. Assuming sand to be cohesionless so that $c = 0$ in the first term on the right side of Eq. (16.6.1), leads to the following form of the equation for sands:

$$q = d_q i_q s_q \gamma D N_q + \frac{1}{2} d_\gamma i_\gamma s_\gamma \gamma B' N_\gamma \tag{16.6.6}$$

The first term of Eq. (16.6.6) relates to the contribution to bearing capacity of the soil *above* the base of the foundation, while the second term relates to the contribution of the material below the base of the foundation. This distinction between the terms is important if the correct unit weight of soil is to be introduced into the equation.

If, for example, the groundwater level is at the base of a footing at a depth D below ground surface, then the effective unit weight γ in the first term is the total unit weight of the soil, while the effective unit weight γ in the second term is the submerged unit weight of the soil. The submerged unit weight may be taken as the total unit weight of soil less the unit weight of water.

Furthermore, if the groundwater level is at a depth greater than B below the bottom of the footing ($D + B$ from the surface), the effect of submergence may be ignored. The effective unit weight for the second term of Eq. (16.6.6) may be obtained by simple proportions if the water level is between the bottom of the footing and a depth of B below the bottom. Similarly, the effective unit weight for the first term of Eq. (16.6.6) may be obtained by simple proportions if the water level is between the ground surface and the depth D.

It is fortunate that the bearing capacity of foundations on sand is only significant for narrow footings [say 1–2 ft (300–600 mm) wide] bearing on loose sand when the water table is near the surface of the ground. If this were not the case, there would be a strong need to define accurately the angle of shearing resistance, ϕ, of the soil. As it is, only approximations are necessary for most problems.

Approximations of ϕ may be made from the results of standard penetration test data or cone penetration data obtained during the program of subsurface exploration. The cone penetration

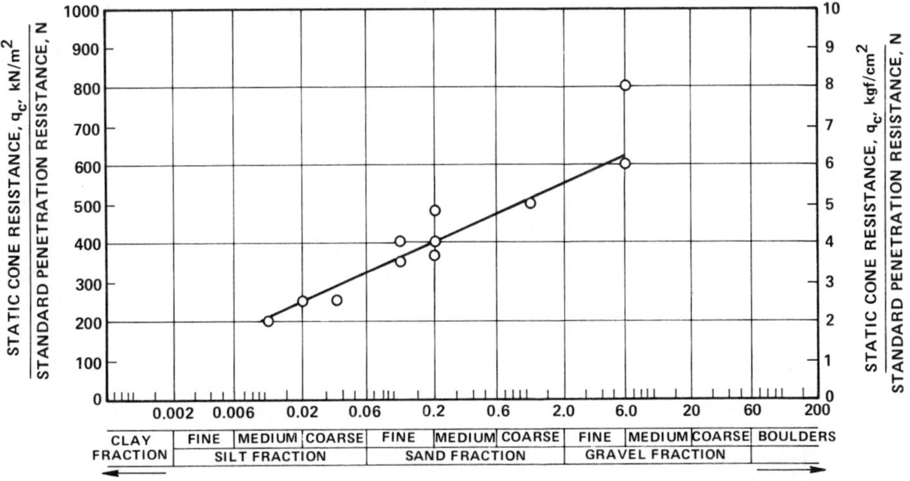

Fig. 16.8 General relationship between static cone resistance/standard penetration test (q_c/N) and particle size.

resistance q_c is obtained using a 60° cone of 1.5-in.² (1000-mm²) cross-sectional area pushed (not driven) into the ground. The standard penetration test value N is the number of blows of a 140-lb (0.62-kN) hammer dropping 30 in. (760 mm) required to drive a standard split-spoon sampler 2-in. OD, 1.4-in. ID (50-mm OD, 35-mm ID) 12 in. (300 mm) into the ground.

In the tabulation given below, cone penetration resistance in kN/m² have been assumed to be approximately equal to 400 times the standard penetration test value N. The ratio of cone resistance in kN/m² to standard penetration N for granular materials is shown in Fig. 16.8. The following approximations to the angle of shearing resistance ϕ are considered reasonable for preliminary design of foundations on sand:

	N	q_c		Estimated ϕ
Description of Sand	blows/12 in. (blows/300 mm)	ksf	kN/m²	degrees
Very loose	0–5	0–40	0–2,000	28
Loose	5–10	40–80	2,000–4,000	30
Medium	10–30	80–240	4,000–12,000	34
Dense	30–50	240–400	12,000–20,000	39
Very dense	above 50	above 400	above 20,000	43

In conventional practice, the allowable bearing capacity of a footing q_a is taken as one-third the ultimate bearing capacity q given by Eq. (16.6.1).

Settlement of Footings on Sand

The settlement of a footing foundation bearing on sand ordinarily is the major concern for design. It has already been stated in Section 16.3.6 that measurements have shown that the maximum differential settlement of a foundation on sand is between 80 and 100% of the maximum total settlement.

Estimates of the settlement of foundations bearing on sand cannot yet be made by theory alone. This is because the load/deformation behavior of sand depends upon such factors as the depth of the point of concern below ground level, the state of packing of the particles (usually defined as the relative density of the sand, but also related to the size of the particles and the grain-size distribution of the material), as well as the stress history of the deposit since its deposition. Because stress history produces anisotropy of stress in sands, it is also not possible to estimate settlements accurately from in-situ tests, such as standard penetration tests or cone resistances which are taken vertically, or from in-situ pressuremeter tests which are taken horizontally. Consequently, a semi-empirical design procedure is commonly followed.

The following rules for design are based upon a combination of settlement observations and theoretical concepts. They pertain to conventional footings founded on sand at depths of 3–6 ft (1–3 m) below grade.

1. The allowable soil bearing pressure q_a in ksf is given as

$$q_a = \frac{1.1sN[1 + 0.4(D/B)]}{B^{1/2}} \qquad (16.6.7)^*$$

where s = settlement in in. considered acceptable
N = the average standard penetration value for a depth $2B$ below the bottom of the footing
B = width of footing in ft, except that this width must equal or exceed 30 in. (0.75 m) as discussed below
D = depth from surface to base of footing and always must equal or exceed 3 ft (1 m)
D/B = ratio of depth of footing to its least width and is to be less than or equal to 2

2. The allowable bearing pressure varies from zero to the value computed at B = 30 in. (0.75 m) when the footing width is smaller than 30 in. (0.75 m), except that the allowable pressure may not exceed one-half the ultimate bearing capacity obtained using Eq. (16.6.1)

3. The doubling of the bearing pressure beneath a footing will double the settlement of the footing, as long as the final load is less than about one-half of the ultimate bearing capacity given by Eq. (16.6.1).

4. Footings subjected to eccentric loads should be proportioned for the width B' shown in Fig. 16.6. and the size then increased to width B for inclusion on design drawings. However, the extreme edge pressure computed as occurring under the eccentric loads assuming the footing and foundation to be elastic should not exceed the allowable bearing capacity of the sand deposit.

5. Increases in water level above those found at the time of the field exploration program may increase the settlement of a footing. The increase in settlement resulting from a rise in water level is approximately given by

$$\frac{S_f}{S_o} = \frac{\gamma'(D + B) + \gamma_w D_{w_o}}{\gamma'(D + B) + \gamma_w D_{w_f}} \qquad (16.6.8)$$

where S_f, S_o = the final and original predicted settlement
γ' = the submerged unit weight of soil which is the total unit weight of soil less the unit weight of water
γ_w = the unit weight of water
D_{w_o}, D_{w_f} = the original and final depths to groundwater, respectively, measured from ground surface or closest ground level adjacent to footing
D = the depth from ground surface or closest ground level adjacent to footing to the bottom of the footing
B = depth equal to the width of footing beneath its base.

6. In order to maintain likely differential settlements within acceptable limits, the heaviest footing should be assumed to bear on the soil with the lowest average N value.

The settlement of foundations on sand that are wider than about 20 ft (6 m), such as foundations for mats, bridge piers, and tanks, requires special discussion. Conventional settlement analyses most often overestimate the magnitude of the settlements of wide footings and predict incorrect patterns of deflection [16.2, 16.11]. Both these errors result, at least in part, from the fact that the modulus of a deep deposit of sand increases approximately with the square root of depth. Thus, at 30 ft (9 m) the modulus is about three times greater than it was at 3 ft (1 m), altering the compressibility of the soil with depth and showing clearly that computations based on a Boussinesq stress distribution, for which the modulus must remain constant throughout, must produce erroneous results. Predicting the settlement of wide footings requires special procedures similar to those described by Schmertmann [16.23]. However, for preliminary purposes, the allowable pressure for mat or pier foundations may be assumed equal to that given by Eq. (16.6.7), assuming an acceptable settlement s of 2 in. (50 mm). This is valid for mat foundations founded at depths greater than 8 ft (2.5 m) and for piers where $D/B > 3$.

---

* To determine q_a in kN/m^2 using Eq. (16.6.7), s must be in mm and B must be in m.

16.6.4 Shallow Foundations on Clay

The design of foundations bearing on clay must take into consideration both the bearing capacity of the material and its settlement (consolidation) under the applied loads. Neither the bearing capacity nor the settlement of foundations on deposits of clay can be predicted with a high degree of certainty. Some of the many reasons for this have been discussed in Section 16.2.1, where the selection of appropriate safety factors is described.

Bearing Capacity

The bearing capacity of foundations on clay is usually estimated on the assumption that ϕ equals zero. This assumption is conservative for all cases in which the current stress is increased by the application of foundation loads, but may be unconservative for cases in which unloading occurs. For example, the strength of clay at the base of an excavation decreases with time as heave or swelling occurs and, therefore, the stability of the bottom of the excavation is reduced with the passage of time.

The assumption that ϕ equals zero is only valid when:

1. The clay is fully saturated so that no volume change occurs immediately on application of load.
2. The strength (shearing resistance) of the clay has been determined from a laboratory test in which no volume change occurs.
3. The analysis of stability is performed using total stresses with no account taken of the groundwater levels.

Under these conditions, and ignoring the effect of the inclination of the load, Eq. (16.6.1) reduces to

$$q = cN_c + \gamma DN_q \tag{16.6.9}$$

where all terms have been defined earlier. The value of c in Eq. (16.6.9) is obtained from an undrained test in which no volume change occurs, is equal to one-half the compressive strength of the soil sample, and is hereafter termed c_u. The value of N_c is obtained from the following approximate relationship:

$$N_c = 5\left(1 + 0.2\frac{D}{B}\right)\left(1 + 0.2\frac{B}{L}\right) \tag{16.6.10}$$

in which D/B is always less than or equal to 2.5. Thus, for a strip footing at the ground surface, for which D/B and B/L are both zero, N_c equals 5, while for the same footing at great depth, where D/B exceeds 2.5, N_c equals 7.5. For the end bearing of a pile where D/B is large and B/L is unity, N_c equals 9.

The value of N_q in Eq. (16.6.9), for the case where ϕ equals zero is unity and the equation reduces to the more common form

$$q = c_uN_c + \gamma D \tag{16.6.11}$$

It should be noted that the term γD is the overburden pressure already existing at the base of the footing, as shown in Fig. 16.6. Therefore, the stress in addition to the overburden pressure at the base of the footing that must be applied to cause catastrophic failure of the footing equals c_uN_c. (The application by the designer of a safety factor of 2 or more to c_u to arrive at the soil pressures allowable for design will avoid the unacceptable creep deformations that occur as catastrophic failure loads are approached.)

Several special conditions must be considered by designers. The first is "local" shear failure at the edge of a foundation, and the second is design when a weak clay stratum underlies a denser stratum.

Local shear failures have been reported at the extreme edges of large foundations, such as mats or storage tank foundations. These failures result from stress concentrations at the perimeter of the structure which arise in response to eccentric loads, wall loads, or the redistribution of load to the edges of rigid foundations. Designers must consider the possibility of local stress concentrations that might lead to a localized, but significant, area of failure. Normal bearing capacity analyses are believed to pertain to analysis of the potential for local failure.

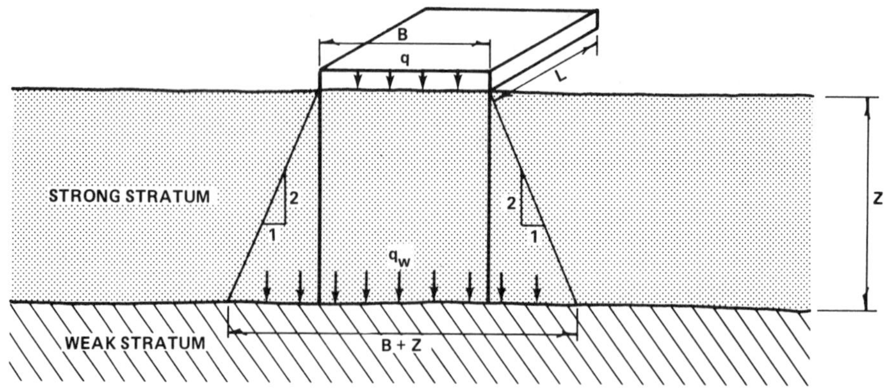

$$q_w = q \cdot \frac{B}{B+Z} \text{ (STRIP FOOTING)}$$

$$q_w = q \cdot \frac{B^2}{(B+Z)^2} \text{ (SQUARE FOOTING)}$$

$$q_w = q \cdot \frac{B \cdot L}{(B+Z)\,(L+Z)} \text{ (RECTANGULAR FOOTING)}$$

Fig. 16.9 Simple distribution of stress.

When a weak layer is found beneath a denser stratum, then it is usually conservative to distribute loads through the dense stratum and perform a bearing capacity analysis at the top of the weak material using the attenuated stress value computed from the distribution. A simple distribution of stress, as shown in Fig. 16.9, is usually sufficient for preliminary analyses. The real distribution is much more complex than suggested in the figure, generally producing a lower stress on the weak stratum than that obtained by the simple analysis. A fuller treatment of the bearing capacity of layered systems has been presented by Mitchell and Gardner [16.15].

Settlement

The prediction of the settlement of a footing foundation on clay requires considerable experience. It is common for engineers using standard procedures to underpredict by 20 to 50% the settlement of foundations on soft clays, and to overpredict by 100% the settlement of foundations on very stiff to hard clays. Part of the problem with the predictions must be related to changes in the properties of the soil which occur as a result of the sampling and testing operation. Additional error is introduced when the actual loads applied to the foundations are uncertain.

Prediction of settlement requires knowledge of the change in stress produced by the foundation, the distribution of this stress throughout the compressible deposit, the thickness of the deposit, and the change in volume associated with the change in stress. The latter is the stress–strain relationship or modulus appropriate for the compressible material. Thus, in its simplest form, the settlement may be expressed as

$$s = \sum \frac{\Delta p}{E_c} H \tag{16.6.12}$$

where s = the total settlement (in. or mm)

Δp = the average change in stress for each stratum of thickness H

H = the thickness of the stratum or portion of the stratum being considered in the calculation in the same units as s

E_c = the modulus of compressibility associated with the material being compressed expressed in the same units as used for Δp.

To compute the total settlement s, the compressible stratum should be divided for computation purposes into layers that are 5 to 10 ft (1.5 to 3 m) thick, but no thicker than one-half the width of

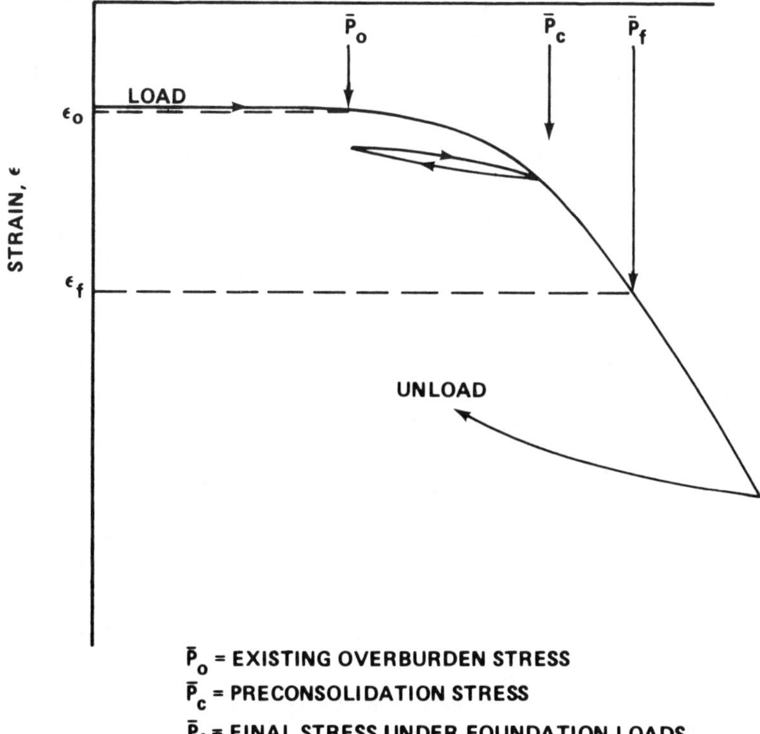

Fig. 16.10 Schematic representation of results of consolidation test.

the footing or foundation whose settlement is being computed. The stress produced by the foundation is then distributed throughout the thickness of the compressible stratum, being certain to include the contribution to the stresses produced by adjacent or nearby footings (the 2 vertical : 1 horizontal distribution shown in Fig. 16.9 is frequently used for this purpose). The stress increase at the center of each layer is determined and the settlements of the several layers are added together using Eq. (16.6.12). Determination of the appropriate modulus E_c is, of course, a major problem. Experience has shown that for heavily overconsolidated hard clays*, it is often sufficiently accurate for preliminary rough estimates of settlement, to assume E_c as equal to $500c_u$, where c_u is the undrained shear strength of the clay. Normally, however, a laboratory test is performed to determine the soil modulus for the range of stress in question.

A schematic representation of the results of such a laboratory test, called a consolidation or oedometer test, is shown as Fig. 16.10. The results of that test can be used directly to estimate settlements by rewriting Eq. (16.6.12) to include the strain appropriate for the increment of stress at the center of each soil layer:

$$s = \Sigma(\epsilon_f - \epsilon_o)H \tag{16.6.13}$$

where ϵ_f is the final strain obtained from the consolidation test (Fig. 16.10) at the final stress at the center of each layer, and ϵ_o is the strain from the consolidation test corresponding to the overburden stress at the center of each layer.

* An overconsolidated soil is one that has been subjected in its geologic past to a stress in excess of the current stress acting on it. The overconsolidation ratio is the ratio of the maximum past pressure at a point to the current pressure at that point. Overconsolidation ratios in excess of 10 or more pertain to hard, heavily overconsolidated soils.

The performance of a laboratory test to obtain a reasonable representation of the correct stress–strain relationship may require the use of special loading sequences. Sample disturbance, for example, is frequently reduced by loading the specimen to about the preconsolidation pressure (shown on Fig. 16.10), unloading the specimen to the overburden pressure, and then reloading well beyond the maximum stress to be applied by the foundation. Other special loading sequences may also be used if appropriate to the problem being studied.

The rate at which settlement will occur and the length of time needed before settlement ceases to be of concern *cannot* be predicted with accuracy. There are many explanations for the inability of the foundation engineer to predict the time required for soil consolidation (i.e., foundation settlement) to occur. Among the explanations are:

1. The rate of settlement of saturated soils is a function of the square of the distance that water, compressed out of the pores of the soil by the loads of the structure, must flow to a drainage face, such as a sand layer. Minor layers of sand, not revealed by the exploration program, can be present in natural deposits and can decrease the time necessary for settlement to occur.

2. The laboratory test, from which estimates of settlement rates in the field are obtained, only permits water to drain vertically. Water flow beneath a foundation can be (and usually is) three-dimensional, which will decrease the time necessary for settlement to occur.

3. The laboratory test specimen is subjected to a condition of uniform stress and uniform strain. The soil beneath a building is subjected to stress levels that usually decrease with depth and, therefore, to nonuniform strain conditions. Nonuniform strain also tends to shorten the time period over which settlement occurs from that obtained in the laboratory test.

4. The soil specimen from which the estimates are made of the likely period over which settlement may occur are individual samples about 1 in. (25 mm) high selected from a long column of soil believed to be "representative" of the compressible materials. This selection process can produce test data that lead to either an overestimate or an underestimate of the likely period of settlement.

Engineers should recognize the uncertainty associated with predicting the time period over which settlement occurs and assess the impact of a range of times on the use of the structure. Generally, the settlement of building foundations occurs over a shorter period of time than predicted from laboratory tests, while the settlement of large areas loaded, for example, by fill to raise the general grade of a site will be predicted reasonably well from the results of these tests.

The time necessary for settlement to occur beneath a structure can be estimated from the results of laboratory tests as

$$t_{\text{field}} = t_{\text{lab}} \frac{H_{\text{field}}^2}{H_{\text{lab}}^2} \tag{16.6.14}$$

where t_{field} = the time for settlement of the structure to occur
t_{lab} = the time for primary settlement to occur in the laboratory
H_{field} = the maximum distance water must travel to reach a drainage face, such as a sand or gravel layer in the soil column beneath a structure (this is one-half the thickness of the compressible stratum if drainage can occur at both the top and bottom of the stratum, or the total thickness if only one-way drainage is likely)
H_{lab} = one-half the thickness of the specimen tested in the most common laboratory test which allows drainage to both top and bottom of the specimen

Soft clays and organic deposits, such as peat, are usually considered to be normally consolidated (believed to never have been subjected to a vertical stress in excess of the current vertical stress). These materials commonly continue to settle at an ever decreasing rate after completion of the primary settlement described above by a process known as secondary compression. The amount of secondary compression as a fraction of the thickness of the clay stratum per tenfold increase in time is known as c_α and can be estimated from the results of laboratory consolidation tests. It can also be approximated from the results of water content determinations for purposes of preliminary estimates.

$$c_\alpha = 0.01 W_c \tag{16.6.15}$$

where W_c is the natural water content of the normally consolidated clay or organic soil expressed as a decimal rather than as a percent.

The magnitude of secondary compression is given approximately by

$$s_{\text{sec}} = H_{\text{total}} c_\alpha \log \frac{t_2}{t_1} \tag{16.6.16}$$

where s_{sec} = the magnitude of the secondary compression
H_{total} = the total thickness of the compressible deposit
t_2 = the time since initial stressing of the compressible deposit for which the total secondary compression is to be computed
t_1 = the time necessary for primary compression to be completed and is equal to t_{field} in Eq. (16.6.14), or the time since primary compression was probably completed if concern is for a material deposited many years ago

Secondary compression is increasingly important as construction on marginal lands increases throughout the world. It can be virtually eliminated by loading (surcharging) a construction area with soil that imposes a stress on the compressible soil at least *twice* that imposed by the building loads, by waiting for primary compression to be complete, and then by removing the soil surcharge. This is often feasible for loads imposed by floor slabs and low-rise structures. It does, however, require considerable forethought because the time required to complete a program of surcharging can be substantial.

16.6.5 Shallow Foundations on Silt

Little has been said thus far about design of foundations bearing on silt deposits. Silts can be particularly difficult materials for designers, especially if the unit density of the soil is less than about 90 pcf (1.4 kg/m^3) and collapse of the soil structure under load must be suspected.

For preliminary design, the designer should design foundations on nonplastic silt as foundations on sand and those on plastic silt as foundations on clay. However, special concern should be given to design of foundations on any silt deposit when the standard penetration value N is 10 or less.

16.7 PRELIMINARY DESIGN OF DEEP FOUNDATIONS

16.7.1 Introduction

Driven piles are a common deep foundation system in most areas of the United States. Drilled (or bored) piers (also called caissons) are more common than piles in areas where deep deposits of clay are encountered and a bearing stratum is within reach of the drilling equipment, or where foundations must be established at depth to avoid many of the problems associated with swelling or collapsing soils. In some areas of the world, the use of small-diameter drilled piers, frequently called bored piles, far exceeds the use of driven piles.

The objective of a deep foundation is to limit the displacements of the structure it supports; no foundation system eliminates all displacements. Therefore, it is important to recognize how displacements arise and how to design a foundation system to produce tolerable displacements.

Clearly, elastic displacements of foundation elements must occur under load. The lower the stress applied to a foundation element, the smaller its elastic displacement. Furthermore, displacement of a foundation element must occur if load is to be transferred from the element to the soil surrounding it or to the bearing stratum beneath its tip. A knowledge of how much displacement of a foundation element is required to transfer loads is critical if the designer is to understand how a structure will perform.

The load-carrying capacity of a deep foundation is derived from two sources; the capacity associated with skin friction (i.e., the interaction of the surface of the foundation element with the soil or rock it penetrates) and the capacity associated with the end bearing of the foundation element [16.12, 16.13, 16.22]. Observations have shown that the full skin frictional capacity of a driven pile or a drilled pier is mobilized at displacements of 0.1 in. (about 3 mm) or less. Observations have also shown that the displacement necessary to mobilize the full end-bearing capacity of a pile or pier in soil or weak rock is about 10% of the diameter of the element. Thus, the tip of a pile about 12 in. (300 mm) in diameter must penetrate about 1.2 in. (30 mm) to mobilize its full end-bearing capacity. A pier 3 ft (1 m) in diameter requires the intolerable displacement of about 4 in. (100 mm) to mobilize end bearing. As a result of the incompatibility of deformation required to mobilize the skin frictional and end-bearing resistances, design of most deep foundations should not take credit for the full end-bearing capacity available. Deep foundations are often designed to mobilize full skin frictional capacity and only that end-bearing capacity available at an acceptable displacement of less than 5% of the base width or 1 in. (25 mm), whichever is less. Sometimes foundations are designed so that the skin frictional capacity is adequate to carry loads at least equal to the design loads applied to the element. With this approach, the factor of safety against ultimate failure is provided by the end-bearing capacity of the pile or pier, when the available end bearing is adequate for that purpose, and large deformations are not expected until a substantial portion of the available end bearing has been mobilized.

If it is believed necessary to determine, by testing, the bearing capacity of the tip of a deep foundation, it is important to recognize that minor displacement of the shaft of the foundation element will mobilize skin frictional resistance and prevent the applied load from being transferred to the tip [16.21]. Therefore, the testing of the capacity of the tip requires that the load transfer of the shaft be eliminated or that the actual load applied to the soil or rock be measured. This is difficult, but possible, to accomplish. It must be done when the bearing capacity of the soil or rock is in question or when settlement of the ground around the deep foundation will eliminate, with time, the load transferred from the shaft to the soil and will cause all the applied load to be transferred to the tip. The latter is a common occurrence when fill is placed over weak soils and piles are driven in the filled area to support structures.

A common practice, and a requirement of most building codes, is for piles to be tested to establish their load-carrying capacity when the design loads exceed some minimum value, frequently about 60 kips (about 300 kN). Below this minimum value, pile-driving formulas are relied upon to indicate that adequate capacity is available. The load–displacement relations obtained from load tests on *individual piles* do not indicate the likely displacements of the groups of piles that will support the structure; interpretation and analyses are required if the behavior of pile groups is to be estimated. This is particularly important for piles bearing in clay and some sand soils because the long-term settlement of these deposits in response to the loads transferred from the pile foundation to the surrounding soil is not indicated by pile load tests.

Information for the design of deep foundations is provided in the remainder of this section.

16.7.2 Friction Piles

Piles that penetrate soil deposits and whose tips bear in a dense (hard) material or a weak rock may be considered as friction piles which have a significant capacity associated with end bearing. This follows from the earlier discussion of the small displacement necessary to mobilize the load-carrying capacity of the pile shaft (0.1 in.) (3 mm), and the much larger displacement necessary to mobilize the load-carrying capacity at the tip (5 to 10% of pile diameter). The discussion of design that follows is divided into consideration of shaft resistance in sand, shaft resistance in clay, end bearing in sand and clay, and end bearing on rock. Implicit in the discussion is recognition that the capacity Q of a single pile is given approximately by

$$Q = \text{skin friction} + \text{end bearing}$$

$$Q = f_s A_s + (c_u N_c + \bar{\sigma}_v N_q) A_b \qquad (16.7.1)$$

where f_s = the skin frictional resistance
A_s = the total area of shaft embedded in the soil
A_b = the cross-sectional area of the base of width B
$\bar{\sigma}_v$ = vertical effective stress at the base of the pile*

All other terms have been previously defined.

Shaft Resistance in Sand

The portion of the ultimate bearing capacity of a pile that is associated with skin friction, f_s, is believed related to the average lateral stress acting on the pile shaft, multiplied by the coefficient of friction between the soil (sand) and the shaft. Thus, it is believed that

$$f_s = K\bar{\sigma}_{\text{av}} \tan \delta \qquad (16.7.2)$$

where K is a coefficient that when multiplied by $\bar{\sigma}_{\text{av}}$, the average vertical effective stress along the pile shaft, gives the average horizontal effective stress, and $\tan \delta$ is the coefficient of friction between the soil and the shaft of the pile.

Direct measurement of the coefficient of friction between pile material and sand by Yoshimi and Kishida [16.29] confirm data collected by other investigators; the coefficient of friction is related to

* $\bar{\sigma}_v = \gamma D$, where γ is the appropriate unit weight of the soil (natural, saturated, or submerged) and D is the depth of the pile tip below the ground surface. It is also equal to the total unit weight of soil, γ_t, multiplied by the depth to groundwater (when groundwater level is above the tip of the pile), plus the submerged unit weight of soil multiplied by the distance between groundwater and the tip of the pile.

the surface roughness of the material, not to the density of the sand deposit, and is approximately as shown below:

Material	Measured Coefficient of Friction, tan δ	Design Value for Piles, tan δ
Steel	0.50 ± 0.1	0.4
Concrete	0.65 ± 0.05	0.6
Wood	0.60 ± 0.05	0.55

The values in the table above are for "normal" material roughness. The values can be substantially lower if unusual surface smoothness is present. The lowest values in the table are the values pertaining to large displacements and are recommended for use in estimating the capacity of driven piles because driven piles undergo many meters of displacement at the pile–soil interface as they are driven into the ground.

Knowledge of the appropriate coefficient of friction, tan δ, is not useful unless the coefficient K can be determined. This coefficient is related at least to the density of the sand deposit and to the shape of the pile section. Limited experimental and analytical data suggest that K for piles with a taper of more than about 1% (1 in./100 in.) is at least 1.5 times the value for a straight-sided pile. The value of K for a straight-sided pile in sand appears to vary between about 0.5 and about 1.0, increasing with increasing sand density. Thus, the function K tan δ can vary between about 0.2 (smooth steel in loose sand) to about 0.7 (rough concrete in dense sand) and achieve values as high as 1.1 if the pile is tapered.

A further complication to estimating shaft capacity stems from the fact that the bottom section of a long pile may not deflect sufficiently to mobilize the full shearing resistance available. This phenomenon has been noted by many researchers, such as Meyerhof [16.12] and has been described analytically as related to the stiffness and length of the pile by Poulos [16.21].

Thus, for present practice, it is believed most appropriate to estimate shaft resistance empirically, rather than by attempting to estimate K tan δ. An empirical approach that is believed to be conservative has been suggested by Meyerhof [16.12]. A modification to his approach, based on the author's experience, is shown as Fig. 16.11. Meyerhof related the available ultimate skin friction of

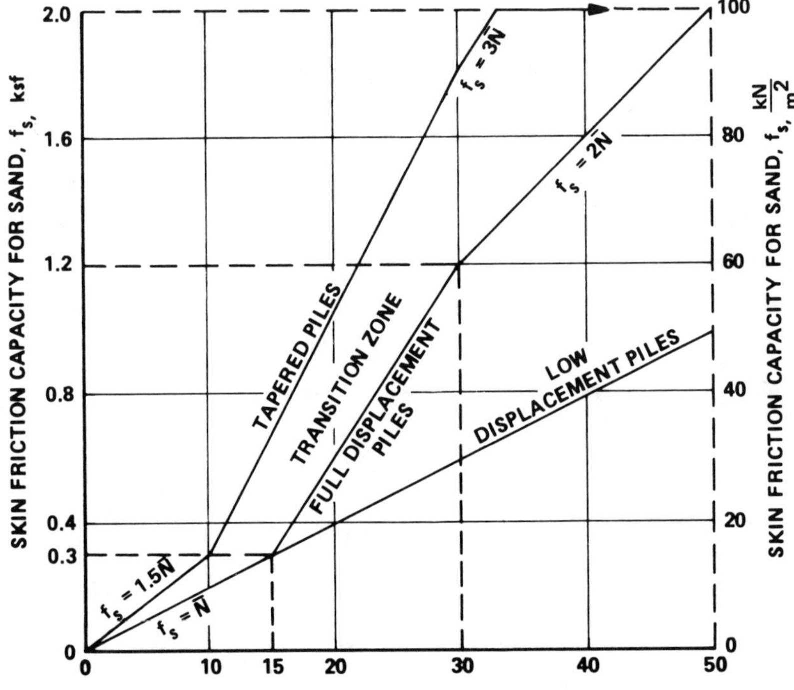

Fig. 16.11 Empirical estimate of skin frictional capacity of piles embedded in sand.

a straight-sided pile to $\overline{N}$, the average standard penetration resistance along the full length of the pile. The relationship $f_s = 2\overline{N}$ kN/m^2 ($\overline{N}$/25 ksf) is appropriate for displacement piles driven into medium dense sands. The relationship $f_s = \overline{N}$ kN/m^2 ($\overline{N}$/50 ksf) is appropriate for low-displacement piles, such as H piles, and for full-displacement piles driven into loose to very loose sands. The values for displacement piles may be multiplied by 1.5 for piles with a uniform taper that is greater than 1%, but less than 3% for a portion of their length. However, the shaft resistance should not be assumed to exceed about 2000 psf (100 kN/m^2), no matter what the sand density or pile length. For design, appropriate safety factors (usually at least 2) must be applied to the estimated values of shaft resistance.

The above rules are believed appropriate for sand deposits that are predominantly quartz and/or other massive materials, but may be inappropriate for calcareous sands or piles driven into cemented materials with large voids, such as coral.

Shaft Resistance in Clay

There has been increasing interest and research since about 1975 about the applicability of Eq. (16.7.2) for prediction of shaft resistances of displacement piles driven into clay. Although a great many insights have come out of the research, a simple design procedure acceptable to a majority of those who have thought deeply about the problem has not evolved. Therefore, a totally empirical approach is suggested here for use in preliminary design. Because the scatter is great between capacities predicted by this approach and those actually measured, the method is not to be relied upon for final design.

The empirical approach relates ultimate skin friction to the undrained shear strength of the clay, c_u, as follows:

$$f_s = \alpha c_u \tag{16.7.3}$$

where α is a coefficient varying between 1.0 and 0.25, as shown in Fig. 16.12.

The use of the coefficient α to relate ultimate skin friction to the undrained shear strength of the clay *prior* to driving the pile was first introduced about 30 years ago, but has lost favor during the past decade. As more data have been gathered, the approximate nature of this approach has become apparent. However, more accurate predictive methods, which are considerably more complex to apply, are not considered appropriate for inclusion here.

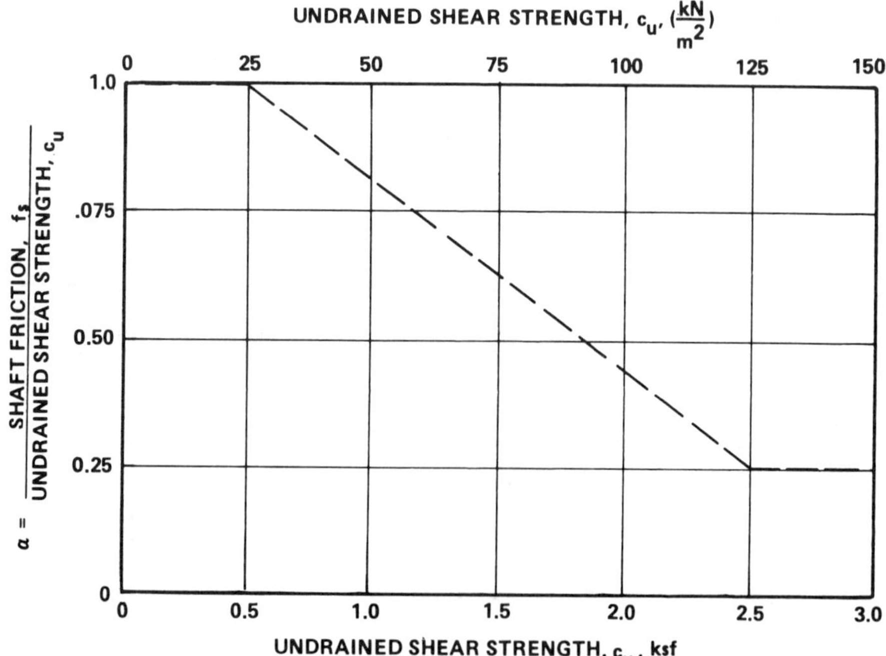

Fig. 16.12 Approximate relationship between α and c_u for preliminary design.

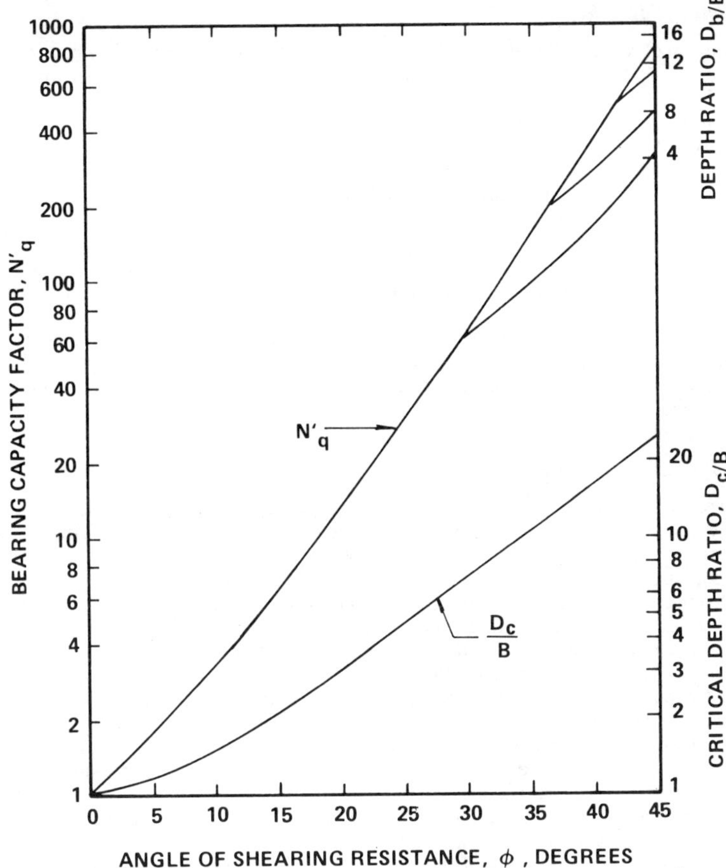

Fig. 16.13 Bearing capacity factors and critical depth for piles in sand.

Tip Resistance in Sand

The unit resistance of the tip of a pile bearing in sand, q_p, increases approximately linearly with depth of the tip below ground surface to a maximum, limiting value of unit point resistance, q_ℓ, as given by

$$q_p = \bar{\sigma}_v N'_q \leq q_l \tag{16.7.4}$$

The limiting unit point resistance, q_l, occurs at the critical depth, the depth at which the ratio of the depth of penetration of the pile tip (D) to the width of the pile tip (B) exceeds between 10 and 20. A reasonable approximation to the critical depth is shown in Fig. 16.13 as a function of the angle of shearing resistance, ϕ, of the sand.

Also shown in Fig. 16.13 are values of the bearing capacity factor N'_q that pertain if the tip of the pile penetrates to a depth at least one-half the critical depth. At intermediate depths, N'_q varies approximately linearly from N_q pertinent for a square footing at ground surface [which can be derived from Fig. 16.7 and Eq. (16.6.5*b*)] to the values shown in Fig. 16.13.

However, the crushing of sand grains at high stresses suggests that the limiting value of the unit point resistance q_l should not exceed about 300 ksf (15,000 kN/m²). As a reasonable approximation, the limiting value of point resistance can be related to the standard penetration test values N by

$$q_l = 8N \text{ ksf} \leq 300 \text{ ksf} \tag{16.7.5}$$ *

* $q_l = 400N$ kN/m² $\leq$ 15,000 kN/m²

Tip Resistance in Clay

Piles bearing in clay rarely develop a significant portion of their capacity in end bearing [the second term in Eq. (16.7.1)]. Therefore, it is common practice, and sufficiently correct for most applications, to obtain the bearing capacity factor N_c for the tip of a pile by using Eq. (16.6.10). Because piles penetrate to depths in excess of 2.5B and because they are ordinarily round or square, Eq. (16.6.10) gives $N_c = 9$.

The term N_q appearing in Eq.(16.7.1) equals 1 when ϕ equals zero. Thus, the unit point resistance q_b in addition to the load already present at the tip for a pile bearing in clay is

$$q_b = 9c_u \tag{16.7.6}$$

where c_u is the average undrained shear strength (one-half the compressive strength) of the clay at the elevation of the pile tip.

The undrained shear strength is used for bearing capacity computations because any drainage of pore water occurring during the loading of the pile will cause the strength of the clay to increase. Therefore, the use of undrained shear strength is a conservative approach to assessing the capacity of the tip of a pile to resist load.

16.7.3 End-Bearing Piles

Piles bearing on or penetrating only a short distance into a strong material, such as rock, are commonly thought of as end-bearing piles. Although some engineers consider a pile that penetrates through relatively weak soils and obtains capacity primarily from the portion that bears in sand as end bearing, such a pile should be considered a friction pile and designed as described earlier.

The capacity of piles bearing on rock is usually governed by the allowable stresses in the pile material, rather than by the bearing capacity of the rock. Only rarely, when piles bear on weak rock, will the capacity be governed by the strength and quality of the rock.

It is common practice to evaluate the capacity of end-bearing piles by load test. When it is considered necessary to assess the load-carrying capacity of the material at the tip of the pile without including the contribution of the soil surrounding the shaft, a casing is used to separate the pile shaft from the soil. Alternatively, to reduce the cost of the load test, the casing is not used, but the pile is subjected to a load that is increased above the required test load by the estimated capacity produced by skin resistance.

End-bearing piles driven into weathered rock, weak rock, or rock with many fractures have been known to exhibit reduced capacity when part of a large group. This can occur for at least two reasons. The first is the fracturing of rock beneath the tip of a pile already driven to its required capacity, by the driving of adjacent piles. To determine if fracturing has occurred, it is prudent to redrive at least some of the piles in a group of end-bearing piles before accepting the group as satisfactory to support the design column loads. In one case, 200-ton end-bearing piles, each driven until they penetrated no more than ¼ in. (6 mm) under the last 10 blows of the pile hammer, penetrated into the rock an additional 10 in. (250 mm) when redriven after completion of initial driving of all piles in the group. All piles in this group had to be redriven three times before the initial resistance to penetration was once again achieved.

The second is the result of the vertical (upward) displacement (heave) of end-bearing piles within a group which, sometimes occurs as the group is driven. The heave of piles within a large group should always be measured as driving occurs when the piles are end bearing. It is sometimes found that the displacement of soil that results from the penetration of a pile into the ground causes nearby piles to heave and become separated from the bearing stratum. When an end-bearing pile has heaved, redriving is necessary if building settlement is to be avoided. In contrast, the vertical displacement of piles achieving their capacity in skin resistance is usually not a problem because their capacity is achieved through friction along the sides of the shaft, not through end bearing.

16.7.4 Capacity of Pile Groups

Individual piles driven in groups are commonly spaced 2.5 to 4 pile diameters center-to-center. The "efficiency"* of such groups is normally about unity. The available (limited) evidence suggests the following for the bearing capacity of pile groups:

1. The ultimate bearing capacity of a pile group is usually equal to the sum of the point loads of the individual piles. However, for piles whose tips bear in clay, the sum of the point loads should not exceed the bearing capacity of the group assuming it to be an equivalent pier consisting of the

---

* Efficiency is the ratio of the group capacity to the sum of the capacities of the individual piles.

piles and the soil mass enclosed by the piles and assuming that failure occurs under undrained conditions [Eq. (16.6.9)] [16.19].

2. The ultimate skin resistance of a pile group is the lesser of the sum of the skin resistances of the individual piles in the group or the skin resistance of the perimeter of the equivalent pier consisting of the piles and the enclosed soil mass.

3. The additional capacity that may be provided by the pile cap bearing on soil is ordinarily not taken into consideration. It is assumed, conservatively, that this capacity might be eliminated at some time during the life of the structure by erosion beneath the cap or by ground settlement.

16.7.5 Settlement of Pile Groups

The likely settlement of an individual pile under the working loads appropriate for a particular structure is usually defined by load tests performed on piles driven at the site of the building. In general, the vertical deflection during a load test of piles considered acceptable for use in supporting a proposed structure rarely will exceed about 0.5 in. (12 mm) when the piles are subjected to the design loads; the vertical deflection may be considerably greater than 0.5 in. (12 mm) when the test piles are subjected to twice the design load.

Experience indicates that, at working loads, individual piles behave as inclusions in an elastic medium. Therefore, the vertical deflection of the individual pile during a load test is equal to the settlement of that pile as if it were a single pile incorporated into the structure, provided there is no compressible zone of soil at some depth beneath the pile tip. The settlement of a group of piles can be estimated by proper use of elastic analysis, as described in detail by Poulos and Davis [16.22] or by empirical methods. Empirical methods are believed most appropriate for preliminary design, but the results of these analyses *must* be verified as part of final design to avoid the possibility that the designer was misled by the approximations inherent in them, or by lack of understanding of the conditions for which they apply.

A first approximation of the settlement of a regular group of piles spaced between two and three pile diameters, center-to-center and without considering any compressible material beneath the pile tips, may be obtained from

$$\frac{\rho_G}{\rho_1} = \frac{1.5 + n}{2.5} \tag{16.7.7}$$

where ρ_G = the group settlement
ρ_1 = the settlement of the test pile at working loads
n = the total number of piles in a group

A simplified, empirical approach to estimating the total settlement of a pile group, and including in the estimate the effects of compressible material beneath the tip of the piles, has been described by Peck, Hanson, and Thornburn [16.18], and verified as reasonable by others. In this approach, the group of piles is replaced by a flexible footing or pier whose base is at a depth equal to two-thirds the average length of the piles in the group and the settlement of the footing is computed using Eq. (16.6.12). The presence of the lower one-third of the piles is ignored and, commonly, load is distributed to the compressible material, as illustrated in Fig. 16.9. This approach is also valid if the pile group is bedded entirely in clay for which analysis no displacements are assumed to occur as a result of any load in the soil above the base of the equivalent pier.

Prediction of the likely settlement of pile groups is of major significance for the design of foundations of structures; all too often, it is tacitly assumed that piles do not settle and/or that the settlement of a group of piles will be no more than the settlement of the individual pile observed in the pile load test.

16.7.6 Special Problems of Driven Pile Foundations

Negative Skin Friction (Dragdown)

When soil surrounding a pile or a pile group settles, the resistance of the relatively rigid piles to the settlement of the ground around them adds load to the piles (negative skin friction). The maximum available negative skin friction is equal to, and can be estimated in the same manner as, positive skin friction (see sections on shaft resistance). Because negative skin friction eliminates (by the downward movement of the soil) the load carried by the pile as positive skin friction, the loads formerly carried by the shaft of the pile are transferred to its tip. This transfer could overstress the material supporting the tip of the pile, causing it to penetrate further into the ground or, perhaps, could overstress the pile material. Therefore, the pile capacity used for design must take into account negative skin friction. Negative skin friction is often a problem when building on marginal lands which have been filled to achieve some desired final grade and are underlain by clay.

It should be recognized that, if ground settlement around a pile foundation is stopped after negative skin friction has developed, for example, because the pumping of groundwater was stopped, the load-carrying capacity of the pile foundation has not been impaired. The reapplication of load to the head of such a pile will cause small pile displacements relative to the soil and will redevelop the positive skin resistance along the pile shaft.

Tension Capacity of Piles

Evidence has accumulated to show that the capacity of a pile to resist tension loads is smaller than its capacity to resist compression loads, with due consideration given to the resistance associated with the tip of the pile. The resistance to tension loads is provided by the shearing resistance along the shaft of the pile, and not by the pile tip.

The available evidence indicates that the shaft of the pile driven into or through a deposit of sand will support only about two-thirds the load in tension that it will support in compression. No satisfactory explanation for this observation is available as yet.

The shaft of a pile driven into or through a deposit of clay will frequently, but not always, be found to support the same load in tension as in compression. Therefore, it is conservative and reasonable to design all piles assuming the ultimate tension capacity to be only two-thirds of the ultimate skin friction capacity in compression.

Lateral Capacity of Piles

The capacity of an individual pile or a group of piles to resist lateral loads can be assessed by analysis, but is best established by the testing of a single pile and by the extrapolation through analytical means of the results of this test to predict the deformations of a pile group. Several concepts should be remembered by designers when assigning lateral capacities to piles:

1. The lateral capacity of a pile is heavily dependent on the strength and characteristics of the soil within a depth below the ground surface of 10 to 20 pile diameters. Therefore, the placement of a thick layer of compacted granular fill over a deposit of weak soil will permit the development of significant resistance to lateral loads.

2. Most piles will be found able to support loads of at least 2 kips (10 kN) with modest deflection. When greater loads are to be applied, testing is usually appropriate. Lateral loads for design ranging between 10 and 20 kips (roughly 50 and 100 kN) can often be validated by tests.

3. The lateral capacity of a single pile determined from a lateral load test is appropriate for an infrequently applied transient loading. Long-term loading, as well as frequent transient loadings, will increase the displacement of the pile head from that observed in the lateral load test by factors that may approach 2. Therefore, in defining an acceptable lateral capacity for a single pile, consideration must be given to the frequency and duration of the loads and the lateral displacement believed acceptable.

4. The lateral capacity of a group of piles is *less* than the number of piles multiplied by the capacity of the single pile. Group displacement under lateral loadings increases with increasing size of the group and may be many times greater than the displacement of the individual pile in the pile test.

5. No credit is ordinarily taken for the fact that the pile cap in which the piles are embedded will also resist the lateral loads. This is to add conservatism to the design, to account for the fact that the pile cap is near the ground surface and can, therefore, only generate a modest passive pressure, and to account for the fact that the contractor's methods in constructing the pile cap, in stripping the forms, and in compacting the backfill around the pile caps are most often not able to be controlled or influenced by the design engineer.

16.7.7 Bored Piles or Drilled Piers

Bearing Capacity

The discussion that follows pertains to bored piles or drilled piers with a ratio of embedded shaft length L to shaft diameter B not less than about 5, and more frequently 10 or greater.

A bored pile or drilled pier differs from a driven pile because (1) the soil at the boundary of the pile shaft has undergone stress relief rather than compression during the installation (excavation) process, and (2) the base of the pile might not be totally free of material loosened by the excavation process before concrete is placed. Concern over the latter issue has restricted the use in the United

States of drilled piers for which the cleanliness of the base cannot be ascertained by visual inspection. There is, however, considerable experience elsewhere in the world to suggest that the use of proper underwater cleanout techniques, such as reverse circulation pumping or airlift, and careful tremie-placement of concrete in shafts 3 to 6 ft (1 to 2 m) in diameter will produce a satisfactory foundation element.

Drilled piers have been installed in or passed through virtually all types of soil and rock. They are most frequently termed bored piles or drilled caissons when installed in soil or weak rock; when installed through soil to bear in (as contrasted with bearing on) hard rock, these foundation elements are referred to as drilled-in caissons or caissons socketed into rock. The design of the shafts of drilled-in caissons is usually governed by the strength of the concrete used in the shaft and/or the bond between the concrete and any strengthening elements (reinforcing bars, H-sections, or pipe sections) placed within the shaft.

Capacity of Bored Piles in Clay

The unit skin frictional capacity, f_s, of a shaft in clay has been given earlier by Eq. (16.7.3):

$$f_s = \alpha c_u$$

or, in terms of the unconfined compressive strength, q_u, of the soil as

$$f_s = \frac{\alpha}{2} q_u \tag{16.7.8}$$

The relationship between $\alpha/2$ and q_u is shown as curve A in Fig. 16.14. This curve is almost, but not quite, identical to that given in Fig. 16.12. However, in Fig. 16.14, the unconfined compressive strength, q_u, which is twice the undrained shear strength, c_u, has been shown on the abscissa and used to normalize the unit skin frictional capacity, f_s.

A safety factor of at least 2 is usually believed appropriate when computing the design capacity of a drilled pier using Eq. (16.7.8). Furthermore, it is prudent to exclude from the calculation of capacity portions of the shaft subjected to uncertain stress and strain conditions. To accomplish this, it is believed appropriate for computation purposes to assume that the length of shaft is reduced by two shaft diameters at the top and by two shaft diameters at the bottom (eliminate from the shaft length the top and bottom two shaft diameters). In addition, if there is a bell at the bottom, the surface area of the bell should not be considered in the capacity computation. Ignoring the surface area of the bell is in addition to the elimination of a length of shaft equal to two shaft diameters just above the top of the bell.

For end bearing, the unit point resistance, q_p, whether or not the end of the pier has a bell, is

$$q_p = 9c_u \tag{16.7.9}$$

It is usual to apply a safety factor of 3 to the end-bearing capacity to account for, in a qualitative manner, the fact that it takes 5 to 10 times more deflection to mobilize full end bearing than to mobilize full skin frictional resistance.

Capacity of Bored Piles in Sand

It is necessary to stabilize the sides of the excavation for the shaft of a bored pile to be installed in a sand deposit in order to prevent collapse of the hole. Rarely can this be done using water as a drilling fluid; instead, it requires the introduction of drilling mud into the excavation. Drilling mud has the advantage of weighing more than water and, therefore, requiring a smaller positive head above the natural groundwater level to maintain the hole open, and will form a filter cake of low permeability on the walls of the shaft, thereby reducing fluid losses during drilling. Drilling mud will also keep the soil cuttings in suspension so that they can be removed as the mud is circulated, thereby avoiding the accumulation of loose sand at the base of the shaft.

The use of drilling mud, however, has two potential disadvantages that must be recognized:

1. Some engineers will argue that the formation of a filter cake at the boundary of the shaft (at the shaft/sand contact) reduces the capacity of the shaft to resist load.
2. Sand cuttings will accumulate in the drilling mud unless special precautions are taken to clean them out and could impair the continuity of the concrete tremie-placed to form the shaft.

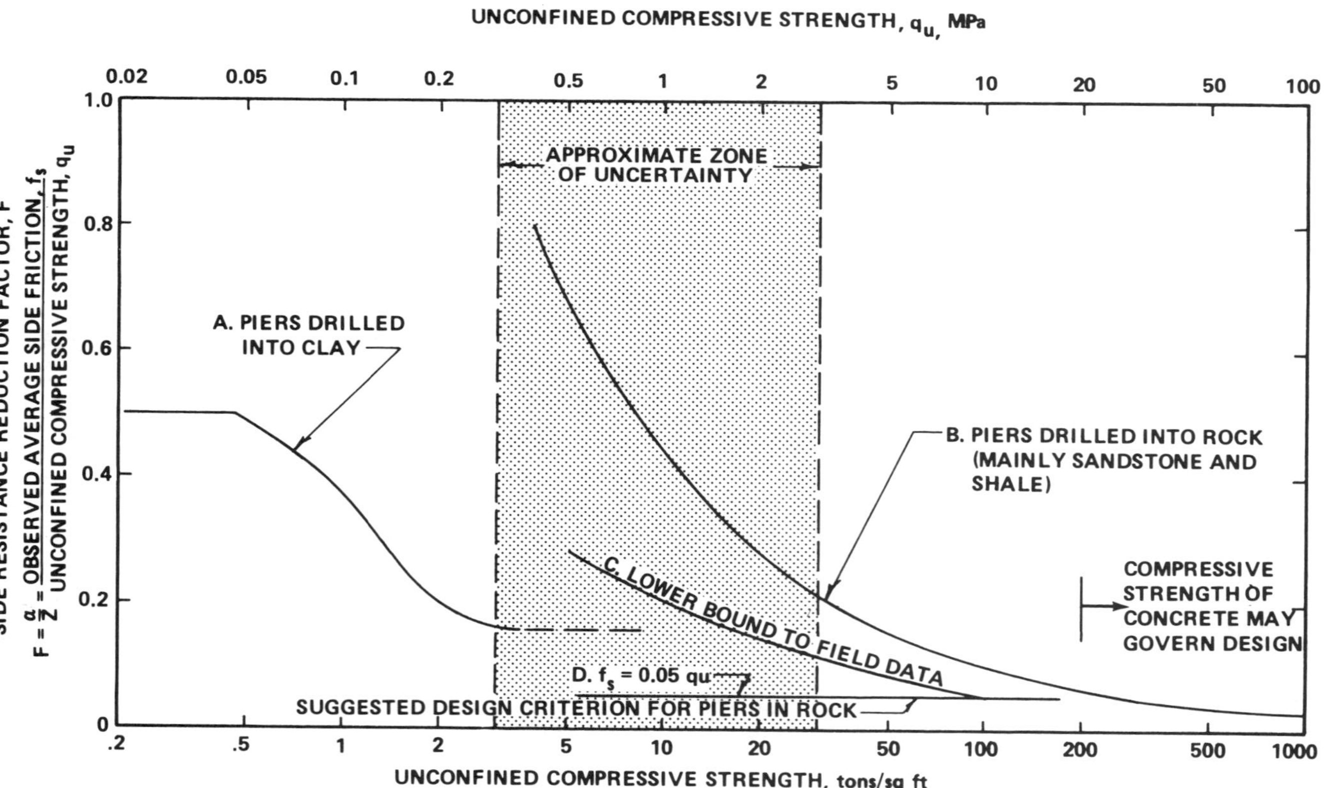

Fig. 16.14 Side friction reduction factor vs. unconfined compressive strength.

The field and laboratory evidence that is available through 1984 indicates that the filter cake does not affect the capacity of the shaft of a pile bored into sand that is not calcareous.* There is substantial evidence to suggest that the combination of a filter cake and soft, calcareous sand can reduce shaft capacity by 50% or more.

Circulation of drilling mud through vibrating screens to remove excess sand from the mud is believed necessary if problems with concrete placement are to be avoided. Problems with the tremie-placement of concrete appear to develop when the sand content of the drilling mud exceeds about 14% (by volume) and the density of the drilling mud exceeds about 1.25.

The unit skin frictional resistance f_s along the shaft of a bored pile in sand can be estimated by either of the following equations:

$$f_s = 0.7 \tan(\phi - 5°)\bar{\sigma}_v \tag{16.7.10a}$$

or

$$f_s = 0.02\bar{N} \text{ ksf} = \bar{N} \text{ kPa} \tag{16.7.10b}$$

where $\bar{\sigma}_v$ = the average vertical effective stress along the shaft
ϕ = the angle of shearing resistance of the sand
$\bar{N}$ = the *average* standard penetration value N over the length of the shaft

The unit skin resistance should not exceed 2000 psf (about 100 kPa). A factor of safety of at least 2 should be used in arriving at the unit skin resistance to be utilized in design.

Shafts drilled in sand usually cannot have a bell formed at the bottom. The drilling mud cannot be made heavy enough to support the overhang produced by the belling operation.

End Bearing. The end-bearing capacity of a bored pile in sand can be very high, but the displacement necessary to mobilize this capacity is normally too high for the structure to accept. The unit end-bearing capacity, q_p, associated with a deflection of the tip of the bored pile of 5% of the pile diameter is approximately

$$q_p = 2N \text{ ksf} = 100N \text{ kPa} \tag{16.7.11}$$

where N is the standard penetration test value at the tip of the pile.

The tip resistance indicated in Eq. (16.7.11) is one-quarter that given by Eq. (16.7.5) (for driven piles) and should be limited to 100 ksf (5 MPa). This accounts, in an approximate manner, for any loosening that may occur at the pile tip during the excavation process and any loose material that may be left at the bottom of the shaft.

Capacity of Bored Piles in Weak Rock

The unit skin frictional capacity f_s of a bored pile in weak rock is related to the average unconfined compressive strength q_u of the rock, and is given by curve *B* shown in Fig. 16.14. The curve was derived mainly from tests on sandstone and shale and is an "average" of many data points.

Curve *C*, also shown in Fig. 16.14, is a lower-bound representation of the data as suggested by Horvath and Kenney [16.9]. Also shown in the figure is curve *D* which is a design curve used in Australia to estimate the unit skin frictional capacity of bored piles in weak rock [16.27]. Curve *D* has the equation

$$f_s = 0.05q_u \tag{16.7.12}$$

The various curves in Fig. 16.14 illustrate that the Australian design methodology may become unconservative when the compressive strength of the rock exceeds about 1500 psi (10 MPa). The figure also indicates that the available unit skin frictional capacity of the pile, which is governed by the strength of the weaker material (rock or concrete), may be related to the strength of the concrete when the rock strength exceeds about 3000 psi (20 MPa). It is quite clear, for example, that the shearing resistance of the concrete governs pile capacity when a drilled-in caisson is installed in competent rock with a compressive strength in excess of about 7000 psi (about 50 MPa).

There is a central zone shown in Fig. 16.14 between compressive strengths of about 40 and 400 psi (0.3 and 3 MPa) for which there is considerable uncertainty about the available side resistance.

---

* Calcareous sands are composed of fragments of coral or sea shells and are found, mainly, off shore.

It is in this range that there is a transition between soil and rock. The curves appear to suggest that the available side resistance may be less for a rock derived from clay and evolving into a shale than for a weak sandstone or, perhaps, limestone. More data are needed to clarify this question and, until it is available, caution is recommended.

Curves *B* and *C* in Fig. 16.14 do not take into account explicitly the benefits associated with the roughness of the surface of the shaft. There is evidence to suggest that substantial improvement in available side resistance occurs if the shaft is rough and the concrete–rock interface is such that slippage must occur through the rock.

In some cases, it is necessary to control the flow of water into a rock socket and the shaft is drilled with drilling mud filling the excavation. The available evidence suggests that the tremie-casting of the pile shaft and the displacing of drilling mud can reduce the available side resistance by a substantial amount if the surface of the shaft is smooth (or, perhaps, is made smooth by the filling of holes in the rock by drilling mud). No effect of drilling mud has been observed when the rock socket is rough.

End Bearing. Bored piles whose length L is greater than 2.5 to 4 pile diameters have been found to continue to increase their load-carrying capacity as displacements increase to as much as 25% of the pile diameter. Yielding of the weak rock appears to occur at unit pressures three to four times greater than the unconfined compressive strength of the rock. Thus, it is common to use, as an allowable unit stress on the tip of the pile, a value as high as one-half the unconfined compressive strength of the rock (assuming rock strength, not concrete strength, governs design). This provides a factor of safety of 6 or more against yielding of the rock at the tip.

It is also common to find that the allowable unit stress at the tip of a pile is governed by a building code with no explicit reference to the compressive strength of the rock. Typical values for tip stress given by codes are:

"Sound" rock	900–1800 psi (6–12 MPa)
"Medium" rock	600–1200 psi (4– 8 MPa)
"Poor" rock	300– 600 psi (2– 4 MPa)

To use these tip strengths, one would expect the compressive strength of the rock to be about twice the values allowed by code.

The designer must recognize that there is a substantial risk associated with assuming the availability of any significant end bearing if the absence of loose material or rock debris at the base of the pile cannot be verified by direct observation. Many designers will assume that debris is present at the base and design bored piles to carry all load in side resistance if the bottom of the pile cannot be inspected adequately.

Settlement

The settlement of individual or groups of drilled piers is believed best estimated using elastic analyses; load testing of these units is not commonly done because of the high loads they carry. The most comprehensive discussion of such analyses has been presented by Poulos and Davis [16.22]. A major difficulty with the use of elastic analyses is the need to determine or assume such parameters as modulus and Poisson's ratio. Thereafter, the distribution of load between the shaft and the tip of the pier must be computed, whereupon the settlement (or vertical displacement) of the top of the pier may be estimated.

Details of these procedures are not considered appropriate for inclusion here.

SELECTED REFERENCES

16.1 ASCE. "Subsurface Investigation for Design and Construction of Foundations of Buildings," *Journal of the Soil Mechanics and Foundation Division,* ASCE, *Part I,* **98,** May 1972 (SM5), 481–490; *Part II,* **98,** June 1972 (SM6), 557–578; *Part III,* **98,** July 1972 (SM7), 749–764; *Part IV,* **98,** August 1972 (SM8), 771–785.

16.2 L. Bjerrum and A. Eggestad. "Interpretation of Loading Test on Sand," *Proceedings of European Conference on Soil Mechanics and Foundation Engineering,* **1,** 1963, 149–203.

16.3 J. B. Burland, B. B. Broms, and V. F. B. deMello. "Behavior of Foundations and Structures," State-of-the-Art Report, *9th International Conference,* Tokyo (2), 1977, #95–546.

16.4 J. B. Burland and C. P. Wroth. "Settlement of Buildings and Associated Damage," State-of-the-Art Review, *Proceedings, Conference on Settlement of Structures,* Cambridge, MA, 1974, 611–654.

16.5 D. J. D'Appolonia. "Effects of Foundation Construction on Nearby Structures," State-of-the-Art Paper, *4th Pan American Conference,* **1,** 1971, 189–236.

16.6 B. Fischhoff, S. Lichtenstein, P. Slovic, S. L. Derby, and R. Keeney. *Acceptable Risk.* New York: Cambridge University Press, 1981.

16.7 Rebecca Grant, John T. Christian, and Eric H. Vanmarcke. "Differential Settlement of Buildings," *Journal of the Geotechnical Engineering Division,* ASCE, **100,** September 1974 (GT9), 973–991.

16.8 A. J. Hendron and C. H. Dowding. "Ground and Structural Response to Blasting," *Proceedings of 3rd Congress of International Society for Rock Mechanics.* Washington, D.C.: National Academy of Sciences, 1974, 1359–1364.

16.9 J. S. Horvath and T. C. Kenny. Discussion, *Journal of the Geotechnical Engineering Division,* ASCE, **108,** July 1982 (GT7), 991–993.

16.10 H. O. Ireland. *Settlement Due to Building Construction in Chicago,* PhD Thesis, University of Illinois, 1955.

16.11 A. C. Meigh. Discussion, *Proceedings of European Conference on Soil Mechanics and Foundation Engineering,* **2,** 1963, 71–72.

16.12 G. G. Meyerhof. "Bearing Capacity and Settlement of Pile Foundations," *Journal of the Geotechnical Engineering Division,* ASCE, **102,** March 1976 (GT2), 195–228.

16.13 G. G. Meyerhof. "Some Recent Research on the Bearing Capacity of Foundations," *Canadian Geotechnical Journal,* **1,** 1, September 1963, 16–26.

16.14 J. K. Mitchell. "Soil Improvement—State-of-the-Art Report," *10th International Conference on Soil Mechanics and Foundation Engineering,* Stockholm, **4,** 1981, 509–566.

16.15 J. K. Mitchell and W. S. Gardner. "In-Situ Measurement of Volume Change Characteristics," *Proceedings, ASCE Specialty Conference on In-Situ Measurements of Soil Properties,* Raleigh, NC, **2,** 1975.

16.16 D. C. Moorhouse and G. L. Baker. "Sand Densification by Heavy Vibratory Compactor," *Proceedings of Conference on Placement and Improvement of Soil to Support Structures,* Cambridge, MA. New York: American Society of Civil Engineers, August 1968, pp. 379–388.

16.17 R. B. Peck. "Deep Excavations and Tunneling in Soft Ground," State-of-the-Art Report, *7th International Conference on Soil Mechanics and Foundation Engineering,* Mexico, 1969, 225–290.

16.18 R. B. Peck, W. E. Hanson, and T. H. Thornburn. *Foundation Engineering,* 2nd ed. New York: Wiley, 1974.

16.19 R. B. Peck, W. E. Hanson, and T. H. Thornburn. *Foundation Engineering.* New York: Wiley, 1953.

16.20 D. E. Polshin and R. A. Tokar. "Maximum Allowable Nonuniform Settlement of Structures," *Proceedings, 4th International Conference on Soil Mechanics and Foundation Engineering,* **1,** 1957, 402–406.

16.21 H. G. Poulos. "The Influence of Shaft Length on Pile Load Capacity in Clays," *Geotechnique,* **XXXII,** 2, June 1982, 145–148.

16.22 H. G. Poulos and E. H. Davis. *Pile Foundation Analysis and Design.* New York: Wiley, 1980.

16.23 J. H. Schmertmann. "Static Cone to Compute Static Settlement over Sand," *Journal of the Soil Mechanics and Foundation Division,* ASCE, **96,** March 1970 (SM3), 1011–1043.

16.24 A. W. Skempton and D. H. MacDonald. "Allowable Settlement of Buildings," *Proceedings, Institution of Civil Engineers,* Part III, **5,** 1956, 727–768.

16.25 H. Sommer. "Recent Findings Concerning Allowable Differential Settlements of Structures—Criteria of Damage," *7th European Conference on Soil Mechanics and Foundation Engineering,* **3,** 1979, 275–280.

16.26 H. E. Wahls. "Tolerable Settlement of Buildings," *Journal of the Geotechnical Engineering Division,* ASCE, **107,** November 1981 (GT11), 1489–1504.

16.27 A. F. Williams, I. W. Johnston, and I. B. Donald. "The Design of Socketed Piles in Weak Rock," *International Conference on Structural Foundations on Rock,* Sydney, Australia, May 1980, 327–347.

16.28 A. H. Wu and D. J. Scheessele. "A Cost-Effective, Shallow Foundation Accommodates Three Feet of Settlement," *Civil Engineering,* March 1982, 65–67.

16.29 Y. Yoshimi and T. Kishida. "A Ring Torsion Apparatus for Evaluating Friction Between Soil and Metal Surfaces," *Geotechnical Testing Journal,* ASTM, **4,** 4, December 1981, 145–152.

CHAPTER 17

COMPUTER-AIDED ANALYSIS AND DESIGN

LeROY A. LUTZ

Vice-President
Computerized Structural Design
Milwaukee, Wisconsin

17.1 INTRODUCTION

The computer has become an essential part of structural analysis and design. Structural analysis and design generally requires a substantial amount of mathematical calculation. The design process often requires examining several alternatives, each needing a series of calculations. Thus there is a definite need for computer assistance. Having a computer available is not enough, however. Having good software for the computer is essential in order to have computer-aided analysis and design capability.

17.1.1 Use of Computers

The computer should be used only when it can be considered beneficial. This can occur if computer use:

1. Is cost effective in providing the answers needed.
2. Provides the accuracy or correctness that may be required.
3. Is necessary to obtain the answer or answers desired within the time constraints imposed.
4. Reduces the chance for making errors.

Once it is decided that it is beneficial to use the computer, there is essentially no limit (except that of the computer capacity and software capability) on what the computer can be used to do. There are three basic aspects to the computer-aided analysis and design process.

Input Phase

In this phase the problem must be described to the computer. Presently, in design practice, by far the majority of all data input is numerical. That is, span identification and properties, joint locations, and loadings are described simply by inputting numbers. However, the use of graphics (pictures) as a means of data entry is increasing.

Creation of the structure geometry or configuration is most amenable to graphics input. Use of a digitizer in a graphics system permits the establishment of numerical locations for the joints being input. Movement of the digitizer wand is used to establish members in a structure. An alternate graphics technique involves the use of light pens on the CRT.

It is also possible to describe loadings in a graphical manner using digitizer capability. An alternate is use of a tabular format which can be presented either with or without graphics. The identification of members with a particular load can be done graphically by referring directly to a plot of the structure.

Creation of element properties can be done in a tabular format with a graphics system, but is usually not a significant improvement over standard input of numbers without graphics. If a library of sections is stored in a program, the identification of sections is a relatively simple process with or without graphics.

Analysis Phase

Presently the vast majority of analysis is linear elastic analysis. That is, the analysis is conducted on the undeformed geometry and the resulting deflections are small. It is assumed that there is no significant difference between the behavior of the deformed structure and the undeformed structure subjected to the same loading.

Programs are available that do consider nonlinear behavior of structures. The first-order analysis described above is extended to a second-order analysis where the member axial forces and relative joint deflections of the structure under load are used in determining the final deflected shape of the structure. Another type of nonlinear analysis is the consideration of inelastic behavior of members. In this type of an analysis, the program steps the loading, changing the member (and the structure) properties as yielding is introduced into these members. Both types of nonlinear analysis could be combined into the same program.

There are also specialty structural analysis programs that do bifurcation type analyses either to evaluate the buckling modes of a particular structure or the vibrational modes. The features of the nonlinear programs can be combined with time-dependent loading to produce a program that combines all the behavior considerations mentioned.

Generally, most analysis programs will use what is considered the stiffness method of analysis to solve the problem. The stiffnesses of all the component elements are evaluated and then combined at the joints to form a stiffness matrix for the entire structure. All structure loads are converted to joint loads. The set of stiffness equations is then solved to determine the joint displacements. The final member forces are then evaluated using the joint displacements. Other analysis techniques such as moment distribution could be used to solve analysis problems; however, if the basic assumptions are the same, the solutions will be identical.

Larger analysis capability is becoming available to the structural engineer because memory core size in computers as well as disk storage is becoming much more affordable. This will permit the typical structural engineer to analyze much larger structures with relative ease.

Output Phase

The output of a structural program can be simply a printout of analysis results or a printout of design results as well. On occasion the analysis output is minimal because the design output is the final desired result needed by the engineer. Today (1986) most output is printed or in graphics format.

Many programs supplement the printed output with graphical output. This is especially true of structural analysis programs where the deflected shape of the structure is plotted for the various load cases. The structure can usually be viewed from any position so that the most informative plot can be displayed. Another type of plot, done less frequently, is a plot of moments, shears, and deflections of individual members of a structure.

For very well-defined problems, the computer program could be directed to actually plot the design output. As an example, for a truss design, the truss could be plotted and the member sections, e.g., WT6 × 20, labeled on the plot. A similar plot could be constructed for a multistory frame structure.

An area of computer use related to the analysis and design process is that of computerized drafting. Although present use of computers in drafting is only a fraction of the drafting volume, its use will increase dramatically in the near future. Although this is not a direct part of the design process, it is an important aspect of the work of an engineering firm.

Initially the computerized analysis–design aspect of engineering will be basically separate from the computerized drafting aspect for the most part. However, in time the analysis–design portion will be tied directly to the drafting portion to produce a fully integrated structural system. This fully integrated structural system is possible now and actually used in some limited areas. Advancements toward completely automated analysis–design–drafting will continue steadily as the software and hardware continue to improve in quality and become increasingly more cost effective.

17.1.2 When to Use Computers

The engineer must make intelligent decisions in the analysis and design process. One of these decisions involves knowing how to use the computer properly. The computer is only a tool to be used, such as a calculator or a slide rule.

Sometimes it is appropriate to do a calculation by hand. Other times it is more appropriate to use a hand-held or desk-top calculator because the answer requires more accuracy or the computation is more correct. If a series of operations are required to obtain a particular result, use of the programmable calculator or computer is probably desirable.

As an example, consider the computation of the section properties for an area composed of a number of fundamental areas and/or special shapes (such as a wide-flange). It is proper to use a computer if I_x, I_y, and I_{xy} are all required, if the area is composed of 10 components, and there is a minicomputer in the next room. It is not proper if only I_x is to be obtained for an area with two components and a large mainframe computer must be accessed to get the solution.

As a second example, consider that a set of stiffeners must be designed for a flat plate subjected to hydrostatic pressure. In making this decision one must compare the ease of writing and debugging a small computer program with the ease of producing a large series of repetitive calculations. It is probably proper to use a computer if the wall has three heights, four widths, with steels of two stress levels and changes in the heights and widths are anticipated such that the calculations might have to be redone. It is probably not proper if there are two heights, two widths, one stress level, and no chance of redesign.

It is becoming increasingly difficult to find a problem that is too small for the computer. This is especially true with the increasing use of mini- and microcomputers which places the accessibility of the computer both physically and economically within reach of almost every structural engineer.

17.2 USE OF COMPUTERS IN PRELIMINARY WORK

The structural engineer is often asked to perform preliminary analysis and design work on a project. Acceptable structural schemes and approximate size of structural members must be obtained in a relatively short period of time.

This preliminary work should be done or directed by an experienced engineer who is able to sort out the important and unimportant. He must know when and how to approximate and simplify the structural model without altering the model behavior significantly.

The preliminary analysis models should follow the points discussed in this section, those raised in Section 17.3, as well as the wealth of preliminary analysis information available in other chapters. Preliminary design should also be done following the guidelines put forth in the preceding chapters.

17.2.1 Preliminary Analysis Models

Structural models used in preliminary analyses should not include components that contribute little to the structural integrity. Nontypical components should be made typical to eliminate model complexity.

The most important aspect in the analysis process is to create a model that considers the behavior desired. The engineer must thoroughly understand structural behavior.

Are the members in the structure subjected primarily to axial loadings or to flexural action? What type of deformation is to be expected in the structure? These questions must be answered before the model is formulated.

A good example of a preliminary analysis modeling problem is that of evaluating whether or not a structure has an appropriate shear wall integrity. There are usually a large number of places where shear walls can exist in a building.

Certain places are selected as appropriate for shear walls. This selection is based on such items as location, length of wall available, and vertical load on the wall (to minimize overturning problems). Several walls that are similar in size should be made identical for simplicity. Other vertical elements will probably be considered not to contribute to the lateral integrity for various reasons.

Unless the size and position of the shear walls is simple, some approximation of the lateral load to each wall must be used to obtain easily the information desired. The accuracy of the answers will depend on the degree of approximation used and the complexity of the structure.

The analysis is probably quite good unless the walls are tied in with a monolithic concrete floor and column system. In this case, the shear wall analysis will likely substantially overestimate the overturning moment on the shear wall. The concrete frame will interact with the shear wall to change appreciably the overall behavior. If this type of interaction is suspected it is usually necessary to do a substantially more complex analysis in order to make the preliminary design decisions.

Another preliminary analysis task might be to size a truss such that a reasonably good estimate of its weight and cost might be obtained. Many engineers would find it necessary to analyze a truss of 50 or so members using a computer in order to obtain the preliminary design. An experienced engineer may simply consider the truss as a beam and find a moment and shear diagram for this beam. From the moment, the chord forces would easily be calculated; from the beam shears, the forces on the diagonals would be obtained. This engineer would then concentrate on doing a preliminary design of the truss components.

In both of these examples, the emphasis is on finding the most direct and simple means of obtaining an analysis. This usually means simplifying but may mean completely eliminating use of the computer in the analysis that is required. Do not get involved in detail that will have little or no bearing on the results. Use the computer as a tool that will be beneficial in the preliminary analysis.

If a preliminary analysis is required for a complex or unusual structure, there is no way to simplify or shortcut the amount of analysis required. In fact, extensive analysis may be required in order to examine all aspects of the structural behavior for complex or unusual structures.

17.2.2 Preliminary Design

Computers can aid in the preliminary design process. Preliminary analyses will provide design moments for beams and design moments and axial loads for columns. Computer programs capable of producing designs for steel or concrete beams and steel or concrete columns can be invaluable in producing quick, accurate member designs.

In addition to the loadings, input values of material properties, lateral bracing position, and effective lengths will lead to a member design, or a series of member designs. A change in the bracing and effective length values will lead to a new series of designs. Many options or alternate designs can be explored quickly with the aid of computer hardware and design software.

17.3 USING COMPUTERS FOR FINAL ANALYSIS

17.3.1 Selecting the Analysis Technique

The analysis technique refers to the analytical assumptions that are incorporated into various analysis programs. Selection of the appropriate program for a particular purpose is not obvious. For some analysis problems there can be a program selection problem. Several examples follow:

Case I. If a continuous beam without columns requires analysis, there are three types of programs that could be used:

1. A program written specifically to analyze continuous beams.
2. A general planar frame program in which a continuous beam can be modeled.
3. A general (three-dimensional) space frame program in which a two-dimensional continuous beam can be modeled.

The ease of use would undoubtedly correspond to the order listed above. All three programs would give the same answer to the problem.

Case II. If a continuous beam line with columns below having continuity with the beam line requires analysis, more analysis program options are possible:

1. A single-story beam line program.
2. A multistory frame program.
3. A general planar frame program.
4. A general space frame program.

In this case, the program to use depends on the analysis technique desired or required. The programs will not all provide identical answers. The first does not consider story drift. Thus the shears developed by the column moments are not balanced. The second considers story drift, but does not consider axial column deformation. The last two permit both sidesway and axial column deformation.

All four programs provide generally acceptable analyses for vertical loads. However, if the column stiffnesses vary significantly and the loading or columns are not close to symmetric, considerably better answers can be obtained with the three program types that permit sidesway or lateral movement of the beam line.

The first program type may not permit hinged bases for the columns or different length columns, whereas the more general programs are more likely to have these features. Should the beams (or columns) be nonprismatic, a program considering nonprismatic sections in the analysis is essential.

Should a lateral load analysis of the beam line with columns below be required, the first program type could not be used. Should analysis for axial deformations of the beams or columns, or column settlement be desired, the general planar or space frame programs would undoubtedly have to be used.

Case III. If a two-dimensional truss were to be analyzed, the following programs should be considered:

1. A general planar truss analysis program.
2. A general planar frame program.
3. A general space frame program.

Practically all trusses are satisfactorily analyzed as members with pinned ends forming triangular configurations. All three program types would provide identical answers provided the frame programs have the feature of being able to pin the ends of members. The truss analysis program would undoubtedly be easier to use because the pinned-end condition is automatic.

However, analysis using the frame programs would be quite accurate simply by modeling the truss with continuity at the joints. Member moments of inertia should either be the actual value or some lesser value (see Section 17.3.2 for further guidelines). Typically the joint moment output would be ignored.

There is a definite advantage to modeling a complex truss using pinned-end members, in that most truss instability can be caught by the program during the analysis. If the truss is analyzed considering joint continuity, the truss instability may be concealed by frame (flexural) behavior.

The frame programs do have the advantage that loadings may be applied to the members as well as the joints. Often the top chord of a truss member is a flexural as well as an axial member. A frame program would permit consideration of the top chord as a continuous beam as well as the top chord of the truss.

Should the truss not be entirely a true truss in that member lines of action do not coincide at a joint (or joints) or should the truss be tied to columns such that it forms a frame, it is probably advisable to use a frame program in which the flexural behavior introduced could be properly modeled.

Most other structures of a two-dimensional nature require the use of a general planar frame analysis which considers both axial and flexural deformation of all members. An analysis program with this capability can be used to model coupled shear walls (wall penetrated by a relatively uniform series of openings) or shear wall–frame interaction type problems. See Fig. 17.1.

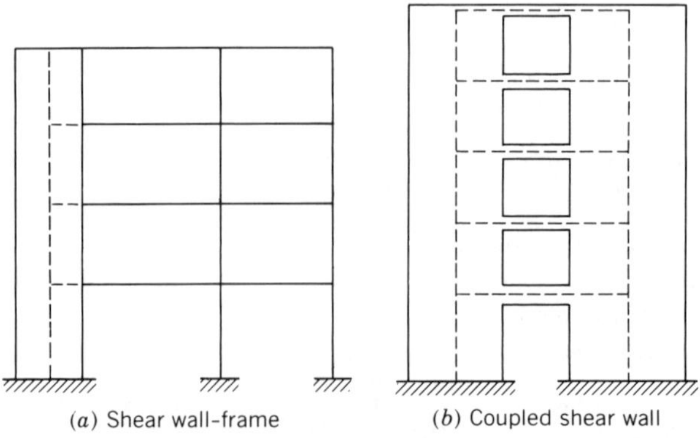

Fig. 17.1 Shear walls as frames.

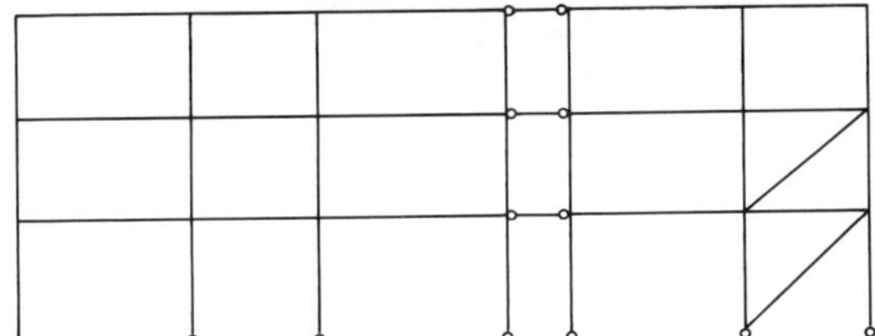

Fig. 17.2 Two linked plane frames.

If the shear walls have a small height-to-depth ratio such that shear deformations become important, reduced effective moments of inertia (reflecting the shear distortion) should be used unless shear deformations are incorporated into the analysis technique used by the program. In this case a shear area for each element must be input or the shear area may be determined automatically from the gross area that is input.

Multistory frame buildings are typically analyzed as a series of planar frames. This assumption, while being fine for analysis of vertical loads, can be substantially in error under lateral loads if the lateral stiffnesses of the various frames differ appreciably. This difference can be caused by diagonal bracing, more and/or larger columns (or wall columns), or in differences in the beam framing between columns.

If the multistory frame is such that there can be considered to be a single displacement (and no rotation) for each floor (usually from symmetry), the frame can be analyzed using a planar multistory or planar general frame program. The two or three representative frames can be considered to be in the same plane and linked to produce the same lateral movement at each floor as shown in Fig. 17.2. All frames would thus be analyzed together as part of a single frame analysis.

Should the floors twist and also translate perpendicular to the direction of load, the building may have to be analyzed as a three-dimensional structure. The three-dimensional (space) frame idealization could be used for this purpose.

The space frame analysis is actually one of many types of three-dimensional analysis techniques available; it is a particular type of finite element program in which the element is a flexural member. Most other finite elements are either surface elements which are used to model structures such as walls, slabs, and shells, or solid elements which are used to model three-dimensional solids such as foundations, dams, and machine parts.

The surface elements are usually either triangular or quadrilateral in shape. The solid elements are usually tetrahedrons or have six quadrilateral faces. The properties of these elements are made to correspond to the surface or solid being modeled.

No attempt is made here to further explain the finite element method. The engineer or designer is referred to the large number of articles and texts already available on finite element methods. Use of the finite element method usually requires the assistance of those experienced in the method. It is not a procedure that every engineer should be expected to learn. Even use of space frame programs requires special training of the engineer.

With complex structures such as domes and shells, the program user must be knowledgeable about the general behavior or might be led to incorrect analytical conclusions. Often nonsymmetrical loads on a dome can lead to a structural response vastly different from that under symmetrical loads. First-order analysis may be inadequate and second-order analysis may reveal large deflections or a buckling analysis may reveal that buckling has occurred. The designer must know when more advanced analysis techniques are necessary.

17.3.2 Proper Modeling of the Structure

The analysis *technique* pertains to a computer program's capability to consider the analytical behavior that is required. Proper *modeling* refers to the ability to describe a structure properly for the computer analysis. The structure must be modeled consistent with the support constraints to be provided and the member connections that are intended.

Support Conditions

Continuous beams are typically considered to have pinned (hinged) or roller supports. For loads perpendicular to the beams it does not matter whether all supports are pinned as axial deformation of the beams is not being considered. There should be one pinned support if a general frame program is being used to prevent a possible analysis solution problem from lack of having horizontal restraint. See Fig 17.3*a*.

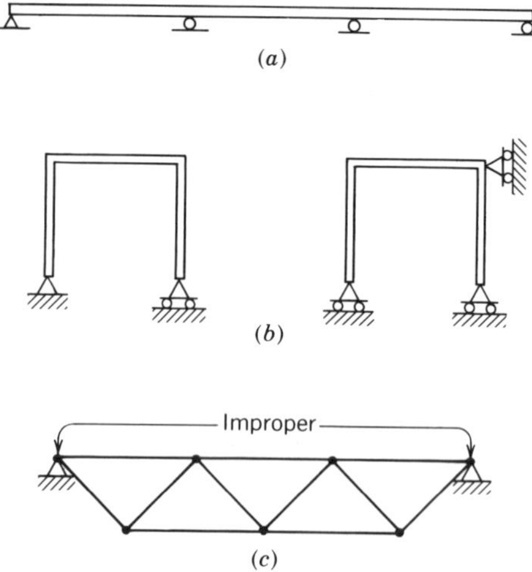

Fig. 17.3 Support conditions.

Planar frames may have either fixed, pinned, or roller supports. Fixed supports provide complete rotational restraint for the columns while the pin or roller provides complete rotational freedom for the column. The pinned and fixed supports provide lateral restraint for columns so that they can take shear from lateral load. A minimum of three restraint components is required to provide static equilibrium for the structural frame. This would be a minimum of one fixed support, one pinned plus one roller support, or three roller supports not situated in the same or parallel lines or arcs. See Fig. 17.3*b*.

Planar trusses that are truss beams must have support conditions that permit axial deformation of all its members, especially its top and bottom chords. Pinned supports must not be used at both ends of a simple span truss beam. See Fig. 17.3*c*. With pinned supports at the top chord, no compression is allowed into this chord as the truss is deflected down by the load.

In a continuous truss beam only one pinned support should be used; the other supports should be on rollers. A truss in the form of a frame may be pinned if no translation of the support is desired. Normally a truss model would not have a fixed joint support. A truss is restrained from rotating by the pinning of a pair of joints.

Three-dimensional trusses require a minimum of three pinned joints in order to provide restraint against the three translational and three rotational degrees-of-freedom of the structure. These pinned joints cannot be colinear, but they can be in the same plane. With roller support conditions considered as well, the possible combination of roller and pin support conditions makes it very difficult to generalize as to what provides a stable condition.

Three-dimensional frames also require restraint of the basic 6 degrees-of-structure-freedom using fixed, pinned, and/or roller supports. The best way to determine whether or not a structure is stable is to examine whether the model can possibly translate unrestrained in any direction or whether there is any axis about which it can rotate without restraint. The ability to do such an examination successfully improves with experience, but this ability can save much delay and computer time.

Symmetry can often be used to cut the size of a problem by a half or more. This not only saves time and money, but allows much larger problems to be done on smaller computer systems.

Joints on the line of symmetry can deflect only in the plan of symmetry so that a roller support on the symmetry plane provides the translational restraint desired. For the frame, rotational restraint is also required. The joint in a frame must not rotate about the two axes that lie on the plane of symmetry. Thus, referring to Fig. 17.4*a*, if Δ_x is restrained, θ_y and θ_z must also be restrained on the line of symmetry.

When a member crosses a line of symmetry in a frame, a joint must be placed at the center of this member on the line of symmetry, and the symmetry constraint conditions are then imposed at this joint. In a pinned-ended truss, when a member crosses the line of symmetry the added joint should be given a pinned support to avoid a member instability, as shown in Fig. 17.4*b*.

A member lying on the plane of symmetry should have its area and moment of inertia about the axis perpendicular to the symmetry plane reduced by 50%. The other moment of inertia and the torsional constant are not used in the analysis, so to avoid confusion, half of all four member properties should be input.

Fig. 17.4 Symmetry support conditions.

Arches and gabled frames are two planar structures that require their supports to take thrust. These structures are normally analyzed with either hinged or fixed supports. Should there be a tie member between supports to take the thrust, this tie should be included in the model and one of the supports should be allowed to translate horizontally to permit tensioning of the tie member. The ramifications of spreading of the supports should be examined. This is especially true if the arch has a low profile and thus is subject to large displacements from spreading.

Domes and pyramids (see also Chapter 13) are three-dimensional structures which employ their own tension rings to resist the thrust from loading. The rings must be permitted to translate radially outward in order to permit the ring(s) to function. However, these structures must also take lateral wind or seismic load. Therefore, radially oriented roller supports with circumferential constraint should be used unless the supports rest on columns in which case the columns should be included in the model. Appropriate column base and circumferential restraint conditions would then be employed.

Inclined roller support conditions can be specified in some structural analysis programs. If they are not part of the program capability they can be simulated by a linkage (pin-ended) member inclined at the appropriate angle, as shown in Fig. 17.5*a*. This works in two- or three-dimensional frame problems. The member can be of any arbitrary length with a relatively high stiffness so as not to compress significantly under load.

Spring supports used to simulate real springs or soil, etc., can be formed from linkages of an arbitrary length with elastic modulus and/or areas specified to equate the spring constant desired. As shown in Fig. 17.5*b*, the spring constant of a linkage is AE/L where A = area, E = elastic modulus, and L = length.

Partial rotational restraint to simulate support restraint between fixed and hinged can be provided by a rotational spring member (usually placed perpendicular to restrained member for clarity); see Fig. 17.5*c*. The rotational spring constant is $3EI/L$ where the far end of the rotational spring is pinned (I = moment of inertia of spring member). Changing the column moment of inertia properties is inadvisable because translational stiffness is altered, which leads to improper results.

Member Interconnection

Members are straight line segments used in a structural analysis model. These member segments may extend from support to support or may be only a portion of what is generally considered as a beam. These members, defined by two joints, may be prismatic or nonprismatic.

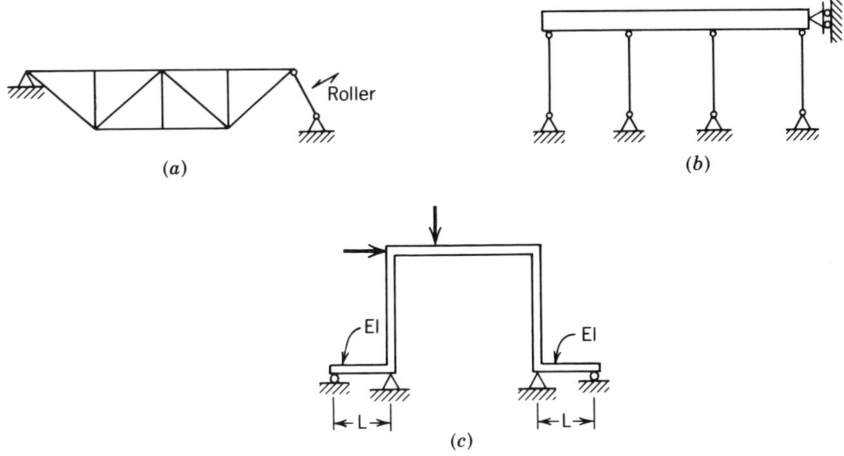

Fig. 17.5 Special support conditions.

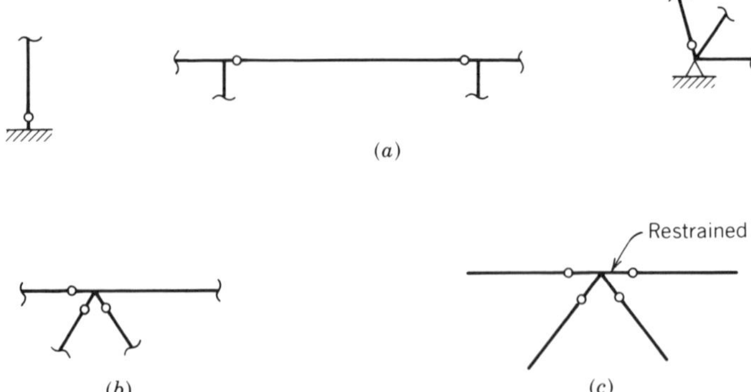

Fig. 17.6 Member hinges.

The length of each member (segment) is based on the curvature of the actual structure as well as numerous other considerations. In the typical structural analysis program the members are considered to be continuous at their interconnections (the joints). This means that the displacement and rotation of a joint specifies the displacement and rotation of all members framing into the joint.

End releases are permitted by many structural analysis programs should continuity not be desired at either end of any member. For most programs the end releases are only rotational releases (not translational). In two-dimensional problems there is one rotational end release component; in three-dimensional problems there are three rotational end release components, two flexural and one torsional.

It is important to remember that end releases are in the member adjacent to the joint. The joint is a separate entity representing the intersection of two or more members, as illustrated in Fig. 17.6*a*. Stiffness equations are written for the joints based on their connection to the members. If a moment is applied to the joint of a two-dimensional structure, this moment must be transferred into at least one of the connecting members in order to achieve joint equilibrium. This means that at least one of the members framing into a joint, such as shown in Fig. 17.6*b*, must *not* have a hinge (end release) at the joint if the joint is free to rotate.

When a frame analysis program is used for analysis of a truss with pin-ended members the temptation is to release the ends of all members. This can be done only when the program provides the capability of restraining the rotation of all joints, as shown in Fig. 17.6*c*. Otherwise one of the members framing into each joint must maintain its continuity with the joint to maintain joint equilibrium.

In three-dimensional structures a decision has to be made as to whether flexural continuity exists about the two principal axes at the member ends. It may exist about one axis, both axes, or neither. In addition, a decision is required on the degree of torsional restraint. One can release only one end of a member torsionally and still maintain torsional stability of the member. It is recommended that no torsional release be used unless it is obvious. Practically any connection of a wide flange to another member will maintain torsional continuity. A tube is usually connected to another tube with a torsionally rigid connection as well. To be essentially torsionally free, a connection such as shown in Fig. 17.7 is required.

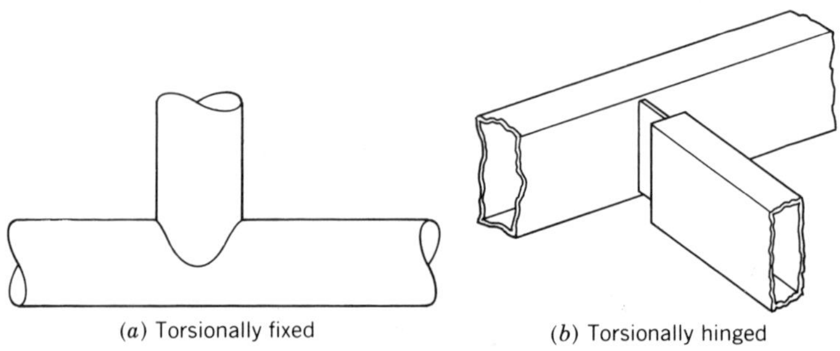

Fig. 17.7 Torsional restraint.

17.3.3 Checking the Analysis

Checking the Input

In those computer programs having graphics capability a plot of the structure can be made. This plot can usually be visualized from several directions. A careful study of these plots can guarantee that the structure geometry is properly described to the computer. Without a graphics plot a careful study of the member lengths, orientations, and joint numbers for each member is needed to determine that the structural geometry is correct.

Very few programs permit easy verification that the support input, member end releases, and member properties are correct. It is most important to prepare carefully the data for input and check what is to be input *prior* to entering any of the data. If this is done, when the data get echo printed, the only check necessary is to verify that the echo print checks the input data form. It is unlikely that one will spend the time to make a thorough check of data during the actual data input stage except for the geometry check discussed above.

The loadings are also difficult to check at the input stage. They should be examined with care during the creation of the model which is prior to the entry of any input data into the program.

Checking the Output

The output should be checked by an experienced engineer who is capable of recognizing what type of response a particular structure should have to the various loadings imposed.

The engineer should have a good idea, based on experience and preliminary design efforts, of what the answer should be before actually doing the analysis. The final analysis is done to obtain a refined answer. If there is a vast difference between the answer and what is expected, the engineer must resolve whether the results contain an error or the program indicated actual behavior which was not anticipated.

The size and location of any large deflections should be noted. All deflections should be scanned to look for possible bad solutions caused by improper modeling of the structure. This check is made easy if the program has ability to produce a deformed geometry plot.

The reactions should be checked to see how they relate to each other and how they relate to the various loading combinations applied. Sum up the vertical and horizontal reactions and compare with the loading that was to have been on the structure.

The shears, axial loads, and moments in the various members should be reviewed and compared with approximate expected values. Check equilibrium at several joints.

17.4 OBTAINING THE DESIGN

The design results can be obtained without use of the computer simply by using the output from an analysis program. Conversely, the design results can be obtained by computer based on user input after the computer analysis phase is completed. The complete design information can be generated by the computer following completion of a computer analysis. These three levels of design using computers are discussed in this section.

17.4.1 Using Analytical Results

A computer program generates an analysis which is used as the basis for designing the various members included in the analysis. There is no basic difference in this design process as compared to the design based on a hand analysis except for the interpretation of the analysis data. The sign convention of the output must be fully understood so that it is clear if the axial load is compressive or tensile and clear which beam face is in compression.

Concerns about effective lengths, lateral bracing, stress level, and material strengths are present as always. The analysis results typically present only numbers based on a fully elastic first-order analysis without regard to stress level, buckling considerations, or level of deflection. It may be necessary to do a more sophisticated analysis before design can proceed. This is especially true when analyzing more complex structures such as arches and domes.

If the design requires member sizes which differ from those assumed in the analysis, then there must be a decision made as to whether a repeat analysis is required. If a rerun is deemed necessary, it is usually fairly simple to do if the analysis program has rerun capability.

17.4.2 Using Interactive Design Procedures

The computer program generates an analysis or part of an analysis and then assists the user in designing the member or members. The computer does the number crunching in the design process and lets the user do the fun part of the design process, the selection of the members and other components.

A good example of this is a program that does the design of a reinforced concrete beam interactively. A particular span in a continuous beam is to be designed. The analysis is done for a particular set of loadings for a continuous beam of constant cross-section. The program has retained all of the analysis results and the apparent critical span is selected for design.

First the material properties are input to the program. The program begins with the flexural design for the positive moment. The user is able to input the cross-section size desired; in response, the area of reinforcing steel and reinforcing ratio is indicated. The user can accept this data and have it printed or can change the cross-section size. The reinforcing bar selection is not made but bar cutoff information is provided for various bar sizes. Bar areas in the negative moment regions are then determined along with corresponding bar cutoff information for the top bars. The designer is able to make the bar selection after all designs are made using the tabulated output.

The program then proceeds to the design of the stirrups. The program uses the shears and the beam cross-section and indicates possible stirrup alternatives at the critical locations. The user then selects the stirrup size and spacing desired at the critical section and subsequent spacings, if desired, farther out in the span. The computer does the drudgery work of evaluating how many stirrups are required at each spacing and where the stirrups end.

The designer still does the actual design using the interactive process. The computer does the analysis and all of the routine number crunching in the design process. The designer does not lose any skills in the process, but actually gains experience in the design process.

17.4.3 Using the Computer's Design

The designer can proceed automatically into the design process after completing the analysis with both analysis and design output being printed in sequence. The design can be semiautomatic whereby the analysis data must be transferred by the user to the design program with the aid of a data storage medium or other type of interface procedure. The analysis data may have to be input directly by the user into the design program.

In the latter two cases the design program is more likely to be a general-purpose program that can accept a multitude of design problems. An example of this might be the concrete beam design program referred to in the previous section, or a steel column design program which could produce member designs for any set of input loadings, effective length, and material properties.

In the fully automated programs the variability of design output is usually more restricted. The material properties and other general design input data are supplied to the program in conjunction with the analysis input. In this type of program either the design alternatives are usually limited so that the design output is generally acceptable to most program users, or the program is specifically written for a particular organization or firm which follows a specific design approach for a specified product.

An example of the first program type is a general flat slab analysis and design program. The analysis procedure and much of the design process follows the criteria of the Building Code of the American Concrete Institute.

The program as part of the design process evaluates the punching shear force at the columns and compares with allowable force values. If the slab thickness input by the user causes the punching shear capacity to be inadequate, a reanalysis and design is required using a larger slab thickness or strength.

The flexural design output indicates required areas of reinforcement at various locations as well as a selection of reinforcement to use (including lengths) at the various locations in each span. It may well be possible to transmit this design information directly to the structural plans if the structure is very regular in layout of columns.

More likely than not several column lines will have to be analyzed in each of the two directions. In this process, approximations are made in tributary width; strip widths vary in each span and in the same span in different analyses, etc. As a result, the designer must review all of the design output and create interpolations between the various designs so that the final floor slab design reflects the consistent use of particular bar sizes, lengths, and spacings throughout the floor for ease of construction.

Unless the computer program has the capability of absorbing the information for an entire floor, it is almost always necessary to have the designer's input in the design process. This is probably good because it educates the designer and provides another means of checking the output of the program.

An example of the second program type is a truss analysis and design program. An organization has a specific type of truss to analyze and follows certain design constraints with regard to use of cross-sections, material strengths, and attachment details. A program can be written which minimizes the input required, creates the truss configuration desired, analyzes, and designs all of the members using the cross-sections that are acceptable.

The user of the program merely has to check the input and output, compare the design with previous designs, if desired, and then transmit the design into shop drawings. It is even possible the production of the shop drawings can be part of the automated process.

17.5 DESIGN PRACTICE

17.5.1 Incorporating Computer Results into Design Calculations

When the engineer is faced with submitting a formal set of structural design calculations for approval, the treatment of the computer calculations is often uncertain. A computer program may have been used to do a portion or all of the analysis involved; it may have done a portion or all of the design involved. The treatment of the computer results depends on the extent of their involvement in the overall analysis and design.

If design calculations are intended to be sample calculations that are representative of a much larger set of calculations that were done for a building, then the engineer has some freedom to pick and choose what is used. If a computer program was used to analyze building frames, one such analysis should be incorporated into the calculations. If the columns and footings were designed using computer programs, then a representative column design and footing design input and output should be included. The analysis and the design examples should interrelate and be tied together by a written explanation or whatever hand calculation might have been used in the design process.

Computer printout should never be just inserted into a set of calculations, and be expected to explain itself. Computer printout should always be referenced in the written portion of the calculations. The purpose for it being included, or the steps taken to develop the input for the computer printout, should be clearly explained prior to referencing the computer printout.

The computer printout should always be complete unless it is simply too voluminous. Never just insert the output; input and output should both be included. If some portion is missing for any reason it should be clearly explained.

Whenever the computer results are used, the location of a particular moment or load must be made clear. Indicate the member, the load case, and the page number in the printout where the particular value was obtained.

When the computer results are very short, such as a section properties printout, the printout can be trimmed and taped or glued directly to the calculation sheet. When the computer results become a full page, insert the page of output and reference as indicated above.

If the computer program is a specialty type program that one might not be familiar with, an explanation of the program is in order. Explain the analytical and design assumptions, and possibly incorporate a sample calculation, a flowchart, or even a listing of the program into the design calculations.

If the computer output is voluminous it may not be practical to include the entire printout. A graphics plot of the input should be adequate to explain the problem modeled. Selected pages of printout containing the critical moments or stresses might suffice as output.

When the structural calculations involve a particular aspect of a building design such as the curtain wall or a skylight, usually the entire set of such calculations is expected rather than just a sample. This would mean that all computer printout pertinent to the analysis and design would have to be included with the calculations unless the entire amount is deemed too voluminous.

17.5.2 Documenting Computer Analysis and Design Results

When the computer program used is not a name program such as STRUDL, users may be asked to verify that the program is satisfactory by documenting the program. This can be done in several ways.

If the computer program is a program sold on the open market, one can use a copy of the program manual with its sample problems and program explanation as a means to providing some credence to the computer output. This can be supplemented with a sample textbook problem with known answers run with the same computer program. Often it is adequate to simply submit the formal name of the program along with the name of the firm that markets the program in order to satisfy the request.

If the computer program used is a specialty program developed and used by one or just several firms, a program listing can be submitted as a means of documentation. If the program is considered to be a private or confidential program, one may simply wish to give a written explanation of the program and its operation or may wish to include a sample hand calculation to verify aspects of the program.

CHAPTER 18

DYNAMIC LOADING—CONCEPTS AND EFFECTS

ROBERT D. EWING

President
Ewing & Associates
Engineering Consultants
Rancho Palo Verdes, California

M. S. AGBABIAN

President
Agbabian Associates
Engineers and Consultants
El Segundo, California
and
Fred Champion Professor of Engineering and Chairman
Department of Civil Engineering
University of Southern California
Los Angeles, California

18.1 INTRODUCTION

Earthquakes, explosive effects (blast and ground shock), wind, and impact are principal sources of dynamic excitation of buildings. Vibrating equipment inside buildings or accidental internal explosions also require an understanding of the dynamic response characteristics of buildings.

Textbooks and manuals should be consulted for descriptions of dynamic forces on structures. There are a number of sourcebooks that will be helpful to the engineer [18.1, 18.2, 18.11, and 18.12]. Also, there are many books, papers, and monographs written on the subject of general dynamics and specialized topics in dynamics; and a partial selection of useful publications is given in the references [18.3, 18.4, 18.5, 18.6, and 18.10]. The authors have freely drawn from the published material and have attempted to give proper credit wherever it was due.

This chapter is designed to meet the needs of the structural engineer for the design and analysis of buildings subjected to dynamic loadings, and it is intended to complement the material given in the other chapters. The reader is expected to be skilled in the use of static analysis procedures, and familiar with the calculus and elementary differential equations.

This chapter focuses on the dynamic characteristics of building systems and the effects of dynamic loadings. Emphasis has been placed on developing structural idealizations of building systems, deriving mathematical models of these idealizations, and determining the effects of dynamic loadings on the components of the buildings from the models. Mathematical derivations are introduced to provide an insight to the dynamic response of structures. For rigorous mathematical treatments, the reader should refer to texts on dynamics, such as Clough and Penzien [18.6] and Norris et al. [18.10].

18.2 TYPES OF LOADING

Earthquake forces on a structure are inertia forces resulting from the shaking of the foundation beneath a structure. Ground motions at the base of a structure are usually obtained from strong motion accelerograph records. Simple procedures have been established to obtain the base shears as functions of the natural period of the structure, the weight, and other structural properties. Equivalent static methods of analysis use the base shear as the equivalent static force on the structure, as described in Chapter 2. However, the ground shock motions at the base of the structure are used directly for dynamic analysis. Figure 18.1 is an example of earthquake ground motions that excite structures. The recorded earthquake motion is the acceleration time history. The velocity and displacement time histories are obtained by integration.

Blast loads resulting from explosions have very short durations, almost instantaneous rise times, and very rapid decays. The blast pressure (overpressure), the reflected pressure from the structure, and the dynamic pressure associated with the air flow behind the shock combine to form a complex

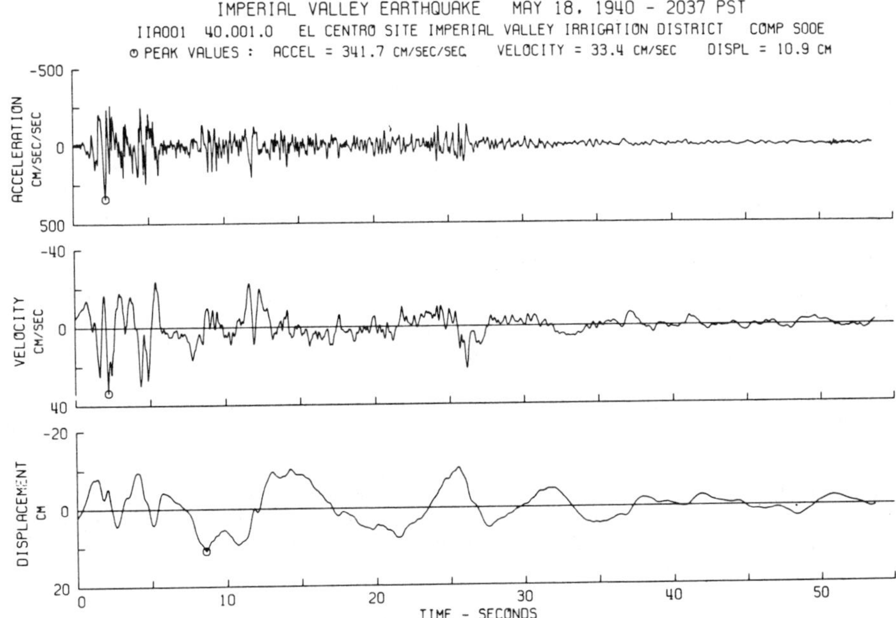

Fig. 18.1 Ground acceleration and integrated ground velocity and displacement curves for a typical earthquake.

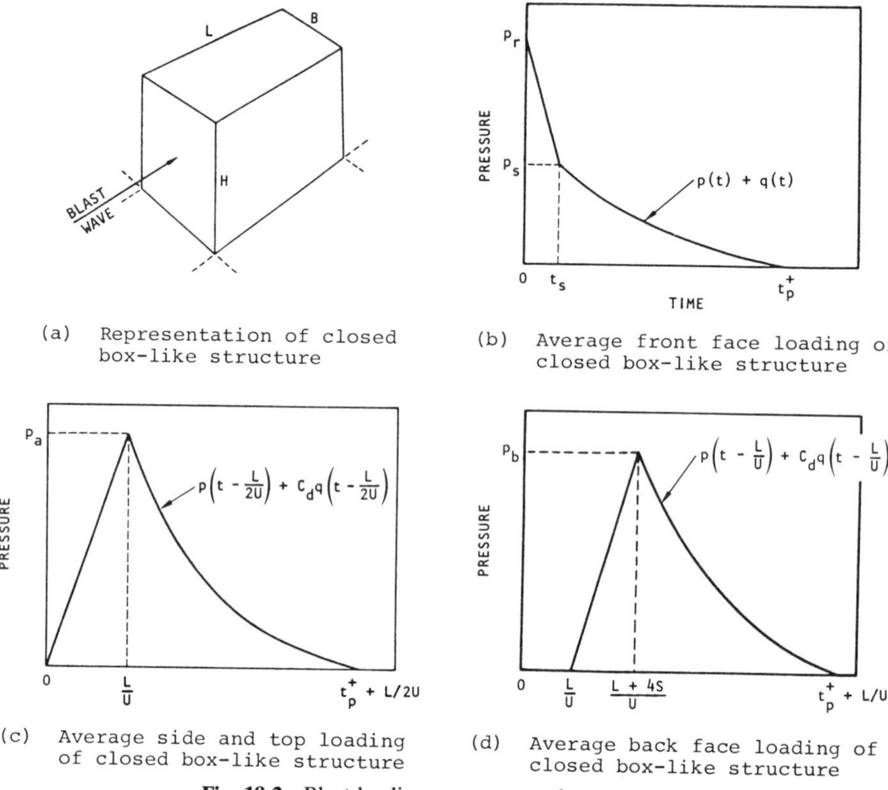

Fig. 18.2 Blast loading on a rectangular structure.

load-time history. Figure 18.2 is an example of a blast loading on the front wall, rear wall, and roof of a rectangular structure. The front wall experiences the combined effects of the reflected pressure and the dynamic pressure, whereas the roof and the rear wall experience the blast and the drag pressure as the airblast envelops the structure.

The airblast time-history is more complicated as the shock front propagates along a curved surface. Figure 18.3 shows how the blast pressure (overpressure) and the dynamic pressure may be combined at any point on an arch to obtain their combined effect. The time intervals t_1, t_2, and t_3 are functions of the ratio of the height of the arch to the velocity of the shock front and the values of p_1, p_2, p_3, and C_d (the drag coefficient) are plotted as a function of the location of a point on the arch with respect to the ground surface.

Wind loading (see Chapter 2) requires a definition of the wind profile along the height of the building and its time variation. In practice, gust response factors are included in design wind pressure formulas, and the wind profile is treated as an equivalent static force on the structure. The dynamic nature of the wind load is sometimes represented as an oscillating vertical profile. The periods of these oscillations may be of the order of minutes and/or seconds, and it is necessary to study the climatological conditions and the roughness of the terrain to develop a time-dependent wind profile. Wind tunnel tests are also needed to define the load patterns on high-rise buildings surrounded by other structures.

Impact forces on buildings are caused by high velocity missiles that are lifted from the ground by tornados and highly turbulent winds. In special cases, such as nuclear power plants, the accidental impact of an airplane has also been considered. The shape, size, and mechanical properties of the missile, its terminal velocity, and the properties of the target (building) will define the load characteristics. The missile upon impact will deform severely or will shatter or pulverize, depending on its mechanical properties. The building will be impulsively loaded and the effect will generally be local. Under large impulsive loads, structural elements in the building may either fracture or undergo large plastic deformations.

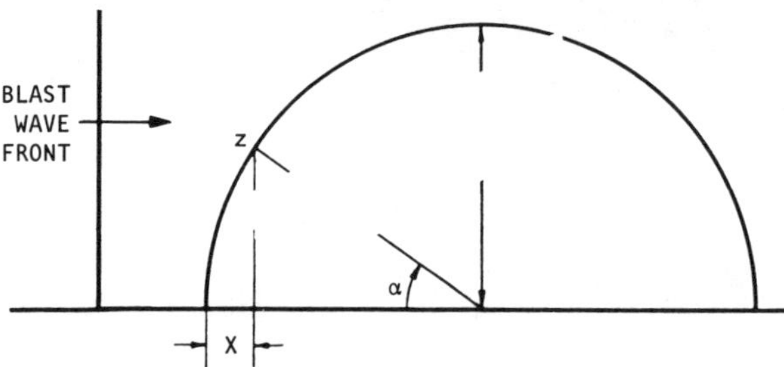

(a) Representation of a typical semi-circular arched structure

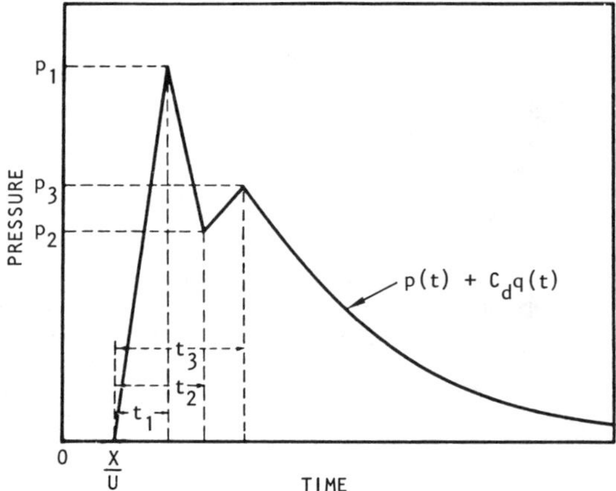

(b) Typical pressure variation at a point on an arched structure subjected to a blast wave

Fig. 18.3 Blast loading on an arched structure.

18.3 DYNAMIC CHARACTERISTICS OF STRUCTURES AND STRUCTURAL COMPONENTS

18.3.1 Degrees-of-Freedom

All structures are continuous systems that have distributed physical properties and, as such, they have an infinite number of degrees-of-freedom (DOF). The dynamic equations of motion involve a system of partial differential equations. However, most of the structural systems of interest to the engineer can be quite accurately represented by idealizations that employ lumped physical properties or lumped parameters. Models of these structural idealizations have a finite number of degrees-of-freedom, and the dynamic equations of motion involve a system of simultaneous ordinary differential equations.

The number of degrees-of-freedom of a lumped parameter structural idealization is equal to the minimum number of coordinates necessary to specify its shape or configuration at any time. A model with one concentrated mass may require as many as six coordinates to define its position (i.e., three translational coordinates and three rotational coordinates) or as few as one coordinate, and is termed a six degree-of-freedom system and a single degree-of-freedom (SDOF) system, respectively. The number of coordinates required per concentrated mass depends on the structural idealization and on the dynamic loading.

18.3.2 Damping

Every structural system has damping of one type or another and the actual damping mechanism depends on the type of materials and their methods of construction. An exact mathematical formulation for the damping is difficult, if not impossible. At best, only mathematical approximations of the actual damping in structural systems can be obtained.

Many of the explicit formulations that have been employed in the past have been special cases of the following damping formulation:

$$f_d(t) = \text{sign}(\dot{x})\ c|\dot{x}|^n \tag{18.3.1}$$

where $f_d(t)$ = damping force
c = constant of proportionality or damping constant
$\dot{x}$ = absolute or relative velocity
n = a positive exponent
$\text{sign}(\dot{x})$ = algebraic sign of the velocity

The case of $n = 0$ corresponds to the formulation associated with Coulomb friction

$$f_d(t) = \text{sign}(\dot{x})\ c \tag{18.3.2}$$

where the damping force is a constant and acts in a direction opposite to the velocity vector. This type of damping is associated with dry frictional resistance that might occur in mechanical joints.

The case of $n = 1$ corresponds to the formulation associated with viscous damping

$$f_d(t) = c\dot{x} \tag{18.3.3}$$

where the damping force is proportional to velocity and acts in a direction opposite to the velocity vector. This type of damping can be associated with the resistance to the motion of a body in a Newtonian fluid, and is considered to be a good approximation for the damping provided by air at low velocities. More important, this formulation is an acceptable approximation for the internal damping provided by most solid materials, and it is fairly conveniently handled mathematically.

The case of $n = 2$ corresponds to the formulation associated with hydraulic damping

$$f_d(t) = \text{sign}(\dot{x})\ c\dot{x}^2 \tag{18.3.4}$$

where the damping force is proportional to the velocity squared and acts in a direction opposite to the velocity vector. This type of damping is usually associated with hydraulic resistance or the damping provided by air at relatively high velocities. Another explicit form of damping is termed structural damping and is given by

$$f_d(t) = \text{sign}(\dot{x})\ g|x| \tag{18.3.5}$$

where x = absolute or relative displacement
g = structural damping constant

In this case, the damping is proportional to displacement and acts in a direction opposite to the velocity vector. This formulation has been used to approximate the internal damping provided by some solid materials, particularly when relative displacement is used.

It has been suggested by Jacobsen [18.7] that most of the damping forces given above can be replaced, for all practical purposes, with equivalent viscous damping forces which result in the same amount of energy dissipation per cycle of vibration. Accordingly, it is the practice of many engineers to use this concept of equivalent viscous damping in their structural models, and this approach is used in this chapter.

18.3.3 Dynamic Characteristics

The dynamic characteristic of an undamped structural model is defined by three interchangeable quantities; namely,

ω_n = natural circular frequency of vibration, rad/sec

f_n = natural frequency of vibration, Hz

T_n = natural period of vibration, sec

where $f_n = \dfrac{\omega_n}{2\pi}$

$$T_n = \frac{1}{f_n}$$

Similarly, the dynamic characteristic of a damped structural model is defined by three interchangeable quantities:

ω_d = damped natural circular frequency of vibration, rad/sec

f_d = damped natural frequency of vibration, Hz

T_d = damped natural period of vibration, sec

The above dynamic characteristics are termed natural quantities because they are natural properties of the model when it is allowed to vibrate freely without any externally applied forcing functions. The relationships between the undamped and damped dynamic characteristics are discussed in subsequent sections of this chapter.

18.4 IDEALIZATION FOR DYNAMIC ANALYSIS

In developing dynamic idealizations of structural systems and their models, the engineer should keep the actual structural system in mind and should model only those features of the structure that are significant in defining its response. This is the most important stage in any dynamic analysis, and there are no universal, foolproof rules or formulas to guide the engineer. The structural idealization must be developed first, and then the mathematical model will naturally follow from the idealization. The engineer should understand the important properties of the structure, the source of the dynamic loading, and how the dynamic loadings affect the structure. Specifically, this requires five steps; namely, (1) determine the spatial distribution of the building mass; (2) determine the spatial distribution and characteristics of the structural system stiffness; (3) determine the spatial distribution and characteristics of the loading; (4) locate and identify the structural components that resist the loading; and (5) determine the load path through the structural system.

Typical examples of structural idealizations and their models are given in the following subsections.

18.4.1 Models for Single-Story Structures

Structural idealizations of single-story structures can vary greatly in complexity, and the models of these idealizations can vary from SDOF systems to multi-degree-of-freedom systems. General types of single-story structures are discussed in this subsection.

A very common situation in single-story structures, and a structure of importance in building design, is one that has a very massive and stiff roof–girder system supported on relatively flexible, lightweight columns. A cross-section of such a structure is shown in Fig. 18.4*a*, a structural idealization is shown in Fig. 18.4*b*, and a mathematical model of the dynamic system is shown in Fig 18.4*c*. In constructing this structural idealization and its corresponding mathematical model, several assumptions have been made, namely:

1. The vertical response of the system to horizontal loads is not important.
2. The columns are essentially fixed at the foundation and at the roof (i.e., the effect of joint rotations is not important).
3. The weight of the column is small compared to the roof.
4. The roof is rigid and massive.
5. The loading and response are completely defined by the lateral deflection of the roof.

Accordingly, the response of the analytical model of Fig. 18.4*c* is completely described by a single coordinate u (the relative deflection between the top and bottom of the columns) and is classified as a single degree-of-freedom system. The equation of motion for the mathematical model which is subjected to a time-dependent forcing function at the roof level and a kinematic motion at the structure foundation (e.g., earthquake-induced motion) is given below

$$m\ddot{u} + c\dot{u} + ku = F(t) - m\ddot{x}_g \tag{18.4.1}$$

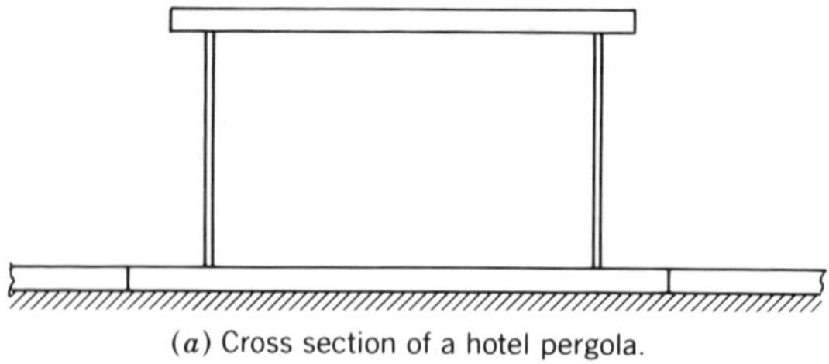

(*a*) Cross section of a hotel pergola.

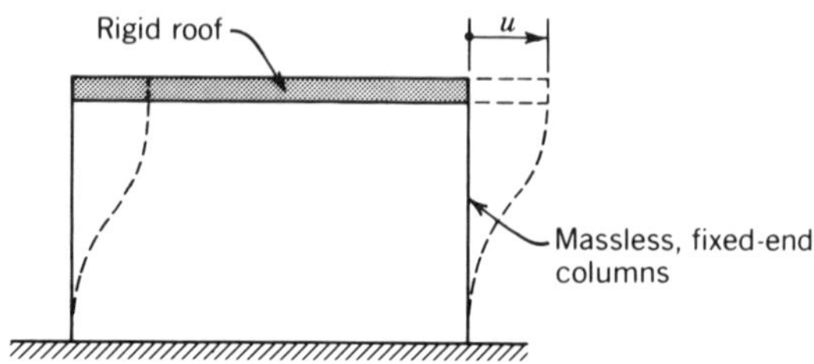

(*b*) Structural idealization of a hotel pergola.

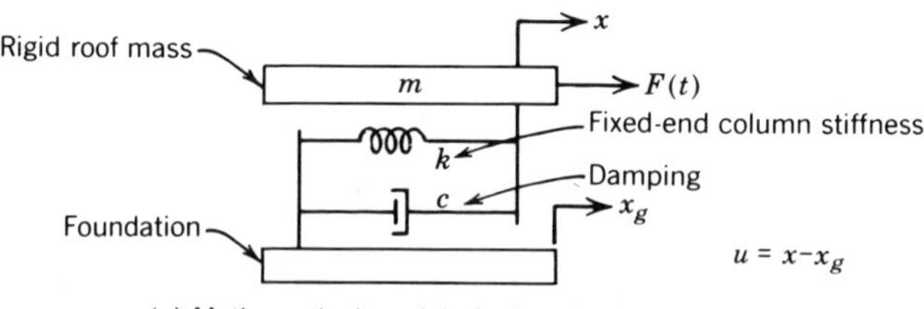

(*c*) Mathematical model of a hotel pergola.

Fig. 18.4 Idealization and model of a single-story structure.

where $u = x - x_g$
x = absolute displacement of the roof
x_g = kinematic displacement of foundation
m = mass of the roof structure
c = damping coefficient
k = fixed-end stiffness of the columns
$F(t)$ = external forcing function

The dynamic characteristic of the structure is its frequency or period of vibration, which can be determined by examining its response in the absence of any forcing functions [$F(t) = 0$ and $\ddot{x}_g = 0$].
Two equations can be written, one without damping and one with damping as given below

$$m\ddot{u} + ku = 0 \tag{18.4.2a}$$

$$m\ddot{u} + c\dot{u} + ku = 0 \tag{18.4.2b}$$

If the mass shown in Fig. 18.4*c* is offset from its equilibrium position by giving it some initial displacement u_0 and/or initial velocity $\dot{u}_0$, the roof will vibrate with a harmonic oscillation. In the undamped case [Eq. (18.4.2*a*)], it can be shown that the time-dependent displacement is given by

$$u(t) = \frac{\dot{u}_0}{\omega} \sin \omega t + u_0 \cos \omega t \tag{18.4.3}$$

where ω = natural circular frequency of vibration, rad/sec
f = natural frequency of vibration, Hz
T = natural period of vibration, sec

Note that

$$\omega = \sqrt{k/m} \tag{18.4.4}$$

It can be shown that the dynamic characteristics of the damped and undamped models are related by

$$\omega_d = \omega\sqrt{1 - \xi^2} \tag{18.4.5}$$

$$f_d = f\sqrt{1 - \xi^2} \tag{18.4.6}$$

$$T_d = \frac{T}{\sqrt{1 - \xi^2}} \tag{18.4.7}$$

$$\xi = \frac{c}{2m\omega} \tag{18.4.8}$$

where ξ = critical damping ratio or damping coefficient

The effect of damping is to decrease the natural frequency of vibration or to increase the natural period of vibration of the model. For critical damping ratios less than 20% ($\xi = 0.2$), a range that includes most structural systems, the effect of damping on the natural period or frequency is small (approximately 2%) and can be neglected. However, damping does have an important effect on structure response in that it can greatly reduce the amplitudes of the responses even for small values of the critical damping ratio.

For a given set of forcing functions [i.e., $F(t)$ and/or $\ddot{x}_g$] Eq. (18.4.1) can be solved in several ways, which are discussed later, to produce a response time history $u(t)$, $\dot{u}(t)$, and $\ddot{u}(t)$ or peak responses. Once the responses have been determined, the horizontal force in the columns can be given by

$$f_h(t) = ku(t) \tag{18.4.9}$$

If the concept of equivalent lateral force is used, the total horizontal base force or base shear and moment at the foundation or base of the structure can be determined from static analysis

$$V_B(t) = f_h(t) = ku(t) \tag{18.4.10}$$

$$M_B(t) = HV_B(t) \tag{18.4.11}$$

where $V_B(t)$ = total horizontal base force or base shear
$M_B(t)$ = total moment at the structure foundation or base
H = height of the roof above the foundation

For the case of the structure of Fig. 18.4*a*, the stiffness k is determined by

$$k = \frac{12EI}{H^3(1 + \phi)} \tag{18.4.12}$$

$$\phi = \frac{12EI}{H^2GA_s} \tag{18.4.13}$$

where E = modulus of elasticity
I = equivalent moment of inertia of the columns
ϕ = shear deformation parameter
G = shear modulus
A_s = effective shear area

For most slender columns, ϕ can be assumed to be zero and shear deformation can be neglected.

Another single-story building of interest is shown in Figs. 18.5 and 18.6. This box or shear wall building is typical of many small commercial or industrial buildings that have concrete or masonry walls and either rigid or flexible roof diaphragms. In the case of a rigid diaphragm (Fig. 18.5), the assumptions made for the structural idealization are:

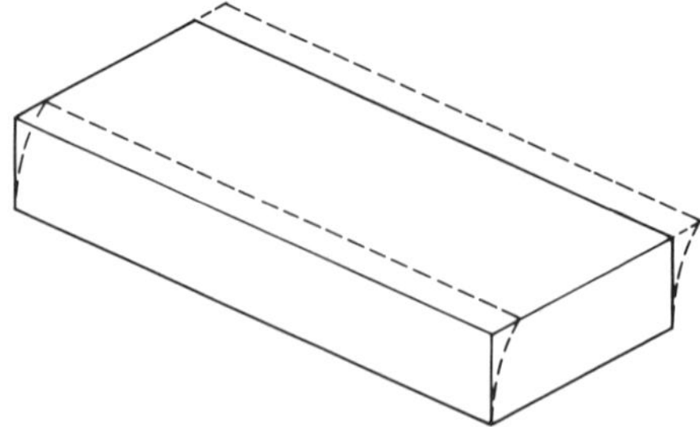

Fig. 18.5 Single-story building with a rigid roof diaphragm.

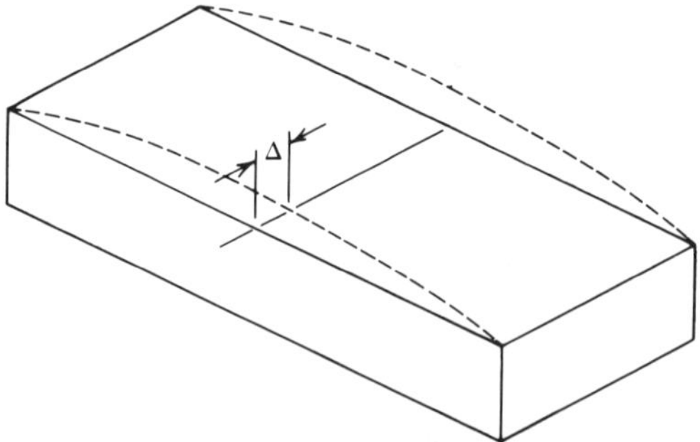

(*a*) Schematic of a single story building showing deflected roof diaphragm.

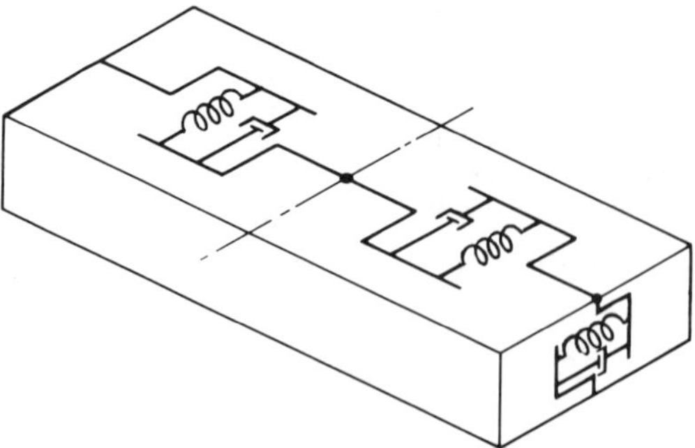

(*b*) Lumped-parameter structural idealization.

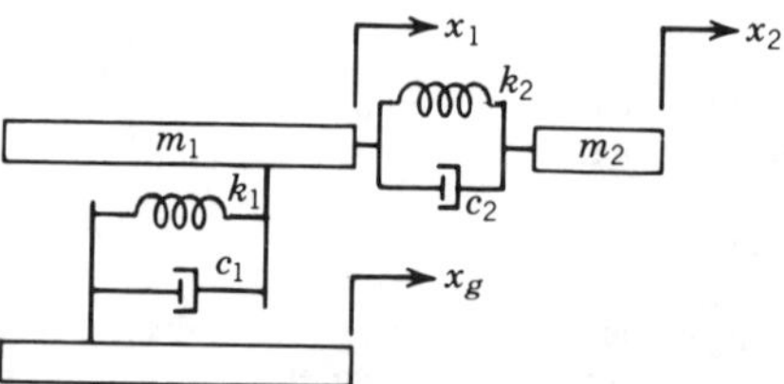

(*c*) Mathematical model of one-story building with a flexible roof diaphragm.

Fig. 18.6 Single-story building with a flexible roof diaphragm.

1. The vertical response of the system to horizontal loads is not important.
2. The end shear walls act as a cantilever (i.e., free to rotate at the roof level).
3. A fraction of the mass of the end shear and side walls can be assumed to be lumped with that of the diaphragm.
4. The roof diaphragm is rigid.
5. The loading and response are completely defined by the lateral deflection of the roof.

Accordingly, the mathematical model is identical to the one shown in Fig. 18.4*c*; and the static and dynamic response of the model is completely described by a single coordinate u (the relative deflection between the base and the roof diaphragm). The equation of motion for the SDOF system is defined by Eq. (18.4.1), where m is the mass of the roof diaphragm and a fraction of the side and end walls, and k is the stiffness of the end shear walls.

The end shear wall stiffness is composed of a flexural and shear component, where

$$k_f = \text{flexural stiffness component}$$

$$= \frac{3EI}{h^3} \text{ (base fixed and top free to rotate)} \tag{18.4.14}$$

$$k_s = \text{shear stiffness component} = \frac{GA_s}{h} \tag{18.4.15}$$

$$k = \frac{k_f k_s}{k_f + k_s} \tag{18.4.16}$$

$$k = \frac{k_f}{1 + \dfrac{k_f}{k_s}} \tag{18.4.17}$$

or

$$k = \frac{k_s}{1 + \dfrac{k_s}{k_f}} \tag{18.4.18}$$

In the case of a flexible diaphragm (Fig. 18.6), the assumptions made for the structural idealization are:

1. The vertical response of the system to horizontal loads is not important.
2. The end shear walls act as cantilevers.
3. The masses of the side walls participate with that of the diaphragm.
4. The diaphragm flexibility influences the system response.
5. The loading and response are defined by the lateral deflection of the top of the end walls and the lateral deflection of the diaphragm.

One possible mathematical model is shown in Figs. 18.6*b* and 18.6*c*, and it is a two degree-of-freedom system. The equations of motion for a two degree-of-freedom system are given below in terms of the relative deflections u_1 and u_2

$$m_1\ddot{u}_1 + (c_1 + c_2)\dot{u}_1 - c_2\dot{u}_2 + (k_1 + k_2)u_1 - k_2u_2 = F_1(t) - m_1\ddot{x}_g \tag{18.4.19a}$$

and

$$m_2\ddot{u}_2 - c_2\dot{u}_1 + c_2\dot{u}_2 - k_2u_1 + k_2u_2 = F_2(t) - m_2\ddot{x}_g \tag{18.4.19b}$$

where $u_1 = x_1 - x_g$
$u_2 = x_2 - x_g$

These equations can be written in matrix form as

$$\begin{bmatrix} m_1 & 0 \\ 0 & m_2 \end{bmatrix}\begin{Bmatrix} \ddot{u}_1 \\ \ddot{u}_2 \end{Bmatrix} + \begin{bmatrix} (c_1 + c_2) & -c_2 \\ -c_2 & c_2 \end{bmatrix}\begin{Bmatrix} \dot{u}_1 \\ \dot{u}_2 \end{Bmatrix} + \begin{bmatrix} (k_1 + k_2) & -k_2 \\ -k_2 & k_2 \end{bmatrix}\begin{Bmatrix} u_1 \\ u_2 \end{Bmatrix} = \begin{Bmatrix} F_1(t) - m_1\ddot{x}_g \\ F_2(t) - m_2\ddot{x}_g \end{Bmatrix} \quad (18.4.20)$$

The normal modes (i.e., natural frequencies) and mode shapes for the undamped system can be obtained from Eqs. (18.4.20) by setting the forcing functions to zero [i.e., $F_1(t) = 0$, $F_2(t) = 0$, and $\ddot{x}_g = 0$] and eliminating the damping terms. These equations reduce to

$$\begin{bmatrix} m_1 & 0 \\ 0 & m_2 \end{bmatrix}\begin{Bmatrix} \ddot{u}_1 \\ \ddot{u}_2 \end{Bmatrix} + \begin{bmatrix} (k_1 + k_2) & -k_2 \\ -k_2 & k_2 \end{bmatrix}\begin{Bmatrix} u_1 \\ u_2 \end{Bmatrix} = \begin{Bmatrix} 0 \\ 0 \end{Bmatrix} \quad (18.4.21)$$

A normal mode of an elastic structural model is identified by a motion in which all degrees-of-freedom vibrate in unison at the same frequency, with each degree-of-freedom reaching its maximum displacement at the same time. A structural model with N degrees-of-freedom possesses N normal modes, each with a distinct frequency (modal frequency) and associated vibratory motion or mode shape. The normal modes of a model provide useful information on its characteristics and a means of determining its response to applied forces and/or kinematic motions. The model defined by Eqs. (18.4.21) has two normal modes, and the response of the degrees-of-freedom, u_1 and u_2, in either normal mode is defined by

$$\begin{Bmatrix} u_1 \\ u_2 \end{Bmatrix} = \begin{Bmatrix} A_1 \sin \omega t \\ A_2 \sin \omega t \end{Bmatrix} \quad (18.4.22a)$$

$$\begin{Bmatrix} \dot{u}_1 \\ \dot{u}_2 \end{Bmatrix} = \omega \begin{Bmatrix} A_1 \cos \omega t \\ A_2 \cos \omega t \end{Bmatrix} \quad (18.4.22b)$$

$$\begin{Bmatrix} \ddot{u}_1 \\ \ddot{u}_2 \end{Bmatrix} = -\omega^2 \begin{Bmatrix} A_1 \sin \omega t \\ A_2 \sin \omega t \end{Bmatrix} = -\omega^2 \begin{Bmatrix} u_1 \\ u_2 \end{Bmatrix} \quad (18.4.22c)$$

where ω is the normal mode frequency and A_1 and A_2 define the mode shape. Substituting Eqs. (18.4.22) into Eqs. (18.4.21), the equations for modal vibration are

$$\begin{bmatrix} -m_1\omega^2 & 0 \\ 0 & -m_2\omega^2 \end{bmatrix}\begin{Bmatrix} u_1 \\ u_2 \end{Bmatrix} + \begin{bmatrix} (k_1 + k_2) & -k_2 \\ -k_2 & k_2 \end{bmatrix}\begin{Bmatrix} u_1 \\ u_2 \end{Bmatrix} = \begin{Bmatrix} 0 \\ 0 \end{Bmatrix}$$

or

$$\begin{bmatrix} (k_1 + k_2 - m_1\omega^2) & -k_2 \\ -k_2 & (k_2 - m_2\omega^2) \end{bmatrix}\begin{Bmatrix} A_1 \\ A_2 \end{Bmatrix} \sin \omega t = \begin{Bmatrix} 0 \\ 0 \end{Bmatrix} \quad (18.4.23)$$

Equations (18.4.23) can be solved for the normal mode frequencies, ω_1 and ω_2, and the mode shape amplitude ratios, $(A_1/A_2)_1$ and $(A_1/A_2)_2$, in the following way.

Equations (18.4.23) are satisfied for any A_1 and A_2 only if the determinant of the matrix is zero. Accordingly, the modal frequencies are obtained from the following quadratic polynomial in ω^2

$$m_1m_2(\omega^2)^2 - (m_1k_2 + m_2k_1 + m_2k_2)\omega^2 + k_1k_2 = 0$$

or

$$(\omega^2)^2 - (p_{11}^2 + p_{21}^2 + p_{22}^2)\omega^2 + p_{11}^2p_{22}^2 = 0 \quad (18.4.24)$$

where

$$p_{ij}^2 = \frac{k_i}{m_j} \quad (18.4.25)$$

and the two normal mode frequencies, ω_1 and ω_2, of the undamped system defined by Eqs. (18.4.21) are obtained by solving Eq. (18.4.24)

$$\omega_1^2, \omega_2^2 = \left(\frac{p_{11}^2 + p_{21}^2 + p_{22}^2}{2}\right) \pm \sqrt{(p_{11}^2 + p_{21}^2 + p_{22}^2)^2 - 4p_{11}^2 p_{22}^2} \tag{18.4.26}$$

The amplitude ratios of the normal modes are obtained from Eqs. (18.4.23) as

$$\frac{A_1}{A_2} = \frac{k_2}{k_1 + k_2 - m_1\omega^2} = \frac{p_{21}^2}{p_{11}^2 + p_{21}^2 - \omega^2} \tag{18.4.27a}$$

or

$$\frac{A_1}{A_2} = \frac{k_2 - m_2\omega^2}{k_2} = \frac{p_{22}^2 - \omega^2}{p_{22}^2} \tag{18.4.27b}$$

where Eqs. (18.4.27a) and (18.4.27b) are equivalent and yield the same displacement amplitude ratios. The exact amplitudes, A_1 and A_2, are not obtainable, since no information has been provided on how the motions are initiated. It is common practice to assign one of the amplitudes to be unity, in which case the modes are termed to be normalized to that particular amplitude.

The amplitude ratios of the two normal modes are obtained from either Eqs. (18.4.27a) or (18.4.27b) by substituting ω_1^2 and ω_2^2 for ω^2. The first normal mode is defined by the frequency ω_1 and the mode shape amplitude ratio of

$$\left(\frac{A_1}{A_2}\right)_1 = \frac{p_{22}^2 - \omega_1^2}{p_{22}^2} \tag{18.4.28}$$

The second normal mode is defined by the frequency ω_2 and the mode shape amplitude ratio of

$$\left(\frac{A_1}{A_2}\right)_2 = \frac{p_{22}^2 - \omega_2^2}{p_{22}^2} \tag{18.4.29}$$

It will be shown that Eqs. (18.4.20) can be solved using the normal mode characteristics obtained above. Moreover, these characteristics can provide information to guide the engineer in determining what constitutes a rigid or a flexible diaphragm. Generally speaking, it can be shown that the diaphragm contributes about 6% to the system response when $p_{22}^2/p_{11}^2 = 8$, and the diaphragm can be considered to be rigid at this and larger ratios.

18.4.2 Models for Multistory Structures

The development of dynamic idealizations and models for multistory structures follows the same procedure as that used for single-story structures. Again, the engineer should only model those features of the structure that are important in defining its response. A very common multistory structure is one that is termed the multistory shear building and is the multistory counterpart of the single-story structure described in Fig. 18.4. A structural idealization/model of this structure is given in Fig. 18.7. For the purpose of this description, a shear building is defined as a structure that has no rotation at the floor levels, and the response of the rigid floors completely determines the response of the building in terms of lateral deformations and horizontal forces in the walls or columns.

This type of structural model is termed a multi-degree-of-freedom system in that more than one degree-of-freedom or coordinate is required to describe its response completely. For a multistory shear building, the number of degrees-of-freedom is equal to the number of stories in the building. The response of this structural model is governed by a set of ordinary differential equations which are most conveniently handled using matrix algebra.

18.5 ANALYSIS FOR DYNAMIC LOADS

18.5.1 Dynamic Response Spectra

Of the available ways to measure the intensity of earthquakes, the response spectrun is the most valuable to the structural engineer. An earthquake response spectrum is obtained from an earth-

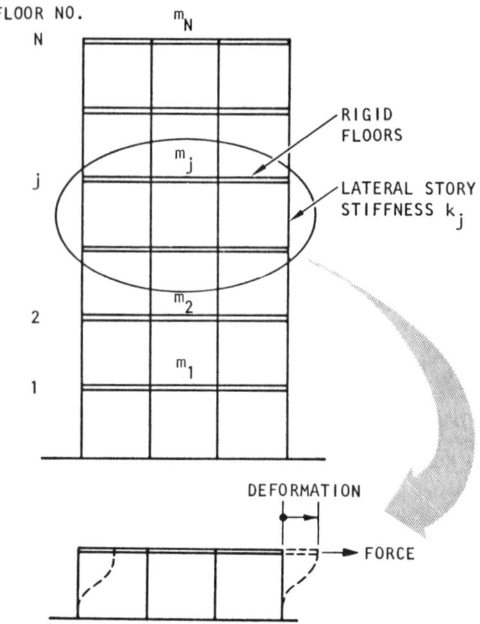

Fig. 18.7 Structural idealization/model of a multistory shear building [18.5].

quake accelerogram (i.e., acceleration-time history) and quantitatively shows the response of a family of viscously damped single degree-of-freedom oscillators to a support motion described by the accelerogram. The analytical model for calculating the oscillator's response is similar to the one shown in Fig. 18.4*c*, and the equation of motion as defined by Eq. (18.4.1) can be written as

$$\ddot{u} + \frac{c}{m}\dot{u} + \frac{k}{m}u = -\ddot{x}_g(t)$$

or

$$\ddot{u} + 2\xi\omega_n\dot{u} + \omega_n^2 u = -\ddot{x}_g(t) \tag{18.5.1}$$

Equation (18.5.1) can be solved using the following integral solutions

$$u(t) = -\frac{1}{\omega_d}\int_0^t \ddot{x}_g(\tau)e^{-\xi\omega_n(t-\tau)}\sin[\omega_d(t-\tau)]\,d\tau \tag{18.5.2}$$

$$\dot{u}(t) = -\int_0^t \ddot{x}_g(\tau)e^{-\xi\omega_n(t-\tau)}\cos[\omega_d(t-\tau)]\,d\tau - \xi\omega_n u(t) \tag{18.5.3}$$

$$\ddot{u}(t) = -2\xi\omega_n\dot{u}(t) - \omega_n^2 u(t) \tag{18.5.4}$$

$$\ddot{x}(t) = \ddot{x}_g(t) - 2\xi\omega_n\dot{u}(t) - \omega_n^2 u(t) \tag{18.5.5}$$

For a given accelerogram and values of ξ and ω_n (or T_n), the relative deflection $u(t)$, relative velocity $\dot{u}(t)$, and relative acceleration $\ddot{u}(t)$, between the oscillator mass and the ground can be obtained as a function of time as given by Eqs. (18.5.2), (18.5.3), and (18.5.4). In addition the absolute acceleration of the oscillator mass can be obtained as a function of time using Eq. (18.5.5). The maximum absolute values of $u(t)$, $\dot{u}(t)$, and $\ddot{x}(t)$ are of special interest and are defined below:

$S_d(T_n, \xi) = |u(t)|_{\max}$ = relative displacement response spectrum or spectral displacement

$S_v(T_n, \xi) = |\dot{u}(t)|_{\max}$ = relative velocity response spectrum or spectral velocity

$S_a(T_n, \xi) = |\ddot{x}(t)|_{\max}$ = absolute acceleration response spectrum or spectral acceleration

These spectral values define the response of a damped, single degree-of-freedom oscillator to a given accelerogram, and help characterize the earthquake and define its intensity. Two additional parameters are also of special interest and are defined below:

$$\mathrm{PS}_v(T_n, \xi) = \omega_n S_d = \left(\frac{2\pi}{T_n}\right) S_d = \text{pseudo-velocity spectrum}$$

$$\mathrm{PS}_a(T_n, \xi) = \omega_n^2 S_d = \left(\frac{2\pi}{T_n}\right)^2 S_d = \text{pseudo-acceleration spectrum}$$

For a given accelerogram and value of ξ, any of these spectral values can be calculated and plotted as a function of T_n or f_n, and the resulting graph is called a response spectrum of the earthquake motion. Typically, these spectra are computed for several values of the critical damping ratio and are all plotted on a single graph. Due to the unique relationship among S_d, PS_v, and PS_a, it is possible to plot them all on a single graph as a function of T_n or f_n using four-way logarithmic axes. In this case, the log of the period and/or frequency is plotted on the abscissa axis, the log of the pseudo-velocity is plotted on the ordinate axis, and the log of pseudo-acceleration and log of the spectral displacement are plotted on axes oriented at 45° to the ordinate and abscissa axes. This type of graph is commonly called a tripartite response spectrum plot, and a typical response spectrum using the tripartite plot is shown in Fig. 18.8.

Response spectra for explosive effects are usually referred to as shock spectra. The method of obtaining shock spectra is the same as for response spectra except that the loads are impulsive, the accelerations are much higher, and the ranges of interest are therefore different than in the case of earthquake spectra.

In blast-resistant design, the structure is allowed to have inelastic deformation. In ductile materials, linear elastic behavior is followed by yielding, and this behavior is idealized with an elasto-plastic resistance curve.

18.5.2 Modal Analysis Method

If the analytical model of a structural idealization of a building system is a multi-degree-of-freedom linear system, the dynamic or quasi-static analysis of the model is best handled by one of the modal analysis methods.

The dynamic equations of motion for a multi-degree-of-freedom system can be written in matrix form as

$$[m]\{\ddot{u}\} + [c]\{\dot{u}\} + [k]\{u\} = \{p\} \tag{18.5.6}$$

It can be shown that the undamped normal modes of the model can be used to uncouple Eq. (18.5.6), and greatly simplify the response calculations. The undamped normal modes of Eq. (18.5.6) are obtained by the solution of the following eigenvalue or characteristic value problem

$$[m]\{\ddot{u}\} + [k]\{u\} = 0 \tag{18.5.7}$$

where harmonic response is invoked by

$$\{\ddot{u}\} = -\omega^2\{u\} \tag{18.5.8}$$

Using Eq. (18.5.8) in Eq. (18.5.7)

$$[[k] - \omega^2[m]]\{u\} = \{0\} \tag{18.5.9}$$

Equation (18.5.9) can be solved in several ways [18.6], one of which is shown in Section 18.4. The solution yields a matrix of eigenvalues (modal frequencies) and eigenvectors (mode shapes),

where $[\phi]$ = modal matrix, a matrix of mode shapes stored by columns
$[\diagdown\omega^2]$ = modal frequencies, a diagonal matrix of modal frequencies corresponding to the mode shapes

For certain forms of the damping matrix $[c]$ the modal matrix uncouples Eq. (18.5.6) by defining normal coordinates η as

IMPERIAL VALLEY EARTHQUAKE MAY 18, 1940 - 2037 PST

IIIA001 40.001.0 EL CENTRO SITE IMPERIAL VALLEY IRRIGATION DISTRICT COMP S00E

DAMPING VALUES ARE 0, 2, 5, 10 AND 20 PERCENT OF CRITICAL

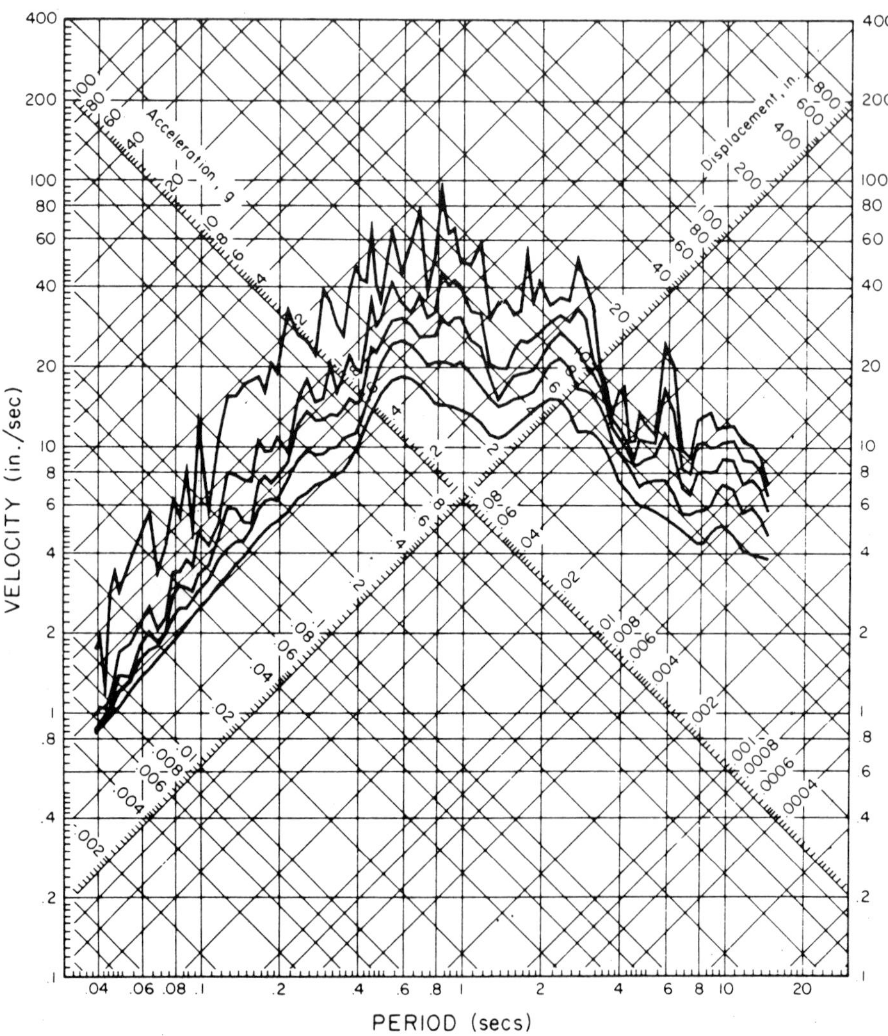

Fig. 18.8 Typical earthquake response spectrum curves—tripartite logarithmic plot [18.5].

$$\left.\begin{aligned}\{u\} &= [\phi]\{\eta\} \\ \{\dot{u}\} &= [\phi]\{\dot{\eta}\} \\ \{\ddot{u}\} &= [\phi]\{\ddot{\eta}\}\end{aligned}\right\} \tag{18.5.10}$$

Using Eqs. (18.5.10) in Eqs. (18.5.6) and premultiplying by the transpose of the modal matrix gives

$$[\phi]^T[m][\phi]\{\ddot{\eta}\} + [\phi]^T[c][\phi]\{\dot{\eta}\} + [\phi]^T[k][\phi]\{\eta\} = [\phi]^T\{p\} \tag{18.5.11}$$

From the orthogonality properties of the normal modes is can be shown that

$$\begin{aligned}[\phi]^T[m][\phi] &= [\diagdown M \diagdown] \\ [\phi]^T[k][\phi] &= [\diagdown K \diagdown] = [\diagdown \omega^2 \diagdown][\diagdown M \diagdown] \\ [\phi]^T[c][\phi] &= [\diagdown C \diagdown] = 2[\diagdown \xi \diagdown][\diagdown \omega \diagdown][\diagdown M \diagdown]\end{aligned} \tag{18.5.12}$$

Using Eqs. (18.5.12) in Eqs. (18.5.11) results in a set of uncoupled equations of motion in η

$$[\diagdown M \diagdown]\{\ddot{\eta}\} + 2[\diagdown \xi \diagdown][\diagdown \omega \diagdown][\diagdown M \diagdown]\{\dot{\eta}\} + [\diagdown \omega^2 \diagdown][\diagdown M \diagdown]\{\eta\} = [\phi]^T\{p\} \tag{18.5.13}$$

where the rth equation is

$$M_r\ddot{\eta}_r + 2\xi_r\omega_r M_r\dot{\eta}_r + \omega_r^2 M_r\eta_r = \{\phi_r\}^T\{p\}$$

or

$$\ddot{\eta}_r + 2\xi_r\omega_r\dot{\eta}_r + \omega_r^2\eta_r = \frac{1}{M_r}\{\phi_r\}^T\{p\} \tag{18.5.14}$$

and

$$M_r = \{\phi_r\}^T[m]\{\phi_r\} \tag{18.5.15}$$

Equations (18.5.13) can be solved for the normal coordinates as described in Section 18.5.1.

18.5.3 Response of Elasto-Plastic Single Degree-of-Freedom Systems

For a triangular pulse, the maximum response of an elasto-plastic SDOF model has been calculated and plotted in a convenient form as shown in Fig. 18.9 (see Ref. 18.8). In this figure, the load parameters are P_m/r_y, the ratio of the peak pressure to peak resistance, and t_d/T, the ratio of the duration of the pulse to the natural period of the SDOF model. The designer must specify the allowable deformation of the structure, or the ductility factor, defined as the ratio of maximum deformation to the deformation at yield. The maximum allowed pressure can then be read for different time ratios t_d/T. For example, from Fig. 18.9, if $x_m/x_y = 7$ and $t_d/T = 1.5$, read $P_m/r_y = 1.4$, that is, the allowed peak pressure is 1.4 times the pressure that initiates yielding. The ratio t_m/T in the figure gives the time t_m when the maximum displacement occurs.

For other than triangular pulse shapes, the reader is referred to Ref. 18.1. For an idealized elasto-plastic (e.g., steel) material, the stress–deformation relationship is shown in Fig. 18.10.

Ductility factor μ of an element is defined as the ratio of maximum deformation to the deformation at yield point. Thus

$$\mu = \frac{\Delta_u}{\Delta_y} \tag{18.5.16}$$

The ductility factor of an element can be obtained from the calculated elastic stress F_u, assuming the same deformation Δ_u is attained, as follows:

$$\mu = \frac{\Delta_u}{\Delta_y} = \frac{F_u}{F_y} \tag{18.5.17}$$

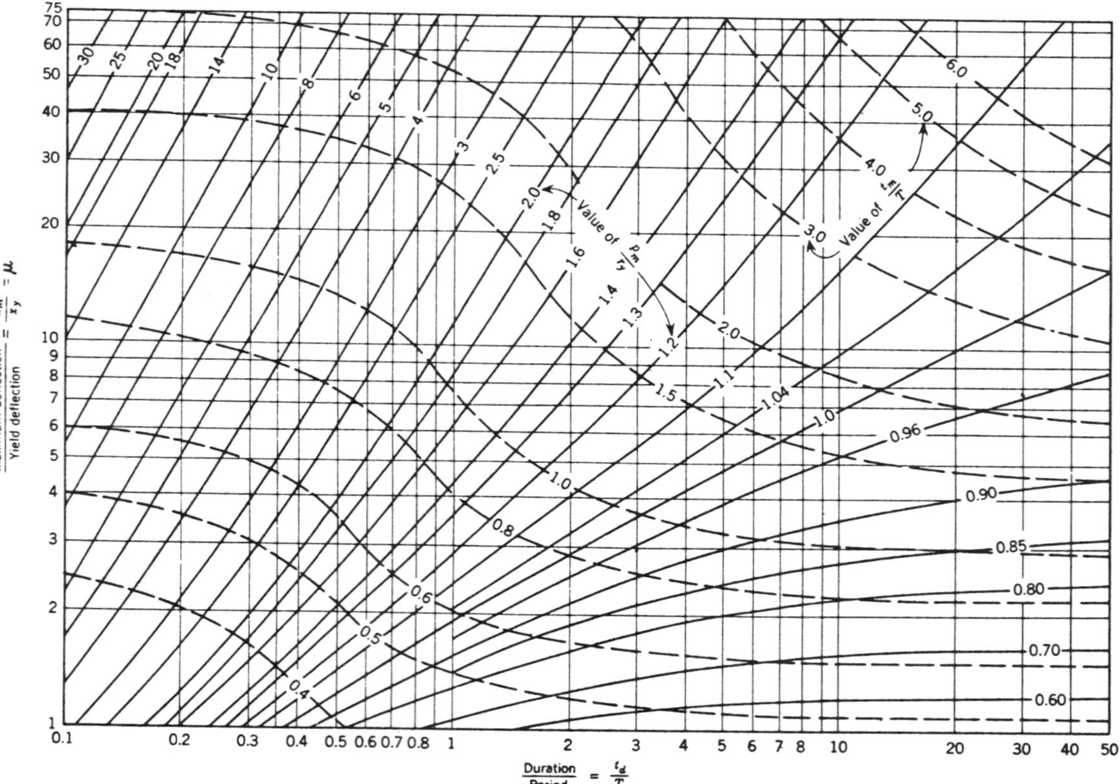

DESIGN CHART FOR INITIAL PEAK TRIANGULAR FORCE PULSE ON ELASTO-PLASTIC SYSTEM

Fig. 18.9 Design chart for initial peak triangular force pulse on elasto-plastic system [18.8].

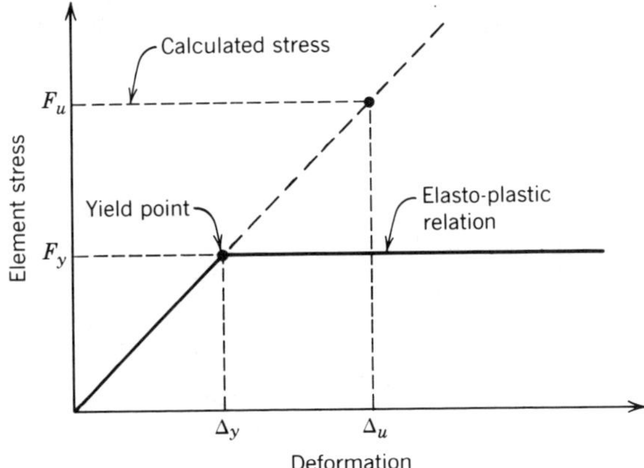

Fig. 18.10 Idealized elasto-plastic material behavior.

Thus the ductility factor is the ratio of the maximum calculated stress and the yield stress of the material.

In blast-resistant design, inelastic spectra are used very frequently, and approximate methods have been developed to obtain inelastic spectra from elastic spectra that are now being used both for earthquake and shock spectra (see Ref. 18.9).

18.6 MODELING OF STRUCTURES

EXAMPLE 18.6.1

The simplest example of a structure is a rigid frame with a stiff roof on flexible columns. This type of structure is often used to introduce the concept of a SDOF system. The mass of the roof is usually much larger than that of the columns, and in modeling the structure, it is customary to neglect the column mass. Figure 18.4*a* shows the structure; Fig. 18.4*b* shows it in its deformed shape. The mathematical model is shown in Fig 18.4*c*. This SDOF model typically illustrates the use of Eq. (18.4.1). The solution of the SDOF equation in turn provides the response of the structure to a base motion or to a force at the roof level.

EXAMPLE 18.6.2

A one-story rectangular structure with a flexible roof may be represented as a two degree-of-freedom system. Figure 18.6 shows such a structure. The end walls and the roof are represented by masses m_1 and m_2, respectively, k_1, c_1 are the stiffness and damping of the end walls, and k_2, c_2 describe the roof stiffness and damping. Since the roof motions are coupled to the out-of-plane motions of the side walls, m_2, k_2, and c_2 should be obtained as equivalent values allowing for this coupling action. The mathematical model is shown in Figs. 18.6*b* and 18.6*c*. The relationships give the equations of motion for a two degree-of-freedom system as described in Section 18.4.1.

EXAMPLE 18.6.3

Figure 18.11*a* is a one-story tilt-up wall building with a wood diaphragm roof. The diaphragm is flexible and its deflection produces out-of-plane bending of the side walls. The end walls are shear walls and earthquake motions are transmitted to the roof in the plane of the shear walls. In this example, it is intended to model the effect of the flexible roof diaphragm on the side walls. A separate model may be used to analyze the shear walls.

Figure 18.11*b* shows the first step in modeling the structure as a lumped parameter model. The diaphragm is modeled as an interconnected spring and dashpot (damping) system transferring their displacements to the side walls through their reaction points that are connected to the wall. The wall is represented as a series of beams spanning from the ground to the roof. The earthquake excitations affecting this model are at the base of the side walls and at the roof level above the end walls. Figure 18.11*c* shows the mathematical model of the diaphragm–wall system. This is a multi-

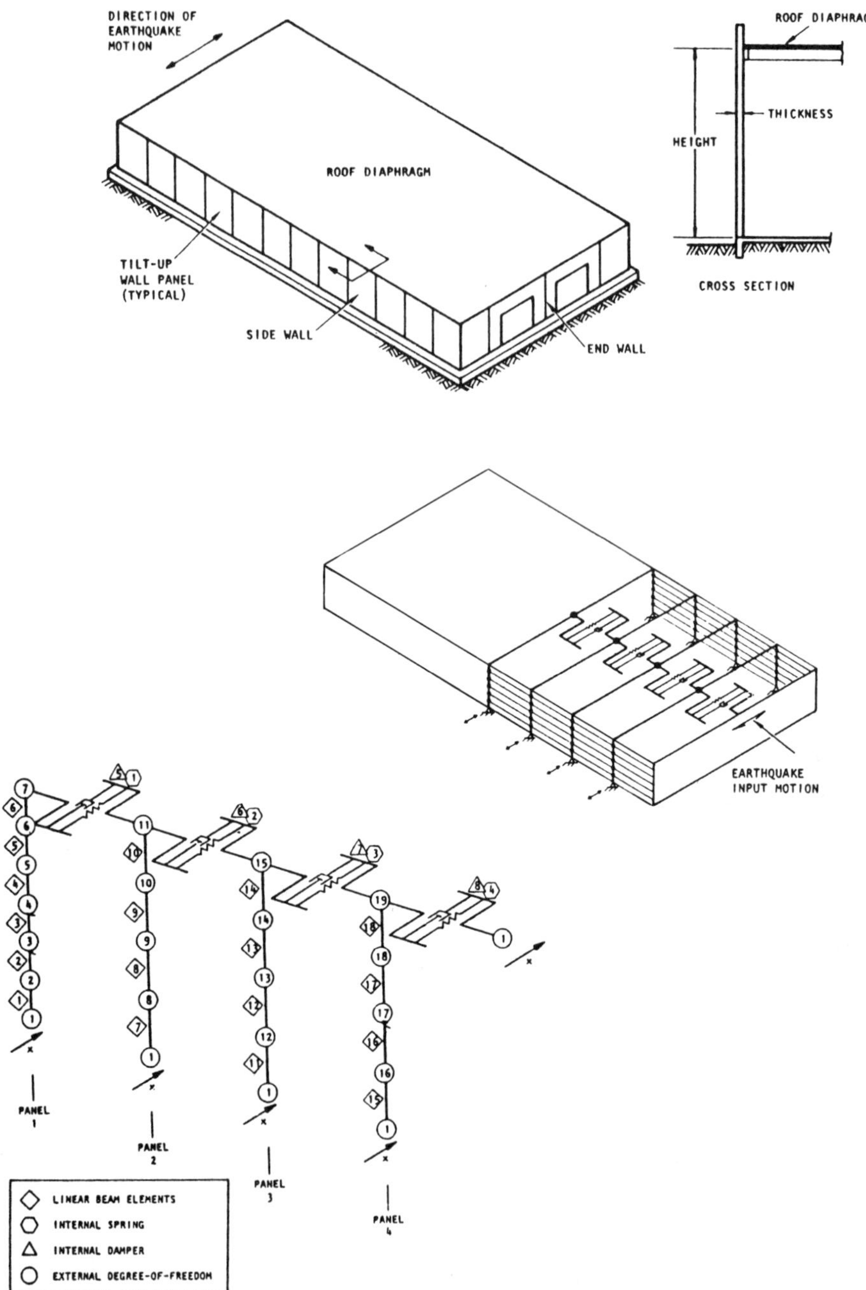

Fig. 18.11 One-story tilt-up wall building with a wood roof diaphragm.

degree-of-freedom system. Computer programs giving solutions to Eqs. (18.5.6) are available in the public domain.

EXAMPLE 18.6.4

Figure 18.12 shows a plan view of a rectangular steel frame building with an atrium. The heavy lines indicate those frames that have moment-resisting connections and are designed to act as the

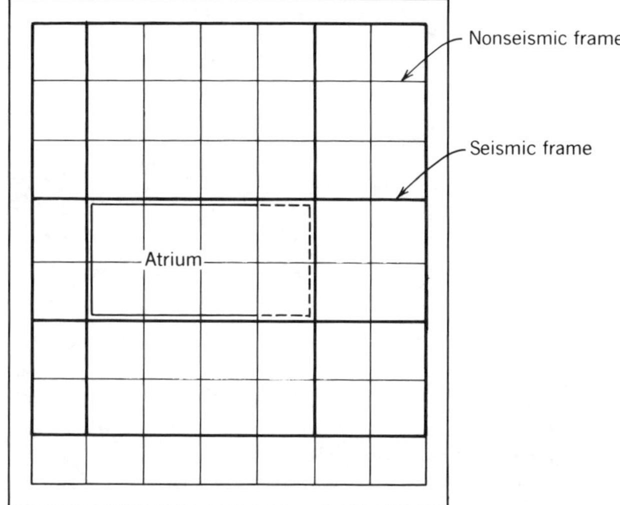

Fig. 18.12 Plan view of a three-story rectangular steel frame building with an atrium.

framework that will resist the earthquake motions. This is a multi-degree-of-freedom system. It can be analyzed as a two-dimensional frame in two directions, the N–S and E–W inputs being applied separately, or it may be analyzed as a three-dimensional frame applying the N–S and E–W inputs simultaneously. Figure 18.13 shows the three-dimensional model representing the entire framework for a finite element analysis. The earthquake excitations are applied at the bases of all the columns. A simpler model may be used by ignoring the influence of all the frames except the frames with moment-resisting connections. This is shown in Fig. 18.14. The advantage of using the more

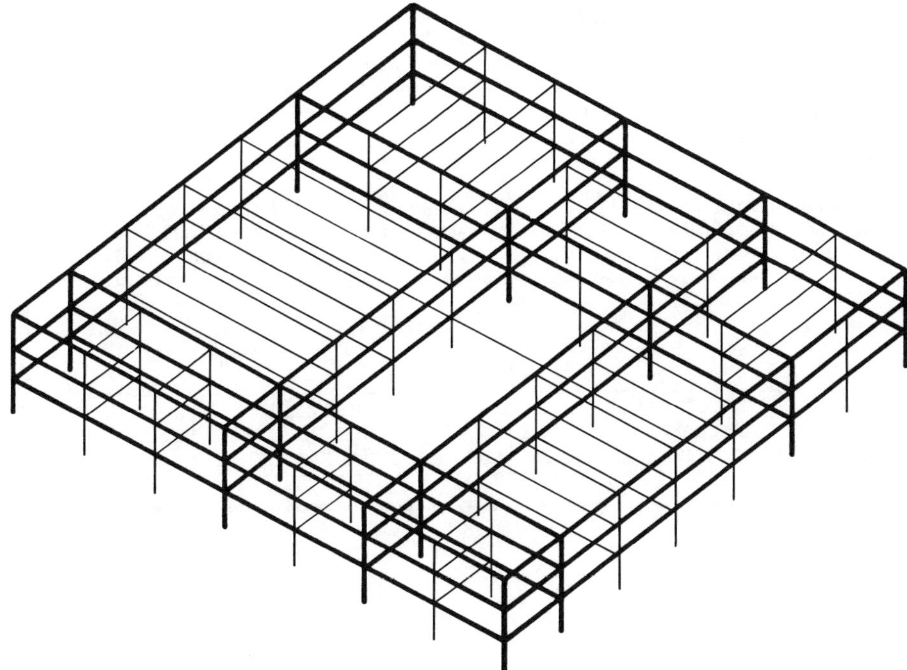

Fig. 18.13 Three-dimensional finite element model including the seismic and nonseismic framework.

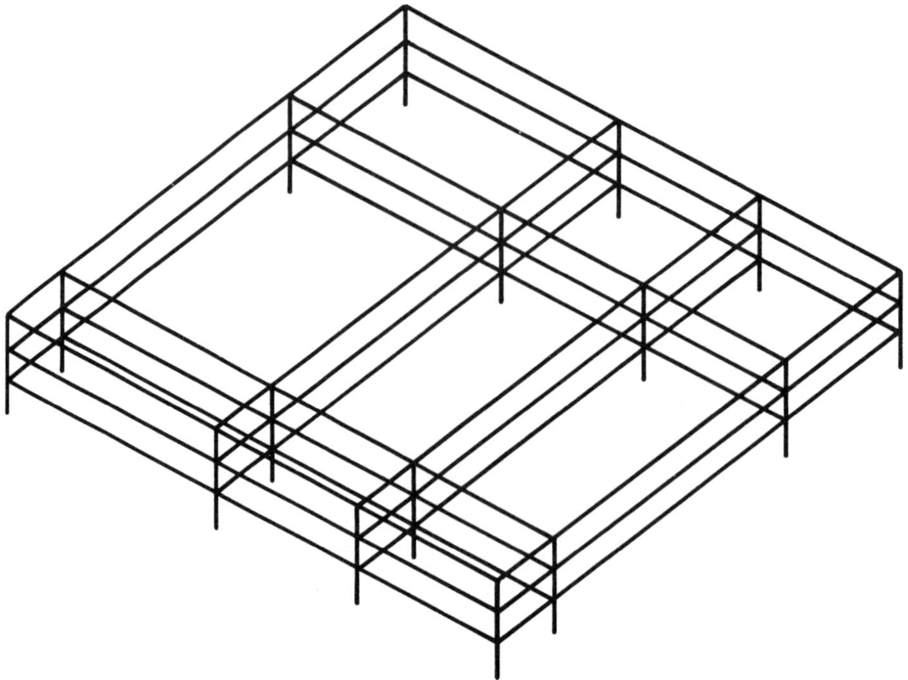

Fig. 18.14 Three-dimensional finite element model including only the seismic framework.

complex frame is in calculating the stresses due to the combined effect of gravity forces and lateral forces. The computer does this automatically. The designer has three choices—two two-dimensional analyses, a simplified three-dimensional analysis, and a more complete three-dimensional model. The availability of the computer allows the analyst to choose any one of these three models, depending on his preference and experience. It should be noted that there will be some differences in the answers obtained from the three methods, but these differences will be small because the asymmetry of the structure is not very large. There is lack of symmetry in the loading distribution around the atrium, but there is symmetry in the number of frames on each side of the atrium.

EXAMPLE 18.6.5

The building shown in Fig 18.15 is a three-story masonry structure with shear walls. The designer is asked if separation of parts A, B, and C by expansion joints is a better design than a fully connected structure. An experienced designer may devise a simple model to make a comparison of the two cases, starting with three masses interconnected with gap-springs. More refined analyses are

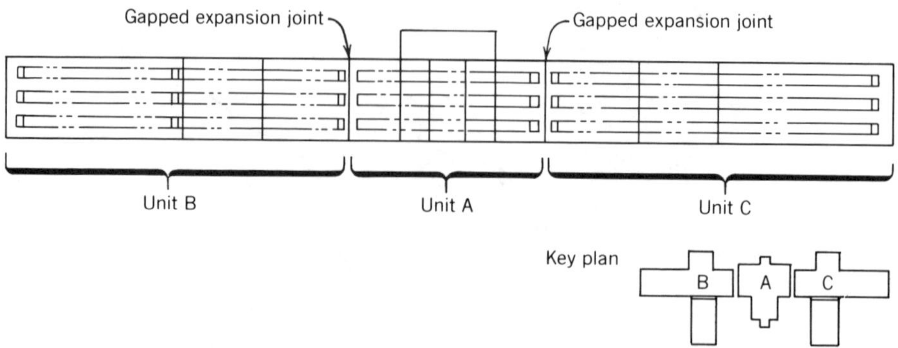

Fig. 18.15 Three-story masonry building with shear walls.

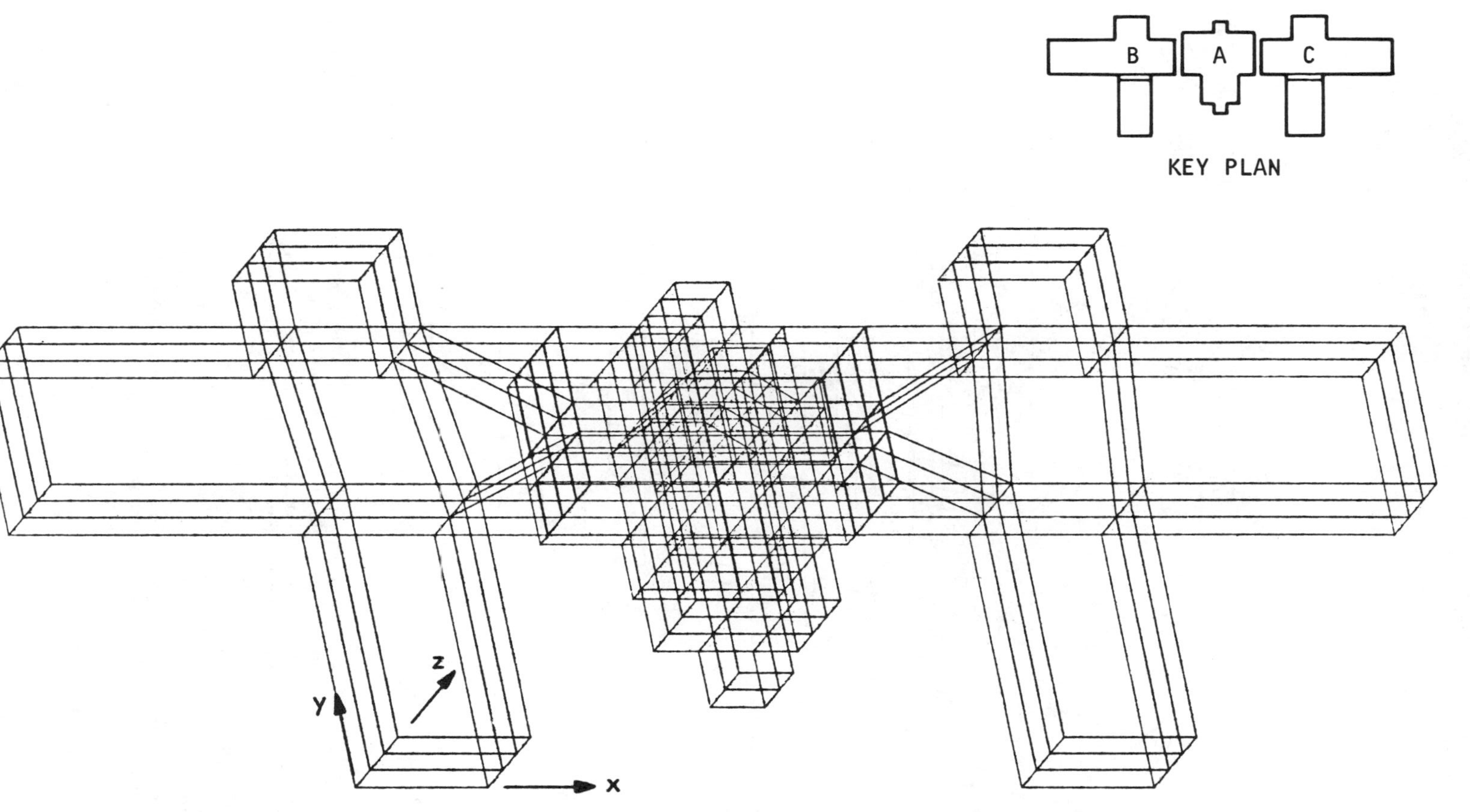

Fig. 18.16 Finite element model of building with segments A, B, and C connected.

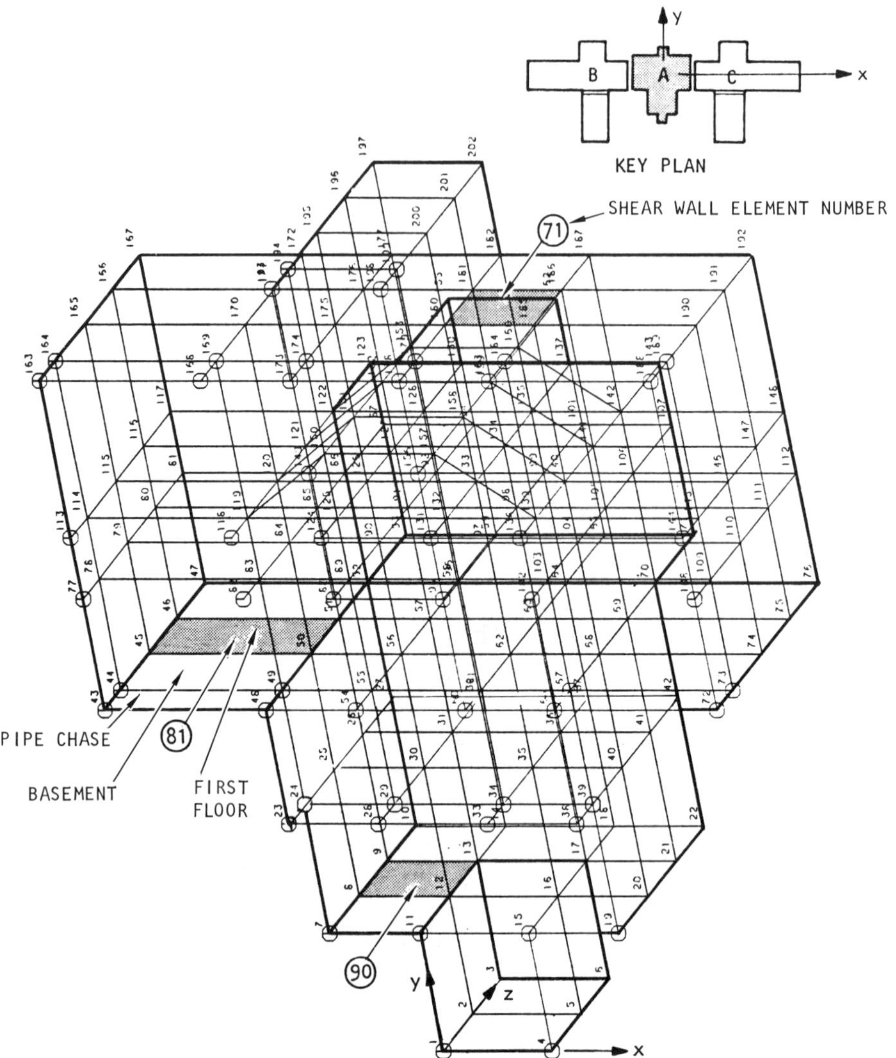

Fig. 18.17 Finite element model of building A without connection to the outside wings.

possible with finite element models that describe the state of stress in critical members of the structure. In this example, the middle structure is analyzed with and without connections to the outside wings. Two models are shown as Figs. 18.16 and 18.17. It is noted that the effect of connecting structures is included in Fig. 18.16, with a less refined mesh of the outside wings since the investigation is focused on the middle building. The analyst then makes a direct comparison between the two cases.

Shear walls and floor slabs are modeled by plate elements, and columns are modeled as bending elements under the effect of lateral forces. The modal response method may be used for the analyses. In this example, calculations by the computer show that the shear walls shown as elements 71, 81, and 90, have calculated stresses that are 30 to 65% higher for the interconnected building than when the three building parts are separated by expansion joints.

SELECTED REFERENCES

18.1 ASCE. *Design of Structures to Resist Nuclear Weapon Effects,* ASCE Manuals and Reports on Engineering Practice No. 42. New York: American Society of Civil Engineers, 1985.

18.2 W. E. Baker. *Explosions in Air*. Austin: Univ. of Texas Press, 1973.

18.3 K. J. Bathe. *Finite Element Procedures in Engineering Analysis*. Englewood Cliffs, NJ: Prentice-Hall, 1982.

18.4 K. J. Bathe and E. L. Wilson. *Numerical Methods in Finite Element Analysis*. Englewood Cliffs, NJ: Prentice-Hall, 1976.

18.5 A. K. Chopra. *Dynamics of Structures,* a Primer. Berkeley, CA: Earthquake Engineering Research Institute, 1980.

18.6 R. W. Clough and J. Penzien. *Dynamics of Structures*. New York: McGraw-Hill, 1975.

18.7 L. S. Jacobsen. "Steady Forced Vibration as Influenced by Damping," *Transactions,* ASME, **52,** 1930, 168–181.

18.8 N. M. Newmark. "An Engineering Approach to Blast Resistant Design," *Transactions,* ASCE, **121,** 1956, 45–64.

18.9 N. M. Newmark and W. J. Hall. *Earthquake Spectra and Design*. Berkeley, CA: Earthquake Engineering Research Institute, 1982.

18.10 Charles H. Norris, Robert J. Hansen, Myle J. Holley, Jr., John M. Biggs, Saul Namyet, and John K. Minami. *Structural Design for Dynamic Loads*. New York: McGraw-Hill, 1959.

18.11 E. Simiu and R. H. Scanlan. *Wind Effects on Structures: An Introduction to Wind Engineering*. New York: John Wiley, 1978.

18.12 R. L. Wiegel (ed.). *Earthquake Engineering*. Englewood Cliffs, NJ: Prentice-Hall, 1970.

CHAPTER 19
STEEL DESIGN—SPECIAL CONSIDERATIONS

HORATIO ALLISON

Structural Engineers
Allison, McCormac & Nickolaus
Rockville, Maryland

19.1 ECONOMIC CONSIDERATIONS

The basic selection of an optimum as opposed to a workable and satisfactory structural system for any particular project will be, at best, an extremely difficult task. The active participation of the entire design team in the evaluation of structural systems is necessary if a truly optimum solution is to be obtained. In steel structures, selection of the type of framing system should include consideration of not only the structural costs, but also the associated costs of other affected components. For instance, the type of fire rating assembly selected for a composite steel framing system may have a great impact on the cost of the heating, ventilating, and air conditioning systems. An assembly consisting of a rated composite floor slab and steel beams and girders with sprayed on fireproofing, may result in an HVAC (heating–ventilating–air conditioning) system which minimizes the use of fire dampers and permits use of the ceiling plenum space.

However, as a practical matter, the final choice of a structural system is often made on the basis of professional experience or personal preference. Hence, even a comprehensive discussion of the subject will not result in the development of an explicit procedure which will lead to a truly optimum solution. However, a number of items relating to overall economy will be discussed.

19.1.1 Price Comparison Methods

One of the most demanding tasks of a designer of steel structures is that of the preparation of meaningful price comparison studies which take into account the appropriate factors involved. Decisions concerning structural systems can materially affect the architecture and economic performance of the final product. Sometimes, the conclusion drawn from a particular serviceability design criterion selected can affect the quality of the product as much or perhaps more than the cost. In any case, the most economical solution seldom will lead to the most satisfactory solution in terms of serviceability. Thus, the final choice must be made on the basis of a combination of cost and quality. It is not possible to present real cost information that will be accurate over an extended period of time. Therefore, the information presented will focus upon the subjects which are germane to the determination of an economical solution.

The Bethlehem Steel Corporation publishes a valuable pamphlet [19.12] which tabulates the cost per foot of structural shapes for the various grades of steel that they produce. The tabulated prices include base price, grade, and section extras. Some fabricators may charge extra handling costs for material other than ASTM A36 steel.

19.1.2 Yield Strength

A review of the costs of rolled shapes for the various grades of structural steel available will quickly reveal that the higher the yield strength, the higher the strength-to-price ratio. The price per foot indicates that the cost of ASTM A572 (Grade 50) steel is only 11% more than ASTM A36.

For beams, the selection of $F_y = 50$ ksi instead of $F_y = 36$ ksi material results in a basic increase of 39% in strength for a basic increase of cost of only 11%. Obviously, when the use of the higher strength steel is appropriate, it should be used for beams and girders. However, serviceability considerations (deflection and/or vibration) may preclude the use of the more cost-effective steel.

The use of higher strength steel in columns is another matter. Due to stability problems, both local and overall, as members become more slender, the use of higher strength steels becomes less efficient. For slenderness ratio KL/r values of about 105 and higher, A36 steel columns would be less costly than columns of Grade 50 steel. However, for a typical office building column supporting an axial load of 1500 kips with a length of 14 ft, the use of A36 steel will require a W14×283 section costing $71.88/ft while the use of Grade 50 steel will require only a W14×211 costing $59.50/ft. In this case, the saving in mill material costs is 17%. For normal tier buildings, the use of high-strength steel columns will result in an appreciable saving of mill material costs, and fabrication costs are no higher for high-strength steel.

19.1.3 Floor Slab Systems

The selection of a floor slab system is not straightforward. A true cost comparison involves considerations other than the structural framing. In all but the smallest buildings, the fire rating requirements are often the governing factor in the determination of the most economical system. Two basic types of floor systems will be discussed, open-web steel joists and composite metal deck floor systems with concrete slabs on steel filler beams.

In an open-web joist system, a floor–ceiling assembly is normally used to achieve the required fire resistance rating (Fig. 19.1). Typically, a rather thin concrete slab, 2–2½ in. (51–64 mm) minimum thickness, plus a 9/16-in. (14-mm) deep metal deck forming material, and a fire-rated acoustical tile or gypsum board ceiling is used with the steel joists and girder beams to form a fire-rated assembly which will satisfy the local building code requirements for fire resistance. The use of this type of floor may not be suitable for large open spaces as the damping characteristics of the system are poor and the performance of the assembly under the influence of vibration induced by foot traffic may be so poor as to be objectionable. The evaluation of this phenomenon is described

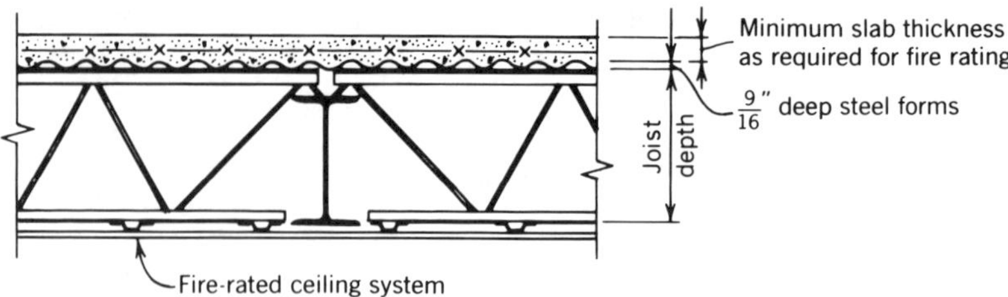

Fig. 19.1 Open-web steel joist floor system section.

in Section 19.2.1. In comparing the real costs of this type of floor system, items other than structural requirements must be considered. For instance, generally the space between the slab and the ceiling may not be used as an air plenum for HVAC systems. This may result in additional ducts for return air. Additionally, penetrations through the ceiling membrane are closely controlled in order to ensure that the integrity of the fire resistance system will not be compromised. Many more fire dampers may be necessary than would be required for a system not dependent upon the floor–ceiling assembly. Items that must be considered when comparing open-web joist systems with other systems include:

1. Cost of concrete slab in place.
2. Cost of metal forms.
3. Cost of erected steel frame.
4. Cost evaluation of HVAC installation (see Chapter 4 for more details).
5. Additional cost of fire-rated ceiling.

On the other hand, when comparing alternative open-web joist systems, the cost items to be considered are reduced to simply the cost of the steel frame.

Many modern office buildings utilize a composite metal deck floor slab in conjunction with composite designed structural steel floor beams and girders (Fig. 19.2). The slab system consists of a reinforced concrete slab placed on a composite metal deck. Comments on the design philosophies utilized with composite construction are given here, with more details provided in Chapters 20 and 23. A number of options must be considered in the selection of a floor deck system. These include:

1. Slab design.
2. Metal deck design.
3. Beam and/or girder design.

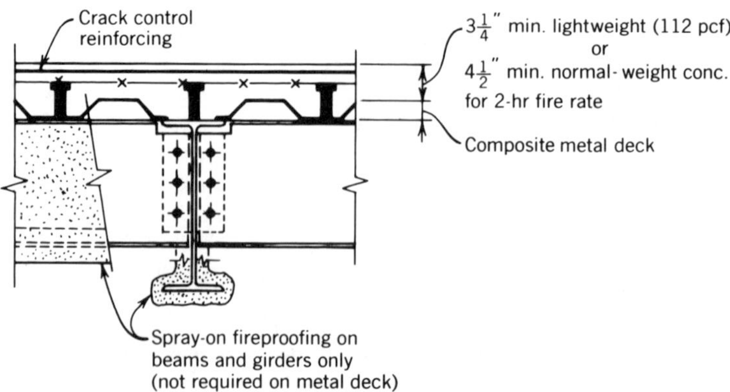

Fig. 19.2 Composite metal deck floor system section.

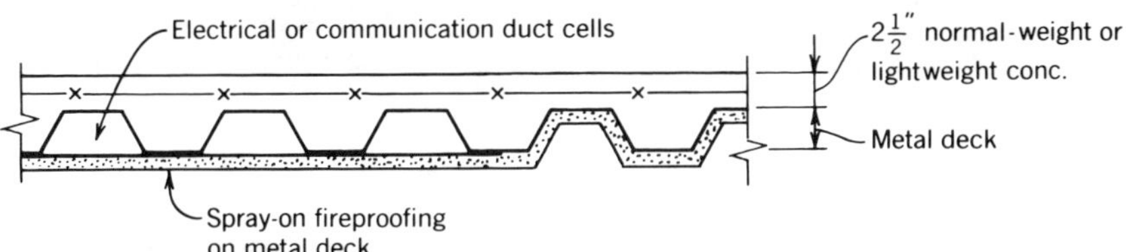

Fig. 19.3 Under floor duct composite metal deck section.

Several items influence the selection of the concrete slab depth and decking material. If the space between the ceiling and the floor slab is not used as a plenum for the HVAC system, a floor–ceiling assembly may be used to obtain the required fire resistance rating. Undoubtedly, this combination will prove to be the most economical solution structurally. If an electrical and/or communication distribution system is to be installed in the floor, generally a 2½-in. (64-mm) minimum thickness lightweight concrete slab will be used. Since the portions of the slab which contain the ducts must be sprayed with a fireproofing membrane, generally, the entire floor area must also be sprayed. For typical office building use, the installation of the HVAC system is generally simpler and less expensive if the fire rating is not dependent upon the ceiling. Thus, more often than not, the assembly consists of a concrete slab of sufficient thickness on a composite metal deck that does not require additional fireproofing, and steel girder and filler beams that require a sprayed-on fireproofing membrane. Typically, a 2-hr fire rating may be obtained using a 3¼-in. (83-mm) minimum thickness lightweight concrete slab (112 pcf; 1100 N/m^3) or a 4½-in. (114-mm) minimum thickness normal-weight concrete slab placed on a composite metal deck without any other fireproofing membrane. The increased volume of concrete (30%) for the 4½-in. (114-mm) slab will more than offset the added cost for the lightweight concrete (see Fig. 19.2). Also, for a typical office building, the total dead load of the floor system is increased by about 40% and the total design load on the columns and foundations is increased by about 20% if the 4½-in. (114-mm) normal-weight concrete slab is used. More often than not, the lighter slab is selected on the basis that the increased floor stiffness and better vibration characteristics of the heavy slab do not justify the extra cost.

The selection of a composite metal deck is straightforward. Filler beam spacing is the major factor in the selection of metal deck. Decks available include 1½-, 2-, and 3-in. depths (38-, 51-, and 76-mm) of various thicknesses. A depth of 3 in. (76 mm) is the maximum permitted by the *AISC Specification* [19.6]. Generally 2- or 3-in. (51- or 76-mm) decks are used for floors as they are the most efficient. The key to the selection of the economical system is to choose the deck with the minimum depth and thickness which is suitable. Thus, for beam spacings up to about 9 ft (2.7 m), a 2-in. (50-mm) deep deck may be used and a 3-in. (76-mm) deep deck for larger spans. Often, composite metal deck slabs are blended with the underfloor duct distribution systems for both communications and power distribution (Fig. 19.3).

Composite floor systems can be designed as totally unshored during the placement of concrete, generally with no penalty insofar as cost. However, wet load deflection of the noncomposite steel beams tends to be so high that an acceptably level floor cannot be obtained without the addition of a significant amount of concrete. Thus, where the appearance of sagging floors is objectionable, shored construction may be the proper choice for both appearance and economy. One should be aware of the fact that if shored construction is used, when the props are removed, it is likely that a crack will form over girders supporting filler beams; therefore, crack control slab reinforcement should be used over girders (see Section 19.2).

When comparing costs of composite floor systems with open-web joist systems, items to be considered should include:

1. Cost of concrete slab in place.
2. Cost of composite metal deck.
3. Cost of erected steel frame.
4. Cost of spray-on fireproofing.
5. Cost evaluation of HVAC installation.

Figure 19.4 illustrates a wood-formed composite concrete slab. While in the past, this type of construction was common, it is now seldom used as the composite metal deck slabs have proved to be considerably less costly. However, formed slabs are often used where the exposure to weather and/or corrosive atmosphere will be encountered, as for example, in parking structures and swimming pool enclosures.

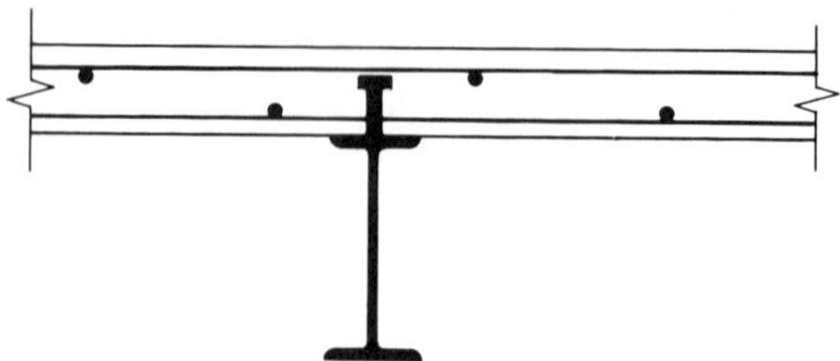

Fig. 19.4 Wood formed concrete slab composite beam section.

19.1.4 Type of Member

There are a number of design alternatives which may be considered in the selection of individual members and framing schemes. The alternative selections for filler beams and girders include noncomposite, composite, rigid connections, and perhaps semi-rigid connections. In addition to the above choices different yield strength steels may be considered.

Figure 19.5 is a part plan of a typical speculative office building bay. The alternatives for the filler beams (designated B-1) include:

1. W21×44 – A36 Simple span
2. W18×35 – A36 Composite
3. W16×31 – A36 Composite
4. W16×26 – F_y = 50 Composite

Since the fabrication and erection costs of all of the alternatives are identical, a basis of cost comparison will be simply the cost of mill material plus the cost of the studs for the composite beams.

The selection of a girder beam can be much more complicated. Assuming that lateral loads are not a factor, girder beam choices are likely to be:

1. Simple span (AISC Type 2).
2. Composite.
3. Rigid frame (AISC Type 1).

The selection of simple span and composite girders is no different than for filler beams. As a practical matter, a girder with a rigid moment connection will be the same size or perhaps one size removed from that of a composite girder. Thus, the cost differential between the two will be the

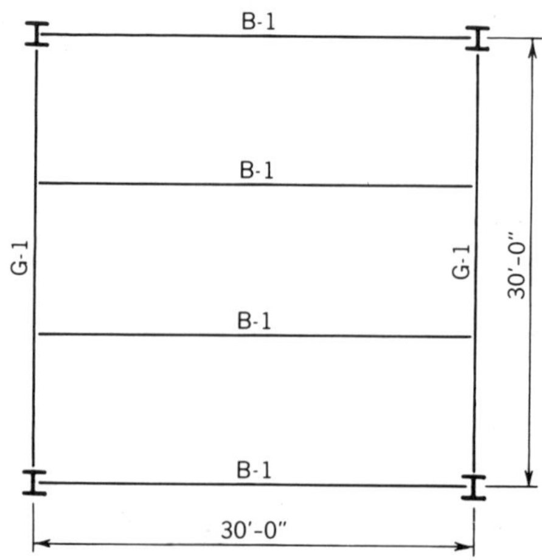

Fig. 19.5 Part plan of typical office building bay.

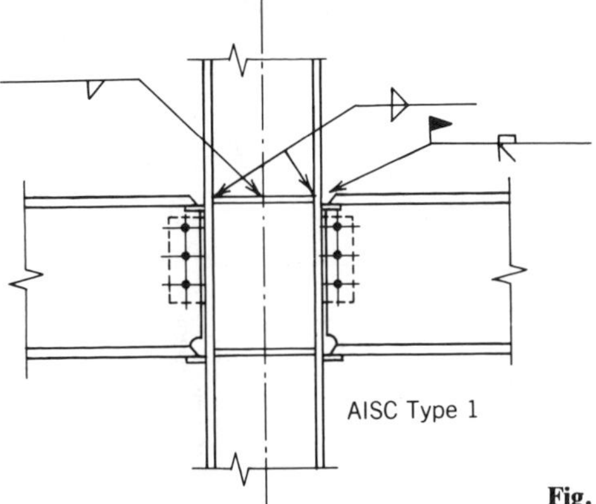

Fig. 19.6 Continuous girder connection detail.

difference in mill material plus the cost of studs vs. the cost of the continuous connection. In tall buildings, the effect of the rigidly connected girders on the column size (due to girder-induced gravity load moments) will be small. However, in low-rise buildings where column sizes are relatively small, continuous girders can significantly increase the cost of columns by increasing their weight and also creating the need for column stiffeners (Fig. 19.6). It should be noted that column stiffeners should not be arbitrarily specified as they increase costs significantly.

If lateral loads are considered, the use of continuous beams is often the economical solution. If a moment-resistant frame is required for lateral load resistance, the girder beams should be utilized for lateral resistance rather than the filler beams.

For the more sophisticated design, other options are available such as stub-girder (Fig. 19.7) or a combination design of composite and plastic analyses. Bjorhovde and Zimmerman [19.13] provide an up-to-date review of design procedures used and an excellent bibliography on the subject. The combination design of composite–plastic girders (Fig. 19.8) will result in very small girders. The full moment capacity of a composite steel beam is generally more than 140% of the steel beam alone. Thus for a composite–plastic girder, the total design moment capacity available for the steel section is about 2.5 times that available for use as a simple span beam. This type of an analysis, while not prohibited, is not specifically addressed by the *AISC Specification* [19.6].

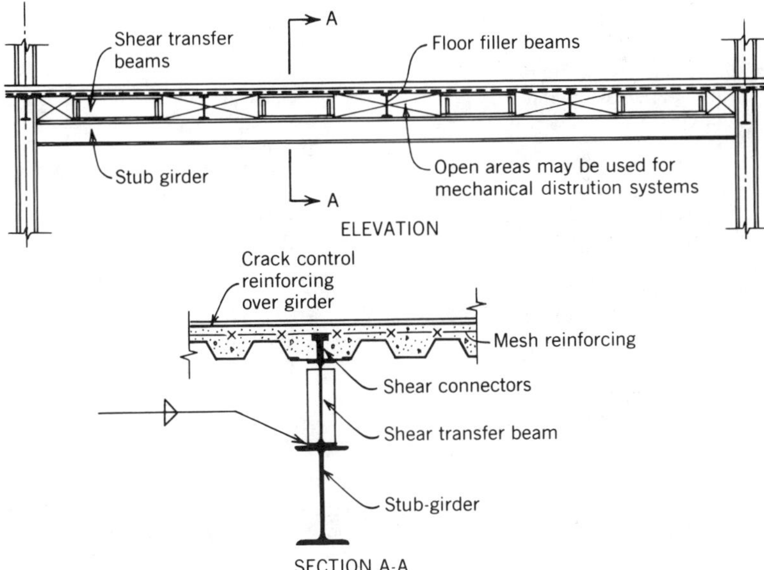

Fig. 19.7 Stub-girder.

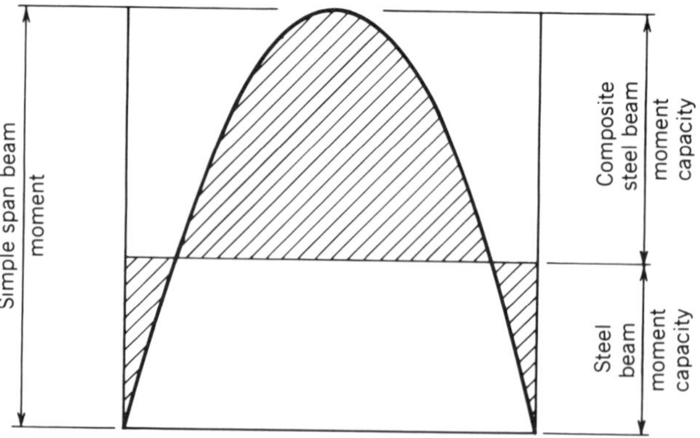

Fig. 19.8 Composite–plastic girder moment diagram.

19.1.5 Selection of Floor Beams or Girders

The seemingly simple task of selecting a floor beam or girder requires consideration of a number of factors, all of which affect the cost. Among the considerations are cost, vibration, deflection, and type of connection, type of construction, etc. As an example, the selection process for a filler beam in an office building bay (see Fig. 19.5) is reviewed:

1. W21×44 A36 simple span:

Mill material cost 30(11.51)	= \$345.30 (\$1.15 psf)
Wet load deflection	= 0.39 in.
Fully composite live load (LL) deflection	= 0.16 in.
Noncomposite LL deflection	= 0.57 in.

2. W18×35 A36 composite beam:

Mill material cost 30(9.07)	= \$272.10
Stud cost 10(1.50)	= 15.00
Total	= \$287.10 (\$0.96 psf)
Wet load deflection	= 0.64 in.
Fully composite LL deflection	= 0.24 in.
Partially composite LL deflection	= 0.37 in.

3. W16×31 A36 composite beam:

Mill material cost 30(8.03)	= \$240.90
Stud cost 16(1.50)	= 24.00
Total	= \$264.90 (\$0.88 psf)
Wet load deflection	= 0.87 in.
Fully composite LL deflection	= 0.31 in.
Partially composite LL deflection	= 0.40 in.

4. W16×26 (F_y = 50 ksi) composite beam:

Mill material cost 30(7.46)	= \$233.80
Stud cost 10(1.50)	= 15.00
Total	= \$248.50 (\$0.80 psf)
Wet load deflection	= 1.08 in.
Fully composite LL deflection	= 0.36 in.
Partially composite LL deflection	= 0.57 in.

As is discussed in Section 19.2.1 the performance of the four beams with respect to foot-traffic-induced vibrations is virtually identical. Even though the simple span beam is designed noncomposite and the filler beams are designed as partially composite, it is assumed that at service loads, they will act as fully composite. In all cases, live load deflections meet the generally accepted limit of $L/360$. Unless the beams are shored, the dead load deflection of the lighter beams may be objectionable. If extra concrete is added to level the floor, the extra cost and extra weight of the concrete should be considered. The best choice from an economic standpoint is probably the W16 × 26 (F_y = 50 ksi) beam shored to limit dead load deflection. However, this will probably result in the formation of a crack in the slab over the supporting girders at each end. This condition creates another problem that is discussed in the design of composite members (Section 19.4.3).

19.1.6 Bay Size

The selection of the bay size for any type building is a most difficult task as a myriad of factors must be considered. For an office building, the selection of bay size is a most important decision as it may critically affect the economic function (rental rates) of the finished product as well as the initial cost. On the other hand, the bay size in a residential building needs to be no larger than required to accommodate the desired plan layout. Thus, the bay size in a residential building may be selected on the basis of the least structural cost. In some types of construction, the quantity of material required is a direct ratio of the span length resulting in higher costs for longer spans and vice-versa. In steel structures, this is not necessarily true. For example, an office building with proposed bay sizes of 24 ft (7.3 m) square and 30 ft (9.1 m) square is shown in Figs. 19.9 and 19.10. The various options available for the selection of member size considering serviceability (quality) are considered in Section 19.4. For this example, a middle-quality selection will be used for comparison of bay size costs. Many engineers would select the least costly members, and rightly so, as their performance in service will nearly duplicate the heavier sections. An analysis using the least costly members will yield similar results. Only items that materially affect the comparative costs will be considered. For instance, only the cost of the mill material will be considered for beams and

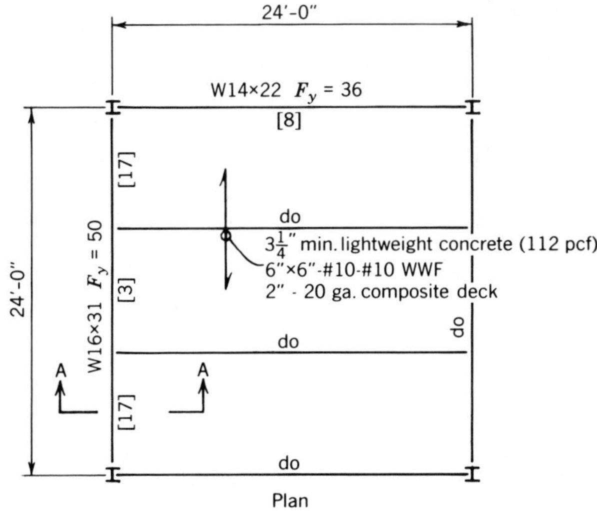

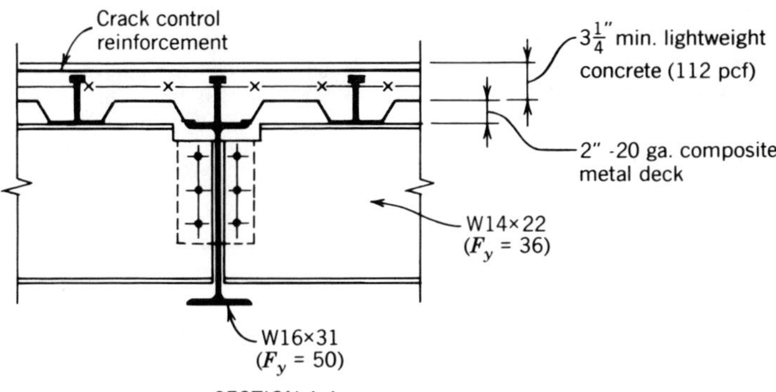

Fig. 19.9 Office building plan—24-ft square bay.

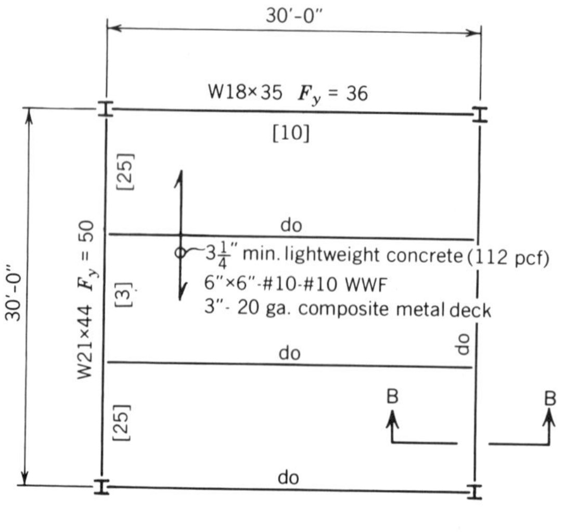

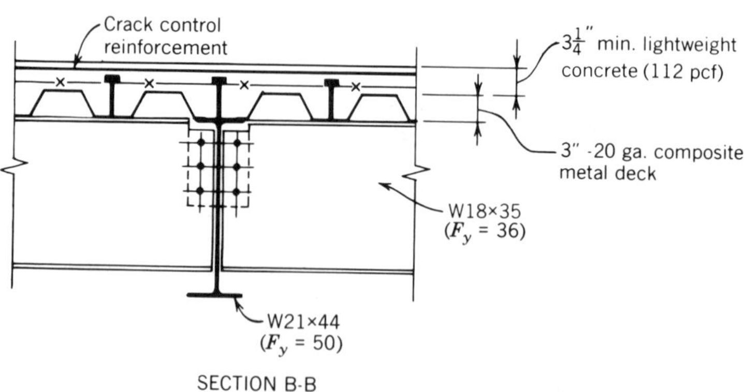

Fig. 19.10 Office building plan—30-ft square bay.

girders. In this example, the fabrication and erection effort required for both bay sizes is identical. Thus, for practical purposes, the only difference in cost for the members should be only a function of the cost of the material. That is not to say that the final price paid will not involve a judgment on the part of the fabricator which considers the cost per ton furnished rather than the cost of material plus labor. The difference in the cost of the 2-in. (51-mm) deep and 3-in. (76-mm) deep metal decks should be included in the consideration. The difference in the cost of the lightweight concrete fill is small but can be included. The effect of the bay size on foundation cost will not likely be significant for a multistory building. On the other hand, when dealing with bay size analyses for single story buildings, if deep foundations are required, the foundation cost may be a most significant factor.

Table 19.1 Costs for 24 × 24 ft Bay

Item	Number	Length	Unit Cost	Cost
W14×22 (F_y = 36 ksi)	3	24 ft	$5.75	$414
Studs (6)	3		1.50	27
W16×31 (F_y = 50 ksi)	1	24	8.90	214
Studs (37)	1		1.50	56
			Total cost	$711
			sq ft cost	$1.23

Table 19.2 Costs for 30 × 30 ft Bay

Item	Number	Length	Unit Cost	Cost
W18×35 (F_y = 36 ksi)	3	30 ft	$ 9.07	$ 816
Studs (10)	3		1.50	45
W21×44 (F_y = 50 ksi)	1	30	12.74	382
Studs (52)	1		1.50	78
			Total cost	$1321
			sq ft cost	$1.47

Tables 19.1 and 19.2 show the fabricated steel costs with an allowance to account for the different number of studs. Unit cost of mill material is taken from *Cost Per Foot of Structural Shapes* [19.12]. Unit cost of studs is taken as $1.50 each. This number will vary from location to location. However, the actual number is not usually significant. For instance in this example, the 24 × 24 ft (7.3 × 7.3 m) bay requires 0.095 studs/sq ft while the 30 × 30 ft (9.1 × 9.1 m) bay requires 0.091 studs/sq ft. So far, costs of erection have been ignored. Each of the examples has the same number of pieces in each bay. It is possible that the larger bay filler beams will have one more bolt per connection or six more bolts per bay. Thus, theoretically, except for the possibility of six additional bolts, the erection costs should nearly be the same. Assuming the above rationale to be true one must include some consideration of erection costs. For this example, select an erection cost of $300/ton for the 24 × 24 ft (7.3 × 7.3 m) bay. The total weight is 2328 lb or 1.16 tons/bay which is $349/bay or $0.61/sq ft. Assuming the actual cost of erection per bay will be the same for the 30 × 30 ft (9.1 × 9.1 m) bay, the price per ton will be $349/2.24 tons which is $156/ton or $0.39/sq ft. Obviously, other factors affecting the overall cost of erection have not been considered, but nevertheless, it is just as obvious that the erection of five pieces (one column, one girder, and three filler beams) costs less per sq ft for a 30 × 30 ft bay than for a 24 × 24 ft bay. For this example, assume that the indicated sq ft price will be half the difference or $0.50/sq ft. Taking this difference of erection costs into consideration the indicated price differential is only $0.06/sq ft.

Table 19.3 indicates the comparative costs for the 24-ft and 30-ft bay sizes with allowances made for erection costs. Again, these prices are not meant in any way to represent job prices but only differential prices between different bay sizes. If an allowance of $0.10 extra for 3-in. (76-mm) deep deck instead of 2-in. (51-mm) deep deck is made, the increase in price for the larger bay size is only a modest $0.23/sq ft. The indicated difference ($0.23/sq ft) is so small as to be meaningless in the cost of the final structure. That is not to say that this will normally be the case. Comparative costs per sq/ft can vary greatly for various load and span combinations.

In the foregoing example, it was demonstrated that for all practical purposes, the cost of the structural frame for 30-ft square bay was the same as for a 24-ft square bay. The implications drawn from this information can greatly affect the economic productivity of the building. The larger bay size will result in a building with approximately 40% fewer interior columns. The larger open spaces produce a structure that will be more desirable for leasing purposes.

19.1.7 Wind Bracing

Buildings of virtually any size and height should be rationally designed for an appropriate wind load. In years gone by, buildings had masonry exterior facades and interior partitions. Many modern office buildings have no masonry partitions at all, nor other components that can materially contribute to the ability of the structure to resist drift. Thus it would appear that all buildings without adequate wind shear capacity built into their facades should be analyzed for wind loads.

For low-rise buildings, the lateral loads can easily be resisted by a traditional bracing system, a moment-resisting frame, or perhaps even masonry-infilled panels in the facade or core. When a bracing system has a low aspect ratio (height-to-depth) of about 5 or 6, a traditional bracing system

Table 19.3 Comparative Costs—24 × 24 ft and 30 × 30 ft Bays

Bay Size	24 × 24 ft	30 × 30 ft
Structural steel and studs	$1.23	$1.47
Erection	$0.61	$0.50
	$1.84/sq ft	$1.97/sq ft

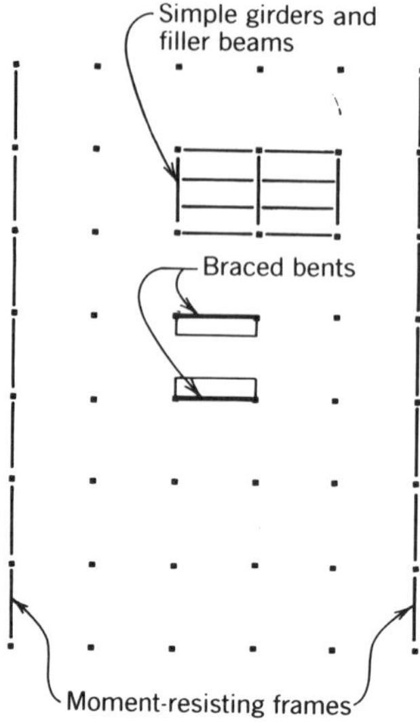

Fig. 19.11 Combination braced bent and moment-resisting frame.

will be an economical choice. When higher aspect ratios for bracing systems are necessary due to architectural constraints, it is not an unusual condition that very large net uplift loads occur in the columns. This condition should be avoided if possible as it results in complicated and expensive construction details. For small buildings, the use of end plate connections for a moment-resisting frame often works or one might use a combination of both bracing and moment-resisting frame. Figures 19.11, 19.12, and 19.13 show possible plans with solutions for moment-resisting frames, braced frames, and combinations.

The number of members in a moment-resisting frame should be minimized as the cost for each rigid connection is substantial. The system in Fig. 19.11 utilizes a combination of bracing bents and moment-resisting frames to resist lateral loads. The system in Fig. 19.12 uses moment-resisting frames only as does the scheme indicated in Fig. 19.13. The selection of a scheme such as that shown in Fig. 19.12 most likely will not be as economical as that shown in Fig. 19.13 because of the cost of the larger number of rigid connections.

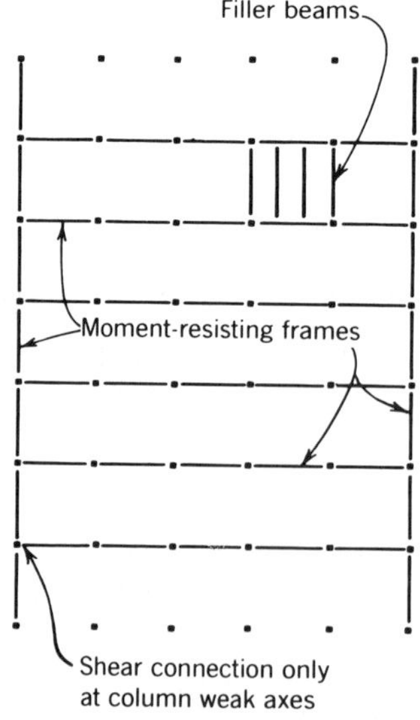

Fig. 19.12 Moment-resisting frames.

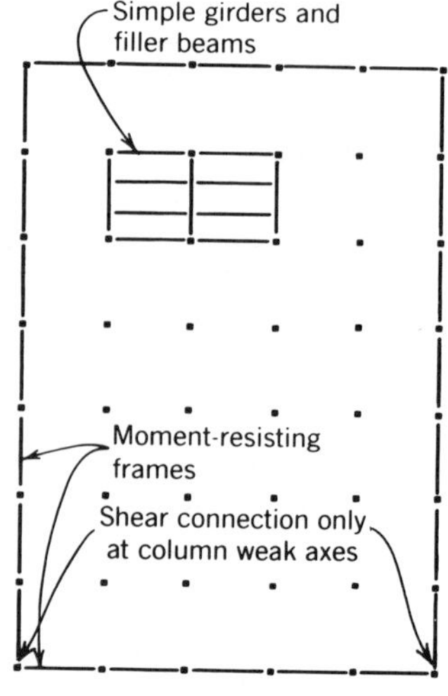

Fig. 19.13 Moment-resisting frames at perimeter only.

Where possible, moment connections to the weak axis of wide-flange columns should be avoided. The fabrication of this type of connection is difficult as the alignment of the parts must be more precise than the usual fabrication tolerances. Also, research has shown that if not carefully designed such connections may lack a desirable ductility level [19.18]. Thus, where members in moment-resisting frames frame into the weak axis of columns, it may be desirable to provide shear connections only. Another factor that may affect the structural layout is that if possible, members in moment-resisting frames should be girders instead of beams. Their larger initial size provides a greater initial moment capacity for resisting lateral loads. The taller the building, the less likely that a traditional braced frame can be economically used. Unless bracing frame aspect ratios are kept low, very high uplift forces are encountered which are extremely difficult to handle from a design standpoint and often will result in costly foundation installations. However, for the taller building the use of a combination of a bracing frame and a moment-resisting frame is likely to be an excellent choice. The introduction of hat and belt truss levels enhances the efficiency of a bracing system. The use of combination frames is discussed in Section 19.7.3.

19.1.8 Connections

The proper design and specification of connections for steel structures can be a significant factor in the overall economy of the design. Often fabricators of structural steel (justly or unjustly) accuse structural engineers of being overly conservative in their requirements for simple shear connections while at the same time neglecting to specify adequately the requirements for special and/or critical connections.

Shear Connections

Prior to the general acceptance of the use of high-strength bolts (circa 1950), most beam-to-girder connections were made with double angles shop riveted to the beam web and field riveted or bolted to the supporting girder (see Fig. 19.14). Since that time, a number of alternate methods of connecting members has evolved. As shop riveting was phased out, the rivets were replaced with shop-welded angles and field bolts (Fig. 19.15). Another method that is widely used is shown in Fig. 19.16. The shear tabs permit the erection of the beams to the girder to be independent of each other; thus the erection is easier, quicker, and safer. Automated beam line fabricating machines are capable of producing steel beams with any variety of holes punched in the webs or flanges of most beam sections ordinarily used. Some fabricators using these machines have returned to the type of connection shown in Fig. 19.14. The only difference is that the double angles are now shop bolted to the beams, and gages and edge distances are greater. The introduction of various types of automated equipment in fabricating shops has resulted in the need for the engineer to be willing to accept the type of connection that any particular shop finds to be the most economical to produce.

The use of double angle connections (Fig. 19.14 and 19.15) is favored by many engineers as being the "best" method insofar as conforming to the standards of the industry and having stood the test of time. However, from the viewpoint of safety during the erection process, other methods may be desirable. In order to erect the second beam, at one point in time, the end of the first beam will be supported by only a single drift pin. This has led to accidents during the erection process. Figure 19.16 illustrates a connection that uses shear tabs welded directly to the girder web.

Moment Connections

As with shear connections, where possible, an engineer should provide adequate information for detailing and allow adequate flexibility so that the fabricator may provide a type of connection

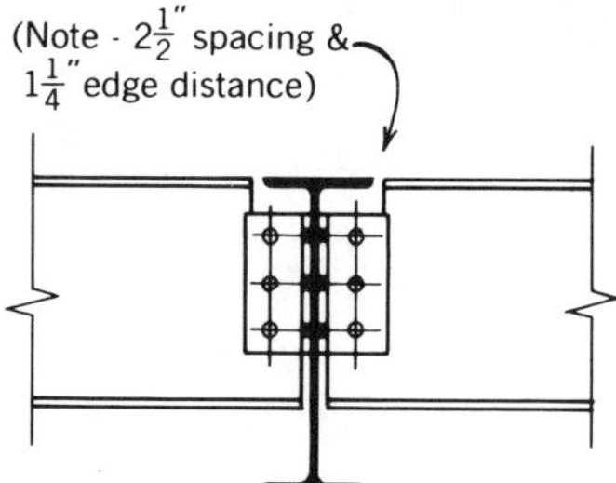

Fig. 19.14 Bolted or riveted shear connection detail.

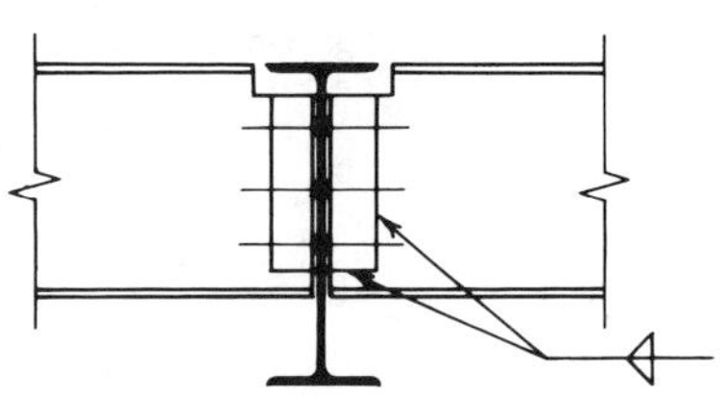

Fig. 19.15 Bolted and welded shear connection detail.

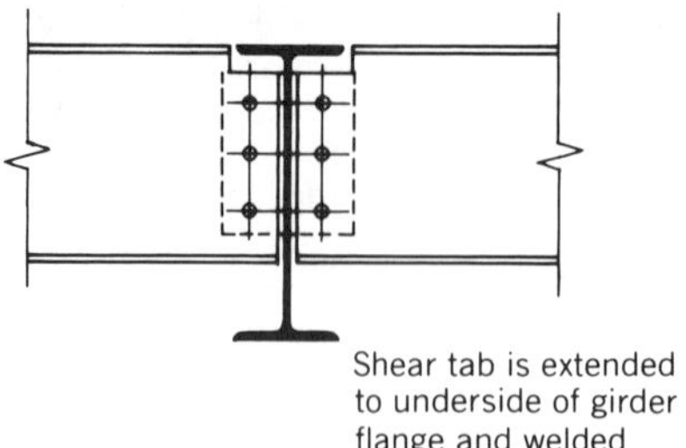

Fig. 19.16 Bolted shear tab connection detail.

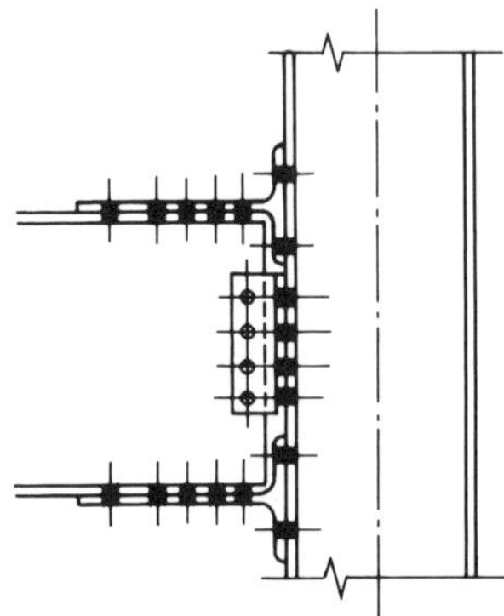

Fig. 19.17 Bolted tee stub connection. This type of connection is not recommended for new buildings.

which is suitable as well as economical. For instance, Fig. 19.17 illustrates the type of moment connection which was widely used before welding was generally accepted by the industry. This type of connection is not now widely used. Current specification requirements for prying force as well as the expense of drilling thick material makes this type of connection uneconomical. Today, most shops would prefer to produce an alternative connection which may utilize field welding (Fig. 19.18). Another alternative which is practical for relatively light sections, is the use of end plates (Fig. 19.19). The indiscriminate specifying of stiffeners in the column webs for moment connections can be extremely wasteful and uneconomical.

An engineer should provide the fabricator with adequate information such that satisfactory connections can be detailed. If shear capacities greater than those indicated by the beam tables based on uniform load in the *AISC Manual* [19.4] are required, such requirements should be given. In particular, for composite construction, some indication of the required shear capacity is essential.

Thus, if the engineer is to provide the client with the most economical job which meets all of the design requirements, the engineer may indicate the type of typical connections preferred, however,

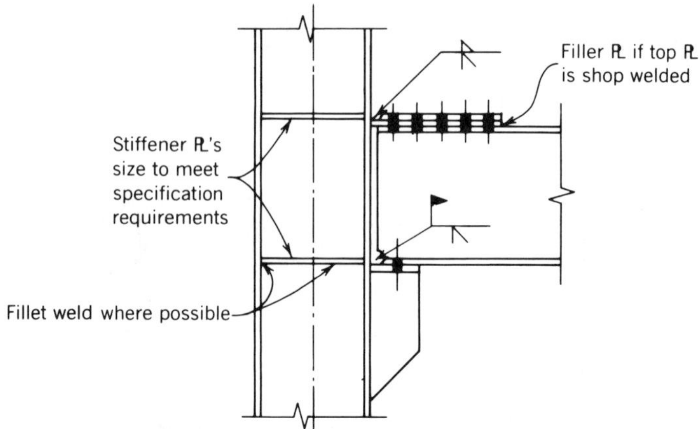

Fig. 19.18 Welded moment connection.

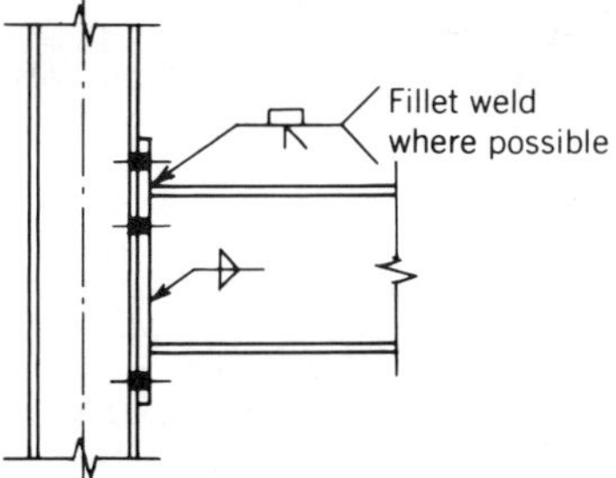

Fig. 19.19 End plate connection. Note that prying forces must be considered in bolt design.

it should be clearly stated that alternative connections will be considered. For the owner's protection and the engineer's peace of mind, approval prior to bidding may be required. To arbitrarily rule out slotted holes, oversize holes, threads excluded from the shear plane, one-sided connections, etc., is clearly a disservice to the client. Connections can be engineered—hence, connections should be engineered.

19.2 SERVICEABILITY CRITERIA

The establishment of any serviceability criteria should be essentially a choice of quality versus cost. In some instances, various codes will provide minimum requirements which may or may not result in the desired degree of quality. In other instances, the decision is solely the responsibility of the building design team.

19.2.1 Floor Systems

Two separate serviceability criteria should be considered in the selection of floor system members—deflection and vibration.

Deflection

The 1978 *AISC Specification* [19.6] limits live load deflections of beams and girders supporting plastered ceilings to a maximum of the span length divided by 360 ($L/360$). The basis of the original limit is unknown to the author. Apparently it was the result of the experience of the early designers of steel structures. As construction methods changed, the limit was liberalized. While the $L/360$ limit appears to function well in so far as protecting building components from deterioration due to deflection, it does not necessarily control vibration to the extent desired for human occupancy.

Vibration

Buildings with large open spaces such as found in shopping malls and office building loft areas are susceptible to objectionable transient vibrations caused by foot traffic. The present state of art of evaluating this problem is at best marginal. Acceptability criteria have been established by a number of different groups. Unfortunately, under some circumstances, the various criteria conflict as to whether a system is or is not acceptable. A review of the state of art is given by Murray [19.31]. Included in this paper is the following proposed acceptability criterion:

$$D \geq 35A_0 f + 2.5 \tag{19.2.1}$$

where D = damping, %
A_0 = amplitude, in.
f = frequency, cps

Methods for computing the frequency and amplitude are given by Murray [19.32] and Galambos [19.23].

Some designers choose to use lower yield stress steel and/or simple span (noncomposite) floor beams with the intent of providing stiffer, more vibration-free floor systems. Unfortunately, their assumption is wrong with regard to vibration. Typically, the performance of a floor system with regard to acceptability of occupancy-induced floor vibrations is not a direct function of steel yield or beam depth-to-span ratios.

For illustration, the floor beams in Sections 19.1.5 and 19.4.1 have been analyzed using the methods given by Murray [19.31, 19.32] and are shown on Plates 19.1, 19.2, and 19.3 in the Appendix.

The present state of the art does not permit one to determine accurately the damping characteristics (i.e., % of critical damping or damping factor) of a floor system. The Canadian Standards Association [19.15] permits values of 3% for bare floors; 6% for finished assemblies (ceiling, mechanical equipment, finish, furniture, etc.); and 13% for partitioned areas. Murray [19.32] suggests values of 1–3% for bare floors (1% for thin lightweight slabs, 3% for thick normal-weight slabs) and 1–3% for ceilings (1% for hung ceilings, 3% for ceilings attached to the structure). Additionally, ductwork and mechanical equipment can provide for 1–10% damping (depending upon the amount) and 10–20% for partitions, depending upon the amount. Obviously, with this large range of possibilities, any determination is, at best, not very reliable. Murray further recommends that if the estimated damping in a floor is less than 8–10% an analysis should be made.

The damping factor for a typical loft space office building or shopping mall probably will fall below the recommended lower limit. Using the above values, minimum damping will be on the order of 2% for bare floor and 1% for ceiling and perhaps 1% for open building areas; in all likelihood, an analysis should be made. For the illustrations, a damping factor of 4% has been assumed.

The frequency of a simply supported beam is

$$f = 1.57 \left[\frac{gEI_t}{WL^3}\right] \tag{19.2.2}$$

where g = 386 in./sec^2
E = modulus of elasticity, ksi
I_t = transformed moment of inertia, in.^{4}
W = total weight supported by beam, kips
L = beam span, in.

The amplitude of a simple beam is given by

$$A_{0t} = (\text{DLF})_{\max} \left[\frac{L^3}{80EI_t}\right] \tag{19.2.3}$$

where $(\text{DLF})_{\max}$ = maximum dynamic load factor which is plotted in Fig. 19.20.

An important parameter affecting the vibration phenomenon is the effective area of slab actively participating in vibration. Galambos [19.23] includes a method of determining the number of effective open-web steel joists to be included in an analysis. Murray [19.32] includes a graph (Fig. 19.21) from which the effective number of beams can be determined as a function of the beam span, moment of inertia, beam spacing, and slab depth.

Further clouding the issue of vibration analysis, is the fact that there is no universally accepted criterion for performance. In addition to the proposed criterion of Murray, the CSA/PBS Guide floor rating is calculated. This criterion is based on the work of Wiss and Parmelee [19.39].

$$R = 5.08 \left[\frac{fA_0}{D^{0.217}}\right]^{0.265} \tag{19.2.4}$$

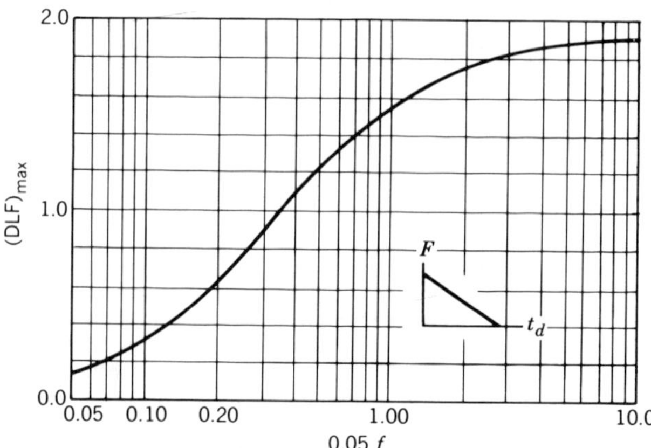

Fig. 19.20 Maximum dynamic load factor.

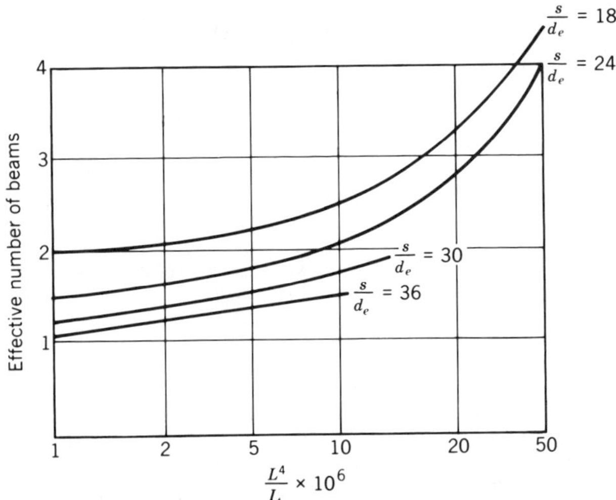

Fig. 19.21 Effective number of beams (from Murray [19.32]).

where A_0 = amplitude, in.
D = damping, %
R = the mean rating response
f = frequency, cps

Response ratings are interpreted as follows:

R = 1.0 Vibration is imperceptible.
R = 2.0 Vibration is barely perceptible.
R = 2.5 Maximum acceptable value of R.
R = 3.0 Vibration is distinctly perceptible.
R = 4.0 Vibration is strongly perceptible.
R = 5.0 Severe vibration.

The illustrations will be for an A36 simple span beam and A36 and $F_y = 50$ composite beams (see Plates 19.1, 19.2, and 19.3 in the Appendix, page 620). For floor vibration analyses, beams act compositely with the slab whether or not they have adequate shear connectors. The slab thickness to be used is an equivalent thickness of concrete (based on the weight of the concrete per sq ft) plus the steel deck. Assumed loads will include 10 psf for people and furnishings. In the computation of R values, a damping factor equal to 4.0 is used.

In all of the examples the R ratings range from 2.03 to 2.08. These ratings indicate that the floor is acceptable with vibrations that will be perceptible.

Using Murray's recommendations, the minimum damping required ranges from 3.98 to 4.13. Thus for open office spaces, normal composite floors are likely to fall into a category of barely acceptable insofar as foot-traffic-induced vibrations are concerned. It can be seen from these examples that it is often a waste of money to select lower-yield strength steel or simple span noncomposite beams in an effort to provide more vibration-free floor systems.

19.2.2 Drift

Since the early days of skyscraper buildings, designers have sought to ensure the occupants creature comfort; that is, a task of the designer is to limit vibration and/or motion to the extent that building movement is tolerable in the view of the occupants. Early designers sought to achieve this goal by limiting deflection to certain specified limits. Today's designers of very tall buildings recognize that the problem to be solved is to limit the perceptibility of motion rather than building deflection per se. This topic is also treated in Chapters 10 and 11.

Two early guides for the lateral load design of tall buildings were "Designing Wind Bracing for Skyscrapers" [19.9] published in 1931 and "Final Report, Wind Bracing in Steel Buildings" [19.10] published in 1940. The 1940 final report set forth criteria which recommended a deflection limit of

$L/500$, or a D.I. = 0.002. Often the term Deflection Index (D.I.) is used to define the drift limit or drift of a bare building frame. It is simply the ratio of the building deflection to the height.

In the period of the 1940s to the mid-1960s, bare frame deflections were computed for building code wind loads which were much lower than those in use today. Also, deflections were computed by hand using approximate methods which often neglected the affect of axial strains of members on the deflection. When comparing the building deflections of the 1940s to those of today, it must be recognized that most older buildings had much stiffer cladding. The exterior walls were likely to be masonry and perhaps even the interior partitions might be masonry. Thus, the effect of the cladding would be substantially greater in the damping of vibration and the stiffening of a building frame than the type of light curtain walls often used today.

To further cloud the issue of drift, engineers seldom publicly provide information which will allow one to compare designs from one era to another or for that matter from one project to another. In *Design of Steel Structures* by Gaylord and Gaylord [19.24], an example of the wind drift calculations for Lever House is included. Lever House was built as the corporate headquarters for Lever Brothers Company and fronts on Park Avenue in New York City. Lever House was one of the first (if not the first) totally glass-enclosed buildings to be built. The designers selected a target drift index of 0.0025. It is interesting to note that the target index is 25% greater than the maximum recommended by the ASCE Committee [19.10]. Even though the Lever House was very slender and by that day's standard had light cladding, the engineering firm, which was one of the noted firms in New York City, selected the higher drift index. This would lead one to speculate that perhaps the committee's recommendation may not have represented the actual limits selected by experienced practicing consultants at that time.

With the information provided in the example, it was possible to prepare the data required to analyze the frame using a stiffness method computer program. The frame has been analyzed for four separate loading conditions:

1. Original wind loads with original frame member areas multiplied by a factor of 100.
2. Original wind loads with original frame.
3. *BOCA Code* [19.14] wind loads with original frame. (This wind load is similar to ANSI-A58 [19.8], 50-yr storm, Exposure B for 80-mph fastest mile.)
4. ANSI-A58 wind loads with original frame using 90-mph fastest mile, Exposure A.

Loading condition No. 1 had the member areas increased by a factor of 100 in order to negate the effects of axial strain and thus compare the deflections using the computer with the original hand calculations. From the information available, it is not clear how the wind loads were handled in the upper stories above the roof. Therefore, only the stories for which the wind bent member properties were known were included in the analysis. Overall for loading condition No. 1, the agreement was good. The drift index at the top was computed to be 0.0027 compared to the 0.0025 target. At the sixth floor, the story deflection by the computer was 0.380 in. (10 mm) compared to the original 0.375 in. (10 mm). This example will be used to compare the practice of the earlier times to those of today.

The second loading condition uses the original wind loads and frame members with their original areas. The overall drift index increased to 0.0031. This indicates that the axial strains of the frame members contributes about 15% additional deflection.

The third loading condition, *BOCA Code* wind loads, using an 80-mph fastest mile with the original frame resulted in a drift index of 0.0052. This is more than twice the original design deflection calculation.

The fourth loading conditions is an ANSI A58.8 Exposure A with 90-mph fastest mile (which may be considered appropriate to New York City). The resulting deflection (drift) index is 0.0042.

A comparison of these deflections is shown in Table 19.4. The difference in the magnitudes of deflections is so large that one might question the validity of limits established under a completely

Table 19.4 Lever House Frame Deflections

Loading Condition	Frame	D.I.
Original	Modified	0.0027
Original	Original	0.0031
BOCA (80 mph)	Original	0.0052
ANSI (90 mph) (Exposure A)	Original	0.0042

different set of circumstances. When one considers the large differences in calculated deflections due to wind loading and design techniques, one may well conclude that the original limit of 0.0020 is too conservative.

For buildings over 20 to 25 stories in height, the consideration of motion perceptibility by the occupants should be the determining factor in the design of a wind bracing system. Unfortunately, no official or semiofficial guidelines for building performance with regard to motion perception by the occupants have been widely distributed or accepted. An easy to understand discussion of the state of art appeared in an ASCE–IABSE Tall Building Conference reprint "Human Response to Tall Building Wind-Induced Motion" [19.35] by Reed, Hansen, and Vanmarcke. In terms of serviceability, a drift or motion perception due to acceleration criterion should be selected with due regard to the subject building's intended use. Resistance to drift or perceptible motion is a factor of quality. Thus, a selection of drift index of 0.0025 under the influence of a 25-yr storm may be appropriate for a speculative office building. On the other hand, it may be completely inappropriate for a hospital, library, or any other type of high-quality building project. In the end, a designer of tall buildings must make the decision of quality based on the published information available as well as personal experience.

One way to handle the problem of choosing the proper drift index may be to select a rather conservative value (say 0.0020) to control drift using a 10- or 25-yr storm wind load. However, the frame design for strength should be made using a 50- or 100-yr storm wind load. One may also wish to consider the problem of frame stability when choosing a service load drift limit (see Section 19.8).

19.2.3 Roofs

Two serviceability criteria should be considered in the selection of roof systems members—deflection and ponding.

Deflection

The *AISC Specification* simply restricts the maximum live load deflection to 1/360 of the span for members supporting plastered ceilings. This results in approximate span maximum depth ratios as shown in Table 19.5. When the span-to-depth ratio is high, it is likely that ponding will control.

Ponding

Ponding of flat roof steel-framed structures can be a serious design problem. The best way to eliminate the ponding problem is to provide a modest roof slope, perhaps ¼ in. per ft (20 mm per m) or more. Deflections due to initial loads in two-way steel roof structures form concave-shaped ponds between the high points at the columns. The initial loads may include structure dead load plus additional accumulated live load (ice or snow). If the roof structure does not have adequate stiffness, additional liquid load (rain) may not act as the assumed uniform load. That is, when additional water load in the ponded areas causes more vertical deflection than the depth of water causing the deflection, the roof structure is susceptible to catastrophic collapse due to ponding. AISC-1.13.3 [19.6] contains a relatively simple procedure for dealing with ponding in steel structures. Often open-web steel joists are used in large flat roof systems. Moments of inertia can be found in the publication "Structural Design of Steel Joist Roofs to Resist Ponding Loads" [19.25].

19.3 THE *AISC SPECIFICATION*—ITS USE AND FUNCTION

The AISC *Specification for the Design, Fabrication, and Erection of Structural Steel for Buildings* [19.6] is exactly what the title implies. The specification explicitly sets minimum requirements for design of steel members. Except in a few cases, items concerning analysis and serviceability are

Table 19.5 Roof Beam Span-to-Depth Ratio Limitations

$\frac{\text{Dead Load}}{\text{Total Load}}$ Ratio	$\frac{\text{Span}}{\text{Depth}}$ Ratio
0.3	24
0.4	28
0.5	33

not discussed. In the instances where serviceability considerations or analysis methods are discussed, the reasons appear to fall into two categories: one, frame stability and roof ponding analysis methods for overall safety; two, deflection limitations and span-to-depth ratios to control vibration. In the case of floor systems subjected to foot-induced vibration, it can be shown that neither the deflection criterion nor the span-to-depth ratio criterion nor the combination of both will necessarily produce a satisfactory structure. In some cases, the specification is not clear-cut, that is, the real meaning of the provision is hidden in the verbiage. In other cases, the specification is clear-cut, but often designers of structures fail to recognize the basic engineering principles involved in their design.

AISC-1.8.1 [19.6] contains the following:

> *General stability shall be provided for the structure as a whole and for each compression element. Design consideration should be given to significant load effects resulting from the deflected shape of the structure or of individual elements of the lateral load resisting systems, including the effects on beams, columns, bracing, connections, and shear walls.*

The implications of the foregoing statement are either overlooked or misunderstood by many members of the design profession.

In the design of unbraced frames, many designers simply assume that columns are to be designed using effective length factors K as described in Section 1.8 of the Commentary on the Specification. The Commentary also explains that in some cases it may be possible that this method may not provide adequate capacity to resist the secondary moments induced by the frame's drift under the influence of lateral loads. The opening sentences in Section 1.8.1 infer that the designer should be cognizant of these requirements. Many are not. This matter is discussed further in Section 19.8.

For instance, the requirements for the bracing of compression members are straightforward. However, the specification does not address the problem of the details of bracing and/or braced members. If it is necessary to brace a compression member eccentrically as shown in Fig. 19.22, one must not only provide a brace with an L/r value not exceeding 200 but also a brace with sufficient stiffness to restrain the member from any tendency to translate or rotate. Often, a value of 2% of the vertical load acting in any direction is used for the forces in the design of restraining bracing system.

AISC-1.10.11 [19.6] is one short seemingly innocuous sentence which can be and has been overlooked by engineers. The sentence simply states "at points of support, beams, girders and trusses shall be restrained against rotation about their longitudinal axis." Figure 19.23 illustrates a condition that can easily become unstable if rotation is not restrained. A small accidental eccentricity in the beam, perhaps due to rolling, can cause a secondary moment which leads to the formation of a plastic hinge in the web, in turn leading to the collapse of the structure.

Thus, in using the *AISC Specification,* one must combine not only knowledge of the provisions and their meanings, but, one must also develop and use a sense of engineering judgment which will result in the selection of adequate and meaningful analyses that fit the problem at hand. Detailed treatment of basic design of steel structures is available in many standard textbooks, such as those of Salmon and Johnson [19.45], McCormac [19.46], and Kuzmanovic and Willems [19.47]. Thus, only a brief summary of basic design of steel elements is included in this chapter.

19.4 BEAMS

The design of steel beams is a simple task. For the most part, beams are designed by using the linear elastic theory developed more than 150 years ago. The permitted stresses are determined not only

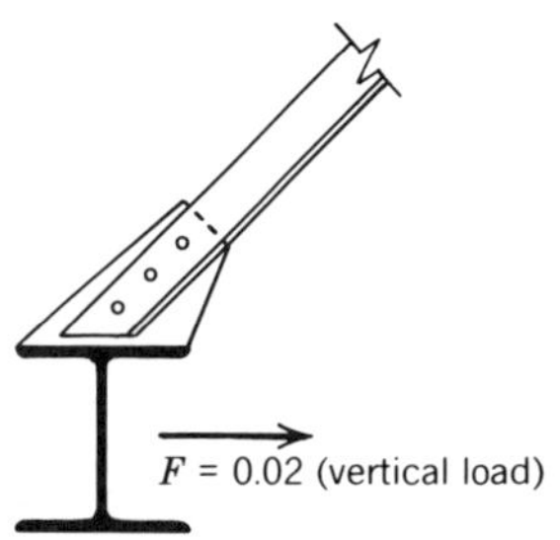

Fig. 19.22 Eccentrically braced column.

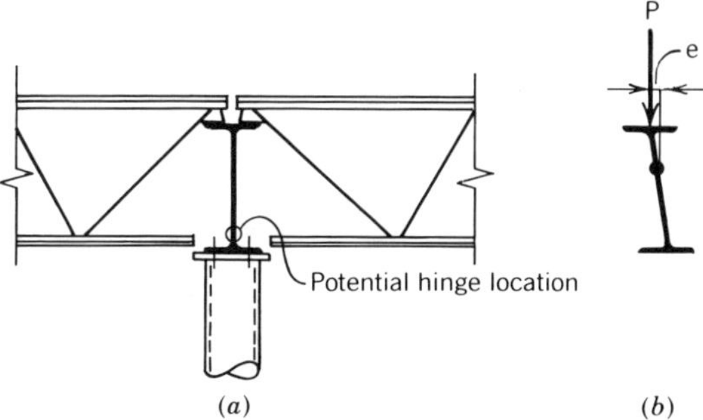

Fig. 19.23 Inadequate beam restraint at support.

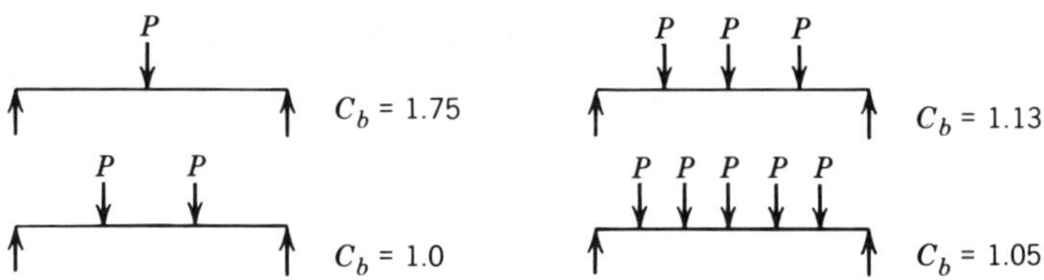

Fig. 19.24 C_b values for beams with uniformly spaced concentrated loads.

by the properties of the steel but also the properties of the specific shape and conditions of its bracing. The *AISC Specification* is quite specific in its requirement and this information is not repeated. However, alternative design methods for unbraced beams, composite beams, and cantilever beam systems are discussed.

19.4.1 Laterally Unbraced Beams

The design of laterally unbraced beams does not have to be an onerous task for the designer. If an unbraced beam is duplicated a number of times in a structure, a careful, complete analysis should be made. However, for an isolated unbraced beam, it may be questionable whether such an effort is warranted.

At present the *AISC Specification* permits a bending stress of $0.66F_y$ for compact sections up to a length tabulated as L_c and $0.60F_y$ between the lengths tabulated L_c and L_u in the *AISC Manual* [19.4]. As an unbraced length is increased beyond L_u, the allowable moment capacity is reduced. For a simple span beam braced only at the ends, the use of the curves in the *AISC Manual* is the quickest method of selecting a beam.

In certain instances where the factor C_b does not equal 1.0, it is improper to use the curves in the *AISC Manual* as the result may be too conservative. The value of factor C_b directly affects the permitted bending stress. Figure 19.24 indicates the C_b values for beams with uniformly spaced concentrated loads which are laterally braced at load points. For a beam with either a single load or a uniform load, braced at the centerline, C_b is equal to 1.75. A C_b value of this magnitude may affect the capacity of a given member materially and should be considered. On the other hand, for a beam with two uniformly spaced concentrated loads the value of C_b is 1.0. In fact, any beam with an even number of uniformly spaced loads and/or braced points will have a C_b value equal to 1.0. A beam with three equally spaced loads has a C_b value equal to 1.13. The average increment of S_x or Z_x in an economy table of steel beam properties is about 13%. Thus, only for beams with a single or perhaps three uniformly spaced loads and/or bracing points should the value of C_b materially affect the size of the member. For cantilevers C_b should be taken as 1.0 unless a detailed analysis is made [19.27].

The *AISC Specification* permits the stress for laterally unbraced beams to be the greater of the values determined as a function of L/r_T, using AISC Formulas (1.5-6a) or (1.5-6b), and as a function of Ld/A_f using AISC Formula (1.5-7). More often than not Formula (1.5-7) results in the higher value. Therefore, for a single unbraced beam, one may be justified in considering the use of Formula (1.5-7) only. If this is the case, it can be shown that

$$\frac{ML_b}{C_b} = \frac{1000S_x}{12(d/A_f)} \tag{19.4.1}$$

where M = maximum moment, ft-kips
L_b = maximum unbraced length, ft

with other factors as defined in the *AISC Specification*. The factor

$$F_{br} = \frac{1000S_x}{12(d/A_f)} \tag{19.4.2}$$

is unique for each section. Thus one may use the tabulated values as shown in Table 19.6 to find a satisfactory laterally unbraced beam. To select an unbraced beam, one multiplies the bending moment by the laterally unbraced length divided by C_b to find a required minimum value for the factor F_{br} (see Plate 19.4).

Table 19.6 Unbraced Beam Selection Table (Partial Example)

F_y = 36 ksi						F_y = 50 ksi			
L_c (ft)	M_c (ft-kips)	L_u (ft)	M_u (ft-kips)	Shape	Factor	L_c (ft)	M_c (ft-kips)	L_u (ft)	M_u (ft-kips)
8.7	280	12.4	255	W21×68	3128	7.4	385	8.9	350
8.7	254	11.2	231	W21×62	2556	7.4	349	8.1	318
6.9	222	9.4	202	W21×57	1872	5.9	305	6.7	278
6.9	189	7.8	176	W21×50	1321	5.6	260	6.0	236
6.6	163	7.0	148	W21×44	963	4.7	224	5.9	204
8.0	216	13.3	196	W18×60	2594	6.8	297	9.6	270
7.9	197	12.1	179	W18×55	2144	6.7	270	8.7	246
7.9	178	11.0	162	W18×50	1760	6.7	244	7.9	222
6.4	158	9.4	143	W18×46	1332	5.4	217	6.8	197
6.4	137	8.2	124	W18×40	1005	5.4	188	5.9	171
6.3	115	6.7	105	W18×35	692	4.8	158	5.6	144
7.5	162	12.7	147	W16×50	1849	6.3	223	9.1	203
7.4	145	11.4	132	W16×45	1492	6.3	200	8.2	182
7.4	129	10.2	118	W16×40	1190	6.3	178	7.4	162
7.4	113	8.8	103	W16×36	892	6.3	155	6.7	141
5.8	94.4	7.1	85.8	W16×31	602	4.9	130	5.2	118
5.6	76.8	6.0	69.8	W16×26	387	4.0	106	5.1	96
7.1	109	11.5	99.3	W14×38	1126	6.1	150	8.3	137
7.1	97.2	10.2	88.4	W14×34	888	6.0	134	7.3	122
7.1	84.0	8.7	76.4	W14×30	655	6.0	116	6.5	105
5.3	70.6	7.0	64.2	W14×26	446	4.5	97.1	5.1	88.2
5.3	58.0	5.6	52.7	W14×22	295	4.1	79.8	4.7	72.5

$$F_{br} = \frac{1000S}{12(d/A_f)}$$

19.4.2 Cantilever Beam Systems

The design of cantilever beam systems can be an arduous task if proper design aids are not available. AISC-1.3.2 requires that snow loads shall be applied in any probable arrangement. Thus, the designer must exercise judgment in making the decision for the load arrangement. Some designers believe it prudent to use a maximum load of dead load plus live load and a minimum load of dead load plus half live load.

In designing cantilever beam systems, there are two *AISC Specification* requirements which may be easily overlooked. AISC-1.5.1.4.5 indicates that the effective length of a cantilever braced against twisting only at the support may be taken as the actual length. Further, AISC-1.10.11 requires rotational restraint at supports. Often the cantilever portions of the beams will have to have their bottom flanges braced or their capacity reduced due to their unbraced length.

Tables 19.7 through 19.9 tabulate coefficients that can be used to determine maximum moments and cantilever lengths for systems with three equally spaced concentrated loads, four equally spaced concentrated loads, and for uniform loads for all equal spans. Coefficients for the three concentrated load and four concentrated load tables are from Mehta [19.30].

These tables should be used for the preliminary selection of member sizes and cantilever lengths. Both maximum positive and negative moments are very sensitive to small changes in cantilever length, and, as a practical matter, the use of the exact theoretical lengths may not be desirable. One should remember to check the unbraced length between the spacing of the joists or purlins as often it will exceed the L_c and perhaps L_u lengths for the girder.

Table 19.7 Cantilever Beam Coefficients—Three Equal Loads

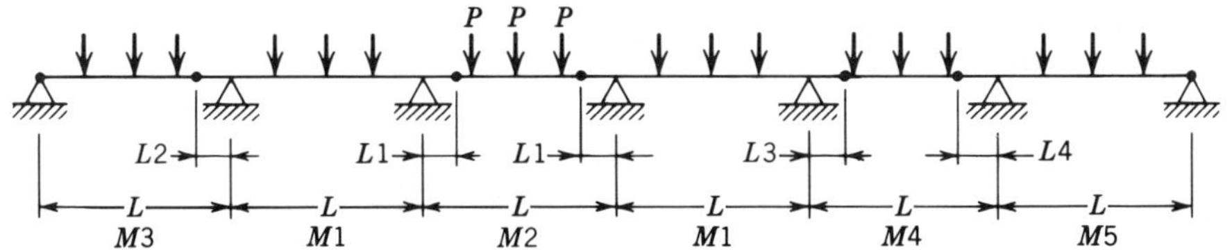

Cantilever Length Coefficients

Dead Load / Total Load	$L1$	$L2$	$L3$	$L4$
0.250	0.30	0.21	0.33	0.36
0.375	0.24	0.19	0.27	0.31
0.500	0.22	0.18	0.24	0.27
0.625	0.20	0.17	0.22	0.24
0.750	0.18	0.16	0.20	0.23
1.000	0.17	0.14	0.17	0.21

Moment Coefficients

Dead Load / Total Load	$M1$	$M2$	$M3$	$M4$	$M5$
0.250	0.40	0.11	0.30	0.08	0.45
0.375	0.36	0.14	0.32	0.11	0.43
0.500	0.33	0.17	0.34	0.14	0.40
0.625	0.30	0.19	0.35	0.16	0.38
0.750	0.28	0.22	0.36	0.18	0.36
1.000	0.25	0.25	0.38	0.21	0.33

For $\frac{\text{Dead Load}}{\text{Total Load}} = \frac{\text{DL}}{\text{TL}} = 0.50$,
find $L1 = 0.22L$ and $M1 = 0.33PL$.

19.4.3 Composite Beams

The use of steel beams designed to act compositely with the concrete floor slab has proved to be an economical method of construction. The major advantage of the system is the economics. For the relatively small cost of installing the stud connectors, the moment capacity of the steel section will be increased by a factor generally ranging between 40% and 75%. Figures 19.25 and 19.26 are cross-sections through a filler beam and a girder. Chapter 23 provides extended treatment of composite construction.

The *AISC Specification* contains explicit requirements for the design parameters involved in the selection of composite beams. Design examples for both a typical filler and girder are shown on Plates 19.5 through 19.9. The nomenclature for the composite design examples is repeated below:

A_c = effective concrete flange area = bt, sq in.
A = steel section area, sq in.
b = effective concrete flange width, in.
D = overall depth of composite section, in.
d = depth of steel beam, in.
E_s = modulus of elasticity of steel, ksi
E_c = modulus of elasticity of concrete, ksi
F_y = steel yield strength, ksi
f'_c = concrete strength, ksi
H_s = stud length after installation, in.

Table 19.8 Cantilever Beam Coefficients—Four Equal Loads

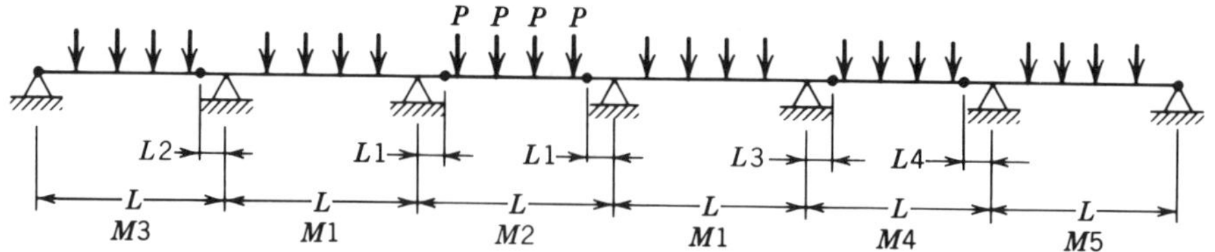

Cantilever Length Coefficients

Dead Load / Total Load	$L1$	$L2$	$L3$	$L4$
0.250	0.28	0.19	0.30	0.33
0.375	0.24	0.18	0.26	0.30
0.500	0.20	0.17	0.23	0.27
0.625	0.18	0.16	0.20	0.25
0.750	0.17	0.15	0.18	0.23
1.000	0.15	0.13	0.16	0.20

Moment Coefficients

Dead Load / Total Load	$M1$	$M2$	$M3$	$M4$	$M5$
0.250	0.48	0.13	0.41	0.09	0.55
0.375	0.44	0.17	0.43	0.13	0.52
0.500	0.40	0.20	0.44	0.16	0.50
0.625	0.37	0.23	0.45	0.19	0.48
0.750	0.34	0.26	0.46	0.22	0.46
1.000	0.30	0.30	0.48	0.25	0.43

For $\dfrac{\text{Dead Load}}{\text{Total Load}} = \dfrac{\text{DL}}{\text{TL}} = 0.5,$
find $L1 = 0.2L$ and $M1 = 0.4PL$.

h_r = metal deck depth, in.
I_{eff} = effective moment of inertia of partial composite section, in.^{4}
I_s = moment of inertia of steel section, in.^{4}
I_{tr} = moment of inertia of composite section, in.^{4}
Q = shear connector capacity, kips
MD = wet load moment for unshored construction, ft-kips
ML = dry load moment for unshored construction, ft-kips
N = number of studs to develop required horizontal shear
N_r = number of studs in one rib of metal deck
n = modular ratio, E_s/E_c
S_{eff} = effective section modulus of partial composite section, in.^{3}
S_s = section modulus of steel section, in.^{3}
S_{tr} = effective section of fully composite section, in.^{3}
t = concrete flange thickness, in.
V_h = total horizontal shear between point of zero moment and maximum, kips
V'_h = horizontal shear between point of zero moment and maximum moment for partial composite construction, kips
w_r = width of steel deck rib, in.
$\phi = A_c/nA_s$

The design of a composite beam is performed using a simple elastic analysis. Two methods are used in the example. The filler beam analysis utilizes formulae developed to find the neutral axis of

Table 19.9 Cantilever Beam Coefficients—Uniform Loads

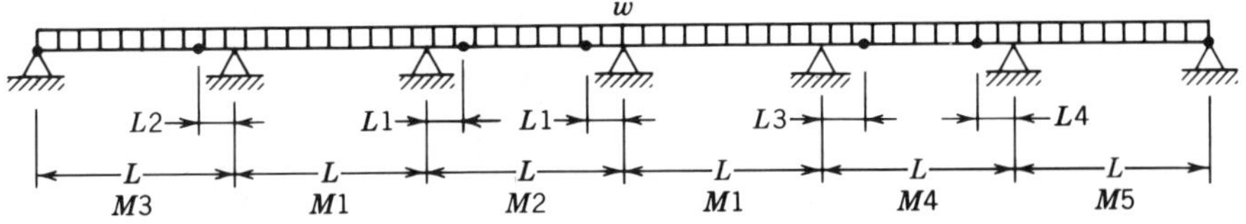

Cantilever Length Coefficients

Dead Load / Total Load	$L1$	$L2$	$L3$	$L4$
0.250	0.28	0.20	0.29	0.31
0.375	0.24	0.18	0.25	0.28
0.500	0.21	0.17	0.23	0.26
0.625	0.19	0.15	0.20	0.24
0.750	0.17	0.14	0.19	0.23
1.000	0.14	0.12	0.16	0.20

Moment Coefficients

Dead Load / Total Load	$M1$	$M2$	$M3$	$M4$	$M5$
0.250	0.100	0.025	0.080	0.019	0.112
0.375	0.091	0.034	0.084	0.027	0.106
0.500	0.084	0.042	0.087	0.033	0.101
0.625	0.077	0.048	0.090	0.038	0.097
0.750	0.072	0.054	0.092	0.043	0.093
1.000	0.063	0.063	0.096	0.051	0.086

For $\dfrac{\text{Dead Load}}{\text{Total Load}} = \dfrac{\text{DL}}{\text{TL}} = 0.5$,
find $L1 = 0.21L$ and $M1 = 0.084wL^2$.

the section and the girder method uses a traditional method for the determination of the section properties.

A 3¼-in. (83-mm) thick lightweight slab on 3-in. (76-mm) deep metal deck is selected in order to provide a fire rating economically. The design strength of the concrete is selected to be $f'_c = 3000$ psi (21 MPa). It should be noted that for some fire-rating systems, higher concrete strengths may be required. The concrete weight of 112 pcf is selected since that is the maximum weight permitted by many fire-rated assemblies. The 3-in. (76-mm) deep deck is chosen as the economical selection for

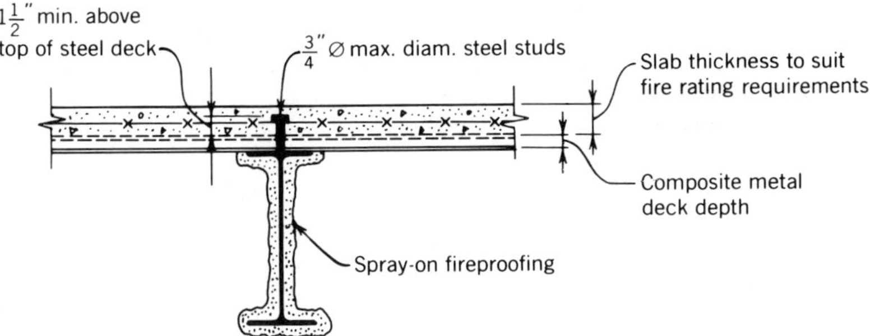

Fig. 19.25 Typical composite beam. It may be prudent to provide mild steel top reinforcing when moving loads will be present.

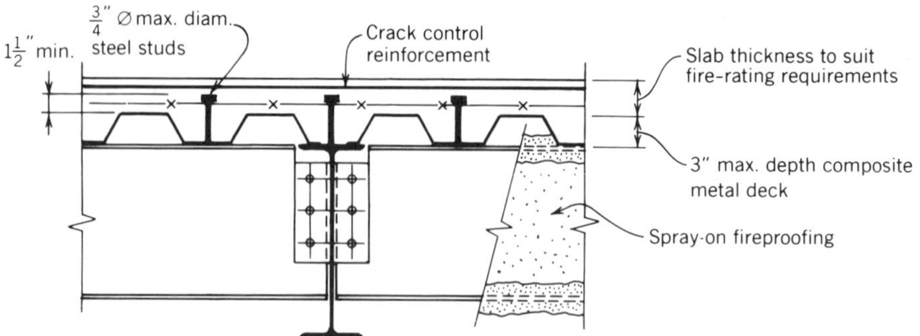

Fig. 19.26 Composite girder with crack control reinforcement.

the 10-ft (3-m) span. The gage of the deck will be selected as the lightest that will not require shoring between beams.

The design of a composite section is not a direct design procedure. A trial section must be picked for an analysis to verify the section's suitability. In the design of a composite section, the greater the depth of the composite deck, and the greater the slab thickness, the greater the efficiency of the composite section. Thus, the selection of a 3-in. (76-mm) deck instead of a 2-in. (50-mm) deck helps the efficiency of the section not only for strength considerations, but also for vibration and deflection considerations.

Generally, concrete stresses are not a problem in composite construction. There are some exceptions, notably long span girders supporting long-span filler beams with relatively thin slabs; such as, a clear span parking structure. The use of composite construction for typical steel beams and girders increases the efficiency by varying amounts depending upon the configuration of the section. For beams with formed slabs directly on the top flange to beams with 3-in. (76-mm) deep composite decks, the efficiency factor (S_{tr}/S_s) varies from about 1.4 to 1.8. In the example, a factor of 1.6 was selected and resulted in the proper initial trial section.

The decision of whether to shore beams or not to shore beams in composite construction poses a dilemma which must be considered. If one chooses unshored construction, wet load deflections of both beams and girders are likely to be large. In this example, combined wet load deflections of the beams and girders result in a total centerline beam deflection of 1.97 in. (50 mm). With deflections this large, the slab cannot be placed "flat" without an adjustment in the design to account for the added weight. The beams and girders may be ordered cambered. This is likely to produce a floor that will be humped in the center of the panels. One may reduce the dead load deflection problem to a minimum by the use of shores to prop the filler beams at their quarter points. This will reduce the total centerline deflection to 0.64 in. (16 mm). This value should be acceptable as it is practically unobservable to the general public. The use of shoring has one serious drawback—it will virtually guarantee the formation of a substantial crack in the slab over the girder. Results of tests of composite members reflecting this condition if they exist at all are certainly not widespread. For that matter, results of tests of beams with composite slabs which are loaded as they are in service are unknown to the author. Thus, it is possible that the tests on which designs are based do not reflect the conditions of actual use of these materials.

In view of the foregoing, it would seem that for shored construction it would be prudent to provide some crack control reinforcement near the top of the slab as shown in Fig. 19.26. Many engineers feel that it would be prudent to specify crack control reinforcement over all girders. In addition to the need for crack control reinforcement over girder beams, some designers feel that the normally specified wire fabric is not sufficient reinforcement to assure composite action as indicated by the testing done to date.

19.5 COMPRESSION MEMBERS AND BEAM-COLUMNS

The *AISC Specification* requirements for the design of axially loaded compression members are explicitly simple, with one exception, the determination of the effective length factors K_x and K_y when frame stability is involved. This subject is discussed in Section 19.8. Values for the permitted compressive stress may be obtained directly from *AISC Specification,* (Appendix A) Tables 3-36 and 3-50 for steels having yield strengths of 36 and 50 ksi (248 and 345 MPa), respectively. Additionally, column load tables in the *AISC Manual* have loads tabulated for various lengths for all shapes normally used as columns.

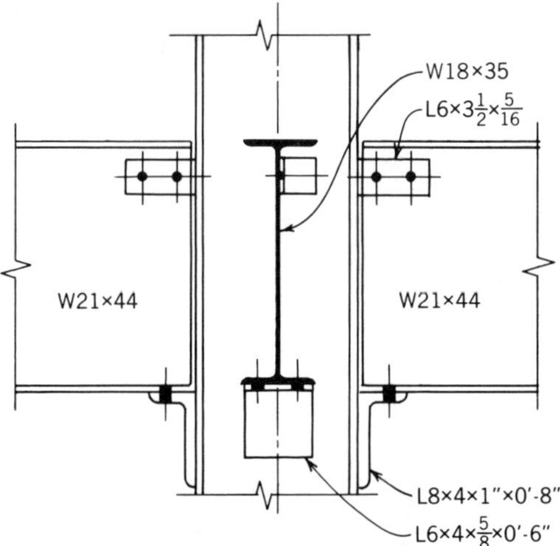

Fig. 19.27 Column with seated connections.

Values for the effective length factor *K* for individual columns are illustrated in Table C1.8.1 in the Commentary of the Specification. Building columns that are subjected to axial loads only, for example, depend on other members to provide stability, and should be designed with the effective length factor *K* equal to unity.

In high-rise buildings, the design of axially loaded columns is simply a matter of load tabulation and section selection. If floor beams and girders have seated connections (Fig. 19.27), some minor eccentricity will be introduced which must be considered. Thus, in some instances, a simple column will have to be designed as a beam-column.

Figure 19.28 is an idealized plan of a typical interior column at a typical floor. This is the condition which will be considered in the following design example. Plate 19.10 is a tabulation of the column loads for a 12-story building. Live load reductions have been taken as permitted by the *BOCA Basic Building Code* [19.14]. From AISC-1.6.1,

$$C_m = \left(0.6 - 0.4\,\frac{M_1}{M_2}\right) \geq 0.4 \tag{19.5.1}$$

where M_1/M_2 is the ratio of smaller to larger moment in the plane of bending under consideration. The sign of M_1/M_2 is negative for single curvature bending, positive for reverse curvature bending.

The loading used is conservative as the occurrence of the maximum and minimum reactions as shown is mutually exclusive under the influence of uniform loading. Further, the occurrence of full checkerboard loading which would cause uniform single curvature bending is highly unlikely. However, unlike the loading condition, the single curvature bending condition will have to be considered. For both major and minor axis bending, the ratio M_1/M_2 will be -1.0 (Fig. 19.29). Thus,

$$C_m = 0.6 - 0.4(-1.0) = 1.0$$

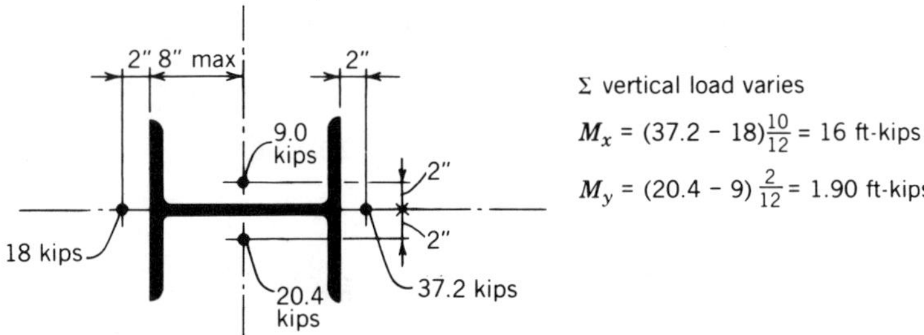

Fig. 19.28 Idealized column cross-section having loading caused by checkerboard loading on building.

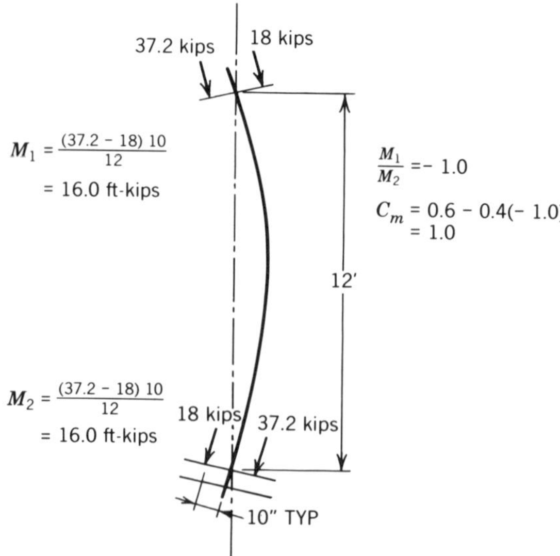

Fig. 19.29 Checkerboard loaded column—schematic loading diagram.

Sections may be easily selected for this type of column using the information in the *AISC Manual*. In this case, the selection of the column is based primarily on the axial load, that is, the effect of the moments is relatively small. Using the AISC procedure,

$$P_{\text{eff}} = P_0 + M_x m(C_m/0.85) + M_y mU(C_m/0.85) \tag{19.5.2}$$

where P_{eff} = required tabular load, kips
P_0 = actual axial load, kips
M_x = strong axis moment, ft-kips
M_y = weak axis moment, ft-kips
m = factor taken from *AISC Manual* (Table B, p. 3-10)
U = factor taken from column load table

For the top lift column, from the column load tabulation (Plate 19.10) one finds $P_0 = 270$ kips. Using the *AISC Manual* Column Load Tables for a W14×48 ($F_y = 50$ ksi) and $KL = 12$, the tabular load results give $P = 381$ kips and $U = 4.39$. From Table B then select $m = 1.7$.

A safe increment of load to be added to the actual column axial load may be computed as follows:

$$P_{\text{add}} = M_x m(1/C_m) + M_y mU(1/C_m)$$

$$P_{\text{add}} = 16(1.7)(1/0.85) + 1.9(1.7)(4.39)(1/0.85)$$

$$= 32 + 17 = 49 \text{ kips}$$

or, the minimum required tabular load 270 + 49 = 319 kips. At the bottom, the column load is $P_0 =$ 1368 kips and $U = 2.31$.

A safe increment of load to add to the bottom column may be computed:

$$P_{\text{add}} = 16(1.7)(1/0.85) + 1.9(1.7)(2.31)(1/0.85)$$

$$= 32 + 9 = 41 \text{ kips}$$

Thus, from the top column to the bottom column, the range of loads to be added to the actual axial load to find an equivalent axial load varies from 41 to 49 kips. Notice that the only variation is that of U as it reflects the variation of the depth-to-width ratios and the associated properties. Keeping this in mind, the selection of the proper section from the column load table will generally be obvious. For the top lift, P_{eff} will range between

$$P_{\text{eff}} = 270 + 49 = 319 \text{ kips maximum}$$

$$= 270 + 41 = 311 \text{ kips minimum}$$

Obviously, a W14×61, $F_y = 50$, will be required as the U value for a W14×53, $F_y = 50$, is virtually the same as for the W14×48. Therefore, a W14×61 is selected.

The second lift from the top will have a range

$$P_{\text{eff}} = 591 + 49 = 640 \text{ kips maximum}$$

$$= 591 + 41 = 632 \text{ kips minimum}$$

Select the W14×90, $F_y = 50$, $P_{\text{tab}} = 689$ kips.

The third lift from the top will have a range

$$P_{\text{eff}} = 913 + 49 = 962 \text{ kips maximum}$$

$$= 913 + 41 = 954 \text{ kips minimum}$$

Select the W14×132, $F_y = 50$, $P_{\text{tab}} = 1101$ kips.

The bottom lift column will have a

$$P_{\text{eff}} = 1234 + 41 = 1275 \text{ kips}$$

Select the W14×176, $F_y = 50$, $P_{\text{tab}} = 1368$ kips.

It must be realized that these sections are initial selections and their suitability must be checked using AISC Formulas (1.6-1a) and (1.6-1b) or (1.6-2) if appropriate. The *AISC Manual* includes a method to check the formulas using additional tabulated factors. However, the use of the formula in the specification is not difficult with a pocket computer. A sample calculation for the top lift column is shown on Plate 19.11.

It appears that the initial selection may be overly conservative. However, the next lightest section (W14×53) is inadequate. A check of all of the other trial selections will reveal that the initial selections were correct. Early editions of the *AISC Specification* handled the design of beam-columns with the very simple interaction formula

$$\frac{f_a}{F_a} + \frac{f_b}{F_b} \leq 1.0 \tag{19.5.3}$$

Under certain circumstances a similar formula may be used in lieu of Formulas (1.5-1a) and (1.5-1b) of the current *AISC Specification* for all the W-shapes in the *AISC Manual*. The modified formula takes the form

$$\frac{f_a}{F_a} + \frac{f_{bx}}{F_{bx}} \leq C1 \tag{19.5.4}$$

where the nomenclature is as identified in the *AISC Specification*. The following restrictions apply:

$$f_{by} = 0$$

$$C_m = 0.85$$

$$K_x \leq 1.0$$

Values for $C1$ are given in Table 19.10.

The initial values for $C1$ were calculated by substituting values of f_{bx} as determined by

$$f_{bx} = \left(1.0 - \frac{f_a}{F_a}\right) F_{bx} \tag{19.5.5}$$

into AISC Formulas (1.5-1a) and (1.5-1b) for values of f_a/F_a varying from 0.15 to 0.95 in steps of 0.01 and KL/r_y values from 0 to 120 in steps of 1.0. $C1$ was computed as the inverse of the sum of the two factors and checked by computing a new value

$$f_{bx} = \left(C1 - \frac{f_a}{F_a}\right) F_{bx} \tag{19.5.6}$$

Table 19.10 Values for C1—Modified Interaction Formula

$$\frac{f_a}{F_a} + \frac{f_{bx}}{F_{bx}} \leq C1$$

where $f_{by} = 0$, $C_m = 0.85$, and $K_x \leq 1.0$

$\frac{KL}{r_y}$	For $F_y = 36$ ksi	For $F_y = 50$ ksi
0– 60	1.0	0.99
60– 80	0.98	0.97
80–100	0.95	0.95

The new values for f_{bx} were then substituted back into the AISC Formulas to check the validity of the $C1$ values.

19.6 CONNECTIONS

With the adoption of the 1978 *AISC Specification* [19.6] and the subsequent issue of the *AISC Manual* [19.4], the entire procedure for the design and detailing of ordinary shear connections for beams has been drastically changed. Research at Lehigh and other universities indicated that the shear capacity in bearing of high-strength bolts was much higher than the values being used prior to the adoption of the 1978 *AISC Specification*. The *Guide to Design Criteria for Bolted and Riveted Joints* by Fisher and Struik [19.20] provided much of the basis for the *Specification for Structural Joints Using ASTM A325 or A490 Bolts* [19.36]. This Specification was adopted by AISC for use with the 1978 Specification. Provisions are included which allow the use of long and short slotted holes. Subsequently, in 1980, the bolt specification was updated. The major change was the elimination of the use of the calibrated wrench method as an approved method of installation of high-strength bolts.

The large increase in capacity for bolts in bearing-type connections has introduced two factors which heretofore were not generally considered critical in the design of connections. In prior specifications, minimum edge distances and spacings were specified for various size connectors and edge conditions. With the higher capacity bolts, bolt capacities are often determined by edge distance, end distance, and/or spacing. Figure 19.30 illustrates the critical distances L_v and L_h for a double angle connection. If the old standard end distance or edge distance L_h of 1¼ in. (32 mm) is used, a web thickness of over ½ in. (13 mm) is required for an A36 steel beam in order to develop the full double shear capacity of an A325 high-strength bolt in a bearing-type connection. Thus, for most normal beam sections used in light steel framing, connection capacities may well be determined by end and/or edge distances. Table 19.11 presents tabulation of the connecting material thicknesses required to develop the full strength of bolts in bearing-type connections for 3-in. spacing and various end and/or edge distances for A36 ($F_u = 58$ ksi, 400 MPa) and $F_y = 50$ ksi ($F_u = 65$ ksi, 448 MPa). The second factor which is new to the *AISC Specification* is a requirement that for end connections where the top flange is coped, "block shear" shall be limited to $0.30F_u$. The critical perimeter for block shear is the distance ($L_h + L_d$) as shown in Fig. 19.31. A conservative

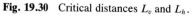

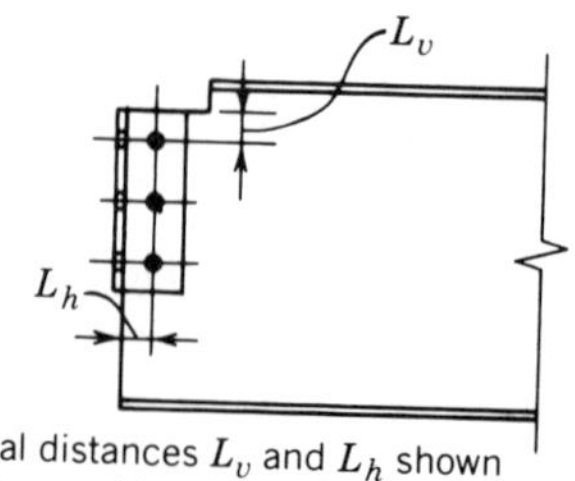

Fig. 19.30 Critical distances L_v and L_h.

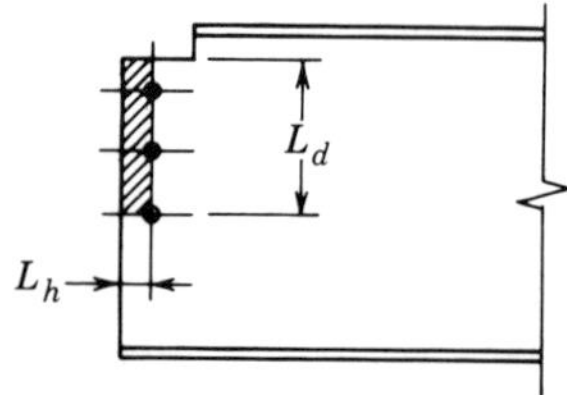

Fig. 19.31 Block shear critical perimeter.

Table 19.11 Minimum Material Thickness to Develop Full Bolt Capacities

Bolt Designation		Shear Capacity (kips)	3-in. Spacing*		1¼-in. Edge Distance**		1½-in. Edge Distance**		2-in. Edge Distance**	
			A36	$F_y = 50$	A36	$F_y = 50$	A36	$F_y = 50$	A36	$F_y = 50$
¾-in. diam. A325	S.S. D.S.	9.3 18.6	0.124 0.247	0.110 0.221	0.257 0.513	0.229 0.458	0.214 0.428	0.191 0.382	0.160 0.320	0.143 0.286
⅞-in. diam. A325	S.S. D.S.	12.6 25.3	0.172 0.345	0.153 0.308	0.348 0.698	0.311 0.623	0.290 0.582	0.258 0.519	0.217 0.436	0.194 0.389
¾-in. diam. A490	S.S. D.S.	12.4 24.7	0.138 0.262	0.147 0.293	0.342 0.681	0.305 0.608	0.285 0.568	0.254 0.507	0.214 0.426	0.191 0.380
⅞-in. diam. A490	S.S. D.S.	16.8 33.7	0.229 0.459	0.204 0.410	0.463 0.930	0.414 0.830	0.386 0.775	0.345 0.691	0.290 0.581	0.258 0.518

Notes: Threads included in shear planes.

F_u = 58 ksi for A36 and F_u = 65 ksi for F_y = 50 ksi.

$$* \; t = \frac{2P}{(s - d/2)F_u} \qquad \text{[from AISC Formula (1.16-1)]}$$

$$** \; t = \frac{2P}{LF_u} \qquad \text{[from AISC Formulas (1.16-2) and (1.16-3)]}$$

where t = thickness of connecting or connected material, in.
P = bolt capacity, kips
s = bolt spacing, in.
d = nominal bolt diameter + 1/16 in., in.
L = minimum of L_h or L_v, in.
L_h = edge distance, in.
L_v = end distance, in.
F_u = specified minimum tensile strength of connecting or connected material, ksi

value for block shear capacity is equal to $0.30F_u(L_h + L_d)t_w$. Alternatively, the capacity may be computed as $0.30F_uL_dt_w + 0.50F_uL_ht_w$. With the introduction of these factors into the *AISC Specification,* the checking of simple shear connection details is no longer a simple matter. There are literally an unlimited number of combinations of edge distances, end distances, beam web thicknesses, and connecting material thicknesses which can be selected. To minimize the time required to check connections one may choose to use predesigned connections which are tabulated in "Predesigned Bolted Framing Angle Connections" [19.3]. If connections other than the predesigned ones which utilize the high bolt capacities are used, computations must be made by either the designer or the fabricator.

The use of single plate shear connections (Fig. 19.32) either with simple beams or fully continuous members with welded flanges has become widespread. In the case of simple beams, no published procedure was available for the design of this type of connection until the 1981 publication of "Design Aids for Single Plate Framing Connections" by Young and Disque [19.40]. The critical factor controlling the capacity of this type of connection is the ability of the connection to accommodate the rotation of the beam end. In order to provide adequate rotation capacity this method limits the thicknesses of the connection plate and the beam web or requires the use of short slotted holes. The limit on material thickness is made to allow some ductile deformation in the upper and lower bolt hole areas. The procedure results in weld quantities which are seemingly excessive. However, the welds are proportioned to allow plowing of the bolts in the plate without the welds up-zipping. In addition, the method requires greater edge distance than the minimums given in the *AISC Specification,* and the method also limits the span-to-depth ratios for beams.

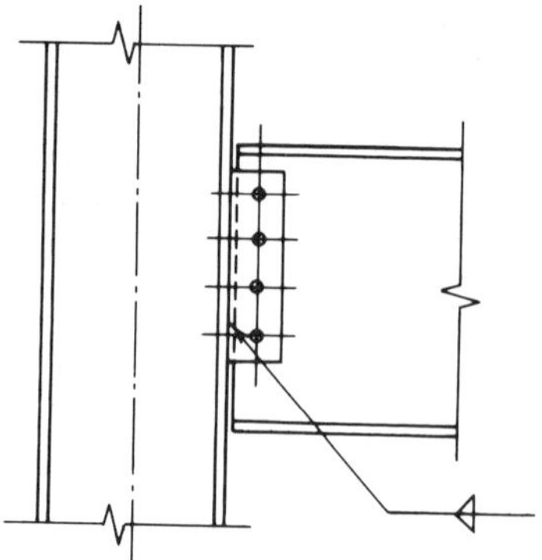

Fig. 19.32 Single plate framing connection.

The design of bolts in friction-type connections now specifies shear values for nine different classes of contact surfaces. Allowable shear values for the various classes of contact surface finish vary from a low of 16.5 ksi (114 MPa) to a high of 30 ksi (207 MPa) for A325 bolts. In any event, the allowable bolt capacity is limited by the lower of the two values for bearing-type or friction-type connection. In ordinary building construction bearing-type connections are generally used for ordinary shear connections. Although it is unlikely that a wind frame in a building will be subjected to the number of cyclic loadings that would require the consideration of fatigue problems, friction-type connections are often specified for use in these members.

Many types of moment-resisting connections have been developed to suit the needs for both fully rigid frame design (*AISC Specification* Type 1 Construction) and wind frame moment connections in simple framed buildings (*AISC Specification* Type 2 Construction). AISC Type 3, semi-rigid framing, is not widely used. See Figs. 19.18, 19.19, and 19.33.

It is important that design engineers either properly design or, as a minimum, provide adequate information for the proper design of moment connections. Too often, design drawings simply require that stiffener plates are to be installed to match beam flange size and/or area. Such a requirement not only can increase steel fabrication costs appreciably, in some instances it may lead to weld shrinkage problems, including lamellar tearing. The shop installation of thick stiffener plates requiring full penetration welds in heavy columns leads to the conditions that are often associated with lamellar tearing (see Sections 6.12 and 19.9). To minimize these problems and to produce more economical structures, stiffener plates should be proportioned as required by the *AISC Specification*. Where weld sizes are practical, fillet welds should be used to attach stiffener plates to columns as this will tend to minimize the stresses locked-in due to the welding process. (Refer to Chapter 6 for more specifics on welding-related considerations.) The weld between the stiffeners and the column web need only be designed to carry the shear force resulting from the

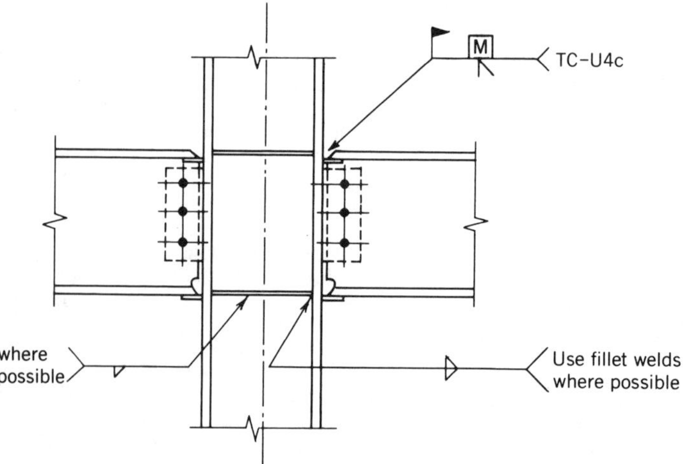

Fig. 19.33 Typical welded moment connection. To minimize potential lamellar tearing problems and reduce costs, stiffeners and welds should be designed for applied loads.

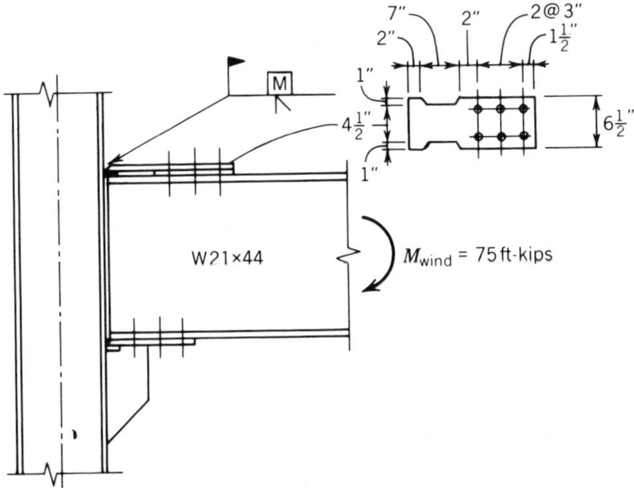

Fig. 19.34 Typical wind connection.

unbalanced moments on the opposite sides of the column. Often, a fillet weld on one side of the stiffener will be adequate.

The design of stiffeners using the *AISC Manual* is simple and quick. The *AISC Specification* requirements for stiffeners are based on the factored flange forces of the connected beams. For fully rigid frames, it is customary to use the full moment capacity of the connected girders to proportion stiffeners. For simple framed buildings with wind connections on selected girders, the flange force used to design the stiffeners should be the maximum factored force that the connected flange or plate can deliver.

The *AISC Specification* requires that wind connections in Type 2 Construction be designed so that connections have adequate inelastic rotation capacity to prevent overstress of the fasteners and welds under combined gravity and wind loads. An error in design which should be avoided in the design of Type 2 wind connections is the specifying of moment connections that are "too strong". Connections that have excessive moment capacity may possibly induce excessive gravity load moments in light columns. Many designers detail wind connections so that the top flange plate may elongate inelastically under the influence of gravity and wind loads (Fig. 19.34). In any case, the moment capacity should be limited by the connecting plate size, not the fasteners or welds, and the stiffeners should be proportioned for the maximum force that the plate can deliver. The design of the top connection plate shown in Fig. 19.34 is illustrated on Plate 19.12.

One should note that an arbitrary increase in the plate thickness may result in an increase in the number of bolts required in the connection to satisfy the requirements of AISC-1.2. It also should be noted that for columns with wind connections on one side only, the column should be designed for a moment not less than the smaller of either the connection moment capacity or the moment resulting from gravity plus wind loading (Fig. 19.35). The latter condition may well be the critical loading in the upper stories of the tier building.

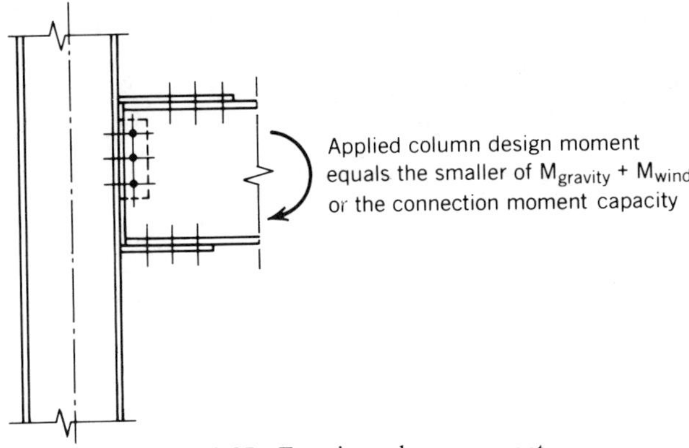

Fig. 19.35 Exterior column moment.

AISC-1.15.5 requires for stiffeners

$$A_{st} = \frac{P_{bf} - F_{yc}t(t_b + 5k)}{F_{yst}} \tag{19.6.1}$$

$$d_c \leq \frac{4100t^3\sqrt{F_{yc}}}{P_{bf}} \tag{19.6.2}$$

$$t_f \geq 0.4\sqrt{\frac{P_{bf}}{F_{yc}}} \tag{19.6.3}$$

where F_{yc} = column yield stress, ksi
F_{yst} = stiffener yield stress, ksi
k = distance from outer face of column to toe of fillet, in.
P_{bf} = factored flange force, kips
t = thickness of column web, in.
t_b = thickness of beam flange or moment connection plate, in.
d_c = column web depth clear of fillets, in.
t_f = thickness of column flange, in.

If the value for A_{st} in Eq. (19.6.1) is positive, or the column web depth d_c criterion is not met, a column web stiffener is required opposite the beam compression flange. If the column flange thickness t_f criterion is not met, a column web stiffener is required opposite the beam tension flange. As a practical matter, if stiffeners are required, they are generally placed opposite both tension and compression flanges. In the event that the value of A_{st} is negative or very small and stiffeners are required, they are proportioned in accordance with AISC-1.15.5.4 and 1.9.1.2.

In the column load tables in the *AISC Manual*, four of the tabulated values may be used to determine easily all stiffener requirements. Equation (19.6.1) may be rearranged to show that A_{st} = 0 when

$$[F_{yc}t(5k) + F_{yc}tt_b] > P_{bf} \tag{19.6.4}$$

One can observe that the values of $5k$ and t_b are added to compute the resistance of the column web to crippling. Also, it can be noted that for the very light columns, k values are seldom less than 1 in. and that beam flange thicknesses of 1 in. or more occur only for very heavy beams. Thus one can conclude that the affect of t_b in the quantity $(5k + t_b)$ is normally less than 10%. This fact can simplify the use of the values in the *AISC Manual*. Four values are tabulated in the *AISC Manual* "Column Load Tables," P_{w0}, P_{wi}, P_{wb}, P_{fb}. The first two values are associated with Eq. (19.6.1):

$$P_{w0} = F_{yc}t(5k) \text{ kips}$$

$$P_{wi} = F_{yc}t_b \text{ kips/in.}$$

P_{wb} and P_{fb} are associated with Eqs. (19.6.3) and (19.6.2) respectively, and represent the maximum factored flange force permitted for a column section without stiffeners for the compression and tension flange forces, respectively.

For an example, using the tabulated values in *AISC Manual* "Column Load Tables" and a factored flange force of 46.5 kips resulting from a 1/4 × 8 in. moment plate, and A36 8-in. wide-flange columns, for W8×31:

P_{w0} = 48 kips, therefore $A_{st} = 0$

P_{wb} = 93 kips, O.K.

P_{fb} = 43 kips, stiffener required at beam tension flange

From AISC-1.15.5.4,

$$\text{minimum stiffener width} = \frac{\text{flange width}}{3} - \frac{\text{column web}}{2}$$

$$= 8/3 - t_{wc}/2$$

Use 3-in.-wide stiffener.

From AISC-1.9.1.2,

$$t_{\min} = \frac{3\sqrt{36}}{95} = 0.19 \text{ in.}$$

Use 1/4 × 3 in. plate stiffeners.

For W8×35 and heavier 8-in. sections, P_{w0}, P_{wb}, and P_{fb} are larger than P_{bf}; therefore, no stiffeners are required.

In the event that the factored flange force is larger than P_{w0}, one may compute $(P_{w0} + P_{wi}t_b)$. If this value is less than the factored flange force, A_{st} may be computed:

$$A_{st} = [P_{bf} - (P_{w0} + P_{wi}t_b)]/F_{yst} \tag{19.6.5}$$

where F_{yst} = stiffener plate yield stress

Alternatively, one may compute with little significant error

$$A_{st} = (P_{bf} - P_{w0})/F_{yst} \tag{19.6.6}$$

A list of do's for connections on the structural design drawings might include the following:

1. Record reactions for all composite beams.
2. Record reactions for all beams with concentrated loads.
3. Record reactions for each beam whose reaction is greater than $W_c/(2L)$ where W_c is the recorded value in the beam tables in the *AISC Manual*.
4. If high bolt values are to be used, require the use of AISC preengineered connections [19.3] or that complete design calculations be submitted by the fabricator.
5. Column web stiffener requirements should be indicated on the structural drawings. In the event that proportioning of stiffeners is to be a responsibility of the fabricator, proper information for their design should be included on the structural drawings.
6. Wind connection design should be included on the structural drawings. In some cases, knowledge of wind connection details is essential to column design.

19.7 LATERAL FORCE RESISTANCE SYSTEMS

Finding an efficient economical bracing system for a multistory building offers the greatest challenge to the structural engineer. The design of bracing systems for buildings of all sizes presents the engineer an opportunity to use imaginative, innovative design. There are basically three types of frame bracing to be considered: braced frames, unbraced frames, and combinations of the two. Even the modern "tube" designed buildings consist of essentially either braced frames and/or moment-resisting frames. However, in many "tube" designs, shear deformations can significantly contribute to the building deflection.

The actual design of members in bracing systems is different than that of other members in a building frame. There are beams, columns, and beam-columns. However, there are additional considerations except for bracing systems with low aspect ratios. The selection of member sizes is generally based upon deflection rather than strength considerations. In low-rise buildings, members in the upper stories may be minimum sizes or perhaps controlled by gravity or combined loads. However, these members, while greater than optimum size, generally will not affect the total deflection significantly. Thus, the key to an efficient design system is the proper selection of the initial trial member sizes. Topics to be discussed include braced frames, unbraced frames, combinations of the two, and frame stability considerations.

19.7.1 Braced Frames

The use of trussed bracing frames in moderately tall buildings (up to 20 stories) is often the best overall solution to the lateral load design problem. In very tall buildings, it is unlikely that a reasonable aspect ratio (building height/bracing bent length) for an interior bracing bent can be accommodated in the building architectural plan. In recent years, a number of tall buildings using bracing members built into their facades have been constructed. Placement of the bracing in the facade can result in a very efficient bracing system. The aspect ratio of the system can become the ratio of the building height to width.

Basically, there are two types of braced bents, X-braced (Fig. 19.36*a*) and K-braced (Figs. 19.36*b*,*c*). X-braced bents are less efficient and more difficult to properly design than K-braced

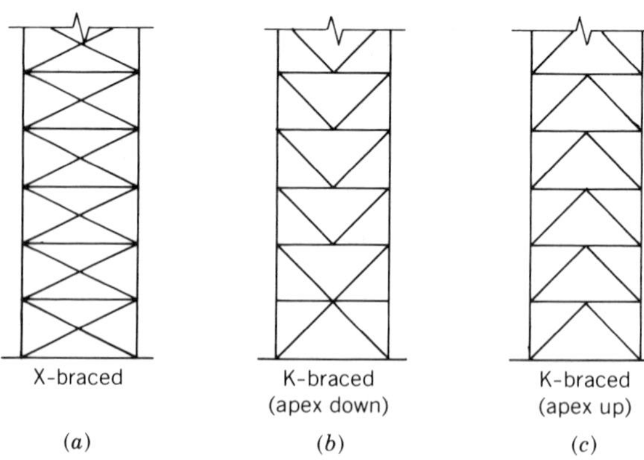

Fig. 19.36 Bracing bents.

bents. In a typical X-braced bent, both diagonals are designed to carry the entire story shear which results in a duplication of material. Geometrically, the K-braced bent is more efficient resulting in less deflection for equivalent areas of members. X-braced frames must be carefully designed to account for column strain due to live or dead loads applied after the installation of the bracing members. Bracing members must be preloaded in some manner so that they will be effective when lateral loads are applied or the initial deflection before they become effective must be considered (Fig. 19.38). There are an unlimited number of bracing schemes which can be used (Fig. 19.37). However, for the more complicated systems, the final analysis must be made using a computer.

For most bracing bents, many of the members except for some in the upper stories are likely to be proportioned to control deflection rather than the load effects. Thus, the solution to the design problem is one of selecting the member size on the basis of deflection with a subsequent check for strength. The AISI *Plastic Design of Braced Multi-Story Steel Frames* [19.7] contains a description of a complete procedure for the design of braced frames for plastically designed buildings. Even though information pertains to plastically designed frames, it can easily be used for allowable stress design. Methods are included for the selection of initial member sizes and for the calculation of the bracing bent deflection.

In order to reduce overall costs, the number of braced bents should be kept to a minimum. This results in the concentration of the special fabrication and erection processes to a fewer number of pieces of material. It also results in the concentration of lateral load effects in a fewer number of members. Traditionally many engineers have designed K-braced bents with apex of the V pointed down so that the gravity load was tension, thus reducing the compression under the influence of wind loads. Often at the bottom story, the apex position was reversed (see Fig. 19.36*b*). However, a K-braced frame is much more efficient if the apex of all of the Vs is up (see Fig. 19.36*c*). Figure 19.39 illustrates the deflected shape of a bent with the apex down. In a typical braced bent, in the upper stories, the deflection is largely a result of the accumulated rotation of the frame due to the column strains in each story. With the apex down, rotation in any story includes the effect of column strain in that story due to the total shear in that story. Additionally, deflection occurs within the story due to column strain. With the apex up, the column strain in any story is a function of the story shear in the story above. Additionally, there is no deflection in the story due to the column strain in that story.

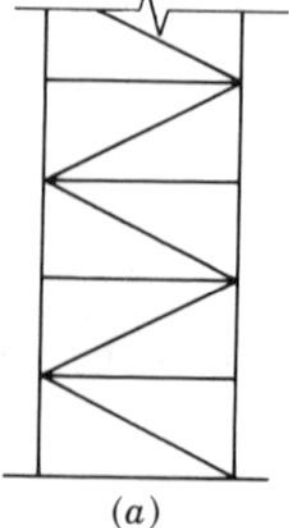

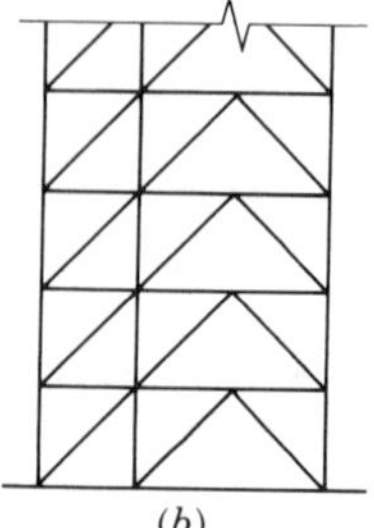

Fig. 19.37 Bracing systems.

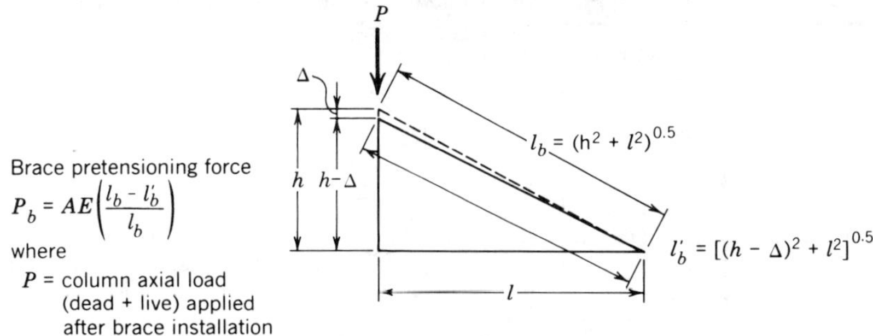

Fig. 19.38 X-brace pretensioning force.

Figure 19.40 depicts the deflected shape of a K-braced bent. As it would be for a cantilever beam, the slope of the truss is ever increasing. The actual floor-to-floor deflection increases from bottom to top. Thus, if the individual floor-to-floor deflections are critical, a pure braced frame may not be appropriate.

If wind loads were uniform from bottom to top, the stress distribution in bracing frame members would be primarily a function of the aspect ratio of the frame. Tables 19.12 and 19.13 contain optimum stress levels for members in K-braced frames with a drift index of 0.0025 using *BOCA Code* [19.14] wind loading distribution for both 10- and 20-story buildings. The apex of the bracing members points up. A uniform story height of 12 ft–4 in. (3.8 m) was assumed for all frames. Aspect ratios (height-to-truss depth) of 5, 7.5, and 10 are tabulated. The stress level values were obtained using a computer program which uses the deflection algorithm contained in *Plastic Design of Braced Multi-Story Steel Frames* [19.7]. The program selects initial areas for all members and computes the system deflection. A small increment of area is then added to each member individually and a deflection calculation is made. A permanent increment of area is added to the member that reduces the deflection the most. Small initial areas are selected so that all members will have several increments of area added. The tables are meant to be used to assist in the initial selection of member sizes to be used in an analysis. If a different target drift index (D.I.) is desired, the stress levels may be modified by the ratio of the desired D.I. to 0.0025 (D.I. required/0.0025). For different aspect ratios or different numbers of stories, values may be interpolated.

The deflection of bracing frames is not too sensitive to the changing of stress levels from one location to another. By examining the values of R_C, R_G, and R_B shown in the story rotation table on Plate 19.16, one can observe that in the upper-half of the total height the accumulation of column strains contributes more than 75% of the total story rotations. Therefore, where possible one should use the optimum size columns in the lower stories. If the bottom lift column sizes are reduced below the optimum, the next lift or lifts should be oversized to control the total story rotation due to column strain in the upper stories. On the other hand, if architectural constraints prevent the use of optimum brace sizes in the lower stories, providing that strength considerations are not a problem, brace sizes in lower stories may be reduced and those in the upper stories increased.

Plates 19.13 through 19.17 illustrate a method of calculation utilizing Table 19.12 for a small K-braced bent with the apex of the V's pointed up. The method is similar to that presented in *Plastic Design of Braced Multi-Story Steel Frames* [19.7]. After initial member sizes are selected, a strength design check for both gravity and wind plus gravity loading conditions should be made.

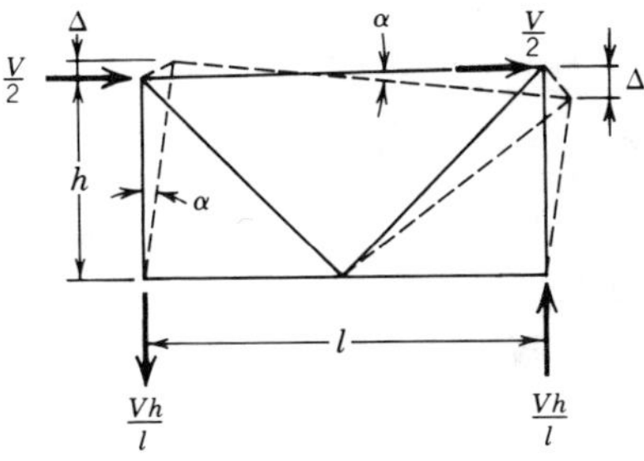

Fig. 19.39 K-bracing—apex down.

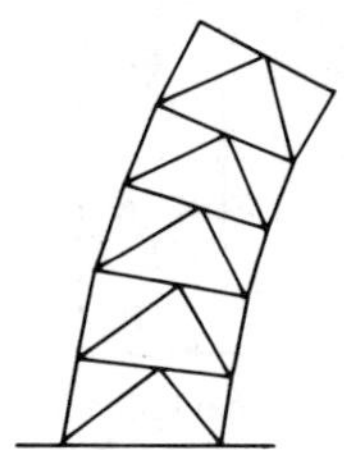

Fig. 19.40 K-braced bent—deflected shape.

Table 19.12 K-Bracing Member Stress Distribution for 10-Story Building (H = 123.33 ft)

Floor	Member	Aspect Ratio (H/D)		
		10	7.5	5
9–10	Column Girder Brace	2.3 9.6 4.4	3.5 9.2 5.9	4.6 8.5 6.6
7–8	Column Girder Brace	3.6 11.6 7.3	4.6 12.1 9.2	5.6 12.3 12.3
5–6	Column Girder Brace	4.8 12.8 8.0	6.0 14.5 11.0	6.7 13.4 14.7
3–4	Column Girder Brace	5.7 13.9 9.6	7.2 14.5 14.5	8.0 16.0 16.0
1–2	Column Girder Brace	7.2 13.9 9.6	8.6 17.3 14.5	10.0 16.0 16.0

Tabulated values of member bar stresses (ksi) due to wind loads to produce a maximum bracing system deflection index of 0.0025 using BOCA wind loading distribution.

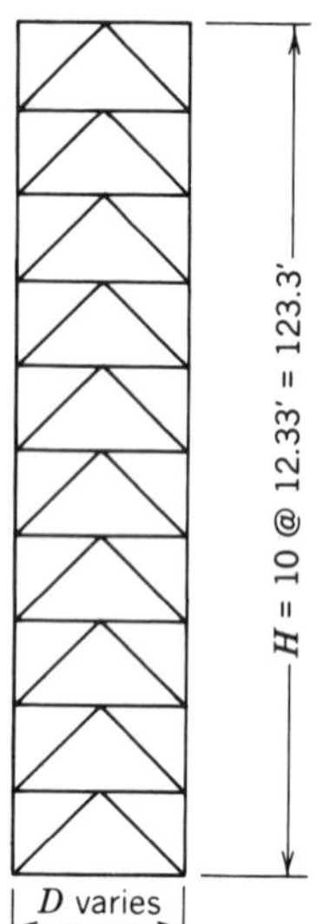

Table 19.13 K-Bracing Member Stress Distribution for 20-Story Building (H = 246.67 ft)

Floor	Member	Aspect Ratio (H/D)		
		10	7.5	5
17–20	Column Girder Brace	2.2 4.0 4.3	3.4 4.6 4.6	3.9 4.2 5.0
13–16	Column Girder Brace	3.5 5.9 6.8	4.5 7.4 7.1	5.1 6.5 7.8
9–15	Column Girder Brace	4.4 6.9 8.4	5.9 8.9 9.4	6.4 8.6 11.2
5–8	Column Girder Brace	5.8 8.0 9.6	7.0 10.2 12.2	7.7 11.2 11.2
1–4	Column Girder Brace	6.6 9.2 9.6	7.8 11.2 12.2	8.5 11.7 11.2

Tabulated values of member bar stresses (ksi) due to wind loads to produce a maximum bracing system deflection index of 0.0025 using BOCA wind loading distribution.

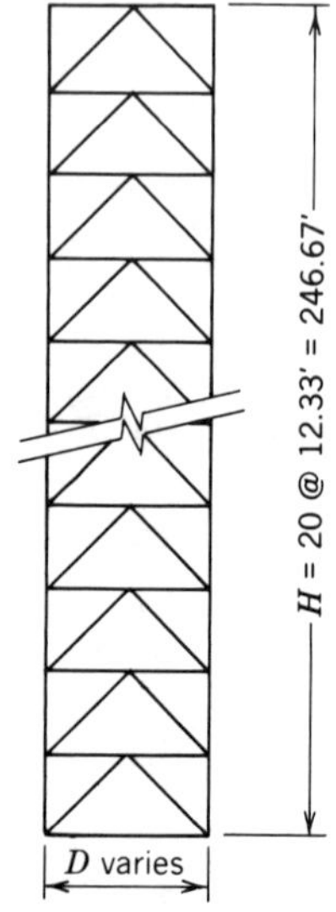

This step is omitted in the design example. The resulting maximum D.I. of 0.0024 is very close to the target D.I. of 0.0025. Plate 19.18 is an abbreviated calculation for a K-braced bent with the apex of the V's pointed down. The maximum deflection of the bent is increased by an astonishing 33% simply by reversing the direction of the bracing.

19.7.2 Unbraced Frames

The design of unbraced frames is not as straightforward as the design of braced frames. Most designers keep the number of such "wind frames" to a minimum. This strategy also reduces the number of expensive moment connections to a minimum. As a result, girder member sizes are often determined by drift considerations rather than strength considerations. Often, column sizes proportioned for strength will be larger than the optimum size required for drift.

A basic procedure for the design of an unbraced frame is as follows:

1. Gravity load design (to find approximate sizes required for gravity loads only).
2. Perform a simple portal wind analysis.
3. Determine the approximate minimum member sizes for combined load effects of lateral plus gravity loads.
4. Adjust member sizes to control drift.
5. Perform final frame analysis.
6. Make a final member selection and design.

The key to success is to perform Steps 1 through 4 in such a manner that Steps 5 and 6 do not need repeating.

The gravity load analysis may be approximate. In most cases in the final design, girder stiffnesses will increase relative to the column stiffnesses as determined by a gravity load analysis. Thus, accurate determination of gravity column moments is very difficult. A preliminary girder size may be selected by modifying its fixed-end moments by a factor of 12/11. End column moments may be determined by making a single distribution of the girder fixed-end moments between the two columns and the girder in proportion to their relative stiffnesses. One may elect to use an end column moment equal to 40% of the girder fixed-end moment. The determination of interior column moments is similar to that used for the exterior columns except the difference between the total load and dead load fixed-end moments is distributed between the columns and girders. One may wish to perform the above distributions for both a light and heavy column section in order to establish a range of moments. To repeat, at this point the accuracy of the values obtained is not critical; the values will be used only to select initial column sizes. In the final design, except for a few upper stories, it is likely that gravity column moments will be a small percentage of the total moments due to gravity plus lateral loads. The lateral load analysis may be performed using a simple portal analysis. Interior column moments are equal to the total story shear divided by the number of bays multiplied by half the story height. Exterior column moments are equal to one-half of the value of the interior column moments. Girder moments are equal to the average of the interior column moments above and below.

If uniform girder and column stiffness values (I/L) are maintained, the simple portal analysis is valid for nonuniform bents. If the geometry of any bent is such that it is not practical to maintain uniform stiffness values, a modified portal analysis may be used as illustrated in Fig. 19.41. Minimum member sizes are then selected which will satisfy the strength requirements for both gravity and gravity plus lateral load loading conditions.

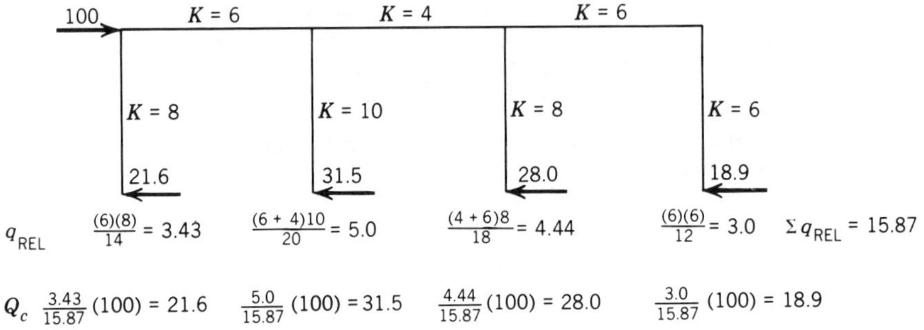

Fig. 19.41 Modified portal analysis.

Once minimum member sizes for strength are known, an analysis for the control of deflection is performed. A general formula for story deflection is

$$\Delta = \frac{Q}{\dfrac{12E}{h^2} \sum_{1}^{m} \left[\dfrac{I_c/h}{1 + 2\left(\dfrac{I_c/h}{\Sigma I_b/L_b}\right)} \right]} \tag{19.7.1}$$

where Δ = story deflection, in.
Q = average story shear in stories above and below, kips
E = modulus of elasticity, ksi
h = story height, in.
L_b = beam length, in.
I_c = column moment of inertia, in.^{4}
I_b = beam moment of inertia, in.^{4}
m = number of rows of columns

Solving such an equation is quite burdensome and also is not required. Note that the sum of the stiffness values for both beams and columns in any story is used in the determination of deflection. Thus, if one can determine the average stiffness value required for each member in the story, the total stiffness values in any story will be the average values multiplied by the number of bays. For the following method, exterior column moments of inertia are assumed to be one-half of the interior column moments of inertia.

Equation (19.7.1) may be rearranged as shown in Eqs. (19.7.2), (19.7.3), and (19.7.4).

$$K = \frac{2X_1}{n} \tag{19.7.2}$$

$$K_b = \frac{X_1 K_c}{nK_c - X_1} \tag{19.7.3}$$

$$K_c = \frac{X_1}{n - X_1/K_b} \tag{19.7.4}$$

$$X_1 = \frac{Qh^2}{12E\Delta} \tag{19.7.5}$$

where K = average I_b/L_b and I_c/L_c required for specified Δ
K_b = average I_b/L_b required for specified Δ and K_c
K_c = average I_c/h required for specified Δ and K_b
n = number of bays

Knowing the required member sizes for load effects, one may compare those stiffnesses with the stiffnesses indicated by Eqs. (19.7.2), (19.7.3), and (19.7.4). It is unlikely that the column stiffnesses K_c will be small relative to the girder stiffnesses, therefore, one may expect that girder stiffnesses should be adjusted. Assuming that both column and girder stiffnesses needed to be increased, one may concentrate the increase for the member that is most efficient (least total weight) in increasing total stiffness. More often than not, it will be more efficient to increase girder stiffnesses. Since the preliminary selection of the column sizes was not very precise, again assuming that an increase in the stiffness of the entire system is required, an initial increase in column stiffness may be advisable so that when the final analysis is made, often column sizes will not have to be adjusted for strength requirements.

Once the preliminary selection of member sizes is completed, a final analysis of the frame and subsequent strength check for all members should be made.

19.7.3 Combination Systems

Due to the different deflection characteristics of moment-resisting and braced frames (Fig. 19.42), the combination of the two in a single system results in a very efficient system. The braced bent deflection characteristic is similar to that of a cantilever beam. At the bottom, the floor-to-floor deflections are very large. On the other hand, the moment-resisting frame deflections are approxi-

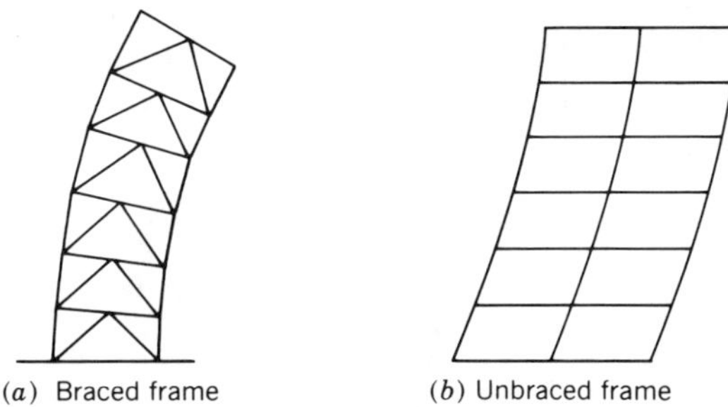

Fig. 19.42 Frame deflection characteristics.

mately uniform, if anything perhaps a little larger at the bottom. Thus, in the combination system, near the bottom the braced frame will restrict deflections to a small value and at the top, the moment-resisting frame tends to restrain the deflection of the braced frame. If combination braced and moment-resisting frames are to be used, a computer analysis of the frame must be used.

Another method of dramatically increasing the efficiency of a bracing system is the use of hat and belt trusses. Figure 19.43 illustrates the effect of introducing a hat truss. The addition of the hat truss tends to reduce the slope of the cantilever truss. Figure 19.44 illustrates a bracing system which was used in a mid-rise building. In this example, a rather shallow braced bent (aspect ratio = 12.4) was placed adjacent to the elevator shaft. The system was enhanced with the addition of two additional bracing members in the mechanical level at the top. The deflection of the braced bent alone was 4.26 in. (108 mm) (D.I. = 0.0030). The addition of the two extra bracing members reduced the deflection to 2.66 in. (68 mm) (D.I. = 0.0019). Thus, the addition of two light members reduces the deflection by 38%. More striking, the reduction in deflection was accomplished with the addition of 1000 lb of steel to the 25,000-lb bracing system. Thus, the introduction of truss web members at the top of the building (perhaps in a mechanical space) is an excellent method of increasing the efficiency of a bracing system. In taller buildings which require an intermediate mechanical equipment level, the introduction of a belt truss at the intermediate level in addition to the hat truss will prove effective in reducing the angular rotation of the braced truss, thus materially reducing deflection (Fig. 19.45). References 19.29 and 19.38 discuss this matter more fully.

19.8 FRAME STABILITY

The subject of overall stability of multistory frames is perhaps the least understood and least well-documented subject with which the building designer must deal. The *AISC Specification* simply requires that general stability be provided for individual members and structures as a whole. Further, the specification states

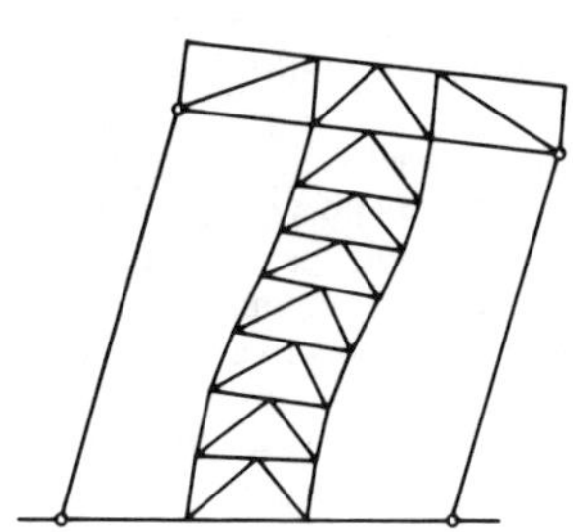

Fig. 19.43 Bracing frame with hat truss.

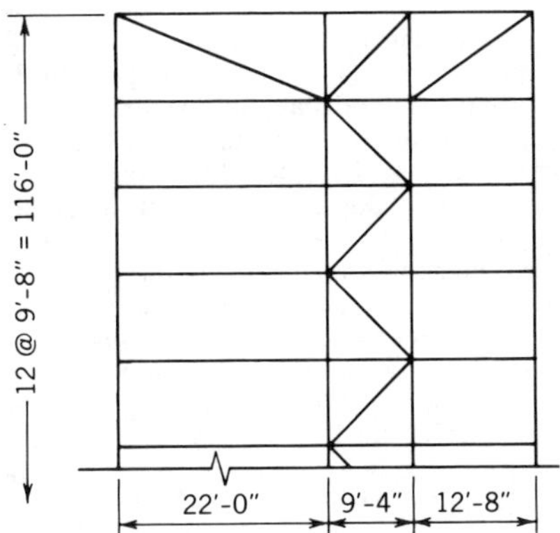

Fig. 19.44 Elevator shaft bracing frame.

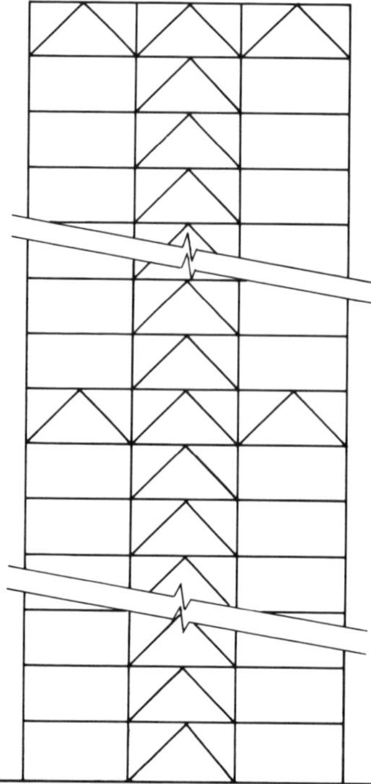

Fig. 19.45 Bracing frame with hat and belt truss.

Design consideration should be given to significant load effects resulting from the deflected shape of the structure or of individual elements of the lateral load resisting system, including the effects on beams, columns, bracing, connections, and shear walls.

These seemingly innocuous statements effectively place the responsibility for overall frame stability directly on the shoulders of the structural engineer. The *AISC Specification* and *Commentary* do not contain sufficient background information for the exclusive use of it in the design of multistory unbraced frames. More detailed discussions on the subject of frame stability may be found in the *Structural Stability Research Council, Guide to Stability Design Criteria for Metal Structures* [19.27] and in Chapter 4 of *Structural Design of Tall Steel Buildings* [19.16]. Topics to be discussed in this section include braced frames and P–Δ and K-factor methods for unbraced frames.

19.8.1 P–Δ Action

The basic aim in devising methods and design procedures for overall frame stability really revolves around the problem of how to cope with the required stabilizing forces which result from deflections due to lateral loads and/or erection deviations (see also Chapters 10 and 11). Figure 19.46 is an illustration of two columns with pinned ends. If for any reason the top of the column deflects horizontally with respect to its base, either a stabilizing force H or a moment is required for stability. In building frames, horizontal deflections due to lateral loads and/or initial out-of-plumbness also result in a need for these stabilizing forces.

In buildings, the stabilizing forces must be provided by additional frame moment capacity or bracing bent capacity. These forces are often referred to as fictitious sway forces or P–Δ forces. This phenomenon can be simply illustrated by an idealized subassemblage consisting of two half-story columns and one-half span beam (Fig. 19.47). In this case, the column and girder moment magnifications due to the initial frame deflection are $P\Delta/2$ and $P\Delta$, respectively, as shown in Fig. 19.47*b*. In an analysis this moment magnification can be represented by an equivalent fictitious sway force equal to $P\Delta/h$. The application of the fictitious sway force $P\Delta/h$ results in an additional deflection for which a new deflection and sway force can be computed, as shown in Fig. 19.47*c*. Thus, a second-order P–Δ analysis consists of computing successive deflections and sway forces. Providing the frame is not unstable, within a few cycles, the sum of added sway forces should

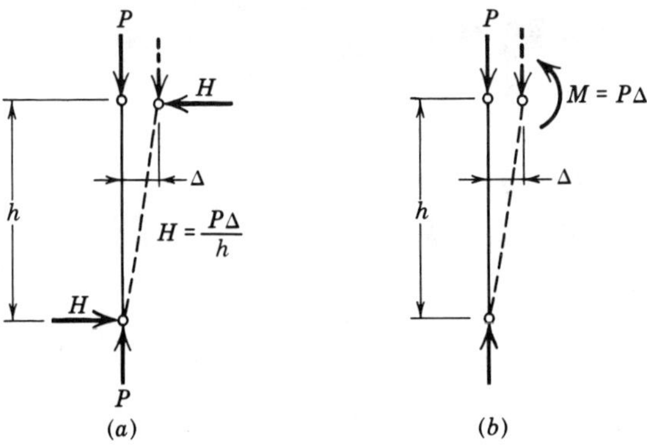

Fig. 19.46 P–Δ stabilizing force.

converge to a constant value. An example of a simple P–Δ analysis is shown in Fig. 19.48. Figure 19.48*b* illustrates a simple iteration process which may be used for single or multistory frames. However, in lieu of the iterative process, a simple amplification factor may be used. The factor $A1$ is a ratio of initial frame deflection to final frame deflection after a second-order analysis. The amplification factor, as given by Iffland and Birnstiel [19.26], may be stated as follows:

$$A1 = \frac{1}{1 - \left(\frac{\Sigma P}{\Sigma H}\right)\left(\frac{\Delta}{h}\right)} \tag{19.8.1}$$

where $A1$ = amplification factor applied to first-order analysis deflection to obtain second-order analysis deflections
ΣP = sum of vertical loads in story
ΣH = sum of horizontal loads in story
Δ = first order analysis story deflection
h = story height

This factor may be used in the design of unbraced frames.

19.8.2 Unbraced Frames

The *SSRC Guide* [19.27] discusses three methods for the design of unbraced steel frames in some detail. The three methods are discussed here with the emphasis placed on the practicability for normal office design procedures.

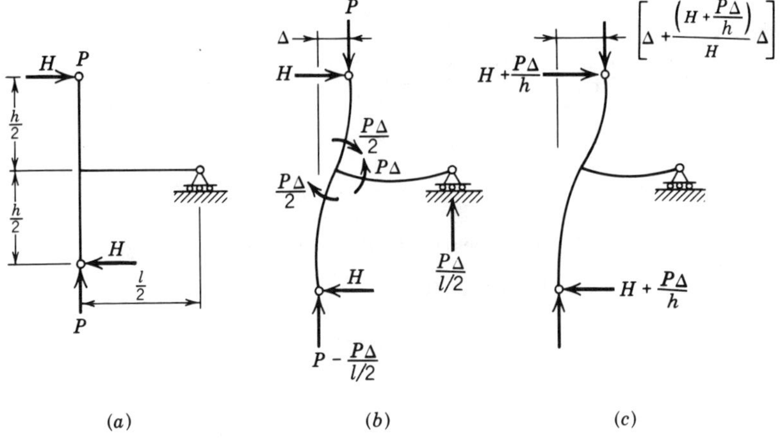

Fig. 19.47 Idealized subassemblage.

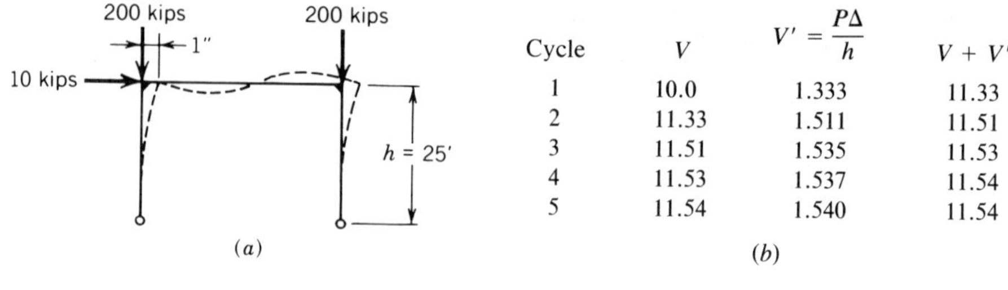

Cycle	V	$V' = \frac{P\Delta}{h}$	$V + V'$
1	10.0	1.333	11.33
2	11.33	1.511	11.51
3	11.51	1.535	11.53
4	11.53	1.537	11.54
5	11.54	1.540	11.54

$$A_1 = \frac{1}{1 - \frac{400}{10}\left(\frac{1.0}{2.5(12)}\right)} = 1.154$$

(c)

Fig. 19.48 P–Δ analysis.

The simplest procedure is those frames that fall into the category of "frames that can be designed without considering frame stability." Some building frames that are designed to control drift will be within requirements as stated by the *SSRC Guide* [19.27]. They are:

1. All columns must be designed in accordance with the *AISC Specification* with $K = 1.0$. The value for C_m will be equal to $(0.6 - 0.4M_1/M_2) \geq 0.4$.
2. The maximum values for f_a/F_a or $f_a/0.60F_y$ shall be 0.75.
3. The maximum in-plane L/r value shall not exceed 35.
4. The bare frame first-order deflection index shall be limited to

$$\frac{\Delta}{h} < \frac{1}{7}\frac{\Sigma V}{\Sigma P} \tag{19.8.2}$$

where Δ = story deflection
h = story height
ΣV = sum of story shears
ΣP = sum of story vertical loads

The requirement for a maximum in-plane L/r ratio of 35 effectively prohibits weak axis bending in columns if the stability problem is to be ignored. Table 19.14 tabulates the maximum permitted story heights for the normal column shapes. As can be observed, the height limitation is not too restrictive. Equation (19.8.2) may be rearranged in the form

$$P_{\min} > 7(\text{D.I.})(D)(\text{PCF}) \tag{19.8.3}$$

where $P_{\min}$ = minimum average design wind pressure, psf
D.I. = deflection index ratio of story deflection to story height
D = depth of structure parallel to wind direction
PCF = total weight of building with appropriate live load reduction expressed in pcf

Table 19.14 Maximum Story Heights

Nominal Column Depth (in.)	Maximum Story Height (ft)
8	10
10	12
12	15
14	17

Table 19.15 Minimum Design Wind Pressures (per 100 ft of Building Height) for Frames in Which Stability Considerations May Be Ignored

Building Weight	10.0 pcf			12.5 pcf		
D.I.	0.0020	0.0025	0.0030	0.0020	0.0025	0.0030
Wind load	14.0	17.5	21.0	17.5	22.9	26.3

Note: See Section 19.8.2 for design limitations.

Several relationships regarding the stability of buildings are well illustrated in Eq. (19.8.3):

1. The minimum design wind pressure is a direct function of the building weight. As the weight of a building increases, the additional in-plane strength requirement increases.
2. The minimum design wind pressure is a direct function of the depth of the structure perpendicular to the wind. Again, the greater the weight of the structure to be laterally supported, the greater the in-plane strength requirement.
3. The greater the permitted first-order analysis deflection, the greater the additional in-plane strength requirement.

The above relationships are considerations that cannot be ignored no matter what method of analysis is used. Table 19.15 tabulates the minimum design wind pressures for a range of initial frame deflections and building weights for frames in which stability considerations may be omitted providing the foregoing design limitations are followed. The wind pressures are tabulated for a 100-ft depth of building parallel to the load. The building weights of 10.0 pcf (1570 N/m^3) to 12.5 pcf (1970 N/m^3) encompass a range into which most multistory buildings fall. From the table one may find the minimum design wind pressure for a 150-ft (45.7-m) square building with a building weight of 10.0 pcf (1570 N/m^3) and a design deflection index of 0.0020 as follows: The average wind pressure for the story and the stories above must equal (150/100)(14) = 21.0 psf. On the other hand, a 12.5-pcf building with a D.I. of 0.0030 would require a (150/100)(26.3) = 39.4-psf average wind pressure.

19.8.3 *P–Δ* Designed Unbraced Frames

The use of the *P–Δ* method for design of unbraced frames is philosophically attractive as it is completely rational and can be easily understood by the designer. At this time, the AISC permits the use of a *P–Δ* analysis for allowable stress design. Several items affecting the use of the *P–Δ* method need to be further evaluated. Research has shown that in many instances, actual live loads in buildings are much lower than those presently specified in building codes. Adoption of ANSI A58.1-82 *Minimum Design Loads and Other Structures* [19.8] results in a substantial decrease in column loads for most structures. This reduction will materially reduce *P–Δ* loadings for office buildings. At this time, neither the AISC nor the SSRC comment on the effect of the facade of a building in the reduction of real frame deflection. As a result, we take the bare frame drift, calculate *P–Δ* loads, and then add a "safety factor" for deflections that will not actually occur. This procedure may seem to be and probably is overly conservative; however at this time, no other course of action has been accepted by the profession.

If a building is slender and/or the lateral loading is high, and the members are proportioned to resist drift, the whole issue of second-order analysis may be mute. That is, initial stresses due to lateral loads may be so low that secondary effects are tolerable. On the other hand, a long narrow building subjected to a low level of lateral loading parallel to the long dimension will require a consideration of secondary effects due to deflection. Older building codes often permitted the omission of lateral load analysis for buildings with a depth-to-height ratio greater than a specified minimum. In these cases, frame stability was provided by facade and/or interior partitioning. However, in buildings with light cladding and open space planning, *P–Δ* analysis provides a rational method for the evaluation of the problem. Purportedly, at least one well-known building has suffered difficulties because of this loading condition.

The *SSRC Guide* presents a design method for *P–Δ* analysis but refrains from recommending its use. No value for the load factor $\bar{F}$ is recommended. Iffland and Birnstiel [19.26] point out that the value $\bar{F}$ used in Canadian practice, where *P–Δ* procedure is commonly used, is 1.67. However, at this time, their recommendation is that $\bar{F}$ be taken as 2.0. A building frame that is subjected to a reasonable level of lateral loads is not likely to experience critical frame stability problems under the influence of gravity loads only. It has been recommended that a minimum P–Δ load determined by a deflection of $h/500$ should be applied to all frames. In most buildings the effect of axial load in reducing column stiffness is not a factor that needs to be considered. However, if the quantity $(PL^2/EI)^{0.5}$ is greater than 0.9, an analysis considering this phenomenon should be made [19.22].

Story shears to be used including the P–Δ forces may be expressed as follows:

$$H = V + V' \tag{19.8.4}$$

$$V' = \frac{P\Delta}{h}\bar{F}(1 - A1)V \tag{19.8.5}$$

$$H = \frac{P\Delta}{h}\bar{F}(A1)V \tag{19.8.6}$$

where H = total story shear including P–Δ sway forces
V = total story shear excluding P–Δ forces
V' = story shear due to secondary effects
P = sum of vertical loads in story (not the sum of all loads above the story)
Δ = first-order story deflection
h = story height
$\bar{F}$ = load factor (recommended value = 2)

$$A1 = \text{amplification factor} = \frac{1}{1 - \left(\frac{\Sigma P}{\Sigma H}\right)\left(\frac{\Delta}{h}\right)}$$

Two factors, the ratio of the wind load per sq ft of loaded facade to the building weight per sq ft of facade and the initial frame deflection determine the magnitude of the amplification of a P–Δ analysis. The amplification factor may be restated:

$$A1 - \frac{1}{1 - Y1(\text{D.I.})} \tag{19.8.7}$$

where $Y1$ = (depth)(PCF)/P
depth = depth of building parallel to wind force, ft
PCF = average weight of building, per cu ft
P = average design wind pressure
D.I. = ratio of story deflection to story height (drift index)

Table 19.16 gives values for the amplification factor $A1$ for various ratios of building weights to wind loads per square foot of facade and initial frame deflection indices. One may observe that for a 100-ft (30-m) square building that weighs an average of 10 pcf (1600 N/m^3) with an initial D.I. of 0.0025 under an average of a 20-psf wind load, that is, $Y1 = 10(100)/20 = 50$, the increase in loading due to secondary affects with a factor safety ($\bar{F} = 1.$) is a modest 14%. If a value of 2 is used for $\bar{F}$, the increase will be 28%. However, for a 240-ft (70-m) square 12.5-pcf (1970-N/m^3) building with an initial D.I. of 0.0025 under the influence of a 20-psf wind load, the increase with $\bar{F} = 1.0$ is 60%. If a value of 2.0 is used for $\bar{F}$, the increase will be an astonishing 120%.

A design method that includes the P–Δ affect may or may not be a direct design procedure. If member sizes are selected to control drift, a direct design may result in first order lateral load analysis stresses which are low.

When using the P–Δ method, frame members must be designed in accordance with the *AISC Specification* provisions. The *SSRC Guide* [19.27] states that effective length factor K may be taken as unity. However, it may be prudent to calculate the value for the critical buckling strength F_a using the effective length factor K as determined by the *AISC Specification*. Values for F_e may be determined with $K = 1.0$. The factor C_m may be taken as $C_m = (0.6 - 0.4M_1/M_2) \geq 0.4$. In most instances, a value for C_m of 0.85 will be conservative. However, if wind loads are low and the potential for single curvature bending in the column is high, a higher value may be required.

A simple design procedure for a P–Δ analysis follows:

1. Determine minimum member sizes for the gravity and gravity plus lateral load conditions including drift control considerations.
2. Determine frame drift Δ.
3. If deflection considerations clearly control, go to Step 9.

Table 19.16 Tabulated Values for $A1$ for Eq. (19.8.1)

$Y1$	Initial Frame D.I. 0.0020	0.0025	0.0030
300	2.50	4.00	10.0
275	2.22	3.20	5.71
250	2.00	2.67	4.00
225	1.82	2.29	3.08
200	1.67	2.00	2.50
175	1.54	1.78	2.11
150	1.43	1.60	1.82
125	1.33	1.45	1.60
100	1.25	1.33	1.43
75	1.18	1.23	1.29
50	1.11	1.14	1.18
25	1.05	1.07	1.08

Note:

$$Y1 = \frac{\text{(building depth)(building weight, pcf)}}{\text{average wind pressure, psf}}$$

$A1$ = amplification factor due to P–Δ forces

4. Compute the amplification factor or from Table 19.16 select the value for $A1$ using the initial deflection and compute P–Δ forces for a deflection equal to $\Delta + \Delta(A1 - 1.)(0.9)$.
5. Select member sizes for strength using story shears $H = V + V'\overline{F}$ as determined in Steps 4 or 7.
6. Determine revised frame drift using story shears not including P–Δ effects. If new amplified deflection is less than or equal to that as determined by Step 4, the design is complete.
7. Compute P–Δ forces as in Step 4 for the drift computed in Step 6.
8. Go to Step 5.
9. Select member sizes to control drift to a specified limit.
10. Determine frame drift.
11. Check member stresses for story shears $H = V + V'$. If initial member sizes are adequate, the design is complete.
12. Increase member sizes to satisfy strength requirements.

In the above procedure, if the increase in member size significantly increases the frame stiffness one may wish to continue an iteration process to select the optimum member size as follows:

13. Determine frame drift.
14. Check member sizes for story shears $H = V + V'$.
15. Reduce member sizes to the strength level required.
16. Go to Step 10.

19.8.4 *K*-Factor Designed Unbraced Frames

At the present time, the *AISC Specification* includes the concept of the effective length factor K for the design of unbraced frames. Other rational methods of analysis are acceptable; however, no other method is described in the *AISC Specification* or *Commentary,* where a brief explanation is given. A complete discussion of the use of the K-factor method is contained in the *SSRC Guide* [19.27].

The need for magnification of girder moments as well as column moments was illustrated in Fig. 19.47. The *SSRC Guide* notes that the *K*-factor method of design ignores the need for girder moment magnification. The AISC carefully couches the requirement in Section 1.8.1 which states

> *Design consideration should be given to significant load effects resulting from the deflected shape of the structure or of individual elements of the lateral load resisting system, including the effects on beams, columns, bracing, connections and shear walls.*

Many studies have indicated that building frames designed to a drift limit using *K* factors are satisfactory when analyzed for ultimate strength. However, there is no explanation as to why girder moment magnification is not necessary.

As a practical matter, in most moment-resisting frames, girder sizes are determined on the basis of required stiffness for deflection control and often columns need be sized for strength considerations only. Thus, the *K*-factor method would appear to be valid only when girder sizes have been increased to control working (service) load drift and thereby possess adequate reserve capacity to resist secondary moments due to deflection. It may be prudent for a designer to give consideration to girder moment magnification in situations where secondary effects may be significant.

The *Commentary* contains a method of determining column effective length factors based upon the ratio of the sum of column to girder stiffnesses at the ends of the columns. An accompanying nomograph is used to select *K* factors. Eight idealized assumptions are made which may seldom be obtainable in a real building. Two of the assumptions that may be difficult to justify include the following:

1. All joints are rigid. This item infers that stabilizing forces for columns not participating in the rigid frame in question must be supported by some other means.
2. The stiffness parameter $L\sqrt{PE/I}$ must be equal for all columns. This likely will not be true in most designs.

Regarding Item 1, ordinarily it is impossible to include all building columns in any rigid frame providing frame stability. Regarding Item 2, the stiffness parameter may be nearly equal only if column depths and widths are equal.

Assuming diaphragm action of the floor slab with symmetrical framing and loading and thus no floor rotation, the *SSRC Guide* [19.27] suggests that columns not participating in frame stability may be included in an analysis in the following manner. All columns participating in frame action may be designed using an amplification factor equal to $1/[1 - (\Sigma f_a/\Sigma F'_{ex})]$ in the interaction equation. In the above $\Sigma F'_{ex}$ is equal to the sum of the critical (Euler) stresses for only those columns participating in the frames providing stability. Columns not participating in frame stability may be designed for the case of sidesway prevented or $K = 1.0$. As the quantity $(\Sigma f_a/\Sigma F'_e)$ approaches unity, the amplification factor approaches infinity. Thus, for a frame with a small percentage of the columns participating in the stability frames, this method is inappropriate.

The *SSRC Guide* [19.27] also suggests that the situation of a moment-resisting frame providing lateral stability for columns other than those in the frame may be handled in still another manner. Figure 19.49 is an idealistic representation of a building frame which depends upon moment-resisting frames located in the exterior bents to provide lateral load resistance. In addition, the frames must also provide frame stability forces for columns not participating in the moment-resisting frames. In addition to the external lateral loads, P–Δ loads equal to $(P\Delta/h)$ for each column are added to the external loads in the directions indicated by the arrows.

Other methods of computing *K* factors and/or making stability analyses for frames are available for use by the engineer. Zweig and Kahn [19.44] presented a method for the design of large industrial buildings. In order to provide stability for large areas, alternative columns are rotated 90° so that half of the columns are available to resist lateral loads and/or provide frame stability in the two orthogonal directions.

In 1971 Joseph Yura [19.41] introduced the concept of inelastic *K* factors. A considerable amount of discussion followed by Adams, Johnston, and Zweig with response by Yura [19.43][19.42]. The subsequent discussion by Disque [19.17] showed a direct design method utilizing inelastic *K* factors. The use of this method can result in a significant reduction in the value of *K*. The design procedure is simple:

1. Select a trial column size.
2. Calculate column axial stress $f_a = P/A$.
3. Compute the stiffness reduction factor $S_r = 0.60F_y/F'_e$ or f_a/F'_e.
4. Calculate inelastic $G = S_r G_{\text{elastic}}$.
5. Determine K from nomograph using $G_{\text{inelastic}}$.

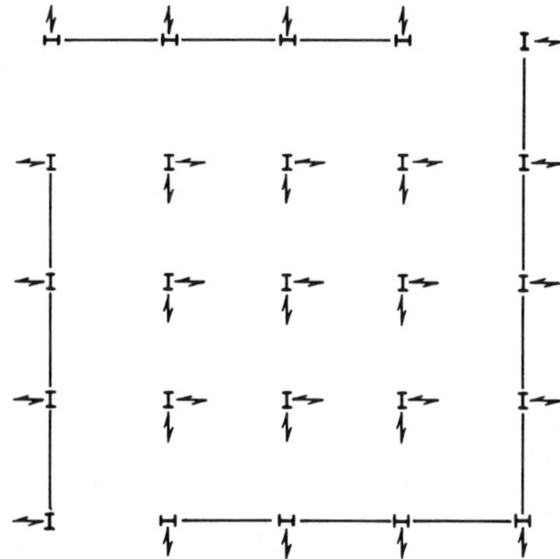

Fig. 19.49 Plan showing destabilizing force.

6. Calculate KL/r and determine F_a.
7. If $F_a > f_a$, column is adequate.

The use of a stiffness reduction factor equal to $0.60F_y/F'_e$ will result in a more conservative stiffness reduction factor. The value for F'_e in Step 3 above is that value associated for an F_a equal to f_a in Step 2. In the Disque paper, stiffness reduction factors equal to $0.60F_y/F'_e$ are tabulated for $F_y = 36$ ksi and $F_y = 50$ ksi. The use of this method may well reduce the G values, and the effect of F_a values may be relatively small.

19.8.5 Braced Frame Stability

The *AISC Specification* has no specific requirement for the consideration of secondary effects due to deflection for allowable stress design. For plastic design, the effect of deflection is required to be included in any analysis.

Most designers assume that secondary effects due to deflection are not serious for braced frames. From Tables 19.12 and 19.13 one can observe that bar stresses for bracing frames are quite low. Thus, the magnification of stress due to deflection will also be quite low. Even though the magnification of moment may be low it may be prudent to apply the magnification factor $\bar{F}(A1)$ [Eq. (19.8.1)] to the wind load forces when checking members for strength.

19.9 FATIGUE AND BRITTLE FRACTURE

With a few exceptions, fatigue and/or brittle fracture are subjects that normally need not be considered in the design of ordinary buildings. Exceptions may include, but are not limited to, crane runways, equipment supports, parts of structures supporting traffic decks, and perhaps for seismic loadings. However, there is a need for structural engineers to have a fundamental understanding of some of the basic factors involved in the phenomena. A good review of the state of art as applicable to buildings is contained in the *Tall Buildings Monograph* [19.16].

Brittle Fracture

Prior to World War II, ships were built utilizing riveted and bolted connections. Once the war effort was under way, in order to save time and money, the first large number of all-welded ships were built. A number of the vessels failed by brittle failure cracking which split the ships entirely into two separate pieces. As a result of these failures, an extensive research program was undertaken at the U.S. Naval Research Laboratory to explore the brittle fracture problem. In "Fracture Analysis Diagram Procedures for the Safe Engineering Design of Steel Structures" Pellini and Puzak [19.34] discuss the fundamental concepts of fracture safe design. A later paper [19.33] provides a more detailed explanation of the fracture mechanics involved. Rolfe and Barsom [19.37] provide a complete text on the subject of fracture mechanics.

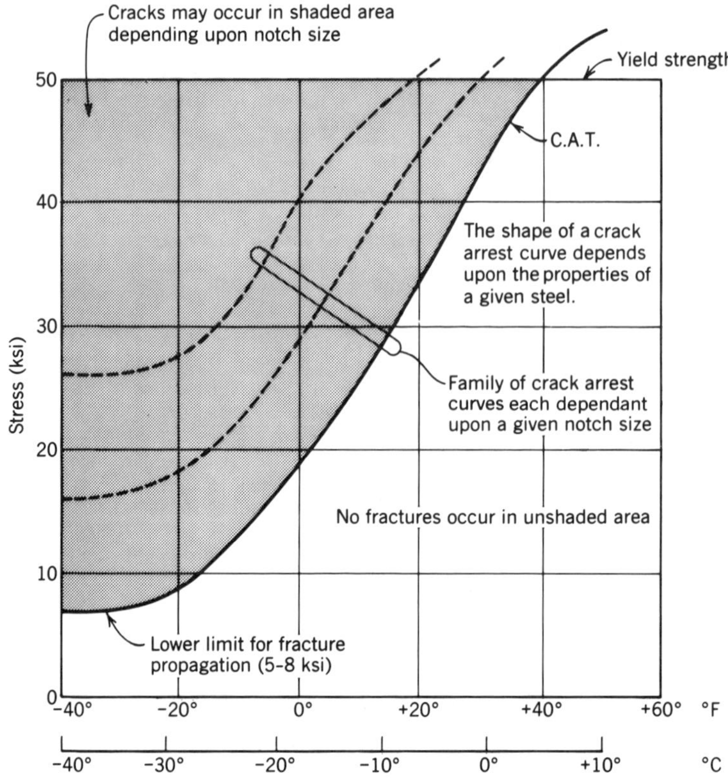

Fig. 19.50 Crack arrest temperature transition curve.

Brittle fracture cracks are spontaneous in that they propagate at velocities of several thousand feet per second. The fractures are square with little evidence of inelastic strain except for shear lips at the free surfaces of the crack. Three conditions must be satisfied for a brittle fracture to occur at conventional levels of stress. There must be a flaw such as a sharp notch or crack present, the stress level must be high enough to develop a small strain at the notch tip, and the ambient service temperature must be low enough for cleavage fractures of the metal crystals to occur at the crack tip. Figure 19.50 illustrates a typical crack arrest curve. The characteristics of the crack arrest curve depend upon the properties of a given steel. A lower limit stress for crack propagation ranges from 5 to 8 ksi (34 to 55 MPa) for various steels. As a flaw size decreases, the stress level at which a brittle fracture will occur increases. A family of crack arrest curves for varying flaw sizes may be constructed as shown in Figure 19.50. Thus, it can be shown that for a given level of stress, there is a corresponding notch size at which brittle fractures may occur. A flaw may be any discontinuity or stress raiser which induces abnormal strains at any point. A complete catastrophic failure of a liberty ship has been attributed to a small flaw caused by an arc strike. Other failures have been attributed to small fatigue-induced cracks which grew to a critical size.

The ability of a steel to resist brittle fracture may be measured by its notch toughness. Notch toughness may be defined as the ability of the material to absorb energy when loaded dynamically in the presence of a flaw. For most building structures, the required notch toughness is usually found in the normally specified steels.

For most building applications, the problem of brittle fracture is of no concern. That is not to say that under some circumstances the possibility of brittle fracture should not be given some consideration. For example, bridge crane runways are often operated in a manner that can produce impact loadings at relatively low ambient temperatures. There have been instances of brittle fracture of connecting elements which contained fabricated notches of sufficient magnitude for crack propagation. If a building designer feels that it is necessary or prudent to provide brittle fracture protection, one may use as a guide the provisions contained in *Standard Specification for Highway Bridges* [19.1].

Fatigue

Fatigue, as with brittle fracture, is ordinarily of little concern to the building designer. The *AISC Specification* [19.6] is specific in its design requirements concerning fatigue loading. The fatigue life of a member is primarily determined by the number of cycles of loading, the stress range to which a member is to be subjected, and the initial size of the flaw (notch size or crack) in the member. The AISC—Appendix B controls all three parameters in the following manner:

1. There are four loading conditions based on the expected number of loading cycles.
2. The flaw size determination is made by the selection of a stress category. Stress categories are selected from a table describing various types of structural elements. The elements are illustrated for easy determination of stress category.
3. A permissible stress range (or stress amplitude) is tabulated for each combination of the four loading conditions and six stress categories.

The least severe loading condition is for 20,000 to 100,000 cycles of loading. The 20,000 cycles is the equivalent of two applications a day for a period of 25 yr. Thus it can be seen that very few building components will be subject to fatigue-loading design constraints. Supports for elevators are likely to be subjected to a sufficient number of cycles to require fatigue-loading consideration. However, the stress amplitudes are likely to be low enough to meet the requirements for the worst loading condition. In any case, it will be prudent to eliminate slot welds, plug welds, intermittent fillet welds, or other high-stress category conditions for elevator beam supports.

Bolts and threaded parts used in bearing-type connections subjected to cyclic loading may be designed without regard to fatigue loading insofar as the strength of the fasteners themselves. However, the base material itself may have limitations placed on the permissible amplitude of stress. High-strength bolts loaded in direct tension require special attention in both the design and construction phases. It is important to note that the high level of permissible stresses in high-strength bolts loaded in tension is based on their being pretensioned as required by the *AISC Specification.* Improperly tightened high-strength bolts have led to catastrophic failures in the past. If high-strength bolts are loaded in direct tension, it may be prudent to specify that direct load indicators of some type be used so as to ensure proper pretensioning. For bolts loaded in tension including prying action, the basic allowable stresses may be used for connections subjected to less than 20,000 cycles of loading. If connections are subjected to more than 20,000 cycles of loading, the allowable stress may be reduced depending upon the number of cycles and the ratio of prying action stress to total tensile stress. Thus, it is prudent to design bolts that are cyclically loaded in a manner that will result in low prying action stresses. The use of bolts or threaded parts other than A325 or A490 bolts is not recommended for fatigue-loading situations.

19.10 LAMELLAR TEARING

Lamellar tearing (see also Section 6.12) is a subject not well understood by many structural engineers. The problem generally occurs in highly restrained welded joints with material thicknesses of ¾ to 1 in. (19 to 25 mm) or thicker. However, instances involving thinner material have been observed. It appears that high-strength low-alloy steels are more prone to lamellar tearing than are carbon steels. Material in this section is taken in large part from two references. The first, "An Evaluation of Factors Significant to Lamellar Tearing" by Kaufman, Pense, and Stout [19.28] relates to the selection of proper fabrication procedures, welding processes, and techniques. The paper presents results of numerous tests made using high-strength low-alloy steels. The results may not be directly applicable to A36 carbon steel. The second paper, "Commentary on Highly Restrained Welded Connections" [19.2] focuses on the design consideration involved in reduction of restraint and thus the risk of lamellar tearing. Kaufman et al. [19.28] note the problem is multifaceted. Material properties, welding processes, welding techniques, and the design of the welds and their profiles are all major factors in the problem.

Lamellar tears occur as a result of decohesion of nonmetallic inclusions in the base metal under the influence of through-thickness strains caused by shrinkage of the deposited weld metal. Cooling shrinkage of molten weld metal as deposited is inherent in the welding process and cannot be altered. The initial welding passes at the root of the weld (Fig. 19.51) solidify and then offer restraint to the shrinkage of the subsequent passes and thus strains are induced perpendicular to the thickness of plate. Weld shrinkage strains may be many times as great as yield stress level strains. Thus it is the lack of ductility of the steel in the direction of the thickness of the plate that causes the susceptibility to lamellar tears. Beyond the elastic limit, steel stressed in the rolling direction and transverse direction (see Fig. 19.52) demonstrates great ductility. Unfortunately, steel loaded in the through-thickness direction does not possess this quality. In the through-thickness direction, strains due to weld shrinkage may cause decohesion and lamellar tearing.

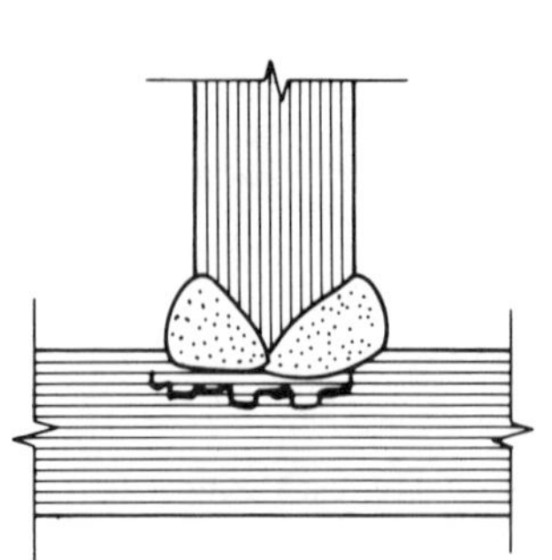

Fig. 19.51 Lamellar tear cross-section.

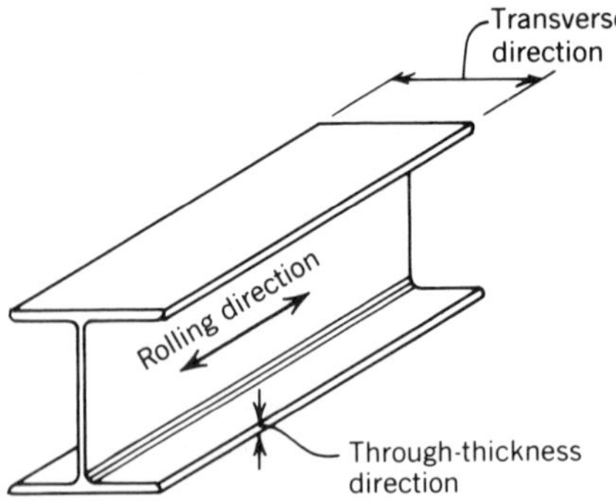

Transverse direction

Rolling direction

Through-thickness direction

Fig. 19.52 Rolling direction nomenclature.

19.10.1 Welding Procedure Considerations

Kaufman et al. [19.28] provide guidelines for reducing the risk of lamellar tearing, based on tests on ASTM A588 and A572, Grade 50 steel. It is generally considered that the high-strength low-alloy steels are more likely to have lamellar tearing problems than A36 carbon steel. Four welding processes were used in the tests [19.28], gas metal arc (GMAW), submerged arc (SAW), shielded metal arc (SMAW), and flux core metal arc (FCAW). Chapter 6 discusses these processes in detail.

The results indicate that generally the GMAW and SAW processes consistently produced more tear-resistant welds. It is assumed that the presence of hydrogen in the arc atmosphere of these processes resulted in their lower overall ratings.

Preheating of test specimens for the GMAW, SMAW, and FCAW processes increased tear resistance of each. However, for the SMAW and FCAW the increase was much greater, perhaps because of the higher potential for the presence of hydrogen in the arc atmosphere. Local preheating will be beneficial for applications where the cooling strain will not add to the final contraction strains in the joint. It was also demonstrated that lower strength electrodes led to improved tear resistance as the lower strength weld metal will take a larger share of the total shrinkage strain. Unfortunately, appropriate low strength electrodes are not available for all processes and types of steel.

The importance of careful quality control procedures insofar as assuring the absence of moisture in the arc atmosphere can greatly influence the tear resistance of a welded connection. The AWS *Structural Welding Code* [19.11] requires that certain electrodes be kept in heating ovens after the seal of their container is broken. If the electrodes are not protected, moisture is absorbed into the coating and provides a source for hydrogen in the arc atmosphere. This point is dramatically illustrated in the Kaufman et al. [19.28] tests. For the GMAW process with 1% H_2O introduced into the argon, critical welding restraint levels were reduced by about 40%. Tests were made using E7018 low-hydrogen electrodes in three conditions, oven dry (100° C), humidified, and with preheat. For comparison, tests were made with E7010 electrodes which have a high hydrogen potential. Using the oven dry test as a base, the critical restraint levels were 20% lower for the humidified electrode. On the other hand, preheating increased the restraint level by 70%. As could be expected, the use of the non-low-hydrogen E7010 electrodes degraded the critical restraint level to that of the humidified E7018 electrodes.

Kaufman et al. [19.28] as well as the AISC indicate that controlled peening as described in AWS D1.1 [19.11] may be beneficial in reducing the stresses associated with weld shrinkage.

Buttering (Fig. 19.53) is a technique that can be used to greatly reduce the risk of lamellar tears in through-thickness welds. Tests using a ¼-in. (6-mm) deep buttering layer deposited using E7018 low-hydrogen electrodes with the joint being completed using the GMAW process indicate that the presence of the buttering layer significantly increases the tear resistance.

In summary, to reduce the risk of lamellar tears, some or all of the following welding techniques may be considered:

1. Use GMAW or SAW processes rather than SMAW or FCAW processes.
2. Take appropriate steps to ensure that hydrogen levels in the arc atmosphere are kept to a minimum. For the SMAW this means that the electrodes must be kept as required by AWS D1.1, that is, after the container is opened, electrodes must be kept in an oven to ensure that hydrogen-producing moisture is not absorbed by the electrodes.

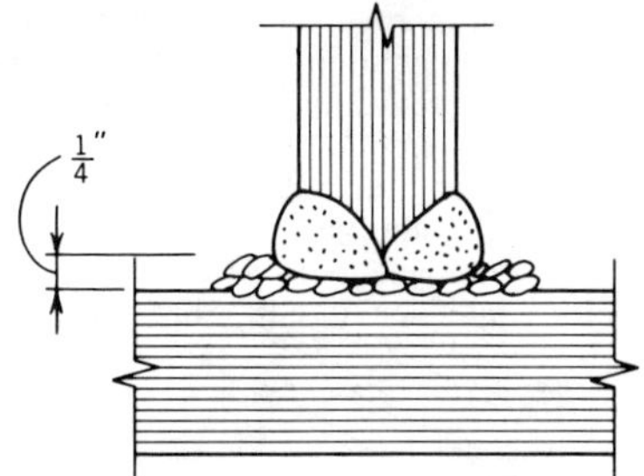

Fig. 19.53 Buttering layer.

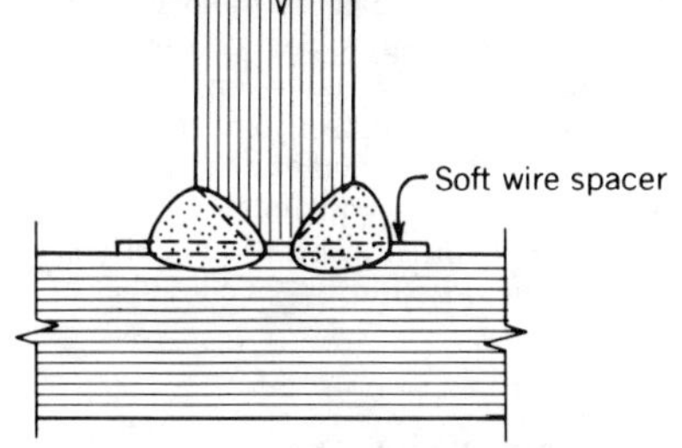

Fig. 19.54 Tee joint with wire spacers.

3. Where cooling strains will not add to the final contraction strains in the joint, preheating is effective in increasing lamellar tear resistance, especially for the SMAW and FCAW processes.
4. Buttering of the material subjected to through-thickness strains is effective in increasing tear resistance. The buttering should be made in two layers of not less than ¼ in. (6 mm) in thickness.

19.10.2 Design Procedure Considerations

The structural designer can help reduce the risk of lamellar tearing problems by using common sense in designing through-thickness welded connections that have the minimum amount of weld shrinkage and/or strain to which the through-thickness material is subjected. In some applications it is necessary to use joints that build in high restraint. In these cases, welding procedures should be considered which reduce the risk of tearing.

Close fit-up in a fabricating shop is normally considered to be a measure of excellence. However, in the case of through-thickness welds, this may not be the case. Care should be taken to try to develop details that permit weld shrinkage to occur with minimum locked-in stresses. For instance, the use of soft wire spacers to ensure that direct bearing of the plates themselves will not increase locked-in shrinkage stresses is one way of reducing the risk of lamellar tearing (Fig. 19.54).

Other steps that a structural designer may take to reduce the risk of lamellar tearing include:

1. Omit stiffener plates in column troughs at girder connections where stiffeners are not required.
2. Where stiffener plates are required, they should be sized as required by the *AISC Specification.*
3. Welds should be proportioned only for the applied loads.
4. Use fillet welds where possible if appropriate.

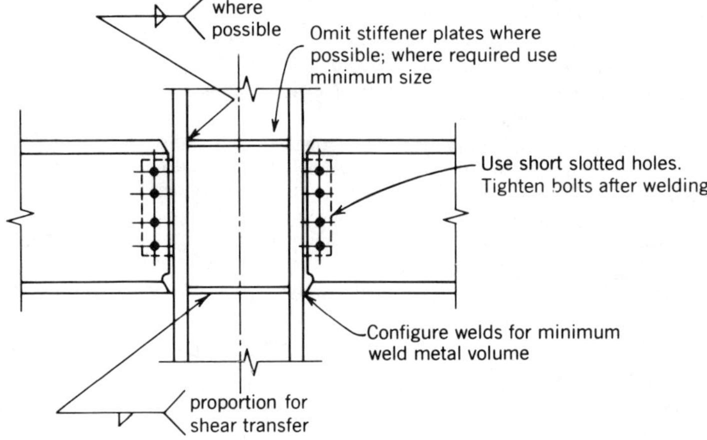

Fig. 19.55 Girders to column detail.

Often designers are prone to specify unneeded, oversized, and/or overwelded stiffener plates in welded girder to column connections. As stated earlier, an important factor in the reduction of the risk of tearing is the reduction of the total restraint in the joint produced by weld shrinkage. Experience has shown that joints involving heavy plates with large volumes of weld metal are those that are most likely to experience lamellar tearing problems. Consideration of the factors indicated in Fig. 19.55 should be included in the design. To that end, when designing girder connections, one should proportion stiffeners in accordance with the minimum requirements in AISC-1.15.5. The *AISC Commentary* illustrates a number of joint details which are intended to reduce high restraint.

SELECTED REFERENCES

19.1 AASHTO. *Standard Specifications for Highway Bridges*, 13th ed. Washington, D.C.: The American Association of State Highway and Transportation Officials, 1983.

19.2 AISC. "Commentary on Highly Restrained Welded Connections," *Engineering Journal*, AISC, **10**, 3(Third Quarter), 1973, 61–73.

19.3 AISC. "Predesigned Bolted Framing Angle Connections," *Engineering Journal*, AISC, **19**, 1(First Quarter), 1982, 1–11.

19.4 AISC. *Manual of Steel Construction*, 8th ed. Chicago: American Institute of Steel Construction, 1980.

19.5 AISC. *Proposed Load and Resistance Factor Design Specification for Structural Steel Buildings* (September 1, 1983). Chicago: American Institute of Steel Construction, 1983.

19.6 AISC. *Specification for the Design, Fabrication and Erection of Structural Steel for Buildings* (November 1, 1978). Chicago: American Institute of Steel Construction, 1978.

19.7 AISI. *Plastic Design of Braced Multi-Story Steel Frames*. New York: American Iron and Steel Institute, 1968.

19.8 ANSI. *American National Standard Minimum Design Loads for Buildings and Other Structures* (ANSI-A58.1). New York: American National Standards Institute, 1982.

19.9 ASCE. "Designing Wind Bracing for Skyscrapers," *Civil Engineering*, ASCE, **1**, May 1931, p. 700.

19.10 Subcommittee 31, Committee on Steel, Structural Division. Final Report, "Wind Bracing in Steel Buildings," *Transactions*, ASCE, **105**(1940), 1713–1739.

19.11 AWS. *Structural Welding Code* (D1.1). Miami, Florida: American Welding Society, 1985.

19.12 *Cost Per Foot of Structural Shapes*. Bethlehem, PA: Bethlehem Steel Corporation, November 12, 1983.

19.13 Reidar Bjorhovde and T. J. Zimmerman. "Some Aspects of Stub Girder Design," *Engineering Journal*, AISC, **17**, 3(Third Quarter), 1980, 54–69.

19.14 BOCA. *BOCA Basic Building Code/1978*, 7th ed. Homewood, Illinois: Building Officials and Code Administrators International, Inc., 1977.

19.15 CSA. *Steel Structures for Buildings* (proposed Appendix G, Std. S16-69), Rexdale, Ontario, Canada: Canadian Standards Association, 1969.

19.16 Council on Tall Buildings and Urban Habitat. *Structural Design of Tall Steel Buildings*. New York: American Society of Civil Engineers, 1979.

19.17 Robert O. Disque. "Inelastic *K*-factor for Column Design," *Engineering Journal*, AISC, **10**, 2(2nd Quarter), 1973, 33–35.

19.18 George C. Driscoll and Lynn S. Beedle. "Suggestions for Avoiding Beam-To-Column Web Connection Failure," *Engineering Journal*, AISC, **19**, 1(1st Quarter), 1982, 16–19.

19.19 Bruce Ellingwood, Theodore V. Galambos, James G. MacGregor, and C. Allin Cornell. *Development of a Probability Based Load Criterion for American National Standard A58* (NBS Special Publication 577). Washington, D.C.: U.S. Department of Commerce, National Bureau of Standards, June 1980.

19.20 John W. Fisher and John H. A. Struik. *Guide to Design Criteria for Bolted and Riveted Joints*. New York: Wiley, 1974.

19.21 T. V. Galambos. "History of Steel Beam Design," *Engineering Journal*, AISC, **14**, 4(Fourth Quarter), 1977, 141–147.

19.22 Theodore V. Galambos. *Structural Members and Frames*. Englewood Cliffs, NJ: Prentice-Hall, 1968.

19.23 T. V. Galambos. "Vibration of Steel Joist—Concrete Floors," Technical Digest No. 5. Myrtle Beach, South Carolina: Steel Joist Institute. (Suite A, 1205 48th Avenue North, Myrtle Beach, SC 29577)

19.24 Edwin H. Gaylord and Charles N. Gaylord. *Design of Steel Structures*. New York: McGraw-Hill, 1957.

19.25 J. E. Heinzerling. "Structural Design of Steel Joist Roofs to Resist Ponding Loads," Technical Digest No. 3. Myrtle Beach, South Carolina: Steel Joist Institute, May 1971.

19.26 J. S. B. Iffland and C. Birnstiel. "Stability of Tall Structures," presented at AISC National Engineering Conference, Dallas, Texas, 1981 (not published).

19.27 Bruce G. Johnston, ed. *Structural Stability Research Council, Guide to Stability Design Criteria for Metal Structures,* 3rd ed. New York: Wiley, 1976.

19.28 E. J. Kaufman, A. W. Pense, and R. D. Stout. "An Evaluation of Factors Significant to Lamellar Tearing," *Welding Journal,* March 1981, 43-s–49-s.

19.29 J. W. McNabb and B. B. Muvdi. "Drift Reduction Factors for Belted High-Rise Structures," *Engineering Journal,* AISC, **12,** 3(Third Quarter), 1975, 88–91.

19.30 P. B. Mehta. "A Graphical Method for Design of Cantilevered and Suspended Beams," *Engineering Journal,* AISC, **18,** 2(Second Quarter), 1981, 54–57.

19.31 Thomas M. Murray. "Acceptability Criterion for Occupant Induced Floor Vibrations," *Engineering Journal,* AISC, **18,** 2(Second Quarter), 1981, 62–69.

19.32 Thomas M. Murray. "Design to Prevent Floor Vibrations," *Engineering Journal,* AISC, **12,** 3(Third Quarter), 1975, 82–87.

19.33 W. S. Pellini. "Evolution of Principles for Fracture Safe Design of Steel Structures," Report 6957, U.S. Naval Research Laboratory, September 1977.

19.34 W. S. Pellini and P. P. Puzak. "Fracture Analysis Diagram Procedures for the Safe Engineering Design of Steel Structures," Report 5920, U.S. Naval Research Laboratory, March 1963.

19.35 J. W. Reed, R. J. Hansen, and E. H. Van Marke. "Human Response to Tall Buildings Wind-Induced Motion," ASCE-IABSE International Conference Preprints: Reports Vol. II-17, ASCE, August 1972, 59.

19.36 Research Council on Riveted and Bolted Structural Joints of the Engineering Foundation (Research Council on Structural Connections). *Specification for Structural Joints Using ASTM A325 and A490 Bolts*. 1980.

19.37 S. T. Rolfe and J. M. Barsom. *Fracture and Fatigue Control in Structures—Applications of Fracture Mechanics*. Englewood Cliffs, NJ: Prentice-Hall, 1977.

19.38 B. S. Taranath. "Optimum Belt Truss Locations for High-Rise Structures," *Engineering Journal,* AISC, **11,** 1(First Quarter), 1974, 18–21.

19.39 John F. Wiss and Richard A. Parmelee. "Human Perception of Transient Vibrations," *Journal of the Structural Division,* ASCE, **100,** April 1974 (ST4), 773–787.

19.40 Ned W. Young and Robert O. Disque. "Design Aids for Single Plate Framing Connections," *Engineering Journal,* AISC, **18,** 4(Fourth Quarter), 1981, 129–148.

19.41 Joseph A. Yura. "The Effective Length of Columns in Unbraced Frames," *Engineering Journal,* AISC, **8,** 2(April), 1971, 37–42.

19.42 Joseph A. Yura. Discussion of "The Effective Length of Columns in Unbraced Frames," *Engineering Journal,* AISC, **9,** 4(October), 1972, 167–169.

19.43 Joseph A. Yura, Bruce G. Johnston, Peter F. Adams, and Alfred Zweig. Discussion of "The Effective Length of Columns in Unbraced Frames," *Engineering Journal,* AISC, **9,** 1(January), 40–48.

19.44 Alfred Zweig and Albert Kahn. "Buckling Analysis of One-Story Frames," *Journal of the Structural Division,* ASCE, **94,** September 1968 (ST9), 2107–2134.

19.45 Charles G. Salmon and John E. Johnson. *Steel Structures: Design and Behavior,* 2nd ed. New York: Harper & Row, 1980.

19.46 Jack C. McCormac. *Structural Steel Design,* 3rd ed. New York: Harper & Row, 1981.

19.47 Bogdan O. Kuzmanovic and Nicholas Willems. *Steel Design for Structural Engineers,* 2nd ed. Englewood Cliffs, NJ: Prentice-Hall, 1983.

APPENDIX

Plate 19.1 Floor Beam Vibration Analysis

Loads

Live load	10
$3\frac{1}{4}$-in. lightweight concrete + deck	45
Beam	5
Ceiling	3
Fireproofing + misc.	2
	65 psf

$$E_c = (115)^{1.5}(33)\sqrt{3500} = 2.41 \times 10^6 \text{ psi}$$

$$n = E_s/E_c = 29 \times 10^6/(2.41 \times 10^6) = 12$$

Case I W21×44, L = 30 ft, Spacing = 10 ft.

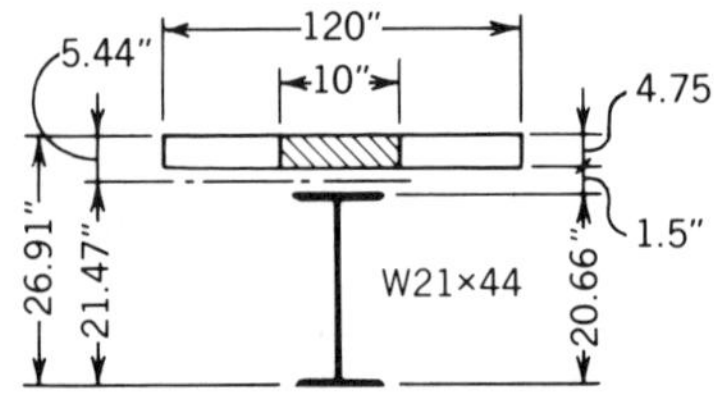

Find y_b:

Concrete, 10(4.75) =	47.50(24.54)	= 1165
Beam	13.0 (20.66/2)	= 134
	60.5 sq in.	1299 in.^{3}

$$y_b = 1299/60.5 = 21.47 \text{ in.}$$

Moment of inertia:

$$I_t = 10(4.75)^3/12 + 47.5(24.54 - 21.47)^2 + 843 + 13(21.47 - 20.66/2)^2$$

$$= 2993 \text{ in.}^4$$

$$W = 65(10)(30)/1000 = 19.5 \text{ kips}$$

$$\text{Frequency } f = 1.57\sqrt{\frac{gEI}{WL^3}} = 1.57\sqrt{\frac{386(29 \times 10^3)(2993)}{19.5(30 \times 12)^3}} = 9.53$$

Dynamic load factor:

$$0.05f = 0.05(9.53) = 0.48$$

From Fig. 19.20, $(\text{DLF})_{\max} = 1.2$

Plate 19.2 Floor Beam Vibration Analysis

Find single beam amplitude:

$$A_{0t} = (\text{DLF})_{\max}[L^3/80EI_t)]$$

$$A_{0t} = 1.2\{(30 \times 12)^3/[80(29 \times 10^3)2993]\} = 0.0081$$

Find N_{eff}, the number of effective beams:

$$s/d_e = 120/4.75 = 25.3$$

where s = beam spacing
d_e = equivalent slab thickness

$$L^4/I_t = (30 \times 12)^4/2993 = 5.61 \times 10^6$$

From Fig. 19.21, find $N_{\text{eff}} = 1.80$

Find maximum amplitude of the system:

$$A_0 = A_{0t}/N_{\text{eff}} = 0.0081/1.8 = 0.0045$$

Find "R" rating [from Eq. (19.2.4)] using $D = 4.0$

$$R = 5.08[9.53(0.0045)/(4.0)^{0.217}]^{0.265} = 2.04$$

Find $D_{\min}$:

$$D_{\min} = 35A_0f + 2.5 = 35(0.0045)(9.53) + 2.5 = 4.0$$

Plate 19.3 Floor Beam Vibration Analysis

Case II

$$L = 30 \text{ ft, Spacing} = 10 \text{ ft}$$

$$\text{W18}\times 35,\ I_{tr} = 1967 \text{ in.}^4$$

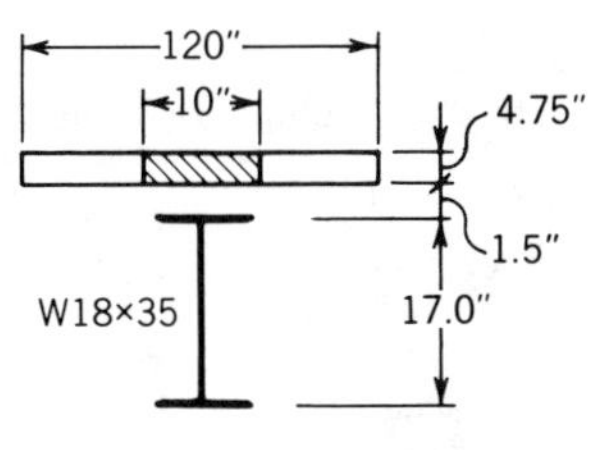

$$f = 1.57\sqrt{\frac{389(29 \times 10^3)1967}{19.5(30 \times 12)^3}} = 7.75$$

$$0.05f = 0.05(0.388); \qquad (\text{DLF})_{\max} = 1.05$$

$$s/d_e = 120/4.75 = 25.26 \qquad L^4/I_{tr} = 8.54 \times 10^6$$

$$N_{\text{eff}} = 1.8$$

$$A_0 = 1.05\left[\frac{(30 \times 12)^3}{80(29 \times 10^3)1967}\right]\frac{1}{1.8} = 0.0060$$

$$R = 5.08[7.75(0.006)/(4)^{0.217}]^{0.265} = 2.08$$

$$D_{\min} = 35(0.006)7.75 + 2.5 = 4.13$$

Case III

$$L = 30 \text{ ft, Spacing} = 10 \text{ ft}$$

$$\text{W16}\times 26,\ I_{tr} = 1294 \text{ in.}^4$$

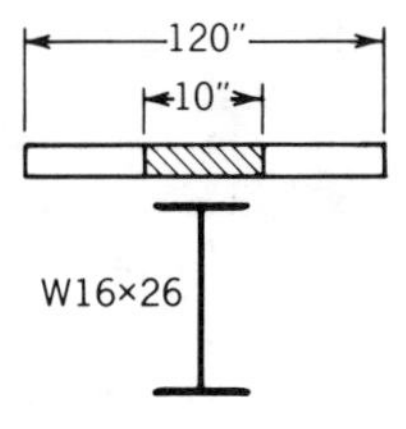

$$f = 6.29$$

$$0.05f = 0.31; \qquad (\text{DLF})_{\max} = 0.92$$

$$s/d_e = 25.26 \qquad L^4/I_{tr} = 12.98 \times 10^6$$

$$N_{\text{eff}} = 2.15$$

$$A_0 = 0.0067$$

$$R = 2.03$$

$$D_{\min} = 3.98$$

Plate 19.4 Laterally Unbraced Beam Design

Example No. 1

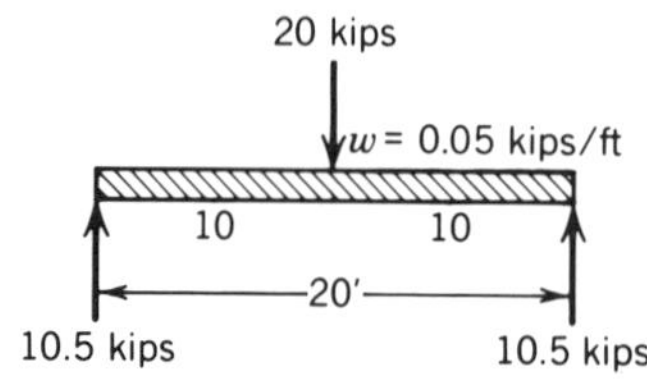

A36 beam braced at midspan:

$$M_{max} = 0.05(20)^2/8 + 20(20)/4 = 102.5 \text{ ft-kips}$$

$$F_{br} \text{ reqd} = M_{max}L/C_b = 102.5(10)/1.75 = 587$$

From Table 19.6, find

W16×36	M_{max} = 103 ft-kips	OK
W18×35	M_{max} = 105 ft-kips	OK

A36 beam unbraced except at ends:

$$\text{Factor } F_{br} \text{ required} = 102.5(10)/1.0 = 1025$$

From Table 19.6, find

W16×40	OK
W18×46	OK

Plate 19.5 Composite Design

Loads	
Live load	100
$3\frac{1}{4}$-in. lightweight concrete + 3-in. deck	46
Miscellaneous	5
Ceiling	4
Steel	5
	160 psf

$$R_{max} = 23\left(1 + \frac{60}{100}\right) = 37\%$$

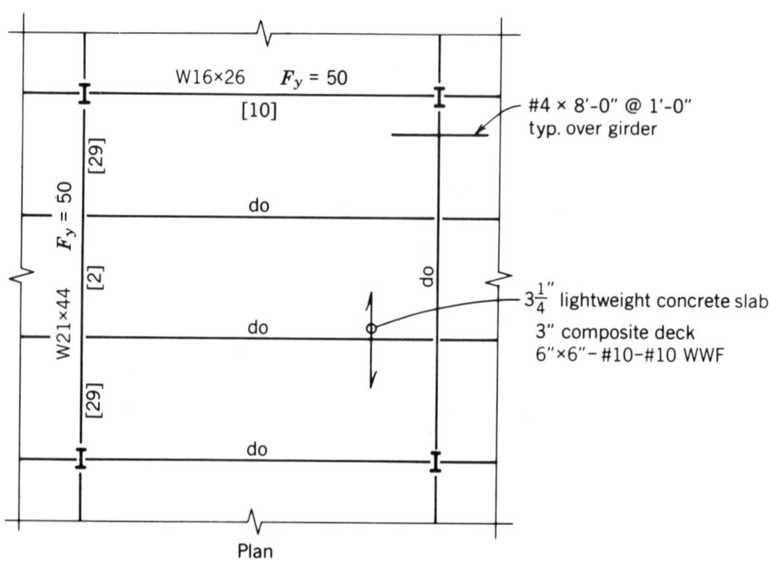

Typical filler beam, $L = 30$ ft, Spacing = 10 ft

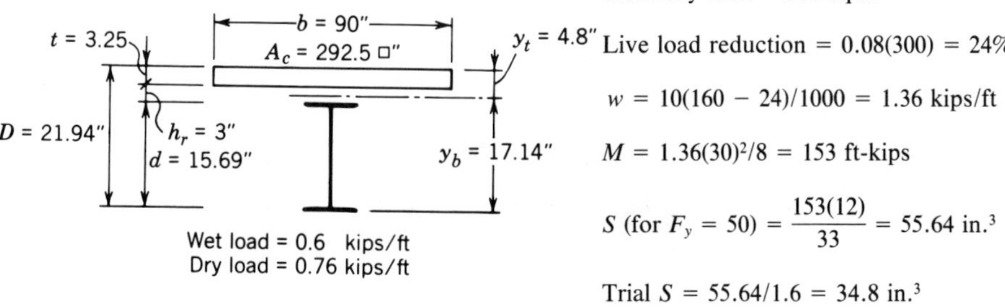

Tributary area = 300 sq ft

Live load reduction = 0.08(300) = 24%

$w = 10(160 - 24)/1000 = 1.36$ kips/ft

$M = 1.36(30)^2/8 = 153$ ft-kips

$$S \text{ (for } F_y = 50) = \frac{153(12)}{33} = 55.64 \text{ in.}^3$$

Trial $S = 55.64/1.6 = 34.8$ in.^{3}

Try W16×26, $F_y = 50$ ksi, $A = 7.68$ sq in., $I_x = 301$ in.^{4}, $S_x = 38.4$ in.^{3}, $d = 15.69$ in.

Use structural lightweight concrete (112 pcf) with $f'_c = 3000$ psi

$$E_c = (112)^{1.5}(33)\sqrt{3000}/1000 = 2140 \text{ ksi}$$

$$n = E_s/E_c = 29 \times 10^3/2140 = 13.5, \text{ say } 13$$

Determine b: Use the smallest of

$$30/4 = 7.5 \text{ ft (90 in.)} \qquad \textit{Controls}$$

$$2(8)6.25 + 5.5 = 105.5 \text{ in.}$$

$$\text{Spacing} = 120 \text{ in.}$$

$$y_b = \frac{d}{2} + \frac{d + 2h_r + t}{2}\left(\frac{\phi}{\phi + 1}\right)$$

where $\phi = \dfrac{bt}{nA_s} = \dfrac{A_c}{nA_s}$

Plate 19.6 Composite Design

Filler beam

$$\phi = \frac{292.5}{13(7.68)} = 2.93$$

$$y_b = \frac{15.69}{2} + \frac{15.69 + 2(3) + 3.25}{2}\left(\frac{2.93}{3.93}\right) = 7.84 + 9.30 = 17.14 \text{ in.}$$

$$I_{tr} = I_s + A_s\left(y_b - \frac{d}{2}\right)^2 + \frac{A_c}{n}\left(D - \frac{t}{2} - y_b\right)^2 + \frac{A_c}{n}\left(\frac{t^2}{12}\right)^\dagger$$

$$I_{tr} = 301 + 7.68\left(17.14 - \frac{15.69}{2}\right)^2 + \frac{292.5}{13}\left(21.94 - \frac{3.25}{2} - 17.14\right)^2 + \frac{292.5(3.25)^2}{13(12)}$$

$$= 301 + 664 + 227 + 20 = 1212 \text{ in.}^4$$

$$S_b = S_{tr} = 1212/17.14 = 70.71 \text{ in.}^3$$

$$S_t = 1212/4.80 = 252.5 \text{ in.}^3$$

† One may substitute the value of $2n$ for n for long term deflections.

Plate 19.6 (*Continued*)

$$f_s = M(12)/S_b = 153(12)/70.71 = 25.97 \text{ ksi}$$

$$f_c = M(12)/S_t n = 153(12)/(252.5 \times 13) = 0.56 \text{ ksi}$$

Determine V_h: Use the smaller of

$$0.85 f'_c A_c/2 \quad \text{or} \quad A_s F_y/2$$

$$0.85(3.5)(292.5)/2 = 435 \text{ kips} \quad \text{or } 7.68(50)/2 = 192 \text{ kips}$$

$$V_h = 192 \text{ kips}$$

Find V'_h for partial composite design:

$$V'_h = V_h \left(\frac{S_{\text{reqd}} - S_s}{S_{tr} - S_s}\right)^2$$

$$V'_h = 192 \left(\frac{55.64 - 38.4}{70.71 - 38.4}\right)^2 = 54.66 \text{ kips}$$

For ¾-in. diam. headed stud, $f'_c = 3.5$ ksi, $w_c = 112$ pcf

$$q = 11.5(0.84) = 9.66 \text{ kips}$$

$$\text{Number required} = 54.66/9.66 = 5.66, \text{ or a total of } 12$$

$$S_{\text{eff}} = S_s + \sqrt{\frac{V'_h}{V_h}}\left(S_{tr} - S_s\right) = 38.4 + \sqrt{\frac{6(9.66)}{192}}\,(70.71 - 38.4) = 56.2 \text{ in.}^3$$

Plate 19.7 Composite Design

Deflections (Partial Composite Section)

$$I_{\text{eff}} = I_s + \sqrt{\frac{V'_h}{V_h}}\,(I_{tr} - I_s)$$

$$I_{\text{eff}} = 301 + \sqrt{\frac{6(9.66)}{192}}\,(1212 - 301) = 802 \text{ in.}^4$$

For unshored construction,

$$S_{tr\,\max} = \left(1.35 + 0.35\,\frac{M_L}{M_D}\right) S_s \qquad (M_D \text{ is wet load})$$

$$= [1.35 + 0.35(0.85/0.51)]38.4 = 74.24 > 56.2 \qquad \text{OK}$$

$$\Delta_{\text{wet load}} = \left(\frac{0.51}{1.36}\right)\frac{153(30)^2}{161(301)} = 1.07 \text{ in.}$$

$$\Delta_{\text{dry load}} = \left(\frac{0.85}{1.36}\right)\frac{153(30)^2}{161(802)} = 0.67 \text{ in.}$$

For shored construction,

$$\Delta_{\text{DL}} = \left(\frac{0.60}{1.36}\right)\frac{153(30)^2}{161(802)} = 0.47 \text{ in.}$$

$$\Delta_{LL} = \left(\frac{0.76}{1.36}\right) \frac{153(30)^2}{161(802)} = 0.60 \text{ in.}$$

Check stud efficiency for steel deck:

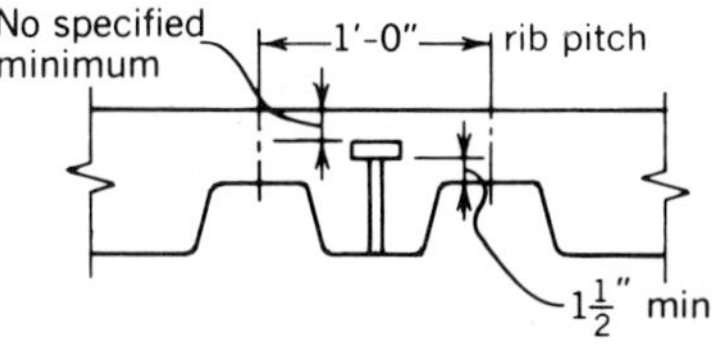

For 12 studs total, one stud per rib is the maximum

$$H_{s\,\min} = \frac{\sqrt{N_r}h_r^2}{0.85w_r} + h_r = \frac{\sqrt{1}(3)^2}{0.85(6)} + 3 = 4.76 \text{ in.}$$

Use 5-in. headed studs.

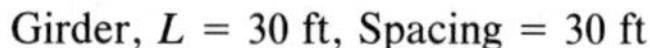

Plate 19.8 Composite Design

Girder, L = 30 ft, Spacing = 30 ft

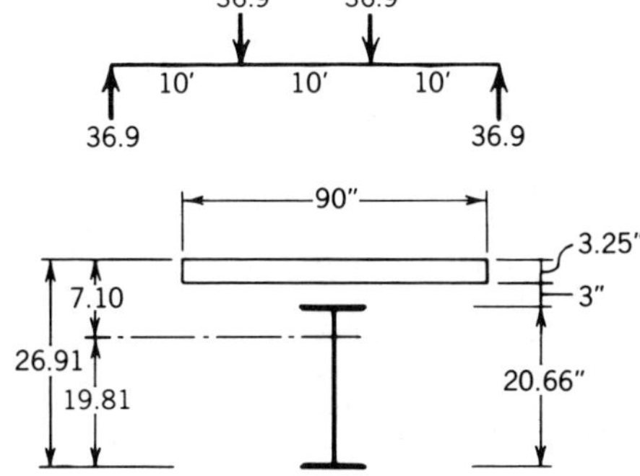

Area = 20(30) = 600 sq ft

R = 600(0.08) > 37%

P = 10(30)(160 − 37)/1000 = 36.9 kips

M = 36.9(10) = 369 ft-kips

$$\text{Required } S = \frac{369(12)}{33} = 134.2 \text{ in.}^3$$

$$S_{\text{trial}} = \frac{134.2}{1.6} = 84 \text{ in.}^3$$

Try W21×44, A = 13.0 sq in., I_x = 843 in.^{4}, S_x = 81.6 in.^{3}, d = 20.66 in., n = 13

Find y_b by taking moments about bottom of steel beam:

Beam	13.0 (20.66/2)	= 134.29
Slab, 3.25(90)/13 =	22.5 (26.91 − 3.25/2)	= 568.91
	A = 35.5 sq in.	Ay = 703.20 in.^{3}

$$y_b = \frac{703.20}{35.5} = 19.81 \text{ in.}$$

Find I_{tr}:

$$\text{Beam} = 843 + 13(19.81 - 20.66/2)^2 = 2011$$

$$\text{Slab} = \frac{90(3.25)^3}{13(12)} + 22.5\left(26.91 - \frac{3.25}{2} - 19.81\right)^2 = 694$$

$$I_{tr} = 2705 \text{ in.}^4$$

S_b = 2705/19.81 = 136.5 in.^{3}

S_t = 2705/7.10 = 381.0 in.^{3}

f_s = 369(12)/136.5 = 32.44 ksi OK

f_c = 369(12)/[(381.0)13] = 0.89 ksi

Plate 19.9 Composite Construction

For unshored construction,

$$S_{tr\,max} = [1.35 + 0.35(0.63/0.51)]81.6 = 145.4 \text{ in.}^3 \qquad \text{OK}$$

Deflections (Fully Composite Section)

Wet load deflections:

$$\Delta \text{ at load} = \left(\frac{51}{123}\right)\frac{369(30)^2}{181(843)} = 0.90 \text{ in.}$$

$$\Delta \text{ at centerline} = 0.90(181/158) = 1.03 \text{ in.}$$

Dry load deflections:

$$\Delta \text{ at load} = \left(\frac{63}{123}\right)\frac{369(30)^2}{181(2708)} = 0.35 \text{ in.}$$

$$\Delta \text{ at centerline} = 0.35(181/158) = 0.40 \text{ in.}$$

For shored construction,

$$\Delta_{\text{DL}} \text{ at load} = \left(\frac{60}{123}\right)\frac{369(30)^2}{181(2708)} = 0.33 \text{ in.}$$

$$\Delta_{\text{DL}} \text{ at centerline} = 0.33(181/158) = 0.38 \text{ in.}$$

$$\Delta_{\text{LL}} \text{ at load} = 0.33(63/60) = 0.35 \text{ in.}$$

$$\Delta_{\text{LL}} \text{ at centerline} = 0.38(63/60) = 0.40 \text{ in.}$$

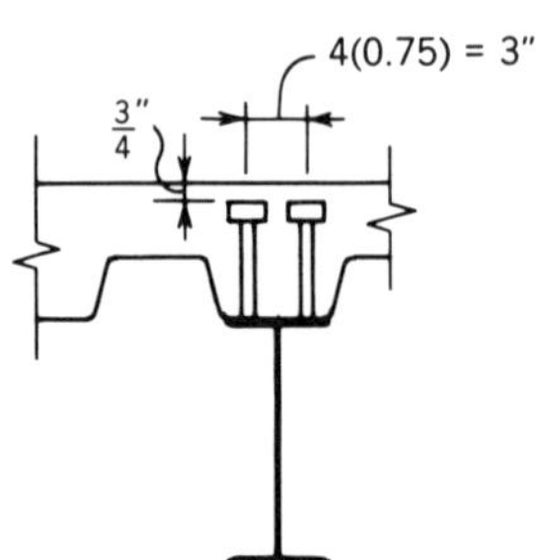

$$Vn = (50/2)13 = 325 \text{ kips}; \qquad q = 9.66 \text{ kips}$$

$$N = \frac{325}{9.66} = 34$$

Total studs = 34(2) + 2 = 70
at minimum spacing of $4\frac{1}{2}$ in.

Use 6 rows of 2 studs + 19 single studs.

$$H_{s\,min} = \frac{h_r^2}{0.6w_r} + h_r = \frac{(3)^2}{0.6(6)} + 3 = 5.5 \text{ in.}$$

Plate 19.10 Column Load Tabulation

Loads

Roof		Floor	
Live load	30 psf	Live load	100 psf
Dead load	30 psf	Dead load	55 psf
Total	60 psf	Total	155 psf

$$\text{Maximum reduction, } R_{max} = 23\left(1 + \frac{55}{100}\right) = 36\%$$

Typical interior column, A = 900 sq ft

Floor	Load P	% Reduction	Additional Dead Load	Total P	ΣP	Initial Size	Final Size
Roof	54	0	2	56	56	W14×61 $F_y = 50$	
12	139.5	32.4	2	107.1	163		
11					270		
10					377	W14×90 $F_y = 50$	
9					484		
8					591		
7					699	W14×132 $F_y = 50$	
6					806		
5					913		
4					1020	W14×61 $F_y = 50$	
3					1127		
2	139.5	32.4	2	107.1	1234		

Plate 19.11 Typical Beam-Column Design

Try W14×61, $F_y = 50$ ksi

$$P = 270 \text{ kips} \qquad M_x = 16 \text{ ft-kips} \qquad M_y = 1.9 \text{ ft-kips} \qquad L = 12 \text{ ft}$$

From the *AISC Manual* property tables for W sections:

$$A = 17.9 \text{ sq in.}, \; S_x = 92.2 \text{ in.}^3, \; S_y = 21.5 \text{ in.}^3$$

$$r_x = 5.98 \text{ in.} \qquad r_y = 2.45 \text{ in.}$$

From the *AISC Manual* Column Load Tables,

$$L_c = 9.0 \text{ ft}, \; L_u = 15.5 \text{ ft} \qquad \text{Thus,} \quad F_{bx} = 0.60(50) = 30 \text{ ksi}$$

$$F_{by} = 0.75(50) = 37.5 \text{ ksi}$$

$$KL/r_x = 1(144)/5.98 = 24 \qquad KL/r_y = 1(144)/2.45 = 59$$

From Table 3.5 and Table 9 of *AISC Specification*—Appendix A,

$$F_a = 22.89 \text{ ksi} \qquad F'_{ex} = 259.26 \text{ ksi} \qquad F'_{ey} = 42.9 \text{ ksi}$$

Check AISC Formulas (1.6-1a) and (1.6-1b),

$$f_a = \frac{P}{A} = \frac{270}{17.9} = 15.08 \text{ ksi}$$

$$f_{bx} = \frac{M}{S_x} = \frac{16(12)}{92.2} = 2.08 \text{ ksi} \qquad f_{by} = \frac{M}{S_y} = \frac{1.9(12)}{21.5} = 1.06 \text{ ksi}$$

$$\frac{f_a}{F_a} + \frac{C_m f_{bx}}{(1 - f_a/F'_{ex})F_{bx}} + \frac{C_m f_{by}}{(1 - f_a/F'_{ey})F_{by}} \leq 1.0 \qquad (1.6\text{-}1a)$$

Plate 19.11 (*Continued*)

$$\frac{15.08}{22.89} + \frac{1.0(2.08)}{(1 - 15.08/259)30} + \frac{1.0(1.06)}{(1 - 15.08/42.9)37.5} = 0.78 < 1.0 \quad \text{OK}$$

$$\frac{f_a}{0.60F_y} + \frac{f_{bx}}{F_{bx}} + \frac{f_{by}}{F_{by}} \leq 1.0 \qquad (1.6\text{-}1b)$$

$$\frac{15.08}{30} + \frac{2.08}{30} + \frac{1.06}{37.5} = 0.60 < 1.0 \quad \text{OK}$$

Plate 19.12 Wind Connection Top Plate Design

Refer to Fig. 19.34 for sketch.
Use A36 detail material and A325 bolts in a friction-type connection.

$$\text{Wind moment} = 75 \text{ ft-kips}$$

$$\text{Minimum flange force} = \frac{M_{\text{wind}}}{0.95d_b} = \frac{75(12)}{0.95(21)} = 45.11 \text{ kips}$$

$$\text{Minimum net area required} = \frac{P_{f1}}{0.50F_u(1.3)} = \frac{45.11}{0.50(58)(1.3)} = 1.20 \text{ sq in.}$$

$$\text{Minimum gross area required} = \frac{P_{f1}}{0.60F_y(1.3)} = \frac{45.11}{22(1.3)} = 1.58 \text{ sq in.}$$

$$\text{Minimum plate width} = \text{bolt gage} + 2(\text{edge distance})$$

$$= 3.5 + 2(1.5) = 6.5 \text{ in.}$$

Minimum plate thickness, $1.2/[6.5 - 2(0.875)] = 0.253$ in.

$1.58/6.5 = 0.243$ in.

Use 5⁄16 × 6½ plate. $A_{\text{gross}} = 2.037$ sq in.

$$\text{Minimum width at reduced section} = 1.58/0.373 = 4.24 \text{ in.}$$

Use 4½-in. width at reduced section.

$$\text{Maximum bolt shear force} = 4.5(0.373)(22)(1.3) = 48 \text{ kips}$$

For ¾-in. diam. A325 bolts in a friction-type connection, and with "clean mill scale" surfaces of contact, *AISC Manual* Tables I-D, I-E, and I-F give:

Shear capacity, standard holes	=	7.7 kips
Bearing capacity, ⅜-in plate	=	16.3 kips
Edge distance, ⅜-in. plate, 0.375(43.5)	=	16.4 kips

$$\text{Minimum number of bolts} = \frac{48}{1.3(7.7)} = 4.8 \text{ bolts*}$$

Use 6–¾-in diam. A325 bolts in standard holes ("clean mill scale").

* Note: Number of bolts proportioned using plate capacity.

Plate 19.13 K-Braced Frame Design Example

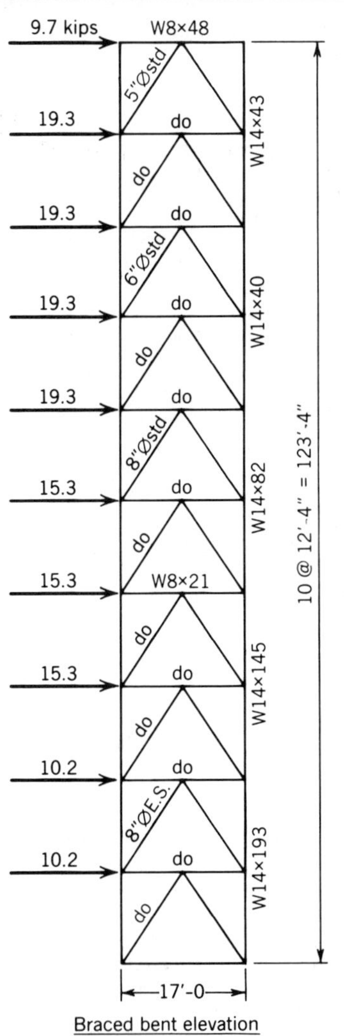

Braced bent elevation

Loads:

Roof

LL	30 psf	
DL	35 psf	
Total	65 psf	

Wall loads
25 psf of wall

Floor

LL	100 psf
LL RED(max)	37 psf
Reduced LL	63 psf
DL	67 psf
Total	130 psf

Wind load
BOCA Code: 80 mph fastest mile

Building Description:

127 ft–0 in. square

10 stories @ 12 ft–4 in = 123 ft–4 in. height

Two symmetrical placed K-braced bents with V's up

Aspect ratio = 10(12.33)/17 = 7.25

Select initial member sizes on basis of stresses given in Table 19.12 for aspect ratio = 7.5

Plate 19.14 K-Braced Frame Design Example, Member Force Determination (V's Up)

Story	(1) Elevation (ft)	(2) Q_F (psf)	(3) H (kips)	(4) ΣH (kips)	(5) P_1 (kips)	(6) P_{C1} (kips)	(7) P_G (kips)	(8) P_B (kips)
10	123.33	19	9.7	9.7	0	0	4.9	8.6
9	111.00	19	19.3	29.0	7.1	7.1	14.5	25.5
8	98.67	19	19.3	48.3	21.0	28.1	24.2	42.6
7	86.33	19	19.3	67.6	35.1	63.2	33.8	59.5
6	74.00	19	19.3	86.9	49.0	112.2	43.5	76.6
5	61.67	15	15.3	102.2	63.1	175.3	51.1	89.9
4	49.33	15	15.3	117.5	74.1	249.4	58.8	103.5
3	37.00	15	15.3	132.8	85.3	334.7	66.4	116.9
2	24.67	10	10.2	143.0	96.3	431.0	71.5	125.8
1	12.33	10	10.2	153.2	103.7	534.7	76.6*	134.6

* Girder force for V's down = 5.1 kips.

(1) Floor elevation above ground
(2) Q_{Fi} = effective velocity pressure
(3) H_i = story wind shear = $Q_F C_p$ (tributary area)/1000
$= Q_F(0.8 + 0.5)(127/2)(12.33)/1000 = 1.02Q_F$
(4) ΣH_i = sum of story wind shears above
(5) P_{1i} = column vertical load increment = $(\Sigma H_i + 1)h/L = 0.725(\Sigma H_i + 1)$
(6) P_{C1i} = column vertical load = ΣP_1 above
(7) P_{Gi} = girder axial load = $\Sigma H/2$
(8) P_{Bi} = brace axial load = $P_G\sqrt{(0.5L)^2 + h^2}/(0.5L) = 1.76P_G$

Plate 19.15 K-Braced Design Example, Initial Member Selection

Story	Member	Load	Stress*	Area	Section	Area
10	Column	21.0	3.5	6.0	W14×40	11.8
9	Girder	14.5	9.2	1.58	W8×18	5.26
	Brace	25.5	5.9	4.32	5-in. diam. stnd†	4.30
8	Column	45.7	4.6	9.93	W14×40	11.8
7	Girder	29.0	12.1	2.40	W8×18	5.26
	Brace	51.1	9.2	5.55	6-in. diam. stnd	5.58
6	Column	143.8	6.0	23.97	W14×82	24.1
5	Girder	47.3	14.5	3.26	W8×18	5.26
	Brace	83.3	11.0	7.57	8-in. diam. stnd	8.4
4	Column	292.1	7.2	40.57	W14×145	42.7
3	Girder	62.6	14.5	4.32	W8×21	6.16
	Brace	110.2	14.5	7.60	8-in. diam. stnd	8.4
2	Column	482.9	8.6	56.15	W14×193	55.8
1	Girder	71.5	17.3	4.13	W8×21	6.16
	Brace	130.2	14.5	8.98	8-in. diam. E.S.‡	12.8

* Initial stresses from Table 19.12. Use aspect ratio = 7.5.
† standard weight pipe
‡ extra strong pipe

Combined stress may control for girder:

Assume girder $w = 12.5(1.6) = 2.0$ kips/ft.

$$M = 2(8.5)^2/8 = 18.1(0.9) = 16.3 \text{ ft-kips}$$

$L_x = 8.5$ ft; $L_y = 0$; for W8's, $KLr_x = 8.5(12)/3.4 = 30$

$$F_a = 19.9 \text{ ksi}$$

Assume $F_{bx} = 24$ ksi;

Using $f_a/F_a + f_{bx}/F_{bx} = 0.95(1.3)$; thus $f_a = (1.24 - f_{bx}/F_{bx})F_a$

For W8×18: $f_{bx} = 16.3(12)/15.2 = 12.9$ ksi

$$f_a = (1.24 - 12.9/24)19.9 = 14.0; \quad P_{all} = 14.0(5.26) = 73.8 \text{ kips}$$

For W8×21: $f_{bx} = 16.3(12)/18.2 = 10.8$ ksi

$$f_a = (1.24 - 10.8/24)19.9 = 15.8; \quad P_{all} = 15.8(6.16) = 97.3 \text{ kips}$$

Note: Gravity load and gravity plus wind load strength analysis omitted in this example.

Plate 19.16 K-Braced Frame Design Example, Story Rotations (V's Up)

Story	(1) A_C	(2) P_{C1}	(3) α_a ($\times 10^5$)	(4) R_C ($\times 10^5$)	(5) A_G	(6) P_G	(7) R_G ($\times 10^5$)	(8) A_B	(9) P_B	(10) R_B ($\times 10^5$)
10	11.8	0	0	260	5.26	4.9	2	4.3	8.6	11
9	11.8	7.1	3.0	257	5.26	14.5	7	4.3	25.5	34
8	11.8	28.1	11.9	245	5.26	24.2	11	5.58	42.6	56
7	11.8	63.2	26.8	218	5.26	33.8	15	5.58	59.5	79
6	24.1	112.2	23.3	195	5.26	43.5	20	8.4	76.6	67
5	24.1	175.3	36.4	159	5.26	51.1	22	8.4	89.9	79
4	42.7	294.4	34.5	124	6.16	58.8	23	8.4	103.5	91
3	42.7	334.7	39.2	85	6.16	66.4	26	8.4	116.9	103
2	56.8	431.0	37.9	47	6.16	71.5	28	12.8	125.8	73
1	56.8	534.7	47.1	0	6.16	76.6*	30*	12.8	134.6	78

* For V's down $P_G = 5$; therefore $R_G = 12$.

(1) A_{Ci} = column area from initial member size selection
(2) P_{Ci} = column force from member force determination

(3) α_{ai} = column story rotation $= \dfrac{P_{Ci}}{A_{Ci}}\left(\dfrac{2h}{EL}\right) = \dfrac{P_{Ci}}{A_{Ci}}\left[\dfrac{2(12.33)}{29{,}000(17)}\right] = \dfrac{P_{Ci}}{A_{Ci}}(5.0 \times 10^{-5})$

(4) R_{Ci} = total story rotation due to column strain $= \Sigma\alpha_{ai}$
(5) A_{Gi} = girder area
(6) P_{Gi} = girder force

(7) R_{Gi} = girder story rotation $= \dfrac{P_G}{A_G}\left(\dfrac{0.5L}{Eh}\right) = \dfrac{P_G}{A_G}(2.37 \times 10^{-5})$

(8) A_B = brace area
(9) P_{Bi} = brace load

(10) R_{Bi} = brace story rotation $= \dfrac{P_B}{A_B}\left(\dfrac{2L_B^2}{EhL}\right) = \dfrac{P_B}{A_B}(7.38 \times 10^{-5})$

Plate 19.17 K-Braced Frame Design Example, Story Rotations, Deflections, and Deflection Indexes (D.I.) (V's Up)

Story	(1) R_C (×10⁵)	(2) R_G (×10⁵)	(3) R_B (×10⁵)	(4) ΣR (×10⁵)	(5) Story Δ	(6) $\Sigma\Delta$	(7) D.I.
10	260	2	11	273	0.0337	0.302	0.0024
9	257	7	34	298	0.0368	0.268	0.0024
8	245	11	56	312	0.0385	0.231	0.0023
7	218	15	79	312	0.0385	0.193	0.0022
6	195	20	67	282	0.0348	0.154	0.0021
5	159	23	79	261	0.0322	0.120	0.0019
4	124	23	91	238	0.0294	0.087	0.0018
3	85	26	103	214	0.0264	0.058	0.0016
2	47	28	73	148	0.0183	0.032	0.0013
1	0	30	78	108	0.0133	0.013	0.0011

(1) R_C = story rotation due to column strain
(2) R_G = story rotation due to girder strain
(3) R_B = story rotation due to brace strain
(4) ΣR = sum of story rotations
(5) Story Δ = story deflection = $\Sigma R(h)$
(6) $\Sigma\Delta$ = sum of story deflections below
(7) D.I. = deflection index = $\Sigma\Delta/\Sigma h$

Plate 19.18 K-Braced Frame Design Example, Story Rotations, Deflections, and Deflection Indexes (D.I.) (V's Down)

Story	A_C	P_{C2}	α_a (×10⁵)	R_C (×10⁵)	R_G (×10⁵)	R_B (×10⁵)	ΣR (×10⁵)	Story Δ	$\Sigma\Delta$	D.I.
10	11.8	7.1	3.0	371	2	11	384	0.047	0.394	0.0032
9	11.8	28.1	11.9	368	7	34	409	0.050	0.347	0.0031
8	11.8	63.2	26.8	356	11	56	423	0.052	0.297	0.0030
7	11.8	112.2	47.5	329	15	79	423	0.052	0.245	0.0028
6	24.1	175.3	36.4	281	20	67	368	0.045	0.193	0.0026
5	24.1	294.4	61.1	245	23	79	347	0.043	0.148	0.0024
4	42.7	334.7	39.2	184	23	91	298	0.037	0.105	0.0021
3	42.7	431.0	50.5	145	26	103	274	0.034	0.068	0.0018
2	56.8	534.7	47.1	94	28	73	195	0.024	0.034	0.0014
1	56.8	534.7	47.1	0	2	78	80	0.010	0.010	0.0008

CHAPTER 20

COLD-FORMED STEEL CONSTRUCTION

GEORGE WINTER

Professor of Engineering, Emeritus*
Cornell University
Ithaca, New York

* Deceased 1982.

20.1 INTRODUCTION TO COLD-FORMED STEEL STRUCTURES

20.1.1 Members, Types, and Applications

Cold-formed steel structural members are made from carbon or low-alloy sheet or strip steel by cold-rolling in sets of special rolling machines. To a lesser degree, they are also made in press-brakes, mostly for producing relatively small quantities of identical shapes. Thicknesses range most frequently from 0.018 to 0.13 in. (0.46 to 3.3 mm), although sheets are available from 0.012 to 0.24 in. (0.30 to 6.1 mm). There is also limited use of cold-forming of plate and bar steel from about ¼ to 1.0 in. (6.3 to 25.4 mm) for producing structural shapes in configurations different from those of standard hot-rolled sections.

The chief practical difference between cold-formed and hot-rolled steel shapes is the fact that the former can be mass produced in a great variety of specialized shapes since the cold-forming process is easily adapted to produce most any reasonable configuration. In contrast, hot-rolled sections are generally available only in a relatively small number of standardized shapes. Correspondingly, cold-formed steel members are used for a considerable variety of functions. They can be divided into two categories: framing members and surface members.

Figure 20.1 shows a number of current cold-formed framing members. Their primary use is in two types of buildings and structures: The first application is in structures in which hot-rolled steel main framing is combined with cold-formed secondary structural framing shapes and surface members. Examples include medium- to wide-span single-story buildings with single- or multispan portal-type transverse main framing and all other steel members cold-formed such as roof purlins, girts, studs, chords of open-web joists, etc. Such framing is used for industrial mill buildings, large storage buildings, covered sports and recreation centers, large supermarkets, and retail malls. Multistory office and apartment buildings are often fabricated from hot-rolled main framing with all other steel shapes cold-formed. The second class of structures includes those made entirely of cold-formed steel members. Examples include one- to three- and even four-story office, institutional, or apartment buildings, occasional single-family dwellings, small to medium size single-story industrial and commercial buildings such as industrial plants, shopping centers, single- and two-story schools, agricultural structures for storage and other uses, and industrial and commercial storage racks and rack buildings, from minor ones to large installations up to 150 ft (46 m) high.

A number of cold-formed surface members are shown in Fig. 20.2. In contrast to framing members, surface members are load-carrying shapes which also constitute useful surfaces, chiefly floors, roofs, and walls. These can be bearing walls or, more often, curtain walls in which wall panels resist the wind loads. Correspondingly, while framing members are optimized to produce maximum strength and stiffness per unit weight of steel, surface members must satisfy a

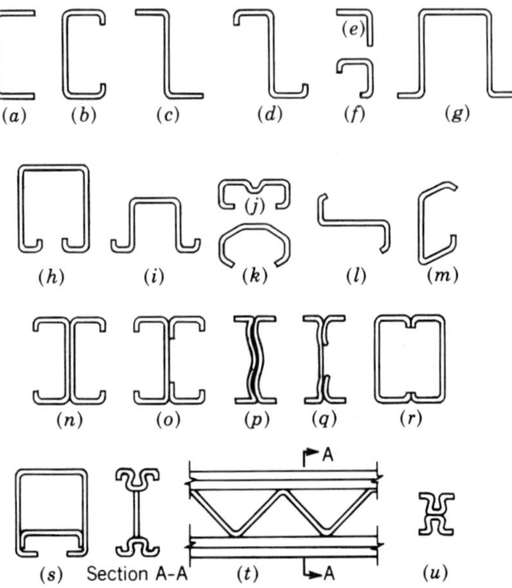

Fig. 20.1 Cold-formed steel framing members. Courtesy of IABSE.

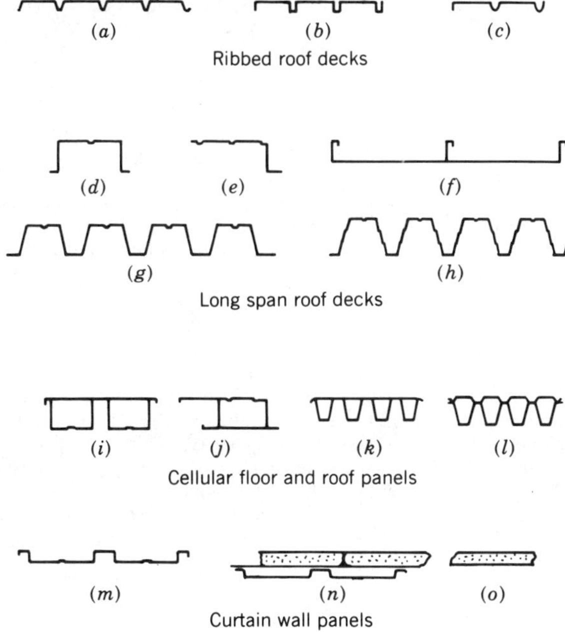

Fig. 20.2 Cold-formed steel surface members. Courtesy of IABSE.

number of additional functional requirements. Among these are coverage, that is, the amount of surface area provided by one unit of deck or panel; specific requirements to permit floor electrification and/or conduction of hot and cold air (see Section 4.12); provision for joint action with other materials, such as floor concrete and wall or roof insulation; and aesthetic requirements in the frequent case where the steel wall panels constitute the permanent, visible exterior of the building, either as curtain walls or, in smaller structures, as load-carrying walls. Use of metal roof decking is also covered in Section 12.4.2.

There are several important and frequent special uses of panels and decks. Roofs and floors when made of appropriately interconnected panels, act as *shear diaphragms,* transmitting in-plane forces such as wind resultants to properly braced vertical planes, such as shear walls. When appropriately shaped to ensure reliable bond, floor and roof steel panels act not only during construction as working surfaces and formwork for the slab concrete, but also as permanent positive slab reinforcement, resulting in *steel-deck-reinforced composite slabs.*

Because of the sizable number of requirements and roles of decks and panels, the designer of an individual structure usually will not design custom panels and decks and then have them fabricated in press-brakes or, for large quantities, in cold-rolls. Instead, the designer will select appropriate shapes and sizes from the catalogues of the many firms which manufacture a great variety of floor, roof, and wall panels. In addition to providing details of section shape, size, and thickness, these catalogues contain elaborate load tables which are calculated according to the latest edition of the *AISI Specification for the Design of Cold-Formed Steel Structural Members,** Ref 20.2, often supported by load tests. These catalogues often also show typical framing and connection details and other technical information.

Figures 20.3, 20.4, and 20.5 show several typical buildings where framing and surfaces consist in part of cold-formed components.

20.1.2 Design Basis

In the United States, the design of cold-formed members and assemblies is governed by the *AISI Specification* [20.2]. This specification is distinct from the *AISC Specification for the Design, Fabrication and Erection of Structural Steel for Buildings* [20.1] which governs the design of hot-

* Hereafter referred to as *AISI Specification.*

Fig. 20.3 Cold-formed steel roof and wall panels.

rolled steel members and assemblies. A separate design specification for cold-formed construction is needed for several reasons:

1. Hot-rolled construction is generally confined to a limited number of rolled shapes or members fabricated from plate. In contrast, inspection of Figs. 20.1 and 20.2 shows the great variety of differently shaped cold-formed members of an almost unlimited multitude of forms, most of which could not be analyzed by the methods of the *AISC Specification*.

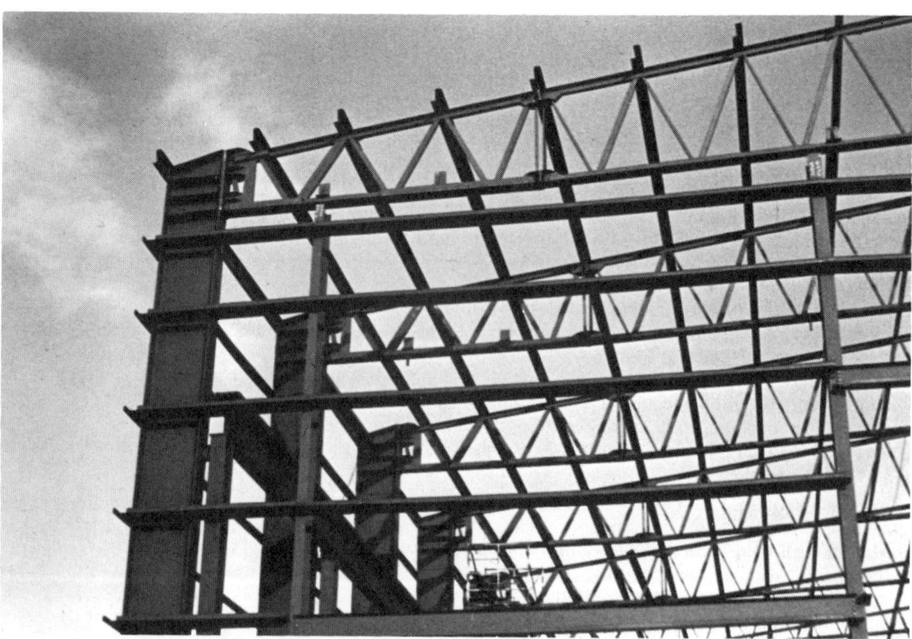

Fig. 20.4 Cold-formed steel purlins and girts.

Fig. 20.5 Cold-formed steel floor decking for composite slab.

2. Most of the cold-formed members consist of flat elements (flanges, webs, etc.) with large width-to-thickness ratios; this calls for consideration of plate buckling and of post-buckling strength.
3. Most of these members are singly symmetrical or even asymmetrical, which calls for appropriate consideration of torsional stresses, deformations, and buckling.
4. The possibility of both local plate buckling and overall member buckling requires consideration of possible interaction of local and overall buckling.
5. The presence of cold-formed rounded corners joining flat elements of equal thickness is quite different from corner fillets joining unequal thickness elements of hot-rolled shapes.
6. The distribution of residual stresses and strain hardening caused by cold-forming differs from that of the residual stresses caused by cooling of hot-rolled shapes.
7. Apart from these matters, which refer to members, the types and ranges of connections in cold-formed construction are different from those in hot-rolled assemblies, chiefly because of the generally much smaller thicknesses of the elements of cold-formed members.

It is because of these features that the *AISI Specification* had to be developed for cold-formed construction, based on extensive experimental and analytical research. This Specification consists of 48 large-size pages. In the framework of this handbook it is possible, of course, to deal only with a fraction of the content, limited to the most important and basic features. The Specification is part of the *AISI Cold-Formed Steel Design Manual,* Ref. 20.3, likewise published by the American Iron and Steel Institute. The manual also contains a Commentary, Supplementary Information, Illustrative Examples, and Charts and Tables, all of which greatly clarify and simplify the work of the structural designer.

Designers who utilize load tables published by producers of cold-formed steel members and components should make sure that these tables are based on the latest edition of the *AISI Specification* and/or on load tests carried out in the manner prescribed in that Specification. Additional details on design and applications are given in Ref. 20.8.

20.1.3 Materials

Members of cold-formed construction are mostly produced from sheet and strip steels. However, the Specification applies not only to the usual, thin-walled members, but also to cold-formed shapes up to 1 in. thick. Such members with thicknesses exceeding those of the sheet and strip range can be rolled from structural steel plates or bars. However, this is the exception. Correspondingly, Table 20.1 gives designations and main properties of ASTM sheet and strip steels. For the infre-

Table 20.1 ASTM Standard Sheet and Strip Steels

ASTM Designation and Scope	Product	Grade	F_y ksi (min)	F_u ksi (min)	Percent Elongation in 2 in. (min)
A446-83					
This specification covers zinc-coated (galvanized) steel sheet of structural quality in coils and cut lengths. Sheet of this quality can be produced in six grades, A through F, with the base metal mechanical requirements shown. Structural quality galvanized sheet is produced with any of the types of coating in the latest revision of ASTM Specification A525.	Sheet	A	33	45	20
		B	37	52	18
		C	40	55	16
		D	50	65	12
		E	80	82	—
		F	50	70	12
A570-84a					
This specification covers hot-rolled carbon steel sheet and strip of structural quality in cut lengths or coils. This material is available up to maximum thickness of 0.2299 in. (5.8 mm) except as limited by Specification A568.	Sheet and strip	30	30	49	21 to 25
		33	33	52	18 to 23
		36	36	53	17 to 22
		40	40	55	15 to 21
		45	45	60	13 to 19
		50	50	65	11 to 17
A606-75 (1981)					
This specification covers high-strength, low-alloy, hot- and cold-rolled sheet and strip in cut lengths or coils, for structural purposes, where savings in weight or added durability are important. These steels have enhanced atmospheric corrosion resistance and are supplied in two types: Type 2 having corrosion resistance at least two times that of plain carbon steel and Type 4 having corrosion resistance at least four times that of plain carbon steel.	Sheet and strip	Hot-rolled —As rolled cut lengths	50	70	22
		Hot-rolled —As rolled coils	45	65	22
		Hot-rolled —Annealed or normalized	45	65	22
		Cold-rolled	45	65	22
A607-83					
This specification covers high-strength, low-alloy columbium and/or vanadium hot-rolled and cold-rolled steel sheet and strip in either cut lengths or coils, for structural purposes where greater strength and savings in weight are important. This material is available in six strength levels as shown. Atmospheric corrosion resistance of these steels without copper specified is equivalent to plain carbon steel. With copper specified, the atmospheric corrosion resistance is twice that of plain carbon steel.	Sheet and strip	45	45	60	Hot-rolled 25 Cold-rolled 22
		50	50	65	Hot-rolled 22 Cold-rolled 20
		55	55	70	Hot-rolled 20 Cold-rolled 18
		60	60	75	Hot-rolled 18 Cold-rolled 16
		65	65	80	Hot-rolled 16 Cold-rolled 15
		70	70	85	14
A611-84					
This specification covers cold-rolled carbon structural steel sheet, in cut lengths or coils. It includes five strength levels designated as Grades A through E with yield points from 25 to 80 ksi, as shown.	Sheet	A	25	42	26
		B	30	45	24
		C	33	48	22
		D	40	52	20
		E	80	82	—

Table 20.1 *(Continued)*

ASTM Designation and Scope	Product	Grade	F_y, ksi (min)	F_u, ksi (min)	Percent Elongation in 2 in. (min)
A715-81 This specification covers high-strength low-alloy, hot-rolled steel sheet and strip having improved formability when compared with steels covered by Specifications A606 and A607. The product is furnished in cut lengths or coils and is available in four strength levels, Grades 50, 60, 70, and 80 (corresponding to minimum yield point) and in seven types (according to chemical composition). Not all grades are available in all types. The steel is intended for applications where higher strength, savings in weight, improved formability, and weldability are important.	Sheet and strip	50	50	60	22 to 24
		60	60	70	20 to 22
		70	70	80	18 to 20
		80	80	90	16 to 18

Structural ASTM Specifications for Steel Shapes, Plates and Bars

A36-81a	Carbon steel	Yield point 36 ksi
A242-81	High-strength low-alloy steel	Yield points 50 and 46 ksi
A441-81	High-strength low-alloy corrosion resistant steel	Yield points 50 and 46 ksi
A572-82	High-strength low-alloy steels	Yield points 42 to 65 ksi
A588-82	High-strength low-alloy corrosion resistant steel	Yield point 50 ksi
A529-82	Carbon steel	Yield point 42 ksi

quently used plate and bar steels, only the pertinent ASTM designations and yield points are listed. The various ASTM Specifications should be consulted directly when more information on properties of these steels is needed.

20.2 MEMBER DESIGN

20.2.1 Basic Design Stresses

The basic design stresses in the *AISI Specification* apply to situations where attainment of the yield stress is regarded as limiting the strength of the member. This does not apply to cases of local or overall buckling in its various forms, stress concentrations in connections, etc. The basic design stress so defined is

$$F = 0.60F_y \tag{20.2.1}$$

where F_y is the minimum specified yield point of the particular steel. This basic design stress applies to the net section of tension members, and to tension and compression at the extreme fibers of flexural members. In all other cases allowable stresses are generally smaller than F because of the effects of buckling and other influences. Equation (20.2.1) thus defines the basic factor of safety as $1/0.60 = 1.67$.

Table 20.2 Basic Design Stresses, F, ksi

Yield Point, F_y	Basic Tension or Compression Stress, F
33	20
40	24
45	27
50	30
60	36
70	42
Other	$0.60\ F_y$

If yielding is caused by shear rather than normal stress, the basic design stress for shear is

$$F_v = (2/3)F = 0.40F_y \tag{20.2.2}$$

Again, this is the allowable shear stress, for instance, in the web of a beam, provided that other influences, such as local shear buckling of thin webs, do not require a reduction of the allowable shear stress below F_v.

In general, then, the basic design stresses depend on the yield point F_y of the steel before forming. However, the cold-forming operations involved in creating structural shapes from flat sheet or strip produce local strain hardening which increases the yield point above the virgin value F_y. The *AISI Specification* provides three methods for determining this increased yield point F_{ya}. When this has been done by any of these methods, the basic design stress may be based on the higher yield point produced by cold-work: that is, in Eq. (20.2.1), F_y may be replaced by the larger value F_{ya}. This holds only for Eq. (20.2.1) that is, for tension and compression stresses. The effect of cold-work on yielding in shear has not been sufficiently investigated and consequently Eq. (20.2.2) is valid regardless of possible effects of cold-work.

For easy reference, Table 20.2 gives the basic design stresses for the virgin yield points of the most frequently used ASTM steels.

Inspection of Table 20.1 shows that two ASTM Specifications include steels with yield points as high as 80 ksi. However, the ductilities of these steels are considerably smaller than those of lower, more customary strengths. The special conditions for their use are specifically spelled out in the *AISI Specification*.

20.2.2 Calculation of Section Properties

Calculation of section properties of the frequently quite complicated cold-formed shapes (see Figs. 20.1 and 20.2) can be fairly tedious. They may be simplified in various ways. For one thing, one can assume the areas of the individual constituent elements (flanges, webs, corners, etc.) to be concentrated in their midlines. Then one needs to calculate only the properties (length, moment of inertia, etc.) of these linear elements, multiply them by the sheet thickness, and then obtain by summation the properties of the entire cross-section. This is known as the "linear method" which is discussed in some detail in Part III of the *AISI Design Manual* [20.3]. Further, Fig. 20.6, taken from this Manual, gives the properties of the most frequent individual line elements of which such cross-sections are composed, thus further simplifying calculations.

In certain types of compression elements the actual area is replaced by a reduced effective area, which is then used to determine effective section properties. This is discussed below. One must then distinguish between properties of actual, full sections and of reduced, effective sections.

20.3 THIN COMPRESSION ELEMENTS

When flat compression elements, for example, compression flanges of beams, are sufficiently thick, failure will be initiated when the yield stress is reached in the flange. However, when the thickness is a small fraction of the width w, local buckling will occur before yielding is reached. Hence, the flat-width to thickness ratio w/t is the main structural characteristic of thin compression elements in regard to local buckling. Large w/t ratios result in local buckling at stresses below the yield point. However, initiation of such local plate buckling in general does not mean collapse of the element and of the member of which it is a part.

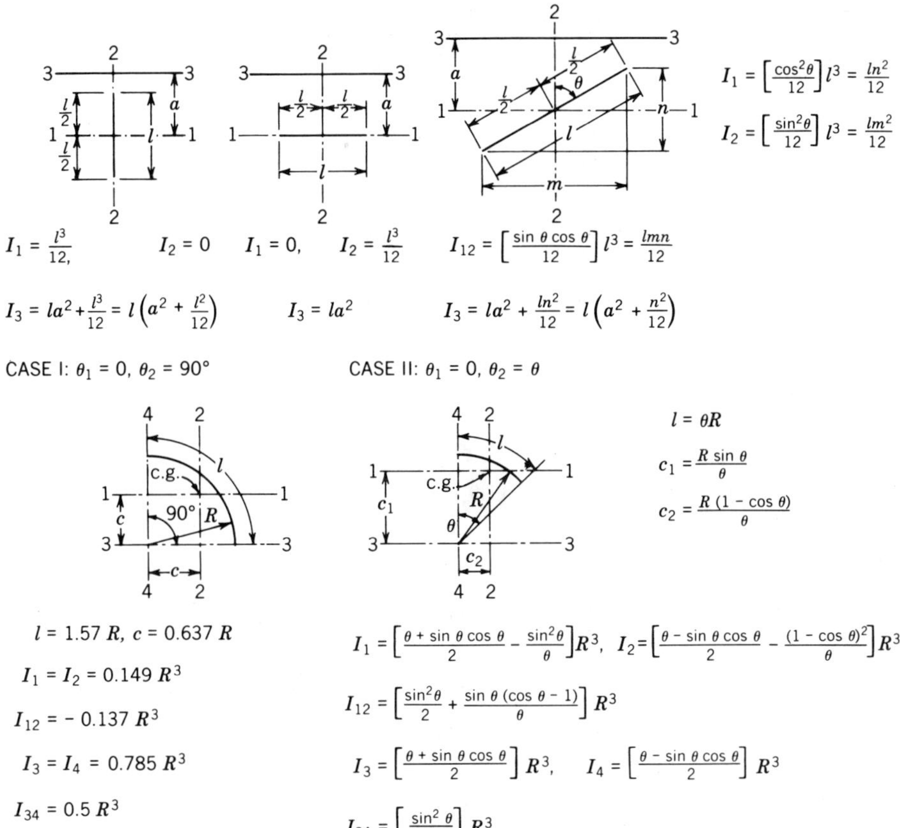

$I_1 = \frac{l^3}{12}$, $I_2 = 0$
$I_3 = la^2 + \frac{l^3}{12} = l\left(a^2 + \frac{l^2}{12}\right)$
$I_1 = 0$, $I_2 = \frac{l^3}{12}$
$I_3 = la^2$
$I_{12} = \left[\frac{\sin\theta\cos\theta}{12}\right] l^3 = \frac{lmn}{12}$
$I_3 = la^2 + \frac{ln^2}{12} = l\left(a^2 + \frac{n^2}{12}\right)$
$I_1 = \left[\frac{\cos^2\theta}{12}\right] l^3 = \frac{ln^2}{12}$
$I_2 = \left[\frac{\sin^2\theta}{12}\right] l^3 = \frac{lm^2}{12}$
CASE I: $\theta_1 = 0$, $\theta_2 = 90°$
CASE II: $\theta_1 = 0$, $\theta_2 = \theta$
$l = \theta R$
$c_1 = \frac{R\sin\theta}{\theta}$
$c_2 = \frac{R(1-\cos\theta)}{\theta}$
c.g.
90°
$l = 1.57\,R$, $c = 0.637\,R$
$I_1 = I_2 = 0.149\,R^3$
$I_{12} = -0.137\,R^3$
$I_3 = I_4 = 0.785\,R^3$
$I_{34} = 0.5\,R^3$
$I_1 = \left[\frac{\theta + \sin\theta\cos\theta}{2} - \frac{\sin^2\theta}{\theta}\right]R^3$, $I_2 = \left[\frac{\theta - \sin\theta\cos\theta}{2} - \frac{(1-\cos\theta)^2}{\theta}\right]R^3$
$I_{12} = \left[\frac{\sin^2\theta}{2} + \frac{\sin\theta(\cos\theta - 1)}{\theta}\right]R^3$
$I_3 = \left[\frac{\theta + \sin\theta\cos\theta}{2}\right]R^3$, $I_4 = \left[\frac{\theta - \sin\theta\cos\theta}{2}\right]R^3$
$I_{34} = \left[\frac{\sin^2\theta}{2}\right]R^3$

Fig. 20.6 Properties of straight and curved line elements. Courtesy of AISI.

Figure 20.7 shows two beams with thin flanges. One of these flanges is supported, that is, stiffened at both of its longitudinal edges, and the other only at one edge. The former is known as a *stiffened compression element,* and the latter as an *unstiffened compression element.* When the compression stress exceeds the plate buckling stress, the flanges will buckle and distort as shown. However, further deformation is counteracted by so-called membrane stresses. Thus, in the stiffened element of Fig. 20.7*a* the waving caused by longitudinal compression clearly causes tension in the transverse fibers which are curved, that is, elongated by the buckling wave action. This transverse tension, the so-called membrane stress, counteracts unlimited growth of the wave distortions and causes the flange to continue resisting increasing compression beyond the stress which caused first buckling. This is known as post-buckling strength and is utilized in the design of such elements. Unstiffened elements, when buckled, distort as shown in Fig. 20.7*b*. Here, however, the transverse fibers are supported only along one of the longitudinal edges. Hence, membrane tension cannot develop in the manner of Fig. 20.7*a*. However, even here membrane stresses counteract unlimited waving, but they are mostly shear stresses which are much less effective than tension stresses. Hence, unstiffened compression elements show considerably less post-buckling strength than stiffened elements.

20.3.1 Stiffened Compression Elements

The slight waving of the compression element in its post-buckled shape (Fig. 20.7*a*) causes the compression stresses, which were uniformly distributed prior to plate buckling, to redistribute themselves. The portions of the element at or close to the stiffened edges remain more nearly straight than the portions along mid-flange. Therefore, these more curved portions shed part of their compression stresses to the straighter portions near the edges. The resulting nonuniform stress distribution is shown in Fig. 20.8. Of course, the total compression force in the flange in its

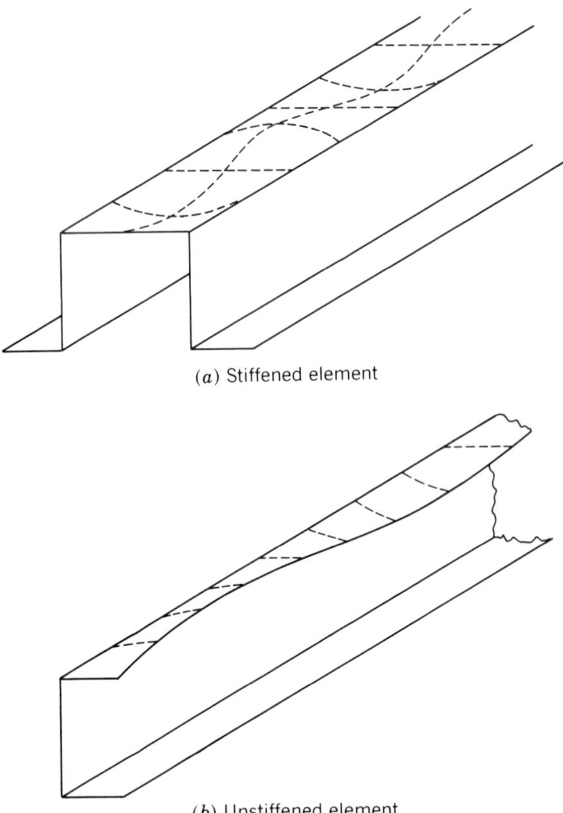

Fig. 20.7 Buckling of thin compression elements. Courtesy of AISI.

post-buckling state is the area under the stress curve times the flange thickness t. In order to simplify design calculations, this nonuniform stress distribution is replaced by two equal rectangles with the same maximum edge stress and the same combined area as those of the actual nonuniform stress distribution across the real width, w, by replacing it by the effective width, b, distributed as shown. This means that for calculating flexural stresses, the actual full area of the cross-section, of width w, is replaced by the effective area of width b, but of the same maximum stress f, as shown on the lower part of Fig. 20.8.

The *AISI Specification* stipulates that stiffened compression elements with w/t ratios up to $(w/t)_{\text{lim}} = 171/\sqrt{f}$ are fully effective (i.e., $b = w$). For compression elements with w/t exceeding this value, the effective width b is calculated from

$$\frac{b}{t} = \frac{253}{\sqrt{f}}\left[1 - \frac{55.3}{(w/t)\sqrt{f}}\right] \tag{20.3.1}$$

This equation is used when calculating allowable loads, for which purpose the safety factor is built into the equation. When determining deflections, $(w/t)_{\text{lim}} = 237/\sqrt{f}$ and the equation for the effective width b becomes

$$\frac{b}{t} = \frac{326}{\sqrt{f}}\left[1 - \frac{71.3}{(w/t)\sqrt{f}}\right] \tag{20.3.2}$$

Apart from the simple shape of Fig. 20.7*a*, there are many other sections that employ stiffened compression elements. A very frequent case occurs when the element is supported by a web only along one longitudinal edge, whereas the other edge is stiffened by a special edge stiffener. On Fig. 20.1, the shapes (*b*), (*d*), (*f*), (*n*), and (*o*) are examples of sections incorporating edge stiffeners. In addition to edge stiffeners, intermediate stiffeners are frequently employed between webs, particu-

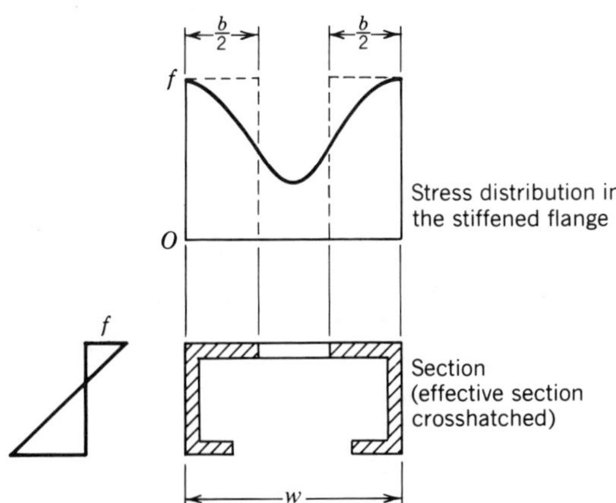

Fig. 20.8 Effective width concept for post-buckling behavior.

larly in panels and decks. Some of these, which further increase economy, are illustrated in Fig. 20.2, in the panel shapes (*d*), (*e*), (*g*), (*h*), and (*j*). The *AISI Specification* contains detailed provisions on stiffeners for compression elements, both edge stiffeners and intermediate stiffeners, which define the minimum moment of inertia I for a stiffener to be effective, and a number of other related requirements.

Finally, the effective width of compression elements in closed square or rectangular tubes is specified to be somewhat larger than for open sections, because of the formers' greater rigidity of shape.

20.3.2 Unstiffened Compression Elements

Because one of the two longitudinal edges of unstiffened elements is unsupported and therefore free to wave, such elements buckle at stresses considerably lower than stiffened elements. Also, the fact that only one edge is stiffened generally results in considerably larger buckling wave distortions than in stiffened elements. It is for this reason that the *AISI Specification* does not apply the effective width concept to unstiffened elements. Rather, it stipulates allowable compression stresses F_c which decrease with increasing w/t, which prevents failure by local buckling and reduces distortions at design loads.

These allowable stresses are:

$$\text{For } w/t \le 63.3/\sqrt{F_y}: \qquad F_c = 0.60F_y \tag{20.3.3}$$

$$\text{For } 63.3/\sqrt{F_y} \le w/t \le 144/\sqrt{F_y}: \qquad F_c = F_y[0.767 - 0.00264(w/t)\sqrt{F_y}] \tag{20.3.4}$$

$$\text{For } 144/\sqrt{F_y} \le w/t < 25: \qquad F_c = 8000/(w/t)^2 \tag{20.3.5}$$

$$\text{For } 25 < w/t < 60: \qquad F_c = 19.8 - 0.28(w/t) \tag{20.3.6}$$

$$\text{except that for angle struts} \qquad F_c = 8000/(w/t)^2 \tag{20.3.7}$$

The fact that effective widths are specified for stiffened elements and reduced allowable stresses for unstiffened elements makes for some complications and inconsistencies in the present edition of the *AISI Specification*. It may be assumed that when research on this rather complex topic is completed, the Specification will apply the effective width concept also to unstiffened elements.

20.4 WEBS OF FLEXURAL MEMBERS

Webs of beams, panels, or other flexural members are subject to three types of actions: shear stresses caused by the external shear forces; bending stresses caused by the external bending moments; and web crippling caused by concentrated loads or reactions which must be locally resisted and absorbed, unless special web stiffeners are provided. These three actions can occur singly or, more often, in combination. Thus, at most cross-sections along the span of a beam there will be both a shear force V and a bending moment M, which will cause corresponding shear and bending stresses in the web.

The depth-to-thickness ratio h/t is the main variable that affects the strength and stability of webs in regard to shear, bending, crippling, or combinations thereof. In the expression h/t, h = clear distance between flanges, and t = web thickness. The *AISI Specification* sets the following upper limits for h/t:

1. For unreinforced webs: $(h/t)_{max} = 200$.
2. For webs with transverse stiffeners that satisfy the special stiffener requirements of the Specification:
   a. When bearing stiffeners are used at reactions and concentrated loads, $(h/t)_{max} = 260$.
   b. When bearing stiffeners and intermediate stiffeners are used, $(h/t)_{max} = 300$.

When a web consists of two or more sheets, such as in an I-shape made by connecting two channels back-to-back, the h/t ratio shall be computed for the individual sheets.

20.4.1 Shear Stresses in Webs

The shear strength of thin webs depends on the h/t ratio, the support conditions along the top and bottom edges, the presence or absence of adequate transverse stiffeners, and, of course, the yield strength of the steel. The majority of cold-formed flexural members have webs without stiffeners except, perhaps, at points of support, and with webs longitudinally supported by flanges along top and bottom edges.

The *AISI Specification* gives the allowable shear stress F_v on webs without or with transverse stiffeners as:

$$\text{for} \quad h/t \le 237\sqrt{k_v/F_y}: \qquad F_v = 65.7\,\frac{\sqrt{k_v/F_y}}{(h/t)} \le 0.40F_y \tag{20.4.1}$$

$$\text{for} \quad h/t > 237\sqrt{k_v/F_y}: \qquad F_v = \frac{15{,}600k_v}{(h/t)^2} \tag{20.4.2}$$

The value of the shear buckling coefficient k_v depends on the absence or presence of transverse stiffeners.

For webs without transverse stiffeners: $k_v = 5.34$

For webs with adequate transverse stiffeners:

$$k_v = 4.00 + 5.34/(a/h)^2 \qquad \text{when} \quad a/h \le 1.0$$

$$k_v = 5.34 + 4.00/(a/h)^2 \qquad \text{when} \quad a/h > 1.0$$

where a = distance between transverse stiffeners. For transverse stiffeners to be adequate, that is, to remain straight and to confine the web buckling waves to the flat shear panels between stiffeners, they must satisfy requirements for minimum stiffener area and minimum stiffener moment of inertia, which are separately stipulated in the *AISI Specification*.

20.4.2 Bending Stresses and Combined Bending and Shear Stresses in Webs

The compression portions of thin webs, that is, the portions between the neutral axis and the outer compression fiber, are subject to buckling and develop post-buckling strength in the same manner as compression flanges. The allowable web compression stress for such web buckling is calculated for the cross-section consisting of the effective area of the compression flange and the full area of web and tension flange. For stiffened compression flanges, the effective width is calculated as discussed in Section 20.3.1. For unstiffened compression flanges, the *AISI Specification* stipulates that the effective flange areas shall be assumed as full areas times $F_c/0.60F_y$. (For F_c, see Section 20.3.2.)

Then the maximum allowable web bending compression stress, F_{bw}, for beams with stiffened compression flanges is

$$F_{bw} = [1.21 - 0.00034(h/t)\sqrt{F_y}](0.60F_y) \le 0.60F_y \tag{20.4.3}$$

and for beams with unstiffened flanges,

$$F_{bw} = [1.26 - 0.00051(h/t)\sqrt{F_y}](0.60F_y) \le 0.60F_y \tag{20.4.4}$$

The necessity of separate equations reflects the fact that the support provided to the web by a stiffened compression flange is more effective than that from an unstiffened compression flange.

There are many situations in flexural members where large bending moments occur at the same locations as large shear forces. Examples are support regions of cantilevers, interior supports of continuous beams, and often at concentrated loads in the center portion of the span. For such situations of *combined shear and bending stresses in webs,* the *AISI Specification* provides interaction equations

$$\left(\frac{f_{bw}}{F_{bw}}\right)^2 + \left(\frac{f_v}{F_v}\right)^2 \le 1.0 \tag{20.4.5}$$

For webs with adequate intermediate stiffeners,

$$0.6\left(\frac{f_{bw}}{F_{bw}}\right) + \left(\frac{f_v}{F_v}\right) \le 1.3 \tag{20.4.6}$$

when $f_{bw}/F_{bw} > 0.5$ and $f_v/F_v > 0.7$. For values smaller than 0.5 and 0.7, respectively, interaction need not be considered.

In these equations, f_{bw} and f_v are, respectively, the bending stress and the shear stress permissible in the simultaneous presence of the other stress.

20.4.3 Web Crippling

When concentrated loads or reactions are applied to the flanges over short bearing lengths, the high local intensity of the compression stress transmitted to the web may cripple the web in the manner shown in Fig. 20.9. Such web crippling can be affected by a variety of conditions. Thus in Fig. 20.9*a* the bottoms of the webs are restrained against rotation to a higher degree than those in Fig. 20.9*b*, which results in a higher web crippling strength in the former than the latter. Further, concentrated loads are mostly applied either to the top or to the bottom flange, but there are situations, mostly at supports, where two loads opposite to each other are applied at the same location. Also, such concentrated loads may act either somewhere along the span, or at or near the free end of a beam. In the former case the load in the web will spread in both directions, giving a higher crippling strength than for a similar load applied, for instance, at the end of a cantilever. The main variables or parameters that affect web crippling are the bearing length ratio N/t, the depth ratio h/t, and the yield point F_y. Here N = length of bearing of concentrated load or reaction and, as before, h = clear depth of web between flanges, and t = thickness.

In the absence of a general theory of web crippling, and because of the great variety of conditions, the AISI provisions on allowable concentrated loads and reactions are based on many hundreds of tests. The resulting design formulas were obtained by systematic evaluation of these tests, such as by regression analysis. In consequence, the formulas are fairly complex and contain a

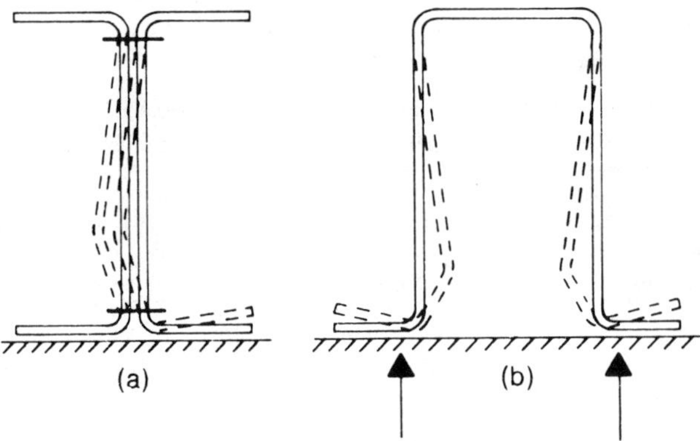

Fig. 20.9 Web crippling effects.

great many constants in order to produce satisfactory fit. The reader is advised to consult the Specification directly when faced with the web crippling problem. Part V of the *AISI Cold-Formed Steel Design Manual* [20.3] contains a total of nine charts on web crippling, which greatly facilitate design.

When bending moments are present at the locations of concentrated loads, or reactions, provision must be made for *combined bending and web crippling*. For this situation the *AISI Specification* provides interaction equations similar to those for other combined actions, such as shear and bending in webs, compression and bending in beam-columns, and others.

If the concentrated loads or reactions are larger than the specified allowable values for the given webs (h, t, etc.), *bearing stiffeners* must be provided at the points of these local loads. Such bearing stiffeners must be detailed so that the concentrated loads or reactions are applied directly to the bearing stiffeners. If not, each stiffener shall be fitted accurately to the flat portion of the flange to ensure direct load transfer into the end of the stiffener. The Specification provides two design equations for bearing stiffeners. One safeguards against yielding and crushing at the ends of the stiffener; the other prevents column-buckling in a direction perpendicular to the web. These provisions take account of the fact that part of the load will be transferred by shear from the stiffener to the adjacent portions of the web. The magnitude of these portions is specified and depends on whether the stiffener is located at an interior support or concentrated load, or whether it is located at an end support of the beam. In the former case the load spreads into the web to both sides of the stiffener, in the latter case only to one side.

The preceding discussion on thin compression elements and on webs of flexural members has dealt with thin elements of which cold-formed, thin-walled shapes are composed and in particular with those elements which differ in their performance and design from the corresponding components of hot-rolled members. The next several sections of this chapter cover not individual flat elements, but cold-formed members composed of such elements and the manner in which the behavior of those members differs from that of hot-rolled or fabricated steel members of ordinary cross-sectional dimensions and shapes.

20.5 DESIGN OF FLEXURAL MEMBERS

Section properties and allowable moments are calculated differently, depending on whether the compression flange or flanges of flexural members are composed of unstiffened or of stiffened elements. Thus shapes (a), (c), (e), (p), and (q) of Fig. 20.1, when used as beams, have unstiffened compression flanges, shapes (b), (g), (n), (o), (s), and others have stiffened compression flanges, and flexural calculations for the two types of shapes differ somewhat.

The section properties of beams with unstiffened compression flanges are calculated in the usual manner since, for such members, the entire unreduced cross-section is regarded as effective (see Section 20.3.2). However, if the flat-width ratio w/t of the compression flange exceeds $63.3/\sqrt{F_y}$, the allowable stress must be reduced from $0.60F_y$ to the smaller values given in Section 20.3.2, which depend on the actual w/t values.

20.5.1 Effective Width of Compression Flanges

On the other hand, as pointed out in Section 20.3.1, for stiffened elements with $(w/t) > 171/\sqrt{f}$, the effective width is smaller than the full width, and is given by Eqs. (20.3.1) and (20.3.2). This is shown in Fig. 20.8. The section properties, therefore, must be calculated for the shaded portion shown on Fig. 20.8, that is, for the full area of tension flanges and webs, and the reduced, effective area of the compression flange. Now, the effective width b given in Eqs. (20.3.1) and (20.3.2) is seen to depend on the compression stress f in that flange. In this respect, two situations must be distinguished. For one, the compression stress may be known beforehand. This is the case if one wants to calculate the allowable bending moment on the section and if the distance from the neutral axis to the compression flange is greater than that to the tension flange (a frequent case). In that case the stress f in the compression flange is the allowable stress $f = 0.60F_y$ and one has to calculate the effective width b for that stress. Once the effective width b is known, the section properties I, S, etc., are calculated in the usual manner.

In other situations the compression stress may not be known beforehand and so one cannot calculate a final value for the effective width b. This is the case when the distance from the neutral axis is greater to the tension flange than to the compression flange. Also, it is always the case when one wants to calculate deflections at stresses smaller than the allowable. In both these situations, at the outset the effective width b for calculating the moment of inertia or the section modulus is not known. But as long as b is not known, one cannot calculate the compression stress f which is needed for calculating b, and subsequently I, S, etc. In such situations iteration is used. One first assumes a reasonable looking compression stress, f, then calculates b, I, S, etc., and checks whether the assumption was correct. If not, one or two iterations usually lead to a sufficiently accurate value of b.

Once the adequacy of the chosen cross-section has been verified for the given bending moment and, if desired, for deflection, the web strength for the various modes must be ascertained as described in Section 20.4. This includes web crippling and, when needed, the provision of bearing stiffeners at reactions or concentrated loads.

20.5.2 Lateral Buckling Considerations

The features just discussed generally suffice to ensure the adequacy of the beam design. The possibility of *lateral buckling of beams* generally needs to be considered only in special cases. That is, in most situations the structural assembly supported by the beam, say the floor supported by the joist, has sufficient in-plane rigidity to brace the joist and prevent it from deflecting laterally out of its original position. While this is true in the finished structure, during construction situations often arise where flexural members are temporarily unbraced laterally, over part or all of their length. This, when not watched, has often led to structural distress during construction. Also, there are types of cold-formed structures, such as storage racks, where the horizontal girts that carry the stored goods are not continuously connected to rigid horizontal components, such as floors. Such members can buckle laterally unless prevented from doing so by special, appropriately spaced lateral braces. Also, in the frequent case of free-standing columns or studs subject both to axial loads and bending moments, the carrying capacity is calculated from interaction equations which require the analysis of the particular member as a laterally unbraced beam on the one hand and as an axially loaded column on the other.

The following provisions for laterally unbraced beams in the *AISI Specification* apply to I-shapes symmetrical about the plane of the web and to symmetrical channel shapes. Allowable stresses on such beams depend on the slenderness parameter L^2S_{xc}/dI_{yc} where

L = unbraced length, in.
S_{xc} = compression section modulus of the entire section, in.^{3}
d = depth of section, in.
I_{yc} = moment of inertia of compression portion about centroidal axis of entire section, parallel to web, in.^{4}

Denoting for brevity the slenderness parameter L^2S_{xc}/dI_{yc} by η, the specified allowable stresses F_b are,

when $0.36\pi^2EC_b/F_y < \eta < 1.8\pi^2EC_b/F_y$,

$$F = \frac{2}{3}F_y - \frac{F_y^2\eta}{5.4\pi^2EC_b} \tag{20.5.1}$$

and when $\eta \geq 1.8\pi^2EC_b/F_y$,

$$F_b = 0.6\pi^2EC_b/\eta \tag{20.5.2}$$

Here C_b is a coefficient that reflects the influence of the distribution of bending moments along the span. It can be conservatively taken as unity or more accurately as

$$C_b = 1.75 + 1.05(M_1/M_2) + 0.3(M_1/M_2)^2 \leq 2.3 \tag{20.5.3}$$

where M_1 is the smaller and M_2 the larger of the two bending moments at the ends of the unbraced length, and where M_1/M_2 is positive for reverse curvature over that length and negative for single curvature. If the bending moment at any point within the unbraced length is larger than those at the ends, $C_b = 1$. Also, for members subject to combined axial force and bending, $C_b = 1$.

For the rarer case of point-symmetrical Z-shapes bent about the centroidal axis perpendicular to the web, the *AISI Specification* contains special, similar provisions which should be consulted in this situation.

20.5.3 Lateral Bracing of Channels and Z-Beams

Channels and Z's are the simplest two-flange shapes that can be cold-formed from a single steel sheet. For this reason they are very frequently used as flexural members, for example, as floor joists, roof purlins, etc. I-shapes, produced by spot-welding two channels back to back, have the advantage of double symmetry, but require the additional welding. When unbraced channels or Z's are loaded in the plane of the web, they deflect not only vertically, but they also twist and deflect transversely. For the channel this is so because its centroid and its shear center do not coincide.

This causes twisting moments and resulting lateral deflections. In Z-sections, on the other hand, the principal axes are inclined to the web, that is, to the plane of loading. This causes lateral deflections and also resulting rotations. Hence, when channels or Z's are used as flexural members, they must be adequately braced to prevent such twisting and lateral deflections or, at least, to limit these movements to very small, permissible amounts. This can be done in various ways, depending on the specific structural situation. Most frequently channels and Z's are used to support floor panels, roof decks, or other covering material. In most cases these covering materials rest on the compression flanges of the beams. If the individual panels of the covering material are adequately interconnected so that they can resist in-plane forces, they form a rigid diaphragm. The channel or Z tends to deflect and twist but is prevented from doing so if it is adequately connected to these covering diaphragms. The forces transmitted from the beams to the covering material through the connections accumulate in the diaphragm. This means that the interconnections between the individual panels of the diaphragm must be strong enough to resist these accumulated beam connection forces. It also means that the diaphragm itself must be so connected to the main framing of the structure or to other supports, so that these accumulated forces are safely transmitted to that framing and adequately resisted by it.

Although the wording of the *AISI Specification* in this respect is somewhat ambiguous, it prescribes two measures to provide safety for such assemblies of channels or Z's plus covering materials: For one, the flange not braced by covering material must be furnished with individual braces of appropriate spacing and strength, the same as both flanges of channels or Z-beams which are not braced by covering material. Second, tests shall be carried out on the proposed assemblies which must show a strength at least equal to ⅔ of the required design capacity. Generally, this testing requirement is not onerous since, in most cases, large areas are covered with repetitious, identical beam and covering arrangements. At least to date the test requirement is necessary because many as yet not fully explored factors affect the strength of such assemblies. For more details, see Ref. 20.3, Parts II, III, and IV.

When neither flange is connected to continuous bracing by covering material, special braces of appropriate spacing and strength must be provided to both flanges. Such situations occur during erection, but also in permanent configurations, such as in storage rack structures, special roof systems, etc. According to the *AISI Specification,* braces shall be attached to both flanges at the ends of the beam and at intervals no larger than one-quarter of the span. If the beam carries not only uniform loading but also a concentrated load larger than one-third of the total load, an additional brace must be provided at the location of such a concentrated load. The braces must be arranged to prevent tipping of the beam at the supports and lateral deflection at all other brace locations. The Specification, which should be consulted on this matter, provides a relatively simple way to calculate the lateral forces these braces must be able to resist.

20.5.4 Inelastic Reserve Capacity of Flexural Members

For hot-rolled shapes the *AISI Specification* permits so-called plastic design under certain conditions. This method recognizes the fact that a structural member, say a beam, does not fail when yielding starts in the extreme fiber. In fact, if the width-to-thickness ratios of the component parts (flanges, webs) are sufficiently small, the member will fail only when yielding has spread over the entire section, in tension on one side and in compression on the other side of the neutral axis. This occurs, of course, at a significantly higher load than initial yielding. The *AISI Specification* defines the maximum width-to-thickness ratios for which such a design is permissible. Shapes with larger ratios may buckle locally before complete plastification has been obtained.

In cold-formed steel construction made of sheet or strip steel the width-to-thickness ratios are generally much larger than those permitted for plastic design. Nevertheless, in many cases the actual strength of a member, for example, a beam, is significantly larger than that load or moment which causes initial yielding in the extreme fiber. This is particularly so if the neutral axis is located such that first yielding occurs in the tension flange; this is much more frequent in unsymmetrical cold-formed sections than in hot-rolled shapes which are generally symmetrical about the neutral axis parallel to the flanges. If yielding occurs first in the tension flange, inelastic stress redistribution will occur at least until the compression flange also starts yielding. If, in addition, the w/t ratio of the compression flange is sufficiently small, further plastification may occur prior to failure.

Correspondingly, the *AISI Specification* makes provisions for utilizing this inelastic reserve capacity of flexural members, provided the necessary conditions for safe partial plastification are satisfied. These conditions are:

1. The member is not subject to twisting or lateral buckling.
2. The effect of cold-forming is not included in determining the yield point F_y.
3. The ratio of the depth of the compressed portion of the web to its thickness does not exceed $190/\sqrt{F_y}$.

4. The h/t ratio of the entire web does not exceed $640/\sqrt{F_y}$.
5. The shear force does not exceed $0.35F_y$ times the web area.
6. The angle between web and flange is at least 70°.

When these conditions are satisfied, the design moment shall not exceed $0.75M_y$ nor $0.60M_u$ where M_u is the moment causing in a stiffened flange a compression strain $C_y e_y$. Here $C_y = 3$ for w/t less than $190/\sqrt{F_y}$, $C_y = 1$ for w/t greater than $221/\sqrt{F_y}$, with linear transition between these two values of w/t. For unstiffened compression elements, for multiple-stiffened compression elements, and for compression elements with edge stiffeners, $C_y = 1$. No limit is placed on the maximum tensile strain.

Considerable economy can be achieved by utilizing the inelastic reserve capacity. For appropriately shaped sections, gains in design moments up to about 25% are possible.

20.6 DESIGN OF COMPRESSION MEMBERS

20.6.1 Design Considerations

The design of thin-walled, cold-formed compression members must provide safety against several possible modes of failure: simple flexural column buckling; local buckling of the component elements; a combination of local and flexural column buckling; torsional–flexural buckling; and a combination of local and torsional–flexural buckling. In contrast, hot-rolled or fabricated structural steel columns, with their large thicknesses, generally need to be checked only for flexural buckling. This difference is caused by a combination of various influences: the smaller thicknesses t of cold-formed members; the larger width-to-thickness ratios w/t; and the fact that the most economical shapes produced by cold-forming are not doubly symmetrical like hot-rolled W- or M-shapes, but are singly symmetrical, point symmetrical, or occasionally nonsymmetrical. The most frequent ones are channels, angles, Z-sections, C-sections, and hat-sections. It is such singly symmetrical or nonsymmetrical shapes which, unless braced against twisting, are subject to torsional–flexural buckling under axial load.

If such sections are used, it is possible in most cases to provide bracing in a manner that will eliminate torsion. For instance, when wall studs are appropriately connected on both flanges to covering material (wallboard or the like), twisting cannot occur. If wallboard and connections are sufficiently rigid, even a single layer may provide sufficient bracing. Compression chords of roof trusses often can be so connected to purlins or directly to roof sheathing as to eliminate twist. Similar bracing is often possible for the posts of storage racks. Such bracing greatly increases the load capacity of the members. Only when singly symmetrical or unsymmetrical members are not braced against twisting, or when such bracing is provided only at large spacing, is it necessary to calculate torsional–flexural buckling loads.

The *AISI Specification* gives provisions for calculating these torsional–flexural buckling loads. Such calculations are quite lengthy and complex. Parts III, Supplementary Information and V, Charts and Tables of Ref. 20.3 provide extensive and valuable design aids for torsional–flexural buckling calculations. For this reason, only the much more frequent and simple compression member design for flexural buckling and for combined local and flexural buckling will be presented in this handbook.

20.6.2 Compression Members Subject to Flexural Buckling

In the *AISI Specification,* the design equations *for axially loaded compression members subject to flexural buckling with and without local buckling are:*

$$\text{For } \frac{KL}{r} < \frac{C_c}{\sqrt{Q}}: \quad F_{a1} = \frac{12}{23}QF_y - \frac{3(QF_y)^2}{23\pi^2 E}\left(\frac{KL}{r}\right)^2$$

$$= 0.522QF_y - \left[\frac{QF_y KL/r}{1494}\right]^2 \tag{20.6.1}$$

$$\text{For } \frac{KL}{r} \geq \frac{C_c}{\sqrt{Q}}: \quad F_{a1} = \frac{12\pi^2 E}{23(KL/r)^2}$$

$$= \frac{151{,}900}{(KL/r)^2} \tag{20.6.2}$$

In these equations, Q, discussed below, reflects the influence of local buckling, if any. It is easily seen that for $Q = 1$ (no local buckling) the two equations are identical with those of the *AISC Specification,* except that the latter uses a variable safety factor for the first of the two equations, while the above AISI equations use a constant safety factor $23/12 = 1.92$ for the entire range of KL/r.

In these equations

$C_c = \pi\sqrt{2E/F_y}$
F_{a1} = allowable average compression stress under concentric loading on full, unreduced area of cross-section, ksi
E = 29,500 ksi
K = effective length factor
L = unbraced length of member, in.
r = radius of gyration of full, unreduced cross-section, in.
F_y = yield point, ksi
Q = local buckling factor determined as follows

If all elements of the section are stiffened (e.g., Figs. 20.1*b*, 20.1*j*):

$$Q = \frac{A_{\text{eff}}}{A_{\text{full}}} \tag{20.6.3}$$

If all elements of the section are unstiffened (e.g., Fig. 20.1*e*):

$$Q = \frac{F_c}{F} \tag{20.6.4}$$

If the section is composed of both stiffened and unstiffened elements (e.g., Figs. 20.1*a*, 20.1*ψ*):

$$Q = \frac{F_c}{F}\frac{A_{\text{eff}}}{A_{\text{full}}} \tag{20.6.5}$$

where F_c = allowable compression stress for the unstiffened element with the largest w/t ratio [Eqs. (20.3.3) to (20.3.6)]
A_{eff} = reduced effective area of all stiffened elements plus unreduced area of unstiffened elements, if any

The effective width for calculating A_{eff} is determined from Eq. (20.3.1), using $f = F$ in Eq. (20.6.3) and $f = F_c$ in Eq. (20.6.5).

The Design Chart of Fig. 20.10, reproduced from the *AISI Cold-Formed Steel Design Manual,* permits the direct determination of the allowable axial compression stress F_{a1} for any combination of slenderness ratio KL/r, yield point F_y, and local buckling factor Q. The chart applies only to flexural column buckling and not to torsional–flexural buckling.

20.6.3 Torsional–Flexural Buckling of Axially Loaded Shapes

In regard to *singly symmetrical or point-symmetrical axially loaded shapes* it was mentioned previously that they may be subject to *torsional–flexural buckling,* without or with local buckling. If possible, they should be braced against twisting, since torsional–flexural buckling (compared to flexural buckling) leads to significant loss of carrying capacity in the given member.

The *AISI Specification* presents the following two equations for torsional–flexural buckling:

$$\text{for} \quad \sigma_{TFO} > 0.5QF_y: \qquad F_{a2} = 0.522QF_y - \frac{(QF_y)^2}{7.67\sigma_{TFO}} \tag{20.6.6}$$

$$\text{for} \quad \sigma_{TFO} \le 0.5QF_y: \qquad F_{a2} = 0.522\sigma_{TFO} \tag{20.6.7}$$

Here Q is the same local buckling factor as in Eqs. (20.6.3), (20.6.4), and (20.6.5) and σ_{TFO} is the elastic torsional–flexural buckling stress under concentric loading. For point-symmetrical sections this quantity is replaced by σ_t, but the equations (20.6.6) and (20.6.7) are the same. The equations for σ_{TFO} and σ_t are quite complicated and lengthy and involve a great many parameters. They will not be reproduced here; instead the reader is referred to Art. 3.6.1.2 of the *AISI Specification.* For *unsymmetrical shapes* no explicit methods are given and the designer is advised to use either "rational analysis" or tests of the specific shapes being used.

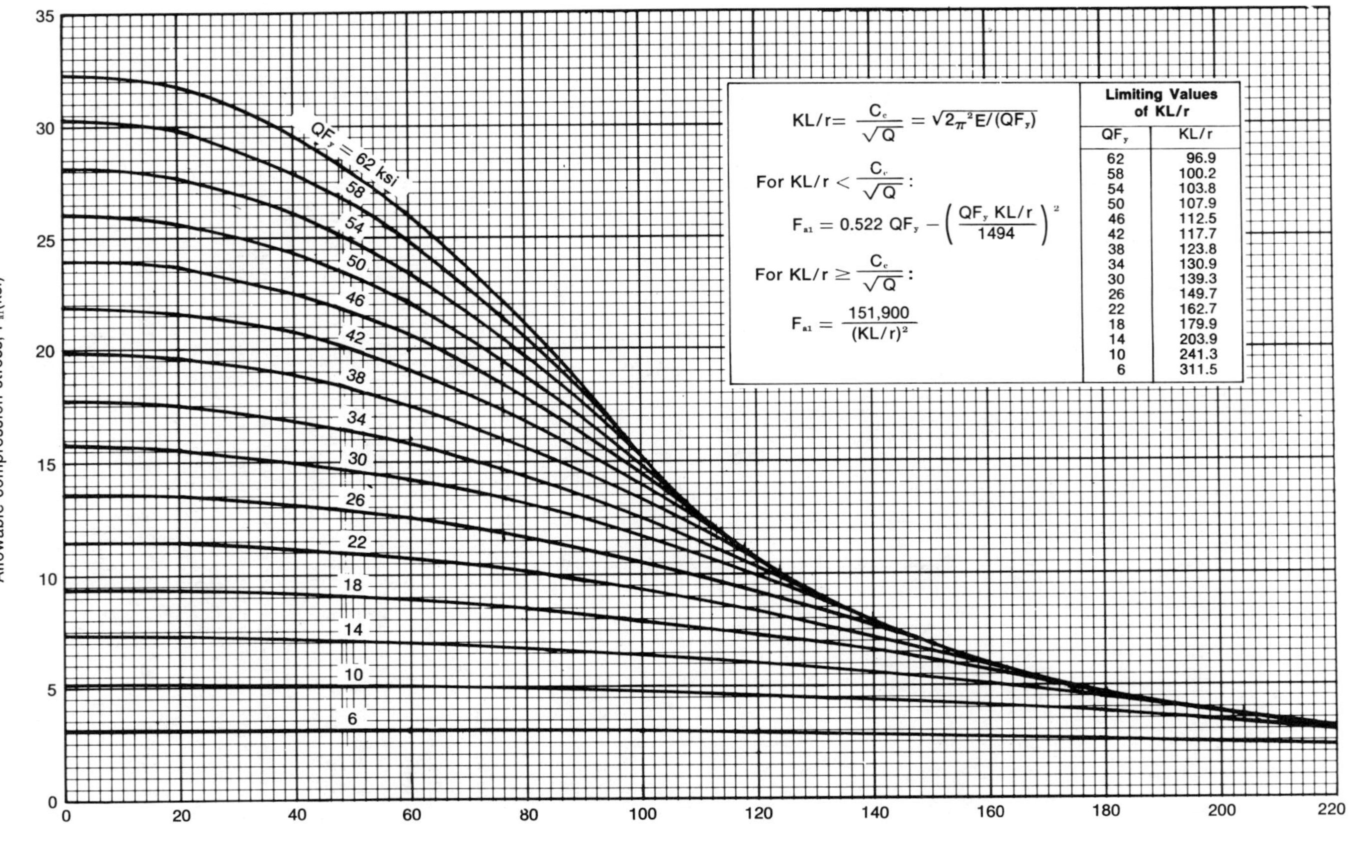

Limiting Values of KL/r	
QF_y	KL/r
62	96.9
58	100.2
54	103.8
50	107.9
46	112.5
42	117.7
38	123.8
34	130.9
30	139.3
26	149.7
22	162.7
18	179.9
14	203.9
10	241.3
6	311.5

Fig. 20.10 Axially loaded compression members—allowable average compression stress under concentric loading, F_{a1}, for shapes not subject to torsional–flexural buckling. Courtesy of AISI.

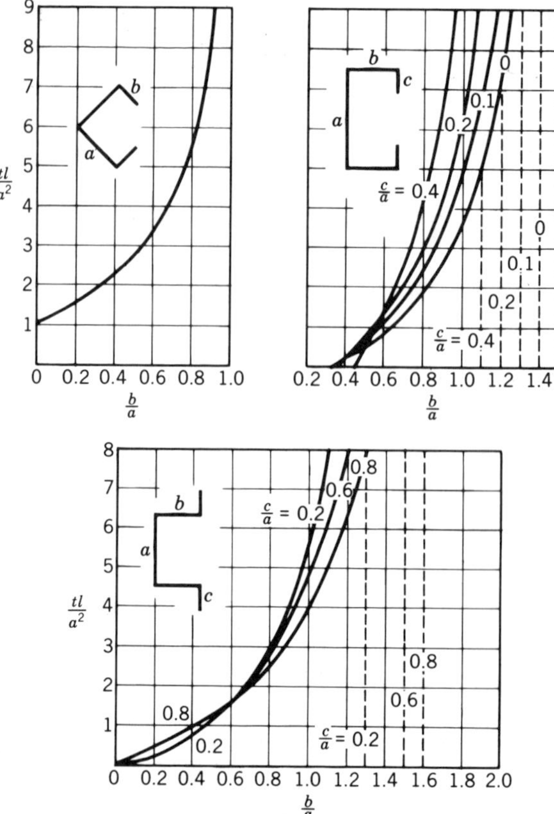

Fig. 20.11 Buckling mode curves.

In many cases the practical problem is to place braces at sufficiently close spacing to avoid torsional–flexural buckling and the corresponding reduction in load capacities. Figure 20.11 permits one to do this for the most frequent cases: angles, channels or Z's, and hats, all without and with edge stiffeners. In these charts, L is the length of the member between points that are torsionally braced, t is the thickness, and the other dimensions are as shown. Any point that falls to the left of the appropriate c/a curve defines a situation where the member will buckle flexurally; any point to the right defines a situation where buckling will be torsional–flexural. Hence, these charts permit (for the most frequent shapes) the determination of the maximum bracing distance which will avoid torsional–flexural buckling.

For these same six types of sections, Part V of Ref. 20.3 contains elaborate charts that greatly facilitate the design of such members if they are in fact subject to torsional–flexural buckling, that is, if they are not braced sufficiently closely to prevent this buckling mode.

20.7 COMBINED AXIAL COMPRESSION AND BENDING

For shapes that are not subject to torsional–flexural buckling, the interaction equations for combined axial compression and bending are the same in the AISI as in the *AISC Specification,* with some minor differences in notations. These equations are:

$$\frac{f_a}{F_{a1}} + \frac{C_{mx} f_{bx}}{\left(1 - \dfrac{f_a}{F'_{ex}}\right) F_{bx}} + \frac{C_{my} f_{by}}{\left(1 - \dfrac{f_a}{F'_{ey}}\right) F_{by}} \le 1.0 \tag{20.7.1}$$

$$\frac{f_a}{F_{ao}} + \frac{f_{bx}}{F_{b1x}} + \frac{f_{by}}{F_{b1y}} \le 1.0 \tag{20.7.2}$$

For relatively small axial forces, when $f_a/F_{a1} \le 0.15$, the following formula may be used in lieu of Eqs. (20.7.1) and (20.7.2):

$$\frac{f_a}{F_{a1}} + \frac{f_{bx}}{F_{bx}} + \frac{f_{by}}{F_{by}} \le 1.0 \tag{20.7.3}$$

Most of the notations in these equations are the same as in the *AISC Specification,* except:

F_{ao} = allowable compression stress under axial loading, for $L = 0$
F_{a1} = allowable compression stress under axial loading for actual L
F_b = allowable compression bending stress for bending only
F_{b1} = allowable compression bending stress for bending only, when lateral buckling is not possible

The design of members under combined stress which may be subject to torsional–flexural buckling is complicated by the fact that the centroid and the shear center of the cross-section do not coincide. This situation creates several possibilities; the point of application of the eccentric load may be located on the side of the centroid opposite that of the shear center, or between centroid and shear center, or on the side of the shear center opposite that of the centroid. The *AISI Specification* gives detailed design methods for all three conditions, to which the reader is referred when faced with one of these relatively rare situations.

20.8 WALL STUDS

Wall studs are compression members which are typical mostly to cold-formed steel and to timber construction. They are relatively closely spaced vertical members primarily carrying gravity load and to which wallboard is attached either along one or both flanges. Thus, the assembly of studs and wall material or sheathing constitutes the complete wall construction. The studs are mostly channel-, C-, or Z-shaped. They are braced by the wall materials. This makes for much lighter members than if they were free standing. However, there are also free-standing studs in use without wall material, such as in storage racks. In that case intermediate bracing along the length of the stud is provided by the horizontal rack members, often combined with special braces.

As shown in Fig. 20.12 the studs are placed with the minor axis perpendicular to the wall. They are prevented from buckling in the direction of the wall by the in-plane rigidity of the wallboard or sheathing. A table in the *AISI Specification* gives shear rigidities and limits of shear strain for most important types of wallboard.

For studs enclosed between two identical layers of sheathing, the following three situations must be checked in design:

1. One must verify whether buckling between fasteners, in the direction of the wall, is prevented. This is done by applying regular column formulas with the distance L equal to the fastener spacing.
2. One must determine the axial load capacity for overall buckling out of the plane of the wall. This is a flexural or torsional-flexural type of buckling which includes adjacent portions of the wallboards.
3. Finally, one must check whether the wallboard is not too brittle, that is, whether the permissible in-plane shear strain of the wallboards is sufficient to accommodate the prebuckling deformations of the stud. Corresponding shear strain values are also tabulated in the Specification.

The actual computations prescribed in the Specification are quite involved even for the simplest case of identical sheathing on both faces. They become even more complex if the cladding on both

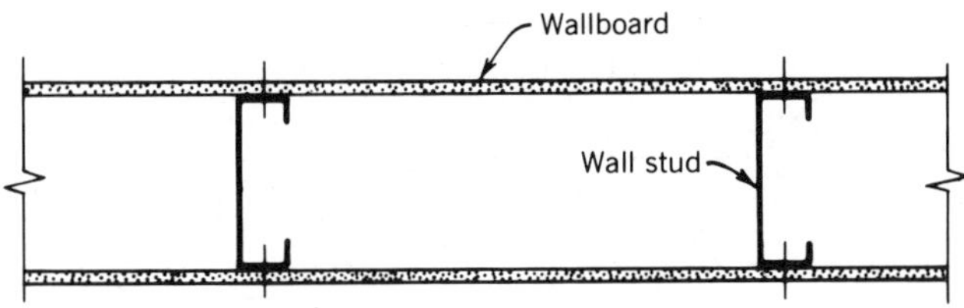

Fig. 20.12 Wall studs braced by connected wallboard.

sides is not identical, or if cladding is provided on one plane of flanges only. Fortunately, computer programs are available for use for such designs.

The situation of combined stress, that is, compression plus bending, occurs frequently in wall studs, for example, in exterior walls exposed to wind loads. The *AISI Specification* uses interaction equations for this case, similar to those previously discussed for free-standing columns.

20.9 CONNECTIONS

In cold-formed steel construction, many different types of connections are used in a great variety of situations. Thus, in fabricating plants, components are connected on a mass production basis to form finished members, for instance, connecting two channels back to back to form an I-shape, or tip to tip to form a box-section. Members are frequently assembled also in the fabricating plant into subassemblies, for example, of trusses, walls, or floors. Further, members and/or subassemblies are connected on the site to other cold-formed members or assemblies to produce parts or all of the finished structure. Alternately, cold-formed components are connected on the site to the main hot-rolled or fabricated structural steel members, such as floor joists being connected to girders or wall assemblies to members of the main steel framing (columns, wall beams, etc.).

It is this great variety of connection situations and of cold-formed shapes that are to be connected which makes for the fact that many different kinds of connections and connectors are in use. The *AISI Specification* contains provisions only for welded and for bolted connections. Apart from these most frequent types, other methods of connection find wide application. Among these are: cold rivets, such as blind rivets for application from one side only, high-shear rivets, explosive rivets, and others; screws, mostly self-tapping and of various shapes; metal stitching done by tools similar to ordinary office staplers; and connecting by upsetting, using special clinching tools which draw the sheets into interlocking projections, particularly in panel construction.

Many of these special types of connections are proprietary. Information on their strength is obtained by tests, either by the proprietor or by the user, or both. General, specification-type information is available only for welded and for bolted connections, chiefly in the *AISI Specification,* but also from the American Welding Society [20.4] and the Research Council on Structural Connections [20.6]. In addition, welded connections where the thinnest connected part is over 0.18 in. (4.6 mm) thick or bolted connections with thinnest part at least 3/16 in. (5 mm) thick, are to be made in accordance with the *AISC Specification* [20.1].

20.9.1 Welded Connections

The types and shapes of welded connections in cold-formed steel in most cases are different from those employed in regular, heavy steel construction. This comes not only from the smaller thickness but also from the shapes peculiar to cold-formed members, such as the curved corners resulting from the cold-rolling process. Most welding is of the arc-welding type, although resistance spot welds are used in the fabricating process, when appropriate. The weld types briefly described below are all arc-welds. Chapter 6 treats welding in considerable detail.

Arc spot welds are used for welding sheet steel to thicker supporting members. Weld washers, as shown in Fig. 20.13, are used when the sheet thickness is less than 0.028 in. (0.7 mm). For this and all subsequently described types of arc welds, the *AISI Specification* gives not only ways to calculate allowable loads, but it also specifies minimum edge distances, positions during the welding process, and other essentials.

Arc seam welds, Fig. 20.14, are used when larger allowable loads are required than can be developed by circular arc spot welds. Fillet welds are employed for connecting sheet to sheet or

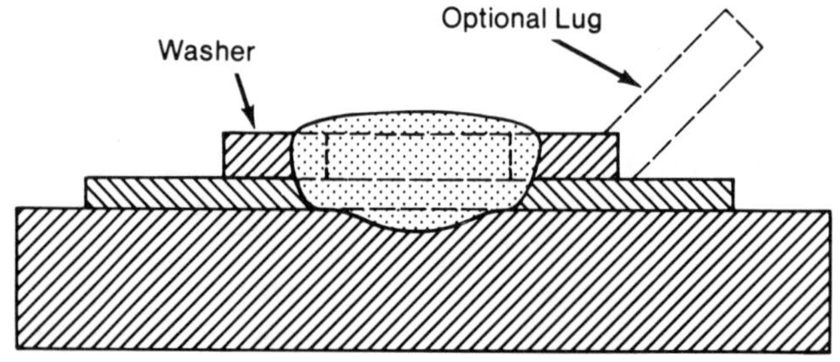

Fig. 20.13 Arc spot weld using washer. Courtesy of AISI.

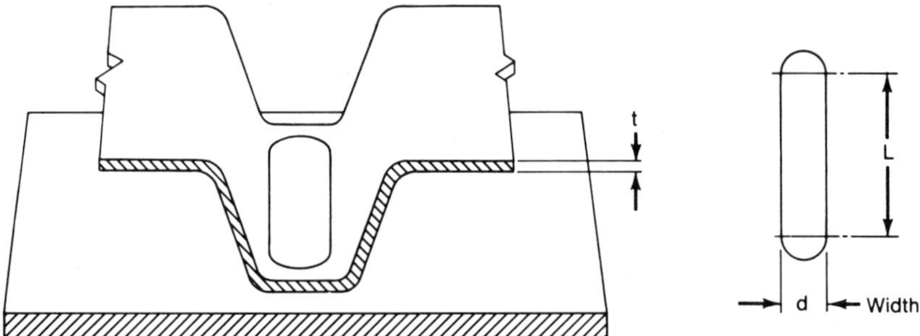

Fig. 20.14 Arc seam welds—sheet to supporting member in flat position. Courtesy of AISI.

sheet to a thicker steel member (Fig. 20.15). The use of groove welds is dictated by the corner configurations resulting from the cold-forming process. Flare bevel-groove welds connect a rounded edge to a flat sheet or plate, while flare V-groove welds connect two rounded edges, Figs. 20.16*a* and 20.16*b*.

The *AISI Specification* provides allowable loads and other information (such as allowable edge distances) for all these types of welds. In addition it gives a table of allowable shear strength for resistance spot welds.

20.9.2 Bolted Connections

Bolts, nuts, and washers in cold-formed steel construction shall conform to one of the following specifications: ASTM A307, Type A (carbon steel); ASTM A325 (high-strength bolts); ASTM A354, Gr. BD (quenched-and-tempered alloy steel for bolt diameters smaller than ½ in. [13 mm]); ASTM A449 (quenched-and-tempered steels); and ASTM A490 (quenched-and-tempered alloy steel).

The *AISI Specification* gives detailed information on minimum spacings and edge distances; allowable tension stresses on the net section for various thickness ranges with and without washers; allowable bearing stresses on the contact surfaces for various thickness ranges and types of joints; and allowable shear stresses on bolts made of the various ASTM bolt steels.

20.9.3 Special Situations

Apart from the provisions on welded and on bolted connections just described, the *AISI Specification* deals with two special situations related to connections.

One is the manner of connecting two channels to form an I-section, a frequently utilized cold-formed shape. When (see Fig. 20.1*n*) such an I-section is used as a compression member, the maximum longitudinal distance between connections is so specified that with adequate safety, it will prevent the individual channel from buckling perpendicular to the web over the free length between connections. On the other hand, when such an I-shape is used as a flexural member, the connections in the web near top and bottom flange must be strong enough to prevent the two channels from breaking away from each other by rotating around their respective shear-centers, and also must prevent the individual channels from distorting excessively between adjacent connections even if the connection strength itself is adequate.

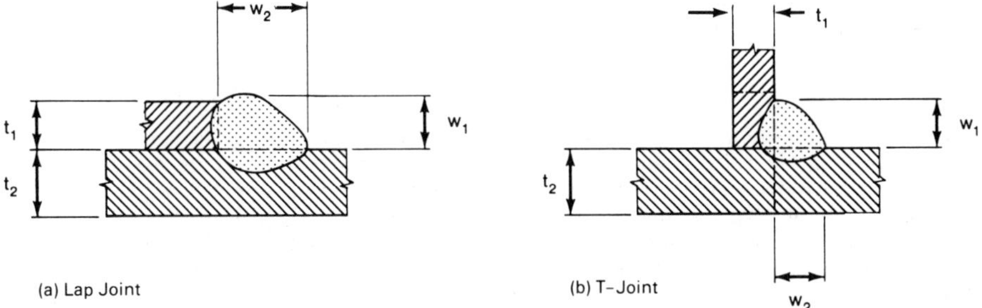

Fig. 20.15 Fillet welds. Courtesy of AISI.

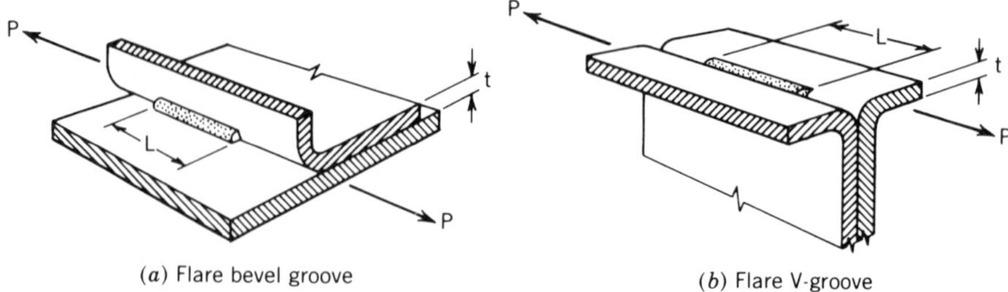

(*a*) Flare bevel groove (*b*) Flare V-groove

Fig. 20.16 Shear in flare groove welds. Courtesy of AISI.

The other special situation refers to cross-sectional shapes in which compression elements, such as compression flanges, are not monolithic parts of the section but represent separate cover plates or sheets. A typical example is a box section obtained by closing the opening of a U-shape by a cover sheet attached to the top edge of each web of the U at definite intervals. The Specification provides three criteria for determining the spacing of the connections of such compression elements (e.g., the cover plates) to the stiffening elements (e.g., the webs).

This brief explanatory listing of the content of the *AISI Specification* in regard to connections is intended to guide the reader to the type of information which is formed in the Specification. The amount of necessary information there included is far too extensive to permit its detailed and complete reproduction in this handbook.

20.10 TESTS FOR SPECIAL CASES

The cold-forming process, by cold-rolling or by press-braking, permits an almost limitless variety of shapes to be produced. They may consist of flat elements with rounded corners, of curved portions, and of any combination thereof. Load-carrying capacities and deflections of most usual shapes and even of many new shapes in the process of development can be calculated by the provisions and methods of the *AISI Specification*. There are, however, and always will be members and assemblies of such unusual shape that they cannot be analyzed by the methods of the Specification. Yet, this should not prevent their use if adequate safety is assured. Clearly, this can be done for such components only by test. Correspondingly, the Specification makes detailed provisions for such load tests and their evaluation. They are applicable only to members that cannot be calculated. In any situation which is covered by the Specification, (elements, members, assemblies, connections, details), design must be made by the Specification and determination by test is not permissible.

The Specification provides that at least three identical specimens be tested. Then, for a member or assembly designed for a dead load D and a live load L, the minimum load carrying capacity R obtained by tests shall be at least

$$R \geq 1.5D + 2L \tag{20.10.1}$$

Also, the load R at which distortions would interfere with proper functioning of the member or assembly in structure shall not be less than

$$R \geq D + 1.5L \tag{20.10.2}$$

Finally, if the load-carrying capacity turns out to be limited by connection failure, the minimum carrying capacity by test shall be

$$R \geq 2.5D + 2.5L \tag{20.10.3}$$

These provisions are not applicable to tests on specimens which can be calculated according to the Specification, so-called confirmatory tests. If such confirmatory tests are made, they shall simply demonstrate a safety factor no smaller than that implied in the Specification for the type of structural action involved (compression, bending, shear, etc.).

Another situation where testing may be necessary is in the utilization of the strain hardening which results from the cold-forming process (see Section 20.2.1). If design is to be based on the increased steel strength (over and above the virgin yield point) obtained by cold-forming, then this increased steel strength must be proved by test. The Specification gives detailed instructions on this type of testing for mechanical properties. It also gives provisions for determining the properties

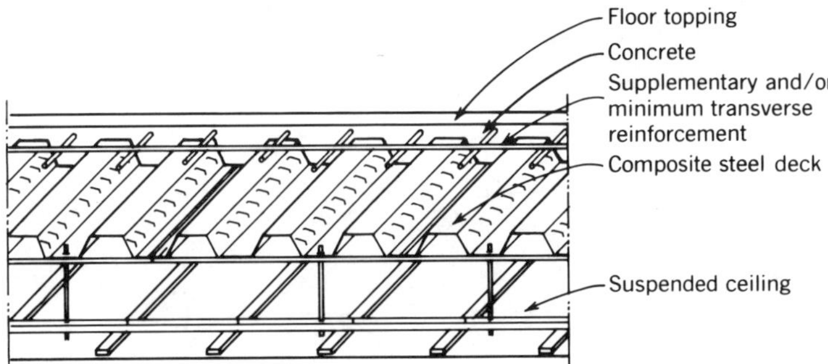

Fig. 20.17 Typical building floor construction utilizing cold-formed steel decking.

of virgin steel when steels are used other than those listed in the pertinent ASTM Specifications (see Section 20.1.3).

20.11 STEEL-DECK-REINFORCED COMPOSITE CONCRETE SLABS

One long-established and widely used application of cold-formed steel decks is in steel-deck-reinforced composite concrete slabs. In this type of construction the steel-deck panels serve both as forms during the construction process, and as positive reinforcement in the finished floor or roof slabs. Such slabs are used primarily in buildings where the main framing is of structural steel, chiefly in multistory buildings. Composite decks are also discussed in Sections 19.1.3 and 23.8.7.

Many types of steel deck are used for that purpose, all of them proprietary. Figures 20.17 and 20.18 show two of these types. Figure 20.17 shows the use of a trapezoidal deck. Embossments are provided on the inclined webs which have a very essential function: They ensure a greatly increased bond between concrete and steel deck compared to smooth steel deck surfaces. In this respect they play the role of the ribs on the surface of reinforcing bars. Since the deck constitutes the tensile reinforcement of the slab, such reliable bond is absolutely essential. The figure also shows nonstructural features, that is, the floor topping and a suspended ceiling.

Figure 20.18 shows another type of reinforcing deck in which increased bond resistance is provided by transverse wires welded to the top surface of the deck. In this application, some of the trapezoidal decks are replaced by cellular decks which serve as ducts for conduits for electrical current and for phone and other cables. Other cells in the same floor serve as air ducts for heating and air conditioning (see Chapter 4). Furthermore, it has been shown that such composite decks can be made to act in the other direction in a composite manner with the supporting steel floor beams. Figure 20.18 shows the shear connectors needed for that purpose. Reference 20.1 makes specific provisions for this use of composite slabs in composite action with the floor beams.

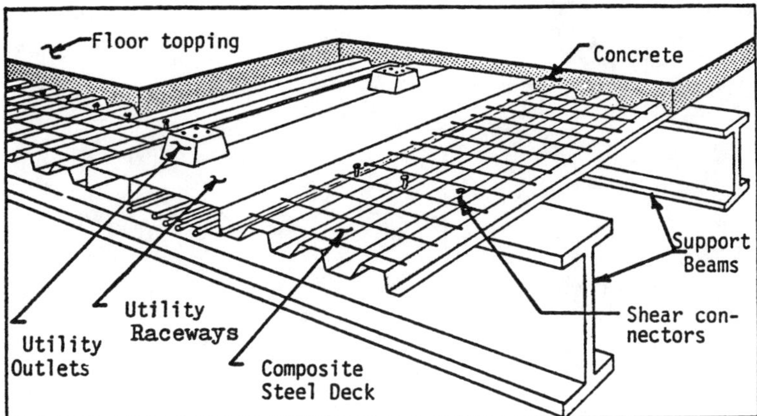

Fig. 20.18 Typical building floor construction utilizing cold-formed steel decking with composite support beams.

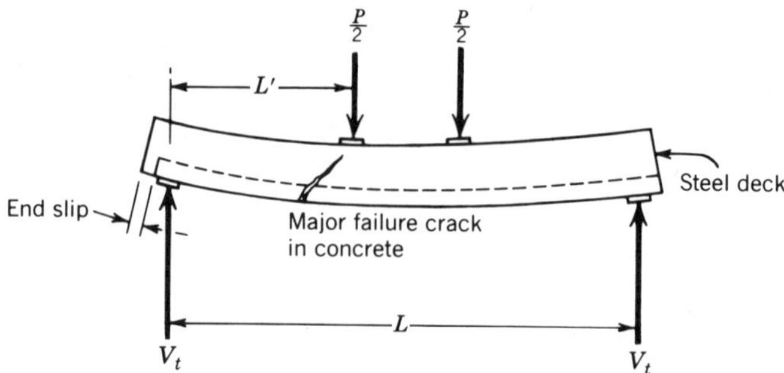

Fig. 20.19 Typical shear-bond failure for composite slab with decking.

The load-carrying capacity of such deck-reinforced slabs may be limited by their flexural strength in much the same way as for ordinary reinforced concrete slabs. Another type of resistance, however, is that known as shear-bond strength. That is, the bond on the deck–concrete interface may be lost, usually in the immediate vicinity of a major shear crack, and this loss of bond then usually propagates to the end of the slab, causing failure. This is shown on Fig. 20.19, where it can be seen that at the left end the concrete is horizontally displaced relative to the end of the deck.

This decisive shear-bond strength can be determined only by test. This is so chiefly because of the great variety of shapes of decks and of shear-transfer devices (variously shaped embossments, transverse wires, shape of the deck proper) produced by the various manufacturers. It is, in general, the manufacturer's responsibility to make the necessary tests and provide load tables for his products. The manner of doing this at this writing is in the process of standardization.

Tests are generally made with the two-point loading shown on Fig. 20.19. For the same deck shape, a variety of shear spans L' and effective slab depths d are tested. It has been shown [20.5, 20.7] that the nominal shear-bond strength per unit width of composite slab can be determined from

$$V_{nb} = \left(k\sqrt{f_c'} + m\,\frac{\rho d}{L'}\right) bd/s \tag{20.11.1}$$

where d = effective slab depth from top of concrete to centroid of steel deck cross-section
b = unit width, generally 12 in.
L' = shear span in test (see Fig. 20.19) = ¼ of span length for uniformly loaded spans
V_{nb} = nominal shear-bond strength per unit width
$\rho = A_s/bd$ = reinforcement ratio
k, m = constants determined for each individual type of deck from test plots (Fig. 20.20)
s = center-to-center spacing of shear transfer devices

It is generally sufficient to test two groups of dimension combinations as shown on Fig. 20.20, and connect them by a straight line.

As previously mentioned, the determination in this manner of composite slab strength is usually carried out by the manufacturer. However, the particular way of testing and evaluating has been

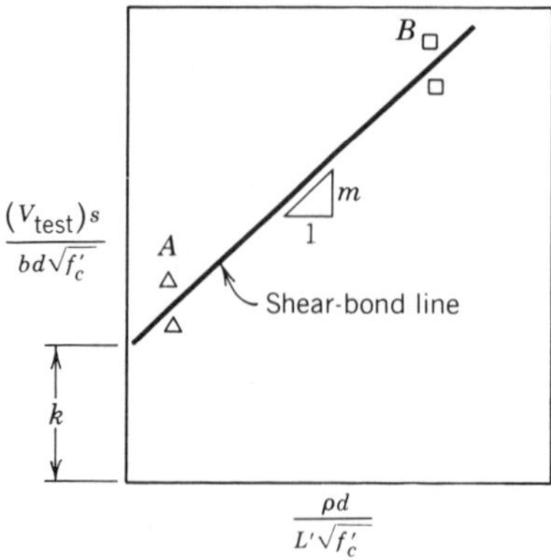

Fig. 20.20 Plot of test data for typical shear-bond failure for composite slab with steel decking.

developed only recently. The user of this type of construction may wish, or be asked by building officials, to verify the load capacity assumed in the design. It is for this reason that a brief account of such testing has been included here. In the foreseeable future "Criteria for the Design and Construction of Composite Steel Deck Slabs" will be officially published by ASCE. The criteria will be quite elaborate and will provide detailed directions for design, construction, testing, and other features.

SELECTED REFERENCES

20.1 AISC. *Specification for the Design, Fabrication and Erection of Structural Steel for Buildings, Effective November 1, 1978, with Commentary*. Chicago, Illinois: American Institute of Steel Construction, 1978.

20.2 AISI. *Specification for the Design of Cold-Formed Steel Structural Members*. Washington, D.C.: American Iron and Steel Institute, September, 1980.

20.3 AISI. *Cold-Formed Steel Design Manual, Part I, Specification; Part II, Commentary; Part III, Supplementary Information; Part IV, Illustrative Examples;* and *Part V, Charts and Tables*. Washington, D.C.: American Iron and Steel Institute, November, 1982.

20.4 AWS. *Welding Sheet Steel in Structures,* AWS D1.3-78. Miami, FL: American Welding Society, 1978.

20.5 Max L. Porter and Carl E. Ekberg. "Design Recommendations for Steel Deck Floor Slabs," *Journal of the Structural Division,* ASCE, **102,** December 1976 (ST11), 2121–2136.

20.6 RCSC. *Specification for Structural Joints Using ASTM A325 or A490 Bolts*. Research Council on Structural Connections of the Engineering Foundation, 1980.

20.7 Arthur H. Nilson and George Winter. *Design of Concrete Structures,* 10th ed. New York: McGraw-Hill, 1986.

20.8 Wei-Wen Yu. *Cold-Formed Steel Structures,* 2nd ed. New York: Wiley, 1985.

CHAPTER 21
PRESTRESSED CONCRETE DESIGN

ROBERT F. MAST

Chairman
ABAM Engineers Inc.
Federal Way, Washington

21.1 GENERAL

21.1.1 What Is Prestressed Concrete?

The term "prestressed" indicates that internal stress is applied prior to the application of gravity or other loads. Prestressing of concrete is done by tensioning steel against the concrete, thus pre-compressing the concrete. This avoids or minimizes cracking at service loads. Early attempts to prestress concrete with ordinary mild steel failed, because the long-term effects of creep and shrinkage caused the prestressing to be lost. Modern prestressing techniques overcome this difficulty through use of very high strength steel. With such steels, a reasonably large amount of residual prestress may be retained throughout the service life of a member after allowing for losses.

There are two primary methods of prestressing. In one method, called *pretensioning,* steel strands are stretched to high tension between two bulkheads anchored in the ground. Concrete is cast around the tensioned strands. After the concrete has reached sufficient strength, the strands are cut or otherwise released at the ends of the member, transferring the prestressing force to the concrete. Alternatively, in the *post-tensioning* method, concrete is cast around flexible metal ducts, creating tubular voids inside the concrete. Steel tendons are placed in these ducts and anchored at each end of the member. The steel is tensioned with jacks after the concrete has hardened.

Both methods of tensioning members are called *prestressing* because internal stress is applied prior to application of loads. The only real difference between pre- and post-tensioning is that the steel is tensioned before or after casting of the concrete, respectively. There is no word "post-stressing", although one occasionally hears this word substituted for post-tensioning.

Prestressing is often associated with precasting. The pretensioning method is usually used when the members are precast, that is, cast somewhere other than their final position in a structure. While post-tensioning is more commonly applied to cast-in-place members, it also proves useful for precast members, sometimes in conjunction with pretensioning. Precast members need not be prestressed; small precast members are often conventionally reinforced, as are most precast columns. Although precasting and prestressing are often used in conjunction with one another, this is not always so.

Some confusion exists as to whether the terms "reinforced concrete" and "reinforcement" include prestressed concrete and prestressing steel, or whether these terms specifically refer to concrete and steel that are not prestressed. The definitions given in *Building Code Requirements for Reinforced Concrete,* ACI 318-83,* [21.1] are as follows:

---

* Hereafter in this chapter, this document will be referred to as the *ACI Code*.

Reinforced Concrete. Concrete reinforced with no less than the minimum amount required by this code, prestressed or nonprestressed, and designed on the assumption that the two materials act together in resisting forces.

Reinforcement. Material that conforms to ACI-3.5, excluding prestressing tendons unless specifically included.

These definitions state that while "reinforced concrete" includes prestressed concrete, "reinforcement" excludes prestressing steel and refers only to conventional mild steel reinforcement. For the purposes of this chapter, "conventional reinforcement" and "conventionally reinforced concrete" are used to specifically refer to unstressed reinforcement and unstressed members, respectively.

21.1.2 Why Use Prestressed Concrete?

When prestressed concrete was initially introduced, it was hailed as a new material quite different from conventionally reinforced concrete, one which could produce concrete members free of tensile stresses and cracking. Prestressing can be used to negate primary longitudinal tensile stresses at service dead and live loads. However, tensile stresses exist in the transverse directions, caused by shear, bearing, and anchorage stresses. Few, if any, prestressed members are prestressed in all directions and so tension exists in prestressed members. Furthermore, it is not necessary nor economically desirable to limit longitudinal tensions to zero in most members.

It is much more useful to think of prestressed concrete as conventionally reinforced concrete with a few added features, a view reflected in the 1983 edition of the *ACI Code* and *Commentary* [21.2]. A major motivation for using prestressed concrete is economy of materials. Prestressing steel has a useful tensile strength about four times that of conventional reinforcement and a basic material cost generally less than four times that of reinforcement. Therefore, it is a more economical material in terms of cost per kip of supported load. High-strength steels cannot advantageously be used in conventionally reinforced concrete, because the high steel strains cause excessive crack widths. This difficulty is overcome by prestressing the steel, which removes the excessive strain during the prestressing operation and allows use of high steel stresses without resulting large crack widths.

Prestressing does not make concrete members stronger than does conventional reinforcement. Higher-strength concretes are obtainable in factory precasting operations and precast prestressed members are often made with higher strength concrete. The additional strength of such members is due to the higher strength concrete, not the prestressing.

Prestressing can greatly improve the durability of concrete members by eliminating most cracking at service loads. When combined with precasting, the better quality concrete obtained in a controlled factory operation may also significantly enhance durability.

Prestressing is very valuable in controlling deflections of concrete members. Because it produces deflections of opposite direction to those produced by the loads, prestressing significantly reduces the total deflection compared to a nonprestressed member of identical size. By controlling cracking, prestressing may also cause the gross cross-section of the concrete to be effective, whereas in conventionally reinforced concrete the effective moment of inertia of the cracked section is about half that of the gross section. These advantages encourage designers to use relatively large span-to-depth ratios in prestressed members. This often causes deflections to be a problem, but the cause is the large span-to-depth ratios used.

Probably the greatest deflection problems in prestressed concrete are those of excessive camber or upward deflection. This is caused by the members having too much prestressing for optimum serviceability. Newer codes have given the designer more latitude to proportion the prestressing force based on deflection considerations rather than solely on stress considerations.

21.1.3 Precasting of Concrete Members

Precasting and prestressing often are used in conjunction with one another, particularly when pretensioning is used. There are many advantages to precasting. These include savings in formwork, speed of construction, and minimized need for on-site labor. There are also disadvantages. Analysis is more complex because the stresses in the member must be considered during various stages of construction. Precast members must be connected together when they are placed in the structure. When precast members are abutted to cast-in-place concrete, the connections are often simple, similar to the construction joints used in cast-in-place construction. When precast members are abutted to other precast members, however, a family of connection designs is necessary for proper behavior of the structure.

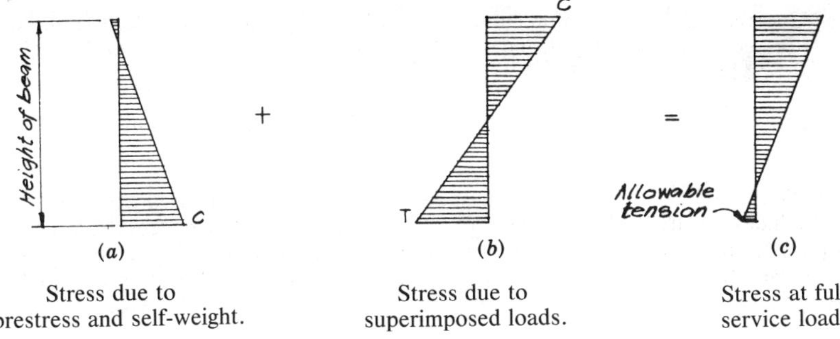

Fig. 21.1 Internal stresses in a prestressed member.

21.1.4 The Basic Principle of Prestress Design

Flexure in a simple span beam produces tension in the bottom fiber. Prestressing may be used to provide an initial compression in the bottom fiber, so that when the superimposed loads are applied, the resulting tensile *change* in stress results primarily in relaxation of precompression in the bottom fiber. A small net tensile stress at full service load is usually permissible. Figure 21.1 shows the flexural stresses graphically.

The stresses in the bottom fiber are given by the following fundamental equation (see Section 21.11 for notation):

$$\frac{P}{A} + \frac{Pe}{S_b} - \frac{M_g}{S_b} - \frac{M_d}{S_{b(c)}} - \frac{M_\ell}{S_{b(c)}} = f_{t\ \text{allow}} \qquad (21.1.1)$$

The first three terms represent the stresses due to prestress and self-weight and are shown graphically in Fig. 21.1*a*. The following two terms represent the stresses due to superimposed dead and live loads, shown in Fig. 21.1*b*. The resultant of all these stresses produces the stress under full service load, shown in Fig. 21.1.*c*. Usually, a small tensile stress is permissible at full service load.

The above discussion concerns bottom fiber stress, for this normally controls in the design of (simple span) prestressed beams. Top fiber stresses must also be checked, of course. Similar considerations apply to continuous beams, but "bottom fiber" should be read "tension side" which is normally the bottom at midspan, but the top at the support.

The bending stress due to self-weight is included with the prestress in Fig. 21.1*a*, rather than with the other superimposed load stresses shown in Fig. 21.1*b*. The application of the prestress force tends to lift the beam from the form. Consequently, the beam "feels" the effects of the prestress force and the self-weight bending moment *simultaneously* as the prestressing force is applied. This is a very important principle. Normally, the beam need not be checked for stresses due to the prestress alone, but rather for stresses due to the simultaneous action of prestress and self-weight bending. Often, the self-weight of a prestressed member may be supported "for free" since the self-weight bending moment may be counteracted by additional eccentricity of the prestressing force.

21.1.5 Scope of This Chapter

It is the intention of this chapter to treat the topics of prestressed concrete relating to building design that are outside the scope of an undergraduate course on prestressed concrete design. The emphasis here is toward the special concerns of the practicing engineer. For extended treatment of the basics of prestressed concrete, the reader is referred to the books by Lin and Burns [21.20], Nilson [21.21], Naaman [21.22], and Libby [21.23].

21.2 COMPARISON OF PRESTRESSED AND CONVENTIONALLY REINFORCED CONCRETE

As previously stated, prestressed concrete should be thought of as conventionally reinforced concrete, with some special features. Ideally, the same design equations should apply to both prestressed and conventionally reinforced concrete design, simply by setting the prestressing term equal to zero in the equations. While the trend of the *ACI Code* is in this direction, the goal has not been fully accomplished. There are a number of considerations unique to prestressed concrete and these are covered in a specific prestressed concrete chapter in the *ACI Code*. The great bulk of the *ACI Code,* however, applies equally to both prestressed and conventionally reinforced concrete.

By definition, a prestressed member has internal stress established prior to application of dead and live loads. A very important consequence is that the stress in a prestressed member is not proportional to the load. It often happens that the highest stresses in the member occur in the unloaded state and the stresses at full design load are low by comparison. This is in contrast to most wood, steel, and conventionally reinforced members in which stress is proportional to load. Such conventional members are normally checked either at service load or at ultimate load, but not both. The proportionality of stress and load allows the behavior at one load level to be inferred from the computations done at the other load level. However, prestressed members, for which stress and load are not proportional, require that analyses be made at both service and ultimate (factored) load levels. Without analyses at both load levels, one might design members which give satisfactory performance at one load level but prove unsatisfactory at the other.

21.2.1 Flexure

The analysis of the ultimate strength of prestressed and conventionally reinforced members is almost identical. The main difference is that the prestressing steel has no flat yield plateau similar to that for conventional reinforcement. As a result, determination of the steel stress at ultimate load is somewhat more complex. *ACI Code* Formulas (18-3), (18-4), and (18-5) may be used to determine the steel stress at ultimate load, or, for bonded members, the stress may be determined from a strain compatibility analysis using the stress–strain curve for the steel. Such strain compatibility analyses show that initial prestressing has only a minor effect on the steel stress at ultimate strength. In most cases involving the use of T-beams, the steel percentages are quite low and identical members with stressed and unstressed high-strength prestressing steel will have approximately equal ultimate strengths. (They will have markedly different performance at service load, however.)

Figure 21.2 compares internal stress conditions existing at service load and at ultimate strength in prestressed and conventionally reinforced members. At ultimate strength, there is very little difference in the internal conditions of the two types of members. At service load, however, the difference is quite significant. The neutral axis is much farther from the face of maximum compression in the prestressed member than in the conventionally reinforced member. The concrete strain in the bottom fiber is much smaller in the case of the prestressed member. While the reinforced member will generally be cracked, the prestressed member generally will not. The strain gradient in the prestressed member is approximately half that of the reinforced member. Since deflections are proportional to the strain gradient, the deflection is also half or less that for a prestressed member.

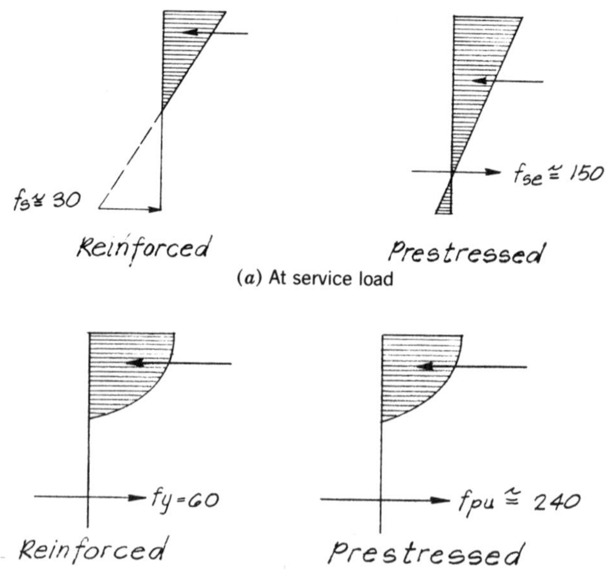

Fig. 21.2 Internal stresses in reinforced and prestressed members.

21.2.2 Deflections

Under dead load, a nonprestressed member will have a downward deflection. The effects of prestressing, however, normally override those of dead load and upward deflection (camber) in prestressed members at dead load is common. At full service load, the shape of the stress diagram indicates a strain gradient that would produce a downward deflection. Service live loads are normally of a transient nature, however, whereas prestress and dead loads are permanently applied. Creep effects from years of application of permanent prestress loads may produce upward deflections in excess of the downward deflection from the live loads. Thus, prestressed members often have an upward deflection even at full service load. This has been a problem. Current codes permit proportioning the prestressing force to achieve a more desirable deflection at service load. The resulting bottom fiber tension at service load is now permissible.

Uncracked prestressed members have smaller deflections than conventionally reinforced members of the same size. However, after cracking, prestressed members deflect at a much more rapid rate than conventionally reinforced members. After cracking, the steel area is very important in determining the moment of inertia of the cracked section. Prestressed members have low steel areas and are subject to rapid increase of deflection after cracking. Deflections of prestressed members at ultimate load are large. This is desirable from the standpoint of ductility because it provides ample warning of overloads. It is obviously undesirable to have large deflections under loads anticipated during normal service.

21.2.3 Shear

Ideally, prestressed and conventionally reinforced members would be analyzed for shear using a common method that would treat the prestressing force as an additional load. Such a goal has not yet been achieved in the *ACI Code*. There is an equation (11-4) for analyzing a nonprestressed member with axial compression, but it gives substantially lower results than the equations for prestressed concrete.

Prestressing has three favorable effects on shear strength. The longitudinal compression from prestressing reduces principal tensions caused by shearing stresses. The prestressing delays the formation of flexural cracks and thus delays the interaction between flexure and shear. Part of the shear may be carried by the upward component of the tension in a sloping tendon. Of these effects, delay of flexural cracks is the most important for members of ordinary proportions. Formula (11-11) of the *ACI Code* is used to analyze the interaction of shear and bending in prestressed members. Unfortunately, this equation is complicated to use because the location of the critical section is not readily apparent and checks must be made at several points along the span. Formula (11-13) of the *ACI Code* is a simplified method for checking the effects of prestressing on the principal tension at the neutral axis.

In conventionally reinforced concrete design, separate equations comparable to (11-11) and (11-13) do not exist. Rather, a single equation, (11-3), is used to cover both the effects of flexure and of principal tension. Formula (11-3) generally gives shear values of half or less those given by the equations for prestressed concrete. Because it is very simple to use, it is sometimes used to analyze prestressed members, albeit conservatively.

The equations cited above are used to determine the shear strength attributable to the concrete. The difference between the factored shear and that attributed to the concrete must be resisted by reinforcement. The design of shear reinforcement (stirrups) is the same for both prestressed and conventionally reinforced concrete. (See Section 11.5.6 of the *ACI Code*.)

21.2.4 Compression

Prestressing adds to the compressive stress in the concrete. With the following important exceptions, compression members are normally not prestressed.

Piles

Piles, normally thought of as compression members, must resist substantial tensile stresses during handling and driving. For this reason, most concrete piles are now prestressed. Prestress levels of 600–800 psi (4.1–5.5 MPa) are common, although experience has shown that using higher prestress levels (up to 1200 psi) (8.3 MPa) will permit heavier driving and higher pile capacities. Prestressing makes possible the handling of very long slender piles, as shown in Fig. 21.3.

Handling

Nominal requirements rather than stress often determine column sizes in low- and medium-rise buildings. It is sometimes desirable to precast columns in fairly long lengths and prestressing may be used to improve the handling characteristics of such members. This is also true of walls.

Fig. 21.3 Handling of long slender prestressed piles. Courtesy of Jones Photo Co.

Resistance to Buckling

Prestressing improves the buckling characteristics of long, slender compression members because it delays the onset of cracking and the reduction in moment of inertia associated with cracking. See Refs. 21.3 and 21.16.

21.2.5 Reversible Stresses

Free-standing ("flagpole") columns are subject to reversible stresses from lateral wind loads. Although such members are called columns, they function more as "vertical beams" for the bending stress usually dominates the compressive stress. Normally, it is more advantageous to use conventional reinforcement rather than prestressing in such members. Figure 21.4 shows internal stress diagrams for conventionally reinforced and prestressed members subject to reversible stresses. While conventional reinforcement works both in tension and compression, prestressing steel works only in tension. Furthermore, since the stresses are reversible, the centroid of the prestressing force must be located at the center of the member, significantly reducing the lever arm of the steel force.

Building members do not normally receive an explicit analysis for fatigue. It is generally assumed that when the stresses and other provisions of the *ACI Code* are followed, the structure will be

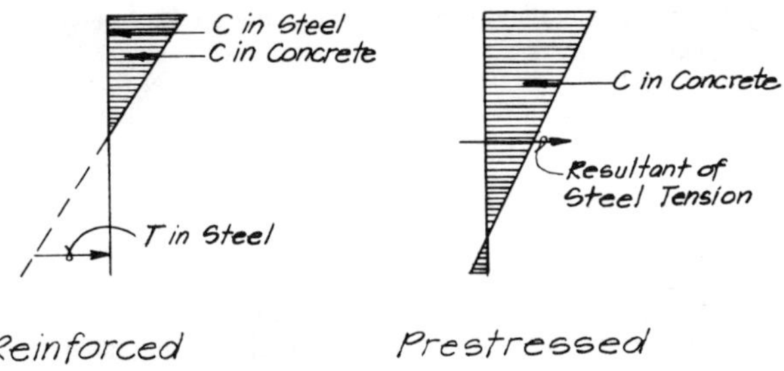

Fig. 21.4 Members subjected to reversible stress.

satisfactory for fatigue loads encountered in ordinary applications. Where fatigue loading is known to be an important consideration, however, such as wave loadings on offshore structures, prestressed concrete may be very advantageously used to resist fatigue. If prestressing is used to eliminate flexural tension, and if concrete stresses are limited to $0.45f'_c$ or less, the concrete will have an essentially unlimited fatigue life. This is also true for the prestressing tendons since the stress range in the prestressing steel associated with the concrete cycling through the stress limits just described is a small percentage of the steel's ultimate strength.

21.2.6 Combinations of Conventional Reinforcement and Prestressed Reinforcement

For members subject to reversals of load, conventional reinforcement has the great advantage of working in both tension and compression. But, prestressing a member can improve shear resistance, reduce crack widths, reduce deflections, and improve durability. Why not combine the two methods of reinforcement within the same member? This is sometimes done and a member of improved performance characteristics should result, but it is difficult to produce such designs following the *ACI Code*. If the member is classed as a prestressed member, it must be checked for tension at service loads. This requires the design to be primarily a prestressed one, although small amounts of conventional reinforcement may be advantageously used to supplement the prestress and improve the post-cracking behavior. Future developments in concrete research may pave the way for more extended use of the two techniques in combination.

21.3 MATERIALS

A wide variety of concretes, steels, and beam sections are available throughout the United States. The following comments are offered as general guidelines for the selection of materials. The designer is encouraged to consult local suppliers and to use materials locally available, whether or not they conform to these general guidelines.

21.3.1 Concrete Strengths

Concrete strengths of 5000 psi (34 MPa) are available throughout the United States for precasting operations and for field-casting of important members such as girders and columns. The strength at release of prestress is an important parameter in the design of prestressed members. Factory-produced members are often produced on a 24-hr cycle. This allows a maximum of about 16 hr for curing prior to release of prestress. A concrete strength at release, f'_{ci}, of 3500 psi (24 MPa) is commonly specified for pretensioned members. An f'_{ci} of 4000 psi (28 MPa) is obtainable in most areas, but should only be specified if it is actually required by the design. For post-tensioned members, a minimum strength of 4000 psi (28 MPa) at the time of application of post-tensioning is normally specified.

Higher strengths are common in certain areas. In some states a 28-day strength of 6000 psi (41 MPa) and a release strength of 4500 or 5000 psi (approx. 30 or 35 MPa) are common. Even higher strengths, 7500–9000 psi (approx. 50–65 MPa) at 28 days, are possible in certain localities for special projects. Local practices and conditions should be determined prior to specifying these high strengths.

21.3.2 Lightweight Concrete

Lightweight concrete can be advantageously used to reduce shipping and handling weights of large precast members, as well as the dead load of the structure. Because local availability and cost of high-quality lightweight concrete suitable for prestressed members vary considerably, both should be determined before specifying. It should also be remembered that the stiffness of lightweight concrete is approximately half that of normal-weight concrete and that the design of many precast prestressed members is controlled by stiffness. If one compares truly equivalent (in stiffness as well as in strength) lightweight and normal-weight concrete members, the equivalent lightweight member will be somewhat deeper.

21.3.3 Prestressing Materials

The most commonly used prestressing material is seven-wire strand of either 0.5- or 0.6-in. nominal diameter with a tensile strength of 270,000 psi (1860 MPa). The same material is the standard in metric countries, although it is called 12.7- or 15.2-mm strand and the strength is rounded off to a metric equivalent. The same strands used for pretensioning are also widely used for post-tensioning. Additionally, button-headed solid wires and solid bars are sometimes used for post-tensioning. Solid bars have the advantage of being easily coupled and anchored and thus are often used for applications requiring short tendons or coupled tendons, such as in segmental construction.

21.3.4 Reinforcement

Grade 60 reinforcement with a yield strength of 60 ksi (414 MPa) is now the universal standard in the United States. Because Grade 60 bars may not be weldable, designers sometimes specify Grade 40 for welded applications. Still, many suppliers will furnish Grade 60 when Grade 40 is called for. If welding of reinforcing bar is contemplated, it should only be done in accordance with strict quality control procedures to ensure that the reinforcing bar being used is truly weldable.

21.4 DESIGNING THE WHOLE BUILDING

Most textbooks concentrate on the design of a building's individual members. If one designs each and every member in a building, has he designed the building? Of course not! Even though each and every member has been designed, there is still the interaction between members to consider. When designing columns, for instance, one often assumes the columns are braced at each floor. Braced to what, another column? A column that needs bracing is hardly fit to brace another. Are the columns braced by shear walls? Not if there is an expansion joint between the column and the shear wall.

One must design the building first, then design the members. "Designing the building" involves making a number of fundamental decisions about building behavior and member interaction.

1. *Lateral Loads.* Are lateral loads resisted by frame action or by shear walls? Or perhaps are they resisted by frames in one direction and shear walls in the other? This decision will have great consequences on the design of members and their connections.

2. *Stability Under Vertical Loads.* This consideration is closely related to resistance to lateral loads because the assumption for stability under vertical loads must be consistent with that for lateral loads. If lateral loads are assumed to be resisted by frame action, one cannot assume that the frame is braced for vertical loads. Conversely, if lateral loads are resisted by shear walls, the columns normally may be assumed to be braced. A force is required to brace a column against P–Δ effects, and this force imposes an added load upon the shear wall.

3. *Diaphragms and Their Connections.* If lateral loads are assumed to be resisted by shear walls and columns assumed to be braced, then floor and roof diaphragms must be provided. The diaphragm connections must have sufficient strength to transfer the lateral and bracing loads from their point of application to the bracing element. What about volume change effects? How are the deformations due to creep and shrinkage accounted for? If each member is assumed to creep and shrink independently of another, this may be inconsistent with the need to have an interconnected diaphragm.

4. *Compatibility of Member and Connection Behavior.* If the member is assumed to be simple span, the connections must allow for the end rotations associated with simple span behavior. If the member is assumed to be unrestrained, the connections must allow for its shortening due to creep and shrinkage.

5. *Compatibility of Load Assumptions from One Member to the Next.* If the designer of a floor or roof unit assumes that the member is supported at the center of the bearings, the designer of the supporting member should not assume that the load is centered on that member when the bearings of the supported units are not centered on the supporting member.

6. *Construction Sequence.* The construction sequence can have a major effect on member design. It may determine what loads are carried on the bare beam and what loads are carried on a composite section. It may also determine what loads are carried on a simple span system and what

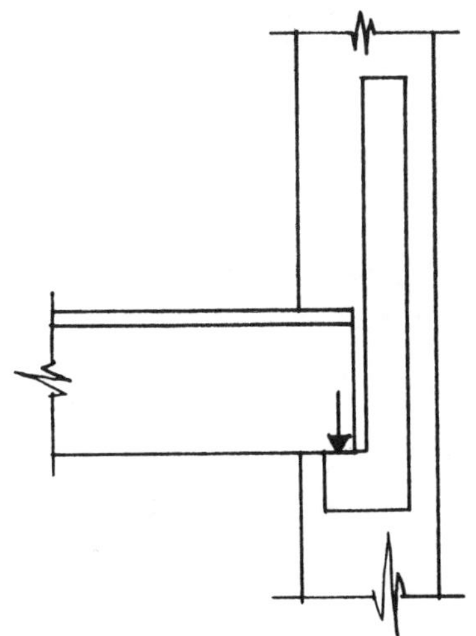

Fig. 21.5 Double tee, spandrel, and column.

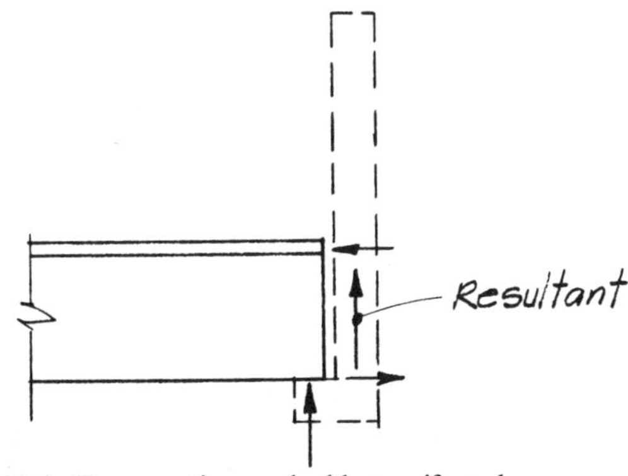

Fig. 21.6 Forces acting on double tee, if resultant is at centerline of spandrel.

loads are carried on a continuous system. These questions are discussed at greater length later in this chapter.

The importance of designing the overall building prior to beginning member connection design can best be illustrated by an example. Consider the detail shown in Fig. 21.5. A double-tee floor is supported by a spandrel beam which is in turn supported by a column. The double tee would normally be designed as a simple span member with the span length being the center to center of bearings. The spandrel designer must therefore assume that the applied loads are at the center of the double-tee bearing, not at the center of the spandrel. One might argue that the double tee could easily span center to center of spandrel since the additional span is only a few inches. This would be true only if the end connection design of the double tee accounted for the horizontal tension at the bearing associated with such an assumption (see Fig. 21.6). If the double tee is designed assuming supports at the center of the bearing, the spandrel and its connection to the column should be designed for the torsional moments consistent with this assumption.

Consider the columns. If the spandrel is designed for a combination of vertical load plus torsion, the column design should not be assumed to be concentrically loaded but rather eccentrically loaded, as shown in Fig. 21.7. Does this mean that the total load on the column produces a moment diagram for the column shown in Fig. 21.8? Not necessarily; equal and opposite effects may be generated by the column on the other side of the building. Of course, there are often certain areas of the building where, due to asymmetry, equal and opposite forces are not present. The column may be braced by shear walls. If the column is braced, either by shear walls or by equal and opposite moments on the other side of the building, then the moment diagram for the column will look more like that shown in Fig. 21.9. It should be noted that if the column is braced by shear walls, bracing forces will be delivered to the shear walls.

The previous illustration should make it clear that the design of the individual elements—double tees, spandrels, columns, and shear walls—cannot be delegated to individual designers without first determining how these members interact with one another. This information must then be communicated to the designers of individual members and their connections.

21.5 LATERAL LOAD RESISTANCE

21.5.1 Comparison of Prestressed and Cast-in-Place Reinforced Concrete Construction

Frame action is commonly used to resist lateral loads in cast-in-place reinforced concrete construction. Girders are often framed into all four sides of a column and continuity is easily obtained by passing reinforcing steel through the column. This four-way beam connection at the column presents difficulties in precast buildings since, without corbels, there is not enough room to make a four-way connection between beam ends. Even with corbels, a four-way connection is undesirable. When precasting columns, it is desirable to have corbels only on the top or bottom of the column as

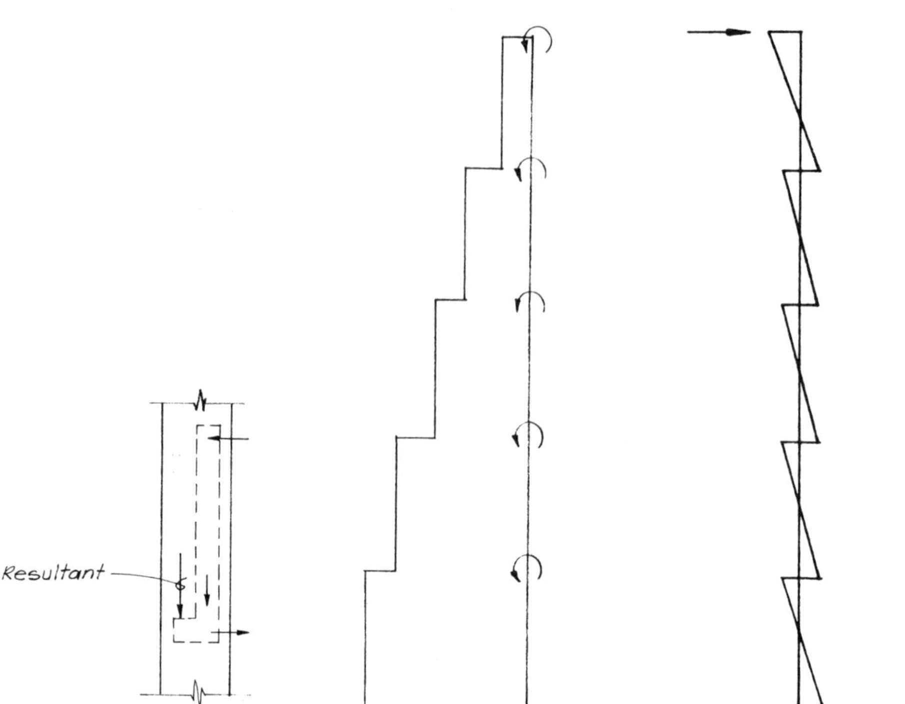

Fig. 21.7 Reaction to column.

Fig. 21.8 Unbraced column moment diagram.

Fig. 21.9 Braced column moment diagram.

it is cast because having corbels projecting from the sides requires penetration of the side forms. Figure 21.10 shows a precast framing system which avoids four-way connections at the columns.

Shear walls are commonly used for lateral load resistance in precast construction. Concrete walls are often present in cast-in-place construction and their contribution is sometimes ignored. Since walls are generally much stiffer than frames, they will resist most of the lateral load. Designing the shear walls to resist all the lateral loads relieves the designer of the connection problems in the beam and girder frames associated with designing for lateral loads.

21.5.2 Column Stability

When designing columns, an important distinction must be made between columns braced and unbraced against sidesway. The effective length of braced columns is equal to or less than the actual length, whereas the effective length of unbraced columns is always greater than the actual length. Assumptions regarding braced vs. unbraced column design must be consistent with assumptions regarding lateral load resistance. If columns are assumed to be braced, then bracing elements (usually shear walls) must be provided and used to resist lateral loads. Conversely, if columns are designed as unbraced, then lateral loads will normally also be resisted by the columns acting as a part of a frame and the P–Δ effects from lateral loads must be considered in the column design for both vertical and lateral loads.

It is valid and fairly common to design a building using shear wall bracing in one direction and beam-column frames for bracing in the other direction. In such cases, the designer must use two different column bracing assumptions when designing for moments about the two axes of the column.

21.5.3 Low-Rise Buildings

Most low-rise precast buildings are designed using shear walls for lateral load resistance (see Chapter 7). This greatly simplifies the beam-to-column connections. Often the lateral resistance can

Fig. 21.10 Precast beam and girder framing.

be provided with walls which are required for other purposes. Less common than shear walls is use of diagonal bracing to resist lateral loads in low-rise buildings. The diagonal bracing may be either steel or concrete. If of steel, it should normally be fireproofed to preserve the fire ratings available to the concrete structure.

Low-rise buildings are sometimes designed using free-standing "flagpole" columns. This is particularly useful in very large one-story buildings such as warehouses and industrial buildings in which there may be a long distance between shear walls. In this system, the columns may be designed fixed at the base with no moment resistance at the top. Each column is then assumed to take its share of lateral load as well as vertical load. This design approach is simple because moment connections are not required at the top of the column. It must be remembered, however, that the effective length of such a "flagpole" column is at least twice its actual length and this must be accounted for in the column stability design.

21.5.4 High-Rise Buildings

Precast prestressed high-rise buildings may also utilize shear walls to resist lateral loads. In high-rise buildings, overturning moments are significant. This is particularly true where end walls, which have relatively small vertical load, are used as the primary lateral load resistance. The overturning problem may be mitigated by use of coupling beams to couple shear walls together and thus broaden the base available for overturning resistance. The design of coupling beams to resist seismic loads is a special art (see Ref. 21.15). Another way of mitigating the overturning problem is to design for interaction between shear walls and frames. When this is done, it will often be found that most of the shear in the lower stories is resisted by the shear walls, whereas most of the overturning moment is resisted by the frame. In such designs, one cannot properly distinguish between braced and unbraced columns. The design should be done in accordance with Section 10.10 of the *ACI Code,* with factored lateral loads applied to the combined shear wall and frame utilizing effective stiffnesses applicable to the deformations at ultimate load. Extensive discussion of high-rise building design is given in Chapters 10 and 11.

High-rise apartments and hotels are often rectangular in plan, with many walls in the transverse direction and few walls in the longitudinal direction. Such buildings may be designed using shear walls in the transverse direction and frame action in the longitudinal direction. For wind design, the forces in the longitudinal direction are quite small and this approach is quite effective. Where seismic loads must also be considered, the longitudinal forces are larger and well-designed frames are needed to resist the longitudinal seismic loads.

21.5.5 Shear Wall and Diaphragm Design

Knowledge of frame design is more common than knowledge of shear wall design. Many good books are available on frame analysis and design, but few on shear wall design. Perhaps this is because shear wall design appears simple and checking the shear in a shear wall is no big design problem. Do not be so easily led astray; there is much more to the design of shear walls, as shown in Chapters 7, 24, and 25.

There must be a way for the load to get to the shear wall. Wind and seismic loads are applied to the building as distributed loads and these loads must be collected and delivered to the shear wall. This requires the use of floor and roof diaphragms. Sufficiently strong diaphragms do not occur automatically in precast buildings. Members must be properly interconnected and an adequate connection must be provided to transfer the load into the wall. The shear stress in the diaphragm at the shear wall is often much higher than the average shear stress in the diaphragm. Diaphragm design also requires considering bending moments as well at shear. For instance, whereas shear walls are commonly placed at the ends of the building, the wind or seismic load is uniformly distributed. The diaphragm must function as a horizontal beam spanning perhaps hundreds of feet. In precast diaphragms, this bending strength must be designed into the connections.

The design of a shear wall also involves design for bending moments as well as shear. The horizontal loads transferred from the diaphragm to shear wall create overturning moments. Designing for these moments can be more difficult than designing for shear. This is particularly so at the footing, where only the gravity loads on the shear wall are available to resist overturning. Sometimes grade beams or other devices must be used to increase the size of the base available to resist overturning. Sliding at the footing must also be checked. If the gravity loads acting on the shear wall are small, sliding resistance may be inadequate. In such cases, the shear wall footing may be interconnected to other footings through floor slabs or grade beams.

The *PCI Design Handbook* [21.17], Chapter 4.8, is a good treatise on diaphragm design. Although this chapter deals with seismic design, the principles are equally applicable to diaphragms resisting wind loads.

21.6 FRAMING FOR VERTICAL LOADS

21.6.1 Framing Plans

For post-tensioned cast-in-place construction, framing plans may be similar to those used in conventionally reinforced cast-in-place construction. One must provide for continuity of tendons and clearance for jacks at the tendon ends. A post-tensioned structure will shorten during the post-tensioning operation and will continue to shorten thereafter due to creep effects. Thus, the layout of rigid elements such as shear walls and stair and elevator shafts must be handled with care. Problems of restraint by rigid members may be overcome by locating the rigid elements near the center of the building, by using pour strips to allow motion during the post-tensioning operation, or by detailing a flexible joint around the rigid elements.

Whereas the framing plan for a post-tensioned structure may be similar to that for a cast-in-place concrete building, the framing plan for a precast pretensioned structure will usually be quite different. The following general rule should be used as a guide in selecting a precast framing system.

Make the connections simple and minimize the number of precast elements.

"Simple" in the above rule means uncomplicated, but not necessarily simple span. Application of this rule often results in precast framing systems quite different from those used in cast-in-place construction.

One strategy for making connections simple is to alternate precast and cast-in-place members within the structure. The connections are then similar to construction joints in cast-in-place construction. Where shears are high, shear transfer may be effected through the use of roughened surfaces and shear-friction reinforcement. (See Section 11.7 of the *ACI Code*.) One common method of alternating precast and cast-in-place construction is to use prestressed girders in conjunction with an otherwise cast-in-place structure (Fig. 21.11). This is a very common method of construction for bridges which also finds some application in buildings in which the use of prestressed girders is advantageous on moderately long spans. Another commonly used combined precast and cast-in-place system is the "soffit beam" construction illustrated in Fig. 21.12. In this construction, a wide precast (usually pretensioned) soffit beam is supported on shores and precast hollow-core floor units are placed on the soffit beam. The space between the ends of the precast floor units is then concreted, forming a composite beam with the soffit. While this technique of

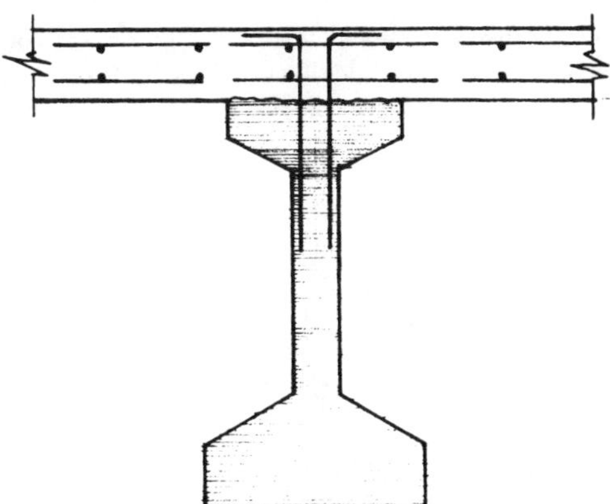

Fig. 21.11 Cast-in-place deck on precast beam.

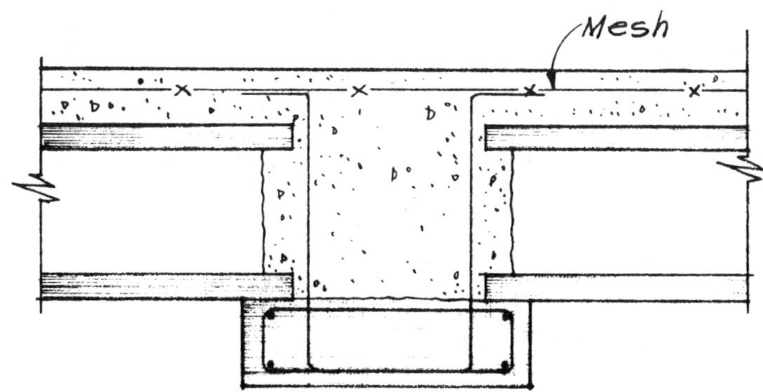

Fig. 21.12 Hollow-core slabs on soffit beam.

combining precast and cast-in-place construction is quite useful, it requires formwork and shoring, and thus some of the advantages of all-precast construction are lost.

In construction having mostly precast members, the application of the general rule requires one to avoid, if possible, having beams and girders in both directions. Placing beams in one direction only obviously helps to minimize the number of precast elements and avoids complex three- and four-way connections at columns. Of course, if beams are used in one direction only, lateral stability in the other direction must be provided by shear walls or other means.

Floor (and roof) units are the most commonly used precast and pretensioned members. A "floor unit" is a member which is abutted to adjacent members to cover an entire floor area without an intermediate space to be formed and cast. Floor units are commonly covered with a topping slab, although this is not always the case. The two most common types of floor units are hollow-core slabs and double tees. Hollow-core slabs are used on short to medium spans (15–50 ft) (4.6–15 m) and double tees are used on medium to long spans (30–70 ft) (9–20 m). Hollow-core slabs are made in 4- and 8-ft (1.2- and 2.4-m) widths. Double tees are made in 8- and 10-ft (2.4- and 3.0-m) widths. Precast solid planks, channel slabs, and single tees are also used as floor units. Solid planks are used on short spans, single tees on long spans, and channel slabs are often made to order for either short or long spans.

Floor units require essentially continuous support. This support may be provided by bearing walls or by beams. Often a spandrel beam proportioned primarily for architectural considerations may be utilized as a structural support for the floor units. The most common framing plan is to run the floor units in one direction and the beams or walls in the other direction. Changes in the direction of framing within a floor plan should be avoided if at all possible. Where bays are rectangular in shape, it is often best to run the precast floor unit in the long direction and the supporting beam or girder in the short direction. The cost increase of longer-span floor units is more than offset by savings in the shorter-span girders. Also, the depth of the floor unit and girder are more in proportion to one another when the building is framed in this manner. Note that this practice is somewhat different from the common cast-in-place construction practice in which longer-span members generally support the larger area.

21.6.2 Simple Span Construction

Simple span construction seems to be the "natural" way to build with precast units. Simple span construction is simple to design, simple to build, and requires only vertical load support at the connections (although lateral loads must be resisted somehow, usually by shear walls). Creep and shrinkage forces may be minimized by allowing a small amount of motion at each joint, thus avoiding the accumulation of forces and deflections caused by creep and shrinkage. In general, simple span construction is an economical way to use precast pretensioned members.

There are some disadvantages to simple span construction, however. Simple span construction implies movement at the member ends. The member ends will rotate under the effects of creep and live load, and this may lead to horizontal separation at the top or bottom of the members. Creep and shrinkage deformations will also cause such separations. These motions may detract from the serviceability of the structure. This is particularly true of floors but may also be a problem on roofs.

A more serious problem with simple span statically determinate construction is the possibility of "progressive collapse". If members are not well interconnected, a situation may arise in which the failure of one member leads to a progressive collapse of other members and total destruction of the structure. The layman's term for this is a "house of cards". Protecting against progressive collapse will generally involve interconnection of members. Lateral load resistance will also usually require interconnection of members. In precast structures, lateral loads are most commonly resisted by shear walls and the floors and roofs must function as diaphragms to transfer the loads to the walls. Thus, a simple span structure must have end connections which are flexible enough to allow rotation and other movements at the end of the beams, but strong enough to transmit lateral loads and resist progressive collapse.

21.6.3 Continuity

Continuity has several advantages to the performance of structures. It eliminates movements at the beam ends, thus giving better serviceability. It reduces moments, thus increasing structural capacity. It significantly reduces deflections. Continuous structures have good diaphragms for resisting lateral loads and avoiding hammering damage at joints during earthquakes. Continuity also has its disadvantages. Design and construction are both more complicated. When a structure is designed to be continuous, there are often a few members that cannot be made continuous because of discontinuities in the structure (stair openings, etc.). The additional cost of making members continuous often outweighs the structural advantages of continuity. When structures are made continuous, creep and shrinkage strains accumulate throughout the length of the continuous members and must be accounted for.

The "natural" way to design a "monolithic" cast-in-place structure is as a continuous structure because this preserves the monolithic nature of cast-in-place construction. This also applies to cast-in-place post-tensioned structures. Even when precasting is used in conjunction with post-tensioning, full continuity may often be readily obtained. For precast pretensioned structures, however, extra design and construction efforts are required to obtain continuity.

21.6.4 Partial Continuity

There is no universal definition of partial continuity although, intentionally or not, many precast structures behave as partially continuous structures.

Structures may be designed to be partially continuous. A common practice is to design the structure as a simple span structure for most dead loads, but to make the structure continuous for live loads. This may be done by adding reinforcement to a cast-in-place topping or deck slab (Fig. 21.13). This is an economical way of obtaining continuity and improving serviceability. It is often not realized that substantial positive moment caused by creep cambering and/or temperature differentials may develop at the support. This positive moment must be allowed for, either by splicing a substantial amount of bottom reinforcement or by providing beam bearing connections which will allow for horizontal motion.

Another type of partial continuity occurs in a structure which is designed totally as a simple span structure but in which some continuity reinforcement is provided. An example of this can be seen in Fig. 21.12. The mesh in the topping provides some continuity for the hollow-core slabs even though the designer probably would not consider it in the hollow-core slab design. It is important to

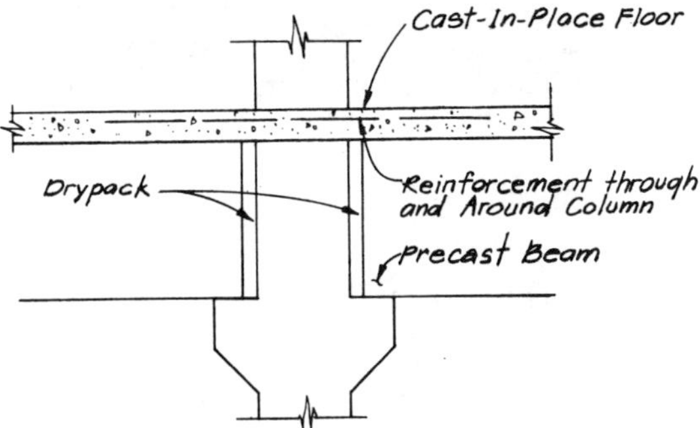

Fig. 21.13 Continuity by reinforcement in deck.

recognize that live loads will create negative moments which may cause the mesh to yield and significant cracking may thus develop in the slab. To avoid this, reinforcing bars may be added over the support to resist the calculated continuity moment from live load only. Another type of partial continuity is illustrated in Fig. 21.14. In this case, no attempt is made to provide moment continuity; rather, continuity is provided only for diaphragm action. In this case, beam end rotations will cause horizontal movements at the beam end bearings and this must be allowed for in the connection design. Also, it must be recognized that beam end rotations will cause some inelastic behavior in a topping slab.

There is no generally accepted method for the shear design of a prestressed member made continuous with conventional reinforcement. Many designers design for shear as though the member were a simple span precast member. This is not always conservative since the interaction of shear and negative moment could reduce the shear strength. Another approach is to design the negative moment region for shear assuming it is a reinforced concrete section and neglecting the prestress effect. This procedure is conservative, in some cases too conservative to be of practical use. In cases where it gives reasonable amounts of shear reinforcement, however, it is certainly a simple and safe approach.

21.6.5 Methods of Obtaining Continuity

A common method of obtaining continuity is to add reinforcing steel over the support in a deck or topping slab. This is a simple method but it must be remembered that this creates only negative moment resistance, not positive moment resistance. Reinforcing bars projecting from precast members are sometimes spliced to obtain continuity using either welding or mechanical splices. Such splices generally have low fatigue resistance and should be used with caution if large fluctuating loads are anticipated. Lapped splices are used infrequently because the required lap lengths are difficult to accommodate. Lapped or welded rebar splices are sometimes used to create positive moment capacity between the ends of the precast girders, in order to resist creep cambering. Prestressing strands may also be projected, lapped, and hooked to create positive moment capacity.

Continuity is readily obtained by post-tensioning in either precast or cast-in-place construction. Long continuous tendons may be used or short tendons may be provided over the supports. Another method is to use crossed tendons (Fig. 21.15). The analysis of moments generated by crossed tendons is complex. It may be done using the methods described in Section 21.9.

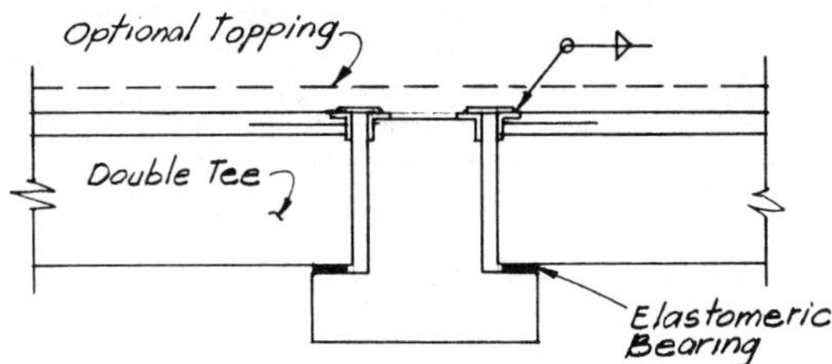

Fig. 21.14 Continuity for diaphragm action only.

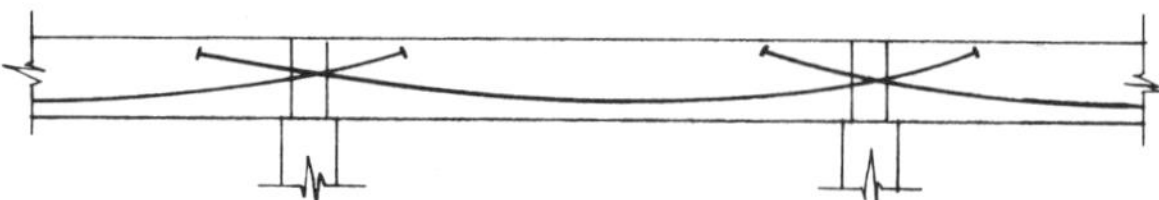

Fig. 21.15 Crossed post-tensioned tendons.

21.7 COMPOSITE CONSTRUCTION

When should composite action be utilized in the design of structural members? The answer varies with the type of member under consideration. The following guidelines are offered to supplement coverage in Chapter 23.

21.7.1 Topping Slab and Floor Unit

When a topping slab is cast on a precast floor unit, composite action may almost always be utilized. The contact surface should be clean, free of laitance, and intentionally roughened. When these conditions are met, the *ACI Code* (Section 17.5.2.1) allows 80 psi (0.55 MPa) of shear on the contact surface without steel ties. With topping slabs of normal thickness (about 2½ in.) (64 mm), the shear will almost always be less than this limit. The provision of ties is not practical in the case of topping slabs of usual thicknesses, and is normally not necessary. One word of caution—topping slabs are often interlaced with electrical conduits and other embedded penetrations. As a result, some designers prefer to neglect the structural value of the topping altogether. An alternate approach is to specify a concrete strength for the topping that is substantially in excess of the strength assumed in the design calculations. For instance, one might specify 4000-psi (28-MPa) concrete but use 2500 psi (17 MPa) in the design calculations. Since conduits should normally be limited to less than half the slab thickness, this would create an allowance for such embedments in the design.

21.7.2 Cast-in-Place Slab on Precast Girders

When a cast-in-place slab is cast on a precast girder, the girder should almost always be designed as a composite girder. All that is normally required to obtain composite action is to leave the top surface of the girder rough and to extend the stirrups into the slab. The requirements of Section 17.5.2.3 of the *ACI Code* must be checked but they will normally be satisfied without difficulty.

It should be noted that in a composite design of a precast girder with a cast-in-place slab, the weight of the slab may be applied either to the composite section or to the bare girder, depending on the method of construction. If the slab forms are supported from below or if the girder is shored, the weight of the slab will be applied to the girder when the supports are removed, that is, after the slab has hardened. Therefore, the weight of the slab will be applied to the composite section. If, however, the slab form is supported by the girder and the girders are not shored, then the weight of the slab will be applied to the bare noncomposite girder.

21.7.3 Precast Floor Units on Girders

The decision of whether or not to take advantage of composite action depends on the configuration involved.

Hollow Core Slabs on Girders or Soffit Beams

Refer to Fig. 21.12. In such cases, the girder may normally be made composite with the concrete between the slabs by simply extending the stirrups and providing a roughened surface on the top of the girder. The topping above the slab may be included in the composite section to a width eight times the topping thickness on each side. The previous comments about using a reduced topping strength to allow for conduits, etc., also apply here, although the calculations may be simplified by using a reduced flange width instead of a reduced concrete strength. If careful grouting of the joints between the hollow-core members can be assured, the top flange of the hollow-core member might also be considered to be a part of the composite section. A simplified approach is to assume that the concrete strength of the hollow-core flange is the same as that of the topping. This allows for less than perfect grouting of the hollow-core joints. In this case, the "$8t$" rule (*ACI Code,* Section 8.10.2) may be applied using t as the sum of the thickness of the topping and the top flange of the hollow core.

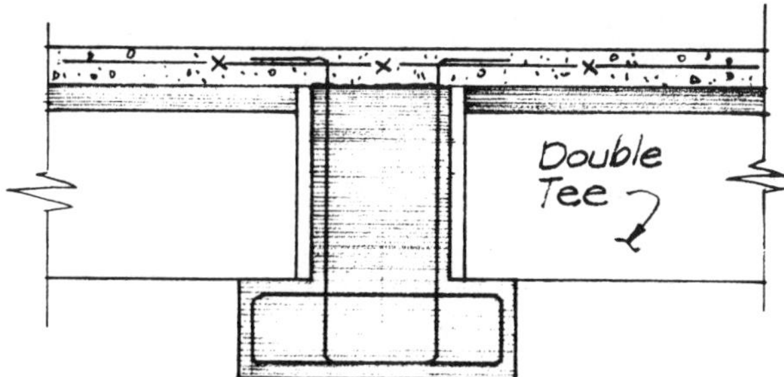

Fig. 21.16 Double tees on inverted tee beams.

Double Tees or Single Tees

Double or single tees are commonly used in conjunction with inverted T- and L-shaped beams (Fig. 21.16). In such a case, the beam may be assumed to be composite with a topping subject to the previously suggested reductions in topping strength or effective flange width to account for conduits, etc. The double- or single-tee flange should not be assumed as a part of the composite section since the joints between flanges are not grouted.

When double or single tees rest on top of precast beams (Fig. 21.17), composite action is virtually never utilized. The difficulty of adequately connecting the floor units to the girder outweighs the advantages that might be gained through composite action.

21.7.4 To Shore or Not to Shore?

Shoring is sometimes used in composite construction in order to cause more of the load to be applied to the composite section. The use of shoring has virtually no effect on the ultimate strength of the member, but it has a significant effect on the stresses and deflections at service load. Shoring also determines whether certain loads are applied to a simple span beam or a continuous beam. The function of shoring is to delay the application of the weight of a concrete slab or topping until such time as that concrete element has hardened and is able to resist load. Thus, the use of shoring can cause the weight of the slab or topping to be applied to a composite section, and also to a continuous section if continuity is obtained through reinforcement embedded in the slab or topping. It is generally beneficial to apply as much load as possible to the composite section since it is a stronger section. It may or may not be beneficial to apply more load to a continuous member, since the T-beam section normally available to resist positive moments at midspan is potentially much stronger than the rectangular section available to resist negative moments at the support.

The cost of shoring must be compared to the potential benefit. The cost is very much determined by the height of the shores. In building construction, where story heights range from 8 to 12 ft (2.4–3.7 m), wood 4 × 4's or 6 × 6's may commonly be used. When the reshoring of lower floors is not

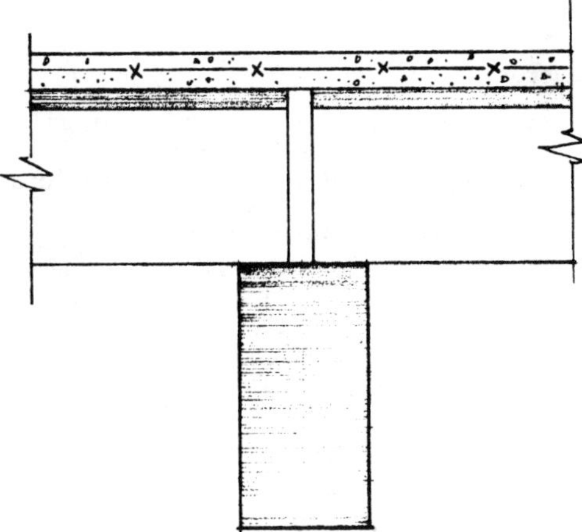

Fig. 21.17 Double tees on precast beams.

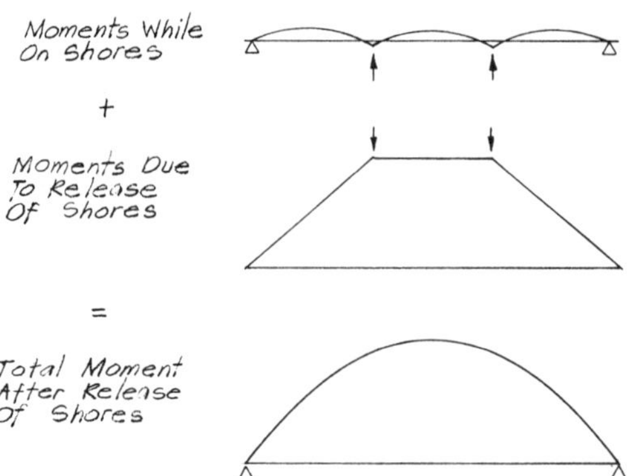

Fig. 21.18 Moments in a shored member.

required, the cost of shoring is relatively minimal when the shore size is no larger than 6 × 6. The decision to shore is also made on the basis of the desired structural behavior. Bridges, which normally have a clear height of 15 ft (5 m) or more, are almost never shored. At these heights, simple wood members are not sufficient and more complex shoring such as scaffold towers is required. Shoring may be useful in certain special situations, such as stretching a span beyond normal limits, but in general, shoring is not used in precast bridge girder construction.

Shoring may also be used to control differential camber in precast/prestressed floor and roof units. Differential camber is common in such units because the net camber is the difference of the upward deflection from prestressing and the downward deflection from self-weight. This net camber is the small difference of two larger numbers; thus, the variability of the difference is large. Differential camber may be negated either by shoring up the lower members or by weighting the high members. It is much easier to apply a ton of upward force with a shore than to apply a ton of downward force with dead weight. Thus, shoring is often used to eliminate differential camber. Of course, the amount of upward deflection should be consistent with the desired final position of the members.

It is commonly assumed that a shored member is fully supported by the shores, and that the loads supported by the shores are applied to a composite (and sometimes continuous) member after the shores are removed. This is only approximately true. A closer examination will show that, while shored, the bare beam functions as a multiple-span beam with negative moments applied to the bare beam at the shoring points and positive moments in between. When the shores are removed, the effect upon the structure is to apply downward loads to the composite structure equal to the loads supported by the shores (Fig 21.18). If the beam is shored at one-third or closer intervals, the effect of the downwardly applied concentrated loads is close to being the same as a uniform load.

21.8 CONNECTIONS

21.8.1 Types of Connection Behavior

Connection behavior may be classified as hard, soft, or semi-rigid. Hard connections involve direct contact of steel and/or concrete at the mating surfaces. Hard connections are useful for transmitting high loads, such as in a column-to-column connection. Hard connections must allow for tolerances of the fit of the precast elements. Tolerances may be absorbed through the use of a filler material in the joint such as grout or epoxy. Sometimes the mating surfaces are so confined with steel that the tolerances may be taken up by local yielding of steel and/or confined concrete. A special type of hard connection is the match-cast connection. In this type of connection, one mating surface is cast against the other previously hardened surface using a bond breaker. Alignment keys or lugs are provided, and the pieces are match-marked and reassembled in the same sequence that they are cast. This results in a theoretically perfect fit. Sometimes match-cast joints are assembled dry. More often, a very thin film of epoxy is used in the joint.

Soft connections usually involve the use of an elastomer between the mating surfaces. The elastomer performs several important functions. It allows for tolerances between the mating surfaces. It also allows for a moderate amount of relative rotation in the case of simple span designs. Additionally, through shear deformation, the elastomer may allow for changes in the length of the horizontal members caused by creep and shrinkage.

Ideally, connections would be designed totally hard or totally soft. In reality, some connections must be designed to accommodate both types of behavior. Such connections are called semi-rigid (or semisoft). An example is the connection at the end of a precast floor or roof member designed as a simple span member for vertical loads. Such a connection must be soft to accommodate the creep and shrinkage shortening of the member and the end rotations associated with simple span behavior. However, when lateral load resistance is through shear walls, a diaphragm connection must be provided at the end of the member capable of transmitting lateral loads to the shear walls. The connection must be reasonably hard to accomplish this. Semi-rigid connections are normally accomplished through the use of steel weldments or reinforcing which can yield under the permanently applied long-term deformation loads and yet transmit the short-term lateral loads.

21.8.2 Loads on Connections

Connections must be designed for three types of loads—vertical, lateral, and deformation. Design for vertical loads is relatively straightforward. Compression may be resisted either by concrete or steel. Tension must be resisted totally by steel. Shear may be resisted by shear-friction. Problems encountered in the design for vertical load are twofold. The vertical load may not be uniform; this may be caused by tolerances at the bearing surfaces, or it may be caused by deformations such as end rotations of simple span members. Also, lateral and deformation loads may cause a weakening of the vertical load capacity.

The forces associated with lateral load design may be readily determined in the case of wind loads (see Chapter 2). Although the design codes give methods for determination of earthquake forces, it must be remembered that earthquake loads are really inertial loads resulting from displacements and that the forces given by the codes are artificial. Lateral loads and deformations usually produce a combined effect on the connection.

21.8.3 Deformation Loads

The term "deformation loads" is a misnomer. Deformations are not loads, but deformations do produce loads that are very important to connection design. The loads produced by deformations are dependent upon the stiffnesses of the connections and the connected elements. Deformation loads may largely be relieved through the use of soft connections. It is seldom feasible to restrain members fully against the loads caused by deformations. When hard connections are used, there is little or no relief within the connection and thus deformation loads must normally be relieved through the elasticity of the connected members.

Deformations from creep, shrinkage, and temperature can produce axial shortening of members. Additionally, deformations from creep and thermal gradients can produce end rotations. Connections are sometimes made rigid for end rotations, in which case the positive moments caused by deformations must be accounted for. The loads caused by axial deformations are normally determined by the flexibility of the columns or other supporting members.

21.8.4 Shear-Friction

Connection design must consider three types of stresses: compression, tension, and shear. Compression stresses may be carried directly through the concrete. Tension stresses should be resisted by reinforcement. Shear-friction provides a method for resisting shearing stresses through the use of tensile reinforcement which mobilizes "frictional" stresses across a crack in concrete. The key assumption of the shear-friction method is that the concrete may crack in the most unfavorable manner and reinforcement must be provided across this potential crack to resist all tension and shear (through shear-friction) across the crack. (See Refs. 21.8, 21.11, 21.12, and 21.13).

21.8.5 Connection References

It would be impractical to describe all the various types of connections and their design methods here. Rather, references are offered for guidance in connection design. The primary reference for connection design is the *PCI Design Handbook* [21.17]. Part 6 of that handbook is the most comprehensive guide to connection design available at this time. A second important reference is *Connections for Precast Prestressed Concrete Buildings,* by L. D. Martin and W. J. Korkosz [21.5]. This reference is a comprehensive survey of the various connection designs presently in use.

In addition to the two basic references cited above, Ref. 21.7, 21.9, 21.10, and 21.14 will be helpful in the design of specific types of connections.

21.9 POST-TENSIONING

Post-tensioning has a wide variety of applications in prestressed concrete construction. Post-tensioning is sometimes used as a substitute for pretensioning when stressing beds of sufficient capacity are not available. This is a simple application of post-tensioning and the analysis of such members is similar to that for pretensioning. Provided the tendons are grouted, the end result is virtually identical to pretensioning.

The basic reference for the design of post-tensioned structures is the *Post-Tensioning Manual* [21.18] published by the Post-Tensioning Institute.

Post-tensioning may be used in more complex structures for which pretensioning would be impractical. These include continuous members, curved structures, shells, and other three-dimensional structures. Such structures provide interesting and challenging problems of analysis and design. In a statically indeterminate structure, "secondary moments" are created and the effective location of the post-tensioning force is not coincident with the steel tendons.

Post-tensioned structures of any degree of complexity may be analyzed by considering the forces applied to the concrete structure by the steel tendons as a system of external loads, and using the same techniques that would be used to analyze the effects of gravity and other loads. This section will be devoted to the applications of this fundamental technique for the analysis of the effects of post-tensioning.

21.9.1 Simple Illustration of the Method

Consider a simple span beam with prestressing draped at the one-third points, as shown in Fig. 21.19. The prestressing tendons exert forces on the ends of the beam where the tendons are anchored. Also, upward forces are applied at the points where the tendon is deflected. Wherever a tendon deviates from a straight line, a force is applied to the concrete. And, of course, forces are applied to the concrete at all anchorage points. Technically, the forces applied at the deflection points have a horizontal component. In many cases, this component may be ignored. A very important principle is that all of the forces applied by the steel tendons to the concrete must be in equilibrium with each other. This may be understood by considering a free body of the steel tendon itself. The forces applied by the concrete to the steel tendon are equal and opposite to those applied by the tendon to the concrete. The tendon must be in static equilibrium. Therefore, the forces applied by the tendon to the concrete must also be in equilibrium. If a simplifying assumption is used and the upward forces at the deflection points are regarded as being vertical, then the downward force at the anchor points should be of the same magnitude and the horizontal force at the anchor points should be equal to the force in the tendon. In order to produce a consistent set of loads, the horizontal loads are computed as P, the force in the tendon, and the vertical loads as $P \tan \Delta\theta$ where $\Delta\theta$ is the angle change at the deflection points. (This is equivalent to assuming that $\sin \Delta\theta = \tan \Delta\theta$ and $\cos \Delta\theta = 1$, which is a reasonable assumption for small angles.)

The analysis of the beam shown in Fig. 21.19 indicates that the trajectory of the post-tensioning force exactly follows that of the tendon and the results may seem trivial. This is so for simple statically determinate beams; however, treating the forces due to post-tensioning as external loads has great value in the analysis of more complex structures.

21.9.2 Statically Indeterminate Structures

Consider the two-span continuous post-tensioned beam shown in Fig. 21.20. In this case, the upward and downward forces are determined by the same methods as in the previous example. However, the moments produced by the vertical forces are not a carbon copy of the tendon profile. In this case, the vertical forces are applied to a statically indeterminate structure and the moments and reactions are determined by the elastic properties of the structure as well as by the geometry of the tendon profile. The moments corresponding to the eccentricity of the tendon profile are sometimes called the "primary moments" and the additional moments resulting from the statically indeterminate nature of the structure called "secondary moments". For instance, in Fig. 21.20 the primary moment at the interior support is 100 ft-kips (136 kN·m), computed from Pe. The secondary moment is therefore 157 − 100 = 57 ft-kips (77 kN·m). This concept has use in analyzing statically indeterminate beams. When considering more complex structures, however, it is more

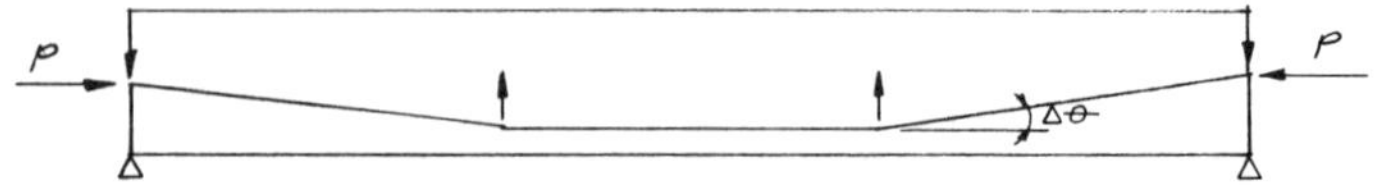

Fig. 21.19 Prestressing treated as external loads.

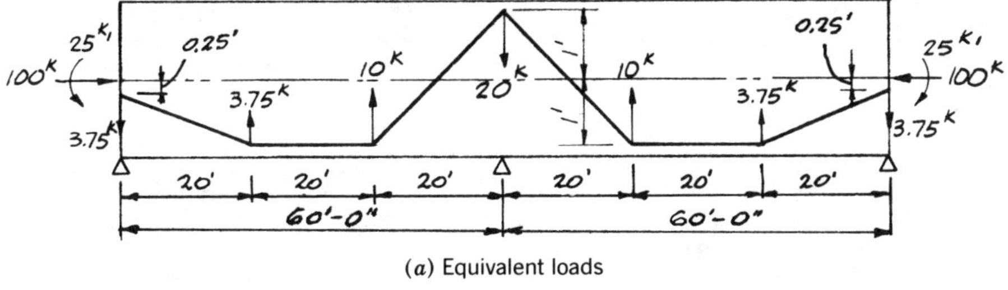

(*a*) Equivalent loads

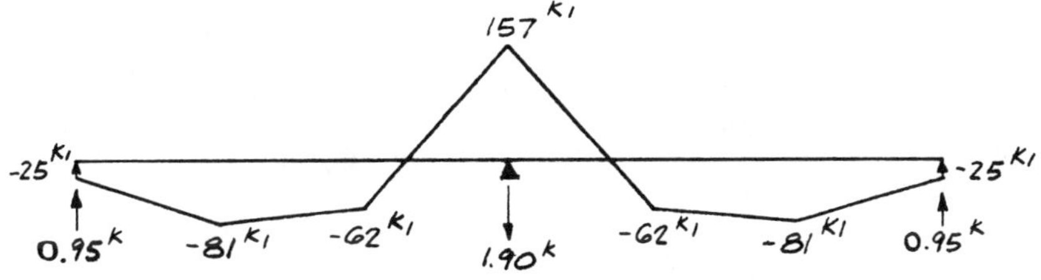

(*b*) Moments due to prestressing

Fig. 21.20 Prestressing of a statically indeterminate structure.

useful to think of the total moments and stresses resulting from the prestressing forces without attempting to distinguish between primary and secondary stresses.

The post-tensioning profile shown in Fig. 21.20 is a simplified one used for purposes of this illustration. A more realistic profile is that shown in Fig. 21.21. Parabolic tendons which create uniform vertical loads are used instead of point deflections of the tendons. It would be tempting to think of the tendons as composed of two parabolas with a sudden angle change at the interior

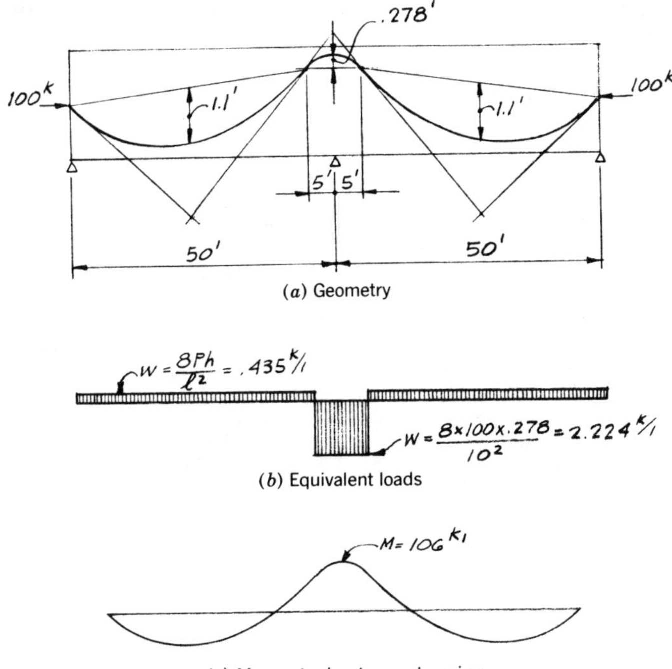

(*a*) Geometry

(*b*) Equivalent loads

(*c*) Moments due to prestressing

Fig. 21.21 Curved tendons.

support. To be realistic, however, a short concave downward section must be introduced at the interior support and this should be accounted for in the analysis. Figures 21.21*b* and *c* show the loads applied by the tendons to the concrete and the moments resulting from these loads.

If the tendon shape is sufficiently complex so that it cannot be readily analyzed as uniform loads resulting from parabolic profiles, it may be more practical to regard the curved tendon profile as a series of chords. The tendon may be divided into 10 or 20 increments per span with each increment being a chord. The forces applied by the tendon to the concrete at each deflection point may then be calculated.

It is also feasible to account for frictional forces in the analysis of the effects of post-tensioning. Once the tendon has been subdivided into chords of one-tenth or one-twentieth of the span, the normal forces (i.e., the forces applied by the tendon to the concrete) may be calculated and the frictional forces will be a percentage of the normal force, determined by the coefficient of friction. Of course, the frictional force will affect the tendon force at the next deflection point and thus the normal and frictional forces at that point as well. When doing computations involving frictional force, the simplifying assumption of constant horizontal force cannot be used and therefore both horizontal and vertical components of the normal force and frictional force must be computed at each deflection point. It is especially important to verify that equilibrium is satisfied by all of the input forces prior to running the analysis. Moment equilibrium, as well as force equilibrium, must be satisfied by the input forces.

21.9.3 More Complex Structures

Structures of virtually any degree of complexity, including shell structures and three-dimensional structures, may be analyzed by the equivalent load method. The loads applied to the concrete structure by each tendon are computed and checked. Again, it is essential that the equivalent loads for each tendon satisfy equilibrium. For the general case, this means that the sums of the X, Y, and Z forces and the X, Y, and Z moments must each be zero. After the equivalent loads have been computed and checked, they may then be applied to the model used for the structural analysis. For complex structures, this would normally be a finite element computer model. The output from this analysis will give the total stresses caused by the effects of prestress. Note that in this case the concept of "secondary moments" becomes meaningless. The distribution of axial force, shear, and moment on the complex structure will be different from that which would be obtained by simply tracing the prestress force in each tendon along its trajectory.

The equivalent load method is very powerful. For complex three-dimensional structures, it is probably the only method of analysis that will give the correct results for the stresses due to the prestressing operations.

21.9.4 Curved Prestressed Members

Curved members may readily be prestressed using the post-tensioning method. This comes as a surprise to many. Some feel that the curved concrete member will buckle under the effects of the compression caused by the prestressing force. Others feel that when the tendon is stressed it will tend to straighten out and straighten the concrete member in the process. In reality, these two effects offset one another and prestressed members of moderate curvature may be analyzed for flexural stresses as though they were straight members. However, torsional stresses are higher in curved members and the analysis of shear and torsion must be done considering the curvature of the member.

Figure 21.22 shows the equilibrium of a curved prestressed concrete member. The steel tendon applies forces to the concrete member which cause the compression force to follow the curvature of the member. For statically determinate members, the compression in the concrete and the tension in the steel will be equal. Therefore, if the radius of curvature of the cgc (center of gravity of the concrete cross-section) and cgs (center of gravity of the prestressing steel force) are the same, the inward forces resulting from the curvature of the steel tendon will be exactly sufficient to cause the compression forces in the concrete to curve to the same radius. This may be readily demonstrated by analyzing a curved member using the equivalent load method.

Fig. 21.22 Curved members.

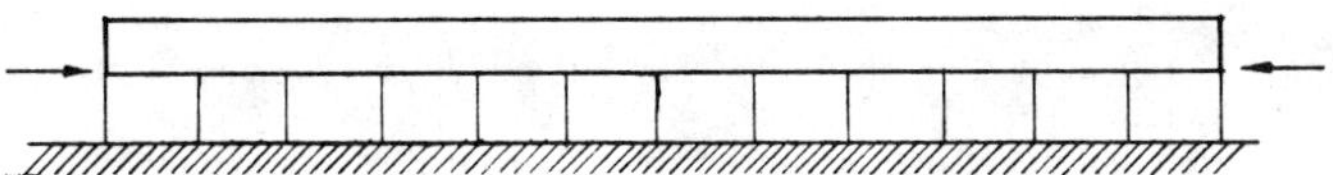

Fig. 21.23 Beam continuous over many supports.

The equivalent load method may also be used for more complex three-dimensionally curved indeterminate structures.

21.9.5 Multiple-Span Structures

Consider a spandrel beam continuous over many supports as shown Fig. 21.23. If this spandrel is post-tensioned eccentrically with a straight tendon, what will be the distribution of stresses in the interior spans of the spandrel beam? The stresses will not be those determined by $P/A + Mc/I$ with M equal to the prestressing force multiplied by the eccentricity of the tendon. If there were a constant moment in the spandrel proportional to the eccentricity of the prestressed tendon, there would have to be overall curvature and deflection of the spandrel caused by the moment. But the many columns supporting the spandrel require that the vertical deflection at each column be virtually zero. This in turn implies that the distribution of stress in the spandrel must be essentially P/A, without moment. What happens is that the columns at each end of the structure provide vertical forces that create moments equal and opposite to the moment caused by the eccentricity of the prestressing force. The equivalent load method may be used to analyze the structure. Figure 21.24 shows the results of such an analysis. The moments induced in the spandrel by the eccentric prestressing die out rapidly and may be neglected beyond the first two spans.

21.9.5 Straight Tendons Combined with Variable Location of Cross-Section Center of Gravity

The previous section would seem to indicate that, regardless of the tendon location, the stress due to prestressing in a multiple-span continuous member must be P/A. Obviously, it is desirable to provide eccentric prestressing in order to counterbalance moments in the members. One way to accomplish this is by draping the prestressing tendons. This is an effective method in unbonded prestressed slabs in which the amount of deflection is very small and the friction is low because of antifriction coatings. However, in multiple-span beams the friction caused by draping tendons throughout many spans makes this technique impractical. This problem may be solved by using straight tendons combined with a variable-depth concrete member which has a variable height of the cgc. An example of this is the hollow box girders with parabolic soffit used for the Walt Disney World monorail in Orlando, Florida (Fig. 21.25). In such a case the equivalent loads applied by the prestressing forces are very simple since the tendons are straight (in straight beams), but the cgc line of the concrete is curved and the section properties of the concrete cross-section also vary throughout the span. Such a structure may be analyzed by dividing the span into one-tenth or one-twentieth increments. The section properties may then be approximated for each increment. It is essential that the variable height of the cgc line be accounted for in the model used for the analysis. The results of such an analysis are shown (with vertical scale exaggerated) in Fig. 21.26.

21.9.7 Special Problems in Post-Tensioning

Whenever a tendon is curved, an inward radial force equal to P/R is produced (where P is the tendon force and R is the radius of curvature). This force must be resisted locally by the concrete, sometimes with the aid of auxiliary reinforcement. In straight beams, the radial force is normally directed inward toward the center of the concrete cross-sections, as in the case illustrated in Fig. 21.21. There are times, however, when a curved tendon may produce a significant radial force directed outward toward a free edge of the concrete. This most commonly occurs when tendons are "splayed" (i.e., given a sharp curve near an anchor, in order to avoid interferences). In such cases,

Fig. 21.24 Moments in a beam continuous over many supports.

Fig. 21.25 Curved continuous post-tensioned beams of variable cross-section. Copyright Walt Disney Productions 1983.

the radius may be on the order of 10 to 25 ft, producing a large P/R radial force. Tendons on the inside face of a curved beam also produce a radial force directed toward the free edge of the concrete.

There exists no recognized standard for computing the resistance of the concrete cross-section to an outward radial force. An analogy may be made to the pullout of a group of headed studs. See Figure 6.5.3 of the *PCI Design Handbook*. A 45° wedge is assumed to pull out, and an ultimate tension of $4\lambda\sqrt{f_c'}$ is assumed to act on the 45° planes. The ultimate resistance per unit length may thus be calculated as

$$\phi P_c = (\phi 4\lambda\sqrt{f_c'})(2\ell_e)$$

For $\lambda = 1$ (normal-weight concrete), $\phi = 0.85$, and $f_{ci}' = 4000$, this reduces to

$$\phi P_c = (0.43 \text{ ksi})(\ell_e)$$

The resistance computed above is based on reliance on the (brittle) tensile strength of concrete, and thus must be used with caution. The use of steel "hairpins" (Fig. 21.27) sized for the total radial force is recommended whenever the factored radial force exceeds one-fourth of the above amount, and is required if it exceeds one-half the above amount.

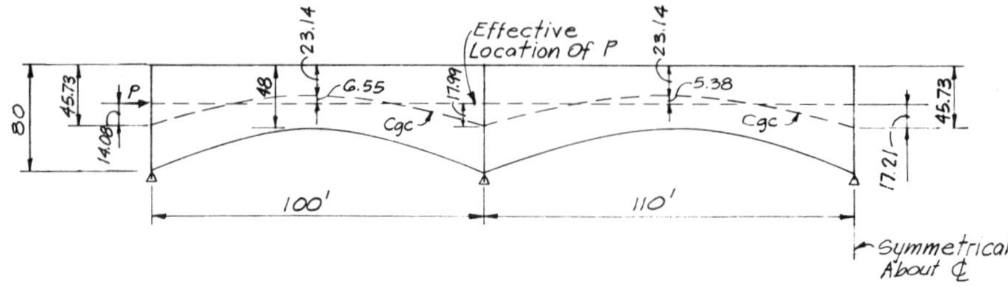

Fig. 21.26 Beams of variable cross-section.

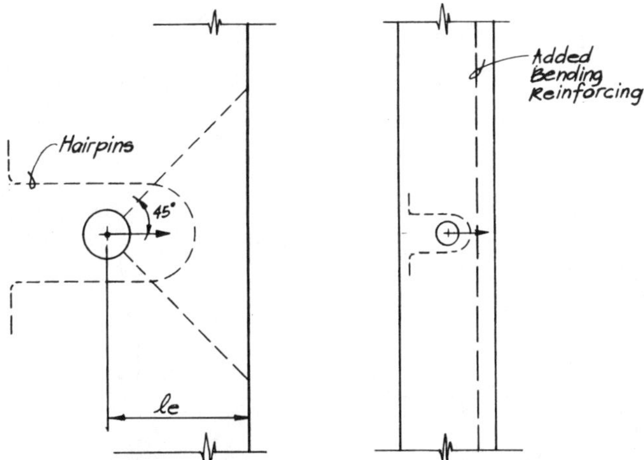

Fig. 21.27 Resistance to radial forces.

The radial force may produce other effects, such as bending in the web of a curved girder. The web should be checked for this effect, considering it to be a beam spanning between the top and bottom slab, and reinforcement should be added to the web for the local bending effect, if necessary.

21.10 DESIGN EXAMPLE

The following design example is not intended to be a treatise on the analysis of a prestressed beam. Before a beam may be thoroughly analyzed, many selections and decisions must be made. This often leads to a design process of repeated trial and analysis. This process may be made more efficient by intelligent choices and selections in the preliminary design. The prime purpose of this example is to offer some guidelines on making these choices and selections.

The widespread use of pocket calculators has led to the practice of carrying calculations to several places beyond the decimal point. This is totally unwarranted. Loads and member properties are known to two significant figures at most. It is quite proper to periodically round off excess (in)significant figures.

A remark about sign conventions is necessary. The traditional sign convention for stresses is that tension is positive. Since most stresses in prestress design are compressive, this convention would cause one to work with mostly negative stresses. Many authors of works on prestressing have thus reversed the sign convention, calling compression positive. This convention is used in this example. When doing longhand calculations, the best way to avoid confusion is to write *c* or *t* after the stress. This is not practical with computerized calculations, for which + and − must be used.

21.10.1 Problem Statement

Design the floor system for a six-story office building. The building is 150 ft (46 m) wide, and will have three 50-ft (15-m) wide bays. See Fig. 21.28. Column spacing in the longitudinal direction is not fixed, although the architect would prefer a spacing of at least 15 ft (4.5 m). Total superimposed load is 100 psf (4800 N/m^2), all treated as live load (without reduction, for purposes of this example problem). Lateral loads are resisted by shear walls; the floor system will be designed for vertical load only.

The desired clear span of 50 ft (15 m) will favor a prestressed, precast member. A cast-in-place post-tensioned member is also a possibility.

Select a framing system and a longitudinal column spacing. Design the 50-ft (15-m) span members.

21.10.2 Selection of Framing System

The selection of a framing system is interrelated with the type and spacing of support provided. If continuous support (i.e., bearing walls) is provided, a precast, prestressed floor unit is a good

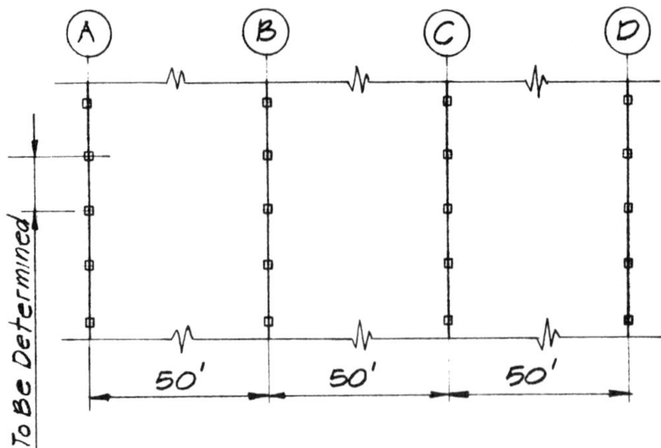

Fig. 21.28 Partial plan of example problem.

solution, as shown in Fig. 21.29. Double tees and hollow-core slabs are useful on a 50-ft (15 m) span.

Single tees, 10 ft (3 m) wide, are sometimes supported directly on columns; see Fig. 21.30. This is a simple, straightforward solution, if a 10-ft (3-m) column spacing can be tolerated.

A beam and slab configuration works well on beam spacings of 15–20 ft (4.5–6.0 m); see Fig. 21.31. The beam may be precast and pretensioned, or cast-in-place and post-tensioned. The slab is normally cast-in-place, although a thin precast plank is sometimes made composite with a cast-in-place topping.

Floor units (double tees or hollow-core plank) may be supported by concrete girders (Fig. 21.32). The addition of girders increases the cost compared to the floor-unit-on-bearing-wall framing system described previously. Still, this is a good framing system where a wider column spacing is desired. Economy results when a spandrel required for architectural reasons may also serve as the structural support at the exterior walls.

A beam and floor unit framing system (Fig. 21.33) is also a possibility. In this system, the beams span the long direction and the floor units the short direction. This system requires a greater depth of construction than the one described previously. It may also be more costly compared to the previous system, as the additional cost of beams may be more than the cost saving in the shorter span floor units.

A beam and girder system (Fig. 21.34) is often used in cast-in-place construction. However, it should be avoided in precast construction. The large lineal footage of beam adds to the cost and the three- and four-way connections increase the complexity and cost of the connections.

The precast/prestressed beam and cast-in-place slab, using a 15-ft (4.5 m) beam and column spacing, is selected for this design example. This is a good solution. It may or may not be the best solution in an individual case. It is selected for this design example because it illustrates a variety of design considerations which make it a useful design example.

21.10.3 Selection of Materials and Modes of Behavior

Pretensioned members are normally controlled by the concrete strength f'_{ci} at release of prestress. An f'_{ci} of 4500 psi (31 MPa) is chosen for this design example. Using an f'_{ci} of 4500 psi, an f'_c of 6000

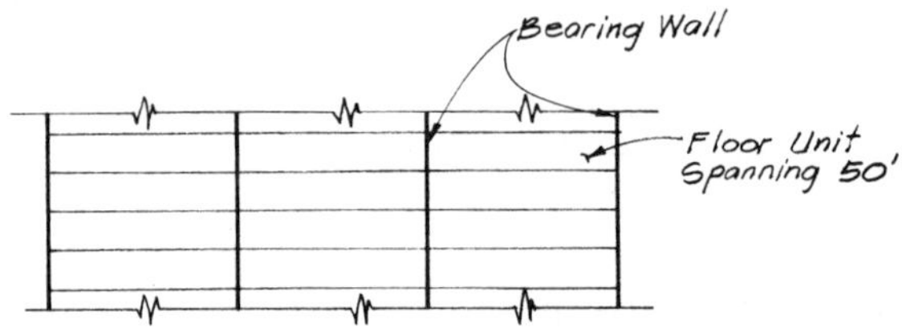

Fig. 21.29 Floor units on continuous support.

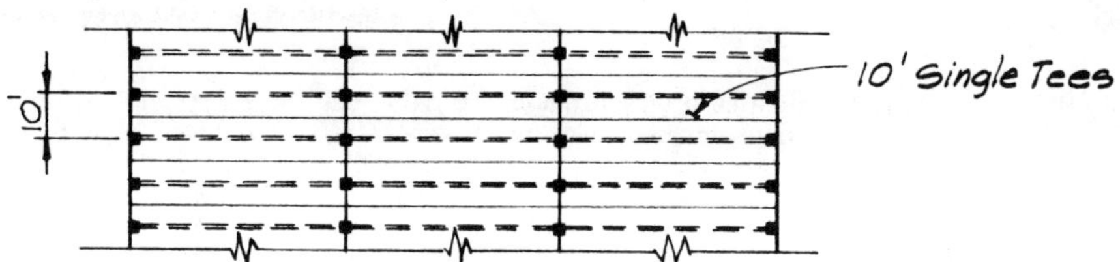

Fig. 21.30 Floor units on columns.

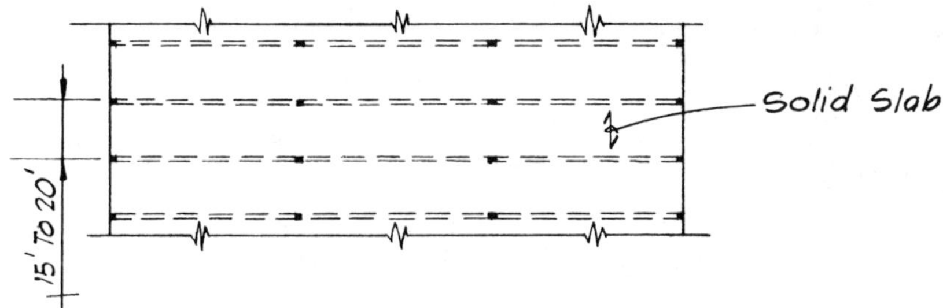

Fig. 21.31 Beam and solid slab.

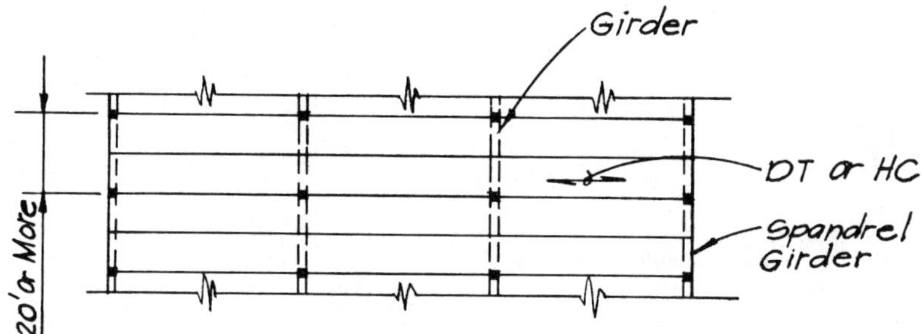

Fig. 21.32 Floor units on girders.

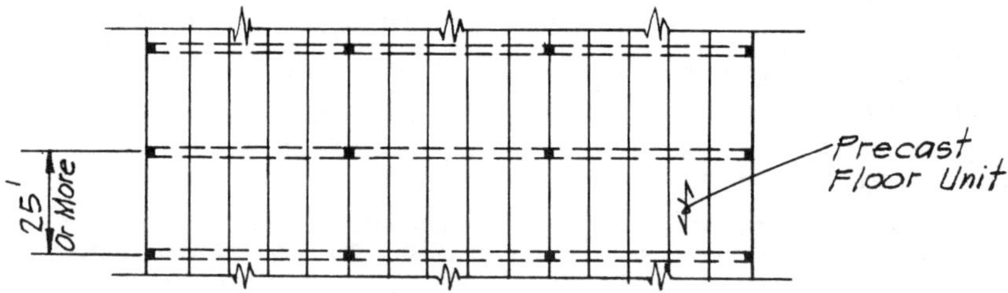

Fig. 21.33 Beam and floor unit.

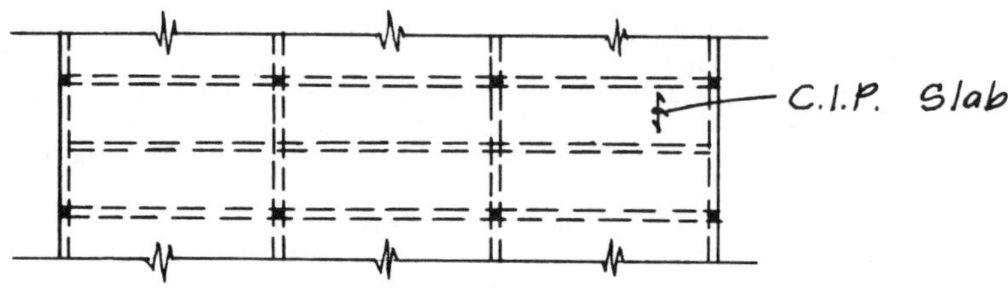

Fig. 21.34 Beam and girder.

psi (41 MPa) at 28 days would be a reasonable choice. Note that these strengths may not be obtainable in all localities. Minimum obtainable strengths are f'_{ci} = 3500 psi and f'_c = 5000 psi (24 and 34 MPa).

Concrete strengths of 3500–4000 psi (24–28 MPa) are readily obtainable for the field-cast concrete. A strength of 3000 psi (21 MPa) will be used in the design calculations, but 3500 or 4000 psi (24 or 28 MPa) will be specified for construction. This will provide a margin for conduits, sleeves, etc., in the slab.

Normal-weight concrete will be used for both beam and slab, as there is no specific reason to use lightweight concrete. For weight computations, the beam concrete will be assumed to weigh 160 pcf, since high-quality concrete is often a little heavier than 150 pcf. For the slab, the conventional weight of 150 pcf will be used. Grade 60 reinforcement will be used. This is the national standard.

The prestressing material used will be ½-in. Type 270K, stress-relieved, seven-wire strand. This is the most commonly used prestressing material in the United States. The strand has a cross-sectional area of 0.1531 sq in., and a tensile strength of 41.3 kip/strand, or 270 ksi (1860 MPa).

The commonly used stresses are as follows:

$$\begin{array}{lll} \text{jacking stress} = 0.7f'_s & = 189 \text{ ksi} & = 28.90 \text{ kip/strand} \\ \text{losses} & = \underline{\ 35 \text{ ksi}\ } & \\ \text{final stress} & = 154 \text{ ksi} & = 23.55 \text{ kip/strand} \end{array}$$

As of the writing of this book (1986), these are the most commonly used stresses. However, the market is changing. A newer material, low-relaxation strand, is becoming more widely used. It has the same cross-sectional area and tensile strength as stress-relieved strand. Higher jacking and final stresses may be used with low-relaxation strand [21.6].

Composite action will reduce beam size and improve beam behavior. The only cost is normally that of extending the beam stirrups, and possibly adding a few stirrups. When a cast-in-place slab is used with a precast beam, it is almost always made composite. Thus, utilize composite action for this example.

Continuity is easy to obtain by placing reinforcement in the cast-in-place slab. For very little cost, the service load behavior of the beam may be significantly improved. Thus, make the beam continuous.

It would be fairly easy to shore the beams (or to support the slab forms from the floor below) when casting the slab. This would cause the weight of the slab to be applied to the continuous span, resulting in large negative moments at the beam ends. This is undesirable, for the T-section at midspan can resist large positive moments, whereas the rectangular section available at the support has less capacity for negative moments. If the slab forms are supported by the unshored beam, then the slab weight will be applied to the simple span, resulting in a better proportion of the total positive and negative moments. Thus, do not shore the beam.

For a 50-ft span, a rectangular beam might be the best choice. I-beams are more efficient, but also more costly per unit volume. Typically, I-beams would be a better choice for longer spans.

Should a tensile stress greater than $6\sqrt{f'_c}$ psi ($0.5\sqrt{f'_c}$ MPa) be utilized? (See ACI 18.4.2.c.) Only if necessary to control camber. Size the beam based on $6\sqrt{f'_c}$ tension, and then adjust the tension later, if necessary.

21.10.4 Selection of Beam Size

To size a simple steel beam, one computes the total bending moment and divides it by the allowable stress, obtaining the required section modulus. For a prestressed composite beam, the procedure is fundamentally the same, but it is more complex. First, the superimposed dead and live load moments are computed as M_d = *3308 in.-kips and* M_ℓ = 2159 in.-kips (the computation is given in Section 21.10.5). The self-weight bending moment, M_g, need not be estimated at this time. The allowable compressive stress at the time the prestress force is released is $0.60f'_{ci}$ = 0.60(4500) = 2700 psi. This stress will decrease approximately 20% with time, due to prestress losses. Thus, the compressive stress in the bottom fiber available to resist the tensile stresses from the applied moments is 80% of 2700 psi, or 2160 psi. The "code allowable" final stress is $0.45f'_c$ = 0.45(6000) = 2700 psi. But, the practical maximum final bottom fiber compression is 2160 psi, as governed by initial stresses, reduced to final values. The total range of stress Δf caused by superimposed loads is from the maximum bottom fiber compression, after losses, of 2160 psi to the allowable tension of 465 psi, giving a total stress range of 2625 psi.

Having the maximum permissible stress range, may one divide the moment by this stress range to find the required section modulus? There is one other consideration. There are two different section modulii—one for the bare beam and one for the composite beam. To reduce the number of unknowns to one, a relationship between the two section modulii must be established. The composite section modulus will normally have a value of 1.5 to 2 times that of the bare beam. Small beams have the higher ratio, for the composite action adds more, percentage-wise, to a small beam.

For this 50-ft span beam, assume $S_{bc} = 1.8S_b$. The required section modulus may now be found from the fundamental equation (21.1.1):

$$\frac{M_d}{S_b} + \frac{M_\ell}{S_{bc}} = \frac{P}{A} + \frac{Pe}{S_b} - \frac{M_g}{S_b} - f_t = \Delta f$$

Substituting $S_{bc} = 1.8S_b$, and solving for S_b,

$$S_b = \frac{M_d + M_\ell/1.8}{\Delta f}$$

$$= \frac{3308 + 2159/1.8}{2.625 \text{ ksi}}$$

$$= 1717 \text{ in.}^3$$

Select a 12 × 30 beam. $S = bd^2/6 = 1800$ in.^{3}.

Note: Since a 12 × 30 rectangular beam appears to be sufficient, it is doubtful that an I-beam would be more cost effective than this relatively small rectangular beam.

21.10.5 Flexural Design

Section properties for the bare (noncomposite) 12 × 30 beam are

$$A = 360 \text{ sq in.}$$

$$I = 27{,}000 \text{ in.}^4$$

$$S_t = S_b = 1800 \text{ in.}^3$$

For the composite beam, one must consider that the precast beam and the in situ slab have different modulii of elasticity. The modulus of elasticity, E_c, may be computed as $57{,}000\sqrt{f'_c}$ (ACI-8.5.1). For the precast beam, $E_c = 57{,}000\sqrt{6000} = 4{,}415{,}000$; and, for the slab, $E_c = 57{,}000\sqrt{3000} = 3{,}122{,}000$. The slab E_c is 0.707 of the beam E_c. This may be accounted for by transforming the slab width to 0.707 of the actual width b. The actual slab width used in the flexural calculations may be computed (ACI-8.10.2) as the beam stem width plus eight times the slab thickness on each side, giving a total width b of 92 in. The transformed width is 0.707 × 92 in. = 65 in. See Fig. 21.35. The transformed section properties are computed as follows:

$$A_c = 685 \text{ sq in.}$$

$$I_c = 79{,}990 \text{ in.}^4$$

$$y_{tc} = 11.70 \text{ in.}$$

$$S_{tc} = 6838 \text{ in.}^3/0.707 = 9670 \text{ in.}^3$$

$$y_{bc} = 23.30 \text{ in.}$$

$$S_{bc} = 3432 \text{ in.}^3$$

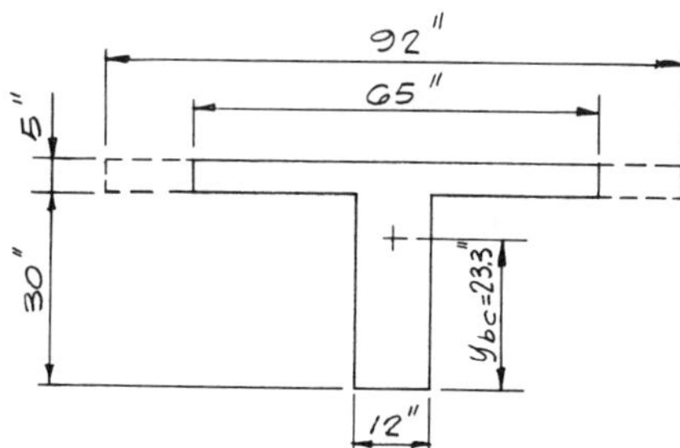

Fig. 21.35 Composite section.

Note that the top (of slab) composite section modulus computed for the transformed composite section is divided by the modular ratio, 0.707, to "untransform" it in order to give a true stress of the actual 92-in.-wide section. It is also useful to compute the composite section modulus applicable to the interface (i.e., the top of the precast beam) in order to compute stresses there.

$$y_{ic} = 6.70 \text{ in.}$$

$$S_{ic} = 11{,}940 \text{ in.}^3$$

The span length L may be taken as the center-to-center of bearing span, which is approximately 48.5 ft. Although a literal reading of ACI-8.7.2 and 8.7.3 might suggest a more complex procedure for determining live load moments on the continuous span, it is sufficiently accurate in most cases to use the same center-to-center span as used in computing simple span dead load moments.

Live load moments are computed from a frame analysis. Many computer programs are available for this purpose; a simple noncomputer method is given in Ref. 21.4. The results of a frame analysis give the following maximum live moments:

$$M^+ = 0.051\, wL^2 \qquad M^- = 0.091 wL^2$$

These are combined maxima for exterior and interior spans. For simplicity, the same moments are used for all three spans.

Service load moments and the bottom fiber stresses resulting from the moments are tabulated in Table 21.1.

Table 21.1 Service Load Moments and Stresses

	w (kip/ft)	Moment Coefficient	Moment (in.-kips)	S_b or S_{bc} (in.^{3})	Stress (psi)
Girder (160 pcf)	0.400	$\frac{1}{8}\, wL^2$	1411	1800	784
Slab (150 pcf)	0.938	$\frac{1}{8}\, wL^2$	3308	1800	1838
Live load (100 psf)	1.500	$0.051\, wL^2$	2159	3432	629
				$\Sigma[M/S_{b(c)}] =$	3251

The required final prestressing force P may be computed from the fundamental equation (21.1.1). All quantities in this equation except P and e have been determined. To find P, a limit must be set on e. There are two possibilities. The eccentricity may be limited by the physical dimensions of the beam, or e may be limited by excessive tension in the top fibers of the beam.

Case 1—Eccentricity Limited by Beam Dimensions

The prestressing strands must be within the beam, with adequate concrete cover. This requirement normally limits the maximum e to y_b less about 3 in. In large beams, the maximum e may be several inches less than y_b. This limitation normally applies on long (60-ft and over) spans.

Case 2—Eccentricity Limited by Top Tension

The maximum permissible top tension (if any) is selected. The required bottom fiber compression is then computed from the known loads. Given the top and bottom fiber stresses, the stress f_{cg} at the centroid is found. Then P is found from $f_{cg} = P/A$.

If Case 1 is assumed to control, top tension must be checked once P and e are found. Conversely, if Case 2 is assumed to control, it must be verified that the required e can be obtained with a reasonable strand pattern within the beam cross-section.

Since the 48.5-ft span is of intermediate length, both cases will be checked. For checking Case 1, assume tht $e = y_b - 3$ in. $= 12$ in. This fixes e, and the fundamental equation may be rearranged and solved for P.

$$P = \frac{\Sigma[M/S_{b(c)}] + f_{t\,\text{allow}}}{1/A + e/S} \tag{21.10.1}$$

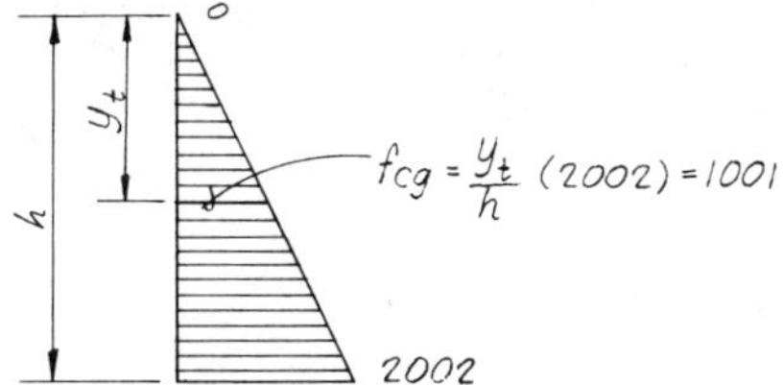

Fig. 21.36 Required stress at midspan.

The quantity $\Sigma[M/S_{b(c)}]$ is the sum of the bottom fiber stresses due to *all* the applied loads including self-weight. Equation (21.10.1) may now be solved for P.

$$P = \frac{3251 - 465}{1/360 + 12/1800}\left(\frac{1 \text{ kip}}{1000 \text{ lb}}\right) = 295 \text{ kips}$$

Top tension at midspan must now be checked. As discussed in Section 21.1.4, the effects of self-weight are normally applied concurrently with those of prestress. Top tension is, therefore, computed for the combined effects of prestress and self-weight.

$$f_t = \frac{P}{A} - \frac{Pe}{S_t} + \frac{M_g}{S_t}$$

$$= \frac{295{,}000}{360} - \frac{295{,}000(12)}{1800} + 784$$

$$= -782 \text{ (782 psi tension)}$$

The *ACI Code* places no explicit limit on top tension (if sufficient reinforcement is provided); however, high top tensions at midspan are undesirable. Therefore, Case 2 will control.

In applying Case 2 to find P and e, the approach is to determine both the top and bottom fiber stresses at midspan. Knowing these stresses, the stress at the centroid may be determined. Then, P may be found from the simple relationship $P = f_{cg}(A)$.

Small (say 200-psi or less) top tension at midspan does little harm. But assume zero top tension for purposes of computing P. The required bottom fiber compression under the combined effects of prestress and self-weight is equal to $M_d/S_b + M_\ell/S_{bc} + f_{t\,\text{allow}}$, or 1838 + 629 − 465 = 2002 psi. Then, f_{cg} may be computed. See Fig. 21.36. In this case, f_{cg} is simply half of the required bottom compression, or 1001 psi. It should be noted that when the precast beam is unsymmetrical (i.e., $y_t \neq y_b$), f_{cg} must be computed using smaller triangles in the stress diagram of Fig. 21.36.

The required prestress force is $P = f_{cg}(A) = 1.001\ (360) = 360.4$ kips. The required number of strands is 360.4/23.55 = 15.3 strands. Use 15 strands, and permit a small top tension. The final prestress force P is 23.55(15) = 353.3 kips. The eccentricity e may be found by solving the fundamental equation for e:

$$e = \frac{S_b}{P}\left[\Sigma\left(\frac{M}{S_{b(c)}}\right) + f_{t\,\text{allow}} - \frac{P}{A}\right]$$

$$= \frac{1800}{353{,}300}\left[3251 - 465 - \frac{353{,}300}{360}\right)$$

$$e = 9.20 \text{ in.}$$

Strand patterns are selected as shown in Fig. 21.37. Ten straight strands and five harped (draped) strands are used to give the required eccentricity at midspan, and to keep the prestressing force within the kern at the end.

Stresses at midspan may now be reviewed. For checking the stresses at time of release, assume that the initial force P_i is 15% greater than the final force.

$$P_i = 1.15P = 406.2 \text{ kips}$$

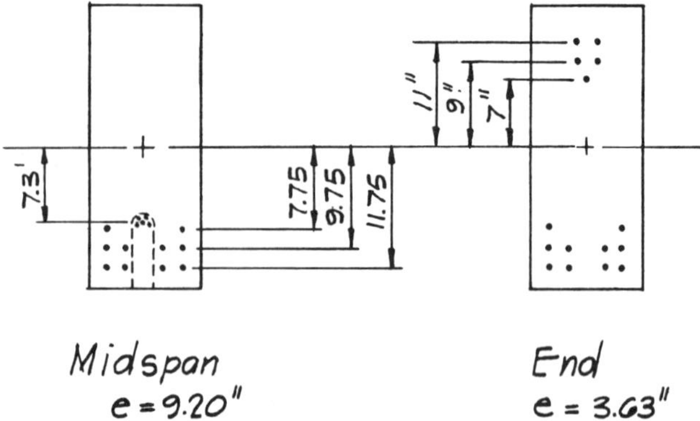

Fig. 21.37 Strand patterns.

Stresses at midspan release are tabulated in Table 21.2. The top tension is less than the code limit beyond which reinforcement is required. Use a minimum of 2-#4 bars in the top of the beam.

The stresses at design load are tabulated in Table 21.3. All stresses are within the allowable. The "stress at delivery" (line 4) is computed assuming all prestress losses have taken place at the time of delivery. This is a simplifying assumption, and is sufficiently accurate for most purposes.

It is recommended that a table similar to Table 21.3 be prepared for most, if not all, beam designs.

Table 21.2 Midspan Stresses (psi) at Release

	Top	Bottom
1. P/A = 406,200/360	1128	1128
2. Pe/S = 406,200(9.20)/1800	−2076	2076
3. M_g/S = 1,411,000/1800	784	−784
4. Stress at release	−164	2420
5. Allowable stress ($3\sqrt{f'_{ci}}$; $0.6f'_{ci}$)	−201 OK	2700 OK

Table 21.3 Final Midspan Stresses (psi)

	Top of Slab	Top of Beam	Bottom
1. P/A = 353,300/360		981	981
2. Pe/S = 353,300 (9.20)/1800		−1806	1806
3. M_g/S = 1,411,000/1800		784	−784
4. Stress at delivery		−41	2003
5. M_d/S = 3,308,000/1800		1838	−1838
6. Dead load stress		1797	165
7. M_ℓ/S_c = 2,159,000/9670; 11,940; 3432	223	181	−629
8. Stress at design condition	223	1978	−464
9. Allowable stress ($0.45f'_c$; $6\sqrt{f'_c}$)	1350 OK	2700 OK	−465 OK

Table 21.4 Factored Shears and Moments

	w (kip/ft)	LF	w_u (kip/ft)	V_u (kip)	M_u^+ (in.-kips)	M_u^- (in.-kips)
Girder	0.400	1.4	1.873	45.4	6609	
Slab	0.938	1.4				
Live load	1.500	1.7	2.550	61.8	3671	−6550
Sum			4.423	107.2	10,280	−6550

It provides a summary of the stress history of the beam and will show any possible problems during the various stages of loading.

Factored shears and moments are tabulated in Table 21.4.

The computation of the ultimate strength M_u (corresponding to ϕM_n, the design strength) is similar to that for a conventionally reinforced beam, except that the determination of the steel stress f_{ps} at ultimate load is more complex. Formula (18-3) of the *ACI Code* may be used:

$$f_{ps} = f_{pu}\left\{1 - \frac{\gamma_p}{\beta_1}\left[\rho_p \frac{f_{pu}}{f'_c} + \frac{d}{d_p}(\omega - \omega')\right]\right\}$$

Since there will be only nominal longitudinal reinforcement, ω and ω' may be taken as zero. First, compute the steel reinforcement ratio ρ_p.

$$\rho_p = \frac{A_{ps}}{bd} = \frac{15(0.1531)}{92(29.2)} = 0.000855$$

Since the compressive stress at ultimate load will be within the composite slab, use the properties of the slab for b and f'_c.

The factor γ_p is 0.4 for stress-relieved strand, and β_1 is 0.85 for $f'_c = 3000$ psi. Thus,

$$f_{ps} = 270{,}000\left[1 - \frac{0.4}{0.85}\left(0.000855\,\frac{270{,}000}{3000}\right)\right]$$

$$f_{ps} = 260{,}000 \text{ psi}$$

The ultimate steel tension, the lever arm, and the ultimate moment are then computed.

$$T_n = A_{ps}f_{ps} = 2.30(260) = 598 \text{ kips}$$

$$C_n = 0.85f'_c ba = 0.85(3)(92)a = 234.6\,a$$

$$a = \frac{598}{234.6} = 2.55 \text{ in.}$$

$$M_n = T_n(d - a/2) = 598(29.2 - 1.27) = 16{,}700 \text{ in.-kips}$$

$$\phi M_n = 0.9(16{,}700) = 15{,}030 \text{ in.-kips} > [M_u = 10{,}280 \text{ in.-kips}] \qquad \text{OK}$$

Note that the provided strength ϕM_n is substantially in excess of the factored moment M_u. This often happens in composite beams, for the service level loads are resisted partly by the bare beam, whereas at ultimate, all loads are resisted by the stronger composite section.

The section at the support is designed as a conventionally reinforced beam. When the steel percentage is less than half of balanced, the compression in the beam from prestress may normally be ignored in the ultimate strength computation [21.19].

Assume

$$a = 4 \text{ in.} \quad \text{and} \quad d = 35 - 3 = 32 \text{ in.}$$

Then

$$d - a/2 = 32 - 2 = 30 \text{ in.}$$

$$C_n = T_n = \frac{\text{Required } M_n}{d - a/2} = \frac{M_u/\phi}{d - a/2} = \frac{6550/0.9}{30}$$

$$= 242.5 \text{ hips}$$

$$\text{Required } A_s = \frac{T_n}{f_y} = \frac{242.5}{60} = 4.04 \text{ sq in.}$$

Use 3-#11, A_s = 4.68 sq in.

Check a,

$$a = \frac{T_n}{(0.85f'_c)b} = \frac{242.5}{(5.1)(12)} = 3.96 \text{ in.} \quad \text{OK}$$

Moments at the various loading stages are plotted in Fig. 21.38. The prestress moment is of opposite sign to the gravity load moments, but it is plotted on the same side of the base line so that it may readily be compared to the other moments. The shaded area represents the difference between the prestress and dead load moments. Since the difference is predominately positive, the long-term creep deflections will tend to be downward.

The moment diagram shows that the positive moments are considerably larger than the negative moments. This is desirable, since the T-beam section for positive moment is stronger than the rectangular sections available for negative moment. The inflection point is reasonably close to the support, minimizing the required length of the reinforcement required for negative moment. The desirable proportions of positive and negative moment were brought about by the decision not to shore, which caused the dead load of the slab to create positive moment only.

21.10.6 Deflections

The computation of deflections in prestressed members is an art. The final deflection is the difference of the upward deflection due to prestress and the downward deflection due to gravity loads. This net difference can be sensitive to small changes in assumptions used to compute the component deflections. The computation must consider six factors:

1. The history of loading on the beam.
2. The proper prestressing force to use in deflection calculations.
3. The proper choice of modulus of elasticity.
4. The use of multipliers for long-term creep effects.
5. The moment of inertia.
6. The choice of deflection coefficient.

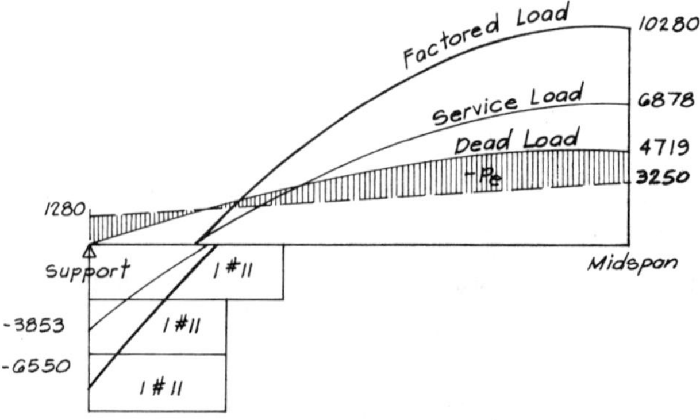

Fig. 21.38 Moment diagrams at various loading stages.

The loading history occurs in three main stages. Prestress (pretensioning) and self-weight are permanently applied at an age of one day. The dead load of the slab is permanently applied at an age of a few weeks or months. The live load is intermittently applied at a later time.

For deflection calculations, the use of a prestress force 10% above the final force level is recommended. This is an average of initial and final forces, weighted toward the initial.

The modulus of elasticity used in deflection calculations is derived from the concrete strength at the time the load is applied. Prestress and self-weight are applied when f'_{ci} = 4500 psi, and $E_{ci} = 57{,}000\sqrt{4500} = 3{,}824{,}000$ psi. The slab and the live load are applied when f'_c = 6000 psi, and $E_c = 57{,}000\sqrt{6000} = 4{,}415{,}000$ psi. The gain in strength after 28 days is normally ignored.

The proper choice of multipliers for long-term creep is the most difficult part of the deflection computation. The *PCI Design Handbook* [21.17], page 4-46, gives some recommendations. These apply to simple span beams. Continuous composite beams are not as free to deflect, once they are made continuous. Thus, for the continuous beam, a multiplier of 2 is used for deflections due to prestress and self-weight, and 1.35 for deflections due to the slab weight. No multiplier is used for the live load deflection.

The bare beam moment of inertia is used for all deflections except live load. The composite moment of inertia is used for computing the live load deflection.

Deflection calculations may be simplified by the use of a deflection coefficient C_Δ such that

$$\Delta = C_\Delta \frac{ML^2}{EI} \tag{21.10.2}$$

where M is the midspan moment.

The various moments for the different loadings have already been determined. The coefficient C_Δ depends on the shape of the moment diagram, as shown in Fig. 21.39. Note that all coefficients for simple span moments vary from 1/8 to 1/12, with the parabolic case being approximately 1/10. Coefficients for other shapes may usually be estimated with sufficient accuracy. The coefficient for the continuous case is approximate; the exact coefficient depends on the ratio of midspan to end moments.

The deflection computations are as follows, using Eq. (21.10.2) (L = 48.5 ft).

Prestress

$$P = 1.1(353.3) = 388.6 \text{ kips}$$

Divide the moment curve into two parts, as shown in Fig. 21.39*e*.

$$M_1 = Pe_{\text{end}} = 388.6(3.63) = 1411 \text{ kip-in.}$$

$$M_2 = M_{\text{msp}} - M_{\text{end}} = Pe_{\text{msp}} - M_{\text{end}} = 388.6(9.20) - 1411 = 2164 \text{ kip-in.}$$

$$\Delta_1 = C_\Delta \frac{ML^2}{EI} C_c = \frac{1}{8}\frac{1411(48.5)^2}{3824(27{,}000)}(144)(2) \qquad 1.15 \text{ in.} \uparrow$$

$$\Delta_2 = \frac{1}{12}\frac{2164(48.5)^2}{3824(27{,}000)}(144)(2) \qquad \underline{1.18 \text{ in.}} \uparrow$$

$$\Delta_{ps} = 2.33 \text{ in.} \uparrow$$

Self-Weight

$$\Delta_g = \frac{5}{48}\frac{1411(48.5)^2}{3824(27{,}000)}(144)(2) \qquad \underline{0.96 \text{ in.}} \downarrow$$

$$\text{camber} = 1.37 \text{ in.} \uparrow$$

Dead Load of Slab

$$\Delta_d = \frac{5}{48}\frac{3308(48.5)^2}{4415(27{,}000)}(144)(1.35) \qquad \underline{1.32 \text{ in.}} \downarrow$$

$$\Delta \text{ at full dead load} = 0.05 \text{ in.} \uparrow$$

Live Load

$$\Delta_\ell = 0.07\frac{2159(48.5)^2}{4415(79{,}990)}(144) \qquad \underline{0.14 \text{ in.}} \downarrow$$

$$\Delta \text{ at service load} = 0.09 \text{ in.} \downarrow$$

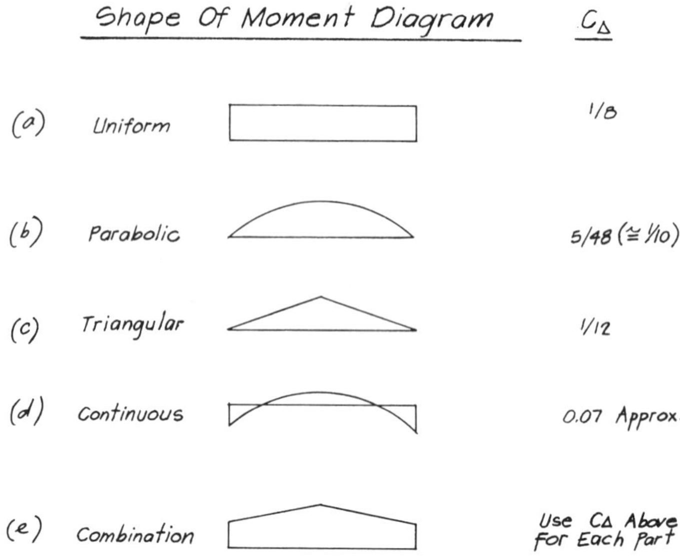

Fig. 21.39 Moment diagrams for deflection calculations.

How accurate is this computation? The accuracy cannot be estimated as a percentage of the final result, which is virtually zero. The result is the difference of several component deflections, whose absolute values sum to 4.75 in. To estimate the accuracy of the computation, one should take a percentage of this sum. The accuracy of the computation might be ±15% of 4.75 in. = ±¾ in., which is several times the net predicted deflection in this case.

21.11 NOTATION

An attempt has been made to make the notation consistent with the *ACI Code* and the *PCI Handbook.*

21.11.1 Main Symbols

a depth of compressive stress block
A area
b width
C compressive force
C_c creep multiplier
C_Δ deflection coefficient
d depth to centroid of steel force
e eccentricity of prestressing force
E modulus of elasticity
f stress
h overall thickness of member
I moment of inertia
ℓ_e embedment length
L span
M bending moment
P prestressing force (final after all losses, unless subscripted)
P_c tensile pullout strength of concrete
R radius of curvature
S section modulus

T tensile force
V shear force
w load per unit length
β_1 see *ACI Code,* 10.2.7.3, $\beta_1 = 0.85$ for 3000-psi concrete
γ a factor equal to 0.4 for stress-relieved strand
Δ change in stress or deflection
ρ steel ratio
Σ sum
ϕ capacity reduction factor
ω see *ACI Code,* Section 18.0

21.11.2 Subscripts

allow allowable
b bottom
c concrete, when applied to stresses
composite, when applied to section properties
cg center of gravity
d superimposed dead load
e effective, when applied to stress
embedment, when applied to length
end end
g self-weight
i initial (at release of prestress), when applied to stresses
interface, when applied to section properties
ℓ live load
msp midspan
n nominal strength as defined in Chapter 2 of the *ACI Code*
p prestress
reqd required
s steel
t tension, when applied to stresses
top, when applied to section properties
u ultimate (factored) load
y yield
1, 2 used for component parts

21.11.3 Superscripts

$'$ ultimate or cylinder strength
$+$ positive (moment)
$-$ negative (moment)

SELECTED REFERENCES

21.1 ACI Committee 318. *Building Code Requirements for Reinforced Concrete (ACI 318-83).* Detroit: American Concrete Institute, 1983.

21.2 ACI Committee 318. *Commentary on Building Code Requirements for Reinforced Concrete (ACI 318-83).* Detroit: American Concrete Institute, 1983.

21.3 Arthur R. Anderson and Saad E. Moustafa. "Ultimate Strength of Prestressed Concrete Piles and Columns," *ACI Journal, Proceedings,* **67,** August 1970, 620–635.

21.4 *Continuity in Concrete Building Frames,* 4th ed. Skokie, Ill.: Portland Cement Association, 1959.

21.5 L. D. Martin and W. J. Korkosz. *Connections for Precast Prestressed Concrete Buildings Including Earthquake Resistance.* Technical Report No. 2. Chicago: Prestressed Concrete Institute, 1982.

21.6 Leslie D. Martin and Donald L. Pellow. "Low-Relaxation Strand—Practical Applications in Precast Prestressed Concrete," *PCI Journal,* **28,** July/August 1983, 84–101.

21.7 Alan H. Mattock. "Design for Reinforced Concrete Corbels," *PCI Journal,* **21,** May/June 1976, 18–42.

21.8 A. H. Mattock. "Shear Transfer in Concrete Having Reinforcement at an Angle to the Shear Plane," *Shear in Reinforced Concrete,* SP-42. Detroit: American Concrete Institute, 1974, pp. 17–42.

21.9 Alan H. Mattock and Timothy C. Chan. "Design and Behavior of Dapped-End Beams," *PCI Journal,* **24,** November/December 1979, 28–45.

21.10 Alan H. Mattock, K. C. Chen, and K. Soongswang. "The Behavior of Reinforced Concrete Corbels," *PCI Journal,* **21,** March/April 1976, 52–77.

21.11 Alan H. Mattock and Neil M. Hawkins. "Shear Transfer in Reinforced Concrete—Recent Research," *PCI Journal,* **17,** March-April 1972, 55–75.

21.12 Alan H. Mattock, L. Johal, and H. C. Chow. "Shear Transfer in Reinforced Concrete with Moment or Tension Acting Across the Shear Plane," *PCI Journal,* **20,** July/August 1975, 76–93.

21.13 Alan H. Mattock, W. K. Li, and T. C. Wang. "Shear Transfer in Lightweight Reinforced Concrete," *PCI Journal,* **21,** January/February 1976, 20–39.

21.14 Saad E. Moustafa. "Effectiveness of Shear-Friction Reinforcement in Shear Diaphragm Capacity of Hollow-Core Slabs," *PCI Journal,* **26,** January/February 1981, 118–132.

21.15 Thomas Paulay. "An Elasto-Plastic Analysis of Coupled Shear Walls," *ACI Journal, Proceedings,* **67,** November 1970, 915–922.

21.16 PCI Committee on Prestressed Concrete Columns. "Recommended Practice for the Design of Prestressed Concrete Columns and Bearing Walls," *PCI Journal,* **21,** November/December 1976, 16–45.

21.17 *PCI Design Handbook,* 3rd ed. Chicago: Prestressed Concrete Institute, 1985.

21.18 *Post-Tensioning Manual,* 3rd ed. Phoenix, AZ: Post-Tensioning Institute, 1981.

21.19 Ferdinand S. Rostasy. "Connections in Precast Concrete Structures—Continuity in Double-T Floor Construction," *PCI Journal,* **7,** August 1962, 18–48.

21.20 T. Y. Lin and Ned H. Burns. *Design of Prestressed Concrete Structures,* 3rd ed. New York: Wiley, 1981.

21.21 Arthur H. Nilson. *Design of Prestressed Concrete*. New York: Wiley, 1978.

21.22 Antoine E. Naaman. *Prestressed Concrete Analysis and Design—Fundamentals*. New York: McGraw-Hill, 1982.

21.23 James R. Libby. *Modern Prestressed Concrete,* 2nd ed. Princeton, N.J.: Van Nostrand Reinhold, 1977.

CHAPTER 22
REINFORCED CONCRETE DESIGN

JACOB S. GROSSMAN

Vice-President
Robert Rosenwasser Associates, P.C.
Consulting Engineers
New York, New York

22.1 INTRODUCTION

This chapter focuses on special topics of crucial importance in buildings designed and constructed of placed reinforced concrete. Considering these special topics in the early stages of design, while options are available for the selection of the building system, will permit the practitioner to select the best system and design approach. This chapter is intended to provide information needed and not usually available in standard textbooks, or other standards and manuals, to help simplify the process of design and construction of large structures for which "exact" solutions are not possible or practical.

22.2 PRELIMINARY TOPICS OF IMPORTANCE

22.2.1 The Practitioner and the Market Place

Upon passing the professional engineering examinations, a structural engineer proudly embosses his PE seal and stands ready to certify to all that he intends to provide safe and serviceable structures. Soon, this newly certified professional is made aware that in addition to being safe and serviceable, the structures must be economical and comply with a schedule that allows introduction of the newly constructed building at the optimum time for marketability.

The owner requires the consulting team input prior to final determination as to whether the project should proceed. The practitioner will, therefore, find that rock-bottom fees are a distinct asset to the undecided owner.

If the owner is a builder of considerable experience and knows the capabilities of his consultants from previous projects, he may commit himself to the project much before the completion of the design concept. The design team, usually a conglomerate of several firms, including architects, mechanical engineers, structural engineers, soil consultants, and sometimes with the expert supervision of a construction management team, is then under the added pressure of producing design drawings, in stages, to allow the project to proceed.

Since market conditions change suddenly and often the best time to introduce a building into the market place was yesterday, a structural consulting firm has little time to provide the owner with an initial set of plans to allow him to start excavation and proceed to the foundation. This construction period is then utilized by the total design team to conclude the design process. This obviously puts added pressure on the structural consultant whose final design for the foundation must be completed prior to other consultant's final input. The structural engineer must, therefore, also have intimate knowledge of the needs of the mechanical and architectural teams in order to anticipate their requirements. The above described process ("fast tracking") has increasingly become the normal process of construction as interest rates, (and therefore time), become an important, if not the most important, consideration in new construction.

22.2.2 Professional Liability

The practicing engineer provides professional assurance that the structure is designed in accordance with the minimum requirements of the local building codes, and in accordance with the best available practices and utilizing the proper materials to assure that the structure is not only safe to enter but also is serviceable to live in. The engineer's liability and exposure to lawsuit varies from state to state. The extent of liability is not fully defined and is presently being tested in several important cases. The statute of limitations also varies and may not exist in some locations or under some conditions. A lengthy discussion of this subject will not be attempted except to point out the present trend to diversify "one-owner" structures into "conglomerate of owners" structures (condominiums, for example) where owners are not necessarily also the builders. While the owner–builder knows the reality of the construction business and is generally tolerant, such is not the case with buyer–owners, some of whom have threatened legal action often for petty transgressions in the construction process.

Liability, therefore, becomes an important factor within the practicing professions for which they have not and are not yet being compensated. Ventures into new types of design or construction are now hindered much more than in the past because the design team cannot continue and explore new designs with as much freedom as in the past. It is likely that new innovations in construction will be set back considerably unless lawsuits can be limited to legitimate cases only.

22.2.3 Compliance with Codes

The bleak reality described above is made even bleaker when the practicing engineer, whose seal and signature makes him liable possibly for the life of the structure, attempts to comply with all Code requirements. Dismissing for now the need to discuss the local and national codes, and focusing attention on codes such as *Building Code Requirements for Reinforced Concrete* (ACI 318-83) [22.1] and its related standards giving the minimum requirements for design and construction of concrete buildings, one finds that here, too, archaic provisions can easily put the practitioner on the defensive and possibly indicate liability for noncompliance.

Codes are written by "experts" in their profession. The societies (such as American Concrete Institute) that promulgate such codes usually seek independence from government legislation in order to produce a document for which the profession self-certifies the product. This independence from government legislation unfortunately attracts liability to such societies. Therefore, their standards are often written with an eye to limiting their liability. The design profession and the manufacturers of the product, on the other hand, attempt to insert provisions which will absolve them of added liability. This vicious cycle may produce some archaic provisions which are either contradictory or made unclear purposely (to allow alternate solutions or methods of analysis) and often cannot be adhered to or even understood by the practitioner. Such requirements may, for example, dictate elastic analysis (*ACI Code* [22.1], Sec. 8.3) for a product (concrete) which is nonhomogeneous and, therefore, cannot be "elastic". The *ACI Code* requires consideration of (but is not limited to) such items as creep, shrinkage, foundation settlement, unforeseen construction loads, wind loads, earthquake loads, and temperature variation for structures of all sizes.

Cast-in-place reinforced concrete structures are statically indeterminate because of the inherent continuity between structural members provided in monolithic construction. Some structures have several thousand structural members, all interconnected and therefore influencing one another. The engineer must provide safe and serviceable structures. If problems occur the engineer may be liable unless all related code provisions have been complied with. This cannot be realistically done. The practitioner, therefore, must know the limitations of the product at hand, understand one's own "sphere of knowledge", enlarge upon it, and keep the design well within such boundaries of knowledge so that exposure to lawsuits is diminished. The engineer should also carry professional liability insurance.

22.2.4 Designing Within One's "Sphere of Knowledge"

The previous comments may discourage both newcomers and presently practicing professionals. What is needed is to improve the situation by reducing the number of parameters that may adversely influence the design. One's "sphere of knowledge" of the topic must constantly enlarge. An old proverb relates that if knowledge equals the volume of a sphere, then ignorance (or areas yet to be explored) is equal to the exposed surface of the sphere.

Therefore, the larger the volume of knowledge, the greater the area of ignorance or areas where exploration is yet necessary. One may argue then against the need to undertake research to enlarge one's knowledge. However, the inner "core" of this knowledge is where one's practice should be concentrated. This core allows the engineer to discount parameters of minor influence and concentrate on those of importance. Regarding design of concrete buildings, a safe economical structure must be designed within the allotted time, hopefully at a profit, without "dotting all the i's and crossing all the t's" and hopefully, without being subject to a liability suit.

It must be stressed that to succeed in design practice, one needs to develop a large measure of "intuition", (a good eye for problem areas which usually is the by-product of experience). The beginner is therefore cautioned to consult the experienced designer prior to utilizing any of the material in this chapter.

In order to provide simple provisions for design, the number of parameters influencing the design must be minimized. Therefore, one must first define the boundaries of knowledge, the "state of the art", and stay well within it for the particular structure being designed. To help define "safe" design zones, the following example is used.

Second-order Analysis as an Example of "Safe" Design Zone

Column failure may be categorized into those columns failing due to "material" failure and those failing because of "instability" developed due to large deformation (sway) according to MacGregor and Hage [22.2], as illustrated in Fig. 22.1. The moment capacity is a function of the deflection index h_s/Δ (story height h_s to lateral deflection Δ). As Δ increases, the loss of moment capacity due to P–Δ moments (so called secondary moments) reduces the residual capacity of the column. Loss of stiffness due to additional cracking of the structural member also occurs thus increasing the effect of the secondary moments. Such a loss of stiffness is nonlinear, complicating the problem. When the "stability" type of failure controls, the analysis will be even more complicated and time-consuming. Iterative procedures are required and the resulting columns will not be economical.

MacGregor and Hage [22.2] indicate that "stability"-type failures were not observed when the deflection index (under service loads) was less than $h_s/300$ nor were stability failures observed in columns having

$$\frac{h_s}{r} < \left(90 - 50\frac{P_u}{P_0}\right) \leq 45 \tag{22.2.1}$$

where P_u is the factored axial load, and P_0 is the nominal strength under axial compression. These limits, when considered for the majority of structures designed today, do not represent undue restrictions on the practitioner. Serviceability of the nonstructural elements and the comfort of the human occupants usually dictate a more severe deflection index. Very slender columns are usually not needed. Selection of columns with, say, $h_s/c < 12$ (c is least lateral dimension), and of a deflection index of, say, $h_s/500$ to control the design will allow the practitioner to develop economical structures, to utilize the materials efficiently, and avoid the need to spend considerable design time on reinforced concrete columns that may not be stable. Columns that achieve their "material" strength require little iteration and have relatively small strength reduction due to slenderness or P–Δ effects. The P–Δ effect reduces the strength of the structure to support both gravity and lateral

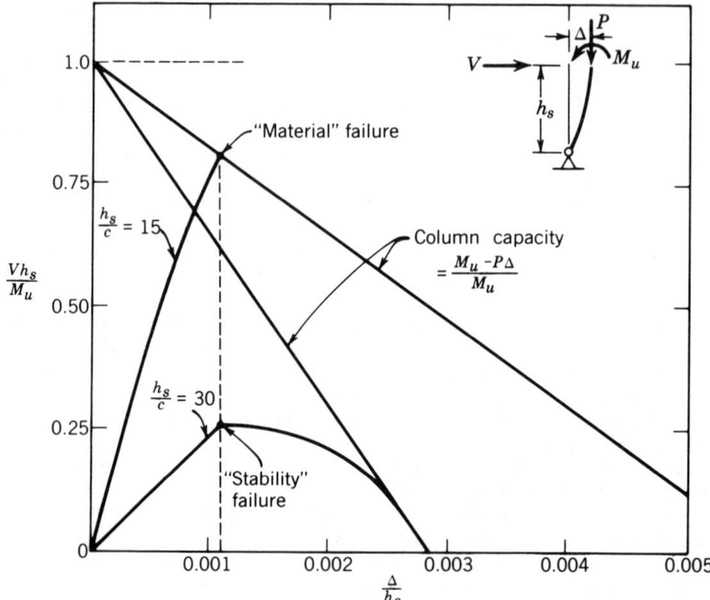

Fig. 22.1 Comparison of column "material" and "stability" failure modes (adapted from MacGregor and Hage [22.2]).

loads. Such reductions are not excessive when sways are small and in some such cases the P–Δ effect may be ignored altogether.

Economy in design time, economy in use of material, and uppermost, safety, will be enhanced dramatically when the designer stays well within his sphere of knowledge, where design requirements are simple, and structures are both safer and more economical to construct.

22.2.5 Office Design Practices (Past, Present, Future)

With the introduction in the last quarter of the century of electronic computers and with scientific data more readily disseminated in a much more orderly fashion, the design process in consulting firms is today quite different from that of even recent history of the late 50s and early 60s. The slide rule and Smoley's tables have become collector's items. The engineer has more time to review and analyze the "whole structure" since less time is devoted to the need for the repetitive, time-consuming, and often boring process of the design of the individual elements. With large-capacity computers available and with cheaper mini- and microcomputers flooding the market, the future is even more science-fiction fashioned and is already taking the form of the various consulting teams, dipping into a common data base to develop a structure from conceptual owner/architect dream to final detailing and scheduling with *any input item introduced only once*. The efficiency of the electronic age can be best measured by the ratio of input required to output obtained. Having to describe geometry only once (say, by the architect), sizes only once (by the structural engineer), penetrations for ductwork, pipes, etc. only once (by the mechanical engineer), could provide the common input base from which all trades could extract and supplement as the design process continues and evolves to the desired structure. This is the future of our profession which is at present in the initial formative stages. Extensions are being made possible by rapid development of graphic capabilities and increasingly compatible equipment for all software. At present, individual efforts by individual firms are preparing the various software necessary for the various trades. Soon all trades will be linked by compatible small or large computers via telephone lines.

The technical know-how for each of the individual trades is now being codified by experts. However, good practical designs require more than the ability to compute at the speed of light. "Experience" is, therefore, of utmost necessity and this "human element" must find its niche in this somewhat overwhelming projection of the future—otherwise we will lose it for all time. This human element of the design aspect cannot easily be programmed and thus may eventually disappear to be replaced by trial and error, no longer time consuming on the speedy computer. It will be interesting in time to evaluate the pros and cons the future will bring.

Present-Day Design Office Practice—Nursing the Computer

Most design firms today are using computers to some extent. It is probable that by the 1990s, more than 90% of design firms will rely on printed output to do at least part of their work. This period of evolution may cause great consternation among executives who were brought up and educated in the precomputer era. However, firms that have not already started the process of "computerizing" their practice are fast approaching obsolescence.

The correct sequence is to attempt to program the manual process the design firm follows. That is, establish the "flowchart" of the design process one step at a time, and then program the process, a part at a time, starting with the simplest or the most time consuming of the manual steps. Existing programs, either those in public domain or those that can be bought, can sometimes be modified and incorporated into this process. In any case, the end result must be an integrated process in which any input item is entered only once. Limitations of equipment and availability of funds to purchase must be initially reviewed by management in order to select the best and most suitable means. Time-sharing has been suitable in the 1970s and may still be suitable for developing one's own programs.

A programmer hired for the job ideally should have at least some engineering background, but by working closely with the experts of the firm it is possible for the programmer without intricate knowledge of engineering to develop practical programs. The language used is often erroneously selected to save computer time. The language used should be the one which the design team can easily learn since eventually the services of the programmer will no longer be required and someone from the design team must be able to modify, improve, and extend programs as required. Developing the program incrementally allows the necessary process of using parts as one goes along and then more easily verifying the output.

The size of the project to be computed will determine the required equipment and to what extent simplified assumptions are necessary so that production is not too costly nor too time consuming. Time-sharing, which limits size of programs that can be compiled, sometimes can be used as a yardstick to help curtail the individual steps taken. A method to link information between the several individual programs can be devised by using a common direct access file. Fragmented output is important to help detect errors and irrational results. Check points (flags) of input items against common values must be established early to help debug input as well. Such a procedure in computerizing the design process is the best way to assure continuation of the ''human element'' and to help develop and improve the firm's practice. Simplified—but safe—procedures are often needed to curtail costs or to await larger computer capacities certain to be available in the near future.

Development of office practices in the described manner is a slow process but should result in reduced expenditures, provide a thorough review of the firm's procedures, and allow for better uniformity of product, for checking of results, and be a great aid to production. Slow periods are better utilized to improve and expand the firm's programs.

22.2.6 Advantages of Fragmented Analysis

With the development of computer software the practitioner has new tools which can be used to great advantage but also can be detrimental to safe design. Again, experience must be exercised when one reviews the often enormous printout a few lines of input may generate.

One must know what it is the computer is spewing out before one can trust such output. Most software programs available today will analyze a structure using theories of elastic behavior for fragmented loading. Fragmented analysis is a correct step to follow. Gravity load output, followed by lateral loads acting alone, followed by temperature effects, etc., is the proper sequence. In addition, a combined loading output is available. The novice will do well to examine each of the separate load outputs so that the final placement of steel reinforcement will be appropriate even when all assumed design parameters are *not* at work.

The designer must also realize that the time sequence is altered. That is, a structure is constructed a floor at a time and dead load of the structure is incrementally added. Effects of creep and shrinkage are therefore a function of varied loading situations, and estimated foundation settlement producing shears and moments is not instantaneous (such as may be assumed in the computer analysis where the whole structure materializes instantly and all external loads and anticipated settlements occur simultaneously).

''Exact'' solutions using theory of elasticity, or finite element analysis, may cloak results with a dress of respectability, which is often unwarranted for concrete structures, and may sometimes be truly misleading and therefore dangerous. A quick check of ''old'' simple methods should accompany strange looking output. When in doubt, a backup analysis comparing the use of, say, ''finite'' joint size with another in which the size of the joint is ignored may help detect irrational results. A few stiff joints and members in a structure can indicate moments and shears that are impossible to develop, while reducing considerably the requirements from the structure's remaining elements.

The practioner must exercise intuition. A good engineer must almost know the answer before the design is made and must practice to develop this intuitive reflex. Unfortunately, as stated before, complex code requirements forcing reliance on computer output more and more tend to dull such instincts. A conscientious designer will look suspiciously at any output and attempt to find at least one irrational result in each review of design requirements. This writer is fond of recalling and repeating the story about the proud programmer displaying new software for a word processor which translates from any one language to another. Satisfying a request from one amazed bystander, he translates from English into Chinese the sentence ''out of sight—out of mind''. Displaying proudly the printed results, he is confronted with the skepticism of the audience. No one can read Chinese. What to do? ''No problem,'' says the resourceful programmer. We will translate back into English to verify. The result—''Blind, idiot''—was, as computers go, accurate enough.

22.2.7 Steel vs. Concrete—Design Approaches

Design of structural steel historically had tedious requirements for the design of the individual members but provided simple requirements for the design of the whole structure. This came about because the designer had control over the fixity of the connection of floor members to columns. Therefore, highly statically indeterminate structures were made statically determinate and simpler to analyze with a single predesign decision about column–beam fixity.

Cast-in-place reinforced concrete structures, where monolithic construction is inherently provided, do not allow such simplifications; hence, statically indeterminate analysis was required for solution, with various degrees of design success. It is the opinion of this writer that both research and analysis were set back many years because of the ''monolithic construction of homogeneous material'' concept assigned to concrete structures. Although much of it has been removed over the years, the code requirements regarding statically indeterminate analysis still remain and only slowly does the profession chip away at such misconceptions. Hardy Cross, by providing easier means to analyze statically indeterminate structures utilizing elastic theories (such as moment distribution and column analogy methods), may have set the concrete industry back. Field observations which were the mainstay source of providing design requirements were then often discarded, and replaced by requirements for theoretical solution in code provisions.

It is encouraging to note that practitioners are again attempting to increase input to code provisions, but much more effort is needed before completely "rational" codes come about. Meanwhile, the designer must face provisions which require full compliance with elastic theories on the one hand and which are countered by others that require none (e.g., on two-way slab systems design by the Equivalent Frame Method vs Direct Design Method of Chapter 13 of the *ACI Code* [22.1]). This conflict causes irrational disjointed code provisions. The time will surely come when the "fixity" syndrome to nonhomogeneous material will disappear from the codes in favor of guidelines for ductility which will allow local yielding while controlling cracking—thus maintaining shear strength. Ductility will allow redistribution of moments and therefore help permit continuous concrete structures to be recognized as having capabilities similar to those of steel structures, where simplicity of design is provided by allocated fixity of the connections and by designing the required capacity into the connection. Experience has taught the practitioner the value of ductility coupled with redundancy. For example, prior to general knowledge of how to design for torsion in spandrel beams (initially introduced into the 1971 *ACI Code*) many practitioners overlooked the problem by following simple practical rules such as use of closed stirrups and by providing alternate route (redundancy) to the unbalanced moments causing torsion. The requirements of the 1971 *ACI Code* were later softened in 1977 to allow such redistribution but only after torsional cracking occurred. The practitioner may take issue with assigning to torsional stiffness the primary deciding role of where and how loads are to be carried to supports. Field observations indicate that concrete structures (because of creep and shrinkage and possibly other relaxation tendencies of the material) usually attempt to support loads in either a "catenary" action or at least approach a simple span condition. The deflected shape of overloaded concrete members (with, however, excess shear capacity) just prior to flexural collapse can verify this tendency. It is quite apparent that concrete members are able and "want" to redistribute in order to reduce the "unbalanced" moments they support if another route (redundancy) is provided.

Torsional moments are "unbalanced" moments and are therefore difficult to develop in laboratories and real structures when alternate routes are available to support the loads. The practitioner should be able to exercise greater freedom to decide how the loads are to be supported. Other examples can also be cited of practices altered not because of field observations indicating the need to increase safety but because the theoreticians have found the way to codify the theory.

How can the lessons of the past be used? The answer is quite simple; by directing research to provide the necessary assurances that legitimate practices will not be discredited without "due process" and by confining the majority of experiments to be within the design boundaries (the inner "core" of actual design practice. Research to produce structures that are constructable and serviceable and aesthetically pleasing to the eye will generally also lead to structures having better safety.

22.2.8 Is It Possible to Simplify and Provide Safe and Economical Structures?

Simplicity in approach to design problems based on experience has many advantages and can produce economical as well as safe structures. The following remarks in this chapter are to be read with the skeptical and critical eye hopefully developed by reading the above. The responsible engineer must also utilize his own knowledge in order to develop new trends in design for the future. Direction from knowledgeable responsible engineers is necessary to obtain research verification of new design practices, meaningful to the profession *within practical design boundaries* in order to release the engineering profession from the confinements of overly elaborate theories developed for homogeneous materials which may not be applicable to nonhomogeneous materials such as concrete.

Needless to say, some or all that is written in this section may be controversial and reflects one engineer's—better yet, one practicing office's—experience, and the reader will do well to question and become satisfied that within one's own practice the shortcuts or simplified assumptions to be discussed later can properly be used. The remainder of this chapter attempts to describe the boundaries required for such a practice (used successfully in the design of several hundred buildings) in addition to the special requirements for (1) ductility (to allow load redistribution); (2) redundancy (provision for more than one path for possible load transfer); and (3) measures to prevent progressive collapse—all at minimal (if any) extra cost to the owner.

22.3 THE FOUR DESIGN CONSIDERATIONS: STRENGTH, SERVICEABILITY, CONSTRUCTABILITY, AND AESTHETIC INTEGRITY

A concrete structure designed and constructed in the United States is governed by the minimum provisions of the *ACI Code* as sometimes augmented by the local ordinance's general building code of which the *ACI Code* forms a part.

A review of the *ACI Code* will reveal that most of the design provisions dictate minimum strength (safety) requirements. ACI-9.5 and ACI-10.6 are orphans in that they do not relate to strength but rather to serviceability. ACI-9.5 involves control of deflections and ACI-10.6 gives crack control provisions. Indirectly both of these sections also enhance safety. Excessive deflection can be detrimental to a structure susceptible, for example, to ponding of rain water. Such ponding will increase the loads sustained and cause an increase in deflections which may further aggravate the situation to the point of failure. Similarly, cracking exposes reinforcement to rusting, causing concrete to spall, thereby destroying the bond of concrete to steel. Excessive cracking can also precipitate sudden (nonductile) shear failure.

The primary objective of the designer is to comply with the *ACI Code's* minimum requirements for strength; however, serviceability must also be satisfactory. Strength requirements are adequately discussed in standard textbooks such as Wang and Salmon [22.3]. The reader should be thoroughly familiar with the topics of reinforced concrete design treated in undergraduate courses. Experienced designers know they do not provide serviceability by providing merely the required strength.

In practice, neither strength nor serviceability is the criterion that commonly establishes the minimum size of structural members. These two design considerations are only a part of the design review. The designer responsibility does not end with the selection of members; an efficient and trouble-free construction process must be developed. In order to facilitate the erection of the nonstructural elements, deflections must be controlled during the construction stage. The "constructability" aspect of design cannot be overlooked.

Certain structures are more intolerant of design/construction deformations than others. For example, "as built" conditions, including deformations occurring up to the time of installation, can greatly influence the ease of installing prefabricated cladding that must fit between floors. On the other hand, on-site-placed brick cladding can be easily adjusted to even large field errors and excessive deformations. The designer must therefore be aware of the non-*ACI Code* "constructibility" needs of the nonstructural elements, in addition to the *ACI Code* "serviceability" and "strength" considerations.

An exposed structural member may also require added attention to assure that it remains pleasing in appearance during its life-long service. An excessively sagging exposed structural member can cause doubt about its strength and be a constant embarrassment to the designer who neglected to consider its "aesthetic integrity". Deeper members, more steel reinforcement, better concrete, and a more suitable construction process are often the result of demands for "constructability" and "aesthetic integrity" design considerations. The *ACI Code* does not (and should not) provide for the constructability and aesthetic integrity of the structure. This must be the responsibility of the design/construction/owner team.

To summarize, members must be selected to satisfy the following requirements using *actual load level* for any deflection-related computations:

1. Strength.
2. Serviceability.
3. Constructability.
4. Aesthetic integrity.

Computation of deflections of concrete members requires the review and estimate of so many parameters that accurate prediction is unlikely. Some of the more important reasons for this uncertainty and the difficulty in accurately predicting deflections are:

1. Concrete is not a homogeneous material and it cracks at relatively low tensile stress due to applied loads and/or restraint to shrinkage.
2. Concrete creeps and shrinks with time: therefore, another dimension, time, enters into the computation process (e.g., when will cladding be installed?)
3. The magnitude of the actual load causing deflection in monolithic construction is highly indeterminate and in two-way systems is influenced by the deflection of each supporting member.
4. Construction loads and construction procedures greatly influence the behavior of the concrete member. Initial load applied at an early age, such as in high-rise construction where freshly placed floors are supported by concrete only a few days old, can greatly increase the degree of initial cracking, as well as the magnitude of long-time creep and shrinkage deformations. A much observed phenomenon in high-rise construction is the better performance of the lowest several floors. These floors were shored during their early age, down to a rigid nonyielding foundation to help support self-weight and construction loads until the required strength was attained. Upper floors on the other hand received initial support from already

unshored and yielding lower levels. They are therefore required to support part of the self-load and to also participate in the support of the construction above prior to concrete attaining design strength. There is a distinctly observable difference in the realized construction tolerance and total deformations of these two groups of floors.

5. The loading history of the concrete members must be considered in the deflection computations. A member that eventually supports its full live load develops additional cracks. Upon removal of the live loads the measured self-load deflection is larger than the self-load deflection during its "first cycle of loading" prior to live load application. Several full load cycles, especially at an early age, will also adversely influence the long-term creep. This will be the case even if the member has not been subjected to construction loads and construction procedure abuse, and has been allowed to gain full strength prior to application of the service live loads.
6. In addition to the above, temperature, humidity, method of stripping shoring, and curing, etc. also influence the deformation of a concrete member.

It is, therefore, understandable that practitioners find it impossible to determine actual deflections with any degree of confidence. The *ACI Code* provisions have been somewhat cognizant of this problem in providing Table 9.5(a), in which limiting span/depth ratios for one-way construction are given so that the practitioner may (in certain situations) bypass the requirement to compute deflections to ascertain serviceability. The design aid is, however, limited in scope. It is applicable only to "members not supporting or attached to partitions or other construction likely to be damaged by large deflections."

Design aids that adequately provide the serviceability requirements need to consider such factors as the load level, member size and shape, material, time lag between removal of the shores and the application of the nonstructural finishes, the ratio of live to dead load, the loading history (i.e., is this the "first cycle of loading" or several cycles later?), the construction process, and construction loads. Table 9.5(a) does not directly consider these parameters. As may be expected, Table 9.5(a) has been shown [22.4] to not comply with the deflection limits requirements of Table 9.5(b), nor to provide safe haven from deflection problems in actual construction.

When architectural considerations limit member size, it may be necessary to increase the reinforcement ratio in order to increase the effective moment of inertia I_e of the member. The member will then be overly reinforced for other than strength considerations. In other situations, it is desirable to limit strength in order to prevent larger lateral forces from concentrating in the member. Ductility must therefore be detailed into such a member.

All aforementioned considerations, as well as perhaps others, must be reviewed by the designer. Design for serviceability, particularly deflection control, has been treated in several references [22.3–22.10]. Most designers, however, are less familiar with procedures of controlling deflections; nor do many of them correctly understand the serviceability requirements of the *ACI Code*. These topics are reviewed in detail in this section. Included in the treatment are a summary of *ACI Code* provisions, a presentation of simplified equations for effective moment of inertia I_e without the need to compute I_{cr}, an approximate deflection computation procedure, and a unified approach for computing the minimum depth of a concrete member which will not only comply with the *ACI Code* but also will satisfy both the need for "constructability" of the nonstructural finishes and the need for "aesthetic integrity" of the member during its service life.

The critical deflections that establish the required minimum depth for each of the design considerations are reviewed, including a review of typical construction loads and construction procedures, as well as an indication of how these important parameters can be used to better estimate actual field-observed deflections.

The minimum depth of a concrete member that will provide for flexural requirements and also for adequate ductility to allow the simplified design approach for gravity loads (see Section 22.4.2) is discussed in Section 22.7.2.

The reader may question the need to emphasize "serviceability", "constructability", and "aesthetic integrity" in the design of concrete structures. This design approach will, however, result in member sizes that give deflection behavior well within the "inner core" of the state of the art (see Section 22.2.4) and the resulting structure will therefore be safe, serviceable, and economical.

22.3.1 ACI Code Provisions for Serviceability—Their Complexity and Shortcomings

The serviceability requirements of the *ACI Code* are intended to assure that a concrete member subjected to flexure has "adequate stiffness to limit deflections or any deformations that may adversely affect strength or serviceability of a structure at service loads." (ACI-9.5.1).

Deflection limits to assure a serviceable structure are provided in ACI-Table 9.5(b). The footnote of Table 9.5(b), however, indicates that the limits provided do not safeguard against ponding. Ponding must be checked by "suitable calculations of deflection, including the added deflections due to ponded water, considering the long-time effects of all sustained loads, camber, construction tolerances and reliability of provisions for drainage.'' Table 9.5(b) therefore provides only the limits to assure that nonstructural elements attached to the structure (not likely to be damaged by large deflections) are not damaged by the ensuing deflections of the structural members. The structure can therefore become ''serviceable'' for service loads.

The *ACI Code* provisions requiring serviceable structures, if not complied with, attach legal liability to the designer. Thus, it is not surprising that much of the design profession has been upset with Code provisions to ascertain serviceability. Such dissatisfaction accelerated after the introduction of the 1971 Code. In that Code the effective moment of inertia I_e of the structural member was introduced [currently 1983 ACI Formula (9-7)].

The effective moment of inertia is given by

$$I_e = \left(\frac{M_{cr}}{M_a}\right)^3 I_g + \left[1 - \left(\frac{M_{cr}}{M_a}\right)^3\right] I_{cr} \leq I_g \tag{22.3.1}$$

as developed by Branson [22.9] to consider the bilinear behavior of a varying moment of inertia decaying with each added load increment because of increased amount of cracking. Equation (22.3.1) represents a single value that may be used in place of a variable moment of inertia for a simply supported span. By contrast, with Eq. (22.3.1), the 1963 ACI Code had indicated that $I_e = I_{cr}$ when the reinforcement ratio ρ exceeded $500/f_y$. For lower reinforcement ratio I_e was to be taken equal to the gross moment of inertia I_g.

The *ACI Code* and *Commentary* [22.11] also note that the proper way to estimate deflections to ascertain the serviceability limits of Table 9.5(b) is to assume ''first cycle of loading'' and therefore to consider I_e at each load level separately. To compute the live load deflection $(\Delta_i)_L$ one must compute separately (1) the immediate total load deflection $(\Delta_i)_{D+L}$ using $(M_a)_{D+L}$ in Eq. (22.3.1), and (2) the immediate deflection for the dead load portion $(\Delta_i)_D$ using $(M_a)_D$ in Eq. (22.3.1). The immediate live load deflection is then obtained as the net difference between these two computed values. Since for the first cycle of loading $(I_e)_D$ is larger than $(I_e)_{D+L}$, the computed net live load deflection is larger, and the immediate dead load deflection is smaller, than their respective proportions of total load.

To consider the long-time effects of creep and shrinkage, a multiplier is applied to the immediate dead load deflection $(\Delta_i)_D$ to obtain the additional time-dependent deflection due to creep and shrinkage. The long-time creep and shrinkage multiplier can be estimated for each separately [22.6] or arrived at collectively [22.1] as

$$\lambda = \frac{\xi}{1 + 50\rho'} \tag{22.3.2}$$

where ξ is a time-dependent factor estimated to equal 2.0 five years after construction. For intermediate time periods lower values can be obtained from ACI-9.5.2.5 or from Fig. 9.2 [22.11].

The total deflection Δ_t due to sustained loads (including creep and shrinkage) and live loads is

$$\Delta_t = (1 + \lambda)(\Delta_i)_D + [(\Delta_i)_{D+L} - (\Delta_i)_D] \tag{22.3.3}$$

The first term represents the sum of the immediate plus long term deflection due to sustained loads. The net value $[(\Delta_i)_{D+L} - (\Delta_i)_D]$ represents the computed live load deflection.

The critical deflection Δ_s to ascertain serviceability requirements of ACI-Table 9.5(b) is the following:

1. When live load deflection alone is considered:

$$\Delta_s = (\Delta_i)_L = [(\Delta_i)_{D+L} - (\Delta_i)_D] \tag{22.3.4}$$

2. Where sustained loads deflection occurring after installation of nonstructural elements is to be added to the live load deflection:

$$\Delta_s = \lambda_n(\Delta_i)_D + [(\Delta_i)_{D+L} - (\Delta_i)_D] \tag{22.3.5}$$

The time-dependent factor λ_n can be obtained by subtracting from λ given by Eq. (22.3.2) at 5 yr the estimated value of λ at the time nonstructural finishes are to be installed. If, for example, this period is estimated to occur not earlier than 6 months after construction of the supporting member,

$$\lambda_n = \frac{2.0}{1 + 50\rho'} - \frac{1.2}{1 + 50\rho'} = \frac{0.8}{1 + 50\rho'}$$

where ρ' is the compression reinforcement ratio provided (either required or not required for strength). The value 1.2 is the long-term multiplier λ at 6 months (ACI-9.5.2.5).

The complexity of the procedure to ascertain serviceability is apparent from the previous discussion. Solution by iteration is required in all design cases. In order to estimate I_e a depth is selected and the required reinforcement determined. The neutral axis and I_{cr} can then be computed. I_e is then arrived at by Eq. (22.3.1) for several load levels, such as dead plus live load, dead load, and other intermediate loads (including sustained live loads). Trial-and-error solutions often require design aids and preferably computers to do the job adequately.

ACI Committee 318 recognized the complexity of the deflection requirements shortly after the 1971 Code was introduced. A Committee 318 Task Group for Code Simplification was formed and its first charge was to simplify the serviceability provisions. The results of this still on-going simplification effort were first published by Grossman [22.4]. The author in response to discussion of the paper [22.5] also added means to incorporate "contructability" and "aesthetic integrity" design considerations with the same initial design effort used to ascertain serviceability. A unified approach to all four design considerations (strength, serviceability, constructability, and aesthetic integrity) is therefore now available.

The influence of construction loads and construction procedures on the deflection of concrete members has not to date been addressed by the *ACI Code*. The result is that practitioners attempting to use the Code procedures are often dismayed by the results. The treatment herein, derived from the observation of actual construction practices, recognizes construction loads and procedures as *major* parameters influencing deformations [22.4, 22.5, 22.12–22.14].

Note, however, that while most construction is influenced by construction loads and procedures, some concrete members may well be protected from these influences. Members designed to support large future loads (such as transfer girders intercepting columns supporting high-rise structures) may not be influenced by construction abuses nor will they complete "first cycle of loading" until the total structure, a floor at a time, is completed. The suggestions of the next sections are therefore developed in such a way as to allow either the inclusion or the exclusion of construction loads and procedures and "first cycle of loading" as design parameters in the computation of deflections.

Readers may be surprised to find that construction loads and procedures actually permit simplification of the design process. Equations followed by (a) include provisions for designing concrete members when construction loads and procedures should be considered.

The treatment in the following sections is essentially the procedure of Refs. 22.4 and 22.5.

22.3.2 Simplified Equations to Estimate Effective Moment of Inertia I_e

The effective moment of inertia I_e was developed [22.9] in the form of Eq. (22.3.1) to estimate the bilinear behavior of a concrete member as it transforms from an uncracked section (when supporting a light load $M_a < M_{cr}$), to a fully cracked section (when loads are heavy $M_a/M_{cr} > 3.5\pm$) for which $I_e = I_{cr}$.

Figure 22.2 from Grossman [22.4] shows how I_e/I_g varies with M_a/M_{cr} for many similar single span uniformly loaded beams, each of which was reinforced for a total service load level preselected to induce a particular M_a/M_{cr} load ratio. The first beam was designed to support a light load producing a service moment M_a at midspan equal to the bending moment causing initial flexural cracks; therefore $M_a/M_{cr} = 1.0$. The other beams were each designed to support a progressively larger M_a/M_{cr} ratio. All beams were identical except for reinforcement ratio ρ.

The curve of Fig. 22.2 may be approximated [22.4] as follows:

1. When $\dfrac{M_a}{M_{cr}} \le 1.6$

$$\frac{I_e}{I_g} = \left(\frac{M_{cr}}{M_a}\right)^2 \le 1.0 \tag{22.3.6}$$

2. When $1.6 < \dfrac{M_a}{M_{cr}} \le 10$

$$\frac{I_e}{I_g} = 0.1\left(\frac{M_a}{M_{cr}}\right) K \sqrt{\frac{145}{w_c}} \tag{22.3.7}$$

I_e/I_g when computed by Eq. (22.3.6) or (22.3.7) need not be smaller than $0.35K\sqrt{145/w_c}$, nor smaller than value attained at a maximum service moment capacity, $M_a = \phi M_n[(D + L)/(1.4D + 1.7L)]$.

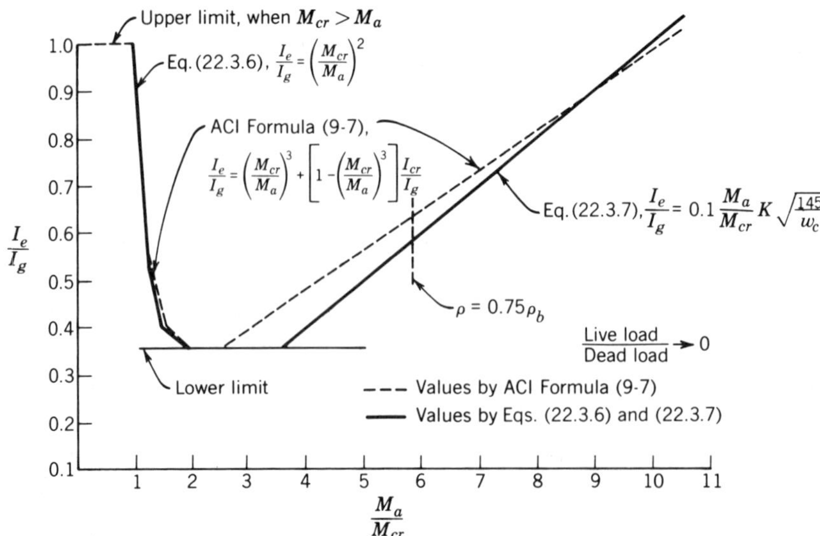

Fig. 22.2 I_e/I_g values for normal-weight concrete beams, supporting total service load moment M_a, computed according to *ACI Code* Formula (9-7) and by simplified Eqs. (22.3.6) and (22.3.7) (from Grossman [22.4]).

In Eqs. (22.3.6) and (22.3.7),

$$K = \frac{d}{0.9h}\left[\frac{1}{0.4 + \left(\dfrac{1.4M_a}{\phi M_n}\right)\left(\dfrac{f_y}{100{,}000}\right)}\right] \tag{22.3.8}$$

M_a = maximum unfactored moment in member at load level for which deflection is being computed
M_n = nominal moment strength
w_c = unit weight of concrete, pcf*
f_y = yield stress of steel reinforcement, psi†

There are three distinct load-level zones indicated in Fig. 22.2. (For the purpose of this discussion, assume that f_y = 60,000 psi and w_c = 145 pcf (stone concrete) to set $K \approx 1.0$). The first load-level zone ($1.0 < M_a/M_{cr} < 1.6$) coincides with that of lightly loaded beams and slabs, often reinforced with only slightly more than temperature steel requirements. The stiffness of concrete members in this zone is due mainly to the tension in the concrete: the low amount of reinforcement provided is of secondary importance in estimating I_e. This load level can have large variations between computed and observed deflections. As can be seen in Fig. 22.2, a small increase in loading quickly reduced I_e from I_g (for a beam carrying a moment $M_a = M_{cr}$) to about 40% of I_g for a similar beam but proportionally reinforced to support a moment $M_a = 1.6\ M_{cr}$. It is therefore understandable that determination of deflection cannot be exact in this load zone, and indeed it is not. Observation of actually constructed structural members (mostly one-way slabs in this load level) indicates that large variations in deformations can occur in seemingly identical members. Clearly then construction loads and construction procedures (such as method of shoring, stripping, and curing) have a significant influence on the behavior of these lightly reinforced members.

The second load-level zone of Fig. 22.2 can be described as a flat lower limit to $I_e \approx 0.35\ I_g$. All beams having total service loads causing ($1.6 < M_a/M_{cr} < 3.5$) will have this lower limit of I_e. In this loading zone Eq. (22.3.1) indicates that the increase in reinforcement (required for beams supporting larger loads) approximately compensates for the reduction in tension stiffening of the concrete

* Ratio $145/w_c$ in Eq. (22.3.7) becomes $2330/w_c$ when w_c is in kg/m³.
† Ratio $f_y/100{,}000$ in Eq. (22.3.8) becomes $f_y/689$ when f_y is in MPa.

due to increase in cracking. Thus I_e remains a fairly constant ratio of I_g for the many beams in this load-level zone.

The third load-level zone in Fig. 22.2 (when $M_a/M_{cr} > 3.5$) is a zone where beams support heavy loads. The large quantities of reinforcement in these beams increase I_{cr} of the section. Since large loads cause many cracks to develop, the tension in the concrete is of little significance in I_e and its value essentially equals I_{cr} of the member. Equation (22.3.7) approximates I_e/I_g in a simple manner and for this load-level zone it also approximates I_{cr}/I_g.

It is of interest to note that Eqs. (22.3.6) and (22.3.7) (with their lower limits) provide I_e values based directly on the total load level (M_a/M_{cr}) the member is supporting. This should greatly simplify the computation process especially during the preliminary stage when iteration to arrive at a final size is required. *This can be done without the need to compute the reinforcement, neutral axis, or* I_{cr} *of the member.* Final refinements using K by Eq. (22.3.8) can be applied to the final size.

Correction factor K by Eq. (22.3.8) allows for adjustments for shallow and deep members, and when f_y is other than 60,000 psi. Via the ratio $(1.4M_a/\phi M_n)$ the equation includes the live-to-dead-load ratio. Members having reinforcement in excess of design requirements (referred to here as *over-reinforced* members) may also be analyzed. (ϕM_n is the design strength of the section.) The lower limit to I_e is computed at a service load level equal to ϕM_n divided by an average overload factor, even if this load level is higher than one anticipates the member needs to support. This situation may arise when the member is more heavily reinforced than strength requires solely to help build up I_e. I_e at the actual load level must be compared against a lower limit obtained by using a load level $M_a = \phi M_n(D + L)/(1.4D + 1.7L)$, and the largest I_e value selected.

When one needs to evaluate I_e for the "first cycle of loading" at intermediate load levels, such as dead load acting alone, several considerations must be reviewed if the simplified equations are to be used. Simplified Eqs. (22.3.6) and (22.3.7) were developed to estimate I_e at the total service load level. The accuracy indicated [22.4] is that in 97% of over 13,000 members reviewed (representing 486 permutations of size, shape, and material) values by the simplified equations were within ±20% of those by Eq. (22.3.1). For intermediate load levels, the correction factor K provides a similar adjustment to that obtained for "over-reinforced" members. I_e at intermediate load levels cannot be lower than I_e at total load level $(M_a)_{D+L}$ or I_e at load level corresponding to ϕM_n. Accuracy, however, may still suffer (or improve) at intermediate load levels as compared with total load level predictions.

Figure 22.3 shows the accuracy of Eqs. (22.3.6) and (22.3.7) for "first cycle of loading" at intermediate load levels for three of the many beams which form (at total service load) a curve similar to the curve in Fig. 22.2. One beam from each of three distinct load-level loads in Fig. 22.2 was selected. Each of the beams is reinforced to support its total service load. At total service load: $M_a/M_{cr} = 1.6$ for beam #1; $M_a/M_{cr} = 2.5$ for beam #2; and $M_a/M_{cr} = 5.5$ for beam #3. For beam #1, Eq. (22.3.6) is shown in Fig. 22.3 to give good agreement with values of I_e by Eq. (22.3.1) also at all intermediate load levels.

For beam #2, the lower limit of $I_e/I_g = 0.35K$ controls and provides good agreement for a fair range of intermediate load levels here too. Note that K is a variable obtained separately for each of

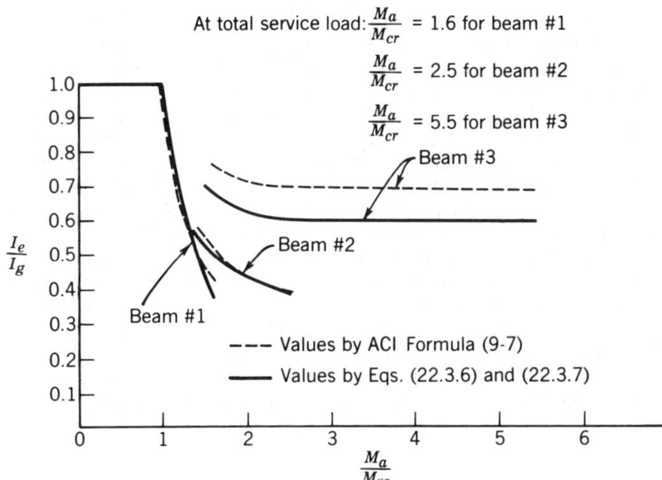

Fig. 22.3 I_e/I_g values at intermediate load levels for "first cycle of loading" computed by *ACI Code* Formula (9-7) and by simplified Eqs. (22.3.6) and (22.3.7) (from Grossman [22.5]).

Table 22.1 Approximate Expressions* for $Q = (I_e)_{D+L}/(I_e)_D$

At Intermediate Load Level	At Full Service Load	
$A = (M_a)_D/M_{cr}$	$B = (M_a)_{D+L}/M_{cr}$	$Q = (I_e)_{D+L}/(I_e)_D \leq 1.0$
$A \leq 1$	$B \leq 1$	1.0
$A \leq 1$	$B \leq 1.6$	$1/B^2$
$1.0 < A < 1.6$		E^2
$A \leq 1$	$1.6 < B < 3.5$	0.35
$1.0 < A < 1.6$		$0.35A^2$
$1.6 < A < 3.5$		$0.4 + 0.6E$
$A \leq 1$	$B \geq 3.5$	$0.1B \leq 1.0$
$1.0 < A < 1.6$		$0.1A^2B \leq 1.0$
$1.6 < A < 3.5$		$0.28B(0.4 + 0.6E) \leq 1.0$
$A \geq 3.5$		1.0

$(M_a)_D$ = service load moment at intermediate load level (including construction loads).
$(M_a)_{D+L}$ = total service load moment.
$E = (M_a)_D/(M_a)_{D+L}$.
* From Grossman [22.4].

the intermediate load levels. When at intermediate load level M_a is less than $1.6M_{cr}$, Eq. (22.3.6) needs to also be considered and may provide a larger value of I_e. The larger value always controls.

For beam #3, at relatively high load level, I_e by Eq. (22.3.7) provides also an estimate of I_{cr}, and I_e at the total load level is the lower limit also for all high–intermediate load levels. At low–intermediate load levels ($M_a/M_{cr} < 3.5$) a larger I_e value can sometimes be realized by using the lower limit of $0.35K$ [or by Eq.(22.3.6) if intermediate load level $M_a/M_{cr} < 1.6$].

$(I_e)_D$ will reduce in value and approach $(I_e)_{D+L}$ after one or more load cycles (self-load to total load) have occurred. From a practical standpoint several load cycles and/or construction loads and procedures negate the need to compute I_e at intermediate load levels [22.4, 22.5, 22.12, 22.13, 22.14] even though literal application of the *ACI Code* might dictate otherwise. The "first cycle of loading" obviously loses significance for the serviceability check in actual construction whenever nonstructural elements are attached after the structural member has undergone load cycles due to construction loads which often can exceed the design live load. When, however, there is a need to compute I_e at intermediate load levels for the "first cycle of loading" the simplified equations can provide a fair estimate.

Another quick procedure is to estimate Q, the $(I_e)_{D+L}/(I_e)_D$ ratio for "first cycle of loading", using Table 22.1 (taken from reference 22.5), which was evaluated using Eqs. (22.3.6) and (22.3.7). "Over-reinforced" members should be treated with total service load assumed to be some intermediate load of an even larger service load corresponding to ϕM_n.

22.3.3 Approximate Deflection Computation

Once I_e is estimated the immediate deflection Δ_i can be computed using the elastic equation for maximum deflection of cantilevers and the midspan deflection of simple and continuous beams as follows:

$$\Delta_i = \frac{5}{48} K_1 \frac{M_a \ell^2}{E_c I_e} \tag{22.3.9}$$

where M_a is the support moment of cantilevers and the midspan moment for simple and continuous beams. K_1 is 12/5 for cantilevers, 1 for simple spans, and $(1.2 - 0.2M_0/M_a)$ for continuous spans, where M_0 is the simple span moment at midspan. The use of only the midspan section I_e (rather than the average of I_e at ends and midspan) is considered satisfactory for prismatic continuous members. The ACI Commentary-9.5.2.4 indicates that somewhat improved results may be obtained using weighted average section properties.

The immediate total load deflection section is therefore equal to

$$(\Delta_i)_{D+L} = \frac{5}{48} K_1 \frac{(M_a)_{D+L}\ell^2}{E_c(I_e)_{D+L}} \tag{22.3.10}$$

and the immediate intermediate load deflection

$$(\Delta_i)_D = \frac{5}{48} K_1 \frac{(M_a)_D\ell^2}{E_c(I_e)_D} \tag{22.3.11}$$

Substituting $Q = (I_e)_{D+L}/(I_e)_D$,

$$(\Delta_i)_D = \frac{5}{48} K_1 \frac{(M_a)_D\ell^2}{E_c(I_e)_{D+L}} Q \tag{22.3.12}$$

with Q estimated using Eq. (22.3.1), Eq. (22.3.6), and Eq. (22.3.7), or from Table 22.1. To simplify, $(\Delta_i)_D$ may be obtained directly from total load immediate deflection, and letting $D/(D + L) = (M_a)_D/(M_a)_{D+L}$

$$(\Delta_i)_D = (\Delta_i)_{D+L} \left(\frac{D}{D + L}\right) Q \tag{22.3.13}$$

and the immediate live load deflection taken as the net difference (to account for $(I_e)_D > (I_e)_{D+L}$):

$$(\Delta_i)_L = (\Delta_i)_{D+L} \left[1 - \left(\frac{D}{D + L}\right) Q\right] \tag{22.3.14}$$

Note that $(M_a)_D$ and $(M_a)_{D+L}$ are service loads (not factored) and D and L are dead and live service loads, respectively.

When construction loads, or several load cycles, cause cracking to reduce $(I_e)_D$ to the lower value of $(I_e)_{D+L}$ (therefore $Q = 1.0$), Eqs. (22.3.13) and (22.3.14) can be simplified [22.4]

$$(\Delta_i)_D = (\Delta_i)_{D+L} \left(\frac{D}{D + L}\right) \tag{22.3.13a}$$

$$(\Delta_i)_L = (\Delta_i)_{D+L} \left(\frac{L}{D + L}\right) \tag{22.3.14a}$$

It is this writer's opinion, based on observed results of actual construction and based on present-day construction procedures, that the simplified equations will more accurately describe deflections realized in actual construction, except in special situations such as extremely oversized members, transfer girders, etc.

EXAMPLE 22.3.1

Given a simply supported T-section that spans 20 ft (6.1 m) and must support a sensitive partition above it, as shown in Fig. 22.4. The beam also spans over another sensitive partition, which in turn is supported on a nonyielding foundation. The contractor indicates that the partitions will be installed not earlier than three months after the beam has been cast: (a) Determine the required clearance under the beam so that there will be no contact with the partition below. (b) Determine whether or not the beam will satisfy the deflection limit of ACI-Table 9.5(b) for the supported partition; that is, $\Delta \leq \ell/480$ for the sum of live load plus creep and shrinkage deflections. (c) Compare results for a similar beam which is to be constructed at one of the top levels of a high-rise structure where construction loads and procedures are expected to influence results adversely.

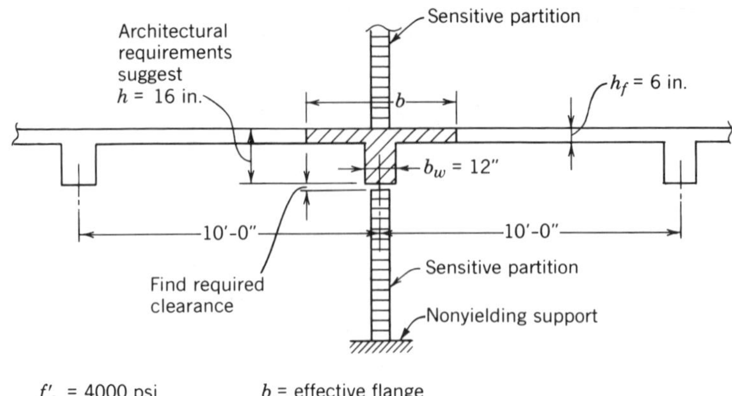

f'_c = 4000 psi
f_y = 60,000 psi
w_c = 145 pcf
Live load = 100 psf
Finishes = 25 psf

b = effective flange $= \frac{\ell}{4} =$ 5'-0"
Assume d = 12.5 in.

Fig. 22.4 Simply supported T-section for Example 22.3.1.

Solution

(a) *Compute loads and moments.*

dead load D: slab	$\frac{6}{12}(150)$ =	75 psf	
finishes	=	25	
est. beam weight	=	15	
		115 psf	(1.4) = 161
live load L:		100	(1.7) = 170
totals		215 psf (service)	331 psf (factored)

$$M_u = 0.331(10)(20)^2/8 = 165.5 \text{ ft-kips } (224.6 \text{ kN}\cdot\text{m})$$

$$(M_a)_D = 0.115(10)(20)^2/8 = 57.5 \text{ ft-kips } (78.0 \text{ kN}\cdot\text{m})$$

$$(M_a)_{D+L} = 0.215(10)(20)^2/8 = 108 \text{ ft-kips } (146.6 \text{ kN}\cdot\text{m})$$

Assume beam is reinforced so that $\phi M_n = M_u$.

(b) *Compute* $(I_e)_{D+L}$. The cracking moment M_{cr} may be computed using the *ACI Design Handbook* [22.15] deflection design aids.

$$\text{for } \frac{b}{b_w} = \frac{60}{12} = 5.0 \quad \text{and} \quad \frac{h_f}{h} = \frac{6}{16} = 0.375$$

$$\text{and } K_{cr} \quad \text{(DEFLECTION 1.1)} = 1.69$$

$$K_{crt} \quad \text{(DEFLECTION 1.2)} = 1.4$$

$$M_{cr} = b_w K_{cr} K_{crt} = 12(1.69)(1.4) = 28.4 \text{ ft-kips } (38.5 \text{ kN}\cdot\text{m})$$

Using DEFLECTION 3 from Ref. 22.15,

$$I_g = K_{14}\left(\frac{b_w h^3}{12}\right) = 1.92\left(\frac{12(16)^3}{12}\right) = 7860 \text{ in.}^4$$

$$\frac{(M_a)_{D+L}}{M_{cr}} = \frac{108}{28.4} = 3.80; \qquad \text{use Eq. (22.3.7)}$$

$$\frac{I_e}{I_g} = 0.1\left(\frac{M_a}{M_{cr}}\right) K \sqrt{\frac{145}{w_c}} \geq 0.35K \sqrt{\frac{145}{w_c}}$$

$$= 0.1(3.80)K(1.0) \geq 0.35K(1.0) = 0.38K$$

Using Eq. (22.3.8) for K,

$$K = \frac{12.5}{0.9(16)}\left[\frac{1}{0.4 + \dfrac{1.4(108)}{165.5}\left(\dfrac{60{,}000}{100{,}000}\right)}\right] = 0.92$$

$$(I_e)_{D+L} = 0.38KI_g = 0.38(0.92)7860 = 2748 \text{ in.}^4$$

(c) *Compute* $(I_e)_D$. Assume partitions represent only a small percentage of the 25 psf attributed to finishes. For "first cycle of loading" (and neglecting construction loads and procedures),

$$\frac{(M_a)_D}{M_{cr}} = \frac{57.5}{28.4} = 2.02; \qquad \text{use Eq. (22.3.7)}$$

$$\frac{I_e}{I_g} = [0.1(2.02)K = 0.20K] < 0.35K$$

Thus,

$$\frac{I_e}{I_g} = 0.35K$$

where

$$K = \frac{12.5}{0.9(16)}\left[\frac{1}{0.4 + \dfrac{1.4(57.5)}{165.5}(0.6)}\right] = 1.25$$

$$\frac{I_e}{I_g} = 0.35(1.25) = 0.44$$

$$(I_e)_D = 0.44I_g = 0.44(7860) = 3440 \text{ in.}^4$$

(d) *Compute immediate total load deflection.* Using Eq. (22.3.10)

$$(\Delta_i)_{D+L} = \frac{5}{48}(1.0)\frac{108(20)^2 1728}{3600(2748)} = 0.79 \text{ in. (20 mm)}$$

(e) *Compute* $(\Delta_i)_D$ *and* $(\Delta_i)_L$. Using Eq. (22.3.13),

$$Q = \frac{(I_e)_{D+L}}{(I_e)_D} = \frac{2748}{3440} = 0.80$$

$$(\Delta_i)_D = 0.79\left(\frac{115}{215}\right)0.80 = 0.34 \text{ in. (8.6 mm)}$$

Compare with results using Eq. (22.3.11),

$$(\Delta_i)_D = \frac{5}{48}(1.0)\frac{57.5(20)^2 1728}{3600(3440)} = 0.33 \text{ in. (8.4 mm)}$$

Alternatively, the multiplier Q may be estimated from Table 22.1 without the need to compute $(I_e)_D$. Enter Table 22.1 with

$$A = \frac{(M_a)_D}{M_{cr}} = 2.02 \qquad B = \frac{(M_a)_{D+L}}{M_{cr}} = 3.80$$

and

$$E = \frac{(M_a)_D}{(M_a)_{D+L}} = \frac{57.5}{108} = 0.53$$

Obtain for $(1.6 < A < 3.5)$ and $B > 3.5$

$$Q = 0.28B(0.4 + 0.6D) \leq 1.0$$

$$= 0.28(3.80)[0.4 + 0.6(0.53)] = 0.77$$

This is approximately the same as actual $Q = 0.80$. Finally, using the theoretical formulas

$$(\Delta_i)_L = (\Delta_i)_{D+L} - (\Delta_i)_D = 0.79 - 0.33 = 0.46 \text{ in. } (11.7 \text{ mm})$$

Alternately, using Eq. (22.3.14),

$$(\Delta_i)_L = 0.79\left[1 - \left(\frac{115}{215}\right)(0.8)\right] = 0.45 \text{ in. } (11.4 \text{ mm})$$

The results are in close agreement.

(f) *Assume the simplified equations, Eq. (22.3.13a) and (22.3.14a), are appropriate since several construction load cycles have occurred prior to installation of the frangible partitions, that is, assume Q approaches* 1.0. Then,

$$(\Delta_i)_D = 0.79\left(\frac{115}{215}\right) = 0.42 \text{ in. } (10.7 \text{ mm})$$

$$(\Delta_i)_L = 0.79\left(\frac{100}{215}\right) = 0.37 \text{ in. } (9.4 \text{ mm})$$

(g) *Compute the net long-term creep and shrinkage multiplier λ_n.* Assume there is no compression reinforcement. From ACI-9.5.2.5,

$$\begin{aligned} \xi \text{ at 5 yr} &= 2.0 \\ -(\xi \text{ at 3 months}) &= -\underline{1.0} \\ \text{net } \xi &= 1.0 \end{aligned}$$

$$\lambda_n = \frac{net\ \xi}{1 + 50\rho'} = \frac{1.0}{1 + 0} = 1.0$$

(h) *Compute sustained load deflection occurring after installation of frangible partitions.* Using Eq. (22.3.5) for "first cycle of loading",

$$\Delta_s = \lambda_n(\Delta_i)_D + [(\Delta_i)_{D+L} - (\Delta_i)_D]$$

$$= 1.0(0.33) + (0.79 - 0.33) = 0.79 \text{ in. } (20 \text{ mm})$$

Therefore, the required clearance below the beam above the frangible partition is 0.79 in. (say ⅞ in.).

(i) *Check ACI-Table 9.5(b) requirements for the partition supported on top of the beam.*

$$\Delta_s = 0.79 \text{ in.} > \left[\frac{\ell}{480} = \frac{20(12)}{480} = 0.50 \text{ in.}\right]$$

The beam will not satisfy the deflection limit of ACI-Table 9.5(b). Note that the deflection check was made without having to compute the required reinforcement. Another iteration is now required using a larger size beam, longer waiting period before placing partitions, or with compression reinforcement. However, compare first results by simplified equations assuming $(I_e)_D = (I_e)_{D+L}$.

(j) *Compare results using approximate equations, Eqs. (22.3.13a) and (22.3.14a), assuming that $Q \approx 1$.*

$$\Delta_s = 1.0(0.42) + (0.79 - 0.42) = 0.79 \text{ in. (20 mm)}$$

This value agrees with the critical deflection for serviceability obtained in (h) where Q was assumed equal to 0.80. Note that when $\lambda_n = 1.0$, $Q \neq 1.0$ does not affect the result, and the simplified equations assuming $Q = 1.0$ could always be used. Even if λ_n is not 1.0 the results are only slightly affected. For example, assuming $\lambda_n = 1.5$ (say, partitions are installed at an earlier time), the "exact" expression, Eq. (22.3.5), gives

$$\Delta_s = 1.5(0.33) + (0.79 - 0.33) = 0.96 \text{ in. (24 mm)}$$

The approximate solution assuming $Q = 1.0$ gives

$$\Delta_s = 1.5(0.42) + (0.79 - 0.42) = 1.0 \text{ in. (25 mm)}$$

which predicts a result 4% higher than the "exact" solution.

If the multiplier $\lambda_n = 0.5$ (say partitions are installed at a later time), Eq. (22.3.5) gives

$$\Delta_s = 0.5(0.33) + (0.79 - 0.33) = 0.63 \text{ in. (16 mm)}$$

The approximate solution assuming $Q = 1.0$ gives

$$\Delta_s = 0.5(0.42) + (0.79 - 0.42) = 0.58 \text{ in. (15 mm)}$$

giving a result about 8% too low.

This writer concludes that for the deflection serviceability check, the assumption that $(I_e)_D \approx (I_e)_{D+L}$ is sufficiently accurate even for the case where "first cycle of loading" criteria are required.

22.3.4 Direct Estimate of $h_{\min}$ to Satisfy Serviceability, Constructability, and Aesthetic Integrity of the Structural Member

The approximate equations for the effective moment of inertia I_e may be used to develop a procedure to obtain a minimum thickness $h_{\min}$ which will approximately satisfy ACI-Table 9.5(b).

When a beam is heavily loaded, that is, $M_a/M_{cr} > 3.5$, Eq. (22.3.7) applies for I_e/I_g and also provides an estimate of I_{cr}/I_g. For heavily loaded beams $I_e = I_{cr}$. Thus, using Eq. (22.3.7) as the starting point

$$\frac{I_e}{I_g} = 0.1\left(\frac{M_a}{M_{cr}}\right) K \sqrt{\frac{145}{w_c}} \qquad [22.3.7]$$

for the development of an $h_{\min}$ expression, the uniformly loaded beam will have a total immediate deflection (dead plus live load), $(\Delta_i)_{D+L}$ as given by Eq. (22.3.9),

$$(\Delta_i)_{D+L} = \frac{5}{48} K_1 \frac{M_a \ell^2}{E_c I_e} \qquad [22.3.9]$$

where

$$E_c = 33(w_c^{1.5})\sqrt{f_c'} \approx 57{,}620\sqrt{f_c'}\left(\frac{w_c}{145}\right)^{1.5} \qquad (22.3.15)$$

When the cross-section is rectangular,

$$M_{cr} = 7.5\sqrt{f_c'}\left(\frac{bh^2}{6}\right) K_2 \qquad (22.3.16)$$

where $K_2 = 1.0$ for normal-weight concrete; 0.85 for sand-lightweight concrete; and 0.75 for all-lightweight concrete. With slight approximation, therefore, $K_2 = \sqrt{145/w_c}$.

Assuming the member is to be heavily reinforced such that Eq. (22.3.7) controls I_e, substitution of Eqs. (22.3.7) and (22.3.16) into Eq. (22.3.9) with $K_2 = \sqrt{145/w_c}$ and $I_g = bh^3/12$ gives

$$(\Delta_i)_{D+L} = \frac{5}{48} K_1 \left[\frac{\ell^2}{57{,}620\sqrt{f_c'}(w_c/145)^{1.5}}\right] \frac{7.5\sqrt{f_c'}(bh^3/6)\sqrt{w_c/145}}{0.1K(bh^3/12)\sqrt{145/w_c}} \qquad (22.3.17)$$

which gives

$$(\Delta_i)_{D+L} = \frac{\ell}{3688}\left(\frac{\ell}{h}\right)\frac{K_1}{K}\sqrt{\frac{145}{w_c}} \tag{22.3.18}$$

In general, Eq. (22.3.18) may be written

$$(\Delta_i)_{D+L} = \frac{\ell}{C}\left(\frac{\ell}{h}\right)\frac{K_1}{K}\sqrt{\frac{145}{w_c}} \tag{22.3.19}$$

where for the example above using simplified I_e values, $C \approx 3700$ for a rectangular cross-section.

Since Eq. (22.3.7) gives only approximate values of I_e [within $\pm 20\%$ of I_e from Eq. (22.3.1)] a parametric computer study [22.4] was made, using Eq. (22.3.1), of 486 simple span ($K_1 = 1$) varied beams to obtain a value for C corresponding to a load level requiring the use of $\rho = 0.75\rho_b$ as follows:

$$C = \frac{\ell}{(\Delta_i)_{D+L}}\left(\frac{\ell}{h}\right)\left(\frac{1}{K}\right)\sqrt{\frac{145}{w_c}} \tag{22.3.20*}$$

The study included variations in size and shape of cross-section, concrete strength, and steel yield stress. The appropriate value of C was found to be about 4000 for rectangular sections and somewhat larger for T-sections.

For simplicity, Grossman [22.4] recommended taking $C = 4320$ when $\rho = 0.75\rho_b$ because this value is a whole multiple of all the deflection limit coefficients of ACI-Table 9.5(b); that is, 4320 is evenly divisible by 180, 240, 360, and 480.

Therefore, for heavily loaded ($\rho \approx 0.75\rho_b$) one-way nonprestressed simply supported members, the total load immediate deflection $(\Delta_i)_{D+L}$ may be estimated as

$$(\Delta_i)_{D+L} = \frac{\ell}{4320}\left(\frac{\ell}{h}\right)\frac{K_1}{K}\sqrt{\frac{145}{w_c}} \tag{22.3.21}$$

Values of C larger than 4320 were also established [22.4] for members less heavily loaded and therefore less heavily reinforced (see Fig. 22.5).

---

* For w_c in kg/m³, $145/w_c$ becomes $2330/w_c$.

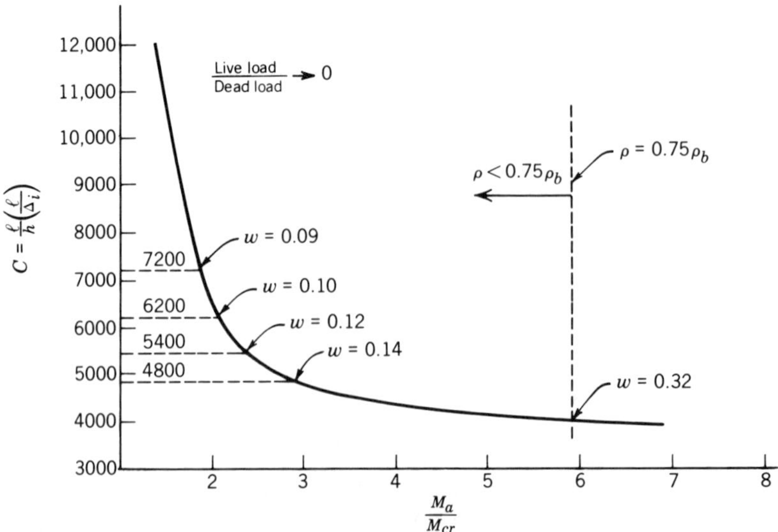

Fig. 22.5 Values of $C = (\ell/h)(\ell/\Delta_i)$ for normal-weight concrete beams having $d/h = 0.9$ and $f_y =$ 60,000 psi (from Grossman [22.4]).

When the total loading dictates $\rho < 0.75\rho_b$, a smaller member may be used if the deflection limit of ACI-Table 9.5(b) remains the same. Figure 22.5 from Grossman [22.4] shows the variation of C with load level in terms of M_a/M_{cr}. Tables 22.2 through 22.7 provide values of C for corresponding ρ ratios of the several variables. The deflections obtained by Eq. (22.3.20) will be the same for various beams, each supporting a different load, when the C value used corresponds to the required ρ value (for the given beam size) computed according to *ACI Code* strength design.

For T-sections, guidelines are more difficult to establish than for rectangular beams, because of the wide range of values for flange width b and flange thickness h_f. Grossman [22.4] determined that by using three different combinations of member variables, sufficient data would be available to draw general conclusions. The three are as follows:

1. T-sections, with $h_f/h = 0.1$ and $b_w/b = 0.1$.
2. T-sections, with $0.2 \leq h_f/h \leq 0.5$ and $b_w/b = 0.1$.
3. Rectangular sections ($h_f/h = 1.0$ and $b_w/b = 1.0$).

For each concrete strength f'_c, the tables provide values for the three different shapes of member. Concrete strengths of 3000, 4000, and 5000 psi are tabulated. Note that the tables have juxtaposed the commonly used 60,000-psi steel with 40,000-psi steel to provide the designer with an indication as to what degree of improvement can be had simply by overly reinforcing the member (as if only 40,000-psi yield stress steel is available). To obtain this evaluation, the designer must not overlook the need for the additional adjustment required by K [computed by Eq. (22.3.8)] since the tables have been prepared by computing C values based on Eq. (22.3.20).

Use of the tables is straightforward. The ρ values for flanged members for which $0.1 < b_w/b < 1.0$ can be obtained by straight-line interpolation between either one of the flanged beam groups and the rectangular beam group. For example, a beam for which $b_w/b = 0.3$ and $h_f/h = 0.1$ requires interpolation between ρ for $b_w/b = 0.1$ of the first group and ρ for $b_w/b = 1.0$ for the third group. For a similar beam but with $0.2 \leq h_f/h \leq 0.5$, interpolation between ρ values in the second and third groups will provide the correct result. (See Examples 22.3.2 and 22.3.3.)

For each of the three groups, values of ρ are provided for shallow ($d/h = 0.79$), average ($d/h = 0.9$), and deep ($d/h = 0.95$) members. For other d/h ratios a straightline interpolation is appropriate. For flanged members the ρ values are computed using the effective flange width b.

Reference 22.4 also provides (see Table 22.8) a very condensed table which can serve all the permutations of member shape, size, and material studied. Table 22.8 indicates, however, limiting $\omega = \rho f_y/f'_c$ values rather than the more readily applicable ρ values in Tables 22.2 through 22.7. Table 22.8 generally can provide conservative C values suitable when the serviceability of the member is reviewed. When determination of deflections is necessary better accuracy is realized when the designer uses Tables 22.2 through 22.7 directly.

EXAMPLE 22.3.2

Determine the coefficient C appropriate for a rectangular cross-section requiring $\rho = 0.0050$. Use $f_y = 60{,}000$ psi, $w_c = 145$ pcf, $D = 3L$, $d/h = 0.90$, and $f'_c = 3000$ psi.

Solution

Using Table 22.2 for $f_y = 60{,}000$ psi, $w_c = 145$ pcf, $D = 3L$, the third group of the table with $b_w/b = 1.0$ and $h_f/h = 1.0$ gives for $f'_c = 3000$ psi and $d/h = 0.90$, five different values of C depending on ρ.

C =	4800	5400	6200	7200	8600
ρ =	0.0076	0.0061	0.0052	0.0047	0.0041

Interpolate for $\rho = 0.0050$. Find $C = 6440$.

Alternatively, use the less accurate but condensed Table 22.8. Compute

$$\omega = \frac{\rho f_y}{f'_c} = \frac{0.0050(60{,}000)}{3000} = 0.10$$

For a rectangular section having $f_y = 60$ ksi no interpolation is required since $C = 5400$ is given for $\omega = 0.10$. The C value is somewhat conservative (i.e., too low).

Table 22.2 Coefficient C for Eq. (22.3.20) For Members Loaded to Require $\rho = 0.75\rho_b$ and Others Less Heavily Loaded to Require $\rho < 0.75\rho_b$*

Type of Member					$f_y = 60{,}000$ psi; $w_c = 145$ psf; $D = 3L$							$f_y = 40{,}000$ psi; $w_c = 145$ psf; $D = 3L$						
					@ $\rho = 0.75\rho_b$		Maximum ρ when $C =$					@ $\rho = 0.75\rho_b$		Maximum ρ when $C =$				
f'_c		b_w/b	h_f/h	d/h	C	$\frac{M_a}{M_{cr}}$	4800	5400	6200	7200	8600	C	$\frac{M_a}{M_{cr}}$	4800	5400	6200	7200	8600
$f'_c = 3000$ psi	Flanged Section	0.1	0.1	0.79	4451	12.9	0.0039	0.0020	0.0013	0.0011	0.0009	4655	13.3	0.0084	0.0051	0.0026	0.0017	0.0014
		0.1	0.1	0.90	4453	15.2	0.0037	0.0017	0.0010	0.0008	0.0007	4682	15.6	0.0078	0.0048	0.0023	0.0013	0.0011
		0.1	0.1	0.95	4465	16.4	0.0035	0.0016	0.0009	0.0007	0.0006	4670	17.0	0.0076	0.0046	0.0022	0.0012	0.0009
		0.1	≥0.2	0.79	4260	21.4	0.0047	0.0020	0.0013	0.0011	0.0009	4477	21.8	0.0121	0.0065	0.0028	0.0018	0.0014
		0.1	≥0.2	0.90	4281	25.3	0.0047	0.0018	0.0010	0.0008	0.0007	4502	25.8	0.0116	0.0064	0.0025	0.0014	0.0011
		0.1	≥0.2	0.95	4303	27.2	0.0047	0.0017	0.0009	0.0007	0.0006	4536	27.6	0.0113	0.0063	0.0024	0.0012	0.0010
	Rect.	1.0	1.0	0.79	4179	4.4	0.0097	0.0080	0.0070	0.0062	0.0056	4246	4.9	0.0163	0.0123	0.0103	0.0091	0.0079
		1.0	1.0	0.90	4129	5.6	0.0076	0.0061	0.0052	0.0047	0.0041	4227	6.3	0.0141	0.0099	0.0079	0.0069	0.0059
		1.0	1.0	0.95	4116	6.3	0.0069	0.0054	0.0046	0.0041	0.0036	4231	7.0	0.0136	0.0090	0.0070	0.0060	0.0052
$f'_c = 4000$ psi	Flanged Section	0.1	0.1	0.79	4254	14.9	0.0044	0.0022	0.0015	0.0012	0.0011	4409	15.3	0.0092	0.0058	0.0030	0.0020	0.0016
		0.1	0.1	0.90	4250	17.6	0.0041	0.0019	0.0012	0.0010	0.0008	4414	18.1	0.0085	0.0054	0.0027	0.0015	0.0012
		0.1	0.1	0.95	4277	18.8	0.0040	0.0018	0.0010	0.0008	0.0007	4419	19.5	0.0082	0.0052	0.0025	0.0014	0.0011
		0.1	≥0.2	0.79	4063	24.8	0.0051	0.0023	0.0015	0.0013	0.0011	4212	25.2	0.0132	0.0072	0.0031	0.0020	0.0016
		0.1	≥0.2	0.90	4110	28.9	0.0051	0.0020	0.0012	0.0010	0.0008	4254	29.6	0.0126	0.0071	0.0028	0.0016	0.0012
		0.1	≥0.2	0.95	4115	31.4	0.0051	0.0019	0.0011	0.0009	0.0007	4288	32.0	0.0124	0.0071	0.0028	0.0014	0.0011
	Rect.	1.0	1.0	0.79	3963	5.0	0.0108	0.0090	0.0079	0.0070	0.0064	3994	5.6	0.0176	0.0138	0.0117	0.0103	0.0090
		1.0	1.0	0.90	3934	6.4	0.0085	0.0069	0.0059	0.0053	0.0047	3986	7.2	0.0153	0.0110	0.0090	0.0078	0.0067
		1.0	1.0	0.95	3926	7.3	0.0077	0.0062	0.0052	0.0047	0.0041	3987	8.1	0.0146	0.0101	0.0080	0.0069	0.0059
$f'_c = 5000$ psi	Flanged Section	0.1	0.1	0.79	4093	16.5	0.0048	0.0025	0.0017	0.0014	0.0012	4202	16.9	0.0101	0.0064	0.0033	0.0022	0.0018
		0.1	0.1	0.90	4112	19.3	0.0045	0.0021	0.0013	0.0011	0.0009	4218	20.0	0.0093	0.0059	0.0029	0.0017	0.0014
		0.1	0.1	0.95	4108	21.0	0.0044	0.0020	0.0012	0.0009	0.0008	4241	21.6	0.0090	0.0057	0.0028	0.0015	0.0012
		0.1	≥0.2	0.79	3930	27.4	0.0055	0.0025	0.0017	0.0014	0.0012	4039	27.9	0.0142	0.0078	0.0035	0.0023	0.0018
		0.1	≥0.2	0.90	3962	32.3	0.0055	0.0022	0.0013	0.0011	0.0009	4088	32.8	0.0136	0.0077	0.0031	0.0018	0.0014
		0.1	≥0.2	0.95	3979	34.5	0.0055	0.0021	0.0012	0.0010	0.0008	4085	35.0	0.0134	0.0077	0.0031	0.0016	0.0012
	Rect.	1.0	1.0	0.79	3844	5.3	0.0118	0.0100	0.0087	0.0078	0.0071	3861	5.9	0.0189	0.0152	0.0129	0.0114	0.0100
		1.0	1.0	0.90	3819	6.8	0.0093	0.0077	0.0065	0.0059	0.0053	3853	7.6	0.0165	0.0121	0.0099	0.0087	0.0075
		1.0	1.0	0.95	3819	7.6	0.0084	0.0068	0.0057	0.0052	0.0046	3854	8.5	0.0157	0.0111	0.0088	0.0077	0.0066

* For flanged sections having $b_w/b > 0.1$, linear interpolation may be used between ρ values at $b_w/b = 0.1$ and $b_w/b = 1.0$ (rectangular section). ρ is the value for the member in its final size. For flanged sections $\rho = A_s/bd$.

Table 22.3 Coefficient C for Eq. (22.3.20) For Members Loaded to Require $\rho = 0.75\rho_b$ and Others Less Heavily Loaded to Require $\rho < 0.75\rho_b$*

Type of Member					$f_y = 60{,}000$ psi; $w_c = 145$ psf; $D = L$							$f_y = 40{,}000$ psi; $w_c = 145$ psf; $D = L$						
					@ $\rho = 0.75\rho_b$		Maximum ρ when $C =$					@ $\rho = 0.75\rho_b$		Maximum ρ when $C =$				
f'_c		b_w/b	h_f/h	d/h	C	$\frac{M_a}{M_{cr}}$	4800	5400	6200	7200	8600	C	$\frac{M_a}{M_{cr}}$	4800	5400	6200	7200	8600
$f'_c = 3000$ psi	Flanged Section	0.1	0.1	0.79	4538	12.3	0.0044	0.0022	0.0014	0.0011	0.0010	4786	12.6	0.0098	0.0058	0.0030	0.0019	0.0015
		0.1	0.1	0.90	4532	14.5	0.0041	0.0020	0.0011	0.0009	0.0007	4792	15.0	0.0091	0.0053	0.0027	0.0015	0.0011
		0.1	0.1	0.95	4561	15.6	0.0040	0.0019	0.0010	0.0008	0.0006	4817	15.9	0.0090	0.0051	0.0026	0.0013	0.0010
		0.1	≥0.2	0.79	4345	20.4	0.0055	0.0023	0.0014	0.0012	0.0010	4587	20.8	0.0137	0.0075	0.0032	0.0019	0.0015
		0.1	≥0.2	0.90	4389	23.8	0.0055	0.0021	0.0011	0.0009	0.0007	4623	24.3	0.0132	0.0074	0.0030	0.0015	0.0011
		0.1	≥0.2	0.95	4388	25.9	0.0054	0.0020	0.0010	0.0008	0.0007	4647	26.4	0.0126	0.0073	0.0030	0.0014	0.0010
	Rect.	1.0	1.0	0.79	4276	4.2	0.0105	0.0085	0.0073	0.0065	0.0059	4360	4.6	0.0180	0.0132	0.0110	0.0096	0.0083
		1.0	1.0	0.90	4212	5.3	0.0083	0.0065	0.0055	0.0049	0.0043	4341	5.9	0.0159	0.0108	0.0084	0.0072	0.0062
		1.0	1.0	0.95	4211	6.0	0.0076	0.0058	0.0048	0.0043	0.0038	4340	6.7	0.0154	0.0099	0.0075	0.0064	0.0055
$f'_c = 4000$ psi	Flanged Section	0.1	0.1	0.79	4332	14.2	0.0049	0.0025	0.0016	0.0013	0.0011	4515	14.6	0.0101	0.0065	0.0034	0.0021	0.0017
		0.1	0.1	0.90	4340	16.7	0.0046	0.0022	0.0013	0.0010	0.0008	4520	17.3	0.0094	0.0060	0.0031	0.0017	0.0013
		0.1	0.1	0.95	4350	18.0	0.0044	0.0021	0.0011	0.0009	0.0007	4541	18.5	0.0090	0.0058	0.0029	0.0015	0.0011
		0.1	≥0.2	0.79	4158	23.5	0.0060	0.0026	0.0016	0.0013	0.0011	4325	23.9	0.0148	0.0083	0.0036	0.0022	0.0017
		0.1	≥0.2	0.90	4192	27.7	0.0059	0.0024	0.0013	0.0010	0.0009	4379	28.1	0.0141	0.0082	0.0034	0.0017	0.0013
		0.1	≥0.2	0.95	4203	29.7	0.0059	0.0023	0.0011	0.0009	0.0008	4386	30.5	0.0138	0.0081	0.0034	0.0016	0.0012
	Rect.	1.0	1.0	0.79	4046	4.8	0.0116	0.0096	0.0083	0.0074	0.0067	4099	5.4	0.0193	0.0148	0.0124	0.0109	0.0095
		1.0	1.0	0.90	4014	6.2	0.0091	0.0073	0.0062	0.0056	0.0049	4090	6.9	0.0169	0.0120	0.0096	0.0083	0.0071
		1.0	1.0	0.95	4010	6.9	0.0083	0.0065	0.0055	0.0049	0.0043	4092	7.7	0.0164	0.0110	0.0085	0.0073	0.0062
$f'_c = 5000$ psi	Flanged Section	0.1	0.1	0.79	4193	15.6	0.0054	0.0028	0.0018	0.0015	0.0012	4313	16.0	0.0110	0.0071	0.0038	0.0024	0.0019
		0.1	0.1	0.90	4200	18.4	0.0050	0.0025	0.0014	0.0011	0.0009	4334	18.8	0.0101	0.0066	0.0034	0.0019	0.0014
		0.1	0.1	0.95	4197	19.9	0.0048	0.0024	0.0012	0.0010	0.0008	4342	20.6	0.0098	0.0064	0.0033	0.0017	0.0013
		0.1	≥0.2	0.79	4012	26.1	0.0064	0.0028	0.0018	0.0015	0.0013	4150	26.5	0.0158	0.0090	0.0040	0.0025	0.0020
		0.1	≥0.2	0.90	4041	30.7	0.0064	0.0026	0.0014	0.0011	0.0010	4189	31.3	0.0152	0.0089	0.0038	0.0019	0.0015
		0.1	≥0.2	0.95	4063	33.0	0.0064	0.0025	0.0013	0.0010	0.0009	4213	33.5	0.0149	0.0088	0.0037	0.0017	0.0013
	Rect.	1.0	1.0	0.79	3931	5.0	0.0126	0.0105	0.0091	0.0082	0.0074	3963	5.6	0.0206	0.0162	0.0136	0.0121	0.0105
		1.0	1.0	0.90	3904	6.5	0.0100	0.0081	0.0069	0.0062	0.0055	3956	7.2	0.0181	0.0131	0.0105	0.0092	0.0078
		1.0	1.0	0.95	3897	7.3	0.0090	0.0072	0.0061	0.0055	0.0048	3954	8.1	0.0175	0.0121	0.0094	0.0081	0.0069

* For flanged sections having $b_w/b > 0.1$, linear interpolation may be used between ρ values at $b_w/b = 0.1$ and $b_w/b = 1.0$ (rectangular section). ρ is the value for the member in its final size. For flanged sections $\rho = A_s/bd$.

Table 22.4 Coefficient C for Eq. (22.3.20) For Members Loaded to Require $\rho = 0.75\rho_b$ and Others Less Heavily Loaded to Require $\rho < 0.75\rho_b$*

Type of Member					$f_y = 60{,}000$ psi; $w_c = 120$ psf; $D = 3L$							$f_y = 40{,}000$ psi; $w_c = 120$ psf; $D = 3L$						
					@ $\rho = 0.75\rho_b$		Maximum ρ when $C =$					@ $\rho = 0.75\rho_b$		Maximum ρ when $C =$				
f'_c		b_w/b	h_f/h	d/h	C	$\frac{M_a}{M_{cr}}$	4800	5400	6200	7200	8600	C	$\frac{M_a}{M_{cr}}$	4800	5400	6200	7200	8600
$f'_c = 3000$ psi	Flanged Section	0.1	0.1	0.79	4464	15.2	0.0044	0.0025	0.0013	0.0010	0.0008	4564	15.6	0.0079	0.0054	0.0030	0.0017	0.0013
		0.1	0.1	0.90	4464	17.9	0.0041	0.0023	0.0010	0.0007	0.0006	4577	18.4	0.0073	0.0050	0.0028	0.0014	0.0010
		0.1	0.1	0.95	4480	19.2	0.0040	0.0022	0.0009	0.0007	0.0005	4578	20.0	0.0071	0.0048	0.0027	0.0013	0.0008
		0.1	≥0.2	0.79	4265	25.3	0.0060	0.0027	0.0013	0.0010	0.0008	4375	25.7	0.0120	0.0074	0.0035	0.0017	0.0013
		0.1	≥0.2	0.90	4300	29.6	0.0058	0.0027	0.0011	0.0008	0.0006	4411	30.3	0.0114	0.0072	0.0035	0.0014	0.0010
		0.1	≥0.2	0.95	4326	32.0	0.0057	0.0027	0.0010	0.0007	0.0005	4459	32.6	0.0110	0.0070	0.0035	0.0013	0.0009
	Rect.	1.0	1.0	0.79	4138	5.1	0.0088	0.0069	0.0060	0.0053	0.0047	4130	5.7	0.0148	0.0111	0.0090	0.0077	0.0067
		1.0	1.0	0.90	4117	6.6	0.0074	0.0055	0.0046	0.0040	0.0036	4127	7.4	0.0137	0.0092	0.0071	0.0059	0.0051
		1.0	1.0	0.95	4113	7.4	0.0069	0.0049	0.0041	0.0036	0.0034	4130	8.2	0.0134	0.0086	0.0064	0.0054	0.0051
$f'_c = 4000$ psi	Flanged Section	0.1	0.1	0.79	4225	17.5	0.0050	0.0028	0.0015	0.0011	0.0009	4271	18.0	0.0089	0.0061	0.0035	0.0019	0.0015
		0.1	0.1	0.90	4233	20.8	0.0046	0.0026	0.0012	0.0009	0.0007	4302	21.4	0.0082	0.0056	0.0032	0.0016	0.0011
		0.1	0.1	0.95	4252	22.4	0.0044	0.0025	0.0011	0.0008	0.0006	4312	23.0	0.0079	0.0054	0.0031	0.0015	0.0010
		0.1	≥0.2	0.79	4040	29.1	0.0065	0.0030	0.0015	0.0011	0.0010	4079	29.7	0.0132	0.0082	0.0040	0.0020	0.0015
		0.1	≥0.2	0.90	4089	34.1	0.0064	0.0030	0.0012	0.0009	0.0007	4131	34.6	0.0126	0.0080	0.0039	0.0016	0.0011
		0.1	≥0.2	0.95	4100	36.9	0.0063	0.0030	0.0011	0.0008	0.0006	4166	37.6	0.0122	0.0078	0.0039	0.0015	0.0010
	Rect.	1.0	1.0	0.79	3913	5.9	0.0097	0.0078	0.0068	0.0060	0.0054	3862	6.6	0.0163	0.0124	0.0102	0.0088	0.0077
		1.0	1.0	0.90	3901	7.6	0.0081	0.0062	0.0052	0.0046	0.0040	3860	8.5	0.0149	0.0104	0.0080	0.0068	0.0057
		1.0	1.0	0.95	3900	8.5	0.0076	0.0056	0.0046	0.0040	0.0035	3882	9.6	0.0146	0.0096	0.0072	0.0060	0.0051
$f'_c = 5000$ psi	Flanged Section	0.1	0.1	0.79	4048	19.4	0.0054	0.0030	0.0016	0.0012	0.0010	4056	19.9	0.0098	0.0067	0.0038	0.0022	0.0016
		0.1	0.1	0.90	4083	22.7	0.0050	0.0029	0.0013	0.0010	0.0008	4088	23.3	0.0090	0.0062	0.0036	0.0018	0.0012
		0.1	0.1	0.95	4073	24.4	0.0049	0.0028	0.0012	0.0008	0.0007	4091	25.3	0.0087	0.0060	0.0034	0.0016	0.0011
		0.1	≥0.2	0.79	3891	32.3	0.0070	0.0033	0.0017	0.0013	0.0011	3907	32.8	0.0143	0.0090	0.0044	0.0022	0.0017
		0.1	≥0.2	0.90	3924	37.9	0.0069	0.0033	0.0014	0.0010	0.0008	3941	38.5	0.0137	0.0088	0.0043	0.0018	0.0013
		0.1	≥0.2	0.95	3931	40.6	0.0068	0.0033	0.0012	0.0009	0.0007	3951	41.5	0.0132	0.0086	0.0043	0.0017	0.0011
	Rect.	1.0	1.0	0.79	3796	6.2	0.0105	0.0086	0.0075	0.0067	0.0060	3727	6.9	0.0176	0.0136	0.0113	0.0098	0.0086
		1.0	1.0	0.90	3785	8.0	0.0089	0.0068	0.0058	0.0051	0.0045	3729	8.9	0.0161	0.0114	0.0089	0.0075	0.0064
		1.0	1.0	0.95	3785	8.9	0.0083	0.0061	0.0051	0.0045	0.0039	3727	10.0	0.0157	0.0106	0.0080	0.0067	0.0056

* For flanged sections having $b_w/b > 0.1$, linear interpolation may be used between ρ values at $b_w/b = 0.1$ and $b_w/b = 1.0$ (rectangular section). ρ is the value for the member in its final size. For flanged sections $\rho = A_s/bd$.

Table 22.5 Coefficient C for Eq. (22.3.20) For Members Loaded to Require $\rho = 0.75\rho_b$ and Others Less Heavily Loaded to Require $\rho < 0.75\rho_b$*

Type of Member					$f_y = 60{,}000$ psi; $w_c = 120$ psf; $D = L$							$f_y = 40{,}000$ psi; $w_c = 120$ psf; $D = L$						
					@ $\rho = 0.75\rho_b$		Maximum ρ when $C =$					@ $\rho = 0.75\rho_b$		Maximum ρ when $C =$				
f'_c		b_w/b	h_f/h	d/h	C	$\frac{M_a}{M_{cr}}$	4800	5400	6200	7200	8600	$C =$	$\frac{M_a}{M_{cr}}$	4800	5400	6200	7200	8600
$f'_c = 3000$ psi	Flanged Section	0.1	0.1	0.79	4548	14.5	0.0048	0.0028	0.0014	0.0010	0.0008	4684	14.8	0.0087	0.0059	0.0034	0.0018	0.0013
		0.1	0.1	0.90	4551	17.1	0.0045	0.0026	0.0012	0.0008	0.0006	4692	17.6	0.0080	0.0054	0.0032	0.0015	0.0010
		0.1	0.1	0.95	4570	18.3	0.0043	0.0025	0.0011	0.0007	0.0006	4701	18.8	0.0079	0.0052	0.0031	0.0014	0.0009
		0.1	≥0.2	0.79	4352	24.0	0.0067	0.0032	0.0014	0.0010	0.0009	4471	24.4	0.0132	0.0083	0.0041	0.0019	0.0014
		0.1	≥0.2	0.90	4400	28.1	0.0065	0.0031	0.0012	0.0008	0.0007	4532	28.6	0.0124	0.0080	0.0041	0.0016	0.0011
		0.1	≥0.2	0.95	4408	30.4	0.0064	0.0031	0.0011	0.0007	0.0006	4560	31.1	0.0121	0.0078	0.0040	0.0015	0.0009
	Rect.	1.0	1.0	0.79	4229	4.9	0.0095	0.0074	0.0064	0.0055	0.0050	4239	5.4	0.0163	0.0120	0.0096	0.0082	0.0071
		1.0	1.0	0.90	4200	6.3	0.0080	0.0059	0.0048	0.0042	0.0037	4231	7.0	0.0153	0.0100	0.0076	0.0062	0.0053
		1.0	1.0	0.95	4202	7.0	0.0076	0.0053	0.0043	0.0038	0.0035	4238	7.8	0.0151	0.0095	0.0068	0.0057	0.0051
$f'_c = 4000$ psi	Flanged Section	0.1	0.1	0.79	4306	16.7	0.0054	0.0031	0.0016	0.0012	0.0010	4375	17.2	0.0096	0.0067	0.0039	0.0021	0.0015
		0.1	0.1	0.90	4322	19.7	0.0050	0.0029	0.0013	0.0009	0.0007	4399	20.4	0.0088	0.0062	0.0036	0.0018	0.0012
		0.1	0.1	0.95	4334	21.4	0.0048	0.0028	0.0012	0.0008	0.0006	4432	22.0	0.0085	0.0059	0.0035	0.0016	0.0010
		0.1	≥0.2	0.79	4137	27.7	0.0073	0.0035	0.0016	0.0012	0.0010	4203	28.1	0.0145	0.0092	0.0046	0.0022	0.0016
		0.1	≥0.2	0.90	4179	32.6	0.0071	0.0035	0.0014	0.0009	0.0008	4267	33.2	0.0137	0.0089	0.0046	0.0019	0.0012
		0.1	≥0.2	0.95	4190	34.8	0.0070	0.0035	0.0013	0.0008	0.0007	4252	35.5	0.0133	0.0087	0.0045	0.0017	0.0011
	Rect.	1.0	1.0	0.79	3996	5.6	0.0104	0.0083	0.0072	0.0063	0.0057	3966	6.3	0.0177	0.0133	0.0109	0.0093	0.0081
		1.0	1.0	0.90	3982	7.2	0.0088	0.0066	0.0055	0.0048	0.0042	3959	8.0	0.0164	0.0112	0.0086	0.0071	0.0060
		1.0	1.0	0.95	3982	8.1	0.0083	0.0060	0.0049	0.0042	0.0037	3958	9.0	0.0161	0.0105	0.0077	0.0063	0.0053
$f'_c = 5000$ psi	Flanged Section	0.1	0.1	0.79	4146	18.4	0.0059	0.0034	0.0018	0.0013	0.0011	4154	18.8	0.0105	0.0073	0.0043	0.0024	0.0017
		0.1	0.1	0.90	4166	21.7	0.0055	0.0032	0.0015	0.0010	0.0008	4216	22.3	0.0097	0.0068	0.0040	0.0020	0.0013
		0.1	0.1	0.95	4172	23.3	0.0053	0.0031	0.0014	0.0009	0.0007	4180	23.8	0.0093	0.0066	0.0039	0.0018	0.0012
		0.1	≥0.2	0.79	3968	30.8	0.0079	0.0038	0.0018	0.0013	0.0011	4007	31.2	0.0157	0.0100	0.0051	0.0024	0.0018
		0.1	≥0.2	0.90	3997	36.1	0.0077	0.0038	0.0015	0.0010	0.0008	4038	36.8	0.0149	0.0097	0.0050	0.0021	0.0014
		0.1	≥0.2	0.95	4024	38.7	0.0075	0.0038	0.0014	0.0009	0.0007	4053	39.3	0.0144	0.0095	0.0050	0.0019	0.0012
	Rect.	1.0	1.0	0.79	3877	5.9	0.0113	0.0091	0.0079	0.0070	0.0063	3827	6.6	0.0191	0.0146	0.0120	0.0103	0.0090
		1.0	1.0	0.90	3865	7.6	0.0096	0.0072	0.0061	0.0053	0.0047	3823	8.5	0.0176	0.0123	0.0095	0.0079	0.0067
		1.0	1.0	0.95	3863	8.5	0.0089	0.0066	0.0054	0.0047	0.0041	3850	9.6	0.0173	0.0115	0.0085	0.0070	0.0059

* For flanged sections having $b_w/b > 0.1$, linear interpolation may be used between ρ values at $b_w/b = 0.1$ and $b_w/b = 1.0$ (rectangular section). ρ is the value for the member in its final size. For flanged sections $\rho = A_s/bd$.

Table 22.6 Coefficient C for Eq. (22.3.20) For Members Loaded to Require $\rho = 0.75\rho_b$ and Others Less Heavily Loaded to Require $\rho < 0.75\rho_b$*

Type of Member					$f_y = 60{,}000$ psi; $w_c = 95$ psf; $D = 3L$							$f_y = 40{,}000$ psi; $w_c = 95$ psf; $D = 3L$						
					@ $\rho = 0.75\rho_b$		Maximum ρ when $C =$					@ $\rho = 0.75\rho_b$		Maximum ρ when $C =$				
f'_c		b_w/b	h_f/h	d/h	C	$\frac{M_a}{M_{cr}}$	4800	5400	6200	7200	8600	C	$\frac{M_a}{M_{cr}}$	4800	5400	6200	7200	8600
$f'_c = 3000$ psi	Flanged Section	0.1	0.1	0.79	4371	17.2	0.0045	0.0029	0.0015	0.0009	0.0007	4342	17.7	0.0073	0.0053	0.0034	0.0018	0.0012
		0.1	0.1	0.90	4389	20.4	0.0042	0.0027	0.0014	0.0007	0.0006	4389	21.1	0.0066	0.0049	0.0031	0.0017	0.0009
		0.1	0.1	0.95	4415	22.0	0.0040	0.0026	0.0013	0.0007	0.0005	4409	22.6	0.0064	0.0047	0.0030	0.0016	0.0008
		0.1	≥0.2	0.79	4181	28.6	0.0065	0.0037	0.0016	0.0010	0.0008	4126	29.0	0.0112	0.0076	0.0044	0.0020	0.0012
		0.1	≥0.2	0.90	4241	33.6	0.0063	0.0036	0.0015	0.0008	0.0006	4214	34.1	0.0104	0.0073	0.0043	0.0019	0.0009
		0.1	≥0.2	0.95	4244	36.1	0.0061	0.0036	0.0015	0.0007	0.0005	4238	36.9	0.0101	0.0071	0.0042	0.0019	0.0008
	Rect.	1.0	1.0	0.79	4049	5.8	0.0083	0.0063	0.0053	0.0046	0.0041	3920	6.5	0.0136	0.0101	0.0080	0.0067	0.0058
		1.0	1.0	0.90	4038	7.4	0.0073	0.0052	0.0041	0.0036	0.0034	3935	8.3	0.0129	0.0088	0.0064	0.0054	0.0051
		1.0	1.0	0.95	4042	8.4	0.0071	0.0048	0.0038	0.0033	0.0033	3957	9.4	0.0128	0.0085	0.0059	0.0050	0.0050
$f'_c = 4000$ psi	Flanged Section	0.1	0.1	0.79	4102	19.9	0.0051	0.0033	0.0017	0.0011	0.0008	4033	20.5	0.0082	0.0060	0.0038	0.0021	0.0014
		0.1	0.1	0.90	4131	23.3	0.0047	0.0031	0.0016	0.0008	0.0006	4004	23.9	0.0075	0.0055	0.0036	0.0019	0.0011
		0.1	0.1	0.95	4132	25.3	0.0045	0.0030	0.0015	0.0008	0.0006	4085	26.2	0.0072	0.0053	0.0034	0.0018	0.0009
		0.1	≥0.2	0.79	3943	33.0	0.0072	0.0041	0.0018	0.0011	0.0009	3866	33.5	0.0124	0.0085	0.0049	0.0023	0.0014
		0.1	≥0.2	0.90	3977	38.7	0.0069	0.0041	0.0017	0.0009	0.0007	3889	39.3	0.0117	0.0082	0.0048	0.0022	0.0011
		0.1	≥0.2	0.95	3993	41.8	0.0068	0.0040	0.0017	0.0008	0.0006	3944	42.6	0.0113	0.0080	0.0048	0.0021	0.0010
	Rect.	1.0	1.0	0.79	3805	6.7	0.0091	0.0071	0.0060	0.0052	0.0047	3634	7.5	0.0149	0.0113	0.0091	0.0077	0.0066
		1.0	1.0	0.90	3799	8.6	0.0081	0.0059	0.0047	0.0040	0.0035	3676	9.7	0.0142	0.0098	0.0073	0.0060	0.0050
		1.0	1.0	0.95	3814	9.7	0.0078	0.0054	0.0042	0.0036	0.0034	3671	10.8	0.0141	0.0095	0.0067	0.0055	0.0051
$f'_c = 5000$ psi	Flanged Section	0.1	0.1	0.79	3945	21.8	0.0056	0.0036	0.0019	0.0012	0.0009	3836	22.4	0.0090	0.0066	0.0042	0.0023	0.0015
		0.1	0.1	0.90	3944	25.9	0.0052	0.0034	0.0017	0.0009	0.0007	3861	26.6	0.0083	0.0061	0.0040	0.0021	0.0012
		0.1	0.1	0.95	3956	27.8	0.0050	0.0033	0.0017	0.0008	0.0006	3823	28.5	0.0080	0.0059	0.0038	0.0020	0.0011
		0.1	≥0.2	0.79	3742	36.6	0.0078	0.0045	0.0020	0.0012	0.0010	3642	37.2	0.0135	0.0094	0.0054	0.0026	0.0016
		0.1	≥0.2	0.90	3797	43.0	0.0075	0.0045	0.0019	0.0010	0.0007	3706	43.7	0.0128	0.0090	0.0053	0.0024	0.0012
		0.1	≥0.2	0.95	3781	45.9	0.0074	0.0044	0.0019	0.0009	0.0006	3677	46.9	0.0124	0.0088	0.0053	0.0024	0.0011
	Rect.	1.0	1.0	0.79	3681	7.1	0.0099	0.0078	0.0067	0.0058	0.0052	3512	7.9	0.0162	0.0124	0.0100	0.0085	0.0073
		1.0	1.0	0.90	3676	9.0	0.0088	0.0064	0.0052	0.0044	0.0039	3507	10.1	0.0154	0.0107	0.0081	0.0066	0.0056
		1.0	1.0	0.95	3684	10.1	0.0085	0.0059	0.0047	0.0040	0.0034	3550	11.4	0.0153	0.0104	0.0074	0.0059	0.0050

* For flanged sections having $b_w/b > 0.1$, linear interpolation may be used between ρ values at $b_w/b = 0.1$ and $b_w/b = 1.0$ (rectangular section). ρ is the value for the member in its final size. For flanged sections $\rho = A_s/bd$.

Table 22.7 Coefficient C for Eq. (22.3.20) For Members Loaded to Require $\rho = 0.75\rho_b$ and Others Less Heavily Loaded to Require $\rho < 0.75\rho_b$*

Type of Member					$f_y = 60{,}000$ psi; $w_c = 95$ psf; $D = L$							$f_y = 40{,}000$ psi; $w_c = 95$ psf; $D = L$						
					@ $\rho = 0.75\rho_b$		Maximum ρ when $C =$					@ $\rho = 0.75\rho_b$		Maximum ρ when $C =$				
f'_c		b_w/b	h_f/h	d/h	C	$\frac{M_a}{M_{cr}}$	4800	5400	6200	7200	8600	C	$\frac{M_a}{M_{cr}}$	4800	5400	6200	7200	8600
$f'_c = 3000$ psi	Flanged Section	0.1	0.1	0.79	4454	16.4	0.0048	0.0032	0.0017	0.0010	0.0008	4441	16.9	0.0078	0.0057	0.0037	0.0021	0.0013
		0.1	0.1	0.90	4473	19.2	0.0044	0.0029	0.0015	0.0008	0.0006	4470	20.0	0.0070	0.0052	0.0034	0.0019	0.0010
		0.1	0.1	0.95	4499	21.0	0.0043	0.0028	0.0015	0.0007	0.0005	4532	21.7	0.0068	0.0050	0.0033	0.0018	0.0009
		0.1	≥0.2	0.79	4286	27.2	0.0071	0.0041	0.0018	0.0010	0.0008	4282	27.6	0.0121	0.0083	0.0049	0.0023	0.0013
		0.1	≥0.2	0.90	4336	32.0	0.0068	0.0041	0.0018	0.0008	0.0006	4357	32.6	0.0112	0.0079	0.0048	0.0022	0.0010
		0.1	≥0.2	0.95	4343	34.2	0.0066	0.0040	0.0018	0.0007	0.0005	4316	34.8	0.0109	0.0077	0.0047	0.0022	0.0009
	Rect.	1.0	1.0	0.79	4133	5.5	0.0089	0.0067	0.0056	0.0048	0.0043	4032	6.2	0.0148	0.0108	0.0085	0.0071	0.0060
		1.0	1.0	0.90	4124	7.1	0.0080	0.0055	0.0044	0.0037	0.0035	4032	7.9	0.0142	0.0096	0.0069	0.0056	0.0051
		1.0	1.0	0.95	4124	7.9	0.0078	0.0051	0.0039	0.0034	0.0034	4039	8.9	0.0141	0.0092	0.0063	0.0051	0.0050
$f'_c = 4000$ psi	Flanged Section	0.1	0.1	0.79	4181	18.9	0.0054	0.0036	0.0019	0.0011	0.0009	4101	19.4	0.0087	0.0064	0.0042	0.0024	0.0015
		0.1	0.1	0.90	4233	22.3	0.0050	0.0033	0.0018	0.0009	0.0007	4176	22.9	0.0080	0.0059	0.0039	0.0021	0.0011
		0.1	0.1	0.95	4208	23.8	0.0048	0.0032	0.0017	0.0008	0.0006	4141	24.7	0.0077	0.0057	0.0038	0.0021	0.0010
		0.1	≥0.2	0.79	4027	31.4	0.0078	0.0046	0.0021	0.0012	0.0009	3984	32.0	0.0134	0.0093	0.0055	0.0026	0.0015
		0.1	≥0.2	0.90	4059	37.0	0.0075	0.0045	0.0020	0.0009	0.0007	4020	37.6	0.0125	0.0089	0.0054	0.0025	0.0012
		0.1	≥0.2	0.95	4070	39.5	0.0073	0.0045	0.0020	0.0008	0.0006	3977	40.1	0.0122	0.0086	0.0053	0.0025	0.0011
	Rect.	1.0	1.0	0.79	3884	6.4	0.0098	0.0076	0.0064	0.0055	0.0049	3737	7.1	0.0162	0.0121	0.0096	0.0081	0.0069
		1.0	1.0	0.90	3882	8.2	0.0088	0.0062	0.0050	0.0042	0.0036	3744	9.1	0.0155	0.0107	0.0078	0.0063	0.0052
		1.0	1.0	0.95	3886	9.2	0.0085	0.0058	0.0044	0.0038	0.0034	3763	10.3	0.0154	0.0103	0.0071	0.0057	0.0050
$f'_c = 5000$ psi	Flanged Section	0.1	0.1	0.79	4017	20.8	0.0060	0.0040	0.0021	0.0013	0.0010	3943	21.4	0.0096	0.0071	0.0046	0.0026	0.0016
		0.1	0.1	0.90	4002	24.5	0.0055	0.0037	0.0019	0.0010	0.0008	3932	25.3	0.0088	0.0066	0.0043	0.0024	0.0013
		0.1	0.1	0.95	4046	26.6	0.0053	0.0036	0.0019	0.0009	0.0007	3978	27.3	0.0085	0.0063	0.0042	0.0023	0.0011
		0.1	≥0.2	0.79	3800	34.8	0.0085	0.0050	0.0023	0.0013	0.0010	3694	35.4	0.0146	0.0102	0.0061	0.0029	0.0017
		0.1	≥0.2	0.90	3851	40.8	0.0082	0.0050	0.0022	0.0010	0.0008	3780	41.6	0.0137	0.0097	0.0059	0.0028	0.0013
		0.1	≥0.2	0.95	3898	43.9	0.0080	0.0049	0.0022	0.0009	0.0007	3812	44.6	0.0133	0.0095	0.0059	0.0028	0.0012
	Rect.	1.0	1.0	0.79	3765	6.7	0.0107	0.0083	0.0070	0.0061	0.0054	3591	7.5	0.0175	0.0133	0.0106	0.0090	0.0077
		1.0	1.0	0.90	3758	8.6	0.0095	0.0069	0.0055	0.0047	0.0040	3639	9.7	0.0168	0.0117	0.0086	0.0070	0.0058
		1.0	1.0	0.95	3775	9.7	0.0092	0.0063	0.0049	0.0041	0.0035	3632	10.8	0.0166	0.0113	0.0079	0.0062	0.0052

* For flanged sections having $b_w/b > 0.1$, linear interpolation may be used between ρ values at $b_w/b = 0.1$ and $b_w/b = 1.0$ (rectangular section). ρ is the value for the member in its final size. For flanged sections $\rho = A_s/bd$.

Table 22.8 Approximate Coefficient C for Eq. (22.3.31a) for Members Having $\omega \leq 0.75\rho_b f_y/f'_c$

$C = 4320$ (when $\rho \cong 0.75\rho_b$)		Maximum ω* when C = 4800	5400	6200	7200	> 30,000
Type of Member	f_y (ksi)					
Flanged section** $b_w/b = 0.1$ $h_f/h = 0.1$	40	0.08	0.06	0.03	0.02	Uncracked sections
	60	0.07	0.04	0.02	0.01	
Flanged section** $b_w/b = 0.1$ $h_f/h \geq 0.2$	40	0.13	0.08	0.04	0.02	
	60	0.10	0.05	0.03	0.01	
Rectangular section ($b_w/b = 1.0$) ($h_f/h = 1.0$)	40	0.17	0.12	0.09	0.07	
	60	0.14	0.10	0.08	0.06	

* ω is equal to $\rho f_y/f'_c$ (of the member in its final size).
** For flanged sections with $b_w/b > 0.1$, linear interpolation may be used between ω values at $b_w/b = 0.1$ and $b_w/b = 1.0$ (rectangular section). ρ for flanged sections computed as $\rho = A_s/bd$.

EXAMPLE 22.3.3

Determine the coefficient C appropriate for a flanged beam requiring $\rho = 0.0060$, $f_y = 60{,}000$ psi, $w_c = 145$ pcf, $D = L$, $h_f/h \geq 0.2$, $d/h = 0.95$, $f'_c = 4000$ psi, and $b_w/b = 0.3$.

Solution

Using Table 22.3 for $f_y = 60{,}000$ psi, $w_c = 145$ pcf, second ($b_w/b = 0.1$ and $h_f/h \geq 0.2$) and third ($b_w/b = 1.0$ and $h_f/h = 1.0$) groups that must be used to interpolate for $b_w/b = 0.3$. For $f'_c = 4000$ psi and $d/h = 0.95$, read

	$C =$	4800	5400	6200	7200	8600
$b_w/b = 0.1$; $h_f/h \geq 0.2$	$\rho =$	0.0059	0.0023	0.0011	0.0009	0.0008
rectangular section	$\rho =$	0.0083	0.0065	0.0055	0.0049	0.0043
Interpolate for $b_w/b = 0.3$	$\rho =$	0.0064	0.0032			

Finally, interpolate for $\rho = 0.0060$. Find $\underline{C = 4875.}$

Alternatively, using the less accurate but condensed Table 22.8 compute

$$\omega = \frac{\rho f_y}{f'_c} = \frac{0.0060(60{,}000)}{4000} = 0.090$$

Then, using the second and third groups for $f_y = 60$ ksi,

	$C =$	4800	5400	6200	7200
$b_w/b = 0.1$; $h_f/h \geq 0.2$	$\omega =$	0.10	0.05	0.03	0.01
rectangular section	$\omega =$	0.14	0.10	0.08	0.06
Interpolate for $b_w/b = 0.3$	$\omega =$	0.109	0.061		

Finally, interpolate for $\omega = 0.090$. Find $C = 5040$. In this case a slightly unconservative (i.e., too high) value for C is obtained.

The reader must realize that great accuracy in interpolation for C values is not necessary. Determination of deflections resulting from error of any 10% in estimating an unconservative value of C will cause *computed* deflection to be 10% higher. Such accuracy in computing deflections is neither warranted nor possible to match in actual construction.

Once the coefficient C has been determined, the immediate total load deflection may be estimated from Eq. (22.3.19) using K computed from Eq. (22.3.8). The immediate dead or live load deflection may then be estimated using Eqs. (22.3.13) and (22.3.14). When $Q = (I_e)_{D+L}/(I_e)_D \approx 1$, simplified Eqs. (22.3.13a) and (22.2.14a) can be used.

General equations for the minimum thickness $h_{\min}$ required to satisfy a limiting deflection such as from ACI-Table 9.5(b), and for any other critical deflection Δ were also proposed by Grossman [22.4, 22.5]. Equation (22.3.19) provides the immediate total dead plus live load deflection of a member. Once $(\Delta_i)_{D+L}$ is known the immediate dead load $(\Delta_i)_D$ and live load $(\Delta_i)_L$ deflections can be computed by Eqs. (22.3.13) and (22.3.14), or from the corresponding simplified Eqs. (22.3.13a) and (22.3.14a) when $Q = (I_e)_{D+L}/(I_e)_D \approx 1.0$.

A critical deflection Δ_s for serviceability can then be established using Eqs. (22.3.4) or (22.3.5). Any other critical deflection Δ to ascertain "constructability" or "aesthetic integrity" can also be computed based on information regarding immediate deflections $(\Delta_i)_D$ and $(\Delta_i)_L$ of the beam. Let

$$C_d = \frac{\Delta}{(\Delta_i)_{D+L}} \tag{22.3.22}$$

The critical deflection Δ based on a given depth h can now be computed by multiplying both sides of Eq. (22.3.19) by C_d, which gives

$$\Delta = \frac{\ell}{C}\left(\frac{\ell}{h}\right)(C_d)\left(\frac{K_1}{K}\right)\sqrt{\frac{145}{w_c}} \tag{22.3.23}$$

Rearranging Eq. (22.3.23), a minimum depth $h_{\min}$ to ensure that a predetermined critical deflection Δ is not exceeded can also be established as follows:

$$h_{\min} = \frac{\ell}{C}\left(\frac{\ell}{\Delta}\right)(C_d)\left(\frac{K_1}{K}\right)\sqrt{\frac{145}{w_c}} \tag{22.3.24*}$$

where C = coefficient obtainable from Tables 22.2 through 22.7; may be taken as 4320 for heavily loaded beams having $\rho \approx 0.75\rho_b$. Value may also be obtained from condensed Table 22.8 to ascertain serviceability.

Δ = limiting deflection, such as given by ACI-Table 9.5(b). The ratio $(\ell/\Delta)_{\min}$ = 180, 240, 360, or 480 from ACI-Table 9.5(b) depending on the loading combinations considered, such as live load only or live load plus creep and shrinkage. The requirements of ACI-Table 9.5(b) for serviceability are given in Table 22.9 in a format more suitable for use with Eq. (22.3.24). To ascertain "constructability" or "aesthetic integrity", Δ is equal to the total computed critical deflection, using Eq. (22.3.23), resulting from a preselected depth h. Alternatively, a minimum depth $h_{\min}$, using Eq. (22.3.24), can be obtained to ensure that a preselected critical deflection Δ is not exceeded.

K = coefficient from Eq. (22.3.8) to provide adjustments to depth, considering material used, L/D and d/h ratios, and for sections reinforced more heavily than required by the loading. M_a in Eq. (22.3.8) is taken as $(M_a)_{D+L}$ at midspan (or at supports for cantilevers). For preliminary design, using f_y = 60,000 psi, K can be taken equal to 1.0. When f_y = 40,000 psi, K can be assumed equal to 1.25.

C_d = $\Delta/(\Delta_i)_{D+L}$ = the ratio of the critical deflection to the total load immediate deflection. See Secs. 22.3.5 through 22.3.8 for values of C_d for various situations.

K_1 = coefficient relating to the support conditions: 1.0 for simply supported, 2.4 for cantilevers and $(1.20 - 0.2M_0/M_a)$ for continuous spans, where M_0 is the simple span total load moment at midspan.

w_c = unit weight of concrete in pcf (use 145 for normal-weight concrete); for w_c in kg/m^3, take $145/w_c$ as $2330/w_c$.

22.3.5 Coefficient C_d for the Critical Deflection

The coefficient C_d in Eqs. (22.3.23) and (22.3.24) equals the ratio of the critical deflection (to be considered) to the immediate deflection caused by the total service load

$$C_d = \frac{\Delta}{(\Delta_i)_{D+L}} \qquad [22.3.22]$$

---

* Grossman [22.4] recommends that $(\ell/C)(\ell/\Delta)C_d$ in Eq. (22.3.24) be taken not less than $(\ell/24)$ to assure that when C_d is small the selected member depth is not too shallow to cause excessive total deflection.

Table 22.9 Minimum Permissible Ratios of Span ℓ to Computed Deflection Δ

Type of Member	Deflection Δ to be considered	$(\ell/\Delta)_{min}$
Flat roots not supporting and not attached to nonstructural elements likely to be damaged by large deflections	Immediate deflection due to live load L	180*
Floors not supporting and not attached to nonstructural elements likely to be damaged by large deflections	Immediate deflection due to live load L	360
Roof or floor construction supporting or attached to nonstructural elements likely to be damaged by large deflections	That part of total defection occurring after attachment of nonstructural elements. Sum of long-time deflection due to all sustained D loads (dead load plus any sustained portion of live load) and immediate deflection due to any additional live load L†	480**
Roof or floor construction supporting or attached to nonstructural elements not likely to be damaged by large deflections		240**

* Limit not intended to safeguard against ponding. Ponding should be checked by suitable calculations of deflection, including added deflections due to ponded water, and considering long-time effects of all sustained loads, camber, construction tolerances, and reliability of provisions for drainage.
** Ratio limit may be lower if adequate measures are taken to prevent damage to supported or attached elements, but shall not be lower than tolerance of nonstructural elements.
† Long-time deflection shall be determined in accordance with ACI-9.5.2.5 or 9.5.4.2 [22.1] but may be reduced by amount of deflection calculated to occur before attachment of nonstructural elements. This amount shall be determined on basis of accepted engineering data relating to time-deflection characteristics of members similar to those being considered.

For each of the design considerations (serviceability, constructability, or the aesthetic integrity) a critical deflection Δ either establishes a minimum required depth, or is the result of the depth selected.

22.3.6 Coefficient C_d for the Serviceability Check by ACI-Table 9.5(b)

For serviceability, $C_d = \Delta_s/(\Delta_i)_{D+L}$ and the limiting span–deflection ratios $[\ell/\Delta)_{min} = 180,360,240,480]$ are obtained from ACI-Table 9.5(b). (See also Table 22.9.)

Case 1

When live load alone is considered $[(\ell/\Delta)_{min} = 180$ or $360]$ the critical deflection ($\Delta = \Delta_s$) equals

$$\Delta_s = (\Delta_i)_L$$

Using Eqs. (22.3.14) and (22.3.22),

$$C_d = \left[1 - \left(\frac{D}{D + L}\right) Q\right] \tag{22.3.25}$$

When Q approaches 1.0, such as the case when constrution loads have acted or several load cycles have already occurred, or if $(M_a)_D > 3.5M_{cr}$, then Eq. (22.3.25) can be simplified. Using Eq. (22.3.14a) and (22.3.22)

$$C_d = \left(\frac{L}{D + L}\right) \tag{22.3.25a}$$

Case 2

When the serviceability check requires in addition to live load deflection inclusion of the long-term creep and shrinkage deflection occurring *after* installation of the nonstructural elements, that is, $(\ell/\Delta)_{\min} = 240$ or 480 from ACI-Table 9.5(b), the critical deflection Δ equals

$$\Delta_s = \lambda_n(\Delta_i)_{D+L} + (\Delta_i)_L \qquad (22.3.26)$$

Using Eqs. (22.3.13), (22.3.14), and (22.3.22) and solving for C_d

$$C_d = \frac{\Delta_s}{(\Delta_i)_{D+L}} = \lambda_n \left(\frac{D}{D+L}\right) Q + \left[1 - \left(\frac{D}{D+L}\right) Q\right] \qquad (22.3.27)$$

Simplifying gives

$$C_d = 1 + (\lambda_n - 1)\left(\frac{D}{D+L}\right) Q \qquad (22.3.28)$$

Equation (22.3.28) accounts for the larger $(I_e)_D$ at the intermediate dead load level. When $Q \approx 1$, Eq. (22.3.28) becomes

$$C_d = \frac{\lambda_n D + L}{D + L} \qquad (22.3.28a)$$

Note that Eqs. (22.3.28) and (22.3.28a) may also be used for Case 1, because when only the live load deflection is considered, $\lambda_n = 0$, giving Eqs. (22.3.25) and (22.3.25a). The net long-term multiplier λ_n is computed using Eq. (22.3.2) as the net difference between λ at 60 months and λ at time nonstructural elements are installed. The use of λ_n discounts the long-term creep and shrinkage deflection that occurred prior to installation of nonstructural elements.

The use of Eq. (22.3.28a) for C_d (which assumed Q equal to 1.0) is recommended [22.4] as sufficiently accurate for serviceability considerations because it generally provides better agreement with field observations. The following reasons are given to justify the simpler approach:

1. Computations for deflections are not accurate to begin with.
2. Construction loads and procedures greatly influence initial cracking so that "first cycle of loading" (for which Q may be much smaller than unity at very low levels only) should not be an important consideration for the (future) serviceability check of the nonstructural elements. Preloadings and/or construction loads and procedures make the simplified approach more accurate.
3. When both dead and live load deflections are considered (using "first cycle loading" criteria) for the serviceability check, an overestimate of $(\Delta_i)_D$ (obtained by using $Q = 1$) will reduce $(\Delta_i)_L$ by the same increment. Both "errors" cancel each other when $\lambda_n \approx 1$ in Eq. (22.3.5) to produce accurate results. When $\lambda_n > 1$, a conservative estimate of $h_{\min}$ will result. When $\lambda_n < 1$, results will not be conservative. However, in this latter case the nonstructural elements are not to be added immediately after construction. With the passing of time, serviceability becomes a lesser problem. In any case, the percent variation is small. For an extreme case where $Q = 0.35$, Ref. 22.4 indicates $h_{\min}$ varies not more than $\pm$ 17%.
4. In practical situations the multiplier λ_n does not vary much from unity. ($\lambda_n = 1.0$ when $\rho' = 0$ and nonstructural elements are applied about 3 months after removal of the shores.) "Accuracy" is therefore maintained even with the simplified approach.

A minimum depth $h_{\min}$ can now be provided which will satisfy any span–deflection limit of ACI-Table 9.5(b) and be based on the load level the member is supporting; the member size, shape, and material used; the long-time creep and shrinkage (sustained load) effects; the ratio of live load to total load; and the time lag between removal of shores and attachment of nonstructural elements likely (or not) to be damaged by large deflections.

Using Eq. (22.3.24) with C_d from Eq. (22.3.28a) where $Q = 1$ gives

$$h_{\min} \geq \frac{\ell}{C}\left(\frac{\ell}{\Delta_{\min}}\right)\left(\frac{\lambda_n D + L}{D + L}\right)\left(\frac{K_1}{K}\right)\sqrt{\frac{145}{w_c}} \qquad (22.3.29a)$$

or when more "accuracy" is desired for "first cycle of loading", using Eq. (22.3.28) for C_d gives

$$h_{\min} \geq \frac{\ell}{C}\left(\frac{\ell}{\Delta_{\min}}\right)\left[1 + (\lambda_n - 1)\left(\frac{D}{D+L}\right)Q\right]\left(\frac{K_1}{K}\right)\sqrt{\frac{145}{w_c}} \tag{22.3.29}$$

22.3.7 Coefficient C_d for Proper Constructability

Case 1

A structural concrete member *not* affected by construction loads and procedures will have a critical deflection Δ at the time nonstructural elements are installed equal to

$$\Delta_c = (\Delta_i)_D + \lambda(\Delta_i)_D = (1 + \lambda)(\Delta_i)_D \tag{22.3.30}$$

where λ is the long-time multiplier at the time nonstructural elements are applied. It may be estimated from Fig. 9.2 in Ref. 22.11 and Eq. (22.3.2). For this condition,

$$C_d = \frac{\Delta_c}{(\Delta_i)_{D+L}} \tag{22.3.31}$$

and using Eq. (22.3.13) in (22.3.30) gives

$$C_d = (1 + \lambda)\left(\frac{D}{D+L}\right)Q \tag{22.3.32}$$

The critical deflection already realized at the time "constructability" is critical is then, using Eq. (22.2.32) for C_d in Eq. (22.3.23),

$$\Delta_c = \frac{\ell}{C}\left(\frac{\ell}{h}\right)\left[(1 + \lambda)\left(\frac{D}{D+L}\right)Q\right]\left(\frac{K_1}{K}\right)\sqrt{\frac{145}{w_c}} \tag{22.3.33}$$

A suitable camber can help reduce this value when deemed necessary to reduce or eliminate this estimated deflection. Caution must be exercised in selecting the camber because of uncertainty resulting from any attempt at computing deflections. It is best to select a size that will indicate Δ_c small enough not to hinder the application of finishes (nonstructural items).

Case 2

A structural concrete member, if cast above and supported by previously cast and deflecting levels, is consequently affected by construction procedures, and will have deflections larger than those indicated by Eq. (22.3.33).

In order to compensate for the additional cracks and creep resulting from the necessity for premature self-support and participation in the support of the freshly cast concrete levels above, Ref. 22.4 suggested using the lowest value of $(I_e)_{D+L}$ to compute the immediate deflection $(\Delta_i)_D$ and to apply the long-time *ACI Code* multipliers to those larger computed deflections. This method, which assumes that $Q = (I_e)_{D+L}/(I_e)_D \approx 1.0$, more accurately predicts the actual deflections apparent in construction.

Sbarounis [22.12], who investigated two-way slab construction loads and procedures, indicated that construction loads were on the order of 80 to 110% of the dead load of the slab itself. The range depended on the number of forms utilized in construction and on the particular stripping and reshoring operation. As result of his investigation, Sbarounis recommended [22.13] use of the secant service load deflection, based on an I_e computed for the maximum load level encountered during construction (i.e., $Q \rightarrow 1.0$).

A creep magnifier of up to 20% larger is also indicated by Branson and Trost [22.16] when concrete members are susceptible to a few load cycles.

Sbarounis [22.13] measured the deformation "permanent set" caused by the lower concrete strengths in slabs required to support load during the construction period. He suggests that for two-way thin slab construction, a "construction magnifier" on the order of 1.5 be used to account for the permanent set and the creep during the first month. This method gives a self-load deflection at 28 days of 2.5 (i.e., 1 + 1.5) times the immediate self-load deflection $(\Delta_i)_D$, which is to be computed using $(I_e)_{D+L}$ (i.e., $Q \approx 1.0$). Deflections computed this way appear to be realistic for many flat plate structures constructed on a two- or three-day cycle.

A larger creep and shrinkage multiplier (after one month) of 2.5 for two-way slabs is also recommended [22.13]. The total multiplier is then equal to 5 (i.e., 1 + 1.5 + 2.5) rather than 3 (i.e., 1 + 2.0), and is recommended in ACI-9.5 [22.1]. This revised multiplier agrees with observed deflections of from four to six times the computed self-load immediate deflection.

For beams, it has been suggested [22.13] that the multipliers given in ACI-9.5 will give acceptable results if they are applied to $(\Delta_i)_D$ that has been obtained using $(I_e)_{D+L}$. This suggestion is in agreement with Grossman [22.4]. When construction loads exceed live loads and adjustments to reinforcement are required in anticipation of such loading, I_e of the larger load level should be used to compute $(\Delta_i)_D$.

It is obviously necessary to adjust Eqs. (22.3.32) and (22.3.33) to accommodate the construction process. Therefore,

$$C_d = (1 + \lambda_c + \lambda)\left(\frac{D}{D + L}\right) \tag{22.3.32a}$$

and

$$\Delta_c = \frac{\ell}{C}\left(\frac{\ell}{h}\right)[1 + \lambda_c + \lambda]\left(\frac{D}{D + L}\right)\left(\frac{K_1}{K}\right)\sqrt{\frac{145}{w_c}} \tag{22.3.33a}$$

where λ_c is an additional multiplier which accounts for the effects of understrength concrete at the time of the application of load, the permanent set built in, and the increased anticipated creep. It is this writer's suggestion that until better evaluation of the construction influence is available λ_c be taken as follows:

for slabs, $\lambda_c \approx 1.0$

for beams, $\lambda_c \approx 0.5$

These proposed values have been adjusted downward to discount first-month long-time creep and shrinkage, which therefore allows the values in ACI-9.5 to remain unchanged.

It is clear, nevertheless, that more research that duplicates construction procedures is needed. Construction loads and procedures are indisputably major parameters which cannot be overlooked whenever an estimate of deflections is desired. (Note that Q is omitted from the revised equations; construction loads justify this simplification.)

22.3.8 Coefficient C_d for "Aesthetic Integrity"

"Aesthetic integrity" may be considered in a manner similar to "constructability". In this case, the total final deflection Δ_f of the sustained loads (that is, immediate dead load plus creep and shrinkage deflection) must be considered.

Case 1

An exposed structural member *not* adversely affected by construction loads and procedures will have a final critical deflection 5 years after construction,

$$\Delta_f = (\Delta_i)_D + \left(\frac{2}{1 + 50\rho'}\right)(\Delta_i)_D = \left[1 + \frac{2}{1 + 50\rho'}\right](\Delta_i)_D \tag{22.3.34}$$

Here, $C_d = \Delta_f/(\Delta_i)_{D+L}$. Using Eqs. (22.3.13) and (22.3.34) gives

$$C_d = \left[1 + \frac{2}{1 + 50\rho'}\right]\left(\frac{D}{D + L}\right) Q \tag{22.3.35}$$

The total deflection of the structural member due to the sustained loads can now be computed by using Eq. (22.3.35) for C_d in Eq. (22.3.23),

$$\Delta_f = \frac{\ell}{C}\left(\frac{\ell}{h}\right)\left(1 + \frac{2}{1 + 50\rho'}\right)\left(\frac{D}{D + L}\right)(Q)\left(\frac{K_1}{K}\right)\sqrt{\frac{145}{w_c}} \tag{22.3.36}$$

A suitable camber can help reduce this value when deemed necessary.

Case 2

If, however, the structural member is supported on previously cast and yielding levels, and is therefore affected by construction loads and construction procedures, the deflection given by Eq. (22.3.36) will be an underestimate of the actual final deflection.

As in the constructability analysis in Sec. 22.3.7, adjustments that account for construction loads can be made. Then,

$$C_d = \left[1 + \lambda_c + \frac{2}{1 + 50\rho'}\right]\left(\frac{D}{D + L}\right) \tag{22.3.35a}$$

and

$$\Delta_f = \frac{\ell}{C}\left(\frac{\ell}{h}\right)\left(1 + \lambda_c + \frac{2}{1 + 50\rho'}\right)\left(\frac{D}{D + L}\right)\left(\frac{K_1}{K}\right)\sqrt{\frac{145}{w_c}} \tag{22.3.36a}$$

Here λ_c is an additional multiplier which accounts for the effects of understrength concrete at time of application of the load. See Sec. 22.3.7 for the proposed values of λ_c presently recommended. In all cases where camber is to be utilized to compensate for anticipated deflections, a realistic projection of actual loads must be used so as to avoid overestimating deflections. Camber much larger than $\ell/300$ should be avoided.

EXAMPLE 22.3.4

For the T-section of Example 22.3.1 (Fig. 22.4) make a direct estimate of minimum thickness $h_{\min}$ required. The beam is constructed in a manner not to be affected by construction load and procedures.

Solution

(a) *Estimate ratios and time-dependent factor* λ_n. For a preliminary depth estimate of 16 in.: $d/h = 12.5/16 = 0.79$; $h_f/h > 0.2$; $b_w/b = 12/60 = 0.2$; and $\lambda_n = 1.0$.

(b) *Compute C.* For a given factored moment $M_u = 166$ ft-kips, the tension steel area required is $A_s = 3.04$ sq in., estimated for $d = 12.5$ in. Then,

$$\rho = \frac{A_s}{bd} = \frac{2.04}{60(12.5)} = 0.0041$$

Enter Table 22.3 and $f_y = 60{,}000$ psi; $w_c = 145$ pcf; $D = L$; and $f'_c = 4000$ psi. First interpolate vertically for $\rho = 0.0041$ between the second and third groups. Use $d/h = 0.79$ line,

	$C =$	4800	5400	6200
$b_w/b = 0.1$, $h_f/h \geq 0.2$	$\rho =$	0.0060	0.0026	0.0016
rectangular section, $b_w/b = 1.0$	$\rho =$	0.0116	0.0096	0.0083
Interpolate for $b_w/b = 0.2$	$\rho =$	0.0066	0.0034	

Then, interpolate for $\rho = 0.0041$ between $C = 4800$ and $C = 5400$. Find $C = 5270$.

(c) *Compute* $h_{\min}$ *using Eq. (22.3.24).*

$$h_{\min} = \frac{\ell}{C}\left(\frac{\ell}{\Delta}\right) C_d \left(\frac{K_1}{K}\right)\sqrt{\frac{145}{w_c}} \qquad [22.3.24]$$

ACI-Table 9.5(b) requires ℓ/Δ not exceed 480 for live load plus creep and shrinkage deflection. For a simply supported beam, $K_1 = 1.0$, and for normal-weight concrete $\sqrt{145/w_c} = 1.0$. Then since $\lambda_n = 1.0$, use simplified Eq. (22.3.28a) to obtain C_d. Thus,

$$C_d = \frac{\lambda_n D + L}{D + L} = \frac{1.0(115) + 100}{215} = 1.0$$

Using Eq. (22.3.8) for K, noting that when a beam is reinforced to the required flexural needs, $1.4M_a/\phi M_n = 1.4(D + L)/(1.4D + 1.7L)$,

$$K = \frac{12.5}{0.9(16)}\left[\frac{1}{0.4 + 1.4(215/331)(0.6)}\right] = 0.92$$

Then,

$$h_{\min} = \frac{20(12)}{5270}(480)(1.0)\left(\frac{1.0}{0.92}\right)1.0 = 23.8 \text{ in.} > 16 \text{ in.}$$

Therefore, the 16-in. deep beam *will not* satisfy serviceability requirements of ACI-Table 9.5(b). This conclusion agrees with the computed result in Example 22.3.1.

The required clearance over the nonstructural sensitive partition below the 16-in. deep beam is, using Eq. (22.3.23) and solving for Δ,

$$\Delta = \frac{\ell}{C}\left(\frac{\ell}{h}\right)C_d\left(\frac{K_1}{K}\right)\sqrt{\frac{145}{w_c}} = \frac{20(12)}{5270}\left(\frac{20(12)}{16}\right)(1.0)\left(\frac{1.0}{0.92}\right)(1.0)$$

$\Delta = 0.74$ in. (Use ⅞ in.) (This value is in close agreement to that computed in (h) of Example 23.3.1.)

(d) *Compute now a required depth to satisfy a serviceability requirement that $\ell/\Delta \geq 480$.* For this solution iteration is required. However, since the deeper member requires less reinforcement, the value of C will increase, as may be noted from Table 22.3. Thus, try $h = 20$ in. (about halfway between the 16-in. first trial and the solution value of 23.5 in.) Assume the moments to change only slightly. Then, $M_u = \phi M_n = 166$ ft-kips and $A_s = 2.3$ sq in. Using $d = 16.5$ in.,

$$\rho = \frac{2.3}{60(16.5)} = 0.0023; \qquad \frac{d}{h} = \frac{16.5}{20} = 0.825$$

From the second group in Table 22.3, interpolate first to obtain ρ values for $d/h = 0.825$:

For $C = 6200$		
Read	$\rho = 0.0016$	for $b_w/b = 0.1$ and $d/h = 0.79$
	$\rho = 0.0013$	for $b_w/b = 0.1$ and $d/h = 0.90$
Interpolate	$\rho = 0.0015$	for $b_w/b = 0.1$ and $d/h = 0.825$
Read	$\rho = 0.0083$	for $b_w/b = 1.0$ and $d/h = 0.79$
	$\rho = 0.0062$	for $b_w/b = 1.0$ and $d/h = 0.90$
Interpolate	$\rho = 0.0076$	for $b_w/b = 1.0$ and $d/h = 0.825$

Then interpolate for $b_w/b = 0.2$,

$$\rho = (0.0076 - 0.0015)\left(\frac{0.2 - 0.1}{1.0 - 0.1}\right) + 0.0015 = 0.0023$$

Since this is equal to the ρ value used, use $C = 6200$.

Then, using Eq. (22.3.8) for K,

$$K = \frac{16.5}{0.9(20)}\left(\frac{1}{0.4 + 1.4(215/331)(0.6)}\right) = 0.97$$

Then, since $\lambda_n = 1.0$, use simplified Eq. (22.3.28a) to obtain $C_d = 1.0$, and therefore,

$$h_{\min} \geq \frac{20(12)}{6200}(480)(1.0)\left(\frac{1.0}{0.97}\right)(1.0) \geq 19.2 \text{ in. (488 mm)}$$

A 20-in. beam will satisfy the $\ell/\Delta \geq 480$ limit.

(e) *Compute adjusted clearance required for future sustained (creep and shrinkage) plus live load deflection under the 20-in. deep beam.* Using Eq. (22.3.23),

$$\Delta = \frac{20(12)}{6200}\left(\frac{20(12)}{20}\right)(1.0)\left(\frac{1.0}{0.97}\right)(1.0) = 0.48 \text{ in. (say ½ in.) (12 mm)}$$

(f) *Estimate the deflection of the beam at the time the partitions are installed.*

$$\lambda \text{ at 3 months} = \frac{1.0}{1 + 50\rho'} = 1.0$$

Although Q may be computed to be less than unity (using steps shown in Example 22.3.1), assume initially that $Q = 1.0$ to obtain most critical deflection. Using Eq. (22.3.33),

$$\Delta_c = \frac{20(12)}{6200}\left(\frac{20(12)}{20}\right)\left[(1 + 1.0)\left(\frac{115}{215}\right) 1.0\right]\left(\frac{1.0}{0.97}\right)(1.0) = 0.5 \text{ in.}$$

Since in this example construction loads and procedures will not affect the beam, verify by more exact computation. Compute $Q = (I_e)_{D+L}/(I_e)_D$ using the procedure explained in Example 22.3.1, step (b). For the 20-in. deep beam,

$$M_{\text{cr}} = 2.64(1.39)12 = 44 \text{ ft-kips}$$

and as explained in Example 22.3.1, step (e), compute

$$A = \frac{(M_a)_D}{M_{\text{cr}}} = \frac{57.5}{44} = 1.31$$

$$B = \frac{(M_a)_{D+L}}{M_{\text{cr}}} = \frac{108}{44} = 2.45$$

Enter Table 22.1 with $A = 1.31$ and $B = 2.45$. Obtain for ($1.0 < A < 1.6$) and ($1.6 < B < 3.5$).

$$Q = 0.35A^2$$

$$Q = 0.35(1.31)^2 = 0.60$$

Using Eq. (22.3.33),

$$\Delta_c = \frac{20(12)}{6200}\left(\frac{20(12)}{20}\right)\left[(1 + 1.0)\frac{115}{215}\, 0.60\right]\left(\frac{1.0}{0.97}\right) 1.0 = 0.3 \text{ in.}$$

Installing a partition above the beam should not represent any difficulty. If the partition below the beam is prefabricated, allow for an additional 0.3-in. reduced clear story height. Construction tolerances must also be considered, in addition to the predetermined clearance required (computed to be about ½ in.) for the expected long-time sustained load plus the live load deflections.

(g) *Review the "aesthetic integrity" of the 20-in. beam.* How much total deformation due to sustained load can be anticipated during the service life of the beam? From ACI-9.5.2.5,

$$\lambda \text{ at 60 months} = \frac{2.0}{1 + 50\rho'} = 2.0$$

For $Q = 0.6$ and using Eq. (22.3.36),

$$\Delta_f = \frac{20(12)}{6200}\left(\frac{20(12)}{20}\right)(1 + 2.0)\left(\frac{115}{215}\right)(0.6)\left(\frac{1.0}{0.97}\right)(1.0) = 0.44 \text{ in.}$$

$$\frac{\ell}{\Delta_f} = \frac{20(12)}{0.44} = 550$$

This is acceptable in most cases. The designer is cautioned against placing too much credence on computed deflections. The foregoing essentially indicates that construction problems relating to deflection are unlikely.

EXAMPLE 22.3.5

For the T-section of Example 22.3.1 (Fig. 22.4) make a direct estimate of the minimum thickness h_{min} required if the beam is constructed at an upper level of a high-rise structure and is supported on previously cast (yielding) levels. Construction loads and procedure *are* therefore affecting the beam. Compare results with those of Example 22.3.4.

Solution

(a) *Compute h_{min} to satisfy serviceability requirements.* Note that since $\lambda_n = 1$ was assumed in Example 22.3.4, serviceability requirements remain the same and $h_{min} = 20$ in.

(b) *Estimate deflection of the beam at the time partitions are installed.* From ACI-9.5.2.5, $\lambda_{3\text{ month}} = 1/(1 + 50\rho') = 1.0$. Assume $\lambda_c = 0.5$ (see Sec. 22.3.6). Using Eq. (22.3.33a),

$$\Delta_c = \frac{20(12)}{6200}\left(\frac{20(12)}{20}\right)[1 + 0.5 + 1.0]\left(\frac{115}{215}\right)\left(\frac{1.0}{0.97}\right)1.0 = 0.65 \text{ in.}$$

This deflection is about twice the deflection in Example 22.3.4, step (f).

(c) *Compute the final sustained loads deflection.* Using Eq. (22.3.36a),

$$\Delta_f = \frac{20(12)}{6200}\left(\frac{20(12)}{20}\right)\left[1 + 0.5 + \frac{2.0}{1 + 50\rho'}\right]\left(\frac{115}{215}\right)\left(\frac{1.0}{0.97}\right)1.0 = 0.9 \text{ in.}$$

This deflection is also twice that of Example 22.3.4 in which the beam was not affected by construction loads and procedures. It appears that a ½-in. camber could be useful.

$$\frac{20(12)}{300} > 0.5 \text{ in.} \qquad \text{OK}$$

Even though the actual deflection is larger, note that in this case the levels directly below also deflect in a similar manner and the net clearance required above the frangible partition can be smaller than that required when the partition below the beam is supported on a nonyielding base (as in Example 22.3.1).

To estimate, however, the required clearance the designer must use a crystal ball to anticipate the variable behavior of the adjacent levels due to the loading, material supplied, and the construction process, each somewhat different even in seemingly identical levels designed for similar occupancy. In addition, the designer must be aware of the need for clearance required due to other influencing parameters such as long-term creep and shrinkage of the supporting columns. See Section 22.5.

22.3.9 Additional Comments on Deflections

A unified procedure which accounts for the construction process, allowing for the considerations for serviceability, constructability, and aesthetic integrity in design has been presented in the previous sections. The equations for I_e, h_{min}, and Δ are presented to allow use with all sizes, shapes, materials, and load levels. Though the equations may appear cumbersome initially, the user will quickly find that in the preliminary analysis, for the common case when $f_y = 60{,}000$ psi and $w_c = 145$ pcf, the coefficient K is approximately 1.0, Q can be set (at least temporarily) equal to 1.0, making the equations simple to use to help estimate quickly initial depth or deflection. An overestimate of deflection will occur when Q is set equal to 1; however, this more accurately reflects actual field observation of members susceptible to construction loads. Such members are constructed by procedures that require them to provide self-support, as well as contribute to support at a very early age of freshly placed floors above.

By setting $C_d = 1$, the immediate total service load deflection $(\Delta_i)_{D+L}$ can be estimated and solving for Δ from Eq. (22.3.23)

$$(\Delta_i)_{D+L} = \frac{\ell}{C}\left(\frac{\ell}{h}\right)K_1 \qquad (22.3.37)$$

The value $C = 4320$ for heavily loaded members cannot be reduced much (see Fig. 22.5) unless the member is very lightly loaded; even when C approaches 7200 there will be only a 40% reduction in depth. A final C value must be selected based on the reinforcement ratio ρ of the member in its final size. However, the range is easily determined to quickly provide by iteration the required size.

Adjustment by coefficient K for size and material, D/L ratio, and over-reinforced members provides the necessary trend to allow the designer to develop the "intuition" needed but often lacking when deformations are of concern.

The need to know actual deformations becomes academic if one is assured that the selected size will satisfy the actual "serviceability". One must realize that deflection computations are at best an exercise in probabilistic prophecy and as such must be reviewed with consideration and understanding of the construction process, the probability of occurrence of anticipated loads, and the variation in modulus of elasticity typical of concrete.

Determination of loads to be used for deflection computation is a dilemma which each designer has to struggle with at one time or another. Obviously for strength the struggle is short as safety is paramount. For serviceability, again, a more severe loading condition and early application of nonstructural elements should be considered. In such cases deeper sections are the result; hence less overall deflection.

When one considers, however, "constructability" and "aesthetic integrity", realistic loading must be used so that under actual service loads the member is deflecting within closer proximity to the anticipated behavior. Cambering the structural member can be useful to allow for constructability of finishes and to reduce the final deflection. Overestimating deflection may produce a result where upward camber is not eliminated. Constructability of finishes can therefore be hindered even more. For example, downward sagging floors can be finished by flash patching to a level floor those areas that have sagged too much. A residual upward camber cannot be leveled in such a manner without causing problems at the circumferential free edges where exterior cladding must be attached. There is some indication (based on field observations) that cambered construction, confined by large (unyielding laterally) supports, deflects less than similar construction uncambered. This variance is probably caused by the extra length required in the cambered (arched) construction which induces small compressive forces to help prevent shrinkage cracks from developing.

For easy application of nonstructural elements, such as exterior cladding or interior floor finishes, the designer must consider carefully the method of application and the manner by which the nonstructural elements are built. Exterior prefabricated cladding which is attached to the face of the structure can accommodate nonlevel floors more easily than prefabricated wall elements which are wedged between two floors (with each deflecting somewhat differently because of such things as the variety in materials supplied, the construction history, or the supported loads). Regarding the construction history, the most severe possible loading on one floor compared to least possible loading on the floor above or below will provide the maximum clearance and the minimum overlap required for detailing of receptors for windows. The depth of the receptor must account for both future reduction or increase in clear story height to prevent the glass from either slipping out of the receptor or being crushed by it.

Preloading to allow more uniformity of behavior between adjacent floors receiving sensitive nonstructural finishes is one means to improve estimate of needs. Construction loads generally provide this measure automatically in high-rise construction. However, even in high-rise construction (as was noted previously) the lower floors which are shored to solid foundation indicate generally a varied behavior from identical upper floors which receive their initial support from more flexible lower floors.

The designer instinct is obviously toward deeper stiffer construction where small deflections can be overestimated or underestimated by a large degree yet not cause any problem because the total deflection is small. However, the trend in concrete construction is toward thinner construction with minimized (or eliminated) form drops (to create beams). It is therefore important to understand the need to provide sufficient size to soft joints and deeper window head receptors. In the evaluation of the deformation requirements of nonstructural elements, it is sometimes necessary to provide some load-bearing capacity in such elements in order to allow equalization of floor deflections under special severe conditions. Often this can be accommodated by masonry construction, but caution is required because of long-term creep and shrinkage in concrete columns (Section 22.5).

22.4 SIMPLIFIED APPROACH TO STIFFNESS COMPUTATIONS

22.4.1 Is There a Need to Simplify?

Designers seem to prefer to use those design processes with which they are familiar; those which they can execute without being re-educated. Old timers who may not have many theoretical tools tend to rationalize old approaches in order to avoid the need to adjust. It is this writer's observation that over the years the voluntary adjustments that have been made to design processes by practitioners were brought about mainly in response to *observed* problems rather than because newer theories that enhanced analytical accuracy were designed. Nevertheless the designer's "need" to obtain better accuracy appears to be proportional to the familiarity one has with so-called more "accurate" methods and the speed by which such computations can be performed. Thus, finite element techniques and computer programs to solve them speedily are making inroads in the profession, resulting in a clash between advocates (those who know how to apply) and opponents (those who find reasons to avoid). This writer's experience indicates that neither group has the upper hand in providing safer and more serviceable structures. It is yet to be proven that overall safer structures result from greater effort to obtain accuracy, utilizing theories which cannot be manually verified (or for which errors cannot be easily detected) and which usually can exhaust either time alloted or capacity to review other design parameters. Finite element solutions that cannot predict development of accidental (or nonaccidental) cracks in nonhomogeneous material such as concrete probably should not be used for actual design except to obtain guidelines (alongside field and laboratory observations) from which better simplified design approaches, or better design boundaries can be drawn. The actual design process must be simple enough to allow an experienced practitioner to spot gross errors using simple manual approximations. It is meaningless to choose accurate design approaches for one design parameter if others cannot also be as accurately reviewed. Structural behavior involves many unpredictable parameters because of such things as construction processes, unknown loads, and variable climate. One may want to consider carefully the need to spend time and money in obtaining accurate theoretical solutions to any of the design parameters involved.

Knowing the scope of the design requirement even for a small project, it is important to be able to develop design approaches which can review *all* important design parameters with equal attention, even though less accurately. To do so, the designer must be familiar with simplified approaches, and this is the target of this chapter. First the weakest "link" must be protected. Ductility with redundancy must be provided in a structure. With ductility and redundancy, "accuracy" is no longer paramount. Simplifications are available and better yet, safety is enhanced.

22.4.2 Simplified Design for Gravity Load Using Ductility and Redundancy

Any designer will soon realize that a structure has a marvelous quality; it can overcome even gross errors in design and construction. This quality is the product of the built-in ductility and the availability of redundancy (multiple load paths) in the structure.

"Ductility" is the ability a structural member has to deform several times the deformation that occurs when the flexural yield capacity of its reinforcement is reached, which allows excessive loads to be distributed either to other structural members, or to other parts of the same member. This must be done without reducing significantly its ability to support the load causing the initial yielding.

The presence of other structural members or other parts of the same member to receive and support the extra redistributed load is called "redundancy". That is, more than one route is available to support the load. Furlong [22.17] states that flexural ductility in beams can be assured if tension reinforcement will begin to yield at curvatures less than 20% as large as curvatures at which concrete will begin to spall. To assure this one must limit the area of tension reinforcement to not more than ⅜ of the area of tension steel in the "balanced" strain condition as defined by ACI-10.3.2. In order to prevent formation of large cracks under maximum service loads, Furlong [22.17] proposes minimum moments that must be developed at various regions along the structural member. According to Furlong [22.17] as long as the assigned factored moments are at least as large as the proportion given by Table 22.10 of the full factored static moment M_0, excessive cracking will not occur at service load. The coefficients in Table 22.10 can safely be used for structural beams in "braced" structures. Unbraced structures were excluded by Furlong since he concluded that the design of a member in unbraced frames required the investigation of all possible combined load failure mechanisms, while in braced structures the member can fail only in the mode of the gravity load mechanism.

The possibility of using a similar coefficient procedure for gravity loads when the member is part of an "unbraced" structure should also be considered. In order to do so the restrictions (boundaries) necessary must be identified.

How can a "braced" structure be distinguished from an "unbraced" one? ACI Commentary-10.11 [22.11] stipulates that when shear walls have six times the cumulative stiffness of all the columns within any given story, the columns can be assumed to be "braced". Obviously the intention here is to allow the shear walls to "brace" the columns and prevent large lateral sways, including the secondary deflection and P–Δ moments, or if such secondary effects do occur to provide the means to counter them. In practice, seldom do we encounter structures that do not sway at all. They may sway only a little, but they still sway. Paradoxically, reliance on shear walls to prevent sway at the lower floors of a high-rise building will generally cause the shear walls to be "braced" (and pulled back) by the frame elements at the top portion of the structure. It is obvious that the intention of the Code is to assure that large secondary moments do not develop and if this is

Table 22.10 Ratio γ of Minimum Factored Design Moment M_u (for *Braced* Frames) to Factored Static Moment M_0 (from Furlong [22.17])

Type of End Restraint	Type of Moment	Beams Loaded Only by One Load at Midspan	All Other Beams
Spans with both ends restrained	Negative	0.37	0.50
	Positive	0.42	0.33
Spans with one end restrained	Negative	0.56	0.75
	Positive	0.50	0.46

Note: At interior supports negative moment reinforcement adequate to resist the larger of the beam moments either side of the support must be extended through the support to serve also as negative moment reinforcement in the adjacent span. *Caution:* γ negative + γ positive ≥ 1.0 must be provided to account for the total static moment.

the case (large sways are prevented) the distinction between "braced" and "unbraced" columns becomes moot.

Therefore, if sways can be limited, simplified methods can be utilized to compute gravity moments. How much sway can be allowed? The ACI Commentary-10.11 indicates that if secondary moments (P–Δ effects) caused by the initial sway do not exceed 5% of the first-order moments, the structure may be considered to be "braced" and therefore the added small increment of 5% can be ignored. Therefore, a stability index Q [22.11] may be used to identify that an apparently "unbraced" structure may be treated as "braced". The indicated requirement [22.11] is

$$Q = \frac{\Sigma P_u \Delta_u}{H_u h_s} \leq \left[0.04 = \frac{1}{25}\right] \tag{22.4.1}$$

where ΣP_u = total factored axial load acting on all columns in a story
Δ_u = relative lateral deflection in a story under factored load
H_u = factored lateral load (shear) in a story
h_s = story height

Alternatively, using service rather than factored loads since drifts are generally computed for service lateral loads, and assuming for the gravity loads in Eq. (9-2) of the *ACI Code* that $P_u = 1.4D + 1.7L \approx 1.5P$

$$Q = 0.75\left[\frac{1.5\Sigma P(1.7\Delta)}{1.7Hh_s}\right] \leq 0.04$$

or

$$Q = \frac{\Sigma P\Delta}{Hh_s} \leq \left[\frac{0.04}{1.125} = \frac{1}{28}\right]$$

For a structure where $\Sigma P/H \approx 50$,

$$\frac{\Delta}{h_s} \leq \left[\frac{1}{28}\left(\frac{1}{50}\right) = \frac{1}{1400}\right]$$

Most structures are designed for sways that do not exceed twice this sway index. The necessary conclusion is that serviceability requirements place most structures within an "inner core" for which secondary effects are small.

The question of what happens to gravity moments in sway structures was extensively reviewed by a recent Reinforced Concrete Research Council (RCRC) project which resulted in a revision to the 1977 *ACI Code*. The 1983 *ACI Code* [22.1] now allows gravity load moments for columns in unbraced structures to be magnified by the "braced" magnification factor rather than by the larger "sway" magnifier as indicated in previous codes. It is therefore now possible [see *ACI Code* Equation (10-6)] to consider gravity-load-induced moments in a "sway" structure as if they are not influenced by the sway. This is so even though it is the gravity loads (ΣP) which in conjunction with the sway Δ cause the secondary moment ($P\Delta$).

The RCRC report indicates that this liberalized procedure can be safely applied also to the flexural floor members that must be designed for the total magnified end moments of the compression members at the joint (ACI-10.11.6).

The Code therefore allows for the consideration of gravity loads as if the structure is not swaying. The result of such an analysis can then be added to results of the lateral loads analysis in which sway is considered and the P–Δ effects do increase the moments and shears due to lateral loads in both compression and flexural members.

The 1983 *ACI Code* [22.1, Chap. 13], in recognizing design limitations of practicing firms, now allows the superposition of results from various analytical models, each suited to analyze a particular load in question. Therefore, to obtain gravity load moments and shears it is permissible to use a suitable analytical model utilizing one set of stiffnesses for the structural members and then to superimpose results from such an analysis to results of another in which a different set of stiffnesses may help estimate lateral load moments and shears more accurately. *Field observations of many structures verify the validity of such assumptions when redundancy coupled with ductility is available.* Because of the many uncertainties generally accompanying either lateral load estimate or the modeling of the structure under lateral load, analysis cannot demand great accuracy for the combined loads. Therefore, if the presence of ductility and redundancy is assured, it is logical to use a simpler model for analysis of the gravity loads.

Lateral load analysis increases the negative moments over the supports, requiring some extra reinforcement in the beam members so that when the extra reinforcement is needed to resist lateral loads, it is available. Meanwhile, this added negative moment reinforcement is also available for the service gravity dead and live loads. Note also that under combined gravity and lateral loads a reversal of moments often occurs at one end of the beam (see Fig. 22.6). Furthermore the negative combined moment at the other end is then numerically larger than the net positive moment region at the end where reversal occurred. The positive moment region of the beam is therefore stiffer than the negative moment region particularly in the case of the downturned web of the beam where a compression flange can be included in stiffness computations in positive moment regions.

A study by Grossman [22.18] of a structure having composite beams with variable moment of inertia under combined gravity and lateral loads identified the occurrence of a redistribution of moments which resulted in an increase in moments at the stiffer end (the positive moment region) and a reduction to moments at the negative moment end of the beam. Total sway of the structure was also found to be reduced.

It appears, therefore, that for the case of combined gravity and lateral load one could benefit in selecting the upper limits to positive moments required by the gravity loads, therefore providing more room for placement of added required lateral load reinforcement in the support (negative moment) regions. Such coefficients for gravity load in sway structures (rounded for easy applica-

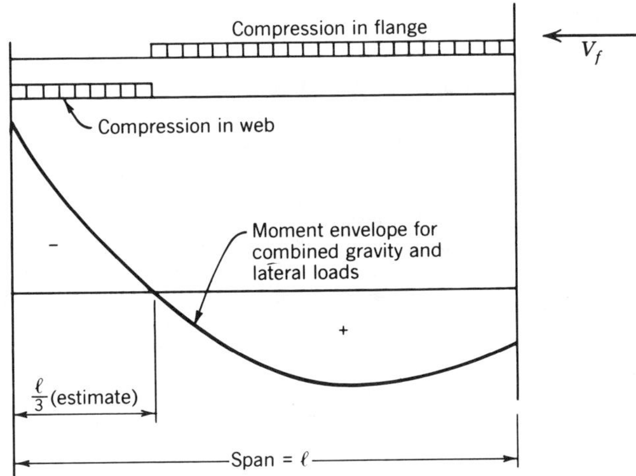

Fig. 22.6 Moment envelope assumption for combined gravity and lateral load analysis.

Table 22.11 Ratio γ of Factored Design Moment M_u (for *Unbraced* Frames) to Factored Static Moment M_0

Type of End Restraint	Type of Moment	Beams Loaded Only by One Load at Midspan	All Other Beams
Continuous interior beams	Negative	0.4	0.5
	Positive	0.6	0.5
Continuous exterior beams	Negative	0.2/0.5	0.2/0.6
	Positive	0.65	0.6
Spans with one end restrained	Negative	0.6	0.67
	Positive	0.7	0.67

* Lateral load moments by a separate analysis are to be added to gravity load moments obtained from this table. Ductility and redundancy must be provided and sway must be restricted (see Section 22.4.3).

tion) are provided in Table 22.11 which is extended also for the common case of exterior beams with integral exterior supports. The result of diverting gravity moments to require added positive moment reinforcement will improve structural behavior under sustained gravity loads by increasing I_e of the positive moment region and therefore reduce gravity load deflections. This small relocation of reinforcement does not reduce construction economy.

Good detailing practices to maintain ductility to allow redistribution of moment dictates that the beam member must have (1) excess shear capacity, (2) continuous top reinforcement (about 25% of bottom reinforcement to be continued along top at midspan will also aid in reducing deflection due to creep and shrinkage), (3) bottom reinforcement continuing into supports and being spliced with adjacent beam reinforcement, and (4) sizes large enough so that $(\rho - \rho') \leq 0.375\rho_b$, (that is, one-half of maximum ρ). The method for computing minimum sizes that respect this limit is explained in Section 22.7.2.

The use of low reinforcement ratios is generally not a hardship in actual design. As was seen in Section 22.3, the size of the member is often dictated by other than strength requirements. Requirements for ductility and redundancy add an insignificant premium to gravity load requirements and provide for design simplicity, simpler constructability, and improved overall factor of safety. In addition future repair and maintenance costs of both the structural and nonstructural elements will also be reduced.

22.4.3 When Can Simplified Assumptions Be Made for Lateral Load Analysis?

The initial selection of allowable sway (the deflection index Δ/h) is the most important parameter in dictating the need for accuracy in stiffness computation for lateral load analysis.

Determining stiffness with accuracy is important in order to judge the structure's sway and dynamic (acceleration-motion perception) characteristics. These are needed in order to make the structure serviceable; that is, to allow the tenant to occupy it comfortably and make sure the nonstructural elements such as partitions and cladding will not require frequent service and repair. Stiffness estimate is also needed to evaluate the distribution of lateral loads to the various members and also to determine the secondary moments and shears caused by the structure's sway (P–Δ effect).

As discussed in Section 22.3, design of a member for gravity load is usually dictated by other than strength requirements. Design for lateral loads is no exception. Sway limits are needed in order to protect the nonstructural elements and to prevent perception of motion from causing discomfort to the occupants. Safe structures are the result of such demands. Limits for serviceability will dictate lower sways than can be tolerated by strength considerations. Considerations for serviceability may allow the use of simplified assumptions for lateral load design, as well as for gravity load design.

Thus, a knowledge of stiffness for lateral loads is necessary:

1. To verify serviceability.
2. To assign lateral loads to the various members.
3. To estimate the secondary P–Δ effect.

Stiffness Assumptions for Serviceability

Concrete structures have larger mass, longer fundamental period of vibration, and greater damping capacity (cracks are, for once, appreciated) than structural steel structures, and hence, their dynamic characteristics of motion are not as pronounced.

An in-depth study of motion in structures is generally outside the scope of this chapter. (Note that along-wind and across-wind design provisions are extensively discussed in the *Supplement to the National Building Code of Canada* [22.19].) Chapters 2, 10, 11, and 19 of this book provide additional treatment of loads and of lateral load and dynamic response.

The wind load on the structure is the result of three separate contributions: one static and two dynamic in nature. The first one is the "mean" static wind load which can be measured quite accurately in wind tunnel testing of rigid models. The second contribution is caused by the fluctuating part of the instantaneous wind pressure. In large slender structures there is a lack of correlation of the wind pressure over the full height of the exterior surfaces so that this dynamic action is small in comparison to the third contributor. The latter is a function of several minutes of wind action rather than the precise wind distribution of any particular instant, and is caused by the inertial forces developed as the structure sways. This inertial force is equal to the product of the mass and the acceleration of the structure. In very tall and very slender structures the dynamic wind loading is often much larger than the mean static loading.

To assess these contributors in a wind tunnel test an aeroelastic model of the structure is constructed. This model must simulate the flexibility, damping, and the mass of the prototype structure in order to provide detailed information on the movement of the structure and the dynamic loads generated by this motion. The construction and testing is time-consuming and expensive.

Lately, a more efficient and economical means to estimate the wind loads has been used in wind tunnel laboratories. The mean and dynamic forces on the building are measured and then dynamic inertial response of the structure is computed. The measured dynamic wind and computed inertial forces are then combined to indicate the peak dynamic force. The peak dynamic force added to the mean force will provide the required design load.

The mean forces and the dynamic wind forces are measured on the entire structure using a lightweight rigid model of the building which is mounted on a high frequency force balance. This simplified method is suitable when torsional effects are small and when the building motion itself does not affect the aerodynamic forces. (A second generation force balance is now being developed and will be more suitable when torsional effects are dominant.) The advantage of early reporting and mathematical projection of results to several alternative structural systems, each with its own mass stiffness and damping, without laboratory testing, make this system attractive to designers.

The frequency of occurrence of wind storms and the acceleration caused by the wind are prime parameters associated with the perception of motion. The acceleration a

$$a = \Delta\left(\frac{2\pi}{T}\right)^2 \tag{22.4.2}$$

is the quantity that induces motion in the structure. Experiments to determine human perception of motion thresholds were based on the natural period T of the structure vs. magnitude of acceleration a and/or the drift Δ caused by the wind forces. In stiff structures, both the fundamental period and the drift are reduced. A sensitive balance must be reached since the human body is more perceptive to motion at certain low fundamental periods.

There can be gleaned from existing concrete structures sufficient evidence to indicate that the limiting sway index recommended in this chapter coupled with larger damping available will produce concrete structures in which the perception of motion is tolerable. An exception to this is the very tall and slender structure which will require in-depth analysis of perception to motion. This subject therefore will not be discussed further here.

Sway limits that effectively reduce the need for repair work to partitions can be more readily measured. Damage to partitions can also be avoided if soft joints are provided. Studies of dry wall partitions by Freeman [22.20] indicate occurrence of initial damage at a sway index of $h_s/400$ (i.e., ¼-in. sway for 8-ft ceiling height). It is apparent, then, that until further research indicates otherwise, a final maximum interstory sway index of $h_s/400$ should be respected in order to satisfy serviceability.

Obviously, it is meaningless to consider sway limits without any indication of how the stiffnesses of the members used in the sway analysis are to be computed. If gross sections are to be used (which will simplify the process immensely), the results of the analysis will be far from realistic. Simplified ("rational") reduced stiffness should therefore be considered. The "rational" reduced stiffnesses should be selected to also eliminate the need for the expensive iteration processes which incrementally match increasing load requirements (caused by the larger sways) which reduce stiffness of the section (nonlinearly) to produce the additional sway increments.

Stiffness Assumptions to Assign Lateral Loads to the Members

Examine first a single member such as a flagpole. Since there is no other means by which any lateral loads reach the foundation except through stressing and deforming this single member, that is, it is statically determinate, it is unnecessary to compute the flagpole properties to determine that *all* the lateral loads must be supported by this member. To make the example more complex, assume two identical vertical members are interlinked by a horizontal member at their top. Again since the members are identical, one can forego stiffness computation to be able to assign lateral forces to each. However, if members are not identical, or if say different gravity loads alter their stiffnesses allowing one member to crack more than the other, and therefore to deflect more, the less stiff member is still linked to the stiffer one. Thus, even though the less stiff member supports a lesser load, both members deflect the same at the link level. To know *exactly* the load and moment carried by each, an iterative series of calculations is necessary.

Next, consider a 60-story structure with 100 (varied in load and shape) columns all interconnected by rigid diaphragms at each floor level. The 6000 "joints" in this structure must translate horizontally the same at each floor level. If lateral torsion is also present, then additional complexities enter the problem.

How can one design such a structure? How can one correctly assume the stiffnesses of each member without having total knowledge and control of where (1) cracks may develop, (2) foundation settlements may occur, or (3) construction processes may influence results? Further, the assignment of the distribution of loads, *even if a correct estimate of the stiffnesses of all members is made,* will be a time-consuming process even for the most powerful computer. To compound the problem some slender structural members absorb shear-mode deformation (frame action) while others may support lateral loads as flexural cantilevers (such as shear walls). Most structures have both shear walls and frame elements and a shear wall–frame interaction analysis to equalize lateral displacements at any level may find some elements (even though the same size) assuming a role of a cantilever at the lower floors and a frame element at an upper floor.

The engineer of the past (precomputer era) did not have such problems with this dilemma. There was no way to solve such a problem. However, many structures were satisfactorily designed and constructed. What made this possible? Again (as for gravity loads) the answer is our Good Samaritan pair, ductility coupled with redundancy. That for structures of old the redundancy was made available by massive masonry partitions and thick brick cladding is not important. Obviously redundancy existed. Today's structures, often clad by glass, must look elsewhere for redundancy. The solution to the problem is still similar. This forgiving characteristic in structures allows one engineer today to totally ignore frame elements in lateral load analysis and to assign all the lateral loads to shear walls, while another engineer may have been wrong in stiffness evaluations, while attempting to consider all structural elements, with both generally producing safe structures. Awareness of this problem is of importance today because redundancy is becoming a "scarce commodity" due to the trends in modern "spacy" design.

Redundancy alone, without ductility, can cause severe problems, especially in monolithic concrete structures. There is a need to assure ductility to allow redistribution without precipitating shear problems. Shear problems can come about when the member cracks, in deforming excessively to allow distribution, and the beam shear capabilities are damaged.

The *ACI Code* appears to provide sufficient ductility details for average structures in mild climates. Where seismic design is required ACI-Appendix A provides some additional ductility requirements for structures in moderate (so-called Zone II) seismicity areas, and larger ductility requirements in areas sensitive to large tremors.

Various low-cost details should be incorporated by the designer who must estimate the need to rely on redundancy and must anticipate the possible parameters that may influence the structure. The ductility requirements for gravity loads discussed in Section 22.4.2 must be reviewed, as well as other requirements of ACI-Appendix A.

Once ductility and redundancy are built into the structure, the designer will find that many design problems he is wondering about, from possibility of a small foundation settlement to the compulsion one has to "exactly" compute stiffnesses of structural members in order to distribute lateral loads to all members as "correctly" as possible, seem less important and are actually fading away.

Thus, based on observation of many safe, existing structures designed by many different design assumptions, there is less need to be exact in stiffness computations in order to distribute lateral loads to the various structural members in structures *where ductility and redundancy are present.*

Stiffness Assumptions to Compute P–Δ Effects

Is an "exact" stiffness evaluation needed to determine the secondary moments (P–Δ effects)? Safe design to balance P–Δ effects always requires assurance of quick convergence to prevent a "mushrooming" sway. Chapters 10 and 11 provide extensive treatment of P–Δ effects.

The secondary moments are caused by displaced gravity loads P when the structure deflects laterally a distance Δ of one end relative to the other, while resisting the lateral loads. This floor-to-floor displacement causes a secondary moment which is equal to P times Δ. Several cycles are usually required to arrive at the final deflection caused by the several incremental displacements caused by each secondary moment. The stiffnesses of the members in the structure must be evaluated so that Δ can be computed.

Stiffnesses of structural members are unfortunately variable, changing with time and service record, and are dependent on construction loads and procedures. Exposures to previous lateral and gravity loads, temperature variances, creep and shrinkage, foundation settlement, magnitude of loads, aging, fatigue and other parameters, including the ability of the designer to make a proper computation, all can provide many answers for the actual stiffness for any particular member.

The need still exists to estimate stiffnesses for determination of P–Δ effects, except perhaps when the P–Δ effects are small. How small is small? The *ACI Commentary* discusses this as reported in Section 22.4.2, wherein it was shown that a structure can be assumed to be "braced" when P–Δ effects do not increase design requirements by more than 5%. The built-in factor of safety is considered to absorb this small reduction in capacity. This indicates that a small stability index ($Q = P_u\Delta_u/H_u h_s$) may be a realistic added boundary which will allow relaxation of demands to stiffness computations.

As an example, assume the designer is not able to compute stiffnesses correctly and it is possible for stiffnesses to be overestimated by say 50%. Therefore, computed story drift Δ is actually 50% larger than computed. To assure that Q is small, initial drift Δ_i must also be kept small. It is customary by many engineers today to limit (for the most critical level) final story sway to, say, $\Delta_f \le h_s/400$ (and the final total structure sway to, say, $\Delta_f \le h/500$). The initial story sway (prior to P–Δ effect causing additional sway) should therefore be kept to a maximum of, say, $h_s/500$. The designer estimates P–Δ effects to increase design requirements by a magnifier δ, given by

$$\delta = \frac{1}{1 - Q} \tag{22.4.3}$$

where using service loads since drifts are usually computed for unfactored Code-prescribed lateral loads,

$$Q = \frac{\Sigma P\Delta_i}{Hh_s} \tag{22.4.4}$$

where ΣP = total service gravity load on the columns in the story
H = external lateral load
V_f = final lateral load in the structure (the external load increased by the P–Δ effects)
h_s = story height
h = total structure height

Thus, the magnified lateral deflection is

$$\Delta_f = \delta\Delta_i = \frac{\Delta_i}{1 - Q} \tag{22.4.5}$$

and the external lateral force becomes

$$V_f = \delta H = \frac{H}{1 - Q} \tag{22.4.6}$$

For high-rise structures having, say,

$$\frac{\Sigma P}{H} \approx 60 \quad \text{and} \quad \frac{\Delta_i}{h_s} \le \frac{1}{500}$$

$$Q = \frac{60}{500} = 0.12$$

Then,

$$\delta = \frac{1}{1 - Q} = \frac{1}{1 - 0.12} = 1.14$$

giving

$$\Delta_f = 1.14\Delta_i$$

$$V_f = 1.14H$$

If one now assumes a 50% error has been made so that Q is 50% higher, then

$$Q = 1.5(0.12) = 0.18$$

$$\delta = \frac{1}{1 - 0.18} = 1.22$$

$$\Delta_f = 1.22\Delta_i$$

$$V_f = 1.22H$$

Designing the structure for the magnified lateral load by the second (reduced) set of stiffnesses requires (1.22/1.14 = 1.07) or 7% more capacity. This is only slightly more than the percentage that the *ACI Code* indicates may be ignored. Yet it is caused by a gross underestimate (50%) of stiffness by the designer. With lower $\Sigma P/H$ values results are improved. Higher $\Sigma P/H$ values are common to structures in mild wind zones in which lower external forces cause smaller initial sways, again improving the results.

In order to limit the possible understrength (caused by stiffness overestimations of 50%) to a maximum of about 10%, it is necessary to limit erroneously computed Q to not exceed 0.15. For this example, this will result in

$$\text{(erroneous)}\ \delta = \frac{1}{1 - 0.15} = 1.176$$

$$\text{(correct)}\quad \delta = \frac{1}{1 - 0.15(1.5)} = 1.290$$

The required strength will then equal (1.290/1.176) = 1.097, or 9.7% understrength.

This understrength in required capacity can be readily accepted since the initial assumption is a gross error of 50% in stiffness on the unconservative side. The load factors required by the *ACI Code* recognize that there may be an even greater magnitude of understrength, even for more readily estimated design loads such as gravity loads.

In any case if the designer attempts to be realistic in estimating stiffnesses, and to limit allowed sway to that required for serviceability and to consider P–Δ effects even if small, it does appear that relaxing demands for stiffness computations and/or computations of sways by using simplistic analytical models are acceptable even in the design of major structures.

Thus, the degree of exactness needed for knowing stiffnesses of structural members in order to estimate P–Δ effects is therefore qualitative. For structures where drift limits are small to start with, simplified assumptions to obtain "rational" stiffnesses can be made. In addition, approximate lateral analyses to determine drifts can be utilized. Practitioners who cannot correctly estimate stiffnesses and those who do not have the physical resources (computing capabilities and programs) to compute drifts "accurately" should establish a small deflection index to start with. An extra small deflection index limit will mean better serviceability to the occupant.

The boundaries that must be observed to allow simplified assumptions are:

1. Deflection index $\Delta_f/h_s \leq 1/400$ (and total structure deflection index $\Sigma\Delta_f/h \leq 1/500$) with loads magnified due to P–Δ effects.
2. Small stability index $Q \leq 0.15$ using a "rational" initial stiffness estimate.
3. Provision of ductility and redundancy in the structure.

Estimation of "rational" initial stiffnesses is discussed in the next section.

22.4.4 "Rational" Stiffnesses of Members for Use in Simplified Lateral Load Analysis

In the previous sections guidelines were provided and special ductility requirements were indicated to allow the designer to use simplified assumptions. The need for such guidelines exists because consideration of all parameters that affect the structure is impossible. Rough estimates can be made of stiffnesses for the structural members commonly participating in the lateral load resistance. Use of these stiffnesses should provide safe structures without excess cost in construction or in design time spent to estimate stiffnesses.

As previously stated, stiffnesses are variable, changing with service record and age of service. The designer must allow for most of the parameters involved, yet the effect of each cannot be traced in a "four-dimensional" structure in which *time* is the fourth dimension during which stiffnesses change. It is meaningless to discuss the deflection index for a structure without identifying also at what load level or at what time in the life of the structure stiffnesses are computed. The designer must also realize that members assumed to behave "elastically" will not do so either immediately upon construction, or as time passes.

An estimate of stiffness must account for effects of cracks, accidental or otherwise, on the member. The efficiency of the "joint" connecting floor members to the supporting columns if frame action is necessary, must also be considered. Structures located in high seismic zones must have the ability to deform into the inelastic range of response without decaying in strength appreciably. This is a safety valve mechanism which allows the structure to increase its "natural period" and provides damping to absorb the seismic inertial forces and to dissipate them. For structures in milder climates which are designed to behave generally "elastically" the estimates of "rational" stiffness for design purposes must therefore assume stiffnesses which can be anticipated just prior to the member reaching yielding of its reinforcement. Since lateral loads are of short duration there is no need to reduce E_c for long-term effects of creep and shrinkage.

Structural elements such as (a) beams and one-way slabs, (b) two-way slabs, (c) columns, (d) shear walls, and (e) stubby columns will be discussed separately. For each type of element the author's guideline suggestion for a "rational" stiffness will be provided, developed over many years of design practice. The suggestions are the results of many discussions with various experts, researchers, consultants, and the review of various publications on this subject. No attempt will be made here to include all references on the subject. The use of these guidelines with the previously established boundaries for using approximate simplified lateral load analysis have produced structures which are behaving well and were economical to build.

22.4.5 Stiffness Assumptions for Beams and One-Way Slabs

In Section 22.3.3, I_e for gravity loads was shown to vary between about 35 and 100% of I_g of the section. I_e was developed for a gravity load parabolic moment diagram and there is little or no experimentation to justify using this value for a lateral load stiffness estimate.

A reinforced concrete beam behaves as a member having variable moment of inertia because along its length cracking usually occurs at service load, and the extent of cracking is proportional to the service load moment. The I_e from Eq. (22.3.1) assumes a cracking pattern caused by uniform loading on a simply supported beam. For a bending moment variation as caused by lateral loading, realistically one must estimate the magnitude variation in moment of inertia in order to obtain the moment due to lateral forces and then revise the moment of inertia variation, iterating until the moment of inertia variation is compatible with the moment variation.

Such a procedure is not practical in the design of actual structures having thousands of such members. The designer therefore must estimate a somewhat conservative value, but not so conservative as to make the structure uneconomical to build. Obviously experiments using whole structures are impractical with the limited resources available to researchers. Theoretical analysis (with some experimental backup) of subassemblies is the only practical basis for stiffness selection. It is necessary to establish values corresponding to member stiffness when first yielding occurs. With redundancy and ductility available, studies of individual members underestimate stiffnesses; however, this added conservatism may be needed for other unforeseen influences ignored when simplified procedures for design are used.

Hage and MacGregor [22.21] extensively reviewed one-way beams and slabs, for various loading combinations, including combined gravity and lateral loads to develop recommendations for a second-order analysis of framed structures. T-section flange effects were also considered. They recommended using 0.5 to 0.6 of I_g for I_e of beams. This value was lowered somewhat to 0.4 in a recent study [22.2]. A similar value appears to be applicable also in one-way slabs. For T-sections values between 1.0 and 1.2 times I_g of the web section alone are recommended [22.21]. The *ACI Commentary* [22.11] recommends 50% of I_g for I_e of beams.

This writer, based on a previous study [22.18], recommends using a weighted average. This average estimates the combined gravity and lateral load moment diagram to cause about one-third of the beam length to have tension on the flange (top) side of the beam (see Fig. 22.6). This weighted average is then reduced to account for cracking and some bar slip in the beam-column joint, and then a weighted average I_e is computed for use with lateral load analysis.

$$I_e = 0.5\left(\frac{1}{3} I_{gw} + \frac{2}{3} I_{gf}\right) \tag{22.4.7}$$

where I_{gw} is the gross moment of inertia for the web only and I_{gf} is the gross moment of inertia of the flanged beam.

The weighted average considers the available ductility which allows redistribution of gravity and lateral moments toward the positive moment region of the member where larger stiffness is available [22.18]. For example, if

$$I_{gf} = 1.5 I_{gw}; \qquad I_e \approx \frac{2}{3} I_{gw}$$

When

$$I_{gf} = I_{gw} \text{ (i.e., one-way slabs)}; \qquad I_e \approx 0.5 I_g$$

22.4.6 Stiffness Assumptions for Flat Slabs and Plates

Flat slabs and in recent years flat plates have been extensively used, especially on the East Coast, in the design of high-rise apartment buildings. In these, architectural constraints did not allow for many shear walls to be incorporated in the design. Engineers, therefore, responding to demands for this economical type of structure, have over the last quarter of a century attempted to estimate the effectiveness of the two-way slab and its contribution to lateral loads. Today the question is still largely unanswered. It is hoped that future research will be directed to solve this intricate dilemma since flat slab construction is now used for all types of structures, including some reaching 600 ft in height and more. Vanderbilt [22.22], under the auspices of the Reinforced Concrete Research Council, undertook the task of accumulating available research and theoretical studies on this subject. It was hoped this research would provide provisions badly needed within the *ACI Code* to design for lateral load on the unbraced frame having a flat plate floor system. Specifically of interest was the possibility of using the Equivalent Frame Method (ACI-13.7) for lateral loads analysis.

Vanderbilt [22.22] accumulated most of the information on this subject and describes the different approaches a design team may select. A considerable part of the report is devoted to the use of the Equivalent Frame Method, concluding that the method is suitable with some adjustment. The *ACI Code* Committee 318 was not in agreement concerning the use of the method. The 1983 *ACI Code* therefore allows various methods including an Equivalent Frame Method as recommended by Vanderbilt. Further treatment of the proposed Equivalent Frame Method for lateral loads is given by Vanderbilt [22.23] and Vanderbilt and Corley [22.24]. Other approaches more suitable for use by a design team with limited resources may also be used. Among such acceptable approaches is one given by Kahn and Sbarounis [22.25] which has been used by this author and many others for nearly 20 years.

The 1983 *ACI Code* therefore is vague on this issue because of the committee's lack of a consensus regarding the appropriate method and ways to stimulate additional research while continuing established practices. It appears that more experiments are necessary, especially for structures which are expected to be constructed within boundaries similar to those already stated in the chapter.

For flat slabs (and plates especially) much of the stiffness of the slabs can be lost if the joint between the column and the slab is not detailed properly. All unbalanced moments (gravity and lateral) must be accounted for within a narrow band—"column head"—which equals a width of $(c_2 + 3h)$, where c_2 is the transverse width of the support and h is the slab thickness. ACI-13.3.4.2 specifies the portion of the unbalanced moment to be transferred by flexure via concentration of reinforcement and the remainder of the unbalanced moment to be transferred by the eccentricity of shear about the centroid of the critical section (ACI-11.12.1.2). This empirical method described in the code provides conservative results but is less accurate then the "beam analogy" method developed more recently by Hawkins, Wong, and Yang [22.26].

A designer can choose between an "equivalent slab width" model and the "transverse–torsional" model to estimate stiffness of two-way slabs [22.22].

Additional discussion of models for use in lateral load frame analysis is presented by Wang and Salmon [22.3]. This writer recommends the "equivalent slab width" model as more straightforward, simpler to use, and more "rational" than the "transverse–torsional" model, especially for flat plate structures.

A review of the state-of-the-art will indicate that no explicit "rational" directions can be provided. Most attempts to provide direction for both the "transverse–torsional" and the "equivalent width" design models are based on elastic plate theory modified and "ironed out" to agree with the meager test results presently available. In this attempt to correlate theory and tests one important parameter—the ability of the connection between the floor member and support to develop the predicted unbalanced moments and shears—is generally not considered. Another often neglected parameter is the availability of redundancy which allows redistribution and better utilization of available capacity. Other simplifications by researchers who must deal with reduced scale models and limited equipment will generally result in lower recommended flexural stiffness values to the slabs. On the other side of the coin, construction procedures and loads may often reduce available stiffness more than can be anticipated by the presence of the service loads. Tests of three-dimensional models that will review also the slab-column connection (considering concentration of reinforcement within the "column-head" c_2/c_1 ratio, presence of capitals, drops, etc.) and also varied panel ℓ_2/ℓ_1, c_2/c_1 aspect ratios are still necessary before any explicit "rational" directions can be provided. This lack is however straining the credibility of the practitioner who must meanwhile continue to design flat slab structures. The writer for many years has used charts [22.25] but more recently has developed the following to obtain the effective slab width $\alpha\ell_2$, for interior panels,

$$\alpha\ell_2 = \left[0.3\ell_1 + \frac{c_1}{\ell_1}\,\ell_2 + \frac{c_2 - c_1}{2}\right] \tag{22.4.8}$$*

and as limits

$$0.2 \le \alpha \le 0.5 \tag{22.4.9}$$

The value of α from Eq. (22.4.8) gives a slightly lower value than obtained from the Khan–Sbarounis charts [22.25]. Equation (22.4.8) also allows for recognition of non-square columns and the transition from two-way to one-way slab action when support width c_2 approaches panel width l_2. In Eq. (22.4.8), the terms are defined as in Chapter 13 of the *ACI Code*.

Equation (22.4.8) appears to provide "rational" estimates of stiffness for two-way slabs to be used in a second-order analysis in structures confined by the drift and stability boundaries previously established. Concentration of reinforcement to develop the unbalanced moment within the "column head" should be allocated so that the ratio of top to bottom reinforcement is about 2 to 1 when reversal of moment occurs on both sides (windward and leeward) of the column and the reinforcement is sized to be developed within column dimension c_1. The net difference between top and bottom reinforcement within the "column head" should not exceed $0.375\rho_b$ to allow for ductile behavior.

Equation (22.4.8) is intended to account for a "normal" amount of accidental cracks developed during the construction process, some unforeseen yielding, and some bar slippage; all are usually observed in average construction. It is applicable only for structures in mild climate zones where lateral wind or seismic forces are low so that the structure is generally behaving elastically. [In moderate seismic zones, use about 75% of the values given by Eqs. (22.4.8) and (22.4.9).] The "joint" efficiency is lumped together with the slab efficiency. Care of the "joint" is tedious and computers are often needed to provide the necessary review. Fortunately many programs are available to the practitioner for flat slabs which can account for the transfer of unbalanced moments. Tables in the *ACI Design Handbook* [22.27] can also be used. Concentrating reinforcement within the "column head" is necessary to improve all aspects of slab behavior and maintain the stiffness.

Slabs in actual structures are usually penetrated by many sleeves and slots. The designer must exercise judgment in further reducing stiffnesses based on location and size of these penetrations. The architect and mechanical engineer should be encouraged not to congregate such penetrations immediately adjacent to supports. Sometimes this is unavoidable and special care must be taken to investigate transfer of unbalanced moment and shear transfer at the column–slab joints.

Effectiveness of shear reinforcement in thin slabs has been demonstrated by Hawkins et al. [22.26] to increase shear capacity and ductility of a flat slab. In critical areas the engineer can provide shear reinforcement. Shear reinforcement is especially useful in structures in seismic zones where assurance of ductility is important to allow the structure to absorb much larger forces than it was designed for.

---

* Laboratory testing now (May 1986) underway at the University of California at Berkeley to study vertical and lateral load resistance of flat-plate construction should, shortly, disclaim or verify the accuracy of this equation.

22.4.7 Stiffness Assumptions for Columns

Column sizes should be established to assure that plastic hinges, should they occur, would be in the floor members and not in the columns ("strong column, weak beam" concept). This will assure that the gravity load transfer mechanism is still available even when larger than anticipated lateral loads cause the structure to behave inelastically.

In general, economical columns are also those that are lightly reinforced. Columns congested with reinforcement should be avoided. Columns having a reinforcement ratio less than 0.02 to 0.03 in high-rise structures are economical and restrict the total sway. The gross moment of inertia I_g of columns can be used for lateral load analysis when the ratio e/c (eccentricity e to lateral dimension c) falls within the kern (no tension). Otherwise, the stiffness of column should be computed as for a flexural member.

The use of h_s/c (story height h_s to lateral dimension c) less than 12 will assure that columns in structures designed within the predescribed boundaries are not penalized excessively for slenderness and will not experience "stability"-type failures (see Section 22.2.4).

22.4.8 Stiffness Assumptions for Shear Walls

Shear walls (see Chapter 7) perform a dual function. They are generally a main lateral load-resisting structural element, supporting loads in a cantilever (flexural) mode if their base is fixed or they are anchored to the foundation. In addition, shear walls, similar to columns, must support the tributary gravity loads.

The gravity loads supported by the shear wall can influence greatly the stiffness of the wall and its capacity to resist overturning moments. Tributary gravity loads, larger than anticipated, will enter the tensile (elongated) end of the wall deforming under lateral loads. This will precipitate an increase in efficiency of the wall to cause larger forces to be resisted by it. This increase cannot be determined easily. In solid coupled walls the compression end, close to the base, will support the great majority of the shear resulting from diagonal strut action developed because of the unyielding foundation and because the compression side is stiffer than the tension side of the wall. It is prudent, therefore, to design the wall for larger-than-computed shears. This will assure ductility to allow redistribution once the tensile capacity of the wall is exhausted.

Shear walls (unless very dominant in number and size, or if part of a shallow structure) usually can be efficient in lateral load participation only for a portion of their height. The rigid construction of the shear wall, when deformed by lateral loads acting close to the base, will tend to deflect the top as dictated by the base curvature. Frame elements (also usually present) generally have sufficient stiffness to "pull back" the shear wall at the top. A familiar shear wall–frame interaction therefore as shown in Fig. 22.7 usually occurs in most high-rise structures. This requires a different effective stiffness for the shear wall at the lower and upper levels. Except for its lower portion where high overturning moments occur the shear wall can be assumed to be uncracked except for accidental temperature, construction, or shrinkage cracks. Therefore, using 80% of I_g for the moment of inertia I_e in this region is probably acceptable for design. Below this level, I_e must be related to gravity loads, overturning moments, and reinforcement provided.

Foundation rotation usually reduces shear wall effectiveness in a manner similar to the wall being cracked. For a "rational" estimate of shear wall stiffness, this writer recommends, for the aforementioned categories of structures, the use of 40% of I_g for the lowest 1/6 of the structure height. Above that level use 80% of I_g. Economy will dictate a reduced size of shear wall at upper levels; in this case, in order to simplify design the shear wall base stiffness (based on 40% I_g) can be used for

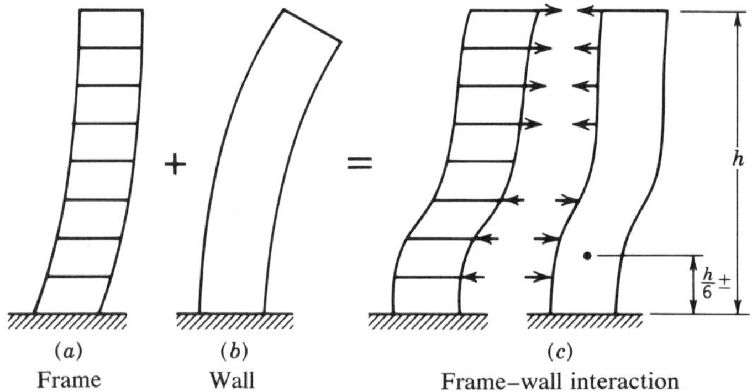

Fig. 22.7 Frame–shear-wall interaction.

the entire height. Results based on using 40% I_g are acceptable for an average structure where shear walls and frames interact to resist lateral loads, with minimal sway. Further adjustments to this initial estimate of stiffness may have to be made if the computed overturning moments cannot be developed at the base of the shear wall.

22.4.9 Stiffness Assumptions for Stubby Columns

Stubby columns are both shear walls and frame elements in the lower levels of high-rise structures. At the upper levels they are able to perform only as part of the frame system.

In reality all columns to some degree are “”stubby” columns. This is the reason why elastic theory dictates shifting of the point of contraflexure in the columns of a frame system upward at the lower couple of floors and downward at the top floors, thus in essence duplicating “shear wall–frame” interaction where all columns can behave both in the cantilever (flexural) mode and the shear (frame) mode simultaneously.

This writer has used a simplified design approach for structural frames in which the point of contraflexure is assumed at midheight for the columns (to allow simplicity in frame analysis) alongside further evaluation of the columns acting also as shear walls to account for the additional moments the columns can develop based on flexural (cantilever) action. This simplified approach was developed to allow large structures having several thousand joints to be analyzed on computers with limited capabilities and yet to maintain the three-dimensional aspect of the project. Therefore, lateral–torsional effects (much more important in many cases than the location of contraflexure points at lower levels) could also be accounted for.

The question often arises regarding when is a vertical member a column and when is it a shear wall. This writer defines a shear wall as an element that can support all tributary gravity and lateral loads, unbraced in the direction of lateral load action. The practitioner using simplified approaches and limited capabilities must decide whether the columns are stiff enough to be treated as a shear wall and to what height.

Each shear wall and “stubby” column are also connected to floor elements (beams and slabs) which develop moments and deform to resist the lateral load action. Therefore all shear walls and “stubby” columns are also part of the “frame” system. Their efficiency in either the “frame” or “cantilever” mode depends on the floor level (height above base) and on the relative stiffness of columns to walls and on the stiffnesses of the floor elements as well. Obviously “exact” solutions are preferred in this situation—but they too, as has been stated before, have their limitations. An approximate process used by the writer is described next.

First determine for the “stubby” columns the height ℓ_s above the base of the structure the column can stand freely (without buckling) and still support its gravity load. For conservatism, assume the gravity load P_u at the base to act at the distance ℓ_s above the base. Using ACI Formula (10-9) [22.1] and taking $EI = E_cI_s/2.5$, I_s required for the stubby column may be obtained from

$$I_s \geq \frac{P_u}{\phi}\frac{(k\ell_s)^2 2.5}{\pi^2 E_c} \approx \frac{P_u \ell_s^2}{3E_c} \tag{22.4.10}$$

In Eq.(22.4.10) the effective length factor k may be taken as 1.0 for a column having a fixed base and rotation fixed (by the frame elements) but translation free at the top. This quick computation determines which columns also participate (and to what floor level) in the cantilever (flexural) lateral resisting mode.

For example, assume the factored load P_u at base equals 1500 kips, and the stubby column size is 12×36 in. and $E_c = 4100$ ksi. Using Eq. (22.4.10),

$$I_s = \frac{1}{12}(12)(36)^3 = 46{,}700 \text{ in.}^4$$

$$\ell_s = \sqrt{\frac{3E_cI_s}{P_u}} = \sqrt{\frac{3(4100)46{,}700}{1500}} = 620 \text{ in.}$$

Thus, ℓ_s is about 50 ft (15.2 m).

In order to allow the stubby column to participate both in “frame” and “shear wall” action the column is “divided” into two parts as shown in Fig. 22.8. The horizontally dashed portion of the column is the “shear wall” part of the column (having stiffness K_{sw}). The diagonally hatched portion is part of the “frame” system and has stiffness K_{cb}. The two parts can now be added to the actual frame elements and to the actual shear wall elements for the determination of “shear wall–frame” interaction. To determine the effective I for each part the following procedure can be used as a yardstick.

The 1963 *ACI Code* indicated that for beams and columns to participate in frame action, their stiffness ratio $(K_{cb} + K_{ct})/(K_{b\ell} + K_{br})$ should not exceed 25 (see Fig. 22.8 for definition of terms). This multiple (25) of the floor member stiffnesses can be subtracted from the stubby column total I_c with the remainder considered a shear wall. Final design of the stubby column will require superimposing moments and shears from both systems.

This relationship $(K_{cb} + K_{ct}) = 25(K_{b\ell} + K_{br})$ is an arbitrary value. For "pure" frame structures (without shear walls) this multiplier can be reduced to say 5, to better compensate for the assumed midheight location of the point of contraflexure in columns at lower floors of high-rise structures.

In structures where large shear walls actually resist most of the lateral loads at the lower levels the multiple may be increased substantially (to say 100) to exclude many of the stubby columns. The stubby (and other) columns' assumed point of contraflexure location is not important in this case because the lateral load moments resisted by the frame elements are generally insignificant when compared to those resisted by the shear walls.

Structures in High Seismicity Zones

For structures that are expected to behave inelastically, such as those in high-risk seismic zones, it is suggested that further reductions in stiffness be taken. Stricter ductility requirements based on ACI-Appendix A must be followed to provide confinement of concrete in columns, beams, and interconnecting joints. Values of stiffnesses between ½ and ⅔ of those indicated previously for lower seismic risk areas will allow for the added decay expected due to inelastic behavior.

Since the lateral loads specified by codes for the design of the structure can be considerably smaller than those corresponding to a linear response of the anticipated earthquake intensity [22.11], as the properly detailed ductile structure responds to strong ground motion, it loses some of its effective stiffness and its capacity to dissipate energy generally increases. In such a case both the lateral earthquake inertial induced forces and accelerations are reduced. (An exception to this will be when the ground motion period matches the natural period of the structure and resonance will cause the structure to absorb larger and larger forces such as occurred in the 1985 Mexico City earthquake.)

Since inelastic (nonlinear) response is anticipated, the stability of the structural members must be investigated at displacement levels larger than obtained by the above estimation of stiffnesses. The *ACI Commentary* [22.11] suggests multiplication by a factor of at least 2 of the computed displacement obtained from a linear analysis using factored lateral loads.

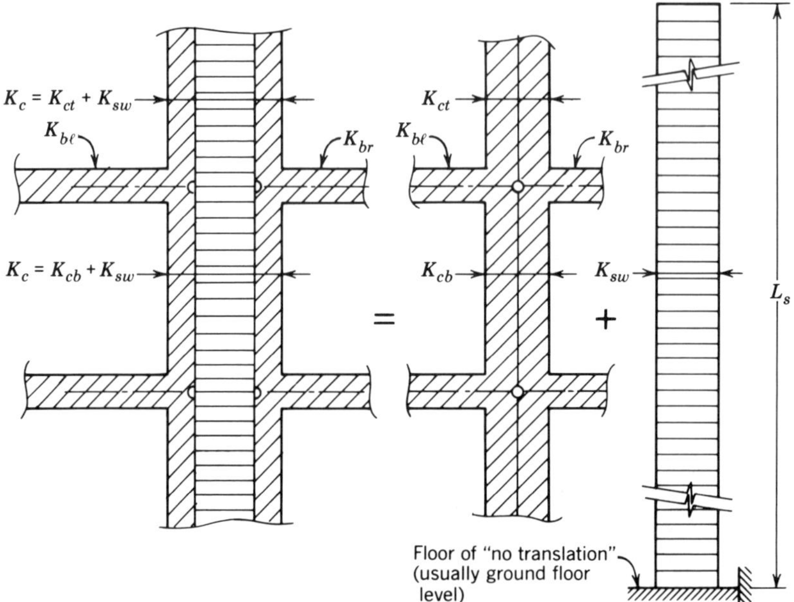

Fig. 22.8 A "stubby" column participating as part of a frame and also as a shear wall in resisting lateral loads.

22.4.10 Approximate Analysis for Large Three-Dimensional Structures for Lateral Loads

It is important to maintain the three-dimensional configuration in the analysis of structures. This is to detect large in-plane torsional moments and shears which result when the center of rigidity (shear center) does not coincide with the centroid of the resisting section through which the centroid of the lateral loading is assumed to act. Often such induced moments and shears, if overlooked, are the cause of major problems. When the structure is very large so that exact solutions are either beyond the capacity of the computer used or are too expensive to obtain, the designer may find it beneficial and safer for the structure to simplify the design procedure but still retain the three-dimensional aspect. This section addresses such a need and presents simplified design approaches that can be used manually or with the help of limited computer capacity.

Simplified Solutions for Frame Structures

The Portal Method, the simplest of all, allows for distribution of lateral loads among the various structural elements in "pure" frame structures. It is probably still used for simple symmetrical structures. In its simplest form this method did not consider relative stiffness. Other methods, almost as simple and easy to apply, have been used extensively with great success. One such method simple enough to use manually and applicable for large unsymmetrical structures is described in the PCA publication, *Continuity in Concrete Building Frames* [22.28]. This publication was a bonanza to the design office of the 1960s because it describes shortcuts to moment distribution procedures (the "two-cycle" method) which saved many firms from bankruptcy. It also describes a simple method for lateral load analysis in which "joint coefficients" are assigned to each column beam connection. The stiffnesses of both vertical and horizontal members are considered, and the "joint efficiency" is a function of both. Efficient "joints" require both vertical and horizontal members to be efficient, otherwise the joint efficiency is controlled by the strength of the weaker member.

The assumption of equal translation to all "joints" at the floor level considered is expressed as a function of the I/ℓ values of the members framing into each joint. A "joint coefficient" can then be computed.

Let the coefficient of any joint be denoted as J_x and the sum of all the coefficients in the floor as ΣJ_x, the share of the total lateral load V_x carried by each joint is equal to $V_x' = V_x J_x / \Sigma J_x$. Simple analysis of a few floors in a structure allows for a reasonable distribution of lateral loads based on stiffness. The plan of the structure is sliced into bents parallel to each major axis and in each "bent" all "joint coefficients" J_x are computed

$$J_x = \frac{K_{cb}(K_{b\ell} + K_{br})}{K_{cb} + K_{ct} + K_{b\ell} + K_{br}} \tag{22.4.11}$$

where K_{ct} and K_{cb} are the column stiffnesses above and below the floor, respectively; and $K_{b\ell}$ and K_{br} are the stiffnesses of the floor members framing into the joint on the windward and leeward sides, respectively (see Fig. 22.9).

This simple approach is shown to distribute lateral loads into each of the "joints" to cause equal lateral translation. Lateral–torsional effects can also be handled. The center of rigidity of all "joint coefficients" is easily determined to allow computation of torsional moments $V_x e_y$ and $V_y e_x$ for both major axes. The moments of inertia $\Sigma J_x y^2$ and $\Sigma J_y x^2$ of the "joint coefficients" about their centroids in both major axes is then computed. Therefore, each "joint" is actually resisting lateral–torsional moments in both major axes. The shear F_x in the supporting column caused by the lateral load V_x (parallel to x direction) and torsional moment $V_x e_y$ is then computed:

$$F_x = V_x' + \frac{(V_x e_y)(J_x y)}{\Sigma(J_x y^2) + \Sigma(J_y x^2)} \tag{22.4.12}$$

and the transverse shear F_y acting simultaneously due to the same torsional moment is

$$F_y = \frac{(V_x e_y)(J_y x)}{\Sigma(J_x y^2) + \Sigma(J_y x^2)} \tag{22.4.13}$$

where in Eqs. (22.4.12) and (22.4.13),

e_y = the torsional eccentricity for lateral load V_x parallel to x axis
V_x = external lateral load parallel to x axis
y = distance to center of rigidity perpendicular to the "joint coefficient" J_x
x = distance to center of rigidity perpendicular to the "joint coefficient" J_y

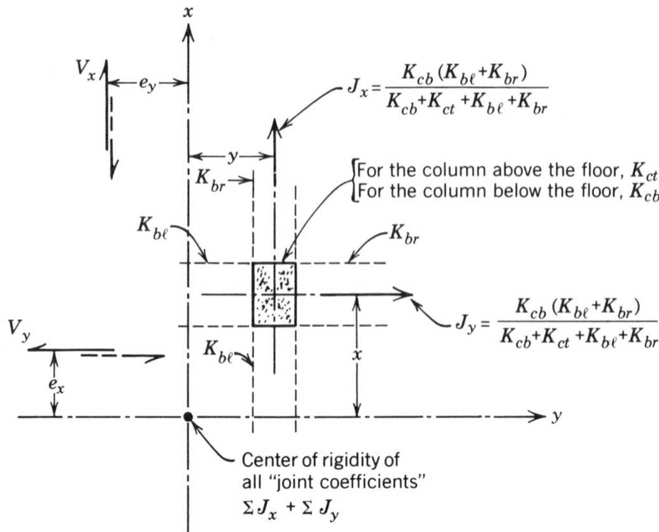

Fig. 22.9 Plan view of a single "joint" of columns and intersecting floor members considered in computation of the individual "joint coefficient." Note that stiffnesses are computed for the columns and floor members in a direction parallel to the x axis to determine J_x and in a direction parallel to the y axis to determine J_y.

J_x = "joint coefficient" when lateral load is in the direction parallel to the x axis
J_y = "joint coefficient" when lateral load is in the direction parallel to the y axis
$V'_x = V_x J_x / \Sigma J_x$

Analogous equations which include the torsional moment $V_y e_x$ in the computations of combined shear F_y, and the transverse component F_x can be developed as well.

Note that F_x and F_y act on the column simultaneously, therefore requiring biaxial moment analysis of the columns. If F_y is small compared to F_x it can be neglected, otherwise the design may be done the same as for uniaxial columns if an enlarged single-axis moment M'_x is used in place of biaxial moment,

$$M'_x \approx 1.1 \sqrt{M_x^2 + \left(\frac{c_x}{c_y} M_y\right)^2} \tag{22.4.14}*$$

where c_x and c_y represent the column dimensions in the x and y directions, respectively.

Equation (22.4.14) can usually provide satisfactory results using uniaxial design charts which are much easier to use than to carry out the design for biaxial bending and compression. The designer can adjust the 1.1 factor based on his experience and range of column sizes and reinforcement used.

When axial deformations can be ignored, initial floor-to-floor sway of frame structures Δ_i can also be estimated in a simplified manner. The moment sway caused by flexure in columns and by the joint rotation due to the flexure of the floor members may be approximated as

$$\Delta_i \approx \frac{Hh_s^2}{12}\left[\frac{1}{E_c \Sigma K_c} + \frac{1}{E_b \Sigma K_b}\right] \tag{22.4.15}$$

where h_s = story height
H = external lateral load
E_c and E_b = modulus of elasticity for columns and beams, respectively
ΣK_c = sum of stiffnesses of all columns, $\Sigma I_c / h_s$
ΣK_b = sum of stiffnesses of all floor beams, $\Sigma I_b / \ell_1$
ℓ_1 = span of beam

* This simplified equation was suggested by Professor R. W. Furlong in a lecture "Design of Reinforced Concrete Columns," sponsored by the Concrete Industry Board, New York City, October 16, 1978.

Simplified assumptions such as (1) taking the point of contraflexure at midheight and (2) shear deformations are small, can be tolerated when K_c and K_b are computed using the assumptions of the preceding section and the initial drift Δ_i [computed by Eq. (22.4.15)] is small. Adjustments to initial drift Δ_i due to axial deformation (so-called chord deformation) can be estimated to increase Δ_i by about 5 to 20%, in "stubby" moderately high-rise structures. The first and last "joint coefficient" in each bent should be reduced by, say, 20% to increase Δ_i indirectly and to duplicate release of lateral moments in the exterior columns due to lengthening and shortening of the windward and leeward columns, respectively. Approximate adjustments may also be made according to McLeod [22.29].

Once Δ_i is computed, the stability index Q can also be computed and if larger than 0.15 the structure will require stiffening, or alternatively, more detailed design procedures must be used (see Section 22.4.3). Next, Δ_i and H are multiplied by the magnifier $\delta = 1/(1 - Q)$ to account for the P–Δ effects.

Slender structures, especially if tall, must be reviewed by more exact means because chord deformation will become a major parameter influencing total drift. Chord deformation caused by the elongation of columns on the windward side and shortening of columns on the leeward side, can be shown to shift moments from exterior bays toward interior bays of a high-rise structure, with increasing intensity at the upper levels. The design wind force in very tall slender structures is to a large extent dynamic in nature, caused mainly by inertial forces which require several minutes of wind action to develop. The building is oscillating about a deformed axis, the result of the mean wind forces about both principal directions. The accumulation (along the length of the column) of the tension or compression forces due to lateral load is not always additive at any one instant. Computer elastic two-dimensional analyses, in which moments, shears, and deformations materialize instantly throughout the structure, may therefore underestimate the efficiency of the frame elements to which large axial tension and compression forces are assigned.

Simplified Design for Structures Having Shear Wall and Frame Components

This common dual system requires division of the lateral loads between the frame elements and the shear wall elements. The floor system can usually provide distribution of loads to all members in a manner that equalizes lateral sway of all members (i.e., "rigid" diaphragm action). A simple process for design is to lump all shear wall elements and all frame elements into two groups, and then to go through the process of mathematically equalizing the translation (sway) of each group at each of the floor levels. This can be done by using design aids [22.25] or by a two-dimensional frame computer analysis.

Shear walls having repetitive penetrations are called *coupled-walls*. The coupled-walls stiffness depends heavily on the connecting beams (or slabs) stiffness. The efficiency of the coupled-wall can be estimated by manual or computer analysis in the following manner: Compare one analysis in which the connecting beams are assumed infinitely stiff, to another, in which the actual (reduced for cracking) stiffness of the connecting beams is considered. The ratio of the deflections obtained from these analyses can assign a reduced "equivalent" stiffness to the coupled-wall prior to lumping it with other walls for the "shear wall–frame" interaction analysis.

At lower levels, the effectiveness of the shear wall will attract most of the lateral load. At mid-levels, frame elements more actively participate and actually "pull back" the shear walls at the upper levels. At lower levels, the frames resist only a small portion of the lateral load and the resulting flexural moments usually will not control the design of the frame members. (For that reason, any simplified assumption that the location of the point of contraflexure is at midheight is less disturbing.) At uppermost levels, smaller in magnitude total lateral load will require less strength from the frame elements. Some adjustments are usually required to moments in columns obtained by simplified assumptions at uppermost levels, to account for shifting downward of the point of contraflexure.

When shear walls are more dominant in lateral load action (because of either great number of shear walls and/or their large size) frame elements tend to show a fairly uniform lateral load requirement, in the order of 20–30% of the total lateral load, throughout the height of the structure. This can greatly simplify design demands. In structures having moderate to small shear walls, large variations in lateral load requirements can be observed for the frame elements, sometimes requiring stiffening of floor members at various critical floors. When flat slabs are used as part of the lateral load system, an increased thickness at midlevels is customary.

Torsional moments in structures where both frame and shear wall elements resist the lateral loads are not easily distributed. The capacity of frame elements to resist torsional moments can be estimated as previously discussed. The capacity of a shear wall to resist torsional moment depends on the shape of the wall, its plan location, and the height above its fixed base. As a result, the effectiveness of the shear wall to resist torques is not uniform at the various levels in the structure.

Frame elements and planar (i.e., rectangular in cross-section) shear walls resist torques only if they are not located at the center of rigidity (shear resistance) of the considered level (neglecting small torques that can be developed about each structural member centroid).

Nonplanar (such as elevator core) shear walls can resist additional torques also in a twisting action about the shear-center of the wall. (Note that the shear-center of a core is not necessarily its geometric area center.) If the core section is closed (or its open end is framed by stiff beams), and each of the wall segments has a large lever distance to the shear-center of the core, then large torques can be developed regardless of the plan location of the core. These torques can oppose the external torque without causing a net lateral displacement to the center of the core. They deform the core by twisting it about its vertical axis (St. Venant mode of action which circulates shear flow around the core) and by warping it (restrained warping mode of action) about its fixed base. The first mode is analogous to shear-mode action and is independent of the distance to the fixed base. The latter is analogous to cantilever-mode action and is dependent on the distance of the level considered from the base of the wall.

The interaction between nonplanar walls and other structural elements in resisting torques is made even more complex because the stiff floor diaphragm connecting all the elements can alter the results of an exact analysis of each individual core. When the wall is pulled back by the frame elements at the upper portion of the structure (Fig. 22.7) it is not clear what torque capacity can be attributed to the core wall, or even how planar walls participate in resisting the torque.

A conservative approach for the structure as a whole is to ignore this added torque capacity of the core and to distribute initially torsional moments assuming all shear walls are planar (linear) elements. In such an analysis the designer should discount torsional shears which reduce the requirements of concentric lateral forces. A separate analysis can then estimate the added stresses in the core.

The interaction of planar shear walls and frame elements in resisting external torques is complex enough in itself. Where exact analysis is not feasible the writer suggests the following procedure: First distribute the external lateral forces along both principal axes into both the frame and shear wall systems, based on the relative stiffness of each. Torsional moments are ignored at this stage. "Shear wall–frame" interaction analysis is used to divide the total external forces, V_x and V_y between these two systems and to compute the relative story drifts, Δ_x, Δ_y, in the x and y directions, respectively. Thereafter, individual shears, V_x' and V_y', are computed based on the relative stiffnesses of all structural members in each of the two systems. Shears in the "stubby" columns are combined from both systems. The center of rigidity (shear resistance) can now be located and eccentricities, e_x and e_y, determined in the normal manner, *but using the individual shears previously computed rather than stiffnesses*.

Torsional moments are then estimated to equal $V_x e_y$ and $V_y e_x$. In order to compute shears induced by the torsional moments, adjust Eqs. (22.4.12) and (22.4.13) by replacing J_x and J_y with V_x'/Δ_x and V_y'/Δ_y, respectively, This substitution is proper since for frame elements, by definition, J_x is proportional to V_x'/Δ_x and J_y is proportional to V_y'/Δ_y. Similarly, the shear wall element stiffness is also proportional to V_x'/Δ_x and V_y'/Δ_y. Δ_x and Δ_y are the incremental lateral deflections between adjacent levels in the x and y directions, respectively, for the dual system of shear wall and frame elements.

Upon substitution and rearrangement, Eqs. (22.4.16) and (22.4.17) are obtained. These equations allow distribution of the torsional moments in dual systems, without the need to use a three-dimensional frame analysis. They respect the relative stiffness of each principal direction and the presence of two separate systems, frames and shear walls, in each of the principal directions. The shears produced by torsional moments are either additive or subtractive to the lateral shears in the columns and shear walls (no reduction, however, is allowed in seismic zones).

$$F_x = V_x' + \frac{(V_x e_y)(V_x' y)}{\Sigma(V_x' y^2) + \left(\dfrac{\Delta_x}{\Delta_y}\right)\Sigma(V_y' x^2)} \tag{22.4.16}$$

$$F_y = \frac{(V_x e_y)(V_y' x)\left(\dfrac{\Delta_x}{\Delta_y}\right)}{\Sigma(V_x' y^2) + \left(\dfrac{\Delta_x}{\Delta_y}\right)\Sigma(V_y' x^2)} \tag{22.4.17}$$

Interaction between the two lateral systems already discussed is sometimes complicated by the presence of other systems. For example, a tube system utilizing closely spaced exterior columns interconnected by either stiff or flexible spandrels is usually encountered together with other frame and shear wall elements. The complexity in design can become enormous and one must usually resort to large computers for interaction analysis. Short cuts for modeling are available [22.30] but one must be careful not to overlook the three-dimensional influence.

In all such analyses stiffnesses are estimated by the designer and entered as input. Where floor member spans are short, the influence of the rigid joints is often a dilemma. Too short a span can overestimate stiffness and cause an overflow of forces to the member, causing it to yield and thus release and redistribute. Most computer programs cannot yet account for this redistribution, and reruns with adjustments are then necessary. Any assumptions by the designer with regard to assigned stiffness must be carried through similarly for all members. Errors are forgiving and erroneous assumptions usually result in overestimate or underestimate of sways but not of the forces resisted by the individual members. Members that can crack because of such items as construction overload or temperature change should be considered cracked (at least partially).

This writer recommends using clear spans to determine stiffnesses for both floor and support members of frames when using the effective moment of inertia I_e. Tension in columns should be avoided if at all possible, because it may create not only foundation anchoring problems but also considerable loss of stiffness. Ideally the designer should attempt to direct the lateral loads in a manner to avoid tension in columns (i.e., gravity loads are dominant). In shear walls, tension of a small magnitude can be allowed at lowest levels. It is sometimes better to reduce shear wall stiffness and increase frame stiffness to avoid large overturning moments in the shear walls. Floor members can be controlled by either reducing stiffness and/or reinforcement so that large tension forces (in excess of gravity loads) are prevented from entering the columns. Most projects require some such "manipulation" and the experienced designer can spot those very stiff floor members and adjust ahead of time, by anticipating and actually implanting provisions for plastic hinges in these members. Again, detailing for ductility is important especially for these cases.

22.5 PROVISION FOR ELASTIC AND TIME-DEPENDENT PHYSICAL CHANGES CAUSED BY CREEP, SHRINKAGE, AND TEMPERATURE CHANGE

22.5.1 Dimensional Changes

The concrete structure begins the process of aging with the first troweling motion. Changes in physical dimensions caused by hydration, creep, and shrinkage are time-dependent; these introduce additional complexities for concrete structures which are not encountered in homogeneous materials such as steel. All materials change dimensions under load. When this occurs instantly, as the load is applied, it is identified as elastic deformation which must be contended with in all structures regardless of the construction material used. In columns, elastic shortening can cause problems unless accounted for. It can manifest itself in installation problems of cladding. It can cause damage to cladding installed at lower levels while the construction of upper levels has not been completed. Avoiding structural distress in high-rise structures due to uneven elastic shortening can be handled generally with ease. A simple procedure is to attempt to equalize unit pressure due to dead (sustained) load so that the elastic shortening is fairly uniform in all supports. In major construction sites where cladding and other nonstructural finishes are placed prior to the completion of top floors, extra joint space must be allowed to discount the yet unrealized elastic shortening due to top floors yet to be constructed and due to finishes and cladding weights above not yet installed.

Thermal effects on cladding and exposed exterior columns can cause distress in both nonstructural and structural elements. When exterior columns are exposed and elongate or shorten because of temperature variance with respect to the interior supports, care in detailing must be made to account for movement (up or down) of the exterior column. Resisting this differential movement altogether can cause more problems. It will require very stiff floor members connecting exterior exposed columns to interior supports to minimize such displacements and very stiff members will develop large moments and shears. The use of very stiff floor members is usually not a practical solution; it is better to make provision for soft joints in cladding and partitions and to allow the displacement to occur without causing damage. In monolithic construction this displacement is resisted even by flexible floor members. Special design considerations for the floor members are discussed [22.31] and are not repeated here. It is recommended that ductility be provided for in the floor members. Different changes in length in adjacent supports will cause redistribution of forces in the structural members. Continuity in reinforcement is a must.

Generally a relative displacement of $\ell/300$ can be tolerated, where ℓ is the distance between a support exposed to the weather and the shielded support (which does not always have to be an interior support: some structures do not have all their exterior columns exposed). If computations indicate that the displacement is likely to exceed $\ell/400$ (the lower threshold of minor partition damage) soft joints to protect the partitions should be provided. If the displacement is to exceed double that value, it is recommended that the architect and owner be advised to shield all columns equally.

The maximum differential displacement obviously occurs at the roof level. The one exception is possibly a sudden foundation cave-in or a delayed uneven foundation settlement which can then cause similar differential settlement for a particular column at all supported levels. Uneven foundation settlement is another cause of concern which must draw attention of the designer to the need to provide ductility and redundancy in the structure. Some soft soils may require design of the foundations based on a uniform soil pressure caused by sustained (dead) loads acting alone. In this case the column supporting the largest live load (as a percent of total load) will establish a reduced soil capacity used to size the footings for all other columns.

The few preventive measures described above usually can be provided at minimal premium to the cost of construction. It must, however, be realized that additional attempts to prevent the physical changes that will occur, or to diminish them considerably, are not practical. The option is to recognize that they will occur and to provide allowance.

Often the differential displacements occur over a long period of time during which creep and shrinkage will diminish the induced stresses (caused by the displacements) considerably. The provisions for ductility will allow for the redistribution needed in the structural elements. Nonstructural elements fitted with soft joints can withstand the associated structural physical changes without the need to participate (adversely) in countering this change. Such soft joints should also consider elastic shortening or elongation caused by the structure swaying while countering temporary lateral wind (or seismic) forces.

22.5.2 Construction Procedures Reduce Problems from Dimensional Changes

Problems in design due to elastic physical changes described above must be recognized for all structures regardless of material used in construction. Concrete structures, however, require more planning due to the long-time creep and shrinkage which cause the floor members to continue to deform (deflect) with time, and cause the columns to also shorten with time, long after all sustained loads are in place.

Two separate design criteria must be considered for the time-dependent effects:

1. How much construction tolerance is necessary to allow easy installation of nonstructural elements?
2. How much tolerance is needed for anticipated future changes in the physical shape of the supporting structure?

Tolerances for installation depend to some degree on the ability of the contractor. This writer's experience indicates that tolerances given by ACI Standard 301 [22.32] are generally not practical because normal construction requires much larger tolerances. When vertical nonstructural elements such as masonry cladding, partitions, and piping are installed, as-built conditions can usually be accommodated. Only future physical changes need be considered. When the nonstructural elements are preassembled such as the case of precast concrete cladding, or light metal and glass cladding, then as-built conditions are of paramount importance. Construction inaccuracies must be considered especially when such cladding is wedged between floors rather than suspended off the spandrel edge.

Since a structure is constructed one floor at a time elastic column shortening at the lower levels is an on-going process. In high-rise structures, often, cladding commences prior to topping of the structure. The designer will attempt to analyze clearances required and indicate on the drawing a soft joint of a certain magnitude. During erection of the cladding some of the anticipated shortening has already taken place. Prefabricated cladding attached to the outside of the spandrel edge, for which a soft joint has been called for on the design drawings, may find itself overriding the top of the high-rise structure by several inches and cause a troublesome time with floor connections because the upper floors are "lower" than they should be. The prudent contractor closely monitors and surveys periodically the different levels, informs the trades and with the help of the design team, adjusts downward the "theoretical" joint sizes provided on the design drawing to discount the portion already consumed.

The "theoretical" joint size in the vicinity of a support should account for: (1) the unrealized elastic shortening; (2) "rocking" due to lateral loads; (3) temperature variances; and (4) future creep and shrinkage of the supports.

In between supports there are additional concerns. Net differences due to variation in deflections (elastic and time-dependent) of the floor members overriding each other must also be accounted for. The individual deflections of the overriding floor members must be estimated. Net maximum and minimum differences due to such things as unequal sizes, unequal loads, and construction procedures can require, for example, deeper window head receptors. As an alternate the cladding must be able to support a small amount of the load in order to equalize deflections where intentional separation is either not sufficient or has not been provided for. Obviously the shorter the spans and the deeper the members the lesser the overall problem. Even if the problem at midspan is solved one cannot overlook the long-term shortening of the columns at all locations in the structure.

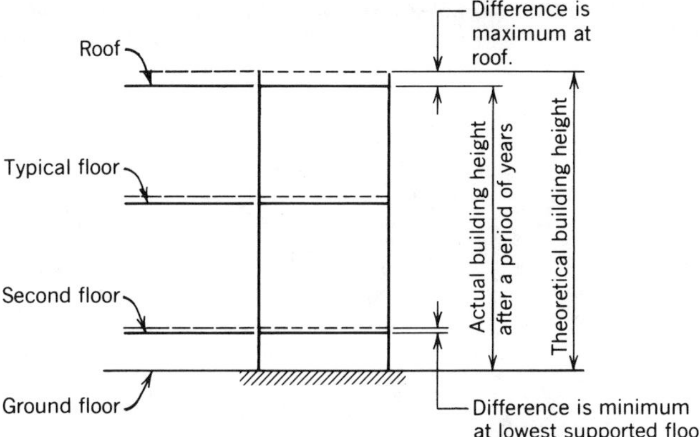

Fig. 22.10 Dimensional changes due to column elastic and time-dependent shortening. From Robert Rosenwasser Associates, P.C.

Figures 22.10, 22.11 and 22.12 indicate what happens to an aging concrete structure and how cladding, piping, and other nonstructural elements must be kept clear of the wedging action of the shrinking structure. Masonry cladding will spall and wedged pipes will burst if proper attention and review is not done by all trades supplying and equipping high-rise concrete structures.

The specific detailing requirements for nonstructural elements are generally easily obtained. The details are already used by mechanical contractors and by curtain wall manufacturers to allow for thermal expansion. The additional criteria for reduction in joint size due to shortening of the columns can therefore be easily accommodated. The problem for manufacturers of nonstructural elements is that they may provide for such joint details but later inhibit the movement by anchoring, say, one long pipe to several structure supports which are many floors apart.

Unless a pipe is slender enough to bend and buckle so that the distance between supports can shorten then the pipe may burst. Rigid piping should be supported on one level only and be allowed to freely move with the supporting floor. The pipe must of course be also guided laterally throughout its length.

When nonstructural elements engage several floor levels without provisions to allow for future shortening of the distance between the supports, such members must be reviewed for load-carrying

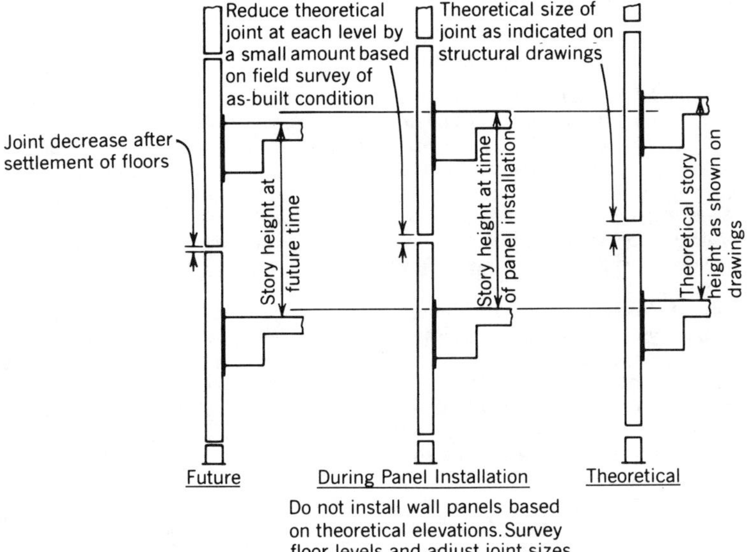

Fig. 22.11 Dimensional change effects on panel installation. From Robert Rosenwasser Associates, P.C.

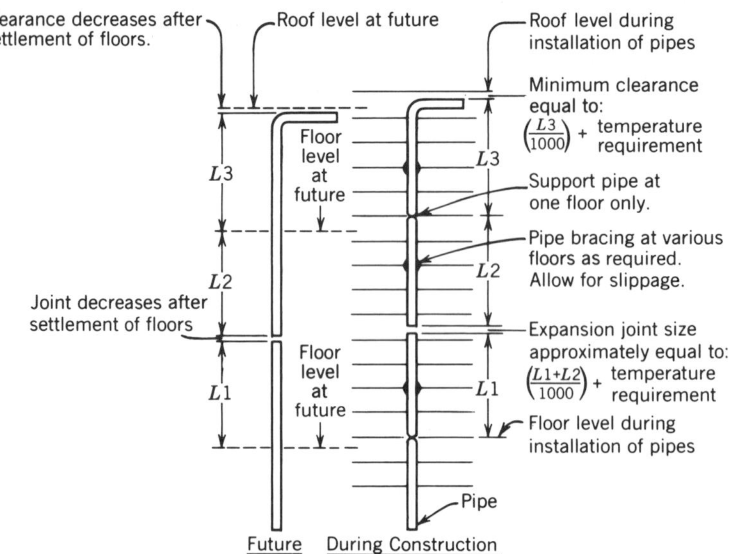

Fig. 22.12 Dimensional change effects on pipe installation. From Robert Rosenwasser Associates, P.C.

capacities because they will eventually participate as structural members. Until the transferred loads exceed the nonstructural members load-bearing capacities no problem is realized. It is possible, however, to cause damage to the structural members supporting overriding stiff load-bearing nonstructural elements. For example, boiler room flues have (in the not-too-distant past) been constructed with fire brick lining attached to masonry construction forming the four sides of the flue. It was customary to allow such flues to be free-standing to the foundations as high as they could freely stand. At a higher level, however, the structural members were reframed to support the walls of the flue. Distress is observed in many such structural members as they and the supported portion of the flue attempt to relocate downward along with the lower portion of the structure, while being resisted by the upward thrust of the free-standing portion of the flue which is independent of the structure and does not shorten similarly.

22.5.3 Cladding-Related Considerations

Good quality cladding (such as masonry, brick, stone, solid exterior walls) have been known to participate in support of both gravity and lateral loads in many structures. Often they become the main structural elements. In the last several decades there has been a shift toward cavity wall construction. In this type of construction, the outer brick wythe is separated by a couple of inches of air space from the inner wythe of masonry block. The inner wythe is usually wedged between the concrete floors and the outer wythe is supported on steel angle lintels. Such masonry construction has shown signs of distress when used as part of buildings that are essentially concrete buildings. Details on masonry walls and cladding are given in Chapter 24, and the general problem of facades is covered in Chapter 27.

The causes of cladding distress are numerous. The primary causes are (1) poor quality control; (2) use of glazed brick having moisture absorption tendencies that cause it to expand when wet; (3) missing ties across the cavity; and (4) missing inserts and even missing lintels. The main culprit, however, seems to be lack of adequate vertical separation between the cladding and the structure which shortens with time. Even where soft joints were specified they were often not constructed in a proper manner, such as being grouted at the outer edge to "hide" the soft material behind, which then results in compressive action of the structure concentrated at the extreme edge of the outer wythe making the problem worse.

Figure 22.13 is an office detail for a brick cladding soft joint intended to prevent accidental grouting of the outer edge of the soft joint. Such grouting worsens the problem because then all the compressive force is concentrated at the exterior edge. Such foolproof details are important since quality control greatly deteriorates once the concrete frame has been topped and the structural engineer no longer inspects the project.

Corrective measures to reopen closed soft joints in older structures to prevent imminent problems may require more than one such corrective application. Upon reopening of such soft joints, loads which were formally supported by the wedged masonry cladding are then transferred to the concrete columns causing additional elastic and time-dependent shortening (creep shortening will commence only with the application of the load) which often necessitates another such adjustment to the joints.

Horizontal joints must be also accompanied by vertical joints to allow for horizontal temperature expansion and to disengage as much as possible the cladding from participating in carrying lateral load. Such vertical joints should be located adjacent to corners and should not exceed 30 to 40 ft in spacing.

Another more recent development in the construction of cavity walls is the "eyebrow" extension of the concrete edge to provide better (than a loose lintel) support to the outer wythe. Since the inner wythe of cavity wall construction seldom indicates signs of distress it is reasoned that providing similar supports also in the outer wythe may solve the problem. Since separation between each floor level confines the displacement to the function of a single-story height only and since brick and block have moderate compression strengths the cavity wall can behave in a better manner and participate more as a load-bearing member. Such construction performs better but has not been in service sufficiently long to categorically state that it is the optimum situation. There is little doubt, however, that an "eyebrow" support with an ample size soft joint at the underside of the slab above will allow the cladding to behave well with the passage of time. Concrete well-vibrated into the exposed edge is needed to avoid honeycomb that may eventually cause spalling. The reinforcement should be placed not less than $1\frac{1}{2}$ in. clear of the exposed edge. The top surface should be sloped slightly to prevent accumulation of moisture, as shown in Fig. 22.14.

22.5.4 Size of the Horizontal Joint

The size of the required joint between supported and supporting elements depends on the sensitivity and the load-bearing capacity of the supported element. When the stresses induced (into both supporting and supported members) can be tolerated no joints are necessary. Where the supported elements (or the supporting elements) cannot withstand the induced stresses, then a joint size sufficient to reduce the induced stresses to tolerable levels is necessary.

The problem is so complex that it is seldom possible to determine "exactly" the size of the horizontal joint, because many of the parameters are not within the control of the designer. It is

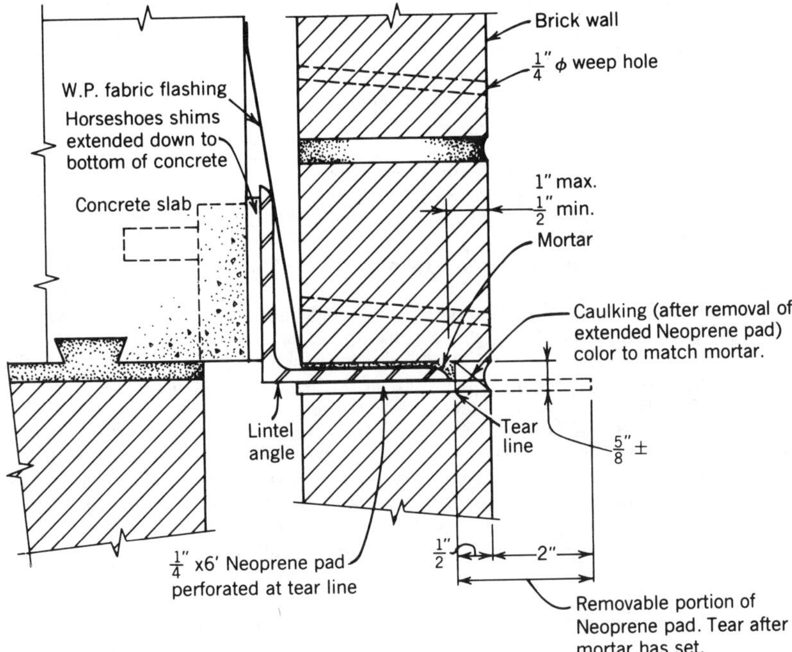

Fig. 22.13 Exterior masonry wall compression detail. From Robert Rosenwasser Associates, P.C.

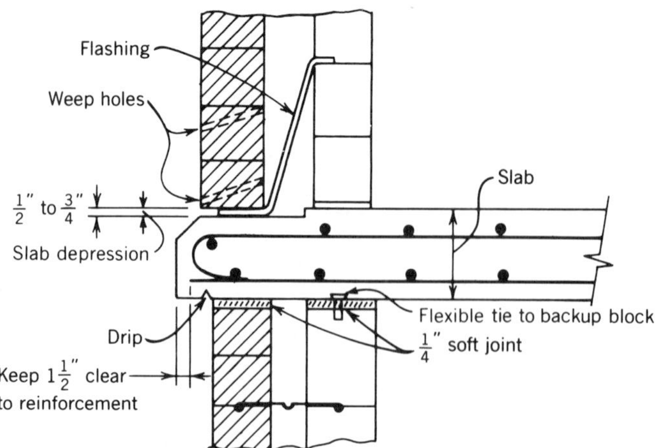

Fig. 22.14 Concrete "eyebrow" detail.

therefore important to rely on past experience (which also requires almost a lifelong observation period). Once a cladding is selected for a structure and the cladding requires a soft joint then in the consideration for a size of the joint, "time" is probably the most important parameter. When will the supported elements be installed? How much of the anticipated shortening and deflection (elastic and time-dependent) is already discounted?

The method of construction, including proper initial separation, use of time, and having an undamaged (in construction) supporting member, is the next important consideration. A proper analysis must consider all parameters causing elastic and time-dependent physical changes in the structure. Any guidelines if provided may apply only to the practices of the office from which this information is obtained. Long-term creep and shrinkage of columns depends on many parameters (such as stress level, materials, and shape). Different practices provide for different parameters and each design firm must establish its own criteria. The story shortening due to long-term creep and shrinkage can vary only slightly, however. The accumulation of small increments over many floor levels can result in a measured value of several inches. For example, a 40-story building with each incremental floor height shortened by only 1⁄16 in. will exhibit over 3 in. of total shortening at the roof level, in addition to elastic shortening of a similar magnitude. If shortening of the forms under the weight of wet concrete is also included it is easy to see that the owner is "buying" less than the plan dimensions indicate.

Common construction practices often allow adjustments during the construction period to accommodate this loss in height by making periodic level readings from a fixed base. Some contractors do not follow this procedure and therefore lose height due to the construction process (this is an important item when cladding is prefabricated). The design firm should require field measurements to establish actual dimensions for prefabricated cladding prior to the fabrication of such cladding.

Figure 22.13 indicates a guideline size of soft joint selected by a design firm having a common practice to utilize oversized columns (i.e., low stress and low reinforcement ratio). This soft joint size is for the solid masonry portion of the wall. Receptors over windows require larger clearance over the glazing (size is a function of the width of the window with the masonry on each side providing low load-bearing capacity when necessary).

Since brick cladding is constructed at the site there is a need to anticipate only the future shortening. Precast cladding will, however, require a "theoretical" joint size larger than the one used during attachment of the cladding (at which time a part of the shortening already has occurred). Generally ½- to ¾-in. "net theoretical" joint size (discounting the non-compressible portion of the sealant) is ample for most high-rise structures. Presurvey of constructed floor levels will indicate the reduced gap at installation. The contractor may also build in a small increment of extra height to compensate for a portion of the anticipated loss.

Wedged prefabricated cladding must have receptors large enough to allow for field errors as well. Field measurements of as-built conditions are required prior to fabrications, otherwise the construction process will be hindered by many delays caused by prefabricated cladding (in accordance to design drawings) not properly fitting.

22.6 SELECTING A FRAMING SYSTEM AND CONSTRUCTION MATERIALS

22.6.1 Economic Considerations

The architectural scheme developed to satisfy the owner requirements generally dictates a framing system which interferes the least with the architect's needs and which can economically provide the structural skeleton required for safety and serviceability within a budget dictated by the market place. The structural system, usually hidden by cladding, ceiling, and flooring, can sometimes replace the finishes at a penalty (requiring better exposed surfaces to suit). This added premium can be weighed against the saving realized by the replaced finishes.

Structures in which the mechanical system does not require false ceilings for distribution (such as apartment buildings) need a structural system where unsightly construction (such as beam drops at conspicuous locations within living quarters) does not have to be hidden by a false ceiling. A false ceiling adds height to the structure, which also means such things as extra wind loads to be resisted, and higher partitions.

Structures, such as office structures, where extensive electrical distribution is required will benefit by having floor distribution cells integrally built. Such cell closures can sometimes also provide the structural strength required to span the floor beams. Frequently, a required building function will easily dictate the best-suited framing system.

Availability of materials will sometimes offset an otherwise clear choice in favor of an alternate framing system using more readily available materials. Availability of qualified contractors can dictate a framing alternate which is only second best but can be built by local contractors. Future flexibility for alterations and additions can also help the selection of a suitable framing system. Familiarity of the design team with one system or another predetermines the type of structure they usually select and which they can easily and economically design. The owner preference based on past experience can also dictate selection of a framing system. Most framing systems can be built with a variety of materials. Present-day cost and availability, often fluctuating, must be considered prior to selecting a system. Sometimes, as pending changes are imminent (such as a possible mill strike), changing to another construction material or another system is required during the design stage.

22.6.2 Time

As discussed previously, consideration of time is an important parameter in design decisions. What material is best suited and what framing system will expedite the construction? Two- and three-day construction cycles can save construction time, reduce construction loan cost, and may put the structure in use at a more appropriate time. A selection of a proper framing system must consider availability of materials that must be preordered and possible advantages of time-saving procedures even at an added cost of materials. A project that is "fast-tracked" must use materials readily available without the need to preorder several months in advance.

Formwork for the placed concrete generally constitutes between 40 and 60% of the total labor and material cost of the structural system. Simplifying the formwork influences favorably the construction cycle. Flat slabs and especially flat plates are recognized to be the most cost-efficient floor systems and should be used whenever possible.

22.6.3 Materials

Repetitive use of forms at each floor level depends on proper care and handling of the forms by the construction crew. Between 10 and 12 uses are optimum. To use a two- or three-day cycle requires devising a means to reuse the forms as quickly as possible. A floor having 8000–10,000 sq ft will require 2 sets of floor forms for a three-day cycle, and 2¼ sets for a two-day cycle. A method of "preshoring" prior to reshoring which will allow the fewest sets of forms and yet maintain a short cycle of construction is explained in Section 22.7. Only well-qualified and trained contractors utilizing early (24-hr) cylinder break with better concrete (4000 psi minimum) should attempt to use this system. The difficulty is the early partial loading of "green" concrete that may cause added cracking and larger initial deflections. The advantage is that the partial self-loading of "green" concrete floor during the process of reshoring eliminates a large build-up of construction loads on the floor below (having concrete almost as "green"). Cold weather is a cause for concern and quality control of the preshoring stripping and reshoring operations is necessary.

To reuse forms as many times as possible and to avoid delays required by patching and adjusting, the designer should consider all form "drops" and maintain where possible such drops constant even when the floor thicknesses vary with service load. Column sizes also should be kept uniform for about 10 lifts or more in high-rise construction to minimize patching of forms needed each time a column is diminished in size.

Economical selection of bar sizes should permit large orders of fewer bar sizes. Considerations of development length and end anchorage of the bars (in addition to crack control requirements) will necessitate a detailed analysis, with further reduction in number of different bar sizes when a final review is made. It is also important to maintain similarity in reinforcement layout on the many consecutive typical floors to allow the construction workers to gain familiarity and avoid errors. The inspector's job is also made easier and more efficient.

Since gravity loads may be constant for many floor levels but lateral load requirements can vary from floor to floor, uniformity of bar reinforcement can be maintained for the gravity loads and reinforcement added for lateral loads scheduled to match the needs of the different floors. Do not revise reinforcement at each floor. Group several floors with common lateral load requirements. Congestion of bar reinforcement should be avoided at all costs. To provide ductility in floor members and allow the simplification required in the design of large projects the reinforcement ratio ρ should be low. Similarly, an economical column is one that requires no more than 2 or 3% reinforcement at the lowest floor level. Easy placement of noncongested bars with reinforcement beam cages of stirrups or ties saves time. The requirements of the 1983 *ACI Code* [22.1] for design of joints in seismic zones will help alleviate some of the complicated detailing and placing problems of the past. The designer should be aware that economy will be realized where the concrete (and not the reinforcement) is relied upon to do most of the work whenever compression forces are present.

22.6.4 Types of Framing Systems

Subassemblies (the "building blocks") of any framing system, such as slabs, beams, shear walls, diaphragms, "joints", diagonals (compression or tension struts), and columns are well known to designers. Each of these subassemblies can be designed to resist shear (inclined tension), bending moment, axial tension, and axial compression. The shape of each subassembly and the axis of force-resisting action determine how suitable and efficient each is within the framing system. The span-to-depth ratio of each member (with the depth parallel to the acting force direction) will dictate the mode of action of the member in resisting the loads applied. For example, a column may be designed as a cantilevered member (i.e., acting as a shear wall deflecting in a flexure mode) or as a column in a frame where the deflection is predominately a shear mode. Whether or not the column is actually a cantilevered shear wall depends on the span (in this case total height above the "floor of no translation") to its depth ratio (as discussed in Section 22.4).

Any combination of several subassemblies will produce a framing system. The function of the framing system is to carry all loads to the foundation. The efficiency of the framing system under any special architectural configuration determines the economy. Needless to say that the most important function of the framing system is to transfer all loads (gravity and lateral) to the foundation. Under a major catastrophic earthquake this function alone is sufficient. However, when minor climatic events occur the framing system must also be judged by its efficiency to protect the nonstructural elements from major damage or to provide habitable shelter for its inhabitants. That is to say the structure should be serviceable too. Excessive deflection should be prevented and the comfort of the occupants generally ensured against perception of motion.

The major categories of concrete framing systems may be described as:

1. Moment-resisting framing system.
2. Bearing wall framing system.
3. Dual (shear wall and moment-resisting) framing system.
4. Outrigger system.
5. Tube system.
6. Diagonally braced system.

Each framing system must be designed to support the prescribed gravity and lateral loads. The second-order effects of the sway (P–Δ effects), plus the internally induced forces due to creep and shrinkage, temperature variation, uneven foundation settlement, and shortening and lengthening of columns due to lateral load action, must also be investigated. The shorter the structure the smaller the effect of some of the influencing parameters. Building more ductility into the structure means less accuracy is required in the analysis. Where redundancy and ductility are available the structure is safer. The structure must also provide the serviceability requirements of its occupants. A thorough review of subassemblies and framing systems is now available in the literature [22.33]. A brief summary and a few supplementary comments (based on this writer's experience) follow in the next subsections. This topic is also treated in many other chapters of this book (such as 8, 10, 11 and 23).

Moment-Resisting Framing System

Concrete framing systems in which subassemblies deform essentially in a shear mode resist lateral loads in the familiar fashion attributed to structural steel frames. In concrete structures the floor members generally contribute a large portion of the total drift.

The P–Δ effects are more dominant in moment-resisting framing systems. When the floor system is made essentially of slabs alone the familiar "flat slab" structures of the East Coast are recognized. Decisions with regard to effective width of slab contributing to lateral stiffness have to be made by the designer (see Section 22.4). Slabs without drops and without spandrel beams are categorized as "flat plates". They too can contribute considerably to resisting lateral load (depending on a variety of parameters, such as concentration of required reinforcement over supports, size of supports, slab thickness, and spans). Even though slabs individually are weak (compared to beams) collectively when in large numbers they can be moderately strong. Economy is realized because each floor member is harnessed to do its share. When the slab system is called upon to resist a large portion of the lateral load, thicker slabs are generally required.

The addition of spandrel beams can enhance total lateral and torsional stiffness and especially accommodate transfer of both gravity and lateral unbalanced moments at the exterior columns. Even though some high-rise flat slab structures (over 40 stories) have been constructed in the past (in mild weather areas) without shear walls and generally behave well, it is preferable, based on present state-of-the-art, to include shear walls (even if limited to elevator and stair shafts) in order to provide added localized stiffness. This is especially needed at lower levels where atriums and other open spaces require extra story height with diminished floor area to participate in the lateral load resistance.

The weak link of the flat slab system is its punching shear strength. Unbalanced moments (due to gravity and lateral loads) according to ACI-11.12 [22.1] cause increased shears. Shear reinforcement is not common in flat slabs. Sometimes, however, it must be used, especially at exterior columns when there are no spandrel beams. Thin slabs are not conducive to development of shear reinforcement. Large columns, providing excess direct shear capacity are required (in high-rise structures the larger column size is common anyhow). Drop panels and capitals can be used to increase shear strength as well as stiffness of the slabs. A good "rule of thumb" is to design preliminarily for the direct shear not to exceed 50% of the strength so as to allow for the additional requirement of unbalanced moments for transfer. Added shear strength increases ductility of a slab. A more rational analysis ("beam-analogy") [22.26] indicates that greater unbalanced moment can be transferred than indicated by the *ACI Code*.

For long spans the two-way grid (waffle) slab allows for uniform flat forms with extra depth (stiffness) and reduced quantity of material. Added costs associated with the extra steps required in the stripping and forming operations can eliminate most or all material savings except when spans are extremely long.

Bearing Wall Structures

Gravity load bearing walls, generally traversing the long axis of the structure (except at corridors) supporting a one-way floor system (often precast) can provide lateral load resistance limited only by the overturning moment capacity of each wall. Such bearing wall systems are widely used where architectural layout is conducive. The walls can be cast in place, precast, or of masonry construction. Economy accrues when the structural walls provide also the required compartmentization within the structure. Ductility is often added by untensioned strands to ensure against progressive collapse placed in the floor–wall joints. (These strands are placed parallel to the wall to account for one wall panel being damaged or removed by blast and are placed perpendicular to the wall and into precast slab joints to provide continuity over a damaged support [22.34].)

Dual Systems

The most common (by far) framing systems are dual systems which utilize various subassemblies in different ways depending on the location of such subassemblies within the three-dimensional structure.

"Frame–shear-wall" systems utilizing flat slabs and shear walls can be used to great heights providing both economy and architectural freedom. They become, however, less efficient when the aspect ratio (height-to-least-width) is over 5 or 6. In this type of system a separate analysis is required to assign lateral loads to each of the components of the system. The complexity involves matching the lateral drift of shear walls which deflect as cantilevers (flexure mode) with the drift of the columns and floor members which deform in frame action as a predominantly shear mode (see Fig. 22.7). The floor slab is generally assumed to be infinitely stiff so that the shear walls and frames must deflect the same at all levels. The shear walls preferably should be so situated that a large tributary floor area is supported by them and that they are a maximum distance away from other columns. Since creep and shrinkage effects can seldom be made compatible between the large shear walls and the smaller columns it is best to allow for some vertical displacement by providing ductility in the connecting floor members. Less stiffness in the floor members will cause less distress. The shape of the shear walls usually conforms to architectural features. For example, a channel-shape may be used where the shear wall forms the elevator space. The shape can meander to follow partitions with no evidence of great loss in efficiency since the wall is braced at each level by a stiff diaphragm. Occasional openings through the walls can also be permitted unless they are repetitive in which case the strength of the connecting links to participate in the anticipated wall behavior must be investigated. A reduction in efficiency of the total wall to that of a series of smaller individual walls will occur when small connecting links span large openings.

Partially-coupled shear walls are very common. It is often not possible (because of architectural constrictions) to provide link beams sufficiently strong to couple completely the parts of the shear wall. Coupled walls can be made only partially-coupled intentionally to avoid attracting large seismic shear forces into the wall [22.35]. Partially-coupled walls with ductile tie beams allow the designer full flexibility. A predetermined coupled action caused by a prearranged portion of the lateral load (for which the link beams are elastically designed) can be established. When larger moments are attracted to the connecting beams plastic hinges will form and the larger moments will be unable to develop. The leftover lateral load can then be assigned to the dual system with the shear walls assumed uncoupled (i.e., with the use of the smaller individual fragments of the total wall acting individually). The overturning moments and shears resulting from this "frame–shear-wall" interaction analysis can then be combined with moments and shears in the fully-coupled wall caused by the preassigned portion of the lateral load. Care must be exercised that results by such separate analyses will not exceed capacity of the coupled wall (assumed fully-coupled) to draw and retain lateral loads. Separate "frame–shear-wall" analyses are usually required, one assuming the coupled wall is fully coupled and the second using only the individual parts of the uncoupled wall. The link beams can then be reinforced to produce in the coupled wall predetermined overturning moments that fall between the upper and lower values obtained by the separate analyses. Analysis and design of shear walls is treated in more detail in Chapter 7.

Most coupled shear walls, as stated before, actually perform only as partially-coupled walls because of architectural constriction to the size of the connecting links. Where the connecting link depth is about equal to its span, a more efficient means to couple will be provided by diagonally placing the reinforcement in the coupling beam with only a small amount of added top and bottom horizontal and vertical shear reinforcement to allow formation of a plastic hinge if necessary [22.36, 22.37].

In high-rise structures the shear walls are not stubby and therefore should be designed for flexure with, however, excess shear capacity to allow formation of plastic hinge under unforeseen extreme loading. The shear wall capacity in high-rise structures will generally be limited to its overturning capacity based on tributary gravity load available. Rock ties are sometimes used to increase the overturning capacity but are not too reliable since cracks in the bedrock cannot easily be determined and the strength of rock anchors will be significantly lower if the rock contains seams unfavorably oriented so as to diminish the anchorage to the weight of the detached rock. As the wall rotates about its compression flanges, added gravity loads will be diverted to the tension flange to greatly increase the wall capacity to resist lateral loads. It is therefore a good practice to provide excess shear capacity in all shear walls.

In high-rise construction, large shear walls can hinder architectural freedom. Even when the shear walls are large it is seldom that they alone can provide the necessary support for lateral loads throughout the height of the structure. A normal case will indicate a shared assignment where the shear walls support most of the lateral load at the lower one-third of the structure height, the other frame elements support the lateral load (and pull back on the shear wall) at the top third, and both frames and walls contribute to the support of the middle third. The relative stiffness of frames and walls will distort this simple division of load-carrying behavior one way or another. (Stiffer walls will participate to a higher level in lateral load-carrying performance.)

Very stiff walls can produce large distortions at the top of the structure unless frame elements can reduce (at the upper levels) the effects of rotation of the stiff wall at or close to the base of the structure. Stiff walls may also contribute to "jerk" effects (the rate of change of acceleration is known as "jerk") which increase perception of motion.

Outrigger Systems

Outrigger beams or trusses at the top level and/or intermediate other levels (such as at one-third or at midheight of the structure) can be used to enhance the performance of interior shear walls by engaging exterior column loads to counter part of the overturning moment in the shear walls. Stiff spandrels at the outrigger beam levels can help engage more of the exterior columns. Total drift and shear wall bending moments are therefore reduced. Differential shortening between shear walls and the columns must be carefully reviewed. Loss of efficiency of the outrigger beams due to creep and shrinkage will reduce efficiency of this dual system. The most efficient location for outrigger beams is at the one-third level of the structure height.

Tube Systems

Tube systems are frame systems where the vertical elements are placed closely together at the outer perimeter of the structure. The short distance between exterior columns allows for an increase in stiffness of the floor members to produce an outer shell penetrated by many openings but otherwise attempting to act as shear walls in a cantilever (flexure) mode. The "shear-lag" (a measure of the inefficiency of the tube) is a function of the stiffness of the floor members. Even systems having large shear-lags have been shown to be structurally viable in high-rise structures because the lateral load resistance is matched against the gravity loads at the extreme perimeter of the structure and therefore provides the largest leverage possible. Gravity loads are therefore used efficiently to resist the lateral loads.

Diagonally Braced Systems

The trend toward high-rise office construction in concrete has received a big boost by the recent adaptation [22.38, 22.39] to concrete structures of a diagonally braced tube system (see Figs. 22.15, 22.16). This system is a hybrid utilizing truss, tube, and frame action acting together to create an efficient structure. This system appears to remove any practical upper height limit to design in concrete.

The effect of diagonal bracing can be created in a concrete structure by infilling the window openings at alternate locations on consecutive levels. Therefore special sloping forms are not required. "Gang" metal forms having special hinges can be used to advantage during erection and stripping. Two- or three-day cycles are also possible for this type of construction.

The "diagonals" are capable of transferring and distributing both lateral and gravity loads. The shear-lag associated with tube structures can therefore be essentially eliminated. Gravity loads are also being uniformly distributed to equalize creep and shrinkage in all columns engaged by the diagonals. This framing system therefore resists both gravity and lateral loads in a most efficient manner.

The infilled panels should be lightly reinforced orthogonally to prevent formation of tension cracks perpendicular to the diagonal forces. Because of the ability to engage gravity loads also, such diagonal forces will generally be compression forces (minimizing the need for large tension splices). Diagonally placed reinforcement may cause some placing problems but is necessary to protect the "joints" where large forces congregate.

The "diagonals" on different faces of the structure should engage each other at the corners. When the face of the structure is too narrow to accommodate two diagonals crossing each other it is necessary to start and terminate the diagonals on the opposite broad faces at different locations to create the continuity around the corners. (The diagonals on the back face of the structure shown in Fig. 22.16 start at the roof level midway between the two corners.)

22.7 DESIGNING THE STRUCTURE

22.7.1 Preliminary Design Considerations

The most important design decision dictating the structure's economy and viability is made by the design team in the initial review during which the structural framing is selected.

Sometimes, a suitable framing system can be selected only after a lengthy review process during which several systems are partially designed, reviewed for strength and serviceability, and then budgeted. This is frequently necessary in order to assure the owner that the system recommended by the design team is indeed the one best suited for the particular project. Experienced design teams, using the lessons learned from the past, can often select the most suitable framing system with minimal deliberation. Once the system is selected, a review of the particular system with alternate placement of components is still necessary in order to determine the most efficient and economical layout.

Many projects offer multiple occupant use such as an apartment tower, which overrides commercial spaces and includes parking at the subgrade levels. The location of columns becomes an important consideration since each of the spaces has its own optimum column layout. Intercepting columns via transfer girders is expensive and the cost must be weighed against the loss of optimum space incurred when columns located to suit one occupancy are less than ideal for other occupancies at other floor levels.

The size and height of the structure (aspect ratio), as well as magnitude of lateral and gravity loads, are parameters that determine the selection of the material and the framing system to be

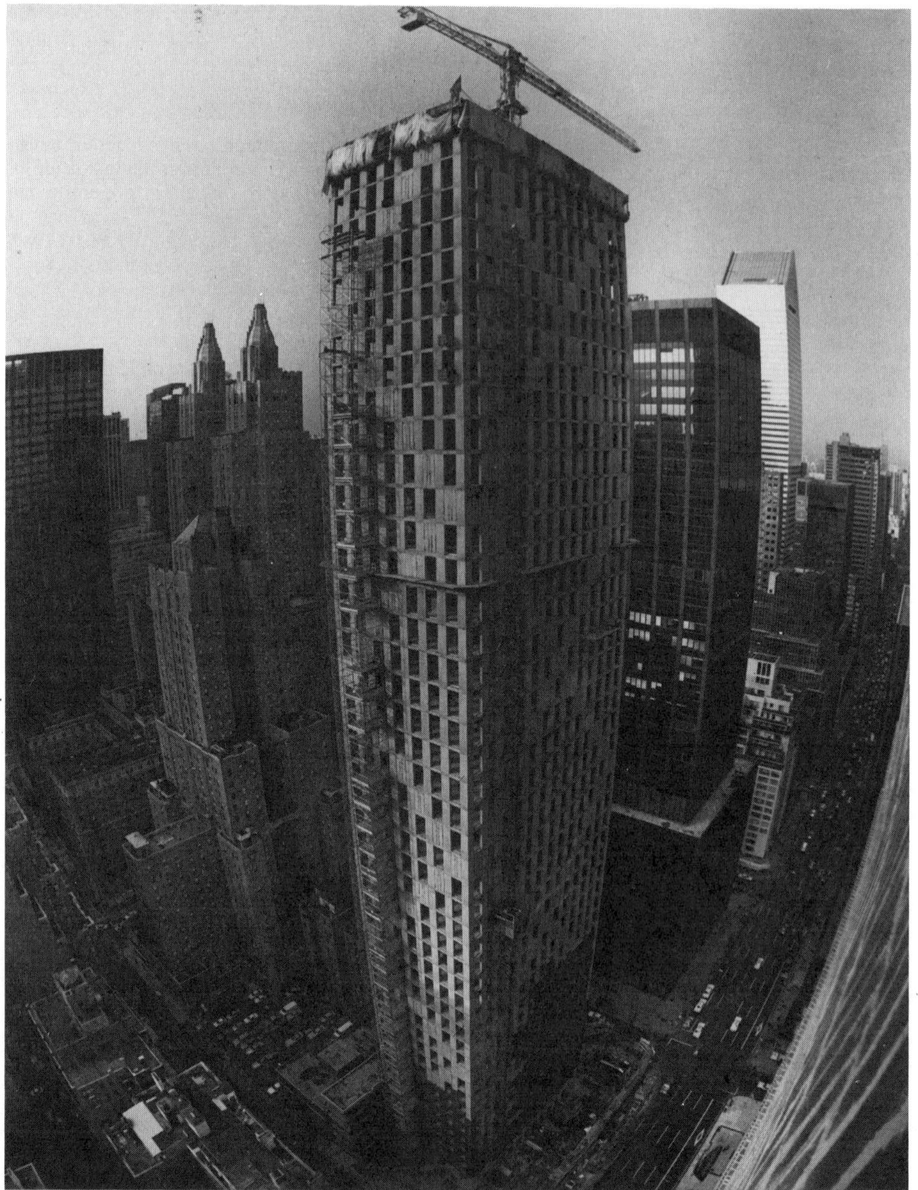

Fig. 22.15 Diagonally braced tube system, 780 3rd Avenue, New York, South and East Walls. Photo courtesy of Robert Rosenwasser Associates, P.C.; photo by Gaston-Doubois, Elmhurst, NY.

used. While it may not be possible to entirely document by rules the method of selecting the materials and framing system, the following guidelines are suggested:

1. The mixing of structural materials, such as structural steel and concrete, should be avoided. If concrete is selected as the major structural material there is usually no need to use structural steel to do part of the job. High strength concrete, now readily available, along with superplasticizers which expedite placement, can provide the designer with the required means to use a single material for the project. The mixing of construction trades and responsibilities is thereby avoided.

Fig. 22.16 Diagonally braced tube system, 780 3rd Avenue, New York, West Wall. Photo courtesy of Robert Rosenwasser Associates, P.C.; photo by Gaston-Doubois, Elmhurst, NY.

2. In multipurpose projects, where different occupant spaces override one another in controlling the framing scheme, it is best to first locate structural supports to suit each of the spaces individually. Once this is done, review the various conflicts and assuage the situation for one by reducing the efficiency of the other. Do not allow the "tail to wag the dog" (i.e., the needs of the dominant space must prevail). If necessary reposition the columns to create better conditions.

3. Encourage discussion with the architect and owner concerning the loss of some space efficiency as a tradeoff for economy. It is during this initial analysis that integrated computer programs, such as "TOWER" [22.40], capable of design and material estimates can help decide the best layout and the cost implication of each of the schemes. The owner can then help make the final selection of the structural system.

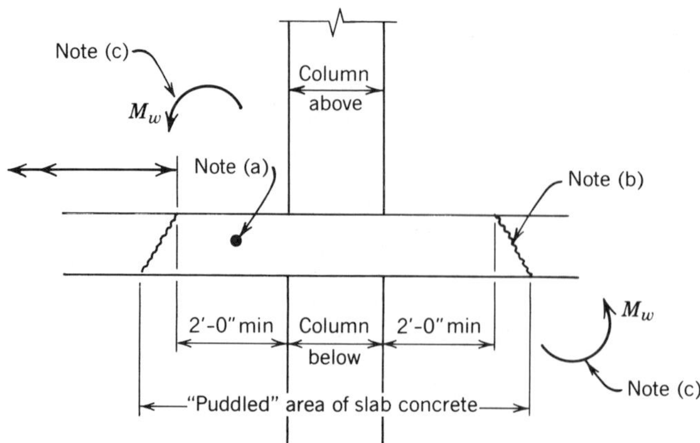

Fig. 22.17 Plurality of concrete mixes in floor construction.
Notes:

(a) The concrete strength in the column above the floor must not exceed by more than 40% the strength of the "puddled" slab concrete.

(b) Cold joint must be avoided. Additional supervision and use of retarders may be required.

(c) When large wind moments M_w cause upward reactions opposite to and larger than gravity load reactions the "puddled" concrete cannot be relied on to transfer the shear to the support.

4. Economical designs in concrete rely on steel reinforcement only to fulfill tension requirements, then allow the concrete to do the work when compression forces are dominant. Sizing the floor members for aesthetic integrity (see Section 22.3.4) often provides this relationship. Flexural members that are designed to require compression reinforcement for strength are generally not properly proportioned. Reinforcement in the compression zone should be added only to enhance ductility and redundancy, as well as to reduce the long-time creep and shrinkage deflection.

The most economical column is the largest that can be tolerated by both the owner and the architect, which can support the required load with the minimum required steel. Oversized columns can be treated as "architectural" columns to reduce the required minimum reinforcement from 1 to ½% of the gross section area. To save in costs of construction and design, slender columns should be avoided. Concrete strength at lower levels of a high-rise structure should be increased to keep the reinforcement to no more than 2 to 3%. Gradual reduction in reinforcement and concrete strength alternating with column size reduction at 10- to 15-floor increments can provide economical columns. At levels where strength or size variance occurs, the column reinforcement should be kept at about 1% of the gross area.

Plurality of concrete design mixes in the floor construction (such as "puddling" for a 2-ft distance around the column with greater concrete strength) should be avoided (see Fig. 22.17). It is more economical and safer to increase the strength of concrete used in the floor member. For example, 7000-psi concrete in columns can be matched with 5000-psi concrete in the floor members so that construction can be carried out with little fear of having cold joints develop too close to the supports. In structures where reversal of moments and shears at supports can occur due to large lateral forces "puddling" should be strictly avoided because shear transfer (in the direction opposite to gravity) is inhibited by the sloping sides of the puddled area. When conditions do require a plurality of design mixes, extra controlled supervision is needed and often dictated by local laws.

5. An orthogonal grid system controlling column layout can provide for economy in both design and construction. Often, however, a rigid alignment curtails the freedom of architectural spaces (such as in many apartment structures). Variable column locations can provide flexibility and gain space and do not necessitate substantially increased costs of construction. The cost of design will be increased, but the prudent engineer will find it a meager argument to use. Uniformity in floor member sizes can still be maintained even with unequal spacing of supports. Steel reinforcement can and should be altered to match the needs of the various span conditions, as well as the varied requirements of the different floor levels.

6. Floor members should be sized to satisfy the "aesthetic integrity" of the member (see Section 22.3.4). In many cases, this type of approach will assure lower reinforcement ratios because of the somewhat increased sizes. If the reinforcement ratio over the supports is kept low, such as $(\rho - \rho') \leq 0.375\rho_b$ (see Section 22.7.2), it is possible to take advantage of simplified design approaches which may allow an even larger distribution of moments than that permitted by ACI-8.4.1 [22.1].

22.7.2 Minimum Depth to Maintain $(\rho - \rho') \leq 0.375\rho_b$

The *ACI Code* [22.1] presently allows up to a $20[1 - (\rho - \rho')/\rho_b]\%$ redistribution of the negative moments calculated by elastic analysis of continuous nonprestressed flexural members. This redistribution is allowed only when $(\rho - \rho') \leq 0.50\rho_b$ (see ACI-8.4.3).

Furlong [22.17] has shown that for braced structures, even larger redistributions are possible when ductility is detailed into the member to limit $(\rho - \rho') \leq 0.375\rho_b$ at supports. Section 22.4.2 extends this concept to "controlled" unbraced structures. To be considered "controlled", sway must be limited so that $P\Delta/Hh_s \leq 0.15$. Within this limit unbraced frames behave similarly to braced frames and the gravity load moments are not affected by the sway.

Table 22.10 for braced structures and Table 22.11 for "controlled" unbraced structures provide simplicity in design for gravity loads without the waste associated with those simplified methods of analysis presently indicated by ACI-8.3.3. The need to use questionable elastic theories in the design of nonhomogeneous materials (specifically concrete) can therefore be avoided in many design situations.

To limit $(\rho - \rho')$ to less than $0.375\rho_b$, the designer may select either to increase the size of the member or to extend more of the compression zone reinforcement through the support to be developed as compression reinforcement so that the net $(\rho - \rho')$ is reduced.

Table 22.12 indicates the limits of $(\rho - \rho')$ which assure that sufficient ductility is provided so that moment redistribution capacity is available.

Since the stress–strain relationship from which the values in Table 22.12 were derived can vary appreciably, it is sufficiently accurate and conservative to use the following expression for all materials to assure that $(\rho - \rho') \leq 0.375\rho_b$. Estimating $(1.4D + 1.7L)/(D + L) \approx 1.55$ gives

$$d \geq 7\sqrt{\frac{M'_a}{b_w}} \tag{22.7.1}^*$$

where M'_a is the net *service* (unfactored) moment (ft-kips) at the support which is resisted by compression in the concrete. The terms d and b_w are the effective beam depth and the beam web width (both in inches), respectively.

EXAMPLE 22.7.1

Determine the minimum depth of a 10-in. wide beam (see Fig. 22.18) constructed at an upper level (construction loads are a factor) which provide sufficient ductility to allow the redistribution of moments. Compare minimum depth obtained to depth obtained by other design requirements. The beam is a part of a lateral load resisting frame. Assume that an area of positive moment reinforcement equal to 50% of the negative moment reinforcement at the support is to be extended continuously over the support. $f'_c = 3000$ psi, $f_y = 60{,}000$ psi, $w_c = 145$ pcf (normal-weight concrete), $L = 100$ psf, and $D = 25 + 75 = 100$ psf.

Solution

(a) *Detail beam for ductility*. Using Table 22.11, for interior continuous beams,

$$-M_a = 0.5M_0$$

$$+M_a = 0.5M_0$$

$$M_0 \text{ (static moment)} = 10(0.1 + 0.1)(30)^2/8 = 225 \text{ ft-kips}$$

$$M_{0u} \text{ (factored static moment)} = 10[1.4(0.1) + 1.7(0.1)](30)^2/8 = 348 \text{ ft-kips}$$

$$-M_a = 0.5(225) = 113 \text{ ft-kips}$$

$$+M_a = 0.5(225) = 113 \text{ ft-kips}$$

* For SI with M'_a in kN·m and b_w and d in mm, Eq. (22.7.1) becomes

$$d \geq 770\sqrt{M'_a/b_w} \tag{22.7.1}$$

Table 22.12 Minimum Effective Depth to Limit $(\rho - \rho') \leq 0.375\rho_b$

Concrete (psi)	3000		4000		5000	
Steel (ksi)	40	60	40	60	40	60
$(\rho - \rho') \leq 0.375\rho_b$	$\leq$1.39%	$\leq$0.80%	$\leq$1.85%	$\leq$1.06%	$\leq$2.13%	$\leq$1.25%
$a_u = M'_u/A_s d$ (at 0.375 ρ_b)	2.67	4.08	2.67	4.08	2.70	4.10
$d \geq \sqrt{\dfrac{M'_u}{a_u(\rho - \rho')b_w}}$ *	$5.2\sqrt{\dfrac{M'_u}{b_w}}$	$5.5\sqrt{\dfrac{M'_u}{b_w}}$	$4.5\sqrt{\dfrac{M'_u}{b_w}}$	$4.8\sqrt{\dfrac{M'_u}{b_w}}$	$4.2\sqrt{\dfrac{M'_u}{b_w}}$	$4.4\sqrt{\dfrac{M'_u}{b_w}}$

Definitions: M'_u (ft-kips) is the net portion of the factored negative moment over the support that is resisted by compression in the concrete. It is assumed that the compression reinforcement (represented by the ratio ρ') present over the support can be stressed to the yield value, which in combination with an equal amount of tension reinforcement in the opposite face will resist the leftover moment $(M_u - M'_u)$. For example, when $\rho' = 25\%$ of ρ, then $M'_u \approx 0.75M_u$. d and b_w are in inches.

* To obtain depth d in mm, for M'_u in kN·m and b_w in mm, coefficients 5.2, 5.5, 4.5, 4.8, 4.2, and 4.4 become 570, 600, 500, 530, 460, and 480, respectively.

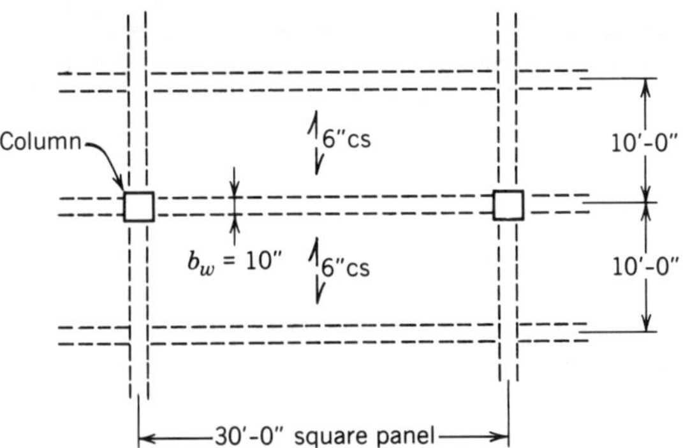

Fig. 22.18 Plan view for Example 22.7.1.

The net moment resisted by compression in the concrete at the support is

$$M'_a = (1 - 0.5)113 = 57 \text{ ft-kips (since at the support } \rho' = 0.50\rho)$$

Using Eq. (22.7.1),

$$d \geq 7\sqrt{\frac{57}{10}} \geq 16.7 \text{ in.}$$

Use $h = d + 3$ in. $\approx$ 20 in. (508 mm).

Verify that $(\rho - \rho') \leq 0.375\rho_b$. See limits in Table 22.12 which indicate that for f'_c = 3000 psi and f_y = 60,000 psi, $(\rho - \rho') \leq 0.80\%$. Find also from Table 22.12,

$$a_u = 4.08 \text{ at } (\rho - \rho') = 0.375\rho_b$$

At the support, the factored negative moment is

$$M_u = 0.5M_{0u} = 0.5(348) = 174 \text{ ft-kips (236 kN·m)}$$

$$A_s = \frac{174}{4.08(17)} = 2.5 \text{ sq in. (1600 mm}^2)$$

Thus, the positive moment reinforcement to be developed at the support is 0.5(2.5) = 1.25 sq in. Then,

$$A_s - A'_s = 2.5 - 1.25 = 1.25 \text{ sq in. (800 mm}^2)$$

$$(\rho - \rho') = \frac{1.25}{10(17)} = 0.0073 \quad \text{or} \quad 0.73\% < 0.80\% \qquad \text{OK}$$

Note that the more accurate expressions in Table 22.12 may also be used for each of the materials listed. At the support, the factored net moment resisted by compression in the concrete is

$$M'_u = (1 - 0.5)175 = 88 \text{ ft-kips (119 kN·m) (since } \rho' = 0.50\rho)$$

and

$$d \geq 5.5\sqrt{\frac{M'_u}{b_w}} \geq 5.5\sqrt{\frac{88}{10}} \geq 16.3 \text{ in.}; \qquad h \approx 20 \text{ in. (508 mm)} \qquad \text{OK}$$

Ductility is therefore assured.

(b) *Review the depth requirements for a serviceability index of 240, which corresponds to nonsensitive partitions being installed three months after construction.* Since this is a panel in an upper floor level of the structure, construction loads and procedures will influence the beam adversely. From ACI-9.5.2.5 compute the net time-dependent magnifier λ_n. Note that the flange sections in the positive moment region will reduce ρ' at the midspan to a small value and

$$\lambda_n = \frac{2}{1 + 50\rho'} - \frac{1}{1 + 50\rho'} \approx 1.0$$

The positive service and factored moments at midspan are (see Table 22.11):

$$+M_a = 0.5M_0 = 0.5(225) = 113 \text{ ft-kips};$$

$$+M_u = 0.5M_{0u} = 0.5(348) = 174 \text{ ft-kips} \qquad A_s \approx 2.5 \text{ sq in.}$$

Compute effective flange width b according to ACI-8.10:

$$b \le \frac{1}{4}(30)12 \quad = 90 \text{ in.} \qquad \underline{\text{controls}}$$

$$b \le 16(6) + 10 = 106 \text{ in.}$$

$$b \le 10(12) \quad = 120 \text{ in.}$$

$$\frac{d}{h} = \frac{17}{20} = 0.85 \qquad \frac{h_f}{h} = \frac{6}{20} = 0.3\ (> 0.2)$$

$$\frac{b_w}{b} = \frac{10}{90} \approx 0.1 \qquad \rho = \frac{2.5}{90(17)} = 0.0017$$

From Table 22.3: $f_y = 60{,}000$ psi; $w_c = 145$ pcf; $f'_c = 3000$ psi; and $D = L$, use second group ($b_w/b =$ 0.1 and $h_f/h = 0.2$):

$$\text{interpolate for } \frac{d}{h} = 0.85 \qquad \left(\text{using } \frac{d}{h} = 0.79 \text{ and } 0.90 \text{ lines}\right)$$

$$C = 5400 \quad \text{for} \quad \rho = 0.0022$$

$$C = 6200 \quad \text{for} \quad \rho = 0.0012$$

Next, interpolate for $\rho = 0.0017$ and estimate $C \approx 5800$. Using Eq. (22.3.8),

$$K = \frac{17}{0.9(20)}\left[\frac{1}{0.4 + 1.4(113)/174(0.6)}\right] = 1.04$$

$$K_1 = \left(1.2 - 0.2\frac{M_0}{M_a}\right) = 1.2 - 0.2(2) = 0.8$$

Using Eq. (22.3.29a) (since construction loads allow the simplified approach), confirm selected depth

$$h_{\min} = 20 \text{ in.} > \frac{30(12)}{5800}(240)\left(\frac{1(100) + 100}{200}\right)\left(\frac{0.80}{1.04}\right)(1) \qquad \text{OK}$$

(c) *Review the aesthetic integrity of the beam.* Assume the owner requires a limit to the total sustained load deflection of about 1-in. maximum. To review the aesthetic integrity it is necessary to anticipate also the final steel reinforcement that the beam may require. Since this beam is part of a frame system which resists lateral loads, any additional reinforcement required due to wind forces must also be computed to enable an assessment of the influence of this added reinforcement on the deformations of the beam under gravity loads. If additional lateral load reinforcement is required then the beam is overly reinforced for the gravity loads alone. This obviously helps to reduce deformation of the beam under gravity loads. These considerations are particularly necessary when a camber is to be built in. If however a camber is not needed, then neglecting the additional stiffness of the beam resulting from the lateral load requirements is a conservative procedure.

Assume initially that only a small amount of reinforcement will be added over the support due to combined gravity and lateral loads for this example. Using Eq. (22.3.36a),

$$\Delta_f = \frac{30(12)}{5800}\left(\frac{30(12)}{20}\right)\left(\overbrace{1 + 0.5 + \frac{2}{1 + 50\rho'}}^{3.5}\right)\left(\frac{100}{200}\right)\left(\frac{0.80}{1.04}\right) 1.0$$

$$= 1.5 \text{ in.} > 1.0 \text{ in. allowed}$$

A slight camber will be required.

(d) *Review the requirements for constructability of the partitions at 3 months.* From ACI-9.5.2.5, $\lambda_{3\text{ month}} = 1.0$. Then compute Δ_c from Eq. (22.3.33a),

$$\Delta_c = \frac{30(12)}{5800}\left(\frac{30(12)}{20}\right)\left(\overbrace{1 + 0.5 + 1.0}^{2.5}\right)\left(\frac{100}{200}\right)\left(\frac{0.8}{1.04}\right)(1.0) = 1.0 \text{ in.}$$

Note that Δ_c can also be directly interpolated from Δ_f, as follows:

$$\Delta_c = \left(\frac{2.5}{3.5}\right)\Delta_f = 1.0 \text{ in.}$$

If the initial estimate of negligible added wind force reinforcement is verified, then a camber of about ¾ in. can be provided ($\ell/300 > 0.75$ in. OK). This will result in a final net deflection due to sustained loads of $1.5 - 0.75 = 0.75$ in. approximately five years after construction. This will satisfy the owner's request to limit total sustained loads deflection to a 1-in. maximum.

If, however, the wind moments are large enough to require substantial increase of reinforcement over the supports the designer must further review the camber requirement and reduce it somewhat due to the additional stiffness the beam acquires. A good measure of prophesy is now required on the part of the designer as to the shape of the gravity moment envelope. However, since larger positive moments have been selected to start with in this simplified gravity load design, the designer may now take advantage of a more favorable gravity moment envelope which will reduce K_1 and increase K to indicate smaller deformations due to the gravity loads.

As was demonstrated in this example, the requirements for ductility need not hinder the designer. The ratio ρ'/ρ at the support can be adjusted so that the member size which assures ductility does not exceed the requirements for the other design considerations. Extending more of the positive moment reinforcement continuously through the support will also reduce the potential for progressive collapse.

22.7.3 Prevention of Progressive Collapse

In Example 22.6.1 the need was indicated to provide some bottom (positive moment) reinforcement continuous over supports to assure ductility. This simple detailing requirement will serve a second purpose of minimizing the possibility of *progressive collapse*.

Progressive collapse may occur when a floor level directly below a damaged floor does not have sufficient strength to withstand the impact load of the falling floor. The result may be that it too collapses, endangering all the floor levels below. Progressive collapse is the result. Progressive collapse can be initiated by any localized failure at any one column location. The collapse can progress horizontally to other supports if they cannot absorb the added shears and moments released by the damaged support. In this case the levels below will also, one at a time, progressively collapse. Hawkins and Mitchell [22.41] have indicated that properly designed and detailed bottom reinforcement capable of developing membrane plate action for maximum loading can terminate horizontal progression of a localized collapse. Since this method can become complex to design into the structure (and possibly also uneconomical) an alternate procedure is recommended [22.41] in which a post-failure shear capacity is maintained by properly anchored, continuous over the support, bottom reinforcement.

Major catastrophic collapses have occurred during construction at which time the floor under construction, regardless of the detailing provided, cannot be prevented from collapsing onto the level below it. It is important, therefore, to consider this extreme case in detailing the reinforcement. It is anticipated that the requirements to prevent vertical progressive collapse from occurring during construction will suffice to prevent progressive collapse from occurring after the structure has been occupied.

The following example will demonstrate the steps necessary to help prevent such a mishap.

EXAMPLE 22.7.2

Assume that the panel of Example 22.7.1 (see Fig. 22.18) at floor level ($N + 1$) was damaged by a blast and collapses onto a similar panel on level N. The panel is supported by an interior column having a tributary area of 900 sq ft. Assume the superimposed load in both panels equals about 30 psf. The dead load of the panels is 100 psf.

Solution

The weight of the collapsing panel equals 900(0.13) = 117 kips. This load should be doubled to account for impact. In addition, the panel at level N must support its own weight. The total load that must be transferred to the interior column at level N equals 2(117) + 117 = 350 kips.

Assume now that all four beams framing into the column at level N have been sufficiently damaged by the collapse of level ($N + 1$) so that they have failed in shear. The panel at level N consequently attempts to slide down around the column. The only restraint to this sliding is the ductile bottom reinforcement (continuous over the support), which is able to deform, prior to shearing off, and to develop large tensile forces resembling a catenary action, which prevents the panel from totally disengaging itself from the column. See Fig. 22.19 for an exaggerated schematic view.

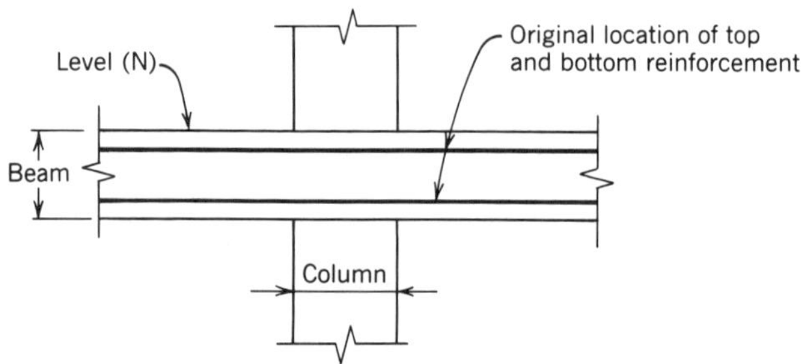

(a) Before collapse of level ($N + 1$)

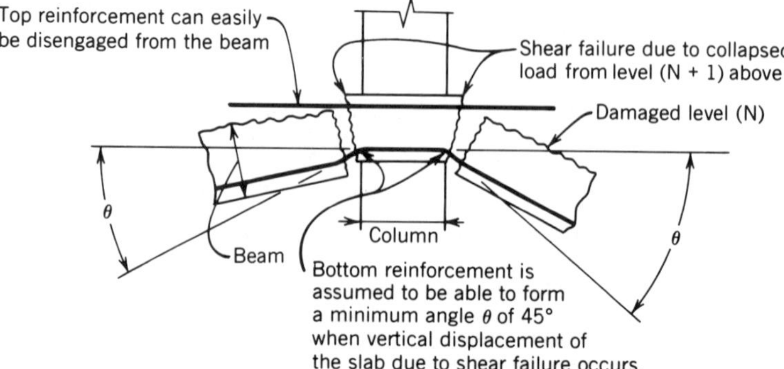

(b) After collapse of level ($N + 1$)

Fig. 22.19 Schematic side views of bottom reinforcement that is capable of developing full tension strength when continuous over or anchored into a support. (For clarity, debris from collapsed level and other details are not shown.)

The reinforcement A_s required to prevent disengagement may now be estimated,

$$A_s = \frac{W}{\phi f_y \sin\theta} \tag{22.7.2}$$

where W is the total weight to be hung (includes an allowance for impact). A_s is the sum of all bottom reinforcement extending from all faces of the column. The reinforcement must be fully anchored in the column and is either continuous into the spans, or is tension spliced with reinforcement which does extend into the spans. The angle θ (see Fig. 22.19) can be taken to equal 45° or more after a small displacement of the damaged slab has occurred in the immediate vicinity of the support. (The small displacement downward at the column faces is due to the shear failure and the reinforcing bars are ductile enough to deform while supported by the column.)

The required reinforcement (taking f_y = 60 ksi) may now be computed,

$$A_s = \frac{350}{0.90(60)(0.707)} = 9.0 \text{ sq in.}$$

At an interior support the draped reinforcement can be tensioned from both sides. When four equal beams frame into the interior column, the required reinforcement in each direction is only 2.25 sq in.

In Example 22.7.1, only 1.25 sq in. were to be continuously extended through the support. The transverse girders should therefore have at least [9.0 − 2(1.25)]/2 = 3.25 sq in. of bottom reinforcement extending continuously over the support.

Note that the top reinforcement over the support cannot provide similar protection against progressive collapse because it can easily be separated through the thin top cover of concrete from the beams.

EXAMPLE 22.7.3

Compute the area of bottom mat reinforcement, continuous over an interior column which supports a 20-ft square flat plate (8-in. thick) panel, required to prevent a progressive collapse. Assume sustained superimposed dead and live loads of 20 psf on both floor levels N and $(N + 1)$. Assume that a construction mishap precipitates the collapse of level $(N + 1)$.

Solution

Compute the loads to be supported. Level $(N + 1)$ loads are

$$\text{8-in. slab} = 100 \text{ psf} + 20 \text{ psf} = 120 \text{ psf}$$

The load should be doubled for the impact effect; thus

$$\begin{aligned} \text{level } N + 1,\quad 2(120) &= 240 \text{ psf} \\ \text{loads on level } N, \quad &= \underline{120 \text{ psf}} \\ \text{total} \quad &= 360 \text{ psf} \end{aligned}$$

$$A_s = \frac{0.36(20)^2}{0.90(60)(0.707)} = 3.8 \text{ sq in.}$$

Thus, the required slab bottom reinforcement to be continuous over the interior column equals 3.8/4 ≈ 1.0 sq in. in each direction. Again, the bottom bars must be long enough to splice with the bottom mat reinforcement. Note that in many flat slab structures the slabs participate in resisting lateral loads and therefore already require continuity of the bottom mat reinforcement to account for the possible reversal of moments due to large lateral forces. The continuous bottom reinforcement also enhances the ductility at the support.

If bottom mat reinforcement is to be made continuous over supports and is to be anchored at discontinuous edges, the weight of the reinforcement required is, as a result, increased by about a tenth of a pound per sq ft of floor area. Simplicity in the placement of reinforcement is now, however, enhanced. Mill-ordered, uncut bars, can now be tension spliced at fewer and more suitable locations (i.e., away from the congested supports and preferably at the one-third points of a span). Furthermore, longer bars of equal length will help placement, eliminate the need to cut (and waste) material, and reduce the extra handling involved. These factors will, in fact, boost the overall economy and will also enhance ductility and safety.

22.7.4 Lateral Loads and Member Sizes

Member sizes selected for gravity loads must often be increased in order to reduce sway and to accommodate the larger flexural requirements due to combined gravity and lateral load factors of ACI-9.2 [22.1].

Increased sizes may be required only in a critical portion of a structure. In dual systems (frame–shear walls) the frame's beam elements may require deeper sections only at the middle one-third of the structure height. Frames without shear walls require more stiffness close to the base of the structure.

Uniformity in sizes should be maintained where possible and necessary variations in sizes should remain constant for at least 10 to 15 consecutive floor levels to allow the efficient use of forms. If it is possible to adjust only the reinforcement and to maintain constant sizes of the floor members, then this should be attempted. Major structures usually do require different stiffnesses at various vertical locations in the structure. Flat slab structures can be adjusted for stiffness with minimal cost attributed to formwork alterations by changing only the slab thickness. In flat slab structures, the columns are much stiffer than the floor members. To increase floor stiffness in order to reduce sway, only the stiffness of the floor members should be increased in order to gain the most from the additional materials used.

Equation (22.4.15) can demonstrate the above argument. Assume that a flat plate structure has columns with stiffnesses about 20 times the stiffness of the floor members. Then,

$$\Delta_i = \frac{Hh_s^2}{12}\left(\frac{1}{E_c\Sigma K_c} + \frac{1}{E_b\Sigma K_b}\right)$$

$$= Z\left(\frac{1}{20} + \frac{1}{1}\right) = Z(1.05)$$

where Z is the product of all other variables in the equation.

Assume now that to gain stiffness only the columns are doubled in stiffness,

$$\Delta_i = Z\left(\frac{1}{40} + \frac{1}{1}\right) = Z(1.025)$$

indicating only an insignificant reduction in sway discounting axial deformation.

However, if the floor member stiffnesses have been doubled,

$$\Delta_i = Z\left(\frac{1}{20} + \frac{1}{2}\right) = Z(0.55)$$

which does indicate a much reduced sway.

Equation (22.4.15) can also be used for preliminary estimation of the slab thickness in flat plate frame structures to shorten the iteration process of the more exact computer analysis that should follow.

EXAMPLE 22.7.4

Estimate the slab thickness to be used for an initial computer input so that the final sway at a critical floor is not more than $h_s/400$. The structure is a 20-story flat plate structure, 160 × 62 ft in plan dimension, which has an average panel size ℓ_2 by ℓ_1 of 20 × 16 ft. The average column size at the midheight of the structure is 16 × 24 in. The story height is 9 ft. There are three slab panels in the 62-ft direction with columns spaced at $\ell_1 = 20$ ft. Assume the critical floor for the frame system is at the midheight of the structure because shear walls provide the major lateral load resistance at the lower levels. The flat slab frame is to support an estimated 260 kips of the total wind load of 360 kips at the tenth floor.

Solution

Compare the stiffness of the columns and slabs. For the columns,

$$K_c = \frac{16(24)^3/12}{9(12)} = 170 \text{ in.}^3$$

Using Eq. (22.4.8), compute the effective slab width $\alpha\ell_2$ for the interior 16-ft wide by 20-ft long slab panel,

$$\alpha\ell_2 = 0.3(20) + \frac{2}{20}(16) + \frac{1.44 - 2.0}{2} = 7.3 \text{ ft}$$

This is about 45% of the interior panel width and is within the limits of 20 and 50% given by Eq. (22.4.9).

Allowing for exterior panels to be less efficient (due to column chord action) and accounting for mechanical penetrations, assume that only about 35% of an average panel width is effective. Thus, an individual average panel stiffness is

$$K_b = \frac{0.35(16)(12)h^3}{12}\left(\frac{1}{20(12)}\right) = 0.023h^3$$

It is apparent that the columns are much stiffer and will contribute only a small increment of the total sway. For this preliminary selection of slab thickness the column contribution to sway can be ignored. (This will be the case for preliminary review of most flat plate high-rise structures.)

Neglecting columns then and rearranging Eq. (22.4.15) to solve for the required slab thickness (and taking centerline to centerline spans to compensate for neglecting the columns contribution to the sway),

$$\Sigma K_b \approx \frac{Hh_s^2}{12E_b\Delta_i} = \frac{L'h^3}{\ell_1(\text{av})(12)} \tag{22.7.3}$$

where $\ell_1(\text{av})$ is the average span ℓ_1 in the direction parallel to the lateral load, and L' is the sum of all the effective $\alpha\ell_2$ widths of the panels present in the structure. In this example the structure is three panels deep and is 160 ft wide; therefore, $L' \approx 3(160)0.35 \approx 168$ ft.

Solving Eq. (22.7.3) for the slab thickness h,

$$h \approx \sqrt[3]{\frac{Hh_s(h_s/\Delta_i)}{E_b[L'/\ell_1(\text{av})]}} \tag{22.7.4}$$

To limit the final story drift Δ_f to not exceed $h_s/400$ it is necessary to allow the anticipated P–Δ effects to increase the initial drift Δ_i. Assuming $Q \leq 0.15$ maximum and taking $\Delta_f = \Delta_i/(1 - Q)$,

$$\frac{h_s}{\Delta_i} = \frac{(h_s/\Delta_f)}{1 - Q} \approx 480$$

Therefore to limit story drift to $h_s/400$ the flat slab thickness must be

$$h \geq \sqrt[3]{\frac{Hh_s(480)}{E_b[L'/\ell_1(\text{av})]}} \tag{22.7.5}$$

For the example, where $H = 260$ kips, $h_s = 9(12) = 108$ in., and $f'_c = 4000$ psi,

$$h \geq \sqrt[3]{\frac{260(108)(480)}{3600(168/20)}} = 7.5 \text{ in.}$$

At midheight of the structure estimate that the total gravity load equals about 17,000 kips. Then,

$$Q = \frac{P\Delta_i}{Hh_s} = \frac{17{,}000(h_s/480)}{360h_s} = 0.10 < 0.15 \qquad \text{OK}$$

The lateral load H will be magnified by

$$\delta = \frac{1}{1 - Q} = \frac{1}{1 - 0.10} = 1.11$$

and the final critical story sway is

$$\Delta_f \approx 1.11\Delta_i = 1.11\frac{h_s}{480} = \frac{h_s}{432} < \frac{h_s}{400} \qquad \text{OK}$$

The preliminary slab thickness of 7.5 in. can now be used for the more exact analysis in which the frame–shear-wall interaction and the effects of chord deformation will be determined. Final adjustments to the floor thickness will then be made.

22.7.5 Construction Procedure

Structures generally are constructed a floor at a time starting with the lowest levels. Some structures, however, such as those with lift slabs, and suspended structures, may take a different route. An important part of the designer's role includes understanding of how the structure will be erected, and providing for the method of construction in the design.

There are many pertinent issues to be considered. What cranes are likely to be used? Climbing cranes require self-support on newly cast floors, and a large penetration must be provided at each level to accommodate the crane body. Sidewalk cranes are limited in reach. Often they require support at street level adjacent to foundation walls (Have the walls been reinforced for extra lateral loads?). Sidewalk cranes may also require support over constructed lower levels. Special foundations, piles, footings, mats, etc. must be anticipated in order to accommodate the large loads and overturning moments exerted by cranes, and be provided for before the foundation contractor has left the job site.

Construction live loads must also be considered. Hoists (material and personnel) attached to the sides of the structure are necessary for high-rise structures to prevent the loss of time moving to and from the higher levels. The location of these hoists generally will require special design considerations, since storage of construction materials to be used for finishes will take the route the hoist provides.

The prudent structural engineer maintains an open line with the contractor to listen to his problems and to help by providing alternate design solutions when needed.

Decisions about field conditions which require revisions must be made immediately so that heavy equipment does not stand idle. The contractor can become the designer's best advocate when the engineer's role is understood, and there will be no sacrifice of safety or serviceability in order to construct the project on time without extra cost.

22.7.6 Two-Day Construction Cycle for High-Rise Structures Using "Preshores"

The forming and stripping especially of flat slab construction have been developed to a high degree of efficiency during the past 25 years. Contractors, with the aid of the structural engineers responsible for quality control, have been able to reduce the construction process to a bare minimum two-day cycle.

The procedure of stripping of primary construction forms using "preshores" has been developed during the past two decades with the aid of this writer's firm in consultation with various leading contractors and with the approval of government agencies supervising housing construction in New York City.

This procedure allows fast cycling with reduced sets of primary forms while maintaining full support of the freshly placed floors during a stripping operation which commences within 24 hr after placement of concrete.

To demonstrate a two-day construction cycle, assume that concrete for a typical floor "N" is placed on Monday. The west side of the floor area is the starting point for placement of concrete. Additional morning and afternoon test cylinders are taken to be field cured overnight in the pour location and tested at 24 hr.

On Tuesday

Step 1 (A.M. Tuesday). Vertical beam faces and column forms are removed. "Preshores" to alternate plywood sheets are installed. (See x locations in Fig. 22.20.) Note that preshores are placed at both ends of the same sheet of plywood and the ribs and stringers supporting the plywood sheets are bypassed (see Fig. 22.21). Preshores are wedged in tight but *not too tight* to cause "lifting" of the "Monday" slab above.

The early morning cylinder results are phoned in. If strength attains 1700-psi minimum, step 2 can proceed but not earlier than noon.

Step 2 (P.M. Tuesday). "Thinning out" operation starts at the west end where concrete was first placed. The contractor removes 50% of the stringers and about 75% of the primary shores (see Fig. 22.22). Preshores are left in place.

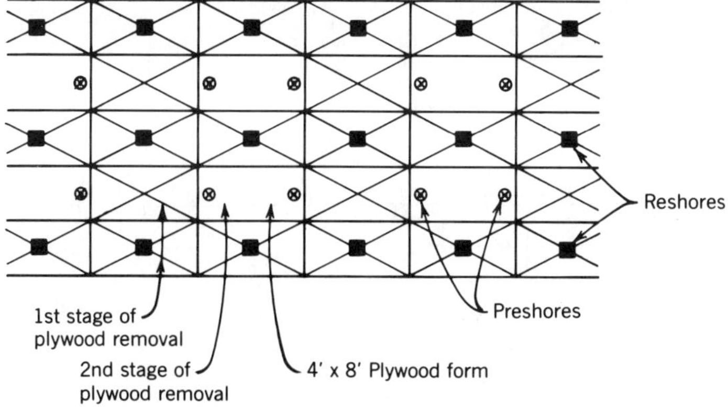

Fig. 22.20 Reflected plan of plywood form material showing schematic location. From Robert Rosenwasser Associates, P.C.

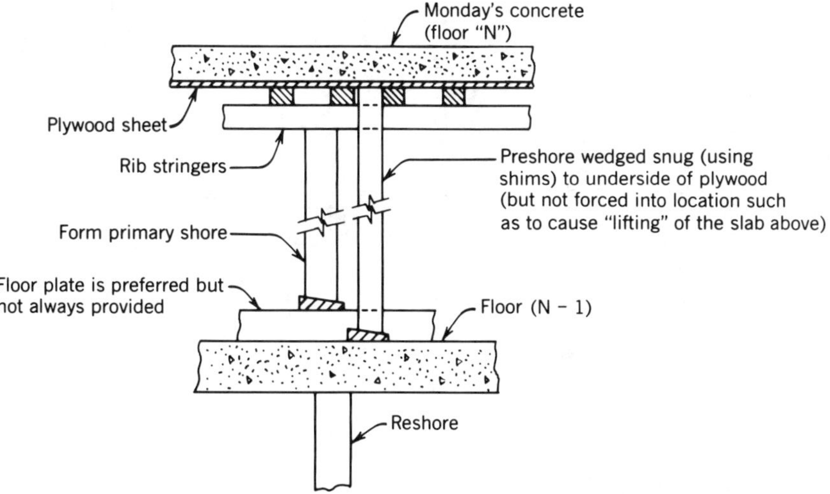

Fig. 22.21 Section view during "thinning out" of primary form shores.

On Wednesday

Step 3 (Early A.M.). Prior to any placement of concrete on level $(N + 1)$, the contractor removes, a small area at a time, all leftover primary shores and plywood sheets not pinned by the preshores. Preshores are left untouched! Simultaneously, reshores are placed and wedged against the bare concrete soffit at 8-ft intervals (see reshores marked at location of "1st stage of plywood removal" in Fig. 22.20).

When a complete bay is stripped of leftover primary shores, and is completely reshored, and while this process is advancing into other bays, the preshores and their pinned plywood sheets are also removed (see "second stage of plywood removal" in Fig. 22.20). Each column is to receive reshores, within 3 ft of column faces, on all four sides.

The stripping and reshoring operation below level N must be well advanced (at least two full bays) of any concrete operation on floor $(N + 1)$. This is necessary to avoid any disturbance in the primary forms supporting level $(N + 1)$ immediately after receiving the placed concrete load and also to keep the crew underneath away from the work area. If the concrete operation catches up with the stripping operation below, such stripping operation is terminated at once and does not proceed until the next day!

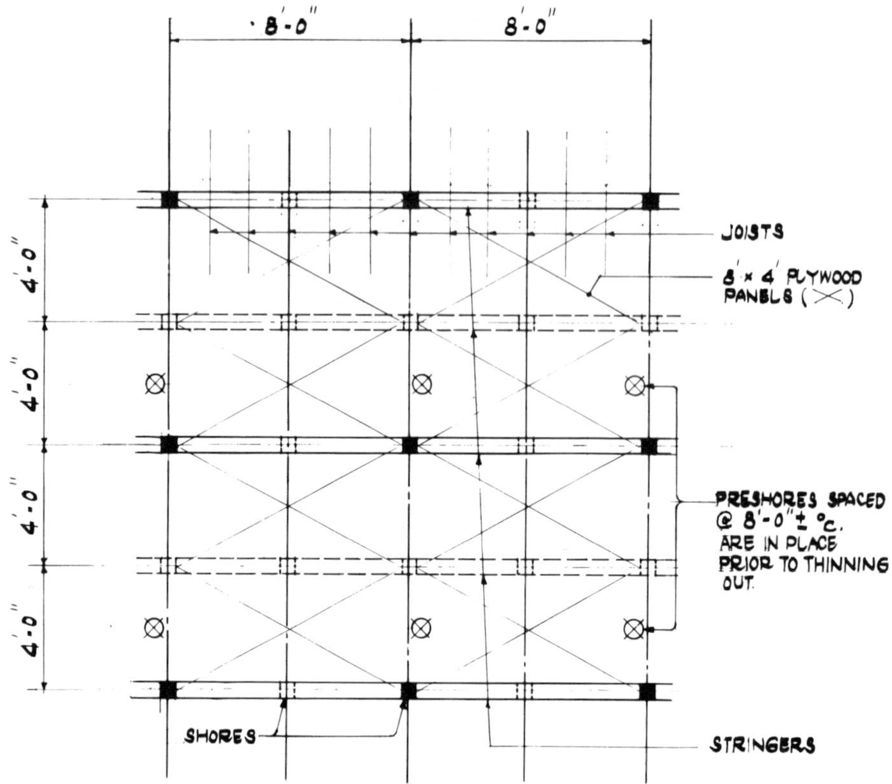

Fig. 22.22 Thinning out of primary form shores.

This construction process utilizes 2¼ sets of forms. The quarter set allows earlier forming start on the west couple of bays Monday afternoon as soon as workers can walk on the freshly placed concrete without leaving footprints. The workers are cautioned to ''hand'' handle all materials (i.e., no crane drops) during this early setup of the primary forms for the next level.

This stripping operation using preshores has been used successfully for nearly two decades, with an excellent safety record. The method has one major advantage over any other forming and stripping procedure. The one-day-old concrete floor is already reshored in a manner similar to the levels directly below it and its supports will remain largely undisturbed for the duration (usually 28 days and not less than 21 days) during which time it is required to support construction loads transferred from the levels above. The distribution of the construction loads is also fairly uniform to all supporting levels and can remain quite low.

Figure 22.23 indicates reshores left in place under ''Monday'' level N undisturbed for about 28 days. It is obvious that a large number of floors (10 in this case) do participate and are ''locked'' into each other to almost equally support the newly constructed floor. Permanent set is induced in all floors supported on yielding levels due to the deformation in the supports while the concrete is still plastic enough not to receive induced stresses. The construction load that must be supported depends on the number of levels participating. Even when reshores below the lowest supporting level are removed somewhat earlier (say at day 21) at least eight levels will participate in the support of the construction of the new level and its construction live loads (estimated at about 50 psf).

Assuming "rigid" reshores and using ACI equation [22.42] for young concrete,

$$(f'_c)_t = \frac{t}{4 + 0.85t}(f'_c)_{28} \tag{22.7.6}$$

where t is the age of concrete in days. It can be shown that when concrete levels remain reshored for 28 days and therefore 10 levels (see Fig. 22.20) participate in the support of the construction the lowest participating level N will support about 11% and the top level ($N + 9$) about 7% of the constructed level ($N + 10$). When reshores are removed after 21 days, level ($N + 2$) will support about 14% and level ($N + 9$) about 9% of the constructed level.

To compute the load that must be supported, assume an 8-in. slab weighing 100 psf, two sets of primary shores to weigh about 20 psf, reshores at 3 psf per floor, and other miscellaneous construction materials at say 10 psf per floor.

The levels supporting construction (see Fig. 22.23) are therefore loaded during the initial month, assuming all 10 floors are "locked in" by the reshores and adjusting for reduced modulus of elasticity of the younger supporting levels, level N can be shown to support about

$$\begin{pmatrix}\text{self-load plus}\\ \text{construction load}\end{pmatrix} = [(100 + 3 + 10)10 + (100 + 50 + 20)]0.11 = 143 \text{ psf}$$

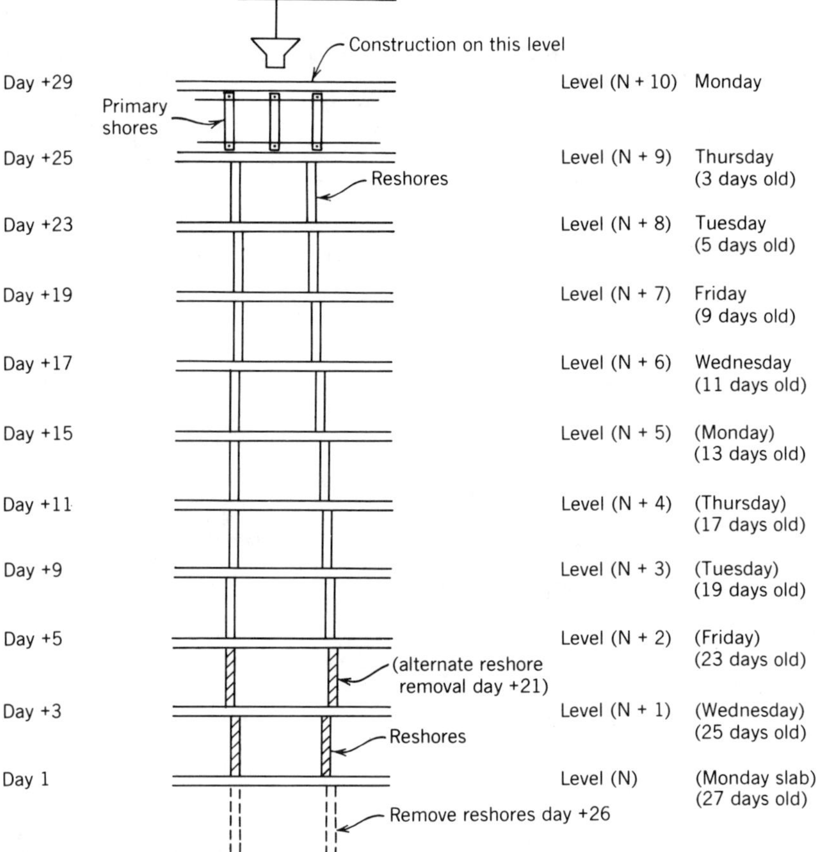

Fig. 22.23 Cross-section of high-rise building. Shores are to yielding levels and preshores *are* used.

When reshores are to be removed after only 21 days [i.e., reshores under level $(N + 2)$ are removed prior to construction of level $(N + 10)$], the self-load plus construction load is still no more than

$$[(100 + 3 + 10)8 + (100 + 50 + 20)]0.14 = 150 \text{ psf}$$

The removal of reshores after only 21 days is sometimes permitted by the engineer when the flat slab is designed as part of the lateral load system and therefore is provided with extra reinforcement and depth above the requirements for gravity load considerations.

In both of the above cases, construction loads do not exceed designated minimum live load! This is a far cry from construction loads of up to 110% of self-load observed [22.13] when another common system (see Fig. 22.24) utilizing three sets of primary forms (but excluding the preshoring procedure) is used.

As was demonstrated, more sets of primary forms do not necessarily improve safety in the construction operation. The margin of safety in the construction procedure described in Fig. 22.24 is perilously small.

The preshoring procedure has some disadvantages that can be minimized with good supervision. Careless "thinning out" operation can topple preshores to cause damage to the day-old slab. Manipulation and removal of primary forms even in a correct manner cannot totally prevent some self-load being carried by the day-old slab. Some cracks will therefore develop.

A more troublesome item is that the new two-day-old slab has already gained enough stiffness to span large distances. It does participate therefore in supporting the freshly placed construction directly above it and must deform to remove the slack in the reshores below it and to account also for the additional compressibility of the reshores under the applied loads. Placement of reshores does not always match location of primary shores; this increases transfer load requirements of the two-day-old slab.

Development of hairline cracks, usually at midspan of large exterior bays, can be observed to occur (even alongside the line of reshores) as the construction load is applied above. Additional reinforcement to help maintain the effective moment of inertia I_e of the slab can be added by the engineer in anticipation of this cracking occurrence.

The reduced construction cycle takes its toll on construction tolerances. Accuracy is not possible when speed of construction is a prime consideration. This is not applicable to construction under

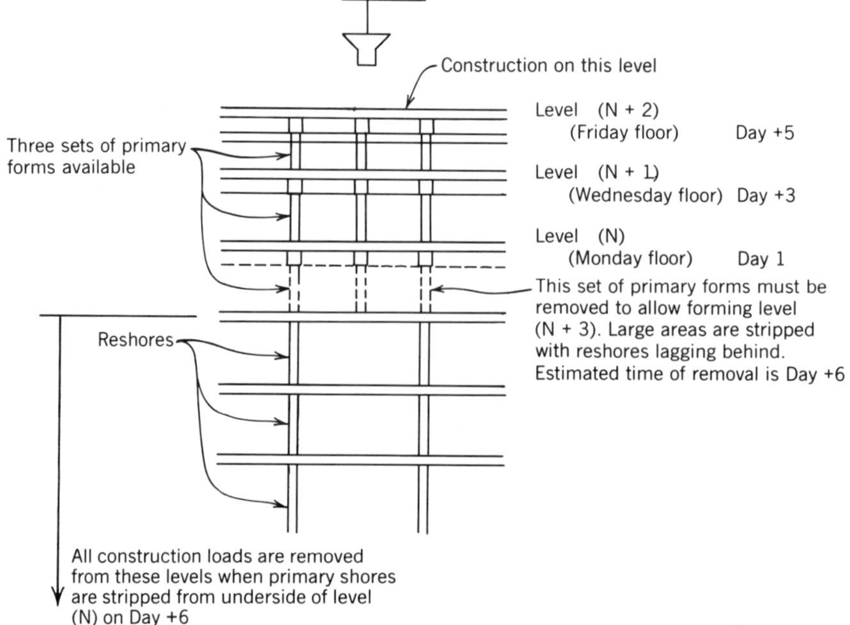

Fig. 22.24 Cross-section of a high-rise building. Shores are to yielding levels and preshores *are not* used.

any forming system which utilizes a two- or three-day cycle. The owner/contractor dialogue must consider this item and its effect on the future application of the nonstructural elements.

It is clear that better control by both the contractor and the engineer is required. The resulting efficient construction method and better safety record is certainly worth the extra effort.

22.8 NOTATION

A_s	area of tension reinforcement
A_s'	area of compression reinforcement
b	effective compression flange width of a beam
b_w	web width of a beam
c_1	size of support measured in the direction of the span for which moments are being determined
c_2	size of support measured in the direction transverse to the span direction for which moments are being determined
C	variable coefficient relating load level to deflection of one-way concrete members. See Eq. (22.3.20) and Tables 22.2 through 22.8.
C_d	Δ/Δ_i
d	distance from extreme compression fiber to centroid of tension reinforcement
D	dead load
E_c	modulus of elasticity of concrete
f_c'	specified compressive strength of concrete
f_y	specified yield strength of reinforcement
h	total height of a structure total depth of a member
h_f	flange thickness for a T-shape
h_s	story height
H	external (service) lateral load
H_u	external (factored) lateral load
I_{cr}	moment of inertia of cracked section transformed to concrete
I_e	effective moment of inertia for computation of deflection
I_g	moment of inertia of gross concrete section about centroidal axis, neglecting reinforcement
I_{gf}	I_g of a flanged member
I_{gw}	I_g of a member excluding flanges
I_s	moment of inertia of a "stubby" column. See Section 22.4.9.
K	correction factor. See Eq. (22.3.8).
K_1	correction factor relating simply supported span. See Section 22.3.3.
ℓ	span
ℓ_s	approximate height above fixed base to which columns can support lateral loads in cantilever action. See Section 22.9.4.
ℓ_1	length of span in direction that moments are being determined, measured center-to-center of support
ℓ_2	length of span transverse to ℓ_1, measured center to center of supports
L	live load
M_a	maximum (unfactored) moment in member at stage deflection is computed
M_{cr}	cracking moment. See ACI-9.5.2.3 [22.1].
M_n	nominal moment strength at section. $A_s f_y(d - a/2)$, where a = depth of equivalent rectangular stress block. See ACI-10.2.7.1 [22.1].
M_0	static moment $= w\ell^2/8$
M_u	factored moment at a section
P	service gravity load
P_u	factored axial load

Q stability index in Section 22.4
$(I_e)_{D+L}/(I_e)_D$ in Section 22.3

r radius of gyration of cross-section of a compression member

V_f magnified external loads

w_c unit weight of concrete

δ magnifier to account for P–Δ effects

Δ computed deflection (drift)

Δ_c computed deflection at stage when nonstructural elements are installed

Δ_f computed final deflection

Δ_i computed immediate deflection
initial (first-order) lateral load deflection

Δ_s computed critical deflection for consideration of serviceability

Δ_u computed deflection for factored loads

$(\Delta_i)_D$ computed immediate deflection caused by dead loads

$(\Delta_i)_{D+L}$ computed immediate deflection caused by total loads

$(\Delta_i)_L$ computed immediate deflection caused by live loads

λ multiplier for additional long-time (creep and shrinkage) deflection as defined by ACI-9.5.2.5 [22.1]

λ_c multiplier for additional deflection caused by construction loads and/or construction procedures

λ_n net difference between λ values computed at the various time periods considered

ρ reinforcement ratio A_s/bd (tension reinforcement)

ρ' reinforcement ratio A_s'/bd (compression reinforcement)

ρ_b reinforcement ratio producing balanced strain conditions. See ACI Code [22.1].

ξ time-dependent factor for sustained loads. See ACI-9.5.2.5 [22.1].

ϕ strength reduction factor. See ACI-9.3 [22.1].

ω $\rho f_y/f_c'$

SELECTED REFERENCES

22.1 ACI. *Building Code Requirements for Reinforced Concrete* (ACI 318-83). Detroit: American Concrete Institute, 1983.

22.2 James G. MacGregor and Sven E. Hage. "Stability Analysis and Design of Concrete Frames," *Journal of the Structural Division,* ASCE, **103,** October 1977 (ST10), 1953–1970.

22.3 Chu-kia Wang and Charles G. Salmon. *Reinforced Concrete Design,* 4th ed. New York: Harper & Row, 1985.

22.4 Jacob S. Grossman. "Simplified Computations for Effective Moment of Inertia I_e and Minimum Thickness to Avoid Deflection Computations," *ACI Journal, Proceedings,* **78,** November–December 1981, 423–439. Disc. **79,** September–October 1982, 413–419.

22.5 Jacob S. Grossman. Closure of Discussion of "Simplified Computations for Effective Moment of Inertia I_e and Minimum Thickness to Avoid Deflection Computations," (*ACI Journal, Proceedings,* **78,** November–December 1981, 423–439), *ACI Journal, Proceedings,* **79,** September–October 1982, 414–419.

22.6 Dan E. Branson. *Deformation of Concrete Structures.* New York: McGraw-Hill International, 1977.

22.7 Dan E. Branson. "Deflections," *Handbook of Concrete Engineering,* 2nd ed., ed. by Mark Fintel. New York: Van Nostrand Reinhold, 1984, Chapter 2.

22.8 Dan E. Branson. "Deflections," *Notes on ACI 318-83 Building Code Requirements for Reinforced Concrete,* 4th ed., ed. by G. B. Neville. Skokie, Ill.: Portland Cement Association, 1984, Chapter 7.

22.9 Dan E. Branson. "Instantaneous and Time-Dependent Deflections of Simple and Continuous Reinforced Concrete Beams," Part 1, Report No. 7, Alabama Highway Research Report, U.S. Bureau of Public Roads, August 1963 (1965). 78 pp.

22.10 Dan E. Branson. "Design for Deflections," *Metric Design Handbook for Reinforced Concrete Elements,* ed. by M. Saatcioglu. Ottawa, Canada: Canadian Portland Cement Association, 1978, Chapter 5, 5-1 to 5-61.

22.11 ACI. *Commentary on Building Code Requirements for Reinforced Concrete* (ACI 318-83). Detroit: American Concrete Institute, 1983.

22.12 John A. Sbarounis. "Multistory Flat Plate Buildings—Construction Loads and Immediate Deflections," *Concrete International,* **6,** February 1984, 70–77.

22.13 John A. Sbarounis. "Multistory Flat Plate Buildings—Effect of Construction Loads on Long-Term Deflections," *Concrete International,* **6,** April 1984, 62–70.

22.14 John A. Sbarounis. "Multistory Flat Plates—Measured and Computed One-Year Deflections," *Concrete International,* **6,** August 1984, 31–35.

22.15 ACI Committee 340. *Design Handbook In Accordance With the Strength Design Method of ACI 318-77.* Vol. 1—*Beams, Slabs, Brackets, Footings, and Pile Caps* [SP-17(81)], 3rd ed. Detroit: American Concrete Institute, 1981.

22.16 Dan E. Branson and Heinrich Trost. "Unified Procedures for Predicting the Deflection and Centroidal Axis Location of Partially Cracked Nonprestressed and Prestressed Concrete Members," *ACI Journal, Proceedings,* **79,** March–April 1982, 119–130.

22.17 Richard W. Furlong. "Design of Concrete Frames by Assigned Limit Moments," *ACI Journal, Proceedings,* **67,** April 1970, 341–353.

22.18 J. Grossman. "Composite Design Cuts Costs," *Modern Steel Construction,* AISC, **IX,** 4th Quarter, 1969.

22.19 *The Supplement to the National Building Code of Canada 1980,* NRCC No. 17724. Ottawa, Ontario: National Research Council of Canada.

22.20 Sigmund A. Freeman. "Racking Tests of High-Rise Building Partitions," *Journal of the Structural Division,* ASCE, **103,** August 1977 (ST8), 1673–1685.

22.21 S. E. Hage and J. G. MacGregor. "The Second-Order Analysis of Reinforced Concrete Frames," Structural Engineering Report No. 49, Department of Civil Engineering, University of Alberta, Edmonton, 1974.

22.22 M. Daniel Vanderbilt. "Equivalent Frame Analysis of Unbraced Reinforced Concrete Buildings for Static Lateral Loads," Structural Research Report No. 36, Civil Engineering Department, Colorado State University, Fort Collins, Colorado, June 1981.

22.23 M. Daniel Vanderbilt. "Equivalent Frame Analysis for Lateral Loads," *Journal of the Structural Division,* ASCE, **105,** October 1979 (ST10), 1981–1998. Disc. **106,** July 1980 (ST7), 1671–1672; **107,** January 1981 (ST1), 245.

22.24 M. Daniel Vanderbilt and W. Gene Corley. "Frame Analysis of Concrete Buildings," *Concrete International,* **5,** December 1983, 33–43.

22.25 Fazlur R. Khan and John A. Sbarounis. "Interaction of Shear Walls and Frames," *Journal of the Structural Division,* ASCE, **90,** June 1964 (ST3), 285–335. Disc. **91,** February 1965 (ST1), 317; **92,** April 1966 (ST2), 389.

22.26 N. M. Hawkins, C. F. Wong, and C. H. Yang, "Slab-Edge Connections Transferring High Intensity Reversing Moments Normal to the Edge of the Slab," Division of Structures and Mechanics, Department of Civil Engineering, University of Washington, Seattle, Washington, May 1978.

22.27 ACI Committee 340. *Design Handbook In Accordance With the Strength Design Method of ACI 318-83.* Vol. 1—*Beams, Slabs, Brackets, Footings, and Pile Caps* [SP-17(84)], 4th ed. Detroit: American Concrete Institute, 1984.

22.28 *Continuity in Concrete Building Frames,* 4th ed. Skokie, Ill.: Portland Cement Association, 1959.

22.29 I. A. MacLeod. "Shear Wall–Frame Interaction," Portland Cement Association, Skokie, Ill., April 1971.

22.30 A. Rutenberg. "Analysis of Tube Structures Using Plane Frame Programs," *Proceedings of Regional Conference on Tall Buildings.* Bangkok, Thailand, 1974, 397–413.

22.31 Mark Fintel and Fazlur R. Kahn. "Effects of Column Creep and Shrinkage in Tall Structures—Analysis for Differential Shortening of Columns and Field Observation of Structures," *Designing for Effects of Creep, Shrinkage, Temperature in Concrete Structures* (SP-27). Detroit: American Concrete Institute, 1971, pp. 95–119.

22.32 ACI Committee 301. *Specifications for Structural Concrete for Buildings* (ACI 301-81). Detroit: American Concrete Institute, 1981.

22.33 Council on Tall Buildings, Committee 21A. *Concrete Framing Systems for Tall Buildings,* Vol. CB, Monograph on Planning and Design of Tall Buildings, ASCE, New York, 1978 (Chap. CB-3).

22.34 *Design and Construction of Large-Panel Concrete Structures—A Design Approach to*

General Structural Integrity, Report No. 4, Portland Cement Association, Prepared for U.S. Department of Housing and Urban Development, Office of Policy Development and Research, October 1977.

22.35 Mark Fintel and S. K. Ghosh. "Explicit Inelastic Dynamic Design Procedure for Aseismic Structures," *ACI Journal, Proceedings,* **79,** March–April 1982, 110–118.

22.36 Mark Fintel and S. K. Ghosh. "Case Study of Aseismic Design of a 16 Story Coupled Wall Structure Using Inelastic Dynamic Analysis," *ACI Journal, Proceedings,* **79,** May–June 1982, 171–179.

22.37 T. Paulay and J. R. Binney. "Diagonally Reinforced Coupling Beams of Shear Walls," *Shear in Reinforced Concrete* (SP-42) Detroit: American Concrete Institute, 1974, pp. 579–598.

22.38 Jacob S. Grossman. "780 Third Avenue: The First High-Rise Diagonally Braced Concrete Structure," *Concrete International,* **7,** February 1985, 53–56.

22.39 Jacob S. Grossman, Mark R. Cruvellier, and Bryan Stafford-Smith. "Behavior, Analysis and Construction of a Braced Tube Concrete Structure", *Concrete International,* **8,** September 1986, 32–42.

22.40 "TOWER," Integrated Computer Program for the Design of High-Rise Concrete Structures, Supervised and developed by the author.

22.41 Neil M. Hawkins and Denis Mitchell. "Progressive Collapse of Flat Plate Structures," *ACI Journal, Proceedings,* **76,** July 1979, 775–808.

22.42 ACI Committee 209. "Prediction of Creep Shrinkage and Temperature Effects in Concrete Structures," *Designing for Creep and Shrinkage in Concrete Structures* (SP-76). Detroit, MI: American Concrete Institute 1982, pp. 193–300.

CHAPTER 23
COMPOSITE CONSTRUCTION

SRINIVASA H. IYENGAR

General Partner
Skidmore, Owings & Merrill
Chicago, Illinois

MOHAMMAD IQBAL

Chief Structural Engineer
Walker Parking Consultants
Elgin, Illinois

23.1 INTRODUCTION

Composite construction as treated here involves the interactive and integral behavior of concrete and structural steel components. The basic principle underlying composite construction is to utilize the most desirable attributes of each material to their best advantage so that the combination may result in a higher order of structural efficiency and cost effectiveness. Elements of composite construction can be broadly classified into two categories: *composite members* and *composite systems*.

Composite members involve interactive behavior of concrete and steel in a single, isolated and specific element of a structure. They generally take the form of steel–concrete beams, trusses, columns, walls, and slabs. In general, plain concrete or concrete with some reinforcement is encased in or appended to structural steel components such as rolled I- or H-shaped sections, trusses, joists, steel plate weldments, and cold-formed metal decks. The most common composite member in building construction is the composite beam-slab arrangement, where a steel beam and a concrete slab are so interconnected with shear connectors that they act together as a unit. Composite slab members with metal deck are also used in conjunction with these composite beams. Other composite members that have been used include concrete-encased steel columns and concrete-filled steel tubular columns. These members can be used as individual members of a structural steel frame building.

A *composite system* consists of an assemblage of reinforced concrete, structural steel, and/or composite members, such that the assembly acts integrally to resist all imposed loads and the system as a whole meets all required structural requirements. The composite system can also be viewed as a combination building system, borrowing component members from either structural steel or reinforced concrete building systems. Therefore, they are termed "mixed steel–concrete systems." During the last decade, several specific combinations have gained common use, particularly for high-rise buildings. They include: (1) composite tubular system which combines an equivalent exterior framed tube and interior steel frame; (2) core-braced composite system, which uses core shear walls and exterior steel frame; and (3) tube-in-tube composite system, which uses an exterior concrete framed tube, interior core shear wall tube, and structural steel floor framing. Some of these systems have been discussed in Chapters 8, 10, 11, and 22.

Several other forms of less-used mixed steel–concrete systems are also possible. They are composite frames, composite claddings, panel-braced frames, and other miscellaneous combinations.

This chapter presents the conceptual basis and design considerations for composite systems including design examples. Similarly, design guidelines are also provided for composite members. The types of connections for mixed steel–concrete systems are also discussed. References 23.12 and 23.14 provide additional detailed information which may be useful in reviewing composite systems.

23.2 HISTORICAL DEVELOPMENT

The history of composite construction is closely linked with the development of reinforced concrete and structural steel construction. Concrete was used primarily as an encasing material in the early 1900s to encase steel columns and beams, and as a fill material on brick or tile arches for floors. In these instances, concrete played a secondary role, providing both fire protection for steel and finished or flat surfaces for columns, beams, and floors. Even though concrete provided some degree of structural composite behavior, it was generally ignored in strength calculations. Solid concrete slabs with encased steel beams were used extensively in the 1940s and 1950s with some composite action permitted for this condition. The development of mechanical shear connections in the late 1950s substantially helped speed up the process of composite beam development. Today, composite beams and trusses with metal deck and shear studs are used extensively for floor systems. The development of sprayed-on fire protection in the 1960s substantially eliminated the composite column for anything but special use, and therefore, its use is limited in practice.

Further developments in reinforced concrete high-rise building systems in the 1960s and 1970s set the stage for mixed steel–concrete types of systems. In this process, elements of concrete systems such as the shear wall and the punched framed tube were recognized as efficient elements to resist wind forces and to provide lateral rigidity. The use of these elements with simple structural steel framing (especially for floor framings) offered advantages of economy and speed. The object of the combination was to utilize the rigidity of concrete for lateral load resistance and the lightness and spannability of steel for floor framing. As the construction methodology of such mixing of systems was established and accepted by the construction industry, their use became widespread, which spurred further refinements and developments. Therefore, mixed steel–concrete systems have gained popularity and acceptance as a viable alternative to either structural steel or reinforced concrete buildings. In current practice, it is almost customary to offer the three alternatives of steel, concrete, and composite systems for cost and system evaluations for any particular high-rise building.

23.3 MIXED STEEL–CONCRETE SYSTEMS

Since there are numerous types of suitable members in both reinforced concrete and structural steel, a variety of practical combinations can be derived that serve and meet the overall criteria of the structure. For this purpose, the components or subsystems of a high-rise building can be classified as: lateral load resisting system, floor framing, floor slabs, and columns or walls resisting only gravity forces.

23.3.1 Lateral Load Resisting Components

Reinforced concrete shear walls represent a very efficient subsystem for use in a mixed form. Their ability to be cast in any plan shape makes them extremely versatile. Their massiveness together with their equivalent cantilever behavior makes them a highly efficient lateral load resisting element. Open or closed, linear or curved shapes can be utilized either inside the building or on its exterior. They form load-bearing enclosures for centralized building cores, stairways, and other shafts of reasonably large size. Since shear walls can be constructed independent of the rest of the system, they can be built before the rest of the structure is built and attached to it at a later time. Closed-form shear wall cores can be effective up to 50 stories.

Exterior concrete framed-tubes which are conceived as equivalent punched-tube forms are also highly efficient for use in the mixed form. The punched-tube is formed by close spacing of columns with wide columns and deep beams giving the appearance of a punched wall. As with shear walls, they are versatile in terms of possible shape modulations. The continuity and rigidity at the joints is obtained from the natural monolithic character of concrete. Architecturally, the concrete of the vertical and horizontal surfaces of the tube can be molded to any reasonable shape to create the desired articulations and can be exposed, if desired. An exterior framed-tube, therefore, performs the dual function of being a structure and a building enclosure. Figure 23.1 shows the elevation of a 37-story composite tubular system which typifies the basic character of the exterior tube. The punched-tube wall generally results in a reduction of glass area on the facade and therefore offers savings in the perimeter air systems because of reduced heating and cooling loads. Another advantage of the closely spaced column form of the framed tube relates to the simplification of window wall details. Since each window is framed all around by the elements of the tube, the window glazing is directly attached to the members by simple gasket details. Concrete framed-tube systems can be effective up to 100 stories or more.

The exterior reinforced concrete tube can be combined with a shear wall core to formulate a tube-in-tube system. The lateral loads then are shared by the two resistive systems in proportion to their relative rigidities. The floor framing can be of steel.

The exterior framed-tube in concrete can be restructured to formulate a modular or bundled tube. The interior tubular lines transform the system into a cellular configuration and the overall system

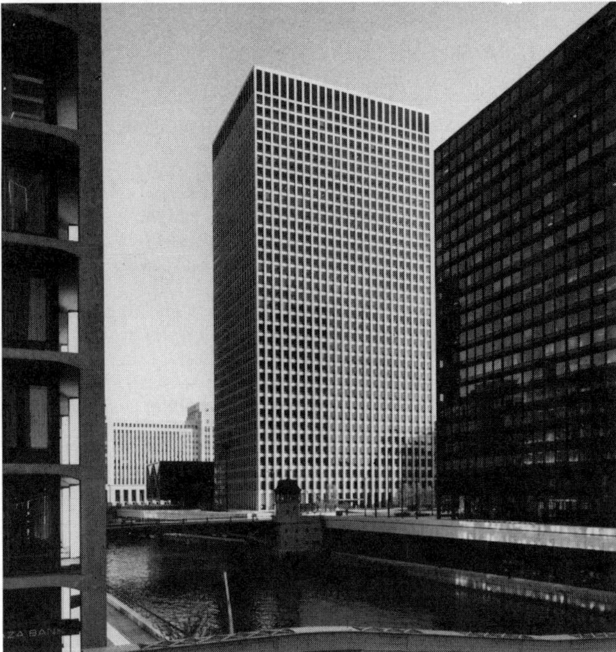

Fig. 23.1 Gateway Three Building, Chicago, Illinois.

Fig. 23.2 25-Story Ohio National Bank, Columbus, Ohio.

efficiency for lateral loads is greatly enhanced. The floor framing can be of steel. Such systems have the potential for being used in ultra-tall buildings. In addition, they offer an organized vocabulary for vertical massing and clustering of buildings shapes. Figure 23.2 shows such a possibility in a 25-story tower.

The moment-resisting Vierendeel steel frame as a lateral load resisting system can be utilized in a mixed form. The frames can be provided either on the exterior or interior of the building. Similarly, an exterior or interior steel shear truss can also be utilized. Both of the above systems can be used in an interacting mode with interior concrete shear walls. Similarly, an exterior framed-tube in steel can be combined with an interior shear wall core as a tube-in-tube system.

The moment-resisting Vierendeel steel frame can be stiffened by concrete encasement as used in steel–reinforced concrete construction in Japan. The moment-resisting steel frame can also be used with concrete shear panel in-fills. Similarly, concrete fascia panels can also be integrated into the structure to provide adequate lateral stiffness. These forms, however, tend to have limited application and are not in common use in the United States.

23.3.2 Floor Framing and Slab Diaphragm Components

Structural steel floor framing consisting of composite or noncomposite beams or trusses whose main function is to support the floor and its gravity load, has a great advantage over other types of floor framing. The benefits are longer spans and lightness, reduction of labor due to prefabrication, faster construction, and the ability to be connected to other concrete or steel elements with simple details. The steel framing may consist of solid-web standard rolled beam sections or open-web steel trusses. The span range is from 30 to 60 ft (10 to 20 m). This aspect of spannability makes this type of floor framing suitable for use with concrete framed-tubes, tube-in-tube, and modular-type systems where a longer span between tubular walls or from the interior tube to the exterior tube is generally desired. The metal deck, often cellularized for floor electrical and power distribution, is commonly used as the slab element with a concrete topping. The metal deck not only supports the floor loads but also serves as a construction platform on steel framing. Composite design for the steel beams as well as for the deck, utilizing the floor slab concrete, is used commonly. In these instances, the metal deck acts as positive reinforcing in the composite slab assembly. The composite beam behavior is established by welding shear studs through the metal deck and the composite action for the deck is established by side embossments on the metal deck or other suitable mechanical devices. Chapter 20 treats metal decking in some detail.

The composite slab and beam arrangements are beneficial in establishing an integral floor diaphragm which is necessary to tie the steel and concrete component parts of the mixed system together. This diaphragm is essential in establishing overall system stability and for transferring wind or earthquake lateral shear forces to respective resistive elements.

23.3.3 Columns

Concrete columns of various types are utilized in the mixed system as vertical elements of concrete framed-tube or other concrete subsystems. Precast concrete columns can also be used as part of a precast framed-tube. Encased steel columns can be part of an encased steel frame or can be used as composite columns in an otherwise concrete building to reduce the size of the concrete columns. Filled steel tube columns have no significant application in the high-rise mixed systems. They are generally confined to low-rise buildings. Steel columns of various types are used in mixed systems. They can be part of a moment-resisting frame braced by walls, panels, or claddings. Steel columns are also utilized as part of a non-moment gravity load subsystem which is braced by shear walls or concrete framed-tubes.

23.3.4 Mixed System Combinations

As a first step, it is essential to examine each system combination for its potential to comply with the basic structural requirements. The integral behavior of all components of the mixed system taken together must satisfy all the requirements of the system. For high-rise buildings, this can be generally stated as that required to satisfy strength, stiffness, stability, and ductility of the system subjected to gravity and lateral loads.

Lateral load resistance is perhaps the most controlling subsystem into and around which are fitted other subsystems required for gravity loads. The lateral load subsystem must therefore: (1) support its share of gravity forces; (2) resist lateral load shears and overturning moments; (3) provide the stiffness required for lateral sway or drift limitations; and (4) provide the stiffness required for overall system stability under gravity loads.

The gravity load subsystems include floor slabs, floor framing, and vertical supports which do not participate directly in lateral load resistance. The lateral load subsystem is generally needed to provide the overall system stability. The floor diaphragm consisting of a floor slab, stiffened by the floor framing, provides the essential link between these two subsystems. The stiffness of the diaphragm and its connections to the subsystems to transfer the in-plane shear forces is of extreme importance.

Investigation of the mixed system to satisfy ductility, strength, and detailed requirements for seismic resistance is essential apart from lateral load resistance. Connection details, in particular those between steel and concrete elements, should undergo critical review for their ability to absorb large deformations produced under seismic conditions.

Construction efficiency of a certain mixed form relates to: (1) level of industrialization and use of prefabricated elements which will minimize shop and field labor; (2) simplicity of connection details between elements; (3) construction sequence and its effect on speed of construction; and (4) feasibility of utilizing larger capacity lifting equipment. The overall construction economy is controlled by such factors as: (1) material and labor availability; (2) availability and capability of facilities to produce prefabricated elements in steel and/or concrete; (3) experience and familiarity with mixed system construction; and (4) labor group requirements associated with different trades.

It is obvious that one has to use considerable judgment in the evaluation of a certain combination. It is clear that what may be efficient and economical under one set of circumstances may be invalid or ineffective under others.

Table 23.1 Mixed Systems Combinations

Systems	Lateral Load Resisting Subsystem	Floor Framing	Columns	Range
Core-braced system (Fig. 23.3)	Closed-form core shear walls	Shear connected composite beam with composite metal deck slab	All columns exterior to core are structural steel	Commonly used. Usual range 20–45 stories. Beyond 45 stories, may not be efficient because of large wall thickness and uplift problems in core.
Core-braced exterior steel frame (Fig. 23.3)	Closed-form core shear walls. Perimeter moment-connected steel frame	Shear connected composite beams with composite metal deck slab	All columns exterior to core are structural steel	Used only when core walls are inadequate to provide lateral resistance. Range 35–55 stories.
Composite framed-tube (Fig. 23.8)	Framed-tube on exterior	Steel composite beams and trusses with composite metal deck	All interior columns are structural steel	Commonly used. Usual range 20–60 stories. High cost effectiveness. Applicable for various plan shapes. High torsional rigidity.
Tube-in-tube system—concrete (Fig. 23.9)	Framed concrete tube on exterior. Closed-form core shear walls	Composite beams framing to exterior tube and interior shear wall	Interior columns outside of core are of structural steel	Used for 45–70 story range
Composite modular tube (Fig. 23.12)	Modular framed tube of concrete	Composite beam and metal deck slab framing	All columns of modular tube are of concrete and the rest structural steel	Up to 80 stories
Tube-in-tube system—concrete/steel	Framed steel tube on exterior. Closed-form core shear walls	Composite beam and metal deck slab framing	All columns exterior to core are structural steel	Used for 40–60 story range
Composite steel frames (Fig. 23.13)	Moment-connected steel frames which are composite with floor slabs	Composite beam and metal deck slab framing	All columns are of structural steel	Up to 35 stories
Panel braced steel frames (Fig. 23.14)	Moment-connected steel frames with interior or exterior panel bracing	Composite beam and metal deck slab framing	All columns are of structural steel	Up to 30 stories

Table 23.1 shows a listing of several feasible system combinations. The identification of different components together with a brief explanation of range and applicability are included. The table can be utilized as a general guide to the forms of mixed systems that have been used or can be considered in design currently. It should be noted that many other forms may also be feasible under specific and special conditions.

23.4 CONCRETE CORE-BRACED SYSTEM

Figure 23.3 shows a common form of concrete core-braced mixed system. For convenience, a rectilinear floor shape with a rectilinear closed-form core is shown, although other shapes are possible. Whatever the shape, the basic concept in this system is that the concrete core provides full lateral resistance. The steel structural framing around the core is designed only for gravity forces using simple shear connections. All steel framing follows the standard details of steel construction. The steel beams are connected to the core walls by a typical corbel detail (Fig. 23.3), by bearing in a wall pocket, or by an anchor shear-plate detail. The diaphragm action is established with the use of a typical arrangement of composite slabs on composite beams. Reinforcing dowels are provided from the core walls to the slab to engage the slab diaphragm. Steel columns outside of the concrete core complete the steel part of the framing. Cladding and window wall details used with a steel building are applicable here.

The core's primary function is to house various centralized building service elements, such as elevators, toilets, closets, etc. And, shear walls are conveniently provided around these elements.

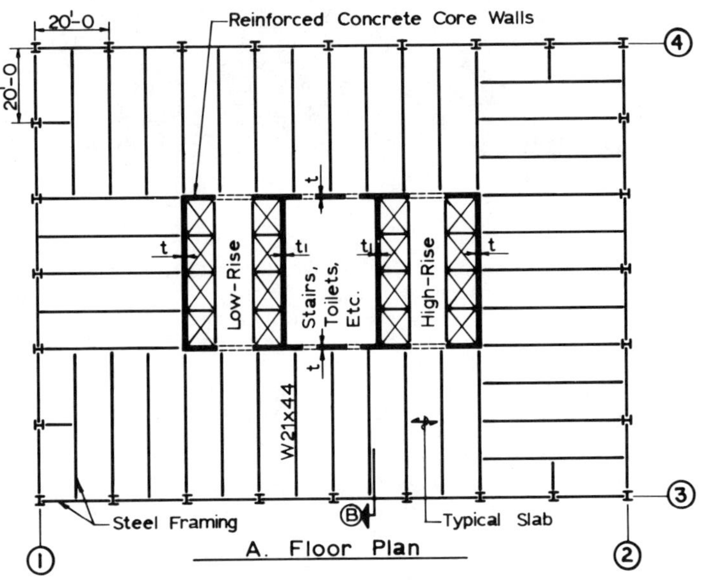

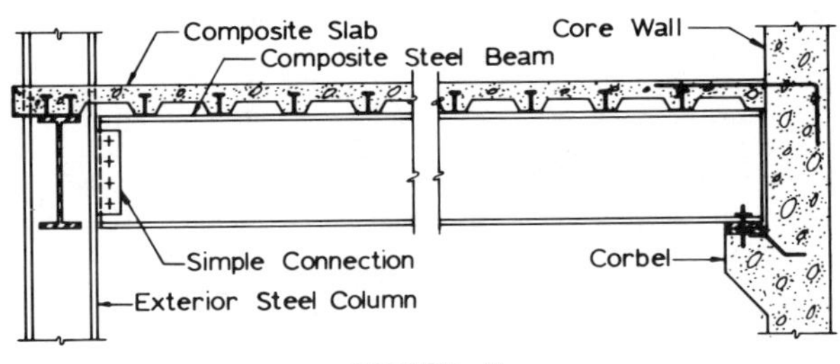

Fig. 23.3 Central core braced system.

Fig. 23.4 Denver Square Tower under construction.

This combination uses a steel framing that is not required to resist lateral loads and therefore, will not involve rigid or semi-rigid connections. During construction, the concrete cores are normally built up to a certain number of stories, after which the steel framing is started and filled around the core. In many recent applications, the core walls have been slip-formed to their entirety before steel framing is started. Figure 23.4 shows a photograph of the system under construction and clearly indicates this sequence of construction. Since the lead time for delivery of structural steel is usually longer than that for reinforcing steel, significant savings in time can be achieved by building the concrete core ahead during the differential lead time period. In situations where the concrete core cannot be built ahead, it is possible to let the steel framing proceed with the aid of small steel columns which are then encased and embedded in concrete core walls built later. This version requires particular coordination of formwork for the construction of shear walls which will be built in a steel frame.

The concrete core-braced system is economical for buildings up to 45 stories in height. For taller structures, the core-braced systems are generally inefficient and ineffective due to the large thicknesses or large plan dimensions required for shear walls.

The shear walls in the core need to be coordinated with the functional requirements of the core. For instance, as one elevator bank terminates, the shear walls for this part may have to be discontinued, thus resulting in an eccentric core which must be considered in core wall analysis and design. The floor framing inside the core can be either cast-in-place concrete, precast concrete, or structural steel. Precast concrete and steel framing offer an efficient alternative to cast-in-place concrete framing since formwork and shoring are not needed.

The concrete core is required to provide the necessary torsional and flexural rigidity and strength with no participation from the steel system to resist lateral loads. The torsional resistance of the core should also be considered for possible asymmetry of the wall system where eccentricities of lateral load with respect to the center of rigidity exist. Conceptually, the core system should be treated as a cellular wall system of a tubular form with punched openings to result in an effective cantilever behavior. Beam interconnections between different wall units are necessary to establish the cantilever behavior.

The plan dimensions of the core should be selected so as to produce optimum flexural stiffness of the cantilever tube for a particular volume of the building. The floor framing span outside the core should be such as to distribute enough gravity loads to the core walls so that their design is controlled by compressive stresses even under wind loads. The geometric location of the core should be selected so as to minimize eccentricities for lateral load.

Simple cantilever wall analysis may be sufficient in most cases. However, if the total form is used as a wall tube, it may be necessary to perform a more sophisticated three-dimensional analysis including the effect of openings. An analysis of this type may also be required to evaluate torsional stresses when the vertical profile of the core wall assembly is unsymmetrical. Such analysis is best performed by use of computerized three-dimensional frame analysis or finite element analysis with the walls simulated as panelized plate elements.

The following is a brief summary of analysis methodologies that can be utilized. If the core consists solely of uncoupled, solid walls, the analysis can be done by simple cantilever beam analysis utilizing the moment of inertia and shear area of the walls. Wind shears can be distributed to each wall on the basis of its relative rigidity. For example, for the core shown in Fig. 23.5 for bending about the *X* axis, the analysis can be done on this basis.

When the walls are connected by link beams, they form coupled shear walls and the distribution of lateral load is no longer based on relative stiffness, since such walls involve a combination of cantilever and shear-frame modes of deformation.

Openings in shear walls generally occur in vertical rows (Fig. 23.5*b*). If the openings are small, their effect in a shear wall may be minor. Larger openings have a more pronounced effect and can result in a system in which frame action is a substantial component. A more detailed discussion is given in Refs. 23.8 and 23.19.

Frame analysis can be performed using computers, recognizing the finite width of the wall system. Infinitely stiff beam stubs from the centerline of the wall to the edges of the actual opening are introduced as shown in Fig. 23.5*c*. Two-dimensional frame analysis can be used for bending about the *Y* axis also as shown in Fig. 23.5*c*. It should be noted that in this type of model, the two

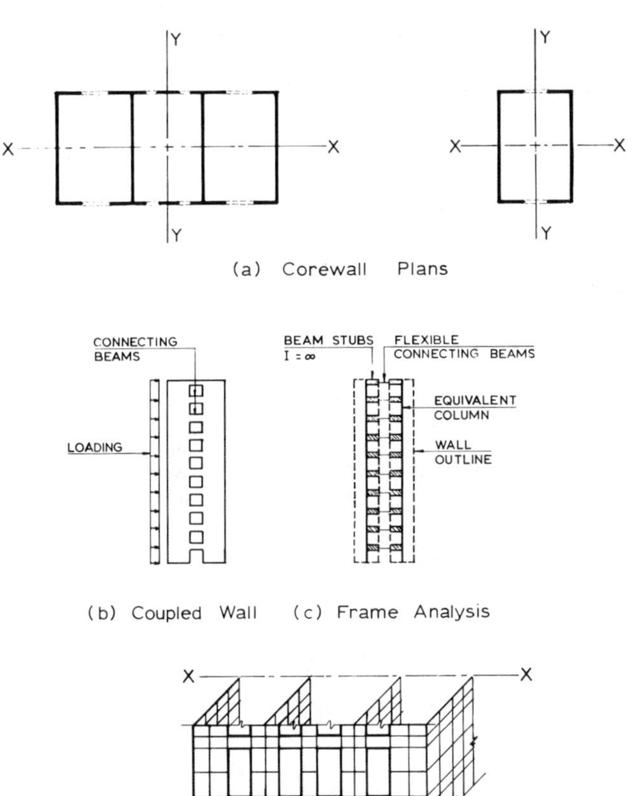

Fig. 23.5 Core wall analysis methods.

link beams are lumped and the end wall properties correspond to a channel, while the middle one corresponds to that of an I-section. If the link beams were to be present in both the X and Y directions, a three-dimensional frame model can similarly be used.

Finite element analysis can also be performed for both two- and three-dimensional models as shown in Fig. 23.5*d*. The wall is divided into a mesh of two-dimensional plate elements and accurate solutions can be obtained by imposing appropriate boundary conditions. The accuracy of the solution depends on the fineness of the mesh and the type of finite element. One can obtain more accurate torsional stresses and deformations by using finite element models.

The steel columns of the system are designed for gravity loads only, using an effective length factor of $K = 1$ on the assumption that the steel frame is braced by the wall system.

The floor diaphragm plays a significant role in terms of providing lateral support to the steel frame which lacks the inherent capacity to resist frame instability. The required bracing force to stabilize a steel column under axial load can generally be computed from $P\Delta/h$, where P is the column axial load, Δ is the maximum expected sway plus erection tolerance, and h is the story height. The diaphragm stresses due to wind forces are generally small and can be readily resisted by the slab. The required $P\Delta/h$ forces are to be considered as additional lateral shears in the wall system and the design should be performed accordingly. Differential shortening between the steel columns and concrete core walls should be considered. The calculations should include shrinkage and creep effect in the concrete core. Elevation corrections to the steel columns are generally necessary to maintain floor levels within reasonable tolerances. See Chapters 10, 11, and 22 for additional material on P–Δ effects, differential shortening, and other effects.

A typical design example for a 39-story office tower is given in Table 23.2 which shows the various structural parameters and is generally self-explanatory. Many such examples of concrete core-braced systems exist in the United States as well as in many other parts of the world [23.12, 23.14].

The concrete core-braced system can be combined with an exterior, moment-connected steel frame, creating a shear-frame interacting system. These systems are generally utilized when the core walls are inadequate to provide all the lateral load resistance. In Fig. 23.3, an interacting core-

Table 23.2 Design Example—Denver Square Tower, Denver, Colorado

No. of stories: 40	Floor area = 810,000 ft/sq ft
Structure height:	507 ft
Floor plan dimensions:	See Figs. 23.3 and 23.4
Office live load:	50 psf
Partition live load:	20 psf
Average wind pressure:	35 psf
Drift limit index:	$H/500$
Translatory period:	5 sec (short direction)
Torsional period:	2 sec

Structural system:

The structural system consists of an exterior steel frame and interior cast-in-place concrete core walls. All lateral loads are resisted by the interior core walls and therefore all the beams on the exterior steel frame are simple connections. The floor framing consists of standard rolled shaped steel beams, all with shear studs to act compositely with the floor slabs.

Slab: 2-in. composite metal deck with 3¼-in. slab
4000-psi structural lightweight concrete.

Steel floor framing:	W21 × 44, A36 typical composite steel–concrete beams @ 10 ft-0 in. o.c.
Core walls:	18 in. up to 25th level 15 in. 25th to roof 12 in. all interior walls 5000 psi normal-weight concrete, ground to 20th floor 4000 psi normal-weight concrete, 21st floor to roof

Material quantities:

Average structural steel	9.5 psf
Corewall steel reinforcement quantity	1.7 psf
Corewall concrete quantity	0.3 cu. ft/sq ft

1 in. = 25.4 mm; 1 ft = 0.3048 m; 1 sq ft = 0.093 m^2; 1 psf = 0.0479 kN/m^2; 1 psi = 0.006895 MPa; 1 cu. ft/sq ft = 0.3048 m^3/m^2.

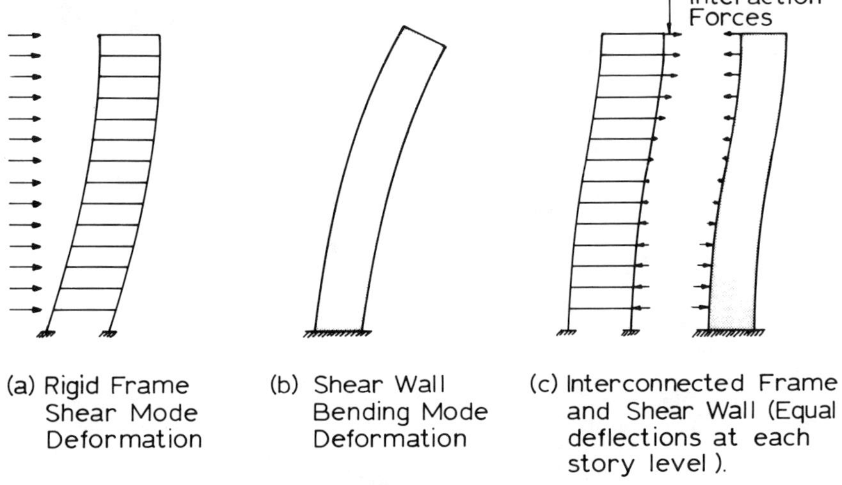

Fig. 23.6 Frame–shear-wall interaction.

braced system is created if the fascia frames along lines 1, 2, 3, and 4 are moment connected. In this case, the fascia frames parallel to the direction of the wind will interact with the core wall. The total drift of the interacting shear wall–frame system is obtained by superimposing the individual modes of deformation as shown in Fig. 23.6. The internal interactive forces reduce the drift of the overall combined system, creating a higher order of stiffness. The interacting system concept has been fully developed for all-concrete buildings and many examples of this construction exist currently. The only difference in the mixed-system concept is that the frame component, as well as floor framing and other interior framing, are in steel.

Table 23.3 shows an interacting system for a 52-story tower. The core walls by themselves would not provide adequate stiffness for wind forces on broad faces of the building. The addition of

Table 23.3 Design Example—Eau Claire Estate Tower, Calgary, Canada

No. of stories: 52	Floor area = 1,350,000 sq ft
Structure height:	730 ft
Floor plan dimensions:	See Fig. 23.7
Office live load:	50 psf
Partition live load:	20 psf
Average wind pressure:	33 psf for deflection; 38 psf for strength
Drift limit index:	$H/500$
Translatory period	
short direction:	5.43 sec
long direction:	4.07 sec
Torsional period:	2.38 sec

Structural system:
Eau Claire Tower utilizes tube-in-tube concept. It has an interior concrete core and an exterior steel frame. Both are connected together by composite trusses. The maximum concrete strength = 6 ksi. See Fig. 23.7 for wall thickness.
Slab: 3-in. metal deck composite with 2½-in. concrete topping

Material quantities:

Average structural steel	14.0 psf
Corewall steel reinforcement quantity	1.8 psf
Concrete quantity	0.37 cu. ft/sq ft

See Table 23.2 for SI units.

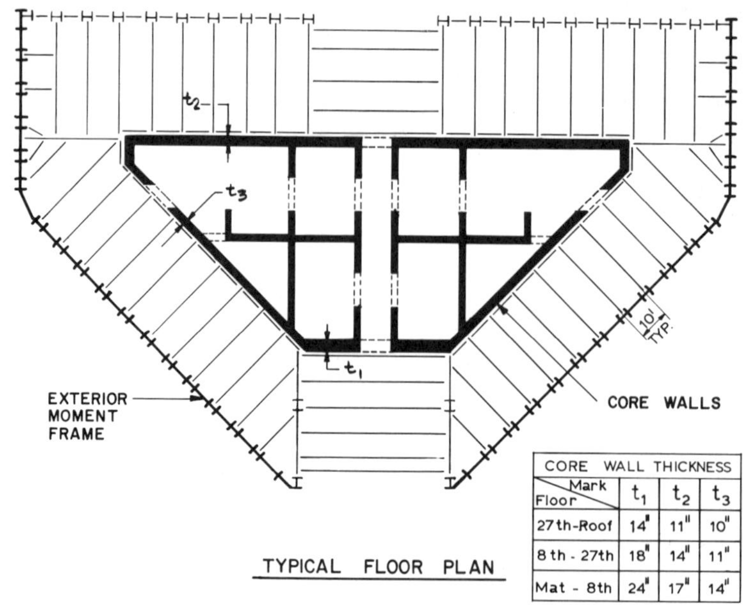

CORE WALL THICKNESS			
Floor \ Mark	t_1	t_2	t_3
27th-Roof	14"	11"	10"
8th - 27th	18"	14"	11"
Mat - 8th	24"	17"	14"

Fig. 23.7 Floor plan of Eau Claire Tower, Calgary, Canada.

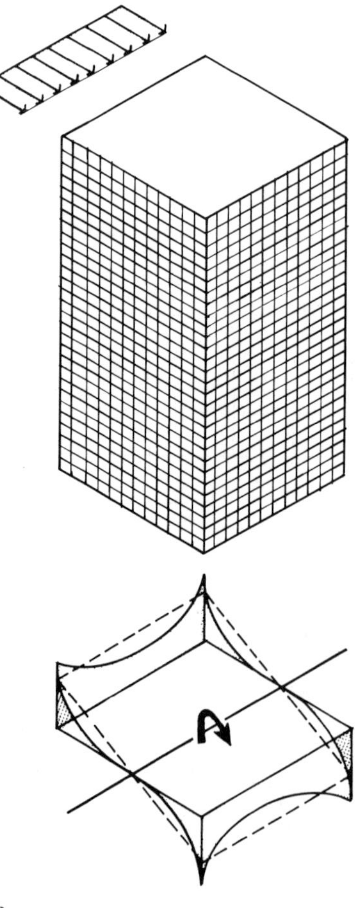

Fig. 23.8 Lateral resistance of tubular system—a schematic view.

steel frames on the sides of the building increased stiffness by 100%. It should be noted that because of the participation of the steel frame, the steel quantities as well as the character of the steel frame have been changed. The table also provides other information which is self-explanatory.

23.5 COMPOSITE FRAMED-TUBES

Composite framed-tubes generally involve the combination of an exterior concrete tubular system with nonrigid, steel framing on the interior. All the wind resistance is derived from the concrete tubular system. Since the system uses the exterior form of the tube as an equivalent cantilever, as shown in Fig. 23.8, it has the potential of being utilized for extremely tall structures up to 100 stories.

The general arrangement is shown in the plan form in Fig. 23.9, which shows the framed-tube on the periphery and steel floor framing and columns on the interior. All the interior framing has nonrigid connections designed for gravity forces only. Composite metal decks with composite steel beams or trusses are used for floor framing as in a steel structure.

These systems are generally constructed with the steel proceeding ahead of the concrete components. Small size exterior steel columns can be used to advance the steel construction which then become encased in the concrete tube as shown in the details of Fig. 23.9. The floor diaphragm is attached by doweling the slab unit into exterior columns and beams. The character of the tubular system is that of a punched wall with wide columns and deep beams. Generally, 30 to 40% of the punched tube wall is covered with columns and beams. The column spacings may vary from 6 ft (2 m) to about 15 ft (5 m), depending on proportions of members. This type of punched tube wall can be constructed with gang forms or slip-forming techniques.

When the composite tubular system was introduced in the mid-1960s, many initial applications were simple prismatic rectilinear shapes, such as the one shown in Fig. 23.9. The ability of the

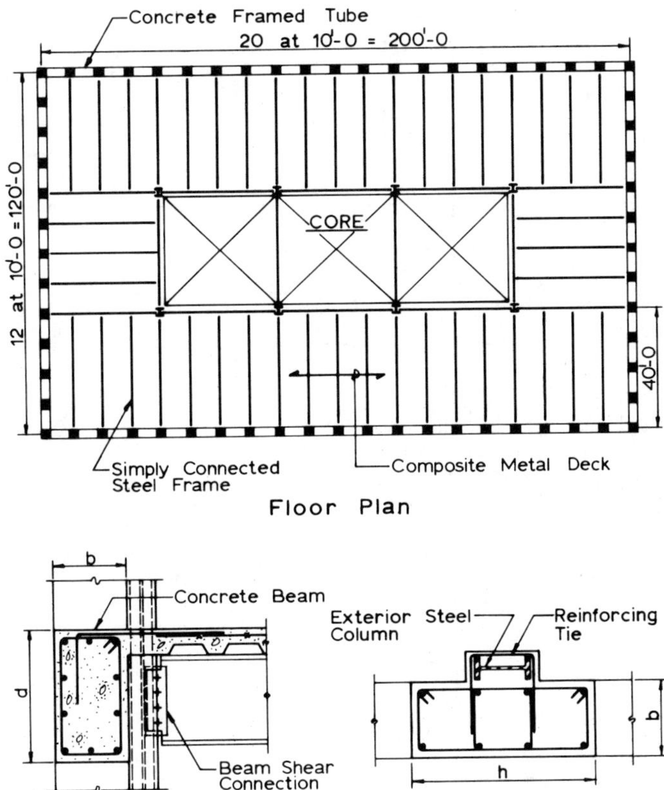

Fig. 23.9 Composite framed tubular system.

Table 23.4 Design Example—Three First National Plaza, Chicago, Illinois

No. of stories: 57	Floor area = 1.5 million sq ft
Structure height:	753 ft
Floor plan dimensions:	See Fig. 23.10
Office live load:	50 psf
Partition live load:	20 psf
Average wind pressure:	30 psf
Drift limit index:	$H/550$
Translatory period:	5.0 sec
Torsional period:	1.5 sec

Structural system:
The 3FNP Tower utilizes composite tubular concept. The entire lateral rigidity is derived from the closely spaced exterior columns connected together through a rigid diaphragm. The maximum concrete strength used in columns was 7 ksi.

Slab: 3-in. composite metal deck with 2½-in. thick 4000-psi lightweight concrete.

Steel floor framing: See Fig. 23.10

Material quantities:

Average structural steel	10.15 psf
Exterior tube steel	3.35 psf
Exterior tube concrete	0.35 cu. ft/sq ft

See Table 23.2 for SI units.

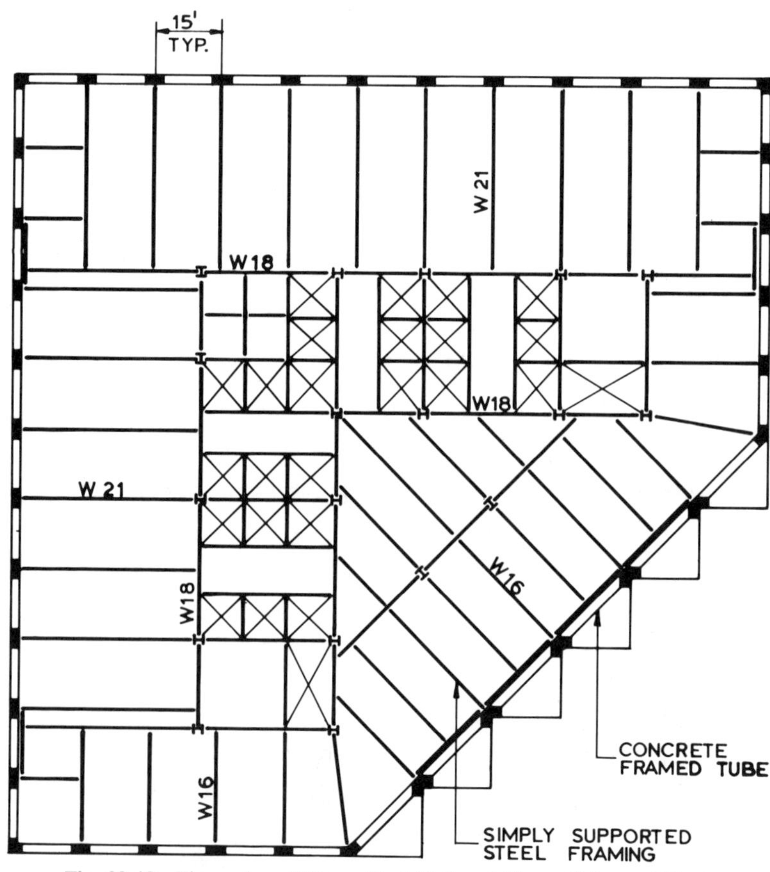

Fig. 23.10 Floor plan of Three First National Plaza, Chicago, Illinois.

exterior concrete tube to conform to various exterior shapes makes it a versatile system and has promoted its use for other shapes. Further, the introduction of interior framed-tube lines to include participation with an interior shear wall has brought about many more possibilities for composite tubular system and a variety of applications can therefore be observed.

Three distinct forms of composite tubular systems can be formulated:

1. Exterior composite tube system involving exterior concrete tube and interior steel framing.
2. Composite tube-in-tube involving exterior concrete tube, interior shear-wall core, and steel floor framing.
3. Modular tubes or bundled tubes involving a combination of a cellular shape with interior framed tube lines. The floor framing is of structural steel.

Design examples for each type are provided in Tables 23.4, 23.5, and 23.6, respectively.

Table 23.5 Design Example—Minneapolis City Center, Minneapolis, Minnesota

No. of stories: 52	Floor area = 1.35 million sq ft
Structure height:	692 ft
Floor plan dimensions	See Fig. 23.11
Office live load:	50 psf
Partition live load:	20 psf
Average wind pressure:	32 psf
Drift limit index:	*H*/500

Structural system:
The tower utilizes tube-in-tube concept. The core walls comprise the interior tube and the exterior concrete columns comprise the exterior tube. Both are connected together by steel framing.

Slab: 3¼-in. lightweight concrete with 2-in. deck electrified

Steel floor framing: W21 × 44 and W21× 50, A36 @ 10 ft-0 in. o.c.

Core walls:	12-in. webs		
	Flanges	18 in. to 4th	6000 psi ground to 16th floor
		16 in. to 30th	5000 psi 16th to 22nd floor
		12 in. to roof	4000 psi 22nd to roof

Material quantities:

Average structural steel	5.7 psf
Corewall steel reinforcement quantity	1.2 psf
Column and spandrel steel reinforcement	2.5 psf
Misc. corbel and slab steel reinforcement	0.5 psf
Concrete quantity	0.52 cu. ft/sq ft

See Table 23.2 for SI units.

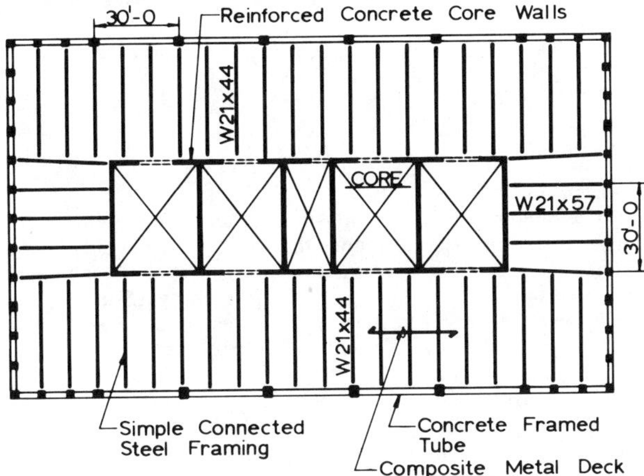

Fig. 23.11 Composite tube-in-tube system—Floor plan.

The choice of a particular structural system is related to the overall geometry, especially the vertical profile and the need for added structural stiffness for extremely tall structures and construction coordination.

The exterior composite tube represents the simplest concept, allows for advance erection of structural steel, and construction coordination is generally simpler because of concrete construction outside or on the periphery of steel construction. However, the rigid discipline of the exterior structure does not allow for exterior profile modifications.

The tube-in-tube composite system does not lend itself to advance erection of steel. Either the core is built first with slip-form techniques and other components follow, or the core, steel, and exterior tube are advanced on a floor-by-floor basis. The advantage in the tube-in-tube system is that added stiffness can be obtained for slender building forms.

The modular tube versions offer the potential for exterior profile modification by terminating individual cellular units when they are not needed for structural resistance. The generalized concept of creating modular tubes of various cellular shapes offers considerable freedom of geometric expressions. The presence of interior tubular frame lines improves the overall efficiency of any structural form. It should, however, be recognized that the presence of interior concrete lines renders the construction cumbersome and may lead to some loss of construction efficiency.

The lateral load analysis generally involves only the framed-tube neglecting nominal participation of the steel part. The lateral loads are resisted by the entire three-dimensional form of the tube, which induces both axial force and bending moment in columns. A preliminary analysis of such structures can be approached on the basis of estimating the cantilever forces and the shear-frame forces. In this analysis, considerable judgment will have to be exercised to allow for the inefficiencies associated with a perforated tubular wall as opposed to a solid wall. A detailed discussion on the estimation of cantilever effectiveness of tubes is presented by Khan and Amin [23.17] which is useful as a practical guide.

The final structural analysis of the framed-tube can be performed by utilizing three-dimensional frame analysis programs which consider column axial deformations. The effect of the large widths and depths of members in terms of stiffness contribution should be included. This can be done by adjusting the member properties by the ratio of clear span to centerline span for beams and columns or by separate evaluation of a one-story model with finite element analysis. Similarly, more refined analysis can be performed by use of the finite element approach on a panelized model of the punched tube.

The design and detailing of the members of the framed-tube are performed using the latest *ACI Code* [23.1]. The exterior concrete column is treated as a composite column including the small

Table 23.6 Design Example—Ohio National Bank Building, Columbus, Ohio

No. of stories: 25	Floor area = 400,000 sq ft
Structure height:	310 ft
Floor plan dimensions:	See Figs. 23.2 and 23.12
Office live load:	50 psf
Partition load:	20 psf
Average wind pressure:	23 psf
Drift limit index:	$H/1000$

Structural system:

The building utilizes composite bundled-tube system. The cluster consists of 6 tubes that are terminated at different levels. The floor framing consists of 60-ft span composite trusses at 15-ft spacing, spanned by a 3-in. cellular deck composite slab. The trusses are attached to concrete tubes by a typical anchor-plate detail.

Material quantities:

Average structural steel	7.0 psf
Tube reinforcement quantity	3.0 psf
Tube concrete quantity	0.4 cu. ft/sq ft

See Table 23.2 for SI units.

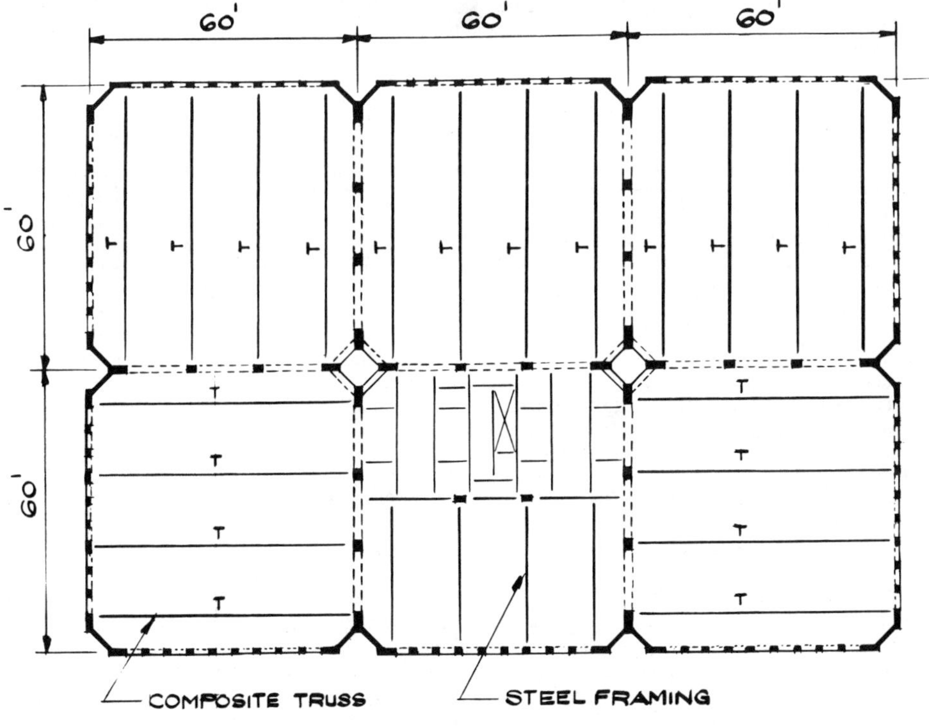

Fig. 23.12 Floor plan of Ohio National Bank, Columbus, Ohio.

steel columns if present. In many cases, the contribution of this steel column is relatively small and therefore it is generally neglected in design.

The floor framing generally consists of composite beams or trusses with a composite floor deck. The design of the composite floor deck is performed following manufacturer's recommendations and test data on proprietary decks. The composite beam/truss design is performed using the *AISC Specification* [23.2] on an allowable stress basis. The steel column design, both interior and exterior, is also performed using the *AISC Specification* [23.2]. Since the concrete framed-tube provides all lateral resistance, an effective length factor of $K = 1$ is used for all steel columns.

23.6 COMPOSITE STEEL FRAMES

The steel frame system, consisting of regularly organized Vierendeel rigid frames in each direction, can be transformed into a composite form by concrete encasement of beams and columns [23.13]. In order to take best advantage of the concrete encasement, the steel part can be reduced and reinforcing steel bars can be added where required. This type of composite steel rigid frame would be similar to the steel–reinforced concrete construction of Japan where light, built-up steel members are encased in concrete and augmented with longitudinal reinforcing steel and ties. This form is labor intensive and, therefore, not used in the United States.

Another form of an unbraced composite steel frame utilizes the normal construction of a rigid frame with unencased composite beams and uses reinforcement in the negative moment areas as shown in the frame plan in Fig. 23.13. Shear studs are provided over the entire beam length to develop the positive and negative regions of the beam. Two types of composite frames of this form are possible. In one type, reinforcements are provided only to control cracks in the negative moment areas, in which case the composite properties are utilized to satisfy only the lateral load stiffness of the frame with the beam section itself capable of resisting the forces from a strength point of view. A second type utilizes the reinforcements for strength development in addition to

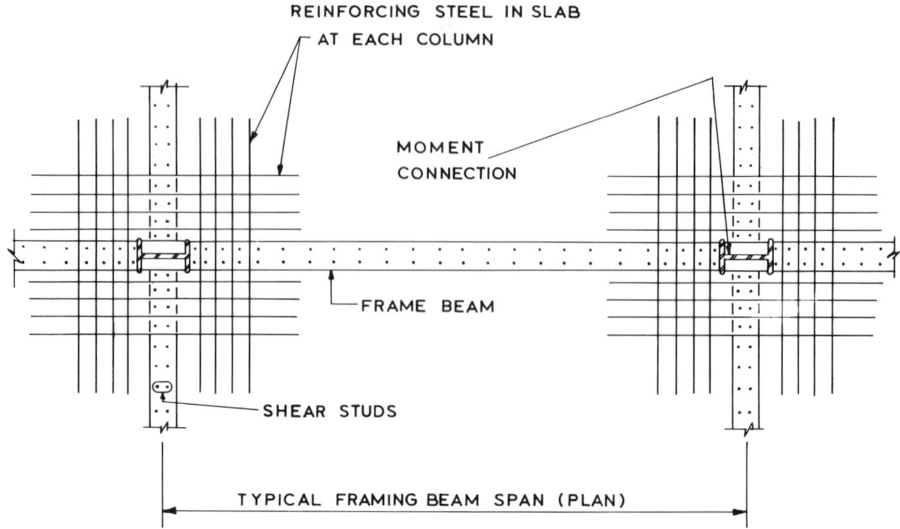

Fig. 23.13 Composite frame plan with additional reinforcement.

stiffness contribution. For the second type to be practically feasible, the spans and the number of stories will have to be reasonably small.

Unencased composite steel frames of the type discussed offer opportunities to utilize floor slabs to contribute to lateral stiffness and they, therefore, improve the general efficiency of a noncomposite rigid steel frame.

23.7 PANEL-BRACED STEEL FRAMES

The panel bracings generally take the form of shear panels infilled or attached to the steel frame, which by virtue of their in-plane rigidity, contribute to the overall lateral load stiffness of the steel frame. These panels serve the basic architectural function of providing building enclosure or interior partitions in the core areas to serve as room dividing partitions in apartment and hotel-type occupancies. Their other function, to perform the structural stiffening function, will result in potential cost savings. In particular, the steel required to achieve drift limits under lateral load is reduced along with simplification of steel details in a typical steel frame structure.

The panels are generally of reinforced concrete and they are attached to the steel frame either on the exterior or interior (Fig. 23.14). Figure 23.14*a* shows some types of concrete panels that can be used on the interior of the steel frame. The panels are precast either at a factory or on the site and they are connected to the frame. The panels may be reinforced with reinforcing bars or light steel shapes, such as angles, channels, or tees. The embedded steel brace, if present, provides for a direct method of transfer of primary forces to the steel frame. The connection of panel with steel frame may be a simple anchor plate equally spaced around the panel, as shown in Fig. 23.14*a*, or a more complex attachment of steel bracing to the steel by means of plates. A logical extension of the interior bracing panel is to use it as architectural cladding enclosing the building and attached to the steel frame in such a way that it could be considered as an exterior panel bracing that resists wind forces (Fig. 23.14*b*). The size of the panel as related to bay dimensions of the steel frame, the extent of vision panels, and the method of attachment of the panel to the frame, all influence the effectiveness of this type of cladding to brace the frame. The transportable size of the precast unit and its lifting weight also are primary factors. In spite of these limitations, considerable potential exists for utilizing the panelized building exterior for structural purposes.

23.8 COMPOSITE MEMBERS

A composite member is comprised of concrete and structural steel elements so interconnected that the component elements act together as a unit. A composite member may be a composite column, composite beam, composite deck, composite truss, or a composite girder. This section summarizes the methods available for design of the composite members.

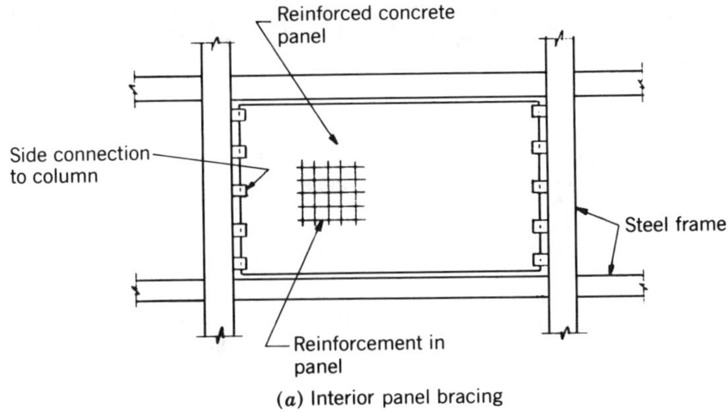

(*a*) Interior panel bracing

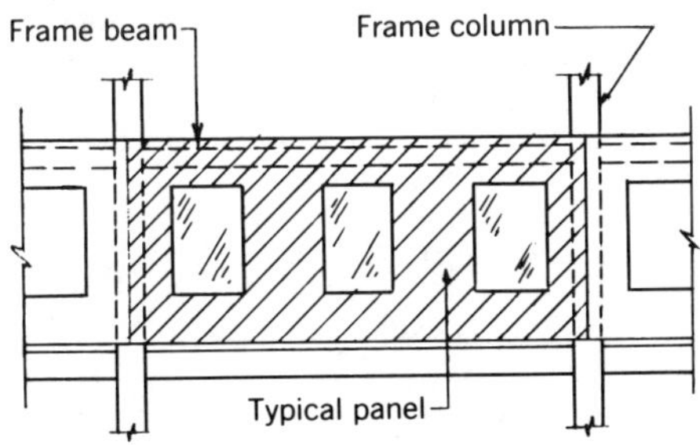

(*b*) Exterior panel bracing

Fig. 23.14 Panel-braced steel frame.

Under vertical and lateral loads the stresses to be accounted for in design within the individual composite elements are principally due to flexure and shear in horizontal members; and compression, flexure, and shear in vertical elements. Both working stress and ultimate strength design methods are used for composite member design.

23.8.1 Composite Columns

A composite column is a compression member that is constructed with load-bearing concrete plus steel in any form different from reinforcement rods. Composite columns may be of two kinds: (a) concrete-filled tubular steel columns, and (b) concrete-encased steel columns, as shown in Figs. 23.15*a* to *d*. Figure 23.15*a* shows a composite column in which the steel tube serves both as form and reinforcement. Figure 23.15*b* shows a steel column encased by plain concrete. Such encasements were used mostly in the 1950s for fireproofing of steel. A very low-strength concrete was used and no attempt was made to utilize the concrete's contribution to strength and stiffness of the column. Low-density, spray-on contact fire protection material has mostly replaced the concrete

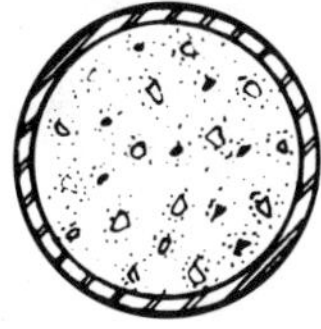

(*a*) Concrete-filled tubular column

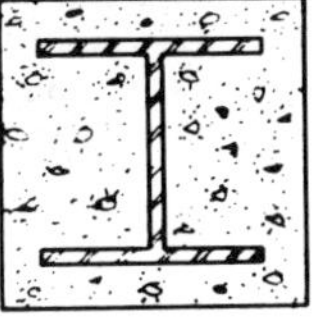

(*b*) Encased steel column

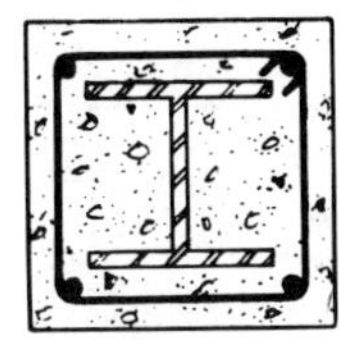

(*c*) Reinforced column with encased steel shape

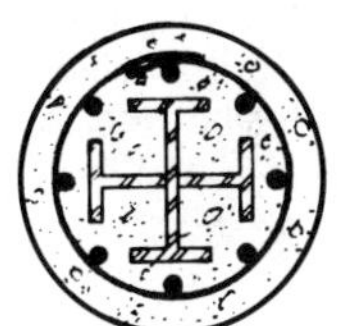

(*d*) Circular composite column

Fig. 23.15 Composite column types.

encasement in North America; however, the concrete encasement may still be used for fire protection outside North America. Figures 23.15*c* and *d* show a column with steel core and additional reinforcing. This allows the steel construction to proceed at the usual structural erection speed after which it is encased in concrete. This form of construction also enhances ductility and shear resistance of reinforced concrete construction in seismic zones.

Because of inelasticity of materials, the mathematical solution to the column problem and method of analysis are relatively complicated. In terms of composite column design, several methods are currently available. A brief review of some pertinent methods is given below.

ACI–UBC Method. Both the *ACI Code* [23.1] and UBC [23.11] stipulate the strength design method for composite columns. The strength of a composite column is computed for the same limiting conditions applicable to ordinary reinforced concrete members. In essence, the ACI–UBC procedure for composite columns is based on an ultimate concrete strain of 0.3%, elastic–plastic steel behavior, linear strain distribution, and disregarding tensile stresses in concrete. Either a rectangular stress block or a parabolic stress distribution can be assumed for concrete in compression. A minimum eccentricity e and a strength reduction factor ϕ are always applied.

The minimum eccentricity provision is desirable for reinforced concrete construction in light of possible accidental eccentricities due to imperfect positioning of members and reinforcement, nonuniformity of materials, and minor discrepancies between assumptions made in the analysis and actual behavior. However, in case of composite construction where structural steel components are prefabricated with controlled tolerance, the ACI minimum eccentricity appears quite conservative.

Comparisons by Furlong [23.10] of these composite column design provisions with test results and AISC [23.2] design provisions for steel columns alone indicated the following:

1. Imposition of minimum eccentricity by the *ACI Code* [23.1] penalizes the column. Therefore, the steel column alone carries more load than the composite column in some instances. This is because the *AISC Specification* [23.2] does not require minimum eccentricity for axial-load-only conditions.
2. Differences in philosophies regarding the strength reduction factor ϕ between steel and concrete column design methodologies accentuate the capacity differences.
3. Application of the creep factor β_d similarly affects the capacity.

Furlong [23.9] published rules for composite columns based on the AISC column design logic, which fit the test data more accurately and reconcile some of the differences that were apparent in the ACI design rules.

The Structural Specifications Liaison Committee (SSLC) and Structural Stability Research Council (SSRC) have attempted to adopt the Furlong approach and an SSRC design proposal [23.25] was published which is briefly reviewed here.

SSRC Design Proposal. The general design requirements are as follows:

1. Steel proportions: at least 4% of gross column sections.
2. Concrete: strength below 3000 psi (21 MPa) is neglected; maximum allowed is 8000 psi (55 MPa).
3. Tie requirements: maximum spacing ≤⅔ the least column dimension; steel area ≥0.007 sq in./in. (0.18 mm²/mm) of bar spacing.

The proposed design approach is based on ultimate strength concepts; however, all design formulae are expressed in allowable stress terms in a form similar to that used in the *AISC Specification* [23.2]. The allowable compressive axial stress F_a on the structural steel area of a composite cross-section is determined from AISC Formula (1.5-1) or (1.5-2), using a modified composite yield stress F_{my} for F_y, a modified composite modulus of elasticity E_m for E, and a radius of gyration r_m for r. The allowable axial compressive force P_a on the composite cross-section is taken as the product of the area of the structural steel shape A_s, and the axial stress F_a.

For concrete-filled pipe or tube:

$$F_{my} = F_y + F_{cr}\frac{A_{cr}}{A_s} + 0.85f'_c\frac{A_{cc}}{A_s} \quad \text{with} \quad F_y \text{ and } F_{cr} \leq 55 \text{ ksi (380 MPa)} \tag{23.8.1}$$

$$E_m = 29{,}000^* + 0.4E_c \frac{A_{cc}}{A_s} \tag{23.8.2}$$

$$r_m = r_s \tag{23.8.3}$$

For concrete-encased structural steel:

$$F_{my} = F_y + 0.7F_{cr} \frac{A_{cr}}{A_s} + 0.6f'_c \frac{A_{cc}}{A_s} \quad \text{with} \quad F_y \text{ and } F_{cr} \leq 55 \text{ ksi (380 MPa)} \tag{23.8.4}$$

$$E_m = 29{,}000^* + 0.2E_c \frac{A_{cc}}{A_s} \tag{23.8.5}$$

$$r_m = r_s, \text{ but not less than } 0.3h_2 \tag{23.8.6}$$

For composite compression members, the allowable flexural stress is as follows:

$$F_b = 0.75F_y \text{ for pipe or tube} \tag{23.8.7}$$

$$F_b = 0.60F_y \text{ for steel shapes} \tag{23.8.8}$$

where A_{cc} = area of concrete effective in composite columns
A_{cr} = area of longitudinal bar reinforcement in a composite column cross-section
A_s = area of steel (shape or tube) in a composite design
F_{cr} = specified yield strength of longitudinal reinforcement in composite column
F_y = specified minimum yield stress of the type of steel being used
f'_c = specified compression strength of concrete
r_s = radius of gyration of the structural shape, pipe, or tube in plane of bending of a composite column
h_2 = overall thickness of a composite column in the plane of bending

The composite columns subjected to biaxial bending in addition to an axial force are designed using AISC Formula (1-6-1a) and a modified section modulus value. Design examples based on this approach are given in Tables 23.7 and 23.8.

A comparison between column capacities reported in laboratory tests and the allowable loads according to these recommendations shows that the average factor of safety for concentrically loaded encased sections was 2.04. The lowest factor of safety (FS) was 1.70. For eccentrically loaded encased column sections, the average FS was 2.02 and the lowest FS was 1.48. No test

* 200,000 MPa in SI units.

Table 23.7 Design of Concentrically Loaded Composite Column

Problem: Design a 20-in. square composite column for an axial load comprised of 700 kips dead load and 300 kips live load for an unsupported length of 15 ft. Use 3500 psi concrete and A36 steel.

Solution:

Try W14 × 82 w/ 8-#8 Grade 60 steel

A_g = 400 sq in.
A_s = 24.1 sq in.
A_{cr} = 6.32 sq in.
A_{cc} = 369.6 sq in.
F_{my} = 78.3 ksi (from Eq. 23.8.4)
E_m = 39343 ksi (from Eq. 23.8.5)
r_y = 6.0 in.
C_c = 99.59
F_a = 42.08 ksi
P_a = 1014 kips > 1000 kips, therefore use W14 × 82 w/ 8-#8 bars

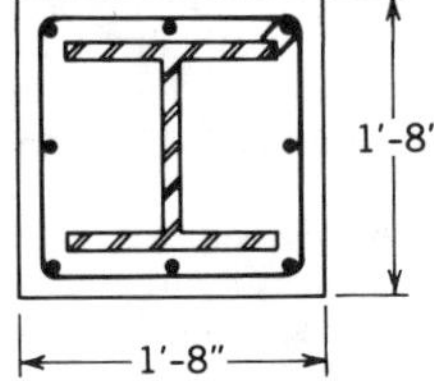

1 in. = 25.4 mm; 1 sq in. = 645 mm²; 1 ksi = 6.895 MPa; 1 kip = 4.448 kN.

Table 23.8 Design of Eccentrically Loaded Composite Column

Problem: Design a 16 × 20 in. rectangular composite column for an axial service load of 500 kips together with a service moment of 180 ft-kips about *x* axis. The unsupported length is 15 ft, $C_m = 0.5$. Use 4000 psi concrete and a 50 ksi steel core shape.

Solution:

Try W8 × 48 w/ 6-#7 Grade 60 bars

$A_g = 320$ sq in.
$A_s = 14.1$ sq in.
$A_{cr} = 3.6$ in.
$A_{cc} = 302.3$ sq in.
$f_a = 35.46$ ksi
$F_{my} = 111.3$ ksi
$E_m = 44458$ ksi
$r_m = 4.8$ in.
$Kl/r = 37.5$
$C_c = 88.8$
$F_a = 55.84$ ksi
$S_m = 79.0$ in.^{3} (from Eq. F in Ref. 23.25)
$f_b = 27.34$ ksi
$F'_e = 162.8$ ksi
$F_{by} = 0.6\ (50) = 30$ ksi

The right-hand side of Eq. (E) in Ref. 23.25 = 0.98 < 1.0 therefore use W8 × 48 w/ 6-#7 bars

See Table 23.7 for SI units.

results are available for biaxially loaded composite column sections. These recommendations are primarily intended for columns of rectangular shapes. The use of this procedure for nonrectangular sections such as those used in some composite systems needs a more rigorous analysis.

Slenderness Effects

Slenderness is generally expressed analytically for columns as a measure of the member flexural stiffness EI/L. In case of composite columns, the contribution of each component toward column stiffness is difficult to establish. The current codes provide only cautious estimates of slenderness effects in composite columns.

It is recognized that the amount of stiffness available for the flexure of concrete contained within a pipe or tube is higher than that which can be anticipated from unconfined concrete. The stability of a steel tube filled with concrete will be influenced much more by the steel tube than by the confined concrete. Conversely, the stability of an encased composite column is influenced more significantly by the concrete than by the steel. The influence of tensile cracking and long-term softening effect of creep reduce effective stiffness of concrete, even when the concrete is confined inside steel tubing. This is reflected in the SSRC proposals.

23.8.2 Composite Beams

Composite beams are of two kinds: (a) steel beams encased in concrete and (b) steel beams connected to the floor slab (or deck) with shear connectors (Fig. 23.16). Concrete encasement has been used to provide fire and corrosion protection. The primary disadvantage of the encased steel beam is that substantial weight is added to the beam without a comparable increase in strength or stiffness. The encased beams are rarely used in high-rise construction and, therefore, their design will not be discussed here.

An often-used type of composite beam is the steel beam with shear connectors embedded in the concrete floor slab. Generally, round steel studs welded to beam flanges are used as shear connectors. Other types of shear connectors are spiral bars, angles, and channels.

As shown in Figs. 23.16*b* and *c*, shear studs are field-welded through the metal deck to the top flange of the beam, thus generating composite action of the concrete slab with the steel deck and beam. The metal deck can be oriented either parallel or perpendicular to the steel beam. The design and construction requirements for composite beam with steel deck are similar to those for composite beams with concrete slab. This type of flooring system is quite popular in North America. The obvious advantage of composite beams is the reduction of structural steel used in beams with resultant economies.

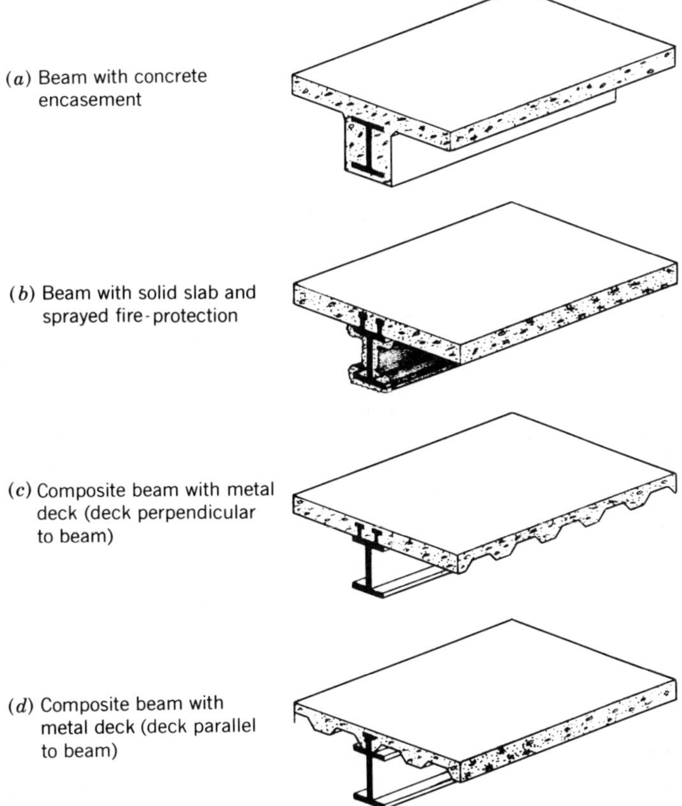

Fig. 23.16 Composite beam types.

Behavior and Design

A composite beam acts as a T-beam and its flexural stress distribution and deflections in the elastic range and plastic moment at ultimate load depend very much on the nature of the shear connection (Fig. 23.17).

Elastic Range

The moment M acting at a section of the composite beam is resisted by

$$M = M_s + M_b + F(z) \tag{23.8.9}$$

where M_s = portion of total moment M resisted by concrete slab
M_b = portion of total moment M resisted by steel beam
z = lever arm between forces F caused by composite action between beam and slab

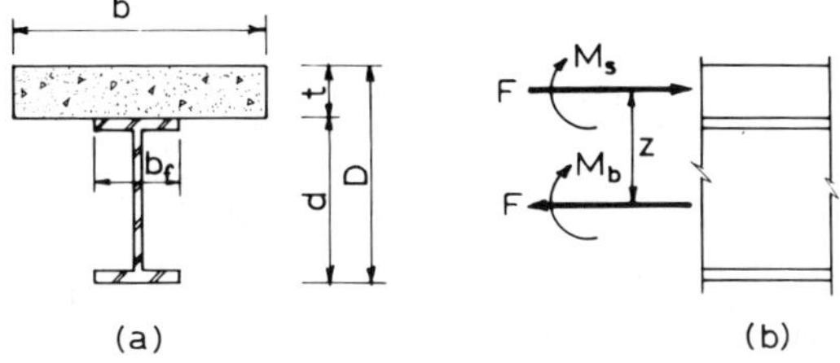

Fig. 23.17 Moment of resistance of composite beam.

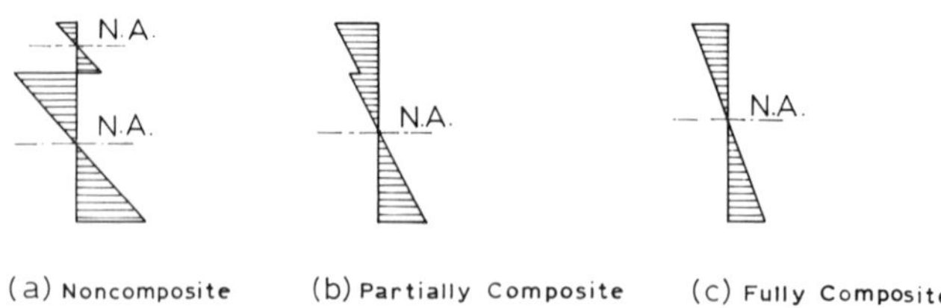

Fig. 23.18 Strain distributions in composite beam.

Figure 23.18 shows the strain distributions for noncomposite, partially composite and fully composite slab–beam arrangements. If no connectors are provided, F will be zero and M will be resisted solely by individual resistance of both the slab and the beam. Slippage will occur at the slab–beam interface. When flexible connectors of moderate stiffness are used, a couple Fz is induced. Some horizontal slip occurs at the interface. The connectors resist the shear force F and constrain the beam and slab to act together in vertical direction. This can produce axial tension in the connectors. Full composite action is achieved with relatively rigid connectors and presence of good bond.

Plastic Range

The ultimate moment capacity of a composite section is dependent upon the yield strength and section properties of the steel beam, slab strength, and interacting capacity of the shear connectors joining the slab to the beam. Provided full interaction exists between beam and slab, the plastic capacity of the composite beam, M_{pc}, is obtained from Eq. (23.8.9) by putting M_b and M_s equal to zero.

The procedure to determine the ultimate capacity depends on whether the neutral axis occurs within the concrete slab or within the steel beam. Figure 23.19 shows the stress distribution for these two cases. Experiments have shown that the plastic moment capacity of a composite beam remains unchanged, whether the steel beam is shored or unshored. The AISC provisions are mostly based on ultimate strength behavior, even though all relationships are adjusted to be in the service load range. A step-by-step process for design of a composite beam is given below. Design examples are available in Refs. 23.7 and 23.23 and in several other textbooks.

Design Procedure

Step 1. Calculate effective concrete flange width. When the slab extends on both sides of the beam, the effective width of concrete flange b_E is taken as the least of the following values:

$$\left.\begin{aligned} b_E &= 16t + b \\ b_E &= L/4 \qquad (L = \text{beam span}) \\ b_E &= \text{center-to-center distance between beams} \end{aligned}\right\} \tag{23.8.10}$$

When the slab is present on only one side of the beam, the effective width of the concrete flange is taken as the least of the following three values:

$$\left.\begin{aligned} b_E &= 6t + b \\ b_E &= L/12 \\ b_E &= \text{center-to-center distance between beams}/2 \end{aligned}\right\} \tag{23.8.11}$$

Step 2. Compute section properties. The transformed area method is used in calculating the properties of a section. Concrete area is transformed into equivalent steel area by using a slab width equal to b_E/n, where n is the modular ratio. Design values for n are listed in Table 23.9. Contribution of concrete slab below neutral axis is neglected. For the beam running perpendicular to deck ribs, concrete below the top of steel deck is neglected in computing the section properties, but for the beam running parallel to the deck ribs it may be included.

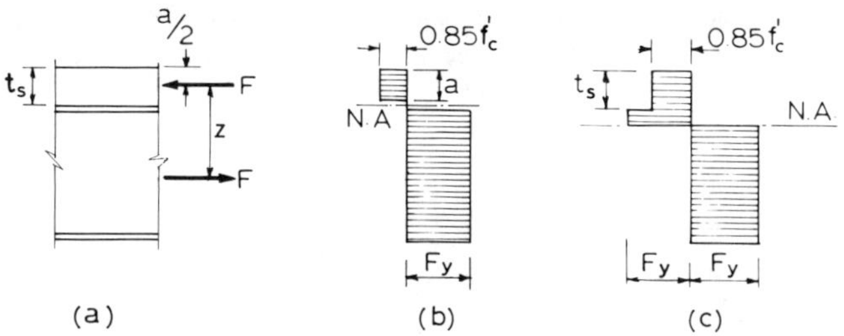

Fig. 23.19 Stress distributions for full plastic moment.

Select section as if shores are to be used; the required composite section modulus, S_{tr}, with reference to the tension fiber is

$$\text{required } S_{tr} = \frac{M_D + M_L}{F_b} \tag{23.8.12}$$

where M_D = the service load moment caused by loads applied prior to the time the concrete has achieved 75% of its required strength

M_L = the service load moment caused by loads applied after the concrete has achieved 75% of its required strength

F_b = allowable service load stress, $0.66F_y$ for positive moment region (where sections are exempt from the "compact section" requirements of AISC-1.5.1.4.1)

The concrete slab and its shear connector attachments are considered to provide adequate lateral support.

Step 3. Determine shoring requirement. To ensure that a designed composite section will perform properly prior to the time the concrete reaches 75% of its strength, stress in the steel beam must be checked. No shoring is needed if the following relation is satisfied:

$$S_{tr}(\text{effective}) \leq \left(1.35 + 0.35 \frac{M_L}{M_D}\right) S_s \tag{23.8.13}$$

where S_s = section modulus of steel beam referred to its tension flange

S_{tr} = section modulus of composite section referred to its tension flange

Step 4. Check steel beam stresses. Check stress on steel beam supporting the loads acting before concrete has hardened.

$$\text{required } S_s = \frac{M_D}{F_b} \tag{23.8.14}$$

where F_b may be $0.66F_y$, $0.60F_y$, or some lower value if adequate lateral support is not provided.

Table 23.9 Design Values for Modular Ratio *n*

f'_c, psi (MPa)	Modular Ratio $n = E_s/E_c$
3000 (21)	9
3500 (24)	8.5
4000 (28)	8
4500 (31)	7.5
5000 (34)	7
6000 (41)	6.5

Step 5. Design shear connectors. The shear connectors (usually headed studs) transfer horizontal shear between the steel beam and the slab/deck. The amount of shear force V_h to be transferred, per one-half beam, is given by the smaller of the two values:

$$\left.\begin{aligned} V_h &= A_s F_y / 2 \\ V_h &= 0.85 f'_c A_c / 2 \end{aligned}\right\} \qquad (23.8.15)$$

The number of the connectors required (N) is obtained by dividing the smaller value of V_h by allowable shear per connector (q):

$$N = \frac{\text{smaller } V_h}{q} \qquad (23.8.16)$$

The allowable shear, q, values are listed in AISC—Table 1.11.4. The reduction factors for connector capacities when using lightweight aggregate concretes are given in AISC—Table 1.11.4A.

Step 6. Partial shear connections. Due to the increasing cost of shear stud construction, full shear connections are rarely used. A 50–75% range provides a more economic construction. Another situation that necessitates partial connection is when studs required for full connection cannot be properly accommodated in the trough of the metal deck. For partial connections, the effective section modulus, S_{eff}, and effective moment of inertia, I_{eff}, are determined as follows:

$$S_{\text{eff}} = S_s + \sqrt{\frac{V'_h}{V_h}}\,(S_{tr} - S_s) \qquad (23.8.17)$$

$$I_{\text{eff}} = I_s + \sqrt{\frac{V'_h}{V_h}}\,(I_{\text{tr}} - I_s) \qquad (23.8.18)$$

where V_h = design horizontal shear for full composite action
V'_h = actual capacity of connectors used ($0.25\ V_h \leq V'_h \leq V_h$)
I_s = moment of inertia of steel beam
I_{tr} = moment of inertia of transformed composite section

23.8.3 Continuous Composite Beams

A continuous beam changes curvature within a span. This situation has been considered traditionally by designing the positive moment region on continuous beams as a composite section and the negative moment region as a noncomposite section (Fig. 23.20). However, some composite action has been known to exist in negative regions. The *AISC Specification* [23.2] permits the use of steel reinforcement that extends parallel to the beam span as part of the composite section. The inclusion of such steel reinforcement enhances negative moment capacity of the beam. When the reinforcing steel in the concrete slab is thus utilized, the force developed by it must be transferred in shear by mechanical shear connectors. The usual lateral buckling provisions of AISC apply to the negative moment regions of continuous composite beams. At ultimate load, the negative moments in continuous beams are often 20–30% higher than the values calculated from simple plastic theory, and are accompanied by a reduction of midspan moment. The redistribution increases the lengths of negative moment regions so that top reinforcement should be continued well beyond calculated points of contraflexure, particularly if shear connectors are bunched near these points.

A review of the steel connection shown in Fig. 23.20 indicates that only the bottom flange of the beam is connected for axial forces, while the forces near the top flange are resisted by the reinforcement in the slab. Vertical shear is resisted by the web connection. A top steel flange connection may also be required if the reinforcement or shear studs cannot be accommodated.

In building frames subjected to combined gravity and lateral load, a negative moment region may develop on one side of an interior column while a positive moment region develops on the other side. The behavior of continuous composite beams under these loading conditions is given in Refs. 23.16, 23.18, and 23.6.

23.8.4 Composite Trusses and Joists

For spans longer than 40 ft (12 m), composite trusses or open-web joists provide an economical solution. The truss geometry permits a relatively open plenum which can be used for mechanical

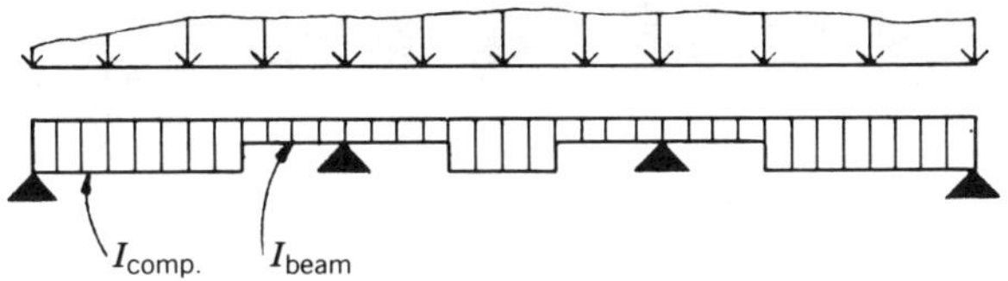

(*a*) Stiffness variation

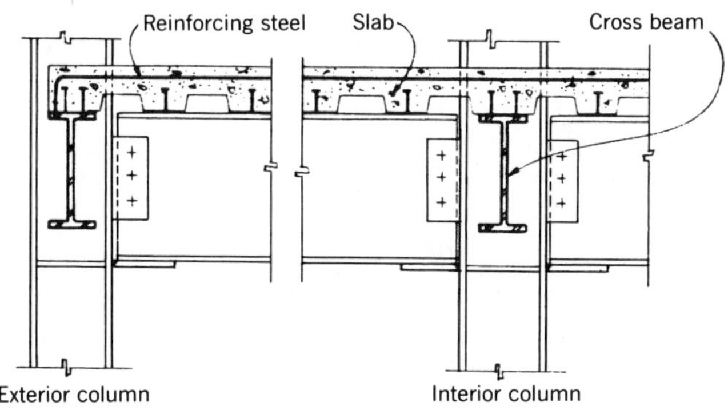

(*b*) Beam-column details

Fig. 23.20 Continuous composite beam details.

distribution systems (Fig. 23.21). Composite trusses with rolled sections, such as tees and angles, have been used quite efficiently for spans up to 80 ft (25 m). The 110-story tall Sears Tower is one example [23.15].

In designing a composite truss, bottom chord and diagonal member sizes are designed on a noncomposite basis, while the top chord is designed on a composite basis. The truss is generally designed such that it acts alone to carry wet concrete on the steel deck and then utilizes the composite top chord for all loads applied after hardening of the concrete. The superposed stresses are then checked against the allowable stresses. The chord between panel points is likely to separate and buckle laterally. This tendency must be guarded against by proper placement of shear studs.

Many composite bar joist systems are also available. Their design is based on tests by manufacturers and involves several different types of shear connectors. The joists are lighter than the equivalent wide-flange structural steel beam, and subsequently they are more flexible in the lateral direction. Since the floor slab is tied to the joist, the lateral buckling of the top chord is prevented.

23.8.5 Stub-Girders

Figure 23.22 shows a typical stub-girder construction. It is fabricated by welding short pieces of floor beams on top of a shallow heavier girder. The length of stub ranges between 5 and 7 ft (1.5 to 2 m), and the distance between the stubs is approximately the same. The entire stub-girder can be made composite with the concrete floor deck. This results in a stiffened girder, continuous secondary floor beams, and openings through which ducts can pass [23.5]. Three structural models have been used in stub-girder analysis: (a) nonprismatic beam model, (b) Vierendeel truss model, and (c) finite element model. The relative merits of each model can be found in Ref. 23.3. Stub-girders are also discussed in Chapter 19.

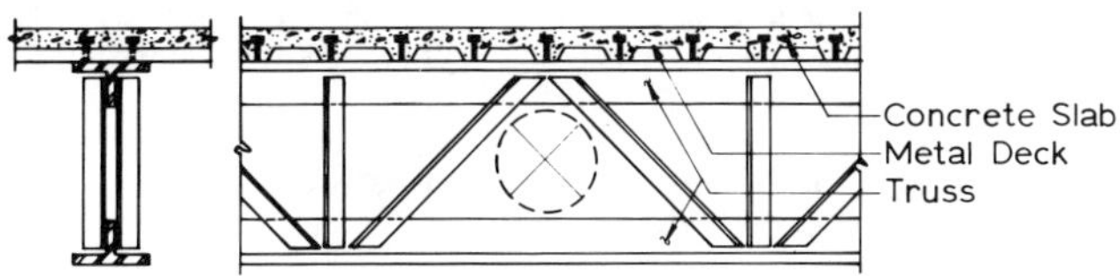

Fig. 23.21 Composite truss.

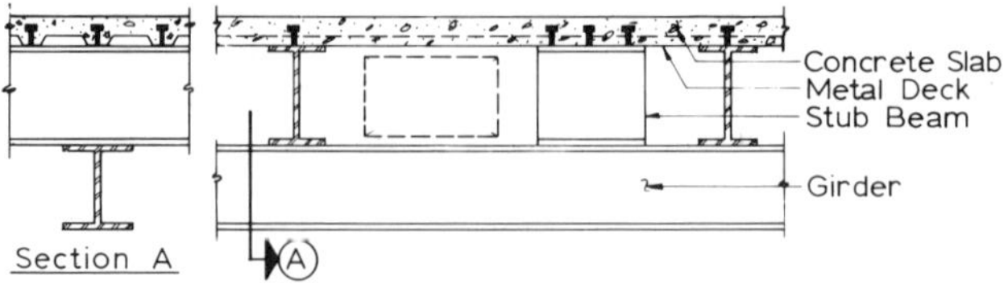

Fig. 23.22 Stub-girder system.

Since its inception in 1972, several buildings have been designed using the stub-girder system. However, the system has not achieved widespread popularity due to the complex design procedure and shoring requirements.

23.8.6 Other Design Considerations for Beams

Creep and Shrinkage

Creep and shrinkage effects and the method of construction need to be considered in computing deflection of a composite beam. When no temporary supports are used during casting and curing of concrete slab, the dead loads are resisted by the steel beam alone and creep effects are smaller. If temporary supports are used during construction, then the dead loads are resisted by the composite section. Since dead loads are sustained, they cause creep. The creep effect may be accounted for by increasing the conventional value of modular ratio. A multiplication factor of 2.9 is generally used.

Vibrations

Vibrations are either steady-state or transient. Steady-state vibrations are usually of no direct concern in structural design because they can and should be eliminated by insulating their source. On the other hand, transient vibrations cannot be eliminated, so the structure must be designed such that the annoyance caused by the transient vibrations is within acceptable limits. The damping characteristics of the structure play a very important role in human reaction to transient vibrations. Murray [23.20] has developed an acceptability criteria for floor vibration which is quite useful in design.

23.8.7 Composite Decks

A composite deck is defined as a metal deck acting compositely with the concrete slab on top to form a composite floor. The steel sheeting or decking performs two functions. First, it acts as tension reinforcement as in a reinforced concrete slab. Second, it acts as a form when the concrete slab is being placed. The sheet is connected to floor beams by means of plug welds, gunned pins, or welded studs. Welded studs permit composite behavior between steel and concrete. Minimum reinforcement is always needed to provide for the effects of temperature and volume changes, and for fire safety reasons. The shear transfer between the deck and concrete is developed by a combination of chemical bond and physical embossments in the deck. The advantages of composite action are: reduction in gage thickness of metal deck, larger load capacity, capability for longer spans, and integral floor diaphragms. Typically, the composite deck is currently used for a slab span range of 6 to 10 ft (1.8 to 3 m) with a 2-in. (50-mm) deep deck and up to 15-ft (4.5-m) span with a 3-in. (76-mm) deep deck.

Metal decks are also utilized for floor electrification, communication lines, and power distribution, as explained in Chapter 4. Header ducts or trench headers which interconnect the cell may be placed in the slab.

Metal Deck Selection

Metal decks are available in different shapes (see Chapter 20). The load-carrying capacity of a metal deck is influenced by depth of cross-section, metal thickness, and shape. The following considerations are necessary in selection of a metal deck:

1. A metal deck is to be strong enough to withstand the wet weight of concrete, preferably without intermediate shoring before composite action is established. Construction loads are also imposed during this phase.

2. The profile of the metal deck should be selected to maximize the shear stud capacity. Height-to-width ratios of the troughs where the shear studs are placed should preferably be 1.75 or more. A larger taper of the sides improves stiffness of the ribs. Side embossments or other mechanical devices provide better tie-down of the concrete to the deck.

3. The choice of metal deck depends on span of the slab and whether floor electrification is desired. For noncellular decks, 2- and 3-in. (50- and 76-mm) deep decks are generally used for spans up to 10 ft (3 m) and 15 ft (4.5 m), respectively. The most commonly used cellular section in the United States is a 3-in. (76-mm) deep, three cell unit with 6-in. (150-mm) wide cells at 8-in. (200-mm) centers. This allows for only narrow ribs about 2 in. (50 mm) in width and is not efficient for shear-stud placement due to a low width-to-height ratio. In many cases, the cellular units are blended with a noncellular part on a modular basis, in which case a relatively wide rib can be selected for the noncellular part where shear studs can be placed. Even in this case, the number of studs that can be accommodated may be quite limited which may result in a partial shear connection. Other cellular configurations with 6-in. (150-mm) cells at 12-in. (300-mm) centers with rib $w/h = 2$ are available, which allows placement of shear studs in the cellular part. Standard header ducts, $4 \times 1\frac{1}{4}$ in. (100×32 mm), completely encased on concrete topping, are more suitable than trench headers, since the former will maintain the continuity of the slab.

4. The metal deck also acts as a diaphragm. The design requirements are discussed in detail in Ref. 23.24.

The thickness of concrete above the metal deck for general and office occupancies typically varies from 2½ to 4 in. (65 to 100 mm). A welded-wire fabric placed directly on top of the steel deck is recommended for shrinkage crack control. The recommended minimum steel area is 0.1% for concrete thickness above the metal deck. A minimum of 6×6 W1.4 $\times$ W1.4 welded-wire fabric is generally used. The concrete thickness may be controlled by fire rating requirements of the slab as contrasted to structural requirements. The choice of lightweight concrete or normal-weight concrete would depend on economics and fire rating considerations. Generally, for high-rise structures, lightweight concrete (usually 110 pcf or 1760 kg/m^3) is more widely used. Another consideration for selection of concrete thickness is adequate projection of shear studs into the slab to maximize stud capacities. The minimum penetration of 1½ in. (40 mm), recommended by various investigators, appears reasonable.

Composite steel decks have also been used in parking garages and industrial floors where the floor is usually subjected to moving wheel loads and/or heavy storage loads. The slab thickness is determined from loading conditions. When the wheel loads are moderate, say 2000 lbs (9 kN) or less, the deck may be designed to act both as form and tension reinforcement. Where considerably greater loads (uniform or concentrated) occur, the metal deck is used to provide slab form only and steel reinforcement is provided to resist flexural tension in concrete. In other words, the composite slab is designed as a reinforced concrete slab to resist all specified load combinations. The test results of several composite slab specimens under static and dynamic loads representing industrial floor loads along with design examples are available in Ref. 23.26. A brief review is given below.

1. The extent to which a concentrated load is distributed in transverse direction depends upon several factors such as: the amount of transverse reinforcement, slab thickness, metal deck properties, slab aspect ratio, etc. A transverse distribution width of 4–5 ft (1.2–1.5 m) is generally used in determining design moments and shears per foot of slab width due to concentrated loads. A punching shear check must always be made where concentrated loads are involved.
2. An industrial floor with a concrete slab having a thickness of 5 in. (130 mm) or more over metal deck may be designed as a one-way continuous slab. Top steel reinforcement should be provided in negative moment regions. Welded-wire fabric mesh can also be used to help resist negative moment and to reduce shrinkage cracks.
3. For composite slabs subjected to long-term static loads, additional deflections due to creep should be considered. The use of top steel reduces the expected creep deflections. The *ACI Code* [23.1] may be used for deflection computations.
4. Fork lift trucks used on industrial floors vary widely in weight, lift capacity, tire type and size, and wheel and axle spacings. Thus, design of floors to resist fork lift truck loads should be done on an individual basis.
5. Pneumatic tire lift trucks provide larger bearing areas and cause less abrasion and spalling problems for concrete surfaces than solid rubber tire vehicles. Thus, where pneumatic tire vehicles can be used, a longer service life for the floor slab will generally result.
6. Impact effect of hard rubber tire vehicles is expected to be more than the impact effect of pneumatic tire vehicles.

Construction Considerations

1. Structural plug welds between the deck [especially a 3-in. (75-mm) deck] and steel improve stiffness characteristics of the beam and prevent premature initiation of rib cracking due to transverse rotation of the rib. Proper location of these welds in coordination with shear stud placement in the same trough is desirable since it tends to reduce the transverse rotation of ribs. It is essential that plug weld qualification procedures be established which will assure strength reliability.

2. Placement of shear studs in prepunched holes should be discouraged because it may lead to poor concrete around the shear studs due to grout leakage through the holes. Further, the prepunched holes reduce the participation of the metal deck in the composite beam stiffness in the working load range. Studs should preferably be welded through the metal deck. The thickness of the metal deck and that of zinc coating for galvanizing determine the reliability of welding through the deck. Stud welding through 18-gage deck with one-half ounce per sq ft of galvanizing has been quite reliably performed for studs up to ¾-in. (19-mm) diameter.

3. For wide troughs, placement of the shear studs to one side of the trough closest to the midspan of the beam appears to improve shear stud capacities. This will reduce the chance of the stud punching through the side of the deck.

23.9 MIXED SYSTEMS CONNECTIONS

Design information concerning standard connections for steel-to-steel joints or concrete-to-concrete joints can be obtained from the AISC, ACI, and PCI standards and various textbooks. Mixed steel–concrete systems require a variety of member connections between steel to concrete or vice-versa, for which information is generally not available in a documented form. Many of these connections have been evolved as a matter of practice or have been developed for particular projects. The following is a brief summary of some of these connections.

23.9.1 Beam End Connections Designed for Gravity Shear Only

Standard bolted or welded shear connections, according to the *AISC Manual,* are used for typical steel composite beams which frame either into a steel column or steel girder. The same type of standard connection is also utilized when the steel beam frames into a steel column which is encased in concrete (Fig. 23.23) or when the steel beam frames into a shear plate embedded in concrete (Fig. 23.24).

Several types of beam end details are shown in Fig. 23.23 for steel beams or trusses framing into a concrete column or wall. These types are required when the steel beam is framing into the core walls, basement walls, precast concrete columns, or panels. Each type is suited for a particular construction method especially related to the organization of formwork. Figure 23.23*a* shows standard T-sections anchored to concrete. Shear studs are designed for shear and axial force and the tension tie resists tension due to moment of gravity load eccentricity. Figure 23.23*b* shows a similar concept with the standard double-angle shear connection welded to an anchored steel plate. In this case, the steel angles come with the beam and are field welded to the steel plate which could be supported on the inside face of formwork during the concrete placement. Figure 23.24*a* shows cast-in-place anchor bolts which are bolted to beam web angles. Allowable shear values for the cast-in-place anchor bolt are given in some codes, such as UBC [23.11]. Shear values for drilled bolts are available from manufacturers. A more common detail is shown in Figs. 23.23*c* or *d* using either a corbel or steel bracket. This type may not be suitable when design for a large axial force in the member is required. The choice of concrete corbel or steel bracket depends entirely on the relative cost. If the concrete member has reinforcing steel at close centers, it may be desirable to eliminate drilled-in anchors. Figure 23.23*e* shows a bearing detail in a pocket in the wall. This is most commonly used for basement wall connections. For connections to heavily reinforced shear walls, this presents some disruption of wall reinforcing which should be evaluated. The pockets are often left open, although in some cases, they are filled in. A bearing-type detail for a truss end connection is shown in Fig. 23.23*f*.

The following is a brief commentary concerning the design of connections:

1. If only shear and eccentricity of the shear are involved, the connection can be designed from available information. For instance, the shear studs in Fig. 23.23*a* can be designed for shear and the ties for tension resulting from eccentricity of the shear. However, if significant tension axial forces are involved, the shear studs will have to resist shear and tension. Some kind of interaction formula needs to be developed for this combination of forces which should be verified by tests. Correlation with the length of the stud should also be considered. A more common connection where large axial tensions are present is represented by the cast-in-place type shown in Fig. 23.24*a*.

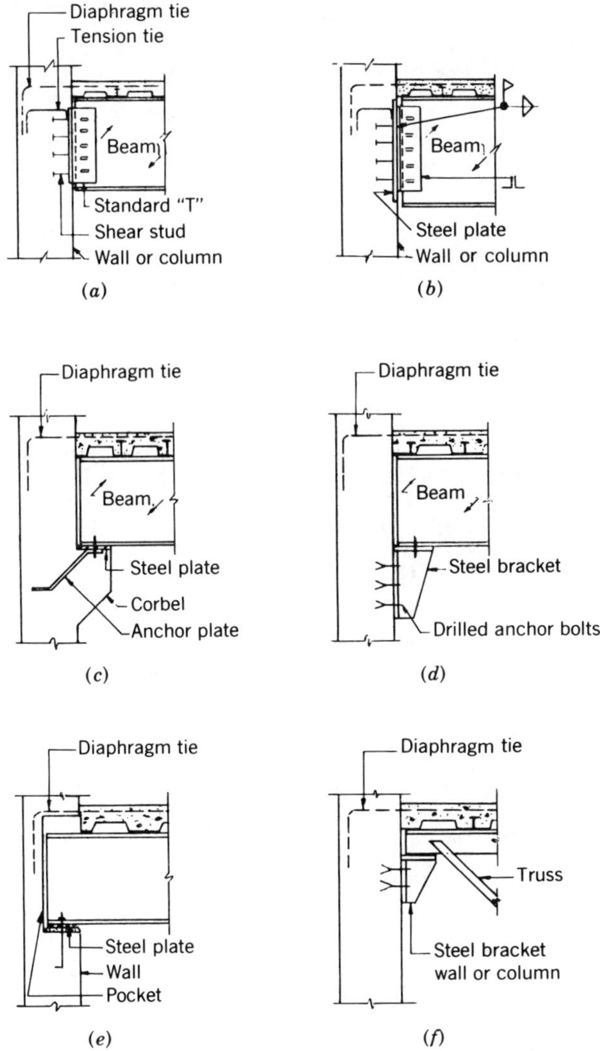

Fig. 23.23 Beam-wall and beam-column connections.

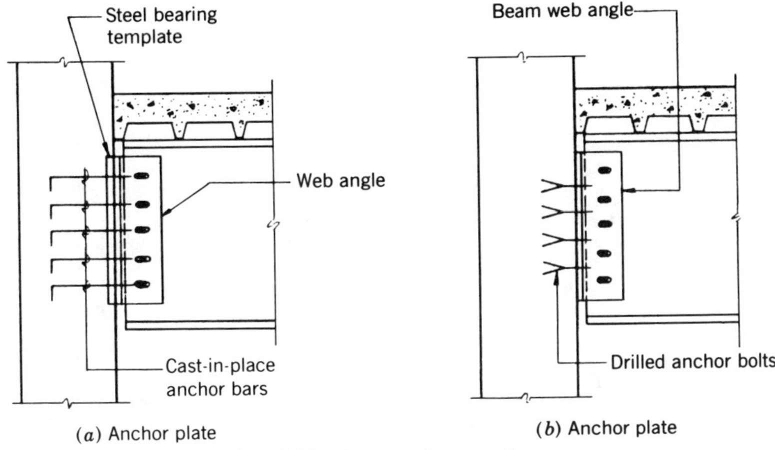

Fig. 23.24 Beam end connections.

2. In most multistory applications, diaphragm attachment to the wall or column is required. This is most conveniently accomplished by providing reinforcing bar dowels from the wall to the diaphragm slab as shown in Fig. 23.23. The reinforcing dowel is stressed only for live loads and for forces in the diaphragm. What is the rotational stiffness of this joint and the force induced in dowels? How should this dowel be proportioned? Some testing for these conditions appears warranted where large shear is present in the diaphragm. This may have to be transferred by an angle detail where the deck is plug welded to the outstanding leg of the angle, while the other leg of the angle is anchor bolted to the wall.

Roeder and Hawkins [23.22] recently investigated the shear and moment behavior of steel beam–concrete wall connections, and have suggested a procedure to determine the moment capacity of a bolted connection similar to that shown in Fig. 23.23*a*. However, the effect of top reinforcement in the metal deck was not included in their experimental work.

23.9.2 Connections for Composite Columns

According to the current version of the *ACI Code* (23.1] and SSRC proposal [23.25], the load transfer to the concrete in a composite column is to be accomplished by elements attached to steel columns which are in direct bearing. Figure 23.25*a* shows a composite column splice for precast encased columns where bearing plates are used for the attachment and, therefore, meet the intent of the *ACI Code*. Other variations may involve small angle members attached to the steel column at various locations to accomplish the bearing transfer as shown in Fig. 23.25*b*. It is not clear if the intent of the Code extends to shear stud connected encased columns as shown in Fig. 23.25*c*. Some investigation is necessary to establish the validity of this connection since it represents a simpler method of attachment. Some investigations are also necessary to establish whether chemical bond between steel and concrete may be counted to accomplish this transfer.

23.9.3 Slab Connections

Typical slab shear connections with beams involving either a solid slab or a metal deck were discussed in Section 23.8. Two types of slab connections involving precast slabs need further investigation. They are shown in Fig. 23.26. In Fig. 23.26*a* a solid precast plank is used between the beams with concrete topping. The topping is to be composite with the plank and also the steel

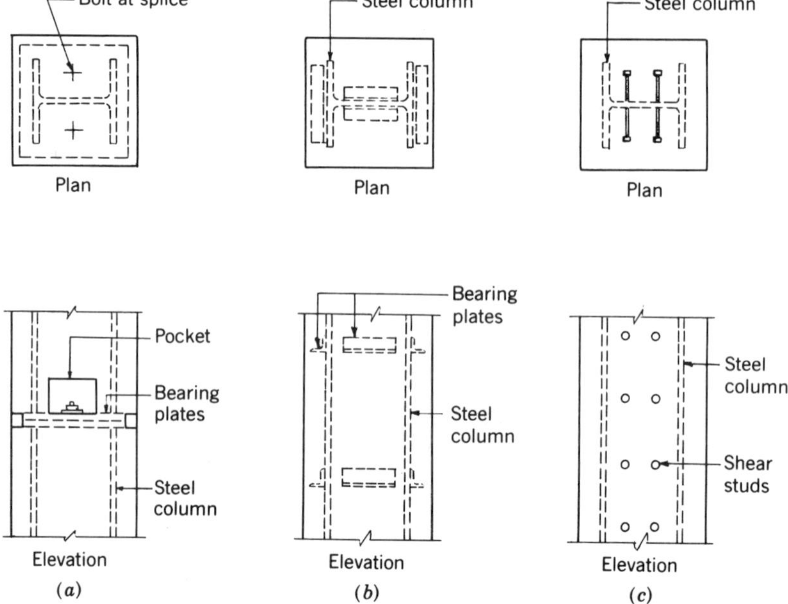

Fig. 23.25 Connections for composite columns.

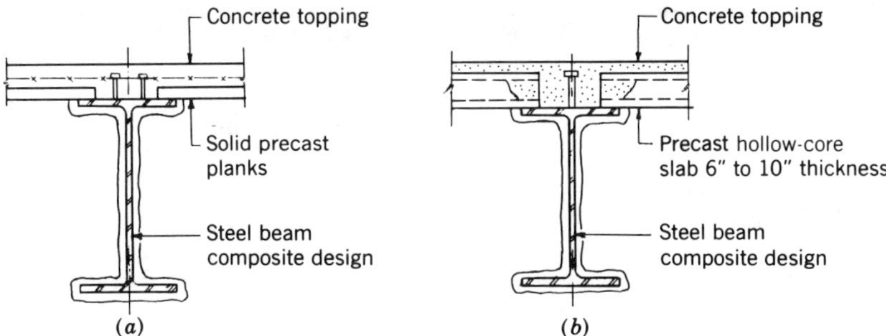

Fig. 23.26 Composite beams with precast slab.

beam. Figure 23.26*b* shows the condition where hollow cellular precast slabs may be used to span the entire bay width. Concrete topping is placed to fill part of the cells. In either case, shear studs are used to establish the composite action.

SELECTED REFERENCES

23.1 ACI Committee 318. *Building Code Requirements for Reinforced Concrete (ACI 318-83).* Detroit: American Concrete Institute, 1983.

23.2 AISC. *Specification for the Design, Fabrication and Erection of Structural Steel for Buildings* (November 1, 1978). Chicago: American Institute of Steel Construction, 1978.

23.3 Reidar Bjorhovde and T. J. Zimmerman. "Some Aspects of Stub-Girder Design," *Engineering Journal,* AISC, **17,** 3(Third Quarter), 1980, 54–69.

23.4 CEB. *International System of Unified Technical Regulations for Structures: Model Code for Structures.* Paris, France: Comite European du Beton, 1977.

23.5 Joseph P. Colaco. "A Stub-Girder System for High-Rise Buildings," *Engineering Journal,* AISC, **9,** 3(July 1972), 89–107.

23.6 Council on Tall Buildings and Urban Habitat. *Monograph on Planning and Design of Tall Buildings.* New York: American Society of Civil Engineers, 1979.

23.7 Phil M. Ferguson. *Reinforced Concrete Fundamentals,* 4th ed. New York: Wiley, 1979.

23.8 Mark Fintel, ed. *Handbook of Concrete Engineering.* New York: Van Nostrand Reinhold, 1974.

23.9 Richard W. Furlong. "AISC Column Design Logic Makes Sense for Composite Columns, Too," *Engineering Journal,* AISC, **13,** 1(First Quarter), 1976, 1–7.

23.10 R. W. Furlong. "A Recommendation: Composite Column Design Rules Consistent With Specifications of the American Institute of Steel Construction," paper presented to the Structural Stability Research Council, Boston, Mass., May 1978.

23.11 ICBO. *Uniform Building Code.* Whittier, CA: International Conference of Building Officials, 1979.

23.12 S. H. Iyengar. "State-of-the-Art Report on Composite or Mixed Steel-Concrete Construction for Buildings." New York: American Society of Civil Engineers, 1977.

23.13 S. H. Iyengar. "System Criteria for Mixed Steel-Concrete Systems," presented at U.S.-Japan Seminar on Mixed Steel-Concrete Construction, Tokyo, Japan, 1978.

23.14 Srinivasa H. Iyengar. "Mixed Steel-Concrete High-Rise Systems," *Handbook of Composite Construction Engineering.* New York: Van Nostrand Reinhold, 1974, Chapter 7.

23.15 S. H. Iyengar and J. J. Zils. "Composite Floor System for Sears Tower," *Engineering Journal,* AISC, **10,** 3(Third Quarter), 1973, 74–81.

23.16 B. Kato. "Cyclic Loading Tests of Composite Beam-to-Column Subassemblages," Progress Report, Laboratory of Steel Structures, Department of Architecture, University of Tokyo, Japan, 1974.

23.17 Fazlur R. Khan and Navinchandra R. Amin. "Analysis and Design of Framed Tube Structures for Tall Concrete Buildings," *Response of Multistory Concrete Structures to Lateral Forces* (SP-36). Detroit: American Concrete Institute, 1973, pp. 39–60.

23.18 L. W. Lu. "Discussion on Composite Steel and Concrete Beams in Japan," *Proceedings,* National Conference on Tall Buildings, Tokyo, Japan, August 1973.

23.19 I. A. MacLeod. "Shear Wall–Frame Interaction—A Design Aid with Commentary," Portland Cement Association, Skokie, Ill., 1971.

23.20 Thomas M. Murray. "Acceptability Criterion for Occupant-Induced Floor Vibrations," *Engineering Journal,* AISC, **18,** 2(Second Quarter), 1981, 62–70.

23.21 *PCI Design Handbook,* 2nd ed. Chicago: Prestressed Concrete Institute, 1978.

23.22 C. W. Roeder and N. M. Hawkins. "Connections Between Steel Frames and Concrete Walls," *Engineering Journal,* AISC, **18,** 1(First Quarter), 1981, 22–29.

23.23 Charles G. Salmon and John E. Johnson. *Steel Structures—Design and Behavior,* 2nd ed. New York: Harper and Row, Publishers, 1980.

23.24 SDI. *Diaphragm Design Manual.* St. Louis, Mo.: Steel Deck Institute, 1981.

23.25 SSRC, Task Group 20, Structural Stability Research Council (S. H. Iyengar, Chairman). "A Specification for the Design of Steel-Concrete Composite Columns," *Engineering Journal,* AISC, **16,** 4(Fourth Quarter), 1979, 101–115.

23.26 "Design Guidelines for Deep Composite Floor Slabs." Pittsburgh, Pa.: H. H. Robertson Company, 1981.

CHAPTER 24
MASONRY

JAMES L. NOLAND

Atkinson-Noland & Associates, Inc.
Consulting Engineers
Boulder, Colorado

24.1 INTRODUCTION

24.1.1 Scope

This chapter addresses the use of masonry for structural purposes, primarily for buildings. Only masonry constructed with clay and concrete units bonded with mortar and grout is discussed because such masonry is the predominant type of masonry in current use for structural applications. The material in this chapter is based upon information from many references. However, the sections on allowable stresses and design are based largely upon the *Uniform Building Code* (UBC) [24.77] and such material from it is included herein with permission.

24.1.2 Typical Structural Applications

Retaining Walls

Masonry retaining walls have been built of three basic types, that is, gravity, cantilever, and counterfort or buttress. Masonry basement or subterranean garage walls are also a form of retaining wall and are essentially flat plates supported on three or four sides.

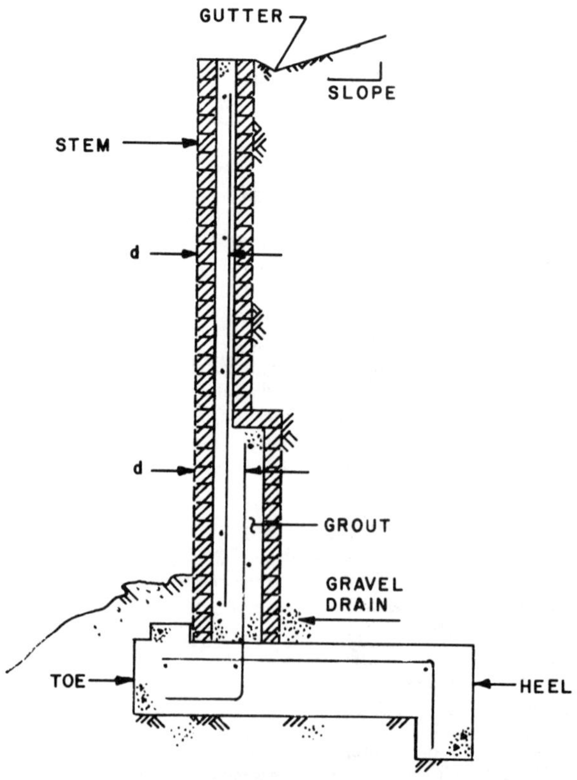

Fig. 24.1 Example of a cantilever masonry retaining wall.

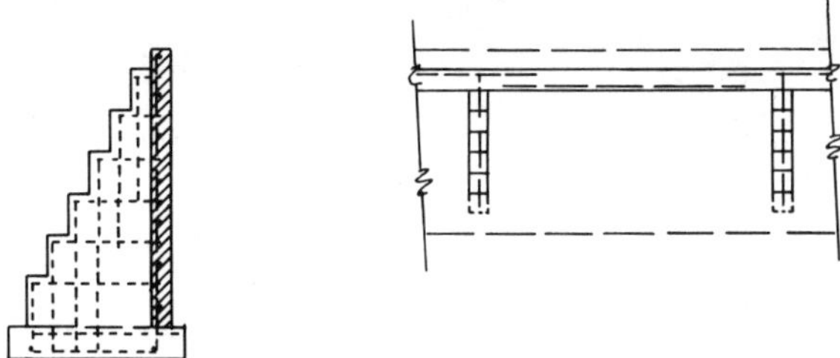

Fig. 24.2 Buttressed retaining wall.

Gravity retaining walls rely upon the mass (weight) of masonry to resist lateral earth pressure and are unreinforced [24.1, 24.50, 24.53] except possibly for light crack control reinforcement or ties. Because the base thickness of such walls is typically one-half to three-quarters of the wall height [24.50], it may be desirable to consider a reinforced wall of the cantilever or buttress type.

Cantilever masonry retaining walls, in cross-section, resemble an inverted T or L. Lateral earth pressures are resisted by the vertical leg (stem) acting as a cantilever fixed at the base and must be reinforced accordingly. Details of construction vary according to the type of masonry unit used, that is, solid or hollow. A representative cross-section of a masonry cantilever wall built with solid units is presented in Fig. 24.1. Design of cantilever masonry walls may be done by basic working stress methods. Design aids are available [24.1, 24.69, 24.73].

Counterfort or buttress masonry retaining walls are usually designed such that the lateral earth pressure is predominantly resisted by the wall spanning horizontally between buttresses or counterforts. The wall is considered to be a continuous flexural member spanning over several supports and reinforced in accordance with the magnitude and direction of bending moment at a given section. This type of wall would be likely to be used for higher walls where cantilever walls may not be practical [24.1, 24.50]. The basic configuration of a masonry buttressed retaining wall is shown in Fig. 24.2.

Load-Bearing Masonry Buildings

Load-bearing masonry buildings are essentially "box" structures in which the walls perform structurally to support vertical compressive loads due to gravity, in-plane shear and flexural loads stemming from wind or earthquake forces, and out-of-plane flexure due to wind, earthquake, or eccentric vertical loads. The walls also serve to enclose space, to provide sound and fire resistance, and often to provide the finished interior and exterior surfaces. Further discussion of this type of structure is presented in Chapter 7.

In plan view, load-bearing masonry buildings exhibit the form of single or multiple, open or closed cells usually rectangular in plan. Single-cell forms are found in single-story buildings used for storage facilities or open-space offices as shown in Fig. 24.3*a*. Motel, office, and commercial facilities up to four stories in height often utilize the "cross-wall" cellular arrangement shown in Fig. 24.3*b* in which the exterior walls may be structurally "open", e.g., glass.

Taller buildings tend to be a compact arrangement of cells formed by load-bearing walls in which the number and arrangement of walls are selected such that a reasonable balance in lateral force resistance is provided in each of two orthogonal directions and stresses due to vertical and lateral forces are relatively uniform [24.30, 24.31, 24.39]. The plan forms depicted by Fig. 24.4 (*a, b,* and *c*) are representative of load-bearing masonry buildings.

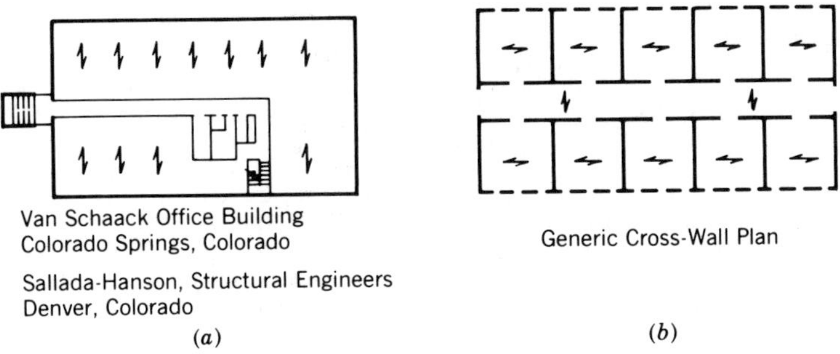

Fig. 24.3 Typical wall arrangements in low-rise masonry buildings.

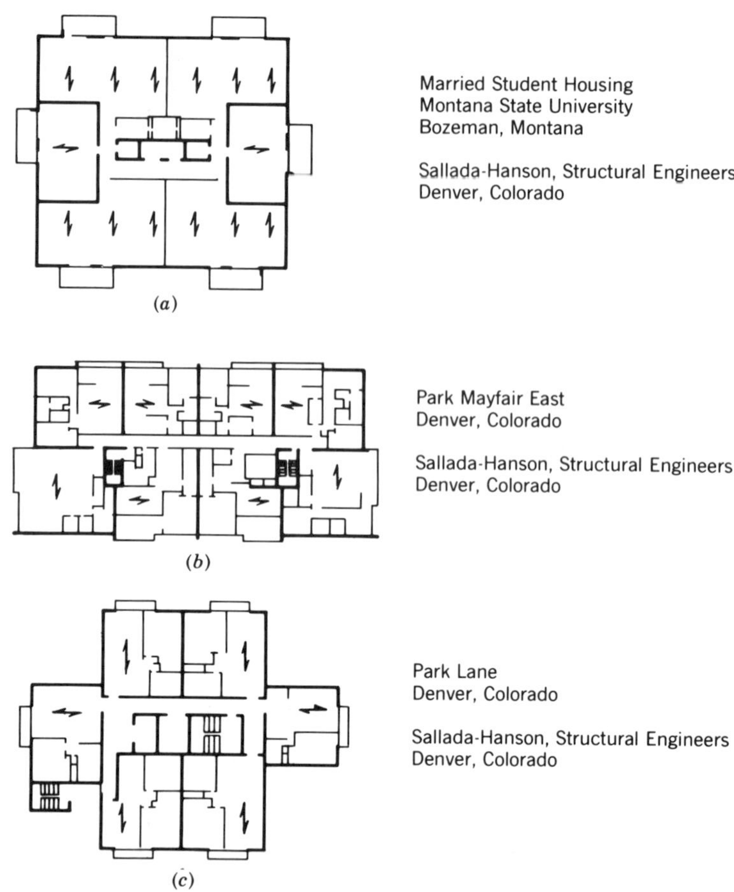

Fig. 24.4 Typical wall arrangements in mid-to-high-rise masonry buildings.

Open space for lobbies, dining areas, etc. at the ground level of masonry buildings may be provided by utilizing frame or partial frame construction [24. 30]. However, the efficiency of load-bearing wall construction is best where the plan form is repeated at each level [24.30, 24.31].

Studies have indicated that load-bearing masonry buildings up to 30 or 40 stories in height are possible [24.15, 24.58]; however, masonry buildings currently in service range from 1 story to just over 20 stories. Figure 24.4 depicts the bearing wall arrangements of actual tall masonry buildings in service. Figure 24.4*a* is the plan form of a building 9 stories in height with a cast-in-place reinforced concrete floor and 10½-in. (270-mm) thick walls. That of Fig. 24.4*b* is of a building 17 stories tall with a precast twin-tee floor system and 10-in. (250-mm) thick walls, and the building of Fig. 24.4*c* is 20 stories high with 10-in. (250-mm) walls.

24.1.3 Definitions

A selected set of definitions of terms commonly associated with masonry design and construction is presented below. The reader is referred to Refs. 24.70 and 24.76, from whence the definitions here are taken or paraphrased, for a complete set of many definitions unique to masonry.

Area, Cross-Sectional. *Gross* area of masonry is that encompassed by the outer periphery of any section. *Net* area is the gross area minus the area of cores (cells) and notches, that is, net area is the actual surface area of a cross-section. *Bedded* area is the area of the surface of masonry in contact with mortar in the plane of the joint.

Bed Joint. The horizontal masonry mortar joint.

Bond. (1) Adhesion between mortar or grout and masonry units or reinforcement. (2) The visual effect created by placement of units in a prescribed pattern. The most common forms are running-bond in which the units in one course overlap the units in the course below, and stack-bond in which units in one course are directly aligned with those in courses below.

Bond Beam. A horizontal reinforced masonry member which is an integral part of a wall and capable of resisting flexural and tensile forces.

C/B Ratio. The ratio of the weight of water absorbed by a masonry unit during immersion in cold water for 24 hr to the weight absorbed during immersion in boiling water for 5 hr. (See *ASTM Specification* C67 [24.64]).

Clay Masonry Unit. A solid or hollow masonry unit formed from clay, shale, fire clay, or a mixture of these and fired in a kiln. Solid clay units, in the shape of rectangular prisms, are commonly referred to as "bricks" and hollow clay units as "hollow bricks".

Cleanout Holes. Openings in one wythe of a multi-wythe masonry wall or cutouts in hollow-unit walls for the purpose of cleaning the space to be grouted and inspection of reinforcement.

Collar Joint. A vertical longitudinal joint between wythes of masonry which is less than ¾ in. (19 mm) in thickness.

Concrete Masonry Unit. A precast solid or hollow masonry unit made with water, portland cement, and suitable aggregate. Hollow concrete units are commonly referred to as "blocks". Solid concrete units are commonly known as "concrete bricks".

Control Joint. A formed, sawed, or tooled vertical groove in masonry to regulate the location and amount of cracking and separation and permit movement due to shrinkage.

Course. A horizontal layer of masonry units.

Core (Cell). An enclosed void in a masonry unit extending between parallel external surfaces.

Effective Width. The width of wall assumed to work, in flexure, with reinforcing bars.

Effective Height. The height of a column or wall used for purposes of determining slenderness effects.

Effective Thickness. The thickness of a column or wall used for purposes of determining slenderness effects.

Effective Area. The cross-sectional area which is considered effective in resisting applied loads.

Face Shell. The side wall of a hollow concrete or clay masonry unit.

Fin. A mortar projection into a grout space.

Grout. A mixture of cementitious materials and aggregate to which sufficient water has been added to produce a pouring consistency without segregation of the constituents.

Head Joint. The vertical mortar joint between the ends of masonry units in the same wythe.

Initial Rate of Absorption (IRA). The weight of water absorbed by a dried clay unit partially immersed in water for 1 min, expressed in grams per 30 sq in. of surface. Tests and calculations should be in accordance with ASTM C67 [24.64].

Joint Reinforcement. Steel reinforcement placed in a bed joint.

Mortar. A mixture of cementitious materials, fine aggregate (sand), and water. It is generally composed of portland cement, hydrated lime, sand, and water.

Masonry, Plain. Masonry in which reinforcement is used only for crack control, that is, reducing the effects of dimensional changes due to variation in moisture content or temperature.

Masonry, Reinforced. Masonry in which reinforcement is embedded in mortar or grout to resist internal forces.

Masonry Unit—Solid. A masonry unit whose net cross-sectional area in every plane parallel to the bearing surface is 75% or more of its gross cross-sectional area measured in the same plane.

Masonry Unit—Hollow. A masonry unit whose net cross-sectional area in every plane parallel to the bearing surface is less than 75% of the gross cross-sectional area.

Prism. A compression test specimen built with masonry units, mortar, and grout if applicable.

Retemper. To moisten and remix mortar to the proper consistency for placing masonry units.

Retentivity. The property of mortar associated with resisting water loss due to evaporation on the board or suction in masonry units.

Tie. A steel wire of C, Z, or rectangular shape, or masonry unit used to tie masonry wythes together, or to connect a masonry element to other elements of a structure.

Wall (Functional). (1) Bearing wall—A wall that is designed to support vertical compressive loads in addition to its own weight. (2) Shear wall—A wall designed to resist in-plane shear forces and moment as well as vertical compressive loads.

Wall (Configuration). (1) Cavity wall—A wall containing a continuous air space between inner and outer wythes, and the wythes are tied together with metal ties or masonry units. (2) Composite wall—A multi-wythe wall in which at least one wythe has strength, stiffness, or other differences affecting structural performance as compared to the other wythe or wythes. (3) Multi-wythe wall—A wall comprised of two or more wythes.

Wythe. A portion of a wall one masonry unit in thickness and composed of masonry.

24.1.4 Differences Between Concrete and Masonry

Masonry and concrete are both considered to be brittle materials with low tensile strength. Compression is the strength property of primary importance for both materials. Flexural and shear capabilities for both materials may be dramatically enhanced by the proper use of steel reinforcement.

Basic differences that can affect design and construction include the following:

Size of Aggregate vs. Size of Masonry Units

Although there are some unique bonding systems in use, masonry for primarily structural purposes is an assemblage of finite-sized, discrete units bonded together with mortar. Concrete is also an assemblage of discrete units, that is, the aggregate bonded together, but the size of the aggregate compared to typical dimensions of concrete elements is such that for most purposes concrete may be considered to be a homogeneous material. Because masonry units are relatively large, because unit and mortar materials properties are often dissimilar, and because of construction site factors, masonry is not a homogeneous material.

Location of Reinforcement

Reinforcement in concrete may be located anywhere within the periphery of a given element with due regard to cover requirements. Reinforcement location in masonry, however, is strongly influenced by the masonry configuration. Reinforcement is normally restricted to cavity spaces, joints, and cores or cells.

Water Content of Mortar and Grout

Mortar and grout strength, as is the case for concrete, is directly affected by the water/cement ratio [24.22]. The basic difference is that for concrete the water/cement ratio *when mixed* is the value affecting strength, however, it is the water/cement ratio *after* the mortar or grout has been in place in masonry which affects strength properties. Masonry units tend to draw water from mortar and grout to an extent dependent on both unit and mortar properties. Ideally, the masonry units absorb enough water from the mortar or grout such that the remaining water content is that associated with maximum strength.

Construction Methods

Concrete construction, whether precast or cast-in-place, relies upon the use of forms, and upon off-site mixing of the concrete in the case of cast-in-place concrete. In contrast, masonry construction does not generally require forms, but may require shores and scaffolds, and mortar is mixed on-site.

Masonry construction is essentially a craft. Units are placed one by one, and quality can be greatly influenced by the skill and care of the mason [24.30, 24.44, 24.53].

Concrete is placed in relatively large quantities and quality is somewhat less dependent upon individual skills.

Design Philosophy

Masonry structural design in the United States is primarily accomplished using working stress concepts [24.66, 24.72, 24.74, 24.77], while concrete design is primarily done using an ultimate strength limit state approach [24.67]. An ultimate strength limit state design approach for masonry is being used in the United Kingdom [24.68] and is under serious consideration in Canada [24.59]. An ultimate strength design procedure has been adopted in the 1985 UBC [24.77] for reinforced slender walls.

24.1.5 Empirical Design vs. Engineered Design

> *. . . until around 1950 load bearing (masonry) walls were proportioned by purely empirical rules which led to excessively thick walls which were wasteful of space and material and took a great deal of time to build. The situation changed in a number of countries after 1950 with the introduction of structural codes of practice which made it possible to calculate the necessary wall thickness and masonry strengths on a more rational basis. These codes of practice were based on research programmes and building experience and, although initially limited in scope, provided a sufficient basis for the design of buildings of twenty stories. [24.31]*

Empirical rules referred to above by Hendry do not rely upon analysis, but are based upon experience and observation and are intended to lead to a conservative design. Empirical rules have been developed which control wall thickness, lateral support, ties, materials, and other factors particularly for unreinforced (plain) masonry [24.25, 24.72, 24.74, 24.75, 24.77].

Engineered design is meant to imply a process of calculating applied loads (stresses) and determining, based on strength and stability considerations, required materials and proportions for a given structure.

Current United States codes, e.g., the *Uniform Building Code* [24.77], governing the design and construction of masonry structures, permit and provide for engineered design for both reinforced and plain masonry structures. Such codes, however, still contain a large number of empirical rules and limitations which must be observed.

24.2 MASONRY UNITS

The material presented here is primarily concerned with the properties of masonry units that affect structural design. Additional information is available in other sources [24.25, 24.30, 24.50, 24.52, 24.53, 24.64].

24.2.1 Clay Masonry Units

Manufacturing Processes

The majority of clay masonry units, both solid and hollow, currently used in the United States are produced by the "stiff-mud" process, also known as the "wire-cut" process. The basic components of the process are: (1) preparation of the clay or clays, (2) mixing with water and additives, if any, (3) extrusion through a die as a continuous ribbon, (4) cutting the clay ribbon into discrete units using steel wire, and (5) controlled firing in which the units are heated to the early stage of incipient vitrification. Peak temperatures attained during the firing sequence are in the 2000°F (1100°C) range [24.32, 24.53].

Solid clay units are also manufactured by molding processes, e.g., the soft mud and dry press. Subsequent to molding the units are dried and fired as in the wire-cut process [24.53].

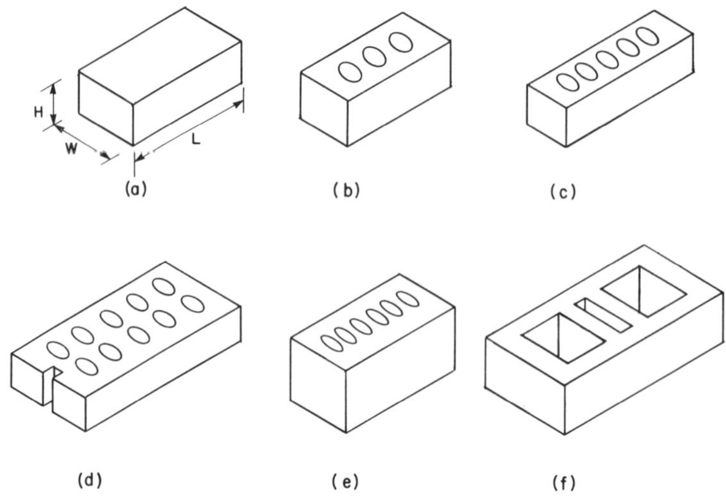

Fig. 24.5 Typical clay masonry units.

Size and Shape

Clay masonry units are available in a wide variety of shapes, sizes, and coring patterns, several of which are illustrated in Fig. 24.5. Figures 24.5*a* through *e* represent "solid clay" units, that is, the net area is 75% or more of the gross area. The width *W* of solid clay units normally ranges from 3 (76 mm) to 4 in. (102 mm), the height *H* from 2¼ in. (57 mm) to 4 in. (102 mm) and the length *L* from 7⅝ in. (194 mm) to 11⅝ in. (295 mm) although larger units have been produced [24.25].

Hollow clay units as shown in Fig. 24.5*f* have been produced in a relatively small number of sizes and core configurations. The shape shown has a length *L* equal to 11⅝ in. (295 mm), a height *H* equal to 3⅝ in. (92 mm), and is available in widths *W* of 3⅝ in. (92 mm), 5⅝ in. (143 mm), and 7⅝ in. (194 mm).

Visual Properties

The color of clay masonry units is determined by the chemical composition, burning intensity, and method of burning control [24.25, 24.53]—factors that also affect the strength of units. The choice of color for aesthetic purposes thus may influence structural performance.

Various textures may be created on the surfaces of clay units that are parallel to the direction of extrusion [24.25, 24.53]. The texture of the cut faces is created by the steel wires used in the cutting process. Surface texture is a factor influencing bond strength between clay units and mortar or grout [24.5, 24.32, 24.38].

Material Properties

Material properties of clay masonry units which can affect structural performance of masonry include: durability, initial rate of absorption, compressive strength, flexural strength, and expansion.

Durability primarily refers to the ability of a masonry unit to withstand environmental conditions, e.g., freeze–thaw action. Clay masonry units have been classified according to their weather-resistant capacities into the following grades: Severe Weathering (SW), Moderate Weathering (MW), and No Weathering (NW), as defined in ASTM C62 [24.83], C216 [24.84], and C652 [24.85].* Durability is evaluated in terms of compressive strength and water absorption measured in accordance with the provisions of ASTM C67 [24.64]. Compressive strength and absorption requirements for each grade are summarized in Table 24.1.

Clay masonry units have a tendency to draw water from mortar or grout with which they are in contact due to a capillary mechanism caused by small pores in the units. This phenomenon is termed the initial rate of absorption (IRA) or suction and has been linked to structural characteris-

---

* Refer also to International Conference of Building Officials, Research Report No. 2730 for alternate requirements applicable to hollow clay units.

Table 24.1 Strength and Absorption Requirements for Clay Unit Grades SW, MW, and NW*

Designation	Minimum Compressive Strength (Brick Flatwise), psi (MPa) Gross Area		Maximum Water Absorption by 5-hr Boiling†, %		Maximum Saturation Coefficient‡	
	Average of 5 Bricks	Individual	Average of 5 Bricks	Individual	Average of 5 Bricks	Individual
Grade SW	3000 (20.7)	2500 (17.2)	17.0	20.0	0.78	0.80
Grade MW	2500 (17.2)	2200 (15.2)	22.0	25.0	0.88	0.90
Grade§ NW	1500 (10.3)	1250 (8.6)	No limit	No limit	No limit	No limit

* Summarized from ASTM C62 [24.83], C216 [24.84], and C652 [24.85].

† Initially immersed for 24 hr in cold water. Five-hr absorption equals the amount of water absorbed after immersion in boiling water for 5 hr expressed as a percentage of the weight of the dry unit.

‡ Saturation coefficient is the ratio of absorption after 24 hr in cold water to the absorption after 5 hr in boiling water.

§ Applies only to a class of masonry units covered by ASTM C62 [24.83].

tics of masonry such as the bond between mortar and unit [24.3, 24.32, 24.38, 24.46, 24.51]. The quality of bond between mortar and unit is a function of properties of each. However, for many often used mortar mixes an IRA† value in the 10–25 range has been observed to be most desirable [24.1, 24.52, 24.53].

Compressive strength of clay masonry units is measured by loading specimens to failure in a direction consistent with the direction of service loading and in accordance with ASTM C67 [24.64]. Compressive strength of units is related to durability as previously discussed, and to compressive strength of masonry [24.3, 24.8, 24.25, 24.30, 24.32, 24.33, 24.36, 24.40, 24.44].

Flexural strength (modulus of rupture) determined in accordance with ASTM C67 [24.64], is basically a measure of the tensile strength of a masonry unit and is somewhat correlated to unit compressive strength.

Clay masonry units immediately after manufacture are extremely dry and expand due to absorption of moisture from the atmosphere. The magnitude of this initial expansion depends on the characteristics of the unit materials, the firing temperature, and the moisture available. Initial expansion is irreversible [24.25, 24.31, 24.35, 24.37]. Additional, but small, amounts of contraction or expansion due to temporary variations in masonry moisture content may occur; a movement coefficient of 0.0002 has been suggested [24.25, 24.31].

As is the case for other materials, clay unit masonry is subject to expansion/contraction due to temperature variations. Expansion coefficients range from 2.8 to 3.9 $\times$ 10^{-6}/°F (5 to 7 $\times$ 10^{-6}/°C) [24.25, 24.31].

The detailing of clay unit facades for moisture control, and for accommodation of volumetric changes and facade/structure interaction, is covered in Chapter 27 (Sections 27.1 and 27.4.2).

24.2.2 Concrete Masonry Units

Ingredients

Concrete masonry units primarily consist of portland cement (ASTM C150 [24.90]) or blended cement (ASTM C595 [24.91]), aggregate (ASTM C33 [24.102] or ASTM C331 [24.103]), and water. Other ingredients such as hydrated lime (ASTM C207, Type S [24.92]), and/or pozzolans (ASTM

† IRA equals the grams of water absorbed by a clay unit in grams per 30 sq in. (19,355 mm²) per min. See Ref. 24.64.

C618 [24.104]) may be used as well as air entraining agents, coloring pigments, ground silica, etc. established as suitable for use in concrete [24.50, 24.53].

Manufacturing Process

Casting of concrete masonry units differs from casting of normal concrete in that the mixture used for concrete is very dry, that is, "no-slump". The mixture is placed into molds and vibrated under pressure for a specified time to obtain compaction. Higher strength units are obtained by subjecting the material to longer compaction periods. Subsequently, the units are removed from the molds and cured either under normal atmospheric conditions or by autoclaving (steam curing) [24.50, 24.53].

Size and Shape

Concrete masonry units are available in a wide variety of sizes and shapes. They may be classified as "hollow" (ASTM C90 [24.86]) or "solid" according to the basic definitions in Section 24.1.3. A type of unit known as "concrete building brick" (ASTM C55 [24.88]) is available which is completely solid as well as "solid units" which are permitted to have up to 25% void area (ASTM C145 [24.87]). Other special units, e. g., varieties of U-shaped units for bond beams and lintels, are also available [24.50, 24.53]. A limited number of concrete masonry configurations presently used are depicted in Fig. 24.6. Hollow units are typically 15⅝ in. (397 mm) long, and either 7⅝ in. (194 mm) or 3⅝ in. (92 mm) high, and 7⅝ in. (194 mm), 5⅝ in. (143 mm), or 3⅝ in. (92 mm) wide. Solid units are typically 7⅝ in. (194 mm) high and are available in several lengths and widths. Concrete "bricks" are normally 3⅝ in. (92 mm) wide, 2¼ in. (57 mm) high, and 7⅝ in. (194 mm) or 15⅝ in. (397 mm) long [24.50, 24.53].

The net concrete cross-sectional areas of most hollow concrete masonry units range from 50 to 70% of the gross area depending on such factors as: unit width, wall (face shell and web) thickness, and core shape. The walls of hollow concrete units taper or are flared and thicker on one bed surface of the unit than the other to enable release from the mold during production. Hence, the net

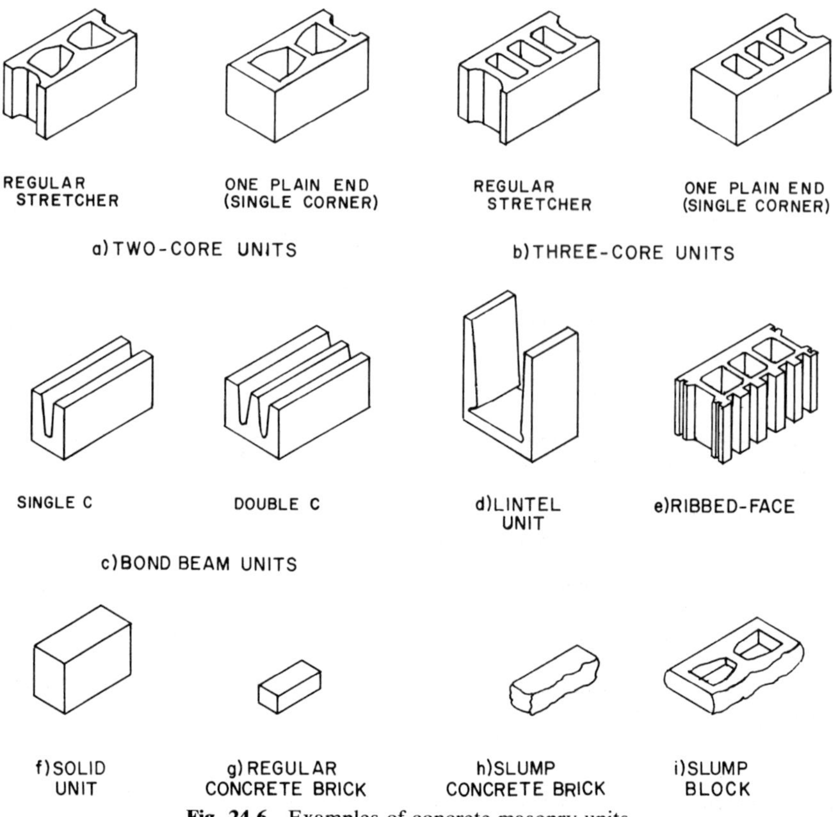

Fig. 24.6 Examples of concrete masonry units.

Table 24.2 Minimum Thickness of Walls of Hollow Load-Bearing Concrete Units*

Nominal Width (W) of Units, in. (mm)	Face-Shell Thickness Min. in. (mm)†	Web Thickness	
		Webs Min. In.† (mm)	Equivalent Web Thickness, Min. in./ft‡ (mm/m)‡
3 (76) and 4 (102)	3/4 (19)	3/4 (19)	1 5/8 (136)
6 (152)	1 (25)	1 (25)	2 1/4 (188)
8 (203)	1 1/4 (32)	1 (25)	2 1/4 (188)
10 (254)	1 3/8 (35) 1 1/4 (32)§	1 1/8 (29)	2 1/2 (209)
12 (305)	1 1/2 (38) 1 1/4 (32)§	1 1/8 (29)	2 1/2 (209)

* Adapted from ASTM C90 [24.86].
† Average of measurements on three units taken at the thinnest point, when measured as described in ASTM C140 [24.101], Sections 15 and 17(b).
‡ Sum of the measured thickness of all webs in the unit, multiplied by 12 and divided by the length of the unit.
§ This face-shell thickness (FST) is applicable where allowable design load is reduced in proportion to the reduction in thickness from basic face-shell thicknesses shown.

concrete cross-sectional area is greater on the top of the unit than the bottom. For structural reasons, ASTM C90 [24.86] stipulates minimum wall thickness for load-bearing concrete masonry units as presented in Table 24.2.

Visual Properties

Color other than the normal concrete gray may be obtained for concrete units by adding pigments into the mix at the time of manufacture or by painting subsequent to use in the field. A variety of surface effects are possible including smooth face, rough (split) face, and fluted, ribbed, recessed, angular, and curved faces [24.50, 24.53], some of which may affect cross-sectional area calculations.

Classifications

Concrete masonry units are produced in several classifications according to recommended use and moisture content [24.50, 24.53].

Grade N hollow (ASTM C90 [24.86]) and solid (ASTM C145 [24.87]) units are "for general use such as in exterior walls below and above grade that may or may not be exposed to moisture penetration or the weather and for interior walls and back-up."

Grade S hollow (ASTM C90 [24.86]) and solid (ASTM C145 [24.87]) units are "limited to use above grade in exterior walls with weather-protective coatings and in walls not exposed to weather."

Grade N concrete brick (ASTM C55 [24.88]) is "for use as architectural veneer and facing in exterior walls and for use where high strength and resistance to moisture penetration and severe frost action are desired."

Grade S concrete brick (ASTM C55 [24.88]) is "for general use where moderate strength and resistance to frost action and moisture penetration are required."

Two types of concrete masonry units are produced within each of the grades N and S, that is:

Type I, Moisture-Controlled Units—Units designated as Type I (Grades N-I and S-I) must conform to the moisture requirements of Table 24.3.

Type II, Nonmoisture Controlled Units—Units designated as Type II need not conform to the requirements of Table 24.3

Material Properties

Material properties of concrete units which can affect structural performance of masonry include: absorption, shrinkage, moisture content, compressive strength, tensile strength, and age.

Table 24.3 Moisture Content Requirements for Type I Concrete Masonry Units*

	Moisture Content Maximum Percent of Total Absorption (Average of Three Units)		
	Humidity Conditions at Job Site or Point of Use		
Linear shrinkage (%)†	Humid‡	Intermediate§	Arid‖
0.03 or less	45	40	35
From 0.03 to 0.045	40	35	30
0.045 to 0.065, max.	35	30	25

* Summarized from ASTM C55 [24.88], C90 [24.86], C145 [24.87].
† Based upon ASTM C426 [24.105].
‡ Average annual relative humidity above 75%.
§ Average annual relative humidity 50 to 75%.
‖ Average annual relative humidity less than 50%.

Absorption of a concrete masonry unit, determined in accordance with ASTM C140 [24.101] is the total amount of water expressed in pcf (kg/m^3) which a dry unit will absorb and is somewhat related to density. Upper limits on absorption are presented in Table 24.4 according to density of the unit material. No limits have been suggested for rate of absorption as have been for clay masonry units; however, the effect of absorption on the quality of mortar joints is the same. Assuming that rate of absorption is related to total absorption, control of the rate may be exercised by limiting the absorption [24.50].

Moisture content is expressed as a percent of the total water absorption possible for a given concrete masonry unit. Dimensional changes of concrete masonry due to changes in unit moisture content from the moisture content when the masonry was built can have serious effects upon the structure depending upon the nature of the boundary conditions and size of a given masonry element. The most common situation that occurs is cracking and other effects due to shrinkage [24.11, 24.50, 24.53] which, in turn, is due to a loss of moisture. Moisture loss is essentially determined by the humidity of the air surrounding a particular masonry element. It should be noted that in modern buildings, humidity conditions applicable to interior elements may be significantly different from those applicable to exterior elements.

Potential shrinkage characteristics of a given unit, determined according to ASTM C426 [24.105], depend upon method of manufacture and materials. Units with low shrinkage potential could be acceptable with a higher moisture content than units with a high shrinkage potential for given humidity conditions. Alternatively, the tolerable moisture content for a given humidity condition will differ depending upon the shrinkage potential of the units. The values in Table 24.3 represent an attempt to equalize drying shrinkage for units of different shrinkage potential considering different humidity conditions.

Compressive strength of concrete masonry units is a measure of unit quality and is related to strength of masonry. It is measured by loading specimens in a direction consistent with the direction of service loading and in accordance with ASTM C140 [24.101]. Factors that affect compressive strength include: water/cement ratio, degree of compaction, and cement control [24.50, 24.53]. Based on an admittedly limited number of tests, indications are that an increase in moisture content, based upon dry unit weight, from 1 to about 4½% (nearly saturated) is associated with a reduction in unit compressive strength of about 15%. A similarly limited number of tests indicates an increase in unit compressive strength, based on net area, of about 10 psi (70 kPa) per day up to approximately 200 days [24.56]. Minimum compressive strength requirements, based on gross cross-sectional area, are summarized in Table 24.4 for the various kinds of units. It should be noted that the values presented are minimums and that test results of concrete units will often exceed these values [24.20, 24.60, 24.61].

Tensile strength of masonry units is a property related to the resistance of concrete masonry to diagonal tensile forces in shear walls and to tensile forces induced by shrinkage of units [24.19, 24.53]. Various methods have been adapted to establish unit tensile strength, including flexural tension, splitting, and direct tension tests [24.17, 24.34, 24.53]. No standard requirements have been established for unit tensile strength.

Table 24.4 Strength and Absorption Requirements for Load-Bearing Concrete Masonry Units*

Type of Unit	ASTM	Grade	Minimum Compressive Strength, psi (MPa), on Average Gross Area		Maximum Water Absorption, pcf (kg/m^3) Average of Three Units Based on Oven-Dry Unit Weight			
					Lightweight Concrete		Medium Weight Less Than 125 to 105 (2000 to (1680)	Normal Weight 125 (2000) or More
			Average of Three Units	Individual Unit	Less than 85 (1360)	Less than 105 (1680)		
Hollow	C90	N	1000 (6.9)	800 (5.5)	—	18 (288)	15 (240)	13 (208)
		S	700 (4.8)	600 (4.1)	20 (320)	—	—	—
Solid	C145	N	1800 (12.4)	1500 (10.4)	—	18 (288)	15 (240)	13 (208)
		S	1200 (8.3)	1000 (6.9)	20 (320)	—	—	—
Brick	C55	N	3500† (24.1)	3000† (20.7)	—	15 (240)	13 (208)	10 (160)
		S	2500† (17.2)	2000† (13.8)	—	18 (288)	15 (240)	13 (208)

* Adapted from ASTM C90 [24.86], C145 [24.87], C55 [24.88].
† Concrete brick tested flatwise.

Detailing of concrete masonry facades for moisture control, volumetric changes, and facade/structure interaction is treated in Sections 27.1 and 27.5.2, Chapter 27.

24.3 MORTAR, GROUT, AND REINFORCEMENT

24.3.1 Mortar

Description

Modern mortar (ASTM C270 [24.89]) is a mixture of cementitious materials, aggregate, and water. Cementitious materials used are (1) portland cement (ASTM C150 [24.90]) or portland blast furnace cement (ASTM C595 [24.91]), and lime (ASTM C207 [24.92]) or (2) masonry cement (ASTM C91 [24.93]). Masonry cement mortar may be and has been used in structural masonry. However, mortar made with portland cement, lime, aggregate (sand), and water is in more general use. While both types of mortar have similar attributes and requirements, the discussion herein applies specifically to conventional mortar made with portland cement.

Function

Mortar serves to bond masonry units together to form a composite structural material, that is, masonry. As such, mortar is a factor in compressive, shear, and flexural tensile strength of masonry. In addition, mortar compensates for dimensional and surface variations of masonry units, resists water and air penetration through masonry, and bonds to metal ties, anchors, and joint reinforcement so that they perform integrally with the masonry [24.1, 24.14, 24.25, 24.31, 24.50, 24.52, 24.53].

Requirements During Construction

Well-constructed masonry may be defined as that in which the mortar is in complete contact with the appropriate surfaces of the masonry units and reinforcement, all joints are filled, and the masonry is dimensionally true. Proper "workability" of mortar, that is, the ease with which a given mortar may be placed and spread on given masonry units is essential in obtaining well-constructed masonry. It is also important that mortar stiffen at a rate which will permit subsequent courses to be placed and supported without causing movement of the masonry below [24.14, 24.25, 24.31, 24.50, 24.52, 24.53].

Workability and rate of stiffening are complex functions of both mortar and unit properties and may be influenced by environmental conditions. These effects are briefly discussed below:

1. *Aggregate (Sand).* Well-graded sand, that is, sand with a uniform distribution of particle sizes, is necessary to produce a workable mortar that is dense and strong in the hardened state. Size and gradation limits shown in Table 24.5 represent current practice. Sand on the finer side of the permitted gradation range will produce a more workable mortar. Such sand, however, results in

Table 24.5 Aggregate Gradation for Masonry Mortar

Sieve Size	Percent Passing*	
	Natural Sand	Manufactured Sand
No. 4 (4.75 mm)	100	100
No. 8 (2.36 mm)	95 to 100	95 to 100
No. 16 (1.18 mm)	70 to 100	70 to 100
No. 30 (600 μm)	40 to 75	40 to 75
No. 50 (300 μm)	10 to 35	20 to 40
No. 100 (150 μm)	2 to 15	10 to 25
No. 200 (75 μm)	—	0 to 10

* Not more than 50% shall be retained between any two sieve sizes nor more than 25% between No. 50 and No. 100 sieve sizes. See ASTM C144 [24.107] for requirements pertaining to masonry with joints thicker than ½ in. (13 mm) or unusually thin joints.

a mortar which requires more water to be workable and is therefore weaker than a mortar made with coarser sand because of the higher water/cement ratio.

Particles of manufactured sand are sharp and angular and tend to produce a less workable mortar than that made with natural sand of rounded particles. More water may be required to obtain adequate workability of mortar made with manufactured sand than that made with natural sand with resulting lower strength due to the higher water/cement ratio [24.10, 24.50, 24.52, 24.53].

2. *Water Retentivity.* Mortar exposed to air tends to lose water by evaporation. Mortar in contact with masonry units tends to lose water to the units because of the suction of the units. Retentivity is the mortar property associated with resistance to such water loss and resultant loss of workability. Lime in mortar improves the water retentivity and workability. Ideally, retentivity of a mortar would be compatible with the suction of the units used and environmental conditions, that is, temperature and humidity, so that adequate workability is maintained. Water content of mortar should be as high as possible consistent with proper workability and suction of the masonry units to maximize bond of the mortar to units [24.14, 24.31, 24.50, 24.53]. Water retentivity is measured by methods described in ASTM C91 [24.93].

3. *Suction and Water Content.* Units with high suction require the use of mortar with high retentivity to prevent excess and rapid water loss and reduced workability. It is noted in ASTM C67 [24.64] and elsewhere [24.53] in the case of clay-unit masonry, that mortar which has stiffened due to water loss because of suction (or drying) results in poor bond, and water permeable joints. It is suggested in ASTM C67 [24.64] that clay masonry units with initial rates of absorption in excess of 30 g/min/30 in.^{2} (30 g/min/19360 mm^2) be wetted prior to placing to reduce suction. Care should be taken to ensure uniformity of wetting.

Flow determined by methods of ASTM C109 [24.94] is a rough measurement of workability, but is not a test amenable to construction sites. No generally accepted procedure has been developed for field measurement of workability; the mason is the best judge [24.14, 24.31].

In general, it may not be possible to specify a mortar, that is, proportions of cement, lime, and sand, which will be optimal for both construction and strength. Priority, in most cases, should be given to specifying a mortar whose characteristics enable masonry to be efficiently and well constructed by normally-skilled masons. A mortar that is workable with the masonry units being used and under site environmental conditions therefore may not yield masonry with the maximum strength properties theoretically possible.

Mortar Proportions

Conventional mortar is described in terms of proportions, by volume, of portland cement, hydrated lime, and aggregate (sand) as given in Table 24.6. Minimum mortar compressive strength of each type, as determined by uniaxial compression tests of 2-in. (50-mm) cube specimens in accordance with ASTM C109 [24.94], is given in Table 24.7.

It should be noted that mortar conforming to the proportion specifications of Table 24.6 may have compressive strength far in excess of the minimum values prescribed in Table 24.7 [24.8, 24.22, 24.41].

Table 24.6 Mortar Proportions by Volume*,†

Type	Parts by Volume of Portland Cement‡	Parts by Volume of Hydrated Lime§
M	1	1/4
S	1	Over 1/4 to 1/2
N	1	Over 1/2 to 1 1/4
O	1	Over 1 1/4 to 2 1/2

* The part by volume of sand measured in a damp, loose condition shall be not less than 2 1/4 and not more than 3 times the sum of the volumes of cement and lime.
† Based on Table 2 of ASTM C270 [24.89].
‡ ASTM C150 [24.90].
§ ASTM C207 [24.92].

Table 24.7 Minimum Compressive Strength of Laboratory Prepared Mortar of 2-in. (50.8 mm) Mortar Cubes*,†

Mortar Type	Average Compressive Strength at 28 Days‡ psi (MPa)
M	2500 (17.2)
S	1800 (12.4)
N	750 (5.2)
O	350 (2.4)

* Based on Table 1 of ASTM C270 [24.89].
† See ASTM C270 [24.89] for additional details.
‡ Average of three laboratory prepared specimens mixed to a flow of 100–115%.

Factors Affecting Mortar Compressive Strength

Mortar compressive strength, typically measured by uniaxial compression of 2-in. (50-mm) cubes in accordance with ASTM C109 [24.94] is a measure of relative mortar quality. Because of several factors, for example, state of stress, water content, and dimensions, the compressive strength of a mortar cube is not directly related to compressive strength of mortar in a masonry joint. The basic factors that affect uniaxial cube compressive strength, however, are essentially those that affect mortar performance in masonry, that is, proportions of portland cement, hydrated lime, and sand (C:L:S), water content, admixtures, air content, mixing time, and sand characteristics [24.10, 24.13, 24.14, 24.22, 24.31, 24.38, 24.40, 24.50, 24.53].

Cube compressive strength:

1. Decreases as the proportion of lime increases with respect to the proportions of cement and sand.
2. Decreases as the ratio of proportion of cement to lime decreases where the proportion of lime plus cement is constant with respect to the proportion of sand.
3. Increases as the proportion of cement increases with respect to the proportion of lime plus sand.
4. Increases as sand gradation becomes coarser.
5. Decreases with increasing water/cement ratio regardless of lime content.
6. Decreases with addition of admixtures, e.g., color agents.
7. Decreases with increasing air content due to admixtures or excessive mixing time.

It should be noted that the reductions in cube strength caused by the factors listed in 1–4 above are basically due to an associated increase in water content required to maintain workability. Hence, the water/cement ratio is the primary factor in cube compressive strength.

Factors Affecting Mortar to Unit Bond

Because mortar not only seals masonry against wind and water penetration, but also binds masonry units together, strength and extent of bond are essential to well-constructed masonry. Two forms of bond strength are important for structural purposes, that is, tensile bond strength and shear bond strength. Tensile bond is required to resist forces perpendicular to a mortar-unit joint while shear bond is required to resist forces parallel to such joints. The factors that affect bond are basically common to both with the exception of the influence of compression on shear bond.

Mortar properties affecting bond include:

1. Cement content—other factors equal, greatest bond strength is associated with high cement content [24.5, 24.12, 24.46, 24.50].
2. Retentivity—bond strength is enhanced if high retentivity mortar is used with high-suction units and low retentivity mortar is used with units of low absorption [24.6, 24.38, 24.51].

3. Water content (flow)—bond is enhanced by using the maximum water content consistent with good workability considering unit properties and environmental conditions [24.6, 24.38, 24.50, 24.51].
4. Air content—bond decreases with increasing air content [24.14].

Unit properties affecting bond include:

1. Surface texture—mortar flows into voids, cracks, and fissures and forms a mechanical attachment to the surface of the unit [24.5, 24.38, 24.50].
2. Suction—for a given mortar, bond strength decreases as unit suction increases. This is perhaps due to the rapid loss of water to the unit on which mortar is placed. The mortar becomes less workable and bond between it and a unit placed on it less reliable [24.38, 24.50, 24.51].

Workmanship factors affecting bond include:

1. Time—the time lapse between spreading mortar on a unit and placing a unit upon that mortar should be minimized to reduce the effects of water loss from mortar due to suction of the unit on which it is placed [24.38, 24.50, 24.51].
2. Movement—movement of units after placing can reduce, if not break, bond between mortar and unit [24.38, 24.50, 24.51, 24.53].
3. Pressure and tapping—units must be placed on mortar with sufficient downward pressure, possibly augmented by tapping, to force the mortar into intimate contact with the unit surface [24.51, 24.53].

Mortar Mixing

Proper mixing is essential to obtain a uniform distribution of materials and the desired workability and strength properties. Mixing by machine is recommended for all but small batches. Procedures for putting ingredients into the mixture vary, but a common trait is to not add cement and water first to avoid lumping of the cement which inhibits blending with the other ingredients.

Mortar should be mixed for a minimum of 3 min (ASTM C270 [24.89]). A maximum time of 10 min is suggested; for some mixes a 5-min maximum may be more appropriate. Longer mixing times tend to reduce mortar strength because of additional water required to maintain workability [24.13].

"Retempering", that is, adding water to mortar taken from the mixer to restore workability is permitted by ASTM C270 [24.89] if done within 2½ hr of initial mixing. This practice should be employed with extreme caution because the water/cement ratio may be altered with attendant loss of strength [24.51]. Different mortars harden at different rates, and the rate is also affected by environmental conditions. In adverse conditions, retempering over a period of 2½ hr may not be advisable [24.53].

24.3.2 Grout

Description

Grout (ASTM C476 [24.106]) is a mixture of cementitious materials and aggregate to which sufficient water has been added to permit the grout to be readily poured into masonry grout spaces without segregation of the materials.

Function

Grout is placed in cavities formed by masonry and bonds to the masonry and any steel reinforcement, ties, and anchors to form a unified composite structure.

Requirements During Construction

Masonry units and grout interact in the same manner as unit–mortar interaction, that is, water is drawn from the grout into the masonry by suction. As in the case of mortar, final strength is a function of water content after suction [24.53].

Proper placement of grout requires that it be sufficiently fluid to be pourable into the grout space available and completely fill the space. The suction of the masonry units (IRA) in the case of clay masonry units will influence the amount of water required in grout. Higher water content is required if masonry units have a high rate of absorption (suction) to reduce the tendency of grout to adhere to the sides of the grout space while it is being poured and thus constricting the space. The converse is true if the units have low suction. Where units have a very high rate of suction, a "grouting aid" admixture may be useful in retarding water loss from grout [24.2, 24.50, 24.53, 24.78]. Water content may be lower for large grout spaces, e. g., least lateral dimension of 4 in. (100 mm), than for small grout spaces, e. g., least lateral dimensions of 2 in. (50 mm) or smaller. Slump, as measured by the standard 12-in. (300-mm) truncated cone test, is typically from 8 in. (200 mm) to 11 in. (280 mm), depending upon fluidity required.

Grout Type and Materials

Grout is identified as "fine" or "coarse" depending on the maximum size of the aggregate used. The proportions of ingredients and aggregate sizes are presented in Table 24.8.

Fine grout should be used where the cavity width in grouted cavity construction is less than 2 in. (50 mm) or the least lateral dimension of cores in grouted hollow-unit construction is less than 3 in. (76 mm). Otherwise coarse grout may be used [24.2, 24.53].

Mixing Grout

Masonry grout may be site-mixed in the same manner as mortar. However, wherever possible grout should be batched and mixed in transit-mix trucks because of better control [24.50, 24.53].

Grout Strength

ASTM does not specify a minimum grout strength. The *Uniform Building Code* [24. 77] requires that the grout have a minimum compressive strength of 2000 psi (14 MPa) as measured by a uniaxial compressive test of a 2 : 1 rectangular prism specimen. UBC Standard 24-22 [24.82] describes the size of specimen and the procedure for preparing it.

Placing Grout

Grout may be poured or pumped into grout spaces according to established procedures [24.2, 24.27, 24.77]. Consolidation is essential to obtaining grout in-place without voids or debonding due to shrinkage. Lack of, or poor, consolidation may cause reduced masonry compressive strength and poor bond of grout to masonry unit. Mechanical vibration has been shown to be a superior method as compared to puddling [24.43, 24.45]. Consolidation should be done soon after placement and again when the excess water has been absorbed from the grout by the masonry units [24.27, 24.71]. It is not always obvious when this occurs. Water migration from grout into the adjacent masonry units can be quite rapid resulting in loss of grout fluidity. Care should be exercised to assure that a second vibration does not cause defects, for example, voids and delamination of grout from reinforcement masonry units.

Table 24.8 Grout Proportions of Ingredients*

Type	Parts by Volume† of Portland Cement or Blended Cement	Parts by Volume‡ of Hydrated Lime or Lime Putty	Aggregate Measured in a Damp, Loose Condition§	
			Fine	Coarse
Fine	1	0 to 1/10	2¼ to 3 times the sum of the volumes of cement and lime	
Coarse	1	0 to 1/10	2¼ to 3 times the sum of the volumes of cement and lime	1 to 2 times the sum of the volumes of cement and lime

* Based on Table 1 of ASTM C476.

† ASTM C150 [24.90] Portland Cement—Types I, IA, II, IIA, III, IIIA.
ASTM C595 [24.91] Blended Cements—Types IS, IS(MS), IS-A, IS-A(MS).

‡ ASTM C207 [24.92] Hydrated Lime, Type S.
ASTM C5 [24.109] Quicklime.

§ ASTM C404 [24.110] Grout Aggregate.

24.3.3 Reinforcement

Description

Masonry may be reinforced with steel to resist tensile forces as in reinforced concrete. Use of deformed bars is normally restricted to #11 size or smaller [24.77]. Ladder or truss bed joint reinforcement is unique to masonry and is primarily used to resist internal forces due to shrinkage or thermally-induced movement.

Steel reinforcement should conform to the requirements of the following ASTM specifications:

1. ASTM A82, *Cold Drawn Steel Wire for Concrete Reinforcement.*
2. ASTM A496, *Deformed Steel Wire for Concrete Reinforcement.*
3. ASTM A615, *Deformed and Plain Billet-Steel Bars for Concrete Reinforcement.*
4. ASTM A616, *Rail-Steel Deformed and Plain Bars for Concrete Reinforcement.*
5. ASTM A617, *Axle-Steel Deformed and plain Bars for Concrete Reinforcement.*
6. ASTM A706, *Deformed Low-Alloy Bars.*

24.3.4 Water

Water used in mixing mortar or grout must be clean and free of deleterious amounts of acids, alkalies, or organic material [24.77].

24.4 BASIC PROPERTIES AND STANDARD TESTS

24.4.1 Compression

Basic Behavior

Masonry loaded in uniform compression tends to fail by tensile cracks formed parallel to the direction of loading [24.7, 24.16, 24.20, 24.30, 24.52, 24.55, 24.72]. The tendency of a brittle material to form cracks parallel to the direction of principal stress can be explained by crack growth theory [24.4, 24.57]. However, the presence of mortar joints, where mortar has a different Poisson's ratio, apparently tends to exaggerate the tendency to tensile splitting [24.4, 24.21, 24.32, 24.57]. This failure mode has been observed in several experimental programs [24.7, 24.9, 24.20, 24.28, 24.30, 24.44, 24.61].

It should be noted that tensile splitting discussed above pertains primarily to solid-unit masonry. The failure mode of ungrouted hollow-unit masonry is affected by the geometry of hollow units. Failure in compression occurs by splitting of the cross-webs with some tensile splitting of the face shells [24.9, 24.41, 24.42, 24.56, 24.60].

The failure mode of grouted hollow-unit masonry is spalling and splitting of the face shells followed by failure of the grout core [24.18, 24.28, 24.47, 24.48].

Factors that have been observed to affect compressive strength of masonry include:

1. Unit strength.
2. Mortar strength.
3. Mortar joint thickness.
4. Coring (degree and pattern).
5. Bond (arrangement of units).
6. Number of wythes.
7. Initial rate of absorption.
8. Workmanship [24.3, 24.4, 24.7, 24.9, 24.13, 24.18, 24.21, 24.28, 24.29, 24.30, 24.32, 24.33, 24.36, 24.42, 24.50, 24.52, 24.53, 24.56].

Standard Test

Compressive strength of masonry may be established by tests of small assemblages, that is, prisms in accordance with ASTM E447 [24.95]. The minimum number of prisms to be tested of a given unit–mortar combination is three.

Prisms may be constructed in stack-bond or in a bonding arrangement that simulates bonding used in the structure, except no structural reinforcement is to be used in the prism. In either case, prisms should be constructed with the same materials, joint thickness, and workmanship used in the structure.

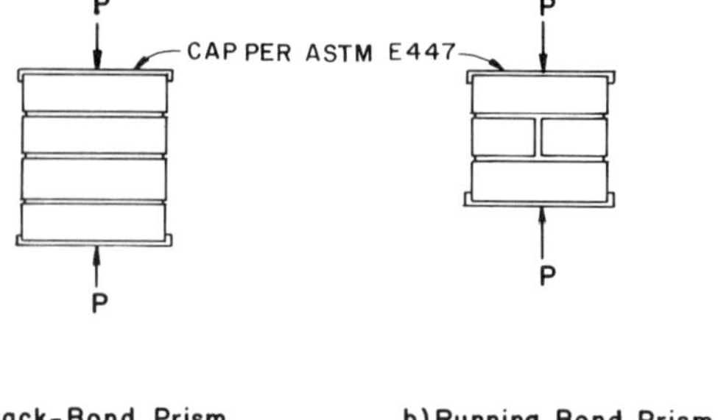

Fig. 24.7 Schematic of prism compression test.

Curing, capping, testing, data recording, and other details are presented in ASTM E447 [24.95]. Especially in very dry environments, it is suggested that measures be taken to prevent a rate of moisture loss from the test specimens higher than would occur in the structure during curing.

The basic test setup is shown in Figs. 24.7*a* and *b*. The first depicts a stack-bond prism and the second depicts a prism built in running-bond typical of masonry in many structures. Note that ASTM E447 [24.95] does not stipulate a number of courses for prisms.

Prism failure stress f'_{mt} is calculated by

$$f'_{mt} = \frac{P}{A} \tag{24.4.1}$$

where A = gross cross-sectional or mortar bedded area depending on the cross-section of the units.

It should be noted that f'_{mt} of hollow-unit prisms mortared on the net area or on the face shells is often calculated based upon gross area. One should therefore be aware of the area basis of f'_{mt} in order to be consistent in its use.

24.4.2 Shear

Basic Behavior

Because masonry is an assemblage of discrete units and mortar, two forms of shear strength exist, that is:

1. Sliding or joint shear—the strength, in bond, between mortar and units which resists relative movement of adjacent units in a direction parallel to the mortar joint between them.
2. Diagonal tension—tensile strength of masonry which resists tensile stresses that exist in a beam or shear wall at some angle to the direction of shear [24.30, 24.52].

In the case of shear walls, where shear is normally considered to be a horizontal force parallel to the bed joints, sufficient bond between mortar and units must exist in order for diagonal tension strength to be developed. Otherwise failure occurs in stepwise fashion along a diagonal in the plane of the wall [24.41, 24.52, 24.54, 24.80]. It has been shown experimentally that joint shear strength is increased by compression across the joint [24.12, 24.24, 24.29, 24.41, 24.46, 24.52, 24.79, 24.81] in a Coulomb friction manner.

Standard Test

Diagonal tension (shear) of 4 × 4 ft (1.22 × 1.22 m) masonry panels may be established by loading the panels in compression along one diagonal. Failure occurs in tension perpendicular to the diagonal. The basic setup and failure mode are shown in Fig. 24.8. Detailed requirements are presented in ASTM E519 [24.96].

The value of P at failure is converted to an equivalent shear stress S_s by

$$S_s = \frac{0.707P}{A} \tag{24.4.2}$$

where A = average of the gross areas (for solid-unit masonry) or net areas (for hollow-unit masonry) of the two continguous upper sides of the specimen, sq in. (or mm^2).

The "racking" test described in ASTM E72 [24.97] (Section 14) has been used to measure diagonal tensile strength of 8 × 8 ft (2.44 × 2.44 m) wall specimens. However, hold-down forces induced by the test fixture complicate the state of stress. This test has essentially been replaced by the test described above.

24.4.3 Tension

Basic Behavior

Masonry is essentially a brittle material that tends toward sudden failure in tension. In structures, tension stresses occur in flexural members and as diagonal tension in piers and walls that carry shear forces. Expansion and contraction of masonry due to moisture or temperature changes may also induce tensile stresses.

Tension strength (in unreinforced masonry) is developed primarily by the shear bond and tensile bond between masonry units and mortar. If bond is sufficient, then failure will occur by tensile splitting through mortar and units. However, in the case of out-of-plane flexure, e.g., in walls subjected to wind pressure, failure normally occurs as bed joint–masonry unit delamination [24.5, 24.20, 24.60, 24.72].

Standard Test

Diagonal tensile strength may be determined or established by loading masonry panels in compression along a diagonal as discussed in Section 24.4.2 and shown in Fig. 24.8.

Tension strength between mortar and unit may be established by the crossed-brick couplet test in the case of solid units in accordance with ASTM C952 [24.98]. A schematic of each setup is shown in Fig. 24.9.

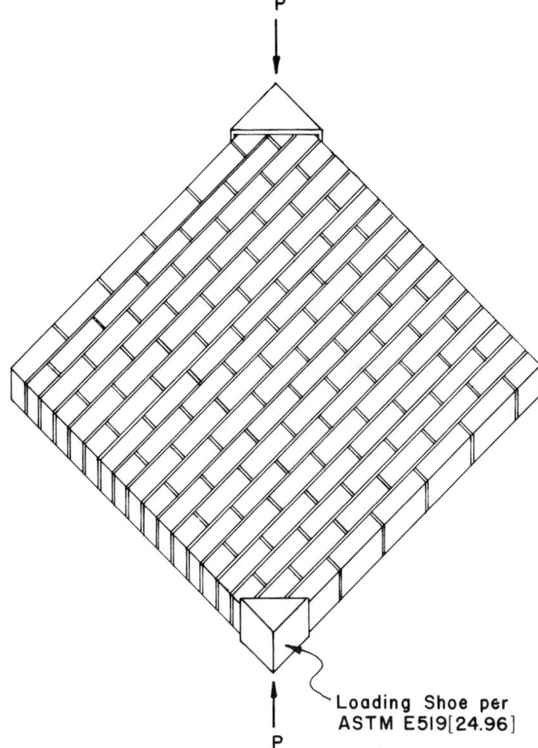

Fig. 24.8 Schematic of diagonal tension test.

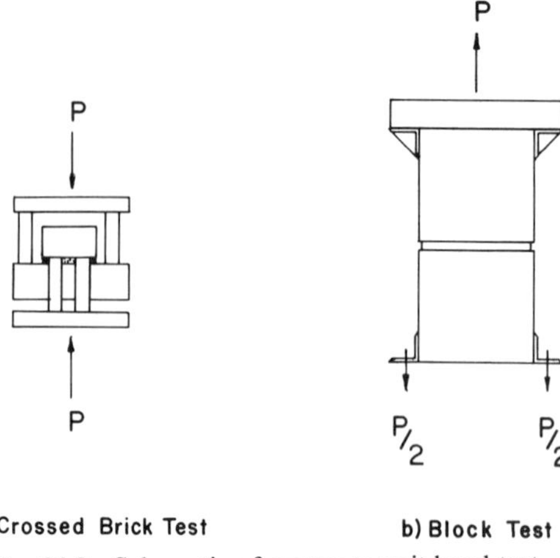

Fig. 24.9 Schematic of mortar-to-unit bond test.

Tension test results typically exhibit significant variability [24.7, 24.12, 24.30]; hence, a large number of tests may be necessary to establish sufficient statistical confidence.

24.4.4 Flexure

Basic Behavior

The flexural capacity of unreinforced masonry walls depends either upon the tensile bond between units or upon the shear-bond of overlapping units depending on the direction of flexure and type of construction as depicted in Fig. 24.10. Flexure that induces shear-bond stresses between overlapping units may be limited by shear-bond strength or by flexural tensile unit strength [24.38, 24.52].

Flexural capacity of reinforced masonry is essentially limited by masonry compressive strength or by tensile strength of the reinforcement. Compression reinforcement can add to flexural strength in beams [24.1, 24.53] particularly if it is confined. Vertical reinforcement in shear walls at or near the ends is primarily used to provide tensile strength for in-plane reversible moment. Failure of slender shear walls in flexure is characterized by progressive damage to the masonry at the compression toes followed by buckling of the unconfined vertical reinforcement [24.48, 24.49].

Tests of prisms and short walls under eccentric compression indicate that at failure the maximum compressive stress, calculated on the assumption of linear elastic behavior, exceeds ultimate uniaxial compressive stress by a factor on the order of $4/3$ [24.20, 24.60].

Standard Test

Bond strength in a direction perpendicular to the bed joint may be established by third-point or uniform loading of stack-bond specimens as shown in Fig. 24.11. ASTM E518 [24.99] provides requirements for materials, specimen preparation and configuration, testing, and calculations. Extreme care is required in handling flexural bond test specimens.

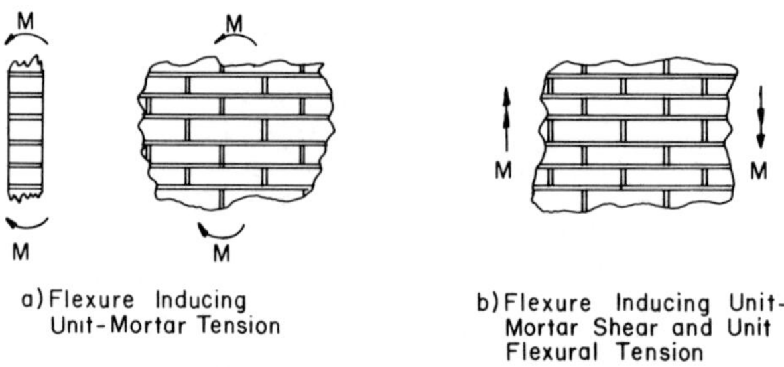

Fig. 24.10 Masonry wall out-of-plane flexure.

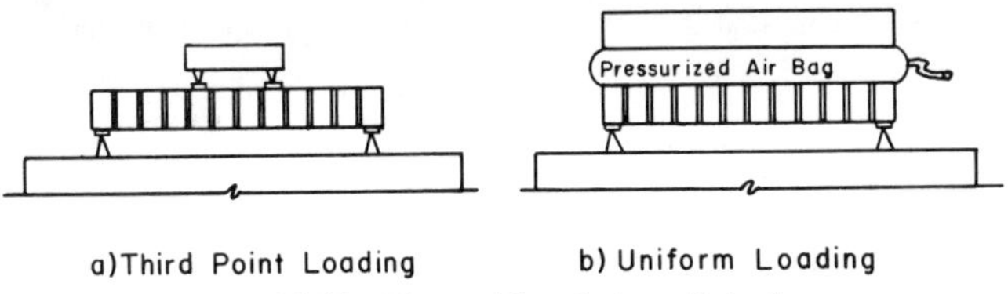

Fig. 24.11 Flexural bond strength test.

An alternate standard method, the bond wrench test, is under consideration by ASTM C15, Manufactured Masonry Units. The proposed standard is entitled "Method for Measurement of Masonry Flexural Bond Strength."

24.4.5 Modulus of Elasticity

Basic Behavior

The modulus of elasticity of masonry is a nonlinear relationship between stress and strain. The shape of the stress–strain curve is basically parabolic, similar to such curves for concrete. However, there appears to be a tendency at very low stress levels toward an increasing value of tangent modulus with an increase in stress producing an S-shaped curve [24.30, 24.52] as shown in Fig. 24.12. The initial stiffening phenomenon has been attributed to breaking down and subsequent compaction of the mortar as load increases and may be more pronounced with lower strength mortars, but has been observed in clay-unit masonry built with higher strength mortar [24.30]. It may also be due to "take-up" in the testing machine used; the influence of this may be avoided by measuring strain between points on the masonry specimen rather than between platens. Factors which influence modulus of elasticity E_m of masonry are primarily the modulus of elasticity of the component materials, that is, mortar and unit, and as implied above, stress level [24.52].

Because of the nonlinearity of the stress–strain relationship, the designer should be aware of the value of E which applies. In many applications it is possible that dead load stress is sufficient to achieve the initial stiffening represented by the lower portion of the curve of Fig. 24.12, thus justifying use of the inner portion which is often approximately linear.

Several U.S. codes and standards permit calculation of the modulus of elasticity E_m of masonry by

$$E_m = 1000 f'_m, \text{ not to exceed } 3{,}000{,}000 \text{ psi } (20{,}700 \text{ MPa}) \tag{24.4.3}$$

where f'_m = design compressive strength of masonry

It is assumed that the value of E_m thus obtained corresponds to the approximately linear portion of the masonry stress–strain curve.

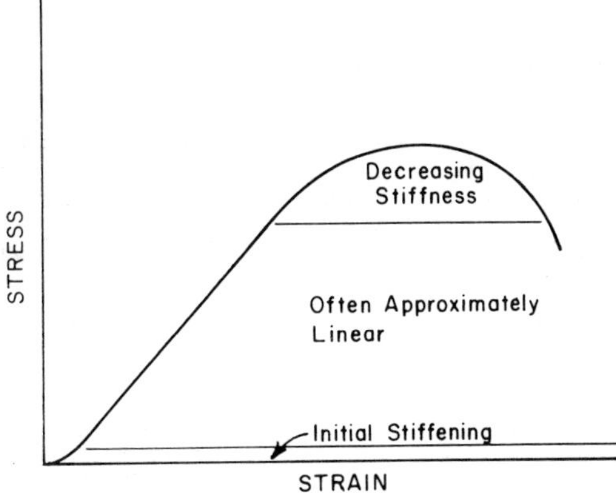

Fig. 24.12 Masonry stress–strain curve.

Experimental data indicate that the correlation between E_m *and* f'_m is rather weak and that

$$E_m = 700f'_m \text{ (approx.)} \tag{24.4.4}$$

for clay-unit (brick) masonry and

$$E_m = 850f'_m \text{ to } 1000f'_m \text{ (approx.)} \tag{24.4.5}$$

for concrete-unit masonry [24.52, 24.60, 24.61, 24.62].

Standard Test

Modulus of elasticity of masonry, that is, the stress–strain relationship, may be obtained by appropriately instrumenting compression specimens, e.g., prisms, in accordance with ASTM E111, "Test for Young"s Modulus at Room Temperatures." Experimental evidence indicates that modulus obtained from tests of flat-end prisms corresponds well to modulus of full scale walls [24.60, 24.61, 24.62].

24.5 ALLOWABLE STRESSES

24.5.1 Basic Considerations

Basis of Design

Structural masonry is designed in United States on the basis of "working stress", that is, stresses at design service loads are limited to values considerably less than ultimate. Stresses at and below design service loads are calculated assuming linear elastic behavior of masonry.

Source of Allowable Stresses

Allowable stresses for design of structural masonry in the United States are presented in several codes and standards maintained throughout the country [24.65, 24.66, 24.74, 24.75, 24.77]. At the time of this writing no national consensus code/standard is available. The applicable code/standard for a given design is that which has been adopted by the governing administrative entity, e.g., the building department of a city or county. It is therefore necessary for the designer to be familiar with the applicable code/standard for a given design and to be aware that differences exist between the various codes/standards.

Basis of Allowable Stresses

Masonry may be classified as (1) unreinforced-unengineered, (2) unreinforced-engineered, and (3) reinforced-engineered. Allowable stresses may be found for each category in one or more of the current codes/standards.

Allowable stresses for the first category are usually empirical numerical values while those for the second and third categories are primarily based upon the design compressive strength of masonry, f'_m, which may be assumed based on unit and mortar properties, established by test, or based on field experience.

Assumed Values of f'_m

Values of f'_m to be used as a basis for allowable stresses may be assumed based on flat-wise compressive strength of the masonry units to be used. Some codes/standards provide values that also depend upon the type of mortar to be used. Unit compressive strength should be established according to the applicable provisions of ASTM C67 [24.64], and ASTM C140 [24.101]. Representative values of f'_m which may be assumed are presented in Table 24.9. The designer should be aware that assumed values may be different between the various codes/standards.

Establishment of f'_m by Test

Ultimate compressive strength of masonry to be used as a basis for design allowable stresses may be established by compressive tests of prisms (see Section 24.4.1). Twenty-eight day strength is usually the basis for f'_m and tests are usually conducted on specimens at that age. Some codes, e.g., Refs. 24.75 and 24.77, permit testing at 7 days of age if the relationship between 7- and 28-day strength is known. Others, e.g., Refs. 24.62 and 24.66, allow f'_m to be based upon 28-day strength *or* at another specified age. The number of tests required ranges from 3 [24.62, 24.66, 24.75] to 5 [24.77].

Table 24.9 Assumed Design Strength of Masonry, f'_m (psi) (MPa)*

Compressive Strength of Units—psi†	Mortar Type‡	
	M or S	N
14,000 or more (97)	5,300 (37)	4,400 (30)
12,000 (83)	4,700 (32)	3,800 (26)
10,000 (69)	4,000 (28)	3,300 (23)
8,000 (55)	3,350 (23)	2,700 (19)
6,000 (41)	2,700 (19)	2,200 (15)
4,000 (28)	2,000 (14)	1,600 (11)
2,000 (14)	1,500 (10)	1,100 (8)
1,000 (7)	800 (6)	600 (4)

* Assumed assemblage compressive strength is gross area strength when using solid units and net area strength when using hollow units. Table based on Table 24-D, *Uniform Building Code,* 1985 [24.77].

† Compressive strength of solid units is based on gross area. Compressive strength of hollow units is based on minimum net area. Values may be interpolated.

‡ Mortar for unit masonry, proportion specification, as specified in Table 24.6. These values apply to portland cement-lime mortars without added coloring or air-entraining materials.

The concrete masonry code [24.66] restricts the height-to-thickness ratio of prisms to the range 1.33–3.0. The average failure stress, f'_{mt} [see Eq. (24.4.1)], from tests of such prisms is used directly as the value of f'_m. Other codes/standards require that the failure stress, f'_{mt}, be "corrected" to correspond to a standard slenderness ratio, that is,

$$f'_m = f'_{mt} \times \text{(correction factor)} \tag{24.5.1}$$

Correction factors from the UBC [24.77] are summarized in Table 24.10.

Table 24.10 Prism Correction Factors [24.77]

h/t_p of prism*:	1.5	2.0	2.5	3.0	4.0	5.0
Factor:	0.86	1.0	1.04	1.07	1.15	1.22

* h = height of prism, t_p = least lateral dimension.

f'_m *Based on Field Experience*

The UBC [24.77] permits f'_m to be based on an existing record, approved by the building official, of at least 30 prism tests. The prisms must have been built with similar mortar, units, and grout (if any) to those to be used for the job at hand. The value of f'_m which may be used is 75% of the average of the previous tests.

24.5.2 Allowable Stresses

The provisions presented in this section are taken from the UBC [24.77]. The allowable stresses herein must be reduced by one-half if job quality control provisions do not include requirements for special inspection as prescribed in Sections 306 and 2405 of the UBC and do not include prism testing prior to and during construction. The sample size of prisms tested as a basis for design is five or more, and the sample size of prisms tested during construction is three. At least one field sample per 5000 square feet of wall area is required and at least one sample for the job.

Compressive Stress F_a—Axial

A. Unreinforced Walls and Columns, Reinforced Walls

$$F_a = 0.20 f'_m \left[1 - \left(\frac{h'}{42t}\right)^3\right] \tag{24.5.2}$$

B. Reinforced Columns

$$P_a = (0.20 f'_m A_e + 0.65 A_s F_{sc}) \left[1 - \left(\frac{h'}{42t}\right)^3\right] \tag{24.5.3}$$

$$F_a = \frac{P_a}{A_e} \tag{24.5.4}$$

Compressive Stress F_b—Flexural

$$F_b = 0.33 f'_m \leq 2000 \text{ psi (14 MPa) maximum} \tag{24.5.5}$$

Tensile Stress F_t—Walls in Flexure

The allowable tensile stress for walls in flexure of masonry elements or members without tensile reinforcement using portland cement and hydrated lime Types M or S mortar shall not exceed the values which follow. For Types M and S masonry cement mortars, the values shall be reduced by 50% for clay units and 25% for concrete units. For Type N mortar, values shall be reduced by 25%.

Values for tension normal to head joints are for running bond; no tension is allowed across head joints in stack bond masonry. These values shall not be used for horizontal flexural members such as beams, girders, or lintels.

A. Tension F_t normal to bed joints, psi (kPa):

	Clay Units	Concrete Units
Solid Units	36 (250)	40 (280)
Hollow Units	22 (150)	25 (170)

B. Tension F_t normal to head joints, psi (kPa):

	Clay Units	Concrete Units
Solid Units	72 (500)	80 (550)
Hollow Units	45 (310)	50 (340)

Shear Stress F_v—Flexural Members

A. No shear reinforcement,

$$F_v = 1.0(f'_m)^{1/2} \leq 50 \text{ psi (340 kPa) maximum} \tag{24.5.6}$$

Exception: For a distance of one-sixteenth the clear span beyond the point of inflection the maximum stress shall be 20 psi (140 kPa).

B. Shear reinforcement designed to take entire shear force,

$$F_v = 3.0(f'_m)^{1/2} \leq 150 \text{ psi (1 MPa) maximum} \tag{24.5.7}$$

Shear Stress F_v—Shear Walls

A. Unreinforced Masonry,

Clay Units

$$F_v = 0.3(f'_m)^{1/2} \leq 80 \text{ psi (550 kPa) maximum} \tag{24.5.8}$$

	M or S Mortar	N Mortar
Concrete units	34 psi (230 kPa)	23 psi (160 kPa)

The allowable shear stress in unreinforced masonry may be increased by $0.2 f_{md}$, where f_{md} is the compressive stress in the masonry due to dead load only.

B. With in-plane flexural reinforcement present,

$$\frac{M}{Vd} < 1; \quad F_v = \frac{1}{3}\left(4 - \frac{M}{Vd}\right)(f'_m)^{1/2} \le \left(80 - 45\frac{M}{Vd}\right) \text{ psi maximum} \quad \left(550 - 310\frac{M}{Vd}\right) \text{ kPa} \tag{24.5.9}$$

$$\frac{M}{Vd} \ge 1; \quad F_v = 1.0(f'_m)^{1/2} \le 35 \text{ psi (240 kPa) maximum} \tag{24.5.10}$$

C. Shear reinforcement designed to take all the shear,

$$\frac{M}{Vd} < 1; \quad F_v = \frac{1}{2}\left(4 - \frac{M}{Vd}\right)(f'_m)^{1/2} \le \left(120 - 55\frac{M}{Vd}\right) \text{ psi maximum} \quad \left[830 - 380\frac{M}{Vd}\right] \text{ kPa} \tag{24.5.11}$$

$$\frac{M}{Vd} \ge 1; \quad F_v = 1.5(f'_m)^{1/2} \le 75 \text{ psi (520 kPa) maximum} \tag{24.5.12}$$

Bearing Stress F_{br}

A. On full area,

$$F_{br} = 0.26 f'_m \tag{24.5.13}$$

B. On one-third area or less,

$$F_{br} = 0.38 f'_m \tag{24.5.14}$$

This increase applies only when the least distance between the edges of the loaded and unloaded areas is a minimum of one-fourth of the parallel side dimension of the loaded area. The allowable bearing stresses on a reasonably concentric area greater than one-third but less than the full area shall be interpolated between the values of Eqs. (24.5.13) and (24.5.14).

Reinforcing Bond Stress u, psi (kPa)

Plain Bars	60 (410)
Deformed Bars	140 (970)

Allowable Tensile Stress in Reinforcement

A. Deformed bars,

$$F_s = 0.5 f_y \le 24{,}000 \text{ psi (165 MPa) maximum} \tag{24.5.15}$$

B. Wire reinforcement,

$$F_s = 0.4 f_y \le 20{,}000 \text{ psi (138 MPa) maximum} \tag{24.5.16}$$

C. Ties, anchors and smooth bars,

$$F_s = 0.4 f_y \le 20{,}000 \text{ psi (138 MPa) maximum} \tag{24.5.17}$$

Allowable Compressive Stress in Reinforcement

A. Deformed Bars in Columns

$$F_{sc} = 0.4 f_y \le 24{,}000 \text{ psi (165 MPa) maximum} \tag{24.5.18}$$

Table 24.11 Allowable Shear on Bolts for All Masonry Except Unburned Clay Units*

Diameter of Bolt, in. (mm)	Embedment,† in. (mm)	Solid Masonry Shear, lb (kN)	Grouted Masonry Shear, lb (kN)
$\frac{1}{2}$ (13)	4 (100)	350 (1.6)	550 (2.5)
$\frac{5}{8}$ (16)	4 (100)	500 (2.2)	750 (3.3)
$\frac{3}{4}$ (19)	5 (130)	750 (3.3)	1100 (4.9)
$\frac{7}{8}$ (22)	6 (150)	1000 (4.4)	1500 (6.7)
1 (25)	7 (180)	1250 (5.6)	1850‡ (8.2)
$1\frac{1}{8}$ (29)	8 (200)	1500 (6.7)	2250 (10.0)

* Based upon Table 24-J, *Uniform Building Code*, 1985 [24.77].
† An additional 2 in. of embedment shall be provided for anchor bolts located in the top of columns for buildings located in (UBC) Seismic Zones 2, 3, and 4.
‡ Permitted only with not less than 2500 psi (17.2 MPa) units.

B. Deformed bars in flexural members,

$$F_s = 0.5 f_y \leq 24{,}000 \text{ psi (165 MPa) maximum} \tag{24.5.19}$$

Allowable shear loads on bolts are given in Table 24.11.

Combined Compressive Stresses

Members subject to combined axial and flexural stresses shall be designed in accordance with accepted principles of mechanics or in accordance with the following formula:

$$\frac{f_a}{F_a} + \frac{f_b}{F_b} \leq 1 \tag{24.5.20}$$

Allowable Stress Increase

All maximum allowable stresses presented for working stress design of masonry may be increased by one-third when the source of load is wind or earthquake either acting alone or combined with gravity load. No increase is allowed for gravity loads acting alone [24.77].

24.6 RECTANGULAR BEAM DESIGN

24.6.1 Basic Considerations

Masonry beams are implicit in the portions of masonry walls over openings where the load from above must be supported by flexural action. Discrete masonry beams may be used to support floor/roof systems; however, such applications are usually restricted to cases where the visual appearance of masonry is required.

The primary structural factors affecting beam design are ability to carry design vertical load and deflection limitations, both of which are unique to every design.

Nonstructural factors, for example, appearance and resistance to environmental forces may affect choice of materials, proportions, etc., and must be considered. These factors are discussed further in Section 24.7.

24.6.2 Beam Configurations

Masonry beams may be constructed in a number of ways depending, in part, upon the type of masonry unit to be used. Some representative singly reinforced designs are presented in Fig. 24.13. Moment capacity may be increased by use of compression reinforcement.

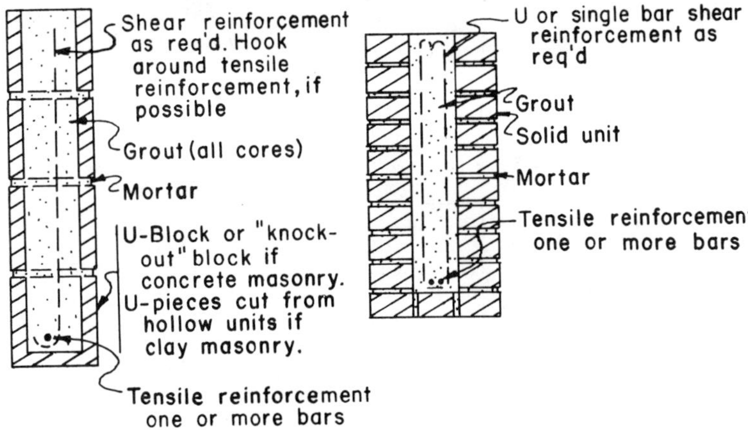

Fig. 24.13 Masonry beam configurations.

Horizontal and vertical reinforcement in building walls enables such walls to act as two-way reinforced plates. More simply, such walls are often designed as if they were a series of parallel one-way beams. Typically, walls are designed to resist out-of-plane forces by spanning vertically. Horizontal spanning capability can often be advantageous, however, and should not be overlooked. Figure 24.14 illustrates representative cross-sections of reinforced walls spanning vertically.

24.6.3 Beam Loading Conditions

Masonry beams are primarily designed to support vertical, in-plane, distributed, and/or concentrated loads. It is possible in some building systems for axial forces to be present as well. Vertical loads produce, in turn, internal shear and moment where moment may be positive or negative depending on support and boundary conditions. Figure 24.15 depicts a generalized loading on a masonry beam element.

24.6.4 Empirical Beam Design Requirements

Masonry codes/standards typically contain design criteria which must be met regardless of loads or other factors (24.65, 24.66, 24.74, 24.75, 24.77]. In some instances exceptions are permitted if the

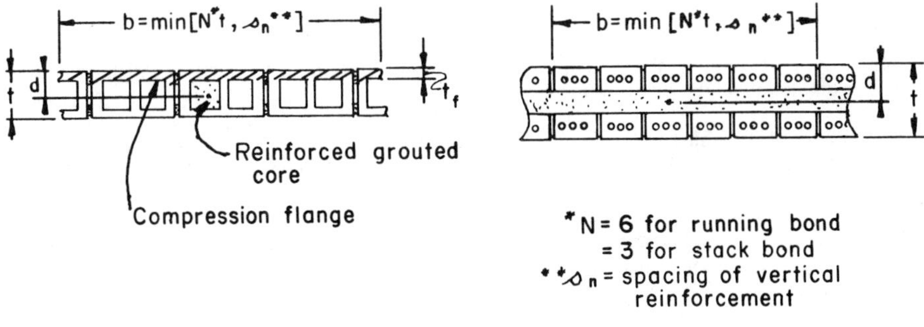

Fig. 24.14 Reinforced wall section.

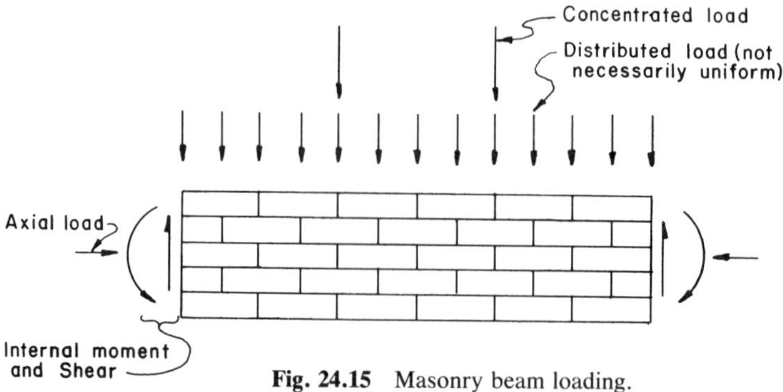

Fig. 24.15 Masonry beam loading.

variation from the requirement can be proven to be structurally acceptable. The empirical criteria pertaining to beam design primarily concern such matters as lateral stability and reinforcement details. The following are found in the UBC [24.77]:

Lateral Stability

The clear distance between lateral supports of a beam should not exceed 32 times the least width of the compression or flange.

Reinforcing Requirements and Details

The maximum size of reinforcing is #11. The maximum area of reinforcement in cells is 6% of the cell area. Where splices occur, the maximum reinforcement area is 12% of the cell area.

Splices

The amount of lap of lapped splices should be sufficient to transfer the allowable reinforcement stress. The length of lap should be at least 30 bar diameters for bars in compression and at least 40 bar diameters for bars in tension.

Spacing and Cover of Longitudinal Reinforcement

The clear distance between parallel bars must be at least the greater of 1 in. (25 mm) or the nominal diameter of the bars except that bars in a splice may be in contact. The clear distance between the surface of a bar and any masonry surface must be at least ¼ in. (6 mm) for fine grout and ½ in. (12 mm) for coarse grout. Cross webs of hollow units may be used as supports for horizontal reinforcement.

Reinforcing bars must be completely embedded in mortar or grout. Cover, which included mortar and/or grout and masonry units, must be at least 1½ in. (38 mm) if the masonry is exposed to weather, 2 in. (50 mm) if exposed to soil, and ¾ in. (19 mm) otherwise.

Anchorage of Flexural Reinforcement

Except at supports or at the free end of cantilevers, every reinforcing bar must extend at the greater of 12 bar diameters or the depth of the beam beyond the point at which it is no longer required to resist stress.

A minimum of one-third of the total reinforcement provided for negative moment at a support shall be extended a distance beyond the extreme position of the point of inflection equal to the greater of:

1. the length sufficient to develop one-half the allowable stress of the reinforcement by bond,
2. one-sixteenth of clear span length, or
3. the depth of the member.

At least one-third of the required positive reinforcement in simple beams, *or* at the freely supported end of continuous beams, shall extend along the same face 6 in. (150 mm) into the support. At least one-fourth of the required positive moment reinforcement at the continuous end of continuous beams shall extend along the same face of the beam into the support at least 6 in. (150 mm).

Compression reinforcement in beams and girders shall be secured by ties or stirrups at least ¼ in. (6 mm) in diameter at a spacing not exceeding the lesser of 48 tie diameters or 16 bar diameters. Such ties or stirrups are required wherever compression reinforcement is required.

Flexural reinforcing bars may not be terminated in a tension zone unless:

1. Shear is less than one-half the shear capacity including capacity of any shear reinforcement,
2. Stirrups in excess of those required are provided each way from the termination point a distance equal to beam depth. Spacing of such stirrups shall be equal to or less than $d/8r_b$, where r_b is the ratio of the area of terminated bars to the total area of bars at the section.
3. The continuing bars provide double the area required for flexure at that point or double the perimeter required for reinforcing bond.

In regions of moment where design tensile stresses in the reinforcement are greater than 80% of the allowable tensile stress, the lap length of splices must be increased by at least 50% of the minimum required length.

24.6.5 Flexural Analysis/Design

Assumptions

The method is based on working stress concepts where:

1. Plane sections remain plane.
2. Materials are linear–elastic.
3. Masonry tensile strength is insignificant.
4. Modulus of elasticity of masonry is uniform throughout the member and remains constant over the working load range.
5. Span is large relative to depth.
6. Individual materials combine to form a homogeneous, isotropic member.

Flexural Relationship—Singly Reinforced Beams

Flexural analysis/design relationships are based upon internal force equilibrium and strain compatibility. Figure 24.16 depicts the assumed stress and strain relationships.

By integrating stress over the area upon which it acts (Fig. 24.16*a*)

$$T = A_s f_s = pbdf_s \tag{24.6.1}$$

$$C = \frac{1}{2} f_b bkd \tag{24.6.2}$$

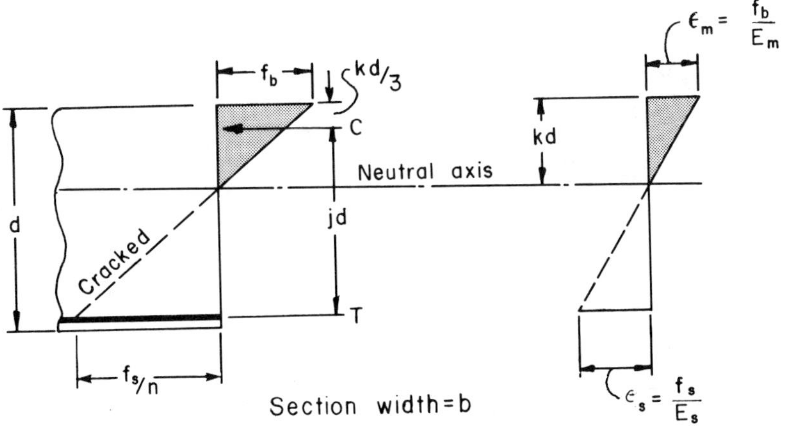

Fig. 24.16 Singly reinforced beam stress and strain diagrams.

Taking moments about the line of action of T and C, respectively,

$$M_m = Cjd = \frac{1}{2} f_b jkbd^2 \tag{24.6.3}$$

$$M_s = Tjd = f_s pjbd^2 = A_s f_s jd \tag{24.6.4}$$

From Eqs. (24.6.3) and (24.6.4),

$$f_s = \frac{M}{pjbd^2} = \frac{M}{A_s jd} \tag{24.6.5}$$

$$f_b = \frac{2M}{bjkd^2} \tag{24.6.6}$$

Based upon conditions of strain compatibility (Fig. 24.16*b*) reinforcement stress is given by

$$f_s = nf_b \frac{(1 - k)}{k} \tag{24.6.7}$$

by rearranging

$$k = \frac{1}{1 + \dfrac{f_s}{nf_b}} \tag{24.6.8}$$

Equating Eqs. (24.6.3) and (24.6.4),

$$k = \frac{2f_s p}{f_b} \tag{24.6.9}$$

Substituting Eq. (24.6.7) into (24.6.9) and solving the resulting quadratic,

$$k = -pn + \sqrt{(pn)^2 + 2pn} \tag{24.6.10}$$

Analysis of a Singly Reinforced Section

The flexural capacity of a section may be determined by:

1. Solving for k using Eq. (24.6.10).
2. Calculating $j = 1 - k/3$.
3. Calculating M_m setting $f_b = F_b$, Eq. (24.6.3).
4. Calculating M_s setting $f_s = F_s$, Eq. (24.6.4).
5. Moment capacity $= \min[M_m, M_s]$.

Alternatively stress in the masonry and reinforcement at a section for a moment M may be determined by:

1. Repeating steps 1 and 2 above.
2. Calculating f_s using Eq. (24.6.5).
3. Calculating f_b using Eq. (24.6.6).

Design of a Singly Reinforced Section

For a beam with dimensions b and d known, the selection of the area of tensile reinforcement is essentially an iterative process. The process can be accelerated by assuming a value of j (which is usually in the neighborhood of 0.90) and calculating the steel ratio, that is, rearranging Eq. (24.6.4) and using F_s for f_s:

$$p_{\text{estimate}} = \frac{M}{F_s(0.90)bd^2} \tag{24.6.11}$$

The estimated value of p and the modular ratio n may then be used to calculate k by Eq. (24.6.10). The moment capacity is then calculated as in the analysis procedure and p adjusted as required so that moment capacity is greater than applied moment.

Balanced Singly Reinforced Design

Balanced flexural design of masonry is defined as a design in which the amount of reinforcement is such that the stress in the masonry and reinforcement are both at their maximum allowable under the design moment. Setting stresses at the maximum allowable in Eq. (24.6.8) the location of the neutral axis for a balanced design is defined by

$$k_b = \frac{1}{1 + \dfrac{F_s}{nF_b}} \tag{24.6.12}$$

Equating internal tension and compression and solving for p_b,

$$p_b = \frac{k_b}{2F_s/F_b} \tag{24.6.13}$$

Substituting for k_b, the "balanced" reinforcement ratio is

$$p_b = \frac{n}{2F_s/F_b(n + F_s/F_b)} \tag{24.6.14}$$

If in a given design, $p < p_b$, reinforcement stress will be at its maximum allowable while the stress in the masonry will be less than its maximum allowable. If $p > p_b$, the converse is true.

Doubly Reinforced Flexural Design

Steel flexural reinforcement may be placed on the compression side to increase flexural capacity and inhibit creep. Such reinforcement, based on linear–elastic analysis, is not as efficient as tensile reinforcement. However, creep tends to increase the stress in compressive reinforcement by an unknown factor [24.23, 24.53]. Unless prohibited by other considerations, it is suggested that the designer attempt to design a suitable flexural member relying upon tensile reinforcement alone. Procedures for design of doubly reinforced flexural sections may be found elsewhere [24.1, 24.23, 24.53].

24.6.6 Flexural Bond

Different values of moment at two different sections of a beam result in a net tension in the reinforcing bars which must be balanced by bond between the bars and the masonry. Flexural bond stress u may be calculated as

$$u = \frac{V}{\sum o\, jd} \tag{24.6.15}$$

Bond stress thus determined must be compared to the maximum allowable stress presented in the applicable code.

If bond is insufficient to develop the required tension in a reinforcing bar, standard hooks or other approved anchorage may be used to augment bond. A standard hook may be considered to develop 7500 psi (52 MPa) and should be either:

1. A complete semicircular turn with a radius of bend at the axis of the bar of at least 3, but not more than 6 bar diameters plus an extension of at least 4 bar diameters but not less than 2½ in. (64 mm) at the free end.
2. A 90° bend with a radius of at least 4 bar diameters plus an extension of 12 bar diameters.

The required development length ℓ_d for deformed reinforcement

$$\ell_d = 0.002 d_b f_s \tag{24.6.16}$$

for bars in tension, and

$$\ell_d = 0.0015 d_b f_s \tag{24.6.17}$$

for bars in compression.

24.6.7 Shear

Assumptions

In addition to the assumptions listed in Section 24.6.5:

1. Shear stress is assumed to be uniformly distributed over the area of the cross-section even though flexural cracking may exist.
2. If shear reinforcement is used, it is designed to carry 100% of the shear at a section.

Design for Shear

The unit shear stress v in a beam is determined by

$$v = \frac{V}{bjd} \tag{24.6.18}$$

If v thus calculated exceeds the maximum allowable shear stress F_v for beams without shear reinforcement, shear reinforcement must be provided. The required area may be determined by the following.

For stirrups:

$$A_v = \frac{Vs}{F_s jd} \tag{24.6.19}$$

For a single bar or group of bars all bent up at the same distance from the support:

$$A_v = \frac{V}{F_s \sin \alpha} \tag{24.6.20}$$

For a series of parallel bars or groups of bars bent up at different distances from the support*:

$$A_v = \frac{Vs}{F_s jd(\sin \alpha + \cos \alpha)} \tag{24.6.21}$$

Shear reinforcement must be spaced horizontally so that every 45° line extending from the mid-depth of the beams to the longitudinal bars is crossed by at least one line of shear reinforcement.

If shear reinforcement is required, it is assumed that it must be capable of carrying 100% of V at the section; however, the stress in the masonry calculated by Eq. (24.6.18) must not exceed F_v for beams with shear reinforcement.

Anchorage of Shear Reinforcement

Single separate bars must be anchored by:

1. Hooking tightly around longitudinal reinforcement through at least 180°.
2. Embedment above or below beam mid-depth on the compression side a distance sufficient to develop the force in the bar.
3. A standard hook plus sufficient embedment length to develop the bar. The effective embedment length may not exceed the distance between beam mid-depth and the tangent of the hook.

---

* Only the center ¾ of the inclined portion of bent-up longitudinal bars can be considered effective. Such bars shall be bent around a pin whose diameter is at least 6 times the bent bar diameter.

The ends of single or multiple U-stirrups may be anchored as described above or by bending through at least 90° around a longitudinal bar and extending 12 stirrup bar diameters beyond the bend. The longitudinal bar diameter should be at least equal to the stirrup bar diameter for this method. All other bends shall be around a longitudinal bar.

24.7 WALL DESIGN

24.7.1 Basic Considerations

Many factors must be considered in the design of a masonry wall and may be broadly classed as "structural" and "nonstructural". While several options of material and configuration may be available to support a given set of loads, the selection of a particular configuration is usually influenced by nonstructural factors. The primary intent of this chapter is to treat wall design in terms of structural considerations; however, nonstructural considerations are briefly reviewed. It should be noted that factors defined as "structural" and "nonstructural" herein may not be exclusive; certain "nonstructural" factors, for example, brick color, may be related to strength properties and therefore affect structural design. Chapter 7 provides more general treatment of wall function and design.

Aesthetics

The visual appearance of walls, that is, color, surface texture and appearance, and geometrical form, is an important attribute. Geometrical form obviously relates to structural capacity. Clay unit color may be related to unit strength as well as mortar color if achieved by additives. Surface texture, if achieved by recessed or protruding units or raked mortar joints, may affect structural performance. Raked joints, for example, could be a source of moisture penetration, material degradation in certain climates, and affect stability [24.1, 24.23, 24.53]. Surface appearance achieved by bonding patterns, for example, stack bond, can have a significant structural effect [24.53, 24.77].

Fire Resistance, Thermal and Acoustical Properties

Depending on wall location, function, and building use, these factors may control material selection and/or thickness. Further information may be found in Refs. 24.1, 24.25, 24.50, 24.53, 24.63.

Environmental Considerations

Resistance to water penetration of exterior masonry walls can directly affect structural design. For example, a wall which could otherwise be solid, may have to be built with provision for internal moisture diversion, for example, a cavity wall. Extreme temperature changes tending to cause relative movement of external walls with respect to internal or other external walls may cause significant structural effects [24.23, 24.25].

Proper detailing of exterior walls for moisture resistance, volumetric changes, and interaction with other structural components is critical in masonry construction. These aspects are dealt with in Chapter 27.

24.7.2 Wall Configurations

Masonry walls may be one unit in thickness (single-wythe) or two or more wythes thick. In some forms, the wythes are separated by a cavity which may or may not be filled with grout. Each wythe is normally of a single type of masonry unit; however, other wythes in a given wall may be of different types of units and of different materials. The choice of a given configuration depends upon pertinent aesthetic, structural, environmental, economic, and functional considerations. Some of the more frequently encountered forms are presented in the following sections.

Multi-Wythe, Solid Unit

This configuration (Fig. 24.17) is typical of unreinforced older brick (solid clay unit) masonry. Current use is normally for low-rise structures.

Multi-Wythe, Brick with Concrete Block

Solid brick as an exterior architectural surface with hollow concrete block backup is widely used for low-rise commercial construction (Fig. 24.18). The brick face is integral with the block backing wythe to form a structural wall. The block wythe may or may not be grouted and reinforced.

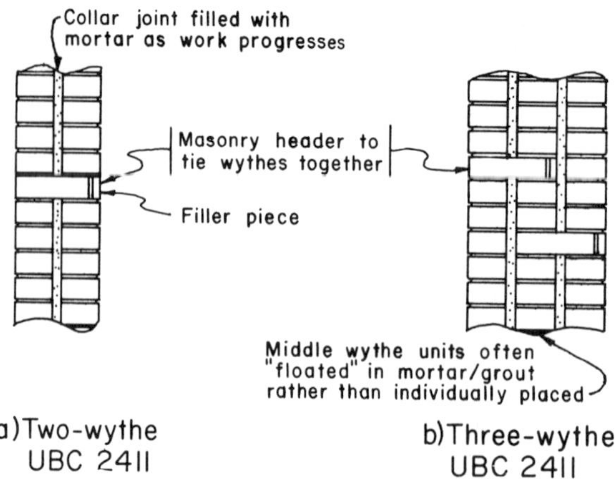

Fig. 24.17 Multi-wythe solid unit wall.

Two-Wythe Cavity

Reinforced and grouted cavity construction (Fig. 24.19) is a means of obtaining a wall with high resistance to in- and out-of-plane flexure and in-plane shear. Vertical reinforcement in walls is not considered to carry vertical loads unless ties are provided (see Section 24.8). This type of wall can provide architectural surfaces on both sides.

The cavity between two wythes may be left empty, filled with insulation, or serve as a space in which reinforcement may be placed and grouted. The outer wythe in the first form is often meant to function as a moisture barrier; flashing is required at vertical intervals to direct the water to the outside. Placing insulation in the cavity is a means of improving thermal performance and is a relatively recent innovation.

Hollow Unit

Hollow clay and concrete masonry units may effectively be used to construct single-wythe walls (Fig. 24.20). Selected or all vertical cores may be grouted and reinforced as required for in- and out-of-plane flexure and in-plane shear. Vertical reinforcement is not considered to carry vertical load unless ties are provided (see Section 24.8). This type of wall can provide architectural surfaces on both sides.

24.7.3 Wall Loading Conditions

Masonry walls may be subjected to a variety of load conditions as components of a building system which must resist overall vertical and lateral loads. Basic loading conditions are depicted in Fig. 24.21. For each of the conditions, dead load (DL) and live load (LL) should be combined in a manner that produces the most critical situation.

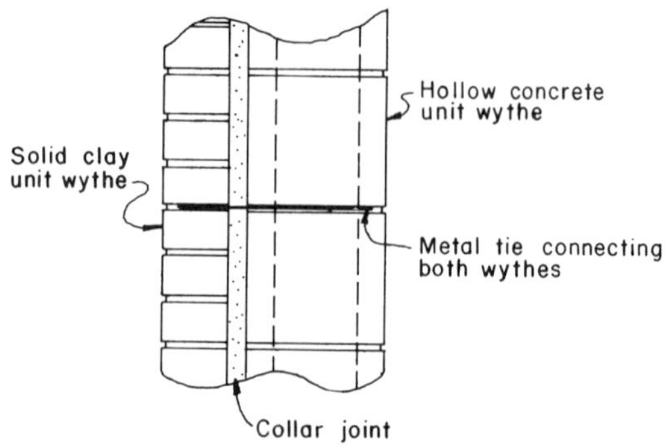

Fig. 24.18 Two-wythe brick face with block backing.

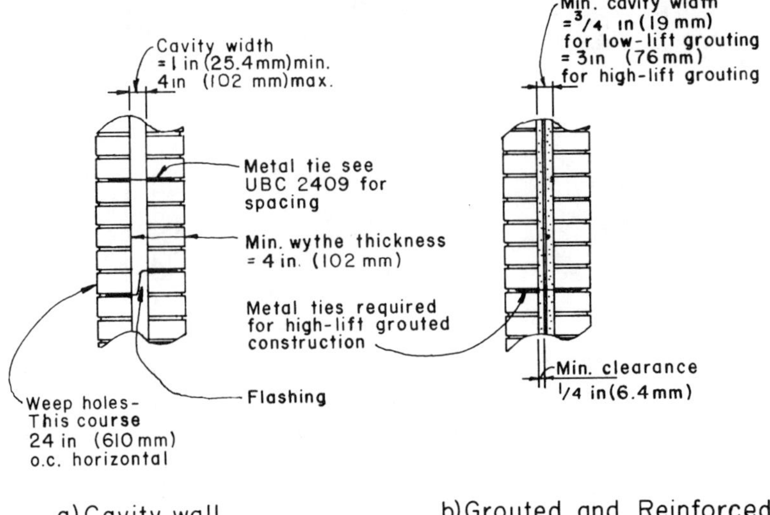

Fig. 24.19 Cavity wall construction.

Figure 24.21*a* represents vertical load on an interior wall where floor spans on each side of the wall are equal. Eccentric vertical load shown in Fig. 24.21*b* is typical of exterior walls where the floor system bears on the inner portion or inner edge of the wall. It is possible for the eccentricities at the top and bottom to be different in magnitude and to be on different sides of the wall centerline, depending on the construction. Moments due to wall–floor interaction must also be considered [24.30, 24.31, 24.52].

The source of lateral load shown in Fig. 24.21*c* could be either wind or earthquake. Which is more severe depends upon the seismic and wind conditions at the building location.

The shear and moment loads of Fig. 24.21*d* may also be due to either wind or earthquake. Again, which is more severe depends upon the seismic and wind conditions at the building location.

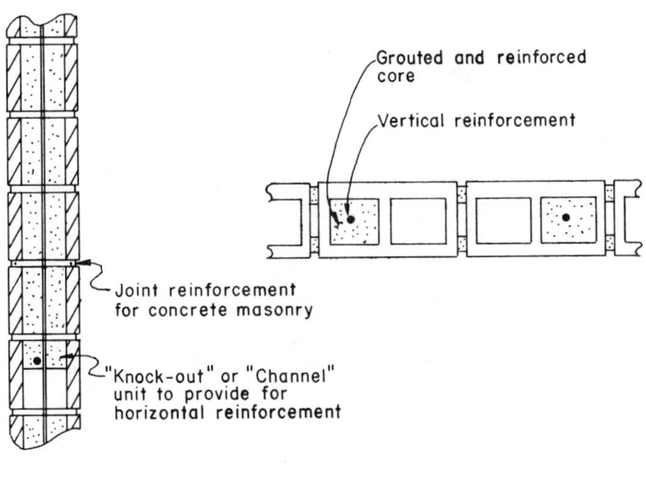

Fig. 24.20 Hollow-unit masonry wall.

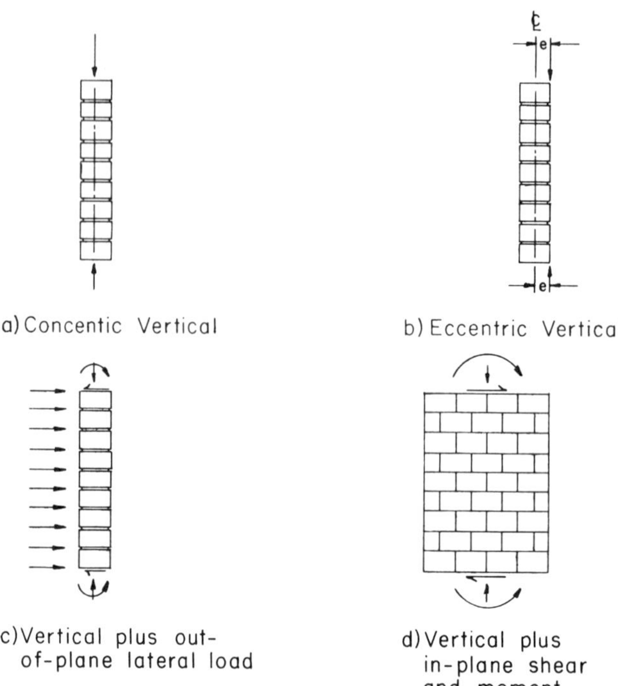

Fig. 24.21 Masonry wall loading conditions.

24.7.4 Empirical Wall Design Requirements

Masonry codes/standards typically contain design criteria which must be met regardless of loads or other factors [24.65, 24.66, 24.74, 24.75, 24.77].

In some instances exceptions are permitted if the variation from the requirement can be proven to be structurally acceptable. The empirical criteria are relevant to structural features such as bonding between wythes of multi-wythe walls, wall thickness, wall slenderness, reinforcement amount, and reinforcement details. The designer should therefore review the code/standard applicable to a given design to identify all pertinent empirical criteria. The designer should be aware that satisfying such criteria does not necessarily assure structural adequacy for all applications; the designer is ultimately responsible. Selected empirical criteria from the UBC [24.77] are presented in the following paragraphs.

Walls Intersecting with Floors and/or Roofs

Integrity of the structural system, of which walls are a part, is greatly enhanced by adequate connection of the components. Walls should be well anchored to floors, roofs or other components which provide stability. Anchorage used as structural connections to transfer forces should be designed accordingly.

Interconnection of Wythes in Multi-wythe Walls

Codes usually require that all wythes of multi-wythe walls be bonded together such that the individual walls act as single structural components. Bonding may be accomplished by grout, transverse masonry units (''headers''), or by corrosion-resistant metal ties on bed joint reinforcement. Interconnection of wythes by the latter method is generally accepted as the preferred method. The UBC (24.77] sets forth empirical requirements for metal ties and joint reinforcement as follows:

> *Metal ties shall be of sufficient length to engage all wythes. The portion of the tie within the wythe shall be completely embedded in mortar or grout. The ends of the ties shall be bent to 90-degree angles with an extension not less than 2 in. (50 mm) long. Ties not completely embedded in mortar or grout between wythes shall be a single piece of metal with each end engaged in each wythe.*

There shall be at least one 3/16-in. (5 mm) diameter metal tie for each 4½ square feet (0.43 m²) of wall area. For cavity walls in which the width of the cavity is greater than 3 in. (76 mm) but not more than 4½ in. (114 mm) at least one 3/16-in. (5 mm) diameter tie for each 3 square feet (0.28 m²) of wall area shall be provided.

Ties in alternate courses shall be staggered; the maximum vertical distance between ties shall not exceed 24 in. (610 mm) the maximum horizontal distance between ties shall not exceed 36 in. (915 mm).

Additional ties spaced not more than 36 in. (915 mm) apart shall be provided around and within 12 in. (305 mm) of the opening.

Metal ties of different size and spacing may be used if they provide equivalent strength between wythes.

Prefabricated joint reinforcement for masonry walls shall have at least one crosswire of at least No. 9 gauge for each 2 square feet (0.19 m²) of wall area. The vertical spacing of the joint reinforcement shall not exceed 16 in. (400 mm). The longitudinal wires shall be thoroughly embedded in the bed joint mortar. The joint reinforcement shall engage all wythes.

Stack Bond Masonry Walls

Each stack-bond wythe of a single or multi-wythe wall must contain horizontal reinforcement of at least two continuous corrosion-resistant steel wires placed in the bed joints with a minimum cross-sectional area of 0. 017 sq in. each (11 mm²). Vertical spacing of such reinforcement must not exceed 16 in. (400 mm).

Cover of Ties, Bolts, and Joint Reinforcement

Cover of metal elements is necessary to provide a measure of environmental protection and bond. The UBC [24.77] requires a minimum of ⅝ in. (16 mm) of mortar between ties and joint reinforcement and any exposed face of a wall. The thickness of grout or mortar between masonry units and metal elements should be at least ¼ in. (6 mm). However, ¼ in. (6 mm) or smaller diameter reinforcement or bolts may be placed in mortar bed joints whose thickness is at least twice the diameter of the metal element.

End Support

Beams, girders, or other concentrated loads on walls should have bearing of at least 3 in. (76 mm) in length on solid masonry at least 4 in. (100 mm) thick, upon an adequate metal bearing plate, or upon a continuous reinforced masonry member projecting at least 3 in. (76 mm) from the face of the wall. Other methods may be shown to be adequate as well. Members supporting masonry should have a bearing length of at least 4 in. (100 mm).

24.7.5. Wall Design—Basic Assumptions

Material Behavior

The assumptions of Section 24.6.5, except number 5, apply to masonry wall design.

Effective Thickness—Single-Wythe Walls [24.77]

The effective thickness of single-wythe masonry walls is assumed to be the specified thickness of the walls.

Effective Thickness—Multi-wythe Walls [24.77]

If the space between wythes is filled with mortar or grout, the effective thickness may be taken as the specified thickness. Otherwise, effective thickness is determined as for cavity walls.

Effective Thickness—Cavity Walls [24.77]

If both wythes are axially loaded, each wythe is assumed to act independently and the effective thickness of each wythe is determined as a single-wythe wall. Where one wythe only is axially loaded, effective thickness of the loaded wythe may be assumed to be the square root of the sum of the squares of the specified thickness of the wythes.

Load Conditions

Masonry walls should be designed for at least the applicable of the load cases depicted in Fig. 24.21.

Effective Height [24.77]

Unless otherwise justified, the effective height of walls and columns is at least the clear height of members laterally supported at their top and bottom in a direction normal to the column or wall axis under consideration. For members not supported at the top normal to the axis considered, the effective height is twice the height of the member above the lower support.

Effective Area [24.77]

Effective area may be based on the minimum bedded area of hollow units, or the gross area of solid units plus any grouted area, or the gross area of grouted hollow units. For hollow unit walls where the cores are aligned perpendicular to the direction of stress, effective area is the lesser of minimum bedded area or minimum cross-sectional area.

Effective Width of Intersecting Walls

Where a wall is intersected by a shear wall and the walls are adequately anchored, the width of the intersected wall, which may be considered effective acting as a flange for purposes of stiffness calculations, shall not exceed six times the thickness of the intersected wall. Only the shear wall shall be assumed to carry shear. The effective width of flange limit may be exceeded if justified.

Flexural Resistance of Cavity Walls [24.77]

Lateral loads acting normal to the plane of the wall may be considered carried by the wythes in proportion to their individual out-of-plane flexural rigidities.

24.7.6 Design of Unreinforced Masonry Walls [24.77]

Axial Stress

Axial stress due to centroidal loads may be computed by assuming uniform distribution over the effective area, that is,

$$f_a = \frac{P}{A_e} \tag{24.7.1}$$

Axial stress shall not exceed the limit given by Eq. (24.5.2).

Flexural Stress

Flexural stress may be calculated by:

$$f_b = \frac{Mc}{I} \tag{24.7.2}$$

Flexural compressive stress shall not exceed the limit given by Eq. (24.5.5).

Combined Stresses

Resultant stresses due to combined flexure and axial load shall not exceed the allowable tensile stress F_t for masonry as provided in Section 24.5.2, where f_a and f_b are of opposite sign. Where f_a and f_b are additive, the limits of Eq. (24.5.20) apply.

Shear Stress

Shear stress may be calculated by the equation:

$$f_v = \frac{V}{A_e} \tag{24.7.3}$$

The stress must not exceed the applicable of values presented in Section 24.5.2.

24.7.7 Design of Reinforced Masonry Walls [24.77]

Axial Stress

Axial stress due to centroidal loads may be computed by assuming uniform distribution over the effective area, that is,

$$f_a = \frac{P}{A_e} \tag{24.7.4}$$

Axial stress should not exceed the limits given by Eq. (24.5.2).

Combined Flexure and Compression or Flexure

Stresses due to combined flexure and compression are limited by Eq. (24.5.20), where f_a is computed by Eq. (24.7.4). Walls subjected to flexure with or without compression must satisfy applicable requirements for flexural design.

Design of walls with an h'/t ratio larger than 30 shall be based on forces and moments based on an analysis of the building where the influence of axial loads and variable moment of inertia on member stiffness and fixed-end moments is considered, along with the effect of deflections on moments and forces and the effects of load duration.

Spacing and Cover of Longitudinal Reinforcement

See Section 24.6.4.

Reinforcement Size and Area

Longitudinal reinforcement may be no larger than #11 bars. The maximum area of reinforcement in cells of hollow unit masonry is 6% of the cell area, but may be 12% where lap splices occur.

Two alternate methods are presented below for analysis/design of sections under combined moment and compression. Both account for the tension relieving effect of the compressive load and are based on accepted principles of mechanics.

Combined Vertical Compression and Flexure—Method A

Method A enables the calculation of required tensile reinforcement for given values of P and M with maximum compressive stress on the masonry limited according to Eq. (24.5.20). The force, stress, and geometry relationships are presented in Fig. 24.22 for the case of compression plus in-plane moment.

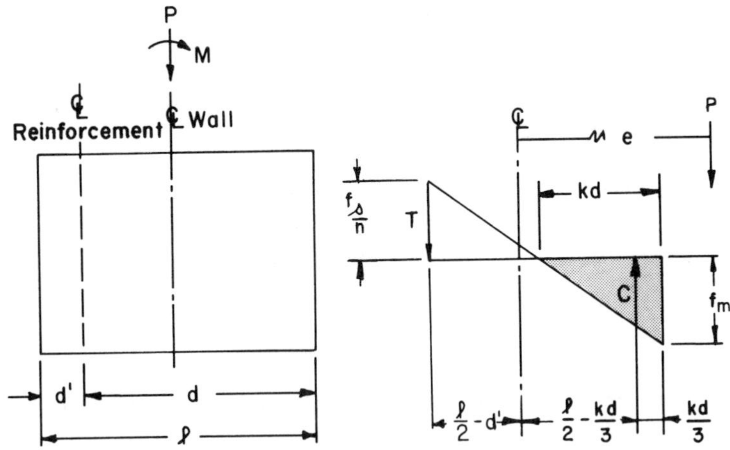

Fig. 24.22 Force/stress/geometry relationships for combined P–M.

Data required are*,†:

Length of wall: ℓ*
Wall thickness: t*
Distance to reinforcement: d
Distance to reinforcement: d'
Compressive load: P
Design compressive masonry strength: f'_m
Modular ratio: $n = E_s/E_m$
Moment: M
Max. allowable reinforcement stress: F_s
Max. allowable masonry compressive stress: F_a
Max. allowable masonry flexural compressive stress: F_b
Wall height: h

With due attention to consistency of units calculate

$$f_a = \frac{P}{A}$$

$$= \frac{P}{\ell t} \text{ for solid walls}$$

From Eq. (24.5.20),

$$f_{b(\max)} = \left(1 - \frac{f_a}{F_a}\right) F_b \tag{24.7.5}$$

which is the maximum allowable flexural compressive stress on a section under axial compressive stress f_a. Now the maximum combined compressive stress is

$$f_m = f_{b(\max)} + f_a \tag{24.7.6}$$

Compute the terms

$$a = \frac{1}{6} t f_m \tag{24.7.7}$$

$$b = -\frac{1}{2} t f_m (\ell - d') \tag{24.7.8}$$

$$c = P\left(\frac{\ell}{2} - d'\right) + M \tag{24.7.9}$$

Solve for the depth of the triangular compression block,

$$kd = \frac{-b - \sqrt{b^2 - 4ac}}{2a} \tag{24.7.10}$$

from which

$$k = \frac{kd}{d}$$

Solve for reinforcement stress:

$$f_s = \left(\frac{1-k}{k}\right) n f_m \leq F_s‡ \tag{24.7.11}$$

* For the case of compression plus out-of-plane flexure, length of wall becomes wall thickness t and wall thickness becomes increment of wall length $\Delta\ell$.
† See Section 24.11, Notation, for definition of terms.
‡ If $f_s > F_s$, try $f_b < f_{b(\max)}$ in Eq. (24.7.5).

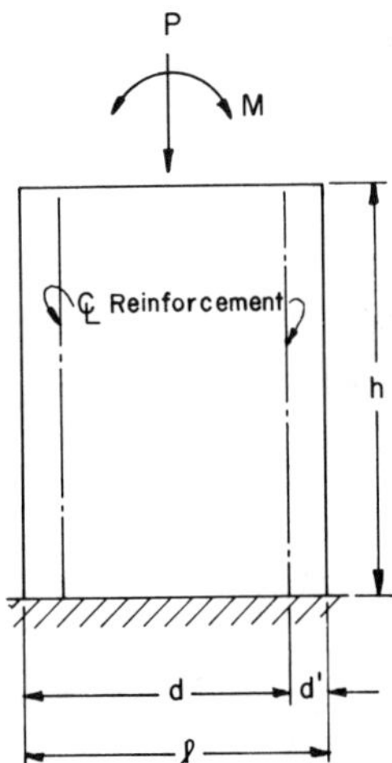

Fig. 24.23 Wall for method A and B examples.

Calculate resultant compressive force in the masonry:

$$C = \frac{1}{2} t(kd)f_m \tag{24.7.12}$$

Calculate the required tensile force in the reinforcement:

$$T = C - P\dagger \tag{24.7.13}$$

Calculate the required area of tensile reinforcement:

$$A_s = \frac{T}{F_s} \tag{24.7.14}$$

Example—Method A [24.1]. Given a wall as shown in Fig. 24.23 where

t = 9.63 in. (245 mm), [nominal thickness = 10 in. (250 mm)]

f'_m = 3000 psi (21 MPa)

h = 14 ft = 168 in. (4.3 m)

n = 10

P = 134 kips (600 kN)

M = 8400 in-kips (950 kN·m) (due to earthquake)

Grade 40 reinforcement [yield stress = 40 ksi (280 MPa)]

Moment is due to earthquake forces, ∴ use 1.33 factor on allowable stresses.

Allowable stresses (for earthquake conditions) are (see Section 24.5.2):

$$F_a = \left\{0.2(3)\left[1 - \left(\frac{168}{40(10)}\right)^3\right](1.33)\right\} = \underline{0.741} \text{ ksi (5.10 MPa)}$$

† If $C - P$ is negative, no flexural tensile reinforcement is required, i.e., the position of P acting at virtual eccentricity $e = M/P$ is coincident with the position of the compressive stress resultant C. In this case specify minimum reinforcement.

$$F_b = \left\{1.33 \text{ min.} \left[0.33 f'_m \le 2000 \text{ psi (13.8 MPa)}\right]\right\} = \underline{1.2} \text{ ksi (8.27 MPa)}$$

$$f_a = \frac{P}{\ell t} = \frac{134}{9.63(96)} = \underline{0.145} \text{ ksi (1 MPa)}$$

$$f_{b(\max)} = \left(1 - \frac{0.145}{0.739}\right)(1.2) = \underline{0.965} \text{ ksi (6.65 MPa)} \qquad \text{[Eq. (24.7.5)]}$$

$$f_m = 0.965 + 0.145 = \underline{1.11} \text{ ksi (7.65 MPa)} \qquad \text{[Eq. (24.7.6)]}$$

Now solve for kd, f_s, C, T, and A_s:

$$a = \frac{1}{6}(9.63)(1.11) = 1.78 \qquad \text{[Eq. (24.7.7)]}$$

$$b = -\frac{1}{2}(9.63)(1.11)(96 - 8) = -470 \qquad \text{[Eq. (24.7.8)]}$$

$$c = 134\left(\frac{96}{2} - 8\right) + 8400 = 13760 \qquad \text{[Eq. (24.7.9)]}$$

$$kd = \frac{470 - [(470)^2 - 4(1.78)(13760)]^{1/2}}{2(1.78)} = \underline{33.54} \text{ in. (852 mm)} \qquad \text{[Eq. (24.7.10)]}$$

$$C = \frac{1}{2}(9.63)(33.54)(1.11) = 179.3 \text{ kips (798 kN)} \qquad \text{[Eq. (24.7.12)]}$$

$$T = 179.3 - 134 = 45.3 \text{ kips (201 kN)} \qquad \text{[Eq. (24.7.13)]}$$

$$k = \frac{33.54}{(96 - 8)} = 0.381$$

$$f_s = \left(\frac{1 - 0.381}{0.381}\right)(10)(1.1) = 17.9 \text{ ksi (123 MPa)} < F_s = 26.6 \text{ ksi} \qquad \text{[Eq. (24.7.11)]}$$

$$A_s = \frac{45.3}{17.9} = 2.53 \text{ sq in. (1630 mm}^2) \qquad \text{[Eq. (24.7.14)]}$$

Use 2 − #10 bars each side for $A_s = 2.54$ in.^{2} (1640 mm^2).

Combined Vertical Compression and Flexure—Method B [24.53]

Method B may be used to develop P–M interaction diagrams for given wall designs. Design P–M combinations located within such envelopes are acceptable and otherwise not.

Data required are:

Length of wall: ℓ
Wall thickness: t
Distance to reinforcement: d'
Modular ratio: $n = E_s/E_m$
Max. allowable axial compressive stress: F_a
Max. allowable flexural compressive stress: F_b
Area of flexural tensile reinforcement: A_s
Max. allowable reinforcement stress: F_s

The process consists of calculating the maximum moment M which can be resisted by a section under a vertical compressive load P such that maximum allowable stresses in the masonry and in the reinforcement are not exceeded. This may be done by calculating P and M associated with various stress distributions beginning with uniform stress and changing with gradually increasing virtual eccentricity of P. The stress distributions shown in Figs. 24.24a–d illustrate states of stress up to incipient tension in the reinforcement. Figure 24.24e illustrates the general state of stress for reinforcement acting. The values of P and M for each are given for solid sections by

$$\left.\begin{aligned}
&\text{(a)} && P = F_a \ell t \\
&\text{(b)} && P = F_a \ell t \\
& && M = \frac{(F_b - F_a)t\ell^2}{6} \\
&\text{(c)} && P = (F_b/2)\ell t \\
& && M = \frac{(F_b/2)t\ell^2}{6} \\
&\text{(d)} && P = \frac{F_b d t}{2} \\
& && M = P\left[\frac{\ell}{2} - \frac{d}{3}\right] \\
&\text{(e)} && P = f_m\left[\frac{at}{2} - \frac{bnA_s}{a}\right] \\
& && M = f_m\left[\frac{at\ell}{4} - \frac{a^2 t}{6} + \frac{nbA_s\ell_s}{a}\right]
\end{aligned}\right\} \quad (24.7.15)$$

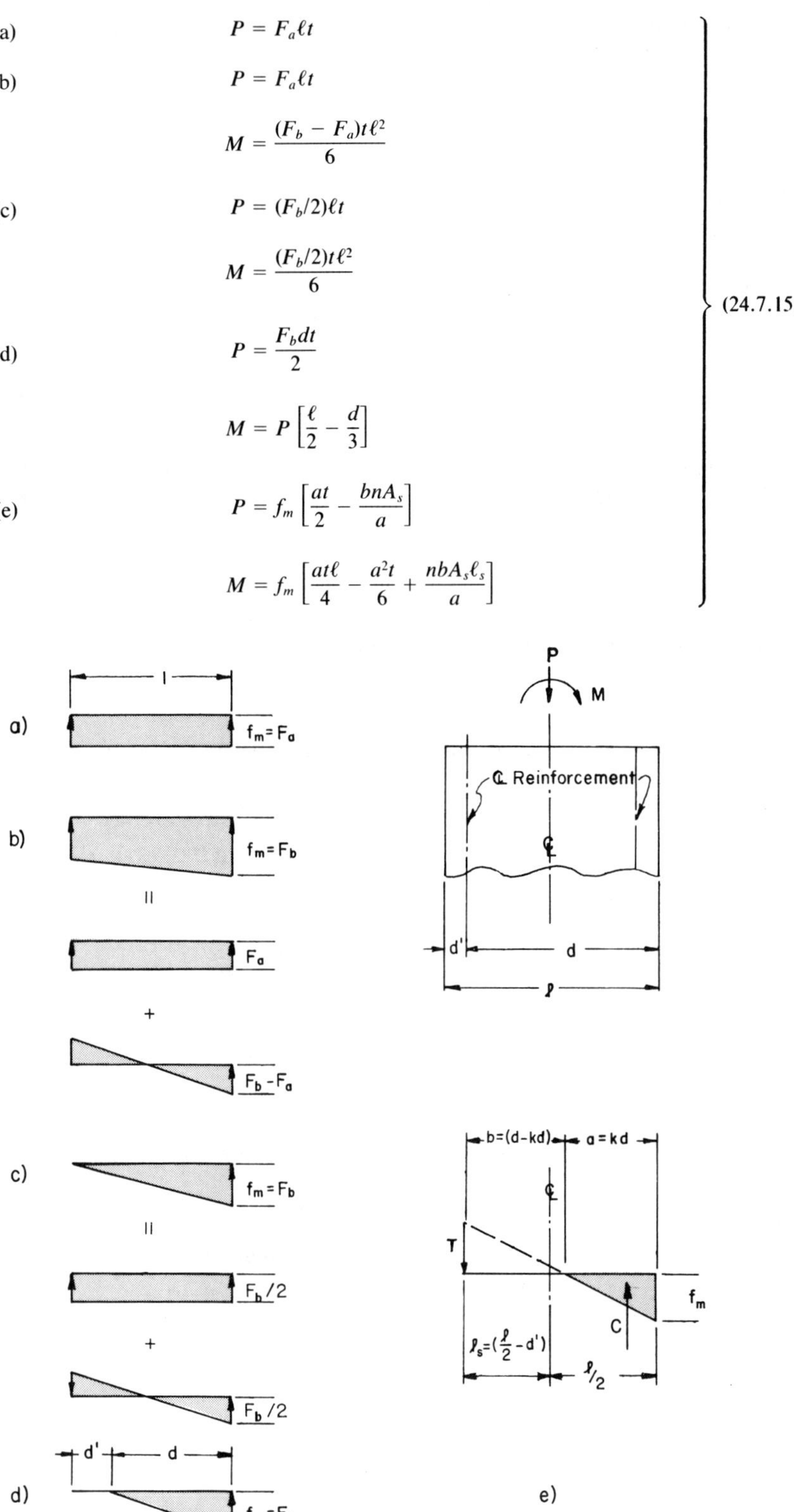

Fig. 24.24 Force/stress/geometry relationships for combined P–M, method B.

Implementing Eqs. [24.7.15(e)] may be done by assuming values of a and b, that is, of kd and $(d - kd)$. For ratios of b/a up to

$$\left(\frac{b}{a}\right)_m = \frac{F_s}{nF_b} \tag{24.7.16}$$

set $f_m = F_b$; f_s is given by

$$f_s = \left(\frac{b}{a}\right) nF_b \leq F_s \tag{24.7.17}$$

At b/a ratios exceeding $(b/a)_m$ given by Eq. (24.7.16), $f_s = F_s$ and

$$f_m = \frac{F_s}{n\left(\frac{b}{a}\right)} \leq F_b \tag{24.7.18}$$

may be substituted in Eqs. [24.7.15(e)].

The value of a, that is, kd, at which $P = 0$ is given by

$$a = \frac{-\left(\frac{2nA_s}{t}\right) + \left[\left(\frac{2nA_s}{t}\right)^2 + \left(\frac{8dnA_s}{t}\right)\right]^{1/2}}{2} \tag{24.7.19}$$

Example—Method B. Method B is applied to the wall of the previous example. Recall that

$$\ell = 96 \text{ in. } (2.44 \text{ m}), \qquad t_{\text{actual}} = 9.63 \text{ in. } (245 \text{ mm})$$

$$d' = 8 \text{ in. } (203 \text{ mm}), \qquad d = 88 \text{ in. } (2.24 \text{ m}), \qquad n = 10$$

$$F_a = 0.741 \text{ ksi } (5.10 \text{ MPa}), \qquad F_b = 1.2 \text{ ksi } (8.27 \text{ MPa})$$

$$A_s = 2.53 \text{ in.}^2 \ (1640 \text{ mm}^2), \qquad F_s = 26.6 \text{ ksi } (183 \text{ MPa})$$

Invoking Eqs. [24.7.15(a–d)],

(a) $P = 683$ kips (3.06 MN)

(b) $P = 683$ kips (3.06 MN)

$M = 568$ ft-kips (770 kN·m)

(c) $P = 555$ kips (2.49 MN)

$M = 740$ ft-kips (1000 kN·m)

(d) $P = 508$ kips (2.28 MN)

$M = 790$ ft-kips (1070 kN·m)

The value of b/a below which $f_s \leq F_s$ is by Eq. (24.7.16)

$$\left(\frac{b}{a}\right)_m = \frac{26.6}{(10)(1.2)} = 2.22$$

or since $a + b = d$:

$$a = 27.33 \quad \text{and} \quad b = 60.67$$

Invoking Eq. [24.7.15(e)] for arbitrary ratios of b/a up to $b/a = 2.22$,

b/a	a	b	P kips (MN)	M ft-kips (kN·m)	f_s ksi (MPa)
0.17	75	13	428 (1.92)	848 (1149)	2.0 (14.5)
0.47	60	28	332 (1.49)	856 (1160)	5.6 (38.6)
0.76	50	38	266 (1.19)	832 (1128)	9.1 (62.7)
1.20	40	48	195 (0.87)	790 (1070)	14.4 (99.3)
1.51	35	53	156 (0.70)	766 (1038)	18.1 (124.8)
2.22	27.33	60.67	90 (0.40)	737 (999)	26.6 (183.4)

Invoking Eq. [24.7.15(e)] for arbitrary ratios of $b/a > 2.22$ note that F_s is limited to the maximum allowable, that is, 26.6 ksi (183 MPa) and f_m is less than 1.2 ksi (8.27 MPa) and is given by Eq. (24.7.18). The minimum depth of the triangular compression zone, that is, the value corresponding to $P = 0$, is by Eq. (24.7.19), $a = 19.07$ in. (484 mm).

b/a	a	b	f_m ksi (MPa)	P kips (MN)	M ft-kips (kN·m)
2.4	25.9	62.1	1.11 (7.7)	70.7 (0.32)	678 (919)
2.8	23.15	64.85	0.95 (6.5)	52.3 (0.23)	581 (787)
3.0	22	66	0.89 (6.1)	26.5 (0.12)	545 (739)
3.3	20.5	67.5	0.81 (5.6)	12.2 (0.05)	500 (678)
3.61	19.07	68.93	0.74 (5.1)	0	462 (626)

For the condition of zero moment, allowable axial load is determined using unfactored allowable stress because the increase of F_a is only permitted when in combination with force due to earthquake or wind [24.77]. Hence

$$P_0 = \frac{0.739}{1.33}(9.63)(96) = 514 \text{ kips } (2.3 \text{ MN})$$

The interaction diagram for the example wall is shown in Fig. 24.25.

Design for In-Plane Shear

Walls that carry in-plane shear V are commonly called "shear walls" although such walls must support their own weight, any imposed floor/roof loads, and out-of-plane forces. Shear stress is assumed uniformly distributed and calculated by

$$f_v = \frac{V}{A_e} \tag{24.7.20}$$

where $A_e = \ell d$ for solid walls
$\quad = $ net area for ungrouted or partially grouted walls

The foregoing also applies to piers, that is, the portion of wall between window openings.

As is the case for beams, if shear reinforcement is used, it must be designed to resist 100% of the shear. Thus if f_v calculated by Eq. (24.7.20) exceeds the maximum allowable shear stress F_v for shear walls without shear reinforcement, shear reinforcement must be provided. The calculated shear stress f_v must not exceed the maximum allowable shear stress in masonry for walls with shear reinforcement. The required area of shear reinforcement is given by

$$A_v{}^* = \frac{Vs}{F_s d} \tag{24.7.21}$$

Allowable shear is established, for walls with or without shear reinforcement, according to the ratio M/Vd where $M =$ the maximum moment due to in-plane shear V, and d is the flexural depth.* For walls (or piers) whose top and bottom are considered fixed against relative in-plane rotation, $M/Vd = h'/2d$ and for cantilever walls (or piers), $M/Vd = h'/d$ as illustrated in Fig. 24.26. Allowable shear between $M/Vd = 0$ and 1 may be determined by linear interpolation.

If the shear V is due to seismic forces in zones 3 or 4, shear walls must be designed for 1.5 times the calculated shear stress. Correspondingly, a 1.33 increase in allowable stress is permitted [24.77].

---

* See Notation, Section 24.11

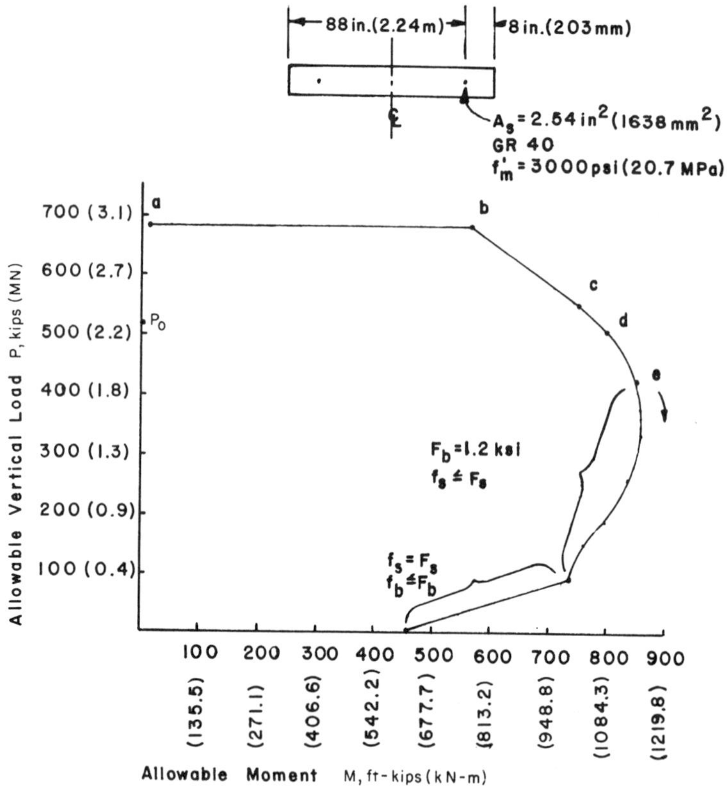

Fig. 24.25 Allowable in-plane moment M—ft-kips (kN·m)—for earthquake conditions.

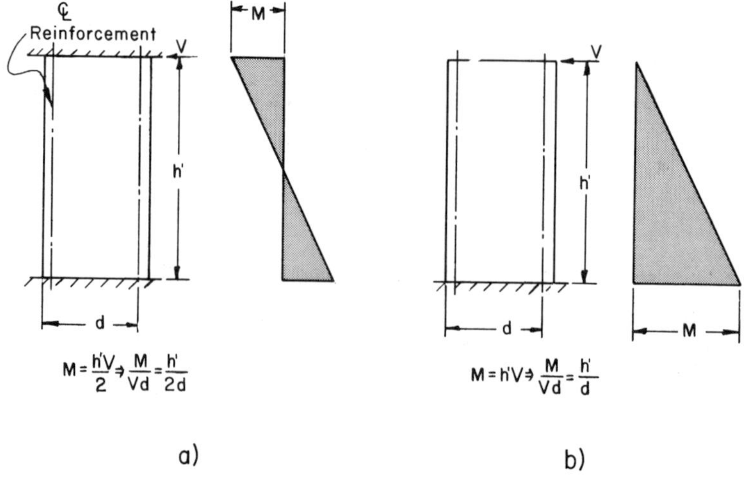

Fig. 24.26 M/Vd in terms of h' and d.

24.8 COLUMN DESIGN

24.8.1 Basic Considerations

Masonry columns are members whose primary structural function is to support concentrated vertical loads and are found in three forms, that is, isolated, contained wholly within the dimensions of a wall (flush wall column), or as a column located in a wall, but projecting beyond one or both vertical surfaces of the wall (pilaster). Architectural requirements may impose limits on column size, location, units used, and other factors that could affect structural performance. Overall building structural system design will determine the magnitude and type of loads to be supported and other factors, for example, fire resistance, durability, height, end conditions, and lateral support.

24.8.2 Column Configurations

The variety of masonry units available and the number of ways in which they can be assembled leads to a very large number of possible column configurations. A few representative types are presented in Fig. 24.27.

24.8.3 Column Loading Conditions

Isolated Column

Isolated columns are primarily intended to carry vertical loads due to the self-weight of the column and vertical load components from beams, trusses, and slabs. Flexural stresses may be produced by eccentric application of such loads, by continuity between columns and supported members, and lateral forces due to wind, earthquake, or blast. Moments may be about either or both horizontal column axes.

Alternatively, pilasters and flush wall columns serve as stiffeners for walls and are often subject to flexural loads due to horizontal wall edge reactions in addition to any flexure caused by eccentric vertical load. Pilasters and flush wall columns normally may be considered to be supported in the plane of the wall. Loading conditions are illustrated in Fig. 24.28.

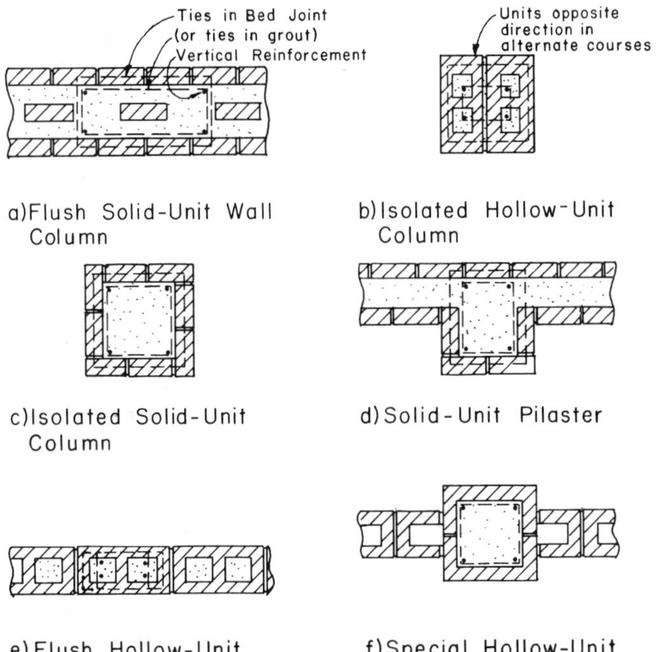

Fig. 24.27 Representative column cross-sections—alternate locations of ties shown.

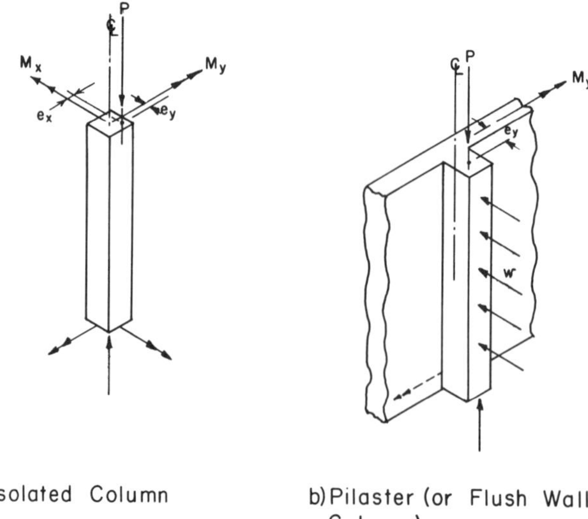

Fig. 24.28 Column loading conditions.

24.8.4 Empirical Column Design Requirements [24.77]

Cover

See Section 24.4.4.

Vertical Reinforcement

Reinforced columns should contain at least 4 longitudinal #3 (nominal 10-mm diameter) bars located in the corners of rectangular columns, equally spaced in circular columns or in circular patterns in rectangular columns. The total area of vertical reinforcement must be 0.005 A_e or more, but not more than 0.04 A_e.

Lateral Ties

Column longitudinal reinforcement must be enclosed by lateral steel ties. Lateral support for longitudinal reinforcement may be provided by enclosing the bars within complete rectangular ties or within a standard hook at the end of a tie as shown in Fig. 24.29.

Corner bars are supported by the corner of complete ties. Alternate longitudinal bars must be supported by hooks on the end of ties and no bar should be further than 6 in. (150 mm) from a laterally supported bar.

Spacing and Cover

The minimum clear distance between parallel bars in a column is 2½ times the diameter of the bars. The requirements of Section 24.6.4 apply except for clear distance requirements.

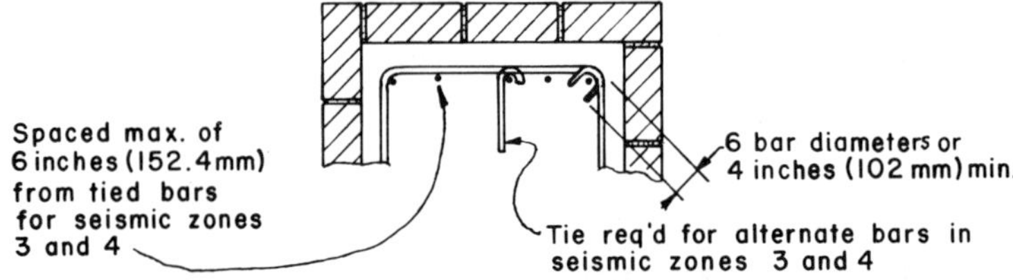

Fig. 24.29 Masonry column tie requirements.

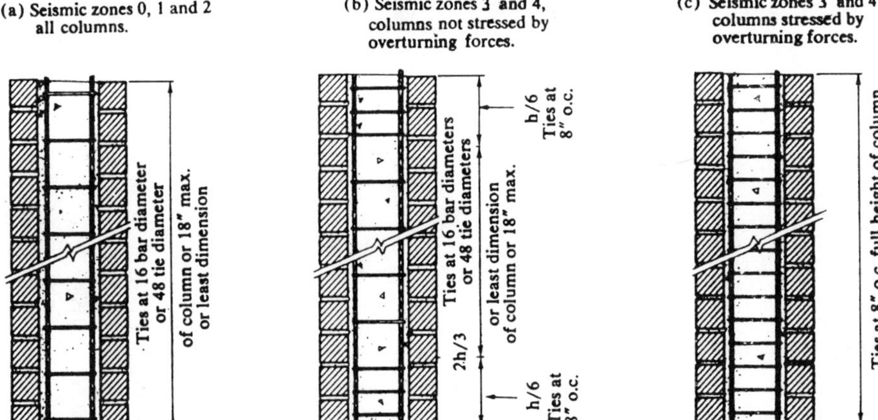

Fig. 24.30 Masonry column tie spacing requirements [24.77] (reprinted from *Reinforced Masonry Engineering Handbook* [24.1] with permission).

Ties must be at least ¼ in. (6 mm) in diameter for #7 or smaller longitudinal bars and be at least #3 for larger longitudinal bars. However, ties smaller than #3 may be used to stabilize #7 or larger longitudinal bars if the total area of such ties is equivalent to the area of #3 ties at their required spacing.

Lateral ties and longitudinal bars must be located at least 1½ in. (38 mm) but not more than 5 in. (130 mm) from the surface of a column. Lateral ties may be placed against longitudinal bars or in the horizontal bed joints if cover requirements stated in Section 24.7.4 are met.

Vertical spacing of lateral ties should not exceed the lesser of 16 longitudinal bar diameters, 48 tie diameters, the least lateral dimension of the column, or 18 in. (450 mm) as shown in Fig. 24.30.

24.8.5 Column Design—Basic Assumptions [24.77]

Material Behavior

The assumptions of Section 24.6.5, except number 5, apply to masonry column design.

Effective Thickness

The effective thickness for rectangular columns is the specified thickness perpendicular to the horizontal axis being considered.

Effective Height

See Section 24.7.5.

Effective Area

See Section 24.7.5.

Load Conditions

Columns should be designed for at least the applicable of load conditions depicted in Fig. 24.28.

24.8.6 Design of Unreinforced Masonry Columns [24.77]

Axial Stress

Stresses due to axial forces applied at the column centroid may be assumed uniformly distributed over the effective area, that is,

$$f_a = \frac{P}{A_e} \tag{24.8.1}$$

Axial stress should not exceed the limits given by Eq. (24.5.2).

Flexural Stress

Flexural stresses may be calculated by:

$$f_b = \frac{Mc}{I} \tag{24.8.2}$$

Flexural compressive stress shall not exceed the limit given by Eq. (24.5.5).

Combined Stresses

Resultant stresses due to combined flexure and axial load shall not exceed the allowable tensile stress F_t for masonry as provided in Section 24.5.2, where f_a and f_b are of opposite sign. Where f_a and f_b are additive, the limits of Eq. (24.5.20) apply.

Shear Stress

Shear stress may be calculated by the equation

$$f_v = \frac{V}{A_e} \tag{24.8.3}$$

The stress must not exceed the applicable of values presented in Section 24.5.2.

24.8.7 Design of Reinforced Masonry Columns

Axial Stress [24.77]

Axial stress due to centroidal loads may be assumed to be uniformly distributed over the effective area, that is,

$$f_a = \frac{P}{A_e} \tag{24.8.4}$$

Because the vertical reinforcement is tied, the allowable load P_a for columns is given in terms of both masonry and reinforcement, that is, by Eq. (24.5.3). The allowable axial stress F_a is given by Eq. (24.5.4).

Combined Axial and Flexural Stresses

Columns subjected to combined axial and flexural stresses must satisfy Eq. (24.5.20), the interaction equation, substituting P/P_a for f/F_a. Columns subjected to bending must satisfy all requirements for flexural design.

An *alternate procedure* developed by NCMA [24.81] based on accepted principles of mechanics which accounts for the effect of compressive load on reinforcement tensile stress and satisfies Eq. (24.5.20) is given below. Refer to Fig. 24.31 for illustration of geometry.

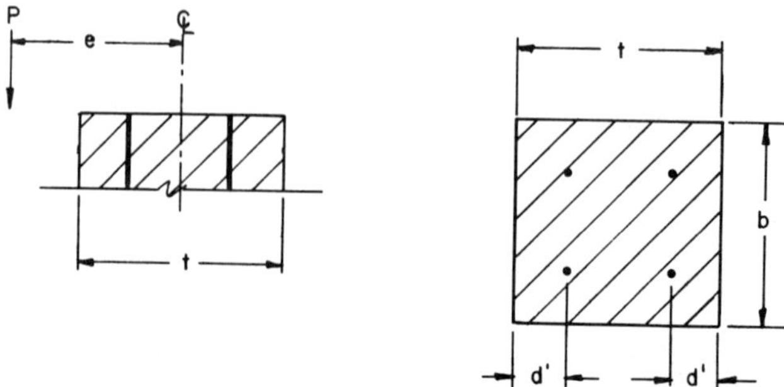

Fig. 24.31 Geometry for NCMA column design method.

The notation in Section 24.10 is applicable. Allowable load P_{ei} is determined for an uncracked section and for a cracked section as limited by masonry stress and reinforcement stresses. The minimum of P_{ei} is then the maximum allowable column load P_e.

1. *Uncracked Section ($0 \le e < t/3$)*

$$P_{e1} = \frac{P_a}{1 + \dfrac{D_1}{F_b}\dfrac{P_a}{A_g}} \tag{24.8.5}$$

where

$$D_1 = \frac{e}{t}\,\frac{6}{\left[1 + 3p_g(n-1)\left(n - \dfrac{2d'}{t}\right)^2\right]} \tag{24.8.6}$$

$$P_a = 0.20\,f'_m A_e + 0.65\,A_s F_{sc} \tag{24.8.7}$$

2. *Cracked Section*

a. *Allowable Load as Limited by Masonry Compressive Stress*

$$P_{e2} = \frac{F_a A_g}{\dfrac{F_b - F_a}{F_b[1 - p_g(n-1)]} + D_2\dfrac{F_a}{F_b}} \tag{24.8.8}$$

where

$$F_a = \frac{P_a/A_g}{1 + p_g(n-1)} \tag{24.8.9}$$

F_b = max. allowable compressive stress in bending [Eq. (24.5.5)]

$$D_2 = \frac{2k}{k^2 + p_g\left[(3n-1)k - \dfrac{d'}{t}(n-1) - n\right]} \tag{24.8.10}$$

k is calculated from:*

$$k^3 + Ak^2 + Bk - C = 0 \tag{24.8.11}$$

where $A = 3\left(\dfrac{e}{t} - \dfrac{1}{2}\right)$

$$B = 3p\left[\frac{2e}{t}(3n-1) + \left(\frac{2d'}{t} - 1\right)(n+1)\right]$$

$$C = \frac{3pd'}{t}\left[\frac{2e}{t}(n+1) + \frac{2d'}{t}(3n+1) - 5n - 1\right] + 3pn\left(\frac{2e}{t} + 1\right)$$

b. *Allowable Load as Limited by Reinforcement Compressive Stress*

$$P_{e3} = \frac{f_m A_g}{D_2} \tag{24.8.12}$$

where

$$f_m = \frac{20{,}000}{(2n-1)\left(1 - \dfrac{d'}{kt}\right)} \tag{24.8.13}$$

---

* See Table 24.12 for values of k for concrete masonry.

Table 24.12 Approximate Values of k (Cracked Section)*

$d'/t = 0.225$	f'_m = 750 psi	e/t						
		0.4	0.5	0.6	0.7	0.8	0.9	1.0
	p_g = 0.005	0.515	0.450	0.410	0.385	0.370	0.355	0.340
	= 0.01	0.575	0.500	0.455	0.430	0.415	0.405	0.395
	= 0.02	0.605	0.530	0.495	0.475	0.455	0.445	0.435
	= 0.03	0.620	0.550	0.515	0.490	0.470	0.460	0.445
	f'_m = 1000 psi	e/t						
		0.4	0.5	0.6	0.7	0.8	0.9	1.0
	p_g = 0.005	0.500	0.435	0.395	0.370	0.355	0.340	0.325
	= 0.01	0.555	0.480	0.445	0.420	0.400	0.390	0.375
	= 0.02	0.585	0.520	0.485	0.460	0.445	0.435	0.425
	= 0.03	0.615	0.540	0.505	0.480	0.465	0.450	0.440
	f'_m = 1500 psi	e/t						
		0.4	0.5	0.6	0.7	0.8	0.9	1.0
	p_g = 0.005	0.490	0.405	0.360	0.340	0.320	0.305	0.295
	= 0.01	0.535	0.460	0.420	0.395	0.375	0.360	0.350
	= 0.02	0.575	0.505	0.470	0.445	0.430	0.415	0.405
	= 0.03	0.595	0.530	0.490	0.470	0.450	0.435	0.425
	f'_m = 2000 psi	e/t						
		0.4	0.5	0.6	0.7	0.8	0.9	1.0
	p_g = 0.005	0.470	0.385	0.340	0.315	0.300	0.285	0.275
	= 0.01	0.520	0.440	0.400	0.375	0.355	0.340	0.330
	= 0.02	0.565	0.490	0.450	0.430	0.410	0.395	0.385
	= 0.03	0.585	0.515	0.480	0.455	0.435	0.420	0.410

* Reprinted from *Design and Construction of Reinforced Concrete Masonry Columns and Pilasters,* courtesy of the National Concrete Masonry Association, Herndon, VA [24.81].

c. *Allowable Load a Limited by Reinforcement Tensile Stress*

$$P_{e4} = \frac{f_m A_g}{D_2} \tag{24.8.14}$$

where

$$f_m = \frac{20{,}000}{n\left(\dfrac{t - d'}{kt} - 1\right)} \tag{24.8.15}$$

D_2 is given by Eq. (24.8.10)
k is determined from Eq. (24.8.11)

3. *Allowable Load P_e*

$$P_e = \min.[P_{e1}, P_{e2}, P_{e3}, P_{e4}] \tag{24.8.16}$$

$d'/t = 0.30$	f'_m = 750 psi	e/t						
		0.4	0.5	0.6	0.7	0.8	0.9	1.0
	p_g = 0.005	0.490	0.435	0.405	0.385	0.375	0.365	0.355
	= 0.01	0.500	0.460	0.430	0.415	0.400	0.395	0.385
	= 0.02	0.520	0.485	0.465	0.450	0.440	0.430	0.425
	= 0.03	0.530	0.495	0.475	0.460	0.450	0.445	0.440
	f'_m = 1000 psi	e/t						
		0.4	0.5	0.6	0.7	0.8	0.9	1.0
	p_g = 0.005	0.480	0.420	0.390	0.370	0.355	0.345	0.340
	= 0.01	0.490	0.450	0.420	0.405	0.390	0.380	0.370
	= 0.02	0.515	0.475	0.455	0.440	0.430	0.420	0.415
	= 0.03	0.525	0.490	0.470	0.455	0.445	0.435	0.430
	f'_m = 1500 psi	e/t						
		0.4	0.5	0.6	0.7	0.8	0.9	1.0
	p_g = 0.005	0.470	0.400	0.365	0.345	0.330	0.320	0.310
	= 0.01	0.485	0.425	0.400	0.380	0.365	0.355	0.345
	= 0.02	0.505	0.465	0.435	0.420	0.410	0.400	0.395
	= 0.03	0.520	0.480	0.460	0.445	0.430	0.425	0.420
	f'_m = 2000 psi	e/t						
		0.4	0.5	0.6	0.7	0.8	0.9	1.0
	p_g = 0.005	0.455	0.380	0.340	0.320	0.305	0.295	0.290
	= 0.01	0.480	0.410	0.380	0.360	0.345	0.335	0.325
	= 0.02	0.500	0.450	0.425	0.405	0.395	0.385	0.380
	= 0.03	0.515	0.470	0.450	0.430	0.420	0.410	0.405

24.9 SPECIAL PROVISIONS IN AREAS OF SEISMIC RISK [24.77]

Applicability

The empirical provisions presented here apply to masonry buildings located in UBC Seismic Zones 2, 3, and 4. These provisions are based upon field experience, the judgement of the code committee, and many qualified reviewers. They are presented herein without editorial modification except for numbering of sections to be consistent with this book.

Special Provisions for Seismic Zone 2

Masonry Structures in Seismic Zone 2 shall comply with the following special provisions.

A. Materials. The following materials shall not be used as part of the structural frame: Type O mortar, masonry cement, plastic cement, nonload-bearing masonry units, and glass block.

B. Wall Reinforcement. Vertical reinforcement of at least 0.20 square inch (130 mm²) in cross-sectional area shall be provided continuously from support to support at each corner, at each side of each opening, at the ends of walls and at a maximum spacing of 4 ft (1.2 m) apart, horizontally throughout the wall.

Horizontal reinforcement not less than 0.2 square inch (130 mm²) in cross-sectional area shall be provided: (1) at the bottom and top of wall openings and shall extend not less than 24 in. (600 mm) nor less than 40 bar diameters past the opening, (2) continuously at structurally connected roof and floor levels and at the top of walls, (3) at the bottom of the wall or in the top of the foundations when dowelled to the wall, (4) at maximum spacing of 10 feet (3 m) unless uniformly distributed joint reinforcement is provided. Reinforcement at the top and bottom of openings when continuous in the wall may be used in determining the maximum spacing specified in Item No. (1) above.

C. Stack Bond. Where stack bond is used, the minimum horizontal reinforcement ratio shall be 0.0007 bt. This ratio shall be satisfied by uniformly distributed joint reinforcement or by horizontal reinforcement spaced not over 4 feet (1.2 m) and fully embedded in grout or mortar.

Special Provisions for Seismic Zones 3 and 4

All masonry structures built in Seismic Zones 3 and 4 shall be designed and constructed in accordance with requirements for Seismic Zone 2 and with the following additional requirements and limitations.

EXCEPTION: One- and two-story structures of Group R, Division 3 and Group M occupancies in Seismic Zone 3 with h'/t not greater than 27 and using running bond construction may be constructed in accordance with the requirements of Seismic Zone 2.

For masonry work without special inspection, the assumed design strength shall be limited to f'_m of 1500 psi (10.3 MPa) for concrete block masonry and 2600 psi (17.9 MPa) for clay unit masonry.

A. Materials. The following materials shall not be used as part of the structural frame: Type N mortar.

B. Wall Reinforcement. All walls shall be reinforced with both vertical and horizontal reinforcement. The sum of the areas of horizontal and vertical reinforcement shall be at least 0.002 times the gross cross-sectional area of the wall, and the minimum area of reinforcement in either direction shall not be less than 0.0007 times the gross cross-sectional area of the wall. The spacing of reinforcement shall not exceed 4 feet (1.2 m). The diameter of reinforcement shall not be less than ⅜ inch (10 mm) except that joint reinforcement may be considered as part or all of the requirement for minimum reinforcement. Reinforcement shall be continuous around wall corners and through intersections. Only horizontal reinforcement which is continuous in the wall or element shall be considered in computing the minimum area of reinforcement. Reinforcement with splices conforming to Section 24.6.4 shall be considered as continuous reinforcement.

C. Column Reinforcement. The spacing of column ties shall not be more than: 8 inches (200 mm) the full height for columns stressed by tensile or compressive axial overturning forces due to the seismic loads of Section 2312; 8 inches (200 mm) for the tops and bottoms of all other columns for a distance of one-sixth of the clear column height, but not less than 18 inches (460 mm) nor the maximum column dimension. Tie spacing for the remaining column height shall be not more than 16 bar diameters, 48 tie diameters or the least column dimension, but not more than 18 inches (460 mm).

D. Stack Bond. Where stack bond is used, the minimum horizontal reinforcement ratio shall be 0.0015 bt. If open-end units are used and grouted solid, then the minimum horizontal reinforcement ratio shall be 0.0007 bt.

E. Minimum Dimension. (i) *Bearing walls.* The nominal thickness of reinforced masonry bearing walls shall not be less than 6 inches (150 mm) except that nominal 4-inch (100 mm) thick load-bearing reinforced hollow clay unit masonry walls with a maximum unsupported height or length to thickness ratio of 27 may be used, provided net area unit strength exceeds 8000 psi (55 MPa), units are laid in running bond, bar sizes do not exceed ½ inch (13 mm) with no more than two bars or one splice in a cell, and joints are flush cut, concave or a protruding V section. Minimum bar coverage where exposed to weather may be 1½ inches (38 mm).

(ii) *Columns.* The least nominal dimension of a reinforced masonry column shall be 12 inches (300 mm) except that if the allowable stresses are reduced to one-half the values given in Section 24.5, the minimum nominal dimension shall be 8 inches (200 mm).

F. Shear Walls. (i) *Design loads.* When calculating shear or diagonal tension stresses, shear walls which resist seismic forces shall be designed to resist 1.5 times the forces required by UBC Section 2312(d)[24.77].

(ii) *Reinforcement.* The portion of the reinforcement required to resist shear shall be uniformly distributed and shall be joint reinforcing, deformed bars, or a combination thereof. The maximum spacing of reinforcement in each direction shall be not less than the smaller of one-half the length or height of the element nor more than 48 in. (1.2 m). Where reinforcement is required to resist 100% of the design shear, special inspection shall be provided.

Reinforcement required to resist in-plane shear shall be terminated with a standard hook or with an extension of proper embedment length beyond the reinforcing at the end of the wall section. The hook or extension may be turned up, down or horizontally. Provisions shall be made not to obstruct grout placement. Wall reinforcement terminating in columns or beams shall be fully anchored into these elements.

(iii) Multi-wythe grouted masonry shear walls shall be designed with consideration of the adhesion bond strength between the grout and masonry units. When bond strengths are not known from previous tests, the bond strength shall be determined by test.

G. Hooks. The term "hook" or "standard hook" as used herein for tie anchorage in Seismic Zones 3 and 4 shall mean a minimum turn of 135° plus an extension of at least 6 bar diameters, but not less than 4 in. (100 mm) at the free end of the bar.

EXCEPTION: Where the ties are placed in the horizontal bed joints, the hook shall consist of a 90-degree bend having a radius of not less than 4 bar diameters plus an extension of 32 bar diameters.

H. Mortar Joints Between Masonry and Concrete. Concrete abutting structural masonry such as at starter courses or at wall intersections not designed as true separation joints shall be roughened to a full amplitude of 1/16 inch (1.6 mm) and shall be bonded to the masonry per the requirements of Section 24.9 as if it were masonry. Unless keys or proper reinforcement are provided, vertical joints as per UBC Section 2407(b)2* shall be considered to be stack bond and the reinforcement as required for stack bond shall extend through the joint and be anchored into the concrete.

24.10 SUMMARY

24.10.1 Current State of Masonry Structural Design and Construction

Present design methodology is based upon linear–elastic, working stress concepts and relies upon a number of empirical requirements. An assumption of isotropic behavior is generally assumed in analyzing system behavior. It is important, however, for the designer to realize that masonry is a complex composite material whose properties may be anisotropic when considering performance at the element level. Allowable flexural tension stresses for unreinforced masonry depend upon the direction of flexure, for example (See Section 24.5.2).

Because masonry is an assemblage of finite-sized units, detail design factors must be considered. Location of reinforcement, for example, must be specified considering constraints imposed by the size and shape of masonry units. Physical layout of masonry elements should consider the finite size of the units so that excessive field cutting and shaping may be avoided.

The quality of masonry is directly related to field construction practice. The designer is often in the awkward position of determining what form of inspection will be required to justify use of "inspected" allowable stresses in design.

24.10.2 Future Trends

For the immediate future, linear–elastic working stress methods will continue to be the predominant basis for masonry structural design. Work has begun to support ultimate strength design methods in the United States. The current UBC [24.77] and ICBO research reports permit limited use of strength methods. However, much remains to be done before a complete method will be available. A significant amount of research is currently in the planning stages which, in addition to recently completed research, should provide data required for a comprehensive strength design method.

---

* In bearing and nonbearing walls, if less than 75% of the units in any transverse vertical plane lap the ends of the units below a distance less than one-half the height of the unit, or less than one-fourth the length of the unit, the wall should be considered laid in stack bond.

24.11 NOTATION

A_e effective area of masonry
A_s effective cross-sectional area of reinforcement in a column or flexural member
A_v area of steel required for shear reinforcement perpendicular to the longitudinal reinforcement
A_s' effective cross-sectional area of compression reinforcement in a flexural member
b effective width of rectangular member or width of flange for T- and I-sections
b' width of web in T- and I-sections
c distance from the neutral axis to extreme fiber
d distance from the compression face of a flexural member to the centroid of longitudinal tensile reinforcement
d' distance from the compression face of a flexural member to the centroid of longitudinal compression reinforcement
d_b diameter of the reinforcing bar
E_m modulus of elasticity of masonry
E_s modulus of elasticity of steel
f_a computed axial compressive stress due to design axial load
f_b computed flexural stress in the extreme fiber due to design bending loads only
f_{md} computed compressive stress in masonry due to dead load only
f_s computed stress in reinforcement due to design loads
f_y tensile yield stress of reinforcement
f_v computed shear stress due to design load
f'_m specified compressive strength of masonry at the age of 28 days
F_a allowable average axial compressive stress for centroidally applied axial load only
F_b allowable flexural compressive stress if members were carrying bending load only
F_{br} allowable bearing stress
F_s allowable stress in reinforcement
F_{sc} allowable compressive stress in column reinforcement
F_t allowable flexural tensile stress in masonry
F_v allowable shear stress in masonry
G shear modulus of masonry
h actual height of a prism
h' effective height of a wall or column
I moment of inertia about the neutral axis of the cross-sectional area
j ratio or distance between centroid of flexural compressive forces and centroid of tensile forces to depth d
k the ratio of depth of the compressive stress in a flexural member to the depth d
ℓ_d required development length of reinforcement
ℓ length of a wall or segment
M design moment
M_c moment capacity of compression steel in a flexural member about the centroid of the tensile force
M_m the moment of the compressive force in the masonry about the centroid of the tensile force in the reinforcement
M_s the moment of the tensile force in the reinforcement about the centroid of the compressive force in the masonry
n modular ratio $= E_s/E_m$
p ratio of the area of flexural tensile reinforcement A_s to the area bd
P design axial load
P_a allowable centroidal axial load for reinforced masonry columns
r_b ratio of the area of bars cut off to the total area of bars at the section
s spacing of stirrups or of bent bars in a direction parallel to that of the main reinforcement
t effective thickness of a wythe, wall, or column
t_p least actual lateral dimension of a prism

u bond stress per unit of surface area of bar

V total design shear force

Σo sum of the perimeters of all the longitudinal reinforcement

α angle between inclined web bars and beam axis

SELECTED REFERENCES

24.1 James E. Amrhein. *Reinforced Masonry Engineering Handbook,* 4th ed. Los Angeles: Masonry Institute of America, 1983.

24.2 J. E. Amhrein and L. L. Thompson. "Methods of Grouting Masonry Walls," *Proceedings of the 5th International Brick Masonry Conference.* McLean, VA: Brick Institute of America, 1982, pp. 516–522.

24.3 G. W. Anderson. "Stack-Bonded Small Specimens as Design and Construction Criteria," *Proceedings of the Second International Brick Masonry Conference.* Stoke-on-Trent, England: British Ceramic Research Association, 1971, pp. 38–43.

24.4 R. H. Atkinson and J. L. Noland. "A Proposed Failure Theory for Brick Masonry in Compression," *Proceedings of the Third Canadian Masonry Symposium,* Paper No. 5, University of Alberta, Edmonton, Canada, June 1983.

24.5 L. R. Baker. *Design of Masonry Panels to Resist Lateral Loads,* Research Report No. CE2/80, Deakin University, Australia, 1980.

24.6 L. R. Baker. "Some Factors Affecting the Bond Strength of Brickwork," *Proceedings of the 5th International Brick Masonry Conference.* McLean, VA: Brick Institute of America, 1982, pp. 84–89.

24.7 J. C. Baur. *Compression and Tension Bond Strength of Small Scale Masonry Specimens,* M.S. Thesis, University of Colorado, Boulder, CO, 1977.

24.8 John Baur, J. L. Noland, and James Chinn. "Compression Tests of Clay-Unit Stackbond Prisms," *Proceedings of the North American Masonry Conference,* Paper No. 25. Denver, CO: The Masonry Society, 1978.

24.9 I. J. Becica. *Behavior of Hollow Concrete Masonry Prisms Under Axial and Bending Loads, and Its Implications on Ultimate Strength Design of Masonry Structures,* Report M80-1, Department of Civil Engineering, Drexel University, Philadelphia, PA, June 1980.

24.10 Delmar L. Bloem. "Effects of Aggregate Grading on Properties of Masonry Mortar," *Symposium on Masonry Testing,* ASTM STP320. Philadelphia: American Society for Testing and Materials, 1963, pp. 67–91.

24.11 R. Copeland. "Shrinkage and Temperature Stresses in Masonry," *ACI Journal, Proceedings,* **28,** No. 8, February 1957, 769–780.

24.12 R. Copeland and Edwin L. Saxer. "Tests of Structural Bond of Masonry Mortars to Concrete Blocks," *ACI Journal, Proceedings,* **61,** November 1964, 1411–1452.

24.13 J. I. Davison. "The Effect of Mixing Time on the Compressive Strength of Masonry Mortars," *Building Research Note ISSN 0701-5232.* Ottawa, Canada: Division of Building Research, National Research Council of Canada, July 1976.

24.14 J. I. Davison. "Mortar Technology," *Proceedings of the 1st Canadian Masonry Symposium,* University of Calgary, 1976, 12–21.

24.15 R. D. Dikkers. "Summary Report of Technical Committee No. 27—Masonry Structures," *Planning and Design of Tall Buildings,* III. New York: American Society of Civil Engineers, 1972, pp. 1103–1114.

24.16 R. D. Dikkers and F. Y. Yokel. "Strength of Brick Walls Subject to Axial Compression and Bending," *Proceedings of the Second International Brick Masonry Conference.* Stoke-on-Trent, England: British Ceramic Research Association, 1971, pp. 125–132.

24.17 Robert G. Drysdale, Ahmad A. Hamid, and Arthur C. Heidebrecht. "Tensile Strength of Concrete Masonry," *Journal of the Structural Division,* ASCE, **105,** July 1979 (ST7), 1261–1276.

24.18 Robert G. Drysdale and Ahmad A. Hamid. "Behavior of Concrete Block Masonry Under Axial Compression," *ACI Journal, Proceedings,* **76,** June 1979, 707–721.

24.19 R. G. Drysdale and A. A. Hamid. "Influence of the Characteristics of the Unit on the Strength of Block Masonry," *Proceedings of the 2nd North American Masonry Conference,* Paper No. 2. Denver, CO: The Masonry Society, 1982.

24.20 S. G. Fattal and L. E. Cattaneo. *Structural Performance of Masonry Walls Under Com-*

pression and Flexure, BSS No. 73. Washington, DC: Center for Building Technology, National Bureau of Standards, 1976.

24.21 A. J. Francis, C. B. Horman, and L. E. Jerrems. "The Effect of Joint Thickness and Other Factors on the Compressive Strength of Brickwork," *Proceedings of the Second International Brick Masonry Conference.* Stoke-on-Trent, England: British Ceramic Research Association, 1971, pp. 31–37.

24.22 D. J. Frey. *Effects of Constituent Proportions on Uniaxial Compressive Strength of 2-Inch Cube Specimens of Masonry Mortars,* M.S. Thesis, University of Colorado, Boulder, CO, 1975.

24.23 J. Glanville. *Engineered Masonry.* Winnipeg, Canada: Cantex Publications, 1983.

24.24 D. G. Grenley and L. E. Cattaneo. "The Effect of Edge Load on the Racking Strength of Clay Masonry," *Proceedings of the Second International Brick Masonry Conference.* Stoke-on-Trent, England: British Ceramic Research Association, 1971, pp. 157–160.

24.25 J. Gross and H. Plummer. *Principles of Clay Masonry Construction,* 2nd ed. (rev.). McLean, VA: Brick Institute of America, 1973.

24.26 P. Haller. "Load Capacity of Brick Masonry," *Designing, Engineering and Constructing with Masonry Products.* Houston: Gulf Publishing Co., 1969, pp. 129–149.

24.27 Robert W. Harrington. "Construction and Quality Control of High-Lift Grouted Reinforced Masonry," *Proceedings of the North American Masonry Conference,* Paper No. 61. Denver, CO: The Masonry Society, 1978.

24.28 G. A. Hegemier, G. Krishnamoorthy, R. O. Nunn, and T. V. Moorthy. "Prism Tests for the Compressive Strength of Concrete Masonry," *Proceedings of the North American Masonry Conference,* Paper No. 18. Denver, CO: The Masonry Society, 1978.

24.29 G. A. Hegemier, S. K. Arya, G. Krishnamoorthy, W. Nachbar, and R. Furgerson. "On the Behavior of Joints in Concrete Masonry," *Proceedings of the North American Masonry Conference,* Paper No. 4. Denver, CO: The Masonry Society, 1978.

24.30 Arnold W. Hendry. *Structural Brickwork.* New York: Halstead Press—A Division of John Wiley & Sons, 1981.

24.31 A. W. Hendry, B. P. Sinha, and S. R. Davies. *An Introduction to Load-Bearing Design.* New York: Halstead Press—A Division of John Wiley & Sons, 1981.

24.32 H. K. Hilsdorf. "Masonry Materials and Their Physical Properties," *Planning and Design of Tall Buildings,* III. New York: American Society of Civil Engineers, 1972, pp. 981–999.

24.33 H. K. Hilsdorf. "Investigation into the Failure Mechanism of Brick Masonry Loaded in Axial Compression," *Designing, Engineering and Constructing with Masonry Products.* Houston: Gulf Publishing Co., 1969, pp. 34–41.

24.34 Thomas A. Holm. "Structural Properties of Block Concrete," *Proceedings of the North American Masonry Conference,* Paper No. 5. Denver, CO: The Masonry Society, 1978.

24.35 J. S. Hosking and H. V. Hueber. "Dimensional Changes Due to Moisture in Bricks and Brickwork," *Symposium on Masonry Testing,* ASTM STP320. Philadelphia: American Society for Testing and Materials, 1963, pp. 107–126.

24.36 J. James, T. McNeilly, and P. Oren. "Predicting the Compressive Strength of Brickwork," *Proceedings of the 5th International Brick Masonry Conference.* McLean, VA: Brick Institute of America, 1982, pp. 334–339.

24.37 E. L. Jessop. "Moisture, Thermal, Elastic and Creep Properties of Masonry," *Proceedings of the 2nd Canadian Masonry Symposium,* Carleton University, Ottawa, Canada, 1980, 505–520.

24.38 L. Kampf. "Factors Affecting Bond of Mortar to Brick," *Symposium on Masonry Testing,* ASTM STP320. Philadelphia: American Society for Testing and Materials, 1962.

24.39 J. J. Kesler. "A Look at Load-Bearing Masonry Design," *Masonry Industry,* Masonry Institute of America, March 1971.

24.40 R. L. Mayes and R. W. Clough. "A Literature Survey—Compressive, Tensile, Bond and Shear Strength of Masonry," *Report No. EERC 75-15,* College of Engineering, University of California, Berkeley, CA, June 1975.

24.41 R. L. Mayes and R. W. Clough. "State of the Art in Seismic Shear Strength of Masonry—An Evaluation and Review," *Report No. EERC 75–21,* College of Engineering, University of California, Berkeley, CA: October 1975.

24.42 D. Miller and J. Noland. "Factors Influencing the Compressive Strength of Hollow Clay Unit Prisms," *Proceedings of the 5th International Brick Masonry Conference.* McLean, VA: Brick Institute of America, 1982, pp. 122–131.

24.43 M. E. Miller, G. A. Hegemier, and R. O. Nunn. "The Influence of Flaws, Compaction and Admixture on the Strength and Elastic Moduli of Concrete Masonry," *Proceedings of the North American Masonry Conference,* Paper No. 17. Denver, CO: The Masonry Society, 1978.

24.44 C. B. Monk, Jr. "A Historical Survey and Analysis of the Compressive Strength of Brick Masonry," *Research Report No. 12*. Geneva, IL: Structural Clay Products Research Foundation, 1967.

24.45 R. O. Nunn, M. E. Miller, and G. A. Hegemier. "Grout-Block Bond Strength," *Proceedings of the North American Masonry Conference,* Paper No. 3. Denver, CO: The Masonry Society, 1978.

24.46 Larry K. Nuss, J. L. Noland, and James Chinn. "The Parameters Influencing Shear Strength Between Clay Masonry Units and Mortar," *Proceedings of the North American Masonry Conference,* Paper No. 13. Denver, CO: The Masonry Society, 1978.

24.47 M. J. N. Priestley. "Ductility of Unconfined and Confined Concrete Masonry Shear Walls," *The Masonry Society Journal,* **1,** 2(July–December), 1981, T28–T39.

24.48 M. J. N. Priestley and D. McG. Elder. "Seismic Behavior of Slender Concrete Masonry Shear Walls," *Report 82-4,* Department of Civil Engineering, University of Canterbury, Christchurch, N.Z., 1982.

24.49 M. J. N. Priestley and D. O. Bridgeman. "Seismic Resistance of Brick Masonry Walls," *Bulletin of the New Zealand Society for Earthquake Engineering,* December 1974, 167–187.

24.50 Frank A. Randall and William C. Panarese. *Concrete Masonry Handbook*. Skokie, IL: Portland Cement Association, 1976.

24.51 T. Ritchie and J. I. Davison. "Factors Affecting Bond Strength and Resistance to Moisture Penetration of Brick Masonry," *Symposium on Masonry Testing,* ASTM STP320. Philadelphia: American Society for Testing and Materials, 1962.

24.52 Sven Sahlin. *Structural Masonry*. Englewood Cliffs: Prentice-Hall, 1971.

24.53 Robert R. Schneider and Walter L. Dickey. *Reinforced Masonry Design*. Englewood Cliffs: Prentice-Hall, 1980.

24.54 J. C. Scrivener. "Static Racking Tests on Concrete Masonry Walls," *Designing, Engineering and Constructing with Masonry Products*. Houston: Gulf Publishing Co., 1969, pp. 185–191.

24.55 J. C. Scrivener and D. Williams. "Compressive Behavior of Masonry Prisms," *Proceedings of the Third Australasian Conference on Mechanics of Structures and Materials,* Auckland, N.Z., August 1971.

24.56 M. Self. *The Structural Properties of Load-Bearing Concrete Masonry,* Department of Civil and Coastal Engineering, Engineering and Industry Experiment Station, University of Florida, March 1974.

24.57 N. G. Shrive. "A Fundamental Approach to the Fracture of Masonry," *Proceedings of the 3rd Canadian Masonry Symposium,* Paper No. 4, University of Alberta, Edmonton, Alberta, June 1983.

24.58 R. J. M. Sutherland. "Structural Design of Masonry Buildings," *Planning and Design of Tall Buildings,* III. New York: American Society of Civil Engineers, 1972, pp. 1035–1056.

24.59 C. Turkstra and J. O. Jinaga. "Towards a Canadian Limit States Masonry Design Code," *Proceedings of the 2nd Canadian Masonry Symposium,* Carleton University, Ottawa, Canada, 1980, 133–142.

24.60 F. Y. Yokel, R. G. Mathey, and R. D. Dikkers. *Strength of Masonry Walls Under Compressive and Transverse Loads,* BSS No. 34. Washington, DC: Center for Building Technology, National Bureau of Standards, 1971.

24.61 F. Y. Yokel, R. G. Mathey, and R. D. Dikkers. *Compressive Strength of Slender Concrete Masonry Walls,* BSS No. 33. Washington, DC: Center for Building Technology, National Bureau of Standards, 1970.

24.62 F. Y. Yokel and S. G. Fattal. "A Failure Hypothesis for Masonry Shearwalls," *NBSIR 75-703,* National Bureau of Standards, Washington, D.C., 1975.

24.63 NCMA. *A Manual of Facts on Concrete Masonry,* An ongoing information series of "TEK" Notes, National Concrete Masonry Association, Herndon, VA.

24.64 ASTM. *Method of Sampling and Testing Brick and Structural Clay Tile* (C67-83). Philadelphia: American Society for Testing and Materials, 1983.

24.65 ANSI. *Building Code Requirements for Reinforced Masonry* [ANSI A41.2(R1970)]. New York: American National Standards Institute, Inc., 1970.

24.66 ACI. *Building Code Requirements for Concrete Masonry Structures* (ACI 531-79). Detroit: American Concrete Institute, 1979.

24.67 ACI. *Building Code Requirements for Reinforced Concrete* (ACI 318-83). Detroit: American Concrete Institute, 1983.

24.68 BSI. *Code of Practice for Structural Use of Masonry* (BS 5628), British Standards Institution, 1978.

24.69 NCMA. "Concrete Masonry Retaining Walls," *TEK No. 4*. Herndon, VA: National Concrete Masonry Association, 1966.

24.70 "Glossary of Masonry Terms," *The Masonry Society Journal,* **1,** 1(January–June), 1981, G13–G21.

24.71 BIA. "High-Lift Grouted Reinforced Masonry," *Technical Note 17D*. McLean, VA: Brick Institute of America, 1972.

24.72 BIA. *Recommended Practice for Engineered Brick Masonry*. McLean, VA: Brick Institute of America, 1969.

24.73 BIA. "Reinforced Brick Masonry Retaining Walls," *Technical Notes No. 76E, F, and G*. McLean, VA: Brick Institute of America, 1965.

24.74 *Standard Building Code*. Birmimgham, AL: Southern Building Code Congress International, Inc., 1979.

24.75 *Standard Building Code Requirements for Masonry Construction* (Document 401-81). Denver, CO: The Masonry Society, 1981.

24.76 *The Masonry Glossary*. Boston: CBI Publishing Co. (for the International Masonry Institute), 1981.

24.77 ICBO. *Uniform Building Code*. Whittier, CA: International Conference of Building Officials, 1985.

24.78 *Report of Investigation of Suconem G. A. (Grout Aid) in Brick and Concrete Block Masonry Test Panels for Super Concrete Emulsions, Ltd.*, Lowry Testing Laboratory, Los Angeles, CA, August 1966.

24.79 "Compressive, Transverse, and Shear Strength Tests of Six and Eight-Inch Walls Built with Solid and Heavy-Duty Hollow Clay Masonry Units," *Research Report No. 16,* Structural Clay Products Institute, McLean, VA, 1969.

24.80 "Compressive, Transverse and Racking Strength Tests of Four-Inch Brick Walls," *Research Report No. 9,* Structural Clay Products Research Foundation, Geneva, IL.

24.81 NCMA. *Design and Construction of Reinforced Concrete Masonry Columns and Pilasters*. McLean, VA: National Concrete Masonry Association, 1960.

24.82 ICBO. *Uniform Building Code Standards*. Whittier, CA: International Conference of Building Officials, 1982.

24.83 ASTM. *Specification for Building Brick (Solid Masonry Units Made from Clay or Shale)* (C62-83). Philadelphia: American Society for Testing and Materials, 1983.

24.84 ASTM. *Specification for Facing Brick (Solid Masonry Units Made from Clay or Shale)* (C216-81). Philadelphia: American Society for Testing and Materials, 1981.

24.85 ASTM. *Specification for Hollow Brick (Hollow Masonry Units Made from Clay or Shale)* (C652-81a). Philadelphia: American Society for Testing and Materials, 1981.

24.86 ASTM. *Specification for Hollow Load-Bearing Concrete Masonry Units* [C90-75(1981)]. Philadelphia: American Society for Testing and Materials, 1981.

24.87 ASTM. *Specification for Solid Load-Bearing Concrete Masonry Units* [C145-75(1981)]. Philadelphia: American Society for Testing and Materials, 1981.

24.88 ASTM. *Specification for Concrete Building Brick* [C55-75(1980)]. Philadelphia: American Society for Testing and Materials, 1980.

24.89 ASTM. *Specification for Mortar for Unit Masonry* (C270-82). Philadelphia: American Society for Testing and Materials, 1982.

24.90 ASTM. *Specification for Portland Cement* (C150-84). Philadelphia: American Society for Testing and Materials, 1984.

24.91 ASTM. *Specification for Blended Hydraulic Cements* (C595-83a). Philadelphia: American Society for Testing and Materials, 1983.

24.92 ASTM. *Specification for Hydrated Lime for Masonry Purposes* [C207-79(1984)]. Philadelphia: American Society for Testing and Materials, 1984.

24.93 ASTM. *Specification for Masonry Cement* (C91-83a). Philadelphia: American Society for Testing and Materials, 1983.

24.94 ASTM. *Test Method for Compressive Strength of Hydraulic Cement Mortars (Using 2-in. or 50-mm Cube Specimens)* (C109-80). Philadelphia: American Society for Testing and Materials, 1980.

24.95 ASTM. *Test Methods for Compressive Strength of Masonry Prisms* (E447-84). Philadelphia: American Society for Testing and Materials, 1984.

24.96 ASTM. *Test Method for Diagonal Tension (Shear) in Masonry Assemblages* (E519-81). Philadelphia: American Society for Testing and Materials, 1981.

24.97 ASTM. *Method for Conducting Strength Tests of Panels for Building Construction* (E72-80). Philadelphia: American Society for Testing and Materials, 1980.

24.98 ASTM. *Test Method for Bond Strength of Mortar to Masonry Units* (C952-76). Philadelphia: American Society for Testing and Materials, 1976.

24.99 ASTM. *Test Method for Flexural Bond Strength for Masonry* (E518-80). Philadelphia: American Society for Testing and Materials, 1980.

24.100 ASTM. *Test Method for Young's Modulus, Tangent Modulus, and Chord Modulus* (E111-82). Philadelphia: American Society for Testing and Materials, 1982.

24.101 ASTM. *Method of Sampling and Testing Concrete Masonry Units* [C140-75(1980)]. Philadelphia: American Society for Testing and Materials, 1980.

24.102 ASTM. *Specification for Concrete Aggregate* (C33-85). Philadelphia: American Society for Testing and Materials, 1985.

24.103 ASTM. *Specification for Lightweight Aggregates for Concrete Masonry Units* (C331-81). Philadelphia: American Society for Testing and Materials, 1981.

24.104 ASTM. *Specification for Fly Ash and Raw or Calcined Natural Pozzolan for Use as a Mineral Admixture in Portland Cement Concrete* (C618-85). Philadelphia: American Society for Testing and Materials, 1985.

24.105 ASTM. *Test for Drying Shrinkage of Concrete Block* [C426-70 (1982)]. Philadelphia: American Society for Testing and Materials, 1982.

24.106 ASTM. *Specifications for Grout for Masonry* (C476-83). Philadelphia: American Society for Testing and Materials, 1983.

24.107 ASTM. *Specification for Aggregate for Masonry Mortar* (C144-84). Philadelphia: American Society for Testing and Materials, 1984.

24.108 ASTM *Specification for Hydrated Lime for Masonry Purposes* [C207-79 (1984)]. Philadelphia: American Society for Testing and Materials, 1984.

24.109 ASTM. *Specification for Quicklime for Structural Purposes* [C5-79 (1984)]. Philaedelphia: American Society for Testing and Materials, 1984.

24.110 ASTM *Specifications for Aggregates for Masonry Grout* [C404-76 (1981)]. Philadelphia: American Society for Testing and Materials, 1981.

CHAPTER 25
WOOD STRUCTURES

ROBERT J. HOYLE, JR.

Mechanical & Structural Engineer
Lewiston, Idaho

25.1 WOOD IN BUILDING DESIGN

Wood has a long record of use as a practical, convenient, and economically competitive building material. It is used successfully throughout the world where timber resources exist. During the past 75 years a large literature on the properties of wood has developed and all manner of wood structural characteristics have been studied and documented to the advantage of the architectural and engineering designer.

Softwood structural timber is a common building construction material. Residential construction accounts for nearly one-half of the market for softwood lumber. About 30% is used for commercial and farm building. Within the building construction field, structural timber uses range from single family units to apartment dwellings, stores, warehouses, office buildings, schools, churches, and industrial structures. It is often used in combination with other common building materials, for floors and roof systems. Temporary structures such as concrete formwork, shoring, and falsework are most economically served by structural wood. Timber piling is an important material for building foundations and port facilities.

The National Forest Products Association (NFPA),* American Plywood Association (APA),† the American Institute of Timber Construction (AITC),‡ and the federal Forest Products Laboratories of the United States§ and Canada¶ possess a wealth of information which is available to the designer.

A particular objective of this chapter is to help designers expand their knowledge of timber design and improve their access to essential engineering information.

Because wood is a traditional building material, there is a tendency to oversimplify the design process and to proceed without an adequate appreciation of wood characteristics important to successful design results. Such characteristics include durability and decay avoidance, moisture relationships and dimensional uniformity, time-related strength effects, creep, stability, variability, and systems of grading, all of unique significance in timber design, which are discussed in this chapter.

25.2 ECONOMIC CONSIDERATIONS

Although the strength and elastic properties of wood are much different than those of other familiar structural materials, wood structures have met the competitive test repeatedly over a long period of time. The combination of material prices, building labor and fabricating costs, insurance expenses,

* National Forest Products Association, 1250 Connecticut Avenue, NW, Washington, DC 20036.
† American Plywood Association, 119 A Street, Tacoma, WA 98401.
‡ American Institute of Timber Construction, 333 W. Hampden Ave., Englewood, CO 80110.
§ USDA Forest Products Laboratory, P.O. Box 5130, Madison, WI 53705.
¶ S. Forintek Canada Corp., 6620 N.W. Marine Drive, Vancouver, BC V6T 1X2; and 800 Montreal Road, Ottawa, ONT K1G 3Z5.

and maintenance and repair outlays considered together lead to decisions about choosing the material for a building structural system. Most wood product prices fluctuate more than those of steel and concrete.

Small projects are supplied from retail building materials outlets at somewhat higher prices than large projects, which may be supplied more directly from mills or wholesale distributors. The warehousing and storage service of distributors is a cost factor that must be recognized in decisions about direct mill buying. Projects with large purchasing lead time can utilize special grades of material with higher strength, or particular form, or more exactly meeting the actual requirements, than the available variety of products in the retail yards.

Whereas residential construction uses common average strength lumber and plywood, truss fabricators and glued laminated timber manufacturers use superior grade materials. The size of the orders and the availability of delivery time both affect the range of choices open to the specifier. Timber specialties found in large commercial and institutional buildings are much less common in residential structures. These include panelized components, prefinished materials, and glued laminated timber of large size and curved shape, all at higher unit cost but with advantages in terms of reduced on-site labor and lower waste in the construction operation. In many cases the wood one buys may still be in the form of a log at the time the order is received.

Sawn lumber is lower in strength and modulus of elasticity than glued laminated beams and is limited in length and cross-sectional size. Glued laminated timber is not so limited but its manufacture involves processing losses and expenses in the mill which are reflected in the price. This is offset somewhat by size uniformity, dimensional stability, and quality. Large sawn timbers must be used unseasoned and the designer must make appropriate allowances for seasoning distortion that will result in a certain percentage of reject or some added labor to fit such material into a structure.

Lumber may be purchased seasoned or unseasoned and since unseasoned lumber may have a price advantage a decision on which to specify is necessary. The advantages of seasoned lumber are reduced shrinkage and distortion during and immediately after construction and less loss and extra building labor in its use. In covered structures where lumber is unlikely to become wet and where its moisture content in use will be well below 15%, the seasoned lumber is most often specified. If it is to be stored for considerable periods of time prior to use, it should be seasoned to preclude development of mold and decay and should be protected from rain during that time. Lumber used in arid regions should definitely be seasoned to prevent development of unsightly gaps between pieces and surface checking. Green or unseasoned lumber may be used for structures in which the moisture content will be high and where the disadvantages of shrinkage and distortion are not a consideration. Some temporary structures fall into this category. When unseasoned lumber is used it should be used as quickly as possible. The price differential is about 10% and is about equal to the degrade loss incurred by drying unseasoned lumber on the job.

Preservatively treated lumber is recommended whenever moisture is likely to be present in permanent structures, as in ground contact and where condensation is likely. If water is likely to accumulate or roof leakage is a definite possibility, preservative treatment is a wise procedure. In marine environments preservative treatment is essential. Treatment should be by a pressure process, not merely dipping. Parts of structures should be precut and drilled prior to treatment if possible because the treatment does not fully penetrate the wood and on-the-job cuts will expose untreated wood. When this is not possible, all cuts and drilled holes should be field treated by soaking or thorough brushing with a suitable preservative. Treating costs are variable according to treatment used. They may be 50–100% of the basic lumber cost so should only be required when necessary.

Fire retardant treatment is useful in meeting certain building code regulations for particular occupancies. It is more useful for light framing than for heavy timber. Plywood sheathing is frequently fire retardant treated when occupancy requires. Fire retardant treatment is more costly than preservative treatment and may equal the cost of the lumber. Fire retardant treatment of lumber and fire protective coating of steel serve similar purposes.

Fabricated wood structural systems are often justified on the basis of construction labor economies, achieving desired sizes, good dimensional stability, and general structural efficiency (as for trusses and I-beams). The load-carrying capacity of fabricated materials is engineered by the manufacturer, a service that saves considerable design time and has value. Specialty products can usually be custom made to exact dimensions and furnished with systems of hardware and bracing. They can be cambered for dead load deflection, which is not possible for sawn lumber.

Particleboard is used to an increasing extent for structural sheathing and is available in exterior grades. Plywood with particleboard core is becoming common. Appropriately grade marked products of this kind are equivalent to plywood of the same grade in nearly all respects. Such particleboard is made with flakes of carefully controlled size and sometimes the orientation of the flakes is controlled. Eventually particleboard structural members may become common. There is no great difference in the cost of plywood and structural particleboard materials.

25.3 TIMBER SPECIFICATIONS AND CODES

Information on design methods and properties are periodically reviewed and updated. To be abreast of these changes the designer should have the latest editions of documents identified in this section.

The *National Design Specification for Wood Construction* (NDS) [25.35] is the accepted resource on structural lumber, glued laminated timber, timber piling, and fastenings. The NDS prescribes design rules and practices to be used by designers of wood structures. A supplement, *Design Values for Wood Construction* [25.36], accompanies the NDS and lists the current grades and design properties of these wood materials.

Properties information used in this chapter is current at the time of writing, but may change as time goes by, owing to shifts in availability of materials, marketing and manufacturing practices, and development of technology.

Designing wood structures is not fundamentally different than designing structures made with other materials. Wood does have some special characteristics that deserve attention. The effect of moisture on properties, shrinkage and expansion due to drying and wetting, creep deformation under sustained load, durability in the presence of wood destroying fungi, thermal expansion characteristics, and insulating properties are some of the special considerations the timber designer should understand.

The American Institute of Timber Construction (AITC), issues the publication, *Glulam Systems* [25.2], containing sample specifications and design practice recommendations for glued laminated timber. AITC also is responsible for the *Timber Construction Manual* [25.4] which contains many timber construction standards of value to the designer.

The *Plywood Design Specification* (PDS) [25.7] from the American Plywood Association (APA) is the definitive resource on plywood properties and grades. APA also publishes four supplements to PDS on curved panels, plywood beams, stressed-skin panels, and sandwich panels. Special publications by APA on design and fabrication of plywood structural systems and concrete formwork are also available.

Wood Preservation: Treating Practices [25.23] is the standard used by federal and state agencies and some private users in procuring treated wood. Another document of value which supplements the federal standard as a guide to the complex variety of treatment procedures is the report, *Selection, Production, Procurement and Use of Preservative-treated Wood* [25.22].

The American Wood Preservers Institute (AWPI), the technical/promotional arm of the wood treating industry, publishes information* on specifying preservative treated and fire retardant treated wood.

The Western Wood Preservers Institute (WWPI) publishes a useful, concise *Architect and Engineers Guide Specifications to the Pressure Treatment of Western Woods* [25.47]. It covers both preservative and fire retardant treated wood. This guide cross-references Federal and American Wood Preservers Association (AWPA) specifications.

American Plywood Association (APA) publishes a brochure and suggested specifications, *Pressure Preserved Plywood* [25.13].

AWPI member plants are not the only suppliers of treated wood. Nearly all treating specifications for structural wood are AWPA and Federal specifications. Quality control and quality marking are essential features of a well-written procurement specification.

Round timber piles for foundations are described in the NDS [25.35] with a table of design properties for four species. For other species the ASTM Method D2899, *Establishing Design Stresses for Round Timber Piles* [25.17], is used to compute design values based on unseasoned clear wood properties from ASTM D2555 [25.16]. Standard sizes of round timber piles are found in ASTM D25 [25.19], *Standard Specifications for Round Timber Piles*.

Round timber poles for buildings and electric utility structures are described in American National Standards Institute (ANSI) 05.1, [25.6] *Specifications and Dimensions for Wood Poles*. That document gives pole size classes and quality requirements and lists the minimum ultimate fiber stresses in bending for all common North American species. It does not give other strength or elastic properties for poles. *The Timber Construction Manual* [25.4] lists design values for bending, compression, and elastic modulus based on ANSI 05.1 [25.6] and ASTM D2899 [25.17]. Another good resource is the American Society of Agricultural Engineers (ASAE) Engineering Practice: ASAE EP388, "Design Properties of Round, Sawn, and Laminated Preservatively Treated Construction Poles" [25.15].

Pole Building Design [25.39], by Donald Patterson, contains design practice discussion and bolt strength values for poles.

---

* These references are in Sweet's File and may be obtained from AWPI, 1651 Old Meadow Road, McLean, VA 22102.

For electric utility structures, design practice is governed by the *National Electrical Safety Code* [25.30]. In this code the design properties are the minimum ultimate bending stress values for the species, and the design practice is to apply overload factors from the Code for the grade of structure required by the Code. Properties are not given in the Code. They can be obtained from other sources.

25.4 LUMBER FOR STRUCTURAL USE

Lumber may be purchased unseasoned (green) or seasoned (dry) with rough sawn or smoothly dressed surfaces. Each of these types of lumber is cut to a nominal size from the unseasoned log and gradually reduced by the processes of drying with attendant shrinkage and planing to uniform final dimensions in the case of dry dressed lumber. Many modern sawmills operate with such precision that they can cut to less than the stated nominal size and dress out seasoned or unseasoned lumber to the national standard final sizes. It is a historical circumstance that lumber is a mass produced commodity rather than a consumer targeted product, and is highly standardized.

The actual sizes of dressed softwood boards and dimension lumber, either dry or green, and green dressed timbers are listed in Table 25.1, in accordance with U.S. Department of Commerce Product Standard PS20-70 [25.34]. Lumber may also be marketed rough sawn and unsurfaced at least ⅛ in. larger than these sizes.

The following terminology is used as related to size classification:

1. *Boards* are lumber of ¾- to 1½-in. nominal thickness.
2. *Dimension lumber* is from 2- to 4-in. nominal thickness.
3. *Timber* is lumber thicker than 4-in. nominal.
4. *Framing lumber* is dimension lumber of 2- to 4-in. nominal width.
5. *Joists and Planks* are dimension lumber 6 in. or more in nominal thickness.
6. *Beams and Stringers* are timber at least 2 in. wider than their nominal thickness.
7. *Posts and Timbers* are timber less than 2 in. wider than their nominal thickness.

Timber is only available in the green condition. The nominal sizes are 5 in. and thicker with the actual dressed size ½ in. less than nominal. These large thicknesses season very slowly, even in a dry kiln so they are rarely sold seasoned.

The dry and green size difference for boards and dimension lumber shown in Table 25.1 is to equalize the size of dry and green lumber at final moisture content in service. This is to provide for a fair measure, accounting for shrinkage, and to simplify design computations. Whether initially dry or green as manufactured, wood placed in any given environment comes to equilibrium at a

Table 25.1 Nominal and Minimum Dressed Sizes of Boards, Dimension Lumber, and Timbers* (Thicknesses apply to all widths. Widths apply to all thicknesses.)

Item	Thicknesses			Face Widths		
		Min. Dressed			Min. Dressed	
	Nom.	Dry	Green	Nom.	Dry	Green
Boards	1 1¼ 1½	¾ 1 1¼	23/32 1 1/32 1 9/32	2 3 4	1½ 2½ 3½	1 9/16 2 9/16 3 9/16
Dimension	2 3 4	1½ 2½ 3½	1 9/16 2 9/16 3 9/16	6 8 10 12 14 16	5½ 7¼ 9¼ 11¼ 13¼ 15¼	5⅝ 7½ 9½ 11½ 13½ 15½
Timbers	5 in. and thicker	Not made dry	½ in. off nominal size	5 in. and wider	Not made dry	½ in. off nominal size

* From Ref. 25.34.

common moisture content. If manufactured to these sizes, either type will arrive at roughly the same final size. As an example, in air at 70° and 65% relative humidity, dry or green wood comes to 12% equilibrium moisture content (*Wood Handbook* [25.43], Table 3-4, page 3-8). The design section properties in Table 25.2 are based on the dry sizes for boards and dimension lumber and the green sizes for timber.

In specifying structural lumber the common designation is S-DRY for seasoned lumber with a specified maximum moisture content of 19% as manufactured and graded. The designation S-GRN means lumber that is either unseasoned or seasoned to some moisture content exceeding 19%. These symbols appear on the grade mark. There is a special drying specification, MC-15, denoting lumber dried to 15% maximum moisture content when graded. It is used for high-quality decking and similar products where shrinkage may be critical to performance. MC-15 lumber is manufactured to the dry dressed sizes in Table 25.1.

Table 25.2 Section Properties for Sawn Lumber and Timber (Adapted from NDS [25.35] Table M-2; Used by Permission)

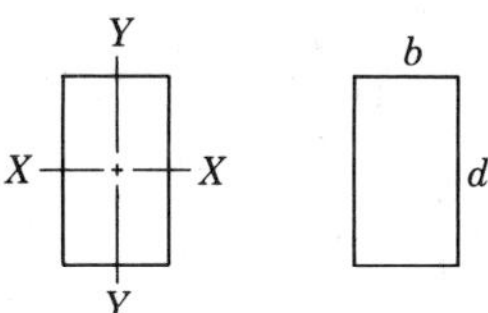

Nominal Size $b \times d$ in.	Standard Dressed Size (S4S) $b \times d$ in.	Area of Section A in.²	X-X Axis: Moment of Inertia I in.⁴	X-X Axis: Section Modulus S in.³	Y-Y Axis: Moment of Inertia I in.⁴	Y-Y Axis: Section Modulus S in.³	Weight, lb/ft: 35 pcf
1 × 3	¾ × 2½	1.875	0.977	0.781	0.088	0.234	0.456
1 × 4	¾ × 3½	2.625	2.680	1.531	0.123	0.328	0.638
1 × 6	¾ × 5½	4.125	10.398	3.781	0.193	0.516	1.003
1 × 8	¾ × 7¼	5.438	23.817	6.570	0.255	0.680	1.322
1 × 10	¾ × 9¼	6.938	49.466	10.695	0.325	0.867	1.686
1 × 12	¾ × 11¼	8.438	88.989	15.820	0.396	1.055	2.051
2 × 3[a]	1½ × 2½	3.750	1.953	1.563	0.703	0.938	0.911
2 × 4[a]	1½ × 3½	5.250	5.359	3.063	0.984	1.313	1.276
2 × 6[a]	1½ × 5½	8.250	20.797	7.563	1.547	2.063	2.005
2 × 8[a]	1½ × 7¼	10.875	47.635	13.141	2.039	2.719	2.643
2 × 10[a]	1½ × 9¼	13.875	98.932	21.391	2.602	3.469	3.372
2 × 12[a]	1½ × 11¼	16.875	177.979	31.641	3.164	4.219	4.102
2 × 14[a]	1½ × 13¼	19.875	290.775	43.891	3.727	4.969	4.831
3 × 4	2½ × 3½	8.750	8.932	5.104	4.557	3.646	2.127
3 × 6	2½ × 5½	13.750	34.661	12.604	7.161	5.729	3.342
3 × 8	2½ × 7¼	18.125	79.391	21.901	9.440	7.552	4.405
3 × 10	2½ × 9¼	23.125	164.886	35.651	12.044	9.635	5.621
3 × 12	2½ × 11¼	28.125	296.631	52.734	14.648	11.719	6.836
3 × 14	2½ × 13¼	33.125	484.625	73.151	17.253	13.802	8.051
3 × 16	2½ × 15¼	38.125	738.870	96.901	19.857	15.885	9.266
4 × 4	3½ × 3½	12.250	12.505	7.146	12.505	7.146	2.977
4 × 6	3½ × 5½	19.250	48.526	17.646	19.651	11.229	4.679
4 × 8	3½ × 7¼	25.375	111.148	30.661	25.904	14.802	6.168
4 × 10	3½ × 9¼	32.375	230.840	49.911	33.049	18.885	7.869
4 × 12	3½ × 11¼	39.375	415.283	73.828	40.195	22.969	9.570
4 × 14	3½ × 13¼	46.375	678.475	102.441	47.340	27.052	11.266
4 × 16	3½ × 15¼	53.375	1,034.418	135.661	54.487	31.135	12.975

Table 25.2 (*Continued*)

Nominal Size $b \times d$ in.	Standard Dressed Size (S4S) $b \times d$ in.	Area of Section A in.^{2}	*X-X* Axis		*Y-Y* Axis		Weight, lb/ft
			Moment of Inertia I in.^{4}	Section Modulus S in.^{3}	Moment of Inertia I in.^{4}	Section Modulus S in.^{3}	35 pcf
6 × 6	5½ × 5½	30.250	76.255	27.729	76.255	27.729	7.352
6 × 8	5½ × 7½	41.250	193.359	51.563	103.984	37.813	10.026
6 × 10	5½ × 9½	52.250	392.963	82.729	131.714	47.896	12.700
6 × 12	5½ × 11½	63.250	697.068	121.229	159.443	57.979	15.373
6 × 14	5½ × 13½	74.250	1,127.672	167.063	187.172	68.063	18.047
6 × 16	5½ × 15½	85.250	1,706.776	220.229	214.901	78.146	20.720
6 × 18	5½ × 17½	96.250	2,456.380	280.729	242.630	88.229	23.394
6 × 20	5½ × 19½	107.250	3,398.484	348.563	270.359	98.313	26.068
6 × 22	5½ × 21½	118.250	4,555.086	423.729	298.088	108.396	28.741
6 × 24	5½ × 23½	129.250	5,948.191	506.229	325.818	118.479	31.415
8 × 8	7½ × 7½	56.250	263.672	70.313	263.672	70.313	13.672
8 × 10	7½ × 9½	71.250	535.859	112.813	333.984	89.063	17.318
8 × 12	7½ × 11½	86.250	950.547	165.313	404.297	107.813	20.964
8 × 14	7½ × 13½	101.250	1,537.734	227.813	474.609	126.563	24.609
8 × 16	7½ × 15½	116.250	2,327.422	300.313	544.922	143.313	28.255
8 × 18	7½ × 17½	131.250	3,349.609	382.813	615.234	164.063	31.901
8 × 20	7½ × 19½	146.250	4,634.297	475.313	684.547	182.813	35.547
8 × 22	7½ × 21½	161.250	6,211.484	577.813	755.859	201.563	39.193
8 × 24	7½ × 23½	176.250	8,111.172	690.313	826.172	220.313	42.839
10 × 10	9½ × 9½	90.250	678.755	142.896	678.755	142.896	21.936
10 × 12	9½ × 11½	109.250	1,204.026	209.396	821.651	172.979	26.554
10 × 14	9½ × 13½	128.250	1,947.797	288.563	964.547	203.063	31.172
10 × 16	9½ × 15½	147.250	2,948.068	380.396	1,107.443	233.146	35.790
10 × 18	9½ × 17½	166.250	4,242.836	484.896	1,250.338	263.229	40.408
10 × 20	9½ × 19½	185.250	5,870.109	602.063	1,393.234	293.313	45.026
10 × 22	9½ × 21½	204.250	7,867.879	731.896	1,536.130	323.396	49.644
10 × 24	9½ × 23½	223.250	10,274.148	874.396	1,679.026	353.479	54.262
12 × 12	11½ × 11½	132.250	1,457.505	253.479	1,457.505	253.479	32.144
12 × 14	11½ × 13½	155.250	2,357.859	349.313	1,710.984	297.563	37.734
12 × 16	11½ × 15½	178.250	3,568.713	460.479	1,964.463	341.646	43.325
12 × 18	11½ × 17½	201.250	5,136.066	586.979	2,217.943	385.729	48.915
12 × 20	11½ × 19½	224.250	7,105.922	728.813	2,471.422	429.813	54.505
12 × 22	11½ × 21½	247.250	9,524.273	885.979	2,724.901	473.896	60.095
12 × 24	11½ × 23½	270.250	10,274.148	1,058.479	2,978.380	517.979	65.686
14 × 16	13½ × 15½	209.250	4,189.359	540.563	3,177.984	470.813	50.859
14 × 18	13½ × 17½	236.250	6,029.297	689.063	3,588.047	531.563	57.422
14 × 20	13½ × 19½	263.250	8,341.734	855.563	3,998.109	592.313	63.984
14 × 22	13½ × 21½	290.250	11,180.672	1,040.063	4,408.172	653.063	70.547
14 × 24	13½ × 23½	317.250	14,600.109	1,242.563	4,818.234	713.813	77.109

25.4.1 Grades and Properties

All structural lumber is stress graded and carries assigned design properties. The common grading method is visual inspection by certified graders supervised by one of the regional grading agencies. Graders sort the lumber according to grades using rules or descriptions published by the agencies and based on ASTM D245 [25.18]. For boards and dimension lumber the grades of the different agencies are nearly identical. For timber there are some variations by agency, but still much similarity.

There are over 60 species, largely coniferous (softwood) woods, graded structurally in North America. Species with similar properties are often grouped for marketing convenience. Hem-fir, for example, is a western hemlock and true fir group. Southern pine is a group of several pines from the southeastern United States. Douglas fir-larch groups two western species. Spruce-pine-fir is a Canadian marketing group often exported to the United States.

The common grade names and size classes are listed in Table 25.3. The properties in this table are for Douglas fir-larch but the grade names are used for all structural softwood species. The "dense" grades are peculiar to Douglas fir-larch and southern pine. Southern pine and California

Table 25.3 Visual Grades of Structural Lumber and Design Values* for Douglas Fir and Larch

Grades	Size Classes	Bending		Tension,	Compression,	Bearing,	Shear,	Modulus of Elasticity,
		Single, psi	Repetitive, psi	psi	psi	psi	psi	psi
Light Framing Grades:	*Dimension*							
Construction	2 to 4 in. thick	1050	1200	625	1150	625	95	1,500,000
Standard	4 in. wide	600	675	350	925	625	95	1,500,000
Utility		275	325	175	600	625	95	1,500,000
Structural light framing grades:	*Dimension*							
Dense Select Structural	2 to 4 in. thick	2450	2800	1400	1850	730	95	1,900,000
Select Structural	2 to 4 in. wide	2100	2400	1200	1600	625	95	1,800,000
Dense No. 1		2050	2400	1200	1450	730	95	1,900,000
No. 1		1750	2050	1050	1250	625	95	1,800,000
Dense No. 2		1700	1950	1000	1150	730	95	1,700,000
No. 2		1450	1650	850	1000	625	95	1,700,000
No. 3 and Stud		800	925	475	600	625	95	1,500,000
Structural Joists & Plank Grades:	*Dimension*							
Dense Select Structural	2 to 4 in. thick	2100	2400	1400†	1650	730	95	1,900,000
Select Structural	5 in. and wider	1800	2050	1200†	1400	625	95	1,800,000
Dense No. 1		1800	2050	1200‡	1450	730	95	1,900,000
No. 1		1500	1750	1000‡	1250	625	95	1,800,000
Dense No. 2		1450	1700	775‡	1250	730	95	1,700,000
No. 2		1250	1450	650‡	1050	625	95	1,700,000
No. 3 and Stud		725	850	375‡,§	675	625	95	1,500,000
Beam & Stringer Grades:	*Timbers*							
Dense Select Structural	5 in. and thicker	1900	—	1100	1300	730	85	1,700,000
Select Structural	Over 2 in. more in thickness	1600	—	950	1100	625	85	1,600,000
Dense No. 1		1550	—	775	1100	625	85	1,600,000
No. 1		1300	—	675	925	625	85	1,600,000
Post & Timber Grades:	*Timbers*							
Dense Select Structural	5 in. and thicker	1750	—	1150	1350	730	85	1,700,000
Select Structural	Not over 2 in. more in thickness	1500	—	1000	1150	625	85	1,600,000
Dense No. 1		1400	—	950	1200	730	85	1,600,000
No. 1		1200	—	825	1000	625	85	1,600,000
Decking Grades:	*Decking*							
Select	T. & G.	—	1200	—	—	625	95	1,800,000
Commercial		—	1650	—	—	625	95	1,700,000

* For use at 19% maximum moisture content, normal duration of load. See Tables 25.4 and 25.5 for adjustments.
† For nominal 8-in., multiply by 0.9; for nominal 10-in. or wider, multiply by 0.8.
‡ For nominal 8-in., multiply by 0.8; for nominal 10-in. or wider, multiply by 0.6.
§ For studs 8-in. and wider, no tension strength is assigned.

Table 25.4 Service Moisture Content Adjustments for Lumber Design Properties (from *NDS Supplement* [25.36])

	F_b	F_t	F_v	$F_{c\perp}$	F_c	E
2- to 4-in. lumber used above 19% moisture content	0.86	0.84	0.97	0.67	0.70	0.97
5-in. and thicker lumber used above 19% moisture content	1.00	1.00	1.00	0.67	0.91	1.00
MC-15 lumber used at 15% moisture content or less	1.08	1.08	1.05	1.00	1.17	1.05
MC-15 decking used above 15% moisture content*	0.79	0.78	0.92	0.67	0.60	0.92

* S-DRY or S-GRN decking used above 19% same as 2- to 4-in. lumber.

redwood are also available in several additional grades which are not described here but can be found in the *NDS Supplement* [25.36] referred to earlier.

The various species of wood have characteristic properties. Among structural softwoods Douglas fir-larch, southern pine, and hem-fir are the densest and strongest. Cedars, spruces, true firs, and other pines have lower densities and strengths. Since each visual stress grade has the same physical description of allowable size of growth defects (knots, grain slope, split, and check) the allowable properties for this grade vary by species. These properties are found in *NDS Supplement* [25.36].

The timber designer must choose a species as well as a grade. The choice will depend on the species that are available in the region where the project will be built, although some species are available anywhere in the United States or Canada.

The allowable properties for Douglas fir-larch in Table 25.3 are for dry use applications, as in most enclosed structures not normally wet, and they are used for either S-DRY or S-GRN lumber. For MC-15 lumber these values may be increased for dry use by factors in Table 25.4.

When wood is used under wet conditions, the Table 25.3 values must be reduced. Use conditions where the moisture content is likely to exceed 19% are "wet-use" conditions. The reduction factors for wet use for Douglas fir and most other species are given in Table 25.4.

The repetitive value properties in Table 25.3 are for use in designing structures with closely spaced parallel members which are well connected and tend to deflect in unison so that the stiffer, stronger pieces assume a higher portion of the load. The criterion is that spacing must not exceed 24 in. and the members be connected by a load-distributing deck of at least the stiffness of ½-in. thick plywood parallel to the face grain or by bridging. Most framed structures meet these requirements. No repetitive values are given for Beams and Stringers or Posts and Timbers because they are seldom so closely spaced. No single-member values are given for decking because decking is laid in parallel courses interconnected by tongues and grooves to provide the required load distribution.

The published design values are for "Normal" load duration, which means the cumulative duration of maximum design load over the life of the structure will not exceed 10 years. Wood can sustain higher loads for short periods of time than for long periods. It is a timber design practice [25.35] to adjust the allowable properties according to the duration of the load, using factors given in Table 25.5. The designer must consider whether the dead load using the 90% factor or higher loads using one of the larger factors will require the larger member and choose accordingly. *These adjustments do not apply to elastic modulus.* In preliminary estimating work it is often sufficient to use the Normal values, unadjusted, and make load duration refinements later.

Table 25.5 Load Duration Adjustments for Lumber Design Properties* (from Appendix B of NDS [25.35])

Permanent Load	Normal (10-yr) Load	Snow (2-month) Load	Hurricane (7-day) Load	Wind and Seismic (½-day) Load	Impact (1-sec) Load
0.9	1.0	1.15	1.25	1.33	2.0

* Adjustments do not apply to elastic modulus (modulus of elasticity).

Table 25.6 Grades and Design Values* for Machine Stress Rated Lumber (adapted from *NDS Supplement* [25.36])

Grade	Bending: Single, psi	Bending: Repetitive, psi	Tension, psi	Compression, psi	Elastic Modulus, psi
1200f—1.2E	1200	1400	600	950	1,200,000
1500f—1.4E	1500	1750	900	1200	1,400,000
1800f—1.6E	1800	2050	1175	1450	1,600,000
2100f—1.8E	2100	2400	1575	1700	1,800,000
2400f—2.0E	2400	2750	1925	1925	2,000,000
2700f—2.2E	2700	3100	2150	2150	2,200,000
3000f—2.4E	3000	3450	2400	2400	2,400,000
3300f—2.6E	3300	3800	2650	2650	2,600,000

	Douglas Fir-Larch	Hem-Fir	Western Hemlock	Pine†	Engelman Spruce	Cedar‡
Bearing (psi)	625	405	410	375	320	425
Shear (psi)	95	75	90	70	70	75

* Values are for use at 19% and less moisture content, normal duration of load, width to 12-in. nominal and 2-in. nominal thickness. See Tables 25.4, 25.5 for adjustments.

† Idaho white, lodgepole, ponderosa, and sugar pines.

‡ Western red and incense cedars.

The *bending strength* values are for members up to 12 in. in depth in the direction of flexure. For depths exceeding 12 in., the design values must be reduced by the size factor

$$C_F = \left(\frac{12}{d}\right)^{1/9} \tag{25.4.1}$$

where d is the depth in inches. This produces a reduction factor of 0.95 for 20 in., 0.9 for 32 in., 0.85 for 52 in., and leveling off at about 0.8 for over 84 in. The adjustment is for uniformly loaded beams with a span-to-depth ratio of 21. For more refined consideration of loading arrangement and span-to-depth ratio see Section 25.8, Beam Design.

Machine stress rated (MSR) lumber is graded using a machine which measures the stiffness of each piece and gives the grader an additional measure of quality to use in sorting decisions. It is used for the manufacture of structural products, notably trusses and to a lesser extent for laminated beam lumber, scaffold plank, and certain industrial column materials. The combination of machine testing and visual inspection afforded by this system results in a more discriminating classification process that identifies the stronger material quite well. The system works because, unlike metals, the elastic modulus of wood is quite variable, but well-correlated with those factors that affect strength, such as density, sloping grain, moisture content, knots, and mismanufacture.

The allowable properties of MSR lumber are not species related for bending, tension, and compression parallel-to-grain, or for elastic modulus. Table 25.6 lists common grades of MSR lumber.

The grade name consists of the design values for bending and elastic modulus. Species identification is still necessary, however, because the compression perpendicular-to-grain (bearing strength) and the shear strength are species characteristics. In general, the machine stress grading system is simple and the species tend to be more easily interchangeable. The particular obstacles to complete species interchangeability are the bearing strength and fastener strengths.

Because of the species density character of wood, Douglas fir-larch, southern pine, and hem-fir provide the grades 2100f-1.8E and better while the other species account for the 1200f-1.2E to 1800f-1.6E grades. This type of lumber is only produced S-DRY and usually only in nominal 2-in. thickness.

25.4.2 Specifying Lumber

Dimension Lumber

Dimension lumber is available in three categories:

1. *Light Framing* (L.F.) widely used by builders for residential construction is nominal 2×4 to 4×4. It is graded to lower stress values than *Structural Light Framing* and is usually selected from span tables [25.37] published by NFPA or found in building codes.
2. *Structural Light Framing* (S.L.F.) for general construction is also in nominal 2×4 to 4×4 sizes, and has higher strength values.
3. *Structural Joists and Planks* (S.J.P.) for general construction is 2×6 to 4×16 nominal size. It is used for roof and floor systems.

All of these grades are specified to length in 2-ft increments and may be obtained either S-DRY or S-GRN.

Typical specifications for dimension lumber would include quantity, size, species, grade, moisture content, and grading agency if the strength value is peculiar to a particular agency.

Examples of such specifications are:

1. 500 pieces 2×4 × 8 ft long Douglas fir-larch, Construction L.F. grade, S-DRY (WCLIB or WWPA) 2666 bd. ft.
2. 330 pieces 2×4 × 14 ft long southern pine, No. 2, S.L.F. grades, S-DRY (SPIB) 3080 bd. ft.
3. 120 pieces 2×8 × 16 ft long hem-fir, No. 1, S.J.P. grade, S-DRY (WWPA or WCLIB) 2560 bd. ft.
4. 200 pieces 2×6 × 12 ft long northern pine, No. 3, S.J.P. grade, S-GRN (NELMA/NHPMA) 2400 bd. ft.

These are merely sample specifications and do not imply availability or seasoning practices for the species.

The board foot tally is used for pricing. It is based on nominal size at 1 in. × 12 in. × 12 in. to the board foot.

Decking

Decking is often ordered random length since it can be laid up in parallel courses with random end joints. Use of MC-15 decking will produce the best looking deck with a minimum amount of gap openings between courses due to low shrinkage. Keeping decking widths below nominal 8 in. will also contribute to a well-fitted deck. The face width of decking is 1 in. off nominal for 6 in. and narrower nominal sizes and 1¼ in. off nominal for widths over 6 in. The actual face width affects coverage. Thus for covering 1000 sq ft of area the ordering information for 2×6 decking would be:

> 2400 lineal ft 2×6 Western Cedar Commercial grade decking MC-15 (WWPA or WCLIB) random length with minimum 6 ft lengths with 70% 10 ft or longer. 2400 bd. ft.

Beams and Stringers and Posts and Timbers

These items are not seasoned and so would be ordered S-GRN typically as follows:

1. 40 pieces 6×14 × 24 ft long Douglas fir-larch, Select Structural B & S grade S-GRN (WWPA or WCLIB) 6720 bd. ft.
2. 30 pieces 6×8 × 16 ft long eastern white pine, No. 1, P & T grade S-GRN (NELMA) 1920 bd. ft.

On these grades it may be necessary to specify "rough sawn" or "dressed four sides" to be sure of obtaining the required finish.

Grading rule writing agencies are:

NELMA	Northeastern Lumber Manufacturers Ass'n, Inc., 4 Fundy Rd., Falmouth, Maine 04105
NHPMA	Northeastern Hardwood and Pine Manufacturers Ass'n, Inc., Northern Building, Green Bay, Wisconsin 54301
NLGA	National Lumber Grading Authority (Canada) 1500/1055 W. Hastings Street, Vancouver, B.C., Canada V6E 2H1

RIS Redwood Inspection Service, 617 Montgomery Street, San Francisco, California 94111
SPIB Southern Pine Inspection Bureau, P.O. Box 846, Pensacola, Florida 32594
WCLIB West Coast Lumber Inspection Bureau, 6980 S.W. Varnes Rd., P.O. Box 23145, Portland, Oregon 97223
WWPA Western Wood Products Ass'n., 1500 Yeon Building, Portland, Oregon 97204

25.5 PLYWOOD FOR STRUCTURAL USE

Plywood is used in both wood- and steel-framed buildings for structural floor, roof, and wall sheathing and to serve as diaphragms and shear walls. It is also used structurally in fabricated components, box, and I-beams, and gussets for trusses and rigid frames.

The *Plywood Design Specification* (PDS) [25.7] is the source of current design values.

Grades

Plywood grades are determined by the species and grade of veneer and the type of adhesive used in its manufacture. Species of wood used in plywood are grouped by stiffness and strength, Group 1 being the highest and Group 5 the lowest. Veneer is graded from highest to lowest by the letter marks N, A, B, C, and D. Structural plywood is almost entirely of the C and D veneer grades. Plywood with veneers of grade higher than C tends to be costly for ordinary structural purposes, but may be used. All plywood containing D grade veneer is classified as "Interior type."

Interior-type plywood bonded with exterior-type glue is the most common structural plywood. The grade name is "C-D Interior with Exterior Glue," often abbreviated CDX. APA designates this as "APA rated sheathing, Exposure 1." If made with interior glue it would be designated "Exposure 2." Plywood *without* D grade veneer and bonded with exterior glue is designated "Exterior type." The Exterior-type designation has two characteristics, veneer grades above D and exterior glue. A typical grade is "C-C Exterior."

Structural plywood grades with design shear values slightly better than "C-D Interior with Exterior Glue," are the two grades "Structural I C-D Interior" and "Structural II C-D Interior."* Each of these grades is made with exterior-type glue only. Structural I is entirely of Group 1 species. Structural II is made with Groups 1, 2, and 3 species. Structural I and II may be obtained with C-C veneer in which case they are Exterior grades. Plywood with D veneers is limited to the Interior classification because D grade veneer does not weather well and should not be permanently exposed to weather even if painted or stained. This is why the Interior designation is required, even if the bonding adhesive is a waterproof exterior-type phenolic resin. Structural I and II plywood are difficult to obtain in small quantities and from retail distributors.

Sizes

Plywood panels are 48 × 96 in. in size with thicknesses varying from 5⁄16 to 1⅛ in. Panels more than 96 in. long are available only by special order.

Properties and Design Values

Plywood section properties are based on the cross-sectional area occupied by the veneers with grain parallel to the direction of the stress. They can not be computed from the width and thickness of the panels. Reference must be made to the PDS [25.7] for section properties. The same grade and thickness of plywood has different section properties for designing parallel to, and perpendicular to, the grain of the face veneers. Table 25.7 lists some section properties from the PDS.

Design values depend on the species group of the face plys, the core and back species, and the veneer thicknesses used in the construction of the panel. These complex combinations are defined in U.S. Department of Commerce Product Standard PS-1 [25.33] to which the manufacturer must conform.

Table 25.8 lists some common structural plywood grades and their design values. These values are used with the section properties for the stress direction and grade for design calculations.

For ordinary roof and floor purposes the first seven grades in Table 25.8 carry an Identification Index as a part of their grade mark. The index indicates the roof and floor panel spans on which they may be used when the face grain is parallel to the span. For example, 32/16 means the panel can span 32 in. on roofs with a 35-psf live load and 16 in. on floors with 100-psf live loads. An index of 24/0 means the panel is suitable for a 24-in. roof span but not for floor construction.

Plyform is concrete form plywood, mill oiled and edge sealed. Such plywood is reusable if

---

* APA designates these as "APA Rated Sheathing, Exposure 1, Structural I (or II)."

Table 25.7 Section Properties for Common Plywood Structural Grades*

				Properties for a 1-ft Width								
				Stress Parallel to Face Grain				Stress Perpendicular to Face Grain				
Grade	Thickness, in.	Identification Index	Effective Thickness for Shear, in.	A Area in.^{2}	I Moment of Inertia in.^{4}	KS Section Modulus in.^{3}	$\frac{Ib}{Q}$ Rolling Shear Constant	A Area in.^{2}	I Moment of Inertia in.^{4}	KS Section Modulus in.^{3}	$\frac{Ib}{Q}$ Rolling Shear Constant	Weight, psf
CS-INT., EXT. GLUE Unsanded	½ ¾	32/16 24/0 48/24 42/20	0.316 0.467	2.500 3.403	0.086 0.243	0.247 0.501	4.189 6.823	1.076 1.632	0.005 0.036	0.057 0.236	2.585 3.717	1.5 2.2
2-4-1 INT., EXT. GLUE Touch sanded	1⅛	—	0.855	4.592	0.653	0.995	9.933	4.120	0.283	0.763	7.452	3.3
STRUCTURAL I & II CC and CD Unsanded	½ ¾	32/16 48/24	0.543 0.747	2.906 4.406	0.091 0.247	0.318 0.573	4.497 7.046	2.325 2.938	0.017 0.085	0.145 0.369	2.574 3.697	1.5 2.2
PLYFORM CLASS I & II Sanded, edge Sealed and oiled	⅝ ¾	— —	0.472 0.589	2.280 2.884	0.129 0.197	0.356 0.452	6.293 7.881	1.627 2.104	0.045 0.093	0.234 0.387	3.922 4.842	1.8 2.2

* Selected from *Plywood Design Specification* [25.7].

Table 25.8 Design Values for Structural Plywood Grades with Exterior Glue for Dry Use—Normal Load Duration

Grades	Thickness, in.	Bending Tension, psi	Compression, psi	Shear ⊥, psi	Shear ‖, psi	Bearing ⊥, psi	Modulus of Rigidity, psi	Modulus of Elasticity, psi
CE INT (EXT GLUE)								
Face Ply Group 1	5/16–3/4	1650	1540	250	53	340	90,000	1,800,000
Face Ply Group 3	5/16–3/4	1200	990	185	53	210	60,000	1,200,000
Face Ply Group 4	5/16–3/4	1110	950	175	53	160	50,000	1,000,000
STRUCTURAL I CD	5/16–3/4	1650	1540	250	75	340	90,000	1,800,000
STRUCTURAL II CD	5/16–3/4	1200	990	185	56	210	60,000	1,200,000
STRUCTURAL I CC	5/16–3/4	2000	1640	250	75	340	90,000	1,800,000
STRUCTURAL II CC	5/16–3/4	1400	1060	185	56	210	60,000	1,200,000
2-4-1 INT (EXT GLUE)	1 1/8	1650	1540	250	53	340	90,000	1,800,000
PLYFORM CLASS I	5/8–3/4	1650	1540	250	53	340	90,000	1,800,000
PLYFORM CLASS II	5/8–3/4	1200	990	185	53	210	60,000	1,200,000

CD INT (EXT GLUE), STRUCTURAL I (CC AND CD), STRUCTURAL II (CC AND CD) are unsanded.
2-4-1 INT (EXT GLUE) is touch-sanded, available tongue and groove.
PLYFORM is sanded and mill oiled and edge sealed.

cleaned and reoiled at intervals. 2-4-1 plywood is a special thick decking product used for roofs on 48-in. spans in many regions. It can be obtained with a tongue-and-groove pattern on its long panel edges, to provide good continuity of performance across panel joints, which is desirable under built-up roofs and vinyl overlayment. Combination subfloor and underlayment panels are described in the PDS [25.7].

Plywood is manufactured at about 8% moisture content. Design values in Table 25.8 are for use at moisture contents up to 16%. At 16% or higher the design values must be reduced using factors given in the PDS. Plywood is specified by grade and thickness, no moisture content specification is required.

25.6 STRUCTURAL GLUED LAMINATED TIMBER

Glued laminated timber (commonly called "glulam") was first produced in Europe and a large body of experience has been developed since its introduction into North America in 1935.

Laminating is a practical way of obtaining long beams of large depth, curved or cambered beams, and arches. Glued laminated members are used as the chords of bowstring trusses and as columns. Large trusses composed of straight members are often of laminated construction and wide glulam panels are used for bridge decks.

It is manufactured by bonding together thin wood laminae using waterproof and heat-resistant adhesives. The laminae are well-seasoned before assembly which greatly reduces shrinkage and improves dimensional stability. The large sections resist heat transfer and will not burn rapidly. Glulam beams may be treated with wood preservatives and fire retardant chemicals to improve their durability.

Quality standards are established by the AITC and the Canadian Institute of Timber Construction (CITC). In the United States, U.S. Department of Commerce Product Standard PS 56-73, *Structural Glued Laminated Timber* [25.32] defines materials, manufacturing practice, and quality requirements. Standard specifications are found in AITC 117 [25.1] for design values. Preservative treatment recommendations are found in AITC 109 [25.5].

Laminating requires a mastery of fabricating technology by the manufacturer. AITC or CITC quality marks should be required by purchasers. Selecting glued laminated timber can easily be done by using the standard combinations, such as indicated in *Glulam Systems* [25.2]. It is useful to indicate design loads and permissible deflections so the manufacturer can confirm the selection. It is not necessary to specify individual lamination quality or arrangement. Laminators have engineering staffs able to work cooperatively with designers.

Design values for common "combinations" are listed in Table 25.9. The values are for Normal load duration as defined in the section on structural lumber. They are for dry-use conditions, in environments where the moisture content will not exceed 16%. At higher moisture contents, (wet-use conditions) property modifications are given in the table.

The bending strength design values are for beams up to and including 12 in. in depth. Deeper beams require a size effect modification. Curved beams also may require a curvature modification. These adjustments of bending strength are described in Sections 25.8 and 25.14 for the design of beams and archs.

Table 25.9 is presented to illustrate typical design values for common use. All values in Table 25.9 are for flexural members with the neutral axis parallel to the glue lines. For bending about an axis perpendicular to the glue lines different design values are used. Combinations especially suited for use as columns or tension members are also made. Complete information is found in the *NDS Supplement* [25.36] and in *Glulam Systems* [25.2].

The term "grade" has a different meaning for glulam beams than for lumber. The term "combination" is analogous to "grade" of lumber. Each glulam combination is available in three different appearance *grades*. The design values for a combination are the same for each of the grades. The three appearance grades are as follows:

1. *Industrial Appearance Grade* is utilitarian. Narrow faces are unsurfaced and may have open knot holes and glue smears.
2. *Architectural Appearance Grade* has a good appearance and is the most commonly specified grade. Surfaces are smooth and clean and the larger knot holes are filled. Exposed edges are rounded (eased).
3. *Premium Appearance Grade* has all knot holes and voids filled, is smoothly surfaced and clean, with all edges eased. It is the superior appearance grade.

Glulam beams are not sold on a board foot basis. Actual size and length are the basis for quotation. Sizes are at the option of buyer and seller but the usual widths are 3⅛, 5⅛, 6¾, 8¾, and 10¾ in. The depth or wide dimension is in multiples of lamination thickness, either ¾ or 1½ in.

Table 25.9 Design Values* for Structural Glued-Laminated Timber, Normal Load Duration, Dry Use (16% Maximum Moisture Content)†

Combination‡ and Species	Bending F_b, psi	Compression Perpendicular-to-Grain: Tension Face $F_{c\perp}$, psi	Compression Perpendicular-to-Grain: Compression Face $F_{c\perp}$, psi	Horizontal Shear F_v, psi	Modulus of Elasticity E, 10^6 psi	Compression Parallel F_c, psi	Tension Parallel F_t, psi
16F,DF	1600	385	385	165	1.5	1550	950
16F,SP	1600	385	385	200	1.4	1550	1000
16F,HF	1600	385	245	155	1.4	1300	875
20F,DP	2000	450	385	165	1.6	1550	1000
20F,SP	2000	450	385	200	1.6	1550	1000
20F,HF	2000	385	245	155	1.5	1350	975
22F,DF	2200	450	385	165	1.7	1500	1050
22F,SP	2200	450	450	200	1.6	1550	1050
22F,HF	2200	385	385	155	1.5	1350	950
24F,DF	2400	450	450	165	1.8	1650	1150
24F,SP	2400	450	450	200	1.8	1700	1150
24F,HF	2400	385	385	155	1.5	1450	1000
Wet-use factors	0.8	0.667	0.667	0.875	0.833	0.73	0.80

* Adapted from Table 5A, *Design Values for Wood Construction* [25.36].

† For beams stressed principally in bending, load perpendicular to wide face of laminations, four or more laminations.

‡ There are many combinations with properties different and sometimes superior to those in this table. Three beams were chosen as typical available combinations. They will be for preliminary design work. The glulam producer will be able to provide better choices for a particular purpose. Combinations 22F and 24F are not available below 15-in. depth. DF = Douglas fir; HF = hem-fir; SP = southern pine.

Glulam members may be obtained paper wrapped for protection from soiling during shipment and construction. Some manufacturers offer custom finishing and nearly all will make specified end cuts, drill holes for connections, and design and furnish connection hardware.

25.7 RELIABILITY OF DESIGN VALUES

Factor of safety (F.S.) is often used as a measure of reliability. It is the ratio of ultimate strength to design value. The ultimate values of all materials are variable and the meaning of factor of safety depends on the probability of occurrence of the ultimate value used in computing the factor of safety.

Table 25.10 presents factors of safety for visually stress graded lumber which may be anticipated at three levels of probability. They were obtained using the following:

1. Clear wood average ultimate properties and their coefficients of variation from ASTM D2555 [25.16].
2. Lumber strength grade ratios and their coefficients of variation from research studies.
3. Load duration characteristics of wood.

The factors of safety given in the 50% column are the average factors of safety. Pieces may be expected to have these F.S. values at least half the time. Since the designer is also interested in F.S. values that will be expected more often than that, columns are presented for the values that would be equaled or exceeded 90 and 99% of the time. For example, in column 1, the factor of safety for bending strength will be at least 1.3, 99% of the time. This table allows the designer to understand the probability of failure at design load correlated with factor of safety.

The lower factors of safety for compression perpendicular-to-grain and for elastic modulus are justified for multiple-member systems in which members share the load in proportion to their elastic moduli. In such systems the stronger and stiffer members carry larger shares of the load.

For the other properties, the values of F.S. are larger and exceed 1.0 with a high probability.

Lumber design values were developed for building construction. When lumber is used in structures significantly different in their framing, or when the consequences of failure are more (or less) serious, the designer should make adjustments to provide the required reliability. Reliability based design procedures are now being considered for adoption by the engineering community for all materials of construction. The F.S. values in the table will serve as an interim basis for considering reliability.

Elastic modulus is the important property for long column design, with buckling as the failure mode. When long columns are not in close proximity to one another, substantial reductions in design elastic modulus should be considered, as much as 50% in the case of independent columns with high dead loads. It is also recommended that the curvature permitted by the grade (as much as ½ in.) be considered and the design be for axial load plus eccentricity. Ordinary stud walls in frame buildings have close column spacing and may be designed without reducing the published elastic modulus.

The F.S. for machine-graded lumber and glued laminated timber is somewhat larger than the Table 25.10 values, because these products are less variable in their strength and elastic properties.

No published information on plywood variability and F.S. is available. Because of the layered construction and the common origin with lumber for basic strength information, plywood factors of safety should equal those of lumber.

Table 25.10 Factors of Safety for Visually Stress Graded Lumber

Property	Factor of Safety with a Probability Tabulated Percentage		
	99%	90%	50%
Bending strength	1.3	1.9	2.8
Tension parallel-to-grain	1.3	1.9	2.8
Compression parallel-to-grain	1.3	1.8	2.6
Shear parallel-to-grain	1.1	1.6	2.6
Compression perpendicular-to-grain	0.4	0.5	1.1
Modulus of elasticity	0.5	0.75	1.0

25.8 BEAM DESIGN

Conventional elastic methods of determining the bending moment, shear, and reactions for loaded beams are used for design. Sizes may be determined on the basis of the design values in the *NDS Supplement*. These published values are sometimes modified to account for size effect, moisture content in use exceeding 19%, and load duration effects.

Joists and rafters may be selected from tables published by NFPA [25.37]. Tables 25.11 and 25.12 are typical span tables for floors and rafters for residential construction. These tables are for particular uniformly distributed loads and deflection limitations. Table 25.11 is for a 40-psf live load and 10-psf dead load with deflections limited to 1/360 of the span under live load. This deflection is considered suitable for floor dynamics in residential use. The dominant property is elastic modulus (E) in Table 25.11.

To use this table to select a suitable species and grade, one enters the table for the required clear span between supports, using the joist spacing for the sheathing to be used. This gives the required elastic modulus and allowable bending stress the rafters must have. Then use a properties table such as Table 25.3 to select a suitable species and grade. Table 25.3 is for Douglas fir-larch, but tables for other species are found in the *NDS Supplement* [25.36], or in the NFPA span table publication [25.37].

EXAMPLE 25.8.1

Choose a floor joist for a 14-ft span with 16-in. spacing between the joists.

Solution

Table 25.11 shows two possible choices:

1. 2×8 requiring $E \geq 2{,}200{,}000$ psi and $F_b \geq 1550$ psi.
2. 2×10, requiring $E \geq 1{,}000{,}000$ psi and $F_b \geq 920$ psi.

Table 25.11 Floor Joists*

40 psf Live Load (All Rooms Except Those Used for Sleeping Areas and Attic Floors)

DESIGN CRITERIA:
Deflection - For 40 lbs. per sq. ft. live load.
Limited to span in inches divided by 360.
Strength - Live load of 40 lbs. per sq. ft. plus dead load of 10 lbs. per sq. ft. determines the required fiber stress value.

JOIST SIZE (IN)	SPACING (IN)	Modulus of Elasticity, "E", in 1,000,000 psi													
		0.8	0.9	1.0	1.1	1.2	1.3	1.4	1.5	1.6	1.7	1.8	1.9	2.0	2.2
2x6	12.0	8-6 720	8-10 780	9-2 830	9-6 890	9-9 940	10-0 990	10-3 1040	10-6 1090	10-9 1140	10-11 1190	11-2 1230	11-4 1280	11-7 1320	11-11 1410
	16.0	7-9 790	8-0 860	8-4 920	8-7 980	8-10 1040	9-1 1090	9-4 1150	9-6 1200	9-9 1250	9-11 1310	10-2 1360	10-4 1410	10-6 1460	10-10 1550
	24.0	6-9 900	7-0 980	7-3 1050	7-6 1120	7-9 1190	7-11 1250	8-2 1310	8-4 1380	8-6 1440	8-8 1500	8-10 1550	9-0 1610	9-2 1670	9-6 1780
2x8	12.0	11-3 720	11-8 780	12-1 830	12-6 890	12-10 940	13-2 990	13-6 1040	13-10 1090	14-2 1140	14-5 1190	14-8 1230	15-0 1280	15-3 1320	15-9 1410
	16.0	10-2 790	10-7 850	11-0 920	11-4 980	11-8 1040	12-0 1090	12-3 1150	12-7 1200	12-10 1250	13-1 1310	13-4 1360	13-7 1410	13-10 1460	14-3 1550
	24.0	8-11 900	9-3 980	9-7 1050	9-11 1120	10-2 1190	10-6 1250	10-9 1310	11-0 1380	11-3 1440	11-5 1500	11-8 1550	11-11 1610	12-1 1670	12-6 1780
2x10	12.0	14-4 720	14-11 780	15-5 830	15-11 890	16-5 940	16-10 990	17-3 1040	17-8 1090	18-0 1140	18-5 1190	18-9 1230	19-1 1280	19-5 1320	20-1 1410
	16.0	13-0 790	13-6 850	14-0 920	14-6 980	14-11 1040	15-3 1090	15-8 1150	16-0 1200	16-5 1250	16-9 1310	17-0 1360	17-4 1410	17-8 1460	18-3 1550
	24.0	11-4 900	11-10 980	12-3 1050	12-8 1120	13-0 1190	13-4 1250	13-8 1310	14-0 1380	14-4 1440	14-7 1500	14-11 1550	15-2 1610	15-5 1670	15-11 1780
2x12	12.0	17-5 720	18-1 780	18-9 830	19-4 890	19-11 940	20-6 990	21-0 1040	21-6 1090	21-11 1140	22-5 1190	22-10 1230	23-3 1280	23-7 1320	24-5 1410
	16.0	15-10 790	16-5 860	17-0 920	17-7 980	18-1 1040	18-7 1090	19-1 1150	19-6 1200	19-11 1250	20-4 1310	20-9 1360	21-1 1410	21-6 1460	22-2 1550
	24.0	13-10 900	14-4 980	14-11 1050	15-4 1120	15-10 1190	16-3 1250	16-8 1310	17-0 1380	17-5 1440	17-9 1500	18-1 1550	18-5 1610	18-9 1670	19-4 1780

Note: The required extreme fiber stress in bending, "F_b", in pounds per square inch is shown below each span.

* From Ref. 25.37, used by permission.

Table 25.12 Low or High Slope Rafters*
30 psf Live Load (Supporting Drywall Ceiling)

DESIGN CRITERIA:
Strength - 15 lbs. per sq. ft. dead load plus 30 lbs. per sq. ft. live load determines required fiber stress.
Deflection - For 30 lbs. per sq. ft. live load. Limited to span in inches divided by 240.

RAFTERS: Spans are measured along the horizontal projection and loads are considered as applied on the horizontal projection.

RAFTER SIZE (IN)	SPACING (IN)	Allowable Extreme Fiber Stress in Bending, "F_b" (psi).														
		500	600	700	800	900	1000	1100	1200	1300	1400	1500	1600	1700	1800	1900
2x6	12.0	7-6 0.27	8-2 0.36	8-10 0.45	9-6 0.55	10-0 0.66	10-7 0.77	11-1 0.89	11-7 1.01	12-1 1.14	12-6 1.28	13-0 1.41	13-5 1.56	13-10 1.71	14-2 1.86	14-7 2.02
2x6	16.0	6-6 0.24	7-1 0.31	7-8 0.39	8-2 0.48	8-8 0.57	9-2 0.67	9-7 0.77	10-0 0.88	10-5 0.99	10-10 1.10	11-3 1.22	11-7 1.35	11-11 1.48	12-4 1.61	12-8 1.75
2x6	24.0	5-4 0.19	5-10 0.25	6-3 0.32	6-8 0.39	7-1 0.46	7-6 0.54	7-10 0.63	8-2 0.72	8-6 0.81	8-10 0.90	9-2 1.00	9-6 1.10	9-9 1.21	10-0 1.31	10-4 1.43
2x8	12.0	9-10 0.27	10-10 0.36	11-8 0.45	12-6 0.55	13-3 0.66	13-11 0.77	14-8 0.89	15-3 1.01	15-11 1.14	16-6 1.28	17-1 1.41	17-8 1.56	18-2 1.71	18-9 1.86	19-3 2.02
2x8	16.0	8-7 0.24	9-4 0.31	10-1 0.39	10-10 0.48	11-6 0.57	12-1 0.67	12-8 0.77	13-3 0.88	13-9 0.99	14-4 1.10	14-10 1.22	15-3 1.35	15-9 1.48	16-3 1.61	16-8 1.75
2x8	24.0	7-0 0.19	7-8 0.25	8-3 0.32	8-10 0.39	9-4 0.46	9-10 0.54	10-4 0.63	10-10 0.72	11-3 0.81	11-8 0.90	12-1 1.00	12-6 1.10	12-10 1.21	13-3 1.31	13-7 1.43
2x10	12.0	12-7 0.27	13-9 0.36	14-11 0.45	15-11 0.55	16-11 0.66	17-10 0.77	18-8 0.89	19-6 1.01	20-4 1.14	21-1 1.28	21-10 1.41	22-6 1.56	23-3 1.71	23-11 1.86	24-6 2.02
2x10	16.0	10-11 0.24	11-11 0.31	12-11 0.39	13-9 0.48	14-8 0.57	15-5 0.67	16-2 0.77	16-11 0.88	17-7 0.99	18-3 1.10	18-11 1.22	19-6 1.35	20-1 1.48	20-8 1.61	21-3 1.75
2x10	24.0	8-11 0.19	9-9 0.25	10-6 0.32	11-3 0.39	11-11 0.46	12-7 0.54	13-2 0.63	13-9 0.72	14-4 0.81	14-11 0.90	15-5 1.00	15-11 1.10	16-5 1.21	16-11 1.31	17-4 1.43
2x12	12.0	15-4 0.27	16-9 0.36	18-1 0.45	19-4 0.55	20-6 0.66	21-8 0.77	22-8 0.89	23-9 1.01	24-8 1.14	25-7 1.28	26-6 1.41	27-5 1.56	28-3 1.71	29-1 1.86	29-10 2.02
2x12	16.0	13-3 0.24	14-6 0.31	15-8 0.39	16-9 0.48	17-9 0.57	18-9 0.67	19-8 0.77	20-6 0.88	21-5 0.99	22-2 1.10	23-0 1.22	23-9 1.35	24-5 1.48	25-2 1.61	25-10 1.75
2x12	24.0	10-10 0.19	11-10 0.25	12-10 0.32	13-8 0.39	14-6 0.46	15-4 0.54	16-1 0.63	16-9 0.72	17-5 0.81	18-1 0.90	18-9 1.00	19-4 1.10	20-0 1.21	20-6 1.31	21-1 1.43

Note: The required modulus of elasticity, "E", in 1,000,000 pounds per square inch is shown below each span.

* From Ref. 25.37, used by permission.

Table 25.3 shows that a 2 × 10, No. 2 SJP Douglas fir-larch with E = 1,700,000 psi and F_b (repetitive) = 1450 psi is adequate. Reference to the *NDS Supplement* [25.36] shows other species in the same size could also be used (eastern hemlock, southern pine, western cedar, hem-fir, and many others).

The span tables [25.37] also are for joists for floor loads of 30 psf, deflection-to-span ratio of 1/360, and ceiling joists for 20 and 10 psf at 1/240 of span deflection.

Design shear strength is never exceeded by members chosen from the span tables.

Rafters

Table 25.12 for rafters is arranged differently because bending strength (F_b) is more often the critical property. The span in the table is the horizontal projection. Repetitive member values of F_b can be used if sheathing is ½ in. or thicker plywood as is usually the case.

The values of F_b from Table 25.3 may be increased 15% if maximum load is caused by snow of two-month cumulative duration during the life of the structure.

Table 25.12 is for 30-psf live load, 15-psf dead load, and a deflection-to-span ratio of 1/240 for live load. Rafter tables for other loadings are available.

EXAMPLE 25.8.2

Choose a rafter from Table 25.12 for a 14-ft horizontal projection span, with 16-in. spacing between rafters.

Solution

There are two possibilities shown in the table:

1. 2×8, requiring $F_b > 1400$ psi and $E > 1{,}100{,}000$ psi.
2. 2×10, requiring $F_b > 900$ psi and $E > 570{,}000$ psi.

Referring to a properties table such as Table 25.3, the following grades would be satisfactory:

2×8, No. 2 SJP Douglas fir-larch, F_b (repetitive) = 1450 psi and E = 1,700,000 psi

2×10, No. 3 SJP Douglas fir-larch, F_b (repetitive) = 850 × 1.15 = 978 psi if load is snow of two months cumulative duration, and E = 1,500,000 psi

Using other properties tables from the *NDS Supplement* [25.36], other species or sizes might be selected.

Shear design strength is not exceeded when the span tables are used.

For commercial, industrial, agricultural, and institutional buildings, other load arrangements and sizes will be of interest. Span tables are found in *Wood Structural Design Data* [25.38] and in *Western Woods Use Book* [25.46]. The member sizes can be calculated on the basis of load and lumber properties for bending strength, shear strength, and elastic modulus.

25.8.1 Size Effect

Bending strength is subject to a size effect. Deep beams have lower bending strength than shallow beams. The values given by NDS [25.36] for dimension lumber (2- to 4-in. nominal thickness) are for use when loaded on edge. When loaded on the wide face, as a plank, the design bending values may be increased using factors given in Table 25.13.

For decking, the published properties in bending assume use with the load applied to the wide face. When decking is loaded on its narrow face, the size effect requires a decrease in the values. The adjustments are given in Table 25.14.

For timbers over 4-in. nominal thickness and for glued laminated beams, the published bending values are for depths up to 12 in., under uniformly distributed load. For deeper members, design bending stress must be reduced by the factor

$$C_F = \left(\frac{12}{d}\right)^{1/9} \tag{25.4.1}$$

where d is the actual depth in inches.

Tension strength parallel-to-grain is also affected by size. The reductions from the basic values are given for dimension lumber in 8- and 10-in. nominal widths in Table 25.15.

Where these size effects are small, they may be ignored in preliminary design work but should be introduced in the final design of the structure. These numerous adjustments are inconvenient for the designer, but are indicative of the behavior of wood. Increase adjustments could be ignored, but decrease adjustments must be considered.

25.8.2 Moisture Content Effects

Wood, whether manufactured to seasoned or unseasoned size standards, usually experiences some further seasoning during and after construction. Also, without regard to moisture content at the time of manufacture, wood arrives at a common final size and moisture content in the structure. This in-place seasoning results in a gain in unit strength and elastic modulus and a loss in size and section properties which compensate. The published design values can be used in conjunction with the S-DRY sizes of dimension lumber for design of structures to be constructed of either seasoned or unseasoned lumber, if used at 19% or lower moisture content.

Table 25.13 Size Effect Adjustments for Bending Strength of Dimension Lumber Loaded on the Wide Face*

Nominal width, in.	Nominal Thickness		
	1-in.	3-in.	4-in.
4	1.10	1.04	1.0
6	1.15	1.09	1.05
8	1.19	1.13	1.08
10	1.22	1.16	1.11
12	1.25	1.18	1.14
14	1.28	1.20	1.16

* Based on *NDS Supplement* [25.36].

Table 25.14 Size Effect Adjustments for Bending Strength of Decking Loaded on Narrow Edge*

Nominal width, in.	Nominal Thickness		
	2-in.	3-in.	4-in.
4	0.93	0.98	1.00
6	0.87	0.93	0.96
8	0.85	0.90	0.93
10	0.82	0.87	0.90
12	0.80	0.85	0.88

* Based on *NDS Supplement* [25.36].

Table 25.15 Size Effect Adjustments for Tension Strength of Dimension Lumber*

Grade	Nominal Width	
	8-in.	10-in. and wider
Select Structural	0.90	0.80
No. 1, 2, and 3 includes dense grades	0.80	0.60

* Based on *NDS Supplement* [25.36].

For structures to be used at service moisture content above 19% the design strength and elastic modulus values are adjusted per Table 25.4. The section properties, however, are those given in Table 25.2.

In the case of large timbers (over 4-in. nominal thickness) which are not available seasoned, the section properties are based on the unseasoned size. These section properties are also listed in Table 25.2. Published design strength and elastic modulus values are for dry-use (19% or less in service) and should be adjusted per Table 25.4 for wet-use (over 19% in service).

S-DRY and S-GRN denote seasoning conditions as manufactured. Dry-use and wet-use denote equilibrium moisture content anticipated under service conditions of the structure.

25.8.3 Load Duration Adjustments

Wood has a higher strength for short duration load than for long duration load. The design stresses in Table 25.2 and the *NDS Supplement* [25.36] are for "Normal" load duration (10-year cumulative duration of load). If the duration of the design load is longer than 10 years, all design strength values are reduced by 10%. On the other hand, if design load duration is less than 10 years, the strength values may be increased. Load duration adjustments as given by NDS [25.35] are listed in Table 25.5.

Residential floors are considered to bear their design load for 10 years out of the structure's total life. In most snow load regions of the United States, the duration of maximum snow load is considered to be two months of the structure's total life. In high snow load regions, the designer should refer to local code practices, and if none exist should make inquiries to determine a suitable duration. It would not be unusual to use the 10-year factor in heavy snow load regions, to err on the safe side. Live loads in non-snow regions are sometimes considered to be seven days.

Maximum wind loads are caused by short duration gusts. Seismic loads are also shocks of short individual event durations. Cumulatively, they are usually less than a half-day of a structure's planned life. Wind and seismic load often cause lateral rather than vertical force on a structure. Wind on low slope roofs often produces uplift rather than downward force, as discussed in Chapter 2.

The designer must make judgments about the nature and probable duration of loads to arrive at reasonable beam designs. A few examples of the use of the load duration factor (LDF) will explain its use.

EXAMPLE 25.8.3

Determine the proper load duration factor (LDF) and design stress condition for a roof with an expected maximum snow load of 35 psf, maximum wind load of 15 psf (down) and dead load (permanent) of 15 psf.

Solution

Several combinations must be examined:

(a) Permanent load has a maximum value of 15 psf and the corresponding LDF is 0.9 from Table 25.5.

(b) Snow load which is maximum at 35 psf is accompanied by the permanent load. For this building, assume the cumulative duration of maximum snow load is two months, a common assumption in the northern states. This gives a two-month load of 50 psf and an LDF of 1.15 from Table 25.5.

(c) Wind load has a maximum value of 15 psf, and it is common to assume the cumulative duration of these high wind gusts is a half-day. In hurricane areas, a cumulative duration of a week might be more suitable, but the half-day is a good judgment (or code requirement). The LDF is 1.33 (Table 25.5). The maximum roof load for this duration is the ever present dead load plus the wind load plus any snow load that might be present. It is often assumed that half the snow load could be present at maximum wind. The load for this half-day cumulative duration is 47.5 psf.

(d) It might also be desirable to consider that wind loads of half the maximum intensity occur more frequently, perhaps a cumulative seven days (LDF = 1.25) accompanied by half the snow load and the dead load, for a seven-day loading of 40 psf.

One of these conditions will result in the largest beam, which must be specified. Since required beam section modulus is directly proportional to load and inversely proportional to *design* stress, the largest value of load/LDF determines the controlling design conditions.

(a) 15/0.9 = 16.66 (b) 50/1.15 = 43.5
(c) 47.5/1.33 = 35.7 (d) 40/1.25 = 32

The beam should be designed to carry 50 psf using 1.15 times the Normal design stress values.

EXAMPLE 25.8.4

Reexamine Example 25.8.3 if the wind load had been 30 psf.

Solution

Under (c) the total load would then be 62.5 psf and under condition (d) of half-wind plus half-snow plus dead load the value would be 47.5 psf. Dividing by the same LDF values previously used, it is found that a 62.5-psf load using 1.33 times Normal design stress would require the largest beam.

Structures with large dead loads and small live loads may be controlled by the dead load.

In preliminary work it may be convenient to use maximum loads and Normal design values, then refine the design after the general picture of structural requirements emerges.

25.8.4 Beam Design Calculation

If span tables are not available, the beam size must be determined by calculation. *Beam strength requirements* are based on maximum expected loads, dead plus live (snow, wind, suspended equipment, etc.). The structure is analyzed to determine the maximum bending moment M, the maximum shear V, and the maximum bearing load R. The required values of section modulus S, area of cross-section A, and bearing area A_b are computed using the elastic homogeneous beam stress relationships

$$S = \frac{M}{F_b} \tag{25.8.1}$$

$$A = \frac{3}{2}\frac{V}{F_v} \tag{25.8.2}$$

$$A_b = \frac{R}{F_{c\perp}} \tag{25.8.3}$$

where F_b, F_v, and $F_{c\perp}$ are the allowable bending, shear, and bearing stresses (i.e., design values).

The allowable stress values should be adjusted for load duration and wet-use if applicable. The allowable bending stress value should be adjusted for size effect, unless it is for dimension lumber loaded on edge.

The allowable shear stress values are for members which may be expected to have some end split due to seasoning. If no end split will be present, the published allowable shear stresses may be increased. Since seasoning split usually occurs only at the ends of pieces, shear stress adjacent to supports located far from the ends may be increased by a factor of 2.0, as indicated in Table 25.16 for no split. This applies usually to shear near supports of beams with overhangs and at intermediate supports of long continuous beams. If beam ends are trimmed off after seasoning, seasoning split may be very limited or even absent. In such cases, increased values of F_v per Table 25.16 are used. NDS suggests using $1.5F_v$ if the distance to the end is five times the member depth.

The shear force V near supports can be reduced by ignoring load on the member located within a distance from the support centerline equal to the beam depth. For moving loads, maximum shear occurs when the load is a distance from the support equal to beam depth.

For calculating bearing area requirements, the full reaction at the support is used. For moving loads, maximum reaction occurs when the load is over the support.

If bearing areas are of very short length, the allowable value of $F_{c\perp}$ (allowable compressive stress perpendicular to the grain) can be increased by the factor

$$\frac{L + 0.375}{L} \tag{25.8.4}$$

Table 25.16 Shear Stress Modification Factors*

Nominal 2-in. Lumber		Nominal 3-in. Lumber	
Length of Split on Wide Face	Factor	Length of Split on Wide Face	Factor
No split	2.00	No split	2.00
½ of wide face	1.67	½ of narrow face	1.67
¾ of wide face	1.50	1 × narrow face	1.33
Wide face	1.33	1½ × narrow face	1.00
1½ × wide face	1.00		

* From *NDS Supplement* [25.36].

where L is the length in inches of bearing surface parallel to the grain. This factor can be used if the bearing area is less than 6 in. long and is at least 3 in. from the end of the member. This increase is based on the fact that to deform a short bearing area it is necessary to deform material immediately adjacent to the bearing surface.

The required *moment of inertia I* for beams is determined using deflection criteria and deflection formulas for the beam load and support arrangement. Table 25.17 gives common deflection criteria. Deflection is likely to be a design limitation for long wood beams.

Floor deflection is limited as shown in Table 25.17. The beam must satisfy two criteria; the live load deflection and one of the other three floor deflection criteria.

The live load deflection limit is to assure the floor will not be unacceptably springy to the occupants and owners. Dead plus live load deflection limit is to preclude an objectionable appearance of sagging. For beams (such as glulam beams and trusses) which can be cambered upward to compensate for dead load creep, the dead plus live load deflection is the second limitation. Beams in temporary structures where creep is of no consequence also use this criterion.

For beams (such as lumber beams) which cannot be cambered, the second deflection criterion is 1.5 dead load plus live load for seasoned lumber, and 2.0 dead load plus live load for unseasoned lumber.

The $L/360$ live load criterion for floors is for spans up to 15 ft. On longer spans, the live load floor criterion is limited to ½ in. Some designers prefer a live load deflection criterion for floors of $L/360$ to 15-ft span, decreasing proportionately to $L/480$ at 24-ft span and not exceeding 0.6 in. at longer spans.

Roof deflection criteria are more liberal than for floors because they do not bear the weight of people and the sensation of spring is not important. Appearance is the main consideration. The criteria for high ceiling structures or those where economy of construction is more important than appearance (industrial and agricultural buildings) are less limiting than for structures under the close scrutiny of the general public (residential, commercial, and institutional buildings). Psychological reaction of occupants is of considerable importance in the choice of deflection criteria.

Roof beams that have attached ceilings of brittle materials like plaster or gypsum are designed according to floor deflection criteria for the practical purpose of avoiding cracks in the ceiling finish.

Table 25.17 Recommended Deflection Limits

	Live Load	Dead + Live Load	1.5 Dead + Live Load	2 Dead + Live Load
Floors	$L/360$	$L/240$	$L/240$	$L/240$
Roofs: Residential, commercial, and institutional buildings	$L/240$	$L/180$	$L/180$	$L/180$
Roofs: Industrial and agricultural buildings	$L/180$	$L/120$	$L/120$	$L/120$
		Cambered members	Seasoned lumber	Unseasoned lumber

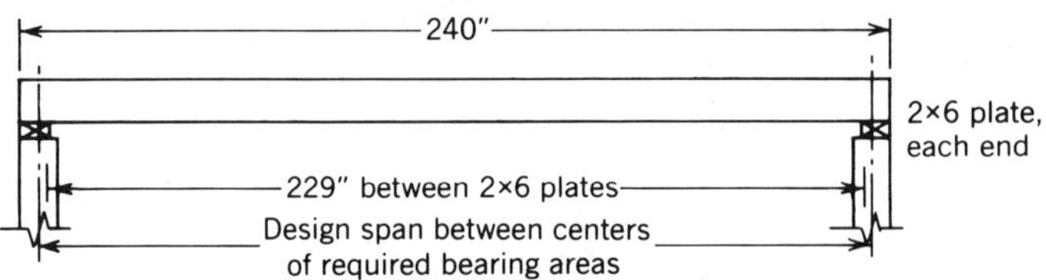

Fig. 25.1 Beam for design Example 25.8.5.

In prefabricated structures, the brittle ceiling finishes may be applied when the components are in assembly fixtures or supported so that dead load is not acting, or may be acting to produce negative deflection. The important consideration is to avoid flexing the brittle layer more than $L/360$.

Creep in wood beams or trusses is assumed to be 50% of dead load deflection for seasoned wood and 100% of dead load deflection for unseasoned wood that seasons after erection (see Table 25.17).

These criteria are basic to beam design. Examples of beam design calculations for a floor and a roof follow.

EXAMPLE 25.8.5

Given that 20-ft long beams rest on nominal 2×6 wall plates as shown in Fig. 25.1, determine the required beam size, grade, and species for a 40-psf live load, 10-psf dead load, and 16-in. beam spacing, using S-DRY lumber in dry-use conditions and "Normal" load duration.

Solution

(a) *Estimate the design span.* The design span is the distance between the centers of the required bearing areas. Since the required bearing areas are unknown at this time, assume bearing length equals the plate width of 5.5 in. and use an estimated design span of 234.5 in.

(b) *Determine the required section modulus S.*

$$S = \frac{wL^2}{8F_b} \qquad (25.8.5)$$

where w = total load, lb/in. of length = (50/12)(16/12) = 5.55
L = span, 234.5 in.
F_b = allowable bending stress for No. 2 SJP Douglas fir-larch (repetitive) = 1450 psi
S = 26.34 in.^{3} = $bh^2/6$, (b = thickness, h = width)

For b = 1.5 in., required h = 10.26 in.
Use a nominal 2×12, h = 11.25 in., S = 31.64 in.^{3}

(c) *Determine the required bearing area A_b.*

$$A_b = \frac{\text{reaction}}{F_{c\perp}} \qquad (25.8.6)$$

where $F_{c\perp}$ = 625 psi for No. 2 SJP Douglas fir-larch
reaction = $wL/2$, where L = total loaded length, 240 in.
= 666 lb
A_b = 1.07 sq in. = 1.5 in. wide by 0.71 in. long

The assumption that bearing length was 5.5 in. was conservative.

$$L = 240 - 11 + 0.71 = 229.71 \text{ in. between centers of bearing}$$

(d) *Determine the shear stress in the 2×12 due to total load.*

$$f_v = \frac{3V}{2A} \qquad (25.8.7)$$

where V = modified vertical shear = $w(L - 2h)/2$ = 575 lb
A = area of nominal 2×12 = 16.875 sq in.
f_v = 51 psi < 95 psi allowable shear for grade.

(e) *Determine deflections under live load, limited to L/360, and under live plus 1.5 dead load, limited to L/240.*

$$\Delta = \frac{5wL^4}{384EI} \tag{25.8.8}$$

where $E = 1.7 \times 10^6$ psi for No. 2 SJP Douglas fir-larch
I = 177.98 in.^{4} for nominal 2×12
L = 229.7 in.

$$w, \text{live} = (40/12)(16/12) = 4.44 \text{ lb/in.}$$

$$\Delta, \text{live} = 0.534 \text{ in.}; \qquad L/360 = 0.638 \text{ in.} > 0.534 \text{ in.}$$

$$\text{w, live plus 1.5 dead load} = (55/12)(16/12) = 6.11 \text{ lb/in.}$$

$$\Delta, \text{live plus dead load} = 0.732 \text{ in.}; \qquad L/240 = 0.957 \text{ in.} > 0.732 \text{ in.}$$

Use No. 2 SJP Douglas fir-larch 2×12 which meets all the criteria.

EXAMPLE 25.8.6

Design a roof beam for a commercial building with a nearly flat roof (pitched only for drainage). It is supported on 2×6 plates on walls as shown in Fig. 25.1, but has a 1.5-ft overhang on each end. The loads are 45 psf for snow of two-month cumulative duration, 20-psf dead load, and wind load is an uplift of less than 45 psf. Beam spacing is 24 in. Beam lumber is unseasoned, will season in place, and use conditions are dry.

Solution

(a) *Determine controlling load for design.*

$$\text{permanent load} = 20 \text{ psf}$$

$$\text{LDF for permanent load} = 0.9; \qquad 20/0.9 = 22.2 \text{ psf}$$

$$\text{two-month load} = 20 + 45 = 65 \text{ psf}$$

$$\text{LDF for two-month load} = 1.15; \qquad 65/1.15 = 56.5 \text{ psf}$$

The total load controls the design.

(b) *Determine required section modulus S.* The load on the overhangs reduces the bending moment. It can be safety ignored in this part of the solution. Snow could blow off the overhang, which is another reason to ignore the overhang load. Use a design span of 234.5 in. as in the previous example. Assume a commonly available species and grade such as No. 2 SJP Douglas fir-larch, for which F_b = 1450 psi repetitive. No moisture content adjustment is necessary for dry-use. A load duration adjustment of 1.15 can be made so F_b = 1668 psi.

$$w = (65/12)(24/12) = 10.83 \text{ lb/in.}$$

$$L = \text{span, } 234.5 \text{ in.}$$

$$S = 44.64 \text{ in.}^3 \text{ [Eq. (25.8.5)]}$$

S for a 3×12 is 52.73 in.^{3} (Table 25.2).

(c) *Determine the required bearing area A_b.*

$$\text{reaction} = wL/2, \text{ where } L = \text{total loaded length} = 276 \text{ in.}$$

$$= 1495 \text{ lb}$$

$$F_{c\perp} = 625(1.15) = 719 \text{ psi}$$

$$A = 1495/719 = 2.07 \text{ sq in. [Eq. (25.8.6)]}$$

$$\text{bearing length} = 2.07/2.5 = 0.83 \text{ in.}$$

The assumption about bearing area-was conservative. The actual design span is 240 − 11 + 0.83 = 229.8 in. Using the new value would not change the choice of a 3×12, No. 2 SJP Douglas fir-larch member.

(d) *Determine the shear stress in the 3×12 due to the total load.*

$$\text{maximum } V = w(L - 2h)/2 = 1123 \text{ lb}$$

$$\text{area of } 3 \times 12 = 28.13 \text{ sq in.}$$

$$f_v = 3V/2A = 60 \text{ psi} < [\text{allowable design shear stress of } 95(1.15) = 109 \text{ psi}]$$

Note that, because of the overhanging ends, there will be no end split at the point of maximum shear and the allowable shear stress could be doubled.

(e) *Determine if beam satisfies the deflection criteria.*

Deflection criteria from Table 25.17 are $L/240$ at 45 psf and $L/180$ at 45 + 2(20) = 85 psf.

$$\Delta = \frac{5wL^4}{384EI} \qquad [25.8.8]$$

where $E = 1.7 \times 10^6$ psi for No. 2 SJP Douglas fir-larch
$I = 296.63$ in.^{4} for 3×12
$L = 229.8$ in. (overhang is ignored)

$$w, \text{ live} = (45/12)(24/12) = 7.5 \text{ lb/in.}$$

$$\Delta, \text{ live} = 0.54 \text{ in.} < L/240 = 0.95 \text{ in.}$$

$$w, \text{ live} + 2.0 \text{ dead load} = (85/12)(24/12) = 14.17 \text{ lb/in.}$$

$$\Delta, \text{ live} + 2.0 \text{ dead load} = 1.02 \text{ in.} < L/180 = 1.28 \text{ in.}$$

Use No. 2 SJP Douglas fir-larch 3×12, which meets all criteria. It should be checked for possible ponding failure, as in the following discussion.

In each of the foregoing examples the features checked were: bending strength, bearing strength, shear strength, and deflection considering creep. It was assumed that the beams support decking at least as stiff as ½-in. plywood and/or are connected by bridging to justify using repetitive values of bending strength.

Ponding

Ponding occurs when the deflection of the roof due to accumulated water increases faster than the water level rises. It occurs most commonly in regions where the design live load is small and the required beam size tends to be small. Such structures will continue to accumulate rain until they fail, if a heavy rainfall occurs. Flat roof buildings should be investigated for possible ponding failure.

It is recommended policy to avoid flat roofs or to provide a drainage pitch equal to ¼ in./ft of horizontal distance from high point to drain. Even so, a ponding calculation is desirable. The recommended practice [25.4] is to compute the deflection of the beam for a 5-psf uniformly distributed load. If that load causes a deflection of ½ in. or more, ponding is a distinct possibility.

Another method of providing for ponding resistance is to design the beams for a magnified bending moment according to the following equation

$$C_p = \frac{1}{1 - (W'L^3/\pi^4 EI)} \tag{25.8.9}$$

where W' is the load in pounds caused by a 5.2-psf uniformly distributed load on the area contiguous to the member and other terms are in consistent units. The beam section modulus must then be increased above that required for design total load by this factor.

EXAMPLE 25.8.7

Check whether ponding may be a problem for the beam of Example 25.8.6.

Solution

(a) Determine the deflection for a 5-psf uniformly distributed load.

$$\Delta = 5wL^4/384EI = 0.06 \text{ in.}$$

This suggests no ponding problem.

(b) Another method is to compute the bending moment magnification factor using Eq. (25.8.9),

$$W' = (229.8/12)(2)(5.2) = 199.2 \text{ lb}$$

$$C_p = 1.052$$

This 5.2% increase in bending stress due to ponding would require a section modulus of 1.052(44.64) = 46.96 in.³. This is less than the section modulus of the 3×12 so the load is not excessive.

The overhanging ends will increase the pond weight by raising the lip of the pond. An exact computation of the overhang effect is complex but a conservative estimate can be obtained by using L = total beam length (276 in.). This gives $C_p = 1.094$ and a required S of 48.8 in.³. This is also less than the section modulus of a 3×12.

Long beams designed for low live loads and large allowable deflections (industrial building deflection criteria) are more likely to present ponding problems. Reference 25.3 is an excellent treatise on roof slope and drainage.

Lateral Support

Beams may require lateral support of the compression edges to prevent instability and buckling. If the compression side is fastened to a laterally rigid deck the beam will be adequately supported. If not, lateral support should be provided by bracing. This may be especially important during erection when the deck is not yet in place and loads of construction materials are on the beams. Thin, deep beams are particularly susceptible to this buckling condition.

According to NDS-3.3.2 [25.35] to preclude buckling the slenderness factor C_s of the beam should not exceed 10. For a single span beam with uniformly distributed load,

$$C_s = \sqrt{\frac{(1.63L + 3h)h}{b^2}} \tag{25.8.10}$$

where L = distance between lateral support points, in.
h = beam depth, in.
b = beam width, in.

If C_s is greater than 10 but less than C_k the bending stress in the beam must be limited to F_b',

$$F_b' = F_b\left[1 - \left(\frac{1}{3}\right)\left(\frac{C_s}{C_k}\right)^4\right] \tag{25.8.11}$$

$$C_k = 0.811\sqrt{\frac{E}{F_b}} \tag{25.8.12}$$

If C_s exceeds C_k the reduced value of beam bending stress is

$$F_b' = 0.438\,\frac{E}{C_s^2} \tag{25.8.13}$$

EXAMPLE 25.8.8

Check the lateral support requirements for the beam in Example 25.8.6.

Solution

(a) The minimum lateral support spacing if the No. 2 SJP Douglas fir-larch 3×12 beam is to be loaded to $F_b = 1668$ psi is obtained by solving for L in Eq. (25.8.10), using $C_s = 10$. L is found to be 13.4 in. A continuous, laterally rigid deck or well-anchored bridging at 13-in. intervals is required.

(b) What is the maximum allowable bending stress if this beam is supported laterally only at its reaction points?

$$C_s = \sqrt{\frac{[1.63(229.8) + 3(11.25)]11.25}{(2.5)^2}} = 27.1 > 10$$

$$C_k = 0.811\sqrt{\frac{1{,}700{,}000}{1668}} = 25.9 < 27.1$$

$$F_b' = \frac{0.438E}{C_s^2} = \frac{0.438(1{,}700{,}000)}{(27.1)^2} = 1010 \text{ psi}$$

The beam bending stress at design load is

$$f_b = \frac{wL^2}{8S} = \frac{10.83(229.8)^2}{8(52.73)} = 1360 \text{ psi}$$

The beam requires lateral support.

(c) Will lateral support at midspan be sufficient? With lateral bracing at ends and center, L = 120 in. and $C_s = 20.3$ which is between 10 and C_k. The maximum allowable bending stress is

$$F_b' = 1668\left[1 - 0.33\left(\frac{20.3}{25.9}\right)^4\right] = 1460 \text{ psi}$$

The actual stress at full load was found to be less than 1460 so bridging at the ends and midspan would be sufficient.

25.8.5 Continuous and Cantilevered Beam Systems

Designing for clear spans between exterior walls without any interior columns is generally desirable in warehouses and industrial or large mercantile buildings and in places of public assembly. However, it is more costly than designing with a system of interior support columns. When intermediate supports are provided, beam sizes are more economical. In floor design, two-span continuous beams are designed for dead load plus live load on one span and dead load plus 50% of live load on the other span.

Beams that are supported by numerous columns to form bays can be arranged to advantage by using a combination of cantilevered members and simple span members as shown in Fig. 25.2 and Table 25.18. By designing for hinged connections at points where zero moment would occur in a continuous beam, a close approximation of long continuous beam performance can be obtained with shorter beams. This is useful in utilizing solid sawn beams of available length, and for glulam timber where length is limited by shipping restrictions.

Table 25.18 provides useful information on moments, shears, reactions, and deflections for typical cantilever beam arrangements.

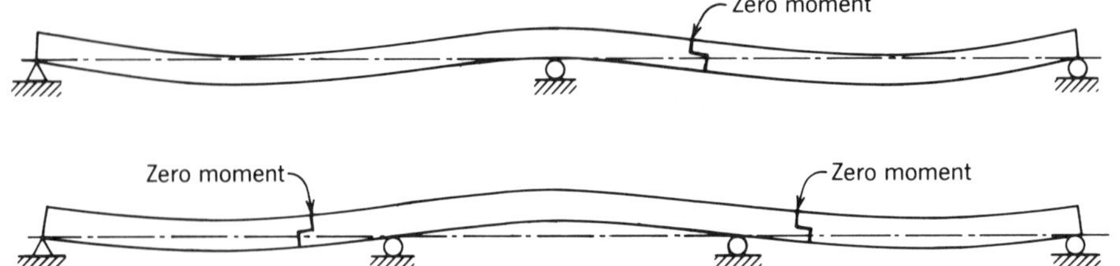

Fig. 25.2 Continuous cantilevered beams (statically determinate).

Table 25.18 Cantilever Beam Coefficients*
(All Spans Equal, Uniformly Distributed Load)

Moment = $M = CwL^2$ Reaction = $R = CwL$

Shear = $V = CwL$ Deflection = $\Delta = \dfrac{CwL^4}{48EI}$

w = loads, lb/ft L = span, ft

Coefficients C		Points							
		A	B	C	D	E	F	G	H
		2 spans A B C D E F Hinge L 0.172L							
	M V R Δ	0.414 0.414	+0.086 −0.370	−0.086 0.586 0.586 1.172	0.414 +0.064	+0.086 −0.306	0.414 0.414		
		3 spans A B C D E Hinge L 0.221L ℄ Sym							
	M V R Δ	0.414 0.414	+0.086 −0.370	−0.086 0.586 0.50 1.086	0.280	+0.039 −0.064			
		4 spans A B C D E F Hinge L 0.2L ℄ Sym L							
	M V R Δ	0.414 0.414	+0.0882 −0.375	−0.080 0.583 0.50 1.08	0.30 +0.085	+0.045 −0.150	+0.080 0.50 0.50 1.0		

Table 25.18 (*Continued*)

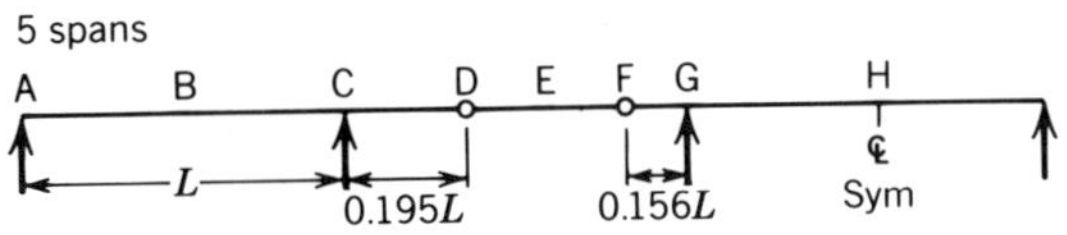

Coefficients C		Points							
		A	B	C	D	E	F	G	H
		5 spans A B C D E F G H *L* 0.195*L* 0.156*L* ℄ Sym							
	M *V* *R* Δ	0.414 0.414	+0.086 −0.375	−0.083 0.583 0.521 1.104	0.329 +0.082	+0.0525 −0.117	0.329 +0.054	−0.0625 0.479 0.50 0.979	+0.0625 −0.250
		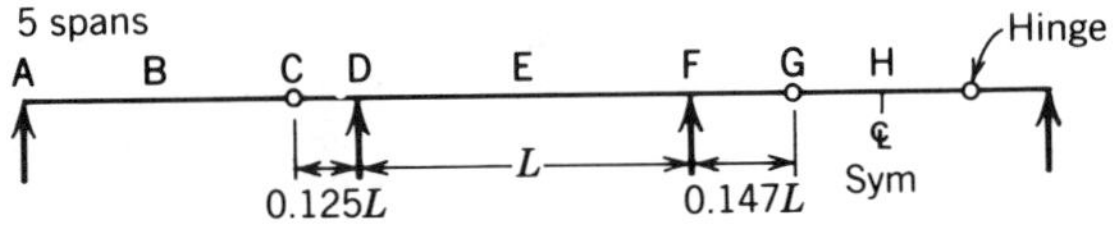 5 spans A B C D E F G H Hinge *L* 0.125*L* 0.147*L* ℄ Sym							
	M *V* *R* Δ	0.4375 0.4375	+0.095 −0.368	0.4375 +0.057	−0.0625 0.5625 0.50 1.0625	+0.0625 −0.250	−0.0625 0.50 0.50 1.0	0.353 +0.055	+0.0625 −0.156
		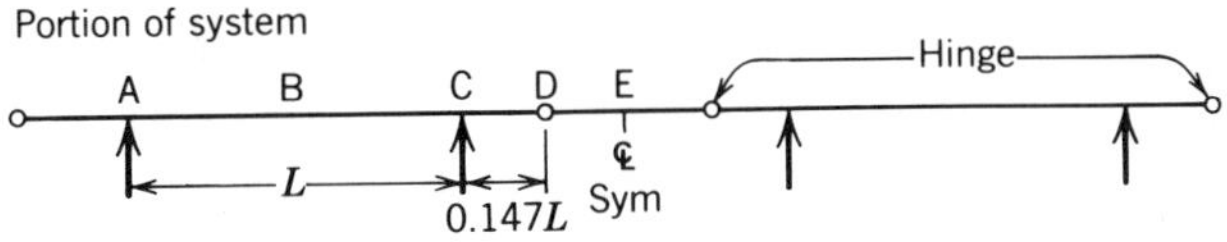 Portion of system A B C D E Hinge *L* 0.147*L* ℄ Sym							
	M *V* *R* Δ	+0.0625	+0.0625 −0.25	−0.0625 0.50 0.50 1.0	0.353 +0.053	+0.0625 −0.156			

* From *Timber Construction Manual* [25.4] used by permission of AITC.

When these cantilever beam arrangements are used it is essential that connections be secure both vertically and longitudinally. The vertical integrity of these designs is usually provided, but the need for longitudinal connection is sometimes overlooked. In seismic and wind load situations the horizontal connections carry load.

25.8.6 Glulam Beams

Selecting a size and strength rating (combination) of glulam beam is similar to selecting dimension lumber or timber. The combinations and their design values have been explained in Section 25.6. The design bending stress values must be reduced by the size factor when the beam depth exceeds 12 in. The load duration factors apply to all design values except elastic modulus.

For glulam timber, the dry-use design values are for service moisture content up to 16%. They are reduced for wet-use conditions, 16% and higher moisture content. Reduction factors are given in the design value tables.

Glulam beams are often cambered to provide for dead load and creep deflection. This camber is usually 1.5 times the dead load deflection. Additional camber may be provided for a drainage pitch

Table 25.19 Size Factor Adjustment*

Span-to-Depth Ratio L/d	Bending Strength Adjustment		
	Uniform Load	Third Point Load	Center Point Load
7	$1.062C_F$	$1.028C_F$	$1.145C_F$
14	$1.023C_F$	$0.990C_F$	$1.103C_F$
21	$1.0C_F$	$0.968C_F$	$1.078C_F$
28	$0.984C_F$	$0.953C_F$	$1.061C_F$
35	$0.972C_F$	$0.941C_F$	$1.043C_F$

$C_F = \left(\frac{12}{d}\right)^{1/9}$ with d = depth in in.

* From *Timber Construction Manual* [25.4].

averaging ¼ in./ft from high point to drain. Camber is usually specified in terms of the rise of the curves from the beam ends, that is, so many inches of camber.

Beams curved in the plane of bending may require a reduction in bending stress, additional to the size factor, depending on the curvature radius. The curvature factor is

$$C_c = 1 - 2000\left(\frac{t}{R}\right)^2 \tag{25.8.14}$$

where t = lamination thickness (usually 1.5 in. or 0.75 in.) and R is the curvature radius, in inches. Beams cambered for dead load creep and drainage almost never have C_c less than 0.99 so the effect can be ignored. But for arches it may become significant.

Curved beams subject to flexure experience radial tension or compression stress. In ordinary cambered beams, these stresses are too small to be significant, but they need to be considered in arch and frame design, as treated in Section 25.14 on arches and domes.

The deflection limitation for glulam beams can be based on the live load deflection criteria recommended for dimension lumber and timber, without regard to dead load deflection, if camber is specified.

The size factor must also be considered for glulam beams,

$$C_F = \left(\frac{12}{d}\right)^{1/9} \tag{25.4.1}$$

is for uniformly loaded simple span beams with a span-to-depth ratio of 21. For other span-to-depth ratios the value of C_F must be modified as in Table 25.19.

EXAMPLE 25.8.9

Design a glulam beam for a school gymnasium with a clear span of 80 ft, a length of 84 ft, and a live load of 50 psf. Dead load will be 15 psf not including the beam weight. Spacing between beams will be 14 ft. The roof will be flat but suitably pitched for good drainage. The species of wood will be southern pine, and use conditions are dry. Maximum load is due to snow of estimated 2-month maximum cumulative duration.

Solution

(a) *Preliminary decisions*. At the outset several facts are not known and must be approximated. These are (1) the size factor, (2) the beam span, and (3) the beam weight. A long span beam on such wide spacing will probably be much deeper than 12 in. so a size factor of 0.86 for a 48-in. depth would be reasonable. The span is the clear distance of 80 ft plus the distance to the center of each bearing area, which might be estimated at 81 ft, as shown in Fig. 25.3.

The beam weight is estimated as the weight of an 8¾ × 48 in. beam at 35 pcf = 102 lb/ft = 8.51 lb/in.

It is also necessary to choose a "combination" and usually the most available one is used, in this case, the 2200-psi bending strength combination 22F, SP from Table 25.9.

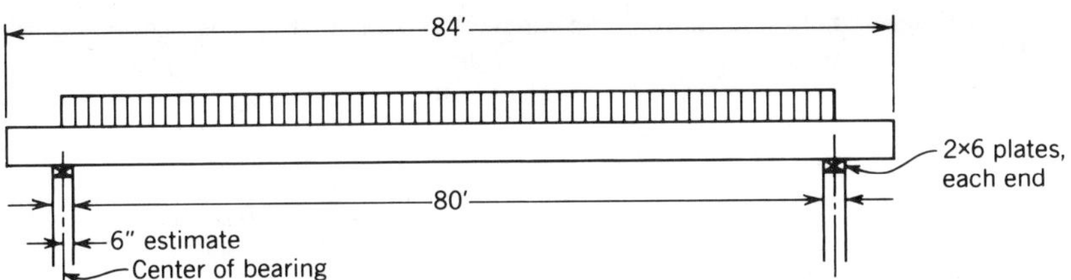

Fig. 25.3 Glulam beam of Example 25.8.9.

(b) *Determine required section modulus.* The negative moment caused by the loaded overhang is ignored,

$$S = \frac{M}{F_b} = \frac{wL^2}{8F_b}$$

$$w = \frac{65(14)}{12} + 8.51 = 84.34 \text{ lb/in.}$$

$$F_b = C_F \text{ (LDF)(design value)} = 0.86\ (1.15)2200 = 2176 \text{ psi}$$

$$S = \frac{84.34(972)^2}{8(2176)} = 4577 \text{ in.}^3$$

(c) *Determine required area for shear.*

$$A = \frac{3V}{2F_v}$$

$$V = \frac{w(L - 2h)}{2} = \frac{84.34(972 - 96)}{2} = 36{,}940 \text{ lb}$$

$$F_v = \text{LDF(Design value)} = 1.15(200) = 230 \text{ psi}$$

$$A = \frac{3(36{,}940)}{2(230)} = 240.9 \text{ sq in.}$$

(d) *Determine required bearing area.* Base the value on reaction under maximum load.

$$A_b = \frac{R}{F_{c\perp}}$$

$$R = \frac{84(12)(84.34)}{2} = 42{,}510 \text{ lb}$$

$$F_{c\perp} = \text{(LDF)(design value)} = 1.15(450) = 518 \text{ psi}$$

$$A_b = \frac{42{,}510}{518} = 82 \text{ sq in.}$$

(e) *Select the member.* A table of section properties for standard sizes of glulam beams is convenient at this point.

An 8¾ × 57 in. beam has $S = 4738$ in.³, $A = 499$ sq in., and $C_F = 0.84$. The weight is 10.1 lb/in. For such a beam the length of the bearing area would be 82/8.75 = 9.37 in. = 0.78 ft. The design span can be 80.78 ft = 969.4 in.

A quick review will now show if the estimates were close enough. Using new values

$$S = \frac{85.93(969.4)^2}{8(2125)} = 4370 \text{ in.}^3$$

$$A = \frac{3(85.93)(969.4 - 114)}{2(2)230} = 240 \text{ sq in.}$$

$$A_b = \frac{84(12)85.93}{2(518)} = 83.6 \text{ sq in.}$$

$$\text{length of bearing area} = \frac{83.6}{8.75} = 9.55 \text{ in.}$$

This is close to previous trial results. A beam 8.75 × 55.5 in. would have the needed section properties for strength.

(f) *Check deflection criteria.* The beam deflection must not exceed $L/240$ at a live load of $(50 \times 14)/12 = 58.33$ lb/in. The required moment of inertia is

$$I = \frac{5wL^3(240)}{384E} = \frac{5(58.33)(969.4)^3(240)}{384(1{,}600{,}000)} = 103{,}800 \text{ in.}^4$$

$$I = 124{,}700 \text{ in.}^4 \text{ for an } 8.75 \times 55.5\text{-in. beam}$$

$$\text{required camber} = 1.5 \text{ dead load deflection} + \text{drainage camber}$$

$$\text{dead load deflection} = \frac{5}{384}\,\frac{(27.6)(969.4)^4}{(1{,}600{,}000)(124{,}700)} = 1.59 \text{ in.}$$

$$\text{drainage camber} = \frac{84}{2}\left(\frac{1}{4}\right) = 10.5 \text{ in.}$$

$$\text{camber} = 1.5(1.59) + 10.5 = 12.9 \simeq 13 \text{ in.}$$

Use 22F-V1 southern pine glulam beam 8.75 in. × 55.5 in. × 84 ft long with 13-in. camber, Architectural Appearance Grade.

Usually the designer has no preference about the species of wood, except insofar as it provides the required performance. In that case, the beam specifications should list the minimum property requirements (F_b, F_c, F_v, $F_{c\perp}$, E) and the laminator will use whatever species are suitable to serve the purpose.

25.8.7 I-Beams

I-beams are manufactured by a number of fabricators who publish span-load tables for use by the designer. These beams have flanges of dimension lumber and webs of plywood. They can be obtained in greater lengths than lumber and serve where lumber of nominal 12-in. or greater depth would be required. I-beams are straight and well seasoned and they are lighter in weight for their load capacity than solid beams. They are 100% usable and can be obtained to the exact length needed for the job.

APA [25.9] publishes a design procedure for I-beams but usually it is more convenient to use the standard product of the fabricator.

25.8.8 Stressed Skin Panels

Roof or floor panels of plywood, with stringers glued to their lower faces, are structurally efficient and usually economical products which provide more load capacity than nailed plywood on joists. They are not available everywhere but in some regions there are specialty fabricators prepared to supply these products.

Stressed skin panels function as T-beams with the plywood contributing to the structural performance much more effectively due to the glue bond between plywood and stringers. They can also be obtained with skins above and below the stringers in which case they function as I-beams.

These types of panelized decking materially reduce the lumber requirements of a floor or roof deck. APA publishes a design procedure [25.10] for stressed skin panels.

25.8.9 Panel-on-Joist Gluing

By gluing floor or roof panels to joists using an elastomeric construction adhesive of an appropriate type, the stiffness of the system can be increased and the deflection reduced. Where the deflection is the limiting design feature, these panel-on-joist systems often permit using a 2×8 instead of a 2×10, or a 2×10 instead of a 2×12. However, if bending stress is the limiting design feature such a size reduction would not be possible unless all panel joints perpendicular to the joists are structurally spliced. Many designs do have a reserve of strength, especially floor systems where deflection is the controlling parameter.

Attachment of these panels to the joists requires an adhesive with a shear modulus G of at least 60 psi in the cured state, and the glue line must have an average thickness of 1⁄32 in. [25.27].

APA publishes information [25.14] on span–load capacities of panel-on-joist glued systems and design methods are published in *Glued Floor Systems* [25.14].

Panel-on-joist systems possess another advantage which may be more important than material saving. The adhesive virtually eliminates floor squeaks. Many builders use adhesives for this purpose exclusively and make no effort to reduce joist sizes beyond those required in ordinary nailed construction. Gluing plus nails at 6- to 12-in. intervals usually does an excellent job on floors.

25.9 COLUMN DESIGN

In building design wood columns are not usually slender. To provide sufficient end surface area for bearing against the side grain of supported wood beams the parallel-to-grain column stress is limited to the bearing stress. Steel bearing plates larger than the column cross section can be used to enlarge the bearing surface and permit more efficient column design. Columns comprise a small part of the total materials cost and can be generously designed.

Design formulas are first presented for the more common rectangular cross-sections. Round column formulas are described in a separate section.

25.9.1 Rectangular Column Formulas

Wood columns are often designed as pin-end members. The degree of end fixity is difficult to evaluate. An assumption of pin-ends, which are free to rotate, is usually conservative.

NDS [25.35] Sec. 3.7.3 describes three classes of wood columns, according to slenderness ratio L/d (length L to least thickness d).

Short columns with slenderness ratios not exceeding 11 are not subject to buckling. The allowable stress for short columns is the parallel-to-grain compression design value F_c modified for load duration and moisture content,

$$\frac{P}{A} < F_c \tag{25.9.1}$$

Long columns for which the failure mode is buckling, depend on the modulus of elasticity and the slenderness ratio for their strength. Parallel-to-grain compression strength is not a factor, except as it affects the lower limit of the long column slenderness ratio range. This limit is defined as

$$K = 0.671\sqrt{\frac{E}{F_c}} \tag{25.9.2}$$

E should be modified for moisture content in use and F_c should be modified for moisture content and load duration.

The long column formula is the well-known Euler formula with a factor of safety of 2.74. Thus the long column allowable stress is

$$F'_c = \frac{\pi^2 EI}{L^2 A}\left(\frac{1}{2.74}\right) = \frac{0.3E}{(L/d)^2} \tag{25.9.3}$$

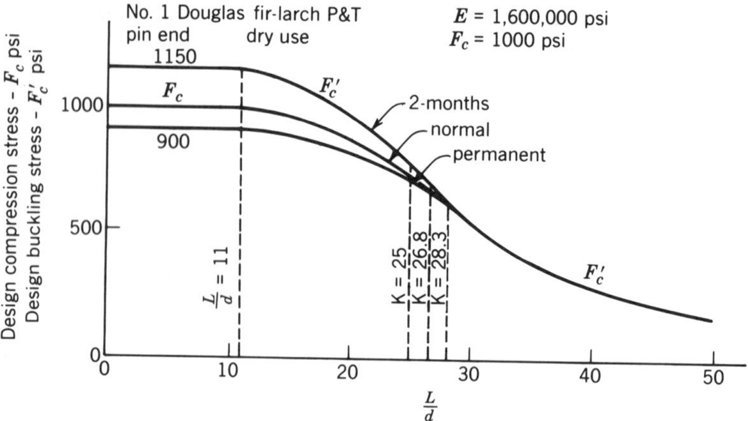

Fig. 25.4 Influence of L/d on column design stress.

The slenderness ratio range for long columns is

$$K \leq \frac{L}{d} \leq 50 \tag{25.9.4}$$

The long column design load is

$$P \leq F_c' A \tag{25.9.5}$$

Intermediate columns have slenderness ratios between 11 and K. Their allowable stress is a function of parallel-to-grain compressive stress and elastic modulus. Their allowable stress (design value) is given by the formula

$$F_c' = F_c \left[1 - \frac{1}{3} \left(\frac{L/d}{K} \right)^4 \right] \tag{25.9.6}$$

K is calculated using Eq. (25.9.2). F_c is modified for load duration and moisture content in both Eqs. (25.9.2) and (25.9.6). E is modified for moisture content but not for load duration.

Figure 25.4 shows the variation of the allowable (design) values through the three ranges of slenderness ratios and their interrelationship to that ratio for a particular grade of lumber.

25.9.2 End Fixity

If the column ends are fixed against rotation and lateral displacement, the elastic curve for the column develops points of inflection (zero moment) as it becomes unstable. The distance between inflection points is called the effective length L_e as shown in Fig. 25.5. By using L_e in place of L in the column formulas of Section 25.9.1 they will apply for the fixed-end condition.

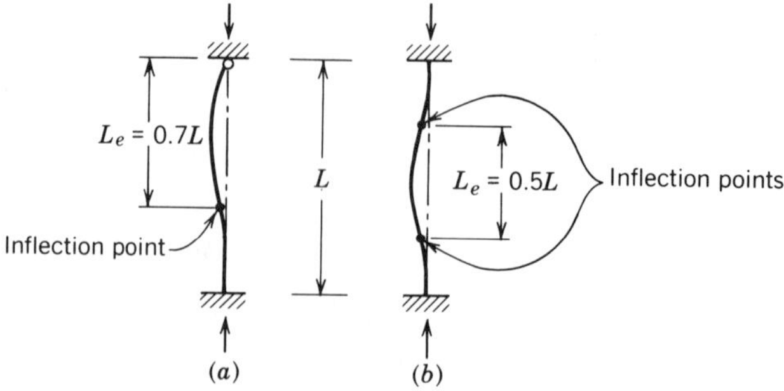

Fig. 25.5 Column effective length. (*a*) Top pin end but laterally fixed; bottom fixed. (*b*) Top and bottom fixed against lateral movement and rotation.

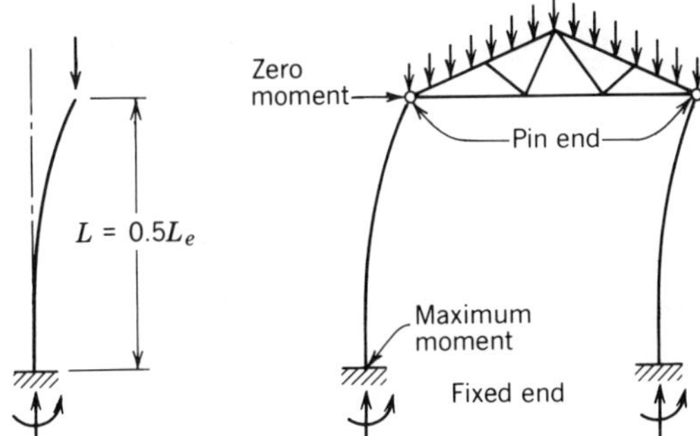

Fig. 25.6 Vertical load on cantilever column or pole.

If one end is fixed and the other is pin-ended but free to displace laterally, as in the case of a cantilever column (Fig. 25.6) carrying top vertical load, the effective length is twice the actual length.

Often columns will have fixed ends supporting a deck without lateral restraint of the deck, as in Fig. 25.7. In that case the deck could shift laterally. The effective length is then $L_e = L$, because the distance shown as $0.5L_e$ has a deflected shape identical to the portion from midheight to the end of a pin-end column.

For a column with lateral restraint at midheight as well as top and bottom and with pin ends, $L_e = L/2$. For a column with fixed ends and lateral restraint at midheight $L_e = 0.35L$ (Fig. 25.8). When different inflection point spacings occur in the same column, the largest one is used as L_e.

There are many possible values of L_e and the designer must assess each specific problem. Also, L_e may differ about the two axes of potential buckling action. Generally a simple pin-end design will be conservative with the exception of the condition in Fig. 25.6 where $L_e = 2L$.

Complete end fixity is rare, occurring mainly when the column is connected to a large footing or set in compacted soil or concrete (in which case the wood must be preservatively treated). Tops connected to trusses or beams are restrained and would be considered partially fixed. In such cases L_e is longer than for full fixity. Some designers increase L_e by 30% for partial fixity. Usually it is better to make a generous estimate than to carry out an elaborate analysis.

25.9.3 Elastic Modulus Variability

The design value of E as usually published is an average value. For individual pieces the E-value varies considerably as shown in Fig. 25.9, a frequency distribution There is a 1% probability that E will be 0.74(E-average) for MSR lumber and glulam, and 0.42(E-average) for dimension lumber and timbers. There is a 5% probability of E = 0.82(E-average) for MSR lumber and glulam, and 0.59(E-average) for dimension lumber and timbers. The long column formula (25.9.3) contains a factor of safety of 2.74 on the average E. This factor of safety will be diminished if the E of the piece is below the published value. For columns in critical structural service, the designer should consider reducing the published value of E for design. A 50% reduction would not be excessive.

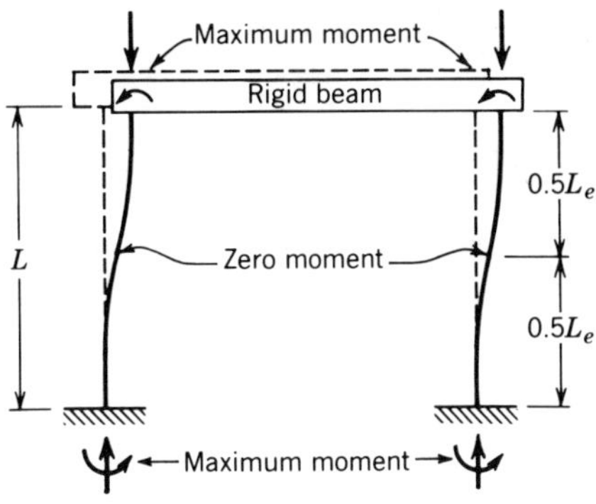

Fig. 25.7 Columns with both ends fixed—top not laterally fixed.

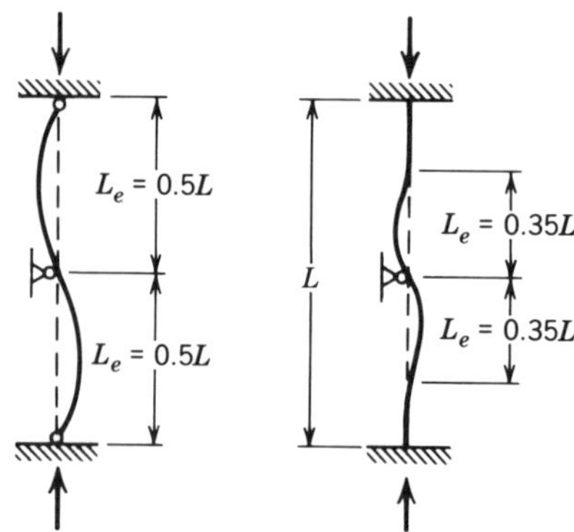

Fig. 25.8 Pin-end and fixed-end columns with lateral support at midheight.

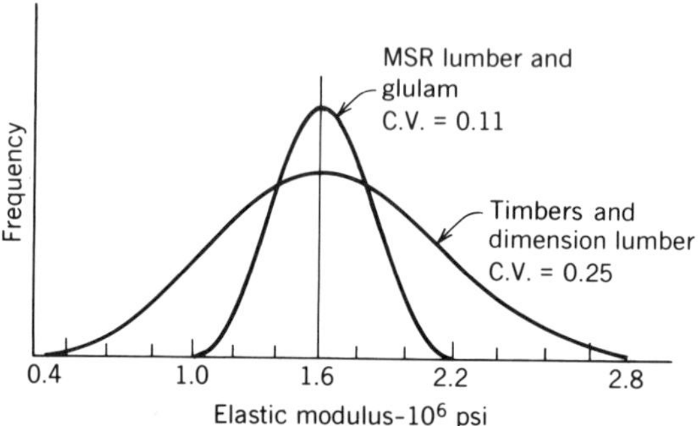

Fig. 25.9 Typical frequency distribution of E for material with an average E value of 1,600,000 psi.

25.9.4 Combined Bending and Axial Loading

Compression members may also carry some bending load. The resulting stresses can be computed by superimposing the bending and axial compression stresses, which gives a reasonable estimate of maximum stress within the elastic limit. Most design stresses would occur within the elastic limit. Examples are columns supporting side loads, with equipment loads attached off center, or with end load applied off center. Top chords of trusses are compression members usually carrying loads normal to their axis as well as forces parallel to their axis. Columns that are not initially straight experience bending moments.

With slenderness ratios in the short column range ($L/d \leq 11$) the interaction equation is

$$\frac{f_c}{F_c} + \frac{f_b}{F_b} < 1.0 \tag{25.9.7}$$

where f_c and f_b are the load-induced axial compression and flexural stresses; F_c and F_b are the allowable design values for the material. These allowable design values are subject to load duration and moisture content adjustment.

With slenderness ratios in the long column range ($K \leq L/d \leq 50$) the interaction equation is

$$\frac{f_c}{F_c'} + \frac{f_b}{F_b - f_c} \leq 1.0 \tag{25.9.8}$$

where F_c' is the allowable buckling stress, Eq. (25.9.3), K from Eq. (25.9.2), and other terms as defined above.

With slenderness ratios in the intermediate range ($11 < L/d < K$) the interaction equation is

$$\frac{f_c}{F'_c} + \frac{f_b}{F_b - Jf_c} \leq 1.0 \tag{25.9.9}$$

where J is an interpolating term which varies from zero to 1 according to the relationship

$$J = \frac{L/d - 11}{K - 11} \tag{25.9.10}$$

The flexural stress f_b may be the consequence of a load applied to the side of the member (such as a piece of equipment) or due to a laterally applied load, either of which could cause a bending moment M in the member.

Flexural stresses in columns may also arise from axial load eccentricity e, if the end load points are not on the column centers, or due to column curvature causing the line of action of the load to be eccentric to the section center of gravity at some points along the column, usually largest near midlength.

For axial load eccentricity NDS [25.35] recommends that f_b be determined as follows:

For short columns: $$f_b = \frac{6Pe}{Ad} \tag{25.9.11}$$

For long columns: $$f_b = \frac{7.5Pe}{Ad} \tag{25.9.12}$$

For intermediate columns: $$f_b = \frac{Pe}{Ad}(6 + 1.5J) \tag{25.9.13}$$

where e is determined by the designer.

Columns may experience bending moment as a result of both side loads and eccentricity, in which case f_b is the sum of flexural stress from each source. In such cases,

$$f_b = \frac{M}{S} + \frac{Pe}{Ad}(6 + 1.5J) \tag{25.9.14}$$

25.9.5 Columns of Circular or Other Than Rectangular Cross-Section

For columns of other than rectangular cross-section, equations are presented in terms of the minimum radius of gyration of the section.

For columns with pin-end conditions (free to rotate) the design formulas are:

Short columns: $\frac{L}{r} \geq 38$

$$P = F_c A \tag{25.9.15}$$

Intermediate columns: $38 < \frac{L}{r} < K_r$

$$K_r = 2.32\sqrt{\frac{E}{F_c}} \tag{25.9.16}$$

$$F'_c = F_c\left[1 - \frac{1}{3}\left(\frac{L/r}{K_r}\right)^4\right] \tag{25.9.17}$$

$$P = F'_c A \tag{25.9.18}$$

Long columns: $K_r \leq \frac{L}{r} \leq 173$

$$F'_c = \frac{3.6E}{(L/r)^2} \tag{25.9.19}$$

$$P = F'_c A \tag{25.9.20}$$

For other end fixity conditions, substitute the effective length L_e for L as explained in Section 25.9.2, in the above equations.

In the above equations the radius of gyration, r is $\sqrt{I/A}$, which for circular sections becomes $D/4$, where D is the diameter.

25.9.6 Spaced Columns

Timber designers often use spaced columns for compression members. Spaced members of this type are paired members separated by spacer blocks at the ends and at midlength. By providing fasteners at the ends which provide shear connection between the members, a degree of end fixity is produced which raises the buckling load of the members.

The degree of end fixity depends on the distance between the column ends and the center of the end connectors or end connector groups in the end spacer blocks. By reference to Fig. 25.10, the degrees of end fixity are defined at two levels.

1. *Condition "a"* is for end connector centroid distances of $L/20$ or less. A fixity factor C_x of 2.5 is characteristic of this condition.
2. *Condition "b"* is for end connector centroid distances of $L/10$ to $L/20$. A fixity factor C_x of 3.0 is characteristic of this end condition.

The design formulas for spaced columns are:

1. Short columns: $\dfrac{L}{d_1} \leq 11$

$$P = F_c A \tag{25.9.21}$$

where A is the cross-sectional area of both members of the spaced column.

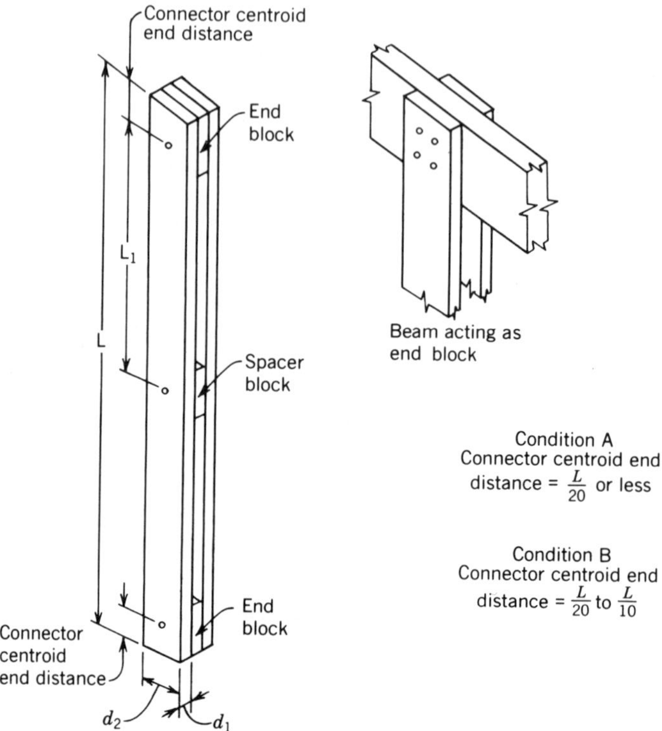

Fig. 25.10 Notation for spaced column.

Table 25.20 End Spacer Block Constants*

L/d_1 of Spaced Column	Group A Douglas Fir Larch Southern Pine (All Dense)	Group B Douglas Fir Larch Southern Pine	Group C Douglas Fir South Hem-Fir Eastern Hemlock Eastern Spruce Ponderosa Pine Lodgepole Pine Redwood (Close Grain)	Group D Eastern White Pine Englemann Spruce Western White Pine Cedars Redwood (Open Grain)
0–11	0	0	0	0
15	38	33	27	21
20	86	73	61	48
25	134	114	94	75
30	181	155	128	101
35	229	195	162	128
40	277	236	195	154
45	325	277	229	181
50	372	318	263	208
55	420	358	296	234
60–80	468	399	330	261

* From NDS [25.35], Table 3A, used by permission.

2. Intermediate columns: $11 < \dfrac{L}{d_1} < K_s$

$$F'_c = F_c\left[1 - \frac{1}{3}\left(\frac{L/d_1}{K_s}\right)^4\right] \tag{25.9.22}$$

$$K_s = 0.671\sqrt{C_x E/F_c} \tag{25.9.23}$$

$$P = F'_c A \tag{25.9.24}$$

3. Long columns: $K_s \le \dfrac{L}{d_1} \le 80$

$$F'_c = \frac{0.3C_x E}{(L/d_1)^2} \tag{25.9.25}$$

$$P = F'_c A \tag{25.9.26}$$

In spaced column design L/d_1 must not exceed 40. Values of F'_c and K_s obtained by the above formulas apply to design for buckling in the d_1 direction.

Spaced columns must also be investigated for buckling capacity in the d_2 direction as solid members (see Section 25.9.1). End fixity conditions that prevail in the d_2 direction should be considered.

When a spaced column has one spacer block between the two end spacer blocks, and when it is located in the middle tenth of the spaced column, the connector in that middle block is not required to possess shear strength. It merely serves to prevent the two column elements from opposite lateral movement, which requires little force.

When a spaced column has two or more spacer blocks located between the end spacer blocks, the distance between these intermediate spacer blocks must not exceed half the distance between the centroids of the end spacer blocks. The connectors in these intermediate spacer blocks should have the same shear strength as required in the end spacer blocks.

The required shear capacity of the connectors in each end block and its contacting column element must be equal to the member area multiplied by the end spacer block constant from Table 25.20. The end spacer block constant depends on member elastic modulus. Species are grouped into categories A, B, C, and D for this purpose (see Table 25.20).

Spacer block size must be at least equal in thickness to column elements and should equal column members in width. The spacer block length must be sufficient for the fastener end distances and spacings specified in Section 25.10 on bolts and connectors.

Connectors are usually split rings or shear plates. Bolts may be used. Bolted glued block-to-member connectors are possible if designed for the required shear strength.

The compression chords of trusses may be designed as spaced members, in which case L is the distance between points of chord lateral support. Care must be taken that the shear connection between blocks and members is sufficient to withstand both end block shear and truss member forces transmitted at the connections.

EXAMPLE 25.9.1

Determine the load which can be used for a spaced column consisting of two 2×6 elements for which $L = 120$ in. and the end block connector centroids are 5½ in. from the column ends. The species is Douglas fir of No. 1 grade, dry use. Maximum load is caused by snow of two-months cumulative duration. Grain of end blocks is parallel to grain of elements.

Solution

(a) The maximum L/d_1 for spaced columns is 80. L/d_1 for this column is $120/1.5 = 80$.

(b) The centroid of the connectors is located 5.5 in. from the end which is $L/21.8$, less than $L/20$, so condition "a" formulas must be used.

(c) $K_s = 0.671\sqrt{2.5E/F_c}$. For 2×6, No. 1 grade Douglas fir, $F_c = 1250$ psi and $E = 1{,}800{,}000$ psi for dry-use. Load duration factor is 1.15.

$$K_s = 0.671\sqrt{\frac{2.5(1{,}800{,}000)}{1.15(1250)}} = 37.5$$

Since $L/d_1 = 80$, this is a long column.

(d) Compute allowable stress F'_c.

$$F'_c = \frac{0.3(2.5)E}{(L/d_1)^2} = \frac{0.75(1{,}800{,}000)}{(80)^2} = 211 \text{ psi}$$

(e) The load per element $= F'_cA = 211(1.5)(5.5) = 1733$ lb. The spaced column load capacity is 3466 lb.

(f) The end block connector shear requirement is based on an end block spacer constant for Group B wood (Table 25.20) for $L/d_1 = 80$. Table 25.20 gives the constant as 399 psi. The shear requirement is $399(1.5)(5.5) = 3921$ lb for each face of the block.

(g) One 2½-in. split ring loaded parallel-to-grain can carry $1.15(2100) = 2415$ lb with a 1.5-in. thick spacer block with connectors in both faces. Two rings will carry 4830 lb, which exceeds the required 3291 lb. Split ring strength properties are explained in Section 25.10.3.

(h) The spaced column is now checked for possible buckling as a pin-end solid column per Section 25.9.1, in the d_2 or 5.5-in. direction. $L/d_2 = 21.8$.

$$K = 0.671\sqrt{\frac{E}{F_c}} = 0.671\sqrt{\frac{1{,}800{,}000}{1.15(1250)}} = 23.74.$$

This is an intermediate solid column.

$$F'_c = F_c\left[1 - \frac{1}{3}\left(\frac{L/d_2}{K}\right)^4\right] \text{ with } F_c = 1.15(1250)$$

$$F'_c = 1438\left[1 - \frac{1}{3}\left(\frac{21.8}{23.74}\right)^4\right] = 1094 \text{ psi}$$

which exceeds the 211 calculated for the spaced column.

(i) The end blocks must be large enough to provide the connector spacing. The 2½-in. split rings will be 68% loaded at 3291 lb. Spacing charts for split ring connectors show a required edge distance of 1.75 in. As Fig. 25.11 shows, this will leave 2 in. between connectors perpendicular to

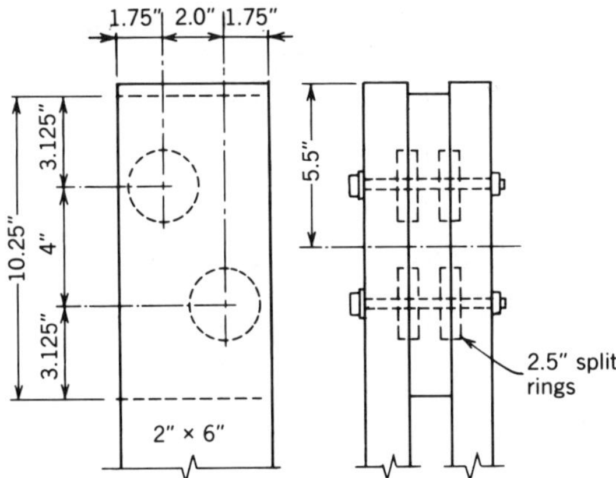

Fig. 25.11 Example of end spacer block connection.

the grain, for which the spacing chart indicates a minimum parallel-to-grain spacing of 4 in. when 68% loaded. The minimum end distance is 3.125 in. when 68% loaded. The spacer block length will be 10.25 in. minimum.

25.10 CONNECTIONS

25.10.1 Nailed Connections

Nailed joints are designed for lateral or withdrawal nail strength using load values such as from NDS [25.35]. The strength of a joint connected by several nails is the sum of the load values of the individual nails comprising the joint.

In this section, load values are given for four types of nails and a special table is included for staple load values. Table 25.21 lists the diameter and length of the nails covered under NDS [25.35]. Common steel wire nails are most widely used for structural work. Steel box nails are smaller in diameter than common nails of the same length, and are used because they drive easily and have less tendency to split the wood. Note in Table 25.21 that the pennyweight nail size is a length designation. Box nails have less lateral and withdrawal strength than common nails of equal length. They bend more easily than common nails and are difficult to drive successfully in the denser woods, especially if the wood is seasoned.

Design load values for nails are given in Table 8.8 in NDS [25.35].

Threaded hardened steel nails are slightly smaller in diameter than common nails (except 6*d* and 8*d* sizes). They have the same load values as common nails for 6*d* to 20*d* lengths. Above 20*d* the threaded nails are not as strong as the common nails. They bend less easily than common nails and tend to provide a stiffer connection. They are useful in nailing predrilled steel plates to wood members.

Common steel wire spikes are larger in diameter than common nails, and are especially useful when long nails are needed. Their load values are larger because of their diameter. They are often used to nail together parallel courses of heavy wood decking, using predrilled holes in the first piece to be penetrated.

When nails are loaded perpendicular to their axis (laterally) as in spliced or lapped joints, the load values in NDS Table 8.8c apply for normal load duration in dry use, whether loaded parallel-to-grain or across the grain. For other load durations nail load values are adjusted using wood load duration factors (see Section 25.8.3).

Nail load values are affected by the moisture content of the wood when the nails are driven as well as by the service moisture content. Adjustment factors for these effects are given in Table 25.22.

It is not generally recommended that nails be driven into the end grain but sometimes this is necessary. Lateral load values are 75% of those for side grain. Withdrawal values in end grain are considered negligible.

Lateral load values depend upon wood density and strength. Wood species are grouped as shown in Table 25.23 for nail strength categories. Lateral load values for nails require a minimum penetration into the piece holding the point as shown in NDS Table 8.8c. For less penetration the values are reduced in proportion to the actual penetration, but never to less than one-third the tabulated design load values.

Table 25.21 Nail and Spike Sizes*

Pennyweight	Length Inches	Wire diameter, inches			
		Box nails	Common wire nails	Threaded hardened-steel nails	Common wire spikes
6*d*	2	0.099	0.113	0.120	–
8*d*	2½	0.113	0.131	0.120	–
10*d*	3	0.128	0.148	0.135	0.192
12*d*	3¼	0.128	0.148	0.135	0.192
16*d*	3½	0.135	0.162	0.148	0.207
20*d*	4	0.148	0.192	0.177	0.225
30*d*	4½	0.148	0.207	0.177	0.244
40*d*	5	0.162	0.225	0.177	0.263
50*d*	5½	–	0.244	0.177	0.283
60*d*	6	–	0.263	0.177	0.283
70*d*	7	–	–	0.207	–
80*d*	8	–	–	0.207	–
90*d*	9	–	–	0.207	–
5/16	7	–	–	–	0.312
3/8	8½	–	–	–	0.375

* From NDS [25.35], Table 8.8A, used by permission.

Three member joints with nails or spikes driven from one side and fully penetrating all three members have design lateral loads one-third greater than tabulated, if side member thickness is one-third the main member thickness or more. If side members equal or exceed main members in thickness, load values are two-thirds greater than tabulated values. These are double-shear fastener arrangements.

For nails or spikes in three-member joints which extend through all three members and three diameters beyond, and are clinched, the design loads can be doubled. This applies for 6d to 12d nails and requires side member thickness of at least ⅜ in. Threaded hardened steel nails can be used at this increased value without clinching. Penetration distance requirements in the member opposite the nail heads must be observed in determining tabulated values.

Side panels of steel can be used, and if designed for adequate nail bearing, the normal nail load values can be increased 25%.

Nails laterally loaded in diaphragms and shear walls have 30% larger design values than tabulated, and may be further adjusted for load duration.

Withdrawal design load values in NDS Table 8.8b depend on wood specific gravity from Table 25.23. These values are in pounds per inch of penetration into the member holding the point. The tabulated values are for normal load duration and may be adjusted. They are for dry lumber used dry and may be adjusted for unseasoned and wet-use conditions according to Table 25.22. These values are for nails driven into side grain. No design load values are reliable for withdrawal of nails from end grain.

Toe nailing is a term used to describe nails driven into the side of a vertical member such as a wall stud at an angle of about 30° to the member axis, with the point penetrating the side grain of a horizontal member. Joists and rafters are often toe nailed to wall plates or headers. Withdrawal loads are two-thirds those for nails in side grain. The value need not be adjusted for wood moisture content or service condition. Lateral load values for toe nails are five-sixths those tabulated for nails in side grain. The member containing the nail head should be loaded toward the side into which the nail is driven.

There are no recommended minimum nail spacing distances. The spacing should be large enough so the nails do not cause splits. If nails cause splits, their design loads as given in these tables are invalid.

Splitting can be avoided by predrilling and for important structural connections this may be

Table 25.22 Fastener Load Modification Factors for Moisture Content*

Type of Fastener	Condition of Wood[1]: At Time of Fabrication	Condition of Wood[1]: In Service	Factor
Timber connectors[2]	Dry	Dry	1.0
	Partially seasoned[3]	Dry	See Note[3]
	Wet	Dry	0.8
	Dry or wet	Partially seasoned or wet	0.67
Bolts or lag screws	Dry	Dry	1.0
	Partially seasoned[3] or wet	Dry	See Table 25.30
	Dry or wet	Exposed to weather	0.75
	Dry or wet	Wet	0.67
Drift bolts or pins—Laterally loaded	Dry or wet	Dry	1.0
	Dry or wet	Partially seasoned or wet, or subject to wetting and drying	0.70
Wire nails and spikes—Withdrawal loads	Dry	Dry	1.0
	Partially seasoned or wet	Will remain wet	1.0
	Partially seasoned or wet	Dry	0.25
	Dry	Subject to wetting and drying	0.25
—Lateral loads	Dry	Dry	1.0
	Partially seasoned or wet	Dry or wet	0.75
	Dry	Partially seasoned or wet	0.75
Threaded, hardened steel nails	Dry or wet	Dry or wet	1.0
Wood screws	Dry or wet	Dry	1.0
	Dry or wet	Exposed to weather	0.75
	Dry or wet	Wet	0.67
Metal plate connectors	Dry	Dry	1.0
	Partially seasoned or wet	Dry or wet	0.8

* Adapted from NDS [25.35], Table 8.1B, used by permission.

[1] Condition of wood definitions applicable to fasteners are:

"Dry" wood has a moisture content of 19% or less.

"Wet" wood has a moisture content at or above the fiber saturation point (approximately 30%).

"Partially seasoned" wood has a moisture content greater than 19% but less than the fiber saturation point (approximately 30%).

"Exposed to weather" implies that the wood may vary in moisture content from dry to partially seasoned, but is not expected to reach the fiber saturation point at times when the joint is under full design load.

"Subject to wetting and drying" implies that the wood may vary in moisture content from dry to partially seasoned or wet, or vice versa, with consequent effects on the tightness of the joint.

[2] For timber connectors, moisture content limitations apply to a depth of ¾ in. from the surface of the wood.

[3] When timber connectors, bolts, or laterally loaded lag screws are installed in wood that is partially seasoned at the time of fabrication but which will be dry before full design load is applied, proportional intermediate values may be used.

Table 25.23 Grouping of Species for Fastener Design†

Grouping for timber connector loads		Grouping for lag screw, drift bolt, nail, spike, wood screw and metal plate connector loads		
Connector load group*	Species of wood	Group	Species of wood	Specific gravity** (G)
Group A	Ash, Commercial White Beech Birch, Sweet & Yellow Douglas Fir-Larch (Dense)*** Hickory & Pecan Maple, Black & Sugar Oak, Red & White Southern Pine (Dense)	Group I	Ash, Commercial White Beech Birch, Sweet & Yellow Hickory & Pecan Maple, Black & Sugar Oak, Red & White	0.62 0.68 0.66 0.75 0.66 0.67
Group B	Douglas Fir-Larch*** Southern Pine Sweetgum & Tupelo	Group II	Douglas Fir - Larch*** Southern Pine Sweetgum & Tupelo Virginia Pine - Pond Pine	0.51 0.55 0.54 0.54
Group C	California Redwood Douglas Fir, South Eastern Hemlock-Tamarack*** Eastern Spruce Hem - Fir*** Lodgepole Pine Mountain Hemlock Northern Aspen Northern Pine Ponderosa Pine**** Ponderosa Pine-Sugar Pine Red Pine**** Sitka Spruce Southern Cypress Spruce-Pine-Fir Western Hemlock Yellow Poplar	Group III	California Redwood Douglas Fir, South Eastern Hemlock Eastern Hemlock - Tamarack*** Eastern Softwoods Eastern Spruce Hem - Fir*** Lodgepole Pine Mountain Hemlock Mountain Hemlock - Hem Fir Northern Aspen Northern Pine Ponderosa Pine*** Ponderosa Pine-Sugar Pine Red Pine**** Sitka Spruce Southern Cypress Spruce-Pine-Fir Western Hemlock Yellow Poplar	0.42 0.48 0.43 0.45 0.42 0.43 0.42 0.44 0.47 0.44 0.42 0.46 0.49 0.42 0.42 0.43 0.48 0.42 0.48 0.46
Group D	Aspen Balsam Fir Black Cottonwood California Redwood, Open grain Coast Sitka Spruce Cottonwood, Eastern Eastern White Pine*** Engelmann Spruce - Alpine Fir Idaho White Pine Northern White Cedar Western Cedars*** Western White Pine	Group IV	Aspen Balsam Fir Black Cottonwood California Redwood, Open grain Coast Sitka Spruce Coast Species Cottonwood, Eastern Eastern White Pine*** Eastern Woods Engelmann Spruce - Alpine Fir Idaho White Pine Northern Species Northern White Cedar West Coast Woods (Mixed Species) Western Cedars*** Western White Pine White Woods (Western Woods)	0.40 0.38 0.33 0.37 0.39 0.39 0.41 0.38 0.38 0.36 0.40 0.35 0.31 0.35 0.35 0.40 0.35

*When stress graded.
**Based on weight and volume when oven-dry.
***Also applies when species name includes the designation "North".
****Applies when graded to NLGA rules.

Note: Coarse grain Southern Pine, as used in some glued laminated timber combinations, is in Group C.

† From NDS [25.35], Table 8.1A, used by permission.

worth the effort. If the drilled hole does not exceed 0.9 nail diameters for Group 1 wood or 0.75 diameters for other species groups, the lateral and withdrawal values in the tables are valid. Avoid driving a row of closely spaced nails in exact alignment along the grain of dense species of wood.

Table 25.24 provides general guidance for nail spacing but is subject to the splitting tendencies of the wood. The spacings are for soft woods.

There is a great variety of staples and special nails for pneumatic nailing machines. International

Table 25.24 Minimum Nail Spacings in Terms of Nail Diameters for Lumber.*

	Holes Not Prebored	Holes Prebored
End distance	20	10
Edge distance	5	5
Perpendicular-to-grain spacing	10	3
Parallel-to-grain spacing	20	10

* From *Wood Technology in Design of Structures* [25.26].

Conference of Building Officials* Research Report 2403 gives design values for a variety of these special fasteners. Tables 25.25 and 25.26 were derived from that report and from the *Wood Handbook* [25.43] and will serve as a guide to lateral and withdrawal loads for several sizes of staples. The values given are for the combined performance of the two prongs of the staples, uncoated. Penetration is not additive for the two prongs. Pneumatic nailers should be adjusted to avoid overdriving the staples. Pneumatic nailers will not draw two members together as effectively as nailing with a hammer, so pieces should be pressed together before the nailer is used.

25.10.2 Bolted Connections

Bolts used in timber engineering usually carry load in double shear (see figure in Table 25.27) or single shear (shear transfer between two pieces). They are used to connect wood-to-wood, and wood-to-steel. The strength is developed by bearing between the bolt and the wood, or when steel hardware is used, between the bolt and the steel. Bolts are usually of ordinary mild steel (ASTM A307) either black or galvanized.

Table 25.27 lists typical bolt design values as published in the NDS [25.35]. These values are for double shear connections between one main member and two side members with side member thickness equal to or exceeding half the main member thickness. These values are for normal load duration in seasoned wood for dry-use service.

Inspection of Table 25.27 shows that bolt design values are given for parallel-to-grain loads, P, and for perpendicular-to-grain loads, Q. For any given bolt diameter the load value increases as the bearing area of the bolt on the wood increases, reaching a maximum at $L = 6d$ for P and $L = 8d$ for Q. For larger bolt bearing areas the load values either remain constant or, in the case of Q, diminish.

* 5360 South Workman Mill Road, Whittier, CA 90601.

Table 25.25 Staples—Lateral Load Design Values*—Normal†

Diameters and penetrations are in inches. Loads are in pounds

Wire gage	16	15	14	13	12
Wire diameter	0.0625	0.072	0.080	0.0915	0.1055
Penetration‡					
10 diameters	0.63	0.72	0.80	0.92	1.06
11 diameters	0.69	0.79	0.88	1.01	1.16
13 diameters	0.81	0.94	1.04	1.19	1.37
14 diameters	0.88	1.01	1.12	1.28	1.48
Design values					
Species Group I§	64	79	93	114	140
Species Group II	52	64	75	92	113
Species Group III	43	52	62	75	93
Species Group IV	34	42	49	60	73

* For staples with 7/16-in. minimum crown width, uncoated. Values are for dry use. Use nail modification factors for wet use.
† Adapted from *Wood Handbook* [25.43].
‡ Penetration in member holding the points must equal or exceed 10*d* for Group I, 11*d* for Group II, 13*d* for Group III, and 14*d* for Group IV.
§ ICBO UBC limits Group I values to loads shown for Group II.

Table 25.26 Staples—Withdrawal Design Values—Normal*,†

Pounds per inch of penetration into side grain of members holding points. Design value not to exceed twice the tabulated value, regardless of penetration.

Specific Gravity G	16 ga. 0.0625 in.	15 ga. 0.072 in.	14 ga. 0.080 in.	13 ga. 0.0915 in.	12 ga. 0.1055 in.
0.75	84	97	108	123	142
0.68	66	76	84	97	111
0.67	63	73	81	92	106
0.66	61	70	78	89	103
0.62	54	60	67	76	88
0.55	39	45	50	57	66
0.54	37	43	47	54	62
0.51	32	37	41	47	54
0.49	29	33	37	42	49
0.48	28	32	36	41	47
0.47	26	30	33	38	44
0.46	25	29	32	37	42
0.45	23	26	29	34	39
0.44	22	25	28	32	37
0.43	21	24	27	31	35
0.42	20	23	26	29	34
0.41	19	22	24	28	32
0.40	17	20	22	25	29
0.39	16	19	20	23	27
0.38	15	18	19	22	25
0.37	14	17	18	20	24
0.36	13	15	17	19	22
0.35	13	14	16	18	21
0.33	11	13	14	16	19
0.31	9	10	12	13	15

For dry-use. Use nail modification factors for wet-use.

* Adapted from *Wood Handbook* [25.43].

Bolt load values are subject of the usual duration of load adjustments for wood. Moisture content adjustments are given in Tables 25.22 and 25.28.

Bolts often bear on the wood at angles other than parallel-to-grain ($\theta = 90°$). Load values for any angle of load-to-grain can be obtained from Fig. 25.12 (see p. 934) which is entered using the ratio of P/Q to obtain the factor N/P where N is the load value at angle θ. If the angle θ of load to grain is 35° and the species is southern pine, from Table 25.27 for a ¾-in. bolt in a 2.5-in. thick main member, P and Q are 2310 and 900 lb, respectively. With $P/Q = 2.57$, $N/P = 0.66$ from Fig. 25.12, and $N = 1525$ lb.

When steel splice plates or hardware are bolted to wood members, the value of P in Table 25.27 is increased by 25%. The value of Q, however, is *not* increased. Furthermore, steel side plate thicknesses need not equal or exceed half of the wood main member thickness. Their thickness must be sufficient to develop the loads carried through the plates.

Bolt holes must be slightly larger than the bolts in wood members to accommodate any shrinkage that may be expected across the grain. This clearance is about 1⁄32- up to ¾-in. bolt size and 1⁄16 in. for larger bolts. Washers should be used whenever steel plates are not used.

When side members are less than half main member thickness, the tabulated bolt loads for a main member twice the thickness of the side member are used. For ¾-in. side members and 2.5-in. main member, P and Q are chosen for a 1.5-in. main member.

Two member (single shear) joint bolt design loads may be obtained from the tables for double shear joints by the following methods:

1. For *both members of equal thickness* use one-half the tabulated design value for the member thickness. If the angle of load-to-grain is not the same for each member, the value for the larger load-to-grain angle shall be used.
2. For *members of unequal thickness* use one-half the tabulated design load for the thicker member, or one-half the tabulated design load for a piece twice the thickness of the thinner member, whichever is less.

EXAMPLE 25.10.1

Illustrate the computation of allowable joint load for several cases of Douglas fir connected by ¾-in. diameter bolts.

Solution

(a) For two members, each 1.5 in. thick and loaded parallel to the grain, from Table 25.27,

$$P = 1420 \text{ lb}; \qquad \text{allowable joint load} = P/2 = 710 \text{ lb}$$

(b) For two members, each 1.5 in. thick, one loaded parallel to the grain and one loaded perpendicular to the grain,

$$P = 1420 \text{ lb}; \qquad Q = 540 \text{ lb}$$

$$\text{allowable joint load} = Q/2 \text{ or } P/2; \qquad Q/2 \text{ controls} = 270 \text{ lb}$$

(c) For two members, each 1.5 in. thick, one loaded parallel-to-grain and the other loaded at 45° to grain,

$$P = 1420 \text{ lb} \qquad Q = 540 \text{ lb}$$

N for $P/Q = 2.62$ and using the factor 0.55 from Fig. 25.12 is 781 lb. Allowable joint load is $N/2$ which is less than $P/2$, and is 390 lb.

(d) For two members of unequal size with one 1.5 in. thick and the other 2.5 in. thick and both loaded parallel-to-grain:

For the 2.5-in. member $P/2 = 1155$ lb.
For the 1.5-in. member $P/2$ for $\ell = 3$ in. is 1315 lb.
Allowable joint load is 1155 lb.

Note that large member is less than twice the small member thickness.

(e) For two members of unequal size with one member ¾ in. thick and the other member 2.5 in. thick and both loaded parallel-to-grain:

For the 2.5-in. member $P/2 = 1155$ lb.
For the ¾-in. member $P/2$ for $\ell = 1.5$ in. is = 710 lb.
Allowable joint load is 710 lb.

Note that large member is more than twice small member thickness.

(f) For two members of unequal thickness with one member 1.5 in. thick and the other 2.5 in. thick (the thicker member is loaded at 90° to the grain and the thinner member parallel-to-grain):

For the 2.5-in. member loaded perpendicular-to-grain, $Q/2 = 450$ lb.
For the 1.5-in. member loaded parallel-to-grain, $P/2$ for $\ell = 3$ in. is 1315 lb.
Allowable joint load is 450 lb.

(g) For the same as (f) but with the thinner member only ¾ in. thick:

For the 2.5-in. member loaded perpendicular-to-grain, $Q/2$ is 450 lb.
For the ¾-in. member loaded parallel-to-grain, $P/2$ for $\ell = 1.5$ in. is 710 lb.
Allowable load is 450 lb.

Although the thinner member is less than half the size of the thicker member, the angle of load to grain in the thicker member makes it control this joint design.

Four member joints are not common. Provisions for their design are given in the NDS [25.35].

Bolted joints with two members arranged so the load is not perpendicular to the bolt axis of one member sometimes occur. Figure 25.13 (see p. 935) shows such a joint. This joint is designated for member thicknesses equal to bolt lengths in each member. For the member in which the load is not

Table 25.27 Bolt Design Values*
Design values, in pounds, on one bolt loaded at both ends (double shear)† for following species.

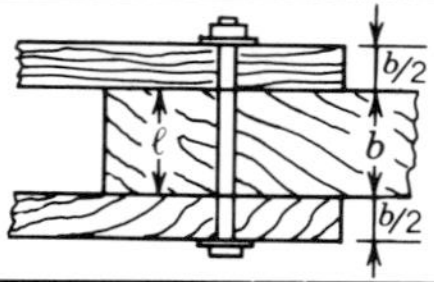

				Douglas Fir-Larch (Dense), Southern Pine (Dense)		California Redwood (Close Grain), Douglas Fir-Larch, Southern Pine, Southern Cypress		Eastern Hemlock-Tamarack, California Redwood (Open Grain), Hem-Fir, Western Hemlock		Spruce-Pine-Fir, Sitka Spruce, Yellow Poplar, Eastern Spruce, Lodgepole Pine		Red Pine, Western White Pine, Ponderosa Pine-Sugar Pine, Eastern White Pine, Balsam Fir, Idaho White Pine		Aspen, Eastern Cottonwood, Engelmann Spruce-Alpine Fir, Northern White Cedar	
Length of Bolt in Main Member ℓ	Diameter of Bolt D	ℓ/D	Projected Area of Bolt $A = \ell \times D$	Parallel to Grain P	Perpendicular to Grain Q	Parallel to Grain P	Perpendicular to Grain Q	Parallel to Grain P	Perpendicular to Grain Q	Parallel to Grain P	Perpendicular to Grain Q	Parallel to Grain P	Perpendicular to Grain Q	Parallel to Grain P	Perpendicular to Grain Q
1½	½	3.00	.750	1100	500	940	430	800	280	680	280	630	190	530	210
	⅝	2.40	.938	1380	570	1180	490	1000	310	850	320	790	210	660	230
	¾	2.00	1.125	1660	630	1420	540	1200	350	1020	350	950	240	800	260
	⅞	1.71	1.313	1940	700	1660	600	1400	380	1190	390	1110	260	930	290
	1	1.50	1.500	2220	760	1890	650	1600	420	1360	420	1260	290	1060	310
2	½	4.00	1.000	1370	670	1170	570	1040	370	900	370	840	250	700	280
	⅝	3.20	1.250	1820	760	1550	650	1330	410	1130	420	1050	290	890	310
	¾	2.67	1.500	2210	840	1890	720	1600	460	1360	470	1260	320	1060	350
	⅞	2.29	1.750	2580	930	2200	790	1870	510	1580	520	1480	350	1240	380
	1	2.00	2.000	2960	1010	2520	870	2130	550	1810	560	1690	380	1420	420
2½	½	5.00	1.250	1480	840	1260	720	1180	460	1080	470	1010	320	840	340
	⅝	4.00	1.563	2140	950	1820	810	1620	520	1410	530	1310	360	1100	390
	¾	3.33	1.875	2710	1060	2310	900	1990	580	1700	590	1580	400	1330	430
	⅞	2.86	2.188	3210	1160	2740	990	2330	630	1980	650	1840	440	1550	480
	1	2.50	2.500	3680	1270	3150	1080	2670	690	2260	700	2110	480	1770	520
3	½	6.00	1.500	1490	1010	1270	860	1210	550	1160	560	1080	380	910	410
	⅝	4.80	1.875	2290	1140	1960	970	1810	620	1640	630	1530	430	1280	470
	¾	4.00	2.250	3080	1270	2630	1080	2340	690	2030	700	1890	480	1590	520
	⅞	3.43	2.625	3770	1390	3220	1190	2790	760	2380	780	2210	520	1860	570
	1	3.00	3.000	4390	1520	3750	1300	3200	830	2710	850	2530	570	2120	620
3½	½	7.00	1.750	1490	1140	1270	980	1210	640	1160	650	1080	440	910	480
	⅝	5.60	2.188	2320	1330	1980	1130	1890	730	1790	740	1670	500	1400	550
	¾	4.67	2.625	3280	1480	2800	1260	2580	810	2310	820	2150	560	1800	610
	⅞	4.00	3.063	4190	1630	3580	1390	3180	890	2760	900	2570	610	2160	670
	1	3.50	3.500	5000	1770	4270	1520	3710	970	3170	990	2950	670	2480	730

4	½	8.00	2.000	1490	1180	1270	1010	1210	700	1160	750	1080	500	910	550
	⅝	6.40	2.500	2330	1510	1990	1290	1900	830	1820	840	1690	570	1420	620
	¾	5.33	3.000	3340	1690	2850	1440	2690	920	2520	940	2340	630	1970	690
	⅞	4.57	3.500	4450	1860	3800	1590	3480	1020	3090	1030	2880	700	2420	760
	1	4.00	4.000	5470	2030	4670	1730	4150	1110	3600	1130	3360	760	2820	830
4½	⅝	7.20	2.813	2330	1640	1990	1400	1890	930	1810	950	1690	640	1420	700
	¾	6.00	3.375	3350	1900	2860	1620	2730	1040	2610	1060	2440	710	2050	780
	⅞	5.14	3.938	4530	2090	3870	1790	3630	1140	3360	1160	3130	790	2630	860
	1	4.50	4.500	5770	2280	4930	1950	4500	1250	3990	1270	3710	860	3120	940
	1¼	3.60	5.625	7980	2670	6820	2280	5940	1460	5080	1490	4740	1000	3980	1100
5½	⅝	8.80	3.438	2330	1650	1990	1410	1900	1010	1810	1110	1690	750	1420	820
	¾	7.33	4.125	3350	2200	2860	1880	2730	1270	2610	1290	2430	870	2040	950
	⅞	6.29	4.813	4570	2550	3900	2180	3720	1400	3560	1420	3320	960	2790	1050
	1	5.50	5.500	5930	2790	5070	2380	4820	1520	4550	1550	4240	1050	3560	1150
	1¼	4.40	6.875	8940	3260	7640	2790	6940	1780	6110	1820	5700	1230	4780	1340
7½	⅝	12.00	4.688	2330	1480	1990	1260	1890	950	1820	1090	1690	730	1420	800
	¾	10.00	5.625	3350	2130	2860	1820	2730	1320	2620	1500	2440	1010	2050	1110
	⅞	8.57	6.563	4560	2840	3890	2430	3720	1730	3560	1900	3310	1280	2780	1400
	1	7.50	7.500	5950	3550	5080	3030	4850	2060	4650	2110	4330	1430	3640	1560
	1¼	6.00	9.375	9310	4450	7950	3800	7580	2430	7260	2480	6770	1670	5680	1830
9½	¾	12.67	7.125	3350	1920	2860	1640	2720	1250	2610	1430	2430	970	2040	1060
	⅞	10.86	8.313	4570	2660	3900	2270	3720	1660	3560	1890	3320	1280	2780	1400
	1	9.50	9.500	5950	3460	5080	2960	4850	2130	4650	2410	4330	1630	3640	1780
	1¼	7.60	11.875	9300	5210	7950	4450	7570	3040	7250	3140	6750	2120	5680	2320
	1½	6.33	14.250	13410	6480	11460	5530	10910	3540	10470	3610	9760	2440	8190	2660
11½	⅞	13.14	10.062	4560	1980	3900	2060	3700	1590	3570	1820	3330	1230	2790	1350
	1	11.50	11.500	5950	3240	5080	2770	4860	2040	4650	2340	4330	1580	3640	1730
	1¼	9.20	14.375	9300	5110	7950	4360	7570	3130	7240	3530	6750	2380	5670	2610
	1½	7.67	17.250	13410	7200	11450	6150	10930	4210	10430	4370	9720	2950	8160	3230
13½	1	13.50	13.500	5960	2410	5100	2530	4850	1970	4670	2270	4350	1530	3650	1680
	1¼	10.80	16.875	9300	4860	7950	4160	7590	3030	7270	3460	6770	2340	5690	2560
	1½	9.00	20.250	13400	7070	11450	6040	10930	4340	10460	4850	9750	3280	8190	3580

* From NDS [25.35], adapted from Table 8.5A, used by permission.
† Three (3) member joint.

Table 25.28 Modification Factors for Laterally Loaded Bolts in Timber Seasoned in Place*

Factors apply when wood is at or above the fiber saturation point (wet) at time of fabrication but dries to a moisture content of 19 percent or less (dry) before full design load is applied. For wood partially seasoned when fabricated, adjusted intermediate values may be used.

Arrangement of bolts or lag screws	Type of splice plate	Modification factor
—One fastener only, or —Two or more fasteners placed in a single line parallel to grain, or —Fasteners placed in two or more lines parallel to grain with separate splice plates for each line	Wood or metal	1.0
—All other arrangements	Wood or metal	0.4

* From NDS [25.35], Table 8.1C, used by permission.

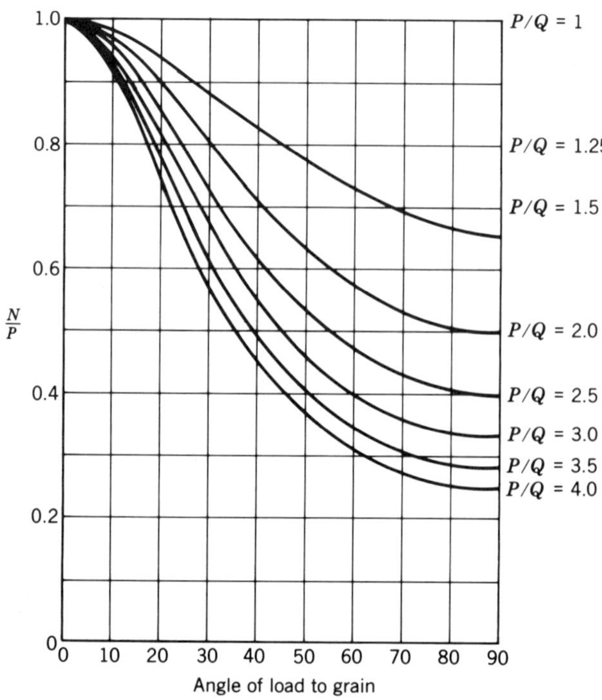

* Based on Hankinson formula, see Appendix J of NDS (25.35).

Fig. 25.12 Adjustment for angle of load to grain.

normal to the bolt axis the component of load normal to the bolt axis is determined, considering the angle of load to grain.

EXAMPLE 25.10.2

Determine the allowable load on the joint in Fig. 25.13 with $\ell_2 = 1.5$ in. and $\ell_1 = 1.06$ in. The species is Douglas fir S-DRY, used dry, and the bolt diameter is ¾ in.

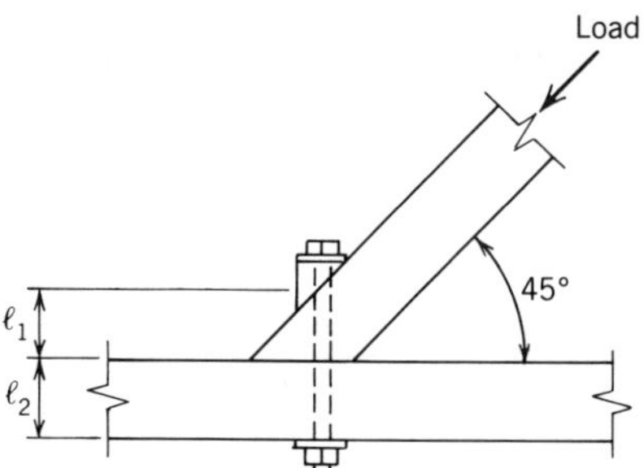

Fig. 25.13 Load at angle to bolt axis.

Solution

The basic P and Q values were obtained in Example 25.10.1, part (a). For the thicker member,

$$P/2 = 710 \text{ lb}$$

For the thinner member, the angle of load to the grain is 45°; thus,

$$P/2 \text{ for } \ell = 2.12 \text{ is about } 1890/2 = 945 \text{ lb}$$

$$Q/2 \text{ for } \ell = 2.12 \text{ is } 720/2 = 360 \text{ lb}$$

Using $P/Q = 2.62$ and a factor of 0.55 from Fig. 25.12, $N/2 = 0.55\ (945) = 520$ lb.

The component of load perpendicular to the bolt axis must not exceed 520 lb, so the load must not exceed 1.414 (520) = 735 lb.

Figure 25.14 summarizes the permissible loads on the bolt and the joint members. Bolted joint connections can be seriously affected by wood shrinkage if there is any possibility of the wood splitting under shrinkage forces which occur after the joint is assembled. This is one reason for the bolt hole clearances of 1/32 and 1/16 in. mentioned previously. Another condition that will lead to shrinkage splitting is the design of joints with two or more rows of bolts with both rows in the same side member. The side member may shrink at a different rate than the main member. Forces can develop to split either of the wood members and this will usually occur at the centerline of the row of bolts. To avoid this, each row should have separate side plates as in Fig. 25.15 if shrinkage can occur after the joint is assembled. If separate side plates are not provided, the joint strength for design must be multiplied by 0.4, which is usually an intolerable bolt strength reduction.

The load is not uniformly distributed among bolts in a row parallel to the direction of load. The load per bolt is higher for bolts at each end of the row than near the center. The degree of load concentration in the end bolts depends on the number of bolts in a row, the relative cross-sectional areas of main and side members, and the main member area. In typical joints where each side member has half the area of the main member, strength reduction for main members 12 sq in. or less in cross-section is 15% for 4 bolts per row and 35% for 8 bolts per row. For steel side plates

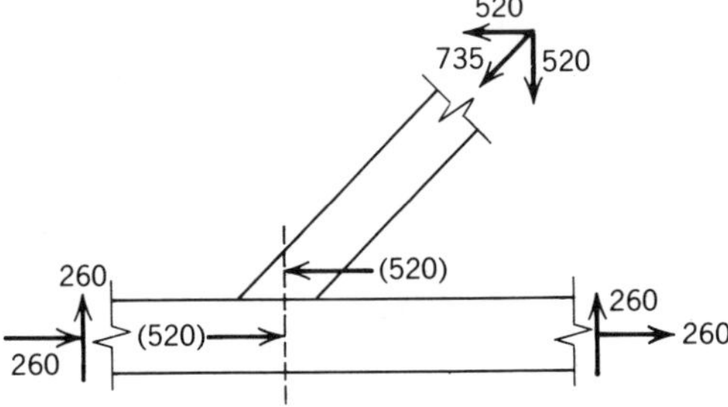

Fig. 25.14 Solution for joint in Example 25.10.2.

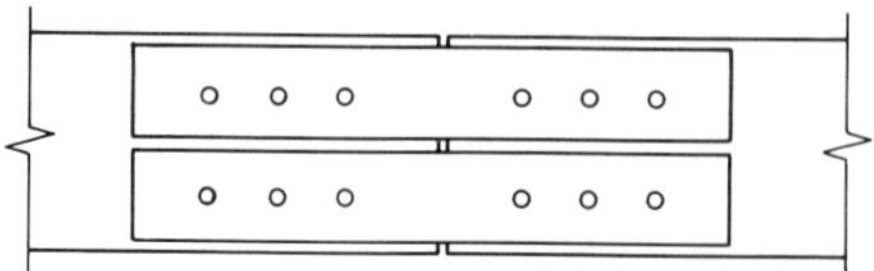

Fig. 25.15 Joint with separate side plates for each row of bolts.

comparable reductions occur. The NDS [25.35] contains tables for assessing the load concentration effect for joints of all configurations. No reduction is required when the number of bolts per row does not exceed two.

The spacing of bolts must conform to certain minimum dimensions for reliable joint performance:

Bolted Joints with Load Parallel-to-Grain: (Fig. 25.16)

1. Spacing between bolt centers in a row parallel-to-grain is 4 bolt diameters.
2. The minimum edge distance of bolt centers is 1.5 bolt diameters.
3. When two or more rows of bolts are used, the spacing between rows shall be at least 1.5 bolt diameters. When bolt length in main member is more than 6 bolt diameters, edge distance shall be one-half the distance between rows or 1.5 diameters, whichever is larger.
4. If the spacing between parallel-to-grain rows of bolts exceeds 5 in., separate splice plates (side members) must be used for each row.
5. The distance from the end of a member to the nearest bolt center shall be at least:
   7 diameters for softwoods in tension;
   5 diameters for hardwoods in tension;
   4 diameters for softwoods or hardwoods in compression.

Bolted Joints with Load Perpendicular-to-Grain: (Fig. 25.17)

1. Spacing between rows loaded perpendicular-to-grain will be determined by spacing requirements in attached members (metal or wood parallel-to-grain).
2. If applied load is less than bolt load values in the attached parallel-to-grain loaded members, the bolt spacing may be reduced in direct proportion.
3. The distance measured parallel-to-grain between rows of bolts loaded perpendicular-to-grain should be at least:
   2.5 diameters when bolt length in main member is 2 diameters;
   5 diameters when bolt length in main member is 6 or more bolt diameters.
   For intermediate values use straight line interpolation.
4. End distance (parallel-to-grain) must not be less than 4 bolt diameters.
5. The distance from bolt centers to the edge toward which the bolt load is acting shall be at least 4 diameters.
6. The distance from bolt centers to the edge away from which the bolt load is acting shall be at least 1.5 diameters.

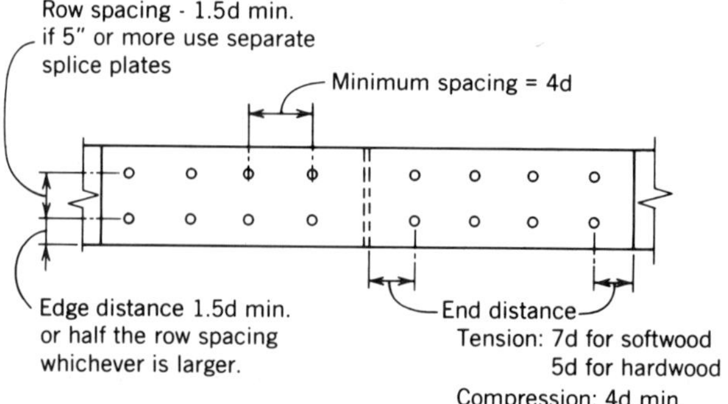

Fig. 25.16 Bolt spacing for parallel-to-grain loading.

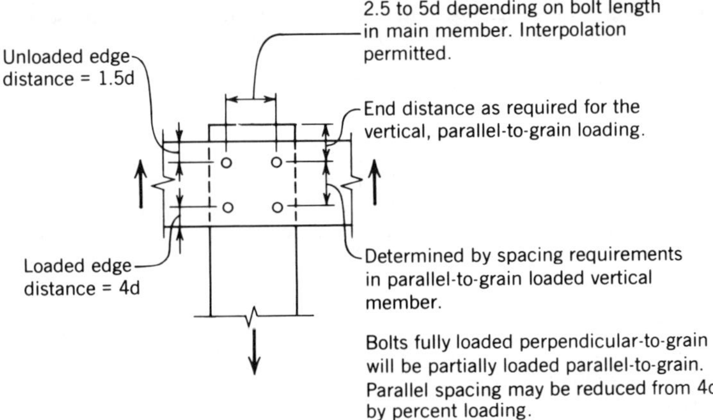

Fig. 25.17 Bolt spacing for perpendicular-to-grain loading.

Staggering of bolts is desirable in members loaded perpendicular-to-grain. The gravity axis of the members and the center of action of bolt groups should coincide. Bolt holes reduce the cross-sectional area of members. The reduced area (net section) is used in determining the ability of the member to carry loads.

If the designer does not have supervision of fabrication, the member strength in tension or compression is taken as the net section multiplied by the allowable tension or compression stress for the grade of lumber.

If the designer does have supervision of fabrication and can control the area of the knots, knot holes, and drilled holes, the design strength of the member is the net section accounting for all of these area losses and allowable clear wood stresses in tension and compression as shown in Table 25.29.

Table 25.29 Allowable Stresses* for Computing the Tension and Compression Strength at the Net Section† of Lumber, for Supervised Fabrication

	Dry Use		Wet Use
Species	Over 4-in. Thick, psi	4-in. Thick or Less, psi	Any Thickness, psi
Douglas fir-larch, dense	1730	2360	1570
Douglas fir-larch	1480	2020	1340
Douglas fir—South	1340	1820	1220
Southern pine, dense	1690	2130	1540
Southern pine	1450	1970	1320
Hem-fir	1220	1670	1110
Spruce-pine-fir	1040	1410	940
Eastern hemlock	1260	1710	1140
Western hemlock	1350	1860	1240
Lodgepole pine	1060	1450	970
Western cedar	1140	1520	1040

Note: Values are listed for common species. For other species see latest NDS [25.35]. For glulam see *Timber Construction Manual* [25.4].

* Table was derived from NDS [25.35].

† Net section is gross section reduced by area occupied by fasteners, knots, knot holes, or other voids.

These values are for normal load duration in dry use. Multiply by 0.8 for wet use. Load duration adjustments for wood should be used for bolt strengths. This method may not be used for glued laminated timber since knots are often hidden.

Staggered bolt holes are bolt holes arranged in parallel rows, not directly opposite one another in the rows, but offset by half the row bolt spacing. All bolt holes located at sections which are not at least four bolt diameters apart, measured parallel-to-grain, shall be considered in the same cross-section for determining net section for unsupervised fabrication.

Design of members containing bolted joints in shear should exclude the area of the member in the unloaded portion of the cross-section. If d_e is the depth of the member from the center of the bolt nearest the unloaded edge, to the loaded edge of the member, shear is calculated as

$$f_v = \frac{3V}{2bd_e} \tag{25.10.1}$$

if the bolts are five times the member full depth from the end of the member. The allowable shear F_v may be 1.5 times the tabulated value for the species and grade of wood member. The shear is calculated as

$$f_v = \frac{3V}{2bd_e}\left(\frac{d}{d_e}\right) \tag{25.10.2}$$

if the bolts are less than five times the full depth from the end of the member. The allowable shear F_v may not exceed the tabulated value for the species and grade of wood.

25.10.3 Split Rings and Shear Plates

Split rings and shear plates are special steel connector devices which have much greater lateral load resistance than bolts. These connectors are manufactured by the Timber Engineering Company (TECO) and are widely available.

A split ring is a steel ring and a bolt (Figs. 25.18*a* and *b*). The ring has a double-tapered cross-section and a radial gap in its circumference. A special grooving tool is used to cut the seat for the

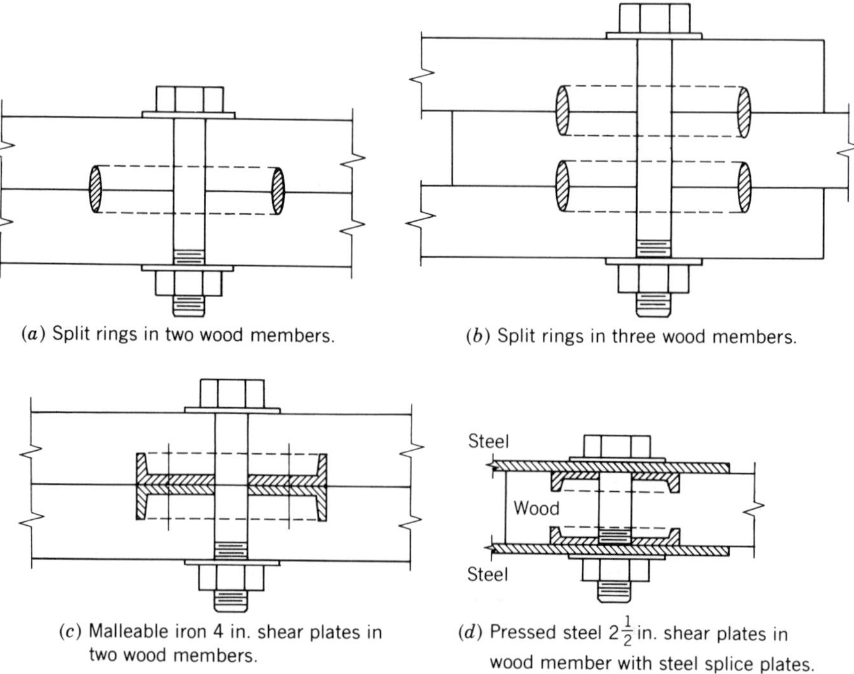

(*a*) Split rings in two wood members.

(*b*) Split rings in three wood members.

(*c*) Malleable iron 4 in. shear plates in two wood members.

(*d*) Pressed steel $2\frac{1}{2}$ in. shear plates in wood member with steel splice plates.

Fig. 25.18 Split rings and shear plates.

split ring in the wood members. This tool also drills the bolt hole. It is used in a drill press or a portable electric drill. The purpose of the bolt is to keep the members drawn firmly together. The split ring develops the lateral resistance. Split ring design values are for the single shear load transfer situation. Split rings are made in two sizes, 2½-in. and 4-in. diameter. Bolt diameters are ½ in. for 2½-in. split rings and ¾ in. for 4-in. split rings. Lag screws may be used instead of bolts.

A shear plate is a steel cup and bolt (Figs. 25.18*c* and *d*). The cup has a bolt hole through its center. There are two sizes, 2⅝-in. and 4-in. diameter. The small size is a pressed steel unit and the large size is malleable iron. Shear plates are used to connect wood to steel, with the shear plate transmitting load to the wood at its periphery and to the bolt at the central hole. The bolt then transmits load to the steel member in bearing. Shear plates are used to transmit column loads to footings through steel cover plates, to connect heel joints of trusses and web members to chords using steel straps or gussets.

Shear plates are also used for wood-to-wood joints which must be demountable. The shear plates are mounted in seats grooved in the wood with central bolt holes. They are fastened in place by nailing through holes provided, to prevent them from becoming detached and lost when the members are disassembled. Two shear plates, back-to-back, bear on the wood at their peripheries and bear on the bolt at their faces.

The 2⅝-in. shear plate uses a ¾-in. bolt and the 4-in. shear plate may be obtained for ¾-in. or ⅞-in. bolts. Lag screws may be used instead of bolts.

The grooving tool for shear plates is of different groove shape than the one for split rings. Grooving tools can be purchased or rented from connector suppliers.

Split rings and pressed steel shear plates are SAE-1010 carbon steel. Malleable iron shear plates are ASTM A47 Grade 32510. They may be black (ASTM A123) or galvanized (ASTM A153).

Design values for split ring connectors given in Table 25.30 are single shear values for normal load duration in S-DRY lumber used dry. Species groups are listed in Table 25.23. Load duration factors should be used. Moisture content adjustment factors are in Table 25.22. Table 25.30 gives lateral load values for loads at 0 and 90° to the grain (axis of member). For other angles of load to grain the chart in Fig. 25.12 can be used.

For piece thicknesses between those in the table, linear interpolation is used. For loaded edge distances between those in Table 25.30 linear interpolation of Q is permitted, between the minimum and maximum loaded edge distances.

Connector spacing terminology is explained in Fig. 25.19. The spacing of split ring and shear plate connectors is obtained from Figs. 25.20 and 25.21. The end and edge distance charts in these figures require a note of explanation. The minimum edge distance from these charts is for parallel-to-grain loading and for the *non-loaded* edge when the angle of load to grain is other than 0°; that is, not parallel-to-grain. In Fig. 25.20, for example, the minimum edge distance is 1¾ in. and is suitable for parallel-to-grain loading (0°). For other angles of load to grain the minimum loaded edge distances are 2⅛ in. at 15°, 2 7/16 in. at 30°, and 2¾ in. at 45° or more. These edge distances apply when the connectors are fully loaded, that is, to 100% of the tabulated values. Reduced minimum loaded edge distances are permitted when less than 100% of the full connector capacity is acting.

End distances for tension members are the same for all angles of load-to-grain, but may be reduced for less than 100% of full connector capacity acting, with the maximum reduction occurring at 62.5% of full connector capacity. End distances for compression members are a function of both load-to-grain angle and percentage of full load, down to 62.5%.

Spacings between connectors measured parallel (0°) and perpendicular (90°) to the member axis (i.e., to grain), are given in these charts for various angles of load-to-grain at full loading and can be interpolated down to 50% of full loading. Interpolation must be along radial lines connecting the 50 and 100% curves for the angle of load-to-grain. This is illustrated in the examples.

In the assembly of structures, connectors are often located in the end-cut surfaces (Figs. 25.22 and 25.23) and secured with lag bolts or pins and hardware. In a square-cut end (Fig. 25.22*a*) the connector load values are 60% of Q. For bevel end-cuts (Fig. 25.22*b*) the load values are calculated using the formulas in Figs. 25.22 and 25.23. The compound bevel end-cut (Fig. 25.23) values require that α be determined, using the bevel angles for each of the member faces. When lag screws are used with the connectors instead of bolts, the lag screw modification factors in Table 25.31 must be used.

The spacing of connectors in end-cut surfaces is the same as perpendicular-to-grain spacing for side-cut surfaces for α from 45° to 90°. If α is less than 45° the spacing of connectors for side-cut surfaces is used.

EXAMPLE 25.10.3

Three 2½-in. split rings are to be used in the wide face of a 2×6. They will be 75% loaded parallel-to-grain. Determine, using Fig. 25.20, the most compact spacing of the connectors.

Table 25.30 Split Ring Design Values*

Design values apply to ONE split ring and bolt in single shear when installed in seasoned wood that will remain dry in service and be subject to normal loading conditions

Split-ring diam. (inches)	Bolt diam. (inches)	Number of faces of piece with connectors on same bolt	Net thickness of piece (inches)	Loaded parallel to grain (0°)					Loaded perpendicular to grain (90°)					
				Minimum edge distance (inches)	Design value per connector unit and bolt (pounds)				Edge distance (inches)		Design value per connector unit and bolt (pounds)			
					Group A woods	Group B woods	Group C woods	Group D woods	Unloaded edge, min.	Loaded-edge (See Fig. 8.4H)	Group A woods	Group B woods	Group C woods	Group D woods
2½	½	1	1 min.	1¾	2630	2270	1900	1640	1¾	1¾ min. 2¾ or more	1580 1900	1350 1620	1130 1350	970 1160
			1½ or more	1¾	3160	2730	2290	1960	1¾	1¾ min. 2¾ or more	1900 2280	1620 1940	1350 1620	1160 1390
		2	1½ min.	1¾	2430	2100	1760	1510	1¾	1¾ min. 2¾ or more	1460 1750	1250 1500	1040 1250	890 1070
			2 or more	1¾	3160	2730	2290	1960	1¾	1¾ min. 2¾ or more	1900 2280	1620 1940	1350 1620	1160 1390
4	¾	1	1 min.	2¾	4090	3510	2920	2520	2¾	2¾ min. 3¾ or more	2370 2840	2030 2440	1700 2040	1470 1760
			1½	2¾	6020	5160	4280	3710	2¾	2¾ min. 3¾ or more	3490 4180	2990 3590	2490 2990	2150 2580
			1-5/8 or more	2¾	6140	5260	4380	3790	2¾	2¾ min. 3¾ or more	3560 4270	3050 3660	2540 3050	2190 2630
		2	1½ min.	2¾	4110	3520	2940	2540	2¾	2¾ min. 3¾ or more	2480 2980	2040 2450	1700 2040	1470 1760
			2	2¾	4950	4250	3540	3050	2¾	2¾ min. 3¾ or more	2870 3440	2470 2960	2050 2460	1770 2120
			2½	2¾	5830	5000	4160	3600	2¾	2¾ min. 3¾ or more	3380 4050	2900 3480	2410 2890	2080 2500
			3 or more	2¾	6140	5260	4380	3790	2¾	2¾ min. 3¾ or more	3560 4270	3050 3660	2540 3050	2190 2630

* From NDS [25.35], Table 8.4A, used by permission.

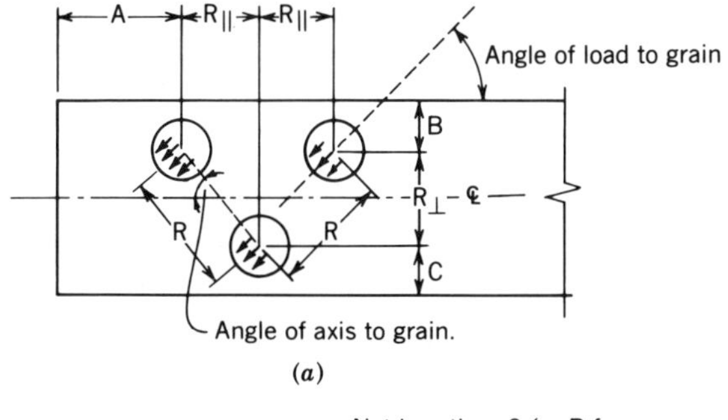

(a)

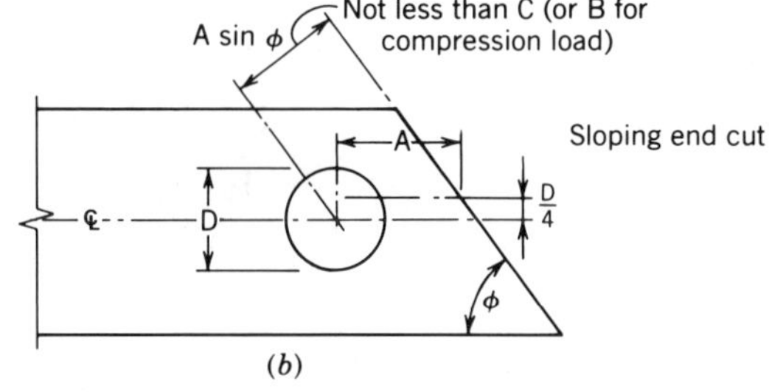

(b)

A = end distance
B = unloaded edge distance
C = loaded edge distance
R = connector spacing
$R_{\parallel}$ = parallel-to-grain spacing
$R_{\perp}$ = perpendicular-to-grain spacing
D = connector diameter, nominal

Fig. 25.19 Connector spacing nomenclature.

Solution

To facilitate interpolation, all spacings read from the charts are expressed in decimal form. The edge distance will be 1.75 in. minimum. This will leave a maximum perpendicular-to-grain spacing of $5.5 - 2(1.75) = 2$ in. To find the parallel-to-grain spacing, locate a radial line on Fig. 25.20, which crosses the horizontal line for 2-in. perpendicular-to-grain spacing at (75-50)/50 or one-half of the length of the radial line subtended by the 50% load curve and the 0° angle of load-to-grain curve. The point of crossing defines the parallel-to-grain spacing of 4 in. The most compact connector spacing is, therefore, 4 in. parallel-to-grain and 2 in. perpendicular-to-grain.

From the end distance chart of Fig. 25.20 the connector end distances for compression or tension are read for 75% of full load. For compression use the 0° angle of load-to-grain line and obtain an end distance of 3.2 in. For tension only one line is given in Fig. 25.20 for all angles of load-to-grain. At 75% of full load this line shows an end distance of 3.625 in.

The connector layout for the most compact spacing arrangement is shown in Fig. 25.24.

EXAMPLE 25.10.4

Two members which intersect at 90° (Fig. 25.25*a*) are to be fastened by two 2½-in. split rings. One member is a 2×8 and will be loaded perpendicular to its axis at the largest allowable value of Q. The other member is a 2×6 and will be loaded parallel to its axis by the load Q. The wood is Group B species. Determine the most compact spacing of connectors.

Solution

(a) The largest allowable value of Q obtained from Table 25.30 for 2½-in. split rings, loaded at 90° to the grain with connectors on one face of a 1½ in. thick piece of Group B wood, is 1940 lb.

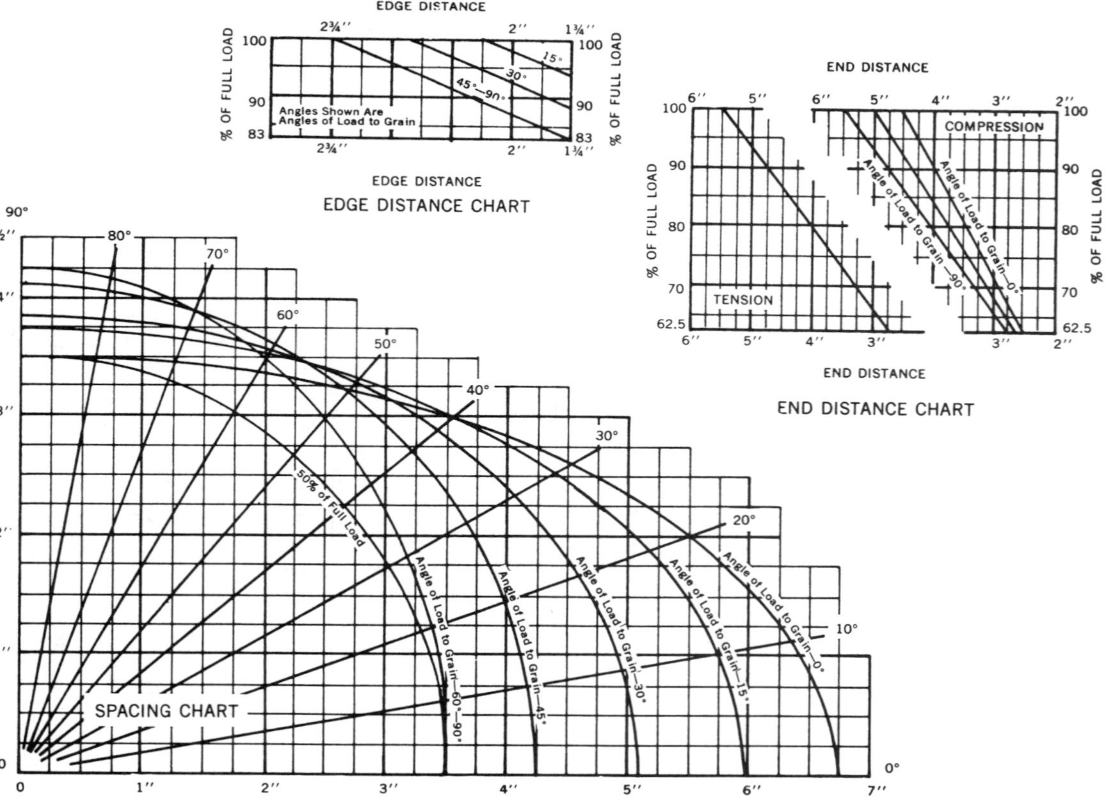

Fig. 25.20 Spacing chart for 2½-in. split rings and 2⅝-in. shear plates (from TECO [25.41] by permission of TECO, Chevy Chase, MD).

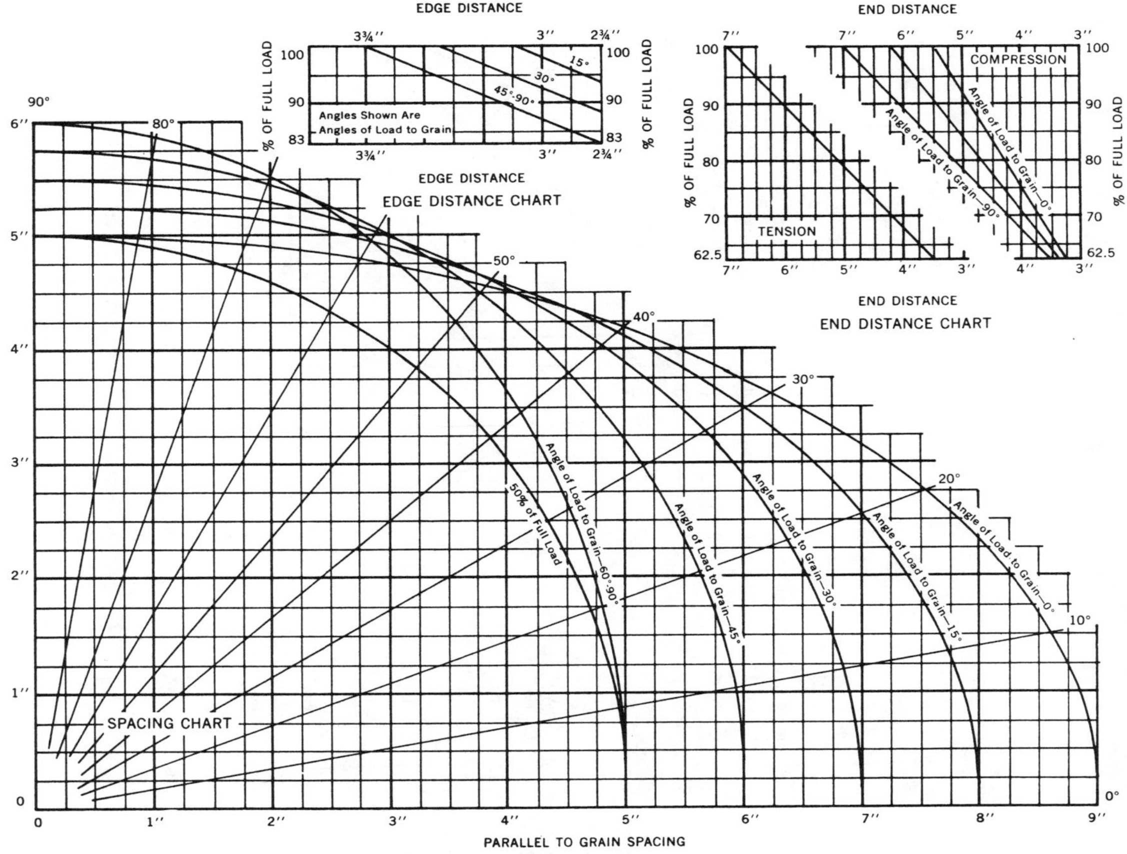

Fig. 25.21 Spacing chart for 4-in. split rings and shear plates (from TECO [25.41], by permission of TECO, Chevy Chase, MD).

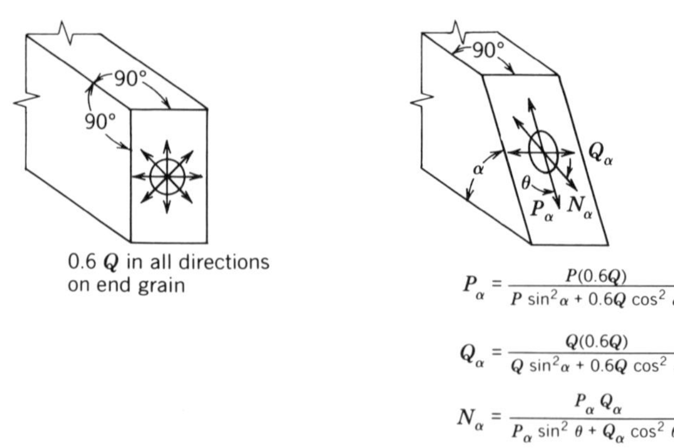

Fig. 25.22 Split rings in square and bevel end cut surfaces.

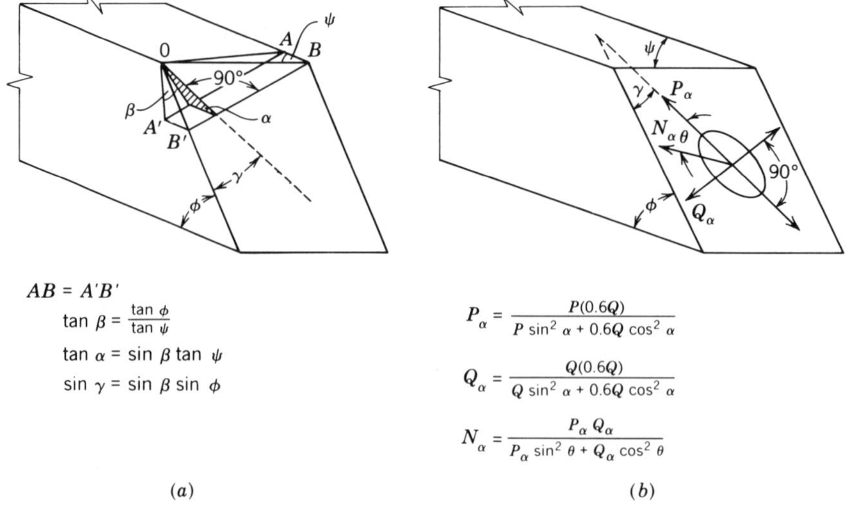

$AB = A'B'$

$$\tan\beta = \frac{\tan\phi}{\tan\psi}$$

$$\tan\alpha = \sin\beta\tan\psi$$

$$\sin\gamma = \sin\beta\sin\phi$$

(a)

$$P_\alpha = \frac{P(0.6Q)}{P\sin^2\alpha + 0.6Q\cos^2\alpha}$$

$$Q_\alpha = \frac{Q(0.6Q)}{Q\sin^2\alpha + 0.6Q\cos^2\alpha}$$

$$N_\alpha = \frac{P_\alpha Q_\alpha}{P_\alpha\sin^2\theta + Q_\alpha\cos^2\theta}$$

(b)

Fig. 25.23 Split rings in compound end cut surfaces.

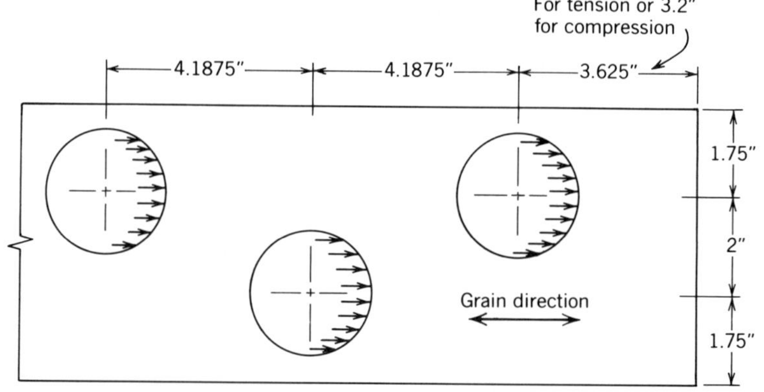

Fig. 25.24 Most compact spacing solution for Example 25.10.3.

Table 25.31 Modification Factors for Connectors Used with Lag Screws*

Factors apply to design values tabulated for connector units used with bolts

Connector size and type	Side plate	Penetration	Penetration of lag screw into member receiving point (Number of shank diameters) — Fastener species group (See Table 8.1A) I	II	III	IV	Modification factor[1]
2-1/2" split ring 4" split ring 4" shear plate	Wood or Metal	Standard	7	8	10	11	1.00
		Minimum	3	3-1/2	4	4-1/2	0.75
2-5/8" shear plate	Wood	Standard	4	5	7	8	1.00
		Minimum	3	3-1/2	4	4-1/2	0.75
2-5/8" shear plate	Metal	Standard and Minimum	3	3-1/2	4	4-1/2	1.00

1. Use straight line interpolation for intermediate penetrations.

* From NDS [25.35], Table 8.4C, used by permission.

(b) For ease of interpolation, spacing values read from the tables and charts will be expressed as decimals. Table 25.30 indicates that for the above value of Q the minimum required loaded edge distance is 2.75 in. and the minimum required unloaded edge distance is 1.75 in. With these edge distances in a 2×8 the perpendicular-to-grain spacing is $7.25 - 2.75 - 1.75 = 2.75$ in. See Fig. 25.25*a*.

(c) The spacing chart, Fig. 25.20, is used to obtain compatible minimum spacing values for connectors in the two principal directions for the 2×8. With the above 2.75-in. perpendicular-to-grain spacing determined, use the 90° angle of load-to-grain curve to obtain a parallel-to-grain spacing of 2.7 in. (The 2.75-in. perpendicular-to-grain spacing and the 2.7-in. parallel-to-grain spacing are coordinates of a point on the 90° angle of load-to-grain curve.) The most compact spacing in the 2×8 that will permit a perpendicular-to-grain connector load of 1940 lb is 2.75 in. perpendicular and 2.7 in. parallel-to-grain.

(d) The 2×6 must have a matching spacing of 2.7 in. perpendicular-to-grain. Will that be permissible at a load of 1940 lb parallel-to-grain? To check this, determine the parallel-to-grain connector load as a percentage of the full parallel-to-grain connector load which could be allowed. According to Table 25.30 for a 2½-in. connector in one face of a 1½-in. thick Group B wood P is 2730 lb. At 1940 lb the connectors would be 1940/2730, or 71% loaded in the 2×6.

(e) From either Table 25.30 or Fig. 25.20, the minimum edge distance for parallel-to-grain loading is 1.75 in. This leaves a perpendicular-to-grain spacing of $5.5 - 1.75 - 1.75 = 2.0$ in. which is less than the 2.7 in. required to match the spacing needs in the 2×8. *Adequate spacing cannot be provided in the 2×6 if Q is to be the maximum permissible value of 1940 lb*. See Fig. 25.25*a*.

EXAMPLE 25.10.5

If the 2×6 in Example 25.10.4 could be replaced by a 2×8 (see Fig. 25.25*b*) would adequate spacing be available for the connector?

Solution

(a) In that case the perpendicular spacing in the parallel-to-grain loaded 2×8 would be $7.25 - 1.75 - 1.75 = 3.75$ in., which is more than adequate to meet the spacing requirements in the perpendicular-to-grain loaded 2×8.

(b) Will this provide sufficient parallel-to-grain spacing for the 71% loaded split rings? It will, because for a perpendiçular-to-grain spacing of 2.7 in. or more, interpolation along radial lines will give a load capacity of 71% or more. Changing the 2×6 to a 2×8 will provide adequate room for the

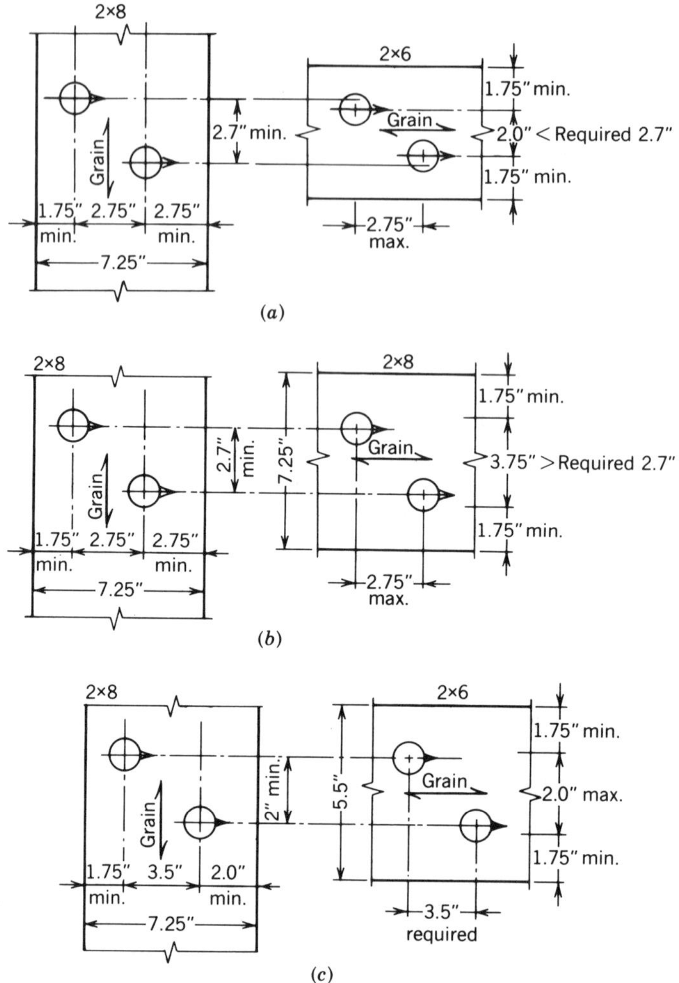

Fig. 25.25 Examples using split ring connectors.

connectors in both pieces. See Fig. 25.25*b*. As long as the required spacing for the perpendicular-to-grain member is provided, matching parallel-to-grain loaded members (with the necessary end and edge distances) will carry the load.

EXAMPLE 25.10.6

If the 2×6 in Example 25.10.4 cannot be changed to a 2×8, what is the largest value of load that can be carried by this joint?

Solution

(a) Table 25.30 shows Q = 1620 lb with a loaded edge distance of 1.75 in. and 1940 lb for a loaded edge distance of 2.75 in. Other values of Q vs. loaded edge distance can be interpolated between these extremes and the value at 2 in. is 1700 lb. This will provide a maximum perpendicular-to-grain spacing in the 2×8 of 7.25 − 1.75 − 2.0 = 3.5 in. See Fig. 25.25*c*.

(b) The parallel-to-grain loaded 2×6 with a spacing of 2 in. perpendicular-to-grain and 3.5 in. parallel-to-grain is represented in Fig. 25.20 by a point that falls on the 30° angle of axis-to-grain radial line. Interpolating between the 50% curve and the 0° angle of load-to-grain curve it is found that the connector may be 66.5% loaded—66.5% of 2730 lb is 1810 lb which exceeds the 1700-lb limitation on the loading of the 2×8. *The joint can be safely loaded to 1700 lb.*

Shear plate lateral load values are also tabulated in the NDS [25.35] in the same way as for split rings. For metal side plates the values may be increased 18, 11, 5, and 0% for species Groups A, B, C, and D. Shear plate spacing requirements are treated in an identical manner as for split rings; that is, using Figs. 25.20 and 25.21.

If lag screws are used instead of bolts, the modification factors for lag screws in Table 25.31 must be used. Lag screw design values are given in the NDS [25.35].

Long rows of connectors develop more highly concentrated loads in the extreme end connectors. The reductions for such arrangements, described in Section 25.10.2 for bolts, also apply to split ring and shear plate connectors.

Design of members containing split ring or shear connector joints in shear should exclude the area of the member in the unloaded portion of the cross-section, as described in Section 25.10.2 for bolted joints. For split rings or shear plates, the dimension d_e is measured from the edge of the connector nearest the unloaded edge of the member to the loaded edge of the member.

25.10.4 Metal Plate Connectors

Metal plates with punched-out teeth are used in the fabrication of light frame trusses. The plates are usually 18 gage galvanized steel. The teeth vary in shape according to the manufacturer but most designs have lateral resistance values of about 100 lbs per square inch of plate surface, depending on the species of wood with which they are used. They have a residual net section of about 50% of the gross plate section.

Toothed metal plates are usually sold to truss fabricators who use the plate manufacturer's truss design, engineering services, and fabricating machinery. Design practices for metal plate connected trusses are described in the *Truss Plate Institute* (TPI) *Specification* [25.42].

Metal plates perforated with holes for nailing are available from hardware manufacturers. They are of the same metal gage thickness as the toothed plates, with lateral load values corresponding to the nail size and spacing.

25.10.5 Glued Gussets and Splices

Gussets and splice plates of plywood are often used for truss member connections or to fabricate structural joints for beams. They are bonded to the members with phenol–resorcinol–formaldehyde resin glue under pressure to create close contact between the surfaces. Pressure may be produced by clamping, or by closely spaced bolts, nails, or wood screws. The shear strength between the plates or gussets and the lumber members is usually taken as the design shear strength of the wood. The plywood properties for designing the plates and gussets are obtained from the PDS [25.7]. Splice plates must have a glued face length approximately 30 times their thickness to develop 75% of their tension and bending strength. This is necessary because the load carried in shear must be transferred to the veneer layers through several cross-ply veneer layers.

25.10.6 Structural End Joints

Glued structural end joints in lumber may be sloping scarf joints or finger joints, as shown in Fig. 25.26. Sloping scarf joint strengths are considered to be equal to the strength of the lumber if the scarf slope is 1 in 8 or less in straight grained wood (1 in 10 if the lumber is to be used above 16% moisture content), PDS [25.7]. The scarfed surface must be smoothly machined and bonded under pressure with a waterproof phenol–resorcinol–formaldehyde adhesive or an equal strength heat resistant and waterproof resin glue.

Structural finger joints are produced by laminators and by companies that specialize in making long lumber from short lengths. The design of the fingers is critical with regard to slope of the sides

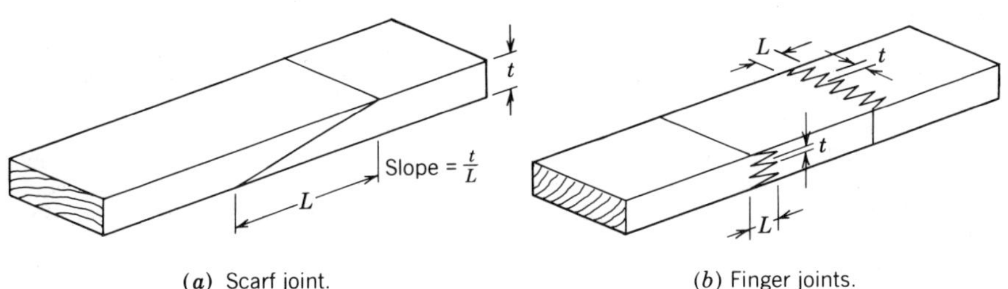

Fig. 25.26 Structural scarf and finger joints.

and thickness of the tips. Quality control agencies provide certification marks for structural end-jointed lumber. Material bearing these certifications may be used at the compression, tension, and bending design values for the lumber.

25.11 DURABILITY

When wood must be used in situations where decay could develop, or when it must be protected from fire, a variety of preservative treatments and fire retardants are used. The cost of such treatments is significant and they are specified only when they appear needed. Wood used in contact with soil (poles, piling, sills, cross-ties, mine timbers) is almost always treated with a preservative. Products for exterior exposure (wood window sash, doors, open decks, advertising signs, timber bridge members, wood boats, truck body parts) should usually be treated unless they can be shielded from the weather by roofing materials or flashings. But if the service conditions will not cause the wood moisture content to exceed 19%, untreated wood will not be destroyed by fungi, insects, or marine organisms.

Most of the wood used in buildings will serve well without treatment, if the designer can avoid features that create moisture traps and condensation. The roofs of extremely large buildings probably cannot be made entirely leak proof, so preservative treatment should be considered, especially if the essential structural support members will be exposed to the leakage.

Insulation and vapor barriers should be designed to exclude moist air from wood at temperatures below the dew point. Attic spaces and crawl spaces should be well ventilated.

Air conditioned or refrigerated buildings may condense water from the outdoor air in summer in humid climates. Humidified buildings (textile mills, furniture plants, wet process buildings, paper mills, steam rooms, laundries, food processing plants and kitchens, livestock buildings, enclosed pools, reservoir roofs) will have high dewpoint air which condenses readily in the winter if it contacts cold roof and wall structures.

Some features of ordinary residential, commercial, and school buildings intended to conserve energy by reducing air infiltration may create conditions favorable to condensation unless given careful design attention.

Wood in direct contact with concrete and masonry should be preservatively treated.

Fungi

Decay can occur when fungi feed on wood. Fungi exist in most natural environments and their spores, carried by the air, are ever present. Fungi will flourish if there is adequate moisture, adequate oxygen, and suitable temperatures. Wood moisture content below 20% is insufficient to produce decay. Oxygen sufficient to cause decay is usually present, except in submerged wood structures. Temperatures between 40 and 115°F are favorable to fungus activity and 70 to 95°F is especially ideal.

Most covered structures with good roofing systems will be too dry for fungi to grow. Dryness is the best protection from decay. When dryness cannot be assured, preservative treatment of the wood will poison the food supply of the fungi and protect the structure.

Structures that are occasionally wet may be expected to have occasional decay. Fungi can lie inactive when the wood is dry and burst into activity during periods of wetness. End grain in the top of poles and posts can absorb water and decay even though most of the pole or post is too dry to decay. Checks, daps, joints, and areas around bolts and nails, can trap water and be nuclei for localized decay, if they are exposed to rain or snow.

Insects

Insects may also destroy wood by building galleries in the timber. They, too, require moisture, and dryness (below 20%) is the best preventative of insect damage. Certain types of termites are capable of transporting water from the ground to dry wood in structures.

Subterranean termites live in the soil and feed on wood in or near the soil. They will construct tubelike structures bringing moisture to otherwise dry wood, making it digestible. They are particularly active in the southern United States. Soil poisoning which must be renewed at intervals is a common method of combatting these insects.

Dry wood termites inhabit wood and feed upon it. They can digest wood with moisture content as low as 10%. Infected structures can be fumigated and treated with toxic powders and liquids. Chemical treatments to prevent termite damage are available and are used on wood located near foundations or in soil contact.

Damp wood termites, as their name suggests, require more moisture to feed on wood. They occur in the United States west of the Rocky Mountains and in the Pacific Northwest. Damp wood termites often damage wood water tanks and fresh water piling.

Carpenter ants do not feed on wood but inhabit it. They often occupy areas that are decayed or decaying. They augment the spread of decay. Fumigation and poisoning by dry or liquid treatments are effective measures.

Whenever insects are discovered in a wood structure it should be inspected carefully to determine the extent of damage. Insects often destroy the entire interior of a member without showing much external evidence of their presence.

Marine borers are crustaceans and mollusks found in salt water where they do enormous damage to untreated wood piling. Marine biology specialists can offer guidance on the types of marine borers living in various waters. Generally heavy impregnation treatment is necessary to give reasonable life to marine piling. The type of preservative should be specified by an expert consultant or an experienced treating organization. These marine organisms sometimes develop immunity to preservatives and recommended treatment changes with time.

The publications *Pressure Preserved Plywood* [25.13], AITC 109 [25.5], and Federal Specification TT-W-571 [25.23] are useful for a suitable wood preservative.

Pressure Preservatives

The basic pressure preservatives are creosote, coal tar–creosote, oil-borne pentacholorophenol treatment, and waterborne salts of various types. Pressure treatment is recommended. Dip and brush treatments are less lasting and do not penetrate side grain well, but are often effective in penetrating end grain by soaking or dipping. Even pressure treatments do not completely penetrate most timber but produce a shell of treated wood on the outside of the timber. Wood should be cut to length, drilled, and dapped before treatment. Fabricating after treatment exposes untreated wood where fungi gain a foothold. When holes or end cuts must be made on treated wood, treatment of the exposed untreated wood by soaking or the injection of preservative liquids or greases into the areas so exposed will reduce decay prospects.

The specification should be chosen upon the advice of a reliable authority such as the American Wood Preservers Institute. It should indicate the preservative, as well as its concentration, penetration, and retention requirements.

Some species of wood which are difficult to penetrate should be incised by puncturing the surface to provide avenues of entry for the preservative. Incising usually does not affect the strength of the timber and preservatives usually have no effect on strength.

Preservatives do not necessarily render wood completely immune to decay but they extend the life immensely, 30 to 50 years or more. Preservatives carried by oil also reduce the ability of weather to moisten the wood and are especially useful for highway bridge treating. They improve the weatherability of these outdoor structures.

Creosote and creosote–oil mixtures are well suited to treatment of heavy timbers, especially if exposed to weather. However, they are not clean treatments and are avoided if contact with clothing is a consideration. They have a distinct odor and are not used in enclosed structures. Wood treated with these preservatives cannot be painted.

Pentachlorophenol in No. 2 fuel oil is used for utility poles and building poles, lumber, and plywood. It is oily and not good in contact with clothing, or materials that will be stained by oil. It cannot be painted.

Pentachlorophenol in more volatile solvents is a clean paintable treatment. Upon evaporation of the solvent the odor is minimal. However, it is not recommended for food storage buildings, bins, silos, etc. Instead, copper-8-quinolinolate is a more suitable treatment.

Pentacholorophenol in liquefied petroleum gas is a clean, dry, paintable treatment which evaporates rapidly following treatment. It is not always recommended for soil contact situations but has performed well in the less humid regions of the country.

Waterborne salts are used for lumber and plywood extensively because they offer certain cost advantages, are not flammable, and are clean and paintable after seasoning. Wood so treated is not very weatherable, but if painted is quite satisfactory. Some waterborne salts will leach in the presence of water. Others are highly leach resistant. Chromated copper arsenate is an effective waterborne salt treatment for marine piling in southern waters

Waterborne preservatives will swell the wood, cause some warping, and twist. Lumber treated with these preservatives must be redried in kilns if the S-DRY lumber specification is essential to the designers purpose. Some waterborne salt solutions will corrode metal fastenings and require the use of galvanized or stainless steel nails.

Fire Retardant Treatments

Fire retardant treatments are used to reduce the flamespread and the contribution of wood to fueling a fire. They are waterborne salt treatments and present problems of wood warping unless carefully dried after treatment.

Fire retardant treatments are used for lumber, plywood, laminating lumber for glulam beams, and decking. Fire retardant salts are often hygroscopic and raise the equilibrium moisture content of the wood, and so may be unsuitable where relative humidity is high. In locations where fire retardants are desired, but where relative humidities are 70% or more, an exterior nonhygroscopic type should be specified.

Fire retardants are not toxic to fungi and do not prevent decay. They do, however, reduce smoke evolution, slow the burning rate of wood exposed to intense heat, and often permit increased building space limits in otherwise limited fire zones as defined in building codes.

Decisions about specifying fire retardants involve balancing the economics of treating cost and insurance rate benefits. The treatments are often desirable for industrial, storage, and retail sales buildings, places of public assembly, construction scaffolding, shopping centers, schools, and churches. They are rarely used in residential structures, except apartment dwellings and nursing homes.

25.12 TRUSSES

Wood trusses are an efficient form of load-carrying structure. They are often used for buildings with masonry walls, supporting plywood or metal roof sheathing. Wood trusses have largely replaced conventional rafter framing of wood frame buildings because of this efficiency. They can be spaced 24 to 48 in. apart to support plywood sheathing or with wood purlins to support metal roof panels. They may be erected rapidly to enclose a building in a minimum amount of time. Within their load and space range, trusses are more economical than sawn lumber and glulam beams.

Wood trusses must be erected with care since they possess little lateral stability and will buckle and overturn unless they are laterally restrained by well-designed bracing. Bracing must be installed to prevent lateral movement of top and bottom chords. This lateral buckling characteristic requires strong emphasis. All bracing specified by the engineer or the truss fabricator must be completely installed before any loads are placed on the trusses. Failures during erection can be caused by gravity loads or by wind.

Trusses should be supported close to the heel joints. The trusses should never be field-modified to make them fit without the advice of a competent structural engineer. Chords should never be cut or drilled to accommodate electrical wiring and utilities.

Lower chord loads should be applied with strict adherence to the design to avoid pulling the lower chords off the trusses. Gypsum ceilings may be nailed directly to lower chords because their weight is then uniformly distributed. Suspended ceilings should be attached to avoid points of excessive concentrated load. Suspended ceiling weight should bear on every truss, not just alternate trusses.

Sprinkler piping and plumbing weight should be carried by the top chords. Mechanical equipment should also be suspended from top chords by well-designed hangers, and never suspended from the lower chords unless they have been designed for the concentrated loading.

Handling and hoisting operations should be carefully supervised to avoid damaging trusses by excessive deflection and shock loads. Lifting cradles and straps are often necessary to avoid handling damage.

Light Frame Trusses

Light frame trusses are manufactured by fabricators using the better grades of dimension lumber. Machine stress rated lumber, visually graded lumber, and veneer laminated lumber are all used in trusses. The common types are metal plate-connected and pin-connected with steel tubular web members. Numerous manufacturers issue truss catalogs with load charts, span-tables, and technical information from which the designer can obtain dimensions and capacity data.

Engineers may not need to design these trusses because design services are available from the metal plate connector manufacturers through the fabricators, including drawings carrying the seal of a licensed engineer. However, when it is necessary to review the design of an existing truss structure Ref. 25.42 is valuable. A detailed discussion of truss design according to the *TPI Specification* [25.42] is given by Hoyle [25.26].

Light frame truss fabricators are licensed by the metal plate connector manufacturers who have engineering staffs and computerized design facilities to provide complete design service. Their designs have the approval of most building code authorities.

Roof trusses may be pitched rafter designs, top or bottom chord bearing parallel chord trusses, scissors trusses, and curved chord designs; there are no "standard" configurations. Camber is usually provided for dead load deflection and creep. Parallel chord trusses are often used for floor systems in buildings.

Light frame trusses are manufactured for spans of 16 to 60 ft and sometimes longer. Load capacities of single light frame trusses of 120 lb/ft are possible. The limitation is usually the web-to-chord capacity at the heel joint.

Several manufacturers fabricate metal web wood trusses and plywood gusset trusses which serve the same general range of performance as the metal plate connected trusses. Plywood gusset truss design data can be obtained from APA, the Midwest Plan Service (Ames, Iowa), and the extension departments of many agricultural colleges.

Heavy Wood Trusses

Wood trusses are also manufactured with split ring connectors and shear plates. They may be designed using the connector data found in Section 25.10. They may be field or shop fabricated and are often large multiple chord trusses. Such trusses may have glued laminated timber chords and steel gusset plates with shear plate connectors at the joints. Spans of 100 ft or more are not uncommon.

Bowstring Trusses

Bowstring trusses using curved glued laminated chords and glulam or sawn timber webs are connected by steel gussets and shear plate connectors or bolts. The buckling strength of the web-to-chord connector plates must receive design attention. Bowstring trusses are highly efficient because the variation of truss depth along the span maintains a nearly uniform chord force. Ideally, the curved chord would be parabolic but in practice it is usually formed as the arc of a circle without sacrificing efficiency. Bowstring trusses impose very low web forces when uniformly loaded. Wind and seismic loads, unbalanced gravity loads, and concentrated loads impose larger web member forces. These trusses are usually spaced 8 to 20 ft, and span 100 ft or more. They are popular for warehouses, industrial buildings, recreational buildings, aircraft hangers, and auditoriums.

Truss Deflection

The deflection of trusses may be estimated by usual elastic methods of structural analysis, the results of which are available in the *Timber Construction Manual* [25.4], giving consideration to joint connector clearances, since much of the deflection relates to the slip of joints.

Deflection limitations for trusses are the same as those for beams, given in Table 25.17. If camber is provided, and if the expected effect of creep is included in the camber specification, the permissible deflection can be assumed to occur under full dead and live loads.

Camber for flat trusses should include a suitable drainage pitch (¼ in./ft from high point to drain) to avoid ponding loads (Section 25.8), and compensate for creep and dead load deflection. The drainage camber is sometimes applied only to the top chord. Large trusses are often cambered for the full dead plus live load, drainage, and creep requirements.

For metal plate connected trusses protected from the elements, creep may be 2.0 to 2.5 times the permanent load deflection. For glued nailed plywood gussets the creep allowance may be 1.5 to 2.0 times permanent load deflection and for unglued, nailed plywood gusset trusses the factor should be about 3.0 according to experience and tests.

25.13 DECKING

Lumber decking is used for roof and floor systems where spans are too long for plywood to serve without excessive deflection. It may also be used for walls of bins or compartments for storage of bulk materials or for its aesthetic and structural qualities in certain architectural styles of building. Typically, decking has tongue and groove edge patterns to make a continuous interlocking joint. Loads placed on one course of decking will receive support from adjoining courses and the pieces will deflect essentially as a panel. Tongue and groove (T & G) patterns are sometimes cut on the ends of the pieces as well as the sides. Such decking is said to be "edge and end matched."

Another form of decking has grooves on both edges. Wood splines are fitted in these grooves to provide the continuity.

Decking is usually made from rough dimension lumber of standard size run to the T & G pattern. The distance across the resulting wide face is less than the standard lumber width. The actual face width of nominal 6 in. and narrower decking is 1 in. less than *nominal;* wider sizes are 1¼ in. less than nominal. The type that is grooved on both edges to receive splines is the full width of dressed dimension lumber, as there is no loss in cutting the groove. When figuring quantities it is necessary to account for the reduced coverage of T & G decking.

Lumber decking nominal thicknesses are 2 and 4 in. Some grade rules list widths nominal 6 in. and wider, while others list nominal 4–12 in. Six- and 8-in. widths are most usual. Wider deck, unless very carefully seasoned, will show unacceptable shrinkage gaps if it shrinks in place, or may buckle if it is tightly laid and then expands in a moist environment (roof leaks).

Nominal 3- and 4-in. thick deck has a double T & G. These sizes can be obtained with holes predrilled through the width along the neutral axis so the courses can be spiked together to draw them up tightly when they are laid.

Decking is basically of two grades, Select and Commercial. Usually it is manufactured S-DRY but some producers, recognizing the inherent problems of shrinkage and unsightly gaps, prefer to supply MC-15 decking (see Section 25.4) for jobs where appearance and good fit are important. Unseasoned decking is a highly unsatisfactory product for buildings and should be avoided.

Decking stress and deflection calculations are made using conventional beam theory, when the lay-up pattern has the decking end joints at the same cross-section in all the courses. But decks that are continuous over several supports are seldom arranged with all the end joints at the same location in adjacent courses. Offsetting the end joints improves the resistance to *lateral loads* by virtue of the interlocking effect. Furthermore, there are economic advantages to using random length decking in terms of availability and price and random lengths are actually easier to install. Random length end joints must be laid according to some simple rules.

Three good interlocking end joint arrangements and the formulas for their design for uniformly distributed loading are given in Fig. 25.27. These formulas are based on evaluative testing and are theoretically conservative if the spacing rules for end joints are observed. The formulas are only valid for four or more equally spaced supports. Single span and two span decks should not have any end joints located between supports, only over supports.

Decks with overhanging ends should have no joints beyond the end support beams. It is good practice to avoid end joints over the end support beams when overhangs exist.

In designing decks the moment of inertia I and the section modulus S are usually computed for a 1-ft wide section. The space between tongues and grooves is considered negligible. The formulas in Fig. 25.27 account for the influence of end joints and the section properties (I and S) are for a unit width section of deck.

For concentrated loads on decks laid in the "Random Rule" (as described below) pattern (Fig. 25.27*c*), deflections should be computed using a continuous beam formula with I reduced by 20%. For concentrated load on the other two lay-up patterns given in Fig. 25.27 the 20% reduction in I will give a slightly high estimate of deflection.

Calculation of strengths under concentrated load should use continuous beam formulas using the moment of inertia for the smallest net section in the region of the highest bending moment. For combined simple and two-span continuous lay-up this will be one-half the gross section and for random and cantilevered pieces intermixed, it will be two-thirds the gross section.

The "Random Rules" for decking lay-up are slightly different for nominal 2-in. and for nominal 3- and 4-in. decking.

For 2-in. sawn, T & G, heavy timber decking, the "Random Rules" are:

1. End joints in adjacent courses must be at least 2 ft apart.
2. End joints in every second course must be at least 6 in. apart, measured in the span direction.
3. End joints not more than 6 in. apart measured along the span must be separated by at least two continuous courses in that area.
4. All pieces must bear on at least one support (no floaters allowed).
5. If the decking is predominantly short pieces, special attention must be given to providing a continuous tie between the supporting beams at intervals of six or seven courses.
6. All pieces must be nailed to the supports which they bear upon. Nailing shall be designed for uplift forces. Provision should be made for expansion if wetting is a factor. On pitched roofs install deck with tongue edge on the high edge.
7. Minimum length pieces shall not be less than 75% of the support spacing.
8. Two-in. decking is secured to beams with at least two nails for nominal 6-in. widths and three nails for nominal 8-in. or wider. One of these nails is driven slantwise just above the tongue. This nail is driven first to draw the courses together. The first two courses on each edge of a deck should contain lengths such that the joints can be distributed to positively fix the beam spacing. The two courses in Fig. 25.27 satisfy this condition.

For 3- and 4-in., sawn, T & G, heavy timber decking, the "Random Rules" are:

1. All pieces shall be nailed to at least one support.
2. Minimum distance between end joints in adjacent courses is 4 ft.
3. Minimum distance, measured along the span, between joints in every second course shall be 6 in.
4. End joints within 6 in. of one another, measured longitudinally, must be separated laterally by two courses without end joints in that area.
5. For 3-in. decking the mixture of lengths must be at least 40% 14 ft or longer, with a minimum of 10% less than 10 ft and no "floaters". For 4-in. decking, the mixture must be 25% or more

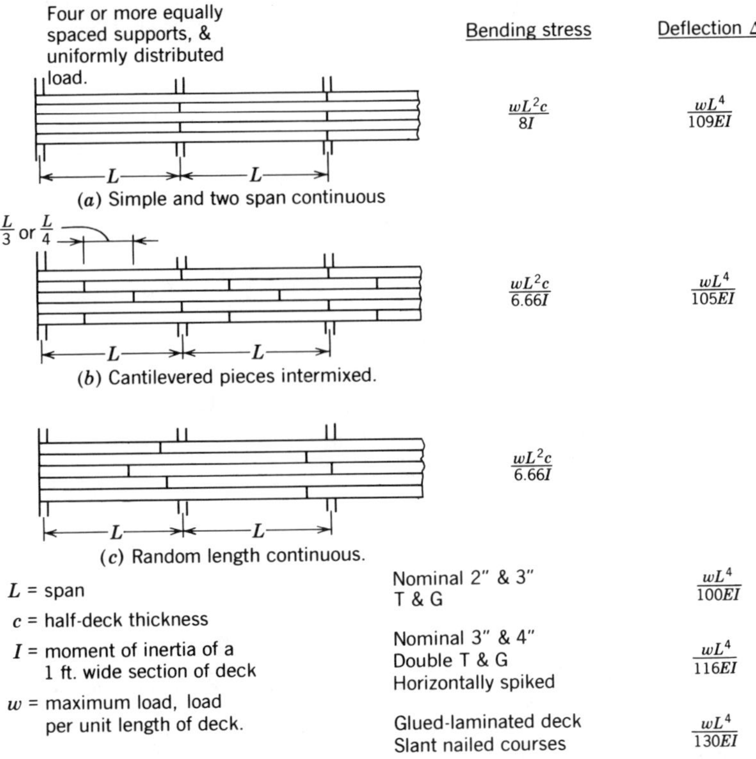

The bending stress formulas are modified for end joints as given.

Fig. 25.27 Decking lay-up arrangements.

16 ft or longer, 50% or more 14 ft or longer, and not more than 10% in the 5- to 10-ft range, and no "floaters". For both thicknesses, a maximum of 10% may be in the 4- and 5-ft range. These percentages are in terms of total number of lineal or board footage in the deck.

Overhangs require careful attention in design. The *Timber Construction Manual* [25.4] contains comprehensive charts for the design of decks with overhangs. Loaded overhangs reduce interior span deflections and seldom raise the bending stresses unless the overhang is very long.

Overhang deflection for two to five span decks with equal overhangs on each end may be computed using the curves given in Fig. 25.28. Coefficient C_1 applied to the computed deflection for the end span of a similar uniformly loaded deck without overhangs, gives the end span deflection of such a deck with overhangs. C_2 applied in the same way gives the overhang deflection. C_2 changes rapidly as overhang length exceeds 30% of the support spacing. The sensitivity to unbalanced loading makes predicting deflections on long overhangs unreliable; thus, overhangs exceeding 30% are not recommended.

Uniformly distributed loads on overhangs and interior spans is not a realistic assumption. Snow can blow off overhangs. Due to exposures, ice can accumulate on overhangs.

To prevent irregular deflection of overhangs, the ends should be laterally connected by a continuous header or strong fascia board.

Glued Laminated Decking

Glued laminated decking is produced by numerous companies. It has the advantages of dryness, straightness, low waste from twist, ease of installation, deep T & G connections, and decorative faces on utilitarian cores and backs. Such decking is usually dried to 12% moisture content and shows little shrinkage after laying. The plywood industry manufactures thick decking plywood for use on 4-ft spans, with T & G edges for continuity.

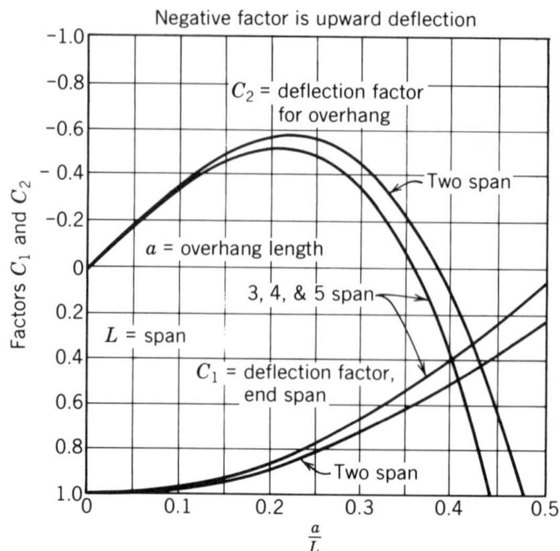

Fig. 25.28 Deflection factors for overhangs (from Ref. 25.4, used by permission of AITC).

Wood decks have good insulating properties. The thermal conductivities of dry decking, in BTU per hr/sq ft/°F, are in the range of 1.1 for southern pine to 0.78 for western red cedar. Figure 25.29 gives the overall coefficient of heat transfer for wood deck constructions with various thicknesses of insulating board for two insulation thermal conductivities. The thermal resistance or *R*-factor is the reciprocal of the U-value in this figure.

For design of glued laminated decking the lay-up rules are generally in accord with those for solid sawn decking of the same thickness. Glulam decking manufacturers (such as Weyerhaeuser Co., Boise Cascade Corp., Potlatch Corp.) issue rules specific to their products.

25.14 ARCHES AND DOMES

Wood arches are usually of glued laminated timber. The oldest glulam building in North America, built in 1935 at Madison, Wisconsin, is a three-hinged arch. Two-hinged arches of a variety of styles are also used. Arches constructed of lumber with glued plywood haunch gussets or with metal plate connectors, and curved Warren truss style barrel arches are also frequently designed.

Sketches of several arch styles are shown in Fig. 25.30. The Tudor arch with tapered leg and rafter portions is a common glulam design. The laminations are continuous on the inside edge and tapered on the outside edge. The depth of cross-section at the tangent points of the column and rafter are usually made equal to simplify construction. The depth at the crown should not be less than the width at that point. The degree of taper should not be large because the strength of the wood at the extreme outer edge where the grain runs out of the face is reduced as the angle is increased. Consultation with a glulam timber manufacturer is always desirable. The short laminations on the outside of the haunch are not considered to contribute to design strength.

The design strength in bending for the curved part is reduced to compensate for the stresses induced by forming. This reduction factor is

$$C_c = 1 - 2000\left(\frac{t}{R}\right)^2 \tag{25.14.1}$$

where R is the radius of curvature and t is the lamination thickness.

The laminations are well-seasoned and are formed cold. Thickness-to-curvature radius should not exceed 1/100 for southern pine and 1/125 for other softwoods. The glulam manufacturers recommend 9.33 ft as the minimum radius of curvature for ¾-in. thick laminations and 27.5 ft for 1½-in. lamination thickness.

Members curved in the plane of flexure (i.e., as an arch) experience compression or tension stress perpendicular to the neutral plane (in the radial direction). This radial stress is tensile when the load puts the inside edge in tension and compressive when it put the inside edge in compression. Radial stress at the neutral plane is calculated by the formula

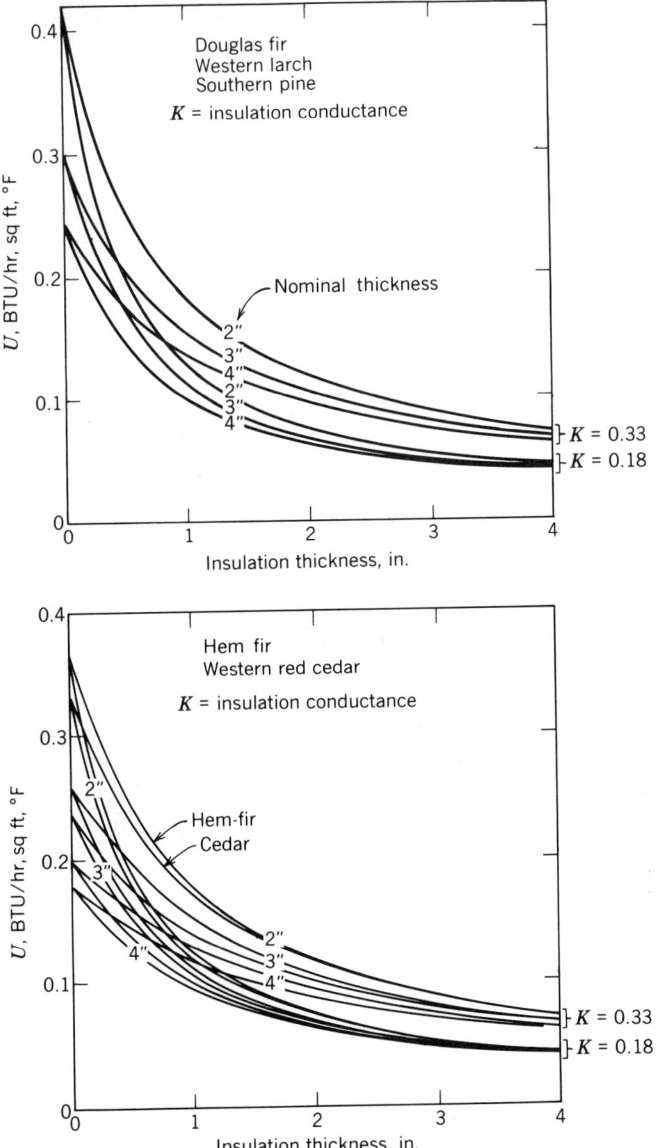

Fig. 25.29 Heat transfer coefficients.

$$f_r = \frac{3M}{2Rbd} \tag{25.14.2}$$

where M is the bending moment, b is the horizontal width, and d is the depth of the section.

Allowable stresses perpendicular to the grain are published in NDS [25.35] for compression but not for tension. Tension stresses perpendicular to the grain for southern pine and California redwood given in the TCM [25.4] and ASTM D2555 [25.16] are one-third of the allowable shear stress parallel to the grain. Experience has suggested a more limited radial tension stress limitation for Douglas fir-larch and hem-fir of 15 psi for all types of load-induced tension perpendicular to grain stresses except those caused by wind or earthquake, in which cases one-third of the shear values may be used. If the 15-psi limit is exceeded, reinforcing bolts are used up to a calculated radial tension of one-third the design shear value.

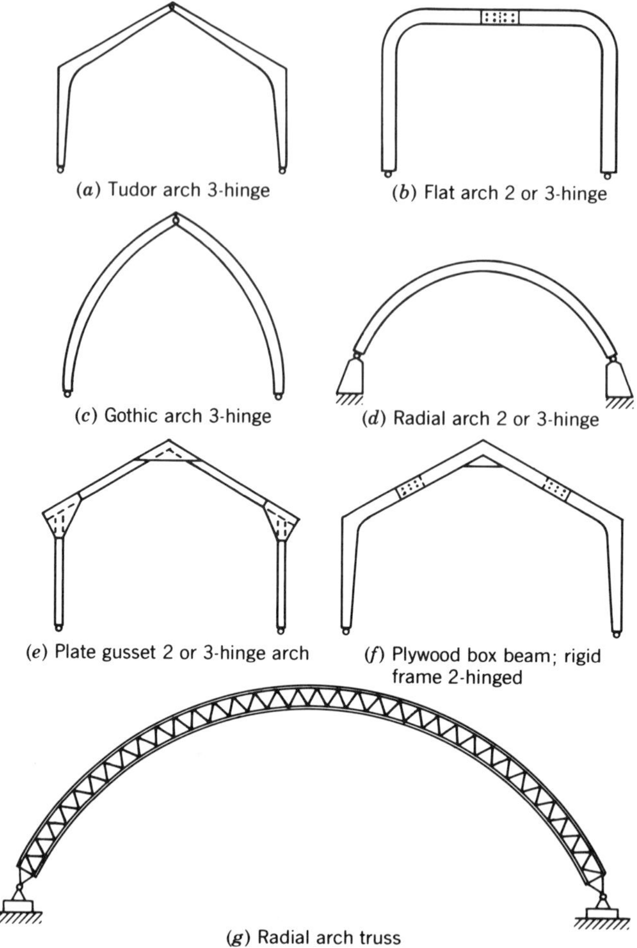

Fig. 25.30 Types of arches.

Depth effect reductions and load duration modifications mentioned in Section 25.6 are applied to glulam arch design.

AITC [25.2] publishes a table of arch dimensions for preliminary design.

25.14.1 Analysis of the Three-Hinged Arch

Arches with pinned connections at the crown and base are three-hinged and are statically determinate. They can be analyzed by any elastic method of structural analysis. Bending moment diagrams can then be constructed for different types of load and combined by superposition to find the most critical maximum moments and their locations. The following loading cases are usually investigated:

1. Dead and live load on the full span.
2. Dead load on the full span and live load on half the span.
3. Dead load on the full span and wind load on the side.

A comprehensive design for arches with tapered leg and rafter portions is lengthy but basically simple, once the structural analysis has been done. Though any method may be used, the equilibrium polygon may be a convenient method for the common cases.

Three-Hinged Arch—Uniformly Distributed Load on Full Span

Equilibrium diagrams represent the bending moment in the rafter portion of a three-hinged arch. For the loading of Fig. 25.31 the rafter has a bending moment at its junction to the leg equal to the product of the horizontal base reaction and the leg length. The uniform load on the rafter opposes this moment so the rafter moment decreases toward the crown, passes through zero where the polygon intersects the rafter axis, becomes positive beyond that point, passes through a positive maximum, then decreases to zero at the crown.

Under this loading the equilibrium diagram is a parabola of the form $y = kx^2$, with its vertex at the crown joint. The bending moment at any point along the rafter axis is equal to the product of the horizontal base reaction and the distance b measured vertically from the equilibrium diagram to the rafter axis. The bending moment at any point in the leg is equal to the distance from the base times the horizontal base reaction.

For any rafter slope and leg height, equations may be written for y to the rafter and y' to the equilibrium diagram. The distance b is $(y' - y)$.

If the rafter should be a circular arch instead of a straight member, the appropriate equation for y' can be used to obtain b.

The reactions for three-hinged arches can be obtained by statics. Figure 25.31 gives the necessary relationships for computing bending moments at any points along the arch axis. Figure 25.32 shows moment diagrams for an arch with $L = 60$ ft, $h = 30$ ft, and $r = 16$ ft.

Three-Hinged Arch—Uniformly Distributed Wind Load on Half-Span

This load situation is depicted in Fig. 25.33. The equilibrium diagram is a parabola with the vertex a distance $h/4$ below the crown and $L/16$ to the right. The horizontal distance between the equilibrium diagram and the arch axis on the loaded side is a measure of the bending moment. The bending moment is the product of b and the vertical reaction at the base.

On the leeward side of this structure the equilibrium diagram is a straight line connecting the crown and the right support. It is useful to note that the slopes of the left and right equilibrium diagrams are equal at the crown.

The reactions are obtained by statics. Figure 25.33 gives the essential information for computing reactions and moments for symmetrical arches. Figure 25.34 shows moment diagrams.

Three-Hinged Arch—Uniformly Distributed Load on One-Half the Span

This load arrangement is necessary in investigating the effect of unbalanced loading. Figure 25.35 shows equilibrium diagrams and the essential equations for determining bending moments and

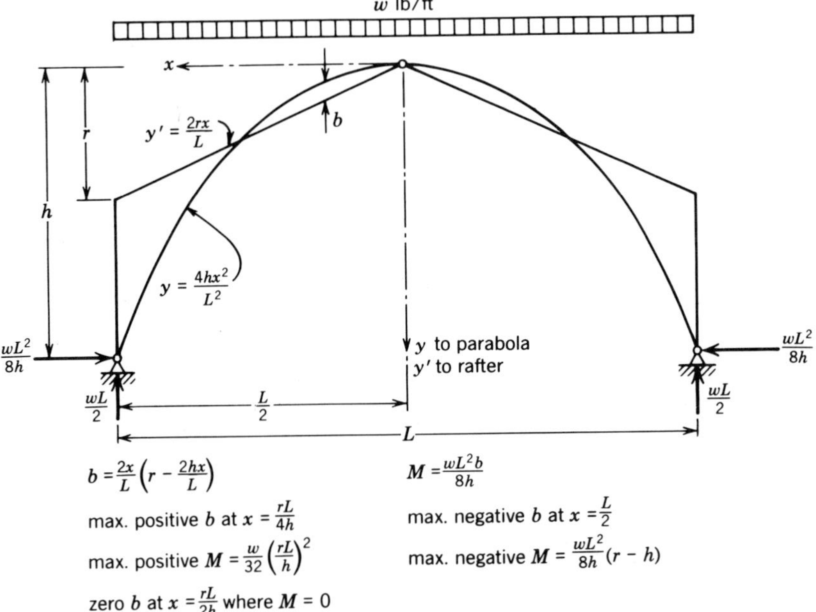

Fig. 25.31 Three-hinged arch with uniformly distributed load on entire span.

Example: For L = 60 ft, h = 30 ft, r = 16 ft

Max. positive moment is $32w$ at x = 8 ft
Max. negative moment is $-210w$ at x = 30 ft
Zero moment is at x = 16 ft
Moment at $x = \frac{wx}{2}(16 - x)$

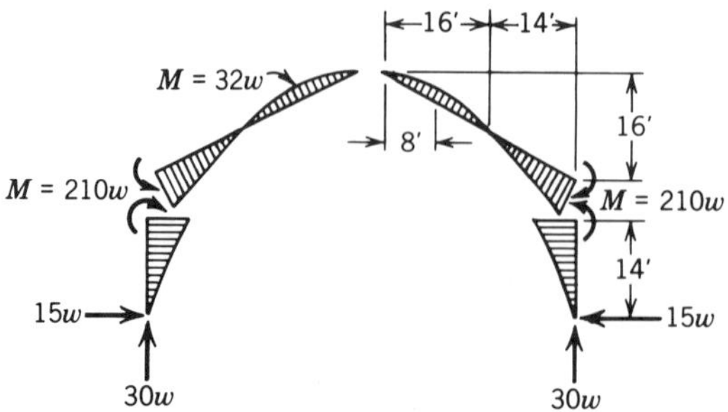

Fig. 25.32 Example for Fig. 25.31.

reactions. In this case the equilibrium diagram has its vertex at $L/8$ to the left of the crown and $h/8$ above the peak. The equilibrium diagrams have equal slopes at the peak. Results for the example structure are shown in Fig. 25.36.

Shear Forces—Axial Forces

The member shear forces may be obtained by recalling that shear at any point is the slope of the moment diagram at that point. The axial thrust at any point can be determined by use of the equations for static equilibrium.

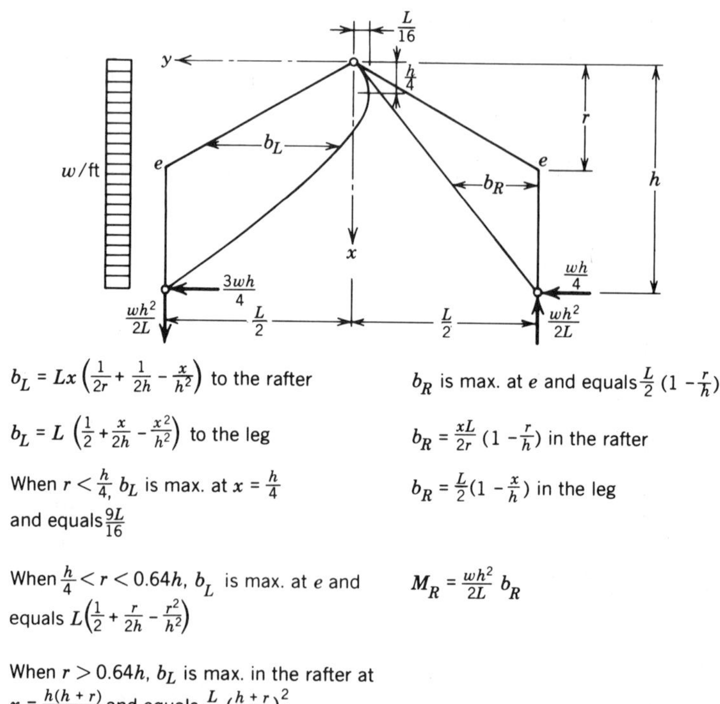

$b_L = Lx\left(\frac{1}{2r} + \frac{1}{2h} - \frac{x}{h^2}\right)$ to the rafter

$b_L = L\left(\frac{1}{2} + \frac{x}{2h} - \frac{x^2}{h^2}\right)$ to the leg

When $r < \frac{h}{4}$, b_L is max. at $x = \frac{h}{4}$ and equals $\frac{9L}{16}$

When $\frac{h}{4} < r < 0.64h$, b_L is max. at e and equals $L\left(\frac{1}{2} + \frac{r}{2h} - \frac{r^2}{h^2}\right)$

When $r > 0.64h$, b_L is max. in the rafter at $x = \frac{h(h+r)}{4r}$ and equals $\frac{L}{16}\left(\frac{h+r}{r}\right)^2$

$M_L = \frac{wh^2}{2L} b_L$

b_R is max. at e and equals $\frac{L}{2}\left(1 - \frac{r}{h}\right)$

$b_R = \frac{xL}{2r}\left(1 - \frac{r}{h}\right)$ in the rafter

$b_R = \frac{L}{2}\left(1 - \frac{x}{h}\right)$ in the leg

$M_R = \frac{wh^2}{2L} b_R$

Fig. 25.33 Three-hinged arch with uniformly distributed wind load on one side.

Example: $L = 60$ ft, $h = 30$ ft, $r = 16$ft, $\frac{h}{4} < r < 0.64h$

$$\max b_L = 60\left(\frac{1}{2} + \frac{16}{60} - \left(\frac{16}{30}\right)^2\right) = 28.933 \text{ ft}$$

$$M_{\max} = \frac{wh^2}{2L}\, b_L = \frac{w(30)^2}{2(60)}\,(28.933) = 217w \text{ on left}$$

$$\max b_R = \frac{60}{2}\left(1 - \frac{16}{30}\right) = 14 \qquad M_{\max} = \frac{wh^2}{2L}\, b_R = \frac{w(30)^2}{2(60)}(14) = 105w$$

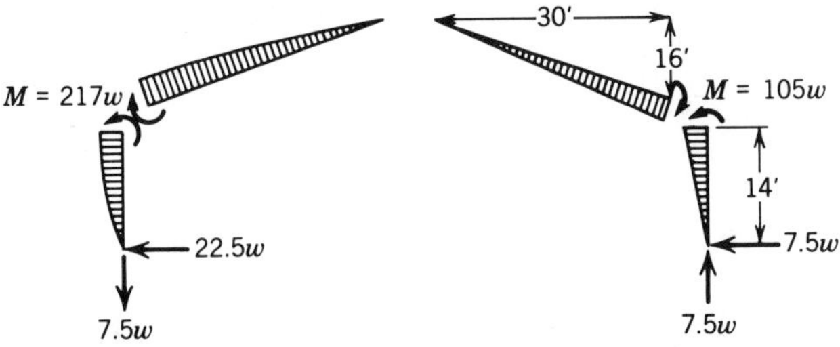

Fig. 25.34 Example for Fig. 25.33.

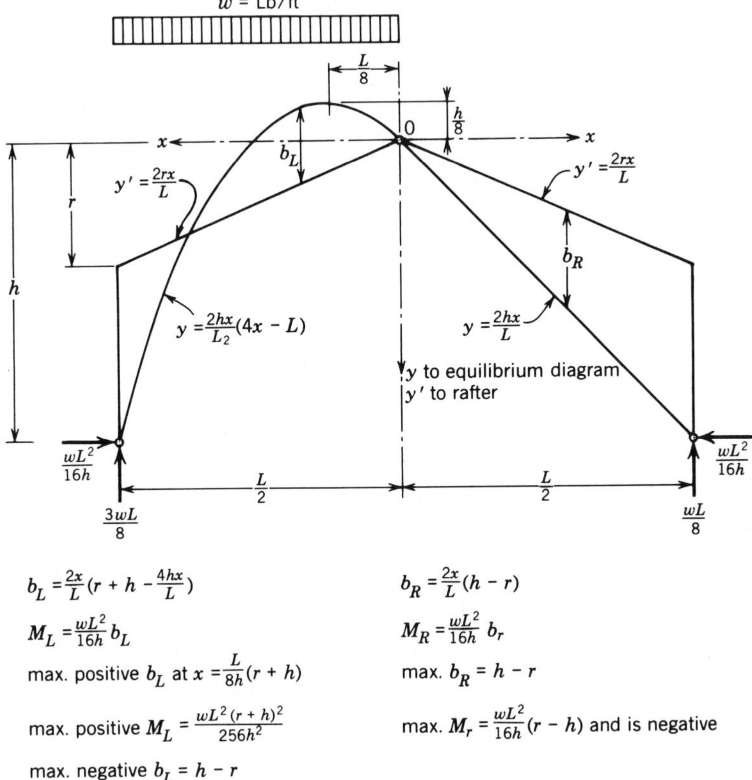

$b_L = \frac{2x}{L}\left(r + h - \frac{4hx}{L}\right)$

$M_L = \frac{wL^2}{16h}\, b_L$

max. positive b_L at $x = \frac{L}{8h}(r + h)$

max. positive $M_L = \frac{wL^2\,(r+h)^2}{256h^2}$

max. negative $b_L = h - r$

max. negative $M_L = \frac{wL^2}{16h}(r - h)$

b_L and $M_L = 0$ at $x = \frac{L}{4h}(r + h)$

$b_R = \frac{2x}{L}(h - r)$

$M_R = \frac{wL^2}{16h}\, b_r$

max. $b_R = h - r$

max. $M_r = \frac{wL^2}{16h}(r - h)$ and is negative

Fig. 25.35 Three-hinged arch with uniformly distributed load on half span.

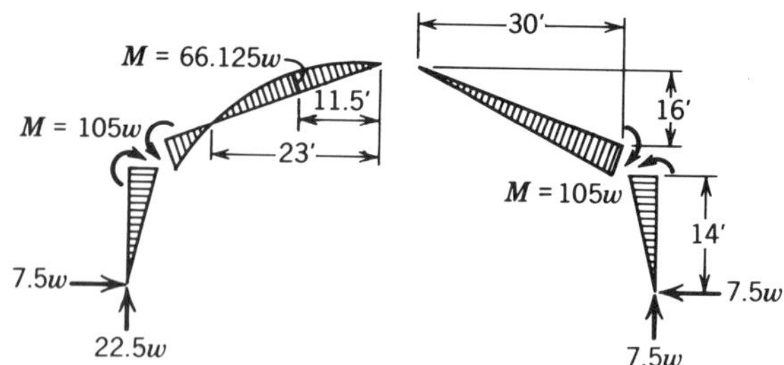

Fig. 25.36 Example for Fig. 25.35.

Member Size

The size of the member is then obtained to provide the required section modulus and area for the strength properties of the arch material. The interaction equations for beams and columns will be used in these calculations.

For the curved haunch portions the radial stress should be checked. Only the laminations which are continuous through the curved portion are included in designing this section. Depth (size) effect factors should be applied to the allowable bending stress for all sections more than 12 in. deep. The curvature factor should be applied to bending stress for the curved portion. Load duration factors should be applied to all properties except elastic modulus.

If the arch has tapered rafter and column members, the usual procedure is to divide the arch into equal segments and superimpose the moments and members forces at the ends of each segment. Member sizing is then determined at regular intervals along the arch.

Lateral Buckling

Arches are susceptible to lateral buckling of the rafter section and should be investigated as in Section 25.8 for beams. Usually the top edge receives lateral support from decking but the portion of the rafter section near the haunch is in compression on the lower edge. Under certain unbalanced loads the lower edge may be in tension. In the vicinity of the haunch the vertical leg and the wall sheathing provide support. The lateral stability requirements for arches generally require a depth-to-width ratio of 5 or less. Special attention should be given to the stability of long, deep, narrow arches.

Column Action

The rafters are subject to axial compression, the more so as they are more steeply pitched. They should be examined in terms of possible column buckling. The major difficulty here is in arriving at effective length values.

Deflections

Arch deflections can be obtained using any elastic method after the member sizes and properties have been determined. Usually deflection will not be the controlling limitation for arches, but it should be investigated before the design is finalized. It is often sufficient to make a rough check, treating the rafters as simple beams; if that does not show objectionable deflection, a more elaborate analysis will be unnecessary.

25.14.2 Analysis of the Two-Hinged Arch

The two-hinged arch is pin-connected at each base and continuous between these points. The pinned connections are restrained from horizontal movement by large foundation buttresses or by horizontal tension ties between the arch ends where they rest on walls or through the building floor system. Two-hinged arches are statically indeterminate. The use of a computer is recommended for this analysis. The *Timber Construction Manual* [25.4] describes a method of analysis suitable for small desk calculators.

Preliminary Design

AITC suggests the following procedure for preliminary design:

1. Using the desired rise and span, compute the radius of curvature using the simple geometric relationship

Table 25.32
Maximum Bending Stress for Two-Hinged Radial Arch*

h/R	f_b, psi
0.2	800
0.3	1125
0.4	1400
0.5	1600
0.6	1775
0.7	1900
0.8	2025
0.9	2100
1.0	2200

* From AITC *Glulam Systems* [25.2].

$$R = \frac{4h^2 + L^2}{8h} \tag{25.14.3}$$

where h is the rise and L the span.

2. Limit the maximum bending moment to

$$M = 1.5wh^2 \text{ in.-lb} \tag{25.14.4}$$

where w is in lb/ft of horizontal projection and h is the rise in ft. Since the dead load is uniformly distributed along the curve, an average value may be used for this estimate.

3. Limit the bending stress to a value based on the ratio h/R from Table 25.32.
4. Choose a section modulus using an assumed size factor C_F of 0.9

$$S = \frac{M}{f_b C_F} \tag{25.14.5}$$

5. Select a standard glued laminated beam size with a depth-to-width ratio of 5 or less. The horizontal reactions at the base will not exceed $wL^2/8h$ for a symmetrical arch.

25.14.3 Analysis of A-Frames

A-frames are really three-hinged arches and can be designed using the methods already described. For preliminary design, AITC [25.2] suggests a method that will serve to rough out the design. This is based on uniformly distributed live dead and wind load as shown in Fig. 25.37. The procedure is as follows:

1. Compute the constants:

$$k_1 = 1/\cos\alpha \tag{25.14.6}$$

k_2 = an allowable bending stress adjustment which accounts for the duration of live load (1.15), a size factor, and the axial contribution to combined stress. Use $k_2 = 0.9$ for preliminary work.

k_3 = an allowable bending stress adjustment for wind load and size factor, assumed to be 1.13 for preliminary purposes

2. Calculate bending moment M_{TL} due to total load:

$$M_{TL} = 0.375 W_{TL} L^2 \text{ in.-lb when } L \text{ is in ft} \tag{25.14.7}$$

where
$W_{TL} = W_{DL} + W_{LL}$ lb/ft
$W_{DL} = (\text{spacing})(\text{DL})\, k_1$ lb/ft
$W_{LL} = (\text{spacing})(\text{LL})$ lb/ft

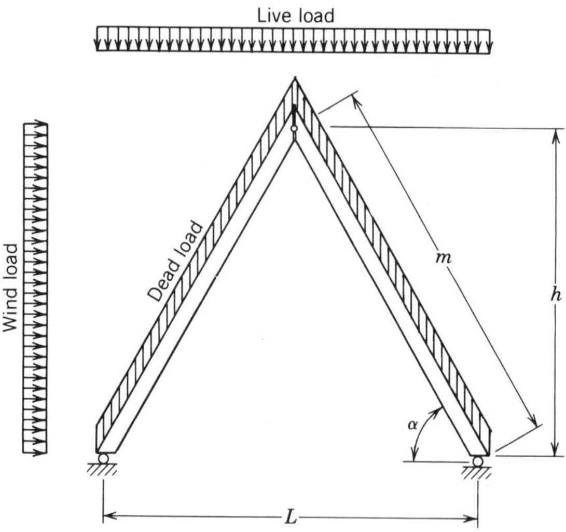

Fig. 25.37 Three-hinged arch, A-frame. Courtesy of American Institute of Timber Construction.

3. Calculate bending moment due to dead load and wind:

$$M_{\mathrm{DL}} = M_{\mathrm{TL}}(W_{\mathrm{DL}}/W_{\mathrm{TL}}) \tag{25.14.8}$$

$$M_{\mathrm{WL}} = 1.5W_{\mathrm{WL}}h^2 \text{ in.-lb when } h \text{ is in ft} \tag{25.14.9}$$

$$M_{\mathrm{DL}} + M_{\mathrm{WL}} = \text{maximum moment under wind loading}$$

4. Determine the required section modulus S for total load:

$$S = \frac{M_{\mathrm{TL}}}{k_2F_b} \text{ in.}^3 \tag{25.14.10}$$

5. Determine the required section modulus S for wind load:

$$S = \frac{M_{\mathrm{DL}} + M_{\mathrm{WL}}}{k_3F_b} \text{ in.}^3 \tag{25.14.11}$$

6. Determine moment of inertia I for a deflection of 1/120th of length of the rafter measured perpendicular to the rafter:

$$I = 75M_{\mathrm{TL}}Lk_1 \text{ in.}^4 \tag{25.14.12}$$

when L is in ft and M_{TL} is in.-lb.

7. Choose a section with the required maximum S and I and the least possible area, but with a depth-to-width ratio of 5 or less. This should produce a close preliminary estimate.

A complete design should be made before the approval of the building drawings.

25.14.4 Other Types of Two- and Three-Hinged Arches

Two- and three-hinged arches made by connecting straight glulam beams together with gussets of thick plywood or steel plates, with bolts, or with timber connectors are sometimes used when the spans are not above about 40 ft. The connection strength and the incursion of the connectors into the beam section at this high stress haunch location are the limiting considerations.

Figure 25.30*f* shows a two-hinged plywood box beam arch with a moment-carrying splice at the approximate location of the inflection point for full span load. Such an arch will probably require steel straps at the outside of the haunches and the crown, with a tension splice below the crown and compression bearing plates at the interior corners of the haunches. The arch could also be a three-hinged design, pin-connected at the crown and without the rafter splice joint.

Radial arch trusses, as shown in Fig. 25.30*g*, are used for spans as long as 400 ft. These usually have been Warren-type pin-connected trusses. They should receive careful design attention as the problems of stability and erection are very substantial.

24.14.5 Durability Considerations

Arches or portions of arches exposed to weather must be treated with a wood preservative, preferably an oil-borne type. Arch foundation connections should be designed to remain dry. They should not rest directly on concrete and connection hardware should be provided with drainage holes to prevent the accumulation of water in contact with the wood. Efforts to seal the ends to exclude water are rarely successful. Only good drainage and ventilation will work. Preservative treatment of the ends at the foundation connections is desirable and should be mandatory if exposed to the weather regularly. Flashings to keep water off the upper surfaces of exposed arches are often unsuccessful, and preservative treatment is recommended for such members. Flashings and coatings on the projecting ends of beams have proven satisfactory if carefully chosen.

Preservative or fire retardant treatment of arches must be done before the gluing and fabricating work because treating cylinders are usually too small to accommodate the completed members.

25.14.6 Segmented Arches

A simple type of wood radial arch which can be made in short segments, mechanically fastened with nails or bolts, is shown in Fig. 25.38. This style of arch is often used for barn rafters, for quonset-style light frame buildings, and for utility and farm structures. It can be constructed with simple tools and assembled in the field.

The segments are sawn to a curve on the outer edge. Gluing, in addition to bolting or nailing, will greatly improve the strength and stiffness of these arches. They are, however, not particularly efficient, because of the many joints and the labor involved in their fabrication.

The literature contains very little information on the actual structural analysis and design of segmented arches. By using conservative allowable stresses and a conservative method of analysis, they can be safely proportioned. Although they are usually installed as two-hinged arches, a simpler analysis as three-hinged arches will be conservative. Segmented arches are usually built to a constant radius of curvature which is sufficiently close to parabolic that the equilibrium diagram

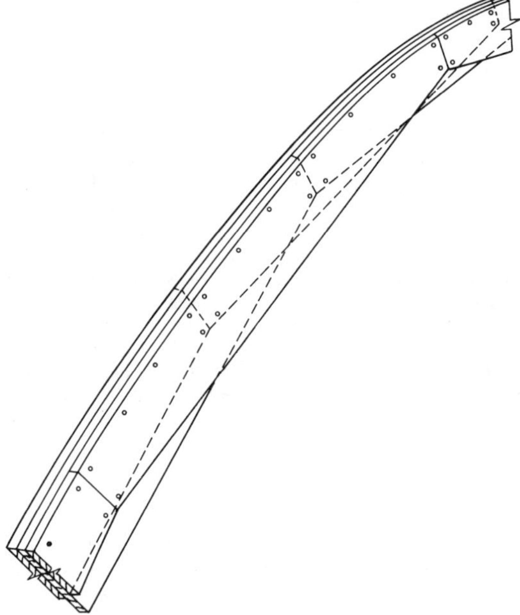

Fig. 25.38 Perspective view of a part of a vertically laminated segmented arch. Courtesy of USDA Forest Products Laboratory, Madison, WI.

for a full uniformly distributed load does not depart very much from the arch axis. This results in minimal bending moments under full span loading. The critical loads will be wind and half-unbalanced snow loads. Their installation as two-hinged arches will reduce the actual maximum bending moments computed by the three-hinged assumption.

Usually segmented arches are more closely spaced than glued laminated arches. Spacing of 24 and 48 in. is not uncommon.

Two special problems arise when designing segmented arches:

1. The transverse joints at the ends of the segments are staggered in the several laminations. Within each lamination these joints will not carry moment and, unless it can be assured that they will bear tightly on one another, they cannot be depended upon to carry thrust. The net section must be carefully determined and mechanical fasteners must be designed to carry the moment at the splices.
2. Sawing to a curve means the stress direction will not be at the angle to the grain for which the allowable stresses are published; thus design requires a suitable reduction in the design properties. The stress direction is parallel to the grain only at the midlength of each segment. Reductions based on expected slope of grain with respect to the stress direction are suggested. If segment lengths do not exceed one-quarter of the radius of curvature a reduction of 50% in allowable bending stress would be reasonable.

25.14.7 Domes

Domes are the largest wood building structures for enclosing space with clear spans, having diameters from 20 to 600 ft (6 to 180 m). They rival steel, aluminum, or concrete shell domes in this respect. They are also well suited to small structures. Wood is well suited to dome construction because of its high strength-to-weight ratio, its insulating quality, and the speed with which wood-fabricated frames can be erected and sheathed over.

Dome architecture extends into the historical past. Domed churches, temples, public buildings (state capitols and universities), and sports arenas are found in many places. Ancient domes often had more "rise" for a given span than is needed for the lightweight wood domes of recent years. The masonry block domes of a few hundred years ago required a nearly semi-circular cross-section. Early wood domes built with mechanical fasteners and without adhesives also had more rise. Modern wood domes are spherical surfaces with a span to rise on the order of 5 or 6.

Wood domes are of two types, the radial-rib arch dome as shown in Fig. 25.39, and the more common triaxial-rib dome shown in Fig. 25.40. They can be designed for heavy loading and many are located in heavy snow load regions.

Radial-Rib Arch Dome

The radial-rib arch dome has radially placed glulam timber arches, continuous from base to crown and terminating in a compression ring at the top and a tension ring at the base. Concentric rings consisting of straight glulam purlins spanning the space between the radial-rib arches are uniformly spaced from the crown ring to the outer periphery of the frame. These purlins are connected by hangers with their top surfaces flush with the top faces of the radial-rib arches. Diagonal cross-bracing in each framing rectangle consists of steel rods with turnbuckles. The entire frame rests on foundation walls or columns around the circumference.

The dome is sheathed with heavy timber decking, plywood, or some specialty sheathing product on the supporting purlins. Insulation is placed either beneath the deck or on its top surface, according to the requirements for thermal design and roofing membrane. Design methods for the radial-rib arch dome are described in the *Timber Construction Manual* [25.4] and in the *Manual of Structural Designs and Engineering Solutions* [25.44]. The design is executed for concentrated loads at each arch–purlin intersection for balanced and unbalanced loading, for wind and seismic forces.

Traxial-Rib Dome

The triaxial-rib dome as shown in Fig. 25.40 differs from the radial-rib arch. The geometry is based on six radial arches placed at 60° angles and connected at the top center by a steel hardware unit. These six arches are great circles of a sphere. Secondary arches, which are lesser circles of the sphere and parallel to the great circles forming the sides of each 60° repeating dome segment, intersect at equal intervals along the radius. The framing generates triangular framed sectors which are stable and require no bracing. These lesser circle glulams are curved to the surface of the sphere with their wide dimension perpendicular to that surface.

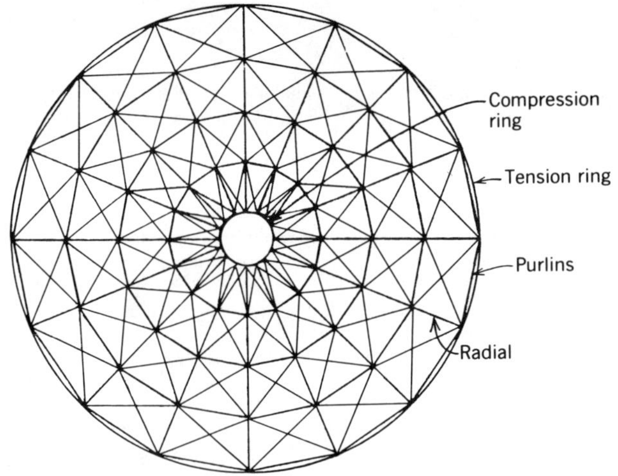

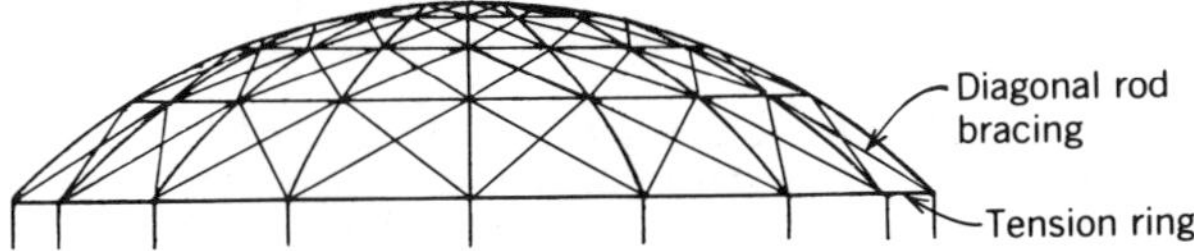

Fig. 25.39 Radial-rib arch dome.

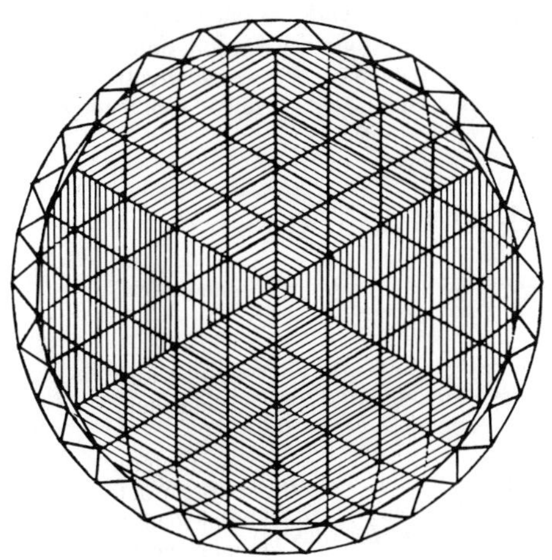

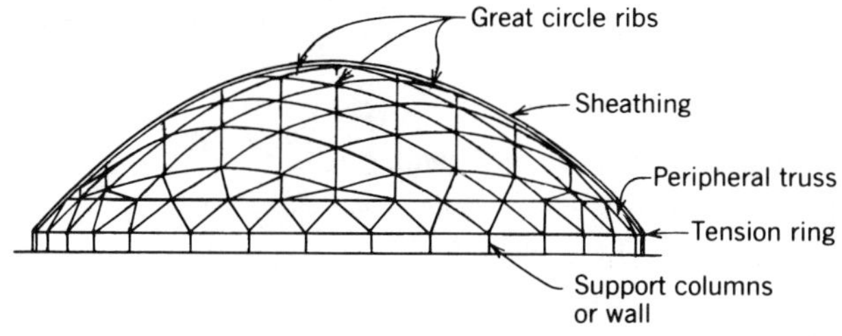

Fig. 25.40 Triaxial-rib dome.

The members frame into steel connecting hardware at every node in the system. All circles are made with short glulam members extending from node to node. The structure is erected from the ground upward.

At the outer periphery of the structure, a steel tension ring to resist the outward thrust of the framing rests on foundation supports (columns or a wall). Immediately above the steel tension ring is a circular truss with its lower chord panel points on the tension ring and its top chord panel points on the surface of the sphere, completely encircling the dome. The framing members (greater and lesser circles of the sphere) are connected to the top chord panel points of this symmetrical truss by special wood framing members.

With the exception of a few pieces at the wall-to-dome junction, all members are of uniform thickness glulam lumber, cambered to the spherical surface.

The spaces within the triangular framing sectors are spanned by cambered wood purlins set perpendicular to the axis of each of the six main sectors, and spaced about 4 ft on centers, connected with joist hangers.

The dome is sheathed with heavy timber decking (2×8 tongue and groove lumber) or with plywood or other types of wood panels. The sheathing is applied in sections and curved to fit the spherical surface according to a predetermined plan. The decking is flat material of uniform width and thickness.

The dome surface is insulated with a suitable thickness of foamed-in-place polyurethane foam and coated with a suitable thickness of sprayed weatherproof membrane. The lower surface of the dome may be covered with an acoustical treatment and a suitable vapor barrier.

The analysis of the structure can be done using a program which solves for forces and joint displacements using the direct stiffness finite element method. From the analysis, all of the forces are obtained, from which members may be sized and checked. The camber for the members is obtained from the geometry of the sphere and the member lengths.

Since the domes are covered structures and the wood moisture content is not likely to exceed 19%, there should be no need to treat the material with a wood preservative.

These are very large roof areas and it would be expected that some leakage would occur from time to time. The foamed insulation should be a type that will not absorb and retain water if the exterior membrane should leak. Attention should also be given to the proper placement of an effective vapor barrier to exclude moist air from contact with cold surfaces which could cause condensation of water, wetting of the insulation and wood, and subsequent decay.

If dryness cannot be reasonably assured, preservative treatment of the sheathing might be considered, but it has not been the usual practice.

25.15 DIAPHRAGMS AND SHEAR WALLS

The functional behavior of diaphragms and shear walls is described in Chapter 7. Wood buildings are often box-type systems, dependent on diaphragm and shear wall action for resisting lateral forces.

Roof, floor, and wall systems designed to carry the vertical load on a building also possess lateral strength and stiffness, useful for resisting wind and seismic load. This lateral resistance can be improved by appropriate connection of the sheathing and support beams or trusses, and the splicing of the boundary members.

Except for small buildings, codes usually require that a lateral force design be made. Pitched roofs resist load both laterally and vertically by diaphragm action. Folded plate roofs depend largely on diaphragm action for both vertical and lateral load resistance. The maximum lateral forces do not usually occur coincident with the maximum vertical loads so structures need not resist both simultaneously.

Any rational design that will resist the lateral forces through frames, the bracing, and the shear strength of sheathing panels can be used. Even though both wind and seismic forces are dynamic, "equivalent" static methods using empirically determined stiffness properties are usually employed.

25.15.1 Diaphragms with Lumber Sheathing Laid Parallel to Diaphragm Boundary

Wood decking or sheathing lumber laid parallel to the panel edges is the simplest type of diaphragm or shear wall construction. This is not a very strong type of construction, but is often adequate where the lateral loads are modest, as indicated in Table 25.33. It can be immensely improved by providing shear connection between the courses or by diagonal bracing.

A simple unbraced diaphragm of this type resists load by the individual flexural stiffness of each course under the action of its share of the total load. Where each course of decking is nailed to the supporting joists, beams, or trusses with more than one nail at each crossing, resisting nail couples

Table 25.33 Horizontal Diaphragm Properties for Douglas Fir or Southern Pine*,†

Type	Web Flexibility Factor F, in./ft/lb	Allowable End Shear V, lb/ft
1-in. straight sheathing	1500×10^{-6}	50
2-in. straight sheathing	1500×10^{-6}	40
1-in. diagonal sheathing single layer	250×10^{-6}	300
2-in. diagonal sheathing single layer	250×10^{-6}	400
1-in. diagonal sheathing double layer	75×10^{-6}	600

* Multiply by 0.82 for Group III and 0.65 for Group IV species.
† From Ref. 25.48.

will develop to counter the applied load. The effect of these couples (Fig. 25.41*b*) on the deflection of the diaphragm is difficult to determine analytically, but is embodied in the empirically derived flexibility factors in Table 25.33. Well-distributed butt joints in the courses have been found [25.31] to have little influence on the lateral deflection of the diaphragm.

The sheathing (or web) component of the in-plane deflection is estimated using the flexibility factor (Table 25.33).

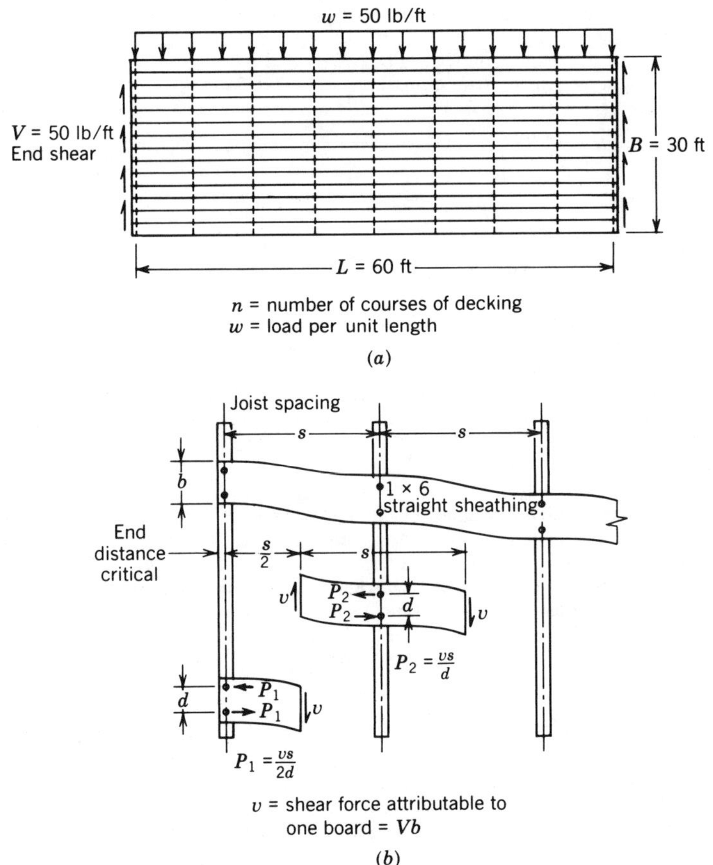

Fig. 25.41 Lumber diaphragm.

EXAMPLE 25.15.1

Compute the diaphragm (in-plane) deflection for a wind loading of 50 lb/ft acting on the panel of Fig. 25.41.

Solution

(a) *Determine the end shear for load perpendicular to the long edge.*

$$V = \frac{wL}{2B} = \frac{50(60)}{2(30)} = 50 \text{ lb/ft} \tag{25.15.1}$$

(b) *Empirical calculation of deflection as a diaphragm.*

$$\Delta_w = q_{av} L_1 F \tag{25.15.2}$$

where q_{av} = average shear in the diaphragm, load per unit length = $V/2$
L_1 = distance from the vertical resisting element (shear wall) to the point at which the deflection is to be determined (in this case, midlength)
F = empirical flexibility factor; 1500×10^{-6} in./ft (for 1-in. or 2-in. straight sheathing, from Table 25.33). Then, upon substitution of the values,

$$\Delta_w = 25(30)(1500 \times 10^{-6}) = 1.13 \text{ in.}$$

(c) *Determine the moment-resisting nail couple.* Refer to Fig. 25.41*b*, which shows a portion of the diaphragm near the left wall. Conservatively assuming inflection points at midspans of s and that no friction between the courses occurs, the shear v per course is

$$v = \frac{Vb}{12} = \frac{50(5.125)}{12} = 21.35 \text{ lb} \tag{25.15.3}$$

A 1×6 T&G board has width b of 5.125 in. and the nail spacing d will be 4 in., giving nail loads of

$$P_2 = \frac{vs}{d} = \frac{21.35(24)}{4} = 128 \text{ lb} \tag{25.15.4}$$

for a joist spacing s of 24 in. The nail load P_1 at the end of the course is lower.

(d) *Determine the required nail size.* Assuming the shear at P_2 is essentially the same as at P_1, nails must be selected to develop 128 lb lateral resistance. In Douglas fir lumber or other Group II species, the normal allowable load for an 8*d* (0.131 × 2.5 in.) common steel nail is 78 lb at a penetration of 1.44 in. or more. This can be increased 1.3 for diaphragm design and 1.33 for load duration (wind or seismic) to 135 lb, which exceeds the required 128 lb [25.32, Secs. 2.2.5.3 and 8.8.5.5]. The boundary members (or flanges) are not considered effective and are not included in the design.

(e) *Determine diaphragm deflection for load perpendicular to the short edge.* This diaphragm deflection Δ is the consequence of slip between the parallel courses of boards. It is the sum of the slip displacements along all course interfaces from the center to the edge courses. This ignores shear deformation of the wood, which will be negligibly small. For wind loading of 50 lb/ft along the 30-ft edge, the resisting shear along the 60-ft edge is

$$V = \frac{wB}{2L} = \frac{50(30)}{2(60)} = 12.5 \text{ lb/ft} \tag{25.15.5}$$

Shear at the course interfaces varies linearly from 12.5 lb/ft at the long edge to zero at the long centerline. The average shear is 6.25 lb/ft.

In this diaphragm there are joints at 2-ft intervals, with two 8*d* nails at each point where a sheathing board crosses a joist. The shear transfer between courses is thus carried by a row of 8*d* nails spaced at 2-ft intervals in each board. The average shear load per nail is

$$6.25 \text{ lb/ft}(2 \text{ ft/nail}) = 12.5 \text{ lb/nail}$$

Allowable nail load values for lateral loading are established on the basis of the load at 0.015-in. lateral deformation [25.43, page 7–6]. For the nail strength given in part (d) of 78 lb, the average deformation of the nails would be

$$\frac{12.5}{78}(0.015) = 0.0024 \text{ in.}$$

The average relative slip of two adjoining courses would be twice this amount, or 0.0048 in. The sum of the interface slips for all courses from the edge to the centerline of the diaphragm is

$$35 \text{ interfaces at an average slip/interface of } 0.0048 = 0.168 \text{ in.}$$

In this example the loads were assumed to be due to wind. The total lateral force due to wind depends on the length of the wall on which the wind is acting. In this case, with one wall twice as long as the other, the total load in the direction of the short axis is twice that along the long axis. If the given 50 lb/ft loading on the long edge had been due to seismic force, the load on the short edge would have twice the intensity or 100 lb/ft. This is because seismic force depends on building mass and is independent of the direction of the action.

In either case, the deflection of the short edge is small compared to the deflection of the long edge.

25.15.2 Lumber Shear Walls

Lumber shear walls with board courses parallel to the sides or the top and bottom edges (Fig. 25.42) are dependent almost entirely on nail resistance for their strength and stiffness. Nail couples develop as previously discussed in regard to Fig. 25.41*b*.

EXAMPLE 25.15.2

Calculate the shear wall deflection for the wall in Fig. 25.42. $V = 50$ lb/ft and $v = 21.35$ lb/board. Use wall studs at 24-in. spacing and nails 4 in. on centers.

Solution

(a) *Determine the resisting nail couple*. Because v, s, and d are the same as in the previous example, the same nail couple will occur ($21.35 \times 24 = 512$ in.-lb) and the same nail load will be present (128 lb/nail).

(b) *Determine the shear deflection*. This deflection may be estimated from the assumed nail slip of 0.015 in./nail. As shown in Fig. 25.43

$$\Delta = \frac{0.03H}{d} \tag{25.15.6}$$

For a wall height of 14 ft this deflection is 1.26 in. The acceptability of a deflection of this amount must be decided by the engineer, based on the tolerable movement of tributary parts of the structure. For a building with wood walls, without attached brittle finishes, this might be acceptable. For a building with masonry walls having attached or brittle finishes it might be unacceptable.

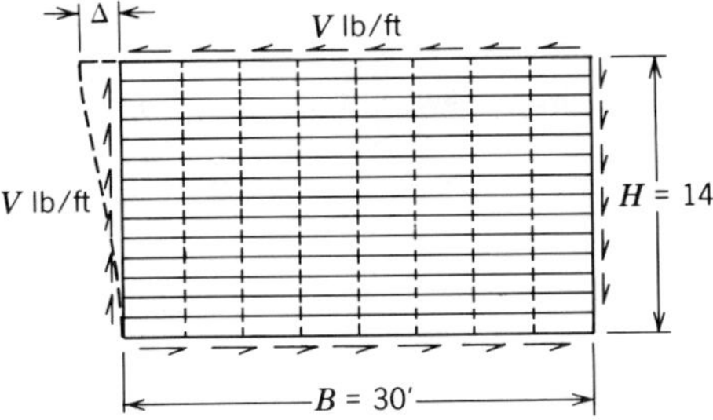

Fig. 25.42 Lumber shear wall.

0.015″ at rated nail load

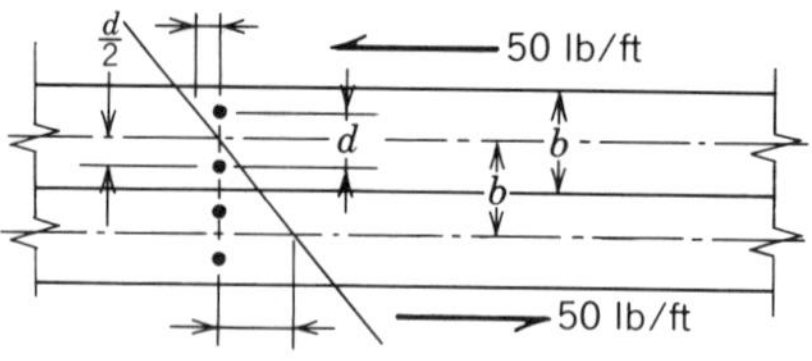

Course slip = $0.015 \frac{b}{d/2}$

Shear wall slip = course slip × number of courses

$$\Delta = 0.015\left(\frac{2b}{d}\right)\frac{H}{b} = 0.03\frac{H}{d}$$

H = Wall height

Fig. 25.43 Shear wall slip.

It is necessary that the vertical wall studs be securely connected to the foundation. Any slip in these connections would cause additional shear deflection of the wall.

Door openings in shear walls can be accounted for considering only the portions of the wall that are full height to be effective. Total shear is then distributed to those parts, each being regarded as deflecting equally since they are tied together at the top. This may call for additional nailing or closer stud spacing.

In the example, nominal 1×6 boards were assumed, using the maximum allowable shear of 50 lb/ft recommended in Table 25.33. For nominal 2-in. lumber only 40 lb/ft is allowed.

New construction should utilize other types of diaphragms, such as plywood sheathed diaphragms, which have greater strength. The engineer must often review old structures constructed of lumber, for which these design methods will be useful.

25.15.3 Diagonally Sheathed, Single-Layer Diaphragms

By laying the lumber at 45° to the diaphragm edges in a single layer, as shown in Fig. 25.44, shear strengths up to 300 to 400 lb/ft (Table 25.33) can be obtained. The diaphragm may be visualized as a thin, deep girder oriented to perform in the horizontal plane. Structurally spliced continuous boundary members function as flanges to carry the bending moment and diagonal sheathing is designed to carry the shear.

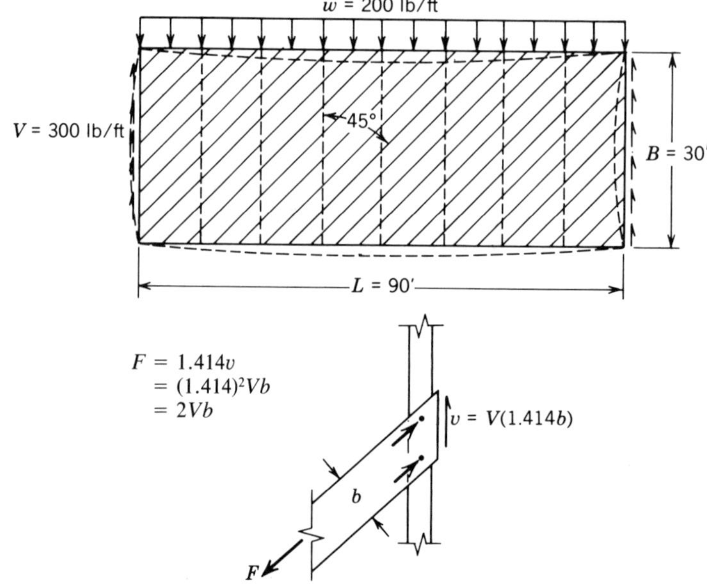

Fig. 25.44 Diagonal lumber diaphragm.

EXAMPLE 25.15.3

Illustrate the design method for a diagonally sheathed diaphragm. For the diaphragm in Fig. 25.44 consider the lateral load of 200 lb/ft as shown.

Solution

(a) *Determine the End Shear.* For this 30 × 90 ft diaphragm the end shear will be

$$V = \frac{wL}{2B} = \frac{200(90)}{2(30)} = 300 \text{ lb/ft}$$

which is the maximum value given in Table 25.33 for single layer diagonally sheathed diaphragms of nominal 1-in. lumber.

(b) *Determine the Forces on the Long Boundary Members.* The long boundary members are the tension (bottom) and compression (top) chords in Fig. 25.44. They resist the bending moment.

$$M = \frac{wL^2}{8} = \frac{200(90)^2}{8} = 202{,}500 \text{ ft/lb}$$

The chord forces are

$$M/B = 202{,}500/30 = 6750 \text{ lb}$$

The chord should be designed to carry this force at the net section of the splice joints in tension and compression. It would be reasonable to design splice joints for lower chord forces when the joints are not near midlength, but usually a single joint design is easier to specify and actually obtain from the builder. The chords will be restrained against buckling by the sheathing and by connections to the walls.

An additional chord strength requirement found in some codes and cited here from the *Uniform Building Code* [25.28] is:

> *Each chord or portion thereof may be considered as a beam loaded with a uniform load per foot equal to 50 percent of the unit shear due to diaphragm action. The load shall be assumed as acting normal to the chord in the plane of the diaphragm. The span of the chord, or portion thereof, shall be the distance between structural members of the diaphragm, such as joists, studs and blocking, which serve to transfer the assumed load to the sheathing.*

The diagonal sheathing at the right will act in tension to resist shear. The maximum tension in a sheathing board will be

$$(1.414)^2 \frac{Vb}{12} = 256 \text{ lb} \qquad (25.15.7)$$

for the example. The diagonal sheathing at the left will be in compression (also 256 lb) diminishing toward the center where the shear is lower. Possible compression buckling of the sheathing should be checked.

(c) *Nailing Requirements.* Using the 8*d* common nail value of 135 lb obtained for the straight parallel course diaphragm (Examples 25.15.1 and 25.15.2) the requirement is two nails. Theoretically the nailing could be reduced toward the center as the shear decreases, but it is always desirable to have two or more nails per crossing to keep the boards flat and develop a resisting couple. Nail end distances should be about 16 nail diameters, indicating the need for nominal 3- or 4-in. edge members, doubled nominal 2-in. lumber, or allowing sheathing to project beyond the boundary members. The designer might also consider using three nails per board at the boundary members to keep the nail loads moderate. Common practice is to use two nails between sheathing and board crossings, and three nails at ends at boundaries and butt joints. There are a minimum of two boards between sheathing board joints on any support and a minimum of two supports between joints in any board.

(d) *Deflection.* The deflection can be determined by superimposing the web deflection, chord flexure, and any chord splice slip which may be anticipated.

$$\text{web deflection} = \Delta_w = q_{av} L_1 F$$

$$= 150(45)(250)(10^{-6}) = 1.69 \text{ in.}$$

$$\text{chord deflection} = \Delta_f = \frac{5wL^4}{384EI}$$

where I is the moment of inertia of the chords about the longitudinal centerline of the diaphragm. For 3×8 Douglas fir chords with $E = 1{,}700{,}000$ psi, chord centroid distance is 178.75 in. and

$$I = 2[2.039 + 10.875(178.75)^2] = 694{,}950 \text{ in.}^4$$

$$\Delta_f = \frac{5(200)(1080)^4}{384(1.7 \times 10^6)(694{,}950)(12)} = 0.25 \text{ in.}$$

The chord splice slip component of diaphragm deflection is

$$\Delta_{cs} = \frac{\Sigma \Delta_c x}{2B} \tag{25.15.8}$$

where Δ_c = slip, one chord splice joint
x = distance from splice to nearest end of diaphragm
B = diaphragm width
$\Sigma\Delta_c x$ = sum of the $\Delta_c x$ product for all the splices in the tension and compression chords

Assume that each chord is made up of six pieces of 15-ft lumber, and there can be $\Delta_c = 0.0625$-in. slip in each of the bolted joints. Thus,

$$\Delta_{cs} = \frac{\Sigma\Delta_c x}{2B} = \frac{0.0625[2(45) + 4(30) + 4(15)]}{2(30)} = 0.28 \text{ in.}$$

The total diaphragm deflection is

$$\Delta_T = \Delta_w + \Delta_f + \Delta_{cs} = 2.22 \text{ in.}$$

Note the deflection of the transverse end members in the sketch in Fig. 25.44. This can be prevented by double sheathing, or by herringbone lay-up as in Fig. 25.45.

25.15.4 Diagonally Sheathed Shear Walls

These shear walls are designed as cantilever beams, fixed at the base. Vertical edge members are assumed to carry the chord stresses due to bending moment and diagonal sheathing to carry the shear.

A 15-ft high shear wall to carry the 300 lb/ft end shear in Example 25.15.3 would have a cantilever beam moment of $VBH = 300(30)(15) = 135{,}000$ ft-lb and chord forces of $VBH/H = 9000$ lb (H = wall height).

Shear wall deflection is computed using the flexibility factor from Table 25.33 with Eq. (25.15.2) to obtain the web component. The cantilever beam component and anchorage and/or chord splice slip components are then added. This procedure is analogous to that for diaphragm deflection discussed in Section 25.15.3.

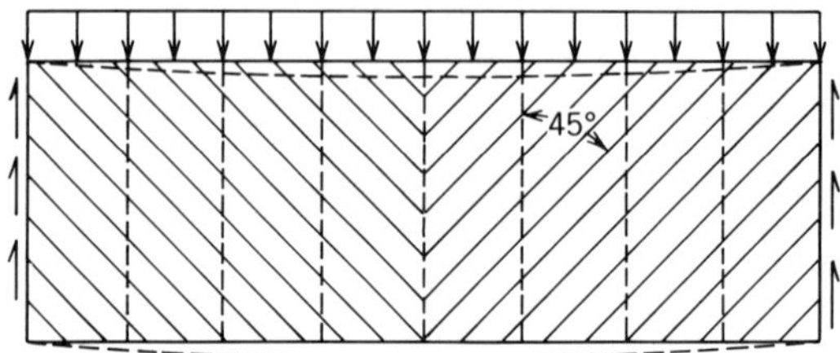

Fig. 25.45 Herringbone pattern lumber sheathed diaphragm.

Sheathing strength and nailing requirements are determined in the manner described for the diaphragm.

Total shear wall design should consider the effect of vertical dead load on wall member forces, by superposition on lateral force requirements.

25.15.5 Double Diagonal Sheathed Diaphragms

This diaphragm using sheathing at 45° to edges, double layer, is stiffer and stronger than those described in the previous sections. The markedly improved stiffness shown in Table 25.33 is due to the interaction between the sheathing layers. Under load the top layer is in tension at all points where the bottom layer is in compression. This results in net reactions of zero perpendicular to boundary members and sheathing connections. There is, of course, axial force on all the boundary members, and the chords act more effectively to carry all the moment.

The bottom sheathing layer is nailed to the joists, rafters, and boundary members, and those nails transfer all the boundary member axial forces. The top layer is nailed to the bottom layer. Each layer carries about half the web shear. Therefore, the strength of the diaphragm is larger if adequate transfer nailing can be provided.

Deflections are calculated using the flexibility factor for the web component, the flange flexure component previously illustrated, and the chord splice slip component.

Double sheathed shear walls are designed in the manner previously described using the appropriate flexibility factors.

Any construction method that will improve the connection strength will improve the performance of this or other diaphragms and shear walls. Flexibility factors would need to be determined for evaluating these diaphragms. Some flexibility factor data can be found in the literature for several types of formed steel and aluminum sheathing for use on wood framing.

Information useful in design of heavy timber decking diaphragms is found in the *Timber Construction Manual* [25.4].

25.15.6 Plywood Diaphragms and Shear Walls

Plywood diaphragms and shear walls are designed on the assumptions that moment is resisted by the chords and shear is resisted by the plywood sheathing.

Chord forces are determined by dividing the bending moment due to lateral load by the chord centroid spacing. The chords must be structurally continuous, which means they must be spliced and the spliced joints must have the capacity to carry the chord tension and compression forces. Since most lumber grades have lower tension than compression strengths, allowable tension controls the chord size and grade.

Chord splices are of two general kinds. Chords made using doubled nominal 2-in. lumber are common and are spliced as shown in Fig. 25.46*a*. In this design both members carry the chord force, except where the end joints occur. The chords must be of a size and grade so one member can carry the chord force at its minimum net section. Each bolt group is designed for half the chord force, because both members carry equal loads for the long distance between bolt groups, that is, half the chord force.

Single chords spliced with wood or steel plates as in Fig. 25.46*b* must transfer all the chord force through the bolted connection and splice plates. Either design will use about the same total number of bolts or connectors. The connector and lumber stresses should be increased for short time load (LDF = 1.33).

All boundary members must be designed to function as both chords and end posts since they may be called upon to serve either purpose depending on direction of wind or seismic load.

The sheathing design to meet the end shear for the diaphragm or shear wall is determined by reference to Tables 25.34 and 25.35. These tables are typical of those published in codes and design guides. They are based on strengths determined by testing and the values are for wind and seismic

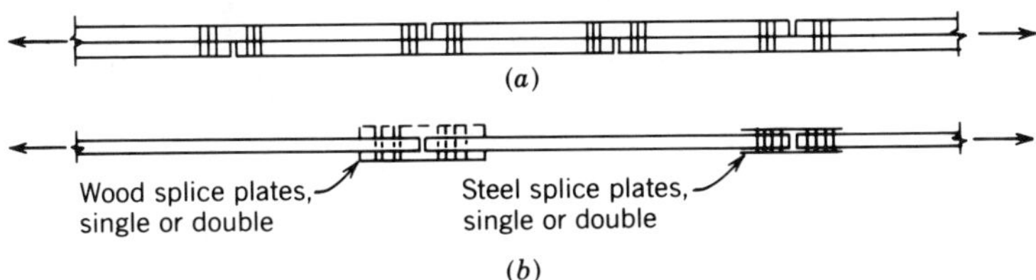

Fig. 25.46 Chord splice designs.

Table 25.34 Recommended Shear for Horizontal Plywood Diaphragms for Wind and Seismic Loading (Values Given in lb/ft) (Courtesy American Plywood Association)

Panel Grade	Common Nail Size	Minimum Nail Penetration in Framing (inches)	Minimum Nominal Panel Thickness (inch)	Minimum Nominal Width of Framing Member (inches)	Blocked Diaphragms				Unblocked Diaphragms	
					Nail Spacing (in.) at diaphragm boundaries (all cases), at continuous panel edges parallel to load (Cases 3 & 4), and at all panel edges (Cases 5 & 6)[b]				Nails Spaced 6″ max. at Supported Edges[b]	
					6	4	2½[c]	2[c]	Case 1 (No unblocked edges or continuous joints parallel to load)	All other configurations (Cases 2, 3, 4, 5 & 6)
					Nail Spacing (in.) at other panel edges (Cases 1, 2, 3 & 4)					
					6	6	4	3		
APA STRUCTURAL I RATED SHEATHING EXP 1 or EXT	6d	1-1/4	5/16	2	185	250	375	420	165	125
				3	210	280	420	475	185	140
	8d	1-1/2	3/8	2	270	360	530	600	240	180
				3	300	400	600	675	265	200
	10d	1-5/8	15/32	2	320	425	640	730[c]	285	215
				3	360	480	720	820	320	240
APA RATED SHEATHING EXP 1, EXP 2 or EXT; APA STRUCTURAL II RATED SHEATHING EXP 1 or EXT; and other APA grades except Species Group 5	6d	1-1/4	5/16	2	170	225	335	380	150	110
				3	190	250	380	430	170	125
			3/8	2	185	250	375	420	165	125
				3	210	280	420	475	185	140
	8d	1-1/2	3/8	2	240	320	480	545	215	160
				3	270	360	540	610	240	180
			7/16	2	255	340	505	575	230	170
				3	285	380	570	645	255	190
			15/32	2	270	360	530	600	240	180
				3	300	400	600	675	265	200
	10d	1-5/8	15/32	2	290	385	575	655[c]	255	190
				3	325	430	650	735	290	215
			19/32	2	320	425	640	730[c]	285	215
				3	360	480	720	820	320	240

[a] For framing of species other than Douglas fir, larch, or Southern pine. (1) Find species group of lumber in Table 25.23. (2) Find shear value from table for nail size, and for Structural I panels (regardless of actual grade). (3) Multiply value by 0.82 for Lumber Group III or 0.65 for Lumber Group IV.

[b] Space nails 12 in. oc along intermediate framing members for roofs, and 10 in. oc for floors.

[c] Framing shall be 3-in. nominal or wider, and nails shall be staggered where nails are spaced 2 in. oc or 2½ in. oc, and where 10*d* nails having penetration into framing of more than 1⅝ in. are spaced 3 in. oc. Exception: Unless otherwise required. 2-in. nominal framing may be used where full nailing surface width is available and nails are staggered.

Notes: Design for diaphragm stresses depends on direction of continuous panel joints with reference to load, not on direction of long dimension of sheet. Continuous framing may be in either direction for blocked diaphragms.

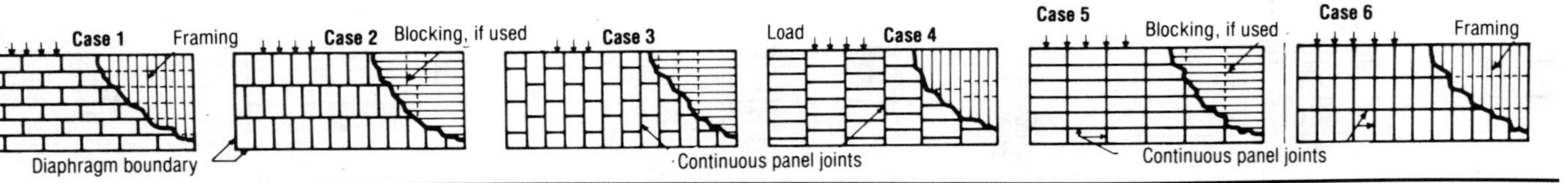

Table 25.35 Recommended Shear for Plywood Shear Walls for Wind or Seismic Loading (Values Given in lb/ft) (Courtesy American Plywood Association)

Panel Grade	Minimum Nominal Panel Thickness (in.)	Minimum Nail Penetration in Framing (in.)	Panels Applied Direct to Framing					Panels Applied Over 1/2″ Gypsum Sheathing				
			Nail Size (common or galvanized box)	Nail Spacing at Panel Edges (in.)				Nail Size (common or galvanized box)	Nail Spacing at Panel Edges (in.)			
				6	4	3	2(e)		6	4	3	2(e)
APA STRUCTURAL I RATED SHEATHING EXP 1 or EXT	5/16	1-1/4	6d	200	300	390	510	8d	200	300	390	510
	3/8	1-1/2	8d	230(d)	360(d)	460(d)	610(d)	10d	280	430	550(e)	730
	7/16			255(d)	395(d)	505(d)	670(d)					
	15/32			280	430	550	730					
	15/32	1-5/8	10d	340	510	665(e)	870	—	—	—	—	—
APA RATED SHEATHING EXP 1, EXP 2 or EXT; APA STRUCTURAL II RATED SHEATHING EXP 1 or EXT; APA panel siding (f) and other APA grades except species Group 5.	5/16 or 1/4(c)	1-1/4	6d	180	270	350	450	8d	180	270	350	450
	3/8			200	300	390	510		200	300	390	510
	3/8	1-1/2	8d	220(d)	320(d)	410(d)	530(d)	10d	260	380	490(e)	640
	7/16			240(d)	350(d)	450(d)	585(d)					
	15/32			260	380	490	640					
	15/32	1-5/8	10d	310	460	600(e)	770	—	—	—	—	—
	19/32			340	510	665(e)	870	—	—	—	—	—
APA panel siding (f) and other APA grades except species Group 5	5/16(c)	1-1/4	Nail Size (galvanized casing) 6d	140	210	275	360	Nail Size (galvanized casing) 8d	140	210	275	360
	3/8	1-1/2	8d	160	240	310	410	10d	160	240	310(e)	410

[a] For framing of other species: (1) Find species group of lumber in Table 25.23. (2) (a) For common or galvanized box nails, find shear value from table for nail size, and for STRUCTURAL I panels (regardless of actual grade). (b) For galvanized casing nails, take shear value directly from table. (3) Multiply this value by 0.82 for Lumber Group III or 0.65 for Lumber Group IV.

[b] All panel edges backed with 2-in. nominal or wider framing. Install panels either horizontally or vertically. Space nails 6 in. oc along intermediate framing members for 3/8-in. and 7/16-in. panels installed on studs spaced 24 in. oc. For other conditions and panel thicknesses, space nails 12 in. oc on intermediate supports.

[c] 3/8-in. or 303- 16 oc is minimum recommended when applied direct to framing as exterior siding.

[d] Shears may be increased to values shown for 15/32-in. sheathing with same nailing provided (1) studs are spaced a maximum of 16 in. oc, or (2) if panels are applied with long dimension across studs.

[e] Framing shall be 3-in. nominal or wider, and nails shall be staggered where nails are spaced 2 in. oc, and where 10d nails having penetration into framing of more than 1 5/8 in. are spaced 3 in. oc. Exception: Unless otherwise required, 2-in. nominal framing may be used where full nailing surface width is available and nails are staggered.

[f] 303- 16 oc plywood may be 11/32-in., 3/8-in., or thicker. Thickness at point of nailing on panel edges governs shear values.

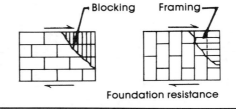

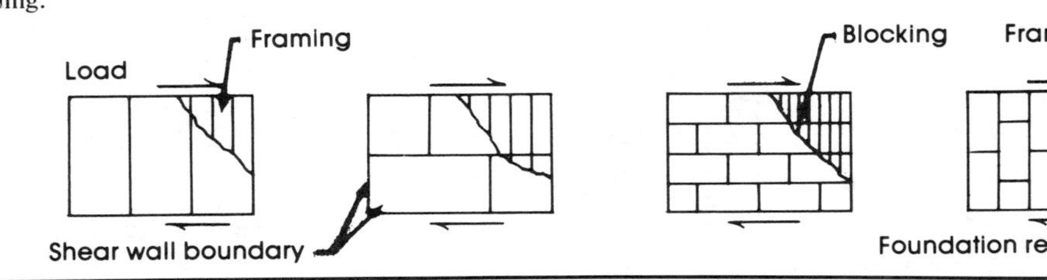

load. If a diaphragm is for use at normal load duration as in a folded plate or box girder, the allowable unit shears are 75% of those in the tables.

Diaphragm allowable unit end shear is dependent on several construction features:

1. Plywood grade and thickness.
2. Framing member thickness and species.
3. Plywood pattern with reference to the direction of load.
4. Nail size and spacing.
5. Presence of blocking under plywood panel edges perpendicular to the supports. (Blocking is a wood member to which panel edges are nailed. It transfers shear and prevents buckling of the plywood.)

Table 25.34 considers all these features.

Diaphragm strength is improved if each plywood panel is nailed to stiff supporting framing at all edges. Panel edges in one direction can be nailed to the structural support members. The other edges can be nailed to wood blocking set between the supports under the panel edges. In roof systems, blocking serves the added purpose of preventing differential vertical panel movement under concentrated loads and prevents damage to the roofing. Plywood panels with tongue and groove edges can be used to prevent differential movement. Tongue and groove edges are not equivalent to diaphragm blocking, however.

Unblocked diaphragms have simple nail spacing requirements; say 6-in. spacing on panel edges over framing, 12-in. spacing on intermediate framing for roofs, and 10-in. spacing for floors. Some plywood panel layouts have edges in a continuous line in one direction. When continuous edges are unblocked and perpendicular to the load direction, the allowable unit end shear is larger than for other patterns, as shown in the two columns at the right side of Table 25.34. Note that Cases 1, 2, and 5 become Cases 3, 4, and 6 when load direction changes by 90°. The performance of unblocked diaphragms is not improved by using closer nail spacing, because the failure mode is buckling along the unblocked and unnailed edges.

Blocked diaphragms can provide large end shear strength if the boundaries (chords) and the continuous panel edges parallel to the load are fastened with more closely spaced nails. Examine Table 25.34 for 6*d* nail size in ⅜-in. CD plywood, noting that the addition of blocking raises the end shear from 165 to 185 lb/ft with 6-in. nail spacing. Increasing the boundary and parallel-to-load panel edge nail spacing to 4 in. raises the end shear to 250 lb/ft. Nail spacing of 2.5 in. in these critical locations raises the end shear to 375 lb/ft and 2-in. spacing gives 420 lb/ft end shear.

Table 25.34 is for Douglas fir and southern pine framing. If Group III or IV species are used, unit end shear is reduced because of the lower nail holding properties.

Structural I plywood is rarely available from distributors and must be ordered well in advance of the need.

To save nailing labor, nail spacing in high unit shear diaphragms can be reduced near the center where the applied unit shear is lower.

Shear wall allowable shears are given in Table 25.35. All edges must be blocked in plywood shear wall construction and all vertical chords should be double and continuous. Shear walls are designed as vertical cantilevers, anchored at the base with moment carried by the chords and shear carried by the plywood. The portions of shear walls above and below door and window openings are considered ineffective. Only the full height parts of the shear wall are relied upon. Shear walls adjacent to openings are tied together over the openings so they will deflect equally. The total shear is shared in proportion to the length of the shear wall segments. Wall openings reduce the effective shear wall length and raise the applied unit end shear in the walls. The vertical load on the shear walls must also be considered in designing chords and vertical wall studs.

The allowable unit end shears for plywood diaphragms are for length-to-width ratios not larger than 4 and for shear walls with height-to-length ratios not exceeding 3.5. Applicable codes should be checked for these permissible ratios. Diaphragms and shear walls within these proportional dimension ratios will have acceptable deflection for wood buildings.

For masonry or concrete wall buildings, the diaphragm deflection must be limited to avoid overstressing the walls in bending. The relative deflection or drift between wall top and bottom is limited to

$$\Delta_p = \frac{100h^2f}{E_w t} \tag{25.15.9}$$

where Δ_p = allowable deflection between adjacent supports of wall
h = height of wall between adjacent horizontal connections
f = allowable flexural compressive stress of wall material

E_w = modulus of elasticity of the wall material
t = wall thickness

25.15.7 Plywood Overlaid Lumber Deck Diaphragms and Shear Walls

A common method of upgrading the performance of decks made with parallel courses of wood decking is to build a plywood diaphragm over the deck. The plywood diaphragm is designed as in Section 25.15.6 and secured to the wood deck with the nailing pattern prescribed for the plywood to obtain the desired end shear. Blocking is not used because the underlying lumber functions as the blocking. In this system the diaphragm behavior of the decking is not counted on and the entire lateral load is regarded as carried by the plywood. The edge rows of decking are spliced to form continuous chords and the ends perpendicular to the lumber courses are fitted with continuous chords secured by equivalent boundary nailing.

25.15.8 Laminated Wood Decking with Elastomeric Adhesive Bonding

Glued laminated wood decking made by laminating three or more boards to form a thick decking product has the virtues of straightness, dryness, and uniformity. The lower face can be of a decorative quality and the core and back may be utilitarian. This decking can provide good end shear values if the courses are bonded together with a structural elastomeric adhesive. The adhesive must be a particular type as approved by the building official (see ICBO Research Committee Recommendation Report No. 1379 [25.29]).

Elastomeric adhesives are glues that cure by chemical polymerization in the presence of moisture from the air. They form thick, elastic bonds of known shear modulus of elasticity and strength, and can be field-applied using caulking guns. In constructing the diaphragm or shear wall, a bead of adhesive is placed on the tongue edge of each course of decking. When the next course is laid and drawn up tightly to the preceding course the adhesive bead forms a glue line of prescribed thickness. The bead may be continuous or intermittent, depending on the end shear requirements of the design.

This system requires that the lumber deck be seasoned to 12% or lower moisture content and that the T & G pattern be cut to provide a clearance for forming a glue line about 1/16 in. thick and 1.25 in. wide.

These diaphragms use the random length continuous system (Fig. 25.27). They require continuous chords to resist the externally applied bending moment as for plywood diaphragms.

EXAMPLE 25.15.4

Illustrate the design of a 50 × 150 ft laminated wood decking diaphragm to resist a lateral load of 250 lb/ft acting perpendicular to the edges due to wind. The decking is 3 11/16 × 5 3/8 in. laminated wood with an elastic modulus of 1,500,000 psi laid in the 150-ft direction. The chords and headers (boundary members) are No. 2 Douglas fir-larch or southern pine, with elastic modulus of 1,700,000 psi and allowable tensile stress of 825 psi, which is less than its allowable compressive stress.

Solution

(a) *Determine end shear, end wall reaction, and glue line fraction required.* With the lateral load on the 150-ft edge the end shear is (250 × 150)/(2 × 50) = 375 lb/ft. The end wall reaction is 18,750 lb. Table 25.36 indicates that this unit end shear can be obtained if the glue line bonds 50% of the length of each course (glue a foot, skip a foot), $p = 0.5$.

(b) *Determine side wall reaction and glue line requirement.* With lateral load acting on the 50-ft edge the end shear is (250 × 50)/(2 × 150) = 41.7 lb/ft for a side wall reaction of 6250 lb. The allowable shear per inch of glue line for the recommended adhesive is 41 lb. The shear load per inch of glue line for this diaphragm is 3.475 when $p = 1$ and 6.95 lb when $p = 0.5$.

The maximum shear on the glue lines when load is on the long edge is 375/12 = 31.25 lb/in. near the end walls. This means that near the diaphragm ends the glue lines should be 75% continuous and near midspan they can be 25% continuous, averaging the required 50% ($p = 0.5$).

(c) *Determine nailing requirement.* Nailing must meet the requirements of wind uplift which depends on support member locations; it is a separate design problem. For diaphragm performance the edge nailing at the boundaries is critical. Along the 50-ft end a resistance to 375 lb/ft end shear is needed. There should be at least two nails per deck piece at this location. The load per nail, for two nails every 5.375 in. will be (5.375/12)(375/2) = 84 lb. Nails of suitable size to resist this lateral force (LDF = 1.33) plus the wind uplift (withdrawal force) must be selected. The same nail spacing is required along the 150-ft boundary at the ends but may be increased toward diaphragm center.

With wind on the 50-ft end the nailing of the 150-ft boundary edge must resist 41.7 lb/ft. Select

Table 25.36 Elastomeric Bonded Diaphragm Constants

Glue Line Fraction of Total Length p	Deflection Constant m	Modulus of Rigidity G, lb/in.	Allowable End Shear V_s, lb/ft	Allowable Glue Line Shear Stress, lb/in.
1.0	0.23	19,300	500	41
0.9	0.32	17,730	475	41
0.8	0.42	16,080	450	41
0.7	0.51	14,430	425	41
0.6	0.60	12,790	400	41
0.5	0.69	11,140	375	41
0.4	0.79	9490	350	41
0.3	0.88	7840	325	41
0.2	0.97	6190	300	41
0.1	1.07	4540	275	41

and space nails to provide this strength which, in this example, will be less demanding than that required by the load on the long edge.

(d) *Design the chords.* Chords are designed to resist the maximum flexural moment of $wL/8$ = 703 ft-kips. With chords spaced 50 ft apart the chord force is 14 kips requiring a chord area of 12.8 sq in. at 825-psi strength of the lumber, increased by LDF = 1.33. A 2×10 has a 13.9-sq in. area. Depending on the net section due to chord splice bolts, 2×10 or larger chords will be needed.

(e) *Compute deflection for loading on the 150-ft edge.* The deflection for uniformly distributed load on the 150-ft edge is

$$\Delta = \left[\frac{5}{384}\frac{wL^4}{EI} + \frac{6}{40}\frac{wL^2}{AG}\right](1 + m) \tag{25.15.10}$$

where w = load on long edge = 20.83 lb/in.
E = elastic modulus of chord = 1.7×10^6 psi
L = diaphragm length = 1800 in.
I = moment of inertia of chords about diaphragm centerline

$$= 2A_c\left(\frac{B}{2}\right)^2 = \frac{1}{2}A_cB^2$$

B = diaphragm width = 600 in.
A_c = chord area = 13.88 sq in.
A = area of deck cross section = 2212 sq in.
G = shear modulus of the construction as given in Table 25.36 = 11,140 psi (relative to the adhesive properties)
m = nonlinear but recoverable deflection constant from Table 25.36 = 0.69

$$\Delta = (0.670 + 0.411)1.69 = 1.83 \text{ in.}$$

(f) *Compute deflection for loading on the 50-ft edge.* The deflection for uniformly distributed load is

$$\Delta = \left[\frac{5wB^4}{384EI} + \frac{1.468wBn}{Lp}10^{-4}\right](1 + m) \tag{25.15.11}$$

where $I = \frac{1}{2}A_cL^2 = 22.49 \times 10^6$ in.^{4}
$n = B/b = 112$
b = decking width = 5.375 in.
p = 0.5 (Table 25.36)

$$\Delta = (0.00092 + 0.2263)(1.69) = 0.38 \text{ in.}$$

When diaphragms and shear walls are designed for the end shear limits in the published tables (such as Tables 25.33, 25.34, and 25.35) the deflections will not exceed limits regarded as acceptable by the building authorities, insofar as the diaphragms or shear wall itself is concerned.

The calculation of deflection is not necessary unless there is some restriction on the displacement of any attached portion of structure (such as an adjoining wall). Engineers often wish to make a deflection estimate and methods are described for this purpose. Should the deflection exceed the engineer's definition of acceptability, the *design* end shear may be increased just to limit deflection at the expected end shear. Such cases occur when brittle materials of modest strength are used for the tributary structure or finish (masonry, gypsum, stucco, or glass, for example).

25.15.9 Metal Sheathed Roof Diaphragms

Wood-framed buildings are often sheathed with steel or aluminum roof panels which are corrugated or formed to pattern which provides a folded plate or trough-like shape to resist vertical loading. The metal panels are fastened to wood purlins which run at 90° to the building rafters or trusses. These panels possess shear resistance to lateral forces on the building. The resistance is characteristic of the construction, being a combined product of the metal sheathing, the wood purlins, their blocking, and connections. The stiffness of the panels is determined by testing to obtain the load-deflection ratio characteristic of that particular panel construction [25.24, 25.25, 25.45]. There are no standard values because of the wide variety of panels, fastener patterns, uses of adhesives to supplement mechanical fasteners, and arrangements of purlins. Some typical values of stiffness and ultimate strength which could be expected from existing metal products are given in Table 25.37.

Buildings that use metal sheathing are usually designed as a series of frames or trusses supported by poles or wood columns. Without the diaphragm effect of the metal roof sheathing, each frame would be designed to bear all the lateral force on its tributary area. This would often require inconvenient knee braces. If the end frames can be designed as rigid shear walls, the roof sheathing must be deformed as the frames deform, thus reducing the moment on the frames. The amount of moment reduction depends on the relative stiffness of frames and roof panels. Moment reduction factors published by Bryan [25.20] are partially reproduced in Table 25.38.

The moment on the middle frame is the greatest and it decreases toward the end shear walls. As the frame-to-roof stiffness ratio decreases, the roof transmits increased portions of the lateral force to the end walls, and the moment carried by the frames diminishes. This is evident in Table 25.38. The end wall shear increases as the frame-to-roof stiffness ratio decreases.

EXAMPLE 25.15.5

Illustrate the design procedure for the building of Fig. 25.47 that is 40 ft wide by 64 ft long consisting of nine frames spaced 8 ft apart, with steel sheathing fastened with screws to wood purlins spaced 24 in. on centers and perpendicular to the frames. The building is to be designed for a lateral wind loading of 20 lb/sq ft.

Solution

(a) *Determine the ratio of lateral load to the deflection at the eave line.* The stiffness of the frames must be determined by elastic analysis. This is done by obtaining the deflection of the frame at the eave line due to any arbitrarily chosen load. The ratio of load to deflection is the frame stiffness k. Assume this has been determined to be 3000 lb/in.

Table 25.37 Stiffness and Strength of Metal Roof Sheathing in Lateral Shear

Panel Description	C, lb/in. of Panel Deflection	Ultimate Strength, lb/ft
Aluminum panels:		
8-ft rafter spacing, 12-ft length		
Screws with or without sidelap gluing	1500	125
Screws, glued sidelap seams, glued to rafters	2700	160
4-ft rafter spacing, 12-ft length		
Screws	4300	240
Screws and glued to rafters	5000	280
Steel panels:		
8-ft rafter spacing, 12-ft length		
Screws with or without sidelap gluing	1800	150
Screws, glued sidelap seams and glued to rafters	3000	200

Table 25.38 Moment Reduction Factor, m

Number of frames includes framed endwalls. From the design example, a 64-ft long building with 8-ft o. c. trusses and poles has nine frames. Frame number 5 is critical. With a k/c ratio of 0.15, multiply restraining force R by 0.59 to get sidesway moments in the center frame of the building.

No. of frames	Frame no.	k/c, ratio of frame to roof stiffness																		
		0.10	0.15	0.20	0.25	0.30	0.40	0.50	0.60	0.70	0.80	0.90	1.00	1.25	1.50	2.00	2.50	3.00	3.50	4.00
		m, moment reduction factor																		
3	2	0.05	0.07	0.09	0.11	0.13	0.17	0.20	0.23	0.26	0.29	0.31	0.33	0.38	0.43	0.50	0.56	0.60	0.64	0.67
4	2	0.09	0.13	0.17	0.20	0.23	0.29	0.33	0.38	0.41	0.44	0.47	0.50	0.56	0.60	0.67	0.71	0.75	0.78	0.80
5	2	0.13	0.18	0.23	0.27	0.30	0.36	0.41	0.45	0.49	0.52	0.55	0.57	0.62	0.66	0.71	0.75	0.78	0.81	0.82
	3	0.17	0.24	0.30	0.35	0.39	0.47	0.53	0.58	0.62	0.66	0.69	0.71	0.77	0.80	0.86	0.89	0.91	0.93	0.94
6	2	0.16	0.22	0.27	0.31	0.35	0.41	0.45	0.49	0.53	0.55	0.58	0.60	0.64	0.68	0.73	0.76	0.79	0.81	0.83
	3	0.24	0.32	0.39	0.45	0.50	0.58	0.64	0.68	0.72	0.75	0.78	0.80	0.84	0.87	0.91	0.93	0.95	0.96	0.96
7	2	0.19	0.25	0.30	0.34	0.38	0.43	0.48	0.51	0.54	0.57	0.59	0.61	0.65	0.68	0.73	0.76	0.79	0.81	0.83
	3	0.29	0.38	0.46	0.52	0.56	0.64	0.69	0.73	0.77	0.79	0.82	0.83	0.87	0.89	0.92	0.94	0.95	0.96	0.97
	4	0.32	0.43	0.51	0.57	0.62	0.70	0.75	0.80	0.83	0.85	0.87	0.89	0.92	0.94	0.96	0.97	0.98	0.99	0.99
8	2	0.21	0.27	0.32	0.36	0.39	0.45	0.49	0.52	0.55	0.58	0.60	0.62	0.65	0.69	0.73	0.77	0.79	0.81	0.83
	3	0.33	0.43	0.50	0.56	0.60	0.67	0.72	0.76	0.79	0.81	0.83	0.85	0.88	0.90	0.93	0.94	0.96	0.96	0.97
	4	0.39	0.50	0.58	0.65	0.69	0.77	0.81	0.85	0.87	0.89	0.91	0.92	0.95	0.96	0.98	0.98	0.99	0.99	0.99
9	2	0.22	0.28	0.33	0.37	0.40	0.45	0.49	0.53	0.55	0.58	0.60	0.62	0.66	0.69	0.73	0.77	0.79	0.81	0.83
	3	0.37	0.46	0.53	0.58	0.63	0.69	0.74	0.77	0.80	0.82	0.84	0.85	0.88	0.90	0.93	0.94	0.96	0.96	0.97
	4	0.45	0.56	0.64	0.70	0.74	0.80	0.84	0.87	0.90	0.91	0.93	0.94	0.95	0.97	0.98	0.99	0.99	0.99	0.99
	5	0.47	0.59	0.67	0.73	0.77	0.84	0.88	0.90	0.92	0.94	0.95	0.96	0.97	0.98	0.99	0.99	1.00	1.00	1.00
10	2	0.23	0.30	0.34	0.38	0.41	0.46	0.50	0.53	0.56	0.58	0.60	0.62	0.66	0.69	0.73	0.77	0.79	0.81	0.83
	3	0.39	0.49	0.55	0.60	0.64	0.70	0.74	0.77	0.80	0.82	0.84	0.85	0.88	0.90	0.93	0.95	0.96	0.96	0.97
	4	0.49	0.60	0.67	0.73	0.77	0.82	0.86	0.89	0.91	0.92	0.93	0.94	0.96	0.97	0.98	0.99	0.99	0.99	0.99
	5	0.54	0.65	0.73	0.78	0.82	0.87	0.91	0.93	0.94	0.96	0.96	0.97	0.98	0.99	0.99	1.00	1.00	1.00	1.00
11	2	0.24	0.30	0.35	0.38	0.41	0.46	0.50	0.53	0.56	0.58	0.60	0.62	0.66	0.69	0.73	0.77	0.79	0.81	0.83
	3	0.41	0.50	0.56	0.61	0.65	0.71	0.75	0.78	0.80	0.82	0.84	0.85	0.88	0.90	0.93	0.95	0.96	0.96	0.97
	4	0.52	0.63	0.69	0.74	0.78	0.83	0.87	0.89	0.91	0.92	0.93	0.94	0.96	0.97	0.98	0.99	0.99	0.99	0.99
	5	0.58	0.69	0.76	0.81	0.85	0.89	0.92	0.94	0.95	0.96	0.97	0.98	0.98	0.99	0.99	1.00	1.00	1.00	1.00
	6	0.60	0.71	0.78	0.83	0.87	0.91	0.94	0.95	0.97	0.97	0.98	0.98	0.99	0.99	1.00	1.00	1.00	1.00	1.00
12	2	0.25	0.31	0.35	0.39	0.41	0.46	0.50	0.53	0.56	0.58	0.60	0.62	0.66	0.69	0.73	0.77	0.79	0.81	0.83
	3	0.43	0.51	0.57	0.62	0.65	0.71	0.75	0.78	0.80	0.82	0.84	0.85	0.88	0.90	0.93	0.95	0.96	0.96	0.97
	4	0.54	0.64	0.71	0.76	0.79	0.84	0.87	0.89	0.91	0.92	0.94	0.94	0.94	0.96	0.97	0.98	0.99	0.99	0.99
	5	0.62	0.72	0.79	0.83	0.86	0.90	0.93	0.95	0.96	0.97	0.97	0.98	0.99	0.99	0.99	1.00	1.00	1.00	1.00
	6	0.65	0.76	0.82	0.87	0.89	0.93	0.95	0.97	0.98	0.98	0.99	0.99	0.99	1.00	1.00	1.00	1.00	1.00	1.00

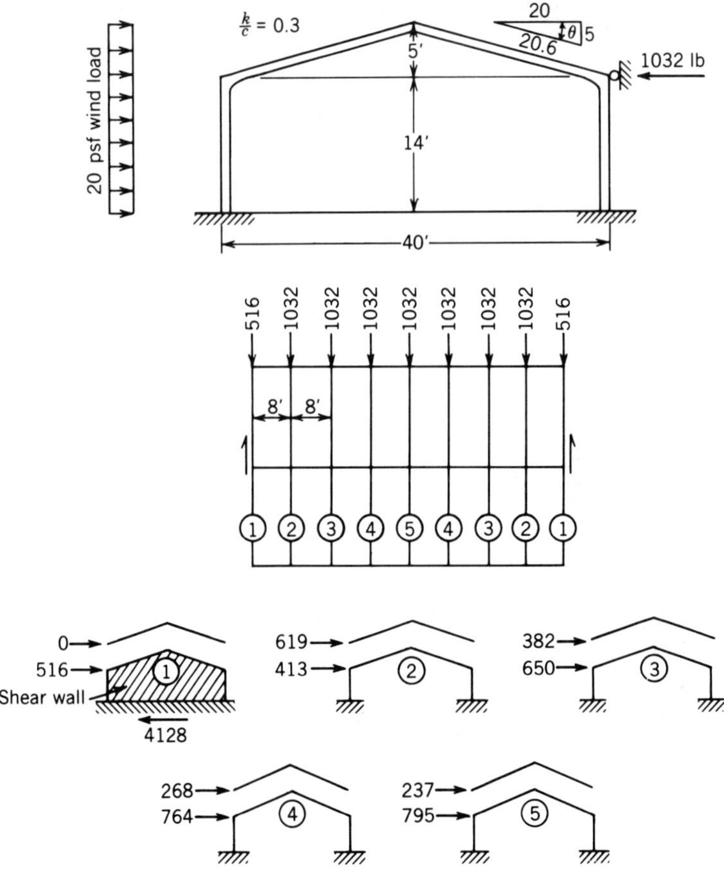

Fig. 25.47 Metal roof diaphragm of Example 25.15.5.

(b) *Determine the reaction at the eave due to wind load.* The frame must also be analyzed to determine the reaction at the eave line due to the wind load tributary to the frame. This is done by placing a roller support at the eave on the leeward side of the building and determining the reaction by any convenient method. The load is 3040 lb and assume the analysis shows the reaction at the eave to be 1032 lb. This is the load that a completely rigid roof system would need to exert to restrain the frame and transmit all the lateral force to the end walls. Realistically, the use of Table 25.38 will give the actual reaction as 1032 times the moment reduction factor m.

(c) *Obtain the actual stiffness provided by the selected diaphragm.* The roof panel length is about 21 ft along each slope. The stiffness of the panel is 1800 lb/in. for a 12-ft panel (see Table 25.37). The stiifness of a 21-ft panel can be obtained using the Luttrell formula [25.25, 25.45] for panel edge load C in shear load per unit length,

$$C = \frac{Et}{2(1 + \nu)(g/p) + K/(Lt)^2} \tag{25.15.12}$$

where L = panel length, in.
E = elastic modulus of elasticity, psi
t = metal thickness, in.
ν = Poisson's ratio, 0.3 for steel
g/p = ratio of panel width measured along the profile to straight line width, usually about 1.3
K = a constant, in.^{4}

The value of C is usually obtained experimentally. Once C is known for a particular panel (1800 lb/in. for a 12-ft long panel of 8-ft width, as in Table 25.37) the value of C for a 21 × 8 ft wide panel is obtained by solving for K from Luttrell's formula using a C value of 1800, then using this K and the 21-ft panel length (252 in.) solving for C_{21}. Thus,

$$1800 = \frac{29{,}000{,}000(0.03)}{2(1.3)(1.3) + K/[144(0.03)]^2}$$

$$K = 8955$$

$$C_{21} = \frac{29{,}000{,}000(0.03)}{2(1.3)(1.3) + 8955/[252(0.03)]^2} = 5435 \text{ lb/in.}$$

Accounting for the roof slope θ,

$$C_{21} \text{ horizontal} = C_{21} \cos\theta = 5125$$

There are two panels, one on each slope making the actual roof stiffness 10,250 lb/in.

(d) *Determine the required panel loads*. The frame-to-roof stiffness ratio will be $k/C = 3000/10250 = 0.293$ or 0.3 and from Table 25.38, $m = 0.77$

The load on the middle frame will be 1032 m = 795 lb. The load on the roof panel must be 1032 − 795 = 237 lb.

In like manner, from Table 25.38 the frame and roof panel loads can be obtained for each frame with results as shown in Fig. 25.47.

(e) *Determine the end shear*. All roof loads are ultimately carried to the shear wall for a maximum diaphragm shear of 237/2 + 268 + 382 + 619 = 1387 lb or 1387 $\cos\theta/(40\cos\theta)$ = 34.7 lb/ft end shear (see Fig. 25.45).

(f) *Determine deflections*. Deflections at the various frames may be obtained from the shear loads carried by the 8-ft roof diaphragm segments.

The stiffness in lb/ft for a 1-in. deflection is 5435/21 = 259.

Segment	Shear load (lb/ft)	Stiffness (lb/ft/in.)	Δ_i (in.)
4–5	2.96	259	0.0115
3–4	9.66	259	0.0374
2–3	19.21	259	0.0742
1–2	34.7	259	0.1340

Total deflection = 0.257 in.

Consider an *extreme case* where the frames possessed zero stiffness. The fixed reaction load at each frame would be 1920 lb and it would all be carried by the roof panels. The end shear would be 6720 lb or 168 lb/ft. The ultimate strength of the steel diaphragm is only 150 lb/ft although this could be increased by 1.3 to 195 lb/ft. This leaves no safety factor, however. The deflections would be:

Segment	Shear load (lb/ft)	Stiffness (lb/ft/in.)	Δ_i (in.)
4–5	24	259	0.093
3–4	72	259	0.278
2–3	120	259	0.463
1–2	168	250	0.649

Total deflection = 1.48 in.

The frames would usually possess some stiffness either due to moment resistance at the eave joint or frame legs cantilevered from the ground line, as in post or pole frame buildings.

(g) *Design of chords*. No mention has been made of chords. It is assumed that the panels are on frames with strong joints at all corners. These corner joints must be designed to carry the eave line

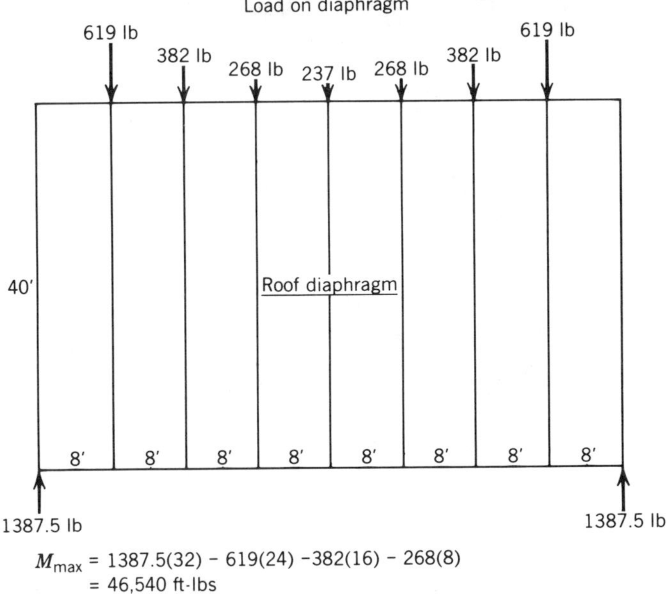

M_{max} = 1387.5(32) − 619(24) −382(16) − 268(8)
= 46,540 ft-lbs

Chord force = $M_{max}/40$ = 1164 lb

Fig. 25.48 Chord force for metal roof diaphragm of Example 25.15.5.

tension and compression, or the building will come apart at these locations. The tension and compression loads can be determined in a manner similar to that for plywood diaphragms, by dividing the maximum horizontal plane moment by the chord centroid distance. As shown in Fig. 25.48 the maximum moment is 46,540 ft/lb and the chord tension or compression is 1164 lb. Continuous chords with well-designed connections are an essential part of this diaphragm design.

25.16 FOLDED PLATES

Wood folded plates are made with plywood skins and lumber or glulam framing members as chords and rafters. This type of design utilizes the skins to carry load to the end supports, in lieu of the beam requirements in ordinary roof systems. The plates function as diaphragms in resisting vertical as well as horizontal loads.

Design depends on the proper analysis of the structure to establish the forces acting parallel and normal to the plates and at the plate junctions. Carney [25.21] and Payne [25.40] are the principal sources of folded plywood plate design procedure.

Folded plate structures usually consist of multiple bays of folded plates, but analysis of a single-bay folded plate, similar in appearance to a simple gabled roof, is a good starting point for explaining the method of analysis.

25.16.1 Single Bay Folded Plate Supported by Posts

Consider the folded plate of Fig. 25.49 supported by posts at the corners. Design is based on a uniformly distributed load acting vertically. The vertical reactions at the posts are each one-quarter of the total load.

The horizontal reaction at the top of each post is a function of the plate slope (H/B). This reaction increases as the slope decreases. This reaction must be resisted by a horizontal tie member between the end posts, or by end walls.

The top edges of the plates bear on one another, horizontally with a force equivalent to these horizontal reactions at the posts, but distributed uniformly along the ridge line.

The forces Z in the plane of the plates are uniformly distributed and load the plate as a deep beam (i.e., diaphragm). This distributed load is carried to the ends of the plates where it is resisted by the posts, or the end walls.

Structurally continuous framing members serve as tension and compression chords in the two edges of each plate. The chord force per plate is $wBL^2/8H$. If one chord member serves two plates, as at the ridge line in Fig. 25.49, the chord force is twice as large, $wBL^2/4H$.

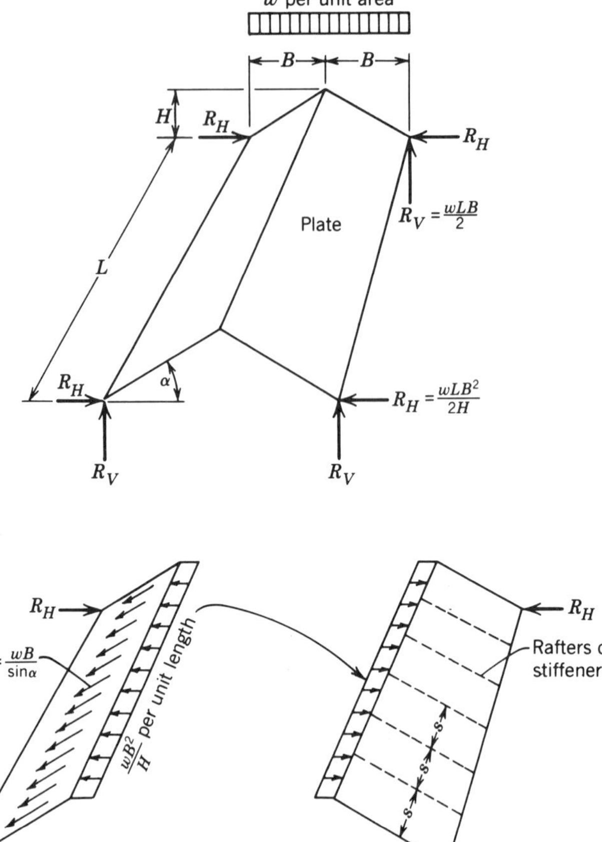

Fig. 25.49 Single bay folded plate (from Ref. 25.21, courtesy of American Plywood Association).

The top edge of the plate is supported against horizontal displacement by the mating plate, but at the lower edge no such continuous edge support exists. If this edge of the plate is to remain unsupported this chord must be designed to resist both the chord tension and a uniformly distributed vertical load of $wB/2$.

The plates usually have rafters extending between their chords, spaced at intervals determined by the span capacity of the plywood. The load for selecting the plywood is the load ($w \cos \alpha$) normal to the roof, and the span s is the distance between rafters. The plywood properties used will depend on whether the plywood face grain is parallel to the span or to the rafters. For plates of narrow width, with face grain along the slope, rafters may be unnecessary to support the plywood but are desirable as stiffeners.

Rafters can be designed for bending strength using a vertical load of ws lb/ft and a span of B,

$$M = \frac{wsB^2}{8} \tag{25.16.1}$$

25.16.2 Single Bay Folded Plate Supported by Side Wall or Beam

The lower edge of the plate is often supported by a beam or a continuous wall. This alters some of the forces just described.

A wall or beam carries half of the load on each plate. The other half is carried to the corner posts by way of the panel end rafters or stiffeners. If end walls are used they carry this load to the foundations. The side walls must resist the lateral force developed between the top edges of the plates, wB^2/H per foot, and may require horizontal ties at intervals along the length of the building at the eaves if the deflection is large enough to distort the wall beyond acceptable limits.

The wall or beam support reduces the in-plane load on the plates by half and Z becomes $wB/(2 \sin \alpha)$. They also reduce the chord forces by half, to $wBL^2/16H$ in the lower edge chord and $wBL^2/8H$ in a ridge chord serving two plates.

The skins are designed for the load acting normal to the slope and to carry the shear between the plate chords. This shear is $wBL/2 \sin \alpha$ when the plate edge is not vertically supported and $wBL/4 \sin \alpha$ when a beam or wall provides support. The unit shear in the skin at the ends of each plate is $wBL/2H$ without lower plate edge support and $wBL/4H$ when that support is provided. The plywood unit shear loads decrease linearly to zero at midlength of the folded plate. Skin-to-plate connections, nailed or glued, should be provided to resist these shears.

Deflections in the plane of the plates are computed for the plates as composite plywood–lumber beams or as diaphragms, depending on the size of the plates. When plate width exceeds the size of a continuous plywood sheet (between chords) the plates should be designed as diaphragms. Lateral ties at the eaves, at intervals along the length of the plates cause the plates to behave as beams on multiple supports, instead of simple beams of length L.

25.16.3 Multiple Folded Plates

A folded plate structure supported by four exterior walls is shown in Fig. 25.50. The gable end walls support the interior folded plates. The side walls support half the load on each exterior folded plate.

The in-plane loads for design of the folded plates are twice as large for the interior plates as they are for the exterior plates.

For the interior plates the vertical load per foot of length is wB and the in-plane load per foot is $wB/\sin \alpha$. The interior plates cause reactions along the top of each gable end wall of $wBL/(2 \sin \alpha)$ parallel to the inclined top edge, and $wBL/2$ vertically upon each top edge.

At the interior ridges the plate chords bear horizontally upon one another with the force wB^2/H per foot. At the interior valleys there are tension forces between the plate chords of wB^2/H per foot.

The chord force at the valleys and interior ridges is $wBL^2/8H$ ($wBL^2/4H$ if one chord serves both plates). This force is compression at the ridges and tension at the valleys.

For the exterior plates supported by the exterior side walls, the in-plane loads are $wB/(2 \sin \alpha)$, since the walls resist half of the vertical load upon the plates. The gable end walls then support half the loads mentioned above for gable end walls under interior plates. The side walls carry a horizontal thrust from the plates equal to $wB^2/2H$ per foot, which must be resisted by cantilever action of the side walls or by horizontal ties between the exterior walls at intervals along the sidewall length.

The exterior ridge chord forces are $wBL^2/16H$, that is, one-half of those for the interior ridges. If a common chord is used for the plates that come together at the exterior ridge their chord load is $3wBL^2/16H$. The exterior edge chord force is $wBL^2/16H$.

The side wall vertical loads are $wB/2$ per foot of length. The gable end wall vertical loads are

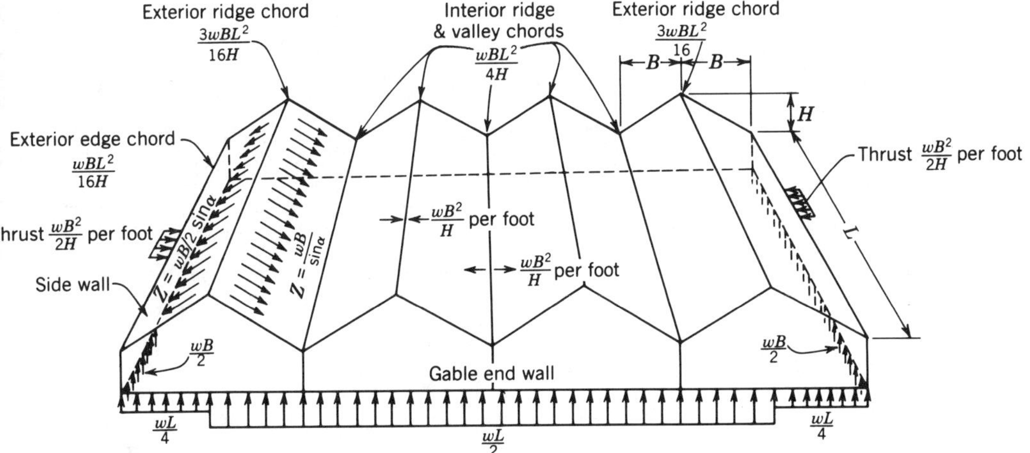

Fig. 25.50 Multiple bay folded plate supported by exterior walls (from Ref. 25.21, courtesy of American Plywood Association).

$wL/2$ between exterior ridges and $wL/4$ from side wall to exterior ridge. These loads add up to $wBLn$, where n is the number of folded plates in the roof.

To carry the load in tension perpendicular-to-the-grain in the valley chords, bolts should be installed, since wood is weak in tension perpendicular-to-grain.

Rafters and plywood skins are designed as previously described for single-bay folded plates (Sections 25.16.1 and 25.16.2).

25.16.4 Lateral Loads

For lateral loading by wind or seismic forces, the structure is designed as a blocked or unblocked diaphragm or as a series of composite beams, each carrying a share of the total lateral force.

25.16.5 Use of Adhesives

Plywood folded plates can be built with nails and bolts or adhesives can be used. When adhesives are used the fabrication should be carried out under the supervision of a well-qualified inspection service to assure that the adhesive bonds are of high quality. Folded plates made with adhesives of the proper type will be much stiffer than nailed plates.

25.17 HYPERBOLIC PARABOLOIDS

Wood hyperbolic paraboloid shells have been designed in a variety of combinations to obtain interesting architectural forms. Some of these are depicted in Fig. 25.51.

The fabrication skills to produce these structures limit their use to structures where the aesthetic characteristics are worth the extra cost. For such cases the wood hyperbolic paraboloid can probably be constructed as economically as one of any other material; perhaps more so. Hyperbolic paraboloids are light in weight. They can be made with lumber or glulam framing and lumber or plywood sheathing, depending on the size requirements of the job.

Fig. 25.51*a* is a simple square building with sloping walls which support the perimeter members. Straight sheathing boards connect points that are equally spaced along the opposite perimeter members. These boards have a slight twist. The distance between perimeter members is shortest for the horizontal boards at the midlength of the perimeter members so the width of some boards will need to be cut to a curve at some point along the width of a layer of boards. Two layers of sheathing boards are needed if lumber is used. If plywood is used, framing members to serve as blocking between plywood panel edges are required and usually they should be structurally spliced. One plywood layer may be adequate.

Lumber layers may be laid parallel to the perimeter members or parallel to the diagonal axes of the building. When laid in the latter manner the layers will be stressed axially in tension (concave face up) and compression (concave face down) to give the best strength and continuity of the shell. To form these layers a falsework of straight members parallel to the perimeter members will be necessary, and can be removed when the shell is finished. The two layers must be fastened together with mechanical fasteners or adhesives, or both, to prevent separation and provide two-way action of the system.

The perimeter members must be shaped to a twisted top surface configuration to fit the sheathing.

Wall support on all sides as in Fig. 25.51*a* carries the load directly to the footings and reduces the axial compression requirements of the perimeter members. This particular design is uncomplicated.

Figure 25.51*b* shows a structure supported at two corners. The load produces tension along the *BD* axis and compression along the *AC* axis. Thrust occurs outward at the support points *A* and *C*. This design may be square or diamond shaped in plan view. The perimeter members must carry the edge forces to the foundation points in axial compression. They are laterally supported by the sheathing but not in the vertical direction.

In Fig. 25.51*c* four hyperbolic paraboloids are arranged in a square plan to make a peaked structure. The perimeter members on two sides are horizontal. On the other sides they lie in a vertical plane. Figure 25.51*d* is a single module used in this design. The pitched members bear upon one another at the module interface, and may be in either tension or compression normal to this joint depending on the state of loading. The joint should therefore be bolted. As in Fig. 25.51*b*, thrust will develop at points *A* and *C* and provision must be made to resist this force with large horizontal beams, tie rods, or a buttress-type foundation. Foundation posts would be appropriately located at *A*, *C*, and *D*. These units should probably be designed for the load applied assuming the structure oriented as in Fig. 25.51*b*. Regardless of how the module is actually oriented, each perimeter member carries one-quarter of the load on the horizontal projection.

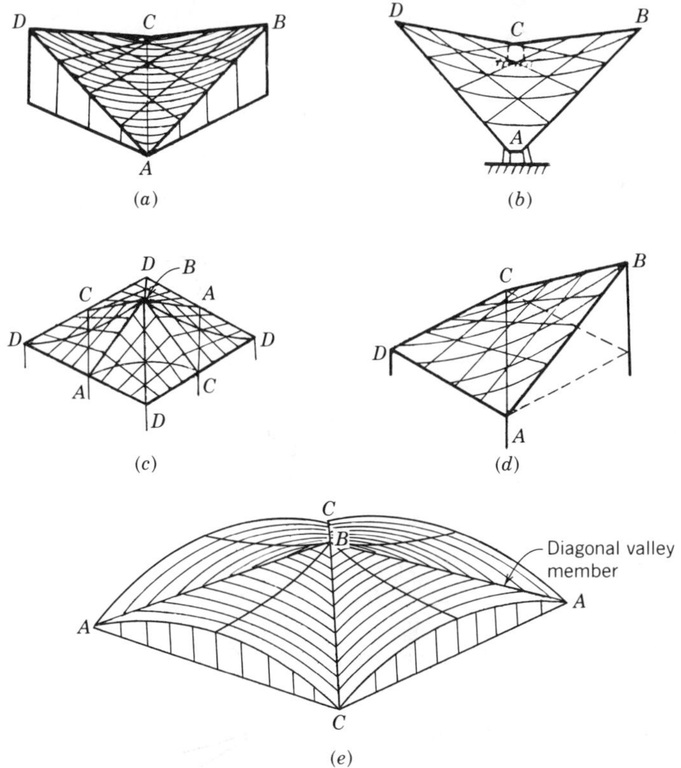

Fig. 25.51 Hyperbolic paraboloids.

Figure 25.51*e* shows four half shells arranged in a square. The diagonal valley members are in compression and exert thrust on the corner foundations. There is also thrust between these valley members. The wall arches are in compression. Sheathing parallel to the walls is in compression while that normal to the walls is in tension.

The design method for the hyperbolic paraboloid shown in Fig. 25.51*b* is very useful. The total load is assumed to be uniformly distributed over the horizontal projection, even though the dead load is not distributed exactly that way. The vertical reactions at *A* and *C* of the diamond-shaped plan in Fig. 25.52 are each half the total load. The structure is covered with a double layer of sheathing boards laid parallel to the major and minor axes. The top layer will be in the L_2 direction and will carry half the load in tension to the perimeter members. The bottom layer in the L_1 direction will carry half the load in compression to the perimeter members. The perimeter members transfer the sheathing forces to the foundation by compression along their axes. Shell tension and compression force components perpendicular to the perimeter members cancel one another while components parallel to these edge members are additive and are resisted by shear. The perimeter members are loaded in axial compression.

The vertical load is uniformly distributed along the horizontal projection of the perimeter members. The thrust at the supports is a function of L_1 and H. The horizontal thrust and vertical reaction are components of the resultant shown in Fig. 25.52.

Each perimeter member supports one-quarter of the total load. The maximum perimeter compression P at the base is computed from half the vertical reaction as in Fig. 25.53. The uniformly distributed shear load is obtained by dividing P by the member length a giving v in lb/ft along the top of the perimeter member.

The maximum sheathing board forces, c and t, acting along the axis of diagonally laid boards may be computed as in Fig. 25.53. These forces are for board widths shown. Stresses can be determined by dividing by the crosssection areas for these board widths.

The sheathing connections at the perimeter should be designed to resist these loads. The sheathing layers should be nailed, stapled, and/or glued together to prevent buckling of the compression layer. End joints in lumber courses should be structurally spliced or dispersed so the continuity will be adequate.

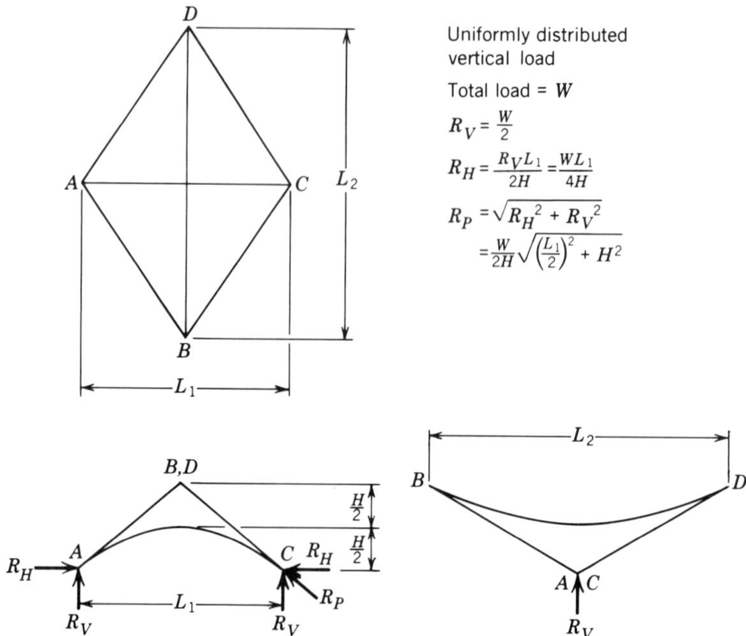

Fig. 25.52 Reactions for hyperbolic paraboloid.

25.18 CURVED ROOF PANELS

Plywood can be formed into curved panels by cold bending over a curved form and pressure gluing to ribs, or by glue laminating to cores of plywood or other core material such as insulating board or paper honeycomb. PDS Supplement No. 1, *Design and Fabrication of Plywood Curved Panels,* [25.8] describes the procedure in detail.

Curved panels with lumber ribs may be made with a topskin, or with top and bottom skins to form hollow core panels. Paper honeycomb or insulated core panels have curved wood boundary strips. Ribs and core can be offset to form tongue and groove edges for making good joints for field connection by nailing. The panels should be constructed with waterproof adhesives of the synthetic resin type which are rigid and creep resistant. Nails alone are not adequate and gluing should be done under clamping pressure to bring the parts into intimate contact for forming quality glued joints. The fabrication should be done by a qualified manufacturer.

Panels that exceed the length of standard plywood sheets should use skins that have been structurally joined by glued splices or scarf joints, designed to carry the stresses.

Curved panels must be designed for structural adequacy with a system of support to carry the vertical reactions and the thrust at the panels ends. Curved panels are not intended to be used as curved "folded plates" to carry load on spans parallel to their axis of curvature.

These panels may be designed as *curved flexural beams* (arches) without horizontal restraint at the reaction points (see Section 25.14). Provision of space to permit horizontal movement at the ends obviates horizontal thrust. If horizontal movement is positively prevented (by tie-rods, for example) the panels become two-hinged arches. These *arch panels* have lower bending movments but high axial compression reaching a maximum at the reactions.

Curved flexural panels develop their highest bending moment at midspan, if the load is uniformly distributed. The panel can be analyzed by treating it as a beam with a span equal to the horizontal projection of the panel. The deflection can be approximated by using a beam deflection formula. This will overestimate the deflection by 10%. Deflections calculated by this method are satisfactory for most designs of this kind.

The increase in span for curved flexural panels under uniformly distributed load is equal to the midspan deflection multiplied by the rise-to-span ratio times 4, for small deflections of the usually acceptable amount.

The span increase to be permitted at the supports can be computed in this way. Maximum axial force in these panels occurs at the support points. At the top of the curve where the moment is largest the axial load is zero under uniformly distributed load. Analysis for other loadings can be based on static equilibrium methods.

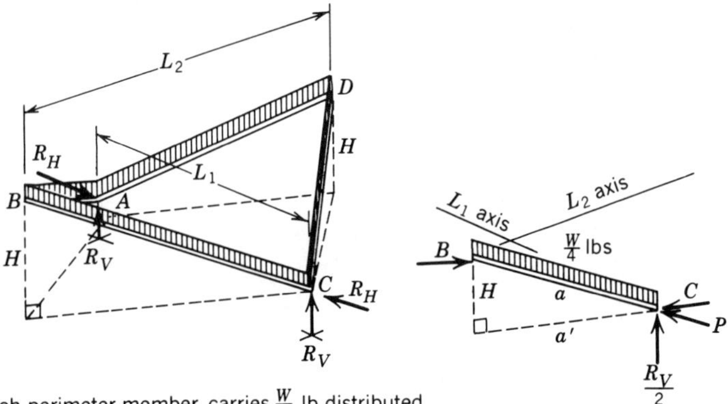

Each perimeter member carries $\frac{W}{4}$ lb distributed uniformly along the length and acting vertically;

W = total load $\quad P = \frac{R_V a}{2H} = \frac{W a}{4H}$

Shear force parallel to perimeter member between sheathing and member is $v = \frac{P}{a}$ per foot.

(*a*)

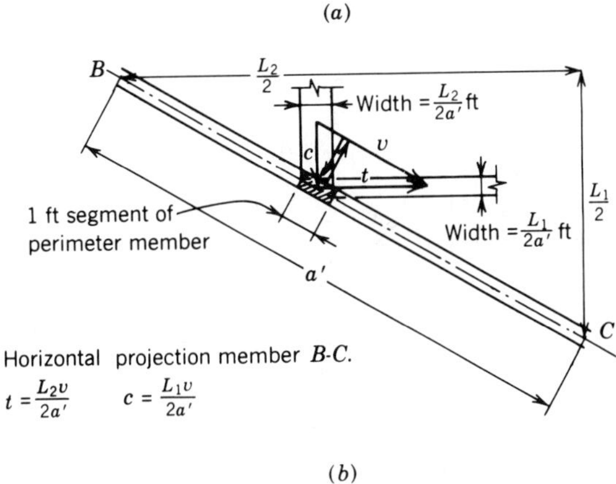

(*b*)

Fig. 25.53 Hyberbolic paraboloid design.

Arch panels are statically indeterminate structures. The analysis can be simplified by using the thrust coefficient graph, Fig. 25.54, to obtain the horizontal reaction caused by concentrated loads. For uniformly distributed loading, convert to 10 or more segments of equal length on the horizontal projection and sum the thrusts due to concentrated loads at the center of each segment. This will give thrust coefficients for load distributed uniformly on the horizontal projections which vary approximately linearly with span-to-rise ratio from 0.22 at $L/h = 2$ to 1.2 for $L/h = 10$.

After determining horizontal reactions (the left and right thrusts must balance), moments along the arch can be obtained by statics.

Maximum moments in arch panels are about 15 to 20% of the midspan maximum moments for curved flexural panels under the same loads, but the arch panels have large axial compressive forces which must be considered by combined stress methods.

Arch panels do not deflect as much as curved flexural panels. An easy method of computing these deflections uses the following steps:

1. Divide the arch into 10 equal segments, measured along the curve.
2. Determine the coordinates of points at middle of each segment.
3. Determine the horizontal and vertical reactions at the supports due to a uniformly distributed load on the horizontal projection, and, using statics, calculate the moment M at each segment midpoint.
4. Using the chart, Fig. 25.54, calculate the values of a moment m at the midpoint of each segment due to a unit load at midspan.

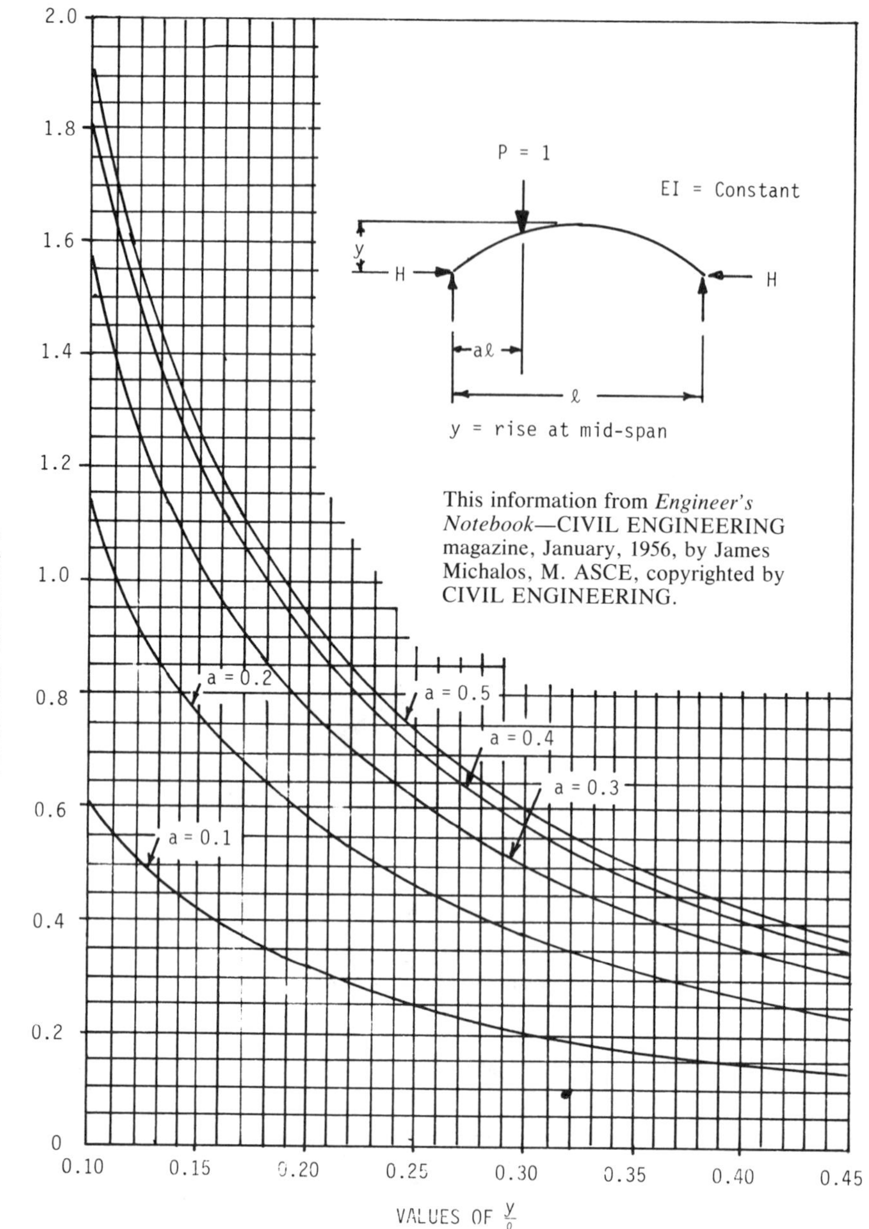

Fig. 25.54 Thrust coefficients for two-hinged circular arch.

5. Obtain the sum of Mm for the 10 segments and use the following equation in which s equals the segment length to obtain the midspan deflection of the arch.

$$\Delta = \frac{s\Sigma Mm}{EI} \tag{25.18.1}$$

The results of such analyses will give deflections which can also be approximated by the equation

$$\Delta = \frac{KwL^4}{EI} \tag{25.18.2}$$

Table 25.39
Deflection Constants for Approximate Midspan Deflection of Arch Panels

$\frac{\text{Rise}}{\text{Span}}$	K
0.1	0.00003
0.2	0.00006
0.3	0.00017
0.4	0.00039
0.5	0.00080

with values of K given in Table 25.39. Use of the table may be faster and sufficiently accurate for most work.

Section properties for these panels are obtained using the transformed section principle. This is illustrated by an example, using ½-in. CD-INT (EXT. GLUE) 32/16 skins with three 2×3 ribs of No. 2 Douglas fir surfaced to 1.375-in. thickness. This is a symmetrical section with the neutral axis at the center. The plywood face grain is parallel to the span and the ribs. Figure 25.55 shows the section and the calculated section properties. I for deflection is the I computed as shown. If skins are thin and clear distance between ribs exceeds about 40 times the plywood thickness, limit the plywood width to $40t$ in computing bending stresses. Otherwise use full plywood width for I for bending stress or deflection.

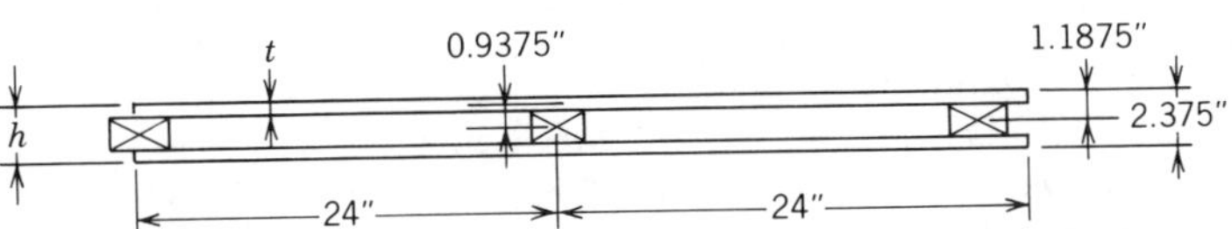

Plywood skins: ½-in. CD INT (EXT GLUE) 32/16
$I_{\parallel} = 0.086$ in.^{4}/ft of width
$A_{\parallel} = 2.5$ sq in./ft
$E = 1{,}200{,}000$ psi

Lumber ribs: Nominal 2×3 in. Douglas fir surfaced to 1.375 in.
$I = 0.5415$ in.^{4}
$A = 3.4375$ sq in.
$E = 1{,}700{,}000$ psi

Transform lumber sections to: $I = (1.7/1.2)0.5415 = 0.7671$ in.^{4}
$A = (1.7/1.2)3.4375 = 4.8556$ sq in.
width $= (1.7/1.2)2.5 = 3.542$ in.

$$I_{\text{plywood}} = 2(4)[0.086 + 2.5(0.9375)^2] = 18.2661 \text{ in.}^4$$

$$I_{\text{lumber}} = 3(0.7671) = 2.283 \text{ in.}^4$$

$$I_{\text{section}} = 18.2661 + 2.283 = 20.5491 \text{ in.}^4$$

$$S_{\text{section}} = 2I/h = 17.3045 \text{ in.}^3$$

$$Q \text{ at glue line} = A_{\text{plywood}}(0.9375) = 9.375 \text{ in.}^3$$

$$Q \text{ at the neutral axis} = A_{\text{plywood}}(0.9375)\ f\ \tfrac{1}{2}A_{\text{lumber}}(1.375/4) = 12.92 \text{ in.}^3$$

If clear distance between ribs is more than $40t$, use only $40t$ of the plywood between ribs in computing I for flexure.

Fig. 25.55 Section properties example for a curved stressed skin panel.

Curved panels experience tension perpendicular to their faces when the bending moment is in a direction that will straighten the curve. This radial tension stress is conservatively estimated by the equation:

$$f_r = \frac{3M}{2Rbh} \tag{25.18.3}$$

where M is the maximum bending moment that straightens the panel, R is the radius of curvature of the panel neutral axis, h is the panel thickness, and b is the width at the neutral axis. It may be preferable to separate the total moment into moments carried by the skin and the ribs and compute radial tension for each element separately.

The allowable radial tension stresses should not exceed one-half the allowable shear stress for lumber, or one-half the allowable rolling shear stress for plywood. For other core materials it should not exceed one-third the allowable shear stress.

For curved sandwich panels using paper honeycomb, foamed plastic, or other lightweight core material, reference should be made to *Design and Fabrication of Plywood Sandwich Panels* [25.11].

Cold bending the plywood and lumber introduces prestress that must be taken into account by reducing the design allowable bending, tension, and compression parallel-to-grain stresses by the curvature factor mentioned under arch design.

The radius of curvature should not be less than 125 times the plywood or lumber rib thickness. A smaller radius may damage the materials during cold bending.

SELECTED REFERENCES

25.1 AITC. *Design Standard Specifications for Structural Glued Laminated Timber of Softwood Species* (AITC 117-79). Englewood, CO: American Institute of Timber Construction, 1979.

25.2 AITC. *Glulam Systems*. Englewood, CO: American Institute of Timber Construction, 1981.

25.3 AITC. *Roof Slope and Drainage for Flat or Nearly-Flat Roofs* (Technical Note No. 5). Englewood, CO: American Institute of Timber Construction, 1978.

25.4 AITC. *Timber Construction Manual,* 3rd ed. New York: Wiley, 1986.

25.5 AITC. *Treating Standard for Structural Timber Framing* (AITC 109-69). Englewood, CO: American Institute of Timber Construction, 1969.

25.6 ANSI. *Specifications and Dimensions for Wood Poles* (ANSI Standard 05.1). New York: American National Standards Institute, 1972. (1430 Broadway, New York, NY 10018.)

25.7 APA. *Plywood Design Specification*. Tacoma, WA: American Plywood Association, 1976.

25.8 APA. *Plywood Design Specification Supplement No. 1, Design and Fabrication of Plywood Curved Panels*. Tacoma, WA: American Plywood Association, 1976.

25.9 APA. *Plywood Design Specification Supplement No. 2, Design and Fabrication of Plywood-Lumber Beams*. Tacoma, WA: American Plywood Association, 1979.

25.10 APA. *Plywood Design Specification Supplement No. 3, Design of Plywood Stressed-Skin Panels*. Tacoma, WA: American Plywood Association, 1976.

25.11 APA. *Plywood Design Specification Supplement No. 4, Design and Fabrication of Plywood Sandwich Panels*. Tacoma, WA: American Plywood Association, 1978.

25.12 APA. *Plywood Diaphragm Construction* (U310G). Tacoma, WA: American Plywood Association, 1978.

25.13 APA. *Pressure Preserved Plywood*. Tacoma, WA: American Plywood Association, 1966.

25.14 APA. *Glued Floor System* (U405). Tacoma, WA: American Plywood Association, 1974.

25.15 ASAE. "Design Properties for Round, Sawn and Laminated Preservatively Treated Construction Poles," *Agricultural Engineers Yearbook,* American Society of Agricultural Engineers, EP 388, St. Joseph, MO, 1981.

25.16 ASTM. *Establishing Clear Wood Strength Values* (D2555). Philadelphia: American Society for Testing and Materials, 1976. (1916 Race Street, Philadelphia, PA 19103.)

25.17 ASTM. *Establishing Design Stresses for Round Timber Piles* (D2899). Philadelphia: American Society for Testing and Materials, 1974.

25.18 ASTM. *Establishing Structural Grades and Related Allowable Properties for Visually Graded Lumber* (D245). Philadelphia: American Society for Testing and Materials, 1977.

25.19 ASTM. *Standard Specifications for Round Timber Piles* (D25). Philadelphia: American Society for Testing and Materials, 1973.

25.20 E. R. Bryan. *The Stressed Skin Design of Steel Buildings*. London: Granada Publishing Ltd., 1973.

25.21 J. M. Carney. *Plywood Folded Plates* (Laboratory Report No. 121). Tacoma WA: American Plywood Association, 1971.

25.22 L. R. Gjovik and R. H. Baechler. *Selection, Production, Procurement and Use of Preservative-Treated Wood* (General Technical Report FPL-15). Madison, WI: USDA Forest Service, Forest Products Laboratory, 1977.

25.23 GSA. *Wood Preservation: Treating Practices* (Federal Specification TT-W-571). Washington, D.C.: General Services Administration, 1974.

25.24 C. T. Hausmann and M. L. Esmay. "The Diaphragm Strength of Pole Buildings," *Transactions,* The American Society of Agricultural Engineers, 1977.

25.25 R. C. Hoagland. *Strength and Stiffness of Screw Fastened Roof Panels for Pole Buildings.* (Masters Thesis). Ames, IO: Iowa State University, 1981.

25.26 R. J. Hoyle, Jr. *Wood Technology in Design of Structures,* 4th ed. Missoula, MT: Mountain Press Publishing Co., 1978.

25.27 R. J. Hoyle, Jr. "Designing Wood Structures Bonded with Elastomeric Adhesives," *Forest Products Journal,* **26,** 3, 28–34, 1976.

25.28 ICBO. *Uniform Building Code.* Pasadena, CA: International Conference of Building Officials. (50 South Los Robles, Pasadena, CA 91101.)

25.29 ICBO. *Research Committee Recommendation Report No. 1379.* Pasadena, CA: International Conference of Building Officials.

25.30 IEEE. *National Electrical Safety Code.* New York: Institute of Electrical and Electronics Engineers, 1981. (345 East 47th Street, New York, NY 10017.)

25.31 Rafik Y. Itani, Houssain M. Morshed, and Robert J. Hoyle, "Experimental Evaluation of Composite Action," *Journal of the Structural Division,* ASCE, **107,** March 1981 (ST3), 551–565.

25.32 NBS. *Structural Glued Laminated Timber* (PS 56-73). Washington, DC: National Bureau of Standards, US Department of Commerce, 1974.

25.33 NBS. *Softwood Plywood, Construction and Industrial* (PS 1-74). Washington, DC: National Bureau of Standards, US Department of Commerce, 1974.

25.34 NBS. *American Softwood Lumber Standard* (PS 20-70). Washington, DC: National Bureau of Standards, US Department of Commerce, 1970.

25.35 NFPA. *National Design Specification for Wood Construction.* Washington, DC: National Forest Products Association, 1986.

25.36 NFPA. *Design Values for Wood Construction* (A Supplement to the National Design Specification for Wood Construction). Washington, DC: National Forest Products Association, 1986.

25.37 NFPA. *Span Tables for Joists and Rafters.* Washington, DC: National Forest Products Association, 1977.

25.38 NFPA. *Wood Structural Design Data.* Washington, DC: National Forest Products Association, 1970.

25.39 D. Patterson, *Pole Building Design.* McLean, VA; American Wood Preservers. (1651 Old Meadow Road, McLean, VA 22101.)

25.40 R. J. Payne. *Plywood Construction Manual.* Vancouver, B.C.: Council of the Forest Industries of British Columbia, 1969.

25.41 TECO. *Design Manual for TECO Timber Connector Construction.* Washington, DC: Timber Engineering Company, 1973. (5530 Wisconsin Ave., Chevy Chase, MD 20815.)

25.42 TPI. *Design Specifications for Light Metal Plate Connected Wood Trusses* (TPI-78). Glenview, IL: Truss Plate Institute, 1978. (1800 Pickwick Ave., Glenview, IL 60025.)

25.43 USDA. *Wood Handbook: Wood as an Engineering Material* (Agricultural Handbook No. 71). Washington, DC: USDA Forest Products Laboratory, 1974. (Available through the Superintendent of Documents, Washington, DC.)

25.44 M. W. Walmer and S. L. Baron. *Manual of Structural Designs and Engineering Solutions.* Englewood Cliffs, NJ: Prentice-Hall, 1972.

25.45 R. N. White, *Diaphragm Action on Aluminum Clad Timber-framed Buildings.* American Society of Agricultural Engineers Paper 78-4501, St. Joseph, MO, 1978.

25.46 WWPA. *Western Woods Use Book.* Portland, OR: Western Wood Products Association, 1973. (1550 Yeon Building, Portland, OR 97204.)

25.47 WWPI. *Architect and Engineers Guide Specifications to the Pressure Treatment of Western Woods.* Portland, OR: Western Wood Preservers Institute. (1021 Yeon Building, Portland, OR 97204.)

CHAPTER 26

ROOFS AND ROOFING

DAVID L. ADLER

Senior Associate
Simpson Gumpertz & Heger Inc.
Consulting Engineers
Arlington, Massachusetts

STEPHEN J. CONDREN

Formerly, Staff Engineer
Simpson Gumpertz & Heger Inc.
Consulting Engineers
Arlington, Massachusetts

26.1 INTRODUCTION

A building is a structure that must provide security and protection from the elements. The weatherproofing of a structure should not be an afterthought of the building designer. If the design process consists of fitting the building envelope around a predetermined interior space and to a previously established framework with little regard to weatherproofing functions, the result will be poor performance of the weatherproofing systems. Proper design of a building requires integration of the weatherproofing systems with all other building elements as the design proceeds.

Materials commonly used in buildings include concrete, steel, aluminum, glass, plastics, masonry, and wood. Each of these materials have different physical properties that affect their behavior and the need to weatherproof them. Their use in combination with one another presents additional sets of design conditions.

The design of building weatherproofing systems should be related to the expected life of the building, system cost and its reliability to prevent water penetration, and the consequences of failure. The designer should provide for repairs and replacement as needed, and for opportunities for future retrofits.

Weatherproofing systems are often victims of cost cutting during design and construction, since many of the important elements of protection are not visible. This is shortsighted, particularly during inflationary times and in consideration of the major direct and consequential damages to building components and contents which can occur from water. Studies of the price of the original installation, operating expenses, and replacement costs will show the wisdom of providing effective weatherproofing systems, with initial cost as a secondary consideration.

Assuming the designer avoids building collapse, the most costly and enervating problems can result from leakage. Leakage is reported to be a leading subject of construction disputes and litigation.

This chapter discusses principles of roof design and construction for consideration by engineers and architects early in the design phase of a project. Proper use of these principles by early attention to design of roofing systems will avoid most of the pitfalls and problems we have seen in our many years of consulting on building leakage problems. As in other elements of building construction, products and methods change. Most, but not all, of these changes are improvements. The authors have tried to present time-proven principles that are not likely to change quickly, but designers and others using this chapter should keep abreast of current industry practices.

Roofing has its own nomenclature; as an aid to the reader a glossary is provided (see Table 26.1).

This chapter has been prepared with invaluable assistance and contributions from the management and staff of Simpson Gumpertz & Heger Inc. Their support for this project is appreciated by the authors. The critical review and suggestions of Mr. Werner H. Gumpertz, the manuscript preparation by Ms. Cynthia B. Topping, and the figures by Mr. Michael J. Louis were particularly helpful to us.

Table 26.1 Glossary of Roofing Terminology

Aggregate Gravel, crushed stone, slag, or mineral granules either (1) embedded in a conventional built-up membrane's bituminous flood coat or (2) applied to a loose-laid roof system as a protective ballast.

Alligatoring Deep shrinkage cracks in bitumen coatings, progressing down from the surface, in smooth-surfaced membrane coatings and sometimes in bare spots of aggregate-surface membranes. It is a consequence of photo-oxidative hardening.

Asphalt Dark brown to black, highly viscous, hydrocarbon produced from the residuum left after the distillation of petroleum, used as the waterproofing agent of a built-up roof.

Backnailing "Blind" (i.e., concealed by overlapping felt) nailing in addition to hot mopping to prevent membrane slippage.

Ballast Aggregate, concrete pavers, or other material designed to prevent wind uplift or flotation of a loose-laid roof system.

Base sheet A felt (often coated) placed as the first (nonshingled) ply in a multi-ply built-up roofing membrane.

Bitumen Generic term for an amorphous, semisolid mixture of complex hydrocarbons derived from petroleum or coal. In the roofing industry there are two basic bitumens: asphalt and coal tar pitch. Before application, they are (1) heated to a liquid state, (2) dissolved in a solvent, or (3) emulsified.

Blister Spongy, humped portion of a roof membrane, formed by entrapped air-vapor mixture under pressure, with the blister chamber located either between felt plies or at the membrane–substrate interface.

Table 26.1 (*Continued*)

Blocking Continuous wood components anchored to the deck at roof perimeters and openings and doubling as cross-sectional fillers and anchorage bases, used in conjunction with nailers.

Brooming Field procedure of pressing felts into a layer of fluid hot bitumen to ensure full embedment and continuous adhesion—that is, elimination of blister-originating voids—of the bitumen film.

Built-up roof membrane Continuous, semi-flexible roof covering of laminations or plies of saturated or coated felts alternated with layers of bitumen, surfaced with mineral aggregate or asphaltic materials.

BUR Abbreviation for built-up roof membrane.

Cap flashing See *Flashing*.

Cap sheet Mineral-surfaced coated felt used as the top ply of a built-up roof membrane; also facers on prefabricated insulation boards.

Coal tar pitch Dark brown to black solid hydrocarbon obtained from the residuum of distilled coke-oven tar, used as the waterproofing agent of dead-level or low-slope built-up roofs.

Coated felt (or base sheet) A felt that previously has been saturated (impregnated with asphalt) and later coated with harder, more viscous asphalt, which increases its resistance to moisture.

Control joint Deliberate joint constructed between building materials or components to permit their free movement without damage to the roofing or flashing system.

Counterflashing See *Flashing*.

Crack Membrane fracture produced by bending, often at a ridge (see *Ridging*).

Cricket Ridge built up in a level valley or perimeter to direct rainwater to a drain.

Curing Final step in the irreversible polymerization of a thermosetting plastic, usually requiring some combination of heat, radiation, and pressure.

Cutback Solvent-thinned bitumen used in cold-process roofing adhesives, flashing cements, and roof coatings.

Dead level Horizontal, or "zero" slope (see **Slope**), in general less than 0.5-in-12.

Deck Structural supporting surface of a roof system.

Double pour Doubling of the flood-coat, graveling-in operation, in theory to provide additional waterproofing integrity to the membrane.

Edge stripping Application of felt strips to cover a joint between flashing and built-up membrane.

Elastomer Macromolecular material that rapidly regains its original shape after release of a light deforming stress.

Emulsion Intimate mixture of bitumen and water, with uniform dispersion of the bitumen globules achieved through a chemical or clay emulsifying agent.

Equiviscous temperature (EVT) Temperature at which asphalt has the correct viscosity (50 − 150 cSt) for hot mopping.

Ethylene-propylene-diene monomer (EPDM) Thermosetting synthetic rubber used in single-ply elastomeric sheet roof membranes.

Expansion joint Structural separation between two building segments, designed to permit free movement without damage to the roof system.

Exposure Tranverse dimension of a felt (or shingle) not overlapped by an adjacent unit in a roof system. Correct felt exposure in a shingled, built-up membrane is computed by dividing the felt width minus 2 in. by the number of plies—for example, for four plies of 36-in.-wide felt, exposure = (36 − 2)/4 = 8½ in.

Fabric Woven cloth of organic or inorganic filaments, threads, or yarns.

Felt Flexible sheet manufactured by interlocking fibers with a binder or through a combination of mechanical work, moisture, and heat.

Fishmouth Membrane defect consisting of an opening in the edge lap of a felt in a built-up membrane or a flexible sheet in single-ply membranes, a consequence of an edge wrinkle.

Flashing Connecting devices that seal membrane joints at walls, expansion joints, drains, gravel stops, and other places where the membrane is interrupted. Base flashing forms the upturned edges of the watertight membrane. Cap or counterflashing shields the exposed edges and joints of the base flashing.

Flashing cement Trowelable, plastic mixture of bitumen and asbestos (or other inorganic) reinforcing fibers and a solvent.

Glaze coat Thin (but continuous) protective coating of bitumen applied to the lower plies or top ply of a built-up membrane when application of additional felts or the flood coat and aggregate surfacing is delayed.

Gravel Coarse, granular aggregate resulting from natural erosion or crushing of rock, used as protective surfacing or ballast on roof systems.

Table 26.1 (*Continued*)

Gravel stop Flanged device, usually metallic, with vertical projection above the roof level, designed to prevent loose aggregate from rolling or washing off the roof and to provide a finished edge detail for the roof.
Holiday Area where interply bitumen mopping or other fluid-applied coating is discontinuous.
Hot stuff or hot Roofer's term for hot bitumen.
Ice dam Drainage obstructive ice formation at eave of snow-covered sloped roof.
Inorganic Comprising matter other than hydrocarbons and derivatives, not of plant or animal origin.
Insulation Rigid boards of insulating material for the purpose of increasing interior comfort, preventing condensation on interior building surfaces, and reducing heating and cooling costs. Insulation helps attenuate stress transfer to a built-up membrane from movement in the structural deck. Board insulation can also provide an acceptable substrate for applying a built-up membrane on a steel deck.
Lap Dimension by which a felt covers an underlying felt in a multi-ply built-up bituminous membrane. Edge lap indicates the transverse cover; end lap indicates the cover at the end of the roll. These terms also apply to single-ply membranes and metal flashings.
Loosely-laid roof system Design concept in which insulation boards and membrane are not anchored to the deck but ballasted by loose aggregate or concrete pavers.
Membrane Flexible or semiflexible roof covering, the waterproofing component of the roof system.
Mineral granules Natural or synthetic aggregate particles, ranging in size from 500-μm (1 $\mu m = 10^{-6}$ m) to ¼-in. diameter, used to surface cap sheets, asphalt shingles, and some cold-process membranes.
Mineral-surfaced sheet Asphalt-saturated felt, coated on one or both sides and surfaced on the weather-exposed side with mineral granules.
Mopping Application of hot, fluid bitumen to substrate or to plies of built-up membrane with a manually wielded mop or a mechanical applicator.
  Solid mopping A continuous coating.
  Strip mopping Mopping pattern featuring parallel mopped bands.
Nailer Wood member bolted or otherwise anchored to a non-nailable deck or wall to provide nailing anchorage of membrane roof felts or flashings.
Neoprene Synthetic rubber (polychloroprene) used in fluid- or sheet-applied elastomeric single-ply membranes or flashing.
Organic Comprising hydrocarbons or their derivatives, or matter of plant or animal origin.
Parting agent Powdered mineral (talc, mica, and so forth) placed on coated felts to prevent adhesion of concentric felt layers in the roll (sometimes called a *releasing agent* or *antistick compound*).
Perlite Aggregate used in lightweight insulating concrete and preformed insulating board, formed by heating and expanding silicaceous volcanic glass.
Permeance Index of a material's resistance to water-vapor transmission.
Phased application Applying the felt plies of a built-up roof or waterproofing membrane in two or more operations, separated by a delay normally of at least 1 day.
Picture framing Rectangular membrane pattern formed over insulation-board joints by ridges or thermal transmission.
Pitch pocket Flanged, open-bottomed metal container placed around a column or other roof-penetrating element and filled with bitumen or plastic cement to seal the joint.
Plasticizer High-boiling-point solvent or softening agent added to a polymer to facilitate processing or increase flexibility or toughness in the manufactured material.
Ply Layer of felt in a built-up roof membrane; a four-ply membrane has at least four plies of felt at any vertical cross-section cut through the membrane.
Polymer Long chain macromolecules produced from monomers for the purpose of increasing tensile strength of sheets used as membranes or flashing.
Polyvinyl chloride (PVC) Thermoplastic polymer, formulated with a plasticizer, used as a single-ply sheet membrane material or liquid coating.
Primer Thin liquid solvent applied to seal a surface, absorb dust, and promote adhesion of subsequently applied materials.
Protected membrane roof (PMR) Roof assembly with insulation on top of the membrane instead of vice versa, as in the conventional roof assembly (also known as an inverted (IRMA) or upside-down roof assembly).
Re-covering Covering an existing roof assembly with a new membrane instead of removing the existing roof system before installing the new membrane.
Reglet Horizontal groove in a wall or other vertical surface adjoining a roof surface for anchoring flashing.

Table 26.1 (*Continued*)

Reroofing Removing and replacing an existing roof system (as opposed to mere recovering); also called tearoff-replacement.

Ridging Membrane defect characterized by upward displacement of the membrane, usually over insulation-board joints (see *Picture framing*).

Roll roofing Coated felts, generally mineral-surfaced, supplied in rolls and designed for use without field-applied surfacing.

Roof system Assembly of interacting components designed to weatherproof, and normally insulate, a building's top surface.

Saddle See *Cricket*.

Saturated felt Felt that has been immersed in hot bitumen.

Scupper Channel through parapet, designed for peripheral drainage of the roof, usually as safety overflow system to limit accumulation of ponded rainwater caused by clogged drains.

Scuttle Curbed opening, with hinged or loose cover, providing access to roof (synonymous with hatch).

Self-healing Property of the least viscous roofing bitumens, notably coal tar pitch, that enables them to seal cracks formed at lower temperatures.

Selvage joint Lapped joint detail for two-ply, shingled roll roofing membrane, with mineral surfacing omitted over a transverse dimension of the cap sheets to improve mopping adhesion. For a 36-in. wide sheet, the selvage (unsurfaced) width is 19 in.

Shark fin Curled felt projecting up through the aggregate surfacing of a built-up membrane.

Shingling Pattern formed by laying parallel felt rolls with lapped joints so that one longitudinal edge overlaps the longitudinal edge of one adjacent felt, whereas the other longitudinal edge underlaps the other adjacent felt (see *Ply*.) Shingling is the normal method of applying felts in a built-up roofing membrane.

Single-ply membrane Membrane, either sheet- or fluid-applied, with only a single layer of material, designed to prevent water intrusion into the building.

Slag Porous aggregate used as built-up bituminous membrane surfacing, comprising silicates and alumino-silicates of calcium and other bases, developed with iron in a blast furnace.

Slippage Relative lateral movement of adjacent felt plies in a built-up membrane. Occurs mainly in sloped roofing membranes, exposing the lower plies, or even the base sheet, to the weather.

Slope Tangent of the angle between the roof surface and the horizontal, in inches per foot. The Asphalt Roofing Manufacturers' Association ranks slopes as follows:
Level: ½-in. maximum
Low slope: over ½ in. up to 1½ in.
Steep slope: over 1½ in.

Smooth-surface roof Built-up roofing membrane surfaced with a layer of hot-mopped asphalt or cold-applied asphalt–clay emulsion or asphalt cutback, or sometimes with an unmopped, inorganic felt.

Softening point Temperature at which bitumen becomes soft enough to flow, as measured by standard laboratory test in which a steel ball falls through a measured distance through a disk made of the tested bitumen.

Solid mopping See *Mopping*.

Split Membrane tear resulting from tensile stress.

Stripping (1) Technique of sealing the joint between base flashing and membrane plies or between metal and built-up membrane with one or two plies of felt or fabric and hot- or cold-applied bitumen. (2) Taping joints between insulation boards or deck units.

Substrate Surface (structural deck, insulation, or vapor retarder) upon which the roof membrane is placed. Also, the deck, vapor retarder, or membrane surface upon which insulation or other roof system component is placed.

Sump Depression in roof deck around drain.

Tearoff Removing a failed roof system down to the structural deck.

Thermal shock factor (TSF) Mathematical expression for calculating the theoretical temperature drop required to split a rigidly held membrane test sample under tensile contractive stress.

Through-wall flashing Water-resistant membrane or material assembly extending through a wall's horizontal cross-section and designed to direct water flow through the wall toward the exterior.

Vapor barrier See *Vapor retarder*.

Vapor retarder Roof component designed to obstruct water vapor flow through a roof (or wall).

Vent Opening designed to convey water vapor, or other gas, from inside a building or building component to the atmosphere.

Table 26.1 (*Continued*)

Vermiculite Aggregate used in lightweight insulating concrete, formed by heating and consequent expansion of mica rock.
Wrinkling See *Ridging*.

26.2 DESIGN DECISIONS AFFECTING ROOFING

Decisions about systems and materials for roofing should be made concurrently with the selection of the shape of the building. A preselected roof system cannot be fitted to a previously designed structural system without inviting compromise, leading to higher costs and lower reliability.

Because of their direct exposure to weather, roofing systems are subjected to an almost endless variety of physical and chemical effects, such as:

1. Solar exposure, including heat and ultraviolet rays.
2. Radiant cooling to space (at night).
3. Thermal cycling, including rapid temperature changes or "thermal shock".
4. Aging.
5. Time-dependent dimensional changes (creep).
6. Continued chemical reactions within the materials.
7. Chemical incompatibility by reaction between different components.
8. Chemical influences from the atmosphere and airborne pollutants (acid rain).
9. Influences of rain, snow, ice, hail, and wind.
10. Dimensional changes of components in response to changes in temperature and moisture.
11. Pressures from water vaporization caused by heat.
12. Movements in the building, particularly the roof deck.
13. Deflections in the building, and creep of structural elements.
14. Influences of roof-top mechanical equipment, including expelled waste materials.
15. Roof-top traffic for maintenance and other "recreational" use of the roof surfaces.
16. Ponded water and root penetration from vegetation.
17. Mechanical abuse and vandalism.

No other building components are exposed to such a variety of conditions. It follows that roofing design must consider all of these influences, and must provide not only proper materials, but a suitable substrate and the assembly of well-balanced and compatible components.

26.2.1 Support Systems

The structural deck supporting the roofing system is supported by a structural framing system. This system must be sufficiently rigid to support safely the roofing system to be selected. The design of the roofing system also must take into account the likelihood of relative (shear) movement between the roofing components (supported by the deck) and the edge flashings (independently supported by the building walls). Even if the building structural elements are tied together, it is prudent to provide support for roofing and flashing from the roof deck only.

In addition, interior walls, not part of the roof deck support system, should not come into contact with the roof deck, and items such as sprinklers and light fixtures should not be hung from the deck (unless specifically considered in the design), because they can affect the expected deflections of deck components.

All roof-top equipment should be supported by the structural framing system, not by the roof deck. All support elements should be designed with proper regard for equipment weight, snow (including drifting where applicable), and wind loads, with proper anchorage and connections. Allowance should be made in the design of the framing for all interior building components that will be hung from the roof structure; these include sprinkler systems, roof drainage systems, conveyors, light fixtures, and HVAC equipment.

Roof decks and support systems are discussed in Section 26.4.

26.2.2 Building Element Intersections

The intersections of roof structural systems with walls require special attention. Perimeter walls should not extend as parapets above the roof, unless required by special circumstances such as local building codes. Parapets present special problems because the exposure of the outside walls to the elements results in severe and uneven thermal loads. Parapets should be fully capped, and covered from the roof system to the cap.

If the deck is rigidly anchored to and supported by the exterior walls, a rigid flashing system can be used. In any other case, a flexible roof-to-wall flashing joint must be designed to accommodate differential movements between the roof and walls.

Similar consideration is required whenever different materials are used for the roof decks (such as at intersections of concrete and steel decks), at changes in direction of supports or deck systems, and at abrupt changes in roof shape (such as re-entrant corners) because stress concentrations that can cause damaging strains on roofing components will occur at these locations.

Where roofs intersect rising walls, the structural deck and wall construction will determine the need to provide expansion joints. If the wall rests on the deck, they will move together, and a fixed flashing can be provided. If the wall and deck have separate supports, a flexible flashing is needed. In any case, it is necessary to design a counterflashing system which is not integral with the wall finish to protect the roofing system below. This allows future roof repairs without disturbing the wall, provides proper construction sequence, and allows flashing to perform independent of wall movements.

Even when commonly supported, the wall may move independently of the structure that supports the roof deck, such as when the wall extends far above the roof line, if it is exposed to weather and differential thermal movements, or if it is without suitable control joints. This condition requires a suitable flexible flashing that can accommodate the differential movement.

26.2.3 Drainage Provisions

All ''horizontal'' building elements should be sloped to shed water. The roof surface itself must drain promptly, and other components such as flashings, tops of walls, and mechanical equipment and its supports must slope so that water (or snow) will not accumulate, or remain, on the surfaces. Water standing on roofing components increases the risk and amount of leakage at even minor defects, accelerates deterioration of materials, allows ice damage, permits penetrating vegetation growth and insect breeding, prevents maintenance, and increases deflection.

Most commercial buildings have ''flat'' roofing systems. For many years the roofing industry has recognized the need to slope these ''flat'' roofs for positive and prompt drainage. A minimum slope of ¼ in./ft ($\approx$20 mm/m) is usually necessary to avoid negative slopes resulting from elastic and creep deflections and from tolerances in normal building construction, unless detailed provisions are made for accurate lower slopes. Water should leave the roof through interior drains. Experience has shown that roof-edge scuppers become blocked by ice, or leak at the intersection between the roof membrane and the metal flashing where water drains against the flashing connections to the membrane. Also, connections may fail due to differential movements of the sheet metal scupper flashing and the roofing materials. For the same reasons, gutters at roof edges are often unsuccessful on low slope roofs.

Structural decks are easily sloped by varying the heights of structural supports. Drains are usually located near columns for convenience and support and concealment of the interior piping. When drains are placed near supports, normal elastic and creep deflection may overcome planned slope, so that the slope becomes ''negative'', resulting in ponding and additional deflections between drains. Drains should be placed between supports unless adequate slope is provided to accommodate structural deflections.

Roof slopes should be continuous; significant changes in slope produce difficult roofing problems. The roofing systems must be discontinuous, but sealed with special flashings across ''breaks'' in slope so strains do not concentrate at the transitions and cause membrane fracture.

26.2.4 Vapor Diffusion and Condensation

The roofing design must consider the presence and migration of water vapor below the roof that may condense within the roofing system; this can result in moisture accumulation and accelerated deterioration of components. Computations based on physical laws of moisture migration and partial vapor pressure will determine the need for special provisions such as vapor retarders and venting spaces below the roof.

The climate in the building locale and the use of the building are also important considerations. Normally, dry occupancy buildings in temperate climates will not require any special vapor control. Buildings for special use such as swimming pools, freezers, or processes with high moisture conditions (paper mills), as well as buildings in cold climates, should be carefully studied for potential condensation problems.

Detailed information about condensation problems and vapor retarders is included in Section 26.3.

26.2.5 Energy Efficiency

Commercial and public buildings require insulation to meet building codes and to provide for energy efficiency of the building envelope. In most cases, the largest area of heat loss (or gain) is the roof. Insulation can be provided inside (ceilings), outside (on the roofs), or in combination of the two. Roof membranes must be divorced from direct contact with the roof deck to avoid damage to the membrane from deck surface irregularities or movements; therefore, at least some insulation is usually provided in conjunction with the roofing membranes to serve the dual purposes of providing energy efficiency and an attenuating layer between the membrane and the roof deck.

Factors that determine the amount and type of insulation include the characteristics of the roofing and structural deck systems, applicable codes, and the heat transmission value desired (*U* factor) for the roof system.

Long-term energy efficiency depends on the integrity of the entire roofing system, and specifically on maintaining the insulation in a dry condition. In addition, the efficiency of some insulation declines with age; this should be considered in the design process.

These topics are discussed in more detail in Section 26.3.

26.2.6 Installation Sequence of Roofing and Other Building Elements

The roofing system should be installed after all contiguous building elements have been completed, because the membrane is easily damaged by construction activities. Building components should not be installed above or adjacent to the roof if this will result in construction traffic or the risk of falling debris over the roofing. If this kind of activity is necessary, temporary protection (such as plywood sheets over the roof surfaces) should be provided.

All roof edge and penetration construction should be in place and complete so flashings can be installed at the same time as the membrane and thus avoid water entry into partially completed work via unfinished edges.

26.2.7 Roof-Top Equipment and Services

A correctly designed and properly installed roofing system can be fatally damaged by roof-top traffic from the installation and maintenance of roof-top equipment. In addition, the equipment supports must allow for sufficient space below the equipment so inspection, maintenance, and repairs can be readily performed to the roofing. A recommended guide [26.3] for heights is published by the National Roofing Contractors Association and is provided as Fig. 26.1. A completely curbed enclosure is the best method to avoid leaks and maintenance problems.

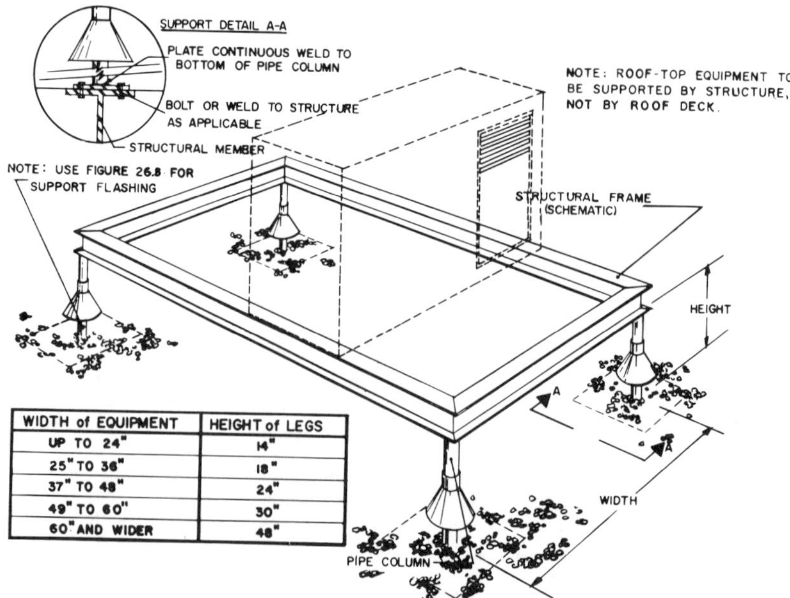

WIDTH of EQUIPMENT	HEIGHT of LEGS
UP TO 24"	14"
25" TO 36"	18"
37" TO 48"	24"
49" TO 60"	30"
60" AND WIDER	48"

Fig. 26.1 Recommended heights of roof-top equipment.

The supports for equipment should be connected to the structural framing system, and should not rest on the roofing or roof deck. Provide supports that can be made watertight with standard flashing details. Do not use "pitch pockets"; they are high-maintenance, low-reliability "flashings".

The coordination of mechanical units and their location, size, and support requirements is one of the design parameters most often neglected by designers. Curbs must be undersized sufficiently to allow for flashing between the mechanical unit and the curb.

26.2.8 Exposures to Unusual Conditions

The designer of the roofing system must consider its exposure to extremes of weather such as heat, cold, wind, and hail, and any other conditions peculiar to the particular environment. Unusual conditions include spills of chemicals, oils, or process dust on the roof surface, and acidic water from "acid rain". Consider the effect that these exposures have on the roof components, together with the need for periodic cleaning of roof surfaces. Design a system that will not deteriorate under unusual conditions, and can be cleaned without roof damage.

In addition, many roofs are subject to almost continuous equipment maintenance service. In this event, enclose equipment in penthouses, construct raised walkways mounted on structural supports, or provide other suitable protection for the roof surface.

26.2.9 Code Requirements

Roof design must comply with applicable building codes. In some cases, codes will restrict the materials that can be used and the method of their installation. Local regulations may require items such as perimeter guard rails and roof access doors, and limit water drainage rates.

In addition to building codes, insurance requirements for fire and wind protection may govern roofing design. The Factory Mutual System, Norwood, MA and Underwriters Laboratory, Chicago, IL both publish such standards. A prudent designer will use these standards as guidelines for minimum design parameters, upgrading them as necessary.

26.2.10 Maintenance and Replacement

The designer should consider that periodic roof maintenance must be performed. While roofing should be designed for minimum maintenance, all roofs require some regularly scheduled care. Select basic roof materials that are readily available, and easily repaired or replaced by competent roofing and sheet metal workers. Design details so that building elements intersecting the roof can be separated and replaced without disturbing the building's exterior finishes such as stucco or siding, and so there is sufficient headroom and work space to perform repairs.

Workers will have to bring equipment to the roof to make repairs and to service roof-top mounted equipment. Safe access should be provided by roof access doors or scuttles, secure ladders, walkways if necessary, and railings where needed. All means of access to roofs should be locked or under security control.

26.2.11 Reroofing Over Existing Roof Systems

The advent of reroofing with single ply membranes that need not be attached to existing built-up roofs has led to installation of many "new roofs" over existing ones. This chapter addresses roof design on structural decks, not overlays of existing systems. Almost exclusively, new roofs are provided only when old ones leak. Thus the moisture content and other problems of an old roof can well be imposed in the new one. In addition, few roof structures are designed to accommodate multiple roof systems.

It is recommended that old roofs be removed before new ones are installed. This permits verification of the deck conditions and assures the opportunity for the new roof to perform effectively for its life expectancy.

26.3 ENERGY CONSERVATION

26.3.1 General Design Principles of Heat Flow

Heat transfer involves the displacement of heat energy from one point to another. There are three modes of heat transfer: conduction, convection, and radiation. Heat transfer by conduction involves the transfer of heat energy through a material from a point of high temperature to a cooler area. The heat is transferred by molecular activity; warmer areas, being more molecularly active, transmit energy to the adjacent less active molecules. Heat transfer by convection involves the physical transport of a warmer mass of material to a cooler area. The heat is then dissipated by conduction from the warmer mass to the surrounding area. Radiant heat transfer involves the emission and transmission of electromagnetic energy from heat source to another area where it is converted back to internal molecular energy and the temperature of the receiving substance increases.

Energy conservation in buildings involves reducing the transfer of heat energy from the interior of the building into the environment or vice versa. If the energy losses (or gains) are reduced, then less energy will have to be provided to maintain the interior temperature, thus reducing operating costs for the building. Heat transfer by convection involves air leaks that transport warm air through the building envelope; slow or stop the air leaks by sealing the building and energy losses by convection are reduced or eliminated. Energy losses in buildings by radiation are low due to the relatively low operating temperatures. The largest source of radiant heat loss is through glass, and can be reduced using blinds or reflective films on glass surfaces. The largest energy losses in buildings occur by conduction through building surfaces. Thermal insulation, placed in the walls and roof, will slow the heat transfer, conserve energy, and reduce operating costs. It is important to note that insulation, no matter how high its thermal resistance, will not stop the conductance of heat as long as a temperature difference exists between the wall surfaces. Heat transfer by conduction can only be reduced.

The equation for steady-state heat flow q through a material is

$$q = A\,\Delta T/R \qquad (26.3.1)$$

where q = heat flow rate, Btu/hr
ΔT = overall temperature difference, °F
R = thermal resistance, °F (hr)(sq ft)/Btu
A = Area, sq ft

The total thermal resistance R_T of a wall or roof assembly per unit area is the sum of the thermal resistance of all components of the assembly

$$R_T = R_1 + R_2 + R_3 \cdots \qquad (26.3.2)$$

The overall coefficient of heat transfer U is defined as the reciprocal of the total thermal resistance R_T of the assembly:

$$U = 1/R_T \qquad (26.3.3)$$

Thus, heat loss q is

$$q = UA\,\Delta T \qquad (26.3.4)$$

The various components, that is, walls, windows, roof, etc., in a building envelope will have different coefficients of heat transfer. The total heat losses are the sum of all the heat losses through the various components:

$$q_t = (U_1A_1 + U_2A_2 + U_3A_3 + \cdots + U_nA_n)\,\Delta T \qquad (26.3.5)$$

26.3.2 Insulation

Thermal insulation slows the transfer of heat energy. Insulations are manufactured as cellular, reflective, granular, or fibrous materials containing voids and interfaces that hinder the transfer of heat energy and provide a product with a high thermal resistance. Still air is a good insulator. Foam insulations entrap air within cells in the material; fibrous materials provide many material interfaces that result in large areas of thin air films which slow heat passage. Other insulations entrap air and fill the available space, preventing air convection. Various base materials with different physical properties are combined with numerous manufacturing processes, producing different configurations and densities, resulting in the many insulation products and properties available on the market.

Thermal conductivity k is a measure of the amount of heat that will pass through 1 sq ft of a 1-in. thickness of a material in 1 hr when a temperature difference of 1 °F exists across the two surfaces of the material. The thermal conductivity property provides a means for comparing the various insulation products available.

26.3.3 Optimization of Thermal Insulation

Thermal insulations slow, but do not stop, the loss of heat through the building envelope. The effectiveness, and benefits, of thermal insulation drop off very rapidly with an increasing thermal resistance. For example, compare the performance of three conditions in a building envelope, $R = 1, 10, 20$ or, converting to the coefficient of heat transfer, $U = 1, 0.1, 0.05$. $U = 1$ allows 1 Btu/[hr(sq ft)(°F)] to pass through the envelope. $U = 0.1$ provides an envelope that reduces heat transmission by 90% but uses 900% more insulation; $U = 0.05$ reduces heat transmission another 5% from the original amount, but requires an additional 1000% more insulation than the $U = 1$ system to do so.

The designer can calculate the optimum amount of thermal insulation based on the environment, present and future cost of fuel, type of fuel, cost and efficiency of heating and cooling system, the useful life of the building and the insulation, the cost of insulation and its installation, and inflation and interest factors [26.3]. This calculation should be performed when the designer considers using insulation quantities in excess of the amount required by building codes.

26.3.4 Special Problems of Thermally Efficient Roofing Systems

Thermally efficient roof systems use an increased quantity of insulation when compared to pre-"energy crisis" conditions. This increase in insulation thickness may result in the following problems:

1. Lowering of impact resistance of the roof system by producing a softer base for the membrane.
2. Decreasing the lateral stability within the system; the lateral stability in a roof system that is properly attached to the structure is provided by the shear resistance of the insulation. A thick insulation layer will deform more in shear than a thin layer when developing the necessary stresses to restrain the working forces in the membrane caused by changes in temperature and moisture conditions. The larger displacements that result may lead to membrane splitting at drains and other points where the membrane is fixed.

The increased insulation quantities in a thermally efficient roof will not cause a significant increase or decrease in the working temperatures of the membrane when compared to less efficient roofs.

26.3.5 Roof Insulation and the Thermal Efficiency of the Building

The roof is only part of the building surface area or envelope separating the interior, controlled conditions from the outside. Heat will transfer from a warm area to a cooler area by conduction and radiation through roof, wall, doors, windows, and floors. Heat transfer also occurs by air convection through cracks and joints in windows, doors, vents, chimneys, and other penetrations through the building envelope. The effect that the insulated roof system has on the total thermal efficiency of the building is dependent on the relative area between the roof and the other building surfaces, and the proportion of overall heat losses shared by conduction, convection, and radiation as follows:

$$Q = q + q_c + q_r \qquad (26.3.6)$$

where Q = overall rate of heat flow
q = heat flow rate by conduction
q_c = heat flow rate by convection
q_r = heat flow rate by radiation

The term for heat loss by conduction can be expressed as the sum of the products of the coefficient of heat transfer for the various components of the building envelope and the area of the component as follows:

$$q = U_1A_1 + U_2A_2 + U_3A_3 + \cdots + U_nA_n \qquad (26.3.7)$$

The contribution of the roof system will be one of the above products.

The effect of the roof insulation on the thermal efficiency of a building can be significant. A large, single-story industrial building will benefit considerably from a properly insulated roof, yet the effect roof insulation would have on the heat loss in a high-rise building may be small.

26.3.6 General Design Principles of Water Vapor Transmission

The process of molecular diffusion occurs continuously in air. The nature and properties of a gas are such that the molecules are widely spaced and in constant rapid random motion. Water vapor behaves as a gas, and will migrate from an area of high molecular concentration to an area of lower concentration.

Water vapor flow through a building envelope depends on the difference in partial vapor pressure (molecular concentration) between the building air and outside air and the combined permeance of the elements of the envelope. In the special case of a roofing system, the permeance of the membrane is very low and for practical purposes it can be considered impermeable. In this case, water vapor flows to the membrane but not past it. Condensation will occur whenever the temperature at the surface of the membrane is below the dew point of the building air. A vapor retarder does not stop the transmission of water vapor, but only slows its passage. Condensation may occur in a roof system, even with a vapor retarder, unless the permeability of the retarder is less than that of the roof membrane.

The rate of condensation (or evaporation) at the membrane, and accumulation (or loss) of water in the system depend on the difference in the partial water vapor pressure between the interior air and the saturation vapor pressure at the underside of the membrane for the condition being investigated. Water condenses in a roof system during the colder season and evaporates during warmer seasons. The designer must investigate the climatic and operating conditions of the building over an annual cycle to determine if condensation will be harmful and if a vapor retarder is needed to reduce levels of condensation to acceptable levels.

26.3.7 Design and Location of Vapor Retarders, Venting Provisions

The rate of vapor flow in a roof system depends on the difference in partial vapor pressures at the membrane, at the interior of the building, and the combined resistances of the roof system components to the transmission of water vapor (vapor flow resistance is the reciprocal of the water vapor permeance of a material) as follows:

$$W = p/r \tag{26.3.8}$$

where W = water vapor flow
p = vapor pressure drop between interior and membrane surface
r = total vapor flow resistance to the membrane

The condensing and drying conditions over an annual cycle are investigated for any harmful effects of water accumulation in the roof system. If a vapor retarder is necessary, it should be provided on the warm side of the roofing system during the condensation period, and be located within the envelope so that the temperature of its surface always remains above the dew point of the interior air.

The decision to vent a roof system is affected by different rationales depending on the objective of the venting. Moisture migrating through a roof system can be slowed by a vapor retarder to a level that no harmful effects occur, or the moisture can be removed by circulating dry air below the roofing. Venting to remove moisture cannot be accomplished without an exchange of air, thus an air space must be available below the roof, such as an attic space, to allow free circulation of dry air. Venting a roof system that does not have a free air space is done only to relieve the pressure from solar heating and expansion of entrapped moisture. Wet roofs cannot be dried, nor can condensation be prevented, by including vents. Vents that penetrate a roof system increase the potential for problems, since they create another "hole" in the roof. Detail the roof system to provide venting at perimeters, at equipment curbs, and other required penetrations, through the system.

Venting-type roof systems are also used on sloping shingled roofs in cold climates to prevent formation of ice dams and subsequent water backup. This is called a "cold roof" design.

26.3.8 Insulation Types

Insulation on roofs is almost always by the use of board stock of rigid insulating material. Natural materials such as glass fiber, wood fiber, and perlite (mineral fiber) have proven to be stable, suitable materials when properly fabricated with cap sheets and binders. Fiberglas conforms well to minor deck irregularities and fiberboard provides a firm base for roofing. Perlite boards are very fragile and dusty, and have low cohesive strength.

Chemical foam insulations (polyurethane, isocyanurate, expanded polystyrene) and composites of these materials are often dimensionally unstable; furthermore, they may not meet local fire codes and may be incompatible with roofing materials. Built-up roofing should not be applied directly to these materials, and their use with other products should be carefully researched for acceptable field applications.

26.4 ROOF DECKS AND SUPPORT SYSTEMS

The structural system used to support the roof deck, and the deck itself, must safely withstand the loadings prescribed by governing building codes. These loadings should be increased by the designer if justified by circumstances such as drifting snow. The properties, behavior, analysis, design, and other pertinent information regarding the types of roof decks and their uses are thoroughly covered in other chapters in this design handbook. Although adequate for structural safety requirements, designing the roof deck to resist structural loading only does not ensure its suitability for the roofing system. General parameters that must be considered when designing roof decks are presented in this section, and specific behavior and properties of the various roof decks that directly affect the selection and performance of the roofing will be reviewed. Also included are precautions, based on our experience, to improve the chances for a successful roofing installation.

26.4.1 Deflections

Deflections will take place in the structural supports and roof decking when these components experience any loading. The deflections fall into two categories: elastic (temporary) and creep (permanent) deflections.

Structural components will return to their original position from an elastic deflection when the imposed load is removed. Creep deflections result from a permanent deformation in the material, which is usually irrecoverable. All materials experience elastic deflections while creep deflections are characteristic of only certain materials and they will be discussed where applicable to the individual materials.

Deflections usually result in problems when they are not considered during the design phase. The major problems that can occur are:

1. Water can pond within a deflected area. When the deck system deflects, it forms a depression between the main supports, forming a dished area that traps water; the additional weight of the water, in turn, causes additional deflection. This condition of cumulative deflections may even result in complete roof collapse if the system is too flexible, as discussed in Section 19.2.3.
2. The roofing system can become delaminated from the deck if the horizontal shear, developed as a result of the deflected shape, exceeds the strength of the bond between components. This is most likely to take place during construction, from local traffic, or from material storage loads.
3. Differential deflections between two adjacent structural components can breach the roofing system. Such deflections can occur from uneven loading or thermal response, and from unequal spans of deck or support components.
4. End rotation of deck components over support beams or interior walls can strain the roofing system over these areas and cause splitting in the membrane.

26.4.2 Moisture and Thermal Movement

Movement of the substrate can damage the roofing system. Such movement may result from the expansion and/or contraction of the roof deck or support system caused by changes in temperature or moisture conditions. Modern roofing systems, efficiently insulated for energy conservation, tend to stabilize the temperature of the structural roof system; but the designer should review each project for potential problems. Wood and wood product components are susceptible to swelling and/or shrinking from moisture changes. The effects of these phenomena are readily controlled by proper detailing and by the use of flexible joints in the building and in the roofing system.

26.4.3 Membrane Stress Concentrations

Bituminous built-up roofing systems are particularly susceptible to the effects of stress concentrations. When construction is complete, the membrane behavior can be compared to a continuous plate, undergoing temperature and moisture fluctuations with changing weather conditions.

When the membrane contracts from a drop in the temperature, or shrinks from a loss of moisture, stresses develop in the membrane. The average magnitude of the stresses is a function of the strains developed due to the restraint imposed on the membrane where it is attached to the structure. If the attachment between the membrane and the structure is uniform, as achieved by good adhesion between all roof system components and the roof deck, then the stresses developed will also be uniform and should remain within the strength limits of the materials.

When the attachment between the components of the roof system is erratic, the induced stresses will be nonuniform and may accumulate in local areas, exceed the strength of the membrane, and cause it to split. Stress concentrations are usually caused by re-entrant corners in the plane of the roof at equipment curbs, drains, vent pipes, courtyards, penthouses, roof perimeter changes, and any other condition that creates a corner or hole in the membrane plate, or otherwise results in nonuniform restraint of the edge of the membrane.

Irregularly shaped roofs, which connect two or more roofs at the same level, will produce increased stresses in the roofing systems at the corners where the roofs intersect, as each roof section responds to changes in temperature and moisture conditions as an individual unit. In this case, roofing problems are likely to occur in the form of membrane splits.

The strength of any roof system is insignificant relative to the structural roof deck components. Unrelieved movements in the substrate will damage the roofing and destroy its waterproofing integrity, particularly if it is a built-up roof membrane with low fracture resistance. If no provisions are made in the membrane to accommodate building movements at a building expansion joint, the accumulation of movements will strain the roofing system beyond its strength limits and the membrane will rupture. Every building expansion joint must be continuous through the roofing system.

There are other less obvious, but often just as critical, features in the roof deck: changes in deck direction or deck materials may transmit concentrated movements into the roofing system, and changes in temperature or applied loads will cause relative movements between the deck components which will be transferred into the roofing system if the roof system design does not accommodate such movements.

26.4.4 Metal Decks

There are two categories of structural metal roof decks. The most popular one is the light-gage, corrugated steel deck, used to support an insulated roofing system. The second type is a structural metal roof; it provides the weatherproof barrier, as well as performing the structural role.

The structural design procedures for corrugated steel decking are not reviewed in this chapter (see Chapter 20). The minimum design requirements that should be included to provide an adequate substrate for a built-up roof are included. This information is extracted from the Factory Mutual Engineering Corp. *Insulated Steel Deck* [26.2] (Fig. 26.2). This guide should be thoroughly understood by the designer and all its recommendations should be followed. The owner's insurance carrier may require, as a minimum, that the steel deck and the roofing system conform to those standards or the owner will pay higher premiums. Criteria for the steel deck include, but are not limited to:

1. Steel must have a minimum 22-gage thickness.
2. Deflection limited under concentrated load at midspan.
3. Deck side lap fasteners are required.
4. Flat top flanges, wide enough for adhesive bonding of insulation.
5. Top flanges shall remain flat relative to each other.
6. No stiffening groove in the top flange.

Gage	Max. Allowable Span ft.-in. (m)		
	Wide Rib	Intermediate Rib	Narrow Rib
22	5′-7″ (1.7)	4′-10″ (1.5)	4′-5″ (1.3)
20	6′-3″ (1.9)	5′-6″ (1.7)	5′-0″ (1.5)
18	7′-7″ (2.3)	6′-7″ (2.0)	5′-10″ (1.8)

The above spans apply to 1½-in. (38-mm) deep deck with 6-in. (150-mm) rib spacing, deck to span continuous over three or more supports.

Fig. 26.2 Maximum recommended spans for metal deck. From Ref. 26.2.

7. Installation must be straight and true, secured with minimum ½-in. (13-mm) puddle welds or mechanical fasteners, 12 in. (300 mm) on center maximum at all supports.
8. Deck sections should be secured to the supports in the ribs, where the sheets lap.

Structural metal roofing should be designed according to the requirements of governing building and material codes. Generally, the structural metal roofing is provided as part of a complete building system. The problems with structural metal roofs are usually associated with the design philosophy that seeks to minimize material and workmanship costs at the expense of the weatherproofing qualities and attachment of the structure. The following lists some of the major considerations with structural metal roofing:

1. Spans should not be excessive; the structural system may meet the design strength requirements for service loads, but often the system is so flexible it cannot be installed without being damaged from construction traffic.
2. The roofs should drain promptly; provide a minimum 3-in-12 (25%) slope. There is no maximum slope for structural metal roofing.
3. Structural metal roofing should always be attached to the structure with concealed fasteners. Never use exposed fasteners such as screws, rivets, or bolts without sealing washers. Thermal movements of the metal will loosen the fasteners or enlarge the holes in the base metal and allow leakage. Metal roofs with exposed fasteners will always require frequent, and often costly, maintenance.
4. Do not rely on seam sealant in end laps or side laps for watertightness of structural metal roofing. Thermal movements of the metal will soon break these seals. The formation of ice dams during winter months may submerge end laps and side joints in the roofing, causing leaks.
5. The roofing must be free to undergo thermal movements. Movements cannot be prevented, therefore, the connections to the structure, and at curbs, penetrations, valleys, and gutters, etc., must provide freedom for such movements.
6. Steel roofing must be protected from corrosive elements in the atmosphere. Many steel products are coated during manufacture, but care must be exercised during installation to protect the finishes. Rusting of steel panel edges and field cuts often produces unsightly staining on the roof surface, and failure at fasteners.

26.4.5 Cast-in-Place Concrete Decks

Cast-in-place concrete decks include both structural and lightweight insulating concrete. Both types can be easily sloped to drains during construction.

Cast-in-place structural concrete provides an excellent substrate for a roofing system, when the following principles are followed:

1. Structural concrete is subject to creep (long-term deflection) under constant load, as described in Section 22.5. The slope in the deck and the concrete support members must be designed to take this into account, or the roof will develop ponding. Structural lightweight concrete (125 pcf) (2000 kg/m^3) experiences creep, over about 5 yr, of approximately 1.5 to 2.0 times the creep of normal-weight concrete (145 pcf) (2320 kg/m^3). In normal-weight concrete, creep amounts to about 2 times the elastic deflection at a constant load.
2. The deck must be relatively smooth and flat (but not level) to allow adhesive bonding of the roof system components. The surface should be no rougher than a broom finish. The surface should not vary from flatness more than ⅛ in. in 3 ft (3.5 mm/m) to ensure full contact between roof insulation boards and the deck surface, without excessive distortion of the insulation board.
3. It is normal for concrete decks to develop small surface cracks from loading and from concrete shrinkage. Roof membranes should not be applied directly to the surface of a concrete deck, as crack movements will break the membrane.

Lightweight insulating concrete using vermiculite or perlite for aggregate is the other type of cast-in-place concrete for decks. These materials are installed "wet"; they are batched with more water than is required to hydrate the cement component. The additional water is included to produce a high slump, so the mix can be pumped into place on the roof and easily worked and finished. Use of these materials for roof decks is very risky because of the high water content of the material. Most of the mix water is entrapped by the application of the roofing and can not escape despite various deck venting schemes in common use. This moisture will either diffuse into roofing materials, causing their deterioration, or freeze, causing mechanical damage to the membrane. If the deck materials dry, the shrinkage that occurs can result in widespread cracking of the deck. The published thermal properties of the material are for dry deck materials; they may never be achieved in practice.

If these deck systems must be used, the following precautions should be followed:

1. Roof insulation boards must be used on top of the fill, separating the fill from the membrane, to ensure the desired heat transmission value and to protect the deck from freezing.
2. Always provide a vapor retarder system that allows horizontal venting between the deck and the roof insulation. Construction details should allow for the release of vapor pressure at the roof perimeters and roof penetrations. Do not add roof vents; the deck cannot be dried by venting.
3. Apply roofing as soon after casting as concrete strength allows. Insulating concretes are very porous and will retain large quantities of water if they are exposed to precipitation.

26.4.6 Precast–Prestressed Concrete Decks

There are a number of precast–prestressed concrete roof systems used for roof decks. The most popular are single and double tee-beams, and cored and channel slabs (see Section 21.7.3). These deck components are usually supported on precast beams and columns or by a steel structure. Tee-beams can span substantial distances, commonly 40 ft (12 m) to greater than 100 ft (30 m). The behavior of the precast roof components can affect the performance of the roof system. The following items should be considered during the design of a precast concrete building:

1. The precast units experience differential movements caused by deflections under loading, changes in temperature, creep deflections, and concrete shrinkage. These movements are often substantial enough to break connections between precast members. The resultant unrestrained movements of individual units can split roofing systems and cause leaks. Provision must be made in the structure and the roofing to accommodate these movements at side connections, at ends of tee units, and over support beams.
2. Variations in concrete unit dimensions which are within fabrication and construction tolerances often result in a roof deck that has an uneven surface for the application of the roofing. Always provide a reinforced structural concrete leveling course to be applied to the roof deck before the application of roofing.
3. When designing for roof drainage, the designer should consider camber and long-term reverse creep deflections, particularly in long-span tee-beams. The camber is an initial upward deflection fabricated into the beam to reduce service deflection. If the camber is not removed by the weight of the beam after erection, the beam will have a high point at midspan, contrary to conventional structural theory. A similar phenomenon may occur with prestressed beams where the concrete, under compression from the prestressed reinforcing, experiences creep resulting in additional upward deflection of the beam. Drains located at midspan will be at a high point, and the drains cannot be located at low points (at tee-beam ends) due to interference from support beams. Sloped reinforced concrete topping must be used to provide positive drainage.

26.4.7 Wood Plank and Plywood Decks

Wood is used in buildings of light construction, and in areas where it is readily available and inexpensive relative to other building systems. Most wood decks today are plywood, although sawn lumber is still used, particularly for buildings with exposed ceilings. Structural design of the wood deck is beyond the scope of this section, but standards to provide for the proper performance of the roofing system are reviewed below. Design is treated in Chapter 25.

A stable roof deck must be provided to preserve the integrity of the roof system. Wood, in both cross and side grain, moves considerably under changes in thermal and moisture conditions, compared to other structural materials. Expected movements must be prevented or accounted for in the design and detailing. Plywood end and side joints must be spaced to allow for such movements, and all sheets must be securely anchored on continuous supports at all edges.

Construct wood decks with lumber that has been properly seasoned to a moisture content not to exceed 19% by weight. Treatments for preservation or fire retardant purposes must be checked for compatibility with the roofing materials. Plywood should be exterior-type and a minimum of ⅝ in. (16 mm) thick. All edges of boards should be continuously supported to prevent differential deflections or edge curling. Support can be provided by wood blocking, tongued and grooved joints, or metal clips for plywood.

Wood decking must be securely fastened to supports. It should have a stable, smooth surface free from knot holes, cracks, and projections; small holes can be covered with sheet metal firmly nailed in place. End joints in boards should be staggered over supports, and boards should be secured with annular ring or hexagonal twist nails with a minimum embedment of 1 to 1.2 in. (25 to 30 mm) into support members. Spacing of nails should not exceed 6 in. (150 mm) on center for plywood edges and sawn lumber, and 12 in. (300 mm) on center for intermediate supports below plywood. All supported end joints of sawn lumber decks should have a minimum of two nails to prevent curling or twisting.

A layer of insulation boards should be nailed to wood decks, and a second layer should be set in asphalt as an attenuating layer between the deck and the membrane, even when interior insulation is provided. If board insulation is not used, bituminous based roofing products should be nailed to wood decks over unsaturated paper with annular ring or spiral-thread nails driven through metal discs, or factory capped nails, at a spacing required by the roofing materials manufacturer. Rosin-sized building paper must be used to separate saturated or coated roofing plies from the wood deck to prevent the asphalt from adhering to the deck surface during warm weather. Normal dimensional changes in the wood decks can split a built-up roof that is adhered to it, especially over deck joints. Unmopped plies below the membrane also serve to prevent bitumen drippage between deck elements, and it is prudent to use them below insulation as well.

Elastomeric roofing systems can be fully adhered to plywood decks but a release tape should be applied to all deck joints prior to the application of the membrane. The release tape will provide an unadhered strip to accommodate dimensional changes in the deck joints so they will not split the membrane.

Wood decks are particularly susceptible to deflections from long-term creep and they must be provided with sufficient slope to prevent ponding, and to reduce the risk for progressive ponding resulting in collapse of the structure.

26.4.8 Cast Gypsum Decks

Cast gypsum decks are constructed of reinforced gypsum concrete spanning between bulb-tee purlins that are supported by steel joists or roof purlins. The gypsum concrete is a mixture of gypsum and wood chips and shavings, mixed with water at the job site and pumped into place on the roof. Formboards of glass fiber, fiberboard, or other porous materials, span between the bulb-tees to provide temporary support for the steel reinforcing mesh and gypsum concrete slurry until the curing process is complete. The gypsum concrete gains strength rapidly by drying of the gypsum. The deck can support traffic and receive roofing soon after casting. Once the deck is cured, the formboards no longer have any structural significance, but they do provide some insulation value.

The roofing should be applied to a gypsum deck as soon as possible after curing, according to the deck manufacturers. The gypsum concrete will still be wet, but it is as dry as it is going to be for some time. Gypsum concrete is very porous. If rained on while left exposed to the weather (to dry), it will absorb far more water than it will lose in a day or two's drying time. After the roofing is applied, the gypsum deck will dry by evaporation of the water downward through the formboards. The wet gypsum must not be permitted to freeze or damage will result to the deck. Roofing systems must be mechanically fastened with special nails to fresh gypsum concrete decks. The material is so wet that adhesives will not bond to its surface.

Fresh gypsum concrete is relatively soft and easily penetrated by nails. Shanks of cut nails, driven through metal discs, will corrode in the gypsum and provide excellent withdrawal resistance. Other special fasteners are capped nails with cones or flanges that flare or spread when driven into the gypsum, forming a wedge to prevent withdrawal.

The roofing membrane should be divorced from the gypsum deck. As the fresh gypsum dries it will experience shrinkage-cracking. The deck will also crack over its supporting bulb-tees under service loading and at carelessly installed reinforcing. Asphalt bonds very effectively to primed, dry gypsum. A bituminous membrane that is nailed directly to gypsum deck may eventually adhere itself through solar heating. The cracks developing in the deck will split a membrane that is adhered to the surface of the gypsum. Use rosin-sized paper and multiple insulation layers to separate the membrane from the deck.

A roof membrane should be protected from wet gypsum concrete with a vapor retarder. The vapor retarder is nailed to the deck and the insulation and membrane can be adhesive-bonded to it. The necessity for a vapor retarder on a reroofing of an existing gypsum deck depends on the occupancy of the building. This is discussed in another section.

26.4.9 Gypsum, Wood Fiber, and Concrete Plank Decks

Roof deck units can be factory made of gypsum, wood fibers, or asbestos fibers with a cementitious binder, and lightweight and foamed concrete. The precautions outlined for precast–prestressed concrete decks should be followed with these systems, in addition to the following:

1. The preformed plank systems are often supported by clipping them to steel framing. This method prevents uplift but may not restrain lateral movement caused by changes in temperature and moisture content; such movements will cause membrane damage.
2. Some manufacturers restrict the allowable humidity levels in buildings using their planks as the materials are unstable and can deteriorate at high humidity conditions.
3. Roofing components are difficult to attach to wood fiber planks because only a special mechanical fastener provides good holding power into the fibrous material. Bolts with toggles are adequate for securing wood blocking, but are not practical for membrane attachment. Special fasteners that have a capped metal tube can be driven into the plank, followed by a nail driven into the tube. The nail point is deflected out of the tube end into the plank, to form a barb which anchors the device into the plank material. This fastener is the only presently adequate fastener, but holding power is low, so close spacing is required. Withdrawal tests for the particular material should be performed and the number of fasteners specified to meet the design requirements. Membranes that are fully or partially bonded to these planks experience splitting associated with plank movements accumulating at their joints.
4. Use two layers of roof insulation boards with offset joints between layers, applied to the precast plank to separate the membrane from the deck; this will often attenuate the joint movement from the membrane and prevent damage. Unmopped rosin-sized paper below the insulation is used to prevent bitumen drippage at fasteners and between units.

26.4.10 Parapets and Rising Walls

Avoid using parapets whenever possible. They cause problems for the roofing and often for the walls themselves. The roofing should be extended to the outside edge of the building and protected with perimeter flashing. Parapets are exposed to the weather from three sides and deteriorate rapidly, often providing a path for water to enter the roofing system and the building. Once the roofing system contains entrapped water, rapid deterioration occurs. If parapets are necessary, the following principles should be followed to minimize problems:

1. Securely anchor parapets to the structure.
2. Provide control joints at one-half the recommended spacing (twice the usual number) for conventional walls. Building and roofing expansion joints should continue through the parapet, without offsets.
3. Provide full metal copings with drip edges and carefully designed transverse joints on the tops of the parapets; slope the top of the copings toward the roof to prevent accumulation of water on the top surface which may enter transverse joints in the coping.
4. Provide roof flashing for the full height of the parapet on the interior, roof side. Include tapered edge strips below the membrane to raise the roof surface and direct water away from the parapet.
5. Provide through-wall flashing above the level of the roof deck to prevent water entry from the exterior side of the parapet; take precautions that the flashing does not provide a slippage plane for the masonry and that transverse joints are secure.
6. Design the roof structure for the live load increase from weight of drifted snow.
7. Provide emergency overflow scuppers to prevent ponding if roof drains become blocked.
8. Use flashings which are compatible with joint sealant in precast and control joints.
9. In cases where parapets are required by codes, consider railings attached to the exterior walls as an alternate design.

Rising walls do not pose all the same problems that parapets do, but there are similarities as follows:

1. The roof structure adjacent to the rising wall must be designed to support the additional weight of drifted snow.
2. It is usually impractical to flash the entire surface of the rising wall; a through-wall flashing over appropriate base flashing is needed, as well as a weep system to drain the wall above the through-wall flashing. The top of the base flashing should be 12 in. (300 mm) above the adjacent membrane surface.
3. Provide tapered edge strips below the membrane to raise the roof surface and direct water away from the wall.
4. Provide for movement within the flashing construction if the wall structure is independent

from the roof structure or the wall is continuous (no control joints) or exposed on the outside.

26.4.11 Expansion and Control Joints

Roof expansion joints should coincide with every building expansion joint. They also subdivide large buildings into relatively smaller buildings, and provide a means of controlling movements from changes in temperature and structural loadings. Roof expansion joint cover flashings must be designed to accommodate anticipated movements while they maintain watertight conditions.

Control joints are used to divide large roof surfaces into smaller roof surfaces and may, but usually do not, coincide with building expansion joints. The roof system is fragile and must not be subjected to high stresses which will breach the system. Control joints should be provided under the following conditions:

1. To ensure that no unbroken roof dimension exceeds 200 ft (60 m) (an arbitrary number based on experience only).
2. To separate a change in deck materials.
3. To separate a change in deck spanning direction.
4. To eliminate stress concentrations caused by re-entrant corners in irregularly shaped roofs or courtyards (the goal is to provide a number of compact rectangular roof areas).
5. To separate additions from a central building.
6. To separate the roofing at a change in occupancy or operating conditions in the building such as a cooler or freezer.
7. To separate "interior" roofs from "exterior" roofs at canopies or unheated sections of the building.

Control joint and expansion joint covers should be securely anchored to blocking; they are not connected directly to the membrane. They must have capability to accommodate maximum anticipated amounts of movements between elements.

26.5 ROOF SYSTEM SELECTION

Select a roofing system that has a history of successful performance in the field. The system should have been successfully used on a similar type of building, with similar deck, in a similar environment.

Designers are often tempted to try new materials, methods, and systems; however, experience has shown that laboratory tests and accelerated weathering tests often do not accurately predict field performance of roofing systems.

Frequently, the designer is presented with a situation requiring the use of a "new" system. Before any unfamiliar system is selected, visit the earliest installations; preferably, they should be at least 5 to 10 years old. Do not make decisions solely on sales department lists of "satisfied customers". A telephone survey, interviewing owners of roofs of at least a few years' standing, has proven to be revealing and most useful. Even this is not certain to provide accurate performance information, as many owners are not aware of the condition of their roof unless it leaks, and incipient problems may be developing. A field trip always provides more accurate information.

If the designer is faced with a project that requires substantial remedial roofing, a replacement roof is probably in order. Factors affecting reroofing decisions are beyond the scope of this chapter; however, in most cases a failed roof should be completely removed, the deck inspected and repaired as needed, and new roofing designed in accordance with the principles for new roofing systems. Experience has shown that deck deterioration cannot be accurately determined from below the roof.

Factors affecting the selection process for a roofing system are discussed in the following sections.

26.5.1 Deck Slopes

Most large industrial and commercial buildings have "flat" roofs. A roof is commonly defined as flat when the slope is from 0 to 3-in-12, but all flat roofs should slope sufficiently to prevent ponding under all circumstances, including deflections from snow loads and creep effects.

The deck slope limits may actually limit or dictate the use of many roofing systems. Flat roofs must be waterproofed with continuous built-up roofs or polymeric sheet membranes. All roofing systems must be constructed with provision for positive and prompt drainage of water.

Roofs that slope 3-in-12, or more, are classified as steep. Continuous roofing systems that can be effective on these slopes include built-up and polymeric membranes. Asphalt and wood shingles, slate, tile, and metal roofing consist of individual components that must be fastened to wood decking; other substrates generally have proven unsuitable for attachment of these products. Appearance is usually important for these roofs. Most of the above materials are available in several colors and textures.

Mineral surfaced roll roofing (asphalt-saturated felt with factory-applied mineral granule surfacing), is a special type of bituminous built-up roof that can be applied with hot asphalt or cold adhesives. It can be used on flat roofs at slopes between 1-in-12 and 3-in-12, but these roofs have not given long-term service compared to conventional built-up roofs. Workmanship is extremely important to assure good results with these products. Improperly applied roll roofing has a tendency to crack and lift at laps.

Cold process roofing includes numerous combinations of fabrics and asphalts or synthetic liquids. Many of these are available in colors of coatings or in granule surfacing. Some systems are reported to be successful in limited geographic areas, but cold adhesive systems have not achieved general industry acceptance at this time.

Polymeric membranes can be loosely laid (attached only at perimeters and penetrations) and held in place by stone, or other ballast, on slopes up to 1½ or 2-in-12. On steeper slopes the polymeric sheets must be fully adhered to the substrate or mechanically fastened with special manufacturers' systems, as the ballast will not remain in place. Workmanship is important on these systems because they are applied in a single layer and their continuity must be perfect.

Metal roofing can be used on flat roofs, but all seams must be soldered. Metal roofing with loose seams requires a minimum slope of 3-in-12. Metal roofs are made with flat, standing, or batten seams. Care must be taken to make proper allowances for thermal movements of metal components, while maintaining watertight seams.

Asphalt shingles may be used on roof slopes of 4-in-12 or steeper. Use on lower slopes, but not less than 2-in-12, is possible when special precautions are taken to avoid water backup. In cold regions where snow and ice dam formation are possible, the minimum slope should be 5-in-12. Asphalt shingle manufacturers allow the use of shingles on slopes down to 2-in-12, but such applications should be restricted to warm climates. The Asphalt Roofing Manufacturers Association publishes standards [26.1] for the design and application of shingle roofs which should be rigidly followed.

Wood shingles are sawn from red cedar logs to have uniform taper; they must be kiln dried before use in order to provide satisfactory results. Cedar shingles are available in several grades and sizes, each best suited to particular job conditions. They are durable, lightweight, and attractive. The shingles may be stained or left to weather naturally. Some building codes require that the wood shingles be treated with a fire retardant before use. A minimum 3-in-12 slope for the roof deck is permissible but steeper slopes are recommended, particularly in snow areas. The steeper the roof, the longer life is expected for the shingles. Cedar shingles are usually applied over solid sheathing, but application on open sheathing is permissible. The Red Cedar Shingles and Handsplit Shake Bureau provides detailed installation specifications which must be rigidly adhered to during installation. Irregular, thick split wood shingles are called shakes; these are decorative but not weathertight and waterproof underlayment is required with their use.

Slate and tile roofing are similar to shingles, but they provide a more durable and aesthetically unique roofing system. Their use should be restricted to slopes of 5-in-12, particularly in areas of strong winds, freezing temperature, and where snowfall is possible. Also, slate or tile roofing is heavier than most other systems and the structure and deck must be designed for the additional weight. Slate and tile can be applied on open sheathing rather than solid decking.

26.5.2 Drainage Provisions

All roofs should be designed to drain water completely and promptly. Water drains naturally from steep roofs and can be collected at edges by gutters and channeled from the building perimeter by exterior conductors (downspouts). Gutters in cold regions promote the formation of ice dams that cause water backup and leakage, and they are often blocked and damaged by snow and ice; therefore, in cold climates, gutters are best omitted to allow snow, ice, and water to leave the roof surface unimpeded. In this event, other provisions are needed to direct water away from the building face.

Flat roofs should always be drained with interior drainage systems. The slope of roofs to the drains should be provided by the structural system; methods of providing slope by tapered insulation are difficult to construct, and use of lightweight fills is not desirable. Tapered systems have often performed poorly, because of inaccuracies in construction leaving flat spots, the difficulty of obtaining smooth, tight joints between insulation boards, and the problems of covering the sloped surface promptly to protect it from weather. Wet fill materials do not dry sufficiently before the roofing is applied and asphaltic concrete is often improperly mixed and compacted. Both systems usually retain water on their substrates in the event of precipitation before the roofing is applied. With exceptionally good installation procedures and careful planning, these methods have been successful in roofing reconstruction projects.

The simplest method to provide slope is to vary the heights of columns to provide slope of the roof deck support framing. Steel decks are easily "warped" during installation to conform to gradual changes in slope. All decks require transverse sloping of valleys to avoid ponding along "flat" valley lines. Cast-in-place concrete decks can be sloped during installation of the deck forms. Precast decks should be sloped with a reinforced concrete topping which can also offset camber and provide a continuous substrate for the roofing system which is unaffected by the differential movement between precast units.

Drains should not be placed near columns (which are often "high" points), unless all deck deflections are accounted for. Drains located at low points will not be hindered by deck deflections between supports. Placing drains in sumped areas can compensate for minor interference which may prevent location of the drains at the actual low points. Reducing roof insulation thickness at the sumps also inhibits freezeup of drains, and the sump improves drainage locally.

Roof drains consist of cast iron bodies, usually with integral sumps, that are clamped onto steel or precast decks at the proper elevation, or cast into concrete decks. Plastic drains are available, but their performance is unproven. Clamping the drain in place is preferable because it allows adjustments and replacement if necessary. Drains should include a separate clamping ring to secure the roofing membrane and flashing, and an expansion sleeve connection to the interior piping system so the drain and its flashing are not displaced by deck deflection or movement of the interior piping. Do not permit flashing turned down into the drains; blockage within drain piping will force water under the flashing into the roofing system.

Drains should not be smaller than 4-in. diameter, to reduce danger of clogging. Cleanouts should be provided at convenient locations as drains may be inadvertently blocked by roofing materials. The size and spacing of drains is based on expected rainfall as shown on Table 26.2, If the roof is adjacent to a higher wall, added capacity may be needed to carry away water from runoff along the wall.

Each roof area should have at least two drains, preferably connected to separate downdrain pipes, so all drainage is not stopped if one drain is blocked. If parapets are present, emergency overflow scuppers or drains are needed to prevent a buildup of water and subsequent leakage and structural overload in the event of drain blockage. Overflows should be located so that accidental flooding does not accumulate water weight in excess of the structural capacity of the roof deck or ponding above flashing height. Local codes may have specific requirements for the overflows.

Gutters that drain steep roofs should be designed following published local area guidelines for sizing. Gutters become filled with debris, water, and ice; they should be of heavy metal with substantial anchorage. The gutters must have adequate slope and transverse expansion joints. Exterior downspouts will freeze in cold climates. The weight of ice can dislodge downspouts; they should be securely fastened to the building, not to the gutter.

When gutters freeze, expect water from the roof to overflow the roof edge, and cascade onto walls, windows, and walks. Design roofs with generous overhangs to minimize damage to the building from these effects. Gutter design is discussed in Section 26.7.

Gutter downspouts should connect to underground water disposal systems with provisions for cleanout. In cold climates where overflow is possible or where gutters are omitted, grade the ground surface to direct the runoff of water away from the building foundation to reduce water seepage problems at foundation walls.

26.5.3 Multiple Use of Roofs

Roof areas may serve as usable space for parking, patios, gardens, and other recreational purposes. Liquid (roller- or brush-applied) and semi-liquid (trowel-applied) systems are available for use with pedestrian and automobile traffic. These systems are usually protected by embedded aggregate surfacing. When the system is protected by pavers, cast-in-place concrete, or bituminous paving, and is not directly exposed, it is classified as a waterproofing system and should be dealt with separately. Most of the roofing design principles apply but effective drainage of the membrane and above the paving is even more essential.

The roofing systems for traffic bearing surfaces are usually applied directly to the substrate, since insulation is too soft to serve as an adequate substrate. The substrate should be cast-in-place concrete or precast units with reinforced structural concrete topping. The concrete should be placed in accordance with the roofing system manufacturer's specifications for surface finish and with respect to the use of curing compounds. Control or expansion joints require careful detailing to avoid random cracking. Membranes of this type are, at best, only as good as the substrate. Prompt maintenance of defects or wear areas is mandatory for prolonging service life.

Table 26.2 Roof Drainage Chart
Interior roof drains should be properly located and sufficient in number and size to drain all accumulated water from the surface of the roof in a 24-hr period. Special consideration should be given to the location of the drains to ensure their usefulness when deflection of the decking may reasonably be expected to occur after its installation.

Area of Roof (in Squares) Drained by Each Leader Based on Rainfall Intensity, by State

State	Diameter of Leader: 3 in. max.–min.	4 in. max.–min.	5 in. max.–min.	6 in. max.–min.
Alabama	22/17	46/36	86/69	135/108
Arizona	88/44	184/92	346/173	540/270
Arkansas	29	61	115	179
California	88/29	184/61	346/115	540/179
Colorado	88/22	184/46	346/86	540/135
Connecticut	44/29	92/61	173/115	270/179
Delaware	44	92	173	270
Florida	29/14	61/30	115/57	179/90
Georgia	29/17	61/36	115/69	179/108
Idaho	88/44	184/92	346/173	540/270
Illinois	29/22	61/46	115/86	179/135
Indiana	29/22	61/46	115/86	179/135
Iowa	29	61	115	179
Kansas	44/17	92/36	173/69	270/108

State	Diameter of Leader: 3 in. max.–min.	4 in. max.–min.	5 in. max.–min.	6 in. max.–min.
Nebraska	44/22	92/46	173/86	270/135
Nevada	88/44	184/92	346/173	540/270
New Hampshire	29	61	115	179
New Jersey	44/22	92/46	173/86	270/135
New Mexico	44/29	92/61	173/115	270/179
New York	44/22	92/46	173/86	270/135
North Carolina	29/14	61/30	115/57	179/90
North Dakota	29/22	61/46	115/86	179/135
Ohio	44/17	92/36	173/69	270/108
Oklahoma	29/22	61/46	115/86	179/135
Oregon	88/44	184/92	346/173	540/270
Pennsylvania	29/22	61/46	115/86	179/135
Rhode Island	29	61	115	179
South Carolina	29/17	61/36	115/69	179/108

Kentucky	29/22	61/46	115/86	179/135
Louisiana	22/17	46/36	86/69	135/108
Maine	44/29	92/61	173/115	270/179
Maryland	22	46	86	135
Massachusetts	44/29	92/61	173/115	270/179
Michigan	44/22	92/46	173/86	270/135
Minnesota	44/22	92/46	173/86	270/135
Mississippi	29/22	61/46	115/86	179/135
Missouri	29/14	61/30	115/57	179/90
Montana	44	92	173	270

South Dakota	44/29	92/61	173/115	270/179
Tennessee	29/22	61/46	115/86	179/135
Texas	29/17	61/36	115/69	179/108
Utah	88/44	184/92	346/173	540/270
Vermont	44/29	92/61	173/115	270/179
Virginia	44/17	92/36	173/69	270/108
Washington	88/44	184/92	346/173	540/270
West Virginia	29	61	115	179
Wisconsin	29/22	61/46	115/86	179/135
Wyoming	88/29	184/61	346/115	540/179

For example, the number of roof leaders required to prevent excessive accumulation of water on a 40-square job in Alabama (depending upon location within the state) would be two or three 3-in. leaders, or one or two 4-in. leaders. 5-in. and 6-in. leaders would not be required.

No roof should have fewer than two roof leaders and the maximum spacing for leaders should be no more than 75 ft in any direction.

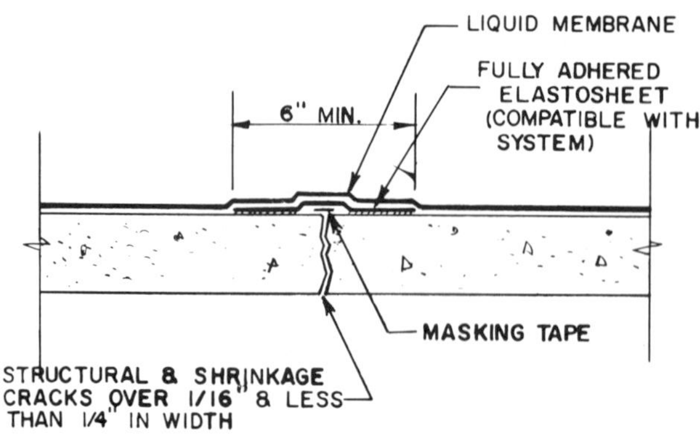

Fig. 26.3 Deck crack.

The surface of the concrete must be sloped to provide positive drainage. The liquid-applied systems are particularly susceptible to deterioration from ponded water, which also creates safety hazards to traffic during freezing weather.

Most surface-applied liquid systems are polyurethane, cured to a rubber-like coating, and have mineral aggregate broadcast to embed in the surface as a wear coat. Some include reinforcing mats or fibers, usually of glass fiber. As these coatings are totally adhered to the concrete, leakage generally is limited to the location of any damage in the coating; however, some leakage water can travel below surface-applied membranes and may even loosen them from the substrate.

Despite manufacturers' claims and the elastic characteristics of the cured systems, they are not able to bridge cracks reliably in the substrate unless the crack is discovered before application and a relief joint constructed. An effective joint for minor cracks [up to ⅛ in. (3 mm)] can be constructed by masking the crack to prevent the system from adhering to the surfaces on each side of the crack, thus creating a bridge of elastic material an inch or two wide (25–50 mm). This type of bridge should be reinforced with sheet flashing materials, but it remains subject to mechanical damage from traffic. This principle is illustrated in Fig. 26.3. If patching or grinding of the concrete surface is necessary, special precautions are needed to provide proper bonding of the liquid system and to prevent lifting of the patch from the concrete.

Problems of these systems include the contamination of surfaces between applications of multi-coat systems, and the formation of blisters in heavy coats of liquid during cure. In addition, as appearance is often an important consideration for these systems, patching and repairs will have color and texture variations that may be unacceptable. As with all materials, verify that the coatings are compatible with the environmental conditions at the project location. Traffic deck surfaces require constant maintenance, particularly at high traffic areas and at spills of fuels, etc.

Special consideration must be given to waterproofing of perimeters and penetrations. Many of the coatings are self-flashing; that is, the coating is continued at edge construction, often with reinforcing or compatible polymeric sheets. Where discontinuities exist, provide for movement between adjacent walls and surfaces in the flashing design. Generally, dissimilar materials such as wood or metal should not be included in these systems.

The diverse nature of coatings makes the above recommendations only guidelines. Each product should be carefully investigated because surface-applied waterproof coatings have had a mixed overall performance record in the field. Unfortunately, there is no other effective method of waterproofing traffic bearing surfaces. Do not use these systems over slabs-on-grade as ground-water pressure and dampness will lift the coating. Once the system is selected, the manufacturer's recommendations should be strictly followed.

26.5.4 Dead Load Capacity of Structure

The weight (load) of the roofing system selected must be included in the dead load capacity of the structure. The conventional built-up roofing system dead load is about 9 psf (430 N/m^2), including 2 psf (96 N/m^2) of insulation boards and 4 psf (190 N/m^2) of aggregate. Acceptable variations in field application, such as differences in bitumen amounts, felt weight, and insulation thickness and type, can result in 15% more or less than this amount. A range of 7 to 10 psf (340 to 480 N/m^2) should be expected. If a white (or specially colored) aggregate is required, an additional 1 or 2 psf (50 to 100 N/m^2) of material is necessary to assure coverage to hide the bituminous surfacing.

The built-up roof system weight can be reduced by use of a smooth surface built-up roofing system that substitutes a liquid (emulsion, asphalt, or aluminized coating) for the aggregate surfacing. This reduces the system weight by 4 to 5 psf (200 to 250 N/m^2), but does not provide the same degree of protection from traffic or weather as a gravel surface.

Polymeric sheet membranes weigh less than 0.5 psf (24 N/m^2); however, if they are loosely laid, 10 to 12 psf (500 to 600 N/m^2) of stone ballast is needed. A ballasted system will weigh about 14 psf (670 N/m^2) and an adhered or mechanically fastened system could weigh less than 2 psf (100 N/m^2) including insulation boards.

Sloped roofs may have relatively lightweight shingles or heavy slate or tile and the structural design must allow for these choices.

In addition, designers should include the probability of multiple roof systems being applied during the building's life, and increase dead loads accordingly. Table 26.3 shows the weights of the commonly used roofing systems.

In many ways, the construction period is the most critical time for the survival of the roof, primarily because of potential physical abuse and the accompanying risk of moisture inclusion during application. In addition, the roof deck, particularly in flat roof design, must be able to support the construction loads during roof application. Roofing contractors' machinery often imposes concentrated loads far in excess of design loads. An example is a moving load of a full wheelbarrow of gravel (400 lb) and a worker (200 lb) at midspan on a steel roof deck. Materials stored on the roof can also cause problems. A pallet (36 × 42 in.) of roofing felt can weigh 25(60) = 1500 lb, resulting in a load of about 150 psf (7 kN/m^2); contractors sometimes stack materials resulting in even heavier loads.

If concrete fill is used to slope roof surfaces, its weight must be added to the dead load.

26.5.5 Deck Characteristics

Deck selection and design is discussed in Section 26.4. The deck system characteristics to be considered in the selection of roofing systems often vary from their structural requirements.

Relative structural flexibility is an important element in roofing design. Although sufficient for structural requirements, a deck (particularly steel deck) may be too flexible for permanent attachment of the roofing system, especially under concentrated loading during construction. Exact standards for deflection limits do not exist, but prudent standards are provided by Factory Mutual Corporation [26.2].

Mechanical "fastening" to "nailable" decks is becoming the accepted industry practice in place of the use of hot bitumen. Cold adhesives have proven to be unsatisfactory. Table 26.4 shows common decks and recommended fastening methods for insulation and roofing.

For corrugated steel decks, insulation or other suitable underlayment such as gypsum board (usually Type "X") or plywood is required to bridge the corrugations and provide a smooth substrate to receive roofing. The width of the deck flutes limits the type and thickness of insulation board; manufacturers' literature should be consulted to determine the allowable span for each thickness of board. Insulation boards are made in multiples of 2 ft (609 mm), so they fit with 6-in. (152 mm) cycle steel deck ribs (flanges). Steel deck installation should be carefully monitored to ensure that these dimensions are held between panels and that panels are aligned so insulation boards will have adequate bearing on all flanges. Flanges may be "off-cycle" at deck side laps requiring cutting insulation boards to maintain adequate bearing.

When using precast concrete decks, the fabrication and erection tolerances between units must be compatible with the requirements for a smooth substrate for the application of the roofing system. Since it is difficult to obtain a sufficiently stable surface smoothness, a reinforced structural concrete topping is usually required. Precast planks must be securely anchored to prevent any lateral movement, which could lead to splits in the membrane, unless provisions are made to accommodate the plank movements.

Wet fill decks such as vermiculite, perlite, or gas-foamed (cellular) insulating concrete are susceptible to damage from precipitation or freezing weather before the roofing system is applied. Deck manufacturers' recommendations for roofing application to commence as soon as possible after pouring do not allow upward drying of the fill. These recommendations generally conflict with roofing manufacturers' recommendations for dry decks for roofing application. If the deck is not "dry" when the roofing is applied, the entrapped moisture will damage built-up roofing systems. The moisture can also result in the membrane freezing to the roof deck and cracking over deck cracks. Polymeric sheet systems may be less sensitive to moisture damage; although the sheet membrane may be relatively unaffected by moisture, the lap seam can be at risk. The degree of dryness may also affect the holding power of mechanical fasteners. Depending on the time of year and project location, the requirements for installation of roofing and insulating concrete are likely to be mutually exclusive, leading to premature roof failure because of entrapped moisture. Some contractors report good results when these decks are vented, but this procedure is not uniformly accepted.

Table 26.3 Characteristic Properties of Built-up Roofing Materials, Felts, and Membranes

Material	ASTM Spec.	Weight, lb/100 sq ft	Coefficient of Thermal Expansion, 10^{-6}/°F		Breaking Load, lb/in.		Load Strain Modulus, 0°F × 10^{-3} lb/in		Vapor Permeance, Perms	Thermal Shock Factor	
			L	T	L	T	L	T		L	T
Roofing Felts											
Asphalt organic felt Type I	D226	13	18	26	30	15	57	36	3	1000	500
Coal tar organic felt	D227	13	19	29	30	15	67	74	4	500	300
Asphalt asbestos felt 15 lb	D250	13	7	14	20	10	80	55	3	1200	700
Asphalt asbestos base felt—uncoated	D250	28			40	20	—	—		—	—
Glass fiber felt Type IV	D2178	7	11	10	44	44	28	22		800	500
Asphalt coated organic felt	D2626	37			35	20	—	—	0.03	—	—
Asphalt coated asbestos felt	D3378	37			30	15	—	—	0.03	—	—
Mineral cap sheet (roll roofing)	D249	74	5	11	—	—	—	—	0.05	—	—
Reinforced asbestos flashing sheet	—	65									
Venting type base felt		70									
Modified Bitumen Products											
Mineral-surfaced sheet		100				130–200			0		600
Plain surfaced sheet		60				130–200			0		
*Roofing Membranes**											
Base + 3 #15 organic—asphalt	—	190	11	21	437	302	4.9	3.9	0	358	231
Base + 3 #15 organic coal tar pitch	—	195	22	30	426	230	6.2	9.2	0	315	84
Asbestos base + 3 #15 asbestos asphalt†	—	170	7	17	317	165	5.9	5.3	0	418	89
Asbestos base + 2 #15 asbestos asphalt†	—	130	9	18	269	152	8.5	5.5	0	208	194
3-ply glass fiber asphalt†	—	115	25	24	304	277	2.1	1.9	0	570	605
4-ply glass fiber asphalt†	—	150	19	29	418	372	5.3	2.4	0	528	532
4 #15 asphalt organic	—	155	22	62	400	233	2.9	1.7	0	618	216
4 #15 coal tar pitch organic	—	160	22	33	477	311	—	—	0	302	202

* For roofs surfaced with aggregate, add:

60 lb for asphalt flood coat surfacing
75 lb for coal tar pitch flood coat surfacing
and:
400 lb for gravel
500 lb for white aggregate
300 lb for blast furnace slag

L = Longitudinal
T = Transverse

† Add: 20 lb for smooth surface asphalt coating.

Wood decking characteristically changes dimension with changes in moisture content. Today, most wood decks are plywood and are considered to be nailable. In some areas of the United States built-up roofing is adhered to plywood decks with hot asphalt, but this continuous attachment is likely to lead to membrane cracking over the plywood panel joints. Some polymeric sheet systems can be adhered to plywood decks if special provisions are made at the panel joints. With wood decks use the following general precautions:

1. Deflections should be limited by the needs of the roofing system, which may be more stringent than called for by structural needs.
2. An initial nailed sheet, sealed at the joints, should always be used to prevent bitumen drippage into the building from bituminous built-up roofs.
3. All wood products used in roofing should be preservative treated and, in the case of plywood, be exterior grade.
4. Plywood should be at least ⅝ in. (16 mm) thick for adequate nail retention.

26.5.6 Appearance

The appearance of flat roofs is seldom a design consideration unless the view from adjacent windows is important. Normally a roof surfaced with neatly applied aggregate is satisfactory.

Roofing aggregates are available in several colors (such as red, gray, and white), depending on local availability, and they may be applied in patterns if desired. Roofing aggregate should be smooth "river bottom" gravel with as little crushed material as possible to avoid damage to the membrane. Before selecting aggregates other than gravel, examine samples for sharp and broken stone that can damage the roof membrane. Gravel punctures in new built-up roofs have been caused by construction traffic. This is true for built-up roof membranes as well as loosely-laid ballasted systems. Loosely-laid membranes can be protected by insulation or by a scrim, but the use of sharp stone could lead to inadvertent damage to the membrane at edges and penetrations, and during application or maintenance operations.

Flat built-up roofs can be smooth surfaced with a thin coating of exposed asphalt or painted with a white or aluminum pigmented cutback bituminous coating. Heavy bitumen coatings should not be used as they will crack (alligator), exposing the membrane to deterioration from the top downward. Several manufacturers offer gray/white mineral surface "cap" sheets in place of other surfacing, but their performance has generally not been satisfactory.

Sloping roofs may be visible from the ground and are frequently an architectural feature of the building. Sloping roof materials are available in a broad range of "standard" colors. Special colors are available in some products at added cost if the roof area is large. Colors vary between manufacturers, and, in some cases, between production lots. Most colors will change and fade with time. Surplus material should be ordered and stored, so that future repairs will not cause major color variations.

Polymeric sheets are manufactured in several colors, but most materials are black or gray. Field-applied color coating (Hypalon) painted onto synthetic rubber sheets is offered by several manufacturers. The coating protects the sheet for longer life but painted surfaces require maintenance and periodic recoating. The coatings can be damaged by ice, hail, and roof-top traffic, and light-colored materials quickly become discolored from atmospheric dirt.

Metals for roofing include copper, aluminum, stainless steel, galvanized-coated steel, and lead-coated copper. All accept paint readily, and some are available with factory-applied coatings, which should be used when painting is necessary. Aluminum can be anodized in a variety of colors if the correct alloys are specified. Copper can be oil-rubbed for a dull dark finish, or "pickled" for the classic greenish patina. Proper metal characteristics must be specified so the materials are suitable for fabrication processes.

Slate, tiles, and wood (cedar) shingles provide a variety of textures and colors. None of these products are truly color-fast when exposed to weather; however, the weathered effect may be visually desirable.

Because of their exposure to sun, rain, dirt, and pollution, all roofing surfaces may stain, streak, fade, or show other signs of deterioration with time. Discoloration will be accelerated in ponded areas, in vicinities with air pollutants, and where light-colored polymeric sheets are used.

26.5.7 Life Cycle Considerations

Roof systems generally have life expectancies of 10 to 50 years. The useful life of built-up roofing systems is habitually judged against a 20-yr standard. This is an artificial basis, resulting from the sales-tool "guarantees" offered by most manufacturers over the past several decades. Some manufacturers provided a 20-yr bond which has been incorrectly interpreted as a guarantee of the service

Table 26.4 Roof Deck and Fastener Data

Type of Deck	Fastener Recommendation
Wood-tongue and groove sheathing	1,3,4,7,8,11
plywood	3,4,7,8,11
Gypsum-poured (1–7 Days)	9
Poured (Dry)	6,12
Precast metal edge plank	4,12
Concrete-poured	
Precast-Alaslab	
Doxplank	
Flexicore	
CPC Channel Slab	
Span-Deck	
Spancrete	
Prestressed concrete "T" or "TT"	
Precast lightweight concrete	
Calsi-Crete	5,6,2
Castilte	5,6,2
Creteplank	5,6,2
Cantilite	5,6,2
Federal Featherweight	5,6,2
CPC Concrete Plank	5,6,2
Lightweight poured concrete	
Perlite	6,10,2
Vermiculite	6,10,2
Zonolite	6,10,2
Cellular concrete	6,10,2
Structural wood fiber	
Fibroplank	5,6,2
Petrical	5,6,2
Permadeck	5,6,2
Fibertex	5,6,2
Asphalt lightweight aggregate	
Dri-Pac	—
All Weathercrete	—

More comprehensive data now available concerning wind action on structures and roof coverings dictates that designers and architects be more aware of possible damage to roofs. Factory Mutual data sheets I-7, I-28, I-47, and I-49 call attention to the fact that some geographical locations are subject to wind conditions that require increased fastening means at critical areas. It is the designer's and architect's responsibility to consider these factors in design of a roofing system.

The table on this page indicates the proper built-up roof specifications for application over various roof decks. Not every roof

Detailed Description of Fasteners & Sources of Supply.			
1 	2 	3 	4 
Roofing nail, 11 or 12 gage, 3/8–7/16 in. diam. head	Insuldeck Loc-Nail, E.G. Building Fasteners Corp.	Roofing nail, annular thread, 11 gage, 3/8 in. diam. head	Roofing nail, spiral thread, 11 gage, 3/8 in. diam. head
		Independent Nail Co. W. H. Maze Co.	
5 	6 	7 	8 
Capped Es-nail, 1 in. cap	Tube-Loc nail, 1 in. diam. cap	Squarehead cap nail, annular thread, 1 in. diam. cap	Squarehead cap nail, spiral thread, 1 in. diam. cap
Es-Products, new	Simplex Nail & Mfg. Co.	Independent Nail Co. Simplex Nail and Mfg. Co.	
9 	10 	11 	12 
Nail-Tite Type A 1¼ in. diam. cap	Zonolite or Mark III	Roofing staple, for power-driven application only	Do-All nail, hardened
Es-Products, New	Es-Products	Bostitch Spotnails Berry-Fast	E.G. Building Fasteners Corp.

deck has been included by trade name. However, the unlisted ones can be readily identified in one of the major groupings of decks and treated as indicated.

Also shown is a table of fastener recommendations for securement of roofing to nailable decks. This chart identifies the fasteners referred to in the table.

As density of decks vary and in many cases more than one type of fastener is specified, field tests should be conducted to determine the most effective fastener.

life of a roofing system. A system should not be selected on the basis of a manufacturer's guarantee. These bonds and guarantees are discussed in more detail in Section 26.15.4.

Field experience and industry surveys present widely different pictures for the life of built-up roofs. If a roof is properly designed, installed, and maintained, a 20- to 30-yr life is not unusual, and many roofs survive longer. Built-up roofs are particularly susceptible to water damage, so proper design (i.e., drainage), workmanship, and regularly scheduled maintenance are critical to their performance.

The selection of a system with proven field performance is also critical; many premature roof failures have resulted from the use of unproven systems and materials.

The life expectancy of roofing systems also depends on good design and workmanship. Examples of useful life expectancy of some materials are:

Built-up roofing	15–30 yr
Asphalt shingles	15–20 yr
Metal roofing (noncorrosive)	40–50 yr
Cedar shingles	35–40 yr
Mineral surface roll roofing	10–12 yr
Tile and slate (except metal accessories)	50–75 yr
Polymeric sheets	(estimated) 5–15 yr

In the case of tile and slate, metal flashings and other adjacent materials such as masonry and wood trim will probably need repair or replacement before the roofing materials. The designer should allow for this in the construction details.

Steep roofing products generally will perform longer than built-up roofing, partly because of their use on sloped roofs that drain rapidly. In general, the steeper the roof slope the longer the roof life. An exception to this may be shingles on southern exposures, where sunlight and heat may cause deterioration earlier than on other elevations.

Steep roof materials (because they must look satisfactory as an architectural feature) and the construction of the roof support system (because of the sloped surface) are usually more costly than flat roof construction. They can be expected to have a longer life because of reduced exposure to moisture. Estimates based on current prices of labor and material in the building's locale will help to determine the roof system selection. The cost figures should be applied to a suitable economic model which includes the cost of money and an inflation factor.

The economic effects of insulation in the roofing system are discussed in Section 26.3.

26.6 BUILT-UP ROOFING AND FLASHING

Traditionally, roofing systems used on flat or moderately sloped roofs are bituminous built-up roofs. They are assembled on the roof by cementing alternate layers of bituminous saturated felts and hot bitumen. The felts provide the "body" or reinforcing for the membrane, and the bitumen provides the waterproofing materials. The resulting composite membrane can have a smooth surface or be protected with mineral aggregate. The membrane should be applied over the smooth, firm substrate of multiple layers of board insulation to attenuate the effect of any deck movements on the membrane.

The built-up roofing is necessarily applied in the presence of atmospheric moisture by a mixture of skilled and unskilled labor. Exacting design and workmanship standards are required to achieve acceptable results because the felts are water sensitive, insulation performance is adversely affected by moisture, and the membrane has thermoplastic characteristics.

This section describes the use of built-up roofing materials. Unless otherwise noted, all the materials described below have ASTM and/or Federal specifications. Table 26.5 is a listing of these, cross-referenced where applicable, to both standards.

26.6.1 Insulation Installation

It is important to provide insulation between all built-up roof systems and their roof decks to isolate the membrane from movements in the roof deck. The selection and types of insulation are discussed in Section 26.3.

Roof insulation is supplied as "boards", 2 × 4 ft or 3 × 4 ft. The boards should be installed in at least two layers. The first layer must be mechanically fastened to the deck wherever it is possible. Mechanical securement is more reliable than either cold or hot adhesives. The joints between all the boards must be staggered in and between layers to provide a continuous "diaphragm" of insulation for a stable base for the roofing.

Hot asphalt is to be used where mechanical fasteners are impractical or the deck is a non-nailable

Table 26.5 Roofing Materials, ASTM and Federal Specifications

	ASTM	Federal
Aggregate	D1863	—
Aluminum coatings (asphalt-based)	D2824	TT-C-1079B
Asphalt (emulsified) for protective coatings	D1227	SS-A-694D
Asphalt (for built-up roofs)	D312	SS-A-00666a
Asphalt (for dampproofing/waterproofing)	D449	SS-A-666D
Asphalt mastic	D491	—
Cap sheets (mineral-surfaced)	D371	—
Caulking compound (Grade I or II)	—	TT-C-598
Cement (lap) for asphalt roll roofing	D3019	—
Cement (roof) plastic, asphalt	D2822	SS-C-153
Coal-tar bitumen	E450	R-P-381A
Coal-tar roof cement	D4022	—
Felts:		
Asphalt-coated/saturated	D3158	—
Asphalt-impregnated (glass mat/felt)	D2178	SS-R-620B
Asphalt-saturated, asbestos	D250	HH-R-590
Asphalt-saturated, organic	D227	HH-R-595B
Insulation		
Blankets and batts	—	HH-1-528B
Boards		
Cellular glass	—	HH-1-551E
Mineral aggregate	—	HH-1-529B
Mineral fiber	C208	HH-1-526C
Perlite	—	HH-1-574B
Polystyrene	—	HH-1-524B
Urethane	—	HH-1-530A
Vermiculite	—	HH-1-585C
Nonbituminous:		
Elastomeric and plastomeric	D3105	—
Neoprene and chlorosulfonated polyethylene synthetic rubber, liquid-applied	D3468	—
Primer (asphalt) for roofing/dampproofing/waterproofing	D41	SS-A-701
Roll roofing:		
Glass mat asphalt (mineral-surfaced)	D3909	—
Organic felt (mineral-surfaced)	D249	SS-R-630D
Organic felt (smooth-surfaced)	D224	—
Rubber:		
Chloroprene rubbers (CR)		
Ethylene-propylene-diene rubbers (EPDM)		
Isobutene-isoprene rubbers (IIR)		
Vulcanized sheeting		
Fabric reinforced	D3254	—
Nonfabric-reinforced	D3253	—
Shingles:		
Mineral-surfaced, Class A	D3018	SS-S-294a
Mineral-surfaced made from glass mat	D3462	—
Mineral-surfaced	D225	—
Sheet (asphalt-saturated/asphalt-coated) asphalt base (for BUR/vapor barriers)	D2626	UU-B-790A
Wood, preservative-treated	D1760	TT-W-571
Woven fabric		
Burlap (saturated with bituminous substances) for membrane waterproofing	D1327	HH-B-0080
Cotton (saturated with bituminous substances) for membrane waterproofing	D173	SS-C-4501
Glass (treated)	D1668	—

material such as concrete. Cold adhesives, popular in years past because of their fire resistance, are not recommended since they have been proven inadequate in field use. Firm anchorage of the insulation is mandatory to the successful performance of the built-up roofing system to provide wind resistance and to prevent stress accumulations in the membrane.

Factory Mutual Bulletin I-28 [26.2] describes the types, numbers, and locations recommended for mechanical fasteners.

When hot asphalt is used as insulation adhesive, the deck (except steel) must be primed to produce an "adhering" surface. Use Type III (steep) asphalt as the adhesive; its relatively high softening point makes it less likely to drip into the building through openings in the deck during and after construction. Steep asphalt provides good adhesion when applied uniformly and when the insulation is fully embedded while the asphalt is in a plastic state. If embedment of the insulation is delayed, the asphalt will cool, solidify, and will not be an effective adhesive. If the asphalt/insulation bond fails due to deck deflection or other causes, it will not re-adhere, leaving the insulation unattached.

The second and succeeding insulation layers should be set in continuous hot moppings of steep asphalt. The joints of these layers must be staggered from board to board, as well as between layers (Fig. 26.4) to eliminate any through-joints in the system between the deck and the membrane. Through-joints provide discontinuities below the membrane where "working" stresses can accumulate, heat leakage can occur, and mechanical damage is more likely.

In some installations of glass fiber insulation, the joints are aligned and taped with glass fiber tape set in asphalt. Owens-Corning Fiberglas Corporation claims that this procedure stabilizes the insulation layer and prevents splitting of the membrane.

The insulation must be installed with firmly butted and level joints. Continuous (unstaggered) joints should be oriented to allow application of the roofing felts across the joints as described in Section 26.6.2. All insulation board edges must be fully supported. On steel decks, insulation boards may have to be cut during installation to ensure adequate support by the deck flanges.

26.6.2 Felts

Roofing felts are made of glass fibers, mineral fibers, wood pulp (paper), waste fibers, binders, and fillers. The felt manufacturing process is similar to paper production. The felt mat is capable of absorbing relatively large amounts of bituminous saturant. For example, ASTM specifications require organic felt (made from waste paper and wood pulp), when saturated with asphalt, to reach at least 2.4 times its unsaturated weight. Other felt materials are made from glass or asbestos fibers, and recently, polyester.

Organic felts are the least expensive and have the longest successful performance record. A disadvantage to their use is that they readily absorb moisture when exposed to rain or a humid atmosphere. The felt strength is dependent on fiber orientation in the manufacturing process; the felts are significantly weaker across the machine direction than along it.

Glass fiber felts are manufactured in several ways, and classified by their mat type: continuous strand or chopped fiber. Most glass fiber roofing felts have been produced only since the late 1970s. A product with acceptably extensive field history is a continuous strand type felt, "Perma Ply-R," manufactured by Owens-Corning Fiberglas Corp. This felt is stronger in tensile strength than organic felt; and is reported to be equally strong in both directions, along and across the machine direction. Because these felts are porous, extra care is needed during installation to ensure full embedment in the hot bitumen layers.

Asbestos felts have been used for constructing roofing membranes since the 1960s and for roofing specialty felts long before that. Their tensile strength is lower than organic felts, particularly across the machine direction. Asbestos membrane roofs have a history of splitting and blistering problems. As a result of these problems, their use has decreased. The use of asbestos felts for built-up roof membranes is not recommended, but they are useful for built-up flashing systems. By 1984 the manufacture and use of asbestos products for roofing in the United States stopped for all practical purposes. Asbestos has been largely supplemented by glass fiber materials for roofing felts.

Asbestos and glass fiber felts are less susceptible to weathering and moisture absorption than organic felts because of their inorganic fibers. For this reason, their use has been promoted by their manufacturers for smooth surfaced membranes. However, water can be absorbed into spaces between fibers, and in the fillers used to make the felts, leading to membrane problems.

Table 26.3 gives the strength of felts in laboratory-tested membranes. Expect lower values from field installations, because of variations in workmanship.

Most of these types of felt are also available as "asphalt-coated felt". A coated felt is usually used as the initial membrane ply over the deck or insulation layer; this type of felt is called a "coated base sheet." Coated base sheets are used to reduce moisture penetration into the membrane from below. Coating the felts does not make them impervious to water absorption; these felts

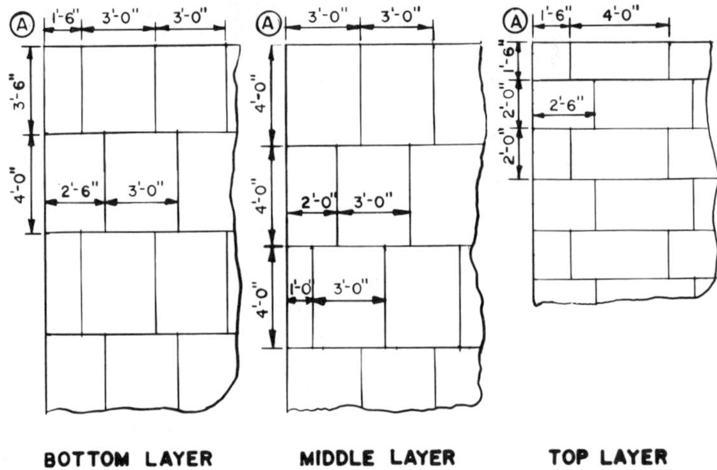

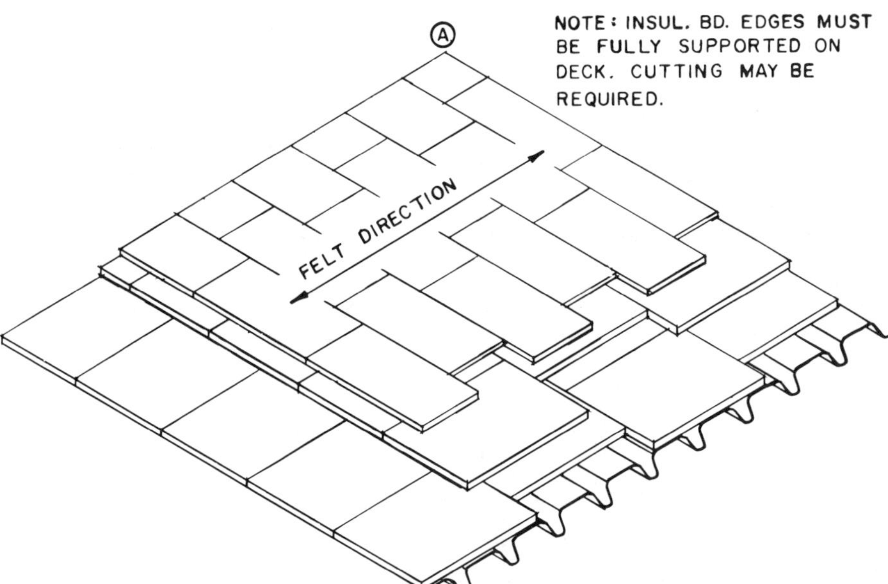

Fig. 26.4 Insulation stagger pattern.

are particularly vulnerable at their edges. Coated felts should never be used in place of saturated felts, except on a 1 for 1 basis.

Similar coated felts have been used as the roofing plies in some systems. Multiple plies of coated felts often contain interply blisters at lap edges because moisture entrapped in the felt escapes at the felt side laps during hot mopping. These felts are stiff and difficult to apply uniformly. Coated felts are usually rolled with separator materials such as sand, talc, or mica dust to prevent sticking in the rolls; if excessive, this material can interfere with bitumen adhesion. These systems are not recommended. When surfaced with mineral granules, the coated felts are used in some systems as the uppermost felt ply for roofs in place of aggregate surfacing.

A coated base sheet is usually used for the first ply of a membrane. This provides an asphalt coating on the underside of the membrane above substrate discontinuities such as at insulation board joints. The underside of the membrane above these areas may not be coated by the mopping asphalt during installation because the hot bitumen runs into the joint. The sheet coating prevents moisture from being absorbed through the underside of the membrane and also reduces wrinkling and blistering. The base sheet is difficult to handle because of its greater thickness. It must be installed using careful workmanship, especially in cold weather, because of the risk of buckling and cracking during unrolling. Full brooming into the hot bitumen is essential.

Felts are applied shingle fashion from the low to high points on the roof slope to prevent water from flowing against the felt side laps. Organic felts are stronger along the machine (long) direction, and should be applied across continuous insulation joints. This practice orients the strongest dimension of the membrane across the weakest link in the insulated substrate.

26.6.3 Bitumens

Bitumen is the adhesive and the waterproofing agent in the built-up roofing membrane. The bitumen adheres the felt plies to each other, the insulation layers together, and in some cases, the insulation to the roof deck. The felts alone are not waterproof; in fact, most are deliberately perforated with venting holes during their manufacture to allow vapor release during their application with hot bitumen.

Roofing bitumen is either coal-tar pitch or asphalt. Asphalt is a refinery product derived from the heavy portion of crude oil. Coal-tar pitch is a by-product of the production of coke from bituminous coal. While they are similar in physical characteristics, these bitumens are very different chemically and should not be mixed. The results of their contact with each other are unpredictable. When the result of these combinations is unfavorable, asphalt softening or coal pitch embrittlement "incompatibility" occurs. Pitch and asphalt should never be allowed to contact one another in roof construction; felts saturated with one bitumen should only be used in combination with the same bitumen. Some manufacturers recommend application of asphalt-coated base sheets and the use of asphaltic built-up flashings with pitch, but this practice should be avoided because the risk of incompatibility always exists.

Both asphalt and pitch are solids at ambient temperatures. Coal-tar has a softening point of about 120°F and is not soluble in water. Its low melt point allows it to flow on roofs during hot weather and to seal small surface defects. Because of its cold-flow characteristics, coal-tar pitch roofs should be limited to slopes of less than 1-in-12 to avoid slippage of the felts and aggregate surfacing. On slopes greater than ½-in-12, the tar felts must be "back nailed" to prevent slippage. Coal-tar pitch is more brittle than asphalt in cold weather, and as such is more likely to lose its bond to metal flashings and to crack in areas experiencing differential movements of materials.

Asphalt products are more commonly used for built-up roofing because they perform as well as coal-tar pitch on low slopes, have less flow on moderate slopes, are less brittle in cold weather, are less costly than coal-tar pitch, are less irritating to workers, and are more readily available. Asphalt felts must be backnailed at slopes of 1-in-12 and greater. The proper grade of mopping asphalt must be used, dependent on the roof slope.

Roofing asphalt is processed from petroleum. The asphalt is asphalt "oxidized" or "blown" by bubbling air through heated asphalt to produce materials with particular melt characteristics. The commonly available types are ASTM Types I, II, III, and on special order, Type IV. Table 26.6 shows the characteristics and uses of these asphalts. These four types increase in "hardness" as their "softening point" increases.

Bitumens are not compatible with some other materials. They should not be used with other coatings, polymeric materials, wood preservatives, and insulations except after thorough investigation for compatibility.

Both asphalt and coal-tar pitch are used to make many other products that are useful in roofing. Products containing solvents are called "cutbacks" and are used, together with fibers and clay extenders, for cold adhesives and coatings. Primers are usually mixtures of asphalts and solvents. Pitch and asphalt are also combined with clays and water as emulsions for weather-resistant surfacing of built-up roofs.

The most widely used "accessory" product is (plastic) roofing cement; it is a solvent-based trowel grade mastic formulated from either asphalt or coal-tar pitch and contains asbestos fibers and fillers. Roofing cement is used for cement in built-up flashing, for bedding metal flashings, and for temporary seals during roof construction. It is available in several grades for workability at hot or cold ambient temperatures. Solvent-based materials are usually not compatible with polymeric products nor with some foam insulations. Their use below impermeable sheets will result in blisters as solvent release from the roofing cement is impeded.

26.6.4 Built-up Flashings

Built-up roof membranes are joined to vertical surfaces such as parapets, roof curbs, and walls with built-up flashing systems.

Table 26.6 Physical Properties of Roofing Asphalts*

Property	Type I		Type II		Type III		Type IV	
	Min.	Max.	Min.	Max.	Min.	Max.	Min.	Max.
Softening point, °C (°F)	57 (135)	66 (151)	70 (158)	80 (176)	85 (185)	96 (205)	99 (210)	107 (225)
Flash point, °C (°F)	225 (437)	—	225 (437)	—	225 (437)	—	225 (437)	—
Penetration, units:								
at 0°C (32°F)	3	—	6	—	6	—	6	—
at 25°C (77°F)	18	60	18	40	15	35	12	25
at 46°C (115°F)	90	180	—	100	—	90	—	75
Ductility at 25°C (77°F), cm	10.0	—	3.0	—	2.5	—	1.5	—
Solubility in trichloroethylene, %	99	—	99	—	99	—	99	—

Slope guidelines: On the slope at which a specific asphalt must be used, the following guidelines are provided:

Type I includes asphalts that are relatively susceptible to flow at roof temperatures with good adhesive and ''self-healing'' properties. They are generally used in slag- or gravel-surfaced roofs on inclines up to 4.17% (½ in./ft) slope.

Type II includes asphalts that are moderately susceptible to flow at roof temperatures. They are generally for use in built-up roof construction on inclines from approximately 4.17% (½ in./ft) slope to 12.5% (1½ in./ft) slope.

Type III includes asphalts that are relatively nonsusceptible to flow at roof temperatures for use in the construction of built-up roof construction on inclines from approximately 8.3% (1 in./ft) slope to 25% (3 in./ft) slope.

Type IV includes asphalts that are generally nonsusceptible to flow at roof temperatures for use in the construction of built-up roofing on inclines from approximately 16.7% (2 in./ft) slope to 50% (6 in./ft) slope. These asphalts may be useful in areas where relatively high year-round temperatures are experienced.

Built-up flashing is constructed of alternate layers of special felts such as asbestos felt, glass fiber reinforced asbestos felt, and roofing cement. A 45° cant strip is installed in the right angle formed by vertical and horizontal surfaces before the flashing is installed. The cant strip may be wood or fiberboard, and it must be firmly nailed or adhered to the horizontal and vertical surfaces. This construction is necessary because the felts cannot bend 90° without cracking, and because it is essential to direct surface water away from the roofing/flashing junction. A tapered edge strip of fiberboard (1½ × 18 in.) (38 × 457 mm) is also used below the cant strip to provide slope away from the flashing.

The flashing may also be installed with Type III (steep) asphalt, but roofing cement is more effective because it produces better adhesion in all weather conditions, assures full contact between sheets, and provides more cushion below the felts. Not all reinforced flashing sheets are compatible with roofing cement solvents; verify this with the sheet manufacturer. If the flashing is applied to concrete or masonry walls, the surfaces are primed before applying the roofing cement. Nail the flashing with capped nails or nails driven through metal washers, along its top edge to prevent it from sliding or falling down. A wood nailer should be provided no less than 8 or more than 12 in. (200 to 300 mm) above the cant strip. Higher flashings should be considered in areas of deep snowfall, but special precautions must be taken to prevent the sheets from sliding (use two-tier construction) and to seal top edges and other roof openings such as at ventilating units. Figure 26.5 illustrates a typical built-up base flashing installation.

Recently, Owens-Corning Fiberglas has marketed glass fiber based felts and fabric for flashing. These are currently recommended for use only with the Perma Ply-R roofing systems.

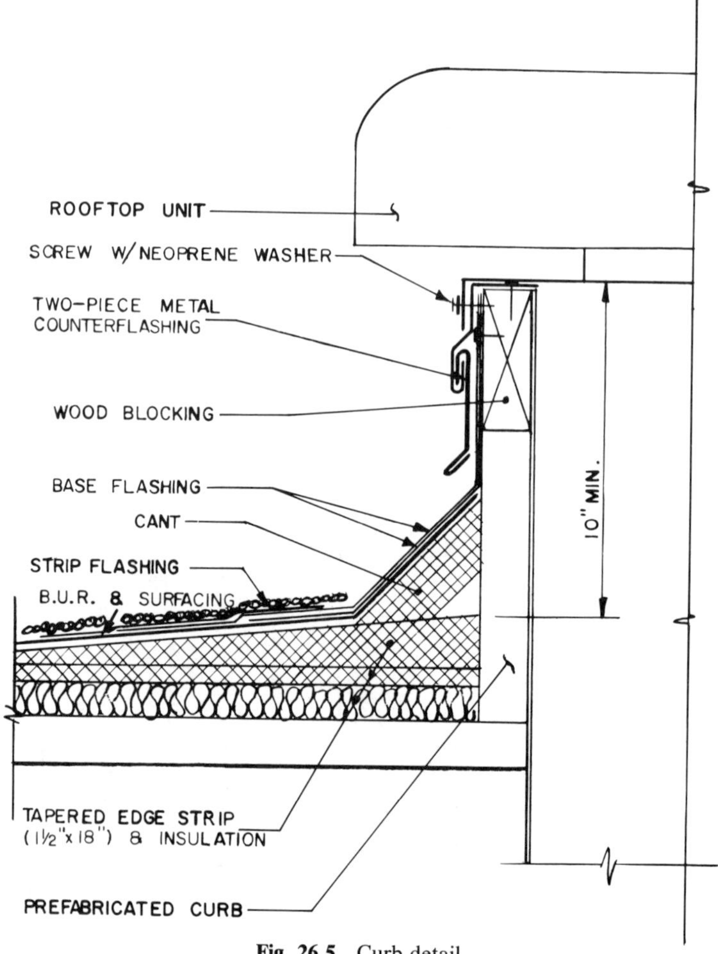

Fig. 26.5 Curb detail.

26.6.5 Surfacings

Built-up roof membranes perform significantly better when protected by aggregate surfacing than when smooth surfaced. Some surfacing is required by most manufacturers for all built-up roofs. Gravel surfacing has four basic functions:

1. Provides ballast against wind uplift.
2. Increases fire resistance capability.
3. Improves weathering characteristics by shielding the membrane from environmental exposure, particularly solar radiation.
4. Protects the roof surface from casual foot traffic.

Some manufacturers of glass fiber felts do not require the use of any surfacing, but the authors recommend it for these roofs as well.

Aggregate surfacing is predominantly commercially graded gravel from ⅜ to ½ in. (10 to 13 mm) diameter. Gravel should be without sharp edges; it should be clean and free of fines and of water (it may be applied damp but not wet). In many areas of the United States only crushed stone is available. If this material must be used, it should be applied with care, so as not to result in damage to the felts during construction and maintenance traffic.

Other aggregates include white marble chips, blast furnace slag (gray), granite chips, and scoria. The marble and granite chips have particles with sharp corners and edges, and roof-top traffic can force the particles through the flood coat into the felts. They should be used with caution. Some aggregates, such as scoria (slag), are too lightweight and friable, and wind erosion readily displaces the aggregate when it is not properly embedded, leaving bare asphalt or felts.

The aggregate is broadcast into a poured coat of 60 lb/100 sq ft of hot asphalt, or 75 lb/100 sq ft of coal-tar pitch. The usual quantity of stone aggregate applied is 400 lb/100 sq ft. Lighter-weight materials such as slag, or scoria, require only 300 lb/100 sq ft for surface coverage due to their large bulk per unit weight; white roof surfacing, if specified, requires at least 500 lb/100 sq ft to cover the black bitumen effectively. Expect about 250 lb/100 sq ft to be firmly embedded in the pour coat; even then, exposed areas such as corners may be subject to wind erosion and require periodic resurfacing.

Excess gravel (over 400 lb/100 sq ft) should be carefully removed from the roof surface and the remaining gravel raked smooth.

When localized depressed roof areas result in ponding, additional gravel and pour coats are sometimes used to displace the water. This is effective only in very small, shallow pond areas and it is not a recommended method for general use.

Perma-Ply R membranes can be protected by mineral surfaced felts called cap sheets. They have not performed well on low slope roofs. Several other reinforced "modified" asphaltic cap sheets have recently appeared on the market; while their performance to date is promising, their long-term performance is unproven in the field.

26.6.6 Standard Specifications for Membranes

Built-up roof membranes should consist of four plies of felts. More do not seem to be necessary based on field experience, and fewer may not provide the strength needed for membranes to perform satisfactorily. The first ply should be a coated base sheet, followed by three plies of Type I asphalt-saturated organic felt. Principal manufacturers of roofing felts publish detailed recommendations for installation of built-up roofing systems. In many cases, the availability of a manufacturer's guarantee or warranty depends on the designer using a published specification. However, a specification should not be used merely because a guarantee is available, particularly if it contains less than four felt plies or if gravel surfacing is not required. The designer can usually upgrade the specification and still qualify for a manufacturer's guarantee, but this should be verified in advance.

Built-up roofing systems specifications are generally classified by substrate (nailable or non-nailable) and by slope. Some manufacturers also list their specifications by geographical areas of the United States: Central and Eastern, Southern, and Far Western. The designer should select a traditional system with proven components and a record of successful field performance of the components in combination with one another, used over a suitable deck and insulation boards.

26.6.7 Flashing Details

Designers must provide details for roofing construction at all roofing terminations, interruptions, and penetrations. The waterproofing systems constructed for this purpose are called "flashings". Most roof leaks occur at improperly designed or built flashings, or at locations where flashings should have been provided. Ideally, roof interruptions and penetrations should be avoided. Terminations at perimeters should be raised slightly above roof level. Flashings (except roof drains) should always be elevated from the roof surface to assure positive drainage from the transition between the flashing and the roof system, and to keep standing water away from the transverse joints of the flashing materials.

A flashing system should be provided at every membrane interruption including, but not limited to:

1. Curbs for mechanical equipment, skylights, and access hatches.
2. Walls of intersecting structures, penthouses, parapets, and changes in building (roof) elevation.
3. Differences in roof deck slope (e.g., 3-in-12, and 4-in-12).
4. Pipe penetrations for sanitary vents, conduits, equipment supports, and flues.
5. Special use penetrations such as lightning protection and antenna equipment.
6. Roof relief or control joints and building expansion joints.
7. Roof drains.

The design of every roof flashing should be reduced to one of the following categories:

1. Curb (including curbed expansion joints), parapets, and equipment curbs.
2. Flange-on-roof (including roof edge flashing, low expansion joint covers, and pipe sleeves).
3. Drains.

For built-up roofing systems, vertical planes such as roof curbs at equipment supports and expansion joints, access scuttles, penthouses, and rising walls, should be flashed with bituminous

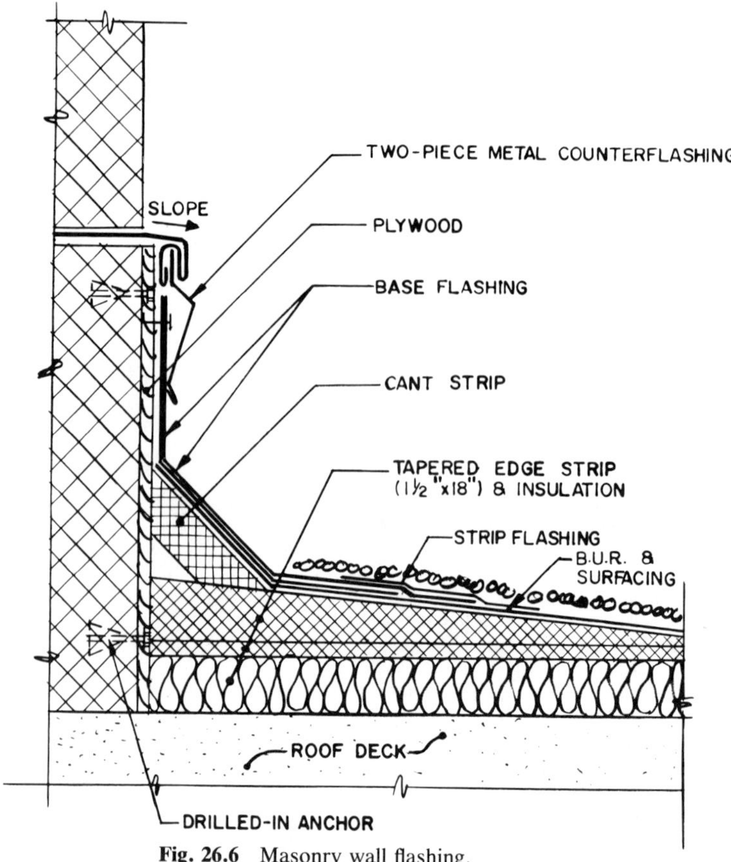

Fig. 26.6 Masonry wall flashing.

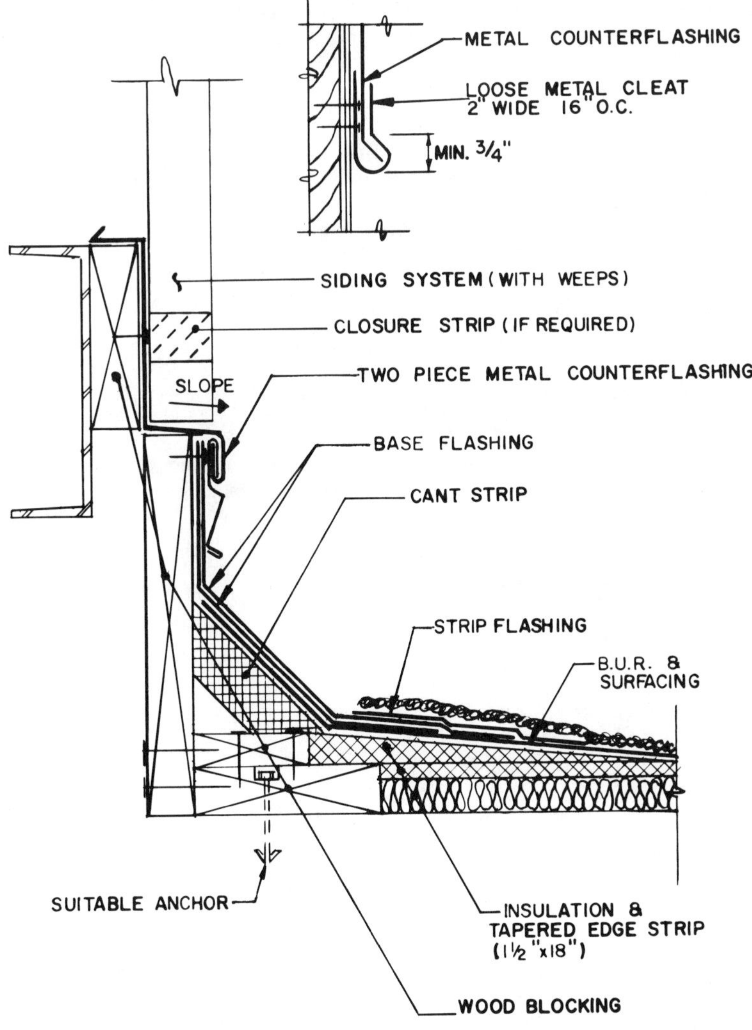

Fig. 26.7 Wall flashing at penthouse.

built-up base flashings, as described in Section 26.6.4. For polymeric membranes, proprietary materials are provided by the basic sheet manufacturer, but most of the same design principles described below apply with polymeric systems. Do not connect polymeric materials directly to bituminous materials.

All base flashings must have separate counterflashings to protect their upper edges and to shed water draining across the base flashing surfaces from above. Counterflashing can be a separate metal or polymeric flashing, an extension of a rising wall flashing, or an apron that is an integral part of the roof top equipment housing.

Building finishes such as window sills, siding, or stucco should not be used for counterflashing. The counterflashing should be able to be independently removed and replaced when necessary for replacement and repair of the roofing and base flashing. Typical details are shown in Figures 26.5, 26.6, and 26.7 for curbs, walls, and penthouses. Also, window and door sills must be raised above the roof surface to provide adequate flashing height.

It would be ideal if all roof-top equipment could be enclosed in penthouses; the equipment can then be maintained without exposing the roof to traffic damage. Some items (such as solar collectors) cannot be enclosed. These items should be easily accessible for maintenance. Roof-top equipment should have clearance between the roof surface and the equipment sufficient for membrane and flashing application and for inspection and maintenance of the roofing components.

Mechanical and electrical services that must penetrate the roof system should be located within the unit's curbed support, or far enough from the flashing system to provide adequate space for independent flashing of the service penetration. Generally, flashing cannot be applied to pipe insulation, but the bare pipe must be flashed and insulation overlaid.

Piping should have minimum length roof-top runs. Long pipe runs expand and contract significantly, making movable roller pipe supports necessary. A fixed support for the roller must be anchored to the structural system and be properly flashed.

Some machinery needs isolation fixtures to prevent vibrations from damaging the flashings. Anchor bolts must be sealed at penetrations and should be placed on vertical faces only.

A convenient method of supporting roof-top equipment is with a curb that encloses the equipment base; this eliminates the need for post-like supports and reduces the height needed for working clearance below post-supported units. The curb must be constructed as a watertight box, and the top should slope to shed water. Services can be run through the roof inside the curb to the unit above. Verify that totally enclosed roof-top units advertised as such are, in fact, permanently weathertight, that they do not pond water on top, and that their housings have perimeter skirts to serve as counterflashing to protect the built-up base flashings on the curbs. Provide additional metal counterflashing to protect the curb base flashing during removal of units during routine maintenance (Fig. 26.5), and additional flashing extensions to permit roof work if a unit cannot be lifted.

Avoid the use of metal base flashing. Experience shows that the differential movement caused by thermal cycling debonds the metal from the bituminous membrane. The different coefficients of thermal expansion of the two materials also lead to membrane splitting at the edges and transverse joints at the metal flashing. Metal base flashings can be successful for small penetrations; they should be well anchored to wood nailers, and raised from the roof surface (for drainage) with tapered blocking.

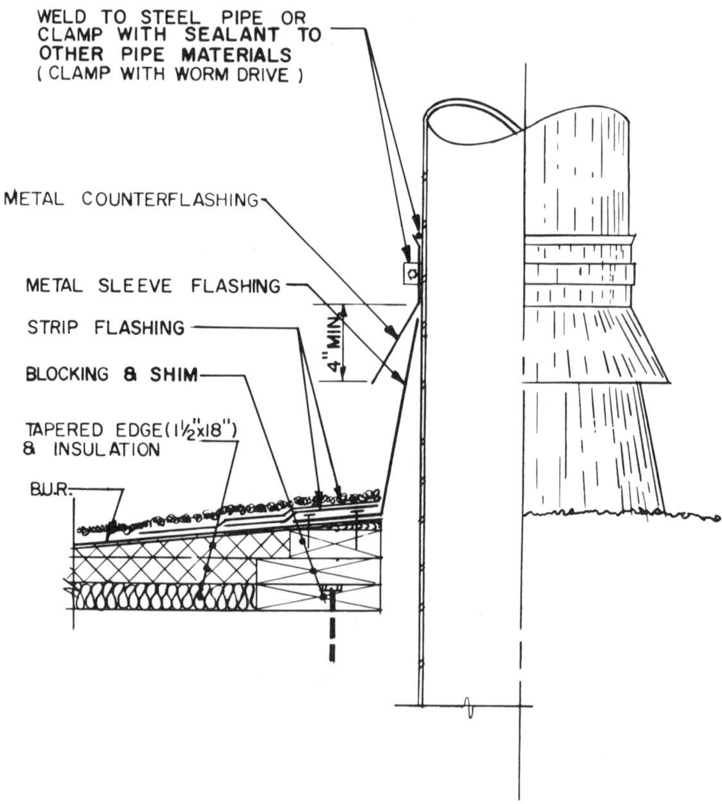

Fig. 26.8 Pipe flashing.

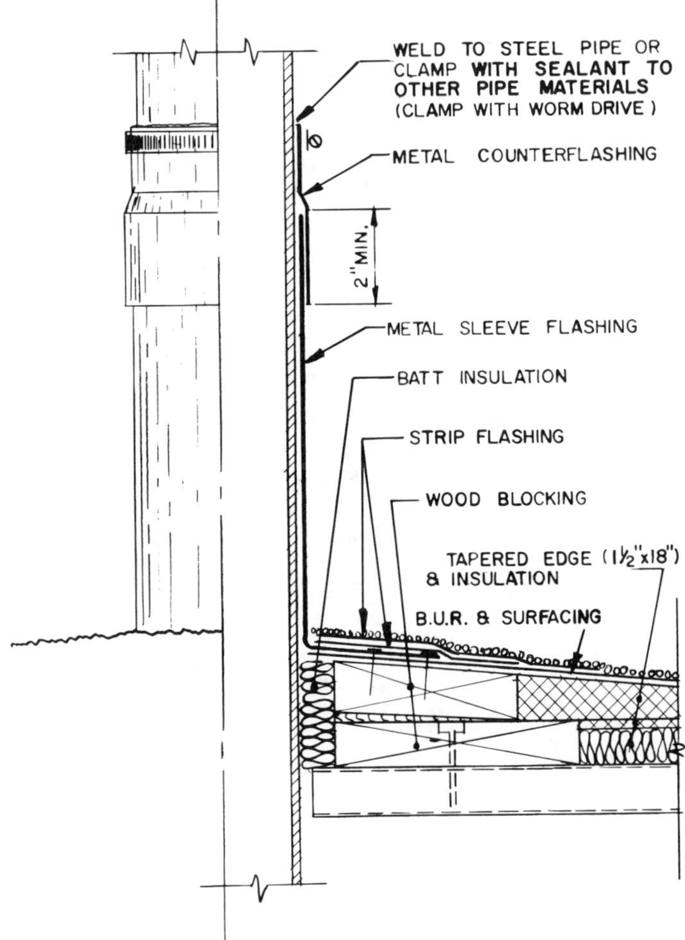

Fig. 26.9 Pipe flashing.

Pipe penetrations are an exception to this rule, as bituminous flashings cannot be formed around the pipes. Pipes, flues, vents, and pipe supports for equipment are flashed with metal tubes soldered to a base flange that is nailed securely to blocking anchored to the roof deck, and integrated into the roofing system with strippings of reinforced flashing felts and roofing cement. The relatively small size of this metal base prevents the magnitude of the thermal movements that would create a problem. The flashing tube is not connected to the pipe; a separate counterflashing, clamped to the pipe, overlaps the base flashing. Figures 26.8 and 26.9 illustrate typical flashings of this type. A special case is flashing of a sanitary vent stack. Unless it is over 2 ft (600 mm) high, the stack is enclosed in a metal tube with flange as described above, but the separate counterflashing turns into the pipe (Fig. 26.10). In areas of exposure to vandalism, sanitary pipes may be fitted with screened tops. The standard pipe flashing detail can simplify an endless number of different conditions by utilizing a pipe column support and special fittings for an antenna lead-in, lightning arrester cable, or guy wire.

Complicated or special shaped penetrations such as multiple pipes or I-beams are sometimes flashed with "pitch-pockets". These are open-top metal boxes filled with bitumen or roofing cement. The filler material is retained by the metal box and forms a seal around the penetration. This type of "flashing" is unreliable because the seal is not permanent and frequent maintenance of the filler is needed. Pitch pockets should not be used. All penetrations should be constructed so their flashing details can be adapted from curb or pipe flashing details. Complicated penetrations or unusual shapes should be modified to fit these design standards.

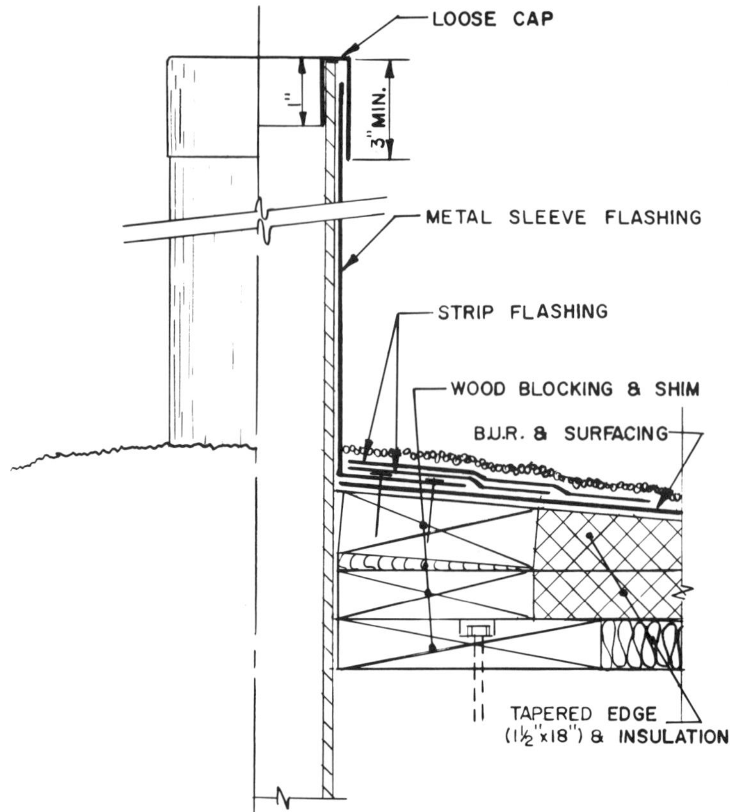

Fig. 26.10 Sanitary vent pipe flashing.

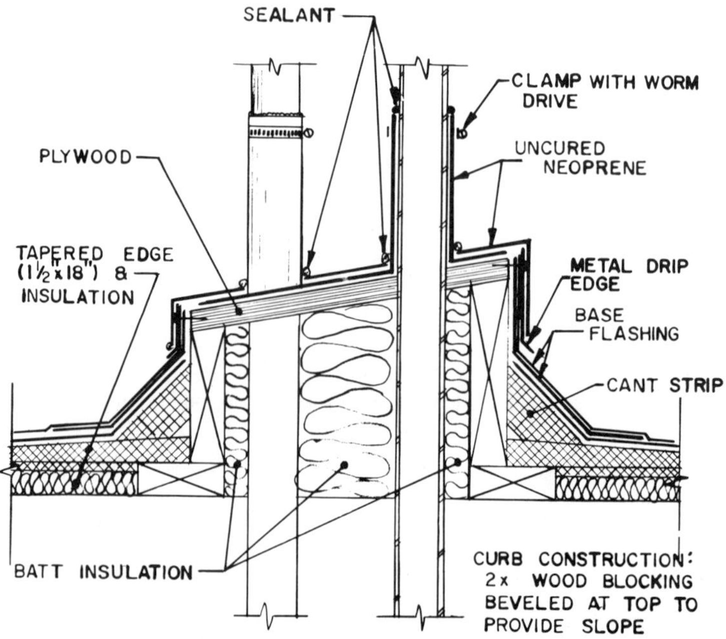

Fig. 26.11 Multiple pipe penetration.

Very small conduits [1 to 1½ in. (25 to 37 mm) diameter] may be counterflashed with polymeric materials, because of the difficulty in forming metals to the small diameters. This design is shown in Figs. 26.11 and 26.12.

Exceptions to these rules are hot pipes such as from heater flues, antenna cables, lightning rod cables, and glass or plastic piping. Special fittings are available for some of these such as antenna lead wires and flues; others will require imaginative detailing, but these same principles discussed above should be followed. Figures 26.13 and 26.14 illustrate some typical solutions.

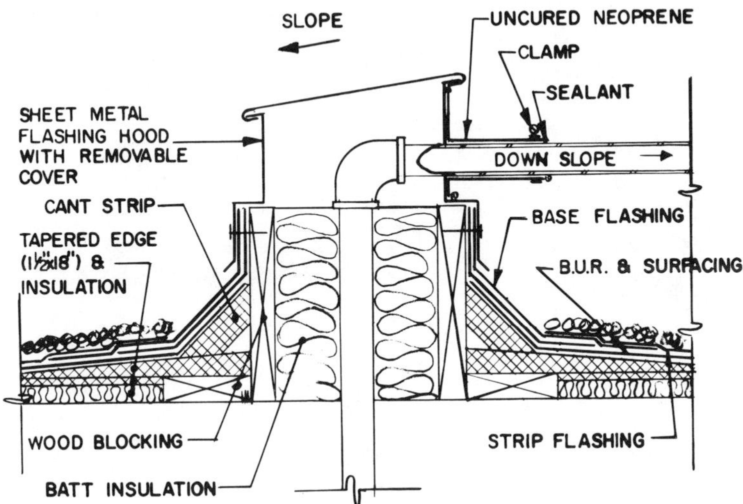

Fig. 26.12 Pipe penetration.

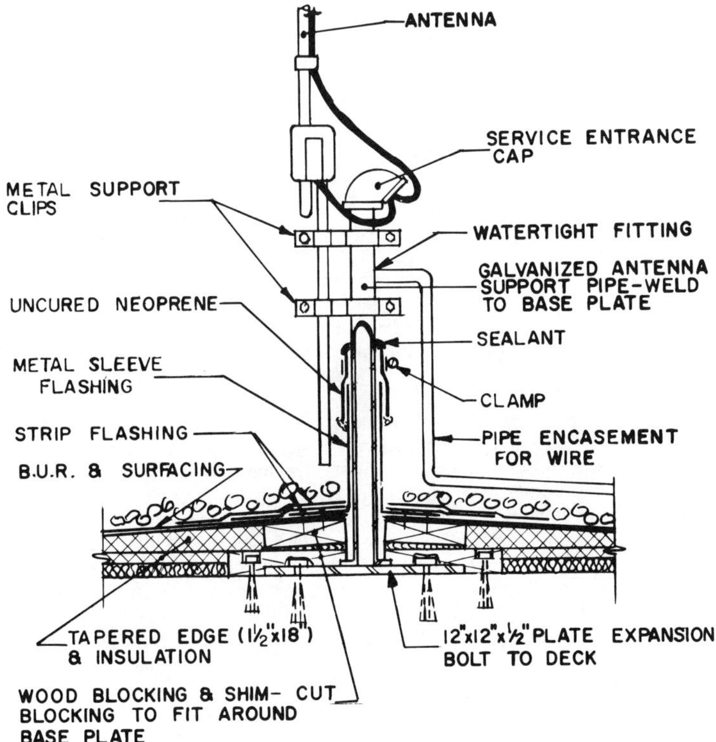

Fig. 26.13 Antenna support.

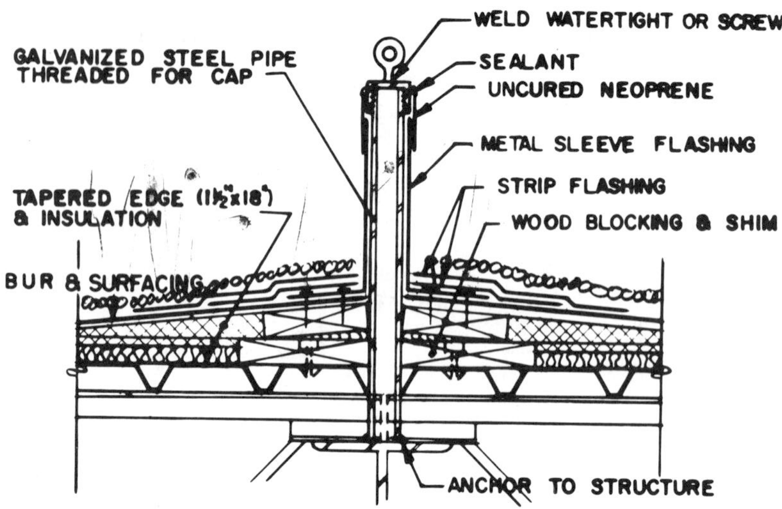

Fig. 26.14 Lead wire detail.

Roof expansion and control joints can be flashed with prefabricated covers of metal-flanged neoprene bellows. The neoprene portion is insulated on its underside by factory-applied foam which prevents condensation and helps to maintain a convex profile for the bellows. The cover may be raised on a curb so that the metal flange serves as counterflashing over the curb's built-up base flashing. Preferably, the cover is installed on raised, tapered blocking similar to perimeter metal flashing. Figures 26.15 and 26.16 illustrate these two methods. The curb type removes the metal flange from the built-up roof system, but exposed fasteners are required at the curb. The roof level design eliminates the exposed nails, but introduces the metal flange into the roofing. The low profile method reduces the height of the assembly and minimizes the amount of water that can accumulate in an isolated roof area in case of drain blockage.

If the expansion joint is at an intersection with a vertical surface, free movement in two dimensions is required. This can be accomplished as shown in Fig. 26.17.

Roof edge flashing ("gravel stop") is generally L-shaped sheet metal, with the horizontal leg (the flange) set on top of the membrane in roofing cement, nailed and cover-stripped, integrating the flashing into the built-up roofing system. The vertical leg is turned down over the building wall as a closure and a fascia, visible on the exterior. The metal is formed in an upward vee at the corner to act as an edge stop, and the horizontal flange is securely nailed to roof blocking that is supported by the deck system. The exterior (lower) edge is formed into a 45° drip edge and hooked and clamped securely over a continuous metal locking strip. There are no exposed fasteners and the metal is free to move along its face. Transverse joints of the metal flashings are loose locked and sealant filled. Joints occur at 8- or 10-ft (2.4 or 3 m) intervals depending on the stock used to fabricate the flashings. Transverse joint spacing should not exceed 10 ft (3 m). This detail is shown in Fig. 26.18. Nails for anchorage are annular thread to prevent nail backout; they should be placed 4 in. (100 mm) on center on the horizontal flange, in two staggered rows, to minimize the effect of the transverse thermal movement of the metal. Inadequately nailed metal flanges result in broken transverse joints and debonding of the strip flashing system from the flange.

An alternate method is shown in Fig. 26.19 This consists of a low perimeter cant covered by the metal. This method has the advantage of allowing the roofing to be completed without installation of the metal; however, the cant flashing does not provide a positive metal to roofing seal and is not a true built-up base flashing system. As such, it is less secure than the method described previously.

All perimeter flashing should be continuous over the edge of the building. Attempts to stop the metal short of the outside of the perimeter by use of reglets and sealants on top of concrete or stone copings are always unsuccessful. Thermal movements of the metal invariably loosen the flashings from the reglet and cause the sealant joints to fail. Water can bypass the reglets at coping cracks and at control and expansion joints and the system inevitably fails.

Roof drains, of necessity, must be level with, or preferably lower than, the roof surface. The standard roof drain must have a clamping ring held in place with bolts to integrate the roofing and flashings with the drain to prevent leakage. Cement the roofing plies onto the drain body below the ring area; a lead sheet flashing is bedded in roof cement, over the plies, and the entire assembly is

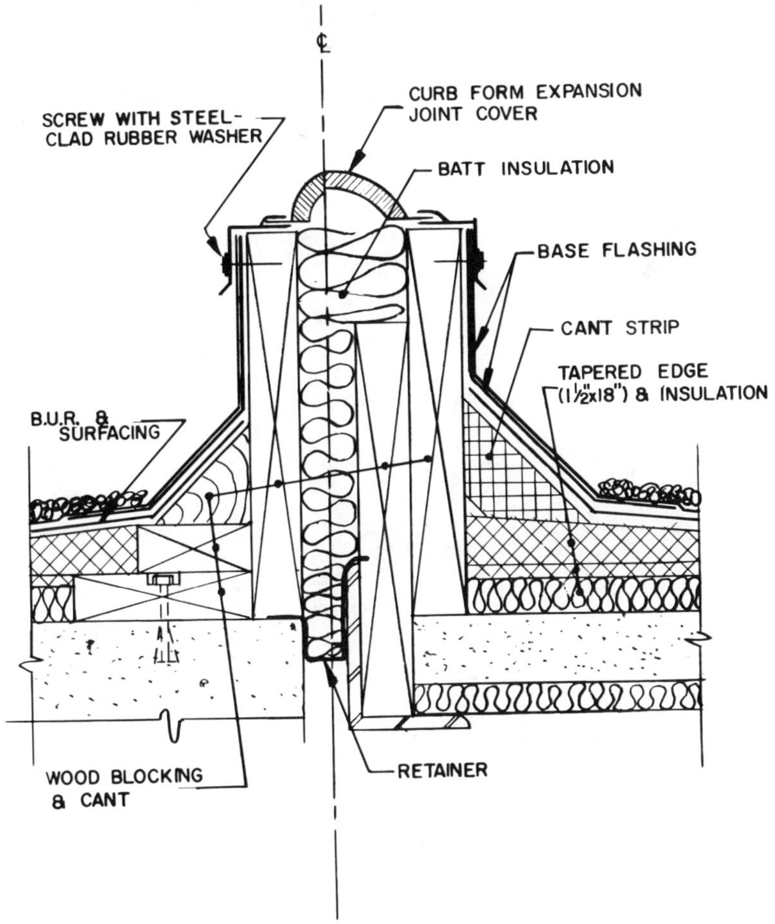

℄
CURB FORM EXPANSION JOINT COVER
SCREW WITH STEEL-CLAD RUBBER WASHER
BATT INSULATION
BASE FLASHING
CANT STRIP
TAPERED EDGE (1½"x18") & INSULATION
B.U.R. & SURFACING
WOOD BLOCKING & CANT
RETAINER

Fig. 26.15 Expansion joint curb.

METAL FLANGED NEOPRENE EXPANSION JOINT COVER
STRIP FLASHING
BATT INSULATION
B.U.R. & SURFACING
RETAINER
SHIM & WOOD BLOCKING
TAPERED EDGE STRIP (1½"x18") & INSULATION

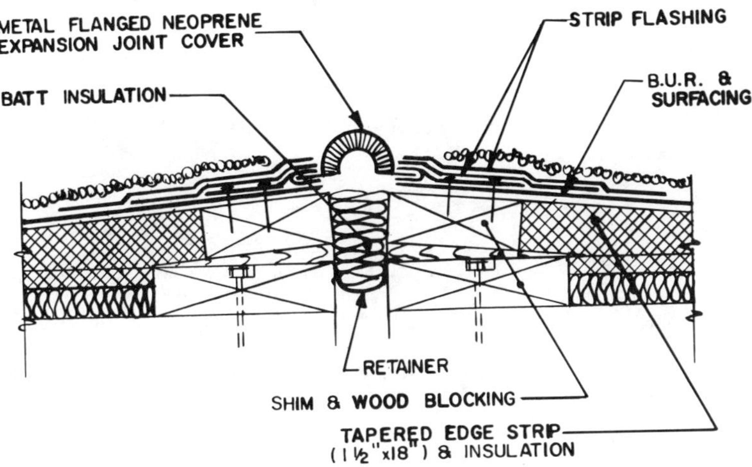

Fig. 26.16 Expansion joint.

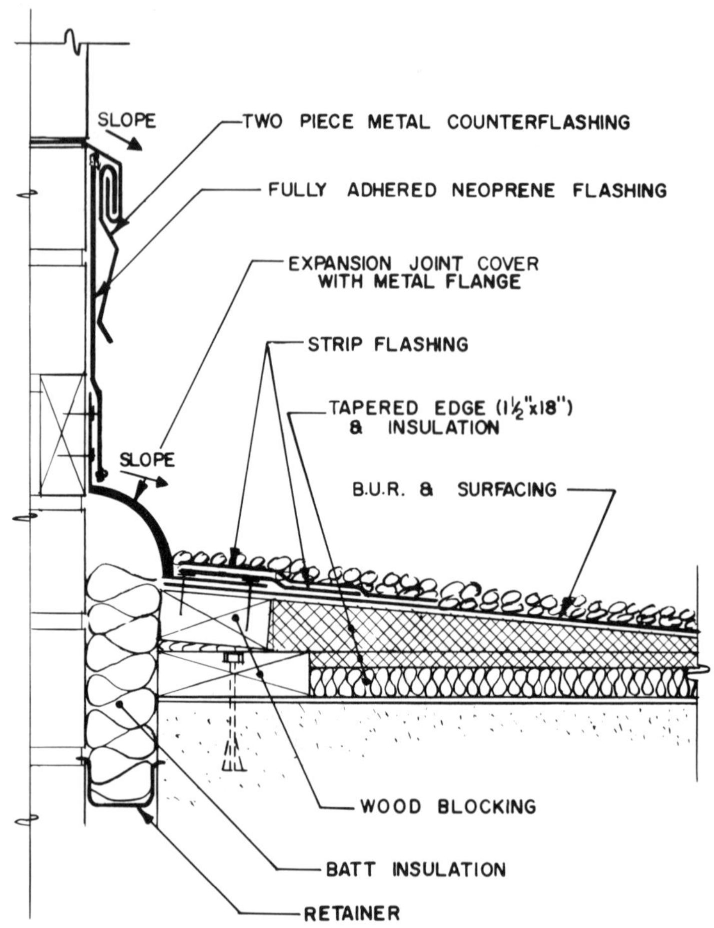

Fig. 26.17 Expansion joint at wall.

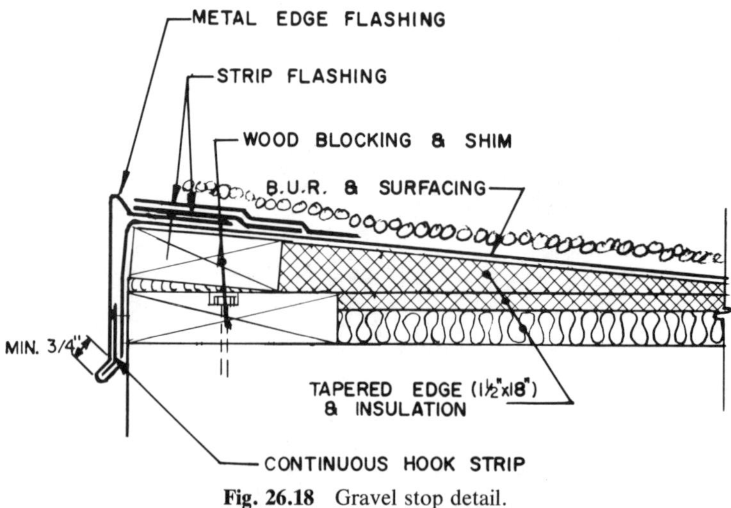

Fig. 26.18 Gravel stop detail.

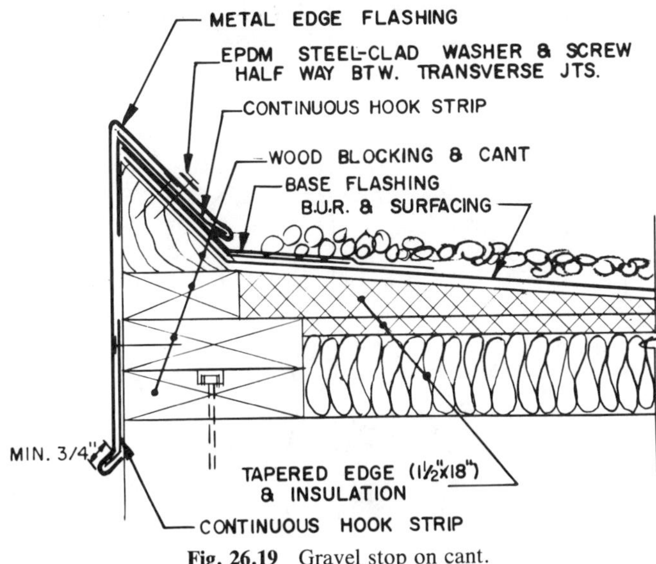

Fig. 26.19 Gravel stop on cant.

clamped into place. All clamping ring bolts must be firmly secured. Felt stripping finishes the detail. A typical drain flashing is shown in Fig. 26.20. The drain detail is the most critical one in any built-up roofing system because it receives the most exposure to water. Impeccable workmanship is required both in the flashing and in the installation of the roof drain itself.

26.6.8 Metal Flashing Materials

Metals should not be included within roofing systems except in unavoidable circumstances. Connections between sheet metal used as flashing and bituminous roofing materials are often the cause of roof leaks because the different thermal movements between the metal and the bituminous roofing materials result in breaks at the transverse joints and in the strip flashings.

The use of sheet metals integrated within a built-up roofing system should be restricted to edge flashings, pipe sleeve flashings, expansion joint covers, and drain flashings. Sheet metal is very effective when used for cap or counterflashing to overlap base flashings, protecting their upper edge from mechanical damage and water penetration. Details for these conditions are shown on preceding pages.

Metals for flashing should have the following characteristics:

1. Low coefficient of thermal movement.
2. Easily formed by standard sheet metal tools.
3. Simply joined in the field.
4. Resistant to corrosion, embrittlement, and fatigue cracking.
5. Compatible with other roofing materials.
6. Moderate in cost.
7. Readily obtained from suppliers.
8. Aesthetically pleasing or able to accept protective and decorative finishes.

Sheet metals that exhibit these characteristics and are commonly used as flashings for roofing include copper, lead, stainless steel, galvanized steel, and aluminum. Their physical characteristics are given in Table 26.7.

To prevent electrochemical corrosion, different metals should not be installed in contact with each other. The corrosion resistant properties of dissimilar metals in contact depend on their relative position on a scale of the galvanic series which is shown in Table 26.8. Metal flashings of different materials that must overlap should always be separated from each other by a suitable

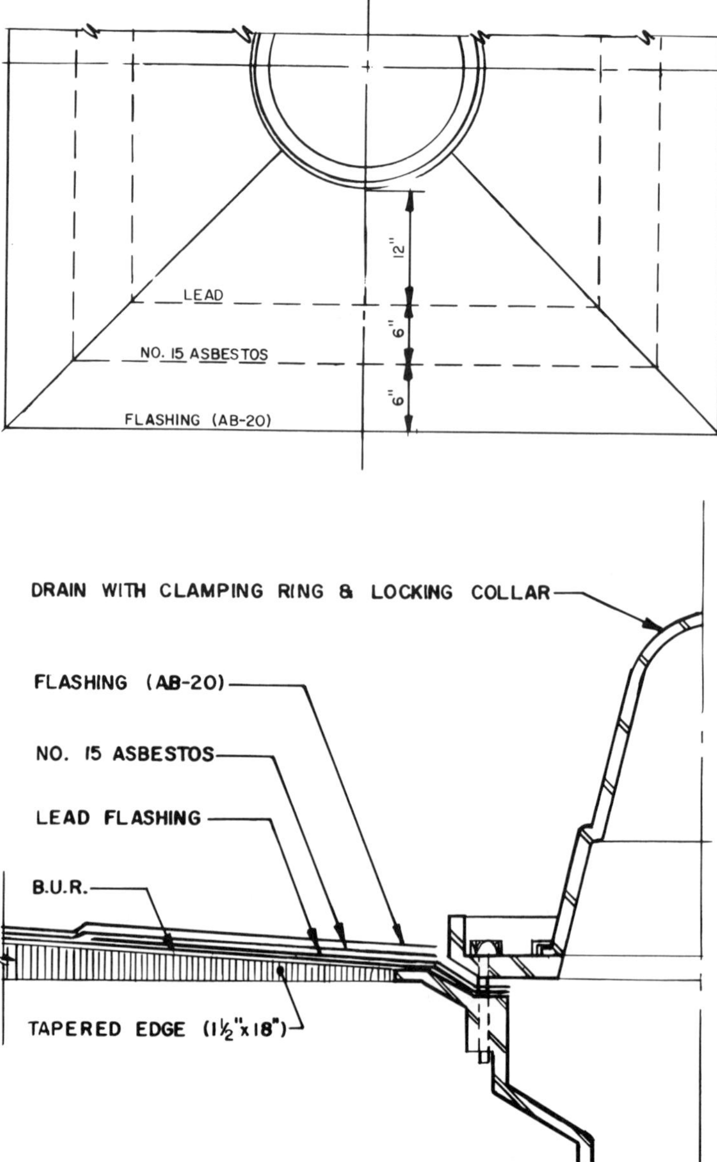

Fig. 26.20 Drain flashing.

coating or layer of roofing felt. Sheet metal flashing should not be placed so that water drains from a metal of higher position on the series over one of lower rank. The further apart the metals are in the series, the greater the corrosion potential.

Because the commodity-type raw materials for sheet metal manufacture are subject to wide price fluctuations, material cost comparisons are not included here.

The most versatile and suitable metal for roof flashing is copper. Copper is easy to fabricate and it is easily soldered, providing a strong, watertight connection readily made in the field. Copper is available lead-coated (standards for coating are ASTM B101, see Ref. 26.6) for soft gray color and because water running over uncoated copper will stain building elements below. The lead coating

Table 26.7 Properties of Metals

Materials	Coefficient of Thermal Expansion, $\times 10^{-6}$/°F	Modulus of Elasticity, 10^6 psi	Tensile Yield Stress at 0.5% Offset, psi	Tensile Strength, psi	Ultimate Tensile Elongation, %	Ductility in Cold Temperature	Corrosion Resistance	Reflection of Radiant Heat	Embrittlement at °F	Field Joints	Comments
Copper	9.4	17	28,000	36,000	30	Good	Good	Fair		Excellent	50/50 solder
Lead-coated copper	9.4	17	28,000	36,000	30	Good	Excellent	Fair		Fair	Unless coating is removed before soldering
Galvanized steel	6.5	29	35,000+	50,000	33	Good	Poor	Fair		Good	50/50 solder
Aluminum-coated steel	6.5	29	35,000+	50,000	30	Good	Poor	Fair		Poor	Cannot solder
Aluminum	13	10	21,000	22,000	8	Good	Good	Fair		Poor	Cannot solder
Stainless steel	9.6	28	42,000	85,000	60	Excellent	Excellent	Good		Fair	60/40 solder
Terne-coated stainless	9.6	28	42,000	85,000	60	Excellent	Excellent	Good		Good	50/50 solder
Monel	8	26	25,000+	70,000	40	Excellent	Excellent	Good		Poor	—
Lead	16	2	1600	4100	43*	Fair	Good	Fair		Excellent	50/50 solder
Terne metal	6.5	29*	33,000	45,000	29		Poor	NA		Good	Must be painted
Zinc	18–20	?	(low)	20,000	42*	Poor	Good	Fair	+45	Good	50/50 solder
Zinc alloy (Zn-Cu-Ti)	11–14	?	25,000	32,000	14–42	Fair	Good	Fair		Good	50/50 solder

* These materials creep under loading.

Table 26.8 Activity Series or Electromotive Force Series (Abbreviated List of Metals)

Metal	Single Electrode Potential	
Lithium	+3.02	Less noble
Potassium	+2.92	
Sodium	+2.71	
Magnesium	+2.34	
Aluminum	+1.67	
Zinc	+0.76	
Chromium	+0.71	
Iron	+0.44	
Cadmium	+0.40	
Nickel	+0.25	
Tin	+0.135	
Lead	+0.125	
Hydrogen	0	
Copper	−0.345	
Mercury	−0.80	
Silver	−0.80	
Platinum	−1.2	
Gold	−1.42	More noble

On the basis of the rate of displacement of hydrogen from hydrochloric acid, metals may be arranged in an activity series. For those metals not affected by hydrochloric acid, other reactions are used. If pieces of two different clean metals are electrically connected and placed in an electrolyte the more active one (higher electrode potential) tends to go into the solution and the less active one tends to be protected from the solution; e.g., zinc protects iron in galvanized sheet. Further any metal will displace from its salt in water solution a metal lying below it in the series. Thus iron will displace copper from a copper salt solution; e.g., an iron nail in a solution of copper sulphate is rapidly coated with copper.

should be removed at joints before soldering to ensure a properly ''sweated'' joint. Both copper and lead coating can be effectively painted. Lead-coated copper should not be used for hook strips as the coating restricts movements of the metal.

Stainless steel is preferred for flashings for its superior strength, stiffness, and moderate coefficient of expansion. Drawbacks are that it is more difficult to fabricate than copper, and while currently less expensive it has minimum scrap value. It is the only metal sure not to cause any streaking or staining on other materials and it is the most resistant material against mechanical erosion. Use dead-soft stainless steel, 2-D dull finish. Stainless steel sheets are also manufactured with ''Terne'' coating. This material uses a lower grade stainless steel and a coating of tin/lead. It has performed well in roofing and flashing sheet installations.

Galvanized steel is used only in areas not subject to corrosive, marine, or hot-humid atmospheres. This material is moderately easy to form and solder, readily painted (if properly primed), and relatively inexpensive. Its life expectancy is less than for copper or stainless steel. It has the important advantage of a significant lower thermal coefficient of expansion than many other flashing metals.

Aluminum is inexpensive, corrosion resistant, and easily color-coated and formed. It cannot be field soldered. Aluminum is relatively weak and has a high coefficient of thermal expansion. Aluminum sheets often develop fatigue cracks around fasteners. As such, it is not recommended for use in roofing or flashing. If aluminum must be used, pay particular attention to selection of the correct alloy and temper; only certain alloys are suitable for forming (bending) without cracking, for particular corrosion resistance, and for color anodizing.

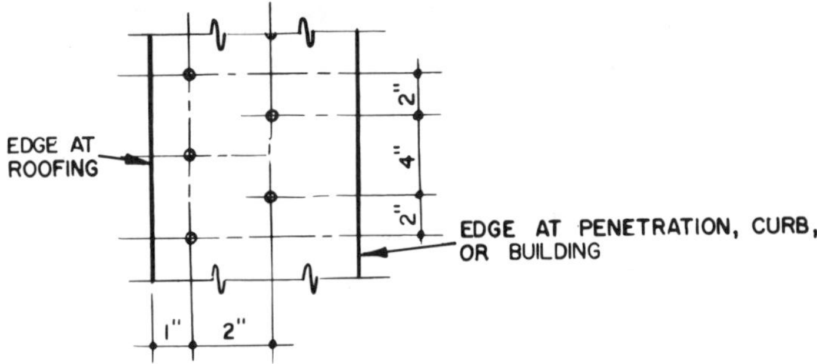

Fig. 26.21 Nailing pattern for metal flanges.

Lead is very easily fabricated but it is too soft for most flashing. Lead creeps in time and may fail by this mode if improperly used. It can be readily formed in drains and provides excellent reinforcing for the roofing felt plies and satisfactory gasketing action in the drain clamping ring. Special alloys are required to minimize creep problems. In substantial lead flashing installations, lead joining should be done by specially trained "lead burners".

All the metal flashing (except lead) must be securely anchored to treated wood blocking, which is in turn anchored to the structure. Use annular-thread nails of the same materials as the metal flashing (except use stainless steel nails for aluminum). The horizontal flanges should be about 4 in. (100 mm) wide, and secured with two rows of nails, staggered (Fig. 26.21) to minimize the effect of thermal movement, which may produce buckling, transverse joint failure, and destruction of the bond between the metal and the built-up roofing. The flange should not extend more than ½ in. beyond the nails.

Set the metal flange in a bed of roofing cement and nail to wood blocking. Strip flash the flange to the roofing with a two-ply felt flashing system of Type I asbestos felt and reinforced flashing sheet. The flashing is raised from the roofing surface and sloped to shed water as shown in the recommended details. Thicknesses of metals are usually in the 0.015–0.040 in. (0.4–1 mm) range depending on the use, shape, exposed dimension, and producer's recommendations.

Excellent design manuals are produced by the manufacturers of copper, stainless steel, aluminum, and by the Sheet Metal and Air Conditioning Contractors National Association. Even these details must be used with caution as they are very general and may not always show the most reliable methods. Factory Mutual Bulletin I-52 describes standards of design for various perimeter conditions as related to wind damage prevention, and prudent designers should follow the recommendations contained therein.

Occasionally it is necessary to combine built-up base flashing with a metal coping cover (such as at parapets), as shown in Fig. 26.22. The coping should be fully covered and the metal fully supported and sloped to drain. Continuous hook strips are used for concealed anchorage on both sides of the parapet. Transverse joints can be made in several ways, but allowance must be made for the thermal movements of the metal. A properly made standing seam joint accomplishes this effectively. Provide expansion joints adjacent to all building corners where maximum movements are expected.

Properly designed metal flashing details use a minimum of soldered joints. Special attention is required to details at flashing terminations such as at walls above the roof line and at intersections with other flashing systems. These conditions are frequently leakage points and they must not be neglected in roof design.

Metal counterflashings should always be two-piece assemblies. This allows other construction trades to work independently and allows for convenient repair or replacement of roofing components. The upper element should be built into the wall or curb construction and firmly anchored. Surface-mounted flashings with reglets or sealant-protected assemblies are unreliable and their use should be avoided. Built-in reglets are not recommended because it is difficult to install them securely, water often bypasses them, and metal usually becomes loose because of differential movements. If it is absolutely necessary to use reglets, the metal must be mechanically anchored and a properly designed sealant joint provided.

Sealants must be compatible with the metal and roofing materials. Follow the sealant manufacturer's directions for joint shape and size, primer, and backer material. All sealants, at best, should be considered hole fillers and not waterproofing. Sealants are not suitable as adhesives between sheet metal laps.

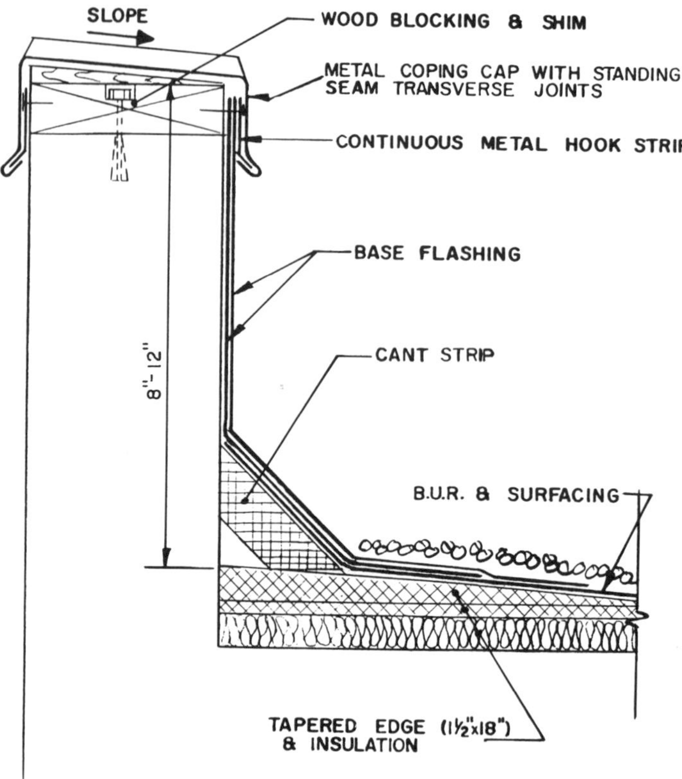

Fig. 26.22 Parapet flashing.

26.7. METAL ROOFING

26.7.1 Materials

Metal in sheet form is an ideal roofing material. Metal roof systems can be used as prominent architectural features of a building, or be purely functional, used solely as a weatherproofing system. Metal, when properly designed and installed, provides a relatively maintenance-free roof system that can provide reliable service for more than 20 years. When a corrosion-resistant metal is used, the life of the roof can easily exceed 50 years with proper maintenance to repair localized areas of wear.

The initial cost of installation of a metal roof is considerably more than for other roofing systems, but may be competitive when life-cycle costs are considered. Metal roofing systems and details vary somewhat from one manufacturer to another, but the designs must follow the basic principles and minimum standards as provided in the *Architectural Sheet Metal Manual* [26.5] by the Sheet Metal and Air Conditioning Contractors National Association, Inc. (SMACNA) or manuals of the various metal producers, or problems are likely to occur. The various metals used for roofing applications are described in the following sections.

26.7.2 Deck Supported Roofing

Deck supported roofing provides weatherproofing as well as architectural effect; it must be supported by a structural roof deck. The roof deck must resist all design loadings and must be adequate to hold the fasteners that secure the metal roofing. Design requirements and details for the various conventional metal roofing systems are provided in the architectural sheet metal manuals, but many of the details in these manuals are not fully in accordance with the recommendations in this chapter and should be used with caution. Specifications and special design requirements are available from the manufacturers of the particular roofing material. The following are general recommendations that should be followed when designing metal roof systems:

1. Use proven standard industry details.
2. Wood decks resist the loads from wind pressure. If an alternate deck system is used, the metal roofing must be designed to provide full support for the resistance of wind loads; full scale testing will be needed before construction.
3. Slope all roofs for rapid drainage, 3-in-12, minimum.
4. Use soldered, flat lock seam roofing where the slope is less than 3-in-12 but in any case not less than ½-in-12; provide expansion battens of a spacing not to exceed 30 ft (9 m) in perpendicular direction. This system uses small sheets that are installed to lift independently for minor thermal changes.
5. For decks other than wood, nailing strips must be provided to receive the fasteners and holding cleats.
6. The deck must be smooth, dry, and covered with a saturated roofing felt. A smooth building paper should be applied over the felt to separate the metal roofing from the roofing felt and allow it to move freely.
7. Use heavier gage sheets for valleys gutters, base flashings, cleats, and hook strips, as required by the design conditions, to prevent local buckling from metal expansion and to resist erosion in areas of increased water flow. At fixed joints riveting and soldering is necessary for heavier metals; where locked, soldered joints can suffice for lighter gages.

26.7.3 Deck Supported Systems: Provision for Thermal Movement

The components of the metal roofing must withstand thermal cycling without buckling. Relief joints must be provided that allow free movement of the components in place, or the component must be fabricated using sheet material of a sufficient gage to prevent buckling under the thermal stresses that will develop when the roofing is fixed. If the material is allowed to buckle, the metal will rapidly fatigue and crack, and the roof will leak. The buckling strength of a component is dependent on its moment of inertia and the modulus of elasticity of the metal. Material strength has no effect on buckling resistance. Design the metal roofing with ample provisions for free movement, thus eliminating the need for an engineering analysis for buckling strength requirements. The transverse joints must also be designed either to resist these movements or to function as expansion joints. Generally, longitudinal joints should be permitted to move unrestrained.

26.7.4 Structural Metal Roofing

Metal roofing systems are available in steel and aluminum serving both as the structural deck and the weatherproofing for the building. Systems are available that use corrugated sheets, standing seams, batten-type seams, or combinations of these systems; the roofing is secured with concealed clips, exposed fasteners, or both. Materials used are noncorrosive, galvanized steel, factory or field painted, or protected with other metallic factory coatings. Some systems use insulated structural sandwich panels.

This type of roofing spans between structural roof beams or purlins. Structural roof systems can be part of a complete building system or can be used for retrofitting existing buildings.

The following design principles provide the best chance for a successful structural metal roof:

1. Do not use structural roofing on a slope less than 3-in-12.
2. There is no limitation for maximum slope for metal roofing, but special precautions must be included when the exposure approaches vertical. The side connection details of many systems remain watertight provided the water does not travel "uphill" to enter the building at side laps in the sheets. Negative pressure within the building and wind pressures can suck or drive water through simple lap joints. The water can easily travel around these laps on walls or mansards, unless a compressible gasket is included to block water entry.
3. Do not use exposed fasteners in the roofing system. Thermal movement in the system will loosen the fasteners or deform the roofing sheet around the fasteners and cause leakage.
4. The method to secure the roof panels must allow for their thermal movements. The roofing should be secured with concealed anchors that restrict all movements to one location only: eave, ridge, or midpoint. All other means of securing the roofing must allow free lateral movement. This procedure will leave the roofing free to accommodate thermal movements.
5. A structural metal roof system that is free to move and accommodate thermal movements will not provide lateral stability to the structural framing.

The structural metal roofing is insulated below the deck. The usual practice is to roll glass fiber batt insulation (that is adhered to a plastic film) across the structural support members before applying the metal roofing. When the metal roofing is installed and secured to the structure, the insulation becomes compressed between the structural membrane and the roof. The thermal efficiency of the compressed insulation is greatly reduced, resulting in strips that conduct heat from the building at a much faster rate than adjacent areas. Snow and ice will melt quickly over the structural supports and ice dams will form. Leakage can occur through unsealed sidelaps that become submerged, so any defects in the roof will be significant. The compressed insulation also results in cooling of the structural supports and potential condensation and water drippage. This deficiency can be avoided by placing a structural insulating material, such as wood, between the roof and the support structure. The insulating value of this component should be approximately equal to those in the rest of the roof to prevent uneven melting of snow.

The intersection of the roofing and side walls in a building must be watertight and airtight to prevent water leakage and heat loss by air leaks. The seals must be secured adequately so they will not become dislodged by thermal movements in the roofing.

Structural metal roofs should be drained to open eaves where water can be collected by external metal gutters secured to the wall or roof. If the roof must drain to valleys or internal gutters, the latter should slope to drains; avoid using metal components to waterproof these areas as thermal movements within the system are very difficult to accommodate. By design, these areas handle large quantities of water so leakage will always be significant when it occurs. These areas can be effectively waterproofed with an polymeric sheet, provided it is protected from the impact of sliding ice and snow and is designed to accommodate the expected movements in the roofing.

26.8 POLYMERIC AND MODIFIED BITUMINOUS SHEET ROOFING

Polymeric sheet roofing describes synthetic rubber and thermoplastic sheets for use as single-ply roofing. In 1981 the National Roofing Contractors Association published a listing of more than 60 available products by over 30 manufacturers. Because each is differently manufactured and marketed, and because each manufacturer publishes their own installation instructions and standards for guarantees, this section is limited to a general discussion of these types of materials. No material should be selected without thorough investigation of prior installations, as well as of the adequacy of the technology of components and of installation methods. In addition, a determination should be made of the producer's commitment to technological and financial support of the system. The authors have reports of single ply membranes with 20 years of successful field performance.

History of the technology of roofing materials has shown that laboratory tests, such as accelerated weathering, have not always accurately predicted field performance. Manufacturers' entries into the roofing market have been nearly equal to their exits. Many "roofing materials of the future" have become "materials of the past" when they were tested by full scale field installations. Too many inexperienced manufacturers have tried to join the rush to single sheet membranes with untested and unproven products, and without being able to give field installations knowledgeable support.

26.8.1 General

Polymeric systems have in common the use of a single ply of flexible sheet material. The range of that material's extensibility is usually advertised to be 300 to 400%. The available "stretch" is dependent on the amount of material that is free to accept the imposed loading, the ambient temperature, and the exposure history (including aging) of the membrane. Thermoplastic sheet membranes may become embrittled and shrink from loss of plasticizer (the component that makes rigid plastics pliable). Loss of plasticizer has caused shrinkage, splitting, and leakage of some membranes within five years of application. Improved formulation and reinforcing are claimed to have reduced this problem. Table 26.9 lists characteristic properties of elastomeric membranes.

Beware of the frequently advertised "breathability" characteristic of polymeric sheets. Claims that the membranes will shed water from above, yet will allow "water vapor" release from below, are misleading. Field tests by the National Research Council of Canada and the U.S. Corps of Engineers have shown these claims to be untrue, or the amount of venting to be negligible, as far as any useful roof system drying is concerned.

Polymeric systems can be loosely-laid or attached. If loosely-laid, the sheet must be ballasted to prevent wind uplift. Alternatively, the sheet can be fully attached with adhesives, or partially attached by mechanical fasteners at suitable intervals.

Except in unusual installations where high wind uplift resistance is necessary, where the structural framing requires a very lightweight system, where slopes are too steep to retain ballast, or where special use of the roof surface is required, all polymeric sheet systems should be loosely-laid

Table 26.9 Characteristic Properties for Sheet and Liquid-Applied Elastomeric Membranes

Material	Thickness, mils	Tensile Strength, psi	Ultimate Elongation, %	Service Temperature °—	Permeability Perms	Lap Seal Methods	Color	Flashings	Comments
Sheet Membranes									
Butyl	30–60	1000	300		0.001	Adhesive	Black, grey	Uncured rubber	
Chlorinated polyethylene (CPE) reinforced		3500		−50–250	0.05	Heat weld	Grey, white	CPE coated metal	
Chlorosulfonated polyethylene (CSPE) reinforced	45	3500		−50–250	0.001	Heat weld	Grey, white	CSPE coated metal	
Ethylene-propylene-diene-monomer (EPDM)	45–60	1400	300	−65–300	0.1–0.007		Black	Uncured rubber	
Modified bitumens		250–1500	300	−40–180	0.007	Adhesive or heat weld	Aluminum, dark grey, black	Self-flashing	
Neoprene (polychloroprene)	32–60	1400–1800	450–700	−40–200	0.007	Adhesive	Black, grey	Uncured neoprene	
Polyvinyl chloride (PVC) plain	45–60	2800	250	−30–160	0.5	Solvent or heat weld	Grey, tan	PVC coated metal	
Polyvinyl chloride (PVC) reinforced	45	2500	330	−20–160	0.5	Solvent or heat weld	Grey, tan, clear, black	PVC coated metal	
Silicone					0.23				
Liquid Applied Membranes									
Acrylic	20	180–500	200–300	−70–180			Optional		
Butyl	15–30	600	250	−55–180					
Hypalon (CSPE)	20–45	800	400	−50–250	0.18–0.25		Optional		
Neoprene	20	700	600	−50–180	0.32–0.46				
Polyvinyl chloride (PVC)	15–30	1200	150–200	−30–160					
Silicone	20	400–600	100–150	−70–350	2.5–5.4				
Urethane (modified)	20–60	100–3000	300–800	−60–200	0.01–0.06				
Polysulfide modified bitumens				−40–180			Black		Hot Applied

and ballasted. This method takes advantage of the flexibility and extensibility of the materials, and the ballast provides protection for the membrane from roof-top traffic, sunlight, and impact damage. A ballasted system is heavier than a conventional roof system, because of the additional stone required. A ballasted system without adequate drainage may result in a shortened life because of ponding water; good drainage is essential.

A polymeric system can also be protected by application of extruded polystyrene insulation between the membrane and the ballast. No other insulation currently available is suitable for this exposed use.

26.8.2 Elastomeric Materials

Typical elastomeric materials are synthetic rubbers, such as EPDM (ethylene-propylene-diene monomer), neoprene (polychloroprene), and Hypalon CSPE or CSM (chlorosulfonated polyethylene) and CPE (chlorinated polyethylene).

Over the years, 30- to 45-mils thick synthetic rubber sheets (all of them thermosetting) have had the best performance records; first installations date from the early 1960s. By the late 1970s, neoprene and EPDM remained as the principally used roofing sheets because of failures of other materials and difficulties with complicated seaming methods.

EPDM has excellent weathering resistance characteristics and the sheet has performed well. Since 1978 the major producer of sheet EPDM roofing has marketed a contact adhesive for seaming. This has greatly simplified the field seaming of individual sheets, but the joint is not as secure as with the previous seaming process using uncured butyl tape and two-part adhesives. Because many sheets are produced with separators (mica or talc) to prevent sticking in the roll, scrupulous cleaning of surfaces before seaming is an absolute necessity. The field history of the contact adhesive is limited, as it dates only from 1978. Under load, seam failure will always occur at one face of the seam by the adhesive peeling cleanly from the sheet. Inspection is required on all field seams; this is a difficult task because the seam can be lifted by finger pressure even when properly made. An advantage of some manufacturer's EPDM sheet is vulcanized seaming for sheets up to 40 by 100 ft (12 by 30 m), which reduces the potential for problems at field seams. However, even factory seams are not always perfect, and they must be field inspected. Some manufacturers have marketed a cleaner–primer for pretreatment of laps before adhesion; at this time, the success of these products is unknown. Currently, technological work by adhesives experts is underway; primers and tape adhered joints are being considered. This is a developing science and the latest data should be carefully checked before use.

Neoprene is expected to have a somewhat shorter useful life than EPDM because it has less resistance to sunlight and ozone. The advantage of neoprene sheet is that a secure field seam can be made using contact adhesive. With proper workmanship the seam is tight; the adhesive bonds well to the neoprene, and it can be inspected without damaging the seam. Field installations of neoprene, 5 to 15 years old, have shown some chalking and embrittlement. When neoprene sheets are ballasted or coated with hypalon paint, they have a longer life expectancy.

Limited marketing of Hypalon sheets for roofing continues, but their use is not recommended until a significant field performance history has been established. They are seamed in an uncured state by either adhesive or heat welds. Hypalon is manufactured as a thermoplastic and then vulcanized or cured making heat welding difficult and unreliable. Aged sheets must be spliced with adhesive after careful cleaning and priming.

Chlorinated polyethylene (CPE) is a specialty roofing membrane product. CPE is a thermoplastic polymer that has properties of an elastomer. CPE resins provide excellent weathering and ozone resistance and remain uncured so their thermoplastic properties are "permanent". CPE has a relative short history of field performance as a roofing membrane. Seams can be heat welded to develop splices that do not rely on adhesives.

All the sheets discussed in this section are synthetic; therefore, their manufacture can vary considerably in formulation and production techniques. Only materials of proven performance, in actual roofing installations by experienced producers and contractors, should be selected for use. Accessories used with different manufacturer's products, such as solvents, primers, and adhesives, are usually not interchangeable. Each system should be designed and constructed as recommended by the sheet manufacturer.

26.8.3 Plastic Materials

Thermoplastic sheet materials, as opposed to elastomers, do not return promptly to their original shape after deformation. They are usually thermoplastic and can be jointed by heating or solvents, without adhesives. Among them are PVC (polyvinyl chloride), CPA (copolymer alloy) and PIB (polyisobutylene).

PVC is a rigid plastic material which is softened by adding a plasticizer. Seaming can be achieved by solvent or heat welding. PVC sheet membranes were developed in Europe and have been used in the United States since the early 1970s. Some early systems failed by embrittlement due to plasticizer migration, shrinkage, and splitting. Some of these problems have been reduced, at least in the short term, by changes in chemical formulations, reinforcing, and increased material thickness, as well as changes in installation techniques, flashings, and seams. Plastomeric roofing sheets are usually 30–60 mils thick. Generally, the seaming of PVC sheets has not been a problem. PVC sheets are sensitive to mechanical damage, including melting from discarded smoking materials. PVC materials are incompatible with bitumens and polystyrene insulating materials. PVC must also be protected from coal tar pitch fumes.

Cautions in the use of synthetic rubbers also apply to PVC in regard to their manufacture and field performance.

Some manufacturers offer reinforced PVC and CPE sheets. Of the presently available plastic sheets only reinforced PVC sheets should be considered for use on roofs. Even these PVC sheets do not have a field history of success sufficient for recommending their general use in roofing; however, their performance to date is promising.

Galvanized steel sheets covered with factory bonded PVC coatings are produced by some sheet producers and can be used for flashings; they permit field bonding of the roofing membrane directly to the metal flashings.

26.8.4 Plastic Modified Bituminous Sheets

Several polymeric sheet products consist of laminates, or composite sheets of polymer modified asphaltic materials bonded to foil, plastic, and glass fiber materials. Seams are achieved by melting or mopping the asphaltic portion, or by torching the underside with a gas (propane) burner to melt the material. This has been shown to present a significant fire hazard and suitable installation precautions are needed. The brief field performance history of these products combined with the many variations of the products' composition make it impossible to present a valid opinion on their use in the United States at this time. Reports of their use in Europe indicate a promising future for some of them. One limitation is that the seams of some of these materials cannot be easily inspected without their destruction, because an inspector cannot see the joining process at the torch point, beyond the lap edge. Advantages claimed by manufacturers include factory control of the membrane; however, one should be very cautious about the use of a single ply bituminous membrane.

26.8.5 Loosely-Laid Polymeric Sheets

The majority of the polymeric sheet membrane roofs are loosely-laid; attachment of the sheet to the roof deck is limited to perimeters and penetrations, including roof drains. Where the membrane is attached, the sheets are fastened, through rubber or perforated metal nailing strips, with capped nails or screws at specified intervals to anchored wood blocking or other structural building components. At roof drains the sheet is adhered to the drain (or metal flashing) using adhesives, and clamped by the drain flashing ring. Ballast is applied directly over the membrane to hold it in place and to protect it from damage. Drainage of surface water remains an important principle for all roofing systems; there is no exception.

The typical ballast is rounded stone, ¾ to 1½ in. diameter [10–15 psf (480–720 N/m^2) is usually needed to cover the membrane], or concrete pavers. If concrete pavers are used, an underlayment (as recommended by the sheet manufacturer) should protect the sheet membrane from abrasion. If extruded polystyrene insulation is used on top of the system as described above, a polyester scrim should be used below the ballast to help spread the ballast load evenly and to prevent flotation of the polystyrene boards. The scrim also retains the stone ballast, preventing it from filtering through insulation joints to the membrane surface where it could damage the membrane or block the drainage channels between insulation boards. When the scrim is used, manufacturers may not insist on rounded stone for ballast; however, some stones may reach the membrane surface inadvertently, and there is risk of membrane puncture from sharp aggregate. Pavers should be carefully designed to resist water absorption and freeze–thaw deterioration (where applicable), and they should be elevated from the membrane surface by pedestals.

The membrane should not be placed directly on the roof deck. Use an insulation layer to attenuate the effects of deck irregularities or movements on the membrane. At least one and preferably two layers of insulation should be provided, with staggered joints. If the layer of insulation directly below the membrane is mechanically fastened, a patch of the membrane material is applied over the head of each fastener to avoid puncture damage to the membrane. Rigid insulation must be used to prevent the weight of ballasting equipment from compressing the insulation around the fixed fastener and driving it through the membrane. The extra membrane patch will not always prevent this, with soft insulations. It is preferable to cover the fasteners with an insulation layer because the through fasteners act at thermal conductors and condensation may occur on them below the roof. Each layer of insulation should be attached as described in Section 26.6.1, to assure tight joints during installation and to limit the risk of the loose boards lifting and stacking on one another.

If fire safety laws permit, closed-cell insulation should be used. Single-ply membranes that are loosely-laid will experience condensation on their underside during cold weather; this moisture may be absorbed by the underlying insulation and results in attachment of the sheet to the substrate by freezing. Use closed-cell insulations such as extruded polystyrene below loosely-laid systems to minimize the degradation of the insulation from moisture; their potential dimensional instability should have little effect on the loosely-laid membrane. Some urethane foam and isocyanurate boards also may be suitable. Low density expanded polystyrenes (''bead boards'') may be too weak to provide a sound base for the membrane, and they can be damaged by seam and lap adhesives and sealers, even through the membrane.

26.8.6 Fully Adhered and Mechanically Attached Polymeric Sheets

When a lightweight roof is needed, when the roof is too steep to retain ballast (over 1.5-in-12) or when the appearance of the sheet is important to the design, an unballasted single-ply membrane system can be used. PVC sheets are usually light gray, but are available from some manufacturers in other colors. EPDM is white or black; and neoprene is black, or grey which must be color-coated in the field with hypalon paint; this coating will have to be renewed at approximately 5-yr intervals.

The seaming procedures for the adhered sheets is the same used for loosely-laid systems. The sheet membrane is fully adhered to the insulation which has been installed as described above, or on a plywood deck. Some insulations and their cap sheets are not compatible with solvent-based contact adhesives, and others may have insufficient resistance to wind uplift. Each manufacturer's recommendations should be carefully followed, and only approved insulations should be used to bond the sheets.

Many manufacturers recently have developed mechanically fastened polymeric sheet systems. Methods of attachment vary considerably; in some cases metal bars are used as hold-down straps, in others, fasteners with special caps are used at seams and the membrane is adhered over the caps. Although these systems perform well in laboratory uplift tests, there is little field performance history available on these systems.

Membranes that are fully adhered or mechanically fastened use narrow sheets [maximum 10 ft (3 m)] to ease field application. Each manufacturer will have their design parameters which should be followed for the system selected.

Adhered or mechanically fastened single-ply membranes are exposed to mechanical damage and are without protection from ice, sunlight, or mechanical damage. Their use should be carefully considered in view of these potential hazards.

26.8.7 Flashings

Flashings for polymeric sheet systems are made of compatible materials, almost always provided by the manufacturer of the basic membrane. The flashing materials are uncured rubber used to flash neoprene or EPDM sheet, and PVC for flashing a PVC sheet. (Uncured EPDM/neoprene blends and EPDM flashings have appeared on the market recently but they have virtually no field record.) Some manufacturers also provide reinforced plastic sheets and plastic coated metals for solvent or heat welding to thermoplastic membrane materials. The general principles presented earlier for flashings used with built-up roofs apply equally to polymeric roof systems. Important principles to follow are listed below:

1. Avoid the use of metal flashings in the roof system.
2. Raise flashings from the roof surface and slope them to shed water.
3. Use cant strips and tapered insulation at base flashings to direct water away from the base of the flashings.
4. Secure base flashings at their upper edges.
5. Provide firm support behind vertical polymeric sheet flashings (which are always to be fully adhered).
6. Cap or seal the upper edges of all flashings.
7. Lap flashings to sheet roofing with (not against) water flow, regardless of manufacturers' claims.
8. Avoid 90° bends of flashings whenever possible. Seams at 90° bends are difficult to fabricate because of the thickness of the sheet and its elasticity (memory). Small openings (fish-mouths) almost always occur at seams made at 90° corners.

9. Do not allow contact of polymeric flashings to bituminous materials, except with specific information from their manufacturer and with documented successful field performance history.

Most relief and control joint conditions can be adequately flashed by providing a loop (slack) in the sheet material for accommodating thermal movement.

At building expansion joints, use a prefabricated polymeric flashing system compatible with the membrane, anchored to wood blocking on each side of the building joint similar to the system shown in Fig. 26.16.

Prefabricated pipe sleeves are available from some manufacturers. They perform well in the field.

Polymeric sheets can be well adhered to a variety of clean building materials. Individual manufacturers have varying requirements for primers and will provide information about compatibility with paints and sealants.

Typical details illustrating the principles discussed above are shown in Figures 26.23, 26.24, and 26.25.

26.8.8 General Considerations

Use a complete system provided by a single manufacturer; do not mix various materials from separate manufacturers.

Consideration must be given to proper storage and shelf life of adhesives, primers, and coatings, and in some cases, to storage and shelf life of the flashing material and the sheet goods themselves. Many of the solvents used are dangerous to breathe in enclosed spaces, are irritating to skin and eyes, and are flammable; they must be handled with caution.

Workmen using sheet goods should have clean clothes and tools, free of bitumens. Special care is necessary to avoid damage to sheets and damage to foam, (and other types) insulation from adhesives and sealants during installation. Careful examination of each square inch of sheet for mechanical damage or factory defects is required. Areas of temporary seals at the end of each work day should have the adhered portions of the sheet cut off to avoid a puckered, "fishmouthed" seam with this material.

Single-ply systems have no safety margin by redundancy, such as multiple-ply systems. However, large areas of roof can be covered quickly, and factory controls can help maintain sheet and factory seam quality.

When the roof system is completed, a polymeric membrane will usually have many "patches". These result from normal repairs required by mechanical damage incurred during handling and installation, factory defects, and repairs to seaming errors (fishmouths, puckers, or solvent swell-

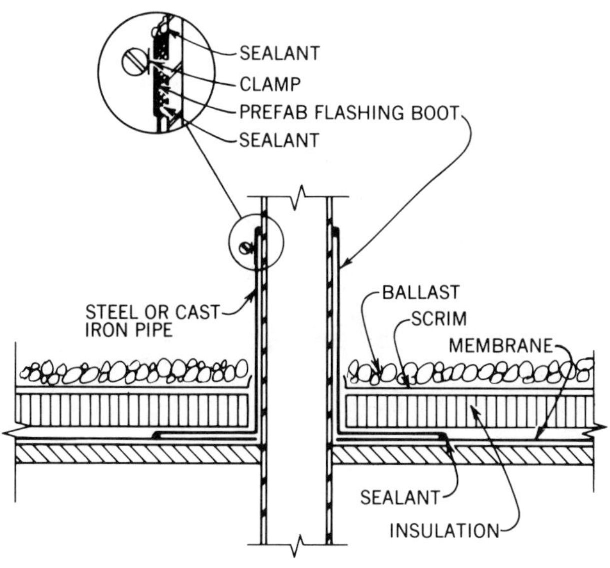

Fig. 26.23 Pipe penetration.

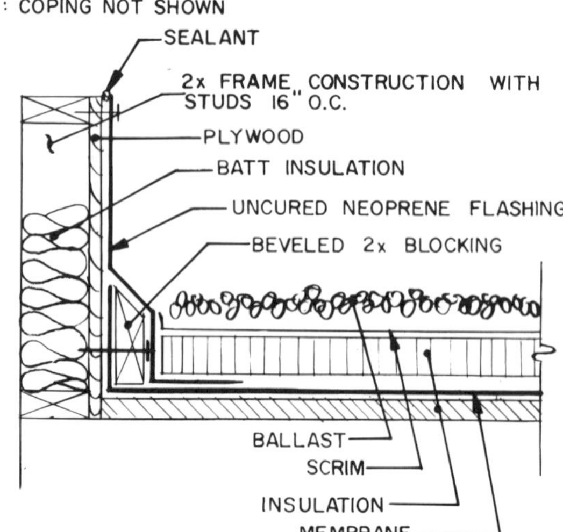

Fig. 26.24 Parapet detail.

ing). The patches are unsightly and may give the impression that the membrane is not "new". There is no criteria established for a number of patches acceptable in a given area, but the designer should be prepared to accept a "reasonable" number of patches as normal for this type of system.

Oils and chemicals vary widely in their effects on polymeric materials; machinery discharges from roof-mounted equipment, acid rain, and wind-borne dirt from nearby processing facilities may result in premature failure of polymeric sheets or affect their ability to be repaired effectively.

Because of the individual characteristics of each polymeric sheet type, verify with the selected manufacturer that their material is compatible with any other materials expected to contact the roof. Most manufacturers list the compatibility of their materials with an extensive list of chemicals, and they are required to make available data sheets of material characteristics with "right-to-know" regulations for material users.

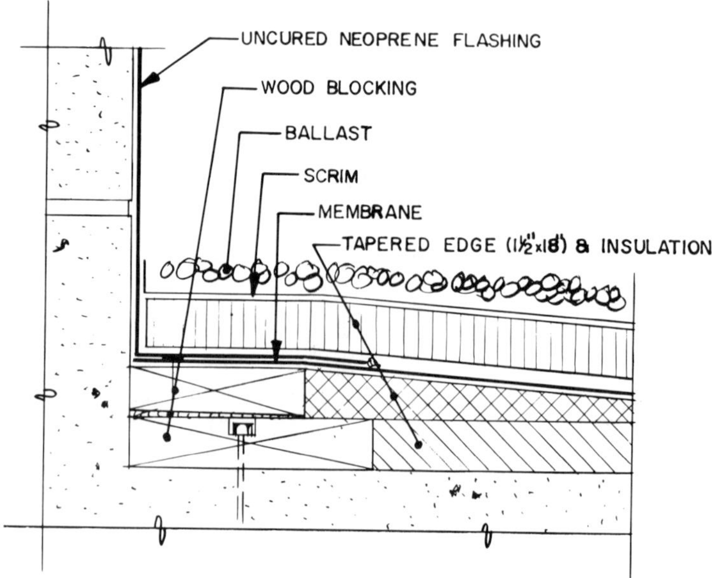

Fig. 26.25 Masonry wall flashing.

26.9 LIQUID-APPLIED POLYMERIC ROOFING

26.9.1 Materials

The polymeric materials presently available for fluid application as a roofing membrane are listed in Table 26.9, along with their relevant properties. Some of the advantages claimed for these materials and systems are low weight, heat and chemical resistance, ease of application to irregular shapes or steep slopes, ease of repair, and resistance to deterioration from moisture. Exposed systems require regular maintenance if they are to provide reasonably acceptable service.

A liquid-applied membrane must be applied to a sound, stable, structural deck, which is compatible with the membrane material. Liquid-applied materials are susceptible to numerous problems when used as roofing membranes:

1. Liquid-applied materials are particularly troublesome when they are required to bridge gaps in the substrate at construction joints or other minor surface cracks. Materials that cure cannot accommodate the strain induced by substrate movements at the crack and will be ruptured; materials that remain uncured, in a liquid or semi-liquid form, will flow into the crack; both conditions result in water leakage.
2. The materials must be applied to a dry substrate to ensure adhesion, or the membrane will blister and peel from the surface. Water vapor escaping during construction can produce pinholes in the membrane.
3. Solvent-based materials that dry too rapidly due to heat exposure from sunlight can become pinholed from solvent release (boiling).
4. Some liquid-applied materials deteriorate rapidly when exposed to sunlight and the weather, and consequently must be protected. The protection system qualifies the system as waterproofing rather than roofing; it must be separated from the membrane and ballasted to hold it in place, adding significant weight to the roofing system. The designer must specify membrane and insulation materials that are compatible. The insulation must not break down during the membrane cure period or during solvent release, if any.
5. Liquid-applied membranes must be protected from traffic and impact, particularly materials that remain uncured and soft.

26.9.2 Deck Preparation

The most critical component in the liquid-applied roofing system is the substrate to which the material will be applied. The success or failure of the membrane is dependent on the designer's understanding of the substrate, and detailing and specifying proper application procedures. Good workmanship must be assured during construction. Workmanship is almost as critical as design, since the best design will not perform adequately if not installed properly; but the best workmanship cannot make an improperly designed system work. (This applies to all roofing work.) The following general principles should be followed, along with the materials manufacturer's requirements for the system, when designing and applying liquid-applied roofing membranes:

1. The deck surface must be clean, free of sharp projections, holes, grease, oil, dirt, paint, curing compounds, or other contaminants; concrete must be unfrozen, free of laitance and loose aggregate, and it must have a smooth trowel or float finish.
2. The membrane should not be applied until all protrusions and projections through the deck are in place, or sleeves are installed and waterproofed.
3. All penetrations, protrusions, projections, and rising walls should be flashed with a polymeric sheet adhered to both the deck and the flashed surface. Leave the sheet unadhered for 1 in. over the joint between the deck and the component.
4. All visible cracks and control joints should be covered with a fully adhered polymeric sheet. Adhesion should be prevented directly over the crack by using a release tape.
5. Expansion joints should be flashed with a preformed polymeric expansion joint cover, adhered to each side of the joint.
6. Prime all surfaces to receive the membrane if required by the system used.
7. Apply the membrane continuously to all deck surfaces and flashings, using a method acceptable to the manufacturer.

26.9.3 Application

The membrane is applied to the required thickness (usually in a single coat) by trowel, squeegee, roller, brush, spray apparatus, or other method acceptable to the membrane manufacturer. This is usually difficult to achieve as all decks have some irregularities resulting in coating thickness irregularities. This is one weakness of the liquid-applied systems. Coverage should be complete over all deck surfaces, and previously flashed expansion joints, control joints, prepared cracks, flashed penetrations, projections and protrusions, wall flashings, and terminations.

The completed membrane should be water tested by flooding prior to the application of the protection course, if any. Be careful not to overload the deck during testing. The waterproofed surface should not be used for any purpose prior to the application of the protection course.

26.10 SHINGLES, SLATE, AND TILE ROOFING

26.10.1 Shingles

Shingles are the most widely used roofing system. They are very reliable when used on roof slopes in excess of 4-in-12, and can be used, with special precautions, to slopes as low as 2-in-12. Shingles are produced from asphalt saturated organic and asbestos felts, glass fiber fabric, cement-asbestos, wood, and sheet metal in a wide variety of shapes, sizes, colors, textures, and styles. Properly produced asphalt shingles are fire and wind resistant and will provide a service life of from 15 to in excess of 20 years. Shingles of metal and cedar can last far longer.

As this is written (1986) the manufacture of asbestos products in the United States has for all practical use, stopped. Shingles made of glass fiber material have begun to dominate the market. Their performance has not yet been extensive enough to evaluate with confidence. Some problems have been reported in cold weather installations, with lower strength of the shingles, and with poor adhesion.

In addition, the use of staples to supplant the traditional nails is common today. Recently, acceptance of this method by shingle manufacturers has been published.

Consult manufacturers' current literature for the types of shingles and application requirements for each type, including slope limitations, underlayment, fasteners, fire rating, wind resistance, and recommended head laps.

Long-term success with shingle roofing requires careful and correct application of the products. General application procedures will not be presented here; they are available from the materials' manufacturer, Asphalt Roofing Manufacturers Association [26.1], and *Architectural Graphic Standards* [26.4]. Many shingle roof failures result because these procedures were not scrupulously followed.

Ventilation of the underside of the roof deck is almost always necessary. Air must be allowed to circulate freely to all areas of the deck to prevent condensation. Venting can be accomplished with eave vents in combination with gable, roof, or ridge vents. The Federal Housing Authority Minimum Property Standards require a ratio of net free ventilation area to ceiling area of 1/150; 1/300 is acceptable, if a vapor barrier is installed on the warm side of the ceiling or approximately half the ventilation area is provided near the ridge.

26.10.2 Slate Roofing

Slate is a natural stone, split along natural cleavage planes, and trimmed to the desired thickness and sizes for roofing. Weatherproofing of slate roofs is obtained by installing the individual slates in a configuration that sheds water similar to other types of shingle roofing. Hence, slate should not be used as the weatherproofing system on roofs that slope less than 5-in-12. Slate roofs are more expensive, heavier, and installation procedures are more critical than for asphalt shingle roofs, but their long-term performance, appearance, fire rating, and wind resistance characteristics are superior to asphalt shingle products.

Table 26.10 shows the slate sizes available. Slates should have a minimum vertical overlap of 3 in. (75 mm) between rows; horizontal lap or head lap requirements depend on the slope of the roof: 4-in. (100 mm) lap for slopes of 5-in-12 to 8-in-12, 3-in. (75 mm) lap for slopes at 8-in-12 to 20-in-12, and a 2-in. (50 mm) lap for slopes of 20-in-12 to vertical or wall siding conditions.

Standard slate colors available are: black, blue black, mottled grey, purple, green, mottled purple and green, purple variegated, and red. The above slate colors are available in unfading or weathering grades. Special colors are also available.

Each slate should have two holes prepunched for nail fastening. Nails should be copper, 1 in. (25 mm) longer than the thickness of the slate. Provide carefully installed asphalt-saturated felt underlayment that ranges from 15-lb saturated felt for commercial standard slate to 65-lb prepared roll roofing for slate that is thicker than ¾ in. (75 mm). Nails must not be overdriven, so slates hang on the nails.

Slate is also used in flat roof applications (sometimes with a concrete setting bed), not for waterproofing but as a traffic-bearing surface. A separate waterproofing membrane is required

Table 26.10 Slate Sizes

Pitched Roof Length, in.	Width, in.	Thicknesses, in.
10	6, 7, 8	3/16, 1/4, 3/8, 1/2, 3/4, 1, 1 1/4, 1 1/2, 1 3/4, 2
12	6, 7, 8, 9, 10	
14	7, 8, 9, 10, 12	
16	8, 9, 10, 12	
18	9, 10, 11, 12	
20	10, 11, 12, 14	
22	11, 12, 14	
24	12, 14	
Flat Roof (Less than 4-in-12) Length, in.	**Width, in.**	**Thicknesses, in.**
6	6, 8, 9	1/4 to 3/8—heavy service
10	6, 7, 8	3/16—ordinary service
12	6, 7, 8	3/4 to 1 1/4—special terraces, walks, etc.

below the traffic deck. Roofing systems described in this chapter are not acceptable for waterproofing.

Valley flashing should be fabricated from sheet metal, and the edges of each intersecting roof should taper outward 1/8-in-12 to provide a valley that is wider at the bottom of the roof slope than at the top. Copper step flashings are interwoven with each slate course at chimneys, dormers, and walls.

Ice and snow often slide from steep slate roof surfaces. Snow and ice guards should be provided over entries or other areas that will be damaged by falling debris. These must be secured firmly and be strong enough to resist the force of the rapidly moving snow and ice.

26.10.3 Concrete and Clay Tiles

Modern concrete and clay tile products, because of their high cost, are used primarily for their architectural appearance. Tile roofs weigh 8.5 to 11.5 psf (400 to 550 N/m^2) and should not be used as the primary roofing on slopes less than 5-in-12. Both clay and concrete tiles are available in a number of architectural styles.

Clay and concrete tile roofs can be applied to decks with solid or spaced sheathing of sufficient thickness for nailing. Barrel-type clay tiles also require battens running parallel to the roof slope for nailing. Nails should be noncorrosive.

Tile and shingle roofs are not airtight and windblown precipitation can bypass the laps in the tiles. Waterproof underlayment such as roofing felt layers must be provided to ensure a leak-free roof. The waterproofing underlayment will also seal the roofing from air leakage which could result in significant heat losses in cold climates.

Valley, wall, eave, and chimney flashings are sheet metal; ridges, hips, and rakes are constructed from molded components to match to the roofing.

Design details and installation procedures are available from the manufacturers; details are included in *Architectural Graphic Standards* [26.4]

26.10.4 Substrate

Shingles, slates, and tiles used for steep roofing must be applied to a nailable substrate, except in special applications where tiles or slates are wired to metal purlins. Wiring of slate and tile to metal purlins, although acceptable for attachment, is costly, and it is difficult to provide a watertight and windproof underlayment.

Specify a wood deck to be installed over a non-nailable deck that is to receive steep roofing. The wood deck should be fastened to wood sleepers that are, in turn, securely anchored to be structural deck. To secure sleepers, use drilled-in steel anchors in concrete; sheet metal screws or toggle bolts in steel decks; and, toggle bolts in wood fiber, gypsum plank, or cast gypsum decks. Provide for condensation problems where necessary.

Wood decks should be well-seasoned tongue and groove sheathing or plywood, with a maximum thickness of ¾ in. (75 mm) for sheathing and ⅝ in. (16 mm) for plywood. Install "ply clips" at all unsupported plywood edges where spacing of supports exceeds 24 in. (600 mm). Badly warped boards or those containing loose knots or other defects should not be used. Minor defects in the deck should be covered with noncorrosive sheet metal patches, nailed to the deck.

In buildings with unusual occupancy moisture conditions and where condensation at the deck is possible, use wood that is preservative treated with chromated copper arsenate under pressure. Minimum net retention of preservative should comply with ASTM D1760 [26.7] for ground contact.

Fasteners for wood to wood connections should be annular ring nails with a length to provide a minimum of 1½-in. embedment into the final piece receiving the nail points. All fasteners used for roofing or deck construction should be hot-dip zinc coated in accordance with ASTM A153 [26.8].

After the wood deck is complete, and prior to the application of the roofing system, apply an underlayment to the entire deck surface. The underlayment provides a dry surface to which the roofing system is applied, acts as a wind stop, serves as an effective water barrier should any water pass through the roofing system, and separates the roofing from the deck preventing any incompatibility reaction or adhesion between the deck and the roofing.

The following are minimum requirements for underlayment for the various systems:

1. Asphalt, cement asbestos, wood or metal shingles, slopes of 4-in-12 or greater: Starting at the eave or low point, install one layer of unperforated 15-lb asphalt-saturated felt over the deck, lap underlayment sheets 2 in. (50 mm) at side laps, 6 in. (150 mm) at end laps, and 6 in. (150 mm) from both sides of hips and ridges. Do not use a vapor retardant material.
2. Square tab asphalt strip shingles with factory applied adhesive for wind resistance on slopes between 2-in-12 and 4-in-12: Starting at the eave or low point install a double underlayment of unperforated 15-lb asphalt-saturated felt. The underlayment should be sealed between layers with roofing cement starting at the eaves and continuing up the slope to cover a point 24 in. (600 mm) inside the interior wall of the building. If self-sealing shingles are not used, cement each tab with a spot of roofing cement.
3. Slate: Starting at the eave or low point install one layer of asphalt saturated felt over the deck. Lap underlayment 2 in. (50 mm) at side laps, 6 in. (150 mm) at end laps, and 6 in. (150 mm) from both sides of hips and ridges. Use 15-lb felt with commercial grade slate; use 30 lb felt with textural roofs, use 30-lb felt for graduated roofs and ¾-in. slate, and use 45-lb, 55-lb, and 65-1b prepared roll roofing for heavier slate.
4. Underlayment for clay and concrete tile roof should provide a watertight roof before the tiles are installed. Use a minimum 30-lb felt applied starting at the low point end, lap all side laps 2 in. (50 mm) and end laps 6 in. and overlap ridges and hips 6 in. (150 mm) from each side.
5. Recently successful use has been reported of composite, modified bitumen sheet materials for underlayment.

26.10.5 Metal Shingles

Metal roof shingles are manufactured from steel and aluminum, prefinished in a variety of colors. In each case the shingles are part of a special interlocking system including substructure and trim elements. Generally, the use of these systems is restricted to steep decorative roofs such as mansards. Some manufacturers permit installation on slopes as low as 3-in-12 but these installations usually have interior gutter systems. Each application should be carefully reviewed with the manufacturer selected for the specific use contemplated.

26.10.6 Attachment of Roofing Units

Shingles are applied with hot-dip zinc coated roofing nails made of 11 or 12 gage wire; the heads are ⅜ to 7⁄16 in. (9.5 to 11.1 mm) in diameter. Nails should be minimum 1¼ in. (32 mm) long for new roofs, but should pass through the roofing and underlayment and penetrate the wood deck a minimum of ¾ in. (19 mm) or full depth of plywood decking.

Do not use staples for applying shingles. The shingles may tear at the staple, resulting in reduced wind uplift resistance. New standards are being developed for the use of staples, but they are as yet unproven.

Nails for wood shingles and shakes should be hot-dip zinc coated, of sufficient length to pass through the roofing and penetrate 1-in. (25 mm) minimum into the decking; special nails are used to prevent excessive split shingles.

Slates, cement-asbestos shingles, and clay and concrete tiles are installed using noncorroding annular ring nails, long enough to have a minimum of 1-in. (25 mm) penetration (or full depth) into the deck.

26.10.7 Details and Flashing

The flashing details of a steep roofing application are critical for its successful performance. Follow the recommendations for decks, minimum slope, underlayment, application, and attachment and the main regions of the roofing system will remain leak-free.

Typical flashing details, such as at ridges, rakes, eaves, valleys, vent pipe penetrations, and for chimneys, vertical walls, dormers, and crickets, are provided in an Asphalt Roofing Manufacturers Association booklet [26.9] for asphalt products, and by other manufacturers' associations for specialty roofing materials.

Details for skylights, clerestories, drainage valleys (interior gutters), and connections between steep and "flat" roofing should have the following features:

1. All water must promptly drain from the roof surfaces. Provide crickets or sufficient slope in drainage areas to prevent standing water.
2. Provide continuous base flashings of mineral surface roll roofing set in plastic roofing cement at all transition points.
3. Do not use asphalt roofing products in valleys where snow accumulations and ice dams may result in standing water.
4. When roofs of different slopes join in a valley, expect the runoff from the steeper roof to flow up the roof with less slope. Provide suitable water diverter flashings.

26.10.8 Drainage

Water from steep roofs is removed using gutters and downspouts, or gutters and interior roof drains. Sizes of gutters and downspouts are dependent upon the roof area, roof slope, expected rainfall intensity, and length and shape of the gutter. In addition, metal gutters require expansion joints which can only be located between outlets. To size gutters and downspouts, the designer should refer to various guidelines and nomographs available, such as one design aid in *Architectural Graphic Standards* [26.4].

Gutters must be sloped to the downspouts to provide prompt drainage, prevent overflow, and to eliminate any standing water, particularly to prevent rot if wood gutters are used. Gutters must be anchored to provide support for their weight and for loads from ladders and accumulations of ice and snow during winter months. Use screws for positive anchorage. The fasteners should not penetrate the eave flashings. The outer gutter edge must be least 1 in. below the eave to prevent water backup. The eaves must be structurally reinforced to support the anticipated loads from water, snow, and ladders. Gutters can be spaced away from eaves to permit safe overflow in the event of ice dams or other blockage.

Gutters can be badly damaged, and even torn off, by ice and snow sliding from roof surfaces during periods of thaw. Locate gutters below the slope, clear of sliding ice and snow or provide a system of snow guards to prevent this from occurring. Snow guards are devices fabricated from sheet metal, wire, or brackets and tubing that are anchored to the roof deck and project from the roof surface. Snow guards hold the accumulated ice and snow, preventing it from sliding off the roof. Spacing and anchorage of the snow guards must be adequate to support the total weight of accumulated ice and snow, or they will be torn from the deck and considerable damage will result to the roofing. Snow guards are not necessary on all steep roofing, but are usually needed on roofs constructed of smooth roofing materials such as metal or slate of any slope and with most other roofing material applied to very steep slopes, in excess of 5-in-12 rise, or when falling ice and snow cannot be tolerated. Ice dams and the resultant water backup must be expected at snow guards, and suitable waterproofing provisions must be made.

26.11 PROPRIETARY SYSTEMS

26.11.1 IRMA

The IRMA (Insulated Roof Membrane Assembly) is a patent protected system of The Dow Chemical Company, using "Styrofoam RM," which consists of extruded, closed-cell, polystyrene foam board insulation over a roofing membrane.

The insulation is applied loosely or adhered to the membrane; in both cases, the boards are held in place with ballast, to counteract buoyancy from submerging the low-density foam insulation.

Ballast can include pavers or large diameter stones (¾ to 1½ in.). Dow markets a developmental interlocking composite panel of Styrofoam topped with latex-concrete.

The protected membrane system isolates the roof membrane from foot traffic, hail, dropped tools, ice, and stresses resulting from extreme thermal cycling. The membrane also acts as a vapor retarder provided the insulation quantities are sufficient to maintain the temperature at the underside of the membrane above the dew point of interior air. The advantage of Styrofoam insulation, as claimed by the manufacturer, is its resistance to moisture, its high compressive strength, and its retention of a high percentage of the original properties after repeated abuse and moisture exposure. The IRMA system is conditionally guaranteed by its manufacturer.

The disadvantages of the protected membrane roof are:

1. The temperature of the membrane drops during rain or melting ice and snow, resulting in an increase in heat transmission; more heat is lost while the roof is draining.
2. A reduction of the membrane temperature may result in condensation, if the temperature at the underside of the membrane drops below the dew point of the interior air and remains at that level for a prolonged period of time.
3. The arrangement of these components in the protected membrane roof results in the retention of moisture at the membrane level. Water trapped between and below insulation boards cannot readily evaporate and dry. The membrane materials must be moisture resistant, or they will deteriorate rapidly because of the increased duration of the wet environment.
4. Dirt and other debris may accumulate between and below insulation boards, providing a condition favorable to vegetation growth; the roots can penetrate the membrane, thus admitting moisture.
5. Styrofoam is not fully resistant to moisture.
6. Excess bitumen may block drainage channels, keeping the roof wet.

The insulation that provides membrane protection will also hinder the detection and repair of any leaks that may develop. To prevent some of these problems, consider the following:

1. The membrane must be designed to accommodate potential structural movements, by being applied loosely or by separating it from the deck with additional insulation.
2. The flashings of penetrations and walls are not protected, and they must be designed to withstand abuse during construction and maintenance, and during service exposure.
3. Provide a continuous filter layer between the insulation and stone ballast to prevent the stones from dropping between insulation joints and accumulating under and lifting the insulation boards.
4. The membrane must drain. Provide adequate slope to the roof decks and design special drain sumps that will not clog easily and that can be cleaned readily by maintenance personnel. Ideally, design an open drainage area above the system below any "drainage layer".
5. Use concrete pavers as ballast; the pavers must be elevated from the Styrofoam to prevent formation of a water film that interferes with normal evaporation. Pavers must be structurally able to support imposed loading in addition to their own weight, have low moisture absorption, and be resistant to deterioration from water immersion and freeze/thaw action.

26.11.2 Styrofoam Protection Systems for Waterproofing

The IRMA system may be defined as a waterproofing system because the membrane is not exposed. Similar systems can be used for waterproofing applications. Follow the design principles for roofing but, even more important, provide positive drainage of the surfaces in waterproofing because the system is subjected almost constantly to moisture and frequently to freeze/thaw cycling.

Specially designed pavers are supported by pedestals to provide drainage space below and serviceable surfacing above. The pavers can be topped with fill or other landscaping. The structural capacity of pavers, insulation, and decks must be checked carefully before using this system.

In addition, drainage is usually needed at the membrane level as well as above the protection course; specially manufactured double drains are available for this purpose, but the drainage holes are often inadequate or become blocked. Figure 26.26 shows a schematic drawing for this type of drain. The drain system must provide for convenient in-service cleaning.

As decks for waterproofing are almost always concrete, the membrane should be loosely laid neoprene sheet; adhered membranes will crack at concrete cracks and built-up membranes cannot be used over the unstable Styrofoam; other available insulations are too soft to use between the membrane and the roof deck.

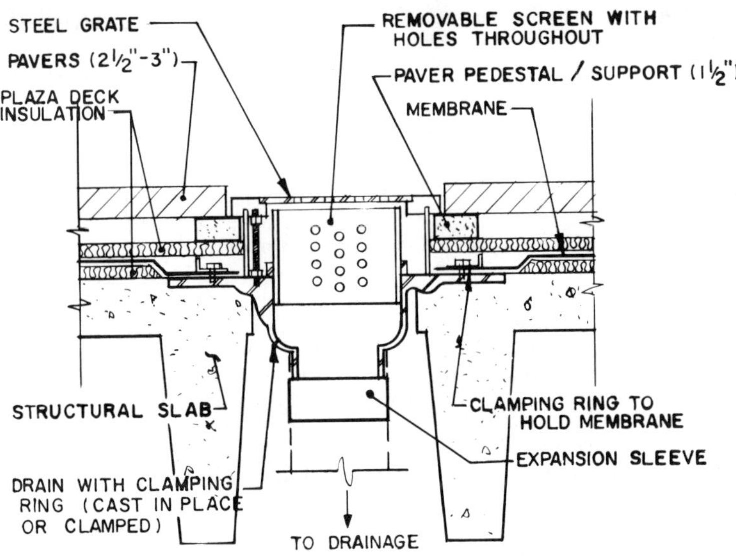

Fig. 26.26 Plaza drain detail.

Styrofoam is also available as "PD" insulation with increased compressive strength for roofs used as traffic spaces.

26.12 PREPARATION OF DOCUMENTS FOR CONSTRUCTION

26.12.1 Use of Specifications

The specifications, in conjunction with the detail drawings, establish the minimum standards and quality for construction of the roofing system. The designer communicates his roof system design, through these documents, to the contractor. Specification documents must be as complete as is necessary to achieve the roof installation that is envisioned. Specification documents can be prepared from two different approaches:

1. Direct instructions to the contractor, combined with detailed drawings which establish the exact system, flashings, and application methods to be used.
2. Performance specifications that outline the requirements for the general type of roofing, insulation value, and system life expectancy. The roofing contractor may be required to provide shop drawings of flashing details and manufacturer's literature for approval.

The use of a performance specification shifts the actual design of the roofing system from the designer to the roofing contractor and his selected suppliers. The responsibilities for the design will remain with the designer, since he has established the design concepts and submissions must be reviewed and approved. Submissions under a performance specification often involve various systems and products which increases the difficulty of evaluating them during bidding and prior to a contract award. A performance specification may be the command "use anything you wish to keep the building dry." More specific requirements must be used to establish the desired level of quality. This is the duty of the designer, not that of the contractor; the designer designs, the contractor builds. However, communication between them is vital to a successful project.

There are often as many opinions of how to weatherproof a building as there are roofing contractors bidding. Do not use this procedure to obtain the least expensive "guaranteed" roof system for your project. It always results in problems.

Specifications and drawings must deal with all aspects of the roofing system and its application. Nothing should be left to the interpretation of the contractor because his interpretation might not always be to the best interest of quality construction. In theory, the specifications are the minimum acceptable standards for construction, but in practice, the specifications become the maximum obtainable standards; adjusted savings almost always are largely a benefit to the contractor.

Guidelines for the development of contract documents for roofing include:

1. All elements of construction must be described in detail, from preparation and acceptance of the roof deck to final cleanup procedures.
2. All materials must be specified. If a particular manufacturer's product is required, state this in the document. "Equal" products are not always equal. The designer must assume the responsibility for this determination; recent court decisions have affirmed this professional duty.
3. Include all workmanship standards, materials qualities, quantities per unit area, step-by-step installation procedures, acceptable weather conditions for work, protection of uncompleted work, circumstances for rejection of work, and the correction of rejected work.
4. Include all required guarantees from the contractor and materials manufacturer.
5. Provide detailed drawings of the construction, including all types of flashings, as outlined in Section 26.12.4.
6. Provide detailed drawings showing the procedure for the intersections of various flashings and building components. Typical flashing details can be confusing and often impossible to provide at nontypical locations. The flashings and materials transitions are the most critical areas on the roof because they are the most difficult to install and frequently they are ignored by all parties until it is too late to construct them properly. These conditions are often sources of leaks because of inattention to details.
7. Provide working tolerances when possible, described to be meaningful to the construction of the work, and enforceable.

26.12.2 Interpretation and Use of Manufacturer's Literature

Roofing manufacturers publish product use information and application specifications and procedures in their catalogues. The information provided, although generally useful, is rarely sufficient to ensure a proper installation of the roof system, since it is restricted to the mention of material quantities, deck and slope requirements, general application procedures, and flashing details for typical conditions. This literature usually provides little, if any, information for variations from typical conditions.

The designer should not be satisfied solely with meeting the specifications of the manufacturer; these specifications should be a starting point. All minimum requirements of the manufacturer's specifications must be met or guarantees may be jeopardized; there is generally no problem with exceeding the manufacturer's specifications but this must be verified in each case.

Roof system manufacturers have made claims about their products' use that are contrary to good roofing practice such as: phasing membrane construction, leaving the roof membrane exposed without the surfacing, or applying the system in the presence of small amounts of ice or snow. Historically, application practices or material uses that vary from the accepted norm have resulted in failures. Approach all claims cautiously; whenever there is any doubt, adhere to known product uses and application procedures.

26.12.3 ASTM and Other Standards

Materials and components specified in the roofing system should conform to accepted standards. The American Society of Testing and Materials and many trade associations develop and publish minimum standards for characteristics and performance of materials, systems, and workmanship. These standards offer the designer an opportunity to specify materials that have known minimum qualities and performance characteristics.

The specifications should require that all materials have imprinted on their label that the product conforms to the appropriate standard. If this is not the case, obtain separate certifications from the manufacturers. This certification ensures that the manufacturer has performed the tests required by the standards and that the product meets the minimum requirements within the standards. The consumer cannot be assured of the quality of products without conducting testing programs to determine the material and performance characteristics of each product. Certification by manufacturers are not always accurate; for critical items, require independent testing for quality standards.

26.12.4 Development of Flashing Details

Flashings are used to provide a watertight connection between the roof membrane and the interruptions and penetrations on the roof surface. Roof drains, vent pipes, electrical conduits, expansion joints, rising walls, gravel stops, skylight and equipment curbs, and equipment supports are among the more common items that require flashings for watertight connections to the roof membrane.

Flashings are fabricated from metal, bituminous membrane, or polymeric materials. Membrane flashings can be connected and sealed to a built-up membrane, and polymeric materials can be connected and sealed to polymeric membranes, metal, concrete, masonry, and wood components. Metal flashings are readily connected to other roofing materials but a permanent seal is difficult to maintain due to differential thermal movements that occur during temperature cycling. Because of this sealing problem, metal flashings are principally designed to divert and to shed water onto other roof surfaces.

The contract documents should include drawings showing all roof-related details and their flashings. All materials must be shown in the sequence which they should be applied. Sections through a detail may provide the required information for simple details but complex flashing should be presented using isometric drawings. With isometric drawings, the various materials can be cut away showing application sequence, attachment procedures, extent of overlap, and procedures at complex intersections for different details. All materials must be identified; all significant dimensions shown, including but not limited to, minimum heights, overlaps, and fastener spacings.

When designing flashing details the following points should be considered:

1. Allow room for proper flashing. Locate all penetrations, curbs, and equipment supports a minimum of 2 ft (600 mm) from walls or other components.
2. Avoid contact between sheet metal and roofing. Do not use metal base flashing where other acceptable material will perform the task.
3. Ensure that all materials are compatible.
4. Raise the flashing around all penetrations (except drains), walls, and perimeters using cedar bevels, cants, and tapered edge strips.
5. Set roof drains below the roof surface and form a sump with tapered insulation strips.
6. Use cant strips at the base of vertical intersections with the roof to provide a transition for the flashing materials.
7. Construct built-up base flashings with a reinforced system, applied with flashing cement and covered with aluminum roof coating for additional protection and improved appearance. The tops of flashing sheets should be secured at 4 in. (100 mm) on center with large head mechanical fasteners or nails driven through metal discs. Fasteners are driven into wood nailers provided specifically for attaching base flashings.
8. Detail metal counterflashings of two-piece construction; be certain they overlap base flashings a minimum of 4 in. (100 mm).
9. Isolate base flashing from walls or parapets that are not tied into the roof structure to prevent differential movements from rupturing the components and causing leakage.
10. Raise expansion joint covers above the general roof surface using wood curbs and metal flashings or wood nailers, tapered edge strips, and elastomeric flashings. Polymeric flashings should only be used where structural axial and shear movements will be relatively small, and when they are not in direct contact with any bituminous material.
11. Limit metal roof flanges to the width of the nailer in the roofing system, and nail them securely with two rows of nails 4 in. (100 mm) on center and with the nails between rows staggered to provide a nail every 2 in. (50 mm). Flanges are placed on top of roofing, and nails should be properly waterproofed with a strip flashing system.
12. Use through-wall flashings in all walls and parapets to prevent water from bypassing surface flashing within the wall.

26.13. THERMODYNAMIC PROPERTIES OF MOIST AIR

The psychrometric chart graphically represents the thermodynamic properties of moist air. Some of the properties shown on the chart are: entropy and humidity ratio coordinates; lines of constant dry bulb, wet bulb, and dew point temperatures; relative humidity; humidity ratio; and volume. If two properties of a given air condition are known, any of the other properties can be easily obtained graphically. The psychrometric chart is frequently used to determine the dew point of interior air to predict condensation potential (see Fig. 26.27).

26.14 CONSTRUCTION PHASE

The roofing design is a critical part of any building project. This chapter has emphasized that the design of roofing must be integrated with other building components early in the design of the building. Previous sections discuss selection and design of the structure, roof deck, roofing system, adjacent construction materials and finishes, and the placement of roof-top equipment. Logically, the construction process must address these elements in the same process as the design. Each

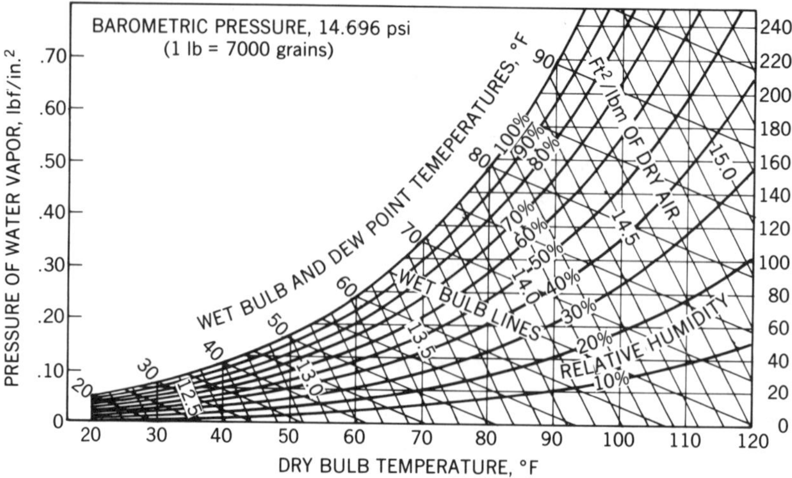

Fig. 26.27 Psychrometric chart.

contractor, subcontractor, and supplier must be aware of the requirements of their performance in the overall roofing construction.

26.14.1 Coordination of Contract Work

A general contractor, a construction manager, or the owner may manage the project; in each instance, similar procedures are to be followed. The timing for each component involved in the roofing is critical to a successful installation. The following checklist is representative of a typical project, but it is not all-inclusive. Before roofing work begins, the following items should be resolved in detail.

Roof Deck Construction, Including Approval by Roof System Manufacturer if Applicable

Check tolerances of precast units, flatness, accuracy and attachment of steel deck ribs, requirements for anchorage, and, in the case of cast decks, any weather restrictions for their installation and protection. Provide deck protection for trades that will use the deck as work surfaces before roofing is installed, as some decks will not sustain construction loads and traffic without damage. Crickets and slopes to be constructed on top of the basic deck should be in place. Deck slopes and deflections must be checked to be certain there will be no ponding; this should generally not be the roofing contractor's responsibility.

Walls and Structures Above the Roof

All construction necessary above the roofs should be in place prior to the roofing work; masonry washdown or treatment should be completed. Exterior walls should be in place, so perimeter flashing can be completed with the roofing work.

Plumbing and Mechanical Services

Roof drains should be in place and piped to outlets; they should never be plugged after roofing is installed. All penetrations, including sanitary vent stacks and all other piping should be through the roof and secured.

Roof-Top Equipment

All units should be approved from dimensioned shop or manufacturers' drawings. Exact sizes are needed to provide properly located and sized supports and curbs with adequate clearance allowance for flashings. Units should be placed before roofing work begins. Expect major problems from the installation of solar collection units that cover large roof areas. Arrange for transportation and installation of large mechanical units so the roof will not be damaged.

Wood Blocking

All wood blocking needed for roofing construction should be in place or ready for immediate installation as the roofing work begins. Any wood preservative or fire-resistant treatments must be compatible with roofing materials.

Control of Construction Traffic

During construction of the roofing, personnel not involved should be restricted from the roof, and from the area immediately below the work if hot bituminous materials are in use. The roofing contractor should be provided with a safe, secure work area, free from interference.

26.14.2 Materials Storage

Before the installation work begins, materials that have been previously approved should be delivered to the job-site and properly stored. Contractors may attempt last-minute substitutions because of alleged unavailability of specified materials; this should not be allowed.

Materials should be stored off the ground (and roof surface) on dunnage, and covered with opaque, breathing tarpaulins. Polyethylene covers are not acceptable as condensation forms on the inner surfaces and the materials become wet.

Some materials are shipped in polyethylene covers. These covers are not suitable for protection during outside storage; they should be removed at the site.

Some roll goods (mineral-surfaced roofing) and canned goods (cements and adhesives) require storage at manufacturer-specified maximum and minimum temperatures, or at least conditioning at these temperatures before use. Ideally, storage of all materials should be in covered (waterproof) trailers. Some materials (such as organic felts and fiberboard insulation) can absorb significant amounts of moisture from a humid environment. In hot humid climates, amounts of material and the time they are stockpiled on site should be kept to a minimum. To avoid damage to the materials, they should be stored where minimum handling is required between storage and installation.

If materials are to be stockpiled on the roof deck, the designer should provide information to the contractor about their placement in relation to allowable deck loads. Only the material for one days work should be stored on the roof.

The contractor should arrange security procedures to safeguard the materials from theft, vandalism, and damage from weather in the event the tarpaulin protection becomes dislodged. Some materials are not immediately and conveniently replaceable; loss by damage, vandalism, or theft could delay the work substantially. Any unprotected materials should be clearly marked and removed from the site.

The party responsible for monitoring the work should examine the stored materials for compliance with contract requirements, and take samples for laboratory testing to assure that they meet the specifications.

Many contracts provide for payment to the contractor for stored materials. In this event, conduct an inventory of the properly stored materials with the contractor. The value of the materials is usually ascertained by certified invoices. When the owner pays for stored materials, their security becomes even more important.

26.14.3 Preconstruction Conference

A properly managed conference should be held in advance of the roofing construction. The conference should be scheduled after initial submittals of materials and details so they can be reviewed at the meeting. Representatives of all interested parties should be present. Table 26.11 is a typical checklist for a preconstruction conference.

The roofing work should begin only when all the items that must precede it have been completed. When the designer's responsibility extends to project coordination, preparation for the roofing work is included. Often, pressures of the owner's or general contractor's schedules have the effect of forcing the roofing contractor to perform the roofing work before the project is ready or under adverse weather conditions. These procedures can defeat even the best design practices and can result in roofing failures.

For scheduling purposes, all parties should understand weather conditions that will limit roofing work and who will make work or no-work decisions. In general, work should not be performed in below-freezing weather or in the presence of any moisture; 40°F is a good minimum guideline for low-temperature work. Wind will also restrict work, particularly if a single-ply sheet membrane is being installed.

Effective communications are a necessity in any successful project, particularly in roofing installations, because of the complex requirements for coordination. All communications should be

Table 26.11 Agenda for Preconstruction Conference On Roofing

1. The following parties should attend:
   Architect (if applicable)
   Owner (representative)
   Project manager and resident engineer/architect
   Roofing contractor, including field superintendent (and foreman)
   Sheet metal contractor (if applicable)
   General contractor (and/or his superintendent)
   Other subcontractors (if appropriate; especially masons, etc., if working above roofs)
   Materials suppliers (if asked by roofing contractor)
2. Contract and bond (ask owner if all items are correct)
   a. unit prices
   b. time scheduling
   c. special items (such as access and elevator use)
   d. progress payment dates and submittal details
3. General and supplementary general conditions
   a. insurance certificates (note forms, amount, hold harmless clauses)
   b. changes in work—procedures
   c. communications—through engineer/architect
   d. progress schedule—submit
   e. schedule of values—submit
   f. payment of materials not in work, or stored off site—establish procedure
4. Special conditions: services, water, power, sanitary, parking, access.
5. Technical sections: an item by item read-through with questions (and answers), particularly:
   a. submittal procedures—review items not yet submitted
   b. function of resident engineer (no jobsite approvals)
   c. material storage
   d. existing roof drain conditions
   e. coordination of work with other trades
   f. testing
   g. early order items (stainless annular ring nails, etc.)
6. Communications: establish communications procedures for observer to clarify or amend specifications if issues arise or serious problems occur, for change order notification, and to clarify who is authorized to direct work stoppage. Include list of who gets daily reports.
7. Safety is the contractor's responsibility; he shall provide safe conditions for monitor to observe work.
8. Drawings and details—page-by-page detail sheet review.
9. Subcontractors: develop coordination through general contractor with subcontractors whose work has bearing on roofing work, such as plumber, carpenter, roof deck applicator, mason, electrician, HVAC contractor.
10. Suppliers—establish duties and notification needed for guarantees and inspections.
11. Summary and questions.
12. Exchange of phone numbers for office and home for emergencies and daily contact.
13. Record meeting minutes and distribute to participants.

through the designer and in writing to maintain project control and avoid misunderstandings. The above procedures for communications should be written into the roofing specifications.

26.14.4 Inspection of the Work

The workmanship of a roofing system is the most important ingredient of a successful installation. The designer cannot, and should not, exercise control through supervision of the contractor's performance of the work. However, the work should be observed full-time by a resident monitor trained in roofing technology, as the first line of defense to assure compliance with the contract documents.

Roofing can not be inspected after construction; each operation covers the preceding one. Some variation is unavoidable in a handmade product; therefore, constant effort to achieve perfection is necessary. If each operation is less than perfectly performed, the cumulative effect becomes the product of the percentages.

The responsibilities of a monitor should be defined in the contract documents and explained at the preconstruction conference. The monitor is to observe the work for compliance with the contract documents, to interpret the documents (when necessary), and to keep all parties (including the contractor) fully and promptly informed in writing. Reports are of little use if they are not distributed until long after improper work has been completed, if they are not read promptly by the designer, owner, and contractor, and if appropriate action is not taken promptly.

The monitor should not have authority to approve materials or details on-site, to direct or stop the work, or to authorize changes (or extra work) to the contract documents.

The designer or owner may assign specific related tasks to the monitor, including coordination of work where building equipment operation or use of spaces by others is involved, assistance to the contractor in preparing quantity estimates of work installed, and limited authorization to direct extra work in accordance with allowance or unit price items in the contract. The monitor should never threaten nonacceptance or disapproval of work or the withholding of payments for work, unless specifically authorized to do so. In general, these functions are best reserved to others in higher positions.

The best procedures and reporting are only effective insofar as the authority (owner or designer) is informed and takes appropriate action.

A successful project results from the combined efforts by all of the parties, who must recognize when prompt action is necessary, and how strong that action should be. The contractor must respond to the directions of the designer in a timely and cooperative fashion, as the work proceeds.

The owner will depend upon the designer for advice, and to evaluate the contractor's response, when appropriate. Constant communication and prompt action when called for are necessary ingredients for success.

26.14.5 Performance of the Work

Roofing work must be performed in accordance with the contract documents, which almost always include references to published information from the manufacturers of the roofing materials and systems.

Specific workmanship procedures are not within the scope of this chapter; the selection of the roofing system will determine partially the applicable workmanship standards. Responsible manufacturers of long-standing reputation publish complete application instructions for their products. In the absence of such detailed standards, the system (material) probably should not be used. Do not vary from the manufacturers' directions unless the procedure is in writing and signed by an authorized technical representative. All manufacturers' "recommendations" should be followed; these documents are usually more stringent than their "requirements".

26.14.6 Testing

Confirm by testing that materials delivered to the site meet specification before inclusion in the construction. Sampling should be of typical and representative materials. The amount of testing performed is a judgment, to some degree based on prior test results and the past performance of familiar materials. Even these sometimes have unusual manufacturing defects or arrive on-site damaged in transit.

Special tests are needed when a dispute arises about materials that have been exposed or improperly installed. Test results are usually for the owner's information and do not relieve the contractor of the responsibility of providing specified materials. Properly labeled materials from reputable sources almost always conform to their label information and only spot checks should be required.

Avoid routine testing of installed work (if monitored); it unnecessarily damages the new roofing.

26.15 USE, MAINTENANCE, AND LONG-TERM PERFORMANCE OF ROOFS

Correctly designed, installed, and maintained roofing systems will provide trouble-free service for more than 20 years. Roofing systems are designed to provide long-term weather protection for a building; however, they can be damaged easily by careless or improper treatment, and are not suitable for use as mechanical room floors, recreational areas, sign platforms, window washing machine traffic pads, storage facilities, or waste drainage transfer.

26.15.1 Use of Roof Surfaces

Many roof systems begin to fail from the day of installation, being subjected to mechanical damage from uncontrolled construction traffic. This problem was discussed in an earlier section. The same principles that apply to construction work also apply to maintenance work. Roof-top traffic should be restricted to only trained personnel with necessary legitimate business on the roof; keep roof access doors and hatches locked. Constant vigilance by the owner is needed because service personnel can gain access to a one or two-story roof with ladders carried by refrigeration, solar, cable TV antenna and telephone equipment servicemen. The inconvenience of going through a security office makes this independent access very enticing to service personnel.

Roof-top security is important for the following reasons:

1. Every equipment service activity in a building should be monitored for reliability, safety, and security reasons. Personnel on roofs should be registered with an inside party. In the event of an accident (personal injury), someone should be aware of their presence, especially in bad weather or at night, when emergency services may be needed.
2. Roof service life will be shortened by abuse of the surface; arrange for assurance that only trained personnel are on the roof.
3. Casual roof-top traffic for sightseeing or sunbathing is hazardous to the roof surface and to the participant who is in danger of falling. The possibility of access for irresponsible persons to endanger themselves should be eliminated.
4. Unsecured roof access doors and hatches are easy ways into buildings for burglars and vandals.
5. Well-intentioned maintenance work can damage roofs and flashings, and void guarantees.
6. Unrepaired damage to the roofing membrane could lead to large-scale roof system degradation requiring costly repairs to correct.
7. Improper use of roofs may void guarantees.

A responsible party should maintain control of roof access including verifying that persons have left the roof safely and resecuring all doors. Check fire safety laws where some doors cannot legally be locked. In these cases an alarm can be activated when the door is opened, alerting security personnel.

26.15.2 Maintenance Schedules

Roofs should be inspected at least twice annually, preferably spring and fall. Roof drains may need to be cleared more often depending on local conditions, such as overhanging trees. New roofs should be viewed during hot weather for the first several years when any blisters will be most apparent, after substantial rain to observe any ponding, and after snow to verify that flashings protect walls and openings from drifting and blowing snow. At sloping and overhanging roof areas, check for damage from falling icicles or water overflow.

Periodic resurfacing treatments (or resaturating) of built-up roofs are generally not recommended. Their beneficial effect on roofs in good condition is questionable, and on roofs in poor condition this kind of treatment will not cure basic defects but is often a detriment, hastening their deterioration. If the membrane is sound, it needs no further help; otherwise, the basic problems must be determined and repaired. Coatings and resurfacers will not "rejuvenate" built-up roofing systems or solve problems associated with blisters, splits, buckles, or tears; they will make reroofing work more difficult and more costly.

Keeping roof-top traffic to an absolute minimum is the most effective preventative step toward assuring good roof performance.

Maintenance of roof-top equipment should be regularly scheduled, so the service personnel can be trained in the particular characteristics of the roofing system and its care. When emergency service is needed, a trained member of the building maintenance staff should accompany equipment service personnel to assure proper care of the roofing and flashing.

Servicing personnel should have written instructions for protecting the roof surface, such as:

1. Instructions for access to the roofs.
2. Limits of loads for roof-top equipment such as liquid gas tanks, replacement motors and tools, and instructions for providing protective underlayment.
3. Precautions for roofing protection such as wheeled carts with large rubber tired wheels and plywood underlayment (no work should be performed or materials transported over unprotected surfaces). Take precautions that plywood is not susceptible to displacement by wind.

4. Precautions against fluid spills onto roof surfaces (and cleanup instructions in case of accidental spills—some materials should not be allowed to enter roof drainage systems).
5. Prohibitions against burning, open flame soldering, or welding on roofs unless prior arrangements have been made for fire watches (inside and outside), extinguishers, and protective blankets. This is particularly true for "new" torched-on modified bitumen sheets.
6. Information about utility services (including fire protection equipment) available on the roof.
7. Directives about whom to notify in case of accidental spills or damages.

26.15.3 Maintenance Personnel

Except in very large organizations, the average maintenance worker cannot usually perform roof repairs. The reasons for this include:

1. Unfamiliarity with roofing.
2. Difficulty of access.
3. Distaste for handling bitumens.
4. Avoidance of exposure to hot and cold temperature extremes.
5. Fear of heights.
6. Lack of special equipment and materials.

Unless trained roofing and sheet metal mechanics are available, roof repairs should not be attempted. This does not mean that roof maintenance should not be continually performed, but it should be generally restricted to the following:

1. Periodic inspection of roofing and flashing.
2. Clearing of debris from roof drains, gutters, and strainers.
3. Emergency patching of holes with cold patching materials promptly followed by arrangements for permanent repairs. The emergency repair materials should be compatible with the roofing system; this requires planning for emergencies.
4. Resecuring loose metal or accessories with screws or nails.

Personnel in charge of maintenance should have available:

1. Copies of guarantees and bonds.
2. Names of approved service organizations to use for repairs.
3. Names of manufacturers of roof system materials.
4. As-built drawings and specifications.
5. Updated records of roof repairs and added roof-top equipment.

Many roof guarantees include provisions that require special notifications when repairs are made, and in some cases repairs may have to be performed by contractors acceptable to all parties. All communications concerning roof repairs should be confirmed in writing to the guarantors.

26.15.4 Bonds, Guarantees, and Records

The term "bonded roof" has been commonly used as a standard in the roofing industry for over 50 years. Designers and owners have depended on roof bonds to assure roof performance and they have used words such as "bonded type" to incorrectly describe standards of roofing systems. In addition, almost every roof carries some form of guarantee. In this discussion the words "guarantee" and "warranty" will be used as equivalent terms. The following information is general; specific instances should be reviewed with legal counsel.

Roof bonds are written assurances that certain limited obligations are undertaken by the manufacturer of the roofing system. Some bonds are also underwritten by a casualty or insurance company. The bond life of 5, 10, 15, 20, or 25 years has no direct relationship to the expected roof life. The terms and conditions of bonds vary from manufacturer to manufacturer and from year to year, as their policies change. The "penal sums" are usually totally inadequate to protect the owner.

In general, bonds state that certain defects in roofs will be repaired by the manufacturer up to a face value of the bond, called the penal sum. The penal sum is usually a nominal value. Some manufacturers offer several levels of penal sum amounts and varying periods of bond enforcement. The cost to the owner for each of these terms is different.

Alternatively, manufacturers offer guarantees that provide for repairs to leaks without limit by a penal sum, although in some cases bond and guarantee are words used interchangeably. There are as many forms for bonds and guarantees as there are roofing materials; therefore, specific examples are not included. Specifiers should exactly describe the guarantee required in the contract specifications, and should verify that it is commercially available with the roofing system to be used. Each guarantee should be carefully studied for its conditions, terms, inclusions, exclusions, and limitations. It is usually best to obtain a simple guarantee; most language in a guarantee restricts the manufacturer's responsibility. Quite often a manufacturer's guarantee specifically omits workmanship problems; this document is simply a material, not a system, guarantee.

Most manufacturers' bonds and guarantees are issued directly to an owner, although the materials are purchased and installed by a contractor. Manufacturers may have a separate agreement with the installing contractor for certain limited repairs, especially for the first two years; many specifically omit responsibility for workmanship defects.

A "personal" guarantee from the contractor should also be obtained. This guarantee should run directly to the owner and should survive the termination of the contract. At least two years, possibly three, is the recommended minimum acceptable guarantee period.

No contract should be considered complete unless all guarantees and bonds have been delivered to the owner, specifying the roof size, location, and date of completion or acceptance as the contract may require. This is often forgotten at the end of a project and can cause problems later when the owner is searching for the guarantee that was never received, or never issued.

Some guarantees and bonds are extra cost items, while others are included in the materials purchase price and are available at "no cost" if requested. "Free" guarantees should always be carefully reviewed in detail, as they may restrict the owner's right of recovery in case of trouble. Bonds and guarantees that have a premium cost are generally poor "insurance" values. The decision to purchase bonded coverage should be carefully reviewed by the designer, and the owner and his attorney. The premium is usually better spent on assuring roof quality. No bond or guarantee will protect against poor workmanship, improper design, or careless use of roofs.

In addition to the above, the following documents should be kept in a safe location and copies should be readily available to those responsible for roof maintenance:

1. As-built drawings including roof opening locations.
2. Specifications including approved changes.
3. Lists and manufacturer's data on each piece of original roof-top equipment as well as any added equipment (include records of utilities to the equipment and the spaces served by the equipment).
4. Names of the manufacturers and distributors of roofing materials.
5. Name of the contractors responsible under all guarantees.
6. Names of parties to be contacted under terms of bonds and guarantees.

All repair and maintenance work, as well as any leaks or roof damage, should be part of a permanent record.

SELECTED REFERENCES

26.1 ARMA. *Residential Asphalt Roofing Manual* (Article 7d). Washington, D.C.: Asphalt Roofing Manufacturers Association, 1980.

26.2 FACTORY MUTUAL. *Insulated Steel Deck* (Bulletin I-28). Norwood, MA: Factory Mutual Engineering Corporation, 1980.

26.3 NRCA. *Good Roofs Save Energy*. IL: National Roofing Contractors Association, 1977.

26.4 Charles G. Ramsey and Harold R. Sleeper. *Architectural Graphic Standards*. New York: Wiley, 1970.

26.5 SMACNA. *Architectural Sheet Metal Manual* (AIA File No. 12-L). IL: Sheet Metal and Air Conditioning Contractors National Association, 1965.

26.6 ASTM. *Specification for Lead-Coated Copper* (B101-83). Philadelphia: American Society for Testing and Materials, 1983.

26.7 ASTM. *Specification for Pressure Treatment of Timber Products* (D1760-83b). Philadelphia: American Society for Testing and Materials, 1983.

26.8 ASTM. *Specification for Zinc Coating (Hot-Dip) on Iron and Steel Hardware* (A153-82). Philadelphia: American Society for Testing and Materials, 1982.

26.9 ARMA. *Manufacture, Selection and Application of Asphalt Roofing and Siding Products*. Washington, D.C.: Asphalt Roofing Manufacturers Association.

CHAPTER 27
BUILDING FACADES

REXFORD L. SELBE

Senior Consultant
Wiss, Janney, Elstner Associates
Northbrook, Illinois

JERRY G. STOCKBRIDGE

Vice-President
Wiss, Janney, Elstner Associates
Northbrook, Illinois

27.1 INTRODUCTION

Although all components of buildings are susceptible to failures resulting from poor design, building facades in recent years have had a disproportionate share of serious problems. Many of these problems have occurred because of inadequate collaboration between the architect and the structural engineer.

Close collaboration normally takes place between the architect and the structural engineer when a building facade performs the structural function of a bearing wall. However, all too often, the architect requests little or no engineering assistance when a facade is a curtain wall.

What the architect must understand is that the assistance of the structural engineer is needed in the design of a curtain wall just as in the design of a bearing wall. First, all facades, even curtain walls, have to perform the structural function of resisting wind and seismic loads, and of supporting their own weight. Second, proper detailing of a curtain wall requires a complete understanding of the anticipated behavior of the structure to which it is attached.

There was a time when only a limited number of facade types were used and safety factors were greater, and the architect could be assured of a successful result by relying only on his knowledge and experience. Today, however, technological progress has greatly widened the architect's area of choice and there are many facade systems in the market only recently developed and of limited proven performance. Even the systems with a proven history of commendable performance and which traditionally have been designed by rule of thumb, frequently are being revised and modified to take advantage of more liberal design criteria. Consequently, architects should solicit the help of a structural engineer to assist in the design of the building facade. Where the architect is not seeking this assistance, the structural engineer is strongly encouraged to get actively involved in educating the architect to the important structural considerations involved in the design of all facades.

The material in this chapter is presented in the following sequence:

The "General Design Considerations" section provides a checklist of some of the more significant items that should be considered in the design of a building facade.

The "General Comparison of Facade Types" section provides data to aid in the selection of facade types potentially suitable to satisfy specific design criteria.

The "Specific Design Considerations" sections highlight important factors which should be considered when designing some of the more commonly used types of building facades.

27.2 GENERAL DESIGN CONSIDERATIONS

The selection of a facade system for a particular building requires a careful comparison of the performance requirements of the building with the performance characteristics of the systems available. The following are some of the more important considerations involved in making the system selection:

1. Aesthetics.
2. Gravity loads.
3. Wind loads.
4. Seismic loads.
5. Thermal resistance.
6. Condensation control.
7. Moisture resistance.
8. Fire resistance.
9. Durability and serviceability.
10. Tolerance compatibility.
11. Volumetric changes.
12. Facade/structure interaction.
13. Cost.
14. Sound control.

27.2.1 Aesthetics

The character of the facade should reflect the function of the building, be sympathetic to adjacent structures, and, possibly, be representative of the region.

It must be remembered, however, that all facade design concepts cannot be executed successfully with all systems and materials that are available. It is essential that design decisions not be dominated by superficial considerations of appearance which are unrelated to performance.

27.2.2 Gravity Loads

Building facades that are designed to carry vertical loads other than their own weight are considered to be bearing walls. Facades that are designed to carry only their own weight are considered to be curtain walls. Because they do not carry other gravity loads, curtain walls generally can be made thinner and lighter than bearing walls. Use of lightweight curtain walls often can lead to a significant reduction in foundation costs, particularly in high-rise buildings.

Building facades that are required to have a large percentage of sizable openings normally do not lend themselves to being constructed as bearing walls. Also, many facades in common use today simply are not structurally capable of functioning as bearing walls irrespective of the layout of the openings.

Table 27.1 includes the approximate weight of some of the more commonly used facades. Table 27.1 also indicates which types of facades normally can be used as bearing walls and those which are normally used only as curtain walls.

27.2.3 Wind Loads

Wind loads are examined in detail in Chapter 2, and in Chapters 10 and 11 the importance of wind loads on tall building behavior is described. In this chapter the particular features of wind loading affecting facade design are presented.

Two aspects of wind loading normally must be considered in the design of buildings. First, there is the overall wind loading which is used to determine the strength and stiffness of the building. Second, there are the localized wind loadings which are used in determining the strength, stiffness, and anchorage of discrete parts and areas of the facade.

Localized wind loadings normally are greater in intensity than overall wind loads, which are the net result of wind loads occurring simultaneously at any one time on all building surfaces. The intensity and direction of localized wind loads will differ considerably on the various surfaces of the facade. They are influenced greatly by the configuration of the building.

Overall wind loads normally only include gusts of about 2 sec or more duration. Localized loads that are instantaneous can be critical and are normally considered. Internal pressures seldom have any influence on the overall wind loading since they normally balance out when summed up. Internal pressures, however, are a factor in determining localized loads. Localized loads on facade walls can be either negative or positive, but in most locations on buildings the critical loads normally will be negative. Overall wind loads may not be of concern for low buildings, but localized loads should be considered in the facade design on buildings of all heights.

Normally, the first step in establishing wind loading will be to determine the "basic wind speed" for the geographical location where the building is situated. Most codes provide maps for this purpose.

Once selected, the basic wind speed normally will be adjusted by a number of factors. The factors vary from code to code, but some which may be included are: type of terrain, building height, internal pressures, corner effects, distribution of operating windows, and height-to-length ratios of facades.

Wind velocity normally increases exponentially with height and produces quite different loading effects on different building facades. Facades on the windward side receive positive pressure while those on the leeward side are subjected to negative pressure. Corners of building facades experience greater pressure than flatter surfaces. Many building codes require that for a distance of one-tenth of the facade width in from the leeward corners, negative pressures must be doubled over the pressures used for the design of the remainder of the facade.

The loads exerted on each building facade are the combined effect of both the external and internal pressures. Internal pressures are influenced by wind, mechanical factors, and in very high buildings, by stack effect. The wind effect on internal pressure will depend on the number, size, and distribution of openings in relation to wind direction. If there are many more operable windows and doors on the windward facade than on the leeward facade, internal pressures will approach those developed on the windward facade. If there are many more operable windows and doors on the leeward facade than the windward facade, internal pressures will approach those of the leeward facade.

Stack effect in high buildings results in positive pressures on the upper floors and negative pressure on the lower floors. This effect is caused by warm air rising in a vertical passage, for

Table 27.1 Typical Facade Systems—Characteristics and Properties

		Approximate Wt, psf (kN/m²)*	Nominal Wall† Thickness, in. (mm)	*U* Value‡ Btu/sq ft/hr/F°	Fire Rating, hr	Sound Transmission Class	Suitable for Use as: Load-Bearing Wall	Suitable for Use as: Curtain Wall	Comments
	MASONRY								
1	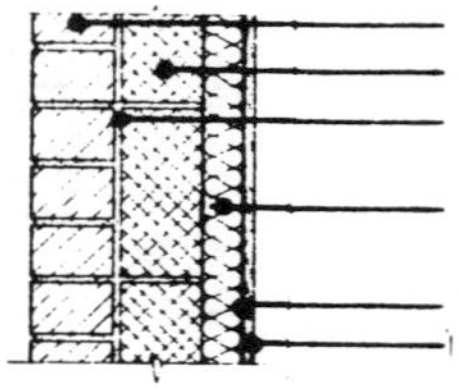 4-in. face brick 4-in. concrete block ⅜-in. filled collar joint 2-in. furring and insulation Vapor barrier ½-in. gypsum panels	89 (4.3)	11.0 (280)	0.27(a) 0.08(b) 0.06(c) 0.10(d) 0.06(e)	4	58	Yes	Yes	Typically has low maintenance requirements, highly durable and abuse resistant Among the heavier walls in common use
2	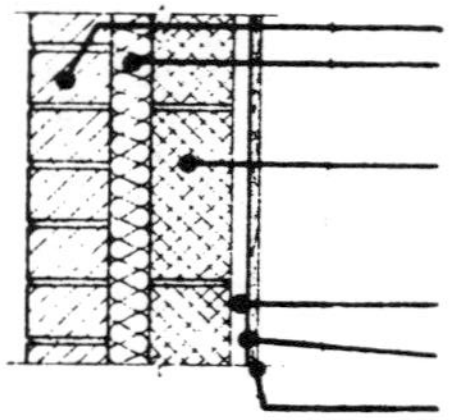 4-in. face brick 2-in. insulation or air space 4-in. concrete block heavy aggregate (see note) Furring J.S. Vapor barrier ½-in. gypsum panels	76 (3.6)	11.0 (280)	0.19(a) 0.07(b) 0.05(c) 0.09(d) — (e)	4	55	Yes	Yes	Typically has low maintenance requirements, highly durable and abuse resistant Among the heavier walls in common use
3	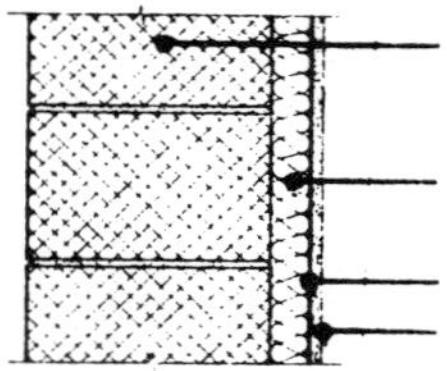 12-in. painted concrete block (lightweight) 2-in. furring and insulation Vapor barrier ½-in. gypsum panels	58 (2.8)	14.5 (368)	0.28(a) 0.08(b) 0.06(c) 0.10(d) 0.06(e)	4	58	Yes	Yes	Normally most economical solid masonry wall Single-wythe masonry walls are normally not as weather resistant as multi-wythe masonry walls.

	CONCRETE								
4	8-in. cast-in-place concrete 2-in. furring and insulation Vapor barrier ½-in. gypsum panels	95 (4.6)	10.5 (267)	0.34(a) 0.08(b) 0.06(c) 0.11(d) 0.07(e)	4	60	Yes	Yes	Normally low maintenance One of the heavier and thicker wall systems Normally only used as a load-bearing and/or shear wall
5	6-in. precast concrete panel 2-in. furring and insulation Vapor barrier ½-in. gypsum panels	72 (3.4)	8.5 (216)	0.36(a) 0.08(b) 0.06(c) 0.11(d) 0.07(e)	4	57	Yes	Yes	Panel size limited by ability to transport and handle
	METAL STUDS								
6	4-in. face brick ½-in. gypsum sheathing 1-in. air space 2½-in. metal studs 2-in. insulation Vapor barrier ½-in. gypsum panel	50 (2.4)	9.0 (229)	0.24(a) 0.08(b) 0.06(c) 0.10(d) 0.08(e)	2	57	Yes	Yes	Often lowest initial cost Seldom used as load bearing over 5 stories in height Thin Lightweight

Table 27.1 (*Continued*)

		Approximate Wt, psf (kN/m²)*	Nominal Wall† Thickness, in. (mm)	*U* Value‡ Btu/sq ft/hr/F°	Fire Rating, hr	Sound Transmission Class	Suitable for Use as: Load-Bearing Wall	Suitable for Use as: Curtain Wall	Comments
	METAL STUDS (*continued*)								
7	¾-in. stucco expanded metal lath ½-in. gypsum sheathing 3½-in. metal studs 3½-in. insulation Vapor barrier ½-in. gypsum panels	12 (0.6)	5.5 (140)	0.35(a) 0.08(b) 0.06(c) 0.10(d) 0.08(e)	2	48	Yes	Yes	Often lowest initial cost Seldom used as load bearing over 5 stories in height Thin Lightweight
	VENEER								
8	2-in. stone veneer e.g., granite, marble 3½-in. mineral fiber insulation Vapor barrier	28 (1.3)	5.4 (140)	N/A(a) 0.09(b) 0.07(c) 0.12(d) 0.07(e)	2§	42	No	Yes	Normally thin. One of the lighter systems Normally contains numerous sealant joints requiring maintenance
9	Aluminum or steel 3½-in. mineral fiber insulation Vapor barrier				2§		No	Yes	Often expensive initial cost Normally supported off the structure by girts, clips, channels, masonry, etc.
	⅛-in. aluminum and insulation	3 (0.14)	3.5 (89)			24			
	⅛-in. steel and insulation	6 (0.29)	3.5 (89)			30			

10	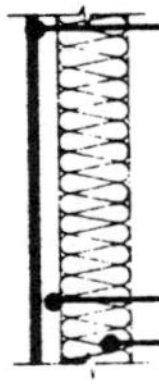 Glass, heat-strengthened, or tempered. Ceramic colored or reflective, clear or tinted 1-in. air space 3½-in. mineral fiber insulation	5 (0.24)	4.5 (114)	0.07(e)	2§	26	No	Yes	
	VISION GLASS								
11	 Annealed or heat strengthened, clear or tinted; reflective, clear or tinted	4 (0.19)	0.25 (6)	1.13	N/A	26	N/A	N/A	One of the lighter systems Thin Low maintenance
12	⅛-, 3/16-, or ¼-in. annealed or heat-strengthened glass ¼- to ½-in. air space ¼-in. with ½-in. air space	8 (0.38)	1 (25)	0.58	N/A	32	N/A	N/A	When the seal of insulating glass fails fogging occurs and replacement is required Often high initial cost Some breakage is normal and provisions for replacement should be made in the design

* Weight of supporting structure, if any, not included.
† Adjust wall thickness for insulation or air space over 2 in.
‡ Insulation:
(a) Air space only
(b) 2-in. polystyrene
(c) 2-in. polyurethane
(d) 2-in. mineral fiber
(e) 3½-in. mineral fiber—(assuming 3½-in. or 3⅝-in. metal studs rather than 2-in. studs).
§ Rating applies only where exterior is the fire side.

example, an elevator shaft. Theoretically, in very cold weather an interior to exterior pressure differential of about 5 psf (240 Pa) could develop in a 400-ft (120-m) high vertical shaft.

27.2.4 Seismic Loads

In regions where local experience or the records of the U.S. Geological Survey show loss of life or damage of buildings resulting from earthquakes, codes normally require buildings and facades to be designed to withstand seismic forces. Facade elements normally are designed to resist forces normal to their vertical surfaces. Ornamentation and appendages on the facade normally are designed for higher loads than the general facade areas and for forces in all directions. Seismic loads are discussed in Chapter 2. Design principles for seismic forces are given in Chapter 19.

27.2.5 Thermal Resistance

The facade must be able to provide sufficient resistance to transfer of heat in order that the interior of the building can be maintained at an acceptable temperature. The heat gain/loss of a building normally is calculated utilizing a steady-state analysis and the thermal resistance of the building envelope. *U* values for some of the more commonly used facades are included in Table 27.1.

The "steady-state" analysis, although very useful, does not take into account the capacity of the heavier building facades to absorb, store, and release heat slowly, helping the building to remain cool during the day and warmer at night. Modifications of thermal calculations to reflect the time lag effect of heavier facades indicate that there will be actual energy cost savings of at least 20% more than the steady-state analysis would indicate.

When detailing connections, wherever possible, one should avoid the development of thermal bridges through the facade which will lead to increased energy loss and increase the potential for condensation in cold weather.

27.2.6 Condensation Control

Condensation is most likely to develop in exterior walls during cold weather when the humidity is significantly higher within the building than outside. Vapor pressure differential, which is a function of humidity difference, is the driving force that causes water vapor to move through the wall. Assuming cold outside conditions and a normal living environment within the building interior, a significant temperature gradient exists through the wall. In the event there is no vapor barrier in the wall, the moisture will migrate outward in vapor form until it is cooled to the point where the air in which it is contained is saturated. This is also called the dew point at which water vapor condenses to liquid water. Condensation will continue to occur until more favorable temperature and humidity conditions prevail. Insulation may become water saturated and ineffective, thus aggravating the condition. Under extremely cold conditions, the condensate may freeze within the wall cavity.

As shown in Fig. 27.1, thermal insulation in the facade walls ideally should be positioned near the outside face and the vapor resistant materials near the inside face to prevent condensation.

27.2.7 Moisture Resistance

Preventing water entry into the building is one of the most important and difficult requirements to achieve in facades. In recent years, water leakage through facades and roofs has been, by far, the most common problem for building owners. In addition to causing discomfort for occupants and damage to building contents, uncontrolled water leakage into facade walls can significantly reduce thermal resistance, as well as cause deterioration, corrosion, and lead to freeze–thaw damage.

Gravity will cause water gaining access into facades to follow paths of least resistance downward until it is absorbed, evaporates, drains out, ponds, or gains access into the interior of the building. The potential for gravity-induced water leakage into facades is normally greatest on horizontal surfaces, such as at sills and balconies, where the water can stand for a long time.

Wind can drive rain laterally through facade openings which are very small. This effect is compounded by pressure differentials sometimes developed between building interior and exterior during high winds. The pressure difference can be sufficient to force water through seals and over flashings which would have been adequate to prevent water penetration resulting from wind pressure alone.

Facades constructed of highly water-absorbent materials under sustained rain exposure can become saturated to the point of becoming wet on the interior face. With this type of facade, a

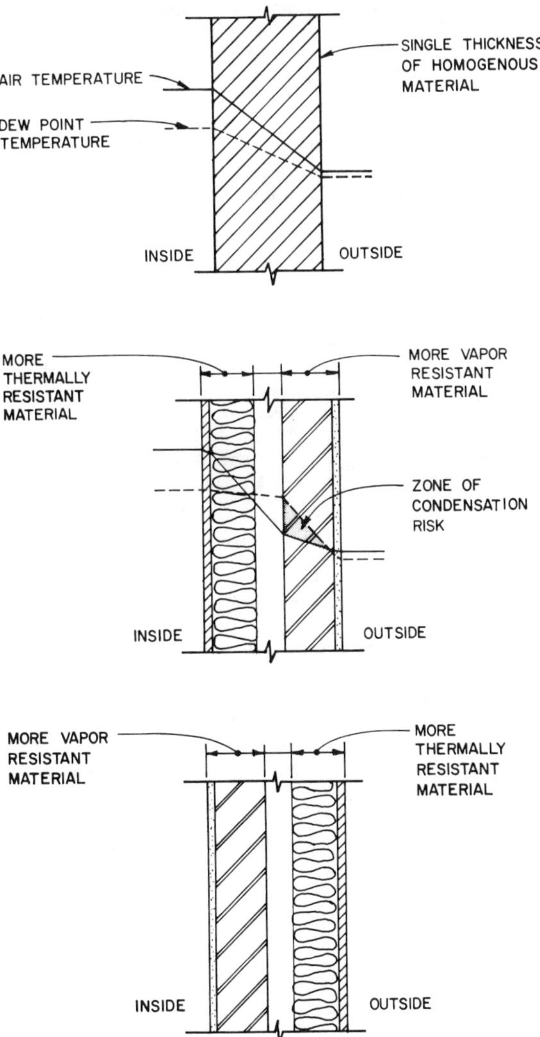

Fig. 27.1 Potential for condensation to develop in facades.

water barrier may be required between the exterior and interior facing of the wall to prevent water penetration into the building.

27.2.8 Fire Resistance

Many building codes specify fire resistive ratings for curtain walls. Interpretations as to stringency of requirements, or whether fire resistive ratings are applicable at all, vary widely within given building code jurisdictions. However, there is a general insistence by most building code officials that gaps occurring between the exterior wall and floor slab be closed off with a noncombustible material. Failure to fill this space can result in the creation of a flue running the complete building height which can lead to catastrophic consequences in certain fire situations.

Facades that act as bearing walls are an integral part of the building structure, and as such, normally are required to meet minimum fire resistive ratings of applicable codes. Facades may be required to provide fire resistive ratings ranging from 1 up to 4 hr. The fire resistive ratings for some of the more commonly used building facades are presented in Table 27.1.

The height to which bearing wall facades can be carried is also normally regulated by code. The most fire resistant facades seldom have any height restriction, while the least may be restricted to as little as two stories.

27.2.9 Durability and Serviceability

Facades made up of newly introduced materials are being found in the market place at an ever-increasing rate. Also, many materials such as natural stones are being used in thinner sections and in combination with other materials as never before. The durability and serviceability of many of these applications have yet to be proven.

Provision for easy removal and replacement should be considered for materials that are susceptible to damage. Elements of facades often are installed from the inside during construction, but because of interior finishes in the completed building, must be replaced from the outside.

27.2.10 Tolerance Compatibility

When detailing building facades, the designer must consider the manufacturing tolerances of the materials used, the tolerances to which the materials can be fabricated or erected, and the tolerances of the structure to which the materials are attached. Manufacturing tolerances for size, straightness, etc. are available for most commonly used facade materials. Fabrication and erection tolerances for facade plumbness, straightness, layout, etc. are also available for most commonly used materials. The fabrication or erection tolerances for the facade normally will be more demanding than for the structure. In buildings where the structural framing is to be left exposed, it may be necessary for the designer to set extremely tight tolerances, recognizing the difficulty in achieving a close fit of other facade elements to the exposed structural elements.

27.2.11 Volumetric Changes

All building materials expand and contract with variations in temperature. The coefficients of thermal expansion for some of the materials commonly used in building facades are presented in Table 27.2.

It is not unusual for dark exterior surfaces of facade walls, in summer, to attain temperatures as high as 140°F (60°C) or more. In winter, in a northern climate, exterior surfaces of facade walls can drop to −20°F (−30°C) or below. Even though the outside surfaces of the facade walls may become very hot or cold, the inside surfaces of the walls, which are normally in a controlled environment, will remain close to 70°F (21°C). Solid facades and facades comprising assemblies that are rigidly connected together, normally should be designed to accommodate movements induced by the mean temperature of the entire exterior facade. Cavity wall facades and facades comprising assemblies that are capable of some independent movement between elements, normally should be designed to accommodate movements induced by the mean temperature of each element that can move independently.

In addition to the thermally induced movement, facades made up of some materials must also be designed to accommodate moisture-induced volumetric changes. Volumetric changes that should be anticipated in clay masonry, concrete masonry, natural stones, and concrete will be discussed later in this chapter.

Dimensional changes in high-rise buildings are described in detail in Section 10.2.5 (thermal effects) and in Section 10.2.6 (axial shortening produced by stresses and by creep).

27.2.12 Facade/Structure Interaction

Provisions normally will be required in facade walls to accommodate the movements of the structure of the building to which they are attached. Facade walls that normally only require control joints to accommodate their own volumetric changes, may very well require some expansion joints at selected locations to accommodate the movements of the building structure to which they are attached.

While facade walls regularly expand and contract in response to changes in outside temperature, those elements within the building which are in a controlled temperature environment will be affected very little. The thermally induced differential movements between the facade and the structure must be considered in the positioning of joints and detailing of connections.

In buildings with concrete frames, the facades must have sufficient horizontal expansion joints to accommodate the creep that will occur in the frame columns. Creep normally will be greater in lightweight concrete frames than in normal-weight concrete frames.

Facade walls must have properly designed joints and connections detailed to accommodate building sway. If not properly detailed, the facade walls in many buildings will be more rigid than the structure to which they are attached. The resulting lateral forces can exceed available facade strength and distress can occur. Sway in high-rise buildings is discussed in Chapters 10, 11, and 22.

Deflection in structural members also should be taken into consideration. The deflection limits set by many codes for beams and slabs are often much greater than the facades supported on them can tolerate without special detailing.

Facade/structure interaction is also treated in Section 10.2.7.

Table 27.2 Representative Thermal Expansion Coefficient Factors

Material	Representative Coefficient of Thermal Expansion, in Millionths (0.000001)/°F
Clay masonry	
Clay or shale brick	3.6
Fire clay brick or tile	2.5
Clay or shale tile	3.3
Concrete masonry	
Dense aggregate	6.2
Cinder aggregate	3.7
Expanded-shale aggregate	5.2
Expanded-slag aggregate	5.5
Stone	
Granite	4.7– 9.0
Limestone	2.4– 4.5
Marble	3.7–12.3
Sandstone	5.0–12.0
Slate	5.4–12.0
Concrete	
Gravel aggregate	6.0
Lightweight, structural	4.5
Metal	
Aluminum	12.8
Bronze	10.1
Stainless steel	9.6
Structural steel	6.7
Wood, parallel to fiber	
Fir	2.1
Maple	3.6
Oak	2.7
Pine	3.6
Wood, perpendicular to fiber	
Fir	32.0
Maple	27.0
Oak	30.0
Pine	19.0
Glass	4.5

27.2.13 Cost

Reliable economic comparisons of facade systems for a particular building are very difficult to establish. Some of the more important factors that should be included in any comprehensive cost comparison include:

1. Material costs.
2. Construction or erection costs.
3. Special equipment costs needed during construction, such as scaffolding, cranes, shoring, and bracing.
4. Influence of facade weight on the cost of other elements in the building.
5. Influence of facade thickness on rentable floor space for the same gross building area.

6. Availability of materials, skilled labor, and special equipment in a particular location, specifically with respect to new systems.
7. Cost considerations associated with speed of erection of the facade and date of occupancy.
8. Maintenance costs over the life of the structure.

27.3 GENERAL COMPARISON OF FACADE TYPES

The selection of a facade requires a careful comparison of the performance requirements of the building with the characteristics of the facade types available.

Space constraints and continuing developments preclude the listing of all building facade types. The facades presented in Table 27.1 are only a few examples of some of the types that are most widely used. Omission of a facade type in no way is intended to infer that it is unable to perform in a satisfactory manner.

The information presented is only representative. Complete information should be obtained for final design decisions from applicable codes, from material manufacturers or, if necessary, by testing.

27.4 SPECIAL DESIGN CONSIDERATIONS FOR CLAY UNIT MASONRY

The more commonly used types of clay units used in the construction of facade walls are: building bricks, face bricks, glazed bricks, hollow bricks, and glazed structural facing tiles. Units used in facade walls should normally conform to Grade MW or SW in the appropriate ASTM standard. When glazed units are used, the manufacturer should be consulted to ensure that his specific materials are suitable for exterior use.

Mortars used in clay unit facades should meet the requirements of the Brick Institute of America (BIA) [27.36]. Types M, S, and N mortars normally are used in facade walls. Type 0 mortar also can be used in facade walls, but only in the limited areas of the country where the walls will not be subjected to freezing action. No single type of mortar is best suited for all purposes, but generally it is desirable to select the lowest strength mortar that is consistent with performance requirements of the project. The greater the compressive strength of the mortar in the facade walls, the more rigid the walls become, and the less they are able to absorb unanticipated movements and forces without cracking.

Type N mortar is a medium strength mortar which is suitable for general use in facade walls, and is specifically recommended when the walls will be subjected to severe weathering. Type S mortar is recommended where maximum flexural strength is required or where somewhat higher compressive strength is required. Type M is only recommended where the highest achievable compressive strength is required.

Masonry cements are not recommended by the BIA for use in clay brick masonry because of the lack of limitations on the type and amount of ingredients and the high air content permitted. The BIA contends that air content in excess of 12% generally will reduce bond strength in clay unit walls. Air contents up to 22% are allowed in masonry cements. Portland cement mortars have an air content of less than 12%.

27.4.1 Structural Design Considerations

Most building codes allow masonry facade walls to be designed either empirically or by rational analysis procedures. Empirical design procedures are contained in ANSI A41.1, *American Standard Building Code Requirements for Masonry* [27.16]. Chapter 24 of this handbook also contains considerable information on masonry design procedures.

Rational design procedures for brick facades are contained in *Building Code Requirements for Engineered Brick Masonry* [27.34]. Brick masonry compressive strengths for rational analysis are based on test results on prisms or clay units using a prescribed formula. Shear strength calculations are based on allowable compressive strengths. Flexural strength calculations are based on the type of mortar used. Higher stresses are allowed in the design when walls are inspected during construction. In some codes, connections in masonry facades may require a minimum strength per linear foot value which is greater than the strength calculated by the rational analysis.

27.4.2 Detailing Considerations

Moisture Control

The two principal approaches employed for preventing the penetration of wind driven rain through clay unit facades are shown in Fig. 27.2. The most effective method is to provide a path for the drainage of water that enters the wall before it reaches the building interior. The second method used to prevent the entry of water is to rely on the facade wall to act as a barrier. Even when a barrier-type wall is solid and a gravity flow path for the water is not defined, it is still recommended that flashings and weep holes be provided to direct out any water that does reach the bottom of the walls. To achieve the maximum resistance to rain penetration in clay unit facades in areas of severe weathering, a drainage wall system should be used. In other locations, either system may be used. The BIA provides a map to determine severities of exposure in terms of wind pressure and annual precipitation.

Careful detailing of flashings is essential to ensure that the strength of the facade walls is not adversely effected. Flashings will not affect wall compressive strengths, but they will substantially affect shear and flexural tensile strengths. Shear strengths normally are dependent on the coefficient of friction between the flashing and the masonry. Flexural tensile strength through flashings should be assumed to be equal to zero.

Accommodating Volumetric Changes

When a clay masonry unit is delivered from the kiln, it is at its minimum volume, and subsequently will absorb moisture from the air and expand. The rate of expansion will be greatest immediately after leaving the kiln, decreasing with passage of time. Normally, about one-third of the expansion will take place in the first month, and more than half will take place in the first year. The amount of moisture expansion will vary from brick to brick, but the BIA indicates that for design purposes, a value of 0.0002 in./in. normally should be sufficient.

Carefully designed and located expansion joints must be provided in facade walls to accommodate volumetric changes. In clay unit masonry where moisture expansion occurs, high temperatures, which also cause expansion, will be of greatest concern. The spacing of expansion joints to a great extent will be influenced by the layout of facade walls and by the location of openings. In general, expansion joints will be required at regularly spaced intervals in long walls, at offsets, at or

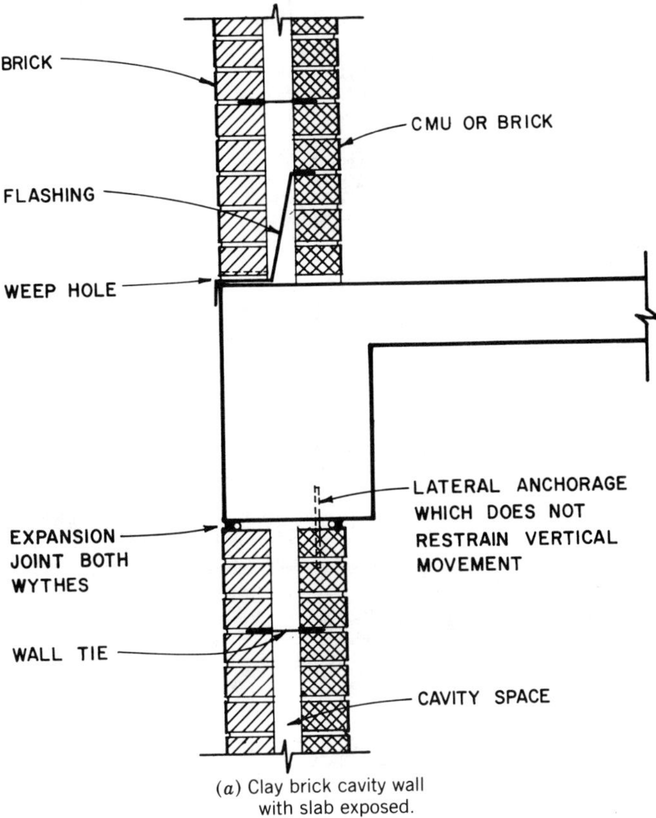

(*a*) Clay brick cavity wall with slab exposed.

Fig. 27.2 Clay brick cavity walls (*continued on following pages*).

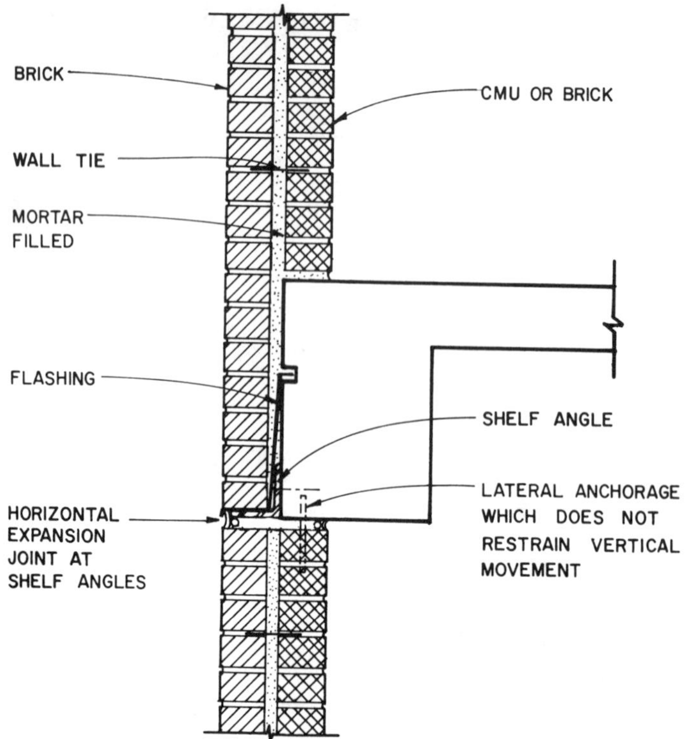

(*b*) Clay brick solid wall with slab concealed.

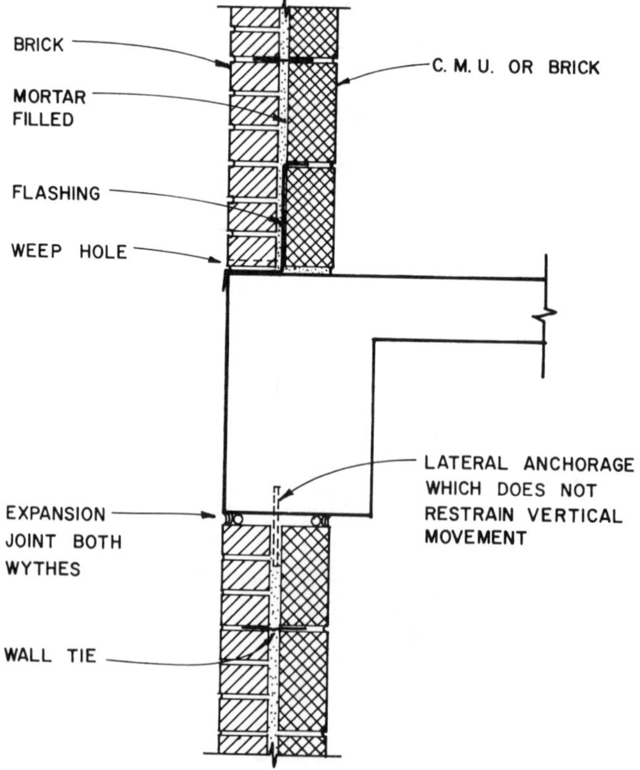

(*c*) Clay brick solid wall with slab exposed.

Fig. 27.2 (continued).

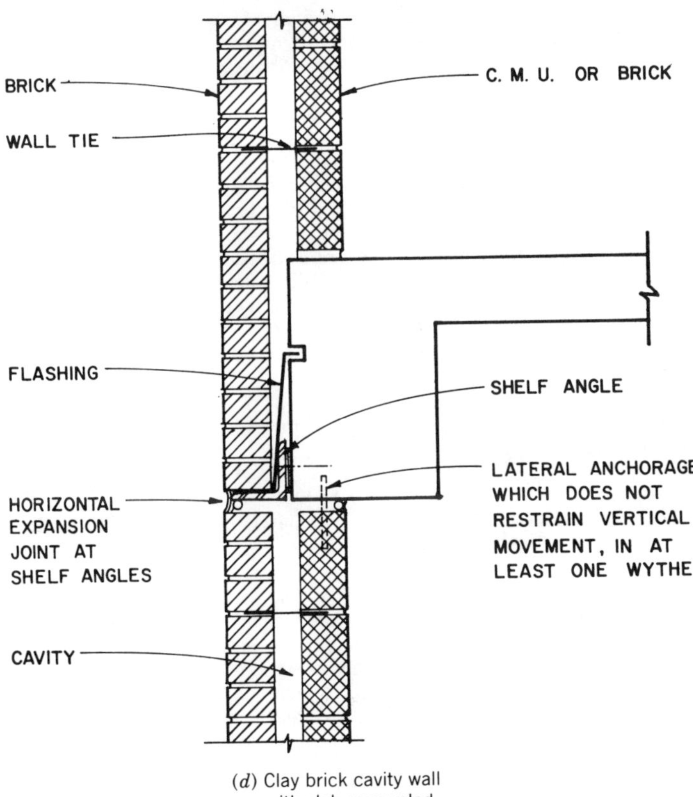

(*d*) Clay brick cavity wall with slab concealed.

Fig. 27.2 (continued).

near corners, and at planes of weakness such as windows and doors. Expansion joints should be carried completely through parapets. Additional expansion joints may be needed in parapets because of the high thermally induced movements which result from exposure on two sides to extreme temperature changes. Reinforced masonry walls normally will require significantly fewer expansion joints than nonreinforced walls. Figure 27.3 illustrates some of the more commonly used expansion joint details.

Accommodating Facade/Structure Interaction

The weight of masonry facade walls on lower buildings may be carried completely on the foundations, with the structure providing only lateral support. In such cases, the accumulation of move-

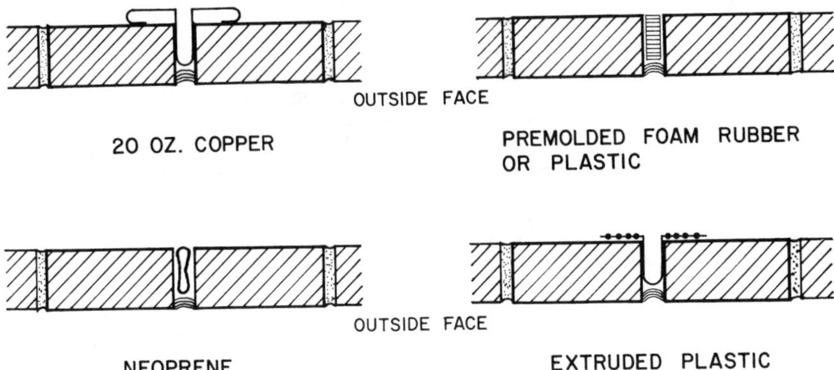

Fig. 27.3 Expansion joint details for clay brick facades.

ments with increased height must be studied very carefully. Details at windows and parapets often must be able to accommodate quite large differential movements between the facade and the structure. Flexible wall anchors may also be required. Interaction of cladding with reinforced concrete structures is discussed in Section 22.5.3.

In higher buildings and buildings with punch-through windows, it is usually desirable to carry the weight of the facade walls at floor lines at regular intervals. Carrying the walls at a greater spacing than every other floor line may require the development of special window and finish details that are capable of accommodating large differential movements. Horizontal expansion joints should be provided under locations of vertical support to allow for vertical expansion of the facade below the joint, deflection of the structural supporting element above the joint, and downward movement of the structural element above the joint due to column shortening. Figure 27.2 shows a few of the more commonly used methods of supporting the facade at floor lines.

To ensure that the structure is free to move, masonry facades should be kept clear of columns as much as possible, particularly near the corners.

27.4.3 Suggested References

References 27.16, 27.20, 27.22, 27.24, 27.32. and 27.34 through 27.37 will be useful in the design of clay unit masonry facades.

27.5 SPECIFIC DESIGN CONSIDERATIONS FOR CONCRETE UNIT MASONRY

Concrete building bricks, hollow load-bearing blocks, and solid load-bearing blocks are regularly used for the construction of concrete masonry facade walls. Concrete units used in facade walls normally should be required to conform to Grade N in the appropriate ASTM specification. Concrete units conforming to Grade S should normally only be used in conjunction with a weather-protection coating. Both sand-gravel units and lightweight concrete blocks may be used in facade walls. Split face block, split face concrete brick, and ribbed concrete brick have become particularly popular in recent years. General information on concrete masonry materials and design approaches are given in Chapter 24.

Mortars used in unreinforced concrete unit facades should meet the requirements of ASTM C270 [27.25]. Type N or S mortars normally are recommended for use in unreinforced facade walls. In bearing walls where higher strengths are required than can be achieved with Type N or S mortar, Type M mortar may be used. In non-load-bearing curtain walls, it seldom will be necessary or desirable to use Type M mortar.

Mortars and grouts for use in reinforced concrete unit facades should meet the requirements of ASTM C476 [27.26], Facade walls containing only joint reinforcement to control cracking should not be considered to be reinforced.

One of the significant differences between the recommendations of ASTM C270 [27.25] and C476 [27.26] for concrete unit masonry and the recommendations of the BIA for clay unit masonry is that ASTM C270 and C476 allow the use of masonry cement. Before specifying masonry cement, however, all other applicable codes should also be checked. Some building codes require that the air content of all mortars not exceed 12%. Mortars made with masonry cement often will not satisfy this requirement.

27.5.1 Structural Design Considerations

Most codes allow concrete unit facade walls to be designed either empirically or based on rational analysis procedures. Empirical design procedures are contained in A41.1, *American Standard Building Code Requirements for Masonry* [27.16].

When special shaped concrete blocks are used, the thickness to be used for calculating the height-to-thickness ratio may be subject to some interpretation. For example, if the rib-faced block in Fig. 27.4 is used in a wall that spans vertically and the projecting ribs are bedded in mortar, it may be considered to be a nominal 8-in. (200-mm) wall. If the ribs are not bedded in mortar or the wall spans horizontally, it should be considered a nominal 6-in. (150-mm) wall.

Rational design procedures are contained in the ACI publication, *Building Code Requirements for Concrete Masonry Structures and Commentary* [27.9]. Concrete unit masonry compressive strengths for rational analysis are based on the results of prism tests or concrete unit test results used in a prescribed formula. Shear and flexural strengths are calculated based on the determined compressive strength. Higher stresses are allowed in the design of walls when they are inspected than when they are not inspected. In some codes, connections in masonry facades may be required to satisfy a minimum strength per linear foot greater than the strength determined by rational analysis. See Chapter 24 for detailed design methods.

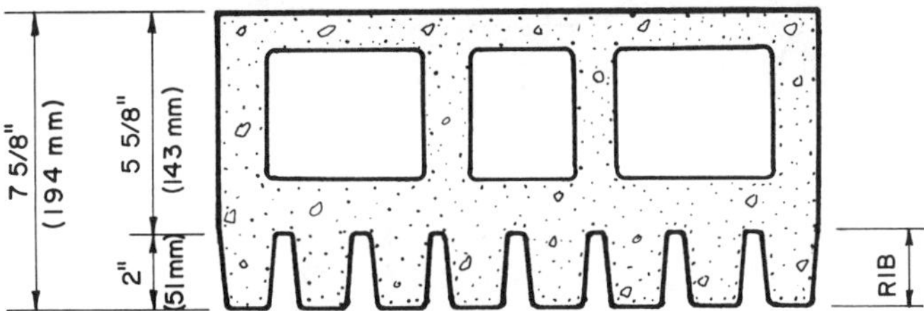

Fig. 27.4 Rib faced concrete masonry unit.

27.5.2 Detailing Considerations

Moisture Control

In concrete unit masonry, as in clay unit masonry, cavity walls are the most effective system for resisting the penetration of rain, however, they are not as commonly used. Single-wythe and double-wythe concrete unit facade walls without cavities for drainage are used widely in all parts of the country. Facade walls, whether they contain a cavity or not, should be constructed with weep holes and flashings. Figure 27.5 shows an example of a cavity wall and a solid wall.

For most geographical locations, the National Concrete Masonry Association (NCMA) [27.45] recommends that a waterproof coating be applied to solid concrete unit masonry walls to reduce water penetration. Several types of waterproof coatings are available. Generally, coatings for concrete masonry are opaque. A filler coat normally is applied first to level out and fill surface pores, and then a finish coat is applied. The most common finish coats are latex, portland cement, or these materials in combination. In most geographical locations, the finish coat will have to be reapplied at least once every 10 years.

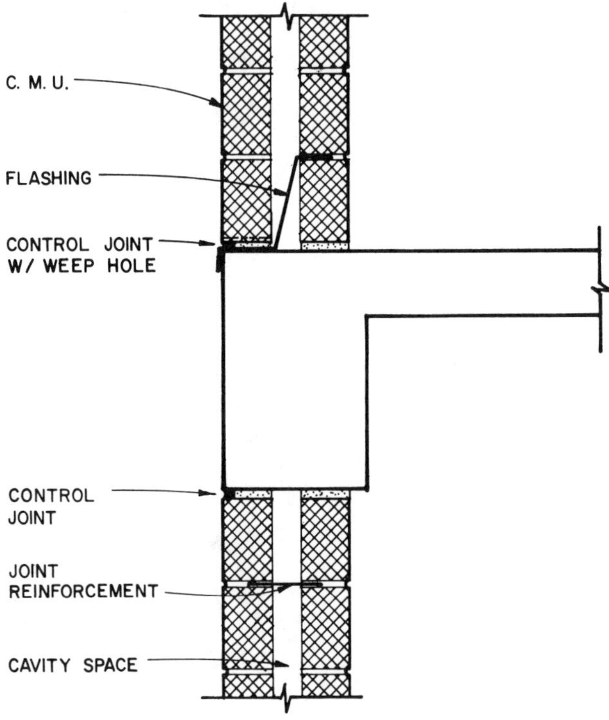

(*a*) Concrete block cavity wall.

Fig. 27.5 Concrete block walls (*continued on next page*).

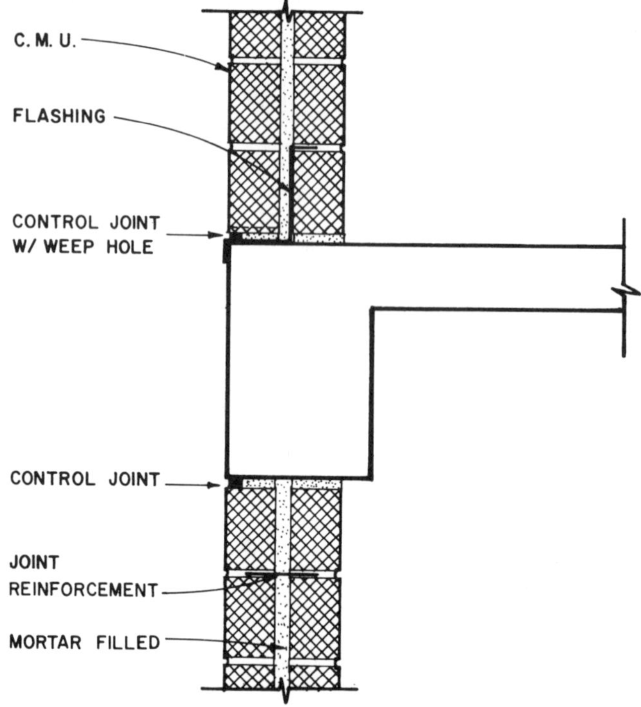

(*b*) Concrete block solid wall.

Fig. 27.5 (continued).

To improve watertightness of solid walls, limitations are also suggested on mortar joint tooling. Concave and V-shape mortar joints are considered to be best. Beaded and weathered joints are less desirable. Flush, raked, extruded, and struck joints are not recommended. Joint tooling is illustrated in Fig. 27.6.

The careful detailing of flashings is essential so as not to compromise wall strength. Flexural, tensile, and shear strength of mortar joints containing flashing cannot be utilized according to ACI 531 [27.9].

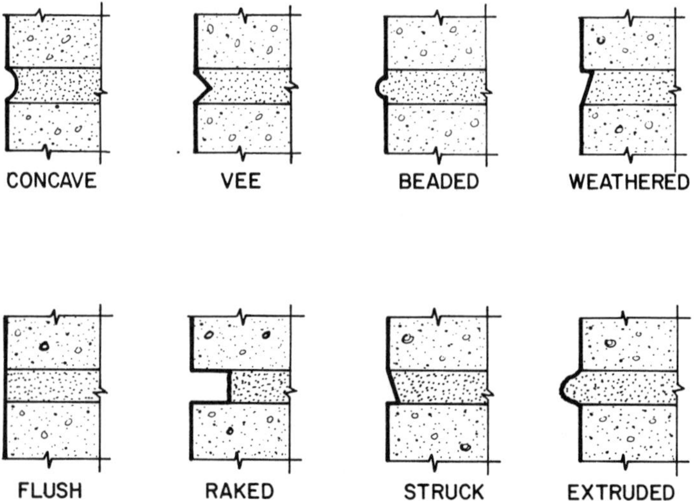

Fig. 27.6 Types of masonry joint tooling.

Accommodating Volumetric Changes

When a concrete unit comes from the mold, it is at its maximum size. Subsequently, it will dry out and shrink. Although there are many factors that affect the drying shrinkage of concrete units, the major controllable factors are the type of aggregate used, the curing procedure, and the method of storage.

Concrete units made with sand and gravel aggregates normally will have substantially less shrinkage than those produced with lightweight aggregates. The amount of drying shrinkage that takes place in concrete units can be reduced by one-third or more by high-pressure steam curing (autoclaving). Because of the cost of energy, however, most plants have stopped autoclaving block. If specified, the designer should make sure that a source of autoclaved blocks is available.

The amount of drying shrinkage which takes place in a concrete block cladding wall can be reduced by drying the blocks before they are laid. Blocks which are dried during production and then kept dry so that their moisture content does not exceed prescribed standards, are called moisture-controlled units. Some codes and standards limit the moisture content of concrete blocks at the time of installation, but normally, moisture-controlled units will only be utilized in facade walls in areas of the country having very dry climates.

In concrete unit masonry, where shrinkage from moisture loss occurs, low temperatures which cause contraction will normally be of greatest concern. To accommodate drying shrinkage and thermally induced dimensional changes in concrete masonry facade walls, control joints normally will be required. Control joints are not expansion joints of the type used in clay brick masonry facade walls. Control joints have no space provided to accommodate expansion in the cladding walls. Control joints are continuous planes of weakness intentionally built into the cladding wall which when used in conjunction with joint reinforcement, control the locations of cracking. They must be able to crack and open while remaining watertight, and be located so that the structural integrity of the masonry is not affected when the crack develops.

Figure 27.7 illustrates some of the more commonly used control joint details. Most of the control joints are first laid up with mortar. After the mortar becomes stiff, the joint is raked out and filled with caulk, at least on the outside face of the wall.

Since there are many possible layouts of facade walls and location and size of openings, judgment is required in determining where control joints should be installed. In general, control joints should be installed at regular spacings in straight walls, at abrupt changes in wall height and thickness, at dissimilar materials, at corners, and at openings.

The spacing of control joints in straight walls without openings will depend upon the height and the amount of horizontal joint reinforcement provided. Joint reinforcement will not prevent crack-

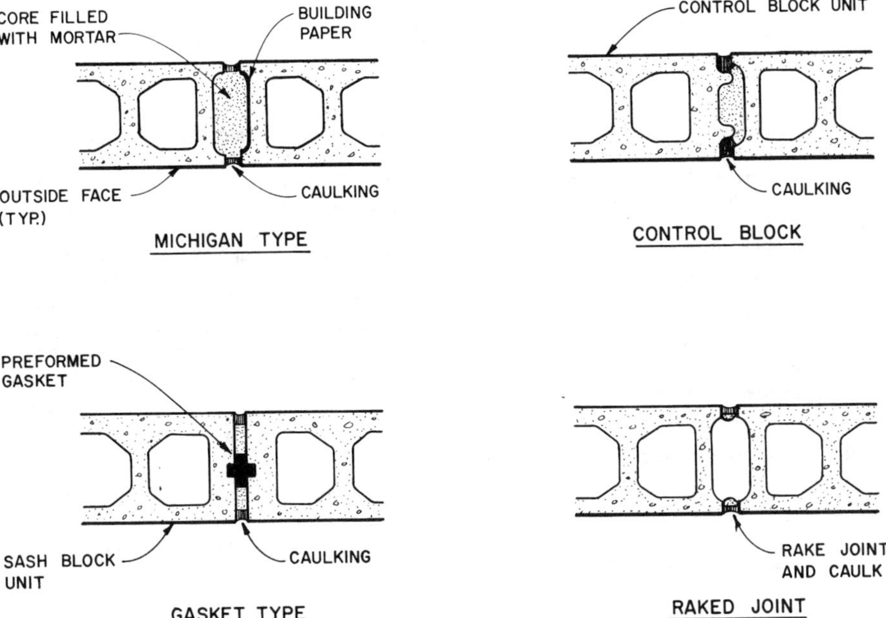

Fig. 27.7 Control joints for concrete unit masonry facades.

Table 27.3 Control Joint Spacing for Moisture-Controlled Type 1 Concrete Masonry Units*

	Vertical Spacing of Joint Reinforcement, in. (mm)			
	None	24 (600)	16 (400)	8 (200)
Expressed as ratio of panel length to height L/H	2	2½	3	4
With panel length, ft (m) not to exceed:	40 (12.3)	45 (13.9)	50 (15.4)	60 (18.5)

* Reproduced from NCMA-TEK Note 53, National Concrete Masonry Association [27.45].

ing in concrete masonry walls between control joints, but will distribute cracks evenly so that they will be very fine and hardly visible. For moisture-controlled units, NCMA recommends the control joint spacings shown in Table 27.3. Joint reinforcement should be discontinuous at control joints. The NCMA has no recommendations for the spacing of control joints in cladding walls constructed of non-moisture-controlled units. For walls constructed of non-moisture-controlled units, the Indiana Concrete Masonry Association recommends a control joint spacing of approximately 25 ft (7.6 m) in walls 8 ft (2.4 m) to 12 ft (3.7 m) high. In cavity walls where both wythes are concrete blocks, the control joints should extend completely through both wythes. Reinforced masonry walls normally will require significantly fewer expansion joints than nonreinforced walls.

Accommodating Facade/Structure Interaction

The facade/structure interaction discussion in Section 27.4.2. on clay unit masonry is also applicable to concrete unit masonry.

27.5.3 Suggested References

References 27.7 and 27.10 through 27.17 are suggested for concrete unit masonry facades.

27.6 SPECIFIC DESIGN CONSIDERATIONS FOR NATURAL STONE MASONRY

A wide variety of natural stones are used in the construction of buildings. Sandstone, limestone, marble, granite, and slate are probably the natural stones most commonly used in facade walls. Natural stones used in facades must be sound and free of cracks or other imperfections that would impair their structural integrity or durability. The stones should conform to appropriate ASTM standards. Not all types and grades of stones are suitable for exterior use, and without a proven record of performance, a new product should be used only after careful study.

Stone facades may be laid up with mortar or each stone may be supported individually. When each stone is supported individually, the joints between stones usually are caulked. When mortar is used, it normally should be Type N or S, conforming to ASTM C270 [27.25]. To prevent staining, white masonry cement is recommended for use with some types of stones. Some building codes require that the air content of mortars not exceed 12%. Mortars made with masonry cement may not satisfy this requirement. To improve watertightness, tuckpointing joints after the stones are set is recommended by some stone suppliers.

27.6.1 Structural Design Considerations

Empirical design procedures sometimes are used for facade walls laid up with mortar. Empirical design procedures for stone masonry are contained in ANSI A41.1 [27.16].

Natural stone facade walls, however, normally are designed based on rational analysis. When individual stones are independently supported with anchors, the flexural strength of stone usually controls the design. The minimum physical properties for some of the more commonly used natural stones are presented in Table 27.4.

Because stone is a natural product, physical properties including strength can vary widely. Representative values may be used for preliminary design purposes, but final designs, with few

Table 27.4 Representative Minimum Physical Properties of Natural Stones in ASTM Specifications (Unless Noted Otherwise)*

	Compressive Strength, psi (MPa)	Modulus of Rupture, psi (MPa)
Sandstone	2000 (13.8)	300 (2.1)
Limestone		
Low density	1800 (12)	400 (2.9)
Medium density	4000 (28)	500 (3.4)
High density	8000 (55)	1000 (6.9)
Indiana*	4000 (28)	700 (4.8)
Marble	7500 (52)	1000 (6.9)
Granite	19,000 (131)	1500 (10.3)
Slate		
Across grain		9000 (62.1)
Along grain		7200 (49.6)

* *Indiana Limestone Handbook* [27.41].

exceptions, should be based on strength values of the actual stone ultimately selected for use in a specific facade. These values can be determined by test or, in some cases, may be available from the stone supplier.

Tests indicate that the flexural strength of some types of natural stones decrease with time. Because of this factor and the variable nature of most natural stone, larger safety factors are recommended for natural stone masonry than for masonry manufactured from materials such as clay or portland cement. Different factors of safety are recommended by the trade associations and suppliers for the different types of stone.

When stones are laid up with mortar, the bond strength between the stone and mortar usually controls the flexural strength of assemblies. Assembly strengths may be determined by test or calculated from conservative design values provided in some codes, for example, *Uniform Building Code* [27.49].

In some codes, connections in masonry facades may be required to satisfy a minimum strength per linear foot which is greater than the strength indicated by rational analysis.

27.6.2 Detailing Considerations

Anchorage

A wide variety of stone anchorage systems are being used today. Thicker stones, 3 in. (75 mm) or 4 in. (100 mm) thick, normally are supported on shelf angles at regular intervals. Figure 27.8 shows an example of one commonly used type of installation.

Thinner stones on the order of 1¼ in. to 2¼ in. (approx. 30 to 60 mm) thick are also sometimes supported directly on shelf angles, but more often they are supported on pieces of stone which are bonded to the back of the stone to provide better bearing, as shown in Fig. 27.9. When the angle connections are set blindly as shown in Fig. 27.9, great care must be taken during construction to ensure that the angles and stones are positioned properly to provide adequate bearing. Thinner stones also are regularly supported on grid strut systems attached to the building structure or in preassembled panels backed up with structural steel sections which are lifted and anchored in place. Figure 27.10 shows a grid strut system. Figure 27.11 shows a preassembled panel.

Moisture Control

The types of anchorage systems frequently used with natural stone facades make the installation of flashings very difficult. Many natural stone trade associations and suppliers indicate that flashings are not necessary with their products. Acceptable water resistance has been achieved in stone facades both with and without flashings. Normally it will be more important that flashing and weepholes be provided in stone facades laid up with mortar than those where the joints between stones are caulked.

It is recommended that a damp-proof coating be applied to the back of some types of stone to prevent discoloration when they are placed on continuous concrete ledges, concrete haunches, continuous steel angles, etc. The same types of stones should not be built tightly against backup materials that contain alkaline or discoloration can also occur. Cement products contain alkaline components.

Accommodating Volumetric Changes

Irreversible expansion occurs in some types of natural stone after being quarried when they are subjected to normal temperature cycling and exposure to moisture. Minimum thicknesses for stone panels are recommended by various trade associations and suppliers for their specific products. Stone panels must be of sufficient thickness to ensure that irreversible temperature/moisture-induced differential expansion of the exposed and unexposed faces does not cause the panels to bow outward with the passage of time.

Expansion joints generally will be required in facade walls to accommodate thermal movements and residual material expansion. The joints should be positioned at regular intervals in long and tall walls, at corners, and at potential planes of weakness such as window and door openings. The actual size and spacing of expansion joints should be designed based on the anticipated movements for the particular stone chosen. The Marble Institute of America [27.42] recommends that in marble facades vertical expansion joints usually be provided at intervals of about 20 ft (6 m) on center, and horizontal expansion joints be provided at least at every other story height.

Accommodating Facade/Structure Interaction

When stones are supported independently and the joints between stones are caulked, differential movement between the facade and structure usually is easily accommodated. When stones are laid up with mortar, the facades must be carefully detailed to accommodate facade/structure interaction. The facade/structure interaction discussion in Section 27.4.2 on clay unit masonry also is applicable to stone masonry set in mortar.

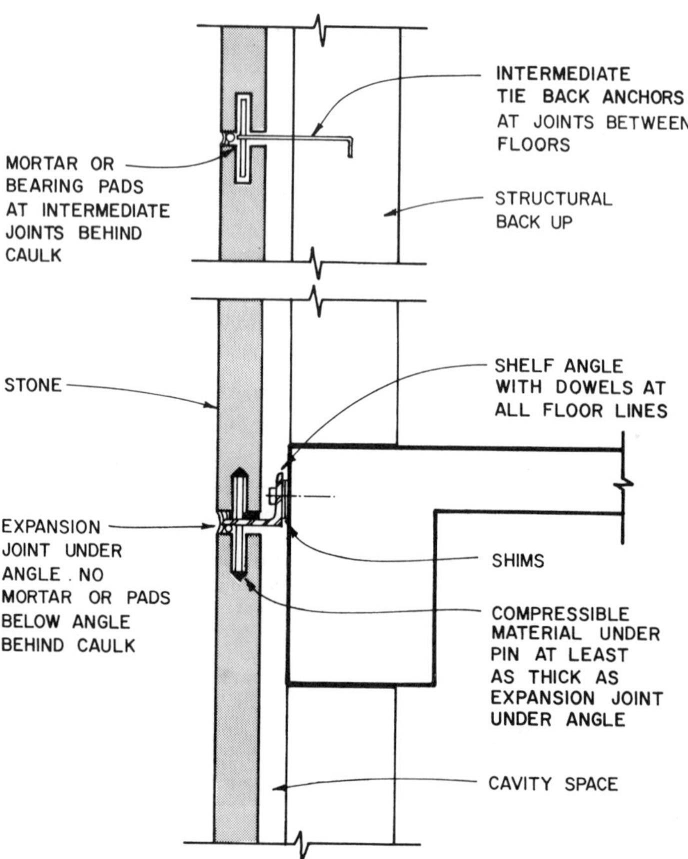

Fig. 27.8 Natural stone wall with thick material.

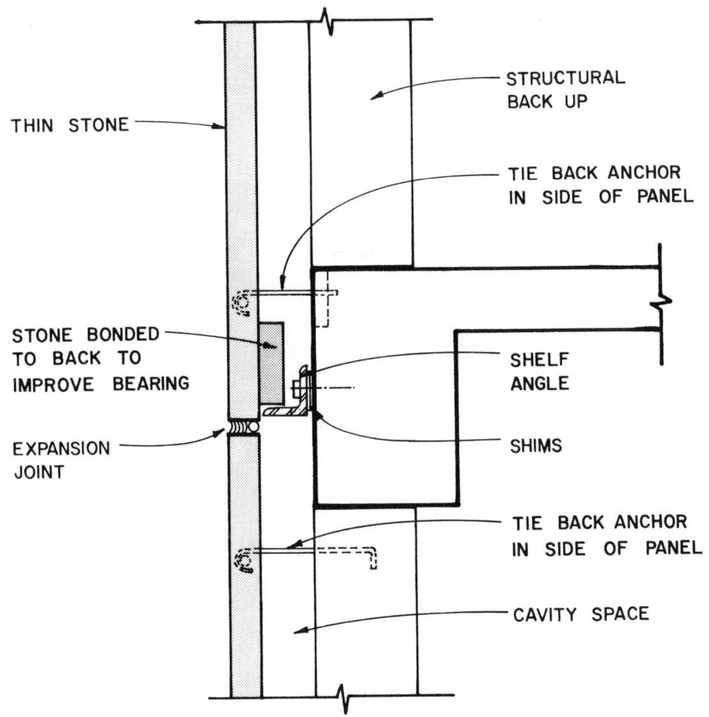

Fig. 27.9 Natural stone wall with thin material.

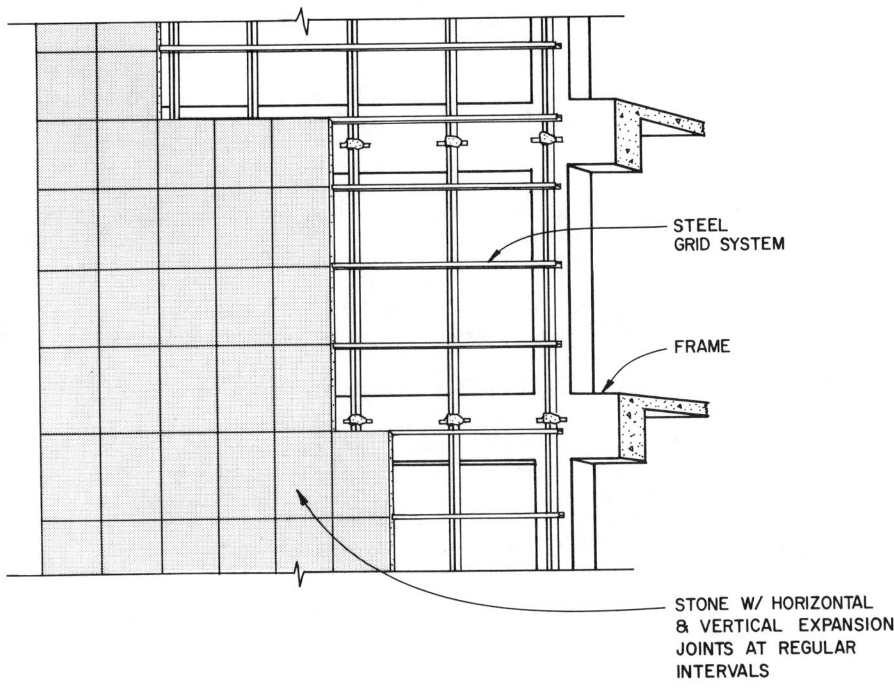

Fig. 27.10 Natural stone wall set on grid strut system.

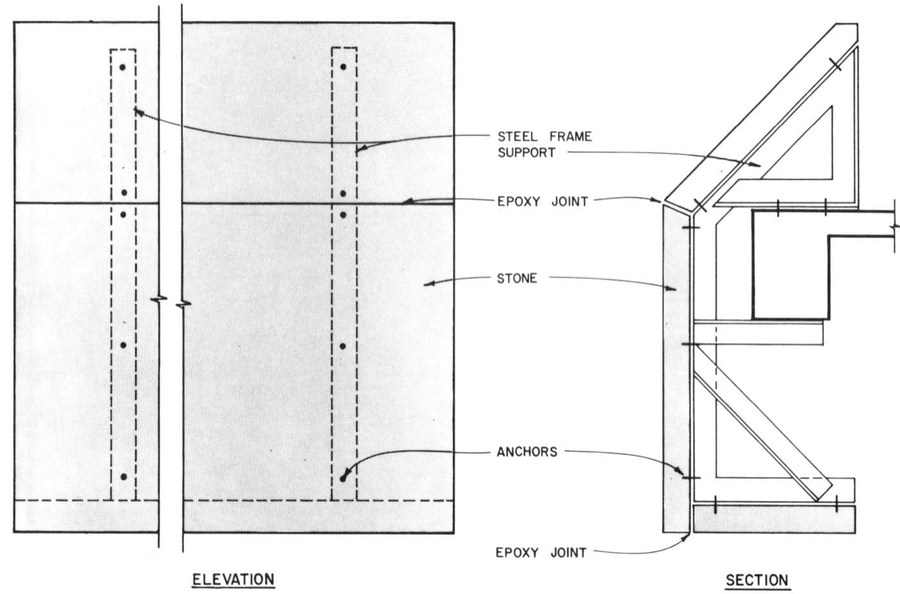

Fig. 27.11 Natural stone wall component—preassembled.

27.6.3 Suggested References

References 27.28 through 27.31, 27.41, 27.42, 27.44, and 27.48 are suggested for natural stone masonry facades.

27.7 SPECIFIC DESIGN CONSIDERATIONS FOR CONCRETE FACADES

Concrete facade walls should conform to the requirements of *Specifications for Structural Concrete for Buildings* [27.3]. Normal-weight or lightweight concrete may be used in facade walls. In smooth finished concrete the color of the cement is a dominant factor. White concrete should be considered for color uniformity. Whether concrete with white or normal cement is used, it should be air-entrained if severe weather exposure is anticipated. Exposed aggregate finishes can be provided in a wide variety of colors and textures. All exposed aggregates should have a proven service record. Some limestones, marbles, and other high calcium materials which are soft and nondurable are not suitable for exposed aggregate surfaces. The desired appearance and performance requirements of the concrete should be specified, but the actual design of the mix normally should be left to the precaster.

As a general rule, reinforcing bars should be kept reasonably small. Small bars closely spaced improve the distribution of thermal stresses and decrease the potential for unsightly cracking.

27.7.1 Structural Design Considerations for Cast-in-Place Concrete

Cast-in-place concrete facade walls are designed based on rational analysis procedures in accordance with the ACI *Building Code Requirements for Reinforced Concrete* [27.5]. Facade walls must be able to accommodate service loads and stresses without an objectionable amount of cracking, spalling, or deflection which would detract from the appearance and serviceability of the building. The *ACI Code* [27.5] limits the flexural tensile stresses in reinforcement in exposed concrete elements, such as facade walls, to keep cracking at a minimum.

The use of ultimate strength design requires careful analysis of all facade elements subject to deflection, including spandrels. Where there are long spans, normally allowable deflections may be visually objectionable, and more stringent criteria may have to be established. Chapter 7 has additional information on concrete wall design.

27.7.2 Detailing of Cast-in-Place Concrete

It is good practice to avoid large, flat, smooth uninterrupted expanses of concrete surface. Facade walls should be divided into manageably sized areas which will provide the contractor freedom in planning the construction operation. Having the ability to reduce the width of lift will accelerate the vertical rate of casting and reduce the tendency to develop cold joints, spatter, and lift lines. Construction joints should be incorporated into the joint patterns or rustification.

Accommodating Volumetric Changes

A cast-in-place concrete facade wall element is at its maximum size when the forming is removed. Subsequently, it will continue to dry and shrink until its moisture content reaches equilibrium.

Horizontal cracking usually is not a problem because the walls are in compression. However, vertical cracking can produce problems if not properly considered. To accommodate drying shrinkage and thermally induced dimensional changes in the horizontal direction, vertical control joints normally will be required. The simplest and most practical method of forming vertical control joints is with rustification strips on one or both faces of the facade walls.

Generally, the maximum amount of reinforcing crossing vertical control joints should not be greater than one-half of the horizontal reinforcement elsewhere in the wall. Additional cover should be provided over the reinforcement in the facade walls to compensate for the depth of control joints.

Vertical control joints should be installed at about 20 ft (6.1 m) on center. Some examples of control joint details in cast-in-place walls are shown in Fig. 27.12

27.7.3 Structural Design of Precast Concrete

Precast concrete facade walls are designed based on rational analysis procedures in accordance with the *ACI Code* [27.5] and the *PCI Manual for Structural Design of Architectural Precast Concrete* [27.46]. Precast concrete facade walls regularly are used as both bearing walls and curtain walls. The sculptural configurations of many wall panels readily lend themselves to carrying vertical loads with little modification.

Rapid strength development which allows early stripping of units is of prime importance to the precaster. A concrete strength of about 5000 psi (35 MPa) normally should be specified for precast panels to give the precaster flexibility in the development of production and erection procedures.

The specification of the in-service loads that the precaster must accommodate is the responsibility of the design engineer. Proper design to prevent excessive stresses during handling and erection of the units is the responsibility of the precaster. Connections in curtain walls should be designed to ensure that each panel supports only its own weight and does not transmit load to the panels below. Braces often will be required behind panels where the connections are not sufficiently spread to effectively resist rotation. Figure 27.13 illustrates some representative examples. In some codes, connections in concrete facade elements are required to satisfy a minimum strength per linear foot that is greater than the strength indicated by rational analysis.

Whether the reinforcing and connections are designed and detailed by the design engineer or left to the precaster is usually optional. If relevant codes permit, and local precasters have demonstrated design experience, there can be advantages in leaving the design of the reinforcing and connections to the precaster. An experienced precaster is in a position to select the details that are best suited to his specific production and erection procedures. If the design and/or detailing responsibilities are transferred from the design engineer to the precaster, the precaster should have a registered engineer on his staff or retain one for the project.

On precast concrete projects, it is essential that the design responsibility of the design engineer and the precaster's engineer be clearly defined in the contract documents.

27.7.4 Detailing of Precast Concrete

Maximum economy and quality is encouraged by designing and detailing precast panels at typical floors which are as repetitious as the design will permit. Special panels will almost always be required at ground level, at mechanical floors, and at the roof. Narrower panels should be considered at these normally higher floors to ease handling and erection. Large size top floor panels, in particular, often are difficult and expensive to erect.

Construction requirements are more likely to control minimum panel thicknesses than structural considerations. The Prestressed Concrete Institute (PCI) [27.46] recommends that the ratio of the unsupported length to minimum thickness of flat panels normally be between 1/20 and 1/40 when conventionally reinforced, or be between 1/30 and 1/60 when prestressed. They also point out that a thickness of at least 3 in. (76 mm) normally is required to achieve adequate concrete cover in conventionally reinforced panels.

When the design permits, closed units are usually preferable to open units. Some examples of open and closed units are shown in Fig. 27.14. Closed units tend to be more rigid and easier to handle. Open units are often delicate and may require temporary stiffeners or strong backs for handling. Some open type units also have the potential to bow or twist during storage. In the completed structure, good joints are often more difficult to achieve when open units are used and where windows occur between open units. Appropriate window frames must be installed to accommodate the glazing.

Moisture Control

The quality of concrete in facades is such that clear waterproofing coatings are seldom required to prevent the entry of water. The sealing of the joints between prefabricated concrete panels is the main concern. The development of watertight joints is particularly critical because flashings frequently are not installed in precast concrete walls because of the difficulty of installation.

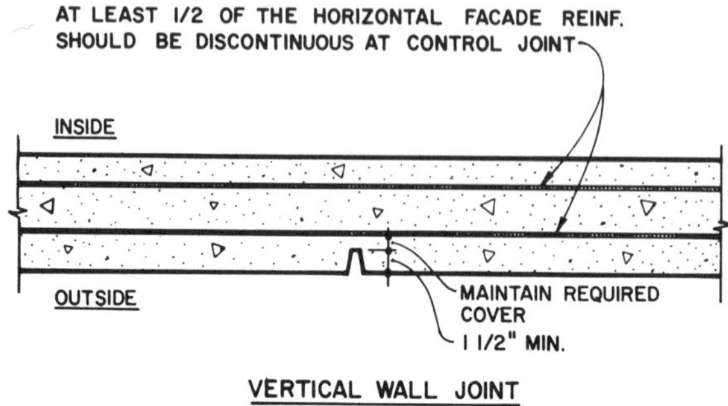

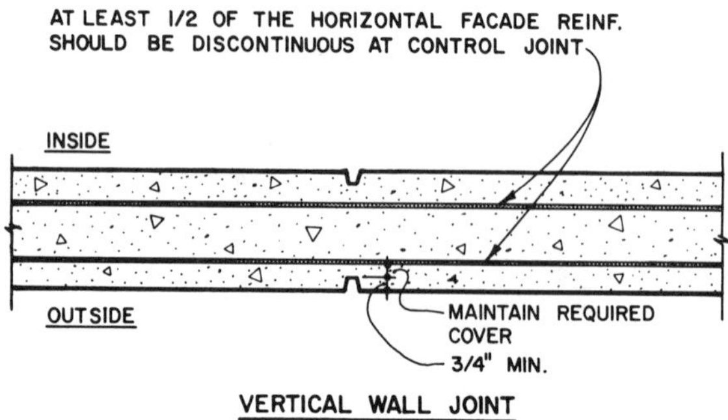

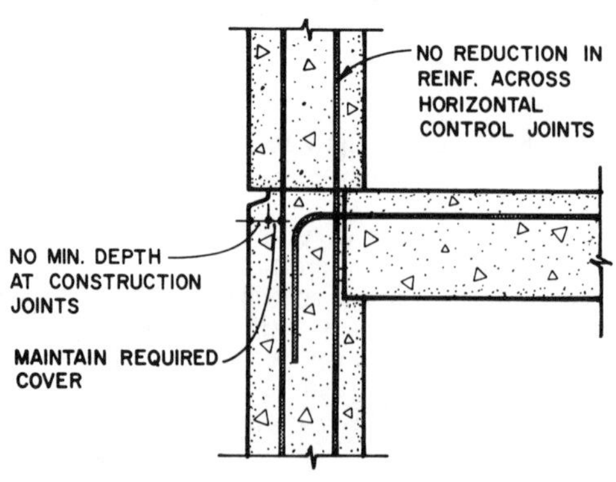

Fig. 27.12 Control joint details for cast-in-place concrete.

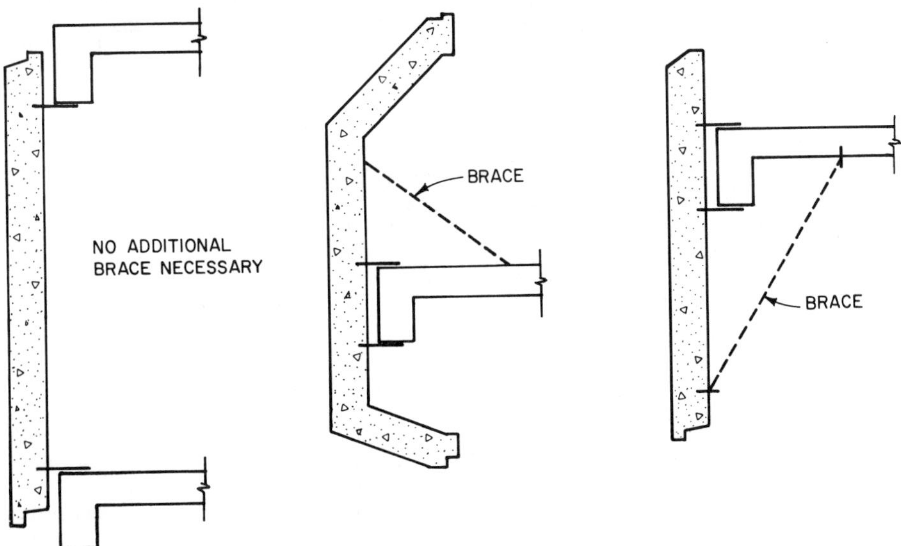

Fig. 27.13 Examples of additional bracing which may be required for precast concrete panels.

Both one- and two-stage joints have been used successfully. Figure 27.15 shows examples of each type of joint. The one-stage joint is typical of those used in most types of building facades which provide only a single line of defense against water entry. More maintenance normally will be required to keep the building watertight with this type.

Two-stage joints provide two lines of defense against the entry of water. Typically, there is a rain barrier at the exterior face of the joint, and a rain and air seal close to the interior face of the joint. The rain barrier stops most of the rain, while the inside seal prevents internal building pressures from adversely affecting the drainage of water which gets past the rain barrier.

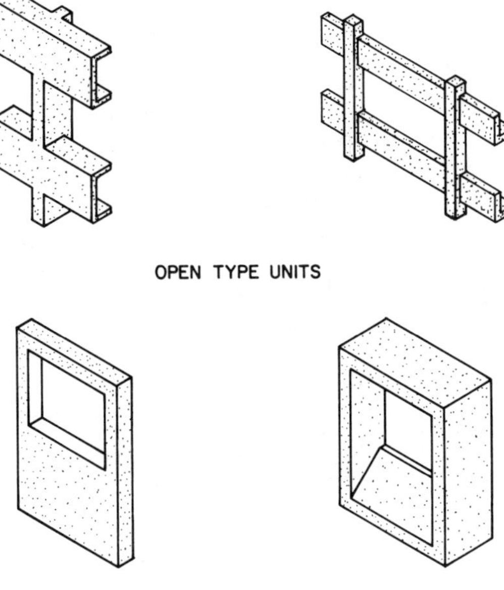

Fig. 27.14 Examples of precast concrete wall panels.

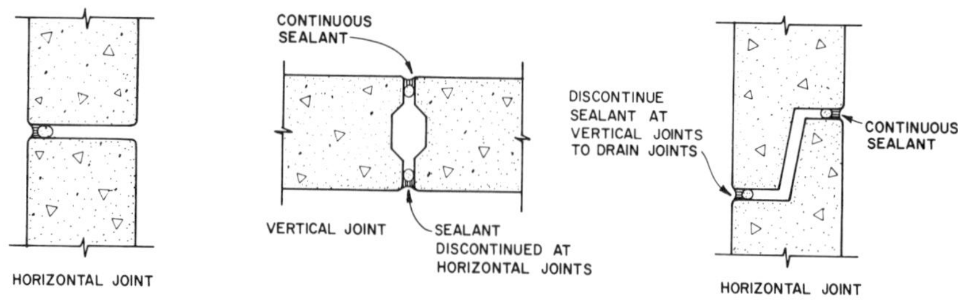

Fig. 27.15 Sealant joint details for precast concrete panels.

Accommodating Volumetric Changes

When a precast concrete panel is removed from its mold, it is at its maximum size, and subsequently shrinks as it dries out.

Precast concrete panels should be designed in such a manner that movements can take place in individual panels with no, or only minimum, effect on adjacent panels. The most common practice for supporting precast panels is to provide vertical/lateral support near the bottom of panels and to provide only lateral support near the top of the panels. To accommodate longitudinal movements near the bottom of the panel, details can be designed to permit the vertical/lateral supports to slide horizontally on neoprene or Teflon bearing pads. Vertical/lateral supports can also be designed with sufficient flexibility in the horizontal direction to accommodate the longitudinal panel movements. To accommodate longitudinal movements near the top of the panel, details can be designed to allow the lateral supports to move both horizontally and vertically by using either slotted holes or oversize holes with nylon or Teflon washers. An example of a precast concrete panel with bearing pad supports at bottom and angles with oversize holes at top is shown in Fig. 27.16.

If the same mix is not used throughout the total thickness of concrete panels, the face mix and backup mix should be reasonably compatible to prevent undue bowing and warping. As a mini-

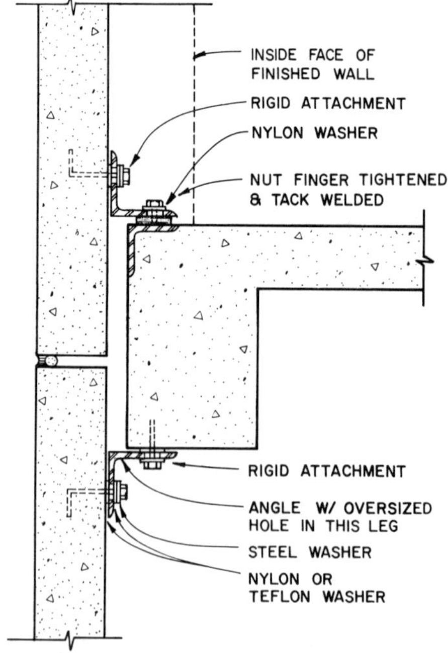

Fig. 27.16 Precast concrete wall.

mum, at least the water–cement and cement–aggregate ratios should be similar. Panels with normal-weight concrete face mixes and lightweight concrete backup mixes should be selected very carefully.

When insulation is required in a facade, it is sometimes incorporated into the precast concrete panels. Both composite and noncomposite concrete sandwich panels are manufactured. The outer concrete layer in noncomposite panels is relatively free to move relative to the inner layer when differences in temperature develop in service. In composite panels where the outer and inner layers are rigidly attached, there is a tendency for bowing to occur when differences in temperature develop. Composite sandwich panels have performed successfully when both layers have been prestressed.

Accommodating Facade/Structure Interaction

When detailing precast concrete facade panels for tall buildings, allowance should be made for frame sway (see Chapters 10 and 11). Attention also should be given to accommodate the deflections and rotations of the spandrel beams which support panels. Excessive deflections of long spandrels can create tapered vertical joints between adjacent panels.

Space should be provided between the facade and the frame to avoid tolerance problems. Facades comprising precast concrete panels and exposed cast-in-place concrete must be detailed with great care because of the differences in achievable tolerances. The totally precast facade, shown in Fig. 27.17, is likely to be more successful than the infill design. The joints in the infill

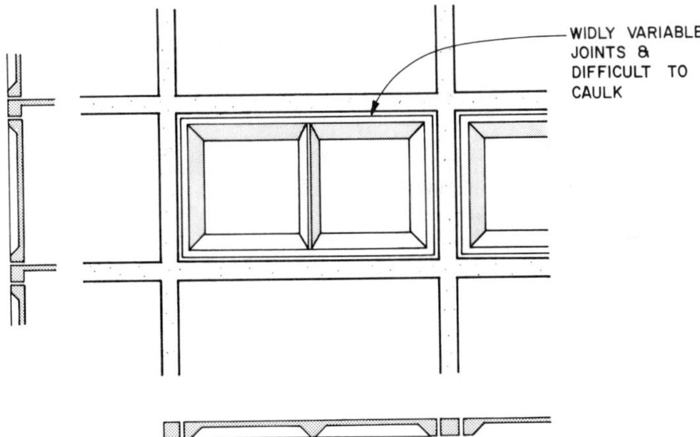

PRECAST & EXPOSED CAST IN PLACE CONCRETE FACADE

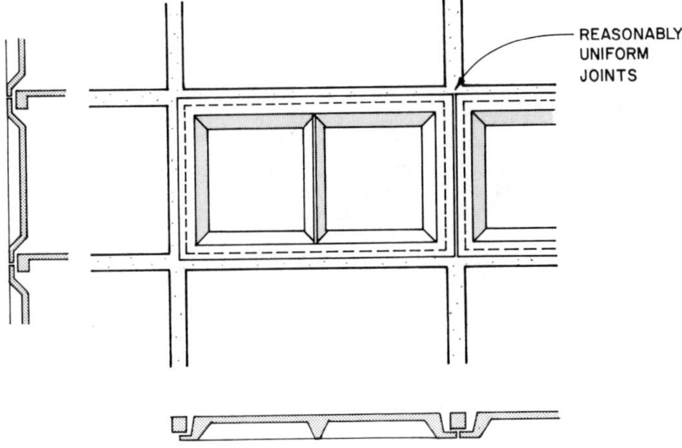

TOTALLY PRECAST CONCRETE FACADE

Fig. 27.17 Precast concrete facade walls.

design are likely to vary in width and uniformity of width, making them very difficult to seal against the entry of water.

27.7.5 Suggested References

References 27.3 through 27.7, 27.10 through 27.13, 27.46, and 27.47 are suggested for precast concrete facades.

27.8 SPECIFIC DESIGN CONSIDERATIONS FOR METAL AND GLASS FACADES

Relatively few tall buildings have facades or curtain walls constructed completely from metal. There is, however, a general class of one- and two-story structures with metal exteriors unbroken except for panel joints and door openings. These are primarily the so-called pre-engineered buildings for which the design and building components are supplied by a manufacturer. The fabricated panels and structural frame components are delivered to franchised contractors who erect the buildings. The exterior metal panels, which generally are fabricated from steel, must meet applicable building code requirements for lateral load and fire resistance. Other than this, the primary design considerations are panel joints and flashing details to prevent water intrusion. Insulation is normally integral to the panel as shipped.

More prevalent is the use of metal in combination with glass. Metal/glass facades are found on all types of structures.

The accommodation of thermal movements is particularly critical in metal curtain wall facades where there are often larger thermal swings than in other types of systems. Maximum ambient temperature change from summer to winter in the northern United States is on the order of 120°F (67°C). Actual temperature readings on metal exteriors substantially exceed this value. Temperatures as high as 175°F (97°C) may be reached on dark colored metals exposed directly to the sun. At the other extreme, metal temperatures may be 10°F (6°C) colder than the surrounding air on a clear winter night. Consequently, from summer to winter, the range of metal temperatures can be greater than that of the air. Except in the most southern states, the temperature swing should be assumed to be at least 180°F (100°C) for design purposes. Where extreme swings are possible, up to 215°F (119°C) may need to be assumed.

Design wind loads may be selected from ANSI Standard A58.1, *Building Code Requirements for Minimum Design Loads in Buildings and Other Structures,* unless preempted by other requirements in the local governing code. The allowable stress for use in design calculations should be that recommended by the Aluminum Association, American Iron and Steel Institute, American Institute of Steel Construction, or the Copper Development Association, as is applicable for the alloy specified.

Normally, the architect should specify a minimum deflection of 1/175 of the clear span and a lesser deflection if this appears desirable. Midspan deflections greater than 1/240 often can be seen under critical light exposure. Reducing deflection often requires the use of additional metal in supports. Consequently, the importance of visual impact vs. added cost must be considered.

Metal curtain walls in large and monumental buildings should normally be required to satisfy a load test. Usually the test is conducted using a uniformly applied load of 1.5 times the design pressure. Deflection is not a criterion. What governs is the ability of the unit to withstand the load without damage or permanent deformation in excess of limits specified.

The concentrated loading imposed on guide rail members by mechanically operating window washing equipment can be a critical consideration. Design information should be obtained from the manufacturer. Usually the load acting normal to the plane of the wall should include the equipment load added to one-third of the inward acting design wind load.

It is very difficult to install a metal curtain wall system that will not have some water leakage through the exterior skin. Consequently, it is recommended that secondary barriers be designed into the assembly to prevent water from entering the interior of the building. Also, a built-in means of collecting and draining the intruding water to the exterior is highly desirable.

It is also common practice to test metal curtain walls for water and air penetration. The most common test method for water penetration is ASTM E331 which is done with a static pressure difference between outer and inner surfaces of the wall. Rain is simulated using a spray rig calibrated to deliver a given rate of water in a uniform pattern. There normally should be no uncontrolled water leakage under a differential static pressure equal to 20% of the inward acting design pressure. The static pressure differential for testing aluminum windows and sliding doors is only 10% of the inward acting design pressure. Reference should be made to Architectural Aluminum Manufacturers Association's *Aluminum Curtain Wall Design Manual* [27.2], Volume I for further information on testing.

Design improvements in composite metal/glass curtain walls have made it possible to meet rigorous performance requirements for thermal efficiency. Both wall and window framing systems are available with thermal breaks which interrupt the highly thermal conductive path characteristics of metals.

Dissimilar metals in contact may create a galvanic cell when exposed to moisture. (See Section 26.6.8). This is an electrochemical phenomenon causing corrosion. Consequently, it is important that fasteners and materials intersecting at joints be compatible from the standpoint of preventing corrosion. The AAMA *Metal Curtain Wall, Window, Storefront and Entrance Guide Specification Manual* [27.1] provides guidelines on fastener selection.

27.8.1 Specific Design Considerations for Aluminum

Aluminum is the most often specified material for both panels and framing when a metal curtain wall system is selected. As described in Chapter 28, the factors leading to this choice include: excellent weatherability, design versatility, wide selection of colors and finishes, ease of fabrication and installation, corrosion resistance, and light weight.

Aluminum curtain walls are broadly classified as either custom or standard—the custom being designed specifically for a given project, whereas the standard employs components and details standardized by the manufacturer.

Both custom and standard walls are further classified according to method of installation. The five basic systems are shown in Fig. 27.18.

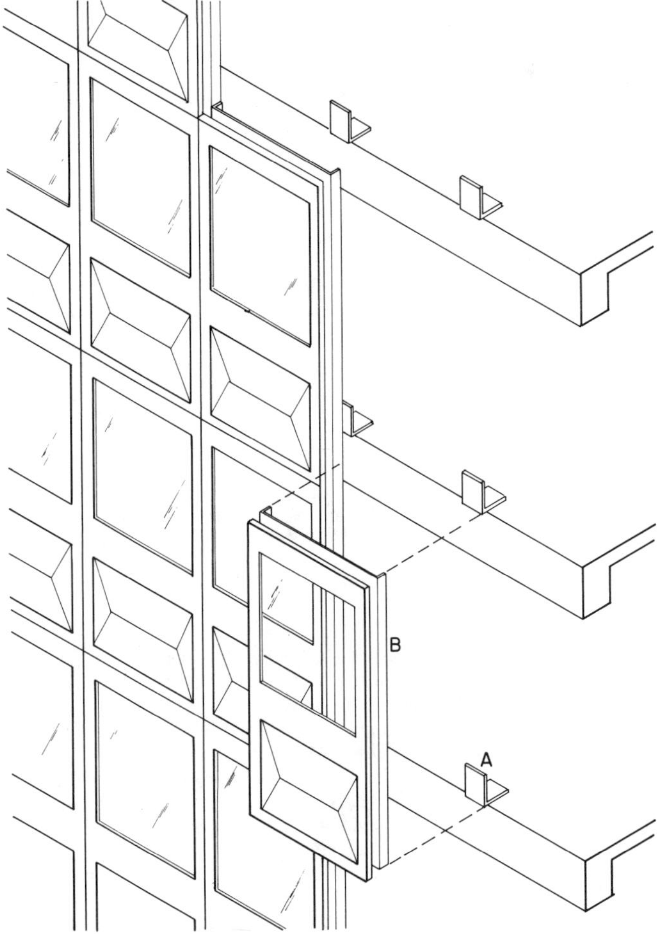

PANEL SYSTEM
A. ANCHOR B. PANEL

Fig. 27.18 Examples of some metal curtain wall systems (*continued on following pages*).

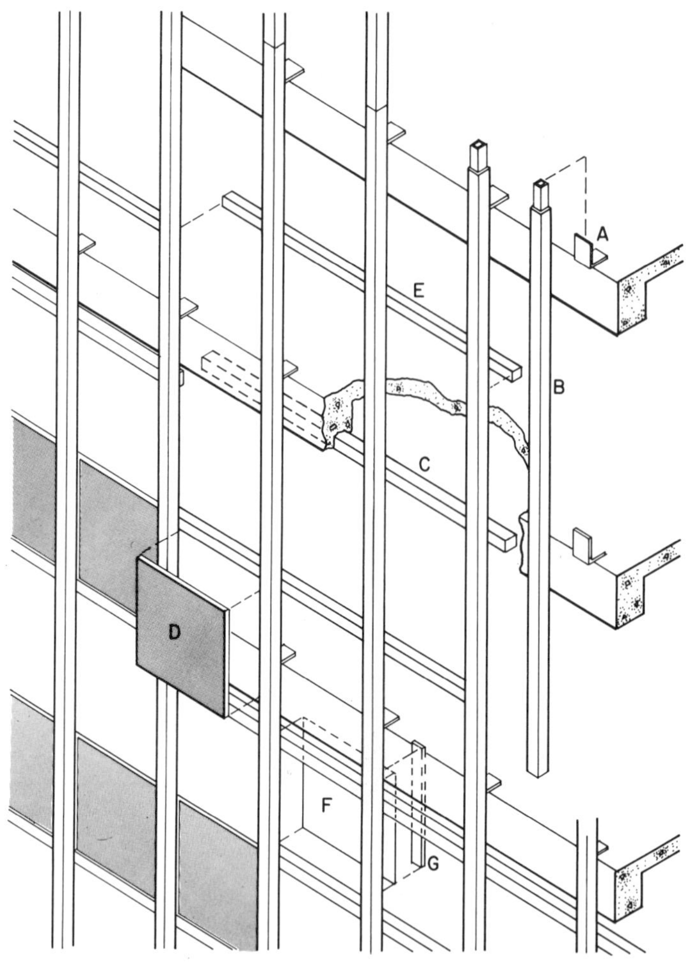

STICK SYSTEM
A. ANCHORS B. MULLION C. HORIZONTAL RAIL D. SPANDREL PANEL
E. HORIZONTAL RAIL (SILL) F. VISION GLASS G. INTERIOR MULLION TRIM.

Fig. 27.18 (continued).

Anodizing, although one of the earliest processes for providing protective coatings to aluminum, is still extensively used for exposed aluminum. Clear anodic finishes as well as integral colors can be obtained from a variety of anodizing and related processes (see Section 28.3.2).

There is also a more recent process known as Integral Color Hardcoat Anodizing. Finishes obtained by this process are designated as AA-A42. They are currently, by far, the most popular selection for architectural aluminum finishing for both interior and exterior use.

Another finish, also designated as AA-A42, is achieved by a process by which metallic compounds are electrolytically deposited in a previously formed oxide coating. The medium bronze to black colors from this method are suitable for exterior use.

Aluminum components of a curtain wall assembly will melt when exposed directly to the flames of a severe fire. Meltdown can occur within 12–15 min after initial flame contact. Glass will break usually after a 3- to 4-min exposure. Although there have been no reported serious consequences resulting from aluminum curtain walls exposed to fire, the designer should keep this limitation of aluminum in mind when designing connections to the structure and in those instances where fire resistive ratings are required. High melt point mineral fiber blankets are available which can provide thermal insulation, as well as excellent fire protection to both aluminum panels and nonvision glass if the design warrants. Because of vulnerability to fire, it obviously is not advisable to incorporate any of the aluminum facade elements as part of the basic structural design.

27.8.2 Specific Design Considerations for Steel

Steel of many types is used for building cladding. Thicknesses normally range from light-gage cold-formed steel up to ¼-in. plate.

The light-gage steel is used in facades in many different ways. For preengineered buildings, panels of standard size and configuration are produced by each manufacturer for their particular building designs. The panels are usually complete with insulation and interior facings, and designed to meet lateral load and fire resistive requirements of the prevailing building codes. Some manufacturers also produce light-gage steel panels to meet specific design requirements of a particular building. Design procedures for cold-formed steel panels are given in Chapter 20.

Another category of panels is made by attaching light-gage steel facings to core materials, such as plastic foam insulation or treated paper honeycomb. The interior facing may be metal, gypsum board, or any one of a number of interior wall materials. These prefabricated panels usually are available both as standard catalog items and in nonstandard sizes and thicknesses.

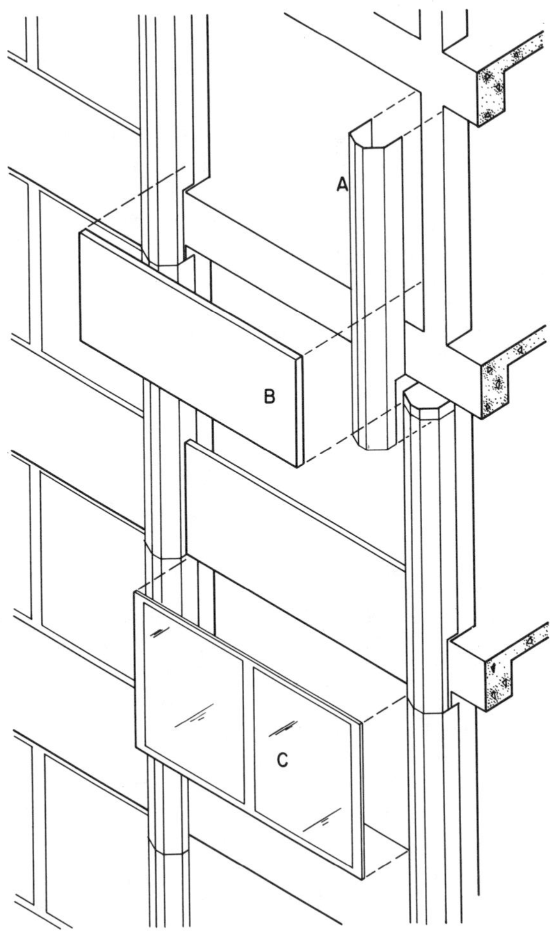

SPANDREL SYSTEM AND COLUMN COVER
A.COLUMN COVER SECTION. B SPANDREL PANEL C.GLAZING INFILL

Fig. 27.18 (continued).

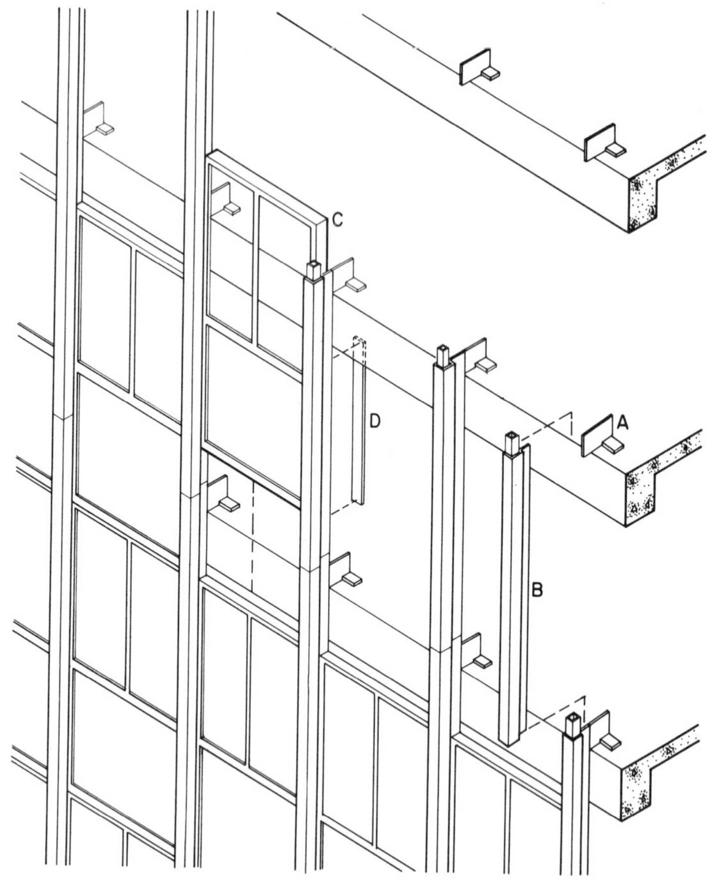

UNIT-AND-MULLION SYSTEM
A. ANCHORS B. MULLION (ONE OR TWO STORY LENGTHS)
C. PRE-ASSEMBLED UNIT (LOWERED INTO PLACE)
D. INTERIOR MULLION TRIM

Fig. 27.18 (continued).

Heavy steel plate up to ¼ in. (6 mm) thick is used for spandrel panels, modular infill panels, or panels connected to the structural frame to stiffen the building. This latter concept of making the panels integral with the structure is a departure from the conventional approach of infilling panels at the exterior of the building or hanging the curtain wall off the floor and structural frame. The advantage of this stressed skin tube concept is that the structural steel framing requirements often can be significantly reduced.

Both steel plate and light-gage steel panels require special primers and finishes to prevent corrosion. It should be established both from test information and actual performance that the coating will not discolor, become dull, or deteriorate for the anticipated use cycle of the building. Fired porcelain enamel finishes normally are applied only to light-gage steel which typically is used as a facing on composite panels.

Stainless steel and weathering steel normally require no special protective finish. Types 302 and 304 stainless are the most commonly used architecturally. These alloys have excellent corrosion resistance, good wearing qualities, high strength, and are easily formed and weldable. Type 316 is similar to 304, but with better corrosion resistance, and should be specified for use in highly corrosive atmospheres. Chapter 29 contains additional information on stainless steel.

27.8.3 Specific Design Considerations for Glass

The designer has many options in glass selection and glazing design to meet functional, safety, and aesthetic requirements for a particular building. The options include annealed, tempered, or heat-

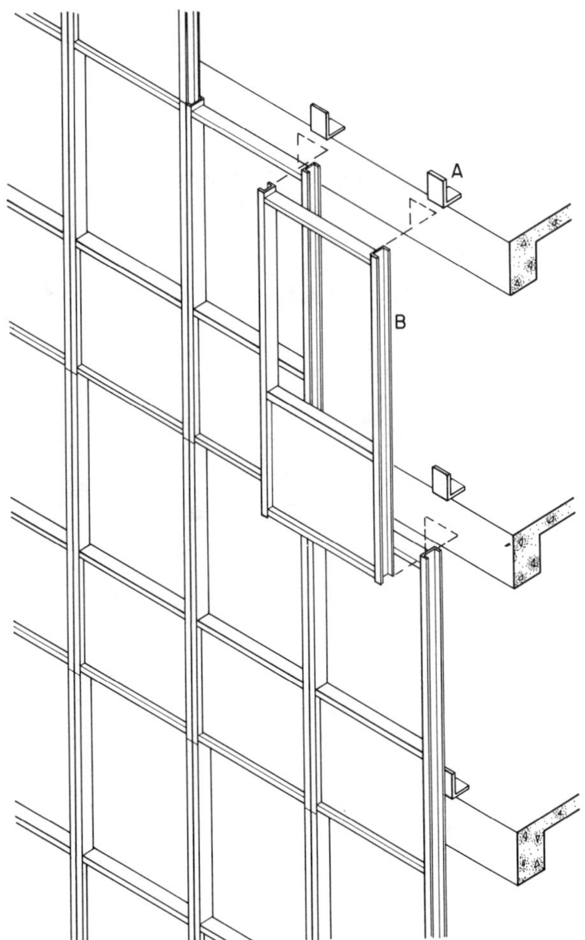

UNIT SYSTEM
A. ANCHOR B. PRE-ASSEMBLED FRAMED UNIT

Fig. 27.18 (continued).

strengthened glass. These in turn can be extended to insulated glass, laminated glass, reflective glass, ceramic colored glass, and sandblasted or etched glass.

The choice for a particular facade design should be based on several considerations including: type and thickness of glass to meet wind load requirements, hazard potential from glass breakage, breakage potential resulting from differential heat gain or loss, thermal insulation requirements, solar gain (greenhouse effect), fire safety, appearance, and cost.

It is normally expected that a certain amount of glass breakage after installation is unavoidable. Statistically, it can be expected even when good practice is followed that up to 10 lights out of every 1000 will break. This usually occurs early in the life of the building. The replacement of glass should be considered by the designer especially when insulating glass is used and there is the potential for widespread replacement when the seals fail.

27.8.4 Suggested References

For aluminum facades, the reader is referred to publications of the Aluminum Association [27.15] and the Architectural Aluminum Manufacturers' Association [27.1, 27.2]. In Ref. 27.1 are listed appropriate ASTM and ANSI standards and federal specifications for:

1. Aluminum alloys.
2. Copper alloys.

3. Carbon steel.
4. High-strength low-alloy steel.
5. Patterned metal sheets.
6. Glass and plastic glazing materials.

Also, this manual lists organizations publishing referenced standards and information relating to metal/glass curtain walls and should be considered a major resource in identifying standards and specifications.

27.9 SPECIFIC DESIGN CONSIDERATIONS FOR STUCCO FACADES

Exterior stucco has a long history of use and remains a popular construction material in warmer areas of the United States. Stucco for building exteriors is comprised of portland cement, hydrated lime, and aggregate—usually sand.

Stucco is applied over some type of metal reinforcing, typically expanded metal lath or a paper-backed wire mesh or metal lath. The metal reinforcing can be used over metal framing, wood framing, or unit masonry. Each of these types of construction calls for special design considerations.

Control joints of suitable design normally are required to break the stucco at intervals of 10 ft (3 m) or less with a maximum area enclosed by control joints not to exceed 100 sq ft (9.3 m^2). In order to function properly, the control joint must interrupt the metal lath or reinforcing mesh. Vertical framing should be doubled up behind control joints, as shown in Fig. 27.19.

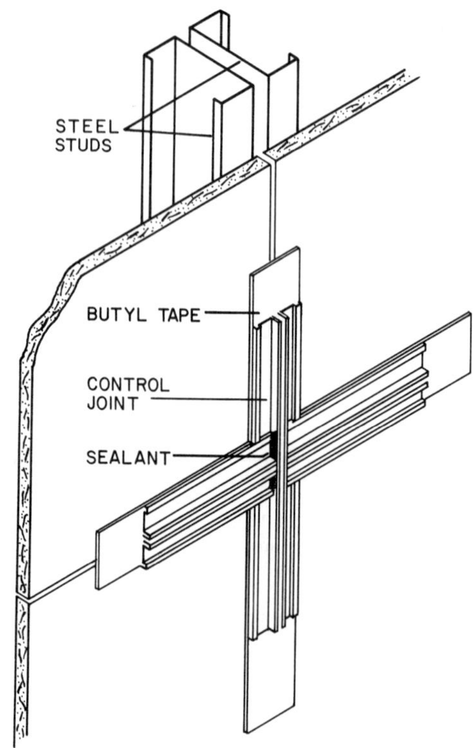

Fig. 27.19 Stucco facade detail.

27.9.1 Metal Framing Backup for Stucco

A stucco/metal stud facade has the features of being noncombustible, lightweight, architecturally versatile, and relatively low cost. Selection of the framing member for a given lateral load condition is determined on the basis of steel thickness, strength of steel, section properties, wall height, and framing spacing. Framing guides showing stud selection and spacing for given lateral load conditions are provided by most metal framing manufacturers and MLSFA.

Although somewhat more costly than other systems, maximum resistance to cracking and water leakage can be achieved by use of gypsum sheathing attached with screws to the steel framing followed by application of lapped 15-lb building felt [or alternatively, sheathing joints sealed with 2-in. (50-mm) wide butyl tape] and 3.4-lb galvanized self-furring lath. Application of ⅞-in. (22-mm) overall thickness of stucco from lath face to stucco surface is normally recommended. Adequate flashing and wall drainage should also be included.

Paper-backed metal lath or wire mesh attached directly to the metal studs is widely used for stucco.

Unless special provision is made to fur out these materials at the stud, there is no opportunity for stucco to penetrate the mesh or lath at the stud. This provides a line of weakness which may result in a vertical crack over the stud. The paper backing of the lath also should be removed at the laps in order to maintain uniform stucco thickness.

27.9.2 Stucco Over Concrete Block

Because of great variability in absorption characteristics of concrete block, differential shrinkage cracks in the stucco may result. Direct application of stucco often is not recommended.

Paper-backed galvanized wire mesh or felt-backed galvanized metal lath may be used, secured to the block with masonry nails. The preferred system is self-furring galvanized 3.4-1b metal lath applied over 15-lb felt.

27.10 SPECIFIC DESIGN CONSIDERATIONS FOR METAL STUD BACKUP WALL

The introduction of light-gage steel framing in conjunction with self-tapping, self-drilling screws has resulted in widespread use of this type of framing system for both load-bearing and non-load-bearing applications.

Load-bearing steel studs, with few exceptions, can be used as an alternate to conventional wood framing in facade walls. When used in load bearing, the practical height limit for light steel facades normally is about five stories. As with any load-bearing exterior wall design, the steel framing components must be selected to meet anticipated axial, tensile, and lateral load requirements. Diagonal steel strap bracing may be used to provide requisite racking resistance. Alternatively, a structural sheathing such as plywood, if permitted by code, suitably fastened to the studs, may be used to provide racking resistance.

Load-bearing studs are typically in a "C" shape. Steel studs are available in a configuration that can be nested allowing them to be used singly, doubled (back-to-back), or in double nested combination.

In addition to load-bearing wall applications, light-gage steel framing is widely used in curtain wall construction, primarily as backup framing for masonry or stucco veneers. When used with masonry, the studs must be rigid enough that cracks do not develop in the masonry when the walls are subjected to lateral loads.

Calculations of structural properties of light-gage steel members are normally based on AISI *Specification for the Design of Cold-Formed Steel Structural Members* [27.14] (See also Chapter 20). However, most manufacturers publish complete load tables which are convenient to use.

Light-gage steel framing is available with a painted or galvanized finish. Painted steel is widely used, since generally it is slightly less in cost than galvanized steel and less troublesome to weld. However, painted steel is more vulnerable to corrosive effects than galvanized steel, and normally its use is not recommended where significant corrosive influences are found. Details are to be found in the MLSFA *Steel Framing Systems Manual* [27.43].

A factor often overlooked in the design of exterior walls with light-gage steel framing is the high thermal conductivity of the framing members. Heat loss through the framing can be significant, although frequently neglected in calculating the *U* factor for the wall.

It is advisable to determine overall conductivity of the wall by test, either with a guarded or calibrated hot box. Alternatively, the *U* factor can be calculated using the procedure shown in the *ASHRAE Handbook* [27.18], Fundamentals, Section 23.3, "Heat Flow Through Panels Containing Metal."

27.11 JOINTS AND SEALANTS

The subject of joints and joint details in building facades, and the selection of sealant(s) to ensure against air or water passage at the joints, is one of great diversity and complexity. For comprehensive treatment of this subject, the reader is referred to "Sealants and Waterproofing," Sweets Division, McGraw-Hill Information Systems.

SELECTED REFERENCES

27.1 AAMA. *Metal Curtain Wall, Window, Storefront and Entrance Guide Specifications Manual.* Des Plaines, IL: Architectural Aluminum Manufacturers' Association, 1976 (2700 River Rd., Suite 118, Des Plaines, IL 60018) 1976.

27.2 AAMA. *Aluminum Curtain Wall Design Guide Manual.* Des Plaines, IL: Architectural Aluminum Manufacturers' Association, 1979.

27.3 ACI. *Specifications for Structural Concrete for Buildings* (ACI 301-72) (Revised 1981). Detroit: American Concrete Institute, 1981.

27.4 ACI. *Guide to Cast-in-Place Architectural Concrete Practices* (ACI 303R-74) (Revised 1982). Detroit: American Concrete Institute, 1982.

27.5 ACI. *Building Code Requirements for Reinforced Concrete* (ACI 318-83). Detroit: American Concrete Institute, 1983.

27.6 ACI. *Commentary on Building Code Requirements for Reinforced Concrete* (ACI 318-83). Detroit: American Concrete Institute, 1983.

27.7 ACI Committee 504. *Guide to Joint Sealants for Concrete Structures* (ACI 504R-77). Detroit: American Concrete Institute, 1977.

27.8 ACI Committee 531. *Specification for Concrete Masonry Construction* (ACI 531.1-81). Detroit: American Concrete Institute, 1981.

27.9 ACI Committee 531. *Building Code Requirements for Concrete Masonry Structures and Commentary* (ACI 531-83). Detroit: American Concrete Institute, 1983.

27.10 ACI Committee 533. "Design of Precast Concrete Wall Panels," *ACI Journal, Proceedings,* **68,** July 1971, 504–513.

27.11 ACI Committee 533. "Selection and Use of Materials for Precast Concrete Wall Panels," *ACI Journal, Proceedings,* **66,** October 1969, 814–822.

27.12 ACI Committee 533. "Fabrication, Handling and Erection of Precast Concrete Wall Panels," *ACI Journal, Proceedings,* **67,** April 1970, 310–340.

27.13 ACI Committee 533. "Quality Standards and Tests for Precast Concrete Wall Panels," *ACI Journal, Proceedings,* **66,** April 1969, 270–275.

27.14 AISI. *Specification for the Design of Cold-Formed Steel Structural Members.* Washington, DC: American Iron and Steel Institute, September 3, 1980 (1000 16th Street N.W., Washington, DC 20036).

27.15 Aluminum Association. *Aluminum Standards and Data.* Washington, DC: Aluminum Association (818 Connecticut Avenue N.W., Washington DC, 20006).

27.16 ANSI. *American Standard Building Code Requirements for Masonry* (A41.1) (National Bureau of Standards Miscellaneous Publication 211). Washington, DC: American National Standards Institute, 1954.

27.17 ANSI. *Safety Performance Specification and Method of Test for Safety Glazing Materials Used in Buildings* (Z97.1-75). Washington, DC: American National Standards Institute, 1975.

27.18 ASHRAE. *ASHRAE Systems Handbook for 1981.* New York: American Society of Heating, Refrigerating & Air Conditioning Engineers, Inc. 1981.

27.19 ASTM. *Specification for Concrete Building Brick* (C55-75) (1980). Philadelphia: American Society for Testing and Materials, 1980.

27.20 ASTM. *Specification for Building Brick (Solid Masonry Units Made from Clay or Shale)* (C62-83). Philadelphia: American Society for Testing and Materials, 1983.

27.21 ASTM. *Specification for Hollow Load-Bearing Concrete Masonry Units* (C90-75) (1981). Philadelphia: American Society for Testing and Materials, 1981.

27.22 ASTM. *Specification for Ceramic Glazed Structural Clay Facing Tile, Facing Brick, and Solid Masonry Units* (C126-82). Philadelphia: American Society for Testing and Materials, 1982.

27.23 ASTM. *Specification for Solid Load-Bearing Concrete Masonry Units* (C145-75) (1981). Philadelphia: American Society for Testing and Materials, 1981.

27.24 ASTM. *Specification for Facing Brick (Solid Masonry Units Made from Clay or Shale)* (C216-81). Philadelphia: American Society for Testing and Materials, 1981.

27.25 ASTM. *Specification for Mortar for Unit Masonry* (C270-82). Philadelphia: American Society for Testing and Materials, 1982.

27.26 ASTM. *Specification for Grout for Reinforced and Nonreinforced Masonry* (C476-83). Philadelphia: American Society for Testing and Materials, 1983.

27.27 ASTM. *Specification for Marble Building Stone (Exterior)* (C503-79). Philadelphia: American Society for Testing and Materials, 1979.

27.28 ASTM. *Specification for Limestone Building Stone* (C568-79). Philadelphia: American Society for Testing and Materials, 1979.

27.29 ASTM. *Specification for Granite Building Stone* (C615-80). Philadelphia: American Society for Testing and Materials, 1980.

27.30 ASTM. *Specification for Sandstone Building Stone* (C616-80). Philadelphia: American Society for Testing and Materials, 1980.

27.31 ASTM. *Specification for Slate Building Stone* (C629-80). Philadelphia: American Society for Testing and Materials, 1980.

27.32 ASTM. *Specification for Hollow Brick (Hollow Masonry Units Made from Clay or Shale)* (C652-81a). Philadelphia: American Society for Testing and Materials, 1981.

27.33 ASTM. *Specification for Sealed Insulating Glass Units* (E774-84a). Philadelphia: American Society for Testing and Materials, 1984.

27.34 BIA. *Building Code Requirements for Engineered Brick Masonry*. McLean, VA: Brick Institute of America, Reprinted March, 1979 (1750 Old Meadow Rd., McLean, VA 22102).

27.35 BIA. *Recommended Practice for Engineered Brick Masonry*. McLean, VA: Brick Institute of America, November, 1969.

27.36 BIA. *Standard Specifications for Portland Cement Lime Mortar for Brick Masonry (M1)*. McLean, VA: Brick Institute of America, November, 1969.

27.37 BIA. *Technical Notes on Brick Construction,* pamphlet series. McLean, VA: Brick Institute of America.

27.38 CPSC. *Safety Standard for Architectural Glazing Materials* (16CPR 1201). Washington, DC: Consumer Product Safety Commission.

27.39 Federal Specification DD-G 451D, *Glass, Float or Plate, Sheet, Figured (Flat for Glazing, Mirrors and Other Uses),* April 25, 1977.

27.40 Federal Specification DD-G 1403C, *Glass, Float, Sheet, Figured, Coated (Heat-Strengthened and Tempered),* September 13, 1983.

27.41 ILI. *Indiana Limestone Handbook*. Bedford, IN: Indiana Limestone Institute of America, Inc. (Stone City Bank Building, Suite 400, Bedford, IN 47421).

27.42 MIA. *Marble Institute of America Manual*. Farmington, MI: The Marble Institute of America (33505 State St., Farmington, MI 48024).

27.43 MLSFA. *Steel Framing Systems Manual*. Chicago: Metal Lath/Steel Framing Association and Association of The Wall and Ceiling Industries International (221 N. LaSalle St., Chicago, IL 60601).

27.44 NBGQA. *Specifications for Architectural Granite*. West Chelmsford, MA: Building Granite Quarries Association, Inc. (c/o H.E. Fletcher Co., West Chelmsford, MA 01863).

27.45 NCMA. *TEK Information Series*. Herndon, VA: National Concrete Masonry Association, continuing series since 1970 (P.O. Box 781, Herndon, VA 22070).

27.46 PCI. *PCI Manual for Structural Design of Architectural Precast Concrete* (MNL-121-77). Chicago: Prestressed Concrete Institute, 1977 (201 North Wells Street, Chicago, IL 60606).

27.47 PCI. *PCI Architectural Precast Concrete*. Chicago: Prestressed Concrete Institute, 1973 (201 North Wells Street, Chicago, IL 60606).

27.48 PSPG. *Specifications for Exterior Structural Slate*. Pen Argyl, PA: The Pennsylvania Slate Producers Guild.

27.49 ICBO. *Uniform Building Code*. Whittier, CA: International Conference of Building Officials, 1982.

CHAPTER 28
ALUMINUM STRUCTURES

ROBERT W. HAUSSLER

Consulting Structural Engineer
P.O. Box 669
Templeton, CA 93465

28.1 TYPICAL APPLICATIONS

28.1.1 Introduction

Aluminum consumption per capita in the United States is higher than that in any other country, having been 57 lb (26 kg) per person per year in 1980, followed by Norway, 53, West Germany, 49, and Japan, 46. Of the total U.S. consumption, the building and construction industries account for 18%. Shipments of aluminum for these industries reached a peak in 1973 of 3.6 billion lb, dropped to 2.2 billion in 1975, rose to 3.2 billion in 1978, and in 1980 was 2.6 billion lb.

28.1.2 Building Hardware

Aluminum is widely used for sash and glass door frames and jambs. This is due mainly to the intricate precision shapes that can be extruded. Also, its weather resistance, dimensional stability, and ease of fabrication add greatly to its desirability for these applications and many others such as hand railings, partition mullions, louvers, door hoods, awnings, screening, and panels for hiding equipment. The beautiful colored finishes obtainable by anodizing have made aluminum quite popular for architectural flourishes such as fasciae, wall panels, ceilings, and many articles of building hardware.

28.1.3 High-Rise Exterior Cladding

Aluminum has been successfully used for forming the exterior panels below the windows of high-rise buildings. In this application it is found to save weight, resist weathering, and be easily fabricated and installed. Section 27.8.1 treats aluminum facades.

28.1.4 Roofing and Siding

Because of its excellent resistance to weathering, aluminum has become a favorite material for residential patio covers. Ease of installation has made it possible for the home owner to erect a prefabricated unit or for a contractor to do the installation with a minimum of labor. These factors plus the fact that the roof or siding needs no further finish or protection than the factory-applied paint have made aluminum an important material for commercial and industrial canopies, roof structures, and light metal buildings; however, at present it is generally more expensive than the equivalent light-gage steel cladding that is protected from weathering with galvanizing or a coating of aluminum or aluminum–zinc alloy.

Despite such competitive products, the endurance of aluminum roofing and siding over a long period of time (such as 20 years or more) under severe climatic conditions without any maintenance requirement has made aluminum a preferred material in areas of heavy snowfall, for farm buildings, and for buildings that are to be owner-occupied for many years.

Aluminum roofing and siding is used structurally in building design in several ways. It spans between purlins or girts and, in some small structures, spans from wall to wall or floor to roof, thus eliminating the need for purlins or girts. Where snow is not a problem, a span of 20 ft (6 m) is often achieved with a profile 2.5 in. (64 mm) or so in depth. In snow areas profiles as deep as 6 in. (150 mm) have been used. When purlins or girts are used the aluminum sheathing can be employed in the design to resist rotation and lateral displacement of these members, and therefore achieve high efficiency with minimal or no bracing. In small structures the sheathing is often used to brace the entire building by diaphragm action. Without going to the expense of full-scale testing, it is common practice to develop a shear resistance in the range of 50 lb/ft (730 N/m) in the sheathing design by calculating bearing stress in the panel lap fasteners.

28.1.5 Mobile Home Cladding and Patio Enclosures

Ease of fabrication, light weight, and weather resistance have made aluminum the most widely used material for mobile home cladding. These factors along with the ease with which intricate extrusions can be made have resulted in extensive use of aluminum for patio enclosure mullions, perimeter members, kickplates, sash, and doors.

28.1.6 Water Storage Tank Covers and Chemical Plants

Corrosion resistance in a humid atmosphere, durability, light weight, and resistance to birds and mice are the factors that have dictated extensive use of aluminum for covering swimming pools, water storage tanks, and certain types of buildings and tanks containing corrosive vapors such as those from sewage treatment sludge.

28.1.7 Nonmagnetic Structures

Aluminum beams, columns, brackets, fasteners, and sheathing have been used in industrial buildings where the magnetic effects of steel could not be tolerated because of the type of work to be performed in the structures.

28.1.8 Buildings at Inaccessible Sites

Major components of astronomical observatories, remote weather reporting stations, microwave repeaters, and other similar buildings where ground transportation to the erection site is not available and a helicopter must be used to carry the components, are often fabricated in aluminum because of its light weight and low maintenance.

28.2 PHYSICAL PROPERTIES

28.2.1 Strength of Various Alloys

Building codes and specifications for aluminum are based primarily on the guaranteed minimum tensile ultimate and tensile and compressive yield strengths of certain alloys, as these quantities are listed by the Aluminum Association under their registration procedure. Other important but less frequently needed quantities may or may not be covered in the registration such as shear and bearing strengths. Table 28.1 lists five alloys widely used for structural members in buildings.

28.2.2 Strength at Various Temperatures

The influence of temperature on the strength of a typical high-strength alloy is shown in Table 28.2 for alloy 6061-T6 assuming 100% strength at 75°F (24°C). Other alloys follow a similar but not identical pattern.

28.2.3 Elastic Properties

As can be seen from Table 28.1, the modulus of elasticity of the various alloys generally used in buildings can be taken as 10,000 ksi (69,000 MPa) for computation of the deflection of members in bending. For axial stress the modulus of elasticity for these alloys is 10,100 ksi (70,000 MPa). Poisson's ratio is generally 0.33 for aluminum alloys. The modulus of rigidity corresponding to these values is 3800 ksi (26,000 MPa) based on Eq. (28.2.1) [28.12].

$$G = \frac{0.5E}{(1 + \nu)} \tag{28.2.1}$$

where G = modulus of rigidity
E = modulus of elasticity
ν = Poisson's ratio

28.2.4 Density and Thermal Properties

Table 28.3 gives typical values of the density and important thermal properties of aluminum alloys. The values of the top four items are for alloy 6061 while the bottom two are for pure aluminum. For roughly estimating the properties of aluminum in comparison with steel, its density and modulus of elasticity are each approximately one-third that of steel. Thermal expansion of aluminum is approximately twice that of steel. However, for steel and aluminum bars of identical dimensions which are constrained from expansion the thermally induced stress in aluminum is approximately two-thirds that in steel.

28.2.5 Machining

The ease with which aluminum can be machined has helped make it a popular item for building construction. The recommended back rake angle for lathe tools is 35° which is more acute than the angle recommended for steel [28.11]. However, in actual practice it is often more convenient to use tools designed for cutting steel or wood because such tools are easily obtained and generally give satisfactory results. Examples are: (1) portable and fixed carbide-tipped circular power saws (wood cutting), (2) band saws (metal cutting), (3) drill bits (metal cutting), and (4) lathe tools ground for steel.

When cutting aluminum with a circular saw it is very important to arrange the workpiece and the saw in such a way that the operator has complete control over their relative position at all times because the pressure at the cutting teeth is generally greater with aluminum than with wood. For a radial arm saw this means that the workpiece and the waste piece must be well secured and the direction of rotation of the saw must be such that it does not tend to pull into the workpiece. Soft alloys tend to form a deposit on the cutting surface of each tooth which can usually be overcome by the use of an ordinary spray lubricant.

28.2.6 Extruding, Forming, and Casting

Aluminum members of complex cross-section can be made by the extrusion process which involves heating a billet in a large press and forcing it through a precision die that has been cut to the desired cross-sectional dimensions. Several shapes that have been produced in quantity are shown in

Table 28.1 Physical Properties [stresses in ksi (MPa)]*

	Alloy				
	3003-H16	3005-H28	6061-T6	6063-T5	6063-T6
	Use				
Property	Sheathing	Structural Sheathing	Beams and Columns	Architectural Components	Structural Mullions
Tensile ultimate strength	24 (165)	31 (214)	38 (262)	22 (152)	30 (207)
Tensile yield strength	21 (145)	27 (186)	35 (241)	16 (110)	25 (172)
Compressive yield strength	18 (124)	25 (172)	35 (241)	16 (110)	25 (172)
Shear ultimate strength	14 (97)	17 (117)	24 (165)	13 (90)	19 (131)
Shear yield strength	12 (83)	16 (110)	20 (138)	9 (62)	14 (97)
Bearing ultimate strength	46 (317)	56 (386)	80 (552)	46 (317)	63 (434)
Bearing yield strength	31 (214)	38 (262)	56 (386)	26 (179)	40 (276)
Modulus of elasticity for deflection	10,000 (69,000)	10,000 (69,000)	10,000 (69,000)	10,000 (69,000)	10,000 (69,000)

* From Ref. 28.7.

Table 28.2 Strength of Alloy 6061-T6*

Temperature		Tensile Ultimate	Tensile Yield
°F	°C	Percent of Strength at 75°F (24°C)	
−320	−196	133	118
−112	−80	109	105
−18	−28	104	103
75	24	100	100
212	100	93	95
300	149	76	78
400	204	42	38
500	260	17	13
600	316	10	7
700	371	7	5

* From Ref. 28.3.

Table 28.3 Density and Thermal Properties of Aluminum

Density*	169 pcf	2700 kg/m³
Specific gravity*	2.70	2.70
Coefficient of thermal expansion*	13.1/°F	23.6/°C
Melting range*	1080–1205°F	582–652°C
Thermal conductivity†	0.292 Btu·in./(s·ft²·°F)‡	152 W/(m·K)§
Specific heat†	0.214	0.214

* From Ref. 28.3.
† From Ref. 28.13.
‡ Btu·inches per second per square foot per degree Fahrenheit.
§ watt per metre kelvin.

Fig. 28.1. These are samples that demonstrate the versatility of the product. In addition, many of the shapes produced in hot-rolled steel are available in aluminum extrusions. As aluminum is extruded, it is also stretched at a uniform rate to secure the remarkable straightness that characterizes this product. After cutting to the desired stock lengths the extrusions are then heat-treated in an oven at carefully controlled temperatures to secure the proper temper. Popular alloys are 6061 and 6063, the properties of which are in Table 28.1.

Sheet aluminum and coil stock are accurately formed into many varied shapes by the use of heavy machinery such as press brakes and roll formers. Examples of such shapes are those produced in cold-formed steel products; in fact, the manufacturer often uses the same dies for forming both steel and aluminum. Most production items use prepainted coil stock and a roll former because of the great speed with minimal labor. In this process the material enters as a flat strip and is progressively formed by precision rollers arranged in stages. The final shape issues from the last stage at high speed and is chopped to the proper length by a "flying" cutoff. Popular alloys are 3003 and 3005, see Table 28.1.

For shapes of irregularly varying cross-section such as brackets, base plates, fittings, and hangers, aluminum can be readily die cast or sand cast. A popular casting alloy is 356.0 which has a tensile strength in T-6 temper of 30.0-ksi (207-MPa) ultimate and 20.0-ksi (138-MPa) yield.

Fig. 28.1 Typical aluminum extrusions.

28.2.7 Welding

Aluminum alloys used for building can be satisfactorily welded by the gas tungsten-arc or gas metal-arc process. Allowable stresses for the portion of a member within 1.0 in. (25 mm) of a weld are required to be reduced. For example, the allowable local short column strength of alloy 6063-T6 drops from 13.5 ksi (93 MPa) in nonwelded members to 6.5 ksi (45 MPa) in the vicinity of a weld [28.7]. A popular filler alloy is 4043 [28.7, Table 7.1.3.1] which has an allowable shear stress in fillet welds of 5 ksi (34 MPa).

28.2.8 Fasteners

Aluminum members are generally bolted to the other portions of a building by the use of plated steel bolts for large forces or plated steel sheet metal screws where the forces are smaller such as in sheathing, sash, and mullions. However, where corrosion is a problem such as at the seashore, stainless steel or aluminum fasteners are recommended. A typical aluminum bolt alloy is 2024-T4 which has an allowable shear stress on the effective area of 16 ksi (110 MPa) and 26-ksi (179-MPa) allowable tensile stress on the root area [28.7].

Sheet metal screws do not have published withdrawal values; however, because they generally enter a member that is considerably thicker than the sheathing this is not usually a problem. Shear of the screw is also not usually considered because the screws are hardened for use with a portable power-driven insertion tool. The over-the-head tearing strength of the sheathing can be calculated by using Eq. (28.2.2) as follows [28.7]:

$$P_t = 0.17tF_{ty} \tag{28.2.2}$$

where P_t = allowable tensile force
t = sheathing thickness
F_{ty} = tensile yield strength of the sheathing

This equation is based on a large number of tests of fasteners with ⅝-in. (16-mm) diameter washers in the valley of sheathing and should be adjusted in proportion to the outside diameter of the washer or bolt head if no washer is used where that diameter differs from ⅝ in. (16 mm). Bearing values (see Table 28.4) for sheet metal screws are generally taken as the same as those permitted for bolts.

Where sheet metal screws are used to fasten two relatively thin sheets together and the fasteners are in tension, it is recommended that tests be performed to determine the joint strength because of the possibility of stripping of the threads in the second sheet. Rivets have been widely used in aircraft work. As a result there are excellent alloys available that are easily driven and harden later at room temperature. However, because of the fact that they generally must be stored in a cold box to keep them soft, they have never reached wide usage for building construction. To overcome this problem and to allow rivets to be inserted from one side without a backup helper, a blind (pop) rivet is used to some extent in building components that resist small loads. These are generally of an alloy that does not harden at room temperature. A typical alloy is 5052-H32 which has an allowable shear stress on the effective area of 8 ksi (54 MPa) [28.7]. These rivets are tubular and have a mandrel that is pulled through the center to form a head on the far end. The installing tool breaks the mandrel when the head is completely formed. The shank of the mandrel is discarded and its shear area is not available for resisting loads. The effective area of these rivets ranges from 59 to 62% of the nominal diameter.

Table 28.4 Allowable Bearing Stresses*

	Allowable Bearing Stress	
Alloy and Temper	ksi	Mpa
3003-H16	19	131
3005-H28	23	159
6061-T6	35	241
6063-T5	16	110
6063-T6	24	165

* From Ref. 28.7.

28.3 CHEMICAL PROPERTIES AND PROTECTIVE COATINGS

28.3.1 Corrosion Resistance

Aluminum alloys vary in corrosion resistance with pure aluminum having very high resistance and those with large amounts of alloying elements, especially copper, having lower resistance. High-strength alloys such as 6061-T6 develop a grey oxidation when used on exterior surfaces that are exposed to dew or rain. Alloy 6063 is less subject to corrosion; however, to preserve the bright finish it is either painted with a clear paint or anodized where exposed to moisture. Roofing and siding panels are normally painted prior to forming. An exception is the sheathing intended for farm applications on short spans where material with a low alloy content is used without paint.

More severe conditions may require further precautions. Exposure to salt air requires special attention to fasteners, using aluminum or stainless steel instead of galvanized steel fasteners. Cadmium plating is superior to galvanizing and works well for steel fasteners in areas away from the seashore. Aluminum in contact with dissimilar metals or porous substances such as wood or masonry requires painting or anodizing. Properly coated aluminum can be embedded in concrete if the choice of alloys is restricted to those with very high corrosion resistance. For water storage, water distribution, and cooling towers, alloys of the 3000, 5000, and 6000 series are recommended [28.1].

28.3.2 Coatings

Anodizing is a surface treatment of aluminum alloys obtained by immersing the finished product as anode in an acid electrolitic bath. The resultant oxide finish is very tough and can be tinted various colors by adding dyes during the process. The length of time of immersion determines the thickness of the oxide coating and thereby the Class, the greatest thickness commercially available being Class I [28.8]. A similar process called chromate treatment provides a high degree of protection against marine and humid environments [28.7].

Painted aluminum coil stock is available in a wide range of colors. The metal is cleaned, etched, treated in a hot phosphoric acid solution, then painted with generally two coats of special enamel and baked. These coils can be formed in a brake or rollformer without the development of cracks or crazing as would occur with anodic coating. If cut, punched, or scratched the base metal quickly forms a thin oxide coating which protects it from further corrosion, thus preventing flaking of the surrounding paint. Matching paint is available for touching up scratches where appearance is important.

28.4 ALLOWABLE STRESSES AND DEFLECTIONS

28.4.1 Factors of Safety

To determine an allowable stress that will reduce the possibility of structural failure of an aluminum member to an acceptable level, the guaranteed minimum ultimate and yield properties are divided by factors of safety [28.7]. The allowable stress for an element of a member that is short and has a low enough width-to-thickness ratio so that buckling is not a problem is the smaller of the values obtained from the factored ultimate and yield strengths using the factors shown in the upper portion of Table 28.5. Factors of safety for long members and thin elements are shown in the lower portion

Table 28.5 Factors of Safety

For Short Members and Thick Elements	
Tensile ultimate	1.95
Tensile, compressive, shear, and bearing yield	1.65
Shear ultimate of rivets, bolts, and fillet welds	2.34
For Long Members and Thin Elements	
Column buckling or crippling of thin elements	1.95
Beam buckling or crippling of thin flange elements	1.65
Shear crippling of beam webs	1.20

of Table 28.5. These are used to reduce by division the ultimate buckling strengths predicted in formulas for the various types of members and elements subject to overall buckling, intermediate buckling, and local buckling (crippling).

The factor of safety for beam elements (1.65) is smaller than that for column elements (1.95) just as it is in the design of cold-formed steel structural members. The reasons for this are, first, that the failure mode of beams is less likely to be catastrophic, thus enabling the occupant to take corrective action before failure, and, second, that the column formulas for overall buckling are based on perfectly straight columns with no local or overall imperfections.

For wind and seismic loads these safety factors are generally reduced 25%.

28.4.2 Buckling Formulas

Tables 28.6 through 28.8 indicate the numerical values of allowable stresses for various types of buckling and crippling, as they pertain to three typical alloys used in building construction [28.7]. To convert to SI units (MPa) stresses in ksi should be multiplied by 6.895. Definitions are as follows:

I_y = moment of inertia of a beam about axis parallel to web, in.^{4}
L = length of compression member between points of lateral support, or twice the length of a cantilever column (except where analysis shows that a shorter length can be used), in.
L_b = length of beam between points at which the compression flange is supported against lateral movement, or length of cantilever beam from free end to point at which the compression flange is supported against lateral movement, in.
r = least radius of gyration of a column, in.
r_y = radius of gyration of a beam (about axis parallel to web), in. (For beams that are unsymmetrical about the horizontal axis, r_y should be calculated as though both flanges were the same as the compression flange.)
S_c = section modulus of a beam, compression side, in.^{3}
t = thickness of flange, plate, web or tube, in. (For tapered flanges, t is the average thickness.)

For additional information see Ref. 28.7 which contains among other items the following buckling information:

1. Formulas and constants for other alloys.
2. Web crippling at concentrated loads or reactions. This is covered by the use of partly empirical formulas involving the web depth, slope, and thickness, the bend radius at the base of the web, the length of bearing, and the compressive yield strength of the material.
3. Intermediate buckling of beam or panel flanges with elastic lateral bracing.
4. Crippling of curved elements of beams.
5. Combined compression and bending with and without shear. This is covered by three simplified interaction formulas which are selected on the basis of the ratio of the bending moment at the center to the maximum bending moment in the span.
6. Stiffener requirements.
7. Weighted average allowable compressive stress.

28.4.3 Distortion of Thin Elements

Sheathing panels and other aluminum members formed of thin sheet or coil may suffer distortion of cross-sectional configuration under load. The generally accepted minimum thickness of structural sheathing is 0.018 in. (0.46 mm). Distortion may be caused by such items as unequal web slopes, web angle flatter than 45°, eccentric loading, unsymmetric configuration, curl of wide flats, placing the load on wide flats, and attempting to transfer load across wide flats. To avoid problems of reduced section modulus or instability of cross-section that may result from distortion, the design of beams and panels less than 0.050 in. (1.3 mm) should include an analysis of the effect of load on the shape of the cross-section. For configurations that are too complex to so analyze, tests should be performed to determine the amount of distortion, and if too great, the allowable load should be based on test results factored down as indicated in the following section [28.7, Appendix B].

In the tables a distinction is made between column elements and beam elements. The difference in the formulas for these two cases is explained partly by the factors of safety employed (see Section 28.4.2 above). The other main difference in the formulas is that those for columns include no allowance for restraint in a torsional mode of failure, while the beam component formulas include partial torsional restraint.

Table 28.6 Allowable Stresses for Building and Similar Type Structures, 3003-H16 (Sheet, Rolled Rod and Bar, Drawn Tube)

Type of Stress	Type of Member or Component	Spec. No.	Allowable Stress, ksi				
TENSION, axial, net section	Any tension member:	1	12.5				
			4.2 (shaded)				
TENSION IN BEAMS, extreme fiber, net section	Rectangular tubes, structural shapes bent about strong axis	2	12.5				
			4.2 (shaded)				
	Round or oval tubes	3	15				
			5 (shaded)				
	Rectangular bars, plates, shapes bent about weak axis	4	17				
			5.8 (shaded)				
BEARING	On rivets and bolts	5	19				
			9.5 (shaded)				
	On flat surfaces and pins	6	12.5				
			4.8 (shaded)				
			Allowable Stress, ksi, Slenderness ≤ S_1	Slenderness Limit, S_1	Allowable Stress, ksi Slenderness Between S_1 and S_2	Slenderness Limit, S_2	Allowable Stress, ksi Slenderness ≥ S_2
COMPRESSION IN COLUMNS, axial, gross section	All columns	7	10	$\frac{L}{r} = 8.8$	$10.5\text{-}0.057\frac{L}{r}$	$\frac{L}{r} = 121$	$\frac{51{,}000}{(L/r)^2}$
			4.2 (shaded)	$\frac{L}{r} = 111$	$10.5\text{-}0.057\frac{L}{r}$	$\frac{L}{r} = 121$	$\frac{51{,}000}{(L/r)^2}$
COMPRESSION IN COMPONENTS OF COLUMNS, gross section	Outstanding flanges and legs	8	10	$\frac{b}{t} = 6.3$	$12.4\text{-}0.38\frac{b}{t}$	$\frac{b}{t} = 22$	$\frac{1{,}970}{(b/t)^2}$
			4.2 (shaded)	---	4.2	$\frac{b}{t} = 22$	$\frac{1{,}970}{(b/t)^2}$
	Flat plates with both edges supported	9	10	$\frac{b}{t} = 20$	$12.4\text{-}0.119\frac{b}{t}$	$\frac{b}{t} = 52$	$\frac{320}{(b/t)}$
			4.2 (shaded)	---	4.2	$\frac{b}{t} = 76$	$\frac{320}{(b/t)}$
	Curved plates supported on both edges, walls of round or oval tubes	10	10	$\frac{R}{t} = 24$	$12.1\text{-}0.43\sqrt{\frac{R}{t}}$	$\frac{R}{t} = 330$	$\frac{3{,}200}{(R/t)(1+\sqrt{R/t}/35)^2}$
			4.2 (shaded)	$\frac{R}{t} = 6.7$	$4.5\text{-}0.116\sqrt{\frac{R}{t}}$	$\frac{R}{t} = 810$	$\frac{3{,}200}{(R/t)(1+\sqrt{R/t}/35)^2}$

Allowable Stresses for BUILDING and Similar Type Structures

3003–H16

Sheet
Rolled Rod and Bar
Drawn Tube

COMPRESSION IN BEAMS, extreme fiber, gross section	Single web beams bent about strong axis	11	11	$\frac{L_b}{r_y} = 25$	$12.4\text{-}0.057\frac{L_b}{r_y}$	$\frac{L_b}{r_y} = 145$	$\frac{87{,}000}{(L_b/r_y)^2}$
			4.2	---	4.2	$\frac{L_b}{r_y} = 144$	$\frac{87{,}000}{(L_b/r_y)^2}$
	Round or oval tubes	12	13	$\frac{R_b}{t} = 49$	$21.4\text{-}1.20\sqrt{\frac{R_b}{t}}$	$\frac{R_b}{t} = 146$	Same as Specification No. 10.
			5	$\frac{R_b}{t} = 88$	$8.0\text{-}0.32\sqrt{\frac{R_b}{t}}$	$\frac{R_b}{t} = 290$	Same as Specification No. 10
	Solid rectangular beams	13	14	$\frac{d}{t}\sqrt{\frac{L_b}{d}} = 18$	$19.5\text{-}0.31\frac{d}{t}\sqrt{\frac{L_b}{d}}$	$\frac{d}{t}\sqrt{\frac{L_b}{d}} = 42$	$\frac{11{,}400}{(d/t)^2(L_b/d)}$
			5.5	---	5.5	$\frac{d}{t}\sqrt{\frac{L_b}{d}} = 46$	$\frac{11{,}400}{(d/t)^2(L_b/d)}$
	Rectangular tubes and box sections	14	11	$\frac{L_bS_c}{I_y} = 165$	$12.4\text{-}0.109\sqrt{\frac{L_bS_c}{I_y}}$	$\frac{L_bS_c}{I_y} = 5720$	$\frac{24{,}000}{(L_bS_c/I_y)}$
			4.2	---	4.2	$\frac{L_bS_c}{I_y} = 5710$	$\frac{24{,}000}{(L_bS_c/I_y)}$
COMPRESSION IN COMPONENTS OF BEAMS, (component under uniform compression), gross section	Outstanding flanges	15	11	$\frac{b}{t} = 8.2$	$14.7\text{-}0.45\frac{b}{t}$	$\frac{b}{t} = 16$	$\frac{120}{(b/t)}$
			4.2	---	4.2	$\frac{b}{t} = 29$	$\frac{120}{(b/t)}$
	Flat plates with both edges supported	16	11	$\frac{b}{t} = 26$	$14.7\text{-}0.141\frac{b}{t}$	$\frac{b}{t} = 52$	$\frac{380}{(b/t)}$
			4.2	---	4.2	$\frac{b}{t} = 90$	$\frac{380}{(b/t)}$
COMPRESSION IN COMPONENTS OF BEAMS, (component under bending in own plane), gross section	Flat plates with compression edge free, tension edge supported	17	14	$\frac{b}{t} = 12$	$19.5\text{-}0.47\frac{b}{t}$	$\frac{b}{t} = 28$	$\frac{4{,}900}{(b/t)^2}$
			5.5	---	5.5	$\frac{b}{t} = 30$	$\frac{4{,}900}{(b/t)^2}$
	Flat plates with both edges supported	18	14	$\frac{h}{t} = 61$	$19.5\text{-}0.090\frac{h}{t}$	$\frac{h}{t} = 108$	$\frac{1{,}050}{(h/t)}$
			5.5	---	5.5	$\frac{h}{t} = 191$	$\frac{1{,}050}{(h/t)}$
	Flat plates with horizontal stiffener, both edges supported	19	14	$\frac{h}{t} = 141$	$19.5\text{-}0.039\frac{h}{t}$	$\frac{h}{t} = 250$	$\frac{2{,}400}{(h/t)}$
			5.5	---	5.5	$\frac{h}{t} = 440$	$\frac{2{,}400}{(h/t)}$
SHEAR IN WEBS, gross section	Unstiffened flat webs	20	7.5	$\frac{h}{t} = 39$	$9.9\text{-}0.061\frac{h}{t}$	$\frac{h}{t} = 108$	$\frac{39{,}000}{(h/t)^2}$
			2.4	---	2.4	$\frac{h}{t} = 127$	$\frac{39{,}000}{(h/t)^2}$
	Stiffened flat webs $a_e = a_1/\sqrt{1+0.7(a_1/a_2)^2}$	21	7.5	$\frac{a_e}{t} = 74$	$13.7\text{-}0.084\frac{a_e}{t}$	$\frac{a_e}{t} = 108$	$\frac{53{,}000}{(a_e/t)^2}$
			2.4	---	2.4	$\frac{a_e}{t} = 149$	$\frac{53{,}000}{(a_e/t)^2}$

WHITE BARS apply to nonwelded members and to welded members at locations farther than 1.0 in. from a weld. **SHADED BARS** apply within 1.0 in. of a weld.

Table 28.7 Allowable Stresses for Building and Similar Type Structures, 6060-T6, -T651, -T6510, -T6511 (Extrusions, Sheet and Plate, Standard Structural Shapes, Rolled Rod and Bar, Drawn Tube, Pipe)

Type of Stress	Type of Member or Component	Spec. No.	Allowable Stress, ksi				
TENSION, axial, net section	Any tension member:	1	19 / 11Φ				
TENSION IN BEAMS, extreme fiber, net section	Rectangular tubes, structural shapes bent about strong axis	2	19 / 11Φ				
	Round or oval tubes	3	24 / 13.5Φ				
	Rectangular bars, plates, shapes bent about weak axis	4	28 / 16Φ				
BEARING	On rivets and bolts	5	34 / 18Φ				
	On flat surfaces and pins	6	23 / 12Φ				
			Allowable Stress, ksi, Slenderness $\le S_1$	Slenderness Limit, S_1	Allowable Stress, ksi Slenderness Between S_1 and S_2	Slenderness Limit, S_2	Allowable Stress, ksi Slenderness $\ge S_2$
COMPRESSION IN COLUMNS, axial, gross section	All columns	7	19	$\frac{L}{r} = 9.5$	$20.2\text{-}0.126\frac{L}{r}$	$\frac{L}{r} = 66$	$\frac{51{,}000}{(L/r)^2}$
			12Φ	---	12Φ	$\frac{L}{r} =$ 65Φ	$\frac{51{,}000}{(L/r)^2}$
COMPRESSION IN COMPONENTS OF COLUMNS, gross section	Outstanding flanges and legs	8	19	$\frac{b}{t} = 5.2$	$23.1\text{-}0.79\frac{b}{t}$	$\frac{b}{t} = 12$	$\frac{1{,}970}{(b/t)^2}$
			12Φ	---	12Φ	$\frac{b}{t} =$ 13Φ	$\frac{1{,}970}{(b/t)^2}$
	Flat plates with both edges supported	9	19	$\frac{b}{t} = 16$	$23.1\text{-}0.25\frac{b}{t}$	$\frac{b}{t} = 33$	$\frac{490}{(b/t)}$
			12Φ	---	12Φ	$\frac{b}{t} =$ 41Φ	$\frac{490}{(b/t)}$
	Curved plates supported on both edges, walls of round or oval tubes	10	19	$\frac{R}{t} = 16$	$22.2\text{-}0.80\sqrt{\frac{R}{t}}$	$\frac{R}{t} = 141$	$\frac{3{,}200}{(R/t)(1 + \sqrt{R/t}/35)^2}$
			12Φ	$\frac{R}{t} =$ 9.0Φ	$13.5\text{-}0.50\sqrt{\frac{R}{t}}$	$\frac{R}{t} = 200$	$\frac{3{,}200}{(R/t)(1 + \sqrt{R/t}/35)^2}$

Allowable Stresses for BUILDING and Similar Type Structures

6061 – T6, – T651, – T6510, – T6511

Extrusions, Sheet and Plate, Standard Structural Shapes, Rolled Rod and Bar Drawn Tube, Pipe

COMPRESSION IN BEAMS, extreme fiber, gross section	Single web beams bent about strong axis	11	21	$\frac{L_b}{r_y} = 23$	$23.9\text{-}0.124\frac{L_b}{r_y}$	$\frac{L_b}{r_y} = 79$	$\frac{87{,}000}{(L_b/r_y)^2}$
			12①	---	12①	$\frac{L_b}{r_y} = 85$①	$\frac{87{,}000}{(L_b/r_y)^2}$
	Round or oval tubes	12	25	$\frac{R_b}{t} = 28$	$39.3\text{-}2.7\sqrt{\frac{R_b}{t}}$	$\frac{R_b}{t} = 81$	Same as Specification No. 10
			14①	$\frac{R_b}{t} = 51$①	$23.9\text{-}1.39\sqrt{\frac{R_b}{t}}$	$\frac{R_b}{t} = 137$	Same as Specification No. 10
	Solid rectangular beams	13	28	$\frac{d}{t}\sqrt{\frac{L_b}{d}} = 13$	$40.5\text{-}0.93\frac{d}{t}\sqrt{\frac{L_b}{d}}$	$\frac{d}{t}\sqrt{\frac{L_b}{d}} = 29$	$\frac{11{,}400}{(d/t)^2(L_b/d)}$
			16①	$\frac{d}{t}\sqrt{\frac{L_b}{d}} = 26$①	$40.5\text{-}0.93\frac{d}{t}\sqrt{\frac{L_b}{d}}$	$\frac{d}{t}\sqrt{\frac{L_b}{d}} = 29$	$\frac{11{,}400}{(d/t)^2(L_b/d)}$
	Rectangular tubes and box sections	14	21	$\frac{L_bS_c}{I_y} = 146$	$23.9\text{-}0.24\sqrt{\frac{L_bS_c}{I_y}}$	$\frac{L_bS_c}{I_y} = 1700$	$\frac{24{,}000}{(L_bS_c/I_y)}$
			12①	---	12①	$\frac{L_bS_c}{I_y} = 2000$①	$\frac{24{,}000}{(L_bS_c/I_y)}$
COMPRESSION IN COMPONENTS OF BEAMS, (component under uniform compression), gross section	Outstanding flanges	15	21	$\frac{b}{t} = 6.8$	$27.3\text{-}0.93\frac{b}{t}$	$\frac{b}{t} = 10$	$\frac{182}{(b/t)}$
			12①	---	12①	$\frac{b}{t} = 15$①	$\frac{182}{(b/t)}$
	Flat plates with both edges supported	16	21	$\frac{b}{t} = 22$	$27.3\text{-}0.29\frac{b}{t}$	$\frac{b}{t} = 33$	$\frac{580}{(b/t)}$
			12①	---	12①	$\frac{b}{t} = 48$①	$\frac{580}{(b/t)}$
COMPRESSION IN COMPONENTS OF BEAMS, (component under bending in own plane), gross section	Flat plates with compression edge free, tension edge supported	17	28	$\frac{b}{t} = 8.9$	$40.5\text{-}1.41\frac{b}{t}$	$\frac{b}{t} = 19$	$\frac{4{,}900}{(b/t)^2}$
			16①	$\frac{b}{t} = 17$①	$40.5\text{-}1.41\frac{b}{t}$	$\frac{b}{t} = 19$	$\frac{4{,}900}{(b/t)^2}$
	Flat plates with both edges supported	18	28	$\frac{h}{t} = 46$	$40.5\text{-}0.27\frac{h}{t}$	$\frac{h}{t} = 75$	$\frac{1{,}520}{(h/t)}$
			16①	---	16①	$\frac{h}{t} = 95$①	$\frac{1{,}520}{(h/t)}$
	Flat plates with horizontal stiffener, both edges supported	19	28	$\frac{h}{t} = 107$	$40.5\text{-}0.117\frac{h}{t}$	$\frac{h}{t} = 173$	$\frac{3{,}500}{(h/t)}$
			16①	---	16①	$\frac{h}{t} = 219$①	$\frac{3{,}500}{(h/t)}$
SHEAR IN WEBS, gross section	Unstiffened flat webs	20	12	$\frac{h}{t} = 36$	$15.6\text{-}0.099\frac{h}{t}$	$\frac{h}{t} = 65$	$\frac{39{,}000}{(h/t)^2}$
			7.5①	---	7.5①	$\frac{h}{t} = 72$①	$\frac{39{,}000}{(h/t)^2}$
	Stiffened flat webs $a_e = a_1/\sqrt{1+0.7(a_1/a_2)^2}$	21	12	---	12	$\frac{a_e}{t} = 66$	$\frac{53{,}000}{(a_e/t)^2}$
			7.5①	---	9.8①	$\frac{a_e}{t} = 84$①	$\frac{53{,}000}{(a_e/t)^2}$

WHITE BARS apply to nonwelded members and to welded members at locations farther than 1.0 in. from a weld. SHADED BARS apply within 1.0 in. of a weld.

① For all thicknesses with filler alloys 5356 or 5556. With filler alloys 4043 or 5554, values apply to metal 3/8 in. or less in thickness; for greater thicknesses multiply allowable stresses by 0.8.

① For all thicknesses with filler alloys 5356 or 5556; for metal 3/8 in. or less in thickness with filler alloys 4043 or 5554.

Table 28.8 Allowable Stresses for Building and Similar Type Structures, 6063-T6 (Extrusions, Pipe)

Allowable Stresses for BUILDING and Similar Type Structures

6063 – T6

Extrusions, Pipe

Type of Stress	Type of Member or Component	Spec. No.	Allowable Stress, ksi	
TENSION, axial, net section	Any tension member:	1	15	6.5
TENSION IN BEAMS, extreme fiber, net section	Rectangular tubes, structural shapes bent about strong axis 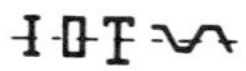	2	15	6.5
	Round or oval tubes 	3	18	8.0
	Rectangular bars, plates, shapes bent about weak axis 	4	20	8.5
BEARING	On rivets and bolts	5	24	13.5
	On flat surfaces and pins	6	16	9.0

Type of Stress	Type of Member or Component	Spec. No.	Allowable Stress, ksi, Slenderness ≤ S_1	Slenderness Limit, S_1	Allowable Stress, ksi Slenderness Between S_1 and S_2	Slenderness Limit, S_2	Allowable Stress, ksi Slenderness ≥ S_2
COMPRESSION IN COLUMNS, axial, gross section	All columns	7	13.5	$\frac{L}{r} = 9.5$	$14.2\text{-}0.074\frac{L}{r}$	$\frac{L}{r} = 78$	$\frac{51{,}000}{(L/r)^2}$
			6.5	---	6.5	$\frac{L}{r} = 89$	$\frac{51{,}000}{(L/r)^2}$
COMPRESSION IN COMPONENTS OF COLUMNS, gross section	Outstanding flanges and legs 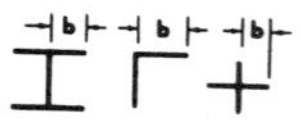	8	13.5	$\frac{b}{t} = 5.7$	$16.1\text{-}0.46\frac{b}{t}$	$\frac{b}{t} = 15$	$\frac{1{,}970}{(b/t)^2}$
			6.5	---	6.5	$\frac{b}{t} = 17$	$\frac{1{,}970}{(b/t)^2}$
	Flat plates with both edges supported 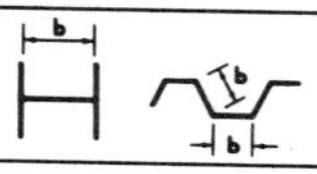	9	13.5	$\frac{b}{t} = 18$	$16.1\text{-}0.144\frac{b}{t}$	$\frac{b}{t} = 39$	$\frac{410}{(b/t)}$
			6.5	---	6.5	$\frac{b}{t} = 63$	$\frac{410}{(b/t)}$
	Curved plates supported on both edges, walls of round or oval tubes 	10	13.5	$\frac{R}{t} = 18$	$15.6\text{-}0.50\sqrt{\frac{R}{t}}$	$\frac{R}{t} = 188$	$\frac{3{,}200}{(R/t)(1+\sqrt{R/t}/35)^2}$
			6.5	$\frac{R}{t} = 10$	$7.2\text{-}0.22\sqrt{\frac{R}{t}}$	$\frac{R}{t} = 510$	$\frac{3{,}200}{(R/t)(1+\sqrt{R/t}/35)^2}$

COMPRESSION IN BEAMS, extreme fiber, gross section	Single web beams bent about strong axis	11	15	$\frac{L_b}{r_y} = 23$	$16.7\text{-}0.073\frac{L_b}{r_y}$	$\frac{L_b}{r_y} = 94$	$\frac{87{,}000}{(L_b/r_y)^2}$
			6.5	– – –	6.5	$\frac{L_b}{r_y} = 116$	$\frac{87{,}000}{(L_b/r_y)^2}$
	Round or oval tubes	12	18	$\frac{R_b}{t} = 33$	$27.7\text{-}1.70\sqrt{\frac{R_b}{t}}$	$\frac{R_b}{t} = 102$	Same as Specification No. 10
			8	$\frac{R_b}{t} = 62$	$12.8\text{-}0.61\sqrt{\frac{R_b}{t}}$	$\frac{R_b}{t} = 206$	Same as Specification No. 10
	Solid rectangular beams	13	20	$\frac{d}{t}\sqrt{\frac{L_b}{d}} = 15$	$27.9\text{-}0.53\frac{d}{t}\sqrt{\frac{Lb}{d}}$	$\frac{d}{t}\sqrt{\frac{L_b}{d}} = 35$	$\frac{11{,}400}{(d/t)^2(L_b/d)}$
			8.5	– – –	8.5	$\frac{d}{t}\sqrt{\frac{L_b}{d}} = 37$	$\frac{11{,}400}{(d/t)^2(L_b/d)}$
	Rectangular tubes and box sections	14	15	$\frac{L_bS_c}{I_y} = 145$	$16.7\text{-}0.141\sqrt{\frac{L_bS_c}{I_y}}$	$\frac{L_bS_c}{I_y} = 2380$	$\frac{24{,}000}{(L_bS_c/I_y)}$
			6.5	– – –	6.5	$\frac{L_bS_c}{I_y} = 3690$	$\frac{24{,}000}{(L_bS_c/I_y)}$
COMPRESSION IN COMPONENTS OF BEAMS, (component under uniform compression), gross section	Outstanding flanges	15	15	$\frac{b}{t} = 7.4$	$19.0\text{-}0.54\frac{b}{t}$	$\frac{b}{t} = 12$	$\frac{152}{(b/t)}$
			6.5	– – –	6.5	$\frac{b}{t} = 23$	$\frac{152}{(b/t)}$
	Flat plates with both edges supported	16	15	$\frac{b}{t} = 24$	$19.0\text{-}0.170\frac{b}{t}$	$\frac{b}{t} = 39$	$\frac{480}{(b/t)}$
			6.5	– – –	6.5	$\frac{b}{t} = 74$	$\frac{480}{(b/t)}$
COMPRESSION IN COMPONENTS OF BEAMS, (component under bending in own plane), gross section	Flat plates with compression edge free, tension edge supported	17	20	$\frac{b}{t} = 9.8$	$27.9\text{-}0.81\frac{b}{t}$	$\frac{b}{t} = 23$	$\frac{4{,}900}{(b/t)^2}$
			8.5	– – –	8.5	$\frac{b}{t} = 24$	$\frac{4{,}900}{(b/t)^2}$
	Flat plates with both edges supported	18	20	$\frac{h}{t} = 51$	$27.9\text{-}0.155\frac{h}{t}$	$\frac{h}{t} = 90$	$\frac{1{,}260}{(h/t)}$
			8.5	– – –	8.5	$\frac{h}{t} = 148$	$\frac{1{,}260}{(h/t)}$
	Flat plates with horizontal stiffener, both edges supported	19	20	$\frac{h}{t} = 118$	$27.9\text{-}0.067\frac{h}{t}$	$\frac{h}{t} = 209$	$\frac{2{,}900}{(h/t)}$
			8.5	– – –	8.5	$\frac{h}{t} = 340$	$\frac{2{,}900}{(h/t)}$
SHEAR IN WEBS, gross section	Unstiffened flat webs	20	8.5	$\frac{h}{t} = 39$	$10.7\text{-}0.056\frac{h}{t}$	$\frac{h}{t} = 78$	$\frac{39{,}000}{(h/t)^2}$
			3.9	– – –	3.9	$\frac{h}{t} = 100$	$\frac{39{,}000}{(h/t)^2}$
	Stiffened flat webs $a_e = a_1/\sqrt{1+0.7(a_1/a_2)^2}$	21	8.5	– – –	8.5	$\frac{a_e}{t} = 79$	$\frac{53{,}000}{(a_e/t)^2}$
			3.9	– – –	3.9	$\frac{a_e}{t} = 117$	$\frac{53{,}000}{(a_e/t)^2}$

WHITE BARS apply to nonwelded members and to welded members at locations farther than 1.0 in. from a weld. SHADED BARS apply within 1.0 in. of a weld.

28.4.4 Tests

Members or systems that cannot be satisfactorily analyzed using the accepted formulas of Ref. 28.7 or the principles of mechanics can generally be used for structural purposes by testing in accordance with the rules given in the same reference. This calls for three identical specimens to be tested and if the results of each do not fall within ±10% of the average, three additional tests are to be performed. The average of the three lowest test results is then reduced by dividing by a factor of safety of 2.00 on live load and 1.50 on dead load except that for wind or seismic load these factors may be reduced 25%. For fastener testing a factor of 2.20 is recommended with a similar reduction for wind or seismic loads.

Where an analysis can be made and the testing is for confirmation of the analysis, the factor of safety generally used is the same as that shown in Table 28.5 for the element that is observed to precipitate the failure; frequently only one test is performed.

28.4.5 Deflection and Ponding

A very important consideration in the design and loading of aluminum structural members is allowance for deflection. Although the allowable stresses are in many cases comparable to steel, the modulus of elasticity is roughly one-third that of steel. This larger flexibility often results in a need to calculate the deflection and occasionally to limit the load of a member. The deflection of sheathing is generally limited to 1/60 of the span, see Ref. 28.2 and Appendix B of Ref. 28.7. For primary structural members of a building it is usually necessary to apply a more severe limit on deflection to assure proper appearance and performance.

Flat or nearly level roof structures utilizing aluminum members should be checked for the possibility of ponding of rainwater or snowmelt. Once a deflection pocket forms, all of the runoff from adjacent roof areas tends to flow into the pocket and make it deeper. This effect can compound and bring failure by overload to the member in question. Such computations of deflection should assume an unlimited supply of water because the adjacent members and roof areas generally give their entire load to the first member to reach a slightly larger deflection. Also, for snow loads the assumption of simultaneous full snow load is recommended. The most effective remedy for a ponding situation is almost always to increase the slope of the roof and supply ample outflow and overflow drains. A method of computing the required slope is available in Ref. 28.10, and Section 19.2.3 treats ponding of steel-framed roofs.

28.5 FUTURE PROSPECTS

The largest factor in the cost of aluminum is the electrical energy used in its production. During the recent energy crisis the industry has made a strong effort to increase the efficiency of energy usage during production, with the result that the rise in cost of aluminum products has not been nearly as great as one would expect. However, the wood and steel industries also have made changes to conserve energy, and this, combined with the general search for economy in the construction industry, has made aluminum a premium material. Because of its many excellent properties, the use of aluminum will continue to expand in the near future, even though this expansion is slower and more selective. For the longer range aluminum could some day be a very extensively used building material because of its abundance in the rocks that form the earth's crust.

SELECTED REFERENCES

28.1 *Aluminum Alloys in Water Storage and Distribution.* Washington, D.C.: The Aluminum Association, Inc. (900 Nineteenth Street, Washington, DC 20006).

28.2 *Aluminum Formed-Sheet Building Sheathing Design Guide.* Washington, D.C.: The Aluminum Association, Inc., 1969.

28.3 *Aluminum Standards and Data,* 8th ed. Washington, D.C.: The Aluminum Association, Inc., 1984.

28.4 *Commentary on Specifications for Aluminum Structures,* Section 1A (Aluminum Construction Manual). Washington, D.C.: The Aluminum Association, Inc., 1982.

28.5 *Engineering Data for Aluminum Structures,* Section 3 (Aluminum Construction Manual), 4th ed. Washington, D.C.: The Aluminum Association, Inc., 1981.

28.6 *Illustrative Examples of Design Based on Specifications for Aluminum Structures,* Section 2 (Aluminum Construction Manual). Washington, D.C.: The Aluminum Association, Inc., 1978.

28.7 *Specifications for Aluminum Structures,* Section 1 (Aluminum Construction Manual), 4th ed. Washington, D.C.: The Aluminum Association, Inc., 1982.

28.8 *Standards for Anodized Architectural Aluminum.* Washington, D.C.: The Aluminum Association, Inc.

28.9 ASTM. *Practice for Chromate Treatments on Aluminum* [B449-67 (1982)]. Philadelphia: American Society for Testing and Materials, 1982.

28.10 Robert W. Haussler. "Roof Deflection Caused by Rainwater Pools," *Civil Engineering,* ASCE, October 1962, 58.

28.11 Rupert LeGrand. *The New American Machinists Handbook.* New York: McGraw-Hill, 1973, pp. 12–26.

28.12 R. A. Millikan, D. Roller, and E. C. Watson. *Principles of Physics,* 2nd Preliminary ed. Pasadena, CA: California Institute of Technology, 1935, p. 426.

28.13 Erik Oberg and F. D. Jones. *Machinery's Handbook,* 11th ed. New York: The Industrial Press, 1943, pp. 1658–1660.

CHAPTER 29

DESIGN FOR STAINLESS STEEL

PAUL T. LOVEJOY

Supervisor, Materials Processing
Allegheny Ludlum Steel Corporation
Research Center
Brackenridge, Pennsylvania

29.1 INTRODUCTION

The optimum functional and economic choice in many structural applications may be a component made of stainless steel. Good aesthetics, excellent durability, low maintenance requirements, and exceptional strength and ductility can all be factors in an engineering decision to use stainless steel. A wide variety of architectural, transit, automotive, and industrial applications can be cited where stainless steel is the rational long-range economic choice. Typical uses of stainless steel in roofing, flashing, coping, fascias, gravel stops, and drainage are described in AISI [29.3].

There are several areas where careful attention is needed in order to get the best use out of a stainless steel. These include the corrosive environment, the mechanical loading, possible fabrication techniques, and the desired visual appearance. Decisions to be made include selection of a specific alloy and manufacturing and joining techniques.

29.2 CORROSION RESISTANCE

In structural applications as well as in most other cases, the best guide to corrosion resistance is prior experience. While test data exist, the use of such information always involves some mathematical or phenomenological extension of short time experience to the longer service times which will be expected of the stainless steel article. For many structural applications an austenitic or ferritic alloy containing 14–18% chromium is likely to be a reasonable choice. If the environment is near the ocean or involves strong chemicals, it may be necessary to utilize one of the molybdenum-containing grades such as T316. As a point of reference, the usual grades of stainless steel are not suitable for use in seawater environments. Stainless boat railings and other nautical parts require periodic cleaning to retain their pleasing appearance. There are proprietary grades which will resist the salt in the ocean situation. Information on these alloys is best sought from a supplier of these alloys. The use of salt to aid snow removal can be a factor of some concern. Historically the greater use of salt in snow belt states forced the automotive use of T434 trim, with a molybdenum addition, over the prior T430 selection which had proven satisfactory for many years.

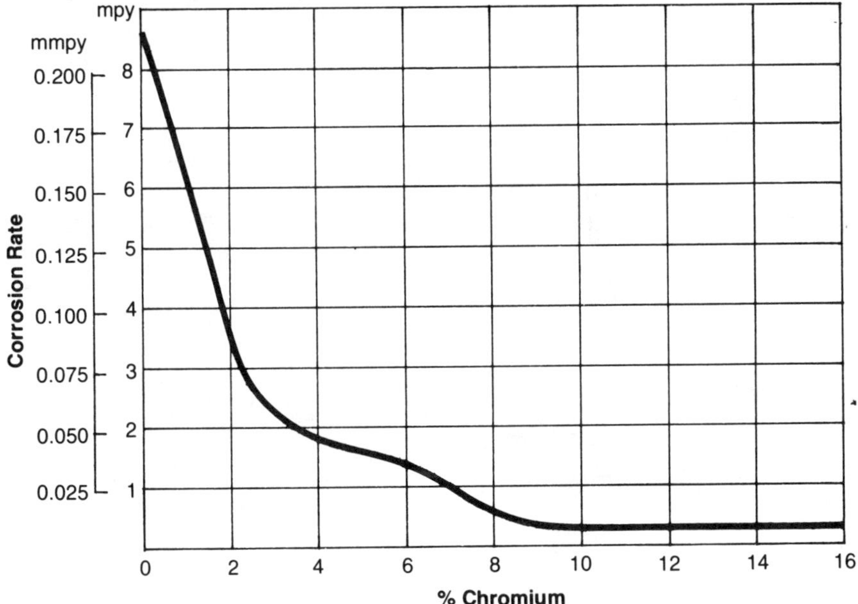

Fig. 29.1 Effect of chromium content on corrosion rate.

Figure 29.1 shows the trend toward reduced metal attack as chromium is alloyed into iron. There are two abrupt reductions in the level of attack and ultimately a plateau of greatly reduced corrosion. Comparing the chemical analyses ranges in Table 29.1 with this figure, it is evident that most grades have more chromium than needed to reach the lowest level of attack for a standardized environment, thus providing an important plus for long-term service.

Once general metal loss or "rust" that might occur in an alternate metal has been overcome by selecting a stainless grade, the designer should give attention to some details that can cause corrosive problems in stainless steel.

29.2.1 Crevice Corrosion

Crevices are known sources of corrosion problems in many alloy systems. The tight crevice can trap and concentrate corrosion agents and effectively raise the intensity of the environment to the point where attack will occur. In the case of stainless steel the exclusion of oxygen can also be a problem since the protective oxide-rich "passive film" cannot form. From a design viewpoint, the considerations are:

1. If the crevice cannot be avoided, keep it as open as possible.
2. Avoid the use of metal/nonmetal joints since these are generally tighter.
3. Consider painting the surrounding areas that are generally cathodic to the crevice.

Note, however, that in structural applications such as stainless steel transit cars, the crevices caused by spot welding are not generally a problem.

An additional benefit in some exterior applications comes from adopting a preference for bold exposures. The washing by rain water will aid in maintaining a uniform appearance.

29.2.2 Stress Corrosion

Unless there is some sustained exposure to aqueous solutions in excess of 150°F (65°C), it is not likely that a structural application would result in a stress corrosion cracking problem. The general requirements for such cracking are:

1. Exposure to some chloride-containing solution at an elevated temperature over 150°F (65°C).
2. An austenitic stainless alloy.

Table 29.1 Chemical Analysis of Selected Stainless Steels

Type	C	Mn	P	S	Si	Cr	Ni	Mo	Other	Tensile Strength, ksi	Tensile Strength, MPa	Yield Strength (0.2% offset), ksi	Yield Strength (0.2% offset), MPa	Elongation in 2 in. (50 mm), %	Hardness (Rockwell)
	Chemical Analysis, % (Max. Unless Noted Otherwise)									Nominal Mechanical Properties (Annealed Sheet Unless Noted Otherwise)					
Austenitic Stainless Steels															
201	0.15	5.50/7.50	0.060	0.030	1.00	16.00/18.00	3.50/5.50		0.25N	95	655	45	310	40	B90
202	0.15	7.50/10.0	0.060	0.030	1.00	17.00/19.00	4.00/6.00		0.25N	90	612	45	310	40	B90
301	0.15	2.00	0.045	0.030	1.00	16.00/18.00	6.00/8.00			110	758	40	276	60	B85
304	0.08	2.00	0.045	0.030	1.00	18.00/20.00	8.00/10.50			84	579	42	290	55	B80
304L	0.030	2.00	0.045	0.030	1.00	18.00/20.00	8.00/12.00			81	558	39	269	55	B79
304N	0.08	2.00	0.045	0.030	1.00	18.00/20.00	8.00/10.50		0.10/0.16N	90	621	48	331	50	B85
305	0.12	2.00	0.045	0.030	1.00	17.00/19.00	10.50/13.00			85	586	38	262	50	B80
309	0.20	2.00	0.045	0.030	1.00	22.00/24.00	12.00/15.00			90	621	45	310	45	B85
310	0.25	2.00	0.045	0.030	1.50	24.00/26.00	19.00/22.00			95	655	45	310	45	B85
316	0.08	2.00	0.045	0.030	1.00	16.00/18.00	10.00/14.00	2.00/3.00		84	579	42	290	50	B79
316L	0.030	2.00	0.045	0.030	1.00	16.00/18.00	10.00/14.00	2.00/3.00		81	558	42	290	50	B79
316N	0.08	2.00	0.045	0.030	1.00	16.00/18.00	10.00/14.00	2.00/3.00	0.10/0.16N	90	621	48	331	48	B85
321	0.08	2.00	0.045	0.030	1.00	17.00/19.00	9.00/12.00		5 × C Ti (min.)	90	621	35	241	45	B80
347	0.08	2.00	0.045	0.030	1.00	17.00/19.00	9.00/13.00		10 × C Cb-Ta (min.)	95	655	40	276	45	B85
Ferritic Stainless Steels															
405	0.08	1.00	0.040	0.030	1.00	11.50/14.50			0.10/0.30Al	65	448	40	276	25	B75
409	0.08	1.00	0.045	0.045	1.00	10.50/11.75			6 × C/0.75Ti	65	448	35	241	25	B75
430	0.12	1.00	0.040	0.030	1.00	16.00/18.00				75	517	50	345	25	B85
434	0.12	1.00	0.040	0.030	1.00	16.00/18.00		0.75/1.25		77	531	53	365	23	B83
436	0.12	1.00	0.040	0.030	1.00	16.00/18.00			5 × C/0.70 Cb + Ta	77	531	53	365	23	B83
446	0.20	1.50	0.040	0.030	1.00	23.00/27.00			0.25N	80	552	50	345	20	B83
Martensitic Stainless Steels															
403	0.15	1.00	0.040	0.030	0.50	11.50/13.00				70	483	45	310	25	B80
410	0.15	1.00	0.040	0.030	1.00	11.50/13.50				70	483	45	310	25	B80

3. Some positive tension stress (possibly quite small).
4. Time, which is influenced by temperature and stress.

The only effective cures are:

1. Lower the temperature, which extends life.
2. Select an alloy type which is not susceptible; either a ferritic stainless or a nickel-base alloy (which is technically not classified as a stainless steel).

29.2.3 Dissimilar Metal Effects

The electrical coupling of stainless steel to other metals of differing nobility can also be a problem, although generally to the other metal. The typical ranking of some common alloys is shown in Fig. 29.2. An everyday example is the accelerated corrosion of carbon steel car bodies adjacent to stainless steel trim which occurred in snow belt states. This problem is avoided by interrupting the electrical circuit with plastic or providing a sacrificial electrode such as aluminum which protects both carbon and stainless steel.

29.2.4 Carbide Precipitation and Intergranular Corrosion

A cursory review of stainless steel literature will show much work in this area, but failures have occurred, even in sophisticated industries. On the other hand, tons of "sensitized" stainless are still working satisfactorily. The key is to understand the mechanism.

Stainless steel is "sensitized" when the contained carbon and chromium combine to form a hard carbide particle. The adjacent metal is left reduced in chromium content and is anodic to the bulk of the metal. This small anode area can then selectively corrode to protect the cathode; but only if the environment is severe enough. In many cases the environment is not sufficiently severe to cause failures. In other cases the carbide network may not be extensive enough so that selective attack around the carbide cannot cause any real internal structural damage to the metal.

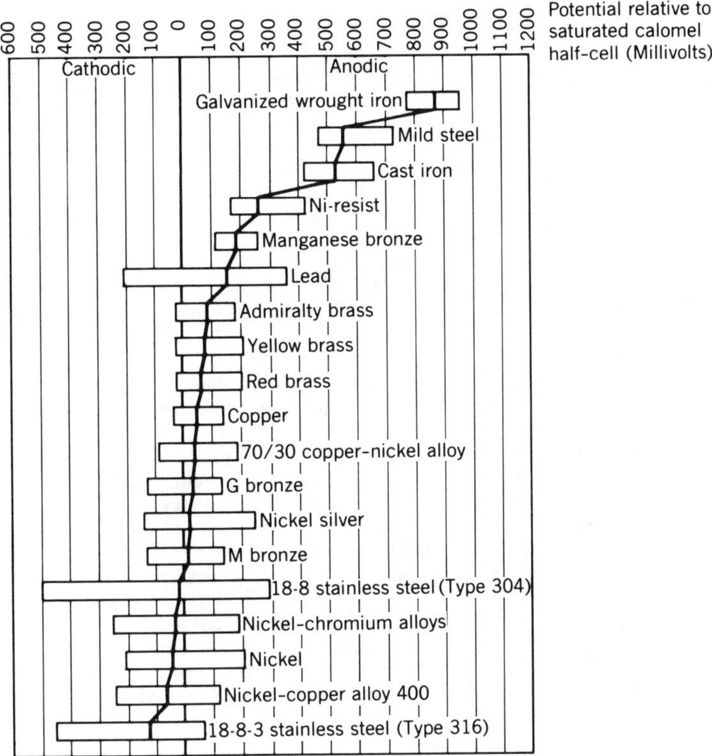

Fig. 29.2 Galvanic series, seawater (LaQue).

In a structural application, welding is the major operation that has sufficient time and temperature for carbide precipitation. However, since the carbide may take minutes to form at 1200°F (650°C), it is obvious that carbides are most likely to precipitate in the heat-affected zones of plates. Lighter gage strip and sheets are unlikely to be held at temperature for the required time. Welds themselves generally cool too quickly.

The most notable exception occurs with welds on the unstabilized ferritic grades such as 430 and 434. Due to faster kinetics of carbide precipitation, in these grades the weld and the heat-affected zone have very poor as-welded corrosion resistance when exposed to any moist exterior environment.

The common solutions to avoid carbide precipitation problems are:

1. Limit the carbon content to 0.03% by weight which results in a 5000 psi (35 MPa) strength penalty, that is, the "L" grades. Some grades, particularly in the Cr–Mn 200 series (such as T-201), may use nitrogen additions to restore strength.
2. Provide a stabilizing or gettering element that will form a carbide in preference to chromium. The examples are T321, T347, and T439.
3. Use low-heat-input welding practices such as resistance or electric arc techniques.
4. In many internal applications if there is not aqueous solution to complete electrical circuits there can be no corrosion in spite of some degree of sensitization.
5. Post-weld annealing is also a possibility.

Corrosion Summary

Experience is the major guide in controlling corrosion. If stainless has worked in similar environments, that information may be used as a reasonable guide to a new application. Many tons of stainless steel continue to resist a variety of environments at low maintenance expense, where the major factor in their selection was experience. In doubtful cases, consult a corrosion engineer or a knowledgeable supplier.

29.3 MECHANICAL PROPERTIES

The broad range in available strengths is a real driving force behind using stainless steel as a structural material. Yield strength levels range from those equivalent to mild steel up to 300,000 psi (2070 MPa). There are examples [29.4] where the high strength and long-range reduced maintenance, coupled with some fabrication and design ingenuity, have made it economically feasible to use stainless instead of some initially less-costly material.

Stainless steels are available in all the usual heat treatment variants that are available in other alloys. That is, there are martensitic stainless grades and age hardening stainless compositions which can be formed and then strengthened. In fact, heat treatment response is one of the major factors utilized to categorize stainless steel into the 200, 300, and 400 series.

For structural purposes, however, the optimum stainless steel is most often found in the temper-rolled versions of the 200 and 300 series. Design guidelines for selecting these and other grades are given in Ref. 29.2. Typical stress–strain curves for annealed 300 series alloys are shown in Fig. 29.3. The large tensile elongation noted in these annealed alloys makes it possible to produce a cold-rolled stainless steel product which will still retain sufficient ductility to permit fabrication into useful parts. The user fabrication process will further strengthen the steel. No subsequent heat treatment is required.

When temper-rolled, or cold-reduced, stainless grades are produced, there is some inevitable anisotropy in mechanical properties produced by the directional rolling operation, as shown in Fig. 29.4. In a structural context it is also important to note that the tangent modulus, which controls buckling, is also anisotropic. Figure 29.5 shows the degree of variation noted in the tangent modulus. Although most stainless applications do not require removal of variation in modulus, a short heat treatment will eliminate both the modulus and strength differences as seen in Table 29.2.

When light-gage stainless steel strip is assembled with sufficient flanges, restraints, and internal supports, it is possible to minimize the amount of metal for support of the applied load, The AISI [29.1] has established guidelines in the form of design manuals for both stainless and carbon steel; the latter topic is covered in Chapter 20 of this handbook. Both of these manuals show that when sheet stock is properly designed it is possible to support considerable loads with lightweight parts designed to have optimum stiffness. Design procedures for the two types of cold-formed materials are quite similar and the treatment provided in Chapter 20 will not be repeated here.

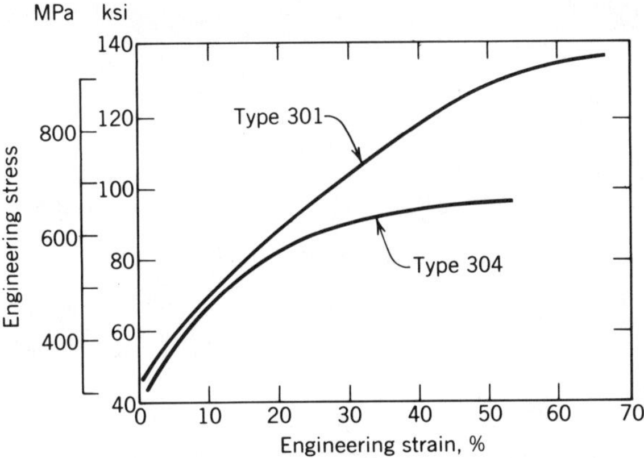

Fig. 29.3 Stress–strain curves for Types 301 and 304.

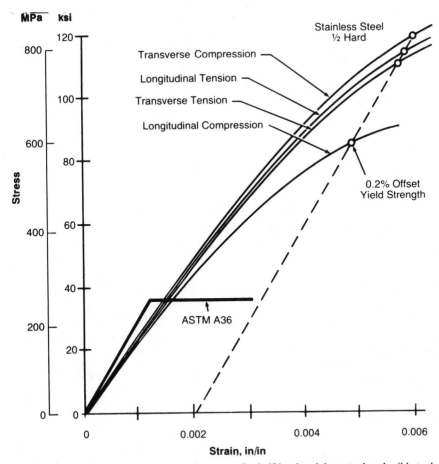

Fig. 29.4 Representative stress–strain curves for half-hard stainless steel and mild steel.

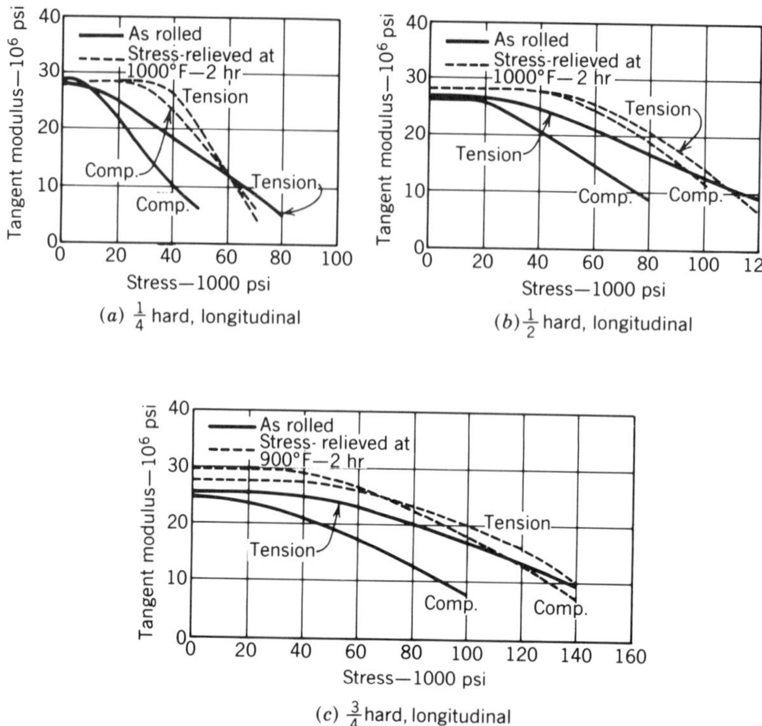

Fig. 29.5 Tangent modulus values for differing hardness.

29.4 JOINING TECHNIQUES

Stainless steels can be joined by a variety of metallurgical and mechanical techniques. In the case of welding, brazing, or soldering, proper procedures as defined by the American Welding Society should be followed. Avoiding high heat input is desirable when welding stainless welds. For a variety of light-gage structures, spot welding techniques have worked well in many cases. Good cleanliness is a desirable practice in all joints.

In the case of either mechanical joints or welds with filler wire it is necessary to select a joining alloy with matching or better corrosion resistance. This avoids sacrificial corrosion of the smaller joint area.

It is an obvious but erroneous thought to conclude that since welding is a high temperature operation (like annealing), the strength of welded joint in a temper-rolled item must necessarily be reduced to the annealed level. The data in Table 29.3 show that butt-welded ¼ and ½ hard T301 alloys do produce strengths higher than in the annealed material. The reason is based on the rapid work hardening of the T301 compositions. As the overall sample or part is strained, the major portion of the deformation is taken up plastically in the small, initially weaker weld and heat-affected zones. Further strain merely causes additional work hardening of these small regions, which is the same mechanism used to develop the initial strength. For parts and welds of normal proportions this mechanism works quite well.

A similar strengthening effect is also experienced in spot welds, as can be seen in Table 29.4. It is noted in this table that high-tensile stainless steel has greater shear strength than annealed stock.

29.5 SURFACE APPEARANCE

Stainless steel is typically recognized by a bright surface finish. In addition to this basic condition, there are a variety of standard mechanical finishes as listed in Table 29.5. Also, most suppliers may have several proprietary finishes. If it is necessary to remove any evidence of welds, a directional ground finish is one of the easiest surfaces to match in the field.

Table 29.2 As Rolled and Stress-Relieved Properties of Type 301

Temper	Condition (As Rolled or 2 hr Stress Relief)	Tension				Compression	
		Tensile Strength, ksi	Yield Strength (0.2% offset), ksi	Elongation, %	E_o, 10^6 psi	Yield Strength (0.2% offset), ksi	E_o, 10^6 psi
			Longitudinal				
Annealed	As rolled	97	36	57	31.0	38	30.2
¼ hard	As rolled	139	80	42	28.0	50	28.2
	800F	132	77	47	28.7	73	28.8
	1000F	134	73	47	29.7	74	30.5
½ hard	As rolled	157	122	23	26.8	90	27.5
	800F	155	128	24	27.9	111	29.2
	1000F	152	120	29	28.0	113	28.8
¾ hard	As rolled	176	142	17	25.8	100	26.5
	800F	181	155	11	27.3	133	27.5
	900F	176	151	11	27.5	133	27.9
Full hard	As rolled	136	160	15	25.2	115	24.6
	850F	199	175	7	28.4	169	27.7
	1000F	168	148	8	27.5	145	29.0
			Transverse				
Annealed	As rolled	93	36	62	30.6	38	30.3
¼ hard	As rolled	139	84	40	28.6	91	28.2
	800F	153	79	48	27.0	84	30.6
	1000F	136	76	47	27.8	79	31.2
½ hard	As rolled	160	123	19	28.1	142	27.5
	800F	157	130	18	28.6	144	29.8
	1000F	151	122	26	27.9	132	29.0
¾ hard	As rolled	180	145	13	27.5	170	27.9
	800F	188	155	10	28.8	176	29.5
	900F	184	154	10	28.3	175	28.5
Full hard	As rolled	204	163	11	28.4	191	29.4
	850F	213	181	5	30.5	209	29.6
	1000F	170	151	11	28.9	179	29.4

Table 29.3 Mechanical Properties of Butt-Welded Type 301 Stainless Steel

		Base Metal Properties (Longitudinal Direction)				Butt-Weld Properties Transverse to Joint		
Temper	Thickness, in.	Ultimate Tensile Strength, psi	Yield Strength, (0.2% offset), psi	Elongation in 2 in., %	Hardness, Rc	Ultimate Tensile Strength, psi	Yield Strength, (0.2% offset), psi	Elongation in 2 in., %
¼ hard	0.026	124,000	84,000	39.5	23	122,000	76,000	24.0
		121,000	83,800	40.0	23	124,000	80,000	29.5
		—	—	—	—	122,000	78,000	26.5
¼ hard	0.125	123,500	83,000	42.0	26	120,700	61,000	32.5
		120,500	82,000	40.5	26	116,000	64,600	28.5
		—	—	—	—	116,500	63,000	32.5
½ hard	0.023	184,000	142,000	23.0	40	143,000	94,700	2.5
		183,500	141,000	24.0	39	144,700	88,900	2.5
		—	—	—	—	139,000	88,000	3.0
½ hard	0.063	—	—	—	—	147,000	89,800	14.0
		—	—	—	—	148,000	102,000	8.5
		—	—	—	—	144,700	94,000	7.5
		—	—	—	—	—	—	—
½ hard	0.105	153,500	145,000	22.0	37	140,000	100,000	8.0
		155,000	146,000	23.5	37	126,700	91,000	10.5
		—	—	—	—	122,800	73,900	10.5
		—	—	—	—	134,900	89,000	8.0
Full hard	0.020	201,500	179,000	18.0	44	134,900	104,800	2.0
		—	—	—	—	133,900	108,000	1.5
		—	—	—	—	138,600	101,900	2.0
Full hard	0.063	192,000	188,500	13.0	43	143,900	86,800	4.5
		193,500	190,500	18.0	43	144,600	87,000	5.5
		—	—	—	—	134,000	83,500	4.5
Full hard	0.094	189,000	183,000	18.5	43	135,000	72,800	8.5
		189,500	182,000	19.5	43	126,800	66,000	9.5
		—	—	—	—	137,500	71,600	8.0

Table 29.4 Spot Welding Stainless Steel

Thickness T of Thinnest Outside Piece (See Notes 1, 2, 3 and 4 Below)	Electrode Diameter and Shape (See Note 5) 20° or OR R = 3″		Net Electrode Force	Weld Time (Single Impulse)	Welding Current (Approx.) Amps		Minimum Contacting Overlap	Minimum Weld Spacing (See Note 6 Below) ℄ to ℄	Diameter of Fused Zone D_ω	Minimum Shear Strength, lb — Ultimate Tensile Strength of Metal		
in.	*D*., in., min.	*d*, in., max.	lb.	Cycles (60/sec)	Tensile Strength Below 150,000 psi	Tensile Strength 150,000 psi and Higher	in.	in.	in. Approx.	70,000 up to 90,000 psi	90,000 up to 150,000 psi	150,000 psi and Higher
0.006	3/16	3/32	180	2	2000	2000	3/16	3/16	0.045	60	70	85
0.012	1/4	1/8	260	3	2100	2000	1/4	1/4	0.076	185	210	250
0.018	1/4	1/8	380	4	3500	2800	1/4	5/16	0.093	320	360	470
0.031	3/8	3/16	650	5	6000	4800	3/8	1/2	0.130	680	800	930
0.044	3/8	3/16	1000	8	8700	7000	7/16	11/16	0.180	1200	1450	1700
0.070	5/8	1/4	1700	12	12300	10000	5/8	1-1/8	0.250	2400	2800	3550
0.125	3/4	3/8	3300	20	18000	15500	7/8	2	0.300	5000	6000	7600

Notes:

1. Types of steel—301, 302, 303, 304, 308, 309, 310, 316, 317, 321, 347, and 349.
2. Material should be free from scale, oxides, paint, grease, and oil.
3. Welding conditions determined by thickness of thinnest outside piece T.
4. Data for total thickness of pileup not exceeding $4T$. Maximum ratio between two thicknesses 3 to 1.
5. Electrode material Class 2 Class 3 or Class 11

	Class 2	Class 3	or	Class 11
Minimum conductivity	75%	45%		30% of copper
Minimum hardness	75	95		98 Rockwell B

6. Minimum weld spacing is that spacing for two pieces for which no special precautions need be taken to compensate for shunted current effect of adjacent welds. For three pieces increase spacing 30%.

Table 29.5 Standard Mechanical Sheet Finishes

Unpolished or Rolled Finishes

No. 1. A rough, dull surface which results from hot rolling to the specified thickness followed by annealing and descaling.

No. 2D A dull finish which results from cold rolling followed by annealing and descaling, and may perhaps get a final light roll pass through unpolished rolls. A 2D finish is used where appearance is of no concern.

No. 2B A bright, cold-rolled finish produced in the same manner as No. 2D finish, except that the annealed and descaled sheet receives a final light roll pass through polished rolls. This is the general-purpose cold-rolled finish that can be used as is, or as a preliminary step to polishing.

Polished Finishes

No. 3 An intermediate polish surface obtained by finishing with a 100-grit abrasive. Generally used where a semifinished polished surface is required. A No. 3 finish usually receives additional polishing during fabrication.

No. 4 A polished surface obtained by finishing with a 120–150 mesh abrasive, following initial grinding with coarser abrasives. This is a general-purpose bright finish with a visible "grain" which prevents mirror reflection.

No. 6 A dull satin finish having lower reflectivity than No. 4 finish. It is produced by Tampico brushing the No. 4 finish in a medium or abrasive and oil. It is used for architectural applications and ornamentation where a high luster is undesirable, and to contrast with brighter finishes.

No. 7 A highly reflective finish that is obtained by buffing finely ground surfaces but not to the extent of completely removing the "grit" lines. It is used chiefly for architectural and ornamental purposes.

No. 8 The most reflective surface, which is obtained by polishing with successively finer abrasives and buffing extensively until all grit lines from preliminary grinding operations are removed. It is used for applications such as mirrors and reflectors.

The surface appearance is developed through the interactions of the rolling, annealing, and pickling operations that occur in the steel mill. Initially the rolls will tend to transfer their surface condition to the stainless surface. Later annealing operations, if done in air, cause oxide formation which is removed during acid treatment. The resulting surface will be rough on a microscopic scale, with a corresponding ability to scatter light. If a protective atmosphere is utilized during annealing, there is no oxide formation and the dense smooth rolled surface is preserved through the bright annealing operation. Multiple rolling operations with bright anneals produce the type of bright surface seen on some automotive trim items.

Another surface appearance parameter has ramifications on design. This is the tendency for large thin panels of stainless to "oil-can". This form of local buckling may have to be controlled by the addition of stiffeners—either through corrugations or an extra reinforcement element. The oil-can effect tends to be more prominent in a brighter finish. The *AISI Structural Design Manual* [29.1] addresses the oil-can problem by suggesting a lower stress when visual appearance is important.

29.6 FABRICATION TECHNIQUES

In most instances, stainless can be fabricated by standard techniques used on other metals. Unlike some other metals, however, there are limited semifabricated shapes available, such as extruded shapes. Most sections must be fabricated from mill-produced sheet, coils, or plate stock. Care should be taken that abraded carbon steel from previous forming is not transferred from the forming devices to the stainless, or rust may appear.

The most economical procedure is high production roll forming of straight cross-sections if sufficient parts are required to amortize the tooling. It is possible to put a slight curvature into a roll-formed part so that it will lie straight under service gravity loads.

For shorter sections, in austenitic 200 and 300 series stainless, it is possible to utilize stretch forming to produce load carrying sections. The high uniform elongation and consequent stretchability in these grades result in a considerable increase in the strength of such a part.

Although not directly a fabrication technique, it should be mentioned that stainless coils are available with various protective coatings from simple paper interleaving to peelable and/or dissolvable organic coatings which protect and lubricate during forming. These coatings can be quite useful in protecting the surface. One point to be remembered is that these coatings generally have a limited life once exposed to sunlight, after which removal can be quite difficult.

29.7 SUMMARY

Although often thought of as a decorative material, or one most often used to resist strong chemicals, the broad class of metals known as stainless steels does in fact have a long history of structural use. There are a variety of alloy types available with strengths up to as much as 300,000 psi complete with long service histories. The combination of strength, low maintenance costs due to excellent corrosion resistance, and modern design techniques can be utilized to give an overall economic advantage to stainless steel in selected structural applications.

SELECTED REFERENCES

29.1 AISI. *Stainless Steel Cold-Formed Structural Design Manual,* 1974 Edition. Washington, DC: American Iron and Steel Institute, 1974 (1000 16th Street, N.W., Washington, DC).

29.2 AISI. *Design Guidelines for the Selection and Use of Stainless Steel.* Washington, DC: American Iron and Steel Institute, 1977.

29.3 AISI. *Moisture Protection—Stainless Steel, Manual 7.* Washington, DC: American Iron and Steel Institute, 1972.

29.4 A. C. Kuentz. *Structural Stainless Steel—Guidelines for Design.* Philadelphia, PA: American Society for Testing and Materials, Special Technical Publication 454.

CHAPTER 30
DESIGN WITH STRUCTURAL PLASTICS

FRANK J. HEGER

Senior Principal
Simpson Gumpertz & Heger Inc.
Arlington, Massachusetts

RICHARD E. CHAMBERS

Principal
Simpson Gumpertz & Heger Inc.
Arlington, Massachusetts

30.1 INTRODUCTION TO STRUCTURAL PLASTICS MATERIALS

30.1.1 Purpose

The purpose of this chapter is to define rationale for and approaches to designing with plastics materials which demonstrate a broad spectrum of engineering properties, and which are in turn affected by magnitude and duration of stress and strain, temperature and environment.

30.1.2 Applications

Although the reader of this chapter may consider as new the use of plastics as structural or semistructural components in buildings, very significant volumes of structural plastics are gaining extensive applications in buildings. Examples are:

1. Pipes for interior drain waste and vent lines, water supply, and foundation drainage, and conduits for electrical cable. Such systems are easily installed and connected, corrosion resistant, and economical.
2. Vessels for containment of industrial corrosive products such as purification chemicals for water and chlorination for swimming pools, and for storage of fuels in aggressive soils. These vessels can be readily fashioned into efficient structural forms, using stiff strong plastics and reinforced plastics.
3. Structural sandwich panels faced with metal skins applied to structural plastic foam cores for exterior building envelopes. The plastic core provides the sole load path for shear in the panels, as well as high levels of thermal resistance.
4. Plastic plumbing fixtures and sanitary ware. These components provide functional, aesthetic, and cost advantages over conventional designs.
5. Clear glazing and translucent panels for skylights, industrial glazing, and other applications that require resistance to vandalism, or shell forms for structural enclosure.
6. Formwork for concrete, particularly in ribbed flat slabs, but also where complex formed shapes are required.
7. Louvers and grilles for lighting, and various HVAC requirements.
8. Air supported tension membrane enclosures.

Plastics are also used as the binder in "polymer concretes" for building panels, in patching mixes for deteriorated concrete, and in structural mortars. Other forms of plastics are used to repair cracks in concrete.

Rubbery plastics are also used as sealants, membranes for waterproofing, roofing, flashing, furniture cushions, beam seats, and vibration mounts. Commercially, these materials are called elastomers, and are excluded from the scope of this chapter (see Chapter 26).

30.1.3 Definition of Plastics Materials

There is no precise description for a plastic material. However, plastics are *high polymers* since they are made up of aggregations of atoms having high molecular weight; they are *organic* since they contain carbon atoms; and they are *synthetic* since they are synthesized in a chemical process.

For the purposes of this chapter, structural plastics are defined as plastics which are relatively stiff at normal temperatures and which are used in applications requiring some level of strength or stiffness. Most plastics used in the building applications listed above can be considered *structural plastics* since they must sustain some level of stress and strain and resist rupture under short-term, long-term, or repeated loads, or under impact.

Structural plastics are usually categorized into two broad divisions, derived mainly from differences in their molecular structure. *Thermoplastics* soften and harden with corresponding increases and decreases in temperature. They may melt or soften into a moldable consistency at high temperatures, and they may become rigid and brittle if temperature is decreased to low enough levels. Different thermoplastics display different degrees of softening and hardening with temperature. *Thermosets,* in contrast, have a cross-linked molecular structure that makes them more resistant to stiffness changes with changes in temperature. Once a thermosetting material is cured or set, it cannot be softened to a point where it can be reformed or molded; likewise, its tendency toward an increase in brittleness at low temperatures is less than with thermoplastics.

30.1.4 Forms of Structural Plastics

Structural plastics may be used in many forms which can be conveniently categorized as follows:

Unreinforced

Basically, a macroscopically homogeneous plastic or blends or alloys of plastic compounds that may contain hard or rubbery particulate fillers, stabilizers, colorants, or other additives to enhance cost effectiveness, durability, appearance, or processing. Usually, but not universally, only thermoplastic materials are used in the unreinforced form. Most thermosetting materials are too brittle to be used without some form of reinforcement.

Reinforced

A composite consisting of plastics material with, or without, the types of additives listed above, and also containing fibers for strengthening and stiffening. In most present building applications, glass fiber is used for reinforcement, although advanced high-performance fibers such as graphite, boron, and aramid may eventually prove cost effective as reinforcements, as they have in the automotive, aerospace, sporting goods, and marine industries. Short-length fibers up to, say ⅛ in.

(3 mm) long, may be used to reinforce both thermoplastic and thermosetting materials. The use of longer fibers, chopped strand up to about 2 in. (50 mm) long, nonwoven and woven fabrics, and continuous filaments, is generally confined to thermosetting materials. Restrictions on fiber length relate mainly to viscosity of the materials during manufacture and also the process, rather than to any specific chemical difference in the thermoplastic and thermosetting materials.

Foams

Most plastics can be formed into foam-like products that can vary in stiffness from rigid to rubbery. Inclusions of gas are responsible for the foam structures. The quantity of gas and plastic can be varied to provide a range of specific gravities from that of the basic plastic resin (usually slightly greater than 1.0) to specific gravities as low as 0.02, depending on material and process. In some processes, the foamed plastic is manufactured with an integral surface layer of solid plastic to form a structure much like bone.

30.2 STRUCTURAL PROPERTIES

The strength and stiffness properties of basic generic plastic materials vary widely depending on the compounding (additives, blends) of the resin. In addition, foam or reinforced types of materials extend the range of plastic structural properties to levels significantly lower and higher, respectively, than that of the parent material.

Figure 30.1 gives specific stress–strain relationships for several forms and types of plastics, demonstrating the wide range of properties available. All of the materials described are used in one form or another in structural or semi-structural applications.

Like other structural materials used in buildings, such as wood and concrete, the stiffness and strength properties of plastics depend on duration of load. Both properties also vary with moisture content and with other environmental influences such as ultraviolet (UV) radiation, various aggressive agents, and temperature. Key considerations in the choice of stiffness and strength values for structural design purposes are presented in this section.

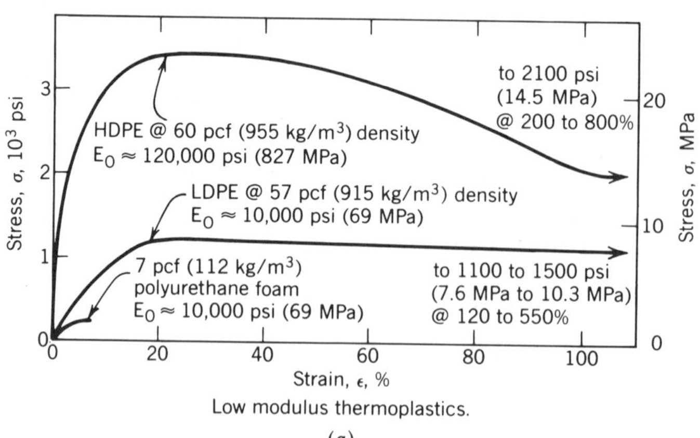

(*a*)

Fig. 30.1 Typical short-term stress–strain curves.

Key:
HDPE = high-density polyethylene
LDPE = low-density polyethylene
PMMA = polymethyl methacrylate
PC = polycarbonate
PVC = polyvinyl chloride
PPO = polyphenylene oxide
TP = thermoplastic

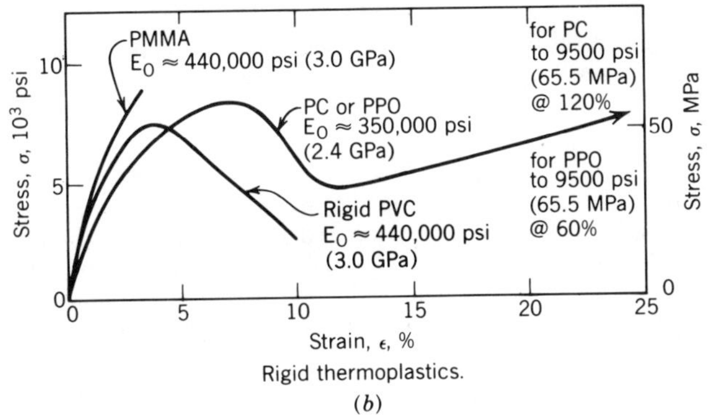

Rigid thermoplastics.
(*b*)

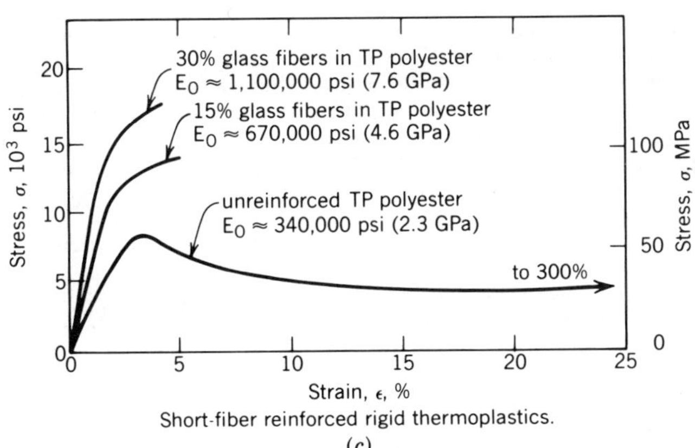

Short-fiber reinforced rigid thermoplastics.
(*c*)

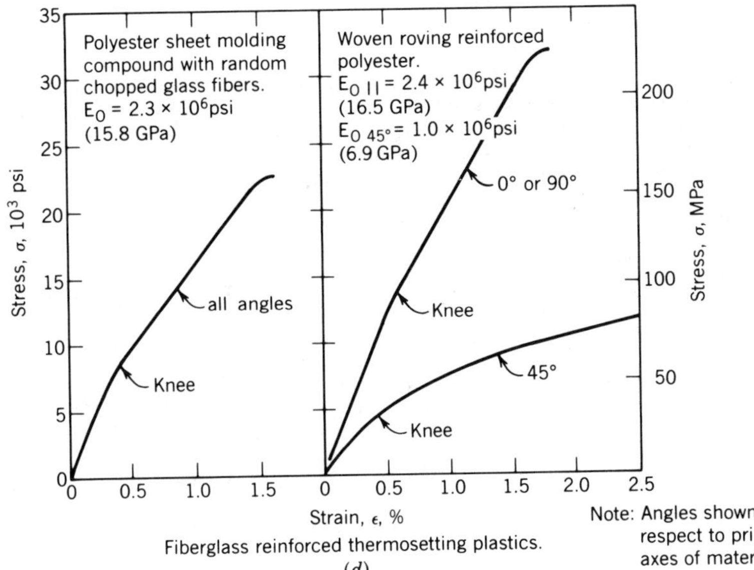

Fiberglass reinforced thermosetting plastics.
(*d*)

Fig. 30.1 (continued).

30.2.1 Stiffness

As illustrated in Fig 30.1, the range of stiffness properties under short-term loads is very large, and it depends on type of material and quantity and orientation of any reinforcement provided. Typical ranges of short-term modulus E_o are as follows:

Low-density rigid foams	1000 psi	7 MPa)
Semi-rigid (leathery) thermoplastics	10,000– 100,000 psi	(70– 700 MPa)
Rigid thermoplastics	100,000– 600,000 psi	(700– 4150 MPa)
Short-fiber fiberglass reinforced plastics	100,000–1,200,000 psi	(700– 8300 MPa)
Fiberglass reinforced plastics	800,000–6,000,000 psi	(5500–41,400 MPa)

There are three key characteristics of stress–strain behavior to be considered in designing plastics for stiffness:

1. Homogeneity of materials properties. Properties of the material may be either isotropic (same in all directions), or directional. Directional effects can result from orientation of the long molecules in thermoplastics, or from fiber alignment and layered construction in reinforced plastics.
2. Nonlinearity of stress–strain behavior. Stress–strain behavior may be nonlinear for many of the softer plastics, and for plastic foams. Stress is usually not proportional to strain up to a well-defined limit, such as the yield point in structural steels. For stiffer thermoplastics and for many reinforced plastics, however, stress–strain behavior is practically linear, at least for the early portion of the stress–strain curve.
3. Time dependence of stress-strain response. Response to stress or strain may be time dependent, or "viscoelastic". Plastics creep under fixed load and relax (i.e., exhibit a decaying stress) under fixed strain.

Because of these characteristics, plastics do not always behave as ideal elastic materials. However, assumptions can be frequently made which provide a sound basis for engineering design, and which lead to a greatly simplified approach to the structural analysis and design process. This approach is discussed below.

Isotropic Materials

Basic short-term stiffness properties required for isotropic plastic materials are the same as for other homogeneous materials such as metals or glass, and are as follows:

elastic modulus—E_0
Poisson's ratio—ν
shear modulus—$G_0 = E_0/2(1 + \nu)$

The above properties are obtained from data taken in short-term stress–strain tests. (See Ref. 30.2 for typical moduli of many types of plastics and reinforced plastics.) Unlike metals used in building applications, these material properties vary significantly with duration of load and with temperature. They also may be affected by other environmental conditions such as moisture, UV, or exposure to various aggressive agents. The effects of these factors will be discussed later.

Oriented Material

In some plastics, and usually in fibrous composites with plastic matrices, stiffness properties vary with direction of applied stress relative to the axes of orientation of molecules, or fibers, or both. The behavior of such anisotropic materials is very complex and complicated relationships are required to define stiffness properties; this is well beyond the scope of this chapter. However, many practical arrangements of oriented fiber reinforced composites yield orthotropic constructions which are *balanced* and *symmetrical*. That is, they are balanced since their principal axes are oriented at 0° and 90° to a reference axis (similar to plywood—with the axes of plies at 0° and 90° to the long direction of the sheet). And, they are also symmetrical about the centroid of the cross-section, since the distribution of plies on one side of the midplane is a mirror image of the opposite side (also similar to plywood—with plies oriented in the same direction, equidistant from the centerline).

When a material or composite is balanced and symmetrical, materials properties can be defined in terms of two orthogonal principal axes, 1 and 2. And for this case, elastic behavior is defined by the following four elastic properties.

1. *Elastic modulus* (E_{011} and E_{022}) along axes 1 and 2, respectively.

2. *Shear modulus* (G_{012}) with shear acting parallel to axes 1 and 2.

3. *Poisson's ratio* (ν_{12}) with deformation along axis 2 resulting from stress applied on axis 1.

4. *Poisson's ratio* (ν_{21}) with deformation along axis 1 resulting from stress applied on axis 2. Note that $\nu_{21} = (E_{022}/E_{011})\nu_{12}$.

In practical structures, the principal axes 1 and 2 frequently coincide with principal axes of structural elements (e.g., parallel to sides of rectangular plates). Furthermore, many common structural elements are thin, and require consideration of properties only in the plane perpendicular to the thickness (e.g., thin plates, thin shells). A thin structural element such as a plate or shell is "specially orthotropic" when its principal material directional properties are parallel to its principal axes, and simplifications are sometimes available for analyses of such structural elements.

With some materials, shear stiffness through the thickness requires special consideration (e.g., foam plastic cores of sandwich panels, and layered composites where transverse stiffness is much less than in-plane stiffness). In this case, the modulus of shearing rigidity, G_{013} or G_{023}, must also be determined.

Linear Behavior

Nonlinear stress–strain behavior is shown schematically for plastics, and for fiber reinforced plastics stressed along principal axes in Fig. 30.2. Viscoelastic limits that define the linear range of behavior for practical purposes are also shown in the figure. The "viscoelastic limit", as proposed in Ref. 30.2 for plastics, is similar to the "elastic limit" used in the design of metal structures. The rationale used in determining these limits is given below:

1. Thermoplastics. For typical short-term stress–strain curves, obtained in a constant strain-rate test, the short-term elastic modulus E_o is the slope of the initial tangent. The intercept of a secant through the origin that has a slope of $0.85E_o$ with the stress–strain curve, is the viscoelastic limit in terms of stress $\bar{\sigma}$ or strain $\bar{\epsilon}$. This limit establishes the bound below which stress is proportional to strain with a maximum error of 15%. This error is assumed to be acceptable for practical design purposes.

2. For long-term properties, an isochronous stress–strain curve is used to characterize behavior at a given time of loading. This curve is obtained from creep tests, where several levels of load or stress are applied to several samples; the resulting strains, measured at a given time of loading,

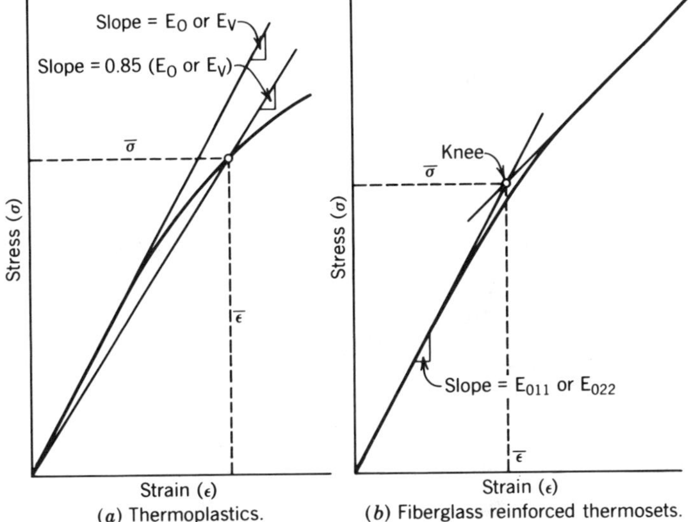

(*a*) Thermoplastics. (*b*) Fiberglass reinforced thermosets.

Note: Curves shown are for either constant strain-rate or isochronous stress-strain relationships.

Fig. 30.2 Stress–strain curves and viscoelastic limits.

are then plotted against the corresponding applied stresses. In this case, the slope of the initial tangent to the curve is the "viscoelastic" modulus E_v. The viscoelastic limits are obtained as above.

3. For most practical reinforced plastics stressed along their principal axes, the stress–strain curve is usually more or less bilinear, and can be idealized as two straight lines which intersect at a "knee". The 15% error criterion above may be applied to establish the viscoelastic limits. However, it is usually desirable to maintain stresses below those occurring at the knee since cracking of the plastic matrix is responsible for the dual modulus.

In highly reinforced plastics with highly directional reinforcements oriented also in the direction of stress, the stress–strain relationship is practically a straight line to failure (i.e., the knee does not develop during loading). Obviously, there is no viscoelastic limit for this case.

As will be discussed subsequently, stress–strain curves are altered by duration or rate of loading. Accordingly, the viscoelastic limits will shift somewhat, depending on loading time. However, the strain at the viscoelastic limit is fairly constant, and is frequently on the order of ¾% for amorphous plastics and 2% for semi-crystalline plastics. These limits should be verified for the specific material used in any final design.

In the case of fiber reinforced plastics, the stress at the knee, or viscoelastic limit, is usually about 25 to 33% of the ultimate tensile strength of the material, which produces a strain at the viscoelastic limit in the range of 0.25 to 0.5%.

Time-Dependent Stiffness

Plastics creep, or continue to deform, with increasing time under load. If, for a given load duration, stresses and strains are held within the viscoelastic limit defined above, creep of plastics can be handled in the same way as in the design of wood and concrete structures. That is, empirical creep factors are applied to account for the effects of load duration. Creep factors that define the relative increase in deformation when load duration is increased from a period of 30 sec to 10 yr are given in Table 30.1. For all materials shown, the creep factor is in the range of 1.2 to 2.4. This suggests that most unreinforced plastics creep by a factor of about 2 to 2.5 in the long term. For fiberglass reinforced plastics having glass fabrics with reasonably straight yarns, the creep factor is reduced to about 1.2 to 1.4 for loads applied in the direction of orientation.

For design with long-term loads, a "reduced modulus" is defined as the short-term modulus divided by an appropriate creep factor (Table 30.1). When applied stress is held constant, the long-term strain is the applied long-term stress divided by the reduced modulus. When applied strain is held constant, the long-term stress is the applied long-term strain multiplied by the reduced modulus.

The above rules of thumb for creep factors apply only for the linear range below the viscoelastic limit, near room temperature. They may be appropriate for other temperatures provided that stresses are below the viscoelastic limit for those temperatures. In the special case of shear modulus for reinforced plastics, orthogonally oriented fibers provide little reinforcement against shear deformation, and a creep factor close to that for the unreinforced resin should be used. Finally, if creep is critical in design, the rules of thumb should be verified on the specific material.

Table 30.1 Creep Factors ($\overline{CF}$) for Viscoelastic Plastics at Room Temperature (Adapted from [30.2])

Thermoplastics		Fiberglass Reinforced Plastics (Thermosetting Matrix)		Laminated Plastics (Thermosetting Matrix)	
Polyethylene	1.9	Polyester resin with:		Melamine/glass fabric	1.2
Polymonochlorotri-fluoroethylene	1.4	181 glass fabric (wet or dry)	1.3	Silicone/glass fabric	1.3
Polyvinyl chloride	1.8	No. 1000 glass fabric (wet or dry)	1.4	Asbestos	2.4
Polystyrene	2.0	Woven roving (dry)	1.9		
		Glass fiber mat (dry)	2.3		
		Epoxy resin with:			
		181 glass fabric (wet or dry)	1.2		

30.2.2 Strength

As illustrated earlier in Fig. 30.1, a wide range of strength properties are provided by plastics and reinforced plastics. As with stiffness, strength properties vary with type of material, and with quantity and orientation of any reinforcement provided. Typical ranges of short-term tensile strength values are as follows:

Low-density rigid foams	25 psi (0.17 MPa)
Semi-rigid (leathery) thermoplastics	1000– 3,000 psi (7–21 MPa)
Rigid thermoplastics	5000–10,000 psi (35–70 MPa)
Short-fiber fiberglass reinforced plastics	3000–20,000 psi (21–140 MPa)
Fiberglass reinforced plastics	8000–60,000 psi (55–415 MPa)

The strength of both thermoplastic and glass fiber reinforced thermosetting materials decreases significantly with increasing duration of load. Conversely, strength increases with increasing loading rate. See Ref. 30.2 for short-term strengths of typical plastics obtained under standard test conditions.

Information on strength decay (regression) with load duration is available from long-term test data on some proprietary plastic compounds, and on all compounds and laminates used in pressure-rated pipe conforming to ASTM standards. The magnitude of strength retained after long periods of sustained stress can be extrapolated from the test data. Specifically, strength retention is defined here as the strength under a load lasting 1 hr, divided by the strength under a load lasting 100,000 hr or 11.4 yr, expressed as a percent.

A comparison of the strength retention of various plastic compounds used for pressure pipe is given in Table 30.2; strength retention of wood is also shown.

The range of strength retention ratios of plastics within the same generic compound (e.g., PVC) is significant, even for those pressure pipe compounds that are designed specifically for good strength retention. As is evident, rules of thumb for strength retention characteristics under long-term loads are inappropriate, unless very conservative, since the variation in this property is so great among different compounds.

The methodology for stress-rating plastic pressure pipe has had the benefit of a substantial record of successful application, and it can be used as the basis for a general strength design approach for plastics and reinforced plastics, as well. ASTM D1598 [30.10] and D2143 [30.11] describe test procedures that involve samples held at several pressure levels to obtain failures in the 10- to 10,000-hr (1.14-yr) time frame. ASTM D2837 [30.12] and D2992 [30.13] define a relationship describing the best-fit straight line through the data points on a log-log plot, which can be rearranged as follows:

$$\sigma_{ut} = \sigma_{uo} t^{-s} \tag{30.2.1}$$

where σ_{ut} = ultimate strength at loading duration t
σ_{uo} = ultimate strength at 1 hr
t = duration of loading, hr
s = material constant, slope of line relating log rupture stress to log time to failure

This equation is used to calculate the extrapolated long-term strength at 100,000 hr (11.4-yr) of loading. See Fig. 30.3.

Table 30.2 Strength Retention Under Sustained Stress (Adapted from [30.2])

Material	Strength Retained @ 11 yr % of 1-hr Strengths*
Wood	68
PVC (polyvinyl chloride)	23 to 71
PE (polyethylene)	54 to 68
ABS (acrylonitrile butadiene styrene)	25 to 57
CAB (cellulose acetate butyrate)	52 to 57

* Strength retention values listed for plastics are for pressure or stress-rated pipe compounds.

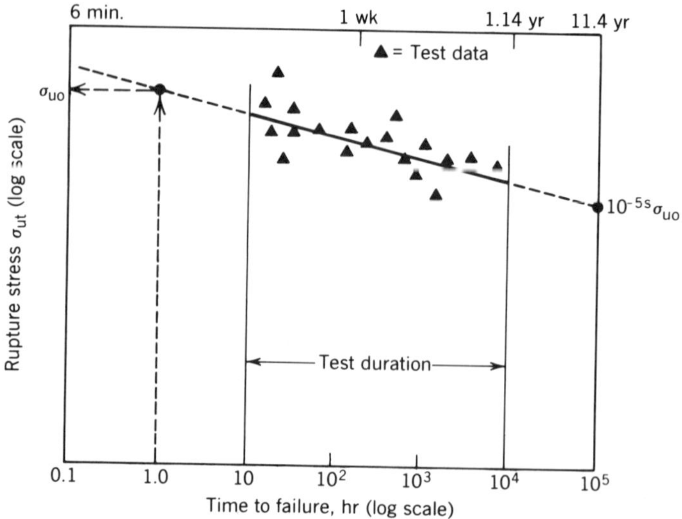

Fig. 30.3 Schematic of strength regression for structural plastic under sustained load.

For design purposes, ultimate strength becomes:

$$\text{short term:} \quad \sigma_{ut} = \sigma_{uo} \tag{30.2.2}$$

$$\text{long term:} \quad \sigma_{ut} = 100{,}000^{-s}\sigma_{uo} \tag{30.2.3}$$

These values are then multiplied by capacity reduction factors to obtain the reduced ultimate strength; this reduced strength is compared to the factored design load, as will be discussed subsequently.

Ductility

Overall, while many plastics are ductile under short-term loads as is shown in Fig. 30.1, they tend to exhibit less ductility under loads of very short and very long durations, and after cyclic loadings. Exposure to aggressive environments and low temperatures may also reduce ductility. And, some thermoplastics (e.g., acrylic plastic) and also fiberglass-reinforced thermosetting composites seldom display much ductility. (Fiberglass-reinforced plastic materials are actually pseudo-brittle as explained in Ref. 30.2.) Either low inherent ductility or potential for loss of ductility must be considered in structural design.

Deleterious Effects on Strength

There are many factors that influence the level of confidence that the ultimate strength obtained as above is representative of ultimate design strength of a specific structure. This depends upon the extent to which the test conditions simulate conditions under which the product is manufactured and used. The following examples illustrate that this can be a crucial consideration in the structural design of plastics:

1. An injection molded plastic product having significant "frozen-in" residual stresses and perhaps orientation of molecules at its surface may self-destruct instantly, without any external load, when contacted by an aggressive agent.
2. An increase in temperature above the test temperature usually increases the slope of the strength regression plot (i.e., *s* increases), and may promote the onset of brittle fracture and unreliable strength predictions. This trend is more pronounced in thermoplastics than in long-fiber fiberglass-reinforced thermosetting materials.
3. Orientation of molecules and/or glass fibers during the manufacturing process can drastically affect properties. Strength and stiffness properties are enhanced in the direction of orientation at the expense of lowering the properties transverse to the orientation.
4. Solar radiation, or radiation from certain types of lamps, can quickly embrittle surfaces of

plastics that are not properly stabilized against such effects. Marked losses in impact strength and creep rupture strength due to radiation are well documented.

As a rule, the deleterious effects on strength illustrated above are aggravated by the combination of process variables and duration of stress, temperature and environment. Experience shows that separate tests for the effects of each variable usually fail to reveal the critical interaction that leads to failure in the service environment. Deleterious effects can be minimized or eliminated by careful selection and evaluation of processes, by review of the track record of a specific plastic compound in other comparable applications, by proper formulation of the plastic for stability in certain environments, and by judicious testing for combined effects as necessary.

Safety Factors

According to the ultimate strength or "limit states" design philosophy, *capacity reduction factors* are applied to ultimate test strengths to account for structural variables and/or unknowns related to *materials*. *Load factors* are applied to design loads to account for variables and unknowns related to the *structure*. The reduced short-term and long-term ultimate strength (stress or strain) of the material is then compared to the factored stress or strain to determine structural adequacy.

Capacity reduction factors for a plastic structure should reflect such important considerations as variations in basic materials strength, further variations introduced by workmanship and manufacturing process, including knit lines and surface orientations, embrittlement and brittle fracture under sustained loads, uncertainties in strength criteria, dimensional variations, known or unknown environmental effects, and the influence of scratches, gouges, and other abuse.

Load factors should reflect characteristics related to the structure such as maximum vs. service or typical load, inaccuracies in stress analysis (e.g., crude estimate vs. detailed finite element analysis), stress due to erection tolerances, and consequences of failure.

30.3 CHOICE OF STRUCTURAL CONFIGURATION

Plastics usually offer the designer wider opportunities than metals for selecting various thin-wall configurations, curved shapes, or composite assemblies that minimize materials or facilitate efficient fabrication. Some of the special concepts available to accomplish these design developments together with illustrations of the supporting engineering theory are outlined in this section and presented later in this chapter. See Ref. 30.2 for more detailed coverage.

Three general characteristics of plastics should be considered when configurations are developed.

Fabrication

Plastics can be fabricated to various shapes by molding, extrusion, or other automatic processes. However, the choices are not unlimited and are closely related to the manufacturing process. First, materials properties may be highly directional. Furthermore, the manufacturing process can alter materials properties significantly, and introduce orientation. Also, the configuration of molded shapes must conform to the size, shape, and thickness limitations of the molding process. Some general rules of design for moldability are discussed in Section 30.4.

Plastics can also be extruded to obtain various tubular and corrugated shapes. Again, materials properties vary with direction relative to the axis of extrusion.

Low Stiffness and Low Stiffness to Strength Ratios

Plastics have low stiffness as compared to metals. They also generally have a low ratio of stiffness to strength. In view of this, criteria related to stiffness often govern the choice of structural configuration and design of sections. These criteria are local buckling of compression elements, overall buckling of compression members, and deflection of bending members. Locked-in compressive stresses resulting from thermal gradients in a molding process may reduce the buckling strength of a component.

Low Ductility and Brittle Fracture in Tension

Plastics may not exhibit short- and long-term ductility that is characteristic of structural metals. Thus, design and analysis usually must be more accurate with plastics, and secondary stresses that arise at structural discontinuities must be accounted for, whereas they are frequently ignored in design with metals that yield. Also, brittle fracture is a significant design consideration with many unreinforced plastics when elements are stressed in tension. In these materials, stress concentrations should be avoided by the use of fillets with generous radii at intersecting elements and by avoiding holes and abrupt changes in thickness. Sometimes components can be interlocked, or connected by adhesives to avoid the use of holes. Locked-in tensile stresses resulting from thermal gradients in a molding process may substantially reduce the usable tensile strength of plastics.

Configurations that take advantage of the moldability of plastics, and also overcome the problems introduced by low stiffness, include: closed or open thin-wall shapes (i.e., tubes, corrugated sections); sandwich construction; membrane action of thin plates; and shells of various shape and section. Some of the special engineering concepts needed for analysis and design of these configurations are presented in Sections 30.5 to 30.9.

30.4 MOLDING DESIGN GUIDE

The requirements of a specific molding process must be accommodated in the configuration of molded components. Some typical considerations are described below.

1. *Draft or taper* is a slight angle relative to direction of opening and closing of a mold. Molding efficiency is facilitated if the designer provides a generous taper. A general guideline for minimum taper angle is 1°; this angle can be reduced to ¼° with some materials and processes. See Ref. 30.2 for minimum draft angles for materials used in compression and injection molding processes.
2. *Fillet radius* defines curvature between two intersecting surfaces; the more generous the radius, the better the flow of material and the less the stress concentration. A minimum radius of 0.6 times part thickness is sometimes used in highly stressed areas to minimize stress concentrations.
3. *Nominal thickness* is the predominant thickness throughout a part. It should be maintained within minimum and maximum limits set by a molding process to allow flow and ejection, and to control shrinkage and locked-in stresses.
4. *Edge stiffening* should be provided to support edges and prevent warping or bowing.
5. *Ribs* are used to reduce the bulk and mass of a part while maintaining strength and stiffness, and to preclude warping of plane surfaces. Ribs should maintain nominal thickness, have simple configurations with fillets, satisfy draft requirements, and have maximum depths and thickness within process limits.
6. *Bosses* are projections from a plane surface to provide for attachments while minimizing the mass of a part.
7. *Undercuts and holes* in exterior surfaces perpendicular to the direction of molding should be avoided, if possible, because they require more complex and expensive molds.
8. *Tolerances* are important design considerations. See Ref. 30.1 for industry standards.

See Ref. 30.2 for more detailed design requirements for molding processes.

30.5 THIN-WALL MEMBERS

Thin-wall sections made by molding, extrusion, or other fabrication processes are cost effective for members subject to bending or compression. Hollow tubular shapes (Fig. 30.4), open ribbed or corrugated configurations (Fig. 30.5), and **I**-, **C**-, **Z**-, **T**-, or **L**-shapes are typical for structural applications.

Elastic beam theory, and elastic axial and lateral buckling theories that are used for metal members are also suitable for the design of thin-wall flexural and compression members. When ribs are comprised of several materials with differing elastic moduli, section properties are determined using the "transformed section" method. This is described in Ref. 30.2.

Control of local buckling is an important design consideration in thin-walled plastic compression elements. Flanges and webs of beams and columns are considered as long plates, with edges either free, or supported without rotational fixity along longitudinal edges by adjacent elements in the section. Equations for basic cases of critical plate buckling stress with specially orthotropic materials are given in Table 30.3. These may also be used for isotropic materials as indicated in the table.

Design methods for columns and beams must take into account both strength and stability of members. Approaches similar to those developed for metal members [30.7] are also used for plastics. These are described in detail in Ref. 30.2 where they are illustrated with design examples.

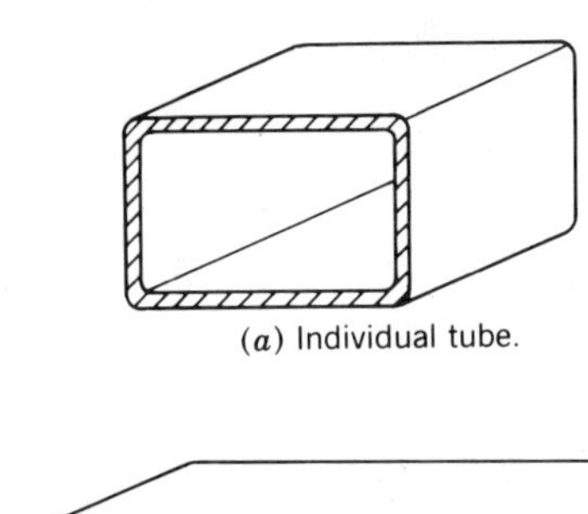
(*a*) Individual tube.

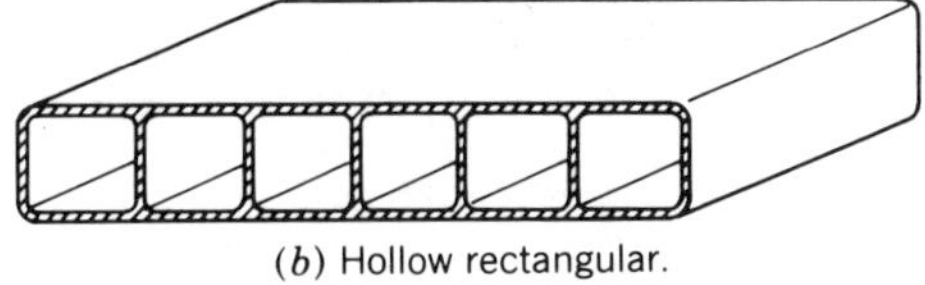
(*b*) Hollow rectangular.

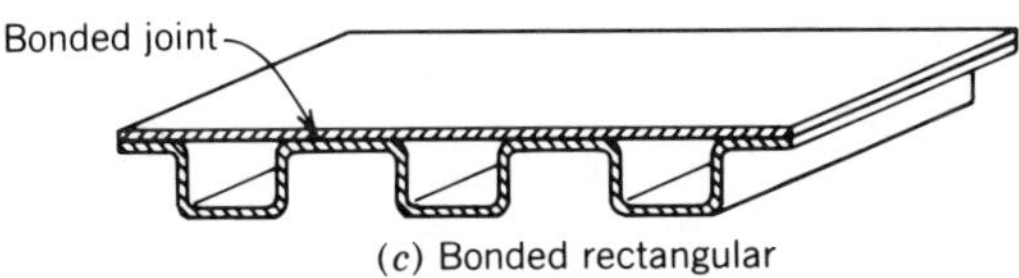

(*c*) Bonded rectangular

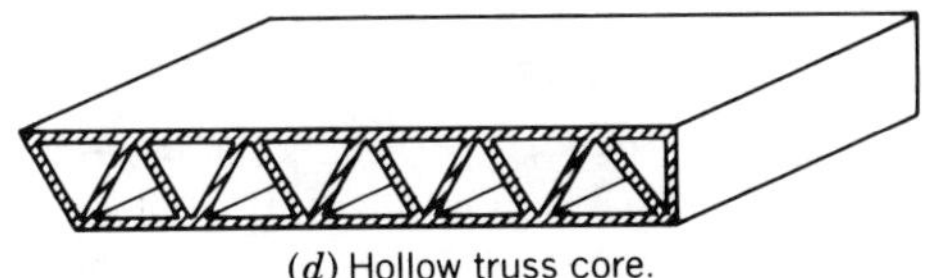
(*d*) Hollow truss core.

Fig. 30.4 Closed thin wall sections.

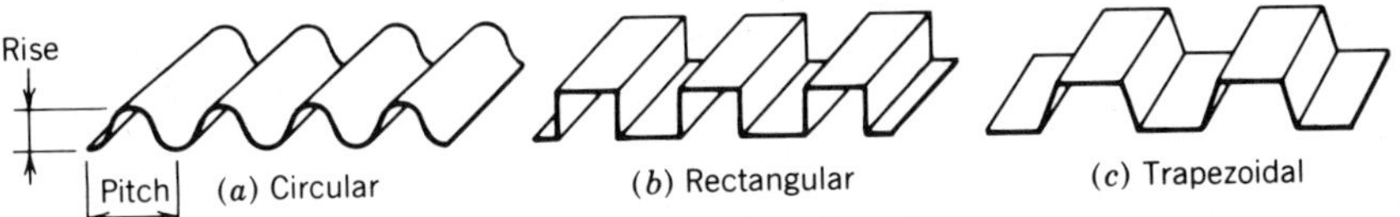

(*a*) Circular (*b*) Rectangular (*c*) Trapezoidal

Fig. 30.5 Open thin wall sections.

30.6 STRUCTURAL SANDWICH MEMBERS

In sandwich construction, thin stiff faces are bonded to a relatively thick, light, and much less stiff core. Typically in building applications, cores are made from foam plastics, end-grain balsa wood, or plastic resin impregnated Kraft paper honeycomb. Overall, highly efficient structural sections can be constructed to provide both strength and deflection control in beams, panels, and plates, and to develop resistance to buckling in compression members.

In building design practice, flat sandwich panels are used extensively for insulating wall and roof components. The high thermal resistance of typical core materials promotes the use of structural sandwich components under conditions of extreme temperature, often requiring consideration of the effects of a thermal gradient through the panel. This is discussed below.

30.6.1 Design for Bending

Sandwich sections (Fig. 30.6) are designed using elastic beam theory, based on the transformed section. For sections with thin stiff faces having cores with low bending stiffness, the simplified approximate section properties together with equations given in Table 30.4 may be used for calculating bending and axial stresses in the facings. For typical cores with low bending stiffness, flexural or axial stresses in the core are neglected.

Table 30.3 Equations for Local Buckling Stress for Thin Wall Plate Elements of Structural Sections

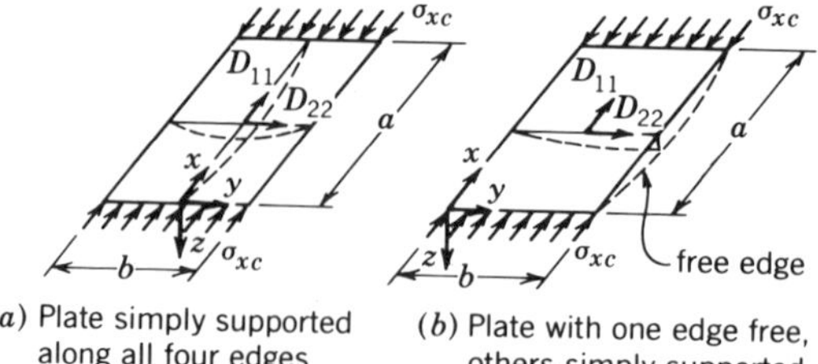

(*a*) Plate simply supported along all four edges (*b*) Plate with one edge free, others simply supported

Typical Use in Thin Wall Section	Isotropic Plate	Orthotropic Plate*
1. Flange of box section beam or column web of **I** or box column [see case (a) above]	$\sigma_{xc} = \dfrac{\pi^2 E t^2}{3(1 - \nu^2)b^2}$	$\sigma_{xc} = \dfrac{2\pi^2}{b^2 t}(\sqrt{D_{11}D_{22}} + D_o)$
2. Flange of **I** section, beam or column [see case (b) above]	$\sigma_{xc} = \dfrac{E t^2}{2(1 + \nu)b^2}$	$\sigma_{xc} = G_{12}\left(\dfrac{t}{b}\right)^2$

* For orthotropic materials,

$$D_{11} = \frac{E_{11}t^3}{12(1 - \nu_{12}\nu_{21})}; \quad D_{22} = \frac{E_{22}t^3}{12(1 - \nu_{12}\nu_{21})}; \quad D_o = \nu_{21}D_{11} + \frac{G_{12}t^3}{6}$$

In order to develop the bending resistance of the sandwich section, the core must have adequate shear strength and stiffness (shear rigidity). The effective section for shear stress is the area of the core. An equation for the approximate maximum shear stress in the core is given in Table 30.4.

Because the shear rigidity of the typical cores described above is fairly low, a sandwich panel will not always behave in strict accordance with elastic beam theory. When the core develops significant shear deformation (Fig. 30.7c) "plane sections before bending" do not "remain plane after bending," and "secondary" effects arise that may prove significant in design of the sandwich. These effects are as follows:

1. Shear deflection increases the total panel deflection from that calculated using conventional elastic beam theory. Equations for "shear" deflection of some common sandwich beam types are given in Table 30.4.

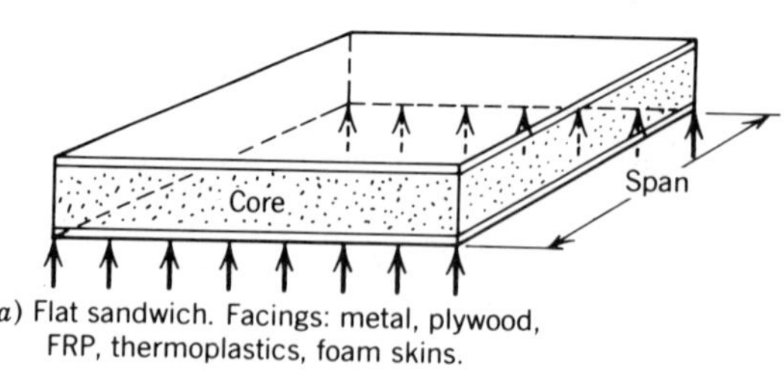

(*a*) Flat sandwich. Facings: metal, plywood, FRP, thermoplastics, foam skins.

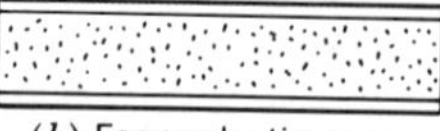

(*b*) Foam plastic core.

(*c*) Plastic impregnated Kraft paper honeycomb core, end grain balsa core.

Fig. 30.6 Sandwich construction.

2. Bending stresses develop within the thickness of the faces. These stresses are in addition to the direct face stresses obtained from the elementary beam theory given in Table 30.4. These additional face bending stresses may be particularly significant in panels subject to concentrated loads, and in panels with thick faces. See Ref. 30.2 for detailed consideration of such secondary bending. These secondary stresses are usually not significant in panels with thin faces subject to distributed loads, and thus, equations are not given here for their determination.

In addition to providing shear strength and rigidity, the core also stiffens the compression face against local buckling. The compression stress that buckles the face (usually by "wrinkling" or dimpling) is a function of the elastic moduli of the face material and the core material in flexure and transverse direct stress, respectively, and of the modulus of shear rigidity of the core. It is also affected by the initial waviness of the face. A lower bound equation for face wrinkling stress σ_{fc} is included in Table 30.4.

30.6.2 Effects of Thermal Gradients

Since typical cores provide good insulation, sandwich panels are frequently subject to significant thermal gradients. These produce "bowing" (deflection) of simply supported panels. The panel deflects toward the side that has the higher temperature. An equation for determining the expected thermal deflection of a simply supported sandwich panel is given in Table 30.4.

Flat sandwich panels that are continuous over one or more supports develop support reactions and shear and bending effects as a result of their tendency to deflect under thermal gradient. The supports restrain the panel from bowing, thereby inducing reactions and shear and bending stresses. This is discussed in Refs. 30.2 and 30.3, where design examples are also given.

When materials that move with moisture change are used as sandwich faces, differential drying will produce effects similar to the above-described thermal gradient. Since the effects are similar, the moisture movement may be estimated simply by establishing an equivalent temperature difference that produces the same effect as the moisture gradient. The use of dissimilar facing materials, or materials that move excessively with moisture change is not recommended for most sandwich panel applications because of the potential of such panels for bowing, and because of the significant stresses that arise from the restraint of bowing.

30.6.3 Optimum Proportions

A wide range of different combinations of face and core thicknesses can be used to satisfy the minimum "required" section modulus s_f and moment of inertia i_f for a given structure. If the costs C_f of a unit volume of facings, and C_c of a unit volume of core are known, the proportions that provide the lowest cost cross-section (called optimum) can be obtained from equations given in Table 30.4. These equations provide a direct design of an optimum sandwich section having thin equal facings. See Ref. 30.2 for a more detailed explanation of the basis of these relations, together with examples of their use in practical design.

30.7 LARGE BOX SECTIONS

Large, hollow reinforced plastic box-shaped cross-sections may be formed by hand lay-up, automated spray-up, or filament winding processes. Such sections are designed using the elastic bending and buckling theories discussed earlier for typical thin-wall shapes. However, the large box shapes may have wide flanges and low ratios of span to width. In this case, flange stresses will vary across the width due to "shear lag". To allow for shear lag, the actual flange area may be replaced by a reduced or "effective" flange area. This may be estimated with the aid of Fig. 30.8*a* and *b*. However, the ratios given in this figure were developed for concrete and steel box girders [30.8], and may require modification for isotropic or orthotropic plastics materials.

30.8 MEMBRANE ACTION OF FLAT PLATES

When thin flat plates are used in structural components, they may derive considerably more strength than indicated by standard formulas for plates that are based on small deflection theory. This behavior is an especially important consideration for thin plastic plates, since they may undergo large deflections due to their inherently low modulus. Thin plates mobilize additional strength as they deflect under load because of membrane action, which becomes important when deflection exceeds about half the plate thickness. The greatest increase in strength and stiffness

Table 30.4 Equations for Analysis and Design of Sandwich Sections with Thin Equal Facings

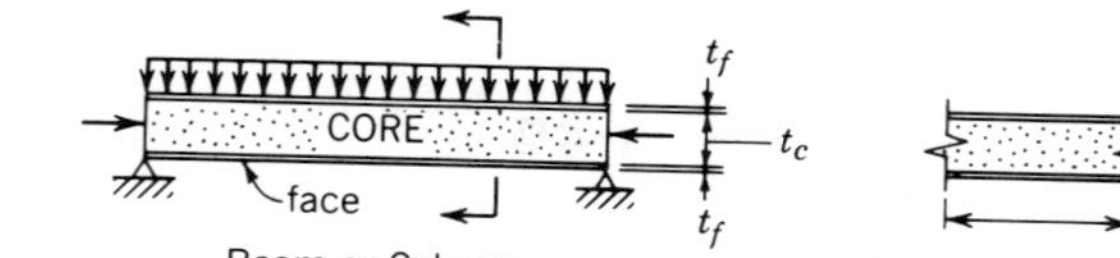

Transformed Section Properties

Bending and axial stress in facings:

Facing area, $a_F = 2t_f$
Core area, $a_c = 0$
Section modulus, $s_f = t_f(t_c + t_f)$
Moment of inertia, $i_f = 0.5t_f(t_c + t_f)^2$

Shear stress in core: Core area, $a_c = t_c$

Stresses

Average tension or compression in facings

$$\sigma_f = \frac{N}{a_f} \pm \frac{M}{s_f} = \frac{N}{2t_f} \pm \frac{M}{t_f(t_c + t_f)}$$

Shear in core

$$\tau_c = \frac{V}{a_c} = \frac{V}{t_c}$$

Maximum facing compression

$\sigma_{fc} = 0.5(E_f E_c G_c)^{1/3}$, or σ_{fu}, whichever is less

Deflection

Load deflection:

Bending portion

$$w_m = \frac{K_m P L^3}{E_f i_f}$$

Shear portion

$$w_v = \frac{K_v P L}{G_c a_c}$$

Thermal deflection:

simply supported member

$$w_t = \alpha_f \frac{(T_1 - T_2)L^2}{8(t_c + t_f)}$$

Notation

t_f = thickness of facing
t_c = thickness of core
a_f = area of two facings, unit width
s_f = section modulus, unit width
i_f = moment of inertia, unit width
a_c = area of core, unit width
σ_f = tension or compression stress in facing
N = axial stress resultant, unit width
M = moment stress resultant, unit width
τ_c = shear stress in core
V = shear stress resultant, unit width
σ_{fc} = local buckling (wrinkling) compression stress in facing
σ_{fu} = ultimate strength of facing
E_f = elastic modulus of facing
E_c = elastic modulus of core
G_c = shear modulus of core
w_m = bending deflection
w_v = shear deflection
w_t = thermal deflection

K_m = coefficient for bending deflection
K_v = coefficient for shear deflection

Loading	K_m	K_v
Simply supported, uniform load P	5/384	1/8
Simply supported, central point load P	1/48	1/4
Fixed ends, uniform load P	1/384	1/8

P = total load on beam
L = span length of beam
α_f = coefficient of thermal expansion of facings
T_1 = temperature of outside face
T_2 = temperature of inside face
C_f = unit volume cost of facing
C_c = unit volume cost of core
s_f^* = section modulus of facings with least cost proportion, based on required i_f
i_f^* = moment of inertia of facings with least cost proportions based on required s_f

Optimum Proportions for Least Panel Cost

Criterion 1:

To provide required i_f	$t_f = \left[\dfrac{C_c^2\, i_f}{2(2C_f - C_c)^2}\right]^{1/3}$
	$t_c = \left(\dfrac{4C_f}{C_c} - 3\right) t_f$
Section modulus provided by these proportions	$s_f^* = t_f(t_c + t_f)$

Criterion 2:

To provide required s_f	$t_f = \left[\dfrac{C_c s_f}{2C_f}\right]^{1/2}$
	$t_c = \dfrac{2C_f t_f}{C_c}$
Moment of inertia provided by these proportions	$i_f^* = 0.5t_f(t_c + t_f)^2$
Which criterion governs?	If $s_f^* \geq$ req'd s_f, Criterion 1 governs If $i_f^* \geq$ req'd i_f, Criterion 2 governs If $s_f^* \leq$ req'd s_f and $i_f^* \leq$ req'd i_f, the proportions that give the required i_f and s_f, simultaneously (Criterion 3) are optimum

Criterion 3:

To provide required i_f and s_f simultaneously	$t_f = \dfrac{s_f^2}{2i_f}; \quad t_c = \dfrac{2i_f}{s_f}$

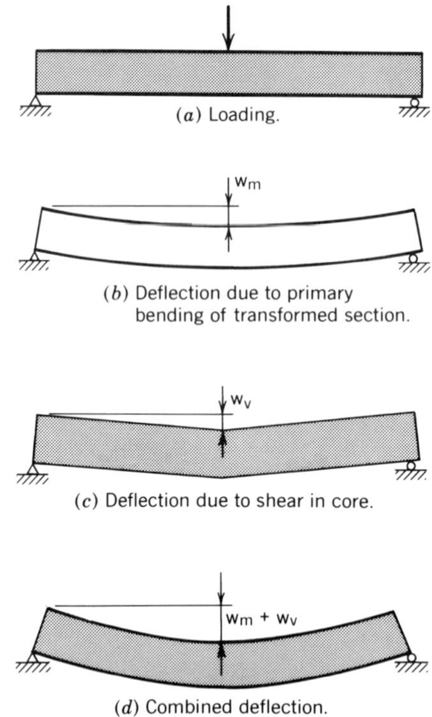

Fig. 30.7 Components of deflection of centrally loaded sandwich beam.

occurs when plate edges are held against lateral translation. Charts and equations for determining deflections and stresses caused by uniformly distributed load on thin rectangular and circular plates having various edge conditions, including membrane action, are given in Ref. 30.2. Alternatively, a conventional bending analysis for plates may be corrected to include membrane effects using the procedure given in Table 30.5 for plates with uniformly distributed loads. Stresses and deflections in plates with only bending considered, and in pure membranes, may also be obtained using equations given in the table. Example calculations for plate deflections and stresses, taking into account membrane action, are included in Ref. 30.2.

30.9 RINGS AND SHELL STRUCTURES

Plastic rings and shells are widely used for pipes, tanks, pressure vessels, skylights, and many other types of components. Pipes are usually analyzed as rings. Unless ends are restrained or closed, uniform radial pressure produces only circumferential direct stresses, and nonuniform circumferential loading produces circumferential bending. However, longitudinal discontinuities, such as thickening at bell and spigot joints, or circumferential ribs, result in secondary shell bending stresses acting in the direction of the longitudinal pipe axis. Frequently, these discontinuity stresses are ignored with metals, but they usually deserve consideration with plastics.

30.9.1 Stress Resultants

See Ref. 30.2 for equations and tables giving an extensive summary of direct, shear, and bending stress resultants in rings under various load distributions.

Thin doubly curved shells and certain singly curved elements can resist distributed loads solely by direct or membrane stress resultants. These are determined by neglecting bending stiffness in such elements and determining the membrane stress resultants using only the laws of statics. Thus, the membrane stress system is statically determinate and independent of elastic moduli and directional variations in stiffness of material.

The ideal support conditions at the edges of a shell are reactions that are equal to the membrane stress resultant required by statics, and with support deformations that are compatible with those occurring in the adjacent membrane. If these conditions are satisfied, undesirable edge bending effects will be avoided. Usually, however, edge conditions do not provide such reactions and deformations, and significant shear and bending stress resultants develop in the vicinity of edges and at other structural discontinuities in the shell. These edge bending stresses depend on the relative stiffness of adjoining discontinuous elements and also are affected by directional variations in material properties. Fortunately, they usually damp out rapidly in directions away from edges and points of discontinuity.

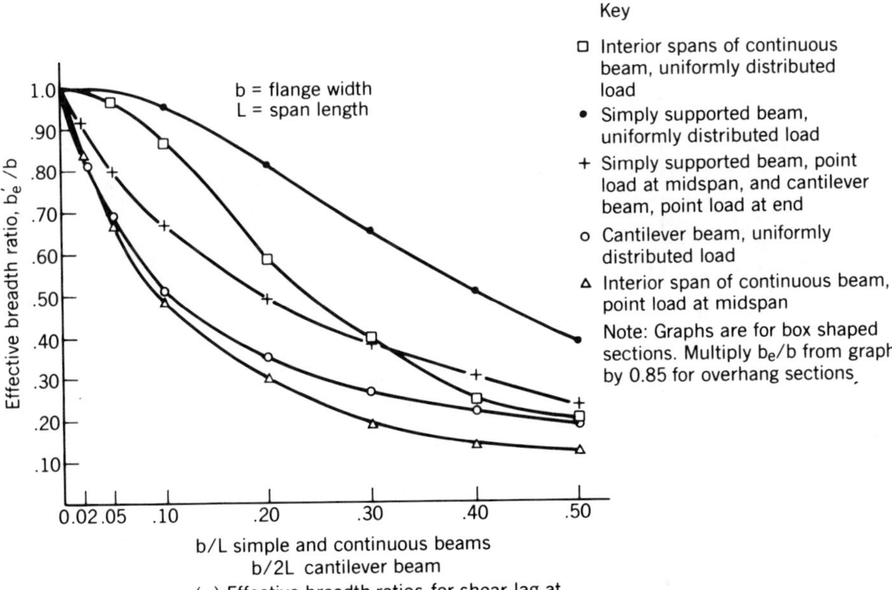

(*a*) Effective breadth ratios for shear lag at sections of maximum moment.

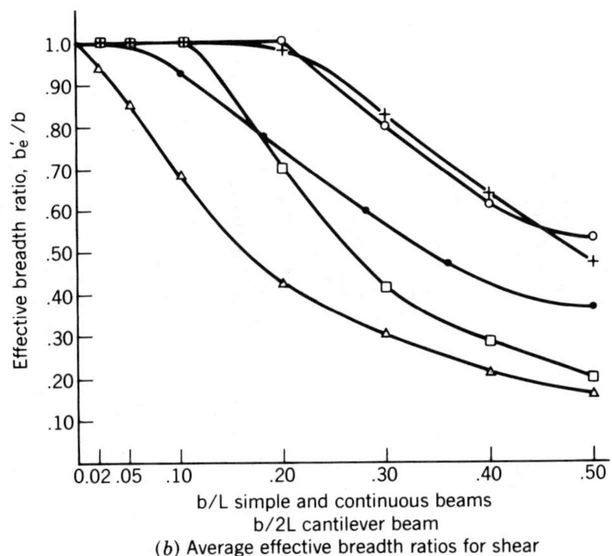

(*b*) Average effective breadth ratios for shear lag at sections for deflection calculations.

Fig. 30.8 Shear lag effects.

Table 30.5 Approximate Deflections and Stresses in Isotropic Plates under Uniformly Distributed Load, Including Membrane Resistance Due to Deflected Shape

	Plate with Edges Not Held		Membrane		Plates with Edges Held
	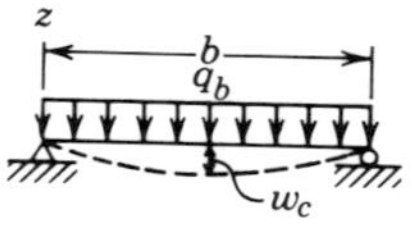	+	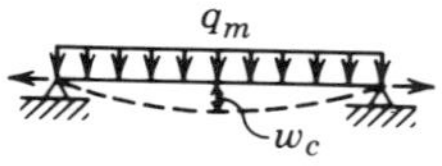	=	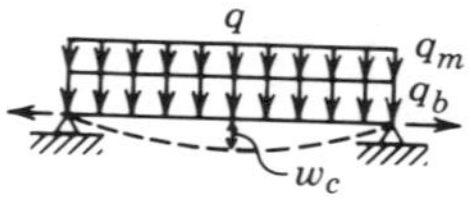
1. Consider lateral load q as comprised of a portion q_b supported by bending action, and a portion q_m supported by membrane action.	q_b	+	q_m	=	q
2. Determine bending and membrane loads in terms of center deflection w_c. First solve for center deflection w_c using (c); then solve for loads q_b and q_m, using (a) and (b). Rectangular plate, simply supported on four sides having lengths a and b, that are held against lateral translation, $\nu = 0.3$, or, circular plate, simply supported on outer periphery having diameter b that is held against lateral translation (See 4 below for C_1 and C_2):	Bending portion of load, q_b (a) $q_b = \frac{C_1 w_c E t^3}{b^4}$		Membrane portion of load, q_m (b) $q_m = \frac{C_2 w_c^3 E t}{b^4}$		Total load, q (c) $q = \frac{w_c E t^3}{b^4}\left[C_1 + C_2\left(\frac{w_c}{t}\right)^2\right]$
3. Considering loads q_b and q_m, solve for maximum stress using equations for bending of plates under lateral load, (d) and for direct stress in pure membranes, (e). Rectangular plate supported on four edges having lengths, a and b, that are held against lateral translation, $\nu = 0.30$, or circular plate, simply supported on outer periphery having diameter a that is held against lateral translation, $\nu = 0.30$ (See 4 below for C_3 and C_4):	(d) $\sigma_b = \frac{C_3 q_b b^2}{t^2}$		(e) $\sigma_t = C_4\left[\frac{q_m^2 b^2 E}{t^2}\right]^{1/3}$		
4. Deflection and stress coefficients for plates with various edge conditions.					

	a/b	C_1 Eq. (a)		C_2 Eq. (b)	C_3 Eq. (d)		C_4 Eq. (e)
		Simply supported edges	Rotationally fixed edges		Simply supported edges	Rotationally fixed† edges	
Rectangular plates	1.0	22.6	72.7	30.0	0.29	0.31	0.25
	1.2	16.2	53.2	24.4	0.38	0.38	0.26
	1.4	13.0	44.2	22.0	0.45	0.44	0.28
	1.6	11.0	39.8	20.0	0.52	0.47	0.29
	1.8	9.8	37.4	18.0	0.57	0.49	0.30
	2.0	9.0	36.0	17.5	0.61	0.50	0.31
	∞*	7.0	35.2	16.0	0.75	0.19	0.32
Circular plates	diameter $= b$	20.9	93.8	55.6	0.31	0.19	0.40

* Long plate, or plate supported on two opposite edges at ends of b.
† Max. stress at edge of plate.

Thus, many shells with regular geometry are analyzed in two steps:

1. Membrane solution.
2. Edge bending solution.

Simplified approximate edge bending analyses are frequently used and usually provide sufficiently accurate results for practical design.

Membrane Analysis

See Ref. 30.2 for summary tables of equations for membrane stresses in cylindrical, spherical, conical, and hypar shells subject to various distributed loads. These are valid for both isotropic and orthotropic materials.

Simplified Edge Bending Analysis

The elastic solution for deformations and shear, bending, and hoop stress resultants in a cylinder subject to axisymmetric edge loads (Fig. 30.9) is frequently used for determining edge bending stress resultants in axisymmetric cylinders, spherical shell, and conical shells subject to axisymmetric loads. The edges of spherical and conical shells are assumed as the edges of a tangent cylinder. Also, the cylindrical shell solutions for these axisymmetric edge effects are the same as the solutions for a concentrated load and moment applied at the end of a beam on an elastic foundation (Fig. 30.9). The hoop tensile or compressive stiffness of the cylinder is equivalent to the foundation modulus of the elastically supported beam.

A shell constant β represents the governing relationship between flexural stiffness in the longitudinal direction and hoop tensile or compressive stiffness. The following equation for β may be used in edge bending analyses of isotropic or orthotropic shells.

$$\beta = \left[\frac{\overline{A}_\theta}{4D_x R^2}\right]^{\frac{1}{4}} \tag{30.9.1}$$

where $\overline{A}_\theta = E_\theta a_\theta$ = stiffness for direct stress in circumferential (θ) direction
$D_x = E_x i_x$ = flexural stiffness in longitudinal (x) direction (see Table 30.3)
R = cylinder radius

Equations are given in Refs. 30.2 and 30.9 for radial deflection, circumferential thrust N_θ, longitudinal bending moment M_x and radial shear Q_x as a function of B and the distance x from the edge of the shell, for the two edge loads, Q_o and M_o, shown in Fig. 30.9. See also Ref. 30.2 for a summary of similar equations for edge bending in spherical shells. For orthotropic materials, the effect of property variation in the x and θ directions is taken into account with the shell factor β as given by Eq. (30.9.1).

30.9.2 Netting Analysis

Netting analysis may be used to design cylinders made from some types of reinforced laminates built up from layers of unidirectional fibers. The objective of this analysis is to provide strength in

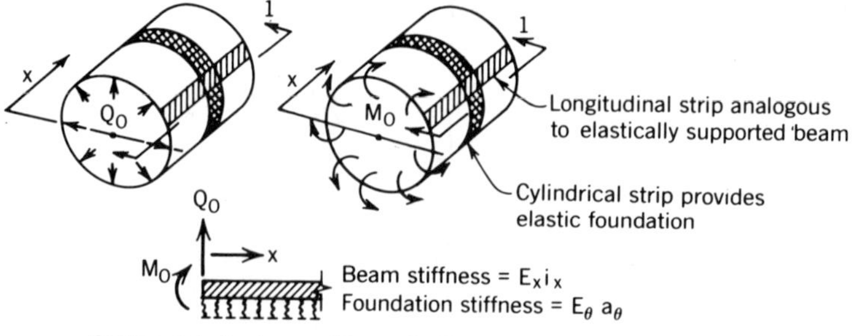

Fig. 30.9 Axisymmetric edge forces on cylindrical shell and analogous "beam-on-elastic foundation."

Table 30.6 Filament Orientation and Strength Based on Netting Analysis

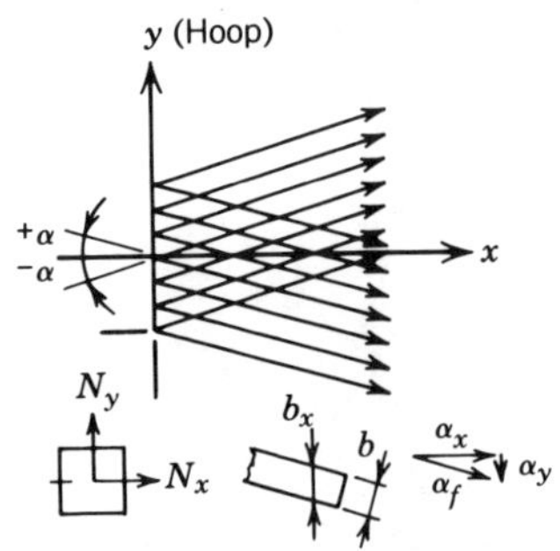

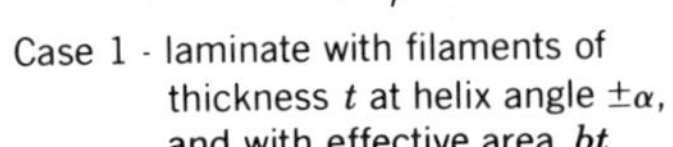

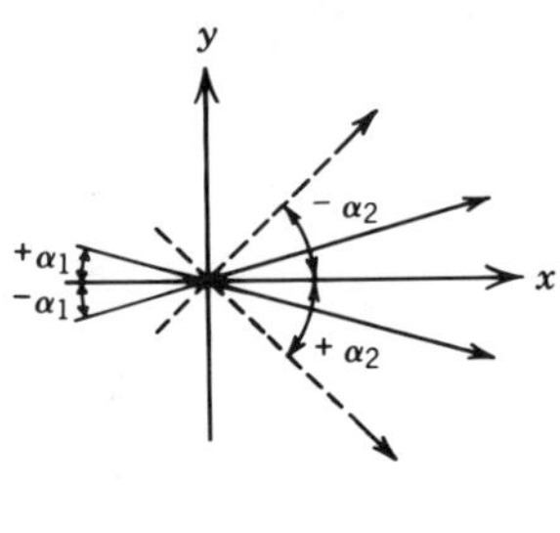

Case 1 - laminate with filaments of thickness t at helix angle $\pm\alpha$, and with effective area bt

Case 2 - general binary laminate with thickness t_1 filaments at $\pm\alpha_1$ and thickness t_2 filaments at $\pm\alpha_2$ and strengths of filaments 1 and 2 are both σ_f

	Expressions for Filament Orientation Angle, α	
	Case 1	Case 2
Filament orientation in terms of helix angle α and equivalent thickness of filaments t_1 and t_2 for membrane stress ratio, N_y/N_x	$\frac{N_y}{N_x} = \tan^2 \alpha$	$\frac{N_y}{N_x} = \frac{t_1 \sin^2 \alpha_1 + t_2 \sin^2 \alpha_2}{t_1 \cos^2 \alpha_1 + t_2 \cos^2 \alpha_2}$
Required strength σ_f in direction of filament	$\sigma_f = \frac{N_x}{t \cos^2 \alpha}$	$\sigma_f = \frac{N_x}{t_1 \cos^2 \alpha_1 + t_2 \cos^2 \alpha_2}$
	Also $\sigma_f = \frac{N_y}{t \sin^2 \alpha}$	Also $\sigma_f = \frac{N_y}{t_1 \sin^2 \alpha_1 + t_2 \sin^2 \alpha_2}$

each principal direction which is just equal to the principal membrane stresses produced by one condition of loading, usually internal pressure. If laminate strength is assumed to be a direct function of fiber strength σ_f in a bidirectional laminate, equations for the required orientation and fiber quantity for various arrangements of filaments in a circular tube or cylindrical pressure vessel are given in Table 30.6. Required fiber quantity is reflected as required thickness of a layer of unidirectional fibers at a particular orientation with a reference axis.

30.9.3 Buckling

Buckling governs the design thickness of most thin plastic shells that are subject to compressive stress resultants. The buckling stress must be equal to or greater than the maximum stress resultant in the same direction times a suitable load (safety) factor.

Buckling resistance of various shell forms may be calculated from one of the following three types of buckling in a cylindrical shell:

1. Longitudinal stress in axially loaded cylinder.
2. Circumferential stress in radially loaded cylinder.
3. Shear stress in torsionally loaded cylinder.

These are illustrated in Figs. 30.10, 30.11, and 30.12 [30.2, 30.4, 30.6]. Equations for the above-described buckling stresses are given in Table 30.7. Equations are given for both isotropic and orthotropic shells. The more general form for orthotropic shells may also be used to evaluate buckling of ribbed shells and sandwich shells with isotropic or orthotropic facings.

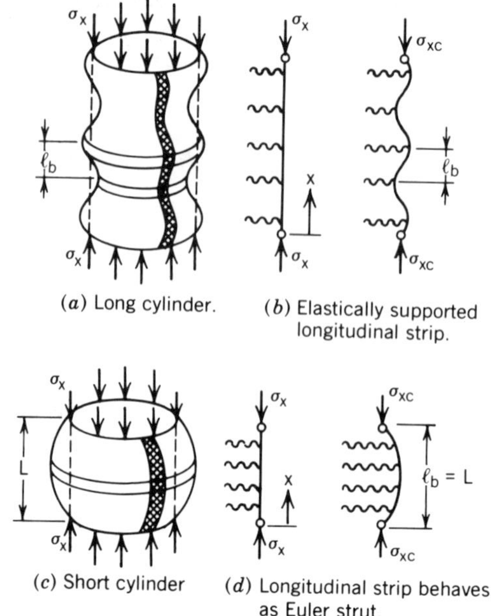

Fig. 30.10 Buckling of longitudinally compressed cylinder.

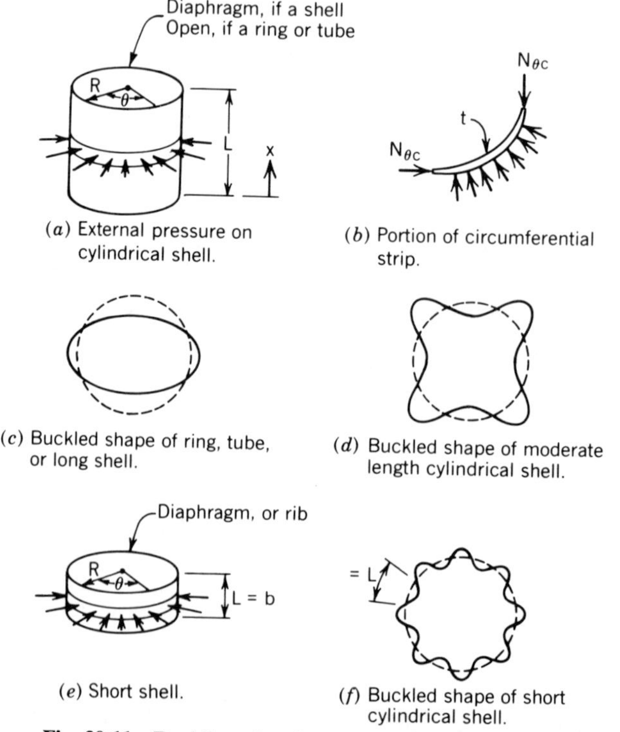

Fig. 30.11 Buckling of radially compressed cylinder.

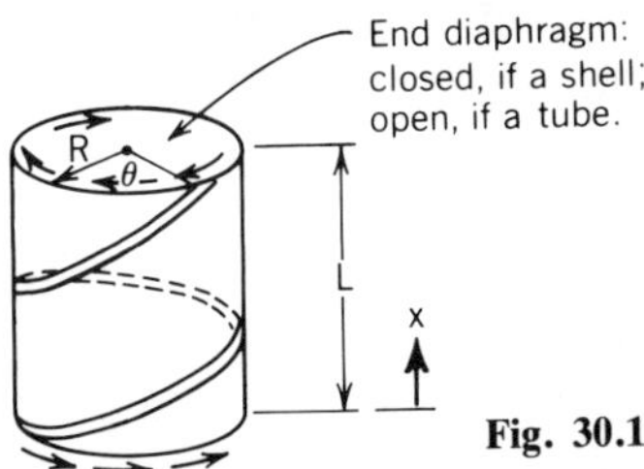

Fig. 30.12 Buckling of torsionally loaded cylinder.

Some of the equations given in the table contain knockdown coefficients k_n that account for the effects of imperfections and deviations from the nominal shape of the cylinder. The knockdown coefficient for longitudinal buckling stress reduces the buckling stress obtained using classical small deflection elastic buckling theory by a factor of 3 to 6 for thin shells. Values for this k_n factor vs. shell R/t are given in Table 30.7. The k_n values for the other types of buckling are either 1.0, or are only slightly less than 1.0, as shown in the table, reflecting the insensitivity of these types of buckling to imperfections.

The equations for longitudinal buckling stress in a cylinder may also be used for determining buckling stress in conical shells [30.2], spherical shells (Table 30.8), and hypar shells [30.2, 30.5]. For these doubly curved shell types, the equivalent cylinder radius is the radius of the shell surface that is perpendicular to the direction of principal compressive stress. When compressive stresses occur in both principal directions, buckling analyses should be made for equivalent cylinders having each of the two principal radii. See Ref. 30.2 for further discussion.

Optimum Sandwich Proportions for Shell Buckling Resistance

The above-described procedure for determining shell buckling stress may also be used to determine the required sectional properties of a sandwich shell. Simple equations for buckling stress in a sandwich shell with equal thin stiff faces (i.e., stiff relative to core stiffness—see Section 30.6) are obtained by substituting $E_f i_f$ for D_x and D_θ and $E_f a_f$ for $\bar{A}_x$ and $\bar{A}_0$ in the shell buckling equations in Tables 30.7 and 30.8, where i_f is the moment of inertia and a_f is the area of a unit width section.

In many cases, the buckling stress is proportional to the cross-sectional property, $\sqrt{i_f a_f}$. The required $\sqrt{i_f a_f}$ property can be obtained with many different sandwich proportions. If the cost C_f of a unit volume of facings and the cost C_c of unit volume of core are known, the proportions of the lowest-cost cross-section (called optimum) are [30.2, 30.6]

$$t_f = \sqrt{\frac{C_c\sqrt{i_f a_f}}{2C_f - C_c}} \qquad (30.9.2)$$

$$t_c = \left(\frac{2C_f}{C_c} - 2\right) t_f \qquad (30.9.3)$$

$$\text{minimum materials cost/unit area} = \sqrt{C_c(2C_f - C_c)\sqrt{i_f a_f}} \qquad (30.9.4)$$

In those cases where buckling is governed by i_f, instead of $\sqrt{i_f a_f}$, minimum cost proportions are obtained using the equations in Table 30.4.

30.10 SELECTION OF MATERIALS AND PROCESS

Plastics and reinforced plastics offer a wide choice of different materials having a broad range of structural properties. See Ref. 30.2 for mechanical property tables for both plastics and plastics-based composite materials. Both unreinforced and fiber reinforced materials are included. Fiber reinforcements are usually glass woven fabrics, mats, continuous filaments, or chopped strands. Advanced fibers such as graphite, aramid, and boron are available, but usually are not cost effective in building applications.

Strengths of some fiberglass reinforced plastics are comparable to metals. Strengths of unreinforced plastics are generally much lower than metals. Stiffness of reinforced and unreinforced plastics are generally lower than metals. Since the elastic modulus of glass reinforcement alone is about 10,000,000 psi (69,000 MPa), the elastic modulus of even highly oriented, unidirectional laminates with high glass content does not exceed about 6,000,000 psi (41,000 MPa). Most practical

Table 30.7 Buckling Stresses and Pressures for Cylindrical Shells

Loading Type	Shell Length	Buckling Stress, σ_c, or Pressure, p_{cr}	Orthotropic, Ribbed, or Sandwich Shells	Isotropic, Uniform Wall, Shell	Buckling Coefficient, C, and Knockdown Coefficient, k_n (for Initial Imperfections and Large Deflections)
Longitudinal See Fig. 30.10	Long or Intermediate	σ_{xc}	$\dfrac{2\sqrt{3}\,C\sqrt{D_x\overline{A}_\theta}}{Ra_x}$	$\dfrac{CEt}{R}$	$C = k_o k_n k_s$ $k_o = \dfrac{1}{\sqrt{3(1-\nu^2)}} \approx 0.6$ $k_n = 1.0 - 0.91\left[1 - \dfrac{1}{e^{0.06\,R/t}}\right] + 1.5\left[\dfrac{R}{L}\right]^2\left[\dfrac{t}{R}\right]$
	Short	σ_{xc}	$\dfrac{k\pi^2 D_x}{a_x L^2}$	$\dfrac{k\pi^2 t^2 E}{12(1-\nu^2)L^2}$	For $100 < R/t < 4000$; $0.2 < R/L < 33$. (For $R/L < 0.2$, neglect term with R/L) where e, the Naperian base, equals 2.7183. $k_s \approx 1.0$, except $k_s < 1.0$ for sandwich shells having cores with low shear rigidity. See Ref. 30.2 for graph for k_s $k = 1.0$ for simply supported circumferential edges and 4.0 for clamped edges
Radial See Fig. 30.11	Long	σ_c	$\dfrac{3D_\theta}{a_\theta R^2}$	$\dfrac{Et^2}{4(1-\nu^2)R^2}$	
		p_{cr}	$\dfrac{3D_\theta}{R^3}$	$\dfrac{Et^3}{4(1-\nu^2)R^3}$	
	Intermediate	$\sigma_{\theta c}$	$\dfrac{5.5k_n(\overline{A}_x)^{1/4}(D_\theta)^{3/4}}{a_\theta L\sqrt{R}}$	$\dfrac{0.855k_n Et}{(1-\nu^2)^{3/4}L\sqrt{R/t}}$	$k_n \approx 0.9$
		p_{cr}	$\dfrac{5.5k_n(\overline{A}_x)^{1/4}(D_\theta)^{3/4}}{LR^{3/2}}$	$\dfrac{0.855k_n Et}{(1-\nu^2)^{3/4}L\sqrt{R/t}}$	

	Short	$\sigma_{\theta c}$	$\frac{2\pi^2}{a_\theta L^2}(\sqrt{D_x D_\theta} + D_o)$*	$\frac{\pi^2 E t^2}{3(1-\nu^2)L^2}$*	See Table 30.3 for D_x, D_θ and D_o (x corresponds to 11 and θ corresponds to 22)
		p_{cr}	$\frac{\sigma_{\theta c} a_\theta}{R}$	$\frac{\pi^2 E t^3}{3(1-\nu^2)L^2 R}$	
Torsion (In-plane shear) See Fig. 30.12	Long	$\tau_{x\theta c}$	$\frac{1.75 k_n (\bar{A}_x)^{1/4}(D_\theta)^{3/4}}{a_{x\theta} R^{3/2}}$	$\frac{0.27 k_n E}{(1-\nu^2)^{3/4}}\left[\frac{t}{R}\right]^{3/2}$	$k_n \approx 0.8$
	Intermediate	$\tau_{x\theta c}$	$\frac{3.46 k_n (\bar{A}_x)^{3/8}(D_\theta)^{5/8}}{a_{x\theta} L^{1/2} R^{3/4}}$	$\frac{0.70 k_n E}{(1-\nu^2)^{5/8}}\left[\frac{t}{R}\right]^{5/4}\left[\frac{R}{L}\right]^{1/2}$	$k_n \approx 0.8$
	Short	$\tau_{x\theta c}$	$\frac{4 k_{x\theta}(D_\theta D_x^3)^{1/4}}{L^2 t}$†	$\frac{k_{x\theta}\pi^2 E t^2}{12(1-\nu^2)L^2}$‡	See Ref. 30.2 for graph for $k_{x\theta}$ ($k_{x\theta} = k_{xy}$ in graph). $k_{x\theta} = 5.34 + (L/\pi R)^2$ for isotropic materials and simply supported circumferential edges.

* For simply supported circumferential edges.
† See Ref. 30.2 for graph for $k_{x\theta}$ ($k_{x\theta} = k_{xy}$ in graph).
‡ $k_{x\theta} = 5.34 + (L/\pi R)^2$ for isotropic materials and simply supported circumferential edges.

Table 30.8 Buckling Stresses and Pressures for Spherical Shells

Buckling Stress σ_c, or Pressure p_{cr}	Orthotropic, Ribbed, or Sandwich Shell	Isotropic Uniform Wall Shell	Buckling Coefficient, C, and Knockdown Coefficient, k_n (for Initial Imperfections and Large Deflections)
$\sigma_{\phi c}$ (for $\sigma_{\theta c}$ use a_θ in place of a)	$\dfrac{2\sqrt{3}\,C\sqrt{D_\phi \bar{A}_\theta}}{Ra_\phi}$	$\dfrac{CEt}{R}$	$C = k_o k_n k_s$ $k_o = \dfrac{1}{\sqrt{3(1-\nu^2)}} \approx 0.6$
p_{cr}	$\dfrac{4\sqrt{3}\,C\sqrt{D_\phi \bar{A}_\theta}}{R^2}$	$\dfrac{2CEt^2}{R^2}$	$k_n = 0.25\left(1 - 0.175\,\dfrac{\phi_k - 20°}{20°}\right)\left(1 - \dfrac{0.07R/t}{400}\right)$ for $20° \le \phi_k \le 60°$ and $400 \le \dfrac{R}{t} \le 2000$ See Ref. 30.2 for other semiempirical expressions for k_n $k_s \approx 1.0$, except $k_s < 1.0$ for sandwich shells having cores with low shear rigidity See Ref. 30.2 for a graph for k_s

$\sigma_{\phi c}$ and $\sigma_{\theta c}$ = buckling stresses in the meridional and circumferential directions, respectively
a_ϕ and a_θ = areas of shell cross section per unit width perpendicular to meridional and circumferential directions, respectively
$\underline{D}_\phi$ = flexural stiffness of a unit width of shell cross section perpendicular to the meridional direction
A_σ = axial stiffness of a unit width of shell cross section perpendicular to the meridional direction
ϕ_k = angle from crown to edge support, degrees

glass reinforced plastics have elastic moduli that range between 800,000 psi (5500 MPa) and 3,500,000 psi (24,000 MPa). Elastic moduli of unreinforced thermoplastics are much lower, ranging from 600,000 psi (4100 MPa) down to less than 100,000 psi (700 MPa). Thus, glass reinforced plastics are most often selected for large components that have substantial structural requirements.

Material selection and process selection are closely related. Mechanical properties often must be related to process. Obviously, component size and configuration are also closely related to manufacturing process. For example, many very efficient molding processes are suitable only for part dimensions below relatively small limiting sizes. Mold costs vary greatly with size and complexity of parts. Obviously, the number of identical components to be made greatly affects the economics of molding, extrusion, or other processes.

The discussion above indicates the complexities that must be considered before structural design standards, comparable to design specifications for steel, aluminum, or reinforced concrete, can be established for specific plastics and reinforced plastics. Thus, only a very limited number of standards are available for special products and applications such as water and sewer pipe. No available standard even begins to provide a basis for general structural design.

In view of the lack of general design standards, rational structural design of plastics components requires a comprehensive understanding of structural behavior, and familiarity with the basic design practices that are available for developing components that will perform adequately for the design life of the structure. Much of the required structural design theory is presented in Ref. 30.2 together with examples of its use for design of plastics structural components. Many of the design procedures suggested in Ref. 30.2 have been taken from design practice for metal members. These should be confirmed or modified with research programs involving tests and evaluations for specific plastics or types of plastics. Until such information is available, relatively high safety factors should be selected and designs involving critical components, or a large number of components, should be proven by tests of prototype structures.

SELECTED REFERENCES

30.1 J. Frados, ed. *Plastics Engineering Handbook,* 4th ed. (Society of the Plastics Industry, Inc.). New York: Van Nostrand Reinhold, 1976.

30.2 F. J. Heger, R. E. Chambers, and A. G. H. Dietz. *Structural Plastics Design Manual* (ASCE Manuals and Reports on Engineering Practice No. 63). New York: American Society of Civil Engineers, Structural Plastics Research Council, 1984.

30.3 F. J. Heger. "Thermal Gradient Deflections and Stresses in Structural Sandwich Insulating Panels," *Proceedings of ASCE Cold Regions Specialty Conference,* Anchorage, Alaska, May, 1978.

30.4 Frank J. Heger. "Design of FRP Fluid Storage Vessels," *Journal of the Structural Division,* ASCE, **96,** November 1970(ST11), 2465–2499.

30.5 F. J. Heger, R. E. Chambers, and A. G. H. Dietz. "On the Use of Plastics and Other Composite Materials for Shell Roof Structures," World Conference on Shell Structures, San Francisco, 1967.

30.6 F. J. Heger. "Design of Reinforced Plastic Shell Structures," Chapter 6 in *Plastics in Building,* edited by I. Skeist. New York: Reinhold, 1966.

30.7 Bruce G. Johnston, ed. *Structural Stability Research Council, Guide to Stability Design Criteria for Metal Structures,* 3rd ed. New York: Wiley, 1976.

30.8 Kevin R. Moffatt and Patrick J. Dowling. "British Shear Lag Rules for Composite Girders," *Journal of the Structural Division,* ASCE, **104,** July 1978(ST7), 1123–1130.

30.9 S. Timoshenko and S. Woinowsky-Krieger. *Theory of Plates and Shells,* 2nd ed. New York: McGraw-Hill, 1972.

30.10 ASTM. *Test Method for Time-to-Failure of Plastic Pipe Under Constant Internal Pressure* (D1598-81). Philadelphia: American Society for Testing and Materials, 1981.

30.11 ASTM. *Test Method for Cyclic Pressure Strength of Reinforced, Thermosetting Plastic Pipe* [D2143-69(1976)]. Philadelphia: American Society for Testing and Materials, 1976.

30.12 ASTM. *Method for Obtaining Hydrostatic Design Basis for Thermosetting Pipe Materials* [D2837-76(1981)]. Philadelphia: American Society for Testing and Materials, 1981.

30.13 ASTM. *Method for Obtaining Hydrostatic Design Basis for Reinforced Thermosetting Resin Pipe and Fittings* [D2992-71(1978)]. Philadelphia: American Society for Testing and Materials, 1978.

INDEX